P9-DCQ-163

BIOCHEMISTRY

BIOCHEMISTRY

THIRD EDITION

LUBERT STRYER
STANFORD UNIVERSITY

W. H. FREEMAN AND COMPANY / NEW YORK

Library of Congress Cataloging-in-Publication Data

Stryer, Lubert.
Biochemistry.

Includes index.
1. Biochemistry. I. Title.
QP514.2.S66 1988 574.19'2 87-36486
ISBN 0-7167-1843-X
ISBN 0-7167-1920-7 (international student ed.)

Printed in the United States of America

6 7 8 9 0 RRD 9 9 8 7 6 5 4 3 2 1 0

To my teachers

Paul F. Brandwein
Daniel L. Harris
Douglas E. Smith
Elkan R. Blout
Edward M. Purcell

Contents

PART III GENERATION AND STORAGE OF METABOLIC ENERGY 313

PART IV BIOSYNTHESIS OF MACROMOLECULAR PRECURSORS 545

PART V GENETIC INFORMATION: storage, transmission, and expression 647

PART VI MOLECULAR PHYSIOLOGY: interaction of information, conformation, and metabolism in physiological processes 887

List of Topics

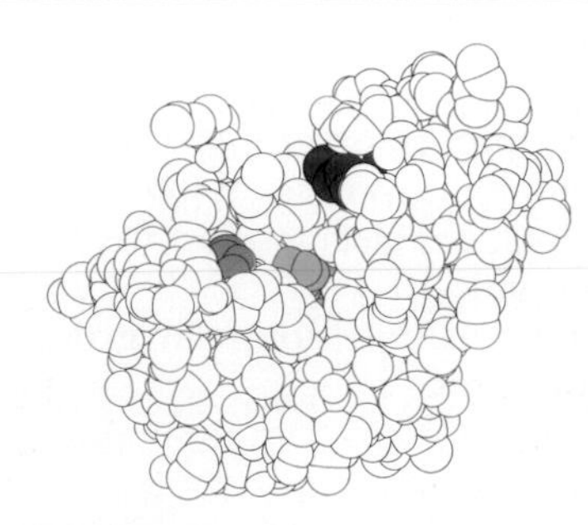

PART III GENERATION AND STORAGE OF METABOLIC ENERGY 313

CHAPTER 13 Metabolism: Basic Concepts and Design 315

CHAPTER 14 Carbohydrates 331

CHAPTER 15 Glycolysis 349

CHAPTER 16 Citric Acid Cycle 373

CHAPTER 17 Oxidative Phosphorylation 397

CHAPTER 18 Pentose Phosphate Pathway and Gluconeogenesis 427

CHAPTER 19 Glycogen Metabolism 449

CHAPTER 20 Fatty Acid Metabolism 469

CHAPTER 21 Amino Acid Degradation and the Urea Cycle 495

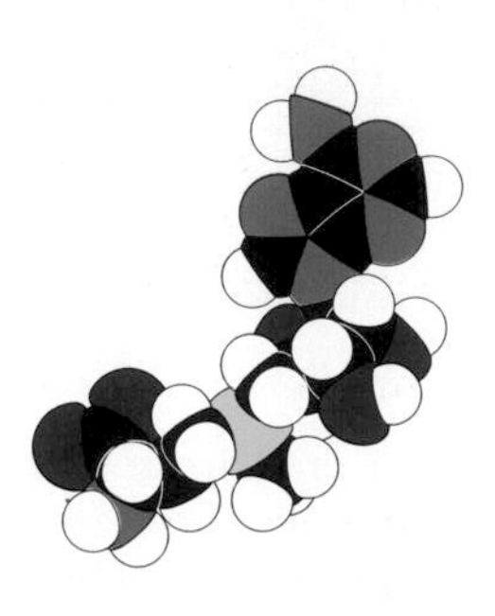

CHAPTER 24 Biosynthesis of Amino Acids and Heme 575

CHAPTER 25 Biosynthesis of Nucleotides 601

CHAPTER 26 Integration of Metabolism 627

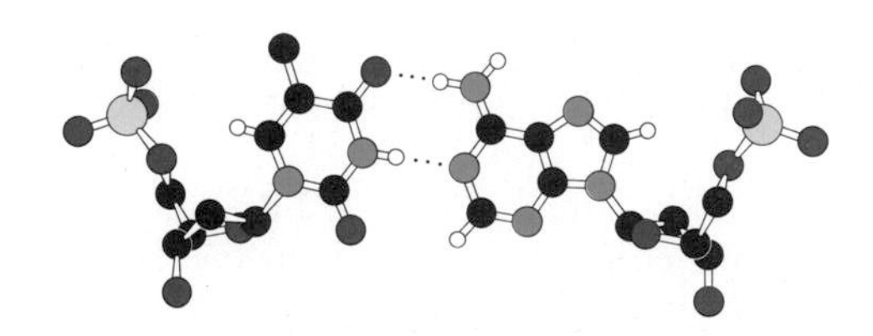

PREFACE

to the Third Edition

Biochemistry has been profoundly transformed by recombinant DNA technology. The genome is now an open book—any passage can be read. The cloning and sequencing of millions of bases of DNA have greatly enriched our understanding of genes and proteins. Indeed, recombinant DNA technology has led to the integration of molecular genetics and protein chemistry. The intricate interplay of genotype and phenotype is now being unraveled at the molecular level. One of the fruits of this harvest is insight into how the genome is organized and its expression is controlled. The molecular circuitry of growth and development is coming into view. The reading of the genome is also providing a wealth of amino acid sequence information that illuminates the entire protein landscape. Scarce proteins can be produced in abundance by transfected cells. Moreover, precisely designed novel proteins can be generated by site-specific mutagenesis to elucidate how proteins fold, catalyze reactions, transduce signals, transport ions, and interconvert different forms of free energy.

Our understanding of molecular evolution also has been greatly enriched by the recombinant DNA revolution. Families and superfamilies of proteins have come into view. Theme and variation at the level of proteins are vivid expressions of the underlying processes of gene duplication and divergence. The genes of complex proteins display the coming together in evolution of exons encoding functional modules. The many recurring structural and mechanistic motifs seen throughout nature testify to the fundamental unity of all forms of life. The discovery of catalytic RNA enables us to envision an RNA world early in the evolution of life, prior to the appearance of DNA and protein. The ubiquity of ribonucleotides in metabolism and the central roles played by them are reflections of their ancient origins—of the early RNA world, when RNA served both as gene and enzyme.

These remarkable advances compel a major change in the way biochemistry is taught. I have altered the architecture of this book to provide a new framework for the exposition of fundamental themes and principles of biochemistry. The book begins with a new part, entitled Molecular Design of Life, that provides an overview of the central molecules of life—DNA, RNA, and proteins—and their interplay. Recombinant DNA technology and other experimental methods for exploring proteins and genes are also presented in this part. This introduction prepares the reader for the detailed consideration of protein structure and function that follows. The teaching of metabolism is likewise enriched by this new organization. The other major structural changes are the addition of a chapter on carbohydrates and one on protein targeting. Many sections of the book have been extensively revised and hundreds of new illustrations have been added. I have tried to preserve the unity of biochemistry as an intellectual discipline. My goal has been to make this powerful language comprehensible and to share its beautiful imagery.

I am grateful to Alexander Glazer, Daniel Koshland, Jr., and Alexander Rich for having encouraged me to write this edition. I would not have embarked on this endeavor had it not been for their warm support and good counsel.

The planning of this edition unexpectedly took place in terrain quite different from Yale, Stanford, and Aspen, where the first two editions took form. In December 1985, my family and I went to Nepal to trek in the Everest region. After two rewarding days in Katmandu, we were on the verge of boarding the plane to Lukla, only to be turned back with the disappointing news that the landing strip was closed because of snow. We then headed for the Annapurna region in four-wheel drive vehicles but had to return a day later because the road vanished in the heavy rain. The inaccessibility of the high Himalayas led to an abrupt change of itinerary. We flew to Bangkok and arrived in 95-degree heat, carrying our parkas and arctic sleeping bags. Instead of hiking at 12,000 feet, we found ourselves at a hotel pool at sea level. This dislocation led to a totally unforeseen benefit. I was able to unhurriedly reflect on the remarkable development of biochemistry since I wrote my last edition. I had the leisure to think and dream and plan this book. Best of all, I was able to share my thoughts with my son Daniel, who was then a senior majoring in human biology, and gain from his insights.

Alexander Glazer, Richard Gumport, Roger Koeppe, James Rawn, Carl Rhodes, and Peter Rubenstein read the entire manuscript. I have benefited greatly from their scholarly and perceptive criticism. Steve Block, Daniel Branton, Simone Brutlag, Carolyn Cohen, Jeffrey Critchfield, Peter Cullis, Russell Doolittle, Marilyn Farquhar, Robert Fletterick, Robert Fox, Michel Goldberg, Jack Griffith, James Hageman, Stephen Harrison, Brian Holl, Leroy Hood, Horace Jackson, Gunther Kohlhaw, Arthur Kornberg, Roger Kornberg, Stephen Kron, Michael Levitt, Bo Malmström, Lynne Mercer, Albert Mildvan, Jeremy Nathans, Christopher Newgard, Marion O'Leary, George Palade, Peter Parham, Frederic Richards, Ed Rock, Gottfried Schatz, Gray Scrimgeour, Paul Sigler, Jeffrey Sklar, James Spudich, Thomas Steitz, Nigel Unwin, Ronald Vale, William Wickner, and Robley Williams also gave valuable advice and help.

The contributors of many striking and informative illustrations are acknowledged in the figure legends. I am also indebted to crystallographers who have deposited the atomic coordinates of their solved structures in the Protein Data Bank, a valuable resource maintained by Brookhaven National Laboratory. Many new figures depicting molecular structure were generated on our departmental molecular graphics computer facility. David Austen and William Hurja helped me use this excellent system.

I was able to concentrate on the writing of this book because my office was in the capable hands of Joanne Tisch. She played a critical role in preparing the manuscript and reading the proofs. Her sensitivity, intelligence, and good spirits lightened my load. The Medline bibliographic retrieval system of the National Library of Medicine greatly facilitated my search of the literature. The staff of the Lane Medical Library and Falconer Biology Library of Stanford University were most helpful in locating books and references.

Andrew Kudlacik edited this manuscript with a fine sense of style and meaning. Mike Suh skillfully integrated word and picture in the design of each page. Susan Moran kept a watchful and discerning eye over many thousands of pages of manuscript, figures, and proofs. I also wish to thank Tom Cardamone and Shirley Baty for many outstanding drawings.

I am grateful to my family for their sustained support of this endeavor, which was more arduous than anticipated. My sons, Michael and Daniel, now embarked on their own careers, cheered me from afar. My wife, Andrea, provided criticism, advice, and encouragement in just the right proportions. I have been nurtured, too, by many who have reached out to express their warmth and interest in continuing this dialogue of biochemistry. I feel very fortunate and privileged to partake in this process at such a wonderful time.

Lubert Stryer

DECEMBER 1987

PREFACE

to the Second Edition

The pace of discovery in biochemistry has been exceptionally rapid during the past several years. This progress has greatly enriched our understanding of the molecular basis of life and has opened many new areas of inquiry. The sequencing of DNA, the construction and cloning of new combinations of genes, the elucidation of metabolic control mechanisms, and the unraveling of membrane transport and transduction processes are some of the highlights of recent research. One of my aims in this edition has been to weave new knowledge into the fabric of the text. I have sought to enhance the book's teaching effectiveness by centering the exposition of new material on common themes wherever feasible and by citing recurring motifs. I have also tried to convey a sense of the intellectual power and beauty of the discipline of biochemistry.

I am indebted to Thomas Emery, Henry Epstein, Alexander Glazer, Roger Kornberg, Robert Martin, and Jeffrey Sklar for their counsel, criticism, and encouragement in the preparation of this edition. Robert Baldwin, Charles Cantor, Richard Caprioli, David Eisenberg, Alan Fersht, Robert Fletterick, Herbert Friedmann, Horace Jackson, Richard Keynes, Sung-Hou Kim, Aaron Klug, Arthur Kornberg, Daniel Koshland, Jr., Samuel Latt, Vincent Marchesi, David Nelson, Garth Nicolson, Vernon Oi, Robert Renthal, Carl Rhodes, Frederic Richards, James Rothman, Peter Sargent, Howard Schachman, Joachim Seelig, Eric Shooter, Elizabeth Simons, James Spudich, Theodore Steck, Thomas Steitz, Judit C.-P. Stenn, Robert Trelstad, Christopher Walsh, Simon Whitney, and Bernhard Witkop also gave valuable advice.

Patricia Mittelstadt edited both editions of this text. I deeply appreciate her critical and sustained contributions. I am indebted to Donna

Salmon for her outstanding drawings. David Clayton, David Dressler, John Heuser, Lynne Mercer, Kenneth Miller, George Palade, Nigel Unwin, and Robley Williams generously provided many fine electron micrographs. Betty Hogan typed the manuscript and played an indispensable role in its preparation. Cary Leiden and Karen Marzotto carefully read the proofs. I also wish to thank Michael Graves for his excellent photographic work.

My wife, Andrea, and my sons, Michael and Daniel, have cheerfully allowed this text to become a member of the family. I am deeply grateful to them for their patience and buoyancy. Andrea provided much advice on style and design, as she did for the first edition.

I have been heartened by the many letters that I have received from readers of the first edition. Their comments and criticisms have enlightened, stimulated, and encouraged me. I look forward to a continuing dialogue with readers in the years ahead.

Lubert Stryer

AUGUST 1980

PREFACE

to the First Edition

This book is an outgrowth of my teaching of biochemistry to undergraduates, graduate students, and medical students at Yale and Stanford. My aim is to provide an introduction to the principles of biochemistry that gives the reader a command of its concepts and language. I also seek to give an appreciation of the process of discovery in biochemistry. My exposition of the principles of biochemistry is organized around several major themes:

1. Conformation—exemplified by the relationship between the three-dimensional structure of proteins and their biological activity
2. Generation and storage of metabolic energy
3. Biosynthesis of macromolecular precursors
4. Information—storage, transmission, and expression of genetic information
5. Molecular physiology—interaction of information, conformation, and metabolism in physiological processes

The elucidation of the three-dimensional structure of proteins, nucleic acids, and other biomolecules has contributed much in recent years to our understanding of the molecular basis of life. I have emphasized this aspect of biochemistry by making extensive use of molecular models to give a vivid picture of architecture and dynamics at the molecular level. Another stimulating and heartening aspect of contemporary biochemistry is its increasing interaction with medicine. I have presented many examples of this interplay. Discussions of molecular diseases such as sickle-cell anemia and of the mechanism of action of drugs such as penicillin enrich the teaching of biochemistry. Finally, I have tried to define several challenging areas of inquiry in biochemistry today, such as the molecular basis of excitability.

In writing this book, I have benefitted greatly from the advice, criticism, and encouragement of many colleagues and students. Leroy Hood, Arthur Kornberg, Jeffrey Sklar, and William Wood gave me invaluable counsel on its overall structure. Richard Caprioli, David Cole, Alexander Glazer, Robert Lehman, and Peter Lengyel read much of the manuscript and made many very helpful suggestions. I am indebted to Frederic Richards for sharing his thoughts on macromolecular conformation and for extensive advice on how to depict three-dimensional structures. Deric Bownds, Thomas Broker, Jack Griffith, Hugh Huxley, and George Palade made available to me many striking electron micrographs. I am also very thankful for the advice and criticism that were given at various times in the preparation of this book by Richard Dickerson, David Eisenberg, Moises Eisenberg, Henry Epstein, Joseph Fruton, Michel Goldberg, James Grisolia, Richard Henderson, Harvey Himel, David Hogness, Dale Kaiser, Samuel Latt, Susan Lowey, Vincent Marchesi, Peter Moore, Allan Oseroff, Jordan Pober, Russell Ross, Edward Reich, Mark Smith, James Spudich, Joan Steitz, Thomas Steitz, and Alan Waggoner.

I am grateful to the Commonwealth Fund for a grant that enabled me to initiate the writing of this book. The interest and support of Robert Glaser, Terrance Keenan, and Quigg Newton came at a critical time. One of my aims in writing this book has been to achieve a close integration of word and picture and to illustrate chemical transformations and three-dimensional structures vividly. I am especially grateful to Donna Salmon, John Foster, and Jean Foster for their work on the drawings, diagrams, and graphs. Many individuals at Yale helped to bring this project to fruition. I particularly wish to thank Margaret Banton and Sharen Westin for typing the manuscript, William Pollard for photographing space-filling models, and Martha Scarf for generating the computer drawings of molecular structures on which many of the illustrations in this book are based. John Harrison and his staff at the Kline Science Library helped in many ways.

Much of this book was written in Aspen. I wish to thank the Aspen Center of Physics and the Given Institute of Pathobiology for their kind hospitality during several summers. I have warm memories of many stimulating discussions about biochemistry and molecular aspects of medicine that took place in the lovely garden of the Given Institute and while hiking in the surrounding wilderness areas. The concerts in Aspen were another source of delight, especially after an intensive day of writing.

I am deeply grateful to my wife, Andrea, and to my children, Michael and Daniel, for their encouragement, patience, and good spirit during the writing of this book. They have truly shared in its gestation, which was much longer than expected. Andrea offered advice on style and design and also called my attention to the remark of the thirteenth-century Chinese scholar Tai T'ung (*The Six Scripts: Principles of Chinese Writing*): "Were I to await perfection, my book would never be finished."

I welcome comments and criticisms from readers.

Lubert Stryer

OCTOBER 1974

BIOCHEMISTRY

On the facing page: Proteins and nucleic acids are the central molecules of life. A complex of DNA bound to the protein that replicates it is shown here. DNA polymerase is shown in blue, one strand of the double-helical DNA template in green, and the other in red. [Courtesy of Dr. Thomas Steitz.]

PART I

MOLECULAR DESIGN OF LIFE

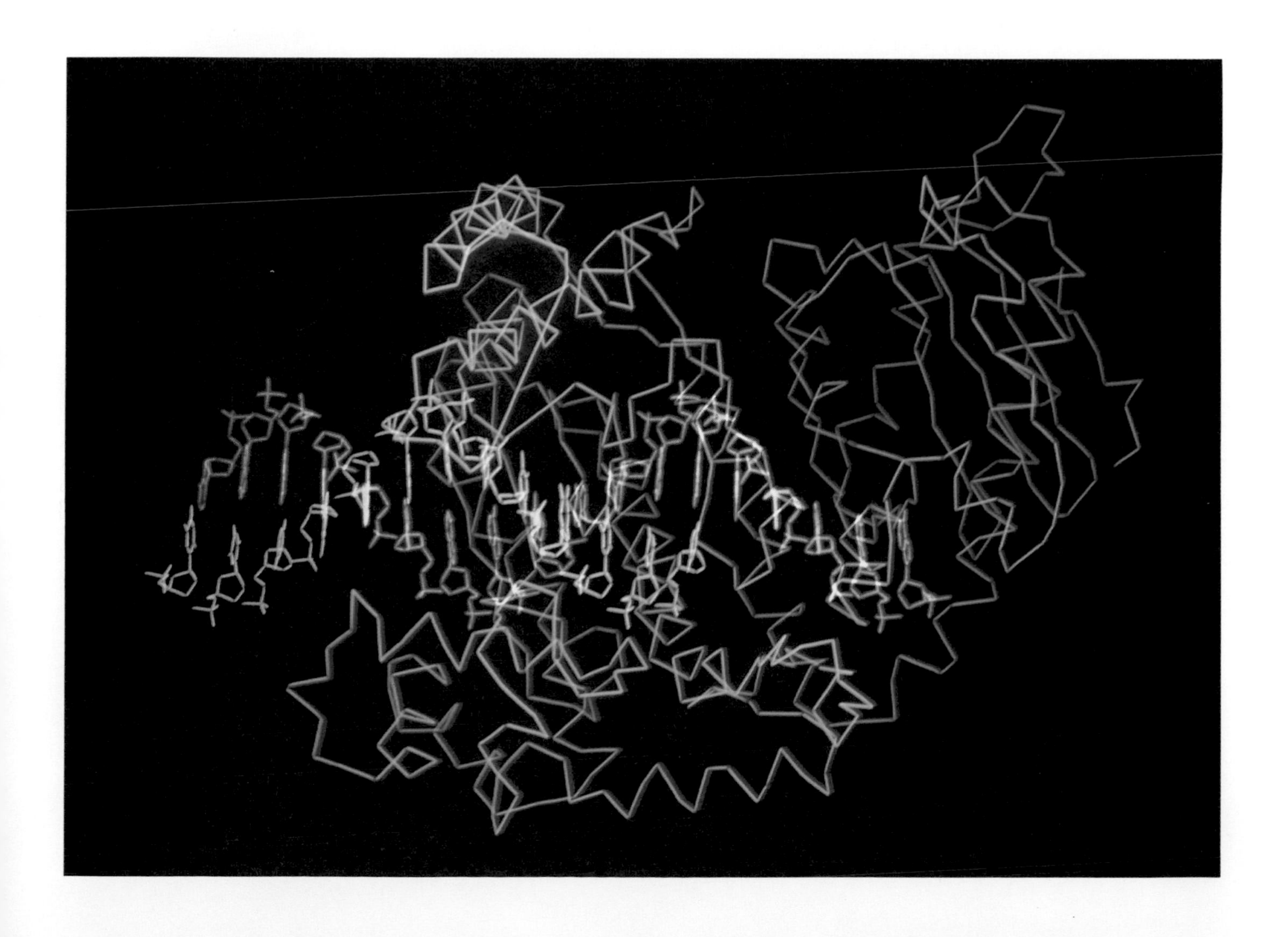

CHAPTER 1

Prelude

Biochemistry is the study of the molecular basis of life. There is much excitement and activity in biochemistry today for several reasons.

First, the chemical bases of many central processes are now understood. The discovery of the double-helical structure of deoxyribonucleic acid (DNA), the elucidation of the flow of information from gene to protein, the determination of the three-dimensional structure and mechanism of action of many protein molecules, the unraveling of central metabolic pathways and energy-conversion mechanisms, and the development of recombinant DNA technology are some of the outstanding achievements of biochemistry.

Second, it is now known that common molecular patterns and principles underlie the diverse expressions of life. Organisms as different as the bacterium *Escherichia coli* and human beings use the same building blocks to construct macromolecules. The flow of genetic information from DNA to ribonucleic acid (RNA) to protein is essentially the same in all organisms. Adenosine triphosphate (ATP), the universal currency of energy in biological systems, is generated in similar ways by all forms of life.

Third, biochemistry is profoundly influencing medicine. The molecular mechanisms of many diseases, such as sickle-cell anemia and numerous inborn errors of metabolism, have been elucidated. Assays for enzyme activity are indispensable in clinical diagnosis. For example, the levels of certain enzymes in serum reveal whether a patient has recently had a myocardial infarction. DNA probes are coming into play in the diagnosis of genetic disorders, infectious diseases, and cancers. Genetically engineered strains of bacteria containing recombinant DNA are producing valuable proteins such as insulin and growth hormone. Furthermore, biochemistry is a basis for the rational design of new drugs. Agriculture, too, is likely to benefit from recombinant DNA technology, which can produce designed changes in the genetic endowment of organisms.

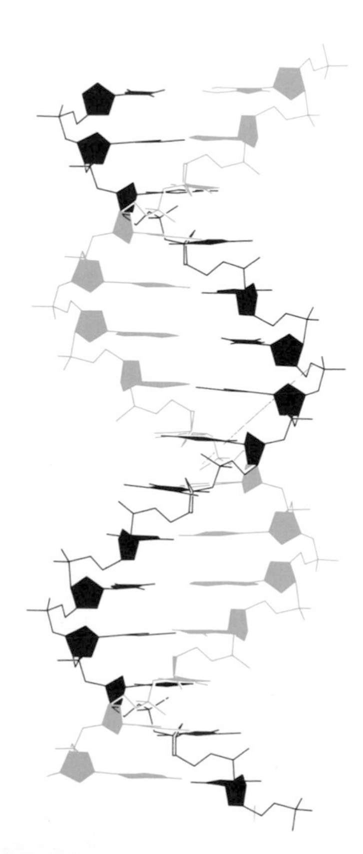

Figure 1-1
Model of the DNA double helix. The diameter of the helix is about 20 Å.

Fourth, the rapid development of powerful biochemical concepts and techniques in recent years has enabled investigators to tackle some of the most challenging and fundamental problems in biology and medicine. How does a fertilized egg give rise to cells as different as those in muscle, the brain, and the liver? How do cells find each other in forming a complex organ? How is the growth of cells controlled? What are the causes of cancer? What is the mechanism of memory? What is the molecular basis of schizophrenia?

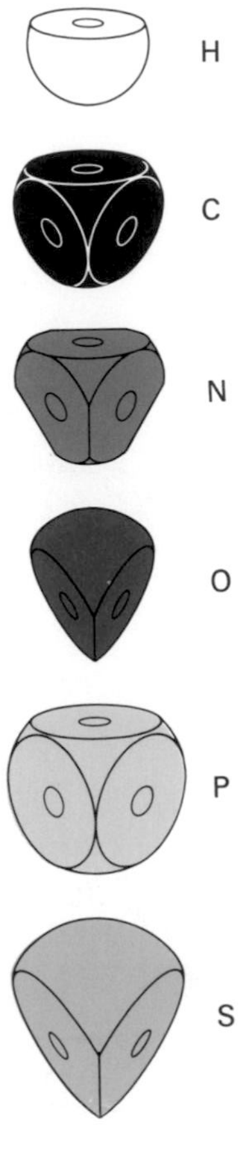

Figure 1-2
Space-filling models of hydrogen, carbon, nitrogen, oxygen, phosphorus, and sulfur atoms.

MOLECULAR MODELS DEPICT THREE-DIMENSIONAL STRUCTURE

The interplay between the three-dimensional structure of biomolecules and their biological function is the unifying motif of this book. Three types of atomic models will be used to depict molecular architecture: space-filling, ball-and-stick, and skeletal. The *space-filling models* are the most realistic. The size and configuration of an atom in a space-filling model are determined by its bonding properties and van der Waals radius (Figure 1-2). The colors of the model atoms are set by convention:

Hydrogen, white	Oxygen, red
Carbon, black	Phosphorus, yellow
Nitrogen, blue	Sulfur, yellow

Space-filling models of several simple molecules are illustrated in Figure 1-3.

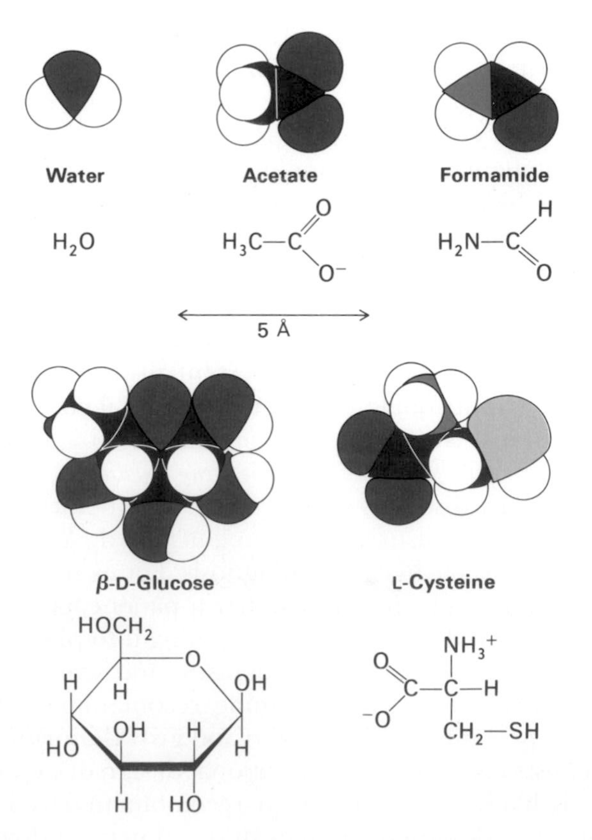

Figure 1-3
Space-filling models of water, acetate, formamide, glucose, and cysteine.

Ball-and-stick models are not as realistic as space-filling models because the atoms are depicted as spheres of radius smaller than the van der Waals radius. However, the bonding arrangement is easier to see because the bonds are explicitly represented by sticks. In an illustration, the taper of a stick tells whether the direction of the bond is in front of the plane of the page or behind it. More of a complex structure can be seen in a ball-and-stick model than in a space-filling model. An even simpler image is achieved with *skeletal models,* which show only the molecular framework. In these models, atoms are not shown explicitly. Rather, their positions are implied by the junctions and ends of bonds. Skeletal models are frequently used to depict large biological macromolecules, such as protein molecules having several thousand atoms. Space-filling, ball-and-stick, and skeletal models of ATP are compared in Figure 1-4.

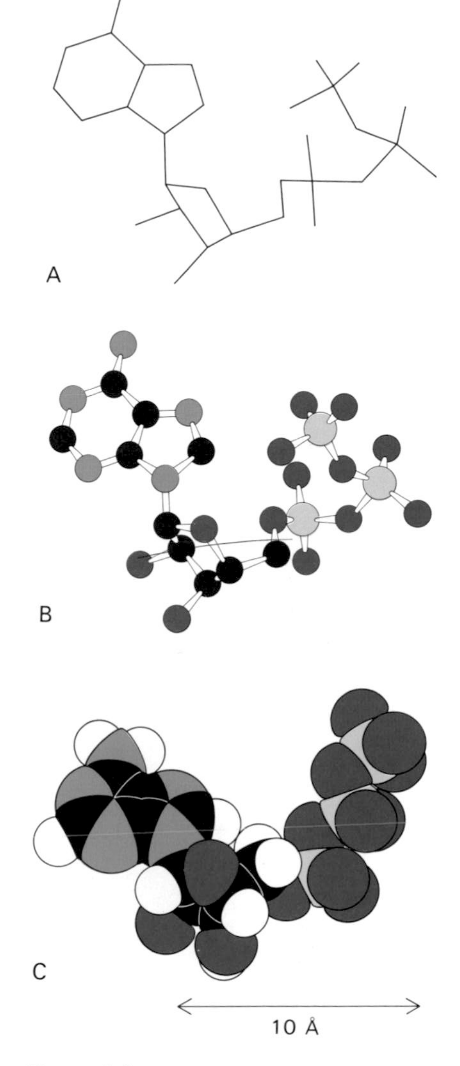

Figure 1-4
Comparison of (A) skeletal, (B) ball-and-stick, and (C) space-filling models of ATP. Hydrogen atoms are not shown in models A and B.

SPACE, TIME, AND ENERGY

In considering molecular structure, it is important to have a sense of scale (Figure 1-5). The angstrom (Å) unit, which is equal to 10^{-10} meter (m) or 0.1 nanometer (nm), is customarily used as the measure of length at the atomic level. The length of a C—C bond, for example, is 1.54 Å. Small biomolecules, such as sugars and amino acids, are typically several angstroms long. Biological macromolecules, such as proteins, are at least tenfold larger. For example, hemoglobin, the oxygen-carrying protein in red blood cells, has a diameter of 65 Å. Another tenfold increase in size brings us to assemblies of macromolecules. Ribosomes, the protein-synthesizing machinery of the cell, have diameters of about 300 Å. The range from 100 Å (10 nm) to 1000 Å (100 nm) also encompasses most viruses. Cells are typically a hundred times as large, in the range of micrometers (μm). For example, a red blood cell is 7 μm (7×10^4 Å) long. It is important to note that the limit of resolution of the light microscope is about 2000 Å (0.2 μm), which corresponds to the size of many subcellular organelles. Mitochondria, the major generators of ATP in aerobic cells, can just be resolved by the light microscope. Most of our knowledge of biological structure in the range from 1 Å (0.1 nm) to 10^4 Å (1 μm) has come from electron microscopy and x-ray diffraction.

The molecules of life are constantly in flux. Chemical reactions in biological systems are catalyzed by enzymes, which typically convert

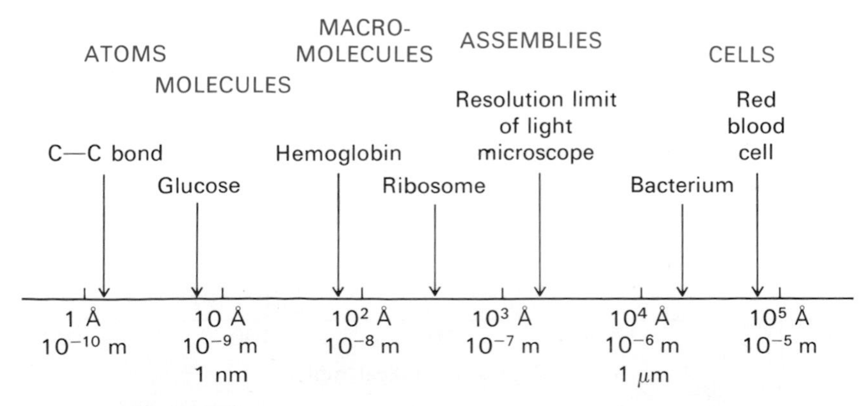

Figure 1-5
Dimensions of some biomolecules, assemblies, and cells.

substrate into product in milliseconds (ms, 10^{-3} s). Some enzymes act even more rapidly, in times as short as a few microseconds (μs, 10^{-6} s). Many conformational changes in biological macromolecules also are rapid. For example, the unwinding of the DNA double helix, which is essential for its replication and expression, is a microsecond event. The rotation of one domain of a protein with respect to another can take place in only nanoseconds (ns, 10^{-9} s). Many noncovalent interactions between groups in macromolecules are formed and broken in nanoseconds. Even more rapid processes can be probed with very short light pulses from lasers. It is remarkable that the primary event in vision—a change in structure of the light-absorbing group—occurs within a few picoseconds (ps, 10^{-12} s) after the absorption of a photon (Figure 1-6). From such brevity to the scale of evolutionary time, biological systems span a broad range. Life on earth arose some 3.5×10^9 years ago, or 1.1×10^{17} s ago.

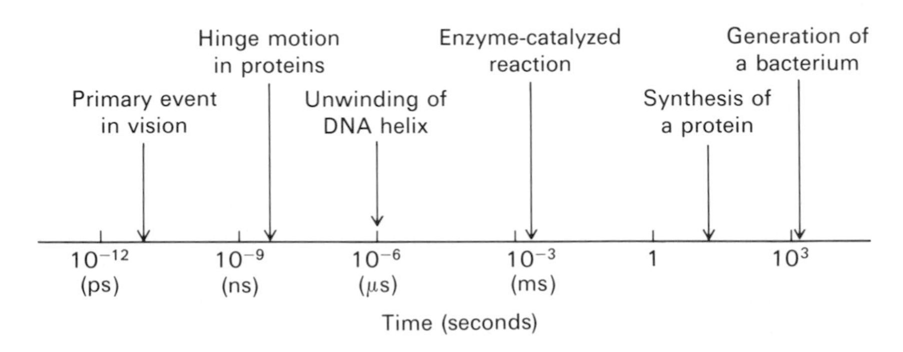

Figure 1-6
Typical rates of some processes in biological systems.

We shall be concerned with energy changes in molecular events (Figure 1-7). The ultimate source of energy for life is the sun. The energy of a green photon, for example, is 57 kilocalories per mole (kcal/mol). ATP, the universal currency of energy, has a usable energy content of about 12 kcal/mol. In contrast, the average energy of each vibrational degree of freedom in a molecule is much smaller, 0.6 kcal/mol at 25°C. This amount of energy is much less than that needed to dissociate covalent bonds (e.g., 83 kcal/mol for a C—C bond). Hence, the covalent framework of biomolecules is stable in the absence of enzymes and inputs of energy. On the other hand, noncovalent bonds in biological systems typically have an energy of only a few kilocalories per mole, so that thermal energy is enough to make and break them. An alternative unit of energy is the joule, which is equal to 0.239 calorie.

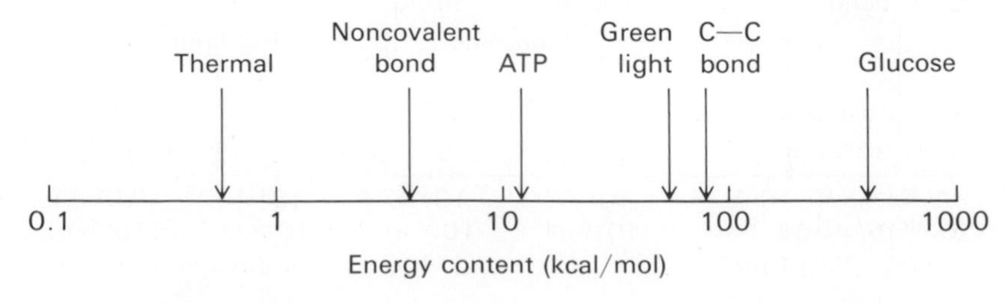

Figure 1-7
Some biologically important energies.

REVERSIBLE INTERACTIONS OF BIOMOLECULES ARE MEDIATED BY THREE KINDS OF NONCOVALENT BONDS

Reversible molecular interactions are at the heart of the dance of life. Weak, noncovalent forces play key roles in the faithful replication of DNA, the folding of proteins into intricate three-dimensional forms, the specific recognition of substrates by enzymes, and the detection of signal molecules. Indeed, all biological structures and processes depend on the interplay of noncovalent interactions as well as covalent ones. The three fundamental noncovalent bonds are *electrostatic bonds, hydrogen bonds,* and *van der Waals bonds*. They differ in geometry, strength, and specificity. Furthermore, these bonds are profoundly affected in different ways by the presence of water. Let us consider the characteristics of each:

1. *Electrostatic bonds*. A charged group on a substrate can attract an oppositely charged group on an enzyme. The force of such an *electrostatic attraction* is given by Coulomb's law:

$$F = \frac{q_1 q_2}{r^2 D}$$

in which q_1 and q_2 are the charges of the two groups, r is the distance between them, and D is the dielectric constant of the medium. The attraction is strongest in a vacuum (where D is 1) and is weakest in a medium such as water (where D is 80). This kind of attraction is also called an ionic bond, salt linkage, salt bridge, or ion pair. The distance between oppositely charged groups in an optimal electrostatic attraction is 2.8 Å.

$-CH_2-C(=O)O^-$

Negatively charged group of a substrate

$H_3\overset{+}{N}-CH_2-$

Positively charged group of an enzyme

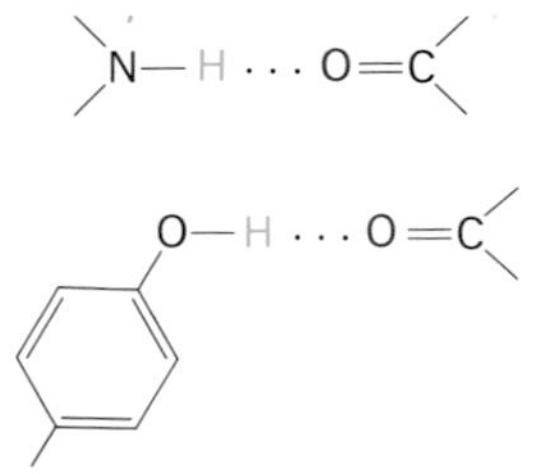

2. *Hydrogen bonds* can be formed between uncharged molecules as well as charged ones. In a hydrogen bond, *a hydrogen atom is shared by two other atoms*. The atom to which the hydrogen is more tightly linked is called the hydrogen donor, whereas the other atom is the hydrogen acceptor. The acceptor has a partial negative charge that attracts the hydrogen atom. In fact, a hydrogen bond can be considered an intermediate in the transfer of a proton from an acid to a base. It is reminiscent of a ménage à trois.

Hydrogen donor, Hydrogen acceptor: —O—H ··· N— (2.88 Å)

Hydrogen donor, Hydrogen acceptor: —N—H ··· O— (3.04 Å)

The donor in a hydrogen bond in biological systems is an oxygen or nitrogen atom that has a covalently attached hydrogen atom. The acceptor is either oxygen or nitrogen. The kinds of hydrogen bonds formed and their bond lengths are given in Table 1-1. The bond energies range from about 3 to 7 kcal/mol. Hydrogen bonds are stronger than van der Waals bonds but much weaker than covalent bonds. The length of a hydrogen bond is intermediate between that of a covalent

O—H ··· O

Strong hydrogen bond

O—H ··· O

Weak hydrogen bond

N—H ··· O=C

O—H ··· O=C

Table 1-1
Typical hydrogen-bond lengths

Bond	*Length* (Å)
O—H···O	2.70
O—H···O$^-$	2.63
O—H···N	2.88
N—H···O	3.04
N$^+$—H···O	2.93
N—H···N	3.10

bond and a van der Waals bond. *An important feature of hydrogen bonds is that they are highly directional.* The strongest hydrogen bonds are those in which the donor, hydrogen, and acceptor atoms are colinear. The α-helix, a recurring motif in proteins, is stabilized by hydrogen bonds between amide (—NH) and carbonyl (—CO) groups (Figure 1-8). Another example of the importance of hydrogen bonding is the DNA double helix, which is held together by hydrogen bonds between bases on opposite strands (Figure 1-1).

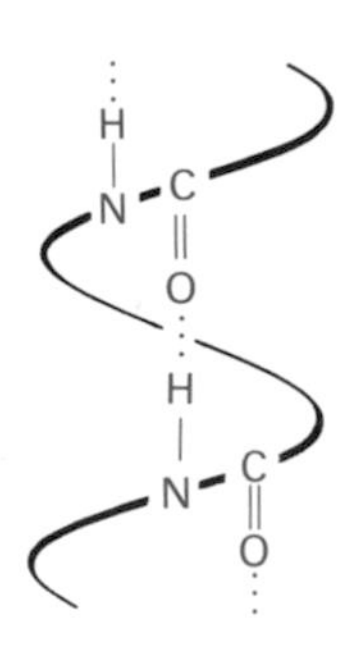

Figure 1-8
Schematic diagram of hydrogen bonding between an amide and a carbonyl group in an α-helix of a protein.

3. *Van der Waals bonds,* a nonspecific attractive force, come into play when any two atoms are 3 to 4 Å apart. Though weaker and less specific than electrostatic and hydrogen bonds, van der Waals bonds are no less important in biological systems. The basis of a van der Waals bond is that the distribution of electronic charge around an atom changes with time. At any instant, the charge distribution is not perfectly symmetric. This transient asymmetry in the electronic charge around an atom encourages a similar asymmetry in the electron distribution around its neighboring atoms. The resulting attraction between a pair of atoms increases as they come closer, until they are separated by the van der Waals *contact distance* (Figure 1-9). At a shorter distance, very strong

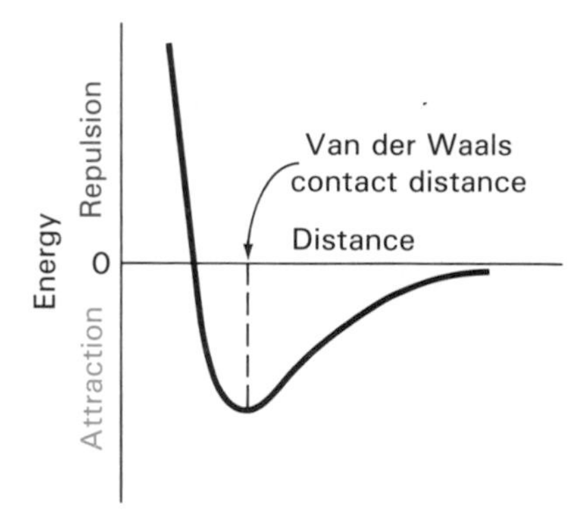

Figure 1-9
Energy of a van der Waals interaction as a function of the distance between two atoms.

Table 1-2
Van der Waals contact radii of atoms (Å)

Atom	*Radius*
H	1.2
C	2.0
N	1.5
O	1.4
S	1.85
P	1.9

repulsive forces become dominant because the outer electron clouds overlap. The contact distance between an oxygen and carbon atom, for example, is 3.4 Å, which is obtained by adding 1.4 and 2.0 Å, the contact radii (Table 1-2) of the O and C atoms.

The van der Waals bond energy of a pair of atoms is about 1 kcal/mol. It is considerably weaker than a hydrogen or electrostatic bond, which is in the range of 3 to 7 kcal/mol. A single van der Waals bond counts for very little because its strength is only a little more than the average thermal energy of molecules at room temperature (0.6 kcal/mol). Furthermore, the van der Waals force fades rapidly when the distance between a pair of atoms becomes even 1 Å greater than their contact distance. It becomes significant only when numerous atoms in one of a pair of molecules can simultaneously come close to many atoms of the other. This can happen only if the shapes of the molecules match. In other words, effective van der Waals interactions depend on *steric complementarity*. Though there is virtually no specificity in a single van der Waals interaction, *specificity arises when there is an opportunity to make a large number of van der Waals bonds simultaneously*. Repulsions between atoms closer than the van der Waals contact distance are as important as attractions for establishing specificity.

THE BIOLOGICALLY IMPORTANT PROPERTIES OF WATER ARE ITS POLARITY AND COHESIVENESS

Water profoundly influences all molecular interactions in biological systems. Two properties of water are especially important in this regard:

1. *Water is a polar molecule.* The shape of the molecule is triangular, not linear, and so there is an asymmetrical distribution of charge. The oxygen nucleus draws electrons away from the hydrogen nuclei, which leaves the region around those nuclei with a net positive charge. The water molecule is thus an electrically polar structure.

− + H O 0.99 Å 105° H

2. *Water molecules have a high affinity for each other.* A positively charged region in one water molecule tends to orient itself toward a negatively charged region in one of its neighbors. Ice has a highly regular crystalline structure in which all potential hydrogen bonds are made (Figure 1-10). Liquid water has a partly ordered structure in which hydrogen-bonded clusters of molecules are continually forming and breaking up. Each molecule is hydrogen bonded to an average of 3.4 neighbors in liquid water, compared with 4 in ice. *Water is highly cohesive.*

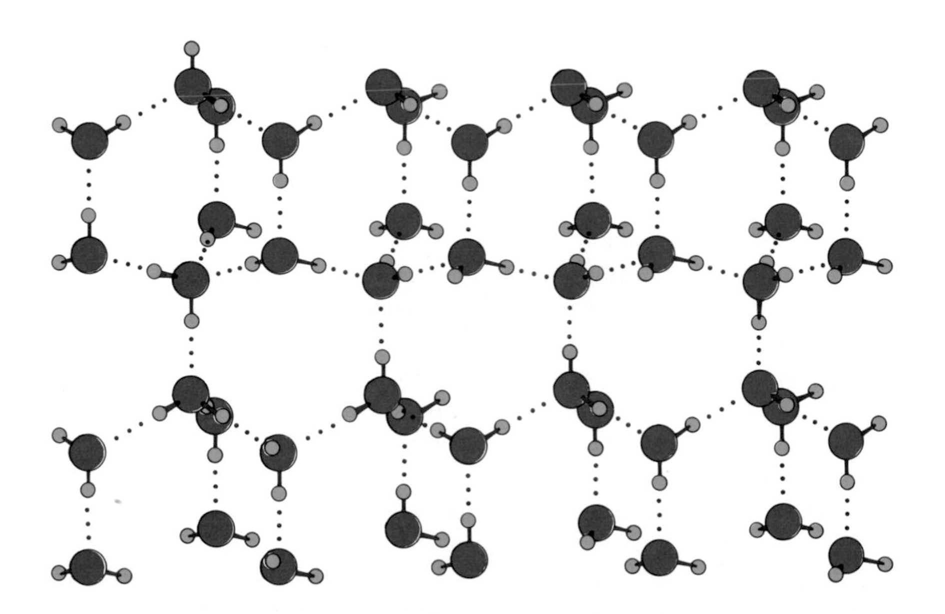

Figure 1-10
Structure of a form of ice. [After L. Pauling and P. Pauling. *Chemistry* (W. H. Freeman, 1975), p. 289.]

WATER SOLVATES POLAR MOLECULES AND WEAKENS IONIC AND HYDROGEN BONDS

The polarity and hydrogen-bonding capability of water make it a highly interacting molecule. Water is an excellent solvent for polar molecules. The reason is that water greatly weakens electrostatic forces and hydrogen bonding between polar molecules by competing for their attractions. For example, consider the effect of water on hydrogen bonding between a carbonyl and an amide group (Figure 1-11). The hydrogen atoms of water can replace the amide hydrogen group as hydrogen-bond donors, and the oxygen atom of water can replace the carbonyl oxygen as the acceptor. Hence, a strong hydrogen bond between a CO and an NH group forms only if water is excluded.

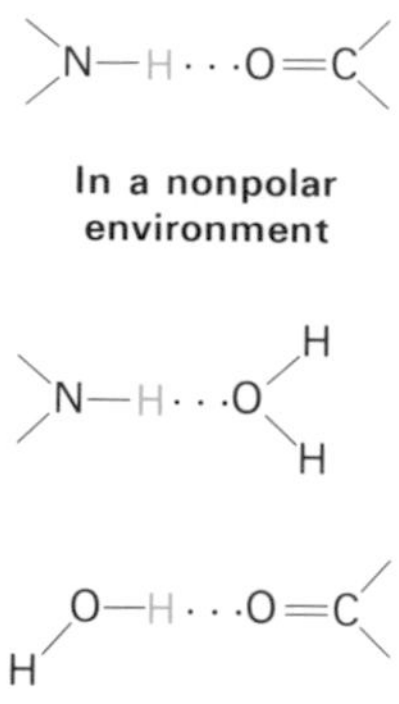

Water diminishes the strength of electrostatic attractions by a factor of 80, the dielectric constant of water, compared with the same interactions in a vacuum. Water has an unusually high dielectric constant (Table 1-3) because of its polarity and capacity to form oriented solvent shells around ions (Figure 1-12). These oriented solvent shells produce electric fields of their own, which oppose the fields produced by the ions. Consequently, electrostatic attractions between ions are markedly weakened by the presence of water.

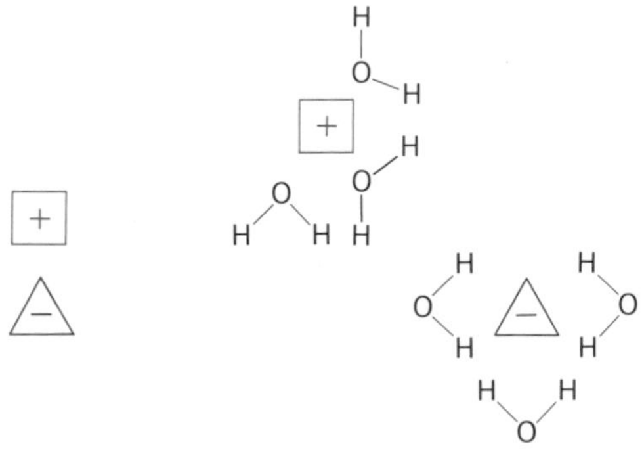

N—H···O=C

In a nonpolar environment

N—H···O(H)H

O(H)—H···O=C

In water

Figure 1-11
Water competes for hydrogen bonds.

Electrostatic interaction in a nonpolar environment

Water surrounds the charged groups and attenuates their interaction

Figure 1-12
Water attenuates electrostatic attractions between charged groups.

Table 1-3
Dielectric constants of some solvents at 20°C

Substance	*Dielectric constant*
Hexane	1.9
Benzene	2.3
Diethyl ether	4.3
Chloroform	5.1
Acetone	21.4
Ethanol	24
Methanol	33
Water	80
Hydrogen cyanide	116

The existence of life on earth depends critically on the capacity of water to dissolve a remarkable array of polar molecules that serve as fuels, building blocks, catalysts, and information carriers. High concentrations of these molecules can coexist in water, where they are free to diffuse and find each other. However, the excellence of water as a solvent poses a problem, for it also weakens interactions between polar molecules. Biological systems have solved this problem by creating water-free microenvironments where polar interactions have maximal strength. We shall see many examples of the critical importance of these specially constructed niches in protein molecules.

HYDROPHOBIC ATTRACTIONS: NONPOLAR GROUPS TEND TO ASSOCIATE IN WATER

The sight of dispersed oil droplets coming together in water to form a single large oil drop is a familiar one. An analogous process occurs at the atomic level: *nonpolar molecules or groups tend to cluster together in water*. These associations are called *hydrophobic attractions*. In a figurative sense, water tends to squeeze nonpolar molecules together.

Let us examine the basis of hydrophobic attractions, which are a major driving force in the folding of macromolecules, the binding of substrates to enzymes, and the formation of membranes that define the boundaries of cells and their internal compartments. Consider the introduction of a single nonpolar molecule, such as hexane, into some water. A cavity in the water is created, which temporarily disrupts some hydrogen bonds between water molecules. The displaced water molecules then reorient themselves to form a maximum number of new hydrogen bonds. This is accomplished at a price: the number of ways of forming hydrogen bonds in the cage of water around the hexane molecule is much fewer than in pure water. The water molecules around the hexane molecule are much more ordered than elsewhere in the solution. Now consider the arrangement of two hexane molecules in water. Do they sit in two small cavities (Figure 1-13A) or in a single larger one (Figure 1-13B)? The experimental fact is that the two hexane molecules come together and occupy a single large cavity. This association releases some of the more ordered water molecules around the separated hexanes. In fact, the basis of a hydrophobic attraction is this enhanced freedom of released water molecules. *Nonpolar solute molecules are driven together in water not primarily because they have a high affinity for each other but because water bonds strongly to itself.*

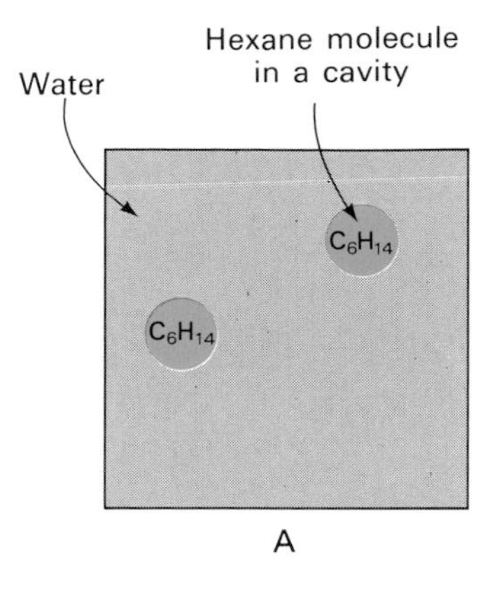

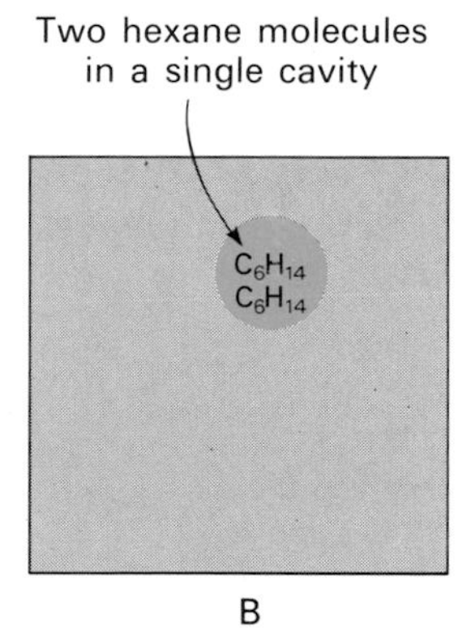

Figure 1-13
A schematic representation of two molecules of hexane in a small volume of water: (A) the hexane molecules occupy different cavities in the water structure, or (B) they occupy the same cavity, which is energetically more favored.

DESIGN OF THIS BOOK

This book has six parts, each having a major theme.

I: Molecular Design of Life

II: Protein Conformation, Dynamics, and Function

III: Generation and Storage of Metabolic Energy

IV: Biosynthesis of Macromolecular Precursors

V: Genetic Information

VI: Molecular Physiology

Part I is an overview of the central molecules of life—DNA, RNA, and proteins—and their interplay. We begin with proteins, which are unique in being able to recognize and bind a remarkably diverse array of molecules. Proteins determine the pattern of chemical transformations in biological systems by catalyzing nearly all of the necessary chemical reactions. We then turn to DNA, the repository of genetic information in all cells. The discovery of the DNA double helix led immediately to an understanding of how DNA replicates. The following chapter deals with the flow of genetic information from DNA to RNA to protein. The first step, called transcription, is the synthesis of RNA, and the second, called translation, is the synthesis of proteins according to instructions given by templates of messenger RNA. The genetic code, which specifies the relation between the sequence of four kinds of bases in DNA and RNA and the twenty kinds of amino acids in proteins, is beautiful in its simplicity. Three bases constitute a codon, the unit that specifies an amino acid. Translation is carried out by the coordinated interplay of more than a hundred kinds of protein and RNA molecules in an organized assembly called the ribosome. Experimental methods for exploring proteins and genes are also presented in Part I. Recombinant DNA technology is introduced here and some examples of its power and generality in analyzing and altering both genes and proteins are given.

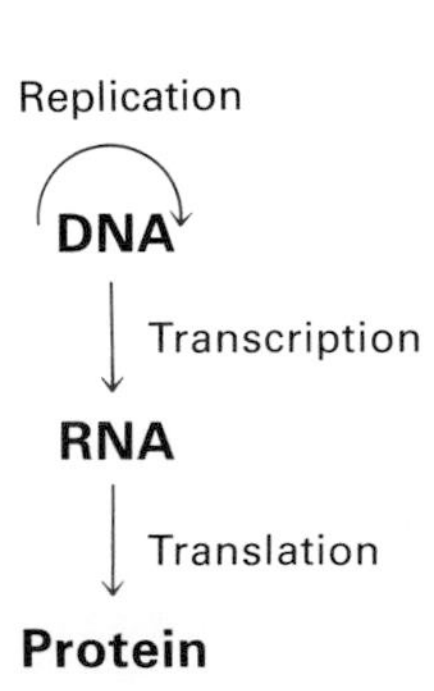

Figure 1-14
Flow of genetic information.

The interplay of three-dimensional structure and biological activity as exemplified by proteins is the major theme of Part II. The structure and function of myoglobin and hemoglobin, the oxygen-carrying proteins in vertebrates, are presented in detail because these proteins illustrate many general principles. Hemoglobin is especially interesting because its binding of oxygen is regulated by specific molecules in its environment. The molecular pathology of hemoglobin, particularly sickle-cell anemia, is also presented. We then turn to enzymes and consider how they recognize substrates and enhance reaction rates by factors of a million or more. The enzymes lysozyme, carboxypeptidase A, and chymotrypsin are examined in detail because the study of them has elucidated many general principles of catalysis. The regulation of enzymatic activity by specific control proteins and other signal molecules is considered next. Rather different facets of the theme of conformation emerge in the chapter on collagen and elastin, two connective-tissue proteins. The final chapter in Part II is an introduction to biological membranes, which are organized assemblies of lipids and proteins. Membranes serve to create compartments and control the flow of matter and information between them.

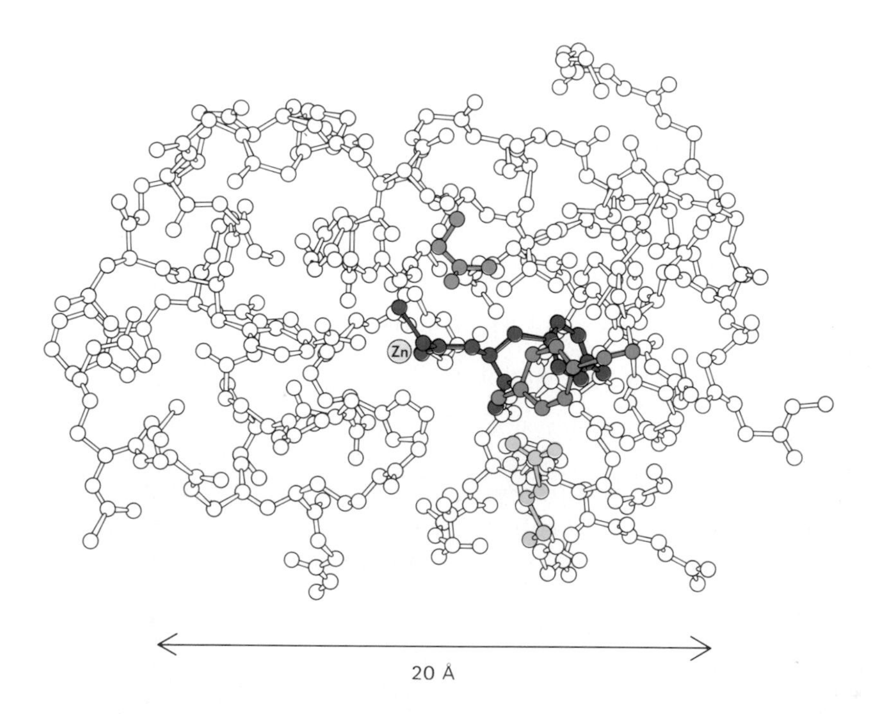

Figure 1-15
Structure of an enzyme-substrate complex. Glycyltyrosine (shown in red) is bound to carboxypeptidase A, a digestive enzyme. Only a quarter of the enzyme is shown. [After W. N. Lipscomb. *Proc. Robert A. Welch Found. Conf. Chem. Res.* 15(1971):141.]

Part III deals with the generation and storage of metabolic energy. First, the overall strategy of metabolism is presented. Cells convert energy from fuel molecules into ATP. In turn, ATP drives most energy-requiring processes in cells. In addition, reducing power in the form of nicotinamide adenine dinucleotide phosphate (NADPH) is generated for use in biosyntheses. The metabolic pathways that carry out these reactions are then presented in detail. For example, the generation of ATP from glucose requires a sequence of three series of reactions—glycolysis, the citric acid cycle, and oxidative phosphorylation. The last two are also common to the generation of ATP from the oxidation of fats and some amino acids, the other major fuels. We see here an illustration of molecular economy. Two storage forms of fuel molecules, glycogen and triacylglycerols (neutral fats), are also discussed in Part III. The concluding topic of this part of the book is photosynthesis, in which the primary event is the light-activated transfer of an electron from one substance to another against a chemical potential gradient. As in oxidative phosphorylation, electron flow leads to the pumping of protons across a membrane, which in turn drives the synthesis of ATP. In essence, life is powered by proton batteries that are ultimately energized by the sun.

Part IV deals with the biosynthesis of macromolecular precursors, starting with the synthesis of membrane lipids and steroids. The pathway for the synthesis of cholesterol, a 27-carbon steroid, is of particular interest because all of its carbon atoms come from a 2-carbon precursor. The reactions leading to the synthesis of selected amino acids and the heme group are then discussed. The control mechanisms in these pathways are of general significance. The biosynthesis of nucleotides, the activated precursors of DNA and RNA, is then considered. The final chapter in this part deals with the integration of metabolism. How are energy-yielding and energy-consuming reactions coordinated to meet the needs of an organism?

The transmission and expression of genetic information constitute the central theme of Part V. The genetic role and structure of DNA were introduced in Part I, as was the flow of genetic information. We now resume our consideration of this theme, enriched with a knowledge of proteins and metabolic transformations. The mechanism of DNA replication and DNA repair are discussed first. An intriguing aspect of DNA replication is its very high accuracy. The processes of genetic recombination and transposition, which produce new combinations of DNA, are then presented. We turn next to transcription and to the processing of nascent transcripts to form functional RNA mole-

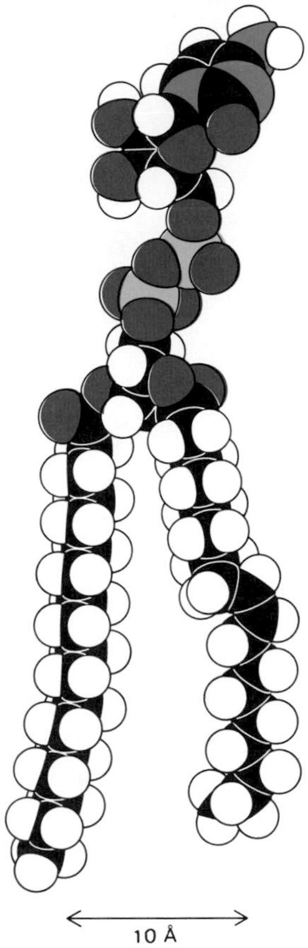

10 Å

Figure 1-16
Model of CDP-diacylglycerol, an activated intermediate in the synthesis of some membrane lipids.

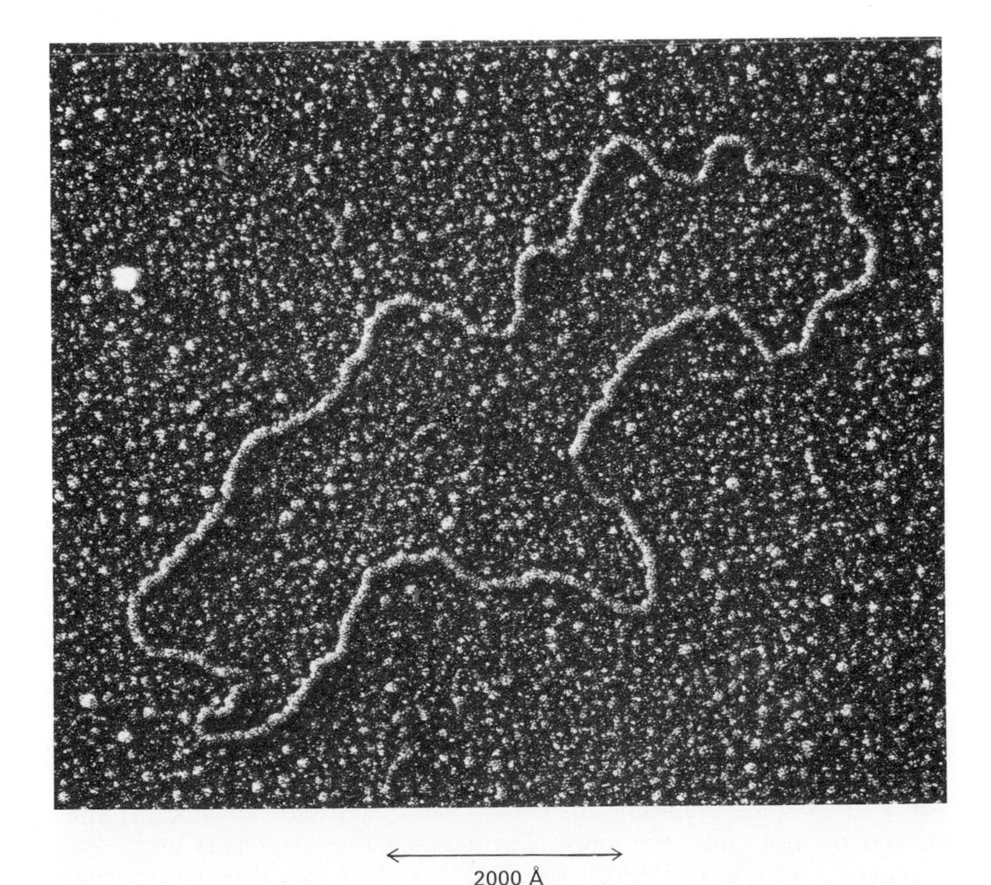

Figure 1-17
Electron micrograph of a DNA molecule. [Courtesy of Dr. Thomas Broker.]

cules. The mechanism of protein synthesis, in which tRNAs, mRNAs, and ribosomes interact, comes next. We then consider how proteins are specifically targeted to many different destinations. The next chapter deals with the control of gene expression in bacteria. The focus here is on the lactose and tryptophan operons of *E. coli*, which are now understood in detail. This is followed by a discussion of the regulation of gene expression in higher organisms—molecules controlling the development of multicellular organisms are now being identified. Virus multiplication and assembly are considered next. Viral assembly exemplifies some general principles of how biological macromolecules form highly ordered structures from a few kinds of building blocks. Viruses have also provided insight into the molecular basis of cancer by revealing the existence of oncogenes, which are altered forms of genes that control cell growth.

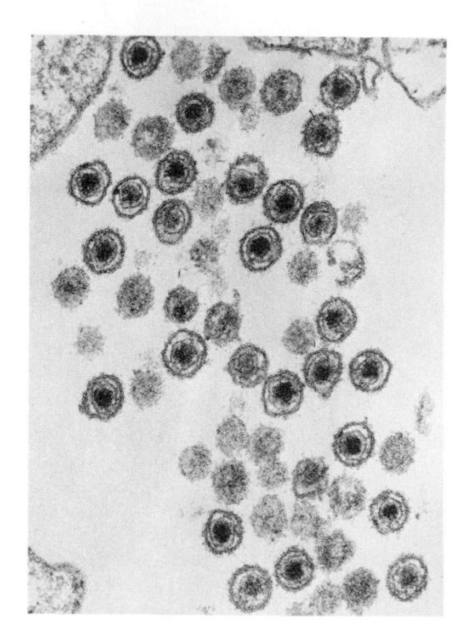

Figure 1-18
Electron micrograph of Rous sarcoma virus. This RNA virus can produce cancer in susceptible hosts. [Courtesy of Dr. Samuel Dales.]

Part VI, entitled "Molecular Physiology," is a transition from biochemistry to physiology. Many of the concepts that were developed earlier in this book are used here, because physiology involves the interplay of genetic information, conformation, and metabolism. We start with the molecular basis of the immune response. How does an organism detect a foreign substance? The next chapter deals with the problem of how the energy of chemical bonds is transformed into coordinated motion—myosin and actin, the major proteins in muscle, have a contractile role in most cells of higher organisms. The transport of ions, such as Na^+, K^+, and Ca^{2+}, and molecules is then considered. Molecular pumps in membranes control the transport of these ions to generate gradients that are at the heart of excitability. We then turn to the molecular basis of the action of hormones and growth factors. Families of receptors and signal-coupling proteins are being discovered, and recurring motifs of signal transduction are becoming evident. The final chapter deals with sensory processes and considers such questions as: How do bacteria detect nutrients in their environment and move toward them? How are action potentials propagated by nerve cells and transmitted across synapses? How is a retinal rod cell triggered by a single photon?

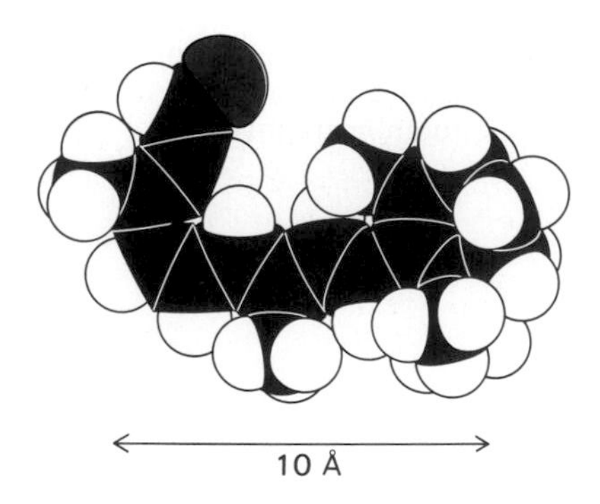

Figure 1-19
Model of 11-*cis*-retinal, the light-absorbing group in rhodopsin. The isomerization of this chromophore by light is the first event in visual excitation.

One of the most satisfying features of biochemistry is that it continually enriches our understanding of biological processes at all levels of organization.

CHAPTER 2

Protein Structure and Function

Proteins play crucial roles in virtually all biological processes. Their significance and the remarkable scope of their activity are exemplified in the following functions:

Protein—
A word coined by Jöns J. Berzelius in 1838 to emphasize the importance of this class of molecules. Derived from the Greek word *proteios*, which means "of the first rank."

1. *Enzymatic catalysis*. Nearly all chemical reactions in biological systems are catalyzed by specific macromolecules called enzymes. Some of these reactions, such as the hydration of carbon dioxide, are quite simple. Others, such as the replication of an entire chromosome, are highly intricate. Enzymes exhibit enormous catalytic power. They usually increase reaction rates by at least a millionfold. Indeed, chemical transformations in vivo rarely proceed at perceptible rates in the absence of enzymes. Several thousand enzymes have been characterized, and many of them have been crystallized. The striking fact is that nearly all known enzymes are proteins. Thus, proteins play the unique role of determining the pattern of chemical transformations in biological systems.

2. *Transport and storage*. Many small molecules and ions are transported by specific proteins. For example, hemoglobin transports oxygen in erythrocytes, whereas myoglobin, a related protein, transports oxygen in muscle. Iron is carried in the plasma of blood by transferrin and is stored in the liver as a complex with ferritin, a different protein.

Figure 2-1
Photomicrograph of a crystal of hexokinase, a key enzyme in the utilization of glucose. [Courtesy of Dr. Thomas Steitz and Dr. Mark Yeager.]

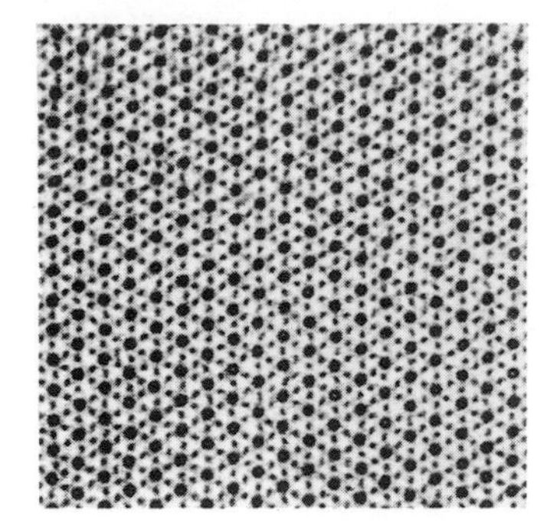

Figure 2-2
Electron micrograph of a cross section of insect flight muscle showing a hexagonal array of two kinds of protein filaments. [Courtesy of Dr. Michael Reedy.]

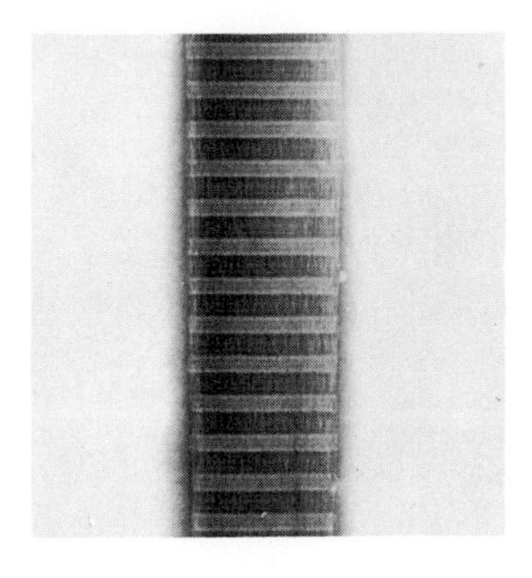

Figure 2-3
Electron micrograph of a fiber of collagen. [Courtesy of Dr. Jerome Gross and Dr. Romaine Bruns.]

3. *Coordinated motion.* Proteins are the major component of muscle. Muscle contraction is accomplished by the sliding motion of two kinds of protein filaments. On the microscopic scale, such coordinated motions as the movement of chromosomes in mitosis and the propulsion of sperm by their flagella also are produced by contractile assemblies consisting of proteins.

4. *Mechanical support.* The high tensile strength of skin and bone is due to the presence of collagen, a fibrous protein.

5. *Immune protection.* Antibodies are highly specific proteins that recognize and combine with such foreign substances as viruses, bacteria, and cells from other organisms. Proteins thus play a vital role in distinguishing between self and nonself.

6. *Generation and transmission of nerve impulses.* The response of nerve cells to specific stimuli is mediated by receptor proteins. For example, rhodopsin is the photoreceptor protein in retinal rod cells. Receptor proteins that can be triggered by specific small molecules, such as acetylcholine, are responsible for transmitting nerve impulses at synapses—that is, at junctions between nerve cells.

7. *Control of growth and differentiation.* Controlled sequential expression of genetic information is essential for the orderly growth and differentiation of cells. Only a small fraction of the genome of a cell is expressed at any one time. In bacteria, repressor proteins are important control elements that silence specific segments of the DNA of a cell. In higher organisms, growth and differentiation are controlled by growth factor proteins. For example, nerve growth factor guides the formation of neural networks. The activities of different cells in multicellular organisms are coordinated by hormones. Many of them, such as insulin and thyroid-stimulating hormone, are proteins. Indeed, proteins serve in all cells as sensors that control the flow of energy and matter.

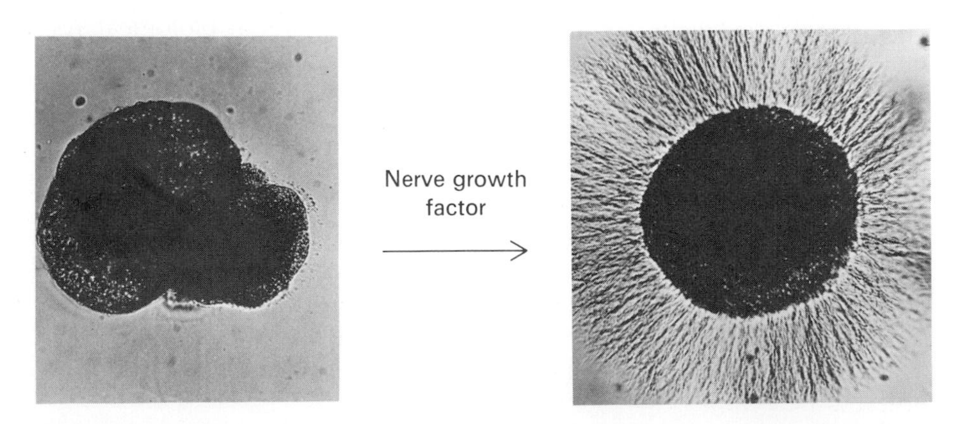

Figure 2-4
Photomicrograph of a ganglion showing the proliferation of nerves after addition of nerve growth factor, a complex of proteins. [Courtesy of Dr. Eric Shooter.]

PROTEINS ARE BUILT FROM A REPERTOIRE OF TWENTY AMINO ACIDS

Amino acids are the basic structural units of proteins. An α-amino acid consists of an amino group, a carboxyl group, a hydrogen atom, and a

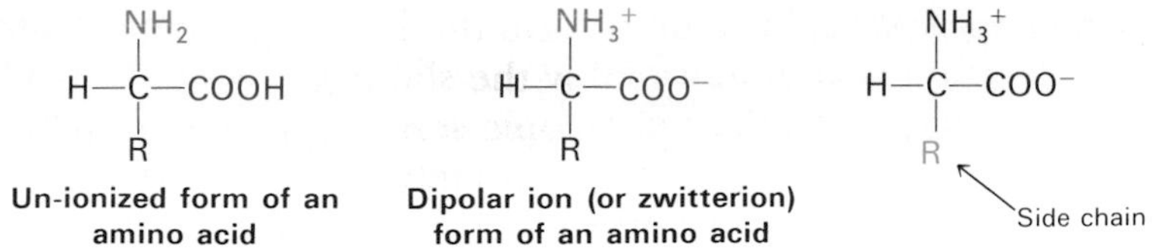

$$H-\underset{R}{\overset{NH_2}{C}}-COOH \qquad H-\underset{R}{\overset{NH_3^+}{C}}-COO^- \qquad H-\underset{R}{\overset{NH_3^+}{C}}-COO^-$$

Un-ionized form of an amino acid — **Dipolar ion (or zwitterion) form of an amino acid** — Side chain

Figure 2-5
Structure of the un-ionized and zwitterion forms of an α-amino acid.

distinctive R group bonded to a carbon atom, which is called the α-carbon because it is adjacent to the carboxyl (acidic) group (Figure 2-5). An R group is referred to as a *side chain* for reasons that will be evident shortly.

Amino acids in solution at neutral pH are predominantly *dipolar ions* (or *zwitterions*) rather than un-ionized molecules. In the dipolar form of an amino acid, the amino group is protonated ($—NH_3^+$) and the carboxyl group is dissociated ($—COO^-$). The ionization state of an amino acid varies with pH (Figure 2-6). In acid solution (e.g., pH 1), the carboxyl group is un-ionized (—COOH) and the amino group is ionized ($—NH_3^+$). In alkaline solution (e.g., pH 11), the carboxyl group is ionized ($—COO^-$) and the amino group is un-ionized ($—NH_2$). For glycine, the pK of the carboxyl group is 2.3 and that of the amino group is 9.6. In other words, the midpoint of the first ionization is at pH 2.3, and that of the second is at pH 9.6. For a review of acid-base concepts and pH, see the Appendix to this chapter.

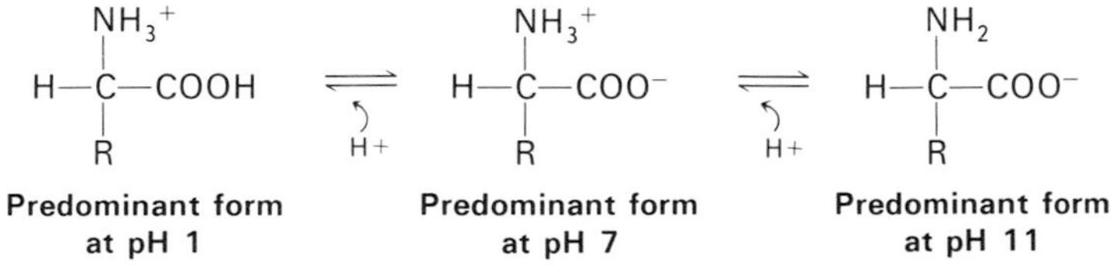

$$H-\underset{R}{\overset{NH_3^+}{C}}-COOH \underset{H^+}{\rightleftharpoons} H-\underset{R}{\overset{NH_3^+}{C}}-COO^- \underset{H^+}{\rightleftharpoons} H-\underset{R}{\overset{NH_2}{C}}-COO^-$$

Predominant form at pH 1 — **Predominant form at pH 7** — **Predominant form at pH 11**

Figure 2-6
Ionization states of an amino acid depend on pH.

The tetrahedral array of four different groups about the α-carbon atom confers optical activity on amino acids. The two mirror-image forms are called the L-isomer and the D-isomer (Figure 2-7). *Only L-amino acids are constituents of proteins.* Hence, the designation of the optical isomer will be omitted and the L-isomer implied in discussions of proteins herein, unless otherwise noted.

Twenty kinds of side chains varying in *size, shape, charge, hydrogen-bonding capacity,* and *chemical reactivity* are commonly found in proteins. Indeed, all proteins in all species, from bacteria to humans, are constructed from the same set of twenty amino acids. This fundamental alphabet of proteins is at least two billion years old. The remarkable range of functions mediated by proteins results from the diversity and versatility of these twenty kinds of building blocks. We shall explore ways in which this alphabet is used to create the intricate three-dimensional structures that enable proteins to carry out so many biological processes.

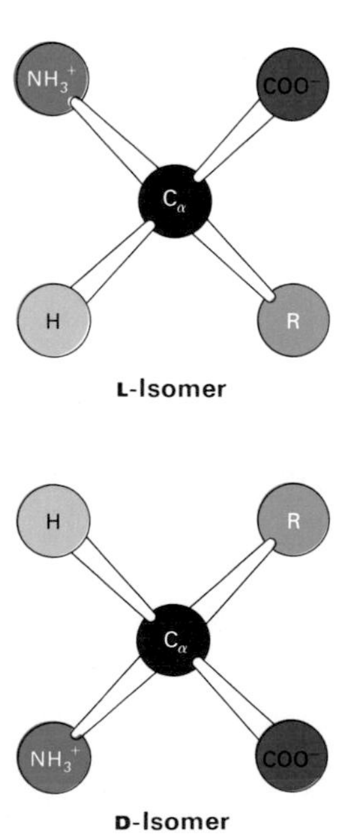

Figure 2-7
Absolute configurations of the L- and D-isomers of amino acids. R refers to the side chain.

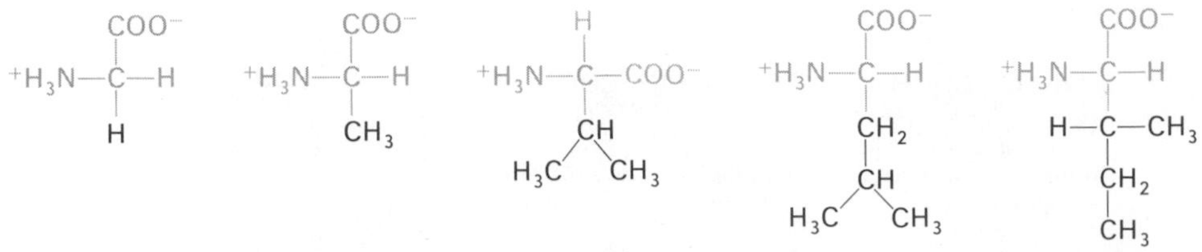

Glycine (Gly, G) | Alanine (Ala, A) | Valine (Val, V) | Leucine (Leu, L) | Isoleucine (Ile, I)

Figure 2-8
Amino acids having aliphatic side chains.

Let us look at this repertoire of amino acids. The simplest one is *glycine*, which has just a hydrogen atom as its side chain (Figure 2-8). *Alanine* comes next, with a methyl group as its side chain. Larger hydrocarbon side chains (three and four carbons long) are found in *valine*, *leucine*, and *isoleucine*. These larger aliphatic side chains are *hydrophobic*—that is, they have an aversion to water and like to cluster. As will be discussed later, the three-dimensional structure of water-soluble proteins is stabilized by the coming together of hydrophobic side chains to avoid contact with water. The different sizes and shapes of these hydrocarbon side chains (Figure 2-9) enable them to pack together to form compact structures with few holes.

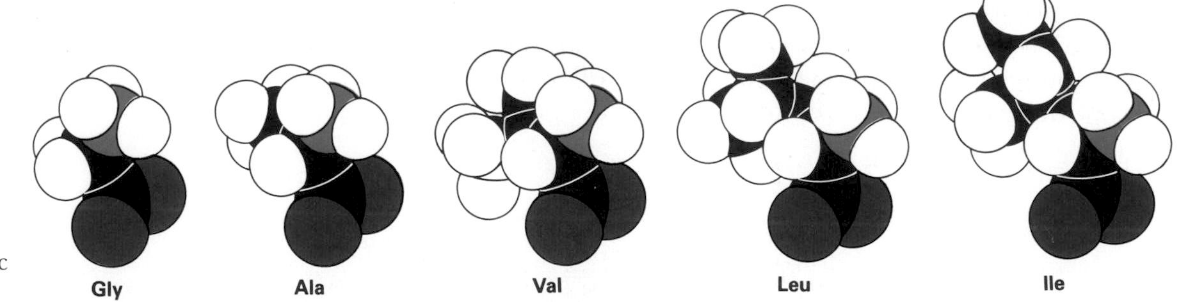

Figure 2-9
Models of aliphatic amino acids

Proline also has an aliphatic side chain but it differs from other members of the set of twenty in that its side chain is bonded to both the nitrogen and α-carbon atoms. The resulting cyclic structure (Figure 2-10) markedly influences protein architecture. Proline, often found in the bends of folded protein chains, is not averse to being exposed to water. Note that proline contains a secondary rather than a primary amino group, which makes it an *imino* acid.

Three amino acids with *aromatic side chains* are part of the fundamental repertoire (Figure 2-11). *Phenylalanine*, as its name indicates, contains a phenyl ring attached to a methylene (—CH_2—) group. *Tryptophan* has an indole ring joined to a methylene group; this side chain contains a nitrogen atom in addition to carbon and hydrogen atoms.

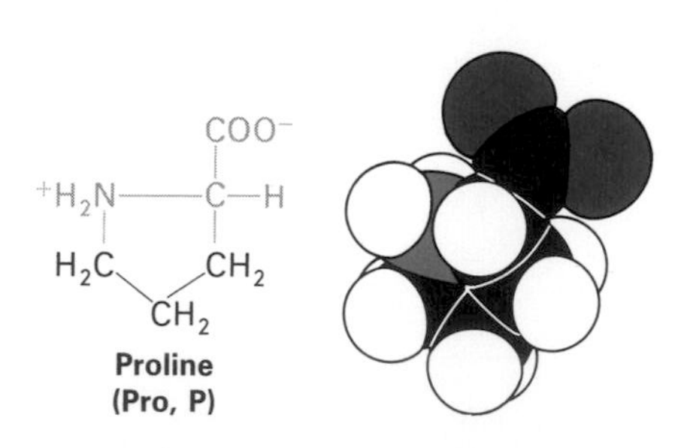

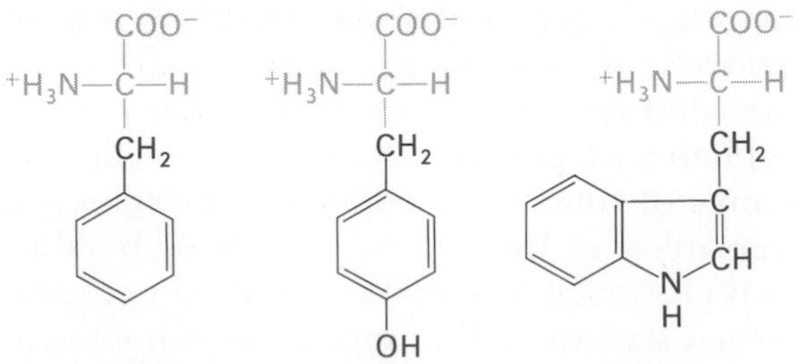

Figure 2-10
Proline differs from the other common amino acids in having a secondary amino group.

Figure 2-11
Phenylalanine, tyrosine, and tryptophan have aromatic side chains.

Phenylalanine (Phe, F) | Tyrosine (Tyr, Y) | Tryptophan (Trp, W)

Phe Tyr Trp

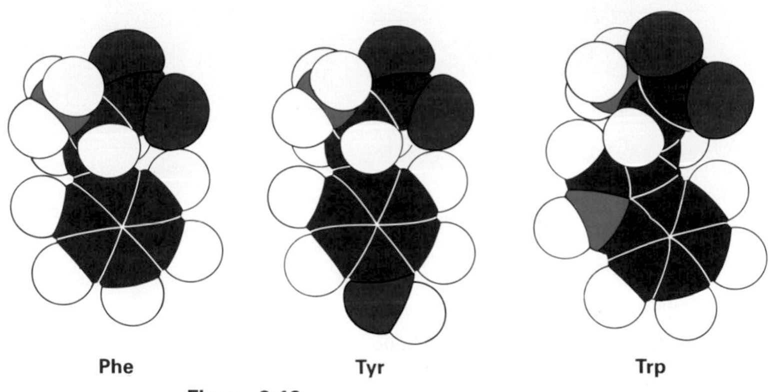

Figure 2-12
Models of the aromatic amino acids.

Phenylalanine and tryptophan are highly hydrophobic. The aromatic ring of *tyrosine* contains a hydroxyl group, which makes tyrosine less hydrophobic than phenylalanine. Moreover, this hydroxyl group is reactive, in contrast with the rather inert side chains of all the other amino acids discussed thus far. The aromatic rings of phenylalanine, tryptophan, and tyrosine contain delocalized pi-electron clouds that enable them to interact with other pi-systems and to transfer electrons.

A *sulfur atom* is present in the side chains of two amino acids (Figure 2-13). *Cysteine* contains a sulfhydryl group (—SH) and *methionine* contains a sulfur atom in a thioether linkage (—S—CH_3). Both of these sulfur-containing side chains are hydrophobic. The sulfhydryl group of cysteine is highly reactive. As will be discussed shortly, cysteine plays a special role in shaping some proteins by forming disulfide links.

COO⁻
+H₃N—C—H
CH₂
SH
COO⁻
+H₃N—C—H
CH₂
CH₂
S
CH₃

Cysteine **(Cys, C)** **Methionine** **(Met, M)**

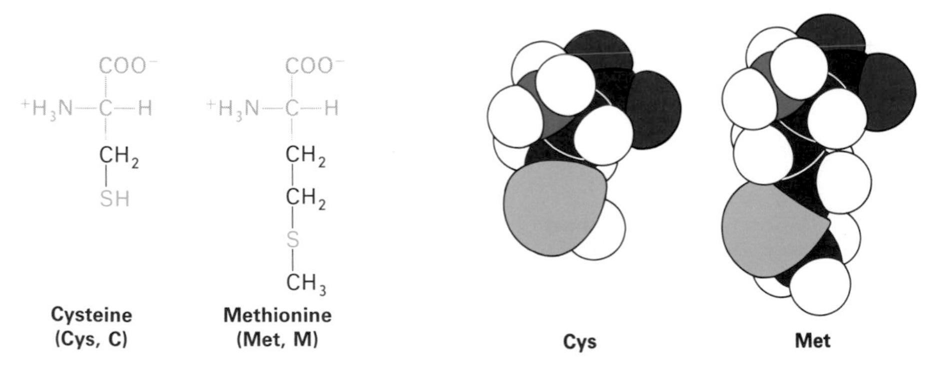

Figure 2-13
Cysteine and methionine have sulfur-containing side chains.

Cys **Met**

Figure 2-14
Models of cysteine and methionine.

Two amino acids, *serine* and *threonine*, contain aliphatic *hydroxyl groups* (Figure 2-15). Serine can be thought of as a hydroxylated version of alanine, and threonine as a hydroxylated version of valine. The hydroxyl groups on serine and threonine make them much more *hydrophilic* (water-loving) and *reactive* than alanine and valine. Threonine, like isoleucine, contains two centers of asymmetry. All other amino acids in the basic set of twenty, except for glycine, contain a single asymmetric center (the α-carbon atom). Glycine is unique in being optically inactive.

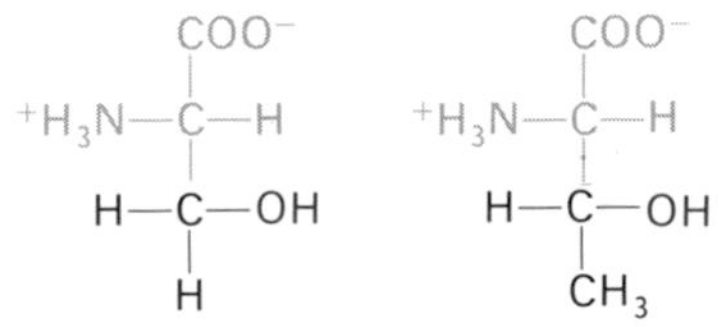

COO⁻
+H₃N—C—H
H—C—OH
H
COO⁻
+H₃N—C—H
H—C—OH
CH₃

Serine **(Ser, S)** **Threonine** **(Thr, T)**

Figure 2-15
Serine and threonine have aliphatic hydroxyl side chains.

We turn now to amino acids with very polar side chains, which render them highly *hydrophilic*. *Lysine* and *arginine* are *positively charged* at neutral pH. *Histidine* can be uncharged or positively charged, depending on its local environment. Indeed, histidine is often found in the active

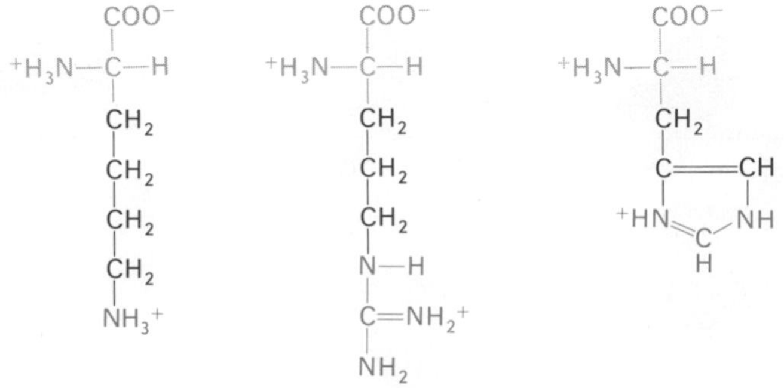

Lysine (Lys, K) | Arginine (Arg, R) | Histidine (His, H)

Figure 2-16
Lysine, arginine, and histidine have basic side chains.

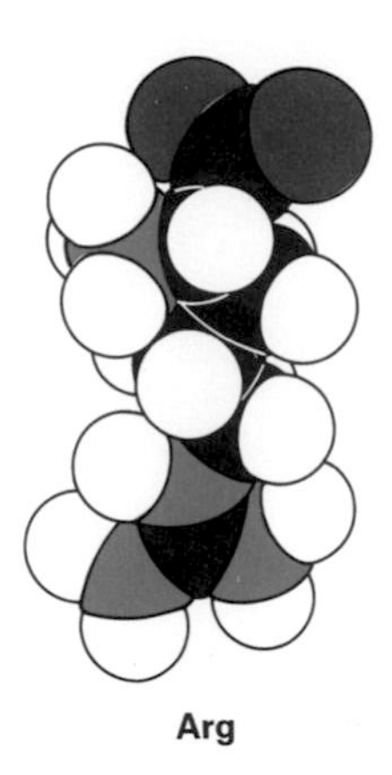

Arg

Figure 2-17
Model of arginine. The planar outer part of the side chain, consisting of three nitrogens bonded to a carbon atom, is called a guanidinium group.

sites of enzymes, where its imidazole ring can readily switch between these states to catalyze the making and breaking of bonds. These *basic amino acids* are depicted in Figure 2-16. The side chains of arginine and lysine are the longest ones in the set of twenty.

The repertoire of amino acids also contains two with *acidic side chains*, *aspartic acid* and *glutamic acid*. These amino acids are usually called *aspartate* and *glutamate* to emphasize that their side chains are nearly always negatively charged at physiological pH (Figure 2-18). Uncharged derivatives of glutamate and aspartate are *glutamine* and *asparagine*, which contain a terminal amide group in place of a carboxylate.

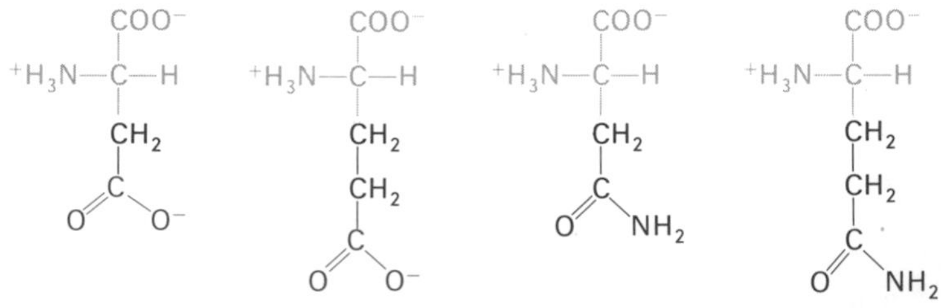

Aspartate (Asp, D) | Glutamate (Glu, E) | Asparagine (Asn, N) | Glutamine (Gln, Q)

Figure 2-18
Acidic amino acids (aspartate and glutamate) and their amide derivatives (asparagine and glutamine).

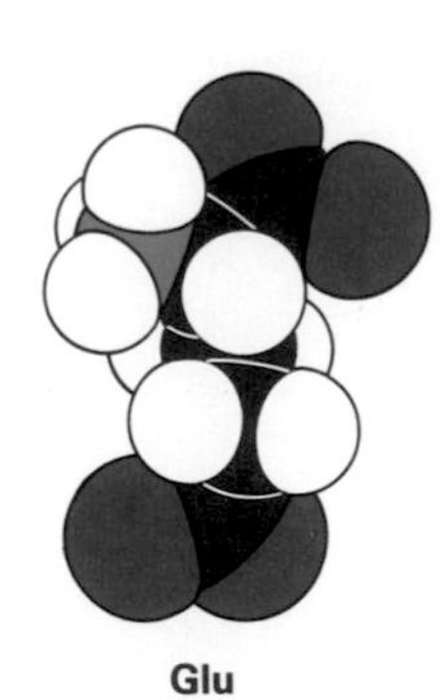

Glu

Figure 2-19
Model of glutamate.

Seven of the twenty amino acids have readily ionizable side chains. Equilibria and typical pK_a values for ionization of the side chains of arginine, lysine, histidine, aspartic and glutamic acids, cysteine, and tyrosine in proteins are given in Table 2-1. Two other groups in proteins, the terminal α-amino group and the terminal α-carboxyl group, can be ionized.

Amino acids are often designated by either a three-letter abbreviation or a one-letter symbol to facilitate concise communication (Table 2-2). The abbreviations for amino acids are the first three letters of their names, except for tryptophan (Trp), asparagine (Asn), glutamine (Gln), and isoleucine (Ile). The symbols for the small amino acids are the first letters of their names (e.g., G for glycine and L for leucine); the other symbols have been agreed upon by convention. These abbreviations and symbols are an integral part of the vocabulary of biochemists.

Table 2-1
pK values of ionizable groups in proteins

Group	Acid ⇌ base + H^+	Typical pK*
Terminal carboxyl	$-COOH \rightleftharpoons -COO^- + H^+$	3.1
Aspartic and glutamic acid	$-COOH \rightleftharpoons -COO^- + H^+$	4.4
Histidine	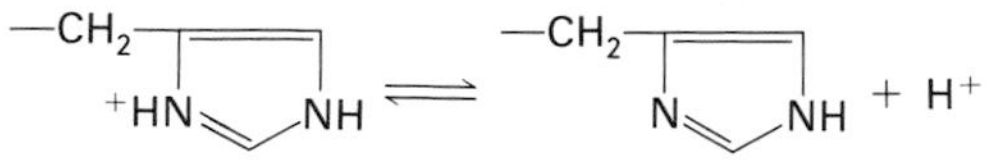 $-CH_2$–imidazolium (^+HN, NH) ⇌ $-CH_2$–imidazole (N, NH) $+ H^+$	6.5
Terminal amino	$-NH_3^+ \rightleftharpoons -NH_2 + H^+$	8.0
Cysteine	$-SH \rightleftharpoons -S^- + H^+$	8.5
Tyrosine	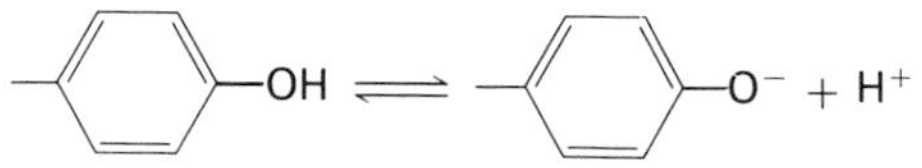 –(phenyl)–OH ⇌ –(phenyl)–O^- $+ H^+$	10.0
Lysine	$-NH_3^+ \rightleftharpoons -NH_2 + H^+$	10.0
Arginine	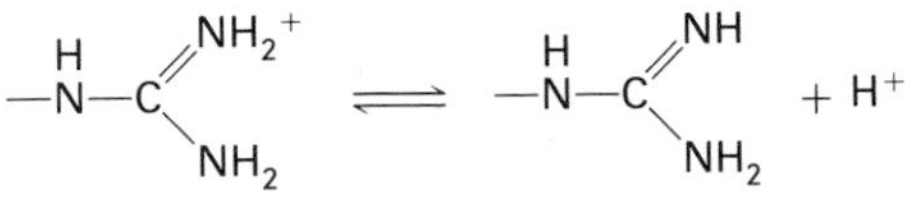 $-N(H)-C(=NH_2^+)NH_2 \rightleftharpoons -N(H)-C(=NH)NH_2 + H^+$	12.0

*pK values depend on temperature, ionic strength, and the microenvironment of the ionizable group.

Table 2-2
Abbreviations for amino acids

Amino acid	Three-letter abbreviation	One-letter symbol
Alanine	Ala	A
Arginine	Arg	R
Asparagine	Asn	N
Aspartic acid	Asp	D
Asparagine or aspartic acid	Asx	B
Cysteine	Cys	C
Glutamine	Gln	Q
Glutamic acid	Glu	E
Glutamine or glutamic acid	Glx	Z
Glycine	Gly	G
Histidine	His	H
Isoleucine	Ile	I
Leucine	Leu	L
Lysine	Lys	K
Methionine	Met	M
Phenylalanine	Phe	F
Proline	Pro	P
Serine	Ser	S
Threonine	Thr	T
Tryptophan	Trp	W
Tyrosine	Tyr	Y
Valine	Val	V

AMINO ACIDS ARE LINKED BY PEPTIDE BONDS TO FORM POLYPEPTIDE CHAINS

In proteins, the α-carboxyl group of one amino acid is joined to the α-amino group of another amino acid by a *peptide bond* (also called an amide bond). The formation of a dipeptide from two amino acids by loss of a water molecule is shown in Figure 2-20. The equilibrium of this reaction lies on the side of hydrolysis rather than synthesis. Hence, the biosynthesis of peptide bonds requires an input of free energy, whereas their hydrolysis is thermodynamically downhill.

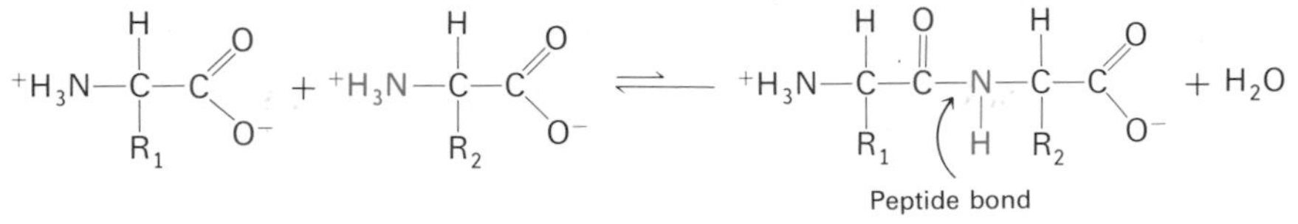

Peptide bond

Figure 2-20
Formation of a peptide bond.

Many amino acids are joined by peptide bonds to form a *polypeptide chain,* which is unbranched (Figure 2-21). An amino acid unit in a polypeptide is called a *residue.* A polypeptide chain has direction because its building blocks have different ends—namely, the α-amino and the α-carboxyl groups. By convention, *the amino end is taken to be the beginning of a polypeptide chain,* and so the sequence of amino acids in a polypeptide chain is written starting with the amino-terminal residue. Thus, in the tripeptide Ala-Gly-Trp (AGW), alanine is the amino-terminal residue and tryptophan is the carboxyl-terminal residue. Note that Trp-Gly-Ala (WGA) is a different tripeptide.

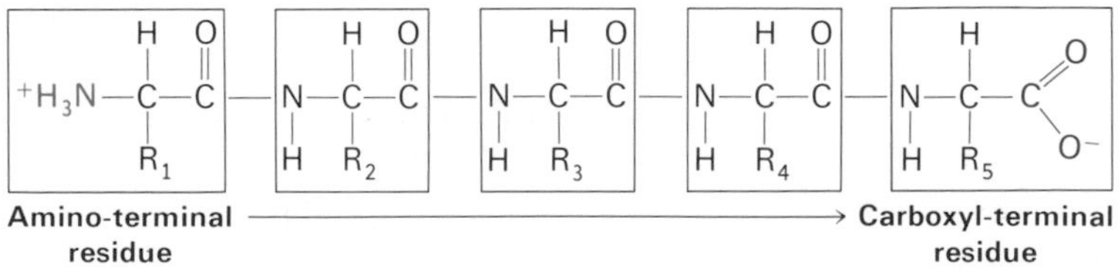

Amino-terminal residue → Carboxyl-terminal residue

Figure 2-21
A pentapeptide. The constituent amino acid residues are outlined. The chain starts at the amino end.

A polypeptide chain consists of a regularly repeating part, called the *main chain,* and a variable part, comprising the distinctive *side chains* (Figure 2-22). The main chain is sometimes termed the backbone. Most natural polypeptide chains contain between 50 and 2000 amino acid residues. The mean molecular weight of an amino acid residue is about 110, and so the molecular weights of most polypeptide chains are between 5500 and 220,000. We can also refer to the mass of a protein, which is expressed in units of daltons; one *dalton* is equal to one atomic mass unit. A protein with a molecular weight of 50,000 has a mass of 50,000 daltons, or 50 kd (kilodaltons).

Dalton—
A unit of mass very nearly equal to that of a hydrogen atom (precisely equal to 1.0000 on the atomic mass scale). Named after John Dalton (1766–1844), who developed the atomic theory of matter.

Kilodalton (kd)—
A unit of mass equal to 1000 daltons.

Figure 2-22
A polypeptide chain is made up of a regularly repeating *backbone* and distinctive *side chains* (R_1, R_2, R_3, shown in green).

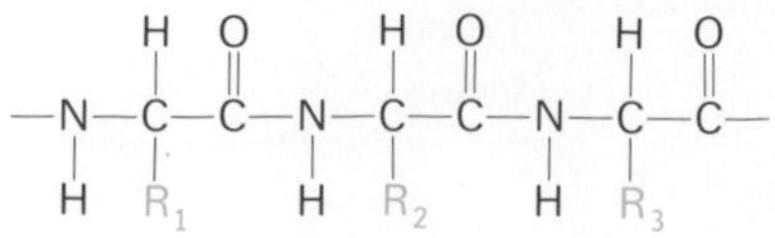

Some proteins contain *disulfide bonds.* These cross-links between chains or between parts of a chain are formed by the oxidation of cysteine residues. The resulting disulfide is called *cystine* (Figure 2-23). Intracellular proteins usually lack disulfide bonds, whereas extracellular proteins often contain several. Nonsulfur cross-links derived from lysine side chains are present in some proteins. For example, collagen fibers in connective tissue are strengthened in this way, as are fibrin blood clots.

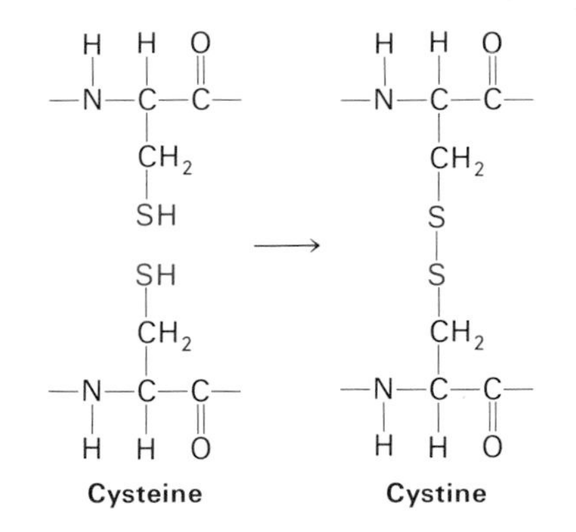

Figure 2-23
A disulfide bridge (—S—S—) is formed from the sulfhydryl groups (—SH) of two cysteine residues. The product is a *cystine* residue.

PROTEINS HAVE UNIQUE AMINO ACID SEQUENCES THAT ARE SPECIFIED BY GENES

In 1953, Frederick Sanger determined the amino acid sequence of insulin, a protein hormone (Figure 2-25). *This work is a landmark in biochemistry because it showed for the first time that a protein has a precisely defined amino acid sequence.* Moreover, it demonstrated that insulin consists only of L-amino acids in peptide linkage between α-amino and α-carboxyl groups. This accomplishment stimulated other scientists to carry out sequence studies of a wide variety of proteins. Indeed, the complete amino acid sequences of more than 2000 proteins are now known. The striking fact is that each protein has a unique, precisely defined amino acid sequence.

A series of incisive studies in the late 1950s and early 1960s revealed that the amino acid sequences of proteins are genetically determined. The sequence of nucleotides in DNA, the molecule of heredity, specifies a complementary sequence of nucleotides in RNA, which in turn specifies the amino acid sequence of a protein (p. 91). In particular, each of the twenty amino acids of the repertoire is encoded by one or more specific sequences of three nucleotides. Furthermore, proteins in all organisms are synthesized from their constituent amino acids by a common mechanism.

Amino acid sequences are important for several reasons. First, knowledge of the sequence of a protein is very helpful, indeed usually essential, in elucidating its mechanism of action (e.g., the catalytic mechanism of an enzyme). Second, analyses of relations between amino acid sequences and three-dimensional structures of proteins are uncovering the rules that govern the folding of polypeptide chains. The amino acid sequence is the link between the genetic message in DNA and the three-dimensional structure that performs a protein's biological function. Third, sequence determination is part of molecular pathology, an emerging area of medicine. Alterations in amino acid sequence can produce abnormal function and disease. Fatal disease, such as sickle-cell anemia, can result from a change in a single amino acid in

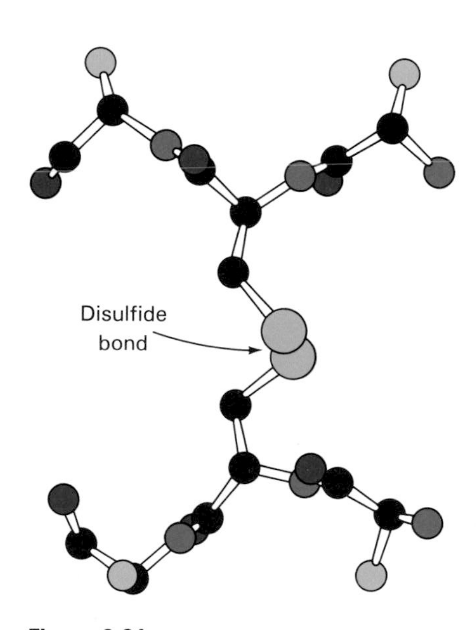

Figure 2-24
Model of a disulfide cross-link.

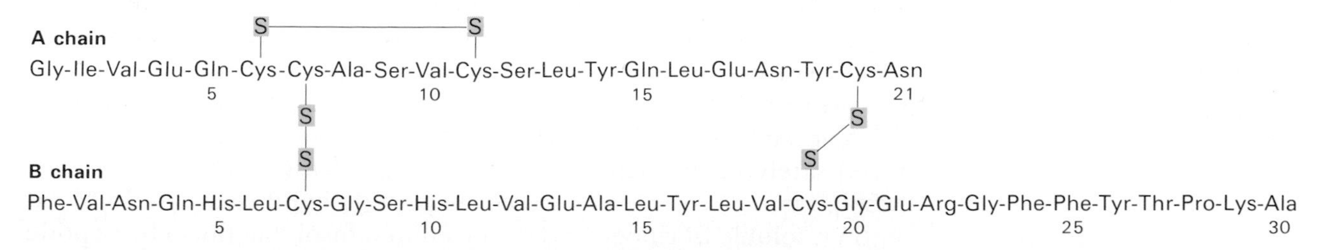

Figure 2-25
Amino acid sequence of bovine insulin.

a single protein. Fourth, the sequence of a protein reveals much about its evolutionary history. Proteins resemble one another in amino acid sequence only if they have a common ancestor. Consequently, molecular events in evolution can be traced from amino acid sequences; molecular paleontology is a flourishing area of research.

PROTEIN MODIFICATION AND CLEAVAGE CONFER NEW CAPABILITIES

The basic set of twenty amino acids can be modified after synthesis of a polypeptide chain to enhance its capabilities. For example, the amino-termini of many proteins are *acetylated*, which makes these proteins more resistant to degradation. In newly synthesized collagen, many proline residues become hydroxylated to form *hydroxyproline* (Figure 2-26). The added hydroxyl groups stabilize the collagen fiber. The biological significance of this modification is evident in scurvy, which results from insufficient hydroxylation of collagen because of a deficiency of vitamin C. Another specialized amino acid produced by a finishing touch is *γ-carboxyglutamate*. In vitamin K deficiency, insufficient carboxylation of glutamate in prothrombin, a clotting protein, can lead to hemorrhage. Most proteins, such as antibodies, that are secreted by cells acquire carbohydrate chains on specific asparagine residues. Many hormones, such as epinephrine (adrenaline), alter the activities of enzymes by stimulating the phosphorylation of the hydroxyl amino acids serine and threonine; *phosphoserine* and *phosphothreonine* are the most ubiquitous modified amino acids in proteins. Growth factors such as insulin act by triggering the phosphorylation of the hydroxyl group of tyrosine to form *phosphotyrosine*. The phosphate groups on these three modified amino acids can readily be removed, enabling them to act as reversible switches in regulating cellular processes. Indeed, some tumor viruses produce cancer by stimulating excessive phosphorylation of tyrosine residues on proteins that control cell proliferation.

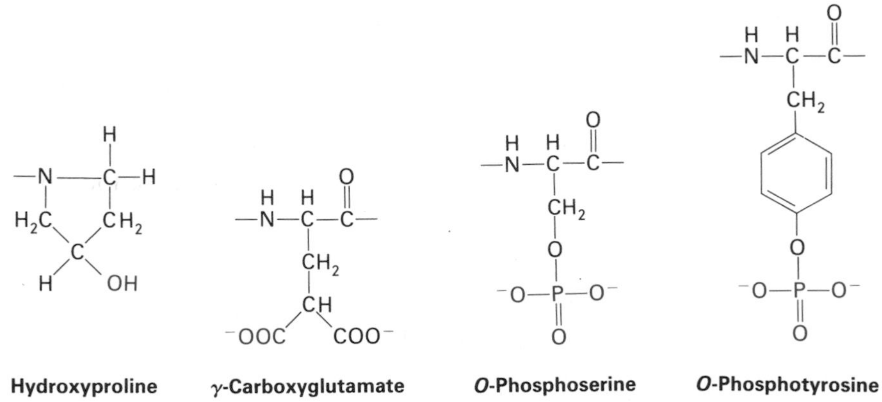

Many proteins are cleaved and trimmed after synthesis. For example, digestive enzymes are synthesized as inactive precursors that can be stored safely in the pancreas. After being released into the intestine, these precursors become activated by peptide bond cleavage. In blood clotting, soluble fibrinogen is converted into insoluble fibrin by peptide bond cleavage. A number of polypeptide hormones that include adrenocorticotropic hormone arise from the splitting of a single large pre-

Hydroxyproline | **γ-Carboxyglutamate** | ***O*-Phosphoserine** | ***O*-Phosphotyrosine**

Figure 2-26
Some modified amino acid residues in proteins: hydroxyproline, γ-carboxyglutamate, *O*-phosphoserine, and *O*-phosphotyrosine. Groups added after the polypeptide chain is synthesized are shown in red.

Phosphoserine

Figure 2-27
Model of phosphoserine.

cursor protein, a molecular cornucopia. The proteins of poliovirus, too, are produced by cleavage of a giant polyprotein precursor. We shall encounter many more examples of modification and cleavage as essential features of protein formation and function. Indeed, these finishing touches account for much of the versatility, precision, and elegance of protein action and regulation.

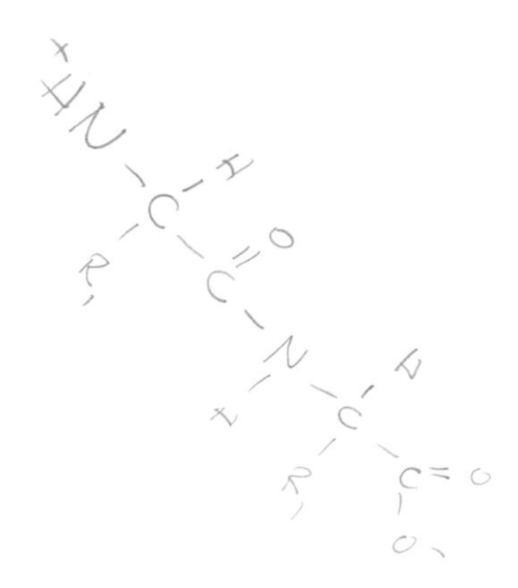

THE PEPTIDE UNIT IS RIGID AND PLANAR

A striking characteristic of proteins is that they have well-defined three-dimensional structures. A stretched-out or randomly arranged polypeptide chain is devoid of biological activity, as will be discussed shortly. *Function arises from conformation,* which is the three-dimensional arrangement of atoms in a structure. Amino acid sequences are important because they specify the conformation of proteins.

In the late 1930s, Linus Pauling and Robert Corey began x-ray crystallographic studies of the precise structure of amino acids and peptides. Their aim was to obtain a set of standard bond distances and bond angles for these building blocks and then use this information to predict the conformation of proteins. One of their important findings was that *the peptide unit is rigid and planar.* The hydrogen of the substituted amino group is nearly always *trans* (opposite) to the oxygen of the carbonyl group (Figure 2-28). There is no freedom of rotation about

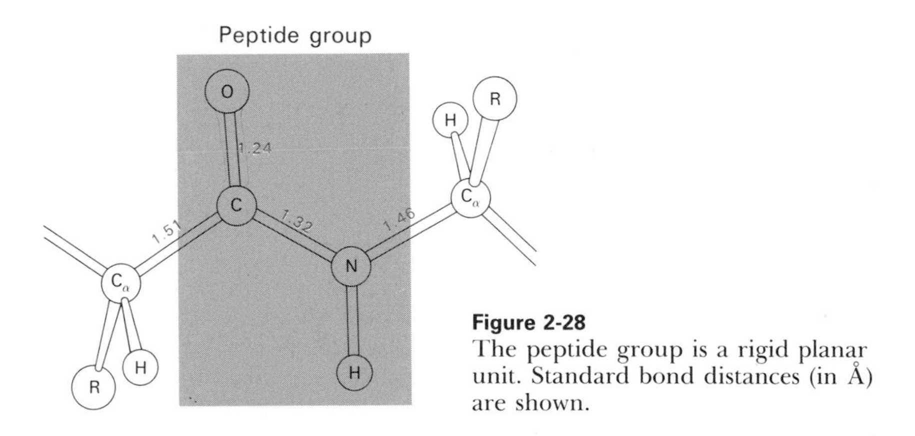

Figure 2-28
The peptide group is a rigid planar unit. Standard bond distances (in Å) are shown.

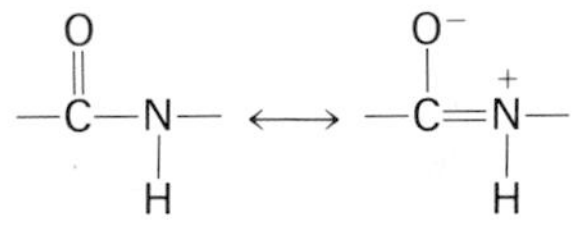

Figure 2-29
The peptide group is planar because the carbon–nitrogen bond has partial double-bond character.

the bond between the carbonyl carbon atom and the nitrogen atom of the peptide unit because this link has partial double-bond character (Figure 2-29). The length of this bond is 1.32 Å, which is between that of a C—N single bond (1.49 Å) and a C=N double bond (1.27 Å). In contrast, the link between the α-carbon atom and the carbonyl carbon atom is a pure single bond. The bond between the α-carbon atom and the peptide nitrogen atom also is a pure single bond. Consequently, *there is a large degree of rotational freedom about these bonds on either side of the rigid peptide unit* (Figure 2-30).

Rigid unit Rigid unit

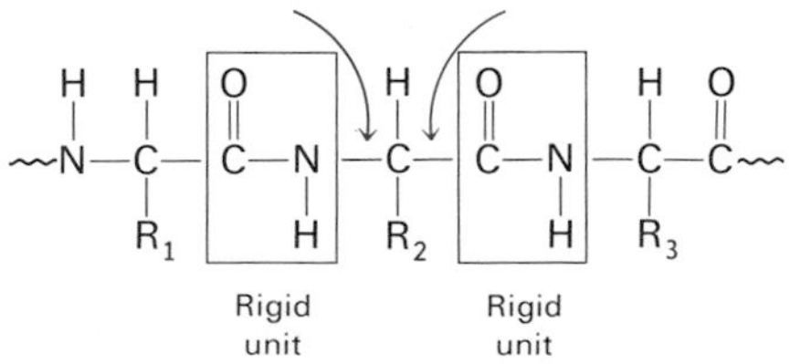

Figure 2-30
There is considerable freedom of rotation about the bonds joining the peptide groups to the α-carbon atoms.

POLYPEPTIDE CHAINS CAN FOLD INTO REGULAR STRUCTURES: THE α HELIX AND β PLEATED SHEETS

Can a polypeptide chain fold into a regularly repeating structure? To answer this question, Pauling and Corey evaluated a variety of potential polypeptide conformations by building precise molecular models. They

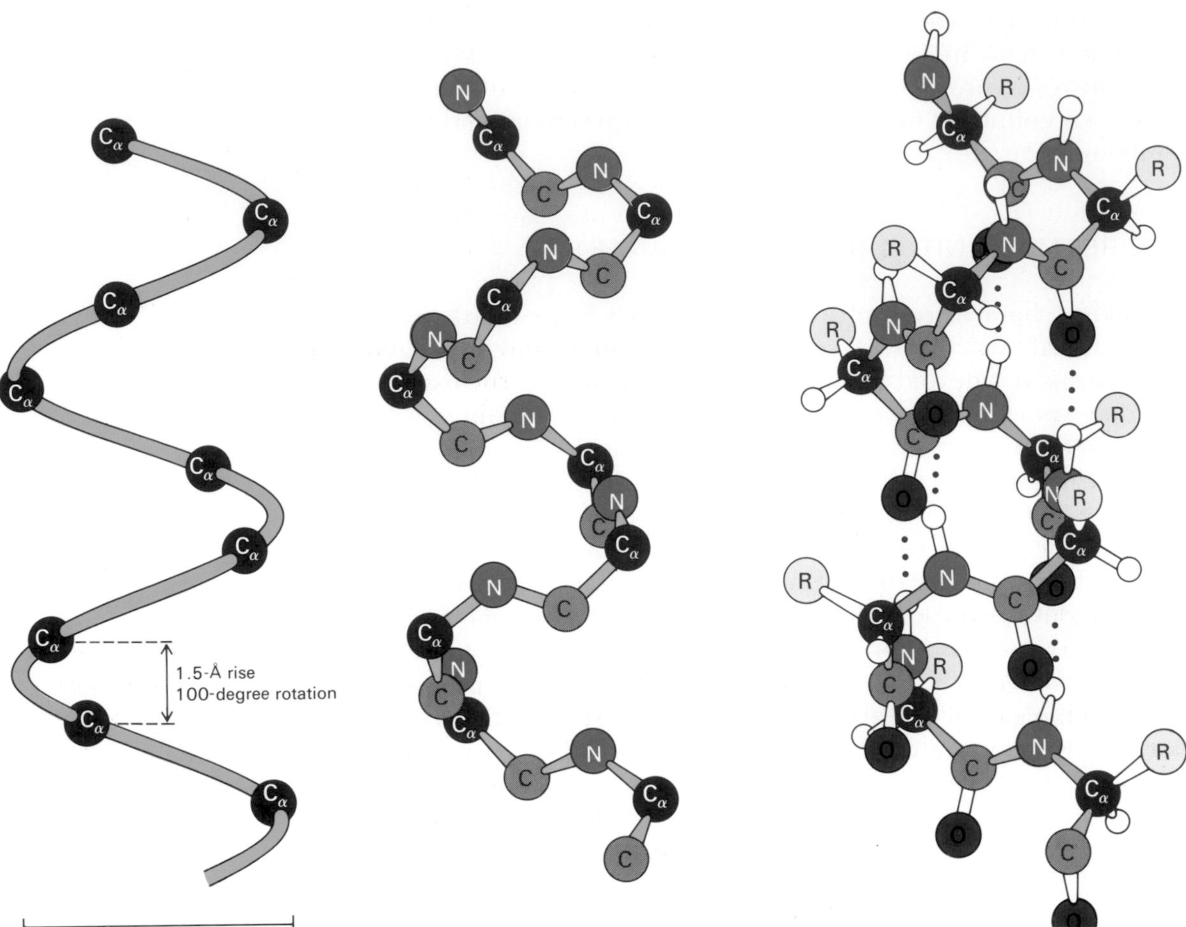

Figure 2-31
Models of a right-handed α helix: (A) only the α-carbon atoms are shown on a helical thread; (B) only the backbone nitrogen (N), α-carbon (C_α), and carbonyl carbon (C) atoms are shown; (C) entire helix. Hydrogen bonds (denoted in part C by red dots) between NH and CO groups stabilize the helix.

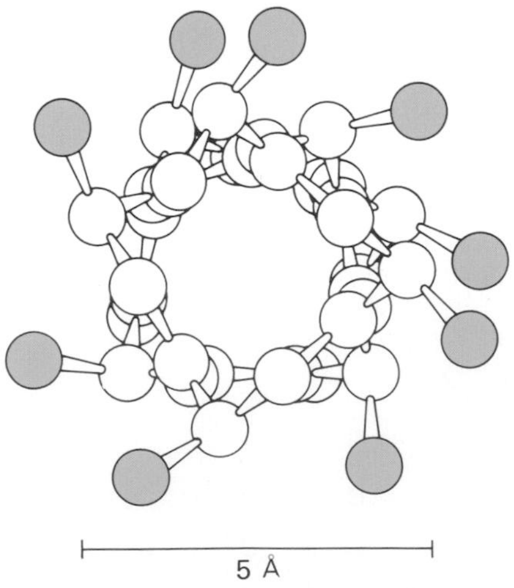

adhered closely to the experimentally observed bond angles and distances for amino acids and small peptides. In 1951, they proposed two periodic polypeptide structures, called the α helix (alpha helix) and the β pleated sheet (beta pleated sheet).

The α helix is a rodlike structure. The tightly coiled polypeptide main chain forms the inner part of the rod, and the side chains extend outward in a helical array (Figures 2-31 and 2-32). The α helix is stabilized by hydrogen bonds between the NH and CO groups of the main chain. The CO group of each amino acid is hydrogen bonded to the NH group of the amino acid that is situated four residues ahead in the linear sequence (Figure 2-33). Thus, *all the main-chain CO and NH groups*

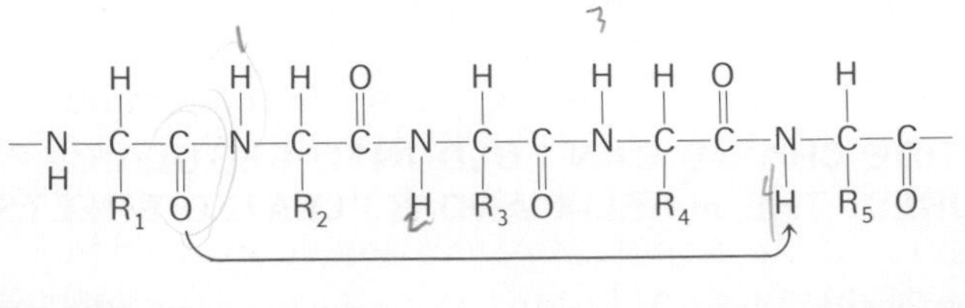

5 Å

Figure 2-32
Cross-sectional view of an α helix. Note that the side chains (shown in green) are on the outside of the helix. The van der Waals radii of the atoms are larger than shown here; hence there is actually almost no free space inside the helix.

Figure 2-33
In the α helix, the CO group of residue n is hydrogen bonded to the NH group of residue (n + 4).

are hydrogen bonded. Each residue is related to the next one by a translation of 1.5 Å along the helix axis and a rotation of 100°, which gives 3.6 amino acid residues per turn of helix. Thus, amino acids spaced three and four apart in the linear sequence are spatially quite close to one another in an α helix. In contrast, amino acids two apart in the linear sequence are situated on opposite sides of the helix and so are unlikely to make contact. The pitch of the α helix is 5.4 Å, the product of the translation (1.5 Å) and the number of residues per turn (3.6). The screw sense of a helix can be right-handed (clockwise) or left-handed (counterclockwise); the α helices found in proteins are right-handed.

Angstrom (Å)—
A unit of length equal to 10^{-10} meter.

$$1\ \text{Å} = 10^{-10}\ \text{m} = 10^{-8}\ \text{cm}$$
$$= 10^{-4}\ \mu\text{m} = 10^{-1}\ \text{nm}$$

Named after Anders J. Ångström (1814–1874), a spectroscopist.

The α helix content of proteins of known three-dimensional structure is highly variable. In some, such as myoglobin and hemoglobin, the α helix is the major structural motif. Other proteins, such as the digestive enzyme chymotrypsin, are virtually devoid of α helix. In most proteins, the single-stranded α helix discussed above is usually a rather short rod, typically less than 40 Å in length. However, the α helical theme is extended in some proteins to much longer rods, as long as 1000 Å (100 nm or 0.1 μm) or more. Two or more such α helices can entwine to form a cable. Such *α helical coiled coils* are found in keratin in hair, myosin and tropomyosin in muscle, epidermin in skin, and fibrin in blood clots. The helical cables in these proteins serve a mechanical role in forming stiff bundles of fibers.

The structure of the α helix was deduced by Pauling and Corey six years before it was actually to be seen in the x-ray reconstruction of the structure of myoglobin. *The elucidation of the structure of the α helix is a landmark in molecular biology because it demonstrated that the conformation of a polypeptide chain can be predicted if the properties of its components are rigorously and precisely known.*

"When we consider that the fibrous proteins of the epidermis, the keratinous tissues, the chief muscle protein, myosin, and now the fibrinogen of the blood all spring from the same peculiar shape of molecule, and are therefore probably all adaptations of a single root idea, we seem to glimpse one of the great coordinating facts in the lineage of biological molecules."

K. Bailey, W. T. Astbury, and K. M. Rudall
Nature, 1943

In the same year, Pauling and Corey discovered another periodic structural motif, which they named the β pleated sheet (β because it was the second structure they elucidated, the α helix having been the first). The β pleated sheet differs markedly from the α helix in that it is a sheet rather than a rod. A polypeptide chain in the β pleated sheet is almost fully extended (Figure 2-34) rather than being tightly coiled as

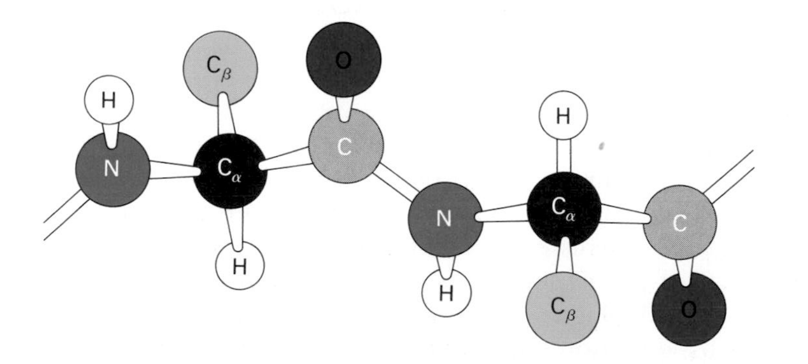

Figure 2-34
Conformation of a dipeptide unit in a β pleated sheet. The polypeptide chain is almost fully stretched out.

in the α helix. The axial distance between adjacent amino acids is 3.5 Å, in contrast with 1.5 Å for the α helix. Another difference is that the β pleated sheet is stabilized by hydrogen bonds between NH and CO groups in *different* polypeptide chains, whereas in the α helix the hydrogen bonds are between NH and CO groups in the *same* polypeptide chain. Adjacent chains in a β pleated sheet can run in the same direction (*parallel β sheet*) or in opposite directions (*antiparallel β sheet*) (Figure

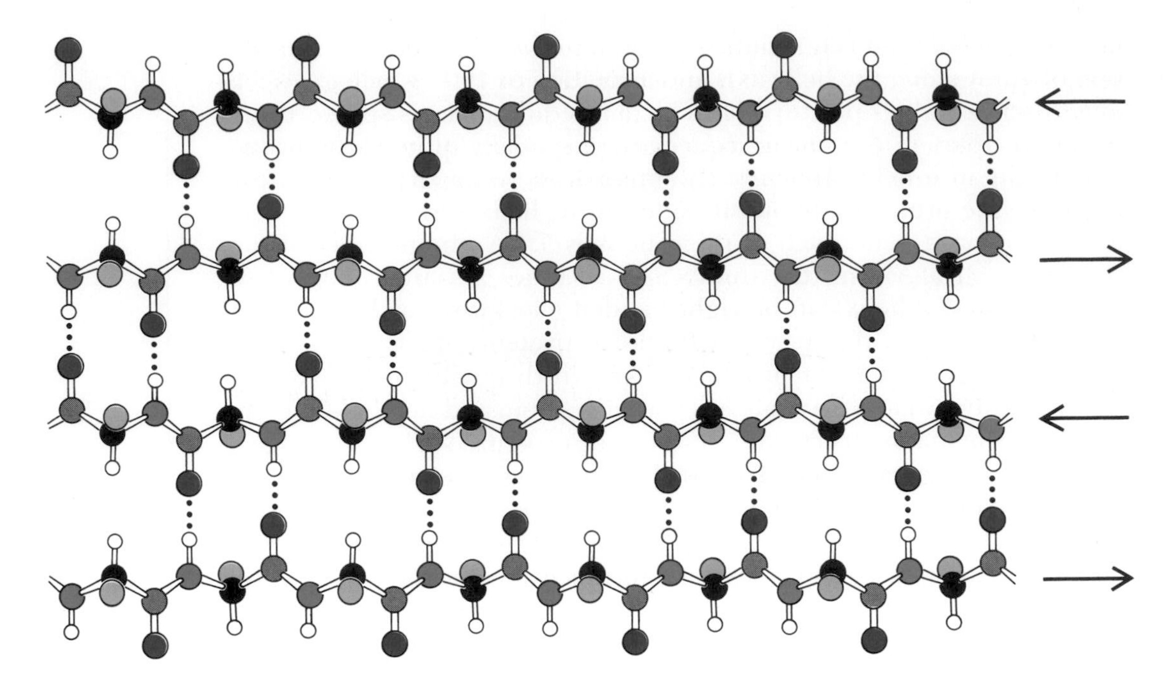

Figure 2-35
Antiparallel β pleated sheet. Adjacent strands run in opposite directions. Hydrogen bonds between NH and CO groups of adjacent strands stabilize the structure. The side chains (shown in green) are above and below the plane of the sheet.

2-35). For example, silk fibroin consists almost entirely of stacks of antiparallel β sheets. Such β-sheet regions are a recurring structural motif in many proteins. Structural units comprising from two to five parallel or antiparallel β strands are especially common.

The *collagen helix,* a third periodic structure, will be discussed in detail in Chapter 11. This specialized structure is responsible for the high tensile strength of collagen, the major protein of skin, bone, and tendon.

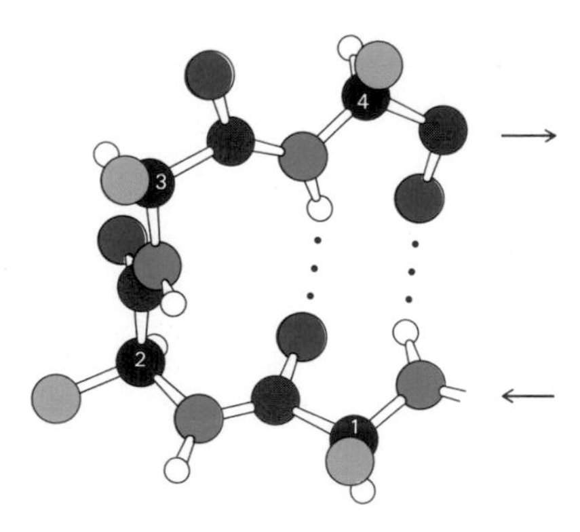

Figure 2-36
Structure of a β-turn. The CO group of residue 1 of the tetrapeptide shown here is hydrogen bonded to the NH group of residue 4, which results in a hairpin turn.

POLYPEPTIDE CHAINS CAN REVERSE DIRECTION BY MAKING β-TURNS

Most proteins have compact, globular shapes due to numerous reversals of the direction of their polypeptide chains. Analyses of the three-dimensional structures of numerous proteins have revealed that many of these chain reversals are accomplished by a common structural element called the *β-turn.* The essence of this hairpin turn is that the CO group of residue n of a polypeptide is hydrogen bonded to the NH group of residue $(n + 3)$ (Figure 2-36). Thus, a polypeptide chain can abruptly reverse its direction. β-Turns often connect antiparallel β strands; hence their name. β-Turns are also known as *reverse turns* or *hairpin bends.*

PROTEINS ARE RICH IN HYDROGEN-BONDING POTENTIALITY

What are the forces that determine the three-dimensional architecture of proteins? As Chapter 1 explained, all reversible molecular interactions in biological systems are mediated by three kinds of forces: *electro-*

static bonds, hydrogen bonds, and *van der Waals bonds.* We have already seen hydrogen bonds between main-chain NH and CO groups at work in forming α helices and β sheets. In fact, side chains of eleven of the twenty fundamental amino acids also can participate in hydrogen bonding. It is convenient to group these residues according to their hydrogen-bonding potentialities:

1. The side chains of tryptophan and arginine can serve as *hydrogen-bond donors only.*

2. Like the peptide group itself, the side chains of asparagine, glutamine, serine, and threonine can serve as *hydrogen-bond donors and acceptors.*

3. The hydrogen-bonding capabilities of lysine (and the terminal amino group), aspartic and glutamic acid (and the terminal carboxyl group), tyrosine, and histidine vary with pH. These groups can serve as both acceptors and donors over a certain range of pH, and as acceptors or donors (but not both) at other pH values, as shown for aspartate and glutamate in Figure 2-37. *The hydrogen-bonding modes of these ionizable residues are pH-dependent.*

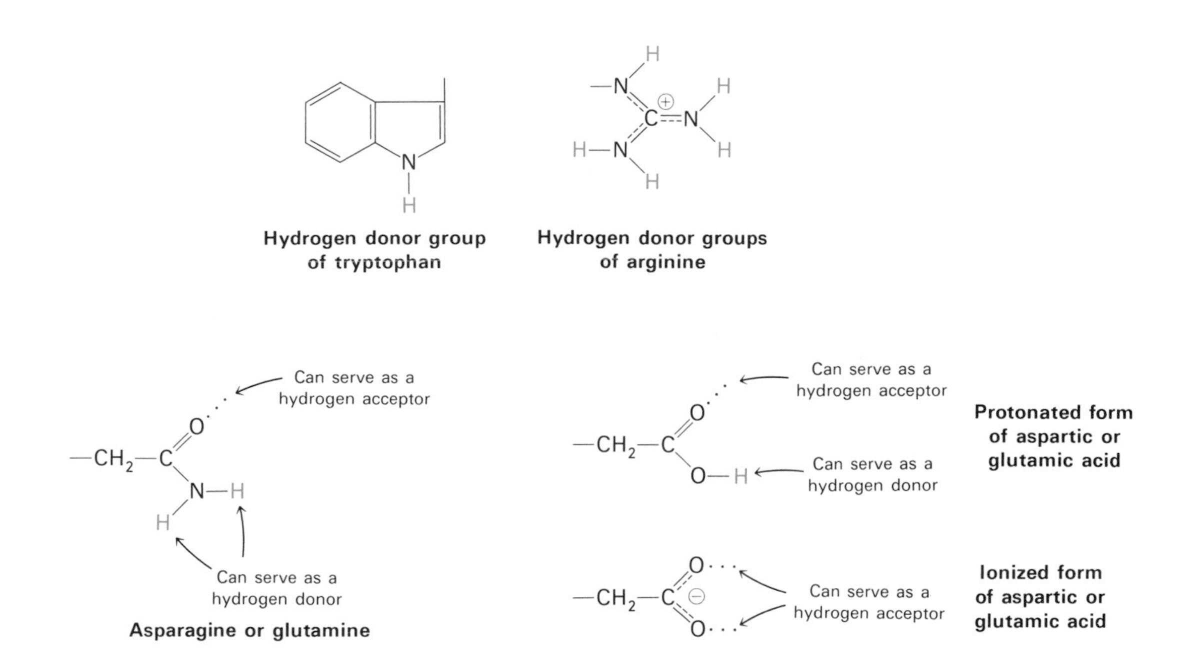

Figure 2-37
Hydrogen-bonding groups of several side chains in proteins.

WATER-SOLUBLE PROTEINS FOLD INTO COMPACT STRUCTURES WITH NONPOLAR CORES

Let us now see how these forces shape the structure of proteins. X-ray crystallographic studies have revealed the detailed three-dimensional structures of more than a hundred proteins. The experimental method of x-ray analysis and some examples of its results will be discussed in later chapters. We begin here with a preview of *myoglobin,* the first protein to be seen in atomic detail.

Myoglobin, the oxygen carrier in muscle, is a single polypeptide chain of 153 amino acids and has a mass of 18 kd. The capacity of myoglobin to bind oxygen depends on the presence of *heme,* a nonpolypeptide *prosthetic* (helper) *group* consisting of protoporphyrin

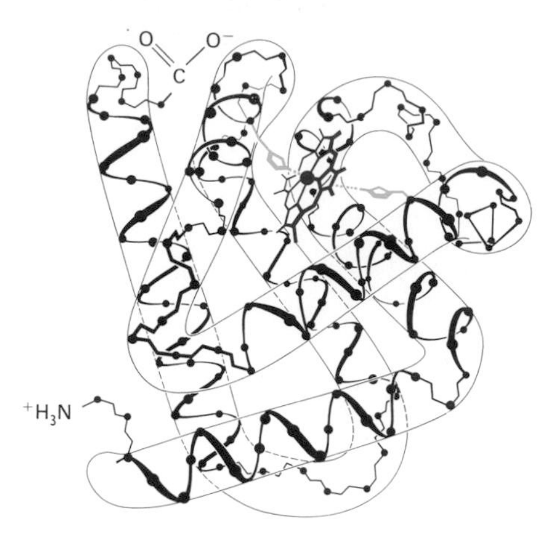

Figure 2-38
Model of myoglobin. Only the α-carbon atoms are shown. The heme group is shown in red and two adjacent histidines in green. [After R. E. Dickerson. In *The Proteins*, H. Neurath, ed., 2nd ed. (Academic Press, 1964), vol. 2, p. 634.]

and a central iron atom. *Myoglobin is an extremely compact molecule with very little empty space inside* (Figure 2-38). Its overall dimensions are 45 × 35 × 25 Å, an order of magnitude less than if it were fully stretched out. *Myoglobin is built primarily of α helices*, of which there are eight. About 70% of the main chain is folded into α helices, and much of the rest of the chain forms turns between helices. Four of the turns contain proline, which disrupts α helices because of its rigid five-membered ring. The folding of the main chain of myoglobin, like that of other proteins, is complex and devoid of symmetry. However, a unifying principle emerges from the distribution of side chains. The striking fact is that *the interior consists almost entirely of nonpolar residues* such as leucine, valine, methionine, and phenylalanine. Polar residues such as aspartate, glutamate, lysine, and arginine are absent from the inside of myoglobin. The only polar residues inside are two histidines, which play critical roles in the binding of heme oxygen. The outside of myoglobin, on the other hand, consists of both polar and nonpolar residues.

This contrasting distribution of polar and nonpolar residues reveals a key facet of protein architecture. In an aqueous environment, protein folding is driven by the strong tendency of hydrophobic residues to be excluded from water. Recall that water is highly cohesive and that hydrophobic groups are thermodynamically more stable when clustered in the interior of the molecule than when extended into the aqueous surroundings (p. 10). *The polypeptide chain therefore folds spontaneously so that its hydrophobic side chains are buried and its polar, charged chains are on the surface.* The fate of the main chain accompanying the hydrophobic side chains is important, too. An unpaired peptide NH or CO markedly prefers water to a nonpolar milieu. The secret of burying a segment of main chain in a hydrophobic environment is to pair all the NH and CO groups by hydrogen bonding. This pairing is neatly accomplished in an α helix or β sheet. Van der Waals bonds between tightly packed hydrocarbon side chains also contribute to the stability of proteins. We can now understand why the repertoire of twenty amino acids contains so many aliphatic ones, differing subtly in size and shape. Nature can choose among them to fill the interior of a protein neatly and thereby maximize van der Waals interactions, which require intimate contact.

Ribonuclease S, a pancreatic enzyme that hydrolyzes RNA, exemplifies a rather different mode of protein folding. This single polypeptide chain of 124 residues is folded mainly into β-sheet strands, in contrast with myoglobin, which contains α helices and lacks β sheets. Ribonuclease, like myoglobin, contains a tightly packed, highly nonpolar interior. This enzyme is further stabilized by four disulfide bonds. The structure of a protein can be symbolized in highly schematic form by depicting β strands as broad arrows, α helices as helical ribbons, and connecting regions as strings. This representation is very useful for concisely representing relations of these elements in proteins, especially large ones, and for detecting structural motifs that recur in different proteins. A ribbon drawing of ribonuclease is shown in Figure 2-39.

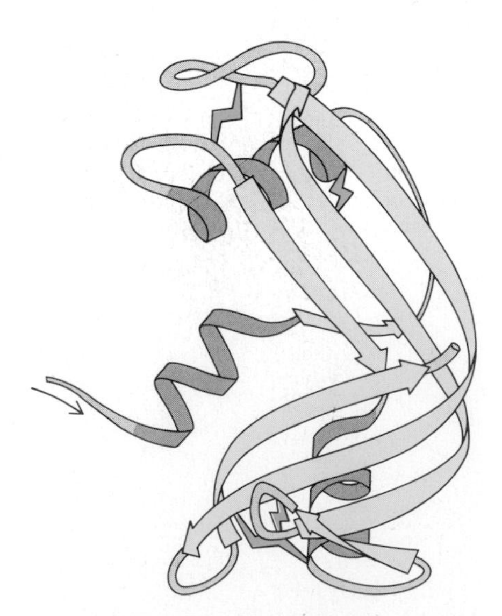

Figure 2-39
Ribbon model of ribonuclease S. Sections of α helix are shown in red, β-sheets in green, and disulfide bridges in yellow. [Courtesy of Dr. Jane Richardson.]

Integral membrane proteins, those that traverse biological membranes, are designed differently from proteins that are soluble in aqueous solution. The permeability barrier of membranes is formed by lipids, which are highly hydrophobic. Thus, the part of a membrane protein that spans this region must have a hydrophobic exterior. As will be discussed in Chapter 12, the transmembrane portion of a membrane protein usually consists of bundles of α helices with nonpolar side chains (such as those of leucine and phenylalanine) facing out from the surface of the protein.

LEVELS OF STRUCTURE IN PROTEIN ARCHITECTURE

Four levels of structure are frequently cited in discussions of protein architecture. *Primary structure* is the amino acid sequence and the location of disulfides, if there are any. The primary structure is thus a complete description of the covalent connections of a protein. *Secondary structure* refers to the spatial arrangement of amino acid residues that are near one another in the linear sequence. Some of these steric relationships are of a regular kind, giving rise to a periodic structure. The α helix, β pleated sheet, and collagen helix are elements of secondary structure. *Tertiary structure* refers to the spatial arrangement of amino acid residues that are far apart in the linear sequence. The dividing line between secondary and tertiary structure is a matter of taste. Proteins containing more than one polypeptide chain exhibit an additional level of structural organization. Each polypeptide chain in such a protein is called a subunit. *Quaternary structure* refers to the spatial arrangement of such subunits and the nature of their contacts (Figure 2-40). The constituent chains of a multisubunit protein can be identical or different. For example, immunoglobulin G, the major antibody molecule in plasma, consists of two L chains and two H chains. The spherical shell of tomato bushy stunt virus, a plant pathogen, is formed from 180 identical coat protein molecules. The interfaces between subunits are often functionally significant. For example, in hemoglobin (consisting of four chains), the subunit interfaces participate in transmitting information between binding sites for O_2, CO_2, and H^+. In antibody molecules, the combining site for antigen is formed by segments of two different kinds of chains.

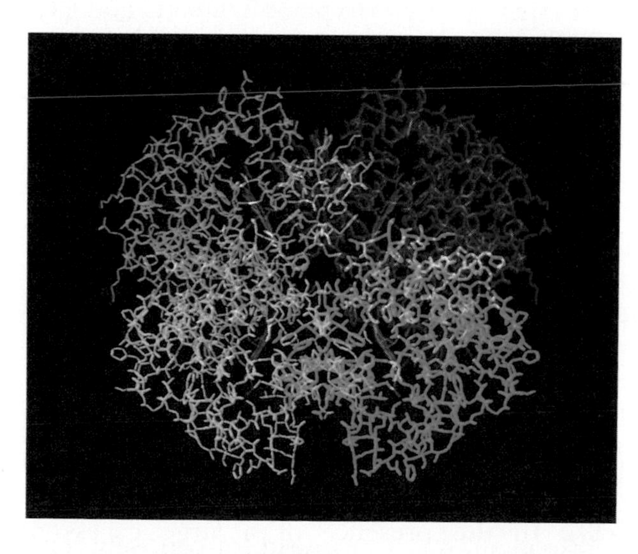

Figure 2-40
Three-dimensional structure of hemoglobin. The four subunits are shown in different colors. Each contains an oxygen-binding heme group (red).

Recent studies of protein conformation, function, and evolution have revealed the importance of two additional levels of organization. *Supersecondary structure* refers to clusters of secondary structure. For example, a β strand separated from another β strand by an α helix is found in many proteins; this motif is called a $\beta\alpha\beta$ unit. It is fruitful to regard supersecondary structures as intermediates between secondary and tertiary structure. Some polypeptide chains fold into two or more compact regions that may be joined by a flexible segment of polypeptide chain, rather like pearls on a string. These compact globular units, called *domains*, range in size from about 100 to 400 amino acid residues. For example, a 25-kd L chain of an antibody is folded into two domains (Figure 2-41). Indeed, these domains resemble one another, which sug-

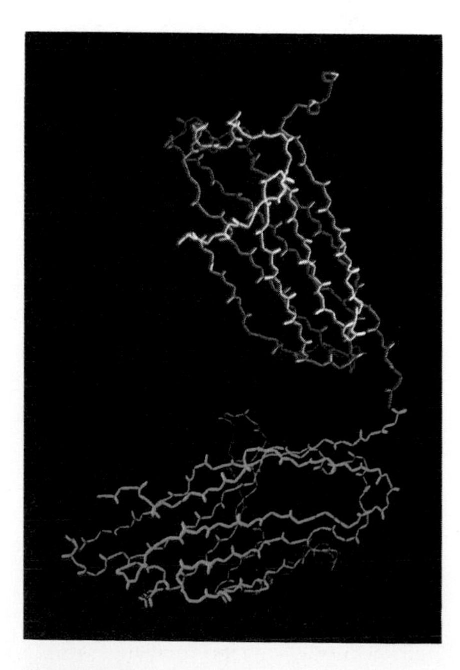

Figure 2-41
The light (L) chain of an antibody molecule consists of two distinct domains.

gests that they arose by duplication of a primordial gene. An important principle has emerged from analyses of genes and proteins in higher eucaryotes: *protein domains are often encoded by distinct parts of genes called exons* (p. 112). In our explorations of genes and proteins, exons and domains will often be at the focal point.

AMINO ACID SEQUENCE SPECIFIES THREE-DIMENSIONAL STRUCTURE

Insight into the relation between the amino acid sequence of a protein and its conformation came from the work of Christian Anfinsen on ribonuclease. As mentioned earlier, ribonuclease is a single polypeptide chain consisting of 124 amino acid residues (Figure 2-42). Its four disul-

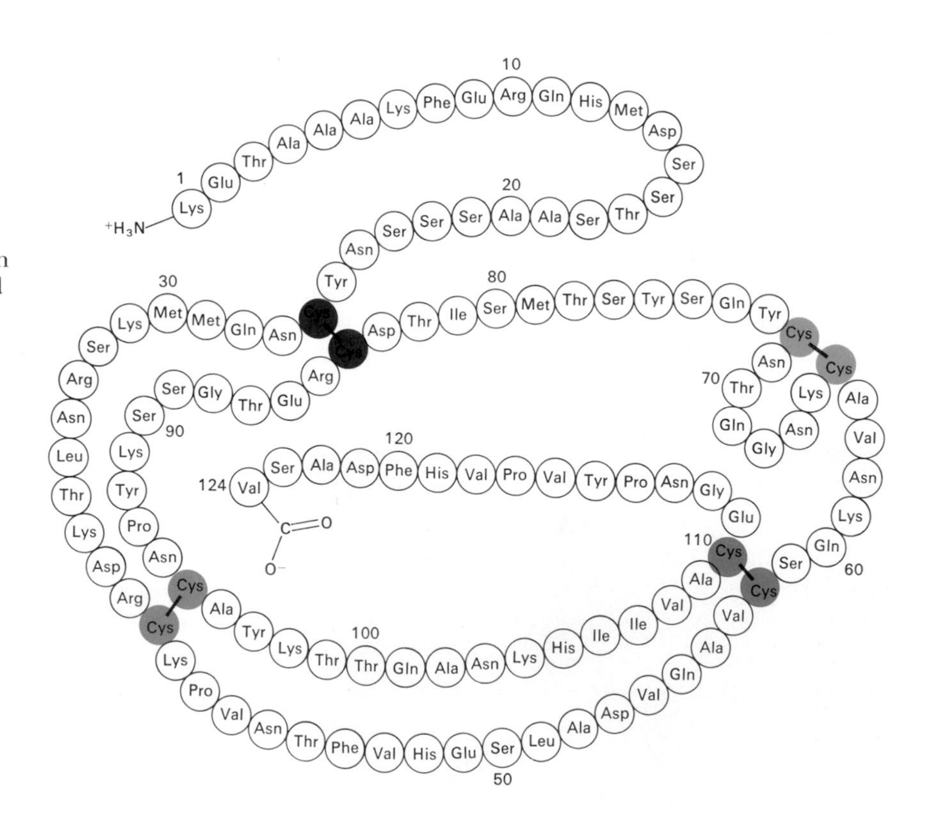

Figure 2-42
Amino acid sequence of bovine ribonuclease. The four disulfide bonds are shown in color. [After C. H. W. Hirs, S. Moore, and W. H. Stein. *J. Biol. Chem.* 235(1960):633.]

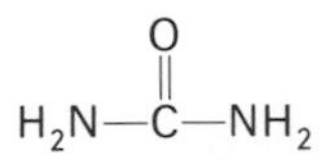

$H_2N—C(=O)—NH_2$
Urea

$H_2N—C(=NH_2^+)—NH_2$ Cl^-
Guanidine hydrochloride

$HO—CH_2—CH_2—SH$
β-Mercaptoethanol

fide bonds can be cleaved reversibly by reducing them with a reagent such as *β-mercaptoethanol,* which forms mixed disulfides with cysteine side chains (Figure 2-43). In the presence of a large excess of β-mercaptoethanol, the mixed disulfides also are reduced, so that the final product is a protein in which the disulfides (cystines) are fully converted into sulfhydryls (cysteines). However, it was found that ribonuclease at 37°C and pH 7 cannot be readily reduced by β-mercaptoethanol unless the protein is partly unfolded by agents such as *urea* or *guanidine hydrochloride.* Although the mechanism of action of

Protein —S—S— (Oxidized protein) —R—SH→ Protein —S—S—R, —SH (Mixed disulfide) —R—SH, R—S—S—R→ Protein —SH, —SH (Reduced protein)

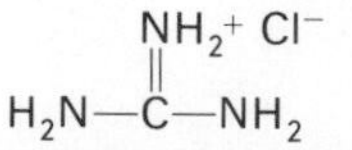

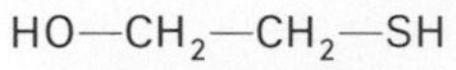

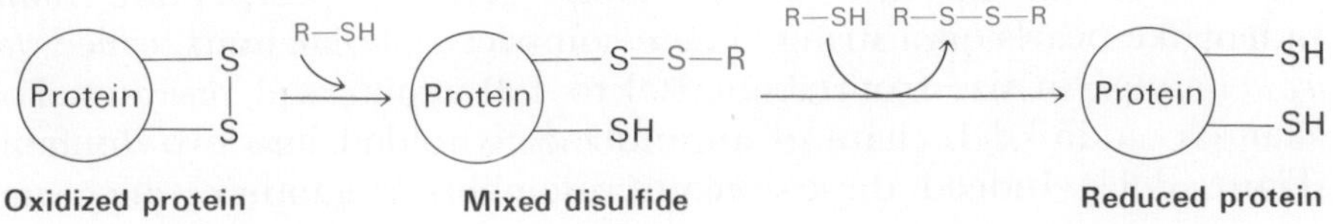

Figure 2-43
Reduction of the disulfide bonds in a protein by an excess of a sulfhydryl reagent such as β-mercaptoethanol.

these agents is not fully understood, it is evident that they disrupt noncovalent interactions. Most polypeptide chains devoid of cross-links assume a *random-coil conformation* in 8 M urea or 6 M guanidine HCl, as evidenced by physical properties such as viscosity and optical rotatory spectra. When ribonuclease was treated with β-mercaptoethanol in 8 M urea, the product was a fully reduced, randomly coiled polypeptide chain *devoid of enzymatic activity*. In other words, ribonuclease was *denatured* by this treatment (Figure 2-44).

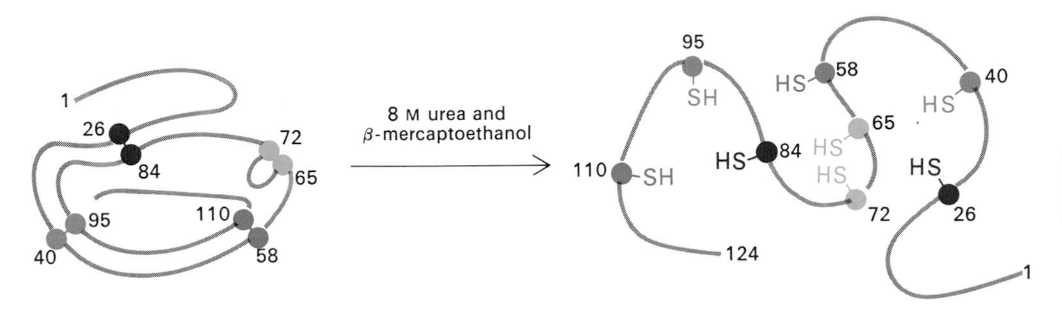

Figure 2-44
Reduction and denaturation of ribonuclease.

Anfinsen then made the critical observation that the denatured ribonuclease, freed of urea and β-mercaptoethanol by dialysis, slowly regained enzymatic activity. He immediately perceived the significance of this chance finding: the sulfhydryls of the denatured enzyme became oxidized by air and the enzyme spontaneously refolded into a catalytically active form. Detailed studies then showed that nearly all of the original enzymatic activity was regained if the sulfhydryls were oxidized under suitable conditions (Figure 2-45). All of the measured physical and chemical properties of the refolded enzyme were virtually identical with those of the native enzyme. These experiments showed that *the information needed to specify the complex three-dimensional structure of ribonuclease is contained in its amino acid sequence*. Subsequent studies of other proteins have established the generality of this central principle of molecular biology: *sequence specifies conformation*.

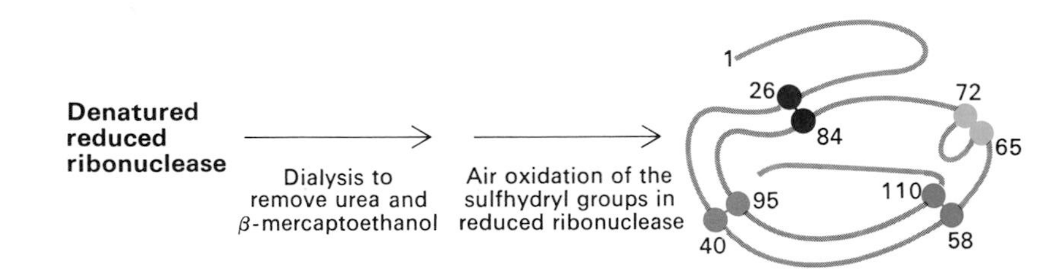

Figure 2-45
Renaturation of ribonuclease.

A quite different result was obtained when reduced ribonuclease was reoxidized while it was still in 8 M urea. This preparation was then dialyzed to remove the urea. Ribonuclease reoxidized in this way had only 1% of the enzymatic activity of the native protein. Why was the outcome of this experiment different from the one in which reduced ribonuclease was reoxidized in a solution free of urea? The reason is that wrong disulfide pairings were formed when the random-coil form of the reduced molecule was reoxidized. There are 105 different ways of pairing eight cysteines to form four disulfides; only one of these combinations is enzymatically active. The 104 wrong pairings have been picturesquely termed "scrambled" ribonuclease. Anfinsen then found that scrambled ribonuclease spontaneously converted into fully active, native ribonuclease when trace amounts of β-mercaptoethanol were added to an aqueous solution of the protein (Figure 2-46). The added

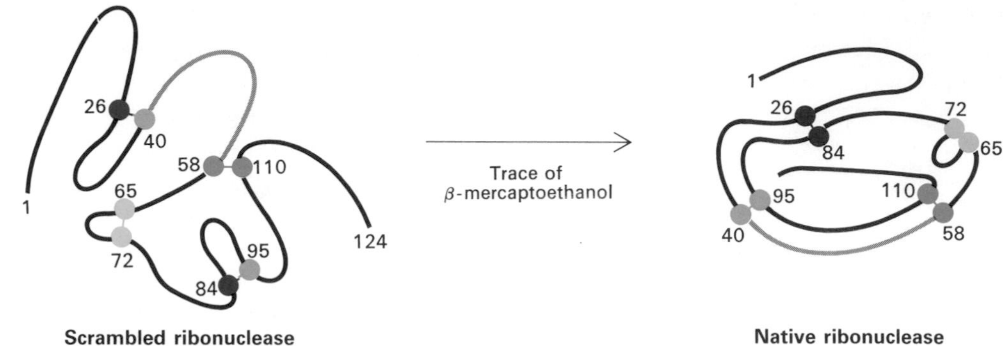

Figure 2-46
Formation of native ribonuclease from scrambled ribonuclease in the presence of a trace of β-mercaptoethanol.

β-mercaptoethanol catalyzed the rearrangement of disulfide pairings until the native structure was regained, which took about ten hours. This process was driven entirely by the decrease in free energy as the scrambled conformations were converted into the stable, native conformation of the enzyme. *Thus, the native form of ribonuclease appears to be the thermodynamically most stable structure.*

Anfinsen (1964) wrote:

> It struck me recently that one should really consider the sequence of a protein molecule, about to fold into a precise geometric form, as a line of melody written in canon form and so designed by Nature to fold back upon itself, creating harmonic chords of interaction consistent with biological function. One might carry the analogy further by suggesting that the kinds of chords formed in a protein with scrambled disulfide bridges, such as I mentioned earlier, are dissonant, but that, by giving an opportunity for rearrangement by the addition of mercaptoethanol, they modulate to give the pleasing harmonics of the native molecule. Whether or not some conclusion can be drawn about the greater thermodynamic stability of Mozart's over Schoenberg's music is something I will leave to the philosophers of the audience.

PROTEINS FOLD BY THE ASSOCIATION OF α-HELICAL AND β-STRAND SEGMENTS

How are the harmonic chords of interaction created in the conversion of an unfolded polypeptide chain into a folded protein? One possibility a priori is that all possible conformations are searched to find the energetically most favored form. How long would such a random search take? Consider a small protein with 100 residues. If each residue can assume three different positions, the total number of structures is 3^{100}, which is equal to 5×10^{47}. If it takes 10^{-13} seconds to convert one structure into another, the total search time would be $5 \times 10^{47} \times 10^{-13}$ seconds, which is equal to 5×10^{34} seconds, or 1.6×10^{27} years! Note that this length of time is a minimal estimate because the actual number of possible conformations per residue is greater than three and the time that it takes to change from one conformation into another is probably considerably longer than 10^{-13} seconds. Clearly, it would take much too long for even a small protein to fold properly by randomly trying out all possible conformations.

How, then, do proteins fold in a few seconds or minutes? The answer is not yet known in detail, but it seems likely that *small stretches of secondary structure serve as intermediates in the folding process.* According to this model, short segments (~15 residues) of an unfolded polypeptide chain flicker in and out of their native α-helical or β-sheet form. These transient structures find each other by diffusion and stabilize each other by forming a complex (Figure 2-47). For example, two α helices, two β strands, or an α helix and a β strand may come together. These $\alpha\alpha$, $\beta\beta$, and $\alpha\beta$ complexes, which are called *folding units,* then act as nuclei to attract and stabilize other flickering elements of secondary structure.

This model is supported by several lines of experimental evidence. The first is that the tendency of a polypeptide to adopt a regular secondary structure depends to a large degree on its amino acid composition. The formation of an α helix is favored by glutamate, methionine, alanine, and leucine residues, whereas β-sheet formation is enhanced by valine, isoleucine, and tyrosine residues. Second, the transition from a random coil to an α helix can occur in less than a microsecond. Thus, short segments of secondary structure can be formed very rapidly. Third, the postulated folding units ($\alpha\alpha$, $\beta\beta$, and $\alpha\beta$ complexes) are very similar to the supersecondary structural motifs discussed earlier (p. 31).

The folding of a polypeptide chain into its native structure is like solving a jigsaw puzzle. A complex puzzle (particularly one without straight borders) is best solved by recognizing component patterns, such as blue sky and green grass. The order in which they are found and assembled is unimportant. *A complex jigsaw puzzle can be solved in many different ways. Likewise, there are many different pathways for the folding of a polypeptide chain.* The analogs of blue sky and green grass are local structures that resemble parts of the final protein, such as the helix-turn-helix and beta-turn-beta supersecondary motifs. In the process of folding, incorrect structures are undoubtedly formed, but they are transient because they do not lead to subsequent productive interactions. In contrast, correctly assembled fragments of structure are likely to persist because they will be stabilized by stereospecific interactions with non-neighboring regions to form a compact structure resembling a native domain. The initial folding units will then undergo structural rearrangements to optimize these tertiary interactions. In short, it seems likely that local folding is followed by long-range interactions and then by local rearrangements to give the final folded state of the protein. The challenge now is to identify the sources of stability of native proteins and detect structural motifs that may begin the folding process and stabilize its intermediate forms.

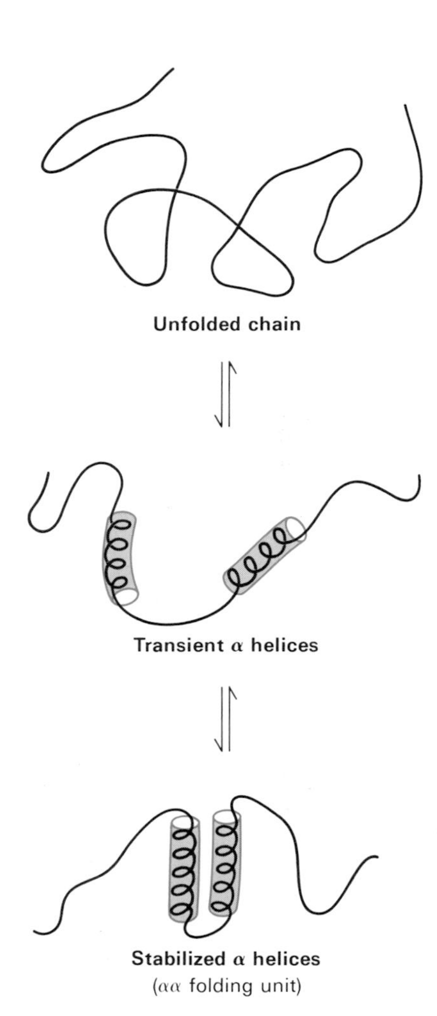

Figure 2-47
Postulated step in protein folding. Two segments of an unfolded polypeptide chain transiently become α helical. These helices are then stabilized by the formation of a complex between the two segments.

PREDICTION OF CONFORMATION FROM AMINO ACID SEQUENCE

Can the conformation of a protein be deduced from its amino acid sequence? Let us begin by considering the polypeptide backbone. Recall that the main chain can rotate on either side of each rigid peptide unit. The amount of rotation at the bond between the nitrogen and α-carbon atoms of the main chain is called phi (ϕ), and the rotation at the one between the α-carbon and carbonyl carbon atoms is called psi

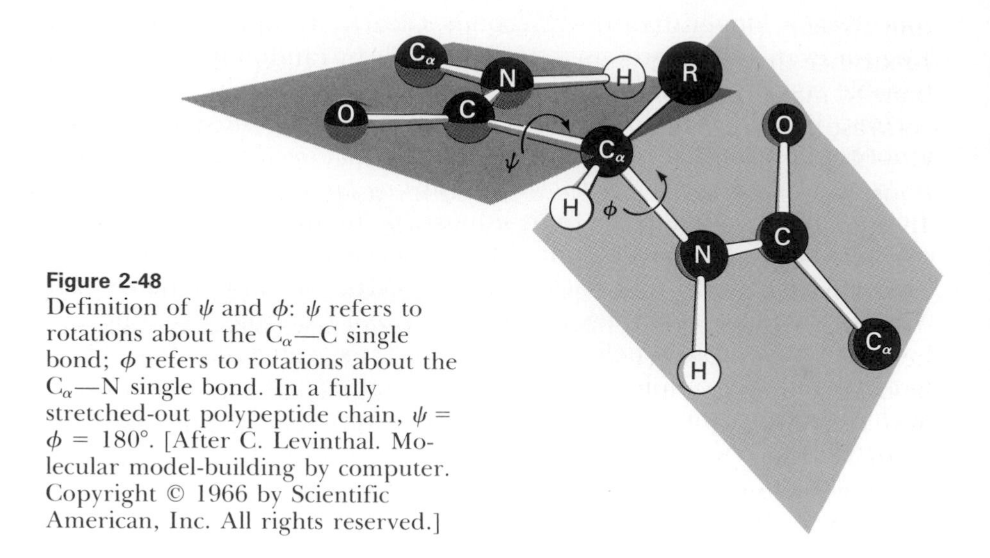

Figure 2-48
Definition of ψ and ϕ: ψ refers to rotations about the C_α—C single bond; ϕ refers to rotations about the C_α—N single bond. In a fully stretched-out polypeptide chain, $\psi = \phi = 180°$. [After C. Levinthal. Molecular model-building by computer. Copyright © 1966 by Scientific American, Inc. All rights reserved.]

(ψ) (Figure 2-48). The conformation of the main chain is completely defined when ϕ and ψ are specified for each residue in the chain. G. N. Ramachandran recognized that a residue in a polypeptide chain cannot have *any* pair of values of ϕ and ψ. Certain combinations are not accessible because of steric hindrance. Allowed ranges of ϕ and ψ can be predicted readily and visualized in steric contour diagrams called *Ramachandran plots*. Such a plot for poly-L-alanine shows three separate allowed ranges (Figure 2-49). In one of them lie the values that produce the antiparallel and parallel β sheets and the collagen helix; in another, those that produce the right-handed α helix; in the third, the left-handed α helix. Though sterically allowed, the left-handed α helix does not occur because it is energetically less favored than the right-handed one. For glycine, these three allowed regions are larger and a fourth

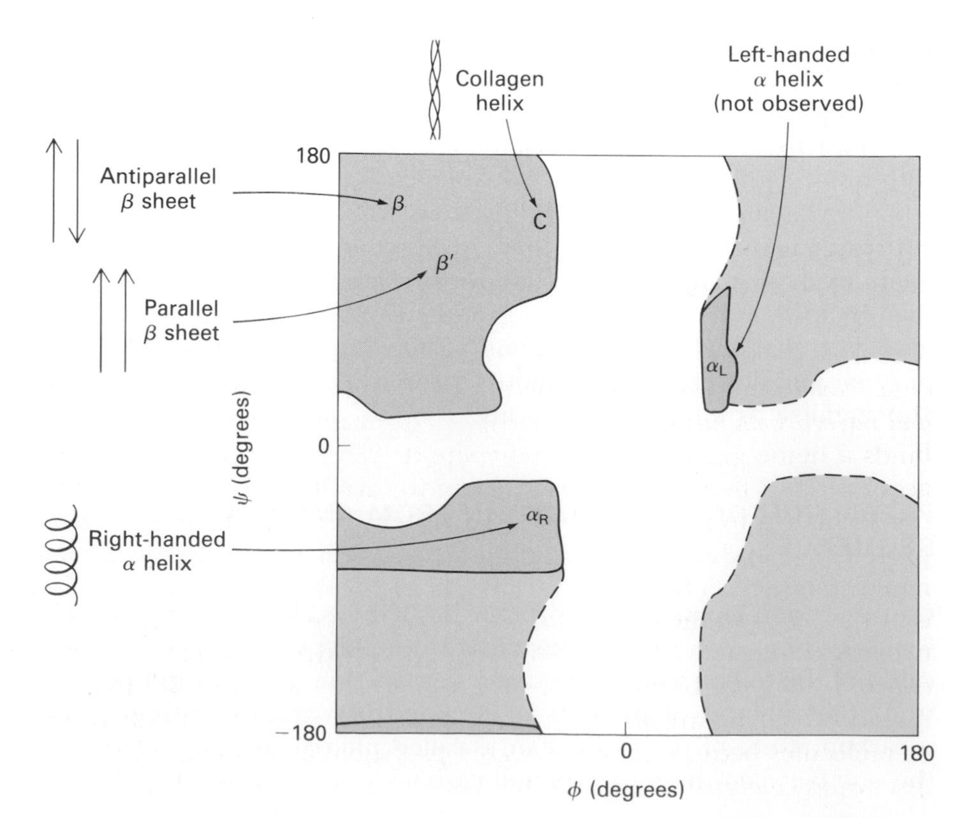

Figure 2-49
Ramachandran plot showing allowed values of ϕ and ψ for L-alanine residues (green regions). Additional conformations are accessible to glycine (yellow regions) because it has a very small side chain.

one appears (Figure 2-49) because a hydrogen atom causes less steric hindrance than a methyl group. Glycine enables the polypeptide backbone to make turns that would not be possible with another residue. In contrast, the five-membered ring of proline prevents rotation about the C_α—N bond, which markedly restricts the range of allowed conformations. The measured values of ϕ and ψ for more than 2500 residues in 13 accurately determined protein structures fit nicely into the regions predicted to be allowed.

Additional insight into protein conformation comes from the finding that amino acid residues have different frequencies of occurrence in α helices, β sheets, and reverse turns (Table 2-3). The formation of an α helix is favored by glutamate, methionine, and leucine. β-Sheet formation is enhanced by valine, isoleucine, and phenylalanine. Reverse turns, on the other hand, are promoted by proline, glycine, aspartate, asparagine, and serine. Studies of the conformation of synthetic polypeptides by Elkan Blout, Gerald Fasman, and Ephraim Katchalski have revealed some of the reasons for these preferences, as have analyses of known three-dimensional structures of proteins. For example, branching at the β-carbon atom (as in valine) tends to destabilize the α helix because of steric hindrance. Serine, aspartate, and asparagine tend to disrupt α helices because their side chains contain a hydrogen-bond donor and acceptors in close proximity to the main chain, where they compete for main-chain NH and CO groups.

Much effort has been devoted to predicting the secondary structure of proteins from amino acid sequence and a knowledge of the different tendencies of residues to occur in α helices, β sheets, and reverse turns. The predicted secondary structure agrees with the actual one for about 60% of the chain of proteins whose structures have been solved by x-ray methods. These are encouraging starts, but it is evident that much remains to be accomplished. A clue to the likely direction of fruitful effort is the finding that a pentapeptide sequence can be part of an α helix in one protein and of a β-sheet region in another protein. Hence, the local amino acid sequence is sometimes not enough to determine secondary structure. *The context in which a peptide segment folds may be crucial.* Powerful experimental approaches such as x-ray crystallography, nuclear magnetic resonance spectroscopy, and recombinant DNA cloning are now being used in concert to solve this fundamental problem.

Table 2-3
Relative frequencies of occurrence of amino acid residues in the secondary structures of proteins

Amino acid	*α helix*	*β sheet*	*β-turn*
Ala	1.29	0.90	0.78
Cys	1.11	0.74	0.80
Leu	1.30	1.02	0.59
Met	1.47	0.97	0.39
Glu	1.44	0.75	1.00
Gln	1.27	0.80	0.97
His	1.22	1.08	0.69
Lys	1.23	0.77	0.96
Val	0.91	1.49	0.47
Ile	0.97	1.45	0.51
Phe	1.07	1.32	0.58
Tyr	0.72	1.25	1.05
Trp	0.99	1.14	0.75
Thr	0.82	1.21	1.03
Gly	0.56	0.92	1.64
Ser	0.82	0.95	1.33
Asp	1.04	0.72	1.41
Asn	0.90	0.76	1.28
Pro	0.52	0.64	1.91
Arg	0.96	0.99	0.88

Source: After T. E. Creighton. *Proteins: Structures and Molecular Properties* (W. H. Freeman, 1983), p. 235.

ESSENCE OF PROTEIN ACTION: SPECIFIC BINDING AND TRANSMISSION OF CONFORMATIONAL CHANGES

The first step in the action of a protein is its binding of another molecule. *Proteins as a class of macromolecules are unique in being able to recognize and interact with highly diverse molecules.* For example, myoglobin tightly binds a heme group when its polypeptide chain is partly folded. The acquisition of heme enables myoglobin to carry out its biological function, which is to reversibly bind O_2. Proteins also combine with other proteins to produce highly ordered arrays, such as the contractile filaments in muscle. The binding of foreign molecules to antibody proteins is at the heart of the capacity of the immune system to distinguish between self and nonself. Furthermore, the expression of many genes is controlled by the binding of proteins that recognize specific DNA sequences. Proteins are able to interact specifically with such a wide range of molecules because they are highly proficient at forming *complementary surfaces and clefts* (Figure 2-50). The rich repertoire of side chains on

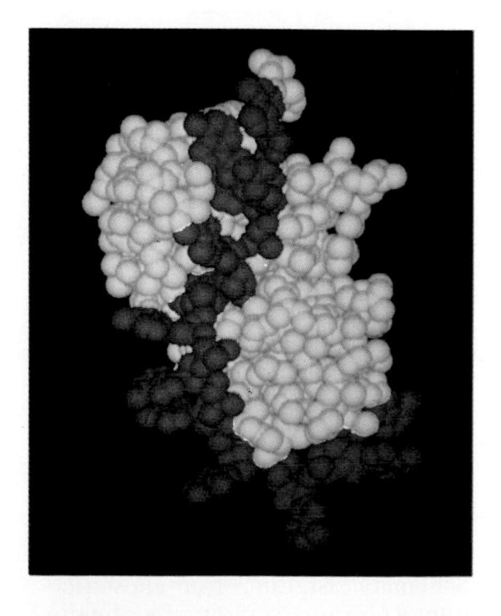

Figure 2-50
Model of ribonuclease (blue) binding an analog of an RNA substrate (red). [Courtesy of Dr. Alex McPherson.]

these surfaces and in these clefts enables proteins to form hydrogen bonds, electrostatic bonds, and van der Waals bonds with other molecules. Moreover, the strength of these interactions and their duration can be precisely controlled.

As was mentioned in the introduction to this chapter, nearly all reactions in biological systems are catalyzed by proteins called enzymes. We can now appreciate why proteins play the unique role of determining the pattern of chemical transformations. *The catalytic power of proteins comes from their capacity to bind substrate molecules in precise orientations and to stabilize transition states in the making and breaking of chemical bonds.* This basic principle can be made concrete by a simple example of enzymatic catalysis, the hydration of carbon dioxide by *carbonic anhydrase.*

$$H_2O + CO_2 \rightleftharpoons HCO_3^- + H^+$$

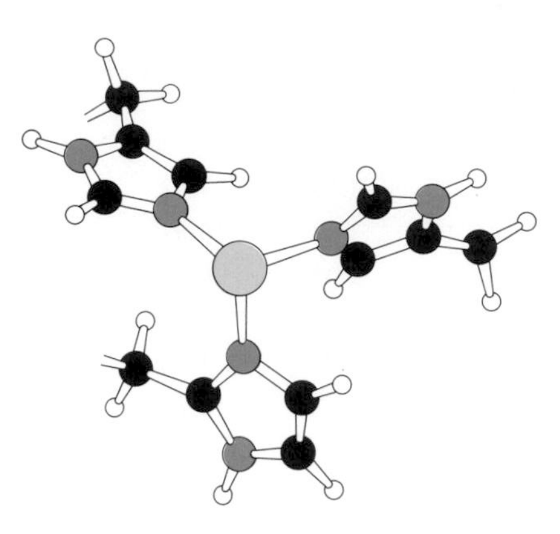

Figure 2-51
Model showing the coordination of a zinc ion (yellow) to three nitrogen atoms of histidine side chains in the catalytic site of carbonic anhydrase.

How does carbonic anhydrase accelerate this reaction by a factor of more than a million? Part of this catalytic enhancement is due to the action of a zinc ion that is coordinated to the imidazole groups of three histidine residues in the enzyme (Figure 2-51). The zinc ion is located at the bottom of a deep cleft some 15 Å from the surface of the protein. Nearby is a group of residues that recognizes and binds carbon dioxide. Water bound to the zinc ion is rapidly converted into hydroxide ion, which is precisely positioned to attack the carbon dioxide molecule bound next to it (Figure 2-52). The zinc ion helps to orient the CO_2 as well as to provide a very high local concentration of OH^-. Carbonic anhydrase, like other enzymes, is a potent catalyst because *it brings substrates into close proximity and optimizes their orientation for reaction.* Another recurring catalytic device is the *use of charged groups to polarize substrates and stabilize transition states.* Some enzymes even form covalent bonds with substrates. We shall consider enzymatic mechanisms in detail in later chapters.

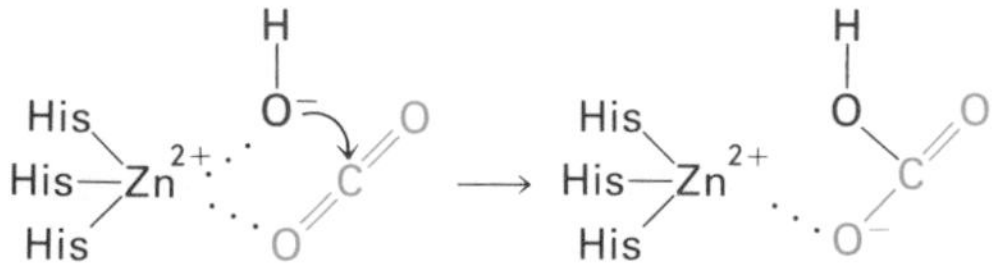

Figure 2-52
Essence of the catalytic mechanism of carbonic anhydrase. Hydroxide ion and carbon dioxide are precisely positioned for the facile formation of bicarbonate by their binding to Zn^{2+}.

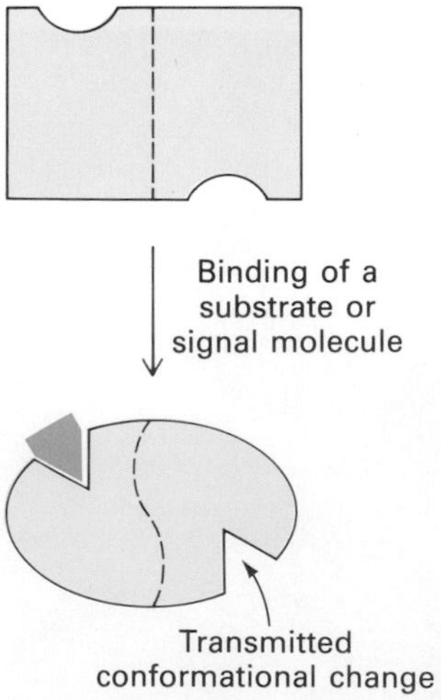

Binding of a substrate or signal molecule

Transmitted conformational change

Figure 2-53
Schematic diagram of an allosteric interaction in a protein. The binding of a small molecule or macromolecule to a site in the protein leads to conformational changes that are propagated to a distant site.

Some of the most interesting and important proteins contain two or more binding sites that communicate with each other. A conformational change induced by the binding of a molecule to one site in a protein can alter other sites more than 20 Å away. Thus, proteins can be built to serve as *molecular switches* to receive, integrate, and transmit signals. Many proteins contain regulatory sites called *allosteric sites* that control their binding of other molecules and alter their catalytic rates (Figure 2-53). For example, the binding of oxygen to the heme groups of hemoglobin is altered by the binding of H^+ and CO_2 to distant sites in the protein. This dependence of oxygen binding on pH and carbon dioxide concentration makes hemoglobin a very efficient oxygen transporter (p. 156). Allosteric control mediated by conformational changes in protein molecules is central to the regulation of metabolism.

Proteins containing pairs of sites that are coupled to each other by conformational changes have the capacity to convert energy from one form to another. Suppose that a protein has a catalytic site that hydrolyzes adenosine triphosphate (ATP) to adenosine diphosphate (ADP), an energetically favored reaction (Figure 2-54). The change from a bound triphosphate to a diphosphate group induces a change at the catalytic site that is transmitted to a different binding site some distance away on the same protein. The role of this second site is to bind another protein when ADP is bound to the first site and to release it when ATP is bound. An enzyme with these properties can function as a molecular motor that converts chemical bond energy into movement, as in muscle contraction (p. 931).

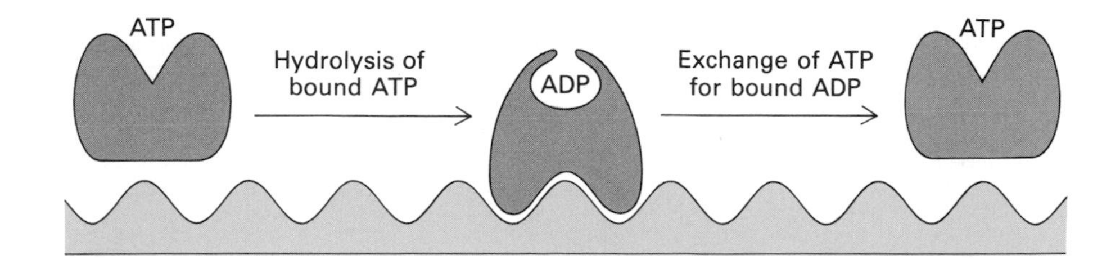

Figure 2-54
The hydrolysis of ATP at the catalytic site of an allosteric protein (red) can increase the affinity of a distant site for binding another protein (green).

SUMMARY

Proteins play key roles in virtually all biological processes. Nearly all catalysts in biological systems are proteins called enzymes. Hence, proteins determine the pattern of chemical transformations in cells. Proteins mediate a wide range of other functions, such as transport and storage, coordinated motions, mechanical support, immune protection, excitability, integration of metabolism, and the control of growth and differentiation.

The basic structural units of proteins are amino acids. All proteins in all species from bacteria to humans are constructed from the same set of twenty amino acids. The side chains of these building blocks differ in size, shape, charge, hydrogen-bonding capacity, and chemical reactivity. They can be grouped as follows: (1) aliphatic side chains—glycine, alanine, valine, leucine, isoleucine, and proline; (2) hydroxyl aliphatic side chains—serine and threonine; (3) aromatic side chains—phenylalanine, tyrosine, and tryptophan; (4) basic side chains—lysine, arginine, and histidine; (5) acidic side chains—aspartic acid and glutamic acid; (6) amide side chains—asparagine and glutamine; and (7) sulfur side chains—cysteine and methionine.

Many amino acids, usually more than a hundred, are joined by peptide bonds to form a polypeptide chain. A peptide bond links the α-carboxyl group of one amino acid and the α-amino group of the next one. Disulfide cross-links can be formed by cysteine residues. Some proteins are covalently modified and cleaved after their synthesis. Proteins have unique amino acid sequences that are genetically determined. The critical determinant of the biological function of a protein is its conformation, which is the three-dimensional arrangement of the atoms of a molecule. Three regularly repeating conformations of polypeptide chains are known: the α helix, the β pleated sheet, and the collagen helix. Segments of the α helix and the β pleated sheet are found in many proteins, as are hairpin β-turns. An important principle is that the amino acid sequence of a protein specifies its three-dimensional structure, as was first shown for ribonuclease. Reduced, unfolded ribonuclease spontaneously forms the correct disulfide pairings and regains full enzymatic activity when oxidized by air after removal of mercaptoethanol and urea. Proteins fold by the association of short polypeptide segments that transiently adopt α-helical or β-sheet forms. The strong tendency of hydrophobic residues to flee from water drives the folding of soluble proteins. Proteins are stabilized by many reinforcing hydrogen bonds and van der Waals interactions as well as by hydrophobic interactions.

Proteins are a unique class of macromolecules in being able to specifically recognize and interact with highly diverse molecules. The repertoire of twenty kinds of side chains enables proteins to fold into distinctive structures and form complementary surfaces and clefts. The catalytic power of enzymes comes from their capacity to bind substrates in precise orientations and to stabilize transition states in the making and breaking of chemical bonds. Conformational changes transmitted between distant sites in protein molecules are at the heart of the capacity of proteins to transduce energy and information.

SELECTED READINGS

Where to start

Doolittle, R. F., 1985. Proteins. *Sci. Amer.* 253(4):88–99. [A lucid overview that emphasizes molecular evolution. Reprinted in *The Molecules of Life*, a collection of readings from *Scientific American*, W. H. Freeman, 1985.]

Goldberg, M. E., 1985. The second translation of the genetic message: protein folding and assembly. *Trends Biochem. Sci.* 10:388–391.

Karplus, M., and McCammon, J. A., 1986. The dynamics of proteins. *Sci. Amer.* 254(4):42–51. [Available as *Sci. Amer.* Offprint 1569.]

Books on protein chemistry

Creighton, T. E., 1983. *Proteins: Structures and Molecular Principles.* W. H. Freeman.

Cantor, C. R., and Schimmel, P. R., 1980. *Biophysical Chemistry.* W. H. Freeman. [Chapters 2 and 5 in Part I and Chapters 20 and 21 in Part III give an excellent account of the principles of protein conformation.]

Fletterick, R. J., Schroer, T., and Matela, R. J., 1985. *Molecular Structure: Macromolecules in Three-Dimensions.* Blackwell Scientific Publications. [A fine introduction to molecular models.]

Schultz, G. E., and Schirmer, R. H., 1979. *Principles of Protein Structure.* Springer-Verlag.

Neurath, H., and Hill, R. L., (eds.), 1976. *The Proteins* (3rd ed.). Academic Press. [A multivolume treatise that contains many fine articles.]

Covalent modification of proteins

Wold, F., 1981. In vivo chemical modification of proteins (post-translational modification). *Ann. Rev. Biochem.* 50:783–814.

Glazer, A. N., DeLange, R. J., and Sigman, D. S., 1975. *Chemical Modification of Proteins.* North-Holland.

Conformation of proteins

Chothia, C., 1984. Principles that determine the structure of proteins. *Ann. Rev. Biochem.* 53:537–572.

Richardson, J. S., 1981. The anatomy and taxonomy of protein structure. *Adv. Protein Chem.* 34:167–339. [A lucid and beautifully illustrated account of three-dimensional architecture, with emphasis on supersecondary structural motifs.]

Folding of proteins

Harrison, S. C., and Durbin, R., 1985. Is there a single pathway for the folding of a polypeptide chain? *Proc. Nat. Acad. Sci.* 82:4028–4030. [The jigsaw-puzzle analogy for protein folding is presented in this incisive article.]

Kim, P. S., and Baldwin, R. L., 1982. Specific intermediates in the folding reactions of small proteins and the mechanism of protein folding. *Ann. Rev. Biochem.* 51:459–490.

Creighton, T. E., 1985. Energetics of protein structure and folding. *Biopolymers* 24:167–182.

Anfinsen, C. B., 1973. Principles that govern the folding of protein chains. *Science* 181:223–230.

Freedman, R. B., Brockway, B. E., and Lambert, N., 1984. Protein disulphide-isomerase and the formation of native disulphide bonds. *Biochem. Soc. Trans.* 12:929–932.

Prediction of protein structure

Blout, E. R., de Lozé, C., Bloom, S. M., and Fasman, G. D., 1960. The dependence of the conformations of synthetic polypeptides on amino acid composition. *J. Amer. Chem. Soc.* 82:3787–3789.

Chou, P. Y., and Fasman, G. D., 1974. Prediction of protein conformation. *Biochemistry* 13:222–244.

Kabsch, W., and Sander, C., 1983. How good are predictions of protein secondary structure? *FEBS Letters* 155:179–182.

PROBLEMS

1. Tropomyosin, a 70-kd muscle protein, is a two-stranded α-helical coiled coil. What is the length of the molecule?

2. Poly-L-leucine in an organic solvent such as dioxane is α-helical, whereas poly-L-isoleucine is not. Why do these amino acids with the same number and kinds of atoms have different helix-forming tendencies?

3. A mutation that changes an alanine residue in the interior of a protein to a valine is found to lead to a loss of activity. However, activity is regained when a second mutation at a different position changes an isoleucine residue to a glycine. How might this second mutation lead to a restoration of activity?

4. An enzyme that catalyzes disulfide-sulfhydryl exchange reactions has been isolated. Inactive scrambled ribonuclease is rapidly converted into enzymatically active ribonuclease by this enzyme. In contrast, insulin is rapidly inactivated by this enzyme. What does this important observation imply about the relation between the amino acid sequence of insulin and its three-dimensional structure?

5. A protease is an enzyme that catalyzes the hydrolysis of peptide bonds of target proteins. How might a protease bind a target protein so that its main chain becomes fully extended in the vicinity of the vulnerable peptide bond?

APPENDIX
Acid-Base Concepts

Ionization of Water

Water dissociates into hydronium (H_3O^+) and hydroxyl (OH^-) ions. For simplicity, we refer to the hydronium ion as a hydrogen ion (H^+) and write the equilibrium as

$$H_2O \rightleftharpoons H^+ + OH^-$$

The equilibrium constant K_{eq} of this dissociation is given by

$$K_{eq} = \frac{[H^+][OH^-]}{[H_2O]} \tag{1}$$

in which the terms in brackets denote molar concentrations. Because the concentration of water (55.5 M) is changed little by ionization, expression 1 can be simplified to give

$$K_w = [H^+][OH^-] \tag{2}$$

in which K_w is the ion product of water. At 25°C, K_w is 1.0×10^{-14}.

Note that the concentrations of H^+ and OH^- are reciprocally related. If the concentration of H^+ is high, then the concentration of OH^- must be low, and vice versa. For example, if $[H^+] = 10^{-2}$ M, then $[OH^-] = 10^{-12}$ M.

Definition of Acid and Base

An acid is a proton donor. A base is a proton acceptor.

$$\text{Acid} \rightleftharpoons H^+ + \text{base}$$

$$\underset{\text{Acetic acid}}{CH_3{-}COOH} \rightleftharpoons H^+ + \underset{\text{Acetate}}{CH_3{-}COO^-}$$

$$\underset{\text{Ammonium ion}}{NH_4^+} \rightleftharpoons H^+ + \underset{\text{Ammonia}}{NH_3}$$

The species formed by the ionization of an acid is its conjugate base. Conversely, protonation of a base yields its conjugate acid. Acetic acid and acetate ion are a conjugate acid-base pair.

Definition of pH and p*K*

The pH of a solution is a measure of its concentration of H^+. The pH is defined as

$$\text{pH} = \log_{10}\frac{1}{[H^+]} = -\log_{10}[H^+] \tag{3}$$

The ionization equilibrium of a weak acid is given by

$$HA \rightleftharpoons H^+ + A^-$$

The apparent equilibrium constant K for this ionization is

$$K = \frac{[H^+][A^-]}{[HA]} \tag{4}$$

The pK of an acid is defined as

$$\text{p}K = -\log K = \log\frac{1}{K} \tag{5}$$

Inspection of equation 4 shows that the pK of an acid is the pH at which it is half dissociated.

Henderson-Hasselbalch Equation

What is the relationship between pH and the ratio of acid to base? A useful expression can be derived from equation 4. Rearrangement of that equation gives

$$\frac{1}{[H^+]} = \frac{1}{K}\frac{[A^-]}{[HA]} \tag{6}$$

Taking the logarithm of both sides of equation 6 gives

$$\log \frac{1}{[H^+]} = \log \frac{1}{K} + \log \frac{[A^-]}{[HA]} \qquad (7)$$

Substituting pH for log $1/[H^+]$ and pK for log $1/K$ in equation 7 yields

$$pH = pK + \log \frac{[A^-]}{[HA]} \qquad (8)$$

which is commonly known as the Henderson-Hasselbalch equation.

The pH of a solution can be calculated from equation 8 if the molar proportion of A^- to HA and the pK of HA are known. Consider a solution of 0.1 M acetic acid and 0.2 M acetate ion. The pK of acetic acid is 4.8. Hence, the pH of the solution is given by

$$pH = 4.8 + \log \frac{0.2}{0.1} = 4.8 + \log 2$$
$$= 4.8 + 0.3 = 5.1$$

Conversely, the pK of an acid can be calculated if the molar proportion of A^- to HA and the pH of the solution are known.

Buffering Power

An acid-base conjugate pair (such as acetic acid and acetate ion) has an important property: it resists changes in the pH of a solution. In other words, it acts as a *buffer*. Consider the addition of OH^- to a solution of acetic acid (HA):

$$HA + OH^- \rightleftharpoons A^- + H_2O$$

A plot of the dependence of the pH of this solution on the amount of OH^- added is called a *titration curve* (Figure 2-55). Note that there is an inflection point in the curve at pH 4.8, which is the pK of acetic acid.

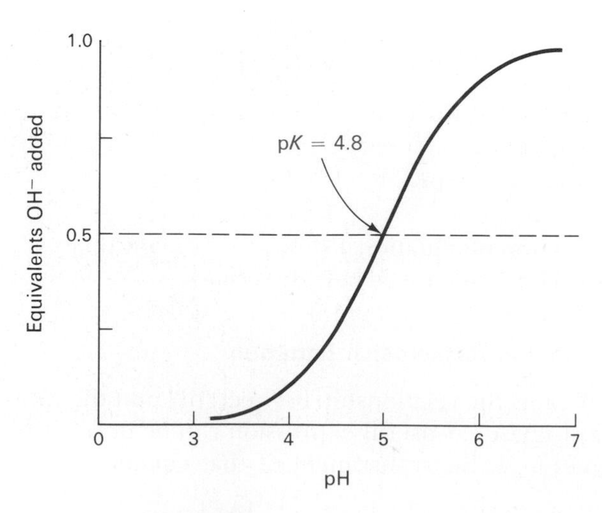

Figure 2-55
Titration curve of acetic acid.

In the vicinity of this pH, a relatively large amount of OH^- produces little change in pH. In general, a weak acid is most effective in buffering against pH changes in the vicinity of its pK value.

p*K* Values of Amino Acids

An amino acid such as glycine contains two ionizable groups: an α-carboxyl group and a protonated α-amino group. As base is added, these two groups are titrated (Figure 2-56). The pK of the α-COOH group is 2.3, whereas that of the α-NH_3^+ group is 9.6. The pK values of these groups in other amino acids are similar. Some amino acids, such as aspartic acid, also contain an ionizable side chain. The pK values of ionizable side chains in amino acids range from 3.9 (aspartic acid) to 12.5 (arginine).

$$^+H_3N{-}CH(R){-}COOH \xrightleftharpoons{OH^-} {}^+H_3N{-}CH(R){-}COO^-$$
$$\xrightleftharpoons{OH^-} H_2N{-}CH(R){-}COO^-$$

Figure 2-56
Titration of the α-carboxyl and α-amino groups of an amino acid.

Table 2-4
pK values of some amino acids

Amino acid	pK values (25°C)		
	α-COOH group	α-NH_3^+ group	Side chain
Alanine	2.3	9.9	
Glycine	2.4	9.8	
Phenylalanine	1.8	9.1	
Serine	2.1	9.2	
Valine	2.3	9.6	
Aspartic acid	2.0	10.0	3.9
Glutamic acid	2.2	9.7	4.3
Histidine	1.8	9.2	6.0
Cysteine	1.8	10.8	8.3
Tyrosine	2.2	9.1	10.9
Lysine	2.2	9.2	10.8
Arginine	1.8	9.0	12.5

Source: After J. T. Edsall and J. Wyman. *Biophysical Chemistry* (Academic Press, 1958), ch. 8.

CHAPTER 3

Exploring Proteins

In the preceding chapter, we saw that proteins play crucial roles in nearly all biological processes—in catalysis, transport, coordinated motion, excitability, and the control of growth and differentiation. This remarkable range of functions arises from the folding of proteins into many distinctive three-dimensional structures that bind highly diverse molecules. One of the major goals of biochemistry is to determine how amino acid sequences specify the conformations of proteins. We also want to learn how proteins bind specific substrates and other molecules, mediate catalysis, and transduce energy and information. An indispensable step in these studies is the purification of the protein of interest.

The three key approaches to analyzing and purifying proteins are electrophoresis, ultracentrifugation, and chromatography. Given a pure protein, it is possible to elucidate its amino acid sequence. The strategy is to divide and conquer, to obtain specific fragments that can readily be sequenced. Automated peptide sequencing and the application of recombinant DNA methods are providing a wealth of amino acid sequence data that are opening new vistas. For seeing beyond primary structure to conformation, for elucidating the precise positions of atoms in proteins, x-ray crystallography is the most powerful technique. The physiological context of a protein also needs to be known to fully understand how it functions. Antibodies are choice probes for locating proteins in vivo and measuring their quantities. The chapter closes with the synthesis of peptides, which makes feasible the synthesis of new drugs and antigens for inducing the formation of specific antibodies.

The exploration of proteins by this array of physical and chemical techniques has greatly enriched our understanding of the molecular basis of life and makes it possible to tackle some of the most challenging questions of biology in molecular terms.

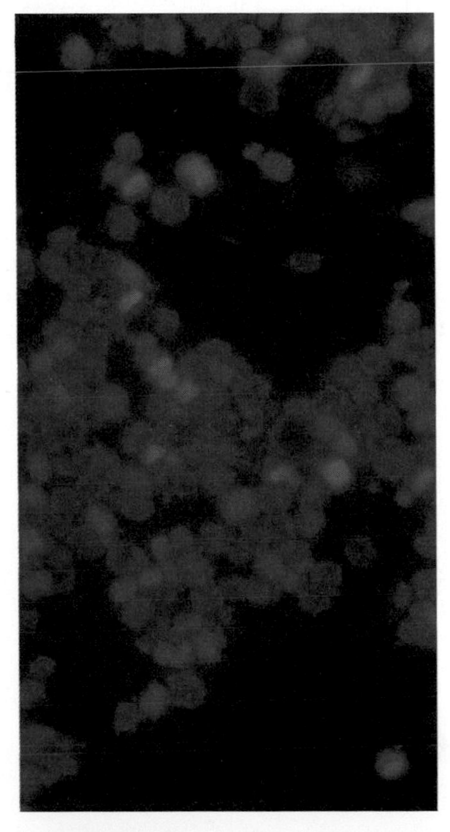

Micrograph of crystals of a light-harvesting protein of a photosynthetic bacterium. [Courtesy of Dr. Alexander Glazer.]

PROTEINS CAN BE SEPARATED BY GEL ELECTROPHORESIS AND DISPLAYED

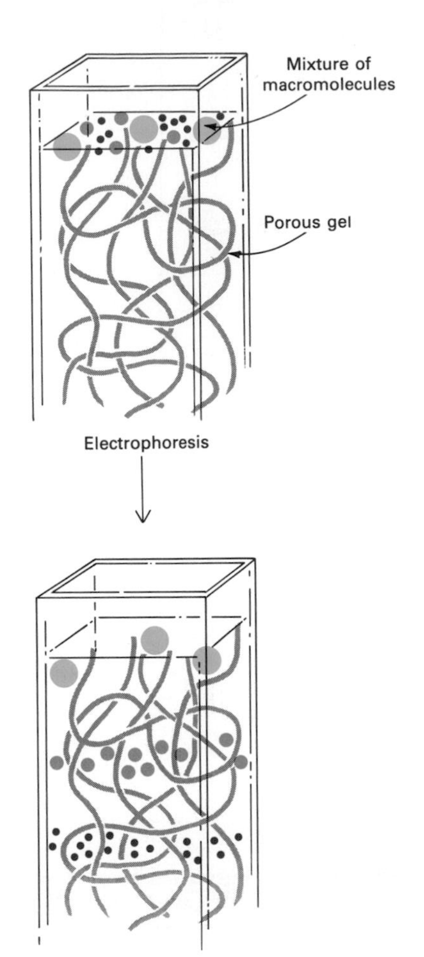

Figure 3-1
Sieving action of a porous polyacrylamide gel.

A molecule with a net charge will move in an electric field. This phenomenon, termed *electrophoresis*, offers a powerful means of separating proteins and other macromolecules, such as DNA and RNA. The velocity of migration (v) of a protein (or any molecule) in an electric field depends on the electric field strength (E), the net charge on the protein (z), and the frictional coefficient (f).

$$v = \frac{Ez}{f} \tag{1}$$

The electric force Ez driving the charged molecule toward the oppositely charged electrode is opposed by the viscous drag fv arising from friction between the moving molecule and the medium. The frictional coefficient f depends on both the mass and shape of the migrating molecule and the viscosity of the medium.

Electrophoretic separations are nearly always carried out in gels rather than in free solution for two reasons. First, gels suppress convective currents produced by small temperature gradients, a requirement for effective separation. Second, gels serve as molecular sieves that enhance separation (Figure 3-1). Molecules that are small compared with the pores in the gel readily move through the gel, whereas molecules much larger than the pores are almost immobile. Intermediate-size molecules move through the gel with various degrees of facility. Polyacrylamide gels are choice supporting media for electrophoresis because they are chemically inert and are readily formed by the polymerization of acrylamide. Moreover, their pore sizes can be controlled by choosing various concentrations of acrylamide and methylenebisacrylamide (a cross-linking reagent) at the time of polymerization (Figure 3-2).

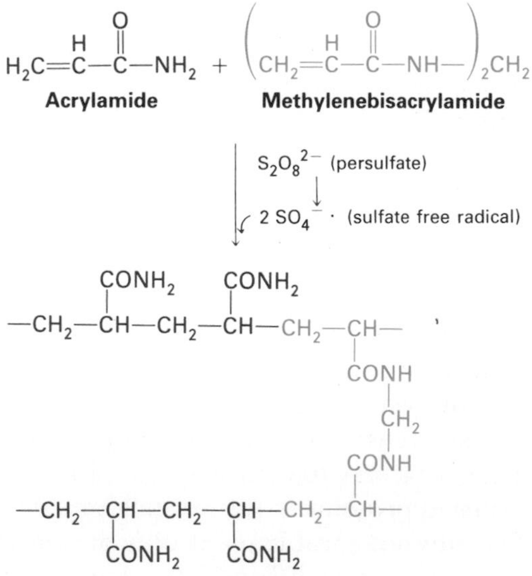

$H_2C{=}CH{-}C({=}O){-}NH_2$ + $(CH_2{=}CH{-}C({=}O){-}NH{-})_2CH_2$

Acrylamide **Methylenebisacrylamide**

$S_2O_8^{2-}$ (persulfate) → $2\ SO_4^{-}\cdot$ (sulfate free radical)

$-CH_2-CH(CONH_2)-CH_2-CH(CONH_2)-CH_2-CH-$
(crosslink: $-CH-$ — CONH — CH_2 — CONH — $-CH-$)
$-CH_2-CH(CONH_2)-CH_2-CH(CONH_2)-CH_2-CH-$

Figure 3-2
Formation of a polyacrylamide gel. The pore size can be controlled by adjusting the concentration of activated monomer (red) and crosslinker (green).

Proteins can be separated largely on the basis of mass by electrophoresis in a polyacrylamide gel under denaturing conditions. The mixture of proteins is first dissolved in a solution of sodium dodecyl sulfate (SDS), an anionic detergent that disrupts nearly all noncovalent interactions in native proteins. Mercaptoethanol or dithiothreitol is also added

$H_3C{-}(CH_2)_{10}{-}CH_2OSO_3^-\ Na^+$
Sodium dodecyl sulfate (SDS)

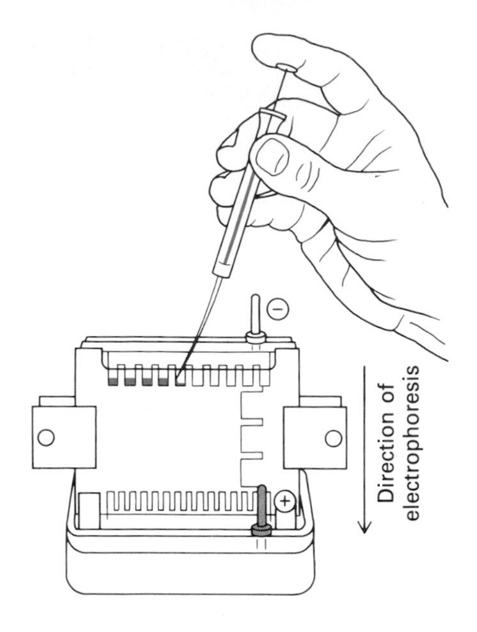

Loading of samples for electrophoresis. Typically, several samples are electrophoresed on one flat polyacrylamide gel. A microliter syringe is used to place solutions of proteins in the wells of the slab. A cover is then placed over the gel chamber and 200 volts are applied. The negatively charged SDS-protein complexes migrate in the direction of the anode, at the bottom of the gel.

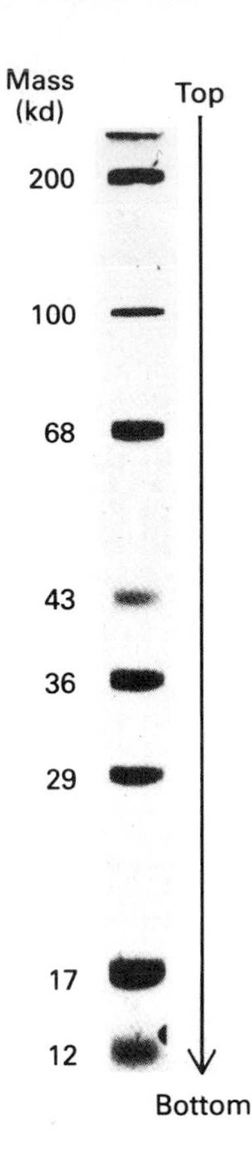

Figure 3-3
Proteins electrophoresed on an SDS-polyacrylamide gel can be visualized by staining with Coomassie blue.

to reduce disulfide bonds. Anions of SDS bind to main chains at a ratio of about one SDS for every two amino acid residues, which gives a complex of SDS with a denatured protein a large net negative charge that is roughly proportional to the mass of the protein. The negative charge acquired on binding SDS is usually much greater than the charge on the native protein; this native charge is thus rendered insignificant. The SDS complexes with the denatured proteins are then electrophoresed on a polyacrylamide gel, typically in the form of a thin verticle slab. The direction of electrophoresis is from top to bottom. Finally, the proteins in the gel can be visualized by staining them with silver or a dye such as Coomassie blue, which reveals a series of bands (Figure 3-3). Radioactive labels can be detected by placing a sheet of x-ray film over the gel, a procedure called autoradiography. *Small proteins move rapidly through the gel, whereas large ones stay at the top, near the point of application of the mixture.* The mobility of most polypeptide chains under these conditions is linearly proportional to the logarithm of their mass (Figure 3-4). This empirical relationship is not obeyed by some proteins; for example, some carbohydrate-rich proteins and membrane proteins migrate anomalously.

SDS-polyacrylamide gel electrophoresis is rapid, sensitive, and capable of a high degree of resolution. Electrophoresis and staining take about a day. As little as 0.1 μg (~2 pmol) of a protein gives a distinct band when stained with Coomassie blue, and even less (~0.02 μg) can be detected with a silver stain. Proteins that differ in mass by about 2% (e.g., 40 and 41 kd, arising from a difference of about ten residues) can usually be distinguished.

Proteins can also be separated electrophoretically on the basis of their relative contents of acidic and basic residues. The *isoelectric point* (pI) of a protein is the pH at which its net charge is zero. At this pH, its electrophoretic mobility is zero because z in equation 1 is equal to zero. For example, the pI of cytochrome c, a highly basic electron-transport protein, is 10.6, whereas that of serum albumin, an acidic protein in blood, is 4.8. Suppose that a mixture of proteins is electrophoresed in a pH gradient in a gel in the absence of SDS. Each protein will move until it reaches a position in the gel at which the pH is equal to the pI of the

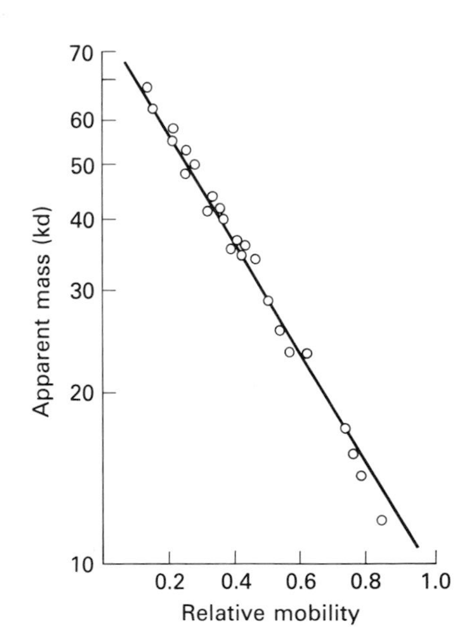

Figure 3-4
The electrophoretic mobility of many proteins in SDS-polyacrylamide gels is proportional to the logarithm of their mass. [After K. Weber and M. Osborn, *The Proteins,* 3rd ed. (Academic Press, 1975), vol. 1, p. 179.]

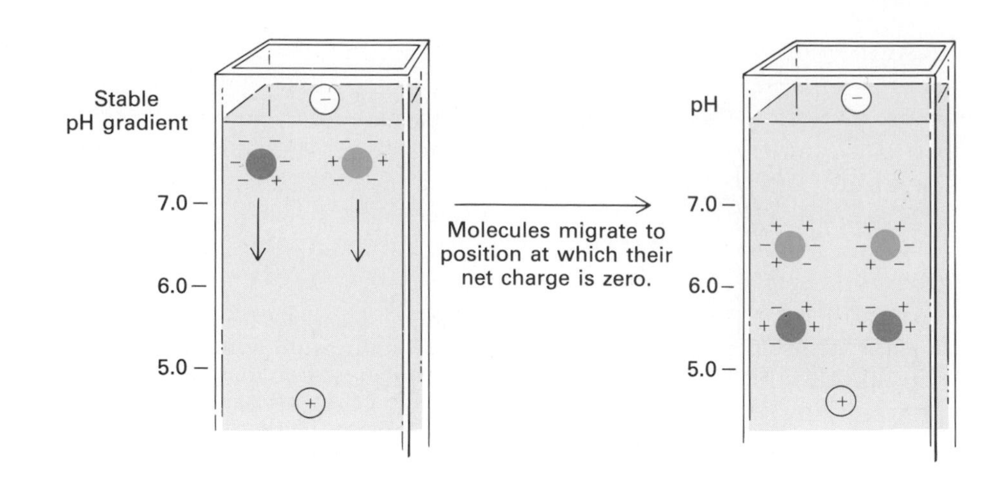

Figure 3-5
Isoelectric focusing separates proteins according to the pH at which their net charge is zero (isoelectric point). A stable pH gradient is produced in the gel before electrophoresing the mixture of proteins.

protein (Figure 3-5). This method of separating proteins according to their isoelectric point is called *isoelectric focusing*. The pH gradient in the gel is formed first by electrophoresing a mixture of *polyampholytes* (small multi-charged polymers) having many pI values. Isoelectric focusing can readily resolve proteins that differ in pI by as little as 0.01, which means that proteins differing by one net charge can be separated.

Isoelectric focusing can be combined with SDS-polyacrylamide gel electrophoresis to obtain very high-resolution separations. A sample is first subjected to isoelectric focusing. This gel is then placed horizontally on top of an SDS-polyacrylamide gel and electrophoresed vertically to yield a two-dimensional pattern of spots. In such a gel, proteins have been separated in the horizontal direction on the basis of isoelectric point, and in the vertical direction on the basis of mass. It is remarkable that more than a thousand different proteins in the bacterium *E. coli* can be resolved in a single experiment by two-dimensional electrophoresis (Figure 3-6).

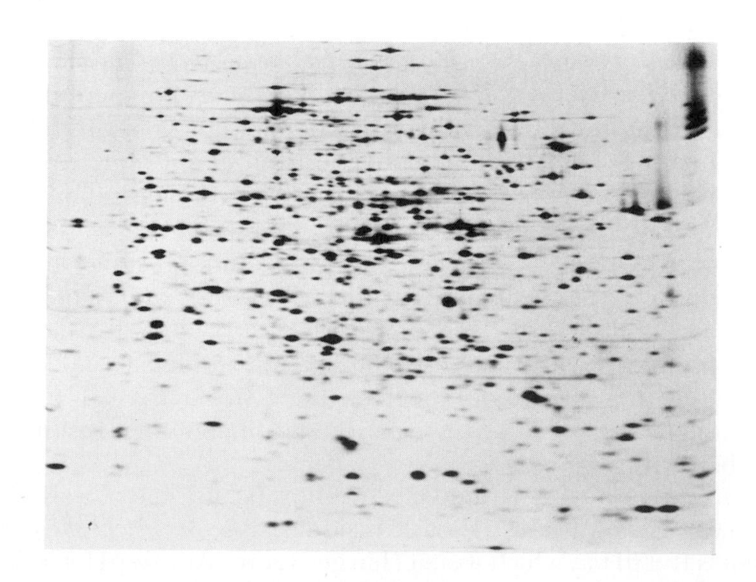

Figure 3-6
Two-dimensional electrophoresis of the proteins from *E. coli*. More than a thousand different proteins from this bacterium have been resolved. These proteins were separated according to their isoelectric pH in the horizontal direction and their apparent mass in the vertical direction. [Courtesy of Dr. Patrick H. O'Farrell.]

PROTEINS CAN BE PURIFIED ACCORDING TO SIZE, CHARGE, AND BINDING AFFINITY

Proteins can readily be visualized and differentiated by the electrophoretic methods described above. These gel techniques can also be used to obtain small quantities (micrograms) of purified polypeptides. However, they do not provide large amounts of purified proteins in their native state. Substantial quantities of purified proteins, of the order of many milligrams, are needed to fully elucidate their three-dimensional structure and their mechanism of action. Several thousand proteins have been purified in active form on the basis of such characteristics as *size, solubility, charge,* and *specific binding affinity*.

Before purifying a protein, various separation methods are tried and their efficiency is evaluated by assaying for a distinctive property of the protein of interest. The assay for an enzyme, for example, is typically a test for its specific catalytic activity. The purification procedure that is chosen is monitored at each step by SDS-polyacrylamide gel electrophoresis; if the procedure is successful, the desired protein becomes increasingly prominent in stained gels. The total amount of protein is measured, too, so that the degree of purification obtained in a particular step can be determined.

Proteins can be separated from small molecules by dialysis through a semipermeable membrane, such as a cellulose membrane with pores (Figure 3-7). Molecules having dimensions significantly greater than the pore diameter are retained inside the dialysis bag, whereas smaller molecules and ions traverse the pores of such a membrane and emerge in the dialysate outside the bag. More discriminating separations on the basis of size can be achieved by the technique of *gel-filtration chromatography* (Figure 3-8). The sample is applied to the top of a column consisting of porous beads made of an insoluble but highly hydrated polymer

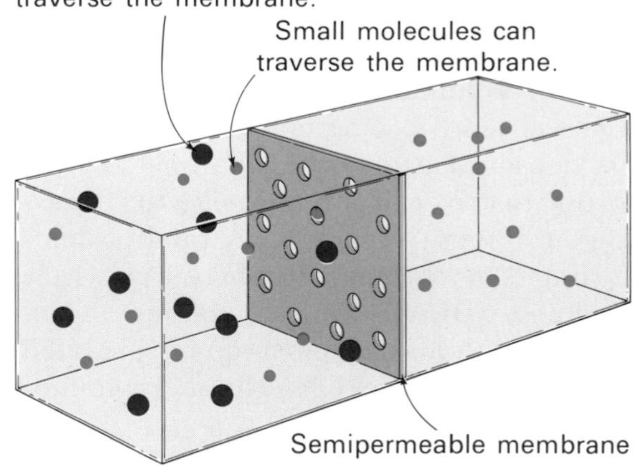
Large molecules cannot
traverse the membrane.
Small molecules can
traverse the membrane.
Semipermeable membrane

Figure 3-7
Separation of molecules on the basis of size by dialysis.

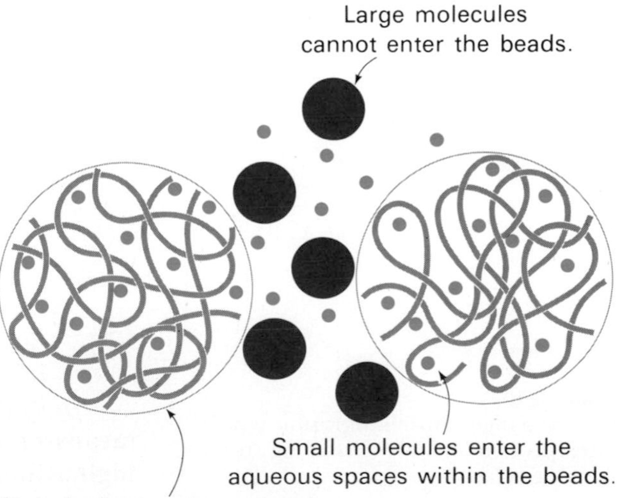
Large molecules
cannot enter the beads.
Small molecules enter the
aqueous spaces within the beads.
Carbohydrate polymer bead

Figure 3-8
Separation of molecules on the basis of size by gel-filtration chromatography.

such as dextran or agarose (which are carbohydrates) or polyacrylamide. Sephadex, Sepharose, and Bio-gel are commonly used commercial preparations of these beads, which are typically 100 μm (0.1 mm) in diameter. Small molecules can enter these beads, but large ones cannot. The result is that small molecules are distributed both in the aqueous solution inside the beads and between them, whereas large molecules are located only in the solution between the beads. *Large molecules flow more rapidly through this column and emerge first because a smaller volume is accessible to them* (Figure 3-9). It should be noted that the order of emergence of molecules from a column of porous beads is the reverse of the order in gel electrophoresis, in which a *continuous* polymer framework *impedes* the movement of large molecules (Figure 3-1). Much larger quantities of protein can be separated by gel-filtration chromatography than by gel electrophoresis, but a price is paid in the lower resolution of gel-filtration chromatography.

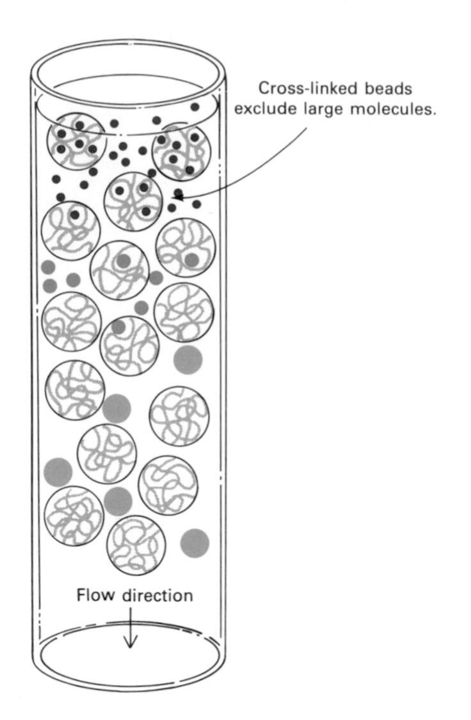

Figure 3-9
Separation of three proteins differing in size by gel-filtration chromatography.

The solubility of most proteins is lowered at high salt concentrations. This effect, called *salting-out*, is very useful, though not well understood. The dependence of solubility on salt concentration differs from one protein to another. Hence, salting-out can be used to fractionate proteins. For example, 0.8 M ammonium sulfate precipitates fibrinogen, a blood-clotting protein, whereas 2.4 M is needed to precipitate serum albumin. Salting-out is also useful for concentrating dilute solutions of proteins.

Proteins can be separated on the basis of their net charge by *ion-exchange chromatography*. If a protein has a net positive charge at pH 7, it will usually bind to a column of beads containing carboxylate groups, whereas a negatively charged protein will not (Figure 3-10). A positively charged protein bound to such a column can then be eluted (released) by increasing the concentration of sodium chloride or another salt in the eluting buffer. Sodium ions compete with positively charged groups on the protein for binding to the column. Proteins that have a low density of net positive charge will tend to emerge first, followed by those having a higher charge density. Factors other than net charge, such as affinity for the supporting matrix, can also influence the behavior of proteins on ion-exchange columns. Negatively charged proteins (anionic proteins) can be separated by chromatography on positively charged diethylaminoethyl-cellulose (DEAE-cellulose) columns. Conversely, positively charged proteins (cationic proteins) can be separated on negatively charged carboxymethyl-cellulose (CM-cellulose) columns.

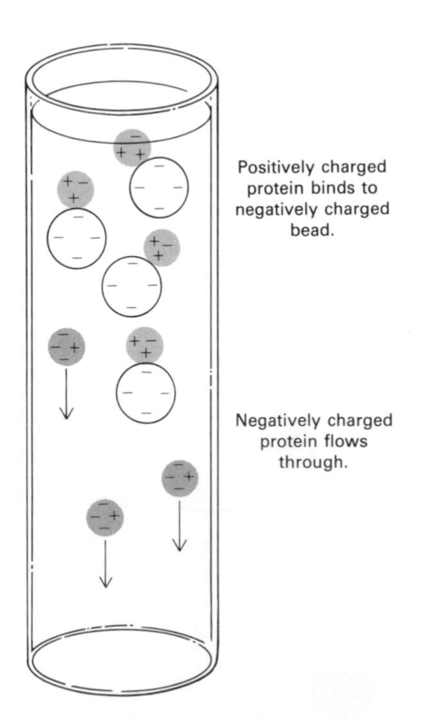

Figure 3-10
Ion-exchange chromatography separates proteins mainly according to their net charge.

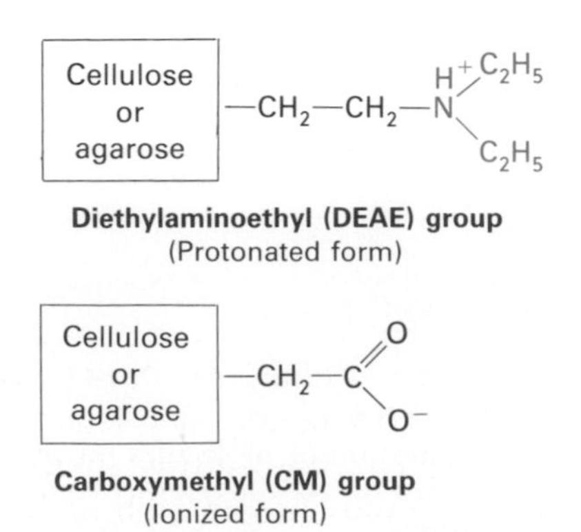

Affinity chromatography is another powerful and generally applicable means of purifying proteins. This technique takes advantage of the high affinity of many proteins for specific chemical groups. For example, the plant protein concanavalin A can be purified by passing a crude extract through a column of beads containing covalently attached glucose residues. Concanavalin A binds to such a column because it has affinity for glucose, whereas most other proteins do not. The bound concanavalin A can then be released from the column by adding a concentrated solution of glucose. The glucose in solution displaces the column-attached glucose residues from binding sites on concanavalin A (Figure 3-11). In general, affinity chromatography can be effectively used to isolate a protein that recognizes group X by (1) covalently attaching X or a derivative of it to a column, (2) adding a mixture of proteins to this column, which is then washed with buffer to remove unbound proteins, and (3) eluting the desired protein by adding a high concentration of a soluble form of X.

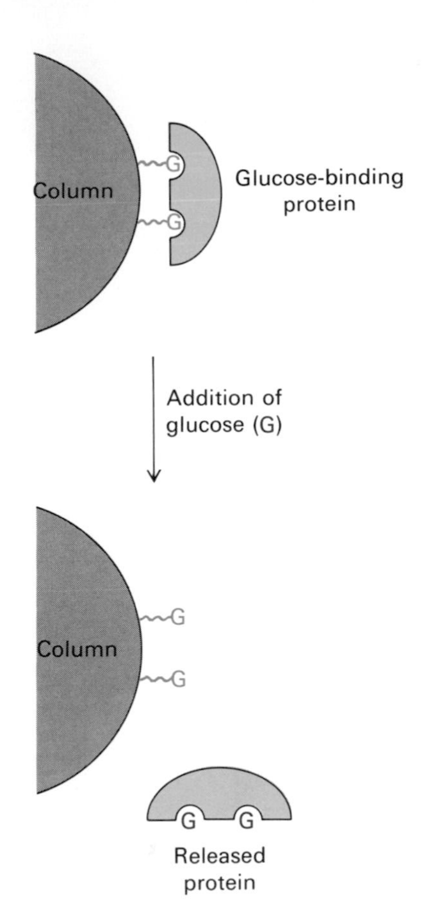

Figure 3-11
Affinity chromatography of concanavalin A (shown in yellow) on a column containing covalently attached glucose residues (G).

ULTRACENTRIFUGATION IS VALUABLE FOR SEPARATING BIOMOLECULES AND DETERMINING MOLECULAR WEIGHTS

Centrifugation is a powerful and generally applicable method for separating and analyzing cells, organelles, and biological macromolecules. A particle moving in a circle of radius r at an angular velocity ω is subject to a centrifugal (outward) field equal to $\omega^2 r$. The *centrifugal force* F_c on this particle is equal to the product of its effective mass m' and the centrifugal field.

$$F_c = m'\omega^2 r = m(1 - \bar{v}\rho)\omega^2 r \tag{2}$$

The effective mass m' is less than the mass m because the displaced fluid exerts an opposing force. This buoyancy factor is equal to $(1 - \bar{v}\rho)$, where $\bar{v}$ is the partial specific volume of the particle and ρ is the density of the solution. A particle moves in this field at a constant velocity v when F_c is equal to the viscous drag vf, where f is the frictional coefficient of the particle. Hence, the migration velocity (sedimentation velocity) of the particle is

$$v = \frac{F_c}{f} = \frac{m(1 - \bar{v}\rho)\omega^2 r}{f} \tag{3}$$

Note that this expression for movement in a centrifugal field is analogous to equation 1 for movement in an electric field.

Equation 3 shows that the sedimentation velocity is directly proportional to the strength of the centrifugal field. Hence, it is possible to define a measure of sedimentation that depends on the properties of the particle and solution but is independent of how fast the sample is spun. The *sedimentation coefficient* s, defined as the velocity divided by the centrifugal field, is equal to

$$s = \frac{v}{\omega^2 r} = \frac{m(1 - \bar{v}\rho)}{f} \tag{4}$$

Sedimentation coefficients are usually expressed in *Svedberg units*. A Svedberg (S) is equal to 10^{-13} seconds. For example, suppose that a 150-kd antibody protein is spun in an ultracentrifuge at a radius of 8 cm at 75,000 revolutions per minute (rpm). The centrifugal field under these conditions is 4.9×10^8 cm s^{-2}, which is about 500,000 times the strength of the earth's gravitational field (g). If the velocity of the protein in this field is 3.4×10^{-4} cm/s, then its sedimentation coefficient is 7S.

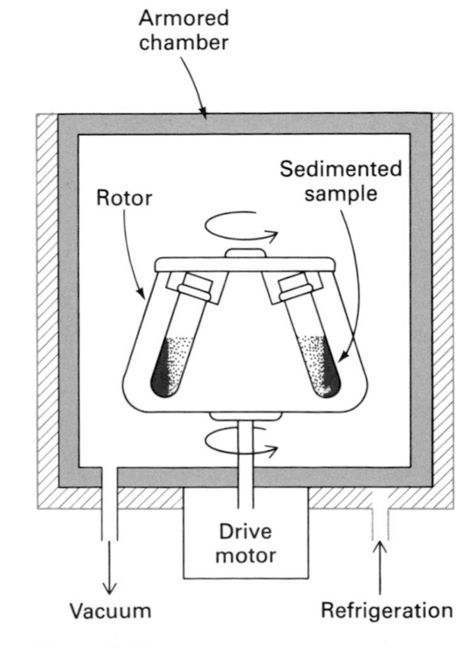

Figure 3-12
Ultracentrifuge rotor containing two sample tubes.

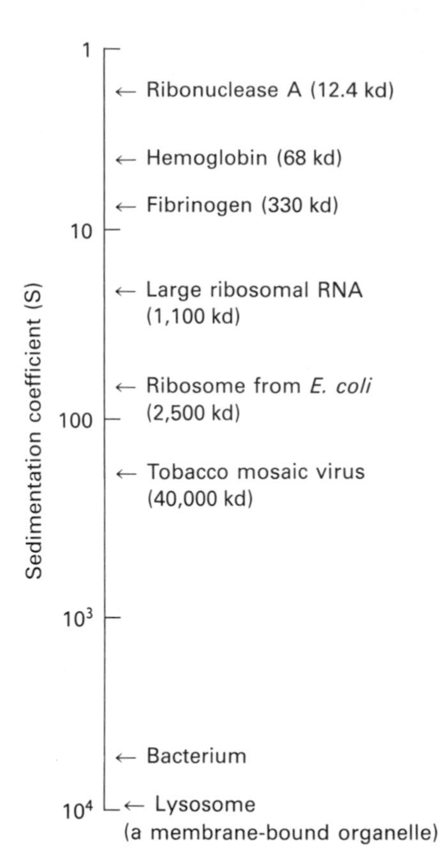

Figure 3-13
Span of S values of biomolecules and cells.

Several important conclusions can be drawn from equation 3:

1. The sedimentation velocity of a particle is proportional to its mass. A 200-kd protein moves at twice the velocity of a 100-kd protein of the same shape and density.

2. A dense particle moves more rapidly than a less dense one because the opposing buoyant force is smaller for a dense particle.

3. Shape, too, is important because it affects the viscous drag. The frictional coefficient of a compact particle is smaller than that of an extended particle of the same mass. A parachutist with a defective unopened parachute falls much more quickly than one with a functioning opened parachute.

4. The sedimentation velocity depends also on the density of the solution (ρ). Particles sink when $\bar{v}\rho < 1$, float when $\bar{v}\rho > 1$, and do not move when $\bar{v}\rho = 1$.

Let us see how centrifugation can be used to separate proteins with different sedimentation coefficients (Figure 3-14). The first step in *zonal centrifugation* (also called *band centrifugation*) is to form a density gradient in a centrifuge tube by mixing different proportions of a low-density solution (such as 5% sucrose) and a high-density one (such as 20% sucrose). The role of the density gradient here is to prevent convective flow. A small volume of the solution containing the mixture of proteins is then layered on top of this density gradient. When the rotor is spun, proteins move through the gradient and separate according to their sedimentation coefficients. Centrifugation is stopped before the fastest protein reaches the bottom of the tube. The separated bands of protein can be harvested by making a hole in the bottom of the tube and collecting drops. The drops can be assayed for protein content and catalytic activity or another functional property. This sedimentation-velocity technique readily separates proteins differing in sedimentation coefficient by a factor of two or more.

The mass (molecular weight) of a protein can be directly determined by *sedimentation equilibrium*, in which a sample is centrifuged at relatively low speed so that sedimentation is counterbalanced by diffusion. Under these conditions, a smooth gradient of protein concentration develops. The dependence of concentration on distance from the rotation axis reveals the mass of the particle. The mass m is given by

$$m = \frac{2kT}{(1 - \bar{v}\rho)\omega^2} \log_e c_2/(c_1(r_2^2 - r_1^2)) \qquad (5)$$

where c_1 and c_2 are the concentrations at distances r_1 and r_2 from the rotation axis, k is Boltzmann's constant, and T is the absolute temperature. *This sedimentation-equilibrium technique for determining mass is rigorous and can be applied under nondenaturing conditions in which the native structure of multimeric proteins is preserved.* In contrast, SDS-polyacrylamide gel electrophoresis (p. 45) provides an *estimate* of the mass of dissociated polypeptide chains under *denaturing* conditions.

Ninhydrin

AMINO ACID SEQUENCES CAN BE DETERMINED BY AUTOMATED EDMAN DEGRADATION

We turn now from the purification of proteins and analysis of their hydrodynamic properties to the elucidation of their amino acid se-

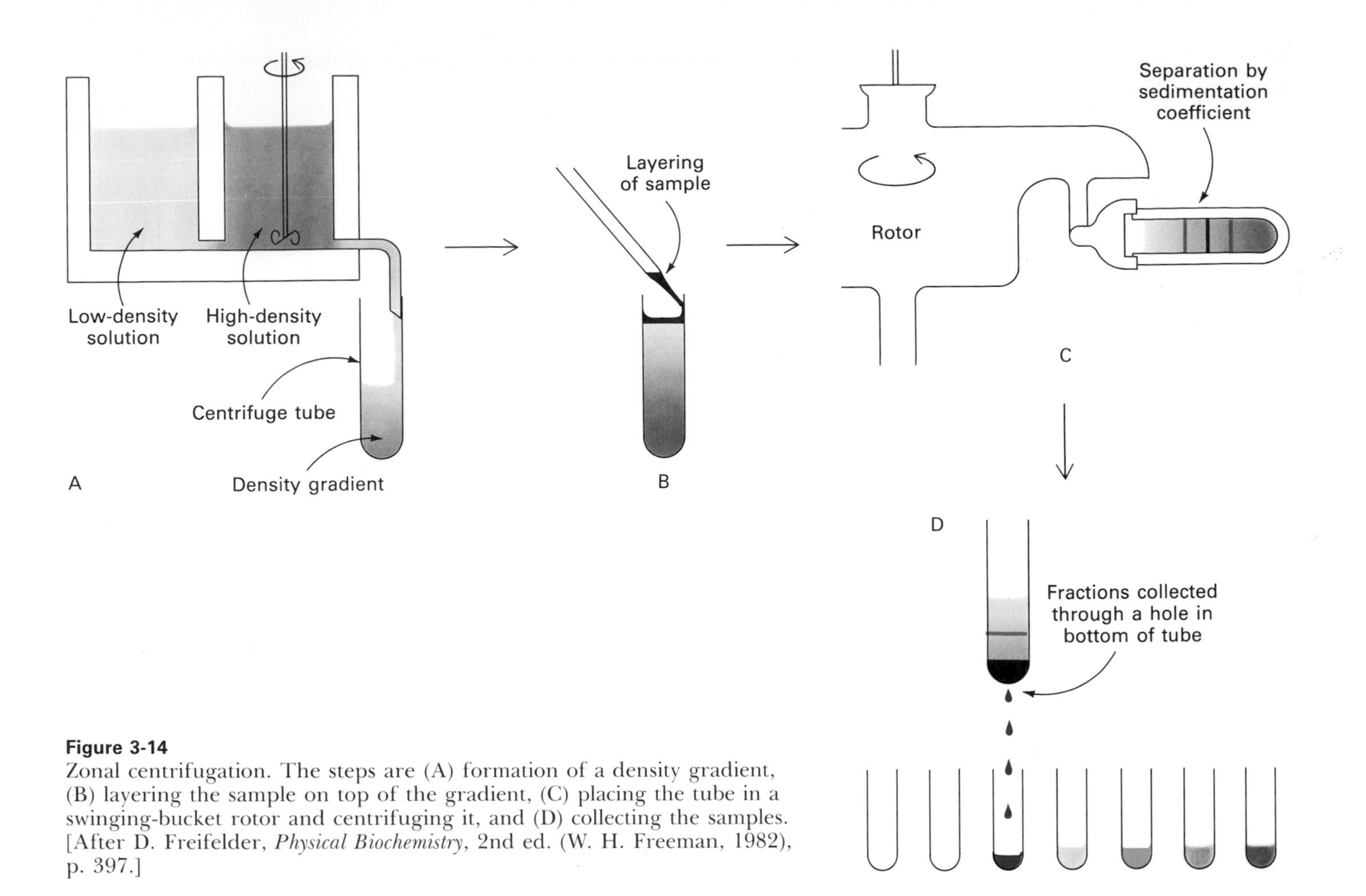

Figure 3-14
Zonal centrifugation. The steps are (A) formation of a density gradient, (B) layering the sample on top of the gradient, (C) placing the tube in a swinging-bucket rotor and centrifuging it, and (D) collecting the samples. [After D. Freifelder, *Physical Biochemistry*, 2nd ed. (W. H. Freeman, 1982), p. 397.]

quences. Let us consider how the sequence of a short peptide, such as

Ala-Gly-Asp-Phe-Arg-Gly

could be established. First, the *amino acid composition* of the peptide is determined. The peptide is hydrolyzed into its constituent amino acids by heating it in 6 N HCl at 110°C for 24 hours. Stanford Moore and William Stein showed that amino acids in hydrolysates can be separated by ion-exchange chromatography on columns of sulfonated polystyrene and quantitated by reacting them with *ninhydrin*. α-Amino acids treated this way give an intense blue color, whereas imino acids, such as proline, give a yellow color. The concentration of amino acid in a solution is proportional to the optical absorbance of the solution after heating it with ninhydrin. This technique can detect a microgram (10 nmol) of an amino acid, which is about the amount present in a thumbprint. As little as a nanogram (10 pmol) of an amino acid can be detected by means of *fluorescamine*, which reacts with the α-amino group to form a highly fluorescent product (Figure 3-15). The identity of the amino acid is revealed by its elution volume, which is the volume of buffer used to remove the amino acid from the column (Figure 3-16). A comparison of the chromatographic patterns of our sample hydrolysate with that of a standard mixture of amino acids would show that the amino acid composition of the peptide is

(Ala, Arg, Asp, Gly_2, Phe)

Fluorescamine

$R{-}NH_2$

Fluorescent amine derivative

Figure 3-15
Reaction of fluorescamine with the α-amino group of an amino acid to form a fluorescent derivative.

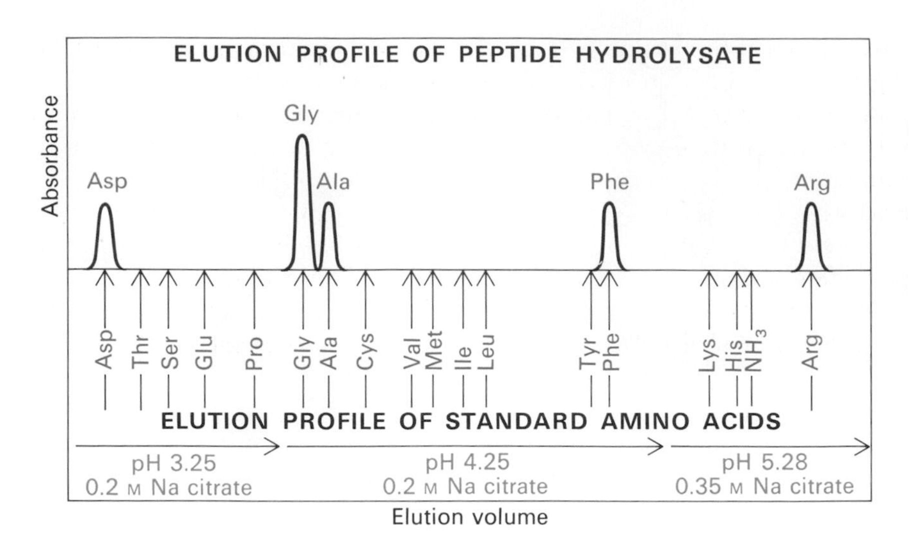

Figure 3-16
Different amino acids in a peptide hydrolysate can be separated by ion-exchange chromatography on a sulfonated polystyrene resin (such as Dowex-50). Buffers of increasing pH are used to elute the amino acids from the column. Aspartate, which has an acidic side chain, is first to emerge, whereas arginine, which has a basic side chain, is the last.

The parentheses denote that this is the amino acid composition of the peptide, not its sequence.

The amino-terminal residue of a protein or peptide can be identified by labeling it with a compound that forms a stable covalent link. *Fluorodinitrobenzene* (FDNB) was first used for this purpose by Frederick Sanger. *Dabsyl chloride* is now commonly used because it forms intensely colored derivatives that can be detected with high sensitivity. It reacts with an uncharged α-NH_2 group to form a sulfonamide derivative that is stable under conditions that hydrolyze peptide bonds (Figure 3-17). Hydrolysis of our sample dabsyl-peptide in 6 N HCl would yield a dabsyl-amino acid, which could be identified as dabsyl-alanine by its

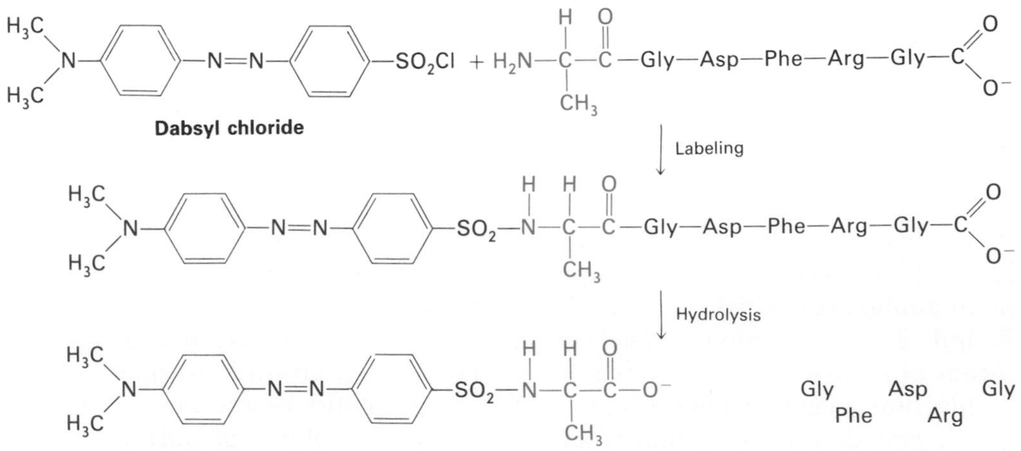

Dabsyl chloride

Labeling

Hydrolysis

Gly Phe Asp Arg Gly

Figure 3-17
Determination of the amino-terminal residue of a peptide. Dabsyl chloride is used to label the peptide, which is then hydrolyzed. The dabsyl-amino acid (dabsyl-alanine in this example) is identified by its chromatographic characteristics.

chromatographic properties. *Dansyl chloride*, another much-used labeling reagent, forms fluorescent sulfonamide derivatives.

Although the dabsyl method for determining the amino-terminal residue is sensitive and powerful, it cannot be used repeatedly on the same peptide because the peptide is totally degraded in the acid-hydrolysis step. Pehr Edman devised a method for labeling the amino-terminal residue and cleaving it from the peptide without disrupting the peptide bonds between the other amino acid residues. The *Edman degradation* sequentially removes one residue at a time from the amino end of a peptide (Figure 3-18). *Phenyl isothiocyanate* reacts with the un-

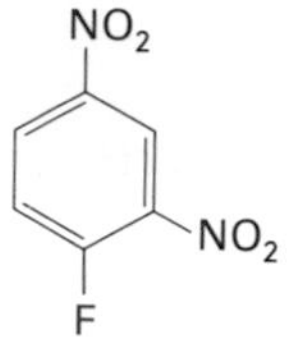

Fluorodinitrobenzene

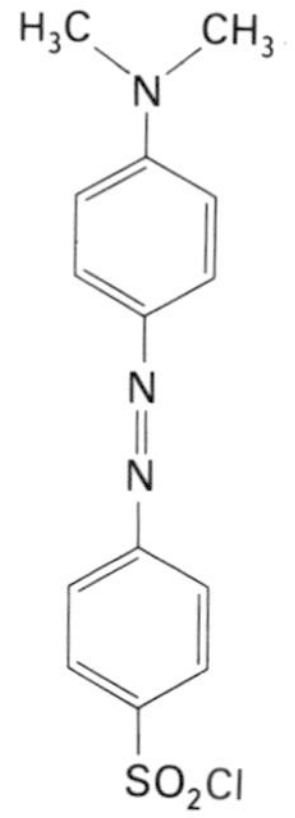

Dabsyl chloride

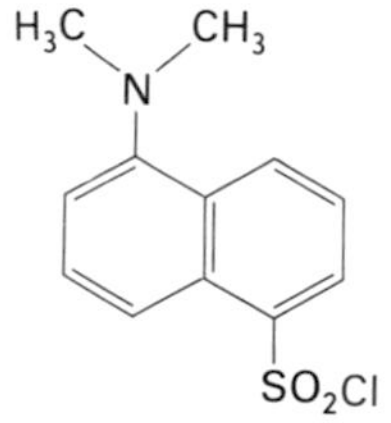

Dansyl chloride

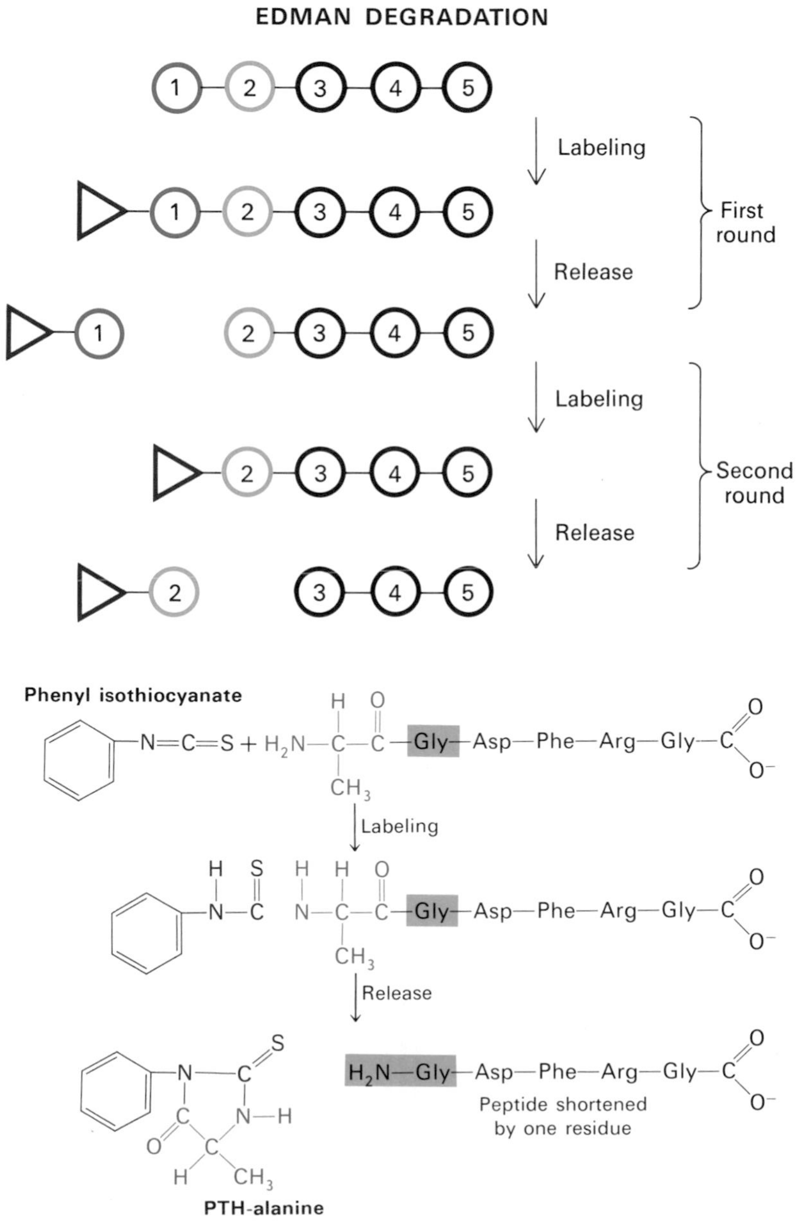

EDMAN DEGRADATION

Labeling
Release
First round
Labeling
Release
Second round

Phenyl isothiocyanate
Labeling
Release
Peptide shortened by one residue
PTH-alanine

Figure 3-18
The Edman degradation. The labeled amino-terminal residue (PTH-alanine in the first round) can be released without hydrolyzing the rest of the peptide. Hence, the amino-terminal residue of the shortened peptide (Gly-Asp-Phe-Arg-Gly) can be determined in the second round. Three more rounds of the Edman degradation reveal the complete sequence of the original peptide.

charged terminal amino group of the peptide to form a phenylthiocarbamoyl derivative. Then, under mildly acidic conditions, a cyclic derivative of the terminal amino acid is liberated, which leaves an intact peptide shortened by one amino acid. The cyclic compound is a phenylthiohydantoin (PTH) amino acid, which can be identified by chromatographic procedures. Furthermore, the amino acid composition of the shortened peptide:

$$(\text{Arg, Asp, Gly}_2\text{, Phe})$$

can be compared with that of the original peptide:

$$(\text{Ala, Arg, Asp, Gly}_2\text{, Phe})$$

The difference between these analyses is one alanine residue, which shows that alanine is the amino-terminal residue of the original peptide. The Edman procedure can then be repeated on the shortened peptide. The amino acid analysis after the second round of degradation is

$$(\text{Arg, Asp, Gly, Phe})$$

showing that the second residue from the amino end is glycine. This conclusion can be confirmed by chromatographic identification of PTH-glycine obtained in the second round of the Edman degradation. Three more rounds of the Edman degradation will reveal the complete sequence of the original peptide.

Analyses of protein structures have been markedly accelerated by the development of *sequenators*, which are automated instruments for the determination of amino acid sequence. In a liquid-phase sequenator, a thin film of protein in a spinning cylindrical cup is subjected to the Edman degradation. The reagents and extracting solvents are passed over the immobilized film of protein, and the released PTH-amino acid is identified by high-performance liquid chromatography (HPLC; Figure 3-19). One cycle of the Edman degradation is carried out in less

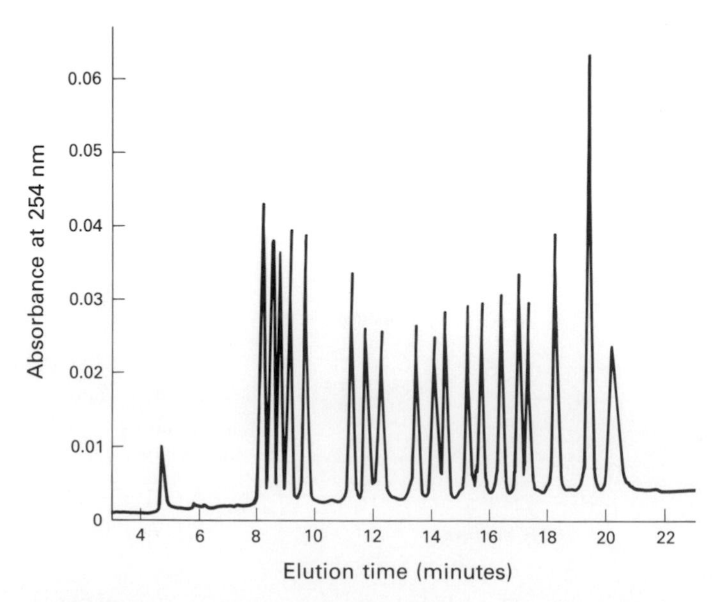

Figure 3-19
PTH-amino acids can be rapidly separated by high-pressure liquid chromatography (HPLC). In this HPLC profile, a mixture of PTH-amino acids is clearly resolved into its components. An individual amino acid can be identified by comparing its profile with this one.

than two hours. By repeated degradations, the amino acid sequence of some fifty residues in a protein can be determined. A recently devised gas-phase sequenator can analyze picomole quantities of peptides and proteins. This high sensitivity makes it feasible to analyze the sequence of a protein sample eluted from a single band of an SDS-polyacrylamide gel.

PROTEINS CAN BE SPECIFICALLY CLEAVED INTO SMALL PEPTIDES TO FACILITATE ANALYSIS

Peptides much longer than about fifty residues cannot be reliably sequenced by the Edman method because not quite all peptides in the reaction mixture release the amino acid derivative at each step. If the efficiency of release of each round were 98%, the proportion of "correct" amino acid released after sixty rounds would be only 0.3 (0.98^{60})—a hopelessly impure mix. This obstacle can be circumvented by specifically cleaving a protein into peptides not much longer than fifty residues. In essence, the strategy is to divide and conquer.

Specific cleavage can be achieved by chemical or enzymatic methods. For example, Bernhard Witkop and Erhard Gross discovered that *cyanogen bromide* (CNBr) splits polypeptide chains only on the carboxyl side of methionine residues (Figure 3-20). A protein that has ten methionines will usually yield eleven peptides on cleavage with CNBr. Highly specific cleavage is also obtained with trypsin, a proteolytic enzyme from pancreatic juice. Trypsin cleaves polypeptide chains on the carboxyl side of arginine and lysine residues (Figure 3-21). A protein

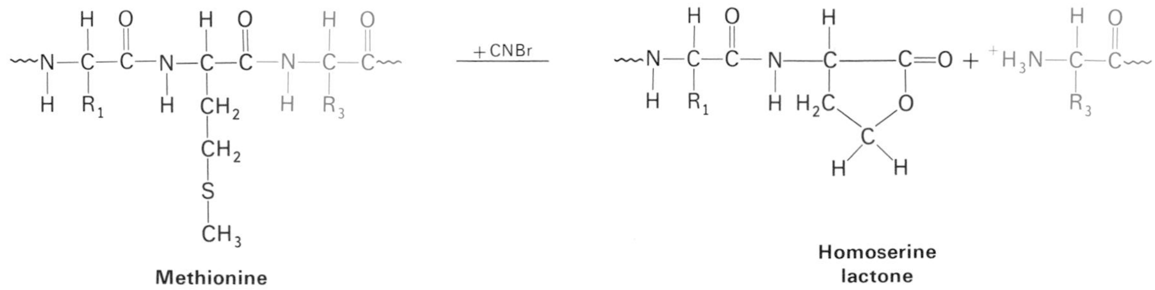

Methionine

+CNBr

Homoserine lactone

Figure 3-20
Cyanogen bromide cleaves polypeptides on the carboxyl side of methionine residues.

Lysine or arginine

$+ H_2O$ Trypsin

Lysine or arginine

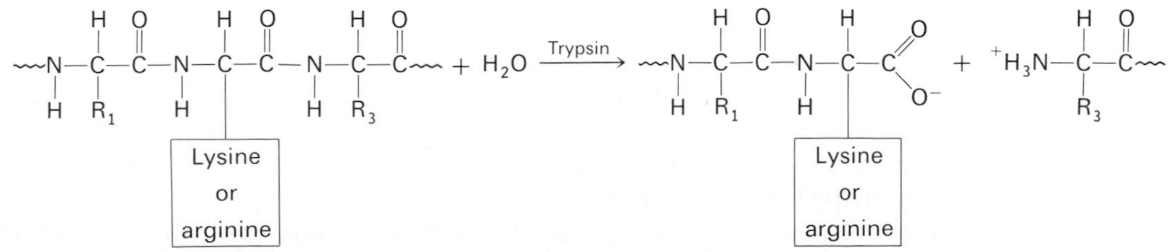

Figure 3-21
Trypsin hydrolyzes polypeptide on the carboxyl side of arginine and lysine residues.

that contains nine lysines and seven arginines will usually yield seventeen peptides on digestion with trypsin. Each of these tryptic peptides, except for the carboxyl-terminal peptide of the protein, will end with either arginine or lysine. Several other ways of specifically cleaving polypeptide chains are given in Table 3-1.

Table 3-1
Specific cleavage of polypeptides

Reagent	Cleavage site
Chemical cleavage	
Cyanogen bromide	Carboxyl side of methionine residues
O-Iodosobenzoate	Carboxyl side of tryptophan residues
Hydroxylamine	Asparagine–glycine bonds
2-Nitro-5-thiocyanobenzoate	Amino side of cysteine residues
Enzymatic cleavage	
Trypsin	Carboxyl side of lysine and arginine residues
Clostripain	Carboxyl side of arginine residues
Staphylococcal protease	Carboxyl side of aspartate and glutamate residues (glutamate only under certain conditions)

The peptides obtained by specific chemical or enzymatic cleavage are separated by chromatography. The sequence of each purified peptide is then determined by the Edman method. At this point, the amino acid sequences of segments of the protein are known, but the order of these segments is not yet defined. The necessary additional information is obtained from *overlap peptides* (Figure 3-22). An enzyme different from trypsin is used to split the polypeptide chain at other linkages. For example, chymotrypsin cleaves preferentially on the carboxyl side of aromatic and some other bulky nonpolar residues. Because these chymotryptic peptides overlap two or more tryptic peptides, they can be used to establish the order of the peptides. The entire amino acid sequence of the polypeptide chain is then known.

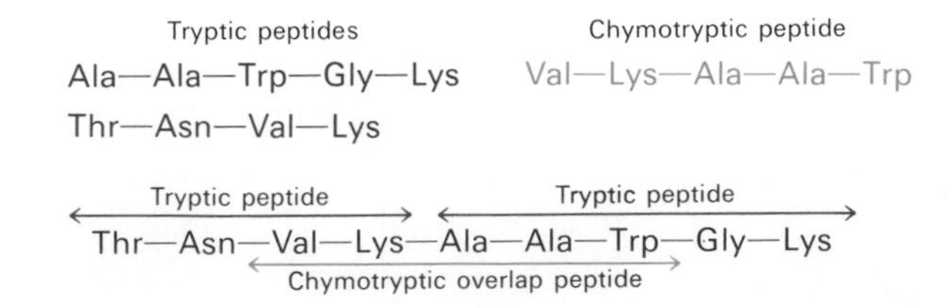

Figure 3-22
The peptide obtained by chymotrypic digestion overlaps two tryptic peptides, which thus establishes their order.

These methods apply to a protein consisting of a single polypeptide chain devoid of disulfide bonds. Additional steps are necessary if a protein has disulfide bonds or more than one chain. For a protein made up of two or more polypeptide chains held together by noncovalent bonds, denaturing agents, such as urea or guanidine hydrochloride, are used to dissociate the chains. The dissociated chains must be separated before sequence determination can begin. Polypeptide chains linked by disulfide bonds are first separated by reduction with β-mercaptoethanol or dithiothreitol. To prevent the cysteine residues

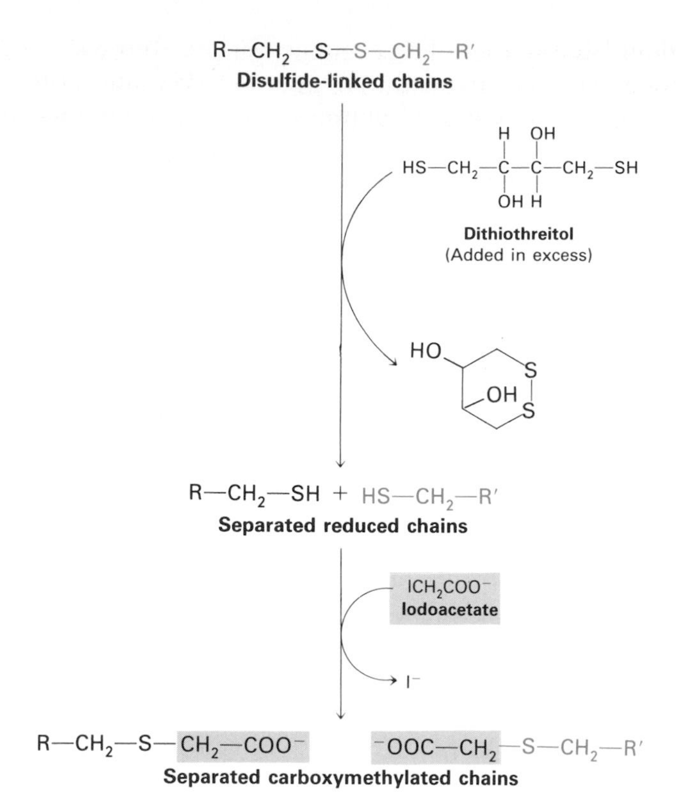

Figure 3-23
Polypeptides linked by disulfide bonds can be separated by reduction with dithiothreitol followed by alkylation.

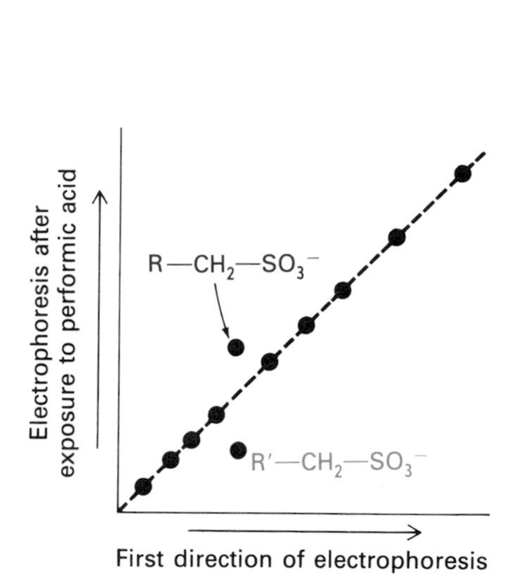

Figure 3-24
Detection of peptides joined by disulfides by diagonal electrophoresis. The mixture of peptides is electrophoresed in the horizontal direction before treatment with performic acid, and then in the vertical direction.

from recombining, they are then alkylated with iodoacetate to form stable *S*-carboxymethyl derivatives (Figure 3-23).

The positions of disulfide bonds can be determined by a *diagonal electrophoresis* technique (Figure 3-24). First, the protein is specifically cleaved into peptides under conditions in which the disulfides stay intact. The mixture of peptides is electrophoresed, and the resulting sheet is exposed to vapors of performic acid, which cleaves disulfides and converts them into cysteic acid residues. Peptides originally linked by disulfides are now independent and also more acidic because of the formation of an SO_3^- group. This mixture is electrophoresed in the perpendicular direction under the same conditions as in the first electrophoresis. Peptides that were devoid of disulfides will have the same mobility as before, and consequently all will be located on a single diagonal line. In contrast, the newly formed peptides containing cysteic acid will usually migrate differently from their parent disulfide-linked peptide and hence will lie off the diagonal.

Cystine

Performic acid

Cysteic acid

RECOMBINANT DNA TECHNOLOGY HAS REVOLUTIONIZED PROTEIN SEQUENCING

Hundreds of proteins have been sequenced by Edman degradation of peptides derived from specific cleavages. Protein sequence determination is a demanding and time-consuming process. The elucidation of the sequence of large proteins, those with more than 1000 residues, usually requires heroic effort. Fortunately, a complementary experimental approach based on recombinant DNA technology has become available. As will be discussed in Chapter 6, long stretches of DNA can

DNA sequence	Amino acid sequence
GGG	Gly
TTC	Phe
TTG	Leu
GGA	Gly
GCA	Ala
GCA	Ala
AGG	Gly
AAG	Ser
CAC	Thr
TAT	Met
GGG	Gly
GCA	Ala

Figure 3-25
The complete nucleotide sequence of the AIDS (acquired immunodeficiency disease syndrome) virus was determined within a year after the isolation of the virus. A portion of the DNA sequence specified by the RNA genome of the virus is shown here with the corresponding amino acid sequence (deduced from a knowledge of the genetic code).

be cloned and sequenced. The sequence of the four kinds of bases in DNA—adenine (A), thymine (T), guanine (G), and cytosine (C)—directly reveals the amino acid sequence of the protein encoded by the gene or the corresponding messenger RNA molecule (Figure 3-25). The amino acid sequence deduced by reading the DNA sequence is that of the *nascent* protein, the direct product of the translational machinery in ribosomes that links amino acids in a sequence specified by a messenger RNA template.

As discussed previously, many proteins are modified after synthesis (p. 24). Some have their ends trimmed, and others arise by cleavage of a larger initial polypeptide chain. Cysteine residues in some proteins are oxidized to form disulfide links, which can be either within a chain or between polypeptide chains. Specific side chains of some proteins are altered. For example, proteins targeted to membranes usually contain carbohydrate units attached to specific asparagine side chains. Amino acid sequences derived from DNA sequences are rich in information, but they do not disclose such posttranslational modifications. Chemical analyses of proteins themselves are needed to delineate the nature of these changes, which are critical for the biological activities of most proteins. *Thus, DNA sequencing and protein chemical analyses are complementary approaches toward elucidating the structural basis of protein function.* Recombinant DNA technology is producing a wealth of amino acid sequence information at a remarkable rate. The sequences of more than 10^6 residues in about 3000 proteins are now known. At the present pace, the number of known sequences is expected to double in less than two years.

AMINO ACID SEQUENCES PROVIDE MANY KINDS OF INSIGHTS

Amino acid sequences can be highly informative in a variety of ways:

1. *The sequence of a protein of interest can be compared with all other known ones to ascertain whether significant similarities exist. Does this protein belong to one of the established families?* For example, myoglobin and hemoglobin belong to the globin family. Chymotrypsin and trypsin are members of the serine protease family, a clan of proteolytic enzymes that have a common catalytic mechanism based on a reactive serine residue. A search for kinship between a newly sequenced protein and several thousand previously sequenced ones takes about twenty minutes on a personal computer. Quite unexpected results sometimes emerge from such comparisons. For example, a viral protein that produces cancer in susceptible hosts was found to be nearly identical to a normal cellular growth factor (p. 877). This startling finding advanced the understanding of both oncogenic viruses (cancer-producing viruses) and the normal cell cycle. Comparison of amino acid sequences has also revealed that many larger proteins of higher organisms are built of domains that have come together by the fusion of gene segments. Proteins with new properties have arisen from novel combinations of these modules.

2. *Comparison of sequences of the same protein in different species yields a wealth of information about evolutionary pathways.* Genealogical relations between species can be inferred from sequence differences between their proteins, and the time of divergence of two evolutionary lines can be estimated because of the clocklike nature of random mutations. For

example, a comparison of serum albumins of primates indicates that human beings and African apes diverged only five million years ago, not thirty million years ago as was previously thought. These sequence analyses have opened a new perspective on the fossil record and the pathway of human evolution.

3. *Amino acid sequences can be searched for the presence of internal repeats.* Many proteins apparently have arisen by duplication of a primordial gene followed by its diversification. For example, antibody molecules are built of a series of similar domains, each consisting of about 108 residues (Figure 3-26). Each 25-kd light chain of antibodies is constructed from two of these modules, and each 50-kd heavy chain from four of them. The amino acid sequences of proteins express their evolutionary history.

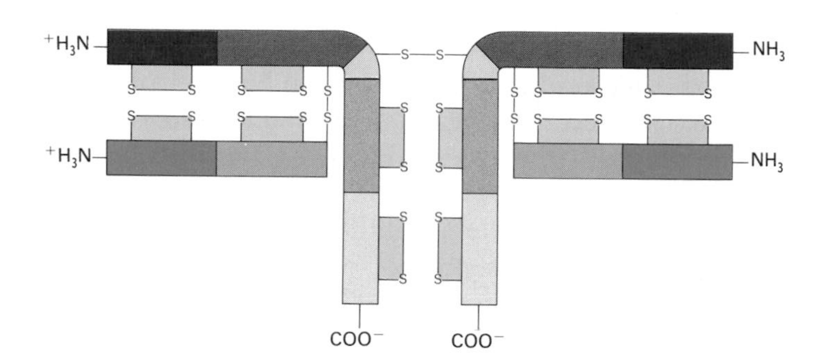

Figure 3-26
Antibody molecules consist of domains that are variations on a common theme produced by gene duplication and diversification. The pattern of disulfide bonds within the domains has been highly conserved.

4. *Amino acid sequences contain signals that determine the destination of proteins and control their processing.* Many proteins destined for export from a cell or for a membrane location contain a signal sequence, a stretch of about twenty hydrophobic residues near the amino terminus. Potential sites for the addition of carbohydrate units to asparagine residues can be identified by finding Asn-X-Ser and Asn-X-Thr in the sequence (X denotes any residue). Pairs of basic residues, such as Arg-Arg, mark potential sites of proteolytic cleavage, as in proinsulin, the precursor of insulin.

5. *Sequence data provide a basis for preparing antibodies specific for a protein of interest.* Specific antibodies can be very useful in determining the amount of a protein, ascertaining its distribution within a cell, and cloning its gene (p. 62).

6. *Amino acid sequences are also valuable for making DNA probes that are specific for the genes encoding the corresponding proteins* (p. 131). Protein sequencing is an integral part of molecular genetics, just as DNA cloning is central to the analysis of protein structure and function.

X-RAY CRYSTALLOGRAPHY REVEALS THREE-DIMENSIONAL STRUCTURE IN ATOMIC DETAIL

The understanding of protein structure and function has been greatly enriched by x-ray crystallography, a technique that can reveal the precise three-dimensional positions of most of the atoms in a protein molecule. Let us consider some basic aspects of this powerful method. First, crystals of the protein of interest are needed. Crystals can often be obtained by adding ammonium sulfate or another salt to a concentrated solution of protein to reduce its solubility. For example, myoglobin crystallizes in 3 M ammonium sulfate (Figure 3-27). Slow salting-out favors the formation of highly ordered crystals instead of amorphous precipitates. Some proteins crystallize readily, whereas others do so only after much effort has been expended in finding the right conditions. Crystallization is an art; the best practitioners have great perseverance and patience, as well as a golden touch. Increasingly large and complex proteins are now being crystallized. For example, polio virus, an 8500-kd complex consisting of 240 protein subunits surrounding an RNA core, has been crystallized and its structure solved by x-ray methods.

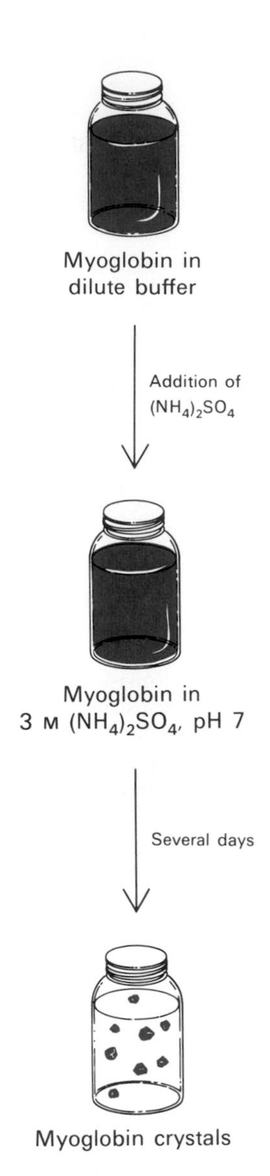

Figure 3-27
Crystallization of myoglobin.

The three components in an x-ray crystallographic analysis are a *source of x-rays*, a *protein crystal*, and a *detector* (Figure 3-28). A beam of x-rays of wavelength 1.54 Å is produced by accelerating electrons against a copper target. A narrow beam of x-rays strikes the protein crystal. Part of it goes straight through the crystal; the rest is *scattered* in various directions. The scattered (or *diffracted*) beams can be detected by x-ray film, the blackening of the emulsion being proportional to the intensity of the scattered x-ray beam, or by a solid-state electronic detector. The basic physical principles underlying the technique are:

1. *Electrons scatter x-rays.* The amplitude of the wave scattered by an atom is proportional to its number of electrons. Thus, a carbon atom scatters six times as strongly as a hydrogen atom.

2. *The scattered waves recombine.* Each atom contributes to each scattered beam. The scattered waves reinforce one another at the film or detector if they are in phase (in step) there and they cancel one another if they are out of phase.

3. *The way in which the scattered waves recombine depends only on the atomic arrangement.*

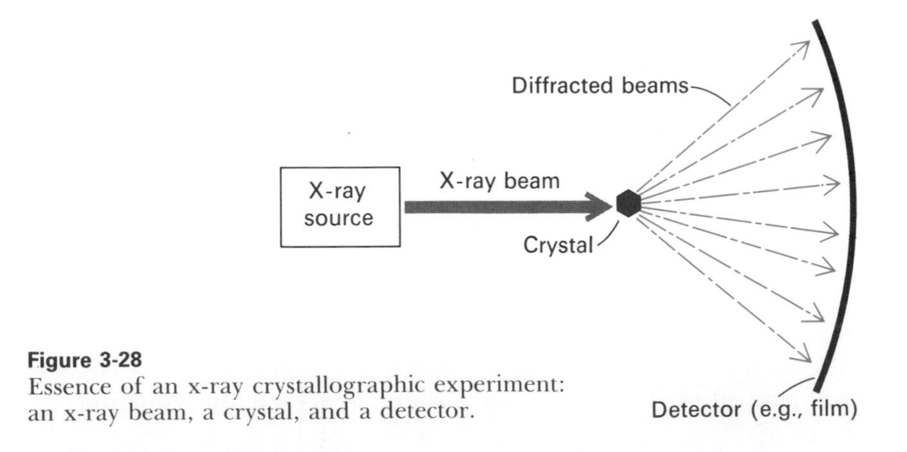

Figure 3-28
Essence of an x-ray crystallographic experiment: an x-ray beam, a crystal, and a detector.

Figure 3-29
X-ray precession photograph of a myoglobin crystal.

The protein crystal is mounted in a capillary and positioned in a precise orientation with respect to the x-ray beam and the film. Precessional motion of the crystal results in an x-ray photograph consisting of a regular array of spots called reflections. The x-ray photograph shown in Figure 3-29 is a two-dimensional section through a three-dimensional array of 25,000 spots. The intensity of each spot is measured. These *intensities* are the basic experimental data of an x-ray crystallographic analysis. The next step is to reconstruct an image of the protein from the observed intensities. In light microscopy or electron microscopy, the diffracted beams are focused by lenses to directly form an image. However, lenses for focusing x-rays do not exist. Instead, the image is formed by applying a mathematical relation called a Fourier transform. For each spot, this operation yields a wave of electron density, whose amplitude is proportional to the square root of the observed intensity of the spot. Each wave also has a *phase*—that is, the timing of its crests and troughs relative to those of other waves. The phase of each wave determines whether it reinforces or cancels the waves contributed by the other spots. These phases can be deduced from the well-understood diffraction patterns produced by heavy-atom reference markers such as uranium or mercury at specific sites in the protein.

The stage is then set for the calculation of an electron-density map, which gives the density of electrons at a large number of regularly spaced points in the crystal. This three-dimensional electron-density distribution is represented by a series of parallel sections stacked on top of each other. Each section is a transparent plastic sheet (or a layer in a computer image) on which the electron-density distribution is represented by contour lines (Figure 3-30), like the contour lines used in geological survey maps to depict altitude (Figure 3-31). The next step is to interpret the electron-density map. A critical factor is the *resolution* of the x-ray analysis, which is determined by the number of scattered intensities used in the Fourier synthesis. The fidelity of the image depends on the resolution of the Fourier synthesis, as shown by the optical analogy in Figure 3-32. A resolution of 6 Å reveals the course of the polypeptide chain but few other structural details. The reason is that polypeptide chains pack together so that their centers are between 5 and 10 Å apart. Maps at higher resolution are needed to delineate groups of atoms, which lie from 2.8 to 4.0 Å apart, and individual atoms, which are between 1.0 and 1.5 Å apart. The ultimate resolution of an x-ray analysis is determined by the degree of perfection of the crystal. For proteins, this limiting resolution is usually about 2 Å.

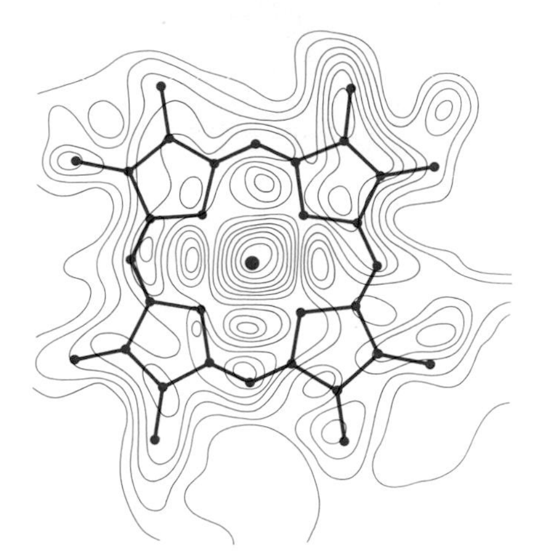

Figure 3-30
Section from the electron-density map of myoglobin showing the heme group. The peak of the center of this section corresponds to the position of the iron atom. [From J. C. Kendrew. The three-dimensional structure of a protein molecule. Copyright © 1961 by Scientific American, Inc. All rights reserved.]

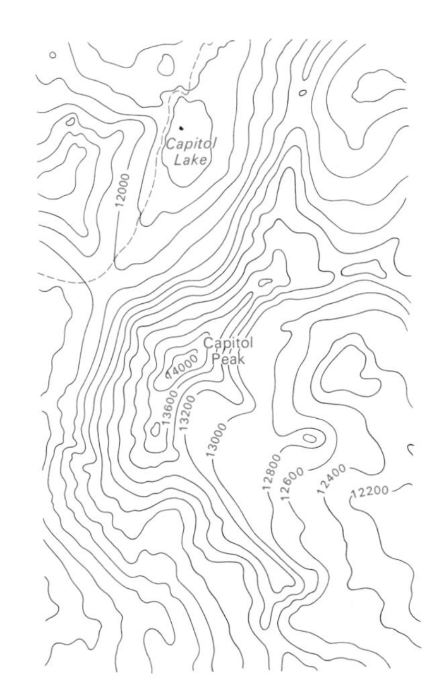

Figure 3-31
Section from a U.S. Geological Survey Map of the Capitol Peak Quadrangle, Colorado.

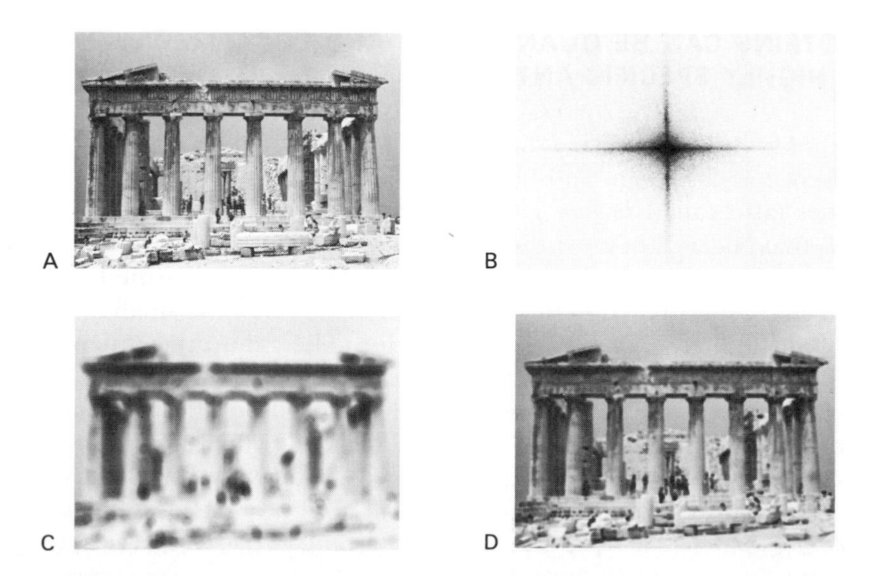

Figure 3-32
Effect of resolution on the quality of a reconstructed image, as shown by an optical analog of x-ray diffraction: (A) a photograph of the Parthenon; (B) an optical diffraction pattern of the Parthenon; (C and D) images reconstructed from the pattern in B. More data were used to obtain D than C, which accounts for the higher quality of the image in D. [Courtesy of Dr. Thomas Steitz (part A) and Dr. David De Rosier (part B).]

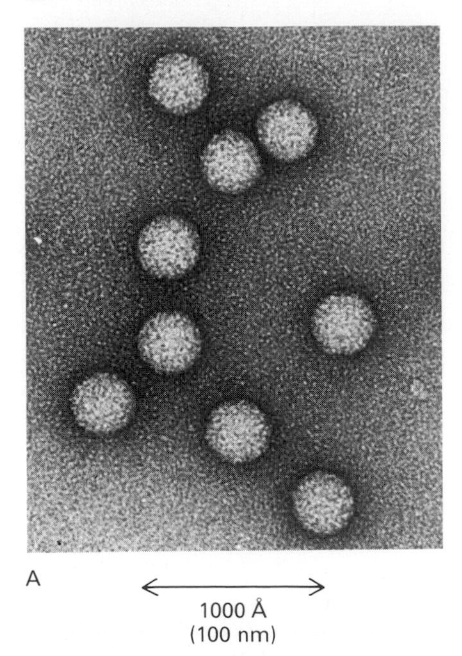

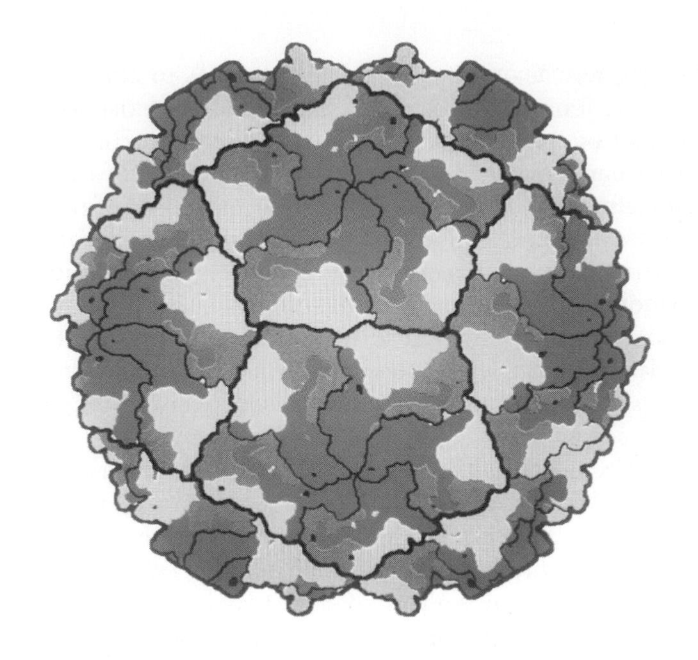

Figure 3-33
Polio virus. (A) Electron micrograph. (B) Model based on x-ray crystallographic analysis. [Part A courtesy of Dr. T. W. Jeng and Dr. Wah Chiu. Part B (computer model and photograph) courtesy of Dr. Arthur J. Olson, Research Institute of Scripps Clinic, © 1987.]

The structures of more than 200 proteins have been elucidated at atomic resolution. Knowledge of their detailed molecular architecture has provided insight into how proteins recognize and bind other molecules, how they function as enzymes, how they fold, and how they evolved. This extraordinarily rich harvest is continuing at a rapid pace and profoundly influencing the entire field of biochemistry. Moreover, x-ray crystallography is being complemented by nuclear magnetic resonance (NMR) spectroscopy, electron microscopy, and electron crystallography, in obtaining increasingly informative views of biomolecules at high resolution. We shall be looking at the three-dimensional structures of proteins and other biomolecules throughout this book and relating the architecture of these molecules to their biological function.

PROTEINS CAN BE QUANTITATED AND LOCALIZED BY HIGHLY SPECIFIC ANTIBODIES

An *antibody* is a protein synthesized by an animal in response to the presence of a foreign substance, called an *antigen* (Chapter 35). Antibodies (also called *immunoglobulins*) have specific affinity for the antigens that elicited their synthesis. Proteins, polysaccharides, and nucleic acids are effective antigens. Antibodies can also be formed to small molecules, such as synthetic peptides, provided that the small molecules are attached to a macromolecular carrier. The group recognized by an antibody is called an *antigenic determinant* (or *epitope*). Animals have a very large repertoire of antibody-producing cells, each producing antibody of a single specificity. An antigen acts by stimulating the proliferation of the small number of cells that were already forming complementary antibody. The major type of antibody in blood plasma is *immunoglobulin G*, a 150-kd protein containing two identical sites for the binding of antigen (Figure 3-34).

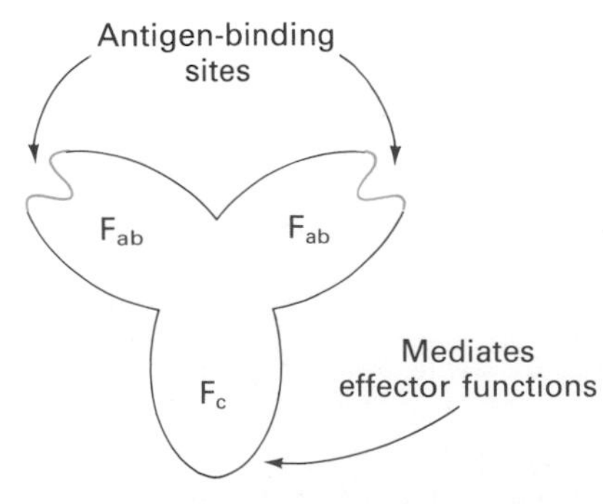

Figure 3-34
Diagram of immunoglobulin G (IgG), the major class of antibody molecules in blood plasma. IgG contains two antigen-binding F_{ab} units and an F_c unit that mediates effector functions such as the lysis of cell membranes.

Antibodies that recognize a particular protein can be obtained by injecting the protein into a rabbit twice, three weeks apart. Blood is drawn from the immunized rabbit several weeks later and centrifuged. The resulting serum, called an *antiserum*, usually contains the desired antibody. The antiserum or its immunoglobulin G fraction can be used directly. Alternatively, antibody molecules specific for the antigen can be purified by affinity chromatography. Antibodies produced in this way are *polyclonal*—that is, they are products of many different populations of antibody-producing cells and hence differ somewhat in their precise specificity and affinity for the antigen. A major advance of recent years is the discovery of a means of producing *monoclonal antibodies* of virtually any desired specificity (p. 896). Monoclonal antibodies, in contrast with polyclonal ones, are homogeneous because they are synthesized by a population of identical cells (a clone). Each such population is descended from a single *hybridoma cell* formed by fusing an antibody-producing cell with a tumor cell that has the capacity for unlimited proliferation.

Closely related proteins can be distinguished by antibodies; indeed, a difference of just one residue on the surface can be detected. Antibodies can be used as exquisitely specific analytic reagents to quantitate the amount of a protein or other antigen. In a *solid-phase immunoassay*, antibody specific for a protein of interest is attached to a polymeric support such as a sheet of polyvinylchloride (Figure 3-35). A drop of cell extract or a sample of serum or urine is laid on the sheet, which is washed after formation of the antibody-antigen complex. Antibody specific for a different site on the antigen is then added, and the sheet is again washed. This second antibody carries a radioactive or fluorescent label so that it can be detected with high sensitivity. The amount of second antibody bound to the sheet is proportional to the quantity of antigen in the sample. The sensitivity of the assay can be enhanced even further if the second antibody is attached to an enzyme such as alkaline phosphatase. This enzyme can rapidly convert an added colorless substrate into a colored product, or a nonfluorescent substrate into an intensely fluorescent product (Figure 3-36). Less than a nanogram (10^{-9} g) of a protein can readily be measured by such an *enzyme-linked immunosorbent assay* (ELISA), which is rapid and convenient. For example, pregnancy can be detected within a few days after conception by immunoassaying urine for the presence of human chorionic gonadotropin (hCG), a 37-kd protein hormone produced by the placenta.

Very small quantities of a protein of interest in a cell or in body fluid can be detected by an immunoassay technique called *Western blotting* (Figure 3-37). A sample is electrophoresed on an SDS polyacrylamide

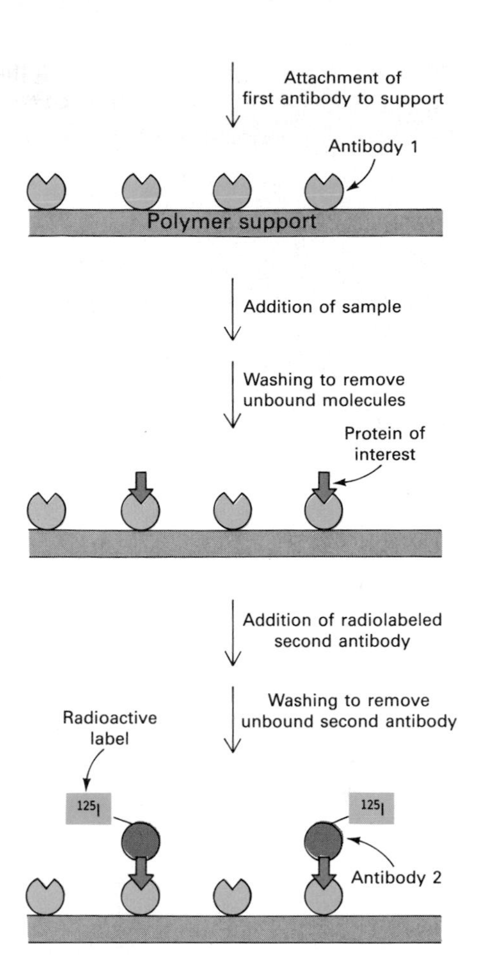

Figure 3-35
Solid-phase immunoassay. The steps are: coupling of specific antibody to a solid support, addition of the sample, washing to remove soluble compounds, and addition of a radiolabeled second antibody specific for a different site on the protein being detected.

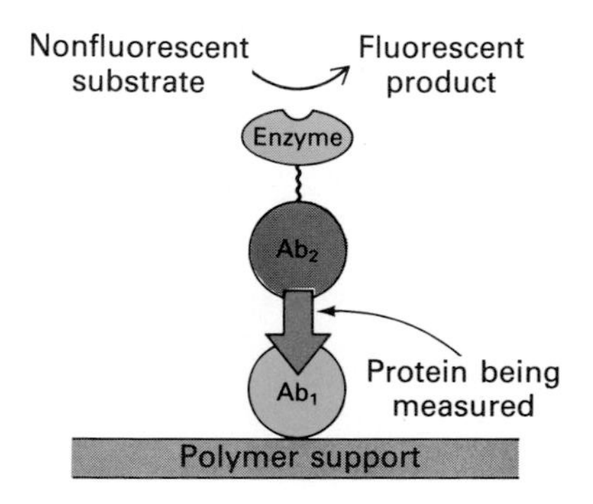

Figure 3-36
Enzyme-linked immunosorbent assay (ELISA). The steps are the same as in the immunoassay described in Figure 3-35 except that an enzyme instead of a radiolabel is attached to the second antibody. An intensely colored or fluorescent compound is formed by the catalytic action of this enzyme.

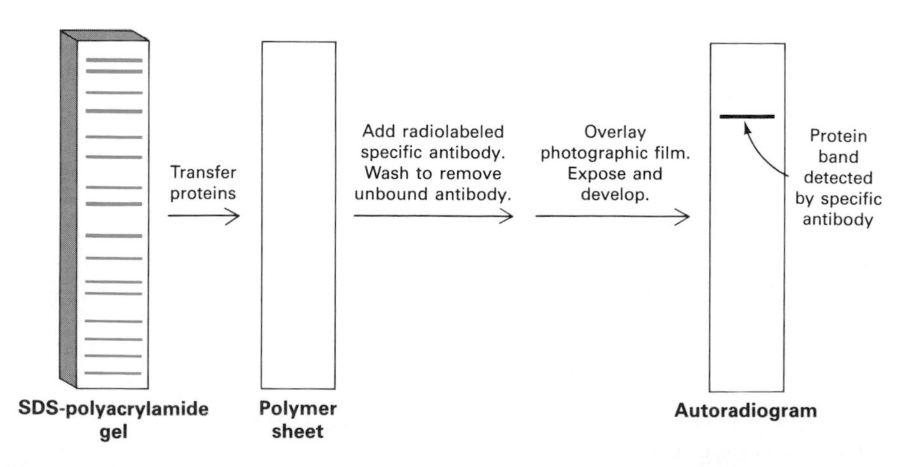

Figure 3-37
Detection of a protein on a gel by Western blotting. Proteins on an SDS-polyacrylamide gel are transferred to a polymer sheet and stained with radioactive antibody. A dark band corresponding to the protein of interest appears in the autoradiogram.

gel. The resolved proteins on the gel are transferred (by blotting) to a sheet to make them more accessible for reaction with a subsequently added antibody that is specific for the protein of interest. The antibody-antigen complex on the sheet then can be detected by rinsing the sheet with a second antibody specific for the first (e.g., goat antibody that recognizes mouse antibody). A radioactive label on the second antibody produces a dark band on x-ray film (an autoradiogram). Alternatively, an enzyme on the second antibody generates a colored product, as in the ELISA method. Western blotting makes it possible to find a protein in a complex mixture, the proverbial needle in a haystack. This technique is used advantageously in the cloning of genes (p. 133).

Antibodies are also valuable in determining the spatial distribution of antigens. Cells can be stained with fluorescent-labeled antibodies and examined by *fluorescence microscopy* to reveal the localization of a protein of interest. For example, arrays of parallel bundles are evident in cells stained with antibody specific for actin, a protein that polymerizes into filaments (Figure 3-38). Actin filaments are constituents of the cytoskeleton, the internal scaffold of cells that controls their shape and movement. The finest resolution of fluorescence microscopy is about 0.2 μm (200 nm or 2000 Å) because of the wavelength of visible light. Finer spatial resolution can be achieved by electron microscopy using antibodies tagged with electron-dense markers. For example, ferritin conjugated to an antibody can readily be visualized by electron microscopy because it contains an electron-dense core of iron hydroxide. Clusters of gold can also be conjugated to antibodies to make them highly visible under the electron microscope. *Immunoelectron microscopy* can define the position of antigens to a resolution of 10 nm (100 Å) or finer (Figure 3-39).

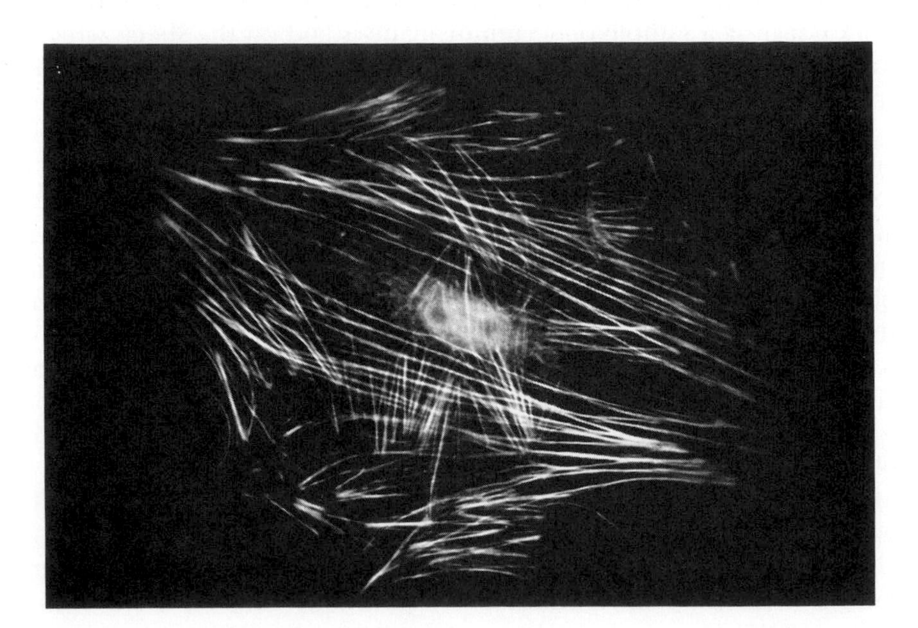

Figure 3-38
Fluorescence micrograph of actin filaments in a cell stained with an antibody specific to actin. [Courtesy of Dr. Elias Lazarides.]

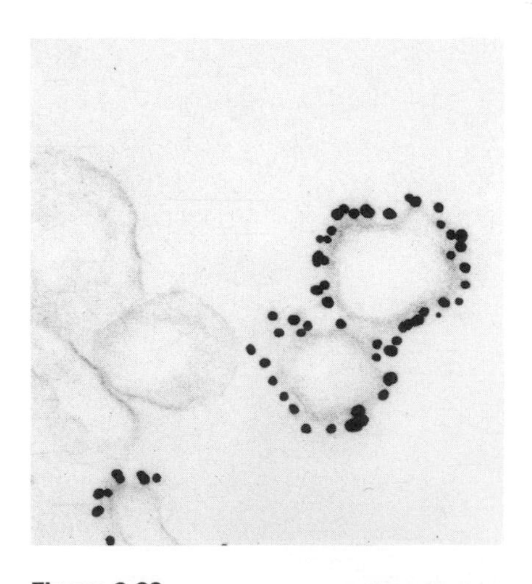

Figure 3-39
The opaque 150 Å (15 nm) diameter particles in this electron micrograph are clusters of gold atoms bound to antibody molecules. These membrane vesicles from the synapses of neurons contain a channel protein that is recognized by the specific antibody. [Courtesy of Dr. Peter Sargent.]

PEPTIDES CAN BE SYNTHESIZED BY AUTOMATED SOLID-PHASE METHODS

Synthesizing peptides of defined sequence is important for several reasons. First, *synthetic peptides can reveal the rules governing the three-dimensional conformation of proteins.* We can ask whether a particular sequence by itself folds into an α-helix, β-strand, or hairpin turn or behaves as a random coil. Second, *many hormones and other signal molecules in nature are peptides*. For example, white blood cells are attracted to bacteria by formylmethionyl peptides that come from the breakdown of bacterial proteins. Synthetic formyl-methionyl peptides have been useful in identifying the cell-surface receptor for this class of peptides. Synthetic peptides can be attached to agarose beads to prepare affinity chromatography columns for the purification of receptor proteins that specifically recognize the peptides. Third, *synthetic peptides can serve as drugs*. Vasopressin is a peptide hormone that stimulates the reabsorption of water in the distal tubules of the kidney, leading to the formation of a more concentrated urine. Patients with diabetes insipidus are deficient in vasopressin (also called antidiuretic hormone), and so they excrete large volumes of urine (more than 5 liters per day) and are continually thirsty because of this massive loss of fluid. This defect can be treated by administering 1-desamino-8-D-arginine vasopressin, a synthetic analog of the missing hormone (Figure 3-40). This synthetic peptide is degraded in vivo much more slowly than vasopressin and, additionally, does not increase the blood pressure. Fourth, *synthetic peptides can serve as antigens to stimulate the formation of specific antibodies*.

fMet peptide

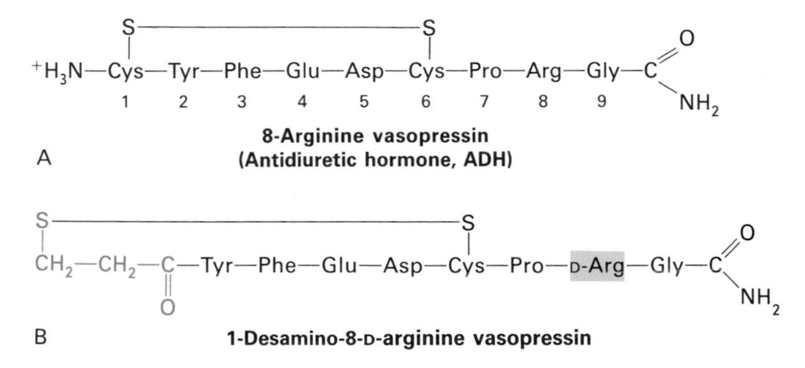

Figure 3-40
Structural formulas of (A) vasopressin (also called antidiuretic hormone), a peptide hormone that stimulates water resorption, and (B) 1-desamino-8-D-arginine vasopressin, a more stable synthetic analog.

Peptides are synthesized by linking an amino group to a carboxyl group that has been activated by reacting it with a reagent such as *dicyclohexylcarbodiimide* (DCC) (Figure 3-41). The attack of a free amino group on the activated carboxyl leads to the formation of a peptide bond and the release of dicylohexylurea. A unique product is formed only if a single amino group and a single carboxyl group are available for reaction. Hence, it is necessary to *block* (protect) all other potentially reactive groups. For example, the α-amino group of the component containing the activated carboxyl group can be blocked with a *tert*-butyloxycarbonyl (*t*-Boc) group. This *t*-Boc protecting group can be subsequently removed by exposing the peptide to dilute acid, which leaves peptide bonds intact.

Dicyclohexylcarbodiimide (DCC)

***t*-Butyloxycarbonyl amino acid (*t*-Boc amino acid)**

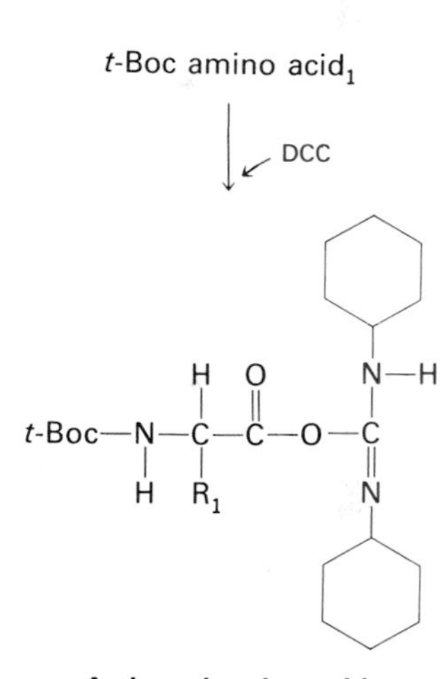

Amino acid$_2$ attached to resin

Dipeptide product attached to resin

\+

Dicyclohexylurea

Figure 3-41
Dicyclohexylcarbodiimide is used to activate carboxyl groups for the formation of peptide bonds.

Peptides can be readily synthesized by a *solid-phase method* devised by R. Bruce Merrifield. Amino acids are added stepwise to a growing peptide chain that is linked to an insoluble matrix, such as polystyrene beads. A major advantage of this solid-phase method is that the desired product at each stage is bound to beads that can be rapidly filtered and washed and so the need to purify intermediates is obviated. All of the reactions are carried out in a single vessel, which eliminates losses due to repeated transfers of products. The carboxyl-terminal amino acid of the desired peptide sequence is first anchored to the polystyrene beads (Figure 3-42). The *t*-Boc protecting group of this amino acid is then removed. The next amino acid (in the protected *t*-Boc form) is added together with dicyclohexylcarbodiimide, the coupling agent. After formation of the peptide bond, excess reagents and dicyclohexylurea are washed away, which leaves the beads with the desired dipeptide prod-

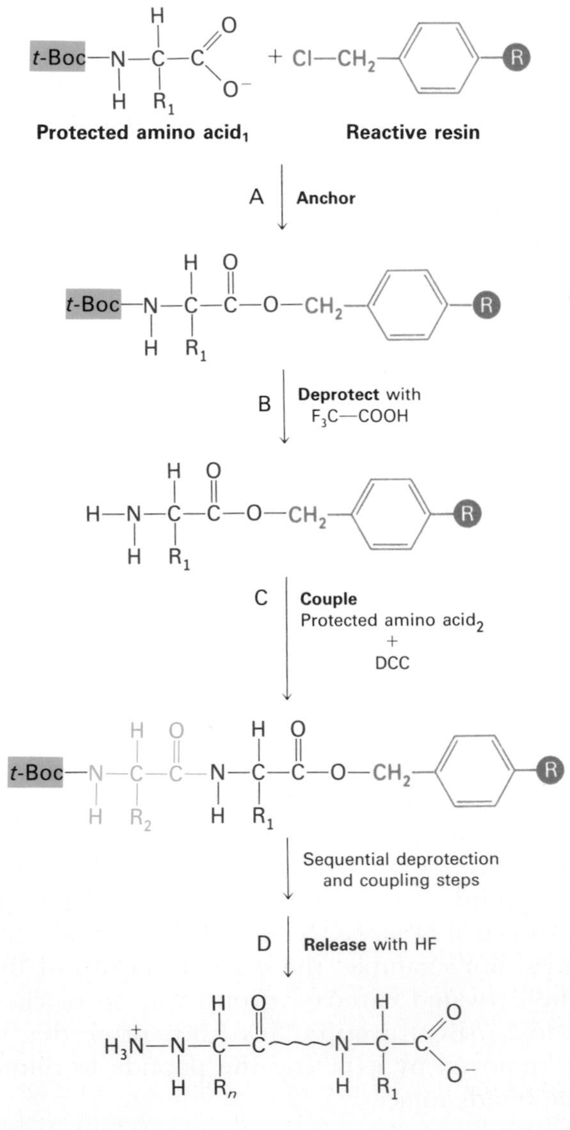

Figure 3-42
Solid-phase peptide synthesis. The steps are (A) anchoring of the C-terminal amino acid to a resin on the surface of polystyrene beads, (B) deprotection of the amino group, (C) addition of the next amino acid (in the protected form) and dicyclohexylcarbodiimide (DCC). Steps (B) and (C) are repeated for each added amino acid. The beads are washed to remove excess reagents and unwanted products after each step. Finally, (D) the completed peptide is released from the resin.

uct. Additional amino acids are linked by the same sequence of reactions. At the end of the synthesis, the peptide is released from the beads by adding HF, which cleaves the carboxyl ester anchor without disrupting peptide bonds. Protecting groups on potentially reactive side chains, such as that of lysine, are also removed at this time. This cycle of reactions can readily be automated, which makes it feasible to routinely synthesize peptides containing about 50 residues in good yield and purity. In fact, Merrifield has synthesized interferons (155 residues) that have antiviral activity and ribonuclease (124 residues) that is catalytically active.

SUMMARY

The purification of a protein is an essential step in elucidating its structure and function. Proteins can be separated from each other and from other molecules on the basis of such characteristics as size, solubility, charge, and binding affinity. SDS-polyacrylamide gel electrophoresis separates the polypeptide chains of proteins under denaturing conditions largely according to mass. Proteins can also be separated electrophoretically on the basis of net charge by isoelectric focusing in a pH gradient. Ultracentrifugation and gel-filtration chromatography resolves proteins according to size, whereas ion-exchange chromatography separates them mainly on the basis of net charge. The high affinity of many proteins for specific chemical groups is exploited in affinity chromatography, in which proteins bind to columns containing beads bearing covalently linked substrates, inhibitors, or other specifically recognized groups.

The amino acid composition of a protein can be determined by hydrolyzing it into its constituent amino acids in 6 N HCl at 110°C. The amino acids can be separated by ion-exchange chromatography and quantitated by reacting them with ninhydrin or fluorescamine. Amino acid sequences can be determined by Edman degradation, which removes one amino acid at a time from the amino end of a peptide. Phenylisothiocyanate reacts with the terminal amino group to form a phenylthiocarbamoyl derivative, which cyclizes under mildly acidic conditions to give a phenylthiohydantoin (PTH)-amino acid and a peptide shorted by one residue. This PTH-amino acid is identified by high-performance liquid chromatography (HPLC). Automated repeated Edman degradations by a sequenator can analyze sequences of about fifty residues. Longer polypeptide chains are broken into shorter ones for analysis by specifically cleaving them with a reagent such as cyanogen bromide, which splits peptide bonds on the carboxyl side of methionine residues. Enzymes such as trypsin, which cleaves on the carboxyl side of lysine and arginine residues, are also very useful in splitting proteins. Recombinant DNA techniques have revolutionized amino acid sequencing. The nucleotide sequence of DNA molecules reveals the amino acid sequence of nascent proteins encoded by them but does not disclose posttranslational modifications. Amino acid sequences are rich in information concerning the kinship of proteins, their evolutionary relations, and diseases produced by mutations. Knowledge of a sequence provides valuable clues to conformation and function.

Polypeptide chains can be synthesized by automated solid-phase methods in which the carboxyl end of the growing chain is linked to an insoluble support. The α-carboxyl group of the incoming amino acid is activated by dicyclohexylcarbodiimide and joined to the α-amino group of the growing chain. Synthetic peptides can serve as drugs and as antigens to stimulate the formation of specific antibodies. They also provide insight into relations between amino acid sequence and conformation.

Proteins can be detected and quantitated by highly specific antibodies. Enzyme-linked immunosorbent assays (ELISA) and Western blots of SDS-polyacrylamide gels are used extensively. Proteins can also be localized within cells by fluorescence microscopy and immunoelectron microscopy using labeled antibodies. X-ray crystallography has revealed the three-dimensional structures of more than a hundred proteins in atomic detail. Knowledge of molecular structure has provided insight into how proteins fold, recognize other molecules, and catalyze chemical reactions.

SELECTED READINGS

Where to start

Moore, S., and Stein, W. H., 1973. Chemical structures of pancreatic ribonuclease and deoxyribonuclease. *Science* 180:458–464.

Hunkapiller, M. W., and Hood, L. E., 1983. Protein sequence analysis: automated microsequencing. *Science* 219:650–659.

Merrifield, B., 1986. Solid phase synthesis. *Science* 232:341–347.

Books on protein chemistry

Creighton, T. E., 1983. *Proteins: Structure and Molecular Properties.* W. H. Freeman.

Cooper, T. G., 1977. *The Tools of Biochemistry*. Wiley. [A valuable guide to experimental methods in protein chemistry and in other areas of biochemistry. Principles of procedures are clearly presented.]

Hirs, C. H. W., and Timasheff, S. N., (eds.), 1983. *Enzyme Structure,* Part I. Methods in Enzymology, vol. 91. Academic Press. [An excellent collection of authoritative articles on amino acid analysis, end-group methods, chemical and enzymatic cleavage, peptide separation, sequence analysis, chemical modification, and active-site labeling. Also see volumes 47–49 in this series.]

Scopes, R., 1982. *Protein Purification: Principles and Practice.* Springer-Verlag.

Langone, J. J., and Van Vunakis, H., 1983. *Immunochemical Techniques,* Part A. Methods in Enzymology, vol. 92. Academic Press.

Physical chemistry of proteins

Cantor, C. R., and Schimmel, P. R., 1980. *Biophysical Chemistry.* W. H. Freeman. [An outstanding exposition of fundamental principles and experimental methods. Part 2 discusses many of the techniques presented in this chapter.]

Freifelder, D., 1982. *Physical Biochemistry: Applications to Biochemistry and Molecular Biology.* W. H. Freeman. [Contains a lucid discussion of ultracentrifugation.]

Amino acid sequence determination

Hunkapiller, M. W., Strickler, J. E., and Wilson, K. J., 1984. Contemporary methodology for protein structure determination. *Science* 226:304–311.

Hewick, R. M., Hunkapiller, M. W., Hood, L. E., and Dreyer, W. J., 1981. A gas-liquid solid phase peptide and protein sequenator. *J. Biol. Chem.* 256:7990–7997.

Konigsberg, W. H., and Steinman, H. M., 1977. Strategy and methods of sequence analysis. *In* Neurath, H., and Hill, R. L., (eds.), *The Proteins* (3rd ed.), vol. 3, pp. 1–178. Academic Press.

Stein, S., and Udenfriend, S., 1984. A picomole protein and peptide chemistry: some applications to the opiod peptides. *Analy. Chem.* 136:7–23.

Sequence comparisons and molecular evolution

Doolittle, R. F., 1981. Similar amino acid sequences: chance or common ancestry? *J. Mol. Biol.* 16:9–16.

Lipman, D. J., and Pearson, W. R., 1985. Rapid and sensitive protein similarity searches. *Science* 227:1435–1441. [Description of an algorithm that searches for similarities between an amino acid sequence and a large database of previously determined sequences. This program can be run on a personal computer.]

Wilson, A. C., 1985. The molecular basis of evolution. *Sci. Amer.* 253(4):164.

X-ray crystallography and NMR spectroscopy

Matthews, B. W., 1977. X-ray structure of proteins. *In* Neurath, H., and Hill, R. L., (eds.), *The Proteins* (3rd ed.), vol. 3, pp. 404–590. Academic Press.

Holmes, K. C., and Blow, D. M., 1965. *The Use of X-ray Diffraction in the Study of Protein and Nucleic Acid Structure.* Wiley-Interscience.

Glusker, J. P., and Trueblood, K. N., 1972. *Crystal Structure Analysis: A Primer.* Oxford University Press. [A lucid and concise introduction to x-ray crystallography in general.]

Kline, A. D., Braun, W., and Wuthrich, K., 1986. Studies by ^{1}H nuclear magnetic resonance and distance geometry of the solution conformation of the α-amylase inhibitor tendamistat. *J. Mol. Biol.* 189:377–382. [Determination of the three-dimensional structure of a small protein by NMR spectroscopy. This important study demonstrates the power of this approach in elucidating conformation.]

PROBLEMS

1. The following reagents are often used in protein chemistry:

CNBr	Dabsyl chloride
Urea	6 N HCl
β-Mercaptoethanol	Ninhydrin
Trypsin	Phenyl isothiocyanate
Performic acid	Chymotrypsin

Which one is the best suited for accomplishing each of the following tasks?
(a) Determination of the amino acid sequence of a small peptide.
(b) Identification of the amino-terminal residue of a peptide (of which you have less than 10^{-7} g).
(c) Reversible denaturation of a protein devoid of disulfide bonds. Which additional reagent would you need if disulfide bonds were present?
(d) Hydrolysis of peptide bonds on the carboxyl side of aromatic residues.
(e) Cleavage of peptide bonds on the carboxyl side of methionines.
(f) Hydrolysis of peptide bonds on the carboxyl side of lysine and arginine residues.

2. What is the ratio of base to acid at pH 4, 5, 6, 7, and 8 for an acid with a p*K* of 6?

3. Anhydrous hydrazine has been used to cleave peptide bonds in proteins. What are the reaction products? How might this technique be used to identify the carboxyl-terminal amino acid?

4. The amino acid sequence of human adrenocorticotropin, a polypeptide hormone, is

Ser-Tyr-Ser-Met-Glu-His-Phe-Arg-Trp-Gly-Lys-Pro-Val-Gly-Lys-Lys-Arg-Arg-Pro-Val-Lys-Val-Tyr-Pro-Asp-Ala-Gly-Glu-Asp-Gln-Ser-Ala-Glu-Ala-Phe-Pro-Leu-Glu-Phe

(a) What is the approximate net charge of this molecule at pH 7? Assume that its side chains have the p*K* values given in Table 2-1 (p. 21) and that the p*K*s of the terminal $—NH_3^+$ and —COOH groups are 7.8 and 3.6, respectively.
(b) How many peptides result from the treatment of the hormone with cyanogen bromide?

5. Ethyleneimine reacts with cysteine side chains in proteins to form *S*-aminoethyl derivatives. The peptide bonds on the carboxyl side of these modified cysteine residues are susceptible to hydrolysis by trypsin. Why?

6. The absorbance *A* of a solution is defined as

$$A = \log_{10} (I_0/I)$$

in which I_0 is the incident light intensity and I is the transmitted light intensity. The absorbance is related to the molar absorption coefficient (extinction coefficient) ϵ (in cm^{-1} M^{-1}), concentration c (in M), and path length l (in cm) by

$$A = \epsilon l c$$

The absorption coefficient of myoglobin at 580 nm is 15,000 cm^{-1} M^{-1}. What is the absorbance of a 1 mg/ml solution across a 1-cm path? What percentage of the incident light is transmitted by this solution?

7. Tropomyosin, a 93-kd muscle protein, sediments more slowly than does hemoglobin (65 kd). Their sedimentation coefficients are 2.6 and 4.31 S, respectively. Which structural feature of tropomyosin accounts for its slow sedimentation?

8. The relative electrophoretic mobilities of a 30-kd protein and a 92-kd protein used as standards on an SDS-polyacrylamide gel are 0.80 and 0.41, respectively. What is the apparent mass of a protein having a mobility of 0.62 on this gel?

9. The relative electrophoretic mobility of a protein on an SDS-polyacrylamide gel decreases from 0.67 to 0.64 on addition of 1 mM dithiothreitol. What is a likely reason for this shift?

10. The gene encoding a protein with a single disulfide bond undergoes a mutation that changes a serine residue into a cysteine residue. You want to find out whether the disulfide pairing in this mutant is the same as in the original protein. Propose an experiment to directly answer this question.

11. A synthetic polypeptide consisting of L-lysine residues is a random coil at pH 7 but becomes α helical as the pH is raised above 10. Account for this pH-dependent conformational transition.

12. Predict the pH dependence of the helix-coil transition of poly-L-glutamate.

13. Glycine is a highly conserved amino acid residue in the evolution of many proteins. Why?

14. Suppose that you are investigating the mechanism of action of vasopressin. You want to isolate the receptor for this hormone. Which chromatographic procedure would you use? How would you carry out the experiment?

15. The sedimentation coefficient of a tetrameric enzyme decreases by 2% on binding substrate. What is your interpretation of this finding.

16. In electrophoresis, the electric force Ez drives a charged molecule toward the oppositely charged electrode. Which parameters correspond to E and z in centrifugation?

17. What is the effect of the shape of a protein on its position relative to the axis of centrifugation in (a) a sedimentation-velocity experiment, and (b) a sedimentation-equilibrium experiment?

18. In SDS-polyacrylamide gel electrophoresis, a 90-kd single-chain protein moves less rapidly than does a 40-kd one. In contrast, the 90-kd protein emerges first from a gel-filtration column. What is the reason for this difference?

19. A complex mixture of proteins from a cell extract is analyzed by Western blotting. A 23-kd and a 57-kd band are stained by one monoclonal antibody, and the same 23-kd but a different 69-kd band are stained by a second monoclonal antibody. Propose a structural basis for this result.

20. The amino-terminal region of myoglobin is not well defined in the electron-density map obtained from an x-ray analysis of this protein. Why?

21. Fluorescence-activated cell sorting (FACS) is a powerful technique for separating cells according to their content of particular molecules. For example, a fluorescent-labeled antibody specific for a cell-surface protein can be used to detect cells containing such a molecule. Suppose that you want to isolate cells that possess a receptor enabling them to detect bacterial degradation products. However, you do not yet have an antibody directed against this receptor. Which fluorescent-labeled molecule would you prepare to identify such cells?

CHAPTER 4

DNA and RNA: Molecules of Heredity

DNA is a very long, threadlike macromolecule made up of a large number of deoxyribonucleotides, each composed of a base, a sugar, and a phosphate group. *The bases of DNA molecules carry genetic information, whereas their sugar and phosphate groups perform a structural role.* This chapter presents the key experiments that revealed that DNA is the genetic material, then describes the DNA double helix. When this structure was discovered, the complementary nature of its two chains immediately suggested that each is a template for the other in DNA replication. DNA polymerases are the enzymes that replicate DNA by taking instructions from DNA templates. These exquisitely specific enzymes replicate DNA with an error frequency of less than one in a million nucleotides. The genes of all cells and many viruses are made of DNA. Some viruses, however, use RNA (ribonucleic acid) as their genetic material. This chapter concludes with examples of the genetic role of RNA in plant viruses and animal tumor viruses.

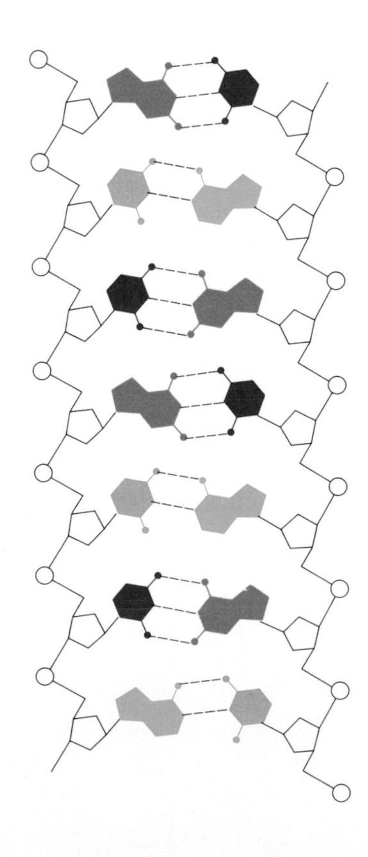

Figure 4-1
Schematic diagram of the structure of DNA. The sugar-phosphate backbone is shown in black, and the purine and pyrimidine bases are shown in color. [After A. Kornberg. The synthesis of DNA. Copyright © 1968 by Scientific American, Inc. All rights reserved.]

DNA CONSISTS OF FOUR KINDS OF BASES JOINED TO A SUGAR-PHOSPHATE BACKBONE

DNA is a polymer of deoxyribonucleotide units. A nucleotide consists of a nitrogenous base, a sugar, and one or more phosphate groups. The sugar in a deoxyribonucleotide is *deoxyribose.* The *deoxy* prefix indicates that this sugar lacks an oxygen atom that is present in ribose, the parent compound. The nitrogenous base is a derivative of *purine* or *pyrimidine.*

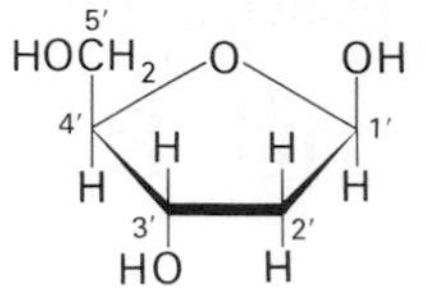

β-D-2-Deoxyribose 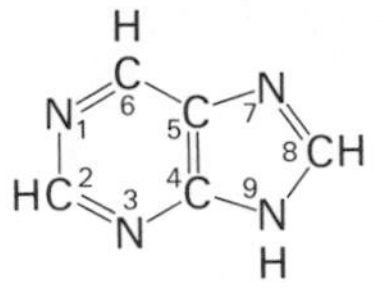 **Purine** 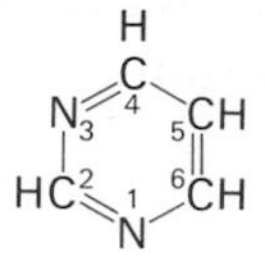 **Pyrimidine**

The purines in DNA are *adenine* (A) and *guanine* (G), and the pyrimidines are *thymine* (T) and *cytosine* (C).

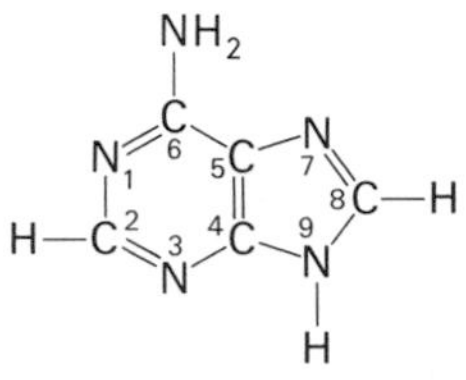

Adenine (A) 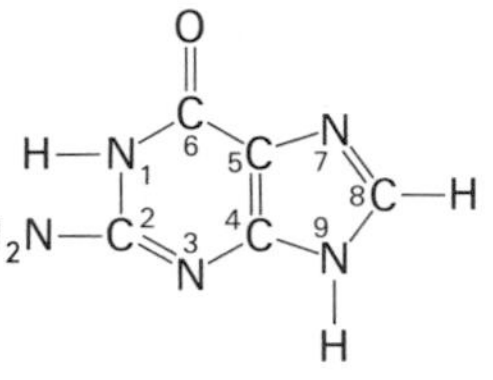 **Guanine (G)** 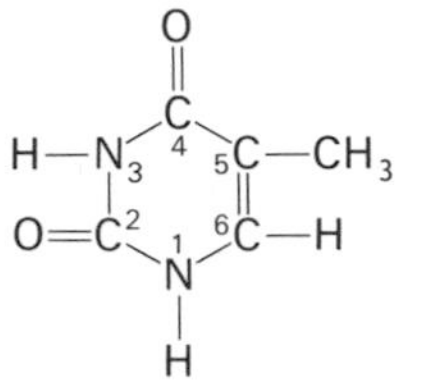 **Thymine (T)** 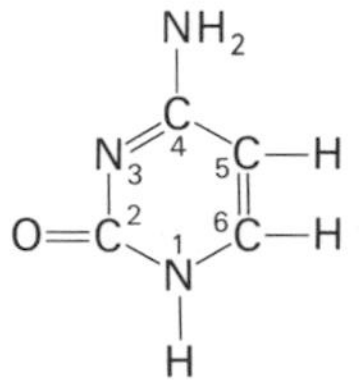 **Cytosine (C)**

In a deoxyribonucleotide, the C-1 carbon atom of deoxyribose is bonded to N-1 of a pyrimidine or N-9 of a purine. The configuration of this *N*-glycosidic linkage is β (the base lies above the plane of the sugar ring). A *nucleoside* consists of a purine or pyrimidine base bonded to a sugar. The four nucleoside units in DNA are called *deoxyadenosine, deoxyguanosine, deoxythymidine,* and *deoxycytidine.* A *nucleotide* is a phosphate ester of a nucleoside. The most common site of esterification in naturally occurring nucleotides is the hydroxyl group attached to C-5 of the sugar. Such a compound is called a *nucleoside 5-phosphate* or a *5′-nucleotide.* For example, *deoxyadenosine 5′-triphosphate* (dATP) is an activated precursor in the synthesis of DNA. A primed number denotes an atom of the sugar, whereas an unprimed number denotes an atom of the purine or pyrimidine ring. The prefix *d* in dATP indicates that the sugar is deoxyribose to distinguish this compound from ATP, in which the sugar is ribose.

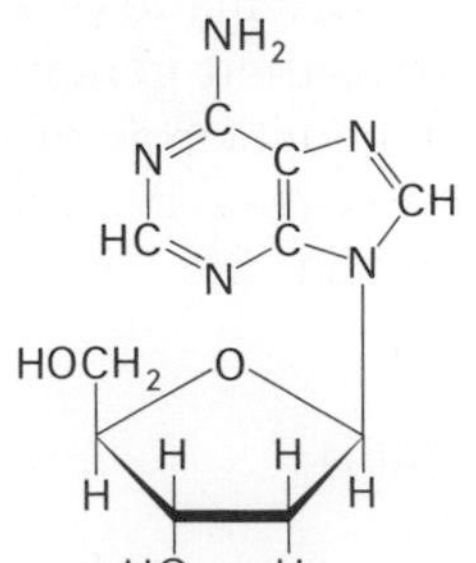

Deoxyadenosine (A nucleoside)

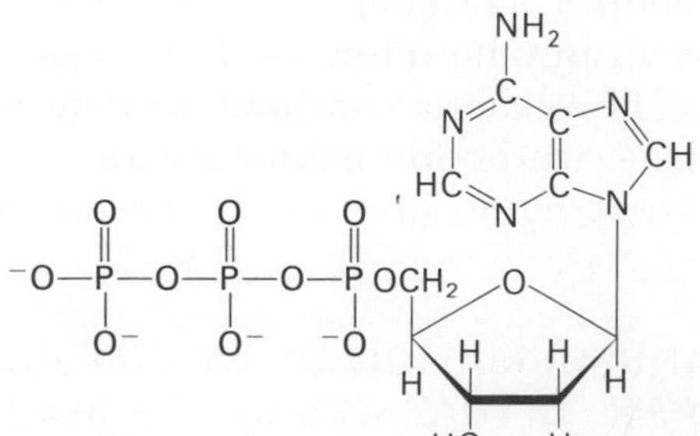

Deoxyadenosine 5′-triphosphate (dATP) (A nucleotide)

The *backbone* of DNA, which is invariant throughout the molecule, consists of deoxyriboses linked by phosphate groups. Specifically, the 3′-hydroxyl of the sugar moiety of one deoxyribonucleotide is joined to the 5′-hydroxyl of the adjacent sugar by a phosphodiester bridge. The *variable part* of DNA is its *sequence of four kinds of bases (A, G, C, and T)*. The corresponding nucleotide units are called *deoxyadenylate, deoxyguanylate, deoxycytidylate,* and *deoxythymidylate.* The structure of a DNA chain is shown in Figure 4-2.

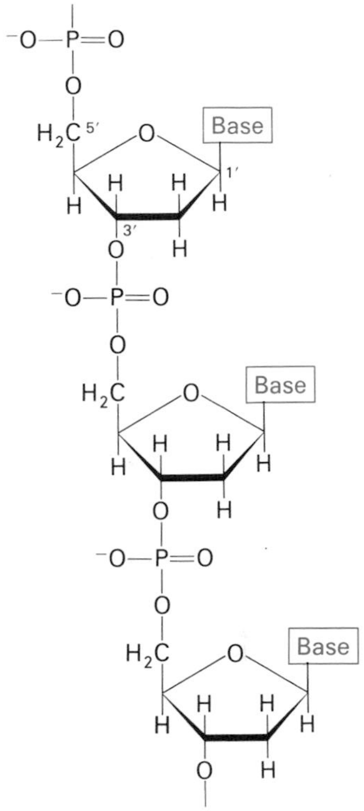

The structure of a DNA chain can be concisely represented in the following way. The symbols for the four principal deoxyribonucleosides are

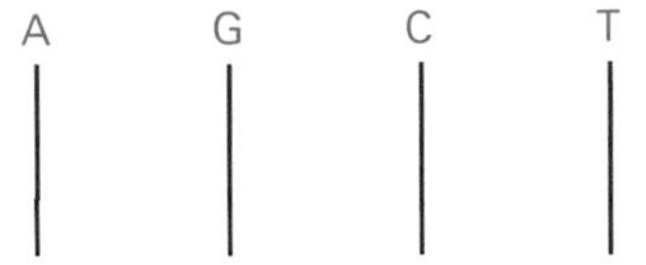

A G C T

The bold line refers to the sugar, whereas A, G, C, and T represent the bases. The Ⓟ within the diagonal line in the diagram below denotes a phosphodiester bond. This diagonal line joins the end of one bold line and the middle of another. These junctions refer to the 5′-OH and 3′-OH, respectively. In this example, the symbol Ⓟ indicates that deoxyadenylate is linked to deoxycytidine by a phosphodiester bridge. Specifically, the 3′-OH of deoxyadenylate is joined through a phosphoryl group to the 5′-OH of deoxycytidine.

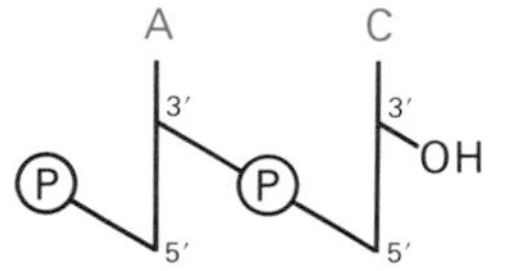

Now suppose that deoxyguanylate becomes linked to the deoxycytidine unit of this dinucleotide. The resulting trinucleotide can be represented by

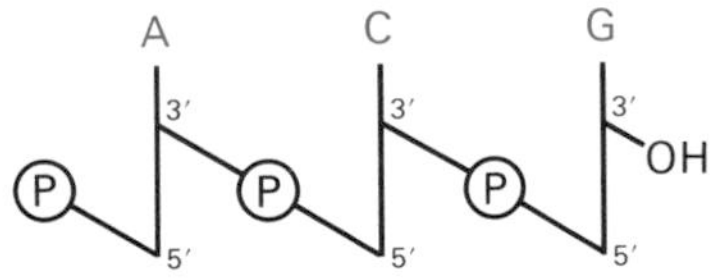

An even more abbreviated notation for this trinucleotide is pApCpG or ACG.

A DNA chain has polarity. One end of the chain has a 5′-OH group and the other a 3′-OH group that is not linked to another nucleotide. By convention, the symbol ACG means that the unlinked 5′-OH group is on deoxyadenosine, whereas the unlinked 3′-OH group is on deoxyguanosine. Thus, *the base sequence is written in the 5′ → 3′ direction.* Recall that the amino acid sequence of a protein is written in the amino → carboxyl direction. Note that ACG and GCA refer to different compounds, just as Glu-Phe-Ala differs from Ala-Phe-Glu.

Figure 4-2
Structure of part of a DNA chain.

TRANSFORMATION OF PNEUMOCOCCI BY DNA REVEALED THAT GENES ARE MADE OF DNA

The pneumococcus bacterium played an important part in the discovery of the genetic role of DNA. A pneumococcus is normally surrounded by a slimy, glistening polysaccharide capsule. This outer layer is essential for the pathogenicity of the bacterium, which causes pneumonia in humans and other susceptible mammals. Mutants devoid of a polysaccharide coat are not pathogenic. The normal bacterium is referred to as the S form (because it forms smooth colonies in a culture dish), whereas mutants without capsules are called R forms (because they form rough colonies). R mutants lack an enzyme needed for the synthesis of capsular polysaccharide.

In 1928, Fred Griffith discovered that a nonpathogenic R mutant could be *transformed* into the pathogenic S form in the following way. He injected mice with a mixture of live R and heat-killed S pneumococci. The striking finding was that this mixture was lethal to the mice, whereas either live R or heat-killed S pneumococci alone were not. The blood of the dead mice contained live S pneumococci. Thus, the heat-killed S pneumococci had somehow transformed live R pneumococci into live S pneumococci. This change was permanent: the transformed pneumococci yielded pathogenic progeny of the S form. It was then found that this R → S transformation can occur in vitro (Figure 4-3). Some of the cells in a growing culture of the R form were transformed into the S form by the addition of a *cell-free extract* of heat-killed S pneumococci. This finding set the stage for the elucidation of the chemical nature of the "transforming principle."

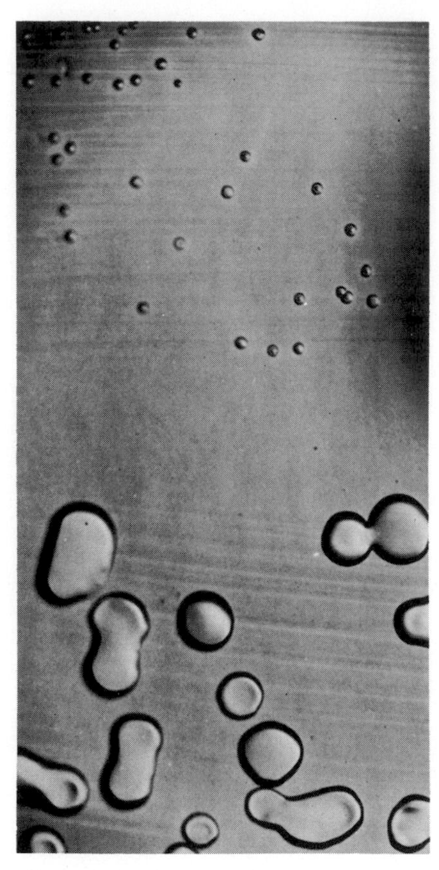

Figure 4-3
Transformation of nonpathogenic R pneumococci (small colonies) to pathogenic S pneumococci (large glistening colonies) by DNA from heat-killed S pneumococci. [From O. T. Avery, C. M. MacLeod, and M. McCarty. *J. Exp. Med.* 79 (1944):158.]

The cell-free extract of heat-killed S pneumococci was fractionated and the transforming activity of its components assayed. In 1944, Oswald Avery, Colin MacLeod, and Maclyn McCarty published their discovery that *"a nucleic acid of the deoxyribose type is the fundamental unit of the transforming principle of Pneumococcus Type III."* The experimental basis for their conclusion was: (1) the purified, highly active transforming principle gave an elemental chemical analysis that agreed closely with that calculated for DNA; (2) the optical, ultracentrifugal, diffusive, and electrophoretic properties of the purified material were like those of DNA; (3) there was no loss of transforming activity upon extraction of protein or lipid; (4) the polypeptide-cleaving enzymes trypsin and chymotrypsin did not affect transforming activity; (5) ribonuclease (known to digest ribonucleic acid) had no effect on the transforming principle; and (6) in contrast, transforming activity was lost following the addition of deoxyribonuclease.

This work is a landmark in the development of biochemistry. Until 1944, it was generally assumed that chromosomal proteins carry genetic information and that DNA plays a secondary role. This prevailing view was decisively shattered by the rigorously documented finding that *purified DNA has genetic specificity.* Avery gave a vivid description of this research and of its implications in a letter that he wrote in 1943 to his brother, a medical microbiologist at another university (Figure 4-4).

Further support for the genetic role of DNA came from the studies of a virus that infects the bacterium *Escherichia coli.* The T2 bacteriophage consists of a core of DNA surrounded by a protein coat. In 1951, Roger Herriott suggested that "the virus may act like a little hypodermic needle full of transforming principles; the virus as such never enters the cell; only the tail contacts the host and perhaps enzymatically cuts a small hole through the outer membrane and then the nucleic acid of the virus head flows into the cell." In 1952, this idea was tested by Alfred Hershey and Martha Chase in the following way. Phage DNA was labeled with the radioisotope ^{32}P, whereas the protein coat was labeled with ^{35}S. These labels are highly specific because DNA does not contain sulfur and the protein coat is devoid of phosphorus. A sample

For the past two years, first with MacLeod and now with Dr. McCarty, I have been trying to find out what is the chemical nature of the substance in the bacterial extract which induces this specific change. The crude extract of Type III is full of capsular polysaccharide, C (somatic) carbohydrate, nucleoproteins, free nucleic acids of both the yeast and thymus type, lipids, and other cell constituents. Try to find in the complex mixtures the active principle! Try to isolate and chemically identify the particular substance that will by itself, when brought into contact with the R cell derived from Type II, cause it to elaborate Type III capsular polysaccharide and to acquire all the aristocratic distinctions of the same specific type of cells as that from which the extract was prepared! Some job, full of headaches and heartbreaks. But at last perhaps we have it.

. . . if we prove to be right—and of course that is a big if—then it means that both the chemical nature of the inducing stimulus is known and the chemical structure of the substance produced is also known, the former being thymus nucleic acid, the latter Type III polysaccharide, and both are thereafter reduplicated in the daughter cells and after innumerable transfers without further addition of the inducing agent and the same active and specific transforming substance can be recovered far in excess of the amount originally used to induce the reaction. Sounds like a virus—may be a gene. But with mechanisms I am not now concerned. One step at a time and the first step is what is the chemical nature of the transforming principle? Some one else can work out the rest. Of course the problem bristles with implications. It touches the biochemistry of the thymus type of nucleic acids which are known to constitute the major part of chromosomes but have been thought to be alike regardless of origin and species. It touches genetics, enzyme chemistry, cell metabolism and carbohydrate synthesis. But today it takes a lot of well documented evidence to convince anyone that the sodium salt of deoxyribose nucleic acid, protein free, could possibly be endowed with such biologically active and specific properties and that is the evidence we are now trying to get. It is lots of fun to blow bubbles but it is wiser to prick them yourself before someone else tries to.

Figure 4-4
Part of a letter from Oswald Avery to his brother Roy, written in May 1943. [From R. D. Hotchkiss. In *Phage and the Origins of Molecular Biology,* J. Cairns, G. S. Stent, and J. D. Watson, eds. (Cold Spring Harbor Laboratory, 1966), pp. 185–186.]

of an *E. coli* culture was infected with labeled phage, which became attached to the bacteria during a short incubation period. The suspension was spun for a few minutes in a Waring Blendor at 10,000 rpm. This treatment subjected the phage-infected cells to very strong shearing forces, which severed the connections between the viruses and bacteria. The resulting suspension was centrifuged at a speed sufficient to throw the bacteria to the bottom of the tube. Thus, the pellet contained the infected bacteria, whereas the supernatant contained smaller particles. These fractions were analyzed for ^{32}P and ^{35}S to determine the location of the phage DNA and the protein coat. The results of these experiments were:

1. Most of the phage DNA was found in the bacteria.

2. Most of the phage protein was found in the supernatant.

3. The blendor treatment had almost no effect on the competence of the infected bacteria to produce progeny virus.

Additional experiments showed that less than 1% of the ^{35}S was transferred from the parental phage to the progeny phage. In contrast, 30% of the parental ^{32}P appeared in the progeny. These simple, incisive

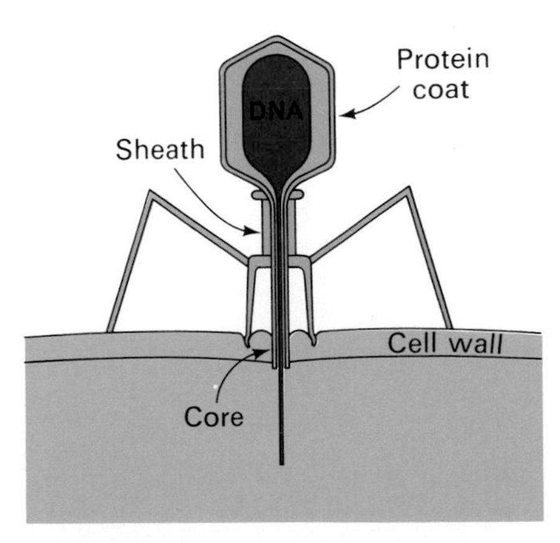

Figure 4-5
Diagram of a T2 bacteriophage injecting its DNA into a bacterial cell. [After W. B. Wood and R. S. Edgar. Building a bacterial virus. Copyright © 1967 by Scientific American, Inc. All rights reserved.]

experiments led to the conclusion that "*a physical separation of the phage T2 into genetic and non-genetic parts is possible. . . .* The sulfur-containing protein of resting phage particles is confined to a protective coat that is responsible for the adsorption of bacteria, and functions as an instrument for the injection of the phage DNA into the cell. This protein probably has no function in the growth of intracellular phage. The DNA has some function. Further chemical inferences should not be drawn from the experiments presented."

The cautious tone of this conclusion did not detract from its impact. The genetic role of DNA soon became a generally accepted fact. The experiments of Hershey and Chase strongly reinforced what Avery, MacLeod, and McCarty had found eight years earlier in a different system. Additional support came from studies of the DNA content of single cells, which showed that in a given species *the DNA content is the same for all cells that have a diploid set of chromosomes. Haploid cells were found to have half as much DNA.*

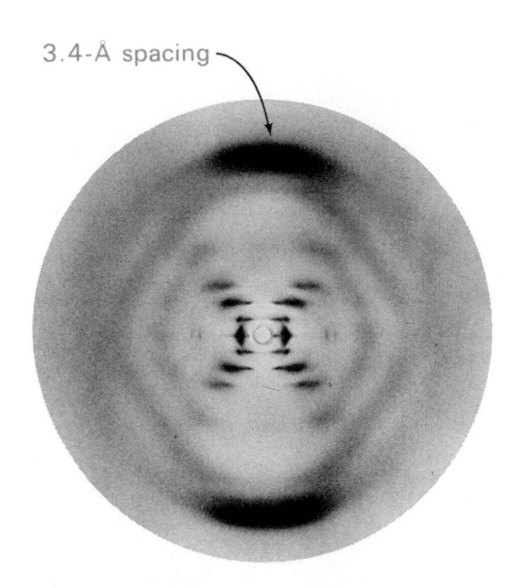

Figure 4-6
X-ray diffraction photograph of a hydrated DNA fiber (form B). The central cross is diagnostic of a helical structure. The strong arcs on the meridian arise from the stack of base pairs, which are 3.4 Å apart. [Courtesy of Dr. Maurice Wilkins.]

THE WATSON-CRICK DNA DOUBLE HELIX

In 1953, James Watson and Francis Crick deduced the three-dimensional structure of DNA and immediately inferred its mechanism of replication. This brilliant accomplishment ranks as one of the most significant in the history of biology because it led the way to an understanding of gene function in molecular terms. Watson and Crick analyzed x-ray diffraction photographs of DNA fibers taken by Rosalind Franklin and Maurice Wilkins and derived a structural model that has proved to be essentially correct. The important features of their model of DNA are:

1. Two helical polynucleotide chains are coiled around a common axis. The chains run in opposite directions (Figure 4-7).

2. The purine and pyrimidine bases are on the inside of the helix, whereas the phosphate and deoxyribose units are on the outside (Figure 4-8). The planes of the bases are perpendicular to the helix axis. The planes of the sugars are nearly at right angles to those of the bases.

3. The diameter of the helix is 20 Å. Adjacent bases are separated by 3.4 Å along the helix axis and related by a rotation of 36 degrees. Hence, the helical structure repeats after ten residues on each chain; that is, at intervals of 34 Å.

4. The two chains are held together by hydrogen bonds between pairs of bases. Adenine is always paired with thymine. Guanine is always paired with cytosine (Figures 4-9 and 4-10).

5. The sequence of bases along a polynucleotide chain is not restricted in any way. *The precise sequence of bases carries the genetic information.*

The most important aspect of the DNA double helix is the specificity of the pairing of bases. Watson and Crick deduced that adenine must pair with thymine, and guanine with cytosine, because of steric and hydrogen-bonding factors. The steric restriction is imposed by the regular helical nature of the sugar-phosphate backbone of each polynucleotide chain. The glycosidic bonds that are attached to a bonded pair of bases are always 10.85

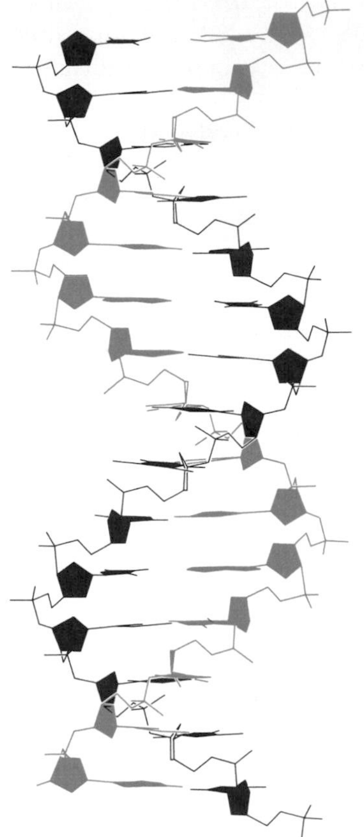

Figure 4-7
Skeletal model of double-helical DNA. The structure repeats at intervals of 34 Å, which corresponds to ten residues on each chain.

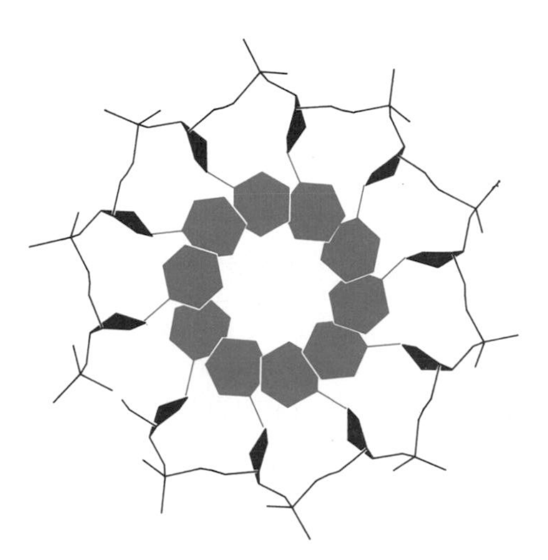

Figure 4-8
Diagram of one of the strands of a DNA double helix, viewed down the helix axis. The bases (all pyrimidines here) are inside, whereas the sugar-phosphate backbone is outside. The tenfold symmetry is evident. The bases are shown in blue and the sugars in red.

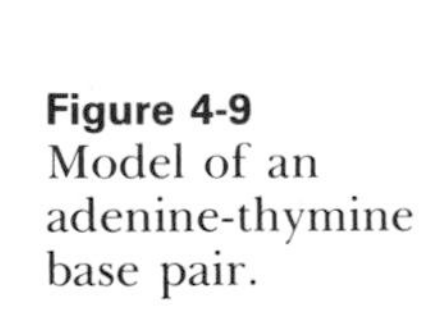

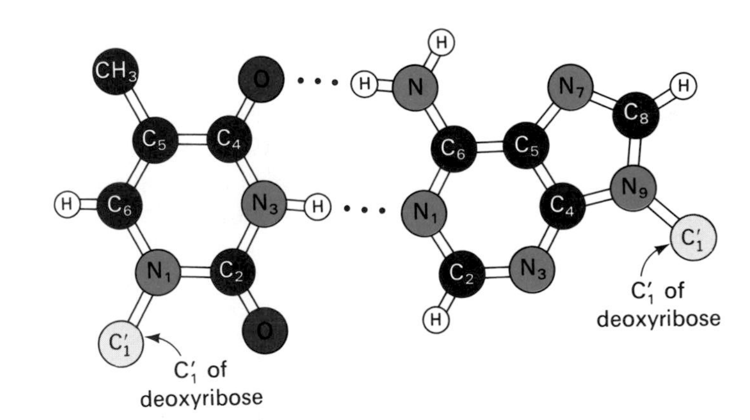

Figure 4-9
Model of an adenine-thymine base pair.

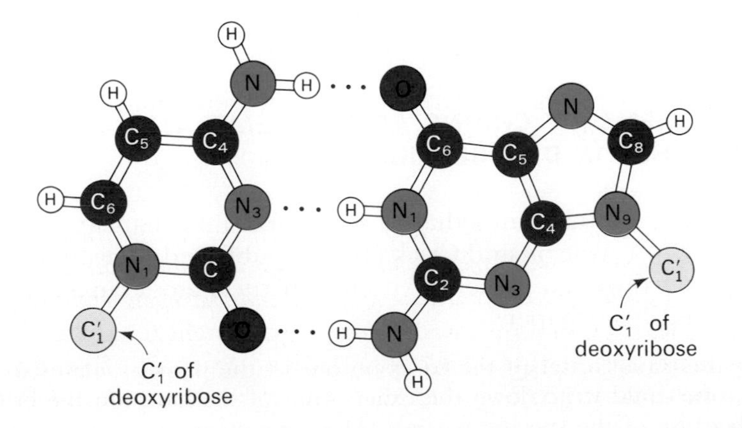

Figure 4-10
Model of a guanine-cytosine base pair.

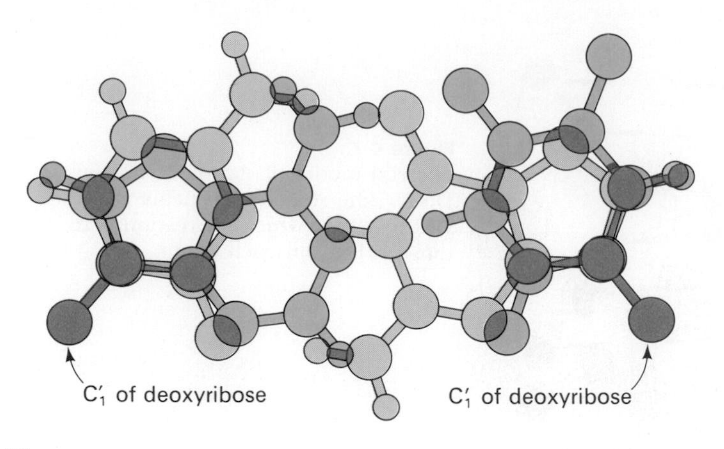

Figure 4-11
Superposition of an A-T base pair (shown in yellow) on a G-C base pair (shown in blue). Note that the positions of the glycosidic bonds and of the C-1′ atom of deoxyribose are identical for the two base pairs.

Å apart (Figure 4-11). A purine-pyrimidine base pair fits perfectly in this space. In contrast, there is insufficient room for two purines. There is more than enough space for two pyrimidines, but they would be too far apart to form hydrogen bonds. Hence, one member of a base pair in a DNA helix must always be a purine and the other a pyrimidine because of steric factors. The base pairing is further restricted by hydrogen-bonding requirements. The hydrogen atoms in the purine and pyrimidine bases have well-defined positions. Adenine cannot pair with cytosine because there would be two hydrogens near one of the bonding positions and none at the other. Likewise, guanine cannot pair with thymine. In contrast, adenine forms two hydrogen bonds with thymine, whereas guanine forms three with cytosine (Figures 4-9 and 4-10). The orientations and distances of these hydrogen bonds are optimal for achieving strong attraction between the bases.

This base-pairing scheme was strongly supported by the results of earlier studies of the base compositions of DNAs from different species. In 1950, Erwin Chargaff found that the *ratios of adenine to thymine and of guanine to cytosine were nearly 1.0 in all species studied.* The meaning of these equivalences was not evident until the Watson-Crick model was proposed. Only then could it be seen that they reflect an essential facet of DNA structure and function—the specificity of base pairing.

The double-helical structure of DNA shown in Figure 4-12 is *the B form of DNA (B-DNA).* As will be discussed in a later chapter (p. 650), DNA can assume different helical forms, such as A-DNA and Z-DNA. Under physiologic conditions, DNA is almost entirely in the Watson-Crick B form.

THE COMPLEMENTARY CHAINS ACT AS TEMPLATES FOR EACH OTHER IN DNA REPLICATION

The double-helical model immediately suggested a mechanism for the replication of DNA. Watson and Crick (1953*b*) published their hypothesis a month after they had presented their structural model in a beautifully simple and lucid paper:

> . . . If the actual order of the bases on one of the pairs of chains were given, one could write down the exact order of the bases on the other one, because of the specific pairing. Thus one chain is, as it were, the

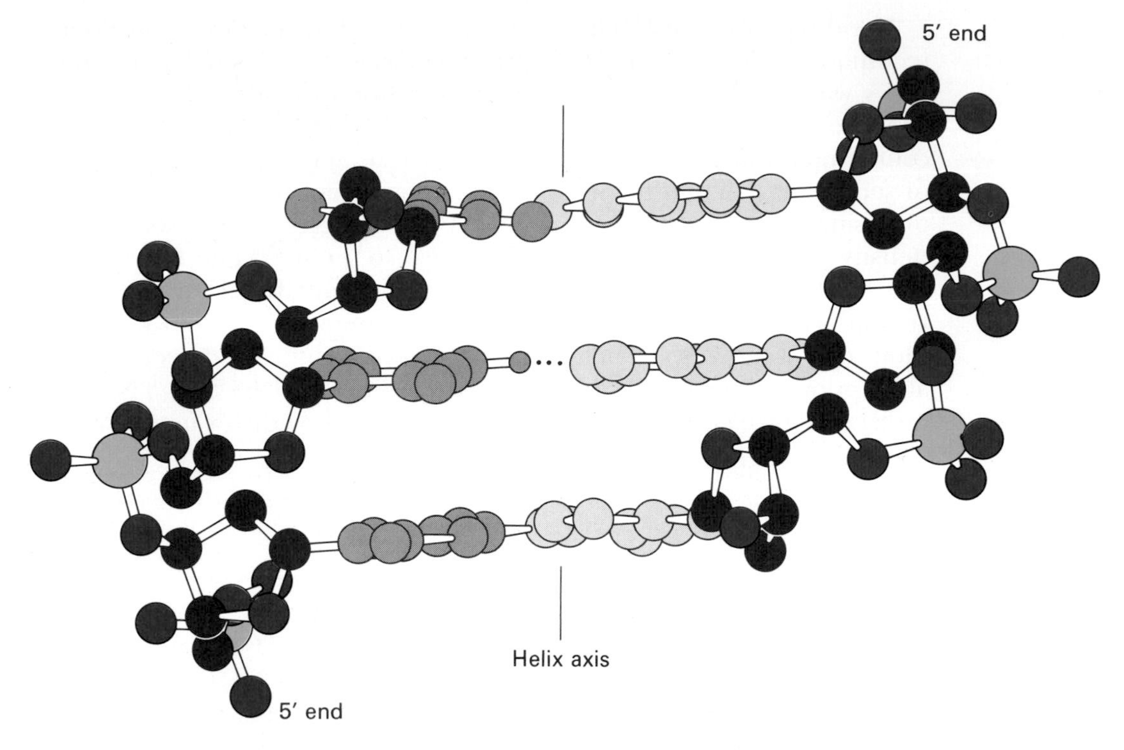

Figure 4-12
Model of a double-helical DNA molecule showing three base pairs (gray). Note that the two strands run in opposite directions.

complement of the other, and it is this feature which suggests how the deoxyribonucleic acid molecule might duplicate itself.

Previous discussions of self-duplication have usually involved the concept of a template, or mould. Either the template was supposed to copy itself directly or it was to produce a "negative," which in its turn was to act as a template and produce the original "positive" once again. In no case has it been explained in detail how it would do this in terms of atoms and molecules.

Now our model for deoxyribonucleic acid is, in effect, a *pair* of templates, each of which is complementary to the other. We imagine that prior to duplication the hydrogen bonds are broken, and the two chains unwind and separate. Each chain then acts as a template for the formation onto itself of a new companion chain, so that eventually we shall have *two* pairs of chains, where we only had one before. Moreover, the sequence of the pairs of bases will have been duplicated exactly.

DNA REPLICATION IS SEMICONSERVATIVE

Watson and Crick proposed that one of the strands of each daughter DNA molecule is newly synthesized, whereas the other is passed on unchanged from the parent DNA molecule. This distribution of parental atoms is called *semiconservative*. A critical test of this hypothesis was carried out by Matthew Meselson and Franklin Stahl. The parent DNA was labeled with ^{15}N, the heavy isotope of nitrogen, to make it denser than ordinary DNA. This was accomplished by growing *E. coli* for many generations in a medium that contained $^{15}NH_4Cl$ as the sole nitrogen source. The bacteria were abruptly transferred to a medium that contained ^{14}N, the ordinary isotope of nitrogen. The question asked was: What is the distribution of ^{14}N and ^{15}N in the DNA molecules after successive rounds of replication?

The distribution of ^{14}N and ^{15}N was revealed by the newly developed technique of *density-gradient equilibrium sedimentation.* A small amount of DNA was dissolved in a concentrated solution of cesium chloride having a density close to that of the DNA (~1.7 g cm^{-3}). This solution was centrifuged until it was nearly at equilibrium. The opposing processes of sedimentation and diffusion created a gradient in the concentration of cesium chloride across the centrifuge cell. The result was a stable density gradient, ranging from 1.66 to 1.76 g cm^{-3}. The DNA molecules in this density gradient were driven by centrifugal force into the region where the solution's density was equal to their own. High-molecular-weight DNA yielded a narrow band that was detected by its absorption of ultraviolet light. A mixture of ^{14}N DNA and ^{15}N DNA molecules gave clearly separate bands because they differ in density by about 1% (Figure 4-13).

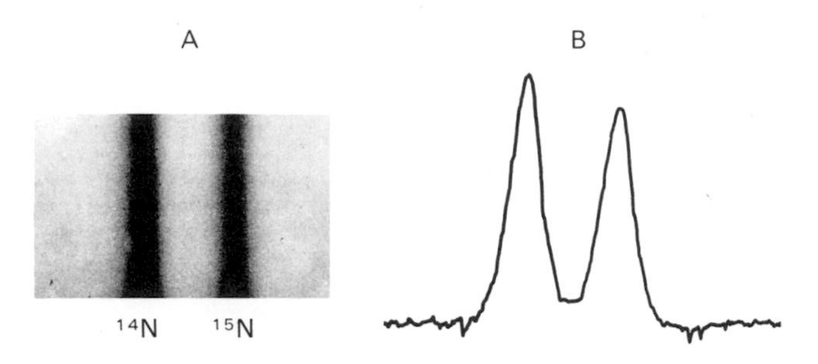

Figure 4-13
Resolution of ^{14}N DNA and ^{15}N DNA by density-gradient centrifugation: (A) ultraviolet absorption photograph of a centrifuge cell; (B) densitometric tracing of the absorption photograph. [From M. Meselson and F. W. Stahl. *Proc. Nat. Acad. Sci.* 44(1958):671.]

DNA was extracted from the bacteria at various times after they were transferred from a ^{15}N to a ^{14}N medium. Analysis of these samples by the density-gradient technique showed that there was a single band of DNA after one generation (Figure 4-14). The density of this band was precisely halfway between those of ^{14}N DNA and ^{15}N DNA. *The absence of ^{15}N DNA indicated that parental DNA was not preserved as an intact unit on replication.* The absence of ^{14}N DNA indicated that all of the daughter DNA molecules derived some of their atoms from the parent DNA. This proportion had to be one-half, because the density of the hybrid DNA band was halfway between those of ^{14}N DNA and ^{15}N DNA.

After two generations, there were equal amounts of two bands of DNA. One was hybrid DNA, the other was ^{14}N DNA. Meselson and Stahl concluded from these incisive experiments *"that the nitrogen of a DNA molecule is divided equally between two physically continuous subunits; that, following duplication, each daughter molecule receives one of these; and that the subunits are conserved through many duplications."* Their results agreed perfectly with the Watson-Crick model for DNA replication (Figure 4-15).

THE DOUBLE HELIX CAN BE REVERSIBLY MELTED

The two strands of a DNA helix readily come apart when the hydrogen bonds between its paired bases are disrupted. This can be accomplished

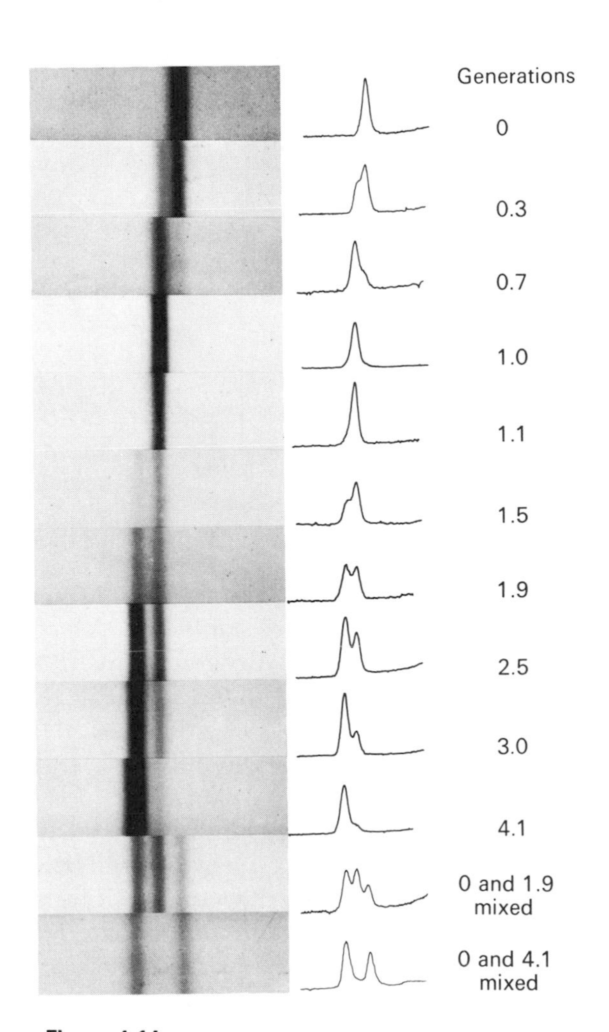

Figure 4-14
Detection of semiconservative replication in *E. coli* by density-gradient centrifugation. The position of a band of DNA depends on its content of ^{14}N and ^{15}N. After 1.0 generation, all of the DNA molecules are hybrids containing equal amounts of ^{14}N and ^{15}N. No parental DNA (^{15}N) is left after 1.0 generation. [From M. Meselson and F. W. Stahl. *Proc. Nat. Acad. Sci.* 44(1958):671.]

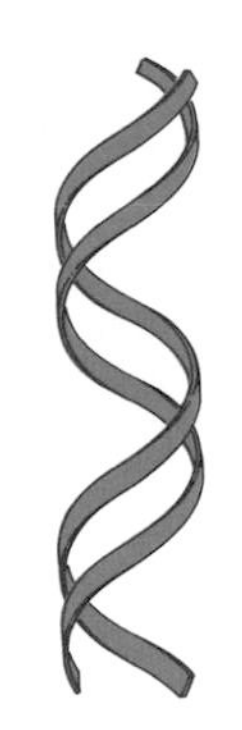

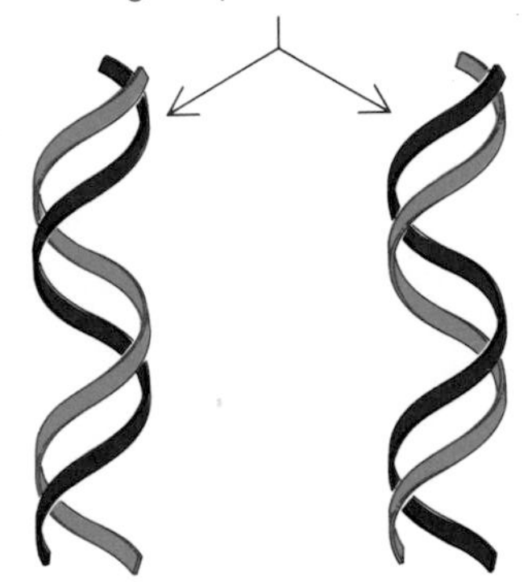

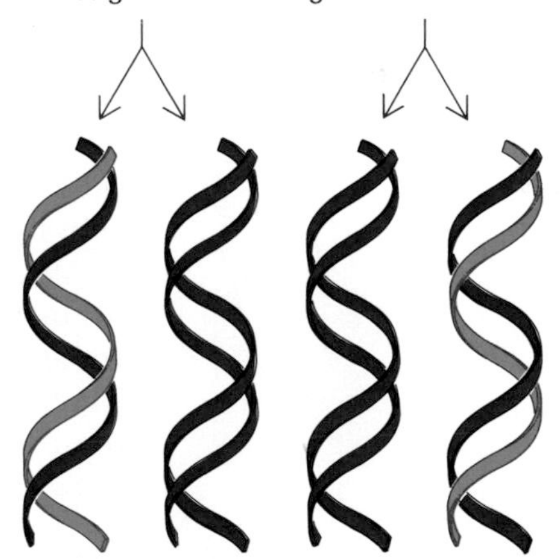

Figure 4-15
Schematic diagram of semiconservative replication. Parental DNA is shown in green and newly synthesized DNA in red. [After M. Meselson and F. W. Stahl. *Proc. Nat. Acad. Sci.* 44(1958):671.]

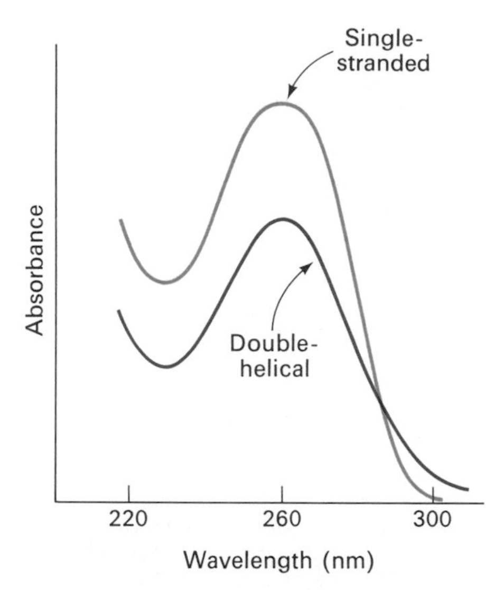

Figure 4-16
The absorbance of a DNA solution at wavelength 260 nm increases when the double helix is melted into single strands.

by heating a solution of DNA or by adding acid or alkali to ionize its bases. The unwinding of the double helix is called *melting* because it occurs abruptly at a certain temperature. The *melting temperature* (T_m) is defined as the temperature at which half of the helical structure is lost. The abruptness of the transition indicates that the DNA double helix is a *highly cooperative structure*, held together by many reinforcing bonds; it is stabilized by the stacking of bases as well as by base pairing. The melting of DNA is readily monitored by measuring its absorbance of light at wavelength 260 nm. The unstacking of the base pairs results in increased absorbance, an effect called *hyperchromism* (Figure 4-16).

The melting temperature of a DNA molecule depends markedly on its base composition. DNA molecules rich in GC base pairs have a higher T_m than those having an abundance of AT base pairs (Figure 4-17). In fact, the T_m of DNA from many species varies linearly with GC

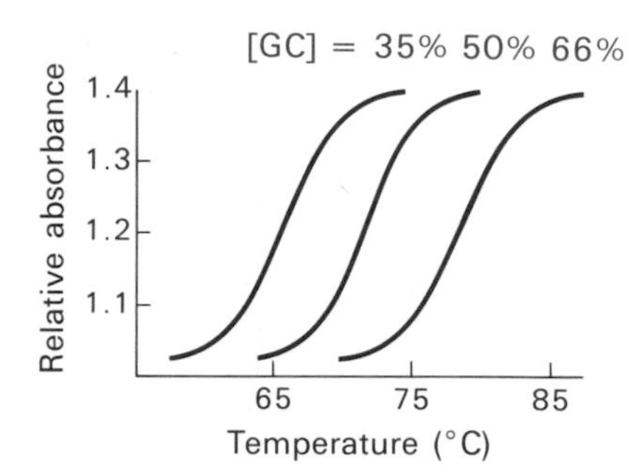

Figure 4-17
DNA melting curves. The absorbance relative to that at 25°C is plotted against temperature. (The wavelength of the incident light was 260 nm.) The T_m is 69°C for *E. coli* DNA (50% GC pairs) and 76°C for *P. aeruginosa* DNA (68% GC pairs).

content, rising from 77°C to 100°C as the fraction of GC pairs increases from 20% to 78%. GC base pairs are more stable than AT pairs because their bases are held together by three hydrogen bonds rather than by two. In addition, adjacent GC base pairs interact more strongly with one another than do adjacent AT base pairs. Hence, *the AT-rich regions of DNA are the first to melt* (Figure 4-18). The double helix is melted in vivo by the action of specific proteins (p. 672).

Separated complementary strands of DNA spontaneously reassociate to form a double helix when the temperature is lowered below T_m. This renaturation process is sometimes called *annealing*. The facility with which double helices can be melted and then reassociated is crucial for the biological functions of DNA.

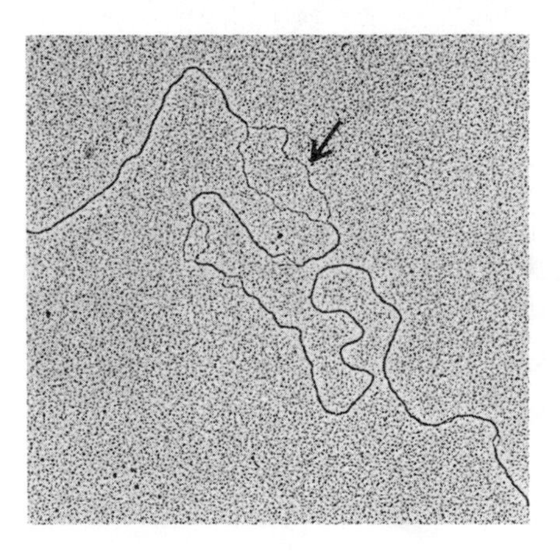

Figure 4-18
Electron micrograph of a DNA molecule partly unwound by alkali. Single-stranded regions appear as loops that stain less intensely than double-stranded segments. These unwound regions are rich in AT base pairs. One of them is marked by an arrow. [From R. B. Inman and M. Schnos. *J. Mol. Biol.* 49(1970):93.]

DNA MOLECULES ARE VERY LONG

A striking characteristic of naturally occurring DNA molecules is their length. DNA molecules must be very long to encode the large number of proteins present in even the simplest cells. The *E. coli* chromosome, for example, is a single molecule of double-helical DNA consisting of four million base pairs. The mass of this DNA molecule is 2.6×10^6 kd. It has a *highly asymmetric shape* when taken out of the cell. The length of *E. coli* DNA is 14×10^6 Å, but its diameter is only 20 Å. The 1.4-mm length of this DNA molecule corresponds to a macroscopic dimension, whereas its width of 20 Å is on the atomic scale. Bruno Zimm found that the largest chromosome of *Drosophila melanogaster* contains a single DNA molecule of 6.2×10^7 base pairs, which has a length of 2.1 cm.

Such highly asymmetric molecules are very susceptible to cleavage by shearing forces. Unless special precautions are taken in their handling, they easily break into segments whose masses are a thousandth of that of the original molecule.

DNA molecules from many bacteria and viruses have been directly visualized by electron microscopy (Figure 4-19). The dimensions of some of these DNA molecules are given in Table 4-1. It should be noted that even the smallest DNA molecules are highly elongate. The DNA from polyoma virus, for example, consists of 5100 base pairs and has a length of 1.7 μm (17,000 Å). Hemoglobin, which is roughly spherical, has a diameter of 65 Å; collagen, one of the longest proteins, has a length of 3000 Å. These comparisons emphasize the remarkable length and asymmetry of DNA molecules.

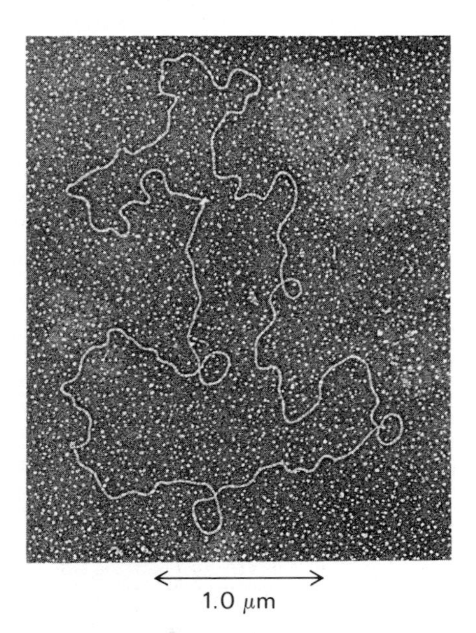

Figure 4-19
Electron micrograph of a DNA molecule from λ bacteriophage (RF II form). [Courtesy of Dr. Thomas Broker.]

Table 4-1
Sizes of DNA molecules

Organism	*Base pairs (in thousands, or kb)*	*Length (μm)*
Viruses		
Polyoma or SV40	5.1	1.7
λ phage	48.6	17
T2 phage	166	56
Vaccinia	190	65
Bacteria		
Mycoplasma	760	260
E. coli	4,000	1,360
Eucaryotes		
Yeast	13,500	4,600
Drosophila	165,000	56,000
Human	2,900,000	990,000

Source: After A. Kornberg, *DNA Replication* (W. H. Freeman and Company, 1980), p. 20.

SOME DNA MOLECULES ARE CIRCULAR AND SUPERCOILED

Electron microscopy has shown that intact DNA molecules from many sources are circular (Figure 4-19). The finding that *E. coli* has a circular chromosome was anticipated by genetic studies that revealed that *the gene-linkage map of this bacterium is circular.* The term "circular" refers to the continuity of the DNA chain, not to its geometrical form. DNA molecules in vivo necessarily have a very compact shape. Note that the length of the *E. coli* chromosome is about a thousand times as long as the greatest diameter of the bacterium.

Not all DNA molecules are circular. DNA from the T7 bacteriophage, for example, is *linear.* The DNA molecules of some viruses, such as the λ bacteriophage, *interconvert between linear and circular forms.* The linear form is present inside the virus particle, whereas the circular form is present in the host cell (see p. 860).

A new property appears in the conversion of a linear DNA duplex into a closed circular molecule. The axis of the double helix can itself be twisted to form a *superhelix.* A circular DNA without any superhelical

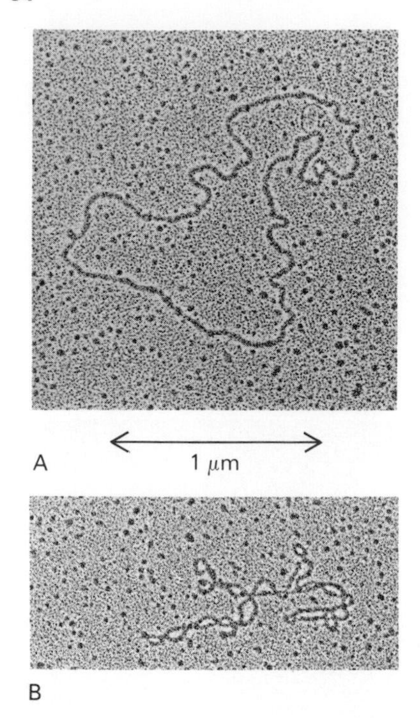

Figure 4-20
Electron micrographs of DNA from mitochondria: (A) relaxed circular form; (B) supercoiled circular form. [Courtesy of Dr. David Clayton.]

turns is known as a *relaxed molecule*. Supercoiling is biologically important for two reasons. First, *a supercoiled DNA has a more compact shape than its relaxed counterpart* (Figure 4-20). Supercoiling is critical for the packaging of DNA in the cell. Second, supercoiling affects the capacity of the double helix to unwind, and thereby affects its interactions with other molecules. These topological features of DNA will be discussed further in a later chapter (p. 660).

DNA IS REPLICATED BY POLYMERASES THAT TAKE INSTRUCTIONS FROM TEMPLATES

We turn now to the molecular mechanism of DNA replication. In 1958, Arthur Kornberg and his colleagues isolated an enzyme from *E. coli* that catalyzes the synthesis of DNA. They named the enzyme DNA polymerase; it is now called *DNA polymerase I* because other DNA polymerases have since been found. DNA replication is mediated by the intricate and coordinated interplay of more than twenty proteins. We focus here on DNA polymerase I to illustrate some new principles.

DNA polymerase I is a 103-kd single polypeptide chain. *It catalyzes the step-by-step addition of deoxyribonucleotide units to a DNA chain:*

$$(\text{DNA})_{n\text{ residues}} + \text{dNTP} \rightleftharpoons (\text{DNA})_{n+1} + \text{PP}_i$$

(The abbreviation dNTP denotes any deoxyribonucleoside triphosphate, and PP_i denotes the pyrophosphate group.) DNA polymerase I requires the following components to synthesize a chain of DNA (Figure 4-21):

1. All four of the activated precursors—the *deoxyribonucleoside 5'-triphosphates dATP, dGTP, dTTP, and dCTP*—must be present. Mg^{2+} ion is also required.

2. DNA polymerase I adds deoxyribonucleotides to the 3'-hydroxyl terminus of a preexisting DNA chain. In other words, a *primer chain* with a free 3'-OH group is required.

3. A *DNA template* is essential. The template can be single- or double-stranded DNA. Double-stranded DNA is an effective template only if its sugar-phosphate backbone is broken at one or more sites.

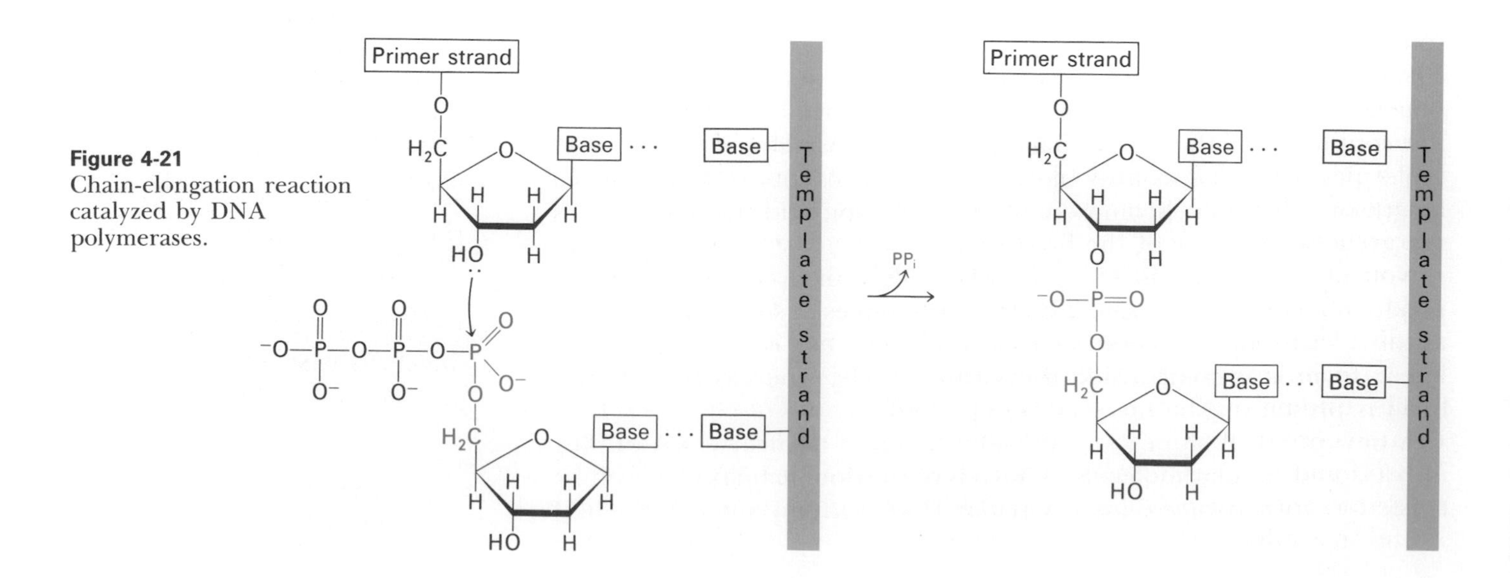

Figure 4-21
Chain-elongation reaction catalyzed by DNA polymerases.

The chain-elongation reaction catalyzed by DNA polymerase is *a nucleophilic attack of the 3′-OH terminus of the primer on the innermost phosphorus atom of a deoxyribonucleoside triphosphate.* A phosphodiester bridge is formed and pyrophosphate is concomitantly released. The subsequent hydrolysis of pyrophosphate by inorganic pyrophosphatase, a ubiquitous enzyme, drives the polymerization forward. *Elongation of the DNA chain proceeds in the 5′ → 3′ direction* (Figure 4-22).

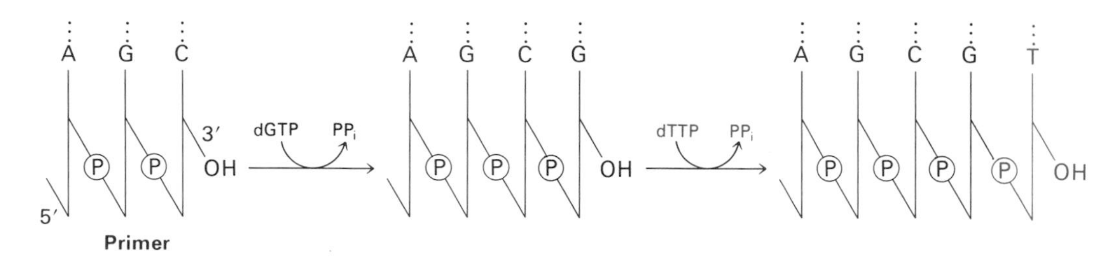

Figure 4-22
DNA polymerases catalyze elongation of DNA chains in the 5′ → 3′ direction.

DNA polymerase catalyzes the formation of a phosphodiester bond only if the base on the incoming nucleotide is complementary to the base on the template strand. The probability of making a covalent link is very low unless the incoming base forms a Watson-Crick type of base pair with the base on the template strand. Thus, DNA polymerase is a *template-directed enzyme.* The enzyme takes instructions from the template and synthesizes a product with a base sequence complementary to that of the template. Indeed, DNA polymerase I was the first template-directed enzyme to be discovered. Another striking feature of DNA polymerase I is that it corrects mistakes in DNA by removing mismatched nucleotides. These properties of DNA polymerase I contribute to the remarkably high fidelity of DNA replication, which has an error rate of less than 10^{-8} per base pair.

SOME VIRUSES HAVE SINGLE-STRANDED DNA DURING PART OF THEIR LIFE CYCLE

Not all DNA is double stranded. Robert Sinsheimer discovered that the DNA in *ϕX174, a small virus that infects* E. coli, *is single stranded.* Several experimental results led to this unexpected conclusion. First, the base ratios of ϕX174 DNA do not conform to the rule that [A] = [T] and [G] = [C]. Second, a solution of ϕX174 DNA is much less viscous than a solution of the same concentration of *E. coli* DNA. The hydrodynamic properties of ϕX174 DNA are like those of a randomly coiled polymer. In contrast, the DNA double helix behaves hydrodynamically as a quite rigid rod. Third, the amino groups of the bases of ϕX174 DNA react readily with formaldehyde, whereas the bases in double helical DNA are virtually inaccessible to this reagent.

The finding of this single-stranded DNA raised doubts concerning the universality of the semiconservative replicative scheme proposed by Watson and Crick. However, it was soon shown that ϕX174 DNA is single stranded for only a part of the life cycle of the virus. Sinsheimer found that infected *E. coli* cells contain a *double-stranded form of ϕX174*

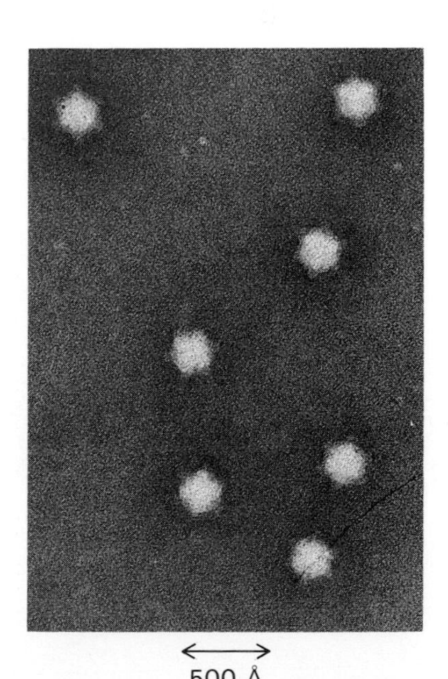

Figure 4-23
Electron micrograph of ϕX174 virus particles. [Courtesy of Dr. Robley Williams.]

DNA. This double-helical DNA is called the *replicative form* because it serves as the template for the synthesis of the DNA of the progeny virus. The generality of the Watson-Crick scheme for replication was reinforced by the finding of this double-stranded viral DNA intermediate.

THE GENES OF SOME VIRUSES ARE MADE OF RNA

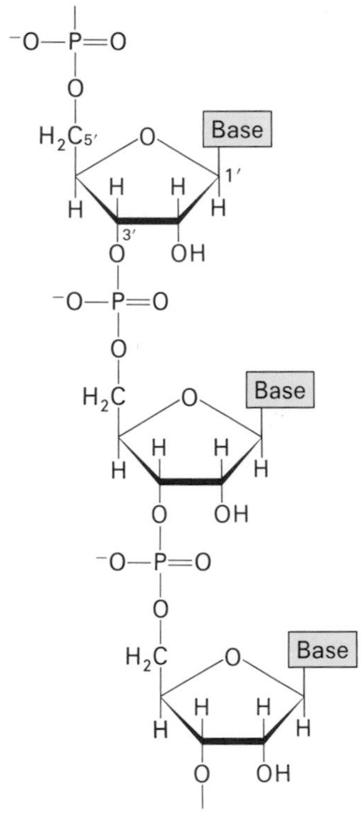

Genes in all procaryotic and eucaryotic organisms are made of DNA. In viruses, genes are made of either DNA or RNA (ribonucleic acid). RNA, like DNA, is a long, unbranched polymer consisting of nucleotides joined by $3' \rightarrow 5'$ phosphodiester bonds (Figure 4-24). The covalent structure of RNA differs from that of DNA in two respects. As indicated by its name, the sugar units in RNA are *riboses* rather than deoxyriboses. Ribose contains a 2′-hydroxyl group not present in deoxyribose. The other difference is that one of the four major bases in RNA is *uracil* (U) instead of thymine (T). Uracil, like thymine, can form a base pair with adenine but it lacks the methyl group present in thymine. RNA molecules can be single-stranded or double-stranded. RNA cannot form a double helix of the B-DNA type because of steric interference by the 2′-hydroxyl groups of its ribose units. However, RNA can adopt a modified double helical form in which the base pairs are tilted about 20 degrees away from the perpendicular to the helix axis, a structure like A-DNA (p. 652).

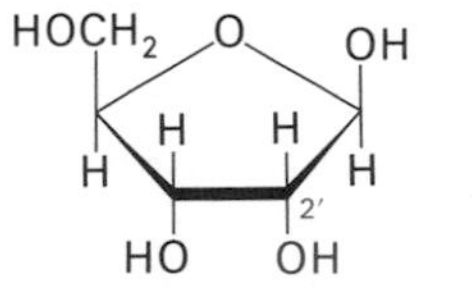

Figure 4-24
Structure of part of an RNA chain.

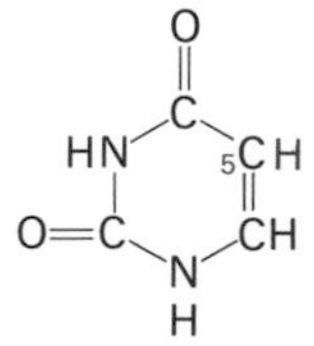

Tobacco mosaic virus, which infects the leaves of tobacco plants, is one of the best-characterized RNA viruses. It consists of a single strand of RNA (6390 nucleotides) surrounded by a protein coat of 2130 identical subunits (see Chapter 34 for a discussion of its structure and assembly). The protein can be separated from the RNA by treatment of the virus with phenol. *The isolated viral RNA is infective, whereas the viral protein is not.* Synthetic hybrid virus particles provide additional evidence that the genetic specificity of the virus resides exclusively in its RNA. A variety of strains of tobacco mosaic virus are known. A synthetic hybrid virus was prepared from the RNA of strain 1 and the protein of strain 2. Another was prepared from the RNA of strain 2 and the protein of strain 1. After infection, *the progeny virus always consisted of RNA and protein corresponding to the RNA in the infecting hybrid virus* (Figure 4-25).

Tobacco mosaic virus replicates in an infected plant cell by first synthesizing a (−) RNA strand that is complementary to the (+) RNA strand in the virus particle. The (−) RNA strand then serves as the template for the synthesis of a large number of (+) RNA strands that become packaged in new virus particles that are released by the cell. These syntheses are catalyzed by RNA polymerases that take instructions from RNA templates (*RNA-directed RNA polymerases*).

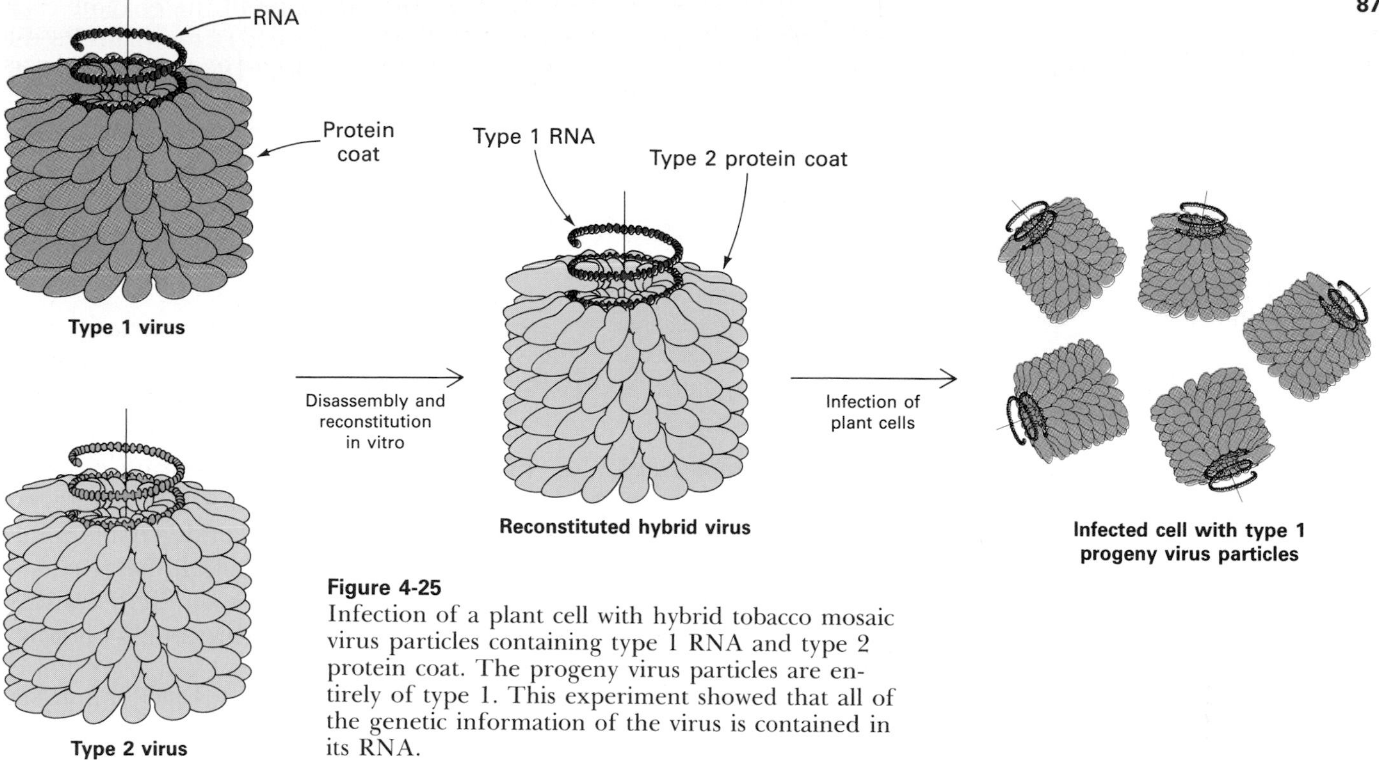

Figure 4-25
Infection of a plant cell with hybrid tobacco mosaic virus particles containing type 1 RNA and type 2 protein coat. The progeny virus particles are entirely of type 1. This experiment showed that all of the genetic information of the virus is contained in its RNA.

RNA TUMOR VIRUSES REPLICATE THROUGH DOUBLE-HELICAL DNA INTERMEDIATES

A number of RNA viruses produce malignant tumors after being injected into susceptible animal hosts. *Rous sarcoma virus* is one of the best-studied members of this group of *RNA tumor viruses*, which contain a single strand of RNA (p. 875). A striking feature of RNA tumor viruses is that they replicate through *DNA* intermediates (Figure 4-26). The RNA of the virus particle, called the (+) strand, is delivered into the host cell. This (+) RNA is the template for the synthesis of a complementary (−) DNA strand by a *reverse transcriptase*, an enzyme that is brought into the cell by the virus particle for this special purpose. Reverse transcriptase is an *RNA-directed DNA polymerase.* In this case, genetic information flows from RNA to DNA, the *reverse* of the normal

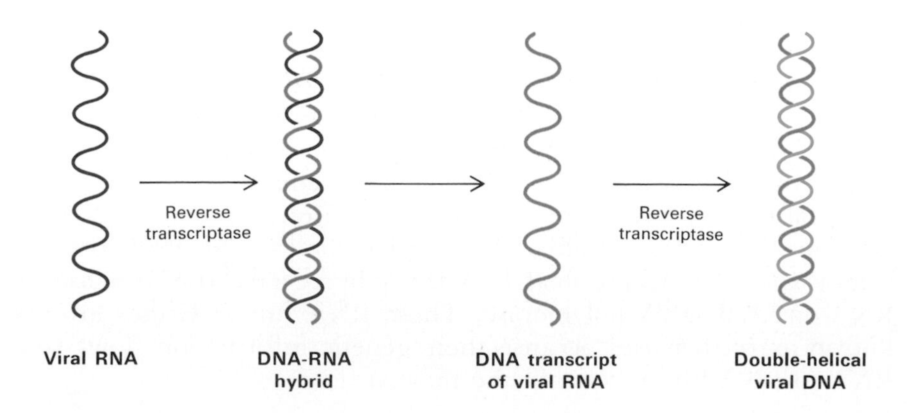

Figure 4-26
RNA tumor viruses replicate through double-helical DNA intermediates. DNA complementary to viral RNA is synthesized by reverse transcriptase, an enzyme brought into the cell by the infecting virus particle.

direction of information transfer (hence the name of the enzyme catalyzing this unusual step). The (−) DNA then serves as a template for the synthesis of (+) DNA. The resulting double-helical DNA version of the viral genome becomes incorporated into the chromosomal DNA of the host and is replicated along with the normal cellular DNA in the course of cell division. At some later time, the integrated viral genome is expressed to form viral (+) RNA and viral proteins, which assemble into new virus particles. RNA tumor viruses are also called *retroviruses* because their genetic information flows from RNA to DNA.

SUMMARY

DNA is the molecule of heredity in all procaryotic and eucaryotic organisms. In viruses, the genetic material is either DNA or RNA. All cellular DNA consists of two very long, helical polynucleotide chains coiled around a common axis. The two strands of the double helix run in opposite directions. The sugar-phosphate backbone of each strand is on the outside of the double helix, whereas the purine and pyrimidine bases are inside. The two chains are held together by hydrogen bonds between pairs of bases. Adenine (A) is always paired with thymine (T), and guanine (G) is always paired with cytosine (C). Hence, one strand of a double helix is the complement of the other. Genetic information is encoded in the precise sequence of bases along a strand. Most DNA molecules are circular. The axis of the double helix of a circular DNA can itself be coiled to form a superhelix. Supercoiled DNA is more compact than is relaxed DNA.

In the replication of DNA, the two strands of a double helix unwind and separate as new chains are synthesized. Each parent strand acts as a template for the formation of a new complementary strand. Thus, the replication of DNA is semiconservative—each daughter molecule receives one strand from the parent DNA molecule. The replication of DNA is a complex process carried out by many proteins, including several DNA polymerases. The activated precursors in the synthesis of DNA are the four deoxyribonucleoside 5′-triphosphates. The new strand is synthesized in the 5′ → 3′ direction by a nucleophilic attack by the 3′-hydroxyl terminus of the primer strand on the innermost phosphorus atom of the incoming deoxyribonucleoside triphosphate. Most important, DNA polymerases catalyze the formation of a phosphodiester bond only if the base on the incoming nucleotide is complementary to the base on the template strand. In other words, DNA polymerases are template-directed enzymes.

Some viruses have single-stranded DNA during a part of their life cycle. The DNA in ϕX174, a small virus that infects *E. coli*, is single stranded. In the infected host, however, a complementary strand is made, to form a double-helical replicative intermediate. The genes of some viruses, such as tobacco mosaic virus, are made of single-stranded RNA. An RNA-directed RNA polymerase mediates the replication of this viral RNA. RNA tumor viruses, such as Rous sarcoma virus, replicate through double-helical DNA intermediates. The RNA of these tumor viruses is transcribed into DNA by reverse transcriptase, an RNA-directed DNA polymerase. These RNA tumor viruses are also known as retroviruses because their genetic information flows from RNA to DNA, the reverse of the normal flow.

SELECTED READINGS

Where to start

Felsenfeld, G., 1985. DNA. *Sci. Amer.* 253(4):58–67. [Also printed in *The Molecules of Life,* an excellent series of articles. W. H. Freeman, 1985.]

Darnell, J. E., Jr., 1985. RNA. *Sci. Amer.* 253(4):26–36. [Also printed in *The Molecules of Life.*]

Dickerson, R. E., 1983. The DNA helix and how it is read. *Sci. Amer.* 249(6):94–111. [Offprint 1545].

Discovery of the major concepts

Avery, O. T., MacLeod, C. M., and McCarty, M., 1944. Studies on the chemical nature of the substance inducing transformation of pneumococcal types. Induction of transformation by a deoxyribonucleic acid fraction isolated from Pneumococcus Type III. *J. Exp. Med.* 79:137–158.

Hershey, A. D., and Chase, M., 1952. Independent functions of viral protein and nucleic acid in growth of bacteriophage. *J. Gen. Physiol.* 36:39–56.

Watson, J. D., and Crick, F. H. C., 1953*a*. Molecular structure of nucleic acid. A structure for deoxyribose nucleic acid. *Nature* 171:737–738.

Watson, J. D., and Crick, F. H. C., 1953*b*. Genetic implications of the structure of deoxyribonucleic acid. *Nature* 171:964–967.

Kornberg, A., 1960. Biologic synthesis of deoxyribonucleic acid. *Science* 131:1503–1508.

Meselson, M., and Stahl, F. W., 1958. The replication of DNA in *Escherichia coli. Proc. Nat. Acad. Sci.* 44:671–682.

Taylor, J. H., (ed.), 1965. *Selected Papers on Molecular Genetics.* Academic Press. [Contains the classic papers listed above.]

DNA structure

Saenger, W., 1984. *Principles of Nucleic Acid Structure.* Springer-Verlag. [An outstanding advanced account of the three-dimensional structure of nucleotides, DNA, and RNA. Contains many excellent illustrations.]

Dickerson, R. E., Drew, H. R., Conner, B. N., Wing, R. M., Fratini, A. V., and Kopka, M. L., 1982. The anatomy of A-, B-, and Z-DNA. *Science* 216:475–485.

Cantor, C. R., and Schimmel, P. R., 1980. *Biophysical Chemistry* (3 vols). W. H. Freeman. [Chapters 3 and 6 (in Part I) and Chapters 22, 23, and 24 (in Part III) give an excellent account of the conformation of nucleic acids.]

DNA replication

Kornberg, A., 1980. *DNA Replication.* W. H. Freeman. [An outstanding and highly readable book. Also see the *1982 Supplement to DNA Replication*, an update.]

Reminiscences and historical accounts

Watson, J. D., 1968. *The Double Helix.* Atheneum. [A lively, personal account of the discovery of the structure of DNA and its biological implications.]

McCarty, M., 1985. *The Transforming Principle: Discovering that Genes Are Made of DNA.* Norton. [A warm and lucid account of one of the major discoveries of this century by a sensitive participant.]

Cairns, J., Stent, G. S., and Watson, J. D., (eds.), 1966. *Phage and the Origins of Molecular Biology.* Cold Spring Harbor Laboratory. [A fascinating collection of reminiscences by some of the architects of molecular biology.]

Olby, R., 1974. *The Path to the Double Helix.* University of Washington Press.

Portugal, F. H., and Cohen, J. S., 1977. *A Century of DNA: A History of the Discovery of the Structure and Function of the Genetic Substance.* MIT Press.

Judson, H., 1979. *The Eighth Day of Creation.* Simon and Schuster.

PROBLEMS

1. Write the complementary sequence (in the standard $5' \rightarrow 3'$ notation) for:
   (a) GATCAA.
   (b) TCGAAC.
   (c) ACGCGT.
   (d) TACCAT.

2. The composition (in mole-fraction units) of one of the strands of a double-helical DNA is [A] = 0.30 and [G] = 0.24.
   (a) What can you say about [T] and [C] for the same strand?
   (b) What can you say about [A], [G], [T], and [C] of the complementary strand?

3. The DNA of a deletion mutant of λ bacteriophage has a length of 15 μm instead of 17 μm. How many base pairs are missing from this mutant?

4. What result would Meselson and Stahl have obtained if the replication of DNA were conservative (i.e., the parental double helix stayed together)? Give the expected distribution of DNA molecules after 1.0 and 2.0 generations for conservative replication.

5. Griffith used heat-killed S pneumococci to transform R mutants. Studies years later showed that double-stranded DNA is needed for efficient transformation and that high temperatures melt the DNA double

helix. Why were Griffith's experiments nevertheless successful?

6. Strains of *Bacillus subtilis* that can be transformed by foreign DNA are termed *competent.* Others, termed *non-competent,* are insusceptible to transformation. How might these strains differ from one another?

7. Bacteriophage M13 infects *E. coli* differently from the way bacteriophage T2 does. The M13 protein coat is removed in the inner membrane of the bacterial cell, where it is sequestered and subsequently used for the envelopment of progeny DNA. Why would M13 have been much less suitable than T2 was for the experiments carried out by Hershey and Chase?

8. Suppose that you want to radioactively label DNA but not RNA in dividing and growing bacterial cells. Which radioactive molecule would you add to the culture medium?

9. Suppose that you want to prepare DNA in which the backbone phosphorus atoms are uniformly labeled with ^{32}P. Which precursors should be added to a solution containing DNA polymerase I and primed template DNA? Specify the position of radioactive atoms in these precursors.

10. A solution contains DNA polymerase I and the Mg^{2+} salts of dATP, dGTP, dCTP, and dTTP. The DNA molecules listed below are added to aliquots of this solution. Which of them would lead to DNA synthesis?
    (a) A single-stranded closed circle containing 1000 nucleotide units.
    (b) A double-stranded closed circle containing 1000 nucleotide pairs.
    (c) A single-stranded closed circle of 1000 nucleotides base paired to a linear strand of 500 nucleotides with a free 3′-OH terminus.
    (d) A double-stranded linear molecule of 1000 nucleotide pairs with a free 3′-OH at each end.

11. Suppose that you want to assay reverse transcriptase activity. If polyriboadenylate is the template in the assay, what should you use as the primer? Which radioactive nucleotide should you use to follow chain elongation?

12. Reverse transcriptase has ribonuclease activity as well as polymerase activity. What is the role of its ribonuclease activity?

13. You have purified a virus that infects turnip leaves. Treatment of a sample with phenol removes viral proteins. Application of the residual material to scraped leaves results in the formation of progeny virus particles. You infer that the infectious substance is a nucleic acid.
    (a) Propose a simple and highly sensitive means of determining whether the infectious nucleic acid is DNA or RNA.
    (b) Is it likely that the virus particle carries an enzyme essential for its replication?

14. Spontaneous deamination of cytosine bases in DNA occurs at low but measurable frequency. Cytosine is converted into uracil by loss of its amino group. After this conversion, which base pair occupies this position in each of the daughter strands resulting from one round of replication? two rounds of replication?

CHAPTER 5

Flow of Genetic Information

We turn now from the storage and transmission of genetic information to its expression. Genes specify the kinds of proteins that are made by cells. However, DNA is not the direct template for protein synthesis. Rather, the templates for protein synthesis are RNA (ribonucleic acid) molecules. This chapter begins with an account of the discovery that a class of RNA molecules called *messenger RNAs* (mRNAs) are the information-carrying intermediates in protein synthesis. Other RNA molecules, such as transfer RNA (tRNA) and ribosomal RNA (rRNA) are part of the protein-synthesizing machinery. All forms of cellular RNA are synthesized by RNA polymerases that take instructions from DNA templates. This process of *transcription* is followed by *translation*, the synthesis of proteins according to instructions given by mRNA templates. Thus, the flow of genetic information in normal cells is

$$\text{DNA} \xrightarrow[\text{transcription}]{} \text{RNA} \xrightarrow[\text{translation}]{} \text{protein}$$

This brings us to the genetic code, the relation between the sequence of bases in DNA (or its mRNA transcript) and the sequence of amino acids in a protein. The code, which is nearly the same in all organisms, is beautiful in its simplicity. A sequence of three bases, called a codon, specifies an amino acid. Codons in mRNA are read sequentially by tRNA molecules, which serve as adaptors in protein synthesis. Protein synthesis takes place on ribosomes, which are complex assemblies of rRNAs and more than fifty kinds of proteins. Newly synthesized proteins contain signals that enable them to be targeted to specific destinations. The last theme considered in this chapter is the interrupted character of most eucaryotic genes, which are mosaics of introns and exons. Both are transcribed, but introns are cut out of newly synthesized RNA molecules, leaving mature RNA molecules with continuous exons. The existence of introns and exons has profound implications for evolution.

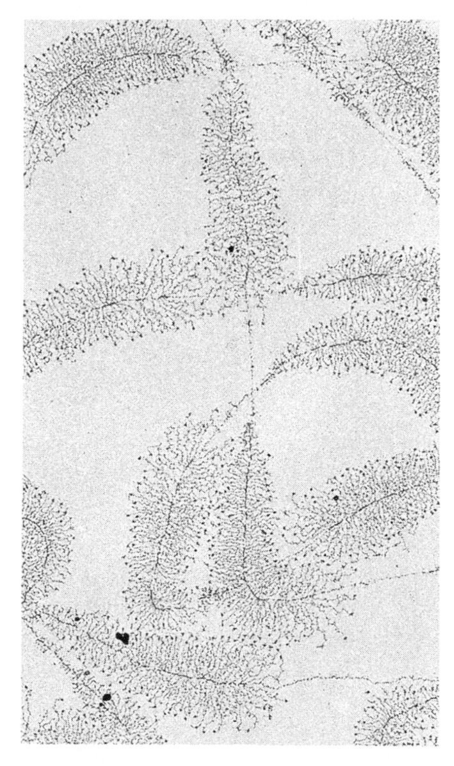

Figure 5-1
Transcription of DNA to form precursors of ribosomal RNA. This electron micrograph shows a tandem array of nascent RNA molecules emerging from a thread of DNA. [From O. L. Miller, Jr., and B. R. Beatty. Portrait of a gene. *J. Cell Physiol.* 74(supp. 1, 1969):225.]

SEVERAL KINDS OF RNA PLAY KEY ROLES IN GENE EXPRESSION

RNA is a long, unbranched macromolecule consisting of nucleotides joined by 3′ → 5′ phosphodiester bonds (Figure 4-24). As the name indicates, the sugar unit in RNA is ribose. The four major bases in RNA are adenine (A), uracil (U), guanine (G), and cytosine (C). Adenine can

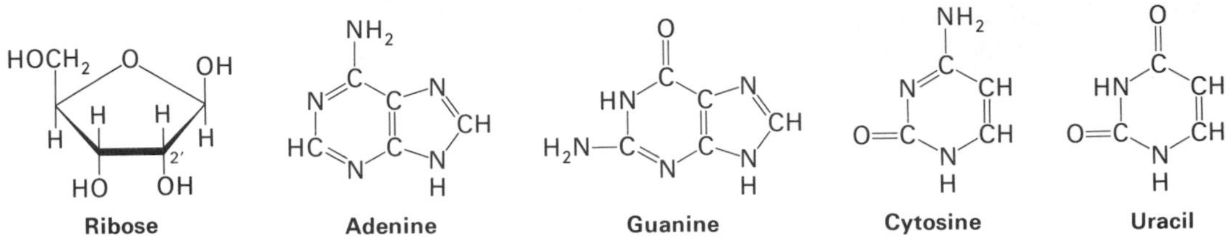

Ribose Adenine Guanine Cytosine Uracil

pair with uracil, and guanine with cytosine. The number of nucleotides in RNA ranges from as few as seventy-five to many thousands. *RNA molecules are usually single stranded*, except in some viruses. Consequently, an RNA molecule need not have complementary base ratios. In fact, the proportion of adenine differs from that of uracil, and the proportion of guanine differs from that of cytosine, in most RNA molecules. However, *RNA molecules do contain regions of double-helical structure that are produced by the formation of hairpin loops* (Figure 5-2). In these regions, A pairs with U, and G pairs with C. The base pairing in RNA hairpins is frequently imperfect. G can also form a base pair with U, but it is less strong than the GC base pair (p. 744). Some of the apposing bases may not be complementary at all, and one or more bases along a single strand may be looped out to facilitate the pairing of the others. The proportion of helical regions in different kinds of RNA varies over a wide range; a value of 50% is typical.

Cells contain several kinds of RNA (Table 5-1). *Messenger RNA* (mRNA) is the template for protein synthesis. An mRNA molecule is produced for each gene or group of genes that is to be expressed. Consequently, mRNA is a very heterogeneous class of molecules. In *E. coli*, the average length of an mRNA molecule is about 1.2 kb. *Transfer*

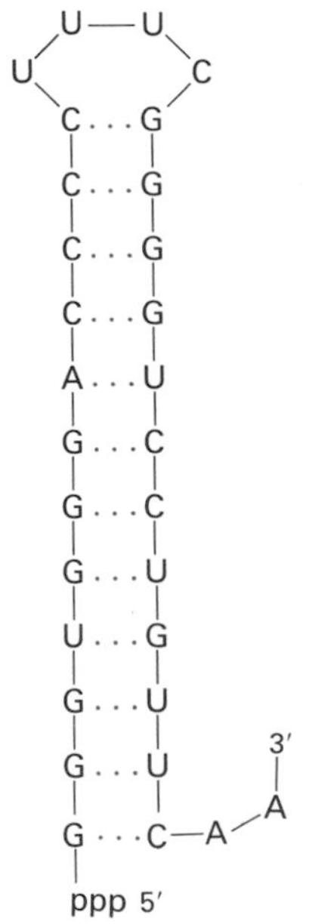

Figure 5-2
RNA can fold back on itself to form double-helical regions.

Table 5-1
RNA molecules in *E. coli*

Type	Relative amount (%)	Sedimentation coefficient (S)	Mass (kd)	Number of nucleotides
Ribosomal RNA (rRNA)	80	23	1.2×10^3	3700
		16	0.55×10^3	1700
		5	3.6×10^1	120
Transfer RNA (tRNA)	15	4	2.5×10^1	75
Messenger RNA (mRNA)	5	Heterogeneous		

RNA (tRNA) carries amino acids in an activated form to the ribosome for peptide-bond formation, in a sequence determined by the mRNA template. There is at least one kind of tRNA for each of the twenty amino acids. Transfer RNA consists of about seventy-five nucleotides (having a mass of about 25 kd), which makes it the smallest of the RNA molecules. *Ribosomal RNA* (rRNA) is the major component of ribosomes, but its precise role in protein synthesis is not yet known. The finding of catalytic RNA (p. 113) makes this question even more intriguing. In *E. coli,* there are three kinds of rRNA, called 23S, 16S, and 5S RNA because of their sedimentation behavior. One molecule of each of these species of rRNA is present in each ribosome. Ribosomal RNA is the most abundant of the three types of RNA. Transfer RNA comes next, followed by messenger RNA, which constitutes only 5% of the total RNA. Eucaryotic cells contain additional small RNA molecules. *Small nuclear RNA* (snRNA) molecules, for example, participate in the splicing of RNA exons. A small RNA molecule in the cytosol plays a role in the targeting of newly synthesized proteins.

FORMULATION OF THE CONCEPT OF MESSENGER RNA

The concept of mRNA was formulated by Francois Jacob and Jacques Monod in a classic paper published in 1961. Because proteins are synthesized in the cytoplasm rather than in the nucleus of eucaryotic cells, it was evident that there must be a chemical intermediate, which they called the structural messenger, specified by the genes. What is the nature of this intermediate? An important clue came from their studies of the control of protein synthesis in *E. coli* (Chapter 32). Certain enzymes in *E. coli,* such as those that participate in the uptake and utilization of lactose, are inducible—that is, the amount of these enzymes increases more than a thousandfold if an inducer (such as isopropylthiogalactoside) is present. The kinetics of induction were very revealing. The addition of an inducer elicited maximal synthesis of the lactose enzymes within a few minutes. Furthermore, the removal of the inducer resulted in the cessation of the synthesis of these enzymes in an equally short time. These experimental findings were incompatible with the presence of stable templates for the formation of these enzymes. Hence, Jacob and Monod surmised that *the messenger must be a very short-lived intermediate.* They then proposed that the messenger should have the following properties:

1. The messenger should be a polynucleotide.

2. The base composition of the messenger should reflect the base composition of the DNA that specifies it.

3. The messenger should be very heterogeneous in size because genes (or groups of genes) vary in length. They correctly assumed that three nucleotides code for one amino acid and calculated that the molecular weight of a messenger should be at least a half million.

4. The messenger should be transiently associated with ribosomes, the sites of protein synthesis.

5. The messenger should be synthesized and degraded very rapidly.

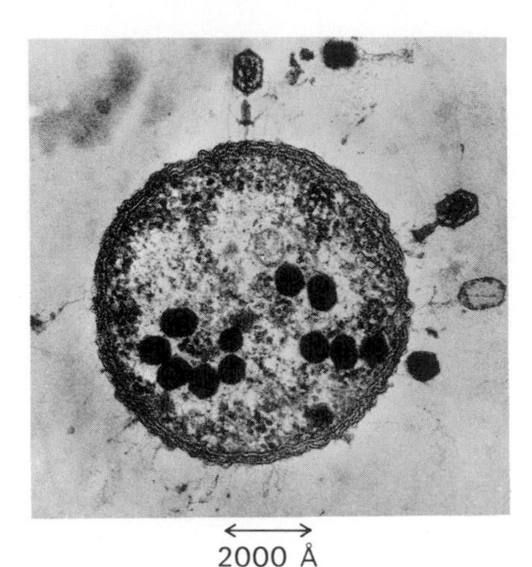

Figure 5-3
Electron micrograph of an *E. coli* cell infected by T2 viruses. [Courtesy of Dr. Lee Simon.]

It was apparent to Jacob and Monod that none of the known RNA fractions at that time met these criteria. Ribosomal RNA, then generally assumed to be the template for protein synthesis, was too homogeneous in size. Also, its base composition was similar in species that had very different DNA base ratios. Transfer RNA also seemed an unlikely candidate for the same reasons. In addition, it was too small. However, there were suggestions in the literature of a third class of RNA that appeared to meet the above criteria for the messenger. In *E. coli* infected with T2 bacteriophage, there was a new RNA fraction of appropriate size that had a very short half-life. Most interesting, the base composition of this new RNA fraction was like that of the viral DNA rather than like that of *E. coli* DNA.

EXPERIMENTAL EVIDENCE FOR MESSENGER RNA, THE INFORMATIONAL INTERMEDIATE IN PROTEIN SYNTHESIS

The hypothesis of a short-lived messenger RNA as the information-carrying intermediate in protein synthesis was tested shortly after the concept was formulated. Sydney Brenner, Francois Jacob, and Matthew Meselson carried out experiments on *E. coli* infected with T2 bacteriophage. Nearly all of the proteins made by the cell after infection are genetically determined by the phage. The synthesis of these proteins is not accompanied by an overall synthesis of RNA. However, a minor RNA fraction with a short half-life appears soon after infection. In fact, this RNA fraction has a nucleotide composition like that of the phage DNA. The fraction seemed optimal for a test of the messenger hypothesis because it appeared with a sudden switch in the kinds of proteins synthesized by the cell, and because neither rRNA nor tRNA are synthesized after infection.

How did this switch in the kinds of proteins made after infection occur? One possibility was that the phage DNA specifies a new set of ribosomes. In this model, genes control the synthesis of specialized ribosomes, and each ribosome can make only one kind of protein. An alternative model, the one proposed by Jacob and Monod, was that ribosomes are nonspecialized structures that receive genetic information from the gene in the form of an unstable messenger RNA. The experiments of Brenner, Jacob, and Meselson were designed to determine whether new ribosomes are synthesized after infection or whether new RNA joins preexisting ribosomes.

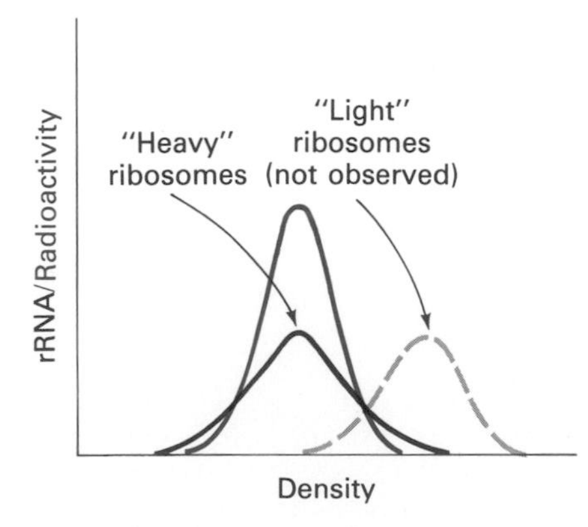

Figure 5-4
Density-gradient centrifugation of ribosomes of normal and T2 phage-infected bacteria. ^{32}P-labeled RNA synthesized *after* infection banded with ribosomes formed prior to infection ("heavy" ribosomes). Ribosomes were not synthesized after infection.

Bacteria were grown in a medium containing heavy isotopes (^{15}N and ^{13}C), infected with phage, and then immediately transferred to a medium containing light isotopes (^{14}N and ^{12}C). Ribosomes synthesized before and after infection could be separated by density-gradient centrifugation because their densities differed ("heavy" and "light," respectively). Furthermore, new RNA was labeled by the radioisotope ^{32}P- or ^{14}C-uracil and new protein by ^{35}S. These experiments showed that

1. *Ribosomes were not synthesized after infection,* as evidenced by the absence of "light" ribosomes.

2. RNA *was* synthesized after infection. Most of the radioactively labeled RNA emerged in the "heavy" ribosome peak. Thus, *most of the new RNA was associated with preexisting ribosomes.* Additional experiments showed that this new RNA turns over rapidly during the growth of phage.

3. The radioisotope ^{35}S appeared transiently in the "heavy" ribosome peak, which showed that *new proteins were synthesized in preexisting ribosomes.*

These experiments led to the conclusion that *ribosomes are nonspecialized structures which synthesize, at a given time, the protein dictated by the messenger they happen to contain.* Studies of uninfected bacterial cells also showed that messenger RNA is the information-carrying link between gene and protein. In a very short time, the concept of messenger RNA became a central facet of molecular biology.

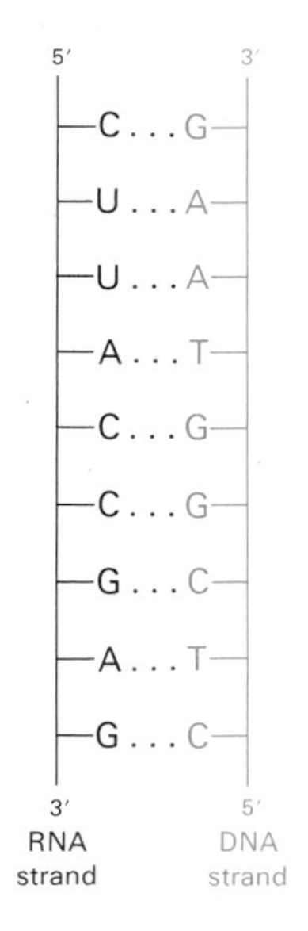

Figure 5-5
An RNA-DNA hybrid can be formed if the RNA and DNA have complementary sequences.

HYBRIDIZATION STUDIES SHOWED THAT MESSENGER RNA IS COMPLEMENTARY TO ITS DNA TEMPLATE

In 1961, Sol Spiegelman developed a new technique called *hybridization* to answer the following question: Is the RNA synthesized after infection with T2 phage complementary to T2 DNA? It was known from the work of Julius Marmur and Paul Doty that heating double-helical DNA above its melting temperature results in the formation of single strands. If the mixture is cooled slowly, these strands reassociate to form a double-helical structure with biological activity. Marmur and Doty also found that double-helical molecules are formed only from strands derived from the same or from closely related organisms. This observation suggested to Spiegelman that a double-stranded DNA-RNA hybrid might be formed from a mixture of single-stranded DNA and RNA if their base sequences were complementary (Figure 5-5). The experimental design was this:

1. A sample of T2 mRNA (the RNA synthesized after infection of *E. coli* with T2 phage) was prepared with a ^{32}P label. T2 DNA labeled with ^{3}H was prepared separately.

2. A mixture of the T2 mRNA and T2 DNA was heated to 100°C, which melted the double-helical DNA into single strands. This solution of single-stranded RNA and DNA was slowly cooled to room temperature.

3. The cooled mixture was analyzed by density-gradient centrifugation. The samples were centrifuged for several days in swinging-bucket rotors. The plastic sample tubes were then punctured at the bottom and drops were collected for analysis.

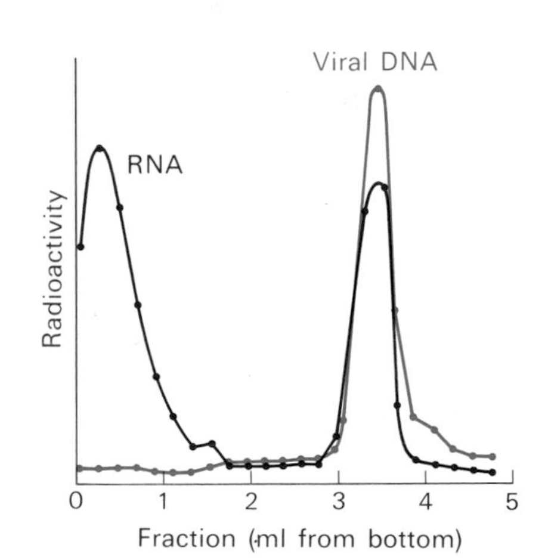

Figure 5-6
The RNA produced after *E. coli* has been infected with T2 phage is complementary to the viral DNA. In this hybridization experiment, RNA was labeled with ^{32}P, whereas T2 DNA was labeled with ^{3}H. The distribution of radioactivity in a cesium chloride density gradient shows that much of the RNA synthesized after infection bands with the T2 DNA. [After S. Spiegelman. Hybrid nucleic acids. Copyright © 1964 by Scientific American, Inc. All rights reserved.]

Three bands were found (Figure 5-6). The densest of these was single-stranded RNA. A second band contained double-helical DNA. A third band consisting of double-stranded DNA-RNA hybrid molecules was present near the DNA band. Thus, T2 mRNA formed a hybrid with T2 DNA. In contrast, T2 mRNA did not hybridize with DNA derived from a variety of bacteria and unrelated viruses, even if their base ratios were like those of T2 DNA. Subsequent experiments showed that the mRNA fraction of uninfected cells hybridized with the DNA derived from that particular organism but not from unrelated ones. These incisive experiments revealed that *the base sequence of mRNA is complementary to that of its DNA template. Furthermore, a powerful tool was developed for tracing the flow of genetic information in cells and for determining whether two nucleic acid molecules are similar.*

RIBOSOMAL RNA AND TRANSFER RNA ARE ALSO SYNTHESIZED ON DNA TEMPLATES

This hybridization technique was then used to determine whether rRNA and tRNA are also synthesized on DNA templates. The formation of RNA-DNA hybrids was detected by a filter assay rather than by density-gradient centrifugation because it is simpler, more sensitive, and much more rapid. Single-stranded RNA passes through a nitrocellulose filter, whereas double-helical DNA and RNA-DNA hybrids are retained by this filter. Various kinds of *E. coli* RNA labeled with ^{32}P were added to unlabeled *E. coli* DNA. These mixtures were heated, slowly cooled, and then filtered through nitrocellulose. The radioactivity retained on the filter was measured. The results were unequivocal: *RNA-DNA hybrids were formed with both rRNA (5S, 16S, and 23S) and tRNA, which shows that complementary sequences for these RNA molecules are present in the* E. coli *genome.*

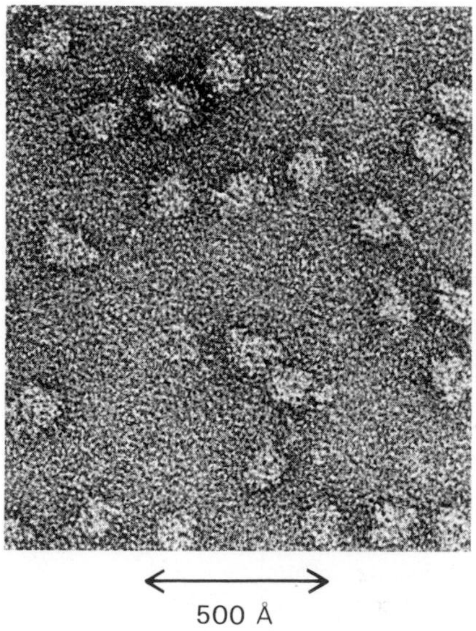

500 Å

Figure 5-7
Electron micrograph of RNA polymerase from *E. coli.* [Courtesy of Dr. Robley Williams and Dr. Michael Chamberlin.]

ALL CELLULAR RNA IS SYNTHESIZED BY RNA POLYMERASES

The concept of mRNA stimulated the search for an enzyme that synthesizes RNA according to instructions given by a DNA template. In 1960, Jerard Hurwitz and Samuel Weiss independently discovered such an enzyme, which they named *RNA polymerase.* The enzyme from *E. coli* requires the following components for the synthesis of RNA:

1. *Template.* The preferred template is *double-stranded DNA.* Single-stranded DNA can also serve as a template. RNA, whether single or double stranded, is not an effective template, nor are RNA-DNA hybrids.

2. *Activated precursors.* All four *ribonucleoside triphosphates*—ATP, GTP, UTP, and CTP—are required.

3. *Divalent metal ion.* Mg^{2+} or Mn^{2+} are effective. Mg^{2+} meets this requirement in vivo.

RNA polymerase catalyzes the initiation and elongation of RNA chains. The reaction catalyzed by this enzyme is

$$(\text{RNA})_{n \text{ residues}} + \text{ribonucleoside triphosphate} \rightleftharpoons (\text{RNA})_{n+1 \text{ residues}} + \text{PP}_i$$

The synthesis of RNA is like that of DNA in several respects (Figure 5-8). First, the direction of synthesis is $5' \rightarrow 3'$. Second, the mechanism of elongation is similar: the 3′-OH group at the terminus of the growing chain makes a nucleophilic attack on the innermost phosphate of the incoming nucleoside triphosphate. Third, the synthesis is driven forward by the hydrolysis of pyrophosphate. In contrast with DNA polymerase, RNA polymerase does not require a primer. Another difference is that the DNA template is fully conserved in RNA synthesis, whereas it is semiconserved in DNA synthesis. Also, RNA polymerase lacks the nuclease capability used by DNA polymerase to excise mismatched nucleotides.

All three types of cellular RNA—mRNA, tRNA, and rRNA—are synthesized in *E. coli* by the same RNA polymerase according to instruc-

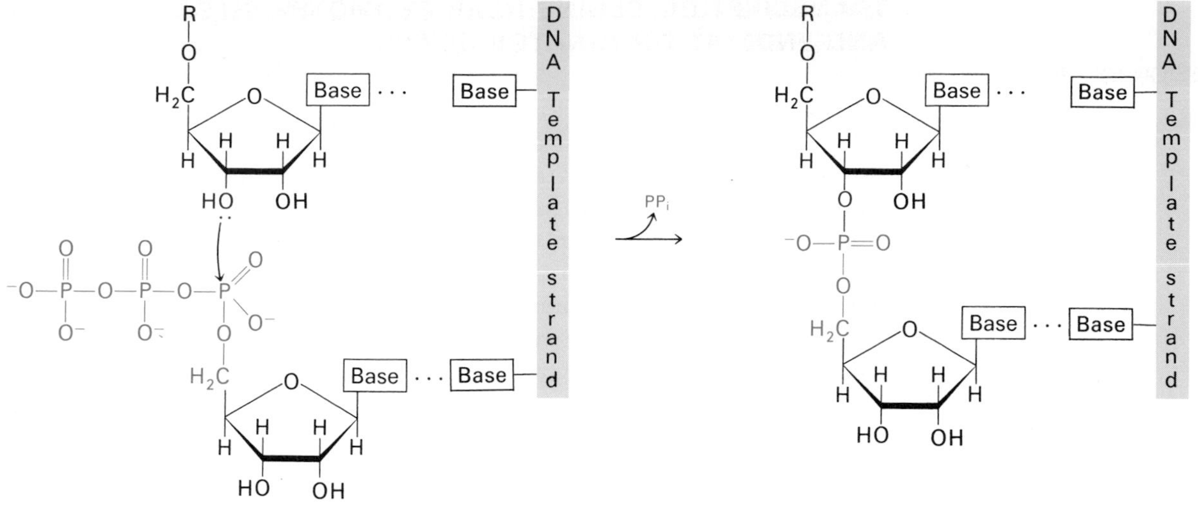

Figure 5-8
Mechanism of the chain-elongation reaction catalyzed by RNA polymerase.

tions given by a DNA template. In mammalian cells, there is a division of labor among several different kinds of RNA polymerases. We shall return to these RNA polymerases in Chapter 29. It is noteworthy that some viruses code for RNA-synthesizing enzymes that are very different from those of their host cells. For example, the RNA replicase that is encoded by Qβ, an RNA phage, is an RNA-dependent RNA polymerase because it takes instructions from an RNA template rather than from a DNA template (Chapter 34). In contrast, the cellular enzymes that synthesize RNA are *DNA-dependent RNA polymerases.*

RNA POLYMERASE TAKES INSTRUCTIONS FROM A DNA TEMPLATE

RNA polymerase, like the DNA polymerases described in the preceding chapter, takes instructions from a DNA template. The earliest evidence was the finding that the *base composition* of newly synthesized RNA is the complement of that of the DNA template strand, as exemplified by the RNA synthesized from single-stranded ϕX174 DNA as template (Table 5-2). Hybridization experiments also revealed that RNA synthesized by RNA polymerase is complementary to its DNA template. The strongest evidence for the fidelity of transcription came from base-sequence studies showing that the RNA sequence is the precise complement of the DNA template sequence (Figure 5-9).

Table 5-2
Base composition of RNA synthesized from a viral DNA template

DNA template (plus strand of φX174)		*RNA product*	
A	0.25	0.25	U
T	0.33	0.32	A
G	0.24	0.23	C
C	0.18	0.20	G

5′-GCGGCGACGCGCAGUUAAUCCCACAGCCGCCAGUUCCGCUGGCGGCAUUUU-3′ **mRNA**

3′-CGCCGCTGCGCGTCAATTAGGGTGTCGGCGGTCAAGGCGACCGCCGTAAAA-5′
5′-GCGGCGACGCGCAGTTAATCCCACAGCCGCCAGTTCCGCTGGCGGCATTTT-3′ **DNA**

Figure 5-9
The base sequence of mRNA is the complement of that of the DNA template strand. The sequence shown here is from the tryptophan operon, a segment of DNA containing the genes for five enzymes that catalyze the synthesis of tryptophan.

TRANSCRIPTION BEGINS NEAR PROMOTER SITES AND ENDS AT TERMINATOR SITES

DNA templates contain regions called *promoter sites* that specifically bind RNA polymerase and determine where transcription begins. In *bacteria*, two sequences on the 5′ (upstream) side of the first nucleotide to be transcribed are important (Figure 5-10). One of them, called the *Pribnow box*, has the consensus sequence TATAAT and is centered at −10 (ten nucleotides on the 5′ side of the first one transcribed, which is denoted by +1). The other, called the *−35 region*, has the consensus sequence TTGACA. The first nucleotide transcribed is usually a purine.

−35 −10 +1
DNA template TTGACA TATAAT
−35 Region
Pribnow box
Start of RNA
A
Procaryotic promoter site

−75 −25 +1
DNA template GGNCAATCT TATA
CAAT box (Sometimes present)
TATA box (Hogness box)
Start of RNA
B
Eucaryotic promoter site

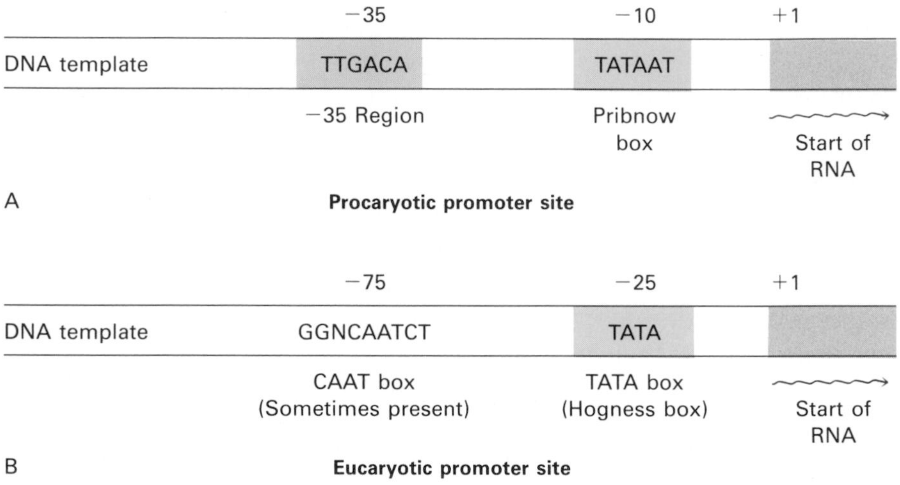

Figure 5-10
Promoter sites for transcription in (A) procaryotes and (B) eucaryotes. Consensus sequences are shown. The first nucleotide to be transcribed is numbered +1. The adjacent nucleotide on the 5′ side is numbered −1.

Consensus sequence—
The base sequences of promoter sites are not all identical. However, they do possess common features, which can be represented by an idealized consensus sequence. Each base in the consensus sequence TATAAT is found in a majority of procaryotic promoters. Nearly all promoter sequences differ from this consensus sequence at only two or fewer bases.

Eucaryotic genes encoding proteins have promoter sites with a TATAAA consensus sequence centered at about −25 (Figure 5-10B). This *TATA box* (also called *Hogness box*) is like the procaryotic Pribnow box except that it is farther upstream. Many eucaryotic promoters also exhibit a *CAAT* consensus sequence centered at about −75. Transcription of eucaryotic genes is further stimulated by *enhancer sequences*, which can be quite distant (up to several kilobases) from the start site, on either its 5′ or its 3′ side.

RNA polymerase proceeds along the DNA template and transcribes one of its strands until a terminator is reached. The termination signal in *E. coli* is a *base-paired hairpin* on the newly synthesized RNA molecule (Figure 5-11). This hairpin is formed by base-pairing of self-complementary sequences that are rich in G and C. Nascent RNA spontaneously dissociates from RNA polymerase when this hairpin is followed by a string of U residues. Alternatively, RNA synthesis can be terminated by the action of *rho*, a protein. Less is known about the termination of transcription in eucaryotes. A more detailed discussion of the initiation and termination of transcription will be given in later chapters. The important point now is that discrete start and stop signals for transcription are encoded in the DNA template.

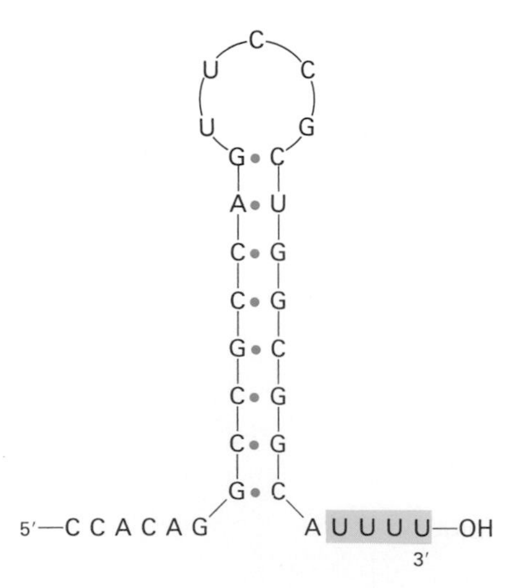

Figure 5-11
Base sequence of the 3′ end of an mRNA transcript in *E. coli*. A stable hairpin structure is followed by a sequence of U residues.

TRANSFER RNA IS THE ADAPTOR MOLECULE IN PROTEIN SYNTHESIS

We have seen that mRNA is the template for protein synthesis. How does it direct amino acids to become joined in the correct sequence? In 1958, Francis Crick wrote:

> One's first naive idea is that the RNA will take up a configuration capable of forming twenty different "cavities," one for the side chain of each of the twenty amino acids. If this were so, one might expect to be able to play the problem backwards—that is, to find the configuration of RNA by trying to form such cavities. All attempts to do this have failed, and on physical-chemical grounds the idea does not seem in the least plausible.

He observed that RNA does not have the knobby hydrophobic surfaces that could distinguish valine from leucine and isoleucine, nor does it have properly positioned charged groups to discriminate between positively and negatively charged amino acid side chains. Crick then proposed an entirely different mechanism for recognition of the mRNA template:

> . . . RNA presents mainly a sequence of sites where hydrogen bonding could occur. One would expect, therefore, that whatever went onto the template in a *specific* way did so by forming hydrogen bonds. It is therefore a natural hypothesis that the amino acid is carried to the template by an adaptor molecule, and that the adaptor is the part which actually fits onto the RNA. In its simplest form one would require twenty adaptors, one for each amino acid.

This highly innovative hypothesis soon became an established fact. *The adaptor in protein synthesis is tRNA.* The structure and reactions of these remarkable adaptor molecules will be considered in detail in Chapter 30. For the moment it suffices to note that tRNA contains an *amino acid attachment site* and a *template-recognition site* (Figures 5-12 and 5-13). A tRNA molecule carries a specific amino acid in an activated form to the site of protein synthesis. The carboxyl group of this amino acid is esterified to the 3′- or 2′-hydroxyl group of the ribose unit at the 3′ end of the tRNA chain. The esterified amino acid may migrate between the 2′- and 3′-hydroxyl groups during protein synthesis. The joining of an amino acid to a tRNA to form an aminoacyl-tRNA is catalyzed by a specific enzyme called an aminoacyl-tRNA synthetase (or activating enzyme). This esterification reaction is driven by ATP. There is at least one specific synthetase for each of the twenty amino acids. The template-recognition site on tRNA is a sequence of three bases called the *anticodon* (Figure 5-13). The anticodon on tRNA recognizes a complementary sequence of three bases on mRNA, called the *codon*.

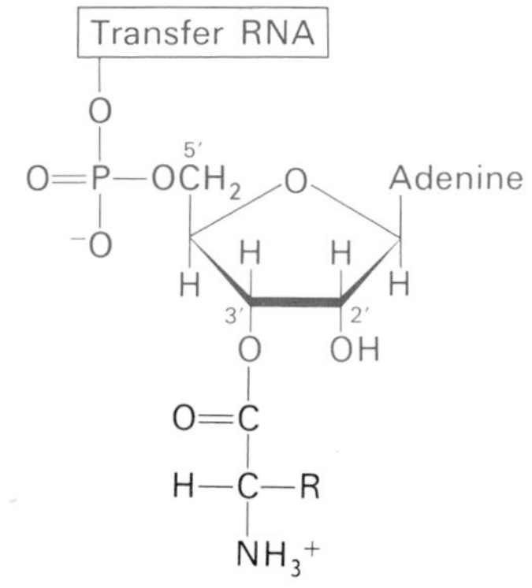

Figure 5-12
Mode of attachment of an amino acid (shown in red) to a tRNA molecule. The amino acid is esterified to the 3′-hydroxyl group of the terminal adenosine of tRNA. A tRNA having an attached amino acid is an aminoacyl-tRNA or a "charged" tRNA, whereas a tRNA without an attached amino acid is "uncharged."

AMINO ACIDS ARE CODED BY GROUPS OF THREE BASES STARTING FROM A FIXED POINT

The *genetic code* is the relation between the sequence of bases in DNA (or its RNA transcripts) and the sequence of amino acids in proteins. Experiments by Francis Crick, Sydney Brenner, and others established the following features of the genetic code by 1961:

1. *What is the coding ratio?* A single-base code can specify only four kinds of amino acids because there are four kinds of bases in DNA. Sixteen kinds of amino acids can be specified by a two-base code

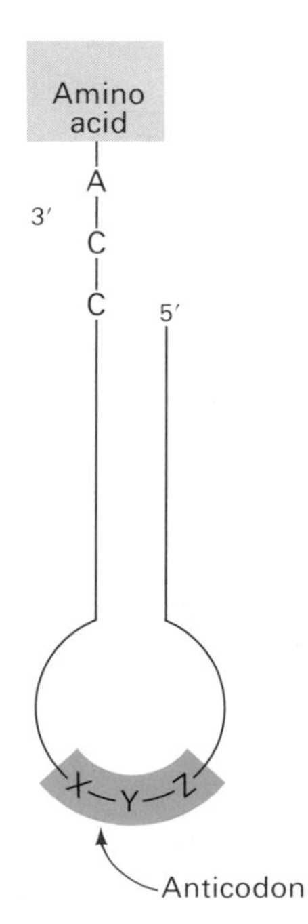

Figure 5-13
Symbolic diagram of an aminoacyl-tRNA showing the amino acid attachment site and the anticodon, which is the template-recognition site.

(4 × 4 = 16), whereas sixty-four kinds of amino acids can be determined by a three-base code (4 × 4 × 4 = 64). Proteins are built from a basic set of twenty amino acids, and so it was evident from this simple calculation that three or more bases are probably needed to specify one amino acid. Genetic experiments then showed that *an amino acid is in fact coded by a group of three bases.* This group of bases is called a *codon.*

2. *Is the code nonoverlapping or overlapping?* In a nonoverlapping triplet code, each group of three bases in a sequence ABCDEF . . . specifies only one amino acid—ABC specifies the first, DEF the second, and so forth—whereas in a completely overlapping triplet code, ABC specifies the first amino acid, BCD the second, CDE the third, and so forth. These alternatives were distinguished by studies of the sequence of amino acids in mutants. Suppose that the base C is mutated to C′. In a nonoverlapping code, only amino acid 1 will be changed. In a completely overlapping code, amino acids 1, 2, and 3 will all be altered by a mutation of C to C′. Amino acid sequence studies of tobacco mosaic virus mutants and abnormal hemoglobins showed that alterations usually affected only a single amino acid. Hence, it was concluded that the *genetic code is nonoverlapping.*

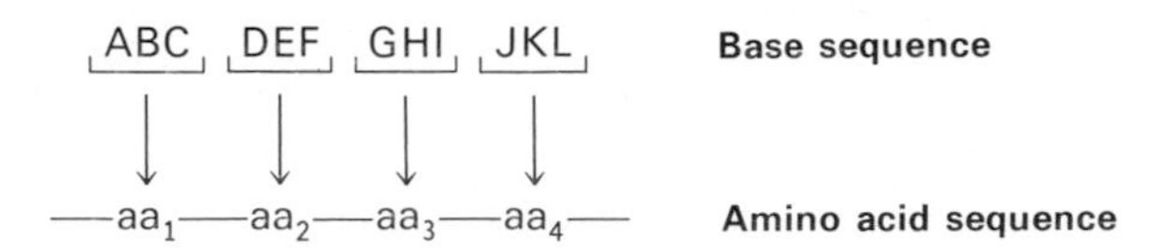

3. *How is the correct group of three bases read?* One possibility a priori is that one of the four bases (denoted as Q) serves as a "comma" between groups of three bases:

. . . QABCQDEFQGHIQJKLQ . . .

This turned out not to be the case. Rather, *the sequence of bases is read sequentially from a fixed starting point.* There are no commas.

Start
↓
ABC | DEF | GHI | JKL | MNO

aa$_1$—aa$_2$—aa$_3$—aa$_4$—aa$_5$

Suppose that a mutation deletes base G:

Start ↓ G (deleted)

ABC | DEF | HIJ | KLM | NOP

aa$_1$—aa$_2$—aa$_3$—aa$_4$—aa$_5$

Normal Altered

The first two amino acids in the resulting polypeptide chain will be normal, but the rest of the base sequence will be read incorrectly because *the reading frame has been shifted* by the deletion of G. Suppose instead that a base Z has been added between F and G:

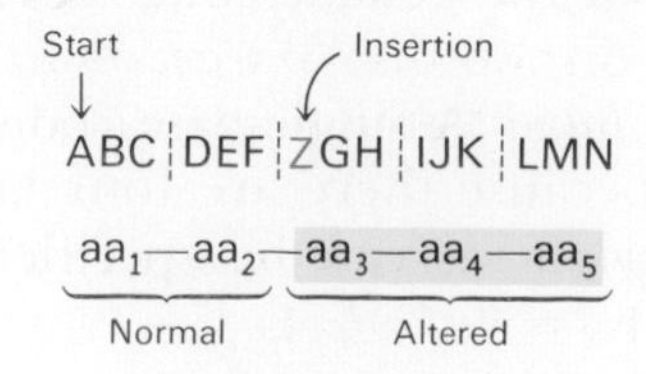

This addition also disrupts the reading frame starting at the codon for amino acid 3. In fact, genetic studies of addition and deletion mutants revealed many of the features of the genetic code.

4. As mentioned earlier, there are sixty-four possible base triplets and twenty amino acids. Is there just one triplet for each of the twenty amino acids or are some amino acids coded by more than one triplet? Genetic studies showed that most of the sixty-four triplets do code for amino acids. Subsequent biochemical studies demonstrated that sixty-one of the sixty-four triplets specify particular amino acids. Thus, *for most amino acids, there is more than one code word.* In other words, the genetic code is *degenerate.*

DECIPHERING THE GENETIC CODE: SYNTHETIC RNA CAN SERVE AS MESSENGER

What is the relation between sixty-four kinds of code words and the twenty kinds of amino acids? In principle, this question can be directly answered by comparing the amino acid sequence of a protein with the corresponding base sequence of its gene or mRNA. However, this approach was not experimentally feasible in 1961 because the base sequences of genes and mRNA molecules were entirely unknown. The breaking of the genetic code then did not seem imminent, but the situation suddenly changed. Marshall Nirenberg discovered that *the addition of polyuridylate (poly U) to a cell-free protein-synthesizing system led to the synthesis of polyphenylalanine.* Poly U evidently served as a messenger RNA. The first code word was deciphered: UUU codes for phenylalanine. This remarkable experiment pointed the way to the complete elucidation of the genetic code.

Let us consider this landmark experiment in more detail. The two essential components were a cell-free system that actively synthesizes protein and a synthetic polyribonucleotide that serves as the messenger RNA. A cell-free protein-synthesizing system was obtained from *E. coli* in the following way. Bacterial cells were gently broken open by grinding them with finely powdered alumina to yield a cell sap. Cell-wall and cell-membrane fragments were then removed by centrifugation. The resulting extract contained DNA, mRNA, tRNA, ribosomes, enzymes, and other cell constituents. Protein was synthesized by this cell-free system when it was supplemented with ATP, GTP, and amino acids. At least one of the added amino acids was radioactive so that its incorporation into protein could be detected. This mixture was incubated at 37°C for about an hour. Trichloroacetic acid was then added to terminate the reaction and precipitate the proteins, leaving the free amino acids in the supernatant. The precipitate was washed and its radioactivity was then measured to determine how much labeled amino acid was incorporated into newly synthesized protein.

A crucial feature of this system is that protein synthesis can be halted by the addition of deoxyribonuclease, which destroys the template for the synthesis of new mRNA (Figure 5-14). The mRNA present at the time of addition of deoxyribonuclease has a short life, and so protein synthesis stops within a few minutes. Nirenberg then found that protein synthesis resumed on addition of a crude fraction of mRNA. Thus, here was a *cell-free protein-synthesizing system that was responsive to the addition of mRNA.*

The other critical component in this experiment was a synthetic polyribonucleotide—namely, poly U. Poly U was synthesized by using *poly-*

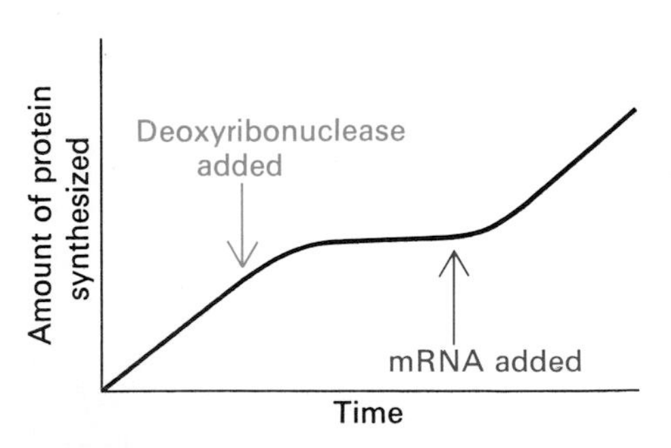

Figure 5-14
Protein synthesis in a cell-free system stops a few minutes after the addition of deoxyribonuclease and resumes following the addition of mRNA.

nucleotide phosphorylase, an enzyme discovered in 1955 by Marianne Grunberg-Manago and Severo Ochoa. This enzyme catalyzes the synthesis of polyribonucleotides from ribonucleoside diphosphates:

$$(\text{RNA})_n + \text{ribonucleoside diphosphate} \rightleftharpoons (\text{RNA})_{n+1} + \text{P}_i$$

The reactions catalyzed by this enzyme and by RNA polymerase are very different. In this reaction, the activated substrates are ribonucleoside diphosphates rather than triphosphates. Orthophosphate is a product instead of pyrophosphate. Hence, this reaction cannot be driven to the right by the hydrolysis of pyrophosphate. Indeed, the equilibrium in vivo is in the direction of RNA degradation, not synthesis. A critical difference is that *polynucleotide phosphorylase does not utilize a template.* The RNA synthesized by this enzyme has a composition dictated only by the ratios of the ribonucleotides present in the incubation mixture and a sequence that is essentially random. For this reason, polynucleotide phosphorylase proved to be a most valuable experimental tool in deciphering the genetic code. Thus, poly U was synthesized by incubating high concentrations of UDP in the presence of this enzyme. Copolymers of two ribonucleotides, say U and A, with a random sequence were prepared by incubating UDP and ADP with this enzyme.

A variety of synthetic ribonucleotides were added to the cell-free protein-synthesizing system, and the incorporation of ^{14}C-labeled L-phenylalanine was measured. The results were striking:

Polyribonucleotide added	*^{14}C counts per minute*
None	44
Poly A	50
Poly C	38
Poly U	39,800

The same experiment was carried out with a different ^{14}C-labeled amino acid in each incubation mixture. It was found that *poly A led to the synthesis of polylysine* and that *poly C led to the synthesis of polyproline.* Thus, three code words were deciphered. The code word GGG could not be deciphered in the same way because poly G does not act as a template, probably because it forms a triple-stranded helical structure. Polyribonucleotides with extensive regions of ordered structure are ineffective templates for protein synthesis.

Code word	*Amino acid*
UUU	Phenylalanine
AAA	Lysine
CCC	Proline

Polynucleotides consisting of two kinds of bases were then used as templates. For example, a random copolymer of U and G contains eight different triplets: UUU, UUG, UGU, GUU, UGG, GUG, GGU, and GGG. With a template containing 0.76 U and 0.24 G, phenylalanine was incorporated to the greatest extent, as expected, because the triplet UUU was the most prevalent one. Valine, leucine, and cysteine came next, followed by tryptophan and glycine. Hence, it was concluded that valine, leucine, and cysteine are specified by codons that contain 2 U and 1 G, whereas tryptophan and glycine are specified by codons that contain 1 U and 2 G. The same type of experiment was carried out with other random copolymers to determine the *compositions* of codons for all twenty amino acids.

Table 5-3
Amino acid incorporation stimulated by a random copolymer of U (0.76) and G (0.24)

Amino acid	*Relative amount incorporated*	*Inferred codon composition*
Phenylalanine	100	UUU
Valine	37	2U, 1G
Leucine	36	2U, 1G
Cysteine	35	2U, 1G
Tryptophan	14	1U, 2G
Glycine	12	1U, 2G

TRINUCLEOTIDES PROMOTE THE BINDING OF SPECIFIC TRANSFER RNA MOLECULES TO RIBOSOMES

The use of mixed copolymers as templates revealed the *composition but not the sequence of codons* corresponding to particular amino acids (except for UUU, AAA, and CCC). Valine is coded by a triplet with 2 U and 1 G, but is it UUG, UGU, or GUU? This question was answered by two entirely different experimental approaches: (1) the use of synthetic polyribonucleotides with an ordered sequence, and (2) the codon-dependent binding of specific tRNA molecules to ribosomes.

In 1964, Nirenberg discovered that *trinucleotides promote the binding of specific tRNA molecules to ribosomes in the absence of protein synthesis.* For example, the addition of pUUU (the *p* indicates that the 5′ end of the trinucleotide is phosphorylated) led to the binding of phenylalanine tRNA, whereas pAAA markedly enhanced the binding of lysine tRNA, as did pCCC for proline tRNA. Dinucleotides were ineffective in stimulating the binding of tRNA to ribosomes. These studies showed that a *trinucleotide (like a triplet in mRNA) specifically binds the particular tRNA molecule for which it is the code word.* A simple and rapid binding assay was devised: tRNA molecules bound to ribosomes are retained by a cellulose nitrate filter, whereas unbound tRNA passes through the filter. The kind of tRNA retained by the filter was identified by using tRNA molecules charged with a particular ^{14}C-labeled amino acid.

All sixty-four trinucleotides were synthesized by organic-chemical or enzymatic techniques. For each trinucleotide, the binding of tRNA molecules corresponding to all twenty amino acids was assayed. For example, it was found that pUUG stimulated the binding of leucine tRNA only, pUGU stimulated the binding of cysteine tRNA only, and pGUU stimulated the binding of valine tRNA only. Hence, it was concluded that the codons UUG, UGU, and GUU correspond to leucine, cysteine, and valine, respectively. For a few codons, no tRNA was strongly bound, whereas, for a few others, more than one kind of tRNA was bound. For most codons, clear-cut binding results were obtained. In fact, about fifty codons were deciphered by this simple and elegant experimental approach.

COPOLYMERS WITH A DEFINED SEQUENCE WERE ALSO INSTRUMENTAL IN BREAKING THE CODE

At about the same time, H. Gobind Khorana succeeded in synthesizing polyribonucleotides with a defined repeating sequence. A variety of copolymers with two, three, and four kinds of bases were synthesized by a combination of organic-chemical and enzymatic techniques. Let us consider the strategy for the synthesis of poly (GUA), for example. This ordered copolymer has the sequence

GUAGUAGUAGUAGUAGUAGUAGUA. . .

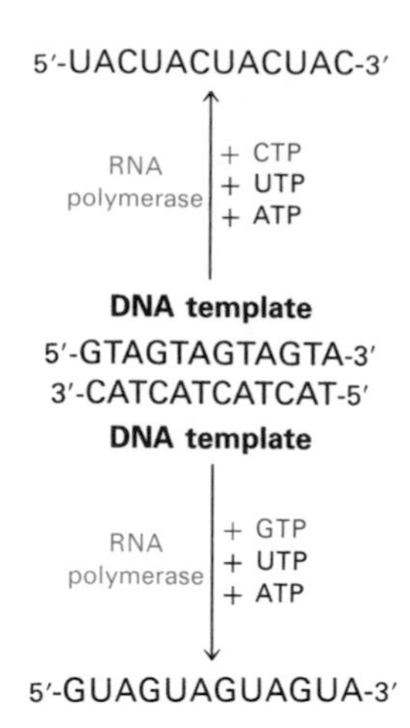

Figure 5-15
One strand or the other of this double-helical DNA template can be selected for transcription by adding just three ribonucleoside triphosphates.

First, Khorana synthesized by organic-chemical methods two complementary deoxyribonucleotides, each with nine residues: $d(TAC)_3$ and $d(GTA)_3$. Partially overlapping duplexes formed on mixing these oligonucleotides then served as templates for the synthesis by DNA polymerase of long repeating double-helical DNA chains. The next step was to obtain long polyribonucleotide chains with a sequence complementary to one of the two DNA strands. The DNA strand to be transcribed could be selected by adding the three complementary ribonucleoside triphosphates. For example, when GTP, UTP, and ATP were added to the incubation mixture, the polyribonucleotide poly (GUA) was synthesized from the poly d(TAC) template strand. The other strand was not transcribed because CTP, one of the required substrates, was missing. Thus, *organic synthesis, followed by template-directed syntheses carried out by DNA polymerase and then RNA polymerase, yielded two long polyribonucleotides having defined repeating sequences* (Figure 5-15).

These regular copolymers were used as templates in the cell-free protein-synthesizing system. Let us examine some of the results. A copolymer consisting of an alternating sequence of two bases P and Q

PQP ¦ QPQ ¦ PQP ¦ QPQ ¦ PQP ¦. . .

contains two codons, PQP and QPQ. The polypeptide product should therefore contain an alternating sequence of two kinds of amino acids (abbreviated as aa_1 and aa_2):

$$aa_1—aa_2—aa_1—aa_2—aa_1—$$

Whether aa_1 or aa_2 is amino-terminal in the polypeptide product depends on whether the reading frame starts at P or Q. When poly (UG) was the template, a polypeptide with an alternating sequence of cysteine and valine was synthesized.

UGU ¦ GUG ¦ UGU ¦ GUG ¦ UGU ¦ GUG

Cys—Val—Cys—Val—Cys—Val

This result unequivocally confirmed the triplet nature of the code and showed that either UGU or GUG codes for cysteine and that the other of these two triplets codes for valine. When this result was considered together with tRNA binding data, it was evident that UGU codes for cysteine and that GUG codes for valine. The polypeptides synthesized in response to several other alternating copolymers of two bases were

Template	*Product*	*Codon assignments*	
Poly (UC)	Poly (Ser-Leu)	UCU ⟶ Ser	CUC ⟶ Leu
Poly (AG)	Poly (Arg-Gln)	AGA ⟶ Arg	GAG ⟶ Gln
Poly (AC)	Poly (Thr-His)	ACA ⟶ Thr	CAC ⟶ His

Now consider a template consisting of three bases in a repeating sequence, poly (PQR). If the reading frame starts at P, the resulting polypeptide should contain only one kind of amino acid, the one coded by the triplet PQR:

Start
↓
PQR┆PQR┆PQR┆PQR┆PQR┆PQR. . .

aa_1—aa_1—aa_1—aa_1—aa_1—aa_1

However, if the reading frame starts at Q, the polypeptide synthesized should contain a different amino acid, the one coded by the triplet QRP:

Start
↓
P┆QRP┆QRP┆QRP┆QRP┆QRP┆QRP. . .

aa_2—aa_2—aa_2—aa_2—aa_2—aa_2

If the reading frame starts at R, a third kind of polypeptide should be formed, containing only the amino acid coded by the triplet RPQ:

Start
↓
PQ┆RPQ┆RPQ┆RPQ┆RPQ┆RPQ┆RPQ. . .

aa_3—aa_3—aa_3—aa_3—aa_3—aa_3

Thus, the expected products are three different homopolypeptides. In fact, this was observed for most templates consisting of repeating sequences of three nucleotides. For example, poly (UUC) led to the synthesis of polyphenylalanine, polyserine, and polyleucine. This result, taken together with the outcome of other experiments, showed that UUC codes for phenylalanine, UCU codes for serine, and CUU codes for leucine. The polypeptides synthesized in response to other templates of this type are given in Table 5-4. Note that poly (GUA) and poly (GAU) each elicited the synthesis of two rather than three homopolypeptides. The reason will be evident shortly.

Table 5-4
Homopolypeptides synthesized using messengers containing repeating trinucleotide sequences

Messenger	*Homopolypeptides synthesized*
Poly (UUC)	Phe, Ser, Leu
Poly (AAG)	Lys, Glu, Arg
Poly (UUG)	Cys, Leu, Val
Poly (CCA)	Gln, Thr, Asn
Poly (GUA)	Val, Ser
Poly (UAC)	Tyr, Thr, Leu
Poly (AUC)	Ile, Ser, His
Poly (GAU)	Met, Asp

Khorana also synthesized several copolymers with repeating tetranucleotides, such as poly (UAUC). This template led to the synthesis of a polypeptide with the repeating sequence Tyr-Leu-Ser-Ile, irrespective of the reading frame:

UAU┆CUA┆UCU┆AUC┆UAU┆CUA┆UCU┆AUC

Tyr—Leu—Ser—Ile—Tyr—Leu—Ser—Ile

Four codon assignments were deduced from this result.

A very different result was obtained when poly (GUAA) was the template. The only products were dipeptides and tripeptides. Why not longer chains? The reason is that one of the triplets present in this polymer—namely, UAA—codes not for an amino acid but rather for the termination of protein synthesis:

GUA┆AGU┆AAG┆UAA┆GUA┆A. . .

Val—Ser—Lys—Stop

Only di- and tripeptides were formed also when poly (AUAG) was the template, because UAG is a second signal for chain termination:

AUA┆GAU┆AGA┆UAG┆AUA┆G. . .

Ile—Asp—Arg—Stop

Now let us look again at Table 5-4. Two rather than three homopolypeptides were synthesized with poly (GUA) as a template because the third reading frame corresponds to the sequence

G ¦ UAG ¦ UAG ¦ UAG . . .
Stop—Stop—Stop

which is a repeating sequence of termination signals. What about poly (GAU)? Again, only two homopolypeptides were synthesized with this template, because the third reading frame corresponds to

GA ¦ UGA ¦ UGA ¦ UGA ¦ U. . .
Stop—Stop—Stop

UGA is yet another signal for chain termination. In fact, *UAG, UAA, and UGA are the only three codons that do not specify an amino acid.*

Khorana's synthesis of polynucleotides with a defined sequence was an outstanding accomplishment. The use of these polymers as templates for protein synthesis, together with Nirenberg's studies of the trinucleotide-stimulated binding of tRNA to ribosomes, resulted in the complete elucidation of the genetic code by 1966, an outcome that would have been deemed a quixotic dream only six years earlier.

Rosetta stone, inscribed in hieroglyphs, demotic, and Greek. [© Archiv/Photo Researchers, Inc.]

MAJOR FEATURES OF THE GENETIC CODE

All sixty-four codons have been deciphered (Table 5-5). Sixty-one triplets correspond to particular amino acids, whereas three code for chain termination. Because there are twenty amino acids and sixty-one triplets that code for them, it is evident that the code is highly *degenerate*. In other words, *many amino acids are designated by more than one triplet.* Only tryptophan and methionine are coded by just one triplet. The other eighteen amino acids are coded by two or more. Indeed, leucine, arginine, and serine are specified by six codons each. Under normal physiological conditions, *the code is not ambiguous:* a codon designates only one amino acid.

Codons that specify the same amino acid are called *synonyms*. For example, CAU and CAC are synonyms for histidine. Note that synonyms are not distributed haphazardly throughout the table of the genetic code (Table 5-5). An amino acid specified by two or more synonyms occupies a single box (unless there are more than four synonyms). The amino acids in a box are specified by codons that have the same first two bases but differ in the third base, as exemplified by GUU, GUC, GUA, and GUG. Thus, *most synonyms differ only in the last base of the triplet.* Inspection of the code shows that XYC and XYU always code for the same amino acid, whereas XYG and XYA usually code for the same amino acid. The structural basis for these equivalences of codons will become evident when we consider the nature of the anticodons of tRNA molecules (p. 743).

What is the biological significance of the extensive degeneracy of the genetic code? One possibility is that *degeneracy minimizes the deleterious effects of mutations.* If the code were not degenerate, then twenty codons would designate amino acids and forty-four would lead to chain termination. The probability of mutating to chain termination would therefore be much higher with a nondegenerate code than with the actual code. It is important to recognize that chain-termination mutations usu-

Table 5-5
The genetic code

First position (5′ end)	Second position				Third position (3′ end)
	U	C	A	G	
U	Phe	Ser	Tyr	Cys	U
	Phe	Ser	Tyr	Cys	C
	Leu	Ser	Stop	Stop	A
	Leu	Ser	Stop	Trp	G
C	Leu	Pro	His	Arg	U
	Leu	Pro	His	Arg	C
	Leu	Pro	Gln	Arg	A
	Leu	Pro	Gln	Arg	G
A	Ile	Thr	Asn	Ser	U
	Ile	Thr	Asn	Ser	C
	Ile	Thr	Lys	Arg	A
	Met	Thr	Lys	Arg	G
G	Val	Ala	Asp	Gly	U
	Val	Ala	Asp	Gly	C
	Val	Ala	Glu	Gly	A
	Val	Ala	Glu	Gly	G

Note: Given the position of the bases in a codon, it is possible to find the corresponding amino acid. For example, the codon 5′ AUG 3′ on mRNA specifies methionine, whereas CAU specifies histidine. UAA, UAG, and UGA are termination signals. AUG is part of the initiation signal, in addition to coding for internal methionines.

ally lead to inactive proteins, whereas substitutions of one amino acid for another are usually rather harmless. *Degeneracy of the code may also be significant in permitting DNA base composition to vary over a wide range without altering the amino acid sequence of the proteins encoded by the DNA.* The [G] + [C] content of bacterial DNA ranges from less than 30% to more than 70%. DNA rich in GC has a higher melting temperature than DNA rich in AT. As might be expected, the DNA of algae residing in hot springs has a high content of GC. DNA molecules with quite different [G] + [C] contents could code for the same proteins if different synonyms were consistently used.

START AND STOP SIGNALS FOR PROTEIN SYNTHESIS

It has already been mentioned that *UAA, UAG, and UGA designate chain termination.* These codons are read not by tRNA molecules but rather by specific proteins called release factors (p. 758). The start signal for protein synthesis is more complex. Polypeptide chains in bacteria start with a modified amino acid—namely, formylmethionine (fMet). A specific tRNA, the initiator tRNA, carries fMet. This fMet-tRNA recognizes the codon AUG or, less frequently, GUG. However, AUG is also the codon for an internal methionine, and GUG is the codon for an internal valine. Hence, the signal for the first amino acid in the polypeptide chain must be more complex than for all subsequent ones. *AUG*

Formylmethionine (fMet)

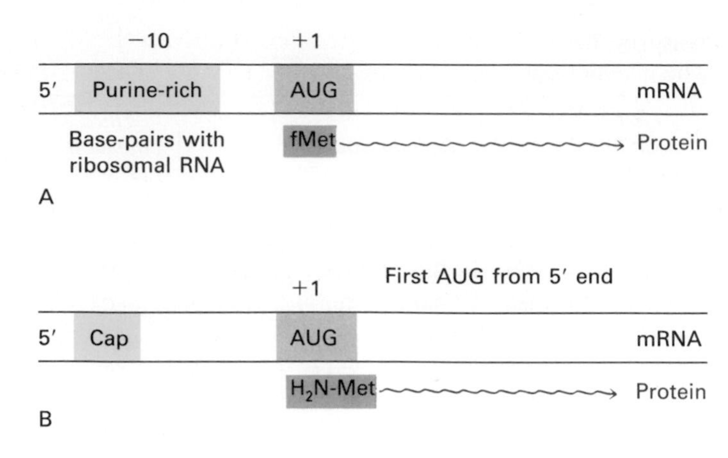

Figure 5-16
Start signals for the initiation of protein synthesis in (A) procaryotes and (B) eucaryotes. In eucaryotic mRNAs the 5′ end, called a *cap*, contains modified bases (Chapter 29).

(or GUG) is part of the initiation signal (Figure 5-16). In bacteria, the initiating AUG (or GUG) is preceded several nucleotides away by a purine-rich sequence that base-pairs with a complementary sequence in a ribosomal RNA molecule (p. 753). In eucaryotes, the AUG closest to the 5′ end of an mRNA is usually the start signal for protein synthesis. This particular AUG is read by an initiator tRNA charged with methionine.

THE GENETIC CODE IS NEARLY UNIVERSAL

The genetic code was deciphered by studies of trinucleotides and synthetic mRNA templates in cell-free systems derived from bacteria. Is the genetic code the same in all organisms? Analyses of spontaneous and specifically designed mutations in viruses, bacteria, and higher organisms have been highly informative. The base sequences of many wild-type and mutant genes are known, as are the amino acid sequences of their encoded proteins. In each case, the nucleotide change in the gene and the amino acid change in the protein are as predicted by the genetic code. Furthermore, mRNAs can be correctly translated by the protein-synthesizing machinery of very different species. For example, human hemoglobin mRNA is correctly translated by a wheat-germ extract. As will be discussed in the next chapter, bacteria efficiently express recombinant DNA molecules encoding human proteins such as insulin. These experimental findings strongly suggested that the genetic code is universal.

However, there was surprise when the sequence of human mitochondrial DNA became known. Human mitochondria read UGA as a codon for tryptophan rather than as a stop signal (Table 5-6). Another difference is that AGA and AGG are read as stop signals rather than as codons for arginine, and AUA is read as a codon for methionine instead of isoleucine. Mitochondria of other species, such as those of yeast, also have a genetic code that differs slightly from the standard one. Mitochondria can have a different genetic code from the rest of the cell because mitochondrial DNA encodes a distinct set of tRNAs. Do any cellular protein-synthesizing systems deviate from the standard genetic code? Recent studies have revealed that ciliated protozoa read AGA and AGG as stop signals rather than as codons for arginine. Thus, *the genetic code is nearly but not absolutely universal.* Variations clearly exist in mitochondria and in species, such as ciliates, that branched off very early in eucaryotic evolution. It is interesting to note that two of the

Table 5-6
Distinctive codons of human mitochondria

Codon	Standard code	Mitochondrial code
UGA	Stop	Trp
UGG	Trp	Trp
AUA	Ile	Met
AUG	Met	Met
AGA	Arg	Stop
AGG	Arg	Stop

codon reassignments in human mitochondria diminish the information content of the third base of the triplet (e.g., both AUA and AUG specify methionine). Most variations from the standard genetic code are in the direction of a simpler code.

Why has the code remained nearly invariant through billions of years of evolution, from bacteria to humans? A mutation that altered the reading of mRNA would change the amino acid sequence of most, if not all, of the proteins synthesized by that particular organism. Many of these changes would undoubtedly be deleterious, and so there would be strong selection against a mutation with such pervasive consequences.

THE SEQUENCES OF GENES AND THEIR ENCODED PROTEINS ARE COLINEAR

Before the genetic code was deciphered, high-resolution gene-mapping studies carried out by Seymour Benzer in the 1950s had revealed that genes are unbranched structures. This important result was in harmony with the established fact that DNA consists of a linear sequence of base pairs. Polypeptide chains also have unbranched structures. The following question was therefore asked: *Is there a linear correspondence between a gene and its polypeptide product?*

Charles Yanofsky approached this problem by using mutants of *E. coli* that produced an altered enzyme molecule. Numerous mutants of the α chain of tryptophan synthetase (an enzyme catalyzing the final step in the synthesis of tryptophan) were isolated and their positions on the genetic map for the α chain were determined by measuring frequencies of recombination. Some of these mutants were very close to each other on the genetic map, whereas others were far apart within the same gene. The next task was to determine the location of the amino acid substitution for each of these mutants. First, the amino acid sequence of the wild-type α chain was determined. Then, the position and nature of the amino acid change in each mutant was identified. *The order of the mutants on the genetic map proved to be the same as the order of the corresponding changes in the amino acid sequence of the polypeptide product.* The same result was subsequently obtained in many other systems. Thus, *genes and their polypeptide products are colinear.*

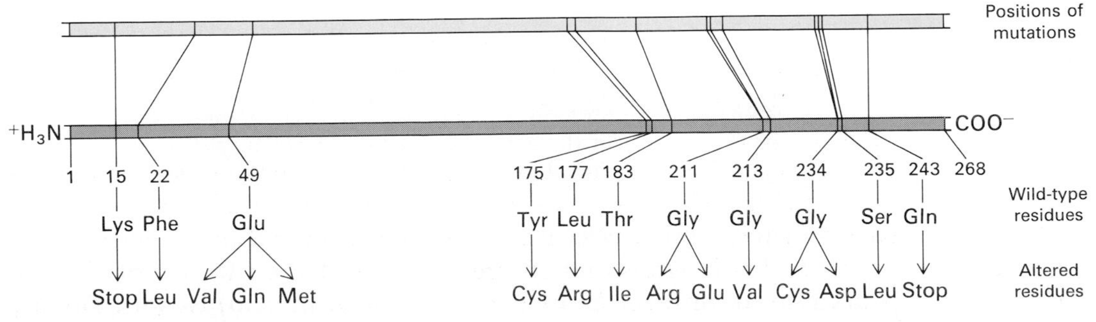

Figure 5-17
Colinearity of the gene and the amino acid sequence of the α chain of tryptophan synthetase. The locations of mutations in DNA (shown in yellow) were determined by genetic-mapping techniques. The positions of the altered amino acids in the amino acid sequence (shown in blue) are in the same order as the corresponding mutations. [After C. Yanofsky. Gene structure and protein structure. Copyright © 1967 by Scientific American, Inc. All rights reserved.]

MOST EUCARYOTIC GENES ARE MOSAICS OF INTRONS AND EXONS

In bacteria, polypeptide chains are encoded by a continuous array of triplet codons in DNA. It was assumed for many years that genes in higher organisms also are continuous. This view was unexpectedly shattered in 1977, when investigators in several laboratories discovered that several genes are *discontinuous*. For example, the gene for the β chain of hemoglobin is interrupted within its amino acid coding sequence by a long *noncoding intervening sequence* of 550 base pairs and a short one of 120 base pairs. Thus, the *β-globin gene is split into three coding sequences.*

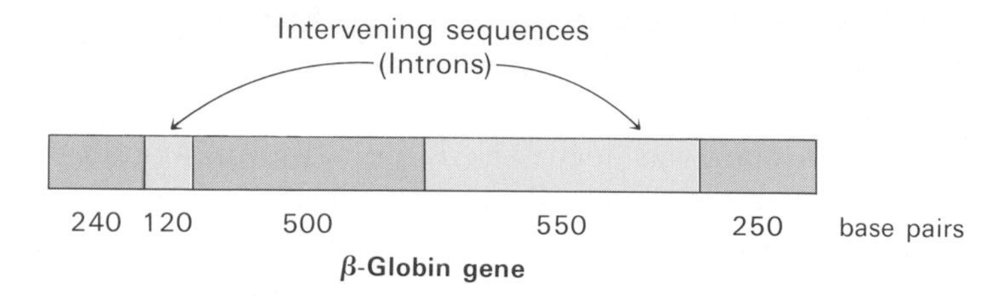

This unexpected structure was revealed by electron-microscopic studies of hybrids formed between β-globin mRNA and a segment of mouse DNA containing a β-globin gene (Figure 5-18). The DNA duplex is

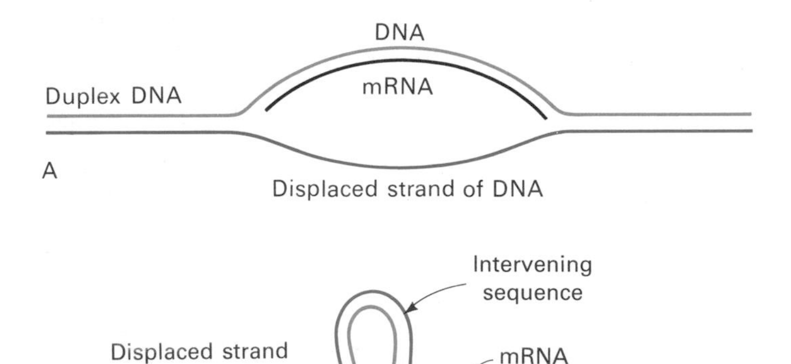

Figure 5-18
Detection of intervening sequences by electron microscopy. An mRNA molecule (shown in red) is hybridized to genomic DNA containing the corresponding gene. (A) A single loop of single-stranded DNA (shown in blue) is seen if the gene is continuous. (B) Two loops of single-stranded DNA (blue) and a loop of double-stranded DNA (blue and green) are seen if the gene contains an intervening sequence.

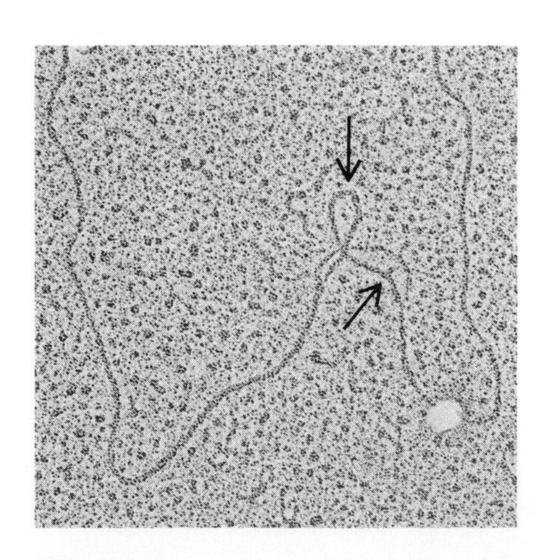

Figure 5-19
Electron micrograph of a hybrid between β-globin mRNA and a fragment of genomic DNA containing the β-globin gene. The thick loops of double-helical DNA that buckle out of the structure are the intervening sequences in DNA that are absent from mRNA (as in part B of Figure 5-18). The upper arrow points to the large intervening sequence, and the lower arrow to the small one. [Courtesy of Dr. Philip Leder.]

partly melted to allow mRNA to hybridize to the complementary strand of DNA. The single-stranded region of DNA then loops out and appears in electron micrographs as a thin line, in contrast with double-stranded DNA or DNA-RNA hybrid regions, which have a thick appearance. If the β-globin gene were continuous, a single loop would be seen. However, three loops are clearly evident in electron micrographs of these hybrids (Figure 5-19), which indicates that the gene is interrupted by at least one stretch of DNA that is absent from the mRNA. The large differences between maps of the β-globin gene and of a reverse transcript of β-globin mRNA showed that the genomic DNA contains untranslated sequences between the coding regions.

At what stage in gene expression are intervening sequences removed? Newly synthesized RNA chains isolated from nuclei are much larger than the mRNA molecules derived from them. In fact, the primary transcript of the β-globin gene contains two untranslated regions. *These intervening sequences in the 15S primary transcript are excised and the coding sequences are simultaneously linked by a precise splicing enzyme to form the mature 9S mRNA* (Figure 5-20). The coding sequences of split genes are called *exons* (for expressed regions), whereas their untranslated intervening sequences are known as *introns.* More generally, introns are sequences that are spliced out in the formation of mature RNA molecules.

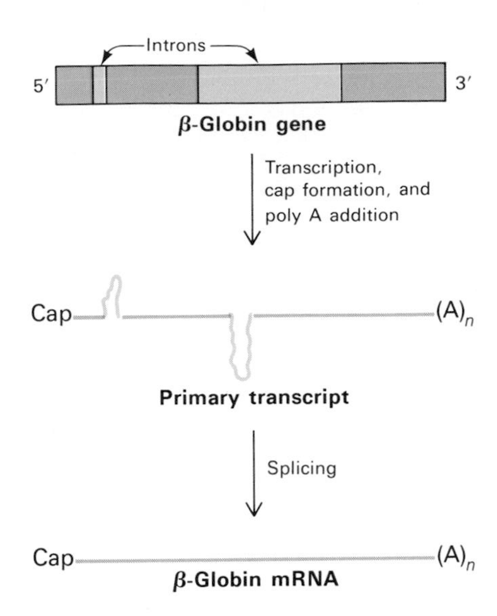

Figure 5-20
Transcription of the β-globin gene and removal of the intervening sequences in the primary RNA transcript. Cap formation and the addition of poly A are discussed in a later chapter (p. 720).

Another split eucaryotic gene is the one for ovalbumin in chickens, which is made up of eight exons separated by seven long introns (Figure 5-21). Even more striking is the collagen gene, which contains more than forty exons. A common feature in the expression of these genes is that their exons are ordered in the same sequence in mRNA as in DNA. Thus, *split genes, like continuous genes, are colinear with their polypeptide products.*

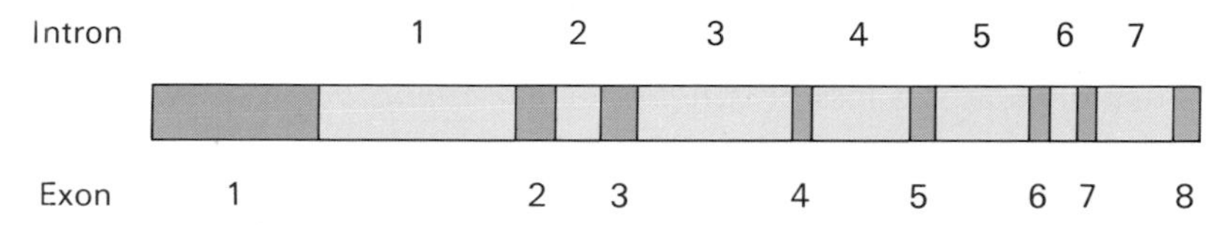

Figure 5-21
Structure of the chick ovalbumin gene. The introns (noncoding regions) are shown in yellow and the exons (translated regions) in blue.

Splicing is a complex operation that is carried out by *spliceosomes,* which are assemblies of proteins and small RNA molecules (p. 724). This enzymatic machinery recognizes signals in the nascent RNA that specify the splice sites. *Introns nearly always begin with GU and end with an AG that is preceded by a pyrimidine-rich tract (Figure 5-22). This consensus sequence is part of the signal for splicing.*

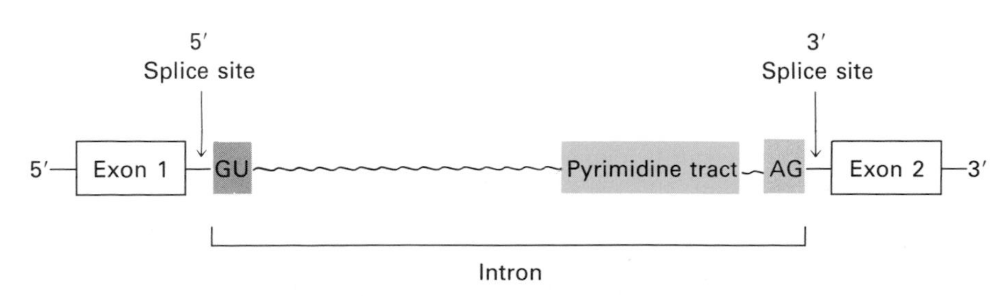

Figure 5-22
Consensus sequence for the splicing of mRNA precursors.

MANY EXONS ENCODE PROTEIN DOMAINS

Most genes of higher eucaryotes, such as birds and mammals, are split. Lower eucaryotes, such as yeast, have a much higher proportion of continuous genes. In eubacteria, such as *E. coli,* no split genes have been found. Have introns been inserted into genes in the evolution of higher organisms? Or have introns been removed from genes to form the streamlined genomes of procaryotes and simple eucaryotes? Comparisons of the DNA sequences of genes encoding proteins that are highly conserved in evolution strongly suggest that *introns were present in ancestral genes and were lost in the evolution of organisms that have become optimized for very rapid growth, such as eubacteria and yeast.* The positions of introns in some genes are at least 10^9 years old. Furthermore, a common mechanism of splicing developed before the divergence of fungi, plants, and vertebrates, as shown by the finding that mammalian cell extracts can splice yeast RNA.

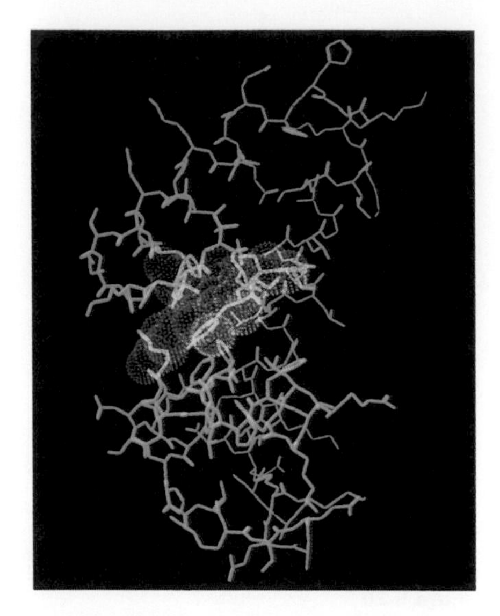

Figure 5-23
The core of myoglobin is encoded by the central exon of its gene. The polypeptide chain encoded by this exon is shown in green and the van der Waals outline of the oxygen-binding heme group in red.

Many exons encode discrete structural and functional units of proteins (Figure 5-23). For example, the central exon of myoglobin and hemoglobin genes encodes a heme-binding region that can reversibly bind O_2 (p. 150). Other exons specify α-helical segments that anchor proteins in cell membranes. An entire domain of a protein may be encoded by a single exon. An attractive hypothesis is that *new proteins arose in evolution by the rearrangement of exons encoding discrete structural elements, binding sites, and catalytic sites.* Exon shuffling is a rapid and efficient means of generating novel genes because it preserves functional units, but allows them to interact in new ways. Introns are extensive regions in which DNA can break and recombine with no deleterious effect on encoded proteins. In contrast, the exchange of sequences between different exons usually leads to loss of function. *Gene duplication* is another means of increasing the genetic potentialities of an organism. The duplicated gene can undergo diversification while the original one continues to serve a vital function. Finally, genes can be altered by *point mutations,* which change a single nucleotide at a time and usually lead to the replacement of a single amino acid residue.

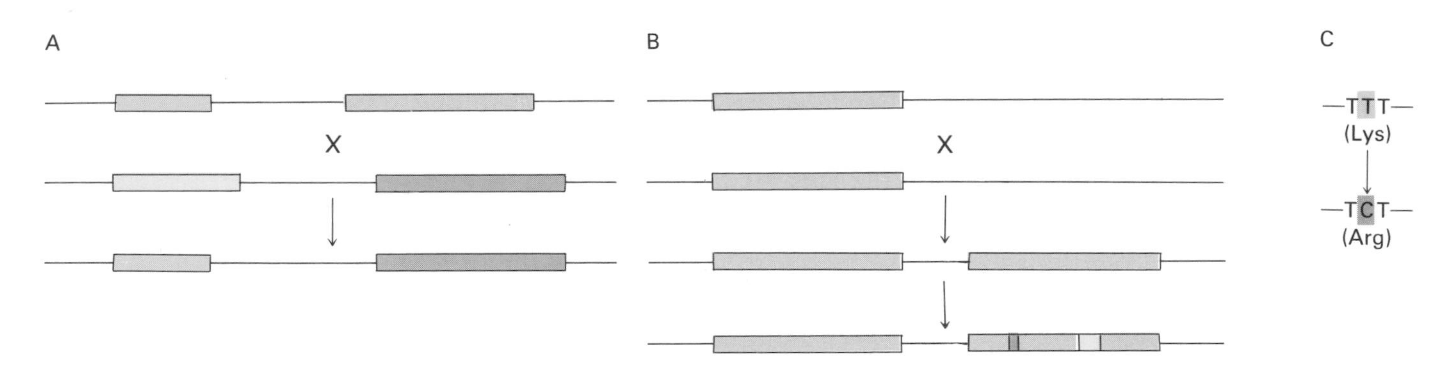

Figure 5-24
The genetic repertoire can be expanded by (A) exon shuffling, (B) gene duplication and diversification, and (C) point mutations.

Another advantage conferred by split genes is the potentiality for generating a series of related proteins by splicing a nascent RNA transcript in different ways. For example, a precursor of an antibody-producing cell forms an antibody that is anchored in the cell's plasma membrane (Figure 5-25). Stimulation of such a cell by a specific foreign antigen that is recognized by the attached antibody leads to cell differentiation and proliferation. The activated antibody-producing cells then splice their nascent RNA transcript in an alternative manner to form soluble antibody molecules that are secreted rather than retained on the cell surface. We see here a clear-cut example of a benefit conferred by the complex arrangement of introns and exons in higher organisms. *Alternative splicing is a means of forming a set of proteins that are variations of a basic motif according to a developmental program.*

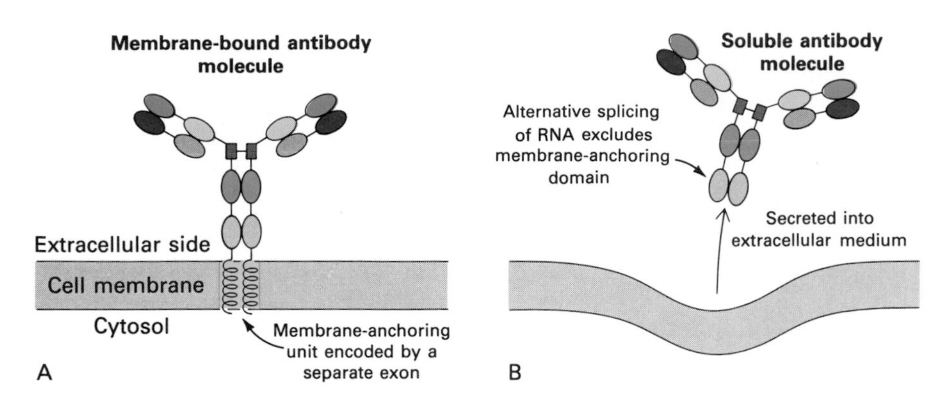

Figure 5-25
Alternative splicing generates mRNAs that are templates for different forms of a protein: (A) a membrane-bound antibody on the surface of a lymphocyte, and (B) its soluble counterpart, exported from the cell. Domains encoded by distinct exons are depicted in different colors. The membrane-bound antibody is anchored to the plasma membrane by a helical segment (highlighted in yellow) that is encoded by its own exon.

RNA PROBABLY CAME BEFORE DNA AND PROTEINS IN EVOLUTION

The flow of genetic information from DNA to RNA to protein depends on the intricate interplay of enzymes and other proteins with nucleic acids. Likewise, the replication of DNA and RNA is mediated by the interaction of polymerases and other proteins with nucleic acid templates. How did nucleic acid molecules early in the evolution of life replicate in the absence of enzymes? A likely solution to this enigma comes from the recent discovery that *RNA molecules as well as proteins can be enzymes.* Thomas Cech found that the precursor of a ribosomal RNA in *Tetrahymena* (a ciliated protozoan) undergoes self-splicing (p. 725). The intron in this precursor RNA molecule is precisely removed by a catalytic activity of the RNA itself. This liberated intron then loses a short 5′-terminal sequence to form a 395-nucleotide RNA molecule that catalyzes the transformation of *other* RNA molecules. This intron-derived RNA catalyzes the cleavage and joining of RNA chains at specific sites without itself being consumed (p. 214). Hence, it is a true enzyme. Protein catalysts, known for a century, are now joined by RNA catalysts (*ribozymes*).

This revolutionary finding enables us to envision an RNA world early in the evolution of life prior to the appearance of DNA and protein. Walter Gilbert has proposed that RNA molecules first catalyzed their own replication and developed a repertoire of enzymatic activities. In the next stage, RNA molecules began to synthesize proteins, which emerged as superior enzymes because their twenty side chains are more versatile than the four bases of RNA. Finally, DNA was formed by reverse transcription of RNA. DNA replaced RNA as the genetic material because its double-helix is a more stable and reliable store of genetic information than is single-stranded RNA. At this point, RNA was left with roles it has retained to this day as the information carrier (mRNA) and adapter (tRNA) in protein synthesis and as components of assemblies that mediate gene expression (e.g., rRNA in ribosomes). The present intricate (indeed baroque) mechanism of information transfer from gene to protein is an ancient epic, which probably began when RNA alone wrote the script, directed the action, and played all the key parts.

SUMMARY

The flow of genetic information in normal cells is from DNA to RNA to protein. The synthesis of RNA from a DNA template is called transcription, whereas the synthesis of a protein from an RNA template is termed translation. Cells contain several kinds of RNA: messenger RNA (mRNA), transfer RNA (tRNA), ribosomal RNA (rRNA), and small nuclear RNA (snRNA). Most RNA molecules are single stranded, but many contain extensive double-helical regions that arise from the folding of the chain into hairpins. The smallest RNA molecules are the tRNAs, which contain as few as 75 nucleotides, whereas the largest ones are some mRNAs, which may have more than 5000 nucleotides. All cellular RNA is synthesized by RNA polymerase according to instructions given by DNA templates. The activated intermediates are ribonucleoside triphosphates. The direction of RNA synthesis is $5' \rightarrow 3'$, like that of DNA synthesis. RNA polymerase differs from DNA polymerase in not requiring a primer. Another difference is that the DNA template is fully conserved in RNA synthesis, whereas it is semiconserved in DNA synthesis. Many RNA molecules are cleaved and chemically modified after transcription.

The base sequence of a gene is colinear with the amino acid sequence of its polypeptide product. The genetic code is the relation between the sequence of bases in DNA (or its RNA transcript) and the sequence of amino acids in proteins. Amino acids are coded by groups of three bases (called codons) starting from a fixed point. Sixty-one of the sixty-four codons specify particular amino acids, whereas the other three codons (UAA, UAG, and UGA) are signals for chain termination. Thus, for most amino acids there is more than one code word. In other words, the code is degenerate. Codons specifying the same amino acid are called synonyms. Most synonyms differ only in the last base of the triplet. The genetic code, which is nearly the same in all organisms, was deciphered after the discovery that the polyribonucleotide poly U codes for polyphenylalanine. Various synthetic polyribonucleotides then were used as mRNAs in cell-free protein-synthesizing systems. Natural mRNAs contain start and stop signals for translation, just as genes do for directing where transcription begins and ends.

Most genes in higher eucaryotes are discontinuous. Coding sequences (exons) in these split genes are separated by intervening sequences (introns), which are removed in the conversion of the primary transcript into mRNA and other functional mature RNA molecules. For example, the β-globin gene in mammals contains two introns. Nascent RNA molecules contain signals that specify splice sites. Split genes, like continuous genes, are colinear with their polypeptide products. A striking feature of many exons is that they encode functional domains in proteins. New proteins probably arose in evolution by the shuffling of exons. Introns may have been present in primordial genes, but were lost in the evolution of such fast-growing organisms as bacteria and yeast. The recent discovery that certain RNA molecules undergo self-splicing and can serve as enzymes suggests that RNA came before DNA and proteins in evolution.

SELECTED READINGS

Where to start

Miller, O. L., Jr., 1973. The visualization of genes in action. *Sci. Amer.* 228(3):34–42. [Available as *Sci. Amer.* Offprint 1267.]

Crick, F. H. C., 1966. The genetic code III. *Sci. Amer.* 215(4):55–62. [Offprint 1052. A view of the code when it was almost completely elucidated.]

Chambon, P., 1981. Split genes. *Sci. Amer.* 244(5):60–71.

Yanofsky, C., 1967. Gene structure and protein structure. *Sci. Amer.* 216(5):80–94. [Available as *Sci. Amer.* Offprint 1074. Presents the evidence for colinearity.]

Books

Darnell, J., Lodish, H., and Baltimore, D., 1986. *Molecular Cell Biology.* Scientific American Books. [Contains an excellent presentation of gene expression in eucaryotes.]

Lewin, B., 1987. *Genes* (3rd ed.). Wiley. [A lucid account of the flow of genetic information in procaryotes and eucaryotes.]

Watson, J. D., Hopkins, N. H., Roberts, J. W., Steitz, J. A., and Weiner, A. M., 1987. *Molecular Biology of the Gene* (4th ed.). Benjamin/Cummings. [Volume 1 of this outstanding work deals with general principles and volume 2 with eucaryotic systems.]

Discovery of messenger RNA

Jacob, F., and Monod, J., 1961. Genetic regulatory mechanisms in the synthesis of proteins. *J. Mol. Biol.* 3:318–356.

Brenner, S., Jacob, F., and Meselson, M., 1961. An unstable intermediate carrying information from genes to ribosomes for protein synthesis. *Nature* 190:576–581.

Hall, B. D., and Spiegelman, S., 1961. Sequence complementarity of T2-DNA and T2-specific RNA. *Proc. Nat. Acad. Sci.* 47:137–146.

Genetic code

Crick, F. H. C., Barnett, L., Brenner, S., and Watts-Tobin, R. J., 1961. General nature of the genetic code for proteins. *Nature* 192:1227–1232.

Khorana, H. G., 1968. Nucleic acid synthesis in the study of the genetic code. In *Nobel Lectures: Physiology or Medicine* (1963–1970), pp. 341–369. American Elsevier (1973).

Nirenberg, M., 1968. The genetic code. In *Nobel Lectures: Physiology or Medicine* (1963–1970), pp. 372–395. American Elsevier (1973).

Crick, F. H. C., 1958. On protein synthesis. *Symp. Soc. Exp. Biol.* 12:138–163. [A brilliant anticipatory view of the problem of protein synthesis. The adaptor hypothesis is presented in this article.]

Woese, C. R., 1967. *The Genetic Code.* Harper & Row.

Crothers, D. M., 1982. Nucleic acid aggregation geometry and the possible evolutionary origin of ribosomes and the genetic code. *J. Mol. Biol.* 162:379–391.

Colinearity of gene and protein

Yanofsky, C., Carlton, B. C., Guest, J. R., Helinski, D. R., and Henning, U., 1964. On the colinearity of gene structure and protein structure. *Proc. Nat. Acad. Sci.* 51:266–272.

Sarabhai, A. S., Stretton, O. W., Brenner, S., and Bolle, A., 1964. Colinearity of gene with polypeptide chain. *Nature* 201:13–17.

Introns, exons, and split genes

Gilbert, W., 1985. Genes-in-pieces revisited. *Science* 228:823–824.

Cochet, M., Gannon, F., Hen, R., Maroteaux, L., Perrin, F., and Chambon, P., 1979. Organization and sequence studies of the 17-piece chicken conalbumin gene. *Nature* 282:567–574.

Tilghman, S. M., Tiemeier, D. C., Seidman, J. G., Peterlin, B. M., Sullivan, M., Maijel, J. V., and Leder, P., 1978. Intervening sequence of DNA identified in the structural portion of a mouse β-globin gene. *Proc. Nat. Acad. Sci.* 75:725–729.

Craik, C. S., Rutter, W. J., and Fletterick, R., 1983. Splice junctions: association with variation in protein structure. *Science* 220:1125–1129.

Padgett, R. A., Grabowski, P. J., Konarska, M. M., Seiler, S., and Sharp, P. A., 1986. Splicing of messenger RNA precursors. *Ann. Rev. Biochem.* 55:1119.

Catalytic activity of RNA

Zaug, A. J., and Cech, T. R., 1986. The intervening sequence RNA of *Tetrahymena* is an enzyme. *Science* 231:470–475.

Altman, S., 1984. Aspects of biochemical catalysis. *Cell* 36:237–239.

Cech, T. R., and Bass, B. L., 1986. Biological catalysis by RNA. *Ann. Rev. Biochem.* 55:599.

Molecular evolution

Wilson, A. C., 1985. The molecular basis of evolution. *Sci. Amer.* 253(4):164.

Gilbert, W., 1986. The RNA world. *Nature* 319:618.

Lewin, R., 1986. RNA catalysis gives fresh perspective on the origin of life. *Science* 231:545–546.

Sharp, P. A., 1985. On the origin of RNA splicing and introns. *Cell* 42:397–400.

Cech, T. R., 1986. A model for the RNA-catalyzed replication of RNA. *Proc. Nat. Acad. Sci. USA* 83:4360–4363.

Marchionni, M., and Gilbert, W., 1986. The triosephosphate isomerase gene from maize: introns antedate the plant-animal divergence. *Cell* 46:133–141.

PROBLEMS

1. Compare DNA polymerase I and RNA polymerase from *E. coli* in regard to each of the following features:
   (a) Activated precursors.
   (b) Direction of chain elongation.
   (c) Conservation of the template.
   (d) Need for a primer.

2. Write the sequence of the mRNA molecule synthesized from a DNA template strand having the sequence

   5′-ATCGTACCGTTA-3′

3. RNA is readily hydrolyzed by alkali, whereas DNA is not. Why?

4. How does cordycepin (3′-deoxyadenosine) block the synthesis of RNA?

5. What amino acid sequence is encoded by the following base sequence of an mRNA molecule? Assume that the reading frame starts at the 5′ end.

   5′-UUGCCUAGUGAUUGGAUG-3′

6. What is the sequence of the polypeptide formed on addition of poly (UUAC) to a cell-free protein-synthesizing system?

7. A protein chemist told a molecular geneticist that he had found a new mutant hemoglobin in which aspartate replaced lysine. The molecular geneticist expressed surprise and sent his friend scurrying back to the laboratory.
   (a) Why was the molecular geneticist dubious about the reported amino acid substitution?
   (b) Which amino acid substitutions would have been more palatable to the molecular geneticist?

8. The RNA transcript of a region of G4 phage DNA contains the sequence 5′-AAAUGAGGA-3′. This sequence encodes three different polypeptides. What are they?

9. Proteins generally have low contents of Met and Trp, intermediate ones of His and Cys, and high ones of Leu and Ser. What is the relation between the number of codons of an amino acid and its frequency of occurrence in proteins? What might be the selective advantage of this relation?

10. A transfer RNA with a UGU anticodon is enzymatically charged with a ^{14}C-labeled cysteine. The cysteine unit is then chemically modified to alanine (using Raney nickel, which removes the sulfur atom of cysteine). The altered aminoacyl-tRNA is added to a protein-synthesizing system containing normal components except for this tRNA. The mRNA added to this mixture contains the following sequence:

    5′-UUUUGCCAUGUUUGUGCU-3′

    What is the sequence of the corresponding radiolabeled peptide?

11. Valine is specified by four codons. How might the relative frequencies of their usage in an alga isolated from a volcanic hot spring differ from those of an alga isolated from an Antarctic bay?

12. The amino acid sequences of a yeast protein and a human protein carrying out the same function are found to be 60% identical. However, the corresponding DNA sequences are only 45% identical. Account for this differing degree of identity.

13. The genes for the green-absorbing and red-absorbing visual pigments mediating human color vision are located next to each other on the X chromosome. These genes are very similar to one another.
    (a) What is the consequence in a male of having only one visual pigment gene on the X chromosome?
    (b) A small proportion of the population has three visual pigment genes adjacent to each other on the X chromosome. Two of them are identical. How might this arise? What is its potential evolutionary significance?

CHAPTER 6

Exploring Genes: Analyzing, Constructing, and Cloning DNA

Recombinant DNA technology, a new approach to exploring the central molecules of life, came into being in the early 1970s. It has revolutionized biochemistry by providing powerful means of analyzing and altering genes and proteins. The genetic endowment of organisms can now be precisely changed in designed ways. Recombinant DNA technology is a fruit of several decades of basic research on DNA, RNA, and viruses. It depends, first, on having enzymes that can cut, join, and replicate DNA and reverse transcribe RNA. Earlier chapters have already discussed DNA polymerases and reverse transcriptase. This chapter begins with restriction enzymes, which cut very long DNA molecules into specific fragments that can be manipulated. The availability of many kinds of restriction enzymes and of DNA ligases, enzymes that join DNA strands, makes it feasible to treat DNA sequences as modules and to move them at will from one DNA molecule to another. Thus, recombinant DNA technology is based on nucleic acid enzymology.

A second foundation is the base-pairing language that mediates nucleic acid recognition. We have already seen that hybridization with complementary DNA or RNA probes is a sensitive and powerful means of detecting specific nucleotide sequences. In recombinant DNA technology, base pairing is used to construct new combinations of DNA as well as to detect particular sequences. This revolutionary technology is also critically dependent on the existence of viruses and detailed knowledge concerning their interplay with susceptible hosts. Viruses are the ultimate parasites. They efficiently deliver their own DNA (or RNA) into hosts, subverting them either to replicate the viral genome and produce viral protein or to incorporate viral DNA into the host genome. Likewise, plasmids, which are accessory chromosomes, have been indispensable in recombinant DNA technology.

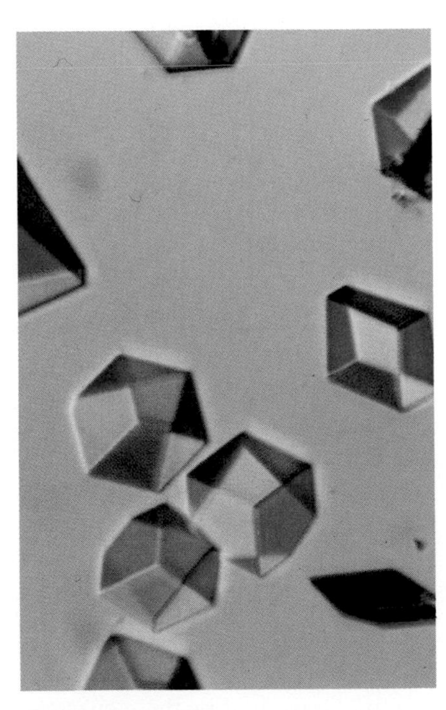

Figure 6-1
Crystals of human insulin produced by bacteria harboring recombinant DNA encoding the hormone. [From R. E. Chance, E. P. Kroeff, and J. A. Hoffmann. In *Insulins, Growth Hormone, and Recombinant DNA Technology*, J. L. Gueriguian, ed. (Raven Press, 1981), p. 77.]

This chapter also introduces some of the benefits of these new methods. For example, the discovery of restriction enzymes led to the development of techniques for the rapid sequencing of DNA. A wealth of information concerning gene architecture, the control of gene expression, and protein structure has come from the sequencing of millions of bases in DNA molecules from viruses, bacteria, and higher organisms. DNA molecules can also be synthesized de novo. The automated solid-phase synthesis of DNA provides highly specific probes and synthetic tailor-made genes. The final part of this chapter deals with the construction and cloning of novel combinations of genes. New genes can be efficiently expressed by host cells, as exemplified by the production of human insulin by bacteria. Moreover, specific mutations can be made in vitro to engineer proteins in designed ways.

Palindrome—
A word, sentence, or verse that reads the same from right to left as it does from left to right.

Radar
Madam, I'm Adam
Able was I ere I saw Elba
Roma tibi subito motibus ibit amor

Derived from the Greek *palindromos*, running back again.

RESTRICTION ENZYMES SPLIT DNA INTO SPECIFIC FRAGMENTS

Restriction enzymes, also called restriction endonucleases, recognize specific base sequences in double-helical DNA and cleave both strands of the duplex at specific places. To biochemists, these exquisitely precise scalpels are marvelous gifts of nature. They are indispensable for analyzing chromosome structure, sequencing very long DNA molecules, isolating genes, and creating new DNA molecules that can be cloned. Restriction enzymes were discovered by Werner Arber, Hamilton Smith, and Daniel Nathans.

Restriction enzymes are found in a wide variety of procaryotes. Their biological role is to cleave foreign DNA molecules. The cell's own DNA is not degraded because the sites recognized by its own restriction enzymes are methylated. The interplay between modification and the action of endonucleases will be considered in a later chapter (Chapter 34). The important point here is that many of them recognize specific sequences of four to eight base pairs and hydrolyze a phosphodiester bond in each strand in this region. A striking characteristic of most of these cleavage sites is that they possess *twofold rotational symmetry*. In other words, the recognized sequence is *palindromic* and the cleavage sites are symmetrically positioned. For example, the sequence recognized by a restriction enzyme from *Streptomyces achromogenes* is

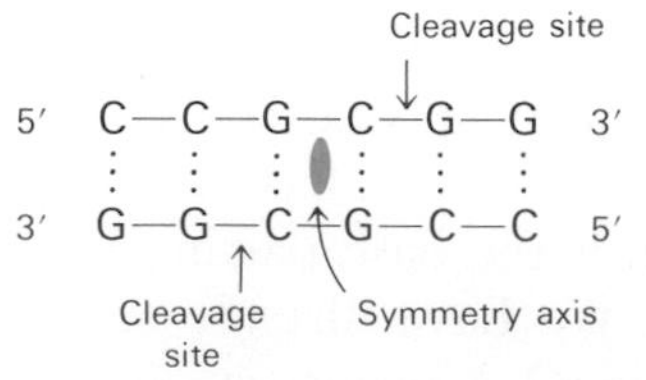

Cleavage site
5′ C—C—G—C—G—G 3′
3′ G—G—C—G—C—C 5′
Cleavage site
Symmetry axis

In each strand, the enzyme cleaves the CG phosphodiester bond on the 3′ side of the symmetry axis.

More than ninety restriction enzymes have been purified and characterized. Their names consist of a three-letter abbreviation for the host organism (e.g., Eco for *E. coli*, Hin for *Haemophilus influenzae*, Hae for *H. aegyptius*) followed by a strain designation (if needed) and a Roman numeral (if more than one restriction enzyme is produced). The specificities of several of these enzymes are shown in Figure 6-2. Note that the cuts may be staggered or even.

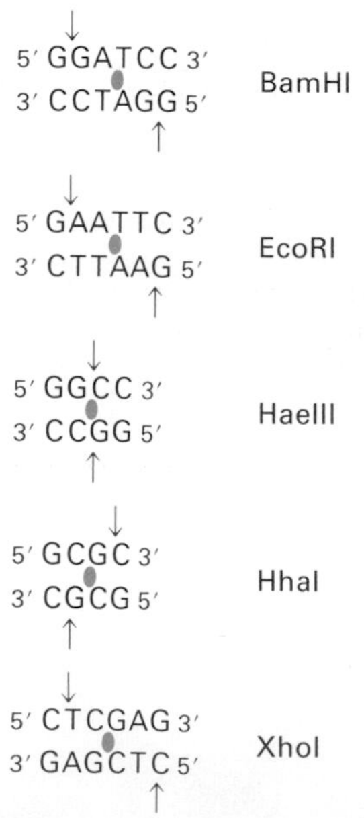

5′ GGATCC 3′
3′ CCTAGG 5′
BamHI

5′ GAATTC 3′
3′ CTTAAG 5′
EcoRI

5′ GGCC 3′
3′ CCGG 5′
HaeIII

5′ GCGC 3′
3′ CGCG 5′
HhaI

5′ CTCGAG 3′
3′ GAGCTC 5′
XhoI

Figure 6-2
Specificities of some restriction endonucleases. The base-pair sequences that are recognized by these enzymes contain a twofold axis of symmetry. The two strands in these regions are related by a 180-degree rotation around the axis marked by the green symbol. The cleavage sites are denoted by red arrows. The abbreviated name of each restriction enzyme is given at the right of the sequence that it recognizes.

Restriction enzymes are used to cleave DNA molecules into specific fragments that are more readily analyzed and manipulated than the parent molecule. For example, the 5.1-kilobase (kb) circular duplex DNA of the tumor-producing SV40 virus is cleaved at one site by EcoRI, four sites by HpaI, and eleven sites by HindIII. A piece of DNA produced by the action of one restriction enzyme can be specifically cleaved into smaller fragments by another restriction enzyme. The pattern of such fragments can serve as a *fingerprint* of a DNA molecule, as will be discussed shortly. Indeed, complex chromosomes containing hundreds of millions of base pairs can be mapped by using a series of restriction enzymes.

Kilobase (kb)—
A unit of length equal to 1000 base pairs of a double-stranded nucleic acid molecule (or 1000 bases of a single-stranded molecule).
One kilobase of double-stranded DNA has a contour length of 0.34 μm and a mass of about 660 kd.

RESTRICTION FRAGMENTS CAN BE SEPARATED BY GEL ELECTROPHORESIS AND VISUALIZED

Small differences between related DNA molecules can be readily detected because their restriction fragments can be separated and displayed by gel electrophoresis. In many types of gels, the electrophoretic mobility of a DNA fragment is inversely proportional to the logarithm of the number of base pairs up to a certain limit. Polyacrylamide gels are used to separate fragments containing up to about 1000 base pairs, whereas more porous agarose gels are used to resolve mixtures of larger fragments (up to about 20 kb). An important feature of these gels is their high resolving power. In certain kinds of gels, fragments differing in length by just one nucleotide out of several hundred can be distinguished. Moreover, entire chromosomes containing millions of nucleotides can now be separated on agarose gels by applying pulsed electric fields in different directions (p. 46). Bands or spots of radioactive DNA in gels can be visualized by autoradiography. Alternatively, a gel can be stained with ethidium bromide, which fluoresces an intense orange when it has been bound to double-helical DNA. A band containing only 50 ng of DNA can readily be seen (Figure 6-3).

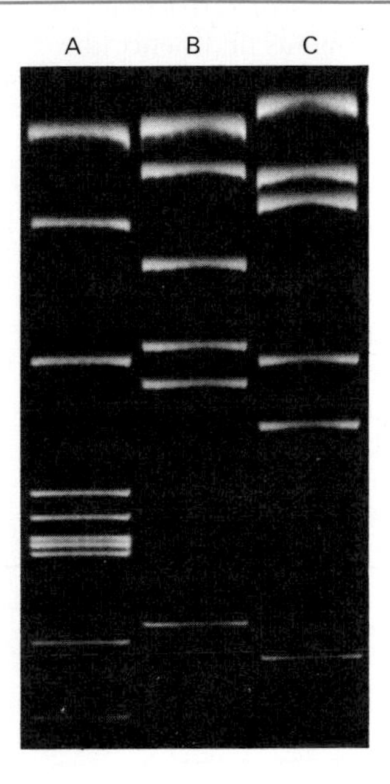

Figure 6-3
Gel electrophoresis pattern showing the fragments produced by cleaving SV40 DNA with each of three restriction enzymes. These fragments were made fluorescent by staining the gel with ethidium bromide. [Courtesy of Dr. Jeffrey Sklar.]

A restriction fragment containing a specific base sequence can be identified by hybridizing it with a labeled complementary DNA strand (Figure 6-4). A mixture of restriction fragments is separated by electro-

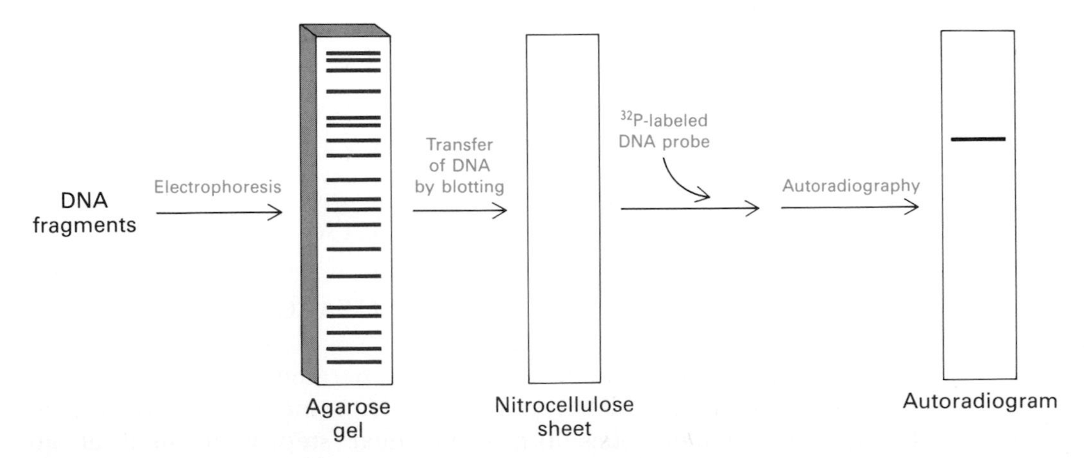

Figure 6-4
Southern blotting. A DNA fragment containing a specific sequence can be identified by separating a mixture of fragments by electrophoresis, transferring them to nitrocellulose, and hybridizing with a ^{32}P-labeled probe complementary to the sequence. The fragment containing the sequence is then visualized by autoradiography.

Restriction-fragment-length polymorphism (RFLP)—Southern blotting can be used to follow the inheritance of selected genes. Mutations within restriction sites change the sizes of restriction fragments and hence the positions of bands in Southern blot analyses. The existence of genetic diversity in a population is termed polymorphism. The detected mutation may itself cause disease or it may be closely linked to one that does. Genetic diseases such as sickle-cell anemia (p. 169), cystic fibrosis, and Huntington's chorea can be detected by RFLP analyses.

phoresis through an agarose gel, denatured to form single-stranded DNA, and transferred to a nitrocellulose sheet. The positions of the DNA fragments in the gel are preserved in the nitrocellulose sheet, where they can be hybridized with a ^{32}P-labeled single-stranded *DNA probe*. Autoradiography then reveals the position of the restriction fragment with a sequence complementary to that of the probe. A particular fragment in the midst of a million others can readily be identified in this way, like finding a needle in a haystack. This powerful technique is known as *Southern blotting* because it was devised by E. M. Southern. Likewise, RNA molecules can be separated by gel electrophoresis, and specific sequences can be identified by hybridization following transfer to nitrocellulose. This analogous technique for the analysis of RNA has been whimsically termed *Northern blotting*. A further play on words accounts for the term *Western blotting*, which refers to a technique for detecting a particular protein by staining with specific antibody (p. 63). Southern, Northern, and Western blotting are also known as DNA, RNA, and protein blotting.

DNA CAN BE SEQUENCED BY SPECIFIC CHEMICAL CLEAVAGE (MAXAM-GILBERT METHOD)

The analysis of DNA structure and its relation to gene expression has also been markedly facilitated by the development of powerful techniques for the sequencing of DNA molecules. The *chemical cleavage method* devised by Allan Maxam and Walter Gilbert starts with a DNA that is labeled at one end of one strand with ^{32}P. Polynucleotide kinase is usually used to add ^{32}P at the 5′-hydroxyl terminus. The labeled DNA is then broken preferentially at one of the four nucleotides. The conditions are chosen so that an average of one break is made per chain. In the reaction mixture for a given base, then, each broken chain yields a radioactive fragment extending from the ^{32}P label to one of the positions of that base, and such fragments are produced for every position of the base. For example, if the sequence is

5′-^{32}P-GCTACGTA-3′

the *radioactive* fragments produced by specific cleavage on the 5′ side of each of the four bases would be

Cleavage at A:	^{32}P-GCT ^{32}P-GCTACGT
Cleavage at G:	^{32}P-GCTAC
Cleavage at C:	^{32}P-G ^{32}P-GCTA
Cleavage at T:	^{32}P-GC ^{32}P-GCTACG

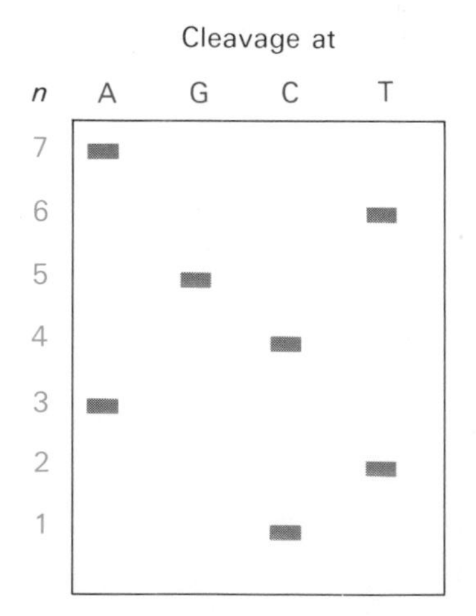

Figure 6-5
Schematic diagram of a gel showing the radioactive fragments formed by specific cleavage of 5′-^{32}P-GCTACGTA-3′ at each of the four bases (A, G, C, and T lanes). The number of nucleotides (n) in a fragment is shown on the left. The base sequence is read in ascending order.

The fragments in each mixture are then separated by polyacrylamide-gel electrophoresis, which can resolve DNA molecules differing in length by just one nucleotide. The next step is to look at an autoradiogram of this gel. In our example (Figure 6-5), the lowest band would be in the C lane, and the next one up in the T lane, followed by one in the A lane. Hence, the sequence of the first three nucleotides is 5′-CTA-3′ (the identity of the G at the 5′ end is not revealed). Reading all seven bands in ascending order gives the sequence 5′-CTACGTA-3′.

Thus, the autoradiogram of a gel produced from four different chemical cleavages displays a pattern of bands from which the sequence can be read directly. In practice, DNA is specifically cleaved by reagents that modify and then remove certain bases from their sugars (Figure 6-6). Purines are damaged by *dimethylsulfate*, which methylates guanine at N-7 and adenine at N-3. The glycosidic bond of a methylated purine is readily broken by heating at neutral pH, which leaves the sugar without a base. Then the backbone is cleaved and the sugar unit eliminated by heating in alkali. When the resulting end-labeled fragments are resolved in a lane on a polyacrylamide gel, a pattern of dark and light bands is seen on the autoradiogram. The dark bands correspond to fragments formed by breakage at guanine, because this base is methylated much more rapidly than is adenine. The lane containing this sample is called the G lane because nearly all the cleavages are at G. In contrast, the glycosidic bond of methylated adenosine is less stable to dilute acid than that of methylated guanosine. Hence, treatment with dilute acid after methylation causes cleavages at both A and G, giving rise to an A + G lane. Comparison of this lane of a gel with a parallel G lane reveals whether cleavage occurred at A or G (Figure 6-7). Cytosine and thymine are split by *hydrazine*. The backbone is then cleaved by *piperidine*, which displaces the products of the hydrazine reaction and catalyzes elimination of the phosphates. The resulting mixture gives rise to a C + T lane, a series of bands of about equal intensity from cleavages at cytosines and thymines. These pyrimidines are distinguished by preparing a C lane, for which hydrazinolysis is carried out in the presence of 2 M NaCl. (This suppresses the reaction with thymine.)

The DNA sequence is then read from the autoradiogram of the gel by comparing the G, A + G, C + T, and C lanes (Figure 6-7). The shortest fragment has the highest electrophoretic mobility and so the 5′ end of the sequence is at the bottom of the gel. Sequences of more than 250 bases can be readily determined by running several gels for different time intervals to resolve all of the labeled fragments.

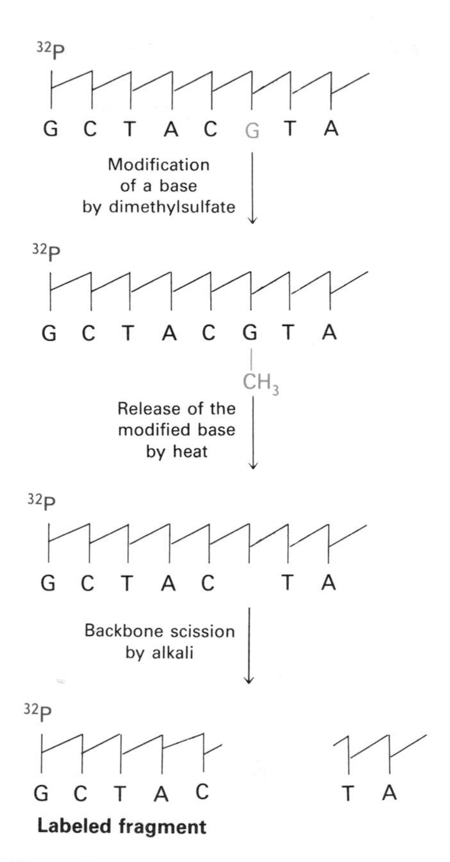

Figure 6-6
Strategy of the chemical-cleavage method for the sequencing of DNA. This particular procedure would produce the fragments visualized in the G lane of a set of gels.

DNA CAN BE SEQUENCED BY CONTROLLED INTERRUPTION OF REPLICATION (SANGER DIDEOXY METHOD)

DNA can also be sequenced by generating fragments through the *controlled interruption of enzymatic replication*, a method developed by Frederick Sanger and his associates. DNA polymerase I is used to copy a particular sequence of a single-stranded DNA. The synthesis is primed by a complementary fragment, which may be obtained from a restriction enzyme digest or synthesized chemically. In addition to the four deoxyribonucleoside triphosphates (radioactively labeled), the incubation mixture contains a *2′,3′-dideoxy analog* of one of them. The incorporation of this analog blocks further growth of the new chain because it lacks the 3′-hydroxyl terminus needed to form the next phosphodiester bond. Hence, fragments of various lengths are produced in which the

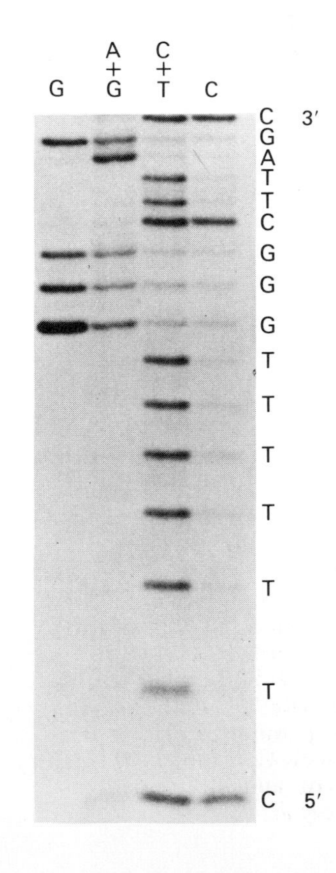

Figure 6-7
Autoradiogram of a gel showing labeled fragments produced by chemical cleavage. The 5′ end of the DNA strand was labeled with ^{32}P. The shortest nucleotide is at the bottom of the gel. Hence, the base sequence is 5′-CTTTTTTGGGCTTAGC-3′. [Courtesy of Dr. David Dressler.]

Base

Ⓟ—Ⓟ—ⓅOCH₂

2′,3′-Dideoxy analog

DNA to be sequenced

3′——GAATTCGCTAATGC————

5′—CTTAA

Primer

DNA polymerase I
dATP, dTTP, dCTP, dGTP
Dideoxy analog of dATP

3′——GAATTCGCTAATGC————

5′—CTTAAGCGATTA

+

3′——GAATTCGCTAATGC————

5′—CTTAAGCGA

New DNA strands are separated and electrophoresed

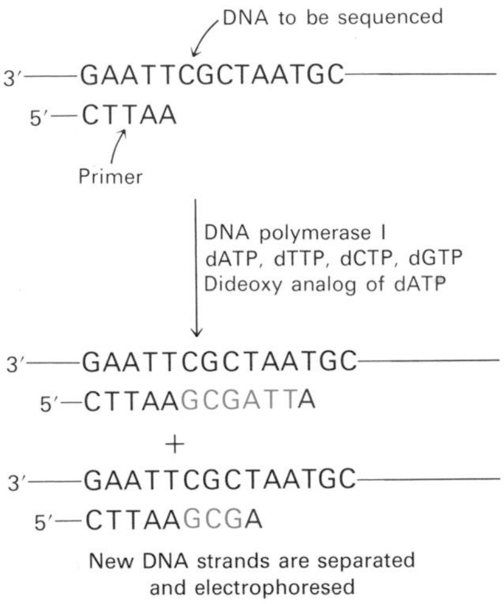

Figure 6-8
Strategy of the chain-termination method for the sequencing of DNA. Fragments are produced by adding the 2′,3′-dideoxy analog of a dNTP to each of four polymerization mixtures. For example, the addition of the dideoxy analog of dATP (shown in red) results in fragments ending in A.

dideoxy analog is at the 3′ end (Figure 6-8). Four such sets of *chain-terminated fragments* (one for each dideoxy analog) are then electrophoresed, and the base sequence of the new DNA is read from the autoradiogram of the four lanes.

The complete sequence of the 5386 bases in ϕX174 DNA was determined in this way by Sanger and co-workers in 1977, just a quarter century after Sanger's pioneering elucidation of the amino acid sequence of a protein. This accomplishment is a landmark in molecular biology because it revealed the total information content of a DNA genome. This tour de force was followed several years later by another one, the determination of the sequence of human mitochondrial DNA, a double-stranded circular DNA molecule containing 16,569 base pairs. It encodes two ribosomal RNAs, 22 transfer RNAs, and 13 proteins. This achievement was followed by the sequencing of the 48,513 base pairs of the DNA of λ phage, a virus that infects *E. coli*. The wealth of information derived from these remarkable accomplishments will be considered in detail in later chapters.

About 5×10^6 bases of DNA have been sequenced in laboratories around the world since the introduction of the Maxam-Gilbert and Sanger methods. All of these studies have used autoradiographic images of gels to ascertain the lengths of DNA fragments generated by chemical cleavage and controlled interruption of replication. Recently, a variant of the dideoxy method has been devised. A fluorescent tag is attached to the oligonucleotide primer—a differently colored one in each of the four chain-terminating reaction mixtures (e.g., a blue emitter for termination at A and a red one for termination at C). The reaction mixtures are combined and electrophoresed together. The separated bands of DNA are then detected by their fluorescence as they pass out the bottom of the tube, and the sequence of their colors directly yields the base sequence (Figure 6-9). Sequences of up to 500 bases can now be determined in this way. An attractive feature of this fluorescence detection method is that it can readily be automated. The sequencing of the en-

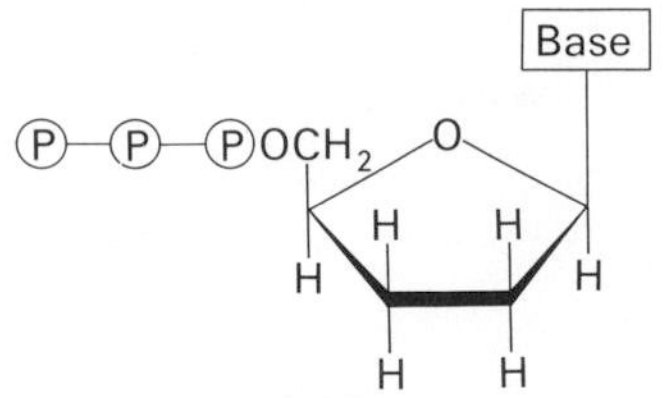

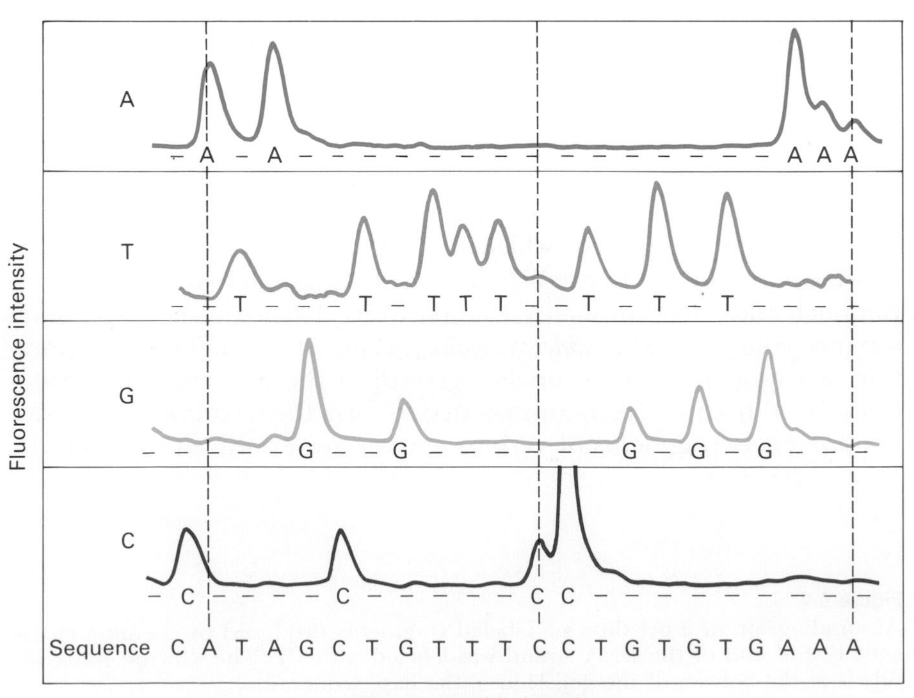

Figure 6-9
Fluorescence detection of oligonucleotide fragments produced by the dideoxy method. Each of the four chain-terminating mixtures is primed with a tag that fluoresces at a different wavelength (e.g., blue for A). The sequence determined by fluorescence measurements at four wavelengths is shown at the bottom of the figure. [From L. M. Smith, J. Z. Sanders, R. J. Kaiser, P. Hughes, C. Dodd, C. R. Connell, C. Heiner, S. B. H. Kent, and L. E. Hood. *Nature* 321(1986):674.]

tire *E. coli* genome (3×10^6 base pairs) has now become feasible. We can even begin to think about determining the sequence of extensive stretches of the human genome, which contains 3×10^9 base pairs.

DNA PROBES AND GENES CAN BE SYNTHESIZED BY AUTOMATED SOLID-PHASE METHODS

DNA strands, like polypeptides (p. 66), can be synthesized by the sequential addition of activated monomers to a growing chain that is linked to an insoluble support. The activated monomers are protonated *deoxyribonucleoside 3′-phosphoramidites* (Figure 6-10). In step 1, the

Protonated phosphoramidite
(The 5′-hydroxyl is blocked by a dimethoxytrityl protecting group.)

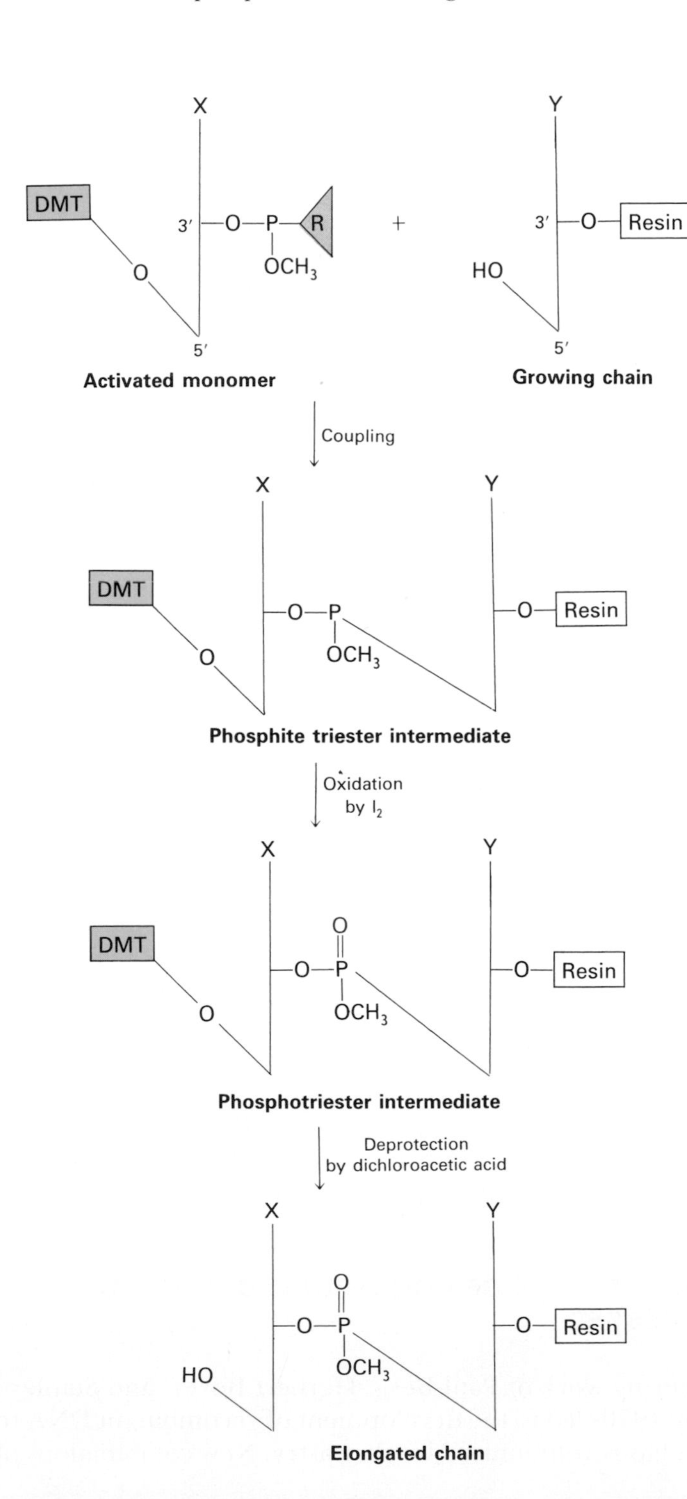

Figure 6-10
Solid-phase synthesis of a DNA chain by the phosphite triester method. The activated monomer added to the growing chain is a deoxyribonucleoside 3′-phosphoramidite containing a DMT (dimethoxytrityl) blocking group on its 5′-oxygen atom.

3′-phosphorus atom of this incoming unit becomes joined to the 5′-oxygen of the growing chain to form a *phosphite triester*. The 5′-OH of the activated monomer is unreactive because it is blocked by a dimethoxytrityl (DMT) protecting group. Likewise, amino groups on the purine and pyrimidine bases are blocked. Coupling is carried out under anhydrous conditions because water reacts with phosphoramidites. In step 2, the phosphite triester (in which P is trivalent) is oxidized by iodine to form a *phosphotriester* (in which P is pentavalent). In step 3, the DMT protecting group on the 5′-OH of the growing chain is removed by addition of dichloroacetic acid, which leaves other protecting groups intact. The DNA chain is now elongated by one unit and ready for another cycle of addition. Each monomer addition cycle takes only about ten minutes and elongates more than 98% of the chains.

This solid-phase approach is ideal for the synthesis of DNA, as it is for polypeptides, because the desired product stays on the insoluble support until the final release step. All of the reactions occur in a single vessel, and excess soluble reagents can be added to drive reactions to completion. At the end of each step, soluble reagents and by-products are washed away from the glass beads that bear the growing chains.

After assembly of the desired DNA chain, the methyl groups protecting the phosphates are removed by addition of thiophenol. The DNA strand is then released from the glass bead by cleavage of the ester bond between the 3′-OH of the terminal nucleoside and the resin that links it to the glass support. This bond is hydrolyzed by the addition of concentrated ammonium hydroxide. Finally, the benzoyl and isobutyryl groups protecting the bases are removed by heating the DNA in ammonium hydroxide. Because elongation is never 100% complete, the new DNA chains are of diverse lengths—the desired chain is the longest one. The sample can be purified by high-performance liquid chromatography or by electrophoresis on polyacrylamide gels. DNA chains up to 100 nucleotides long can readily be synthesized by this automated method.

The ability to rapidly synthesize DNA chains of any selected sequence opens many experimental avenues. For example, an oligonucleotide labeled at one end with ^{32}P can be used to search for a complementary sequence in a very long DNA molecule or even in a genome consisting of many chromosomes. The use of labeled oligonucleotides as *DNA probes* is powerful and general. For example, a DNA probe that is base-paired to a known complementary sequence in a chromosome can serve as the starting point of an exploration of adjacent uncharted DNA. For example, the probe can be used as a *primer* to initiate the replication of neighboring DNA by DNA polymerase. One of the most exciting applications of the solid-phase approach is the *synthesis of new tailor-made genes*. New proteins with novel properties can now be produced in abundance by expressing synthetic genes. *Protein engineering* has become a reality. Moreover, regulatory sequences in DNA can be changed at will to control gene expression.

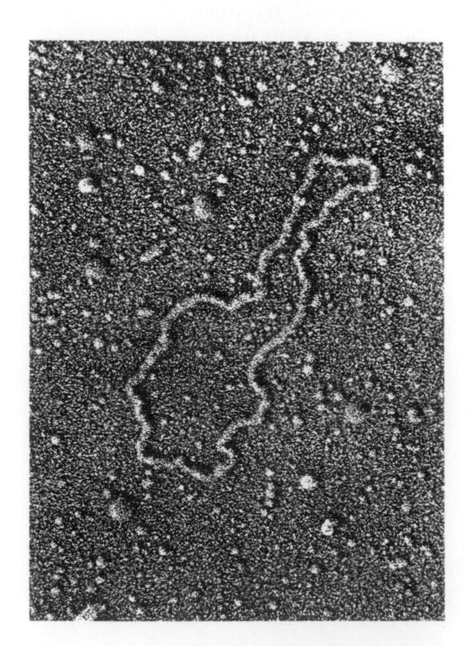

Electron micrograph of pSC101, the first plasmid vector used in the cloning of DNA. [Courtesy of Dr. Stanley N. Cohen.]

NEW GENOMES CAN BE CONSTRUCTED, CLONED, AND EXPRESSED

The pioneering work of Paul Berg, Herbert Boyer, and Stanley Cohen in the early 1970s led to the development of recombinant DNA technology, which has revolutionized biochemistry. New combinations of unrelated genes can be constructed in the laboratory by applying recombinant DNA techniques. These novel combinations can be cloned, amplified manyfold, by introducing them into suitable cells, where they are replicated by the DNA synthesizing machinery of the host. The inserted genes are often transcribed and translated in their new setting. What is most striking is that the genetic endowment of the host can be permanently altered in a designed way.

The major steps in the cloning of DNA are (Figure 6-11):

1. *Construction of a recombinant molecule.* A DNA fragment of interest is covalently joined to a DNA *vector*. The essential feature of a vector is that it can replicate autonomously in an appropriate host. Plasmids (naturally occurring circles of DNA that act as accessory chromosomes) and λ phage (a virus) are choice vectors for cloning in *E. coli.*

2. *Introduction into host cells.* Many bacterial and eucaryotic cells take up naked DNA molecules from the medium. The efficiency of uptake is low (about 1 of 10^6 DNA molecules), but an appreciable proportion of cells can be transformed under appropriate experimental conditions. Mutant bacteria that do not rapidly degrade foreign DNA are often used as host cells. DNA molecules can also be injected into many animal and plant cells. Alternatively, target cells can be infected with virus particles reassembled to harbor the recombinant DNA molecule. In this synthetic viral genome, a gene of interest replaces a segment of viral DNA that is not essential for replication.

Chimeric DNA—
A recombinant DNA molecule containing unrelated genes. From *chimera,* a mythological creature with the head of a lion, the body of a goat, and the tail of a serpent.

"... a thing of immortal make, not human, lion-fronted and snake behind, a goat in the middle, and snorting out the breath of the terrible flame of bright fire. ..."

Iliad (6.179)

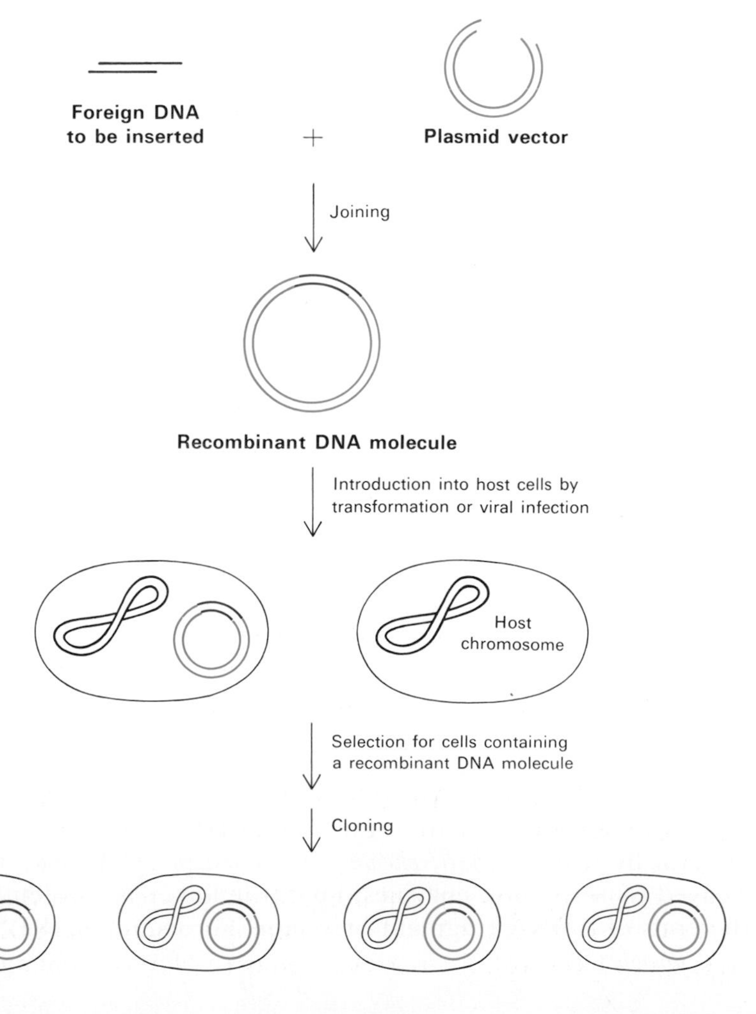

Figure 6-11
Construction and cloning of recombinant DNA molecules.

3. *Selection.* The next step is to determine which cells harbor the recombinant DNA molecule containing the gene of interest. The desired cells can be selected by the presence either of the vector or of the inserted gene itself. For example, some plasmid vectors confer resistance to an antibiotic, which can be used to eliminate the undesired cells. Another approach is to culture individual cells, then test a sample from each clone for the desired DNA sequence by Southern blotting with a labeled complementary probe. An expressed protein can be detected by Western blotting with a specific antibody, or by assaying for a functional property such as the appearance of functional enzymatic activity. Many cell lines containing recombinant DNA are genetically stable.

RESTRICTION ENZYMES AND DNA LIGASE ARE KEY TOOLS IN FORMING RECOMBINANT DNA MOLECULES

Let us begin by seeing how novel DNA molecules can be constructed in the laboratory. The vector in a recombinant DNA experiment can be prepared for splicing by cleaving it at a single specific site with a restriction enzyme. For example, the plasmid pSC101 (a 9.9-kb double-helical circular DNA molecule) is split at a unique site by the EcoRI restriction enzyme. The staggered cuts made by this enzyme produce *complementary single-stranded ends*, which have specific affinity for each other and hence are known as *cohesive ends*. Any DNA fragment can be inserted into this plasmid if it has the same cohesive ends. Such a fragment can be prepared from a larger piece of DNA by using the same restriction enzyme as was used to open the plasmid DNA. The single-stranded ends of the fragment are then complementary to those of the cut plasmid. The DNA fragment and the cut plasmid can be annealed and then joined by *DNA ligase*, which catalyzes the formation of a phosphodiester bond between two DNA chains (Figure 6-12). DNA ligase requires a free OH group at the 3′ end of one DNA chain and a phosphate group at the 5′ end of the other. Furthermore, the chains joined by ligase must belong to double-helical DNA molecules. An energy source, such as ATP or NAD^+ is required for the joining reaction, which will be discussed in Chapter 27.

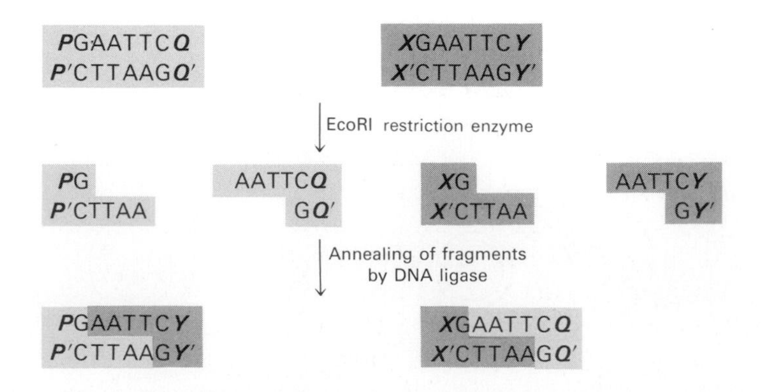

Figure 6-12
Joining of DNA molecules by the cohesive-end method. One of the parental DNA molecules (shown in green) carries genes *P* and *Q* separated by a restriction site, and the other (shown in red) carries *X* and *Y*. One of the recombinant molecules contains *P* and *Y*, and the other contains *Q* and *X*.

This cohesive-end method for joining DNA molecules can be made general by using a *short, chemically synthesized DNA linker* that can be cleaved by restriction enzymes. First, the linker is covalently joined to the ends of a DNA fragment or vector. For example, the 5′ ends of a decameric linker and a DNA molecule are phosphorylated by polynucleotide kinase and then joined by the ligase from T4 phage (Figure 6-13). This ligase can form a covalent bond between blunt-ended (flush-ended) double-helical DNA molecules. Cohesive ends are produced when these terminal extensions are cut by an appropriate restriction enzyme. Thus, *cohesive ends corresponding to a particular restriction enzyme can be added to virtually any DNA molecule.* We see here the fruits of combining enzymatic and synthetic chemical approaches in crafting new DNA molecules.

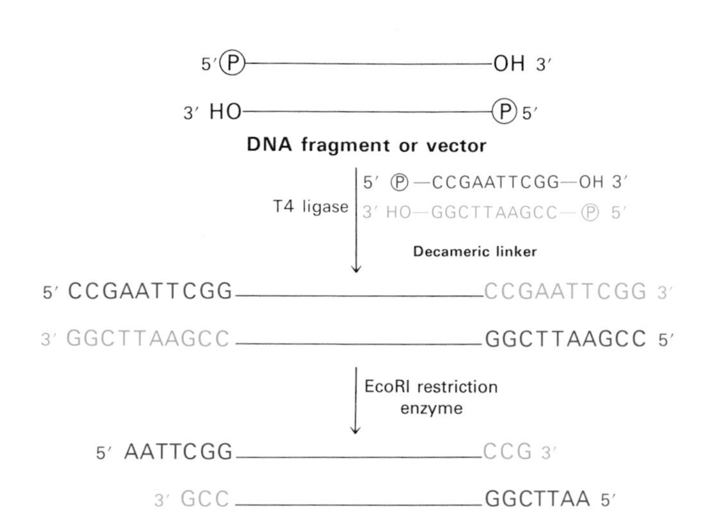

Figure 6-13
Formation of cohesive ends by the addition and cleavage of a chemically synthesized linker.

PLASMIDS AND LAMBDA PHAGE ARE CHOICE VECTORS FOR DNA CLONING IN BACTERIA

Many plasmids and bacteriophages have been ingeniously modified to enhance the delivery of recombinant DNA molecules into bacteria and to facilitate the selection of bacteria harboring them. *Plasmids* are naturally occurring circular duplex DNA molecules ranging in size from two kilobases to several hundred kilobases. They carry genes for the inactivation of antibiotics, the production of toxins, and the breakdown of natural products. These *accessory chromosomes* can replicate independently of the host chromosome. In contrast with the host genome, they are dispensable under certain conditions. A bacterial cell may have no plasmids at all or it may house as many as twenty copies of a plasmid.

One of the most useful plasmids for cloning is pBR322, which contains genes for resistance to tetracycline and ampicillin (an antibiotic like penicillin). This plasmid can be cleaved at a variety of unique sites by different endonucleases, and DNA fragments inserted. Insertion of DNA at the EcoRI restriction site does not alter either of the genes for antibiotic resistance (Figure 6-14). However, insertion at the HindIII, SalI, or BamHI restriction site inactivates the gene for tetracycline resistance, an effect called *insertional inactivation.* Cells containing pBR322 with a DNA insert at one of these restriction sites are resistant to ampicillin but sensitive to tetracycline, and so they can be readily *selected.* Cells that failed to take up the vector are sensitive to both antibiotics, whereas cells containing pBR322 without a DNA insert are resistant to both.

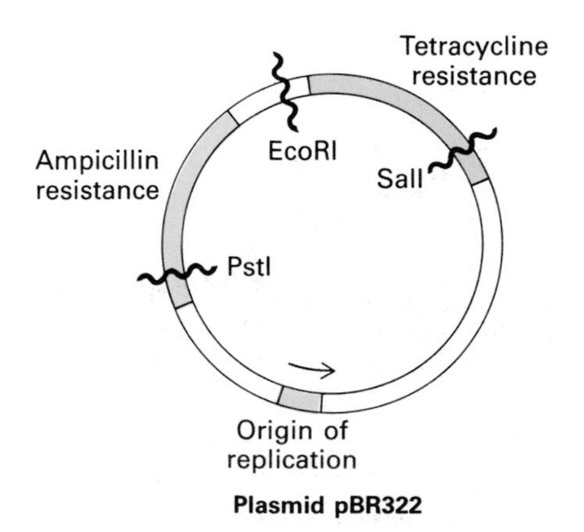

Figure 6-14
Genetic map of pBR322, a plasmid with two genes for antibiotic resistance.

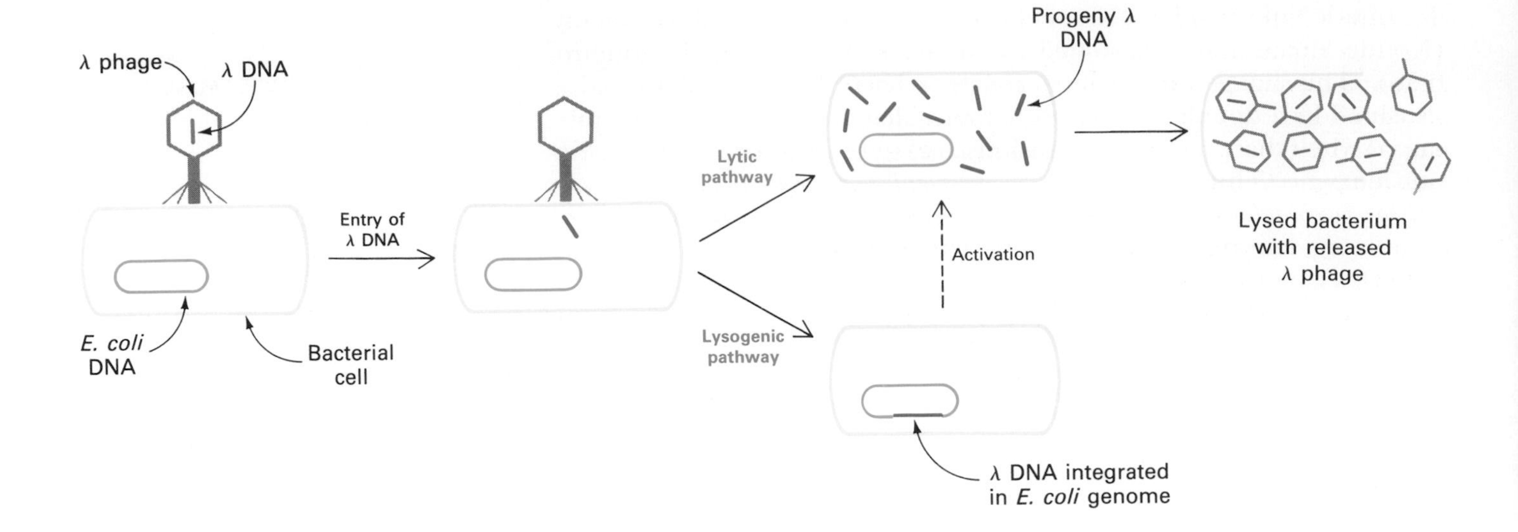

Figure 6-15
Lambda phage can multiply within a host and lyse it (lytic pathway) or its DNA can become integrated into the host genome (lysogenic pathway), where it is dormant until activated.

Lambda (λ) *phage* is another widely-used vector (Figure 6-15). This bacteriophage enjoys a choice of life styles: it can destroy its host or it can become part of its host (Chapter 34). In the *lytic pathway*, viral functions are fully expressed: viral DNA and proteins are quickly produced and packaged into virus particles, which leads to the lysis (destruction) of the host cell and the sudden appearance of about 100 progeny virus particles, or *virions*. In the *lysogenic pathway*, the phage DNA becomes inserted into the host-cell genome and can be replicated together with host-cell DNA for many generations, remaining inactive. Certain environmental changes can trigger the expression of this dormant viral DNA, which leads to the formation of progeny virus and lysis of the host. Large segments of the 48-kb DNA of λ phage are not essential for productive infection and can be replaced by foreign DNA.

Mutant λ phages designed for the cloning of DNA have been constructed. One of the mutants, called λgt-λβ, contains only two EcoRI cleavage sites instead of the five normally present (Figure 6-16). After

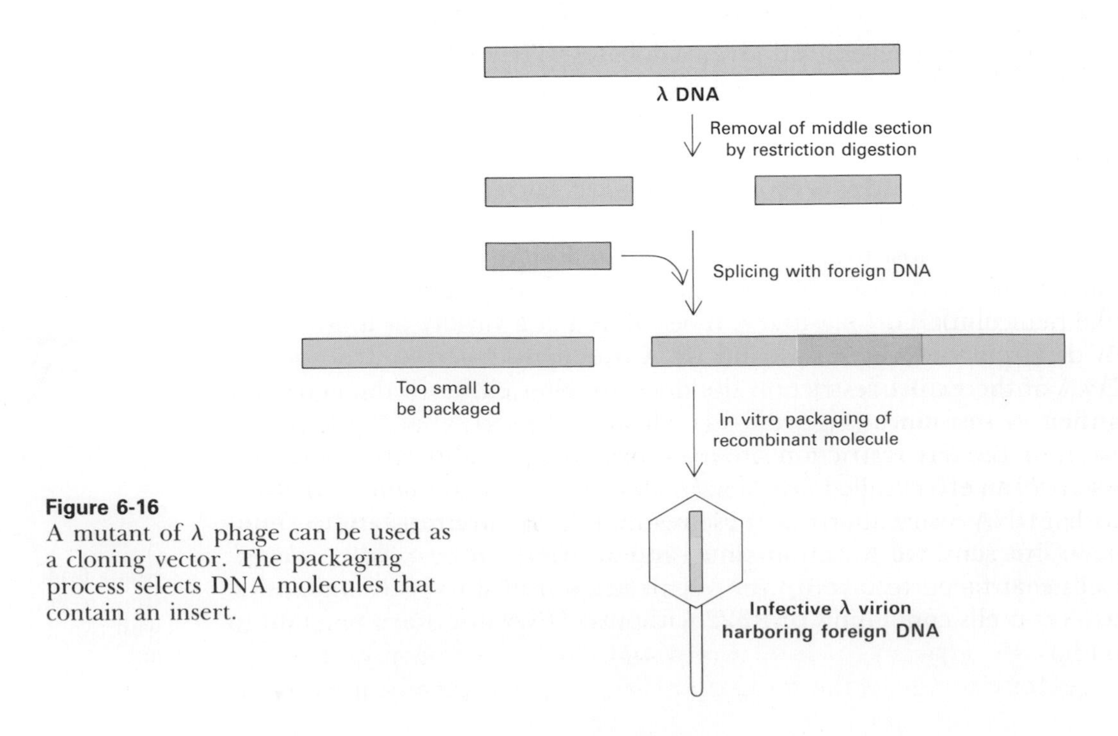

Figure 6-16
A mutant of λ phage can be used as a cloning vector. The packaging process selects DNA molecules that contain an insert.

cleavage, the middle segment of this λ DNA molecule can be removed. The two remaining pieces of DNA have a combined length equal to 72% of a normal genome length. This amount of DNA is too little to be packaged into a λ particle. The range of lengths that can be readily packaged is from 75% to 105% of a normal genome length. However, *a suitably long DNA insert (such as 10 kb) between the two ends of λ DNA enables such a recombinant DNA molecule (93% of normal length) to be packaged.* Nearly all infective λ particles formed in this way will contain an inserted piece of foreign DNA. Another advantage of using these modified viruses as vectors is that they enter bacteria much more easily than do plasmids. A variety of λ mutants have been constructed for use as cloning vectors. One of them, called a *cosmid*, can serve as a vector for large DNA inserts (up to about 45 kb).

M13 phage is another very useful vector for cloning DNA. This filamentous virus is 900 nm long and only 9 nm wide (Figure 6-17). Its 6.4-kb single-stranded circle of DNA is protected by a coat of 2710 identical protein subunits. M13 enters *E. coli* through the bacterial sex pilus, a protein appendage. The single-stranded DNA in the virus particle (called the + strand) is replicated by a double-stranded replicative form (RF) containing + and − strands, much as in ϕX174 (p. 85). Only the + strand is packaged into new virus particles. About a thousand progeny M13 are produced per generation. A striking feature of M13 is that it does not kill its bacterial host. Consequently, large quantities of M13 can be grown and easily harvested (1 g from 10 liters of culture fluid).

Figure 6-17
Electron micrograph of M13 filamentous phage. [Courtesy of Dr. Robley Williams.]

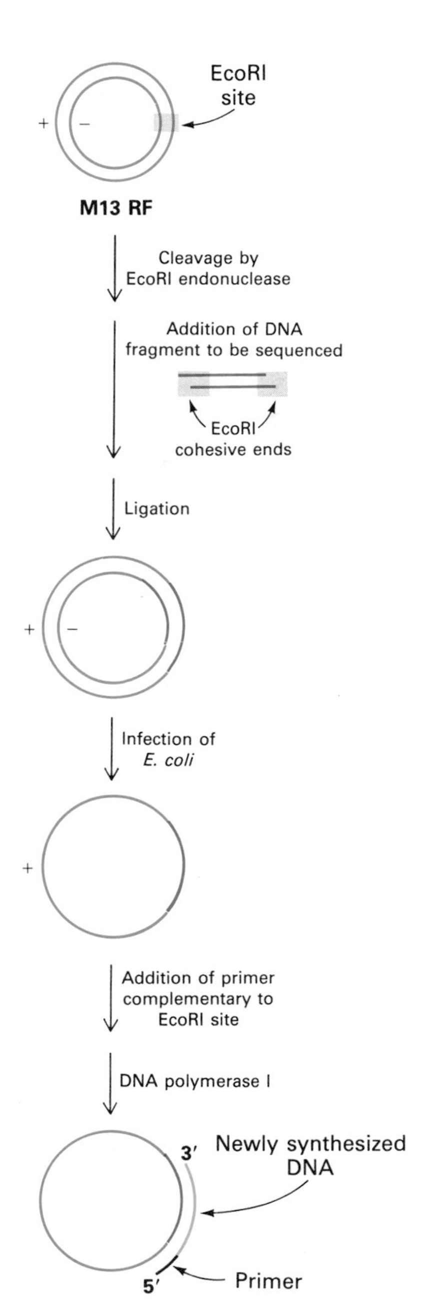

Figure 6-18
Sequencing by the dideoxy method of a DNA fragment inserted into M13 phage DNA. Synthesis of a new strand is primed by an oligonucleotide that is complementary to the restriction sequence adjacent to the inserted DNA.

M13 is prepared for cloning by cutting its circular double-stranded RF at a single site with a restriction enzyme. A double-stranded foreign DNA fragment produced by cleavage with the same restriction enzyme is then ligated to the cut RF (Figure 6-18). The foreign DNA can be inserted into the RF in two different orientations because the ends of both DNA molecules are the same. Hence, half of the new + strands packaged into virus particles will contain one of the strands of the foreign DNA, and half will contain the other strand. Infection of *E. coli* by a single virus particle will yield a large amount of single-stranded M13 DNA containing the same strand of the foreign DNA. The sequence in M13 DNA adjacent to the inserted DNA is known because it is the target for cleavage by the restriction enzyme. Consequently, a synthetic oligonucleotide with a complementary sequence can serve as a primer for dideoxy sequencing of any inserted DNA fragment. M13 is ideal for sequencing but not for long-term propagation of recombinant DNA, because inserts longer than about 1 kb are not stably maintained.

SPECIFIC GENES CAN BE CLONED FROM A DIGEST OF GENOMIC DNA

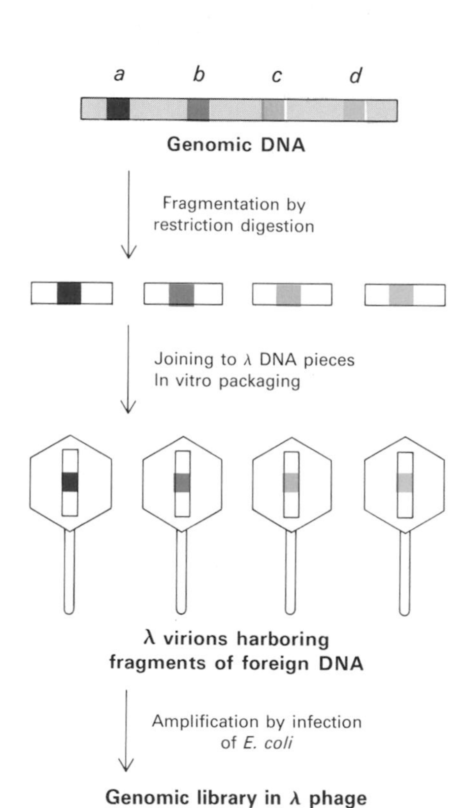

Figure 6-19
Creating a genomic library from a digest of a whole eucaryotic genome.

Ingenious cloning and selection methods have made feasible the isolation of a specific DNA segment several kilobases long out of a genome containing more than 3×10^6 kb. Let us see how a gene that occurs just once in a human genome can be cloned. A sample of the total genomic DNA is first mechanically sheared or partly digested by restriction enzymes into large fragments (Figure 6-19). This nearly random population of overlapping DNA molecules is then separated by gel electrophoresis into a set of fragments about 20 kb long. Synthetic linkers are attached to the ends of these fragments, cohesive ends are formed, and the fragments are then inserted into a vector, such as λ phage DNA, prepared with the same cohesive ends. The in vitro packaging of DNA into virus particles selects for recombinant DNA molecules that contain a large insert. *E. coli* are then infected with these recombinant phages. The resulting lysate contains fragments of human DNA housed in a large number of virus particles. These constitute a *genomic library* because they contain fragments of the entire human genome. Phage can be propagated indefinitely, and so the library can be used repeatedly over long periods.

This genomic library is then screened to find the very small proportion of phage harboring the gene of interest. A calculation shows that a 99% probability of success requires screening about 500,000 clones; hence, a very rapid and efficient screening process is essential. This can be accomplished by DNA hybridization.

A dilute suspension of the recombinant phage is first plated on a lawn of bacteria (Figure 6-20). Where each phage particle has landed and infected a bacterium, a plaque develops on the plate. Then a replica of this "master" plate is made by applying a sheet of nitrocellulose. Infected bacteria and phage DNA released from lysed cells adhere to the sheet in a pattern of spots corresponding to the plaques. Intact bacteria on this sheet are lysed with NaOH, which also serves to denature the DNA so that it becomes accessible for hybridization with a ^{32}P-labeled probe. *The presence of a specific DNA sequence in a single spot on the replica can be detected by using a radioactive complementary DNA or RNA molecule as a probe.* Autoradiography then reveals the positions of spots harboring recombinant DNA. The corresponding plaques are picked out of the intact master plate and grown. A million clones can readily be screened in a day by a single investigator.

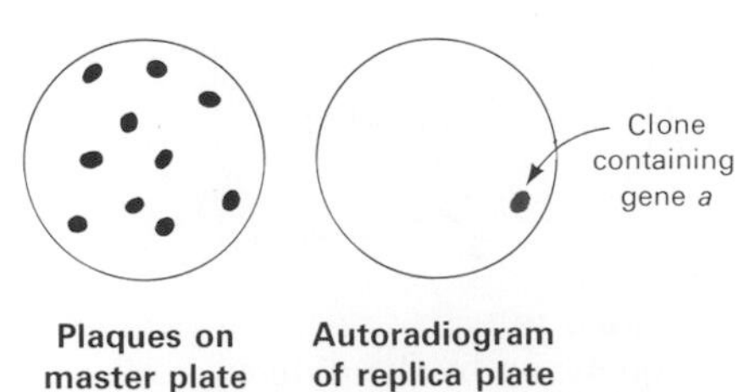

Figure 6-20
Screening a genomic library for a specific gene. Here, a plate is tested for plaques containing gene *a* of Figure 6-19.

This method makes it possible to isolate virtually any gene, *provided that a probe is available.* How does one obtain a specific probe? One approach is to *start with the corresponding mRNA from cells in which it is abundant.* For example, precursors of red blood cells contain large amounts of mRNAs for hemoglobin, and plasma cells are rich in mRNAs for antibody molecules. mRNAs from these cells can be fractionated by size to enrich for the one of interest. As will be described shortly, a DNA complementary to this mRNA can be synthesized in vitro and cloned to produce a highly specific probe.

Alternatively, *a probe for a gene can be prepared if part of the amino acid sequence of the protein encoded by the gene is known.* A problem arises because a given peptide sequence can be encoded by a number of oligonucleotides. Thus, for this purpose, peptide sequences containing tryptophan and methionine are preferred, because these amino acids are specified by a single codon, whereas other amino acid residues have between two and six codons (p. 107). The strategy is to choose a peptide that has a high proportion of tryptophan and methionine. For example, even for the pentapeptide

Trp-Tyr-Met-Cys-Met

there are four possible DNA coding sequences because tyrosine and cysteine can each be specified by either of two codons.

3′ ACC-ATA-TAC-ACA-TAC 5′
ACC-ATG-TAC-ACA-TAC
ACC-ATA-TAC-ACG-TAC
ACC-ATG-TAC-ACG-TAC
Trp-Tyr-Met-Cys-Met

A mixture of all the coding DNA sequences (or their complements) is synthesized by the solid-phase method and made radioactive by phosphorylating their 5′ ends with ^{32}P-orthophosphate. The replica plate is exposed to these probes and autoradiographed to identify any clone with a complementary DNA sequence. Among these, the ones containing the desired gene can be identified by sequencing the recombinant DNA from their plaques to determine whether it matches the known amino acid sequence of the protein of interest.

A typical genomic DNA library consists of DNA fragments about 20 kb long. How can we obtain information about longer stretches of DNA, say 300 kb long? Recall that the fragments in the library are produced by random cleavage of many DNA molecules and so some of the fragments overlap one another. Suppose that a fragment containing region A selected by hybridization with a complementary probe A′ also contains region B (Figure 6-21). A new probe B′ can be prepared by cleaving this fragment and subcloning region B. If the library is screened again with probe B′, new fragments containing region B will be found. Some will contain a contiguous region C. Hence, we now have information about a segment of DNA encompassing regions A, B, and C. This process of subcloning and rescreening is called *chromosome walking*. Long stretches of DNA can be analyzed in this way provided that each of the new probes is complementary to a unique region.

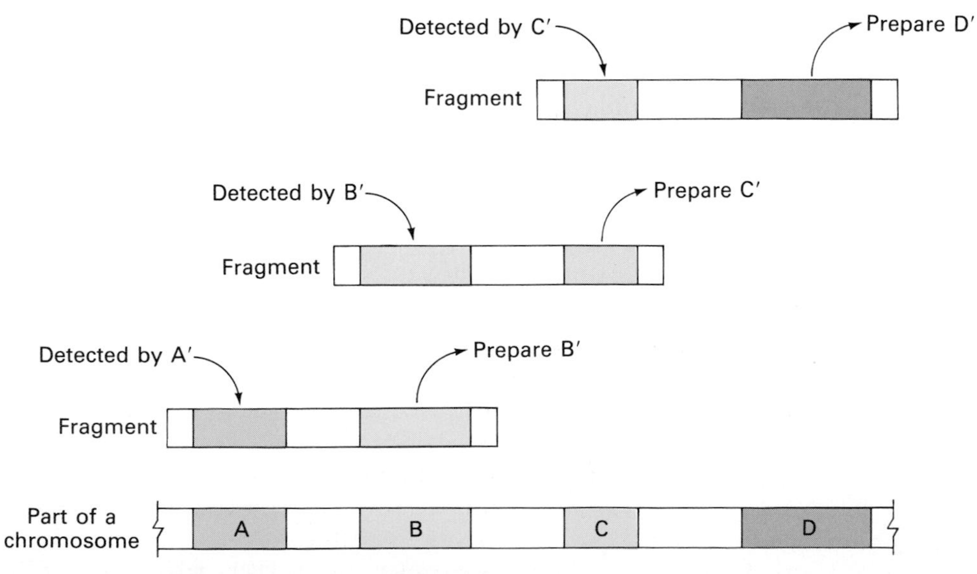

Figure 6-21
Chromosome walking. Long regions of unknown DNA can be explored starting with a known base sequence by subcloning and rescreening.

COMPLEMENTARY DNA (cDNA) PREPARED FROM mRNA CAN BE EXPRESSED IN HOST CELLS

Can mammalian DNA be cloned and expressed by *E. coli*? Recall that most mammalian genes are mosaics of introns and exons. These interrupted genes cannot be expressed by bacteria, which lack the machinery to splice introns out of the primary transcript. However, this difficulty can be circumvented by causing bacteria to take up recombinant DNA that is complementary to mRNA. For example, proinsulin, a precursor of insulin, is synthesized by bacteria harboring plasmids that contain DNA complementary to mRNA for proinsulin (Figure 6-22). Indeed, much of the insulin used today by millions of diabetics is produced by bacteria.

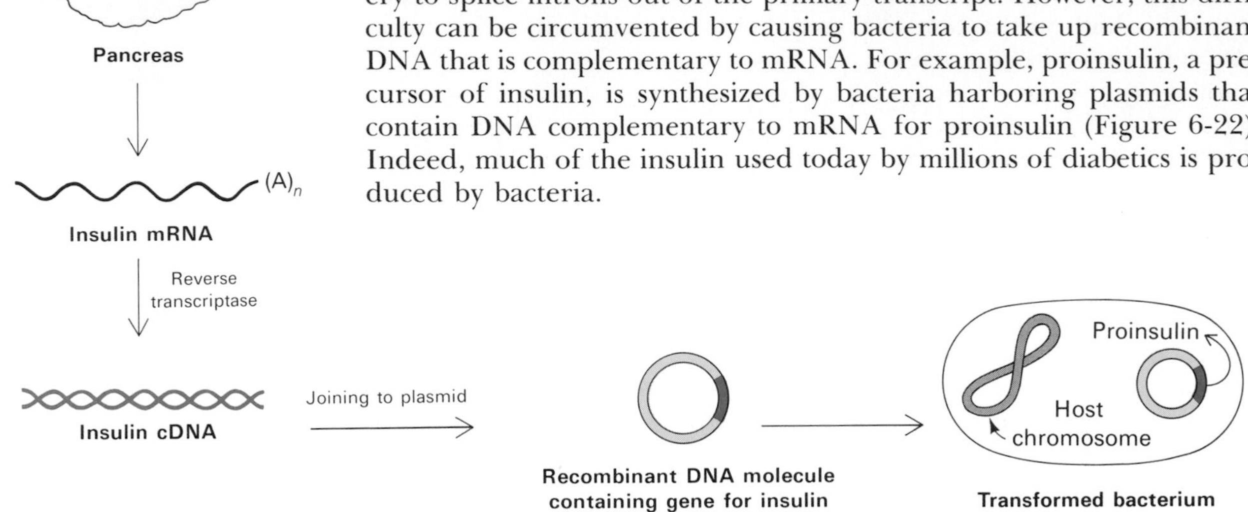

Figure 6-22
Synthesis of proinsulin, a precursor of insulin, by transformed clones of *E. coli.*

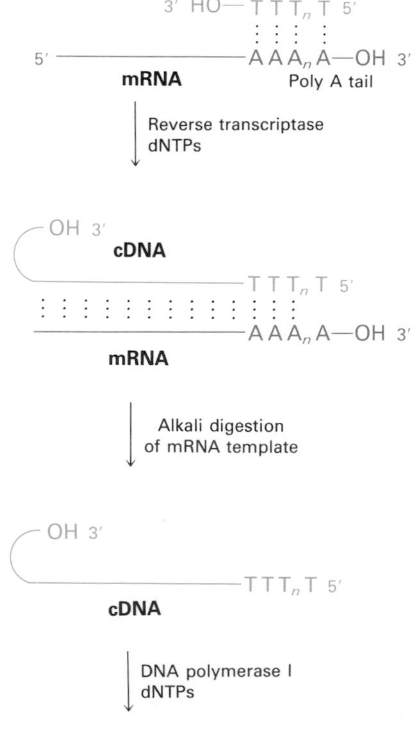

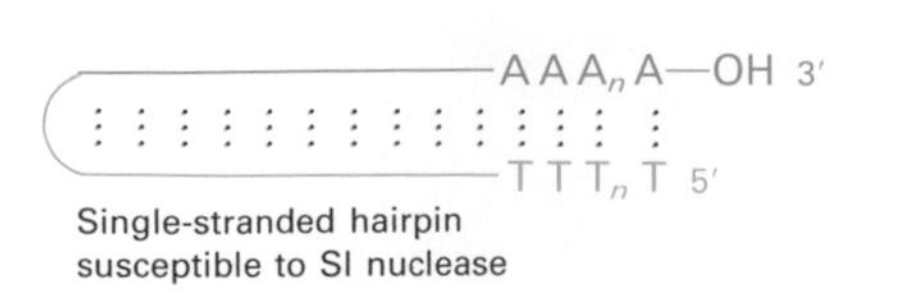

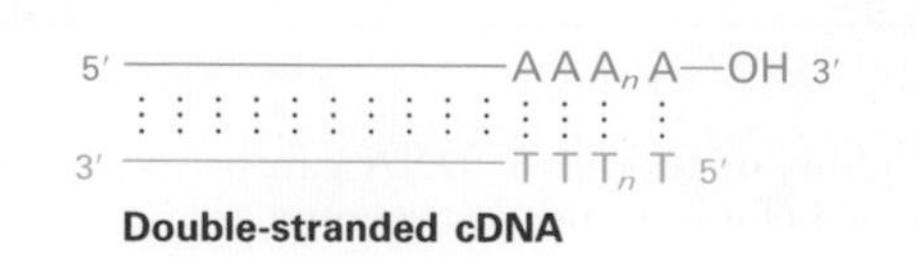

Figure 6-23
Forming a cDNA duplex from mRNA by using reverse transcriptase.

The key to forming complementary DNA (cDNA) is the enzyme *reverse transcriptase*. As was discussed earlier (p. 87), retroviruses use this enzyme to form a DNA-RNA hybrid in the replication of their genomic RNA. Reverse transcriptase synthesizes a DNA strand complementary to an RNA template if it is provided with a primer that is base-paired to the RNA and contains a free 3′-OH group. We can use this enzyme to synthesize DNA from mRNA by providing an oligo-dT primer that pairs with the poly-A sequence at the 3′ end of most eucaryotic mRNA molecules (Figure 6-23). The rest of the cDNA strand is then synthesized in the presence of the four deoxyribonucleoside triphosphates. The RNA strand of this RNA-DNA hybrid is subsequently hydrolyzed by raising the pH. Unlike RNA, DNA is resistant to alkaline hydrolysis. The 3′-end of the newly-synthesized DNA strand then forms a hairpin loop and primes the synthesis of the opposite DNA strand. The hairpin loop is removed by digestion with S1 nuclease, which recognizes unpaired nucleotides. Synthetic linkers can be added to this double-helical DNA for ligation to a suitable vector.

cDNA molecules can be inserted into vectors that favor their efficient expression in hosts such as *E. coli*. Such plasmids or phages are called *expression vectors*. To maximize transcription, the cDNA is inserted into the vector in the correct reading frame near a strong bacterial promoter. In addition, these vectors assure efficient translation by encoding a ribosome-binding site on the mRNA near the initiation codon. *cDNA clones can be screened on the basis of their capacity to direct the synthesis of a foreign protein in bacteria.* A radioactive antibody specific for the protein of interest can be used to identify the colonies of bacteria that harbor the corresponding cDNA vector (Figure 6-24). As before, spots of bacteria on a replica plate are lysed to release proteins, which bind to an applied nitrocellulose filter. A ^{125}I-labeled antibody is added, and autoradiography reveals the location of the desired colonies on the master plate. This *immunochemical screening* approach can be used whenever a protein is expressed and corresponding antibody is available.

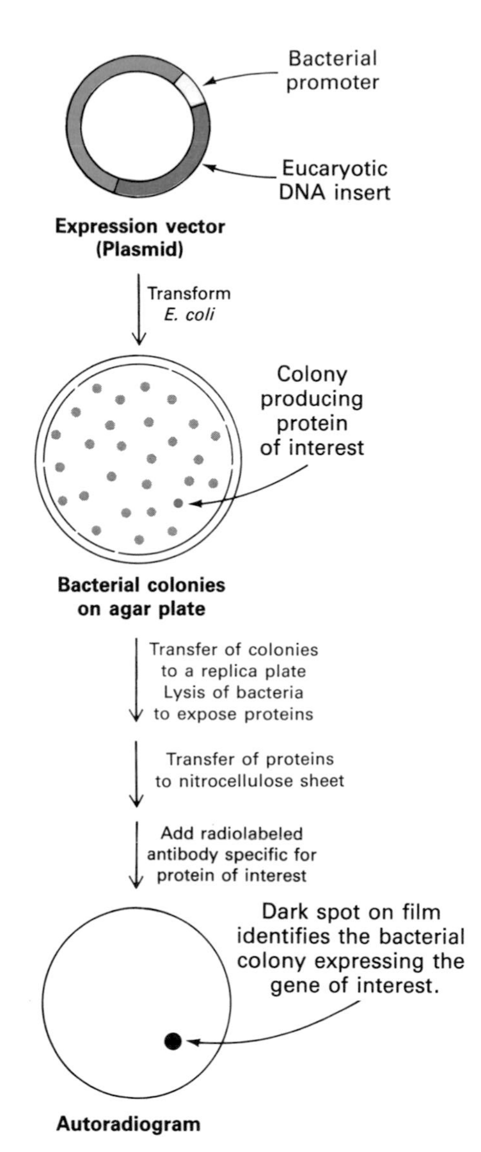

Figure 6-24
Screening of expression vector products by staining with specific antibody.

NEW GENES INSERTED INTO EUCARYOTIC CELLS CAN BE EFFICIENTLY EXPRESSED

Bacteria are ideal hosts for the amplification of DNA molecules. They can also serve as factories for the production of a wide range of procaryotic and eucaryotic proteins. However, posttranslational modifications such as specific cleavages of polypeptides and attachment of carbohydrate units are not carried out by bacteria, because they lack the necessary enzymes. Thus, many eucaryotic genes can be correctly expressed only in eucaryotic host cells. Another motivation for introducing recombinant DNA molecules into cells of higher organisms is to gain insight into how their genes are organized and expressed: How are genes turned on and off in embryological development? How does a fertilized egg give rise to an organism with highly differentiated cells that are organized in space and time? These central questions of biology can now be fruitfully approached because it has become feasible to express foreign genes in mammalian cells as well as in bacteria.

Recombinant DNA molecules can be introduced into animal cells in several ways. In one, foreign DNA molecules precipitated by *calcium phosphate* are taken up by animal cells. A small fraction of the imported DNA becomes stably integrated into the chromosomal DNA. The efficiency of incorporation is low but the method is useful because it is easy to apply. In another method, DNA is injected into cells. A fine-tipped (0.1 μm diameter) glass micropipet containing a solution of foreign DNA is inserted into a nucleus (Figure 6-25). A skilled investigator can inject hundreds of cells per hour. About 2% of injected mouse cells are viable and contain the new gene. In a third method, *viruses* can be used to bring new genes into animal cells. The most effective vectors are

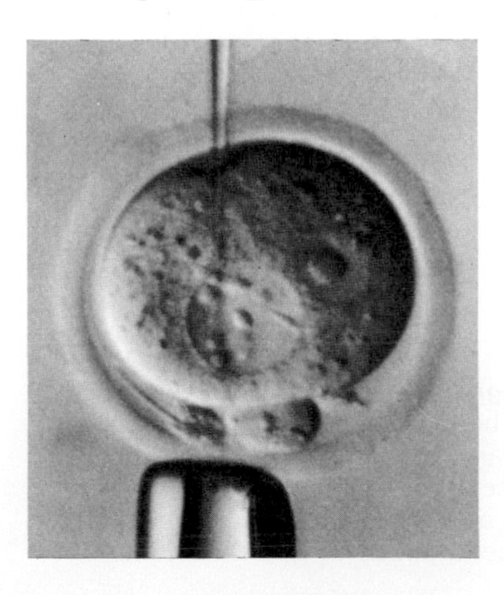

Figure 6-25
Microinjection of cloned plasmid DNA into the male pronucleus of a fertilized mouse egg.

retroviruses (RNA tumor viruses) (Figure 6-26). As discussed earlier (p. 87), these viruses replicate through DNA intermediates, the reverse of the normal flow of information (hence the prefix *retro*). A striking feature of the life cycle of a retrovirus is that the double-helical DNA form of its genome, produced by the action of reverse transcriptase, becomes randomly incorporated into host chromosomal DNA. This DNA version of the viral genome, called proviral DNA, can be efficiently expressed by the host cell and replicated along with normal cellular DNA (Chapter 34). Retroviruses do not usually kill their hosts. Foreign genes have been efficiently introduced into mammalian cells by infecting them with vectors derived from *Moloney murine leukemia virus,* which can accept inserts as long as 6 kb. Some genes introduced by this retroviral vector into the genome of a transformed host cell are efficiently expressed.

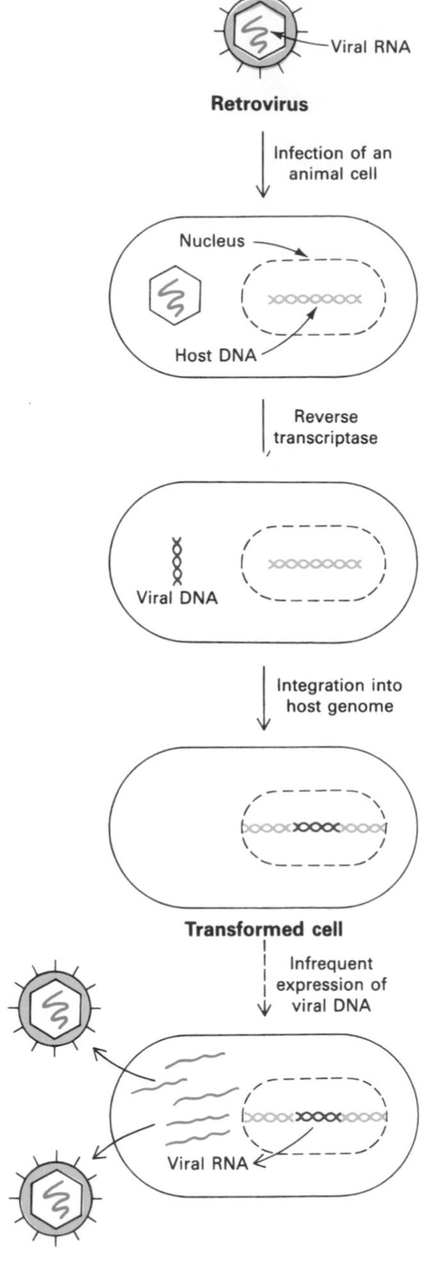

Figure 6-26
Life cycle of a retrovirus.

Genetically engineered giant mice (Figure 6-27) illustrate the expression of foreign genes in mammalian cells. *Growth hormone* (somatotropin), a 21-kd protein, is normally synthesized by the pituitary gland. A deficiency of this hormone produces dwarfism and an excess leads to gigantism. The gene for rat growth hormone was placed next to the mouse metallothionein promoter on a plasmid (Figure 6-28). This promoter is normally located on a chromosome, where it controls the transcription of metallothionein, a cysteine-rich protein that has high affinity for heavy metals. The synthesis of this protective protein by the liver is induced by heavy-metal ions such as cadmium. Several hundred copies of this plasmid were microinjected into the male pronucleus of a fertilized mouse egg, which was then inserted into the uterus of a foster-mother mouse. A number of mice that developed from these microinjected eggs contained the gene for rat growth hormone, as shown by Southern blots of their DNA. These *transgenic mice*, containing multiple copies (~30 per cell) of the rat growth hormone gene, grew much more rapidly than did control mice. The level of growth hormone in these mice was 500 times as high as in normal mice and their body weight at maturity was twice normal. The foreign DNA had been transcribed and its five introns correctly spliced out to form functional mRNA. *These experiments strikingly demonstrate that a foreign gene under the control of a new promoter can be integrated and efficiently expressed in mammalian cells.*

Figure 6-27
Injection of the gene for growth hormone into a fertilized mouse egg gave rise to giant mouse (left), about twice the weight of his sibling (right). [Courtesy of Dr. Ralph Brinster.]

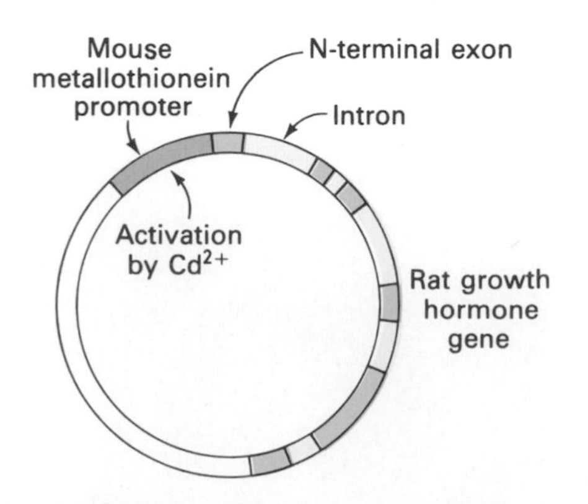

Figure 6-28
The gene for rat growth hormone was inserted into a plasmid next to the metallothionein promoter, which is activated by the addition of heavy metals, such as cadmium ion.

TUMOR-INDUCING (Ti) PLASMIDS CAN BE USED TO BRING NEW GENES INTO PLANT CELLS

The common soil bacterium *Agrobacterium tumefaciens* infects plants and introduces foreign genes into them (Figure 6-29). A lump of tumor tissue called a *crown gall* grows at the site of infection. Crown galls synthesize opines, a group of amino acid derivatives that are metabolized by the infecting bacteria. In essence, the metabolism of the plant cell is diverted to satisfy the highly distinctive appetite of the intruder. The instructions for the synthesis of opines and the switch to the tumor state come from *Ti plasmids* (tumor-inducing plasmids) that are carried by *Agrobacterium*. A small portion of the Ti plasmid becomes integrated into the genome of infected plant cells; this 20-kb segment is called *T-DNA* (transferred DNA).

Ti plasmid derivatives can be used as vectors to deliver foreign genes into plant cells (Figure 6-30). First, a segment of foreign DNA is inserted into the T-DNA region of a small plasmid by restriction enzymes and ligases. This synthetic plasmid is added to *Agrobacterium* colonies harboring naturally occurring Ti plasmids. By recombination, Ti plasmids containing the foreign gene are formed. These Ti vectors hold great promise for exploring the genomes of plant cells and modifying plants to improve their agricultural value and crop yield. However, they are not suitable for transforming all types of plants. Ti-plasmid transfer works with dicots (broad-leaved plants such as grapes) and a few kinds of monocots but not with economically important cereal monocots.

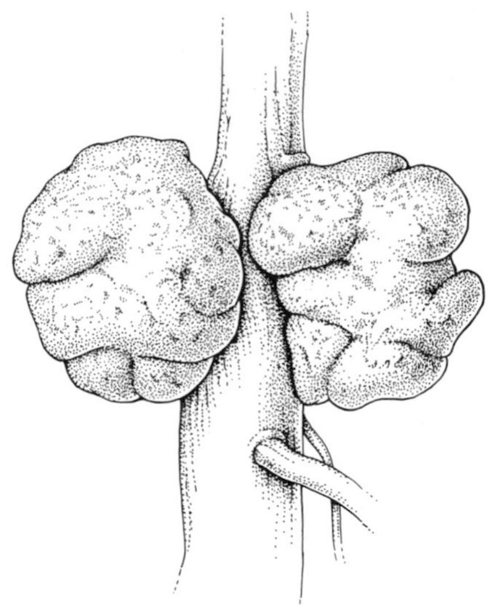

Figure 6-29
Crown gall, a plant tumor, is caused by a bacterium (*Agrobacterium tumefaciens*) that carries a tumor-inducing plasmid (Ti plasmid).

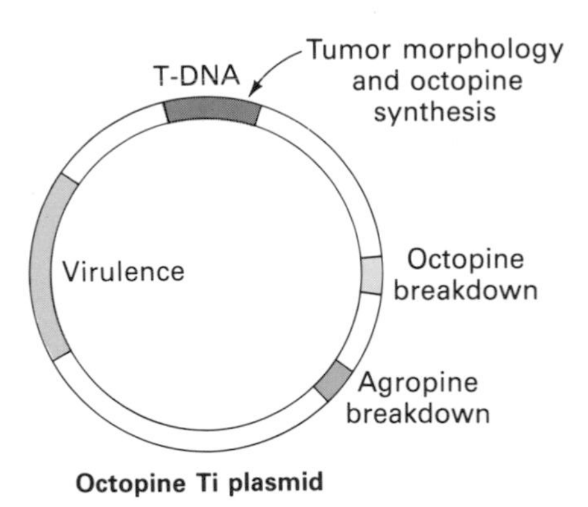

Figure 6-30
Agrobacteria containing Ti plasmids can deliver foreign genes into some plant cells. [After M. Chilton. A vector for new genes into plant. Copyright © 1983 by Scientific American, Inc. All rights reserved.]

Foreign DNA has recently been introduced into cereal monocots as well as dicots by applying intense electric fields, a technique called *electroporation* (Figure 6-31). First, the cellulose wall surrounding plant cells is removed by adding cellulases; this produces *protoplasts*, plant cells with exposed plasma membranes. Electric pulses then are applied to a suspension of protoplasts and plasmid DNA. Because high electric fields make membranes transiently permeable to large molecules, plasmid DNA molecules enter the cells. The cell wall is then allowed to reform, which results in viable plant cells. Maize cells and carrot cells have been stably transformed in this way with plasmid DNA that includes genes for resistance to antibiotics. Moreover, the plasmid DNA is efficiently expressed by the transformed cells.

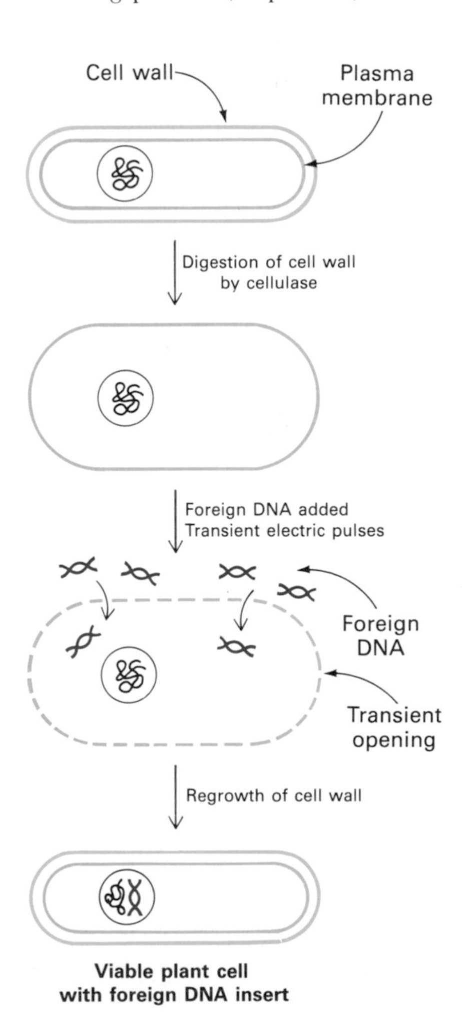

Figure 6-31
Foreign DNA can be introduced into plant cells by electroporation, applying intense electric fields to make their plasma membranes transiently permeable.

NOVEL PROTEINS CAN BE ENGINEERED BY SITE-SPECIFIC MUTAGENESIS

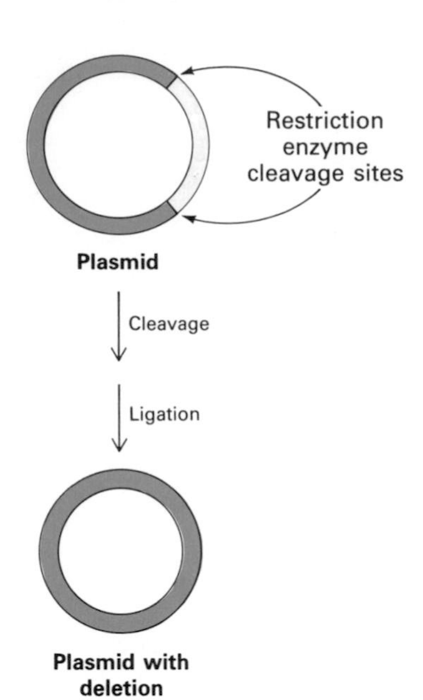

Figure 6-32
Deletions of parts of genes can be produced by cutting a plasmid at a pair of sites with a restriction enzyme (or a pair of enzymes) and religating. A set of smaller deletions can be produced by cutting a plasmid at a single site, digesting the ends to different extents with an exonuclease, and religating.

Much has been learned about genes and proteins by selecting genes from the repertoire offered by nature. In the classic genetic approach, mutations are generated randomly throughout the genome and those exhibiting a particular phenotype are selected. Analysis of these mutants then reveals which genes are altered, and DNA sequencing identifies the precise nature of the changes. *Recombinant DNA technology now makes it feasible to create specific mutations in vitro.* We can construct new genes with designed properties by four kinds of directed changes: *deletion*, *insertion*, *transposition*, and *substitution*.

A specific deletion can be produced by cleaving a plasmid at two sites with a restriction enzyme and religating to form a smaller circle (Figure 6-32). This simple approach usually removes a large block of DNA. A smaller deletion can be made by cutting a plasmid at a single site. The ends of the linear DNA are then digested with an exonuclease that removes nucleotides from both strands. The shortened piece of DNA is then religated to form a circle that is missing a short length of DNA about the restriction site.

Mutant proteins with single amino acid substitutions can readily be produced by *oligonucleotide-directed mutagenesis* (Figure 6-33). Suppose that we want to replace a particular serine residue with cysteine. This mutation can be made if (1) we have a plasmid containing the gene or cDNA for this protein and (2) we know the base sequence in the vicinity of the site to be altered. If the serine of interest is encoded by TCT, we need to change the C to a G to get cysteine, which is encoded by TGT. The key to this mutation is to prepare an oligonucleotide primer that is complementary to this region of the gene except that it contains TGT instead of TCT. The two strands of the plasmid are separated and the primer is then annealed to the complementary strand. (The mismatch of one base pair out of fifteen is tolerable if the annealing is carried out at an appropriate temperature. An attractive feature of nucleic acid hybridization is that its *stringency*—the required closeness of the match—can be experimentally controlled by choice of temperature and ionic strength.) The primer is then elongated by DNA polymerase, and the double-stranded circle is closed by adding DNA ligase. Subsequent replication of this duplex yields two kinds of progeny plasmid, half with the original TCT sequence and half with the mutant TGT sequence. Expression of the plasmid containing the new TGT sequence will produce a protein with the desired substitution of serine for cysteine at a unique site. We will encounter many examples of the use of oligonucleotide-directed mutagenesis to precisely alter regulatory regions of genes and to produce proteins with tailor-made features.

Novel proteins can also be created by splicing gene segments that encode domains that are not associated in nature. For example, a gene for an antibody can be joined to a gene for an enzyme to produce a

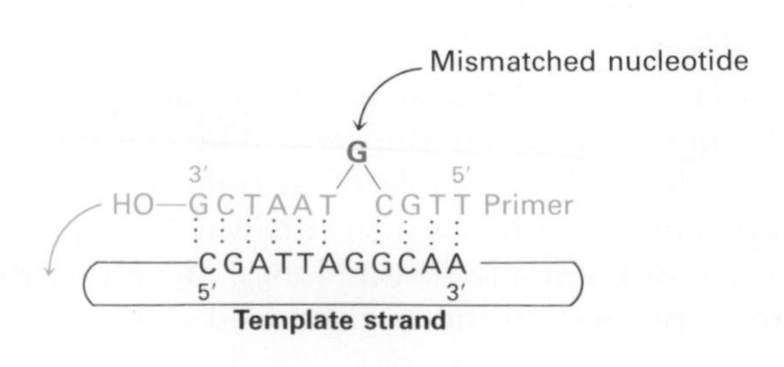

Figure 6-33
Oligonucleotide-directed mutagenesis. A primer containing a mismatched nucleotide is used to produce a desired point mutation.

chimeric protein that could be useful as a therapeutic agent. Moreover, entirely new genes can be synthesized de novo by the solid-phase method. There is much interest in using them to direct the formation of synthetic vaccines that could be safer than conventional vaccines prepared by inactivating pathogenic viruses.

RECOMBINANT DNA TECHNOLOGY HAS OPENED NEW VISTAS

The analysis of the molecular basis of life has been revolutionized by recombinant DNA technology. Complex chromosomes are rapidly being mapped and dissected into units that can be manipulated and deciphered. The amplification of genes by cloning has provided abundant quantities of DNA for sequencing. Genes are now open books that can be read. New insights are emerging, as exemplified by the discovery of introns in eucaryotic genes. Central questions of biology, such as the molecular basis of development, are now being fruitfully explored. The reading of the genome opens a new record of evolution. Biochemists now move back and forth between gene and protein and feel at home in both areas of inquiry.

Analyses of genes and cDNA can reveal the existence of previously unknown proteins, which can be isolated and purified (Figure 6-34A). Conversely, purification of a protein can be the starting point for the isolation and cloning of its gene or cDNA (Figure 6-34B). Very small amounts of protein or nucleic acid suffice because of the sensitivity of recently developed microchemical techniques and the amplification afforded by gene cloning. The powerful techniques of protein chemistry, nucleic acid chemistry, immunology, and molecular genetics are highly synergistic.

New kinds of proteins can be created by altering genes in specific ways. Site-specific mutagenesis opens the door to understanding how proteins fold, recognize other molecules, catalyze reactions, and process information. Large amounts of proteins can be obtained by expressing cloned genes or cDNAs in bacteria. Hormones such as insulin and antiviral agents such as interferon are being produced by bacteria. A

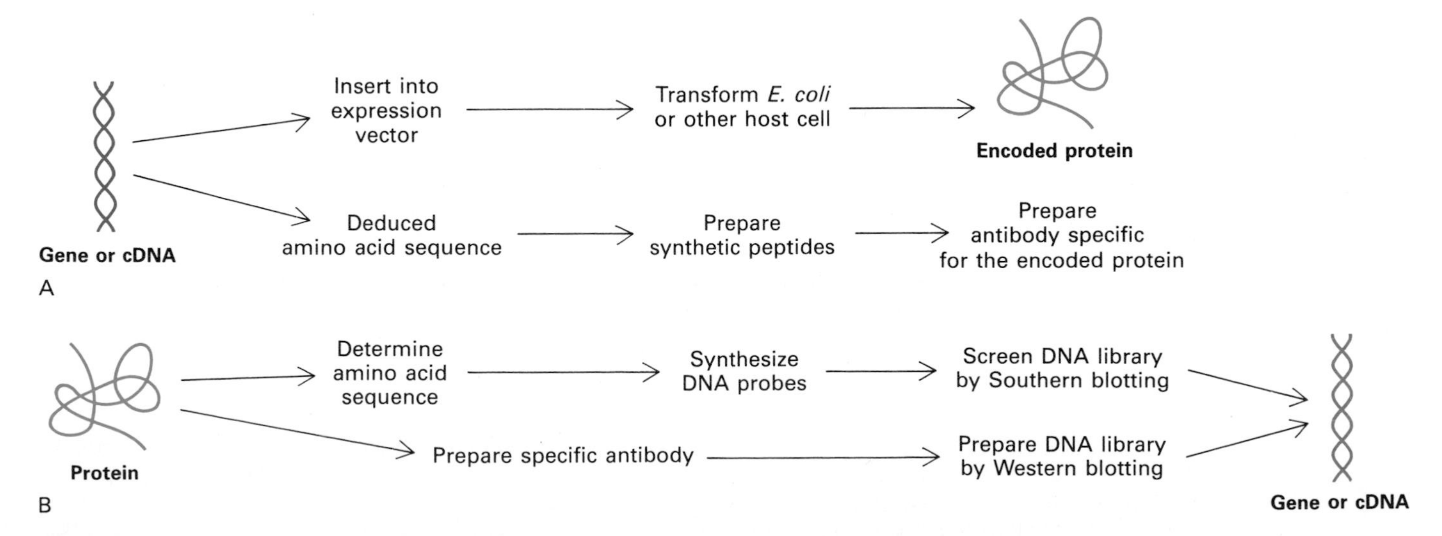

Figure 6-34
The techniques of protein chemistry and nucleic acid chemistry are mutually reinforcing. (A) From DNA (or RNA) to protein, and (B) from protein to DNA.

new pharmacology that will profoundly alter medicine has come into being. Recombinant DNA technology is also providing highly specific diagnostic reagents, such as DNA probes for the detection of genetic diseases, infections, and cancers. Retroviruses are being used to deliver missing genes into mice with genetic diseases, and human gene therapy will probably be initiated in the near future. Agriculture, too, will almost certainly benefit in the near future from the capacity to construct new genes and introduce them into eucaryotic cells. The genetic engineering of plants is likely to lead to crops that are more resistant to drought and have greater nutritional value.

SUMMARY

The recombinant DNA revolution in biology is rooted in the repertoire of enzymes that act on nucleic acids. Restriction enzymes, a group of key reagents, are endonucleases that recognize specific base sequences in double-helical DNA and cleave both strands of the duplex. Specific fragments of DNA formed by the action of restriction enzymes can be separated and displayed by gel electrophoresis. The pattern of these restriction fragments is a fingerprint of a DNA molecule. A DNA fragment containing a particular sequence can be identified by hybridizing it with a labeled single-stranded DNA probe (Southern blotting). Analysis of DNA molecules has also advanced by the development of rapid sequencing techniques. DNA can be sequenced by chemical cleavage at particular bases (Maxam-Gilbert method) or by controlled interruption of replication (Sanger dideoxy method). The fragments produced by either method are separated by gel electrophoresis and visualized by autoradiography of a ^{32}P label at the 5′ end or by fluorescent tags. DNA probes for hybridization reactions, as well as new genes, can be synthesized by the sequential addition of deoxyribonucleoside 3′-phosphoramidites to a growing chain that is linked to an insoluble support. DNA chains a hundred nucleotides long can readily be synthesized by this automated solid-phase method.

New genomes can be constructed in the laboratory, introduced into host cells, and expressed. Novel DNA molecules are made by joining fragments that have complementary cohesive ends produced by the action of a restriction enzyme. DNA ligase joins the ends of DNA chains if they are within a double helix. Plasmids (circular accessory chromosomes) and λ phage are choice vectors for cloning DNA in bacteria. Specific genes can be cloned from a digest of DNA. This genomic library can be screened with a complementary DNA or RNA probe. Alternatively, one can form cDNA from mRNA by the action of reverse transcriptase. cDNA inserted into expression vectors that contain strong promoters is efficiently expressed by bacteria. Foreign DNA can be carried into mammalian cells by a retrovirus or be directly injected. The production of large mice by injecting the gene for rat growth hormone into fertilized mouse eggs shows that mammalian cells can be genetically altered in a designed way. New DNA can be brought into plant cells by the soil bacterium *Agrobacterium tumefaciens*, which harbors Ti (tumor-inducing) plasmids. DNA can also be introduced into cells by applying intense electric fields, which render the cells transiently permeable to very large molecules.

Novel proteins can be engineered by generating specific mutations in vitro. A mutant protein with a single amino acid substitution can be produced by priming DNA replication with an oligonucleotide encoding the new amino acid. The techniques of protein and nucleic acid chemistry are highly synergistic. Investigators now move back and forth between gene and protein with great facility.

SELECTED READINGS

Where to start

Berg, P., 1981. Dissections and reconstructions of genes and chromosomes. *Science* 213:296–303.

Gilbert, W., 1981. DNA sequencing and gene structure. *Science* 214:1305–1312.

Nathans, D., 1979. Restriction endonucleases, simian virus 40, and the new genetics. *Science* 206:903–909.

Sanger, F., 1981. Determination of nucleotide sequences in DNA. *Science* 214:1205–1210.

Books on recombinant DNA technology

Watson, J. D., Tooze, J., and Kurtz, D. T., 1983. *Recombinant DNA: A Short Course.* Scientific American Books. [A highly readable and lucid introduction.]

Old, R. W., and Primrose, S. B., 1985. *Principles of Gene Manipulation: An Introduction to Genetic Engineering.* Blackwell Scientific Publications.

Mantell, S. H., Matthews, J. A., and McKee, R. A., 1985. *Principles of Plant Biotechnology: An Introduction to Genetic Engineering in Plants.* Blackwell Scientific Publications.

Inouye, M., (ed.), 1983. *Experimental Manipulation of Gene Expression.* Academic Press.

Maniatis, T., Fritsch, E. F., and Sambrook, J., 1982. *Molecular Cloning: A Laboratory Manual.* Cold Spring Harbor Laboratory.

Automated synthesis of DNA

Caruthers, M. H., 1985. Gene synthesis machines: DNA chemistry and its uses. *Science* 230:281–285.

Itakura, K., Rossi, J. J., and Wallace, R. B., 1984. Synthesis and use of synthetic oligonucleotides. *Ann. Rev. Biochem.* 53:323–356.

Hunkapiller, M., Kent, S., Caruthers, M., Dreyer, W., Firca, J., Giffin, C., Horvath, S., Hunkapiller, T., Tempst, P., and Hood, L., 1984. A microchemical facility for the analysis and synthesis of genes and proteins. *Nature* 310:105–111.

DNA sequencing

Sanger, F., Nicklen, S., and Coulson, A. R., 1977. DNA sequencing with chain-terminating inhibitors. *Proc. Nat. Acad. Sci.* 74:5463–5467.

Maxam, A. M., and Gilbert, W., 1977. A new method for sequencing DNA. *Proc. Nat. Acad. Sci.* 74:560–564.

Smith, L. M., Sanders, J. Z., Kaiser, R. J., Hughes, P., Dodd, C., Connell, C. R., Heiner, C., Kent, S. B. H., and Hood, L. E., 1986. Fluorescence detection in automated DNA sequence analysis. *Nature* 321:674–679.

Origins of DNA cloning

Jackson, D. A., Symons, R. H., and Berg, P., 1972. Biochemical method for inserting new genetic information into DNA of simian virus 40: circular SV40 DNA molecules containing lambda phage genes and the galactose operon of *Escherichia coli. Proc. Nat. Acad. Sci.* 69:2904–2909.

Lobban, P. E., and Kaiser, A. D., 1973. Enzymatic end-to-end joining of DNA molecules. *J. Mol. Biol.* 78:453–471.

Cohen, S. N., Chang, A., Boyer, H., and Helling, R., 1973. Construction of biologically functional bacterial plasmids in vitro. *Proc. Nat. Acad. Sci.* 70:3240–3244.

Cohen, S. N., 1985. DNA cloning: historical perspectives. *Biogenetics of Neurohormonal Peptides*, pp. 3–14. Academic Press.

Watson, J. D., and Tooze, J., 1981. *The DNA Story.* W. H. Freeman.

Protein production by recombinant bacteria

Pestka, S., 1983. The purification and manufacture of human interferons. *Sci. Amer.* 249(2):36.

Gilbert, W., and Villa-Komaroff, L., 1980. Useful proteins from recombinant bacteria. *Sci. Amer.* 242(4):74–94. [Available as *Sci. Amer.* Offprint 1466.]

Johnson, I. S., 1983. Human insulin from recombinant DNA technology. *Science* 219:632–637.

Introduction of genes into animal cells

Brinster, R. L., and Palmiter, R. D., 1986. Introduction of genes into the germ lines of animals. *Harvey Lectures* 80:1–38. [The production of large mice by injection of the growth hormone gene is discussed in this review.]

Karlsson, S., Humphries, R. K., Gluzman, Y., and Nienhuis, A. W., 1985. Transfer of genes into hematopoietic cells using recombinant DNA viruses. *Proc. Nat. Acad. Sci.* 82:158–162.

Anderson, W. F., 1984. Prospects for human gene therapy. *Science* 226:401–409.

Cepko, C. L., Roberts, B. E., and Mulligan, R. C., 1984. Construction and applications of a highly transmissible murine retrovirus shuttle vector. *Cell* 37:1053–1062.

Friedman, R. L., 1985. Expression of human adenosine deaminase using a transmissible murine retrovirus vector system. *Proc. Nat. Acad. Sci.* 82:703–707.

Introduction of genes into plant cells

Chilton, M-D., 1983. A vector for introducing new genes into plants. *Sci. Amer.* 248(6):50.

Hooykaas, P. J. J., and Schilperoort, R. A., 1985. The Ti-plasmid of *Agrobacterium tumefaciens*: a natural genetic engineer. *Trends Biochem. Sci.* 10:307–309.

Fromm, M. E., Taylor, L. P., and Walbot, V., 1986. Stable transformation of maize after gene transfer by electroporation. *Nature* 319:791–793.

Site-specific mutagenesis

Botstein, D., and Shortle, D., 1985. Strategies and applications of in vitro mutagenesis. *Science* 229:1193–1201.

Myers, R. M., Lerman, L. S., and Maniatis, T., 1985. A general method for saturation mutagenesis of cloned DNA fragments. *Science* 229:242–247.

PROBLEMS

1. An autoradiogram of a gel containing four lanes of DNA fragments produced by chemical cleavage is shown in Figure 6-35. The DNA contained a ^{32}P label at its 5′ end. What is its sequence?

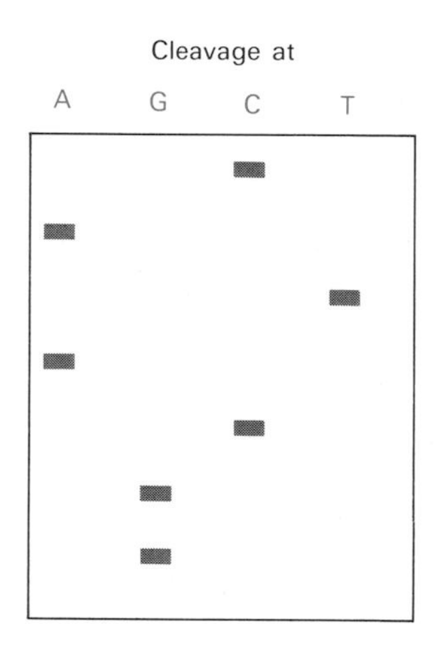

Figure 6-35
Schematic diagram of a gel showing the radioactive fragments formed by specific cleavage of an oligonucleotide.

2. Suppose that you determined the DNA sequence of 5′-GCCATTGCA-3′ by the Sanger dideoxy method. Sketch the gel pattern that revealed the sequence of this oligonucleotide.

3. Ovalbumin is the major protein of egg white. The chicken ovalbumin gene contains eight exons separated by seven introns. Should one use ovalbumin cDNA or ovalbumin genomic DNA to form the protein in *E. coli*? Why?

4. Suppose that a human genomic library were prepared by exhaustive digestion of human DNA with the EcoRI restriction enzyme. Fragments averaging about 4 kb in length would be generated.
   (a) Would this procedure be desirable for cloning large genes? Why?
   (b) Would this procedure be desirable for mapping extensive stretches of the genome by chromosome walking? Why?

5. Sickle-cell anemia arises from a mutation in the gene for the β chain of human hemoglobin. The change from GAG to GTG in the mutant eliminates a cleavage site for the restriction enzyme MstII, which recognizes the target sequence CCTGAGG. These findings form the basis of a valuable diagnostic test for the presence of the sickle-cell gene. Propose a rapid diagnostic procedure for distinguishing between the normal and the mutant gene. Would a positive result with your test prove that the mutant contains GTG in place of GAG?

6. Thomas Cech showed that the ribosomal RNA (rRNA) precursor in the ciliate protozoan *Tetrahymena* can self-splice without binding any *Tetrahymena* protein. He cloned in a plasmid a region of *Tetrahymena* DNA consisting of the intron and flanking sequences present in the precursor RNA. Suggest how this plasmid was used to establish that *Tetrahymena* proteins are not required for the splicing of the ribosomal RNA precursor.

7. The introduction into a bacterial cell of a single M13 RF DNA molecule containing an inserted fragment of foreign DNA yields M13 progeny DNA containing only one strand of the foreign DNA.
   (a) Why is it important to establish which DNA strand is contained in a particular virus particle?
   (b) Suppose that the foreign DNA strand was inserted into M13 DNA using restriction enzyme 1 and that the foreign DNA has an internal site for restriction enzyme 2. How could one determine which strand of foreign DNA is contained in the virus particle?
   (c) Agarose gel electrophoresis can distinguish between single-stranded and double-stranded DNA. How would this method be used to determine whether two phage particles contain the same strand or different strands of foreign DNA? How could an enzyme be used to make this determination?

8. Suppose that a colleague purified from cardiac muscle a 30-residue peptide that modulates calcium transport across membranes. The peptide shows promise as a drug, but unfortunately it is rapidly degraded by proteolytic enzymes in blood. You are intrigued by this problem and decide that it would be worthwhile to use site-specific mutagenesis to engineer a series of modified peptides in the hope that one of them will retain biological activity and be resistant to proteolytic degradation. What information is essential before you can begin to prepare new peptides by molecular genetic techniques? What other information would be helpful in this project?

PART II

PROTEIN CONFORMATION, DYNAMICS, AND FUNCTION

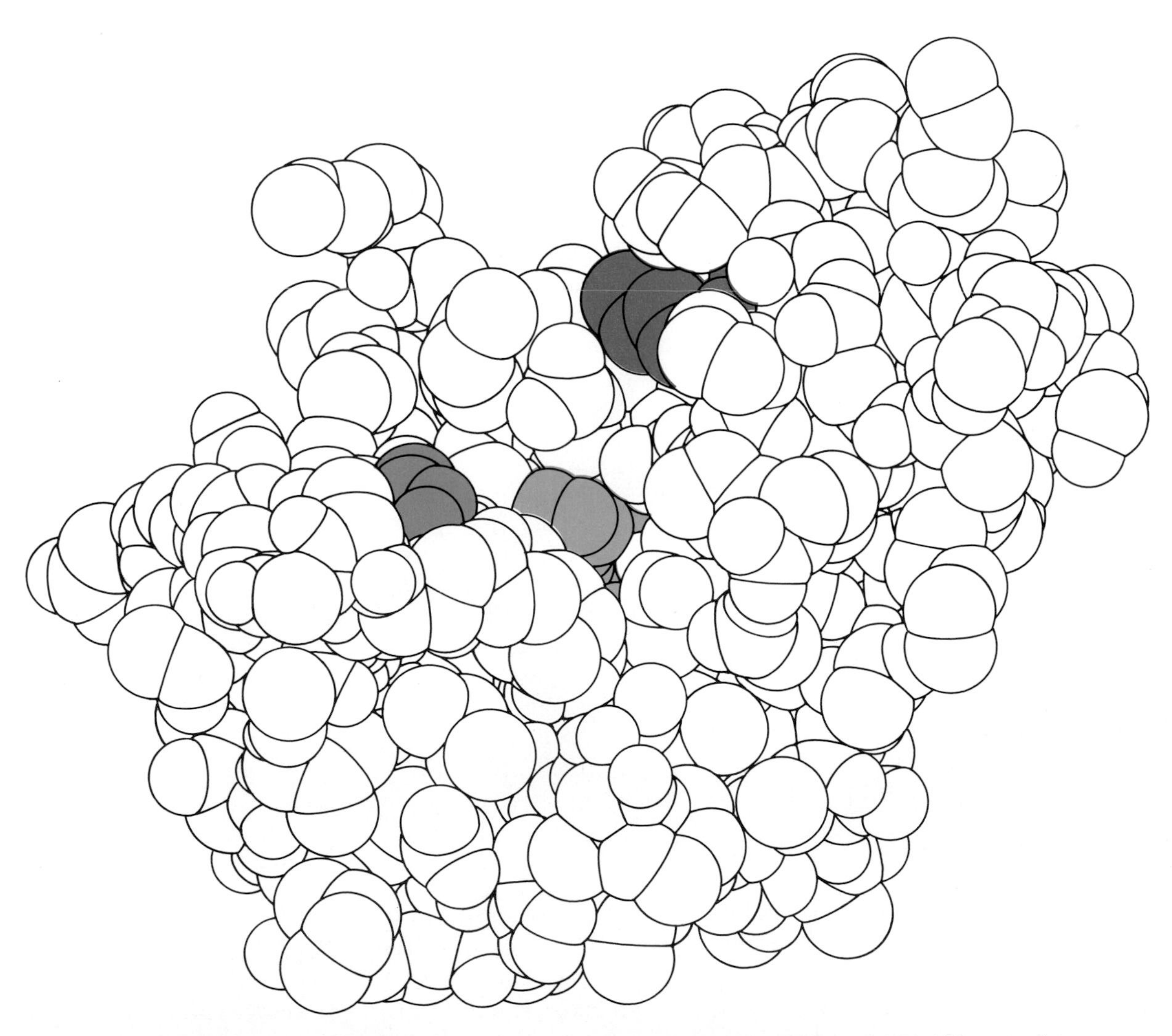

Model of ribonuclease S, an enzyme that hydrolyzes ribonucleic acids. Amino acid residues at the active site that are critical for catalysis are shown in color. The three-dimensional structure of this enzyme was solved by Frederic Richards and Harold Wyckoff. [After a drawing kindly provided by Dr. Frederic Richards, Steven Anderson, and Arthur Perlo.]

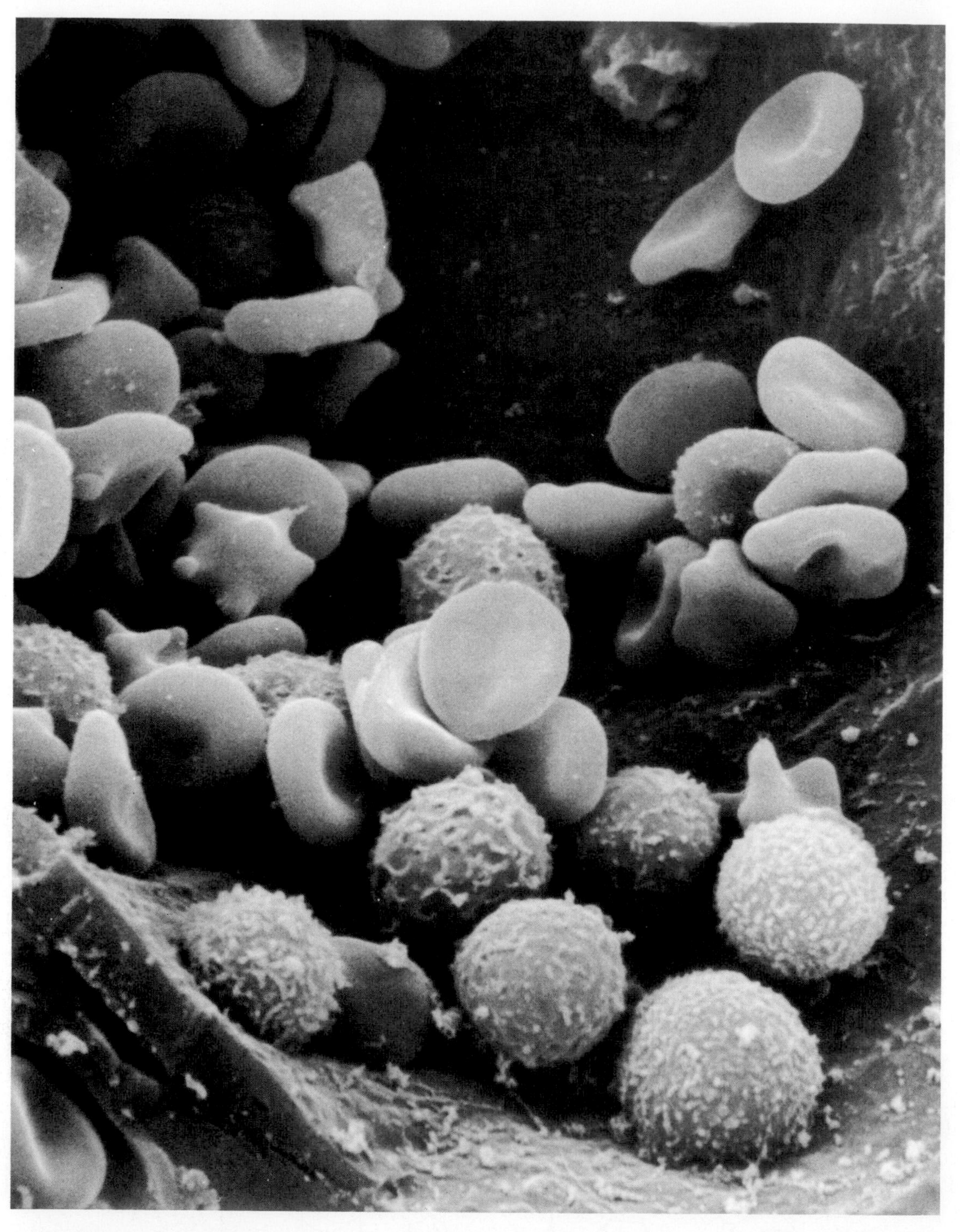

Scanning electron micrograph showing erythrocytes (biconcave-shaped) and leucocytes (rounded) in a small blood vessel. [From R. G. Kessel and R. H. Kardon. *Tissues and Organs*, W. H. Freeman. Copyright © 1979.]

CHAPTER 7

Oxygen-transporting Proteins: Myoglobin and Hemoglobin

The transition from anaerobic to aerobic life was a major step in evolution because it uncovered a rich reservoir of energy. Eighteen times as much energy is extracted from glucose in the presence of oxygen as in its absence. Vertebrates have evolved two principal mechanisms for supplying their cells with a continuous and adequate flow of oxygen. The first is a circulatory system that actively delivers oxygen to cells. The second is the use of *oxygen-carrying molecules* to overcome the limitation imposed by the low solubility of oxygen in water. The oxygen carriers in vertebrates are the proteins *hemoglobin* and *myoglobin*. Hemoglobin, which is contained in red blood cells, serves as the oxygen carrier in blood and also plays a vital role in the transport of carbon dioxide and hydrogen ion. Myoglobin, which is located in muscle, serves as a reserve supply of oxygen and facilitates the movement of oxygen within muscle.

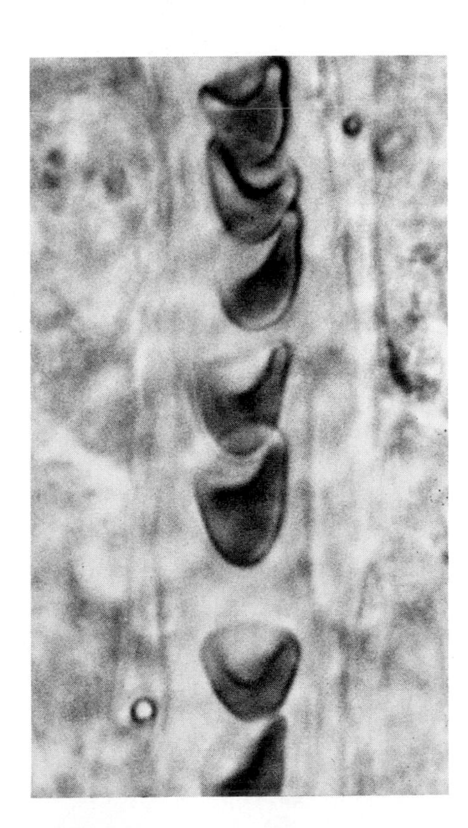

Figure 7-1
Erythrocytes flowing through a small blood vessel. [From P. I. Brånemark. *Intravascular Anatomy of Blood Cells in Man* (Basel: S. Karger AG, 1971).]

Part II of this book begins with myoglobin and hemoglobin because they illustrate many important principles of protein conformation, dynamics, and function. Their three-dimensional structures, known in atomic detail, reveal much about how proteins fold, bind other molecules, and integrate information. The binding of O_2 by hemoglobin is regulated by H^+, CO_2, and organic phosphates. These regulators greatly affect the oxygen-binding properties of hemoglobin by binding to sites on the protein far from where O_2 is bound. Indeed, interactions between separate sites, termed *allosteric interactions*, occur in many proteins. Allosteric effects play a critical role in controlling and integrating molecular events in biological systems. Hemoglobin is the best understood allosteric protein, so that examining its normal structure and function in some detail is rewarding. Furthermore, the discovery of mutant hemoglobins has revealed that disease can arise from a change

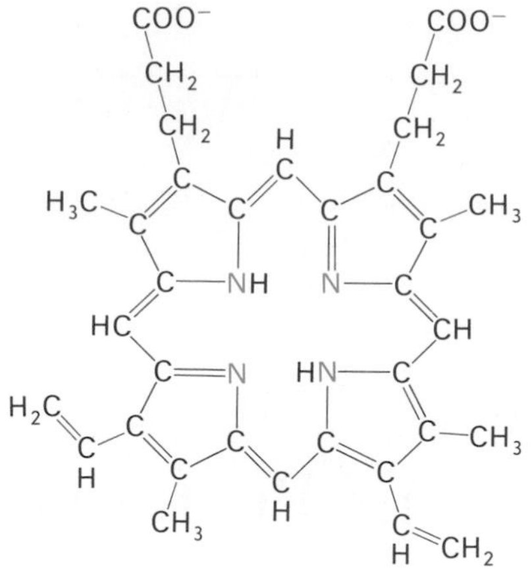

Protoporphyrin IX

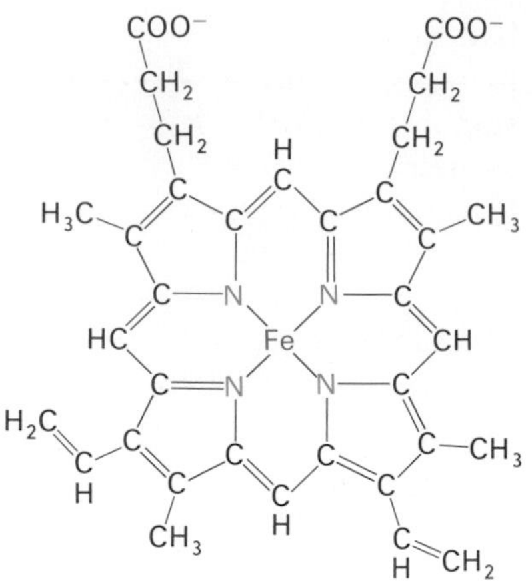

Heme
(Fe-protoporphyrin IX)

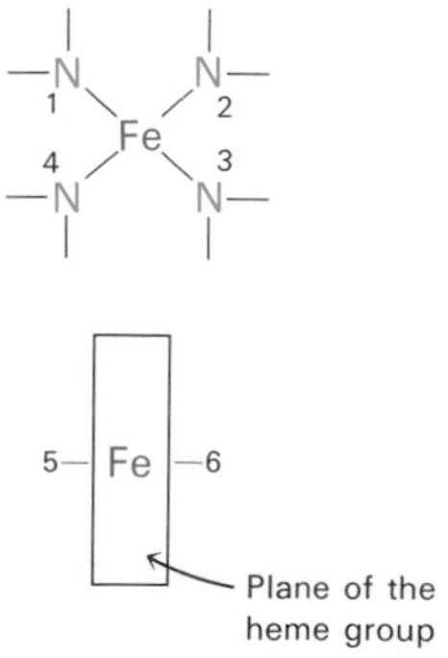

of a single amino acid in a protein. The concept of molecular disease, now an integral part of medicine, came from studies of the abnormal hemoglobin causing sickle-cell anemia. Hemoglobin has also been a rich source of insight into the molecular basis of evolution.

Plane of the heme group

Figure 7-2
The iron atom in heme can form six bonds.

OXYGEN BINDS TO A HEME PROSTHETIC GROUP

The capacity of myoglobin or hemoglobin to bind oxygen depends on the presence of a nonpolypeptide unit, namely, a *heme group*. The heme also gives myoglobin and hemoglobin their distinctive color. Indeed, many proteins require tightly bound, specific nonpolypeptide units for their biological activities. Such a unit is called a *prosthetic group*. A protein without its characteristic prosthetic group is termed an *apoprotein*.

The heme consists of an organic part and an iron atom. The organic part, *protoporphyrin*, is made up of four *pyrrole* rings. The four pyrroles are linked by methene bridges to form a tetrapyrrole ring. Four methyl, two vinyl, and two propionate side chains are attached to the tetrapyrrole ring. These substituents can be arranged in fifteen different ways. Only one of these isomers, called protoporphyrin IX, is present in biological systems.

The iron atom in heme binds to the four nitrogens in the center of the protoporphyrin ring (Figures 7-2 and 7-3). The iron can form two additional bonds, one on either side of the heme plane. These bonding sites are termed the fifth and sixth coordination positions. The iron atom can be in the ferrous (+2) or the ferric (+3) oxidation state, and the corresponding forms of hemoglobin are called *ferrohemoglobin* and *ferrihemoglobin*. Ferrihemoglobin is also called methemoglobin. Only ferrohemoglobin, the +2 oxidation state, can bind oxygen. The same nomenclature applies to myoglobin.

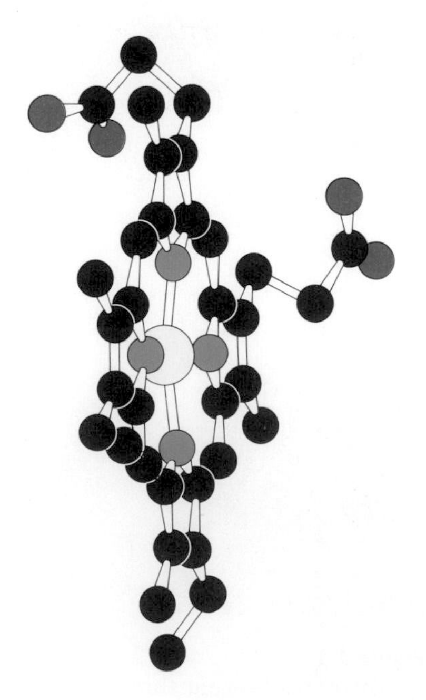

Figure 7-3
The heme group in myoglobin (yellow for Fe, blue for N, red for O, and black for C).

MYOGLOBIN WAS THE FIRST PROTEIN TO BE SEEN AT ATOMIC RESOLUTION

The elucidation of the three-dimensional structure of myoglobin by John Kendrew and of hemoglobin by Max Perutz are landmarks in molecular biology. These studies came to fruition in the late 1950s and showed that x-ray crystallography (p. 60) can reveal the structure of molecules as large as proteins. The largest structure solved before then

had been vitamin B_{12}, which is an order of magnitude smaller than myoglobin and hemoglobin. The determination of the three-dimensional structures of these proteins was a great stimulus to the field of protein crystallography. Kendrew chose myoglobin for x-ray analysis because it is relatively small, easily prepared in quantity, and readily crystallized. It had the additional advantage of being closely related to hemoglobin, which was already being studied by his colleague Perutz. Myoglobin from the skeletal muscle of the sperm whale was selected because it is stable and forms excellent crystals. The skeletal muscle of diving mammals, such as whale, seal, and porpoise, is particularly rich in myoglobin, which serves as a store of oxygen during a dive.

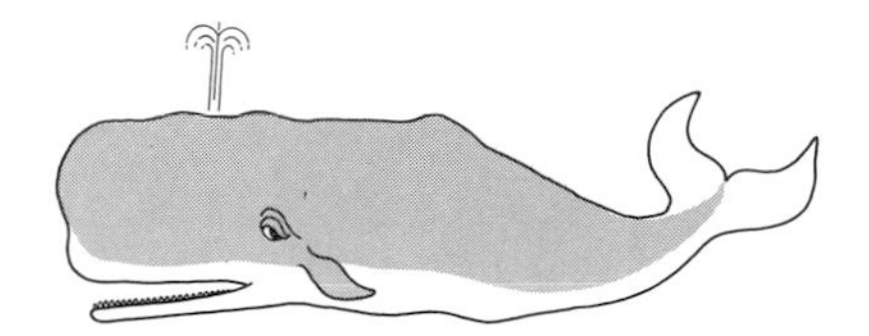

A sperm whale.

In 1957, Kendrew and his colleagues saw what no one had ever seen before: a three-dimensional picture of a protein molecule in all its complexity. The model derived from their Fourier synthesis at 6-Å resolution contained a set of high-density rods of just the dimensions expected for a polypeptide chain. The molecule appeared very compact. Closer examination showed that it consisted of a complicated and intertwining set of these rods, going straight for a distance, then turning a corner and going off in a new direction (Figure 7-4). The location of the iron atom of the heme was also evident because it contains many more electrons than does any other atom in the structure.

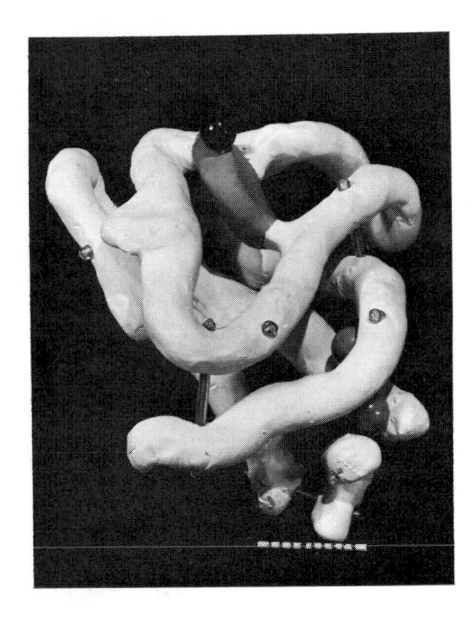

Figure 7-4
Model of myoglobin at low resolution. [Courtesy of Dr. John Kendrew.]

MYOGLOBIN HAS A COMPACT STRUCTURE AND A HIGH CONTENT OF ALPHA HELICES

The high-resolution electron-density map of myoglobin, obtained two years later, contained a wealth of structural detail. The positions of 1200 of the 1260 nonhydrogen atoms were clearly defined to a precision of better than 0.3 Å. The course of the main chain and the position of the heme group are shown in Figure 7-5. Some important features of myoglobin are:

1. Myoglobin is *extremely compact*. The overall dimensions are about 45 × 35 × 25 Å. There is rather little empty space inside.

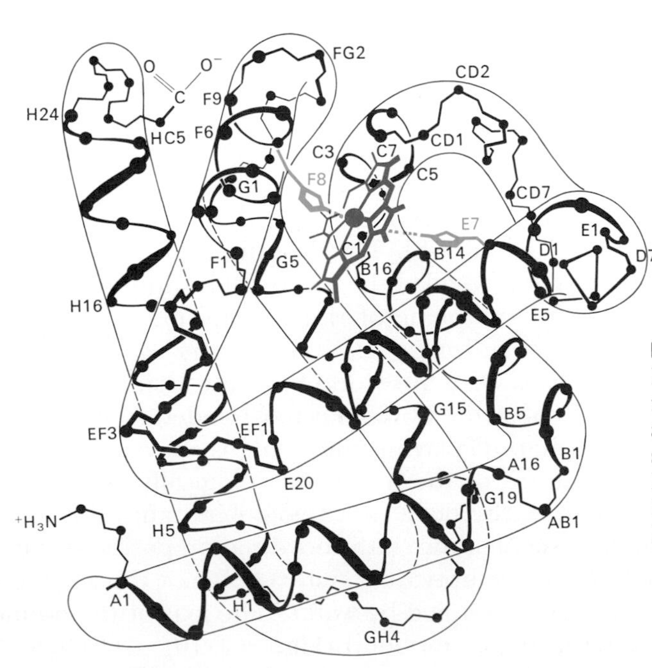

Figure 7-5
Model of myoglobin at high resolution. Only the α-carbon atoms are shown. The heme group is shown in red. [After R. E. Dickerson. In *The Proteins*, H. Neurath, ed., 2nd ed., (Academic Press, 1964), vol. 2, p. 634.]

2. About 75% of the main chain is in an *α-helical conformation.* The eight major helical segments, all right-handed, are referred to as A, B, C, . . . , H. The first residue in helix A is designated A1, the second A2, and so forth (Figure 7-6). Five nonhelical segments lie between helices (named CD, e.g., if located between the C and D helices). Myoglobin has two other nonhelical regions: two residues at the amino-terminal end (named NA1 and NA2) and five residues at the carboxyl-terminal end (named HC1 through HC5).

3. Four of the helices are terminated by a *proline* residue, whose five-membered ring simply does not fit within a straight stretch of α helix.

4. The main-chain peptide groups are *planar,* and the carbonyl group of each is *trans* to the NH. Also, the bond angles and distances are like those in dipeptides and other organic compounds.

5. The inside and outside are well defined. *The interior consists almost entirely of nonpolar residues* such as leucine, valine, methionine, and phenylalanine. In contrast, glutamic and aspartic acids, glutamine, asparagine, lysine, and arginine are absent from the interior of the protein. Residues that have both a polar and a nonpolar part, such as threonine, tyrosine, and tryptophan, are oriented so that their nonpolar portions point inward. The only polar residues inside myoglobin are two histidines, which have a critical function at the binding site. The outside of the protein has both polar and nonpolar residues.

Sequence	Position labels	
Val-Leu-Ser-Glu-Gly-Glu-Trp-Gln-Leu-Val-	NA1 NA2 A1 A2 A3 A4 A5 A6 A7 A8	10
Leu-His-Val-Trp-Ala-Lys-Val-Glu-Ala-Asp-	A9 A10 A11 A12 A13 A14 A15 A16 AB1 B1	20
Val-Ala-Gly-His-Gly-Gln-Asp- Ile -Leu- Ile -	B2 B3 B4 B5 B6 B7 B8 B9 B10 B11	30
Arg-Leu-Phe-Lys-Ser-His-Pro-Glu-Thr-Leu-	B12 B13 B14 B15 B16 C1 C2 C3 C4 C5	40
Glu-Lys-Phe-Asp-Arg-Phe-Lys-His-Leu-Lys-	C6 C7 CD1 CD2 CD3 CD4 CD5 CD6 CD7 CD8	50
Thr-Glu-Ala-Glu-Met-Lys-Ala-Ser-Glu-Asp-	D1 D2 D3 D4 D5 D6 D7 E1 E2 E3	60
Leu-Lys-Lys-His-Gly-Val-Thr-Val-Leu-Thr-	E4 E5 E6 E7 E8 E9 E10 E11 E12 E13	70
Ala-Leu-Gly-Ala- Ile -Leu-Lys-Lys-Lys-Gly-	E14 E15 E16 E17 E18 E19 E20 EF1 EF2 EF3	80
His-His-Glu-Ala-Glu-Leu-Lys-Pro-Leu-Ala-	EF4 EF5 EF6 EF7 EF8 F1 F2 F3 F4 F5	90
Gln-Ser-His-Ala-Thr-Lys-His-Lys- Ile -Pro-	F6 F7 F8 F9 FG1 FG2 FG3 FG4 FG5 G1	100
Ile -Lys-Tyr-Leu-Glu-Phe- Ile -Ser-Glu-Ala-	G2 G3 G4 G5 G6 G7 G8 G9 G10 G11	110
Ile - Ile -His-Val-Leu-His-Ser-Arg-His-Pro-	G12 G13 G14 G15 G16 G17 G18 G19 GH1 GH2	120
Gly-Asn-Phe-Gly-Ala-Asp-Ala-Gln-Gly-Ala-	GH3 GH4 GH5 GH6 H1 H2 H3 H4 H5 H6	130
Met-Asn-Lys-Ala-Leu-Glu-Leu-Phe-Arg-Lys-	H7 H8 H9 H10 H11 H12 H13 H14 H15 H16	140
Asp- Ile -Ala-Ala-Lys-Tyr-Lys-Glu-Leu-Gly-	H17 H18 H19 H20 H21 H22 H23 H24 HC1 HC2	150
Tyr-Gln-Gly	HC3 HC4 HC5	153

Figure 7-6
Amino acid sequence of sperm-whale myoglobin. The label below each residue in the sequence refers to its position in an α-helical region or a nonhelical region. For example, B4 is the fourth residue in the B helix; EF7 is the seventh residue in the nonhelical region between the E and F helices. [After A. E. Edmundson. *Nature* 205(1965):883; H. C. Watson. *Prog. Stereochem.* 4(1969):299–333.]

OXYGEN BINDS WITHIN A CREVICE IN MYOGLOBIN

The heme group is located in a crevice in the myoglobin molecule. The highly polar propionate side chains of the heme are on the surface of the molecule. At physiological pH, these carboxylic acid groups are ionized. The rest of the heme is inside the molecule, where it is surrounded by nonpolar residues except for two histidines. The iron atom of the heme is directly bonded to one of these histidines, namely, residue F8 (Figures 7-7 and 7-8). This histidine, which occupies the fifth coordination position, is called the proximal histidine. The iron atom is about 0.3 Å out of the plane of the porphyrin, on the same side as histidine F8. *The oxygen-binding site is on the other side of the heme plane, at the sixth coordination position.* A second histidine residue (E7), termed the distal histidine, is near the heme but not bonded to it. A section of an electron-density map displaying the heme is shown in Figure 7-9.

The conformations of the three physiologically pertinent forms of myoglobin—deoxymyoglobin, oxymyoglobin, and ferrimyoglobin—are very similar except at the sixth coordination position (Table 7-1). In deoxymyoglobin, it is empty; in oxymyoglobin, it is occupied by O_2; in ferrimyoglobin, it is occupied by water. The axis of the bound O_2 is at an angle to the iron-oxygen bond (Figure 7-10).

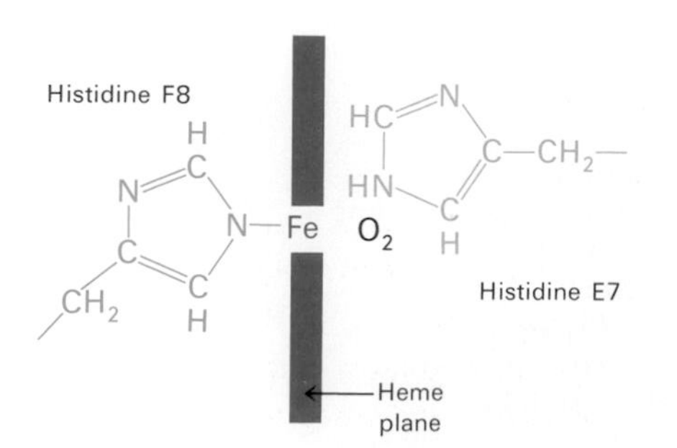

Figure 7-7
Schematic diagram of the oxygen-binding site in myoglobin.

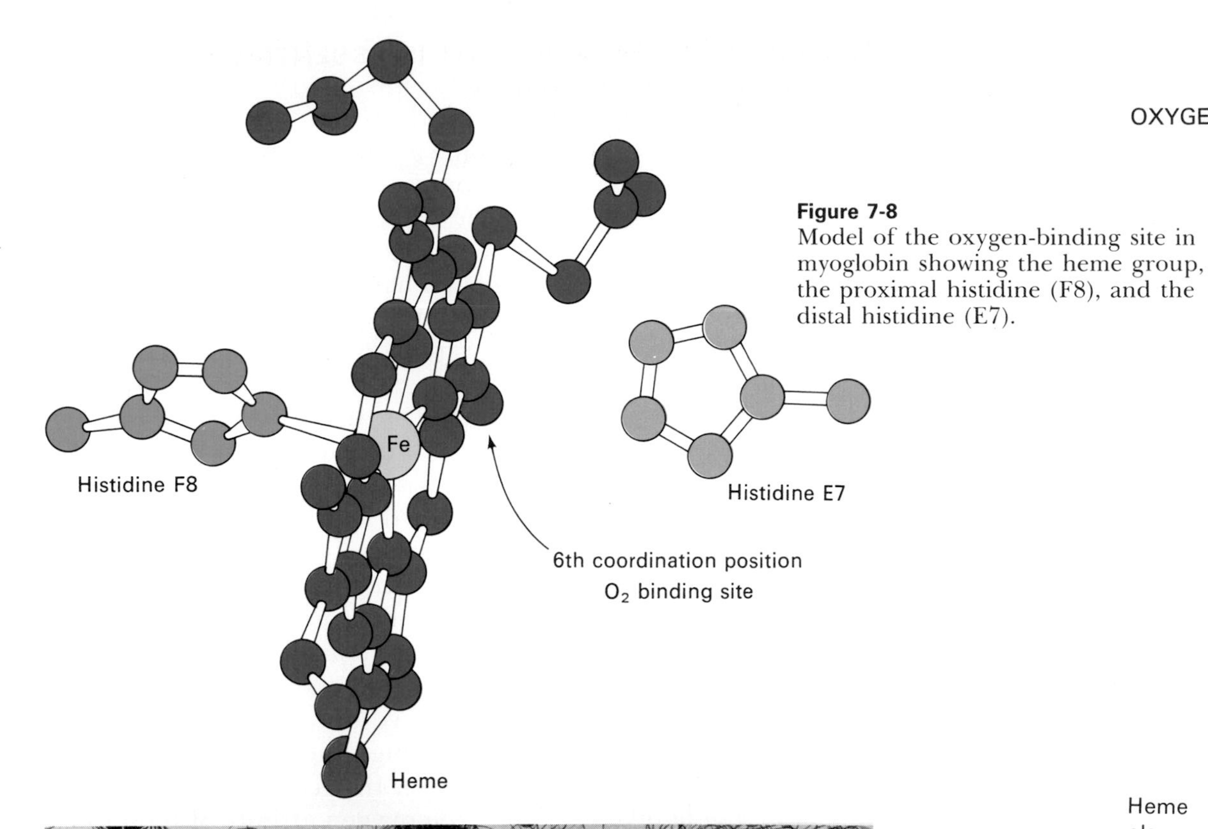

Figure 7-8
Model of the oxygen-binding site in myoglobin showing the heme group, the proximal histidine (F8), and the distal histidine (E7).

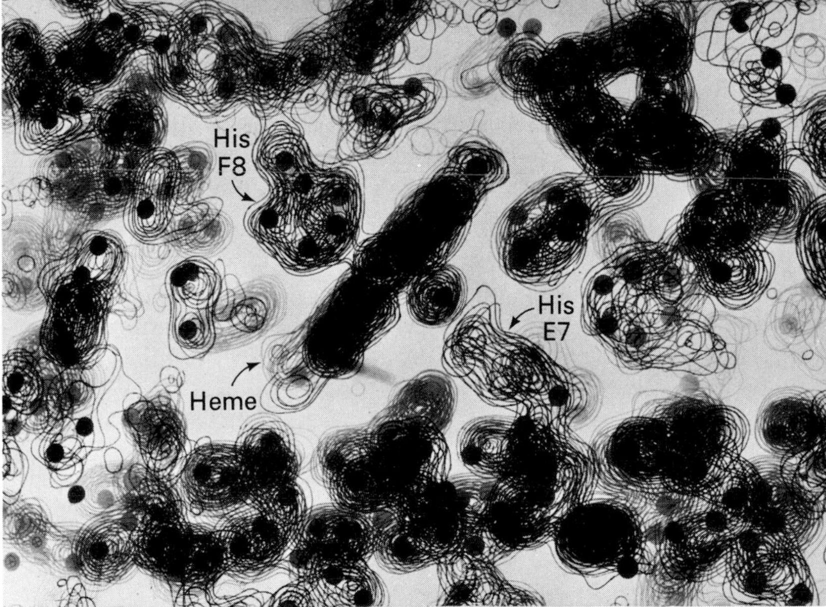

Figure 7-9
Section of an electron-density map of myoglobin near the oxygen-binding site. Electron density extending across the lower part of the map is the E helix. [Courtesy of Dr. John Kendrew.]

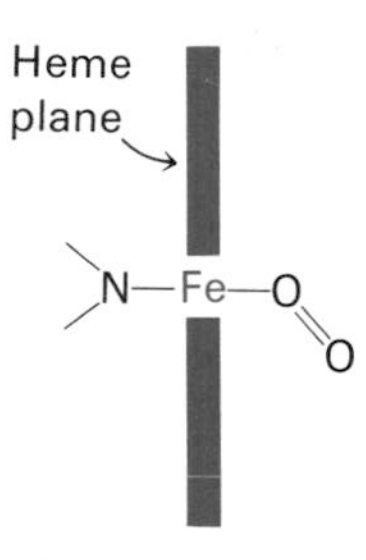

Figure 7-10
Bent, end-on orientation of O_2 in oxymyoglobin. The angle between the O_2 axis and the Fe–O bond is 121 degrees.

Table 7-1
Heme environment

Form	*Oxidation state of Fe*	*Occupant*	
		5th coordination position	*6th coordination position*
Deoxymyoglobin	+2	His F8	Empty
Oxymyoglobin	+2	His F8	O_2
Ferrimyoglobin	+3	His F8	H_2O

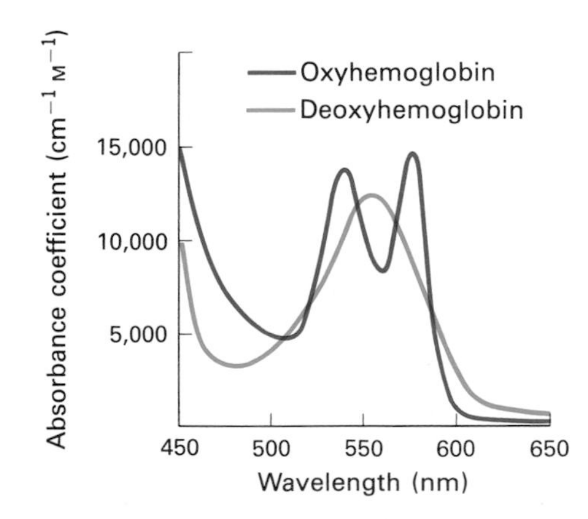

Figure 7-11
Visible absorption spectra of myoglobin and hemoglobin change markedly on binding oxygen. Myoglobin and hemoglobin have very similar visible absorption spectra.

A HINDERED HEME ENVIRONMENT IS ESSENTIAL FOR REVERSIBLE OXYGENATION

The oxygen-binding site comprises only a small fraction of the volume of the myoglobin molecule. Indeed, oxygen is directly bonded only to the iron atom of the heme. Why is the polypeptide portion of myoglobin needed for oxygen transport and storage? The answer lies in the oxygen-binding properties of an isolated heme group. In water, a free ferrous heme group can bind oxygen, but it does so for only a fleeting moment. The reason is that O_2 very rapidly oxidizes the ferrous heme to ferric heme, which cannot bind oxygen. A complex of O_2 sandwiched between two hemes is an intermediate in this reaction. *In myoglobin, the heme group is much less susceptible to oxidation because two myoglobin molecules cannot readily associate to form a heme-O_2-heme complex.* The formation of this sandwich is sterically blocked by the distal histidine and other residues surrounding the sixth coordination site.

The strongest evidence for the importance of steric factors in determining the rate of oxidation of heme comes from studies of synthetic model compounds. James Collman synthesized *picket-fence iron porphyrin complexes* (Figure 7-12) that mimic the oxygen-binding sites of myoglobin and hemoglobin. These compounds have a protective enclosure for binding O_2 on one side of the porphyrin ring, whereas the other side is left unhindered so that it can bind a base. In fact, when the base is a substituted imidazole (like histidine), the oxygen affinity of the picket-fence compound (Figure 7-13) is like that of myoglobin. Furthermore, the picket fence stabilizes the ferrous form of this iron porphyrin and thus enables it to reversibly bind oxygen for long periods. The critical difference between this model compound and free heme is the presence of the picket fence, which blocks the formation of the sandwich dimer.

Thus, myoglobin has created a special microenvironment that confers distinctive properties on its prosthetic group. In general, *the function of a prosthetic group is modulated by its polypeptide environment.* For example, the same heme group has quite a different function in cytochrome *c*, a protein in the terminal oxidation chain in the mitochondria of all aerobic organisms. In cytochrome *c*, the heme is a reversible carrier of electrons rather than of oxygen. Heme has yet another function in the enzyme catalase, where it catalyzes the conversion of hydrogen peroxide into water and oxygen.

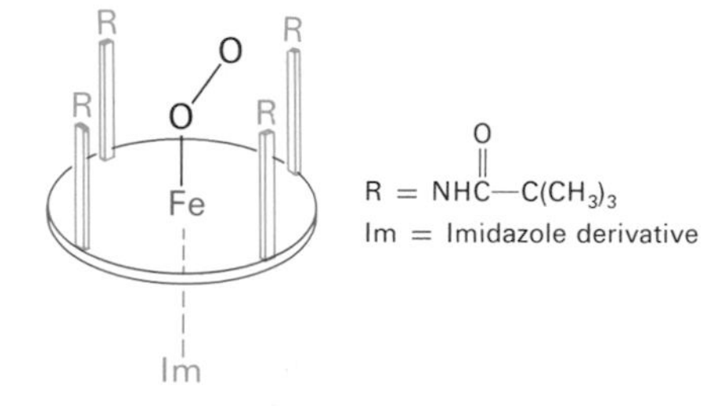

Figure 7-12
Schematic diagram of a picket-fence iron porphyrin with a bound O_2. The picket fence prevents two of these porphyrins from coming together to form the major intermediate in oxidation.

Figure 7-13
Structural formula of a picket-fence iron porphyrin. [After J. Collman, J. I. Brauman, E. Rose, and K. S. Suslick. *Proc. Nat. Acad. Sci.* 75(1978):1053.]

CARBON MONOXIDE BINDING IS DIMINISHED BY THE PRESENCE OF THE DISTAL HISTIDINE

Carbon monoxide is a poison because it combines with ferromyoglobin and ferrohemoglobin and thereby blocks oxygen transport. An isolated heme in solution typically binds CO some 25,000 times as strongly as O_2. However, the binding affinity of myoglobin and hemoglobin for CO is only about 200 times as great as for O_2. How do these proteins suppress the innate preference of heme for carbon monoxide? The answer comes from x-ray crystallographic and infra-red spectroscopic studies of complexes of CO and O_2 with myoglobin and model iron porphyrins. In the very tightly bound complexes of CO with isolated iron porphyrins, the Fe, C, and O atoms are in a linear array (Figure 7-14). In carbonmonoxymyoglobin, by contrast, the CO axis is at an angle to the Fe–C bond. The linear binding of CO is prevented mainly by the steric hindrance of the distal histidine. On the other hand, the O_2 axis is at an angle to the Fe–O bond both in oxymyoglobin and in model compounds. *Thus, the protein forces CO to bind at an angle rather than in line. This bent geometry in the globins weakens the interaction of CO with the heme.*

The decreased affinity of myoglobin and hemoglobin for CO is biologically important. Carbon monoxide was a potential hazard long before the emergence of industrialized societies because it is produced endogenously (within cells) in the breakdown of heme (p. 597). The level of endogenously formed carbon monoxide is such that about 1% of the sites in myoglobin and hemoglobin are blocked by CO, a tolerable degree of inhibition. However, endogenously produced CO would cause massive poisoning if the affinity of these proteins for CO was like that of isolated iron porphyrins. This challenge was solved by the evolution of heme proteins that discriminate between O_2 and CO by sterically imposing a bent and hence weaker mode of binding for CO.

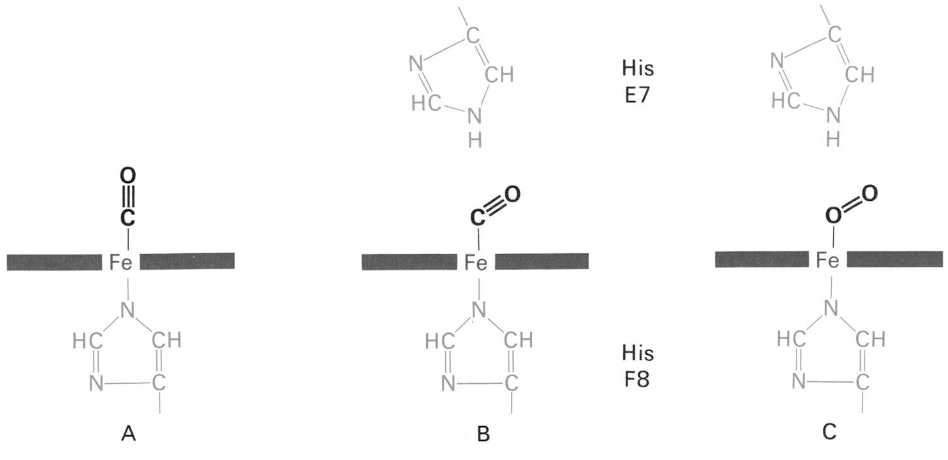

Figure 7-14
Structural basis of the diminished affinity of myoglobin and hemoglobin for carbon monoxide: (A) linear mode of binding of CO to isolated iron porphyrins; (B) bent mode of binding of CO to myoglobin and hemoglobin, in which the distal histidine (E7) prevents CO from binding linearly and so the affinity for CO is markedly reduced; (C) bent mode of binding of O_2 in myoglobin and hemoglobin. Isolated iron porphyrins also bind O_2 in a bent mode.

THE CENTRAL EXON OF MYOGLOBIN ENCODES A FUNCTIONAL HEME-BINDING UNIT

As was discussed in Chapter 5, most eucaryotic genes are mosaics of exons (coding sequences) and introns (noncoding intervening sequences). It was mentioned that exons encode discrete structural and functional units of proteins (p. 112). Myoglobin illustrates this general principle nicely. The gene for myoglobin consists of three exons: an amino-terminal one encoding residues 1 to 30 (NA1 to B2), a central one encoding residues 31 to 105 (B3 to G6), and a C-terminal one encoding residues 106 to 153 (G7 to HC5). A look at the structure of native myoglobin shows that nearly all of the heme-binding site is specified by the central exon (Figure 7-15). The regions encoded by the other two exons make very few contacts with the heme. What are the functional properties of an isolated polypeptide consisting of residues 31 to 105? The answer to this question is not yet known but the properties of another fragment are revealing. Digestion of apomyoglobin with clostripain, an arginine-specific protease, yields a polypeptide containing residues 32 to 139, which corresponds to the central exon and part of the C-terminal one. This fragment binds heme. Furthermore, this complex, called *mini-myoglobin*, binds O_2 and CO reversibly. Indeed, the rates of association and dissociation are nearly the same as those of the intact protein.

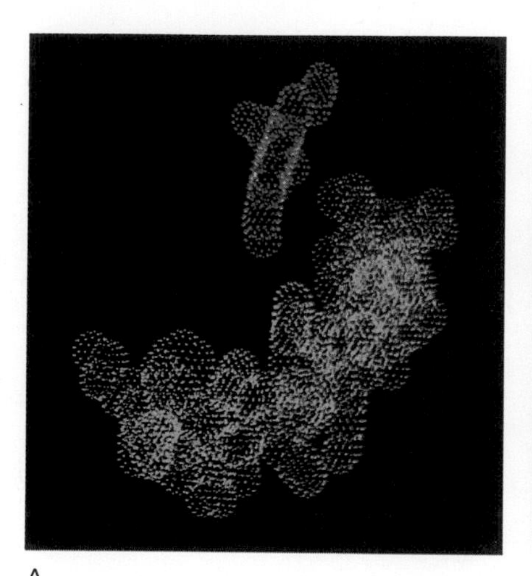
A

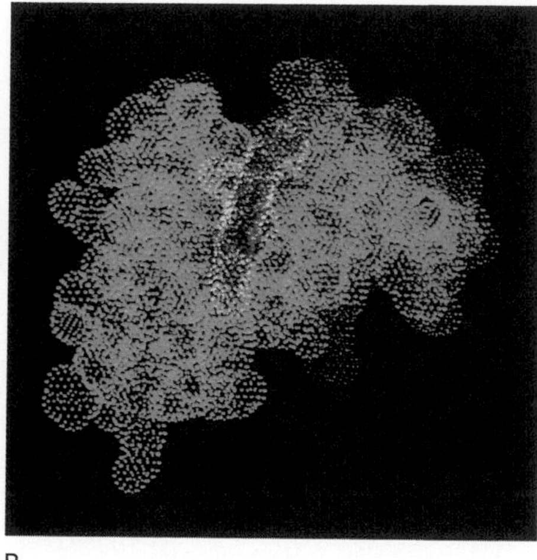
B

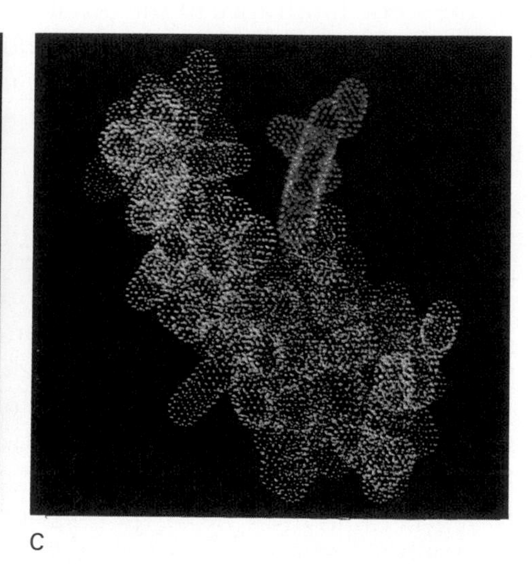
C

Figure 7-15
Region of myoglobin encoded by (A) exon 1, (B) exon 2, and (C) exon 3. The central exon encodes a functional oxygen-binding module.

These findings indicate that the conformation of mini-myoglobin is very similar to that of native myoglobin. Because the amino-terminal exon and the last fourteen residues of the C-terminal exon are not essential for reversible oxygenation, it is evident that the central exon contains much of the information for binding heme and maintaining the native fold of an oxygen-binding protein. The exon-intron organization of the constituent chains of hemoglobin is very similar to that of myoglobin. Amino acid sequence comparisons suggest that the genes for myoglobin and hemoglobin diverged some 700 million years ago (p. 841). Thus, the central exon of these oxygen carriers is an ancient piece of DNA that encoded a functional heme-binding module eons ago.

Table 7-2
Subunits of human hemoglobins

Embryonic hemoglobins	
Hb Gower 1	$\zeta_2\epsilon_2$
Hb Gower 2	$\alpha_2\epsilon_2$
Hb Portland	$\zeta_2\gamma_2$
Fetal hemoglobin	
Hb F	$\alpha_2\gamma_2$
Adult hemoglobins	
Hb A	$\alpha_2\beta_2$
Hb A_2	$\alpha_2\delta_2$

HEMOGLOBIN CONSISTS OF FOUR POLYPEPTIDE CHAINS

We turn now to hemoglobin, the oxygen transporter in erythrocytes. Vertebrate hemoglobins consist of four polypeptide chains, two of one kind and two of another (Table 7-2). The four chains are held together by noncovalent attractions. Each contains a heme group and a single oxygen-binding site. Hemoglobin A, the principal hemoglobin in adults, consists of two alpha (α) chains and two beta (β) chains. Adults also have a minor hemoglobin (~2% of the total hemoglobin) called hemoglobin A_2, which contains delta (δ) chains in place of the β chains of hemoglobin A. Thus, the subunit composition of hemoglobin A is $\alpha_2\beta_2$, and that of hemoglobin A_2 is $\alpha_2\delta_2$.

Fetuses have distinctive hemoglobins. Shortly after conception, fetuses synthesize zeta (ζ) chains (which are α-like chains) and epsilon (ϵ) chains (which are β-like). In the course of fetal life, ζ is replaced by α, and ϵ is replaced by gamma (γ), which is replaced by β (Figure 7-16). The major hemoglobin during the latter two-thirds of fetal life, hemoglobin F, has the subunit composition $\alpha_2\gamma_2$. The α and ζ chains contain 141 residues; the β, γ, and δ chains contain 146 residues. Why do hemoglobins consist of two kinds of chains? Why do fetuses have distinctive hemoglobins? Some answers to these intriguing questions will be given later in this chapter.

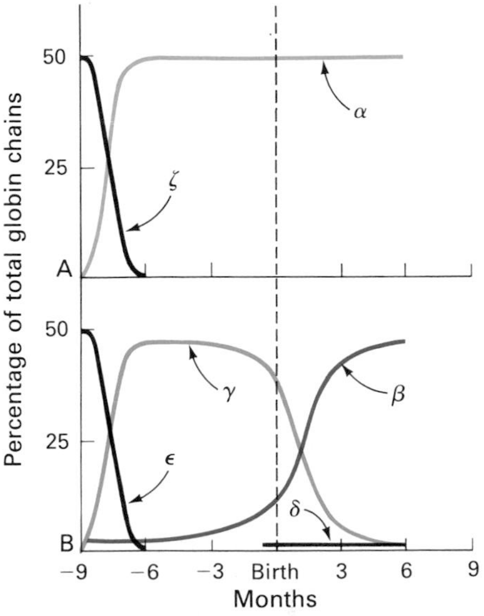

Figure 7-16
Expression of hemoglobin genes in human development. (A) α and ζ genes. (B) β, γ, and δ genes.

X-RAY ANALYSIS OF HEMOGLOBIN

Perutz's elucidation of the three-dimensional structure of hemoglobin, a monumental accomplishment, began in 1936. He left Austria that year to pursue graduate work in England and decided on hemoglobin as his thesis subject. The largest structure that had been solved then was the dye phthalocyanin, which contains 58 atoms. In choosing to tackle a molecule one hundred times as large, as Perutz wrote years later, it was little wonder that "my fellow students regarded me with a pitying smile. . . . Fortunately, the examiners of my doctoral thesis did not insist on a determination of the structure, otherwise I should have had to remain a graduate student for twenty-three years." Fortunately, too, Lawrence Bragg became director of the Cavendish Laboratory in Cambridge at this time and supported the project. In 1912, he and his father were the first to use x-ray crystallography to solve structures. Bragg wrote, "I was frank about the outlook. It was like multiplying a zero probability that success would be achieved by an infinity of importance if the structure came out; the result of this mathematic operation was anyone's guess." Success came in 1959, when Perutz obtained a low-resolution electron-density image of horse oxyhemoglobin, followed several years later by high-resolution maps of both human and horse oxyhemoglobin and deoxyhemoglobin. The three-dimensional structures of human and horse hemoglobins are very similar.

The hemoglobin molecule is nearly spherical, with a diameter of 55 Å. The four chains are packed together in a tetrahedral array (Figure 7-17). The heme groups are located in crevices near the exterior of the

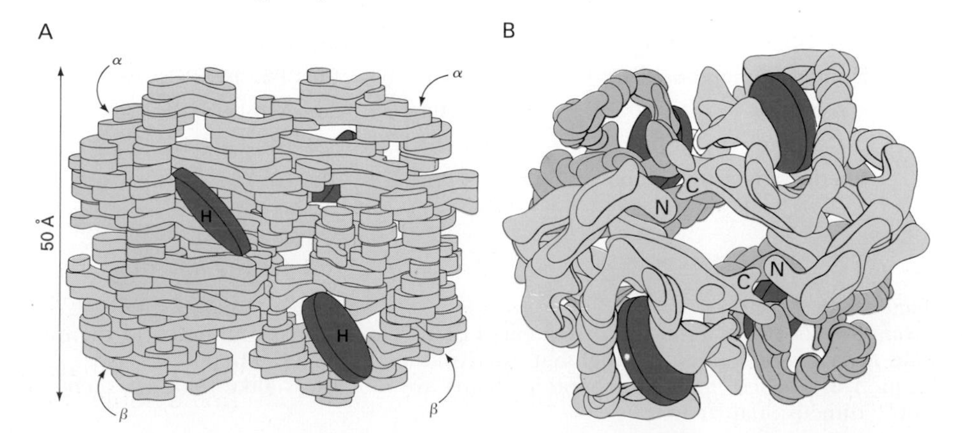

Figure 7-17
Model of hemoglobin at low resolution. The α chains in this model are yellow, the β chains blue, and the heme groups red. View (A) is at right angles to view (B). [After M. F. Perutz. The hemoglobin molecule. Copyright © 1964 by Scientific American, Inc. All rights reserved.]

molecule, one in each subunit. The four oxygen-binding sites are far apart; the distance between the two closest iron atoms is 25 Å. Each α chain is in contact with both β chains. In contrast, there are few interactions between the two α chains or between the two β chains.

THE ALPHA AND BETA CHAINS OF HEMOGLOBIN CLOSELY RESEMBLE MYOGLOBIN

The three-dimensional structures of myoglobin and the α and β chains of human hemoglobin are strikingly similar (Figure 7-18). This close resemblance in the folding of their main chains was unexpected because their amino acid sequences are rather different. In fact, these three chains are identical at only 24 of 141 positions. Hence, quite different amino acid sequences can specify very similar three-dimensional structures (Figure 7-19).

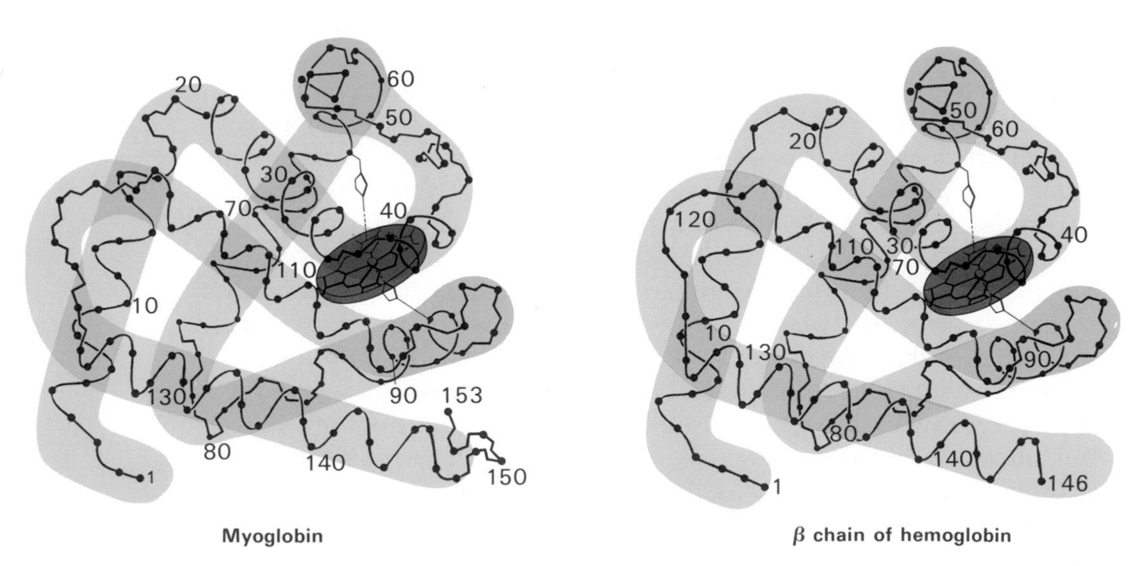

Figure 7-18
Comparison of the conformations of the main chain of myoglobin and the β chain of hemoglobin. The similarity of their conformations is evident. [From M. F. Perutz. The hemoglobin molecule. Copyright © 1964 by Scientific American, Inc. All rights reserved.]

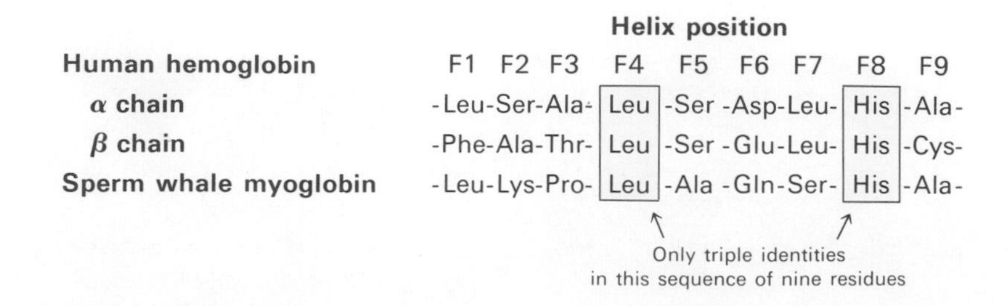

Helix position
Human hemoglobin
F1 F2 F3 F4 F5 F6 F7 F8 F9
α chain
-Leu-Ser-Ala- Leu -Ser -Asp-Leu- His -Ala-
β chain
-Phe-Ala-Thr- Leu -Ser -Glu-Leu- His -Cys-
Sperm whale myoglobin
-Leu-Lys-Pro- Leu -Ala -Gln-Ser- His -Ala-
Only triple identities
in this sequence of nine residues

Figure 7-19
Comparison of the amino acid sequences of sperm whale myoglobin and the α and β chains of human hemoglobin, for residues F1 to F9. The amino acid sequences of these three polypeptide chains are much less alike than are their three-dimensional structures.

It is evident that the three-dimensional form of sperm whale myoglobin and of the α and β chains of human hemoglobin has broad biological significance. In fact, this motif, called the *globin fold*, is common to all known vertebrate myoglobins and hemoglobins. *The intricate folding of the polypeptide chain, first discovered in myoglobin, is nature's fundamental design for an oxygen carrier: it places the heme in an environment that enables it to carry oxygen reversibly.* The genes for myoglobin and for the α, β, and other chains of hemoglobin are variations on a fundamental theme. This family of genes almost certainly arose by gene duplication and diversification.

Hemo—
A prefix from the Greek word meaning "blood"

Myo—
A prefix from the Greek word "muscle"

Globin—
A protein belonging to the myoglobin-hemoglobin family

CRITICAL RESIDUES IN THE AMINO ACID SEQUENCE

The amino acid sequences of hemoglobins from more than sixty species (ranging from lamprey eels to humans) are known. A comparison of these sequences shows considerable variability at most positions. However, nine positions have the same residue in all or nearly all species studied thus far (Table 7-3). These invariant residues are especially important for the function of the hemoglobin molecule. Several of them directly affect the oxygen-binding site. Another invariant residue is tyrosine HC2, which stabilizes the molecule by forming a hydrogen bond between the H and F helices. Glycine B6 is invariant because of its small size: a side chain larger than a hydrogen atom would not allow the B and E helices to approach each other as closely as they do (Figure 7-20). Proline C2 may be essential because it defines one end of the C helix.

The amino acid residues in the interior of hemoglobins vary considerably. However, the change is always of one nonpolar residue for another (as from alanine to isoleucine). Thus, *the striking nonpolar character of the interior of the molecule is conserved*. The nonpolar core is also important in stabilizing the three-dimensional structure of hemoglobin.

In contrast, residues on the surface of the molecule are highly variable. Indeed, few are consistently positively or negatively charged. It might be thought that proline residues would be preserved because of their role as helix-breakers. However, this is not so. Only one proline is invariant, yet the lengths and directions of the helices in all globins are very similar. Obviously, there are other ways of terminating or bending α helices. For example, an α helix can be disrupted by the interaction of the OH group of serine or threonine with a main-chain carbonyl group.

Table 7-3
Invariant amino acid residues in hemoglobins

Position	*Amino acid*	*Role*
F8	Histidine	Proximal heme-linked histidine
E7	Histidine	Distal histidine near the heme
CD1	Phenylalanine	Heme contact
F4	Leucine	Heme contact
B6	Glycine	Allows the close approach of the B and E helices
C2	Proline	Helix termination
HC2	Tyrosine	Cross-links the H and F helices
C4	Threonine	Uncertain
H10	Lysine	Uncertain

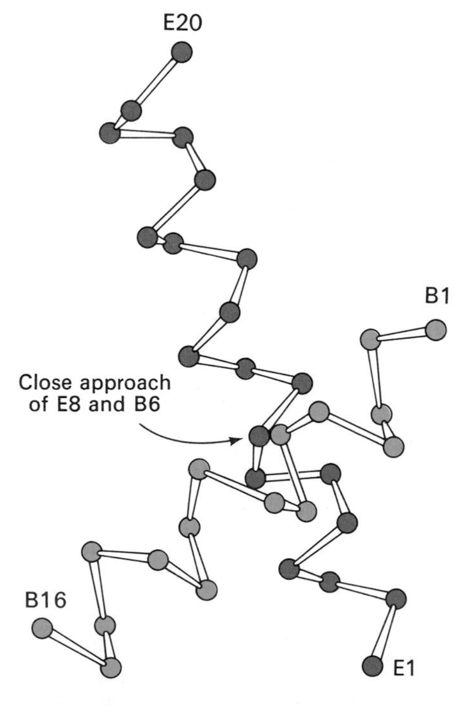

Figure 7-20
Crossing of the B and E helices in myoglobin. Residue B6 is almost invariably glycine because there is no space for a larger side chain.

HEMOGLOBIN IS AN ALLOSTERIC PROTEIN

The α and β subunits of hemoglobin have the same structural design as myoglobin. However, new properties of profound biological importance emerge when different subunits come together to form a tetramer. Hemoglobin is a much more intricate and sentient molecule than is myoglobin. Hemoglobin transports H^+ and CO_2 in addition to O_2. Furthermore, the oxygen-binding properties of hemoglobin are regulated by interactions between separate, nonadjacent sites. Hemoglobin is an allosteric protein, whereas myoglobin is not. This difference is expressed in three ways:

1. The binding of O_2 to hemoglobin enhances the binding of additional O_2 to the same hemoglobin molecule. In other words, O_2 binds cooperatively to hemoglobin. In contrast, the binding of O_2 to myoglobin is not cooperative.

2. The affinity of hemoglobin for oxygen depends on pH, whereas that of myoglobin is independent of pH. The CO_2 molecule also affects the oxygen-binding characteristics of hemoglobin.

3. The oxygen affinity of hemoglobin is further regulated by organic phosphates such as 2,3-bisphosphoglycerate (BPG). The result is that hemoglobin has a lower affinity for oxygen than does myoglobin.

Torr—
A unit of pressure equal to that exerted by a column of mercury 1 mm high at 0°C and standard gravity (1 mm Hg).
Named after Evangelista Torricelli (1608–1647), the inventor of the mercury barometer.

OXYGEN BINDS COOPERATIVELY TO HEMOGLOBIN

The saturation Y is defined as the fractional occupancy of the oxygen-binding sites. The value of Y can range from 0 (all sites empty) to 1 (all sites filled). A plot of Y versus pO_2, the partial pressure of oxygen, is called an *oxygen dissociation curve*. The oxygen dissociation curves of myoglobin and hemoglobin differ in two ways (Figures 7-21 and 7-22). For any given pO_2, Y is higher for myoglobin than for hemoglobin. This means that *myoglobin has a higher affinity for oxygen than does hemoglobin*. Oxygen affinity can be characterized by a quantity called P_{50}, which is the partial pressure of oxygen at which 50% of sites are filled (i.e., at which $Y = 0.5$). For myoglobin, P_{50} is typically 1 torr, whereas for hemoglobin, P_{50} is 26 torrs.

The second difference is that *the oxygen dissociation curve of myoglobin is hyperbolic, whereas that of hemoglobin is sigmoidal*. Let us consider these curves in quantitative terms, starting with the one for myoglobin be-

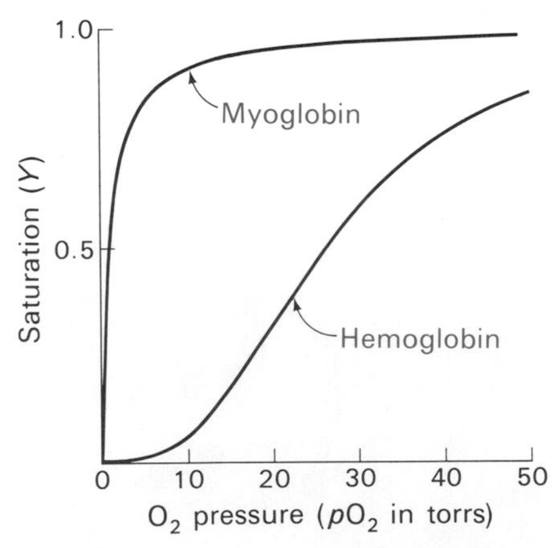

Figure 7-21
Oxygen dissociation curves of myoglobin and hemoglobin. Saturation of the oxygen-binding sites is plotted as a function of the partial pressure of oxygen surrounding the solution.

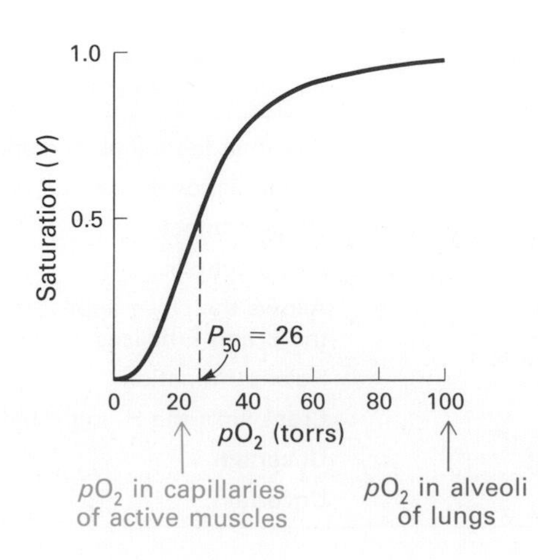

Figure 7-22
Oxygen dissociation curve of hemoglobin. Typical values for pO_2 in the capillaries of active muscle and in the alveoli of the lung are marked on the horizontal axis. Note that P_{50} for hemoglobin under physiological conditions lies between these values.

cause it is simpler. The binding of oxygen to myoglobin (Mb) can be described by a simple equilibrium:

$$MbO_2 \rightleftharpoons Mb + O_2 \tag{1}$$

The equilibrium constant K for the dissociation of oxymyoglobin is

$$K = \frac{[Mb][O_2]}{[MbO_2]} \tag{2}$$

in which $[MbO_2]$ is the concentration of oxymyoglobin, [Mb] is the concentration of deoxymyoglobin, and $[O_2]$ is the concentration of uncombined oxygen, all in moles per liter. Then the saturation Y is

$$Y = \frac{[MbO_2]}{[MbO_2] + [Mb]} \tag{3}$$

Substitution of equation 2 into equation 3 yields

$$Y = \frac{[O_2]}{[O_2] + K} \tag{4}$$

Because oxygen is a gas, it is convenient to express its concentration in terms of pO_2, the partial pressure of oxygen (in torrs) in the atmosphere surrounding the solution. Equation 4 then becomes

$$Y = \frac{pO_2}{pO_2 + P_{50}} \tag{5}$$

Equation 5 plots as a hyperbola. In fact, the oxygen dissociation curve calculated from equation 5, taking P_{50} to be 1 torr, closely matches the experimentally observed curve for myoglobin.

In contrast, the sigmoidal curve for hemoglobin cannot be matched by any curve described by equation 5. In 1913, Archibald Hill showed that the curve obtained from the oxygen-binding data for hemoglobin agrees with the equation derived for the *hypothetical* equilibrium:

$$Hb(O_2)_n \rightleftharpoons Hb + n\ O_2 \tag{6}$$

This expression yields

$$Y = \frac{(pO_2)^n}{(pO_2)^n + (P_{50})^n} \tag{7}$$

which can be rearranged to give

$$\frac{Y}{1 - Y} = \left(\frac{pO_2}{P_{50}}\right)^n \tag{8}$$

This equation states that the ratio of oxyheme (Y) to deoxyheme ($1 - Y$) is equal to the nth power of the ratio of pO_2 to P_{50}. Taking the logarithms of both sides of equation 8 gives

$$\log \frac{(Y)}{1 - Y} = n \log pO_2 - n \log P_{50} \tag{9}$$

A plot of $\log [Y/(1 - Y)]$ versus $\log pO_2$, called a *Hill plot,* approximates a straight line. Its slope n at the midpoint of the binding ($Y = 0.5$) is called the *Hill coefficient.* The value of n increases with the degree of cooperativity; the maximum possible value of n is equal to the number of binding sites.

Myoglobin gives a linear Hill plot with $n = 1.0$, whereas $n = 2.8$ for hemoglobin (Figure 7-23). The slope of 1.0 for myoglobin means that O_2 molecules bind independently of each other, as indicated in equa-

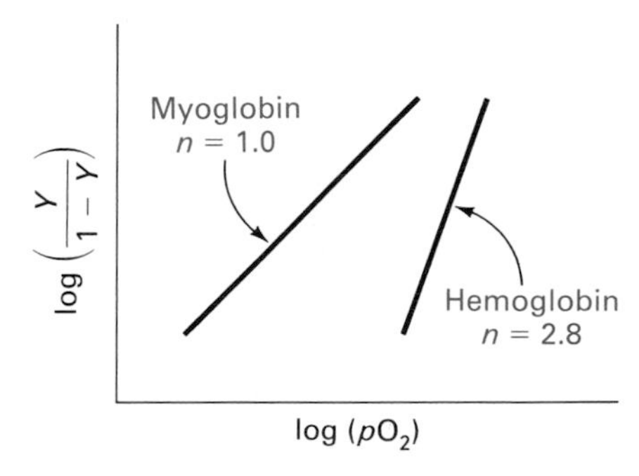

Figure 7-23
Hill plot for the binding of O_2 to myoglobin and hemoglobin. The slope of 2.8 for hemoglobin indicates that it binds oxygen cooperatively, in contrast with myoglobin, which has a slope of 1.0.

tion 1. In contrast, the Hill coefficient of 2.8 for hemoglobin means that *the binding of oxygen in hemoglobin is cooperative*. Binding at one heme facilitates the binding of oxygen at the other hemes on the same tetramer. Conversely, the unloading of oxygen at one heme facilitates the unloading of oxygen at the others. In other words, the heme groups of a hemoglobin molecule communicate with each other. The mechanism of cooperative binding of oxygen by hemoglobin, sometimes called *heme-heme interaction*, will be discussed shortly.

THE COOPERATIVE BINDING OF OXYGEN BY HEMOGLOBIN ENHANCES OXYGEN TRANSPORT

The cooperative binding of oxygen by hemoglobin makes hemoglobin a more efficient oxygen transporter. The oxygen saturation of hemoglobin changes more rapidly with changes in the partial pressure of O_2 than it would if the oxygen-binding sites were independent of each other. Let us consider a specific example. Assume that the alveolar pO_2 is 100 torrs, and that the pO_2 in the capillary of an active muscle is 20 torrs. Let P_{50} = 26 torrs, and take n = 2.8. Then Y in the alveolar capillaries will be 0.98, and Y in the muscle capillaries will be 0.32. The oxygen delivered will be proportional to the difference in Y, which is 0.66. Let us now make the same calculation for a hypothetical oxygen carrier for which P_{50} is also 26 torrs, but in which the binding of oxygen is not cooperative (n = 1). Then $Y_{\text{alveoli}} = 0.79$, and $Y_{\text{muscle}} = 0.43$, and so the difference in Y is equal to 0.36. Thus, *the cooperative binding of oxygen by hemoglobin enables it to deliver 1.83 times as much oxygen under typical physiological conditions as it would if the sites were independent.*

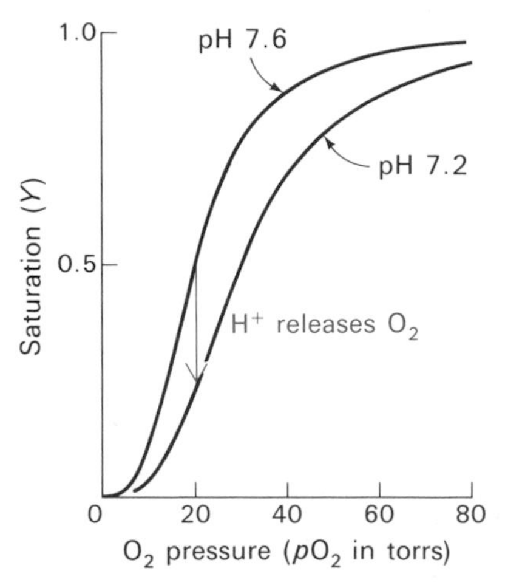

Figure 7-24
Effect of pH on the oxygen affinity of hemoglobin. Lowering the pH from 7.6 to 7.2 results in the release of O_2 from oxyhemoglobin.

H^+ AND CO_2 PROMOTE THE RELEASE OF O_2 (THE BOHR EFFECT)

Myoglobin shows no change in oxygen binding over a broad range of pH, nor does CO_2 have an appreciable effect. In hemoglobin, however, acidity enhances the release of oxygen. In the physiological range, a lowering of pH shifts the oxygen dissociation curve to the right, so that the oxygen affinity is decreased (Figure 7-24). Increasing the concentration of CO_2 (at constant pH) also lowers the oxygen affinity. In rapidly metabolizing tissue, such as contracting muscle, much CO_2 and acid are produced. *The presence of higher levels of CO_2 and H^+ in the capillaries of such metabolically active tissue promotes the release of O_2 from oxyhemoglobin.* This important mechanism for meeting the higher oxygen needs of metabolically active tissues was discovered by Christian Bohr in 1904. The reciprocal effect, discovered ten years later by J. S. Haldane, occurs in the alveolar capillaries of the lungs. The high concentration of O_2 there unloads H^+ and CO_2 from hemoglobin, just as the high concentration of H^+ and CO_2 in active tissues drives off O_2. These linkages between the binding of O_2, H^+, and CO_2 are known as the *Bohr effect* (Figure 7-25).

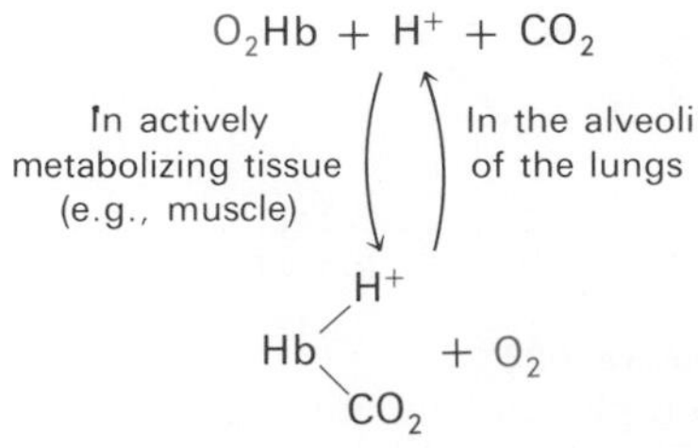

$O_2Hb + H^+ + CO_2$

In actively metabolizing tissue (e.g., muscle)

In the alveoli of the lungs

$Hb(H^+)(CO_2) + O_2$

Figure 7-25
Summary of the Bohr effect. The actual mechanism and stoichiometry are more complex than indicated in this diagram.

BPG LOWERS THE OXYGEN AFFINITY OF HEMOGLOBIN

The oxygen affinity of hemoglobin within red cells is lower than that of hemoglobin in free solution. As early as 1921, Joseph Barcroft wondered, "Is there some third substance present . . . which forms an integral part of the oxygen-hemoglobin complex?" Indeed there is. Reinhold Benesch and Ruth Benesch showed in 1967 that 2,3-bisphosphoglycerate (BPG, also known as 2,3-diphosphoglycerate, DPG) binds to hemoglobin and has a large effect on its affinity for oxygen. This highly anionic organic phosphate is present in human red cells at about the same molar concentration as hemoglobin. In the absence of BPG, the P_{50} of hemoglobin is 1 torr, like that of myoglobin. In its presence, P_{50} becomes 26 torrs (Figure 7-26). Thus, *BPG lowers the oxygen affinity of hemoglobin by a factor of 26, which is essential in enabling hemoglobin to unload oxygen in tissue capillaries.* BPG diminishes the oxygen affinity of hemoglobin by binding to deoxyhemoglobin but not to the oxygenated form.

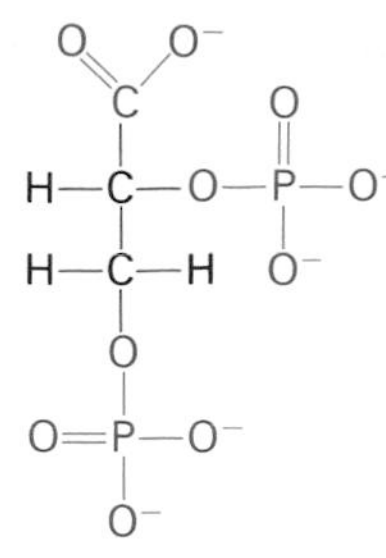

2,3-Bisphosphoglycerate (2,3-Diphosphoglycerate, DPG)

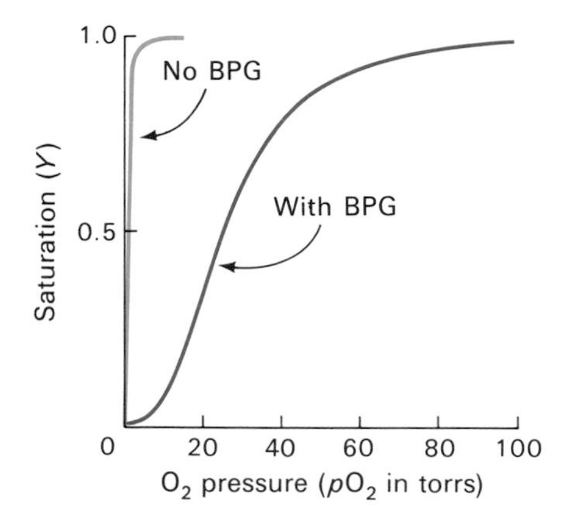

Figure 7-26
2,3-Bisphosphoglycerate (BPG) decreases the oxygen affinity of hemoglobin.

FETAL HEMOGLOBIN HAS A HIGHER OXYGEN AFFINITY THAN MATERNAL HEMOGLOBIN

Fetuses have their own kind of hemoglobin, called hemoglobin F ($\alpha_2\gamma_2$), which differs from adult hemoglobin A ($\alpha_2\beta_2$), as mentioned previously. An important property of hemoglobin F is that it has a higher oxygen affinity under physiological conditions than does hemoglobin A (Figure 7-27). The higher oxygen affinity of hemoglobin F optimizes the transfer of oxygen from the maternal to the fetal circulation. Hemoglobin F is oxygenated at the expense of hemoglobin A on the other side of the placental circulation. The higher oxygen affinity of fetal blood was known for many years, but an understanding of its basis could come only after the discovery of BPG. *Hemoglobin F binds BPG less strongly than does hemoglobin A and consequently has a higher oxygen affinity.* In the absence of BPG, the oxygen affinity of hemoglobin F is actually lower than that of hemoglobin A.

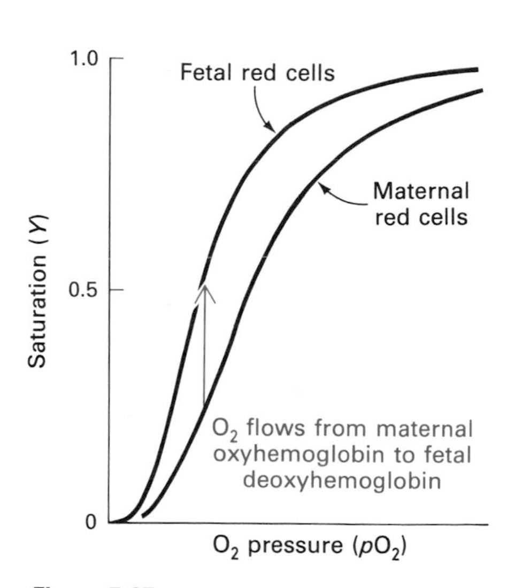

Figure 7-27
Fetal red blood cells have a higher oxygen affinity than do maternal red blood cells. The reason is that, in the presence of BPG, the oxygen affinity of fetal hemoglobin is higher than that of maternal hemoglobin.

SUBUNIT INTERACTIONS ARE REQUIRED FOR ALLOSTERIC EFFECTS

Let us now consider the structural basis of these allosteric effects. Hemoglobin can be dissociated into its constituent chains. The properties of the isolated α chain are very much like those of myoglobin. The α chain by itself has a high oxygen affinity, a hyperbolic oxygen dissociation curve, and oxygen-binding characteristics that are insensitive to pH, CO_2 concentration, and BPG level. The isolated β chain readily forms a tetramer, β_4, which is called hemoglobin H. Like the α chain and myoglobin, β_4 entirely lacks the allosteric properties of hemoglobin and has a high oxygen affinity. In short, *the allosteric properties of hemoglobin arise from interactions between its subunits. The functional unit of hemoglobin is a tetramer consisting of two kinds of polypeptide chains.*

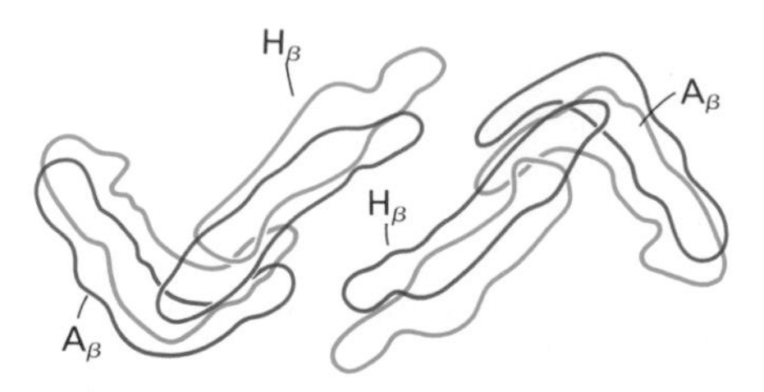

Figure 7-28
Projection of part of the electron-density maps of oxyhemoglobin (shown in red) and deoxyhemoglobin (shown in blue) at a resolution of 5.5 Å. The A and H helices of the two β chains of hemoglobin are shown here. The center of the diagram corresponds to the central cavity of the molecule. It shows one of the conformational changes accompanying oxygenation—a movement of the H helices toward each other. [After M. F. Perutz. *Nature* 228(1970):738.]

THE QUATERNARY STRUCTURE OF HEMOGLOBIN CHANGES MARKEDLY ON OXYGENATION

X-ray crystallographic studies have shown that oxy- and deoxyhemoglobin differ markedly in quaternary structure (Figure 7-28). The oxygenated molecule is more compact. For instance, the distance between the iron atoms of the β chains decreases from 40 to 33 Å on oxygenation. The changes in the contacts between the α and β chains are of special interest. There are two kinds of contact regions between the α and β chains (Figure 7-29). The $\alpha_1\beta_1$ contact (and the identical $\alpha_2\beta_2$ contact) is one type. The other type is the $\alpha_1\beta_2$ contact (and the identical $\alpha_2\beta_1$ contact). In the transition from oxy- to deoxyhemoglobin, large structural changes take place at the $\alpha_1\beta_2$ contact but only small ones at the $\alpha_1\beta_1$ contact. Furthermore, the $\alpha_1\beta_1$ pair rotates relative to the other pair by 15 degrees (Figure 7-30). Some atoms at the interface between these pairs shift by as much as 6 Å.

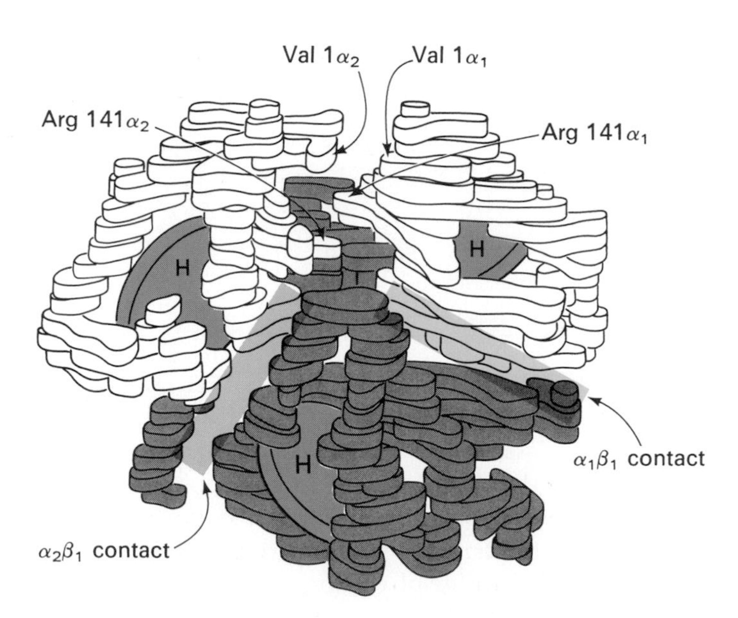

Figure 7-29
Model of oxyhemoglobin at low resolution showing the two kinds of interfaces between α and β chains. The α chains are white, the β chains black. Three hemes (red) can be seen in this view of the molecule. The $\alpha_2\beta_1$ contact region is shown in blue, the $\alpha_1\beta_1$ contact region in yellow. [After M. F. Perutz and L. F. TenEyck. *Cold Spring Harbor Symp. Quant. Biol.* 36(1971):296.]

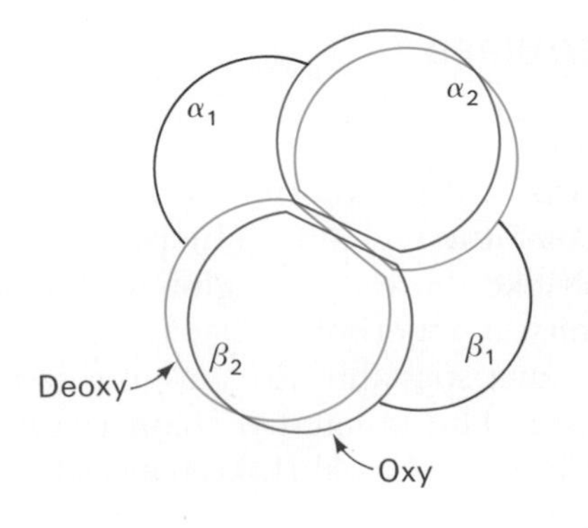

Figure 7-30
Schematic diagram showing the change in quaternary structure on oxygenation. One pair of $\alpha\beta$ subunits shifts with respect to the other by a rotation of 15 degrees and a translation of 0.8 Å. The oxy form of the rotated $\alpha\beta$ subunit is shown in red and the deoxy form in blue. [After J. Baldwin and C. Chothia. *J. Mol. Biol.* 129(1979):192. Copyright by Academic Press, Inc. (London) Ltd.]

In fact, the $\alpha_1\beta_2$ contact region is designed to act as a switch between two alternative structures. The two forms of this dove-tailed interface are stabilized by different sets of hydrogen bonds (Figure 7-31). This interface is closely connected to the heme groups, and so structural changes in it can be expected to affect the hemes. The importance of the interface is reinforced by the finding that most residues in it are the same in all species. Also, almost all mutations in this interface region diminish heme-heme interaction, whereas mutations in the $\alpha_1\beta_1$ interface do not.

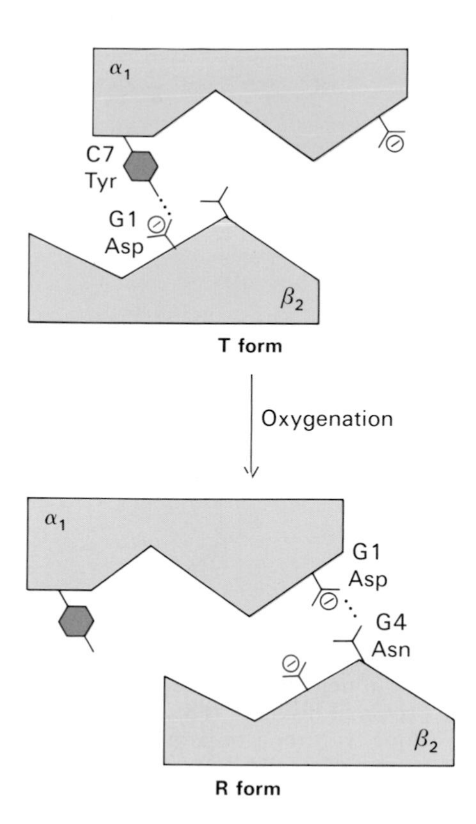

Figure 7-31
The $\alpha_1\beta_2$ interface switches from the T to the R form on oxygenation. The dove-tailed construction of this interface allows the subunits to readily adopt either of the two forms.

DEOXYHEMOGLOBIN IS CONSTRAINED BY SALT LINKS BETWEEN DIFFERENT CHAINS

In oxyhemoglobin, the carboxyl-terminal residues of all four chains have almost complete freedom of rotation. In deoxyhemoglobin, by contrast, these terminal groups are anchored (Figure 7-32), and all of them, as well as the side chains of the C-terminal residues, participate in salt links (electrostatic interactions). Deoxyhemoglobin is a tauter, more constrained molecule than oxyhemoglobin because of these eight salt links. *The quaternary structure of deoxyhemoglobin is termed the T (tense or taut) form; that of oxyhemoglobin, the R (relaxed) form.* The designations R and T are generally used to describe alternative quaternary structures of an allosteric protein, the T form having a lower affinity for the substrate (p. 239).

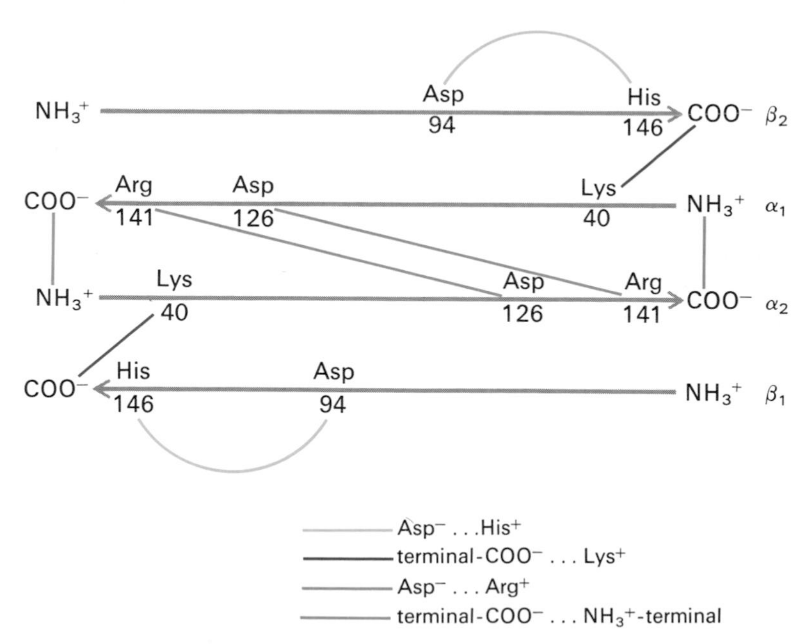

Figure 7-32
Salt links between different subunits in deoxyhemoglobin. These noncovalent, electrostatic cross-links are disrupted on oxygenation.

OXYGENATION MOVES THE IRON ATOM INTO THE PLANE OF THE PORPHYRIN

The conformational changes discussed thus far take place at some distance from the heme. Now let us see what happens at the heme group itself on oxygenation. In deoxyhemoglobin, the iron atom is about 0.4

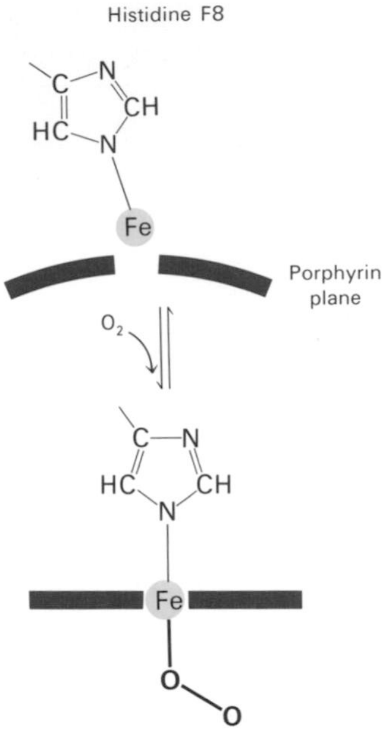

Figure 7-33
The iron atom moves into the plane of the heme on oxygenation. The proximal histidine (F8) is pulled along with the iron atom and becomes less tilted.

Å out of the heme plane toward the proximal histidine, so that the heme group is domed (convex) in the same direction (Figure 7-33). On oxygenation, the iron atom moves into the plane of the porphyrin to form a strong bond with O_2, and the heme becomes more planar. Structural studies of many synthetic iron porphyrins have shown that the iron atom is out of plane in five-coordinated compounds, whereas it is in plane or nearly so in six-coordinated complexes.

MOVEMENT OF THE IRON ATOM IS TRANSMITTED TO OTHER SUBUNITS BY THE PROXIMAL HISTIDINE

How does the movement of the iron atom into the plane of the heme favor the switch in quaternary structure from T to R? The proximal histidine is thought to play a key role in transmitting structural changes from one heme to another. When the iron atom moves into the plane of the heme on oxygenation, it pulls the proximal histidine with it. This movement of histidine F8 results in shifts of the F helix, the EF corner, and the FG corner (Figure 7-34). These conformational changes are transmitted to the subunit interfaces, where they rupture interchain salt links. Consequently, oxygenation shifts the equilibrium between the two quaternary structures to the R form. Thus, *a structural change (oxygenation) within a subunit is translated into structural changes at the interfaces between subunits. The binding of oxygen at one heme site is thereby communicated to parts of the molecule that are far away.*

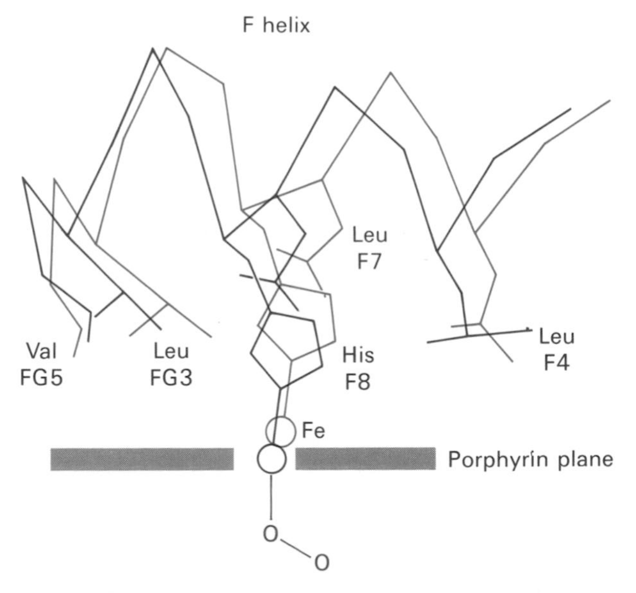

Figure 7-34
Conformational changes induced by the movement of the iron atom on oxygenation. The oxygenated structure is shown in red and the deoxygenated structure in blue. [After J. Baldwin and C. Chothia. *J. Mol. Biol.* 129(1979):192.]

Figure 7-35
Postage-stamp analogy of the oxygenation of hemoglobin. Two perforated edges must be torn to remove the first stamp. Only one perforated edge must be torn to remove the second stamp, and one edge again to remove the third stamp. The fourth stamp is then free.

MECHANISM OF THE COOPERATIVE BINDING OF OXYGEN

The fourth molecule of O_2 binds to hemoglobin some 300 times as tightly as the first one. Why? A postage-stamp analogy is helpful in considering this question (Figure 7-35). Deoxyhemoglobin is a taut molecule, constrained by the eight salt links between its four subunits. Oxygenation does not readily occur unless some of these salt links are broken, enabling the iron atom to move easily into the plane of the heme group. The number of salt links that need to be broken for the binding of an O_2 molecule depends on whether it is the first, second, third, or fourth to be bound. More salt links must be broken to permit the entry of the first O_2 than of subsequent ones. Because energy is required to break salt links, the binding of the first O_2 is energetically less favored than that of subsequent oxygen molecules. In this scheme, the binding affinity for the second and third O_2 molecules is intermediate between that for the first and last. This sequential increase in oxygen affinity would give the observed sigmoidal oxygen-binding curve.

BPG DECREASES OXYGEN AFFINITY BY CROSS-LINKING DEOXYHEMOGLOBIN

BPG binds specifically to deoxyhemoglobin in the ratio of one BPG per hemoglobin tetramer. This is an unusual stoichiometry—an $\alpha_2\beta_2$ protein would be expected to have at least two binding sites for a small molecule. The finding of one binding site immediately suggested that BPG binds on the symmetry axis of the hemoglobin molecule in the central cavity, where the four subunits are near each other (Figure 7-36). Indeed, x-ray analysis confirmed this proposal and showed that the binding site for BPG is constituted by three positively charged residues on *each* β chain: the α-amino group, lysine EF6, and histidine H21. These groups interact with the strongly negatively charged BPG, which carries nearly four negative charges at physiological pH. BPG is stereochemically complementary to this constellation of six positively charged groups facing the central cavity of the hemoglobin molecule (Figure 7-37). BPG binds more weakly to fetal hemoglobin than to hemoglobin A, because residue H21 in fetal hemoglobin is serine instead of histidine.

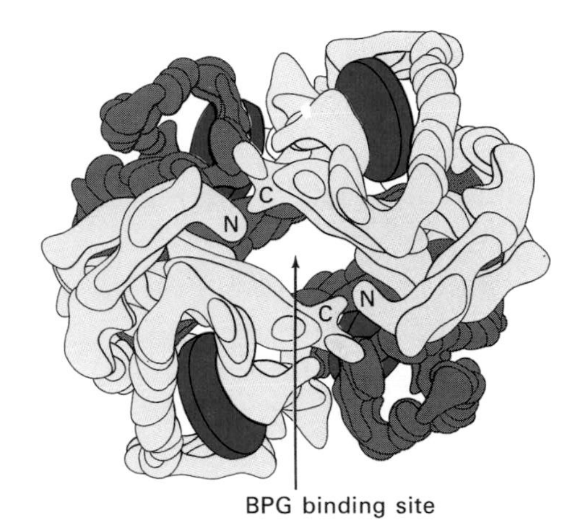

Figure 7-36
The binding site for BPG is in the central cavity of deoxyhemoglobin. [After M. F. Perutz. The hemoglobin molecule. Copyright © 1964 by Scientific American, Inc. All rights reserved.]

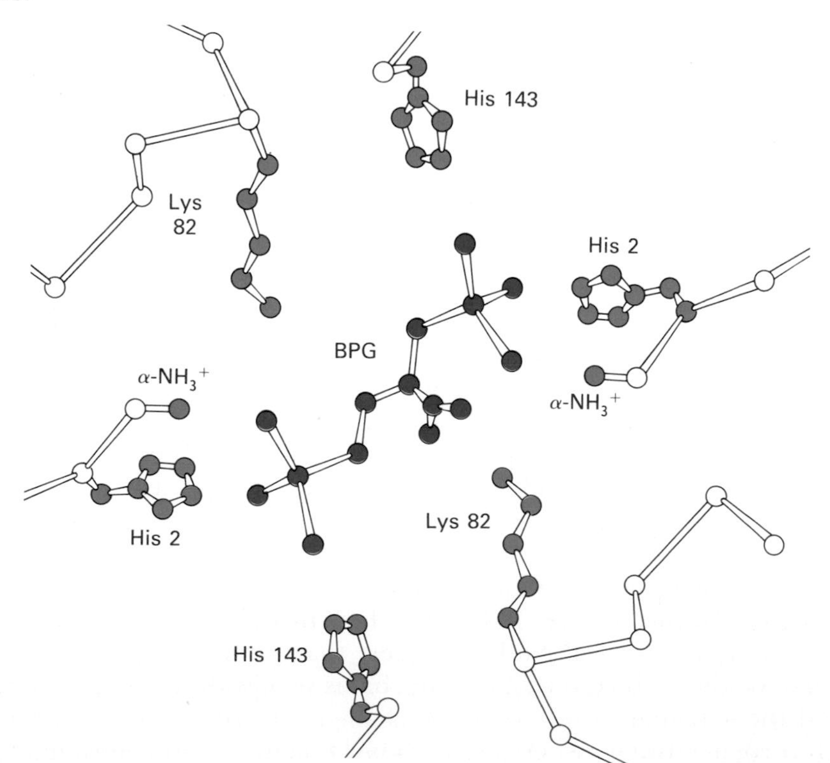

Figure 7-37
Mode of binding of BPG to human deoxyhemoglobin. BPG interacts with three positively charged groups on each β chain. [After A. Arnone. *Nature* 237(1972):148.]

On oxygenation, BPG is extruded because the central cavity becomes too small. Specifically, the gap between the H helices of the β chains becomes narrowed (see Figure 7-28). Also, the distance between the α-amino groups increases from 16 to 20 Å, which prevents them from simultaneously binding the phosphates of a BPG molecule.

The reason why BPG decreases oxygen affinity is now evident. *BPG stabilizes the deoxyhemoglobin quaternary structure by cross-linking the β chains.* In other words, BPG shifts the equilibrium toward the T form. As mentioned previously, the carboxyl-terminal residues of deoxyhemoglobin form eight salt links that must be broken for oxygenation to occur. The binding of BPG contributes additional cross-links that must be broken, and so the oxygen affinity of hemoglobin is diminished.

CO_2 BINDS TO THE TERMINAL AMINO GROUPS OF HEMOGLOBIN AND LOWERS ITS OXYGEN AFFINITY

In aerobic metabolism, about 0.8 equivalents of CO_2 are produced per O_2 consumed. Most of the CO_2 is transported as *bicarbonate,* which is formed within red cells by the action of *carbonic anhydrase:*

$$CO_2 + H_2O \rightleftharpoons HCO_3^- + H^+$$

Much of the H^+ generated by this reaction is taken up by deoxyhemoglobin as part of the Bohr effect. The CO_2 is carried by hemoglobin in the form of *carbamate,* because the un-ionized form of the α-amino groups of hemoglobin can react reversibly with CO_2:

$$R{-}NH_2 + CO_2 \rightleftharpoons R{-}N(H){-}C({=}O){-}O^- + H^+$$

The bound carbamates form salt bridges that stabilize the T form. *Hence, the binding of CO_2 lowers the oxygen affinity of hemoglobin.*

MECHANISM OF THE BOHR EFFECT

About 0.5 H^+ is taken up by hemoglobin for each molecule of O_2 that is released. This uptake of H^+ helps to buffer the pH in active tissues. It indicates that deoxygenation increases the affinities of some sites for H^+. Specifically, the p*K*s of some groups must be *raised* in the transition from oxy- to deoxyhemoglobin; an increase in p*K* means stronger binding of H^+. Which groups have their p*K*s raised? The potential sites and typical p*K* values are given in Figure 7-38. The side-chain carboxylates of glutamate and aspartate normally have p*K*s of about 4. It is unlikely that their p*K*s would be raised to at least 7 or 8, which would be necessary for a group to participate in the Bohr effect. On the other hand, the normal p*K*s of the side chains of tyrosine, lysine, and arginine are usually more than 10, and so it seems unlikely that they could be lowered enough to allow uptake of H^+. Hence, the most plausible candidates are the terminal amino group and the side chains of histidine and cysteine, which normally have p*K* values near 7.

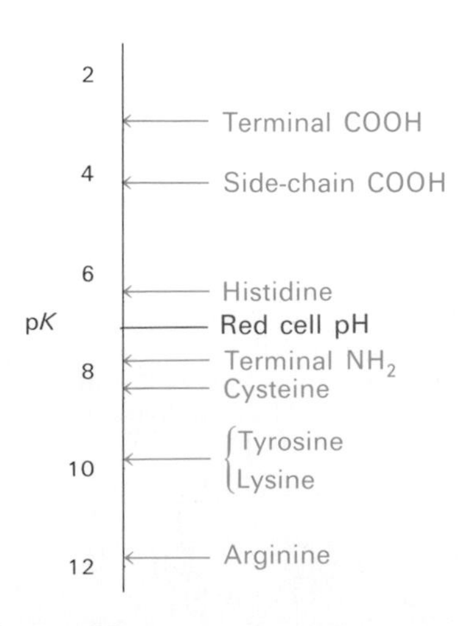

Figure 7-38
Typical p*K*s of some functional groups in proteins. The environment of a particular group can make its actual p*K* higher or lower than the value given here.

X-ray and chemical studies suggest that three groups account for much of the Bohr effect: the side chains of histidines 146β and 122α and the α-amino group of the α chain. In oxyhemoglobin, histidine 146β rotates freely, whereas, in deoxyhemoglobin, this terminal residue participates in a number of interactions. Of particular significance is the interaction of its imidazole ring with the negatively charged aspartate 94 in the FG corner of the same β chain. The close proximity of this negatively charged group enhances the likelihood that the imidazole group binds a proton (Figure 7-39). In other words, the proximity of aspartate 94 raises the pK of histidine 146. Thus, *in the transition from oxy- to deoxyhemoglobin, histidine 146 acquires a greater affinity for H^+ because its local environment becomes more negatively charged.* The immediate environments of the other two groups implicated in the Bohr effect also are more negatively charged in deoxyhemoglobin because of changes of quaternary structure induced by the release of O_2.

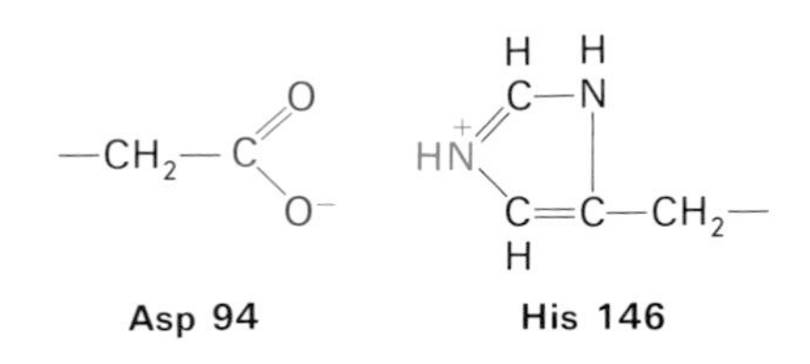

Figure 7-39
Aspartate 94 raises the pK of histidine 146 in deoxyhemoglobin but not in oxyhemoglobin. The proximity of the negative charge on aspartate 94 favors protonation of histidine 146 in deoxyhemoglobin.

COMMUNICATION WITHIN A PROTEIN MOLECULE

We have seen that the binding of O_2, H^+, CO_2, and BPG by hemoglobin are linked. These molecules are bound to spatially distinct sites that communicate with each other by means of conformational changes within the protein. The binding sites are separate because these molecules are structurally very different. *The interplay between these different sites is mediated by changes in quaternary structure.* In fact, every known allosteric protein consists of multiple polypeptide chains. The contact region between two chains can amplify and transmit conformational changes from one subunit to another. An allosteric protein does not have fixed properties. Rather, its functional characteristics are regulated by specific molecules in its environment. Consequently, allosteric interactions have immense importance in cellular function. *In the evolutionary transition from myoglobin to hemoglobin, a macromolecule capable of perceiving information from its environment has emerged.*

SICKLE-SHAPED RED BLOOD CELLS IN A CASE OF SEVERE ANEMIA

In 1904, James Herrick, a Chicago physician, examined a twenty-year-old black college student who had been admitted to the hospital because of a cough and fever. The patient felt weak and dizzy and had a headache. For about a year he had been experiencing palpitation and shortness of breath. On physical examination, the patient appeared rather well developed physically. There was a tinge of yellow in the whites of his eyes, and the visible mucous membranes were pale. His heart was distinctly enlarged. Examination of the blood showed that the patient was markedly anemic. The number of red cells was half of what is normal.

The patient's blood smear contained unusual red cells, which were described by Herrick in these terms: "The shape of the red cells was very irregular, but *what especially attracted attention was the large number of thin, elongated, sickle-shaped and crescent-shaped forms.*" The treatment was supportive, consisting of rest and nourishing food. The patient left the hospital four weeks later, less anemic and feeling much better. However, his blood still exhibited a "tendency to the peculiar crescent-shape in the red corpuscles though this was by no means as noticeable as before."

Figure 7-40
Red cells from the blood of a patient with sickle-cell anemia, as viewed under a light microscope. [Courtesy of Dr. Frank Bunn.]

Herrick was puzzled by the clinical picture and laboratory findings. Indeed, he waited six years before publishing the case history and then candidly asserted that "not even a definite diagnosis can be made." He noted the chronic nature of the disease, and the diversity of abnormal physical and laboratory findings: cardiac enlargement, a generalized swelling of lymph nodes, jaundice, anemia, and evidence of kidney damage. He concluded that the disease could not be explained on the basis of an organic lesion in any one organ. He singled out the abnormal blood picture as the key finding and titled his case report *Peculiar Elongated and Sickle-Shaped Red Blood Corpuscles in a Case of Severe Anemia*. Herrick suggested that "some unrecognized change in the composition of the corpuscle itself may be the determining factor."

SICKLE-CELL ANEMIA IS A GENETICALLY TRANSMITTED, CHRONIC, HEMOLYTIC DISEASE

Figure 7-41
Scanning electron micrograph of an erythrocyte from a patient with sickle-cell anemia. [Courtesy of Dr. Jerry Thornwaite and Dr. Robert Leif.]

Other cases of this disease, called *sickle-cell anemia*, were found soon after the publication of Herrick's description. Indeed, sickle-cell anemia is not a rare disease. It is a significant public health problem wherever there is a substantial black population. The incidence of sickle-cell anemia among blacks is about four per thousand. In the past, it has usually been a fatal disease, often before age thirty, as a result of infection, renal failure, cardiac failure, or thrombosis. Sickled red cells become trapped in the small blood vessels, which impairs the circulation and leads to damage of multiple organs. Sickled cells are more fragile than normal ones. They hemolyze readily and consequently have a shorter life than normal cells, which leads to a severe anemia. The chronic course of the disease is punctuated by crises in which the proportion of sickled cells is especially high. During such a crisis, the patient may go into shock.

Sickle-cell anemia is genetically transmitted. *Patients with sickle-cell anemia are homozygous for an abnormal gene* located on an autosomal chromosome. Offspring who receive the abnormal gene from one parent but its normal allele from the other have *sickle-cell trait. Such heterozygous people are usually not symptomatic.* Only 1% of the red cells in a heterozygote's venous circulation are sickled, in contrast with about 50% in a homozygote. However, sickle-cell trait, which occurs in about one of ten blacks, is not entirely benign. Vigorous physical activity at high altitude, air travel in unpressurized planes, and anesthesia are potentially hazardous to people with sickle-cell trait. The reason will be evident shortly.

THE SOLUBILITY OF DEOXYGENATED SICKLE HEMOGLOBIN IS ABNORMALLY LOW

Herrick correctly surmised the location of the defect in sickle-cell anemia. Red cells from a patient with this disease will sickle on a microscope slide in vitro if the concentration of oxygen is reduced. In fact, the hemoglobin in these cells is itself defective. Deoxygenated sickle-cell hemoglobin has an abnormally low solubility, only 4% of that of normal deoxygenated hemoglobin. A fibrous precipitate is formed when a concentrated solution of sickle-cell hemoglobin is deoxygenated. This precipitate deforms red cells and gives them their sickle shape. Sickle-cell hemoglobin is commonly referred to as *hemoglobin S* (Hb S) to distinguish it from hemoglobin A (Hb A), the normal adult hemoglobin.

HEMOGLOBIN S HAS AN ABNORMAL ELECTROPHORETIC MOBILITY

In 1949, Linus Pauling and his associates examined the physical-chemical properties of hemoglobin from normal people and from those with sickle-cell trait or sickle-cell anemia. Their experimental approach was to search for differences between these hemoglobins by electrophoresis. They found that the isoelectric point (p. 45) of sickle-cell hemoglobin is higher than that of normal hemoglobin in both the oxygenated and the deoxygenated state (Figure 7-42):

	Normal	*Sickle-cell anemia*	*Difference*
Oxyhemoglobin	6.87	7.09	0.22
Deoxyhemoglobin	6.68	6.91	0.23

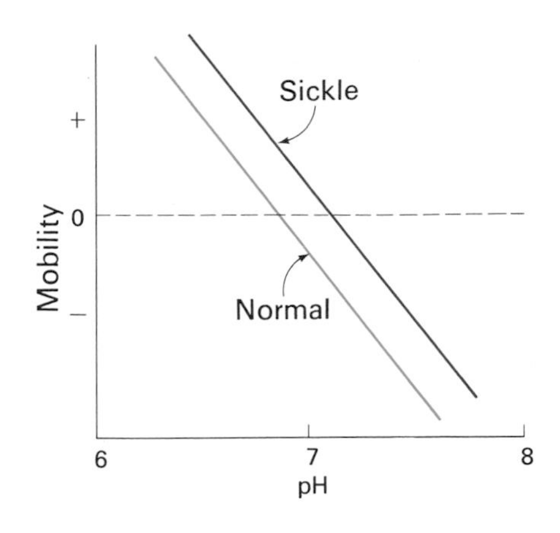

Figure 7-42
Electrophoretic mobility of sickle-cell hemoglobin and of normal hemoglobin as a function of pH. The isoelectric point of a molecule is the pH at which its mobility is 0.

These observations suggested that *there is a difference in the number or kind of ionizable groups in the two hemoglobins*. An estimate was made from the acid-base titration curve of hemoglobin in the neighborhood of pH 7. A change of one pH unit in the hemoglobin solution produces a change of about thirteen charges. The difference in isoelectric pH of 0.23 therefore corresponds to about three charges per hemoglobin molecule. It was concluded that *sickle-cell hemoglobin has between two and four more net positive charges per molecule than does normal hemoglobin*.

Patients with sickle-cell anemia (who are homozygous for the sickle gene) have hemoglobin S but no hemoglobin A. In contrast, people with sickle-cell trait (who are heterozygous for the sickle gene) have both kinds of hemoglobin in approximately equal amounts (Figure 7-43). Thus, Pauling's study revealed "*a clear case of a change produced in a protein molecule by an allelic change in a single gene*." This was the first demonstration of a *molecular disease*.

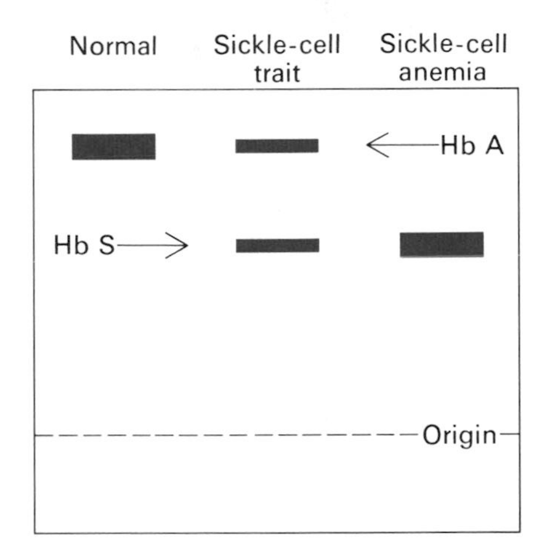

Figure 7-43
Gel electrophoresis pattern at pH 8.6 of hemoglobin isolated from a normal person, from a person with sickle-cell trait, and from a person with sickle-cell anemia.

A SINGLE AMINO ACID IN THE BETA CHAIN IS ALTERED IN SICKLE-CELL HEMOGLOBIN

The precise difference between normal and sickle-cell hemoglobin was identified in 1954, when Vernon Ingram devised a new technique for detecting amino acid substitutions in proteins. The hemoglobin molecule was split into smaller units for analysis, because it was anticipated that it would be easier to detect an altered amino acid in a peptide containing about 20 residues than in a protein ten times as large. Trypsin was used to specifically cleave hemoglobin on the carboxyl side of its lysine and arginine residues. Because an $\alpha\beta$ half of hemoglobin contains a total of 27 lysine and arginine residues, *tryptic digestion* produced 28 different peptides. The next step was to separate the peptides. This was accomplished by a *two-dimensional procedure* (Figure 7-44). The mix-

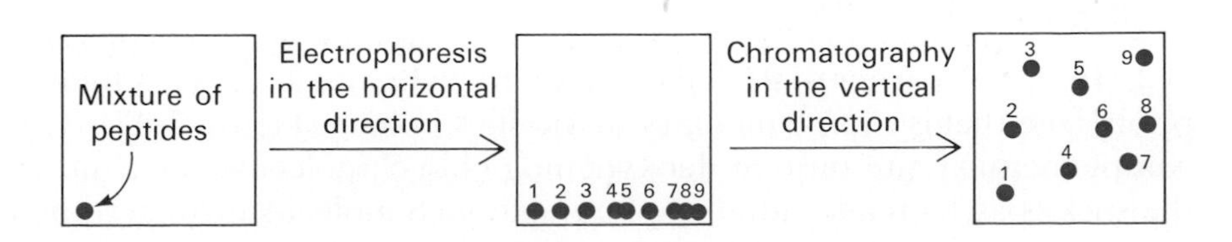

Figure 7-44
Fingerprinting. A mixture of peptides produced by proteolytic cleavage is resolved by electrophoresis in the horizontal direction followed by chromatography in the vertical direction.

> *Chromatography*—
> A term introduced by Mikhail Tswett in 1906 to describe the separation of a mixture of leaf pigments on a calcium carbonate column, like "light rays in the spectrum."
> Derived from the Greek words *chroma*, color, and *graphein*, to draw or write.

ture of peptides was placed in a spot at one corner of a large sheet of filter paper. *Electrophoresis* was first carried out in one direction, followed by *chromatography* in the perpendicular direction. Finally, the peptide spots were made visible by staining the filter paper with ninhydrin. This sequence of steps—selective cleavage of a protein into small peptides, followed by their separation in two dimensions—is called *fingerprinting*.

The fingerprints of hemoglobins A and S were highly revealing (Figure 7-45). When they were compared, *all but one of the peptide spots matched*. The one spot that was different was eluted from each fingerprint and shown to be a single peptide consisting of eight amino acids. Amino acid analysis indicated that this peptide in hemoglobin S differed from the one in hemoglobin A by a single amino acid. Ingram determined the sequence of this peptide and showed that *hemoglobin S contains valine instead of glutamate at position 6 of the β chain:*

Hemoglobin A	Val-His-Leu-Thr-Pro-Glu-Glu-Lys-
Hemoglobin S	Val-His-Leu-Thr-Pro-Val-Glu-Lys-
	β1 2 3 4 5 6 7 8

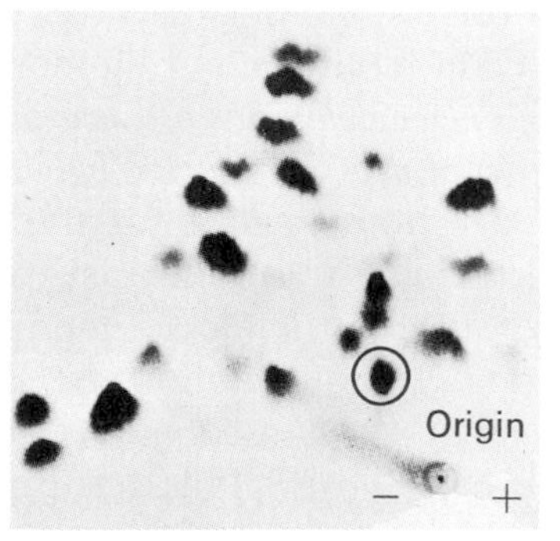

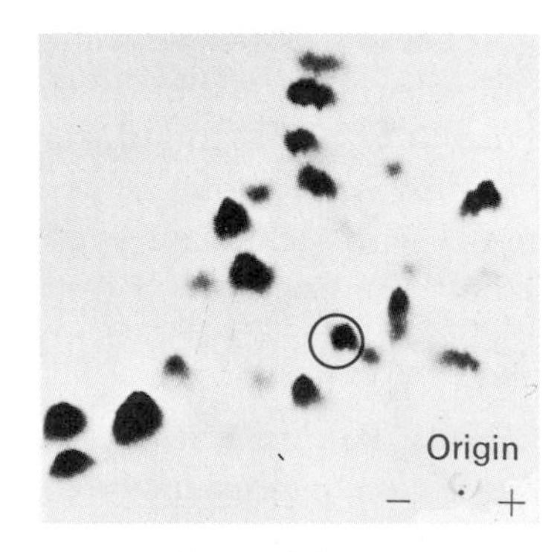

Figure 7-45
Comparison of the ninhydrin-stained fingerprints of hemoglobin A and hemoglobin S. The position of the peptide that is different in these hemoglobins is encircled in red. [Courtesy of Dr. Corrado Baglioni.]

SICKLE HEMOGLOBIN HAS STICKY PATCHES ON ITS SURFACE

The side chain of valine is distinctly nonpolar, whereas that of glutamate is highly polar. The substitution of valine for glutamate at position 6 of the β chains places a nonpolar residue on the outside of hemoglobin S (Figure 7-46). *This alteration markedly reduces the solubility of deoxygenated hemoglobin S but has little effect on the solubility of oxygenated hemoglobin S.* This fact is crucial to an understanding of the clinical picture of sickle-cell anemia and sickle-cell trait.

The molecular basis of sickling can be visualized as follows:

1. The substitution of valine for glutamic acid gives hemoglobin S a sticky patch on the outside of each of its β chains (Figure 7-47). This sticky patch is present on both oxy- and deoxyhemoglobin S and is missing from hemoglobin A.

2. In the EF corner of each β chain of deoxyhemoglobin S is a hydrophobic site that is complementary to the sticky patch (Figure 7-47). The complementary site on one deoxyhemoglobin S molecule can bind to the sticky patch on another deoxyhemoglobin S molecule, which results in the formation of long fibers that distort the red cell.

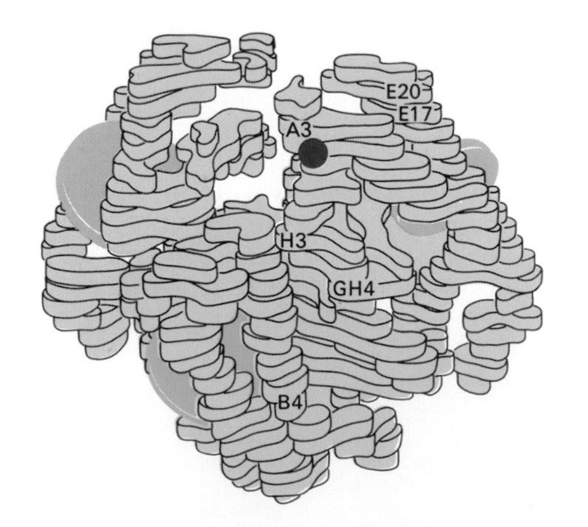

Figure 7-46
The position of the amino acid change in hemoglobin S (glutamate to valine at β6) is marked in red in this model of deoxyhemoglobin. Note that this site is at the surface of the protein. The α chains are shown in yellow, the β chains in blue. [After J. T. Finch, M. F. Perutz, J. F. Bertles, and J. Dobler. *Proc. Nat. Acad. Sci.* 70(1973):721.]

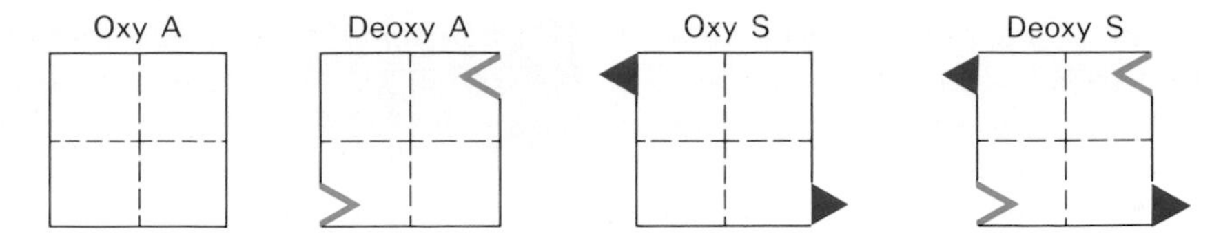

Figure 7-47
The red triangle represents the sticky patch that is present on both oxy- and deoxyhemoglobin S but not on either form of hemoglobin A. The complementary site is represented by an indentation that can accommodate the triangle. This complementary site is present in deoxyhemoglobin S and is probably also present in deoxyhemoglobin A.

3. In oxyhemoglobin S, the complementary site is masked. The sticky patch is present, but it cannot bind to another oxyhemoglobin S because the complementary site is unavailable.

4. Thus, *sickling occurs when there is a high concentration of the deoxygenated form of hemoglobin S* (Figure 7-48).

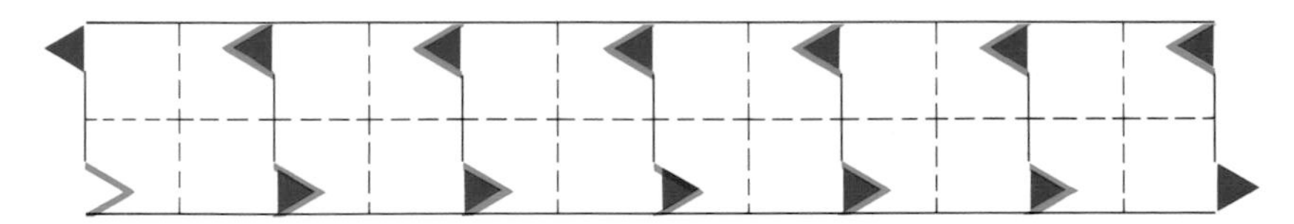

Figure 7-48
Interaction of sticky patches on deoxyhemoglobin S molecules with the complementary sites on other deoxyhemoglobin S molecules forms long strands.

These facts account for several clinical characteristics of sickle-cell anemia. A vicious cycle is set up when sickling occurs in a small blood vessel. The blockage of the vessel creates a local region of low oxygen concentration. Hence, more hemoglobin goes into the deoxy form and so more sickling occurs. A person with sickle-cell trait is usually asymptomatic because not more than half of his hemoglobin is hemoglobin S. This is too low a concentration for extensive sickling at normal oxygen levels. However, if the oxygen level is unusually low (as at high altitude), sickling can occur in such a person.

DEOXYHEMOGLOBIN S FORMS LONG HELICAL FIBERS

As described previously, deoxyhemoglobin S forms fibrous precipitates that deform red cells and give them their sickle shape. The main precipitate seen by electron microscopy consists of fibers having a diameter of 215 Å (Figure 7-49). Each fiber appears to be a fourteen-stranded

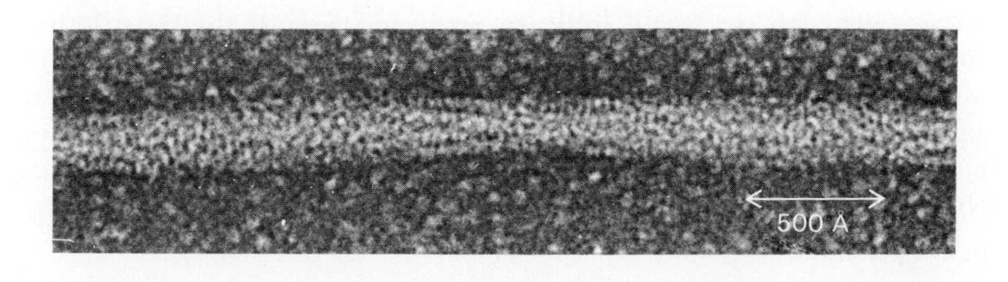

Figure 7-49
Electron micrograph of a negatively stained fiber of deoxyhemoglobin S. [From G. Dykes, R. H. Crepeau, and S. J. Edelstein. *Nature* 272(1978):509.]

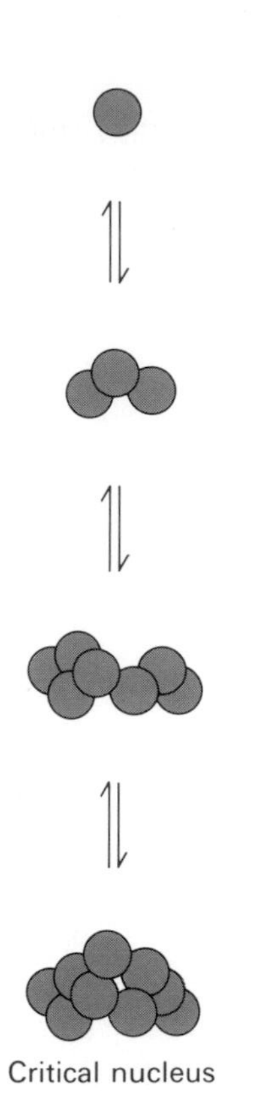

Figure 7-51
Nucleation phase in the formation of deoxyhemoglobin S fibers. The assembly of these nuclei is much slower than their subsequent growth. [After J. Hofrichter, P. D. Ross, and W. A. Eaton. *Proc. Nat. Acad. Sci.* 71(1974):4864.]

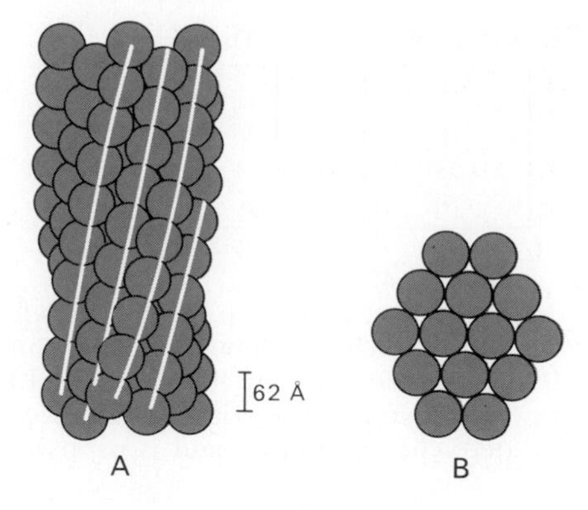

Figure 7-50
Fourteen-stranded helical model of the deoxyhemoglobin S fiber: (A) axial view; (B) cross-sectional view. Each circle represents a hemoglobin S molecule. [After G. Dykes, R. H. Crepeau, and S. J. Edelstein. *Nature* 272(1978):509.]

helix (Figure 7-50). A significant feature of this helix is that each hemoglobin S molecule makes contact with at least eight others. It is evident that the fiber is stabilized by multiple interactions.

What determines whether a red cell becomes sickled during its passage through the capillary circulation, which takes about a second? The most important determinant is the concentration of deoxyhemoglobin S. *The striking experimental finding is that the rate of fiber formation is proportional to the tenth power of the effective concentration of deoxyhemoglobin S. Thus, fiber formation is a highly concerted reaction.* These kinetic data indicate that nucleation is the rate-limiting phase in fiber formation. The fiber grows rapidly once a critical cluster of about ten deoxyhemoglobin S molecules has been formed (Figure 7-51). Such a nucleus may correspond to a major part of one turn of the fourteen-stranded helix. These findings have important clinical implications. They demonstrate that *kinetic,* as well as thermodynamic, factors are important in sickling. A red cell that is supersaturated with deoxyhemoglobin S will not sickle if the lag time for fiber formation is longer than the transit time from the peripheral capillaries to the alveoli of the lungs, where reoxygenation occurs. The very strong dependence of the polymerization rate on the concentration of deoxyhemoglobin S accounts for the fact that people with sickle-cell trait are usually asymptomatic. The concentration of deoxyhemoglobin S in the red cells of these heterozygotes is about half of that in homozygotes, and so their rate of fiber formation is about a thousandfold slower ($2^{10} = 1024$).

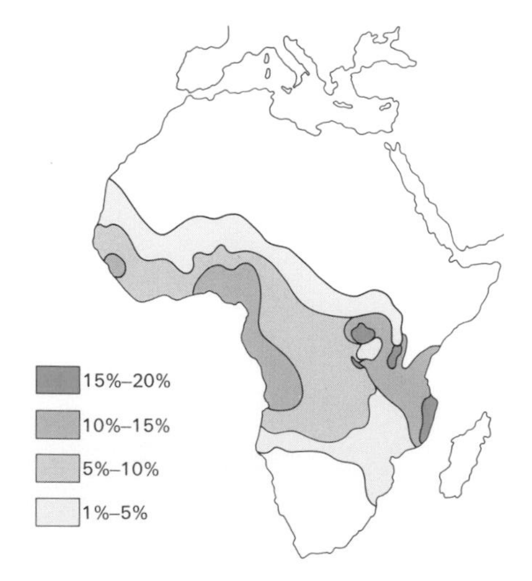

Figure 7-52
Frequency of the sickle-cell gene in Africa. High frequencies are restricted to regions where malaria is a major cause of death. [After A. C. Allison. Sickle cells and evolution. Copyright © 1956 by Scientific American, Inc. All rights reserved.]

HIGH INCIDENCE OF THE SICKLE GENE IS DUE TO THE PROTECTION CONFERRED AGAINST MALARIA

The frequency of the sickle gene is as high as 40% in certain parts of Africa. Until recently, most homozygotes have died before reaching adulthood, and so there must have been strong selective pressures to maintain the high incidence of the gene. James Neel proposed that the heterozygote enjoys advantages not shared by either the normal homozygote or the sickle-cell homozygote. In fact, Anthony Allison found that people with sickle-cell trait are protected against the most lethal form of malaria. The incidence of malaria and the frequency of the sickle gene in Africa are definitely correlated (Figure 7-52). This is a clear-cut example of *balanced polymorphism*—the heterozygote is protected against malaria and does not suffer from sickle-cell disease, whereas the normal homozygote is vulnerable to malaria.

FETAL DNA CAN BE ANALYZED FOR THE PRESENCE OF THE SICKLE-CELL GENE

The substitution of valine for glutamate at $\beta6$ in hemoglobin S results from a change in a single base, T for A. This mutation can readily be detected by cleaving DNA with a restriction enzyme that recognizes the sequence in this immediate region. The target for the restriction endonuclease MstII is the palindromic sequence CCTNAGG (where N denotes any base), which is present in the gene for the β chain of hemoglobin A (β^{A} gene) but not in the one for hemoglobin S (β^{S}). Because of the absence of this target site in the β^{S} gene, complete digestion of the gene by MstII produces a 1.3-kb fragment, corresponding to a 1.1-kb fragment from the β^{A} gene (Figure 7-53A). The fragments in the digested sample of DNA are separated by gel electrophoresis and visualized by Southern blotting (p. 120) with a ^{32}P-labeled DNA probe that is complementary to the 1.1-kb fragment. The 1.3-kb fragment is also stained by this probe because it contains the 1.1-kb sequence. An autoradiogram reveals whether the β^{A} gene, the β^{S} gene, or both are present in the DNA sample (Figure 7-53B).

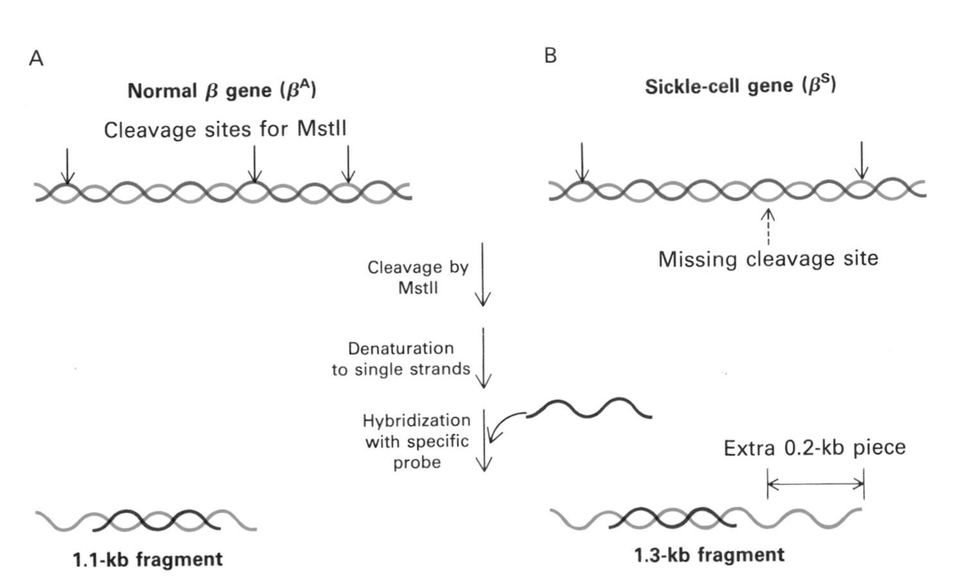

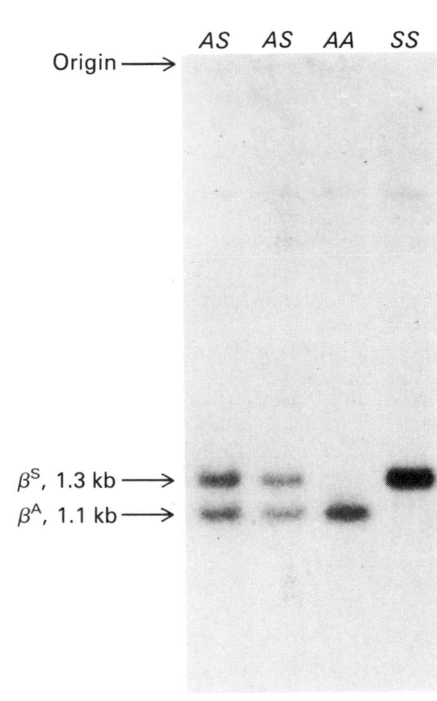

Figure 7-53
Restriction endonuclease method for detecting the sickle-cell gene. (A) Target site in the gene and fragments produced by digestion. (B) Electrophoresis pattern of a digest from parents who are heterozygous for the gene (lanes labeled *AS*), a normal child (*AA*), and a child with sickle-cell anemia (*SS*). [Part B is from Y. W. Kan. In *Medicine, Science, and Society*, K. J. Isselbacher, ed., (Wiley, 1984), p. 297.]

An attractive feature of this restriction enzyme method is that the DNA sample can come from any fetal cell. In contrast, a hemoglobin sample could be obtained only from red blood cells or their precursors. A sample of amniotic fluid is obtained from the fetus more readily and with less potential hazard than are blood-forming cells. DNA analyses can also be performed on biopsy samples of the chorionic villi obtained early in pregnancy, at about eight weeks gestation. These techniques for obtaining and analyzing genomic DNA are generally applicable. The number of genetic diseases that can be detected early in pregnancy by restriction enzyme cleavage of fetal DNA followed by Southern blotting with highly specific probes is increasing rapidly. Parents of fetuses who are at risk can now make informed decisions as to whether to terminate pregnancy.

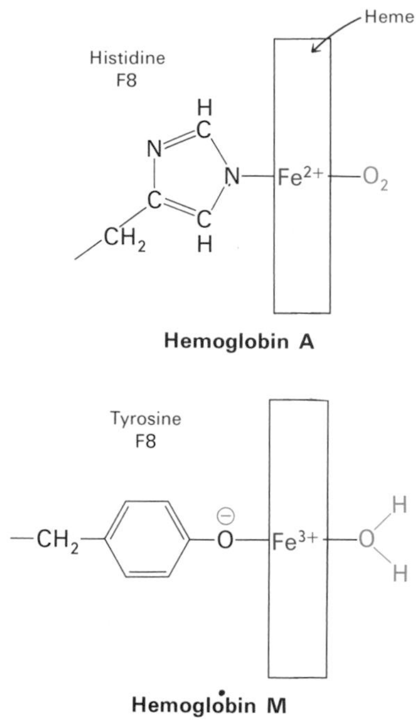

Histidine F8

Heme

Hemoglobin A

Tyrosine F8

Hemoglobin M

Figure 7-54
Substitution of tyrosine for the proximal histidine (F8) results in the formation of a hemoglobin M. The negatively charged oxygen atom of tyrosine is coordinated to the iron atom, which is in the ferric state. Water rather than O_2 is bound at the sixth coordination position.

MOLECULAR PATHOLOGY OF HEMOGLOBIN

More than three hundred abnormal hemoglobins have been discovered by the examination of patients with clinical symptoms and by electrophoretic surveys of normal populations. In northern Europe, one of three hundred persons is heterozygous for a variant of hemoglobin A. The frequency of any one mutant allele is less than 10^{-4}, which is lower by several orders of magnitude than the frequency of the sickle allele in regions where malaria is endemic. In other words, most abnormal hemoglobins do not confer a selective advantage on the person. They are almost always neutral or harmful.

Abnormal hemoglobins are of several types:

1. *Altered exterior.* Nearly all substitutions on the surface of the hemoglobin molecule are harmless. Hemoglobin S is a striking exception.

2. *Altered active site.* The defective subunit cannot bind oxygen because of a structural change near the heme that directly affects oxygen binding. For example, substitution of tyrosine for the histidine proximal or distal to heme stabilizes the heme in the ferric form, which can no longer bind oxygen (Figure 7-54). The tyrosine side chain is ionized in this complex with the ferric ion of the heme. Mutant hemoglobins characterized by a permanent ferric state of two of the hemes are called *hemoglobin M*. The letter M signifies that the altered chains are in the *methemoglobin* (ferrihemoglobin) form. These patients are usually cyanotic. The disease has been seen only in heterozygous form, because homozygosity would almost certainly be lethal.

3. *Altered tertiary structure.* The polypeptide chain is prevented by the amino acid substitution from folding into its normal conformation. These hemoglobins are usually unstable. For example, in hemoglobin Hammersmith, phenylalanine at CD1, adjacent to the heme, is replaced by serine (Figure 7-55). The affinity of this hemoglobin for its heme groups is much lower than normal. Amino acid substitutions at sites far from the heme also can prevent hemoglobin from folding into its normal conformation. An instructive mutant is hemoglobin Riverdale-Bronx, which has arginine in place of glycine at B6. Recall that glycine occupies this position in all known normal myoglobins and hemoglobins (p. 153). This mutant hemoglobin does not fold normally because arginine at B6 is too large to fit into the narrow space between the B and E helices (see Figure 7-20).

4. *Altered quaternary structure.* Some mutations at subunit interfaces lead to the loss of allosteric properties. These hemoglobins usually have an abnormal oxygen affinity. The $\alpha_1\beta_2$ contact region, which changes markedly on oxygenation, is especially vulnerable to effects of mutation.

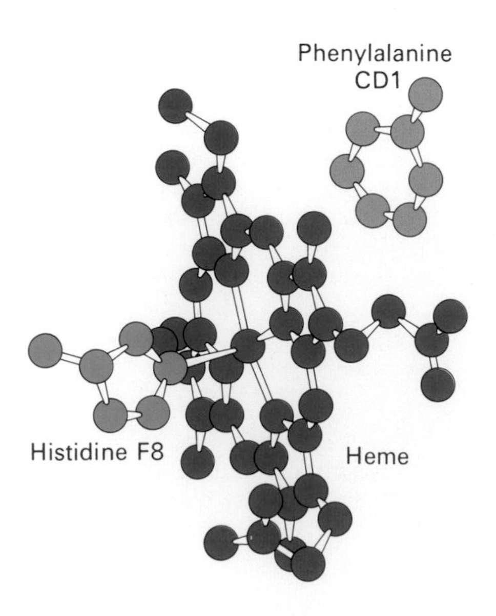

Figure 7-55
Location of phenylalanine CD1, in normal hemoglobin. The aromatic ring of this residue is in contact with the heme. In hemoglobin Hammersmith, serine replaces this phenylalanine residue; this markedly weakens the binding of heme.

THALASSEMIAS ARE GENETIC DISORDERS OF HEMOGLOBIN SYNTHESIS

The genetic diseases considered thus far are ones in which hemoglobin molecules are produced in essentially normal amounts but have impaired function because of the change of a single amino acid residue. The *thalassemias*, a different class of genetic disorders, are characterized by *defective synthesis of one or more hemoglobin chains*. The chain that is affected is denoted by a prefix, as in α-thalassemia. The name of this group of diseases comes from the Greek word "thalassa," meaning "sea," because of the high incidence of some forms of thalassemia among people living near the Mediterranean Sea. Indeed, some 20% of the population in parts of Italy are carriers of a gene for this disease. The geographic distribution of some thalassemia genes parallels that of malaria, which suggests that the heterozygote benefits from the presence of the gene, as in sickle-cell trait.

Thalassemias are produced by many different mutations that lead to the absence or deficiency of a globin chain in a variety of ways:

1. *The gene is missing.*

2. The gene is present but *RNA synthesis or processing is impaired.* For example, a mutation in the TATA box of the promoter site (p. 98) decreases the amount of RNA synthesized. Mutations within introns or near the exon-intron boundary can lead to aberrant splicing of the primary RNA transcript.

3. Globin mRNA is produced in normal amount but it encodes a grossly abnormal protein. For example, mutation of an amino acid codon to a stop codon (say TGG for tryptophan to TGA) will result in *premature termination* of protein synthesis. The deletion or addition of a nucleotide, a *frameshift mutation*, leads to an entirely different amino acid sequence on the distal side of the mutation. Most of these hemoglobin chains are very unstable and are rapidly degraded.

IMPACT OF THE DISCOVERY OF MOLECULAR DISEASES

Analyses of mutations affecting oxygen transport have had a major impact on molecular biology, medicine, and genetics. Their significance is threefold:

1. *They are sources of insight into relations between the structure and function of normal hemoglobin.* Mutations of a single amino acid residue are highly specific chemical modifications provided by nature. They shed light on facets of the protein that are critical for function.

2. *The discovery of mutant hemoglobins has revealed that disease can arise from a change of a single amino acid in one kind of polypeptide chain.* The concept of molecular disease, which is now an integral part of medicine, had its origins in the incisive studies of sickle-cell hemoglobin. The thalassemias have provided striking illustrations of the consequences of aberrant transcription, RNA splicing, and protein synthesis.

3. *The finding of mutant hemoglobins has enhanced our understanding of evolutionary processes.* Mutations are the raw material of evolution; the studies of sickle-cell anemia have shown that a mutation may be simultaneously beneficial and harmful.

On disease and evolution—
"Subjectively, to evolve must most often have amounted to suffering from a disease. And these diseases were of course molecular. The appearance of the concept of good and evil, interpreted by man as his painful expulsion from Paradise, was probably a molecular disease that turned out to be evolution."

From E. Zuckerkandl and L. Pauling. In *Horizons in Biochemistry*, M. Kasha and B. Pullman, eds. (Academic Press, 1964), pp. 189–225.

SUMMARY

Myoglobin and hemoglobin are the oxygen-carrying proteins in vertebrates. Myoglobin facilitates the transport of oxygen in muscle and serves as a reserve store of oxygen, whereas hemoglobin is the oxygen carrier in blood. These proteins contain tightly bound heme, a substituted porphyrin with a central iron atom. The ferrous (+2) state of heme binds O_2, whereas the ferric (+3) state does not.

Myoglobin, a single polypeptide chain of 153 residues (18 kd), has a compact shape. The inside of myoglobin consists almost exclusively of nonpolar residues. About 75% of the polypeptide chain is α helical. The single ferrous heme group is located in a nonpolar niche, which protects it from oxidation to the ferric form. The iron atom of the heme is directly bonded to a nitrogen atom of a histidine side chain. This proximal histidine occupies the fifth coordination position. The sixth coordination position on the other side of the heme plane is the binding site for O_2. The nearby distal histidine diminishes the binding of CO at the oxygen-binding site and inhibits the oxidation of heme to the ferric state. The central exon of the myoglobin gene encodes nearly all of the heme-binding site.

Hemoglobin consists of four polypeptide chains, each with a heme group. Hemoglobin A, the predominant hemoglobin in adults, has the subunit structure $\alpha_2\beta_2$. The three-dimensional structure of the α and β chains of hemoglobin is strikingly similar to that of myoglobin. Yet new properties appear in tetrameric hemoglobin that are not present in monomeric myoglobin. Hemoglobin transports H^+ and CO_2 in addition to O_2. Furthermore, the binding of these molecules is regulated by allosteric interactions, which are interactions between separate sites on the same protein. Indeed, hemoglobin is the best-understood allosteric protein.

Hemoglobin exhibits three kinds of allosteric effects. First, the oxygen-binding curve of hemoglobin is sigmoidal, which means that the binding of oxygen is cooperative. The binding of oxygen to one heme facilitates the binding of oxygen to the other hemes in the same molecule. Second, H^+ and CO_2 promote the release of O_2 from hemoglobin, an effect that is physiologically important in enhancing the release of O_2 in metabolically active tissues such as muscle. Conversely, O_2 promotes the release of H^+ and CO_2 in the alveolar capillaries of the lungs. These allosteric linkages between the bindings of H^+, CO_2, and O_2 are known as the Bohr effect. Third, the affinity of hemoglobin for O_2 is further regulated by 2,3-bisphosphoglycerate (BPG), a small molecule with a high density of negative charge. BPG can bind to deoxyhemoglobin but not to oxyhemoglobin. Hence, BPG lowers the oxygen affinity of hemoglobin. Fetal hemoglobin ($\alpha_2\gamma_2$) has a higher oxygen affinity than adult hemoglobin because it binds BPG less tightly.

The allosteric properties of hemoglobin arise from interactions between its α and β subunits. The T (tense) quaternary structure is constrained by salt links between different subunits, giving it a low affinity for O_2. These intersubunit salt links are absent from the R (relaxed) form, which has a high affinity for O_2. On oxygenation, the iron atom moves into the plane of the heme, pulling the proximal histidine with it. This motion cleaves some of the salt links, and the equilibrium is shifted from T to R. BPG stabilizes the deoxy state by binding to positively charged groups surrounding the central cavity of hemoglobin. Carbon dioxide, another allosteric effector, binds to the terminal amino groups of all four chains by forming readily reversible carbamate linkages. The hydrogen ions participating in the Bohr effect are bound to several pairs of sites that have a more negatively charged environment in the deoxy than in the oxy state.

Gene mutation resulting in the change of a single amino acid in a single protein can produce disease. The best-understood molecular disease is sickle-cell anemia. Hemoglobin S, the abnormal hemoglobin in this disease, consists of two normal α chains and two mutant β chains. In hemoglobin S, glutamate at residue 6 of the β chain is replaced by valine. This substitution of a nonpolar side chain for a polar one drastically reduces the solubility of deoxyhemoglobin S, which leads to the formation of fibrous precipitates that deform the red cell and give it a sickle shape. The resulting destruction of red cells produces a chronic hemolytic anemia. Sickle-cell anemia arises when a person is homozygous for the mutant sickle gene. The heterozygous condition, called sickle-cell trait, is relatively asymptomatic. About one of ten blacks in the United States is heterozygous for the sickle gene, and as many as four of ten in some parts of Africa. This high incidence of a gene that is harmful in homozygotes is due its beneficial effect in heterozygotes—people with sickle-cell trait are protected against the most lethal form of malaria. This is an example of balanced polymorphism. Sickle-cell anemia can be diagnosed in utero by restriction-endonuclease digestion of a sample of fetal DNA, followed by Southern blotting.

Several hundred mutant hemoglobins have been found by studies of the hemoglobin of patients with hematologic symptoms and by surveys of normal populations. Several classes of mutant hemoglobins are known: (1) Most substitutions on the surface of hemoglobin are harmless. Hemoglobin S is a striking exception. (2) Most substitutions near the heme impair the oxygen-binding site. For example, replacement of the proximal or distal histidine with tyrosine locks the heme in the ferric state, which cannot bind oxygen. (3) Many alterations in the interior of the molecule distort the tertiary structure and produce unstable hemoglobins. (4) Many changes at subunit interfaces lead to changes in oxygen affinity and the loss of allosteric properties. Thalassemias, a different class of genetic disorders, are characterized by the absence or defect of a globin chain. Some underlying causes are (1) absence of a globin gene, (2) impaired transcription or defective RNA processing, and (3) premature termination of protein synthesis or a shift in the reading frame.

SELECTED READINGS

Where to start

Kendrew, J. C., 1961. The three-dimensional structure of a protein molecule. *Sci. Amer.* 205(6):96–11. [Available as *Sci. Amer.* Offprint 121.]

Perutz, M. F., 1978. Hemoglobin structure and respiratory transport. *Sci. Amer.* 239(6):92–125. [Available as *Sci. Amer.* Offprint 1413.]

Dickerson, R. E., and Geis, I., 1983. *Hemoglobin: Structure, Function, Evolution and Pathology*. Benjamin/Cummings. [A beautifully illustrated account with a broad perspective.]

Structure of myoglobin and hemoglobin

Fermi, G., and Perutz, M. F., 1981. *Atlas of Molecular Structures in Biology. 2. Haemoglobin and myoglobin.* Clarendon Press, Oxford.

Takano, T., 1977. Structure of myoglobin refined at 2.0 Å resolution. *J. Mol. Biol.* 110:537–584.

Shaanan, B., 1983. Structure of human oxyhaemoglobin at 2.1 Å resolution. *J. Mol. Biol.* 171:31–59.

Fermi, G., Perutz, M. F., Shaanan, B., and Fourme, R., 1984. The crystal structure of human deoxyhaemoglobin at 1.74 Å resolution. *J. Mol. Biol.* 175:159–174.

Model systems

Collman, J. P., 1977. Synthetic models for the oxygen-binding hemoproteins. *Acc. Chem. Res.* 10:265–272. [A highly informative review of picket-fence porphyrins.]

Collman, J. P., Brauman, J. I., Halbert, T. R., and Suslick, K. S., 1976. Nature of O_2 and CO binding to metalloporphyrins and heme proteins. *Proc. Nat. Acad. Sci.* 73:3333–3337. [Presents the structural basis for the decreased binding of carbon monoxide by myoglobin and hemoglobin.]

Interaction of hemoglobin with H^+, CO_2, and BPG

Kilmartin, J. V., 1976. Interaction of haemoglobin with protons, CO_2, and 2,3-diphosphoglycerate. *Brit. Med. Bull.* 32:209–222.

Benesch, R., and Benesch, R. E., 1969. Intracellular organic phosphates as regulators of oxygen release by haemoglobin. *Nature* 221:618–622.

Tyuma, I., and Shimizu, K., 1970. Effect of organic phosphates on the difference in oxygen affinity between fetal and adult human hemoglobin. *Fed. Proc.* 29:1112–1114.

Allosteric mechanism of hemoglobin

Ho, C., (ed.), 1982. *Hemoglobin and Oxygen Binding.* Elsevier.

Perutz, M. F., 1980. Stereochemical mechanism of oxygen transport by haemoglobin. *Proc. Roy. Soc. Lond. Ser. B* 208:135–162.

Baldwin, J., and Chothia, C., 1979. Hemoglobin: the structural changes related to ligand binding and its allosteric mechanism. *J. Mol. Biol.* 129:175–220.

Shulman, R. G., Hopfield, J. J., and Ogawa, S., 1975. Allosteric interpretation of haemoglobin properties. *Quart. Rev. Biophys.* 8:325–420.

Gelin, B. R., Lee, A. W.-M., and Karplus, M., 1983. Hemoglobin tertiary structure change on ligand binding: its role in the cooperative mechanism. *J. Mol. Biol.* 171:489–559.

Friedman, J. M., 1985. Structure, dynamics, and reactivity in hemoglobin. *Science* 228:1273–1280.

Sickle-cell anemia and hemoglobin S

Embury, S. H., 1986. The clinical pathophysiology of sickle-cell disease. *Ann. Rev. Med.* 37:361–376.

Noguchi, C. T., and Schechter, A. N., 1985. Sickle hemoglobin polymerization in solution and in cells. *Ann. Rev. Biophys. Biophys. Chem.* 14:239–263.

Herrick, J. B., 1910. Peculiar elongated and sickle-shaped red blood corpuscles in a case of severe anemia. *Arch. Intern. Med.* 6:517–521.

Pauling, L., Itano, H. A., Singer, S. J., and Wells, I. C., 1949. Sickle cell anemia: a molecular disease. *Science* 110:543–548.

Ingram, V. M., 1957. Gene mutation in human haemoglobin: the chemical difference between normal and sickle cell haemoglobin. *Nature* 180:326–328.

Allison, A. C., 1956. Sickle cells and evolution. *Sci. Amer.* 195(2):87–94. [Available as *Sci. Amer.* Offprint 1065.]

Other genetic disorders of hemoglobin

Embury, S. H., Scharf, S. J., Saiki, R. K., Gholson, M. A., Golbus, M., Arnheim, N., and Erlich, H. A., 1987. Rapid prenatal diagnosis of sickle cell anemia by a new method of DNA analysis. *New Eng. J. Med.* 316:656–661.

Winslow, R. M., and Anderson, W. F., 1983. The hemoglobinopathies. *In* Stanbury, J. B., Wyngaarden, J. B., Fredrickson, D. S., Goldstein, J. L., and Brown, M. S., (eds.), *The Metabolic Basis of Inherited Disease* (5th ed.), pp. 1666–1710. McGraw-Hill.

Kan, Y. W., 1983. The thalassemias. *In* Stanbury, J. B., Wyngaarden, J. B., Fredrickson, D. S., Goldstein, J. L., and Brown, M. S., (eds.), *The Metabolic Basis of Inherited Disease* (5th ed.), pp. 1711–1725.

Weatherall, D. J., and Clegg, J. B., 1982. Thalassemia revisited. *Cell* 29:7–9.

Orkin, S. H., and Kazazian, H. H., Jr., 1984. The mutation and polymorphism of the human β-globin gene and its surrounding DNA. *Ann. Rev. Genet.* 18:131–171.

Honig, G. R., and Adams, J. G., 1986. *Human Hemoglobin Genetics.* Springer-Verlag.

Stamatoyannopoulos, G., Niehaus, A. W., Leder, P., and Majerus, P. W., (eds.), 1986. *Molecular Basis of Blood Diseases.* Saunders.

DNA probe detection of abnormal hemoglobins

Chang, J. C., and Kan, Y. W., 1982. A sensitive new prenatal test for sickle cell anemia. *New Engl. J. Med.* 307:30–32.

Goosens, M., Dumez, Y., Kaplan, L., Lupker, M., Charbet, C., Henrion, R., and Rosa, J., 1983. Prenatal diagnosis of sickle-cell anemia in the first trimester of pregnancy. *New Engl. J. Med.* 309:831–833.

Saiki, R. K., Scharf, S., Faloona, F., Mullis, K. B., Horn, G. T., Erlich, H. A., and Arnheim, N., 1985. Enzymatic amplification of beta-globin genomic sequences and restriction site analysis for diagnosis of sickle cell anemia. *Science* 230:1350–1354.

Bunn, H. F., and Forget, B. G., 1986. *Hemoglobin: Molecular, Genetic and Clinical Aspects.* Saunders.

Relation of exons to structural units of proteins

Go, M., 1981. Correlation of DNA exonic regions with protein structural units in haemoglobin. *Nature* 291:90–92.

Craik, C. S., Buchman, S. R., and Beychok, S., 1981. O_2 binding properties of the product of the central exon of beta-globin gene. *Nature* 291:87–90.

De Sanctis, G., Falcioni, G., Giardina, B., Ascoli, F., and Brunori, M., 1986. Mini-myoglobin: preparation and reaction with oxygen and carbon monoxide. *J. Mol. Biol.* 188:73–76.

PROBLEMS

1. The average volume of a red blood cell is 87 cubic micrometers. The mean concentration of hemoglobin in red cells is 34 g/100 ml.
   (a) What is the weight of the hemoglobin contained in a red cell?
   (b) How many hemoglobin molecules are there in a red cell?
   (c) Could the hemoglobin concentration in red cells be much higher than the observed value? (Hint: Suppose that a red cell contained a crystalline array of hemoglobin molecules 65 Å apart in a cubic lattice.)

2. How much iron is there in the hemoglobin of a 70-kg adult? Assume that the blood volume is 70 ml/kg of body weight and that the hemoglobin content of blood is 16 g/100 ml.

3. The myoglobin content of some human muscles is about 8 g/kg. In sperm whale, the myoglobin content of muscle is about 80 g/kg.
   (a) How much O_2 is bound to myoglobin in human muscle and in that of sperm whale? Assume that the myoglobin is saturated with O_2.
   (b) The amount of oxygen dissolved in tissue water (in equilibrium with venous blood at 37°C) is about 3.5×10^{-5} M. What is the ratio of oxygen bound to myoglobin to that directly dissolved in the water of sperm whale muscle?

4. The equilibrium constant K for the binding of oxygen to myoglobin is 10^{-6} M, where K is defined as

$$K = \frac{[\text{Mb}][\text{O}_2]}{[\text{MbO}_2]}$$

   The rate constant for the combination of O_2 with myoglobin is $2 \times 10^7\ \text{M}^{-1}\ \text{s}^{-1}$.
   (a) What is the rate constant for the dissociation of O_2 from oxymyoglobin?
   (b) What is the mean duration of the oxymyoglobin complex?

5. What is the effect of each of the following treatments on the oxygen affinity of hemoglobin A in vitro?
   (a) Increase in pH from 7.2 to 7.4.
   (b) Increase in pCO_2 from 10 to 40 torrs.
   (c) Increase in [BPG] from 2×10^{-4} to 8×10^{-4} M.
   (d) Dissociation of $\alpha_2\beta_2$ into monomer subunits.

6. The erythrocytes of birds and turtles contain a regulatory molecule different from BPG. This substance is also effective in reducing the oxygen affinity of human hemoglobin stripped of BPG. Which of the following substances would you predict to be most effective in this regard?
   (a) Glucose 6-phosphate.
   (b) Inositol hexaphosphate.
   (c) HPO_4^{2-}.
   (d) Malonate.
   (e) Arginine.
   (f) Lactate.

7. The pK of an acid depends partly on its environment. Predict the effect of the following environmental changes on the pK of a glutamic acid side chain.
   (a) A lysine side chain is brought into close proximity.
   (b) The terminal carboxyl group of the protein is brought into close proximity.
   (c) The glutamic acid side chain is shifted from the outside of the protein to a nonpolar site inside.

8. The concept of linkage is crucial for the understanding of many biochemical processes. Consider a protein molecule P that can bind A or B or both:

$$\begin{array}{ccc} \text{P} & \overset{K_A}{\rightleftharpoons} & \text{PA} \\ K_B \updownarrow & & \updownarrow K_{AB} \\ \text{PB} & \underset{K_{BA}}{\rightleftharpoons} & \text{PAB} \end{array}$$

   The dissociation constants for these equilibria are defined as

$$K_A = \frac{[\text{P}][\text{A}]}{[\text{PA}]} \qquad K_B = \frac{[\text{P}][\text{B}]}{[\text{PB}]}$$

$$K_{BA} = \frac{[\text{PB}][\text{A}]}{[\text{PAB}]} \qquad K_{AB} = \frac{[\text{PA}][\text{B}]}{[\text{PAB}]}$$

   (a) Suppose $K_A = 5 \times 10^{-4}$ M, $K_B = 10^{-3}$ M, and $K_{BA} = 10^{-5}$ M. Is the value of the fourth dissociation constant K_{AB} defined? If so, what is it?
   (b) What is the effect of [A] on the binding of B? What is the effect of [B] on the binding of A?

9. Carbon monoxide combines with hemoglobin to form CO-hemoglobin. Crystals of CO-hemoglobin are isomorphous with those of oxyhemoglobin. Each heme in hemoglobin can bind one carbon monoxide molecule, but O_2 and CO cannot simultaneously bind to the same heme. The binding affinity for CO is about 200 times as great as that for oxygen. Exposure for 1 hour to a CO concentration of 0.1% in inspired air leads to the occupancy by CO of about half of the heme sites in hemoglobin, a proportion that is frequently fatal.
   An interesting problem was posed (and partly solved) by J. S. Haldane and J. G. Priestley in 1935:

   > If the action of CO were simply to diminish the oxygen-carrying power of the hemoglobin, without other modification of its properties, the symptoms of CO poisoning would be very difficult to understand in the light of other knowledge. Thus, a person whose blood is half-saturated with CO is practically helpless, as we

have just seen; but a person whose hemoglobin percentage is simply diminished to half by anemia may be going about his work as usual.

What is the key to this seeming paradox?

10. A protein molecule P reversibly binds a small molecule L. The dissociation constant K for the equilibrium

$$P + L \rightleftharpoons PL$$

is defined as

$$K = \frac{[P][L]}{[PL]}$$

The protein transports the small molecule from a region of high concentration $[L_A]$ to one of low concentration $[L_B]$. Assume that the concentrations of unbound small molecules remain constant. The protein goes back and forth between regions A and B.
(a) Suppose $[L_A] = 10^{-4}$ M and $[L_B] = 10^{-6}$ M. What value of K yields maximal transport? One way of solving this problem is to write an expression for ΔY, the change in saturation of the ligand-binding site in going from region A to B. Then, take the derivative of ΔY with respect to K.
(b) Treat oxygen transport by hemoglobin in a similar way. What value of P_{50} would give a maximal ΔY? Assume that the P in the lungs is 100 torrs, whereas P in the tissue capillaries is 20 torrs. Compare your calculated value of P_{50} with the physiological value of 26 torrs.

11. A hemoglobin with an abnormal electrophoretic mobility is detected in a screening program. Fingerprinting after tryptic digestion reveals that the amino acid substitution is in the β chain. The normal amino-terminal tryptic peptide (Val-His-Leu-Thr-Pro-Glu-Glu-Lys) is missing. A new tryptic peptide consisting of six amino acid residues is found. Valine is the amino terminal residue of this peptide.
(a) Which amino acid substitutions are consistent with these data?
(b) Which single-base changes in DNA sequence could give these amino acid substitutions? The DNA sequence encoding the normal amino-terminal region is GTGCACCTGACTCCTGAG-GAGAAG.
(c) How should the electrophoretic mobility of this hemoglobin compare with those of Hb A and Hb S at pH 8?

12. Does the appearance of a 1.3-kb fragment following digestion with MstII and Southern blotting with a specific probe (p. 169) prove that the DNA sample contains a sickle-cell gene? Could other mutations give the same result?

13. Some mutations in a hemoglobin gene affect all three of the hemoglobins A, A_2, and F, whereas others affect only one of them. Why?

14. Hemoglobin A inhibits the formation of long fibers of hemoglobin S and the subsequent sickling of the red cell upon deoxygenation. Why does hemoglobin A have this effect?

15. Cyanate was a promising antisickling drug until clinical trials uncovered its toxic side effects, such as damage to peripheral nerves. Cyanate carbamoylates the terminal amino groups of hemoglobin. It behaves as a reactive analog of CO_2.

$$R{-}NH_2 + {}^{-}NCO + H^+ \longrightarrow R{-}N(H){-}C({=}O){-}NH_2$$

Terminal amino group — Cyanate — Carbamoylated derivative

How does the change in oxygen affinity from this modification have an antisickling effect?

16. What is the effect of each of the following treatments on the number of H^+ bound to hemoglobin A in vitro?
(a) Increase in pO_2 from 20 to 100 torrs (at constant pH and pCO_2).
(b) Reaction of hemoglobin with excess cyanate (at constant pH).

CHAPTER 8

Introduction to Enzymes

Enzymes, the catalysts of biological systems, are remarkable molecular devices that determine the pattern of chemical transformations. They also mediate the transformation of different forms of energy. The most striking characteristics of enzymes are their *catalytic power* and *specificity*. Furthermore, the actions of many enzymes are regulated. Nearly all known enzymes are proteins. The recent discovery of catalytically active RNA molecules, however, indicates that proteins do not have an absolute monopoly on catalysis.

Proteins as a class of macromolecules are highly effective in catalyzing diverse chemical reactions because of their capacity to specifically bind a very wide range of molecules. By utilizing the full repertoire of intermolecular forces, enzymes bring substrates together in an optimal orientation, the prelude to making and breaking chemical bonds. In essence, they catalyze reactions by stabilizing transition states, the highest-energy species in reaction pathways. By doing this selectively, an enzyme determines which one of several potential chemical reactions actually occurs. Enzymes can also act as molecular switches in regulating catalytic activity and transforming energy because of their capacity to couple the actions of separate binding sites.

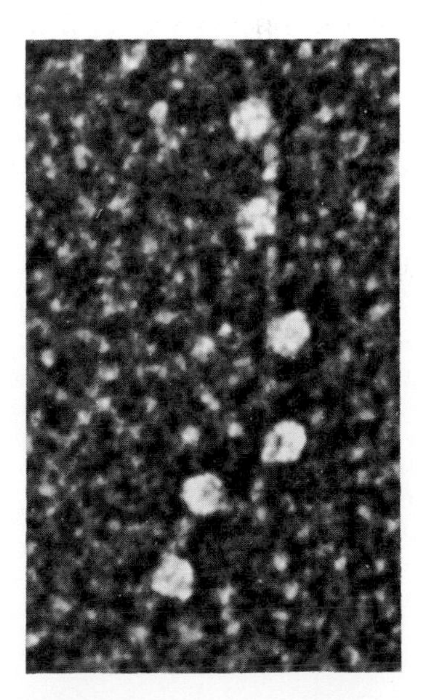

Figure 8-1
Electron micrograph of DNA polymerase I molecules (white spheres) bound to a threadlike synthetic DNA template. [Courtesy of Dr. Jack Griffith.]

ENZYMES HAVE IMMENSE CATALYTIC POWER

Enzymes accelerate reactions by factors of at least a million. Indeed, most reactions in biological systems do not occur at perceptible rates in the absence of enzymes. Even a reaction as simple as the hydration of carbon dioxide is catalyzed by an enzyme, namely, carbonic anhydrase (p. 38). The transfer of CO_2 from the tissues into the blood and then to the alveolar air would be less complete in the absence of this enzyme. In fact, carbonic anhydrase is one of the fastest enzymes known. Each enzyme molecule can hydrate 10^5 molecules of CO_2 per second. This catalyzed reaction is 10^7 times faster than the uncatalyzed one.

ENZYMES ARE HIGHLY SPECIFIC

Enzymes are highly specific both in the reaction catalyzed and in their choice of reactants, which are called *substrates*. An enzyme usually catalyzes a single chemical reaction or a set of closely related reactions. Side reactions leading to the wasteful formation of by-products rarely occur in enzyme-catalyzed reactions, in contrast with uncatalyzed ones. The degree of specificity for substrate is usually high and sometimes virtually absolute.

Let us consider *proteolytic enzymes* as an example. The reaction catalyzed by these enzymes is the hydrolysis of a peptide bond.

$$\sim N(H)-C(H)(R_1)-C(=O)-N(H)-C(H)(R_2)-C(=O)\sim + H_2O \rightleftharpoons \sim N(H)-C(H)(R_1)-COO^- + {}^+H_3N-C(H)(R_2)-C(=O)\sim$$

Peptide → **Carboxyl component** + **Amino component**

Most proteolytic enzymes also catalyze a different but related reaction, namely, the hydrolysis of an ester bond.

$$R_1-C(=O)-O-R_2 + H_2O \rightleftharpoons R_1-COO^- + HO-R_2 + H^+$$

Ester → **Acid** + **Alcohol**

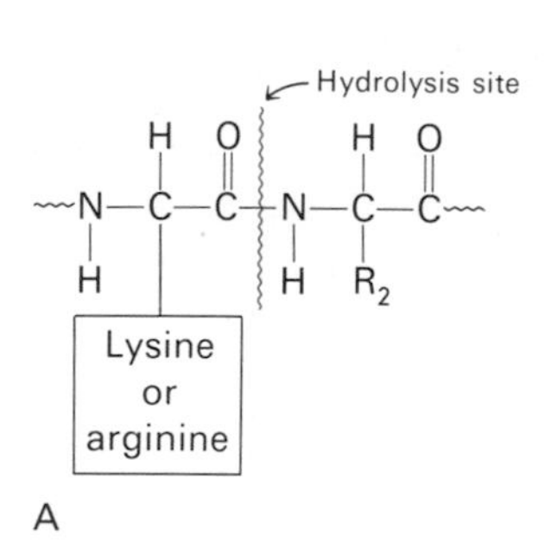

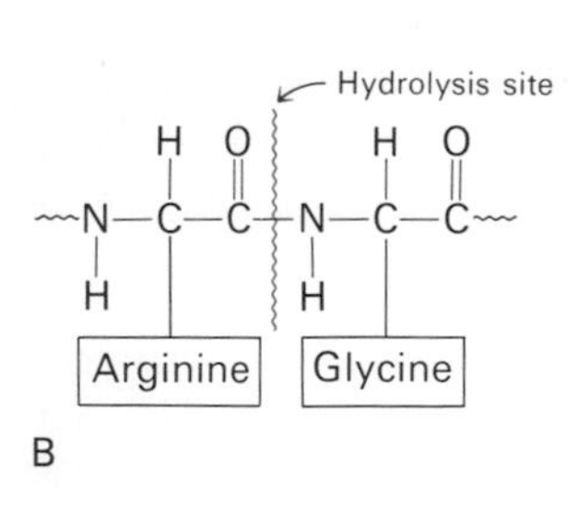

Figure 8-2
Comparison of the specificities of (A) trypsin and (B) thrombin. Trypsin cleaves on the carboxyl side of arginine and lysine residues. Thrombin cleaves Arg-Gly bonds in specific sequences only.

Proteolytic enzymes differ markedly in their degree of substrate specificity. Subtilisin, which comes from certain bacteria, is quite undiscriminating about the nature of the side chains adjacent to the peptide bond to be cleaved. Trypsin, as was mentioned in Chapter 3, is quite specific in that it catalyzes the splitting of peptide bonds on the carboxyl side of lysine and arginine residues only (Figure 8-2A). Thrombin, an enzyme that participates in blood clotting, is even more specific than trypsin. It catalyzes the hydrolysis of Arg-Gly bonds in specific peptide sequences only (Figure 8-2B).

DNA polymerase I, a template-directed enzyme (p. 85), is another highly specific catalyst. The sequence of nucleotides in the DNA strand that is being synthesized is determined by the sequence of nucleotides in another DNA strand that serves as a template. DNA polymerase I is remarkably precise in carrying out the instructions given by the template. The wrong nucleotide is inserted into a new DNA strand less than once in a million times, because DNA polymerase proofreads the nascent product and corrects its mistakes.

THE CATALYTIC ACTIVITIES OF MANY ENZYMES ARE REGULATED

The enzyme that catalyzes the first step in a biosynthetic pathway is usually inhibited by the ultimate product (Figure 8-3). The biosynthesis of isoleucine in bacteria illustrates this type of control, which is called *feedback inhibition*. Threonine is converted into isoleucine in five steps, the first of which is catalyzed by threonine deaminase. This enzyme is inhibited when the concentration of isoleucine reaches a sufficiently high level. Isoleucine inhibits by binding to the enzyme at a regulatory site, which is distinct from the catalytic site. This inhibition is mediated by an *allosteric interaction*, which is reversible. When the level of isoleucine drops sufficiently, threonine deaminase becomes active again, and consequently isoleucine is synthesized once more.

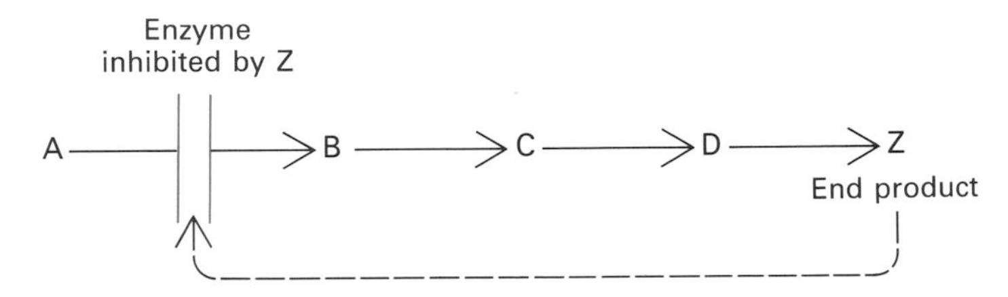

Figure 8-3
Feedback inhibition of the first enzyme in a pathway by reversible binding of the final product.

Figure 8-4
The α-carbon skeleton of calmodulin, a calcium sensor that regulates the activities of many intracellular proteins. The four domains of the protein are shown in different colors. The small circles show bound calcium ions. [Courtesy of Dr. Y. S. Babu and Dr. William J. Cook.]

Enzymes are also controlled by *regulatory proteins*, which can either stimulate or inhibit. The activities of many enzymes are regulated by *calmodulin*, a 17-kd protein that serves as a calcium sensor in nearly all eucaryotic cells (Figure 8-4). The binding of Ca^{2+} to four sites in calmodulin induces the formation of α-helix and other conformational changes that convert it from an inactive to an active form. Activated calmodulin then binds to many enzymes and other proteins in the cell and modifies their activities (p. 989).

Covalent modification is a third mechanism of enzyme regulation. For example, the activities of the enzymes that synthesize and degrade glycogen are regulated by the attachment of a phosphoryl group to a specific serine residue (p. 459). Other enzymes are controlled by the phosphorylation of threonine and tyrosine residues. These modifications are reversed by hydrolysis of the phosphate ester linkage. Specific enzymes catalyze the attachment and removal of phosphoryl and other modifying groups.

Phosphoserine residue **Phosphothreonine residue** **Phosphotyrosine residue**

Some enzymes are synthesized in an inactive precursor form, which is activated at a physiologically appropriate time and place. The digestive enzymes exemplify this kind of control, which is called *proteolytic activation*. For example, trypsinogen is synthesized in the pancreas and is activated by peptide-bond cleavage in the small intestine to form the active enzyme trypsin (Figure 8-5). This type of control is also repeatedly used in the sequence of enzymatic reactions that lead to the clotting of blood. The enzymatically inactive precursors of proteolytic enzymes are called *zymogens*.

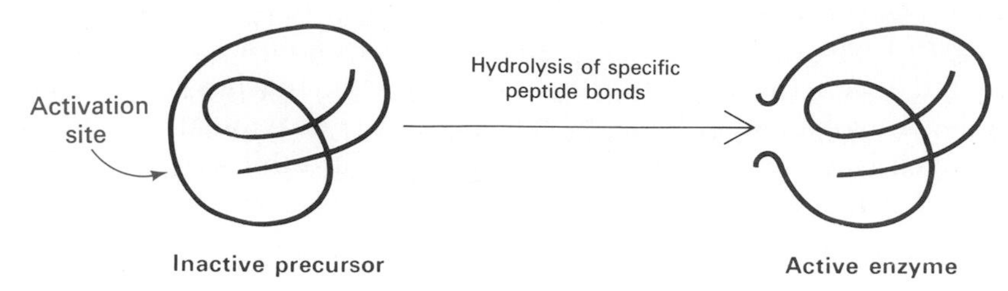

Figure 8-5
Activation of zymogen by hydrolysis of specific peptide bonds.

Figure 8-6
Electron micrograph of sodium-potassium pump molecules in a plasma membrane. These densely packed enzyme molecules catalyze the ATP-driven flux of Na^+ and K^+ out of and into cells. [Courtesy of Dr. Guido Zampighi.]

ENZYMES TRANSFORM DIFFERENT FORMS OF ENERGY

In many biochemical reactions, *the energy of the reactants is converted with high efficiency into a different form.* For example, in photosynthesis, light energy is converted into chemical-bond energy. In mitochondria, the free energy contained in small molecules derived from food is converted into a different currency, the free energy of adenosine triphosphate (ATP). The chemical-bond energy of ATP is then utilized in many different ways. In muscular contraction, the energy of ATP is converted into mechanical energy. Cells and organelles have pumps that utilize ATP to transport molecules and ions against chemical and electrical gradients (Figure 8-6). These transformations of energy are carried out by enzyme molecules that are integral parts of highly organized assemblies.

FREE ENERGY IS THE MOST USEFUL THERMODYNAMIC FUNCTION IN BIOCHEMISTRY

Let us review some key thermodynamic relations. In thermodynamics, a *system* is the matter within a defined region. The matter in the rest of the universe is called the *surroundings. The first law of thermodynamics states that the total energy of a system and its surroundings is a constant.* In other words, energy is conserved. The mathematical expression of the first law is

$$\Delta E = E_B - E_A = Q - W \qquad (1)$$

in which E_A is the energy of a system at the start of a process and E_B at the end of the process, Q is the heat absorbed by the system, and W is the work done by the system. An important feature of equation 1 is that *the change in energy of a system depends only on the initial and final states and not on the path of the transformation.*

The first law of thermodynamics cannot be used to predict whether a reaction can occur spontaneously. Some reactions do occur spontaneously although ΔE is positive (the energy of the system increases). In such cases, the system absorbs heat from its surroundings. It is evident that a function different from ΔE is required. One such function is the *entropy* (S), which is a measure of the *degree of randomness or disorder of a system*. The entropy of a system increases (ΔS is positive) when it becomes more disordered (Figure 8-7). *The second law of thermodynamics states that a process can occur spontaneously only if the sum of the entropies of the system and its surroundings increases.*

$$(\Delta S_{\text{system}} + \Delta S_{\text{surroundings}}) > 0 \text{ for a spontaneous process} \tag{2}$$

Note that the entropy of a system can decrease during a spontaneous process, provided that the entropy of the surroundings increases so that their sum is positive. For example, the formation of a highly ordered biological structure is thermodynamically feasible because the decrease in the entropy of such a system is more than offset by an increase in the entropy of its surroundings.

One difficulty in using entropy as a criterion of whether a biochemical process can occur spontaneously is that the entropy changes of chemical reactions are not readily measured. Furthermore, the criterion of spontaneity given in equation 2 requires that both the entropy change of the surroundings and that of the system of interest be known. These difficulties are obviated by using a different thermodynamic function called the *free energy*, which is denoted by the symbol G (or F, in the older literature). In 1878, Josiah Willard Gibbs created the free-energy function by combining the first and second laws of thermodynamics. The basic equation is

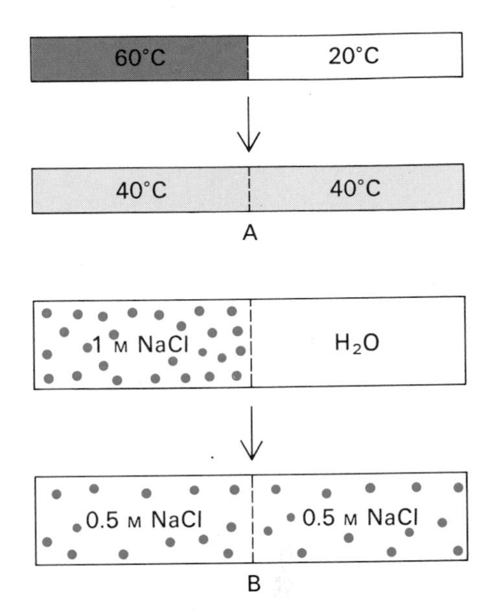

Figure 8-7
Processes that are driven by an increase in the entropy of a system: (A) diffusion of heat; and (B) diffusion of a solute.

$$\Delta G = \Delta H - T\,\Delta S \tag{3}$$

in which ΔG is the change in free energy of a system undergoing a transformation at constant pressure (P) and temperature (T), ΔH is the change in enthalpy of this system, and ΔS is the change in entropy of this system. Note that the properties of the surroundings do not enter into this equation. The enthalpy change is given by

$$\Delta H = \Delta E + P\,\Delta V \tag{4}$$

The volume change, ΔV, is small for nearly all biochemical reactions, and so ΔH is nearly equal to ΔE. Hence,

$$\Delta G \cong \Delta E - T\,\Delta S \tag{5}$$

Thus, the ΔG of a reaction depends both on the change in internal energy and on the change in entropy of the system.

The change in free energy (ΔG) of a reaction, in contrast with the change in internal energy (ΔE) of a reaction, is a valuable criterion of whether it can occur spontaneously.

1. *A reaction can occur spontaneously only if* ΔG *is negative.*

2. A system is at equilibrium and no net change can take place if ΔG is zero.

3. A reaction cannot occur spontaneously if ΔG is positive. An input of free energy is required to drive such a reaction.

Two additional points need to be emphasized here. First, the ΔG of a reaction depends only on the free energy of the products (the final state) minus that of the reactants (the initial state). *The* ΔG *of a reaction is independent of the path (or molecular mechanism) of the transformation.* The mechanism of a reaction has no effect on ΔG. For example, the ΔG for the oxidation of glucose to CO_2 and H_2O is the same whether it occurs by combustion in vitro or by a series of many enzyme-catalyzed steps in a cell. Second, ΔG *provides no information about the rate of a reaction.* A negative ΔG indicates that a reaction can occur spontaneously, but it does not signify whether it will proceed at a perceptible rate. As will be discussed shortly (p. 184), the rate of a reaction depends on the *free energy of activation* ($\Delta G^{\ddagger}$), which is unrelated to ΔG.

STANDARD FREE-ENERGY CHANGE OF A REACTION AND ITS RELATION TO THE EQUILIBRIUM CONSTANT

Consider the reaction

$$\mathrm{A + B \rightleftharpoons C + D}$$

The ΔG of this reaction is given by

$$\Delta G = \Delta G^{\circ} + RT \log_e \frac{[\mathrm{C}][\mathrm{D}]}{[\mathrm{A}][\mathrm{B}]} \tag{6}$$

in which ΔG° is the *standard free-energy change,* R is the gas constant, T is the absolute temperature, and [A], [B], [C], and [D] are the molar concentrations (more precisely, the activities) of the reactants. ΔG° is the free-energy change for this reaction under standard conditions—that is, when each of the reactants A, B, C, and D is present at a concentration of 1.0 M (for a gas, the standard state is usually chosen to be 1 atmosphere). Thus, the ΔG of a reaction depends on the *nature* of the reactants (expressed in the ΔG° term of equation 6) and on their *concentrations* (expressed in the logarithmic term of equation 6).

Units of energy—
A *calorie* (cal) is equivalent to the amount of heat required to raise the temperature of 1 gram of water from 14.5°C to 15.5°C.
A *kilocalorie* (kcal) is equal to 1000 cal.
A *joule* (J) is the amount of energy needed to apply a 1 newton force over a distance of 1 meter. A *kilojoule* (kJ) is equal to 1000 J.
1 kcal = 4.184 kJ

A convention has been adopted to simplify free-energy calculations for biochemical reactions. The standard state is defined as having a pH of 7. Consequently, when H^+ is a reactant, its activity has the value 1 (corresponding to a pH of 7) in equations 6 and 9. The activity of water also is taken to be 1 in these equations. The *standard free-energy change at pH 7,* denoted by the symbol $\Delta G^{\circ\prime}$, will be used throughout this book. The *kilocalorie* will be used as the unit of energy.

The relation between the standard free energy and the equilibrium constant of a reaction can be readily derived. At equilibrium, $\Delta G = 0$. Equation 6 then becomes

$$0 = \Delta G^{\circ\prime} + RT \log_e \frac{[\mathrm{C}][\mathrm{D}]}{[\mathrm{A}][\mathrm{B}]} \tag{7}$$

and so

$$\Delta G^{\circ\prime} = -RT \log_e \frac{[\mathrm{C}][\mathrm{D}]}{[\mathrm{A}][\mathrm{B}]} \tag{8}$$

The equilibrium constant under standard conditions, K'_{eq}, is defined as

$$K'_{\mathrm{eq}} = \frac{[\mathrm{C}][\mathrm{D}]}{[\mathrm{A}][\mathrm{B}]} \tag{9}$$

Substituting equation 9 into equation 8 gives

$$\Delta G^{\circ\prime} = -RT \log_e K'_{\mathrm{eq}} \tag{10}$$

$$\Delta G^{\circ\prime} = -2.303\ RT \log_{10} K'_{\mathrm{eq}} \tag{11}$$

which can be rearranged to give

$$K'_{\mathrm{eq}} = 10^{-\Delta G^{\circ\prime}/(2.303RT)} \tag{12}$$

Substituting $R = 1.98 \times 10^{-3}$ kcal mol^{-1} deg^{-1} and $T = 298°K$ (corresponding to 25°C) gives

$$K'_{eq} = 10^{-\Delta G°'/1.36} \quad (13)$$

when $\Delta G°'$ is expressed in kcal/mol. Thus, the standard free energy and the equilibrium constant of a reaction are related by a simple expression. For example, an equilibrium constant of 10 corresponds to a standard free-energy change of -1.36 kcal/mol at 25°C (Table 8-1).

Let us calculate $\Delta G°'$ and ΔG for the isomerization of dihydroxyacetone phosphate to glyceraldehyde 3-phosphate as an example. This reaction occurs in glycolysis (p. 352). At equilibrium, the ratio of glyceraldehyde 3-phosphate to dihydroxyacetone phosphate is 0.0475 at 25°C (298°K) and pH 7. Hence, $K'_{eq} = 0.0475$. The standard free-energy change for this reaction is then calculated from equation 11:

$$\begin{aligned}\Delta G°' &= -2.303\, RT \log_{10} K'_{eq} \\ &= -2.303 \times 1.98 \times 10^{-3} \times 298 \times \log_{10}(0.0475) \\ &= +1.8 \text{ kcal/mol}\end{aligned}$$

CH_2OH—$C{=}O$—$CH_2OPO_3^{2-}$

Dihydroxyacetone phosphate

⇌

$HC{=}O$—$H{-}C{-}OH$—$CH_2OPO_3^{2-}$

Glyceraldehyde 3-phosphate

Now let us calculate ΔG for this reaction when the initial concentration of dihydroxyacetone phosphate is 2×10^{-4} M and the initial concentration of glyceraldehyde 3-phosphate is 3×10^{-6} M. Substituting these values into equation 6 gives

$$\begin{aligned}\Delta G &= 1.8 \text{ kcal/mol} + 2.303\, RT \log_{10} \frac{3 \times 10^{-6}\ \text{M}}{2 \times 10^{-4}\ \text{M}} \\ &= 1.8 \text{ kcal/mol} - 2.5 \text{ kcal/mol} \\ &= -0.7 \text{ kcal/mol}\end{aligned}$$

This negative value for the ΔG indicates that the isomerization of dihydroxyacetone phosphate to glyceraldehyde 3-phosphate can occur spontaneously when these species are present at the concentrations stated above. Note that ΔG for this reaction is negative although $\Delta G°'$ is positive. *It is important to stress that whether the* ΔG *for a reaction is larger, smaller, or the same as* $\Delta G°'$ *depends on the concentrations of the reactants.* The criterion of spontaneity for a reaction is ΔG, not $\Delta G°'$.

Table 8-1
Relation between $\Delta G°'$ and K'_{eq} (at 25°C)

K'_{eq}	$\Delta G°'$ *(kcal/mol)*
10^{-5}	6.82
10^{-4}	5.46
10^{-3}	4.09
10^{-2}	2.73
10^{-1}	1.36
1	0
10	−1.36
10^{2}	−2.73
10^{3}	−4.09
10^{4}	−5.46
10^{5}	−6.82

ENZYMES CANNOT ALTER REACTION EQUILIBRIA

An enzyme is a catalyst, and consequently it cannot alter the equilibrium of a chemical reaction. This means that an enzyme accelerates the forward and reverse reaction by precisely the same factor. Consider the interconversion of A and B. Suppose that in the absence of enzyme the forward rate constant (k_F) is 10^{-4} s^{-1} and the reverse rate constant (k_R) is 10^{-6} s^{-1}. The equilibrium constant K is given by the ratio of these rate constants:

$$A \underset{10^{-6}\,s^{-1}}{\overset{10^{-4}\,s^{-1}}{\rightleftharpoons}} B$$

$$K = \frac{[B]}{[A]} = \frac{k_F}{k_R} = \frac{10^{-4}}{10^{-6}} = 100$$

The equilibrium concentration of B is 100 times that of A, whether or not enzyme is present. However, it would take more than an hour to approach this equilibrium without enzyme, whereas equilibrium would be attained within a second in the presence of a suitable enzyme. *Enzymes accelerate the attainment of equilibria but do not shift their position.*

ENZYMES ACCELERATE REACTIONS BY STABILIZING TRANSITION STATES

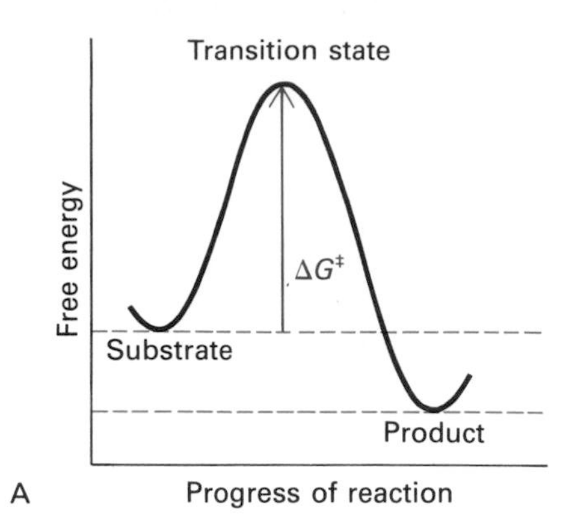

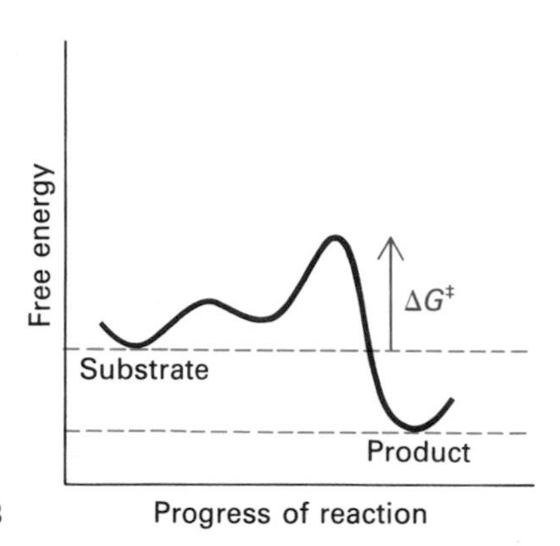

Figure 8-8
Enzymes accelerate reactions by decreasing $\Delta G^{\ddagger}$, the free energy of activation. The free-energy profiles of uncatalyzed (A) and catalyzed (B) reactions are compared.

A chemical reaction of substrate S to form product P goes through a *transition state* $S^{\ddagger}$ that has a higher free energy than either S or P.

$$\underset{\text{Substrate}}{S} \xrightleftharpoons{K^{\ddagger}} \underset{\text{Transition state}}{S^{\ddagger}} \xrightarrow{V} \underset{\text{Product}}{P}$$

The transition state is the most seldom occupied species along the reaction pathway because it has the highest free energy. The *Gibbs free energy of activation*, symbolized by $\Delta G^{\ddagger}$, is equal to the difference in free energy between the transition state and the substrate. The double dagger (‡) denotes a thermodynamic quantity of a transition state.

$$\Delta G^{\ddagger} = G_{S^{\ddagger}} - G_S$$

The reaction rate V is proportional to the concentration of $S^{\ddagger}$, which depends on $\Delta G^{\ddagger}$ because it is in equilibrium with S.

$$[S^{\ddagger}] = [S]e^{-\Delta G^{\ddagger}/RT}$$

$$V = \nu[S^{\ddagger}] = \frac{kT}{h}[S]e^{-\Delta G^{\ddagger}/RT}$$

In these equations, k is Boltzmann's constant and h is Planck's constant. The value of kT/h at 25°C is $6.2 \times 10^{12}\ s^{-1}$. Suppose that the free energy of activation is 6.82 kcal/mol. The ratio $[S^{\ddagger}]/[S]$ is then 10^{-5} (see Table 8-1); we have assumed that $[S] = 1$, and so the reaction rate V is $6.2 \times 10^{7}\ s^{-1}$. A decrease of 1.36 kcal/mol in $\Delta G^{\ddagger}$ results in a tenfold faster V.

Enzymes accelerate reactions by decreasing $\Delta G^{\ddagger}$, the activation barrier. The combination of substrate and enzyme creates a new reaction pathway whose transition-state energy is lower than that of the reaction in the absence of enzyme (Figure 8-8). The essence of catalysis is specific binding of the transition state, as will be discussed in the next chapter.

> "I think that enzymes are molecules that are complementary in structure to the activated complexes of the reactions that they catalyze, that is, to the molecular configuration that is intermediate between the reacting substances and the products of reaction for these catalyzed processes. The attraction of the enzyme molecule for the activated complex would thus lead to a decrease in its energy and hence to a decrease in the energy of activation of the reaction and to an increase in the rate of reaction."
>
> LINUS PAULING
> *Nature* 161:707 (1948)

FORMATION OF AN ENZYME-SUBSTRATE COMPLEX IS THE FIRST STEP IN ENZYMATIC CATALYSIS

Much of the catalytic power of enzymes comes from their bringing substrates together in favorable orientations in *enzyme-substrate* (ES) complexes. The substrates are bound to a specific region of the enzyme called the *active site*. Most enzymes are highly selective in their binding of substrates. Indeed, the catalytic specificity of enzymes depends in part on the specificity of binding. Furthermore, the activities of some enzymes are controlled at this stage.

The existence of ES complexes has been shown in a variety of ways:

1. At a constant concentration of enzyme, the reaction rate increases with increasing substrate concentration until a maximal velocity is reached (Figure 8-9). In contrast, uncatalyzed reactions do not show this saturation effect. In 1913, Leonor Michaelis interpreted the *maximal velocity of an enzyme-catalyzed reaction* in terms of the formation of a discrete ES complex. At a sufficiently high substrate concentration, the catalytic sites are filled and so the reaction rate reaches a maximum. Though indirect, this is the oldest and most general evidence for the existence of ES complexes.

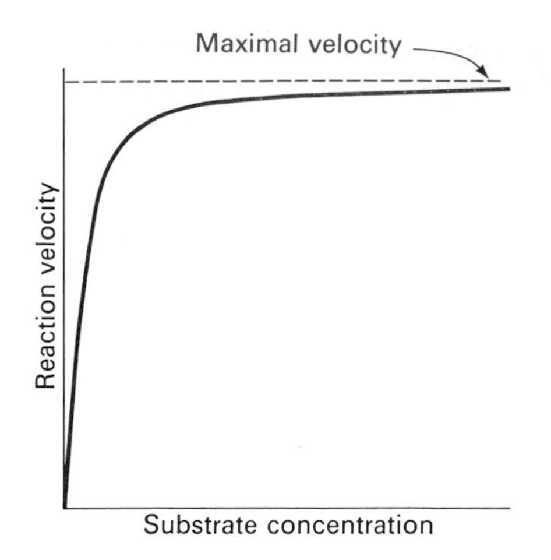

Figure 8-9
Velocity of an enzyme-catalyzed reaction as a function of the substrate concentration.

2. ES complexes have been directly visualized by *electron microscopy*, as in the micrograph of DNA polymerase I bound to its DNA template (see Figure 8-1). *X-ray crystallography* has provided high-resolution images of substrates and substrate analogs bound to the active sites of many enzymes. In the next chapter, we shall take a close look at several of these complexes. Moreover, x-ray studies carried out at low temperatures (to slow reactions down) are providing revealing views of intermediates in enzymatic reactions.

3. The *spectroscopic characteristics* of many enzymes and substrates change upon formation of an ES complex just as the absorption spectrum of deoxyhemoglobin changes markedly when it binds oxygen or when it is oxidized to the ferric state, as described previously (see Figure 7-11, on p. 147). These changes are particularly striking if the enzyme contains a colored prosthetic group. Tryptophan synthetase, a bacterial enzyme that contains a pyridoxal phosphate prosthetic group, affords a nice illustration. This enzyme catalyzes the synthesis of L-tryptophan from L-serine and indole. The addition of L-serine to the enzyme produces a marked increase in the fluorescence of the pyridoxal phosphate group (Figure 8-10). The subsequent addition of indole, the second substrate, quenches this fluorescence to a level lower even than that of the enzyme alone. Thus, fluorescence spectroscopy reveals the existence of an enzyme-serine complex and of an enzyme-serine-indole complex. Other spectroscopic techniques, such as nuclear magnetic resonance and electron spin resonance, also are highly informative about ES interactions.

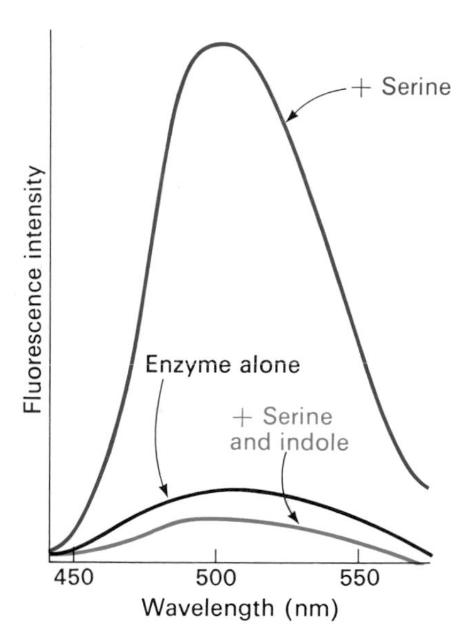

Figure 8-10
Fluorescence intensity of the pyridoxal phosphate group at the active site of tryptophan synthetase changes upon addition of serine and indole, the substrates.

SOME KEY FEATURES OF ACTIVE SITES

The active site of an enzyme is the region that binds the substrates (and the prosthetic group, if any) and contains the residues that directly participate in the making and breaking of bonds. These residues are called the *catalytic groups*. Although enzymes differ widely in structure, specificity, and mode of catalysis, a number of generalizations concerning their active sites can be stated:

1. *The active site takes up a relatively small part of the total volume of an enzyme*. Most of the amino acid residues in an enzyme are not in contact with the substrate. This raises the intriguing question of why enzymes

are so big. Nearly all enzymes are made up of more than 100 amino acid residues, which gives them a mass greater than 10 kd and a diameter of more than 25 Å.

2. *The active site is a three-dimensional entity* formed by groups that come from different parts of the linear amino acid sequence—indeed, residues far apart in the linear sequence may interact more strongly than adjacent residues in the amino acid sequence, as has already been seen for myoglobin and hemoglobin. In lysozyme, an enzyme that will be discussed in more detail in the next chapter, the important groups in the active site are contributed by residues numbered 35, 52, 62, 63, and 101 in the linear sequence of 129 amino acids (Figure 8-11).

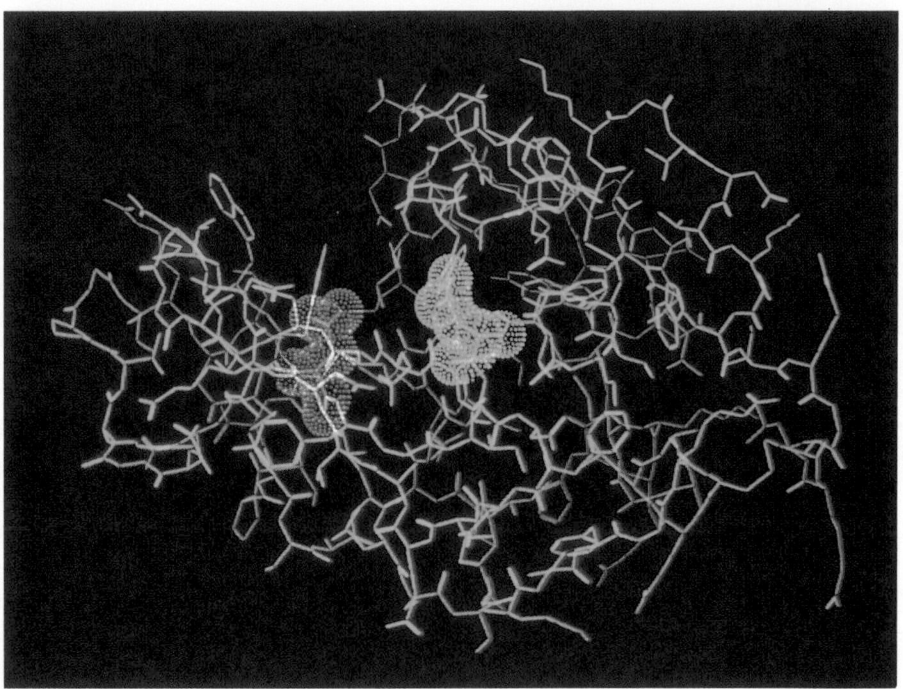

Figure 8-11
Model of lysozyme. The van der Waals surfaces of two catalytically critical residues are shown in color.

3. *Substrates are bound to enzymes by multiple weak attractions.* ES complexes usually have equilibrium constants that range from 10^{-2} to 10^{-8} M, corresponding to free energies of interaction ranging from −3 to −12 kcal/mol. The noncovalent interactions in ES complexes are much weaker than covalent bonds, which have energies between −50 and −110 kcal/mol. As was discussed in Chapter 1 (pp. 7–10), reversible interactions of biomolecules are mediated by electrostatic bonds, hydrogen bonds, van der Waals forces, and hydrophobic interactions. Van der Waals forces become significant in binding only when numerous substrate atoms can simultaneously come close to many enzyme atoms. Hence, the enzyme and substrate should have complementary shapes. The directional character of hydrogen bonds between enzyme and substrate often enforces a high degree of specificity.

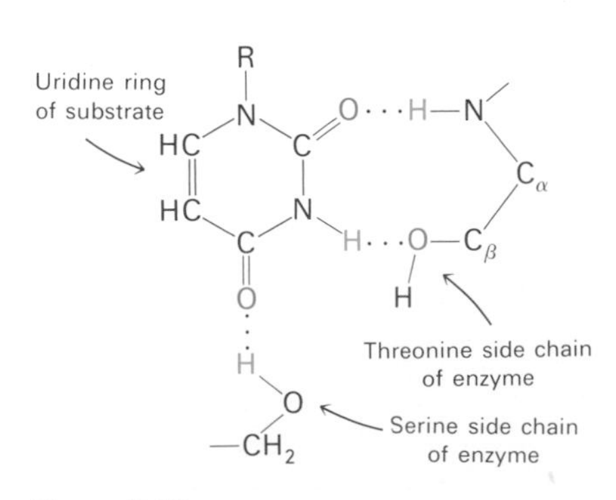

Figure 8-12
Hydrogen-bond interactions in the binding of a uridine substrate to ribonuclease. [After F. M. Richards, H. W. Wyckoff, and N. Allewell. In *The Neurosciences: Second Study Program,* F. O. Schmidt, ed., (Rockefeller University Press, 1970), p. 970.]

4. *Active sites are clefts or crevices.* In all enzymes of known structure, substrate molecules are bound to a cleft or crevice. Water is usually excluded unless it is a reactant. The nonpolar character of much of the cleft enhances the binding of substrate. However, the cleft may also contain polar residues. It creates a microenvironment in which certain of these residues acquire special properties essential for catalysis. The internal positions of these polar residues are biologically crucial exceptions to the general rule that polar residues are exposed to water.

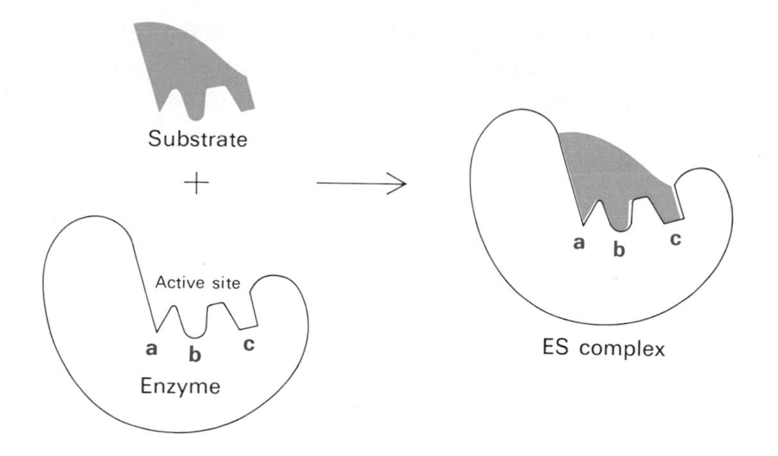

Figure 8-13
Lock-and-key model of the interaction of substrates and enzymes. The active site of the enzyme alone is complementary in shape to that of the substrate.

5. *The specificity of binding depends on the precisely defined arrangement of atoms in an active site.* To fit into the site, a substrate must have a matching shape. Emil Fischer's metaphor of the lock and key (Figure 8-13), expressed in 1890, has proved to be highly stimulating and fruitful. However, it is now evident that the shapes of the active sites of some enzymes are markedly modified by the binding of substrate, as was postulated by Daniel E. Koshland, Jr., in 1958. The active sites of these enzymes have shapes that are complementary to that of the substrate only *after* the substrate is bound. This process of dynamic recognition is called *induced fit* (Figure 8-14).

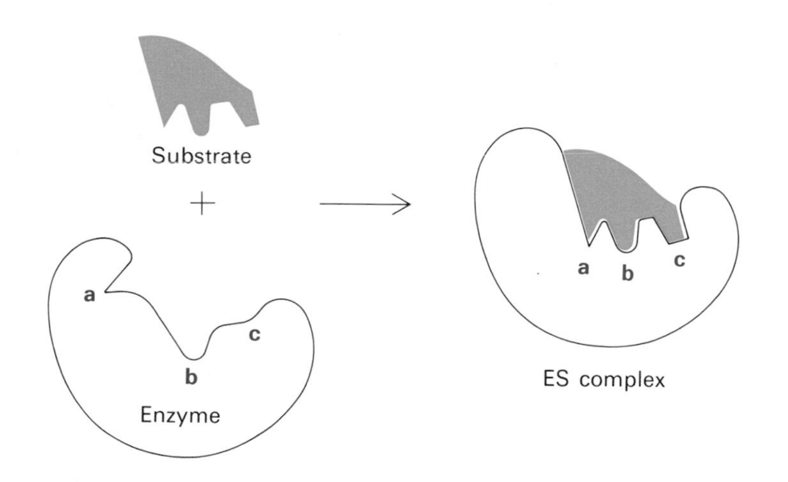

Figure 8-14
Induced-fit model of the interaction of substrates and enzymes. The enzyme changes shape upon binding substrate. The active site has a shape complementary to that of the substrate only after the substrate is bound.

THE MICHAELIS-MENTEN MODEL ACCOUNTS FOR THE KINETIC PROPERTIES OF MANY ENZYMES

For many enzymes, the rate of catalysis, V, varies with the substrate concentration, [S], in a manner shown in Figure 8-15. V is defined as the number of moles of product formed per second. At a fixed concentration of enzyme, V is almost linearly proportional to [S] when [S] is small. At high [S], V is nearly independent of [S]. In 1913, Leonor Michaelis and Maud Menten proposed a simple model to account for these kinetic characteristics. The critical feature in their treatment is that a specific ES complex is a necessary intermediate in catalysis. The model proposed, which is the simplest one that accounts for the kinetic properties of many enzymes, is

$$E + S \underset{k_2}{\overset{k_1}{\rightleftharpoons}} ES \xrightarrow{k_3} E + P \qquad (14)$$

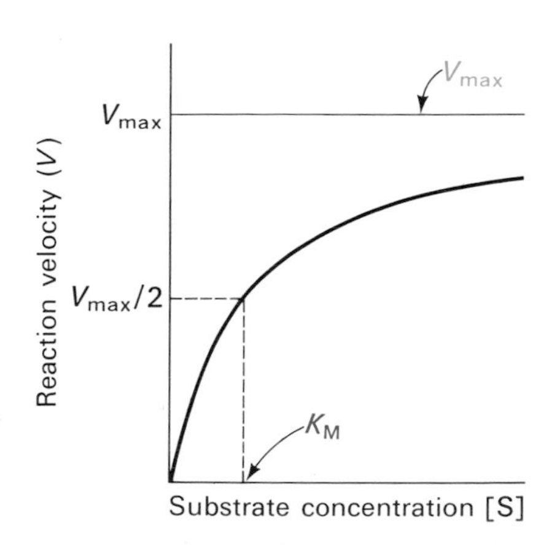

Figure 8-15
A plot of the reaction velocity, V, as a function of the substrate concentration, [S], for an enzyme that obeys Michaelis-Menten kinetics (V_{max} is the maximal velocity and K_M is the Michaelis constant).

An enzyme, E, combines with S to form an ES complex, with a rate constant k_1. The ES complex has two possible fates. It can dissociate to E and S, with a rate constant k_2, or it can proceed to form product P, with a rate constant k_3. It is assumed that almost none of the product reverts to the initial substrate, a condition that holds in the initial stage of a reaction before the concentration of product is appreciable.

We want an expression that relates the rate of catalysis to the concentrations of substrate and enzyme and the rates of the individual steps. The starting point is that the catalytic rate is equal to the product of the concentration of the ES complex and k_3:

$$V = k_3[\mathrm{ES}] \tag{15}$$

Now we need to express [ES] in terms of known quantities. The rates of formation and breakdown of ES are given by

$$\text{Rate of formation of ES} = k_1[\mathrm{E}][\mathrm{S}] \tag{16}$$

$$\text{Rate of breakdown of ES} = (k_2 + k_3)[\mathrm{ES}] \tag{17}$$

We are interested in the catalytic rate under steady-state conditions. In a *steady state,* the concentrations of intermediates stay the same while the concentrations of starting materials and products are changing. This occurs when the rates of formation and breakdown of the ES complex are equal. On setting the right-hand sides of equations 16 and 17 equal,

$$k_1[\mathrm{E}][\mathrm{S}] = (k_2 + k_3)[\mathrm{ES}] \tag{18}$$

By rearranging equation 18,

$$[\mathrm{ES}] = \frac{[\mathrm{E}][\mathrm{S}]}{(k_2 + k_3)/k_1} \tag{19}$$

Equation 19 can be simplified by defining a new constant, K_M, called the *Michaelis constant:*

$$K_M = \frac{k_2 + k_3}{k_1} \tag{20}$$

and substituting it into equation 19, which then becomes

$$[\mathrm{ES}] = \frac{[\mathrm{E}][\mathrm{S}]}{K_M} \tag{21}$$

Now let us examine the numerator of equation 21. The concentration of uncombined substrate, [S], is very nearly equal to the total substrate concentration, provided that the concentration of enzyme is much lower than that of the substrate. The concentration of uncombined enzyme, [E], is equal to the total enzyme concentration, $[E_T]$, minus the concentration of the ES complex.

$$[\mathrm{E}] = [\mathrm{E_T}] - [\mathrm{ES}] \tag{22}$$

On substituting this expression for [E] in equation 21,

$$[\mathrm{ES}] = ([\mathrm{E_T}] - [\mathrm{ES}])[\mathrm{S}]/K_M \tag{23}$$

Solving equation 23 for [ES] gives

$$[\mathrm{ES}] = [\mathrm{E_T}]\frac{[\mathrm{S}]/K_M}{1 + [\mathrm{S}]/K_M} \tag{24}$$

or

$$[\mathrm{ES}] = [\mathrm{E_T}]\frac{[\mathrm{S}]}{[\mathrm{S}] + K_M} \tag{25}$$

By substituting this expression for [ES] into equation 15, we get

$$V = k_3[\mathrm{E_T}]\frac{[\mathrm{S}]}{[\mathrm{S}] + K_\mathrm{M}} \tag{26}$$

The maximal rate, V_{max}, is attained when the enzyme sites are saturated with substrate—that is, when [S] is much greater than K_M—so that $[\mathrm{S}]/([\mathrm{S}] + K_\mathrm{M})$ approaches 1. Thus,

$$V_{max} = k_3[\mathrm{E_T}] \tag{27}$$

Substituting equation 27 into equation 26 yields the Michaelis-Menten equation:

$$V = V_{max}\frac{[\mathrm{S}]}{[\mathrm{S}] + K_\mathrm{M}} \tag{28}$$

This equation accounts for the kinetic data given in Figure 8-15. At very low substrate concentration, when [S] is much less than K_M, $V = [\mathrm{S}]V_{max}/K_\mathrm{M}$; that is, the rate is directly proportional to the substrate concentration. At high substrate concentration, when [S] is much greater than K_M, $V = V_{max}$; that is, the rate is maximal, independent of substrate concentration.

The meaning of K_M is evident from equation 28. When $[\mathrm{S}] = K_\mathrm{M}$, then $V = V_{max}/2$. Thus, *K_M is equal to the substrate concentration at which the reaction rate is half of its maximal value.*

V_{max} AND K_M CAN BE DETERMINED BY VARYING THE SUBSTRATE CONCENTRATION

The Michaelis constant, K_M, and the maximal rate, V_{max}, can be readily derived from rates of catalysis measured at different substrate concentrations if an enzyme operates according to the simple scheme given in equation 14. It is convenient to transform the Michaelis-Menten equation into one that gives a straight line plot. This can be done by taking the reciprocal of both sides of equation 28 to give

$$\frac{1}{V} = \frac{1}{V_{max}} + \frac{K_\mathrm{M}}{V_{max}} \cdot \frac{1}{[\mathrm{S}]} \tag{29}$$

A plot of $1/V$ versus $1/[\mathrm{S}]$, called a *Lineweaver-Burk plot*, yields a straight line with an intercept of $1/V_{max}$ and a slope of K_M/V_{max} (Figure 8-16).

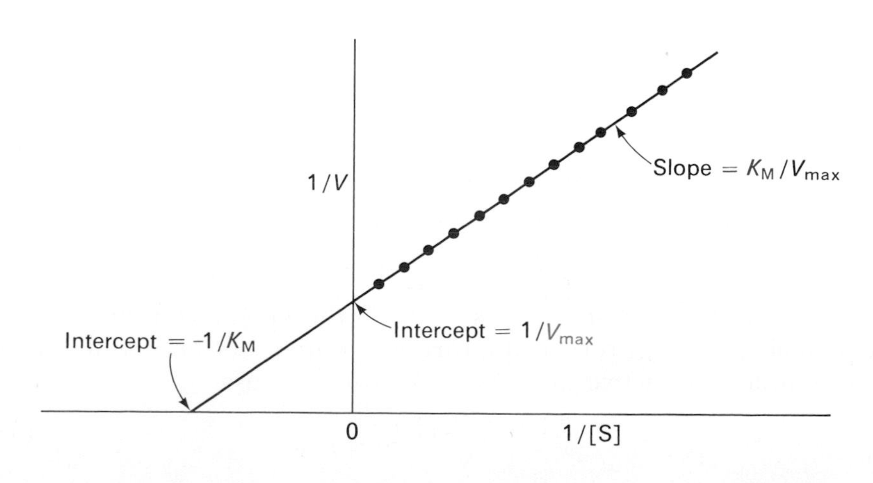

Figure 8-16
A double-reciprocal plot of enzyme kinetics: $1/V$ is plotted as a function of $1/[\mathrm{S}]$. The slope is K_M/V_{max}, the intercept on the vertical axis is $1/V_{max}$, and the intercept on the horizontal axis is $-1/K_\mathrm{M}$.

Table 8-2
K_M values of some enzymes

Enzyme	*Substrate*	K_M (μM)
Chymotrypsin	Acetyl-L-tryptophanamide	5000
Lysozyme	Hexa-*N*-acetylglucosamine	6
β-Galactosidase	Lactose	4000
Threonine deaminase	Threonine	5000
Carbonic anhydrase	CO_2	8000
Penicillinase	Benzylpenicillin	50
Pyruvate carboxylase	Pyruvate	400
	HCO_3^-	1000
	ATP	60
Arginine-tRNA synthetase	Arginine	3
	tRNA	0.4
	ATP	300

SIGNIFICANCE OF K_M AND V_{max} VALUES

The K_M values of enzymes range widely (Table 8-2). For most enzymes, K_M lies between 10^{-1} and 10^{-7} M. The K_M value for an enzyme depends on the particular substrate and also on environmental conditions such as pH, temperature, and ionic strength. The Michaelis constant, K_M, has two meanings. First, K_M is the concentration of substrate at which half the active sites are filled. Once the K_M is known, the fraction of sites filled, f_{ES}, at any substrate concentration can be calculated from

$$f_{ES} = \frac{V}{V_{max}} = \frac{[S]}{[S] + K_M} \qquad (30)$$

Second, K_M is related to the rate constants of the individual steps in the catalytic scheme given in equation 14. In equation 20, K_M is defined as $(k_2 + k_3)/k_1$. Consider a limiting case in which k_2 is much greater than k_3. This means that dissociation of the ES complex to E and S is much more rapid than formation of E and product. Under these conditions ($k_2 \gg k_3$),

$$K_M = \frac{k_2}{k_1} \qquad (31)$$

The dissociation constant of the ES complex is given by

$$K_{ES} = \frac{[E][S]}{[ES]} = \frac{k_2}{k_1} \qquad (32)$$

In other words, *K_M is equal to the dissociation constant of the ES complex if k_3 is much smaller than k_2*. When this condition is met, K_M is a measure of the strength of the ES complex: a high K_M indicates weak binding; a low K_M indicates strong binding. It must be stressed that K_M indicates the affinity of the ES complex only when k_2 is much greater than k_3.

The *turnover number* of an enzyme is *the number of substrate molecules converted into product by an enzyme molecule in a unit time when the enzyme is fully saturated with substrate*. It is equal to the kinetic constant k_3. The maximal rate, V_{max}, reveals the turnover number of an enzyme if the concentration of active sites $[E_T]$ is known, because

$$V_{max} = k_3[E_T] \qquad (33)$$

For example, a 10^{-6} M solution of carbonic anhydrase catalyzes the formation of 0.6 M H_2CO_3 per second when it is fully saturated with substrate. Hence, k_3 is 6×10^5 s^{-1}. This turnover number is one of the largest known. Each round of catalysis occurs in a time equal to $1/k_3$, which is 1.7 microseconds for carbonic anhydrase. The turnover numbers of most enzymes with their physiological substrates fall in the range from 1 to 10^4 per second (Table 8-3).

Table 8-3
Maximum turnover numbers of some enzymes

Enzyme	*Turnover number (per second)*
Carbonic anhydrase	600,000
3-Ketosteroid isomerase	280,000
Acetylcholinesterase	25,000
Penicillinase	2,000
Lactate dehydrogenase	1,000
Chymotrypsin	100
DNA polymerase I	15
Tryptophan synthetase	2
Lysozyme	0.5

KINETIC PERFECTION IN ENZYMATIC CATALYSIS: THE k_{cat}/K_M CRITERION

When the substrate concentration is much greater than K_M, the rate of catalysis is equal to k_3, the turnover number, as described in the preceding section. However, most enzymes are not normally saturated with substrate. Under physiological conditions, the $[S]/K_M$ ratio is typically between 0.01 and 1.0. When $[S] \ll K_M$, the enzymatic rate is much less than k_3 because most of the active sites are unoccupied. Is there a number that characterizes the kinetics of an enzyme under these conditions? Indeed there is, as can be shown by combining equations 15 and 21 to give

$$V = \frac{k_3}{K_M}[E][S] \qquad (34)$$

When $[S] \ll K_M$, the concentration of free enzyme, [E], is nearly equal to the total concentration of enzyme $[E_T]$, and so

$$V = \frac{k_3}{K_M}[S][E_T] \qquad (35)$$

Thus, when $[S] \ll K_M$, *the enzymatic velocity depends on the value of* k_3/K_M *and on [S].*

Are there any physical limits on the value of k_3/K_M? Note that this ratio depends on k_1, k_2, and k_3, as can be shown by substituting for K_M:

$$k_3/K_M = \frac{k_3 k_1}{k_2 + k_3} < k_1 \qquad (36)$$

Thus the ultimate limit on the value of k_3/K_M is set by k_1, the rate of formation of the ES complex. *This rate cannot be faster than the diffusion-controlled encounter of an enzyme and its substrate.* Diffusion limits the value of k_1 so that it cannot be higher than between 10^8 and 10^9 M^{-1} s^{-1}. Hence, the upper limit on k_3/K_M is between 10^8 and 10^9 M^{-1} s^{-1}.

This restriction also pertains to enzymes having more complex reaction pathways than that of equation 14. Their maximal catalytic rate when substrate is saturating, denoted by k_{cat}, depends on several rate constants rather than on k_3 alone. The pertinent parameter for these enzymes is k_{cat}/K_M. In fact, the k_{cat}/K_M ratios of the enzymes acetylcholinesterase, carbonic anhydrase, and triosephosphate isomerase are between 10^8 and 10^9 M^{-1} s^{-1}, which shows that they have attained *kinetic perfection. Their catalytic velocity is restricted only by the rate at which they encounter substrate in the solution.* Any further gain in catalytic rate can come only by decreasing the time for diffusion. Indeed, some series of enzymes are associated into organized assemblies (p. 379) so that the product of one enzyme is very rapidly found by the next enzyme. In effect, products are channeled from one enzyme to the next, much as in an assembly line. Thus, the limit imposed by the rate of diffusion in solution can be partly overcome by confining substrates and products in the limited volume of a multienzyme complex.

ENZYMES CAN BE INHIBITED BY SPECIFIC MOLECULES

The inhibition of enzymatic activity by specific small molecules and ions is important because it serves as a major control mechanism in biological systems. Also, many drugs and toxic agents act by inhibiting enzymes. Furthermore, inhibition can be a source of insight into the mechanism of enzyme action: residues critical for catalysis can often be identified by using specific inhibitors.

Enzyme inhibition can be either reversible or irreversible. An *irreversible inhibitor* dissociates very slowly from its target enzyme because it becomes very tightly bound to the enzyme, either covalently or noncovalently. The action of nerve gases on acetylcholinesterase, an enzyme that plays an important role in the transmission of nerve impulses, exemplifies irreversible inhibition. Diisopropylphosphofluoridate (DIPF), one of these agents, reacts with a critical serine residue at the active site on the enzyme to form an inactive diisopropylphosphoryl enzyme (Figure 8-17). Alkylating reagents, such as iodoacetamide, irreversibly inhibit the catalytic activity of some enzymes by modifying cysteine and other side chains (Figure 8-18).

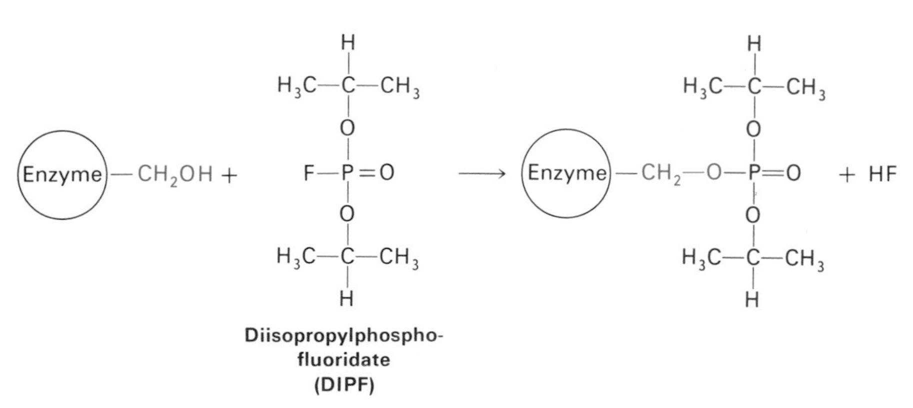

Figure 8-17
Inactivation of chymotrypsin and acetylcholinesterase by diisopropylphosphofluoridate (DIPF).

$$\text{Enzyme}-CH_2SH + ICH_2-\overset{O}{\overset{\|}{C}}-NH_2 \longrightarrow \text{Enzyme}-CH_2-S-CH_2-\overset{O}{\overset{\|}{C}}-NH_2 + HI$$

Iodoacetamide

Figure 8-18
Inactivation of an enzyme with a critical cysteine residue by iodoacetamide.

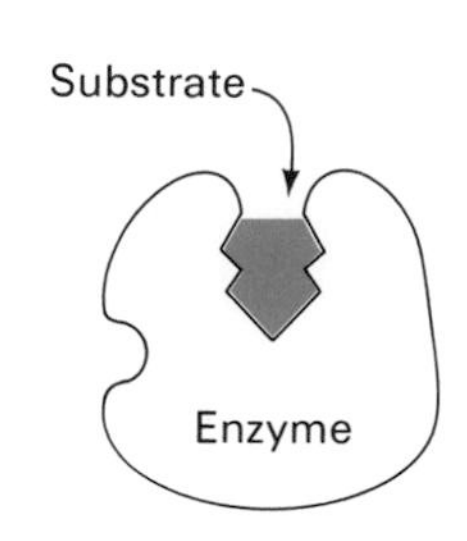

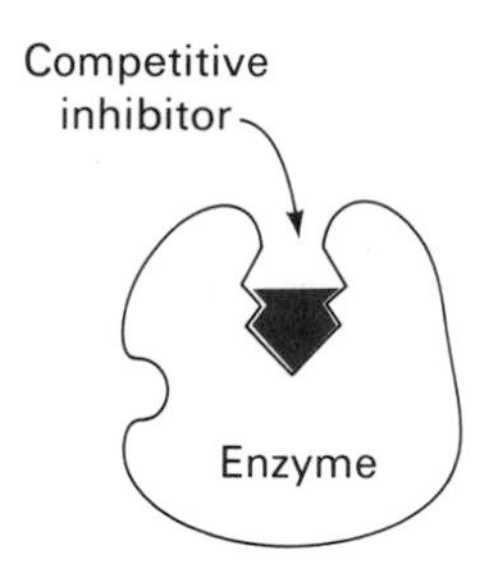

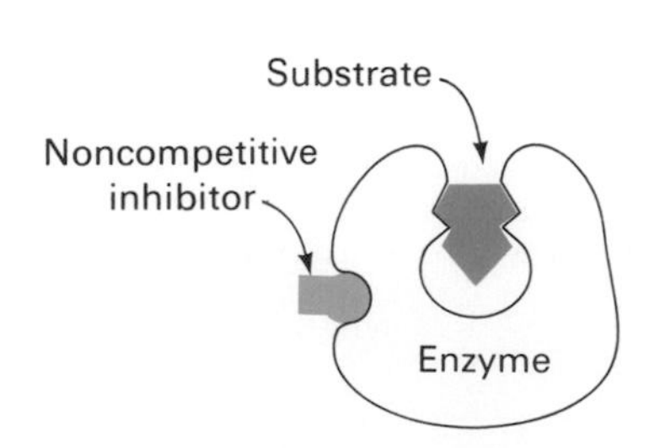

Figure 8-19
Distinction between a competitive inhibitor and a noncompetitive inhibitor: (top) enzyme-substrate complex; (middle) a competitive inhibitor prevents the substrate from binding; (bottom) a noncompetitive inhibitor does not prevent the substrate from binding.

Reversible inhibition, in contrast with irreversible inhibition, is characterized by a rapid dissociation of the enzyme-inhibitor complex. In *competitive inhibition*, the enzyme can bind substrate (forming an ES complex) or inhibitor (EI) but not both (ESI). Many competitive inhibitors resemble the substrate and bind to the active site of the enzyme (Figure 8-19). The substrate is thereby prevented from binding to the same active site. *A competitive inhibitor diminishes the rate of catalysis by reducing the proportion of enzyme molecules bound to a substrate.* A classic example of competitive inhibition is the action of malonate on succinate dehydrogenase, an enzyme that removes two hydrogen atoms from succinate. Malonate differs from succinate in having one rather than two methylene groups. A physiologically important example of competitive inhibition is found in the formation of 2,3-bisphosphoglycerate (BPG, p. 157) from 1,3-bisphosphoglycerate. Bisphosphoglycerate mutase, the enzyme catalyzing this isomerization, is competitively inhibited by even low levels of 2,3-bisphosphoglycerate. In fact, it is not uncommon for an enzyme to be competitively inhibited by its own product because of its structural resemblance to the substrate. Competitive inhibition can be overcome by increasing the concentration of substrate.

In *noncompetitive inhibition,* which is also reversible, the inhibitor and substrate can bind simultaneously to an enzyme molecule. This means that their binding sites do not overlap. A noncompetitive inhibitor acts by decreasing the turnover number of an enzyme rather than by diminishing the proportion of enzyme molecules that are bound to substrate. Noncompetitive inhibition, in contrast with competitive inhibition, cannot be overcome by increasing the substrate concentration. A more complex pattern, called *mixed inhibition,* is produced when an inhibitor both affects the binding of substrate and alters the turnover number of the enzyme.

COO^-—CH_2—CH_2—COO^-
Succinate

COO^-—CH_2—COO^-
Malonate

$O{=}C(-OPO_3^{2-})$—$H{-}C{-}OH$—$H_2C{-}OPO_3^{2-}$
1,3-Bisphosphoglycerate

⇅

$O{=}C(-O^-)$—$H{-}C{-}OPO_3^{2-}$—$H_2C{-}OPO_3^{2-}$
2,3-Bisphosphoglycerate

ALLOSTERIC ENZYMES DO NOT OBEY MICHAELIS-MENTEN KINETICS

The Michaelis-Menten model has greatly affected the development of enzyme chemistry. Its virtues are simplicity and broad applicability. However, the kinetic properties of many enzymes cannot be accounted for by the Michaelis-Menten model. An important group consists of the *allosteric enzymes,* which often display sigmoidal plots (Figure 8-20) of the reaction velocity, V, versus substrate concentration, [S], rather than the hyperbolic plots predicted by the Michaelis-Menten equation (eq. 28). Recall that the oxygen-binding curve of myoglobin is hyperbolic, whereas that of hemoglobin is sigmoidal. The binding of enzymes to substrates is analogous. In allosteric enzymes, the binding of substrate to one active site can affect the properties of other active sites in the same enzyme molecule. A possible outcome of this interaction between subunits is that the binding of substrate becomes cooperative, which would give a sigmoidal plot of V versus [S]. In addition, the activity of allosteric enzymes may be altered by regulatory molecules that are bound to sites other than the catalytic sites, just as oxygen binding to hemoglobin is affected by BPG, H^+, and CO_2.

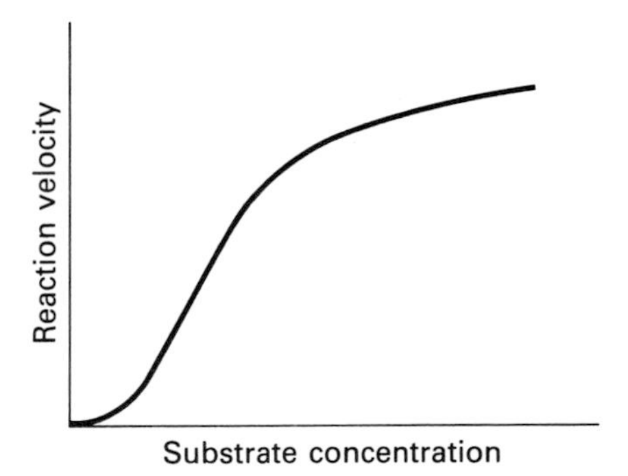

Figure 8-20
Sigmoidal dependence of reaction velocity versus substrate concentration for an allosteric enzyme.

COMPETITIVE AND NONCOMPETITIVE INHIBITION ARE KINETICALLY DISTINGUISHABLE

Let us return to enzymes that exhibit Michaelis-Menten kinetics. Measurements of the rates of catalysis at different concentrations of substrate and inhibitor serve to distinguish between competitive and noncompetitive inhibition. In *competitive inhibition,* the intercept of the plot of $1/V$ versus 1/[S] is the same in the presence and absence of inhibitor, although the slope is different (Figure 8-21). This reflects the fact that V_{max} is not altered by a competitive inhibitor. *The hallmark of competitive inhibition is that it can be overcome by a sufficiently high concentration of substrate.* At a sufficiently high concentration, virtually all the active sites

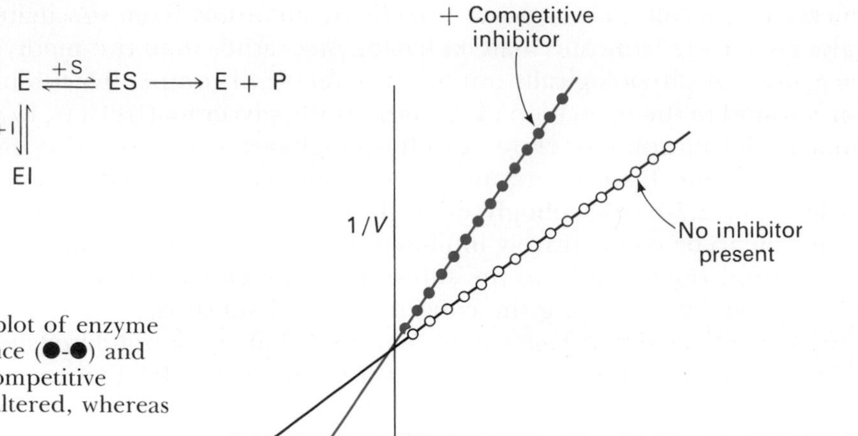

Figure 8-21
A double-reciprocal plot of enzyme kinetics in the presence (●-●) and absence (○-○) of a competitive inhibitor; $V_{\max}$ is unaltered, whereas K_M is increased.

are filled by substrate, and the enzyme is fully operative. The increase in the slope of the $1/V$ versus $1/[S]$ plot indicates the strength of binding of competitive inhibitor. In the presence of a competitive inhibitor, equation 29 is replaced by

$$\frac{1}{V} = \frac{1}{V_{\max}} + \frac{K_M}{V_{\max}}\left(1 + \frac{[I]}{K_i}\right)\left(\frac{1}{[S]}\right) \tag{37}$$

in which [I] is the concentration of inhibitor and K_i is the dissociation constant of the enzyme-inhibitor complex:

$$E + I \rightleftharpoons EI$$

$$K_i = \frac{[E][I]}{[EI]} \tag{38}$$

In other words, the slope of the plot is increased by the factor $(1 + [I]/K_i)$ in the presence of a competitive inhibitor. Consider an enzyme with a K_M of 10^{-4} M. In the absence of inhibitor, $V = V_{\max}/2$ when $[S] = 10^{-4}$ M. In the presence of 2×10^{-3} M competitive inhibitor that is bound to the enzyme with a K_i of 10^{-3} M, the apparent K_M will be 3×10^{-4} M. Substitution of these values into equation 29 gives $V = V_{\max}/4$.

In *noncompetitive inhibition* (Figure 8-22), $V_{\max}$ is decreased to $V^I_{\max}$, and so the intercept on the vertical axis is increased. The new slope,

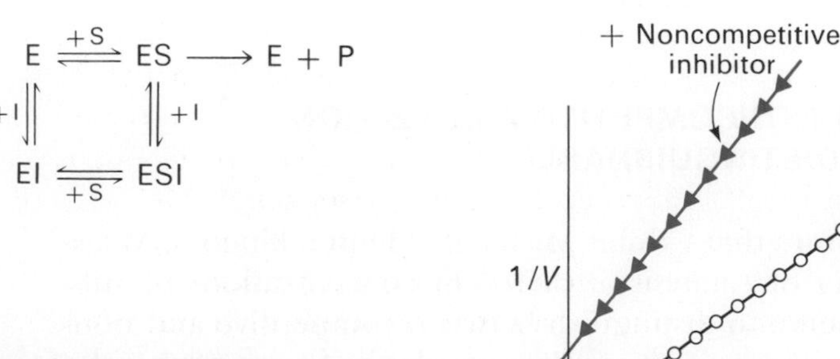

Figure 8-22
A double-reciprocal plot of enzyme kinetics in the presence (▲-▲-▲) and absence (○-○-○) of a noncompetitive inhibitor; K_M is unaltered by the noncompetitive inhibitor, whereas $V_{\max}$ is decreased.

which is equal to K_M/V^I_{max}, is larger by the same factor. In contrast with V_{max}, K_M is not affected by this kind of inhibition. *Noncompetitive inhibition cannot be overcome by increasing the substrate concentration.* The maximal velocity in the presence of a noncompetitive inhibitor, V^I_{max}, is given by

$$V^I_{max} = \frac{V_{max}}{1 + [I]/K_i} \tag{39}$$

ETHANOL IS USED THERAPEUTICALLY AS A COMPETITIVE INHIBITOR TO TREAT ETHYLENE GLYCOL POISONING

About fifty deaths occur annually from the ingestion of ethylene glycol, a constituent of permanent-type automobile antifreeze. Ethylene glycol itself is not lethally toxic. Rather, the harm is done by oxalic acid, an oxidation product of ethylene glycol, because the kidneys are severely damaged by the deposition of oxalate crystals. The first committed step in this conversion is the oxidation of ethylene glycol to an aldehyde by alcohol dehydrogenase (Figure 8-23). This reaction can be effectively inhibited by administering a nearly intoxicating dose of ethanol. The basis for this effect is that *ethanol is a competing substrate and so it blocks the oxidation of ethylene glycol to aldehyde products.* The ethylene glycol is then excreted harmlessly. Ethanol is also used as a competing substrate for treating methanol (wood alcohol) poisoning.

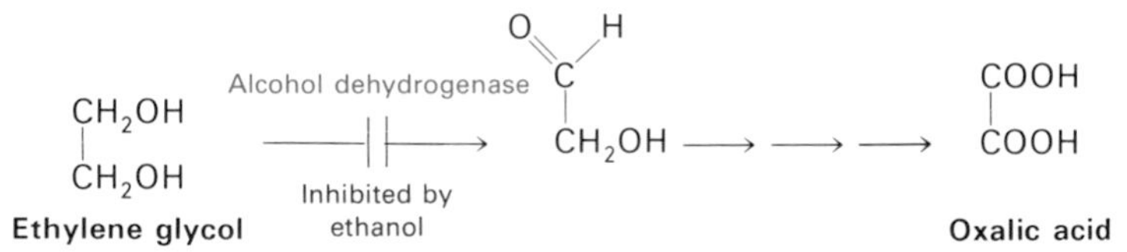

Figure 8-23
Formation of oxalic acid from ethylene glycol is inhibited by ethanol.

PENICILLIN IRREVERSIBLY INACTIVATES A KEY ENZYME IN BACTERIAL CELL-WALL SYNTHESIS

Penicillin was discovered by Alexander Fleming in 1928, when he observed by chance that bacterial growth was inhibited by a contaminating mold (*Penicillium*). Fleming was encouraged to find that an extract from the mold was not toxic when injected into animals. However, on trying to concentrate and purify the antibiotic, he found that "penicillin is easily destroyed, and to all intents and purposes we failed. We were bacteriologists—not chemists—and our relatively simple procedures were unavailing." Ten years later, Howard Florey, a pathologist, and Ernst Chain, a biochemist, carried out an incisive series of studies that led to the isolation, chemical characterization, and clinical use of this antibiotic.

Penicillin consists of a thiazolidine ring fused to a *β-lactam* ring, to which a variable R group is attached by a peptide bond. In benzyl penicillin, for example, R is a benzyl group (Figure 8-24). This structure can undergo a variety of rearrangements, which accounts for the instability

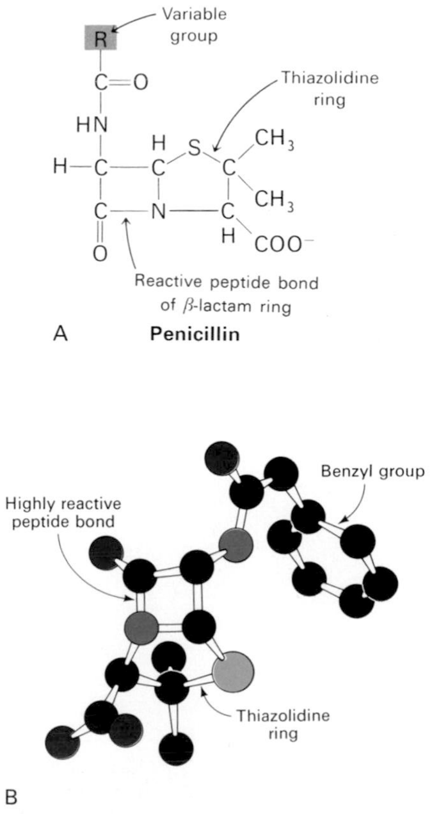

Figure 8-24
(A) Structural formula and (B) model of benzyl penicillin. The reactive site of penicillin is the peptide bond of its β-lactam ring.

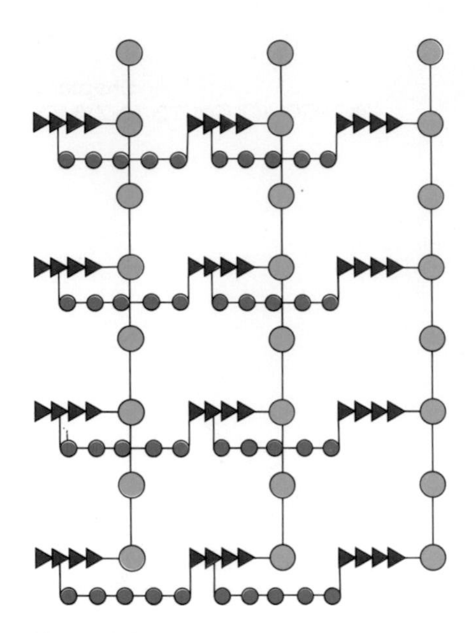

Figure 8-25
Schematic diagram of the peptidoglycan in *Staphylococcus aureus*. The sugars are shown in yellow, the tetrapeptides in red, and the pentaglycine bridges in blue. The cell wall is a single, enormous bag-shaped macromolecule because of extensive cross-linking.

first encountered by Fleming. In particular, the β-lactam ring is very labile. Indeed, this property is closely tied to the antibiotic action of penicillin, as will be evident shortly.

How does penicillin inhibit bacterial growth? In 1957, Joshua Lederberg showed that bacteria ordinarily susceptible to penicillin could be grown in its presence if a hypertonic medium were used. The organisms obtained in this way, called protoplasts, are devoid of a cell wall and consequently lyse when transferred to a normal medium. Hence, it was inferred that penicillin interferes with the synthesis of the bacterial cell wall. The cell-wall macromolecule, called a *peptidoglycan*, consists of linear polysaccharide chains that are cross-linked by short peptides (Figure 8-25). The enormous bag-shaped peptidoglycan confers mechanical support and prevents bacteria from bursting from their high internal osmotic pressure.

In 1965, James Park and Jack Strominger independently deduced that penicillin blocks the last step in cell-wall synthesis, namely the cross-linking of different peptidoglycan strands. In the formation of the cell wall of *Staphylococcus aureus*, the amino group at one end of a pentaglycine chain attacks the peptide bond between two D-alanine residues in another peptide unit (Figure 8-26). A peptide bond is formed between glycine and one of the D-alanine residues, and the other D-alanine residue is released. This cross-linking reaction is catalyzed by *glycopeptide transpeptidase*. Bacterial cell walls are unique in containing D-amino acids, which form cross-links by a mechanism entirely different from that used to synthesize proteins.

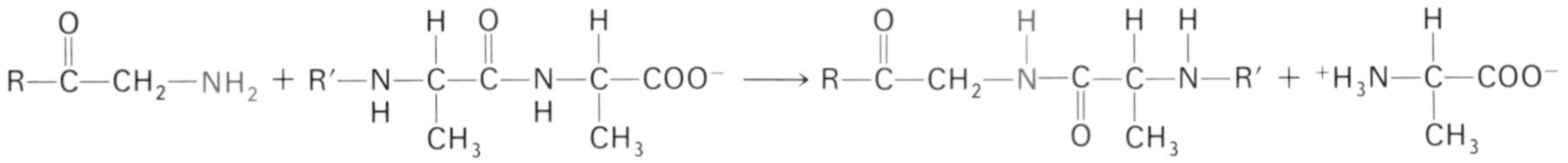

$$\underset{\text{Terminal glycine residue of pentaglycine bridge}}{R{-}\overset{O}{\overset{\|}{C}}{-}CH_2{-}NH_2} + \underset{\text{Terminal D-Ala-D-Ala unit}}{R'{-}\underset{H}{N}{-}\underset{CH_3}{\overset{H}{C}}{-}\overset{O}{\overset{\|}{C}}{-}\underset{H}{N}{-}\underset{CH_3}{\overset{H}{C}}{-}COO^-} \longrightarrow \underset{\text{Gly-D-Ala cross-link}}{R{-}\overset{O}{\overset{\|}{C}}{-}CH_2{-}\overset{H}{N}{-}\underset{O}{\underset{\|}{C}}{-}\underset{CH_3}{\overset{H}{C}}{-}\overset{H}{N}{-}R'} + \underset{\text{D-Ala}}{{}^+H_3N{-}\underset{CH_3}{\overset{H}{C}}{-}COO^-}$$

Figure 8-26
The amino group of the pentaglycine bridge in the *S. aureus* cell wall attacks the peptide bond between two D-Ala residues to form a cross-link.

Penicillin inhibits the cross-linking transpeptidase by the Trojan Horse stratagem. The transpeptidase normally forms an *acyl intermediate* with the penultimate D-alanine residue of the D-Ala-D-Ala-peptide (Figure 8-27). This covalent acyl-enzyme intermediate then reacts with the amino group of the terminal glycine in another peptide to form the cross-link. Penicillin is welcomed into the active site of the transpeptidase because it mimics the D-Ala-D-Ala moiety of the normal substrate. Bound penicillin then forms a covalent bond with a serine residue at

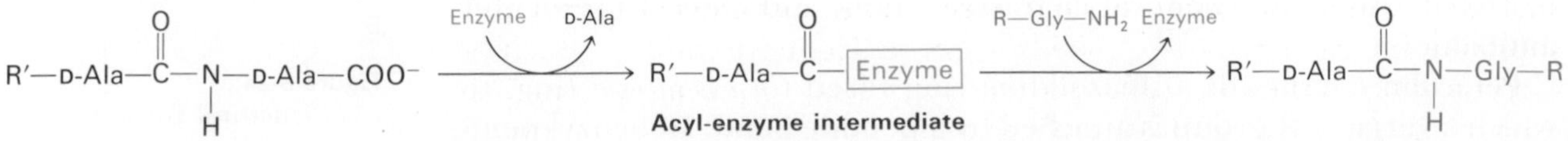

$$R'{-}\text{D-Ala}{-}\overset{O}{\overset{\|}{C}}{-}\underset{H}{N}{-}\text{D-Ala}{-}COO^- \xrightarrow{\text{Enzyme} \quad \text{D-Ala}} \underset{\text{Acyl-enzyme intermediate}}{R'{-}\text{D-Ala}{-}\overset{O}{\overset{\|}{C}}{-}\boxed{\text{Enzyme}}} \xrightarrow{R{-}Gly{-}NH_2 \quad \text{Enzyme}} R'{-}\text{D-Ala}{-}\overset{O}{\overset{\|}{C}}{-}\underset{H}{N}{-}Gly{-}R$$

Figure 8-27
An acyl-enzyme intermediate is formed in the transpeptidation reaction.

Glycopeptide transpeptidase → (Penicillin)

R
|
C=O
|
HN
|
H—C—C(H)—S—C(CH$_3$)(CH$_3$)
|
O=C N(H)—C(H)—COO$^-$
|
O
|
Ser
|
Enzyme

Penicilloyl-enzyme complex
(Enzymatically inactive)

Figure 8-28
Formation of a penicilloyl-enzyme complex, which is indefinitely stable.

the active site of the enzyme (Figure 8-28). *This penicilloyl-enzyme does not react further. Hence, the transpeptidase is irreversibly inhibited.*

Why is penicillin such an effective inhibitor of the transpeptidase? Molecular models show that penicillin resembles acyl-D-Ala-D-Ala, one of the substrates of this enzyme (Figure 8-29). Moreover, the four-membered β-lactam ring of penicillin is strained, which makes it highly reactive. Indeed, the conformation of this part of penicillin is probably very similar to that of the transition state of the normal substrate, a species that interacts strongly with the enzyme. In other words, penicillin is a *transition-state analog*, a striking example of molecular mimicry executed with perfection.

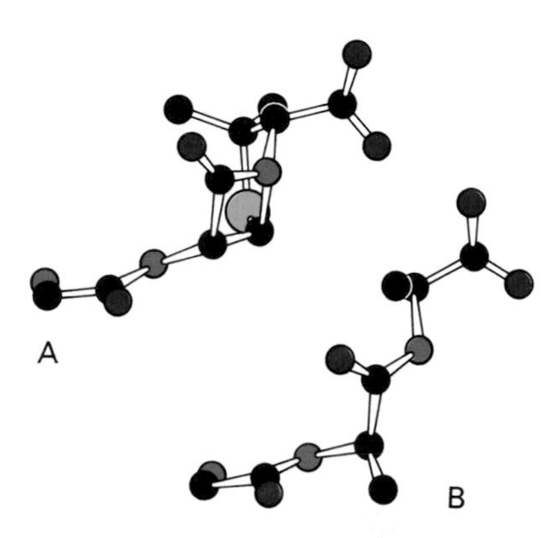

Figure 8-29
The conformation of penicillin in the vicinity of its reactive peptide bond (A) resembles the postulated conformation of the transition state of R—D-Ala—D-Ala (B) in the transpeptidation reaction. [After B. Lee. *J. Mol. Biol.* 61(1971):464.]

SUMMARY

Free energy is the most valuable thermodynamic function for determining whether a reaction can occur and for understanding the energetics of catalysis. Free energy is a measure of the capacity of a system to do useful work at constant temperature and pressure. A reaction can occur spontaneously only if the change in free energy (ΔG) is negative. The ΔG of a reaction is independent of path and depends only on the nature of the reactants and their activities (which can sometimes be approximated by their concentrations). The free-energy change of a reaction that occurs when reactants and products are at unit activity is called the standard free-energy change ($\Delta G°$). Biochemists usually use $\Delta G°'$, the standard free-energy change at pH 7.

The catalysts in biological systems are enzymes, and nearly all of them are proteins. Enzymes are highly specific and have great catalytic power. They enhance reaction rates by factors of at least 10^6. Enzymes do not alter reaction equilibria. Rather, they serve as catalysts by reducing the free energy of activation of chemical reactions. Enzymes accelerate reactions by providing a new reaction pathway in which the transition state (the highest-energy species) has a lower free energy and hence is more accessible than in the uncatalyzed reaction. The first step in catalysis is the formation of an enzyme-substrate complex. Substrates are bound to enzymes at active-site clefts from which water is largely excluded when the substrate is bound. The specificity of enzyme-substrate interactions arises mainly from hydrogen bonding, which is directional, and the shape of the active site, which rejects molecules that do not have a sufficiently complementary shape. The recognition of substrates by enzymes is a dynamic process accompanied by conformational changes at active sites.

The Michaelis-Menten model accounts for the kinetic properties of some enzymes. In this model, an enzyme (E) combines with a substrate (S) to form an enzyme-substrate (ES) complex, which can proceed to form a product (P) or to dissociate into E and S.

$$E + S \underset{k_2}{\overset{k_1}{\rightleftharpoons}} ES \xrightarrow{k_3} E + P$$

The rate V of formation of product is given by the Michaelis-Menten equation:

$$V = V_{max}\frac{[S]}{[S] + K_M}$$

in which V_{max} is the rate when the enzyme is fully saturated with substrate, and K_M, the Michaelis constant, is the substrate concentration at which the reaction rate is half maximal. The maximal rate, V_{max}, is equal to the product of k_3 and the total concentration of enzyme. The kinetic constant k_3, called the turnover number, is the number of substrate molecules converted into product per unit time at a single catalytic site when the enzyme is fully saturated with substrate. Turnover numbers for most enzymes are between 1 and 10^4 per second.

Enzymes can be inhibited by specific small molecules or ions. In irreversible inhibition, the inhibitor is covalently linked to the enzyme or bound so tightly that its dissociation from the enzyme is very slow. For example, penicillin irreversibly inactivates an essential enzyme in the formation of bacterial cell walls by mimicking the normal substrate in the cross-linking reaction. In contrast, reversible inhibition is characterized by a rapid equilibrium between enzyme and inhibitor. A competitive inhibitor prevents the substrate from binding to the active site. It reduces the reaction velocity by diminishing the proportion of enzyme molecules that are bound to substrate. In noncompetitive inhibition, the inhibitor decreases the turnover number. Competitive inhibition can be distinguished from noncompetitive inhibition by determining whether the inhibition can be overcome by raising the substrate concentration.

The catalytic activity of many enzymes is regulated in vivo. Allosteric interactions, which are defined as interactions between spatially distinct sites, are particularly important in this regard. The enzyme catalyzing the first step in a biosynthetic pathway is usually inhibited by the final product. Enzymes are also controlled by regulatory proteins such as calmodulin, which senses the intracellular Ca^{2+} level. Covalent modifications such as phosphorylation of serine, threonine, and tyrosine side chains are a third means of modulating enzymatic activity. The conversion of an inactive precursor protein into an active enzyme by peptide-bond cleavage, a process termed proteolytic activation, is another recurring regulatory device.

SELECTED READINGS

Where to start

Koshland, D. E., Jr., 1973. Protein shape and biological control. *Sci. Amer.* 229(4):52–64. [An excellent introduction to the importance of conformational flexibility for the specificity and regulation of enzyme action. Available as *Sci. Amer.* Offprint 1280.]

Cech, T. R., 1986. RNA as an enzyme. *Sci. Amer.* 255(5):64–75. [An excellent account of the discovery that RNA, as well as proteins, can be an effective catalyst. Available as *Sci. Amer.* Offprint 1575.]

Books on enzymes

Fersht, A., 1983. *Enzyme Structure and Mechanism* (2nd ed.). W. H. Freeman. [A concise and lucid introduction to enzyme action, with emphasis on physical principles.]

Walsh, C., 1979. *Enzymatic Reaction Mechanisms.* W. H. Freeman. [An excellent account of the chemical basis of enzyme action. This book demonstrates that the large number of enzyme-catalyzed reactions in biological systems can be grouped into a small number of types of chemical reactions.]

Bender, M. L., Bergeron, R. J., and Komiyama, M., 1984. *The Bioorganic Chemistry of Enzymatic Catalysis.* Wiley-Interscience.

Dugas, H., and Penney, C., 1981. *Bioorganic Chemistry: A Chemical Approach to Enzyme Action.* Springer-Verlag.

Dixon, M., and Webb, E. C., 1979. *Enzymes* (3rd ed.). Longmans.

Jencks, W. P., 1969. *Catalysis in Chemistry and Enzymology.* McGraw-Hill.

Hiromi, K., 1979. *Kinetics of Fast Enzyme Reactions.* Halsted Press.

Boyer, P. D., (ed.), 1970. *The Enzymes* (3rd ed.). Academic Press. [This multivolume treatise on enzymes contains a wealth of information. Volumes 1 and 2 (available in a paperback edition) deal with general aspects of enzyme structure, mechanism, and regulation. Volume 3 and subsequent ones contain detailed and authoritative articles on individual enzymes.]

Books on thermodynamics

Edsall, J. T., and Gutfreund, H., 1983. *Biothermodynamics: The Study of Biochemical Processes at Equilibrium.* Wiley.

Klotz, I. M., 1967. *Energy Changes in Biochemical Reactions.* Academic Press. [A concise introduction, full of insight.]

Enzyme kinetics and mechanisms

Fersht, A. R., Leatherbarrow, R. J., and Wells, T. N. C., 1986. Binding energy and catalysis: a lesson from protein engineering of the tyrosyl-tRNA synthetase. *Trends Biochem. Sci.* 11:321–325.

Jencks, W. P., 1975. Binding energy, specificity, and enzymic catalysis: the Circe effect. *Advan. Enzymol.* 43:219–410.

Knowles, J. R., and Albery, W. J., 1976. Evolution of enzyme function and the development of catalytic efficiency. *Biochemistry* 15:5631–5640.

Molecular interactions and binding

Richards, F. M., Wyckoff, H. W., and Allewell, N., 1970. The origin of specificity in binding: a detailed example in a protein-nucleic acid interaction. *In* Schmitt, F. O., (ed.), *The Neurosciences: Second Study Program,* pp. 901–912. Rockefeller University Press.

Davidson, N., 1967. Weak interactions and the structure of biological macromolecules. *In* Quarton, G. C., Melnechuk, T., and Schmitt, F. O., (eds.), *The Neurosciences: A Study Program,* pp. 46–56. Rockefeller University Press.

Control of enzymatic activity

Cheung, W. Y., 1982. Calmodulin. *Sci. Amer.* 246(6):62–70.

Monod, J., Changeux, J.-P., and Jacob, F., 1963. Allosteric proteins and cellular control systems. *J. Mol. Biol.* 6:306–329. [A classic paper that introduced the concept of allosteric interactions.]

Neurath, H., 1985. Proteolytic enzymes, past and present. *Fed. Proc.* 44:2907–2913.

Penicillin and other enzyme inhibitors

Waxman, D. J., and Strominger, J. L., 1983. Penicillin-binding proteins and the mechanism of action of β-lactam antibiotics. *Ann. Rev. Biochem.* 52:825–69.

Abraham, E. P., 1981. The beta-lactam antibiotics. *Sci. Amer.* 244:76–86.

Walsh, C. T., 1984. Suicide substrates, mechanism-based enzyme inactivators: recent developments. *Ann. Rev. Biochem.* 53:493–535.

PROBLEMS

1. The hydrolysis of pyrophosphate to orthophosphate is important in driving forward biosynthetic reactions such as the synthesis of DNA. This hydrolytic reaction is catalyzed in *E. coli* by a pyrophosphatase that has a mass of 120 kd and consists of six identical subunits. For this enzyme, a unit of activity is defined as the amount of enzyme that hydrolyzes 10 μmoles of pyrophosphate in 15 minutes at 37°C under standard assay conditions. The purified enzyme has a V_{max} of 2800 units per milligram of enzyme.

   (a) How many moles of substrate are hydrolyzed per second per milligram of enzyme when the substrate concentration is much greater than K_M?

   (b) How many moles of active site are there in 1 mg of enzyme? Assume that each subunit has one active site.

   (c) What is the turnover number of the enzyme? Compare this value with others mentioned in this chapter.

2. Penicillin is hydrolyzed and thereby rendered inactive by penicillinase (also known as β-lactamase), an enzyme present in some resistant bacteria. The mass of this enzyme in *Staphylococcus aureus* is 29.6 kd. The amount of penicillin hydrolyzed in 1 minute in a 10-ml solution containing 10^{-9} g of purified penicillinase was measured as a function of the concentration of penicil-

lin. Assume that the concentration of penicillin does not change appreciably during the assay.

(a) Plot $1/V$ versus $1/[S]$ for these data. Does penicillinase appear to obey Michaelis-Menten kinetics? If so, what is the value of K_M?

(b) What is the value of V_{max}?

(c) What is the turnover number of penicillinase under these experimental conditions? Assume one active site per enzyme molecule.

[Penicillin]	*Amount hydrolyzed (moles)*
0.1×10^{-5} M	0.11×10^{-9}
0.3×10^{-5} M	0.25×10^{-9}
0.5×10^{-5} M	0.34×10^{-9}
1.0×10^{-5} M	0.45×10^{-9}
3.0×10^{-5} M	0.58×10^{-9}
5.0×10^{-5} M	0.61×10^{-9}

3. Penicillinase (β-lactamase) hydrolyzes penicillin. Compare penicillinase with glycopeptide transpeptidase.

4. The kinetics of an enzyme are measured as a function of substrate concentration in the presence and absence of 2×10^{-3} M inhibitor (I).
   (a) What are the values of V_{max} and K_M in the absence of inhibitor? In its presence?
   (b) What type of inhibition is this?
   (c) What is the binding constant of this inhibitor?
   (d) If $[S] = 1 \times 10^{-5}$ M and $[I] = 2 \times 10^{-3}$ M, what fraction of the enzyme molecules have a bound substrate? A bound inhibitor?
   (e) If $[S] = 3 \times 10^{-5}$ M, what fraction of the enzyme molecules have a bound substrate in the presence and absence of 2×10^{-3} M inhibitor? Compare this ratio with the ratio of the reaction velocities under the same conditions.

[S]	*Velocity (μmoles/min)*	
	No inhibitor	*Inhibitor*
0.3×10^{-5} M	10.4	4.1
0.5×10^{-5} M	14.5	6.4
1.0×10^{-5} M	22.5	11.3
3.0×10^{-5} M	33.8	22.6
9.0×10^{-5} M	40.5	33.8

5. The kinetics of the enzyme discussed in problem 4 are measured in the presence of a different inhibitor. The concentration of this inhibitor is 10^{-4} M.
   (a) What are the values of V_{max} and K_M in the presence of this inhibitor? Compare them with those obtained in problem 4.
   (b) What type of inhibition is this?
   (c) What is the dissociation constant of this inhibitor?
   (d) If $[S] = 3 \times 10^{-5}$ M, what fraction of the enzyme molecules have a bound substrate in the presence and absence of 10^{-4} M inhibitor?

[S]	*Velocity (μmoles/min)*	
	No inhibitor	*Inhibitor*
0.3×10^{-5} M	10.4	2.1
0.5×10^{-5} M	14.5	2.9
1.0×10^{-5} M	22.5	4.5
3.0×10^{-5} M	33.8	6.8
9.0×10^{-5} M	40.5	8.1

6. The plot of $1/V$ versus $1/[S]$ is sometimes called a Lineweaver-Burk plot. Another way of expressing the kinetic data is to plot V versus $V/[S]$, which is known as an Eadie-Hofstee plot.
   (a) Rearrange the Michaelis-Menten equation to give V as a function of $V/[S]$.
   (b) What is the significance of the slope, the vertical intercept, and the horizontal intercept in a plot of V versus $V/[S]$?
   (c) Make a sketch of a plot of V versus $V/[S]$ in the absence of an inhibitor, in the presence of a competitive inhibitor, and in the presence of a noncompetitive inhibitor.

7. The hormone progesterone contains two ketone groups. Little is known about the properties of the receptor protein that recognizes progesterone. At pH 7, which amino acid side chains might form hydrogen bonds with progesterone? (Assume that the side chains in the receptor protein have the same p*K*s as in the amino acids in aqueous solution.)

8. Suppose that two substrates A and B compete for an enzyme. Derive an expression relating the ratio of the rates of utilization of A and B, V_A/V_B, to the concentrations of these substrates and their values of k_3 and K_M. (Hint: Express V_A as a function of k_3/K_M for substrate A, and do the same for V_B.) Is specificity determined by K_M alone?

CHAPTER 9

Mechanisms of Enzyme Action

This chapter deals with catalytic strategies used by several well-understood enzymes: lysozyme, ribonuclease A, carboxypeptidase, and chymotrypsin. The three-dimensional structures of these enzymes are known in atomic detail, as is their mode of binding of substrate analogs and inhibitors. The reactions catalyzed are relatively simple—namely, the hydrolysis of glycosidic, phosphodiester, and peptide bonds. Each of these enzymes digests a macromolecular substrate and exemplifies an important category of hydrolytic reactions. The actions of these enzymes illustrate many important principles of catalysis. In particular, we shall see how enzymes facilitate the formation of transition states. Emerging subjects of enzyme research, such as genetically engineered enzymes, catalytic RNA, and catalytic antibodies, will also be presented.

FLEMING'S DISCOVERY OF LYSOZYME

In 1922, Alexander Fleming, a bacteriologist in London, had a cold. He was not one to waste a moment and consequently used his cold as an opportunity to do an experiment. He allowed a few drops of his nasal mucus to fall on a culture plate containing bacteria. He was excited to find some time later that the bacteria near the mucus had been dissolved away, and he thought that the mucus might contain the universal antibiotic he was seeking. Fleming showed that the antibacterial substance was an enzyme, which he named lysozyme—*lyso* because of its capacity to lyse bacteria and *zyme* because it was an enzyme. He also discovered a small round bacterium that was particularly susceptible to lysozyme; he named it *Micrococcus lysodeikticus* because it was a displayer of lysis (*deiktikos* means "able to show" in Greek). Fleming found that tears are a rich source of lysozyme. Volunteers provided tears after they suffered a few squirts of lemon—an "ordeal by lemon." The *St. Mary's*

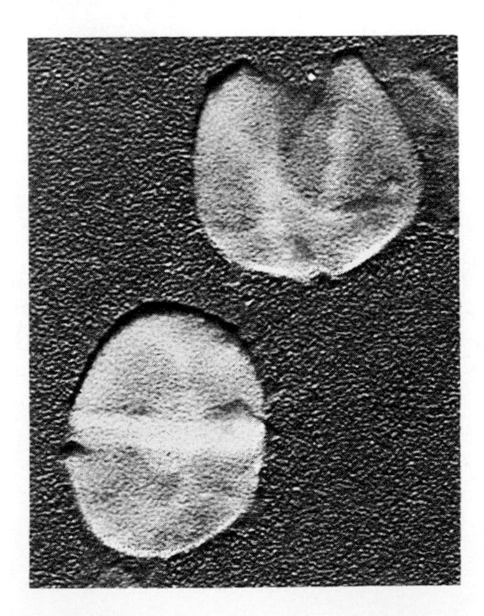

Figure 9-1
Electron micrograph of the isolated cell wall of *Micrococcus lysodeikticus*. [Courtesy of Dr. Nathan Sharon.]

Hospital Gazette published a cartoon showing children coming for a few pennies to Fleming's laboratory, where one attendant administered the beatings while another collected their tears! Fleming was disappointed to find that lysozyme was not effective against the most harmful bacteria. But seven years later, he did discover a highly effective antibiotic: penicillin—a striking illustration of Pasteur's comment that chance favors the prepared mind.

LYSOZYME CLEAVES BACTERIAL CELL WALLS

Lysozyme dissolves certain bacteria by cleaving the polysaccharide component of their cell walls. The function of the cell wall in bacteria is to confer mechanical support. A bacterial cell devoid of a cell wall usually bursts because of the high osmotic pressure inside the cell. We saw in the preceding chapter (p. 196) that penicillin blocks cell-wall synthesis by inactivating a transpeptidase that catalyzes the formation of essential cross-links between peptide units. Let us now focus on lysozyme's target, the polysaccharide portion of cell walls.

The cell-wall polysaccharide is made up of two kinds of sugars: N-*acetylglucosamine* (NAG) and N-*acetylmuramate* (NAM). NAM and NAG are derivatives of glucosamine in which the amino group is acetylated (Figure 9-2). In NAM, a lactyl side chain is attached to C-3 of the sugar ring by an ether bond. In bacterial cell walls, NAM and NAG are joined by *glycosidic linkages* between C-1 of one sugar and C-4 of the other. The oxygen atom in a glycosidic bond can be located either above or below the plane of the sugar ring. In the α configuration, the oxygen is below the plane of the sugar; in the β configuration, it is above (see Chapter 14 for a more detailed discussion of the properties and nomenclature of sugars). All of the glycosidic bonds of the cell-wall polysaccharide have a β *configuration* (Figure 9-3). NAM and NAG alternate in sequence. Thus, the cell-wall polysaccharide is an alternating polymer of NAM and NAG residues joined by $\beta(1 \rightarrow 4)$ glycosidic linkages. Note that polysaccharide chains, like polypeptide and polynucleotide chains, have directionality. Different polysaccharide chains are cross-linked by short peptides that are attached to NAM residues (see Figure 8-25, on p. 196).

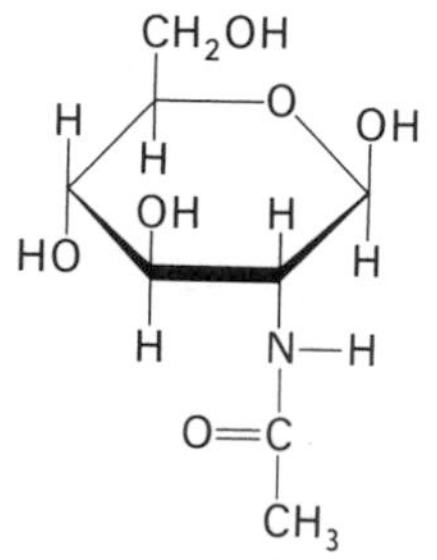

N-Acetylglucosamine (NAG)

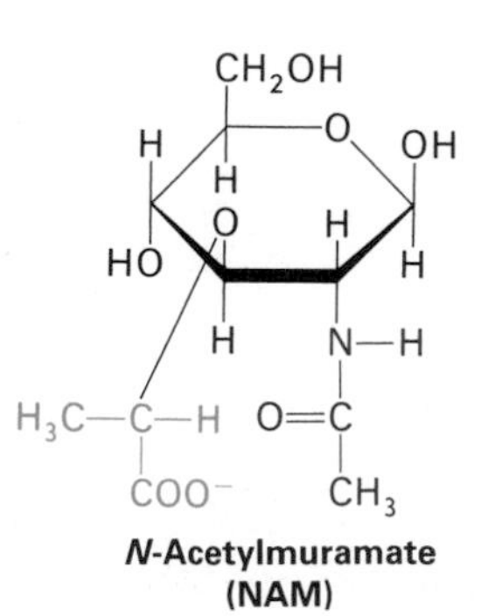

Figure 9-2
Sugar residues in the polysaccharide of bacterial cell walls.

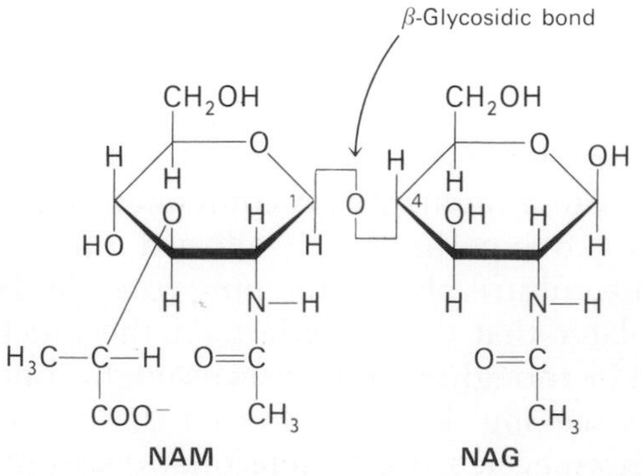

Figure 9-3
NAM is linked to NAG by a $\beta(1 \rightarrow 4)$ glycosidic bond.

Lysozyme, a glycosidase, hydrolyzes the glycosidic bond between C-1 of NAM and C-4 of NAG (Figure 9-4). The other glycosidic bond, between C-1 of NAG and C-4 of NAM, is not cleaved. *Chitin,* a polysaccharide found in the shell of crustaceans, is also a substrate for lysozyme. Chitin consists only of NAG residues joined by $\beta(1 \rightarrow 4)$ glycosidic links.

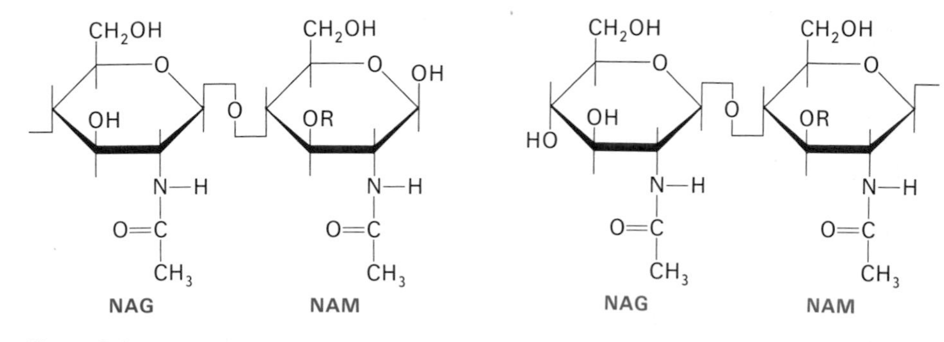

Figure 9-4
Lysozyme hydrolyzes the glycosidic bond (shown in green) between NAM and NAG (R refers to the lactyl group of NAM).

THREE-DIMENSIONAL STRUCTURE OF LYSOZYME

Lysozyme is a relatively small enzyme (14.6 kd). The one obtained from chicken egg white, a rich source, is a single polypeptide chain of 129 residues. This highly stable protein is cross-linked by four disulfide bridges. In 1965, David Phillips and his colleagues determined the three-dimensional structure of lysozyme. Their high-resolution electron-density map was the first for an enzyme molecule. Lysozyme is a compact molecule, roughly ellipsoidal in shape, with dimensions $45 \times 30 \times 30$ Å. The folding of this protein is complex (Figure 9-5). There is much less α helix than in myoglobin and hemoglobin. In a number of regions, the polypeptide chain is in an extended β-sheet conformation (Figure 9-6). The interior of lysozyme, like that of myoglobin and hemoglobin, is almost entirely nonpolar. Hydrophobic interactions play an important role in the folding of lysozyme, as they do for most proteins.

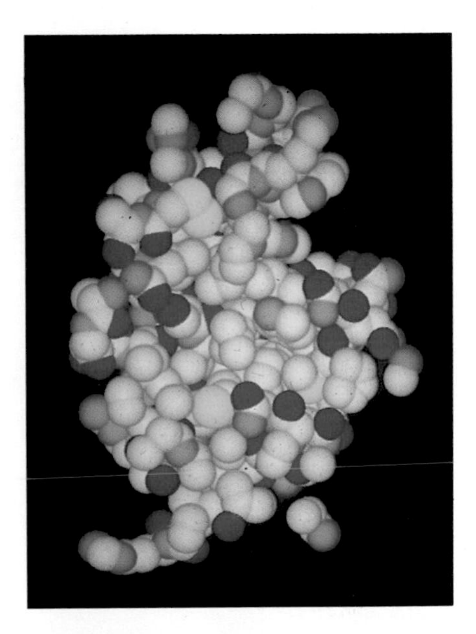

Figure 9-5
Space-filling model of lysozyme.

FINDING THE ACTIVE SITE IN LYSOZYME

Knowledge of the detailed three-dimensional structure of lysozyme did not immediately reveal the catalytic mechanism. Even the location of the active site was not obvious on looking at the electron-density map. Lysozyme does not contain a prosthetic group and thus lacks a built-in marker at its active site, in contrast with such proteins as myoglobin and hemoglobin. The essential information needed to identify the active site, specify the mode of binding of substrate, and elucidate the enzy-

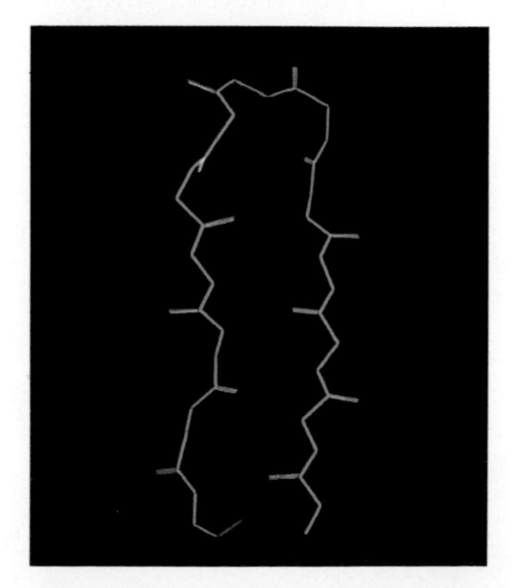

Figure 9-6
A hairpin turn of the main chain of lysozyme.

matic mechanism came from an x-ray crystallographic study of the interaction of lysozyme with inhibitors. When the three-dimensional structure of a protein is known, the mode of binding of small molecules can often be determined quite readily by x-ray crystallographic methods. These experiments are feasible because protein crystals are quite porous. Rather large inhibitor molecules can diffuse in the channels between different protein molecules and find their way to specific binding sites. The electron density corresponding to the additional molecule can be calculated directly from the intensities of the x-ray reflections (using the phases already determined for the native protein crystal) if the crystal structure is not markedly altered. This technique is called the *difference Fourier method*.

Ideally, one would like to use the difference Fourier method to elucidate the structure of an enzyme-substrate (ES) complex undergoing catalysis. However, under ordinary conditions, the conversion of bound substrate into product is much more rapid than the diffusion of new substrate into the crystal, so that ES complexes in the crystal are fleeting and rare. This difficulty can sometimes be circumvented by cooling the crystal (e.g., to −50°C) to slow the catalytic process. This experimental approach is called *cryoenzymology*. Alternatively, much information can be derived from a study of a complex of an enzyme with an unreactive (or very slowly reactive) analog of its substrate. For lysozyme, this was achieved with the trimer of *N*-acetylglucosamine (tri-NAG or NAG_3). Oligomers of *N*-acetylglucosamine consisting of fewer than five residues are hydrolyzed very slowly or not at all. However, they do bind to the active site of the enzyme. Indeed, tri-NAG is a potent competitive inhibitor of lysozyme.

MODE OF BINDING OF A COMPETITIVE INHIBITOR

The x-ray study of the tri-NAG complex with lysozyme showed the location of the active site, revealed the interactions responsible for the specific binding of substrate, and led to the proposal of a detailed enzymatic mechanism. Tri-NAG binds to lysozyme in a cleft at the surface of the enzyme and occupies about half of the cleft. Tri-NAG is bound to lysozyme by hydrogen bonds and van der Waals interactions. Electrostatic interactions cannot occur because tri-NAG lacks ionic groups.

The *hydrogen bonds* between tri-NAG and lysozyme are shown in Figure 9-7. The carboxylate group of aspartate 101 is hydrogen bonded to

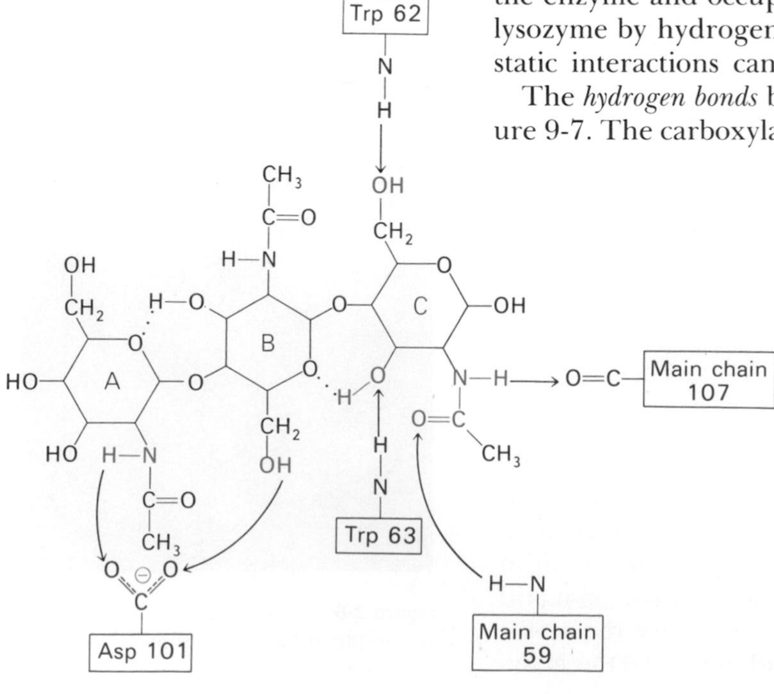

Figure 9-7
Hydrogen bonds between tri-NAG and lysozyme are shown in this highly schematic diagram. The hydrogen-binding groups on tri-NAG are shown in blue; those on lysozyme, in red.

sugar residues A and B. Four hydrogen bonds are made with residue C: the NH of the indole ring of tryptophan 62 is hydrogen bonded to the oxygen attached to C-6 (the ring moves 0.75 Å when tri-NAG binds to the enzyme); the adjacent amino acid residue, tryptophan 63, is similarly hydrogen bonded to the oxygen attached to C-3; and two other hydrogen bonds are formed with peptide groups of the main chain. Tri-NAG also makes many *van der Waals contacts* with the enzyme. For example, sugar residue B fits closely to the indole ring of tryptophan 57.

FROM STRUCTURE TO ENZYMATIC MECHANISM

1. *How would a substrate bind?* The mode of binding of a rapidly cleaved substrate cannot be established directly by x-ray crystallography. However, the structure of an enzyme complex with a competitive inhibitor provides clues to how a substrate binds. The finding that tri-NAG fills only half of the cleft in lysozyme was a very suggestive starting point. It seemed likely that additional sugar residues, which would fill the rest of the cleft, are required for the formation of a reactive ES complex. There was space for three additional sugar residues. This was encouraging, because it was known that the hexamer of *N*-acetylglucosamine (hexa-NAG) is rapidly hydrolyzed by the enzyme.

Three additional sugar residues, named D, E, and F, were fitted into the cleft by careful model building (Figure 9-8). Residues E and F went in nicely, making a number of good hydrogen bonds and van der Waals contacts. However, residue D would not fit unless it was distorted. Its C-6 and O-6 atoms came too close to several groups on the enzyme unless the ring was distorted from its normal conformation, which has the appearance of a chair.

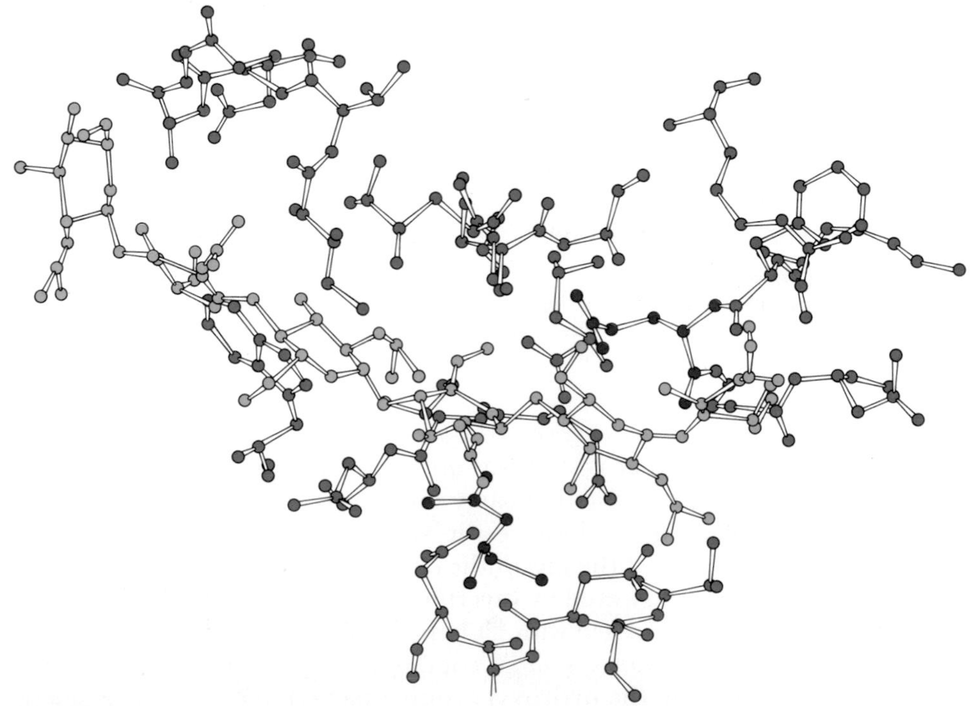

Figure 9-8
Mode of binding of hexa-NAG (shown in yellow) to lysozyme. The locations of sugar residues A, B, and C (left) are those observed in the tri-NAG-lysozyme complex, whereas those of residues D, E, and F (right) are inferred by model building. The two residues shown in red directly participate in catalysis.

2. *Which bond is cleaved?* The rate of hydrolysis of oligomers of *N*-acetylglucosamine increases strikingly as the number of residues is increased from four to five—that is, from NAG_4 to NAG_5 (Table 9-1). The cleavage rate increases further when a sixth residue is added (NAG_6), but it stays the same as the number of residues is increased to eight. This finding is consistent with the crystallographic result showing that the active-site cleft would be fully occupied by six sugar residues.

Table 9-1
Effectiveness of oligomers of *N*-acetylglucosamine as substrates

Substrate	*Relative rate of hydrolysis*
NAG_2	0
NAG_3	1
NAG_4	8
NAG_5	4,000
NAG_6	30,000
NAG_8	30,000

Which bond of hexa-NAG is cleaved? Because tri-NAG is stable, the A–B bond (i.e., the glycosidic bond between the A and B sugar residues) cannot be the site of cleavage. Similarly, the B–C bond cannot be the site of cleavage. A second crucial piece of evidence is that site C cannot be occupied by NAM. NAG fits nicely into site C, but NAM is too large because of its lactyl side chain. The bond cleaved in bacterial cell walls is the NAM-NAG link. Hence, the C–D bond cannot be the site of cleavage if the bacterial cell-wall polysaccharide binds to the enzyme in the same manner as hexa-NAG. The inability of NAM to fit into site C excludes yet another cleavage site: the E–F bond. The cell-wall polysaccharide is an alternating polymer of NAM and NAG, and so NAM cannot occupy site E if it cannot occupy site C.

These observations eliminated the A–B, B–C, C–D, and E–F bonds as possible sites of enzymatic cleavage of the hexamer substrate. Hence, *the D–E bond was the only remaining candidate for the cleavage site* (Figure 9-9).

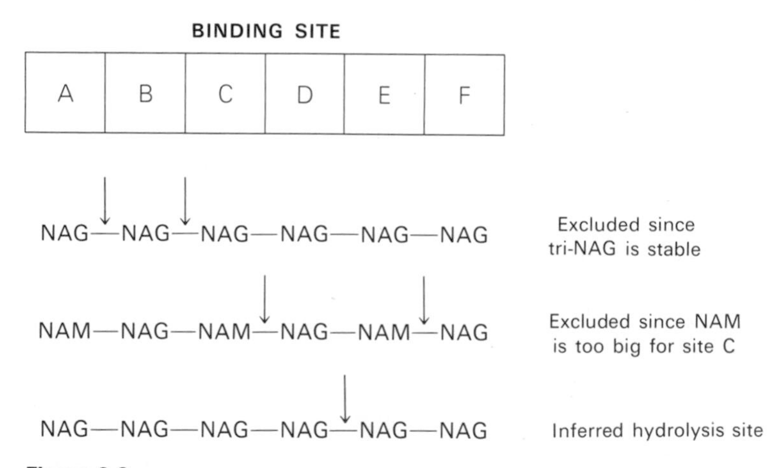

Figure 9-9
Steps in deducing that the glycosidic bond between sugar residues D and E is the one cleaved by lysozyme.

Figure 9-10
Hydrolysis in ^{18}O water showed that lysozyme cleaves the C_1–O bond rather than the O–C_4 bond. (Only the skeletons of the D and E residues are shown here.)

3. *Which groups on the enzyme participate in catalysis?* The inference that it is the D–E bond that is split was an important step toward defining the groups on the enzyme that carry out the hydrolysis reaction. However, it was necessary to localize the site of cleavage even more precisely: on which side of the glycosidic oxygen atom is the bond cleaved? This question was answered by carrying out enzymatic hydrolysis in the presence of water enriched with ^{18}O, the stable heavy isotope of oxygen (Figure 9-10). The sugars isolated contained ^{18}O attached to C-1 of the D sugar. In contrast, the hydroxyl group attached to C-4 of the E sugar contained the ordinary isotope of oxygen. *Hence, the bond split is the one between C-1 of residue D and the oxygen of the glycosidic linkage to residue E.* This experiment illustrates the value of isotopes in elucidating enzy-

matic mechanisms. In the absence of an isotopic marker, it would have been very difficult, if not impossible, to establish the precise site of cleavage.

A search was then made for possible catalytic groups close to the glycosidic bond that is cleaved. A *catalytic group* is one that directly participates in making or breaking covalent bonds. The donation or abstraction of a hydrogen ion is a critical step in most enzymatic reactions. Hence, the most likely candidates are groups that can serve as *proton donors or acceptors*. The only plausible catalytic residues near the glycosidic bond that is cleaved by lysozyme are aspartic acid 52 and glutamic acid 35. The aspartic acid residue is on one side of the glycosidic linkage, and the glutamic acid residue is on the other. These two acidic side chains have markedly different environments. Aspartic acid 52 is in a distinctly polar environment, where it serves as a hydrogen-bond acceptor in a complex network of hydrogen bonds. In contrast, glutamic acid 35 lies in a nonpolar region. Dissociation of a proton from a carboxyl group is less favored in a nonpolar than in a polar environment. Hence, at pH 5, the pH optimum for the hydrolysis of chitin by lysozyme, *aspartic acid 52 is probably in the ionized COO^- form (aspartate), whereas glutamic acid 35 is in the un-ionized COOH form.* The nearest oxygen atom of each of these acid groups is located about 3 Å away from the glycosidic linkage (Figure 9-11).

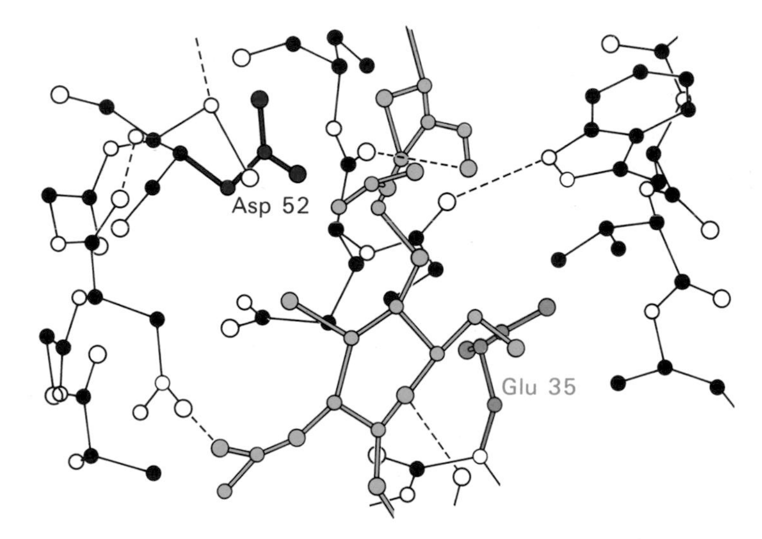

Figure 9-11
Structure of part of the active site of lysozyme. The D and E rings of the hexa-NAG substrate are shown in yellow. The side chains of aspartate 52 (red) and glutamic acid 35 (green) are in close proximity. [After W. N. Lipscomb. *Proc. Robert A. Welch Found. Conf. Chem. Res.* 15(1971):150.]

A CARBONIUM ION INTERMEDIATE IS CRITICAL FOR CATALYSIS

Phillips and his colleagues proposed a detailed catalytic mechanism for lysozyme based on the preceding structural data. The essential steps are:

1. The —COOH group of glutamic acid 35 donates an H^+ to the glycosidic oxygen atom between rings D and E. The transfer of a pro-

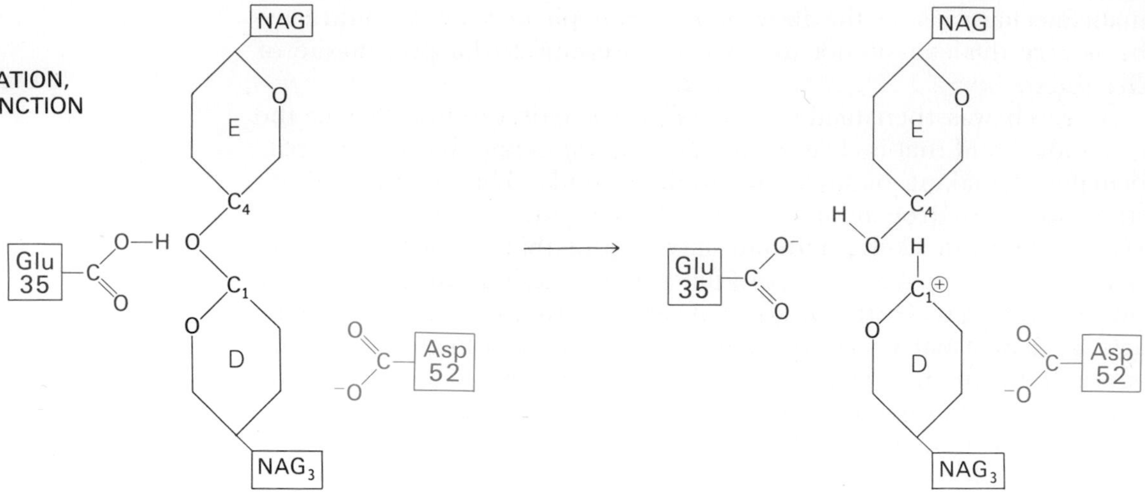

Figure 9-12
The first step in catalysis by lysozyme is the transfer of an H^+ from Glu 35 to the oxygen atom of the glycosidic bond. The glycosidic bond is thereby cleaved, and a carbonium ion intermediate is formed.

ton cleaves the bond between C-1 of the D ring and the glycosidic oxygen atom (Figure 9-12).

2. This creates a positive charge on C-1 of the D ring. This transient species is called a *carbonium ion* because it contains a positively charged carbon atom.

3. The dimer of NAG consisting of residues E and F diffuses away from the enzyme.

4. The carbonium ion intermediate then reacts with OH^- (or H_2O) from the solvent (Figure 9-13). Glutamic acid 35 becomes protonated, and tetra-NAG, consisting of residues A, B, C, and D, diffuses away from the enzyme. Lysozyme is then ready for another round of catalysis.

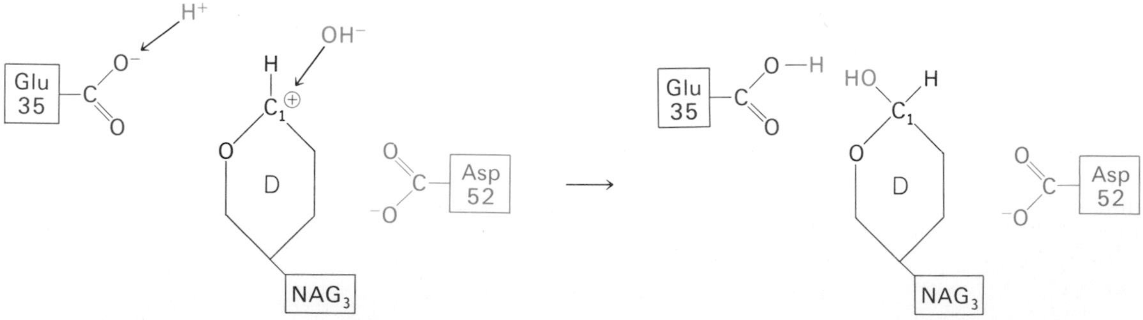

Figure 9-13
The cleavage reaction is completed by the addition of OH^- to the carbonium ion intermediate and H^+ to the side chain of Glu 35.

The critical elements of this proposed catalytic scheme are:

1. *General acid catalysis.* A proton is transferred from glutamic acid 35, which is un-ionized and optimally located 3 Å away from the glycosidic oxygen atom.

2. *Promotion of the formation of the carbonium ion intermediate.* The enzymatic reaction is markedly facilitated by two different factors that stabilize the carbonium ion intermediate:

a. The electrostatic factor is the presence of a negatively charged group 3 Å away from the carbonium ion intermediate. Aspartate 52, which is in the negatively charged carboxylate form, electrostatically stabilizes the positive charge on C-1 of ring D.

b. The geometrical factor is the distortion of ring D (Figure 9-14). Hexa-NAG fits best into the active-site cleft if sugar residue D is distorted out of its customary chair conformation into a half-chair form. This distortion enhances catalysis because the half-chair geometry markedly promotes the formation of the carbonium ion. In the half-chair form, the planarity of carbon atoms 1, 2, and 5 and the ring oxygen atom enables the positive charge to be shared by resonance between C-1 and the ring oxygen atom.

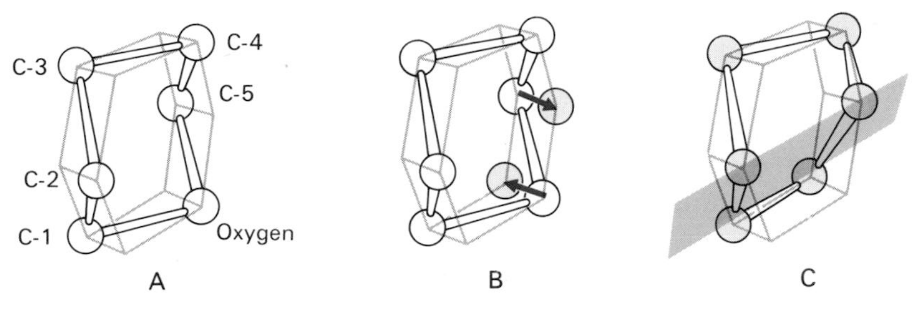

Figure 9-14
Distortion of the D ring of the substrate of lysozyme into a half-chair form: (A) a sugar residue in the normal chair form; (B) on binding to lysozyme, the ring oxygen atom and C-5 of sugar residue D move so that C-1, C-2, C-5, and O are in the same plane, as shown in part C. [After D. C. Phillips. The three-dimensional structure of an enzyme molecule. Copyright © 1966 by Scientific American, Inc. All rights reserved.]

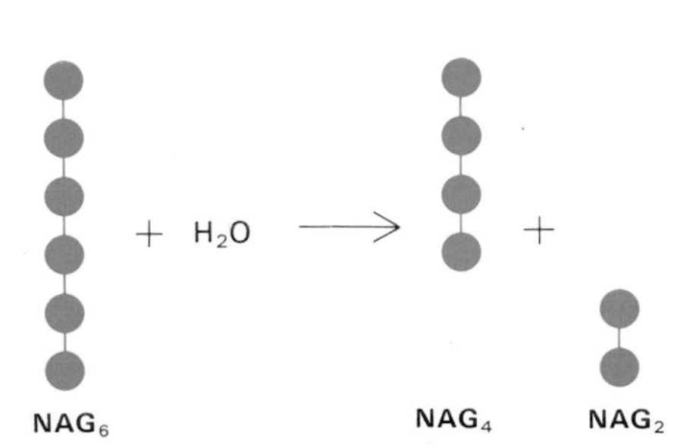

Figure 9-15
Hexa-NAG is hydrolyzed to tetra-NAG and di-NAG.

EXPERIMENTAL SUPPORT FOR THE PROPOSED MECHANISM

The mode of binding of substrate and the mechanism of catalysis proposed on the basis of the crystallographic studies have been tested in a variety of chemical experiments. All of the experimental results thus far support the crystallographic hypothesis. A number of experimental findings are especially pertinent in this regard:

1. *Cleavage pattern.* Hexa-NAG is split into tetra-NAG and di-NAG, which confirms the crystallographic hypothesis that cleavage occurs between the fourth and fifth residues of a hexamer (Figure 9-15).

2. *Binding affinity.* The contributions of each of the six sugar residues to the total free energy of binding to a hexameric substrate have been determined from measurements of binding equilibria (Figure 9-16). The striking finding is that sugar residue D makes a negative contribution to the binding affinity. The cost of binding residue D is about 4 kcal/mol. This result confirms the crystallographic inference that residue D is bound in a strained form. Also pertinent is that residue C makes the largest positive contribution to the binding affinity. The x-ray data show that residue C makes many hydrogen bonds and van der Waals interactions.

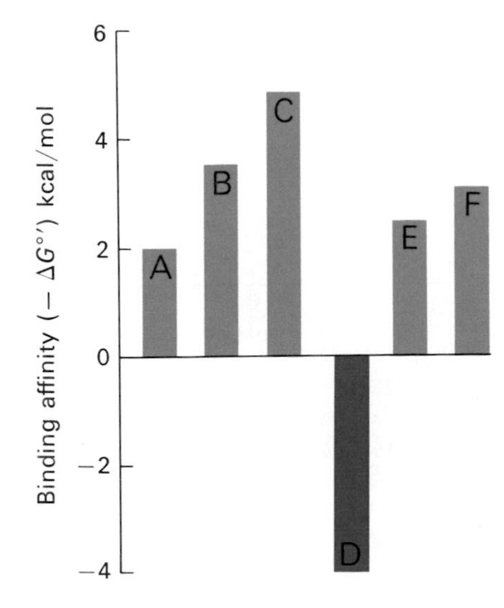

Figure 9-16
Contributions made by each of the six sugars of hexa-NAG to the standard free energy of binding of this substrate. The binding of residue D *costs* free energy because it must be distorted to fit into the active site.

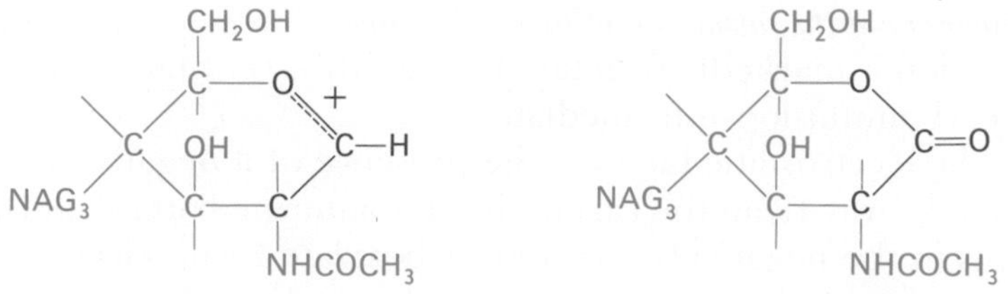

Carbonium ion derivative of tetra-NAG

Lactone analog of tetra-NAG

Figure 9-17
The lactone analog of tetra-NAG resembles the transition-state intermediate in the reaction catalyzed by lysozyme because its D ring has a conformation like that of a half-chair form.

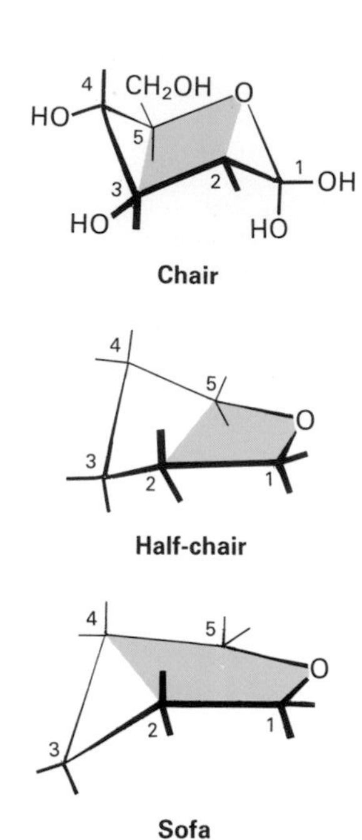

Figure 9-18
Comparison of the chair, half-chair, and sofa conformations of a sugar ring.

3. *Transition-state analogs*. Evidence for the role of distortion of residue D comes from studies of a transition-state analog of the substrate—that is, a compound that has the geometry of the catalyzed transition state *before* it is bound to the enzyme, as well as when it is bound. The D ring of the pure lactone analog of tetra-NAG (Figure 9-17) has a half-chair conformation. When bound to lysozyme, the C-1, C-2, C-4, C-5, and O atoms of the D ring of this analog are coplanar. This sofa conformation (Figure 9-18) is similar to the postulated half-chair form for the transition state. In fact, the binding of this lactone analog of tetra-NAG to subsites A through D of lysozyme is 3600 times as strong as the binding of tetra-NAG itself. This finding suggests that *the distortion of the D ring of a normal substrate could accelerate catalysis several thousandfold.*

4. *Dependence of the catalytic rate on pH*. The rate of hydrolysis of chitin is most rapid at pH 5 (Figure 9-19). The enzymatic activity drops sharply on either side of this optimal pH. The decrease on the alkaline side is due to the ionization of glutamic acid 35, whereas the decrease in rate on the acid side reflects the protonation of aspartate 52. Lysozyme is active only when glutamic acid 35 is un-ionized and aspartate 52 is ionized.

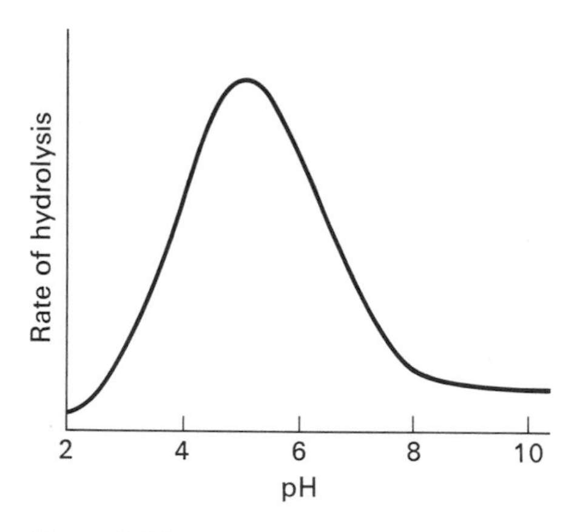

Figure 9-19
The rate of hydrolysis of chitin (poly-NAG) by lysozyme is maximal near pH 5.

5. *Selective chemical modification*. Lysozyme remains active when all of its carboxyl groups except those of glutamic acid 35 and aspartate 52 are esterified. Glutamic acid 35 and aspartate 52 are unmodified if this esterification takes place in the presence of substrate. When the substrate is removed, aspartate 52 also becomes esterified (whereas glutamic acid 35 remains unaltered). The modification of aspartate 52 produces totally inactive enzyme, which supports the proposal that the precise position of the carboxylate group of aspartate 52 is important in stabilizing the carbonium ion intermediate.

6. *Transglycosylation*. Hexa-NAG and di-NAG are formed at a slow rate when tetra-NAG is added to lysozyme (Figure 9-20). This reaction, called a transglycosylation, supports an important feature of the proposed enzymatic mechanism—namely, the binding of a reaction product to sites E and F. In the usual hydrolytic reaction, this species reacts with OH^- or H_2O. In transglycosylation, it reacts with another sugar: ROH. The transglycosylation reaction is specific because the OH^- donor is bound to sites E and F of the active-site cleft. Moreover, the glycosidic bond formed has the β configuration, the same as the normal substrate, which supports the proposed geometry of the catalytic intermediate.

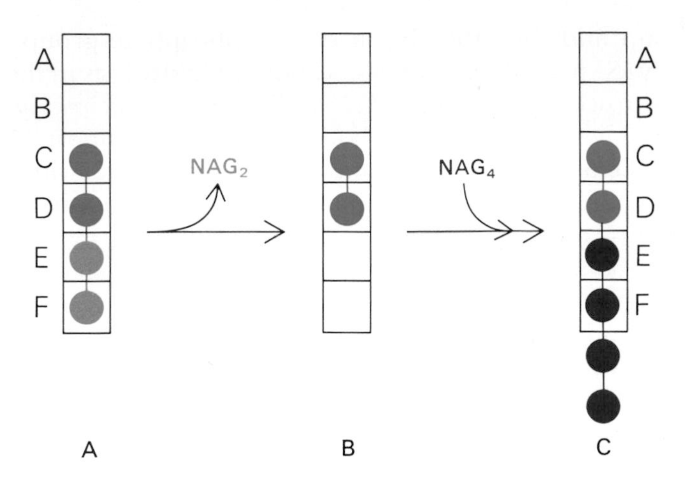

Figure 9-20
Lysozyme catalyzes a slow transglycosylation reaction. NAG_4 (shown in red) adds to bound reaction product NAG_2 (shown in blue in part B) to form NAG_6.

A CYCLIC PHOSPHATE INTERMEDIATE IS FORMED IN THE HYDROLYSIS OF RNA BY RIBONUCLEASE A

Let us now turn to the catalytic action of *ribonuclease A*, a digestive enzyme secreted by the pancreas. The folding of this enzyme (p. 30) and its binding of a uracil base (p. 186) have been mentioned. Ribonuclease A has been intensively studied for many years because it is small (124 residues, 13.7 kd), plentiful, and stable. The enzyme can be cleaved by subtilisin at a single peptide bond to yield *ribonuclease S*, a catalytically active complex consisting of an S-peptide moiety (residues 1–20) and an S-protein moiety (residues 21–124) bound together by multiple noncovalent interactions. Neither the S-peptide nor the S-protein alone is active. The catalytic properties of both ribonuclease A and S are nearly the same, and so we shall refer to them as ribonuclease.

Ribonuclease catalyzes the hydrolysis of phosphodiester bonds in RNA chains. The bonds cleaved are ones between phosphorus and 5′ oxygen atoms (Figure 9-21). One of the new termini formed has a free

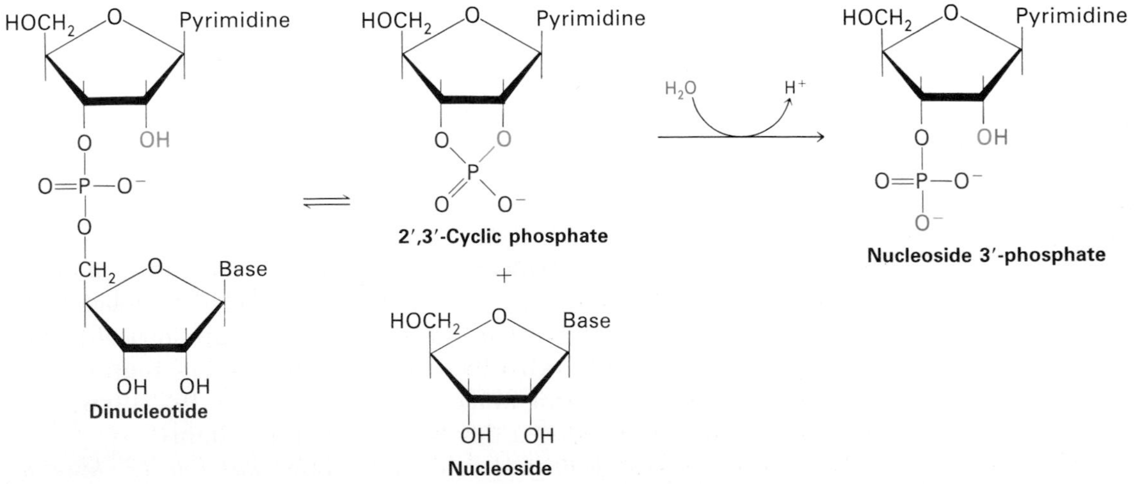

HOCH₂ O Pyrimidine; OH; O=P—O⁻; CH₂ O Base; OH OH
Dinucleotide

⇌

HOCH₂ O Pyrimidine; O O; P; O O⁻
2′,3′-Cyclic phosphate

\+

HOCH₂ O Base; OH OH
Nucleoside

H_2O → H^+

HOCH₂ O Pyrimidine; O OH; O=P—O⁻; O⁻
Nucleoside 3′-phosphate

Figure 9-21
A 2′,3′-cyclic phosphate intermediate is formed in the hydrolysis of RNA by ribonuclease.

Scissile bond—
The bond in a substrate molecule that is cleaved by an enzyme.
From the Latin word *scissus,* meaning "torn asunder." The same root is seen in the word *scissors.*

5′-OH group and the other has a free 3′-phosphate group. Ribonuclease-catalyzed hydrolysis, like base-catalyzed hydrolysis in the absence of enzyme, proceeds through a *2′,3′-cyclic phosphate intermediate.* In the first stage, cleavage of the scissile bond forms a cyclic phosphate terminus and a free 5′-OH. In the second stage, water reacts with the cyclic phosphate to yield a free 3′-phosphate end. The formation of the cyclic phosphate intermediate is reversible, whereas its hydrolysis is virtually irreversible. This cyclic intermediate can readily be isolated because it is formed much more rapidly than it is hydrolyzed.

The nucleotide on the 3′-side of the scissile bond must be a pyrimidine, because a purine ring is too large to be accommodated in the active site without distorting it. A uracil or cytosine ring binds to the active site by multiple, precisely directed hydrogen bonds (p. 186). In contrast, ribonuclease is indifferent to the nature of all other bases in its RNA substrate because its binding to them is primarily electrostatic. A series of nine positively charged lysine and arginine side chains form salt bridges with the negatively charged phosphate backbone of RNA. Single-stranded DNA also binds to ribonuclease, but it is not hydrolyzed because DNA lacks a 2′-hydroxyl group and hence cannot form a 2′,3′-cyclic intermediate. The uncatalyzed hydrolysis of DNA is much slower than that of RNA for the same reason. The absence of 2′-OH groups makes DNA a more reliable store of genetic information.

Much was deduced about the catalytic mechanism of ribonuclease from chemical studies before its three-dimensional structure was visualized. A plot of the catalytic rate versus pH is bell-shaped, like that of lysozyme (see Figure 9-19). Because the pH optimum is about 7, it was proposed that two histidine residues participate in catalysis, one in the basic form and the other in the acidic form. Chemical modification studies supported this notion. Reaction of ribonuclease with iodoacetate, an alkylating reagent, led to the carboxymethylation of the imidazole ring of histidine 119 or histidine 12 but not both in the same enzyme molecule (Figure 9-22). Modification of either histidine side chain inactivated the enzyme. Substrates or competitive inhibitors protected ribonuclease from modification and inactivation by iodoacetate. These findings suggested that *histidines 12 and 119 are near each other in the active site and act in concert as proton donors and acceptors in catalyzing the formation of the cyclic intermediate and its subsequent hydrolysis.*

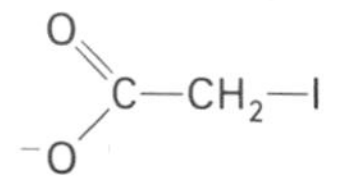

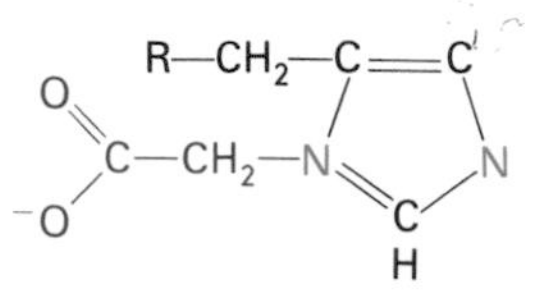

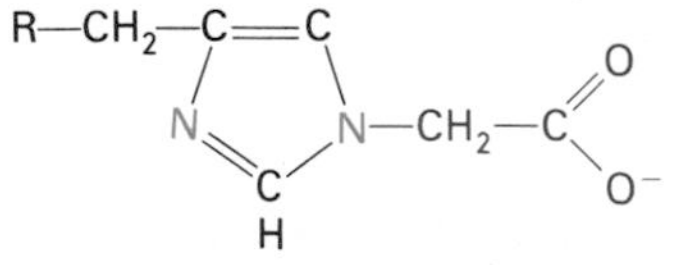

Figure 9-22
Iodoacetate alkylates two histidine residues in ribonuclease. The products are 3-carboxymethyl-histidine 12 and 1-carboxymethyl-histidine 119.

PHOSPHORUS IS PENTACOVALENT IN THE TRANSITION STATE FOR RNA HYDROLYSIS

X-ray crystallographic studies carried out by Frederic Richards and Harold Wyckoff showed that both histidines are at the active site, just where expected. They solved the three-dimensional structure of a complex of ribonuclease with a substrate analog, namely the phosphonate analog of UpC, in which a methylene group (—CH_2—) replaces the oxygen atom in the scissile bond. This analog, though sterically very similar to UpC, cannot be hydrolyzed by ribonuclease. The high-resolution structure of this enzyme-analog complex (Figure 9-23) has been very valuable in formulating a detailed catalytic mechanism.

Three residues play a critical role in catalysis: *histidine 12, histidine 119,* and *lysine 41* (Figure 9-24). The reaction begins with an attack by the 2′-O on the phosphorus atom of the scissile bond, in the following way. First, the un-ionized form of histidine 12 accepts a proton from the 2′-OH, which enhances the nucleophilicity of this oxygen atom. At

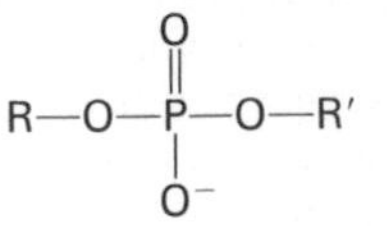

Phosphonate—
A compound containing a C–P bond. Phosphonates can serve as nonhydrolyzable analogs of esters.

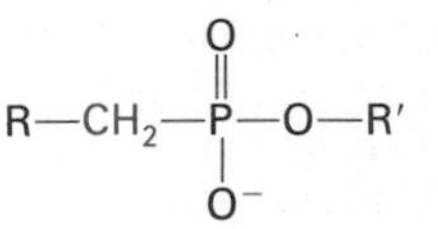

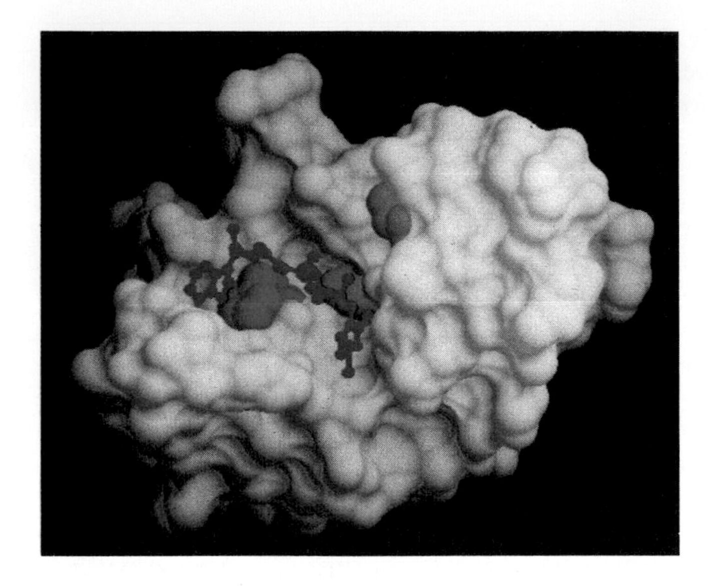

Figure 9-23
Three-dimensional structure of a complex of ribonuclease with a substrate analog. The phosphonate analog of UpC is shown in orange. The side chains shown in color are: histidine 12, green; histidine 119, blue; and lysine 41, pink. [Courtesy of Dr. Frederic Richards, Dr. Harold Wyckoff, and Dr. Art Perlo.]

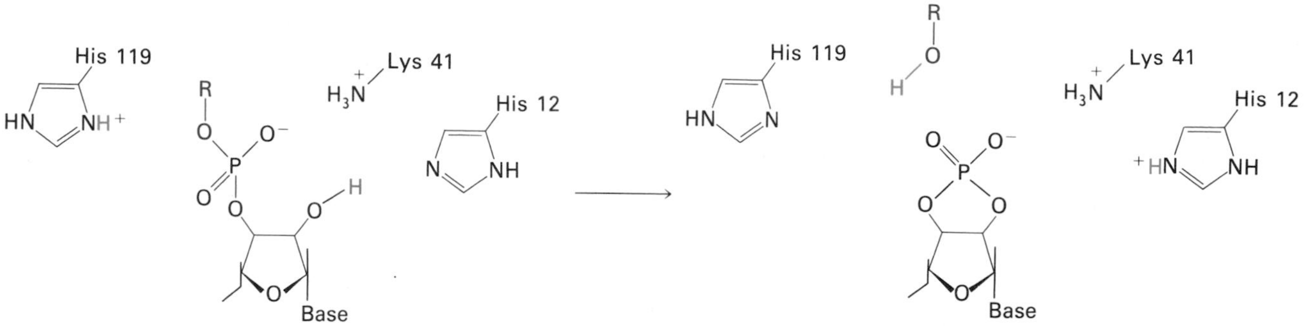

His 119 Lys 41 His 12 R O O^- P O O H Base $\longrightarrow$ His 119 R O H Lys 41 His 12 O O^- P O O Base

Figure 9-24
Proposed mechanism for the formation of the cyclic phosphate intermediate in catalysis by ribonuclease. A pentacovalent transition state is formed. Hydrolysis of the cyclic phosphate intermediate is essentially the reverse of this reaction, with water substituting for ROH.

the same time, the protonated form of histidine 119 begins to donate its proton to the 5′-O, and the 2′-O begins to form a bond with P, which becomes transiently bonded to five oxygen atoms. This pentacovalent transition state is stabilized electrostatically by the nearby positively charged side chain of lysine 41. The bond between P and the 5′-O breaks when the proton from histidine 119 is completely transferred to this oxygen atom. At the same time, a bond between P and the 2′-O becomes fully formed, producing the 2′,3′-cyclic intermediate.

Hydrolysis of this cyclic intermediate, the second stage of the reaction, is nearly a reversal of the first stage. A difference is that water replaces the 5′-O component that was removed. Histidine 12 is now the proton donor and histidine 119 is the proton acceptor. The capacity of histidine to act as either an acid or base at physiological pH accounts for its appearance in the active sites of many enzymes.

The geometry of the pentacovalent transition state is interesting. The tetrahedral geometry of phosphorus in RNA changes to that of a *trigonal bipyramid* when phosphorus becomes pentacovalent. Phosphorus occupies the center, three oxygen atoms lie in the equatorial plane, and two at the apices of the pyramid (Figure 9-25). In the formation of the cyclic intermediate, the outgoing 5′-O is at one apex, and the incoming

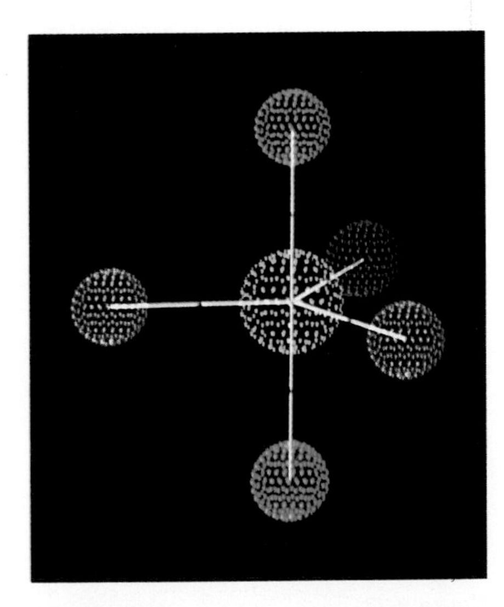

Figure 9-25
Geometry of the pentacovalent transition state of ribonuclease. The central phosphorus atom (yellow) is transiently bonded to five oxygen atoms. Three of them (red) are coplanar with the phosphorus. The oxygen atom of the leaving group (green) is at one apex, and the oxygen atom of the attacking group (blue) is at the other apex of this trigonal bipyramid.

2′-O is at the other. In the hydrolysis of this intermediate, a water O is at one apex and the 2′-O is at the other. *In each stage, one apex is occupied by the attacking nucleophile and the other by the leaving group.* This geometry of attacking and leaving groups is termed *in-line*.

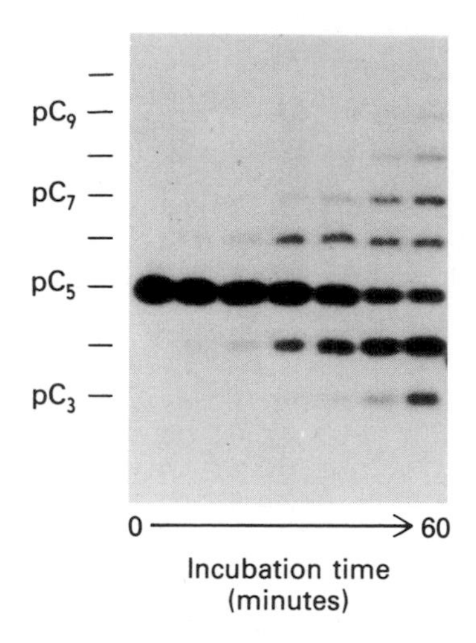

Figure 9-26
Catalysis of the cleavage and elongation of pentacytidylate (C_5) by L19 RNA. The molar ratio of substrate to enzyme was 6. The 5′ end of C_5 was labeled with ^{32}P phosphate so that the reaction products could be visualized by autoradiography. [After A. J. Zaug and T. R. Cech. *Science* 231(1985):471.]

RNA MOLECULES ALSO CAN BE VERY EFFECTIVE ENZYMES

Thousands of proteins have been purified and shown to possess enzymatic activity. It was believed for more than fifty years that all enzymes are proteins. However, recent experiments have revealed that *RNA molecules, too, can be highly active enzymes.* The most striking example is provided by L19 RNA, a shortened form of an intron released by the self-splicing of a ribosomal RNA precursor in *Tetrahymena thermophila*, a protozoan (p. 725). Thomas Cech discovered that L19 RNA catalyzes the cleavage and joining of oligoribonucleotide substrates in a highly specific manner. Pentacytidylate (C_5) is converted by L19 RNA into longer and shorter oligomers (Figure 9-26). Specifically, C_5 is degraded to C_4 and C_3. At the same time, C_6 and longer oligomers are formed. Thus, L19 RNA is both a *ribonuclease* and an *RNA polymerase*.

This RNA enzyme acts much more rapidly on C_6 than on U_6 and not at all on A_6 and G_6. It obeys Michaelis-Menten kinetics with a K_M of 42 μM and a k_{cat} of 0.033 s^{-1} for C_5. The essential need for a 2′-OH in the substrate is demonstrated by the finding that deoxy-C_5 is a competitive inhibitor of C_5 with a K_i of 260 μM. Thus, L19 RNA displays several hallmarks of classic enzymatic catalysis: *a high degree of substrate specificity, Michaelis-Menten saturation kinetics, and susceptibility to competitive inhibition.*

How does this 395-nucleotide RNA molecule act both as a ribonuclease and a polymerase? Its three-dimensional structure is not yet known but some inferences about its catalytic mechanism can be made (Figure 9-27). Pentacytidylate binds to a specific site in L19 RNA. Several of the five C bases of this substrate are probably base-paired to a G-rich sequence of this RNA enzyme. The phosphodiester bond between the fourth and fifth C of bound C_5 is then attacked by the 3′-OH of the terminal G residue of this enzyme. A phosphodiester bond is formed between the terminal G of the enzyme and the terminal C of the substrate to produce a —GpC *covalent intermediate*. C_4 is concomitantly released during this *transesterification reaction*. This highly labile new phos-

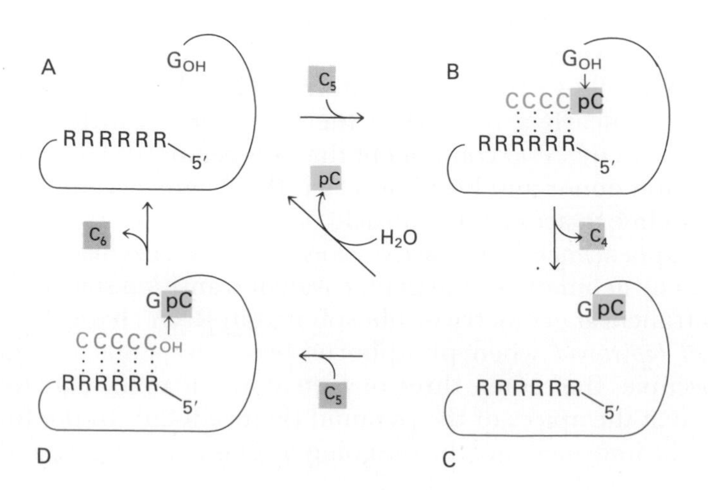

Figure 9-27
Proposed catalytic mechanism of L19 RNA. (A) Enzyme alone. (B) C_5 is hydrogen-bonded to the enzyme, and the terminal pC is linked covalently to the terminal G of the enzyme. The remaining C_4 moiety is free to dissociate from the enzyme. (C) Hydrolysis of GpC restores the enzyme to its original state. (D) Alternatively, the covalently attached pC can be attacked by a second C_5 molecule to form C_6. [After A. J. Zaug and T. R. Cech. *Science* 231(1985):473.]

phodiester bond can be attacked by water to release pC and regenerate the enzyme. This series of three steps accounts for the *ribonuclease* activity of L19 RNA.

Alternatively, the —GpC enzyme intermediate can bind another C_5 molecule, which then attacks —GpC to produce C_6 and regenerate the enzyme. This step accounts for the *polymerase* action of L19 RNA. Thus, the covalent intermediate —GpC can be attacked either by OH^- or by the 3′-OH group of a second substrate. Hydrolysis is favored by a high pH, whereas transesterification to form C_6 is promoted by a high concentration of the C_5 substrate. C_6 formed in this way can be elongated to C_7 and longer oligomers by successive transesterification reactions. It seems likely that the phosphodiester bond of —GpC is unusually reactive because it is activated electrostatically to form a pentacovalent transition state.

How does the catalytic efficacy of this RNA enzyme compare with that of protein enzymes? The k_{cat}/K_M value for L19 RNA is $10^3\ s^{-1}\ M^{-1}$, five orders of magnitude less than that for the most catalytically proficient proteins (p. 191). However, it is similar to that for ribonuclease A. *The rate of hydrolysis of C_5 by this RNA enzyme is about 10^{10} times the uncatalyzed rate. It is evident that RNA molecules, like proteins, can be very effective catalysts.* RNA molecules can form precise three-dimensional structures that bind specific substrates and stabilize transition states. However, RNAs differ from proteins in not being able to form large nonpolar niches. RNAs are also much less versatile than are proteins because they contain only four different building blocks instead of twenty. Hence, most enzymes are proteins, but it must be borne in mind that proteins do not have a monopoly on catalysis. RNA enzymes (sometime called ribozymes) are very well suited for recognizing and transforming single-stranded nucleic acid because they share a common base-pairing language with their substrates. The finding of enzymatic activity in L19 RNA and in the RNA component of ribonuclease P (p. 743), an enzyme that participates in the processing of tRNA precursors, has important evolutionary implications and opens an exciting new area of inquiry.

CARBOXYPEPTIDASE A: A ZINC-CONTAINING PROTEOLYTIC ENZYME

Let us now turn to carboxypeptidase A, a digestive enzyme that hydrolyzes the carboxyl-terminal peptide bond in polypeptide chains. Hydrolysis occurs most readily if the carboxyl-terminal residue has an aromatic or a bulky aliphatic side chain (Figure 9-28).

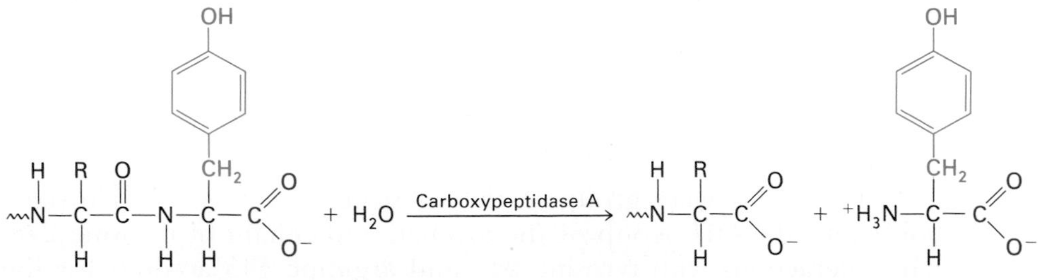

Figure 9-28
Reaction catalyzed by carboxypeptidase A.

The three-dimensional structure of carboxypeptidase A at a resolution of 2 Å was solved by William Lipscomb in 1967. This enzyme is a single polypeptide chain of 307 amino acid residues. Carboxypeptidase A has a compact shape that approximates an ellipsoid of dimensions 50 × 42 × 38 Å. The enzyme contains regions of α helix (38%) and of β pleated sheet (17%). A tightly bound zinc ion is essential for enzymatic activity. This zinc ion is located in a groove near the surface of the molecule, where it is coordinated to a tetrahedral array of two histidine side chains, a glutamate side chain, and a water molecule (Figure 9-29). A large pocket near the zinc ion accommodates the side chain of the terminal residue of the peptide substrate.

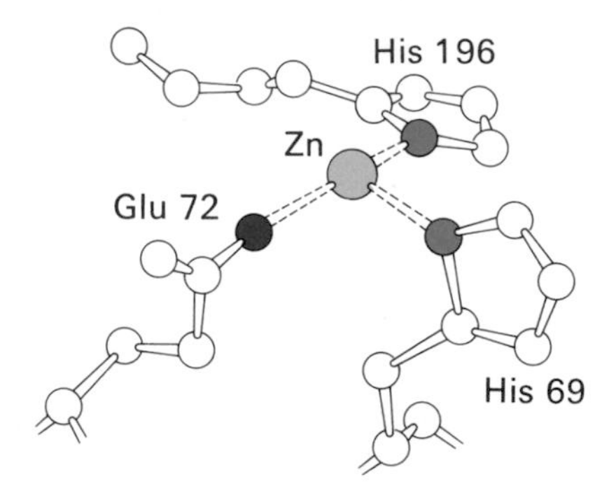

Figure 9-29
A zinc ion is coordinated to two histidine side chains and a glutamate side chain at the active site of carboxypeptidase A. A water molecule coordinated to the zinc is not shown here. [After D. M. Blow and T. A. Steitz. X-ray diffraction studies of enzymes. *Ann. Rev. Biochem.* 39:78. Copyright © 1970 by Annual Reviews, Inc. All rights reserved.]

Two facets of the catalytic mechanism of carboxypeptidase A are particularly noteworthy:

1. *Induced fit.* The binding of substrate is accompanied by many changes in the structure of the enzyme.

2. *Electronic strain.* The enzyme contains a zinc atom and other groups at its active site that induce rearrangements in the distribution of electrons in the substrate, rendering it more susceptible to hydrolysis.

BINDING OF SUBSTRATE INDUCES LARGE STRUCTURAL CHANGES AT THE ACTIVE SITE OF CARBOXYPEPTIDASE A

The structures of carboxypeptidase A complexes with inhibitors and slowly cleaved substrates have provided valuable information concerning the likely geometry of productive enzyme-substrate complexes. Glycyltyrosine, a slowly cleaved substrate, binds to carboxypeptidase A through multiple interactions (Figures 9-30 and 9-31):

1. The negatively charged terminal carboxylate of glycyltyrosine interacts electrostatically with the positively charged side chain of arginine 145.

2. The carboxylate group of the substrate also forms a hydrogen bond with the OH group of the aromatic side chain of tyrosine 248. The interactions with tyrosine 248 and arginine 145 account for the carboxyl-terminal specificity of this protease.

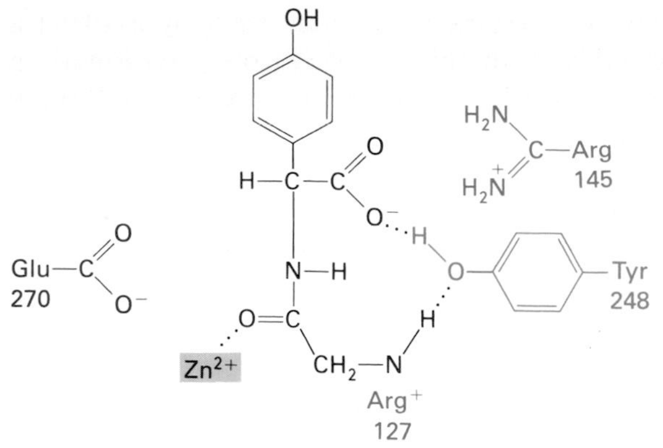

Figure 9-30
Schematic representation of the binding of glycyltyrosine (black) to the active site of carboxypeptidase A.

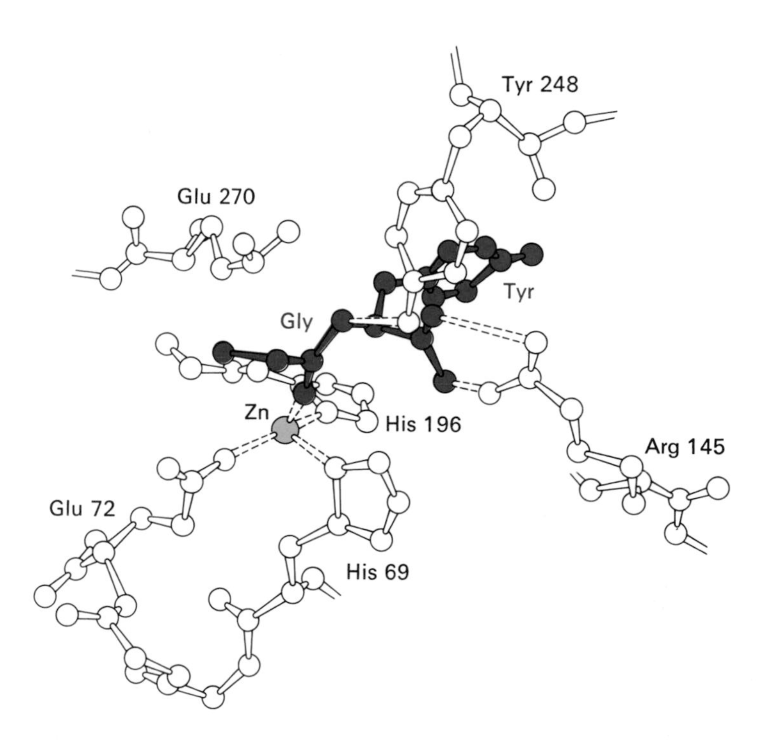

Figure 9-31
Three-dimensional structure of glycyltyrosine at the active site of carboxypeptidase A. Glycyltyrosine, the substrate, is shown in red. [After D. M. Blow and T. A. Steitz. X-ray diffraction studies of enzymes. *Ann. Rev. Biochem.* 39:79. Copyright © 1970 by Annual Reviews, Inc. All rights reserved.]

3. The tyrosine side chain of the substrate binds to a nonpolar pocket in the enzyme. This hydrophobic interaction is the basis for the enzyme's preference for aromatic or bulky aliphatic side chains. Carboxypeptidase B, a related enzyme that is specific for terminal lysine and arginine residues, which are polar, contains an oppositely charged aspartate residue in this pocket.

4. The carbonyl oxygen of the susceptible peptide bond is coordinated to the zinc ion.

5. The terminal amino group of the substrate is hydrogen bonded through an intervening water molecule to the side chain of glutamate 270. This interaction, not found in productive ES complexes, probably accounts for the very slow hydrolysis of glycyltyrosine and other dipeptides with a free amino group.

The binding of glycyltyrosine is accompanied by a structural rearrangement of the active site (Figure 9-32), as envisioned by Daniel Koshland, Jr., in his *induced-fit model* of enzyme action. The guanidinium group of arginine 145 moves 2 Å, as does the carboxylate group of glutamate 270. The binding of the carbonyl group of the substrate to the zinc ion displaces a bound water molecule. At least four other water molecules are displaced from the nonpolar pocket when the tyrosine side chain of the substrate binds there. The largest conformational change is the movement of the phenolic hydroxyl group of tyrosine 248 by 12 Å, a distance equal to about a quarter of the diameter of the

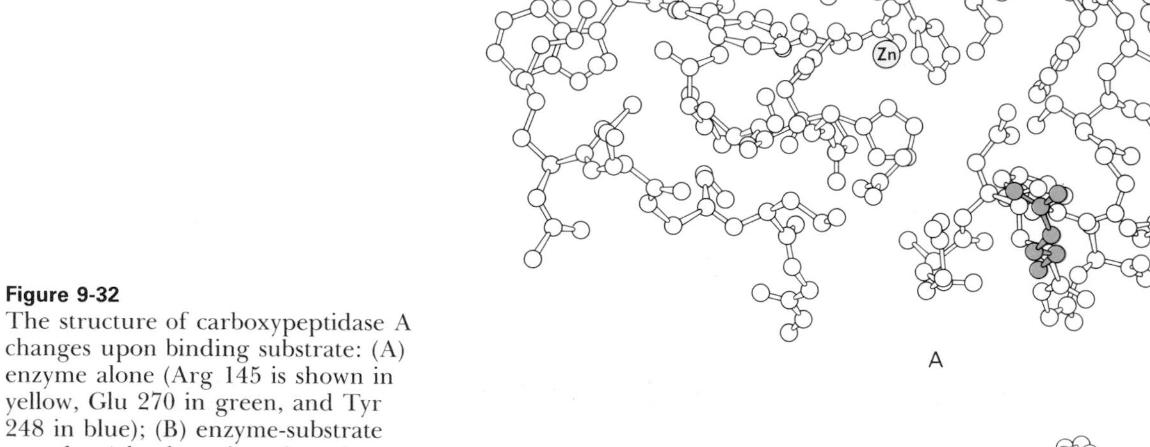

Figure 9-32
The structure of carboxypeptidase A changes upon binding substrate: (A) enzyme alone (Arg 145 is shown in yellow, Glu 270 in green, and Tyr 248 in blue); (B) enzyme-substrate complex (glycyltyrosine, the substrate, is shown in red). [After W. N. Lipscomb. *Proc. Robert A. Welch Found. Conf. Chem. Res.* 15(1971):140–141.]

protein. This motion is accomplished primarily by a facile rotation about a single carbon–carbon bond. The hydroxyl group of tyrosine 248 moves from the surface of the molecule to the vicinity of the terminal carboxylate of the substrate. An important consequence of this motion is that it closes the active-site cavity and extrudes water from it.

CATALYTIC MECHANISM OF CARBOXYPEPTIDASE A

A plausible catalytic mechanism for carboxypeptidase A emerges from x-ray and chemical studies carried out by many investigators. The productive ES complex is postulated to have the structure shown in Figure 9-33. Three groups are thought to be critical for catalysis: the zinc ion, the guanidinium group of arginine 127, and the carboxylate group of glutamate 270. The carbonyl carbon atom of the scissile bond is attacked by the carboxylate of glutamate 270. A *carboxylic-anhydride intermediate* is formed, which makes the scissile bond ripe for attack by a water molecule that is bound to Zn^{2+}. This highly polarized water molecule transfers a proton to the NH group, cleaving the scissile bond. Alternatively, an activated water molecule may directly attack the carbonyl carbon atom of the scissile bond (see problem 11 on page 232.)

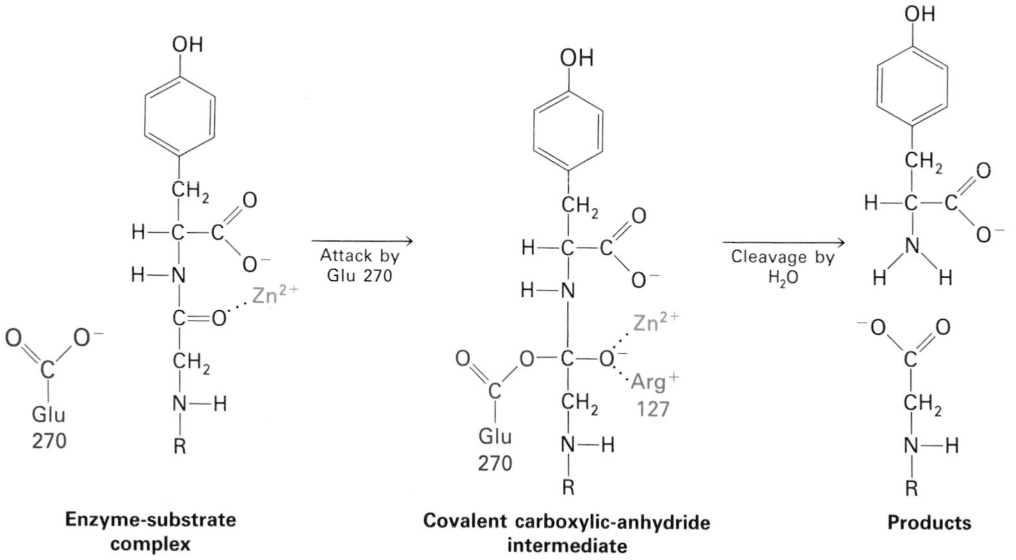

Figure 9-33
A proposed catalytic mechanism for carboxypeptidase A in which Glu 270 directly attacks the carbonyl carbon atom of the scissile bond to form a covalent mixed-anhydride intermediate.

ELECTRONIC STRAIN AND ENZYME FLEXIBILITY ARE KEY FEATURES OF CATALYSIS BY CARBOXYPEPTIDASE A

In carboxypeptidase A, polarization of the carbonyl group by the zinc ion facilitates the attack on the carbon by either glutamate 270 or an activated water molecule. The induction of a dipole is enhanced by the nonpolar environment of the zinc ion, which increases its effective charge. Thus, *carboxypeptidase A accelerates catalysis by inducing electronic strain in its substrate.* Furthermore, the negative charge of the oxygen atom in the tetrahedral intermediate is stabilized by an electrostatic interaction with the positively charged side chain of arginine 127, which has moved to a strategic position nearby.

The need for substrate-induced structural changes in the active site of carboxypeptidase A can now be appreciated. The bound substrate is surrounded on all sides by catalytic groups of the enzyme. This arrangement promotes catalysis for the reasons cited above. It is evident that a substrate could not enter such an array of catalytic groups (nor could a product leave) unless the enzyme were flexible. *A flexible protein provides a much larger repertoire of potentially catalytic conformations than does a rigid one.*

SITE-SPECIFIC MUTAGENESIS IS A POWERFUL METHOD FOR ENGINEERING NEW ENZYMES AND ELUCIDATING MECHANISMS

In Agatha Christie's *Murder on the Orient Express*, Hercule Poirot's task as a detective was compounded by the presence of *so many* plausible suspects. Likewise, the catalytic mechanism of carboxypeptidase A has been difficult to unravel because of the presence of several potential catalytic groups at the active site. For example, the hydroxyl group of tyrosine 248 was postulated to be the proton donor to the NH group of the substrate in the cleavage step. This hypothesis was tested by William Rutter using site-specific mutagenesis. If tyrosine 248 is essential for catalysis, a mutant of this enzyme containing phenylalanine in its place should be inactive. The tyrosine 248 codon (TAT) was converted to that of phenylalanine (TTT) by oligonucleotide-directed mutagenesis (p. 136). The recombinant plasmid containing this gene was inserted into yeast and expressed. The striking finding was that the mutant enzyme had the same k_{cat} value as did the native enzyme but a K_M value that was sixfold higher. These results indicate that tyrosine 248 participates in the binding of substrate but is not essential for catalysis. This incisive experiment illustrates the power of site-specific mutagenesis in delineating the function of a particular residue in a protein.

THREE-DIMENSIONAL STRUCTURE OF CHYMOTRYPSIN, A SERINE PROTEASE

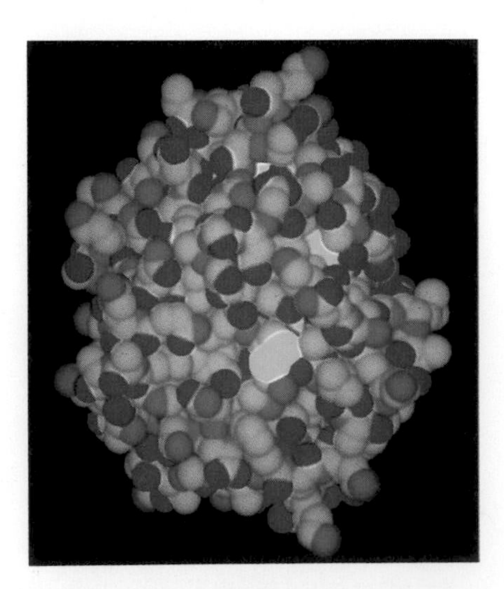

Space-filling model of chymotrypsin. This enzyme is stabilized by several disulfide bonds (yellow).

Chymotrypsin, like carboxypeptidase A, is a digestive enzyme. However, its specificity and catalytic mechanism are different, which makes it rewarding to examine chymotrypsin, too, in some detail. Chymotrypsin has further significance as a member of an important family of proteins, the *serine proteases*. Trypsin, another digestive enzyme, and thrombin, a clotting enzyme, are other members of this pervasive clan.

Chymotrypsin, a 25-kd enzyme, consists of three polypeptide chains connected by two interchain disulfide bonds. As will be discussed in the next chapter, chymotrypsin is synthesized as a single-chain inactive precursor called chymotrypsinogen. The three-dimensional structure of the enzyme was solved at 2-Å resolution (Figure 9-34) by David Blow. The molecule is a compact ellipsoid of dimensions 51 × 40 × 40 Å. Chymotrypsin contains several antiparallel β-pleated sheet regions and little α helix. All charged groups are on the surface of the molecule except for three that play a critical role in catalysis.

The biological role of chymotrypsin is to catalyze the hydrolysis of proteins in the small intestine. Chymotrypsin does not cleave all peptide bonds at a significant rate. Rather, it is selective for peptide bonds on

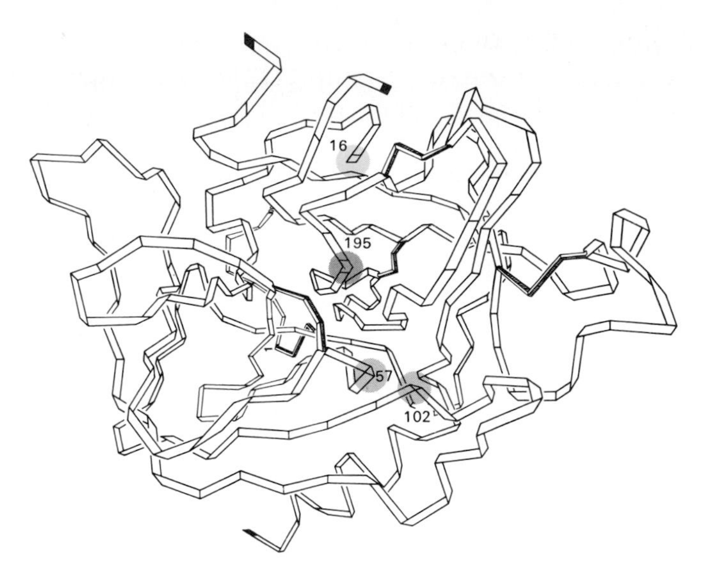

Figure 9-34
Three-dimensional structure of α-chymotrypsin. Only the α-carbon atoms are shown. Catalytically important residues are marked in color. [Courtesy of Dr. David Blow.]

the *carboxyl side* of the *aromatic side chains* tyrosine, tryptophan, and phenylalanine and of large *hydrophobic residues such as methionine*. Chymotrypsin also hydrolyzes *ester bonds*. Although unimportant physiologically, ester-bond hydrolysis is of interest because of its close relationship to peptide-bond hydrolysis (Figure 9-35). Indeed, much of our knowledge of the catalytic mechanism of chymotrypsin comes from studies of the hydrolysis of simple esters.

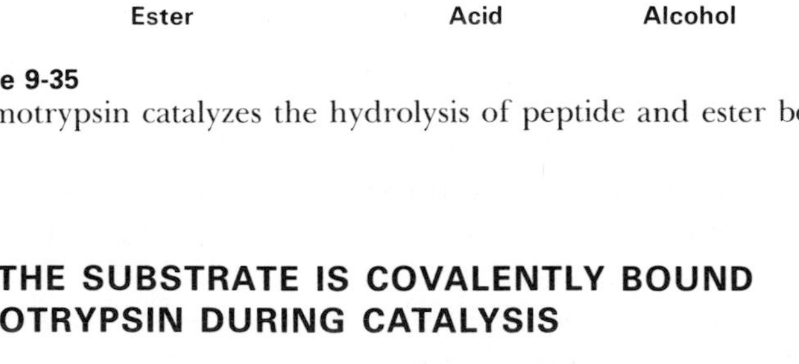

$$R_1{-}C(=O){-}N(H){-}R_2 + H_2O \rightleftharpoons R_1{-}C(=O){-}O^- + {}^+H_3N{-}R_2$$

Peptide — Acid — Amine

$$R_1{-}C(=O){-}O{-}R_2 + H_2O \rightleftharpoons R_1{-}C(=O){-}O^- + HO{-}R_2 + H^+$$

Ester — Acid — Alcohol

Figure 9-35
Chymotrypsin catalyzes the hydrolysis of peptide and ester bonds.

PART OF THE SUBSTRATE IS COVALENTLY BOUND TO CHYMOTRYPSIN DURING CATALYSIS

Chymotrypsin catalyzes the hydrolysis of peptide or ester bonds in two distinct stages. This was first revealed by studies of the kinetics of hydrolysis of *p*-nitrophenyl acetate. When large amounts of enzyme are used, there is an initial *rapid burst* of *p*-nitrophenol product, followed by its formation at a much *slower steady-state rate* (Figure 9-36).

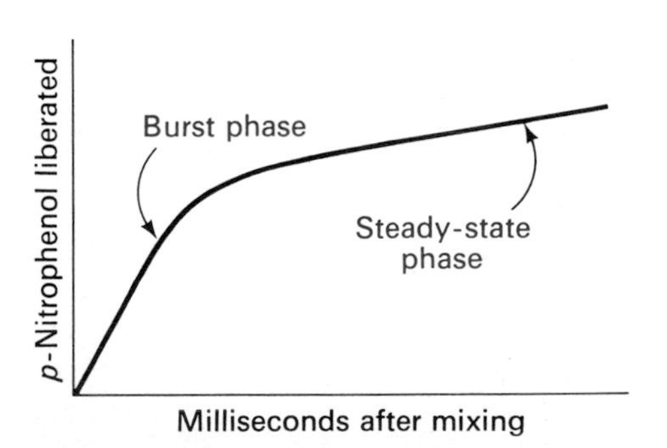

Figure 9-36
Two phases in the formation of *p*-nitrophenol are evident following the mixing of chymotrypsin and *p*-nitrophenyl acetate.

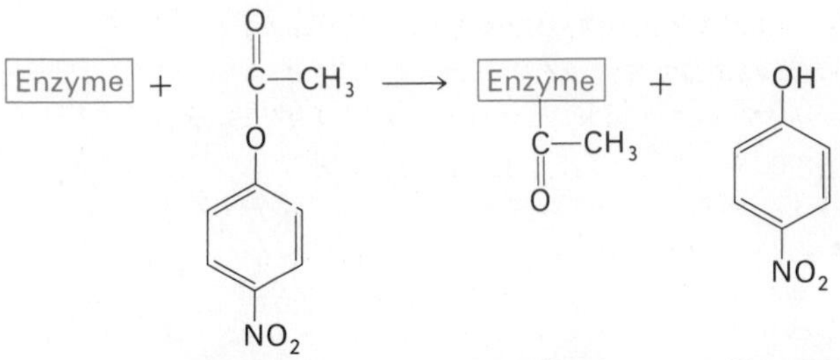

p-Nitrophenyl acetate　Acetyl-enzyme intermediate　*p*-Nitrophenol

Figure 9-37
Acylation: formation of the acetyl-enzyme intermediate in catalysis by chymotrypsin.

The first step is the combination of *p*-nitrophenyl acetate with chymotrypsin to form an enzyme-substrate (ES) complex (Figure 9-37). The ester bond of this substrate is cleaved. One of the products, *p*-nitrophenol, is then released from the enzyme, whereas the acetyl group of the substrate becomes covalently attached to the enzyme. Water then attacks the acetyl-enzyme complex to yield acetate ion and regenerate the enzyme (Figure 9-38). The initial rapid burst of *p*-nitrophenol pro-

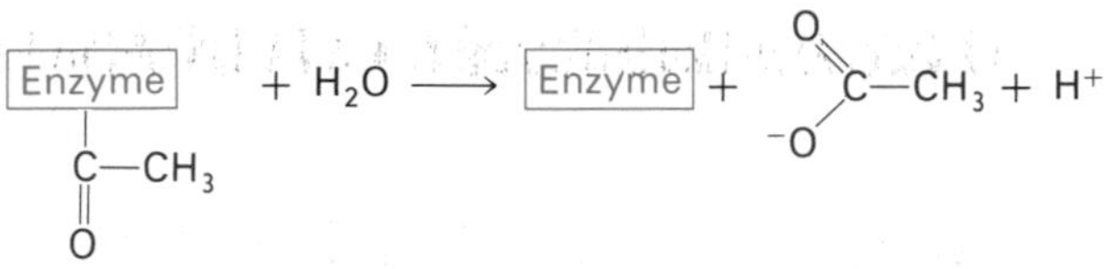

Acetyl-enzyme intermediate　Acetate

Figure 9-38
Deacylation: hydrolysis of the acetyl-enzyme intermediate.

duction corresponds to the formation of the acetyl-enzyme complex. This step is called *acylation*. The slower steady-state production of *p*-nitrophenol corresponds to the hydrolysis of the acetyl-enzyme complex to regenerate the free enzyme. This second step, called *deacylation*, is much slower than the first, so that it determines the overall rate of hydrolysis of esters by chymotrypsin. In fact, the acetyl-enzyme complex is sufficiently stable to be isolated under appropriate conditions. The catalytic mechanism of chymotrypsin can thus be represented by the adjacent scheme, in which P_1 is the amine (or alcohol) component of the substrate, E—P_2 is the covalent intermediate, and P_2 is the acid component of the substrate.

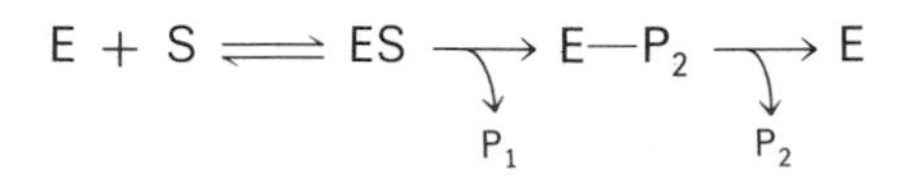

A distinctive feature of this mechanism is the appearance of a covalent intermediate. In the particular reaction discussed above, an acetyl group is covalently bonded to the enzyme. In general, the group attached to chymotrypsin at the E—P_2 stage is an acyl group. Thus, E—P_2 is an *acyl-enzyme intermediate*.

THE ACYL GROUP IS ATTACHED TO AN UNUSUALLY REACTIVE SERINE RESIDUE ON THE ENZYME

The site of attachment of the acyl group was identified following the isolation of E—P_2, which is quite stable at pH 3. The acyl group is linked to the oxygen atom of a specific serine residue, namely, serine

195. This serine residue is unusually reactive. It can be specifically labeled with *organic fluorophosphates,* such as diisopropylphosphofluoridate (DIPF). DIPF reacts only with serine 195 to form an inactive *diisopropylphosphoryl-enzyme complex,* which is indefinitely stable. The remarkable reactivity of serine 195 is highlighted by the fact that the other 27 serine residues in chymotrypsin are untouched by DIPF.

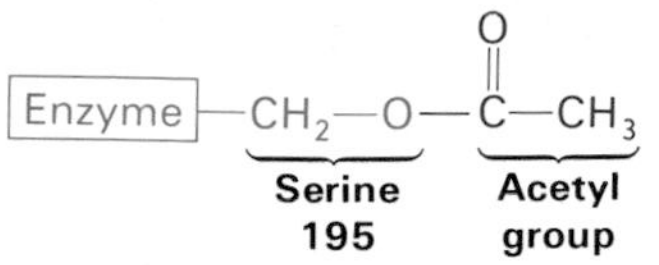

Enzyme—CH_2—O—C(=O)—CH_3
Serine 195 | Acetyl group

Chymotrypsin is not the only enzyme to be inactivated by DIPF. Numerous other proteolytic enzymes, such as trypsin, elastase, thrombin, subtilisin, and acetylcholinesterase react specifically with DIPF and are inactivated. The reaction takes place at a unique serine residue, as in chymotrypsin. Hence, these enzymes are called the *serine proteases.*

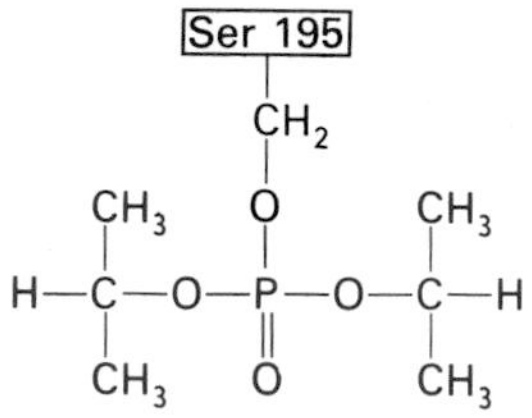

Ser 195 — CH_2 — O — P(=O)(O—CH(CH_3)$_2$)$_2$

Diisopropylphosphoryl derivative of chymotrypsin

DEMONSTRATION OF THE CATALYTIC ROLE OF HISTIDINE 57 BY AFFINITY LABELING

The importance of a second residue in catalysis was shown by *affinity labeling.* The strategy was to react chymotrypsin with a molecule that (1) specifically binds to the active site because it resembles a substrate and then (2) forms a stable covalent bond with a group on the enzyme that is in close proximity. These criteria are met by tosyl-L-phenylalanine chloromethyl ketone (TPCK), whose structure is shown in Figure 9-39. The phenylalanine side chain of TPCK enables it to bind specifically to chymotrypsin. The reactive group in TPCK is the chloromethyl ketone function, which alkylates one of the ring nitrogens of histidine 57. TPCK is positioned to react with this residue because of its specific binding to the active site of the enzyme. The TPCK derivative of chymotrypsin is enzymatically inactive.

Three lines of evidence indicated that histidine 57 is part of the active site. First, the affinity-labeling reaction was highly stereospecific; the D-isomer of TPCK was totally ineffective. Second, the reaction was inhibited when a competitive inhibitor of chymotrypsin, β-phenylpropionate, was present. Third, the rate of inactivation by TPCK varied with pH in nearly the same way as did the rate of catalysis.

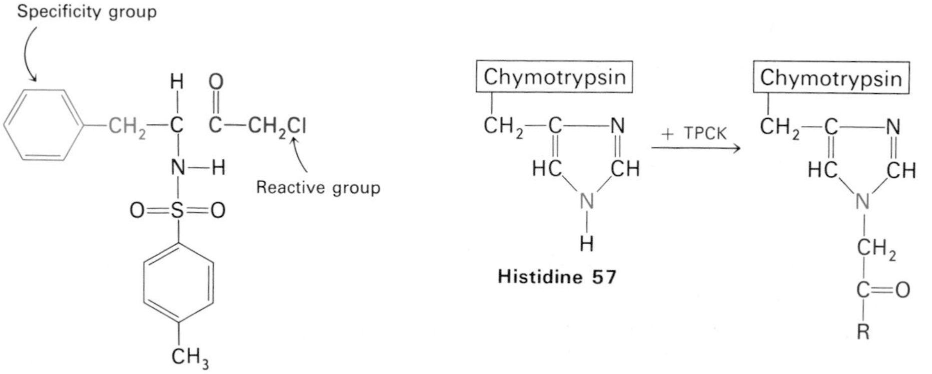

Specificity group
Reactive group
Tosyl-L-phenylalanine chloromethyl ketone (TPCK)

Chymotrypsin
Histidine 57
+ TPCK
Chymotrypsin

Figure 9-39
Affinity labeling of a histidine residue in chymotrypsin by tosyl-L-phenylalanine chloromethyl ketone (TPCK), a reactive substrate analog.

SERINE, HISTIDINE, AND ASPARTATE FORM A CATALYTIC TRIAD IN CHYMOTRYPSIN

The catalytic activity of chymotrypsin depends on the unusual properties of serine 195. A —CH_2OH group is ordinarily quite unreactive under physiological conditions. What makes it so reactive in the active site of chymotrypsin? A convincing explanation has emerged from x-ray studies of the three-dimensional structure of the enzyme. As was foreseen by affinity-labeling studies, histidine 57 is adjacent to serine 195 (Figure 9-40). The carboxylate group of aspartate 102, buried in the protein, also is next to histidine 57. These three residues form a *catalytic triad.*

In the absence of substrate, histidine 57 is unprotonated (Figure 9-41). However, it is poised to accept the proton from the serine 195 —OH group when this oxygen atom carries out a nucleophilic attack on the substrate. It was thought that aspartate 102 in turn became protonated, and so this triad became known as a charge relay network. However, neutron diffraction studies have shown that the proton stays on histidine 57 and that aspartate 102 remains negatively charged. Instead, the role of the $—COO^-$ group of aspartate 102 is to stabilize the positively charged form of histidine 57 in the transition state. In addition, aspartate 102 orients histidine 57 and insures that it is in the appropriate tautomeric form to accept a proton from serine 195.

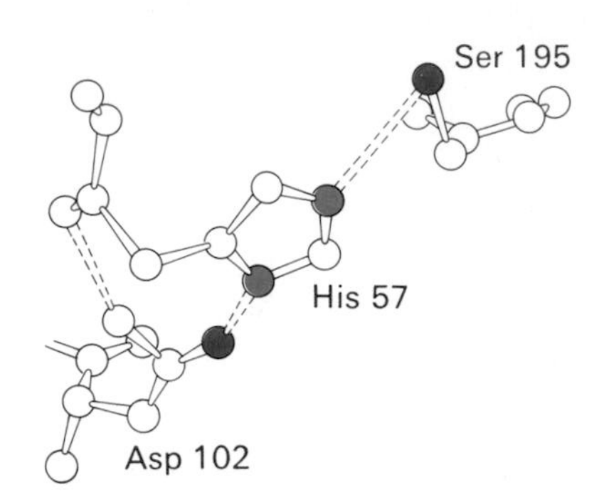

Figure 9-40
Conformation of the serine-histidine-aspartate catalytic triad in chymotrypsin. [After D. M. Blow and T. A. Steitz. X-ray diffraction studies of enzymes. *Ann. Rev. Biochem.* 39(1970):86. Copyright © 1970 by Annual Reviews Inc. All rights reserved.]

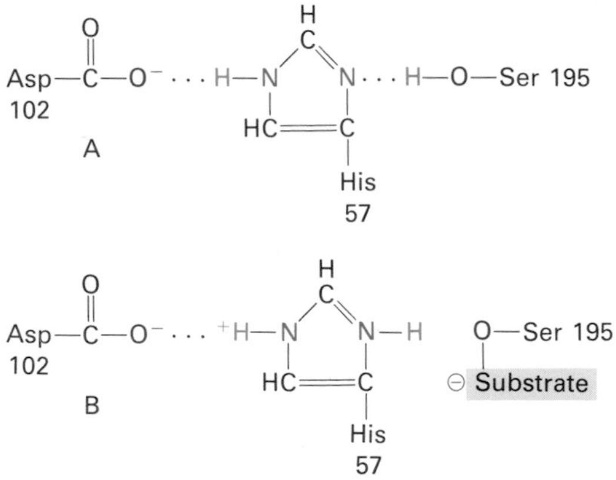

Figure 9-41
Role of the catalytic triad in chymotrypsin: (A) enzyme alone; (B) on addition of a substrate, a proton is transferred from serine 195 to histidine 57. The positively charged imidazole ring is stabilized by electrostatic interaction with negatively charged aspartate 102.

Crystallographic studies of complexes of chymotrypsin with substrate analogs have also shown the location of the site of specific recognition and the likely orientation of the susceptible peptide bond. Formyl-L-tryptophan binds to chymotrypsin with its indole side chain fitted neatly into a pocket near serine 195 (Figure 9-42). This deep cleft accounts for the specificity of chymotrypsin for aromatic and other bulky hydrophobic side chains. Crystallographic analyses of complexes of chymotrypsin and polypeptide substrate analogs show that the main chain of the enzyme and that of the substrate are hydrogen bonded to each other as in an antiparallel β pleated sheet.

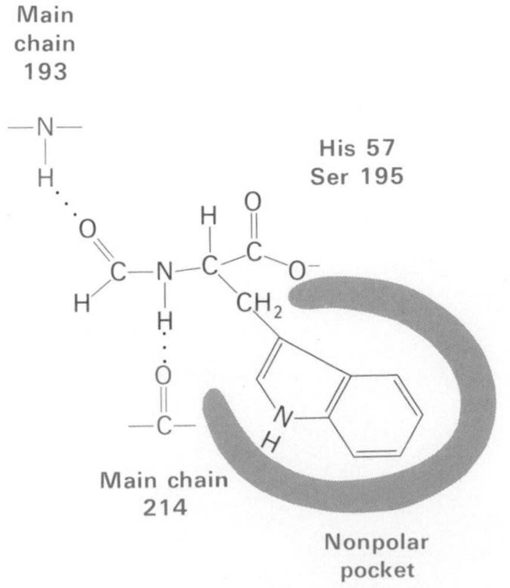

Figure 9-42
Schematic representation of the binding of formyl-L-tryptophan, a substrate analog, to chymotrypsin.

A TRANSIENT TETRAHEDRAL INTERMEDIATE IS FORMED DURING CATALYSIS BY CHYMOTRYPSIN

A plausible catalytic mechanism for chymotrypsin has been deduced from extensive x-ray crystallographic and chemical data. In this mechanism, *histidine 57 and serine 195 participate directly in the cleavage of the susceptible peptide bond of the substrate.* Hydrolysis of the peptide bond starts with an attack by the oxygen atom of the hydroxyl group of serine

195 on the carbonyl carbon atom of the susceptible peptide bond. The carbon–oxygen bond of this carbonyl group becomes a single bond, and the oxygen atom acquires a net negative charge. The four atoms now bonded to the carbonyl carbon are arranged as in a tetrahedron. The formation of this *transient tetrahedral intermediate* from a planar amide group is made possible by hydrogen bonds between the negatively charged carbonyl oxygen atom (called an *oxyanion*) and two main-chain NH groups (Figure 9-43).

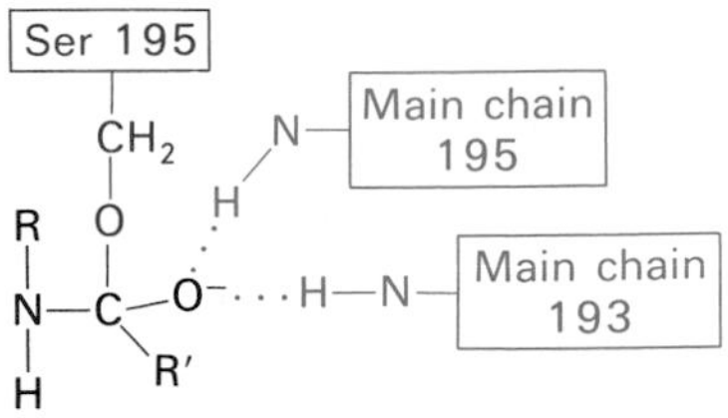

The other essential event in the formation of this tetrahedral transition state is the transfer of a proton from serine 195 to histidine 57 (Figure 9-44). This proton transfer is markedly facilitated by the presence of the catalytic triad. Aspartate 102 precisely orients the imidazole ring of histidine 57 and partly neutralizes the charge that develops on it during the transition state. The proton held by the protonated form of histidine 57 is then donated to the nitrogen atom of the susceptible peptide bond, which thus is cleaved. At this stage, the amine component is hydrogen bonded to histidine 57, whereas the acid component of the substrate is esterified to serine 195. The amine component diffuses away, completing the *acylation stage* of the hydrolytic reaction.

The next stage, *deacylation* (Figure 9-45), begins when a water molecule takes the place occupied earlier by the amine component of the substrate. In essence, *deacylation is the reverse of acylation, with H_2O substi-*

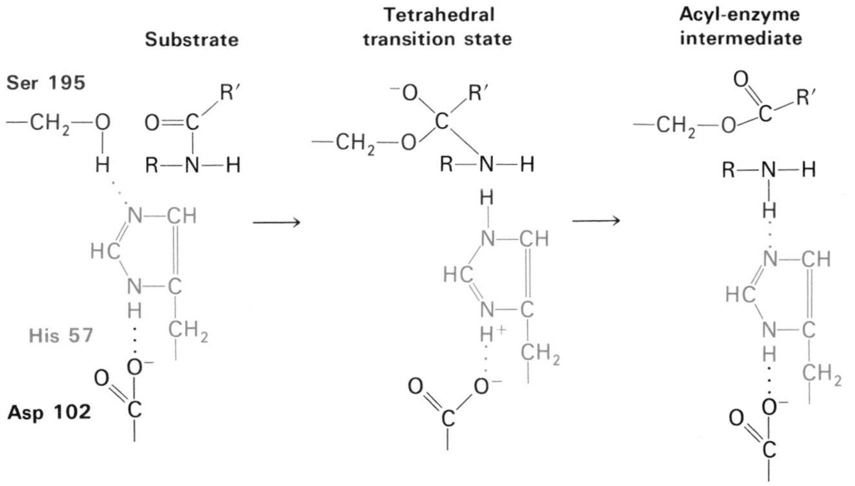

Ser 195 · Main chain 195 · Main chain 193

Figure 9-43
The tetrahedral transition state in the acylation reaction of chymotrypsin. The hydrogen bonds formed by two NH groups from the main chain of the enzyme are critical in stabilizing this species. This site is called the *oxyanion hole*.

Substrate → Tetrahedral transition state → Acyl-enzyme intermediate

Ser 195 · His 57 · Asp 102

Figure 9-44
First stage in the hydrolysis of a peptide by chymotrypsin: *acylation*. A tetrahedral transition state is formed, in which the peptide bond is cleaved. The amine component then rapidly diffuses away, leaving an acyl-enzyme intermediate.

Acyl-enzyme intermediate → Tetrahedral transition state → Acid component of the substrate

His 57 · Asp 102

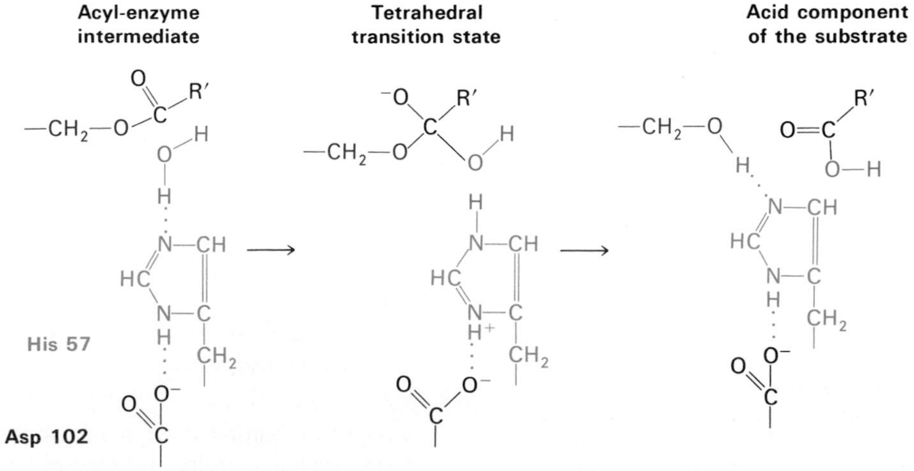

Figure 9-45
Second stage in the hydrolysis of a peptide by chymotrypsin: *deacylation*. The acyl-enzyme intermediate is hydrolyzed by water. Note that deacylation is essentially the reverse of acylation, with water in the role of the amine component of the original substrate.

tuting for the amine component. First, the charge relay network draws a proton away from water. The resulting OH^- ion immediately attacks the carbonyl carbon atom of the acyl group that is attached to serine 195. As in acylation, a transient tetrahedral intermediate is formed. Histidine 57 then donates a proton to the oxygen atom of serine 195, which then releases the acid component of the substrate. This acid component diffuses away and the enzyme is ready for another round of catalysis.

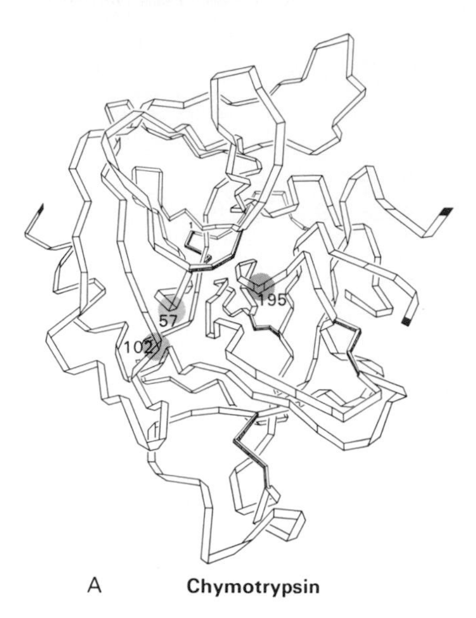

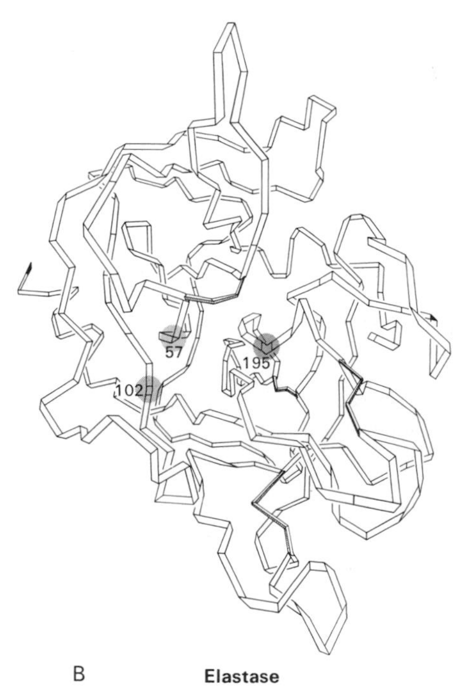

Figure 9-46
Comparison of the conformation of the main chains of (A) chymotrypsin and (B) elastase. The locations of the catalytic triad (residues 102, 57, and 195) are shown in color. [After B. S. Hartley and D. M. Shotton. In *The Enzymes*, P. D. Boyer, ed., 3rd ed., vol. 3 (Academic Press, 1971), p. 362.]

TRYPSIN AND ELASTASE: VARIATIONS ON THE CHYMOTRYPSIN THEME

Trypsin and elastase, two other digestive enzymes, are like chymotrypsin in several respects: (1) About 40% of the amino acid sequences of these three enzymes is identical. The degree of identity is even higher (~60%) for residues located in the interior of the enzymes. (2) X-ray studies have shown that their three-dimensional structures are very similar (Figure 9-46). (3) A serine-histidine-aspartate catalytic triad is present in all three. (4) The serine residue in this triad is modified by fluorophosphates (such as DIPF), causing a loss of catalytic activity. The amino acid sequence around this serine is the same in all three enzymes, Gly-Asp-Ser-Gly-Gly-Pro. (5) All three enzymes have nearly identical catalytic mechanisms. The catalytic triad and oxyanion hole in each promotes the formation of a tetrahedral transition state. A covalent acyl-enzyme intermediate is formed by all three during catalysis. (6) As will be discussed in the next chapter, these enzymes are secreted by the pancreas as inactive precursors that become activated by cleavage of a single peptide bond.

Although similar in structure and mechanism, these enzymes differ markedly in substrate specificity. Chymotrypsin requires an aromatic or bulky nonpolar side chain on the amino side of the scissile bond. Trypsin requires a lysine or arginine residue. Elastase cannot cleave either of these kinds of substrates. Its specificity is directed toward the smaller uncharged side chains. X-ray studies have shown that *these different specificities are due to quite small structural differences in the binding site* (Figure 9-47). In chymotrypsin, a nonpolar pocket serves as a niche for the aromatic or bulky nonpolar side chain. In trypsin, one residue in this pocket is different from chymotrypsin: a serine is replaced by an aspartate. This aspartate in the nonpolar pocket of trypsin can form a strong electrostatic bond with a positively charged lysine or arginine side chain

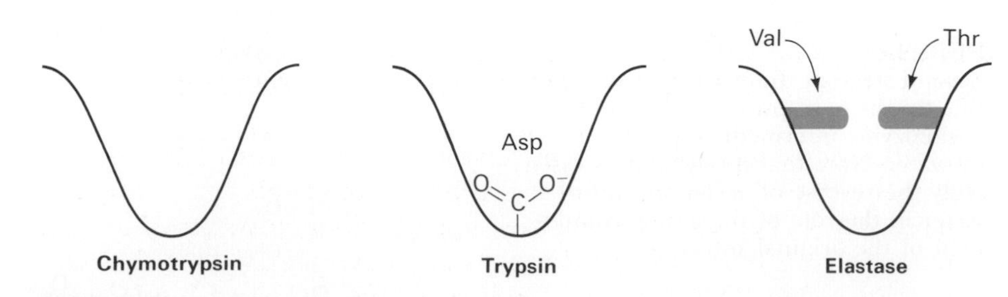

Figure 9-47
A highly simplified representation of part of the substrate-binding site in chymotrypsin, trypsin, and elastase.

of a substrate. In elastase, the pocket does not exist, because the two glycine residues lining it in chymotrypsin are replaced by the much bulkier valine and threonine.

DIVERGENT AND CONVERGENT EVOLUTION OF THE SERINE PROTEASES

Chymotrypsin, elastase, and trypsin have similar overall amino acid sequences because they evolved from a common ancestor. The gene for this common ancestral enzyme probably duplicated many times. Mutations in these duplicates gave rise to present-day proteases that play key roles in digestion, blood clotting, and other physiological processes. The emergence of enzymes of different specificity by duplication and mutation of genes descended from a common ancestor is called *divergent evolution.*

A different kind of evolutionary process is revealed by comparing chymotrypsin with subtilisin, a bacterial enzyme. Subtilisin is also a serine protease. However, the amino acid sequences of chymotrypsin and subtilisin are very different, indicating that they arose independently in evolution. For example, chymotrypsin contains five disulfide bonds, whereas subtilisin has none. The sequences around their active-site serine are quite different:

-Gly-Thr-Ser*-Met-Ala-Ser- (in subtilisin)

-Gly-Asp-Ser*-Gly-Gly-Pro- (in chymotrypsin)

The three-dimensional structures of these enzymes are entirely different. It was therefore surprising to find that *subtilisin has a catalytic triad* that resembles the one in chymotrypsin. Furthermore, subtilisin contains an oxyanion hole, which stabilizes the tetrahedral transition state. The catalytic mechanism of subtilisin is like that of the serine proteases from mammals. However, the residues forming the active site occupy entirely different positions in the amino acid sequences of the mammalian and bacterial enzymes. The functional similarities of these enzymes are probably the consequence of *independent convergent evolution.* Why should the vertebrate pancreatic serine proteases and bacterial subtilisin have attained the same solution by independent, parallel evolutionary processes? There appear to be only a limited number of efficient ways of catalyzing the cleavage of peptide bonds. Studies of several hundred proteolytic enzymes from all types of organisms have revealed only four major types of catalytic mechanisms, with the serine proteases being especially prevalent.

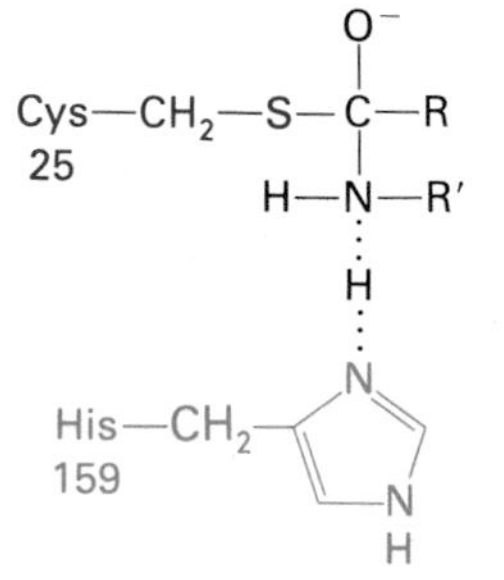

SERINE, ZINC, THIOL, AND CARBOXYL PROTEASES ARE THE MAJOR FAMILIES OF PROTEOLYTIC ENZYMES

Thus far, consideration has been given to two classes of proteolytic enzymes: the *zinc proteases,* exemplified by carboxypeptidase A, and the *serine proteases.* The *thiol proteases* are another widely distributed group. Papain, derived from papaya, contains an active-site cysteine that acts like serine 195 in chymotrypsin. Catalysis proceeds through a thiol ester intermediate and is facilitated by a nearby histidine side chain (Figure 9-48).

Figure 9-48
Transition state in catalysis by papain, a thiol protease. The —SH group of a cysteine residue (shown in red) plays a role like that of the —OH group of serine 195 in chymotrypsin. Another similarity is that the nucleophilicity of the —SH group is enhanced by the adjacent histidine (shown in green), which serves as a proton acceptor.

The other major family of proteolytic enzymes comprises the *carboxyl proteases,* which are also called *acid proteases* because most of them are active only in an acid environment. The best-known member of this family is *pepsin,* the principal protease in gastric juice. The active site of pepsin, a 35-kd protein, contains two aspartate residues. One of them must be ionized and the other un-ionized for the enzyme to be active; the optimum pH is between 2 and 3. Carboxyl proteases with similar structures and enzymatic properties have been isolated from lysosomes, which carry out digestion within cells (p. 787), and from a variety of molds. A common feature of the carboxyl proteases is that they are inhibited by very low concentrations (about 10^{-10} M) of *pepstatin,* a hexapeptide similar to the tetrahedral transition state of normal substrates. Many proteases can be specifically blocked by inhibitors like pepstatin.

ANTIBODIES SPECIFIC FOR TRANSITION STATES ARE CATALYTICALLY ACTIVE

Antibody molecules specifically bind the antigens that stimulated their synthesis (p. 890). Can the very large repertoire of the binding specificities of antibody molecules be tapped to generate novel enzymes? We have seen that the essence of catalysis is the *stabilization of transition states by specific binding interactions.* Hence, it should be feasible to generate catalytically active antibodies by using transition-state analogs as antigens. A promising start in this direction has been made with the preparation of antibodies having esterase activity.

The hydrolysis of the ester shown in Figure 9-49 is expected to occur by attack of a nucleophile (X) on the carbonyl carbon atom of the ester, resulting in the formation of a tetrahedral transition state. The goal was to obtain an antibody that specifically binds this transition state. Mice were immunized with a covalently linked phosphonate analog of this ester. This stable analog probably resembles the transition state for ester hydrolysis. Indeed, a monoclonal antibody specific for this phosphonate analog catalyzed the hydrolysis of the ester. The ratio of the antibody-catalyzed rate to the rate of spontaneous hydrolysis in solution was 1000. This catalytic antibody obeyed Michaelis-Menten kinetics with a K_M of 1.9 μM and a k_{cat} of 2.7×10^{-2} s^{-1}. The k_{cat}/K_M ratio of 1.4×10^4 M^{-1} s^{-1} is three orders of magnitude less than that of enzymes that have achieved kinetic perfection. Ester hydrolysis catalyzed by this

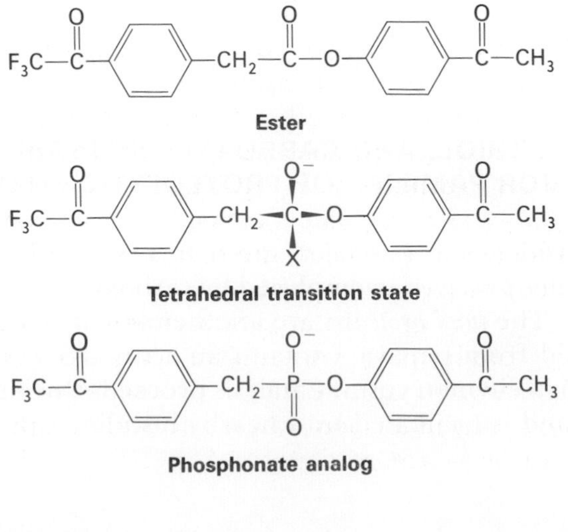

Ester

Tetrahedral transition state

Phosphonate analog

Figure 9-49
Antibodies catalyzing the hydrolysis of the ester were elicited by immunizing mice with the phosphonate analog of the ester. This analog has a geometry like that of the transition state for ester hydrolysis.

antibody was competitively inhibited by addition of the phosphonate analog and displayed a high degree of stereospecificity. It may be feasible to prepare antibodies that catalyze reactions not within the scope of naturally occurring enzymes. For example, catalytic antibodies that cleave viral or tumor-specific proteins could be therapeutically powerful.

SUMMARY

Enzymes can assume conformations that are complementary to the transition states of reactions catalyzed by them. This chapter showed how several well-understood hydrolytic enzymes bind substrates and facilitate the formation of transition states. Lysozyme, a small glycosidase, cleaves bacterial cell walls. Its substrate is an alternating polymer of *N*-acetylglucosamine (NAG) and *N*-acetylmuramic acid (NAM) residues joined by $\beta(1 \rightarrow 4)$ glycosidic linkages. Lysozyme hydrolyzes the glycosidic bond between C-1 of NAM and C-4 of NAG. Oligomers of NAG consisting of six or more residues are also good substrates.

The lysozyme molecule has a cleft to which substrates are bound by many hydrogen bonds and van der Waals interactions. The critical groups in catalysis are the un-ionized carboxyl of glutamic acid 35 and the ionized carboxylate of aspartate 52. Both groups are about 3 Å from the scissile bond between residues D and E of a hexameric substrate. Glutamate 35 cleaves this bond by donating an H^+ to the C-1 of the D ring, which becomes positively charged. This transient species, called a *carbonium ion*, then reacts with OH^- or H_2O from the solvent, and glutamic acid 35 becomes protonated again. Lysozyme is ready for another round of catalysis after the products diffuse away. Catalysis is markedly enhanced by two factors that promote the formation of the carbonium-ion intermediate. The electrostatic factor is the proximity of the negatively charged side chain of aspartate 52. The geometrical factor is the distortion, in the transition state, of ring D into a half-chair form, which enables the positive charge in the carbonium ion to be shared between C-1 and the ring oxygen atom.

Ribonuclease A, a digestive enzyme secreted by the pancreas, hydrolyzes RNA on the 3′ side of pyrimidine residues. The reaction begins with a nucleophilic attack by the 2′-O of ribose on the phosphorus atom of the scissile bond. The un-ionized form of histidine 12 accepts a proton to enhance the nucleophilicity of the 2′-O. The protonated form of histidine 119 simultaneously donates a proton to the 5′-O to cleave the scissile bond and form a *2′,3′-cyclic phosphate intermediate*. In the transition state, phosphorus becomes pentacovalent and acquires the geometry of a trigonal bipyramid, in which three oxygens lie in the equatorial plane and two at the apices. One apex is occupied by the attacking nucleophile (the 2′-O) and the other by the leaving group (the 5′-O). The negatively charged phosphate group in the transition state is stabilized by a salt bridge with the positively charged side chain of lysine 41. Hydrolysis of the cyclic phosphate intermediate, the second stage of the reaction, is essentially a reversal of the first, in which H_2O replaces the 5′OH component that was removed.

The recent discovery of RNA enzymes came as a surprise because it was previously thought that proteins have a monopoly on catalysis. An RNA molecule derived from the intron of a self-splicing RNA from a ciliated protozoan acts as a ribonuclease and an RNA polymerase. This RNA enzyme displays a high degree of specificity, Michaelis-Menten kinetics, and susceptibility to competitive inhibition. RNA enzymes may have preceded protein enzymes in evolution.

Carboxypeptidase A, a proteolytic enzyme, recognizes C-terminal residues that contain an aromatic or bulky aliphatic side chain; it hydrolyzes the adjoining peptide bond. The binding of glycyltyrosine, a substrate analog, induces large structural changes at the active site, which convert it from a water-filled region into a hydrophobic one. Carboxypeptidase A illustrates *induced fit* in catalysis. A noteworthy feature of the active site of this enzyme is a zinc ion, which serves to polarize the carbonyl carbon atom of the scissile bond so that it becomes more vulnerable to nucleophilic attack. This is an example of the induction of *electronic strain* in a substrate. A glutamate residue adjacent to the carbonyl carbon also plays a key role in catalysis. Carboxypeptidase A is a member of the *zinc protease* family.

In chymotrypsin and other *serine proteases*, a highly reactive serine 195 plays a critical role in catalysis. The first stage in the hydrolysis of a peptide substrate is acylation, the formation of a covalent *acyl-enzyme intermediate*, in which the carboxyl component of the substrate is esterified to the hydroxyl group of serine 195. The nucleophilicity of the serine —OH is markedly enhanced by histidine 57, which accepts a proton from serine as serine attacks the carbonyl carbon atom of the substrate. The resulting positively charged histidine is stabilized by electrostatic interaction with negatively charged aspartate 102. Serine, histidine, and aspartate form a *catalytic triad* that is at the heart of the catalytic action of all serine proteases. The negative charge on the *tetrahedral transition state* is also stabilized by hydrogen bonding to two main-chain NH groups in the oxyanion hole. The second stage, deacylation, is in essence a reverse of the first, with H_2O substituting for the amine component. Chymotrypsin, trypsin, elastase, several clotting factors, and other vertebrate serine proteases probably arose from a common ancestral gene. The two other major families of proteolytic enzymes are the *thiol proteases* and the *carboxyl (acid) proteases*.

SELECTED READINGS

Where to start

Phillips, D. C., 1966. The three-dimensional structure of an enzyme molecule. *Sci. Amer.* 215(5):78–90. [A superb article on the three-dimensional structure and catalytic mechanism of lysozyme. Available as *Sci. Amer.* Offprint 1055.]

Stroud, R. M., 1974. A family of protein-cutting proteins. *Sci. Amer.* 231(1):24–88. [Available as *Sci. Amer.* Offprint 1301.]

Neurath, H., 1984. Evolution of proteolytic enzymes. *Science* 224:350–357.

Fersht, A., 1985. *Enzyme Structure and Mechanism.* W. H. Freeman. [Lucid discussions of catalytic mechanisms are given in Chapters 2 and 15.]

Novel enzymes

Zaug, A. J., and Cech, T. R., 1986. The intervening sequence RNA of *Tetrahymena* is an enzyme. *Science* 231:431–475.

Craik, C. S., Largman, C., Fletcher, T., Roczniak, S., Barr, P. J., Fletterick, R., and Rutter, W. J., 1985. Redesigning trypsin: alteration of substrate specificity. *Science* 228:291–297.

Tramontano, A., Janda, K. D., and Lerner, R. A., 1986. Catalytic antibodies. *Science* 234:1566–1570.

Pollack, S. J., Jacobs, J. W., and Schultz, P. G., 1986. Selective chemical catalysis by an antibody. *Science* 234:1570–1573.

Transition-state stabilization

Pauling, L., 1948. Nature of forces between large molecules of biological interest. *Nature* 161:707–709. [A visionary statement of the importance of strain in enzymatic catalysis.]

Leinhard, G. E., 1973. Enzymatic catalysis and transition-state theory. *Science* 180:149–154.

Lysozyme

Imoto, T., Johnson, L. N., North, A. C. T., Phillips, D. C., and Rupley, J. A., 1972. Vertebrate lysozymes. *In* Boyer, P. D., (ed.), *The Enzymes* (3rd ed.), vol. 7, pp. 666–868. Academic Press.

Schindler, M., and Sharon, N., 1976. A transition state analog of lysozyme catalysis prepared from the bacterial cell wall tetrasaccharide. *J. Biol. Chem.* 251:4330–4335.

Ford, L. O., Johnson, L. N., Mackin, P. A., Phillips, D. C., and Tjian, R., 1974. Crystal structure of a lysozyme-tetrasaccharide lactone complex. *J. Mol. Biol.* 88:349–371.

Kelly, J. A., Sielecki, A. R., Sykes, B. D., and James, M. N. G., 1979. X-ray crystallography of the binding of the bacterial cell wall trisaccharide NAM-NAG-NAM to lysozyme. *Nature* 282:875–878.

Weaver, L. H., Grutter, M. G., Remington, S. J., Gray, T. M., Isaacs, N. W., and Mathews, B. W., 1985. Comparison of goose-type, chicken-type, and phage-type lysozymes illustrates that changes occur in both amino acid sequence and three-dimensional structure during evolution. *J. Mol. Evol.* 21:97–111.

Warshel, A., and Levitt, M., 1976. Theoretical studies of enzymatic reactions: dielectric, electrostatic, and steric stabilization of the carbonium ion in the reaction of lysozyme. *J. Mol. Biol.* 103:227–249.

Ribonuclease

Richards, F. M., and Wyckoff, H., 1971. Bovine pancreatic ribonuclease. *In* Boyer, P. D., (ed.), *The Enzymes* (3rd ed.), vol. 4, pp. 647–806. Academic Press.

McPherson, A., Brayer, G., Cascio, D., and Williams, R., 1986. The mechanism of binding of a polynucleotide chain to pancreatic ribonuclease. *Science* 232:765–768.

Borah, B., Chen, C. W., Egan, W., Miller, M., Wlodawer, A., and Cohen, J. S., 1985. Nuclear magnetic resonance and neutron diffraction studies of the complex of ribonuclease A with uridine vanadate, a transition-state analogue. *Biochemistry* 24:2058–2067.

Zinc proteases

Lipscomb, W. N., 1983. Structure and catalysis of enzymes. *Ann. Rev. Biochem.* 52:17–34.

Christianson, D. W., and Lipscomb, W. N., 1987. The complex between carboxypeptidase A and a possible transition state analogue: mechanistic inferences from high resolution x-ray structures of enzyme-inhibitor complexes. *J. Amer. Chem. Soc.* 108:4998–5003.

Monzingo, A. F., and Matthews, B. W., 1984. Binding of *N*-carboxymethyl dipeptide inhibitors to thermolysin determined by x-ray crystallography: a novel class of transition-state analogues for zinc peptidases. *Biochemistry* 23:5724–5729.

Makinen, M. W., Wells, G. B., and Kang, S. O., 1984. Structure and mechanism of carboxypeptidase A. *Advan. Inorg. Biochem.* 6:1–69.

Vallee, B. L., and Galdes, A., 1984. The metallobiochemistry of zinc enzymes. *Advan. Enzymol. Rel. Areas Mol. Biol.* 56:284–430. [A lucid review of the role of zinc in enzymes.]

Sander, M. E., and Witzel, H., 1985. Direct chemical evidence for the mixed anhydride intermediate of carboxypeptidase A in ester and peptide hydrolysis. *Biochem. Biophys. Res. Comm.* 132:681–687.

Gardell, S. J., Craik, C. S., Hilvert, D., Urdea, M. S., and Rutter, W. J., 1985. Site-directed mutagenesis shows that tyrosine 248 of carboxypeptidase A does not play a crucial role in catalysis. *Nature* 317:551–555.

Serine proteases

Steitz, T. A., and Shulman, R. G., 1982. Crystallographic and NMR studies of the serine proteases. *Ann. Rev. Biochem. Biophys.* 11:419–444.

Blow, D. M., 1976. Structure and mechanism of chymotrypsin. *Acc. Chem. Res.* 9:145–152.

Kossiakoff, A. A., and Spencer, S. A., 1981. Direct determination of the protonation states of aspartic acid-102 and histidine-57 in the tetrahedral intermediate of the serine proteases: neutron structure of trypsin. *Biochemistry* 20:6462–6474.

Kraut, J., 1977. Serine proteases: structure and mechanism of catalysis. *Ann. Rev. Biochem.* 46:331–358.

Thiol proteases and acid proteases

Kamphuis, I. G., Drenth, J., and Baker, E. N., 1985. Thiol proteases. Comparative studies based on the high-resolution structures of papain and actinidin, and on amino acid sequence information for cathepsins B and H, and stem bromelain. *J. Mol. Biol.* 182:317–329.

Bradshaw, R. A., and Tang, J., (eds.), 1985. *Molecular Architecture of Proteins and Enzymes*. Academic Press.

PROBLEMS

1. Predict the relative rates of hydrolysis by lysozyme of these oligosaccharides (G stands for an *N*-acetylglucosamine residue, and M for *N*-acetylmuramic acid):
   (a) M-M-M-M-M-M
   (b) G-M-G-M-G-M
   (c) M-G-M-G-M-G

2. Predict on the basis of the data given in Figure 9-16 which of the sugar binding sites, A to F, on lysozyme will be occupied in the most prevalent complex with each of these oligosaccharides (same abbreviations as in problem 1):
   (a) G-G
   (b) G-M
   (c) G-G-G-G

3. Suppose that hexa-NAG is synthesized so that the glycosidic oxygen between its D and E sugar residues is labeled with ^{18}O. Where will this isotope appear in the products formed by hydrolysis with lysozyme?

4. An analog of tetra-NAG containing —H in place of —CH_2OH at C-5 of residue D binds to lysozyme much more strongly than does tetra-NAG. Propose a structural basis for this difference in binding affinity.

5. Compare the coordination of the zinc atom in carboxypeptidase A with that of the iron atom in oxymyoglobin and oxyhemoglobin.
   (a) Which atoms are directly bonded to these metal ions?
   (b) Which side chains contribute these metal-binding groups?
   (c) Which other side chains in proteins are potential metal-binding groups?

6. The transfer of a proton from an enzyme to its substrate is often a key step in catalysis.
   (a) Does this occur in the proposed catalytic mechanisms for ribonuclease A, chymotrypsin, lysozyme, and carboxypeptidase A?
   (b) If so, in each case, identify the proton donor.

7. TPCK is an affinity-labeling reagent for chymotrypsin. It inactivates chymotrypsin by alkylating histidine 57.
   (a) Design an affinity-labeling reagent for trypsin that resembles TPCK.
   (b) How would you test its specificity?

8. Elastase is specifically inhibited by an aldehyde derivative of one of its substrates:

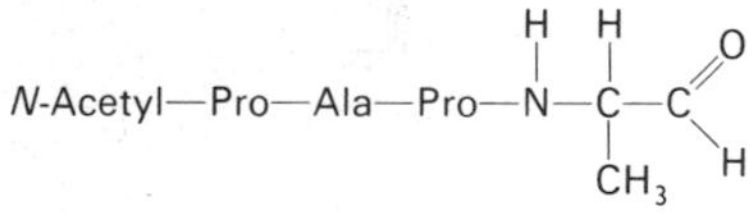

$$N\text{-Acetyl—Pro—Ala—Pro—}\underset{\text{H}}{\text{N}}\text{—}\underset{\text{CH}_3}{\overset{\text{H}}{\text{C}}}\text{—}\overset{\text{O}}{\overset{\|}{\text{C}}}\text{—H}$$

In fact, this aldehyde is an analog of the transition state for catalysis by elastase.

   (a) Which residue at the active site of elastase is most likely to form a covalent bond with this aldehyde?
   (b) What type of covalent link would be formed?

9. Boron acids are another kind of transition-state analog for enzymes that form acyl-enzyme intermediates. Acetylcholinesterase is an enzyme that catalyzes the hydrolysis of the ester bond in acetylcholine:

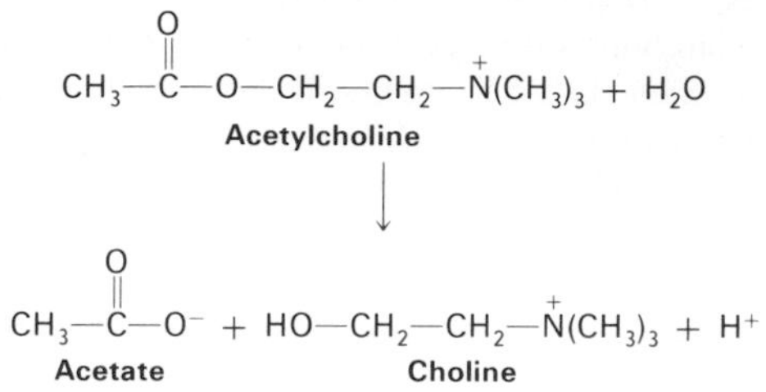

$$CH_3\text{—}\overset{O}{\overset{\|}{C}}\text{—O—}CH_2\text{—}CH_2\text{—}\overset{+}{N}(CH_3)_3 + H_2O$$

Acetylcholine

$$\downarrow$$

$$CH_3\text{—}\overset{O}{\overset{\|}{C}}\text{—}O^- + HO\text{—}CH_2\text{—}CH_2\text{—}\overset{+}{N}(CH_3)_3 + H^+$$

Acetate **Choline**

Acetylcholinesterase is specifically inhibited by this boron analog of acetylcholine:

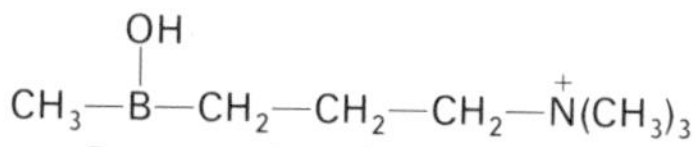

$$CH_3\text{—}\overset{OH}{\overset{|}{B}}\text{—}CH_2\text{—}CH_2\text{—}CH_2\text{—}\overset{+}{N}(CH_3)_3$$

Boron analog of acetylcholine

How might this boron analog be covalently attached to the active site of acetylcholinesterase?

10. List the factors contributing to the catalytic power of enzymes.

11. Thermolysin, like carboxypeptidase A, contains a zinc ion and a glutamate side chain at its active site. X-ray studies suggest that a zinc-bound water molecule directly attacks the carbonyl carbon atom of the peptide substrate. How might Zn^{2+} promote this reaction? How might the carboxylate group of the glutamate residue facilitate the attack on the carbonyl carbon atom?

CHAPTER 10

Control of Enzymatic Activity

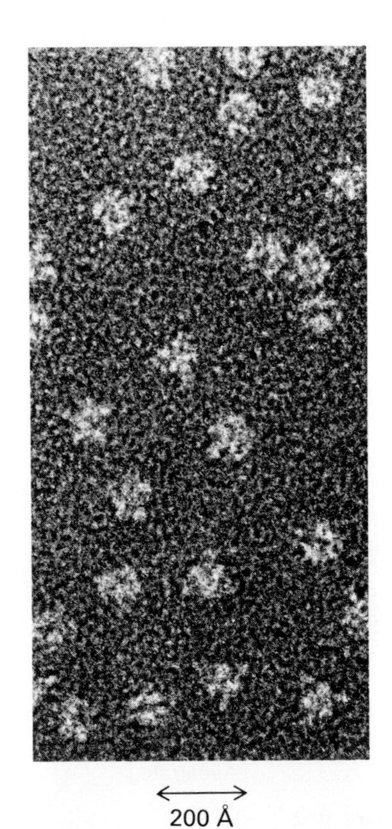

Figure 10-1
Electron micrograph of aspartate transcarbamoylase. [Courtesy of Dr. Robley C. Williams.]

The capacity of proteins to specifically bind virtually any biomolecule and stabilize its transition state endows enzymes with great catalytic power. This chapter considers another remarkable property of enzymes, the precise regulation of their catalytic activity. Their dynamic character, particularly their inherent ability to transmit conformational changes between spatially distinct sites within a molecule, is exploited in many enzymes to make them catalytically active at the right time and in the right place. Four regulatory motifs are found in many enzymes:

1. *Allosteric interactions.* In hemoglobin (Chapter 7), oxygen-binding sites are coupled to each other and to sites that bind H^+, CO_2, and bisphosphoglycerate by conformational changes that are transmitted across subunits. This chapter examines one of the best understood allosteric enzymes, *aspartate transcarbamoylase* from *Escherichia coli.* This feedback-inhibited enzyme carries out the first step in pyrimidine biosynthesis. The presence of separable regulatory and catalytic subunits has made the enzyme a choice object of inquiry. Moreover, its three-dimensional structure is known in atomic detail. The allosteric mechanism of aspartate transcarbamoylase will be discussed in relation to two contrasting models—the concerted and sequential models—of allosteric interactions.

2. *Reversible covalent modification.* The catalytic properties of some enzymes are markedly altered by the covalent attachment of a small group. For example, glycogen phosphorylase (the enzyme that releases sugar units from glycogen, an energy store) is activated by phosphorylation of a specific serine residue (Chapter 19). This reaction is reversed by hydrolysis of the phosphate ester bond. Glutamine synthetase (the

enzyme that synthesizes glutamine from glutamate) from *E. coli* becomes more susceptible to feedback inhibition when specific tyrosine residues acquire AMP units (p. 591). The AMP unit, like nearly all known reversible modifiers, is removed by hydrolysis. As might be expected, enzymes catalyzing covalent modifications are themselves rigorously controlled.

3. *Stimulation and inhibition by control proteins.* The regulatory motifs cited above are mediated by the binding or attachment of small groups. The catalytic activity of some enzymes is controlled in a different way, by the binding of specific stimulatory or inhibitory proteins. For example, some enzymes become active when the intracellular calcium level rises. *Calmodulin,* a ubiquitous regulatory protein in eucaryotes, senses the intracellular concentration of Ca^{2+} and activates many enzymes by binding to them when it contains bound calcium ions. Blood clotting is markedly accelerated by *antihemophilic factor,* a protein that enhances the activity of a serine protease, as will be discussed later in this chapter. Conversely, the catalytic activity of some enzymes is held in check by inhibitory proteins. Visual excitation, for example, is mediated by the reversible release of an inhibitory subunit from an enzyme that cleaves phosphodiester bonds (p. 1034).

4. *Proteolytic activation.* The enzymes controlled by the mechanisms cited above cycle between active and inactive states. A different regulatory motif is used to irreversibly convert an inactive enzyme into an active one. Many enzymes are activated by the hydrolysis of one or a few peptide bonds in inactive precursors called *zymogens* or *proenzymes.* This regulatory mechanism is exemplified by the activation of digestive enzymes such as chymotrypsin. This chapter will also consider the mechanism of blood clotting, a remarkable cascade of zymogen activations. The catalytic activities of digestive and clotting enzymes are switched off by the irreversible binding of specific inhibitory proteins that are irresistible lures to their molecular prey.

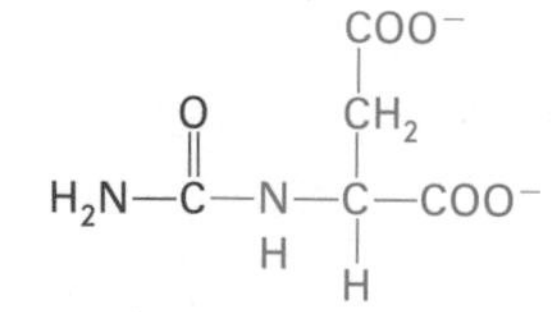

***N*-Carbamoylaspartate**

ASPARTATE TRANSCARBAMOYLASE IS FEEDBACK INHIBITED BY CTP, THE FINAL PRODUCT OF THE PYRIMIDINE PATHWAY

Pyrimidine biosynthesis begins with the formation of N-*carbamoylaspartate* from aspartate and carbamoyl phosphate. This reaction is catalyzed by *aspartate transcarbamoylase* (ATCase), an especially interesting regulatory enzyme. The enzyme from *E. coli* has been studied in detail by many investigators.

John Gerhart and Arthur Pardee discovered that *ATCase is feedback inhibited by cytidine triphosphate (CTP),* the final product of this biosynthetic pathway. Furthermore, they found that the binding of carbamoyl phosphate and aspartate is cooperative, as reflected in the sigmoidal dependence of reaction velocity on substrate concentration (Figure 10-2). Cooperative binding serves to switch on the synthesis of *N*-carbamoylaspartate over a rather narrow range of concentration of substrates. CTP inhibits ATCase by decreasing its affinity for substrates without affecting its V_{max}. The extent of inhibition, which may reach 90%, depends on the concentrations of the substrates. In contrast, *ATP is an activator of ATCase.* It enhances the affinity of the enzyme for its substrates, but also leaves V_{max} unaltered. ATP competes with CTP for binding regulatory sites. Consequently, high levels of ATP prevent CTP from inhibiting the enzyme.

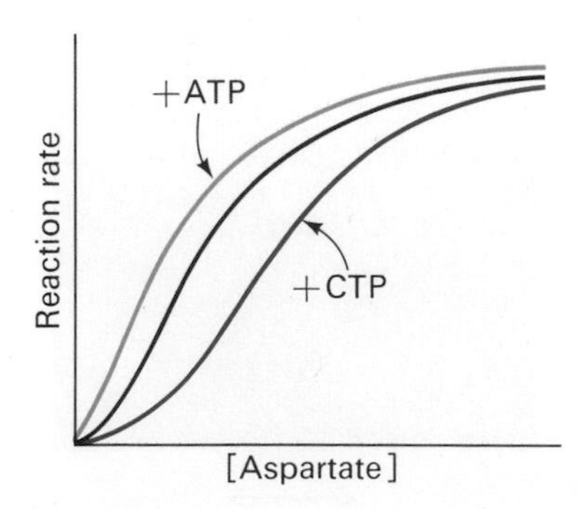

Figure 10-2
Allosteric effects in aspartate transcarbamoylase. ATP is an activator, whereas CTP is an inhibitor. [After J. C. Gerhart. *Curr. Top. Cell Regul.* 2(1970):275.]

The biological significance of the regulation of ATCase by CTP and ATP is twofold. First, activation by ATP signals that energy is available for DNA replication and leads to the synthesis of needed pyrimidine nucleotides. Second, feedback inhibition by CTP assures that *N*-carbamoylaspartate and subsequent intermediates in the pathway are not needlessly formed when pyrimidines are abundant.

ASPARTATE TRANSCARBAMOYLASE CONSISTS OF SEPARABLE CATALYTIC AND REGULATORY SUBUNITS

The regulatory properties of ATCase vanish when the enzyme is treated with a mercurial such as *p*-hydroxymercuribenzoate, which reacts with sulfhydryl groups (Figure 10-3). ATP and CTP no longer have any effect on catalytic activity. Furthermore, the binding of substrates becomes noncooperative. However, the modified enzyme has full catalytic activity. This loss of regulatory properties with retention of enzymatic activity is called *desensitization*.

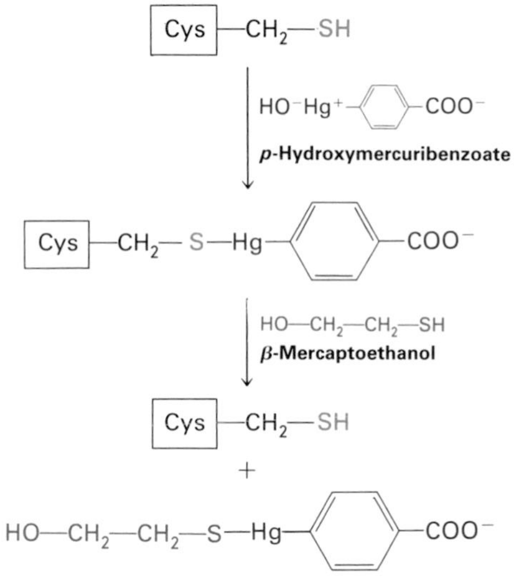

Cys—CH_2—SH

↓ HO–Hg^+–C_6H_4–COO^-
p-Hydroxymercuribenzoate

Cys—CH_2—S—Hg—C_6H_4—COO^-

↓ HO—CH_2—CH_2—SH
β-Mercaptoethanol

Cys—CH_2—SH
+
HO—CH_2—CH_2—S—Hg—C_6H_4—COO^-

Figure 10-3
Reaction of a sulfhydryl group with a mercurial, and its reversal by addition of an excess of β-mercaptoethanol.

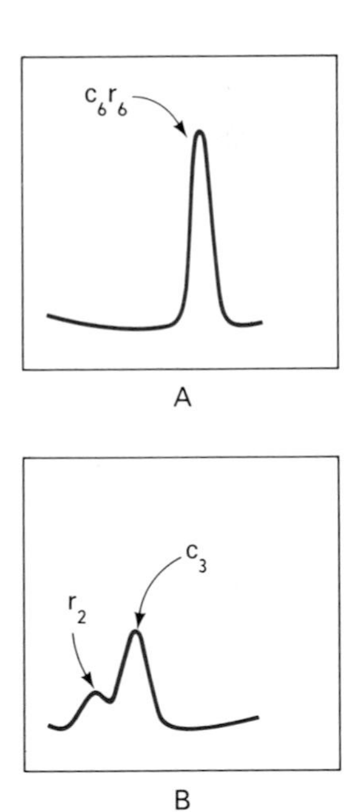

Figure 10-4
Sedimentation velocity patterns of (A) native ATCase and (B) the enzyme dissociated by a mercurial into regulatory and catalytic subunits. [After J. C. Gerhart and H. K. Schachman. *Biochemistry* 4(1965):1054.]

Ultracentrifugation studies carried out by Gerhart and Howard Schachman showed that desensitization of ATCase by mercurials is accompanied by its *dissociation into two kinds of subunits* (Figure 10-4). The sedimentation coefficient of the native enzyme is 11.6S, whereas those of the dissociated subunits are 2.8S and 5.8S. These subunits can be readily separated by ion-exchange chromatography because they differ markedly in charge, or by centrifugation in a sucrose density gradient because they differ in size. *p*-Mercuribenzoate can be removed, after the subunits are separated, by adding an excess of mercaptoethanol (Figure 10-3).

The larger of the subunits, called the *catalytic subunit*, is catalytically active but unresponsive to ATP and CTP. The smaller of the subunits is devoid of catalytic activity but contains a site that can bind CTP or ATP. Hence, it is called the *regulatory subunit*. The catalytic subunit (c_3) consists of three c chains (34 kd each), and the regulatory subunit consists of two r chains (17 kd each). The catalytic and regulatory subunits combine rapidly when they are mixed. The resulting complex has the same structure, c_6r_6, as that of the native enzyme.

$$3\ r_2 + 2\ c_3 \longrightarrow r_6c_6$$

Furthermore, the reconstituted enzyme has the same allosteric properties as those of the native enzyme. Thus, *ATCase is composed of discrete catalytic and regulatory subunits, which interact in the native enzyme to produce its allosteric behavior*. ATCase is a choice object of inquiry because its catalytic and regulatory functions can readily be separated and brought back together.

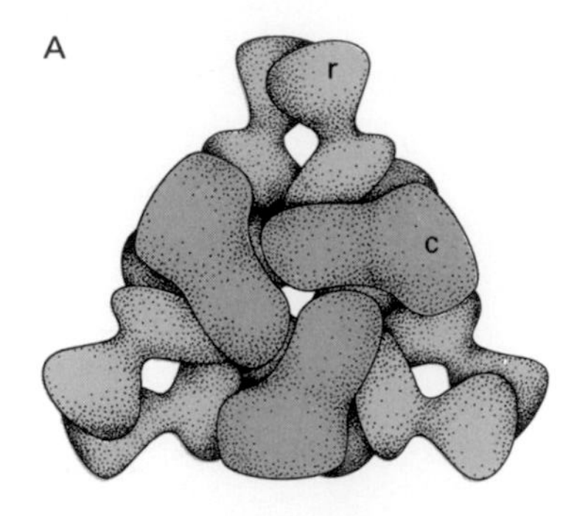

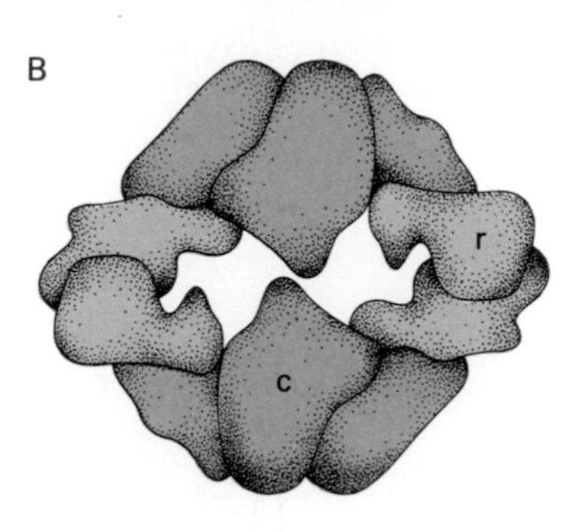

Figure 10-5
Subunit arrangement in aspartate transcarbamoylase. The six catalytic chains (c) are shown in blue and the regulatory chains (r) in red. (A) View looking down the threefold symmetry axis. The positions of the twofold axes are marked. (B) A perpendicular view. The threefold axis is marked. [After K. L. Krause, K. W. Volz, and W. N. Lipscomb. *Proc. Nat. Acad. Sci.* 82(1985):1643.]

THREE-DIMENSIONAL STRUCTURE OF ATCase AND ITS COMPLEX WITH PALA, A BISUBSTRATE ANALOG

William Lipscomb and his coworkers have elucidated the three-dimensional structure of ATCase at a resolution of 2.6 Å. They also have solved the structure of complexes of ATCase with CTP, the allosteric inhibitor, and with a substrate analog. The design of this enzyme is distinctive (Figure 10-5). One of the catalytic trimers (c_3) lies above and the other catalytic trimer lies below an equatorial belt of three regulatory dimers (r_2). The enzyme contains a large central cavity, which is accessible through several channels. The diameter of a sphere encompassing ATCase is about 130 Å, twice the size of hemoglobin.

A view of the path of the main chain of half of the enzyme molecule (c_3r_3) is shown in Figure 10-6. We are looking down the threefold symmetry axis of the molecule. The three identical cr units in this view are related by a rotation of 120 degrees. Each r chain on the periphery consists of two domains. The outer one contains the binding site for CTP and the inner one interacts with two catalytic chains. The inner lobe of the r chain also contains a zinc ion that plays a structural role by coordinating the sulfur atoms of four cysteine residues.

Catalytic site

CTP site

20 Å

Figure 10-6
Three-dimensional structure of half (c_3r_3) of the ATCase molecule. The c chains are shown in blue and the r chains in red. The locations of bound CTP in the regulatory subunits and of PALA in the active sites are marked. The threefold symmetry of ATCase is evident in this view. [After E. R. Kantrowitz, S. C. Patra-Landis, and W. N. Lipscomb. *Trends Biochem. Sci.* 5(1980):152.]

Where is the active site? In the formation of *N*-carbamoylaspartate, carbamoyl phosphate binds first by multiple electrostatic and hydrogen-bond interactions. Aspartate then binds, and its amino group attacks the carbonyl carbon atom of carbamoyl phosphate (Figure 10-7A). A tetrahedral transition state seems likely (Figure 10-7B). *N*-(phosphonacetyl)-L-aspartate (PALA) (Figure 10-7C), which inhibits the enzyme, resembles the two-substrate complex and the transition state. PALA binds tightly to ATCase with a dissociation constant of about 10 nM. This unreactive *bisubstrate analog* has proven to be invaluable in studies of ATCase. X-ray analysis of crystals of ATCase containing PALA shows that each of the six active sites is located near the interface between catalytic chains. Because it has a high negative charge, PALA binds electrostatically to four arginines and one lysine in an active site (Figure 10-8). It is also bound to the enzyme by many hydrogen bonds. An active site is formed by residues belonging to two catalytic chains. Specifically, a serine and a lysine residue from one chain join many residues from the adjacent chain to form a binding pocket for PALA.

The hydrogen bonding of the imidazole side chain of histidine 134 to the carbonyl oxygen atom of PALA is also noteworthy. The carbonyl oxygen atom of carbamoyl phosphate becomes negatively charged when the amino group of aspartate attacks the carbonyl carbon atom. The protonated form of histidine 134 could stabilize this negatively charged atom in the tetrahedral transition state.

A — Bound substrates

B — Transition state

C — *N*-(Phosphonacetyl)-L-aspartate (PALA)

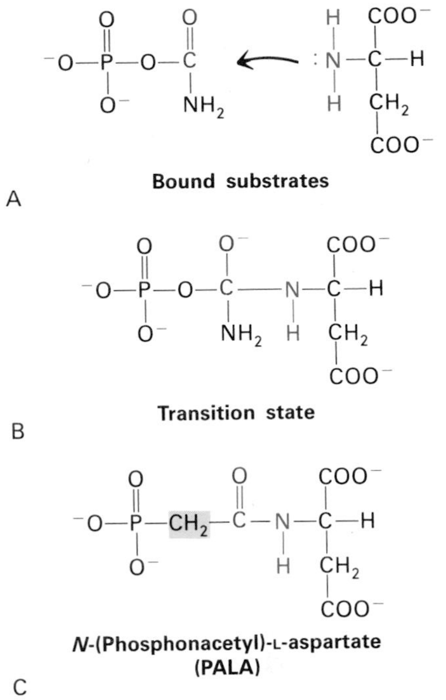

Figure 10-7
(A) Nucleophilic attack of the amino group of aspartate on the carbonyl carbon atom of carbamoyl phosphate. (B) Transition state. (C) PALA resembles the bound pair of substrates and the transition state.

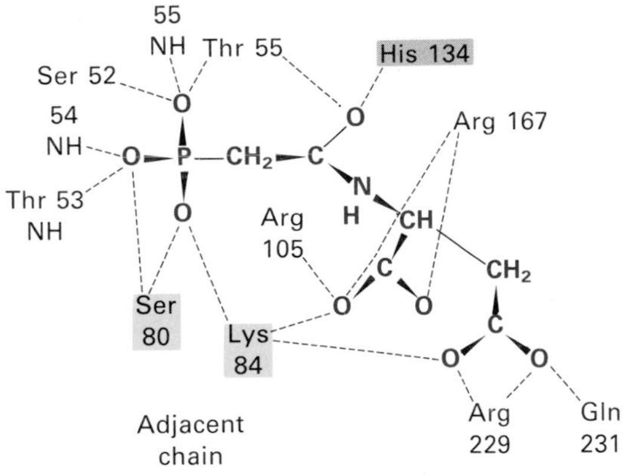

Figure 10-8
Mode of binding of PALA (shown in red) to the catalytic site of ATCase. Not all of the electrostatic and hydrogen bonds formed are shown in this schematic diagram. Histidine 134 (shown in blue) may stabilize the negative charge on the carbonyl oxygen atom in the transition state. This binding site is at the interface of catalytic chains. Residues contributed by the adjacent catalytic chain are marked in yellow. [After K. W. Voltz, K. L. Krause, and W. N. Lipscomb. *Biochem. Biophys. Res. Comm.* 136(1986):823.]

ALLOSTERIC INTERACTIONS IN ATCase ARE MEDIATED BY LARGE CHANGES IN QUATERNARY STRUCTURE

Ultracentrifugation studies showed that the binding of substrate leads to a 3% *decrease* in the sedimentation coefficient of ATCase. This finding indicated that ATCase *expands* on binding substrates. X-ray studies of unliganded ATCase and its complex with PALA have revealed that large changes in quaternary structure accompany the binding of this bisubstrate analog (Figure 10-9). The catalytic trimers move apart 12 Å and turn by 10 degrees. In addition, each regulatory dimer turns 15 degrees about its twofold axis. The binding of PALA also leads to substantial changes in the tertiary structure of each catalytic chain (Figure 10-10). Several loops shift positions, and a lysine and a serine residue are drawn into each active site from the adjacent catalytic chain. In this allosteric transition, the catalytic chains in c_3 are brought closer together to enable them to form optimal active sites.

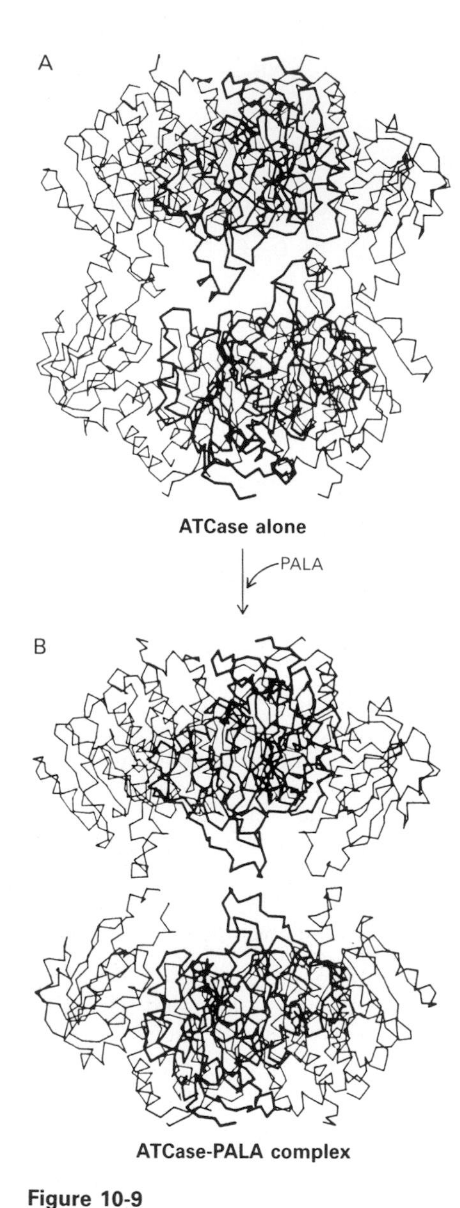

Figure 10-9
Allosteric transition of the catalytic chains of ATCase. (A) Unliganded T state, and (B) liganded R state. The regulatory chains have been omitted for clarity. The catalytic trimers move apart 12 Å and turn 10 degrees in the transition to the high affinity R state. [After K. L. Krause, K. W. Volz, and W. N. Lipscomb. *Proc. Nat. Acad. Sci.* 82(1985):1643.]

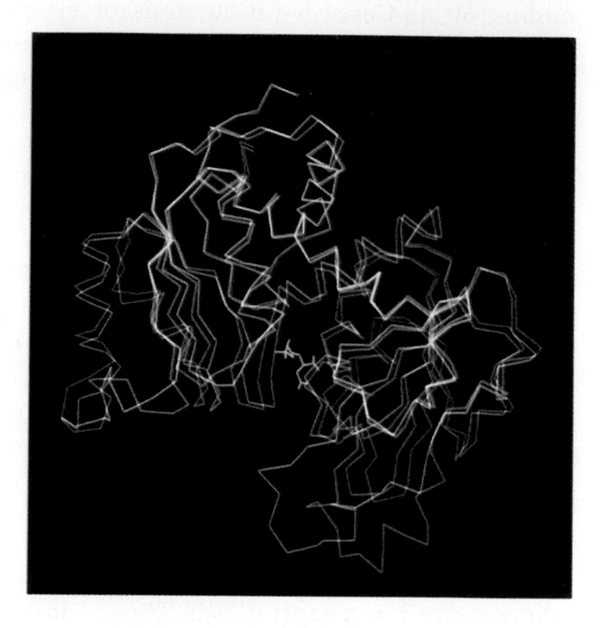

Figure 10-10
Change in tertiary structure of a catalytic chain in the transition from the T state (shown in red) to the R state (shown in blue). [After K. L. Krause, K. W. Volz, and W. N. Lipscomb. *Proc. Nat. Acad. Sci.* 82(1985):1643.]

What is the spatial range of allosteric interactions in ATCase? Active sites in the same c_3 are 22 Å apart. The CTP binding site in the outer lobe of each regulatory subunit is 60 Å away from the nearest active site. Thus, cooperative binding of substrate and feedback inhibition by CTP are mediated by long-range interactions. Indeed, information is transferred from the active sites of one c_3 unit to those of the other c_3 unit. ATCase was reacted with tetranitromethane to form a colored nitrotyrosine group in each of its catalytic chains (Figure 10-11). An essential lysine residue at each of the catalytic sites was also modified to block the binding of substrate. Catalytic trimers from this modified enzyme were then combined with native trimers to form a hybrid enzyme. The binding of substrate to the functional active sites of the native c_3 led to a change in the absorption spectrum of nitrotyrosine residues in the other c_3 unit of this hybrid molecule. This cross-talk experiment

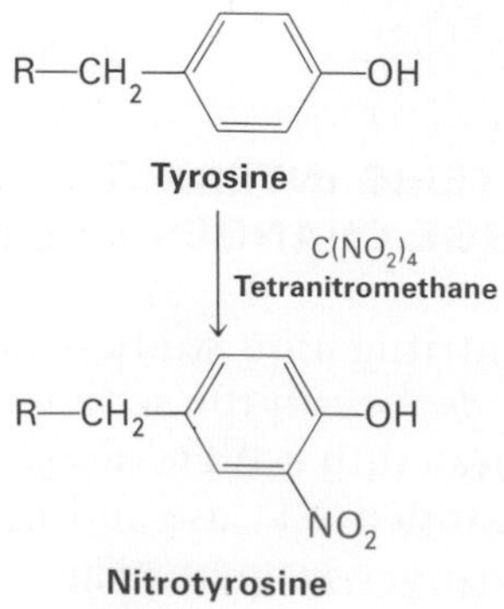

R—CH_2—(ring)—OH
Tyrosine
$C(NO_2)_4$
Tetranitromethane
R—CH_2—(ring)—OH, NO_2
Nitrotyrosine

Figure 10-11
Nitration of tyrosine to form a colored nitrotyrosine group.

demonstrated that *quaternary structural changes resulting from the binding of substrate lead to tertiary structural changes in distant catalytic chains.* Likewise, CTP and ATP induce significant and opposite changes in the spectrum of the nitrotyrosyl groups.

Current work on ATCase centers on two challenging questions: How does the binding of substrates, CTP, and ATP lead to large changes in quaternary structure? In turn, how is the binding affinity of catalytic sites altered by the switching of quaternary structure? The goal now is to delineate the precise pathway by which conformational changes are communicated between distant sites. The three-dimensional structures of the liganded and unliganded forms of ATCase provide a wealth of structural information that will be invaluable in answering these questions. It is already evident that *interfaces between different polypeptide chains play a critical role as molecular switches in mediating allosteric effects in ATCase,* as they do in hemoglobin (p. 158).

THE CONCERTED MODEL FOR ALLOSTERIC INTERACTIONS

It is stimulating and instructive to consider the allosteric mechanism of ATCase, hemoglobin, and other allosteric proteins in terms of general models that make quantitative predictions. An elegant and incisive model was proposed in 1965 by Jacques Monod, Jeffries Wyman, and Jean-Pierre Changeux. Let us apply their approach to an allosteric enzyme made up of two identical subunits, each having one active site. Suppose that a subunit can exist in either of two interconvertible conformations. The R (relaxed) state has a high affinity for substrate, whereas the T (tense) state has a low affinity (Figure 10-12). Recall that these designations were used to describe alternative quaternary structures of hemoglobin (p. 159). They can also be applied to the two quaternary states of ATCase (see Figures 10-9 and 10-10). *An important assumption of this model is that both subunits must be in the same conformational state, so that the symmetry of the dimer is conserved.* Thus, RR and TT are allowed conformations but RT is not.

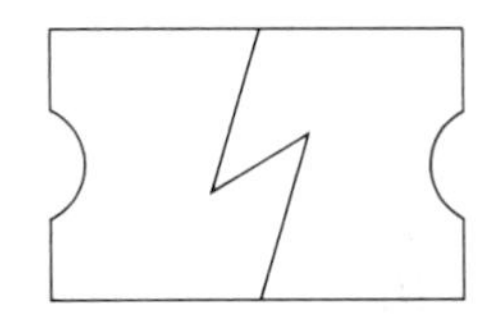

T form
(Low affinity for substrate)

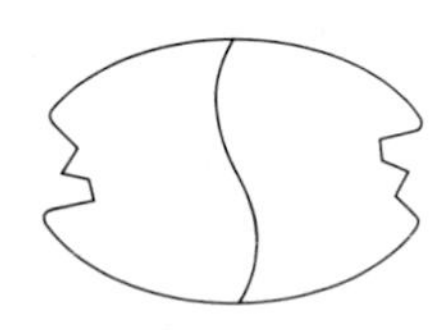

R form
(High affinity for substrate)

Figure 10-12
Schematic representation of the R and T forms of an allosteric enzyme.

In the absence of substrate, the two allowed states are symbolized as T_0 and R_0, and L is the ratio of their concentrations.

$$R_0 \rightleftharpoons T_0 \qquad (1)$$

$$L = \frac{[T_0]}{[R_0]} \qquad (2)$$

For simplicity, let us assume that the substrate does not bind appreciably to the T state. The R state of the dimer can bind one or two substrate molecules; these species are denoted by R_1 and R_2, respectively.

$$R_0 + S \rightleftharpoons R_1 \qquad (3)$$

$$R_1 + S \rightleftharpoons R_2 \qquad (4)$$

In this model, the dissociation constant for a single site, K_R, is the same for the binding of the first and second substrate molecules to the R form of the dimeric enzyme molecule.

$$K_R = \frac{2[R_0][S]}{[R_1]} = \frac{[R_1][S]}{2[R_2]} \qquad (5)$$

The factor of 2 takes into account the fact that a substrate can bind to either of two sites in R_0 to form R_1 and, likewise, the fact that a substrate can be released from either of two sites in R_2 to form R_1.

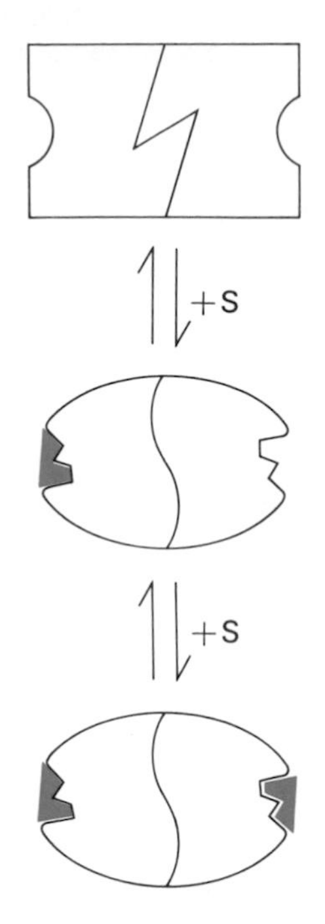

Figure 10-13
Concerted model for the cooperative binding of substrate in an allosteric enzyme. In this example, the T state has negligible affinity for substrate. The low affinity form, TT, switches to the high affinity form, RR, upon binding the first substrate molecule.

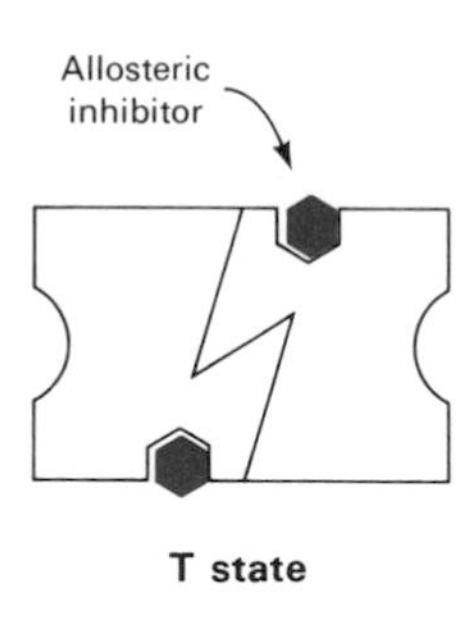

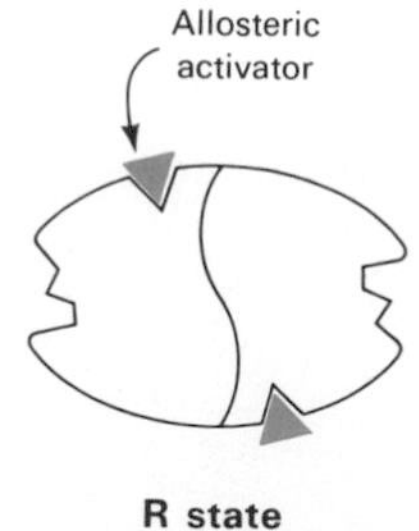

Figure 10-14
In the concerted model, an allosteric inhibitor (represented by a hexagon) stabilizes the T state, whereas an allosteric activator (represented by a triangle) stabilizes the R state.

We want an expression that gives us the *fractional saturation*, Y, which is the fraction of active sites that have a bound substrate, as a function of the substrate concentration.

$$Y = \frac{[\text{occupied sites}]}{[\text{total sites}]} = \frac{[R_1] + 2[R_2]}{2([T_0] + [R_0] + [R_1] + [R_2])} \tag{6}$$

Substituting equations 2 and 5 into equation 6 gives the desired expression for Y:

$$Y = \left(\frac{[S]}{K_R}\right)\frac{1 + [S]/K_R}{L + (1 + [S]/K_R)^2} \tag{7}$$

Let us plot equation 7 with $K_R = 10^{-5}$ M and $L = 10^4$. Such a plot of Y versus [S] is sigmoidal rather than hyperbolic (see Figure 10-15). In other words, *the binding of substrate is cooperative*. If the turnover number per active site is the same for ES complexes in R_1 and R_2, then a plot of reaction velocity versus substrate concentration will also be sigmoidal, because

$$V = YV_{max} \tag{8}$$

Let us look at this binding process (Figure 10-13). In the absence of substrate, nearly all the enzyme molecules are in the T form. Specifically, there is only one molecule in the R form for every 10^4 molecules in the T form, in the above example. The addition of substrate shifts this conformational equilibrium in the direction of the R form, because substrate binds only to the R form. When substrate binds to one site, the other site on the same enzyme molecule must also be in the R form, according to the basic postulate of this model. In other words, the transition from T to R or vice versa is *concerted*. Hence, *the proportion of enzyme molecules in the R form increases progressively as more substrate is added, and so the binding of substrate is cooperative*. When the active sites are fully saturated, all of the enzyme molecules are in the R form.

The effects of allosteric activators and inhibitors can readily be accounted for by this concerted model. An allosteric inhibitor binds preferentially to the T form, whereas an allosteric activator binds preferentially to the R form (Figure 10-14). Consequently, *an allosteric inhibitor shifts the R $\rightleftharpoons$ T conformational equilibrium toward T, whereas an allosteric activator shifts it toward R*. These effects can be expressed quantitatively by a change in the allosteric equilibrium constant, L, which is a variable in equation 7. An allosteric inhibitor increases L, whereas an allosteric activator decreases L. These effects are shown in Figure 10-15, in which

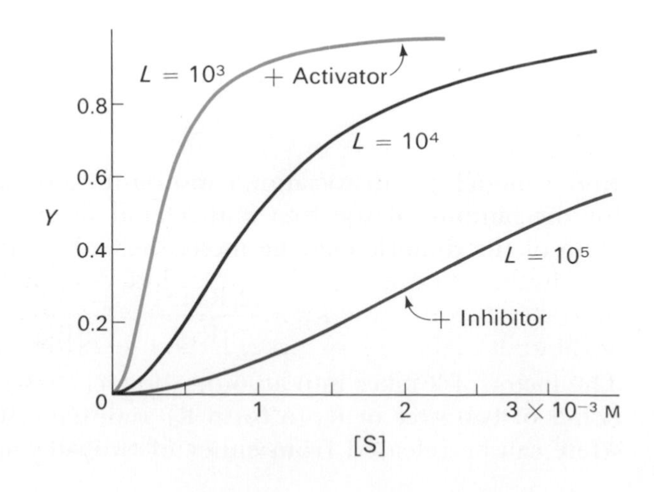

Figure 10-15
Saturation, Y, as a function of substrate concentration, [S], according to the concerted model (equation 7). The effects of an allosteric activator and inhibitor are also shown.

Y is plotted versus [S] for these values of L: 10^3 (activator present), 10^4 (no activator, no inhibitor), and 10^5 (inhibitor present). The fractional saturation, Y, at all values of [S] is decreased by the presence of the inhibitor and increased by the presence of activator. These calculated changes are reminescent of those produced in ATCase by CTP and ATP (see Figure 10-2, on p. 234).

It is useful to define two terms at this point. *Homotropic* effects are those due to allosteric interactions between identical ligands (bound molecules or ions), whereas *heterotropic* effects are those due to interactions between different ligands. In the preceding example, the cooperative binding of substrate to enzyme is a homotropic effect. In contrast, the effect of an activator or inhibitor on the binding of substrate is heterotropic because it is due to interactions between different kinds of molecules. In the concerted model for allosteric interactions, homotropic effects are necessarily positive (cooperative), whereas heterotropic effects can be either positive or negative. For example, the cooperative binding of *N*-carbamoyl phosphate and aspartate by ATCase is a homotropic effect; inhibition by CTP is a negative heterotropic effect; and activation by ATP is a positive heterotropic effect. In the case of hemoglobin (p. 160), the cooperative binding of O_2 is a homotropic effect, and the decreases in oxygen affinity produced by H^+, CO_2, and bisphosphoglycerate (BPG) are negative heterotropic effects.

THE SEQUENTIAL MODEL FOR ALLOSTERIC INTERACTIONS

Allosteric interactions can also be accounted for by a *sequential model*, which has been developed by Daniel Koshland, Jr. The simplest form of this model makes three assumptions:

1. There are only two conformational states (R and T) accessible to any one subunit.

2. The binding of substrate changes the shape of the subunit to which it is bound. However, the conformation of the other subunits in the enzyme molecule are not appreciably altered.

3. The conformational change elicited by the binding of substrate in one subunit can increase or decrease the substrate-binding affinity of the other subunits in the same enzyme molecule.

The sequential model pictures the binding process in an allosteric enzyme to occur as shown in Figure 10-16. The binding is cooperative if the affinity of RT for substrate is greater than that of TT.

This simple sequential model differs from the concerted model in several ways. First, it does not assume an equilibrium between the R and T forms in the absence of substrate. Rather, the conformational transition from T to R is *induced* by the binding of substrate. Second, the conformational change from T to R in different subunits of an enzyme molecule is sequential, not concerted. The hybrid species RT is prominent in the sequential model but excluded in the concerted model. The concerted model supposes that symmetry is essential for the interaction of subunits in oligomeric proteins and therefore requires that it be conserved in allosteric transitions. In contrast, the sequential model assumes that subunits can interact even if they are in different conformational states. Finally, these models differ in that homotropic

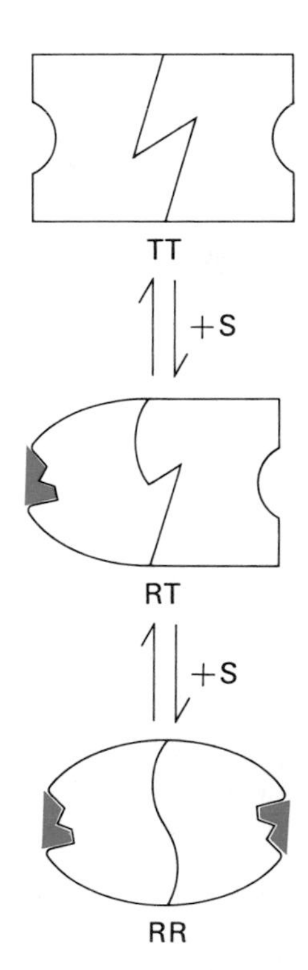

Figure 10-16
Sequential model for the binding of substrate in an allosteric enzyme. The binding is cooperative if the empty active site in RT has a higher affinity for substrate than do the sites in TT.

interactions are necessarily positive in the concerted model but can be either positive or negative in the sequential model. Whether the second substrate molecule can be bound more or less tightly than the first depends on the nature of the distortion induced by the binding of the first substrate molecule.

THE BINDING OF SUBSTRATES TO ATCase LEADS TO A HIGHLY CONCERTED ALLOSTERIC TRANSITION

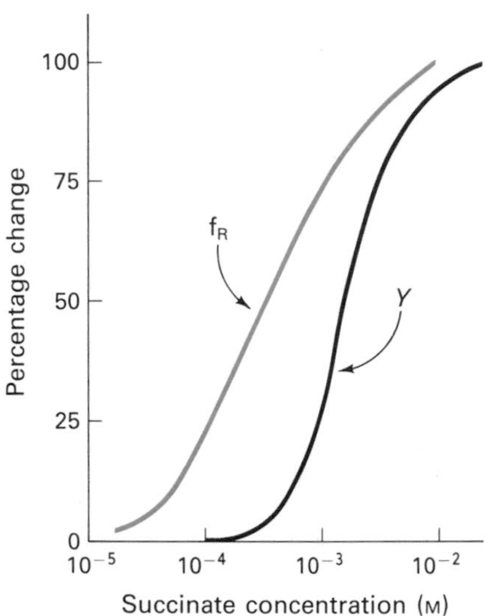

Figure 10-17
Dependence of the fraction of ATCase in the R state (f_R) and the fractional occupancy of substrate-binding sites (Y) on the concentration of succinate, a substrate analog. The change in f_R precedes the change in Y, as predicted by the concerted model. [From M. W. Kirschner and H. K. Schachman. *Biochemistry* 12(1966):2997.]

Which model best accounts for the allosteric properties of ATCase? The concerted model, in contrast with the sequential model, postulates that allosteric transitions are global rather than restricted to the particular subunit bearing a ligand. Specifically, the concerted model predicts that the fraction of catalytic chains in the R state (f_R) is greater than the fraction containing bound substrate (Y). The binding of succinate, an analog of aspartate, to ATCase was detected spectrally, and the T to R transition was monitored by the decrease in sedimentation coefficient caused by the expansion of the enzyme. A decisive result was obtained (Figure 10-17): the change in f_R leads the change in Y as succinate is added, as predicted by the concerted model.

Further evidence concerning the nature of the allosteric transition comes from studies of the effects of inhibitors. As mentioned previously, PALA is a potent inhibitor because it mimics the two substrates of ATCase and occupies the active site. However, low concentrations of PALA *increase* the reaction velocity by *promoting* the binding of substrate. This stimulatory action of an inhibitor is particularly striking in the ATCase-catalyzed breakdown of *N*-carbamoyl phosphate by arsenate, a nonphysiological reaction. On addition of PALA, the reaction rate increases until an average of three PALA are bound per enzyme molecule (Figure 10-18A). This maximal velocity is 17-fold greater than in the absence of PALA. The reaction rate then decreases to nearly zero on adding three more molecules of PALA per enzyme to fill the catalytic sites. The initial slope of this plot reveals that one bound PALA switches the conformation of all six catalytic sites in ATCase. The five not occupied by PALA have been converted to the R state, with high affinity for substrates. Thus, the homotropic allosteric transition in ATCase is *completely concerted* (Figure 10-18B), as predicted by the Monod-Changeux-Wyman model.

The cooperative binding of oxygen to hemoglobin also nicely fits the concerted model. In fact, homotropic effects in most allosteric proteins

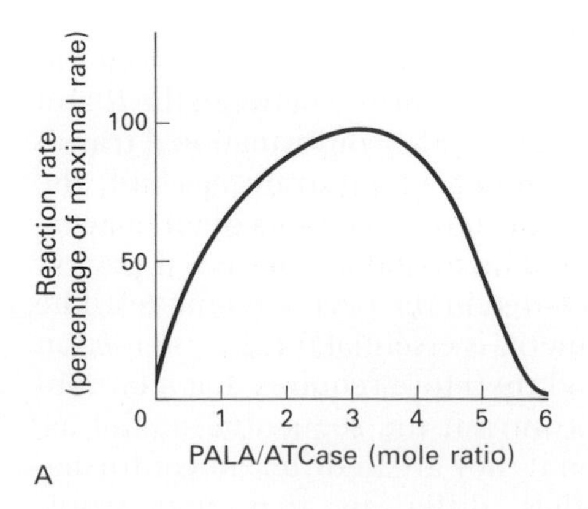

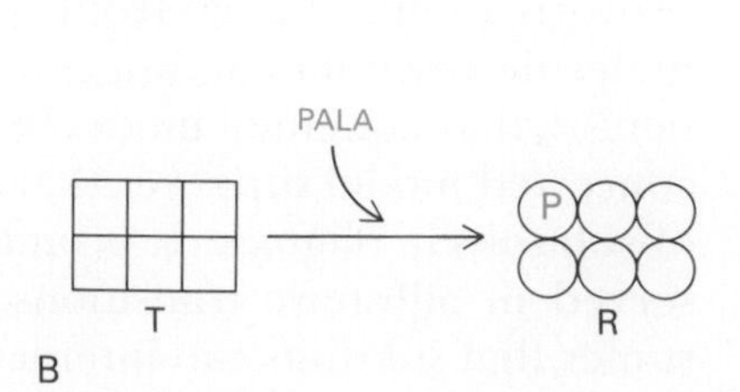

Figure 10-18
(A) Dependence of the reaction rate on the number of PALA bound per ATCase molecule. PALA activates and then inhibits the breakdown of *N*-carbamoyl phosphate by arsenate. [From J. Foote and H. K. Schachman. *J. Mol. Biol.* 186(1985):175–184.] (B) Schematic diagram showing the concerted nature of the transition. A single bound PALA (denoted by P) switches all subunits from the T to the R state.

are best accounted for by the concerted model. In contrast, heterotropic effects are usually better described by the sequential model. In ATCase, homotropic and heterotropic effects appear to be mediated by different allosteric transitions. More than two states need to be postulated to account for the complexity of many allosteric processes.

MANY ENZYMES ARE ACTIVATED BY SPECIFIC PROTEOLYTIC CLEAVAGE

We turn now to a different mechanism of enzyme regulation. Enzymes such as lysozyme acquire full enzymatic activity as they spontaneously fold into their characteristic three-dimensional forms. In contrast, many other enzymes are synthesized as inactive precursors that are subsequently activated by cleavage of one or a few specific peptide bonds. The inactive precursor is called a *zymogen* (or a *proenzyme*). Proteolytic activation, in contrast with allosteric control, occurs just once in the life of an enzyme molecule.

Activation of enzymes and other proteins by specific proteolysis recurs frequently in biological systems. For example:

1. The *digestive enzymes* that hydrolyze proteins are synthesized as zymogens in the stomach and pancreas (Table 10-1).

2. *Blood clotting* is mediated by a cascade of proteolytic activations that assures a rapid and amplified response to trauma.

3. Some protein hormones are synthesized as inactive precursors. For example, *insulin* is derived from *proinsulin* by proteolytic removal of a peptide.

4. The fibrous protein *collagen,* which is present in skin and bone, is derived from *procollagen,* a soluble precursor.

Table 10-1
Gastric and pancreatic zymogens

Site of synthesis	*Zymogen*	*Active enzyme*
Stomach	Pepsinogen	Pepsin
Pancreas	Chymotrypsinogen	Chymotrypsin
Pancreas	Trypsinogen	Trypsin
Pancreas	Procarboxypeptidase	Carboxypeptidase
Pancreas	Proelastase	Elastase

The mechanism of activation of three digestive enzymes—chymotrypsin, trypsin, and pepsin—will be considered first. We shall then turn to blood clotting, which is mediated by a series of proteolytic activations culminating in the conversion of soluble fibrinogen into an insoluble fibrin clot. Proteolytic activation, in contrast with allosteric control, is irreversible. Hence, another means of switching off the activities of these enzymes is needed. Two specific inhibitory proteins that bind very tightly to proteolytic enzymes and block their activity will be discussed.

CHYMOTRYPSINOGEN IS ACTIVATED BY SPECIFIC CLEAVAGE OF A SINGLE PEPTIDE BOND

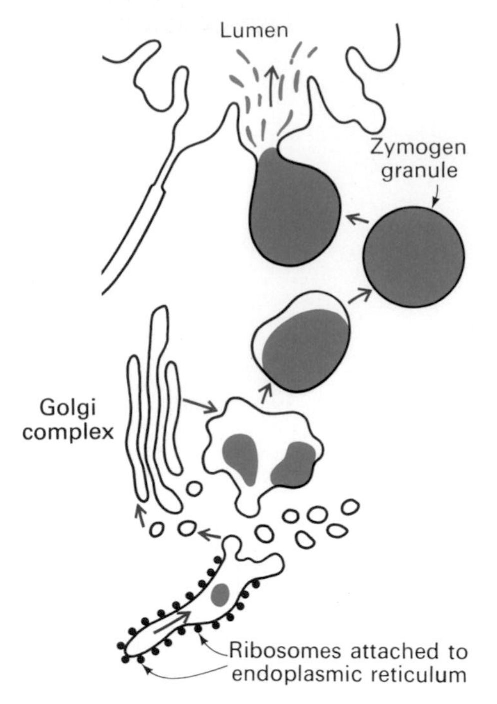

Figure 10-19
Diagrammatic representation of the secretion of zymogens by an acinar cell of the pancreas. [After a drawing kindly provided by Dr. George Palade.]

Chymotrypsin is a digestive enzyme that hydrolyzes proteins in the small intestine. Its inactive precursor, *chymotrypsinogen*, is synthesized in the pancreas, as are several other zymogens and digestive enzymes. Indeed, the pancreas is one of the most active organs in synthesizing proteins. The enzymes and zymogens are synthesized in the acinar cells of the pancreas (Figure 10-19). The proteins travel from the endoplasmic reticulum to the Golgi apparatus, where quantities of them are covered with a membrane made up of lipid and protein. These *zymogen granules* appear in electron micrographs as very dense bodies because they have a high concentration of protein (Figure 10-20). The zymogen granules accumulate at the apex of the acinar cell; when the cell is stimulated by a hormonal signal or a nerve impulse, the granules are secreted into a duct leading into the duodenum

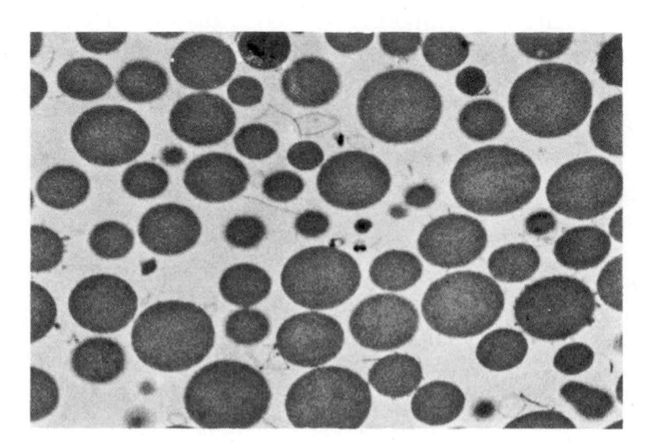

Figure 10-20
Electron micrograph of zymogen granules in an acinar cell of the pancreas. [Courtesy of Dr. George Palade.]

Chymotrypsinogen is a single polypeptide chain consisting of 245 amino acid residues. It is cross-linked by five disulfide bonds. This precursor is virtually devoid of enzymatic activity. It is converted into a fully active enzyme when the peptide bond joining arginine 15 and isoleucine 16 is cleaved by trypsin (Figure 10-21). The resulting active

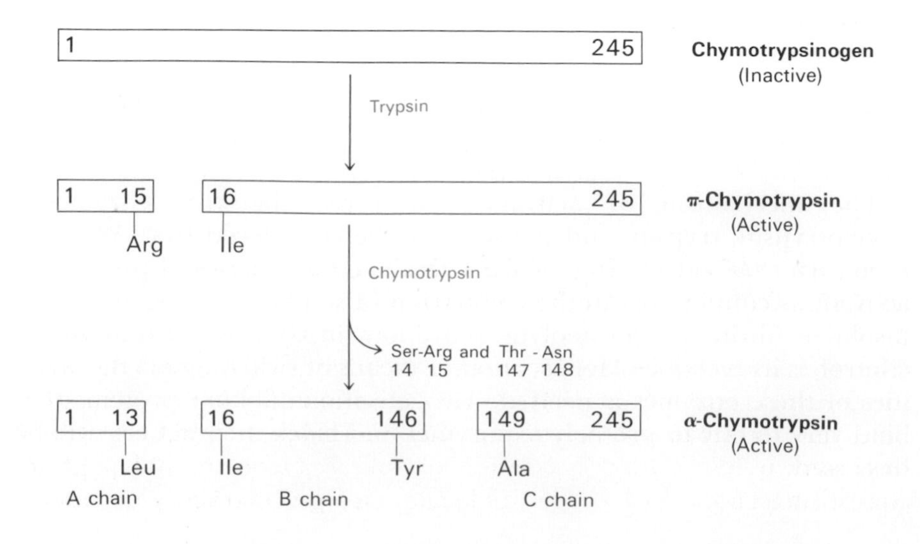

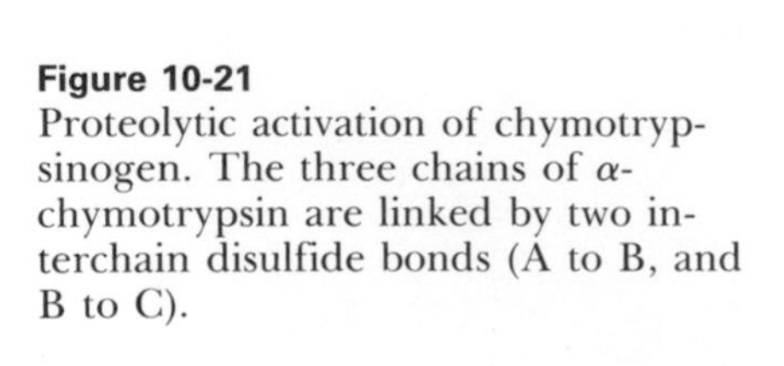

Figure 10-21
Proteolytic activation of chymotrypsinogen. The three chains of α-chymotrypsin are linked by two interchain disulfide bonds (A to B, and B to C).

enzyme, called π-chymotrypsin, then acts on other π-chymotrypsin molecules. Two dipeptides are removed to yield α-chymotrypsin, the stable form of the enzyme. The three resulting chains in α-chymotrypsin remain linked to each other by two interchain disulfide bonds. The additional cleavages made in the conversion of π- into α-chymotrypsin are superfluous, because π-chymotrypsin is already fully active. The striking feature of this activation process is that *cleavage of a single specific peptide bond transforms the protein from a catalytically inactive form into one that is fully active.*

PROTEOLYTIC ACTIVATION OF CHYMOTRYPSINOGEN LEADS TO THE FORMATION OF A SUBSTRATE BINDING SITE

The three-dimensional structure of chymotrypsinogen has been elucidated by Joseph Kraut. The conformational changes resulting in activation have been identified:

1. Hydrolysis of the peptide bond between arginine 15 and isoleucine 16 creates new carboxyl- and amino-terminal groups.

2. The newly formed *amino-terminal group of isoleucine 16 turns inward and interacts with aspartate 194* in the interior of the chymotrypsin molecule (Figure 10-22). Protonation of this amino group stabilizes the active form of chymotrypsin, as shown by the dependence of enzyme activity on pH.

3. This electrostatic interaction between a positively charged amino group and a negatively charged carboxylate ion in a nonpolar region triggers a number of conformational changes. Methionine 192 moves from a deeply buried position in the zymogen to the surface of the molecule in the enzyme, and residues 187 and 193 become more extended. These changes result in the formation of the *substrate specificity site* for aromatic and bulky nonpolar groups. One side of this site is made up of residues 189 through 192. *This cavity for part of the substrate is not fully formed in the zymogen.*

4. The tetrahedral transition state in catalysis by chymotrypsin is stabilized by hydrogen bonds between the negatively charged carbonyl oxygen atom of the substrate and two NH groups of the main chain of the enzyme (see Figure 9-43, on p. 225). One of these NH groups is not appropriately located in chymotrypsinogen, and so *the oxyanion hole is incomplete in the zymogen.*

5. The conformational changes elsewhere in the molecule are very small. Thus, *the switching on of enzymatic activity in a protein can be accomplished by discrete, highly localized conformational changes that are triggered by the hydrolysis of a single peptide bond.*

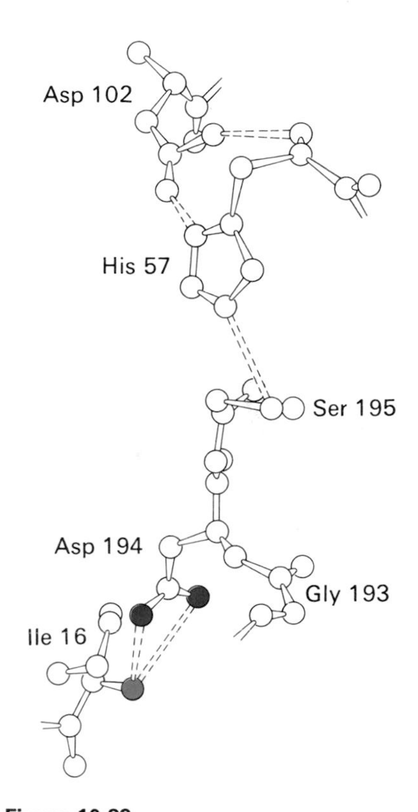

Figure 10-22
Environment of aspartate 194 and isoleucine 16 in chymotrypsin. The electrostatic interaction between the carboxylate of Asp 194 (red) and the α-NH_2 group of Ile 16 (blue) is essential for the activity of chymotrypsin. These groups are adjacent to the charge relay network. [After D. M. Blow and T. A. Steitz. X-ray diffraction studies of enzymes, *Ann. Rev. Biochem.* 39(1970):86. Copyright © 1970 by Annual Reviews Inc. All rights reserved.]

A HIGHLY MOBILE REGION BECOMES ORDERED IN THE ACTIVATION OF TRYPSINOGEN

The structural changes accompanying the activation of trypsinogen are somewhat different from those in the activation of chymotrypsinogen. The independent x-ray analyses of Robert Huber and Robert Stroud have shown that the conformation of four stretches of polypeptide, comprising about 15% of the molecule, changes markedly on activa-

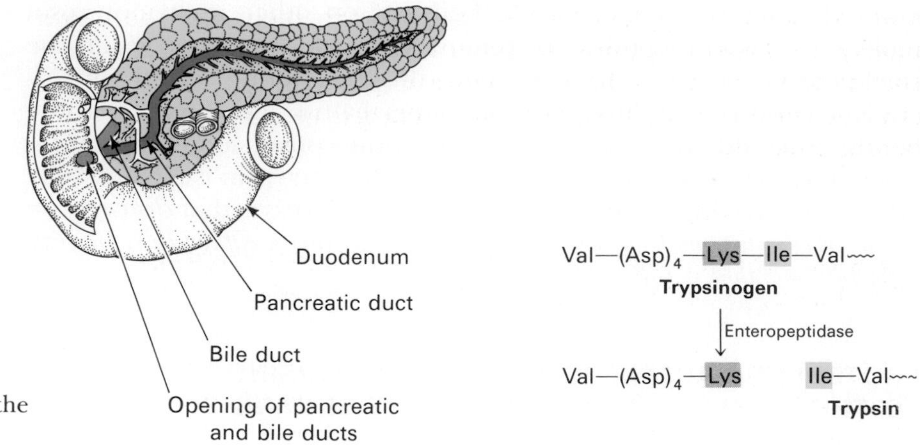

Duodenum

Pancreatic duct

Bile duct

Opening of pancreatic
and bile ducts

Val—$(Asp)_4$—Lys—Ile—Val~~~
Trypsinogen

Enteropeptidase

Val—$(Asp)_4$—Lys   Ile—Val~~~
Trypsin

Figure 10-23
Trypsinogen is activated by enteropeptidase as it enters the duodenum.

tion. *These regions, called the activation domain, are very floppy in the zymogen, whereas they have a well-defined conformation in trypsin.* Furthermore, the oxyanion hole (p. 225) in trypsinogen is too far from histidine 57 to promote the formation of the tetrahedral transition state.

The digestion of proteins in the duodenum requires the concurrent action of several proteolytic enzymes, because each is specific for a limited number of side chains. Thus, the zymogens must be switched on at the same time. Coordinated control is achieved by the action of *trypsin as the common activator of all the pancreatic zymogens*—trypsinogen, chymotrypsinogen, proelastase, and procarboxypeptidase. How is the trypsin produced? The cells that line the duodenum secrete an enzyme, *enteropeptidase,* that hydrolyzes a unique lysine-isoleucine peptide bond in trypsinogen as it enters the duodenum from the pancreas (Figure 10-23). The small amount of trypsin produced in this way activates more trypsinogen and the other zymogens. Thus, *the formation of trypsin by enteropeptidase is the master activation step.*

PEPSINOGEN CLEAVES ITSELF IN AN ACIDIC ENVIRONMENT TO FORM HIGHLY ACTIVE PEPSIN

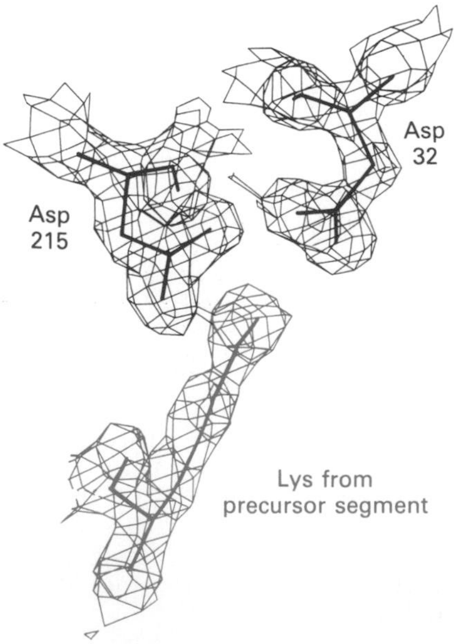

Pepsin digests proteins in the highly acidic environment of the stomach. Its pH optimum of 2 is very unusual for an enzyme. This member of the acid protease family contains two aspartate residues in its catalytic center, one in the —COOH form and the other in the $—COO^-$ form. *Pepsinogen,* the zymogen, contains an amino-terminal precursor segment of 44 residues that is proteolytically removed in the formation of pepsin. This activation occurs spontaneously below pH 5. The key event is the cleavage of the peptide bond between leucine 16 and isoleucine 17 in the precursor segment. The rate of formation of pepsin is independent of the concentration of pepsinogen, which shows that activation is an *intramolecular* process.

Asp 215

Asp 32

Lys from precursor segment

Figure 10-24
The active site of pepsinogen is blocked at neutral pH by an electrostatic interaction between a lysine side chain from the precursor segment (shown in blue) and a critical pair of aspartate carboxylates (shown in red). The polygons depict electron density contour surfaces. [After M. N. G. James and A. R. Sielecki. *Nature* 319(1986):35.]

The precursor segment is highly basic, whereas pepsin is highly acidic. X-ray crystallographic analyses have shown that the active site is fully formed in pepsinogen but is blocked at neutral pH by residues of the precursor segment. Six lysine and arginine side chains of the precursor segment form salt bridges with the carboxylate side chains of glutamate and aspartate residues of the pepsin moiety. Most important, a lysine side chain of the precursor interacts electrostatically with the pair of aspartates at the active site (Figure 10-24). When the pH is lowered, salt bridges between the precursor segment and the pepsin moiety are broken because several carboxylates become protonated. During the ensuing conformational rearrangement, the now exposed catalytic site hydrolyzes the peptide bond between the precursor and pepsin moieties.

PANCREATIC TRYPSIN INHIBITOR BINDS VERY TIGHTLY TO THE ACTIVE SITE OF TRYPSIN

The conversion of a zymogen into a protease by cleavage of a single peptide bond is a precise means of switching on enzymatic activity. However, this activation step is irreversible and so a different mechanism is needed to stop proteolysis. This is accomplished by specific protease inhibitors. For example, *pancreatic trypsin inhibitor,* a 6-kd protein, inhibits trypsin by binding very tightly to its active site. The dissociation constant of the complex is 10^{-13} M, which corresponds to a standard free energy of binding of about −18 kcal/mol. In contrast with nearly all known protein assemblies, this complex is not dissociated into its constituent chains by treatment with 8 M urea or 6 M guanidine hydrochloride.

The reason for the exceptional stability of the complex is that pancreatic trypsin inhibitor is a very effective substrate analog. X-ray analyses have shown that the side chain of lysine 15 of this inhibitor binds to the aspartate side chain in the specificity pocket of the enzyme (Figure 10-25). In addition, there are many hydrogen bonds between the main chain of trypsin and that of its inhibitor, as in an anti-parallel β pleated sheet. This array of hydrogen bonds is like that made by a true substrate. Furthermore, the carbonyl group of lysine 15 and the surrounding atoms of the inhibitor fit snugly in the oxyanion hole of the enzyme. The scissile peptide bond in this complex is not planar; the carbonyl carbon atom is on its way to becoming tetrahedral, a key facet of the transition state (p. 225). Indeed, the peptide bond between lysine 15 and alanine 16 in pancreatic trypsin inhibitor is cleaved, but at a very slow rate. The half life of this trypsin-inhibitor complex is several months. *The inhibitor has very high affinity for trypsin because its structure is almost perfectly complementary to that of the active site of trypsin.* The restricted conformational flexibility of the inhibitor in the binding site contributes to the blocking of catalysis and accounts for the very slow dissociation of the complex.

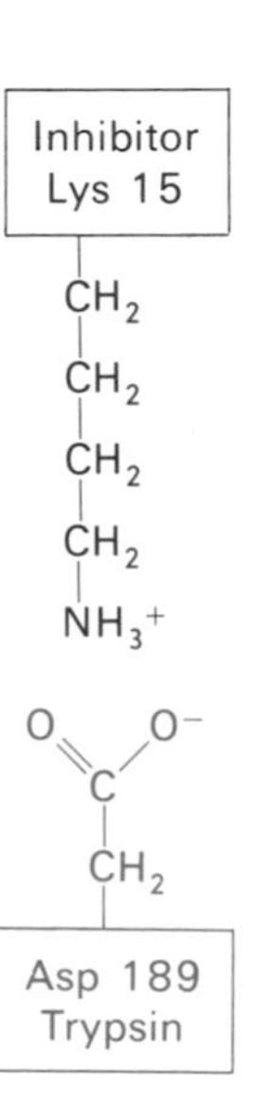

Figure 10-25
A prominent feature of the interaction of pancreatic trypsin inhibitor with trypsin is an electrostatic bond between lysine 15 and aspartate 189. The $-NH_3^+$ group of lysine 15 is also hydrogen bonded to several oxygen atoms in the specificity pocket of trypsin.

INSUFFICIENT α_1-ANTITRYPSIN ACTIVITY LEADS TO DESTRUCTION OF THE LUNGS AND EMPHYSEMA

α_1-Antitrypsin (also called α_1-antiproteinase), a 53-kd plasma protein, protects tissues from digestion by elastase, a secretory product of neutrophils (white blood cells that engulf bacteria). Antielastase would be a more accurate name for this inhibitor, because it blocks elastase more effectively than it blocks trypsin. Like pancreatic trypsin inhibitor, α_1-antitrypsin blocks the action of target enzymes by binding nearly irreversibly to their active site. Genetic disorders leading to deficient α_1-antitrypsin show that this inhibitor is physiologically important. For example, the substitution of lysine for glutamate at residue 53 in the type Z mutant slows the secretion of this inhibitor from liver cells.

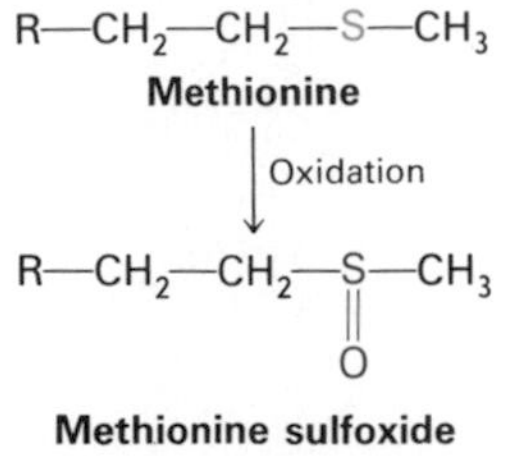

$$R{-}CH_2{-}CH_2{-}S{-}CH_3$$

Methionine

↓ Oxidation

$$R{-}CH_2{-}CH_2{-}\underset{\underset{O}{\|}}{S}{-}CH_3$$

Methionine sulfoxide

Figure 10-26
Oxidation of methionine to the sulfoxide.

Serum levels of the inhibitor are about 15% of normal in individuals homozygous for this defect. The consequence is that unrestrained excess elastase destroys alveolar walls in the lungs by digesting elastic fibers and other connective-tissue proteins.

The resulting clinical condition is called *emphysema* (also known as *destructive lung disease*). A person with emphysema must breathe much harder to exchange the same volume of air because their alveoli are much less resilient than normal. Cigarette smoking markedly increases the likelihood that a type Z heterozygote will develop emphysema. The reason is that smoke oxidizes methionine 358 of the inhibitor (Figure 10-26), a residue essential for binding elastase. Indeed, this methionine side chain of the inhibitor is the bait that selectively traps elastase. The *methionine sulfoxide* oxidation product, in contrast, does not lure elastase, a striking consequence of the insertion of just one oxygen atom into a protein.

CLOTTING OCCURS BY A CASCADE OF ZYMOGEN ACTIVATIONS

Blood clots are formed by a *series of zymogen activations*. In this enzymatic *cascade*, the activated form of one factor catalyzes the activation of the next factor (Figure 10-27). Very small amounts of the initial factors suffice to trigger the cascade because of the catalytic nature of the activation process. The numerous steps yield a *large amplification*, assuring a rapid response to trauma. The clotting cascade illustrates some general

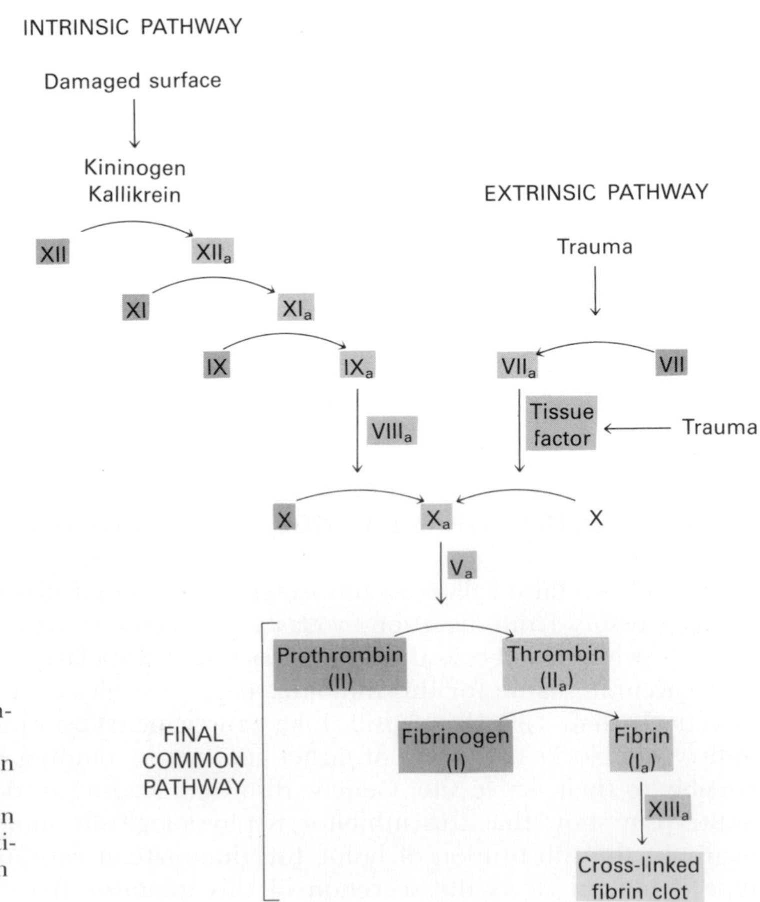

Figure 10-27
Blood clotting cascade. A fibrin clot is formed by the interplay of the intrinsic, extrinsic, and final common pathways. Inactive forms of clotting factors are shown in red and their activated counterparts in green. Stimulatory proteins that are not themselves enzymes are shown in blue. A striking feature of this process is that the activated form of one clotting factor catalyzes the activation of the next factor.

principles of how sequences of reaction are triggered and controlled. Many familiar themes are evident. For example, several of the activated clotting factors are serine proteases.

In 1863, Joseph Lister showed that blood stayed fluid in the excised jugular vein of an ox, but that it rapidly clotted when it was transferred to a glass vessel. This unphysiologic surface activated a reaction sequence that has become known as the *intrinsic clotting pathway*. Clotting can also be triggered by substances that are released from tissues as a consequence of trauma to them. This *extrinsic clotting pathway* and the intrinsic one converge on a common sequence of final steps to form a *fibrin clot*. Recent studies indicate that the intrinsic and extrinsic pathways interact with each other in vivo. Both are needed for proper clotting, as evidenced by clotting disorders caused by a deficiency of a single protein in one of the pathways.

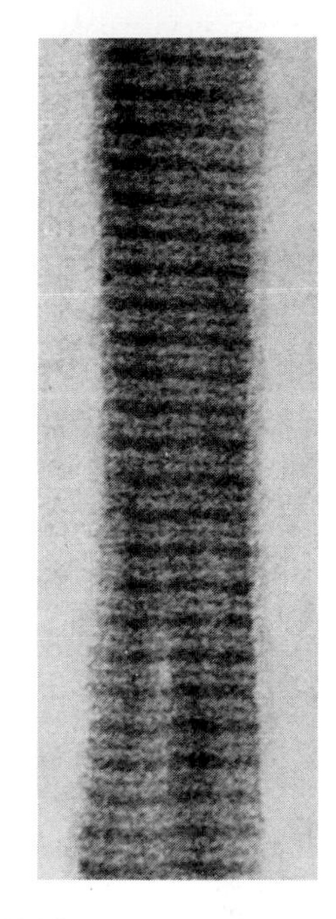

Figure 10-28
Electron micrograph of fibrin. The 230-Å period along the fiber axis is half the length of a fibrinogen molecule. [Courtesy of Dr. Henry Slayter.]

FIBRINOGEN IS CONVERTED BY THROMBIN INTO A FIBRIN CLOT

The best-characterized part of the clotting process is the conversion of fibrinogen into fibrin by thrombin, a proteolytic enzyme. Fibrinogen is made up of three globular units connected by two rods. This 340-kd protein consists of pairs of three kinds of chains: Aα, Bβ, γ. The rod regions are triple-stranded α-helical coiled coils, a recurring motif in fibrous proteins. Thrombin cleaves four *arginine-glycine peptide bonds* in the central globular region of fibrinogen to release an A peptide of 18 residues from each of the two α chains and a B peptide of 20 residues from each of the two β chains. These A and B peptides are called *fibrinopeptides*. A fibrinogen molecule devoid of these fibrinopeptides is called *fibrin monomer* and has the subunit structure $(\alpha\beta\gamma)_2$.

Fibrin monomers spontaneously assemble into ordered fibrous arrays called *fibrin*. Electron micrographs and low-angle x-ray patterns show that fibrin has a periodic structure that repeats every 230 Å (Figure 10-28). Because fibrinogen is about 460 Å long, it seems likely that the fibrin monomers form a half-staggered array (Figure 10-29).

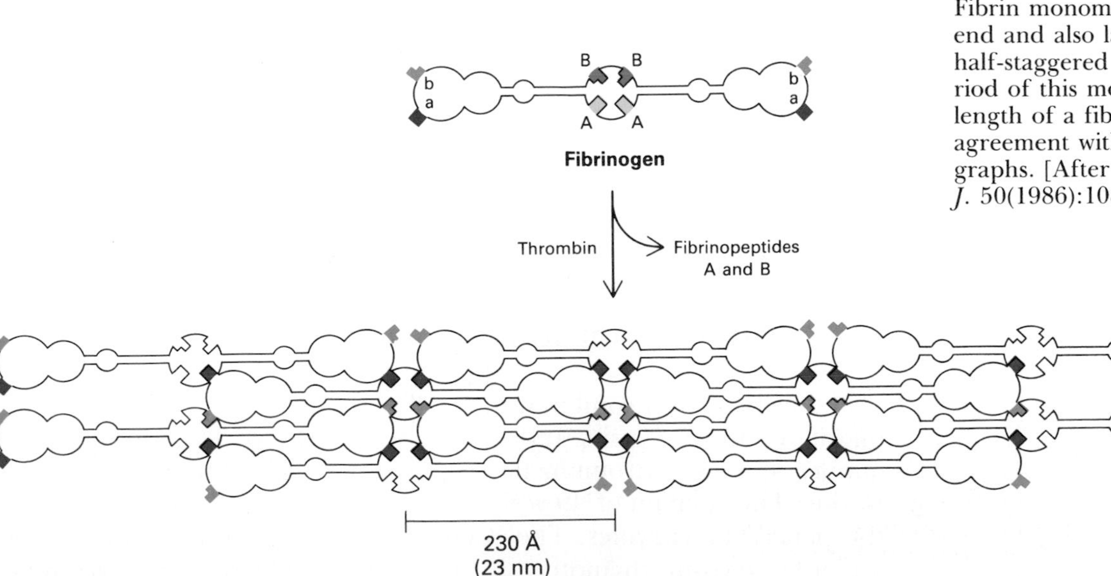

Figure 10-29
Proposed arrangement of fibrin monomers in a fibrin clot. Thrombin cleaves fibrinopeptides A and B from the central globule of fibrinogen. Each of the two end domains of fibrinogen contains sites a and b that are complementary to the newly exposed sites in the central globule. Fibrin monomers associate end-to-end and also laterally to form a half-staggered array. The 230-Å period of this model is half of the length of a fibrinogen molecule, in agreement with electron micrographs. [After J. W. Weisel. *Biophys. J.* 50(1986):1080.]

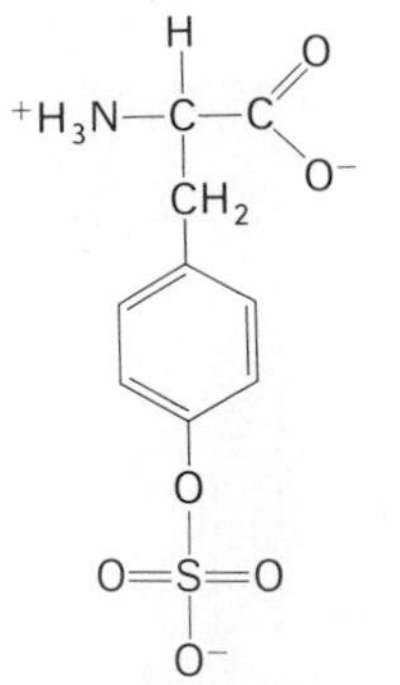

Why do fibrin monomers aggregate, whereas their parent fibrinogen molecules stay in solution? The fibrinopeptides of all vertebrate species studied thus far have a *large net negative charge*. Aspartate and glutamate residues are found in abundance. An unusual negatively charged derivative of tyrosine, namely *tyrosine-O-sulfate,* is found in fibrinopeptide B. The presence of these and other negatively charged groups in the fibrinopeptides probably keeps fibrinogen molecules apart. *Their release by thrombin gives fibrin monomers a different surface-charge pattern, leading to their specific aggregation.* In particular, removal of the fibrinopeptides changes the net charge of the central globular unit from -8 to $+5$. Each of the terminal globular units has a net charge of -4. Thus, electrostatic interactions between the terminal and central globular units probably stabilize the half-staggered array shown in Figure 10-29.

Tyrosine-*O*-sulfate

FIBRIN CLOTS ARE STRENGTHENED BY COVALENT CROSS-LINKS

The clot produced by the spontaneous aggregation of fibrin monomer is stabilized by the formation of covalent cross-links between the side chains of different molecules in the fibrin fiber. In fact, *peptide bonds are formed between specific glutamine and lysine side chains in a transamidation reaction* (Figure 10-30). This cross-linking reaction is catalyzed by an enzyme called Factor XIII_a. Clotting factors are assigned Roman numerals for ease of discussion; the subscript "a" designates an activated factor. Thrombin converts XIII into XIII_a, an active transamidase (also known as fibrin-stabilizing factor or transglutaminase). The peptide cross-links stabilizing fibrin are found in very few other proteins. The importance of this unusual bond in fibrin is evidenced by the finding that patients with a deficiency of Factor XIII have a pronounced tendency to bleed.

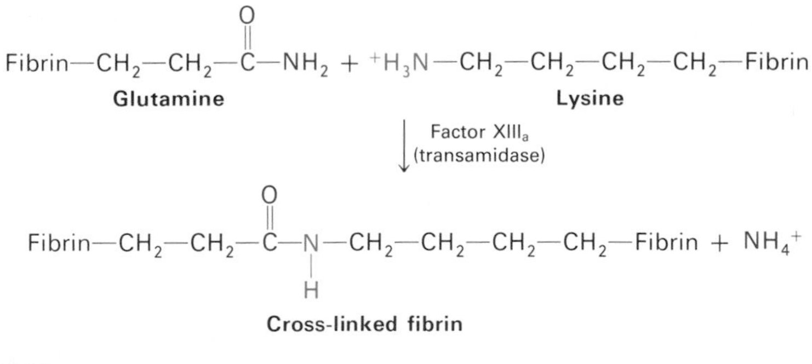

$$\text{Fibrin—CH}_2\text{—CH}_2\text{—C(=O)—NH}_2 + {}^+\text{H}_3\text{N—CH}_2\text{—CH}_2\text{—CH}_2\text{—CH}_2\text{—Fibrin}$$

Glutamine | Lysine

$$\downarrow \text{Factor XIII}_a \text{ (transamidase)}$$

$$\text{Fibrin—CH}_2\text{—CH}_2\text{—C(=O)—N(H)—CH}_2\text{—CH}_2\text{—CH}_2\text{—CH}_2\text{—Fibrin} + \text{NH}_4^+$$

Cross-linked fibrin

Figure 10-30
Fibrin is cross-linked by transamidation. Isopeptide bonds between glutamine and lysine side chains are formed.

THROMBIN IS HOMOLOGOUS TO TRYPSIN

The specificity of thrombin for arginine-glycine bonds suggested that it might resemble trypsin. Indeed it does, as shown by amino acid sequence studies. Thrombin has a mass of 34 kd and consists of two chains. The A chain of 49 residues exhibits no detectable homology to the pancreatic enzymes. The B chain, however, is quite similar in sequence to trypsin, chymotrypsin, and elastase. The sequence around its active-site serine is Gly-Asp-Ser-Gly-Gly-Pro, the same as that in the pancreatic serine proteases. Thrombin, like trypsin, contains a catalytic triad, an oxyanion hole, and an aspartate residue at the bottom of its substrate-binding cleft. This negatively charged group forms an electrostatic bond with a positively charged arginine side chain of its substrate. However, thrombin is much more specific than is trypsin. It cleaves only certain arginine-glycine bonds, whereas trypsin cleaves most peptide bonds following arginine or lysine residues.

Thrombin, like the pancreatic serine proteases, is synthesized as a 66-kd zymogen called *prothrombin*. Proteolytic cleavage of an arginine-threonine bond releases a 32-kd fragment from the amino terminus of prothrombin (Figure 10-31). Cleavage of an arginine-isoleucine bond then yields active thrombin. An ion pair like the one between the positively charged amino group of isoleucine 16 and the negatively charged aspartate 194 in chymotrypsin (see Figure 10-22) stabilizes the active form of thrombin.

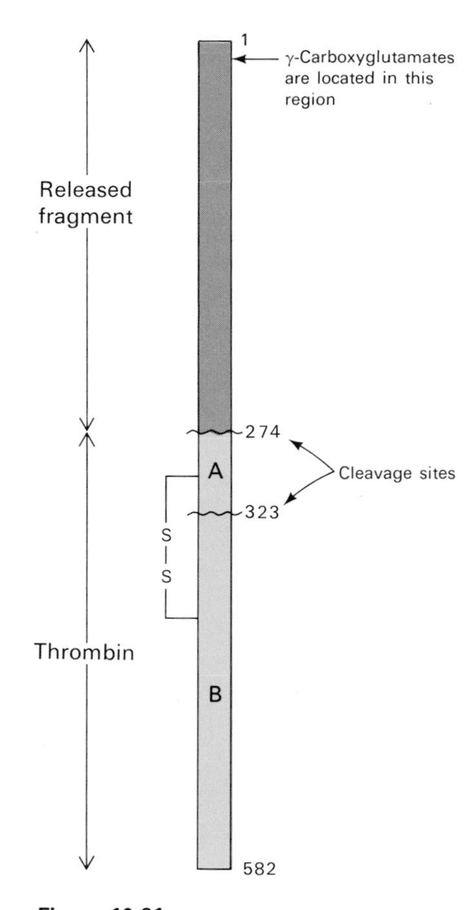

Figure 10-31
Structure of prothrombin. Cleavage of two peptide bonds (Arg 274–Thr 275 and Arg 323–Ile 324) yields thrombin. The released amino-terminal fragment of prothrombin is shown in blue. All of the γ-carboxyglutamate residues are in this fragment. The A and B chains of thrombin, which are joined by a disulfide, are shown in yellow.

VITAMIN K IS REQUIRED FOR THE SYNTHESIS OF PROTHROMBIN AND OTHER CALCIUM-BINDING PROTEINS

Vitamin K (Figure 10-32) has been known for many years to be essential for the synthesis of prothrombin and several other clotting factors. Subsequent studies of the abnormal prothrombin synthesized in the absence of vitamin K or in the presence of vitamin K antagonists, such as dicoumarol, revealed the mode of action of this vitamin. *Dicoumarol* is found in spoiled sweet clover and causes a fatal hemorrhagic disease in cattle fed on this hay. This coumarin derivative is used clinically as an *anticoagulant* to prevent thromboses in patients prone to clot formation. Dicoumarol and such related vitamin K antagonists as *warfarin* also serve as effective rat poisons. Cows fed dicoumarol contain an abnormal prothrombin that does not bind Ca^{2+}, in contrast with normal prothrombin. This difference was puzzling for some time because abnormal prothrombin has the same number of amino acid residues and gives the same amino acid analysis after acid hydrolysis as does normal prothrombin

Fragmentation of normal prothrombin showed that its capacity to bind Ca^{2+} resides in its amino-terminal region. The electrophoretic mobility of an amino-terminal peptide from abnormal prothrombin was then found to be markedly different from that of the correspond-

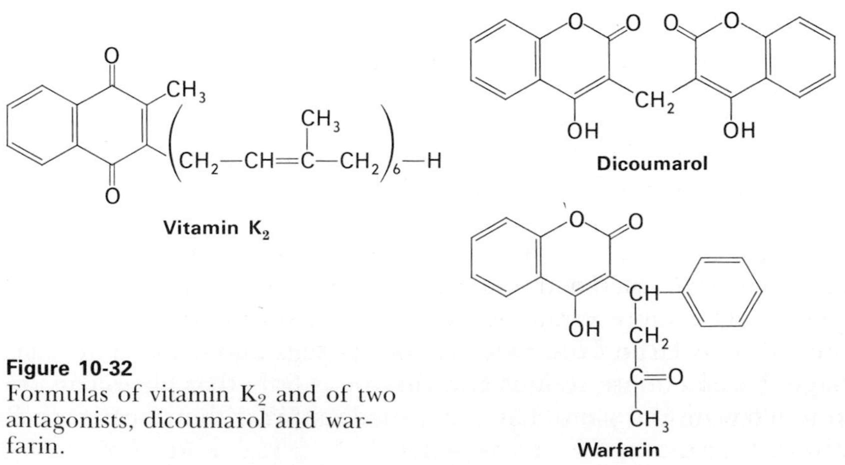

Vitamin K_2

Dicoumarol

Warfarin

Figure 10-32
Formulas of vitamin K_2 and of two antagonists, dicoumarol and warfarin.

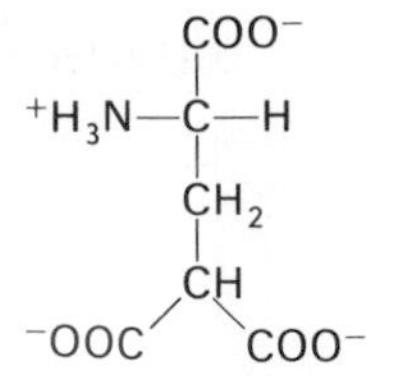

$$
\begin{array}{c}
COO^- \\
| \\
{}^+H_3N{-}C{-}H \\
| \\
CH_2 \\
| \\
CH \\
{}^-OOC \quad COO^-
\end{array}
$$

γ-Carboxyglutamate

ing peptide from the normal molecule. Nuclear magnetic resonance studies of these peptides revealed that normal prothrombin contains *γ-carboxyglutamate,* a previously unknown residue that evaded detection because its second carboxyl group is lost on acid hydrolysis. The abnormal prothrombin formed following administration of anticoagulants lacks this modified amino acid. In fact, the first ten glutamate residues in the amino-terminal region of prothrombin are carboxylated to γ-carboxyglutamate by a vitamin-K-dependent enzyme system.

The vitamin-K-dependent carboxylation reaction converts glutamate, a weak chelator of Ca^{2+}*, into γ-carboxyglutamate, a much stronger chelator.* The binding of Ca^{2+} by prothrombin anchors it to phospholipid membranes derived from blood platelets following injury. The functional significance of the binding of prothrombin to phospholipid surfaces is that it brings prothrombin into close proximity with factor X_a, a serine protease, and factor V, a stimulatory protein that accelerates the activation of prothrombin by more than a factor of 10^4. The amino-terminal fragment of prothrombin, which contains the Ca^{2+}-binding sites, is released in this activation step. Thrombin freed in this way from the phospholipid surface can then activate fibrinogen in the plasma.

Prothrombin is not alone in having γ-carboxyglutamate residues. Specific glutamate residues in Factors VII, IX, and X also become carboxylated to form sites with high affinity for Ca^{2+}. Indeed, this vitamin-K-dependent reaction has a significance extending beyond the clotting cascade. For example, osteocalcin, a calcium-binding protein important in the development of bone, contains γ-carboxyglutamate residues.

HEMOPHILIA REVEALED AN EARLY STEP IN CLOTTING

How is Factor X activated? Biochemical studies of early steps in clotting are more difficult than those of late ones because the early clotting factors are present in only small amounts. The concentration of Factor X in blood is only 0.01 mg/ml, compared with 3 mg/ml for fibrinogen. The concentrations of some of the earlier factors are even lower. Furthermore, these proteins are highly labile. Some of the important breakthroughs in the elucidation of the pathways of clotting have therefore come from studies of patients with bleeding disorders.

Classic hemophilia, the best known clotting defect, is genetically transmitted as a sex-linked recessive characteristic. Heterozygous females are asymptomatic carriers. A famous carrier of this disease was Queen Victoria, who transmitted it to the royal families of Prussia, Spain, and Russia (Figure 10-33).

In 1904, the tsarevich Alexis was born. He was the first male heir to have been born to a reigning Russian tsar since the seventeenth century, which was taken as an omen of hope. Four healthy daughters had previously been born to Tsar Nicholas II and Empress Alexandra, a granddaughter of Queen Victoria. The mood changed six weeks after the birth of Alexis, as reflected in his father's diary: "A hemorrhage began this morning without the slightest cause from the navel of our small Alexis. It lasted with but a few interruptions until evening." The outlook grew more ominous as Alexis started to crawl and toddle, which caused large, blue swellings on his legs and arms. The hemorrhages became more serious and there was little that physicians could do to alleviate the pain. The anguished empress then turned to Ras-

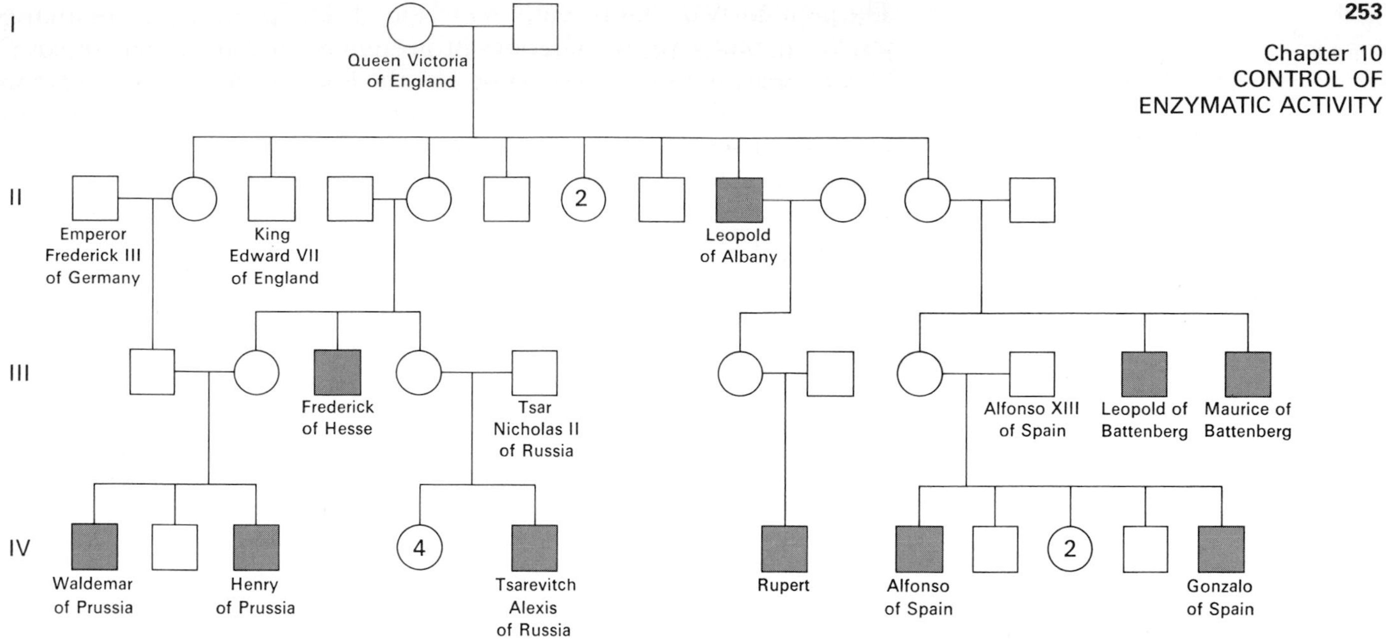

Figure 10-33
Pedigree of hemophilia in the royal families of Europe. All of Queen Victoria's children, but not all individuals in later generations, are included in this diagram. Females are symbolized by circles, normal males by white squares, and hemophilic males by red squares. [After C. Stern. *Principles of Human Genetics*, 3rd ed. (W. H. Freeman and Company). Copyright © 1973.]

putin, who was reputed to be a miracle man. No man knew less about molecular disease, no man ever profited more from it. Rasputin held a position of great power in the Russian court for many years because the empress placed great faith in his healing abilities.

Why did Nicholas and Alexandra marry when it was already known that her brother, nephews, and uncle had hemophilia? In 1873, Grandidier, a French physician, counseled that "all members of bleeder families should be advised against marriage." Indeed, the hereditary nature of the disease was fully appreciated as early as 1803, by John Otto:

> About seventy or eighty years ago, a woman by the name of Smith settled in the vicinity of Plymouth, New Hampshire, and transmitted the following idiosyncrasy to her decendants. . . . It is a surprising circumstance that the males only are subject to this strange affection, and that all of them are not liable to it. . . . Although the females are exempt, they are still capable of transmitting it to their male children.

J. B. S. Haldane has suggested that "kings are carefully protected against disagreeable realities. . . . The hemophilia of the Tsarevich was a symptom of the divorce between royalty and reality."

ANTIHEMOPHILIC FACTOR PRODUCED BY RECOMBINANT DNA TECHNOLOGY IS THERAPEUTICALLY EFFECTIVE

In classic hemophilia, Factor VIII (antihemophilic factor) of the intrinsic pathway is missing or has a markedly reduced activity. VIII stimulates the activation of X by IX_a, another serine protease in the cascade (Figure 10-34).

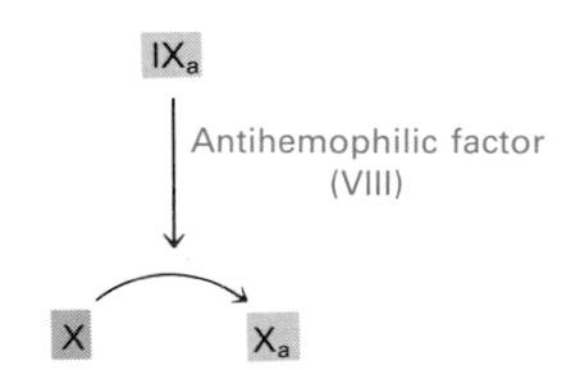

Figure 10-34
Antihemophilic factor (VIII) stimulates the activation of Factor X by IX_a.

The gene for VIII has recently been cloned. The project began with the purification of several milligrams of protein from 25,000 liters of cows' blood, an arduous task necessitated by the low concentration of VIII in plasma (less than 0.05 μg/ml). The amino acid sequences of several peptides from this 330-kd protein were determined; a set of DNA probes could then be synthesized, and the Factor VIII gene was isolated from a human genomic library. A hamster cell that has incorporated this 186-kb gene into its genome secretes large amounts of human Factor VIII.

In the past, hemophiliacs have been treated with transfusions of a concentrated plasma fraction containing VIII, a form of therapy that is expensive and carries the risk of infection by the viruses that cause hepatitis, AIDS, and other diseases. In the near future, purified Factor VIII produced by recombinant DNA technology is likely to replace these plasma concentrates for treating hemophilia. Another benefit of these gene cloning studies is that DNA probes can now be used to diagnose hemophilia in fetuses (as described for sickle-cell anemia, p. 169). These probes have also revealed that classic hemophilia is produced by many kinds of missense and chain-termination mutations in the Factor VIII gene.

THE INTRINSIC AND EXTRINSIC PATHWAYS ARE TRIGGERED BY ABNORMAL SURFACES AND TRAUMA

The intrinsic pathway (Figure 10-27) begins with the activation of Factor XII (Hageman factor) by contact with abnormal surfaces produced by injury. At least two proteins, kininogen and kallikrein, participate in this activation process. XII_a then activates XI, and XI_a in turn activates IX. The last step in the intrinsic pathway is the activation of X by IX_a, a reaction stimulated by antihemophilic Factor (VIII_a). The product of the intrinsic pathway, X_a, then converts prothrombin into thrombin.

The extrinsic pathway seems simpler than the intrinsic one. Trauma to blood vessels releases a lipoprotein called *tissue factor*, which serves as a stimulatory protein. Tissue factor together with VII_a, a serine protease, catalyzes the activation of Factor X. This complex is also a potent activator of Factor IX (Christmas factor) in the intrinsic pathway. Conversely, components of the intrinsic pathway, such as kallikrein, stimulate the extrinsic pathway. Thus, clotting is mediated by the intimate interplay of these pathways.

The roles of clotting factors are summarized in Table 10-2. Seven of them—kallikrein, XII_a, XI_a, IX_a, VII_a, X_a, and thrombin—are *serine proteases*. Four of them—prothrombin, VII, IX, and X—require vitamin K for their synthesis because they contain *γ-carboxyglutamate domains*. These functional modules are encoded by distinct exons. It seems likely that clotting factors evolved by duplication and divergence of an ancestral serine protease gene. Exons encoding domains that target proteins to specific sites (e.g., platelet membranes) became linked to a serine protease exon to form a novel multifunctional protein. Analyses of the sequence and domain structure of clotting proteins strongly support the hypothesis that exon shuffling played a key role in generating proteins with new properties.

Table 10-2
Blood coagulation factors

Factor	*Pathway*	*Function of active form*
Kallikrein	Intrinsic	} Activate XII
Kinin	Intrinsic	
Hageman factor (XII)	Intrinsic	Activates XI
Plasma thromboplastin antecedent (PTA) (XI)	Intrinsic	Activates IX
Christmas factor (IX)	Intrinsic	} Both required to activate X
Antihemophilic factor (VIII)	Intrinsic	
Tissue factor	Extrinsic	} Both required to activate X
Proconvertin (VII)	Extrinsic	
Stuart factor (X)	Common	Activates II
Accelerin (V)	Common	Stimulates activation of II
Prothrombin (II)	Common	Activates fibrinogen (I)
Fibrinogen (I)	Common	FORMS FIBRIN CLOT
Fibrin-stabilizing factor (XIII)	Common	Stabilizes clot by cross-linking fibrin

THROMBIN IS IRREVERSIBLY INHIBITED BY ANTITHROMBIN III

It is evident that the clotting process must be precisely regulated. There is a fine line between hemorrhage and thrombosis. Clotting must occur rapidly yet remain confined to the area of injury. What are the mechanisms that normally limit clot formation to the site of injury? The lability of clotting factors contributes significantly to the control of clotting. The activated clotting factors have a short lifetime because they are diluted by blood flow, removed by the liver, and are degraded by proteases. For example, the stimulatory proteins V_a and VIII_a are digested by protein C, a protease that is switched on by the action of thrombin. Thus, thrombin triggers the deactivation of the clotting cascade in addition to catalyzing the formation of fibrin.

Specific inhibitors of clotting factors are also critical in the termination of clotting. The most important one is *antithrombin III*, a plasma protein that inactivates thrombin by forming an irreversible complex with it. Antithrombin III resembles α_1-antitrypsin except that it inhibits thrombin much more strongly than it inhibits elastase. Antithrombin III also blocks other serine proteases in the clotting cascade, namely, XII_a, XI_a, IX_a, and X_a. The inhibitory action of antithrombin III is enhanced by *heparin*, a negatively charged polysaccharide (p. 276) found in mast cells near the walls of blood vessels and on the surfaces of endothelial cells. Heparin acts as an *anticoagulant* by increasing the rate of formation of irreversible complexes between antithrombin III and the serine protease clotting factors. Antitrypsin and antithrombin are *serpins*, a family of *ser*ine *p*rotease *in*hibitors.

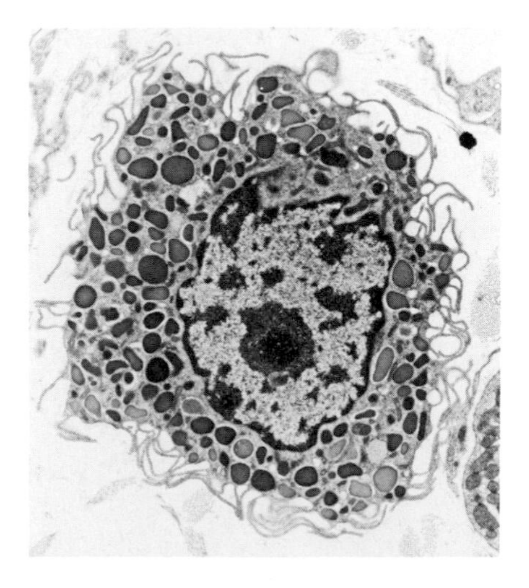

Figure 10-35
Electron micrograph of a mast cell. Heparin and other molecules in the dense granules are released into the extracellular space when the cell is triggered to secrete. [Courtesy of Lynne Mercer.]

The amount of thrombin and antithrombin in blood are precisely balanced to achieve rapid clotting that is limited to the site of injury. This fine balance was disturbed in a fourteen-year-old boy who died of a bleeding disorder because of a mutation in his α_1-antitrypsin, which normally inhibits elastase (p. 247). Methionine 358 in the binding pocket for elastase was replaced by arginine. *This change of a single amino acid residue dramatically altered the specificity of this inhibitory protein. It became an antithrombin because of the lure of arginine 358.* Consequently, his thrombin activity was reduced. α_1-Antitrypsin activity normally increases markedly following injury to counteract excess elastase arising from stimulated neutrophils. During one of these episodes, this patient's thrombin activity dropped to such a low level that a fatal hemorrhage ensued. We see here a striking example of how a change of a single residue in a protein can dramatically alter specificity, an intimation of what probably occurred many times in evolution. This bleeding disorder further illustrates the critical importance of having the right amount of a protease inhibitor.

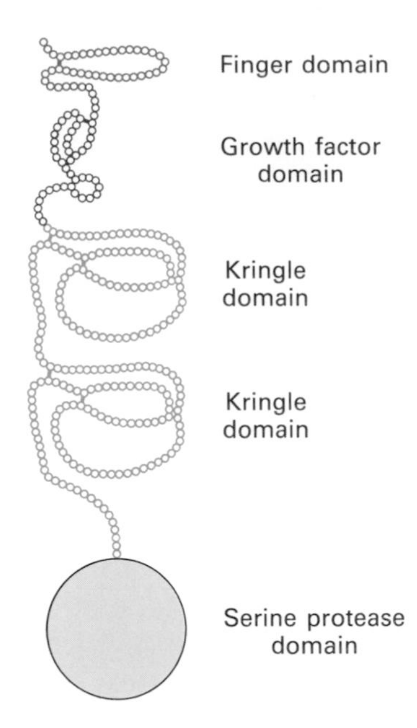

Figure 10-36
Modular structure of tissue-type plasminogen activator (TPA). [After B. A. McMullen and K. Fujikawa. *J. Biol. Chem.* 260(1985):5333.]

FIBRIN CLOTS ARE LYSED BY PLASMIN

Clots are not permanent structures. On the contrary, they are designed to be dissolved when the structural integrity of damaged areas is restored. Fibrin is split by *plasmin*, a serine protease that hydrolyzes peptide bonds in the triple-stranded connector rod regions. We can now see a reason for the porous character of fibrin (see Figure 10-29). The monomer units in this array interact only through the globular regions, which makes the rest of the structure quite open. Plasmin molecules can diffuse through aqueous channels to cut the accessible connector rods.

Plasmin is formed by proteolytic activation of *plasminogen*, an inactive precursor. This conversion is carried out by *tissue-type plasminogen activator* (TPA), a 72-kd protein made up of several types of domains that are found in other proteins: a finger module (found in fibronectin, Chapter 11), a growth factor module (found in proteins that stimulate cell proliferation, Chapter 38), two kringle modules (triply disulfide-bonded units having the shape of a classic Scandinavian pastry), and a serine protease module (Figure 10-36). The kringle modules bind TPA to fibrin clots, where it activates adhering plasminogen. In contrast, free plasminogen is activated very slowly by TPA. The gene for TPA has been cloned and expressed in cultured mammalian cells. Large quantities of fully active TPA produced by recombinant DNA methods are now available. Clinical trials have shown that TPA administered intravenously within an hour of the formation of a blood clot in a coronary artery markedly increases the likelihood of surviving a heart attack.

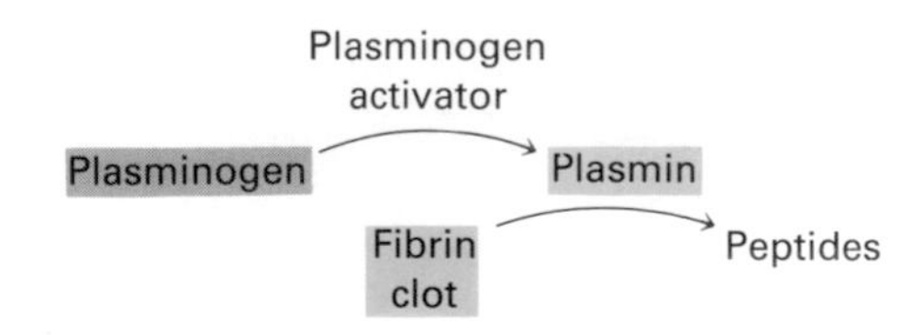

Figure 10-37
Plasmin, formed from plasminogen by the proteolytic action of plasminogen activator, lyses fibrin clots.

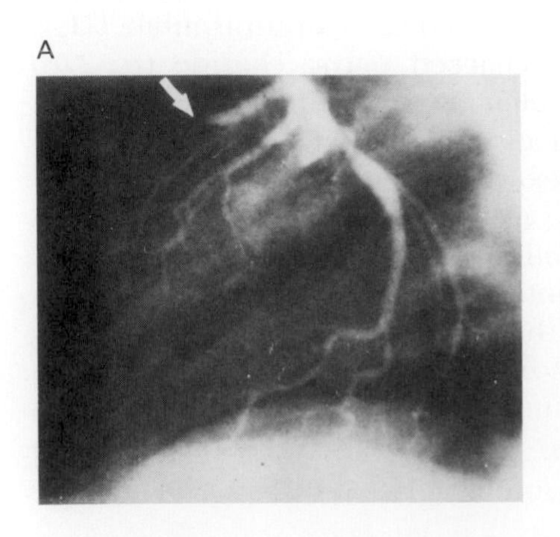

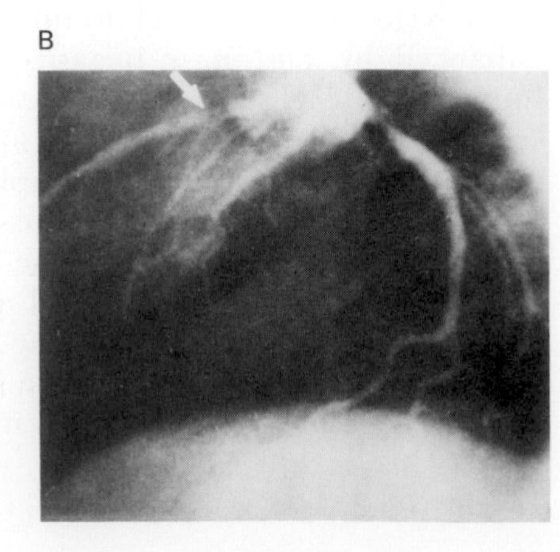

Figure 10-38
Tissue-type plasminogen activator leads to the dissolution of blood clots. X-ray of blood vessels in the heart (A) before and (B) three hours after the administration of TPA. The position of the clot is marked by the arrow in (A). [After F. Van de Werf, P. A. Ludbrook, S. R. Bergmann, A. J. Tiefenbrunn, K. A. A. Fox, H. de Geest, M. Verstraete, D. Collen, and B. E. Sobel. *New Engl. J. Med.* 310(1984):609.]

SUMMARY

The activities of enzymes are controlled by allosteric interactions, proteolytic activation, inhibitory proteins, stimulatory proteins, and reversible covalent modification. Aspartate transcarbamoylase (ATCase), one of the best-understood allosteric enzymes, catalyzes the synthesis of *N*-carbamoyl-L-aspartate, the first intermediate in the synthesis of pyrimidines. ATCase is feedback-inhibited by CTP, the final product of the pathway. This inhibition is reversed by ATP. ATCase consists of separable catalytic (c_3) subunits (which bind the substrates) and regulatory (r_2) subunits (which bind CTP and ATP). The inhibitory effect of CTP, the stimulatory action of ATP, and the cooperative binding of substrates are mediated by large changes in quaternary structure. On binding substrates, the c_3 subunits of the c_6r_6 enzyme move apart and reorient themselves. This allosteric transition is highly concerted, as predicted by the Monod-Changeux-Wyman (MCW) model, which postulates that all of the subunits of an allosteric protein molecule are either in the T (low affinity) or R (high affinity) state. The subunit interfaces are designed to switch between alternative conformations, as in hemoglobin.

The activation of an enzyme by proteolytic cleavage of one or a few peptide bonds is a recurring control mechanism in biological systems. The inactive precursor is called a zymogen (or a proenzyme). Trypsin converts chymotrypsinogen, a zymogen, into active chymotrypsin by hydrolyzing the peptide bond between residues 15 and 16. The newly formed amino-terminal group of isoleucine 16 turns inward and forms an electrostatic bond with aspartate 194 in the interior. This interaction triggers a series of local conformational changes that result in the creation of a pocket for the binding of the aromatic (or bulky nonpolar) side chain of the substrate. In addition, an oxyanion hole that stabilizes the transition state is formed. The activation of trypsinogen by trypsin or enteropeptidase leads to the formation of a binding pocket for the substrate. A molecule of pepsinogen activates itself in acidic media by hydrolyzing a specific peptide bond.

Zymogen activations play a major role in mediating blood clotting. A striking feature of the clotting process is that it occurs by a cascade of zymogen conversions, in which the activated form of one clotting factor catalyzes the activation of the next precursor. Many of the activated clotting factors are serine proteases. Clotting involves the interplay of two reaction sequences, called the intrinsic and extrinsic pathways. They converge on a final common pathway that results in the formation of a fibrin clot. Fibrinogen, a highly soluble molecule in the plasma, is converted by thrombin into fibrin by the hydrolysis of four arginine-glycine bonds. Two A and two B fibrinopeptides are released. The resulting fibrin monomer spontaneously forms long, insoluble fibers called fibrin. The fibrin clot is strengthened by the formation of covalent cross-links as a result of transamidation reactions between specific glutamine and lysine side chains. It seems likely that many clotting proteins evolved by the joining of a serine protease domain to other functional modules. For example, several clotting proteins have a calcium-binding domain containing γ-carboxyglutamate residues formed after translation in a reaction that depends on vitamin K. Zymogen activation is also essential in the lysis of clots. Plasminogen is converted to plasmin, a serine protease that cleaves fibrin, by tissue-type plasminogen activator.

The action of irreversible protein inhibitors is exemplified by α_1-antitrypsin, a plasma protein that inhibits elastase released by neutrophils. A deficiency of this inhibitor leads to the destruction of alveoli in the lung and emphysema. Blood clotting is held in check by antithrombin III, an inhibitor of thrombin and several other serine proteases in the clotting cascade. Antihemophilic factor (Factor VIII), the missing component in hemophilia, illustrates the action of stimulatory proteins. The production of Factor VIII and tissue-type plasminogen activator by recombinant DNA methods demonstrates the power of bringing protein chemistry and molecular genetics together to solve challenging medical problems.

SELECTED READINGS

Where to begin

Kantrowitz, E. R., Pastra-Landis, S. C., and Lipscomb, W. N., 1980. *E. coli* aspartate transcarbamylase. *Trends Biochem. Sci.* 5:124–128 and 150–153.

Neurath, H., 1986. The versatility of proteolytic enzymes. *J. Cell. Biochem.* 32:35–49.

Doolittle, R. F., 1981. Fibrinogen and fibrin. *Sci. Amer.* 245(12):126–135.

Lawn, R. M., and Vehar, G. A., 1986. The molecular genetics of hemophilia. *Sci. Amer.* 254(3):48–65.

Aspartate transcarbamoylase

Gerhart, J. C., 1970. A discussion of the regulatory properties of aspartate transcarbamylase from *Escherichia coli. Curr. Top. Cell Regul.* 2:275–325.

Jacobson, G. R., and Stark, G. R., 1973. Aspartate transcarbamylases. *In* Boyer, P. D., (ed.), *The Enzymes* (3rd ed.), vol. 9, pp. 225–308. Academic Press.

Schachman, H. K., 1974. Anatomy and physiology of a regulatory enzyme: aspartate transcarbamylase. *Harvey Lectures* 68:67–113.

Lahue, R. S., and Schachman, H. K., 1986. Communication between polypeptide chains in aspartate transcarbamoylase. Conformational changes at the active sites of unliganded chains resulting from ligand binding to other chains. *J. Biol. Chem.* 261:3079–3084.

Krause, K. L., Volz, K. W., and Lipscomb, W. N., 1985. Structure at 2.9-Å resolution of aspartate carbamoyltransferase complexed with the bisubstrate analogue *N*-(phosphonacetyl)-L-aspartate. *Proc. Nat. Acad. Sci.* 82:1643–1647.

Volz, K. W., Krause, K. L., and Lipscomb, W. N., 1986. The binding of *N*-(phosphonacetyl)-L-aspartate to aspartate carbamoyltransferase of *Eschericia coli. Biochem. Biophys. Res. Comm.* 136:822–826.

Allosteric models and experimental tests

Monod, J., Wyman, J., and Changeux, J.-P., 1965. On the nature of allosteric transitions: a plausible model. *J. Mol. Biol.* 12:88–118. [Presentation of the concerted model.]

Koshland, D. L., Jr., Nemethy, G., and Filmer, D., 1966. Comparison of experimental binding data and theoretical models in proteins containing subunits. *Biochemistry* 5:365–385. [Presentation of the sequential model for allosteric transitions.]

Foote, J., and Schachman, H. K., 1985. Homotropic effects in aspartate transcarbamylase. What happens when the enzyme binds a single molecule of the bisubstrate analog *N*-phosphonacetyl-L-aspartate? *J. Mol. Biol.* 186:175–184.

Imai, K., 1983. The Monod-Wyman-Changeux allosteric model describes haemoglobin oxygenation with only one adjustable parameter. *J. Mol. Biol.* 167:741–749.

Zymogen activation

Bode, W., and Huber, R., 1986. Crystal structure of pancreatic serine endopeptidases. *In* Desnuelle, P., Sjostrom, H., and Noren, O., (eds.), *Molecular and Cellular Basis of Digestion*, pp. 213–234. Elsevier. [Excellent presentation of the three-dimensional structure and activation mechanism.]

Huber, R., and Bode, W., 1978. Structural basis of the activation and action of trypsin. *Acc. Chem. Res.* 11:114–122.

Stroud, R. M., Kossiakoff, A. A., and Chambers, J. L., 1977. Mechanism of zymogen activation. *Ann. Rev. Biophys. Bioeng.* 6:177–193.

James, M. N. G., and Sielecki, A. R., 1986. Molecular structure of an aspartic proteinase zymogen, porcine pepsinogen, at 1.8 Å resolution. *Nature* 319:33–38.

Blood clotting cascade

Davie, E. W., 1986. Introduction to the blood coagulation cascade and the cloning of blood coagulation factors. *J. Protein Chem.* 5:247–253.

McKee, P. A., 1983. Hemostasis and disorders of blood coagulation. *In* Stanbury, J. B., Wyngaarden, J. B., Fredrickson, D. S., Goldstein, J. L., and Brown, M. S., (eds.), *The Metabolic Basis of Inherited Disease* (5th ed.), pp. 1531–1560. McGraw-Hill.

Gitschier, J., Wood, W. I., Shuman, M. A., and Lawn, R. M., 1986. Identification of a missense mutation in the factor VIII gene of a mild hemophiliac. *Science* 232:1415–1416.

Patthy, L., 1985. Evolution of the proteases of blood coagulation and fibrinolysis by assembly from modules. *Cell* 41:657–663.

Doolittle, R. F., 1984. Fibrinogen and fibrin. *Ann. Rev. Biochem.* 53:195–229.

Weisel, J. W., Stauffacher, C. V., Bullitt, E., Cohen, C., 1985. A model for fibrinogen: domains and sequences. *Science* 230:1386–1391.

Park, C. H., and Tulinsky, A., 1986. Three-dimensional structure of the Kringle sequence: structure of prothrombin fragment 1. *Biochemistry* 25:3977–3982.

Suttie, J. W., 1985. Vitamin K-dependent carboxylase. *Ann. Rev. Biochem.* 54:459–477.

Protease inhibitors and plasmin

Carrell, R., and Travis, J., 1985. α_1-Antitrypsin and the serpins: variation and countervariation. *Trends Biochem. Sci.* 10:20–24.

Travis, J., and Salvesen, G. S., 1983. Human plasma proteinase inhibitors. *Ann. Rev. Biochem.* 52:655–709.

Gadek, J. E., and Crystal, R. G., 1983. α_1-Antitrypsin deficiency. *In* Stanbury, J. B., Wyngaarden, J. B., Fredrickson, D. S., Goldstein, J. L., and Brown, M. S., (eds.), *The Metabolic Basis of Inherited Disease* (5th ed.), pp. 1450–1467. McGraw-Hill.

Carp, H., Miller, F., Hoidal, J. R., and Janoff, A., 1982. Potential mechanism of emphysema: α_1-proteinase inhibitor recovered from lungs of cigarette smokers contains oxidized methionine and has decreased elastase inhibitory capacity. *Proc. Nat. Acad. Sci.* 79:2041–2045.

Owen, M. C., Brennan, S. O., Lewis, J. H., and Carrell, R. W., 1983. Mutation of antitrypsin to antithrombin. *New Engl. J. Med.* 309:694–698. [Case report and molecular analysis of a methionine-to-arginine mutation in α_1-antiproteinase that led to a fatal bleeding disorder.]

PROBLEMS

1. A histidine residue in the active site of aspartate transcarbamoylase is thought to be important in stabilizing the transition state of the bound substrates. Predict the pH dependence of the catalytic rate, assuming that this interaction is essential and dominates the pH-activity profile of the enzyme.

2. A substrate binds 100 times as tightly to the R state of an allosteric enzyme as to its T state. Assume that the MCW model applies to this enzyme.
   (a) By what factor does the binding of one substrate molecule per enzyme molecule alter the ratio of the concentrations of enzyme molecules in the R and T states?
   (b) Suppose that L, the ratio of [T] to [R] in the absence of substrate, is 10^6 and that the enzyme contains four binding sites for substrate. What is the ratio of enzyme molecules in the R state to that in the T state in the presence of saturating amounts of substrate, assuming that the concerted model is obeyed?

3. When very low concentrations of pepsinogen are added to acidic media, how does the half-time for activation depend on zymogen concentration?

4. Suppose that you have just examined a young boy with a bleeding disorder highly suggestive of classic hemophilia (Factor VIII deficiency). Because of the late hour, the laboratory that carries out specialized coagulation assays is closed. However, you happen to have a sample of blood from a classic hemophiliac you admitted to the hospital an hour earlier. What is the simplest and most rapid test you can perform to determine whether your present patient is also deficient in Factor VIII activity?

5. The synthesis of Factor X, like that of prothrombin, requires vitamin K. Factor X also contains γ-carboxyglutamate residues in its amino-terminal region. However, activated Factor X, in contrast with thrombin, retains this region of the molecule. What is a likely functional consequence of this difference between the two activated species?

6. Antithrombin III forms an irreversible complex with thrombin but not with prothrombin. What is the most likely reason for this difference in reactivity?

7. Each of the three types of chains of fibrin contains repeating heptapeptide units (*abcdefg*) in which residues *a* and *d* are hydrophobic. Propose a reason for this regularity.

8. A drug company has decided to prepare by recombinant DNA methods a modified α_1-antitrypsin that will be more resistant to oxidation than is the naturally occurring inhibitor. Which single amino acid substitution would you recommend to them?

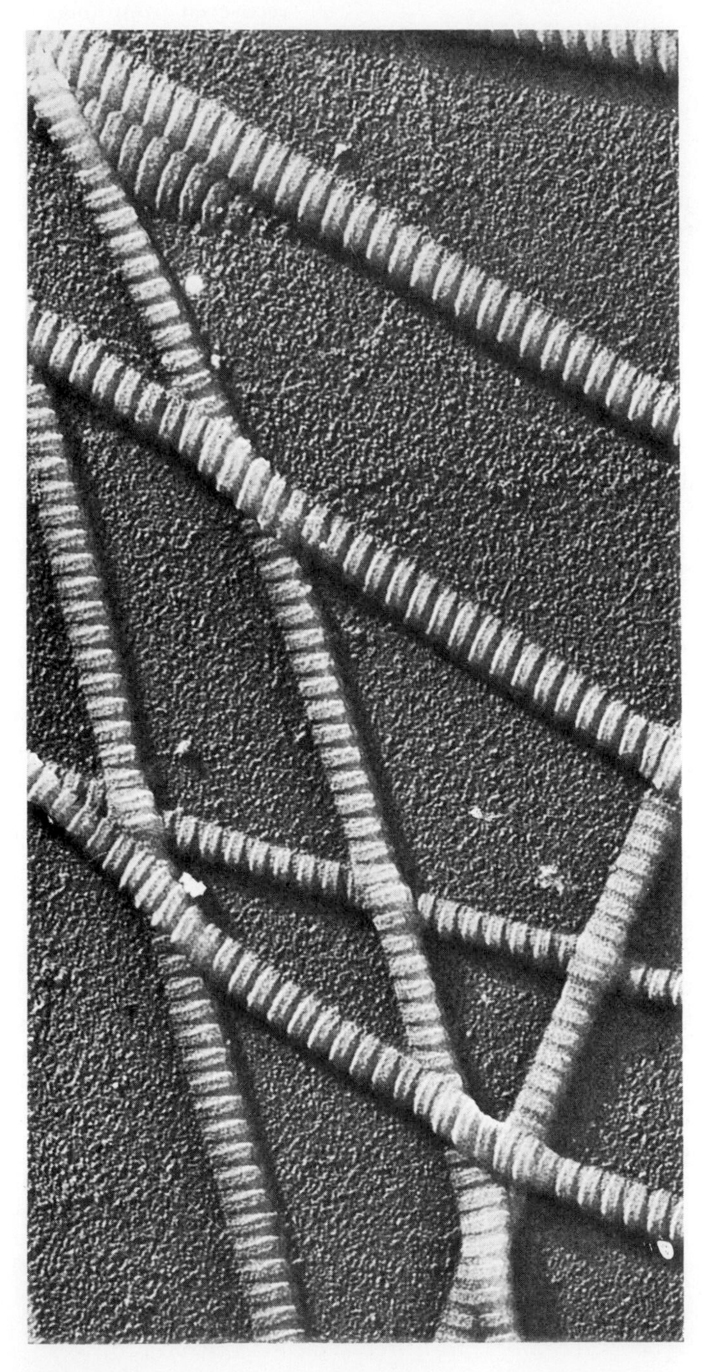

Figure 11-1
Electron micrograph of intact collagen fibrils obtained from skin. The preparation was shadowed with chromium. The period along the fiber axis is 640 Å. [Courtesy of Dr. Jerome Gross.]

CHAPTER 11

Connective-Tissue Proteins

The capacity of proteins for specific binding and catalysis was discussed in the preceding chapters. We turn here to another function of proteins—their structural role. Our focus will be on *collagen*, a family of fibrous proteins present in all multicellular organisms. *Collagen is distinctive in forming insoluble fibers that have a high tensile strength* (Figure 11-1). It is the most abundant protein in mammals, constituting a quarter of their total weight. Collagen is the major fibrous element of skin, bone, tendon, cartilage, blood vessels, and teeth. It is present to some extent in nearly all organs and serves to hold cells together in discrete units. Its basic structural motif is modified to meet the specialized needs of tissues as diverse as bone and cornea. Many variations on a common theme have been characterized. In addition to its structural role in mature tissue, collagen has a directive role in developing tissue.

Collagen—
Derived from the Greek words meaning *to produce glue.*

Two other kinds of connective-tissue proteins will be presented in this chapter: *elastin*, a rubberlike protein in elastic fibers, and *proteoglycans*, very large carbohydrate-rich proteins that form much of the matrix between cells. We shall also consider *fibronectin*, a cell-surface protein that enables cells to interact with the extracellular matrix. Fibronectin plays key roles in cell migration, cell adhesion, and wound healing.

THE BASIC STRUCTURAL UNIT OF COLLAGEN CONSISTS OF THREE CHAINS

The chemical characterization of collagen was at an impasse for many years because of the insolubility of collagen fibers. The breakthrough came when it was found that collagen from the tissues of young animals

Table 11-1
Types of collagen

Type	Composition	Distribution
I	$[\alpha 1(I)]_2\alpha 2$	Skin, tendon, bone, cornea
II	$[\alpha 1(II)]_3$	Cartilage, intervertebral disc, vitreous body
III	$[\alpha 1(III)]_3$	Fetal skin, cardiovascular system, reticular fibers
IV	$[\alpha 1(IV)]_2\alpha 2(IV)$	Basement membrane
V	$[\alpha 1(V)]_2\alpha 2(V)$	Placenta, skin

can be extracted in soluble form because it is not yet extensively cross-linked. The absence of covalent cross-links in immature collagen makes it feasible to extract the basic structural unit, called *tropocollagen*, in intact form. Tropocollagen has a mass of about 285 kd and consists of *three polypeptide chains* of the same size. The composition of the chains depends on the type of collagen (Table 11-1). Type I collagen, the most prevalent species, consists of two chains of one kind, termed $\alpha 1(I)$, and one of another, termed $\alpha 2(I)$. Some types of collagen, such as type II, contain three identical chains. Each of the three strands in collagen consists of about a thousand amino acid residues.

4-Hydroxyproline (Hyp)

5-Hydroxylysine (Hyl)

COLLAGEN HAS AN UNUSUAL AMINO ACID COMPOSITION AND SEQUENCE

The proportion of *glycine* residues in all collagen molecules is nearly one-third, which is unusually high for a protein. In hemoglobin, for example, the glycine content is 5%. *Proline* also is present to a much greater extent in collagen than in most other proteins. Furthermore, collagen contains two amino acids that are found in very few other proteins, namely *4-hydroxyproline* and *5-hydroxylysine*. The amino acid sequence of collagen is remarkably regular: *nearly every third residue is glycine*. Moreover, the sequence glycine-proline-hydroxyproline recurs frequently (Figure 11-2). In contrast, globular proteins rarely exhibit regularities in their amino acid sequences. Two other proteins with regularly repeating sequences are silk fibroin and elastin.

13 -Gly-Pro-Met-Gly-Pro-Ser-Gly-Pro-Arg-
22 -Gly-Leu-Hyp-Gly-Pro-Hyp-Gly-Ala-Hyp-
31 -Gly-Pro-Gln-Gly-Phe-Gln-Gly-Pro-Hyp-
40 -Gly-Glu-Hyp-Gly-Glu-Hyp-Gly-Ala-Ser-
49 -Gly-Pro-Met-Gly-Pro-Arg-Gly-Pro-Hyp-
58 -Gly-Pro-Hyp-Gly-Lys-Asn-Gly-Asp-Asp-

Figure 11-2
Amino acid sequence of part of the $\alpha 1(I)$ chain of collagen. Every third residue is glycine in a region spanning more than a thousand residues.

PROLINE AND LYSINE RESIDUES OF NASCENT CHAINS BECOME HYDROXYLATED

Hydroxyproline and hydroxylysine are not incorporated in the polypeptide chains of collagen by the usual mechanisms of protein synthesis. If ^{14}C-labeled hydroxyproline is administered to a rat, none of the radioactivity appears in the collagen that is synthesized. In contrast, the hydroxyproline in the new collagen is radioactive if ^{14}C-proline is given. Thus, proline is a precursor of the hydroxyproline residues in collagen, whereas exogenous hydroxyproline is not.

Certain proline residues in collagen are converted into hydroxyproline by *prolyl hydroxylase*, an enzyme with a ferrous ion at its active site. The oxygen atom that becomes attached to C-4 of proline comes from O_2. The other oxygen atom of O_2 emerges in succinate, which is formed from α-ketoglutarate, an obligatory substrate in this reaction

(Figure 11-3). Hence, prolyl hydroxylase is a *dioxygenase*. A noteworthy feature of this hydroxylation reaction is the need for a *reducing agent*, such as *ascorbate* (p. 268), to maintain the iron atom in the ferrous state.

This hydroxylation reaction is highly specific. Free proline is not a substrate. Rather, hydroxylation takes place at specific sites on relatively large polypeptide chains before they become helical. *Proline can be hydroxylated at C-4 only if it is situated on the amino side of a glycine residue.* In addition, a few proline residues are hydroxylated at C-3 by a different enzyme. 3-Hydroxyproline, in contrast with 4-hydroxyproline, is located on the carboxyl side of glycine residues.

A small proportion of the lysine residues in collagen become hydroxylated at C-5 by the action of *lysyl hydroxylase*. As in the hydroxylation of proline, molecular oxygen, α-ketoglutarate, and ascorbate are required. Lysines modified in this way are always on the amino side of glycine residues.

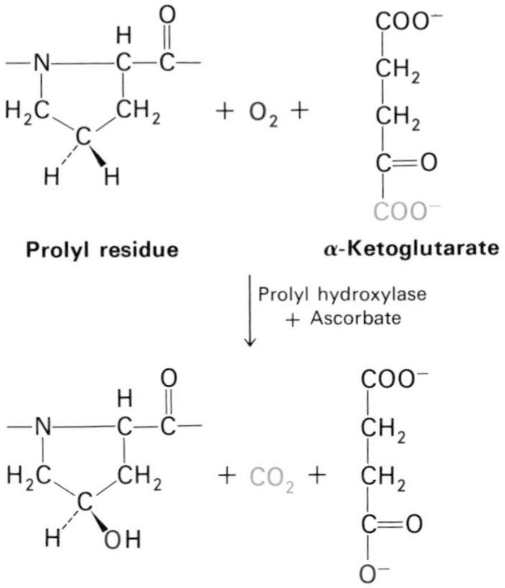

Figure 11-3
Hydroxylation of a proline residue at C-4 by the action of prolyl hydroxylase, an enzyme that activates molecular oxygen.

SUGARS ARE ADDED TO HYDROXYLYSINE RESIDUES

The hydroxylysine residues of collagen are covalently bound to carbohydrate units. A disaccharide of glucose and galactose is commonly found (Figure 11-4). These sugars are attached by the sequential action of *galactosyl transferase* and *glucosyl transferase*. These glycosylating enzymes are specific for hydroxylysine residues in nascent collagen before it becomes helical. The number of carbohydrate units per tropocollagen depends on the tissue. Collagens in fibrils have relatively little carbohydrate, whereas those in sheets are rich in it. For example, tendon tropocollagens (type I) contain 6, whereas those in the lens capsule (type IV) have 110.

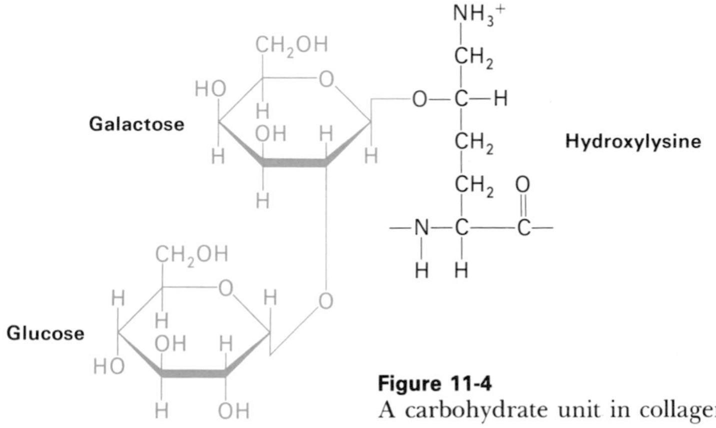

Figure 11-4
A carbohydrate unit in collagen.

COLLAGEN IS A TRIPLE-STRANDED HELICAL ROD

Let us turn to the conformation of the basic structural unit of the type I collagen fiber. Electron microscopic and hydrodynamic studies have shown that *tropocollagen has the shape of a rod 3000 Å long and 15 Å in diameter*. It is one of the longest known proteins. Its elongated character can be appreciated by noting that tropocollagen is sixty times as long as the diameter of chymotrypsin, but its diameter is less than half that of chymotrypsin. Each of the three polypeptide chains is in a helical conformation (Figure 11-5). Furthermore, the three helical strands wind around each other to form a stiff cable. Indeed, the strength of a collagen fiber is remarkable: a load of at least 10 kg is needed to break a fiber 1 mm in diameter.

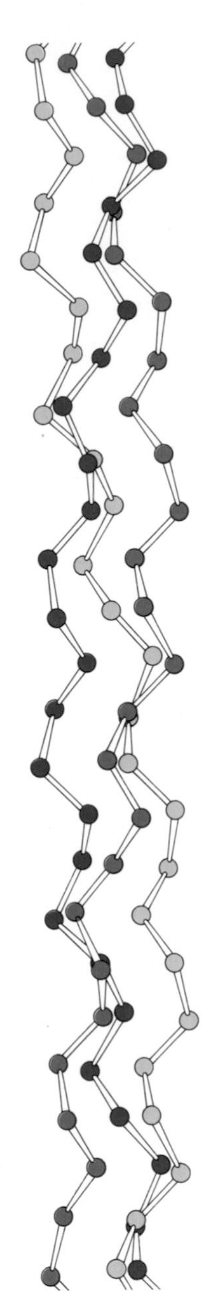

Figure 11-5
Model of the triple-stranded collagen helix. Only the α-carbon atoms are shown.

The helical motif of the individual chains of the triple-stranded collagen cable is nicely illustrated by a model compound, *poly-L-proline*. This synthetic polypeptide has a helical form (Figure 11-6) that is entirely different from the α helix. Hydrogen bonds are absent from the poly-L-proline helix. Instead, *this helix is stabilized by steric repulsion of the pyrrolidone rings of the proline residues*. The pyrrolidone rings keep out of each other's way when the polypeptide chain assumes this helical form (known as the type II *trans* helix), which is much more open than the tightly coiled α helix. The rise per residue along the helix axis of poly-L-proline is 3.12 Å, whereas it is 1.5 Å in the α helix. There are three residues per turn in the poly-L-proline helix.

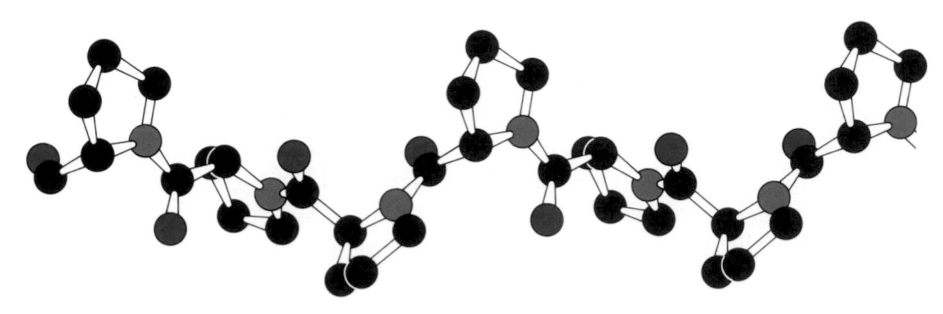

Figure 11-6
Model of a helical form of poly-L-proline (type II *trans* helix). This kind of helix is the basic structural motif of each of the three strands of tropocollagen.

Let us now look at the conformation of a single strand of the triple-stranded collagen helix (Figure 11-7). Each strand has a helical conformation like that shown for poly-L-proline. The three strands wind around each other to form a *superhelical cable* (Figure 11-8). The rise per residue in this superhelix is 2.9 Å and the number of residues per turn is nearly 3.3. *The three strands are hydrogen bonded to each other*. The hydrogen donors are the peptide NH groups of glycine residues, and the hydrogen acceptors are the peptide CO groups of residues on the other chains. The direction of the hydrogen bond is perpendicular to the long axis of the tropocollagen rod. The hydroxyl groups of hydroxyproline residues and bridging water molecules also participate in hydrogen bonding, which stabilizes the triple helix. A space-filling model of the triple-stranded collagen helix is shown in Figure 11-9.

GLYCINE IS CRITICAL BECAUSE IT IS SMALL

We can now understand why glycine occupies every third position in the amino acid sequence of tropocollagen. The inside of the triple-stranded helical cable is very crowded (Figure 11-10). Indeed, *the only residue that can fit in an interior position is glycine*. Because there are three residues per turn of helix, every third residue on each strand must be glycine. The amino acid residue on either side of glycine is located on the outside of the cable, where there is room for the bulky rings of proline and hydroxyproline residues.

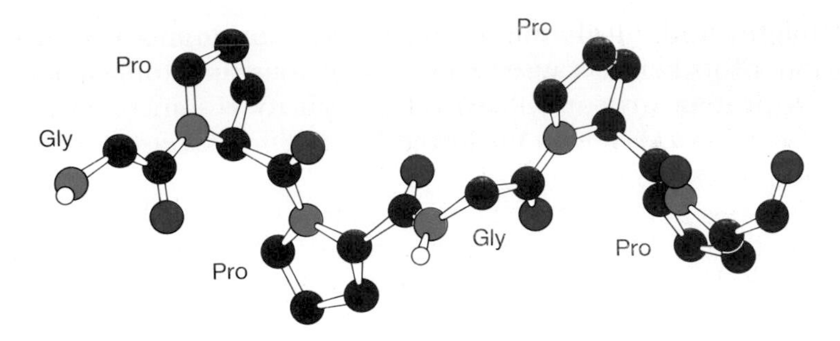

Figure 11-7
Conformation of a single strand of the collagen triple helix. The sequence shown here is Gly-Pro-Pro-Gly-Pro-Pro.

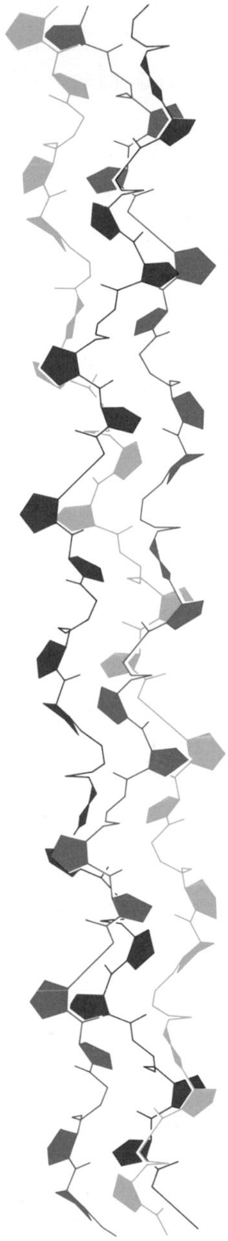

Figure 11-9
A space-filling model of the collagen triple helix. [Courtesy of Dr. Alexander Rich.]

Figure 11-8
Skeletal model of the triple-stranded collagen helix. The repeating sequence shown here is Gly-Pro-Pro.

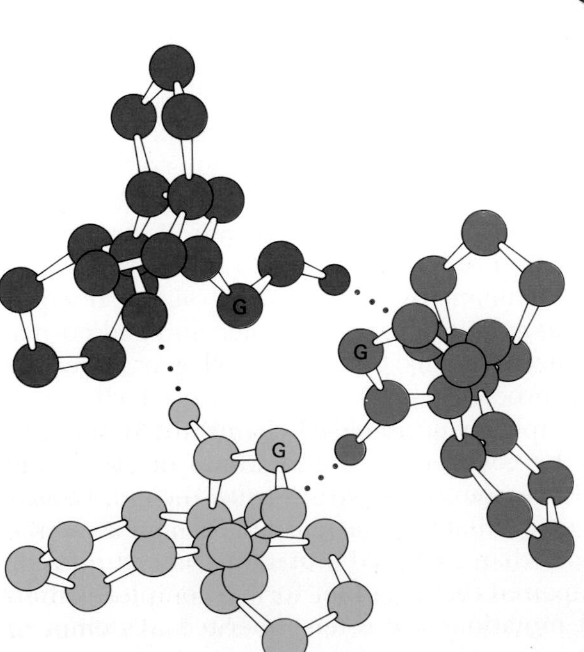

Figure 11-10
Cross section of a model of collagen. Each strand of the triple helix is shown in a different color. Each is hydrogen bonded to the other two strands (• • • denotes a hydrogen bond). The α-carbon atom of a glycine residue in each strand is labeled G. Every third residue must be glycine because there is no space near the helix axis (center) for a larger amino acid residue. Note that the pyrrolidone rings are on the outside.

We might think of glycine as an insignificant amino acid, the least important of its kind because its side chain consists of only a hydrogen atom. In protein structure, however, simplicity is sometimes advantageous. *Glycine is a very important amino acid precisely because it occupies very little space and thereby allows different polypeptide strands to come together.* We have already seen this in myoglobin and hemoglobin, where glycine B6 is invariant because it allows the close approach of the B and E helices (p. 153). In chymotrypsin, the substrate-binding cavity can accommodate a large aromatic group because two of the residues lining the cavity are glycine. In the evolution of cytochrome *c*, an electron carrier, the amino acid residue that has been most conserved is glycine. In collagen, we see once again the critical role of glycine in increasing the range of allowed conformations in the folding of polypeptide chains.

Indeed, *mutation of a single glycine in collagen can be lethal,* as shown by a recent analysis of DNA from an infant afflicted with *osteogenesis imperfecta,* a connective-tissue disorder caused by defective type I collagen. Skeletal deformities arising from multiple fractures due to brittle bones are prominent in this disease. This infant had a normal allele and a mutant allele for the $\alpha 1$(I) chain of collagen. DNA sequencing showed that a single nucleotide was different. The substitution of T for a G changed residue 988 from glycine to cysteine (Figure 11-11). This change disrupted the triple helix near its carboxyl terminus, exposing it to excessive hydroxylation and glycosylation. Consequently, this abnormal collagen was partly unfolded at body temperature and could not form highly ordered fibrillar arrays. Moreover, the secretion of the abnormal chains by the cells that produce them was impaired. We see here another striking example of how a change of a single nucleotide in a genome of 3×10^9 base pairs can produce a fatal disorder.

```
      988
Pro   Gly   Pro  Arg  Gly  Arg  Thr  Gly  Asp  Ala      ]
CCT   GGT   CCT  CGC  GGT  CGC  ACT  GGT  GAT  GCT 3'   ] Normal
       ↓
CCT   TGT   CCT  CGC  GGT  CGC  ACT  GGT  GAT  GCT 3'   ]
Pro   Cys   Pro  Arg  Gly  Arg  Thr  Gly  Asp  Ala      ] Mutant
```

Figure 11-11
Comparison of the DNA sequence and corresponding amino acid sequence of normal and mutant alleles for the pro-$\alpha 1$(I) chain (the precursor of the $\alpha 1$(I) chain) from a patient with osteogenesis imperfecta, a generalized disorder of connective tissue. The mutation of glycine to cysteine caused by substitution of a single base was lethal.

This mutant allele had a deleterious effect because it encoded an aberrant chain that became part of a triple-stranded structure. In classic genetic terms, it was dominant over the normal allele. If equal amounts of normal and mutant $\alpha 1$ chains were formed and had normal affinity for each other, three-fourths of the procollagen molecules would be spoiled because procollagen (I) contains two $\alpha 1$ chains. A mutant allele that led to completely unusable $\alpha 1$ chains would probably be less harmful because the collagen produced would be normal in structure though reduced in amount. *Included mutations*—that is, mutations leading to altered proteins that become part of larger structures—are usually more deleterious than excluded mutations, where the gene product is absent or so impaired that it cannot form a complex with its usual partners. Included mutations are often inherited as dominant traits.

"Nature is nowhere accustomed more openly to display her secret mysteries than in cases where she shows traces of her workings apart from the beaten path; nor is there any better way to advance the proper cause of medicine than to give our minds to the discovery of the usual law of Nature by careful investigation of cases of rarer forms of disease. For it has been found, in almost all things, that what they contain of useful or applicable nature is hardly perceived unless we are deprived of them, or they become deranged in some way."

From a letter written by William Harvey, shortly before his death in 1657. Quoted by Archibald Garrod in his article "The Lessons of Rare Maladies," *Lancet* 1:1055 (1928), and by Victor McKusick in *Heritable Disorders of Connective Tissue,* 4th ed. (Mosby, 1972).

THE STABILITY OF THE COLLAGEN HELIX DEPENDS ON COOPERATIVE INTERACTIONS

If a solution of tropocollagen is heated, large changes in physical properties occur at a characteristic temperature (Figure 11-12). For example, the viscosity of the solution drops sharply, indicating that the molecules have lost their rodlike shape. The altered optical rotatory properties reveal that the helical structure of the individual strands has been destroyed. Thus, thermal motion can overcome the forces that stabilize the triple-stranded helix, yielding a disrupted structure called gelatin, which is a random coil. This structural transition occurs rather abruptly, within a narrow range of temperatures, like the melting of DNA. The abruptness of the transition reveals that the triple-stranded helix is stabilized by *cooperative interactions*. In other words, the helical form is the consequence of *many reinforcing bonds*, each of which is relatively weak. The formation of any one of these stabilizing bonds very much depends on whether adjacent bonds are also made. A zipper is an example of a cooperative structure, as is a DNA molecule.

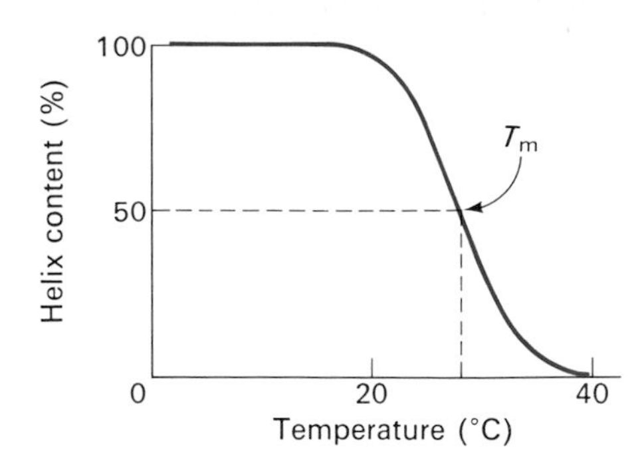

Figure 11-12
Melting curve of solubilized collagen molecules.

The temperature at which half of the helical structure is lost is called the *melting temperature* (T_m). The T_m of tropocollagen is a criterion of the stability of its helical structure. For intact collagen fibers, the comparable index is the *shrinkage temperature* (T_s), at which a fiber shortens markedly. Collagens from different species have different melting temperatures. In fact, the T_m or T_s of a collagen is related to the body temperature of the source species (Table 11-2). Collagens from icefish have the lowest T_m, whereas those from warm-blooded animals have the highest T_m. This difference in thermal stability is correlated with the content of the imino acids (proline and hydroxyproline) in the collagen. *The higher the imino acid content, the more stable the helix.* The imino acid content of collagens increased in the evolution of warm-blooded species from cold-blooded ones.

Table 11-2
Dependence of thermal stability on imino acid content

Source	*Proline plus hydroxyproline (per 1000 residues)*	*Thermal stability (°C)* T_s	T_m	*Body temperature (°C)*
Calf skin	232	65	39	37
Shark skin	191	53	29	24–28
Cod skin	155	40	16	10–14

Studies of the melting temperatures of synthetic polypeptide models of collagen have been sources of insight into the biological significance of the hydroxylation of proline. The T_m of poly-(Pro-Pro-Gly) is 24°C, whereas that of poly-(Pro-Hyp-Gly) is 58°C, which shows that *the triple helix is markedly stabilized by hydroxylation.* The properties of unhydroxylated collagen prepared by incubating tendon cells with α,α'-bipyridyl, an iron chelator that inhibits prolyl hydroxylase, support this conclusion. The unhydroxylated collagen synthesized by these cells does not form a triple helix at 37°C, whereas it rapidly becomes helical when cooled to below 24°C.

DEFECTIVE HYDROXYLATION IS ONE OF THE BIOCHEMICAL LESIONS IN SCURVY

The importance of the hydroxylation of collagen becomes evident in *scurvy*. A vivid description of this disease was given by Jacques Cartier in 1536, when it afflicted his men as they were exploring the Saint Lawrence River:

> Some did lose all their strength, and could not stand on their feet. . . . Others also had all their skins spotted with spots of blood of a purple colour: then did it ascend up to their ankles, knees, thighs, shoulders, arms, and necks. Their mouths became stinking, their gums so rotten, that all the flesh did fall off, even to the roots of the teeth, which did also almost all fall out.

The means of preventing scurvy was succinctly stated by James Lind, a Scottish physician, in 1753:

> Experience indeed sufficiently shows that as greens or fresh vegetables, with ripe fruits, are the best remedies for it, so they prove the most effectual preservatives against it.

Lind urged the inclusion of lemon juice in the diet of sailors. His advice was adopted by the British Navy some forty years later.

Scurvy is caused by a dietary deficiency of ascorbic acid (vitamin C). Primates and guinea pigs have lost the ability to synthesize ascorbic acid and so they must acquire it from their diets. Ascorbic acid, an effective reducing agent (Figure 11-13), maintains prolyl hydroxylase in an active form, probably by keeping its iron atom in the reduced ferrous state. Collagen synthesized in the absence of ascorbic acid is insufficiently hydroxylated and hence has a lower melting temperature. This abnormal collagen cannot properly form fibers and thus causes the skin lesions and blood-vessel fragility that are so prominent in scurvy.

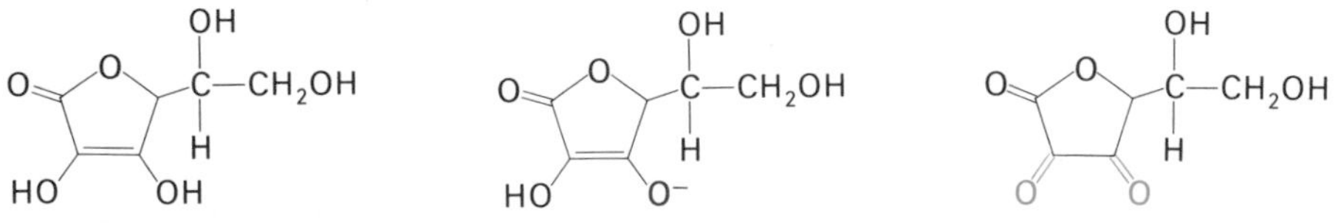

Figure 11-13
Formulas of ascorbic acid (vitamin C) and ascorbate, its ionized form. The pK_a of the acidic hydroxyl group of ascorbic acid is 4.2. Dehydroascorbic acid is the oxidized form of ascorbate.

Ascorbic acid **Ascorbate** **Dehydroascorbic acid**

PROCOLLAGEN IS THE PRECURSOR OF COLLAGEN

The triple-stranded type I collagen helix is rapidly assembled in vivo. In contrast, the process takes days in vitro from a solution of the constituent $\alpha 1$ and $\alpha 2$ chains. Furthermore, the yield is low. Why is there such a disparity between in vitro and in vivo assembly of the tropocollagen helix? The answer is suggested by comparing the refolding of denatured chymotrypsinogen and chymotrypsin. The unfolded zymogen spontaneously refolds into the correct three-dimensional structure, whereas the enzyme does not. The reason is that part of the structure needed to specify the three-dimensional form is missing in chymotrypsin—namely, the two dipeptides that were cleaved in the activation process (p. 244).

It was surmised that the $\alpha 1$ and $\alpha 2$ chains do not spontaneously form the correct tropocollagen structure because part of the necessary information is missing. This is indeed so. *The constituent chains of collagen are synthesized in the form of larger precursors.* The precursor of the $\alpha 1(I)$ chain, called *pro-α1(I)*, has a mass of 140 kd, in contrast with 95 kd for $\alpha 1(I)$. Additional peptides, called *propeptides*, are located at both the amino-terminal and the carboxyl-terminal ends of the pro-$\alpha 1(I)$ chain (Figure 11-14). Similarly, the $\alpha 2(I)$ chain is synthesized as a 140-kd precursor called *pro-α2(I)*. The propeptides of each precursor chain have very different amino acid compositions from the rest of the chain. Propeptides are not rich in glycine, hydroxyproline, and proline. The amino-terminal propeptides of both chains are distinctive in containing intrachain disulfides. Furthermore, *the carboxyl-terminal propeptides are linked by interchain disulfides that are absent from collagen.*

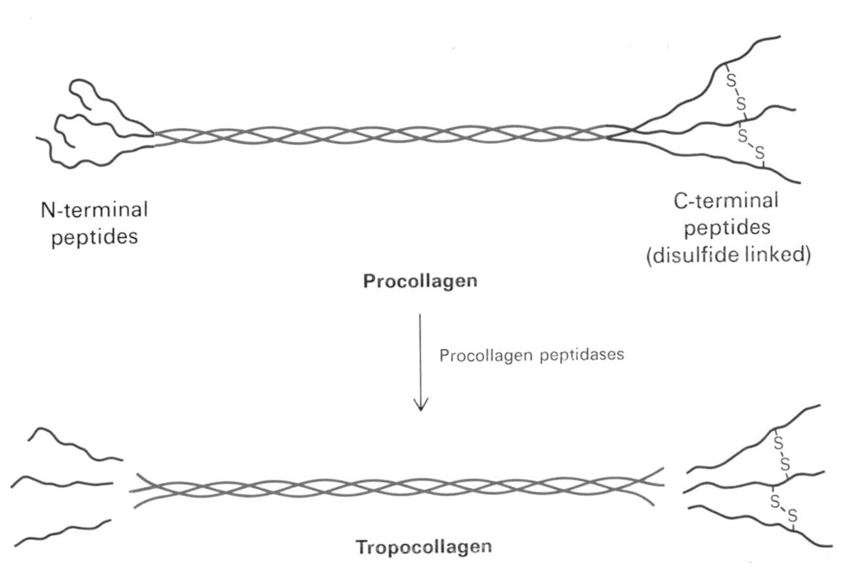

Figure 11-14
Schematic diagram of the conversion of procollagen into collagen by the excision of the amino-terminal peptide (about 15 kd) and the carboxyl-terminal peptide (about 30 kd) from each of the three chains. The carboxyl-terminal regions of the three strands of procollagen are linked by disulfide bonds.

PROPEPTIDES ARE EXCISED OUTSIDE CELLS BY PROCOLLAGEN PEPTIDASES

Procollagen, rather than tropocollagen, is secreted by fibroblasts (Figure 11-15). The pathway of biosynthesis and secretion of exported proteins such as procollagen and the pancreatic zymogens will be discussed in detail in Chapter 31. The propeptides of procollagen are cleaved outside the cell by specific proteases called *procollagen peptidases.* For each type of collagen, there is one for the amino-terminal propeptide and another for the carboxyl-terminal propeptide.

The formation of collagen fibers is analogous to the formation of fibrin fibers. Procollagen is akin to fibrinogen, tropocollagen to fibrin monomer, and the procollagen peptidases to thrombin. In both systems, specific proteolytic cleavages are required for fiber formation. Collagen fiber formation takes place in the extracellular fluid near the cell surface rather than inside the fibroblast because the procollagen peptidases are outside the cell. The propeptides prevent the premature formation of fibers. They probably also facilitate the alignment of the three chains and promote formation of the triple helix. The interchain disulfide bonds may be particularly important in this regard.

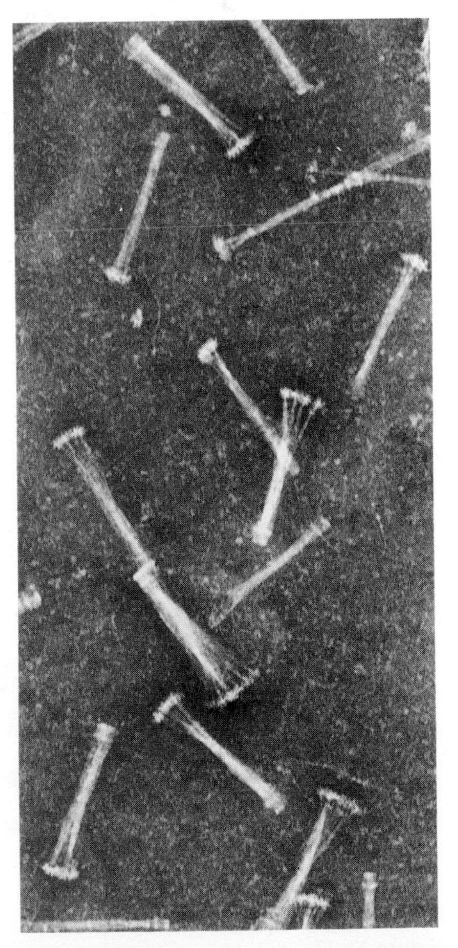

Figure 11-15
Electron micrograph of bundles of procollagen molecules that were secreted into the extracellular medium. [Courtesy of Dr. Jerome Gross.]

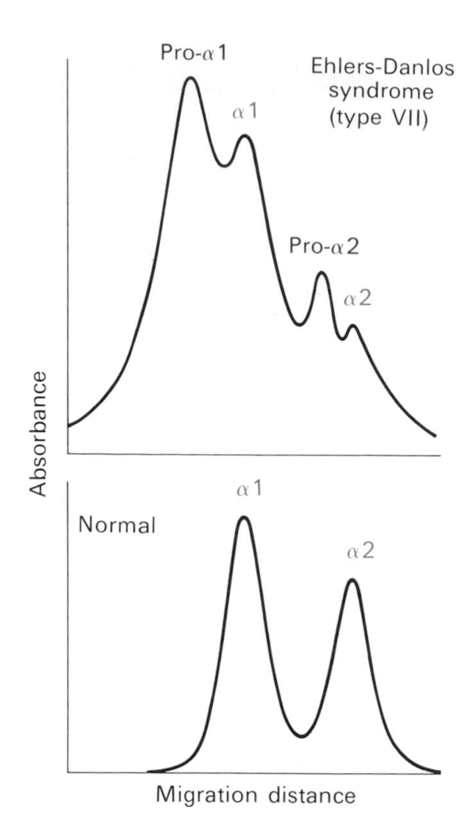

Figure 11-16
Acrylamide gel electrophoresis patterns of skin extracts. Appreciable amounts of procollagen (pro-α1 and pro-α2) are present in acid extracts of skin from patients with type VII Ehlers-Danlos syndrome, whereas none is evident in controls from normal persons. [After J. R. Lichtenstein, G. R. Martin, L. D. Kohn, P. H. Byers, and V. A. McKusick. *Science* 182(1973):299. Copyright 1973 by the American Association for the Advancement of Science.]

Defective removal of propeptides can lead to generalized disorders of connective tissue. For example, patients with the *Ehlers-Danlos syndrome* have stretchable skin, hypermobile joints, and short stature. Patients with one form of this disease have significant amounts of procollagen in extracts of their skin and tendons (Figure 11-16). Furthermore, their fibroblasts have decreased levels of procollagen peptidase. *Dermatosparaxis*, a genetically transmitted (recessive) disease of cattle, is caused by the absence of a procollagen peptidase. The dermis of an afflicted animal is very fragile because it contains disorganized collagen bundles. A significant proportion of its mature type I collagen consists of α1(I) and α2 chains that have retained the amino-terminal propeptides. This disease demonstrates that excision of amino-terminal propeptides is essential for the formation of well-ordered collagen fibers.

COLLAGEN FIBERS ARE STAGGERED ARRAYS OF TROPOCOLLAGEN MOLECULES

Collagen fibers exhibit cross striations every 680 Å (as shown in Figure 11-1), whereas tropocollagen is 3000 Å long. Because the fiber period is smaller than the length of tropocollagen, molecules in adjacent rows cannot be in register. Electron micrographs of stained collagen fibrils indicate that tropocollagens in a row are separated by 400-Å gaps and that adjacent rows are displaced by 680 Å (Figure 11-17). The structure repeats after each five rows. Thus, *the fundamental structural design of a collagen fiber is a staggered array of tropocollagen molecules, an arrangement reminiscent of a musical fugue* (Figures 11-18).

The 400-Å hole between adjacent tropocollagen molecules in a row is important in enabling collagen to become cross-linked after the fiber forms, as will be discussed shortly. This gap may also play a role in *bone formation*. Bone consists of an organic phase, which is nearly entirely collagen, and an inorganic phase, which is calcium phosphate. Specifically, the mineral structure is like that of hydroxylapatite, which has the composition $Ca_{10}(PO_4)_6(OH)_2$. Collagen is required for the deposition of calcium phosphate crystals to form bone. In fact, the initial crystals

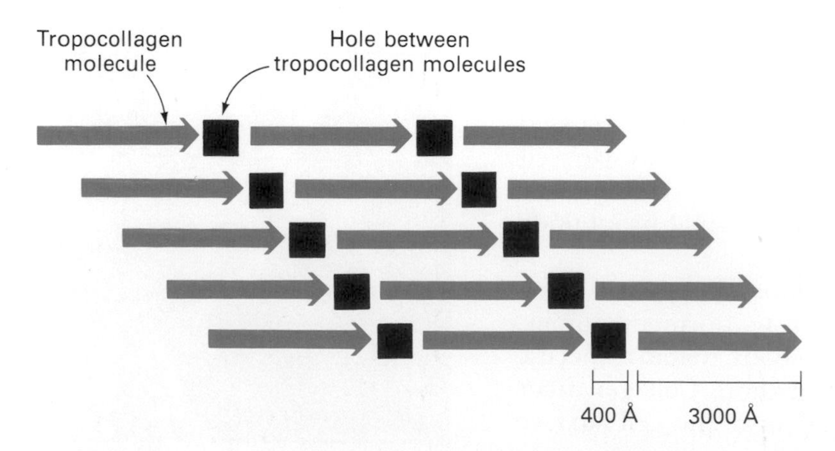

Figure 11-17
A schematic representation of the basic structural design of a collagen fiber. Tropocollagen molecules (shown as blue arrows) form a staggered array. The gaps (represented by red squares) between tropocollagen molecules in a row may be nucleation sites in bone formation.

Figure 11-18
A passage from the Fugue in D Major, Well-tempered Clavier, by J. S. Bach. [From R. Erickson. *The Structure of Music: A Listener's Guide* (Noonday Press, 1955), p. 130.]

are found at intervals of about 680 Å, which is the period of the collagen fibers. It seems likely that the holes between the tropocollagen molecules in a row are *nucleation sites* for calcium deposition.

COLLAGEN FIBERS ARE STRENGTHENED BY CROSS-LINKS

Collagen, like fibrin, is stabilized by the formation of covalent cross-links. Links within a tropocollagen molecule and between different molecules are formed by lysine and hydroxylysine residues. *Lysyl oxidase*, the only enzyme that participates in this process, converts the ϵ-amino group of these residues to an aldehyde. Lysyl oxidase contains a copper ion at its active site. In addition, pyridoxal phosphate (Chapter 21) serves as a prosthetic group in a complex oxidation reaction that utilizes O_2. Lysyl oxidase acts on nascent fibrils outside cells.

Intramolecular cross-links in collagen are derived from lysine side chains. The aldehyde derivatives of two lysine residues undergo an *aldol condensation* (Figure 11-19). In this condensation, the enolate ion derived from one aldehyde adds to the carbonyl group of the other aldehyde. This intramolecular cross-link is formed between lysine residues located in the nonhelical region near the amino terminus.

Lysine residues

Lysyl oxidase

Aldehyde derivatives (Allysine)

Aldol cross-link

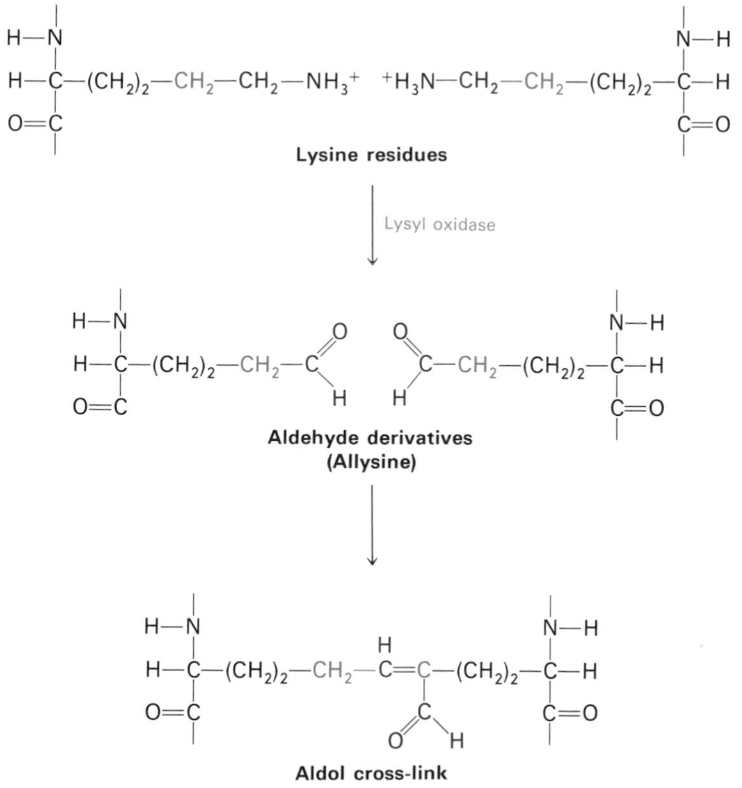

Figure 11-19
Formation of an aldol cross-link from two lysine side chains.

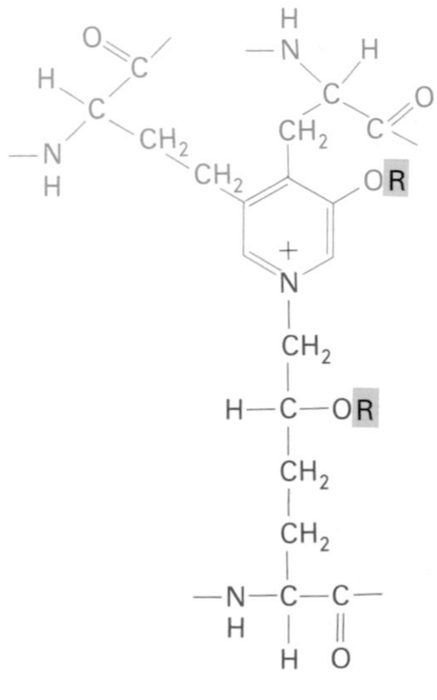

Intermolecular cross-links are formed by the joining of two hydroxylysine residues and one lysine residue. The product is a *hydroxypyridinium cross-link* (Figure 11-20). Four residues in each tropocollagen molecule can participate in these cross-links: a lysine near the amino terminus, a lysine near the carboxyl terminus, and hydroxylysines in helical regions near the ends of the molecule (residues 87 and 930). These

Figure 11-20
Hydroxypyridinium (pyridinoline) cross-link formed by two hydroxylysine residues (shown in blue and red) and a lysine residue (shown in green) in collagen. R is either a hydrogen atom or a carbohydrate unit. Three polypeptide regions are joined by this pyridinium cross-link.

hydroxypyridinium cross-links are formed between residues near the amino terminus of one tropocollagen and the carboxyl terminus of another (Figure 11-21). Each of these contributing residues is located near one of the gaps in a row of tropocollagens. Lysyl oxidase, a 30-kd protein, can fit in this 400-Å hole, so that it can easily initiate cross-linking. However, the subsequent reactions that complete the formation of cross-links are nonenzymatic.

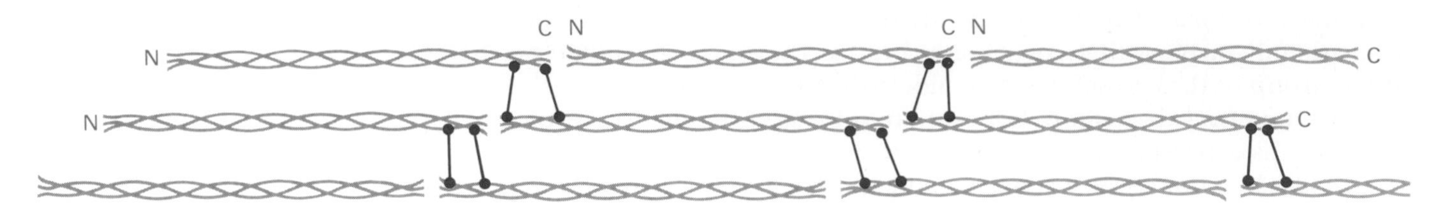

Figure 11-21
Location of cross-links between adjacent tropocollagen molecules in collagen fibers. The amino-terminal region of one molecule is linked to the carboxyl-terminal region of one in an adjacent row.

$N{\equiv}C{-}CH_2CH_2{-}NH_3^+$
β-Aminopropionitrile

The extent and type of cross-linking varies with the physiological function and age of the tissue. The collagen in the Achilles' tendon of mature rats is highly cross-linked, whereas that of the flexible tail tendon is much less so. The importance of cross-linking in conferring mechanical strength on collagen fibers is evident in *lathyrism,* a disease of animals caused by the ingestion of seeds of *Lathyrus odoratus,* the sweet pea. The toxic agent is *β-aminopropionitrile,* which inhibits the transformation of lysyl side chains into aldehydes by forming a tight complex with lysyl oxidase. The collagen in these animals is extremely fragile. Humans with type V Ehlers-Danlos syndrome have hypermobile joints and very extensible skin because of a deficiency of lysyl oxidase.

The steps leading to mature collagen fibers are summarized in Figure 11-22. The nascent polypeptide chains of collagen are hydroxylated and glycosylated before the triple helix forms. Bundles of procollagen are secreted into the extracellular space, where their propeptides are removed by procollagen peptidases. The resulting tropocollagen

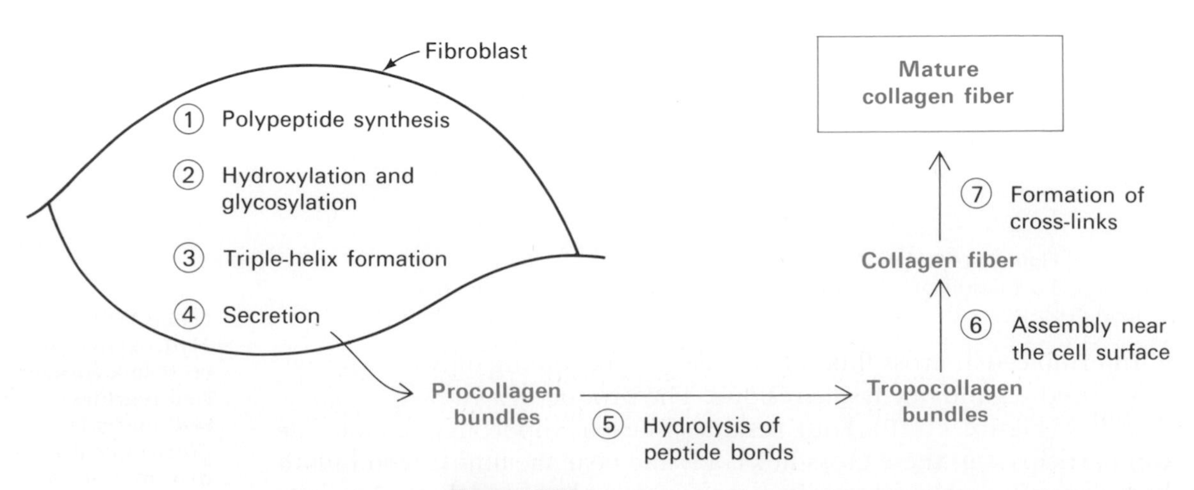

Figure 11-22
Steps in the formation of mature collagen fibers.

molecules assemble in staggered arrays to form collagen fibers. Finally, these fibers are strengthened by cross-links formed from the aldehydes of lysine and hydroxylysine residues.

COLLAGEN GENES CONTAIN MANY SHORT EXONS ENCODING SIX TURNS OF HELIX

The genes for procollagen have an unusual architecture. They contain some fifty exons, most having a length of less than 100 base pairs (Table 11-3). About half of these exons have a length of 54 bp and encode the sequence (Gly-X-Y)$_6$, which corresponds to six turns of helix. Exons precisely twice and three times this size (108 and 162 bp) are also present. Parts of the triple-stranded helix are also encoded by 45- and 99-bp exons, which can be regarded as 54- and 108-bp exons that have lost 9 base pairs. *It seems likely that the triple-helical region of collagen evolved by duplication of an ancestral 54-bp exon.* The encoding of an integral number of turns of helix by each of these exons reinforces the notion that exons correspond to functional units of proteins.

The collagen genes form a large family. Eleven types of collagen encoded by at least eighteen genes have been identified. Why this diversity? Collagen serves a variety of structural roles. It forms ropes and straps in tendons and ligaments, woven sheets in skin, filtration membranes in the glomeruli of kidneys, calcium-reinforced skeletal frameworks in bone and teeth, and many other structural supports throughout the body. The various types of collagen (Table 11-1) differ in how they associate with one another and interact with other molecules. For example, types II, III, and V form fibrils akin to those made by type I collagen. In contrast, type IV collagen (Figure 11-23) forms open nonfibrillar frameworks. Type IV collagen is a major component of membranes surrounding most epithelial cells.

Table 11-3
Sizes of exons in the gene for type II procollagen

Size (bp)	*Encoded sequence*	*Number of exons*
45	(Gly-X-Y)$_5$	5
54	(Gly-X-Y)$_6$	23
99	(Gly-X-Y)$_{11}$	5
108	(Gly-X-Y)$_{12}$	8
162	(Gly-X-Y)$_{18}$	1
188	Nonhelical	1
243	Nonhelical	1
289	Nonhelical	1
667	Nonhelical	1

After W. B. Upholt, C. M. Strom, and L. J. Sandell. *Ann. N.Y. Acad. Sci.* 460 (1985):134.

COLLAGEN IS DEGRADED BY HIGHLY SPECIFIC COLLAGENASES

Collagenases are enzymes that cleave peptide bonds in the triple-helical regions of collagen. Collagen molecules are otherwise very resistant to enzymatic attack. Two types of collagenases are known:

1. *Clostridium histolyticum*, a bacterium that causes *gas gangrene*, secretes a battery of collagenases that split each polypeptide chain of collagen at more than two hundred sites. The bond cleaved by this class of bacterial enzymes is

$$—X\overset{\downarrow}{—}Gly—Pro—Y—$$

These collagenases contribute to the invasiveness of this highly pathogenic clostridium by *destroying the connective-tissue barriers of the host*. The clostridia are unaffected by their secretory product because they are devoid of collagen, which is found only in multicellular eucaryotes.

2. *Tissue collagenases*, the other type, have been found in amphibian and mammalian tissues undergoing growth or remodeling. This enzyme activity was first demonstrated in the tail fin of tadpoles undergo-

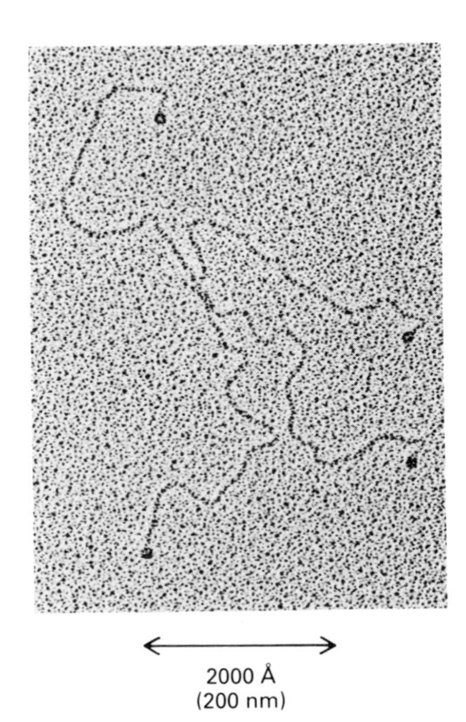

Figure 11-23
Electron micrograph showing four molecules of type IV collagen connected at their N-terminal ends. Each molecule is 400 nm long and has a globular domain at its C-terminal end. Type IV collagen, present in basement membranes, is nonfibrillar, in contrast with type I collagen. [Courtesy of Dr. Klaus Kühn.]

ing metamorphosis (Figure 11-24). Large amounts of collagen are resorbed in that tissue in the course of a few days. Collagenolytic activity was assayed by culturing small pieces of metamorphosing tail fin on a layer of fibrous collagen, which is opaque. The formation of a clear zone within several hours revealed the presence of a high level of collagenase (Figure 11-25). High enzyme activity was then found in mammalian tissue undergoing rapid reorganization, as in the uterus following pregnancy. It is evident that the activity of tissue collagenases must be precisely controlled in timing and magnitude. Human fibroblast collagenase, a 52-kd protein, is formed by proteolytic removal of the amino-terminal segment of procollagenase, the inactive precursor. Collagenase activity is also regulated by specific inhibitory proteins.

Figure 11-24
Collagen is rapidly removed from the tail fin of a tadpole during metamorphosis. [From C. M. Lapiere and J. Gross. In *Mechanisms of Hard Tissue Destruction*, R. F. Sognnaes, ed., p. 665. Copyright 1963 by the American Association for the Advancement of Science.]

The specificity of tadpole collagenase is remarkable. It cleaves tropocollagen across each of its three chains at a single site, between residues 775 and 776 in the sequence of 1056. In the $\alpha 1$ chain, a Gly-Ile bond is split, and in the $\alpha 2$ chain, a Gly-Leu bond. The resulting one-quarter and three-quarter length fragments spontaneously unfold at body temperature and are then cleaved by other proteolytic enzymes. Vertebrate collagenases are metalloproteases containing Ca^{2+} and Zn^{2+}.

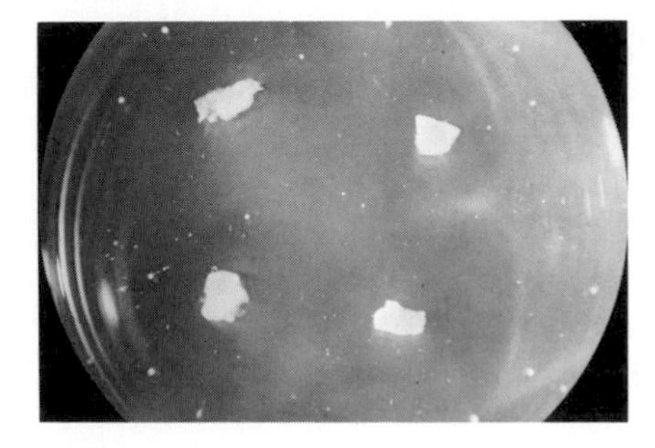

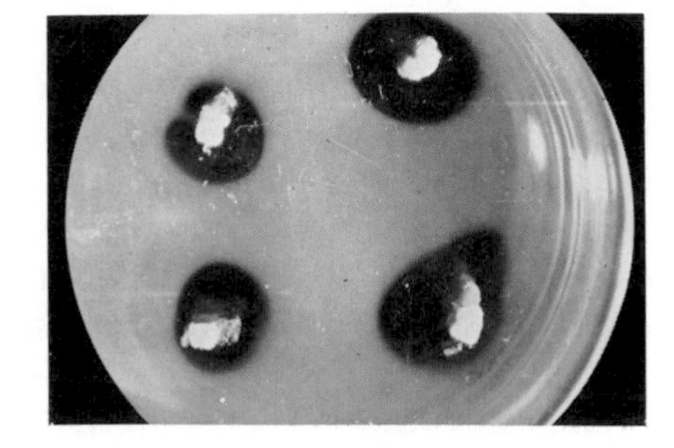

Figure 11-25
Explants from tadpole fin were placed on the surface of a reconstituted collagen gel (left). After 24-hour incubation at 37°C, there was a clear zone around each explant, indicative of collagenase activity (right). [From C. M. Lapiere and J. Gross. In *Mechanisms of Hard Tissue Destruction*, R. F. Sognnaes, ed., p. 681. Copyright 1963 by the American Association for the Advancement of Science.]

ELASTIN IS A RUBBERLIKE PROTEIN IN ELASTIC FIBERS

Elastin is found in most connective tissues in conjunction with collagen and polysaccharides. *It is the major component of elastic fibers, which can stretch to several times their length and then rapidly return to their original size and shape when the tension is released.* Large amounts of elastin are found in the walls of blood vessels, particularly in the arch of the aorta near the heart, and in ligaments. The elastic ligaments prominent in the necks of grazing animals are an especially rich source of elastin. There is relatively little elastin in skin, tendon, and loose connective tissue.

The amino acid composition of elastin is highly distinctive. As in collagen, one-third of the residues are glycine. Also, elastin is rich in proline. In contrast with collagen, elastin contains very little hydroxyproline, no hydroxylysine, and few polar amino acids. Elastin has a very high content of alanine and other nonpolar aliphatic residues.

Mature elastin contains many cross-links, which render it highly insoluble and therefore difficult to analyze. However, *a soluble precursor of*

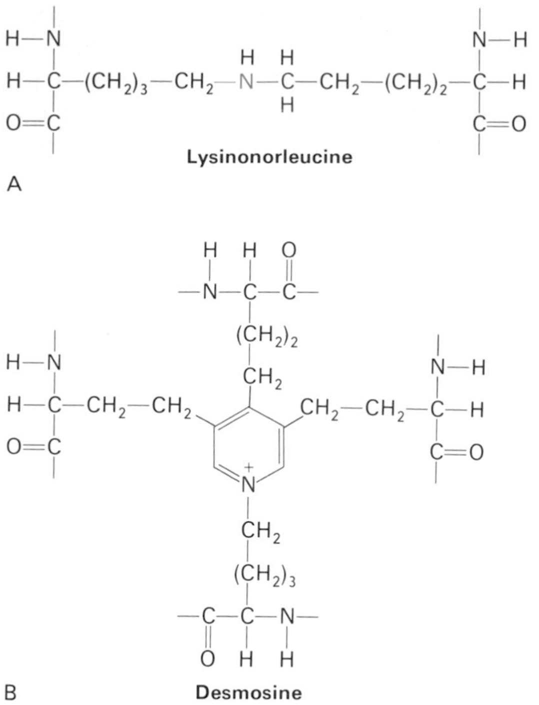

$$\begin{array}{c} \text{H—N} \qquad\qquad\qquad\qquad\qquad\qquad\qquad\qquad \text{N—H} \\ \text{H—C—(CH}_2)_3\text{—CH}_2\text{—N—C—CH}_2\text{—(CH}_2)_2\text{—C—H} \\ \text{O=C} \qquad\qquad\qquad\qquad\qquad\qquad\qquad\qquad \text{C=O} \end{array}$$

Lysinonorleucine

A

$$\begin{array}{c} \text{—N—C—C—} \\ (\text{CH}_2)_2 \\ \text{CH}_2 \\ \text{H—N} \qquad\qquad\qquad\qquad\qquad \text{N—H} \\ \text{H—C—CH}_2\text{—CH}_2 \quad [\text{pyridinium ring, } \text{N}^+] \quad \text{CH}_2\text{—CH}_2\text{—C—H} \\ \text{O=C} \qquad\qquad\qquad\qquad\qquad \text{C=O} \\ \text{CH}_2 \\ (\text{CH}_2)_3 \\ \text{—C—C—N—} \end{array}$$

B **Desmosine**

Figure 11-26
Elastin contains (A) lysinonorleucine and (B) desmosine cross-links. Desmosine is formed from four lysine residues.

-Pro-Gly-Val-Gly-Val-Pro-Gly-Val-Gly-Val-

-Pro-Gly-Val-Gly-Val-Pro-Gly-Val-Ser-Val-

-Pro-Gly-Val-Gly-Val-Pro-Gly-Val-Gly-Val-

Figure 11-27
Part of the amino acid sequence of a soluble precursor of elastin. The repeating nature of this sequence is evident.

elastin called *proelastin* has been isolated from *copper-deficient* pigs. Copper deficiency blocks the formation of aldehydes, which are essential for cross-linking, as in collagen. Amino acid sequence studies of proelastin, which has a mass of about 72 kd, are in progress. Several regions contain lysines separated by two or three alanine residues. Some of the lysine residues in these regions become oxidized by lysyl oxidase, the same copper-containing enzyme that acts on collagen. The resulting aldehyde derivative can condense with an unmodified lysine side chain to form a *lysinonorleucine* cross-link (Figure 11-26A). A different type of cross-link, *desmosine*, is derived from four lysine chains (Figure 11-26B). *These cross-links are likely to be important in enabling elastin fibers to return to their original size and shape after stretching.*

Regions of elastin between cross-links are rich in glycine, proline, and valine. The amino acid sequence of some of these regions displays regularities (Figure 11-27). The sequence Val-Pro-Gly-Val readily forms β-turns. It seems likely that elastin contains helical arrays of such β-turns, a structure known as a *β-spiral* (Figure 11-28). The elucidation of the detailed three-dimensional structure of these regular regions and their relation to the remarkable elasticity of elastin are challenging areas of inquiry. In view of the ubiquity of elastin in arteries, insight into the molecular basis of its action is likely to be important in understanding cardiovascular diseases.

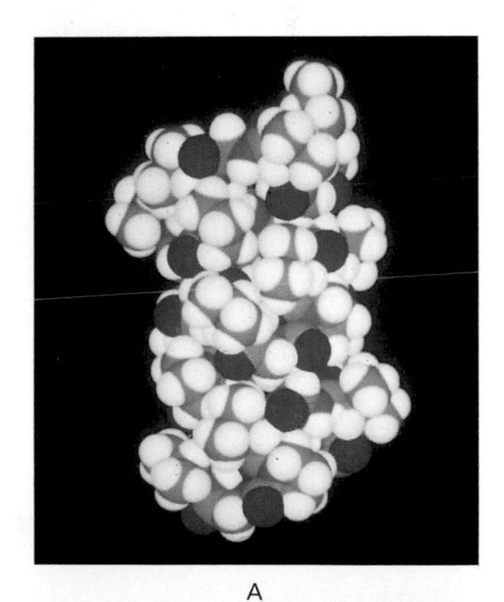

A

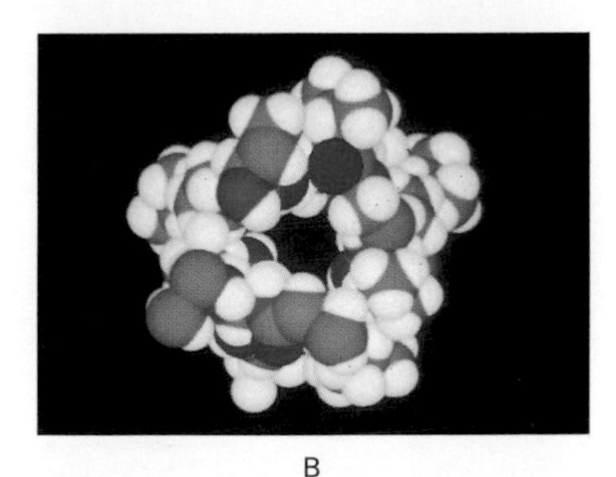

B

Figure 11-28
Model of a β-spiral, a structural motif unique to elastin. (A) Axial view. (B) Cross-sectional view. [Drawn from coordinates kindly provided by Dr. D. W. Urry and D. C. M. Venkatchalam.]

PROTEOGLYCANS FORM THE GROUND SUBSTANCE

Connective tissues are also rich in *proteoglycans*, which consist of units made of polysaccharide (about 95%) and protein (about 5%). These very large polyanions bind water and cations and thereby form the extracellular medium, or *ground substance*, of connective tissue. Proteoglycans are important in determining the viscoelastic properties of joints and of other structures that are subject to mechanical deformation. *Glycosaminoglycans*, the polysaccharide chains in proteoglycans, are made up of *disaccharide repeating units* containing a derivative of an *amino sugar*, either glucosamine or galactosamine. At least one of the sugars in the disaccharide has a *negatively charged* carboxylate or sulfate group. Hyaluronate, chondroitin sulfate, keratan sulfate, heparan sulfate, and heparin are the major glycosaminoglycans (Figure 11-29). Heparan sulfate is like heparin except that it has fewer *N*- and *O*-sulfate groups and more *N*-acetyl groups.

In the proteoglycan from cartilage (Figure 11-30), keratan sulfate and chondroitin sulfate chains are covalently attached to a polypeptide backbone called the *core protein*. About 140 of these proteins are noncovalently bound at intervals of 300 Å to a very long filament of *hyaluronate*. This interaction is promoted by a small *link protein*. The entire complex has a mass of about 2×10^6 daltons and a length of several

Chondroitin 6-sulfate

Keratan sulfate

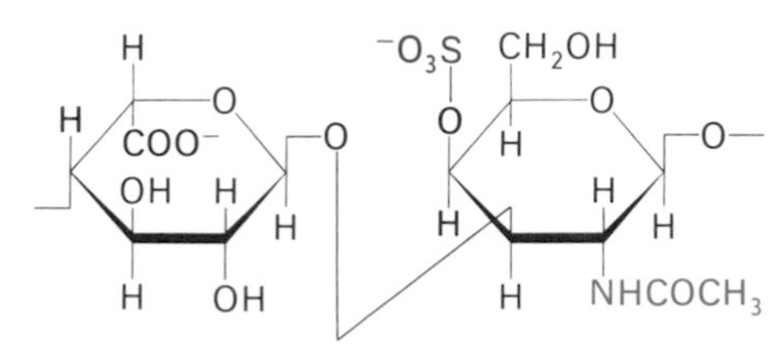

Heparin

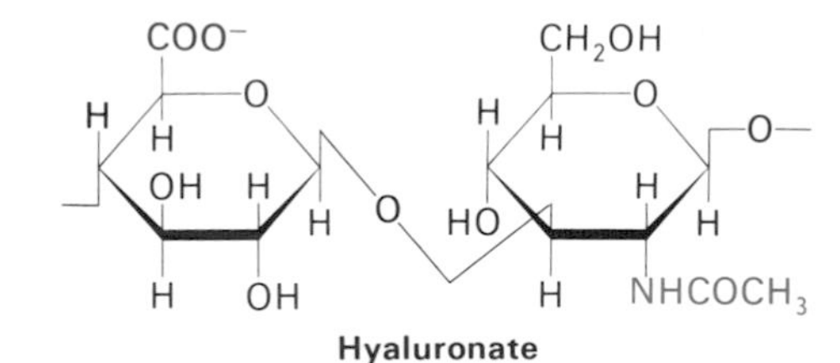

Figure 11-29
Structural formulas of the repeating disaccharide units of some major glycosaminoglycans. The negatively charged groups are shown in red.

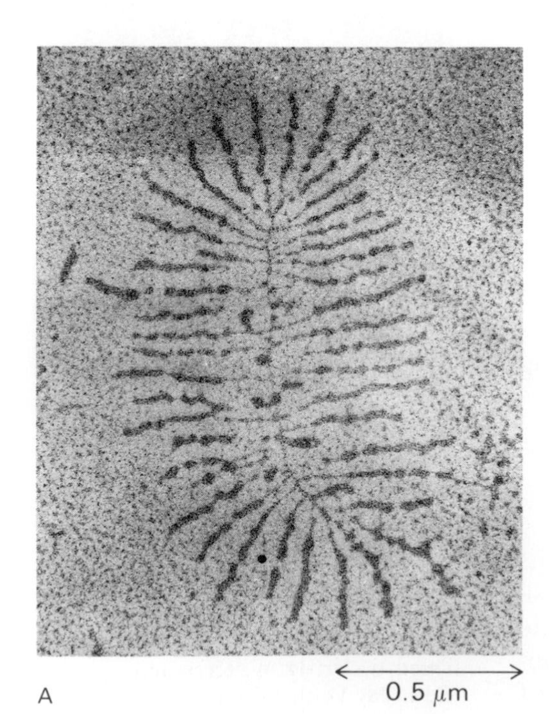

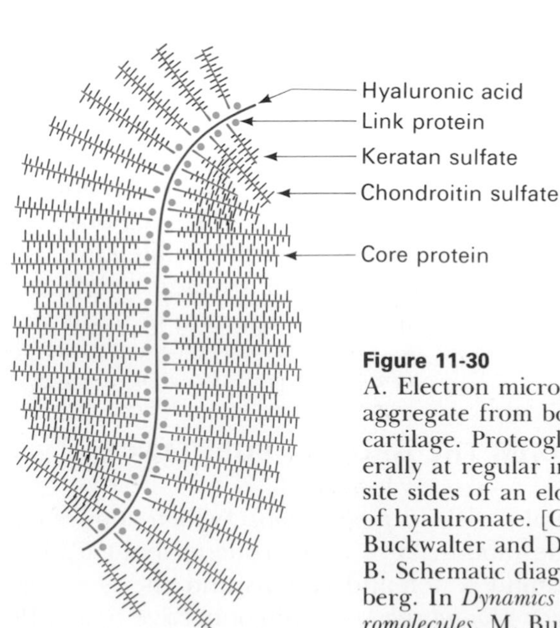

Figure 11-30
A. Electron micrograph of a proteoglycan aggregate from bovine fetal epiphyseal cartilage. Proteoglycan monomers arise laterally at regular intervals from the opposite sides of an elongated central filament of hyaluronate. [Courtesy of Dr. Joseph Buckwalter and Dr. Lawrence Rosenberg.] B. Schematic diagram. [After L. Rosenberg. In *Dynamics of Connective Tissue Macromolecules*, M. Burleigh and R. Poole, eds. (North-Holland, 1975), p. 105.]

microns. Proteoglycan aggregates from some other tissues have a different structure, suggesting that this class of very large biomolecules has several as yet undiscovered functions.

FIBRONECTIN, A CELL-SURFACE PROTEIN, ENABLES CELLS TO INTERACT WITH THE EXTRACELLULAR MATRIX

Cells interact with collagen and other constituents of the extracellular matrix through *fibronectin*, a protein that binds reversibly to the external face of the plasma membrane. Fibronectin consists of two 250-kd polypeptide chains that are linked by a disulfide near their carboxyl termini (Figure 11-31). This highly elongated protein, 600 Å long and 25 Å wide, contains a linear array of domains, each able to specifically bind certain molecules outside the cell. For example, the amino-terminal and carboxyl-terminal domains have high affinity for fibrin. Consequently, fibronectin is incorporated into blood and becomes cross-linked to fibrin by the Factor XIII transamidase that catalyzes the final step in the clotting cascade (p. 250). Fibroblasts and other cells that repair the site of injury adhere to the clot by interacting with the cell-binding region of fibronectin molecules in the clot.

Fibronectin—
An adhesive cell-surface protein.
Derived from the Latin roots *fibra* (fiber) and *nectere* (to bind or connect).

Fibronectin also enables cells to migrate in developing embryos. The cell-binding region of fibronectin binds and releases *integrin*, a complex of proteins that span the plasma membrane. Other domains of the same fibronectin molecule bind and release collagen fibers. Hence, the cells adhere to the extracellular matrix and, in effect, move along tracks of collagen. The heparin-binding domain enables cells to bind to the glycosaminoglycan component of proteoglycans. An intriguing finding is that malignant cells are deficient in fibronectin, which may account in part for their anarchic and invasive properties. *Laminin* is another adhesive glycoprotein in the extracellular matrix. It enables epithelial cells to attach to underlying connective tissue. Specifically, it has high affinity for type IV collagen, a component of basement membranes. Laminin, a very large three-chain protein (~1000 kd) has a cruciform shape (Figure 11-32). Like fibronectin, laminin consists of multiple domains with distinct binding functions.

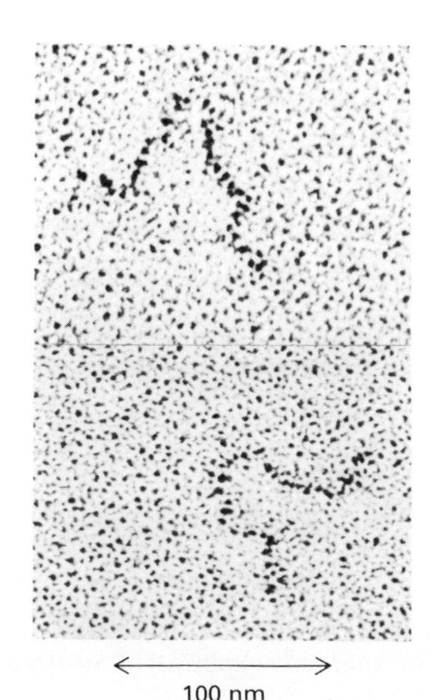

Figure 11-31
Electron micrograph of fibronectin molecules. Each end of a molecule is the amino-terminus of one of the pair of identical subunits. The disulfide-linked pair of carboxyl-terminal regions is located at the center of the molecule. The flexibility of fibronectin is evident. [Courtesy of Dr. Heinz Furthmayr.]

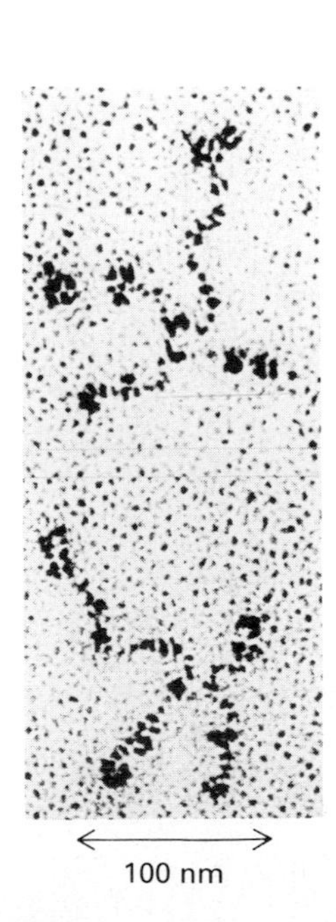

Figure 11-32
Electron micrograph of laminin. [Courtesy of Dr. Heinz Furthmayr.]

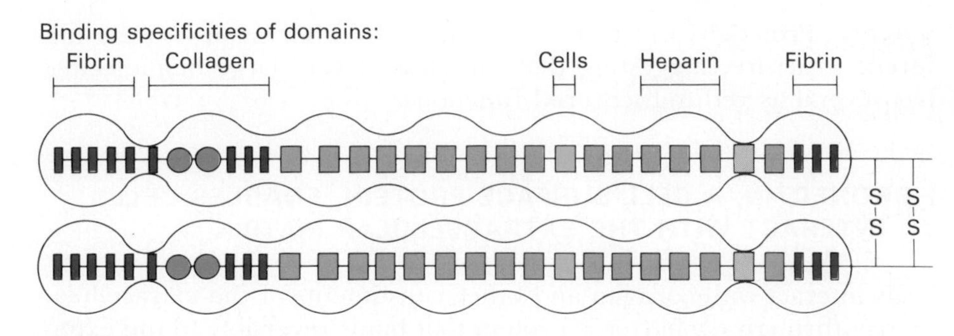

Figure 11-33
Fibronectin consists of domains with partly repeating amino acid sequences. Type I modules are shown in red, type II in blue, and type III in green. A region essential for binding cells is shown in yellow. The function of the unit shown in gray is not yet known. Each of these units is encoded by an exon. [After R. Hynes. Fibronectin. Copyright © 1986 by Scientific American, Inc. All rights reserved.]

The fibronectin gene contains more than fifty exons. Nearly all of these exons are variations of three motifs. In the protein, the fibrin-binding domain at the amino-terminus consists of five repeated type I modules, whereas the collagen-binding domain has three of them and two of type II (Figure 11-33). Type III modules are found in several domains. One of the type III modules in the cell-binding domain contains a sequence

Arg-Gly-Asp-Ser

that binds to a complementary site in integrin. A type III module is missing from plasma fibronectin, a product of liver cells, whereas it is present in fibronectin bound to fibroblasts in connective tissue. Are these closely related fibronectins encoded by different genes? The answer is no. Rather, multiple forms of fibronectin are generated by different patterns of splicing of a single kind of nascent RNA transcript (Figure 11-34). We see here a clear-cut example of how *alternative splicing* produces a family of related proteins differing according to the needs of particular tissues.

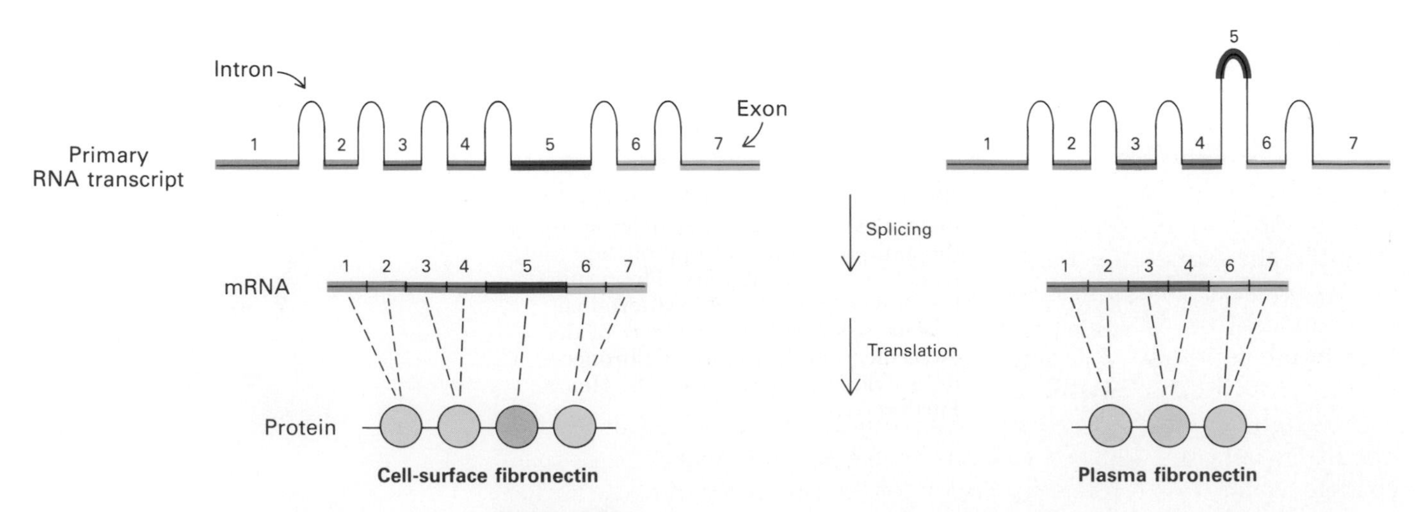

Figure 11-34
Different forms of fibronectin are generated by alternative splicing of an RNA transcript.

SUMMARY

Collagen is a family of proteins with very high tensile strength. It is the major fibrous component of skin, bone, tendon, cartilage, blood vessels, and teeth. Five types of collagens with different tissue distributions have been characterized. The basic structural unit of collagen is tropocollagen, which consists of three strands, each about 1000 residues long. Collagen is unusually rich in glycine and proline and contains hydroxyproline and hydroxylysine, which are found in very few other proteins. The amino acid sequence of collagen is highly distinctive: nearly every third residue is glycine. Tropocollagen is a triple-stranded helical rod, 3000 Å long and 15 Å in diameter. The helical motif of each of its chains is very different from that of an α helix. The collagen triple helix is stabilized by hydrogen bonding between strands and by the steric locking effects of its proline and hydroxyproline residues. The inside of this triple-stranded helix is crowded, and so every third residue in the amino acid sequence must be glycine. Many of the exons in collagen genes encode the sequence $(\text{Gly-X-Y})_6$, which corresponds to six turns of helix. It seems likely that the triple-helical region of collagen evolved by duplication of an ancestral exon of 54 base pairs.

Covalent modifications are important for the synthesis, secretion, and assembly of collagen. The three chains of type I tropocollagen are synthesized as larger precursors called pro-α1(I) and pro-α2. Some proline residues in these procollagen chains are converted into hydroxyproline by prolyl hydroxylase, an enzyme that requires O_2, Fe^{2+}, and α-ketoglutarate for activity; a reducing agent such as ascorbic acid is also needed. Collagen synthesized in the absence of ascorbic acid is less stable because it is insufficiently hydroxylated, which results in scurvy. Lysine is hydroxylated by a different enzyme. Sugars are then attached to hydroxylysine residues on the precursor chains. These hydroxylation and glycosylation reactions take place inside fibroblast cells. Procollagens are secreted into the extracellular medium. Amino-terminal and carboxyl-terminal regions, called propeptides, are then cleaved from these precursors by procollagen peptidase to yield tropocollagen, which spontaneously associates into fibers. Finally, collagen fibers are cross-linked to give them additional strength. Lysyl oxidase converts the amino groups of some lysine and hydroxylysine residues to aldehydes as the first step in cross-linking.

Elastin is an insoluble rubberlike protein in the elastic fibers of connective tissues. These fibers can be reversibly stretched to several times their initial length. Connective tissues such as ligaments and the arch of the aorta contain much elastin. Like collagen, elastin is rich in proline and glycine. Unlike collagen, elastin contains very little hydroxyproline and no hydroxylysine. Its amino acid composition is highly nonpolar. The amino acid sequence of elastin displays certain regularities. For example, the sequence Pro-Gly-Val-Gly-Val recurs frequently. Elastin is synthesized as a soluble precursor, which then becomes cross-linked in a variety of ways. Desmosine, one of these cross-links, is derived from four lysine residues. Aldehyde intermediates participate in the formation of cross-links in elastin, as they do in collagen.

Proteoglycans, the other major macromolecular component of connective tissue, consist of polysaccharide and protein units. The polysaccharide chains, called glycosaminoglycans, are made up of disaccharide repeating units that have a high density of negative charge. Proteoglycans form the ground substance and are important in determining the viscoelastic properties of connective tissues.

Fibronectin, a highly elongated cell-surface protein, enables cells to interact with the extracellular matrix. Each domain of fibronectin can specifically bind certain molecules outside the cell, such as fibrin, collagen, and heparin. In addition, fibronectin has a cell-binding domain. These domains are built of three types of repeating sequences, each encoded by an exon. Alternative splicing produces a variety of fibronectins. This cell-surface protein is important for cell migration in development and for wound healing.

SELECTED READINGS

Where to start

Martin, G. R., Timpl, R., Muller, P. K., and Kühn, K., 1985. The genetically distinct collagens. *Trends Biochem. Sci.* 10:285–287.

Cheah, K. S. E., 1985. Collagen genes and inherited connective tissue disease. *Biochem. J.* 229:287–303.

Caplan, A. I., 1984. Cartilage. *Sci. Amer.* 251(4):84–94. [Contains a discussion of proteoglycans.]

Hynes, R. O., 1986. Fibronectins. *Sci. Amer.* 254(6):42–51. [Available as *Sci. Amer.* Offprint 1571.]

Books

Fleischmajer, R., Olsen, B. R., and Kühn, K., (eds.), 1985. *Biology, Chemistry, and Pathology of Collagen,* vol. 260. *Ann. N. Y. Acad. Sci.* [Contains many excellent papers on collagen structure, genes, and inherited disorders.]

Piez, K. A., and Reddi, A. H., 1984. *Extracellular Matrix Biochemistry.* Elsevier Science.

Hay, E. D., (ed.), 1982. *Cell Biology of Extracellular Matrix.* Plenum.

Collagen: genes and structure

Eyre, D. R., 1980. Collagen: molecular diversity in the body's protein scaffold. *Science* 207:1315–1322.

Chu, M-L., de Wet, W., Bernard, M., Ding, J-F., Morabito, M., Myers, J., Williams, C., and Ramirez, F., 1984. Human pro-α1(I) collagen gene structure reveals evolutionary conservation of a pattern of introns and exons. *Nature* 340:337–340.

Boedtker, H., Finer, M., and Aho, S., 1985. The structure of the chicken α2 collagen gene. *Ann. N. Y. Acad. Sci.* 460:85–116.

Collagen: synthesis and modification

Bornstein, P., Horlein, D., McPherson, J., Sandmeyer, S., and Gallis, B., 1982. Regulation of collagen biosynthesis. *In* Kühn, K., Schoene, H., and Timpl, R. (eds.), *New Trends in Basement Membrane Research,* Raven Press.

Walsh, C., 1979. *Enzymatic Reaction Mechanisms.* W. H. Freeman. [Chapter 16 gives an excellent account of dioxygenases.]

Cardinale, G. J., and Udenfriend, S., 1974. Prolyl hydroxylase. *Advan. Enzymol.* 41:245–300.

Hayaishi, O., (ed.), 1974. *Molecular Mechanisms of Oxygen Activation.* Academic Press.

Berg, R. A., and Prockop, D. J., 1973. The thermal transition of a non-hydroxylated form of collagen: evidence for a role for hydroxyproline in stabilizing the triple-helix of collagen. *Biochem. Biophys. Res. Comm.* 52:115–120.

Hulmes, D. J. S., Bruns, R. R., and Gross, J., 1983. On the state of aggregation of newly secreted procollagen. *Proc. Nat. Acad. Sci.* 80:388–392.

Eyre, D. R., Paz, M. A., and Gallop, P. M., 1984. Cross-linking in collagen and elastin. *Ann. Rev. Biochem.* 53:717–748.

Collagenases

Goldberg, G. I., Wilhelm, S. M., Kronberger, A., Bauer, E. A., Grant, G. A., and Eizen, A. Z., 1986. Human fibroblast collagenase: Complete primary structure and homology to an oncogene transformation-induced rat protein. *J. Biol. Chem.* 261:6600–6605.

Gross, J., and Nagai, Y., 1965. Specific degradation of the collagen molecule by tadpole collagenolytic enzyme. *Proc. Nat. Acad. Sci.* 54:1197–1204.

Elastin

Sandberg, L. B., Soskel, N. T., and Leslie, J. G., 1981. Elastin structure, biosynthesis, and relation to disease state. *New Engl. J. Med.* 304:556–579.

Cleary, E. G., and Gibson, M. A., 1983. Elastin-associated microfibrils and microfibrillar proteins. *Int. Rev. Connect. Tissue Res.* 10:97–209.

Urry, D. W., 1983. What is elastin; what is not. *Ultrastruc. Pathol.* 4:227–251.

Ross, R., and Bornstein, P., 1971. Elastic fibers in the body. *Sci. Amer.* 224(6)44–52. [Available as *Sci. Amer.* Offprint 1225.]

Proteoglycans

Hassell, J. R., Kimura, J. H., and Hascall, V. C., 1986. Proteoglycan core protein families. *Ann. Rev. Biochem.* 55:539.

Hook, M., Kjellen, L., Johansson, S., and Robinson, J., 1984. Cell-surface glycosaminoglycans. *Ann. Rev. Biochem.* 53:847–869.

Fibronectin and laminin

Hynes, R., 1985. Molecular biology of fibronectin. *Ann. Rev. Cell Biol.* 1:67–90.

Odermatt, E., Tamkun, J. W., and Hynes, R. O., 1985. Repeating modular structure of the fibronectin gene: relationship to protein structure and subunit variation. *Proc. Nat. Acad. Sci.* 82:6571–6575.

Engel, J., Odermatt, A., Engel, A., Madri, J. A., Furthmayr, H., Rohde, H., and Timpl, R., 1981. Shapes, domain organization, and flexibility of laminin and fibronectin, two multifunctional proteins of the extracellular matrix. *J. Mol. Biol.* 150:97–120.

Connective-tissue diseases

Pinnell, S. R., and Murad, S., 1983. Disorders of collagen. *In* Stanbury, J. B., Wyngaarden, J. B., Fredrickson, D. S., Goldstein, J. L., and Brown, M. S. (eds.), *The Metabolic Basis of Inherited Disease* (5th ed.), pp. 1425–1449. McGraw-Hill.

Smith, R., 1986. The molecular genetics of collagen disorders. *Clin. Sci.* 71:129–135.

Prockop, D. J., and Kivirikko, K. I., 1984. Heritable diseases of collagen. *New Engl. J. Med.* 311:376–386.

Cohn, D. H., Byers, P. H., Steinmann, B., and Gelinas, R. E., 1986. Lethal osteogenesis imperfecta resulting from a single nucleotide change in one human pro-α1(I) collagen allele. *Proc. Nat. Acad. Sci.* 83:6045–6047.

Major, R. H., (ed.), 1945. *Classic Descriptions of Disease* (3rd ed.). Thomas. [Cartier's description of scurvy is given on page 587.]

Minor, R. R., Sippola-Thiele, M., McKeon, J., Berger, J., and Prockop, D. J., 1986. Defects in the processing of procollagen to collagen are demonstrable in cultured fibroblasts from patients with the Ehlers-Danlos and osteogenesis imperfecta syndromes. *J. Biol. Chem.* 261:10006–10014.

Nishikimi, M., and Udenfriend, S., 1977. Scurvy as an inborn error of ascorbic acid biosynthesis. *Trends Biochem. Sci.* 2:111–112.

PROBLEMS

1. Poly-L-proline, a synthetic polypeptide, can adopt a helical conformation like that of a single strand in the collagen triple helix.
   (a) Poly-L-proline cannot form a triple helix. Why?
   (b) Poly(Gly-Pro-Pro) can form a triple helix like that of collagen. Predict the thermal stability of the triple helix of poly(Gly-Pro-Gly) relative to that of poly(Gly-Pro-Pro).
   (c) Would you expect poly(Gly-Pro-Gly-Pro) to form a triple helix like that of collagen?

2. Consider the amino acid sequence

   Gly-Leu-Pro-Gly-Pro-Pro-Gly-Ala-Pro-Gly

   (a) Which residues are susceptible to hydroxylation at C-4 by prolyl hydroxylase?
   (b) Which peptide bonds are most susceptible to hydrolysis by the collagenase from *Clostridium histolyticum?*

3. Several kinds of covalent cross-links in proteins have been discussed in the preceding chapters. Identify the covalent cross-links, if there are any, in the following proteins.
   (a) Ribonuclease.
   (b) Hemoglobin.
   (c) Fibrin.
   (d) Collagen.
   (e) Elastin.

4. Prolyl hydroxylase can decarboxylate α-ketoglutarate in the absence of a peptidyl prolyl substrate. Ferrous ion, O_2, and ascorbate are needed for this reaction. What does this finding imply about the mechanism of hydroxylation by this enzyme?

5. Collagenases from *Clostridium histolyticum* are inhibited by

   Pro-Pro-Pro-Pro-Pro-Pro-Gly-Pro-Gly-C(=O)H

   The corresponding peptide with a terminal carboxyl group is a much less effective inhibitor.
   (a) Why is the aldehyde a more potent inhibitor?
   (b) Propose a transition state for peptide-bond hydrolysis by this enzyme.

6. The tetrapeptide Arg-Gly-Asp-Ser blocks cell migration during embryonic development. Why?

7. A mutation changes the thermal stability of type I collagen so that the shrinkage temperature and melting temperature become 5°C higher than normal. What is a likely consequence? How might it be deleterious?

8. A mutation leads to a 90% decrease in the activity of proline hydroxylase. What are the likely symptoms of this patient?

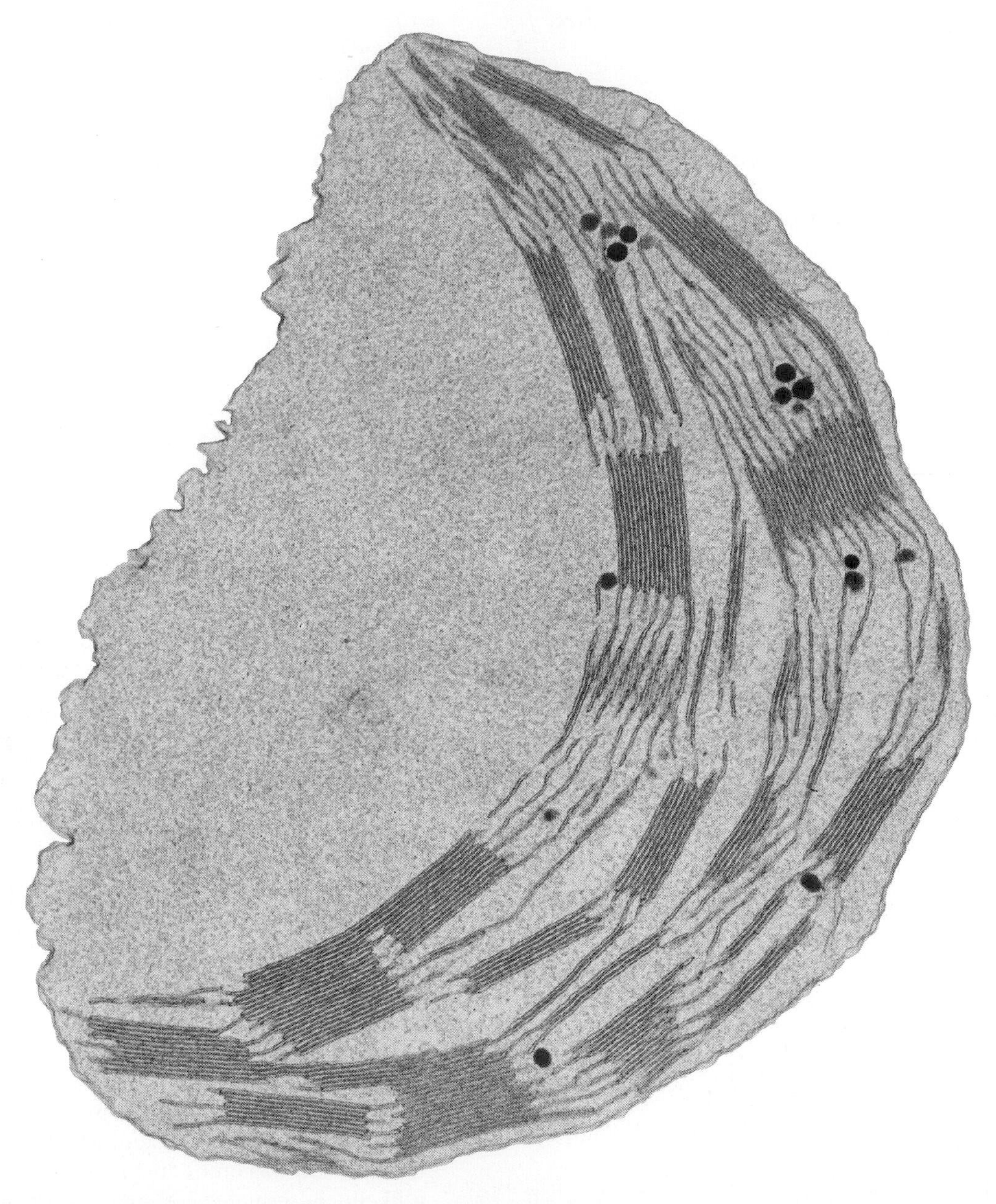

Electron micrograph of a whole chloroplast from a spinach leaf. The stacked thylakoid membranes are the sites of energy conversion in photosynthesis. [Courtesy of Dr. Kenneth Miller.]

CHAPTER 12

Introduction to Biological Membranes

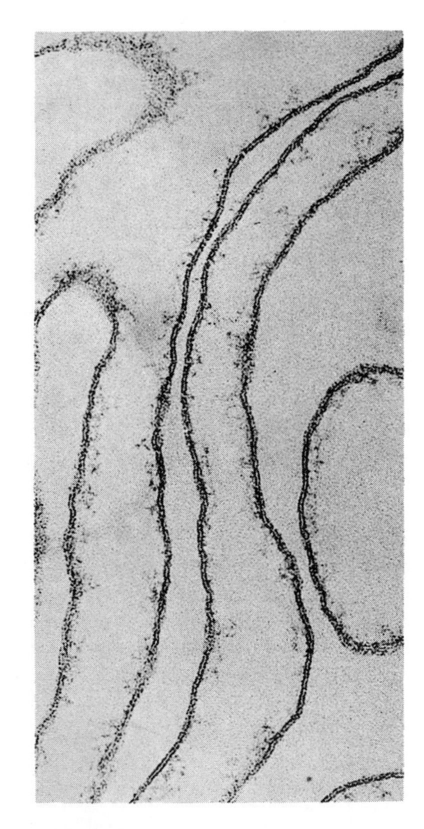

Figure 12-1
Electron micrograph of a preparation of plasma membranes from red blood cells. These membranes are seen "on edge," in cross section. [Courtesy of Dr. Vincent Marchesi.]

We now turn to biological membranes, which are organized sheetlike assemblies consisting mainly of proteins and lipids. The functions carried out by membranes are indispensable for life. Plasma membranes give cells their individuality by separating them from their environment. *Membranes are highly selective permeability barriers* rather than impervious walls because they contain specific molecular *pumps* and *gates*. These transport systems regulate the molecular and ionic composition of the intracellular medium. Eucaryotic cells also contain internal membranes that form the boundaries of organelles such as mitochondria, chloroplasts, and lysosomes. Functional specialization in the course of evolution has been closely linked to the formation of such compartments.

Membranes also control the flow of information between cells and their environment. They contain *specific receptors for external stimuli*. The movement of bacteria toward food, the response of target cells to hormones such as insulin, and the perception of light are processes in which the primary event is the detection of a signal by a specific receptor in a membrane. In turn, *some membranes generate signals*, which can be chemical or electrical. Thus, membranes play a central role in biological communication.

The two most important *energy conversion processes* in biological systems are carried out by membrane systems that contain ordered arrays of enzymes and other proteins. *Photosynthesis*, in which light is converted into chemical-bond energy, occurs in the inner membranes of chloroplasts (Figure 12-2), whereas *oxidative phosphorylation*, in which adenosine triphosphate (ATP) is formed by the oxidation of fuel molecules, takes place in the inner membranes of mitochondria. These and other membrane processes will be discussed in detail in later chapters. This chapter deals with some essential features that are common to most biological membranes.

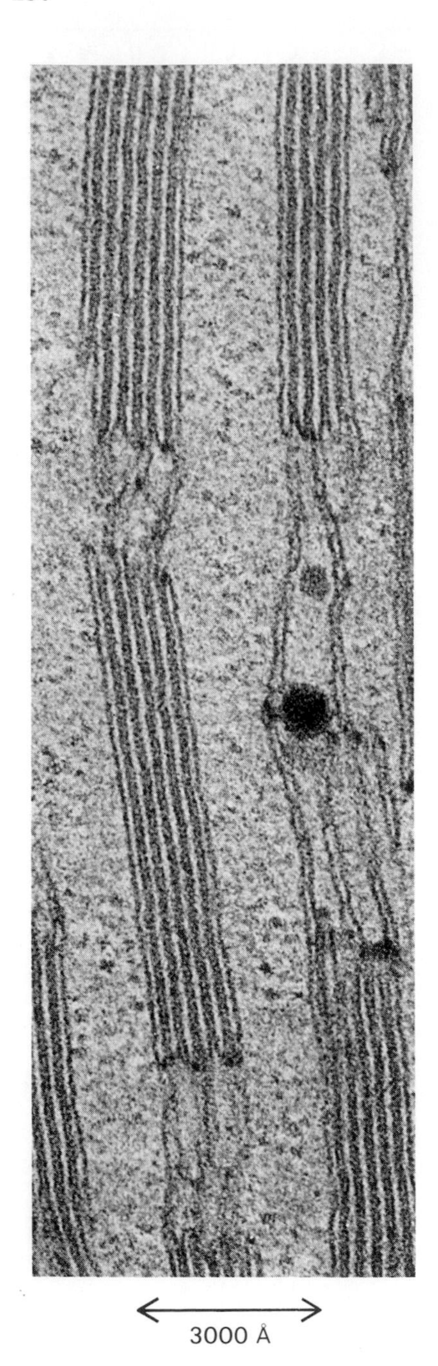

Figure 12-2
Light is converted into chemical-bond energy by photosynthetic assemblies in the thylakoid membranes of chloroplasts. [Courtesy of Dr. M. C. Ledbetter.]

COMMON FEATURES OF BIOLOGICAL MEMBRANES

Membranes are as diverse in structure as they are in function. However, they do have in common a number of important attributes:

1. Membranes are *sheetlike structures*, only a few molecules thick, that form *closed boundaries* between compartments of different composition. The thickness of most membranes is between 60 and 100 Å.

2. Membranes consist mainly of *lipids* and *proteins*. The weight ratio of protein to lipid in most biological membranes ranges from 1:4 to 4:1. Membranes also contain *carbohydrates* that are linked to lipids and proteins.

3. *Membrane lipids are relatively small molecules* that have both a hydrophilic and a hydrophobic moiety. These lipids spontaneously form *closed bimolecular sheets* in aqueous media. These *lipid bilayers* are barriers to the flow of polar molecules.

4. *Specific proteins mediate distinctive functions of membranes*. Proteins serve as pumps, gates, receptors, energy transducers, and enzymes. Membrane proteins are embedded in lipid bilayers, which create suitable environments for their action.

5. Membranes are *noncovalent assemblies*. The constituent protein and lipid molecules are held together by many noncovalent interactions, which are cooperative.

6. Membranes are *asymmetric*. The two faces of a membrane are different.

7. Membranes are *fluid structures*. Lipid molecules diffuse rapidly in the plane of the membrane, as do proteins, unless they are anchored by specific interactions. In contrast, they do not rotate across the membrane. Membranes can be regarded as *two-dimensional solutions of oriented proteins and lipids*.

PHOSPHOLIPIDS ARE THE MAJOR CLASS OF MEMBRANE LIPIDS

Lipids differ markedly from the other groups of biomolecules considered thus far. By definition, lipids are water-insoluble biomolecules that are highly soluble in organic solvents such as chloroform. Lipids have a variety of biological roles: they serve as fuel molecules, highly concentrated energy stores, signal molecules, and components of membranes. The first three roles of lipids will be discussed in Chapters 20 and 23. Here, the concern is with lipids as membrane constituents. The three major kinds of membrane lipids are *phospholipids, glycolipids,* and *cholesterol*.

Let us start with phospholipids, because they are abundant in all biological membranes. Phospholipids are derived from either *glycerol,* a three-carbon alcohol, or *sphingosine,* a more complex alcohol. Phospholipids derived from glycerol are called *phosphoglycerides.* A phosphoglyceride consists of a glycerol backbone, two fatty acid chains, and a phosphorylated alcohol.

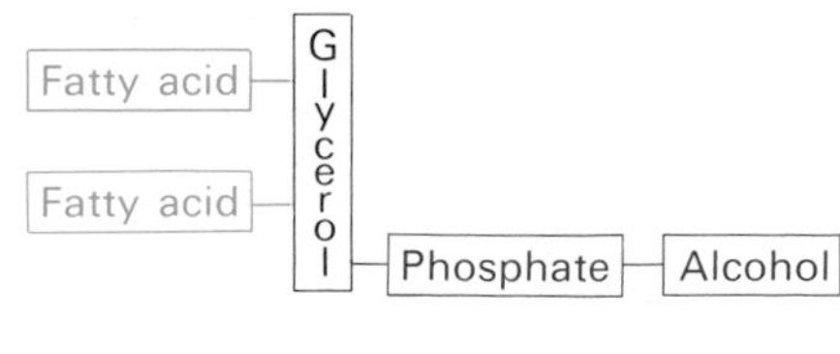

Components of a phosphoglyceride

The *fatty acid chains* in phospholipids and glycolipids usually contain an even number of carbon atoms, typically between 14 and 24. The 16- and 18-carbon fatty acids are the most common ones. In animals, the hydrocarbon chain in fatty acids is unbranched. Fatty acids may be saturated or unsaturated. The configuration of double bonds in unsaturated fatty acids is nearly always *cis.* As will be discussed shortly, the length and the degree of unsaturation of fatty acid chains in membrane lipids have a profound effect on membrane fluidity. The structures of the ionized form of two common fatty acids—palmitic acid (C_{16}, saturated) and oleic acid (C_{18}, one double bond)—are shown in Figure 12-3. They will be referred to as palmitate and oleate to emphasize the fact that they are ionized under physiological conditions. The nomenclature of fatty acids is discussed in Chapter 20.

$$H_3C-(CH_2)_{14}-COO^-$$

Palmitate
(Ionized form of palmitic acid)

$$H_3C-(CH_2)_7-CH{=}CH-(CH_2)_7-COO^-$$

Oleate
(Ionized form of oleic acid)

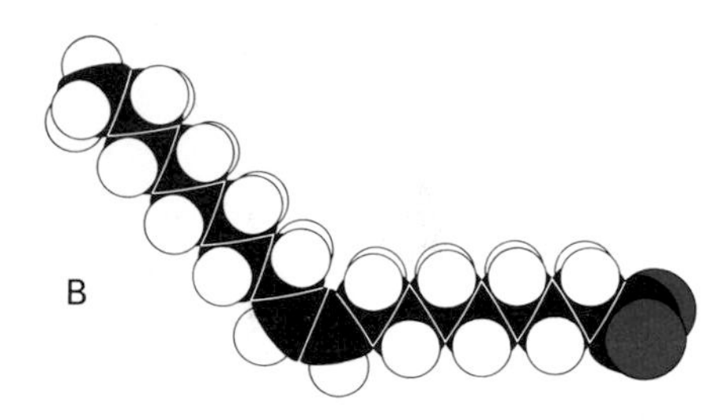

Figure 12-3
Space-filling models of (A) palmitate (C_{16}, saturated) and (B) oleate (C_{18}, unsaturated). The *cis* double bond in oleate produces a bend in the hydrocarbon chain.

In phosphoglycerides, the hydroxyl groups at C-1 and C-2 of glycerol are esterified to the carboxyl groups of two fatty acid chains. The C-3 hydroxyl group of the glycerol backbone is esterified to phosphoric acid. The resulting compound, called *phosphatidate* (or diacylglycerol-3-phosphate), is the simplest phosphoglyceride. Only small amounts of phosphatidate are present in membranes. However, it is a key intermediate in the biosynthesis of the other phosphoglycerides. The absolute configuration of the glycerol 3-phosphate moiety of membrane lipids is shown in Figure 12-4.

A: $_1CH_2OH$; HO—^{2}C—H; $_3CH_2OPO_3^{2-}$

B: CH_2OH; HO—C—H; $CH_2OPO_3^{2-}$

Figure 12-4
Absolute configuration of the glycerol 3-phosphate moiety of membrane lipids: (A) H and OH, attached to C-2, are in front of the plane of the page, whereas C-1 and C-3 are behind it; (B) Fischer representation of this structure. In a Fischer projection, horizontal bonds denote bonds in front, whereas vertical bonds denote bonds behind the plane of the page.

Hydrocarbon chains of fatty acids → R_1—C(=O)—O—CH_2; R_2—C(=O)—O—C—H; H_2C—O—P(=O)(O^-)—O^-

Phosphatidate
(Diacylglycerol 3-phosphate)

The major phosphoglycerides are derivatives of phosphatidate. The phosphate group of phosphatidate becomes esterified to the hydroxyl group of one of several alcohols. The common alcohol moieties of phosphoglycerides are serine, ethanolamine, choline, glycerol, and inositol.

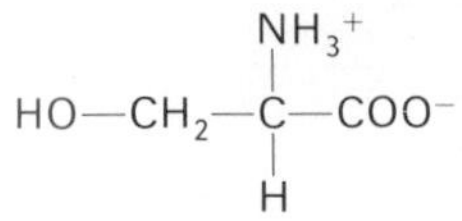

$HO{-}CH_2{-}CH(NH_3^+){-}COO^-$

Serine

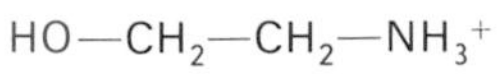

$HO{-}CH_2{-}CH_2{-}NH_3^+$

Ethanolamine

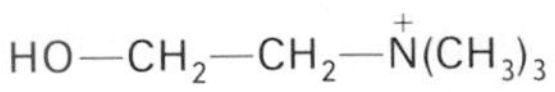

$HO{-}CH_2{-}CH_2{-}\overset{+}{N}(CH_3)_3$

Choline

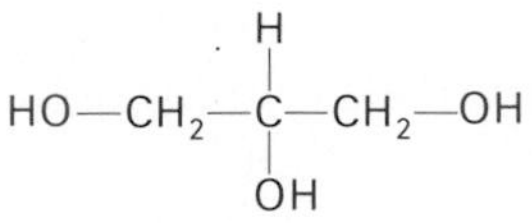

$HO{-}CH_2{-}CH(OH){-}CH_2{-}OH$

Glycerol

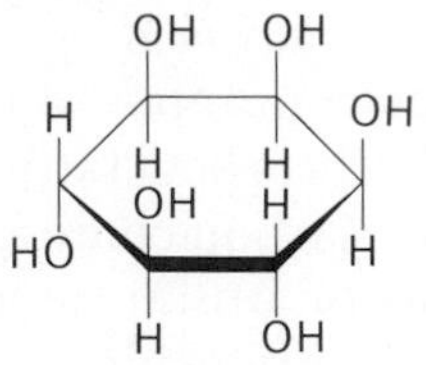

Inositol

Now let us link some of these components to form phosphatidyl choline, a phosphoglyceride found in most membranes of higher organisms.

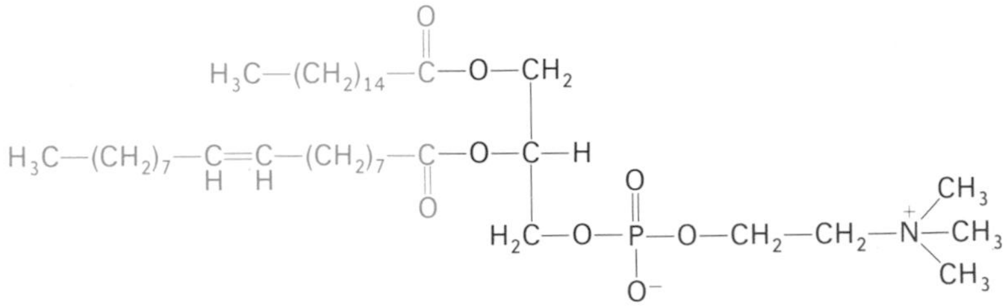

A phosphatidyl choline
(1-Palmitoyl-2-oleoyl-phosphatidyl choline)

The structural formulas of the other principal phosphoglycerides—namely, phosphatidyl ethanolamine, phosphatidyl serine, phosphatidyl inositol, and diphosphatidyl glycerol—are given in Figure 12-5.

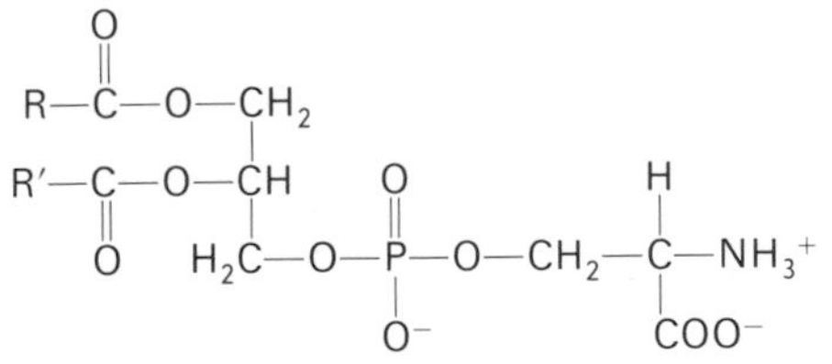

Phosphatidyl serine

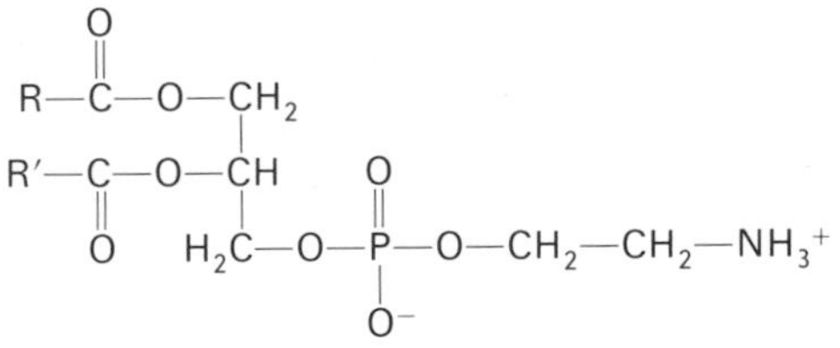

Phosphatidyl ethanolamine

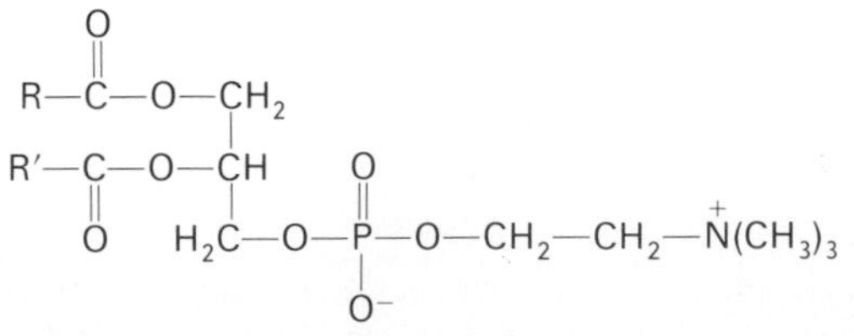

Phosphatidyl choline

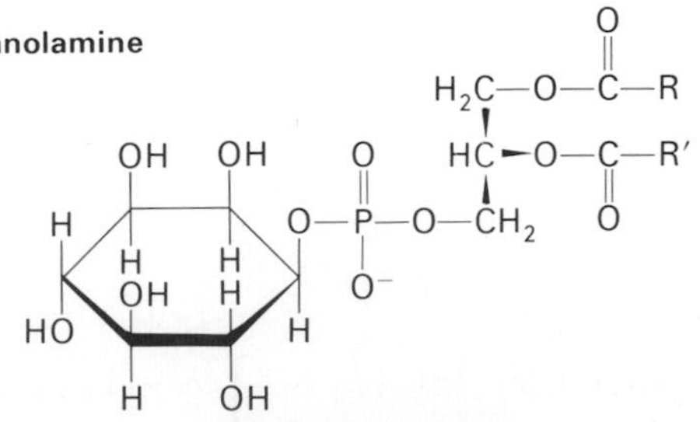

Phosphatidyl inositol

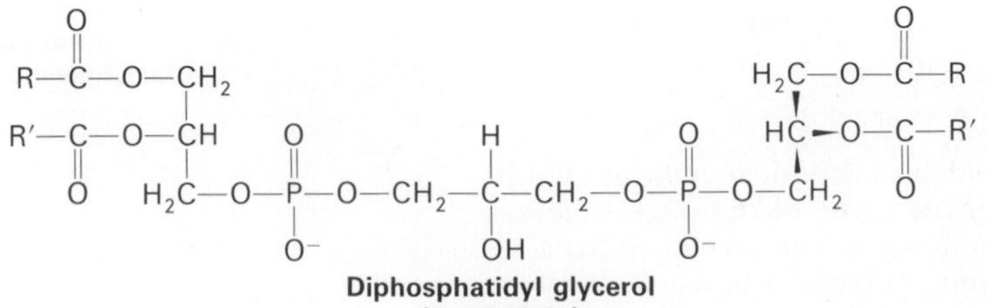

Diphosphatidyl glycerol
(Cardiolipin)

Figure 12-5
Formulas of some phosphoglycerides.

Sphingomyelin is the only phospholipid in membranes that is not derived from glycerol. Instead, the backbone in sphingomyelin is *sphingosine,* an amino alcohol that contains a long, unsaturated hydrocarbon chain. In sphingomyelin, the amino group of the sphingosine backbone is linked to a fatty acid by an amide bond. In addition, the primary hydroxyl group of sphingosine is esterified to phosphoryl choline. As will be shown shortly, the conformation of sphingomyelin resembles that of phosphatidyl choline.

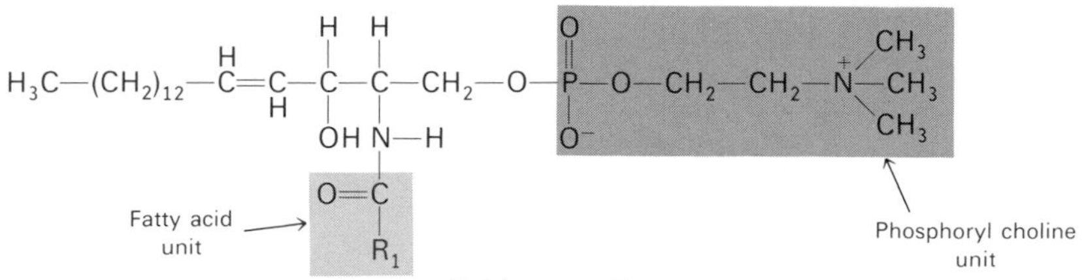

Fatty acid unit
Phosphoryl choline unit

Sphingomyelin

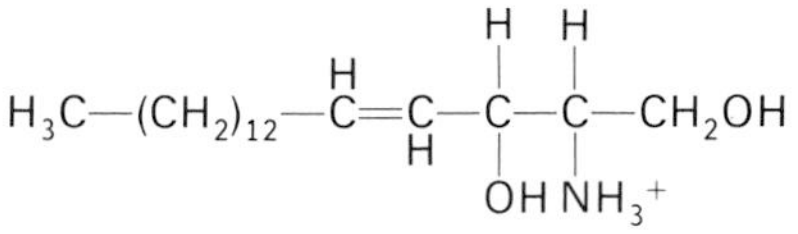

Sphingosine

MANY MEMBRANES ALSO CONTAIN GLYCOLIPIDS AND CHOLESTEROL

Glycolipids, as their name implies, are *sugar-containing lipids.* In animal cells, glycolipids, like sphingomyelin, are derived from sphingosine. The amino group of the sphingosine backbone is acylated by a fatty acid, as in sphingomyelin. Glycolipids differ from sphingomyelin in the nature of the unit that is linked to the primary hydroxyl group of the sphingosine backbone. In glycolipids, one or more sugars (rather than phosphoryl choline) are attached to this group. The simplest glycolipid is *cerebroside,* in which there is only one sugar residue, either glucose or galactose. More complex glycolipids, such as *gangliosides,* contain a branched chain of as many as seven sugar residues.

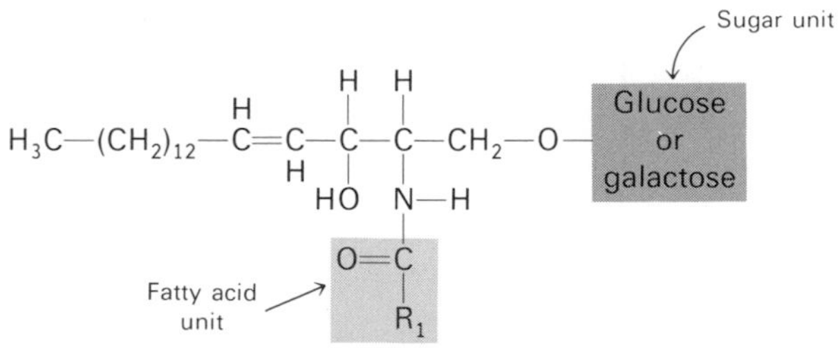

Sugar unit
Glucose or galactose
Fatty acid unit

Cerebroside
(A glycolipid)

Another important lipid in some membranes is *cholesterol.* This sterol is present in eucaryotes but not in most procaryotes. The oxygen atom in its 3-OH group comes from O_2. Cholesterol evolved after the earth's atmosphere became aerobic. The plasma membranes of eucaryotic cells are usually rich in cholesterol, whereas the membranes of their organelles typically have lesser amounts of this neutral lipid.

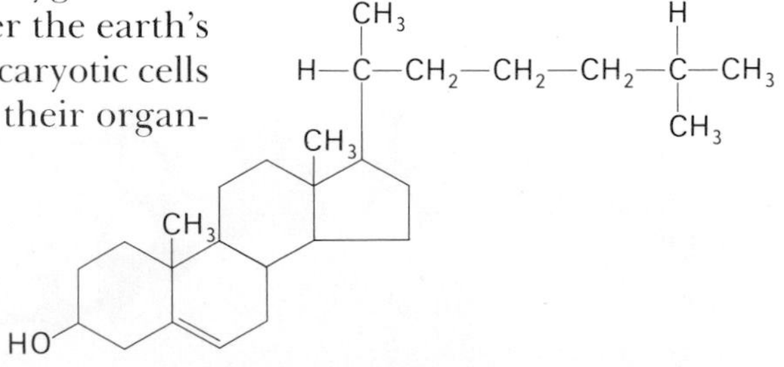

Cholesterol

PHOSPHOLIPIDS AND GLYCOLIPIDS READILY FORM BILAYERS

The repertoire of membrane lipids is extensive, perhaps even bewildering at first sight. However, they possess a critical common structural theme: *membrane lipids are amphipathic molecules.* They contain both a *hydrophilic* and a *hydrophobic* moiety (Table 12-1).

Table 12-1
Hydrophobic and hydrophilic units of membrane lipids

Membrane lipid	*Hydrophobic unit*	*Hydrophilic unit*
Phosphoglycerides	Fatty acid chains	Phosphorylated alcohol
Sphingomyelin	Fatty acid chain and hydrocarbon chain of sphingosine	Phosphoryl choline
Glycolipid	Fatty acid chain and hydrocarbon chain of sphingosine	One or more sugar residues
Cholesterol	Entire molecule except for OH group	OH group at C-3

Let us look at a space-filling model of a phosphoglyceride, such as phosphatidyl choline (Figure 12-6). Its overall shape is roughly rectangular. The two fatty acid chains are approximately parallel to one another, whereas the phosphoryl choline moiety points in the opposite direction. Sphingomyelin has a similar conformation (Figure 12-7). The sugar group of a glycolipid occupies nearly the same position as the phosphoryl choline unit of sphingomyelin. Therefore, the following shorthand has been adopted to represent these membrane lipids. The hydrophilic unit, also called the *polar head group,* is represented by a circle, whereas the hydrocarbon tails are depicted by straight or wavy lines (Figure 12-8).

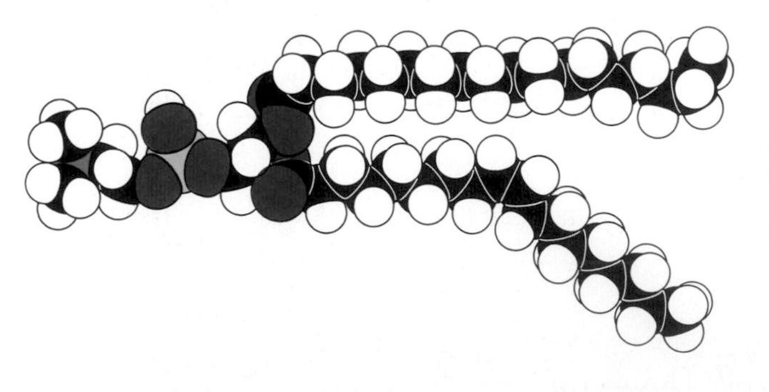

Figure 12-6
Space-filling model of a phosphatidyl choline molecule.

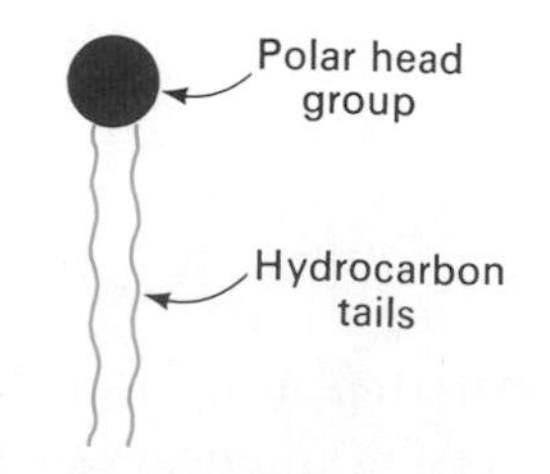

Figure 12-8
Symbol for a phospholipid or glycolipid molecule.

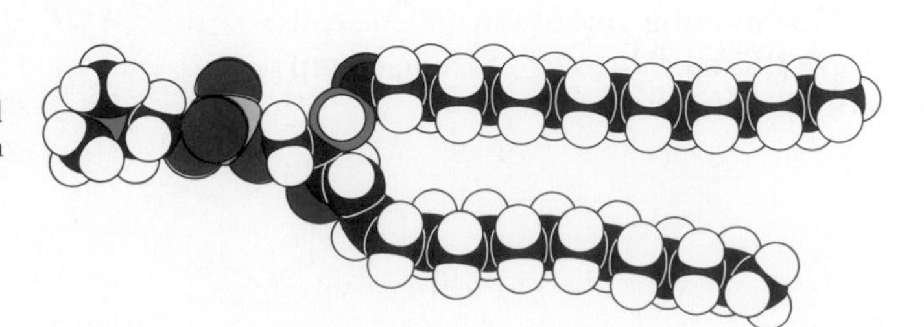

Figure 12-7
Space-filling model of a sphingomyelin molecule.

Now let us consider the arrangement of phospholipids and glycolipids in an aqueous medium. It is evident that their polar head groups will have affinity for water, whereas their hydrocarbon tails will avoid water. These preferences could be satisfied by formation of a *micelle,* a globular structure in which polar head groups are on the surface and hydrocarbon tails are sequestered inside (Figure 12-9). Another arrangement that fulfills both the hydrophilic and hydrophobic preferences of these membrane lipids is a *bimolecular sheet,* also called a *lipid bilayer* (Figure 12-10).

The favored structure for most phospholipids and glycolipids in aqueous media is a bimolecular sheet rather than a micelle. The reason is that their two fatty acyl chains are too bulky to fit into the interior of a micelle. In contrast, salts of fatty acids (such as sodium palmitate, a constituent of soap), which contain only one fatty acyl chain, readily form micelles. The formation of bilayers instead of micelles by phospholipids is of critical biological importance. A micelle is a limited structure, usually less than 200 Å in diameter. In contrast, a bimolecular sheet can have macroscopic dimensions, such as a millimeter (10^7 Å). Phospholipids and glycolipids are key membrane constituents because they readily form extensive bimolecular sheets. Furthermore, these sheets serve as permeability barriers, yet they are quite fluid.

The formation of lipid bilayers is a *self-assembly process.* In other words, the structure of a bimolecular sheet is inherent in the structure of the constituent lipid molecules, specifically in their amphipathic

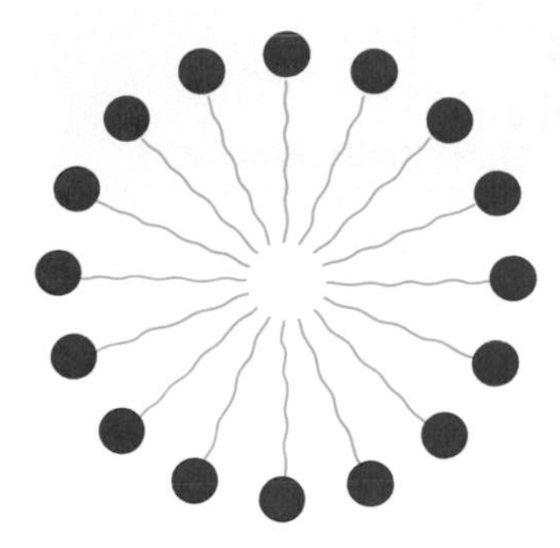

Figure 12-9
Diagram of a section of a micelle formed from ionized fatty acid molecules. Most phospholipids do not form micelles.

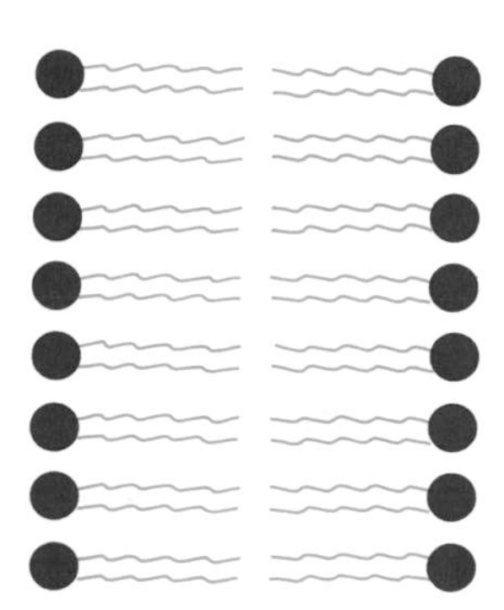

Figure 12-10
Diagram of a section of a bilayer membrane formed from phospholipid molecules.

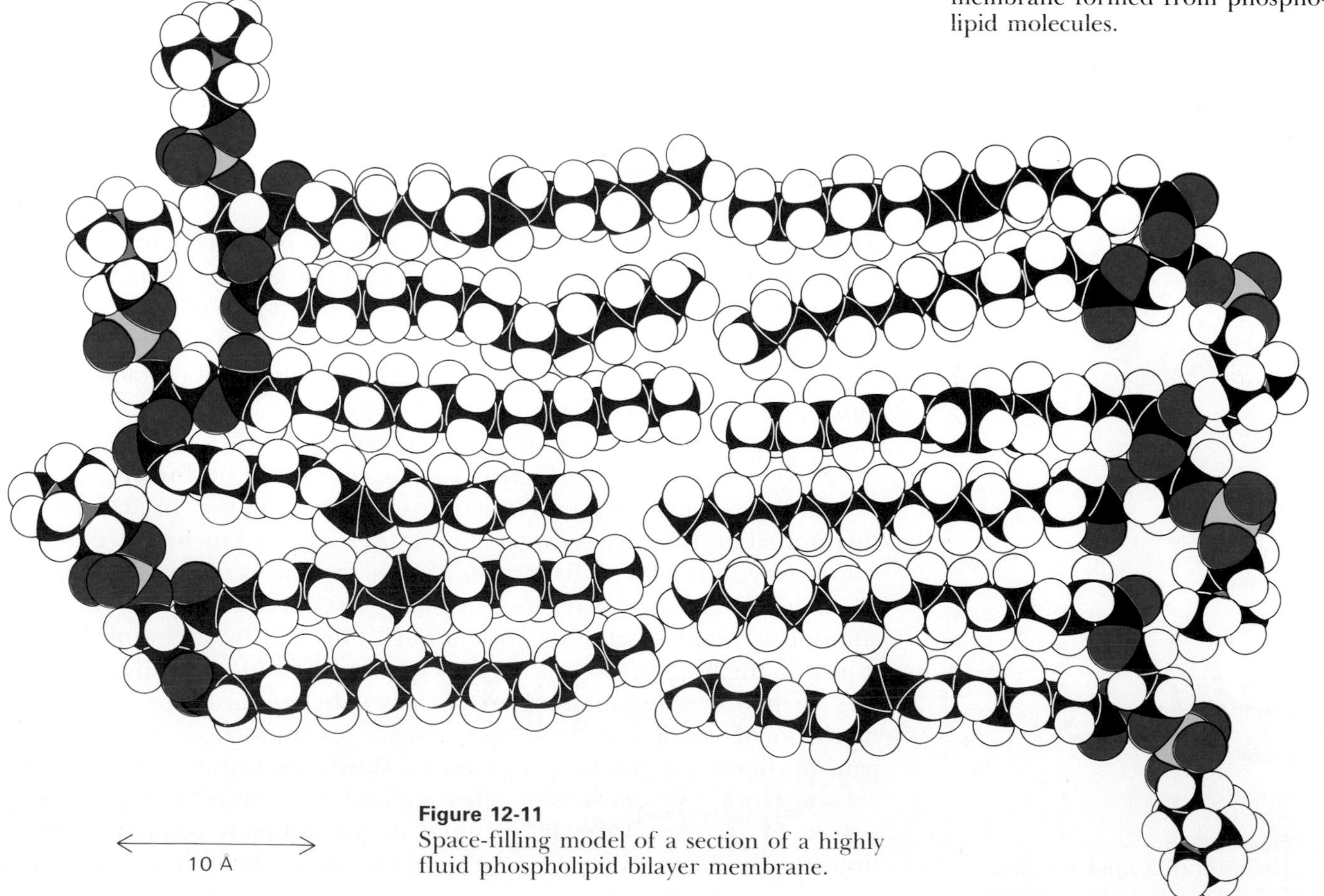

Figure 12-11
Space-filling model of a section of a highly fluid phospholipid bilayer membrane.

character. The formation of lipid bilayers from glycolipids and phospholipids is a rapid and spontaneous process in water. *Hydrophobic interactions are the major driving force for the formation of lipid bilayers.* Recall that hydrophobic interactions also play a dominant role in the folding of proteins in aqueous solution. Water molecules are released from the hydrocarbon tails of membrane lipids as these tails become sequestered in the nonpolar interior of the bilayer. Furthermore, there are *van der Waals attractive forces* between the hydrocarbon tails. These forces favor close packing of the tails. Finally, there are *electrostatic and hydrogen-bonding attractions between the polar head groups and water molecules.* Thus, lipid bilayers are stabilized by the full array of forces that mediate molecular interactions in biological systems.

LIPID BILAYERS ARE NONCOVALENT, COOPERATIVE STRUCTURES

Another important feature of lipid bilayers is that they are *cooperative structures.* They are held together by many *reinforcing, noncovalent interactions.* Phospholipids and glycolipids cluster together in water to minimize the number of exposed hydrocarbon chains. A pertinent analogy is the huddling together of sheep in the cold to minimize the area of exposed body surface. Clustering is also favored by the van der Waals attractive forces between adjacent hydrocarbon chains. These energetic factors have three significant biological consequences: (1) lipid bilayers have an inherent tendency to be *extensive*; (2) lipid bilayers will tend to *close on themselves* so that there are no edges with exposed hydrocarbon chains, which results in the formation of a compartment; and (3) lipid bilayers are *self-sealing* because a hole in a bilayer is energetically unfavorable.

LIPID BILAYERS ARE HIGHLY IMPERMEABLE TO IONS AND MOST POLAR MOLECULES

The permeability of lipid bilayers has been measured in two well-defined synthetic systems: lipid vesicles and planar bilayer membranes. These model systems have been sources of insight into a major function of biological membranes—namely, their role as permeability barriers. The key finding is that lipid bilayers are inherently impermeable to ions and most polar molecules.

Lipid vesicles (also known as *liposomes*) are aqueous compartments enclosed by a lipid bilayer (Figure 12-12). They can be formed by suspending a suitable lipid, such as phosphatidyl choline, in an aqueous medium. This mixture is then *sonicated* (i.e., agitated by high-frequency sound waves) to give a dispersion of closed vesicles that are quite uniform in size. Alternatively, vesicles can be prepared by rapidly mixing a solution of lipid in ethanol with water. This can be accomplished by injecting the lipid through a fine needle into an aqueous solution. Vesicles formed by these methods are nearly spherical in shape and have a diameter of about 500 Å. Larger vesicles (of the order of 10^4 Å, or 1 μm, in diameter) can be prepared by slowly evaporating the organic solvent from a suspension of phospholipid in a mixed solvent system.

Ions or molecules can be trapped in the aqueous compartment of lipid vesicles by forming the vesicles in the presence of these substances

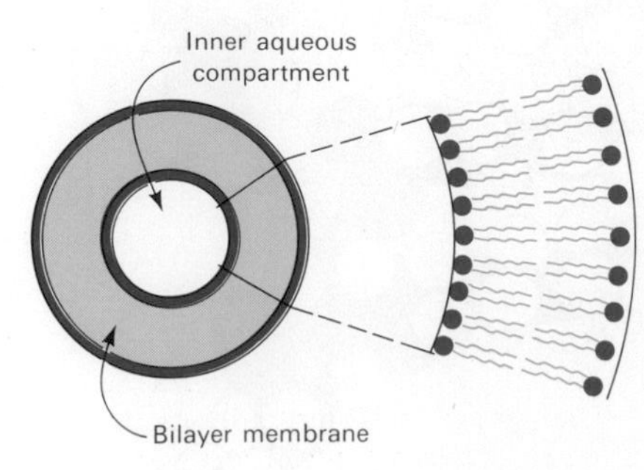

Figure 12-12
Diagram of a lipid vesicle.

(Figure 12-13). For example, if vesicles 500 Å in diameter are formed in a 0.1 M glycine solution, about two thousand molecules of glycine will be trapped in each inner aqueous compartment. These glycine-containing vesicles can be separated from the surrounding solution of glycine by dialysis or by gel-filtration chromatography. The permeability of the bilayer membrane to glycine can then be determined by measuring the rate of efflux of glycine from the inner compartment of the vesicle to the ambient solution. These lipid vesicles are valuable not only for permeability studies. They fuse with the plasma membrane of many kinds of cells and can thus be used to introduce a wide variety of impermeable substances into cells. The selective fusion of lipid vesicles with particular kinds of cells is a promising means of controlling the delivery of drugs to target cells.

Another well-defined synthetic membrane is a *planar bilayer membrane.* This structure can be formed across a 1-mm hole in a partition between two aqueous compartments. Such a membrane is very well suited for electrical studies because of its large size and simple geometry. Paul Mueller and Donald Rudin showed that a large bilayer membrane can be readily formed in the following way. A fine paint brush is dipped into a membrane-forming solution, such as phosphatidyl choline in decane. The tip of the brush is then stroked across a hole (1 mm in diameter) in a partition between two aqueous media. The lipid film across the hole thins spontaneously; the excess lipid forms a torus at the edge of the hole. A planar bilayer membrane consisting primarily of phosphatidyl choline is formed within a few minutes. The electrical conduction properties of this macroscopic bilayer membrane are readily studied by inserting electrodes into each aqueous compartment (Figure 12-14). For example, its permeability to ions is determined by measuring the current across the membrane as a function of the applied voltage.

Permeability studies of lipid vesicles and electrical conductance measurements of planar bilayers have shown that *lipid bilayer membranes have a very low permeability for ions and most polar molecules.* Water is a conspicuous exception to this generalization; it readily traverses such membranes. The range of measured permeability coefficients is very wide (Figure 12-15). For example, Na^+ and K^+ traverse these membranes 10^9 times more slowly than does H_2O. Tryptophan, a zwitterion at pH 7, crosses the membrane 10^3 times more slowly than indole, a structur-

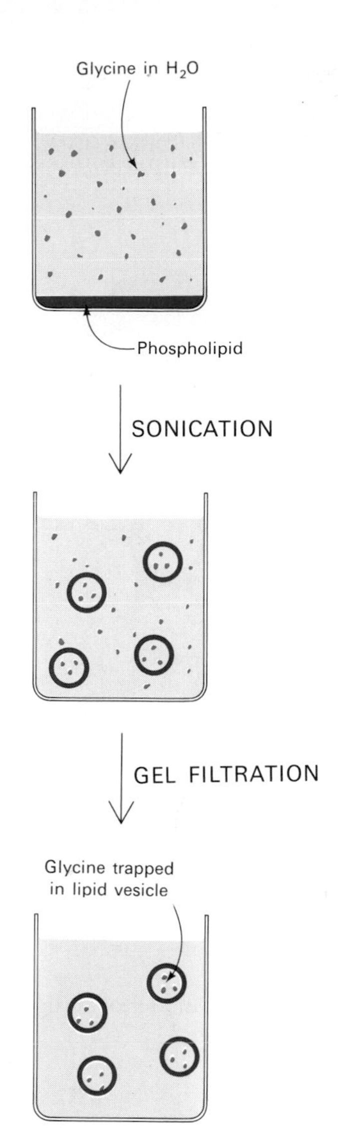

Figure 12-13
Preparation of a suspension of lipid vesicles containing glycine molecules.

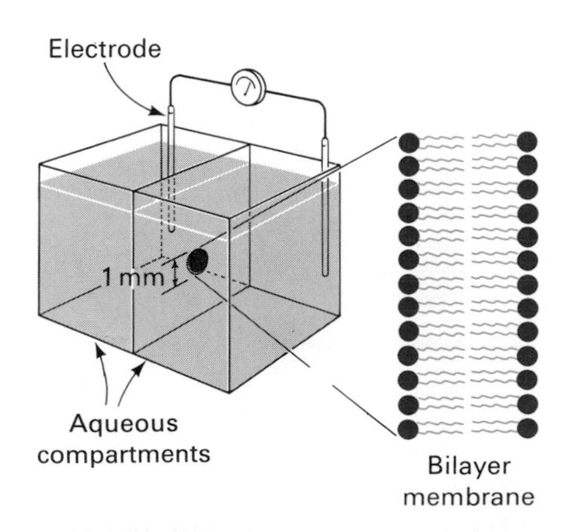

Figure 12-14
Experimental arrangement for the study of planar bilayer membranes. A bilayer membrane is formed across a 1-mm hole in a septum that separates two aqueous compartments.

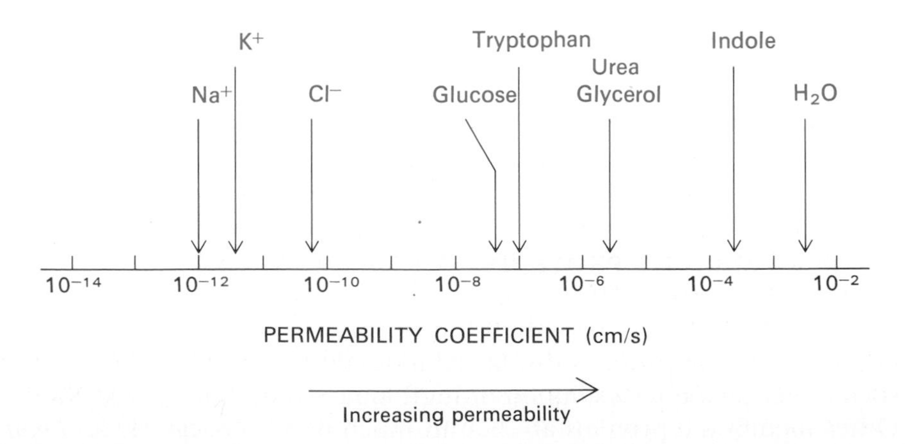

Figure 12-15
Permeability coefficients of some ions and molecules in lipid bilayer membranes.

ally related molecule that lacks ionic groups. *The permeability coefficients of small molecules are correlated with their solubility in a nonpolar solvent relative to their solubility in water.* This relationship suggests that a small molecule might traverse a lipid bilayer membrane in the following way: first, it sheds its solvation shell of water; then, it becomes dissolved in the hydrocarbon core of the membrane; finally, it diffuses through this core to the other side of the membrane, where it becomes resolvated by water. An ion such as Na^+ traverses membranes very slowly because the removal of its coordination shell of water molecules is highly unfavored energetically.

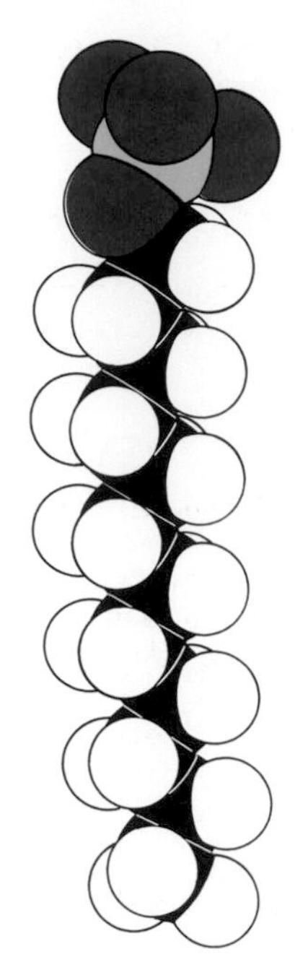

Figure 12-16
Space-filling model of sodium dodecyl sulfate.

PROTEINS CARRY OUT MOST MEMBRANE PROCESSES

We now turn to membrane proteins, which are responsible for most of the dynamic processes carried out by membranes. Membrane lipids form a permeability barrier and thereby establish compartments, whereas *specific proteins mediate nearly all other membrane functions.* Membrane lipids create the appropriate environment for the action of such proteins.

Membranes differ in their protein content. Myelin, a membrane that serves as an insulator around certain nerve fibers, has a low content of protein (18%). Lipid, the major molecular species in myelin, is well suited for insulation. In contrast, the plasma membranes of most other cells are much more active. They contain many pumps, gates, receptors, and enzymes. The protein content of these plasma membranes is typically 50%. Membranes involved in energy transduction, such as the internal membranes of mitochondria and chloroplasts, have the highest content of protein, typically 75%.

The protein components of a membrane can be readily visualized by *SDS polyacrylamide-gel electrophoresis.* The membrane to be analyzed is first solubilized in a 1% solution of sodium dodecyl sulfate (SDS), an amphipathic molecule (Figure 12-16). This detergent disrupts most protein-protein and protein-lipid interactions. The sample is then electrophoresed in an acrylamide gel containing SDS. As discussed earlier (p. 45), the electrophoretic mobility of many proteins in this gel depends on their mass rather than on their net charge in the absence of SDS. The gel electrophoresis patterns of three membranes—the plasma membrane of erythrocytes, the photoreceptor membrane of retinal rod cells, and the sarcoplasmic reticulum membrane of muscle—are shown in Figure 12-17. It is evident that these three membranes have very different protein compositions. In general, *membranes performing different functions contain different proteins.*

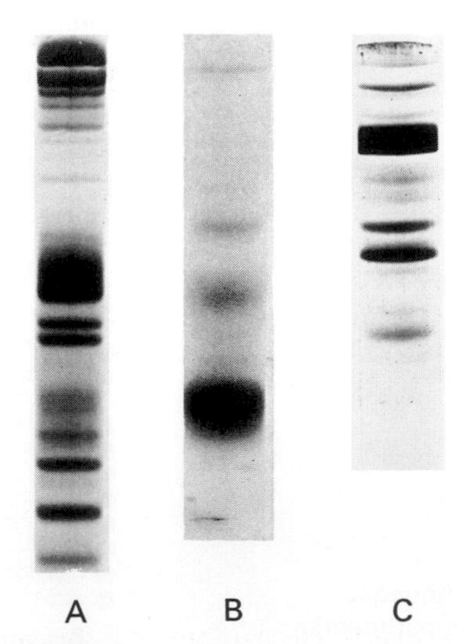

Figure 12-17
SDS-acrylamide-gel patterns of (A) the plasma membrane of erythrocytes, (B) the disc membranes of retinal rod cells, and (C) the sarcoplasmic reticulum membrane of muscle cells. [Courtesy of Dr. Theodore Steck (part A) and Dr. David MacLennan (part C).]

SOME MEMBRANE PROTEINS ARE DEEPLY IMBEDDED IN THE LIPID BILAYER

Some membrane proteins can be solubilized by relatively mild means, such as extraction by a solution of high ionic strength (e.g., 1 M NaCl). Other membrane proteins are bound much more tenaciously and can be solubilized only by using a detergent (see Figure 12-33) or an organic solvent. Membrane proteins can be classified as being either *peripheral* or *integral* on the basis of this difference in dissociability (Figure 12-18). Integral proteins interact extensively with the hydrocarbon chains of membrane lipids and so they can be released only by agents that compete for these nonpolar interactions. In fact, nearly all known integral membrane proteins span the lipid bilayer. In contrast, peripheral proteins are bound to membranes by electrostatic and hydrogen-bond interactions. These polar interactions can be disrupted by adding salts or by changing the pH. Most peripheral membrane proteins are bound to the surfaces of integral proteins, either on the cytosolic or extracellular side of the membrane.

Freeze-fracture electron microscopy is a valuable technique for ascertaining whether proteins are located in the interior of biological membranes. Cells or membrane fragments are rapidly frozen to the temperature of liquid nitrogen. The frozen membrane is then fractured by the impact of a microtome knife. Cleavage usually occurs along a plane in the middle of the bilayer, between its leaflets (Figure 12-19). Hence, extensive regions *within* the lipid bilayer are exposed. These exposed regions can then be shadowed with carbon and platinum, which produces a replica of the interior of the bilayer. The external surfaces of membranes can also be viewed by combining freeze-fracture and deep-etching techniques. First, the interior of a frozen membrane is exposed by fracturing. The ice that covers one of the adjacent membrane surfaces is then sublimed away; this process is termed deep-etching. The combined technique, called *freeze-etching electron microscopy,* provides a view of the interior of a membrane and of both its surfaces. An attractive feature of the freeze-etching technique is that fixatives and dehydrating agents are not required.

Freeze-etching studies provided the first direct evidence for the presence of integral proteins in many biological membranes. The interior of erythrocyte membrane, for example, is dense with globular particles approximately 75 Å in diameter (Figure 12-20). The inside of the sarco-

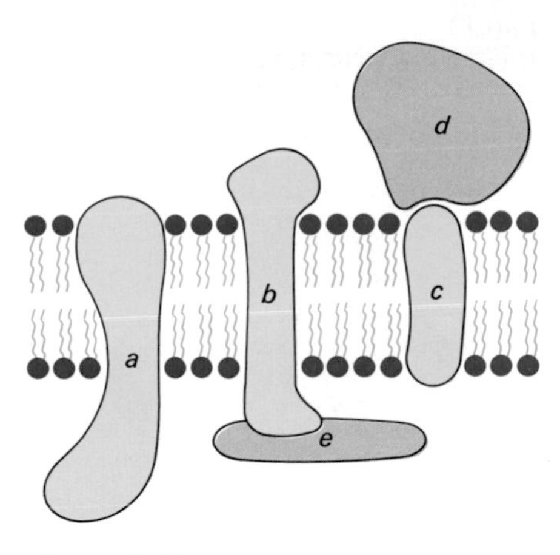

Figure 12-18
Integral membrane proteins (*a, b, c,*) interact extensively with the hydrocarbon region of the bilayer. Nearly all known integral membrane protein traverse the lipid bilayer. Peripheral membrane proteins (*d*) and (*e*) bind to the surface of integral membrane proteins.

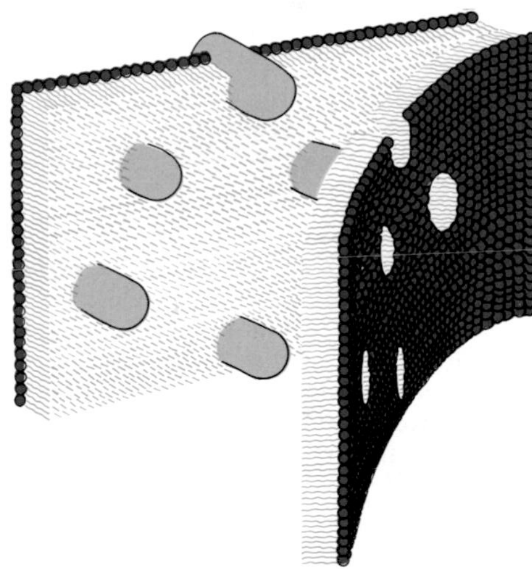

Figure 12-19
Technique of freeze-fracture electron microscopy. The cleavage plane passes through the middle of the bilayer membrane. [After S. J. Singer. *Hosp. Pract.* 8(1973):81.]

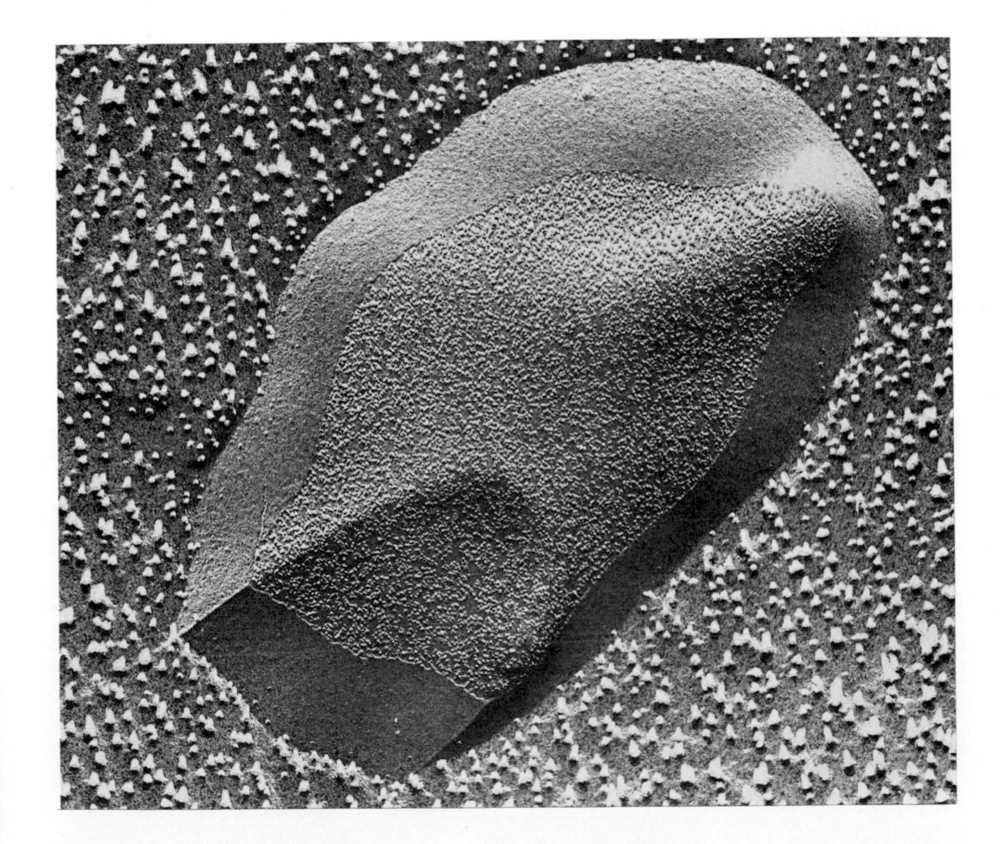

Figure 12-20
Freeze-etch electron micrograph of the plasma membrane of a red blood cell. The interior of the membrane, which has been exposed by fracture of the membrane, is rich in globular particles that have a diameter of about 75 Å. These particles are integral membrane proteins. [Courtesy of Dr. Vincent Marchesi.]

plasmic reticulum membrane is also rich in globular particles. In contrast, synthetic bilayers formed from phosphatidyl choline yield smooth fracture faces. Also, the fracture faces of large areas of myelin membranes are smooth, as might be expected for a relatively inert membrane that serves primarily as an insulator.

LIPIDS AND MANY MEMBRANE PROTEINS DIFFUSE RAPIDLY IN THE PLANE OF THE MEMBRANE

Biological membranes are not rigid structures. On the contrary, lipids and many membrane proteins are constantly in lateral motion. The rapid movement of membrane proteins has been visualized by means of fluorescence microscopy. Human cells and mouse cells in culture can be induced to fuse with each other. The resulting hybrid cell is called a *heterokaryon*. Part of the plasma membrane of this heterokaryon comes from a mouse cell, the rest from a human cell. Do the membrane proteins derived from the mouse and human cells stay segregated in the heterokaryon or do they intermingle? This question was answered by using fluorescent-labeled antibodies as markers that could be followed by light microscopy. An antibody specific for mouse membrane proteins was labeled to show a green fluorescence, and an antibody specific for human membrane proteins was labeled to show a red fluorescence (Figure 12-21). In a newly formed heterokaryon, half of the surface displayed green fluorescing patches, the other half red. However, in less than an hour (at 37°C), the red and green fluorescing patches became completely intermixed. This experiment revealed that a *membrane protein can diffuse through a distance of several microns in approximately one minute*.

Figure 12-21
Diagram showing the fusion of a mouse cell and a human cell, followed by diffusion of membrane components in the plane of the plasma membrane. The green and red fluorescing markers are completely intermingled after several hours.

A more general and quantitative method for measuring the lateral mobility of membrane molecules in intact cells is the *fluorescence photobleaching recovery technique* (Figure 12-22). First, a cell-surface component is specifically labeled with a fluorescent chromophore. A small region of the cell surface (~3 μm^2) is viewed through a fluorescence microscope. The fluorescent molecules in this region are then destroyed by a very intense light pulse from a laser. The fluorescence of

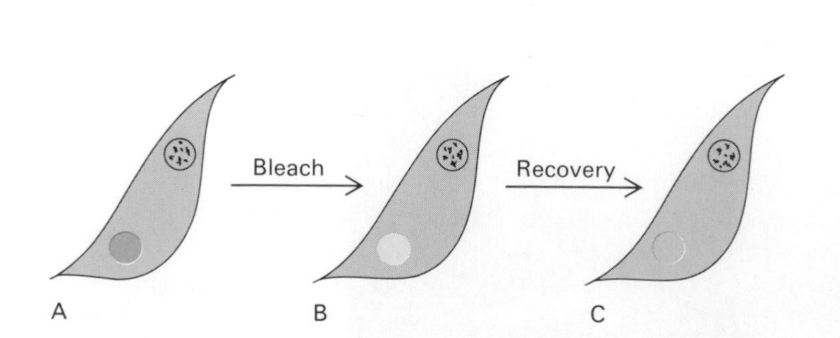

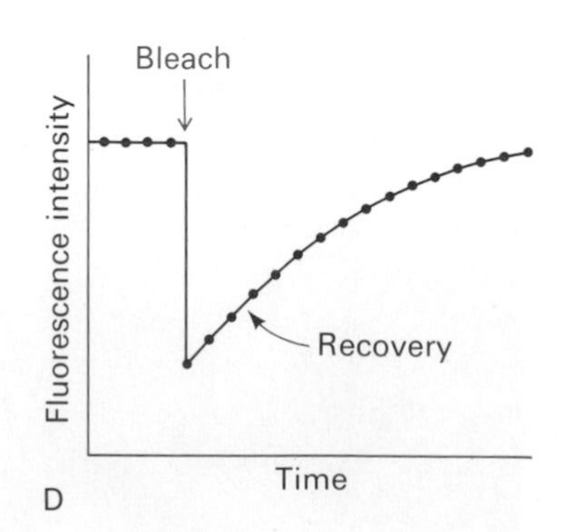

Figure 12-22
Fluorescence photobleaching recovery technique: (A) fluorescence from a labeled cell-surface component in a small illuminated region of a cell; (B) the fluorescent molecules are bleached by an intense light pulse; (C) the fluorescence intensity recovers as bleached molecules diffuse out of the illuminated region and unbleached molecules diffuse into it; the rate of recovery (D) depends on the diffusion coefficient.

this region is subsequently monitored as a function of time using a light level sufficiently low to prevent further bleaching. If the labeled component is mobile, bleached molecules leave and unbleached molecules enter the illuminated region, which results in an increase in the fluorescence intensity. The rate of recovery of fluorescence depends on the lateral mobility of the fluorescent-labeled component, which can be expressed in terms of a diffusion coefficient D. The average distance s (in centimeters) traversed in two dimensions in time t (seconds) depends on D ($cm^2\ s^{-1}$) according to

$$s = (4\ Dt)^{1/2}$$

The diffusion coefficient of lipids in a variety of membranes is about $10^{-8}\ cm^2\ s^{-1}$. Thus, a phospholipid molecule diffuses an average distance of 2×10^{-4} cm, or 2 μm, in 1 s. This means that a *lipid molecule can travel from one end of a bacterium to the other in a second*. The magnitude of the observed diffusion coefficient indicates that the viscosity of the membrane is about one hundred times that of water, rather like that of olive oil.

In contrast, proteins vary markedly in their lateral mobility. *Some proteins are nearly as mobile as lipids, whereas others are virtually immobile*. For example, the photoreceptor protein rhodopsin, a very mobile protein, has a diffusion coefficient of $4 \times 10^{-9}\ cm^2\ s^{-1}$. The rapid movement of rhodopsin is essential for fast signaling. At the other extreme is fibronectin, a peripheral glycoprotein that interacts with the surrounding matrix. For fibronectin, D is less than $10^{-12}\ cm^2\ s^{-1}$. Fibronectin has a very low mobility because it is anchored to actin filaments on the other side of the plasma membrane through *integrin*. This transmembrane protein links the extracellular matrix to the cytoskeleton.

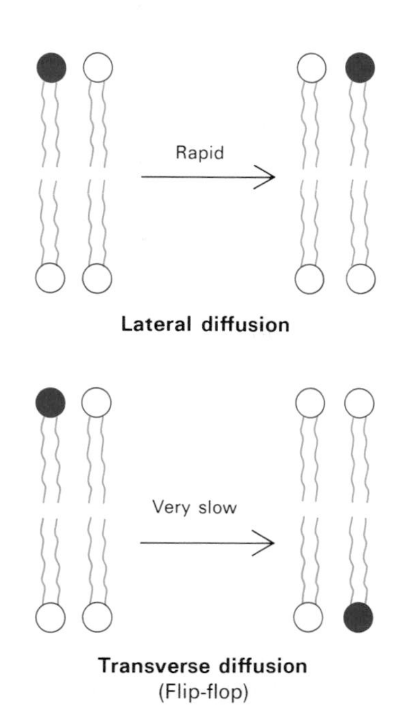

Figure 12-23
Lateral diffusion of lipids is much more rapid than transverse diffusion (flip-flop).

MEMBRANE PROTEINS DO NOT ROTATE ACROSS BILAYERS

The spontaneous rotation of lipids from one face of a membrane to the other is a very slow process, in contrast with their movement parallel to the plane of the bilayer. The transition of a molecule from one membrane surface to the other is called *transverse diffusion*, or *flip-flop*, whereas diffusion in the plane of a membrane is termed *lateral diffusion* (Figure 12-23). The flip-flop of phospholipid molecules in phosphatidyl choline vesicles has been directly measured by electron spin resonance techniques, which showed that *a phospholipid molecule flip-flops once in several hours* (see problem 5, p. 312, for the experimental design). Thus, a phospholipid molecule takes about 10^9 times as long to flip-flop across a membrane as it takes to diffuse a distance of 50 Å in the lateral direction.

The free-energy barriers to the flip-flopping of protein molecules are even larger than for lipids because proteins have more extensive polar regions. In fact, *the flip-flop of a protein molecule has not been observed*. Hence, *membrane asymmetry can be preserved for long periods*.

FLUID MOSAIC MODEL OF BIOLOGICAL MEMBRANES

In 1972, S. Jonathan Singer and Garth Nicolson proposed a fluid mosaic model for the overall organization of biological membranes. The essence of their model is that membranes are *two-dimensional solutions of*

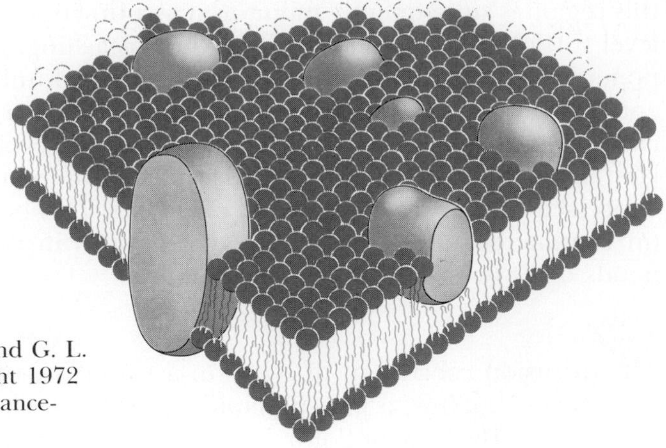

Figure 12-24
Fluid mosaic model. [After S. J. Singer and G. L. Nicolson. *Science* 175(1972):723. Copyright 1972 by the American Association for the Advancement of Science.]

oriented globular proteins and lipids (Figure 12-24). This proposal is supported by a wide variety of experimental observations. The major features of this model are:

1. Most of the membrane phospholipid and glycolipid molecules are arranged in a bilayer. This lipid bilayer has a dual role: it is both a *solvent* for integral membrane proteins and a *permeability barrier*.

2. A small proportion of membrane lipids interact specifically with particular membrane proteins and may be essential for their function.

3. Membrane proteins are free to diffuse laterally in the lipid matrix unless restricted by special interactions, whereas they are not free to rotate from one side of a membrane to the other.

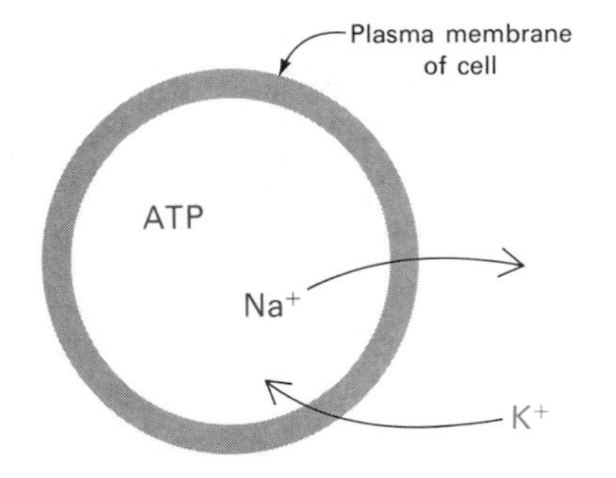

Figure 12-25
Asymmetry of the Na^+-K^+ transport system in plasma membranes.

MEMBRANES ARE ASYMMETRIC

Membranes are structurally and functionally asymmetric. The outer and inner surfaces of *all known biological membranes have different components and different enzymatic activities*. A clear-cut example is the pump that regulates the concentration of Na^+ and K^+ ions in cells (Chapter 37). This transport system is located in the plasma membrane of nearly all cells in higher organisms. The Na^+-K^+ pump assembly is oriented in the plasma membrane so that it pumps Na^+ out of the cell and K^+ into it (Figure 12-25). Furthermore, ATP must be on the inside of the cell to drive the pump. Ouabain, a specific inhibitor of the pump, is effective only if it is located outside.

As will be discussed in detail in Chapter 31, membrane proteins have a unique orientation because they are synthesized and inserted into the membrane in an asymmetric manner. This absolute asymmetry is preserved because membrane proteins do not rotate from one side of the membrane to the other, and because *membranes are always synthesized by growth of pre-existing membranes*. Lipids, too, are asymmetrically distributed as a consequence of their mode of biosynthesis, but this asymmetry is usually not absolute, except in the case of glycolipids. For example, in the membrane of red blood cells, sphingomyelin and phosphatidyl choline are preferentially located in the outer leaflet of the bilayer, whereas phosphatidyl ethanolamine and phosphatidyl serine are mainly in the inner leaflet. Large amounts of cholesterol are present in both leaflets.

MEMBRANE FLUIDITY IS CONTROLLED BY FATTY ACID COMPOSITION AND CHOLESTEROL CONTENT

The fatty acyl chains of lipid molecules in bilayer membranes can exist in an ordered, rigid state or in a relatively disordered, fluid state. In the ordered state, all of the C–C bonds have a *trans* conformation, whereas in the disordered state, some are in the *gauche* conformation (Figure 12-26).

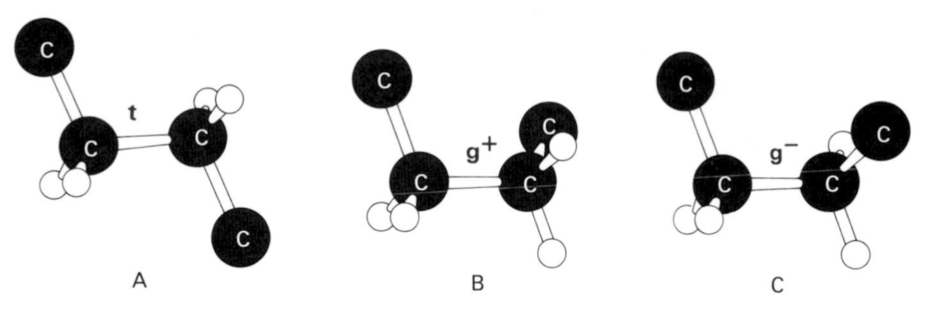

Figure 12-26
Conformation of C–C bonds in fatty acyl chains: (A) *trans* (t) conformation; (B and C) a 120-degree rotation yields a *gauche* (g) conformation, which can be either g^+ (clockwise rotation) or g^- (counterclockwise rotation).

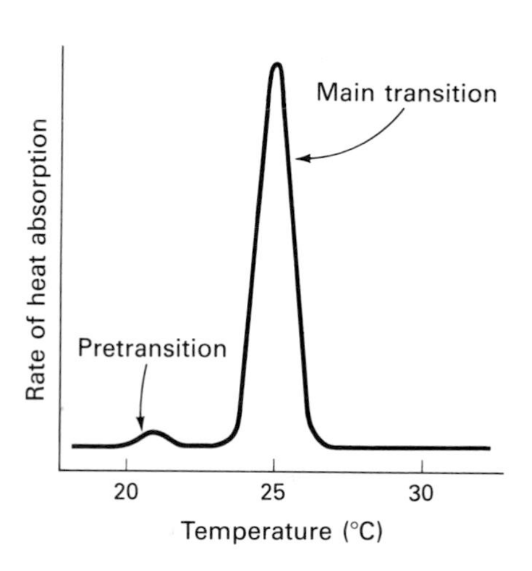

Figure 12-27
Phosphatidyl choline bilayers undergo a phase transition when heated, as detected here by differential scanning calorimetry, which measures the rate of uptake of heat on warming a sample. The small peak (the pretransition) comes from a change in tilt of fatty acyl chains with respect to the bilayer. The major peak arises from a phase transition in which crystalline fatty acyl chains become disordered because of the introduction of kinks.

The transition from the rigid (all *trans*) to the fluid (partly *gauche*) state occurs rather abruptly as the temperature is raised above T_m, the melting temperature (Figure 12-27). *This transition temperature depends on the length of the fatty acyl chains, and on their degree of unsaturation.* The rigid state is favored by the presence of saturated fatty acyl residues because their straight hydrocarbon chains interact very favorably with each other (Figure 12-28A). On the other hand, *a* cis *double bond produces a bend in the hydrocarbon chain. This bend interferes with a highly ordered packing of fatty acyl chains,* and so T_m is lowered (Figure 12-28B). The length of the fatty acyl chain also affects the transition temperature. Long hydrocarbon chains interact more strongly than do short ones. Specifically, each additional —CH_2— group makes a favorable contribution of about −0.5 kcal/mol to the free energy of interaction of two adjacent hydrocarbon chains.

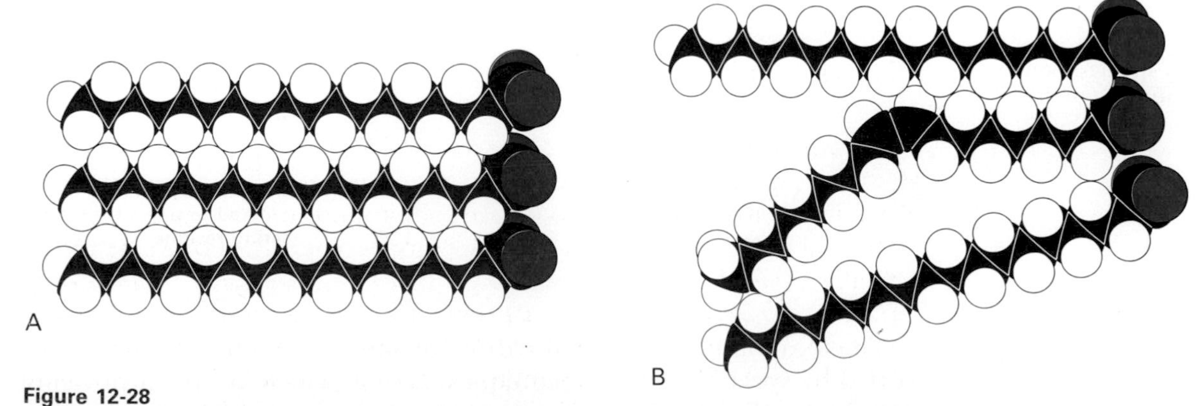

Figure 12-28
The highly ordered packing of fatty acid chains is disrupted by the presence of *cis* double bonds. These space-filling models show the packing of (A) three molecules of stearate (C_{18}, saturated) and (B) a molecule of oleate (C_{18}, unsaturated) between two molecules of stearate.

Procaryotes regulate the fluidity of their membranes by varying the number of double bonds and the length of their fatty acyl chains. For example, the ratio of saturated to unsaturated fatty acyl chains in the *E. coli* membrane decreases from 1.6 to 1.0 as the growth temperature is lowered from 42°C to 27°C. This decrease in the proportion of saturated residues prevents the membrane from becoming too rigid at the lower temperature. *In eucaryotes, cholesterol also is a key regulator of membrane fluidity.* Cholesterol contains a bulky steroid nucleus with a hydroxyl group at one end and a flexible hydrocarbon tail at the other end (Figure 12-29). Cholesterol inserts into bilayers with its long axis perpendicular to the plane of the membrane. The hydroxyl group of cholesterol hydrogen bonds to a carbonyl oxygen atom of a phospholipid head group, whereas the hydrocarbon tail of cholesterol is located in the nonpolar core of the bilayer. Cholesterol prevents the crystallization of fatty acyl chains by fitting between them. In fact, high concentrations of cholesterol abolish phase transitions of bilayers. An opposite effect of cholesterol is to sterically block large motions of fatty acyl chains, which makes membranes less fluid. Thus, *cholesterol moderates the fluidity of membranes.*

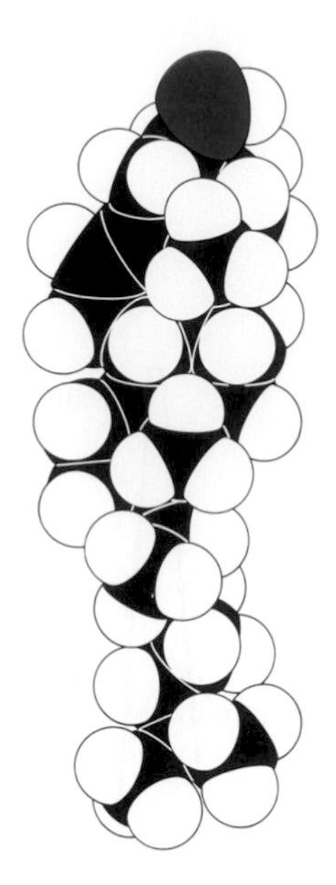

Figure 12-29
Space-filling model of cholesterol.

CARBOHYDRATE UNITS ARE LOCATED ON THE EXTRACELLULAR SIDE OF PLASMA MEMBRANES

Membranes of eucaryotic cells usually have a carbohydrate content of between 2% and 10% contributed by the sugar residues of their *glycolipids* and *glycoproteins*. As mentioned earlier, the glycolipids of higher organisms are derivatives of sphingosine with one or more attached sugar residues. In membrane glycoproteins, sugars are attached either to the amide nitrogen atom in the side chain of asparagine (termed an *N*-linkage) or to the oxygen atom in the side chain of serine or threonine (termed an *O*-linkage). The sugar directly attached to one of these side chains is usually *N*-acetylglucosamine or *N*-acetylgalactosamine (Figure 12-30). The amino acid sequence near an *N*-linked sugar is either Asn-X-Ser or Asn-X-Thr (X is any residue).

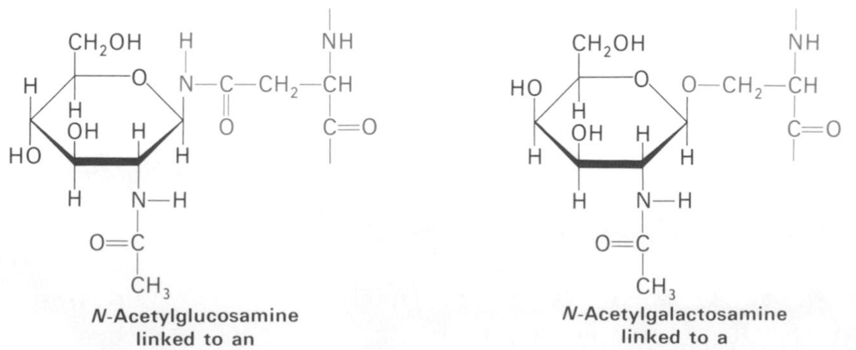

A *N*-Acetylglucosamine linked to an asparagine residue

B *N*-Acetylgalactosamine linked to a serine residue

Figure 12-30
Linkage of a sugar to (A) an asparagine side chain (an *N*-linkage) and (B) a serine side chain (an *O*-linkage).

The location of these carbohydrate groups in membranes can be revealed by specific labeling techniques. *Lectins*, which are plant proteins with high affinity for specific sugar residues, are valuable probes for this purpose. For example, *concanavalin A* binds to internal and nonreducing terminal α-mannosyl residues, whereas *wheat-germ agglutinin* binds to terminal *N*-acetylglucosamine residues. These lectins can be readily seen in electron micrographs if they are conjugated to an electron-dense marker such as ferritin, a 460-kd protein with a central core of about 4000 Fe^{3+} ions. Concanavalin A binds specifically to the outer surface of the erythrocyte membrane and not to the cytosolic surface. Studies of the binding of a wide variety of lectins to many kinds of cell membranes have shown that *sugar residues of membrane glycolipids and glycoproteins are always located on the extracellular side of membranes* (Figure 12-31). The reason for this asymmetry will become evident when we consider how proteins are targeted to membranes (Chapter 31).

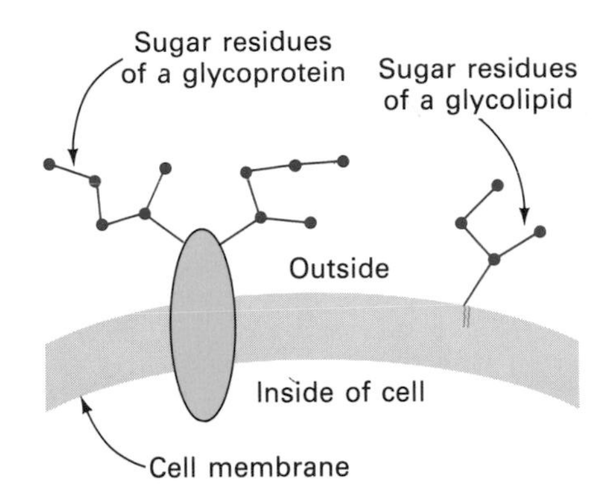

Figure 12-31
Sugar residues of membrane glycoproteins and glycolipids are located on the extracellular surface of mammalian plasma membranes.

Carbohydrate groups may serve to orient glycoproteins in membranes. Because sugars are very hydrophilic, the sugar residues of a glycoprotein or of a glycolipid will strongly prefer to be located near the aqueous surface rather than in the hydrocarbon core. The cost in free energy of inserting an oligosaccharide chain into the hydrocarbon core of a membrane is very high. Consequently, there is a high barrier to the rotation of a glycoprotein from one side of a membrane to the other. The carbohydrate moieties of membrane glycoproteins help to maintain the asymmetric character of biological membranes.

Carbohydrates on cell surfaces may also be important in *intercellular recognition*. The interaction of different cells to form a tissue and the detection of foreign cells by the immune system of a higher organism are examples of processes that depend on the recognition of one cell surface by another. Carbohydrates have the potential for great structural diversity. An enormous number of patterns of surface sugars is possible because (1) different monosaccharides can be joined to each other through any of several hydroxyl groups, (2) the C-1 linkage can have either an α or a β configuration, and (3) extensive branching is possible. Indeed, many more different oligosaccharides can be formed from four sugars than oligopeptides from four amino acids.

FUNCTIONAL MEMBRANE SYSTEMS CAN BE RECONSTITUTED FROM PURIFIED COMPONENTS

Peripheral membrane proteins can be removed from a membrane by changing the pH and salt concentration. The integral membrane proteins in this stripped preparation are then released from their interactions with each other and with membrane lipids by adding a detergent. An ideal detergent has three properties: (1) it should dissociate the protein of interest from other membrane components; (2) it should not unfold the protein; and (3) it should be readily removable following purification. The balance is a delicate one. For example, sodium

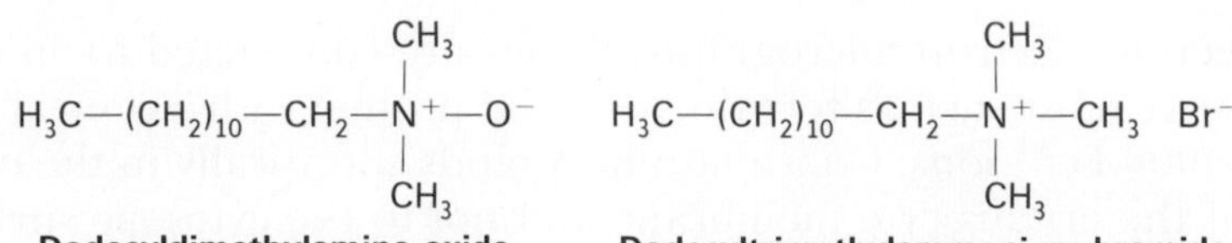

Dodecyldimethylamine oxide (DDAO, Ammonyx LO)

Dodecyltrimethylammonium bromide (DTAB)

Polyoxyethylene *p*-*t*-octyl phenol
(Triton X series)

Sodium cholate

Cholamidopropyldiethyl ammoniopropane sulfonate (CHAPS)

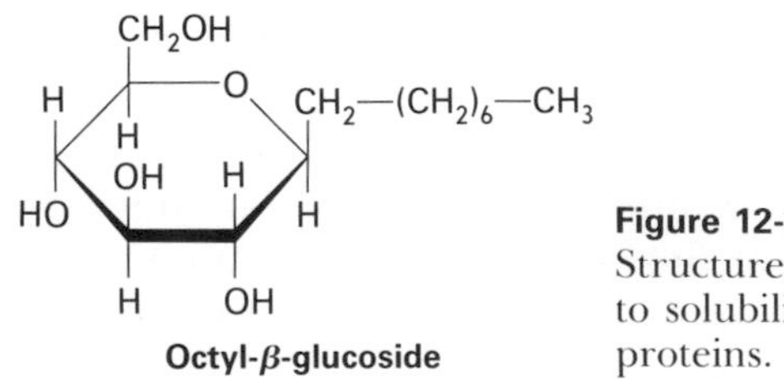

Octyl-β-glucoside

Figure 12-32
Structures of some detergents used to solubilize and purify membrane proteins.

Intact membrane

Detergent

Detergent-solubilized proteins

Figure 12-33
Solubilization of integral membrane proteins by the addition of detergent.

dodecyl sulfate is highly effective in dissociating proteins from other membrane components, but it denatures them and is hard to remove. Gentler detergents are used, such as sodium cholate and octyl glucoside (Figure 12-32). An early step in purifying and reconstituting a membrane protein is to test a series of detergents to determine which one is best suited for the particular task.

An excess of detergent is added so that not more than one protein molecule is present in a detergent micelle (Figure 12-33). The components of this mixture of proteins, phospholipids, and detergent are then separated by chromatographic and sedimentation methods similar to those used to purify water-soluble proteins. For example, ion-exchange chromatography is effective when nonionic (e.g., Triton X-100), or zwitterionic (e.g., DDAO) detergents are used (see Figure 12-32). Some purified membrane proteins in detergent solution are functionally active. For example, rhodopsin, a photoreceptor protein from the rod cells of the retina, has the same 500-nm absorption band

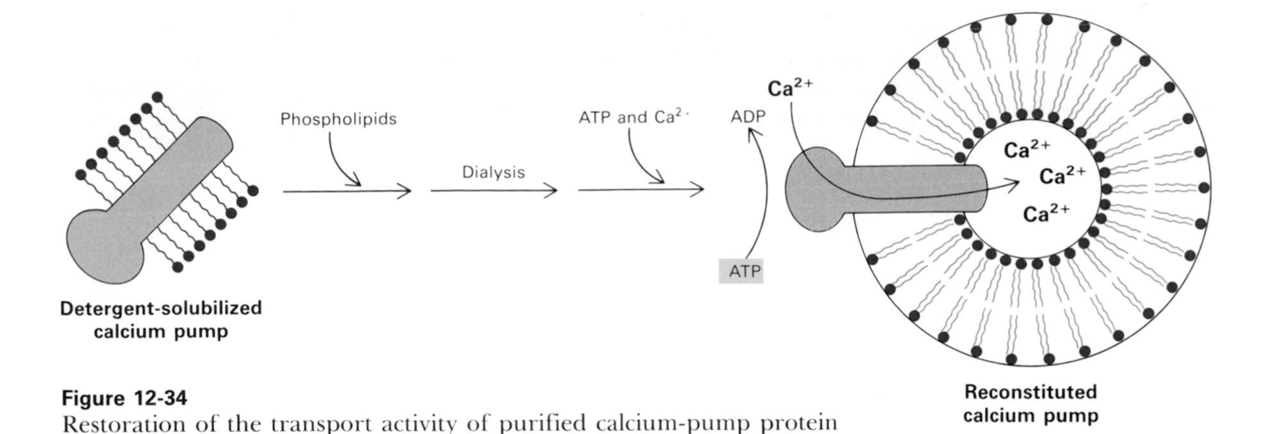

Figure 12-34
Restoration of the transport activity of purified calcium-pump protein (Ca^{+2}-ATPase).

in detergents as in the retinal disk membrane. Furthermore, the prosthetic group of this receptor protein undergoes the same structural change on illumination in both environments. Calsequestrin, a calcium-binding protein from the sarcoplasmic reticulum of muscle, retains this ion-binding property when solubilized in detergent solution.

However, some of the most important properties of membrane proteins are expressed only in a membrane. For example, activity of membrane channels can be detected only in an intact membrane. Likewise, the effect of a transmembrane voltage or pH gradient can be measured only when the protein of interest is located in a membrane separating two compartments. These *vectorial properties* of membrane proteins, which are at the heart of their biological importance, become manifest only when they are situated in a membrane. Hence, the second stage in the study of a membrane protein is to return it to the bilayer (Figure 12-34). This can be accomplished by adding synthetic phospholipids to the purified protein in detergent solution and dialyzing away the detergent. Phospholipid vesicles containing the protein form spontaneously as the concentration of detergent decreases during dialysis. An important difference between such a reconstituted membrane and the native one is that the proteins in the vesicle are usually oriented randomly. However, vectorial functions can be assayed by the addition of substrates and ions to one side of the membrane only. For example, vesicles containing a calcium pump protein (called the Ca^{2+} ATPase) accumulate Ca^{2+} in their inner aqueous space if ATP, the energy source, and Ca^{2+} are added to the outside (Figure 12-34). As will be discussed in detail in Chapters 37 to 39, *the reconstitution of functionally active membrane systems from purified components is a powerful experimental approach to the elucidation of membrane processes.*

THE RED-CELL MEMBRANE CONTAINS A VARIETY OF PERIPHERAL AND INTEGRAL PROTEINS

Erythrocytes have been choice objects of inquiry in studies of membranes because of their ready availability and relative simplicity. They lack organelles and thus have only a single membrane, the plasma membrane. Nearly all of the cytoplasmic contents of these cells can be released by osmotic hemolysis to give *ghosts,* which are quite pure plasma membranes. An SDS-polyacrylamide-gel pattern of this membrane preparation is shown in Figure 12-35. More than ten bands are

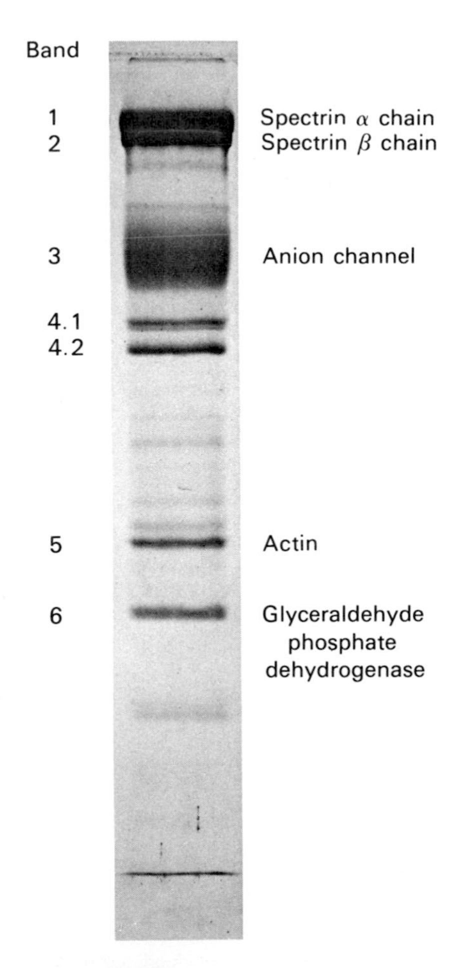

Figure 12-35
SDS-polyacrylamide-gel pattern of the erythrocyte membrane. This gel was stained with coomassie blue. [Courtesy of Dr. Vincent Marchesi.]

evident when this gel is stained with coomassie blue. They are numbered starting at the top of the gel (greatest apparent mass) and referred to as band 1, band 2, and so forth. Staining with the periodic acid–Schiff (PAS) reagent displays several bands that are rich in carbohydrate and are not strongly stained by coomassie blue.

Periodic acid–Schiff reagent (PAS)—
Detection of carbohydrate units by periodate oxidation, which generates aldehydes, followed by addition of a dye containing an amino or hydrazide group.

R—CH(OH)—CH(OH)—R′

Periodate

R—CHO OHC—R′

Schiff's reagent

R—CH=N—Dye Dye—N=CH—R′

What are the locations of these proteins in the red-cell membrane? The proteins corresponding to bands 1, 2, 4.1, 4.2, 5, and 6 can be extracted from the membranes by altering the ionic strength or pH of the medium, and so they are peripheral proteins. Furthermore, they are unaffected when intact red cells or sealed ghosts are incubated with a variety of proteolytic enzymes. In contrast, they are extensively digested when leaky ghosts are treated with proteases. *Hence, these peripheral proteins are located on the cytoplasmic face of the red-cell membrane.*

The identity and function of most of the major peripheral membrane proteins are now known. Bands 1 and 2 are the dimeric and monomeric forms of *spectrin*, a protein that forms extended filamentous networks. Spectrin interacts with other proteins to *stabilize and regulate the shape of the red-cell membrane*, as will be discussed shortly. Band 4.1 helps to link spectrin filaments to the membrane. Band 5 is actin, a key protein in muscle contraction and cell motility (p. 927). Band 6 is glyceraldehyde 3-phosphate dehydrogenase, an enzyme that participates in the generation of metabolic energy from the breakdown of sugars (p. 354).

In contrast with these peripheral membrane proteins, band 3 and all four PAS-positive bands can be dissociated from the red-cell membrane only by detergents or organic solvents. Hence, they are integral membrane proteins. This inference is reinforced by freeze-fracture electron microscopy (see Figure 12-20), which shows that some red-cell membrane proteins are deeply embedded in the hydrocarbon region of the membrane. The band 3 protein constitutes the *anion channel* that enables bicarbonate to be exchanged for chloride in the buffering of red cell pH. The PAS bands are different forms of *glycophorin*, a family of carbohydrate-rich proteins.

GLYCOPHORIN, A TRANSMEMBRANE PROTEIN, FORMS A CARBOHYDRATE COAT AROUND RED CELLS

Glycophorin A is a single polypeptide chain with sixteen attached oligosaccharide units. Indeed, 60% of the mass of this glycoprotein is carbohydrate, and so its name (derived from the Greek words meaning "to carry sugar") is apt. The abundance of carbohydrate in glycophorin is the reason that it stains intensely with the PAS reagent. The *PAS-1 band* is a dimer of glycophorin A held together by noncovalent interactions that resist disruption by sodium dodecyl sulfate.

Vincent Marchesi determined the amino acid sequence of glycophorin A, the first integral membrane protein to be sequenced. Using several experimental approaches, he found that glycophorin A traverses the plasma membrane and that all molecules point in the same direction. Ferritin-labeled antibodies specific for a determinant near the amino-terminus labeled the extracellular side of the membrane, whereas those specific for the carboxyl-terminal region labeled the cytosolic side. Proteolytic digestion and chemical modification studies showed that glycophorin consists of three parts: (1) an amino-terminal region containing all of the carbohydrate units, which is located on the extracellular face of the membrane; (2) a hydrophobic middle region, which is buried within the hydrocarbon core of the membrane; and (3) a carboxyl-terminal region rich in polar and ionized side chains, which is exposed on the cytosolic face of the red-cell membrane (Figure 12-36). The hydrophobic membrane-spanning region is α-helical.

Glycophorin A contains about 100 sugar residues arranged in 16 units. Fifteen of them are attached to the hydroxyl group of serine or threonine residues (*O*-linked), and one of them is linked to the amide NH_2— of an asparagine side chain (*N*-linked). These carbohydrate units are rich in sialic acid (Chapter 14), a negatively charged sugar. Glycophorins give red cells a very hydrophilic, anionic coat. Though much is known about the structure and topography of glycophorin, its function is an enigma heightened by the finding that individuals devoid of glycophorin A are healthy.

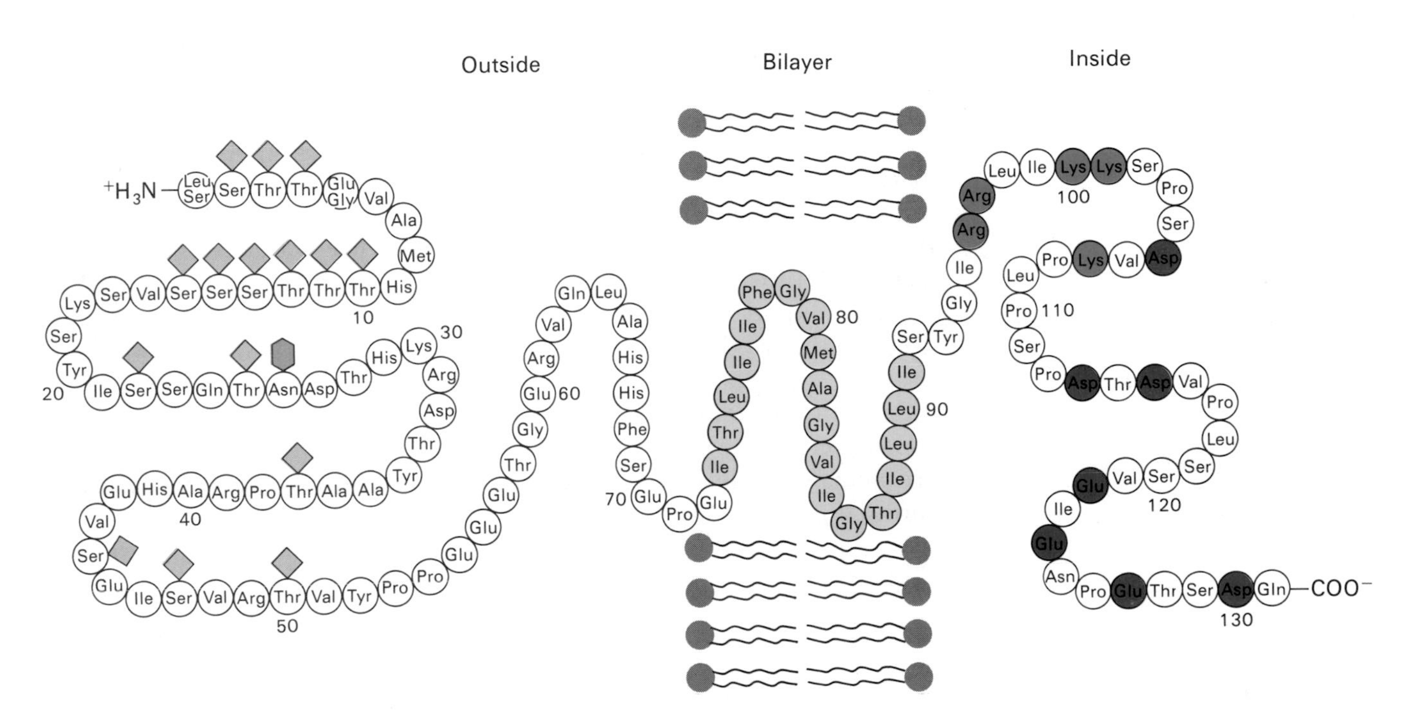

Figure 12-36
Amino acid sequence and transmembrane disposition of glycophorin A from the red-cell membrane. The fifteen *O*-linked carbohydrate units are shown in light green and the *N*-linked in dark green. The hydrophobic residues (yellow) buried in the bilayer form a transmembrane α helix. The carboxyl-terminal part of the molecule, located on the cytosolic side of the membrane, is rich in negatively charged (red) and positively charged (blue) residues. [Courtesy of Dr. Vincent Marchesi.]

TRANSMEMBRANE HELICES CAN BE ACCURATELY PREDICTED FROM AMINO ACID SEQUENCES

The presence of a highly nonpolar sequence of 19 residues in glycophorin (shown in yellow in Figure 12-36) first suggested that this segment of the protein traverses the lipid bilayer and is folded into an α helix. Studies of synthetic polypeptides had shown that α helices are more stable in nonpolar media than in water, which competes for hydrogen bonding with main-chain NH and CO groups. One approach to identifying transmembrane helices is to ask whether a postulated helical segment prefers to be in a hydrocarbon milieu or in water. Specifically, we want to calculate the free-energy change when a helical segment is transferred from the interior of a membrane to water. Free-energy

Table 12-2
Polarity scale for identifying transmembrane helices

Amino acid residue	*Transfer free energy (kcal/mol)*
Phe	3.7
Met	3.4
Ile	3.1
Leu	2.8
Val	2.6
Cys	2.0
Trp	1.9
Ala	1.6
Thr	1.2
Gly	1.0
Ser	0.6
Pro	−0.2
Tyr	−0.7
His	−3.0
Gln	−4.1
Asn	−4.8
Glu	−8.2
Lys	−8.8
Asp	−9.2
Arg	−12.3

Note: The free energies are for the transfer of an amino acid residue in an α helix from the membrane interior (assumed to have a dielectric constant of 2) to water. After D. M. Engelman, T. A. Steitz, and A. Goldman. *Ann. Rev. Biophys. Biophys. Chem.* 15(1986):330.

estimates for amino acid residues are given in Table 12-2. For example, the transfer of a poly-L-arginine helix from the interior of a membrane to water would be highly favorable (−12.3 kcal/mol per arginine in the helix), whereas the transfer of a poly-L-phenylalanine helix would be unfavorable (+3.7 kcal/mol per phenylalanine in the helix).

The hydrocarbon core of membranes is typically 30 Å wide, which can be traversed by an α helix consisting of 20 residues. We can take the amino acid sequence of a protein and calculate the free-energy change from transferring a hypothetical α helix of residues 1 to 20 from the membrane interior to water. The same calculation can be made for residues 2 to 21, 3 to 22, and so forth until we reach the end of the sequence. The span of 20 residues chosen for this calculation is called the window. For glycophorin A, a plot of the free-energy change versus the position of the first residue in the window shows a clear-cut maximum of +40 kcal/mol when residue 70 is the first in the window (Figure 12-37). This *hydropathy plot* correctly identifies the transmembrane α helix of glycophorin. *In general, a peak of +20 kcal/mol or more in a hydropathy plot based on a window of 20 residues indicates that a polypeptide segment could be a membrane-spanning α helix.* By this criterion, glycophorin has only one membrane-spanning helix, in agreement with experimental findings. Indeed, hydropathy plots correctly predicted the positions of all eleven transmembrane helices in a photosynthetic reaction center (p. 309). It should be noted, however, that a peak in a hydropathy plot does not prove that a segment is a transmembrane helix. Even soluble proteins may have such highly nonpolar regions. Conversely, some membrane proteins have amphipathic transmembrane helices that escape detection by these plots. However, it is clear that hydropathy plots are very valuable in suggesting how membrane proteins fold and in stimulating experiments to test such models.

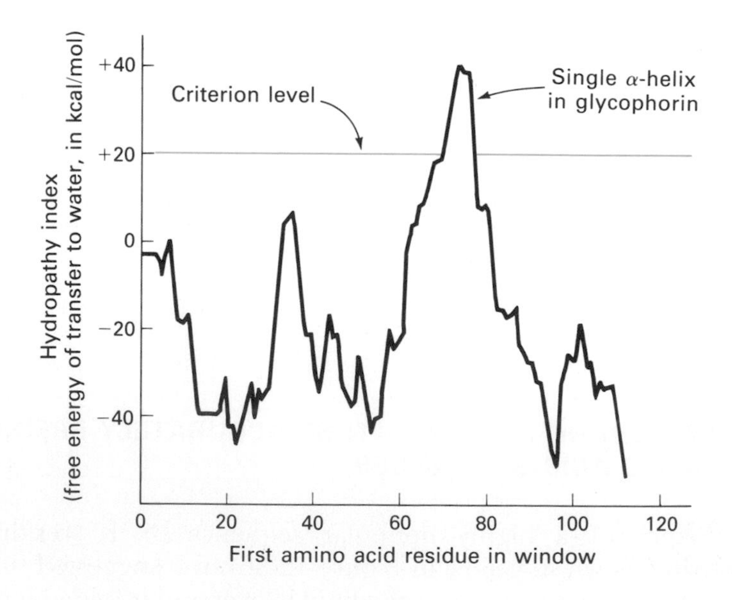

Figure 12-37
Hydropathy plot for glycophorin. The free energy for transferring a helix of 20 residues from the membrane to water is plotted as a function of the position of the first residue of the helix in the sequence of the protein. Peaks of greater than +20 kcal/mol in hydropathy plots are indicative of potential transmembrane helices. [After D. M. Engelman, T. A. Steitz, and A. Goldman. Identifying nonpolar transbilayer helices in amino acid sequences of membrane proteins. *Ann. Rev. Biophys. Biophys. Chem.* 15(1986):343. Copyright © 1986 by Annual Reviews, Inc. All rights reserved.]

A CHANNEL FOR ANIONS IS FORMED FROM SEVERAL MEMBRANE-SPANNING α-HELICES

The red-cell membrane contains an anion channel that plays a key role in the transport of CO_2 by blood and in the buffering of red cell pH. Carbon dioxide entering red cells in tissue capillaries is converted into

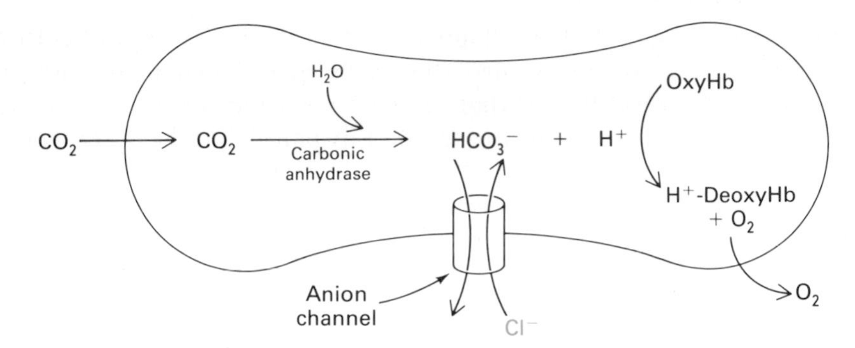

Figure 12-38
Anion channels in erythrocytes play an important role in the transport of CO_2 by enabling HCO_3^- to be exchanged for Cl^-.

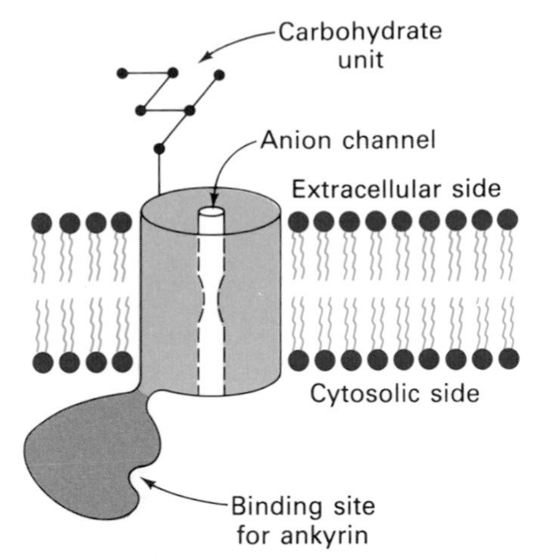

Figure 12-39
Schematic diagram of the domain structure of the anion channel of erythrocytes. The amino-terminal domain (shown in red) binds ankyrin, which in turn binds spectrin. The carboxyl-terminal domain contains a carbohydrate unit and the pore for anion exchange across the membrane.

carbonic acid, H_2CO_3, by the action of carbonic anhydrase (Figure 12-38). Carbonic acid then dissociates into H^+ and HCO_3^-. The released proton is taken up by deoxyhemoglobin as part of the Bohr effect (p. 156). Bicarbonate leaves the red cell through the anion channel in exchange for Cl^-. In the pulmonary circulation, this process is reversed. The *anion channel* (also known as the *band 3 protein* or the *anion exchanger*) is very abundant, comprising about a third of the total membrane protein. Each red cell contains some 10^6 copies of this integral membrane protein, which is a single chain of 929 residues (106 kd).

Proteolytic digestion patterns of intact red cells, leaky ghosts, and inside-out vesicles have shown that the anion channel traverses the red-cell membrane and that all molecules of this protein point in the same direction. Like other integral membrane glycoproteins, its carbohydrate units are located on the extracellular side of the membrane. The protein consists of two domains, which can be separated by proteolytic cleavage (Figure 12-39). The cloning and sequencing of the cDNA for this protein have provided additional valuable information. A rather hydrophilic *amino-terminal domain* (420 residues) is located on the cytosolic side of the membrane. It contains binding sites for ankyrin, a protein linking the plasma membrane to the underlying membrane skeleton, and for several cytosolic proteins. Removal of this amino-terminal domain has no effect on anion exchange activity. The *carboxyl-terminal domain* (509 residues) mediates the exchange of bicarbonate and chloride ions. It traverses the bilayer, as expected for a channel protein. Indeed, hydropathy plots show that this domain may contain as many as twelve membrane-spanning α helices. The mechanism of anion exchange by this channel will be discussed in Chapter 37.

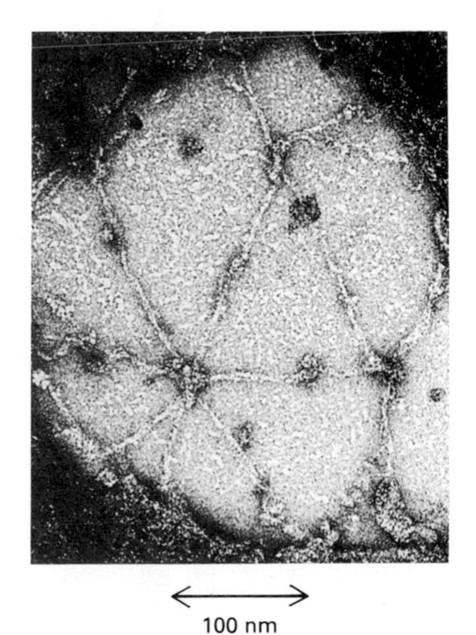

Figure 12-40
Electron micrograph of a portion of the erythrocyte membrane skeleton that has been spread tenfold to display the hexagonal and pentagonal network of spectrin and other protein molecules. [From T. J. Byers and D. Branton. *Proc. Nat. Acad. Sci.* 82(1985):6153.]

SPECTRIN FORMS A MEMBRANE SKELETON THAT ENABLES ERYTHROCYTES TO RESIST STRONG SHEARING FORCES

An erythrocyte is exposed to powerful shearing forces and undergoes large changes in shape during its lifetime of 120 days. It travels some 300 miles in this period, much of it through narrow passageways. The mechanical stability and resilience of the erythrocyte membrane come from a partnership between the plasma membrane and an underlying meshwork called the *membrane skeleton* (Figure 12-40). The major constituent of this infrastructure is *spectrin*, which consists of a 260-kd α

chain and a 225-kd β chain (Figure 12-41). Amino acid and cDNA sequence analyses have revealed that each spectrin chain consists of repeats of a 106-residue unit that folds into a triple-stranded α-helical coiled coil. These two chains are aligned in parallel and twisted about each other to form a flexible rod that is about 1000 Å (100 nm) long. Two of these αβ units join end to end to form a tetramer with a length of 200 nm.

Spectrin does not bind directly to the plasma membrane. Instead, the linkage is through *ankyrin* and *protein 4.1* (the band 4.1 protein), at different sites. Ankyrin forms a cross-link between the β chain of spectrin and the amino-terminal cytosolic domain of the anion channel (Figure 12-42). Protein 4.1, which consists of two nearly identical chains, promotes the binding of actin filaments to the carboxyl-terminal portions of both spectrin chains. Protein 4.1 also links this spectrin-actin complex to the cytosolic face of glycophorin. Several spectrin tetramers insert at each protein 4.1 junction complex to form a continuous meshwork underlying the plasma membrane. Protein 4.1 binds to glycophorin only if a specific phospholipid, phosphatidyl inositol 4,5-bisphosphate, is linked to this integral membrane protein. This finding suggests that the attachment of the spectrin meshwork to the membrane is modulated by changes in the concentration of this phospholipid. Indeed, phosphatidyl inositol 4,5-bisphosphate participates in signal transduction in many kinds of cells (Chapter 38).

The importance of spectrin is strikingly evident in a genetic disorder of mice resulting in a hemolytic anemia. The erythrocytes of these homozygous mice are spherical rather than biconcave. They are extremely fragile and spontaneously lose membrane vesicles (Figure 12-43). *These* sph *(spherocytic) mice have only about 5% of the normal amount of spectrin.* Introduction of normal spectrin into *sph* erythrocytes cures their defect. Defects in spectrin, protein 4.1, and other components of the membrane skeleton are now being found in patients with genetic diseases leading to abnormal erythrocyte shape and diminished mechanical stability.

Another intriguing facet of the erythrocyte membrane skeleton is the presence of similar proteins in other tissues. *Fodrin* in epithelial cells closely resembles spectrin. *Synapsin-1*, a protein once thought to be specific to neurons, is homologous to protein 4.1. *Actin*, a component of contractile assemblies, is present in nearly all eucaryotic cells. *It seems likely that membrane skeletons like the one formed by spectrin in erythrocytes play key roles in altering and stabilizing the shapes of many kinds of cells.*

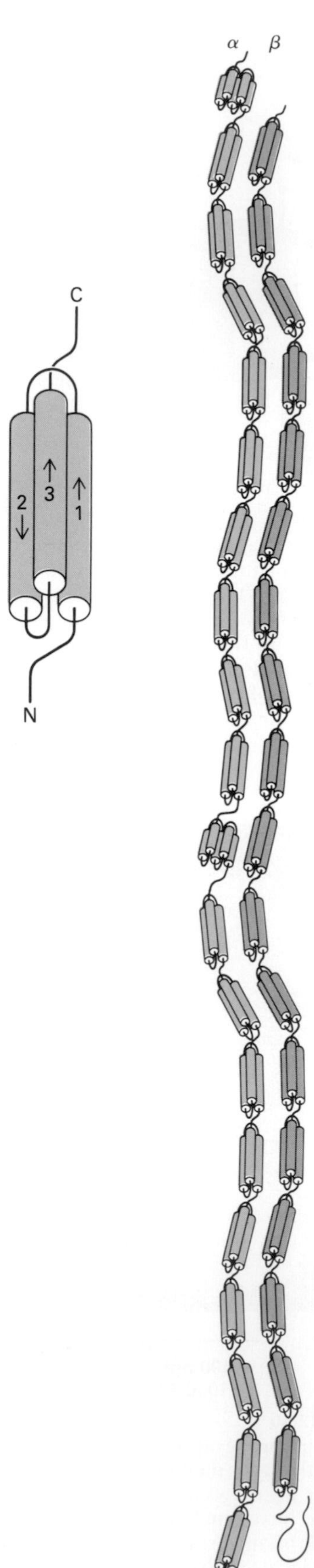

Figure 12-41
Schematic diagram of spectrin. The α chain is shown in blue and the β chain in red. The repeating segment of 106 residues folds into a triple-stranded α-helical bundle. These segments are flexibly joined in both chains. [Courtesy of Dr. Vincent Marchesi.]

ELECTRON-MICROSCOPIC AND X-RAY ANALYSES OF CRYSTALLINE MEMBRANE PROTEINS ARE HIGHLY REVEALING

One of the goals of membrane research is to obtain high-resolution images of membrane systems. X-ray crystallographic studies of proteins such as hemoglobin, lysozyme, and chymotrypsin have provided insight

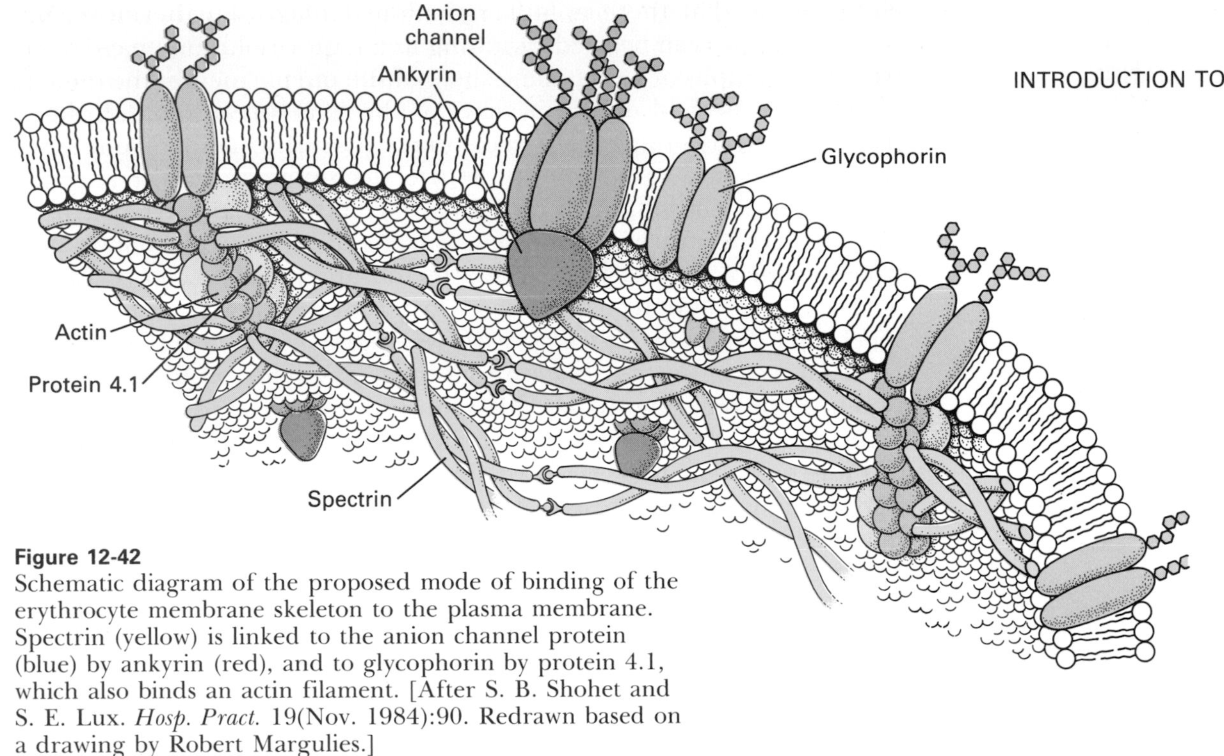

Figure 12-42
Schematic diagram of the proposed mode of binding of the erythrocyte membrane skeleton to the plasma membrane. Spectrin (yellow) is linked to the anion channel protein (blue) by ankyrin (red), and to glycophorin by protein 4.1, which also binds an actin filament. [After S. B. Shohet and S. E. Lux. *Hosp. Pract.* 19(Nov. 1984):90. Redrawn based on a drawing by Robert Margulies.]

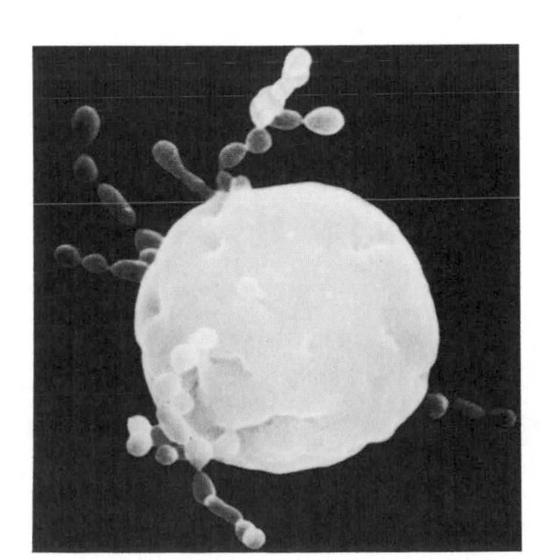

Figure 12-43
Scanning electron micrograph of an erythrocyte from an *sph* mouse deficient in spectrin. These fragile cells spontaneously lose portions of their plasma membrane in the form of vesicles. [Courtesy of Dr. Stephen Shohet.]

into how water-soluble proteins fold, bind other molecules, and catalyze reactions. We would also like to have high-resolution views of membrane channels, pumps, and receptors. An important start has been made with membrane proteins that give rise to two-dimensional crystals, ordered lattices in the plane of the membrane. These crystalline sheets are suitable for structural analysis by electron microscopy, as was first shown for the *purple membrane* of *Halobacterium halobium* (a salt-loving bacterium). This specialized region of the cell membrane contains *bacteriorhodopsin,* a 25-kd protein, which converts the energy of light into a transmembrane proton gradient that is used to synthesize ATP (Chapter 37). Crystalline sheets with diameters as large as 1 μm can be isolated. From a crystalline array of about 20,000 bacteriorhodopsin molecules, an image can be obtained with a very low dose of electrons, so that there is little radiation damage. Furthermore, unstained samples can be used, resulting in a high-resolution image. Electron micrographs of a crystalline sheet of the purple membrane titled at various angles yield different views of the structure projected onto a plane. The next stage is to recombine the information from some twenty electron micrographs by Fourier techniques.

Using this method, in 1975, Richard Henderson and Nigel Unwin reconstructed a three-dimensional image of the purple membrane, which yielded the first picture of a membrane protein (Figure 12-44). Their 7-Å resolution map showed that *bacteriorhodopsin contains seven closely packed α helices extending nearly perpendicular to the plane of the membrane for most of its 45-Å width.* Lipid bilayer regions fill the spaces between the protein molecules. Electron microscopic studies of ordered arrays of the acetylcholine receptor channel (Chapter 37) and of cell-cell junctions (Chapter 37) have shown that these membrane proteins also contain a cluster of closely packed transmembrane α helices. In fact, it seems likely that the transmembrane segment of nearly all integral membrane proteins consists of one or more α helices.

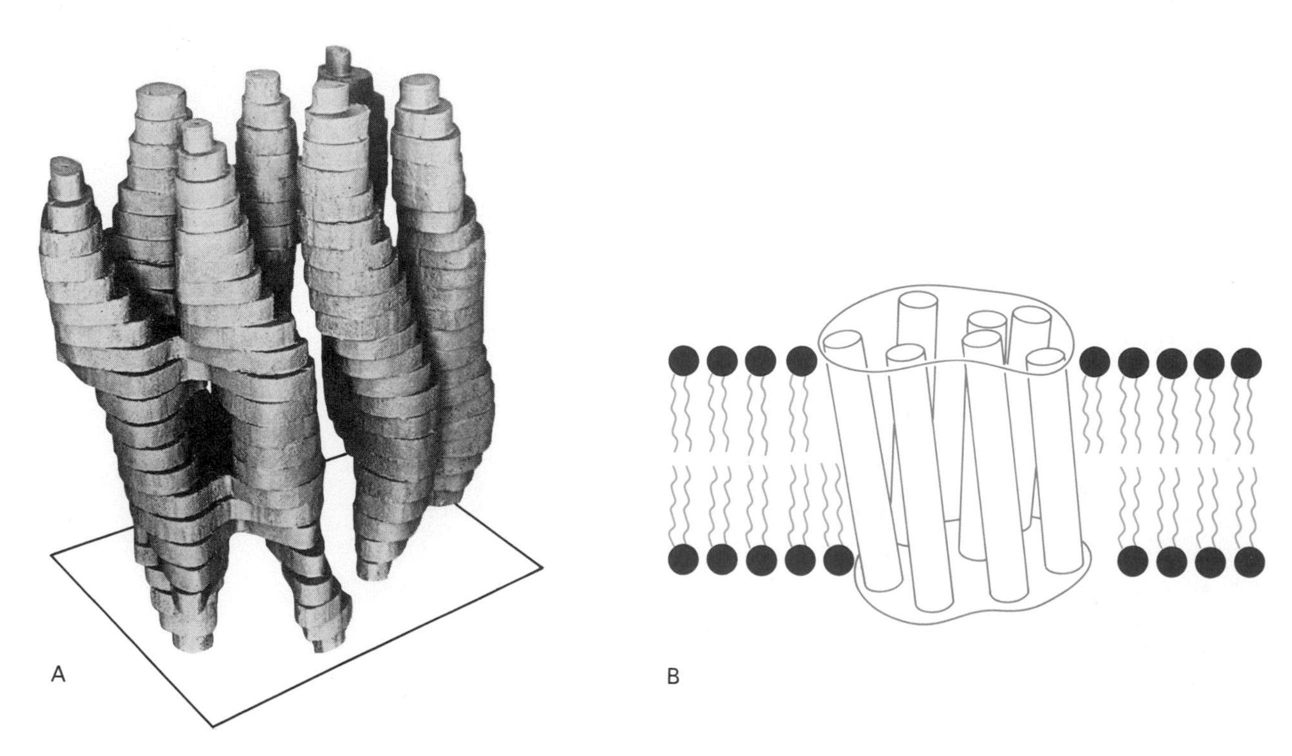

Figure 12-44
A. Model of bacteriorhodopsin constructed from a 7-Å three-dimensional map. B. Interpretative diagram showing the arrangement of α-helical segments in the lipid bilayer. The connections between these helices are not yet known. [Courtesy of Dr. Richard Henderson and Dr. Nigel Unwin.]

The photosynthetic reaction center from a purple bacterium (*Rhodopseudomonas viridis*) has been crystallized in three-dimensions from a detergent solution and its structure has been solved by x-ray analysis at 3-Å resolution. This landmark study by Johann Deisenhofer, Hartmut Michel, and Robert Huber provides a highly detailed and informative image of an energy-converting membrane assembly (Figure 12-45). Moreover, it demonstrates that membrane proteins, like water-soluble proteins, can form highly ordered three-dimensional crystals that are

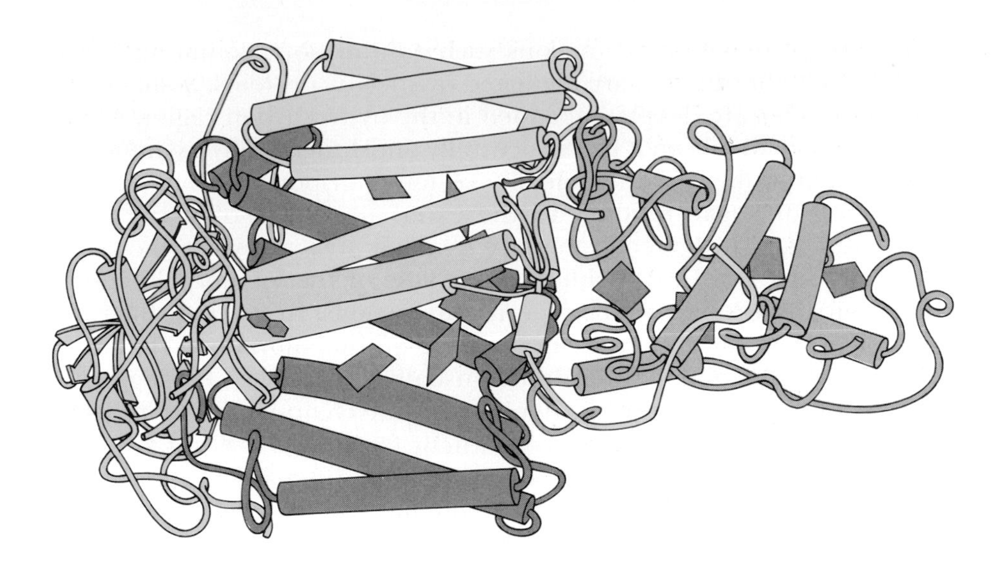

Figure 12-45
Three-dimensional structure of the photosynthetic reaction center from a purple bacterium. Three of the four proteins (shown in yellow, green, and red) traverse the lipid bilayer and consist largely of transmembrane α helices. The prosthetic groups of this energy-converting protein complex are shown in gray. [After a drawing kindly provided by Dr. Jane Richardson. Based on atomic coordinates provided by Dr. Johann Deisenhofer, Dr. Hartmut Michel, and Dr. Robert Huber.]

amenable to x-ray analysis at high resolution. The 150-kd reaction center consists of four polypeptide chains, of which three traverse the membrane. The membrane-spanning segments are α-helical, as in bacteriorhodopsin. The function of the reaction center is to convert light into a separation of charge, a process mediated by its twelve prosthetic groups and an iron atom. We shall return to this remarkable assembly when we consider the primary energy-converting step in photosynthesis (p. 532).

SUMMARY

Biological membranes are sheetlike structures, typically 75 Å wide, that are composed of protein and lipid molecules held together by noncovalent interactions. Membranes are highly selective permeability barriers. They create closed compartments, which may be entire cells or organelles within a cell. Pumps and gates in membranes regulate the molecular and ionic compositions of these compartments. Membranes also control the flow of information between cells. For example, many membranes contain receptors for hormones such as insulin. Furthermore, membranes are intimately involved in such energy conversion processes as photosynthesis and oxidative phosphorylation.

The major classes of membrane lipids are phospholipids, glycolipids, and cholesterol. Phosphoglycerides, a type of phospholipid, consist of a glycerol backbone, two fatty acid chains, and a phosphorylated alcohol. The fatty acid chains usually contain between 14 and 24 carbon atoms; they may be saturated or unsaturated. Phosphatidyl choline, phosphatidyl serine, and phosphatidyl ethanolamine are major phosphoglycerides. Sphingomyelin, a different type of phospholipid, contains a sphingosine backbone instead of glycerol. Glycolipids are sugar-containing lipids derived from sphingosine. A common feature of these membrane lipids is that they are amphipathic molecules. They spontaneously form extensive bimolecular sheets in aqueous solutions because they contain both a hydrophilic and a hydrophobic moiety. These lipid bilayers are highly impermeable to ions and most polar molecules, yet they are quite fluid, which enables them to act as a solvent for membrane proteins.

Distinctive membrane functions such as transport, communication, and energy transduction are mediated by specific proteins. Some membrane proteins are deeply imbedded in the hydrocarbon region of the lipid bilayer. Membranes are structurally and functionally asymmetric, as exemplified by the directionality of ion transport systems and the restriction of sugar residues to the external surface of mammalian plasma membranes. Membranes are dynamic structures in which proteins and lipids diffuse rapidly in the plane of the membrane (lateral diffusion), unless restricted by special interactions. In contrast, the rotation of proteins and lipids from one face of a membrane to the other (transverse diffusion, or flip-flop) is usually very slow. The degree of fluidity of a membrane partly depends on the chain length of its lipids and the extent to which their constituent fatty acids are unsaturated.

The erythrocyte membrane, one of the most intensively studied and best understood membrane systems, contains two abundant transmembrane proteins. Glycophorin forms a carbohydrate coat around red cells. The anion channel (band 3 protein) mediates the exchange of bicarbonate and chloride ions. These integral membrane proteins are linked by the band 4.1 protein and ankyrin to a flexible meshwork consisting mainly of spectrin. This membrane skeleton enables erythrocytes to resist strong shearing forces.

Three-dimensional images of membrane proteins can be reconstructed from electron micrographs of two-dimensional crystalline arrays. Bacteriorhodopsin, a light-driven proton pump in the plasma membrane of halobacteria, contains seven transmembrane α helices. X-ray analyses of three-dimensional crystals of the photosynthetic reaction center of a purple bacterium has revealed the atomic structure of this energy-converting assembly of four proteins.

SELECTED READINGS

Where to start

Bretscher, M. S., 1985. The molecules of the cell membrane. *Sci. Amer.* 253(4):100–108.

Unwin, N., and Henderson, R., 1984. The structure of proteins in biological membranes. *Sci. Amer.* 250(2):78–94.

Singer, S. J., and Nicolson, G. L., 1972. The fluid mosaic model of the structure of cell membranes. *Science* 175:720–731.

Books

Vance, D. E., and Vance, J. E., (eds.), 1985. *Biochemistry of Lipids and Membranes.* Benjamin/Cummings. [Contains many excellent articles on lipid structure, dynamics, metabolism, and assembly into membranes.]

Racker, E., 1985. *Reconstitutions of Transporters, Receptors, and Pathological States.* Academic Press. [A delightful, personal account of the reconstitution of membrane assemblies. Many strategies and examples are presented.]

Houslay, M. D., and Stanley, K. K., 1982. *Dynamics of Biological Membranes.* Wiley. [An excellent account of the physical properties of membranes].

Robertson, R. N., 1983. *The Lively Membranes.* Cambridge University Press. [A highly readable primer with many illustrations of lipids.]

Tanford, C., 1980. *The Hydrophobic Effect: Formation of Micelles and Biological Membranes,* 2nd ed. Wiley-Interscience. [A lucid statement of thermodynamic aspects of membrane structure.]

Membrane lipids and sugars

Yeagle, P. L., 1985. Cholesterol and the cell membrane. *Biochem. Biophys. Acta* 822:267–287.

Thompson, T. E., and Tillack, T. W., 1985. Organization of glycosphingolipids in bilayers and plasma membranes of mammalian cells. *Ann. Rev. Biophys. Biophys. Chem.* 14:361–386.

Lis, H., and Sharon, N., 1986. Lectins as molecules and as tools. *Ann. Rev. Biochem.* 55:35–67.

Op den Kamp, J. A. F., 1979. Lipid asymmetry in membranes. *Ann. Rev. Biochem.* 48:47–71.

Membrane dynamics

Elson, E. L., 1986. Membrane dynamics studied by fluorescence correlation spectroscopy and photobleaching recovery. *Soc. Gen. Physiol. Ser.* 40:367–383.

Frye, C. D., and Edidin, M., 1970. The rapid intermixing of cell surface antigens after formation of mouse-human heterokaryons. *J. Cell Sci.* 7:319–335.

Kornberg, R. D., and McConnell, H. M., 1971. Inside-outside transitions of phospholipids in vesicle membranes. *Biochemistry* 10:1111–1120.

Cone, R. A., 1972. Rotational diffusion of rhodopsin in the visual receptor membrane. *Nature New Biol.* 236:39–43.

Poo, M., and Cone, R. A., 1974. Lateral diffusion of rhodopsin in the photoreceptor membrane. *Nature* 247:438–441.

Raff, M. C., 1976. Cell-surface immunology. *Sci. Amer.* 234(5):30–39. [A lucid and well-illustrated article showing how antibodies have contributed to our understanding of cell-surface structure and dynamics. Available as *Sci. Amer.* Offprint 1338.]

Erythrocyte membrane

Marchesi, V. T., 1985. The cytoskeletal system of red blood cells. *Hosp. Pract.* 20(11):113–131.

Bennett, V., 1985. The membrane skeleton of human erythrocytes and its implications for more complex cells. *Ann. Rev. Biochem.* 54:273–304.

Marchesi, V. T., 1985. Stabilizing infrastructure of cell membranes. *Ann. Rev. Cell Biol.* 1:531–561.

Shohet, S. B., and Lux, S. E., 1984. The erythrocyte membrane skeleton: biochemistry and pathophysiology. *Hosp. Pract.* 19(10):77–83 and 19(11):89–108.

Shen, B. W., Josephs, R., and Steck, T. L., 1986. Ultrastructure of the intact skeleton of the human erythrocyte membrane. *J. Cell Biol.* 102:997–1006.

Byers, T. J., and Branton, D., 1985. Visualization of the protein associations in the erythrocyte membrane skeleton. *Proc. Nat. Acad. Sci.* 82:6153–6157.

Kopito, R. R., and Lodish, H. F., 1985. Primary structure and transmembrane orientation of the murine anion exchange protein. *Nature* 316:234–238.

Reconstitution of membrane assemblies

Madden, T. D., 1986. Current concepts in membrane protein reconstitution. *Chem. Phys. Lipids* 40:207–222.

Montal, M., Darszon, A., and Schindler, H., 1981. Functional reassembly of membrane proteins in planar lipid bilayers. *Quart. Rev. Biophys.* 14:1–79.

Helenius, A., Sarvas, M., and Simons, K., 1981. Asymmetric and symmetric membrane reconstitution by detergent elimination. *Eur. J. Biochem.* 116:27–31.

Fleischer, S., and Packer, L., (eds.), 1974. *Methods in Enzymology*, vol. 32. Academic Press. [Contains excellent articles on experimental methods in membrane research, including procedures for the preparation of model and reconstituted membranes.]

Ostro, M. J., 1987. Liposomes. *Sci. Amer.* 256(1):102–110. [Potential applications of liposomes for drug delivery are discussed.]

Structure of membrane proteins

Eisenberg, D., 1984. Three-dimensional structure of membrane and surface proteins. *Ann. Rev. Biochem.* 53:595–623.

Engelman, D. M., Steitz, T. A., and Goldman, A., 1986. Identifying non-polar transbilayer helices in amino acid sequences of membrane proteins. *Ann. Rev. Biophys. Biophys. Chem.* 15:321–353.

Amos, L. A., Henderson, R., and Unwin, P. N. T., 1982. Three-dimensional structure determination by electron microscopy of two-dimensional crystals. *Prog. Biophys. & Molec. Biol.* 39:183–231.

Henderson, R., and Unwin, P. N. T., 1975. Three-dimensional model of purple membrane obtained by electron microscopy. *Nature* 257:28–32. [The first view of a membrane protein.]

Unwin, P. N. T., and Henderson, R., 1975. Molecular structure determination by electron microscopy of unstained crystalline specimens. *J. Mol. Biol.* 94:425–440. [Presentation of the experimental strategy for determining the structure of membrane proteins that form highly ordered two-dimensional arrays.]

Deisenhofer, J., Epp, O., Miki, N., Huber, R., and Michel, H., 1984. X-ray structure analysis of a membrane protein complex. Electron density map at 3-Å resolution and a model of the chromophores of the photosynthetic reaction center from *Rhodopseudomonas viridis*. *J. Mol. Biol.* 180:385–398.

Branton, D., 1966. Fracture faces of frozen membranes. *Proc. Nat. Acad. Sci.* 55:1048–1056.

PROBLEMS

1. How many phospholipid molecules are there in a 1-μm^2 region of a phospholipid bilayer membrane? Assume that a phospholipid molecule occupies 70 $Å^2$ of the surface area.

2. Bacterial phospholipids contain cyclopropane fatty acid residues.

$$H_3C-(CH_2)_5-\underset{H}{C}\overset{CH_2}{-}\underset{H}{C}-(CH_2)_9-C\begin{matrix}\nearrow O\\ \searrow O^-\end{matrix}$$

Predict the effect of the cyclopropane ring on the packing of hydrocarbon chains in the interior of a bilayer. Would these cyclopropane fatty acid groups tend to make a membrane more or less fluid?

3. What is the average distance traversed by a membrane lipid in 1 μs, 1 ms, and 1 s? Assume a diffusion coefficient of 10^{-8} cm^2/s.

4. The diffusion coefficient D of a rigid spherical molecule is given by

$$D = \frac{kT}{(6\pi\eta r)}$$

in which η is the viscosity of the solvent, r is the radius of the sphere, k is the Boltzman constant (1.38×10^{-16}

erg/deg), and T is the absolute temperature. What is the diffusion coefficient at 37°C of a 100-kd protein in a membrane that has an effective viscosity of 1 poise (1 poise = 1 erg s/cm^3)? What is the average distance traversed by this protein in 1 μs, 1 ms, and 1 s? Assume that this protein is an unhydrated, rigid sphere of density 1.35 g/cm^3.

5. R. D. Kornberg and H. M. McConnell (1971) investigated the transverse diffusion (flip-flop) of phospholipids in a bilayer membrane by using a paramagnetic analog of phosphatidyl choline, called *spin-labeled phosphatidyl choline*.

$H_3C{-}(CH_2)_{14}{-}C(=O){-}O{-}CH_2$

$H_3C{-}(CH_2)_{14}{-}C(=O){-}O{-}CH$

$H_2C{-}O{-}P(=O)(O^-){-}O{-}CH_2{-}CH_2{-}N^+(CH_3)_2{-}$(tetramethylpiperidine ring)${-}N{-}O$

The nitroxide (NO) group in spin-labeled phosphatidyl choline gives a distinctive paramagnetic resonance spectrum. This spectrum disappears when nitroxides are converted into amines by reducing agents such as ascorbate.

Lipid vesicles containing phosphatidyl choline (95%) and the spin-labeled analog (5%) were prepared by sonication and purified by gel-filtration chromatography. The outside diameter of these liposomes was about 250 Å. The amplitude of the paramagnetic resonance spectrum decreased to 35% of its initial value within a few minutes of the addition of ascorbate. There was no detectable change in the spectrum within a few minutes after the addition of a second aliquot of ascorbate. However, the amplitude of the residual spectrum decayed exponentially with a half-time of 6.5 hr. How would you interpret these changes in the amplitude of the paramagnetic spectrum?

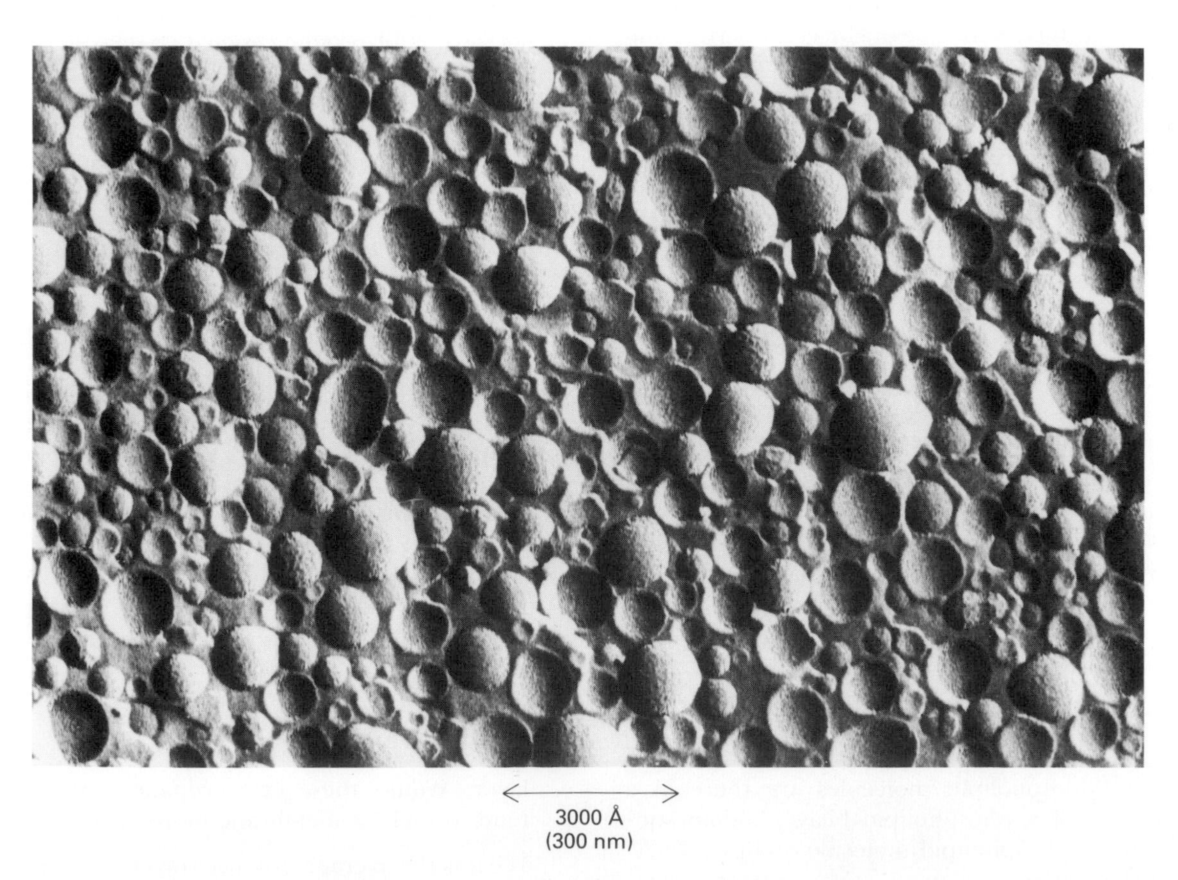

Freeze-fracture electron micrograph of unilamellar phospholipid vesicles. [Courtesy of Dr. P. R. Cullis.]

PART III

GENERATION AND STORAGE OF METABOLIC ENERGY

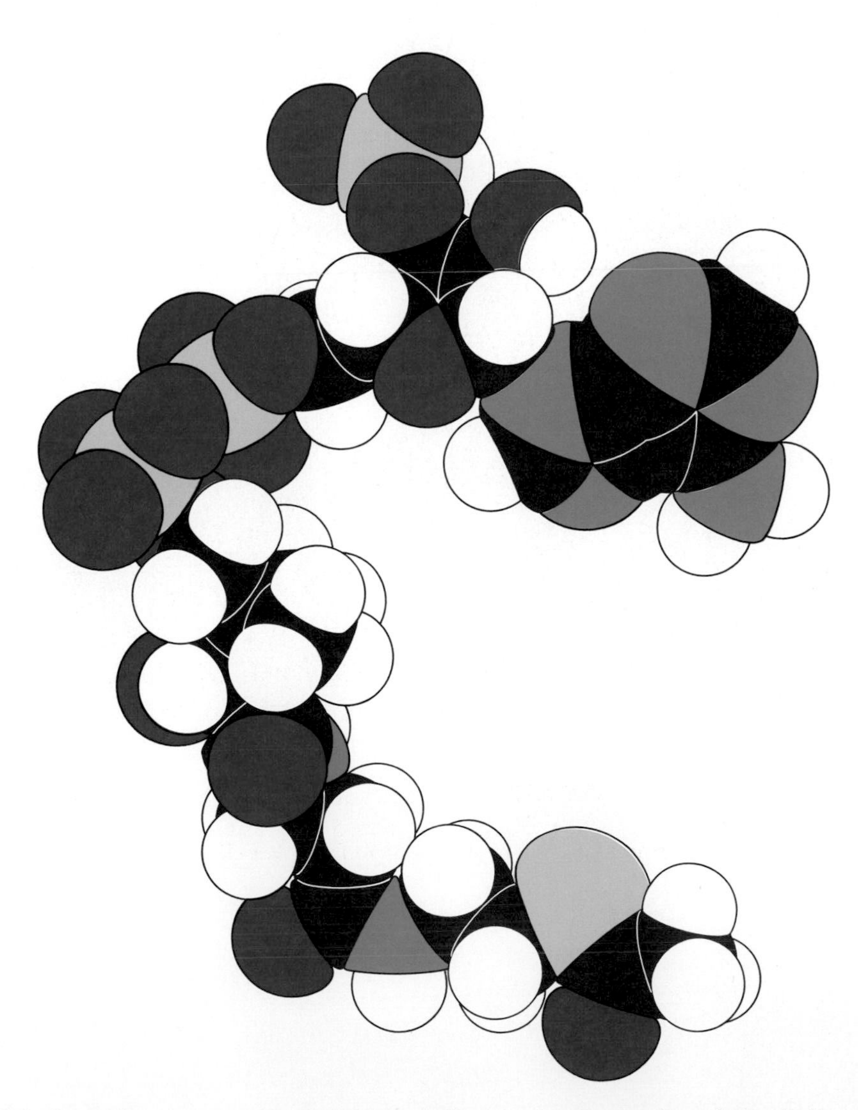

Model of acetyl coenzyme A, a key intermediate in the generation of metabolic energy.

CHAPTER 13

Metabolism: Basic Concepts and Design

The concepts of conformation and dynamics developed in Part II—especially those dealing with the specificity and catalytic power of enzymes, the regulation of their catalytic activity, and the formation of compartments by membranes—enable us to turn now to two major questions of biochemistry:

1. *How do cells extract energy and reducing power from their environments?*
2. *How do cells synthesize the building blocks of their macromolecules?*

These processes are carried out by a highly integrated network of chemical reactions, which are collectively known as *metabolism.*

More than a thousand chemical reactions take place in even as simple an organism as *E. coli.* The array of reactions may seem overwhelming at first glance. However, closer scrutiny reveals that metabolism has a *coherent design containing many common motifs.* The number of reactions in metabolism is large, but the number of *kinds* of reactions is relatively small. The mechanisms of these reactions are usually quite simple. For example, a double bond is often formed by dehydration of an alcohol. Furthermore, a group of about a hundred molecules plays a central role in all forms of life. Moreover, metabolic pathways are regulated in common ways. The purpose of this chapter is to introduce some general principles and motifs of metabolism.

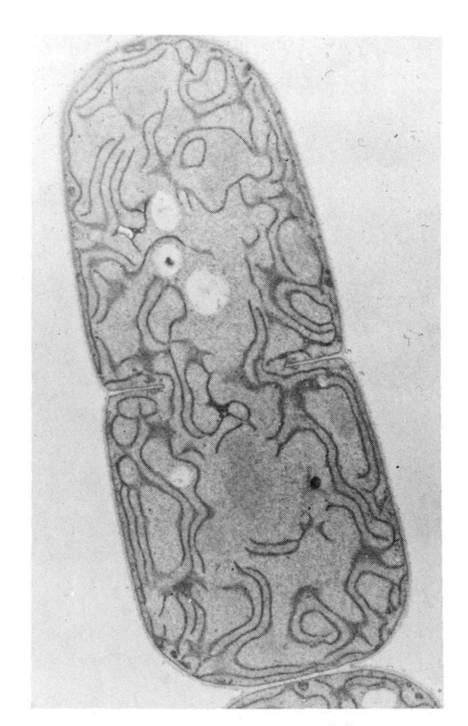

Figure 13-1
Electron micrograph of a cyanobacterium (blue-green alga), a photosynthetic procaryote. The dark-staining membranes contain the photosynthetic assemblies. [Courtesy of Dr. Thomas Giddings and Dr. L. Andrew Staehelin.]

A THERMODYNAMICALLY UNFAVORABLE REACTION CAN BE DRIVEN BY A FAVORABLE ONE

As was discussed earlier (p. 180), free energy is the most useful thermodynamic function in biochemistry. A reaction can occur spontaneously only if ΔG, the change in free energy, is negative. Recall that ΔG for the formation of products C and D from substrates A and B is given by

$$\Delta G = \Delta G^\circ + RT \log_e \frac{[C][D]}{[A][B]}$$

Thus, the ΔG of a reaction depends on the *nature* of the reactants (expressed by the ΔG° term, the standard free-energy change) and on their *concentrations* (expressed by the logarithmic term). The standard free-energy change at pH 7 is denoted by $\Delta G^{\circ\prime}$.

An important thermodynamic fact is that *the overall free-energy change for a chemically coupled series of reactions is equal to the sum of the free-energy changes of the individual steps.* Consider the reactions

$$\begin{array}{ll} A \rightleftharpoons B + C & \Delta G^{\circ\prime} = +5 \text{ kcal/mol} \\ B \rightleftharpoons D & \Delta G^{\circ\prime} = -8 \text{ kcal/mol} \\ \hline A \rightleftharpoons C + D & \Delta G^{\circ\prime} = -3 \text{ kcal/mol} \end{array}$$

Under standard conditions, A cannot be spontaneously converted into B and C because ΔG is positive. However, the conversion of B into D under standard conditions is thermodynamically feasible. Because free-energy changes are additive, the conversion of A into C and D has a $\Delta G^{\circ\prime}$ of -3 kcal/mol, which means that it can occur spontaneously under standard conditions. Thus, *a thermodynamically unfavorable reaction can be driven by a thermodynamically favorable reaction that is coupled to it.* In this example, the reactions are coupled by the common intermediate B. Reactions can also be coupled by conformational changes transmitted through protein molecules and by the flow of ions across membranes. We shall encounter many examples of energy coupling in metabolism.

ATP IS THE UNIVERSAL CURRENCY OF FREE ENERGY IN BIOLOGICAL SYSTEMS

Living things require a continual input of free energy for three major purposes: the performance of mechanical work in muscle contraction and other cellular movements, the active transport of molecules and ions, and the synthesis of macromolecules and other biomolecules from simple precursors. The free energy used in these processes, which maintain an organism in a state that is far from equilibrium, is derived from the environment. *Chemotrophs* obtain this energy by the oxidation of foodstuffs, whereas *phototrophs* obtain it by trapping light energy. Part of the free energy derived from the oxidation of foodstuffs and from light is transformed into a special form before it is used for motion, active transport, and biosyntheses. In most processes, this special carrier of free energy is *adenosine triphosphate* (ATP). The central role of ATP in energy exchanges in biological systems was perceived by Fritz Lipmann and by Herman Kalckar in 1941.

ATP is a nucleotide consisting of an adenine, a ribose, and a triphosphate unit (Figure 13-2). The active form of ATP is usually a complex of ATP with Mg^{2+} or Mn^{2+}. In considering the role of ATP as an energy carrier, we can focus on its triphosphate moiety. *ATP is an energy-*

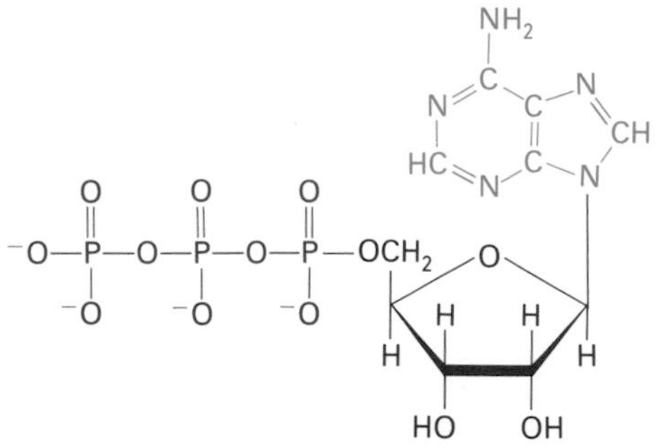

Adenosine triphosphate (ATP)

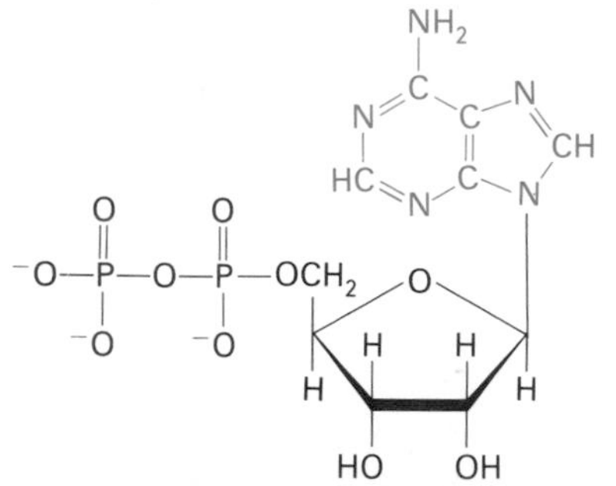

Adenosine diphosphate (ADP)

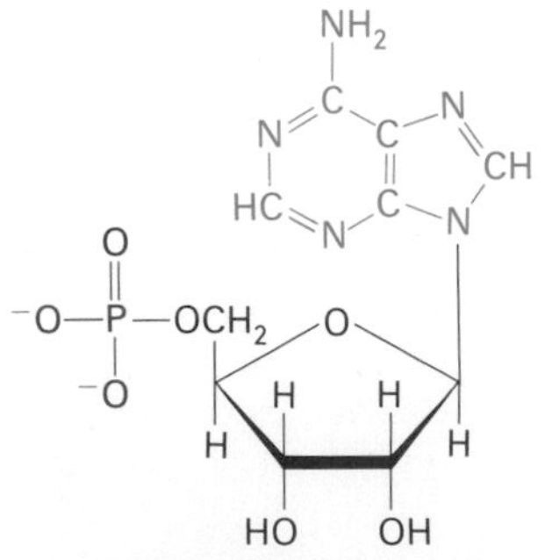

Adenosine monophosphate (AMP)

Figure 13-2
Structure of ATP, ADP, and AMP. These adenylates consist of adenine (blue), a ribose (black), and a tri-, di-, or monophosphate unit (red). The innermost phosphorus atom of ATP (the one bonded to the ribose unit) is designated P_α, the middle one P_β, and the outermost one P_γ.

rich molecule because its triphosphate unit contains two phosphoanhydride bonds. A large amount of free energy is liberated when ATP is hydrolyzed to adenosine diphosphate (ADP) and orthophosphate (P_i) or when ATP is hydrolyzed to adenosine monophosphate (AMP) and pyrophosphate (PP_i).

$$ATP + H_2O \rightleftharpoons ADP + P_i + H^+ \qquad \Delta G^{\circ\prime} = -7.3 \text{ kcal/mol}$$

$$ATP + H_2O \rightleftharpoons AMP + PP_i + H^+ \qquad \Delta G^{\circ\prime} = -7.3 \text{ kcal/mol}$$

The $\Delta G^{\circ\prime}$ for these reactions depends on the ionic strength of the medium and on the concentrations of Mg^{2+} and Ca^{2+}. We shall use a value of −7.3 kcal/mol. Under typical cellular conditions, the actual ΔG for these hydrolyses is approximately −12 kcal/mol.

ATP, AMP, and ADP are interconvertible. The enzyme adenylate kinase (also called myokinase) catalyzes the reaction:

$$ATP + AMP \rightleftharpoons ADP + ADP$$

The free energy liberated in the hydrolysis of ATP is harnessed to drive reactions that require an input of free energy, such as muscle contraction. In turn, ATP is formed from ADP and P_i when fuel molecules are oxidized in chemotrophs or when light is trapped by phototrophs. *This ATP-ADP cycle is the fundamental mode of energy exchange in biological systems.*

Some biosynthetic reactions are driven by nucleotides that are analogous to ATP, namely guanosine triphosphate (GTP), uridine triphosphate (UTP), and cytidine triphosphate (CTP). The diphosphate forms of these nucleotides are denoted by GDP, UDP, and CDP, respectively. Enzymes catalyze the transfer of the terminal phosphoryl group from one nucleotide to another, as in the reactions

$$ATP + GDP \rightleftharpoons ADP + GTP$$

$$ATP + GMP \rightleftharpoons ADP + GDP$$

ATP IS CONTINUOUSLY FORMED AND CONSUMED

ATP serves as the principal *immediate donor of free energy* in biological systems rather than as a long-term storage form of free energy. In a typical cell, an ATP molecule is consumed within a minute following its formation. *The turnover of ATP is very high.* For example, a resting human consumes about 40 kg of ATP in 24 hours. During strenuous exertion, the rate of utilization of ATP may be as high as 0.5 kg per minute. Motion, active transport, signal amplification, and biosyntheses can occur only if ATP is continuously regenerated from ADP (Figure 13-3). Phototrophs harvest the free energy in light to generate ATP, whereas chemotrophs form ATP by the oxidation of fuel molecules.

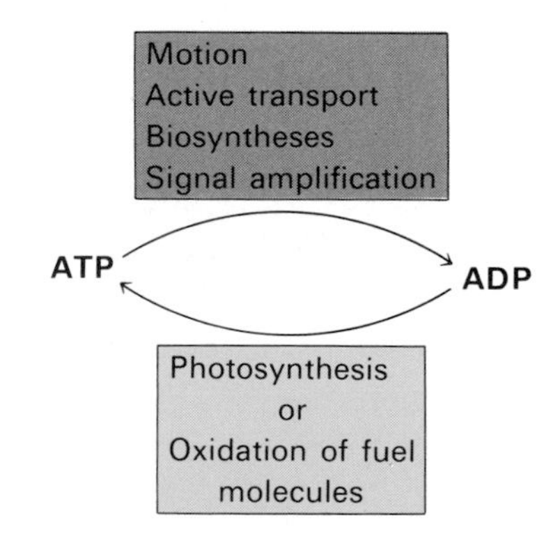

Figure 13-3
The ATP-ADP cycle is the fundamental mode of energy exchange in biological systems.

STRUCTURAL BASIS OF THE HIGH GROUP-TRANSFER POTENTIAL OF ATP

Let us compare the standard free energy of hydrolysis of ATP with that of a phosphate ester, such as glycerol 3-phosphate:

$$ATP + H_2O \rightleftharpoons ADP + P_i + H^+ \qquad \Delta G^{\circ\prime} = -7.3 \text{ kcal/mol}$$

$$\text{Glycerol 3-phosphate} + H_2O \rightleftharpoons \text{glycerol} + P_i \qquad \Delta G^{\circ\prime} = -2.2 \text{ kcal/mol}$$

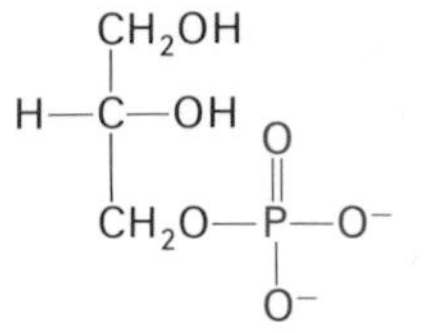

$$\begin{array}{c} CH_2OH \\ | \\ H-C-OH \quad O \\ | \qquad\quad \| \\ CH_2O-P-O^- \\ | \\ O^- \end{array}$$

Glycerol 3-phosphate

The magnitude of $\Delta G^{\circ\prime}$ for the hydrolysis of glycerol 3-phosphate is much smaller than that of ATP. This means that ATP has a stronger tendency to transfer its terminal phosphoryl group to water than does glycerol 3-phosphate. In other words, ATP has a higher *phosphate group-transfer potential* than does glycerol 3-phosphate.

What is the structural basis of the high phosphate group-transfer potential of ATP? The structures of both ATP and its hydrolysis products, ADP and P_i, must be examined to answer this question because $\Delta G^{\circ\prime}$ depends on the *difference* in free energies of the products and reactants. Two factors prove to be important in this regard: *electrostatic repulsion* and *resonance stabilization*. At pH 7, the triphosphate unit of ATP carries about four negative charges. These charges repel each other strongly because they are in close proximity. *The electrostatic repulsion between these negatively charged groups is reduced when ATP is hydrolyzed. The other factor contributing to the high group-transfer potential of ATP is that ADP and P_i enjoy greater resonance stabilization than does ATP*. For example, orthophosphate has a number of resonance forms of similar energy (Figure 13-4). In contrast, the terminal portion of ATP has fewer significant resonance forms per phosphate group. Forms of the type shown in Figure 13-5 are unlikely to occur because the two phosphorus atoms compete for electron pairs on oxygen. Furthermore, there is a positive charge on an oxygen atom adjacent to a positively charged phosphorus atom, which is electrostatically unfavorable.

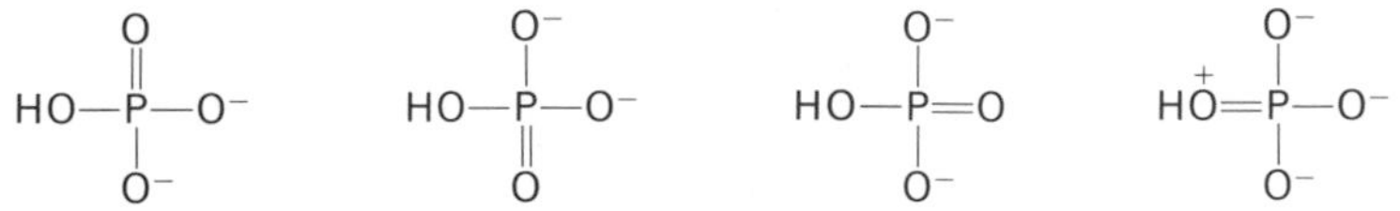

Figure 13-4
Significant resonance forms of orthophosphate.

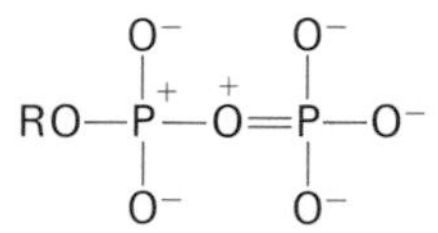

HO—P(=O)(—O⁻)—O⁻ ; HO—P(—O⁻)(=O)—O⁻ ; HO—P(—O⁻)=O (O⁻) ; HO^+=P(—O⁻)(—O⁻)—O⁻

Figure 13-5
An improbable resonance form of the terminal portion of ATP.

RO—P(O⁻)(O⁻)$^+$—O^+=P(O⁻)(O⁻)—O⁻

Various other compounds in biological systems have a high phosphate group transfer potential. In fact, some of them, such as phosphoenolpyruvate, acetyl phosphate, and phosphocreatine (Figure 13-6), have a higher group-transfer potential than does ATP. This means that phosphoenolpyruvate can transfer its phosphoryl group to ADP to form ATP. Indeed, this is one of the ways in which ATP is generated in the breakdown of sugars. It is significant that ATP has a group-transfer potential that is intermediate among the biologically important phosphorylated molecules (Table 13-1). *This intermediate position enables ATP to function efficiently as a carrier of phosphoryl groups.*

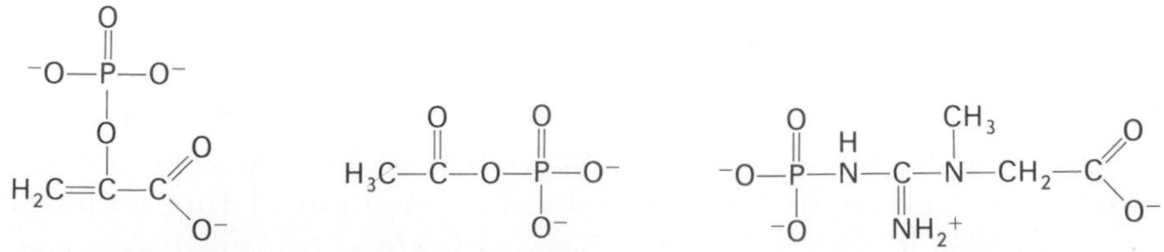

Figure 13-6
Some compounds with a higher phosphate group transfer potential than that of ATP.

Phosphoenolpyruvate — $H_2C=C(OPO_3^{2-})-COO^-$

Acetyl phosphate — $H_3C-C(=O)-O-PO_3^{2-}$

Phosphocreatine — $^-O_3P-NH-C(=NH_2^+)-N(CH_3)-CH_2-COO^-$

ATP is often called a high-energy phosphate compound and its phosphoanhydride bonds are referred to as high-energy bonds. There is nothing special about the bonds themselves. *They are high-energy bonds in the sense that much free energy is released when they are hydrolyzed,* for the reasons given above. Lipmann's term "high-energy bond" and his symbol ~P (squiggle P) for a compound having a high phosphate group-transfer potential are vivid, concise, and useful notations. In fact, Lipmann's squiggle did much to stimulate interest in bioenergetics.

ATP HYDROLYSIS SHIFTS THE EQUILIBRIA OF COUPLED REACTIONS BY A FACTOR OF 10^8

An understanding of the role of ATP in energy coupling can be enhanced by considering a chemical reaction that is thermodynamically unfavorable without an input of free energy. Suppose that the standard free energy of the conversion of A into B is +4 kcal/mol.

$$\mathrm{A} \rightleftharpoons \mathrm{B} \qquad \Delta G^{\circ\prime} = +4 \text{ kcal/mol}$$

The equilibrium constant K'_{eq} of this reaction at 25°C is related to $\Delta G^{\circ\prime}$ by

$$K'_{eq} = \frac{[\mathrm{B}]_{eq}}{[\mathrm{A}]_{eq}} = 10^{-\Delta G^{\circ\prime}/1.36} = 1.15 \times 10^{-3}$$

Thus, A cannot be spontaneously converted into B when the molar ratio of B to A is equal to or greater than 1.15×10^{-3}. However, A can be converted into B when the [B]/[A] ratio is higher than 1.15×10^{-3} if the reaction is coupled to the hydrolysis of ATP. The new overall reaction is

$$\mathrm{A} + \mathrm{ATP} + \mathrm{H_2O} \rightleftharpoons \mathrm{B} + \mathrm{ADP} + \mathrm{P_i} + \mathrm{H^+} \qquad \Delta G^{\circ\prime} = -3.3 \text{ kcal/mol}$$

Its standard free-energy change of −3.3 kcal/mol is the sum of $\Delta G^{\circ\prime}$ for the conversion of A into B (+4 kcal/mol) and for the hydrolysis of ATP (−7.3 kcal/mol). The equilibrium constant of this coupled reaction is

$$K'_{eq} = \frac{[\mathrm{B}]_{eq}}{[\mathrm{A}]_{eq}} \cdot \frac{[\mathrm{ADP}]_{eq}[\mathrm{P_i}]_{eq}}{[\mathrm{ATP}]_{eq}} = 10^{3.3/1.36} = 2.67 \times 10^2$$

At equilibrium, the ratio of [B] to [A] is given by

$$\frac{[\mathrm{B}]_{eq}}{[\mathrm{A}]_{eq}} = K'_{eq}\frac{[\mathrm{ATP}]_{eq}}{[\mathrm{ADP}]_{eq}[\mathrm{P_i}]_{eq}}$$

The ATP-generating system of cells maintains the [ATP]/[ADP][$\mathrm{P_i}$] ratio at a high level, typically of the order of 500. For this ratio,

$$\frac{[\mathrm{B}]_{eq}}{[\mathrm{A}]_{eq}} = 2.67 \times 10^2 \times 500 = 1.34 \times 10^5$$

which means that the hydrolysis of ATP enables A to be converted into B until the [B]/[A] ratio reaches a value of 1.34×10^5. This equilibrium ratio is strikingly different from the value of 1.15×10^{-3} for the reaction A ⇌ B that does not include ATP hydrolysis. In other words, the coupled hydrolysis of ATP has changed the equilibrium ratio of B to A by a factor of about 10^8.

We see here the thermodynamic essence of ATP's action as an energy-coupling agent. *Cells maintain a high level of ATP by using oxidizable substrates or light as sources of free energy. The hydrolysis of an ATP molecule in a coupled reaction then changes the equilibrium ratio of products to reactants by a very large factor, of the order of 10^8*. More generally, the hydrolysis of n ATP molecules changes the equilibrium ratio of a coupled reaction (or sequence of reactions) by a factor of 10^{8n}. For example, the hydrolysis of three ATP molecules in a coupled reaction changes the equilibrium ratio by a factor of 10^{24}. Thus, *a thermodynamically unfavorable reaction sequence can be converted into a favorable one by coupling it to the hydrolysis of a sufficient number of ATP molecules*. It should also be emphasized that A

Table 13-1
Free energies of hydrolysis of some phosphorylated compounds

Compound	*$\Delta G^{\circ\prime}$ (kcal/mol)*
Phosphoenolpyruvate	−14.8
Carbamoyl phosphate	−12.3
Acetyl phosphate	−10.3
Creatine phosphate	−10.3
Pyrophosphate	−8.0
ATP (to ADP)	−7.3
Glucose 1-phosphate	−5.0
Glucose 6-phosphate	−3.3
Glycerol 3-phosphate	−2.2

Note: Phosphoenolpyruvate has the highest phosphate group transfer potential of the compounds listed.

and B in the preceding coupled equation may be interpreted very generally, not only as different chemical species. For example, A and B may represent *different conformations of a protein,* as in muscle contraction. Alternatively, A and B may refer to the *concentrations of an ion or molecule on the outside and inside of a cell,* as in the active transport of a nutrient. The chapters ahead will deal with many remarkable mechanisms of energy coupling in biological processes.

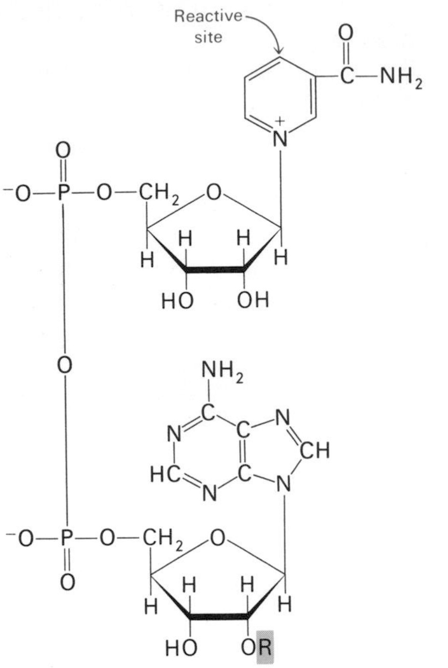

Figure 13-7
Structure of the oxidized form of nicotinamide adenine dinucleotide (NAD^+) and of nicotinamide adenine dinucleotide phosphate ($NADP^+$). In NAD^+, R = H; in $NADP^+$, R = PO_3^{2-}.

NADH AND $FADH_2$ ARE THE MAJOR ELECTRON CARRIERS IN THE OXIDATION OF FUEL MOLECULES

Chemotrophs derive free energy from the oxidation of fuel molecules, such as glucose and fatty acids. In aerobic organisms, the ultimate electron acceptor is O_2. However, electrons are not transferred directly from fuel molecules and their breakdown products to O_2. Instead, these substrates transfer electrons to special carriers, which are either *pyridine nucleotides* or *flavins.* The reduced forms of these carriers then transfer their high-potential electrons to O_2 by means of an electron-transport chain located in the inner membrane of mitochondria. ATP is formed from ADP and P_i as a result of this flow of electrons. This process, called *oxidative phosphorylation* (Chapter 17), is the major source of ATP in aerobic organisms. Alternatively, the high-potential electrons derived from the oxidation of fuel molecules can be used in biosyntheses that require *reducing power* in addition to ATP.

Nicotinamide adenine dinucleotide (NAD^+) is a major electron acceptor in the oxidation of fuel molecules (Figure 13-7). The reactive part of NAD^+ is its nicotinamide ring. *In the oxidation of a substrate, the nicotinamide ring of NAD^+ accepts a hydrogen ion and two electrons, which are equivalent to a hydride ion. The reduced form of this carrier is called NADH.*

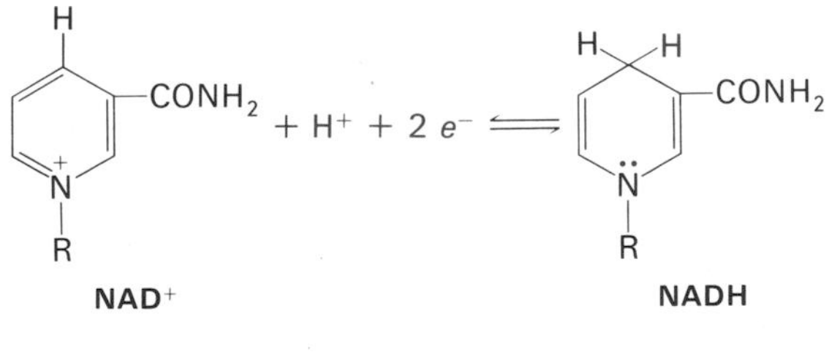

$$NAD^+ + H^+ + 2\,e^- \rightleftharpoons NADH$$

NAD^+ is the electron acceptor in many reactions of the type

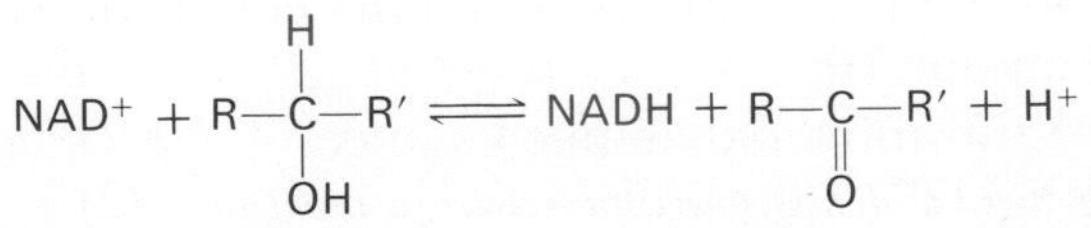

$$NAD^+ + R{-}CH(OH){-}R' \rightleftharpoons NADH + R{-}C({=}O){-}R' + H^+$$

In this dehydrogenation, one hydrogen atom of the substrate is directly transferred to NAD^+, whereas the other appears in the solvent. Both electrons lost by the substrate are transferred to the nicotinamide ring.

The other major electron carrier in the oxidation of fuel molecules is *flavin adenine dinucleotide* (Figure 13-8). The abbreviations for the oxidized and reduced forms of this carrier are FAD and $FADH_2$, respectively. FAD is the electron acceptor in reactions of the type

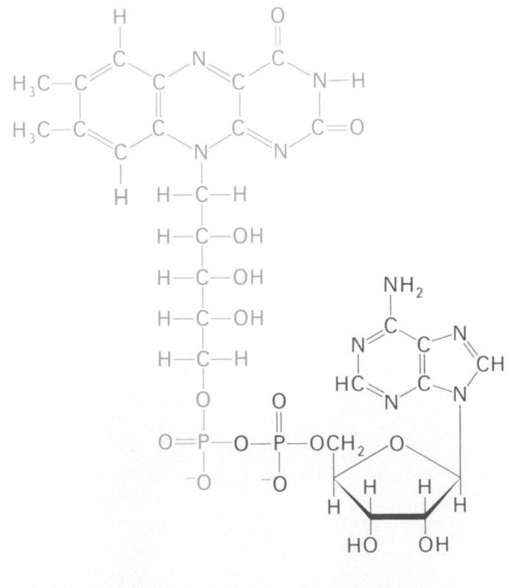

Figure 13-8
Structure of the oxidized form of flavin adenine dinucleotide (FAD). This electron carrier consists of a flavin mononucleotide (FMN) unit (shown in green) and an AMP unit (shown in red).

$$FAD + R-\underset{H}{\overset{H}{C}}-\underset{H}{\overset{H}{C}}-R' \rightleftharpoons FADH_2 + R-\underset{H}{C}=\underset{H}{C}-R'$$

The reactive part of FAD is its isoalloxazine ring (Figure 13-9). FAD, like NAD^+, can accept two electrons. These electron carriers and flavin mononucleotide (FMN), an electron carrier related to FAD, will be discussed further in Chapter 17.

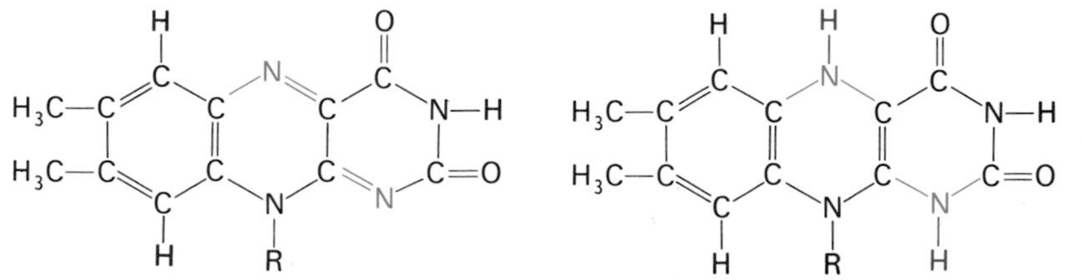

Oxidized form (FAD)

Reduced form ($FADH_2$)

Figure 13-9
Structures of the reactive parts of FAD and $FADH_2$.

NADPH IS THE MAJOR ELECTRON DONOR IN REDUCTIVE BIOSYNTHESES

In most biosyntheses, the precursors are more oxidized than the products. Hence, reductive power is needed in addition to ATP. For example, in the biosynthesis of fatty acids, the keto group of an added C_2 unit is reduced to a methylene group in several steps. This sequence of reactions requires an input of four electrons.

$$R-CH_2-\overset{O}{\overset{\|}{C}}-R' + 4\,H^+ + 4\,e^- \longrightarrow R-CH_2-CH_2-R' + H_2O$$

The electron donor in most reductive biosyntheses is NADPH, the reduced form of nicotinamide adenine dinucleotide phosphate ($NADP^+$, Figure 13-7). NADPH differs from NADH in that the 2′-hydroxyl group of its adenosine moiety is esterified with phosphate. NADPH carries electrons in the same way as NADH. However, *NADPH is used almost exclusively for reductive biosyntheses, whereas NADH is used primarily for the generation of ATP*. The extra phosphate group on NADPH is a tag that directs this reducing agent to discerning biosynthetic enzymes. The biological significance of the distinction between NADPH and NADH will be discussed later (p. 427).

It is important to note that NADH, NADPH, and $FADH_2$ react slowly with O_2 in the absence of catalysts. Likewise, ATP is hydrolyzed slowly (in times of many hours or even days) in the absence of a catalyst. These molecules are kinetically quite stable in the face of a large thermodynamic driving force for reaction with O_2 (in the case of the electron carriers) and H_2O (in the case of ATP). *The stability of these molecules in the absence of specific catalysts is essential for their biological function because it enables enzymes to control the flow of free energy and reductive power.*

COENZYME A IS A UNIVERSAL CARRIER OF ACYL GROUPS

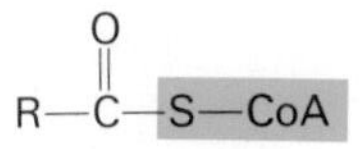

$R-C(=O)-S-CoA$

Acyl CoA

$H_3C-C(=O)-S-CoA$

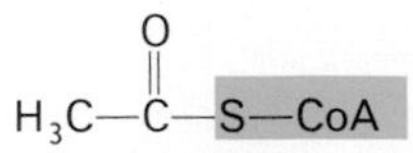

Acetyl CoA

Coenzyme A is another central molecule in metabolism. In 1945, Lipmann found that a heat-stable cofactor was required in many enzyme-catalyzed acetylations. This cofactor was named *coenzyme A* (CoA), the A standing for *acetylation*. It was isolated, and its structure was determined several years later (Figure 13-10). The terminal sulfhydryl group in CoA is the reactive site. Acyl groups are linked to CoA by a thioester bond. The resulting derivative is called an *acyl CoA*. An acyl group often linked to CoA is the acetyl unit; this derivative is called *acetyl CoA*. The $\Delta G^{\circ\prime}$ for the hydrolysis of acetyl CoA has a large negative value:

$$\text{Acetyl CoA} + H_2O \rightleftharpoons \text{acetate} + \text{CoA} + H^+ \qquad \Delta G^{\circ\prime} = -7.5 \text{ kcal/mol}$$

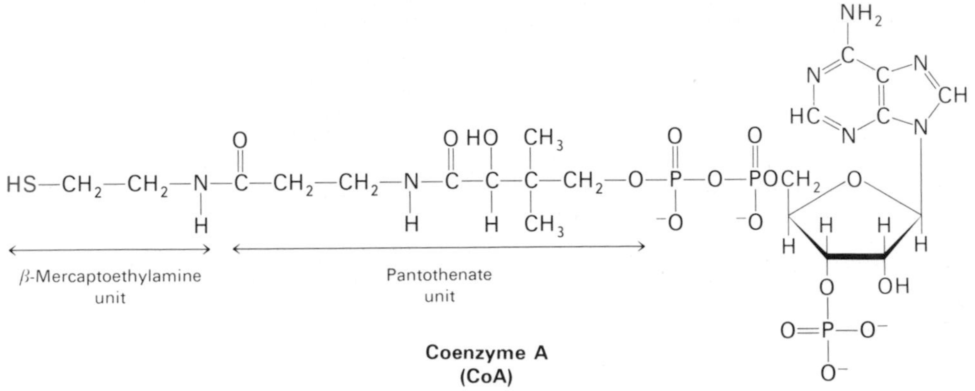

$HS-CH_2-CH_2-NH-C(=O)-CH_2-CH_2-NH-C(=O)-CH(OH)-C(CH_3)_2-CH_2-O-P(=O)(O^-)-O-P(=O)(O^-)-OCH_2$-(ribose 3'-phosphate)-adenine

β-Mercaptoethylamine unit

Pantothenate unit

Coenzyme A (CoA)

Figure 13-10
Structure of coenzyme A (CoA).

The hydrolysis of a thioester is thermodynamically more favorable than that of an oxygen ester because the double-bond character of the C–O bond does not extend significantly to the C–S bond. Consequently, *acetyl CoA has a high acetyl group transfer potential*. Acetyl CoA carries an activated acetyl group, just as ATP carries an activated phosphoryl group.

We shall encounter other carriers of activated groups in our consideration of metabolism. Several of them are listed in Table 13-2. These

Table 13-2
Some activated carriers in metabolism

Carrier molecule	*Group carried in activated form*
ATP	Phosphoryl
NADH and NADPH	Electrons
$FADH_2$	Electrons
$FMNH_2$	Electrons
Coenzyme A	Acyl
Lipoamide	Acyl
Thiamine pyrophosphate	Aldehyde
Biotin	CO_2
Tetrahydrofolate	One-carbon units
S-Adenosylmethionine	Methyl
Uridine diphosphate glucose	Glucose
Cytidine diphosphate diacylglycerol	Phosphatidate
Nucleoside triphosphates	Nucleotides

carriers mediate the interchange of activated groups in a wide variety of biochemical reactions. Indeed, they have very similar roles in all forms of life. Their universal presence is one of the unifying motifs of biochemistry.

MOST WATER-SOLUBLE VITAMINS ARE COMPONENTS OF COENZYMES

Lipmann has commented that "doctors like to prescribe vitamins and millions of people take them, but it requires a good deal of biochemical sophistication to understand why they are needed and how the organism uses them." Vitamins are organic molecules that are needed in small amounts in the diets of higher animals. These molecules serve nearly the same roles in all forms of life, but higher animals have lost the capacity to synthesize them. Vitamins can be grouped according to whether they are soluble in water or in nonpolar solvents. The *water-soluble vitamins* are vitamin C and a series known as the vitamin B complex. Vitamin C is a reducing agent in hydroxylation reactions, as in the conversion of proline residues of collagen to hydroxyproline (p. 263). The vitamin B series (Figure 13-11) are components of coenzymes (Table 13-3). For example, riboflavin (vitamin B_2) is a precursor of FAD, and pantothenate is a component of coenzyme A.

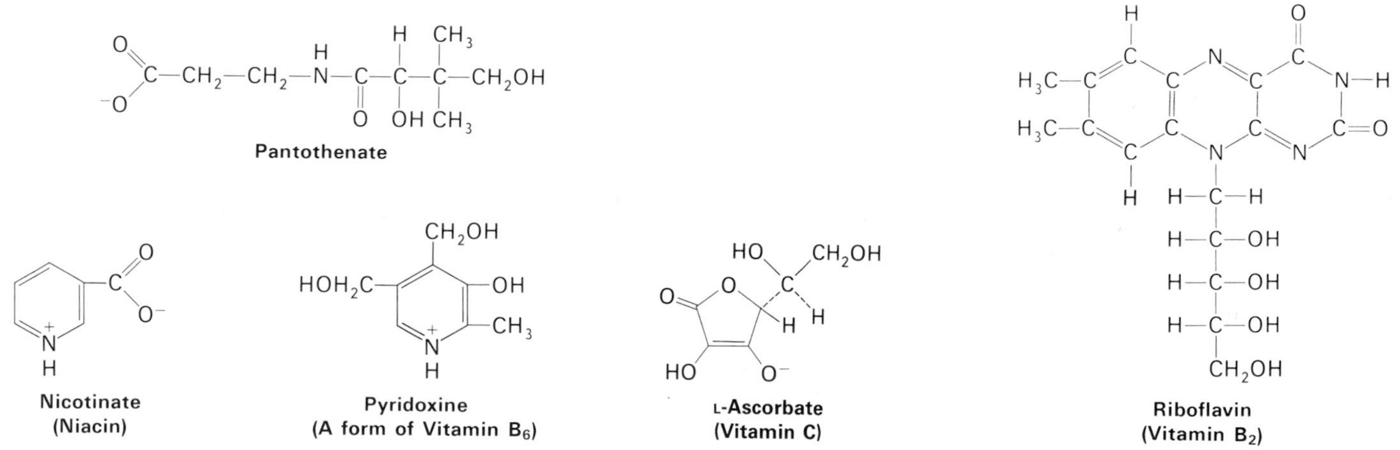

Pantothenate

Nicotinate (Niacin)

Pyridoxine (A form of Vitamin B_6)

L-Ascorbate (Vitamin C)

Riboflavin (Vitamin B_2)

Figure 13-11
Structures of some water-soluble vitamins.

Table 13-3
Coenzyme derivatives of some water-soluble vitamins

Vitamin	*Coenzyme derivative*
Thiamine (vitamin B_1)	Thiamine pyrophosphate
Riboflavin (vitamin B_2)	Flavin adenine dinucleotide and flavin mononucleotide
Nicotinate (niacin)	Nicotinamide adenine dinucleotide
Pyridoxine, pyridoxal, and pyridoxamine (vitamin B_6)	Pyridoxal phosphate
Pantothenate	Coenzyme A
Biotin	Covalently attached to carboxylases
Folate	Tetrahydrofolate
Cobalamin (vitamin B_{12})	Cobamide coenzymes

Much is also known about the molecular actions of *fat-soluble vitamins*, which are designated by the letters A, D, E, and K (Figure 13-12). Vitamin K, which is required for normal blood clotting (K stands for *k*oagulation), participates in the carboxylation of glutamate residues to γ-carboxyglutamate (p. 252). Vitamin A (retinol) is the precursor of retinal, the light-absorbing group in visual pigments (p. 1029). A deficiency of this vitamin results in night blindness. Furthermore, young animals require vitamin A for growth. The metabolism of calcium and phosphorus is regulated by a hormone that is derived from vitamin D (p. 569). A deficiency in vitamin D impairs bone formation in growing animals. Infertility in rats is a consequence of vitamin E (α-tocopherol) deficiency. This vitamin protects unsaturated membrane lipids from oxidation.

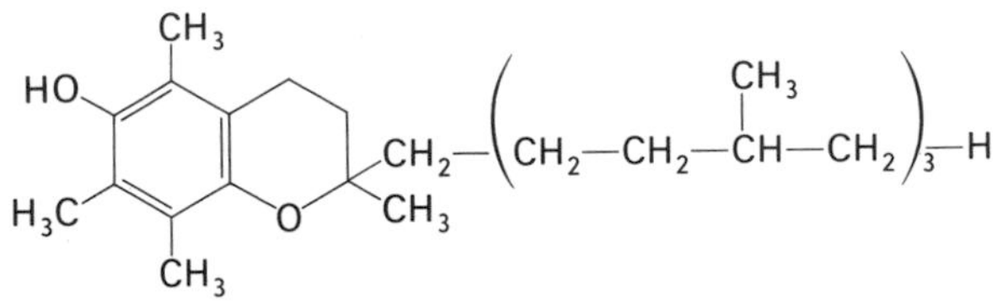

α-Tocopherol (Vitamin E)

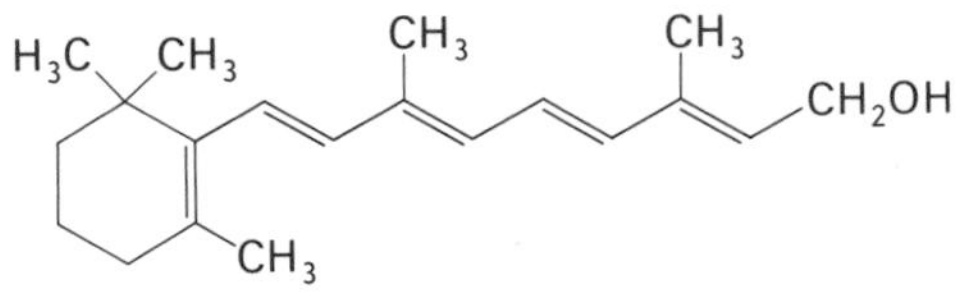

Retinol (Vitamin A)

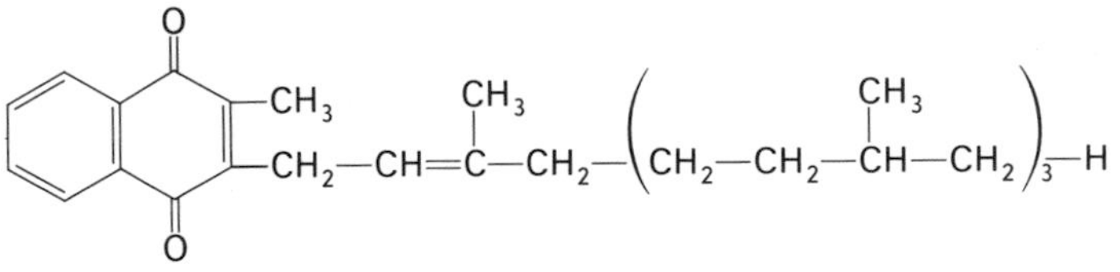

Vitamin K_1

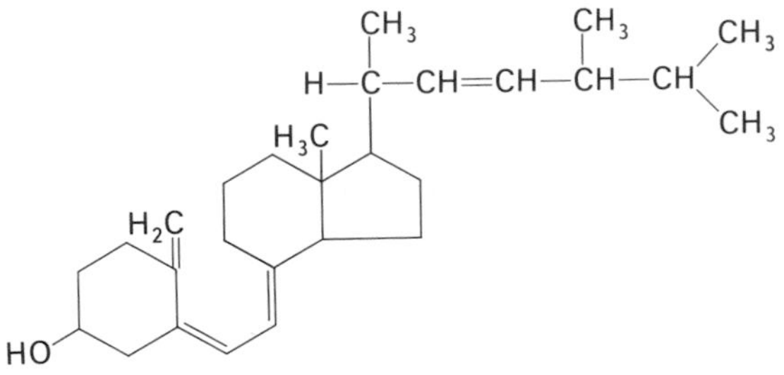

Calciferol (Vitamin D_2)

Figure 13-12
Structures of some fat-soluble vitamins.

STAGES IN THE EXTRACTION OF ENERGY FROM FOODSTUFFS

Let us take an overview of the process of energy generation in higher organisms before considering it in detail in subsequent chapters. Hans Krebs described three stages in the generation of energy from the oxidation of foodstuffs. *In the first stage, large molecules in food are broken down into smaller units.* Proteins are hydrolyzed to their twenty kinds of constituent amino acids, polysaccharides are hydrolyzed to simple sugars such as glucose, and fats are hydrolyzed to glycerol and fatty acids (Figure 13-13). No useful energy is generated in this phase.

In the second stage, these numerous small molecules are degraded to a few simple units that play a central role in metabolism. In fact, most of them—sugars, fatty acids, glycerol, and several amino acids—are converted into the acetyl unit of acetyl CoA. Some ATP is generated in this stage, but the amount is small compared with that obtained from the complete oxidation of the acetyl unit of acetyl CoA.

The third stage consists of the citric acid cycle and oxidative phosphorylation, which are the final common pathways in the oxidation of fuel molecules. Acetyl CoA brings acetyl units into this cycle, where they are completely oxidized to CO_2. Four pairs of electrons are transferred to NAD^+ and FAD for each acetyl group that is oxidized. Then, ATP is

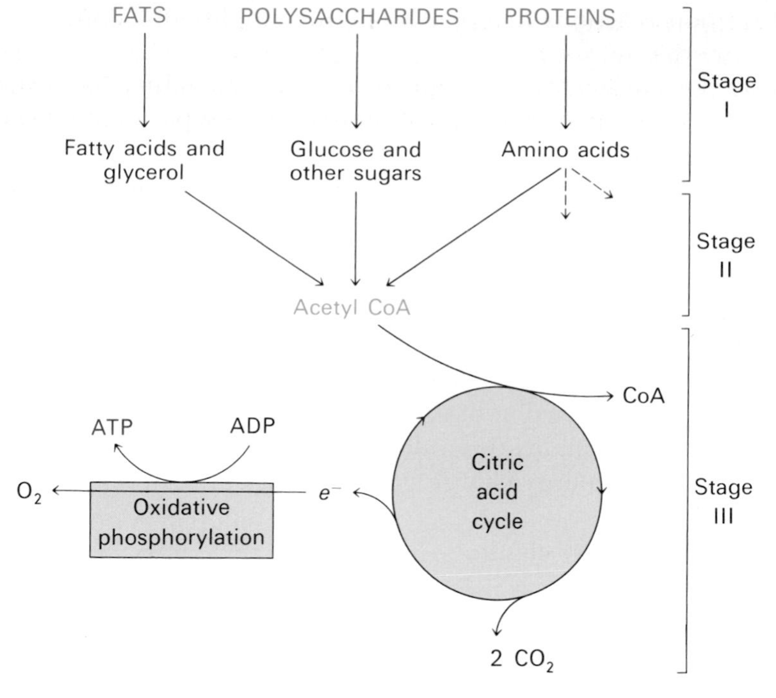

Figure 13-13
Stages in the extraction of energy from foodstuffs.

generated as electrons flow from the reduced forms of these carriers to O_2 in a process called oxidative phosphorylation. Most of the ATP generated by the degradation of foodstuffs is formed in this third stage.

METABOLIC PROCESSES ARE REGULATED BY A VARIETY OF MECHANISMS

Even the simplest bacterial cell has the capacity for carrying out more than a thousand interdependent reactions. It is evident that this complex network must be rigorously regulated. Furthermore, metabolic control must be flexible, because the external environment of cells is not constant. Studies of a wide range of organisms have shown that there are a number of mechanisms for the control of metabolism. It must be emphasized that the central metabolic pathways have been almost fully elucidated, but knowledge concerning their regulation is still in its infancy. Few problems in biochemistry today are as intellectually challenging and important.

Metabolism is regulated in many ways. Control of the *amounts* of certain enzymes, a major mechanism, has been extensively studied in bacteria. Regulation of the *rate of synthesis* of β-galactosidase and other proteins needed for the utilization of lactose is a classic example, which will be discussed in detail in Chapter 32. In bacteria, gene expression is regulated primarily at the level of transcription. The *rate of degradation* of some enzymes is controlled as well. The *catalytic activities* of certain enzymes are regulated to control metabolic fluxes. *Reversible allosteric control* is especially important. For example, the first reaction in many biosynthetic pathways is allosterically inhibited by the ultimate product of the pathway. The inhibition of aspartate transcarbamoylase by cytidine triphosphate (p. 234) is a well-understood example of *feedback inhibition*. The activities of some enzymes are also modulated by *reversible covalent modification*. For example, glycogen phosphorylase, the enzyme catalyzing the breakdown of a storage form of sugar, is activated by phosphorylation of a particular serine residue when glucose is scarce (p. 459). Hormones such as epinephrine trigger cascades of covalent modification reactions that lead to highly amplified changes in metabolic patterns.

An important general principle of metabolism is that *biosynthetic and degradative pathways are almost always distinct.* This separation is necessary for energetic reasons, as will be evident in subsequent chapters. It also facilitates the control of metabolism. In eucaryotes, metabolic regulation and flexibility are also enhanced by *compartmentation.* For example, fatty acid oxidation occurs in mitochondria, whereas fatty acid synthesis occurs in the cytosol (the soluble part of the cytoplasm). Compartmentation segregates these opposed reactions.

Many reactions in metabolism are controlled by the *energy status* of the cell. One index of the energy status is the *energy charge,* which is proportional to the mole fraction of ATP plus half the mole fraction of ADP, given that ATP contains two anhydride bonds, whereas ADP contains one. Hence, the energy charge is defined as

$$\text{Energy charge} = \frac{[\text{ATP}] + \frac{1}{2}[\text{ADP}]}{[\text{ATP}] + [\text{ADP}] + [\text{AMP}]}$$

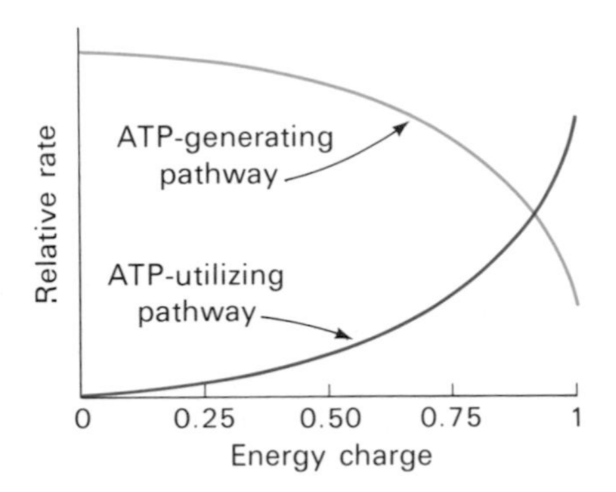

Figure 13-14
Effect of the energy charge on the relative rates of a typical ATP-generating (catabolic) pathway and a typical ATP-utilizing (anabolic) pathway.

The energy charge can have a value ranging from 0 (all AMP) to 1 (all ATP). Daniel Atkinson has shown that *ATP-generating pathways are inhibited by a high energy charge, whereas ATP-utilizing pathways are stimulated by a high energy charge.* In plots of the reaction rates of such pathways versus the energy charge, the curves are steep near an energy charge of 0.9, where they usually intersect (Figure 13-14). It is evident that the control of these pathways is designed to maintain the energy charge within rather narrow limits. In other words, *the energy charge, like the pH of a cell, is buffered.* The energy charge of most cells is in the range of 0.80 to 0.95. An alternative index of the energy status is the *phosphorylation potential,* which is defined as

$$\text{Phosphorylation potential} = \frac{[\text{ATP}]}{[\text{ADP}][\text{P}_i]}$$

The phosphorylation potential, in contrast with the energy charge, depends on the concentration of P_i and is directly related to the free energy available from ATP.

NUCLEAR MAGNETIC RESONANCE SPECTROSCOPY REVEALS METABOLIC EVENTS IN INTACT ORGANISMS

A century ago, Thomas Huxley said "What an enormous revolution would be made in biology, if physics or chemistry could supply the physiologist with a means of making out the molecular structure of living tissues comparable to that which spectroscopy affords to the inquirer into the nature of the heavenly bodies." Recent advances in nuclear magnetic resonance (NMR) spectroscopy are bringing this hope closer to realization. Metabolic processes in skeletal muscle, heart, and brain of intact organisms can now be explored noninvasively by NMR methods. This technique is based on the fact that certain atomic nuclei, such as the naturally occurring isotope of phosphorus (^{31}P), are intrinsically magnetic (Table 13-4). Their magnetic moment can take either of two orientations when an external magnetic field is applied (Figure 13-15). The energy difference between these states is proportional to the strength of the imposed field. The state with the moments aligned along the field has the slightly lower energy of the two, and so it is

Table 13-4
Biologically important nuclei giving NMR signals

Nucleus	*Natural abundance (% by weight of the element)*
^{1}H	99.984
^{2}H	0.016
^{13}C	1.108
^{14}N	99.635
^{15}N	0.365
^{17}O	0.037
^{23}Na	100.0
^{25}Mg	10.05
^{31}P	100.0
^{35}Cl	75.4
^{39}K	93.1

slightly more populated (by a factor of the order of 1.00001 in a typical experiment). A transition from the lower to the upper state occurs when a nucleus absorbs electromagnetic radiation of appropriate frequency. This resonance frequency ν_0 of an isolated nucleus is

$$\nu_0 = \frac{\gamma B_0}{2\pi}$$

where B_0 is the strength of the steady magnetic field and γ is a constant (called the magnetogyric ratio) for a given nucleus. For example, the resonance frequency for ^{31}P in a 100 kilogauss (10 tesla) field is 172 megahertz (MHz), which lies in the radiofrequency region of the spectrum. A plot of the energy absorbed versus frequency would show a peak at 172 MHz.

NMR is a very informative technique because the local magnetic field is not identical to the applied field B_0 for all nuclei in the sample. Depending on the particular arrangement of bonds and other atoms around a nucleus, it is more or less shielded from the applied field. Consequently, nuclei in different chemical environments absorb energy at slightly different resonance frequencies, an effect termed the *chemical shift*. These separations are expressed in fractional units (parts per million, ppm) relative to a standard compound. For example, the chemical shifts of the phosphorus atom in $H_2PO_4^-$ and HPO_4^{2-} differ by 2.4 ppm. However, a single peak is observed from solutions of orthophosphate because $H_2PO_4^-$ and HPO_4^- interconvert rapidly. But the position of this peak depends on the ratio of these species, so that it serves as an indicator of intracellular pH (Figure 13-16).

The ^{31}P NMR spectra of the forearm muscle of a human subject before and during exercise are shown in Figure 13-17. Five peaks are evident in this strikingly simple spectrum from a complex organ. Three of them arise primarily from the α, β, and γ phosphorus atoms of ATP. The other two come from the phosphorus atoms of phosphocreatine and orthophosphate. ADP and other phosphate compounds do not contribute appreciably to these spectra because they are present at

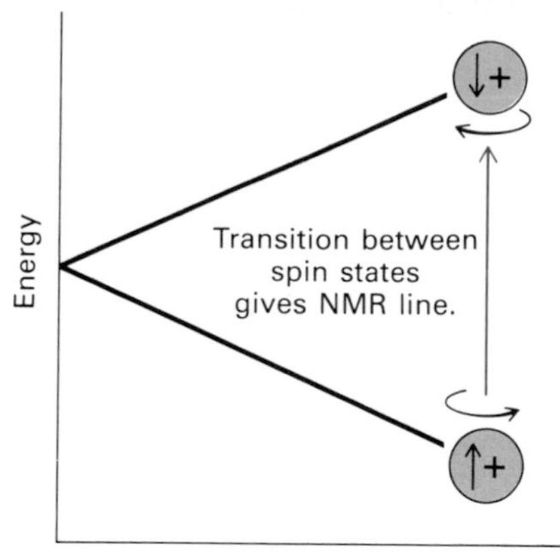

Figure 13-15
Essence of NMR spectroscopy. The energies of the two orientations of a nucleus of spin $\frac{1}{2}$ (such as ^{31}P and ^{1}H) depend on the strength of the applied magnetic field. An oscillating magnetic field of appropriate frequency (ν_0) can induce a transition from the lower to the upper level.

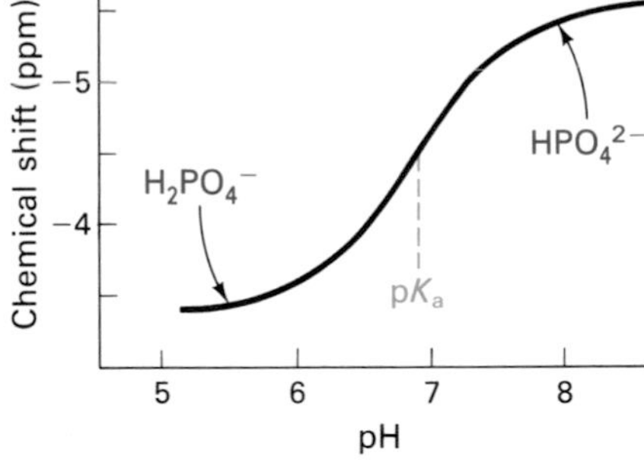

Figure 13-16
The chemical shift of the ^{31}P NMR line of orthophosphate depends on pH. The shifts are expressed relative to the position of the phosphocreatine resonance. [After D. G. Gadian, G. K. Radda, R. E. Richards, and P. J. Seeley. In *Biological Applications of Magnetic Resonance*, R. G. Shulman, ed., (Academic Press, 1979), p. 475.]

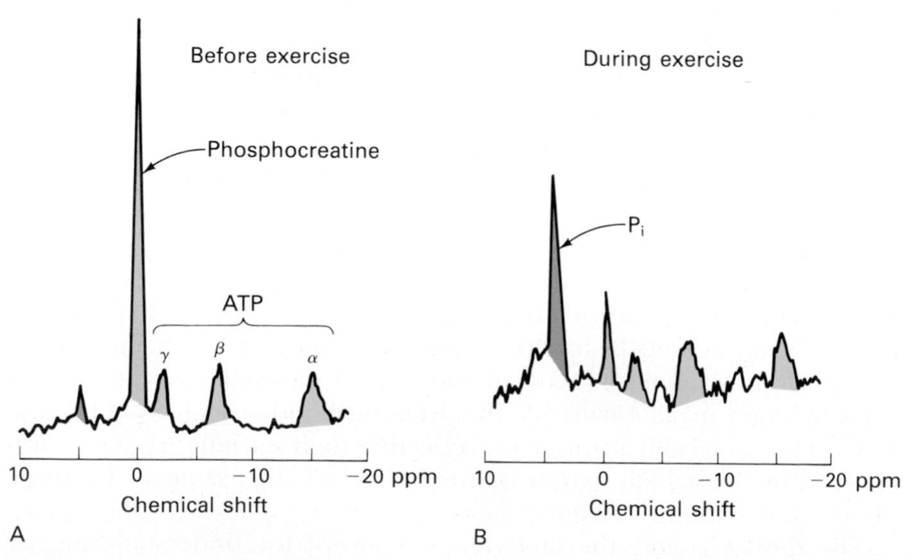

Figure 13-17
Effect of exercise on the level of ATP, phosphocreatine, and orthophosphate in the forearm muscle of a human subject. (A) ^{31}P NMR spectrum before exercise, and (B) after 19 minutes of exercise. Spectrum A was collected in 256 seconds and spectrum B in 64 seconds. Note that the three phosphorus atoms in ATP have different chemical shifts. [After G. K. Radda. *Science* 233(1986):641.]

much lower concentration or are not free to rotate. Slowly rotating nuclei such as the phosphorus atoms of nucleotides tightly bound to proteins and those of phospholipid bilayers and nucleic acids have very broad lines that raise the background of such NMR spectra but do not appear as discrete peaks. The spectrum observed after 19 minutes of exercise shows that *the amount of phosphocreatine has decreased markedly, whereas that of orthophosphate has increased markedly. In contrast, the ATP level stayed nearly constant.* This study provides valuable quantitative information concerning intracellular pH and the flux of phosphoryl groups from phosphocreatine to ATP. In subsequent chapters, we shall see how NMR spectroscopy is being used to obtain insight into control mechanisms and metabolic disorders. NMR is bringing biochemistry, physiology, and clinical medicine closer together.

THE CENTRAL ROLE OF RIBONUCLEOTIDES IN METABOLISM REFLECTS THEIR ANCIENT ORIGINS

Many of the central molecules of metabolism in all forms of life are ribonucleotides. Why do activated carriers such as ATP, NADH, $FADH_2$, and coenzyme A contain adenosine phosphate units? A likely explanation is that RNA came before proteins and DNA in evolution (p. 113). The earliest catalysts most probably were RNA molecules. These ribozymes recruited non-RNA units such as the isoalloxazine ring to form coenzymes that serve as efficient carriers of activated electrons and chemical units, a function not readily performed by RNA itself. We can picture the adenine ring of $FADH_2$ binding to a uracil unit in a niche of a ribozyme by base pairing. When proteins replaced RNA as the major catalysts to achieve greater versatility, the ribonucleotide coenzymes stayed essentially unchanged because they were already well suited to their metabolic roles. The nicotinamide unit of NADH, for example, can readily transfer electrons irrespective of whether the adenine unit interacts with a base in a ribozyme or with amino acid residues in a protein enzyme. That molecules and motifs of metabolism are common to all forms of life testifies to their common origin and to the retention of functioning modules over billions of years of evolution. Our understanding of metabolism, like that of other biological processes, is enriched by inquiry into how these beautifully integrated patterns of reactions came into being.

SUMMARY

Cells extract energy from their environment and convert foodstuffs into cell components by a highly integrated network of chemical reactions called metabolism. Most of the central molecules of metabolism are the same in all forms of life. Ribonucleotides such as ATP and NADH are especially prominent, reflecting their ancient origins. Moreover, many metabolic patterns are essentially the same in bacteria, plants, and animals.

The most valuable thermodynamic concept for understanding the energetics of metabolism is free energy. A reaction can occur spontaneously only if the change in free energy (ΔG) is negative. A thermodynamically unfavorable reaction can be driven by a thermodynamically favorable one. ATP, the universal currency of energy in biological systems, is an energy-rich molecule because it contains two phosphoanhydride bonds. The repulsion between the negatively charged phosphate groups is reduced when ATP is hydrolyzed. Also, ADP and P_i are stabilized by resonance more than is ATP. The hydrolysis of ATP shifts the equilibrium of a coupled reaction by a factor of about 10^8.

The basic strategy of metabolism is to form ATP, NADPH, and macromolecular precursors. ATP is consumed in muscle contraction and other motions of cells, active transport, and biosyntheses. NADPH, which carries two electrons at a high potential, provides reducing power in the biosynthesis of cell components from more-oxidized precursors. ATP and NADPH are continuously generated and consumed.

There are three stages in the extraction of energy from foodstuffs by aerobic organisms. In the first stage, large molecules are broken down into smaller ones, such as amino acids, sugars, and fatty acids. In the second stage, these small molecules are degraded to a few simple units that have a pervasive role in metabolism. One of them is the acetyl unit of acetyl CoA, a carrier of activated acyl groups. The third stage of metabolism is the citric acid cycle and oxidative phosphorylation, in which ATP is generated as electrons flow to O_2, the ultimate electron acceptor, and fuels are completely oxidized to CO_2.

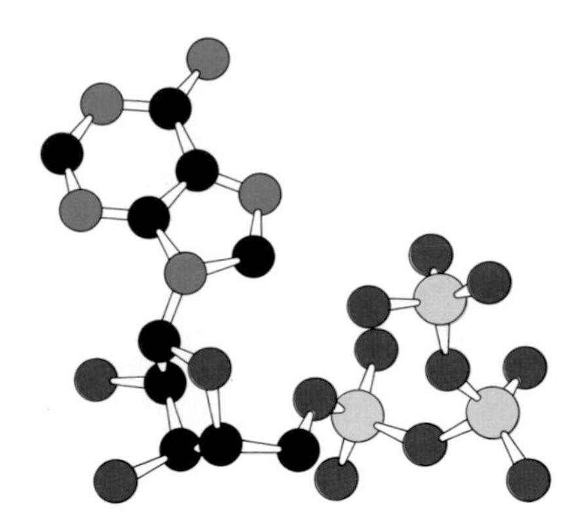

Figure 13-18
Model of adenosine triphosphate (ATP).

Metabolism is regulated in a variety of ways. The amounts of some critical enzymes are controlled by regulation of the rate of protein synthesis and degradation. In addition, the catalytic activities of some enzymes are regulated by allosteric interactions (as in feedback inhibition) and by covalent modification. Compartmentation and distinct pathways for biosynthesis and degradation also contribute to metabolic regulation. The energy charge, which depends on the relative amounts of ATP, ADP, and AMP, plays a role in metabolic regulation. A high energy charge inhibits ATP-generating (catabolic) pathways, whereas it stimulates ATP-utilizing (anabolic) pathways. The level of ATP and other phosphorus-containing molecules in living organisms can be monitored noninvasively by nuclear magnetic resonance.

SELECTED READINGS

Overviews of metabolism

Krebs, H. A., and Kornberg, H. L., 1957. *Energy Transformations in Living Matter.* Springer-Verlag. [Includes a valuable appendix by K. Burton containing thermodynamic data.]

Wood, W. B., 1974. *The Molecular Basis of Metabolism.* Unit 3 in Biocore. McGraw-Hill.

Thermodynamics

Edsall, J. T., and Gutfreund, H., 1983. *Biothermodynamics: The Study of Biochemical Processes at Equilibrium.* Wiley. [A concise account with many informative examples.]

Klotz, I. M., 1967. *Energy Changes in Biochemical Reactions.* Academic Press. [A concise introduction, full of insight.]

Hill, T. L., 1977. *Free Energy Transduction in Biology.* Academic Press.

Alberty, R. A., 1968. Effect of pH and metal ion concentration on the equilibrium hydrolysis of adenosine triphosphate to adenosine diphosphate. *J. Biol. Chem.* 243:1337–1343.

Jencks, W. P., 1970. Free energies of hydrolysis and decarboxylation. *In* Sober, H. A., *Handbook of Biochemistry* (2nd ed.), pp. J181–J186. Chemical Rubber Co.

Regulation of metabolism

Newsholme, E. A., and Start, C., 1973. *Regulation in Metabolism.* Wiley. [An excellent treatment of metabolic regulation in mammals.]

Atkinson, D. E., 1977. *Cellular Energy Metabolism and Its Regulation.* Academic Press. [Contains a detailed account of the concept of energy charge and of other aspects of metabolic control.]

Erecińska, M., and Wilson, D. F., 1978. Homeostatic regulation of cellular energy metabolism. *Trends Biochem. Sci.* 3:219–223. [Review of the regulatory role of the phosphorylation potential in mitochondria.]

Nuclear magnetic resonance spectroscopy

Shulman, R. G., 1983. NMR spectroscopy of living cells. *Sci. Amer.* 248(1):86–93.

Radda, G. K., 1986. The use of NMR spectroscopy for the understanding of disease. *Science* 233:640–645.

Gadian, D. G., 1982. *Nuclear Magnetic Resonance and Its Applications to Living Systems.* Oxford University Press.

Jardetzky, O., and Roberts, G. C. K., 1981. *NMR in Molecular Biology.* Academic Press.

Chance, B., 1983. A noninvasive biochemical assay and imaging of animal and human tissues by optical and nuclear magnetic resonance techniques. *Proc. Amer. Phil. Soc.* 127:1–25.

Avison, M. J., Hetherington, H. P., and Shulman, R. G., 1986. Applications of NMR to studies of tissue metabolism. *Ann. Rev. Biophys. Biophys. Chem.* 15:377.

Radda, G. K., 1986. Control of bioenergetics: from cells to man by phosphorus NMR spectroscopy. *Biochem. Soc. Trans.* 14:517–525.

Evolution of metabolism

Broda, E., 1975. *The Evolution of the Bioenergetic Process.* Pergamon.

Baldwin, J. E., and Krebs, H., 1981. The evolution of metabolic cycles. *Nature* 291:381–382.

Yeh, W. K., and Ornston, L. N., 1980. Origins of metabolic diversity: substitution of homologous sequences into genes for enzymes with different catalytic activities. *Proc. Nat. Acad. Sci. USA* 77:5365–5369.

Historical aspects

Kalckar, H. M., (ed.), 1969. *Biological Phosphorylations.* Prentice-Hall. [A valuable collection of many classic papers on bioenergetics.]

Fruton, J. S., 1972. *Molecules and Life.* Wiley-Interscience. [Perceptive and scholarly essays on the interplay of chemistry and biology since 1800. Metabolism and bioenergetics are among the topics treated in detail.]

Lipmann, F., 1971. *Wanderings of a Biochemist.* Wiley-Interscience. [Contains reprints of some of the author's classic papers and several delightful essays.]

PROBLEMS

1. What is the direction of each of the following reactions when the reactants are initially present in equimolar amounts? Use the data given in Table 13-1.
   (a) ATP + creatine $\rightleftharpoons$ phosphocreatine + ADP
   (b) ATP + glycerol $\rightleftharpoons$ glycerol 3-phosphate + ADP
   (c) ATP + pyruvate $\rightleftharpoons$ phosphoenolpyruvate + ADP
   (d) ATP + glucose $\rightleftharpoons$ glucose 6-phosphate + ADP

2. What information do the $\Delta G^{\circ\prime}$ data given in Table 13-1 provide about the relative rates of hydrolysis of pyrophosphate and acetyl phosphate?

3. Consider the reaction

   ATP + pyruvate $\rightleftharpoons$ phosphoenolpyruvate + ADP

   (a) Calculate $\Delta G^{\circ\prime}$ and K'_{eq} at 25°C for this reaction, using the data given in Table 13-1.
   (b) What is the equilibrium ratio of pyruvate to phosphoenolpyruvate if the ratio of ATP to ADP is 10?

4. Calculate $\Delta G^{\circ\prime}$ for the isomerization of glucose 6-phosphate to glucose 1-phosphate. What is the equilibrium ratio of glucose 6-phosphate to glucose 1-phosphate at 25°C?

5. The formation of acetyl CoA from acetate is an ATP-driven reaction:

   Acetate + ATP + CoA $\rightleftharpoons$ acetyl CoA + AMP + PP_i

   (a) Calculate $\Delta G^{\circ\prime}$ for this reaction, using data given in this chapter.
   (b) The PP_i formed in the above reaction is rapidly hydrolyzed in vivo because of the ubiquity of inorganic pyrophosphatase. The $\Delta G^{\circ\prime}$ for the hydrolysis of PP_i is −8 kcal/mol. Calculate the $\Delta G^{\circ\prime}$ for the overall reaction. What effect does the hydrolysis of PP_i have on the formation of acetyl CoA?

6. The pK of an acid is a measure of its proton group-transfer potential.
   (a) Derive a relation between ΔG° and pK.
   (b) What is the ΔG° for the ionization of acetic acid, which has a pK of 4.8?

7. What is the common structural feature of ATP, FAD, NAD^+, and CoA?

8. Fibrinogen contains tyrosine-*O*-sulfate. Propose an activated form of sulfate that could react in vivo with the aromatic hydroxyl group of a tyrosine residue in a protein to form tyrosine-*O*-sulfate.

9. Two chemical species give rise to a single peak in an NMR spectrum if their rate of interconversion is fast compared with the difference between their resonance frequencies. For example, orthophosphate gives a single peak at pH values at which both $H_2PO_4^-$ and HPO_4^{2-} are present (see Figure 13-16). The chemical shifts of these forms of orthophosphate differ by 2.4 ppm relative to a standard compound that resonates at 129 MHz.
   (a) Calculate the difference between the resonance frequencies of these species.
   (b) What is the minimum interconversion rate of these species?
   (c) What is the minimum rate constant for the association of H^+ with HPO_4^{2-}?

CHAPTER 14

Carbohydrates

Before embarking on our journey through metabolism, let us take an overview of carbohydrates, one of the four major classes of biomolecules. We have already considered the other three—proteins, nucleic acids, and lipids. Carbohydrates are aldehyde or ketone compounds with multiple hydroxyl groups. They make up most of the organic matter on earth because of their multiple roles in all forms of life. First, carbohydrates serve as *energy stores, fuels, and metabolic intermediates.* Starch in plants and glycogen in animals are polysaccharides that can be rapidly mobilized to yield glucose, a prime fuel for the generation of energy. ATP, the universal currency of free energy, is a phosphorylated sugar derivative, as are many coenzymes. Second, ribose and deoxyribose sugars form part of the *structural framework of DNA and RNA.* The conformational flexibility of these sugar rings is important in the storage and expression of genetic information. Third, polysaccharides are *structural elements in the cell walls of bacteria and plants, and in the exoskeletons of arthropods.* In fact, cellulose, the main constituent of plant cell walls, is the most abundant organic compound in the biosphere. Fourth, carbohydrates are *linked to many proteins and lipids.* For example, the sugar units of glycophorin give red cells a highly polar anionic coat. Recent studies suggest that carbohydrate units on cell surfaces are key participants in cell-cell recognition during development.

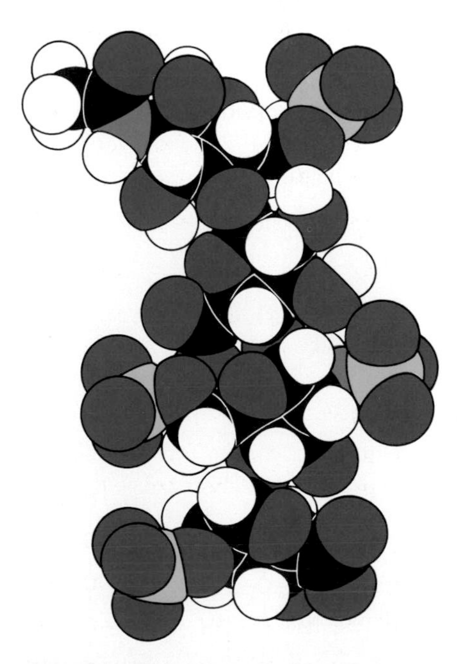

Figure 14-1
Model of heparan sulfate, a carbohydrate that participates in cell-cell recognition.

MONOSACCHARIDES ARE ALDEHYDES OR KETONES WITH MULTIPLE HYDROXYL GROUPS

Monosaccharides, the simplest carbohydrates, are aldehydes or ketones that have two or more hydroxyl groups; their empirical formula is $(CH_2O)_n$. The smallest ones, for which $n = 3$, are glyceraldehyde and dihydroxyacetone. They are *trioses*. Glyceraldehyde is also an *aldose* because it contains an aldehyde group, whereas dihydroxyacetone is a *ketose* because it contains a keto group.

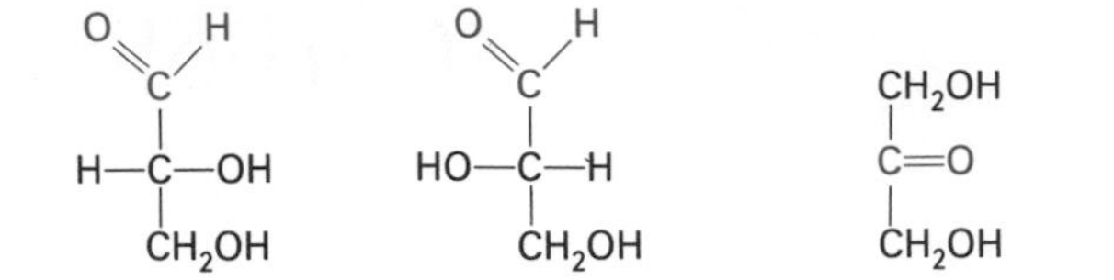

D-Glyceraldehyde
(An aldose)

L-Glyceraldehyde
(An aldose)

Dihydroxyacetone
(A ketose)

Glyceraldehyde has a single asymmetric carbon. Thus, there are two stereoisomers of this three-carbon aldose, D-glyceraldehyde and L-glyceraldehyde. The prefixes D and L designate the absolute configuration. Recall that in a Fischer projection of a molecule, atoms joined to an asymmetric carbon atom by horizontal bonds are in front of the page, and those joined by vertical bonds are behind (Figure 14-2).

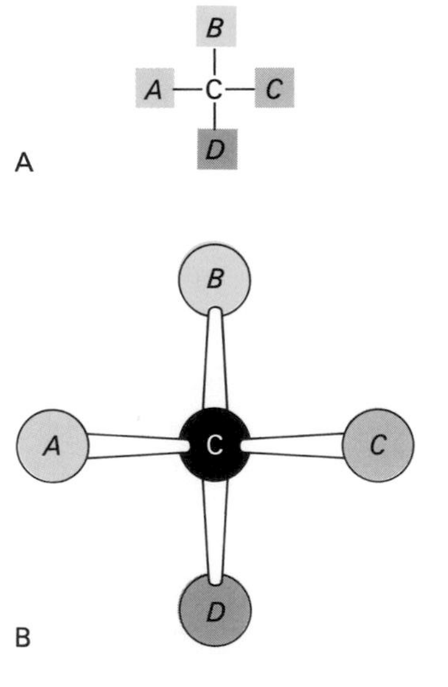

Figure 14-2
(A) Fischer representation of a tetrahedral carbon atom with substituents *A*, *B*, *C*, and *D*, and (B) a model showing the stereochemistry denoted by this projection.

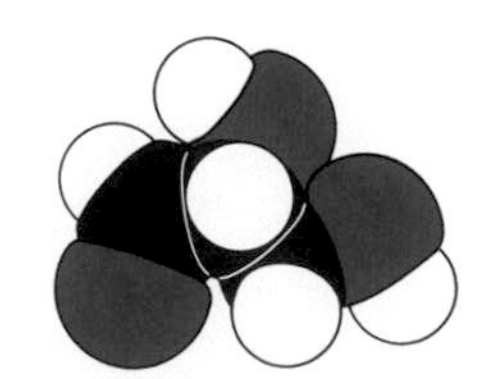

Figure 14-3
Model showing the absolute configuration of D-glyceraldehyde.

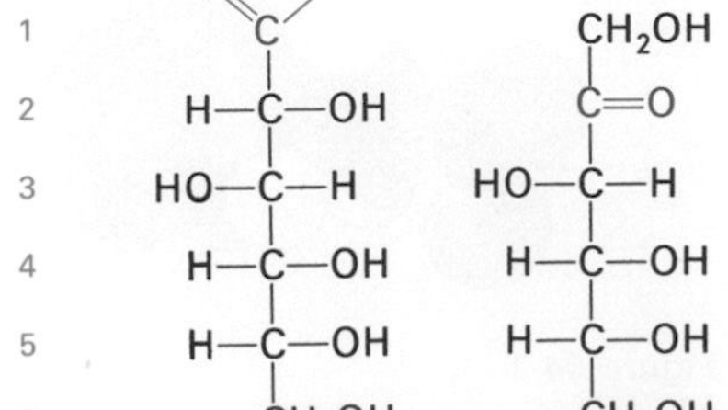

Sugars with 4, 5, 6, and 7 carbon atoms are called *tetroses, pentoses, hexoses,* and *heptoses*. Two common hexoses are D-*glucose* (an aldose) and D-*fructose* (a ketose). For sugars with more than one asymmetric carbon atom, the symbols D and L refer to the absolute configuration of the asymmetric carbon farthest from the aldehyde or keto group. These hexoses belong to the D-series because their absolute configuration at C-5 is the same as that in D-glyceraldehyde.

D-Glucose
(An aldose)

D-Fructose
(A ketose)

In general, a molecule with n asymmetric centers and no plane of symmetry has 2^n stereoisomeric forms. For aldotrioses, $n = 1$, and so there are 2 stereoisomers, D- and L-glyceraldehyde. They are *enantiomers* (mirror images) of each other. Addition of an HCOH group gives four aldotetroses because $n = 2$. Two of them are D-sugars and the

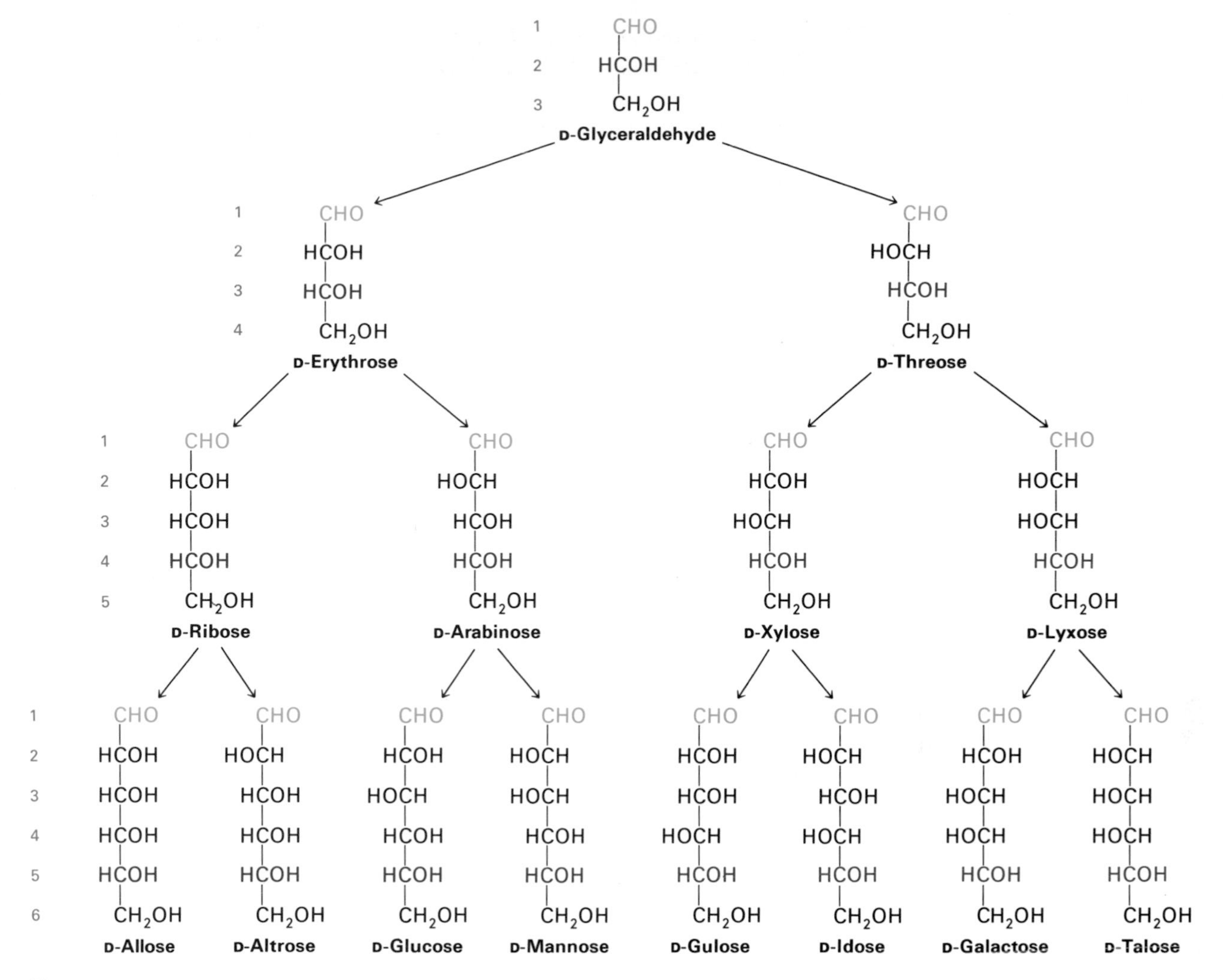

Figure 14-4
Stereochemical relations of D-aldoses containing three, four, five, and six carbon atoms. These sugars are D-aldoses because they contain an aldehyde group (shown in green) and have the configuration of D-glyceraldehyde at their farthest asymmetric center (shown in red).

other two are the enantiomeric L-sugars. Let us follow the D-sugar series (Figure 14-4). One of these four-carbon aldoses is D-erythrose and the other is D-threose. They have the same configuration at C-3 (because they are D-sugars) but opposite configurations at C-2. They are *diastereoisomers*, not enantiomers, because they are not mirror images of one other.

The five-carbon aldoses have three asymmetric centers, which give 8 (2^3) stereoisomers, 4 in the D-series. D-*Ribose* belongs to this group. The six-carbon aldoses have four asymmetric centers, and so there are 16 (2^4) stereoisomers, 8 in the D-series. D-Glucose, D-mannose, and D-galactose are abundant six-carbon aldoses. Note that D-glucose and D-mannose differ only in configuration at C-2. D-Sugars differing in configuration at a single asymmetric center are *epimers*. Thus, D-glucose and D-mannose are epimers at C-2; D-glucose and D-galactose are epimers at C-4. Emil Fischer's elucidation in 1891 of the configuration of D-glucose was a remarkable achievement that greatly stimulated the whole field of organic chemistry.

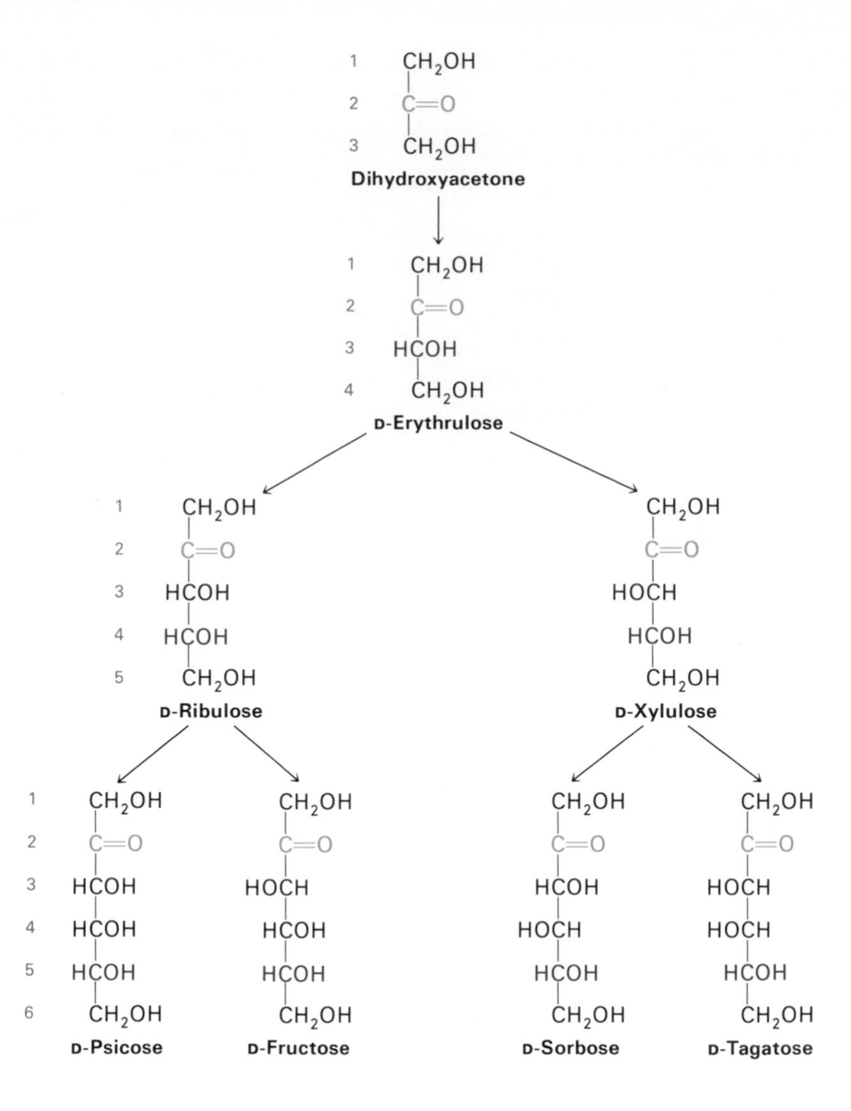

Figure 14-5
Stereochemical relations of D-ketoses containing three, four, five, and six carbon atoms. These sugars are D-ketoses because they contain a keto group (shown in green) and have the configuration of D-glyceraldehyde at their farthest asymmetric center (shown in red).

The stereochemical relations of D-ketoses are shown in Figure 14-5. Dihydroxyacetone, the simplest of these sugars, is optically inactive. D-Erythrulose is the sole four-carbon D-ketose because ketoses have one fewer asymmetric center than do aldoses with the same number of carbon atoms. Hence, there are two five-carbon and four six-carbon D-ketoses. D-*Fructose* is the most abundant ketohexose.

PENTOSES AND HEXOSES CYCLIZE TO FORM FURANOSE AND PYRANOSE RINGS

The predominant forms of glucose and fructose in solution are not open chains. Rather, the open-chain forms of these sugars cyclize into rings. In general, an aldehyde can react with an alcohol to form a *hemiacetal.*

$$R{-}CHO + HOR' \rightleftharpoons R{-}CH(OH){-}OR'$$

Aldehyde Alcohol Hemiacetal

The C-1 aldehyde in the open-chain form of glucose reacts with the C-5 hydroxyl group to form an *intramolecular hemiacetal.* The resulting six-membered sugar ring is called *pyranose* because of its similarity to pyran.

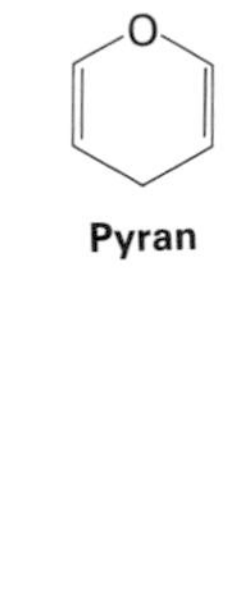
Pyran

Reducing group

D-Glucose
(Open-chain form)

α-D-Glucopyranose

β-D-Glucopyranose

Similarly, a ketone can react with an alcohol to form a *hemiketal.*

$$RR'C{=}O + HOR'' \rightleftharpoons RR'C(OR'')(OH)$$

Ketone **Alcohol** **Hemiketal**

The C-2 keto group in the open-chain form of fructose can react with the C-5 hydroxyl group to form an *intramolecular hemiketal.* This five-membered sugar ring is called *furanose because of its similarity to furan.*

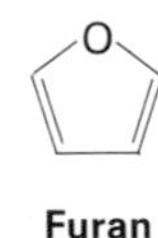
Furan

D-Fructose

α-D-Fructofuranose
(A ring form of fructose)

These structural formulas of glucopyranose and fructofuranose are Haworth projections. In such a projection, the carbon atoms in the ring are not explicitly shown. The approximate plane of the ring is perpendicular to the plane of the paper, with the heavy line on the ring closest to the reader.

An additional asymmetric center is created when glucose cyclizes. Carbon-1, the carbonyl carbon atom in the open-chain form, becomes an asymmetric center in the ring form. Two ring structures can be formed: α-D-glucopyranose and β-D-glucopyranose. *The designation α*

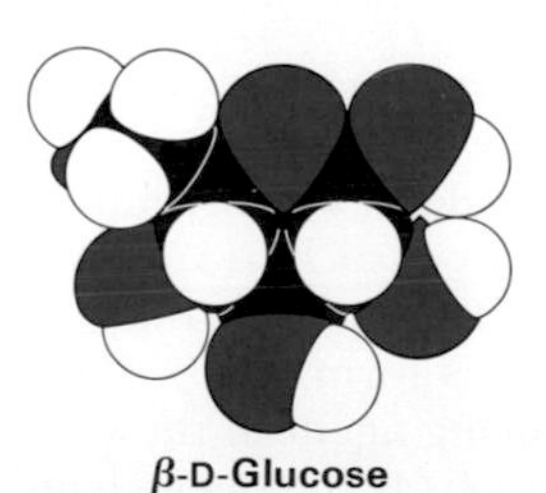

Figure 14-6
Model of β-D-glucopyranose.

means that the hydroxyl group attached to C-1 is below the plane of the ring; β means that it is above the plane of the ring. The C-1 carbon is called the anomeric carbon atom, and so the α and β forms are *anomers.*

The same nomenclature applies to the furanose ring form of fructose, except that α and β refer to the hydroxyl groups attached to C-2, the anomeric carbon atom. Fructose also forms pyranose rings. In fact, the pyranose form predominates in fructose itself, whereas the furanose form is the principal one for most of its derivatives.

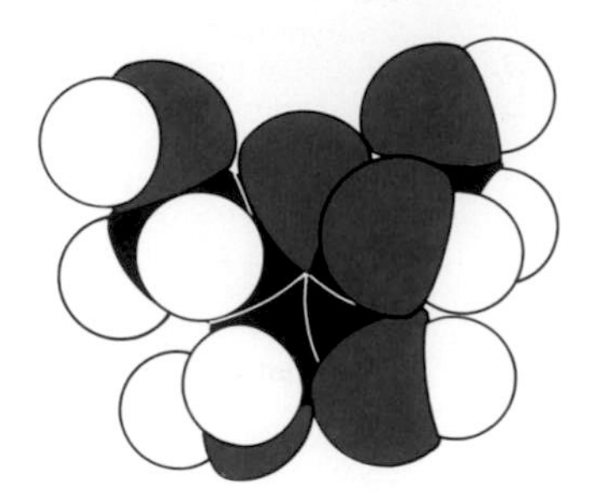

Figure 14-7
Model of β-D-fructofuranose.

α-D-Fructofuranose **β-D-Fructofuranose**

α-D-Fructopyranose **β-D-Fructopyranose**

Five-carbon sugars such as D-ribose and D-deoxyribose form furanose rings, as was exemplified by the structure of these units in RNA and DNA.

β-D-Ribofuranose **β-D-2-Deoxyribofuranose**

In water, α-D-glucopyranose and β-D-glucopyranose interconvert through the open-chain form of the sugar. This interconversion was detected many years ago by *optical rotation* (the rotation of polarized light), a spectroscopic technique that is sensitive to molecular asymmetry. The specific rotation $[\alpha]_D$ is defined as the observed rotation of light of wavelength 589 nm (the D line of a sodium lamp) passing through 10 cm of a 1 g/ml solution of a sample. The specific rotations of the α and β anomers of D-glucose are +112 degrees and +18.7 degrees. When a crystalline sample of either anomer is dissolved in water, $[\alpha]_D$ changes with time until an equilibrium value of +52.7 degrees is attained. This change, called *mutarotation,* results from the formation of an equilibrium mixture containing about one-third α anomer and two-thirds β anomer. Very little of the open-chain form of glucose is present (<1%). Likewise, the α and β anomers of both the pyranose and furanose forms of fructose interconvert through the open-chain form. Some cells contain *mutarotases* that accelerate the interconversion of anomeric sugars. We shall use the terms glucose and fructose to refer to the equilibrium mixture of the open-chain and ring forms of these sugars.

Reducing sugars—
Sugars containing a free aldehyde or keto group reduce indicators such as cupric ion (Cu^{2+}) complexes to the cuprous form (Cu^{+}). The reducing agent in these reactions is the open-chain form of the aldose or ketose.

The six-membered pyranose ring cannot be planar because of the tetrahedral geometry of its saturated carbon atoms. Instead, pyranose rings adopt *chair* and *boat* conformations (Figure 14-8). The substituents of the ring carbon atoms are of two types: *axial* and *equatorial*. Axial bonds are nearly perpendicular to the average plane of the ring, whereas equatorial bonds are nearly parallel to this plane. As can be seen from Figure 14-8 (or even better, from actual molecular models in your hands), axial substituents emerge above and below the average plane of the ring, whereas equatorial substituents emerge at the periphery. Axial substituents other than hydrogen atoms sterically hinder each other if they emerge on the same side of the ring. In contrast, equatorial substituents have much more room. *The chair form of β-D-glucopyranose depicted in this figure is the predominant one because all axial positions are occupied by hydrogen atoms. The bulkier —OH and —CH_2OH groups emerge at the periphery with little steric hindrance.* In contrast, the boat form is sterically hindered.

Furanose rings, like pyranose rings, are not planar. They can be puckered so that four atoms are nearly coplanar and the fifth is about 0.5 Å away from this plane. This conformation is called an *envelope form* because the structure resembles an opened envelope with the back flap raised (Figure 14-9). In the ribose moiety of most biomolecules, either C-2 or C-3 is out of plane on the same side as C-5. These conformations are called $C_{2'}$-*endo* and $C_{3'}$-*endo*. As will be discussed further in Chapter 27, the sugars in RNA are in the $C_{3'}$-*endo* form, whereas the sugars in the Watson-Crick DNA double helix are in the $C_{2'}$-*endo* form. Furanose rings can interconvert rapidly between different conformational states. They are more flexible than pyranose rings, which may account for their selection as components of RNA and DNA.

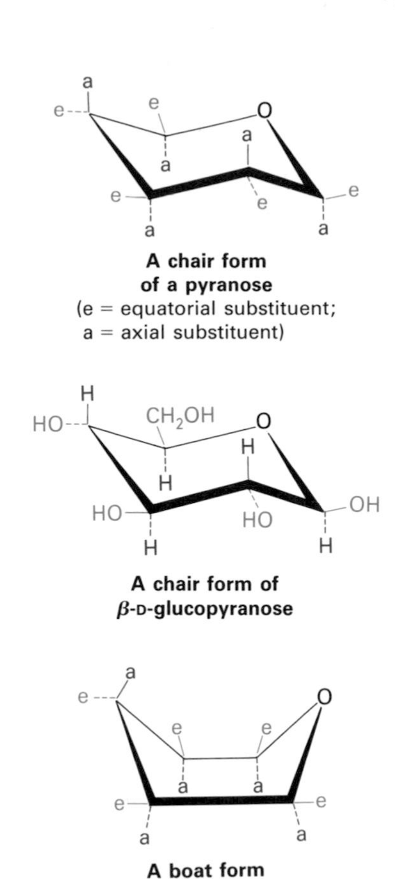

Figure 14-8
Chair and boat conformations of pyranose rings. The chair form is energetically more favorable.

SUGARS ARE JOINED TO ALCOHOLS AND AMINES BY GLYCOSIDIC BONDS

When glucose is warmed in anhydrous methanol containing HCl, its anomeric carbon atom reacts with the hydroxyl group of the alcohol to form two acetals, *α-methylglucoside* and *β-methylglucoside*. HCl facilitates removal of the —OH group from the anomeric carbon. This is an example of acid catalysis.

CH_2OH O H H OH H HO OCH_3 H OH — *O*-Glycosidic bond

α-D-Methylglucoside

CH_2OH O OCH_3 H H OH H HO H H OH

β-D-Methylglucoside

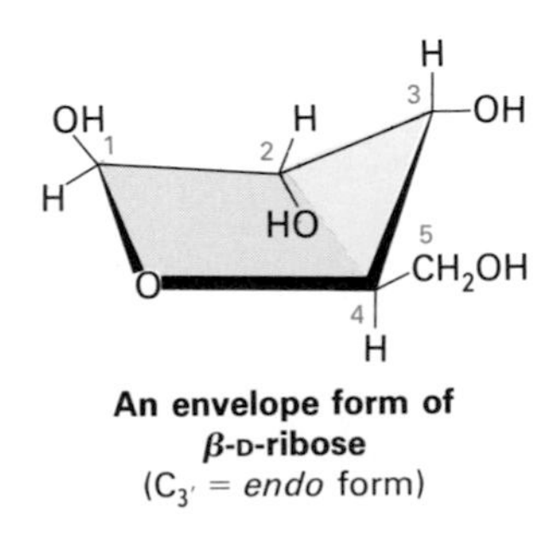

Figure 14-9
An envelope form of β-D-ribose.

The new bond between C-1 of glucose and the oxygen atom of methanol is called a *glycosidic bond*—specifically, an *O*-glycosidic bond. Sugars can be linked to each other by *O*-glyosidic bonds to form disaccharides and polysaccharides. In cellulose, for example, D-glucose residues are joined by glycosidic linkages between C-1 of one sugar and the hydroxyl oxygen atom of C-4 of an adjacent sugar. The glycosidic bonds

β-Glycosidic bond

Cellobiose
(Glucose-β(1→4)-glucose)

N-Glycosidic bond

Adenosine

in cellulose have a β configuration. In other words, the bond emerging from C-1 lies above the plane of the ring when viewed in the standard orientation. Hence, glucose units in cellulose are joined by $\beta(1 \rightarrow 4)$ glycosidic linkages, which can be concisely denoted by the abbreviation β1,4. Recall that *N*-acetylmuramate and *N*-acetylglucosamine sugars in bacterial cell wall polysaccharides also are joined by β1,4 linkages (p. 202).

The anomeric carbon atom of a sugar can be linked to the nitrogen atom of an amine by an *N*-glycosidic bond. The crucial biological importance of this type of glycosidic linkage is evident in such central biomolecules as nucleotides, RNA, and DNA. *N*-glycosidic linkages in virtually all naturally-occurring biomolecules have the β configuration.

PHOSPHORYLATED SUGARS ARE KEY INTERMEDIATES IN ENERGY GENERATION AND BIOSYNTHESES

Phosphorylated sugars are another important class of derivatives. In the next chapter, we shall see that the first step in glycolysis, the breakdown of glucose to obtain energy, is its conversion to *glucose 6-phosphate*, a phosphate ester of the C-6 hydroxyl group. The transfer of a phosphoryl group from ATP to glucose is catalyzed by hexokinase. Several subsequent intermediates in this metabolic pathway, such as dihydroxyacetone phosphate and glyceraldehyde 3-phosphate, are phosphorylated sugars. In fact, one of the strategies of glycolysis is to form three-carbon intermediates that can transfer their phosphate groups to ADP to achieve a net synthesis of ATP.

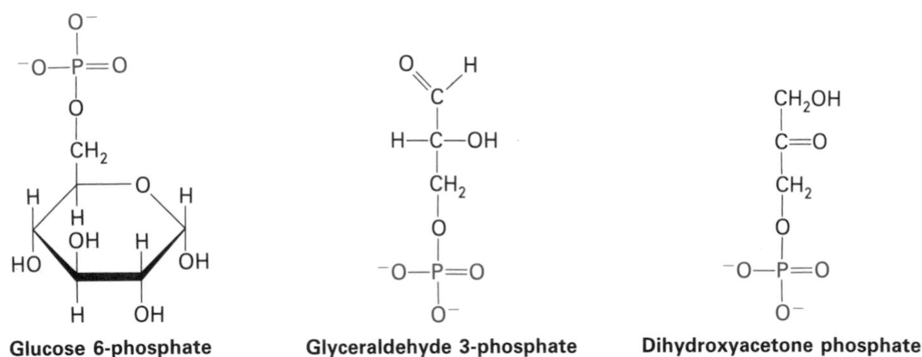

Glucose 6-phosphate **Glyceraldehyde 3-phosphate** **Dihydroxyacetone phosphate**

Phosphorylation also serves to make sugars anionic. The p*K* values of a sugar phosphate group are about 2.1 and 6.8. Hence, at an intracellular pH of 7.4, the net charge of a sugar phosphate such as glucose 6-phosphate is about −1.8. Such a group can have strong electrostatic interactions with the active site of an enzyme. The negative charge contributed by phosphorylation also prevents these sugars from spontaneously crossing lipid bilayer membranes. Phosphorylation helps to retain biomolecules inside cells, an effect whimsically characterized as the importance of being ionized.

Another function of phosphorylation is the *creation of reactive intermediates* for the formation of *O*- and *N*-glycosidic linkages. For example, a multiply phosphorylated derivative of ribose plays key roles in the biosyntheses of purine and pyrimidine nucleotides (p. 602). The pyro-

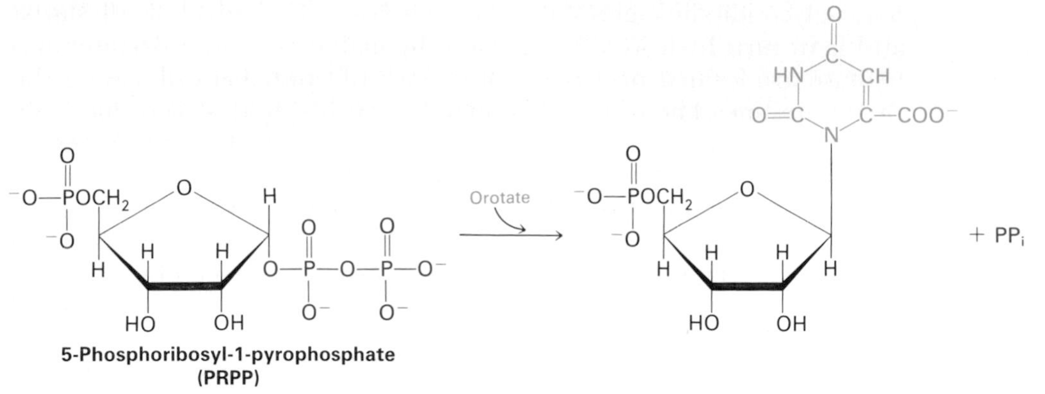

5-Phosphoribosyl-1-pyrophosphate (PRPP)

Orotate

$+ PP_i$

Figure 14-10
The N-glycosidic linkage of pyrimidine nucleotides is formed by displacement of the pyrophosphate group of PRPP, the activated intermediate.

phosphate group of 5-phosphoribosyl-1-pyrophosphate is displaced by the nitrogen atom of a free pyrimidine (orotate) to form a pyrimidine nucleotide (orotidylate) (Figure 14-10).

SUCROSE, LACTOSE AND MALTOSE ARE THE COMMON DISACCHARIDES

Disaccharides consist of two sugars joined by an O-glycosidic bond. Three highly abundant disaccharides are sucrose, lactose, and maltose (Figure 14-11). *Sucrose* (common table sugar) is obtained commercially from cane or beet. The anomeric carbon atoms of a glucose and a fructose residue are in α-glycosidic linkage in sucrose. *Consequently, sucrose lacks a free reducing group (an aldehyde end), in contrast with most other sugars.* The hydrolysis of sucrose to glucose and fructose is catalyzed by *sucrase* (also called *invertase*). *Lactose*, the disaccharide of milk, consists of galactose joined to glucose by a β1,4 glycosidic linkage (Figure 14-12). Lactose is hydrolyzed to these sugars by *lactase* in humans (by *β-galactosidase* in bacteria). In *maltose*, two glucose units are joined by an

Sucrose
(Glucose-α(1→2)-fructose)

Lactose
(Galactose-β(1→4)-glucose)

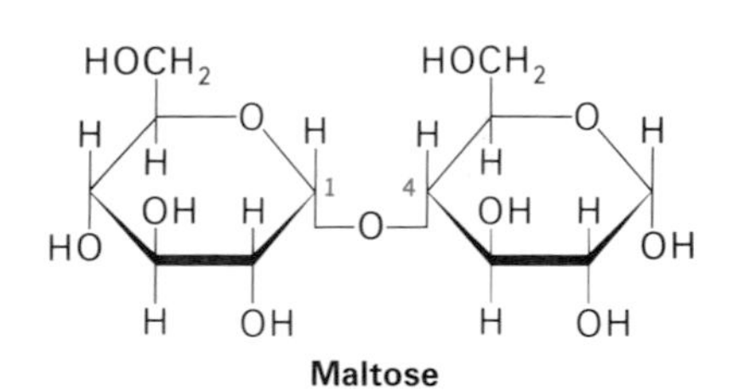

Maltose
(Glucose-α(1→4)-glucose)

Figure 14-11
Formulas of three abundant disaccharides: sucrose, lactose, and maltose. The α configuration of the anomeric carbon atom of maltose and lactose is shown here.

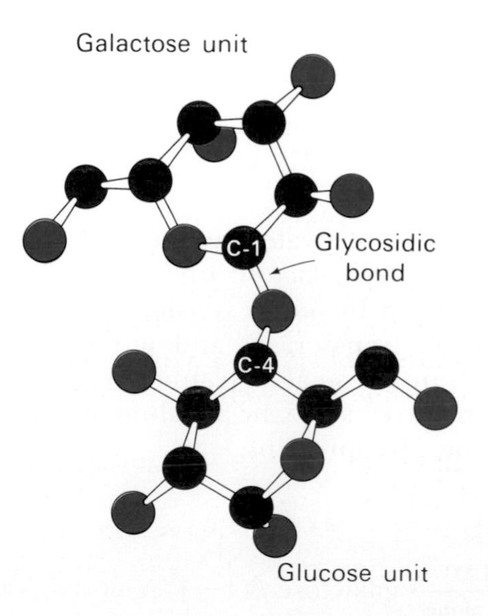

Figure 14-12
Model of lactose.

α-1,4-glycosidic linkage. Maltose comes from the hydrolysis of starch and is in turn hydrolyzed to glucose by maltase. Sucrase, lactase, and maltase are located on the outer surface of epithelial cells lining the small intestine. These cells have many fingerlike folds called microvilli that markedly increase their surface area for digestion and absorption of nutrients.

Sucrose is synthesized by the transfer of glucose from *uridine diphosphate glucose* (*UDP-glucose*), an activated form of the sugar, to fructose 6-phosphate (Figure 14-13). Sucrose 6-phosphate formed by this reaction is then hydrolyzed to sucrose. The UDP group attached to C-1 of glucose activates that carbon atom, just as the pyrophosphate group of phosphoribosylpyrophosphate prepares the ribose unit for condensation with a pyrimidine (p. 339).

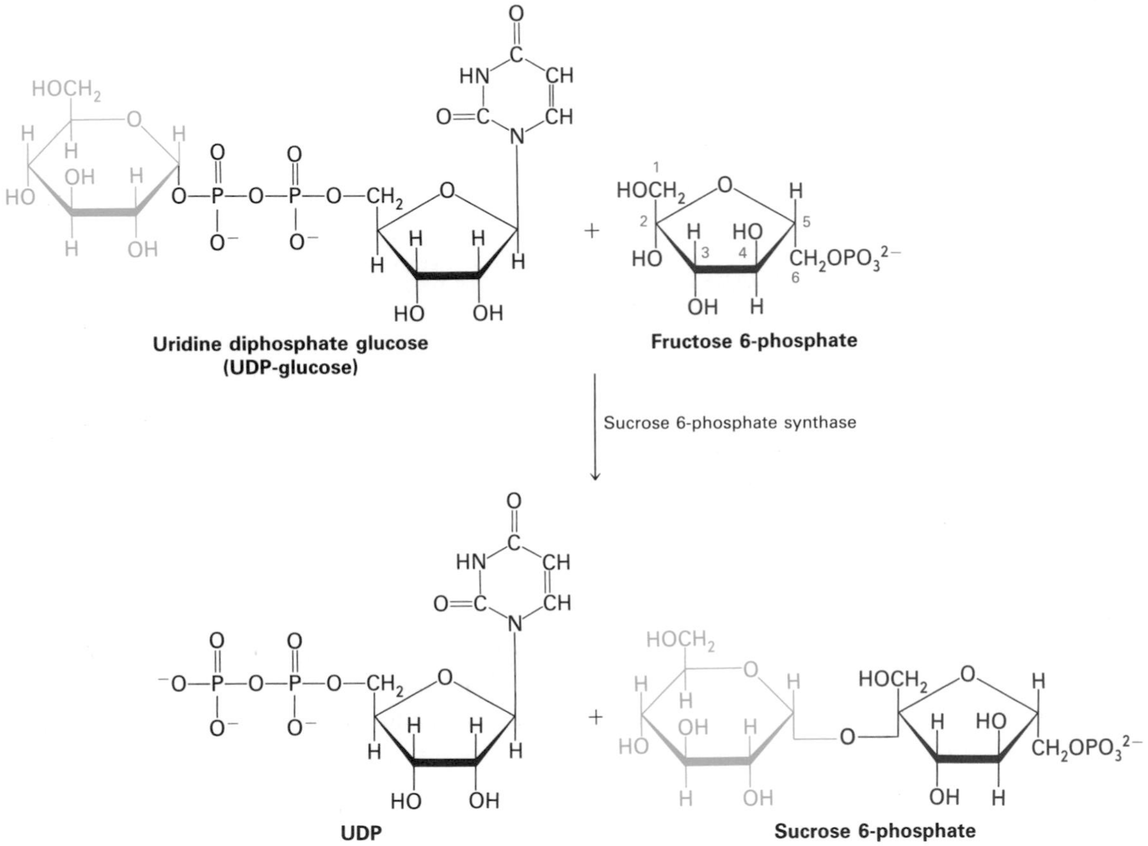

Uridine diphosphate glucose (UDP-glucose) + **Fructose 6-phosphate**

Sucrose 6-phosphate synthase

UDP + **Sucrose 6-phosphate**

Figure 14-13
Uridine diphosphate glucose is the activated intermediate in the synthesis of sucrose.

A UDP-sugar also serves as the activated donor in the synthesis of lactose. *UDP-galactose* transfers its galactose unit to glucose to form lactose. This reaction is catalyzed by *lactose synthase*, which consists of a catalytic subunit (galactosyl transferase) and a modifier subunit (α-lactalbumin). Galactosyl transferase alone cannot synthesize lactose. It has a different role, which is to catalyze the attachment of galactose to *N*-acetylglucosamine units on glycoproteins.

$$\text{UDP-Galactose} \xrightarrow[\text{catalytic subunit alone}]{N\text{-acetyl-glucosamine} \quad \text{UDP}} \text{galactose-}\beta(1\rightarrow 4)\text{-}N\text{-acetylglucosamine}$$

The modifier subunit alters the specificity of the catalytic subunit so that galactose is transferred to glucose rather than to *N*-acetylglucosamine.

$$\text{UDP-Galactose} \xrightarrow[\substack{\text{catalytic subunit} \\ + \\ \text{modifier subunit}}]{\text{glucose} \quad \text{UDP}} \text{galactose-}\beta(1\rightarrow4)\text{-glucose}$$

The level of the modifier subunit is under hormonal control. During pregnancy, the catalytic subunit is formed in the mammary gland but little modifier subunit is synthesized. At the time of birth, a marked increase in the level of the hormone prolactin leads to the synthesis of large amounts of modifier subunit, which joins the catalytic subunit to form active lactose synthetase.

UDP-glucose is the activated intermediate in the synthesis of glycosidic bonds between sugar units in glycogen, as will be discussed in Chapter 19 (p. 455). In the synthesis of starch by plants, adenosine diphosphoglucose (ADP-glucose) is the activated donor of glucose. Likewise, ADP-glucose is the activated intermediate in the synthesis of cellulose by some plants. Other plants utilize cytidine diphosphoglucose (CDP-glucose), or guanosine diphosphoglucose (GDP-glucose) for this purpose. The recurring theme is that *nucleoside diphosphate sugars are the activated intermediates in nearly all biosyntheses of glycosidic linkages between sugars.*

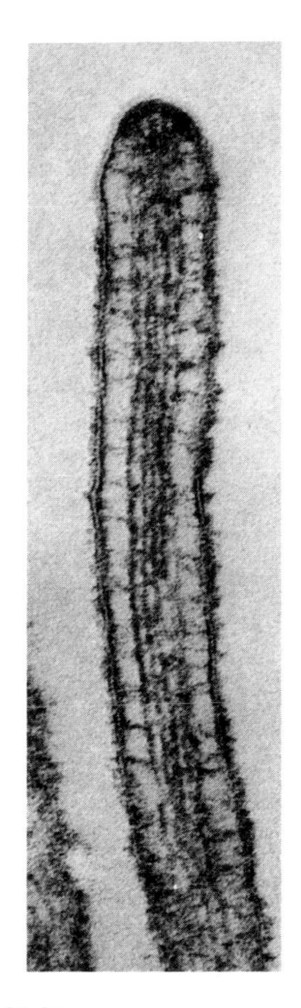

Figure 14-14
Electron micrograph of a microvillus projecting from an intestinal epithelial cell. Lactase and other enzymes that hydrolyze carbohydrates are present on the outer face of the plasma membrane. The filaments inside the microvillus contain actin, a contractile protein. [From M. S. Mooseker and L. G. Tilney. *J. Cell Biol.* 67(1975):725.].

MOST ADULTS ARE INTOLERANT OF MILK BECAUSE THEY ARE DEFICIENT IN LACTASE

Nearly all infants and children are able to digest lactose. In contrast, a majority of adults in certain population groups are deficient in lactase, which makes them intolerant of milk. In a lactase-deficient adult, lactose accumulates in the lumen of the small intestine after ingestion of milk because there is no mechanism for the uptake of this disaccharide. The large osmotic effect of the unabsorbed lactose leads to an influx of fluid into the small intestine. Hence, the clinical symptoms of lactose intolerance are abdominal distention, nausea, cramping, pain, and a watery diarrhea. Lactase deficiency appears to be inherited as an autosomal recessive trait and is usually expressed in adolescence or young adulthood. The prevalence of lactase deficiency in human populations varies greatly. For example, 3% of Danes are deficient in lactase, compared with 97% of Thais. Human populations that do not consume milk in adulthood generally have a high incidence of lactase deficiency, which is also characteristic of other mammals. The capacity of humans to digest lactose in adulthood seems to have evolved since the domestication of cattle some ten thousand years ago.

GLYCOGEN, STARCH, AND DEXTRAN ARE MOBILIZABLE STORES OF GLUCOSE

Animal cells store glucose in the form of glycogen. As will be discussed in detail in Chapter 19, glycogen is a very large, branched polymer of glucose residues. Most of the glucose units in glycogen are linked by α-1,4-glycosidic bonds. The branches are formed by α-1,6-glycosidic

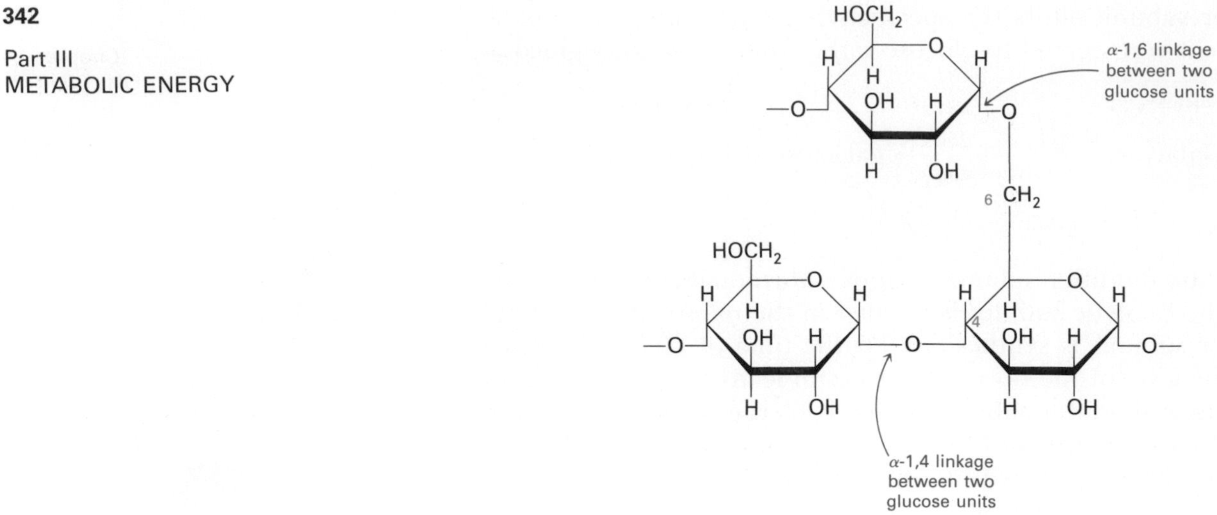

Figure 14-15
A branch in glycogen is formed by an α-1,6-glycosidic linkage.

bonds, which occur about once in ten units (Figure 14-15). These branches serve to increase the solubility of glycogen and make its sugar units more readily mobilized.

The nutritional reservoir in plants is *starch*, of which there are two forms. *Amylose*, the unbranched type of starch, consists of glucose residues in α-1,4 linkage. *Amylopectin*, the branched form, has about one α-1,6 linkage per thirty α-1,4 linkages, and so it is like glycogen except for its lower degree of branching.

More than half of the carbohydrate ingested by humans is starch. Both amylopectin and amylose are rapidly hydrolyzed by *α-amylase*, which is secreted by the salivary glands and the pancreas. α-Amylase hydrolyzes internal α-1,4 linkages to yield *maltose*, *maltotriose*, and *α-dextrin*. Maltose consists of two glucose residues in α-1,4 linkage (p. 339), and maltotriose consists of three such residues. α-Dextrin is made up of several glucose units joined by an α-1,6 linkage in addition to α-1,4 linkages. Maltose and maltotriose are hydrolyzed to glucose by *maltase*, whereas α-dextrin is hydrolyzed to glucose by *α-dextrinase*. Malt contains a different enzyme called *β-amylase*, which hydrolyzes starch into maltose by sequential removal of disaccharide units from nonreducing ends.

Dextran, a storage polysaccharide in yeasts and bacteria, also consists only of glucose residues, but differs from glycogen and starch in that they are joined almost exclusively by α-1,6 linkages. Occasional branches are formed by α-1,2, α-1,3, or α-1,4 linkages, depending on the species.

CELLULOSE, THE MAJOR STRUCTURAL POLYMER OF PLANTS, CONSISTS OF LINEAR CHAINS OF GLUCOSE UNITS

Cellulose, the other major polysaccharide of plants, serves a structural rather than a nutritional role. In fact, *cellulose is the most abundant organic compound in the biosphere*, comprising more than half of all the organic carbon. Some 10^{15} kg of cellulose are synthesized and degraded on earth each year! Cellulose is an unbranched polymer of glucose residues joined by β-1,4 linkages. The β configuration allows cellulose to

Cellulose
(β-1,4-linkages)

Figure 14-16
Schematic diagram showing the conformation of cellulose. The structure is stabilized by hydrogen bonds between adjacent glucose units in the same strand. In fibrils of cellulose, hydrogen bonds are formed between different strands as well.

form very long straight chains (Figure 14-16). Each glucose residue is related to the next by a rotation of 180 degrees, and the ring oxygen atom of one is hydrogen bonded to the 3-OH group of the next. Fibrils are formed by parallel chains. The α-1,4 linkages in glycogen and starch produce a very different molecular architecture. A hollow helix is formed instead of a straight chain. These differing consequences of the α and β linkages are biologically important. *The straight chain formed by β linkages is optimal for the construction of fibers having a high tensile strength. In contrast, the open helix formed by α linkages is well suited to forming an accessible store of sugar.*

Mammals do not have cellulases and therefore cannot digest wood and vegetable fibers. However, some ruminants harbor cellulase-producing bacteria in their digestive tracts and thus can digest cellulose. Fungi and protozoa also secrete cellulases. In fact, the digestion of wood by termites depends on the presence of protozoa in their guts, in a mutually beneficial association.

The exoskeletons of insects and crustacea contain *chitin,* which consists of *N*-acetylglucosamine residues in β-1,4 linkage. Chitin forms long straight chains that serve a structural role. Thus, chitin is like cellulose except that the substituent at C-2 is an acetylated amino group instead of a hydroxyl group.

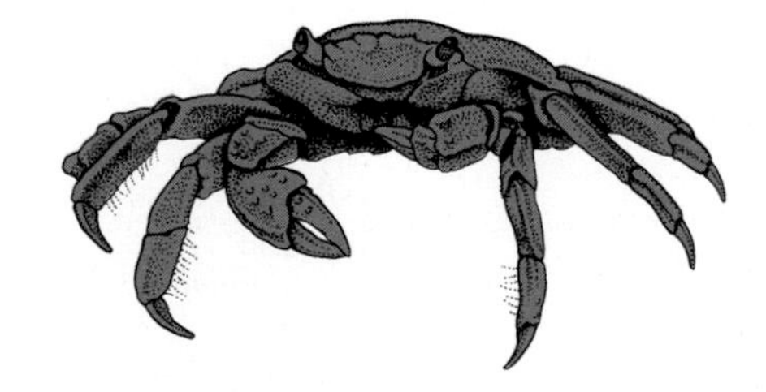

A Sally-lightfoot crab. The exoskeleton of this arthropod is rich in chitin, the second most abundant biopolymer on earth.

OLIGOSACCHARIDES ARE ATTACHED TO INTEGRAL MEMBRANE PROTEINS AND MANY SECRETED PROTEINS

The discussion of membranes in Chapter 12 noted that integral membrane proteins contain covalently attached carbohydrate units, oligosaccharides, on their extracellular face. Many secreted proteins, such as antibodies and clotting factors, also contain oligosaccharide units. These carbohydrates are attached to either the side-chain oxygen atom of serine or threonine residues by *O*-glycosidic linkages or to the side chain nitrogen of asparagine residues by *N*-glycosidic linkages. *N*-linked oligosaccharides contain a *common pentasaccharide core* consisting of three mannose and two *N*-acetylglucosamine residues (Figure 14-17). Additional sugars are attached to this common core in many different ways to form the great variety of oligosaccharide patterns found in glycoproteins. Two examples are shown in Figure 14-17. In the *high mannose type*, additional mannose residues are linked to the core. In the *complex type*, *N*-acetylglucosamine, galactose, sialic acid, and L-fucose residues are built on the core. Fucose is a hexose derived from mannose. Sialic acid (*N*-acetylneuraminate), also derived from mannose, is a 9-carbon sugar with a carboxylate group. The formulas of sugars commonly found in oligosaccharide units of glycoproteins are given in Figure 14-18.

Abbreviations for sugars

Fuc	Fucose
Gal	Galactose
GalNAc	*N*-Acetylgalactosamine
Glc	Glucose
GlcNAc	*N*-Acetylglucosamine
Man	Mannose
Sia	Sialic acid
NAN	*N*-Acetylneuraminate (sialic acid)

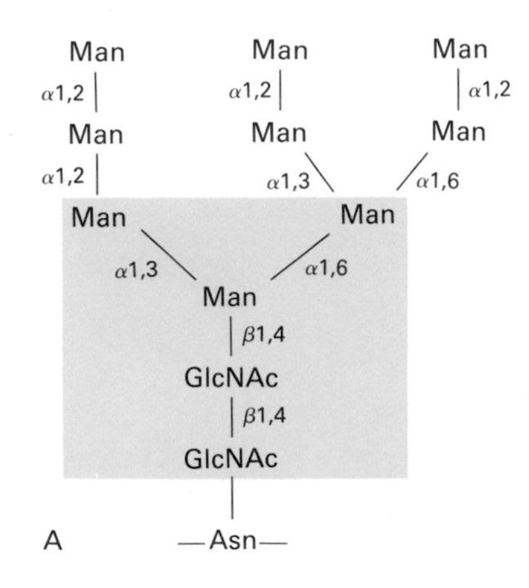

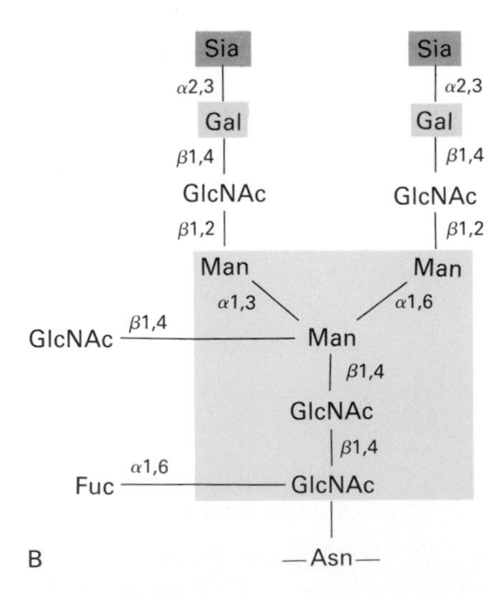

Figure 14-17
N-linked oligosaccharide units in glycoproteins contain a common core (shown in yellow) of three mannose and two *N*-acetylglucosamine residues. Additional sugars are added to form many different patterns, of which only two are shown here. (A) High mannose type. (B) Complex type. [After R. Kornfeld and S. Kornfeld. *Ann. Rev. Biochem.* 54(1985):633.]

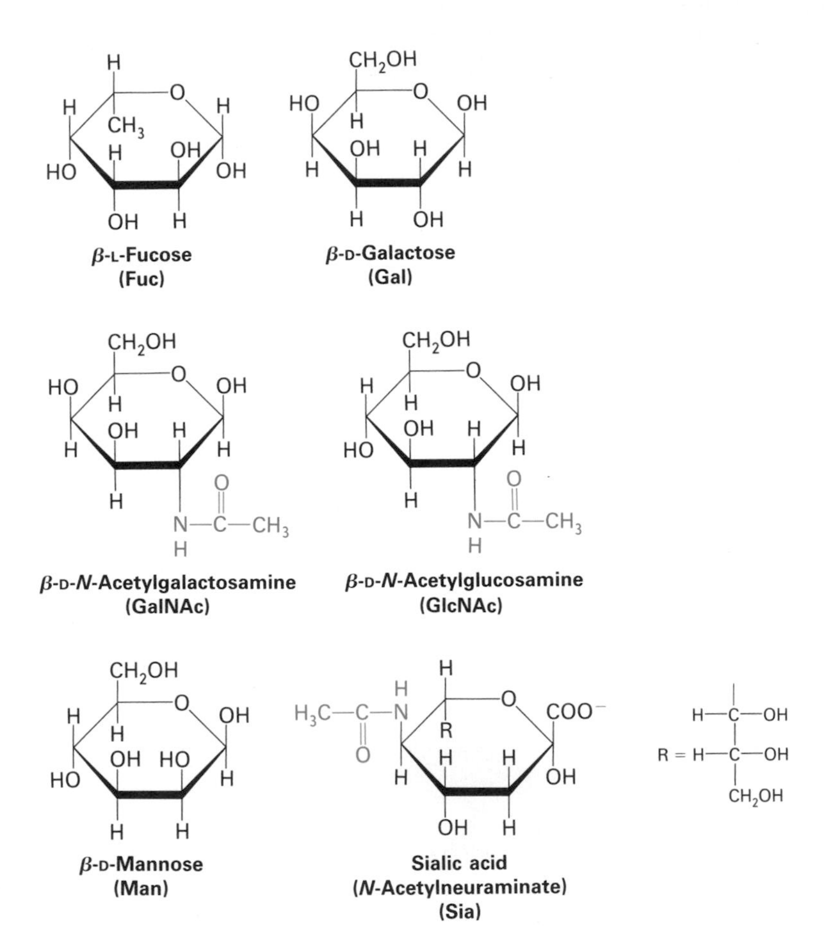

Figure 14-18 Sugar residues commonly found in glycoproteins.

CARBOHYDRATES PARTICIPATE IN MOLECULAR TARGETING AND CELL-CELL RECOGNITION

The diversity and complexity of the carbohydrate units of glycoproteins suggest that they are rich in information and are functionally important. Nature does not construct complex patterns when simple ones suffice. Cellulose and starch, for example, are built from glucose units, nearly all joined by the same type of glycosidic linkage. Glycoproteins, by contrast, contain five kinds of residues joined by many kinds of glycosidic linkages. Why all this intricacy and diversity? A definitive answer to this question cannot yet be given but some clues concerning the biological roles of oligosaccharide units are being uncovered.

The removal of glycoproteins from the blood is accomplished by surface receptor proteins on liver cells. The best understood of these carbohydrate-binding proteins is the *asialoglycoprotein receptor*. Many newly synthesized glycoproteins, such as immunoglobulins and peptide hormones, contain carbohydrate units with terminal sialic acid residues (Figure 14-17). In the course of hours or days, depending on the particular protein, terminal sialic acid residues are removed by sialylases on the surface of blood vessels. The exposed galactose residues of these trimmed proteins are detected by the asialoglycoprotein receptors in the plasma membranes of liver cells (Figure 14-19). The complex of the asialoglycoprotein and its receptor is then internalized by the liver cell (a process called endocytosis, p. 786) to remove the trimmed glycoprotein from the blood. *In essence, these oligosaccharide units mark the passage of time, to indicate when the proteins carrying them should be taken out of circulation.* The rate of removal of sialic acid from glycoproteins is controlled by the structure of the protein itself. Hence, proteins can be designed to have lifetimes ranging from a few hours to many weeks, depending on physiological needs.

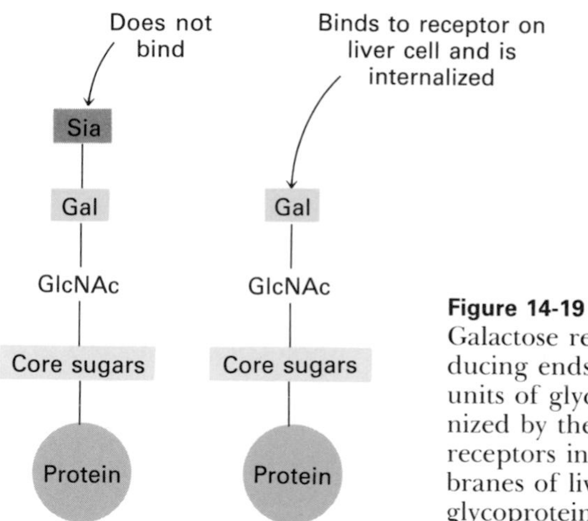

Figure 14-19
Galactose residues at the nonreducing ends of oligosaccharide units of glycoproteins are recognized by the asialoglycoprotein receptors in the plasma membranes of liver cells. The bound glycoproteins are then internalized and removed from the blood.

Plants contain specific carbohydrate-binding proteins called *lectins*. For example, *concanavalin A* (from the jackbean) binds to internal and nonreducing terminal α-mannosyl residues. *Wheat-germ agglutinin*, *peanut lectin*, and *phytohemagglutinin* (from red kidney bean) recognize disaccharide or oligosaccharide units (Figure 14-20). All known lectins contain two or more binding sites for carbohydrate units, which accounts for their ability to agglutinate (cross-link) erythrocytes and other cells. Lectins are very useful probes of cell surfaces because of their capacity to recognize specific oligosaccharide patterns. What is their physiological role? A hint comes from the finding that a lectin participates in binding a nitrogen-fixing bacterium (*Rhizobium trifolii*) to the surface of the root hairs of clover. This lectin cross-links receptors on the cell wall of root hairs to bacterial capsular polysaccharides and lipopolysaccharides. Bacteria, too, contain lectins. The adherence of *E. coli* to epithelial cells of the gastrointestinal tract is mediated by bacterial lectins that recognize oligosaccharide units on the surface of target cells.

Carbohydrates have also been implicated as mediators of cell-cell interactions in animals. The adhesion of neurons in the development of the nervous

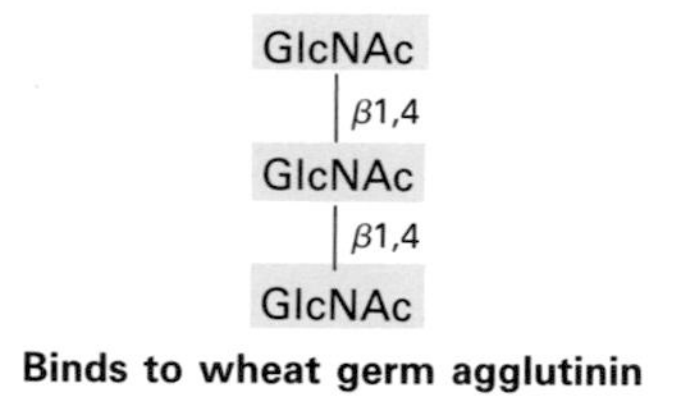

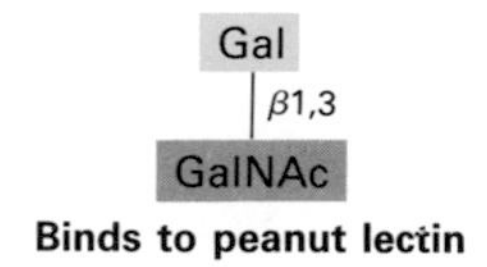

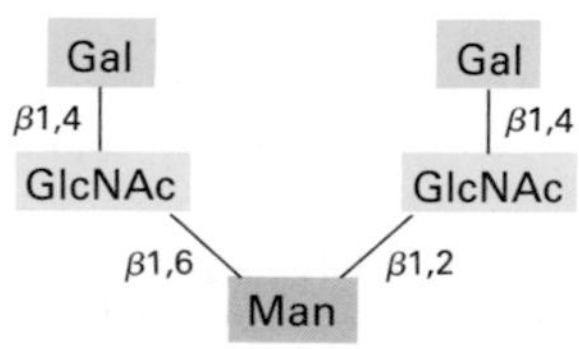

Figure 14-20
Sugar units recognized by wheat germ agglutinin, peanut lectin, and phytohemagglutinin.

system is mediated in part by a neural cell adhesion molecule (N-CAM). An N-CAM molecule on one neuron binds to its counterpart on another neuron to establish contact and trigger the initiation of a permanent liaison between them. Studies of the binding of unattached retinal cells to a monolayer of the same kind of cells suggest that a cell-surface proteoglycan—heparan sulfate (see Figure 14-1)—participates in this process. Adhesion appears to depend on the binding of N-CAM to heparan sulfate as well as to an N-CAM molecule on the apposed cell. The role of carbohydrates on cell surfaces and of the extracellular matrix in morphogenesis is a fascinating area of inquiry. The deciphering of the carbohydrate code will undoubtedly open new vistas in biology.

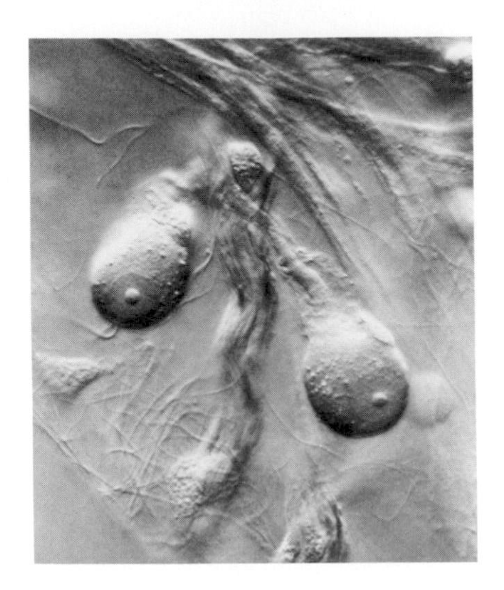

Figure 14-21
Light micrograph showing neurons interacting in the development of the nervous system. Cell-surface carbohydrates have been implicated in the formation of adhesions between interacting cells. [Courtesy of Dr. Peter Sargent.]

Sulfated iduronate | **Bis-sulfated glucosamine** | **Glucuronate** | **Sulfated *N*-acetylglucosamine**

A tetrasaccharide from heparan sulfate

SUMMARY

Carbohydrates are aldehydes or ketones with two or more hydroxyl groups. Aldoses are carbohydrates with an aldehyde group (as in glyceraldehyde and glucose), whereas ketoses contain a keto group (as in dihydroxyacetone and fructose). A sugar belongs to the D series if the absolute configuration of its asymmetric carbon farthest from the aldehyde or keto group is the same as that of D-glyceraldehyde. Nearly all naturally occurring sugars belong to the D-series. The C-1 aldehyde in the open-chain form of glucose reacts with the C-5 hydroxyl group to form a six-membered pyranose ring. The C-2 keto group in the open-chain form of fructose reacts with the C-5 hydroxyl group to form a five-membered furanose ring. Pentoses such as ribose and dexoyribose also form furanose rings. An additional asymmetric center is formed at the anomeric carbon atom (C-1 in aldoses and C-2 in ketoses) in these cyclizations. The hydroxyl group attached to the anomeric carbon atom is below the plane of the ring (viewed in the standard orientation) in the α anomer, whereas it is above the ring in the β anomer. Not all of the atoms in the rings lie in the same plane. Rather, pyranose rings usually adopt the chair conformation, and furanose rings the envelope conformation.

Sugars are joined to alcohols and amines by glycosidic bonds from the anomeric carbon atom. For example, *O*-glycosidic bonds link sugars to one another in disaccharides and polysaccharides. *N*-glycosidic bonds link sugars to purines and pyrimidines in nucleotides, RNA, and DNA. Phosphate-sugar esters such as glucose 6-phosphate are important metabolic intermediates. Another function of phosphorylation is the creation of reactive intermediates for the formation of *O*- and *N*-glycosidic linkages. Nucleoside diphosphate sugars (such as UDP-glucose) are the reactive intermediates in the synthesis of nearly all glycosidic bonds between sugars.

Sucrose, lactose, and maltose are the common disaccharides. Sucrose (common table sugar), obtained from cane or beet, consists of glucose and fructose joined by an α-glycosidic linkage between their anomeric carbons. Lactose (in milk) consists of galactose joined to glucose by a β-1,4 linkage. Maltose (from starch) consists of two glucoses joined by an α-1,4 linkage. Starch is a polymeric form of glucose in plants, and glycogen serves a similar role in animals. Most of the glucose units in starch and glycogen are in α-1,4 linkage. Glycogen has more branch points formed by α-1,6 linkages than does starch, which makes glycogen more soluble. Cellulose, the major structural polymer of plant cell walls, consists of glucose units joined by β-1,4 linkages. These β linkages give rise to long straight chains that form fibrils with high tensile strength. In contrast, the α linkages in starch and glycogen lead to open helices, in keeping with their roles as mobilizable energy stores.

The oligosaccharide units on integral membrane proteins are linked either to the side-chain oxygen atom of serine or threonine residues or the side-chain nitrogen of asparagine residues. *N*-linked oligosaccharides contain a common core consisting of three mannose and two *N*-acetylglucosamine residues. Additional sugars are attached to this core to form diverse patterns. Carbohydrates have been implicated in molecular targeting and cell-cell recognition. The asialoglycoprotein receptor of liver cells removes from circulation glycoproteins that have lost their terminal sialic acid residues, exposing their galactose residues. Plants contain many soluble carbohydrate-binding proteins called lectins. The binding of a nitrogen-fixing microorganism to the root hairs of a plant is mediated by a lectin that recognizes cell-surface sugars on the partners in this symbiotic relationship. Carbohydrates have also been implicated in cell-cell adhesion during the development of animals, as exemplified by the role of a cell-surface proteoglycan in the aggregation of retinal cells.

SELECTED READINGS

Where to start

Sharon, N., 1980. Carbohydrates. *Sci. Amer.* 245(5):90–116. [An interesting introduction to the diverse roles of carbohydrates in nature.]

Steer, C. J., and Ashwell, G., 1986. Hepatic membrane receptors for glycoproteins. *Prog. Liver Dis.* 8:99–123.

Feizi, T., and Childs, R. A., 1985. Carbohydrate structures of glycoproteins and glycolipids as differentiation antigens, tumour-associated antigens and components of receptor systems. *Trends Biochem. Sci.* 10:24–29.

Books on carbohydrate structure and chemistry

Rees, D. A., 1977. *Polysaccharide Shapes.* Wiley. [A concise and lucid account of the conformation of sugars and polysaccharides.]

Davison, E. A., 1967. *Carbohydrate Chemistry.* Holt, Rinehart, and Winston.

Preiss, J., (ed.), 1980. *The Biochemistry of Plants. Volume 3. Carbohydrates: Structure and Function.* Academic Press. [Contains many fine articles on the structure, biosynthesis, and function of carbohydrates in plants.]

Pigman, W. W., and Horton, D., (eds.), 1972. *The Carbohydrates: Chemistry and Biochemistry.* Academic Press.

Carbohydrate-binding proteins

Quiocho, F. A., 1986. Carbohydrate-binding proteins: tertiary structures and protein-sugar interactions. *Ann. Rev. Biochem.* 55:287–315.

Lis, H., and Sharon, N., 1986. Lectins as molecules and as tools. *Ann. Rev. Biochem.* 55:35–67.

Plant cell walls

McNeil, M., Darvill, A. G., Fry, S. C., and Albersheim, P., 1984. Structure and function of the primary cell walls of plants. *Ann. Rev. Biochem.* 53:625–663.

Albersheim, P., and Darvill, A. G., 1985. Oligosaccharins. *Sci. Amer.* 253(3):58–64. [These fragments of the plant cell walls control functions such as development and defense against disease.]

Glycoproteins

Ivatt, R. J., (ed.), 1984. *The Biology of Glycoproteins*. Plenum.

Kornfeld, R., and Kornfeld, S., 1985. Assembly of asparagine-linked oligosaccharides. *Ann. Rev. Biochem.* 54:631–664.

Labat-Robert, J., Timpl, R., and Ladislas, R., (eds.), 1986. *Structural Glycoproteins in Cell-Matrix Interactions*. Karger.

Carbohydrates in recognition processes

Barondes, S. H., 1984. Soluble lectins: a new class of extracellular proteins. *Science* 223:1259–1264.

Cole, G. J., Loewy, A., and Glaser, L., 1986. Neuronal cell-cell adhesion depends on the interactions of N-CAM with heparin-like molecules. *Nature* 320:445–447.

Analyses of carbohydrates

Sweeley, C. C., and Nunez, H., 1985. Structural analysis of glycoconjugates by mass spectrometry and nuclear magnetic resonance spectroscopy. *Ann. Rev. Biochem.* 54:765–801.

Barker, R., and Serianni, A. S., 1986. Carbohydrates in solution: studies with stable isotopes. *Acc. Chem. Res.* 19:307–313. [Review of ^{13}C NMR studies.]

PROBLEMS

1. Indicate whether each of the following pairs of sugars consists of anomers, epimers, or an aldose-ketose pair:
   (a) D-glyceraldehyde and dihydroxyacetone.
   (b) D-glucose and D-mannose.
   (c) D-glucose and D-fructose.
   (d) α-D-glucose and β-D-glucose.
   (e) D-ribose and D-ribulose.
   (f) D-galactose and D-glucose.

2. Glucose and other aldoses are oxidized by an aqueous solution of a silver-ammonia complex (Tollens' test). What are the reaction products?

3. Glucose reacts slowly with hemoglobin and other proteins to form covalent compounds. Why is glucose reactive? What is the nature of the adduct formed?

4. Compounds containing hydroxyl groups on adjacent carbon atoms undergo carbon–carbon bond cleavage when treated with periodate ion (IO_4^-). How can this reaction be used to distinguish between pyranosides and furanosides?

5. Does the oxygen atom attached to C-1 in methylglucoside come from glucose or methanol?

6. Identify the four sugars shown below.

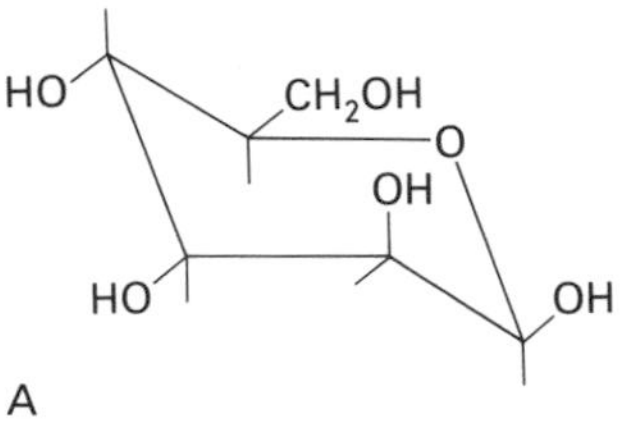

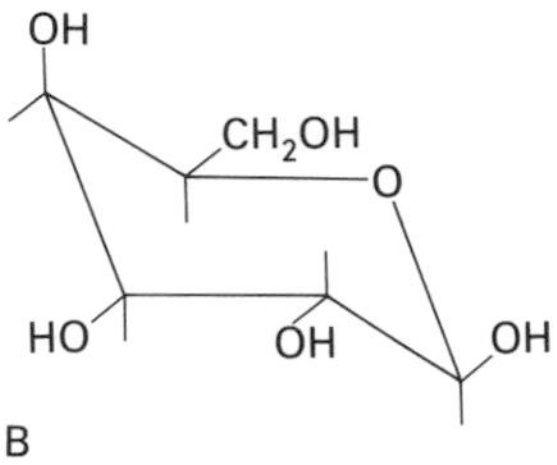

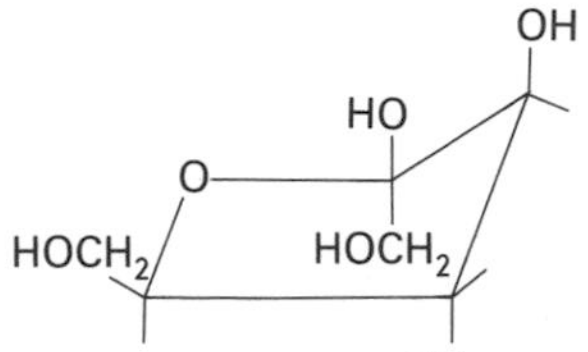

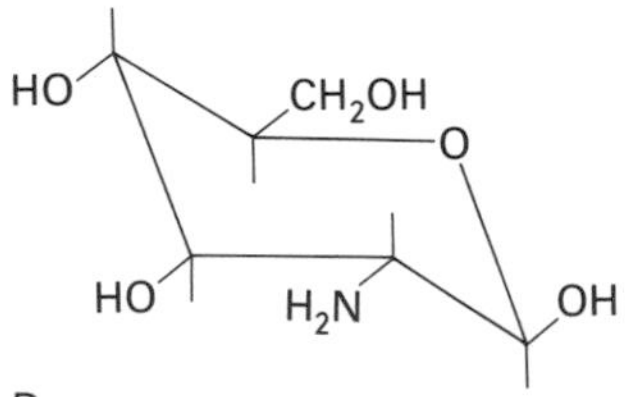

7. A trisaccharide unit of a cell-surface glycoprotein is postulated to play a critical role in mediating cell-cell adhesion in a particular tissue. Design a simple experiment to test this hypothesis.

CHAPTER 15

Glycolysis

We begin our consideration of the generation of metabolic energy with glycolysis, a nearly universal pathway in biological systems. *Glycolysis is the sequence of reactions that converts glucose into pyruvate with the concomitant production of ATP.* In aerobic organisms, glycolysis is the prelude to the citric acid cycle and the electron-transport chain, which together harvest most of the energy contained in glucose. Under aerobic conditions, pyruvate enters mitochondria, where it is completely oxidized to CO_2 and H_2O. If the supply of oxygen is insufficient, as in actively contracting muscle, pyruvate is converted into lactate. In some anaerobic organisms, such as yeast, pyruvate is transformed into ethanol. The formation of ethanol and lactate from glucose are examples of fermentations.

Glycolysis—
Derived from the Greek words *glycos,* sugar (sweet), and *lysis,* dissolution.

The elucidation of glycolysis has a rich history. Indeed, the development of biochemistry and the delineation of this central pathway went hand in hand. A key discovery was made by Hans Buchner and Eduard Buchner in 1897, quite by accident. They were interested in manufacturing cell-free extracts of yeast for possible therapeutic use. These extracts had to be preserved without using antiseptics such as phenol, and so they decided to try sucrose, a commonly used preservative in kitchen chemistry. They obtained a startling result: sucrose was rapidly fermented into alcohol by the yeast juice. The significance of this finding was immense. *The Buchners demonstrated for the first time that fermentation could occur outside living cells.* The accepted view of their day, asserted by Louis Pasteur in 1860, was that fermentation is inextricably tied to living cells. The chance discovery of the Buchners refuted this vitalistic dogma and opened the door to modern biochemistry. Metabolism became chemistry.

Fermentation—
An ATP-generating process in which organic compounds act as both donors and acceptors of electrons. Fermentation can occur in the absence of O_2. Discovered by Pasteur, who described fermentation as "la vie sans l'air" (life without air).

The next important contribution was made by Arthur Harden and William Young in 1905. They added yeast juice to a solution of glucose and found that fermentation started almost immediately. However, the rate of fermentation soon decreased markedly unless inorganic phosphate was added. Furthermore, they found that the added inorganic phosphate disappeared in the course of fermentation, and so they inferred that *inorganic phosphate was incorporated into a sugar phosphate.* Harden and Young isolated a hexose diphosphate, which was later shown to be fructose 1,6-bisphosphate. They found that yeast juice lost its activity if it was dialyzed or heated to 50°C. However, the inactive dialyzed juice became active when it was mixed with inactive heated juice. Thus, activity depended on the presence of two kinds of substances: a heat-labile, nondialyzable component (called *zymase*) and a heat-stable, dialyzable fraction (called *cozymase*). We now know that "zymase" consists of a number of enzymes, whereas "cozymase" consists of metal ions, adenosine triphosphate (ATP), adenosine diphosphate (ADP), and coenzymes such as nicotinamide adenine dinucleotide (NAD^+).

Enzyme—
A term coined by Friedrich Wilhelm Kühne in 1878 to designate catalytically active substances that had previously been called ferments. Derived from the Greek words *en*, in, and *zyme*, leaven.

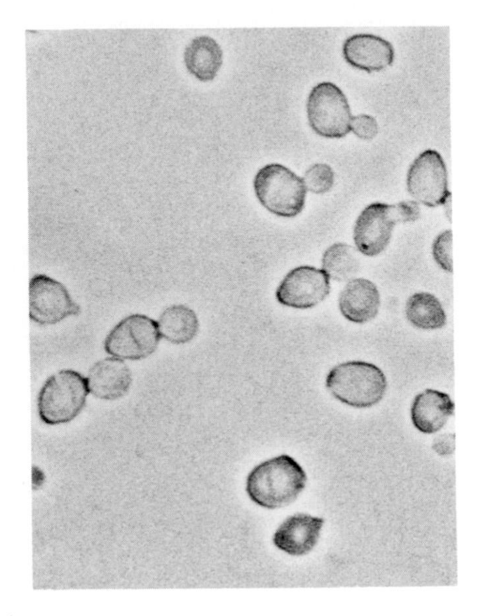

Figure 15-1
Light micrograph of yeast cells. [Courtesy of Dr. Randy Schekman.]

Studies of muscle extracts carried out several years later showed that many of the reactions of lactic fermentation were the same as those of alcoholic fermentation. *This was an exciting discovery because it revealed an underlying unity in biochemistry.* The complete glycolytic pathway was elucidated by 1940, largely because of the contributions made by Gustav Embden, Otto Meyerhof, Carl Neuberg, Jacob Parnas, Otto Warburg, Gerty Cori, and Carl Cori. Glycolysis is sometimes called the Embden-Meyerhof pathway.

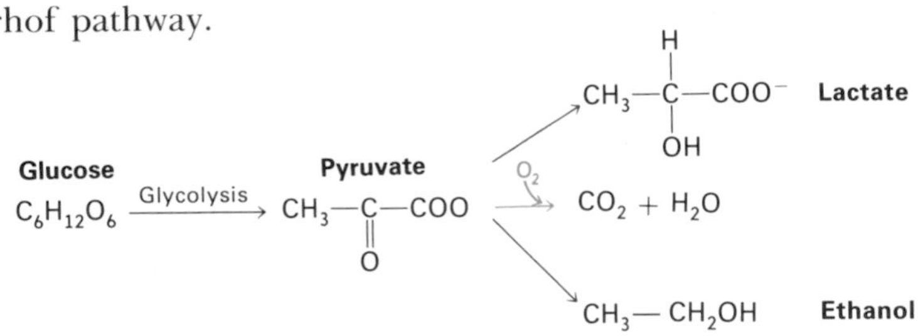

Glucose $C_6H_{12}O_6$ $\xrightarrow{\text{Glycolysis}}$ **Pyruvate** $CH_3{-}C({=}O){-}COO^-$

→ $CH_3{-}CH(OH){-}COO^-$ **Lactate**

$\xrightarrow{O_2}$ $CO_2 + H_2O$

→ $CH_3{-}CH_2OH$ **Ethanol**

Figure 15-2
Some fates of glucose.

AN OVERVIEW OF KEY STRUCTURES AND REACTIONS

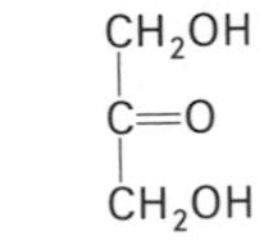

Learning the sequence of events in a metabolic pathway is easier with a firm grasp of the structures of the reactants and an understanding of the types of reactions taking place. The intermediates in glycolysis have either six carbons or three carbons. The *six-carbon units* are derivatives of *glucose* and *fructose*. The *three-carbon units* are derivatives of *dihydroxyacetone, glyceraldehyde, glycerate,* and *pyruvate.*

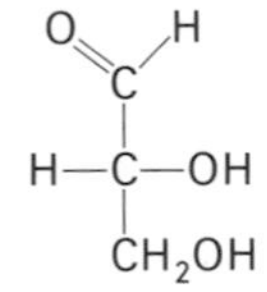

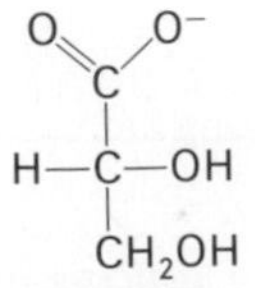

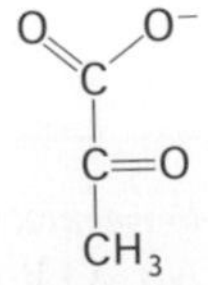

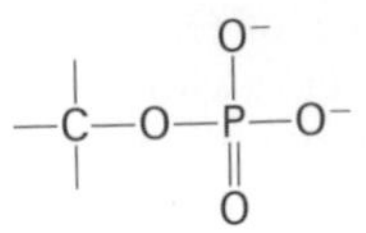

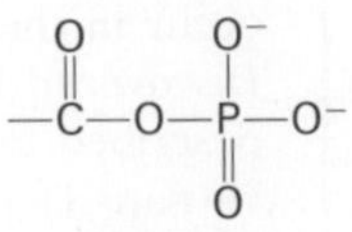

$CH_2OH{-}C({=}O){-}CH_2OH$ **Dihydroxyacetone**

$HC({=}O){-}CH(OH){-}CH_2OH$ **Glyceraldehyde**

$^-OOC{-}CH(OH){-}CH_2OH$ **Glycerate**

$^-OOC{-}C({=}O){-}CH_3$ **Pyruvate**

$-C{-}O{-}PO_3^{2-}$ **Ester**

$-C({=}O){-}O{-}PO_3^{2-}$ **Anhydride**

All intermediates between glucose and pyruvate are *phosphorylated.* The phosphoryl groups in these compounds are linked as either *esters* or *anhydrides*. Now let us look at some of the kinds of reactions that occur in glycolysis:

1. *Phosphoryl transfer*. A phosphoryl group is transferred from ATP to a glycolytic intermediate, or vice versa.

$$R{-}OH + ATP \rightleftharpoons R{-}O{-}P(O)(O^-){-}O^- + ADP + H^+$$

2. *Phosphoryl shift.* A phosphoryl group is shifted within a molecule from one oxygen atom to another.

$$R{-}CH(OH){-}CH_2O{-}PO_3^{2-} \rightleftharpoons R{-}CH(OPO_3^{2-}){-}CH_2OH$$

3. *Isomerization.* A ketose is converted into an aldose, or vice versa.

$$\underset{\text{Ketose}}{HOCH_2{-}C({=}O){-}R} \rightleftharpoons \underset{\text{Aldose}}{H{-}C({=}O){-}CH(OH){-}R}$$

4. *Dehydration.* A molecule of water is eliminated.

$$H{-}\overset{|}{C}{-}\;\;H{-}\overset{|}{C}{-}OH \rightleftharpoons {-}C{=}C{-}H + H_2O$$

5. *Aldol cleavage.* A carbon–carbon bond is split in a reversal of an aldol condensation.

$$R{-}C({=}O){-}CH(OH){-}CH(OH){-}R' \rightleftharpoons R{-}C({=}O){-}CH_2OH + H{-}C({=}O){-}R'$$

FORMATION OF FRUCTOSE 1,6-BISPHOSPHATE FROM GLUCOSE

We now start our journey down the glycolytic pathway. The reactions in this pathway take place in the cell cytosol. The first stage, which is the conversion of glucose into fructose 1,6-bisphosphate, consists of three steps: a phosphorylation, an isomerization, and a second phosphorylation reaction. *The strategy of these initial steps in glycolysis is to trap the substrate in the cell and form a compound that can be readily cleaved into phosphorylated three-carbon units.* Energy is subsequently extracted from the three-carbon units.

$$\text{Glucose} \longrightarrow \longrightarrow \longrightarrow \text{Fructose 1,6-bisphosphate}$$

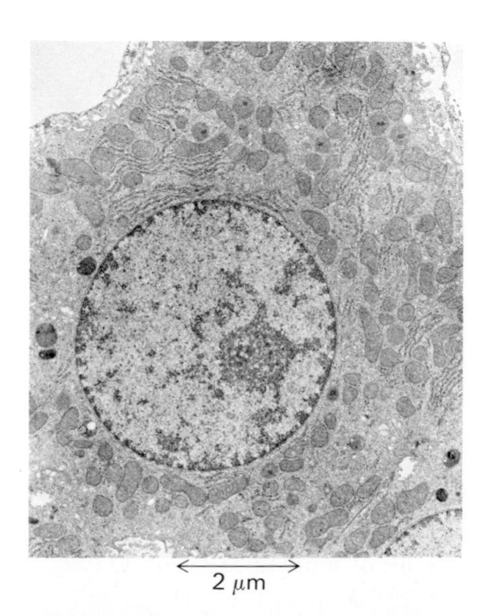

Figure 15-3
Electron micrograph of a liver cell. Glycolysis takes place in the cytosol. [Courtesy of Dr. Anne Hubbard.]

Glucose enters most cells by a specific transport protein and has one principal fate: *it is phosphorylated by ATP to form glucose 6-phosphate*. The transfer of the phosphoryl group from ATP to the hydroxyl group on C-6 of glucose is catalyzed by *hexokinase*.

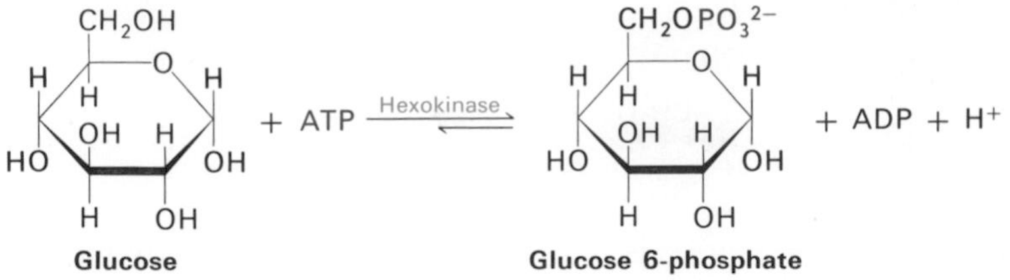

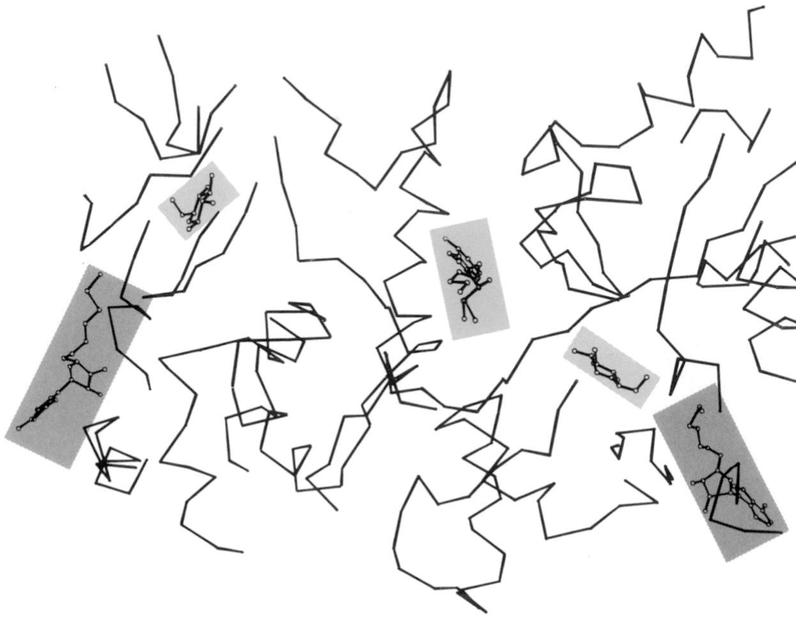

$$\text{Glucose} + \text{ATP} \xrightleftharpoons{\text{Hexokinase}} \text{Glucose 6-phosphate} + \text{ADP} + \text{H}^+$$

Figure 15-4
Schematic diagram of the α-carbon backbone of yeast hexokinase. The two identical subunits of this dimeric enzyme are not arranged symmetrically. Glucose (shaded green) and ATP (shaded red) are bound to the catalytic site of each monomer. An additional ATP molecule (shaded blue) is bound at the interface between subunits. [Courtesy of Dr. Thomas Steitz.]

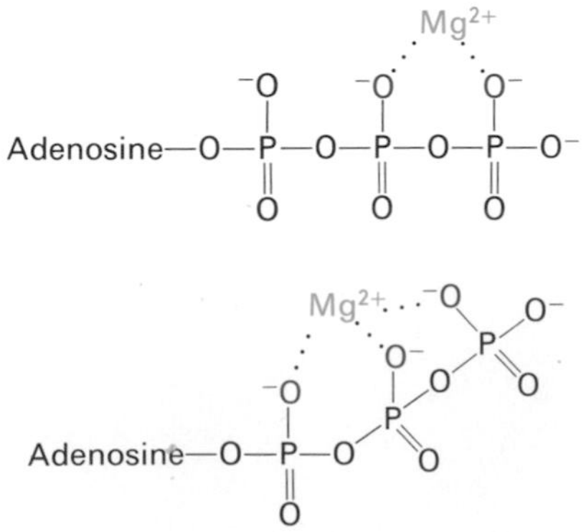

Phosphoryl transfer is a basic reaction in biochemistry. An enzyme that catalyzes the transfer of a phosphoryl group from ATP to an acceptor is called a *kinase*. Hexokinase, then, catalyzes the transfer of a phosphoryl group from ATP to a variety of six-carbon sugars (*hexoses*), such as glucose and mannose. *Hexokinase, like all other kinases, requires* Mg^{2+} *(or another divalent metal ion such as* Mn^{2+}*) for activity*. The divalent metal ion forms a complex with ATP. The structures of two possible Mg^{2+}-ATP complexes are shown at the left.

Figure 15-5
Modes of binding Mg^{2+} to ATP.

The next step in glycolysis is the *isomerization of glucose 6-phosphate to fructose 6-phosphate*. The *six-membered pyranose ring* of glucose 6-phosphate is converted into the *five-membered furanose ring* of fructose 6-phosphate. Recall that the open-chain form of glucose has an aldehyde group on C-1, whereas the open-chain form of fructose has a keto group on C-2. The aldehyde on C-1 reacts with the hydroxyl group on C-5 to form the pyranose ring, whereas the keto group on C-2 reacts with the C-5 hydroxyl to form the furanose ring. Thus, the isomerization of glucose 6-phosphate to fructose 6-phosphate is a *conversion of an aldose into a ketose*.

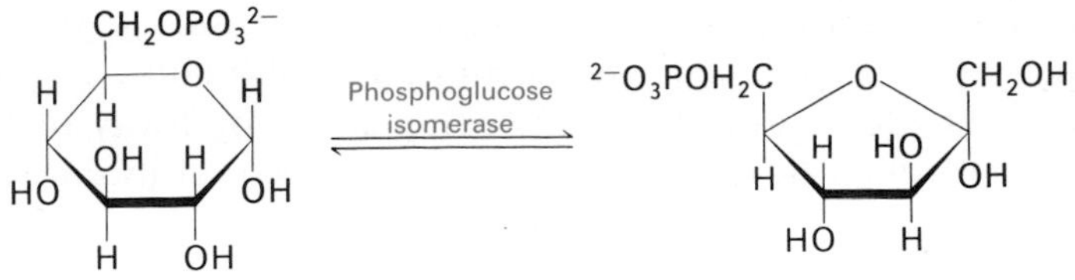

Glucose 6-phosphate ⇌ (Phosphoglucose isomerase) Fructose 6-phosphate

The open-chain representations of these sugars show the essence of this reaction.

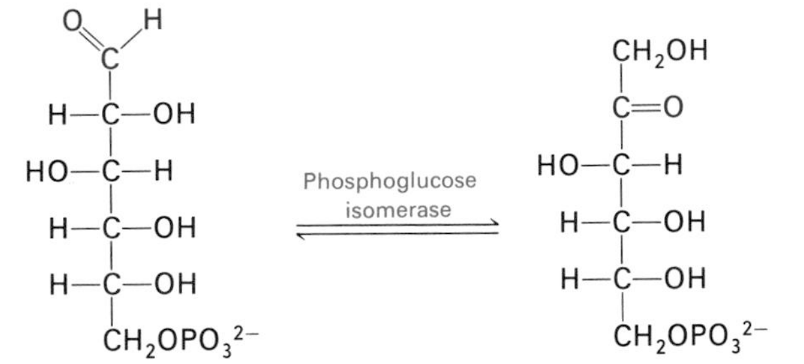

Glucose 6-phosphate (An aldose) ⇌ (Phosphoglucose isomerase) Fructose 6-phosphate (A ketose)

A second phosphorylation reaction follows the isomerization step. *Fructose 6-phosphate is phosphorylated by ATP to fructose 1,6-bisphosphate.* This compound was formerly known as fructose 1,6-diphosphate. *Bis*-phosphate means two separate phosphate groups, whereas *di*phosphate (as in adenosine diphosphate) means two joined phosphate groups. Hence, the name fructose 1,6-bisphosphate is preferable.

Fructose 6-phosphate + ATP → (Phosphofructokinase) Fructose 1,6-bisphosphate + ADP + H^+

This reaction is catalyzed by *phosphofructokinase,* an allosteric enzyme. The pace of glycolysis is critically dependent on the level of activity of this enzyme, which is allosterically controlled by ATP and several other metabolites (p. 359).

FORMATION OF GLYCERALDEHYDE 3-PHOSPHATE BY CLEAVAGE AND ISOMERIZATION

The second stage of glycolysis consists of four steps, starting with the splitting of fructose 1,6-bisphosphate into *glyceraldehyde 3-phosphate* and *dihydroxyacetone phosphate.* The remaining steps in glycolysis involve three-carbon units rather than six-carbon units.

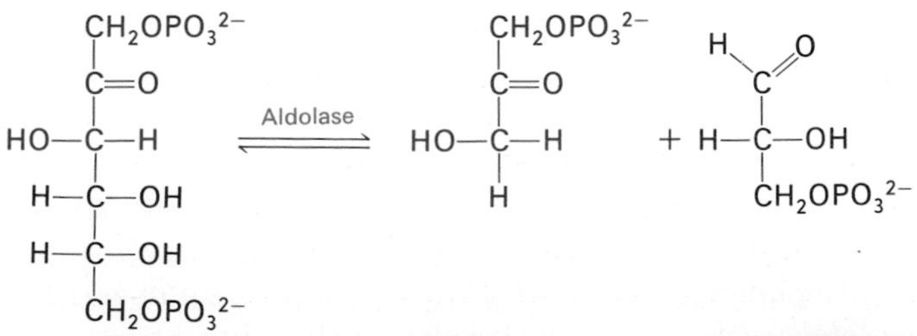

Fructose 1,6-bisphosphate ⇌ (Aldolase) Dihydroxyacetone phosphate + Glyceraldehyde 3-phosphate

This reaction is catalyzed by *aldolase*. This enzyme derives its name from the nature of the reverse reaction, an aldol condensation.

Glyceraldehyde 3-phosphate is on the direct pathway of glycolysis. Dihydroxyacetone phosphate is not, but it can be readily converted into glyceraldehyde 3-phosphate. These compounds are isomers: dihydroxyacetone phosphate is a ketose, whereas glyceraldehyde 3-phosphate is an aldose. The isomerization of these three-carbon phosphorylated sugars is catalyzed by *triose phosphate isomerase* (Figure 15-6). This reaction is rapid and reversible. At equilibrium, 96% of the triose phosphate is dihydroxyacetone phosphate. However, the reaction proceeds readily from dihydroxyacetone phosphate to glyceraldehyde 3-phosphate because of efficient removal of this product.

Aldol condensation—
The combination of two carbonyl compounds (e.g., an aldehyde and a ketone) to form an aldol (a β-hydroxy-carbonyl compound).

Ketone + **Aldehyde** ⇌ aldol

Triose phosphate isomerase: **Dihydroxyacetone phosphate** (A ketose) ⇌ **Glyceraldehyde 3-phosphate** (An aldose)

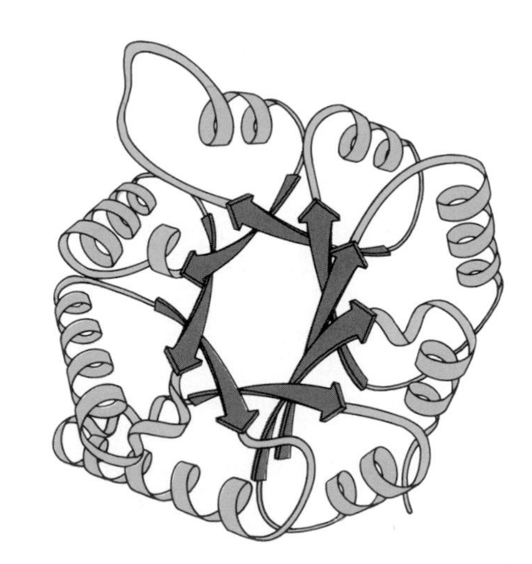

Figure 15-6
Triose phosphate isomerase consists of a central core of eight parallel β strands (red) surrounded by eight α helices (green). Connecting regions are shown in yellow. This structural motif, called an $\alpha\beta$ barrel, is also found in one of the domains of pyruvate kinase. [After a drawing kindly provided by Dr. Jane Richardson.]

Thus, two molecules of glyceraldehyde 3-phosphate are formed from one molecule of fructose 1,6-bisphosphate by the sequential action of aldolase and triose phosphate isomerase.

Glyceraldehyde 3-phosphate + NAD^+ + P_i ⇌ (Glyceraldehyde 3-phosphate dehydrogenase) **1,3-Bisphosphoglycerate (1,3-BPG)** + NADH + H^+

ENERGY CONSERVATION: PHOSPHORYLATION IS COUPLED TO THE OXIDATION OF GLYCERALDEHYDE 3-PHOSPHATE

The preceding steps in glycolysis have transformed one molecule of glucose into two molecules of glyceraldehyde 3-phosphate. No energy has yet been extracted. On the contrary, two molecules of ATP have been invested thus far. We come now to a series of steps that harvest some of the energy contained in glyceraldehyde 3-phosphate. The initial reaction in this sequence is the *conversion of glyceraldehyde 3-phosphate into 1,3-bisphosphoglycerate* (1,3-BPG), a reaction catalyzed by *glyceraldehyde 3-phosphate dehydrogenase*. In the earlier literature, 1,3-BPG was known as 1,3-diphosphoglycerate (1,3-DPG).

A *high-energy phosphate compound* is generated in this oxidation-reduction reaction. The aldehyde group at C-1 is converted into an *acyl phosphate,* which is a *mixed anhydride* of phosphoric acid and a carboxylic acid. The energy for the formation of this anhydride, which has a high phosphate group-transfer potential, comes from the oxidation of the aldehyde group. Note that C-1 in 1,3-BPG is at the oxidation level of a carboxylic acid. The mechanism of this complex reaction, which couples oxidation and phosphorylation, will be discussed later in this chapter (p. 366).

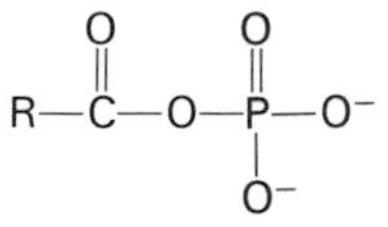

$$R{-}C(=O){-}O{-}P(=O)(O^-){-}O^-$$

Acyl phosphate

FORMATION OF ATP FROM 1,3-BISPHOSPHOGLYCERATE

In the next step, the high phosphoryl transfer potential of 1,3-BPG is used to generate ATP. Indeed, this is the first ATP-generating reaction in glycolysis. *Phosphoglycerate kinase* catalyzes the transfer of the phosphoryl group from the acyl phosphate of 1,3-BPG to ADP. ATP and 3-phosphoglycerate are the products.

Thus, the outcomes of the reactions catalyzed by glyceraldehyde 3-phosphate dehydrogenase and phosphoglycerate kinase are:

1. Glyceraldehyde 3-phosphate, an aldehyde, has been oxidized to 3-phosphoglycerate, a carboxylic acid.
2. NAD^+ has been reduced to NADH.
3. ATP has been formed from P_i and ADP.

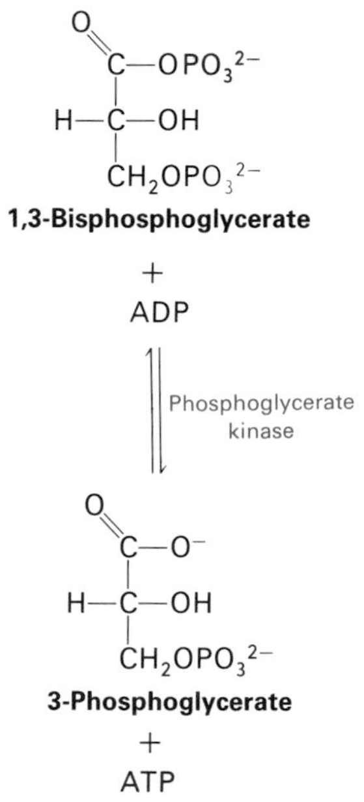

$$O{=}C(OPO_3^{2-}){-}CH(OH){-}CH_2OPO_3^{2-}$$

1,3-Bisphosphoglycerate

+ ADP

⇌ Phosphoglycerate kinase

$$O{=}C(O^-){-}CH(OH){-}CH_2OPO_3^{2-}$$

3-Phosphoglycerate

+ ATP

FORMATION OF PYRUVATE AND THE GENERATION OF A SECOND ATP

In the last stage of glycolysis, 3-phosphoglycerate is converted into pyruvate, and a second molecule of ATP is formed.

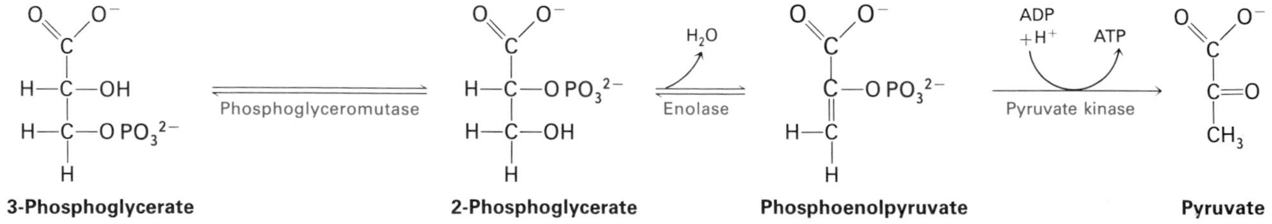

3-Phosphoglycerate $[{}^-OOC{-}CH(OH){-}CH_2OPO_3^{2-}]$ ⇌ (Phosphoglyceromutase) **2-Phosphoglycerate** $[{}^-OOC{-}CH(OPO_3^{2-}){-}CH_2OH]$ ⇌ (Enolase; H_2O released) **Phosphoenolpyruvate** $[{}^-OOC{-}C(OPO_3^{2-}){=}CH_2]$ → (Pyruvate kinase; ADP + H^+ → ATP) **Pyruvate** $[{}^-OOC{-}C(=O){-}CH_3]$

The first reaction is a rearrangement. The position of the phosphoryl group shifts in the conversion of *3-phosphoglycerate into 2-phosphoglycerate,* a reaction catalyzed by *phosphoglyceromutase.* In general, a mutase is an enzyme that catalyzes the intramolecular shift of a chemical group, such as a phosphoryl group.

In the second reaction, an *enol* is formed by the dehydration of 2-phosphoglycerate. *Enolase* catalyzes the formation of *phosphoenolpyruvate.* This dehydration reaction markedly elevates the group-transfer potential of the phosphoryl group. An *enol phosphate* has a high phosphoryl-transfer potential, whereas the phosphate ester of an ordinary alcohol has a low one. The reasons for this difference will be discussed later (p. 368).

In the last reaction, *pyruvate* is formed, and ATP is generated concomitantly. The virtually irreversible transfer of a phosphoryl group from phosphoenolpyruvate to ADP is catalyzed by *pyruvate kinase.*

ENERGY YIELD IN THE CONVERSION OF GLUCOSE INTO PYRUVATE

The net reaction in the transformation of glucose into pyruvate is

$$\text{Glucose} + 2\ P_i + 2\ \text{ADP} + 2\ \text{NAD}^+ \longrightarrow 2\ \text{pyruvate} + 2\ \text{ATP} + 2\ \text{NADH} + 2\ H^+ + 2\ H_2O$$

Thus, *two molecules of ATP are generated in the conversion of glucose into two molecules of pyruvate*. A summary of the steps in which ATP is consumed or formed is given in Table 15-1. Recall that two three-carbon units are formed from fructose 1,6-bisphosphate. The reactions of glycolysis are summarized in Table 15-2 and Figure 15-7.

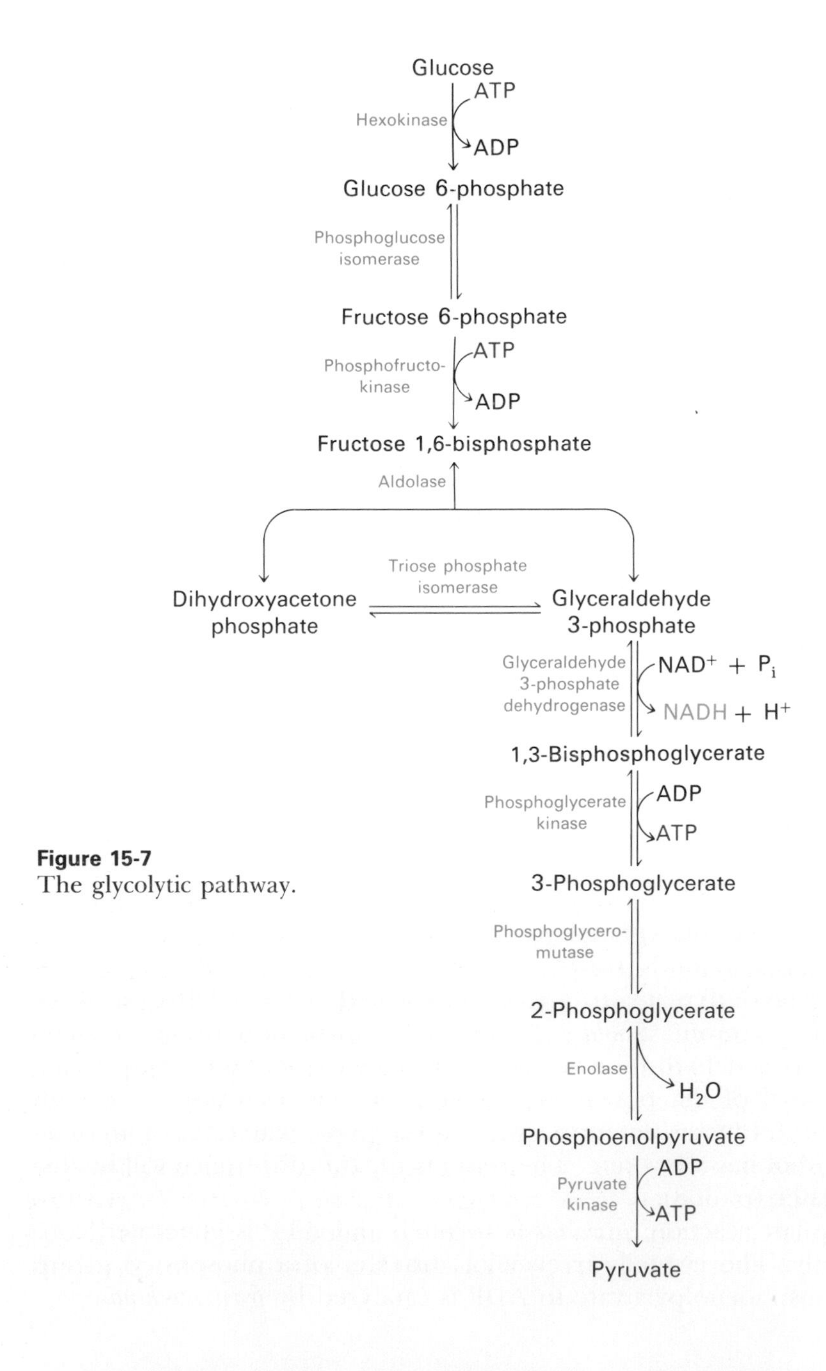

Figure 15-7
The glycolytic pathway.

Table 15-1
Consumption and generation of ATP in glycolysis

Reaction	ATP change per glucose
Glucose ⟶ glucose 6-phosphate	−1
Fructose 6-phosphate ⟶ fructose 1,6-bisphosphate	−1
2 1,3-Bisphosphoglycerate ⟶ 2 3-phosphoglycerate	+2
2 Phosphoenolpyruvate ⟶ 2 pyruvate	+2
Net	+2

Table 15-2
Reactions of glycolysis

Step	Reaction	Enzyme	Type*	ΔG°′	ΔG
1	Glucose + ATP ⟶ glucose 6-phosphate + ADP + H^+	Hexokinase	a	−4.0	−8.0
2	Glucose 6-phosphate ⇌ fructose 6-phosphate	Phosphoglucose isomerase	c	+0.4	−0.6
3	Fructose 6-phosphate + ATP ⟶ fructose 1,6-bisphosphate + ADP + H^+	Phosphofructokinase	a	−3.4	−5.3
4	Fructose 1,6-bisphosphate ⇌ dihydroxyacetone phosphate + glyceraldehyde 3-phosphate	Aldolase	e	+5.7	−0.3
5	Dihydroxyacetone phosphate ⇌ glyceraldehyde 3-phosphate	Triose phosphate isomerase	c	+1.8	+0.6
6	Glyceraldehyde 3-phosphate + P_i + NAD^+ ⇌ 1,3-bisphosphoglycerate + NADH + H^+	Glyceraldehyde 3-phosphate dehydrogenase	f	+1.5	−0.4
7	1,3-Bisphosphoglycerate + ADP ⇌ 3-phosphoglycerate + ATP	Phosphoglycerate kinase	a	−4.5	+0.3
8	3-Phosphoglycerate ⇌ 2-phosphoglycerate	Phosphoglyceromutase	b	+1.1	+0.2
9	2-Phosphoglycerate ⇌ phosphoenolpyruvate + H_2O	Enolase	d	+0.4	−0.8
10	Phosphoenolpyruvate + ADP + H^+ ⟶ pyruvate + ATP	Pyruvate kinase	a	−7.5	−4.0

Note: $\Delta G^{\circ\prime}$ and ΔG are expressed in kcal/mol. ΔG, the actual free-energy change, has been calculated from $\Delta G^{\circ\prime}$ and known concentrations of reactants under typical physiological conditions.

*Reaction type: (a) Phosphoryl transfer (b) Phosphoryl shift (c) Isomerization (d) Dehydration (e) Aldol cleavage (f) Phosphorylation coupled to oxidation

ENTRY OF FRUCTOSE AND GALACTOSE INTO GLYCOLYSIS

Let us consider how two other abundant sugars—fructose and galactose—are funneled into the glycolytic pathway. Recall that the hydrolysis of sucrose (common table sugar) yields fructose and glucose, and that hydrolysis of lactose (milk sugar) gives galactose and glucose. Fructose itself is present in many foods (e.g., honey). A typical daily intake of fructose is 100 grams. Much of it is metabolized by the liver, using the *fructose 1-phosphate pathway* (Figure 15-8). The first step is the phosphorylation of fructose to fructose 1-phosphate by fructokinase. Fructose 1-phosphate is then split into glyceraldehyde and dihydroxyacetone phosphate. This aldol cleavage is catalyzed by a specific fructose 1-phosphate aldolase. Glyceraldehyde is then phosphorylated to glyceraldehyde 3-phosphate by triose kinase so that it too can enter glycolysis.

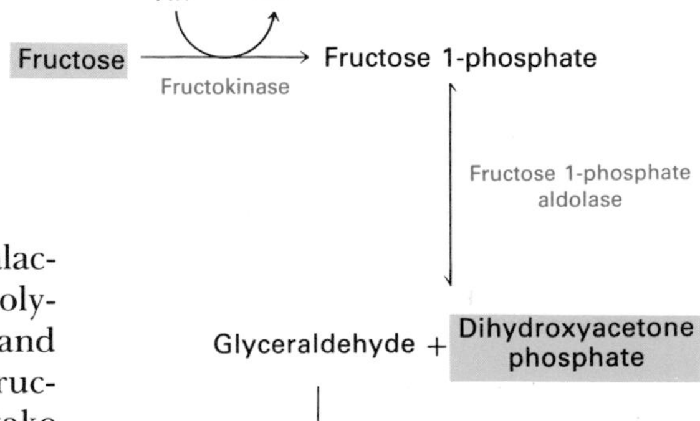

Figure 15-8
Entry of fructose into glycolysis.

Alternatively, fructose can be phosphorylated to fructose 6-phosphate by hexokinase. However, the affinity of hexokinase for glucose is twenty times as high as it is for fructose. Little fructose 6-phosphate is formed in the liver because it has a high glucose level. In contrast, adipose tissue has much more fructose than glucose. Hence, the formation of fructose 6-phosphate is not competitively inhibited to an appreciable extent, and most of the fructose in adipose tissue is metabolized through fructose 6-phosphate.

Galactose is converted into glucose 6-phosphate in four steps. The first reaction in the *galactose-glucose interconversion pathway* is the phosphorylation of galactose to galactose 1-phosphate by *galactokinase.*

$$\text{Galactose} + \text{ATP} \longrightarrow \text{galactose 1-phosphate} + \text{ADP} + \text{H}^+$$

Galactose 1-phosphate then acquires a uridyl group from uridine diphosphate glucose (UDP-glucose), an intermediate in the synthesis of glycosidic linkages (p. 340). The products of this reaction, which is catalyzed by *galactose 1-phosphate uridyl transferase,* are UDP-galactose and glucose 1-phosphate.

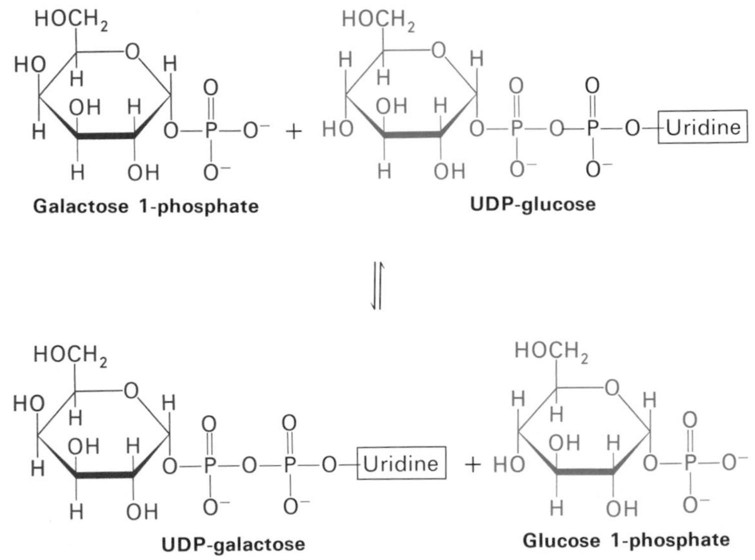

Galactose 1-phosphate + UDP-glucose ⇌ UDP-galactose + Glucose 1-phosphate

Galactose is then epimerized to glucose while attached to UDP. The configuration of the hydroxyl group at C-4 is inverted by *UDP-galactose-4-epimerase,* which contains a tightly bound NAD^+. Fluorescence studies have shown that this NAD^+ is transiently reduced to NADH during catalysis. NAD^+ probably accepts the hydrogen atom attached to C-4 of the sugar, and a 4-keto sugar intermediate is formed.

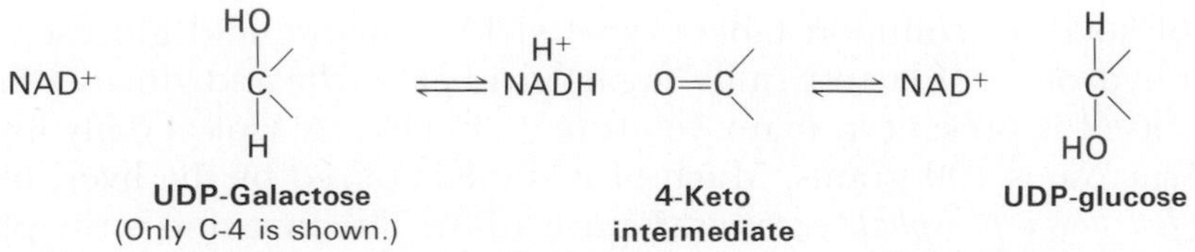

NAD^+ + UDP-Galactose (Only C-4 is shown.) ⇌ H^+ + NADH + 4-Keto intermediate ⇌ NAD^+ + UDP-glucose

NADH then transfers the same hydrogen to the other side of C-4 to form the other epimer. The sum of the reactions catalyzed by galactokinase, the transferase, and the epimerase is

$$\text{Galactose} + \text{ATP} \longrightarrow \text{glucose 1-phosphate} + \text{ADP} + \text{H}^+$$

Note that UDP-glucose is not consumed in the conversion of galactose to glucose because it is regenerated from UDP-galactose by the epimerase. This reaction is reversible, and the product of the reverse direction is also important. *The conversion of UDP-glucose into UDP-galactose is essential for the synthesis of galactosyl residues in complex polysaccharides and glycoproteins if the amount of galactose in the diet is inadequate to meet these needs.*

Finally, glucose 1-phosphate is isomerized to glucose 6-phosphate by *phosphoglucomutase*. We shall return to this reaction when we consider the synthesis and degradation of glycogen, which proceeds through glucose 1-phosphate (p. 454).

GALACTOSE IS HIGHLY TOXIC IF THE TRANSFERASE IS MISSING

The absence of galactose 1-phosphate uridyl transferase causes *galactosemia,* a severe disease that is inherited as an autosomal recessive trait. The metabolism of galactose in persons who have this disease is blocked at galactose 1-phosphate. Afflicted infants fail to thrive. Vomiting or diarrhea occurs when milk is consumed, and enlargement of the liver and jaundice are common. Furthermore, many galactosemics become mentally retarded. The blood galactose level is markedly elevated, and galactose is found in the urine. The absence of the transferase in red blood cells is a definitive diagnostic criterion.

Galactosemia is treated by the exclusion of galactose from the diet. A galactose-free diet leads to a striking regression of virtually all of the clinical symptoms, except for mental retardation, which may not be reversible. Continued galactose intake may lead to death in some patients. *The damage in galactosemia is caused by an accumulation of toxic substances, rather than by the absence of an essential compound.* Patients are able to synthesize UDP-galactose from UDP-glucose because their epimerase activity is normal. One of the toxic substances is *galactitol,* which is formed by reduction of galactose. The presence of aldose reductase in the lens of the eye causes galactitol to accumulate there, which leads to the entry of water and the development of cataracts.

Galactose

Aldose reductase ($NADPH + H^+ \rightarrow NADP^+$)

Galactitol

PHOSPHOFRUCTOKINASE IS THE KEY ENZYME IN THE CONTROL OF GLYCOLYSIS

The glycolytic pathway has a dual role: it degrades glucose to generate ATP and it provides building blocks for synthetic reactions, such as the formation of long-chain fatty acids. The rate of conversion of glucose into pyruvate is regulated to meet these two major cellular needs. *In metabolic pathways, enzymes catalyzing essentially irreversible reactions are potential sites of control.* In glycolysis, the reactions catalyzed by hexokinase, phosphofructokinase, and pyruvate kinase are virtually irreversible; hence, they would be expected to have regulatory as well as catalytic roles. In fact, all three enzymes are control sites.

Phosphofructokinase is the most important control element in the glycolytic pathway of mammals. The enzyme from liver (a 340-kd tetramer) is *inhibited by high levels of ATP,* which lowers the affinity of the enzyme for fructose 6-phosphate. A high concentration of ATP converts the hyperbolic binding curve for fructose 6-phosphate into a sigmoidal one (Figure 15-9). This allosteric effect is elicited by the binding of ATP to a

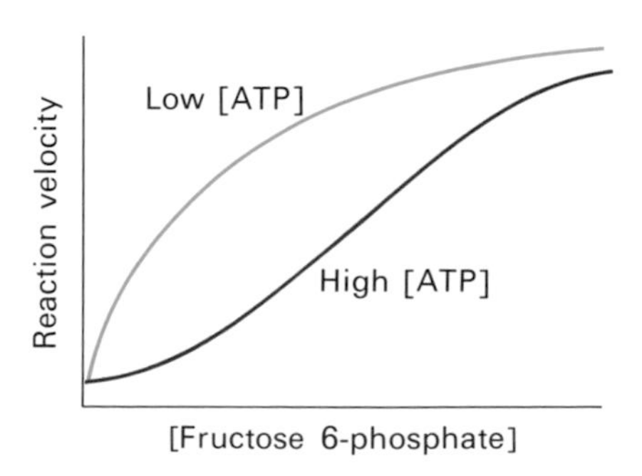

Figure 15-9
Allosteric regulation of phosphofructokinase. A high level of ATP inhibits the enzyme by decreasing its affinity for fructose 6-phosphate. AMP diminishes and citrate enhances the inhibitory effect of ATP.

highly specific regulatory site that is distinct from the catalytic site. The inhibitory action of ATP is reversed by AMP, and so *the activity of the enzyme increases when the ATP/AMP ratio is lowered*. In other words, *glycolysis is stimulated as the energy charge falls*. Phosphofructokinase is also inhibited by H^+, which prevents excessive formation of lactate and a precipitous drop in blood pH (acidosis).

Glycolysis also furnishes carbon skeletons for biosyntheses, and so phosphofructokinase should also be regulated by a signal indicating whether building blocks are abundant or scarce. Indeed, *phosphofructokinase is inhibited by citrate*, an early intermediate in the citric acid cycle (p. 374). A high level of citrate means that biosynthetic precursors are abundant and so additional glucose should not be degraded for this purpose. Citrate inhibits phosphofructokinase by enhancing the inhibitory effect of ATP.

A new regulator of glycolysis was discovered in 1980 by Henri-Géry Hers and Emile Van Schaftingen. They found that *β-D-fructose 2,6-bisphosphate*, a previously unknown metabolite, is a potent activator of phosphofructokinase. Fructose 2,6-bisphosphate (F-2,6-BP) activates phosphofructokinase in liver by increasing its affinity for fructose 6-phosphate and diminishing the inhibitory effect of ATP (Figure 15-10). In essence, F-2,6-BP is an allosteric activator that shifts the conformational equilibrium of this tetrameric enzyme from the T state to the R state.

$^{2-}O_3POH_2C$ — O — CH_2OH; H, HO, H, OPO_3^{2-}, OH, H

Fructose 2,6-bisphosphate (F-2,6-BP)

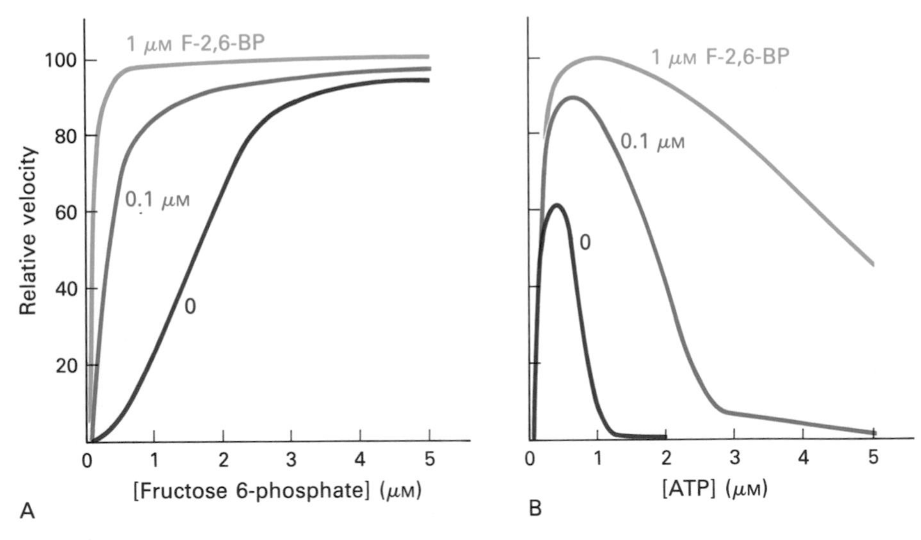

Figure 15-10
Phosphofructokinase is activated by fructose 2,6-bisphosphate. (A) The sigmoidal dependence of velocity on substrate concentration becomes hyperbolic in the presence of 1 μM fructose 2,6-bisphosphate, and (B) the inhibitory effect of ATP is reversed. [After H.-G. Hers and E. Van Schaftingen. *Proc. Nat. Acad. Sci.* 78(1981):2862.]

Fructose 2,6-bisphosphate is formed by the phosphorylation of fructose 6-phosphate, in a reaction catalyzed by phosphofructokinase 2 (PFK2), a different enzyme from phosphofructokinase (PFK). F-2,6-BP is hydrolyzed to fructose 6-phosphate by a specific phosphatase, fructose bisphosphatase 2 (FBPase2). The striking finding is that *both PFK2 and FBPase2 are present in a single 53-kd polypeptide chain*, which thus may be called a *tandem enzyme* (Figure 15-11). Several other enzymes important in metabolic regulation also contain opposing catalytic activities on the same chain (p. 389). Fructose 6-phosphate accelerates the synthesis of F-2,6-BP and inhibits its hydrolysis. Hence, *an abundance of fructose 6-phosphate leads to a higher concentration of F-2,6-BP, which in turn stimulates phosphofructokinase.* Such a process is called *feed-forward stimulation.*

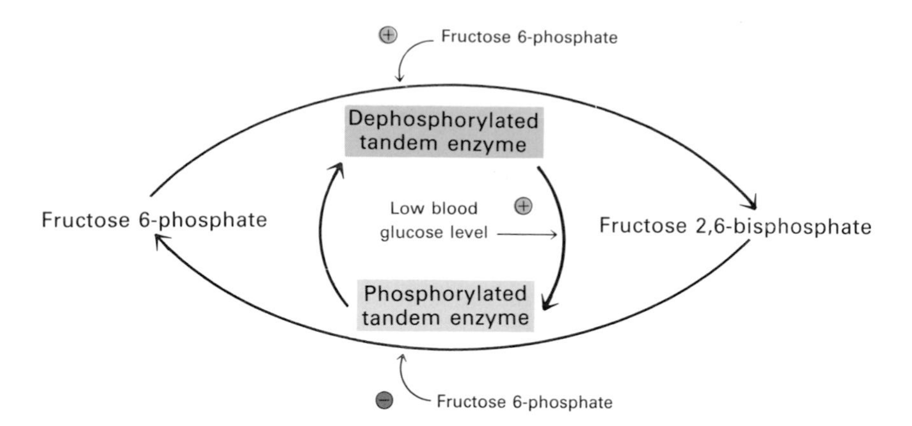

Figure 15-11
Control of the synthesis and degradation of fructose 2,6-bisphosphate. Fructose 6-phosphate accelerates the formation of F-2,6-BP and inhibits its hydrolysis. A low blood glucose level leads to a higher level of the phosphorylated tandem enzyme and hence to a lower level of F-2,6-BP.

Furthermore, the activities of PFK2 and FBPase2 are reciprocally controlled by *phosphorylation of a single serine residue.* When glucose is scarce, a rise in the blood level of the hormone glucagon triggers a cascade leading to the phosphorylation of this bifunctional enzyme. This covalent modification activates FBPase2 and inhibits PFK2, lowering the level of F-2,6-BP. Conversely, when glucose is abundant, the enzyme loses its attached phosphate group, which leads to a rise in the level of F-2,6-BP and the consequent acceleration of glycolysis. We shall return to this remarkable regulatory device when we consider the integration of carbohydrate metabolism (p. 630).

HEXOKINASE AND PYRUVATE KINASE ALSO PARTICIPATE IN REGULATING THE PACE OF GLYCOLYSIS

Hexokinase is inhibited by glucose 6-phosphate. The level of fructose 6-phosphate increases when phosphofructokinase is blocked. In turn, the level of glucose 6-phosphate rises because it is in equilibrium with fructose 6-phosphate. *Hence, the inhibition of phosphofructokinase leads to the inhibition of hexokinase.* Nevertheless, glucose is phosphorylated into glucose 6-phosphate in liver even when the glucose 6-phosphate level is high because of the presence of *glucokinase,* which differs from hexokinase. Glucokinase has a high K_M for glucose, and so it is effective only when glucose is abundant. The role of glucokinase is to provide glucose 6-phosphate for the synthesis of glycogen, a storage form of glucose (p. 449). The high K_M of glucokinase in the liver gives brain and muscle first call on glucose when its supply is limited.

Why is phosphofructokinase rather than hexokinase the pacemaker of glycolysis? The reason becomes evident on noting that glucose 6-phosphate is not only a glycolytic intermediate. Glucose 6-phosphate can also be converted into glycogen or it can be oxidized by the pentose phosphate pathway (p. 428) to generate NADPH. The first irreversible reaction unique to the glycolytic pathway, called the *committed step,* is the phosphorylation of fructose 6-phosphate to fructose 1,6-bisphosphate. Thus, it is highly appropriate for phosphofructokinase to be the primary control site in glycolysis. *In general, the enzyme catalyzing the committed step in a metabolic sequence is the most important control element in the pathway.*

Pyruvate kinase, the enzyme catalyzing the third irreversible step in glycolysis, controls the outflow from this pathway. This stop yields ATP and pyruvate, a central metabolic intermediate that can be oxidized further or used as a building block. Three forms of pyruvate kinase (a tetramer of 55-kd subunits) have been found in mammals: the L type predominates in liver, the M type in muscle and brain, and the A type in other tissues. These variations on a common theme, called *isoenzymes* or *isozymes*, have essentially the same architectural plan and catalytic mechanism but differ in how they are regulated. The L isozyme binds phosphoenolpyruvate cooperatively. *ATP allosterically inhibits this isozyme to slow glycolysis when the energy charge is high* (Figure 15-12). Alanine (synthesized in one step from pyruvate, p. 504) also allosterically inhibits the enzyme, in this case to signal that building blocks are abundant. Fructose 1,6-bisphosphate, the product of the preceding irreversible step in glycolysis, activates pyruvate kinase to enable it to keep pace with the oncoming high flux of intermediates. The catalytic properties of the L isozyme are also controlled by reversible phosphorylation. When the blood glucose level is low, glucagon triggers a cascade that increases the proportion of phosphorylated pyruvate kinase, the less active form. *The role of this hormone-triggered phosphorylation, like that of the tandem enzyme controlling the levels of fructose 2,6-bisphosphate, is to prevent the liver from consuming glucose when it is more urgently needed by brain and muscle* (p. 442). The M isozyme, in contrast, is not reversibly phosphorylated, whereas the A isozyme is intermediate between M and L in its susceptibility to control by covalent modification. We shall encounter other examples of how isoenzymes contribute to the metabolic diversity of different organs.

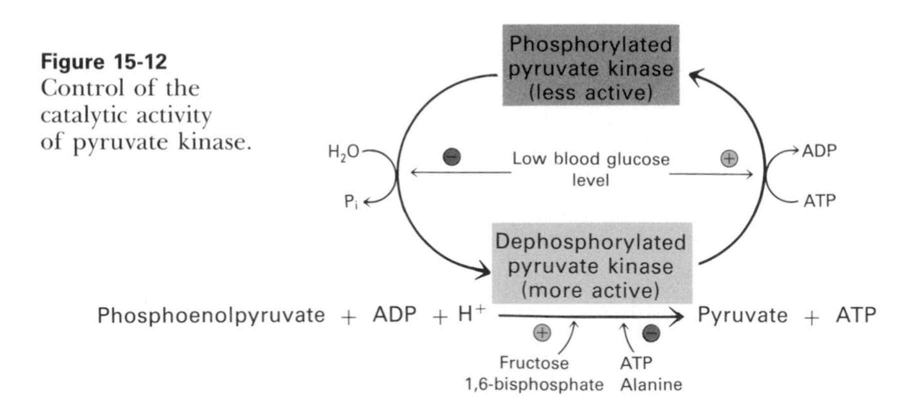

Figure 15-12
Control of the catalytic activity of pyruvate kinase.

PYRUVATE CAN BE CONVERTED INTO ETHANOL, LACTATE, OR ACETYL COENZYME A

The sequence of reactions from glucose to pyruvate is very similar in all organisms and in all kind of cells. In contrast, the fate of pyruvate in the generation of metabolic energy is variable. Three reactions of pyruvate are considered here.

1. *Ethanol* is formed from pyruvate in yeast and several other microorganisms. The first step is the decarboxylation of pyruvate:

$$\text{Pyruvate} + H^+ \longrightarrow \text{acetaldehyde} + CO_2$$

This reaction is catalyzed by pyruvate decarboxylase, which contains thiamine pyrophosphate as a coenzyme. Thiamine pyrophosphate is the coenzyme in a variety of decarboxylases (p. 380).

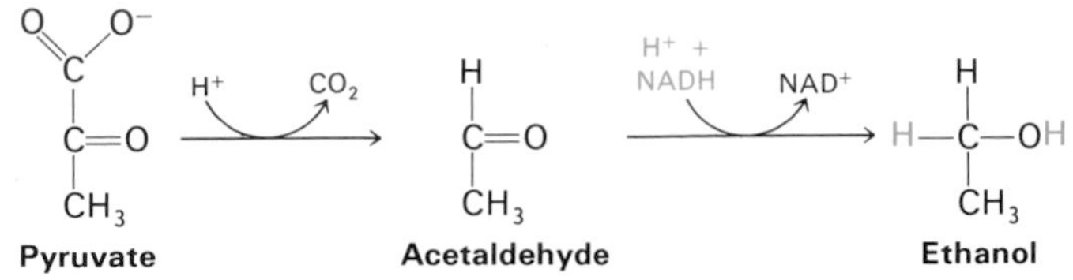

$$\text{Pyruvate} \xrightarrow{H^+ \;\; CO_2} \text{Acetaldehyde} \xrightarrow{H^+ + NADH \;\; NAD^+} \text{Ethanol}$$

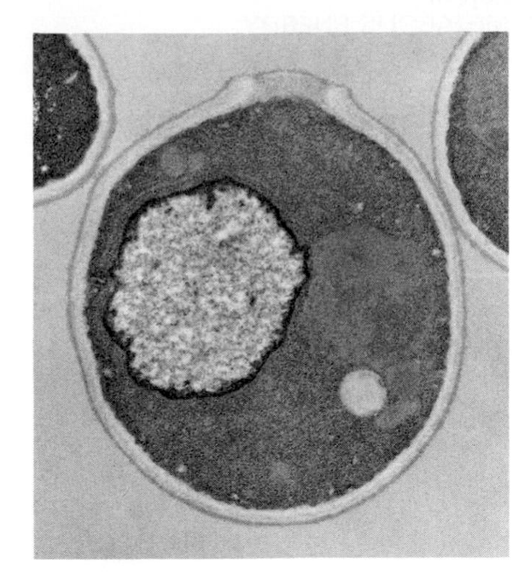

Figure 15-13
Electron micrograph of a yeast cell. [Courtesy of Lynne Mercer.]

The second step is the reduction of acetaldehyde to ethanol by NADH, in a reaction catalyzed by *alcohol dehydrogenase.*

$$\text{Acetaldehyde} + \text{NADH} + H^+ \rightleftharpoons \text{ethanol} + \text{NAD}^+$$

The active site of alcohol dehydrogenase contains a zinc ion that is coordinated to the sulfur atoms of two cysteine residues and a histidine nitrogen atom. As in carboxypeptidase A (p. 216), Zn^{2+} polarizes the carbonyl group of the substrate to stabilize the transition state (Figure 15-14).

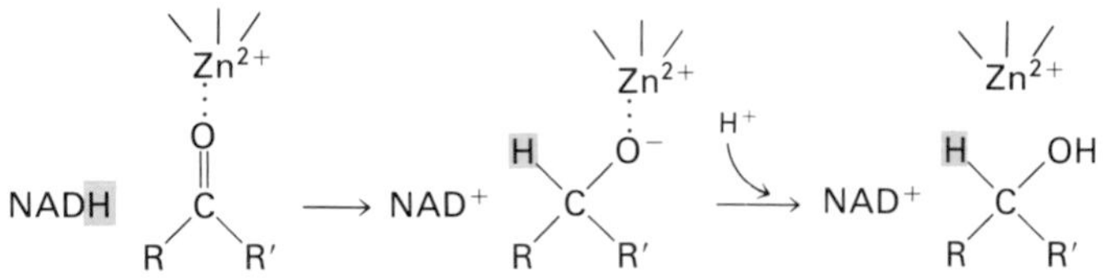

Figure 15-14
Catalytic mechanism of alcohol dehydrogenase.

The conversion of glucose into ethanol is called *alcoholic fermentation.* The net result of this anaerobic process is

$$\text{Glucose} + 2\ P_i + 2\ \text{ADP} + 2\ H^+ \longrightarrow 2\ \text{ethanol} + 2\ CO_2 + 2\ \text{ATP} + 2\ H_2O$$

It is important to note that NAD^+ and NADH do not appear in this equation. Acetaldehyde is reduced to ethanol so that NAD^+ is regenerated for use in the reaction catalyzed by glyceraldehyde 3-phosphate dehydrogenase. Thus, there is no net oxidation-reduction in the conversion of glucose into ethanol.

2. *Lactate* is normally formed from pyruvate in a variety of microorganisms. The reaction also occurs in the cells of higher organisms when the amount of oxygen is limiting, as in muscle during intense activity. The reduction of pyruvate by NADH to form lactate is catalyzed by *lactate dehydrogenase:*

$$\text{Pyruvate} + \text{NADH} + H^+ \xrightleftharpoons{\text{Lactate dehydrogenase}} \text{L-Lactate} + \text{NAD}^+$$

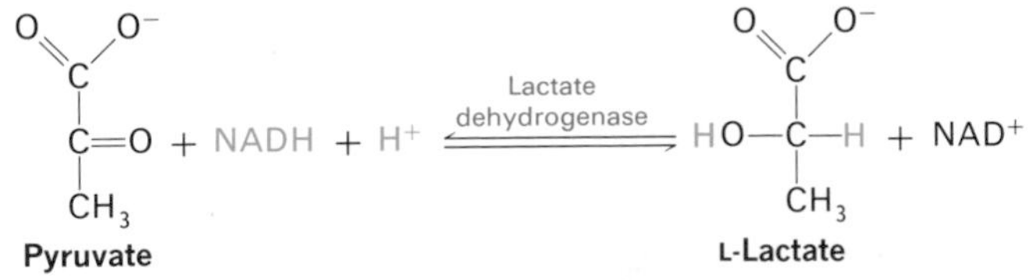

The overall reaction in the conversion of glucose into lactate is

$$\text{Glucose} + 2\ P_i + 2\ \text{ADP} \longrightarrow 2\ \text{lactate} + 2\ \text{ATP} + 2\ H_2O$$

As in alcoholic fermentation, there is no net oxidation-reduction. The NADH formed in the oxidation of glyceraldehyde 3-phosphate is consumed in the reduction of pyruvate. *The regeneration of NAD^+ in the reduction of pyruvate to lactate or ethanol sustains the continued operation of glycolysis under anaerobic conditions.* If NAD^+ were not regenerated, glycolysis could not proceed beyond glyceraldehyde 3-phosphate, which means that no ATP would be generated. In effect, the formation of lactate by aerobic organisms buys time, as will be discussed in Chapter 18.

3. Only a small fraction of the energy of glucose is released in its anaerobic conversion into lactate (or ethanol). Much more energy can be extracted aerobically by means of the citric acid cycle and the electron-transport chain. The entry point to this oxidative pathway is *acetyl coenzyme A* (acetyl CoA), which is formed inside mitochondria by the oxidative decarboxylation of pyruvate:

$$\text{Pyruvate} + NAD^+ + \text{CoA} \longrightarrow \text{acetyl CoA} + CO_2 + \text{NADH}$$

This reaction, which is catalyzed by the pyruvate dehydrogenase complex, will be discussed in detail in the next chapter. The NAD^+ required for this reaction and for the oxidation of glyceraldehyde 3-phosphate is regenerated when NADH ultimately transfers its electrons to O_2 through the electron-transport chain in mitochondria.

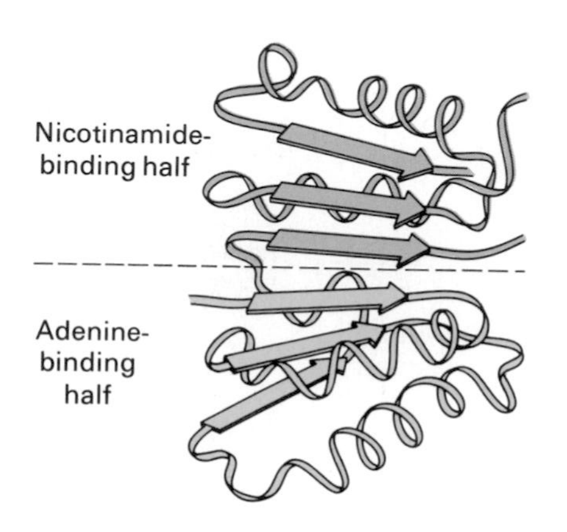

Figure 15-15
Schematic diagram of the NAD^+-binding region in dehydrogenases. The nicotinamide-binding half (shaded green) is structurally similar to the adenine-binding half (shaded yellow) of the site. [After M. G. Rossmann, A. Liljas, C.-I. Brändén, and L. J. Banaszak. In *The Enzymes,* 3rd ed., vol. 10 (Academic Press, 1975), p. 68.]

THE BINDING SITE FOR NAD^+ IS VERY SIMILAR IN A VARIETY OF DEHYDROGENASES

Lactate dehydrogenase from skeletal muscle, a 140-kd tetramer, and alcohol dehydrogenase, an 84-kd dimer, have quite different three-dimensional structures. However, their binding sites for NAD^+ are strikingly similar (Figure 15-15). The NAD^+-binding region is made up of four α helices and six strands of parallel β sheets. The conformations of NAD^+ bound to lactate dehydrogenase and to alcohol dehydrogenase also are nearly the same. The adenosine moiety of NAD^+ is bound in a hydrophobic crevice. In contrast, the nicotinamide unit is bound so that the reactive side of the ring is in a polar environment, whereas the other side makes contact with hydrophobic residues of the enzyme. The bound NAD^+ has an extended conformation (Figure 15-16). The

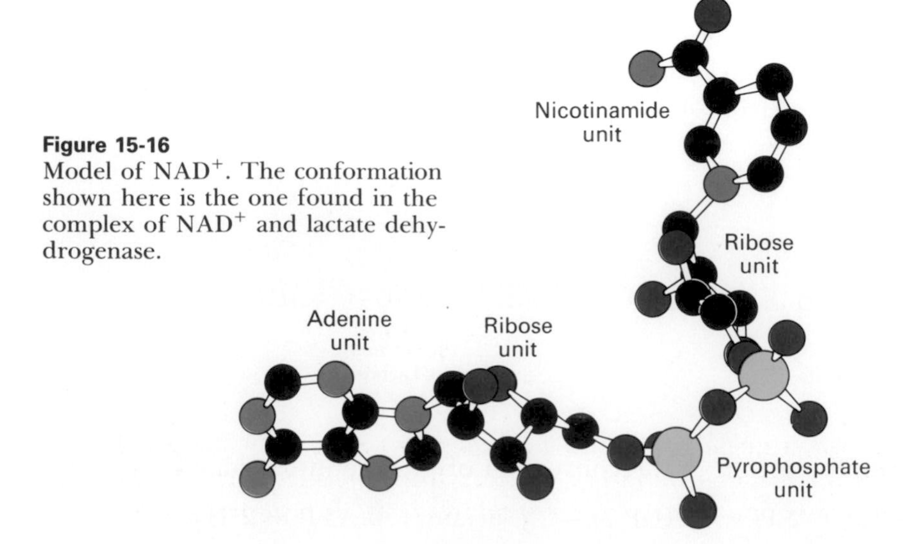

Figure 15-16
Model of NAD^+. The conformation shown here is the one found in the complex of NAD^+ and lactate dehydrogenase.

three-dimensional structures of glyceraldehyde 3-phosphate dehydrogenase and malate dehydrogenase (an enzyme of the citric acid cycle, p. 378) are also known at high resolution. Their NAD^+-binding sites are very similar to those in lactate dehydrogenase and alcohol dehydrogenase. *It seems likely that the NAD^+-binding region common to these four enzymes is a fundamental structural motif of NAD^+-linked dehydrogenases.*

GLUCOSE INDUCES A LARGE CONFORMATIONAL CHANGE IN HEXOKINASE

X-ray crystallographic studies of yeast hexokinase have revealed that the binding of glucose leads to a large conformational change in the enzyme. Hexokinase consists of two lobes, which come together when glucose is bound (Figure 15-17). Glucose induces a 12-degree rotation of one lobe with respect to the other, resulting in movements of the polypeptide backbone of as much as 8 Å. The cleft between the lobes closes and the bound glucose becomes surrounded by protein, except for its 6-hydroxymethyl group.

The closing of the cleft in hexokinase is a striking example of the role of *induced fit* in enzyme action, as originally proposed by Koshland (p. 218). The glucose-induced structural changes are likely to be significant in two ways. First, the environment around the glucose becomes much more nonpolar, which encourages the transfer of phosphoryl from ATP. Second, the embracing of glucose by hexokinase enables the enzyme to discriminate against H_2O as a substrate. If hexokinase were rigid, a water molecule occupying the binding site for the —CH_2OH of glucose would attack the α-phosphate of ATP. In other words, a rigid kinase would necessarily be an ATPase as well as a kinase. This undesirable activity is prevented by making hexokinase enzymatically active only when glucose closes the cleft. It is interesting to note that pyruvate kinase, phosphoglycerate kinase, and phosphofructokinase also contain clefts between lobes that close when substrate is bound. *Substrate-induced cleft closing* is likely to be a general feature of kinases.

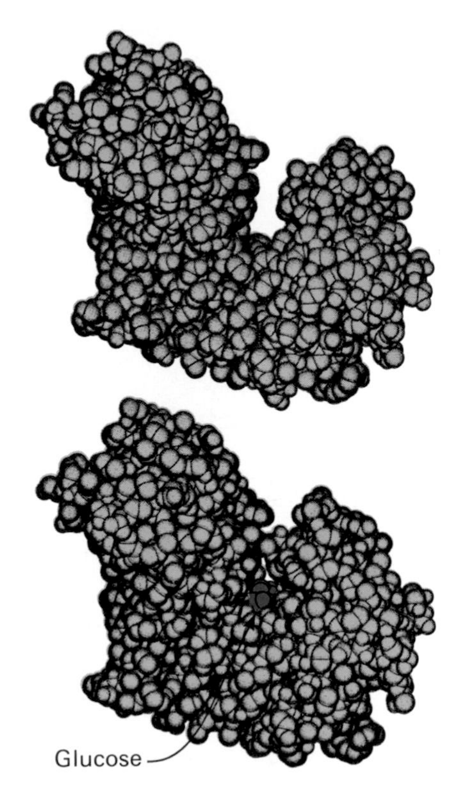

Figure 15-17
The conformation of hexokinase changes markedly on binding glucose (shown in red). The two lobes of the enzyme come together and surround the substrate. [Courtesy of Dr. Thomas Steitz.]

ALDOLASE FORMS A SCHIFF BASE WITH DIHYDROXYACETONE PHOSPHATE

Let us now turn to aldolase, which catalyzes the condensation of dihydroxyacetone phosphate and glyceraldehyde 3-phosphate to form fructose 1,6-bisphosphate. First, dihydroxyacetone phosphate forms a protonated Schiff base with a specific lysine residue in the active site of animal aldolases.

$$\mathrm{E{-}NH_2} + \mathrm{O{=}C(CH_2OPO_3^{2-})(CH_2OH)} + \mathrm{H^+} \rightleftharpoons \mathrm{E{-}\overset{+}{N}H{=}C(CH_2OPO_3^{2-})(CH_2OH)} + \mathrm{H_2O}$$

Protonated Schiff base

In this reaction, a nucleophile (the amino group) attacks the carbonyl group to form a tetrahedral intermediate, which then dehydrates. *The resulting protonated Schiff base plays a critical role in catalysis. It promotes the formation of the enolate anion of dihydroxyacetone phosphate by serving as an electron sink.*

$$\text{E—}\overset{H^+}{N}\text{=C(CH}_2\text{OPO}_3^{2-})\text{—CH(OH)—H} \rightleftharpoons \text{E—}\overset{H^+}{N}\text{=C(CH}_2\text{OPO}_3^{2-})\text{—}\overset{\ominus}{C}\text{H—OH} + H^+$$

Enolate anion

The enolate anion then adds to the aldehyde group of glyceraldehyde 3-phosphate to form the protonated Schiff base of fructose 1,6-bisphosphate.

$$\text{E—}\overset{H^+}{N}\text{=C(CH}_2\text{OPO}_3^{2-})\text{—HO—}\overset{\ominus}{C}\text{—H} \quad + \quad \text{H—}\overset{\delta+}{C}\text{=O}^{\delta-}\text{—H—C—OH—CH}_2\text{OPO}_3^{2-}$$

Enolate anion **Glyceraldehyde 3-phosphate**

Figure 15-18
Photomicrograph of crystals of aldolase.
[Courtesy of Dr. David Eisenberg.]

This Schiff base is deprotonated and hydrolyzed to yield fructose 1,6-bisphosphate and the regenerated enzyme.

$$\text{E—}\overset{H^+}{N}\text{=C(CH}_2\text{OPO}_3^{2-})\text{—R} \xrightarrow{H^+} \text{E—N=C(CH}_2\text{OPO}_3^{2-})\text{—R} \xrightarrow{H_2O} \text{O=C(CH}_2\text{OPO}_3^{2-})\text{—R} + \text{E—NH}_2$$

$$R = \text{—CH(OH)—CH(OH)—CH(OH)—CH}_2\text{OPO}_3^{2-}$$

The pathway for the cleavage of fructose 1,6-bisphosphate is simply the reverse of the one for its formation.

A THIOESTER IS FORMED IN THE OXIDATION OF GLYCERALDEHYDE 3-PHOSPHATE

A different kind of covalent enzyme-substrate intermediate is formed in glyceraldehyde 3-phosphate dehydrogenase. This enzyme catalyzes the oxidative phosphorylation of its aldehyde substrate.

$$\text{Glyceraldehyde 3-phosphate} + P_i + NAD^+ \longrightarrow \text{1,3-BPG} + NADH + H^+$$

The conversion of an aldehyde into an acyl phosphate entails both *oxidation* and *phosphorylation*. Oxidation requires the *removal of a hydride ion* $(:H^-)$, which is a hydrogen nucleus and two electrons. There is a large barrier to the removal of a hydride ion from an aldehyde because of the dipolar character of the carbonyl group: the carbon atom of the carbonyl group already has a partial positive charge. The removal of the hydride ion is facilitated by making the carbon atom less positively charged. This is accomplished by the *addition of a nucleophile*, represented by X^- in the following equation:

$$H^+ + \overset{\delta-}{O}\text{=}\overset{\delta+}{C}(R)\text{—H} + X^- \longrightarrow \text{HO—C(R)(X)—H} \longrightarrow \text{O=C(R)—X} + \underset{\text{Hydride ion}}{:H^-} + H^+$$

The hydride ion readily leaves the addition compound because the carbon atom no longer carries a large positive charge. Furthermore, some of the free energy of the oxidation is preserved in the acyl intermediate. Addition of orthophosphate to this acyl intermediate yields an acyl phosphate, which has a high group-transfer potential.

$$R{-}\overset{O}{\overset{\|}{C}}{-}X + {}^{-}O{-}\overset{O}{\overset{\|}{\underset{OH}{\underset{|}{P}}}}{-}O^{-} \longrightarrow R{-}\overset{O}{\overset{\|}{C}}{-}O{-}\overset{O}{\overset{\|}{\underset{O^{-}}{\underset{|}{P}}}}{-}O^{-} + XH$$

Now let us see how glyceraldehyde 3-phosphate dehydrogenase carries out these steps (Figure 15-19). *The nucleophile X^{-} is the sulfhydryl group of a cysteine residue at the active site.* The aldehyde substrate reacts with the ionized form of this sulfhydryl group to form a hemithioacetal. The next step is the *transfer of a hydride ion to a molecule of NAD^{+} that is tightly bound to the enzyme.* The products of this reaction are the reduced coenzyme NADH and a thioester. *This thioester is an energy-rich intermediate,* corresponding to the acyl intermediate mentioned earlier. NADH dissociates from the enzyme, and another NAD^{+} binds to the active site. Orthophosphate then attacks the thioester to form 1,3-BPG, an energy-rich phosphate. A crucial aspect of the formation of 1,3-BPG from glyceraldehyde 3-phosphate is that a thermodynamically unfavorable reaction, the formation of an acyl phosphate from a carboxylate, is driven by a thermodynamically favorable reaction, the oxidation of an aldehyde.

$$R{-}\overset{O}{\overset{\|}{C}}{-}H + NAD^{+} + H_2O \rightleftharpoons R{-}\overset{O}{\overset{\|}{C}}{-}O^{-} + NADH + 2\,H^{+}$$

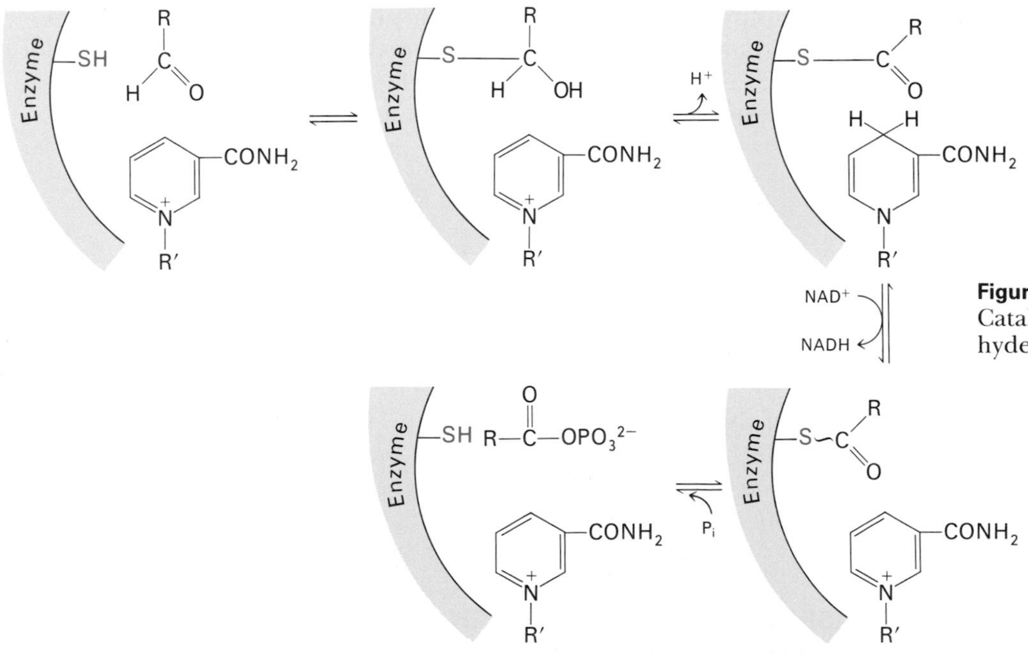

Figure 15-19
Catalytic mechanism of glyceraldehyde 3-phosphate dehydrogenase.

These two reactions are *coupled by the thioester intermediate,* which preserves much of the free energy released in the oxidation reaction. We see here the *use of a covalent enzyme-bound intermediate as a mechanism of energy coupling.*

ARSENATE, AN ANALOG OF PHOSPHATE, ACTS AS AN UNCOUPLER

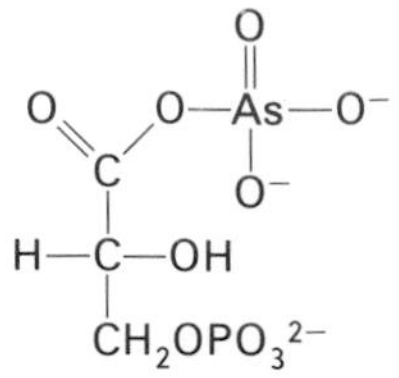

$$O=C(-O-As(=O)(-O^-)-O^-)-CH(OH)-CH_2OPO_3^{2-}$$

1-Arseno-3-phosphoglycerate

Arsenate (AsO_4^{3-}) closely resembles P_i in structure and reactivity. In the reaction catalyzed by glyceraldehyde 3-phosphate dehydrogenase, arsenate can replace phosphate in attacking the energy-rich thioester intermediate. The product of this reaction, 1-arseno-3-phosphoglycerate, is unstable, in contrast with 1,3-bisphosphoglycerate. 1-Arseno-3-phosphoglycerate and other acyl arsenates are very rapidly and spontaneously hydrolyzed. Hence, the net reaction in the presence of arsenate is

$$\text{Glyceraldehyde 3-phosphate} + NAD^+ + H_2O \longrightarrow \text{3-phosphoglycerate} + NADH + 2\ H^+$$

Note that glycolysis proceeds in the presence of arsenate but that the ATP normally formed in the conversion of 1,3-bisphosphoglycerate into 3-phosphoglycerate is lost. Thus, *arsenate uncouples oxidation and phosphorylation by forming a highly labile acyl arsenate.* One likely reason for the choice of phosphorus over arsenic in the evolution of biomolecules is the greater kinetic stability of its energy-rich compounds.

ENOL PHOSPHATES HAVE A HIGH GROUP-TRANSFER POTENTIAL

Because it is an *acyl phosphate,* 1,3-BPG has a high group-transfer potential. A different kind of high-energy phosphate compound is formed several steps later in glycolysis. Phosphoenolpyruvate, an *enol phosphate,* is formed by the dehydration of 2-phosphoglycerate. The $\Delta G^{\circ\prime}$ of hydrolysis of a phosphate ester of an ordinary alcohol is -3 kcal/mol, whereas that of phosphoenolpyruvate is -14.8 kcal/mol. Why does phosphoenolpyruvate have such a high phosphate group-transfer potential? The answer is that the reaction does not stop with the enol formed upon transfer of the phosphoryl group. The enol undergoes a conversion into a ketone—namely, pyruvate.

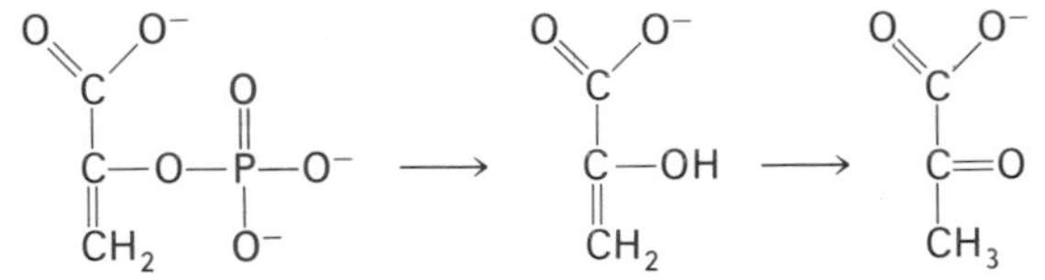

$$\text{Phosphoenolpyruvate} \longrightarrow \text{Enolpyruvate} \longrightarrow \text{Pyruvate}$$

Phosphoenolpyruvate **Enolpyruvate** **Pyruvate**

The $\Delta G^{\circ\prime}$ of the enol $\rightarrow$ ketone conversion is very large, of the order of -10 kcal/mol. Thus, the *high phosphate group-transfer potential of phosphoenolpyruvate arises primarily from the large driving force of the subsequent enol $\rightarrow$ ketone conversion.*

Table 15-3
Typical concentrations of glycolytic intermediates in erythrocytes

Intermediate	μM
Glucose	5000
Glucose 6-phosphate	83
Fructose 6-phosphate	14
Fructose 1,6-bisphosphate	31
Dihydroxyacetone phosphate	138
Glyceraldehyde 3-phosphate	19
1,3-Bisphosphoglycerate	1
2,3-Bisphosphoglycerate	4000
3-Phosphoglycerate	118
2-Phosphoglycerate	30
Phosphoenolpyruvate	23
Pyruvate	51
Lactate	2900
ATP	1850
ADP	138
P_i	1000

After S. Minakami and H. Yoshikawa, *Biochem. Biophys. Res. Comm.* 18(1965):345.

2,3-BISPHOSPHOGLYCERATE, AN ALLOSTERIC EFFECTOR OF HEMOGLOBIN, ARISES FROM 1,3-BISPHOSPHOGLYCERATE

We have seen that fructose 2,6-bisphosphate, a regulatory molecule, arises from a glycolytic intermediate. Another regulatory molecule coming from this pathway is 2,3-bisphosphoglycerate (2,3-BPG), a controller of oxygen transport in erythrocytes. It decreases the oxygen affinity of hemoglobin by stabilizing the deoxygenated form of hemoglobin (p. 161). Red blood cells have a high concentration of 2,3-BPG, typically 4 mM, in contrast with most other cells, which have only trace amounts. The synthesis and degradation of 2,3-BPG are a detour from the glycolytic pathway (Figure 15-20). *Bisphosphoglycerate mutase* converts 1,3-BPG into 2,3-BPG, which is hydrolyzed to 3-phosphoglycerate by *2,3-bisphosphoglycerate phosphatase*. Both reactions are nearly irreversible. In the mutase reaction, 2,3-BPG is a potent competitive inhibitor of its own formation.

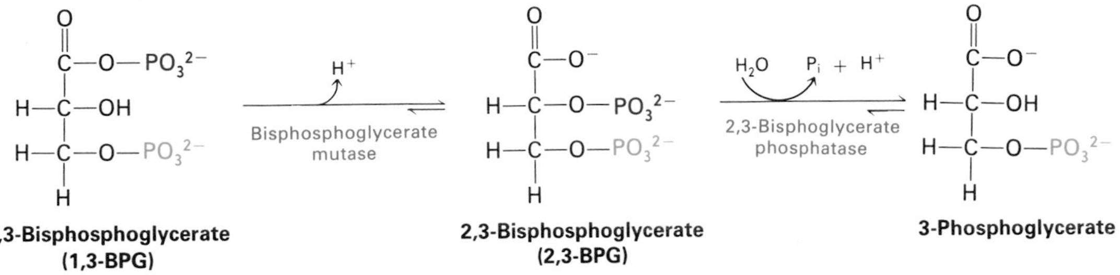

1,3-Bisphosphoglycerate (1,3-BPG) $\xrightarrow[\text{Bisphosphoglycerate mutase}]{H^+}$ 2,3-Bisphosphoglycerate (2,3-BPG) $\xrightarrow[\text{2,3-Bisphosphoglycerate phosphatase}]{H_2O \quad P_i + H^+}$ 3-Phosphoglycerate

Figure 15-20
Synthesis and degradation of 2,3-bisphosphoglycerate.

This mutase reaction has an interesting mechanism. *3-Phosphoglycerate* is an obligatory participant although it does not appear in the overall stoichiometry. The mutase binds 1,3-BPG and 3-phosphoglycerate simultaneously. In this ternary complex, a phosphoryl group is transferred from the 1-position of 1,3-BPG to the 2-position of 3-phosphoglycerate (Figure 15-21).

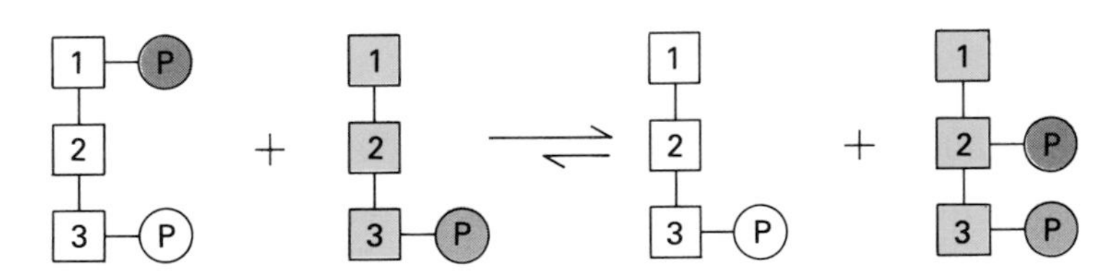

Figure 15-21
3-Phosphoglycerate participates in the conversion of 1,3-bisphosphoglycerate into 2,3-bisphosphoglycerate.

PHOSPHOGLYCERATES INTERCONVERT THROUGH AN ENZYME-BOUND 2,3-BPG INTERMEDIATE

Phosphoglyceromutase, the enzyme catalyzing the interconversion of 3-phosphoglycerate and 2-phosphoglycerate, requires a catalytic amount of 2,3-BPG to be active. 2,3-BPG donates either its 2- or 3-phosphoryl group to a histidine residue at the active site to generate the active form of the enzyme (E-P). The isomerization reaction begins with the binding of 3-phosphoglycerate to the phosphorylated enzyme

(Figure 15-22). Phosphohistidine transfers its phosphoryl group to the 2-OH of the bound substrate to form 2,3-BPG. The 3-phosphoryl group is then transferred to the histidine to generate 2-phosphoglycerate.

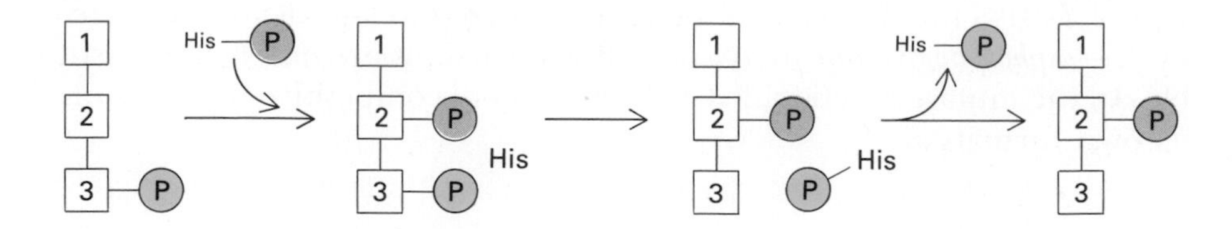

Figure 15-22
A phosphorylated histidine residue in the active site of phosphoglyceromutase plays a key role in the interconversion of 3-phosphoglycerate and 2-phosphoglycerate.

We see here a role of 2,3-bisphosphoglycerate in basic metabolism. It seems likely that 2,3-BPG was present eons before it was recruited by red cells for the purpose of controlling the affinity of hemoglobin for oxygen.

SUMMARY

Glycolysis is the set of reactions that converts glucose into pyruvate. In aerobic organisms, glycolysis is the prelude to the citric acid cycle and the electron-transport chain, where most of the free energy in glucose is harvested. The ten reactions of glycolysis occur in the cytosol. In the first stage, glucose is converted into fructose 1,6-bisphosphate by a phosphorylation, an isomerization, and a second phosphorylation reaction. Two molecules of ATP are consumed per molecule of glucose in these reactions, which are the prelude to the net synthesis of ATP. In the second stage, fructose 1,6-bisphosphate is cleaved by aldolase into dihydroxyacetone phosphate and glyceraldehyde 3-phosphate, which are readily interconvertible. Glyceraldehyde 3-phosphate is then oxidized and phosphorylated to form 1,3-BPG, an acyl phosphate with a high phosphate-transfer potential. 3-Phosphoglycerate is then formed as an ATP is generated. In the last stage of glycolysis, phosphoenolpyruvate, a second intermediate with a high phosphate-transfer potential, is formed by a phosphoryl shift and a dehydration. Another ATP is generated as phosphoenolpyruvate is converted into pyruvate. There is a net gain of two molecules of ATP in the formation of two molecules of pyruvate from one molecule of glucose.

The electron acceptor in the oxidation of glyceraldehyde 3-phosphate is NAD^+, which must be regenerated for glycolysis to proceed. In aerobic organisms, the NADH formed in glycolysis transfers its electrons to O_2 through the electron-transport chain, which thereby regenerates NAD^+. Under anaerobic conditions, NAD^+ is regenerated by the reduction of pyruvate to lactate. In some microorganisms, NAD^+ is normally regenerated by the synthesis of lactate or ethanol from pyruvate. These two processes are of fermentations.

The glycolytic pathway has a dual role: it degrades glucose to generate ATP and it provides building blocks for the synthesis of cellular components. The rate of conversion of glucose into pyruvate is regulated to meet these two major cellular needs. Under physiological conditions, the reactions of glycolysis are readily reversible except for the ones catalyzed by hexokinase, phosphofructokinase, and pyruvate kinase. Phosphofructokinase, the most important control element in glycolysis, is inhibited by high levels of ATP and citrate, and it is activated by AMP and fructose 2,6-bisphosphate. In liver, this bisphosphate signals that glucose is abundant. Hence, phosphofructokinase is active when the need arises for either energy or building blocks. Hexokinase is inhibited by glucose 6-phosphate, which accumulates when phosphofructokinase is inactive. Pyruvate kinase, the other control site, is allosterically inhibited by ATP and alanine, and it is activated by fructose 1,6-bisphosphate. Hence, pyruvate kinase is maximally active when the energy charge is low and glycolytic intermediates accumulate. Pyruvate kinase, like the tandem enzyme controlling the level of fructose 2,6-bisphosphate, is regulated by reversible phosphorylation. A low level of glucose in the blood promotes the phosphorylation of liver pyruvate kinase, which diminishes its activity and so decreases glucose consumption in the liver.

SELECTED READINGS

Where to start

Boiteux, A., and Hess, B., 1981. Design of glycolysis. *Phil Trans. R. Soc. Lond. B* 293:5–22. [A stimulating presentation of the regulation of glycolysis. Published in a symposium volume entitled *The Enzymes of Glycolysis: Structure, Activity, and Evolution,* which brings together x-ray crystallography, enzymology, and evolutionary biology.]

Sols, A., 1981. Multimodulation of enzyme activity. *Curr. Top. Cell. Regul.* 19:77–101. [An interesting discussion of allosteric regulation of metabolism, with emphasis on the control of glycolysis.]

Fothergill-Gilmore, L. A., 1986. The evolution of the glycolytic pathway. *Trends Biochem. Sci.* 11:47–51.

Books

Ochs, R. S., Hanson, R. W., and Hall, J., 1985. *Metabolic Regulation.* Elsevier. [A collection of essays originally published in *Trends in Biochemical Sciences* (*TIBS*), a monthly journal that is highly readable and informative. *TIBS* reports many interesting recent developments in metabolism. I also savor the "Textbook Errors" section of *TIBS*.]

Fersht, A., 1985. *Enzyme Structure and Mechanism* (2nd ed.). W. H. Freeman. [Contains succinct accounts of the catalytic and regulatory mechanisms of numerous enzymes participating in metabolism.]

Newsholme, E. A., and Start, C., 1973. *Regulation in Metabolism.* Wiley. [Chapters 3 and 6 provide excellent accounts of the control of carbohydrate metabolism.]

Structure and enzymatic mechanism

Anderson, C. M., Zucker, F. H., and Steitz, T. A., 1979. Space-filling models of kinase clefts and conformation changes. *Science* 204:375–380. [A well-illustrated and lucid article showing that kinases generally have a deep cleft that closes on binding substrate.]

Rose, I. A., 1981. Chemistry of proton abstraction by glycolytic enzymes (aldolase, isomerases, and pyruvate kinase). *Phil. Trans. R. Soc. Lond. B* 293:131–144.

Knowles, J. R., and Albery, W. J., 1977. Perfection in enzyme catalysis: the energetics of triosephosphate isomerase. *Acc. Chem. Res.* 10:105–111.

Straus, D., Raines, R., Kawashima, E., Knowles, J. R., and Gilbert, W., 1985. Active site of triosephosphate isomerase: *in vitro* mutagenesis and characterization of an altered enzyme. *Proc. Nat. Acad. Sci. USA* 82:2272–2276.

Termonia, Y., and Ross, J., 1981. Oscillations and control features in glycolysis: analysis of resonance effects. *Proc. Nat. Acad. Sci.* 78:3563–3566.

Boyer, P. D., (ed.), 1972. *The Enzymes* (3rd ed.). Academic Press. [Volumes 5 through 9 contain authoritative reviews of each of the glycolytic enzymes. Alcohol dehydrogenase and lactate dehydrogenase are discussed in Volume 10.]

Regulation of glycolysis

Goldhammer, A. R., and Paradies, H. H., 1979. Phosphofructokinase: structure and function. *Curr. Top. Cell Regul.* 15:109–141.

Hers, H.-G., and Van Schaftingen, E., 1982. Fructose 2,6-bisphosphate two years after its discovery. *Biochem. J.* 206:1–12.

Van Schaftingen, E., and Hers, H.-G., 1986. Purification and properties of phosphofructokinase 2/fructose 2,6-bisphosphatase. *Eur. J. Biochem.* 159:359–365.

el-Maghrabi, M. R., Correia, J. J., Heil, P. J., Pate, T. M., Cobb, C. E., and Pilkis, S. J., 1986. Tissue distribution, immunoreactivity, and physical properties of 6-phosphofructo-2-kinase/fructose-2,6-bisphosphatase. *Proc. Nat. Acad. Sci.* 261:8793–8798.

Engstrom, L., 1978. The regulation of liver pyruvate kinase by phosphorylation-dephosphorylation. *Curr. Top. Cell Regul.* 13:29–52.

Ottaway, J. H., and Mowbray, J., 1977. The role of compartmentation in the control of glycolysis. *Curr. Top. Cell Regul.* 12:108–195.

Clark, M. G., and Lardy, H. A., 1975. Regulation of intermediary carbohydrate metabolism. *Intern. Rev. Sci. (Biochem).* 5:223–266.

Genetic diseases

Stanbury, J. B., Wyngaarden, J. B., Fredrickson, D. S., Goldstein, J. L., and Brown, M. S., (eds.), 1983. *The Metabolic Basis of Inherited Disease* (5th ed.). McGraw-Hill. [Chapters 5 and 7 deal with disorders of fructose and galactose metabolism. Deficiencies of glycolytic enzymes in erythrocytes are discussed in Chapter 73.]

Evolution of glycolytic enzymes

Muirhead, H., 1983. Triose phosphate isomerase, pyruvate kinase, and other α/β-barrel enzymes. *Trends Biochem. Sci.* 8:326–330.

Marchionni, M., and Gilbert, W., 1986. The triosephosphate isomerase gene from maize: introns antedate the plant-animal divergence. *Cell* 46:133–141.

Historical aspects

Kalckar, H. M., (ed.)., 1969. *Biological Phosphorylations: Development of Concepts*. Prentice-Hall. [Contains many of the classical papers on glycolysis.]

Fruton, J. S., 1972. *Molecules and Life: Historical Essays on the Interplay of Chemistry and Biology*. Wiley-Interscience. [Includes a meticulously documented account of the elucidation of the nature of fermentation and how it led to enzyme chemistry.]

PROBLEMS

1. Sucrose is commonly used to preserve fruits. Why is glucose not well suited for preserving foods?

2. Glucose labeled with ^{14}C at C-1 is incubated with the glycolytic enzymes and necessary cofactors.
   (a) What is the distribution of ^{14}C in the pyruvate that is formed? (Assume that the interconversion of glyceraldehyde 3-phosphate and dihydroxyacetone phosphate is very rapid compared with the subsequent step.)
   (b) If the specific activity of the glucose substrate is 10 mCi/mM, what is the specific activity of the pyruvate that is formed?

3. Write a balanced equation for the conversion of glucose into lactate.
   (a) Calculate the standard free-energy change of this reaction using the data given in Table 15-2 (p. 357) and the fact that $\Delta G^{\circ\prime}$ is -6 kcal for the reaction

   $$\text{Pyruvate} + \text{NADH} + \text{H}^+ \rightleftharpoons \text{lactate} + \text{NAD}^+$$

   (b) What is the free-energy change ($\Delta G'$, not $\Delta G^{\circ\prime}$) of this reaction when the concentrations of reactants are: glucose, 5 mM; lactate, 0.05 mM; ATP, 2 mM; ADP, 0.2 mM; and P_i, 1 mM?

4. What is the equilibrium ratio of phosphoenolpyruvate to pyruvate under standard conditions when [ATP]/[ADP] = 10?

5. What are the equilibrium concentrations of fructose 1,6-bisphosphate, dihydroxyacetone phosphate, and glyceraldehyde 3-phosphate when 1 mM fructose 1,6-bisphosphate is incubated with aldolase under standard conditions?

6. 3-Phosphoglycerate labeled uniformly with ^{14}C is incubated with 1,3-BPG labeled with ^{32}P at C-1. What is the radioisotope distribution of the 2,3-BPG that is formed on addition of BPG mutase?

7. Xylose has the same structure as glucose except that it has a hydrogen atom at C-5 in place of a hydroxymethyl group. The rate of ATP hydrolysis by hexokinase is markedly enhanced by the addition of xylose. Why?

8. Oxygen transport can be affected in genetic disorders of glycolysis in red cells.
   (a) How are glycolysis and oxygen transport linked?
   (b) How is oxygen affinity altered by a deficiency of hexokinase?
   (c) How is oxygen affinity altered by a deficiency of pyruvate kinase?

9. The intravenous infusion of fructose into healthy volunteers leads to a two- to fivefold increase in the level of lactate in the blood, a far greater increase than that observed following the infusion of the same amount of glucose.
   (a) Why is glycolysis more rapid following the infusion of fructose?
   (b) Fructose has been used in place of glucose for intravenous feeding. Why is this use of fructose unwise?

10. A catalytic amount of glucose 1,6-bisphosphate is required for the interconversion of glucose 1-phosphate and glucose 6-phosphate by phosphoglucomutase. Propose a catalytic mechanism to account for this observation.

11. Aldolases in procaryotes contain a tightly bound divalent metal ion that is essential for catalysis. Propose a catalytic function for this metal ion.

12. Procaryotic aldolases are inactivated by ethylenediaminetetraacetate (EDTA), a chelator of divalent metal ions, whereas animal aldolases discussed in this chapter are inactivated by sodium borohydride. Account for this difference.

CHAPTER 16

Citric Acid Cycle

The preceding chapter considered the glycolytic pathway, in which glucose is converted into pyruvate. Under aerobic conditions, the next step in the generation of energy from glucose is the oxidative decarboxylation of pyruvate to form acetyl coenzyme A (acetyl CoA). This activated acetyl unit is then completely oxidized to CO_2 by the *citric acid cycle,* a series of reactions that is also known as the *tricarboxylic acid cycle* or the *Krebs cycle.* The citric acid cycle is the *final common pathway for the oxidation of fuel molecules*—amino acids, fatty acids, and carbohydrates. Most fuel molecules enter the cycle as acetyl CoA. The cycle also provides intermediates for biosyntheses. In eucaryotes, the reactions of the citric acid cycle occur inside mitochondria, in contrast with those of glycolysis, which occur in the cytosol.

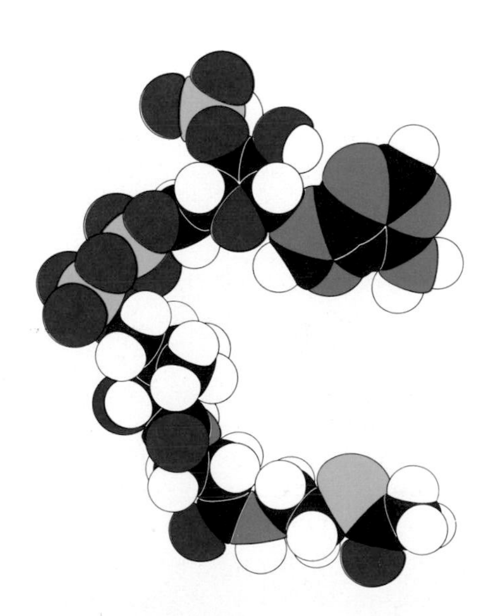

Figure 16-1
Model of acetyl CoA.

FORMATION OF ACETYL COENZYME A FROM PYRUVATE

The oxidative decarboxylation of pyruvate to form acetyl CoA, which occurs in the mitochondrial matrix, is the link between glycolysis and the citric acid cycle:

$$\text{Pyruvate} + \text{CoA} + \text{NAD}^+ \longrightarrow \text{acetyl CoA} + \text{CO}_2 + \text{NADH}$$

This irreversible funneling of the product of glycolysis into the citric acid cycle is catalyzed by the *pyruvate dehydrogenase complex.* This very large multienzyme complex is a highly integrated array of three kinds of enzymes, which will be discussed in detail later in this chapter (p. 379).

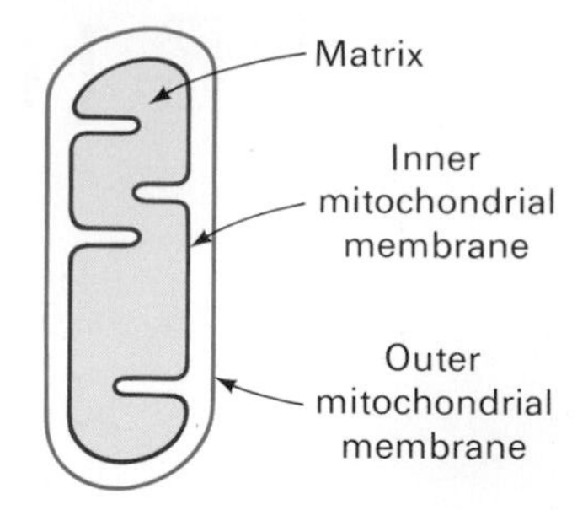

Figure 16-2
Schematic diagram of a mitochondrion. The oxidative decarboxylation of pyruvate and the sequence of reactions in the citric acid cycle occur within the mitochondrial matrix.

AN OVERVIEW OF THE CITRIC ACID CYCLE

The overall pattern of the citric acid cycle is shown in Figure 16-3. A four-carbon compound (oxaloacetate) condenses with a two-carbon acetyl unit to yield a six-carbon tricarboxylic acid (citrate). An isomer of citrate is then oxidatively decarboxylated. The resulting five-carbon compound (α-ketoglutarate) is oxidatively decarboxylated to yield a four-carbon compound (succinate). Oxaloacetate is then regenerated from succinate. Two carbon atoms enter the cycle as an acetyl unit and two carbon atoms leave the cycle in the form of two molecules of CO_2. An acetyl group is more reduced than CO_2, and so oxidation-reduction reactions must take place in the citric acid cycle. In fact, there are four such reactions. Three hydride ions (hence, six electrons) are transferred to three NAD^+ molecules, whereas one pair of hydrogen atoms (hence, two electrons) is transferred to a flavin adenine dinucleotide (FAD) molecule. These electron carriers yield eleven molecules of adenosine triphosphate (ATP) when they are oxidized by O_2 in the electron-transport chain. In addition, one high-energy phosphate bond is formed in each round of the citric acid cycle itself.

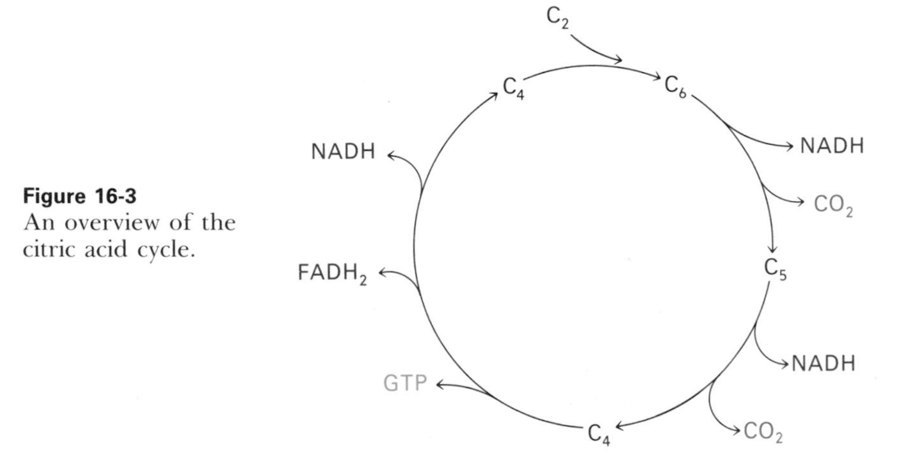

Figure 16-3
An overview of the citric acid cycle.

OXALOACETATE CONDENSES WITH ACETYL COENZYME A TO FORM CITRATE

The cycle starts with the joining of a four-carbon unit, oxaloacetate, and a two-carbon unit, the acetyl group of acetyl CoA. Oxaloacetate reacts with acetyl CoA and H_2O to yield citrate and CoA.

$$\underset{\textbf{Oxaloacetate}}{\mathrm{O{=}C{-}COO^-}\atop \mathrm{H_2C{-}COO^-}} + \underset{\textbf{Acetyl CoA}}{\mathrm{O{=}C{-}CH_3}\atop \mathrm{S{-}CoA}} + H_2O \longrightarrow \underset{\textbf{Citrate}}{\mathrm{HO{-}C(CH_2{-}COO^-)_2{-}COO^-}} + HS{-}CoA + H^+$$

This reaction, which is an aldol condensation followed by a hydrolysis, is catalyzed by *citrate synthase* (originally called condensing enzyme). Oxaloacetate first condenses with acetyl CoA to form *citryl CoA*, which is then hydrolyzed to citrate and CoA. Hydrolysis of citryl CoA pulls the overall reaction far in the direction of the synthesis of citrate.

$$\begin{array}{l} \quad\; O \\ \quad\; \| \\ CH_2\text{—}C\text{—}S\text{—}CoA \\ | \\ HO\text{—}C\text{—}COO^- \\ | \\ CH_2\text{—}COO^- \end{array}$$

Citryl CoA

CITRATE IS ISOMERIZED INTO ISOCITRATE

Citrate must be isomerized into isocitrate to enable the six-carbon unit to undergo oxidative decarboxylation. The isomerization of citrate is accomplished by a *dehydration* step followed by a *hydration* step. The result is an interchange of an H and an OH. The enzyme catalyzing both steps is called *aconitase* because cis-*aconitate* is an intermediate.

$$\begin{array}{c} COO^- \\ | \\ H\text{—}C\text{—}H \\ | \\ {}^-OOC\text{—}C\text{—}OH \\ | \\ CH_2 \\ | \\ COO^- \end{array} \overset{H_2O}{\rightleftharpoons} \begin{array}{c} COO^- \\ | \\ H\text{—}C \\ \| \\ {}^-OOC\text{—}C \\ | \\ CH_2 \\ | \\ COO^- \end{array} \overset{H_2O}{\rightleftharpoons} \begin{array}{c} COO^- \\ | \\ H\text{—}C\text{—}OH \\ | \\ {}^-OOC\text{—}C\text{—}H \\ | \\ CH_2 \\ | \\ COO^- \end{array}$$

Citrate — ***cis*-Aconitate** — **Isocitrate**

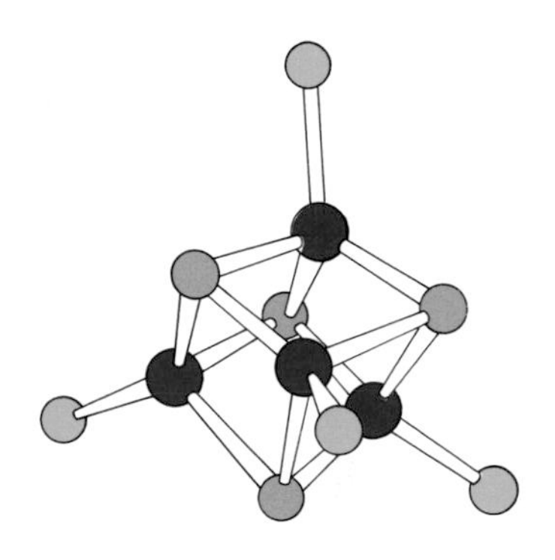

Figure 16-4
An iron-sulfur cluster in aconitase. Each of the four iron atoms (red) in this 4Fe-4S cubic array is bonded to three inorganic sulfides (green) and a cysteine sulfur atom (yellow). [After H. Beinert and A. J. Thomson. *Arch. Biochem. Biophys.* 222(1983):358.]

Aconitase contains iron atoms that are not bonded to a heme group. Rather, its four iron atoms are complexed to four inorganic sulfides and four cysteine sulfur atoms (Figure 16-4). This Fe-S cluster binds citrate and participates in dehydrating and rehydrating the bound substrate. Proteins with Fe-S clusters are known as *iron-sulfur proteins* or *nonheme iron proteins*.

ISOCITRATE IS OXIDIZED AND DECARBOXYLATED TO α-KETOGLUTARATE

We come now to the first of four oxidation-reduction reactions in the citric acid cycle. The oxidative decarboxylation of isocitrate is catalyzed by *isocitrate dehydrogenase:*

$$\text{Isocitrate} + NAD^+ \rightleftharpoons \alpha\text{-ketoglutarate} + CO_2 + NADH$$

The intermediate in this reaction is oxalosuccinate, an unstable β-keto acid. While bound to the enzyme, it loses CO_2 to form α-ketoglutarate. The rate of formation of α-ketoglutarate is important in determining the overall rate of the cycle, as will be discussed later (p. 390).

$$\begin{array}{c} COO^- \\ | \\ CH_2 \\ | \\ H\text{—}C\text{—}COO^- \\ | \\ H\text{—}C\text{—}OH \\ | \\ COO^- \end{array} \xrightarrow{NAD^+ \;\; NADH + H^+} \begin{array}{c} COO^- \\ | \\ CH_2 \\ | \\ H\text{—}C\text{—}COO^- \\ | \\ C\text{=}O \\ | \\ COO^- \end{array} \xrightarrow{H^+ \;\; CO_2} \begin{array}{c} COO^- \\ | \\ CH_2 \\ | \\ H\text{—}C\text{—}H \\ | \\ C\text{=}O \\ | \\ COO^- \end{array}$$

Isocitrate — **Oxalosuccinate** — **α-Ketoglutarate**

SUCCINYL COENZYME A IS FORMED BY THE OXIDATIVE DECARBOXYLATION OF α-KETOGLUTARATE

The conversion of isocitrate into α-ketoglutarate is followed by a second oxidative decarboxylation reaction, the formation of succinyl CoA from α-ketoglutarate:

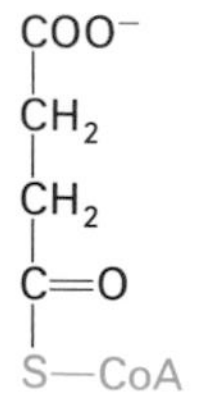

$$\alpha\text{-Ketoglutarate} + \text{NAD}^+ + \text{CoA} \longrightarrow \text{succinyl CoA} + \text{CO}_2 + \text{NADH}$$

This reaction is catalyzed by the *α-ketoglutarate dehydrogenase complex,* an organized assembly consisting of three kinds of enzymes. *The mechanism of this reaction is very similar to that of the conversion of pyruvate into acetyl CoA* (p. 379).

COO^-
CH_2
CH_2
$C{=}O$
$S{-}CoA$

Succinyl CoA

A HIGH-ENERGY PHOSPHATE BOND IS GENERATED FROM SUCCINYL COENZYME A

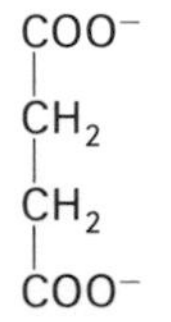

COO^-
CH_2
CH_2
COO^-

Succinate

The succinyl thioester of CoA has an energy-rich bond. The $\Delta G^{\circ\prime}$ for hydrolysis of succinyl CoA is about -8 kcal/mol, which is comparable to that of ATP (-7.3 kcal/mol). *The cleavage of the thioester bond of succinyl CoA is coupled to the phosphorylation of guanosine diphosphate* (GDP):

$$\text{Succinyl CoA} + \text{P}_i + \text{GDP} \rightleftharpoons \text{succinate} + \text{GTP} + \text{CoA}$$

This readily reversible reaction ($\Delta G^{\circ\prime} = -0.8$ kcal/mol) is catalyzed by *succinyl CoA synthetase.* It is the only step in the citric acid cycle that directly yields a high-energy phosphate bond. GTP itself is used as a phosphoryl donor in protein synthesis (p. 754) and signal-transduction processes (p. 979). Alternatively, its γ phosphate group can readily be transferred to adenosine diphosphate (ADP) to form ATP, in a reaction catalyzed by *nucleoside diphosphokinase.*

$$\text{GTP} + \text{ADP} \rightleftharpoons \text{GDP} + \text{ATP}$$

REGENERATION OF OXALOACETATE BY OXIDATION OF SUCCINATE

Reactions of four-carbon compounds constitute the final stage of the citric acid cycle (Figure 16-5). Succinate is converted into oxaloacetate in three steps: an oxidation, a hydration, and a second oxidation reaction. Oxaloacetate is thereby regenerated for another round of the cycle, as energy is trapped in the form of $FADH_2$ and NADH.

Succinate is oxidized to fumarate by *succinate dehydrogenase.* The hydrogen acceptor is FAD rather than NAD^+, which is used in the other

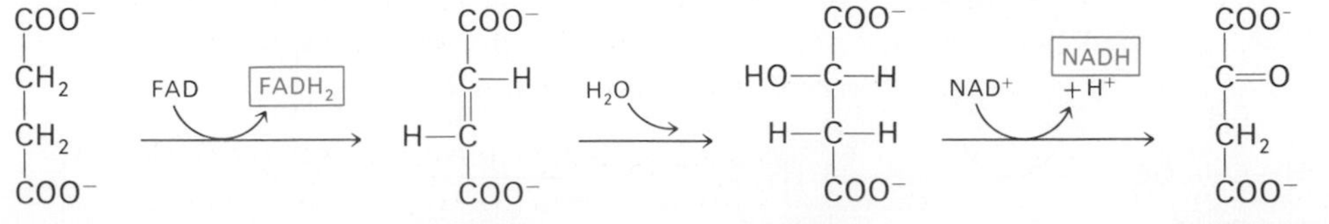

Figure 16-5
Final stage of the citric acid cycle: from succinate to oxaloacetate.

three oxidation reactions in the cycle. FAD is the hydrogen acceptor in this reaction because the free-energy change is insufficient to reduce NAD^+. FAD is nearly always the electron acceptor in oxidations that remove two hydrogen atoms from a substrate. In succinate dehydrogenase, the isoalloxazine ring of FAD is covalently attached to a histidine side chain of the enzyme (denoted E-FAD).

$$\text{E-FAD} + \text{succinate} \rightleftharpoons \text{E-FADH}_2 + \text{fumarate}$$

$$\text{FAD} + \text{R—CH}_2\text{—CH}_2\text{—R}' \rightleftharpoons \text{FADH}_2 + \text{R—CH=CH—R}'$$

Succinate dehydrogenase, like aconitase, is an *iron-sulfur protein* (also called a *nonheme iron protein*). Indeed, succinate dehydrogenase contains three different kinds of iron-sulfur clusters, 2Fe-2S (two irons bonded to two inorganic sulfides), 3Fe-4S, and 4Fe-4S. We shall consider the role of these iron-sulfur clusters in the electron transfer reactions of oxidative phosphorylation (p. 403) and photosynthesis (p. 525). Succinate dehydrogenase, consisting of a 70-kd and a 27-kd subunit, differs from other enzymes in the citric acid cycle in being an integral part of the inner mitochondrial membrane. In fact, *succinate dehydrogenase is directly linked to the electron-transport chain.* The $FADH_2$ produced by the oxidation of succinate does not dissociate from the enzyme, in contrast with NADH produced in other oxidation reactions. Rather, two electrons from $FADH_2$ are transferred directly to FeS clusters of the enzyme. The ultimate acceptor of these electrons is molecular oxygen, as will be discussed in the next chapter.

The next step in the cycle is the hydration of fumarate to form L-malate. *Fumarase* catalyzes a stereospecific *trans* addition of H and OH, as shown by deuterium-labeling studies. The OH group adds to only one side of the double bond of fumarate; hence, only the L-isomer of malate is formed.

Fumarate

D_2O

L-Malate

Finally, malate is oxidized to form oxaloacetate. This reaction is catalyzed by *malate dehydrogenase*, and NAD^+ is again the hydrogen acceptor.

$$\text{Malate} + \text{NAD}^+ \rightleftharpoons \text{oxaloacetate} + \text{NADH} + \text{H}^+$$

STOICHIOMETRY OF THE CITRIC ACID CYCLE

The net reaction of the citric acid cycle is

$$\text{Acetyl CoA} + 3\ \text{NAD}^+ + \text{FAD} + \text{GDP} + \text{P}_i + 2\ \text{H}_2\text{O} \longrightarrow 2\ \text{CO}_2 + 3\ \text{NADH} + \text{FADH}_2 + \text{GTP} + 2\ \text{H}^+ + \text{CoA}$$

The reactions that give this stoichiometry (Figure 16-6 and Table 16-1) are recapitulated on the next two pages.

1. Two carbon atoms enter the cycle in the condensation of an acetyl unit (from acetyl CoA) with oxaloacetate. Two carbon atoms leave the cycle in the form of CO_2 in the successive decarboxylations catalyzed by isocitrate dehydrogenase and α-ketoglutarate dehydrogenase. As will be discussed shortly, the two carbon atoms that leave the cycle are different from the ones that entered in that round.

2. Four pairs of hydrogen atoms leave the cycle in the four oxidation reactions. Two NAD^+ molecules are reduced in the oxidative decarboxylations of isocitrate and α-ketoglutarate, one FAD molecule is reduced in the oxidation of succinate, and one NAD^+ molecule is reduced in the oxidation of malate.

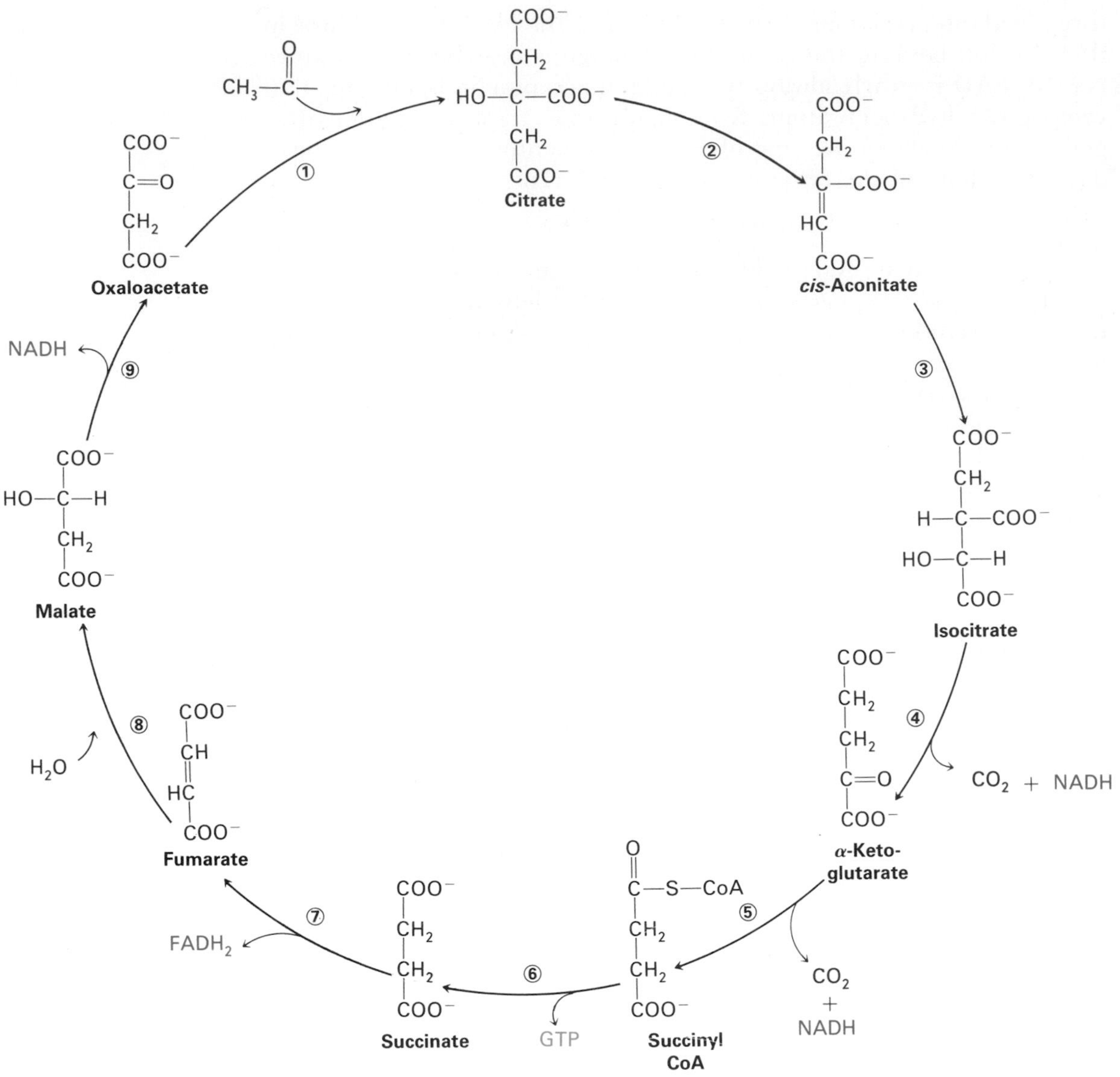

Figure 16-6
Citric acid cycle.

Table 16-1
Citric acid cycle

Step	*Reaction*	*Enzyme*	*Prosthetic group*	*Type**	$\Delta G^{\circ\prime}$
1	Acetyl CoA + oxaloacetate + H_2O ⟶ citrate + CoA + H^+	Citrate synthetase		a	−7.5
2	Citrate ⇌ *cis*-aconitate + H_2O	Aconitase	Fe-S	b	+2.0
3	*cis*-Aconitate + H_2O ⇌ isocitrate	Aconitase	Fe-S	c	−0.5
4	Isocitrate + NAD^+ ⇌ α-ketoglutarate + CO_2 + NADH	Isocitrate dehydrogenase	Lipoic acid FAD	d + e	−2.0
5	α-Ketoglutarate + NAD^+ + CoA ⇌ succinyl CoA + CO_2 + NADH	α-Ketoglutarate dehydrogenase complex	TPP	d + e	−7.2
6	Succinyl CoA + P_i + GDP ⇌ succinate + GTP + CoA	Succinyl CoA synthetase		f	−0.8
7	Succinate + FAD (enzyme-bound) ⇌ fumarate + $FADH_2$ (enzyme-bound)	Succinate dehydrogenase	FAD Fe-S	e	~0
8	Fumarate + H_2O ⇌ L-malate	Fumarase		c	−0.9
9	L-Malate + NAD^+ ⇌ oxaloacetate + NADH + H^+	Malate dehydrogenase		e	+7.1

*Reaction type: (a) Condensation (b) Dehydration (c) Hydration (d) Decarboxylation (e) Oxidation (f) Substrate-level phosphorylation

3. One high-energy phosphate bond (in the form of GTP) is generated from the energy-rich thioester linkage in succinyl CoA.

4. Two water molecules are consumed: one in the synthesis of citrate by the hydrolysis of citryl CoA, the other in the hydration of fumarate.

If we look ahead, the NADH and $FADH_2$ formed in the citric acid cycle are oxidized by the electron-transport chain (Chapter 17). ATP is generated as electrons are transferred from these carriers to O_2, the ultimate acceptor. Three ATP are formed for each NADH in the mitochondrion, whereas two ATP are generated per $FADH_2$. Note that only one high-energy phosphate bond per acetyl unit is directly formed in the citric acid cycle. Eleven more high-energy phosphate bonds are generated when three NADH and one $FADH_2$ are oxidized by the electron-transport chain.

Molecular oxygen does not participate directly in the citric acid cycle. However, the cycle operates only under aerobic conditions because NAD^+ and FAD can be regenerated in the mitochondrion only by the transfer of electrons to molecular oxygen. *Glycolysis has both an aerobic and an anaerobic mode, whereas the citric acid cycle is strictly aerobic.* Recall that glycolysis can proceed under anaerobic conditions because NAD^+ is regenerated in the conversion of pyruvate into lactate.

PYRUVATE DEHYDROGENASE COMPLEX: AN ORGANIZED ENZYME ASSEMBLY

We now turn to some reaction mechanisms. The oxidative decarboxylation of pyruvate in the formation of acetyl CoA is catalyzed by the *pyruvate dehydrogenase complex,* an organized assembly of three kinds of enzymes (Table 16-2). The net reaction catalyzed by this complex is

$$\text{Pyruvate} + \text{CoA} + \text{NAD}^+ \longrightarrow \text{acetyl CoA} + \text{CO}_2 + \text{NADH}$$

The mechanism of this reaction is more complex than might be suggested by its stoichiometry. *Thiamine pyrophosphate* (TPP), *lipoamide,* and *FAD* serve as catalytic cofactors, in addition to CoA and NAD^+, the stoichiometric cofactors.

Table 16-2
Pyruvate dehydrogenase complex of *E. coli*

Enzyme	*Abbreviation*	*Number of chains*	*Prosthetic group*	*Reaction catalyzed*
Pyruvate dehydrogenase component	E_1	24	TPP	Oxidative decarboxylation of pyruvate
Dihydrolipoyl transacetylase	E_2	24	Lipoamide	Transfer of the acetyl group to CoA
Dihydrolipoyl dehydrogenase	E_3	12	FAD	Regeneration of the oxidized form of lipoamide

There are four steps in the conversion of pyruvate into acetyl CoA. First, pyruvate is *decarboxylated* after it combines with TPP. This reaction is catalyzed by the *pyruvate dehydrogenase component* of the multienzyme complex.

$$\text{Pyruvate} + \text{TPP} \longrightarrow \text{hydroxyethyl-TPP} + \text{CO}_2$$

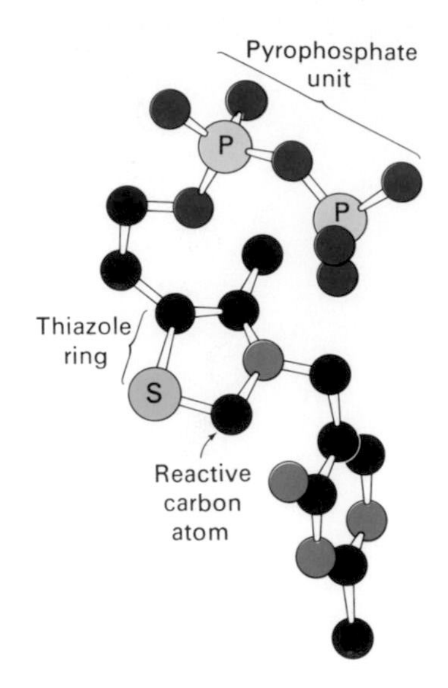

Figure 16-7
Model of thiamine pyrophosphate.

A key feature of TPP, the prosthetic group of the pyruvate dehydrogenase component, is that the carbon atom between the nitrogen and sulfur atoms in the thiazole ring is much more acidic than most =CH— groups.

Dissociable proton

Reactive carbon atom

NH_2 H C—S N C $C—CH_2—N^+$ O⁻ O⁻ $H_3C—C$ C N $C{=}C—CH_2—CH_2—O—P—O—P—O^-$ CH_3 O O

Thiamine pyrophosphate (TPP)

It ionizes to form a *carbanion*, which readily adds to the carbonyl group of pyruvate.

$H^+ + {}^{\delta-}O{=}C^{\delta+}$ (^-O, O, C, CH_3) $+$ $\ominus C$ ($N^+—C—CH_3$, R, S—C, R′) $\longrightarrow$ $HO—C—C$ (^-O, O, C, CH_3, $N^+—C—CH_3$, R, S—C, R′)

Pyruvate **Carbanion of TPP** **Addition compound**

The positively charged ring nitrogen of TPP then acts as an electron sink to stabilize the formation of a negative charge, which is necessary for decarboxylation.

$HO—C—C$ (O^-, O, C, CH_3, R, $N^+—C—CH_3$, S—C, R′) $\xrightarrow{CO_2}$ $HO—C{=}C$ (R, $N—C—CH_3$, CH_3, S—C, R′) $\longleftrightarrow$ $HO—\overset{\ominus}{C}—C$ (R, $N^+—C—CH_3$, CH_3, S—C, R′)

Addition compound **Resonance forms of ionized hydroxyethyl-TPP**

Protonation then gives *hydroxyethyl thiamine pyrophosphate*.

Second, the hydroxyethyl group attached to TPP is *oxidized* to form an acetyl group and concomitantly *transferred to lipoamide*. The oxidant in this reaction is the disulfide group of lipoamide, which is converted into the sulfhydryl form. This reaction, also catalyzed by the pyruvate dehydrogenase component, yields *acetyllipoamide*.

OH $H_3C—C\ominus$ C S $N^+—R$ C═C R′ CH_3 $+$ S—S CH_2 C H R CH_2 $\longrightarrow$ $\ominus$ C S $N^+—R$ C═C R′ CH_3 $+$ CH_3 C═O HS S CH_2 C H R CH_2

Hydroxyethyl-TPP (Ionized form) **Lipoamide** **Carbanion of TPP** **Acetyllipoamide**

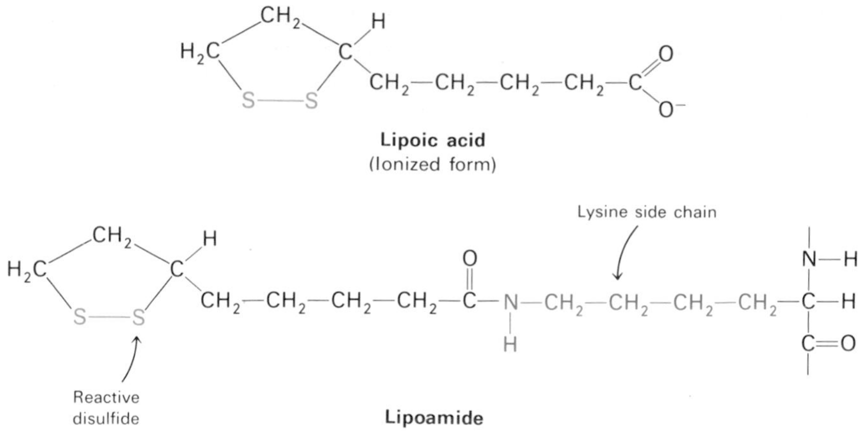

Lipoic acid (Ionized form): H_2C–CH_2–$CH(H)$ ring closed by S–S, with side chain $CH_2-CH_2-CH_2-CH_2-C(=O)O^-$

Lipoic acid
(Ionized form)

Lysine side chain

$CH_2-CH_2-CH_2-CH_2-C(=O)-N(H)-CH_2-CH_2-CH_2-CH_2-C(H)(N-H)(C=O)$

Reactive disulfide

Lipoamide

Figure 16-8
Structures of lipoic acid and lipoamide. Lipoic acid is covalently attached to a specific lysine side chain of dihydrolipoyl transacetylase. Note that this prosthetic group is at the end of a long, flexible chain, which enables it to rotate from one active site to another in the enzyme complex.

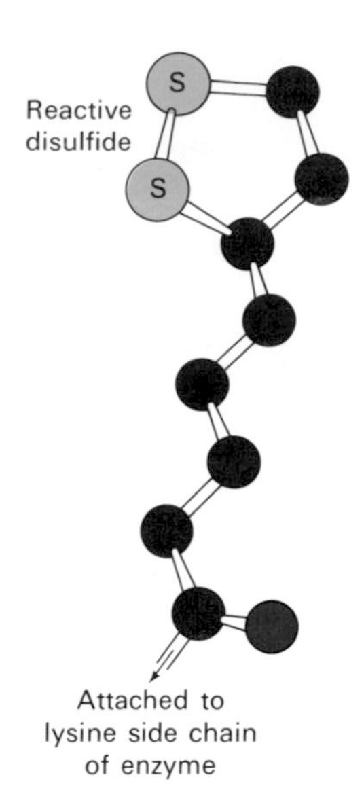

Figure 16-9
Model of the lipoyl part of lipoamide.

Third, *the acetyl group is transferred from acetyllipoamide to CoA to form acetyl CoA*. Dihydrolipoyl transacetylase catalyzes this reaction. The energy-rich thioester bond is preserved as the acetyl group is transferred to CoA.

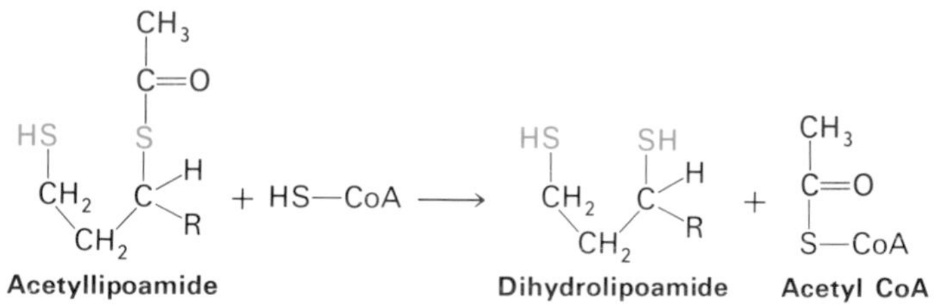

$$\text{Acetyllipoamide} + \text{HS–CoA} \longrightarrow \text{Dihydrolipoamide} + \text{Acetyl CoA}$$

Fourth, *the oxidized form of lipoamide is regenerated by dihydrolipoyl dehydrogenase*. A hydride ion is transferred to an FAD prosthetic group of the enzyme and then to NAD^+.

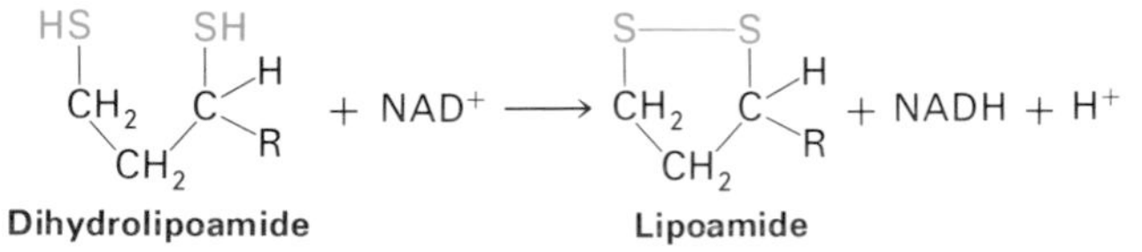

$$\text{Dihydrolipoamide} + NAD^+ \longrightarrow \text{Lipoamide} + NADH + H^+$$

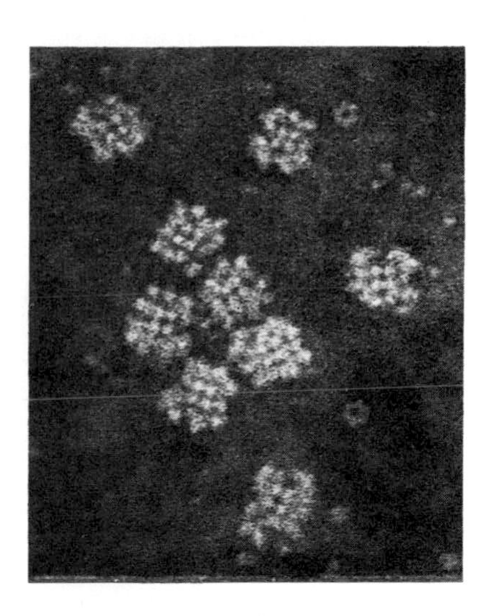

Figure 16-10
Electron micrograph of the pyruvate dehydrogenase complex from *E. coli*. [Courtesy of Dr. Lester Reed.]

The studies of Lester Reed have been sources of insight into the structure and assembly of the pyruvate dehydrogenase complex. The enzyme complex from *E. coli* has been studied intensively. It has a mass of about 4600 kd and consists of 60 polypeptide chains. A polyhedral structure with a diameter of about 300 Å is evident in electron micrographs (Figure 16-10). *The transacetylase polypeptide chains are the core of the pyruvate dehydrogenase complex*. The pyruvate dehydrogenase units and lipoyl dehydrogenase units are bound to the outside of this transacetylase core (Figure 16-11).

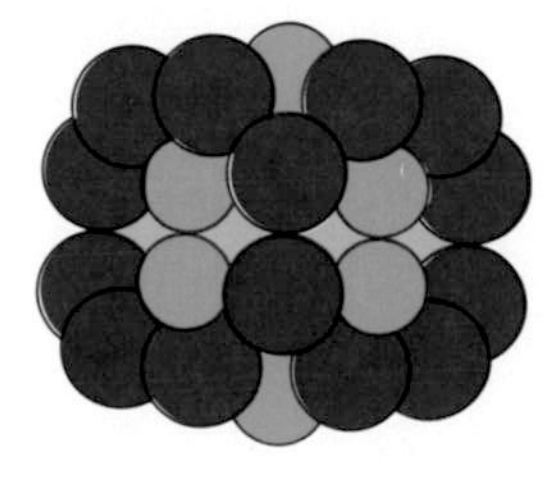

Figure 16-11
Model of the pyruvate dehydrogenase complex from *E. coli*. The transacetylase core (E2) is shown in yellow, the pyruvate dehydrogenase component (E1) in red, and dihydrolipoyl dehydrogenase (E3) in green. [After a drawing kindly provided by Dr. Lester Reed.]

The constituent polypeptide chains of the complex are held together by noncovalent forces. At alkaline pH, the complex dissociates into the pyruvate dehydrogenase component and a subcomplex of the other two enzymes. The transacetylase can then be separated from the dehydrogenase at neutral pH in the presence of urea. *These three enzymes spontaneously associate to form the pyruvate dehydrogenase complex* when they are mixed at neutral pH in the absence of urea, which suggests that the native enzyme complex may be formed by a self-assembly process.

The structural integration of three kinds of enzymes makes possible the coordinated catalysis of a complex reaction (Figure 16-12). All of the intermediates in the oxidative decarboxylation of pyruvate are tightly bound to the complex. The close proximity of one enzyme to another *increases the overall reaction rate* and *minimizes side reactions*. The activated intermediates are transferred from one active site to another by the lipoamide prosthetic group of the transacetylase. The attachment of the lipoyl group to the ϵ-amino group of a lysine residue on the transacetylase provides a flexible arm for the reactive ring. This *14-Å molecular string* enables the lipoyl moiety of a transacetylase subunit to interact with the thiamine pyrophosphate unit of an adjacent pyruvate dehydrogenase subunit and with the flavin unit of an adjacent dihydrolipoyl dehydrogenase. Furthermore, the lipoyl moieties of the multienzyme complex can interact with each other, forming an *interacting network of reactive groups*. The net charge on the lipoyl moiety during its cycle of transformation is 0, -1, or -2, if the sulfhydryl groups are fully ionized. This change in net charge may provide the driving force for the directed movement of the lipoyl group.

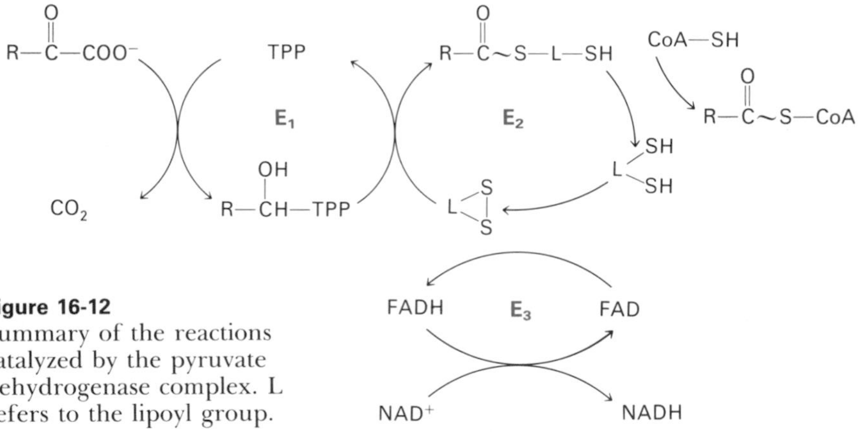

Figure 16-12
Summary of the reactions catalyzed by the pyruvate dehydrogenase complex. L refers to the lipoyl group.

VARIATION ON A MULTIENZYME THEME: THE α-KETOGLUTARATE DEHYDROGENASE COMPLEX

The oxidative decarboxylation of α-ketoglutarate closely resembles that of pyruvate, also an α-keto acid:

$$\alpha\text{-Ketoglutarate} + \text{CoA} + \text{NAD}^+ \longrightarrow \text{succinyl CoA} + \text{CO}_2 + \text{NADH}$$

$$\text{Pyruvate} + \text{CoA} + \text{NAD}^+ \longrightarrow \text{acetyl CoA} + \text{CO}_2 + \text{NADH}$$

The same cofactors are participants: TPP, lipoamide, CoA, FAD, and NAD^+. In fact, *the oxidative decarboxylation of α-ketoglutarate is catalyzed by an enzyme complex that is structurally similar to the pyruvate dehydrogenase complex*. There are three kinds of enzymes in the α-ketoglutarate dehydrogenase complex: an α-ketoglutarate dehydrogenase component (E_1'), a transsuccinylase one (E_2'), and a dihydrolipoyl dehydrogenase one (E_3'). Again, E_1' binds to E_2', and E_2' binds to E_3', but E_1' does not bind directly to E_3'. Thus, *transsuccinylase* (like transacetylase) *is the core of the complex.*

The α-ketoglutarate dehydrogenase component and transsuccinylase are different from the corresponding enzymes in the pyruvate dehydrogenase complex. However, *the dihydrolipoyl dehydrogenase parts* of the two complexes are identical.

It was noted earlier that chymotrypsin, trypsin, thrombin, and elastase are homologous enzymes. Here we see that the pyruvate and α-ketoglutarate dehydrogenase complexes are *homologous enzyme assemblies.* The structural and mechanistic motifs coordinate catalysis in the entrée to the citric acid cycle are used again later in the cycle.

BERIBERI IS CAUSED BY A DEFICIENCY OF THIAMINE

Beriberi, a neurologic and cardiovascular disorder, is caused by a dietary deficiency of thiamine (also called vitamin B_1). The disease has been and continues to be a serious health problem in the Far East because rice, the major food, has a rather low content of thiamine. The problem is exacerbated if the rice is polished, because only the outer layer contains appreciable amounts of thiamine. Beriberi is also occasionally seen in alcoholics who are severely malnourished. The disease is characterized by neurologic and cardiac symptoms. Damage to the peripheral nervous system is expressed in terms of pain in the limbs, weakness of the musculature, and distorted skin sensation. The heart may be enlarged and the cardiac output inadequate.

Which biochemical processes might be affected by a deficiency of thiamine? *TPP is the prosthetic group of three important enzymes: pyruvate dehydrogenase, α-ketoglutarate dehydrogenase, and transketolase.* Transketolase transfers two-carbon units from one sugar to another; its role in the pentose phosphate pathway will be discussed in Chapter 18. *The common feature of enzymatic reactions utilizing TPP is the transfer of an activated aldehyde unit.* In beriberi, the levels of pyruvate and α-ketoglutarate in the blood are higher than normal. The increase in the level of pyruvate in the blood is especially pronounced after ingestion of glucose. A related finding is that the activities of the pyruvate and α-ketoglutarate dehydrogenase complexes in vivo are abnormally low. The low transketolase activity of red cells in beriberi is a reliable diagnostic indicator of the disease.

> *Beriberi—*
> A vitamin-deficiency disease first described in 1630 by Jacob Bonitus, a Dutch physician working in Java:
>
> "A certain very troublesome affliction, which attacks men, is called by the inhabitants Beri-beri (which means sheep). I believe those, whom this same disease attacks, with their knees shaking and the legs raised up, walk like sheep. It is a kind of paralysis, or rather Tremor: for it penetrates the motion and sensation of the hands and feet indeed sometimes of the whole body."

CITRATE SYNTHASE UNDERGOES A LARGE CONFORMATIONAL CHANGE ON BINDING OXALOACETATE

As mentioned earlier, citrate synthase catalyzes the condensation of oxaloacetate and acetyl CoA to form citryl CoA, which is then hydrolyzed to citrate and CoA.

$$\underset{\textbf{Oxaloacetate}}{O{=}C(COO^-)\text{—}H_2C\text{—}COO^-} + \underset{\textbf{Acetyl CoA}}{H_3C\text{—}C({=}O)\text{—}S\text{—}CoA} \longrightarrow \underset{\textbf{Citryl CoA}}{HO\text{—}C(COO^-)(CH_2\text{—}C({=}O)\text{—}S\text{—}CoA)(CH_2\text{—}COO^-)} \xrightarrow{H_2O \quad CoA} \underset{\textbf{Citrate}}{HO\text{—}C(COO^-)(CH_2\text{—}COO^-)(CH_2\text{—}COO^-)}$$

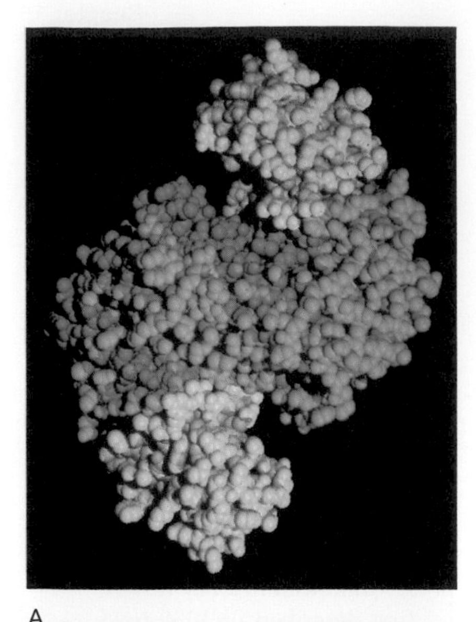

A

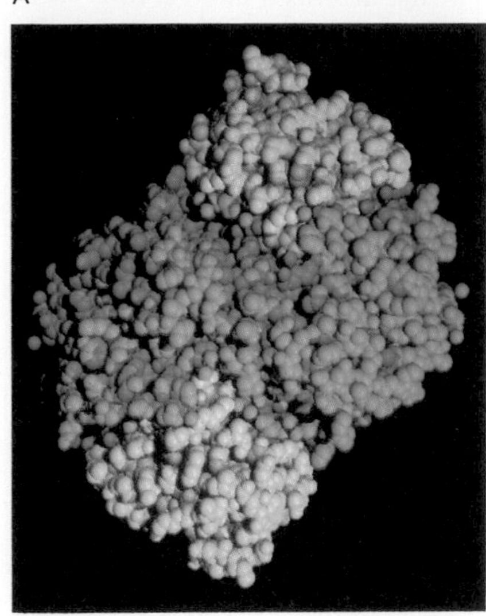

B

Figure 16-13
Citrate synthase undergoes a large conformational change on binding oxaloacetate. (A) Open form of enzyme alone. (B) Closed form of the liganded enzyme. [After S. J. Remington, G. Wiegand, and R. Huber. *J. Mol. Biol.* 158(1982):111.]

Mammalian citrate synthase consists of two identical 49-kd subunits. Each of the two independent active sites is located at the interface between the large and small domain of a subunit. X-ray crystallographic studies of citrate synthase and its complexes with several substrates and inhibitors have revealed that the enzyme undergoes large conformational changes during catalysis. Oxaloacetate binds first, followed by acetyl CoA. The reason for ordered binding is that *oxaloacetate induces a major structural rearrangement leading to the creation of a binding site for acetyl CoA.* The open form of the enzyme observed in the absence of ligands is converted into a closed form (Figure 16-13) by the binding of oxaloacetate. In each subunit, the small domain rotates 18 degrees relative to the large domain. *Movements as large as 15Å are produced by the rotation of α helices elicited by quite small shifts of side chains around bound oxaloacetate.* This conformational transition is reminiscent of the cleft closure in hexokinase induced by the binding of glucose (p. 365) and the subunit rearrangment in hemoglobin caused by the binding of O_2 (p. 158).

The synthase catalyzes the condensation reaction by bringing the substrates into close proximity, orienting them, and polarizing certain bonds. Two histidine residues play critical roles. One of them interacts with the carbonyl oxygen atom of oxaloacetate to make the carbon atom more susceptible to attack. The other histidine promotes the removal of a proton from the methyl carbon atom of acetyl CoA to form an enolate ion; this species attacks the carbonyl carbon atom of oxaloacetate to form a C–C bond.

His$^+$ $^{\delta-}$O=C$^{\delta+}$(CH_2COO^-)(COO^-) ← $HC(H)$=C(O^-)—S—CoA His$^+$

Oxaloacetate **Enolate anion of acetyl CoA**

The newly formed citryl CoA induces additional structural changes in the enzyme. The active site becomes completely enclosed, and an aspartate residue is recruited to enable a trapped water molecule to hydrolyze the thioester bond. CoA leaves the enzyme, followed by citrate, and the enzyme returns to the open conformation.

Citrate synthase is designed so that it does not hydrolyze acetyl CoA, which would be wasteful. First, acetyl CoA is not bound to the enzyme until oxaloacetate is bound and ready for condensation. Second, the aspartate residue crucial for hydrolysis of the thioester linkage is not brought into proximity until citryl CoA is formed. As with hexokinase (p. 365), *induced fit prevents undesirable side reactions.*

SYMMETRIC MOLECULES MAY REACT ASYMMETRICALLY

Let us follow the fate of a particular carbon atom in the citric acid cycle. Suppose that oxaloacetate were labeled with ^{14}C in the carboxyl carbon farthest from the keto group. Analysis of the α-ketoglutarate formed would show that none of the radioactive label had been lost. Decarboxylation of α-ketoglutarate would then yield succinate devoid of radioactivity. All of the label would be in the released CO_2.

$$\mathrm{^{-}OOC{-}C({=}O){-}CH_2{-}COO^{-}} \xrightarrow{\mathrm{CH_3{-}C{=}O}} \mathrm{^{-}OOC{-}CH_2{-}C(OH)(COO^{-}){-}CH_2{-}COO^{-}} \longrightarrow \mathrm{^{-}OOC{-}CH_2{-}CH(COO^{-}){-}CH(OH){-}COO^{-}} \longrightarrow \mathrm{^{-}OOC{-}CH_2{-}CH_2{-}C({=}O){-}COO^{-}} \longrightarrow \mathrm{^{-}OOC{-}CH_2{-}CH_2{-}COO^{-}} + \mathrm{CO_2}$$

Path 1

The finding that *all* of the label emerges in the CO_2 came as a surprise. Citrate is a symmetric molecule. Consequently, it was assumed that the two $—CH_2COO^-$ groups in it would react identically. Thus, for every citrate undergoing the reactions shown in Path 1, it was thought that another citrate molecule would react as shown in Path 2. If so, then only *half* of the label should have emerged in the CO_2.

$$\mathrm{^{-}OOC{-}C({=}O){-}CH_2{-}COO^{-}} \xrightarrow{\mathrm{CH_3{-}C{=}O}} \mathrm{^{-}OOC{-}CH_2{-}C(OH)(COO^{-}){-}CH_2{-}COO^{-}} \longrightarrow \mathrm{^{-}OOC{-}CH_2{-}CH(COO^{-}){-}CH(OH){-}COO^{-}} \longrightarrow \mathrm{^{-}OOC{-}CH_2{-}CH_2{-}C({=}O){-}COO^{-}} \longrightarrow \mathrm{^{-}OOC{-}CH_2{-}CH_2{-}COO^{-}} + \mathrm{CO_2}$$

Path 2
(Does not occur)

Table 16-3
Commonly used radioisotopes

Isotope	*Half-life*
^{3}H	12.26 years
^{14}C	5730 years
^{22}Na	2.62 years
^{32}P	14.28 days
^{35}S	87.9 days
^{42}K	12.36 hours
^{45}Ca	163 days
^{59}Fe	45.6 days
^{125}I	60.2 days
^{203}Hg	46.9 days

The interpretation of these experiments, which were carried out in 1941, was that citrate (or any other symmetric compound) could not be an intermediate in the formation of α-ketoglutarate because of the asymmetric fate of the label. This interpretation seemed compelling until Alexander Ogston incisively pointed out in 1948 that it is a fallacy to assume that the two identical groups of a symmetric molecule cannot be distinguished: "On the contrary, it is possible that *an asymmetric enzyme which attacks a symmetrical compound can distinguish between its identical groups* . . . the asymmetrical occurrence of isotope in a product cannot be taken as conclusive evidence against its arising from a symmetrical precursor."

Let us examine Ogston's assertion. For simplicity, consider a molecule in which two hydrogen atoms, a group X, and a different group Y are bonded to a tetrahedral carbon atom. Let us label one hydrogen A, the other B. Now suppose that an enzyme binds three groups of this substrate—X, Y, and H—at three complementary sites. Can H_A be distinguished from H_B? Figure 16-14 shows X, Y, and H_A bound to three points on the enzyme. In contrast, X, Y, and H_B cannot be bound to this active site; two of these three groups can be bound, but not all three. Thus, *H_A and H_B will have different fates.*

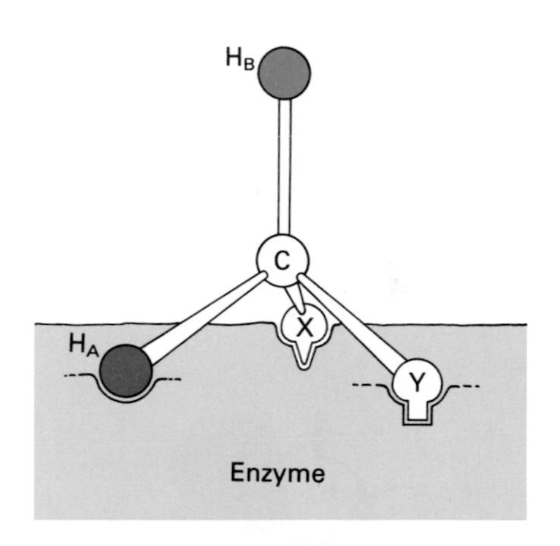

Figure 16-14
H_A and H_B are sterically not equivalent if the substrate $CXYH_2$ is bound to the enzyme at three points.

It should be noted that H_A and H_B are sterically not equivalent even though the molecule $CXYH_2$ is optically inactive. Similarly, the $—CH_2COO^-$ groups in citrate are sterically not equivalent even though citrate is optically inactive. *The symmetry rules that determine whether a compound has indistinguishable substituents are different from those that determine whether it is optically inactive:* (1) a molecule is optically inactive if it can be superimposed on its mirror image; (2) a molecule has indistinguishable substituents only if these groups can be brought into coincidence by a rotation that leaves the rest of the structure invariant.

Sterically nonequivalent groups such as H_A and H_B will almost always be distinguished in enzymatic reactions. The essence of the differentiation of these groups is that the enzyme holds the substrate in a specific orientation. Attachment at three points, as depicted in Figure 16-14, is a readily visualized way of achieving a particular orientation of the substrate, but it is not the only means of doing so.

> *Chirality*—
> "I call any geometrical figure, or any group of points, *chiral,* and say it has *chirality,* if its image in a plane mirror, ideally realized, cannot be brought to coincide with itself."
>
> Lord Kelvin (1893)
>
> Derived from the Greek word *cheir* for hand.

The terms *chiral* and *prochiral* are now extensively used to describe the stereochemistry of molecules. A *chiral* molecule has handedness and hence is optically active. A *prochiral* molecule (or center within a molecule), such as citrate or $CXYH_2$, lacks handedness and is optically inactive, but its identical substituents are distinguishable, and it can become chiral in one step. A prochiral molecule (such as $CXYH_AH_B$) is transformed into a chiral one ($CXYZH_B$) when one of its identical atoms or groups (H_A in this example) is replaced. The prefixes R and S are used to unambiguously designate the configuration of chiral and prochiral centers, as described in the appendix to this chapter (p. 395).

HYDROGEN IS STEREOSPECIFICALLY TRANSFERRED BY NAD⁺ DEHYDROGENASES

In the 1950s, Birgit Vennesland, Frank Westheimer, and their associates carried out some elegant experiments on the stereospecificity of hydrogen transfer by NAD^+ dehydrogenases. Ethanol labeled with two deuterium atoms at C-1 was the substrate in a reaction catalyzed by alcohol dehydrogenase. They found that the reduced coenzyme contained one atom of deuterium per molecule, whereas the other deuterium atom was in acetaldehyde. Thus, none of the deuterium was lost to the solvent, which shows that deuterium was directly transferred from the substrate to NAD^+.

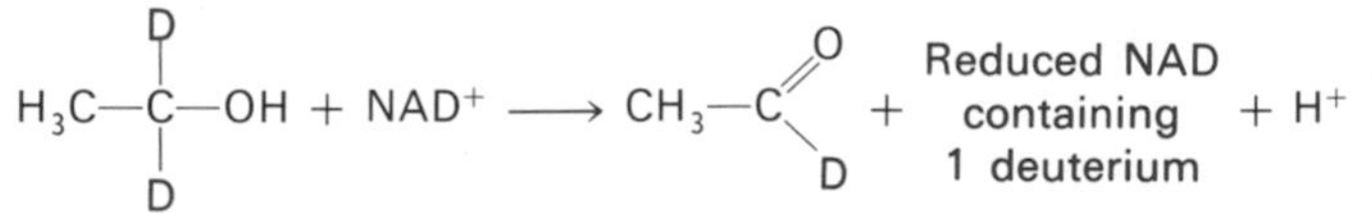

$$H_3C-CD_2-OH + NAD^+ \longrightarrow CH_3-C(=O)D + \begin{array}{c}\text{Reduced NAD}\\ \text{containing}\\ \text{1 deuterium}\end{array} + H^+$$

The deuterated reduced coenzyme formed in this reaction was then used to reduce acetaldehyde. The striking result was that *all of the deuterium was transferred from the coenzyme to the substrate.* None remained in NAD^+.

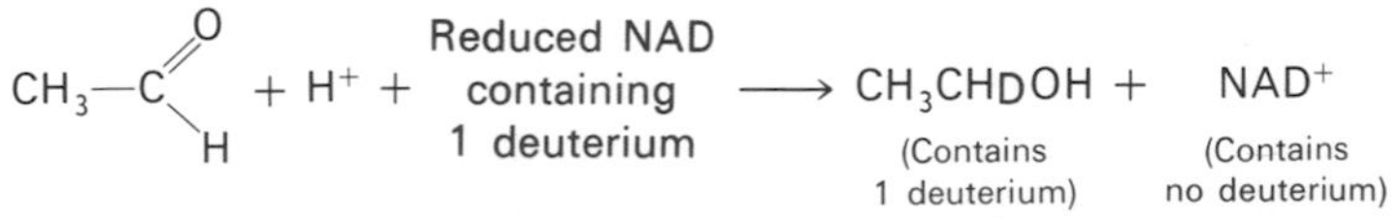

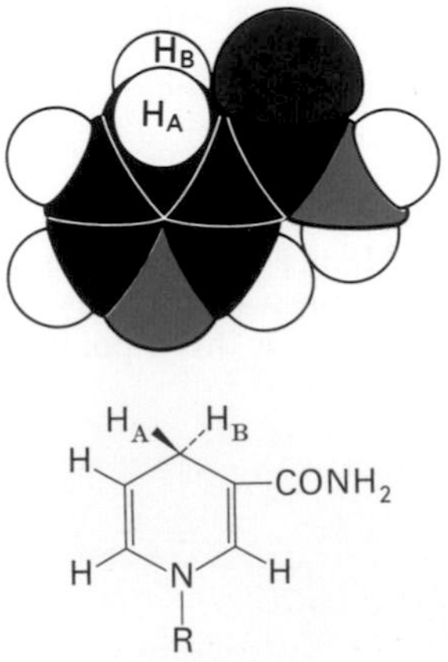

$$CH_3-C(=O)H + H^+ + \begin{array}{c}\text{Reduced NAD}\\ \text{containing}\\ \text{1 deuterium}\end{array} \longrightarrow \underset{\text{(Contains 1 deuterium)}}{CH_3CHDOH} + \underset{\text{(Contains no deuterium)}}{NAD^+}$$

These reactions revealed that the transfer catalyzed by alcohol dehydrogenase is stereospecific. *The positions occupied by the two hydrogen atoms at C-4 in NADH are not equivalent* (Figure 16-15). One of them (H_A) is in front of the nicotinamide plane, the other (H_B) is behind. In other words, C-4 is a *prochiral center.* Alcohol dehydrogenase, a chiral reagent, distinguishes between positions A and B at C-4. Deuterium is transferred from deuterated ethanol to position A only:

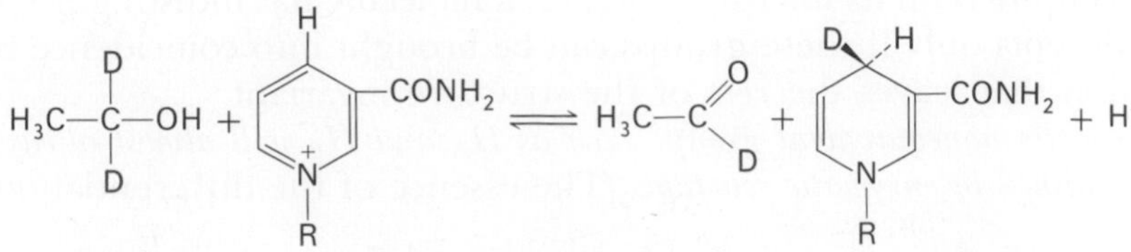

$$H_3C-CD_2-OH + \text{NAD}^+ \rightleftharpoons H_3C-C(=O)D + \text{NAD(4-D}_A\text{)H} + H^+$$

HB HA

NADH

Figure 16-15
Model and formula of the nicotinamide moiety of NADH showing that H_A and H_B are on opposite sides of the ring. In the RS nomenclature, H_A is *pro-R* and H_B is *pro-S*.

In the reverse reaction, the deuterium atom at position A is removed and directly transferred to acetaldehyde.

Some dehydrogenases, such as glyceraldehyde phosphate dehydrogenase, transfer hydrogen to position B. Thus, *there are two classes of NAD^+ (and $NADP^+$) dehydrogenases: A-stereospecific and B-stereospecific.* A comparison of the three-dimensional structures of NAD^+ dehydrogenases shows that the difference in stereospecificity of the A- and B-type enzymes arises from a 180-degree rotation of the nicotinamide ring with respect to the adjacent ribose unit. The result of this 180-degree flip is that opposite sides of the nicotinamide ring are exposed and hence reactive in A- and B-type dehydrogenases. As might be expected, all known dehydrogenases are stereospecific. When a dehydrogenase reacts with a range of substrates, the stereospecificity of hydrogen transfer is the same for all. The conservation of NAD^+-binding sites in the course of evolution is emphasized by the finding that the stereospecificity of a particular dehydrogenase is independent of species (e.g., yeast and horse alcohol dehydrogenase have the same stereospecificity). Furthermore, the stereospecificity of a particular reaction is the same for NAD^+ and $NADP^+$ in those enzymes that can use either coenzyme.

THE CITRIC ACID CYCLE IS A SOURCE OF BIOSYNTHETIC PRECURSORS

Thus far, discussion has focused on the citric acid cycle as the *major degradative pathway for the generation of ATP*. The citric acid cycle also *provides intermediates for biosyntheses* (Figure 16-16). For example, a majority of the carbon atoms in porphyrins come from *succinyl CoA*. Many of the amino acids are derived from *α-ketoglutarate and oxaloacetate*. The biosyntheses of these compounds will be discussed in subsequent chapters. The important point now is that *citric acid cycle intermediates must be replenished if any are drawn off for biosyntheses*. Suppose that oxaloacetate is converted into amino acids for protein synthesis. The citric acid cycle will cease to operate unless new oxaloacetate is formed, because acetyl CoA cannot enter the cycle unless it condenses with oxaloacetate. Even though oxaloacetate is used catalytically, a minimal level must be maintained to allow the cycle to function. How is oxaloacetate replenished? Mammals lack the enzymatic machinery for the net conversion of acetyl

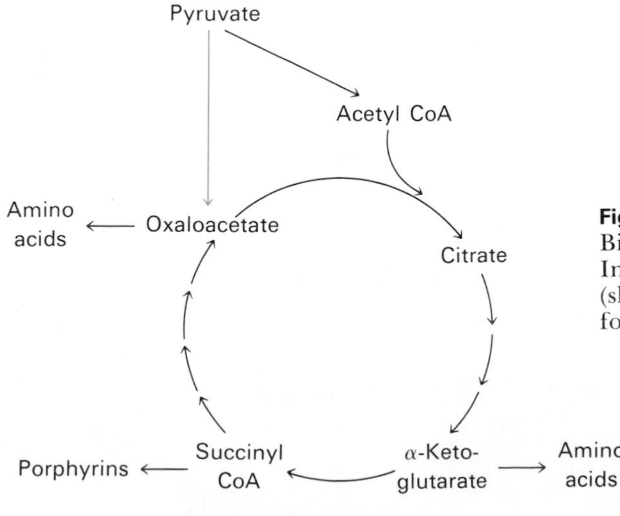

Figure 16-16
Biosynthetic roles of the citric acid cycle. Intermediates drawn off for biosyntheses (shown by red arrows) are replenished by the formation of oxaloacetate from pyruvate.

CoA into oxaloacetate or another citric acid cycle intermediate. Rather, oxaloacetate is formed by the carboxylation of pyruvate, in a reaction catalyzed by pyruvate carboxylase (p. 441).

$$\text{Pyruvate} + CO_2 + \text{ATP} + H_2O \longrightarrow \text{oxaloacetate} + \text{ADP} + P_i + 2\ H^+$$

The carboxylation of pyruvate is an example of an *anaplerotic reaction* (from the Greek, to "fill up").

THE GLYOXYLATE CYCLE ENABLES PLANTS AND BACTERIA TO GROW ON ACETATE

Many bacteria and plants are able to grow on acetate or other compounds that yield acetyl CoA. They make use of a metabolic pathway that converts two-carbon acetyl units into four-carbon units (succinate) for energy production and biosyntheses. This reaction sequence, called the *glyoxylate cycle*, bypasses the two decarboxylation steps of the citric acid cycle. Another key difference is that two molecules of acetyl CoA enter per turn of the glyoxylate cycle, compared with one in the citric acid cycle. The glyoxylate cycle (Figure 16-17), like the citric acid cycle,

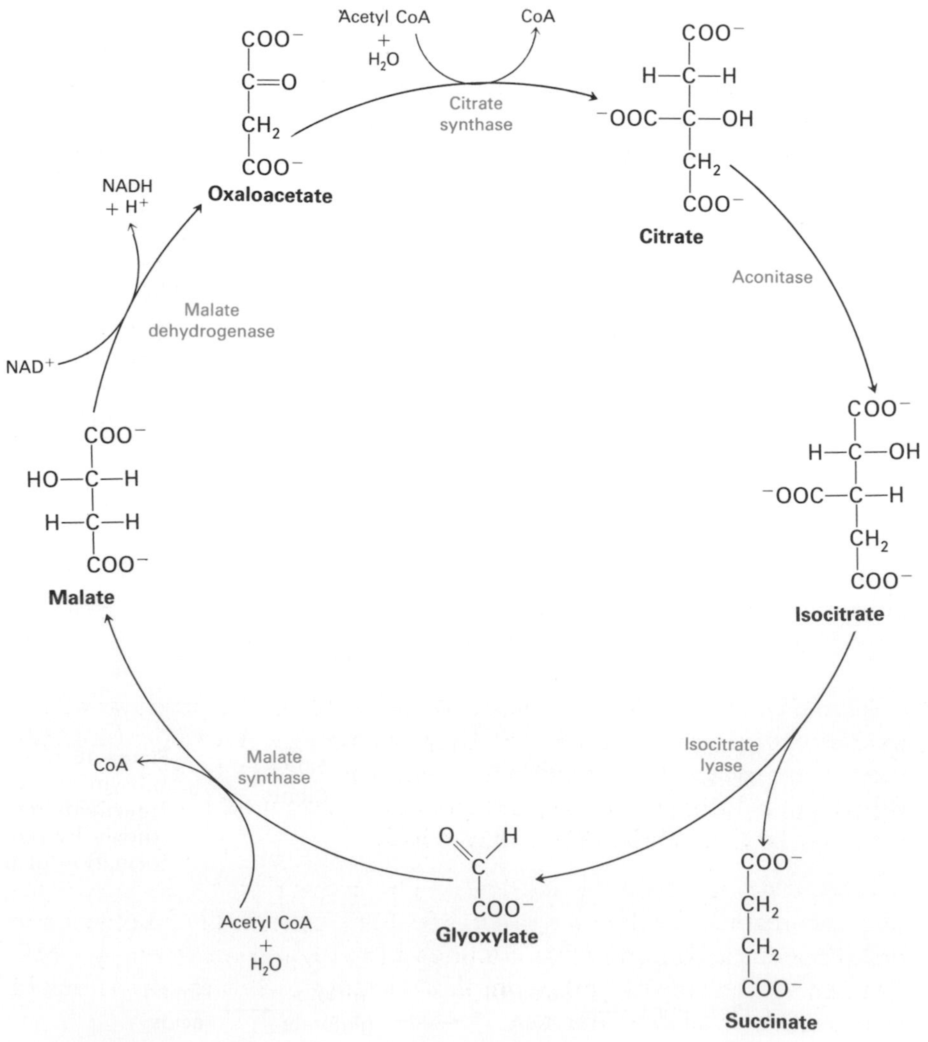

Figure 16-17
Glyoxylate cycle of plants and bacteria. The reactions of this cycle are the same as those of the citric acid cycle except for the ones catalyzed by citrate lyase and malate synthase.

begins with the condensation of acetyl CoA and oxaloacetate to form citrate, which is then isomerized to isocitrate. Instead of being decarboxylated, isocitrate is cleaved by *isocitrate lyase* into succinate and glyoxylate. The subsequent steps regenerate oxaloacetate from glyoxylate. Acetyl CoA condenses with glyoxylate to form malate, in a reaction catalyzed by *malate synthase*, which resembles citrate synthase. Finally, malate is oxidized to oxaloacetate, as in the citric acid cycle. The sum of these reactions is

$$2 \text{ Acetyl CoA} + NAD^+ + 2\ H_2O \longrightarrow \text{succinate} + 2 \text{ CoA} + NADH + 2\ H^+$$

In plants, these reactions occur in organelles called *glyoxysomes.*

Bacteria and plants synthesize acetyl CoA from acetate and CoA by an ATP-driven reaction that is catalyzed by *acetyl CoA synthetase.*

$$\text{Acetate} + \text{CoA} + \text{ATP} \longrightarrow \text{acetyl CoA} + \text{AMP} + PP_i$$

Pyrophosphate is then hydrolyzed to orthophosphate, and so 2 ~P are consumed in the activation of acetate. We shall return to this type of activation reaction in protein synthesis, where it is used to link amino acids to transfer RNAs (p. 735).

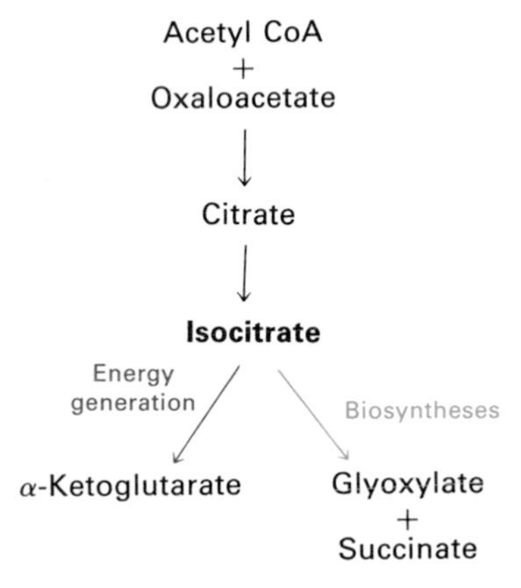

Isocitrate has two major fates in some bacteria and plants. When energy is needed, it is oxidatively decarboxylated to α-ketoglutarate. When energy is abundant, isocitrate is split into succinate and glyoxylate. The citric acid cycle and the glyoxylate cycle compete for isocitrate at this key branch point. In times of plenty (i.e., high levels of ATP), *isocitrate dehydrogenase is switched off by phosphorylation, which serves to funnel isocitrate into the glyoxylate pathway for the formation of biosynthetic intermediates.* The kinase that phosphorylates isocitrate dehydrogenase and the phosphatase that removes the phosphate group from the modified enzyme are located on the same polypeptide chain, like the kinase-phosphatase tandem enzyme regulating the level of fructose 2,6-bisphosphate in the control of glycolysis (p. 361).

CONTROL OF THE PYRUVATE DEHYDROGENASE COMPLEX

In animals, the formation of acetyl CoA from pyruvate is a key irreversible step in metabolism because *they are unable to convert acetyl CoA into glucose.* The oxidative decarboxylation of pyruvate to acetyl CoA commits the carbon atoms of glucose to two principal fates: oxidation to CO_2 by the citric acid cycle, with the concomitant generation of energy, or incorporation into lipid (p. 480). As might be expected of an enzyme at a critical decision point in metabolism, the activity of the pyruvate dehydrogenase complex is stringently controlled:

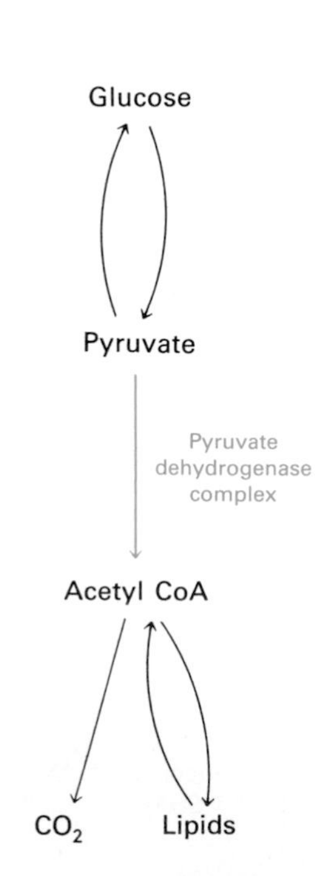

1. *Inhibition by products.* Acetyl CoA and NADH, the products of the oxidation of pyruvate, inhibit the enzyme complex. Acetyl CoA inhibits the transacetylase component, whereas NADH inhibits the dihydrolipoyl dehydrogenase component. These inhibitory effects are reversed by CoA and NAD^+, respectively.

2. *Feedback regulation by nucleotides.* The activity of the enzyme complex is controlled by the *energy charge* (p. 326). Specifically, the pyruvate dehydrogenase component is inhibited by GTP and activated by AMP. In this way, the activity of the complex is reduced when the cell is rich in immediately available energy.

3. *Regulation by reversible phosphorylation.* The complex becomes enzymatically inactive when a specific serine residue of the pyruvate dehydrogenase component is phosphorylated by a specific kinase using ATP. Phosphorylation is enhanced by high ratios of ATP/ADP, acetyl CoA/CoA, and NADH/NAD^+, whereas it is inhibited by pyruvate. The enzyme complex becomes active again when the phosphoryl group is hydrolyzed by a specific phosphatase. Dephosphorylation is enhanced by a high level of Ca^{2+}. Insulin also stimulates dephosphorylation to accelerate the conversion of pyruvate into acetyl CoA, and thus the conversion of glucose into pyruvate. We see in pyruvate dehydrogenase another important example of reversible phosphorylation as a regulatory device and will return to this theme when we consider glycogen synthesis and degradation.

CONTROL OF THE CITRIC ACID CYCLE

The rate of the citric acid cycle is also precisely adjusted to meet the cell's needs for ATP. The availability of NAD^+ and FAD signals that the energy charge is low. *The synthesis of citrate from oxaloacetate and acetyl CoA is an important control point in the cycle.* ATP is an allosteric inhibitor of citrate synthase. The effect of ATP is to increase the K_M for acetyl CoA. Thus, as the level of ATP increases, less of this enzyme is saturated with acetyl CoA and so less citrate is formed.

A second control point is isocitrate dehydrogenase. This enzyme is allosterically stimulated by ADP, which enhances its affinity for substrates. The binding of isocitrate, NAD^+, Mg^{2+}, and ADP is mutually cooperative. In contrast, NADH inhibits isocitrate dehydrogenase by directly displacing NAD^+. ATP, too, is inhibitory.

A third control site in the citric acid cycle is α-ketoglutarate dehydrogenase. Some aspects of its control are like those of the pyruvate dehydrogenase complex, as might be expected from their structural homology. α-Ketoglutarate dehydrogenase is inhibited by succinyl CoA and NADH, the products of the reaction it catalyzes. Also, α-ketoglutarate dehydrogenase is inhibited by a high energy charge. *In short, the funneling of two-carbon fragments into the citric acid cycle and the rate of the cycle are reduced when the cell has a high level of ATP.* This control is achieved by a variety of complementary mechanisms at several sites (Figure 16-18).

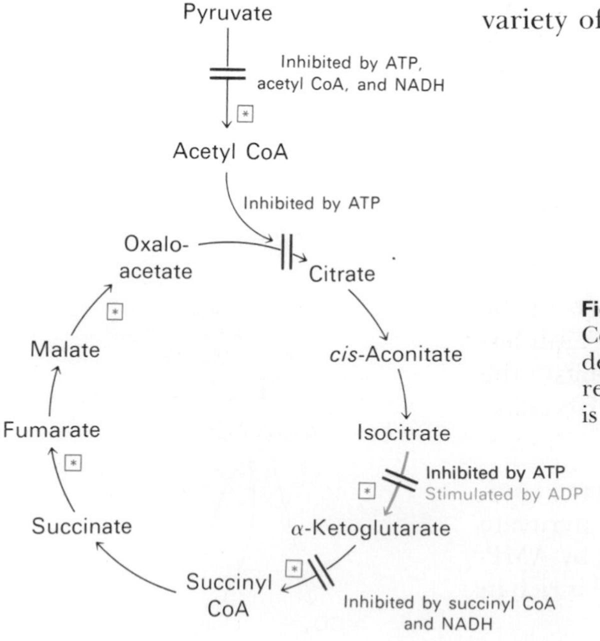

Figure 16-18
Control of the citric acid cycle and the oxidative decarboxylation of pyruvate: ⊡ indicates steps that require an electron acceptor (NAD^+ or FAD) that is regenerated by the respiratory chain.

KREBS' DISCOVERY OF THE CYCLE

"I have often been asked how the work on the tricarboxylic acid cycle arose and developed. Was the concept perhaps due to a sudden inspiration and vision?" Hans Krebs replied that "it was of course nothing of the kind, but a very slow evolutionary process, extending over some five years beginning (as far as I am involved) in 1932." Krebs first studied the rate of oxidation of a variety of compounds by slices of kidney and liver. The substances chosen were ones that might be expected to be intermediates in the oxidation of foodstuffs. The idea behind these experiments was that such intermediates would be rapidly oxidized and thereby identified. The important finding was that *citrate,* succinate, fumarate, and acetate were very readily oxidized in various tissues.

A critical contribution was made by Albert Szent-Györgyi in 1935. He studied the oxidation of various substances by suspensions of minced pigeon-breast muscle. This very active flight muscle has an exceptionally high rate of oxidation, which facilitated the experimental studies. Szent-Györgyi found that the addition of certain C_4-dicarboxylic acids increased the uptake of O_2 far more than could be caused by their direct oxidation. *This catalytic stimulation of respiration was obtained with succinate, fumarate, and malate.*

The next breakthrough was the elucidation of the biological pathway of oxidation of citrate by Carl Martius and Franz Knoop in 1937. They showed that citrate is isomerized to isocitrate by way of *cis*-aconitate and that isocitrate is oxidatively decarboxylated to α-ketoglutarate. It had already been known that α-ketoglutarate can be oxidized to succinate. *Thus, their discovery revealed the pathway from citrate to succinate.* This finding came at a propitious moment, for it enabled Krebs to interpret the recent observation that *citrate catalytically enhanced* the respiration of minced muscle.

COO^-—CH_2—CH_2—COO^-

Succinate

COO^-—CH_2—COO^-

Malonate

Additional critical information came from the use of *malonate, a specific inhibitor of succinate dehydrogenase.* Malonate is a competitive inhibitor of this enzyme because it closely resembles succinate. It had been known for some time that *malonate poisons respiration.* Krebs reasoned that succinate dehydrogenase must therefore play a key role in respiration. This was reinforced by the finding that succinate accumulated when citrate was added to malonate-poisoned muscle. Furthermore, succinate also accumulated when fumarate was added to malonate-poisoned muscle. The first of these experiments showed that the pathway from citrate to succinate is physiologically significant. The second experiment revealed that there is a pathway from fumarate to succinate distinct from the reaction catalyzed by succinate dehydrogenase.

Succinate + E—FAD → (Inhibited by malonate) → Fumarate + E—$FADH_2$

Krebs then discovered that citrate is readily formed by a muscle suspension if oxaloacetate is added. *The discovery of the synthesis of citrate from oxaloacetate and a substance derived from pyruvate or acetate enabled Krebs to formulate a complete scheme.* His postulated tricarboxylic acid cycle suddenly provided a coherent picture of how carbohydrates are oxidized. Many experimental facts, such as the catalytic enhancement of respiration by succinate and other intermediates, fell neatly into place. It is noteworthy that the citric acid cycle was not the only metabolic cycle, or the first, to be discovered by Krebs. Six years earlier, he had shown that urea is synthesized by a cyclic metabolic pathway called the ornithine cycle (p. 500). Thus, the concept of a cyclic metabolic pathway had already been fully recognized by Krebs when he pondered the data and designed the experiments that led to the proposal of the citric acid cycle.

THE CITRIC ACID CYCLE WAS SELECTED IN EVOLUTION BECAUSE OF ITS HIGH ENERGY YIELD

Why did the citric acid cycle arise in the course of evolution? One can envision a simpler sequence of reactions (Figure 16-19) that oxidizes acetate to CO_2 without first joining it to oxaloacetate:

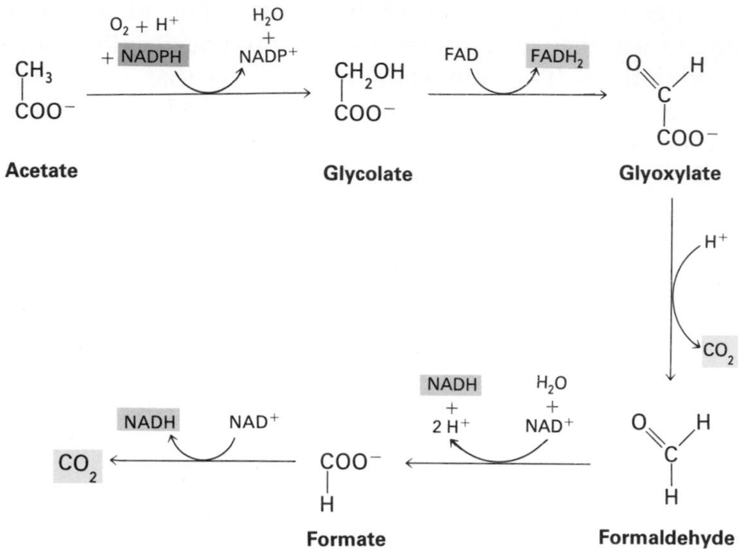

Figure 16-19
A hypothetical direct pathway for the oxidation of acetate to CO_2. The energy yield of this pathway is half of that of the citric acid cycle.

This direct pathway did not evolve because it is energetically less efficient than the citric acid cycle. The first step, the oxidation of acetate to glycolate, cannot be coupled to an energy-yielding dehydrogenation. Instead, molecular oxygen must be used to convert the methyl group of acetate into a hydroxymethyl group. The other oxygen atom of O_2 must be reduced to water by a reductant such as NADPH. Monooxygenases (p. 511) catalyze this type of reaction.

$$H_3C\text{—}COO^- + O_2 + NADPH + H^+ \longrightarrow HOH_2C\text{—}COO^- + H_2O + NADP^+$$

The consumption of an NADPH corresponds to an input of 3 ATP; subsequent steps in this direct pathway could generate 1 $FADH_2$ and 2 NADH, which yield 8 ATP by oxidative phosphorylation (p. 397). Hence, the net yield of this pathway is 5 ATP. In contrast, oxidation of acetate by the citric acid cycle yields 10 ATP because 2 ~P are consumed in forming acetyl CoA from acetate (p. 389), and 12 ATP are generated in oxidizing acetyl CoA to CO_2. Thus, *the ATP yield of the citric acid cycle is more than twice that of the direct pathway.*

Clearly, much more energy can be extracted from an acetyl unit by joining it to a carrier such as oxaloacetate instead of attacking it directly with oxygen. The reason for a reaction cycle is also evident: *the carrier must be regenerated.* If oxaloacetate were not regenerated, a *kilogram* of it would be needed daily to oxidize acetyl units arising from foods taken in by a person on a 2500 kcal diet. The cost of synthesizing such a vast amount of oxaloacetate would far exceed the energy derived from oxidizing acetyl units attached to it. Why the citric acid cycle in particular? It seems likely that the initial steps of the cycle—the synthesis of citrate, the isomerization into isocitrate, and the oxidative decarboxylation to α-ketoglutarate—evolved long before the appearance of oxygen in the atmosphere. Most significant, coenzyme A is an ancient molecule, a ribonucleotide. The first step of the citric acid cycle may well be a motif, a remembrance of the early RNA world.

SUMMARY

The citric acid cycle is the final common pathway for the oxidation of fuel molecules. It also serves as a source of building blocks for biosyntheses. Most fuel molecules enter the cycle as acetyl CoA. The link between glycolysis and the citric acid cycle is the oxidative decarboxylation of pyruvate to form acetyl CoA. In eucaryotes, this reaction and those of the cycle occur inside mitochondria, in contrast with glycolysis, which occurs in the cytosol.

The cycle starts with the condensation of oxaloacetate (C_4) and acetyl CoA (C_2) to give citrate (C_6), which is isomerized to isocitrate (C_6). Oxidative decarboxylation of this intermediate gives α-ketoglutarate (C_5). The second molecule of CO_2 comes off in the next reaction, in which α-ketoglutarate is oxidatively decarboxylated to succinyl CoA (C_4). The thioester bond of succinyl CoA is cleaved by P_i to yield succinate, and a high-energy phosphate bond in the form of GTP is concomitantly generated. Succinate is oxidized to fumarate (C_4), which is then hydrated to form malate (C_4). Finally, malate is oxidized to regenerate oxaloacetate (C_4). Thus, two carbon atoms from acetyl CoA enter the cycle, and two carbon atoms leave the cycle as CO_2 in the successive decarboxylations catalyzed by isocitrate dehydrogenase and α-ketoglutarate dehydrogenase. In the four oxidation-reductions in the cycle, three pairs of electrons are transferred to NAD^+ and one pair to FAD. These reduced electron carriers are subsequently oxidized by the electron-transport chain to generate eleven molecules of ATP. In addition, one high-energy phosphate bond is directly formed in the citric acid cycle. Hence, a total of twelve high-energy phosphate bonds are generated for each two-carbon fragment that is completely oxidized to H_2O and CO_2.

The citric acid cycle operates only under aerobic conditions because it requires a supply of NAD^+ and FAD. These electron acceptors are regenerated when NADH and $FADH_2$ transfer their electrons to O_2 through the electron-transport chain, with the concomitant production of ATP. Consequently, the rate of the citric acid cycle depends on the need for ATP. The regulation of three enzymes in the cycle is also important for control. A high energy charge diminishes the activities of citrate synthase, isocitrate dehydrogenase, and α-ketoglutarate dehydrogenase. The irreversible formation of acetyl CoA from pyruvate is another important regulatory point. The activity of the pyruvate dehydrogenase complex is controlled by product inhibition, feedback regulation by nucleotides, and reversible phosphorylation. These mechanisms complement each other in reducing the rate of formation of acetyl CoA when the energy charge of the cell is high.

SELECTED READINGS

Where to start

Baldwin, J. E., and Krebs, H., 1981. The evolution of metabolic cycles. *Nature* 291:381–382. [A concise essay, full of insight, with emphasis on the citric acid cycle.]

Weitzman, P. D. J., 1981. Unity and diversity in some bacterial citric acid cycle enzymes. *Advan. Microbiol. Physiol.* 22:185–244.

Srere, P. A., 1987. Complexes of sequential metabolic enzymes. *Ann. Rev. Biochem.* 56:89–124.

Reviews

Wieland, O. H., 1983. The mammalian pyruvate dehydrogenase complex: structure and regulation. *Rev. Physiol. Biochem. Pharmacol.* 96:123–170.

Beeckmans, S., 1984. Some structural and regulatory aspects of citrate synthase. *Eur. J. Biochem.* 16:341–351.

Singer, T. P., and Johnson, M. K., 1985. The prosthetic groups of succinate dehydrogenase: 30 years from discovery to identification. *FEBS Lett.* 190:189–198.

Books

Lowenstein, J. M., (ed.), 1969. *Citric Acid Cycle: Control and Compartmentation.* Marcel Dekker.

Gottschalk, G., 1986. *Bacterial Metabolism* (2nd ed.). Springer-Verlag. [An excellent account of the metabolic diversity of bacteria and the richness of their energy-generating pathways.]

Stereospecificity

Popják, G., 1970. Stereospecificity of enzymic reactions. *In* Boyer, P. D., (ed.), *The Enzymes* (3rd ed.), vol. 2, pp. 115–215. Academic Press. [An excellent review containing a discussion of the stereochemistry of the citric acid cycle.]

Ogston, A. G., 1948. Interpretation of experiments on metabolic processes using isotopic tracer elements. *Nature* 162:963.

Bentley, R., 1969. *Molecular Asymmetry in Biology,* vols. 1 and 2. Academic Press. [These volumes contain a wealth of information about stereospecificity in biochemical reactions. Chapter 2 of Volume 1 discusses nomenclature and Chapter 4 discusses prochirality.]

Enzyme structure and mechanism

Packman, L. C., Hale, G., and Perham, R. N., 1984. Repeating functional domains in the pyruvate dehydrogenase complex of *Escherichia coli. EMBO J.* 3:1315–1319.

Oliver, R. M., and Reed, L. J., 1982. Multienzyme complexes. *In* Harris, R., (ed.), *Electron Microscopy of Proteins,* vol. 2, pp. 1–48. Academic Press.

Hackert, M. L., Oliver, R. M., and Reed, L. J., 1983. A computer model analysis of the active-site coupling mechanism in the pyruvate dehydrogenase multienzyme complex of *Escherichia coli. Proc. Nat. Acad. Sci.* 80:2907–2911.

Wiegand, G., and Remington, S. J., 1986. Citrate synthase: structure, control, and mechanism. *Ann. Rev. Biophys. Biophys. Chem.* 15:97–117.

Chothia, C., and Lesk, A. M., 1985. Helix movements in proteins. *Trends Biochem. Sci.* 10:116–118. [A well-illustrated analysis of domain closure in citrate synthase.]

Telser, J., Emptage, M. H., Merkle, H., Kennedy, M. C., Beinert, H., and Hoffman, B. M., 1983. ^{17}O electron nuclear double resonance characterization of substrate binding to the 4Fe-4S cluster of reduced active aconitase. *J. Biol. Chem.* 261:4840–4846.

Barnes, S. J., and Weitzman, P. D., 1986. Organization of citric acid cycle enzymes into a multienzyme cluster. *FEBS Lett.* 201:267–270. [Fumarase, malate dehydrogenase, citrate synthase, aconitase and isocitrate dehydrogenase are loosely associated in a multienzyme cluster.]

Robinson, J. B., Jr., Inman, L., Sumegi, B., and Srere, P. A., 1987. Further characterization of the Krebs tricarboxylic acid cycle metabolon. *J. Biol. Chem.* 262:1786–1790.

Regulation

Reed, L. J., Damuni, Z., and Merryfield, M. L., 1985. Regulation of mammalian pyruvate and branched-chain α-keto acid dehydrogenase complexes by phosphorylation-dephosphorylation. *Curr. Top. Cell Regul.* 27:41–49.

Williamson, J. R., and Cooper, R. H., 1980. Regulation of the citric acid cycle in mammalian systems. *FEBS Lett.* 117(Suppl.):K73–K85.

Randle, P. J., 1978. Pyruvate dehydrogenase complex: meticulous regulator of glucose disposal in animals. *Trends Biochem. Sci.* 3:217–219.

LaPorte, D. C., and Koshland, D. E., 1982. A protein with kinase and phosphatase activities involved in regulation of tricarboxylic acid cycle. *Nature* 30:458–460.

Discovery of the citric acid cycle

Krebs, H. A., and Johnson, W. A., 1937. The role of citric acid in intermediate metabolism in animal tissues. *Enzymologia* 4:148–156.

Krebs, H. A., 1970. The history of the tricarboxylic acid cycle. *Perspect. Biol. Med.* 14:154–170.

Krebs, H. A., and Martin, A., 1981. *Reminiscences and Reflections.* Clarendon Press.

APPENDIX
The RS Designation of Chirality

The absolute configuration of any chiral center can be unambiguously specified using the RS notation introduced in 1956 by Robert Cahn, Christopher Ingold, and Vladimir Prelog. Consider the chiral compound CHFClBr (Figure 16-20A). The first step in using the RS designation is to assign a *priority sequence* to the four substituents by applying the rule that *an atom with a higher atomic number has a higher priority than an atom with a lower atomic number*. Hence, the priority sequence of these four substituents is (a) Br, (b) Cl, (c) F, and (d) H, with Br having the highest priority (a). The next step is to *orient the molecule so that the group of lowest priority (d) points away from the viewer*. For CHFClBr, this means that the molecule should be oriented so that H is away from us (behind the plane of the page in Figure 16-20B). *The final step is to ask whether the path from a to b to c under these conditions is clockwise or counterclockwise*. If it is clockwise (right-handed), the configuration is R (from the Latin *rectus*, right). If it is counterclockwise (left-handed), the configuration is S (from the Latin *sinister*, left). The configuration of the stereoisomer of CHFClBr shown in Figure 16-20 is R.

Now let us consider the RS designation of alanine (Figure 16-21A). The four atoms bonded to the α-carbon are N, C, C, and H. The priority sequence of the methyl carbon and the carboxyl carbon is determined by going outward to the next set of atoms. *It is useful to note the following priority sequence for biochemically important groups:* —SH (highest), —OR, —OH, —NHR, $—NH_2$, —COOR, —COOH, —CHO, $—CH_2OH$, $—C_6H_5$, $—CH_3$, —T, —D, —H (lowest). Thus, the priority sequence of the four groups attached to the α-carbon of alanine is (a) $—NH_3^+$, (b) $—COO^-$, (c) $—CH_3$, and (d) —H. The next step is to orient L-alanine so that its lowest priority group (—H) is behind the plane of the page (Figure 16-21B). The path from (a) $—NH_3^+$ to (b) $—COO^-$ to (c) $—CH_3$ is then counterclockwise (left-handed), and so L-alanine has an S configuration.

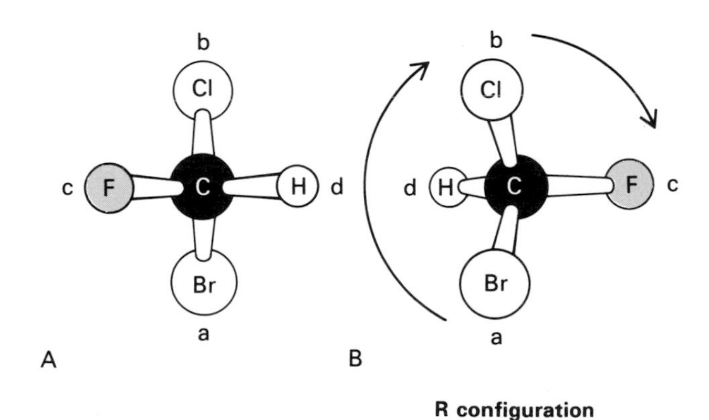

Figure 16-20
The stereoisomer of CHFClBr shown here has an R configuration.

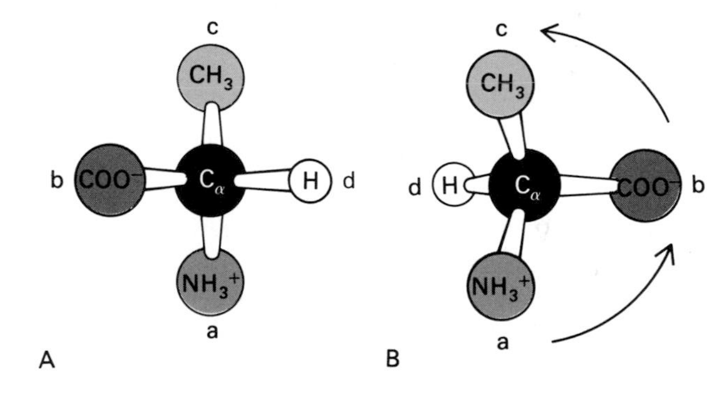

Figure 16-21
L-Alanine has an S configuration.

PROBLEMS

1. What is the fate of the radioactive label when each of the following compounds is added to a cell extract containing the enzymes and cofactors of the glycolytic pathway, the citric acid cycle, and the pyruvate dehydrogenase complex? (The ^{14}C label is printed in red.)

(a) $H_3C—C(=O)—COO^-$

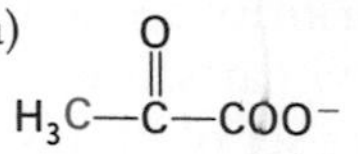

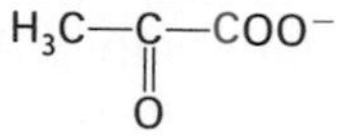

(b) $H_3C—C(=O)—COO^-$

(c) $H_3C—C(=O)—COO^-$

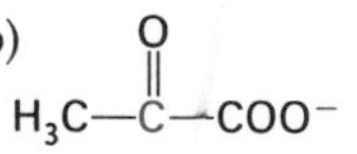

(d) $H_3C—C(=O)—S—CoA$

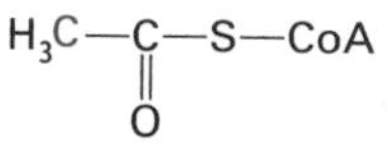

(e) Glucose 6-phosphate labeled at C-1.

2. Is it possible to get *net synthesis* of oxaloacetate by adding acetyl CoA to an extract that contains only the enzymes and cofactors of the citric acid cycle ?

3. What are the relative concentrations of citrate, isocitrate, and *cis*-aconitate at equilibrium? (Use the data given in Table 16-1.)

4. What is the $\Delta G°'$ for the complete oxidation of the acetyl unit of acetyl CoA by the citric acid cycle?

5. A sample of deuterated reduced NAD was prepared by incubating $H_3C—CD_2—OH$ and NAD^+ with alcohol dehydrogenase. This reduced coenzyme was added to a solution of 1,3-BPG and glyceraldehyde 3-phosphate dehydrogenase. The NAD^+ formed by this second reaction contained one atom of deuterium, whereas glyceraldehyde 3-phosphate, the other product, contained none. What does this experiment reveal about the stereospecificity of glyceraldehyde 3-phosphate dehydrogenase?

6. Thiamine thiazolone pyrophosphate binds to pyruvate dehydrogenase about 20,000 times as strongly as thiamine pyrophosphate, and it competitively inhibits the enzyme. Why?

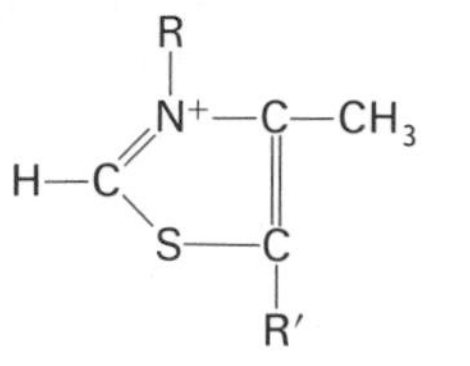

TPP

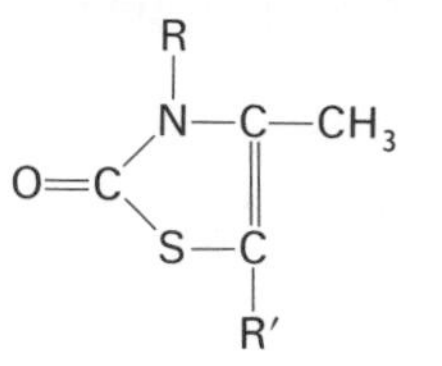

Thiazolone analog of TPP

7. The oxidation of malate by NAD^+ to form oxaloacetate is a highly endergonic reaction under standard conditions ($\Delta G°' = +7$ kcal/mol). The reaction proceeds readily from malate to oxaloacetate under physiological conditions because the steady-state concentrations of the products are low compared with those of the substrates. Assuming a $[NAD^+]/[NADH]$ ratio of 8 and a pH of 7, what is the lowest [malate]/[oxaloacetate] ratio at which oxaloacetate can be formed from malate?

8. Bacteria called methophiles (or methanotrophs) can use methane as a fuel. Propose an energy-conserving reaction sequence for converting methane into CO_2. What is the likely ATP yield of this pathway?

9. Propose a reaction mechanism for the condensation of acetyl CoA and glyoxylate in the glyoxylate cycle of plants and bacteria.

CHAPTER 17

Oxidative Phosphorylation

The NADH and $FADH_2$ formed in glycolysis, fatty acid oxidation, and the citric acid cycle are energy-rich molecules because each contains a pair of electrons having a high transfer potential. When these electrons are donated to molecular oxygen, a large amount of free energy is liberated, which can be used to generate ATP. *Oxidative phosphorylation is the process in which ATP is formed as electrons are transferred from NADH or $FADH_2$ to O_2 by a series of electron carriers.* This is the major source of ATP in aerobic organisms. For example, oxidative phosphorylation generates 32 of the 36 molecules of ATP that are formed when glucose is completely oxidized to CO_2 and H_2O. Some salient features of this process are:

1. Oxidative phosphorylation is carried out by *respiratory assemblies* that are located in the *inner membrane of mitochondria*. The citric acid cycle and the pathway of fatty acid oxidation, which supply most of the NADH and $FADH_2$, are in the adjacent mitochondrial matrix.

2. The oxidation of NADH yields 3 ATP, whereas the oxidation of $FADH_2$ yields 2 ATP. Oxidation and phosphorylation are coupled processes.

3. Respiratory assemblies contain numerous electron carriers, such as the cytochromes. The step-by-step transfer of electrons from NADH or $FADH_2$ to O_2 through these carriers leads to the pumping of protons

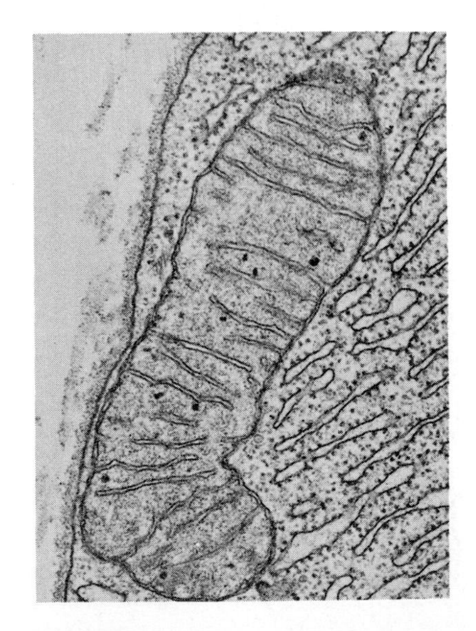

Figure 17-1
Electron micrograph of a mitochondrion. [Courtesy of Dr. George Palade.]

out of the mitochondrial matrix. A proton-motive force is generated consisting of a pH gradient and a transmembrane electric potential. ATP is synthesized when protons flow back to the mitochondrial matrix through an enzyme complex. Thus, *oxidation and phosphorylation are coupled by a proton gradient across the inner mitochondrial membrane* (Figure 17-2).

Respiration—
An ATP-generating process in which an inorganic compound (such as O_2) serves as the ultimate electron acceptor. The electron donor can be either an organic compound or an inorganic one.

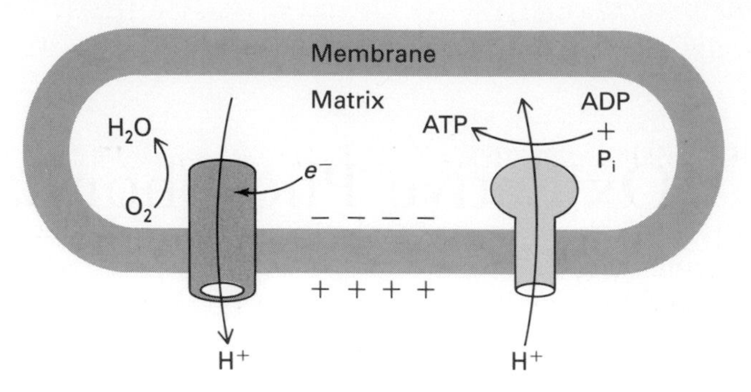

Figure 17-2
Oxidation and ATP synthesis are coupled by transmembrane proton fluxes.

OXIDATIVE PHOSPHORYLATION IN EUCARYOTES OCCURS IN MITOCHONDRIA

Mitochondria are oval-shaped organelles, typically about 2 μm in length and 0.5 μm in diameter. Techniques for isolating mitochondria were devised in the late 1940s. Eugene Kennedy and Albert Lehninger then discovered that *mitochondria contain the respiratory assembly, the enzymes of the citric acid cycle, and the enzymes of fatty acid oxidation.* Electron microscopic studies by George Palade and Fritjof Sjöstrand revealed that mitochondria have two membrane systems: an *outer membrane* and an extensive, highly folded *inner membrane.* The inner membrane is folded into a series of internal ridges called *cristae.* Hence, there are two compartments in mitochondria: the *intermembrane space* between the outer and inner membranes, and the *matrix,* which is bounded by the inner membrane (Figure 17-3). Oxidative phosphorylation takes place in the inner mitochondrial membrane, in contrast with most of the reactions of the citric acid cycle and fatty acid oxidation, which occur in the matrix.

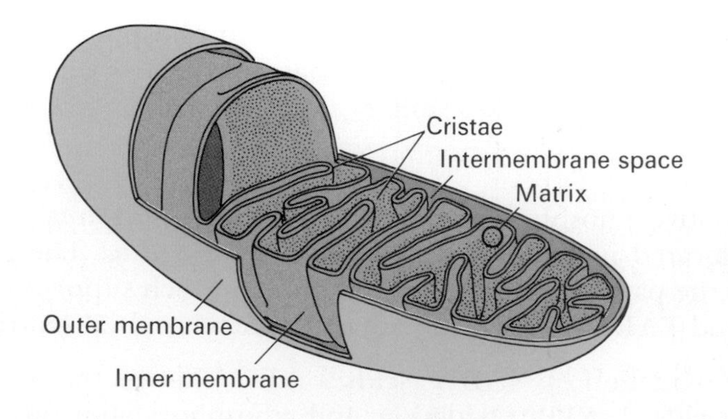

Figure 17-3
Diagram of a mitochondrion. [After *Biology of the Cell* by Stephen L. Wolfe. © 1972 by Wadsworth Publishing Company, Inc. Belmont, California 94002. Adapted by permission of the publisher.]

The outer membrane is quite permeable to most small molecules and ions because it contains many copies of *porin,* a transmembrane protein with a large pore. In contrast, the inner membrane is intrinsically impermeable to nearly all ions and polar molecules. Specific protein carriers transport molecules such as ADP and long-chain fatty acids across the inner mitochondrial membrane.

In procaryotes, the respiratory assemblies and ATP-synthesizing complex are located in the cytoplasmic membrane, the inner of two membranes. The outer membrane of bacteria, like that of mitochondria, is permeable to most small metabolites because of the presence of porin. It seems likely that mitochondria evolved from bacteria that lived symbiotically inside primitive cells.

REDOX POTENTIALS AND FREE-ENERGY CHANGES

In oxidative phosphorylation, the *electron-transfer potential* of NADH or $FADH_2$ is converted into the *phosphate-transfer potential* (phosphoryl potential) of ATP. We need quantitative expressions for these forms of free energy. The measure of phosphate-transfer potential is already familiar to us: it is given by $\Delta G^{\circ\prime}$ for hydrolysis of the phosphate compound. The corresponding expression for the electron-transfer potential is E_0', the *reduction potential* (also called the redox potential or oxidation-reduction potential).

The reduction potential is an electrochemical concept. Consider a substance that can exist in an oxidized form X and a reduced form X^-. Such a pair is called a *redox couple* (Figure 17-4). The reduction potential

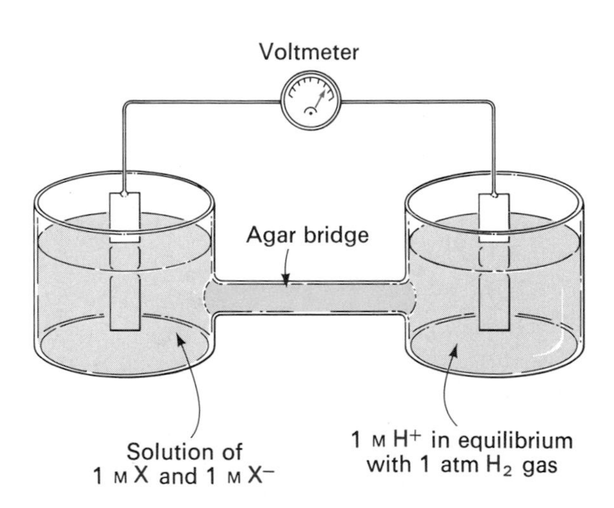

Figure 17-4
Apparatus for the measurement of the standard oxidation-reduction potential of a redox couple.

of this couple can be determined by measuring the electromotive force generated by a sample *half-cell* connected to a *standard reference half-cell.* The sample half-cell consists of an electrode immersed in a solution of 1 M oxidant (X) and 1 M reductant (X^-). The standard reference half-cell consists of an electrode immersed in a 1 M H^+ solution that is in equilibrium with H_2 gas at 1 atmosphere pressure. The electrodes are connected to a voltmeter, and electrical continuity between the half-cells is established by an agar bridge. Electrons then flow from one half-cell to the other. If the reaction proceeds in the direction

$$X^- + H^+ \longrightarrow X + \tfrac{1}{2}\,H_2$$

the reactions in the half-cells are

$$X^- \longrightarrow X + e^-$$

$$H^+ + e^- \longrightarrow \tfrac{1}{2}\,H_2$$

Thus, electrons flow from the sample half-cell to the standard reference half-cell, and consequently the sample-cell electrode is negative with respect to the standard-cell electrode. *The reduction potential of the X:X^- couple is the observed voltage at the start of the experiment* (when X, X^-, and H^+ are 1 M). *The reduction potential of the H^+:H_2 couple is defined to be 0 V* (volts).

The meaning of the reduction potential is now evident. A negative reduction potential means that a substance has lower affinity for electrons than does H_2, as in the above example. A positive reduction potential means that a substance has higher affinity for electrons than does H_2. These comparisons refer to standard conditions, namely 1 M oxidant, 1 M reductant, 1 M H^+, and 1 atmosphere H_2. Thus, *a strong reducing agent (such as NADH) has a negative reduction potential, whereas a strong oxidizing agent (such as O_2) has a positive reduction potential.*

The reduction potentials of many biologically important redox couples are known (Table 17-1). This table is like those presented in chemistry texts except that a hydrogen ion concentration of 10^{-7} M (pH 7) instead of 1 M (pH 0) is the standard state adopted by biochemists. This difference is denoted by the prime in E_0'. Recall that the prime in $\Delta G^{\circ\prime}$ denotes a standard free-energy change at pH 7.

Table 17-1
Standard reduction potentials of some reactions

Oxidant	*Reductant*	n	E_0' *(V)*
Succinate + CO_2	α-Ketoglutarate	2	− 0.67
Acetate	Acetaldehyde	2	− 0.60
Ferredoxin (oxidized)	Ferredoxin (reduced)	1	− 0.43
2 H^+	H_2	2	− 0.42
NAD^+	NADH + H^+	2	− 0.32
$NADP^+$	NADPH + H^+	2	− 0.32
Lipoate (oxidized)	Lipoate (reduced)	2	− 0.29
Glutathione (oxidized)	Glutathione (reduced)	2	− 0.23
Acetaldehyde	Ethanol	2	− 0.20
Pyruvate	Lactate	2	− 0.19
Fumarate	Succinate	2	0.03
Cytochrome *b* (+3)	Cytochrome *b* (+2)	1	0.07
Dehydroascorbate	Ascorbate	2	0.08
Ubiquinone (oxidized)	Ubiquinone (reduced)	2	0.10
Cytochrome *c* (+3)	Cytochrome *c* (+2)	1	0.22
Fe (+3)	Fe (+2)	1	0.77
$\frac{1}{2}$ O_2 + 2 H^+	H_2O	2	0.82

Note: E_0' is the standard oxidation-reduction potential (pH 7, 25°C) and n is the number of electrons transferred. E_0' refers to the partial reaction written as

$$\text{Oxidant} + e^- \longrightarrow \text{reductant}$$

The free-energy change of an oxidation-reduction reaction can be readily calculated from the difference in reduction potentials of the reactants. For example, consider the reduction of pyruvate by NADH:

(a) Pyruvate + NADH + H^+ $\rightleftharpoons$ lactate + NAD^+

The reduction potential of the NAD^+:NADH couple is -0.32 V, whereas that of the pyruvate:lactate couple is -0.19 V. By convention, reduction potentials refer to partial reactions written as reductions: oxidant + e^- → reductant. Hence,

(b) Pyruvate + 2 H^+ + 2 e^- ⟶ lactate $E'_0 = -0.19$ V
(c) NAD^+ + H^+ + 2 e^- ⟶ NADH $E'_0 = -0.32$ V

Subtracting reaction c from reaction b yields the desired reaction a and a $\Delta E'_0$ of +0.13 volt. Now we can calculate the $\Delta G^{\circ\prime}$ for the reduction of pyruvate by NADH. The standard free-energy change $\Delta G^{\circ\prime}$ is related to the change in reduction potential $\Delta E'_0$ by

$$\Delta G^{\circ\prime} = -nF\Delta E'_0$$

in which n is the number of electrons transferred, F is the energy change as a mole of electrons falls through a potential of one volt (23.06 kcal V^{-1} mol^{-1}), $\Delta E'_0$ is in volts, and $\Delta G^{\circ\prime}$ is in kilocalories per mole. Note that $\Delta G^{\circ\prime}$ is the amount of *free energy* that can be obtained per mole from a transformation, whereas $\Delta E'_0$ is the difference in *potential* between two states. Hence, $\Delta E'_0$ must be multiplied by nF to determine the free energy yield. For the reduction of pyruvate, $n = 2$ and so

$$\begin{aligned}\Delta G^{\circ\prime} &= -2 \times 23.06 \times 0.13 \\ &= -6 \text{ kcal/mol}\end{aligned}$$

A *positive* $\Delta E'_0$ (but a *negative* $\Delta G^{\circ\prime}$) signifies an *exergonic reaction* under standard conditions.

THE SPAN OF THE RESPIRATORY CHAIN IS 1.14 VOLTS, WHICH CORRESPONDS TO 53 KCAL

The driving force of oxidative phosphorylation is the electron-transfer potential of NADH or $FADH_2$ relative to that of O_2. Let us calculate the $\Delta E'_0$ and $\Delta G^{\circ\prime}$ of the oxidation of NADH by O_2. The pertinent partial reactions are

(a) $\frac{1}{2}$ O_2 + 2 H^+ + 2 e^- ⟶ H_2O $E'_0 = +0.82$ V

(b) NAD^+ + H^+ + 2 e^- ⟶ NADH $E'_0 = -0.32$ V

Subtracting reaction b from reaction a yields

(c) $\frac{1}{2}$ O_2 + NADH + H^+ ⇌ H_2O + NAD^+ $\Delta E'_0 = +1.14$ V

The free energy of oxidation of this reaction is then given by

$$\begin{aligned}\Delta G^{\circ\prime} &= -nF\Delta E'_0 = -2 \times 23.06 \times 1.14 \\ &= -52.6 \text{ kcal/mol}\end{aligned}$$

THE RESPIRATORY CHAIN CONSISTS OF THREE ENZYME COMPLEXES LINKED BY TWO MOBILE ELECTRON CARRIERS

Electrons are transferred from NADH to O_2 through a chain of three large protein complexes called *NADH-Q reductase, cytochrome reductase,* and *cytochrome oxidase* (Figure 17-5 and Table 17-2). Electron flow within these complexes, which pierce the inner mitochondrial membrane, leads to the pumping of protons across the membrane. As will be discussed shortly, the electron-carrying groups in these enzymes are *flavins, iron-sulfur clusters, hemes,* and *copper ions.* Electrons are carried from NADH-Q reductase to cytochrome reductase, the second complex of the chain, by the reduced form of *ubiquinone,* a hydrophobic

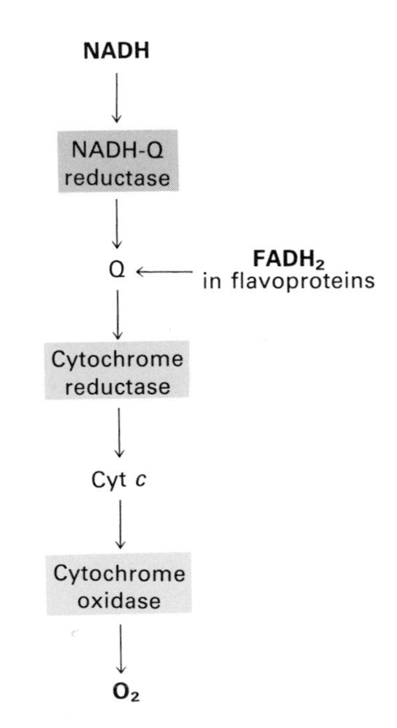

Figure 17-5
Sequence of electron carriers in the respiratory chain. Protons are pumped by the three complexes shown in color.

Table 17-2
Components of the mitochondrial electron-transport chain

Enzyme complex	Mass (kd)	Subunits	Prosthetic group	Location of binding sites		
				Matrix side	Hydrocarbon core	Cytosolic side
NADH-Q reductase	850	25	FMN Fe-S	NADH	Q	
Succinate-Q reductase	140	4	FAD Fe-S	Succinate	Q	
Cytochrome reductase	250	9	Heme *b*-562 Heme *b*-566 Heme c_1 Fe-S		Q	Cyt *c*
Cytochrome *c*	13	1	Heme *c*			Cyt c_1 Cyt *a*
Cytochrome oxidase	160	8	Heme *a* Heme a_3 Cu_A and Cu_B			Cyt *c*

Sources: J. W. DePierre and L. Ernster, *Ann. Rev. Biochem.* 46(1977):215, and Y. Hatefi, *Ann. Rev. Biochem.* 54(1985):1015.
Note: All components are integral membrane proteins, except for cytochrome *c*, which is a water-soluble peripheral membrane protein, and coenzyme Q, which is a lipid-soluble quinone.

quinone. Ubiquinone also carries electrons from $FADH_2$ (produced, for example, by the oxidation of succinate in the citric acid cycle) to cytochrome reductase. Cytochrome *c*, a small protein, shuttles electrons from cytochrome reductase to cytochrome oxidase, the final component in the chain.

THE HIGH-POTENTIAL ELECTRONS OF NADH ENTER THE RESPIRATORY CHAIN AT NADH-Q REDUCTASE

The electrons of NADH enter the chain at *NADH-Q reductase* (also called *NADH dehydrogenase*), a large enzyme (850 kd) consisting of some 25 polypeptide chains. Two electrons are transferred from NADH to the *flavin mononucleotide* (FMN) prosthetic group of this enzyme to give the reduced form, $FMNH_2$.

$$\text{NADH} + \text{H}^+ + \text{FMN} \longrightarrow \text{FMNH}_2 + \text{NAD}^+$$

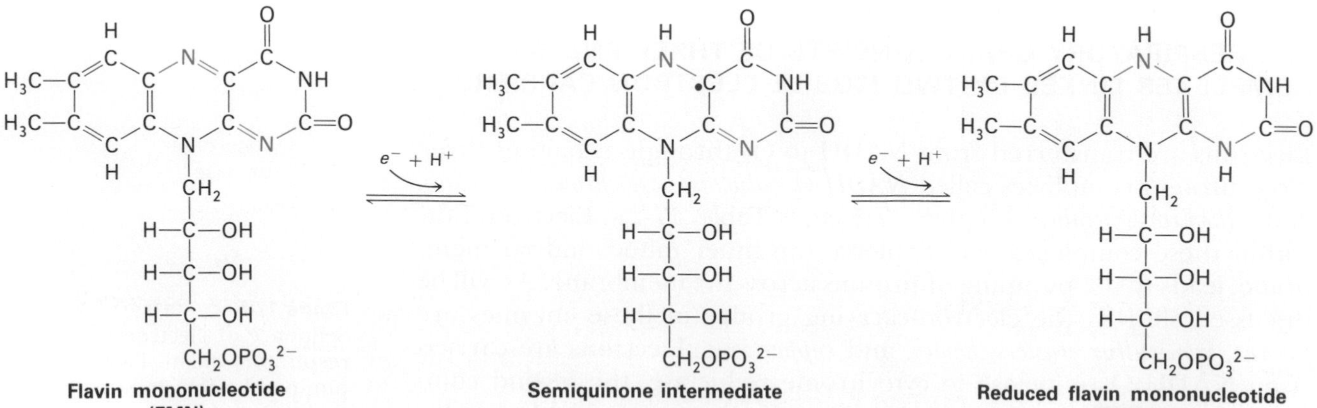

$e^- + H^+$ $e^- + H^+$

Flavin mononucleotide (FMN)   **Semiquinone intermediate**   **Reduced flavin mononucleotide ($FMNH_2$)**

FMN can also accept one electron (or $FMNH_2$ can donate one electron) by forming a semiquinone radical intermediate.

Electrons are then transferred from $FMNH_2$ to a series of *iron-sulfur clusters* (abbreviated as Fe-S), the second type of prosthetic group in NADH-Q reductase. Fe-S clusters in *iron-sulfur proteins* (also called nonheme iron proteins) play a critical role in a wide range of reduction reactions in biological systems. Several types of Fe-S clusters are known (Figure 17-6). In the simplest kind, a single iron atom is tetrahedrally coordinated to the sulfhydryl groups of four cysteine residues of the protein. A second kind, denoted by [2Fe-2S], contains two iron atoms and two inorganic sulfides, in addition to four cysteine residues. A third type, designated as [4Fe-4S], contains four iron atoms, four inorganic sulfides, and four cysteine residues. NADH-Q reductase contains both [2Fe-2S] and [4Fe-4S] clusters. Iron atoms in these FeS complexes cycle between Fe^{2+} (reduced) or Fe^{3+} (oxidized) states.

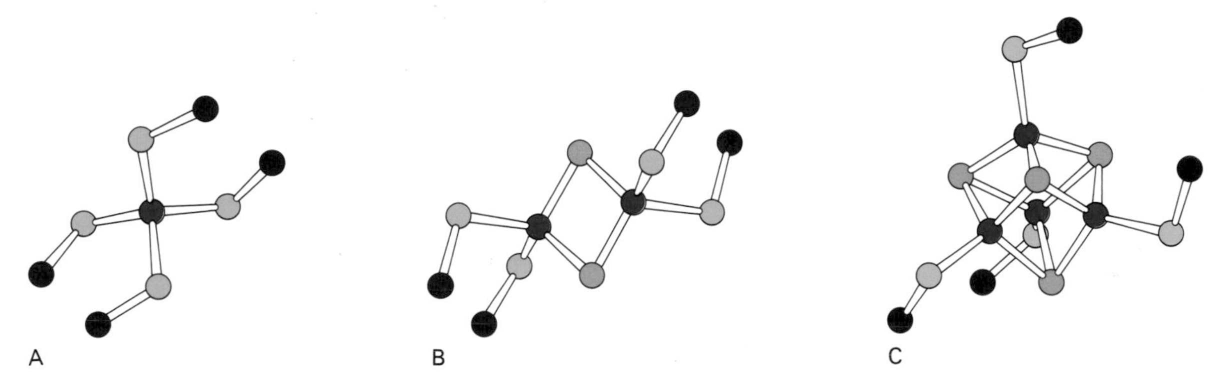

Figure 17-6
Molecular models of iron-sulfur complexes: (A) cluster containing one Fe; (B) [2Fe-2S] cluster; (C) [4Fe-4S] cluster. Iron atoms are shown in red, cysteine sulfur atoms in yellow, and inorganic sulfur atoms in green. [Drawn from atomic coordinates of model compounds kindly provided by Dr. Jeremy Berg.]

Electrons in the iron-sulfur clusters of NADH-Q reductase are then shuttled to *coenzyme Q*, also known as *ubiquinone* (Q) because it is ubiquitous in biological systems. Q is a quinone derivative with a long isoprenoid tail. The number of isoprene units in Q depends on the species. The most common form in mammals contains ten isoprene units (Q_{10}) (Figure 17-7). For simplicity, we will omit the subscript from this abbreviation. Ubiquinone is reduced to a *free-radical semiquinone* by the uptake of a single electron. Reduction of this enzyme-bound intermediate by a second electron yields *ubiquinol* (QH_2).

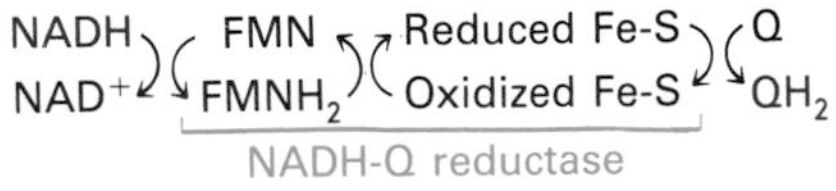

NADH → NAD^+; FMN → $FMNH_2$; Reduced Fe-S → Oxidized Fe-S; Q → QH_2
NADH-Q reductase

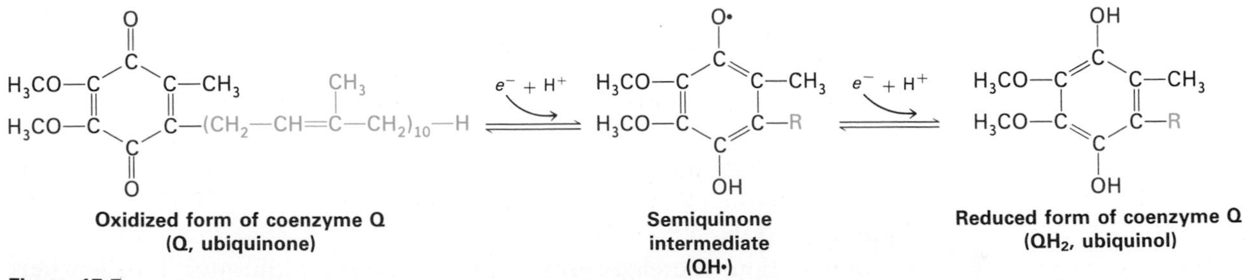

$e^- + H^+$; $e^- + H^+$

Oxidized form of coenzyme Q (Q, ubiquinone)

Semiquinone intermediate (QH•)

Reduced form of coenzyme Q (QH_2, ubiquinol)

Figure 17-7
Ubiquinone (Q) is reduced to ubiquinol (QH_2) through a semiquinone intermediate (QH ·).

UBIQUINOL (QH_2) IS THE ENTRY POINT FOR ELECTRONS FROM $FADH_2$ OF FLAVOPROTEINS

Recall that $FADH_2$ is formed in the citric acid cycle in the oxidation of succinate to fumarate by succinate dehydrogenase, part of the *succinate-Q reductase complex*, an integral membrane protein of the inner mitochondrial membrane. Thus, the newly formed $FADH_2$ does not leave the complex, but rather its electrons are transferred to Fe-S centers and then to Q for entry into the electron-transport chain. Likewise, the $FADH_2$ moieties of *glycerol phosphate dehydrogenase* (p. 417) and *fatty acyl CoA dehydrogenase* (p. 475) transfer their high-potential electrons to Q to form QH_2, the reduced state. The isoprenoid tail makes Q highly nonpolar, which enables it to diffuse rapidly in the hydrocarbon phase of the inner mitochondrial membrane. Q is the only electron carrier in the respiratory chain that is not permanently bound or covalently attached to a protein. Succinate-Q reductase complex and other enzymes that transfer electrons from $FADH_2$ to Q are not proton pumps, in contrast with NADH-Q reductase, because the free-energy change is too small. Consequently, less ATP is formed from the oxidation of $FADH_2$ than from NADH (p. 418).

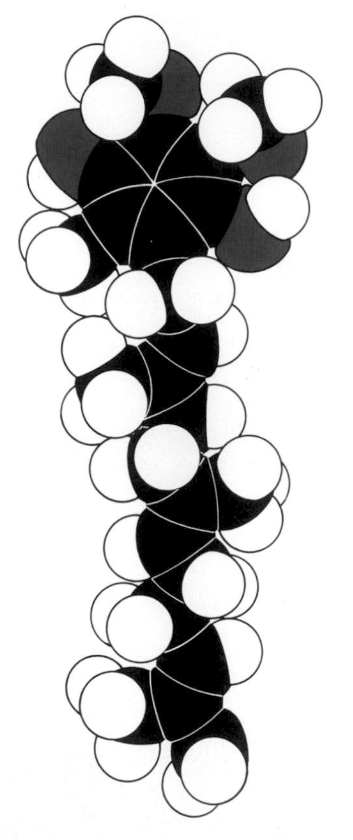

Figure 17-8
Model of ubiquinol. Only three of the ten isoprenoid units are shown. These five-carbon units make the molecule very nonpolar.

ELECTRONS FLOW FROM UBIQUINOL TO CYTOCHROME *c* THROUGH CYTOCHROME REDUCTASE

The second of the three proton pumps in the respiratory chain is *cytochrome reductase* (also called ubiquinol–cytochrome *c* reductase). *A cytochrome is an electron-transferring protein that contains a heme prosthetic group.* The key role of cytochromes in respiration was discovered in 1925 by David Keilin. Their iron atoms alternate between a reduced ferrous (+2) state and an oxidized ferric (+3) state during electron transport. The function of cytochrome reductase is to catalyze the transfer of electrons from QH_2 to cytochrome *c*, a water-soluble protein, and to concomitantly pump protons across the inner mitochondrial membrane.

Q / QH· ⇄ Cyt *b* (+2) / Cyt *b* (+3) ⇄ QH· / QH_2 ⇄ Fe-S (+2) / Fe-S (+3) ⇄ Cyt *c* (+3) / Cyt *c* (+2) ⇄ Cyt *c* (+2) / Cyt *c* (+3)

Cytochrome reductase

Cytochrome reductase itself contains two types of cytochromes, named *b* and c_1. The reductase also contains an Fe-S protein and several other polypeptide chains. The prosthetic group of cytochromes *b*, c_1, and *c* is iron-protoporphyrin IX, the same *heme* as in myoglobin and hemoglobin (p. 144). The hemes of cytochromes *c* and c_1, in contrast with those in *b*, are covalently attached to the protein (Figure 17-9). The linkages are thioethers formed by the addition of the sulfhydryl groups of two cysteine residues to the vinyl groups of the heme.

R—CH=CH_2
Vinyl group of the heme
+
HS—CH_2—R′
Cysteine residue of the protein
↓
CH_3
R—CH—S—CH_2—R′
Thioether linkage

Ubiquinol transfers one of its two high-potential electrons to the Fe-S cluster in the reductase. This electron is then shuttled sequentially to cytochrome c_1 and cytochrome *c*, which carries it away from this complex. This one-electron transfer converts ubiquinol (QH_2) into the semiquinone (QH ·).

What about the electron residing in the semiquinone? This is where cytochrome *b* with its two heme groups enters the act. Its two identical hemes have different electron affinities and spectral properties because

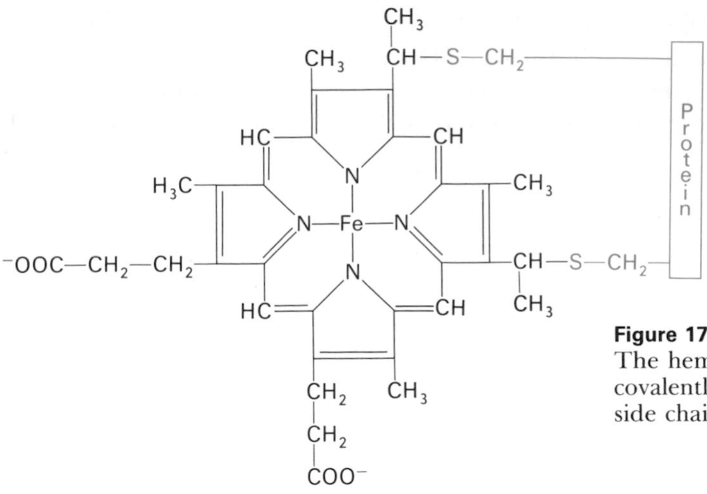

Figure 17-9
The heme in cytochromes c and c_1 is covalently attached to two cysteine side chains.

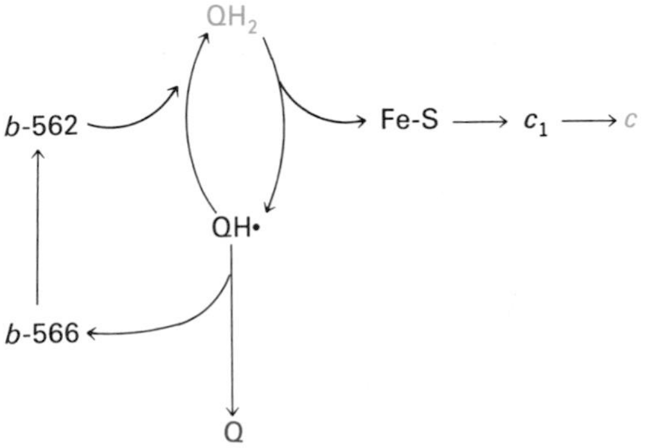

Figure 17-10
Pathway of electron transfer in cytochrome reductase: from QH_2 to an Fe-S cluster to cytochrome c_1 and then to cytochrome c. The hemes of cytochrome b in this complex, oxidize-reduce two molecules of $QH\cdot$ into QH_2 and Q.

they are in different polypeptide environments. One of them is called b-566 and the other b-562 because they absorb light maximally at 566 and 562 nm. $QH\cdot$ transfers its electron to b-566, which then reduces b-562. In turn, b-562 reduces $QH\cdot$ to QH_2. Thus, the net reaction catalyzed by cytochrome b is

$$QH\cdot + QH\cdot \xrightarrow{\text{cyt } b} QH_2 + Q$$

In essence, the cytochrome b component of the reductase is a recycling device that enables a two-electron carrier (ubiquinol) to interact with a one-electron carrier (Fe-S cluster).

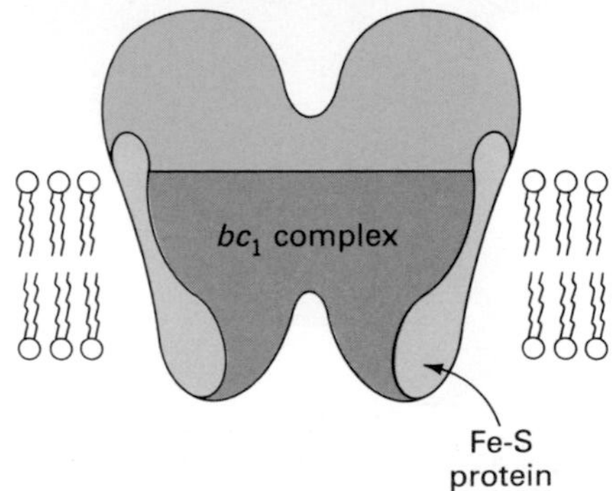

Figure 17-11
A schematic diagram of the structure of cytochrome reductase in a reconstituted bilayer membrane. [Courtesy of Dr. Hanns Weiss and Dr. Kevin Leonard.]

CYTOCHROME OXIDASE CATALYZES THE TRANSFER OF ELECTRONS FROM CYTOCHROME *c* TO O_2

Cytochrome oxidase, the last of the three proton-pumping assemblies of the respiratory chain, catalyzes the transfer of electrons from reduced cytochrome c to molecular oxygen, the final acceptor.

$$4\text{ Cyt } c\ (+2) + 4\ H^+ + O_2 \longrightarrow 4\text{ cyt } c\ (+3) + 2\ H_2O$$

Four electrons are funneled into O_2 to completely reduce it to H_2O and concomitantly pump protons from the matrix to the cytosolic side of the inner mitochondrial membrane. The term *cytosolic* is used to refer to the side of the inner membrane opposite the matrix side because it is freely accessible to most small molecules in the cytosol. The outer mitochondrial membrane has pores that enable most metabolites to pass through.

This reaction is carried out by a complex of at least eight subunits, of which three (called subunits I, II, and III) are encoded by the mitochondrion's own genome. Cytochrome oxidase contains two *heme A* groups and two copper ions. Heme A differs from the heme in cytochrome c and c_1 in that a formyl group replaces one of the methyl groups, and a hydrocarbon chain replaces one of the vinyl groups (Figure 17-13). The hemes, though chemically identical, have different properties because they are located in different parts of cytochrome oxidase. One of them is called heme a and the other heme a_3. Likewise, the two copper ions, called Cu_A and Cu_B are distinct because they are bound differently by the protein. Heme a is close to Cu_A in subunit II, and heme a_3 is next to Cu_B in subunit I. The oxidation-reduction units of cytochrome oxidase known as *cytochromes* a and a_3 are located in subunits I and II.

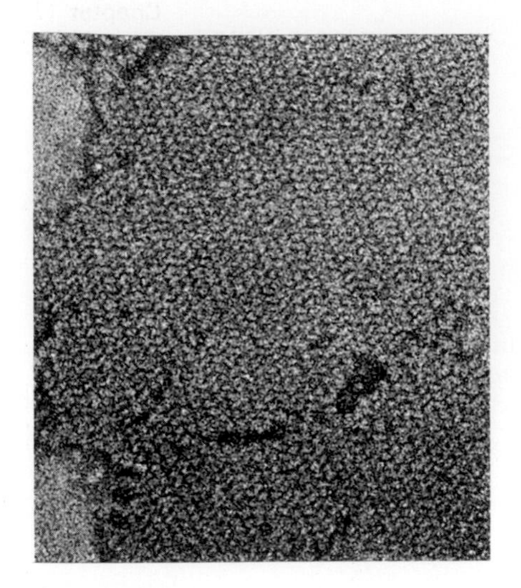

Figure 17-12
Electron micrograph of a two-dimensional crystalline array of cytochrome oxidase. [Courtesy of Dr. Steven Fuller and Dr. Roderick Capaldi.]

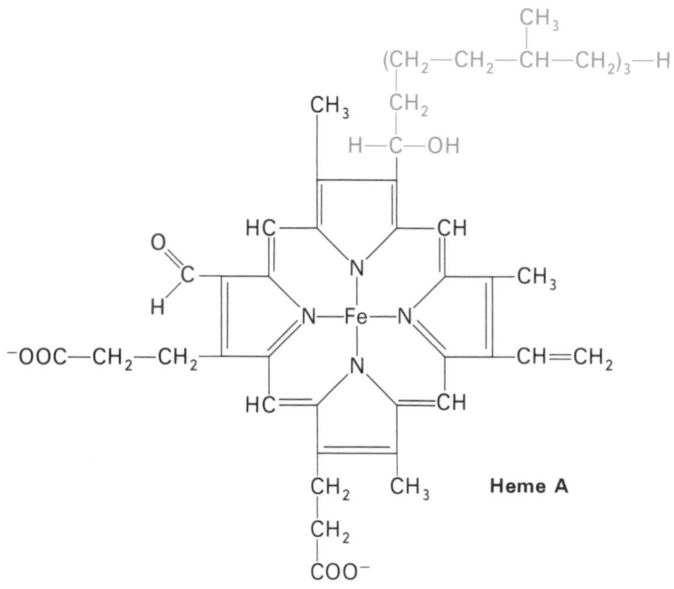

Heme A

Figure 17-13
Heme A, which is present in cytochrome oxidase, differs from other hemes in having a formyl group (red) and a long hydrophobic side chain (green).

Ferrocytochrome c donates its electron to the heme a-Cu_A cluster. An electron is then transferred to the heme a_3-Cu_B cluster, where O_2 is reduced in a series of steps to two molecules of H_2O.

Molecular oxygen is an ideal terminal electron acceptor. Its high affinity for electrons provides a large thermodynamic driving force for oxidative phosphorylation. Moreover, O_2, in contrast with other strong electron acceptors (such as F_2), reacts very slowly unless activated by a catalyst. *However, danger lurks in the reduction of O_2.* The transfer of four electrons leads to safe products (two molecules of H_2O), but partial reduction generates highly hazardous compounds. In particular, *superoxide anion*, a highly destructive compound (p. 422), is formed by the transfer of a single electron to O_2.

$$O_2 + e^- \rightleftharpoons O_2^{\cdot -}$$

Superoxide anion

The strategy for the safe reduction of O_2 is clear: *the catalyst must not release partly-reduced intermediates.* Cytochrome oxidase meets this crucial criterion by binding O_2 between the Fe^{2+} and Cu^+ ions of its a_3-Cu_B center. Each metal then donates an electron to O_2 to convert it to a

Figure 17-14
Structure of cytochrome oxidase in a reconstituted bilayer membrane. This image was reconstructed at low resolution from electron micrographs of a two-dimensional crystalline array. [Courtesy of Dr. Roderick Capaldi.]

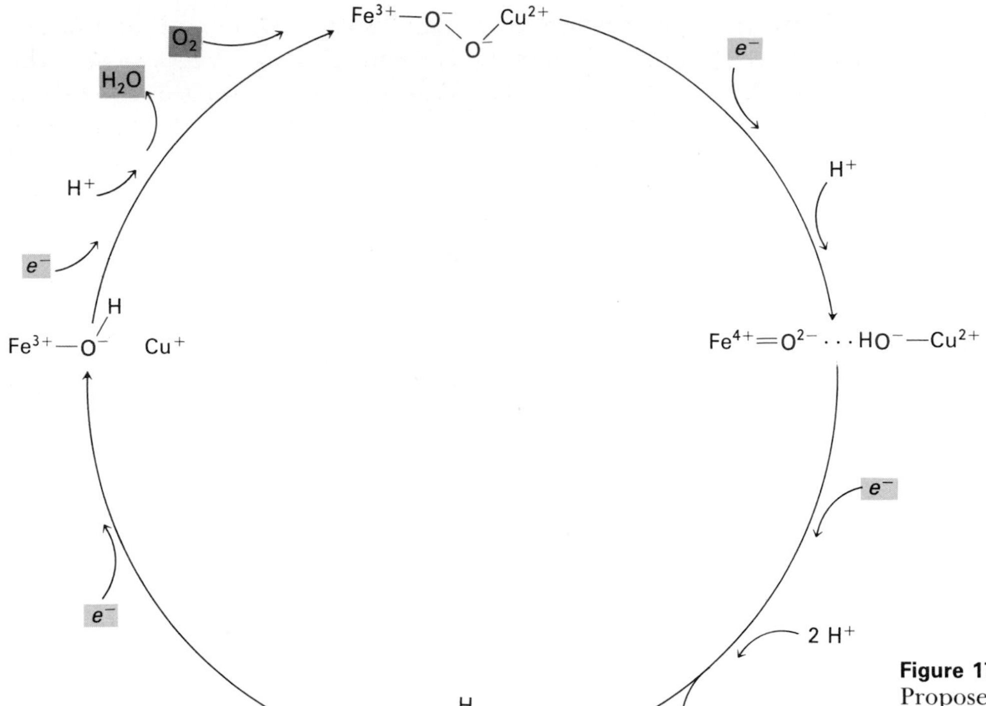

Figure 17-15
Proposed reaction cycle for the four-electron reduction of O_2 by cytochrome oxidase.

dianion (Figure 17-15). The input of an electron from the *a*-Cu_A center then leads to the formation of a *ferryl* intermediate in which iron is in the +4 state. Water is formed and released following acceptance of a second electron, which leaves OH^- bound to Fe^{3+}. A third electron serves to reduce cupric ion to the +1 state. In the last steps of this cycle, an electron and O_2 enter to yield H_2O and form a bound oxygen dianion. Protons enter with electrons in three of the four steps of the cycle.

THREE-DIMENSIONAL STRUCTURE OF CYTOCHROME *c*

Cytochrome *c* is the only electron-transport protein that can be separated from the inner mitochondrial membrane by gentle treatment. The solubility of this peripheral membrane protein in water has facilitated its purification and crystallization. Indeed, much more is known about the structure of cytochrome *c* than of any other electron-carrying protein.

Cytochrome *c* from beef heart consists of a single polypeptide chain of 104 amino acid residues and a covalently attached heme group. The three-dimensional structures of the ferrous and ferric forms of cytochrome *c* have been elucidated at nearly atomic resolution by Richard Dickerson (Figure 17-16). The protein is roughly spherical, with a diameter of 34 Å. The heme group is surrounded by many tightly packed hydrophobic side chains. The iron atom is bonded to the sulfur atom of a methionine residue and to the nitrogen atom of a histidine residue (Figure 17-17). The hydrophobic character of the heme environment makes the reduction potential of cytochrome *c* more positive (corresponding to a higher electron affinity) than that of the same heme

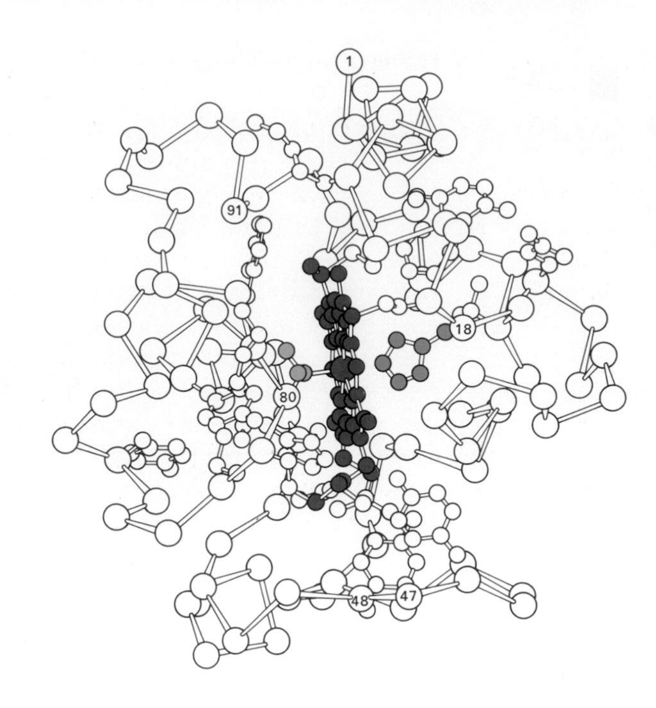

Figure 17-16
Three-dimensional structure of reduced cytochrome *c* from tuna. The heme group (red), methionine 80 (green), histidine 18 (blue), and the α-carbon atoms are shown. [After T. Takano, O. B. Kallai, R. Swanson, and R. E. Dickerson. *J. Biol. Chem.* 248(1973):5244.]

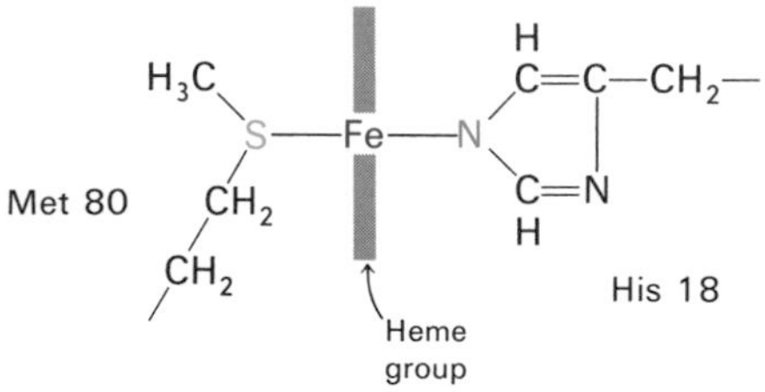

H_3C S—Fe—N $C{=}C{-}CH_2{-}$ $C{=}N$
Met 80 CH_2 CH_2
Heme group
His 18

Figure 17-17
The iron atom of the heme group in cytochrome *c* is bonded to a methionine sulfur atom and a histidine nitrogen atom.

complex in an aqueous milieu. It is energetically more costly to remove an electron from the heme in cytochrome *c* than from a heme in water because the dielectric constant near the iron atom is lower in cytochrome *c*.

The overall structure of the molecule can be characterized as a shell one residue thick surrounding the heme. Hydrophobic side chains make up the innermost part of the shell. The main chain comes next, followed by charged side chains on the surface. There is very little α helix and no β pleated sheet. In essence, the polypeptide chain is wrapped around the heme. Residues 1 to 47 are on the histidine 18 side of the heme (called the right side), whereas 48 to 91 are on the methionine 80 side (the left side). Residues 92 to 104 extend back across the heme to the right side.

ELECTROSTATIC INTERACTIONS ARE CRITICAL IN THE DOCKING OF CYTOCHROME *c* WITH ITS REACTION PARTNERS

As discussed previously, cytochrome *c* carries electrons from cytochrome reductase complex (the second proton-pumping complex) to cytochrome oxidase (the third). How does cytochrome *c* interact with its reductase and then with its oxidase? An important clue came from the distribution of charged residues on the surface of this highly basic protein. Cytochrome *c* molecules from all species studied thus far have clusters of lysine side chains around the heme crevice on one face of the protein (the front of Figure 17-16). This array of positive charges on the surface of cytochrome *c* is central to the recognition and binding of the reductase and oxidase, which are negatively-charged. For example, the interaction of cytochrome *c* with cytochrome oxidase is impaired if lysine 13 is modified. Furthermore, polylysine competes with cytochrome *c* in binding to the oxidase and reductase. *Nearly every collision between cytochrome* c *and its reaction partners is a productive encounter because electrostatic forces bring complementary positively and negatively charged groups into apposition.*

How does cytochrome *c* accept an electron from its reductase, which it then donates to its oxidase? A priori, two kinds of mechanisms are possible. An electron could be transferred between heme groups in different proteins by a relay of aromatic side chains. Alternatively, an electron could be directly transferred from one heme to another. It is important to note that an electron carried by a heme is not confined to its iron atom. Rather, it is partly delocalized over the entire conjugated π-electron network of the heme. Hence, an electron can be transferred from one heme to another if their edges are sufficiently close (less than about 8 Å apart) and if their planes are nearly parallel. The direct mechanism for electron transfer seems more likely because the free energy required to form a free-radical anion of an aromatic side chain is very high. Furthermore, one of the edges of the heme group of cytochrome *c* (the front edge in Figure 17-16) is accessible for direct electron transfer.

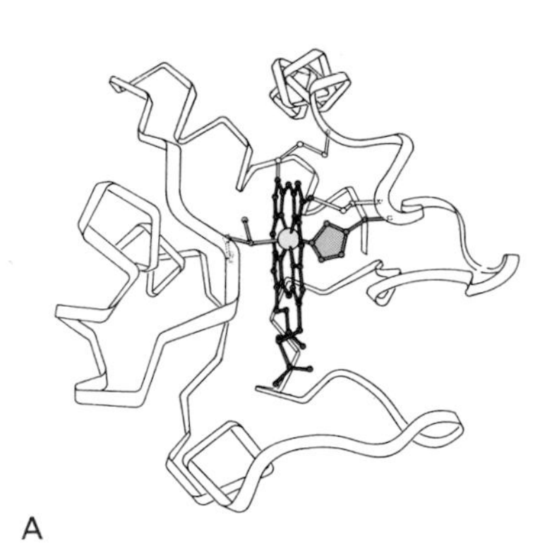

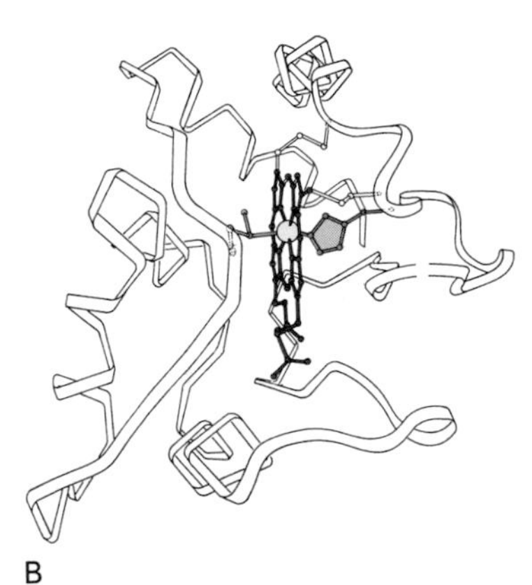

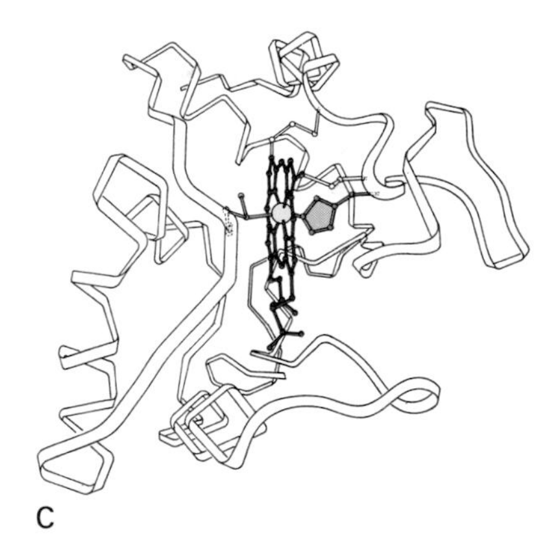

Figure 17-18
Conservation of the three-dimensional structure of cytochrome *c* in evolution is exemplified by the similarity in conformation of (A) cytochrome *c* from tuna-heart mitochondria, (B) cytochrome c_2 from *Rhodospirillum rubrum*, a photosynthetic bacterium, and (C) cytochrome c_{550} from *Paracoccus dentrificans*, a denitrifying bacterium. [After F. R. Salamme. Reproduced, with permission, from the *Annual Review of Biochemistry*, Volume 46. © 1977 by Annual Reviews, Inc.]

THE CONFORMATION OF CYTOCHROME *c* HAS REMAINED ESSENTIALLY CONSTANT FOR A BILLION YEARS

Cytochrome *c* is present in all organisms having mitochondrial respiratory chains: plants, animals, and eucaryotic microorganisms. This electron carrier evolved more than 1.5 billion years ago, before the divergence of plants and animals. Its function has been conserved throughout this period, as evidenced by the fact that *the cytochrome* c *of any eucaryotic species reacts in vitro with the cytochrome oxidase of any other species tested thus far.* For example, wheat-germ cytochrome *c* reacts with human cytochrome oxidase. A second criterion of the conservation of function is that the reduction potentials of all cytochrome *c* molecules studied are close to +0.25 V. Third, the absorption spectra of cytochrome *c* molecules of many species are virtually indistinguishable. Fourth, some procaryotic cytochromes, such as cytochrome c_2 from a photosynthetic bacterium and cytochrome c_{550} from a denitrifying bacterium, closely resemble cytochrome *c* from tuna-heart mitochondria (Figure 17-18).

The amino acid sequences of cytochrome *c* from more than eighty widely ranging eucaryotic species have been determined by Emil Smith, Emanuel Margoliash, and others. The striking finding is that *26 of 104 residues have been invariant for more than one and a half billion years of evolution.* The reasons for the constancy of many of these residues are evident now that the three-dimensional structure of the molecule is known. As might be expected, the heme ligands methionine 80 and histidine 18 are invariant, as are the two cysteines that are covalently bonded to the heme. A sequence of 11 residues from number 70 to number 80 is nearly the same in all cytochrome *c* molecules. Many hydrophobic residues in contact with the heme are invariant. Furthermore, most of the glycine residues in cytochrome *c* have been preserved. As discussed earlier (p. 264), glycine is important because it is small. The compact folding of the polypeptide chain requires the presence of glycine at certain sites. Several invariant lysine and arginine residues are located in the positively charged clusters on the surface of the molecule. One of these clusters interacts with cytochrome *c* reductase, and another with cytochrome oxidase.

OXIDATION AND PHOSPHORYLATION ARE COUPLED BY A PROTON-MOTIVE FORCE

Thus far, consideration has been given to the flow of electrons from NADH to O_2, an exergonic process:

$$\text{NADH} + \tfrac{1}{2}\,\text{O}_2 + \text{H}^+ \rightleftharpoons \text{H}_2\text{O} + \text{NAD}^+ \qquad \Delta G^{\circ\prime} = -52.6 \text{ kcal/mol}$$

This free energy of oxidation is used to synthesize ATP.

$$\text{ADP} + \text{P}_i + \text{H}^+ \rightleftharpoons \text{ATP} + \text{H}_2\text{O} \qquad \Delta G^{\circ\prime} = +7.3 \text{ kcal/mol}$$

The synthesis of ATP is carried out by a molecular assembly in the inner mitochondrial membrane. This enzyme complex (p. 414) has been called the *mitochondrial ATPase* or *H^+-ATPase* because it was discovered through its catalysis of the hydrolytic reaction. *ATP synthase*, its other name, emphasizes its actual role in the mitochondrion.

How is the oxidation of NADH coupled to the phosphorylation of ADP? It was first suggested that electron transfer leads to the formation of a covalent high-energy intermediate that serves as the precursor of ATP. This chemical-coupling hypothesis was based on the mechanism of ATP formation in glycolysis (e.g., the formation of 1,3-bisphosphoglycerate as a high-energy intermediate, p. 355). An alternative proposal was that the free energy of oxidation is trapped in an activated protein conformation, which then drives the synthesis of ATP. Investigators in many laboratories tried for several decades to isolate these putative energy-rich intermediates, but none were found.

A radically different mechanism, *the chemiosmotic hypothesis,* was postulated by Peter Mitchell in 1961. He proposed that electron transport and ATP synthesis are coupled by a proton gradient across the inner mitochondrial membrane rather than by a covalent high-energy intermediate or an activated protein conformation. In his model, the transfer of electrons through the respiratory chain leads to the pumping of protons from the matrix to the other side of the inner mitochondrial membrane. The H^+ concentration becomes higher on the cytosolic side, and an electric potential with that side positive is generated (Figure 17-19). *Mitchell postulated that this proton-motive force drives the synthesis of ATP by the ATPase complex. In essence, the primary energy-conserving event in this model is the movement of protons across the inner mitochondrial membrane.*

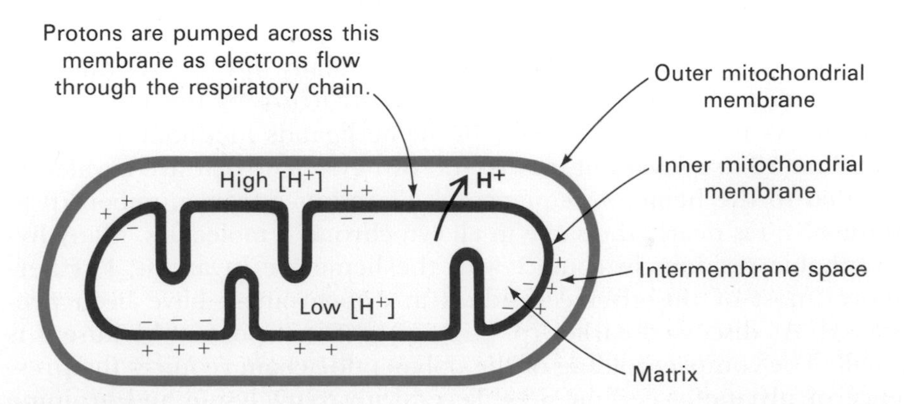

Figure 17-19
Electron transfer through the respiratory chain leads to the pumping of protons from the matrix to the cytosolic side of the inner mitochondrial membrane. The pH gradient and membrane potential constitute a *proton-motive force* that is used to drive ATP synthesis.

Mitchell's highly innovative hypothesis that oxidation and phosphorylation are coupled by a proton gradient is now supported by a wealth of evidence:

1. A proton gradient across the inner mitochondrial membrane is generated during electron transport. The pH outside is 1.4 units lower than inside, and the membrane potential is 0.14 V, the outside being positive. The proton-motive force Δp (in volts) consists of a membrane-potential contribution (E_m) and a H^+ concentration-gradient contribution (ΔpH). In the following equation, R is the gas constant, T is the absolute temperature, and F is the energy change as a mole of electrons passes through an electric potential of one volt.

$$\Delta p = E_m - \frac{2.3\ RT}{F}\Delta pH = E_m - 0.06\ \Delta pH$$
$$= 0.14 - 0.06(-1.4) = 0.224\ V$$

This total proton-motive force of 0.224 V corresponds to a free energy of 5.2 kcal per mole of protons.

2. ATP is synthesized when a pH gradient is imposed on mitochondria or chloroplasts (p. 527) in the absence of electron transport.

3. Bacteriorhodopsin, a purple-membrane protein from halobacteria, pumps protons when illuminated (p. 962). Synthetic vesicles containing this bacterial protein and a purified ATPase from beef-heart mitochondria synthesize ATP when illuminated (Figure 17-20). In this key experiment, carried out by Walther Stoeckenius and Efraim Racker, bacteriorhodopsin replaces the respiratory chain, which shows that the respiratory chain and ATP synthase are biochemically separate systems, linked only by a proton-motive force.

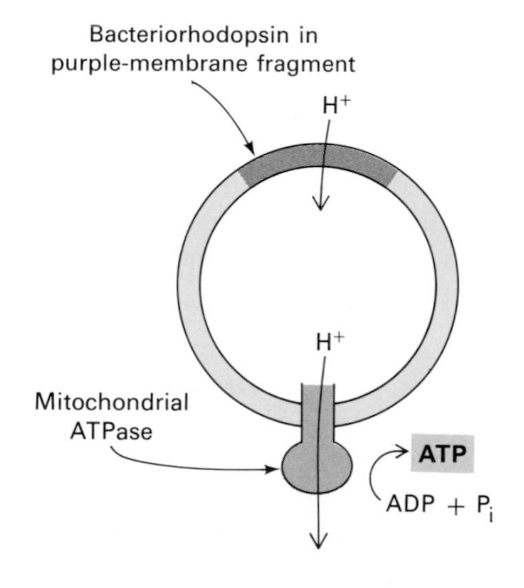

Figure 17-20
ATP is synthesized when reconstituted membrane vesicles containing bacteriorhodopsin (a light-driven proton pump) and ATP synthase are illuminated. The orientation of ATP synthase in this reconstituted membrane is the reverse of that in the mitochondrion.

4. Submitochondrial particles formed by the sonication of mitochondria (see Figure 17-23) have been very informative in showing that the respiratory chain and ATP synthase are vectorially organized. These particles are like miniature inner mitochondrial membranes turned inside out. Hence, both surfaces of the inner mitochondrial membrane are experimentally accessible: the cytosolic side in intact mitochondria, and the matrix side in submitochondrial particles. Studies using proteolytic enzymes, specific antibodies, lectins, and other reagents that do not

cross the inner mitochondrial membrane have shown that each of the producers and consumers of the proton-motive force—NADH-Q reductase, cytochrome reductase, cytochrome oxidase, and ATP synthase—spans the membrane and is asymmetrically oriented.

5. A closed compartment is essential for oxidative phosphorylation. ATP synthesis coupled to electron transfer does not occur in soluble preparations or in membrane fragments lacking well-defined inside and outside compartments.

6. Substances that carry protons across the inner mitochondrial membrane dissipate the proton gradient and uncouple oxidation and phosphorylation (p. 419).

A PROTON-MOTIVE FORCE IS GENERATED BY THREE ELECTRON-TRANSFER COMPLEXES

As was mentioned earlier, protons are pumped at three sites as electrons flow through the respiratory chain from NADH to O_2 (see Figure 17-5): *site 1* is the NADH-Q reductase complex; *site 2* is cytochrome reductase; and *site 3* is cytochrome oxidase. The proton gradient generated at each site by the flow of a pair of electrons can be used to synthesize about one molecule of ATP. These sites have been identified in several ways:

1. *Comparison of the ATP yield from the oxidation of several substrates.* The oxidation of NADH yields about three ATP, whereas the oxidation of succinate yields about two ATP. The electrons from $FADH_2$ enter the electron-transport chain through QH_2, bypassing the first proton-pumping site (NADH-Q reductase) because $FADH_2$ is a less powerful electron donor than is NADH. Only one ATP is formed when ascorbate, an artificial substrate, is oxidized, because its electrons enter at cytochrome *c*, which is at a lower potential than is QH_2. The *P:O ratio*, defined as the number of molecules of inorganic phosphate incorporated into organic form per atom of oxygen consumed, is a frequently used index of oxidative phosphorylation. P:O ratios of about 3, 2, and 1 have been observed for the oxidation of NADH, succinate, and ascorbate, respectively, under a particular set of experimental conditions.

2. *Thermodynamic estimates.* The $\Delta G^{\circ\prime}$ for electron transfer from NADH to the lowest-energy Fe-S center in NADH-Q reductase is −12 kcal/mol; from cytochromes *b* to c_1 in cytochrome reductase, −10 kcal/mol; and from cytochrome *a* to O_2 in cytochrome oxidase, −24 kcal/mol. Each of these oxidation-reduction reactions is sufficiently exergonic to generate a proton-motive force sufficient to drive the synthesis of ATP under standard conditions ($\Delta G^{\circ\prime}$ = −7.3 kcal/mol).

3. *Specific inhibition of electron flow.* The generation of a proton gradient by site 1 is prevented by *rotenone* and *amytal*, which specifically inhibit electron transfer within the NADH-Q reductase complex (Figure 17-21). In contrast, these inhibitors do not interfere with the oxidation of succinate, because the electrons of this substrate enter the electron-transport chain through QH_2, beyond this block. *Antimycin A* blocks electron flow from reduced *b*-562 to QH·. Hence, this antibiotic inhibits electron flow between cytochromes *b* and c_1, and so it prevents proton pumping by site 2 (cytochrome reductase). This block can be bypassed by the addition of ascorbate, which directly reduces cytochrome

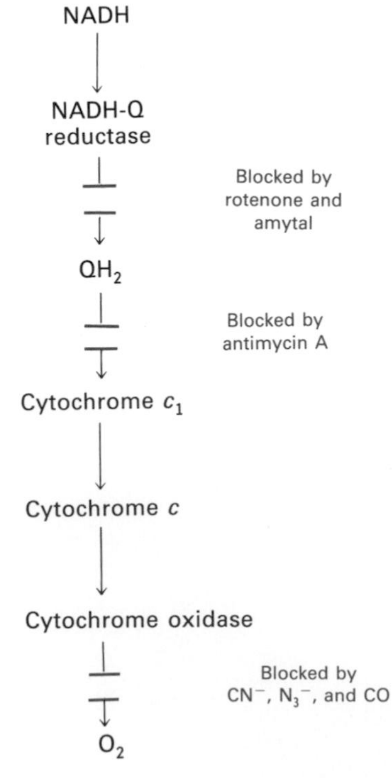

Figure 17-21
Sites of action of some inhibitors of electron transport.

c. Electron flow from cytochrome *c* to O_2 generates a proton-motive force by site 3 (cytochrome oxidase). Finally, electron flow in cytochrome oxidase can be blocked by CN^-, N_3^-, and CO. Cyanide and azide react with the ferric form of heme a_3, whereas carbon monoxide inhibits the ferrous form.

The sites of action of these inhibitors were revealed by the *crossover technique*. Britton Chance devised elegant spectroscopic methods for determining the proportions of the oxidized and reduced forms of each carrier. This is feasible because each carrier has a characteristic absorption spectrum for its oxidized and reduced forms (Figure 17-22). The addition of an electron-transport inhibitor changes the proportion of the oxidized and reduced forms of each carrier. For example, addition of antimycin A causes the carriers between NADH and QH_2 to become more reduced, and those between cytochrome *c* and O_2 to become more oxidized. Hence, it can be concluded that antimycin A inhibits the reduction of c_1 by QH_2, because this step is the *crossover point* after which the electron carriers are more oxidized than in the normal state.

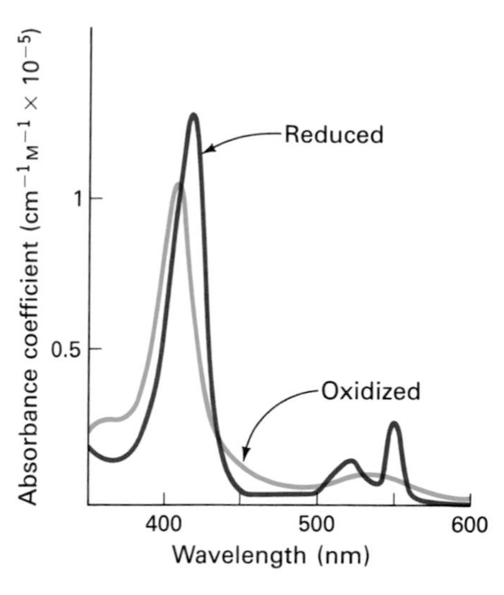

Figure 17-22
Absorption spectra of the oxidized and reduced forms of cytochrome *c*.

4. *Generation of a proton-motive force by reconstituted vesicles*. Each of the three proton-pumping complexes has been reconstituted in synthetic phospholipid vesicles, together with ATP synthase. Addition of oxidizable substrates to these vesicles leads to the generation of a proton-motive force that is sufficient for the synthesis of one ATP per electron pair. For example, the addition of ferrocytochrome *c* to vesicles containing cytochrome oxidase leads to the extrusion of about one proton per electron transferred.

Where do the translocated protons come from? One possibility is that the pumped protons are those *directly* participating in the chemical reactions catalyzed by the three sites. For example, when two electrons are donated by NADH to the NADH-Q reductase complex, two H^+ are involved—one from NADH itself and one from the solvent. Alternatively, the translocated protons may arise *indirectly* by *conformational interactions* between the catalytic site and proton-binding sites in different regions of the enzyme complex. The stoichiometries of the electron-transfer reactions show that only one proton per electron, or two per energy-conserving site, can be pumped by the direct mechanism. However, the number of H^+ per site has been observed to be between 3 and 4, which does not fit this direct mechanism. It seems likely that *electron transfer through each of the energy-conserving sites produces transmitted conformational changes that eject protons from the matrix to the cytoplasmic side of the membrane*. Recall that the protons involved in the Bohr effect in hemoglobin come from sites in the molecule distant from the heme group (p. 162). One of the major challenges now is to obtain high-resolution images of these proton pumps to delineate the precise mechanism by which electron flow causes the directed movement of protons.

ATP IS SYNTHESIZED BY AN ENZYME COMPLEX MADE OF A PROTON-CONDUCTING F_0 UNIT AND A CATALYTIC F_1 UNIT

We turn now to the utilization of the proton-motive force to synthesize ATP. The enzyme catalyzing this process, *ATP synthase*, appears in electron micrographs of submitochondrial particles, small membranous compartments prepared by sonic disruption of the inner mitochondrial membrane. ATP synthase is seen as spherical projections from the membrane surface (Figure 17-23). In intact mitochondria, these 85-Å diameter projections are on the matrix side of the inner mitochondrial membrane. In 1960, Efraim Racker found that these knobs can be removed by mechanical agitation. The stripped submitochondrial particles can transfer electrons through their electron-transport chain, but they can no longer synthesize ATP. In contrast, the separated 85-Å spheres do catalyze the hydrolysis of ATP. Most interesting, Racker found that the addition of these ATPase spheres to the stripped submitochondrial particles restored their capacity to synthesize ATP. These spheres are referred to as F_1 (or *coupling factor 1*). The normal role of the F_1 unit is to catalyze the synthesis of ATP. The ATPase activity exhibited by solubilized F_1 (in the absence of a proton gradient) is the reverse of its normal action. F_1 consists of five kinds of polypeptide chains with the stoichiometry $\alpha_3\beta_3\gamma\delta\epsilon$, and it has a mass of 380 kd (Table 17-3).

The other major unit of ATP synthase is F_0, a hydrophobic segment that spans the inner mitochondrial membrane. *F_0 is the proton channel of the complex.* It consists of four kinds of polypeptide chains. The 8-kd chain, of which there are six per F_1, probably forms the transmembrane

Table 17-3
Components of the mitochondrial ATP-synthesizing complex

Subunits	Mass (kd)	Role	Location
F_1	380	Contains catalytic site for ATP synthesis	Spherical headpiece on matrix side
α	56		
β	52		
γ	34		
δ	14		
ϵ	6		
F_0	25	Contains proton channel	Transmembrane
	21		
	12		
	8		
F_1 inhibitor	10	Regulates proton flow and ATP synthesis	Stalk between F_0 and F_1
Oligomycin-sensitivity-conferring protein (OSCP)	23		
Fc_2 (F_6)	8		

Sources: J. W. DePierre and L. Ernster, *Ann. Rev. Biochem.* 46(1977):216, and Y. Kagawa, in L. Ernster (ed.), *Bioenergetics* (Elsevier, 1984), p. 151.

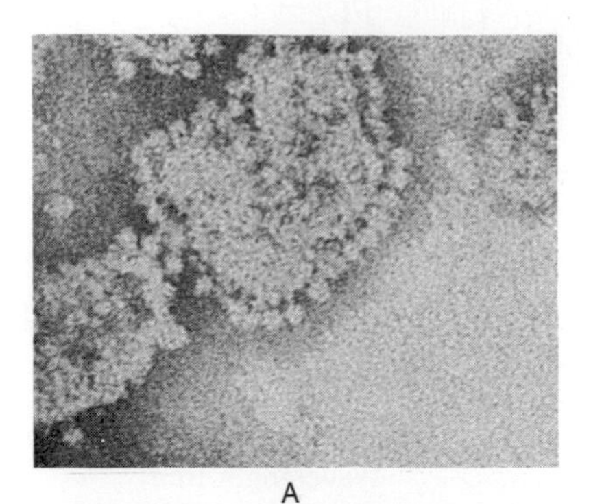
A

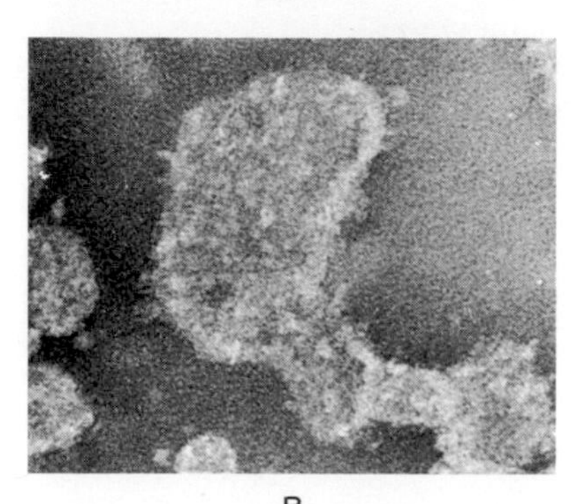
B

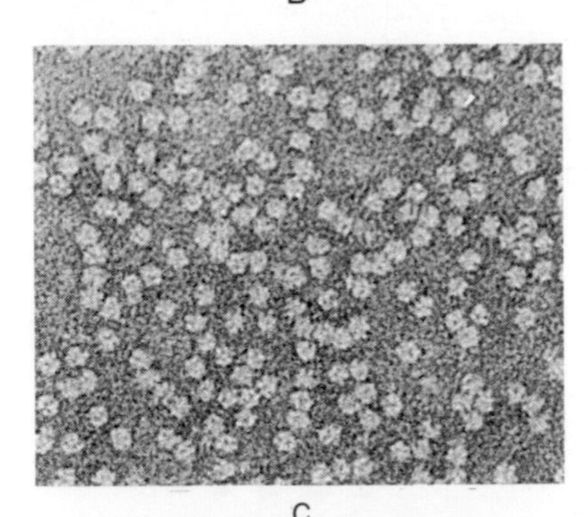
C

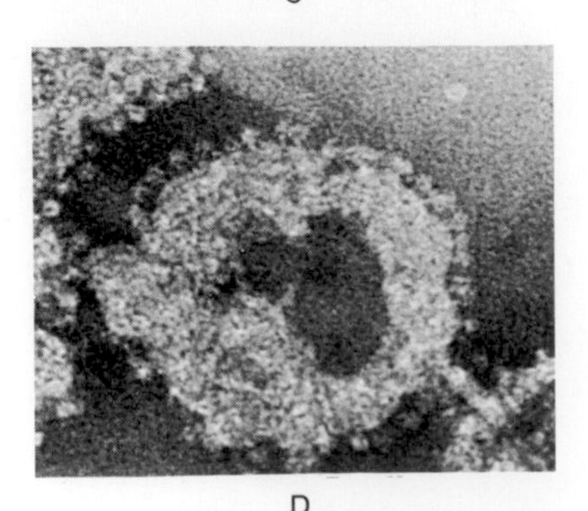
D

Figure 17-23
Electron micrographs of (A) a submitochondrial particle showing F_1 projections on its surface, (B) a submitochondrial particle treated with urea, which has removed the F_1 projections, (C) isolated F_1 subunits, and (D) a reconstituted submitochondrial particle formed by adding F_1 to stripped membranes. The particle shown in B can transfer electrons to O_2 but it cannot form ATP. The reconstituted particle shown in D carries out oxidative phosphorylation. [Courtesy of Dr. Efraim Racker.]

pore for protons. The stalk between F_0 and F_1 contains several other proteins (Table 17-3). One of them renders the complex sensitive to *oligomycin*, an antibiotic that blocks ATP synthesis by interfering with the utilization of the proton gradient. *ATP synthase* is also known as *F_0F_1-ATPase* or *H^+-ATPase*.

Bacterial ATP synthase, like the mitochondrial one, contains a hydrophilic F_1 unit and a hydrophobic F_0 unit. Bacterial F_1 also has an $\alpha_3\beta_3\gamma\delta\epsilon$ structure. F_0 consists of three kinds of chains—a, b, and c—present in a molar ratio of about 1:2:10. The bacterial synthase is simpler than the mitochondrial one in that F_0 is directly linked to F_1 rather than through a connecting stalk. In *E. coli*, the genes for the eight types of subunits of ATP synthase are next to each other (Figure 17-25). This gene cluster, called the *unc* operon, has been cloned and sequenced. The catalytic sites are on the β chains. The homologous α chains also contain binding sites for ATP and ADP. The flow of protons from F_0 to F_1 is controlled by the γ subunit, which serves as a gate. The amino acid sequence of subunit c suggests that it contains two transmembrane regions. It seems likely that a proton channel is formed by the association of multiple copies of subunit c.

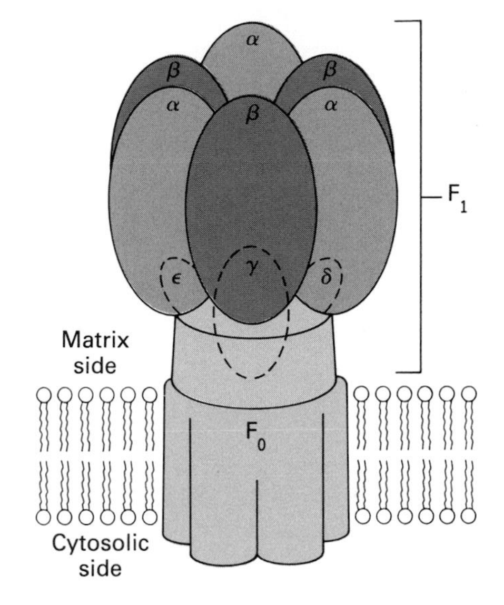

Figure 17-24
Schematic diagram of ATP synthase showing the proton-conducting F_0 unit (green) and the ATP-synthesizing F_1 unit (red). [After F. M. Harold. *The Vital Force: A Study of Bioenergetics* (W. H. Freeman, 1986), p. 238.]

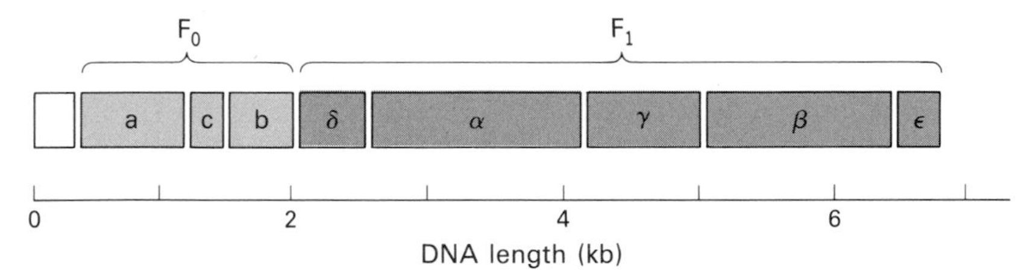

Figure 17-25
Arrangement of genes encoding the subunits of ATP synthase in *E. coli*. This cluster of genes is called the *unc* operon (*unc* stands for uncoupled).

PROTON FLOW THROUGH ATP SYNTHASE LEADS TO THE RELEASE OF TIGHTLY BOUND ATP

ATP synthase catalyzes the formation of ATP from ADP and orthophosphate:

$$ADP^{3-} + P_i^{2-} + H^+ \rightleftharpoons ATP^{4-} + H_2O$$

It should be noted that the actual substrates are the Mg^{2+} complexes of ADP and ATP, as in all known phosphoryl transfer reactions with these nucleotides. A terminal oxygen atom of ADP attacks the phosphorus atom of P_i to form a pentacovalent intermediate, which then dissociates into ATP and H_2O (Figure 17-26). The attacking oxygen atom of ADP

$$\underset{\textbf{ADP}}{R{-}O{-}\overset{O}{\underset{O^-}{P}}{-}O{-}\overset{O}{\underset{O^-}{P}}{-}O^-} + \underset{\textbf{P}_i}{{}^-O{-}\overset{O^-}{\underset{O}{P}}{-}OH} + H^+ \longrightarrow \underset{\textbf{Pentacovalent intermediate}}{R{-}O{-}\overset{O}{\underset{O^-}{P}}{-}O{-}\overset{O}{\underset{O^-}{P}}{-}O{-}\overset{^-O\;\;O^-}{\underset{O^-}{P}}{-}O{\cdots}H^+} \longrightarrow \underset{\textbf{ATP}}{R{-}O{-}\overset{O}{\underset{O^-}{P}}{-}O{-}\overset{O}{\underset{O^-}{P}}{-}O{-}\overset{O}{\underset{O^-}{P}}{-}O^-} + H_2O$$

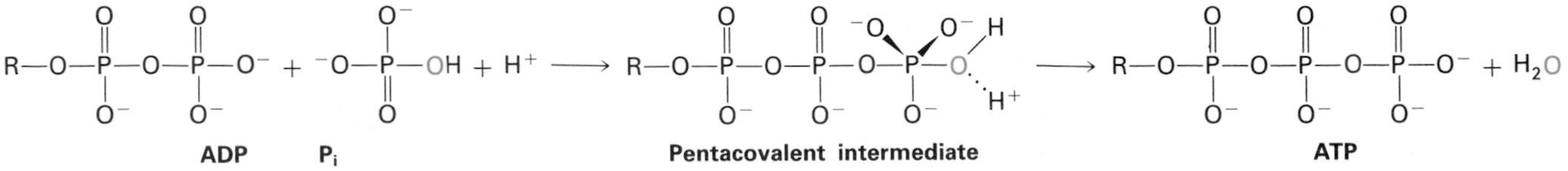

Figure 17-26
A pentacovalent phosphoryl intermediate is formed in the synthesis of ATP. The attacking oxygen atom of ADP and the leaving oxygen atom of P_i are at opposite apices of a trigonal bipyramid.

and the departing oxygen atom of P_i occupy the apices of a trigonal bipyramid. This *in-line* geometry of attacking and leaving groups is also seen in the catalytic mechanism of ribonuclease A (p. 213).

How does the flow of protons drive the synthesis of ATP? One possibility a priori is that the energized protons flowing through F_0 are funneled to the catalytic site on F_1, where they remove an oxygen from P_i to shift the equilibrium toward ATP synthesis. However, isotopic exchange experiments unexpectedly revealed that *enzyme-bound ATP forms readily in the absence of a proton-motive force.* When ADP and P_i were added to ATP synthase in $H_2{}^{18}O$, ^{18}O became incorporated into P_i. The isotopic label was acquired by P_i by the hydrolysis of ATP by $H_2{}^{18}O$.

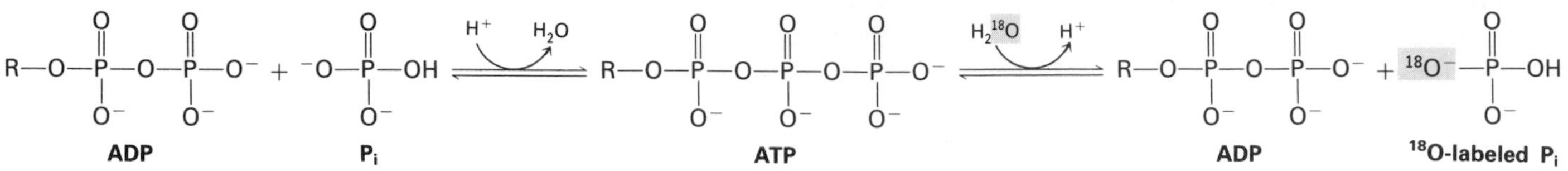

Figure 17-27
Isotope exchange experiment showing that enzyme-bound ATP is formed from ADP and P_i in the absence of a proton-motive force.

The rate of incorporation of ^{18}O into P_i showed that about equal amounts of bound ATP and ADP are in equilibrium at the catalytic site, even in the absence of a proton gradient. However, ATP does not leave the catalytic site unless protons flow through the enzyme. Paul Boyer showed that *the role of the proton gradient is not to form ATP but to release it from the enzyme.* He also found that the nucleotide-binding sites of this enzyme interact with each other. The binding of ADP and P_i to one site promotes the release of ATP from another. In other words, ATP synthase exhibits *catalytic cooperativity.*

Boyer has proposed a *binding-change mechanism* for proton-driven ATP synthesis. In this model (Figure 17-28), the three catalytic β subunits are intrinsically identical but are not functionally equivalent at any particular moment. One catalytic site is in the O form, which is open and has very low affinity for substrates. The second is in the L form, which binds them loosely and is catalytically inactive. The third is in the T form, which binds them tightly and is active. Consider an enzyme molecule with ATP bound to the T site. ADP and P_i then bind to the L site. Energy input by proton flux converts the T site into an O site, the L site into a T site, and the O site into an L site. These conversions permit ATP to be released from the new O site and enable another molecule of ATP to be formed from ADP and P_i at the new T site (former L site). Protons flow from the F_0 to the F_1 side of membrane only when O, L, and T interconvert. These conformational transitions are linked, most likely by changes in subunit interactions.

Figure 17-28
Binding-change mechanism for ATP synthase. The three catalytic sites cycle through three conformational states: O (open), L (loose binding), and T (tight binding). Proton flux through the synthase drives this interconversion of states. The essence of this proposed mechanism is that proton flux leads to the release of tightly-bound ATP. [After R. L. Cross, D. Cunningham, and J. K. Tamura. *Curr. Top. Cell. Regul.* 24(1984):336.]

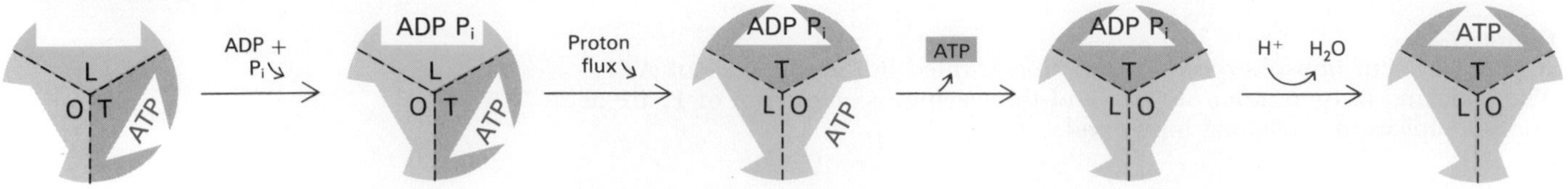

Why are the β chains functionally nonequivalent at a given time? The three $\alpha\beta$ chains interact with just one γ, δ, and ϵ chain. Hence, at a particular moment, the three β chains are in different environments, which produce distinct O, L, and T sites. It is not known how proton flow leads to a switching of sites.

ATP synthase is driven by proton-motive force, which is the sum of contributions by the pH gradient and the membrane potential, as expressed in the equation on page 411. A pH gradient, for example, could cause aspartate residues to be in the —COOH form at pH 4.5 (when facing the cytosolic side) and in the —COO$^-$ form at pH 7.5 (when facing the matrix side). Then protonation on one side and deprotonation on the other could drive a reaction cycle unidirectionally to achieve net ATP synthesis. How would membrane potential achieve the same end? Suppose that the membrane potential is 0.18 V (the cytosolic side positive) and the pH is 7.5 on both sides. The concentration of H^+ within the F_0 channel will not be uniform because H^+ will be attracted to the matrix side of the membrane, which is negative relative to the other side (Figure 17-29). The potential difference of 0.18 V leads to a 1000-fold higher concentration of H^+ at the F_0-F_1 junction than at the entrance of the F_0 channel. Thus, *a positive membrane potential leads to ATP synthesis by producing a high local concentration of H^+ at the gate between F_0 and F_1*; specifically, 0.18 V produces the same concentration gradient as does a 3 pH unit difference between the two sides. Recent studies raise the possibility that protons pumped by the respiratory complexes (the proton source) may diffuse preferentially to ATP synthase (the proton sink), rather than equilibrate with the medium in the intermembrane space.

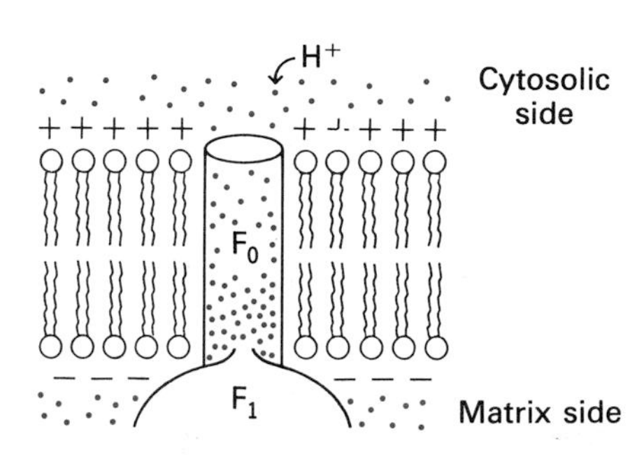

Figure 17-29
A positive membrane potential leads to a high local concentration of H^+ at the junction of the F_0 and F_1 subunits of ATP synthase.

ELECTRONS FROM CYTOPLASMIC NADH ENTER MITOCHONDRIA BY SHUTTLES

The inner mitochondrial membrane is quite impermeable to NADH and NAD$^+$. But NADH is formed by glycolysis in the cytoplasm, in the oxidation of glyceraldehyde 3-phosphate, and NAD$^+$ must be regenerated for glycolysis to continue. How then does cytoplasmic NADH get oxidized by the respiratory chain? The solution is that *electrons from NADH,* rather than NADH itself, are carried across the mitochondrial membrane. One carrier is *glycerol 3-phosphate,* which readily traverses the outer mitochondrial membrane. The first step in this shuttle (Figure 17-30) is the transfer of electrons from NADH to dihydroxyacetone

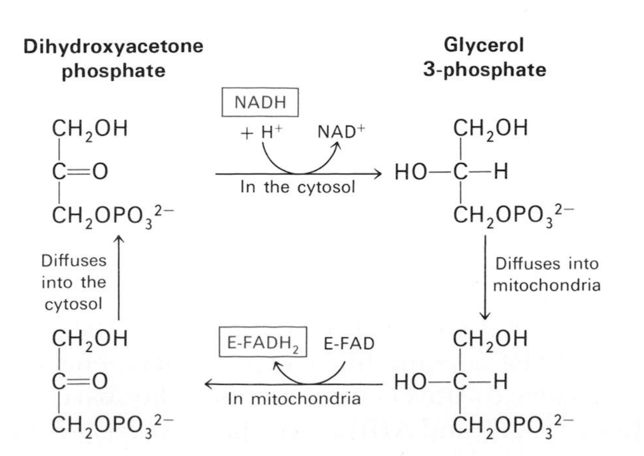

Figure 17-30
Glycerol phosphate shuttle.

phosphate to form glycerol 3-phosphate. This reaction, catalyzed by glycerol 3-phosphate dehydrogenase, occurs in the cytosol. Glycerol 3-phosphate is reoxidized to dihydroxyacetone phosphate on the outer surface of the inner mitochondrial membrane—an electron pair from glycerol 3-phosphate is transferred to the FAD prosthetic group of the mitochondrial glycerol dehydrogenase. This enzyme differs from its cytosolic counterpart: it uses FAD rather than NAD^+ as the electron acceptor and it is a transmembrane protein. The dihydroxyacetone phosphate formed in the oxidation of glycerol 3-phosphate then diffuses back into the cytosol to complete the shuttle. The net reaction is

$$\underset{\text{Cyto-plasmic}}{\text{NADH}} + H^+ + \underset{\text{Mito-chondrial}}{\text{E-FAD}} \longrightarrow \underset{\text{Cyto-plasmic}}{\text{NAD}^+} + \underset{\text{Mito-chondrial}}{\text{E-FADH}_2}$$

The reduced flavin inside the mitochondria transfers its electrons to the electron carrier Q, which then enters the respiratory chain as QH_2. *Consequently, two rather than three ATP are formed when cytoplasmic NADH transported by the glycerol phosphate shuttle is oxidized by the respiratory chain.* At first glance, this shuttle may seem to waste one ATP per cycle. This lower yield arises because FAD rather than NAD^+ is the electron acceptor in mitochondrial glycerol 3-phosphate dehydrogenase. The use of FAD enables electrons from cytoplasmic NADH to be transported into mitochondria against an NADH concentration gradient. The price of this transport is one ATP per two electrons. This glycerol phosphate shuttle is especially prominent in insect flight muscle, which can sustain a very high rate of oxidative phosphorylation.

In heart and liver, electrons from cytoplasmic NADH are brought into mitochondria by the *malate-aspartate shuttle,* which is mediated by two membrane carriers and four enzymes. Electrons are transferred from NADH in the cytoplasm to oxaloacetate, forming malate, which traverses the inner mitochondrial membrane and is then reoxidized by NAD^+ in the matrix to form NADH. The resulting oxaloacetate does not readily cross the inner mitochondrial membrane, and so a transamination reaction is needed to form aspartate, which can be transported to the cytosolic side. The net reaction of the malate-aspartate shuttle is

$$\underset{\text{Cyto-plasmic}}{\text{NADH}} + \underset{\text{Mito-chondrial}}{\text{NAD}^+} \rightleftharpoons \underset{\text{Cyto-plasmic}}{\text{NAD}^+} + \underset{\text{Mito-chondrial}}{\text{NADH}}$$

This shuttle, in contrast with the glycerol phosphate shuttle, is readily reversible. Consequently, NADH can be brought into mitochondria by the malate-aspartate shuttle only if the $NADH/NAD^+$ ratio is higher in the cytosol than in the mitochondrial matrix. No energy is consumed by the malate-aspartate shuttle in transferring electrons from NADH into the mitochondrial respiratory chain, and so three ATP are synthesized per NADH transferred.

THE ENTRY OF ADP INTO MITOCHONDRIA IS COUPLED TO THE EXIT OF ATP BY THE ATP-ADP TRANSLOCASE

ATP and ADP do not diffuse freely across the inner mitochondrial membrane. Rather, a specific transport protein, the *ATP-ADP translocase* (also called the adenine nucleotide carrier), enables these highly charged molecules to traverse this permeability barrier. Most important, the flows of ATP and ADP are coupled. *ADP enters the mitochondrial*

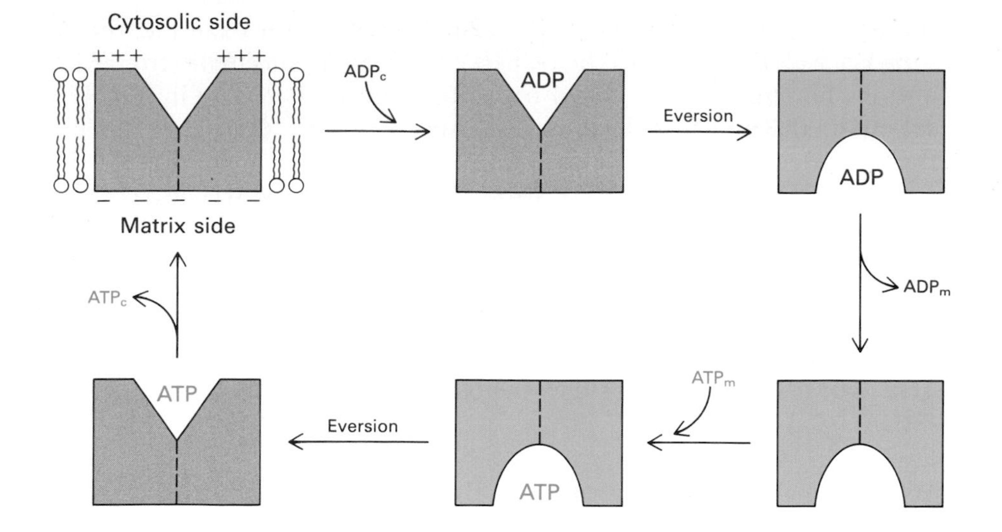

Figure 17-31
The mitochondrial ATP-ADP translocase catalyzes the coupled entry of ADP and exit of ATP into and from the matrix. The reaction cycle is driven by membrane potential. The actual conformational change corresponding to eversion of the binding site could be quite small.

matrix only if ATP exits, and vice versa. The reaction catalyzed by the translocase is

$$ADP_c^{3-} + ATP_m^{4-} \longrightarrow ADP_m^{3-} + ATP_c^{4-}$$

in which the subscript c denotes the cytosolic side and m denotes the matrix side of the inner mitochondrial membrane.

The ATP-ADP translocase is a dimer of identical 30-kd subunits. It contains a single nucleotide binding site that alternately faces the matrix and cytosolic sides of the membrane. ATP and ADP (both devoid of Mg^{2+}) are bound with nearly the same affinity. *In the presence of a positive membrane potential, the rate of eversion of the binding site from the matrix to the cytosolic side is more rapid for ATP than for ADP because ATP has one more negative charge.* Hence, ATP is transported out of the matrix about 30 times more rapidly than is ADP, which leads to a higher phosphoryl potential on the outer side than on the matrix side. The translocase does not evert at an appreciable rate unless a nucleotide is bound. This design feature assures that the entry of ADP into the matrix is precisely coupled to the exit of ATP. The other side of the coin is that *the membrane potential is decreased by the exchange of ATP for ADP, which results in a net transfer of one negative charge out of the matrix.* ATP-ADP exchange is energetically expensive; about a quarter of the energy yield from electron transfer by the respiratory chain is consumed to regenerate the membrane potential that is tapped by this exchange process.

The translocase is abundant, constituting about 14% of the membrane protein. It is specifically inhibited by very low concentrations of *atractyloside*, a plant glycoside, or of *bongkrekic acid*, an antibiotic from a mold. Atractyloside binds to the translocase when its nucleotide site faces outward, whereas bongkrekic acid binds when this site faces the matrix. Oxidative phosphorylation stops soon after either inhibitor is added, showing clearly that the ATP-ADP translocase is essential.

MITOCHONDRIA CONTAIN NUMEROUS TRANSPORT SYSTEMS FOR IONS AND METABOLITES

The ATP-ADP translocase is only one of many mitochondrial transport systems. The inner mitochondrial membrane contains a variety of transporters for ions and charged metabolites. For historical reasons, these transmembrane proteins are called carriers. For example, the *dicarboxylate carrier* mediates the transport of malate, succinate, fumarate, and P_i. The *tricarboxylate carrier* transports citrate. Pyruvate in the cytosol enters the mitochondrial matrix in exchange for OH^- by means of the *pyruvate carrier*. Glutamate and aspartate are exchanged by the *glutamate carrier,* which can also transport OH^-. Mitochondria also contain a *calcium ion transport system*. The proton-motive force generated by electron transfer rather than ATP is the immediate source of free energy for the accumulation of Ca^{2+} in the mitochondrial matrix.

THE COMPLETE OXIDATION OF GLUCOSE YIELDS FROM 36 TO 38 ATP

We can now calculate how many ATP are formed when glucose is completely oxidized (Table 17-4). If the glycerol phosphate shuttle is used, the overall reaction is

$$\text{Glucose} + 36\ \text{ADP} + 36\ P_i + 36\ H^+ + 6\ O_2 \longrightarrow 6\ CO_2 + 36\ \text{ATP} + 42\ H_2O$$

The *P:O ratio is 3* because 36 ATP are formed and 12 atoms of oxygen are consumed. If the malate-aspartate shuttle is used to bring electrons

Table 17-4
ATP yield from the complete oxidation of glucose

Reaction sequence	*ATP yield per glucose*
Glycolysis: glucose into pyruvate (in the cytosol)	
Phosphorylation of glucose	−1
Phosphorylation of fructose 6-phosphate	−1
Dephosphorylation of 2 molecules of 1,3-BPG	+2
Dephosphorylation of 2 molecules of phosphoenolpyruvate	+2
2 NADH are formed in the oxidation of 2 molecules of glyceraldehyde 3-phosphate	
Conversion of pyruvate into acetyl CoA (inside mitochondria)	
2 NADH are formed	
Citric acid cycle (inside mitochondria)	
2 molecules of guanosine triphosphate are formed from 2 molecules of succinyl CoA	+2
6 NADH are formed in the oxidation of 2 molecules each of isocitrate, α-ketoglutarate, and malate	
2 $FADH_2$ are formed in the oxidation of 2 molecules of succinate	
Oxidative phosphorylation (inside mitochondria)	
2 NADH formed in glycolysis; each yields 2 ATP (assuming transport of NADH by the glycerol phosphate shuttle)	+4
2 NADH formed in the oxidative decarboxylation of pyruvate; each yields 3 ATP	+6
2 $FADH_2$ formed in the citric acid cycle; each yields 2 ATP	+4
6 NADH formed in the citric acid cycle; each yields 3 ATP	+18
NET YIELD PER GLUCOSE	+36

from cytosolic NADH to the respiratory chain, a total of 38 ATP is formed per molecule of glucose oxidized. Most of the ATP, 32 out of 36 (or 34 out of 38), is generated by oxidative phosphorylation.

The overall efficiency of ATP generation is high. The oxidation of glucose yields 686 kcal under standard conditions:

$$\text{Glucose} + 6\ O_2 \longrightarrow 6\ CO_2 + 6\ H_2O \qquad \Delta G^{\circ\prime} = -686\ \text{kcal}$$

The free energy stored in 36 ATP is 263 kcal, because $\Delta G^{\circ\prime}$ for the hydrolysis of ATP is -7.3 kcal. Hence, the thermodynamic efficiency of ATP formation from glucose is 263/686, or 38%, under standard conditions.

THE RATE OF OXIDATIVE PHOSPHORYLATION IS DETERMINED BY THE NEED FOR ATP

Under most physiological conditions, electron transport is tightly coupled to phosphorylation. *Electrons do not usually flow through the electron-transport chain to O_2 unless ADP is simultaneously phosphorylated to ATP.* Oxidative phosphorylation requires a supply of NADH (or other source of electrons at high potential), O_2, ADP, and P_i. The most important factor in determining the rate of oxidative phosphorylation is the *level of ADP*. The rate of oxygen consumption by a tissue homogenate increases markedly when ADP is added and then returns to its initial value when the added ADP has been converted into ATP (Figure 17-32).

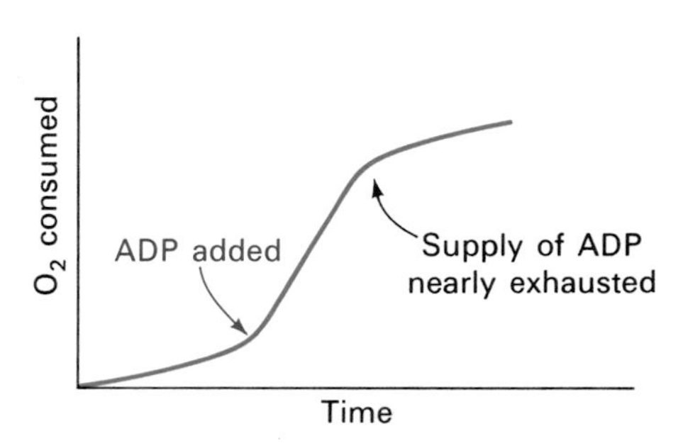

Figure 17-32
Respiratory control. Electrons are transferred to O_2 only if ADP is concomitantly phosphorylated to ATP.

The regulation of the rate of oxidative phosphorylation by the ADP level is called *respiratory control*. The level of ADP likewise affects the rate of the citric acid cycle because of its need for NAD^+ and FAD. The physiological significance of this regulatory mechanism is evident. The ADP level increases when ATP is consumed, and so oxidative phosphorylation is coupled to the utilization of ATP. *Electrons do not flow from fuel molecules to O_2 unless ATP needs to be synthesized.* We see here another example of the regulatory significance of the energy charge.

THE PROTON GRADIENT CAN BE SHORT-CIRCUITED TO GENERATE HEAT

The tight coupling of electron transport and phosphorylation is disrupted by 2,4-dinitrophenol (DNP) and some other acidic aromatic compounds (Figure 17-33). These substances carry protons across the inner mitochondrial membrane. In the presence of these uncouplers, electron transport from NADH to O_2 proceeds normally, but ATP is not formed by the mitochondrial ATPase because the proton-motive force across the inner mitochondrial membrane is dissipated. The loss of respiratory control leads to increased oxygen consumption and oxidation of NADH. In contrast, DNP has no effect on phosphorylations involving high-energy chemical intermediates (e.g., 1,3-bisphosphoglycerate). DNP and other uncouplers are very useful tools in metabolic studies because of their specific effect on oxidative phosphorylation.

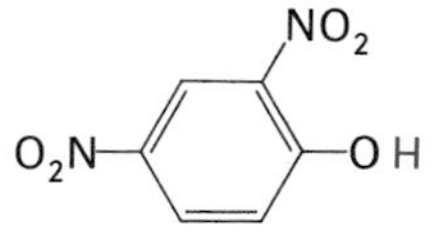

2,4-Dinitrophenol (DNP)

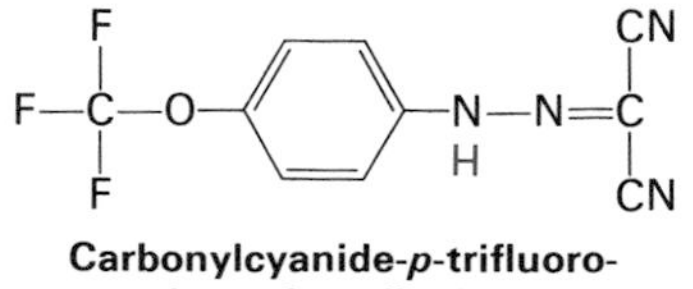

Carbonylcyanide-*p*-trifluoromethoxyphenylhydrazone

Figure 17-33
Formulas of two uncouplers of oxidative phosphorylation. These lipid-soluble substances can carry protons across the inner mitochondrial membrane. The dissociable proton is shown in red.

The uncoupling of oxidative phosphorylation can be biologically useful. *It is a means of generating heat to maintain body temperature in hibernating animals, some newborn animals (including humans), and in mammals adapted to cold.* Brown adipose tissue, which is very rich in mitochondria, is specialized for this process of *thermogenesis*. The inner mitochondrial membrane of these mitochondria contains a large amount of *thermogenin* (also called the uncoupling protein), a dimer of 33-kd subunits that resembles the ATP-ADP translocase. Thermogenin forms a pathway for the flow of protons from the cytosol to the matrix. In essence, *thermogenin generates heat by short-circuiting the mitochondrial proton battery.* This dissipative proton pathway is activated by free fatty acids liberated from triacylglycerols in response to hormonal signals. The skunk cabbage uses a similar mechanism to heat its floral spikes, increasing the evaporation of odoriferous molecules that attract insects to fertilize its flowers.

TOXIC DERIVATIVES OF O_2 SUCH AS SUPEROXIDE RADICAL ARE SCAVENGED BY PROTECTIVE ENZYMES

As was mentioned earlier, the one-electron reduction of O_2 yields $O_2^- \cdot$, *superoxide anion*, a highly reactive and destructive radical. Cytochrome oxidase and other proteins that reduce O_2 have been designed not to release $O_2^- \cdot$. However, a small amount of superoxide anion is unavoidably formed, as in the oxidation of a ferroheme (Fe^{2+}) group of hemoglobin to ferriheme (Fe^{3+}) by O_2. Protonation of superoxide anion yields *hydroperoxyl radical* ($HO_2 \cdot$), which can react spontaneously with another superoxide anion to form *hydrogen peroxide* (H_2O_2).

$$O_2 \xrightarrow{e^-} \underset{\text{Superoxide anion}}{O_2^- \cdot} \xrightarrow{H^+} \underset{\text{Hydroperoxyl radical}}{HO_2 \cdot} \xrightarrow{HO_2 \cdot \quad O_2} \underset{\text{Hydrogen peroxide}}{H_2O_2}$$

Superoxide anion can be scavenged by *superoxide dismutase*, an enzyme present in all aerobic organisms, which catalyzes the conversion of two of these radicals into hydrogen peroxide and molecular oxygen.

$$O_2^- \cdot + O_2^- \cdot \xrightarrow[\text{superoxide dismutase}]{2\,H^+} H_2O_2 + O_2$$

The active site of the cytosolic enzyme in eucaryotes contains a copper ion and a zinc ion coordinated to the side chain of a histidine residue (Figure 17-34). The negatively charged superoxide is guided electrostatically to a very positively charged catalytic site at the bottom of a channel. $O_2^- \cdot$ binds to Cu^{2+} and the guanido group of an arginine residue. An electron is transferred from superoxide to cupric ion to form Cu^+ and O_2, which is released. A second superoxide enters the active site and binds to Cu^+, arginine, and H_3O^+. The bound $O_2^- \cdot$ acquires an electron from Cu^+ and two protons from its binding partners to form H_2O_2 and regenerate the Cu^{2+} state of the enzyme.

The hydrogen peroxide formed by superoxide dismutase and by the uncatalyzed reaction of hydroperoxy radicals is scavenged by *catalase*, a ubiquitous heme protein that catalyzes the dismutation of hydrogen peroxide into water and molecular oxygen.

$$H_2O_2 + H_2O_2 \xrightarrow{\text{catalase}} 2\,H_2O + O_2$$

Peroxidases, also heme enzymes, catalyze an analogous reaction in which hydrogen peroxide is reduced to water by a reductant (AH_2).

$$H_2O_2 + AH_2 \xrightarrow{\text{peroxidase}} 2\,H_2O + A$$

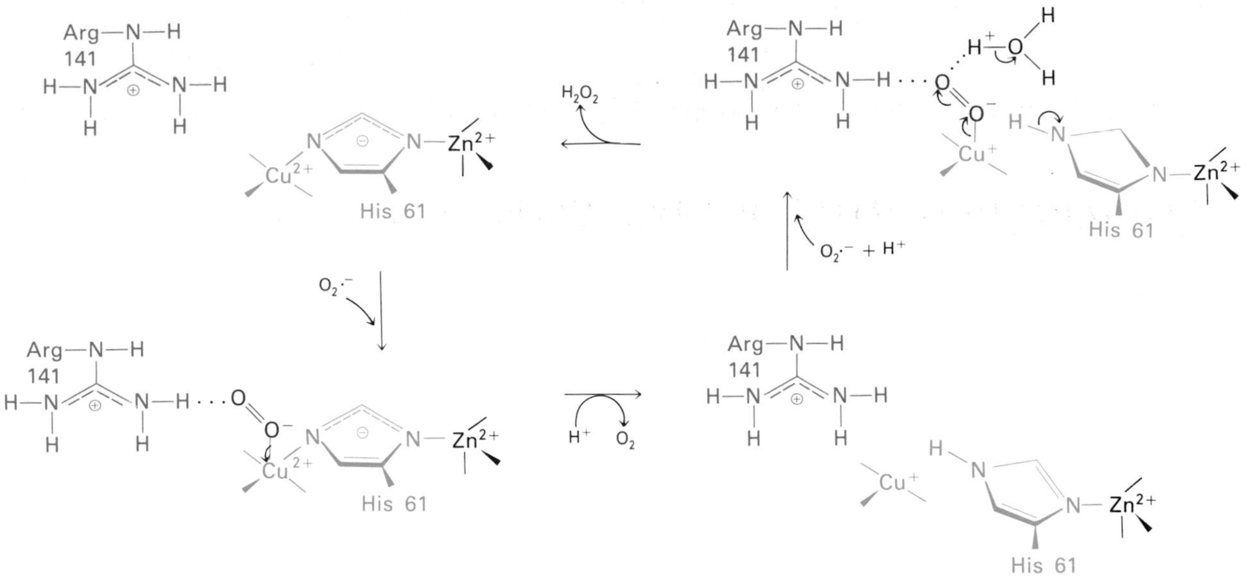

Figure 17-34
Proposed catalytic mechanism of superoxide dismutase. A copper ion at the active site cycles between +1 and +2 oxidation states. The histidine side chain coordinated to Cu^{2+} during part of the cycle is also bonded to Zn^{2+}. The other key participant is an arginine side chain. [After J. A. Tainer, E. D. Getzoff, J. S. Richardson, and D. C. Richardson. *Nature* 306(1983):286.]

POWER TRANSMISSION BY PROTON GRADIENTS: A CENTRAL MOTIF OF BIOENERGETICS

The main concept presented in this chapter is that mitochondrial electron transfer and ATP synthesis are linked by a transmembrane proton gradient. ATP synthesis in bacteria and chloroplasts (p. 527) also is driven by proton gradients. In fact, proton gradients power a variety of energy-requiring processes such as the active transport of Ca^{2+} by mitochondria, the entry of some amino acids and sugars into bacteria, the rotation of bacterial flagella, and the transfer of electrons from NADH to NADPH. Proton gradients can also be used to generate heat, as in hibernation. It is evident that *proton gradients are a central interconvertible currency of free energy in biological systems* (Figure 17-35). Mitchell has noted that the transduction of energy by proticity is marvellously simple and effective because all that is required is a thin topologically closed insulating lipid membrane between two aqueous proton-conductor phases.

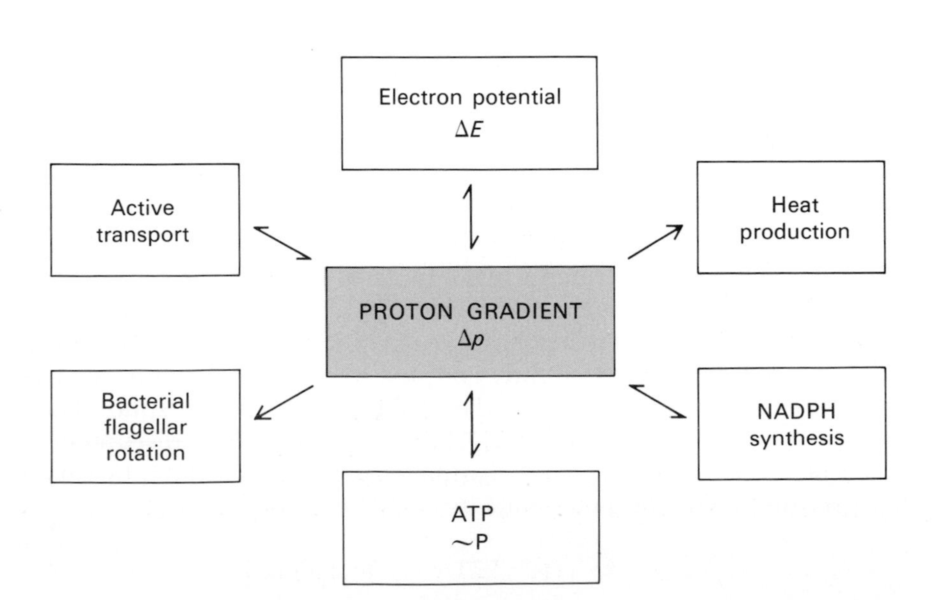

Figure 17-35
The proton gradient is an interconvertible form of free energy.

SUMMARY

In oxidative phosphorylation, the synthesis of ATP is coupled to the flow of electrons from NADH or $FADH_2$ to O_2 by a proton gradient across the inner mitochondrial membrane. Electron flow through three asymmetrically oriented transmembrane complexes results in the pumping of protons out of the mitochondrial matrix and the generation of a membrane potential. ATP is synthesized when protons flow back to the matrix through a channel in an ATP-synthesizing complex, called ATP synthase (also known as F_0F_1-ATPase or H^+-ATPase). Oxidative phosphorylation exemplifies a fundamental theme of bioenergetics: the transmission of free energy by proton gradients.

The electron carriers in the respiratory assembly of the inner mitochondrial membrane are flavins, iron-sulfur complexes, quinones, heme groups of cytochromes, and copper ions. Electrons from NADH are transferred to the FMN prosthetic group of NADH-Q reductase, the first of three complexes. This reductase also contains Fe-S centers. The electrons emerge in QH_2, the reduced form of ubiquinone (Q). This highly mobile hydrophobic carrier transfers its electrons to cytochrome reductase, a complex that contains cytochromes *b* and c_1 and an Fe-S center. This second complex reduces cytochrome *c*, a water-soluble peripheral membrane protein. Cytochrome *c*, like Q, is a mobile carrier of electrons, which it then transfers to cytochrome oxidase. This third complex contains cytochromes *a* and a_3 and two copper ions. A heme iron and a copper ion in this oxidase transfer electrons to O_2, the ultimate acceptor, to form H_2O.

The flow of electrons through each of these complexes leads to the pumping of protons from the matrix side to the cytosolic side of the inner mitochondrial membrane. A proton-motive force consisting of a pH gradient (cytosolic side acidic) and a membrane potential (cytosolic side positive) is generated. The flow of protons back to the matrix side through ATP synthase drives ATP synthesis. The enzyme complex consists of a hydrophobic F_0 unit that conducts protons through the membrane and a hydrophilic F_1 unit that catalyzes ATP synthesis alternately at three active sites. Protons flowing through ATP synthase release tightly bound ATP.

The flow of two electrons through each of the three proton-pumping complexes generates a gradient sufficient to synthesize one molecule of ATP. Hence, three ATP are formed per NADH oxidized in the mitochondrial matrix, whereas only two ATP are made per $FADH_2$ oxidized because its electrons enter the chain at QH_2, after the first proton-pumping site. Also, only two ATP are generated by the oxidation of NADH in the cytosol because one ATP is lost when the glycerol phosphate shuttle carries these electrons into mitochondria. The entry of ADP into the mitochondrial matrix is coupled to the exit of ATP by the ATP-ADP translocase, a transporter driven by membrane potential. Thirty-six ATP are generated when a molecule of glucose is completely oxidized to CO_2 and H_2O; two more ATP are formed if the malate-aspartate shuttle is used instead of the glycerol phosphate shuttle. Electron transport is normally tightly coupled to phosphorylation. NADH and $FADH_2$ are oxidized only if ADP is simultaneously phosphorylated to ATP. This coupling, called respiratory control, can be disrupted by uncouplers such as DNP, which dissipate the proton gradient by carrying protons across the inner mitochondrial membrane.

SELECTED READINGS

Where to start

Hinkle, P. C., and McCarty, R. E., 1978. How cells make ATP. *Sci. Amer.* 238(3):104–123. [Available as *Sci. Amer.* Offprint 1383.]

Mitchell, P., 1979. Keilin's respiratory chain concept and its chemiosmotic consequences. *Science* 206:1148–1159. [Mitchell reviews the evolution of the chemiosmotic concept in this Nobel lecture.]

Dickerson, R. E., 1980. Cytochrome *c* and the evolution of energy metabolism. *Sci. Amer.* 242(3):137–153.

Nicholls, D. G., 1982. *Bioenergetics: An Introduction to the Chemiosmotic Theory.* Academic Press. [A concise and lucid account of the experimental basis of our current understanding of oxidative phosphorylation.]

Books

Harold, F. M., 1986. *The Vital Force: A Study of Bioenergetics.* W. H. Freeman. [A sensitive and scholarly account of bioenergetics emphasizing the development of ideas. Chapter 7 deals with mitochondria and oxidative phosphorylation.]

Ernster, L., (ed.), 1984. *Bioenergetics.* Elsevier. [Contains many excellent articles on mitochondrial and photosynthetic electron transport, ATP synthesis, metabolite transport systems, and thermogenic mitochondria.]

Tzagoloff, A., 1982. *Mitochondria.* Plenum. [An interesting and handsomely illustrated account of mitochondrial structure, biogenesis, genetics, oxidative phosphorylation, and ATP synthesis.]

Spiro, T. G., (ed.), 1982. *Iron-Sulfur Proteins.* Wiley-Interscience.

Racker, E., 1976. *A New Look at Mechanisms in Bioenergetics.* Academic Press. [A lively, personal account of electron transport and phosphorylation.]

Electron transport

Hatefi, Y., 1985. The mitochondrial electron transport and oxidative phosphorylation system. *Ann. Rev. Biochem.* 54:1015–1069.

Wikstrom, M., and Saraste, M., 1984. The mitochondrial respiratory chain. In L. Ernster, (ed.), *Bioenergetics*, pp. 49–94. Elsevier.

Casey, R. P., 1984. Membrane reconstitution of the energy-conserving enzymes of oxidative phosphorylation. *Biochim. Biophys. Acta* 768:319–347.

Slater, E. C., 1983. The Q cycle, an ubiquitous mechanism of electron transfer. *Trends Biochem. Sci.* 8:239–242.

Dickerson, R. E., 1972. The structure and history of an ancient protein. *Sci. Amer.* 226(4):58–72. [A beautifully illustrated account of the conformation and evolution of cytochrome *c*. Offprint 1245.]

Margoliash, E., and Bosshard, H. R., 1983. Guided by electrostatics, a textbook protein comes of age. *Trends Biochem. Sci.* 8:316–320.

Naqui, A., Chance, B., and Cadenas, E., 1986. Reactive oxygen intermediates in biochemistry. *Ann. Rev. Biochem.* 55:137.

Wikstrom, M., Saraste, M., and Penttila, T., 1985. Relationships between structure and function in cytochrome oxidase. *In* A. N. Martonosi, (ed.), *The Enzymes of Biological Membranes*, vol. 4, pp. 111–148. Plenum.

ATP synthesis

McCarty, R. E., 1985. H^+-ATPases in oxidative and photosynthetic phosphorylation. *BioScience* 35:27–30.

Kagawa, Y., 1984. Proton motive ATP synthesis. *In* L. Ernster, (ed.), *Bioenergetics*, pp. 149–186. Elsevier.

Walker, J. E., Saraste, M., and Gay, N. J., 1984. The *unc* operon: nucleotide sequence, regulation, and structure of ATP-synthase. *Biochim. Biophys. Acta* 768:164–200.

Gressert, M. J., Myers, J. A., and Boyer, P. D., 1982. Catalytic site cooperativity of beef heart mitochondrial F_1 adenosine triphosphatase. *J. Biol. Chem.* 257:12030–12038.

Cross, R. L., Cunningham, D., and Tamura, J. K., 1984. Binding change mechanism for ATP synthesis by oxidative phosphorylation and photophosphorylation. *Curr. Top. Cell. Regul.* 24:335–344.

Vignais, P. V., and Lunardi, J., 1985. Chemical probes of the mitochondrial ATP synthesis and translocation. *Ann. Rev. Biochem.* 54:977–1014.

Chance, B., Leigh, J. S., Jr., Kent, J., McCully, K., Nioka, S., Clark, B. J., Maris, J. M., and Graham, T., 1986. Multiple controls of oxidative metabolism in living tissues as studied by phosphorus magnetic resonance. *Proc. Nat. Acad. Sci.* 83:9458–9462.

Translocators

Klingenberg, M., 1985. Principles of carrier catalysis elucidated by comparing two similar membrane translocators from mitochondria, the ADP/ATP carrier and the uncoupling protein. *Ann. N. Y. Acad. Sci.* 456:279–288.

Vignais, P. V., Block, M. R., Boulay, F., Brandolin, G., and Lauquin, G. J.-M., 1985. Molecular aspects of structure-function relationships in mitochondrial adenine nucleotide carrier. In G. Bengha, (ed.), *Structure and Properties of Cell Membranes*, vol. II, pp. 139–179. CRC Press.

Superoxide dismutase

Fridovich, I., 1986. Superoxide dismutases. *Advan. Enzymol.* 58:61–97.

Tainer, J. A., Getzoff, E. D., Richardson, J. S., and Richardson, D. C., 1983. Structure and mechanism of Cu,Zn superoxide dismutase. *Nature* 306:284–287.

Historical aspects

Mitchell, P., 1976. Vectorial chemistry and the molecular mechanics of chemiosmotic coupling: power transmission by proticity. *Trans. Biochem. Soc.* 4:399–430.

Racker, E., 1980. From Pasteur to Mitchell: a hundred years of bioenergetics. *Fed. Proc.* 39:210–215.

Keilin, D., 1966. *The History of Cell Respiration and Cytochromes.* Cambridge University Press.

Kalckar, H. M., (ed.), 1969. *Biological Phosphorylations: Development of Concepts.* Prentice-Hall. [A collection of classical papers on oxidative phosphorylation and other aspects of bioenergetics.]

Fruton, J. S., 1972. *Molecules and Life: Historical Essays on the Interplay of Chemistry and Biology.* Wiley-Interscience. [Includes an excellent account of cellular respiration, starting on page 262.]

PROBLEMS

1. What is the yield of ATP when each of the following substrates is completely oxidized to CO_2 by a mammalian cell homogenate? Assume that glycolysis, the citric acid cycle, and oxidative phosphorylation are fully active.
   (a) Pyruvate.
   (b) NADH.
   (c) Fructose 1,6-bisphosphate.
   (d) Phosphoenolpyruvate.
   (e) Glucose.
   (f) Dihydroxyacetone phosphate.

2. (a) Write an equation for the oxidation of reduced glutathione (see p. 400) by O_2. Calculate $\Delta E_0'$ and $\Delta G^{\circ\prime}$ for this reaction using the data in Table 17-1.
   (b) What is $\Delta E_0'$ and $\Delta G^{\circ\prime}$ for the reduction of oxidized glutathione by NADPH?

3. What is the effect of each of the following inhibitors on electron transport and ATP formation by the respiratory chain?
   (a) Azide.
   (b) Atractyloside.
   (c) Rotenone.
   (d) DNP.
   (e) Carbon monoxide.
   (f) Antimycin A.

4. The addition of oligomycin to mitochondria markedly decreases both the rate of electron transfer from NADH to O_2 and the rate of ATP formation. The subsequent addition of DNP leads to an increase in the rate of electron transfer without changing the rate of ATP formation. What does oligomycin inhibit?

5. Compare the $\Delta G^{\circ\prime}$ for the oxidation of succinate by NAD^+ and by FAD. Use the data given in Table 17-1, and assume that E_0' for the FAD/$FADH_2$ redox couple is nearly 0 V. Why is FAD rather than NAD^+ the electron acceptor in the reaction catalyzed by succinate dehydrogenase?

6. The immediate administration of nitrite is a highly effective treatment for cyanide poisoning. What is the basis for the action of this antidote? (Hint: Nitrite oxidizes ferrohemoglobin to ferrihemoglobin.)

7. For a proton-motive force of 0.2 V (matrix negative), what is the maximum [ATP]/[ADP][P_i] ratio compatible with ATP synthesis? Calculate this ratio three times, assuming that the number of protons translocated per ATP formed is 2, 3, and 4 and that the temperature is 25°C.

8. The standard oxidation-reduction potential for the reduction of O_2 to H_2O is given as 0.82 V in Table 17-1. However, the value given in textbooks of chemistry is 1.23 V. Account for this difference.

9. Suppose that the mitochondria of a patient oxidized NADH irrespective of whether ADP was present. The P:O ratio for oxidative phosphorylation by these mitochondria was less than normal. Predict the likely symptoms of this disorder.

10. ATPγS, a slowly-hydrolyzed analog of ATP, can be used to probe the mechanism of phosphoryl transfer reactions. Chiral ATPγS has been synthesized containing ^{18}O in a specific γ position and ordinary ^{16}O elsewhere in the molecule. Hydrolysis of this chiral molecule by ATP synthase in ^{17}O-enriched water yields inorganic [^{16}O,^{17}O,^{18}O]thiophosphate having the absolute configuration shown in Figure 17-36. In contrast, hydrolysis of this chiral ATPγS by a calcium-pumping ATPase from muscle gives thiophosphate of the opposite configuration. What is the simplest interpretation of these data?

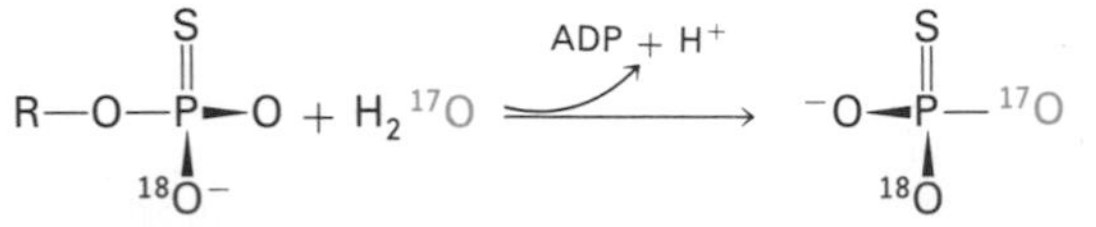

$$\text{R—O—P(=S)(}^{18}\text{O}^-\text{)—O} + \text{H}_2{}^{17}\text{O} \xrightarrow{\text{ADP} + \text{H}^+} {}^-\text{O—P(=S)(}^{18}\text{O)—}^{17}\text{O}$$

ATPγS — Thiophosphate

Figure 17-36
Stereochemistry of the hydrolysis of chiral ATPγS by ATP synthase.

11. Conduction of protons by the F_0 unit of ATP synthase is blocked by modification of a single side chain by dicyclohexylcarbodiimide. What are the most likely targets of action of this reagent? How might you use site-specific mutagenesis to determine whether this residue is essential for proton conduction?

CHAPTER 18

Pentose Phosphate Pathway and Gluconeogenesis

The preceding chapters on glycolysis, the citric acid cycle, and oxidative phosphorylation were primarily concerned with the generation of ATP, starting with glucose as a fuel. We now turn to the generation of a different type of metabolic energy—reducing power. Some of the high-potential electrons of fuel molecules must be conserved for biosynthetic purposes rather than transferred to O_2 to generate ATP. *The currency of readily available reducing power in cells is NADPH.* The phosphoryl group on C-2 of one of the ribose units of NADPH distinguishes it from NADH. As mentioned previously (p. 321), there is a *fundamental distinction between NADPH and NADH in most biochemical reactions. NADH is oxidized by the respiratory chain to generate ATP, whereas NADPH serves as an electron donor (hydride ion donor) in reductive biosyntheses.* This chapter also deals with the synthesis of glucose from noncarbohydrate precursors, by a process called *gluconeogenesis*.

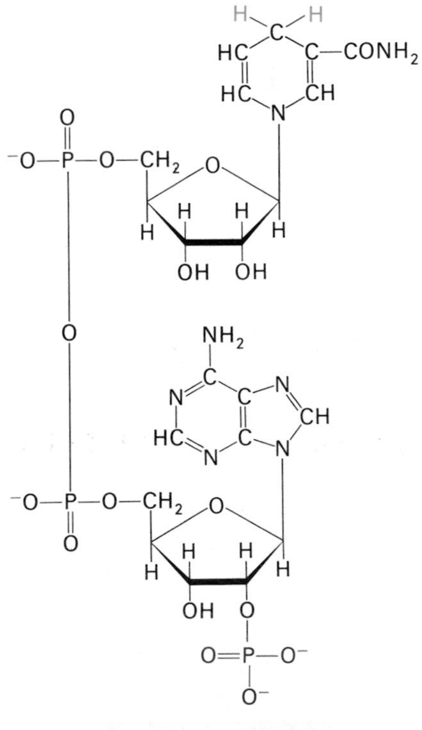

Reduced nicotinamide adenine dinucleotide phosphate (NADPH)

THE PENTOSE PHOSPHATE PATHWAY GENERATES NADPH AND SYNTHESIZES FIVE-CARBON SUGARS

In the pentose phosphate pathway, NADPH is generated when glucose 6-phosphate is oxidized to ribose 5-phosphate. This five-carbon sugar and its derivatives are components of such important biomolecules as ATP, CoA, NAD^+, FAD, RNA, and DNA.

$$\text{Glucose 6-phosphate} + 2\ NADP^+ + H_2O \longrightarrow \text{ribose 5-phosphate} + 2\ NADPH + 2\ H^+ + CO_2$$

The pentose phosphate pathway also catalyzes the interconversion of three-, four-, five-, six-, and seven-carbon sugars in a series of nonoxidative reactions. All of these reactions occur in the cytosol. In plants, part of the pentose phosphate pathway also participates in the formation of hexoses from CO_2 in photosynthesis (p. 537).

The pentose phosphate pathway is sometimes called *the pentose shunt, the hexose monophosphate pathway*, or *the phosphogluconate oxidative pathway*. Glucose 6-phosphate dehydrogenase, the first enzyme in the pathway, was discovered by Otto Warburg in 1931, and the complete pathway was subsequently elucidated by Fritz Lipmann, Frank Dickens, Bernard Horecker, and Efraim Racker.

TWO NADPH ARE GENERATED IN THE CONVERSION OF GLUCOSE 6-PHOSPHATE INTO RIBULOSE 5-PHOSPHATE

The pentose phosphate pathway starts with the dehydrogenation of glucose 6-phosphate at C-1, a reaction catalyzed by *glucose 6-phosphate dehydrogenase* (Figure 18-1). This enzyme is highly specific for $NADP^+$; the K_M for NAD^+ is about a thousand times as great as that for $NADP^+$. The product is *6-phosphoglucono-δ-lactone,* which is an intramolecular ester between the C-1 carboxyl group and the C-5 hydroxyl group. The next step is the hydrolysis of 6-phosphoglucono-δ-lactone by a specific *lactonase* to give *6-phosphogluconate*. This six-carbon sugar is then oxidatively decarboxylated by *6-phosphogluconate dehydrogenase* to yield *ribulose 5-phosphate*. $NADP^+$ is again the electron acceptor.

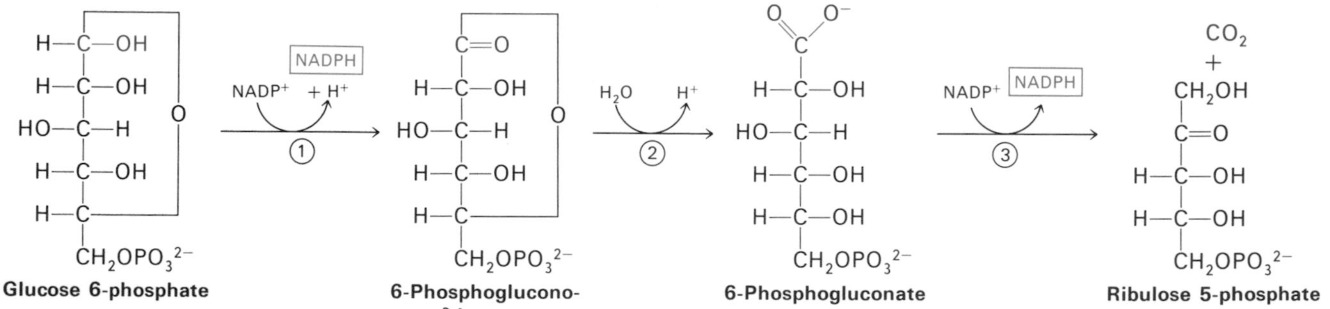

Glucose 6-phosphate → (1) $NADP^+$ → NADPH + H^+ → 6-Phosphoglucono-δ-lactone → (2) H_2O → H^+ → 6-Phosphogluconate → (3) $NADP^+$ → NADPH → Ribulose 5-phosphate + CO_2

Figure 18-1
Oxidative branch of the pentose phosphate pathway. These three reactions are catalyzed by glucose 6-phosphate dehydrogenase, lactonase, and 6-phosphogluconate dehydrogenase.

RIBULOSE 5-PHOSPHATE IS ISOMERIZED TO RIBOSE 5-PHOSPHATE THROUGH AN ENEDIOL INTERMEDIATE

The final step in the synthesis of ribose 5-phosphate is the isomerization of ribulose 5-phosphate by *phosphopentose isomerase*.

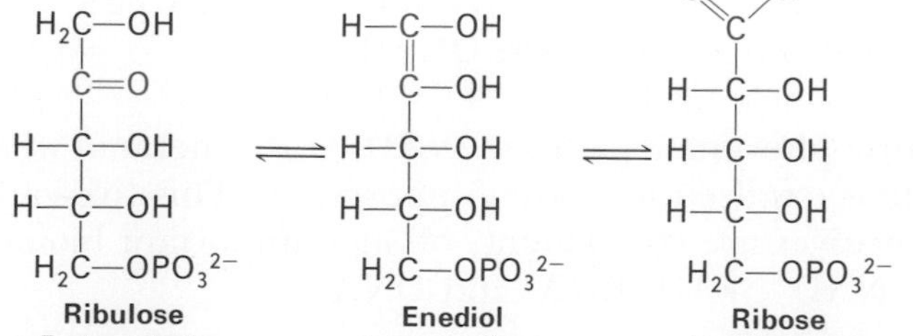

Ribulose 5-phosphate ⇌ Enediol intermediate ⇌ Ribose 5-phosphate

This reaction is similar to the glucose 6-phosphate → fructose 6-phosphate and to the dihydroxyacetone phosphate → glyceraldehyde 3-phosphate reactions in glycolysis. *All three ketose-aldose isomerizations proceed through an enediol intermediate.*

THE PENTOSE PHOSPHATE PATHWAY AND GLYCOLYSIS ARE LINKED BY TRANSKETOLASE AND TRANSALDOLASE

The preceding reactions yield two NADPH and one ribose 5-phosphate for each glucose 6-phosphate oxidized. However, many cells need NADPH for reductive biosyntheses much more than ribose 5-phosphate for incorporation into nucleotides and nucleic acids. In that case, ribose 5-phosphate is converted into glyceraldehyde 3-phosphate and fructose 6-phosphate by *transketolase* and *transaldolase. These enzymes create a reversible link between the pentose phosphate pathway and glycolysis* by catalyzing these three reactions:

$$C_5 + C_5 \xrightleftharpoons{\text{transketolase}} C_3 + C_7$$

$$C_7 + C_3 \xrightleftharpoons{\text{transaldolase}} C_4 + C_6$$

$$C_5 + C_4 \xrightleftharpoons{\text{transketolase}} C_3 + C_6$$

The net result of these reactions is the *formation of two hexoses and one triose from three pentoses.*

The essence of these reactions is that *transketolase transfers a two-carbon unit, whereas transaldolase transfers a three-carbon unit.* The sugar that donates the two- or three-carbon unit is always a ketose, whereas the acceptor is always an aldose.

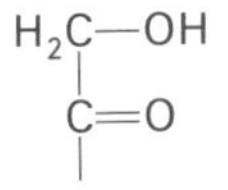

H$_2$C—OH / C=O

Transferred by transketolase

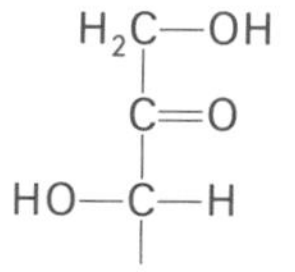

H$_2$C—OH / C=O / HO—C—H

Transferred by transaldolase

The first of the three reactions linking the pentose phosphate pathway and glycolysis is the formation of *glyceraldehyde 3-phosphate* and *sedoheptulose 7-phosphate* from two pentoses.

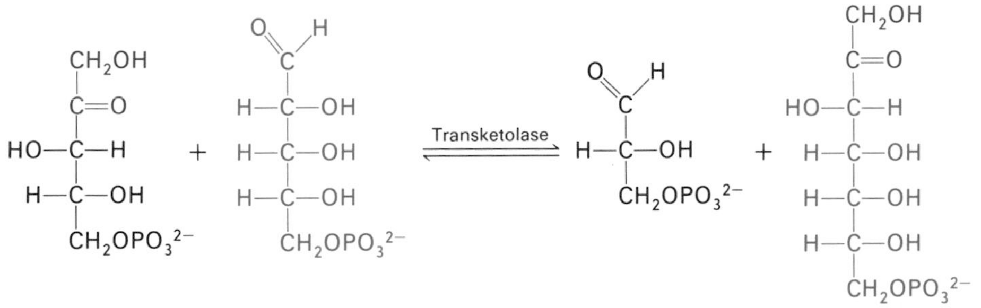

Xylulose 5-phosphate (CH$_2$OH—C=O—HO—C—H—H—C—OH—CH$_2$OPO$_3^{2-}$) + Ribose 5-phosphate (HC=O—H—C—OH—H—C—OH—H—C—OH—CH$_2$OPO$_3^{2-}$) ⇌ (Transketolase) Glyceraldehyde 3-phosphate (HC=O—H—C—OH—CH$_2$OPO$_3^{2-}$) + Sedoheptulose 7-phosphate (CH$_2$OH—C=O—HO—C—H—H—C—OH—H—C—OH—H—C—OH—CH$_2$OPO$_3^{2-}$)

The donor of the two-carbon unit in this reaction is xylulose 5-phosphate, which is an epimer of ribulose 5-phosphate. A ketose is a substrate of transketolase only if its hydroxyl group at C-3 has the configuration of xylulose rather than ribulose. Ribulose 5-phosphate is converted into the appropriate epimer for the transketolase reaction by *phosphopentose epimerase.*

Ribulose 5-phosphate (CH$_2$OH—C=O—H—C—OH—H—C—OH—CH$_2$OPO$_3^{2-}$) ⇌ (Phosphopentose epimerase) Xylulose 5-phosphate (CH$_2$OH—C=O—HO—C—H—H—C—OH—CH$_2$OPO$_3^{2-}$)

Glyceraldehyde 3-phosphate and sedoheptulose 7-phosphate then react to form *fructose 6-phosphate* and *erythrose 4-phosphate*. This synthesis of a four-carbon sugar and a six-carbon sugar is catalyzed by *transaldolase*.

Sedoheptulose 7-phosphate + Glyceraldehyde 3-phosphate $\xrightleftharpoons{\text{Transaldolase}}$ Erythrose 4-phosphate + Fructose 6-phosphate

In the third reaction, transketolase catalyzes the synthesis of *fructose 6-phosphate* and *glyceraldehyde 3-phosphate* from erythrose 4-phosphate and xylulose 5-phosphate.

Xylulose 5-phosphate + Erythrose 4-phosphate $\xrightleftharpoons{\text{Transketolase}}$ Glyceraldehyde 3-phosphate + Fructose 6-phosphate

The sum of these reactions is

2 Xylulose 5-phosphate + ribose 5-phosphate $\rightleftharpoons$
2 fructose 6-phosphate + glyceraldehyde 3-phosphate

Xylulose 5-phosphate can be formed from ribose 5-phosphate by the sequential action of phosphopentose isomerase and phosphopentose epimerase, and so the net reaction starting from ribose 5-phosphate is

3 Ribose 5-phosphate $\rightleftharpoons$
2 fructose 6-phosphate + glyceraldehyde 3-phosphate

Thus, excess ribose 5-phosphate formed by the pentose phosphate pathway can be completely converted into glycolytic intermediates.

THE RATE OF THE PENTOSE PHOSPHATE PATHWAY IS CONTROLLED BY THE LEVEL OF $NADP^+$

The first reaction in the oxidative branch of the pentose phosphate pathway, the dehydrogenation of glucose 6-phosphate, is essentially irreversible. In fact, this reaction is rate-limiting under physiological conditions and serves as the control site. The most important regulatory factor is the level of $NADP^+$, the electron acceptor in the oxidation of glucose 6-phosphate to 6-phosphogluconolactone. Also, NADPH com-

Table 18-1
Pentose phosphate pathway

Reaction	Enzyme
OXIDATIVE BRANCH	
Glucose 6-phosphate + $NADP^+$ $\longrightarrow$ 6-phosphoglucono-δ-lactone + NADPH + H^+	Glucose 6-phosphate dehydrogenase
6-Phosphoglucono-δ-lactone + H_2O $\longrightarrow$ 6-phosphogluconate + H^+	Lactonase
6-Phosphogluconate + $NADP^+$ $\longrightarrow$ ribulose 5-phosphate + CO_2 + NADPH	6-Phosphogluconate dehydrogenase
NONOXIDATIVE BRANCH	
Ribulose 5-phosphate $\rightleftharpoons$ ribose 5-phosphate	Phosphopentose isomerase
Ribulose 5-phosphate $\rightleftharpoons$ xylulose 5-phosphate	Phosphopentose epimerase
Xylulose 5-phosphate + ribose 5-phosphate $\rightleftharpoons$ sedoheptulose 7-phosphate + glyceraldehyde 3-phosphate	Transketolase
Sedoheptulose 7-phosphate + glyceraldehyde 3-phosphate $\rightleftharpoons$ fructose 6-phosphate + erythrose 4-phosphate	Transaldolase
Xylulose 5-phosphate + erythrose 4-phosphate $\rightleftharpoons$ fructose 6-phosphate + glyceraldehyde 3-phosphate	Transketolase

petes with $NADP^+$ in binding to the enzyme. The ratio of $NADP^+$ to NADPH in the cytosol of a liver cell from a well-fed rat is about 0.014, several orders of magnitude lower than the ratio of NAD^+ to NADH, which is 700 under the same conditions. The marked effect of the $NADP^+$ level on the rate of the oxidative branch ensures that NADPH generation is tightly coupled to its utilization in reductive biosyntheses. The nonoxidative branch of the pentose phosphate pathway is controlled primarily by the availability of substrates.

THE FLOW OF GLUCOSE 6-PHOSPHATE DEPENDS ON THE NEED FOR NADPH, RIBOSE 5-PHOSPHATE, AND ATP

Let us follow the fate of glucose 6-phosphate in four different situations:

1. *Much more ribose 5-phosphate than NADPH is required.* Most of the glucose 6-phosphate is converted into fructose 6-phosphate and glyceraldehyde 3-phosphate by the glycolytic pathway. Transaldolase and transketolase then convert two molecules of fructose 6-phosphate and one molecule of glyceraldehyde 3-phosphate into three molecules of ribose 5-phosphate by a reversal of the reactions described earlier. The stoichiometry of this mode (Figure 18-2A) is

$$\text{5 Glucose 6-phosphate} + \text{ATP} \longrightarrow \text{6 ribose 5-phosphate} + \text{ADP} + H^+$$

2. *The needs for NADPH and ribose 5-phosphate are balanced.* The predominant reaction under these conditions is the formation of two NADPH and one ribose 5-phosphate from glucose 6-phosphate by the oxidative branch of the pentose phosphate pathway. The stoichiometry of this mode (Figure 18-2B) is

$$\text{Glucose 6-phosphate} + 2\ NADP^+ + H_2O \longrightarrow \text{ribose 5-phosphate} + 2\ NADPH + 2\ H^+ + CO_2$$

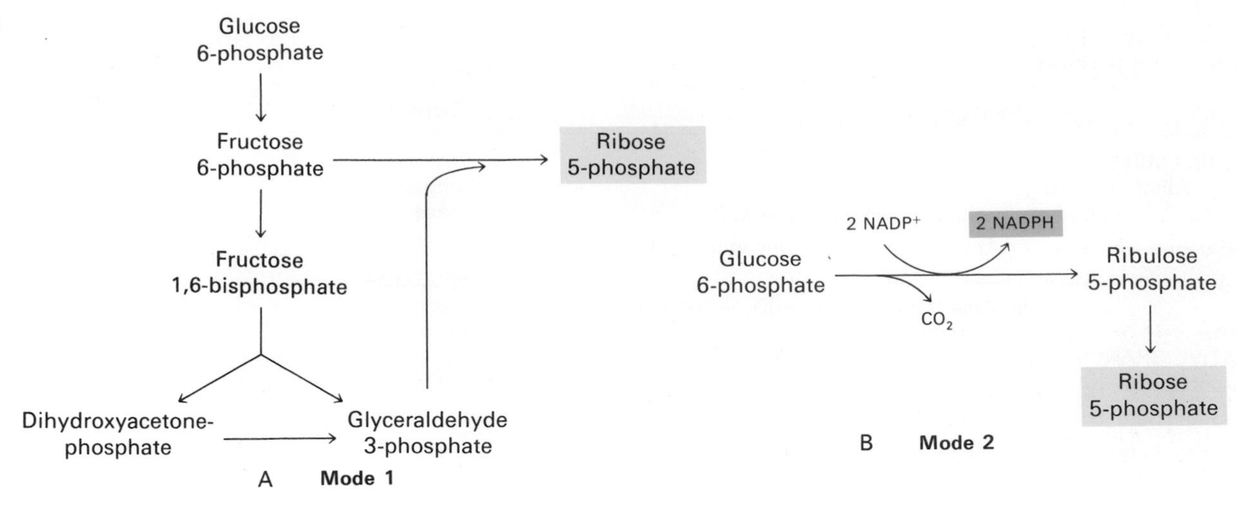

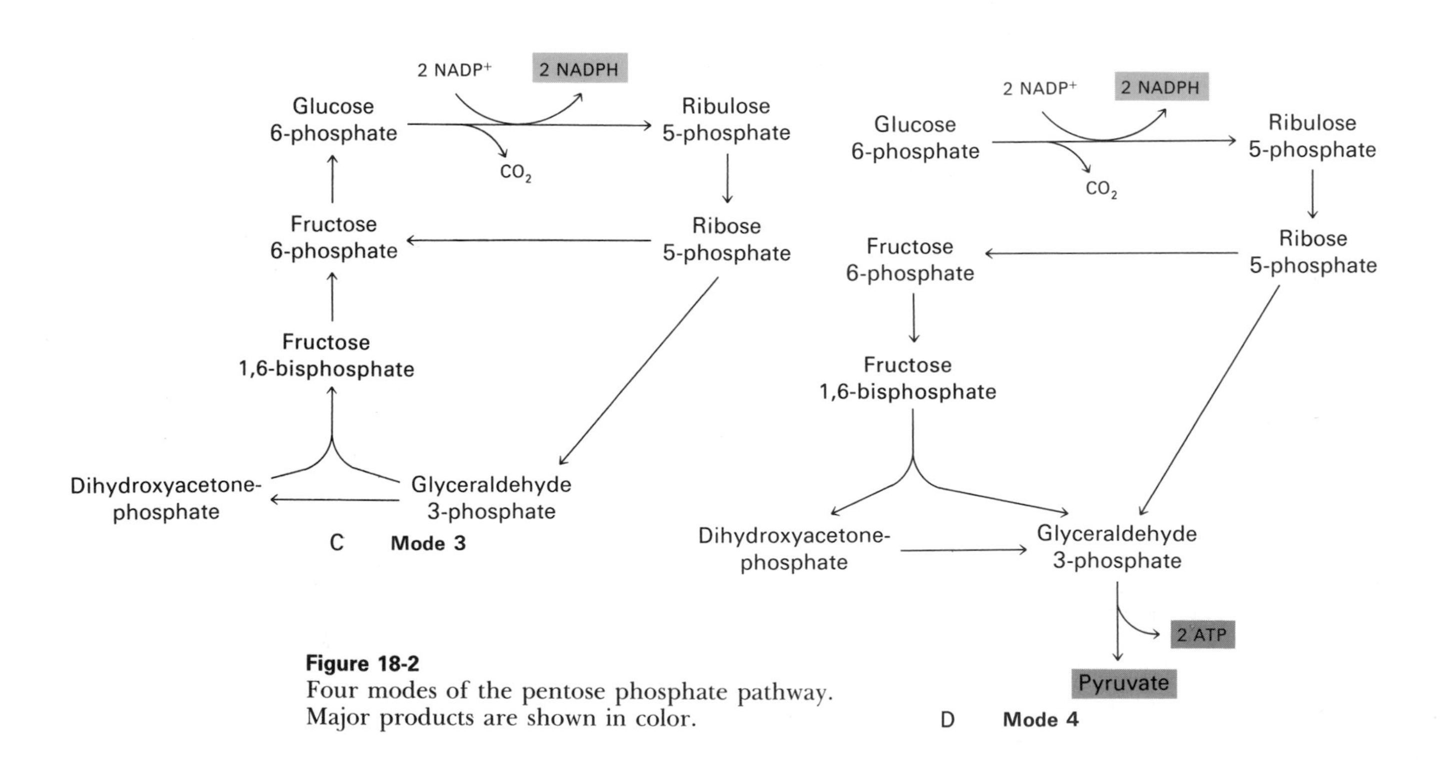

Figure 18-2
Four modes of the pentose phosphate pathway. Major products are shown in color.

3. *Much more NADPH than ribose 5-phosphate is required; glucose 6-phosphate is completely oxidized to CO_2.* Three groups of reactions are active in this situation. First, two NADPH and one ribose 5-phosphate are formed by the oxidative branch of the pentose phosphate pathway. Then, ribose 5-phosphate is converted into fructose 6-phosphate and glyceraldehyde 3-phosphate by transketolase and transaldolase. Finally, glucose 6-phosphate is resynthesized from fructose 6-phosphate and glyceraldehyde 3-phosphate by the gluconeogenic pathway (discussed later in this chapter). The stoichiometries of these three sets of reactions (Figure 18-2C) are

$$6 \text{ Glucose 6-phosphate} + 12 \text{ NADP}^+ + 6 \text{ H}_2\text{O} \longrightarrow 6 \text{ ribose 5-phosphate} + 12 \text{ NADPH} + 12 \text{ H}^+ + 6 \text{ CO}_2$$

$$\text{6 Ribose 5-phosphate} \longrightarrow \text{4 fructose 6-phosphate} + \text{2 glyceraldehyde 3-phosphate}$$

$$\text{4 Fructose 6-phosphate} + \text{2 glyceraldehyde 3-phosphate} + H_2O \longrightarrow \text{5 glucose 6-phosphate} + P_i$$

The sum of these reactions is

$$\text{Glucose 6-phosphate} + 12\ NADP^+ + 7\ H_2O \longrightarrow 6\ CO_2 + 12\ NADPH + 12\ H^+ + P_i$$

Thus, the equivalent of a glucose 6-phosphate can be completely oxidized to CO_2 *with the concomitant generation of NADPH.* The essence of these reactions is that *ribose 5-phosphate produced by the pentose phosphate pathway is recycled into glucose 6-phosphate* by transketolase, transaldolase, and some of the enzymes of the gluconeogenic pathway.

4. *Much more NADPH than ribose 5-phosphate is required; glucose 6-phosphate is converted into pyruvate.* Alternatively, ribose 5-phosphate formed by the oxidative branch of the pentose phosphate pathway can be converted into pyruvate (Figure 18-2D). Fructose 6-phosphate and glyceraldehyde 3-phosphate derived from ribose 5-phosphate enter the glycolytic pathway rather than reverting to glucose 6-phosphate. In this mode, *ATP and NADPH are concomitantly generated, and five of the six carbons of glucose 6-phosphate emerge in pyruvate:*

$$\text{3 Glucose 6-phosphate} + 6\ NADP^+ + 5\ NAD^+ + 5\ P_i + 8\ ADP \longrightarrow \text{5 pyruvate} + 3\ CO_2 + 6\ NADPH + 5\ NADH + 8\ ATP + 2\ H_2O + 8\ H^+$$

Pyruvate formed by these reactions can be oxidized to generate more ATP or it can be used as a building block in a variety of biosyntheses.

THE PENTOSE PHOSPHATE PATHWAY IS MUCH MORE ACTIVE IN ADIPOSE TISSUE THAN IN MUSCLE

Radioactive-labeling experiments can provide estimates of how much glucose 6-phosphate is metabolized by the pentose phosphate pathway and how much is metabolized by the combined action of glycolysis and the citric acid cycle. For example, one aliquot of a tissue homogenate is incubated with glucose labeled with ^{14}C at C-1, and another with glucose labeled with ^{14}C at C-6. The radioactivity of the CO_2 produced by the two samples is then compared. The rationale of this experiment is that only C-1 is decarboxylated by the pentose phosphate pathway, whereas C-1 and C-6 are decarboxylated to the same extent when glucose is metabolized by the glycolytic pathway, the pyruvate dehydrogenase complex, and the citric acid cycle. The reason for the equivalence of C-1 and C-6 in the latter set of reactions is that glyceraldehyde 3-phosphate and dihydroxyacetone phosphate are rapidly interconverted by triose phosphate isomerase.

This experimental approach has shown that *the activity of the pentose phosphate pathway is very low in skeletal muscle, whereas it is very high in adipose tissue.* These findings support the idea that a major role of the pentose phosphate pathway is to generate NADPH for reductive biosyntheses. Large amounts of it are consumed by adipose tissue in the reductive synthesis of fatty acids from acetyl CoA (p. 486).

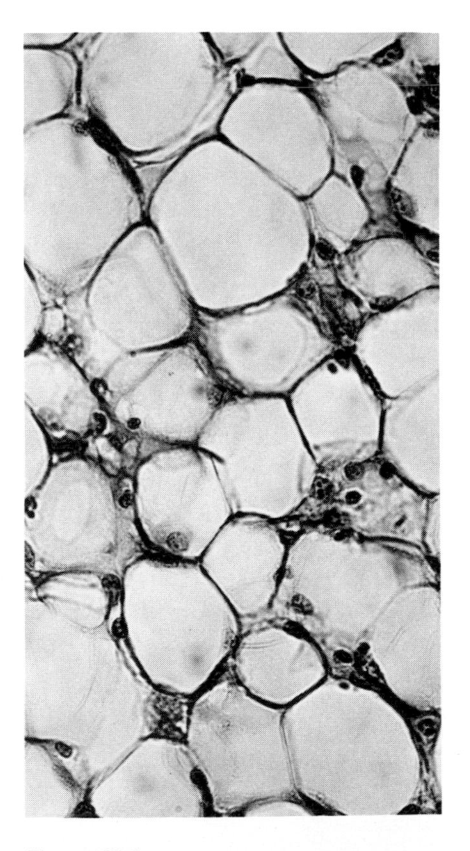

Figure 18-3
Light micrograph of adipose tissue.

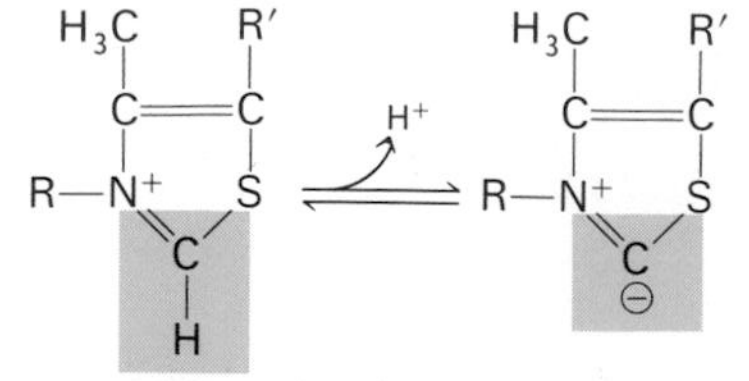

TPP, THE PROSTHETIC GROUP OF TRANSKETOLASE, TRANSFERS ACTIVATED ALDEHYDES

(TPP)
Thiamine pyrophosphate

Carbanion

Transketolase contains a tightly bound thiamine pyrophosphate (TPP) as its prosthetic group. This prosthetic group has been encountered before, in the decarboxylation of pyruvate by the pyruvate dehydrogenase complex (p. 380). The mechanism of catalysis of transketolase is similar in that an *activated aldehyde unit is transferred to an acceptor.* The acceptor in the transketolase reaction is an aldose, whereas in the pyruvate dehydrogenase reaction it is lipoamide. In both reactions, the site of addition of the keto substrate is the *thiazole ring* of the prosthetic group. The C-2 carbon atom of bound TPP readily ionizes to give a *carbanion.* The negatively charged carbon atom of this reactive intermediate adds to the carbonyl group of the ketose substrate (e.g., xylulose 5-phosphate, fructose 6-phosphate, and sedoheptulose 7-phosphate).

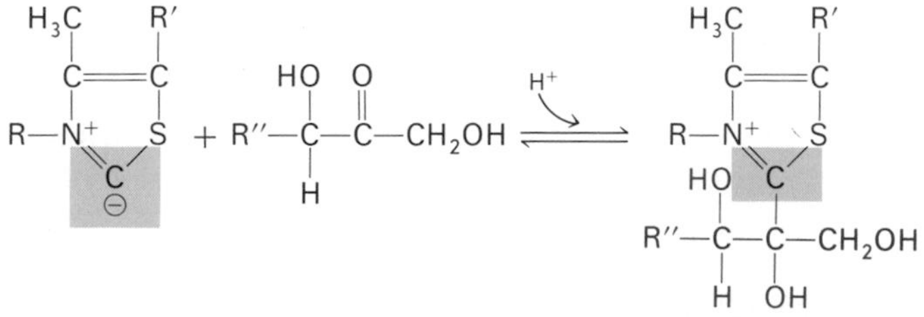

Carbanion | Ketose substrate | Addition compound

This addition compound loses its R—CHOH moiety to yield a negatively charged, *activated glycoaldehyde* unit. The positively charged nitrogen in the thiazole ring acts as an *electron sink* to promote the development of a negative charge on the activated intermediate.

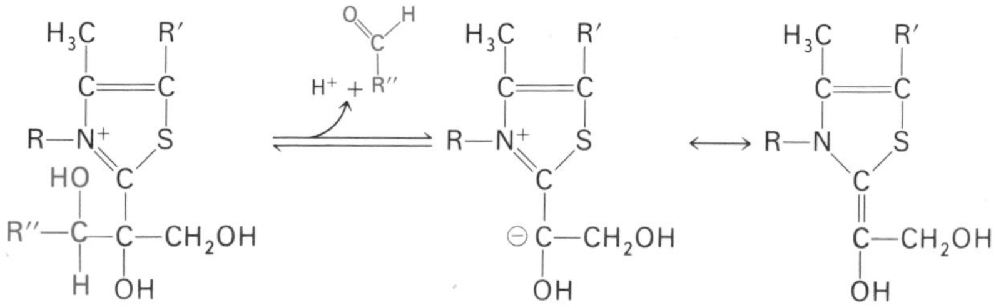

Addition compound

Resonance forms of the activated glycoaldehyde unit

The carbonyl group of a suitable aldehyde acceptor then condenses with the activated glycoaldehyde unit to form a new ketose, which is released from the enzyme.

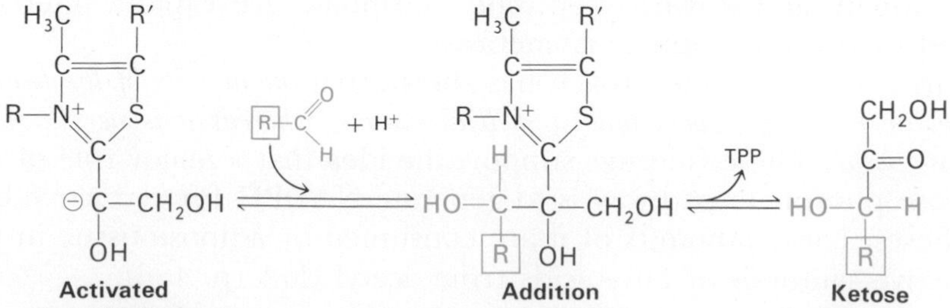

Activated glycoaldehyde

Addition compound

Ketose product

TRANSKETOLASE THAT IS DEFECTIVE IN TPP BINDING CAN CAUSE A NEUROPSYCHIATRIC DISORDER

The *Wernicke-Korsakoff syndrome*, a striking neuropsychiatric disorder, is caused by a lack of thiamine in the diet of susceptible persons. This disease is characterized by paralysis of eye movements, abnormal stance and gait, and markedly deranged mental function. In particular, a patient with this syndrome is disoriented and has a severely impaired memory. Only a small minority of alcoholics and other chronically malnourished persons develop this disorder. Also, its incidence is much higher among Europeans than among non-Europeans on thiamine-deficient diets. These observations suggest that genetic factors may be important determinants of whether a thiamine-deficient person develops the Wernicke-Korsakoff syndrome. Studies of transketolase from cultured fibroblasts show that this is in fact the case. *Transketolase from patients with the Wernicke-Korsakoff syndrome binds thiamine pyrophosphate tenfold less avidly than does the enzyme from normal persons.* Two other thiamine-dependent enzymes, pyruvate dehydrogenase and α-ketoglutarate dehydrogenase, are normal in this disorder. The abnormality in transketolase becomes clinically evident only when the level of thiamine pyrophosphate is too low to saturate the enzyme. This is a clear-cut example of the interplay between genetic and environmental factors in the production of disease. The Wernicke-Korsakoff syndrome also demonstrates vividly how a reduction in the activity of a single enzyme can have profound neurologic and behavioral consequences.

ACTIVATED DIHYDROXYACETONE IS CARRIED BY TRANSALDOLASE AS A SCHIFF BASE

Transaldolase transfers a three-carbon *dihydroxyacetone* unit from a ketose donor to an aldose acceptor. Transaldolase, in contrast with transketolase, does not contain a prosthetic group. Rather, *a Schiff base is formed between the carbonyl group of the ketose substrate and the ε-amino group of a lysine residue at the active site of the enzyme.* This kind of covalent enzyme-substrate (ES) intermediate is like that formed in fructose bisphosphate aldolase in the glycolytic pathway (p. 365).

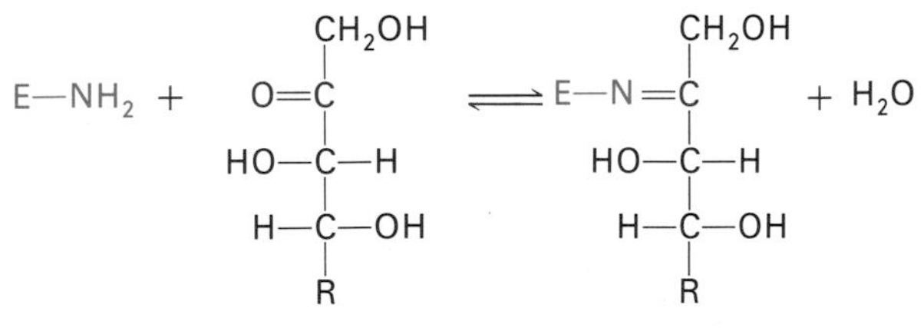

$$\text{E—NH}_2 + \begin{array}{c} \text{CH}_2\text{OH} \\ | \\ \text{O=C} \\ | \\ \text{HO—C—H} \\ | \\ \text{H—C—OH} \\ | \\ \text{R} \end{array} \rightleftharpoons \begin{array}{c} \text{CH}_2\text{OH} \\ | \\ \text{E—N=C} \\ | \\ \text{HO—C—H} \\ | \\ \text{H—C—OH} \\ | \\ \text{R} \end{array} + \text{H}_2\text{O}$$

Ketose substrate **Schiff base**

The Schiff base becomes protonated, the bond between C-3 and C-4 is split, and an aldose is released.

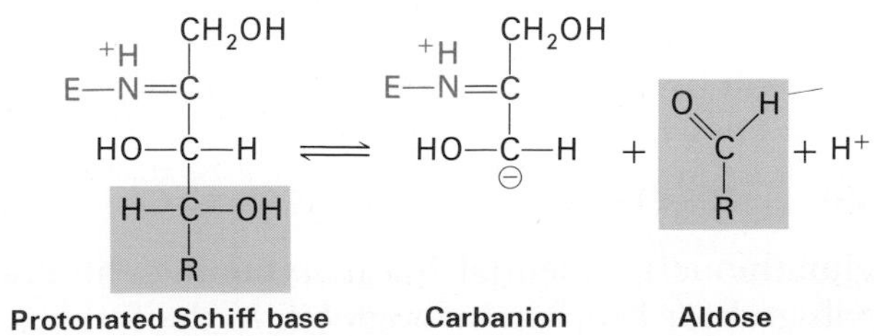

$$\begin{array}{c} {}^{+}\text{H} \quad \text{CH}_2\text{OH} \\ | \\ \text{E—N=C} \\ | \\ \text{HO—C—H} \\ | \\ \text{H—C—OH} \\ | \\ \text{R} \end{array} \rightleftharpoons \begin{array}{c} {}^{+}\text{H} \quad \text{CH}_2\text{OH} \\ | \\ \text{E—N=C} \\ | \\ \text{HO—C—H} \\ \ominus \end{array} + \begin{array}{c} \text{O} \quad \text{H} \\ \diagdown\!\!\!\diagup \\ \text{C} \\ | \\ \text{R} \end{array} + \text{H}^+$$

Protonated Schiff base **Carbanion** **Aldose**

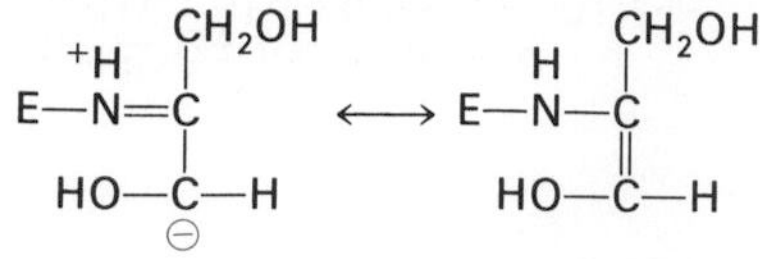

The negative charge on the Schiff base carbanion moiety is stabilized by resonance. *The positively charged nitrogen atom of the Schiff base acts as an electron sink.* This nitrogen atom plays the same role in transaldolase as does the thiazole-ring nitrogen in transketolase.

The Schiff base carbanion is stable until a suitable aldose becomes bound. The dihydroxyacetone moiety then reacts with the carbonyl group of the aldose. The ketose product is released by hydrolysis of the Schiff base.

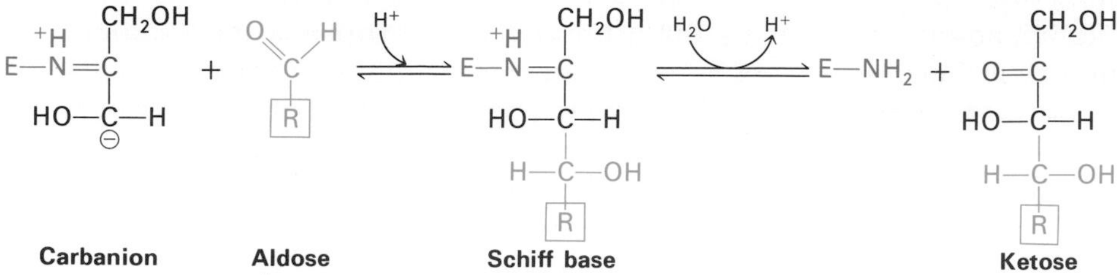

Resonance forms of the Schiff base carbanion

Carbanion + Aldose substrate $\rightleftharpoons$ Schiff base $\rightleftharpoons$ $E{-}NH_2$ + Ketose product

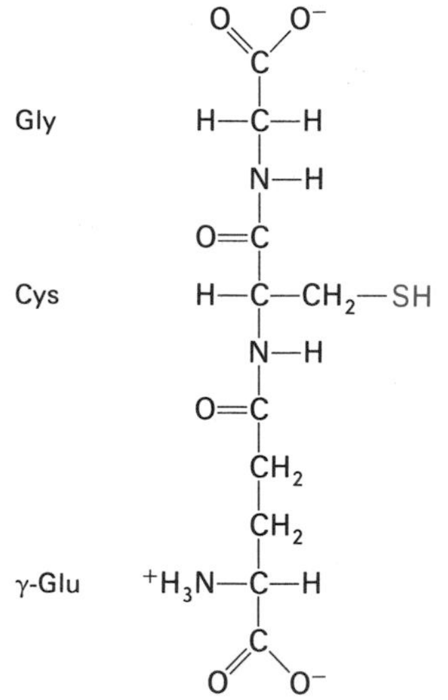

Reduced glutathione (γ-Glutamylcysteinylglycine)

GLUCOSE 6-PHOSPHATE DEHYDROGENASE DEFICIENCY CAUSES A DRUG-INDUCED HEMOLYTIC ANEMIA

An antimalarial drug, pamaquine, was introduced in 1926. Most patients tolerated the drug well, but a few developed severe symptoms within a few days after therapy was started. The urine turned black, jaundice developed, and the hemoglobin content of the blood dropped sharply. In some cases, massive destruction of red blood cells caused death.

The basis of this *drug-induced hemolytic anemia* was elucidated in 1956. The primary defect is a *deficiency in glucose 6-phosphate dehydrogenase in red cells.* The pentose phosphate pathway is the only source of NADPH in red cells because they lack mitochondria, and so the production of NADPH by them is markedly diminished by this deficiency. The major role of NADPH in red cells is to reduce the disulfide form of *glutathione* to the sulfhydryl form. This reaction is catalyzed by *glutathione reductase.*

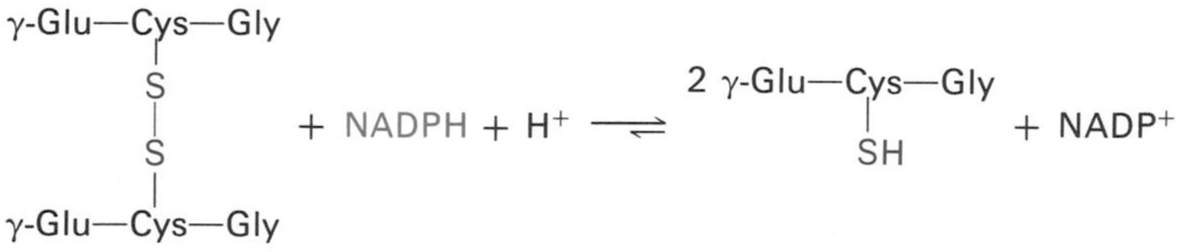

Oxidized glutathione + NADPH + H^+ $\rightleftharpoons$ 2 Reduced glutathione + $NADP^+$

The reduced form of glutathione, a tripeptide with a free sulfhydryl group, serves as a *sulfhydryl buffer* that maintains the cysteine residues of hemoglobin and other red-cell proteins in the reduced state. The ratio of the reduced form of glutathione (GSH) to the oxidized form (GSSG) is normally about 500. The reduced form also plays a role in detoxification by reacting with hydrogen peroxide and organic peroxides, as will be discussed in a later chapter (p. 592).

$$2\ GSH + R{-}O{-}OH \longrightarrow GSSG + H_2O + ROH$$

Reduced glutathione is essential for maintaining the normal structure of red cells and for keeping hemoglobin in the ferrous state. Cells with a lowered level of reduced glutathione are more susceptible to hemolysis for reasons that are not yet fully understood. Drugs such as pamaquine may distort the surface of red cells in the absence of reduced glutathione, which would make them more liable to destruction and removal by the spleen. These drugs also increase the rate of formation of toxic peroxides, which are normally eliminated by reaction with reduced glutathione.

Glucose 6-phosphate dehydrogenase deficiency, a sex-linked trait, is not a rare disease. Female heterozygotes have two populations of red cells: one has normal enzymatic activity, whereas the other is deficient. The glucose 6-phosphate dehydrogenase in most other organs is specified by a different gene, located on an autosome. The incidence of the most common (type A) deficiency of glucose 6-phosphate dehydrogenase, characterized by a tenfold reduction in enzymatic activity in red cells, is 11% among black Americans. This high frequency suggests that the deficiency may be advantageous under certain environmental conditions. Indeed, *glucose 6-phosphate dehydrogenase deficiency in red cells seems to protect a person from falciparum malaria.* The parasites causing this disease require reduced glutathione and the products of the pentose phosphate pathway for optimal growth. Thus, glucose 6-phosphate dehydrogenase deficiency and sickle-cell trait are independent mechanisms of protection against malaria, which accounts for the high frequency of their genes in malaria-infested regions of the world.

The occurrence of glucose 6-phosphate dehydrogenase deficiency clearly demonstrates that *atypical reactions to drugs may have a genetic basis.* This inherited enzymatic deficiency is relatively benign until certain drugs are administered. We see here once again the interplay of heredity and environment in the production of disease. Galactosemia, hereditary fructose intolerance, and phenylketonuria are other striking examples.

GLUTATHIONE REDUCTASE TRANSFERS ELECTRONS FROM NADPH TO OXIDIZED GLUTATHIONE BY MEANS OF FAD

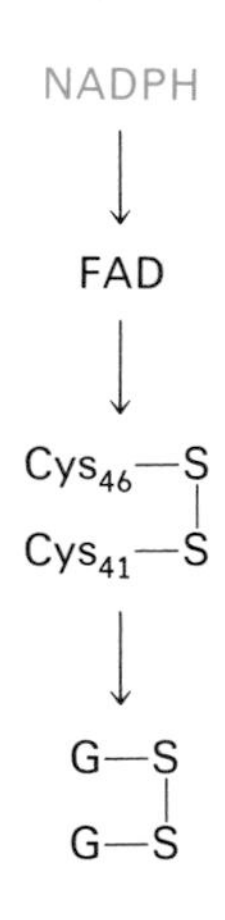

The regeneration of reduced glutathione is catalyzed by *glutathione reductase*, a dimer of 50-kd subunits. The electrons from NADPH are not directly transferred to the disulfide bond in oxidized glutathione. Rather, they are transferred from NADPH to a tightly bound flavin adenine dinucleotide (FAD), then to a disulfide bridge between two cysteine residues in the subunit, and finally to oxidized glutathione. Each subunit consists of three structural domains: an FAD-binding domain, an NADPH-binding domain, and an interface domain (Figure 18-4). The FAD domain and $NADP^+$ domain resemble each other and

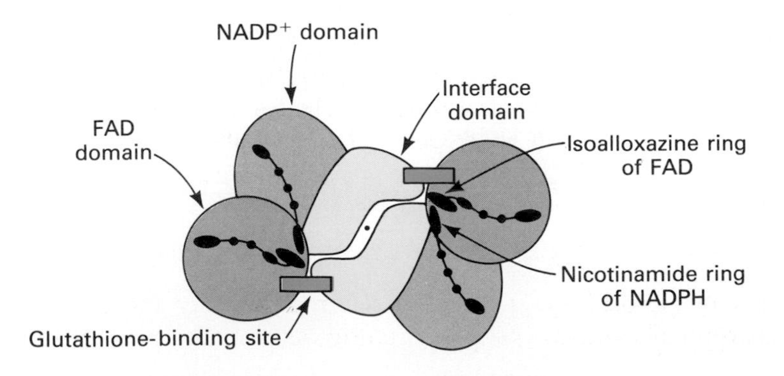

Figure 18-4
Schematic diagram of the domain structure of glutathione reductase. Each subunit in this dimeric enzyme consists of an $NADP^+$ domain, an FAD domain, and an interface domain. Glutathione is bound to the FAD domain of one subunit and the interface domain of another. [After G. E. Schultz, R. H. Schirmer, W. Sachsenheimer, and E. F. Pai, *Nature* 273(1978):123.]

are similar to nucleotide-binding domains in other dehydrogenases. FAD and $NADP^+$ are bound in an extended form, with their isoalloxazine and nicotinamide rings next to each other. It is interesting to note that the binding site for oxidized glutathione is formed by the FAD domain of one subunit and the interface domain of the other subunit.

GLUCOSE CAN BE SYNTHESIZED FROM NONCARBOHYDRATE PRECURSORS

We now turn to the *synthesis of glucose from noncarbohydrate precursors,* by a process called *gluconeogenesis.* This metabolic pathway is very important because the brain is highly dependent on glucose as the primary fuel. Erythrocytes, too, require glucose. The daily glucose requirement of the brain in a typical adult is about 120 g, which accounts for most of the 160 g of glucose needed by the whole body. The amount of glucose present in body fluids is about 20 g, and that readily available from glycogen, a storage form of glucose (p. 449), is approximately 190 g. Thus, the direct glucose reserves are sufficient to meet the needs for glucose for about a day. In a longer period of starvation, glucose must be formed from noncarbohydrate sources for survival. Gluconeogenesis is also important during periods of intense exercise.

The *gluconeogenic pathway converts pyruvate into glucose.* Noncarbohydrate precursors of glucose enter the pathway chiefly at pyruvate, oxaloacetate, and dihydroxyacetone phosphate (Figure 18-5). The major noncarbohydrate precursors are *lactate, amino acids,* and *glycerol.* Lactate is formed by active skeletal muscle when the rate of glycolysis exceeds the metabolic rate of the citric acid cycle and the respiratory chain (p. 363). Amino acids are derived from proteins in the diet and, during starvation, from the breakdown of proteins in skeletal muscle (p. 640). The hydrolysis of triacylglycerols (p. 473) in fat cells yields glycerol and fatty acids. Glycerol is a precursor of glucose, but animals cannot convert fatty acids into glucose, for reasons that will be discussed later (p. 482).

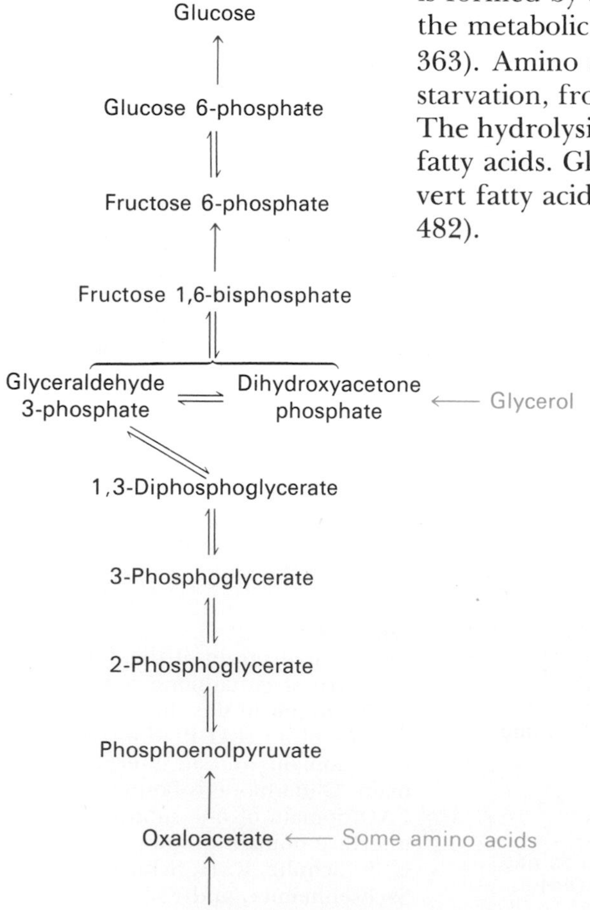

Figure 18-5
Pathway of gluconeogenesis. The distinctive reactions of this pathway are denoted by red arrows. The other reactions are common to glycolysis. The enzymes of gluconeogenesis are located in the cytosol, except for pyruvate carboxylase (in mitochondria) and glucose 6-phosphatase (membrane-bound inside the endoplasmic reticulum). The entry points for lactate, glycerol, and amino acids are shown.

The major site of gluconeogenesis is the *liver*. Gluconeogenesis also occurs in the cortex of the *kidney*, but the total amount of glucose formed there is about one-tenth of that formed in the liver because of the kidney's smaller mass. Very little gluconeogenesis takes place in the brain, skeletal muscle, or heart muscle. Rather, *gluconeogenesis in the liver and kidney helps to maintain the glucose level in the blood so that brain and muscle can extract sufficient glucose from it to meet their metabolic demands.*

GLUCONEOGENESIS IS NOT A REVERSAL OF GLYCOLYSIS

In glycolysis, glucose is converted into pyruvate; in gluconeogenesis, pyruvate is converted into glucose. However, gluconeogenesis is not a reversal of glycolysis. Several different reactions are required because the thermodynamic equilibrium of glycolysis lies far on the side of pyruvate formation. The actual ΔG for the formation of pyruvate from glucose is about -20 kcal/mol under typical cellular conditions (p. 357). Most of the decrease in free energy in glycolysis takes place in the three essentially irreversible steps catalyzed by hexokinase, phosphofructokinase, and pyruvate kinase.

$$\text{Glucose} + \text{ATP} \xrightarrow{\text{hexokinase}} \text{glucose 6-phosphate} + \text{ADP}$$

$$\text{Fructose 6-phosphate} + \text{ATP} \xrightarrow{\text{phosphofructokinase}} \text{fructose 1,6-bisphosphate} + \text{ADP}$$

$$\text{Phosphoenolpyruvate} + \text{ADP} \xrightarrow{\text{pyruvate kinase}} \text{pyruvate} + \text{ATP}$$

In gluconeogenesis, these virtually irreversible reactions of glycolysis are bypassed by the following new steps:

1. *Phosphoenolpyruvate is formed from pyruvate by way of oxaloacetate.* First, pyruvate is carboxylated to oxaloacetate at the expense of an ATP. Then, oxaloacetate is decarboxylated and phosphorylated to yield phosphoenolpyruvate, at the expense of a second high-energy phosphate bond.

$$\text{Pyruvate} + CO_2 + \text{ATP} + H_2O \longrightarrow \text{oxaloacetate} + \text{ADP} + P_i + 2\ H^+$$

$$\text{Oxaloacetate} + \text{GTP} \rightleftharpoons \text{phosphoenolpyruvate} + \text{GDP} + CO_2$$

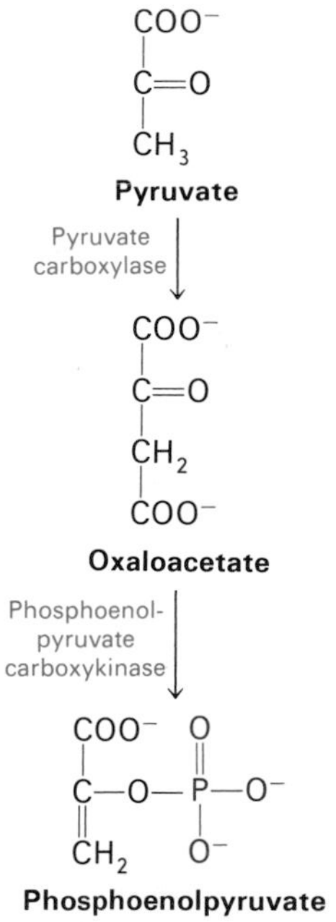

The first reaction is catalyzed by *pyruvate carboxylase,* and the second by *phosphoenolpyruvate carboxykinase*. The sum of these reactions is

$$\text{Pyruvate} + \text{ATP} + \text{GTP} + H_2O \longrightarrow \text{phosphoenolpyruvate} + \text{ADP} + \text{GDP} + P_i + 2\ H^+$$

This two-step pathway for the formation of phosphoenolpyruvate from pyruvate is thermodynamically feasible, because $\Delta G^{\circ\prime}$ is $+0.2$ kcal/mol in contrast with $+ 7.5$ kcal/mol for the reaction catalyzed by pyruvate kinase. This much more favorable $\Delta G^{\circ\prime}$ results from the input of an additional high-energy phosphate bond.

COO^- — C=O — CH_3
Pyruvate
↓ Pyruvate carboxylase
COO^- — C=O — CH_2 — COO^-
Oxaloacetate
↓ Phosphoenol-pyruvate carboxykinase
COO^- — C(—O—P(=O)(—O^-)—O^-)=CH_2
Phosphoenolpyruvate

2. *Fructose 6-phosphate is formed from fructose 1,6-bisphosphate by hydrolysis of the phosphate ester at C-1.* Fructose 1,6-bisphosphatase catalyzes this exergonic hydrolysis.

$$\text{Fructose 1,6-bisphosphate} + H_2O \longrightarrow \text{fructose 6-phosphate} + P_i$$

3. *Glucose is formed by hydrolysis of glucose 6-phosphate,* in a reaction catalyzed by glucose 6-phosphatase.

$$\text{Glucose 6-phosphate} + H_2O \longrightarrow \text{glucose} + P_i$$

Glucose 6-phosphate is transported by a specific carrier protein from the cytosol to the lumen of the endoplasmic reticulum, where it is hydrolyzed by *glucose 6-phosphatase,* a membrane-bound enzyme. Glucose and P_i are transported back to the cytosol. Glucose 6-phosphatase is not present in brain and muscle; hence, glucose cannot be formed by these organs.

Table 18-2
Enzymatic differences between glycolysis and gluconeogenesis

Glycolysis	*Gluconeogenesis*
Hexokinase	Glucose 6-phosphatase
Phosphofructokinase	Fructose 1,6-bisphosphatase
Pyruvate kinase	Pyruvate carboxylase Phosphoenolpyruvate carboxykinase

Biotin

BIOTIN IS A MOBILE CARRIER OF ACTIVATED CO_2

The finding that mitochondria can form oxaloacetate from pyruvate led to the discovery of pyruvate carboxylase by Merton Utter in 1960. This enzyme is of especial interest because of its catalytic and allosteric properties. Pyruvate carboxylase contains a covalently attached prosthetic group, *biotin,* which serves as a *carrier of activated CO_2.* The carboxyl terminus of biotin is linked to the ϵ-amino group of a specific lysine residue by an amide bond.

Lysine residue of enzyme

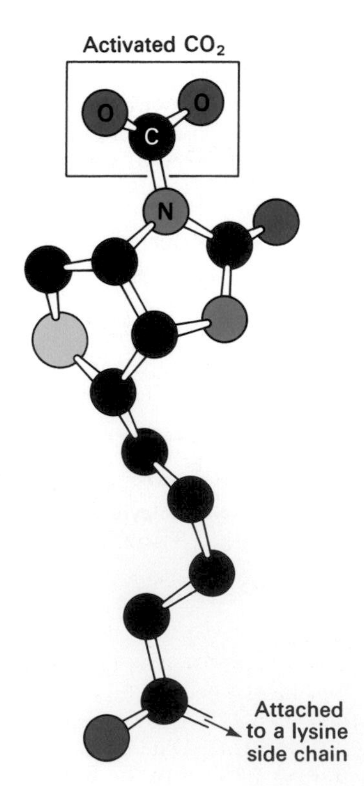

Figure 18-6
Molecular model of carboxybiotin.

Note that biotin is attached to pyruvate carboxylase by a *long, flexible chain* like that of lipoamide in the pyruvate dehydrogenase complex. The carboxylation of pyruvate occurs in two stages:

$$\text{Biotin-enzyme} + \text{ATP} + HCO_3^- \xrightleftharpoons[\text{acetyl CoA}]{Mg^{2+}} CO_2 \sim \text{biotin-enzyme} + \text{ADP} + P_i$$

$$CO_2 \sim \text{biotin-enzyme} + \text{pyruvate} \xrightleftharpoons{Mn^{2+}} \text{biotin-enzyme} + \text{oxaloacetate}$$

The carboxyl group in the carboxybiotin-enzyme intermediate is bonded to the N-1 nitrogen atom of the biotin ring. This carboxyl group is *activated*. The $\Delta G^{\circ\prime}$ for its cleavage

$$CO_2 \sim \text{biotin-enzyme} + H^+ \longrightarrow CO_2 + \text{biotin-enzyme}$$

is −4.7 kcal/mol, which enables carboxybiotin to transfer CO_2 to acceptors without the input of additional free energy.

Carboxybiotin-enzyme intermediate

The activated carboxyl group is then transferred from carboxybiotin to pyruvate to form oxaloacetate. The long, flexible link between biotin and the enzyme enables this prosthetic group to rotate from one active site of the enzyme (the ATP-bicarbonate site) to the other (the pyruvate site).

Pyruvate + Enzyme–Biotin–COO^- ⇌ **Oxaloacetate** + Enzyme–Biotin

PYRUVATE CARBOXYLASE IS ACTIVATED BY ACETYL CoA

The activity of pyruvate carboxylase depends on the presence of acetyl CoA. *Biotin is not carboxylated unless acetyl CoA (or a closely related acyl CoA) is bound to the enzyme.* The second partial reaction is not affected by acetyl CoA. The allosteric activation of pyruvate carboxylase by acetyl CoA is an important physiological control mechanism. Oxaloacetate, the product of the pyruvate carboxylase reaction, is both a stoichiometric intermediate in gluconeogenesis and a catalytic intermediate in the citric acid cycle. *A high level of acetyl CoA signals the need for more oxaloacetate.* If there is a surplus of ATP, oxaloacetate will be consumed in gluconeogenesis. If there is a deficiency of ATP, oxaloacetate will enter the citric acid cycle upon condensing with acetyl CoA.

Thus, not only is pyruvate carboxylase important in gluconeogenesis, but it also plays a *critical role in maintaining the level of citric acid cycle intermediates*. These intermediates need to be replenished because they are consumed in some biosynthetic reactions, such as heme synthesis. This role of pyruvate carboxylase is termed *anaplerotic*, meaning to fill up.

OXALOACETATE IS SHUTTLED INTO THE CYTOSOL AND CONVERTED INTO PHOSPHOENOLPYRUVATE

Pyruvate carboxylase is a mitochondrial enzyme, whereas the other enzymes of gluconeogenesis are cytoplasmic. Oxaloacetate, the product of the pyruvate carboxylase reaction, is transported across the mitochondrial membrane in the form of *malate*. Oxaloacetate is reduced to malate inside the mitochondrion by an NADH-linked malate dehydrogenase. Malate is transported by a carrier across the mitochondrial membrane and is reoxidized to oxaloacetate by an NAD^+-linked malate dehydrogenase in the cytosol.

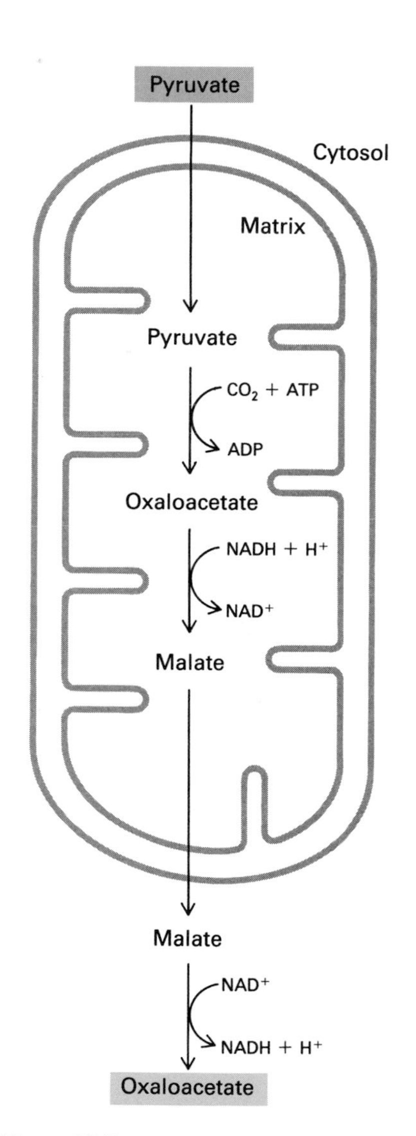

Figure 18-7
Oxaloacetate utilized in the cytosol for gluconeogenesis is formed in the mitochondrial matrix by carboxylation of pyruvate. Oxaloacetate leaves the mitochondrion in the form of malate, which is reoxidized to oxaloacetate in the cytosol.

Oxaloacetate is simultaneously *decarboxylated* and *phosphorylated* by phosphoenolpyruvate carboxykinase in the cytosol.

$$\underset{\textbf{Oxaloacetate}}{{}^{-}OOC{-}C({=}O){-}CH_2{-}COO^{-}} + GTP \rightleftharpoons \underset{\textbf{Phosphoenolpyruvate}}{{}^{-}OOC{-}C({=}CH_2){-}O{-}PO_3^{2-}} + CO_2 + GDP$$

The CO_2 that was added to pyruvate by pyruvate carboxylase comes off in this step. In fact, the phosphorylation reaction is made energetically feasible by the concomitant decarboxylation. *Decarboxylations often drive reactions that would otherwise be highly endergonic.* This device is used in the citric acid cycle and the pentose phosphate pathway, and we shall see it used in fatty acid synthesis (p. 486).

SIX HIGH-ENERGY PHOSPHATE BONDS ARE SPENT IN SYNTHESIZING GLUCOSE FROM PYRUVATE

The stoichiometry of gluconeogenesis is

$$\begin{aligned} &2\ \text{Pyruvate} + 4\ \text{ATP} + 2\ \text{GTP} + 2\ \text{NADH} + 6\ H_2O \longrightarrow \\ &\quad \text{glucose} + 4\ \text{ADP} + 2\ \text{GDP} + 6\ P_i + 2\ \text{NAD}^+ + 2\ H^+ \\ &\qquad \Delta G^{\circ\prime} = -9\ \text{kcal/mol} \end{aligned}$$

In contrast, the stoichiometry for the reversal of glycolysis is

$$\begin{aligned} &2\ \text{Pyruvate} + 2\ \text{ATP} + 2\ \text{NADH} + 2\ H_2O \longrightarrow \\ &\quad \text{glucose} + 2\ \text{ADP} + 2\ P_i + 2\ \text{NAD}^+ \\ &\qquad \Delta G^{\circ\prime} = +20\ \text{kcal/mol} \end{aligned}$$

Note that *six* high-energy phosphate bonds are used to synthesize glucose from pyruvate in gluconeogenesis, whereas only *two* molecules of ATP are generated in glycolysis in the conversion of glucose into pyruvate. Thus, the extra price of gluconeogenesis is four high-energy phosphate bonds per glucose synthesized from pyruvate. The four extra high-energy phosphate bonds are needed to turn an energetically unfavorable process (the reversal of glycolysis, $\Delta G^{\circ\prime} = +20$ kcal/mol) into a favorable one (gluconeogenesis, $\Delta G^{\circ\prime} = -9$ kcal/mol). Another way of looking at this energetic difference between glycolysis and gluconeogenesis is to recall that the input of an ATP equivalent changes the equilibrium constant of a reaction by a factor of about 10^8 (p. 319). Hence, the input of four additional high-energy bonds in gluconeogenesis changes the equilibrium constant by a factor of 10^{32}, which makes the conversion of pyruvate into glucose thermodynamically feasible.

GLUCONEOGENESIS AND GLYCOLYSIS ARE RECIPROCALLY REGULATED

Gluconeogenesis and glycolysis are coordinated so that one pathway is relatively inactive while the other is highly active. If both sets of reactions were highly active at the same time, the net result would be the hydrolysis of four ~P (two ATP plus two GTP) per reaction cycle. Both glycolysis and gluconeogenesis are highly exergonic under cellular conditions, and so there is no thermodynamic barrier to such cycling. However, the activities of the distinctive enzymes of each pathway are controlled so that both pathways are not highly active at the same time. The rate of glycolysis is also determined by the concentration of glucose, and the rate of gluconeogenesis by the concentrations of lactate and other precursors of glucose.

The interconversion of fructose 6-phosphate and fructose 1,6-bisphosphate is a key point of control (Figure 18-8). AMP stimulates phosphofructokinase (p. 359), whereas it inhibits fructose 1,6-bisphosphatase. Citrate has the opposite effect. These enzymes are also

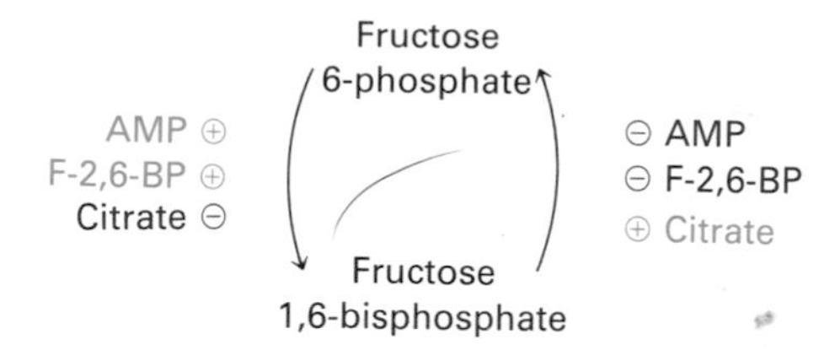

reciprocally controlled by fructose 2,6-bisphosphate, a signal molecule derived from fructose 6-phosphate (p. 360). When glucose is abundant, a hormone-triggered cascade leads to a high level of fructose 2,6-bisphosphate (F-2,6-BP), which stimulates phosphofructokinase and inhibits fructose 1,6-bisphosphatase. Glycolysis is thereby accelerated and gluconeogenesis is diminished. During starvation, the level of F-2,6-BP drops, which decreases the activity of phosphofructokinase and increases that of the phosphatase. Consequently, fructose 1,6-bisphosphate is converted into fructose 6-phosphate to generate glucose. Thus, *fructose 2,6-bisphosphate plays an important role in determining whether glucose is to be degraded or synthesized.*

Pyruvate kinase (p. 362) and pyruvate carboxylase (p. 441) are also reciprocally regulated. Fructose 1,6-bisphosphate stimulates and ATP inhibits pyruvate kinase, whereas acetyl CoA stimulates and ADP inhibits pyruvate carboxylase. Phosphoenolpyruvate carboxykinase also is inhibited by ADP, when the energy charge of a cell is low. Hence, pyruvate flows to phosphoenolpyruvate and gluconeogenesis is favored when the cell is rich in fuel molecules and ATP.

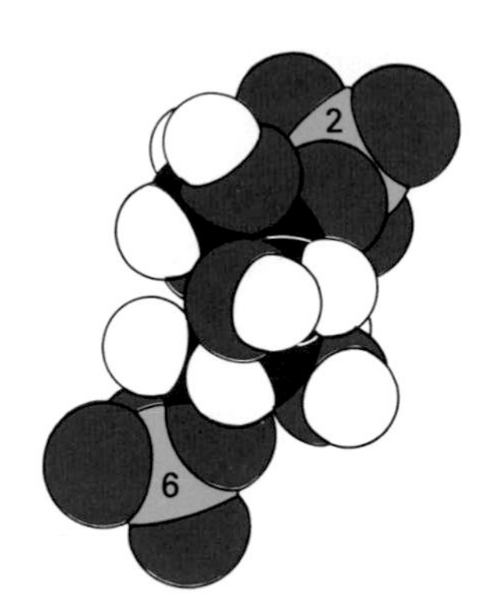

Fructose 2,6-bisphosphate

Figure 18-8
Model of fructose 2,6-bisphosphate, a key regulator of glycolysis and gluconeogenesis.

SUBSTRATE CYCLES AMPLIFY METABOLIC SIGNALS AND PRODUCE HEAT

A pair of reactions such as the phosphorylation of fructose 6-phosphate to fructose 1,6-bisphosphate and its hydrolysis back to fructose 6-phosphate is called a *substrate cycle*. As already mentioned, both reactions are not simultaneously fully active in most cells because of reciprocal allosteric controls. However, isotope labeling studies have shown that fructose 6-phosphate is phosphorylated to fructose 1,6-bisphosphate during gluconeogenesis. A limited degree of cycling also occurs in other pairs of opposed irreversible reactions. This cycling was regarded as an imperfection in metabolic control, and so substrate cycles have sometimes been called *futile cycles*. However, it now seems likely that substrate cycles are biologically important. One possibility is that *substrate cycles amplify metabolic signals*. Suppose that the rate of conversion of A into B is 100 and of B into A is 90, giving an initial net flux of 10. Assume that an allosteric effector increases the A $\rightarrow$ B rate by 20%

to 120 and reciprocally decreases the B → A rate by 20% to 72. The new net flux is 48, and so a 20% change in the rates of the opposing reactions has led to a 380% increase in the net flux. In the example shown in Figure 18-9, this amplification is achieved by the hydrolysis of ATP.

The other potential biological role of substrate cycles is the *generation of heat produced by the hydrolysis of ATP*. A striking example is provided by bumblebees, which have to maintain a thoracic temperature of about 30°C in order to fly. A bumblebee is able to maintain this high thoracic temperature and forage for food even when the ambient temperature is only 10°C because phosphofructokinase and fructose diphosphatase in its flight muscle are simultaneously highly active. This fructose bisphosphatase is not inhibited by AMP, which suggests that the enzyme is specially designed for the generation of heat. In contrast, the honeybee has almost no fructose bisphosphatase activity in its flight muscle and consequently cannot fly when the ambient temperature is low. Excessive heat production caused by too high a rate of cycling between fructose 6-phosphate and fructose 1,6-bisphosphate can occur. Halothane, an anesthetic, produces this condition, called *malignant hyperthermia,* in a susceptible strain of pigs.

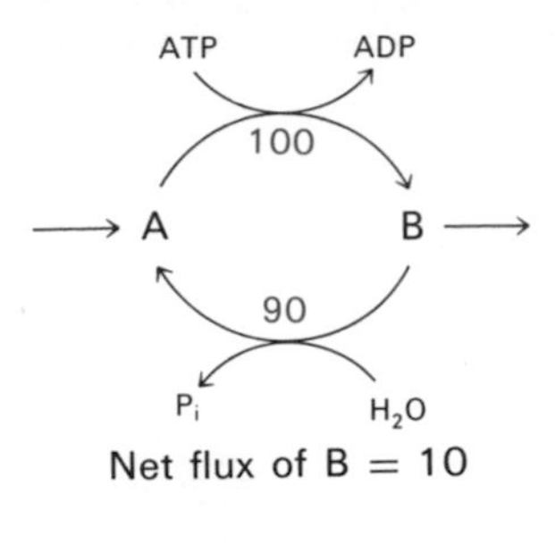

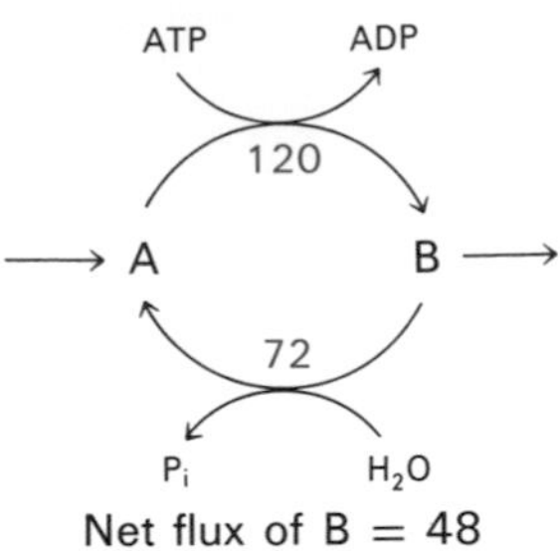

Figure 18-9
An ATP-driven substrate cycle operating at two different rates. A small change in the rates of the two opposing reactions results in a large change in the *net* flux of product B.

LACTATE AND ALANINE FORMED BY CONTRACTING MUSCLE ARE CONVERTED INTO GLUCOSE BY THE LIVER

The major raw materials of gluconeogenesis are lactate and alanine produced by active skeletal muscle and erythrocytes. The rate of production of pyruvate by glycolysis exceeds the rate of oxidation of pyruvate by the citric acid cycle in contracting skeletal muscle under anaerobic conditions, as during vigorous exercise. Under these conditions, moreover, the rate of formation of NADH by glycolysis is greater than the rate of its oxidation by the respiratory chain. Continued glycolysis depends on the availability of NAD^+ for the oxidation of glyceraldehyde 3-phosphate. The accumulation of both NADH and pyruvate is reversed by lactate dehydrogenase, which oxidizes NADH to NAD^+ as it reduces pyruvate to lactate.

$$\underset{\textbf{Pyruvate}}{CH_3-C(=O)-COO^-} + NADH + H^+ \rightleftharpoons \underset{\textbf{Lactate}}{CH_3-CH(OH)-COO^-} + NAD^+$$

Lactate is a dead end in metabolism. It must be converted back into pyruvate before it can be metabolized. The only purpose of the reduction of pyruvate to lactate is to regenerate NAD^+ so that glycolysis can proceed in active skeletal muscle and erythrocytes. Note that the conversion of glucose to lactate does not involve a net oxidation-reduction. Rather, one of the carbon atoms in lactate is more oxidized than in glucose, and another is more reduced. *The formation of lactate buys time and shifts part of the metabolic burden from muscle to liver.*

The plasma membrane of most cells is highly permeable to lactate and pyruvate. Both substances diffuse out of active skeletal muscle into the blood and are carried to the liver. Much more lactate than pyruvate is carried because of the high $NADH/NAD^+$ ratio in contracting skeletal muscle. The lactate that enters the liver is oxidized to pyruvate, a reaction favored by the low $NADH/NAD^+$ ratio in the cytosol of liver cells. Pyruvate is then converted into glucose by the gluconeogenic pathway in liver. Glucose then enters the blood and is taken up by skeletal muscle. Thus, *liver furnishes glucose to contracting skeletal muscle, which derives ATP from the glycolytic conversion of glucose into lactate. Glucose is then synthesized from lactate by the liver. These reactions constitute the Cori cycle* (Figure 18-10). Recent studies have shown that alanine, like lactate, is a major precursor of glucose. In muscle, alanine is formed from pyruvate by transamination (p. 496); the reverse reaction occurs in the liver.

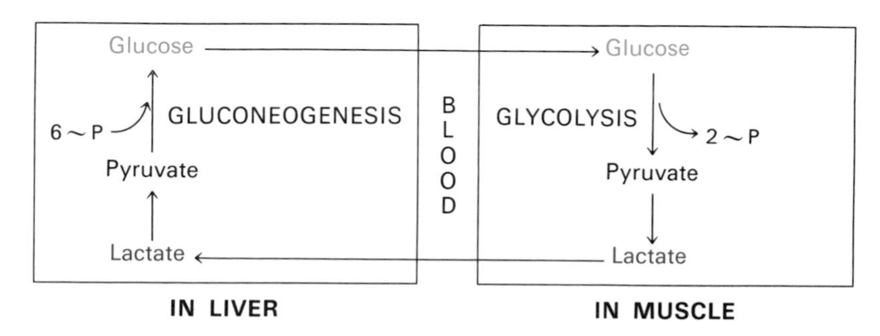

Figure 18-10
The Cori cycle. Lactate formed by active muscle is converted into glucose by the liver. This cycle shifts part of the metabolic burden of active muscle to the liver.

The interconversions of pyruvate and lactate are facilitated by differences in the catalytic properties of lactate dehydrogenase enzymes in different tissues. Lactate dehydrogenase is a tetramer of two kinds of 35-kd subunits: the H type predominates in the heart, and the homologous M type in skeletal muscle and the liver. These subunits associate to form five types of tetramers: H_4, H_3H, H_2M_2, H_1M_3, and M_4. These species are called *isozymes* (or isoenzymes). The H_4 isozyme (type 1) has higher affinity for substrates than does the M_4 isozyme (type 5). They also differ in that H_4 is allosterically inhibited by high levels of pyruvate, whereas M_4 is not. The other isozymes have intermediate properties depending on the ratio of the two kinds of chains. It has been proposed that H_4 is designed to oxidize lactate to pyruvate, which is then utilized as a fuel by the heart. The aerobic metabolism of the heart enables it to funnel pyruvate into the citric acid cycle. In contrast, M_4 is optimized to operate in the reverse direction, to convert pyruvate to lactate to allow glycolysis to proceed under anaerobic conditions. We see here an example of how gene duplication and divergence generate a series of homologous enzymes that foster metabolic cooperation between organs.

SUMMARY

The pentose phosphate pathway generates NADPH and ribose 5-phosphate in the cytosol. NADPH is used in reductive biosyntheses, whereas ribose 5-phosphate is used in the synthesis of RNA, DNA, and nucleotide coenzymes. The pentose phosphate pathway starts with the dehydrogenation of glucose 6-phosphate to form a lactone, which is hydrolyzed to give 6-phosphogluconate and then oxidatively decarboxylated to yield ribulose 5-phosphate. $NADP^+$ is the electron acceptor in both of these oxidations. The last step is the isomerization of ribulose 5-phosphate (a ketose) to ribose 5-phosphate (an aldose). A different mode of the pathway is active when cells need much more NADPH than ribose 5-phosphate. Under these conditions, ribose 5-phosphate is converted into glyceraldehyde 3-phosphate and fructose 6-phosphate by transketolase and transaldolase. These two enzymes create a reversible link between the pentose phosphate pathway and glycolysis. Xylulose 5-phosphate, sedoheptulose 7-phosphate, and erythrose 4-phosphate are intermediates in these interconversions. In this way, twelve NADPH can be generated for each glucose 6-phosphate that is completely oxidized to CO_2.

Only the nonoxidative branch of the pathway is active when much more ribose 5-phosphate than NADPH needs to be synthesized. Under these conditions, fructose 6-phosphate and glyceraldehyde 3-phosphate (formed by the glycolytic pathway) are converted into ribose 5-phosphate without the formation of NADPH. Alternatively, ribose 5-phosphate formed by the oxidative branch can be converted into pyruvate through fructose 6-phosphate and glyceraldehyde 3-phosphate. In this mode, ATP and NADPH are generated, and five of the six carbons of glucose 6-phosphate emerge in pyruvate. The interplay of the glycolytic and pentose phosphate pathways enables the levels of NADPH, ATP, and building blocks such as ribose 5-phosphate and pyruvate to be continuously adjusted to meet cellular needs.

Gluconeogenesis is the synthesis of glucose from noncarbohydrate sources, such as lactate, amino acids, and glycerol. Several of the reactions that convert pyruvate into glucose are common to glycolysis. Gluconeogenesis, however, requires four new reactions to bypass the essential irreversibility of the corresponding reactions in glycolysis. Pyruvate is carboxylated in mitochondria to oxaloacetate, which in turn is decarboxylated and phosphorylated in the cytosol to phosphoenolpyruvate. Two high-energy phosphate bonds are consumed in these reactions, which are catalyzed by pyruvate carboxylase and phosphoenolpyruvate carboxykinase. Pyruvate carboxylase contains a biotin prosthetic group. The other distinctive reactions of gluconeogenesis are the hydrolyses of fructose 1,6-bisphosphate and glucose 6-phosphate, which are catalyzed by specific phosphatases. The major raw materials for gluconeogenesis by the liver are lactate and alanine produced from pyruvate by active skeletal muscle.

Gluconeogenesis and glycolysis are reciprocally regulated so that one pathway is relatively inactive while the other is highly active. Phosphofructokinase and fructose 1,6-bisphosphatase are key control points. Fructose 2,6-bisphosphate, an intracellular signal molecule present at higher levels when glucose is abundant, activates glycolysis and inhibits gluconeogenesis by regulating these enzymes. Pyruvate kinase and pyruvate carboxylase are regulated by other effectors so that both are not maximally active at the same time. A moderate degree of simultaneous activity of opposing enzymes, called substrate cycling, can be beneficial by amplifying metabolic signals.

SELECTED READINGS

Where to start

Horecker, B. L., 1976. Unravelling the pentose phosphate pathway. *In* Kornberg, A., Cornudella, L., Horecker, B. L., and Oro, J., (eds.), *Reflections on Biochemistry,* pp. 65–72. Pergamon.

Hers, H.-G., and Hue, L., 1983. Gluconeogenesis and related aspects of glycolysis. *Ann. Rev. Biochem.* 52:617–653.

Books and reviews

Wood, T., 1985. *The Pentose Phosphate Pathway.* Academic Press.

Pontremoli, S., and Grazi, E., 1969. Hexose monophosphate oxidation. *Compr. Biochem.* 17:163–189.

Pontremoli, S., and Grazi, E., 1968. Gluconeogenesis. *In* Dickens, F., Randle, P. J., and Whelan, W. J., (eds.), *Carbohydrate Metabolism and Its Disorders,* vol. 1, pp. 259–295. Academic Press.

Meister, A., and Anderson, M. E., 1983. Glutathione. *Ann. Rev. Biochem.* 52:711–760.

Enzymes and reaction mechanisms

Horecker, B. L., 1964. Transketolase and transaldolase. *Compr. Biochem.* 15:48–70.

Williams, J. F., 1980. A critical examination of the evidence for the reactions of the pentose phosphate pathway in animal tissues. *Trends Biochem. Sci.* 5:315–320. [Proposes that liver contains a pentose phosphate pathway differing from the classic one presented in this chapter.]

Landau, B. R., and Wood, H. G., 1983. A critical examination of the evidence for the reactions of the pentose phosphate pathway in animal tissues. *Trends Biochem. Sci.* 8:292–296. [A critical review of the evidence indicating that the classic pathway operates in all mammalian tissues.]

Melendez-Hevia, E., and Isidoro, A., 1985. The game of the pentose phosphate cycle. *J. Theor. Biol.* 117:251–263. [What is the simplest way of interconverting five- and six-carbon sugars? The pentose phosphate pathway is analyzed here in terms of a mathematical game of optimization. Read this article to learn whether nature uses the simplest possible scheme.]

Scrutton, M. C., and Young, M. R., 1972. Pyruvate carboxylase. *In* Boyer, P. D., (ed.), *The Enzymes* (3rd ed.), vol. 6, pp. 1–35. Academic Press.

Wood, H. G., and Barden, R. E., 1977. Biotin enzymes. *Ann. Rev. Biochem.* 46:385–414.

Thieme, R., Pai, E. F., Schirmer, R. H., and Schultz, G. E., 1981. Three-dimensional structure of glutathione reductase at 2 Å resolution. *J. Mol. Biol.* 152:763–782.

Meister, A., 1983. Selective modification of glutathione metabolism. *Science* 220:472–477.

Regulation

Newsholme, E. A., and Start, C., 1973. *Regulation in Metabolism.* Wiley. [Chapter 6 deals with the control of glycolysis and gluconeogenesis in liver and kidney cortex.]

Soling, H. D., and Willms, B., (eds.), 1971. *Regulation of Gluconeogenesis.* Academic Press.

Newsholme, E. A., Chaliss, R. A. J., and Crabtree, B., 1984. Substrate cycles: Their role in improving sensitivity in metabolic control. *Trends Biochem. Sci.* 9:277–280.

Katz, J., and Rognstad, R., 1976. Futile cycles in the metabolism of glucose. *Curr. Top. Cell Regul.* 10:238–287.

Nordlie, R. C., 1974. Metabolic regulation by multifunctional glucose-6-phosphatase. *Curr. Top. Cell Regul.* 8:33–111.

Ureta, T., 1978. The role of isozymes in metabolism: A model of metabolic pathways as the basis for the biological role of isozymes. *Curr. Top. Cell Regul.* 13:233–250.

Genetic diseases

Beutler, E., 1983. Glucose 6-phosphate dehydrogenase deficiency. *In* Stanbury, J. B., Wyngaarden, J. B., Fredrickson, D. S., Goldstein, J. L., and Brown, M. S., (eds.), *The Metabolic Basis of Inherited Disease* (5th ed.), pp. 1629–1653. McGraw-Hill.

Luzzatto, L., Usanga, E. A., and Reddy, S., 1969. Glucose 6-phosphate dehydrogenase deficient red cells: Resistance to infection by malarial parasites. *Science* 164:839–842.

Blass, J. P., and Gibson, G. E., 1977. Abnormality of a thiamine-requiring enzyme in patients with Wernicke-Korsakoff syndrome. *New Engl. J. Med.* 297:1367–1370.

PROBLEMS

1. What is the stoichiometry of the synthesis of ribose 5-phosphate from glucose 6-phosphate without the concomitant generation of NADPH? What is the stoichiometry of the synthesis of NADPH from glucose 6-phosphate without the concomitant formation of pentose sugars?

2. Glucose labeled with ^{14}C at C-6 is added to a solution containing the enzymes and cofactors of the oxidative branch of the pentose phosphate pathway. What is the fate of the radioactive label?

3. Which reaction in the citric acid cycle is most analogous to the oxidative decarboxylation of 6-phosphogluconate to ribulose 5-phosphate? What kind of enzyme-bound intermediate is formed in both reactions?

4. Ribose 5-phosphate labeled with ^{14}C at C-1 is added to a solution containing transketolase, transaldolase,

phosphopentose epimerase, phosphopentose isomerase, and glyceraldehyde 3-phosphate. What is the distribution of the radioactive label in the erythrose 4-phosphate and fructose 6-phosphate that are formed in this reaction mixture?

5. Avidin, a 70-kd protein in egg white, has a very high affinity for biotin. In fact, it is a highly specific inhibitor of biotin enzymes. Which of the following conversions would be blocked by the addition of avidin to a cell homogenate?
   (a) Glucose $\longrightarrow$ pyruvate.
   (b) Pyruvate $\longrightarrow$ glucose.
   (c) Oxaloacetate $\longrightarrow$ glucose.
   (d) Glucose $\longrightarrow$ ribose 5-phosphate.
   (e) Pyruvate $\longrightarrow$ oxaloacetate.
   (f) Ribose 5-phosphate $\longrightarrow$ glucose.

6. Design a chemical experiment to identify the lysine residue that forms a Schiff base at the active site of transaldolase.

7. What ratio of NADPH to $NADP^+$ is required to sustain [GSH] = 10 mM and [GSSG] = 1 mM? Use the redox potentials given on page 400.

8. During fetal development, the proportion of H chains in lactate dehydrogenase increases and that of M chains decreases. Propose a selective advantage of this shift in isozyme pattern during development.

9. What are the likely consequences of a genetic disorder rendering fructose 1,6-bisphosphatase in liver less sensitive to regulation by fructose 2,6-bisphosphate?

CHAPTER 19

Glycogen Metabolism

Glycogen is a *readily mobilized storage form of glucose*. It is a very large, branched polymer of glucose residues (Figure 19-1). Most of the glucose residues in glycogen are linked by α-1,4-glycosidic bonds. The branches are created by α-1,6-glycosidic bonds, of which there is one in about ten residues. Recall that α-glycosidic linkages form open helical polymers, whereas β linkages produce nearly straight strands that form structural fibrils, as in cellulose (p. 343).

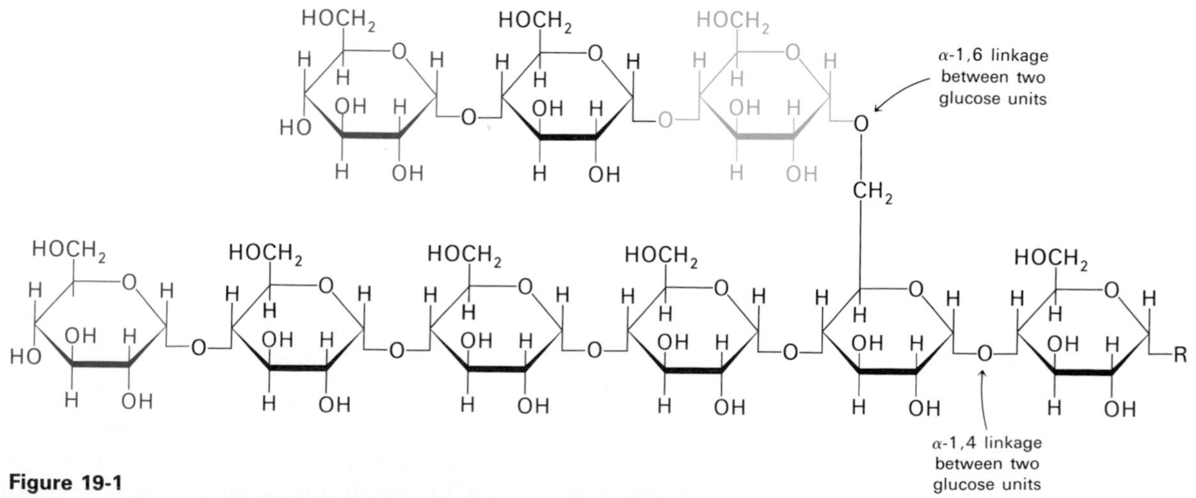

Figure 19-1
Structure of two outer branches of a glycogen particle. The residues at the nonreducing ends are shown in red. The residue that starts a branch is shown in green. The rest of the glycogen molecule is represented by R.

The presence of glycogen greatly increases the amount of glucose that is immediately available between meals and during muscular activity. The amount of glucose in the body fluids of an average 70-kg man has an energy content of only 40 kcal, whereas the total body glycogen has an energy content of more than 600 kcal, even after an overnight fast. The two major sites of glycogen storage are the liver and skeletal muscle. The concentration of glycogen is higher in the liver than in muscle, but more glycogen is stored in skeletal muscle because of its much greater mass. Glycogen is present in the cytosol in the form of granules with diameters ranging from about 100 to 400 Å. This range in size reflects the fact that glycogen molecules do not have a unique size; a typical distribution is centered on a mass of several thousand kilodaltons. Glycogen granules have a dense appearance in electron micrographs (Figure 19-3). They contain the enzymes that catalyze the synthesis and degradation of glycogen and some of the enzymes that regulate these processes. However, a glycogen granule differs from a multienzyme complex (such as the pyruvate dehydrogenase complex) in that the bound enzymes are not present in defined ratios. Also, a glycogen granule is structurally less organized than a multienzyme complex.

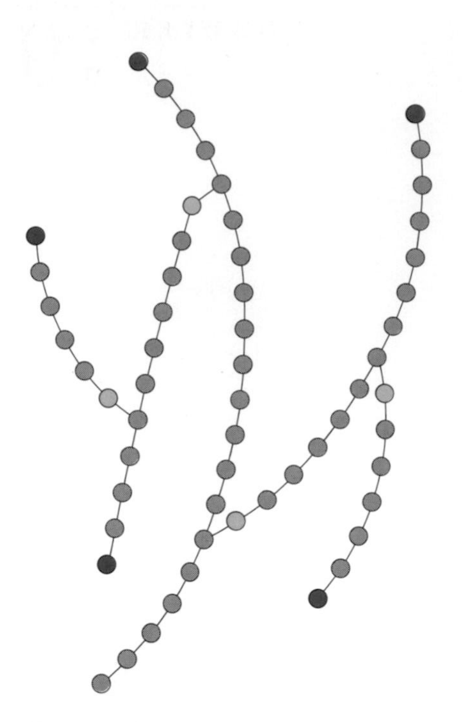

Figure 19-2
Diagram of a cross section of a glycogen molecule. (Residues are differentiated by the same colors as in Figure 19-1.)

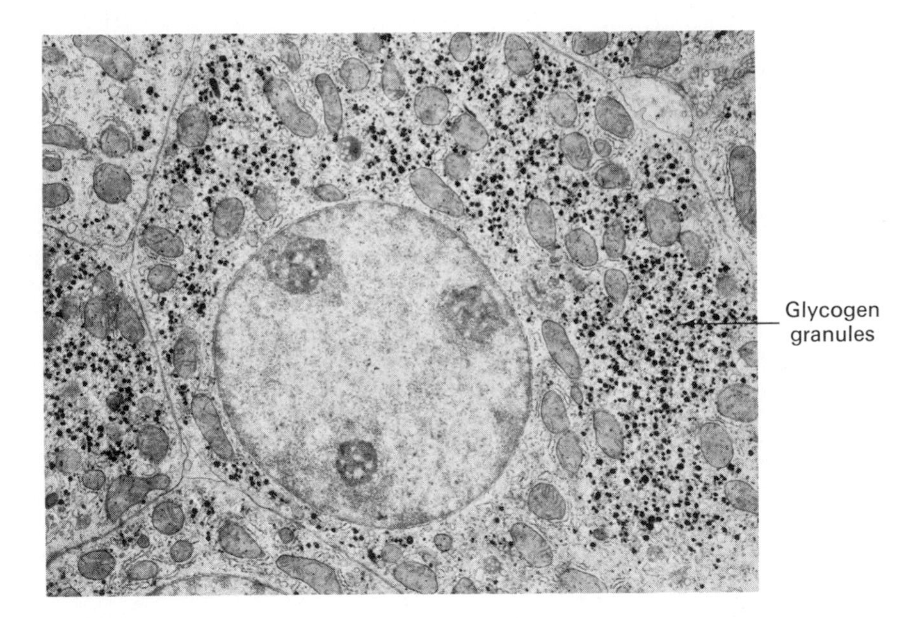

Figure 19-3
Electron micrograph of a liver cell. The dense particles in the cytoplasm are glycogen granules. [Courtesy of Dr. George Palade.]

The synthesis and degradation of glycogen are considered here in some detail for several reasons. First, these processes are important because they *regulate the blood glucose level* and provide a *reservoir of glucose* for strenuous muscular activity. Second, the synthesis and degradation of glycogen occur by *different reaction pathways,* which illustrates an important principle of biochemistry. Third, the hormonal regulation of glycogen metabolism is mediated by mechanisms that are of general significance. The *role of cyclic adenosine monophosphate* (cyclic AMP) in the coordinated control of glycogen synthesis and breakdown is well understood and is a source of insight into the action of hormones in a variety of other systems. The enzymes of glycogen metabolism are regulated by *reversible phosphorylation*, a recurring control device in all biological systems. Fourth, a number of inherited enzyme defects resulting in impaired glycogen metabolism have been characterized. Some of these *glycogen storage diseases* are lethal in infancy, whereas others have a relatively mild clinical course.

PHOSPHORYLASE CATALYZES THE PHOSPHOROLYTIC CLEAVAGE OF GLYCOGEN INTO GLUCOSE 1-PHOSPHATE

The pathway of glycogen breakdown was elucidated by the incisive studies of Carl Cori and Gerty Cori. They showed that glycogen is *cleaved by orthophosphate* to yield a new kind of phosphorylated sugar, which they identified as *glucose 1-phosphate*. The Coris also isolated and crystallized *glycogen phosphorylase*, the enzyme that catalyzes this reaction.

Phosphorolysis—
The cleavage of a bond by orthophosphate (in contrast with *hydrolysis*, which refers to cleavage by water).

$$\underset{(n \text{ residues})}{\text{Glycogen}} + P_i \rightleftharpoons \text{glucose 1-phosphate} + \underset{(n-1 \text{ residues})}{\text{glycogen}}$$

Phosphorylase catalyzes the sequential removal of glycosyl residues from the nonreducing end of the glycogen molecule. The glycosidic linkage between C-1 of the terminal residue and C-4 of the adjacent one is split by orthophosphate. Specifically, the bond between the C-1 carbon atom and the glycosidic oxygen atom is cleaved by orthophosphate, and the α configuration at C-1 is retained.

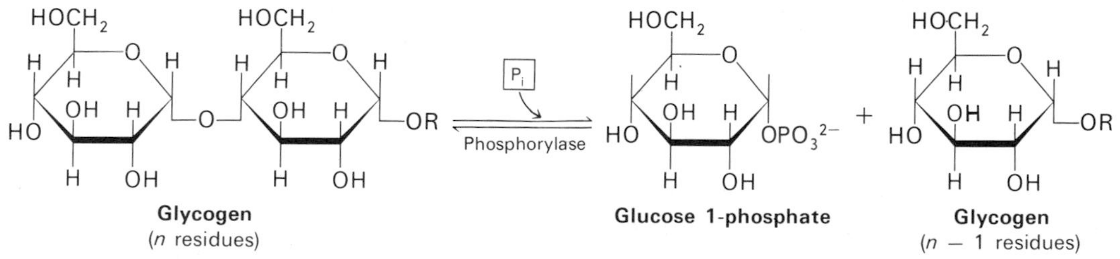

Glycogen (n residues) + P_i ⇌ (Phosphorylase) Glucose 1-phosphate + Glycogen ($n - 1$ residues)

The reaction catalyzed by phosphorylase is readily reversible in vitro. At pH 6.8, the equilibrium ratio of orthophosphate to glucose 1-phosphate is 3.6. The $\Delta G^{\circ\prime}$ for this reaction is small because a glycosidic bond is replaced by a phosphate ester bond that has a nearly equal transfer potential. However, phosphorolysis proceeds far in the direction of glycogen breakdown in vivo because the $[P_i]/[\text{glucose 1-phosphate}]$ ratio is usually greater than 100.

The phosphorolytic cleavage of glycogen is energetically advantageous because the released sugar is phosphorylated. In contrast, a hydrolytic cleavage would yield glucose, which would have to be phosphorylated at the expense of an ATP to enter the glycolytic pathway. An additional advantage of phosphorolytic cleavage for muscle cells is that glucose 1-phosphate, ionized under physiological conditions, cannot diffuse out of the cell, whereas glucose can. We see here an example of the importance of a metabolite's being ionized.

PYRIDOXAL PHOSPHATE PARTICIPATES IN THE PHOSPHOROLYTIC CLEAVAGE OF GLYCOGEN

Both the glycogen substrate and the glucose 1-phosphate product have an α configuration at C-1. This retention of configuration is a valuable clue to the catalytic mechanism of phosphorylase. A direct attack of phosphate on C-1 of a sugar would invert the configuration at this carbon. The finding that the glucose 1-phosphate formed has an α rather than β configuration strongly suggests that an even number of steps (most simply, two) are required. A *carbonium ion intermediate* seems most likely, as in the lysozyme-catalyzed hydrolysis of bacterial cell-wall polysaccharides (p. 208).

A second clue to the catalytic mechanism of phosphorylase is its requirement for *pyridoxal 5′-phosphate* (PLP), a derivative of pyridoxine (vitamin B_6, p. 323). The aldehyde group of this coenzyme forms a Schiff base with a specific lysine side chain of the enzyme (Figure 19-4). X-ray diffraction and nuclear magnetic resonance studies suggest that the reacting orthophosphate group lies between the 5′-phosphate group of PLP and the glycogen substrate (Figure 19-5). The 5′-phosphate of PLP acts in tandem with orthophosphate, probably by serving as a proton donor and acceptor (as a general acid-base catalyst). Orthophosphate (in the HPO_4^{2-} form) donates a proton to O-4 of the departing glycogen chain and simultaneously acquires a proton from PLP. The carbonium ion intermediate formed in this step is then attacked by orthophosphate to form α-glucose 1-phosphate. We shall see in Chapter 21 (p. 496) that PLP acts quite differently in other enzymes. The special challenge faced by phosphorylase is to cleave glycogen phosphorolytically rather than hydrolytically to gain one ~P. This requires that water be excluded from the active site, hence the special role of PLP as a general acid-base catalyst.

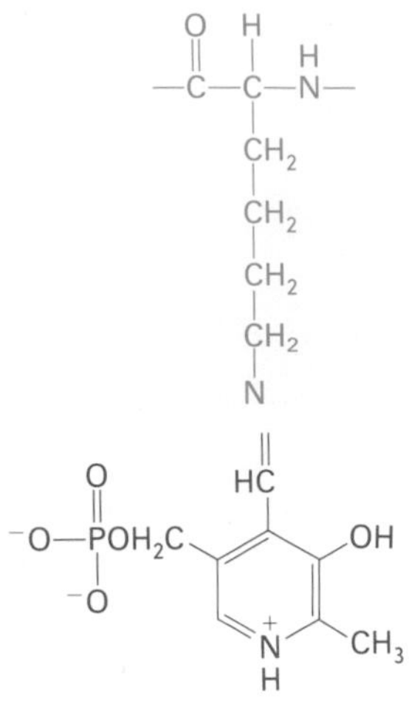

Figure 19-4
Pyridoxal phosphate (red) forms a Schiff base with a lysine residue at the active site of phosphorylase.

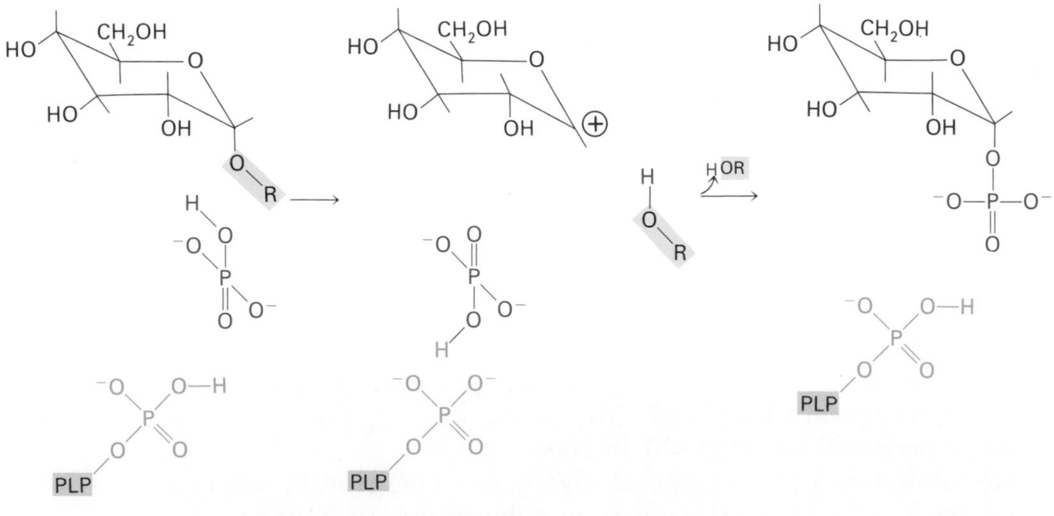

Figure 19-5
Proposed catalytic mechanism of glycogen phosphorylase. PLP denotes pyridoxal phosphate linked as a Schiff base to the active site. The products are glucose 1-phosphate and HOR, the glycogen chain containing one less glucosyl unit. [After P. J. McLaughlin, D. I. Stuart, H. W. Klein, N. G. Oikonomakos, and L. N. Johnson. *Biochemistry* 23(1984):5871.]

A DEBRANCHING ENZYME ALSO IS NEEDED FOR THE BREAKDOWN OF GLYCOGEN

Glycogen is degraded to a limited extent by phosphorylase alone. However, the α-1,6-glycosidic bonds at the branch points are not susceptible to cleavage by phosphorylase. Indeed, phosphorylase stops cleaving

α-1,4 linkages when it reaches a terminal residue four away from a branch point. The action of phosphorylase on two outer branches of a glycogen particle is shown in Figure 19-6. Five α-1,4-glycosidic bonds on one branch and three on the other are cleaved by phosphorylase. Cleavage by phosphorylase stops at this stage because terminal residues *a* and *d* are four away from the branch point *h*. A new enzymatic activity is required. A *transferase shifts a block of three glycosyl residues from one outer branch to the other*. The α-1,4-glycosidic link between *c* and *z* is broken, and a new α-1,4 link between *c* and *d* is formed. This transfer exposes residue *z* to the action of a third degradative enzyme, an α-1,6-glucosidase, which is also known as the *debranching enzyme*. This enzyme hydrolyzes the α-1,6-glycosidic bond between residues *z* and *h*.

Thus, the transferase and the debranching enzyme (α-1,6-glucosidase) convert the branched structure into a linear one, which paves the way for further cleavage by phosphorylase. The hydrolysis of *z* renders all of the residues *a* through *l* susceptible to phosphorylase. It is interesting to note that the transferase and the α-1,6-glucosidase are parts of the same enzyme and, indeed, that a single 160-kd polypeptide chain contains both active sites.

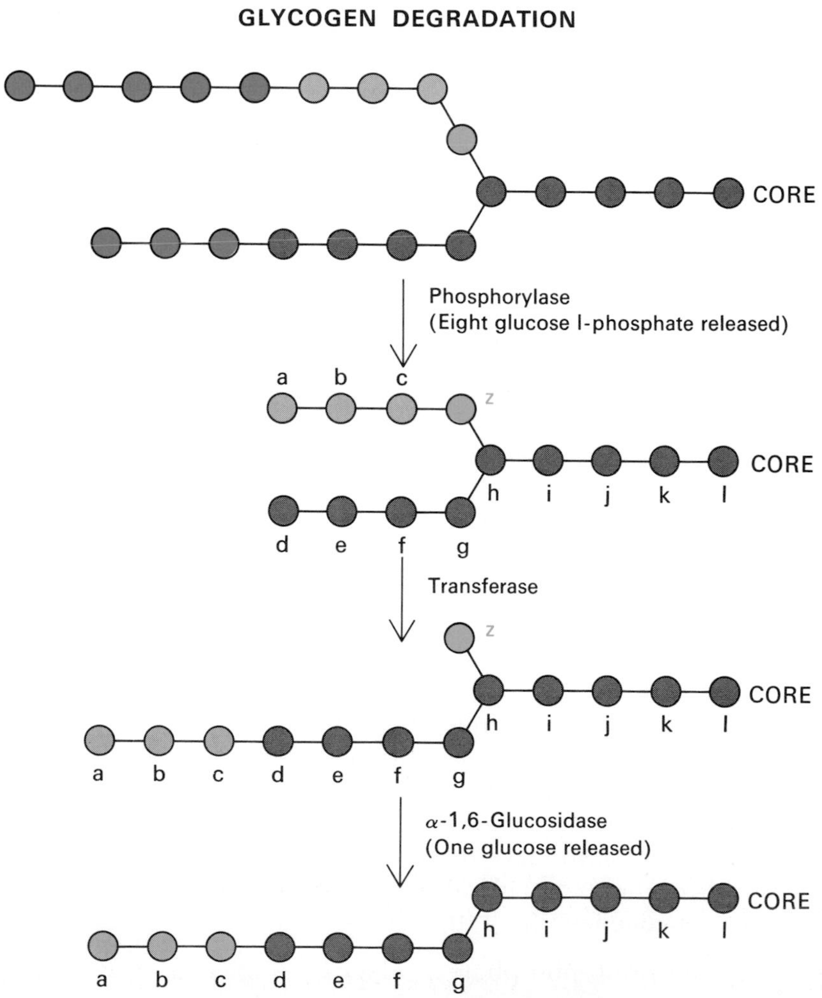

GLYCOGEN DEGRADATION

CORE

Phosphorylase
(Eight glucose l-phosphate released)

a b c z
CORE
h i j k l
d e f g

Transferase

z
CORE
h i j k l
a b c d e f g

α-1,6-Glucosidase
(One glucose released)

CORE
h i j k l
a b c d e f g

Figure 19-6
Steps in the degradation of glycogen.

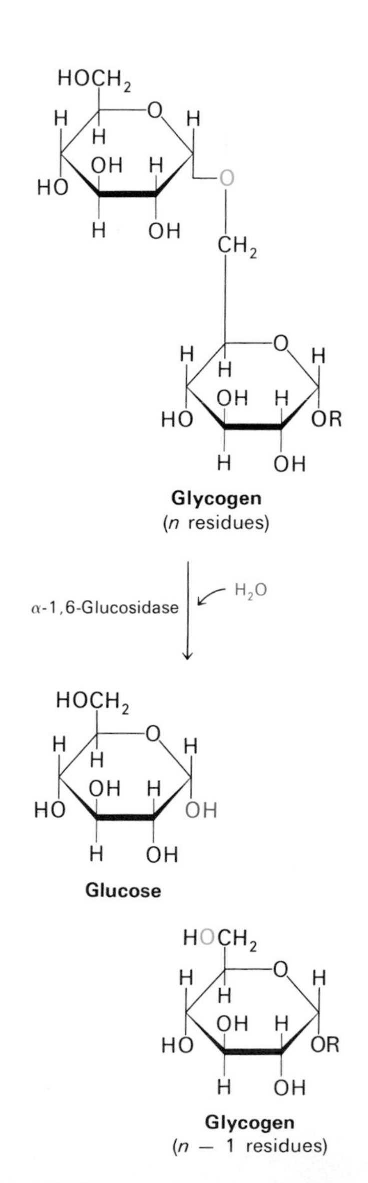

Figure 19-7
Hydrolysis of glycogen at an α-1,6 branch point.

PHOSPHOGLUCOMUTASE CONVERTS GLUCOSE 1-PHOSPHATE INTO GLUCOSE 6-PHOSPHATE

The glucose 1-phosphate formed in the phosphorolytic cleavage of glycogen is converted into glucose 6-phosphate by *phosphoglucomutase*. The equilibrium mixture contains 95% glucose 6-phosphate. The catalytic site of an active enzyme molecule contains a phosphorylated serine residue. In catalysis, this phosphoryl group is probably transferred to the hydroxyl group at C-6 of glucose 1-phosphate to form glucose 1,6-bisphosphate. This intermediate then transfers its C-1 phosphoryl group to the serine residue at the active site of the enzyme, resulting in glucose 6-phosphate and regeneration of the phosphoenzyme.

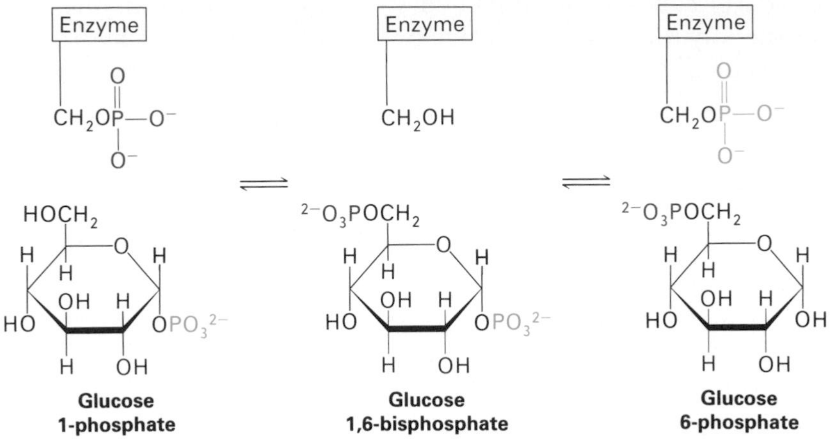

Glucose 1-phosphate ⇌ **Glucose 1,6-bisphosphate** ⇌ **Glucose 6-phosphate**

These reactions are similar to those of *phosphoglyceromutase*, a glycolytic enzyme (p. 370). The role of 2,3-bisphosphoglycerate (2,3-BPG) in the interconversion of 2-phosphoglycerate and 3-phosphoglycerate is like that of glucose 1,6-bisphosphate in the interconversion of the phosphoglucoses. A phosphoenzyme intermediate participates in both reactions.

The phosphoryl group on phosphoglucomutase is slowly lost by hydrolysis. It is restored by phosphoryl transfer from glucose 1,6-bisphosphate, which is formed from glucose 1-phosphate and ATP in a reaction catalyzed by phosphoglucokinase.

LIVER CONTAINS GLUCOSE 6-PHOSPHATASE, A HYDROLYTIC ENZYME ABSENT FROM MUSCLE

A major function of the liver is to maintain a relatively constant level of glucose in the blood. The liver releases glucose into the blood during muscular activity and in the intervals between meals. The released glucose is taken up primarily by the brain and by skeletal muscle. Phosphorylated glucose, in contrast with glucose, cannot readily diffuse out of cells. The liver contains a hydrolytic enzyme, *glucose 6-phosphatase*, that enables glucose to leave that organ. This enzyme, located on the lumenal side of the smooth endoplasmic reticulum membrane, is essential for gluconeogenesis (p. 440).

$$\text{Glucose 6-phosphate} + H_2O \longrightarrow \text{glucose} + P_i$$

Glucose 6-phosphatase is also present in the kidneys and intestine, but *it is absent from muscle and the brain*. Consequently, glucose 6-phosphate is retained by muscle and the brain, which need large amounts of this fuel for the generation of ATP. In contrast, glucose is not a major fuel for the liver. Rather, the liver stores and releases glucose primarily for the benefit of other tissues (see p. 638).

GLYCOGEN IS SYNTHESIZED AND DEGRADED BY DIFFERENT PATHWAYS

The reaction catalyzed by glycogen phosphorylase is readily reversed in vitro, because the $\Delta G^{\circ\prime}$ for the elongation of glycogen by glucose 1-phosphate is -0.5 kcal/mol. In fact, the Coris were able to synthesize glycogen from glucose 1-phosphate using phosphorylase and a branching enzyme. However, a number of subsequent experimental observations indicated that glycogen is synthesized in vivo by a different pathway. First, the reaction catalyzed by phosphorylase is at equilibrium when the $[P_i]$/[glucose 1-phosphate] ratio is 3.6 at neutral pH, whereas this ratio in cells is usually greater than 100. Hence, the phosphorylase reaction in vivo must proceed in the direction of glycogen degradation. Second, hormones that lead to an increase in phosphorylase activity always elicit glycogen breakdown. Third, patients who lack muscle phosphorylase entirely (in McArdle's disease, see p. 466) are able to synthesize muscle glycogen.

In 1957, Luis Leloir and his coworkers showed that glycogen is synthesized by a different pathway. The glycosyl donor is uridine diphosphate glucose (UDP-glucose) rather than glucose 1-phosphate. *The synthetic reaction is not a reversal of the degradative reaction:*

$$\text{Synthesis: Glycogen}_n + \text{UDP-glucose} \longrightarrow \text{glycogen}_{n+1} + \text{UDP}$$

$$\text{Degradation: Glycogen}_{n+1} + \text{P}_i \longrightarrow \text{glycogen}_n + \text{glucose 1-phosphate}$$

We now know that *biosynthetic and degradative pathways in biological systems are almost always distinct.* Glycogen metabolism provided the first example of this important principle. *Separate pathways afford much greater flexibility, both in energetics and in control.* The cell is not at the mercy of mass action; glycogen can be synthesized despite a high ratio of orthophosphate to glucose 1-phosphate.

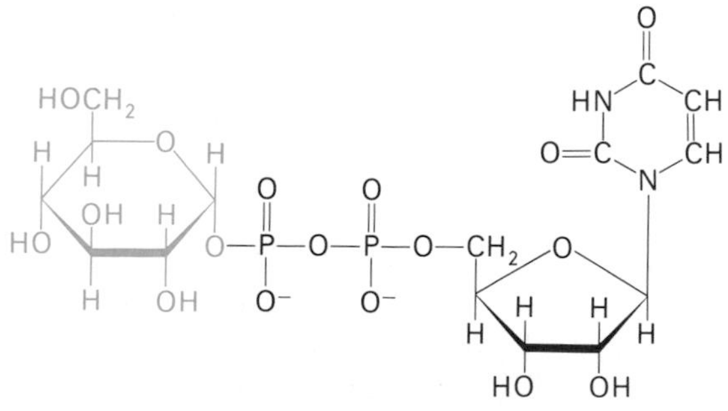

Uridine diphosphate glucose (UDP-glucose)

UDP-GLUCOSE IS AN ACTIVATED FORM OF GLUCOSE

UDP-glucose, the glucose donor in the biosynthesis of glycogen, is an *activated form of glucose,* just as ATP and acetyl CoA are activated forms of orthophosphate and acetate, respectively. The C-1 carbon atom of the glucosyl unit of UDP-glucose is activated because its hydroxyl group is esterified to the diphosphate moiety of UDP.

UDP-glucose is synthesized from glucose 1-phosphate and uridine triphosphate (UTP) in a reaction catalyzed by *UDP-glucose pyrophosphorylase*. The pyrophosphate liberated in this reaction comes from the outer two phosphoryl residues of UTP.

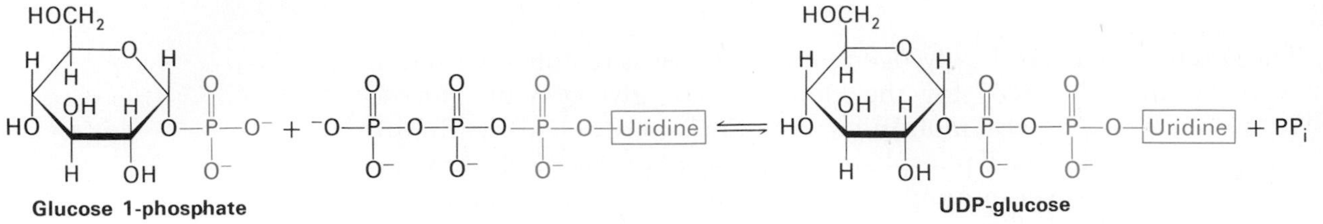

Glucose 1-phosphate + UTP ⇌ UDP-glucose + PP_i

Glucose 1-phosphate   UDP-glucose

This reaction is readily reversible. However, pyrophosphate is rapidly hydrolyzed in vivo to orthophosphate by an inorganic pyrophosphatase. The essentially irreversible hydrolysis of pyrophosphate drives the synthesis of UDP-glucose.

$$\text{Glucose 1-phosphate} + \text{UTP} \rightleftharpoons \text{UDP-glucose} + PP_i$$
$$PP_i + H_2O \longrightarrow 2\ P_i$$
$$\text{Glucose 1-phosphate} + \text{UTP} + H_2O \longrightarrow \text{UDP-glucose} + 2\ P_i$$

The synthesis of UDP-glucose exemplifies a recurring theme in biochemistry: *many biosynthetic reactions are driven by the hydrolysis of pyrophosphate*. Another aspect of this reaction has broad significance. Nucleoside diphosphate sugars serve as glycosyl donors in the biosynthesis of many disaccharides and polysaccharides (p. 340).

GLYCOGEN SYNTHASE CATALYZES THE TRANSFER OF GLUCOSE FROM UDP-GLUCOSE TO A GROWING CHAIN

New glucosyl units are added to the nonreducing terminal residues of glycogen. The activated glucosyl unit of UDP-glucose is transferred to the hydroxyl group at a C-4 terminus of glycogen to form an α-1,4-glycosidic linkage. In elongation, UDP is displaced by the terminal hy-

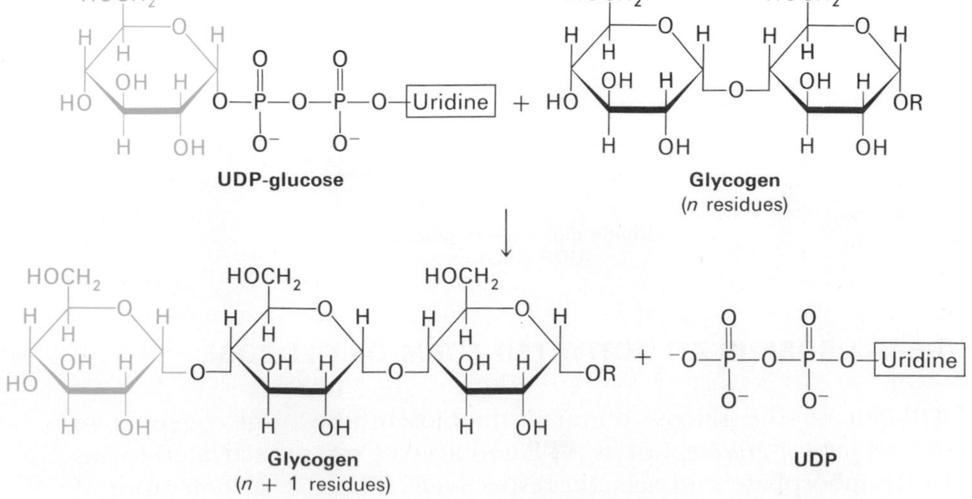

UDP-glucose + Glycogen (n residues) ⟶ Glycogen (n + 1 residues) + UDP

droxyl group of the growing glycogen molecule. This reaction is catalyzed by *glycogen synthase,* which can add glucosyl residues only if the polysaccharide chain already contains more than four residues. Thus, glycogen synthesis requires a *primer*. This priming function is carried out by a protein containing an oligosaccharide of α-1,4-glucose units attached to the phenolic oxygen atom of a tyrosine residue.

Synthase—
An enzyme catalyzing a synthetic reaction in which two units are joined without the direct participation of ATP (or another nucleoside triphosphate).

A BRANCHING ENZYME FORMS α-1,6 LINKAGES

Glycogen synthase catalyzes only the synthesis of α-1,4 linkages. Another enzyme is needed to form the α-1,6 linkages that make glycogen a branched polymer. *Branching is important because it increases the solubility of glycogen.* Furthermore, branching creates a large number of terminal residues, which are the sites of action of glycogen phosphorylase and synthase. Thus, *branching increases the rate of glycogen synthesis and degradation.*

Branching occurs after a number of glucosyl residues are joined in α-1,4 linkage by glycogen synthase. A branch is created by the breaking of an α-1,4 link and the formation of an α-1,6 link: this reaction is different from debranching. A block of residues, typically seven in number, is transferred to a more interior site. The *branching enzyme* that catalyzes this reaction is quite exacting. The block of seven or so residues must include the nonreducing terminus and come from a chain at least eleven residues long. In addition, the new branch point must be at least four residues away from a preexisting one.

GLYCOGEN IS A VERY EFFICIENT STORAGE FORM OF GLUCOSE

What is the cost of converting glucose 6-phosphate into glycogen and back into glucose 6-phosphate? The pertinent reactions have already been described, except for reaction 5 below, which is the regeneration of UTP. UDP is phosphorylated by ATP in a reaction catalyzed by *nucleoside diphosphokinase.*

(1) Glucose 6-phosphate $\longrightarrow$ glucose 1-phosphate
(2) Glucose 1-phosphate + UTP $\longrightarrow$ UDP-glucose + PP_i
(3) $PP_i + H_2O \longrightarrow 2\ P_i$
(4) UDP-glucose + $\text{glycogen}_n \longrightarrow \text{glycogen}_{n+1}$ + UDP
(5) UDP + ATP $\longrightarrow$ UTP + ADP

Sum: Glucose 6-phosphate + ATP + $\text{glycogen}_n + H_2O \longrightarrow$
glycogen_{n+1} + ADP + $2\ P_i$

Thus, one high-energy phosphate bond is spent in incorporating glucose 6-phosphate into glycogen. The energy yield from the breakdown of glycogen is highly efficient. About 90% of the residues are phosphorolytically cleaved to glucose 1-phosphate, which is converted at no cost into glucose 6-phosphate. The other 10% are branch residues, which are hydrolytically cleaved. One ATP is then used to phosphorylate each of these glucose molecules to glucose 6-phosphate. The complete oxidation of glucose 6-phosphate yields about thirty-seven molecules of ATP and storage consumes slightly more than one ATP per glucose 6-phosphate, and so *the overall efficiency of storage is nearly 97%.*

CYCLIC AMP IS CENTRAL TO THE COORDINATED CONTROL OF GLYCOGEN SYNTHESIS AND BREAKDOWN

The existence of separate pathways for the synthesis and degradation of glycogen means that they must be rigorously controlled. ATP would be wastefully hydrolyzed if both sets of reactions were fully active at the same time. In fact, *glycogen synthesis and degradation are coordinated so that glycogen synthase is nearly inactive when phosphorylase is fully active, and vice versa.*

Glycogen metabolism is profoundly affected by specific hormones. *Insulin,* a polypeptide hormone (p. 23), increases the capacity of the liver to synthesize glycogen. The mechanism of action of insulin is not yet fully understood. High levels of insulin in the blood signal the fed state, whereas low levels signal the fasted state (p. 637). Much more is known about the mode of action of *epinephrine* and *glucagon,* which have effects opposite to those of insulin. Muscular activity or its anticipation leads to the release of epinephrine by the adrenal medulla. *Epinephrine* markedly stimulates glycogen breakdown in muscle and, to a lesser extent, in liver. The liver is more responsive to *glucagon,* a polypeptide hormone that is secreted by the α cells of the pancreas when the blood sugar level is low. This hormone increases the blood sugar level by stimulating the breakdown of glycogen in the liver.

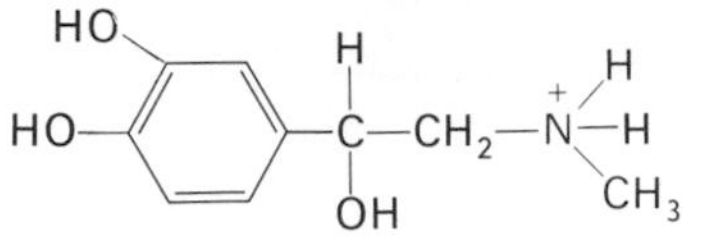

Epinephrine

$^{+}H_3N$-His-Ser-Glu-Gly-Thr-Phe-Thr-Ser-Asp-Tyr- 10

-Ser-Lys-Tyr-Leu-Asp-Ser-Arg-Arg-Ala-Gln- 20

-Asp-Phe-Val-Gln-Trp-Leu-Met-Asn-Thr-COO^- 29

Glucagon

Earl Sutherland discovered that the action of epinephrine and glucagon metabolism is mediated by *cyclic AMP*. This discovery led to the recognition that cyclic AMP is ubiquitous in all eucaryotes and plays a key role in controlling biological processes (see Chapter 38). The synthesis of this regulatory molecule from ATP is catalyzed by *adenylate cyclase,* an enzyme associated with the plasma membrane. The synthesis of cyclic AMP is accelerated by the subsequent hydrolysis of pyrophosphate.

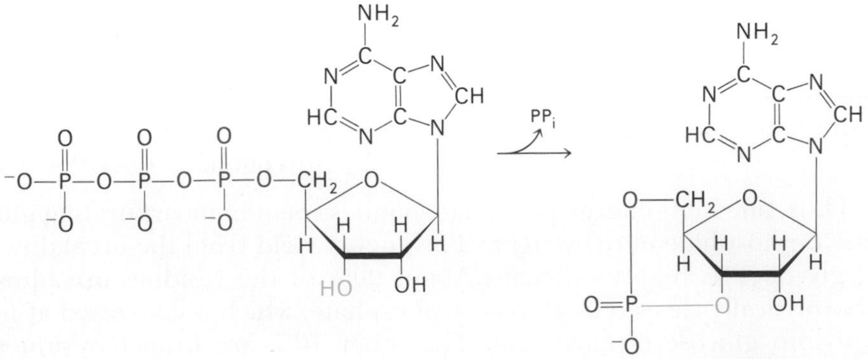

ATP **Cyclic AMP**

Epinephrine and glucagon do not enter their target cells. Rather, they bind to specific receptor proteins on the plasma membrane. These hormone-receptor complexes then activate adenylate cyclase through a signal-coupling protein (p. 979). The increased intracellular level of cyclic AMP triggers a series of reactions that activate phosphorylase and inhibit glycogen synthase. We shall now consider the structural basis of the control of these enzymes, and then turn to the reaction cascade that links cyclic AMP to these crucial enzymes of glycogen metabolism.

PHOSPHORYLASE IS ACTIVATED BY PHOSPHORYLATION OF A SPECIFIC SERINE RESIDUE

Skeletal muscle phosphorylase exists in two interconvertible forms: an *active* phosphorylase *a* and a usually *inactive* phosphorylase *b* (Figure 19-8). The enzyme is a dimer of 97-kd subunits. Phosphorylase *b* is converted into phosphorylase *a* by the phosphorylation of a single serine residue (serine 14) in each subunit.

$$\underset{\text{Ser—OH}}{\boxed{\text{Phosphorylase } b}} + \text{ATP} \longrightarrow \underset{\text{Ser—O—PO}_3^{2-}}{\boxed{\text{Phosphorylase } a}} + \text{ADP} + \text{H}^+$$

This covalent modification is catalyzed by a specific enzyme, *phosphorylase kinase,* which was discovered by Edmond Fischer and Edwin Krebs. Phosphorylase *a* is deactivated by a specific phosphatase that hydrolyzes the phosphoryl group attached to serine 14.

Muscle phosphorylase *b* is active only in the presence of high concentrations of AMP, which acts allosterically—it binds to the nucleotide binding site and alters the conformation of phosphorylase *b*. ATP acts as a negative allosteric effector by competing with AMP. Glucose 6-phosphate also inhibits phosphorylase *b*, primarily by binding to the AMP site. Under most physiological conditions, *phosphorylase* b *is inactive because of the inhibitory effects of ATP and glucose 6-phosphate.* In contrast, *phosphorylase* a *is fully active,* irrespective of the levels of AMP, ATP, and glucose 6-phosphate. The proportion of active enzyme is determined primarily by the rates of phosphorylation and dephosphorylation. In resting muscle, nearly all of the enzyme is in the inactive *b* form. During exercise, the elevated level of AMP leads to the activation of phosphorylase *b*. As described in the next section, increased levels of epinephrine and electrical stimulation of muscle result in the formation of the active *a* form.

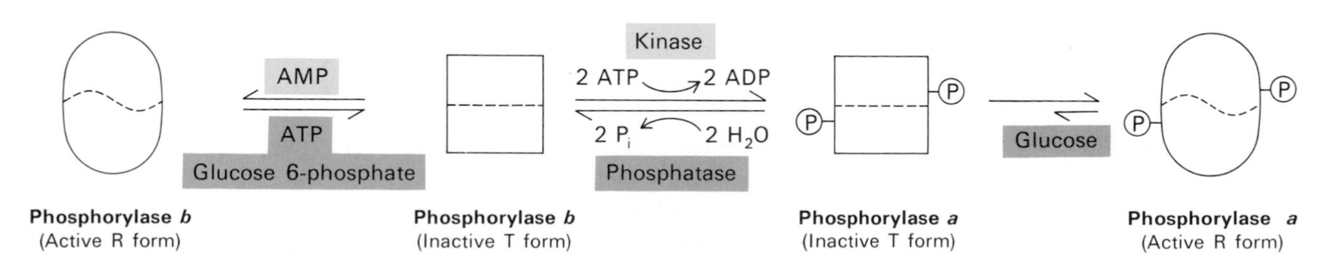

Figure 19-8
Schematic diagram of the control of glycogen phosphorylase in skeletal muscle. The enzyme can adopt a catalytically inactive T (tense) conformation or an active R (relaxed) conformation. The R $\rightleftharpoons$ T equilibrium of phosphorylase *a* is far on the side of the active R state, unless the level of glucose is high. In contrast, phosphorylase *b* is mostly in the inactive T state, unless the level of AMP is high and the levels of ATP and glucose 6-phosphate are low. Under most physiological conditions, the proportion of active enzyme is determined by the rates of phosphorylation and dephosphorylation.

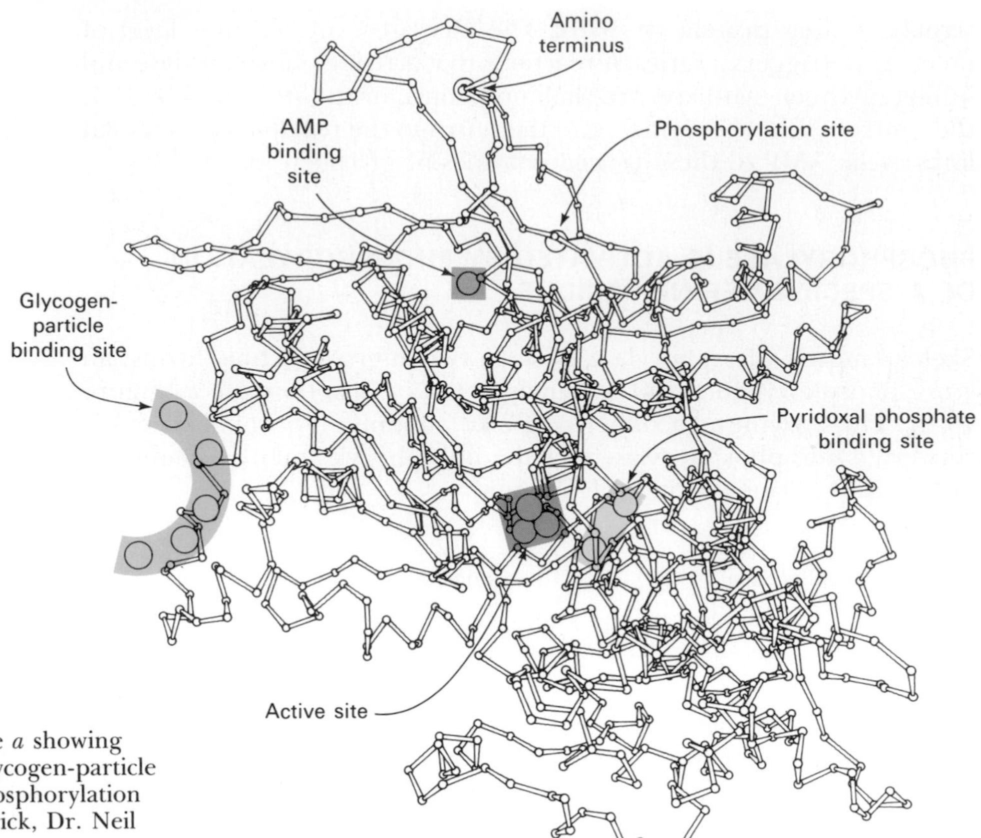

Figure 19-9
α-Carbon diagram of phosphorylase *a* showing the locations of the catalytic site, glycogen-particle binding site, allosteric sites, and phosphorylation site. [Courtesy of Dr. Robert Fletterick, Dr. Neil Madsen, and Dr. Peter Kasvinsky.]

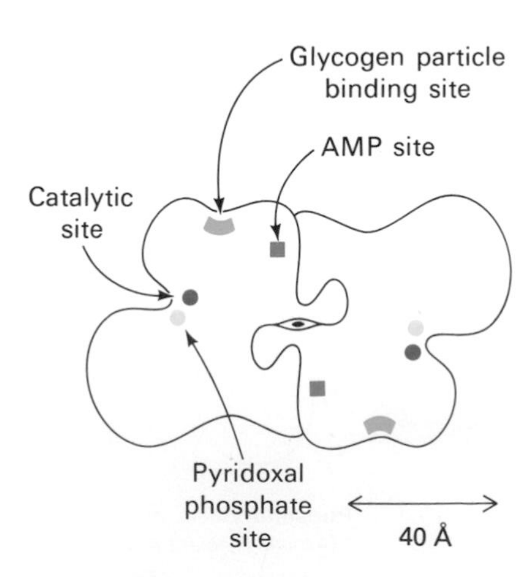

Figure 19-10
Schematic diagram of the phosphorylase *b* dimer viewed down the molecular twofold axis. [After I. T. Weber, L. N. Johnson, K. S. Wilson, D. G. R. Yeates, D. L. Wild, and J. A. Jenkins, *Nature* 274(1978):433.]

STRUCTURAL CHANGES AT THE SUBUNIT INTERFACE ARE TRANSMITTED TO THE CATALYTIC SITES OF PHOSPHORYLASE

X-ray crystallographic studies of the *a* and *b* forms of glycogen phosphorylase from muscle are sources of insight into the catalytic and control mechanisms of this key enzyme in metabolism. The 841 amino acid residues of the monomeric subunit are compactly folded into three structural domains (Figure 19-9): an *amino-terminal domain* (480 residues) with a *glycogen-binding unit* (60 residues), and a *carboxyl-terminal domain* (361 residues). The *catalytic site* is located in a deep crevice formed by residues from the two domains. This shielding of the active site from the aqueous milieu is expected to favor phosphorolysis over hydrolysis. *Pyridoxal phosphate* is positioned at the active site so that its phosphate group is next to the attacking orthophosphate. Phosphorylase also contains a *glycogen-binding site* that is distinct from the catalytic site. This site, which is about 30 Å away from the catalytic site, is important in attaching the enzyme to the glycogen particle. The large separation between this attachment site and the catalytic site enables the enzyme to phosphorolyze many terminal residues without having to dissociate and reassociate after each catalytic cycle.

Phosphorylase contains several *allosteric control sites*. Purines such as caffeine bind to an inhibitory site 12 Å from the active site. Glucose, the major physiological inhibitor of liver phosphorylase *a* (p. 464), binds to the catalytic site differently from the way substrate does, shifting the allosteric equilibrium from the R to the T state. AMP, an allosteric activator of phosphorylase *b*, binds to a site next to the interface between

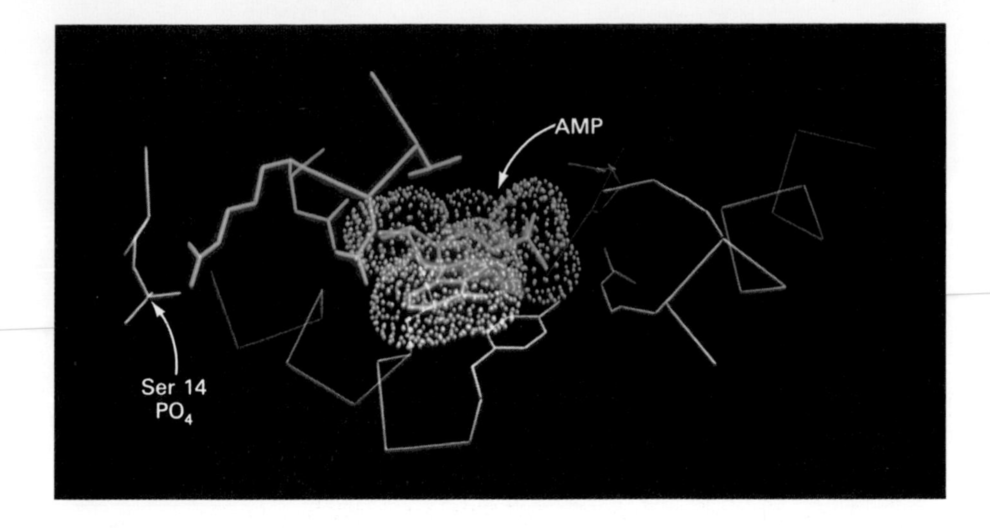

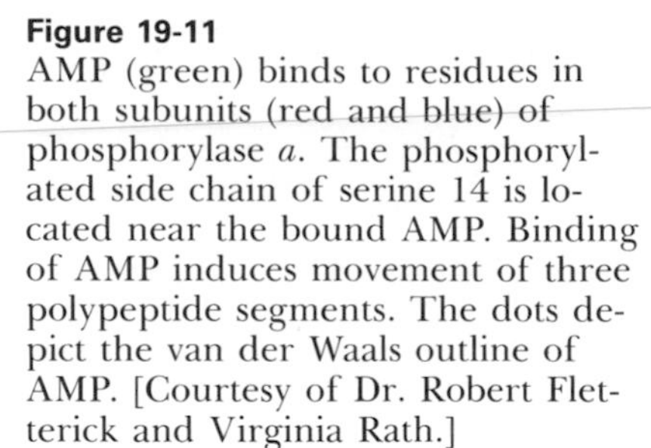

Figure 19-11
AMP (green) binds to residues in both subunits (red and blue) of phosphorylase *a*. The phosphorylated side chain of serine 14 is located near the bound AMP. Binding of AMP induces movement of three polypeptide segments. The dots depict the van der Waals outline of AMP. [Courtesy of Dr. Robert Fletterick and Virginia Rath.]

subunits, far from the catalytic site and the glycogen-binding site. Each of the adenine, ribose, and phosphate moieties of AMP binds a distinct segment of polypeptide chain (Figure 19-11). AMP also makes contact with two residues from the other subunit. ATP, which blocks activation by AMP, binds to the same site but in a different manner. However, ATP does not induce activation because its two additional phosphoryl groups prevent it from moving the three polypeptide segments the way AMP does. The binding of AMP alters the contacts between subunits at their interface and leads to long-range structural changes throughout the dimer. Most important, both catalytic sites, which are more than 30 Å away from the AMP sites, are switched to the catalytically active state. We see here, as in hemoglobin (p. 158) and aspartate transcarbamoylase (p. 237), the critical role of subunit interfaces in transmitting allosteric interactions.

The *phosphorylation site* in the covalent conversion of phosphorylase *b* into *a* is serine 14, which is strategically located at the subunit interface (Figure 19-12). Phosphorylation of this serine leads to a dramatic change in the conformation of the 19 amino-terminal residues: in *b*, they are highly mobile and lack a defined structure, whereas in *a* they have a precise conformation and interact with other residues of both subunits. For example, the negatively charged phosphoryl group on serine 14 interacts electrostatically with two positively charged arginine side chains, one on each chain. These structural changes, like those induced by AMP, are transmitted throughout the protein, and in particular, to both catalytic sites. The flexibility of the amino-terminal region in the *b* form is reminiscent of the very floppy activation domain in trypsinogen, which adopts a well-defined conformation on conversion into trypsin (p. 245).

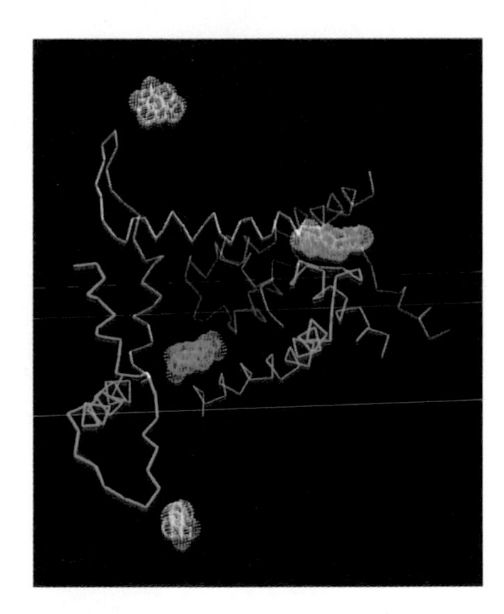

Figure 19-12
Structural links between the AMP-binding sites and the catalytic sites of phosphorylase *a*. The two subunits are shown in red and blue. The two glucose molecules (yellow) mark the catalytic sites. The two AMP molecules at the allosteric sites are shown in green. [Courtesy of Dr. Robert Fletterick and Virginia Rath.]

PHOSPHORYLASE KINASE IS ACTIVATED BY PHOSPHORYLATION AND CALCIUM ION

Phosphorylase kinase, the enzyme that activates phosphorylase by catalyzing the phosphorylation of serine 14, is a very large protein. The subunit composition of the kinase in skeletal muscle is $(\alpha\beta\gamma\delta)_4$, and its mass is 1200 kd. This kinase is under dual control. It is converted from *a low-activity form into a high-activity one by phosphorylation*. The enzyme catalyzing the activation of phosphorylase kinase is a component of the hormone-cyclic-AMP system, as will be discussed shortly. Phosphorylase kinase can be partly activated in a different way, by Ca^{2+} levels of the order of 10^{-6} M. Its δ subunit is calmodulin, a calcium-binding protein that regulates many enzymes in eucaryotes (p. 988); it undergoes a structural change on binding Ca^{2+}. This mode of activation of the kinase is important in muscle, where contraction is triggered by the release of Ca^{2+} (p. 933). Thus, *glycogen breakdown and muscle contraction are coordinated by a transient increase in the Ca^{2+} level in the cytosol.*

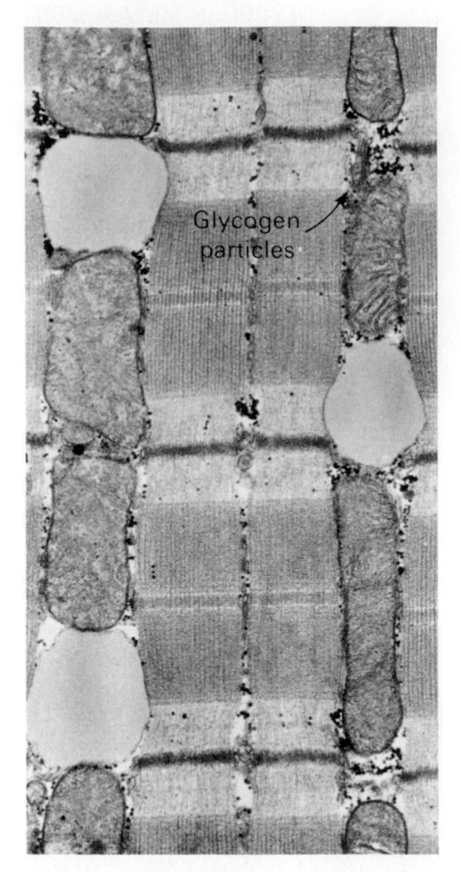

Figure 19-13
Electron micrograph showing the localization of glycogen particles in cardiac muscle. [Courtesy of Dr. Don W. Fawcett.]

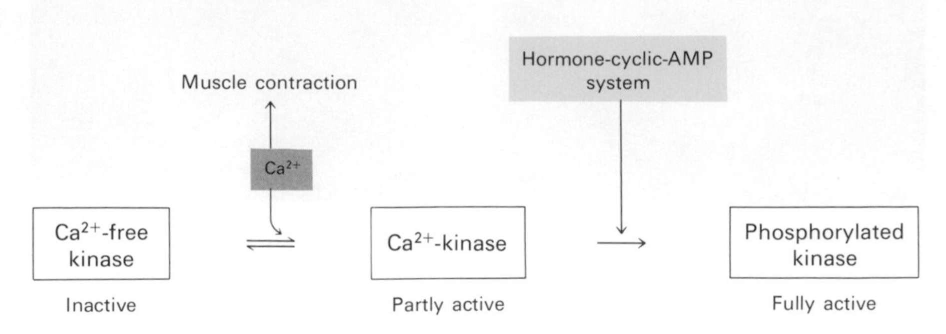

GLYCOGEN SYNTHASE IS INACTIVATED BY THE PHOSPHORYLATION OF A SPECIFIC SERINE RESIDUE

The synthesis of glycogen is closely coordinated with its degradation. The activity of glycogen synthase, like that of phosphorylase, is regulated by covalent modification. Phosphorylation converts glycogen synthase *a* into a usually inactive *b* form. The phosphorylated *b* form requires a high level of glucose 6-phosphate for activity, whereas the dephosphorylated *a* form is active in its presence or absence. Thus, *phosphorylation has opposite effects on the enzymatic activities of glycogen synthase and phosphorylase.*

A REACTION CASCADE CONTROLS THE PHOSPHORYLATION OF GLYCOGEN SYNTHASE AND PHOSPHORYLASE

We turn now to the link between the hormones that affect glycogen metabolism and the phosphorylation reactions that determine the activities of glycogen synthase and phosphorylase. In skeletal muscle, the reaction sequence is (Figure 19-14):

1. Hormones such as epinephrine and glucagon bind to receptors in the plasma membrane of target cells and trigger the activation of adenylate cyclase (p. 977).

2. Adenylate cyclase in the plasma membrane catalyzes the formation of cyclic AMP from ATP.

3. The increased intracellular level of cyclic AMP activates a *protein kinase*. This kinase is inactive in the absence of cyclic AMP. The binding of cyclic AMP allosterically stimulates the protein kinase (p. 982).

4. This protein kinase phosphorylates both phosphorylase kinase and glycogen synthase. *The phosphorylation of both enzymes is the basis of the coordinated regulation of glycogen synthesis and breakdown.* Phosphorylation by this cyclic-AMP–dependent kinase *switches on phosphorylase* (by acti-

vating phosphorylase kinase) and simultaneously *switches off glycogen synthase* (directly). Glycogen synthase is also phosphorylated by phosphorylase kinase, which further assures that glycogen is not being synthesized while it is being degraded. Several other kinases also act on glycogen synthase.

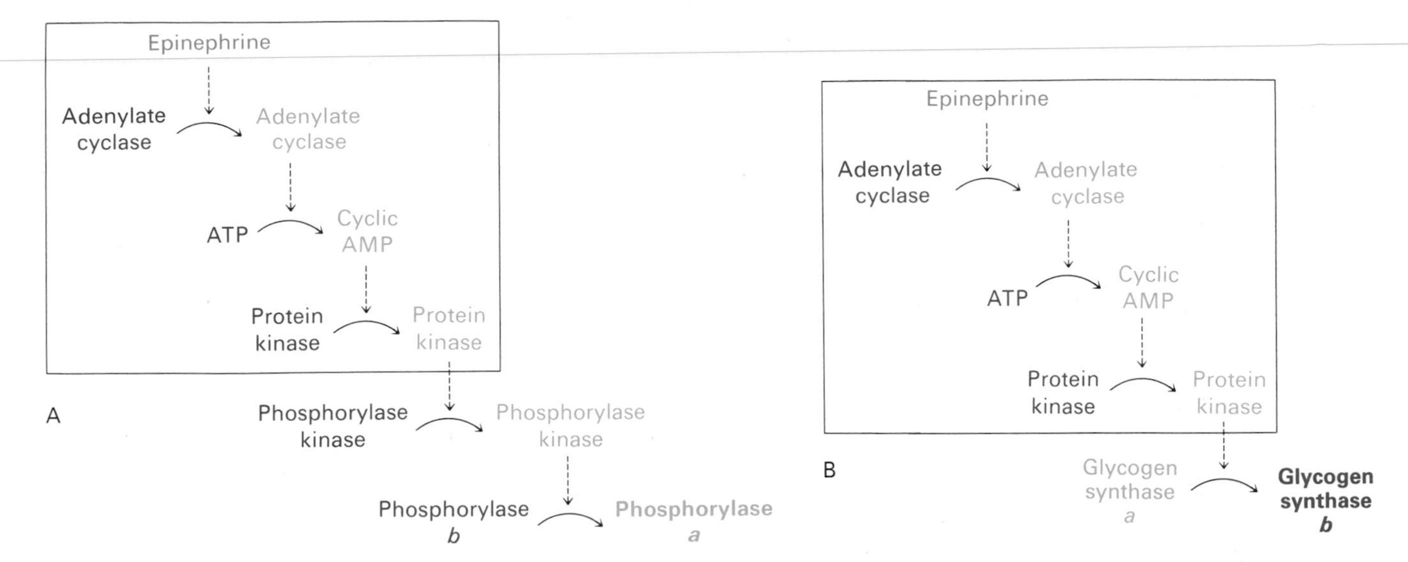

Figure 19-14
Reaction cascades for the control of glycogen metabolism: (A) glycogen degradation; (B) glycogen synthesis. Inactive forms are shown in red, and active ones in green. The sequence of reactions leading to the activation of the protein kinase is the same for the regulation of glycogen degradation and synthesis. Phosphorylase kinase also inactivates glycogen synthase by phosphorylating it.

PHOSPHATASES REVERSE THE REGULATORY EFFECTS OF KINASES

The changes in enzymatic activity produced by phosphorylation can be reversed by the hydrolytic removal of the phosphoryl group. For example, the conversion of phosphorylase *a* into *b* is catalyzed by *protein phosphatase 1*.

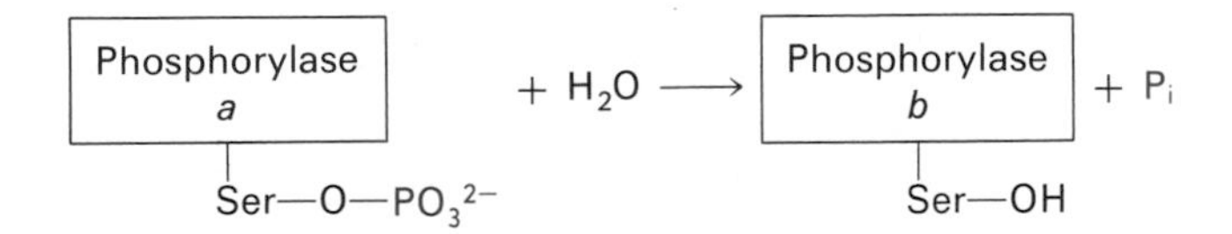

Phosphatase 1 also catalyzes the hydrolysis of the phosphoryl group on the active form of phosphorylase kinase, which renders this kinase inactive. Moreover, the same phosphatase also removes the phosphoryl group from glycogen synthase *b* to convert it into the much more active *a* form. This is yet another molecular device for coordinating glycogen synthesis and breakdown.

Three other phosphatases, called *2A*, *2B*, and *2C*, acting on the enzymes of glycogen metabolism have been isolated. Their activities are regulated. The catalytic action of protein phosphatase 1 is blocked by the phosphorylated form of *inhibitor 1* but not by its dephosphorylated form. The degree of phosphorylation of inhibitor 1 is under hormonal control. Cyclic AMP induces the phosphorylation of inhibitor 1 by activating a protein kinase. Thus, when glycogen degradation is switched on by cyclic AMP, the accompanying phosphorylation of inhibitor 1 keeps phosphorylase in the active *a* form. Conversely, *insulin* decreases the amount of phosphorylated inhibitor 1, unleashing the catalytic activity of phosphatase 1. *The consequent activation of glycogen synthase and deactivation of phosphorylase leads to a build-up of the glycogen store.*

The signal arising from cyclic AMP can also be switched off. The phosphodiester bond in cyclic AMP is hydrolyzed by a specific phosphodiesterase to form AMP, which does not activate the protein kinase. This highly exergonic reaction has a $\Delta G^{\circ\prime}$ of -11.9 kcal/mol. Cyclic nucleotide phosphodiesterases are inhibited by methylxanthines such as *caffeine* and *theophylline*. These compounds prolong the responses to cyclic AMP by slowing the degradation of this intracellular messenger.

Cyclic AMP

H_2O

AMP $+ H^+$

THE REACTION CASCADE AMPLIFIES THE HORMONAL SIGNAL

The enzymatic cascade in the control of glycogen metabolism is analogous to the proteolytic cascade in blood clotting (see p. 248). In both processes, the enzymatic cascade provides a *high degree of amplification.* There are three enzyme-catalyzed control stages in glycogen breakdown and two in glycogen synthesis. If glycogen phosphorylase and synthase were directly regulated by the binding of epinephrine, more than a thousand times as much epinephrine would be required to elicit glycogen breakdown.

GLYCOGEN METABOLISM IN THE LIVER REGULATES THE BLOOD GLUCOSE LEVEL

The control of the synthesis and degradation of glycogen in the liver is central to the regulation of the blood glucose level. The concentration of glucose in the blood normally ranges from about 80 to 120 mg per 100 ml (4.4 to 6.7 mM). The liver senses the concentration of glucose in the blood and takes up or releases glucose accordingly. The amount of liver phosphorylase *a* decreases rapidly when glucose is infused (Figure 19-15). After a lag period, the amount of glycogen synthase *a* increases, which results in the synthesis of glycogen. In fact, *phosphorylase* a *is the glucose sensor in liver cells.* The binding of glucose to phosphorylase *a* shifts the allosteric equilibrium from the R form to the T form (see Figure 19-8), which *exposes the phosphoryl group on serine 14 to hydrolysis by the phosphatase.* It is significant that phosphatase 1 binds tightly to phosphorylase *a* but acts catalytically only when glucose induces it to adopt the T form.

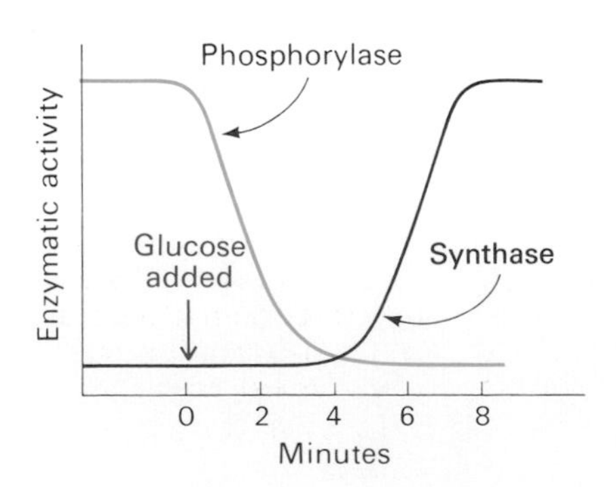

Figure 19-15
The infusion of glucose leads to the inactivation of phosphorylase, followed by the activation of glycogen synthase. [After W. Stalmans, H. De Wulf, L. Hue, and H.-G. Hers, *Eur. J. Biochem.* 41(1974):127.]

How does glucose activate the synthase? Recall that the same phosphatase acts on phosphorylase and glycogen synthase. Phosphorylase *b*, in contrast with the *a* form, does not bind the phosphatase. Consequently, the conversion of phosphorylase *a* into *b* is accompanied by the *release of the phosphatase, which is then free to activate glycogen synthase.* Removal of the phosphoryl group of inactive synthase *b* converts it into the active *a* form. Initially, there are about ten phosphorylase *a* molecules per phosphatase. Hence, the activity of the synthase begins to increase only after most of phosphorylase *a* is converted into the *b* form (Figure 19-15). This remarkable glucose-sensing system depends on three key elements: (1) communication between the serine phosphate and the allosteric site for glucose; (2) the use of the same phosphatase to inactivate phosphorylase and activate the synthase; and (3) the binding of the phosphatase to phosphorylase *a* to prevent premature activation of the synthase.

A VARIETY OF GENETICALLY DETERMINED GLYCOGEN-STORAGE DISEASES ARE KNOWN

The first glycogen-storage disease was described by Edgar von Gierke in 1929. A patient with this disease has a huge abdomen caused by a *massive enlargement of the liver*. There is a pronounced *hypoglycemia* between meals. Furthermore, the blood glucose level does not rise on administration of epinephrine and glucagon. An infant with this glycogen storage disease may have convulsions because of the low blood glucose level.

The enzymatic defect in von Gierke's disease was elucidated by the Coris in 1952. They found that *glucose 6-phosphatase was missing from the liver of a patient with this disease*. This was the first demonstration of an inherited deficiency of a liver enzyme. The liver glycogen is normal in structure but present in abnormally large amounts. The absence of glucose 6-phosphatase in the liver causes hypoglycemia because glucose cannot be formed from glucose 6-phosphate. This phosphorylated sugar does not leave the liver because it cannot traverse the plasma membrane. There is a compensatory increase in glycolysis in the liver, leading to a high level of lactate and pyruvate in the blood. Patients who have von Gierke's disease also have an increased dependence on fat metabolism.

A number of glycogen storage diseases have been characterized (Table 19-1). The Coris elucidated the biochemical defect in another

Table 19-1
Glycogen storage diseases

Type	*Defective enzyme*	*Organ affected*	*Glycogen in the affected organ*	*Clinical features*
I VON GIERKE'S DISEASE	Glucose 6-phosphatase	Liver and kidney	Increased amount; normal structure.	Massive enlargement of the liver. Failure to thrive. Severe hypoglycemia, ketosis, hyperuricemia, hyperlipemia.
II POMPE'S DISEASE	α-1,4-Glucosidase (lysosomal)	All organs	Massive increase in amount; normal structure.	Cardiorespiratory failure causes death, usually before age 2.
III CORI'S DISEASE	Amylo-1,6-glucosidase (debranching enzyme)	Muscle and liver	Increased amount; short outer branches.	Like type I, but milder course.
IV ANDERSEN'S DISEASE	Branching enzyme (α-1,4 $\longrightarrow$ α-1,6)	Liver and spleen	Normal amount; very long outer branches.	Progressive cirrhosis of the liver. Liver failure causes death usually before age 2.
V McARDLE'S DISEASE	Phosphorylase	Muscle	Moderately increased amount; normal structure.	Limited ability to perform strenuous exercise because of painful muscle cramps. Otherwise patient is normal and well developed.
VI HERS' DISEASE	Phosphorylase	Liver	Increased amount	Like type I, but milder course.
VII	Phosphofructokinase	Muscle	Increased amount; normal structure.	Like type V.
VIII	Phosphorylase kinase	Liver	Increased amount; normal structure.	Mild liver enlargement. Mild hypoglycemia.

Note: Types I through VII are inherited as autosomal recessives. Type VIII is sex-linked.

glycogen storage disease (type III), which cannot be distinguished from von Gierke's disease (type I) by physical examination alone. In type III disease, the structure of liver and muscle glycogen is abnormal and the amount is markedly increased. Most striking, the outer branches of the glycogen are very short. *Patients having this type lack the debranching enzyme* (α-1,6-glucosidase), and so only the outermost branches of glycogen can be effectively utilized. Thus, only a small fraction of this abnormal glycogen is functionally active as an accessible store of glucose.

A defect in glycogen metabolism confined to muscle is found in McArdle's disease (type V). *Muscle phosphorylase activity is absent,* and the patient's capacity to perform strenuous exercise is limited because of painful muscle cramps. The patient is otherwise normal and well developed. Thus, effective utilization of muscle glycogen is not essential for life. Phosphorus-31 nuclear magnetic resonance studies of these patients have been very informative. The pH of skeletal muscle cells of normal individuals drops during strenuous exercise because of the production of lactic acid. In contrast, the muscle cells of patients with McArdle's disease become more alkaline during exercise (because of the breakdown of phosphocreatine, p. 935) without the accumulation of glycolytic intermediates. This finding shows that the defect is prior to the first step in glycolysis. NMR studies have also shown that the painful cramps in this disease are correlated with high levels of ADP. NMR spectroscopy is a valuable noninvasive technique for assessing dietary and exercise therapy for this disease.

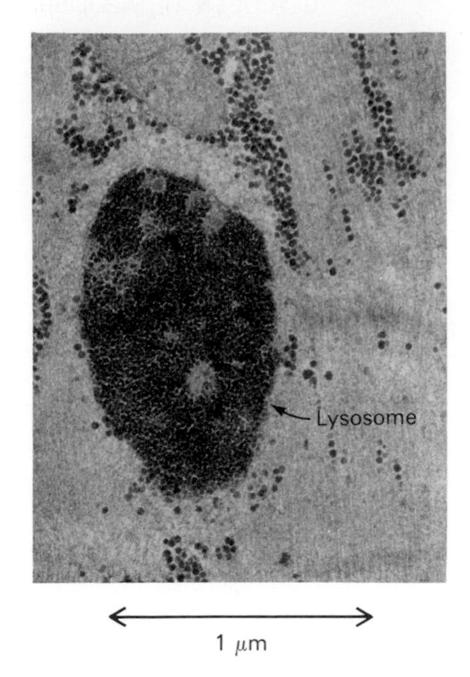

Figure 19-16
Electron micrograph of skeletal muscle from an infant with type II glycogen storage disease (Pompe's disease). The lysosomes are engorged with glycogen because of a deficiency in the α-1,4-glucosidase. This defect in the degradation of glycogen is confined to lysosomes. The amount of glycogen in the cytosol is normal. [From H.-G. Hers and F. Van Hoof (eds). *Lysosomes and Storage Diseases* (Academic Press, 1973), p. 205.]

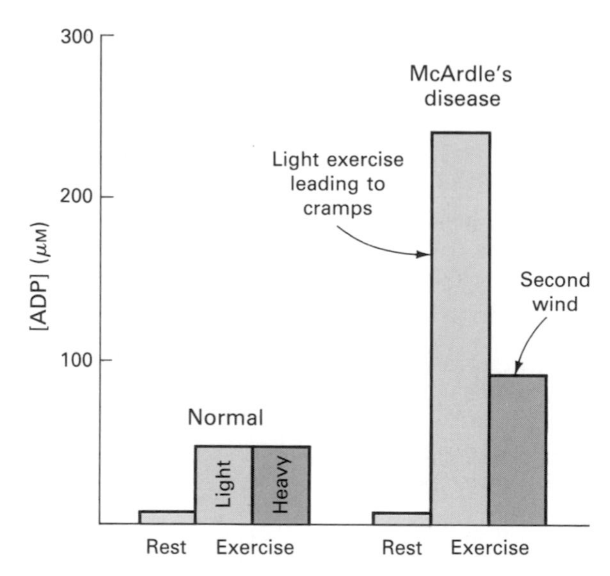

Figure 19-17
NMR study of human arm muscle. The level of ADP during exercise increases much more in a patient with McArdle's glycogen-storage disease than in normal controls. [After G. K. Radda. *Biochem. Soc. Trans.* 14(1986):522.]

SUMMARY

Glycogen, a readily mobilized fuel store, is a branched polymer of glucose residues. Most of the glucose units in glycogen are linked by α-1,4-glycosidic bonds. At about every tenth residue, a branch is created by an α-1,6-glycosidic bond. Glycogen is present in large amounts in muscle and liver, where it is stored in the cytoplasm in the form of hydrated granules. Most of the glycogen molecule is degraded to glucose 1-phosphate by the action of phosphorylase. The glycosidic linkage between C-1 of a terminal residue and C-4 of the adjacent one is split by orthophosphate to give glucose 1-phosphate, which can be reversibly converted into glucose 6-phosphate. Pyridoxal phosphate, a derivative of vitamin B_6, participates in the phosphorolytic cleavage of glycogen. Branch points are degraded by the concerted action of an oligosaccharide transferase and an α-1,6-glucosidase. The latter enzyme (also known as the debranching enzyme) catalyzes the hydrolysis of α-1,6 linkages, yielding free glucose. Glycogen is synthesized by a different pathway. UDP-glucose, the activated intermediate in glycogen synthesis, is formed from glucose 1-phosphate and UTP. Glycogen synthase catalyzes the transfer of glucose from UDP-glucose to the C-4 hydroxyl group of a terminal residue in the growing glycogen molecule. A branching enzyme converts some of the α-1,4 linkages into α-1,6 linkages.

Phosphorylase, a dimeric enzyme, is regulated by allosteric effectors and reversible covalent modifications. Phosphorylase *b*, which is usually inactive, is converted into active phosphorylase *a* by the phosphorylation of a single serine residue in each subunit. The *b* form can also be activated by the binding of AMP. The *a* form in the liver is inhibited by glucose. The AMP binding sites and phosphorylation sites are located at the subunit interface. Conformational changes induced by AMP binding and phosphorylation are transmitted over distances of more than 30 Å to the catalytic sites and other binding sites.

Glycogen synthesis and degradation are coordinated by an amplifying cascade of reactions. Glycogen synthase is inactive when phosphorylase is active, and vice versa. Epinephrine and glucagon stimulate glycogen breakdown and inhibit its synthesis by increasing the intracellular level of cyclic AMP, which activates a protein kinase. Glycogen synthase is then inactivated by phosphorylation. Phosphorylase, also a substrate for this kinase, becomes more active when it is phosphorylated. Hence, cyclic AMP leads to the mobilization of glucose from glycogen. The kinase that activates phosphorylase is also stimulated by Ca^{2+}, leading to the mobilization of glucose from glycogen during muscle contraction. The phosphoryl groups on these enzymes are removed by the same phosphatase. Glycogen synthase and phosphorylase are also regulated by noncovalent allosteric interactions. In fact, phosphorylase is a key part of the glucose-sensing system of liver cells. Glycogen metabolism exemplifies the central role of reversible phosphorylation in the regulation of biological processes.

SELECTED READINGS

Where to start

Cohen, P., 1983. Protein phosphorylation and the control of glycogen metabolism in skeletal muscle. *Phil. Trans. R. Soc. Lond. B* 302:13–25.

Fischer, E. H., 1983. Cellular regulation by protein phosphorylation. *Bull. Instit. Pasteur* 81:7–31.

Fletterick, R. J., 1983. Glycogen phosphorylase: plasticity and specificity in ligand binding. *Proc. Robert A. Welch Found. Conf. Chem. Res.* 27:173–220.

Jenkins, J. A., Johnson, L. N., Stuart, D. I., Stura, E. A., Wilson, K. S., and Zanotti, G., 1981. Phosphorylase: control and activity. *Phil. Trans. R. Soc. Lond. B* 293:23–41.

Books and reviews

Boyer, P. D., and Krebs, E. G., (eds.), 1986. *The Enzymes*, vol. 17. Academic Press. [This volume of the series, entitled *Control by Phosphorylation*, contains several excellent articles on glycogen metabolism. Also see volume 18 of this series for critical reviews of control by reversible phosphorylation.]

Geddes, R., 1986. Glycogen: a metabolic viewpoint. *Biosci. Rep.* 6:415–428.

Preiss, J., Yung, S. G., and Baeker, P. A., 1983. Regulation of bacterial glycogen synthesis. *Mol. Cell Biochem.* 57:61–80.

Hers, H.-G., 1976. The control of glycogen metabolism in the liver. *Ann. Rev. Biochem.* 45:167–190.

Fletterick, R. J., and Madsen, N. B., 1980. The structures and related functions of phosphorylase. *Ann. Rev. Biochem.* 49:31–61.

Newsholme, E. A., and Start, C., 1973. *Regulation in Metabolism.* Academic Press. [Chapter 4 deals with the regulation of glycogen metabolism.]

Structure of glycogen phosphorylase

Lorek, A., Wilson, K. S., Sansom, M. S., Stuart, D. I., Stura, E. A., Jenkins, J. A., Zanotti, G., Hajdu, J., and Johnson, L. N., 1984. Allosteric interactions of glycogen phosphorylase *b*: A crystallographic study of glucose 6-phosphate and inorganic phosphate binding to di-imidate-crosslinked phosphorylase *b*. *Biochem. J.* 218:45–60.

Sprang, S. R., Goldsmith, E. J., Fletterick, R. J., Withers, S. G., and Madsen, N. B., 1982. Catalytic site of glycogen phosphorylase: structure of the T state and specificity for alpha-D-glucose. *Biochemistry* 21:5364–5371.

Withers, S. G., Madsen, N. B., Sprang, S. R., and Fletterick, R. J., 1982. Catalytic site of glycogen phosphorylase: structure changes during activation and mechanistic implications. *Biochemistry* 21:5372–5382.

Weber, I. T., Johnson, L. N., Wilson, K. S., Yeates, D. G. R., Wild, D. L., and Jenkins, J. A., 1978. Crystallographic studies on the activity of glycogen phosphorylase *b*. *Nature* 274:433–437.

Hwang, P. K., and Fletterick, R. J., 1986. Convergent and divergent evolution of regulatory sites in eukaryotic phosphorylases. *Nature* 324:80–84.

Catalytic mechanism of glycogen phosphorylase

Klein, H. M., and Helmreich, E. J. M., 1985. The role of pyridoxal 5′-phosphate in phosphorylase catalysis. *Curr. Top. Cell. Regul.* 26:281–294.

Klein, H. W., Im, W. J., and Palm, D., 1986. Mechanism of the phosphorylase reaction. Utilization of D-*gluco*-hept-1-enitol in the absence of primer. *Eur. J. Biochem.* 157:107–114.

Withers, S. G., Madsen, N. B., and Sykes, B. D., 1981. Active form of pyridoxal phosphate in glycogen phosphorylase. Phosphorus-31 nuclear magnetic resonance investigation. *Biochemistry* 20:1748–1756.

Withers, S. G., Madsen, N. B., Sykes, B. D., Takagi, M., Shimomura, S., and Fukui, T., 1981. Evidence for direct phosphate-phosphate interaction between pyridoxal phosphate and substrate in the glycogen phosphorylase catalytic mechanism. *J. Biol. Chem.* 256:10759–10762.

Genetic diseases

Howell, R. R., and Williams, J. C., 1983. The glycogen storage diseases. *In* Stanbury, J. B., Wyngaarden, J. B., Fredrickson, D. S., Goldstein, J. L., and Brown, M. S. (eds.), *The Metabolic Basis of Inherited Disease* (5th ed.), pp. 141–166. McGraw-Hill.

Ross, B. D., Radda, G. K., Gadian, D. G., Rocker, G., Esiri, M., and Falconer-Smith, J., 1981. Examination of a case of suspected McArdle's syndrome by ^{31}P NMR. *New Engl. J. Med.* 304:1338–1342.

Historical aspects

Cori, C. F., and Cori, G. T., 1947. Polysaccharide phosphorylase. In *Nobel Lectures: Physiology or Medicine (1942–1962)*, pp. 186–206. American Elsevier (1964).

Leloir, L. F., 1971. Two decades of research on the biosynthesis of saccharides. *Science* 172:1299–1302.

PROBLEMS

1. Write a balanced equation for the formation of glycogen from galactose.

2. Write a balanced equation for the formation of glucose from fructose in the liver.

3. A sample of glycogen from a patient with liver disease is incubated with orthophosphate, phosphorylase, the transferase, and the debranching enzyme. The ratio of glucose 1-phosphate to glucose formed in this mixture is 100. What is the most likely enzymatic deficiency in this patient?

4. Suggest an explanation for the fact that the amount of glycogen in type I glycogen-storage disease (von Gierke's disease) is increased.

5. Crystals of phosphorylase *a* grown in the presence of glucose shatter when a substrate such as glucose 1-phosphate is added. Why?

6. Cyclic-AMP–dependent kinase phosphorylates the β subunits of muscle phosphorylase kinase rapidly, activating it. The α subunits of this enzyme are then phosphorylated at a slow rate, making both the α and β subunits susceptible to the action of phosphatase I. What might be the functional significance of the slow phosphorylation of the α subunits?

7. Predict the major consequences of each of the following mutations:
   (a) Loss of the gene for the cyclic-AMP–dependent protein kinase in muscle.
   (b) Loss of the gene for calmodulin in muscle.
   (c) Loss of the gene for inhibitor 1 of protein phosphatase 1.

CHAPTER 20

Fatty Acid Metabolism

We turn now from the metabolism of carbohydrates to that of fatty acids, a class of compounds containing a long hydrocarbon chain and a terminal carboxylate group. Fatty acids have three major physiological roles. First, *they are building blocks of phospholipids and glycolipids.* These amphipathic molecules are important components of biological membranes, as was discussed in Chapter 12. Second, fatty acid derivatives serve as *hormones* (p. 991) and *intracellular messengers* (p. 985). Third, *fatty acids are fuel molecules.* They are stored as *triacylglycerols,* which are uncharged esters of glycerol. Triacylglycerols are also called neutral fats or triglycerides.

$$H_3C{-}(CH_2)_7{-}\overset{H}{C}{=}\overset{H}{C}{-}(CH_2)_7{-}\overset{O}{\overset{\|}{C}}{-}O{-}CH \quad \begin{array}{l} CH_2{-}O{-}\overset{O}{\overset{\|}{C}}{-}(CH_2)_{14}{-}CH_3 \\ CH_2{-}O{-}\underset{O}{\underset{\|}{C}}{-}(CH_2)_{16}{-}CH_3 \end{array}$$

A triacylglycerol

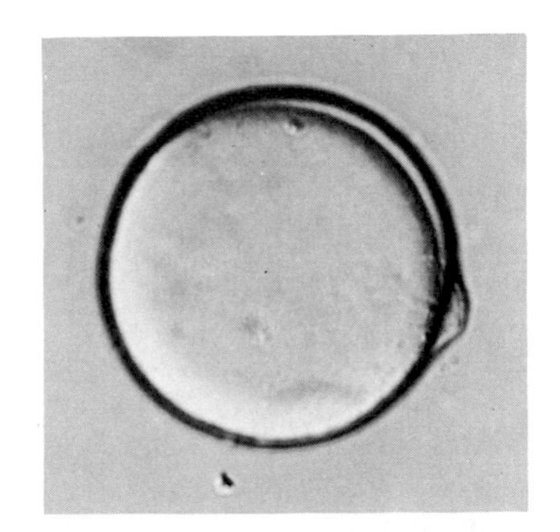

Figure 20-1
Photomicrograph of a fat cell. A large globule of fat is surrounded by a thin rim of cytoplasm and a bulging nucleus. [Courtesy of Dr. Pedro Cuatrecasas.]

NOMENCLATURE OF FATTY ACIDS

The systematic name for a fatty acid is derived from the name of its parent hydrocarbon by the substitution of *oic* for the final *e*. For example, the C_{18} saturated fatty acid is called octadecanoic acid because the parent hydrocarbon is octadecane. A C_{18} fatty acid with one double bond is called octadec*enoic* acid; with two double bonds, octadeca*dienoic* acid; and with three double bonds, octadeca*trienoic* acid. The symbol 18:0 denotes a C_{18} fatty acid with no double bonds, whereas 18:2 signifies that there are two double bonds.

Fatty acid carbon atoms are numbered starting at the carboxyl terminus:

$$\underset{\omega}{H_3C}-(CH_2)_n-\overset{3}{\underset{\beta}{CH_2}}-\overset{2}{\underset{\alpha}{CH_2}}-\overset{1}{C}(=O)-OH$$

Carbon atoms 2 and 3 are often referred to as α and β, respectively. The methyl carbon atom at the distal end of the chain is called the ω carbon. The position of a double bond is represented by the symbol Δ followed by a superscript number. For example, *cis*-Δ^9 means that there is a *cis* double bond between carbon atoms 9 and 10; *trans*-Δ^2 means that there is a *trans* double bond between carbon atoms 2 and 3. Alternatively, the position of a double bond can be denoted by counting from the distal end, with the ω carbon atom (the methyl carbon) as number 1. An ω-3 fatty acid, for example, has the structure shown at the left. Fatty acids are ionized at physiological pH, and so it is appropriate to refer to them according to their carboxylate form: for example, palmitate or hexadecanoate.

ω-Carbon atom: CH_3
CH_2
ω-3 double bond: $C-H$ = $C-H$
CH_2
R
COO^-

An ω-3 fatty acid

FATTY ACIDS VARY IN CHAIN LENGTH AND DEGREE OF UNSATURATION

Fatty acids in biological systems (Table 20-1) usually contain an even number of carbon atoms, typically between 14 and 24. The 16- and 18-carbon fatty acids are most common. The hydrocarbon chain is almost invariably unbranched in animal fatty acids. The alkyl chain may be saturated or it may contain one or more double bonds. The configuration of the double bonds in most unsaturated fatty acids is *cis*. The double bonds in polyunsaturated fatty acids are separated by at least one methylene group.

Table 20-1
Some naturally occurring fatty acids in animals

Number of carbons	*Number of double bonds*	*Common name*	*Systematic name*	*Formula*
12	0	Laurate	*n*-Dodecanoate	$CH_3(CH_2)_{10}COO^-$
14	0	Myristate	*n*-Tetradecanoate	$CH_3(CH_2)_{12}COO^-$
16	0	Palmitate	*n*-Hexadecanoate	$CH_3(CH_2)_{14}COO^-$
18	0	Stearate	*n*-Octadecanoate	$CH_3(CH_2)_{16}COO^-$
20	0	Arachidate	*n*-Eicosanoate	$CH_3(CH_2)_{18}COO^-$
22	0	Behenate	*n*-Docosanoate	$CH_3(CH_2)_{20}COO^-$
24	0	Lignocerate	*n*-Tetracosanoate	$CH_3(CH_2)_{22}COO^-$
16	1	Palmitoleate	*cis*-Δ^9-Hexadecenoate	$CH_3(CH_2)_5CH{=}CH(CH_2)_7COO^-$
18	1	Oleate	*cis*-Δ^9-Octadecenoate	$CH_3(CH_2)_7CH{=}CH(CH_2)_7COO^-$
18	2	Linoleate	*cis,cis*-Δ^9,Δ^{12}-Octadecadienoate	$CH_3(CH_2)_4(CH{=}CHCH_2)_2(CH_2)_6COO^-$
18	3	Linolenate	all *cis*-Δ^9,Δ^{12},Δ^{15}-Octadecatrienoate	$CH_3CH_2(CH{=}CHCH_2)_3(CH_2)_6COO^-$
20	4	Arachidonate	all *cis*-Δ^5,Δ^8,Δ^{11},Δ^{14}-Eicosatetraenoate	$CH_3(CH_2)_4(CH{=}CHCH_2)_4(CH_2)_2COO^-$

The properties of fatty acids and of lipids derived from them are markedly dependent on their chain length and on their degree of saturation. Unsaturated fatty acids have a lower melting point than saturated fatty acids of the same length. For example, the melting point of stearic acid is 69.6°C, whereas that of oleic acid (which contains one *cis* double bond) is 13.4°C. The melting points of polyunsaturated fatty acids of the C_{18} series are even lower. Chain length also affects the melting point, as illustrated by the fact that the melting temperature of palmitic acid (C_{16}) is 6.5 degrees lower than that of stearic acid (C_{18}). *Thus, short chain length and unsaturation enhance the fluidity of fatty acids and of their derivatives* (p. 296).

TRIACYLGLYCEROLS ARE HIGHLY CONCENTRATED ENERGY STORES

Triacylglycerols are highly concentrated stores of metabolic energy because they are *reduced* and *anhydrous*. The yield from the complete oxidation of fatty acids is about 9 kcal/g, in contrast with about 4 kcal/g for carbohydrates and proteins. The basis of this large difference in caloric yield is that fatty acids are much more highly reduced. Furthermore, triacylglycerols are very nonpolar and so they are stored in a nearly anhydrous form, whereas proteins and carbohydrates are much more polar and hence more highly hydrated. In fact, a gram of dry glycogen binds about two grams of water. *Consequently, a gram of nearly anhydrous fat stores more than six times as much energy as a gram of hydrated glycogen,* which is the reason that triacylglycerols rather than glycogen were selected in evolution as the major energy reservoir. Consider a typical 70-kg man, who has fuel reserves of 100,000 kcal in triacylglycerols, 25,000 kcal in protein (mostly in muscle), 600 kcal in glycogen, and 40 kcal in glucose. Triacylglycerols constitute about 11 kg of his total body weight. If this amount of energy were stored in glycogen, his total body weight would be 55 kg greater.

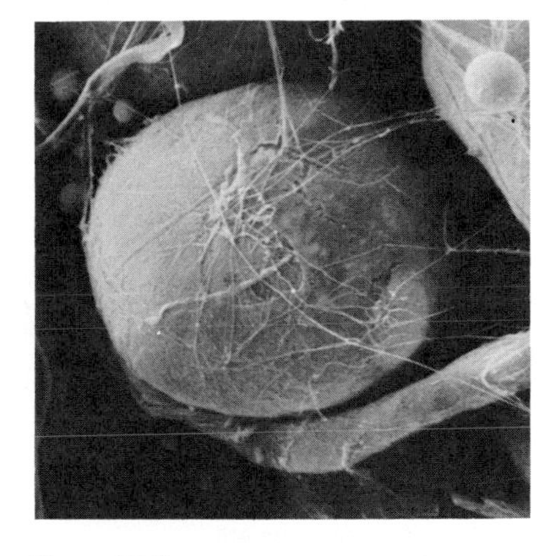

Figure 20-2
Scanning electron micrograph of a fat cell. [From *Tissues and Organs*, by Richard G. Kessel and Randy H. Kardon. Copyright © 1979. W. H. Freeman and Company.]

In mammals, the major site of accumulation of triacylglycerols is the cytoplasm of *adipose cells* (*fat cells*). Droplets of triacylglycerol coalesce to form a large globule, which may occupy most of the cell volume. Adipose cells are specialized for the synthesis and storage of triacylglycerols and for their mobilization into fuel molecules that are transported to other tissues by the blood.

TRIACYLGLYCEROLS ARE HYDROLYZED BY CYCLIC-AMP–REGULATED LIPASES

The initial event in the utilization of fat as an energy source is the hydrolysis of triacylglycerol by lipases.

$$\underset{\text{Triacylglycerol}}{R_2\text{—C(=O)—O—CH}(\text{CH}_2\text{—O—C(=O)—}R_1)(\text{CH}_2\text{—O—C(=O)—}R_3)} + 3\,H_2O \xrightarrow{\text{Lipases}} \underset{\text{Glycerol}}{\text{HO—CH}(\text{CH}_2\text{OH})_2} + \underset{\text{Fatty acids}}{R_1\text{—COO}^- + R_2\text{—COO}^- + R_3\text{—COO}^-} + 3\,H^+$$

The activity of adipose-cell lipase is regulated by hormones. Epinephrine, norepinephrine, glucagon, and adrenocorticotropic hormone stimulate the adenylate cyclase of adipose cells. The increased level of cyclic adenosine monophosphate (cyclic AMP) then stimulates a protein kinase, which activates the lipase by phosphorylating it. Thus, epinephrine, norepinephrine, glucagon, and adrenocorticotropic hormone cause lipolysis. *Cyclic AMP is a second messenger in the activation of lipolysis in adipose cells,* which is analogous to its role in the activation of glycogen breakdown (p. 462). In contrast, insulin inhibits lipolysis.

Glycerol formed by lipolysis is phosphorylated and oxidized to dihydroxyacetone phosphate, which in turn is isomerized to glyceraldehyde 3-phosphate. This intermediate is on both the glycolytic and the gluconeogenic pathways. Hence, glycerol can be converted into pyruvate or glucose in the liver, which contains the appropriate enzymes. The reverse process can occur by the reduction of dihydroxyacetone phosphate to glycerol 3-phosphate. Hydrolysis by a phosphatase then gives glycerol. Thus, glycerol and glycolytic intermediates are readily interconvertible.

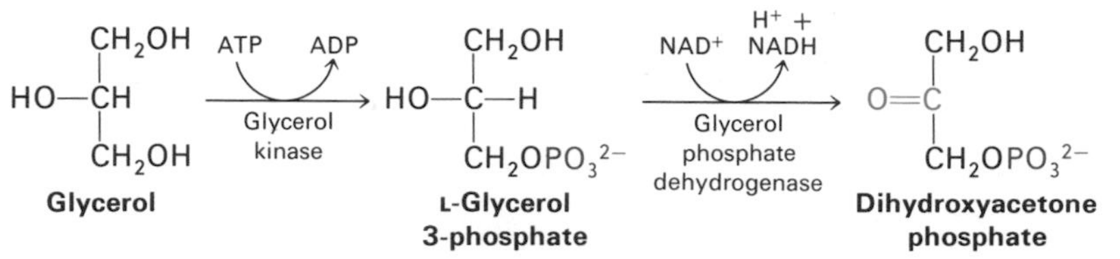

$$\underset{\textbf{Glycerol}}{HO{-}CH(CH_2OH)_2} \xrightarrow[\text{Glycerol kinase}]{ATP \rightarrow ADP} \underset{\textbf{L-Glycerol 3-phosphate}}{HO{-}C(H)(CH_2OH){-}CH_2OPO_3^{2-}} \xrightarrow[\text{Glycerol phosphate dehydrogenase}]{NAD^+ \rightarrow H^+ + NADH} \underset{\textbf{Dihydroxyacetone phosphate}}{O{=}C(CH_2OH){-}CH_2OPO_3^{2-}}$$

FATTY ACIDS ARE DEGRADED BY THE SEQUENTIAL REMOVAL OF TWO-CARBON UNITS

In 1904, Franz Knoop made a critical contribution to the elucidation of the mechanism of fatty acid oxidation. He fed dogs straight-chain fatty acids in which the ω-carbon atom was joined to a phenyl group. Knoop found that the urine of these dogs contained a derivative of phenylacetic acid when they were fed phenylbutyrate. In contrast, a derivative of benzoic acid was formed when they were fed phenylpropionate. In fact, phenylacetic acid was produced whenever a fatty acid containing an even number of carbon atoms was fed, whereas benzoic acid was formed whenever a fatty acid containing an odd number was fed (Figure 20-3). Knoop deduced from these findings that *fatty acids are degraded by oxidation at the β-carbon.* These experiments are a landmark in biochemistry because they were the first to use a synthetic label to elucidate reaction mechanisms. Deuterium and radioisotopes came into biochemistry several decades later.

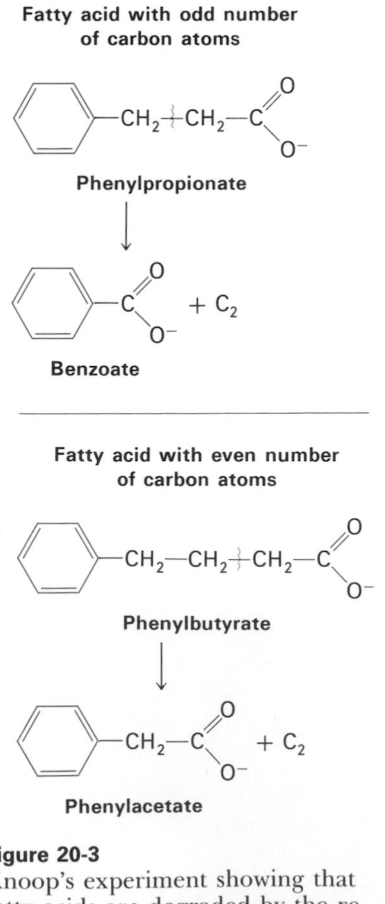

Fatty acid with odd number of carbon atoms

$$C_6H_5{-}CH_2{+}CH_2{-}COO^- \;(\textbf{Phenylpropionate}) \rightarrow C_6H_5{-}COO^- \;(\textbf{Benzoate}) + C_2$$

Fatty acid with even number of carbon atoms

$$C_6H_5{-}CH_2{-}CH_2{+}CH_2{-}COO^- \;(\textbf{Phenylbutyrate}) \rightarrow C_6H_5{-}CH_2{-}COO^- \;(\textbf{Phenylacetate}) + C_2$$

Figure 20-3
Knoop's experiment showing that fatty acids are degraded by the removal of two-carbon units.

FATTY ACIDS ARE LINKED TO COENZYME A BEFORE THEY ARE OXIDIZED

Eugene Kennedy and Albert Lehninger showed in 1949 that fatty acids are oxidized in mitochondria. Subsequent work demonstrated that they are activated before they enter the mitochondrial matrix. Adenosine triphosphate (ATP) drives the formation of a thioester linkage between the carboxyl group of a fatty acid and the sulfhydryl group of CoA. This activation reaction occurs on the outer mitochondrial membrane, where it is catalyzed by *acyl CoA synthetase* (also called fatty acid thiokinase).

$$R{-}C(=O){-}O^- + ATP + HS{-}CoA \rightleftharpoons R{-}\overset{O}{\overset{\|}{C}}{-}S{-}CoA + AMP + PP_i$$

Paul Berg showed that the activation of a fatty acid occurs in two steps. First, the fatty acid reacts with ATP to form an *acyl adenylate*. In this mixed anhydride, the carboxyl group of a fatty acid is bonded to the phosphoryl group of AMP. The other two phosphoryl groups of the ATP substrate are released as pyrophosphate. The sulfhydryl group of CoA then attacks the acyl adenylate, which is tightly bound to the enzyme, to form acyl CoA and AMP.

$$R{-}\overset{O}{\overset{\|}{C}}{-}O{-}\overset{O}{\overset{\|}{\underset{O^-}{\underset{|}{P}}}}{-}O{-}\text{Ribose}{-}\text{Adenine}$$

Acyl adenylate (Acyl-AMP)

$$\underset{\textbf{Fatty acid}}{R{-}C(=O){-}O^-} + ATP \rightleftharpoons \underset{\textbf{Acyl adenylate}}{R{-}\overset{O}{\overset{\|}{C}}{-}AMP} + PP_i$$

$$R{-}\overset{O}{\overset{\|}{C}}{-}AMP + HS{-}CoA \rightleftharpoons \underset{\textbf{Acyl CoA}}{R{-}\overset{O}{\overset{\|}{C}}{-}S{-}CoA} + AMP$$

These partial reactions are freely reversible. In fact, the equilibrium constant for the sum of these reactions is close to 1.

$$R{-}COO^- + CoA + ATP \rightleftharpoons \text{acyl CoA} + AMP + PP_i$$

One high-energy bond is broken (between PP_i and AMP) and one high-energy bond is formed (the thioester linkage in acyl CoA). How is this reaction driven forward? The answer is that pyrophosphate is rapidly hydrolyzed by a pyrophosphatase.

$$R{-}COO^- + CoA + ATP + H_2O \longrightarrow \text{acyl CoA} + AMP + 2\ P_i + 2\ H^+$$

This makes the overall reaction irreversible because two high-energy bonds are consumed, whereas only one is formed. We see here another example of a recurring theme in biochemistry: *many biosynthetic reactions are made irreversible by the hydrolysis of inorganic pyrophosphate.*

Another motif recurs in this activation reaction. The enzyme-bound acyl adenylate intermediate is not unique to the synthesis of acyl CoA. *Acyl adenylates are frequently formed when carboxyl groups are activated in biochemical reactions.* For example, amino acids are activated for protein synthesis by a similar mechanism (p. 734).

CARNITINE CARRIES LONG-CHAIN ACTIVATED FATTY ACIDS INTO THE MITOCHONDRIAL MATRIX

Fatty acids are activated on the outer mitochondrial membrane, whereas they are oxidized in the mitochondrial matrix. Long-chain acyl CoA molecules do not readily traverse the inner mitochondrial membrane, and so a special transport mechanism is needed. Activated long-chain fatty acids are carried across the inner mitochondrial membrane by *carnitine*, a zwitterionic compound formed from lysine. The acyl group is transferred from the sulfur atom of CoA to the hydroxyl group of carnitine to form *acyl carnitine*. This reaction is catalyzed by *carnitine acyltransferase I*, which is located on the cytosolic face of the inner mitochondrial membrane.

$$R-\overset{O}{\overset{\|}{C}}-S-CoA + H_3C-\overset{CH_3}{\underset{CH_3}{\overset{+}{N}}}-CH_2-\overset{H}{\underset{OH}{C}}-CH_2-C\begin{smallmatrix}O\\O^-\end{smallmatrix} \rightleftharpoons HS-CoA + H_3C-\overset{CH_3}{\underset{CH_3}{\overset{+}{N}}}-CH_2-\overset{H}{\underset{O-\overset{}{C}(=O)-R}{C}}-CH_2-C\begin{smallmatrix}O\\O^-\end{smallmatrix}$$

Acyl CoA **Carnitine** **Acyl carnitine**

Acyl carnitine is then shuttled across the inner mitochondrial membrane by a translocase (Figure 20-4). The acyl group is transferred back to CoA on the matrix side of the membrane. This reaction, which is catalyzed by carnitine acyltransferase II, is thermodynamically feasible because the *O*-acyl link in carnitine has a high group-transfer potential. Finally, carnitine is returned to the cytosolic side by the translocase, in exchange for an incoming acylcarnitine.

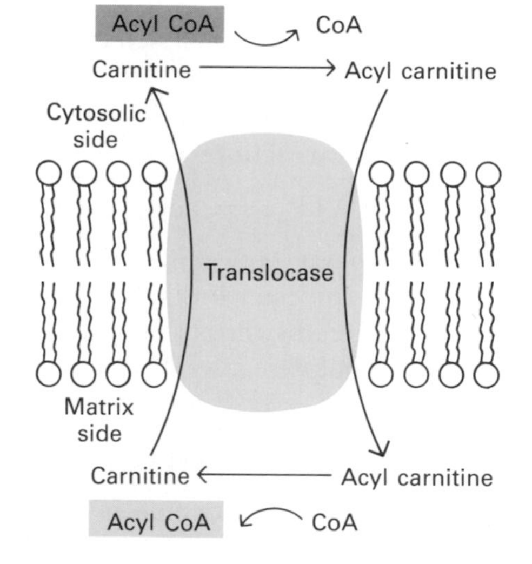

Figure 20-4
The entry of acyl carnitine into the mitochondrial matrix is mediated by a translocase. Carnitine returns to the cytosolic side of the inner mitochondrial membrane in exchange for acyl carnitine.

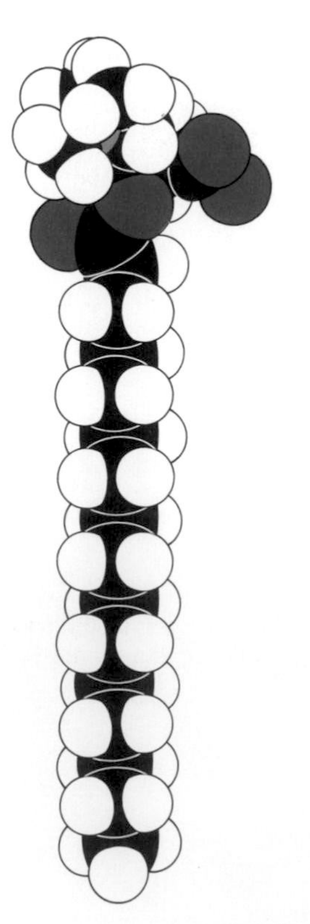

Molecular model of palmitoyl carnitine.

A defect in the transferase or translocase, or a deficiency of carnitine, might be expected to impair the oxidation of long-chain fatty acids. Such a disorder has in fact been found in identical twins who had had aching muscle cramps since early childhood. The aches were precipitated by fasting, exercise, or a high-fat diet; fatty acid oxidation is the major energy-yielding process in these three states. The enzymes of glycolysis and glycogenolysis were found to be normal. Lipolysis of triacylglycerols was normal, as evidenced by a rise in the concentration of unesterified fatty acids in the plasma after fasting. Assay of a muscle biopsy showed that the long-chain acyl CoA synthetase was fully active.

Furthermore, medium-chain (C_8 and C_{10}) fatty acids were normally metabolized. It is known that carnitine is not required for the permeation of medium-chain acyl CoAs into the mitochondrial matrix. This case report demonstrates in a striking way that *the impaired flow of a metabolite from one compartment of a cell to another can cause disease.*

ACETYL CoA, NADH, AND $FADH_2$ ARE GENERATED IN EACH ROUND OF FATTY ACID OXIDATION

A saturated acyl CoA is degraded by a recurring sequence of four reactions: oxidation by flavin adenine dinucleotide (FAD), hydration, oxidation by NAD^+, and thiolysis by CoA (Figure 20-5). The fatty acyl chain is shortened by two carbon atoms as a result of these reactions, and $FADH_2$, NADH, and acetyl CoA are generated. David Green, Severo Ochoa, and Feodor Lynen made important contributions to the elucidation of this series of reactions, which is called the *β-oxidation pathway*.

The first reaction in each round of degradation is the *oxidation* of acyl CoA by an acyl CoA dehydrogenase to give an enoyl CoA with a *trans* double bond between C-2 and C-3.

$$\text{Acyl CoA} + \text{E-FAD} \longrightarrow trans\text{-}\Delta^2\text{-enoyl CoA} + \text{E-FADH}_2$$

As in the dehydrogenation of succinate in the citric acid cycle, FAD rather than NAD^+ is the electron acceptor because the ΔG of this reaction is insufficient to drive the reduction of NAD^+. Electrons from the $FADH_2$ prosthetic group of the reduced acyl CoA dehydrogenase are transferred to a second flavoprotein called electron-transferring flavoprotein (ETF). In turn, ETF donates electrons to *ETF:ubiquinone reductase*, an iron-sulfur protein. Ubiquinone is thereby reduced to ubiquinol, which delivers its high-potential electrons to the second proton-pumping site of the respiratory chain (p. 404). Consequently, two ATP are generated from $FADH_2$ formed in this dehydrogenation step, as in the oxidation of succinate to fumarate.

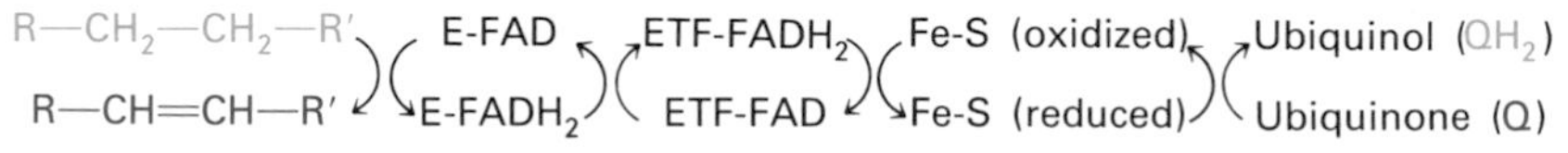

The next step is the *hydration* of the double bond between C-2 and C-3 by enoyl CoA hydratase.

$$trans\text{-}\Delta^2\text{-Enoyl CoA} + H_2O \rightleftharpoons \text{L-3-hydroxyacyl CoA}$$

The hydration of the enoyl CoA is stereospecific, as are the hydrations of fumarate and aconitate. Only the L-isomer of 3-hydroxyacyl CoA is formed when the *trans*-Δ^2 double bond is hydrated. The enzyme also hydrates a *cis*-Δ^2 double bond, but the product then is the D-isomer. We shall return to this point shortly in considering how unsaturated fatty acids are oxidized.

The hydration of enoyl CoA is the prelude to the second *oxidation* reaction, which converts the hydroxyl group at C-3 into a keto group and generates NADH. This oxidation is catalyzed by L-3-hydroxyacyl CoA dehydrogenase, which is absolutely specific for the L-isomer of the hydroxyacyl substrate.

$$\text{L-3-Hydroxyacyl CoA} + NAD^+ \rightleftharpoons \text{3-ketoacyl CoA} + \text{NADH} + H^+$$

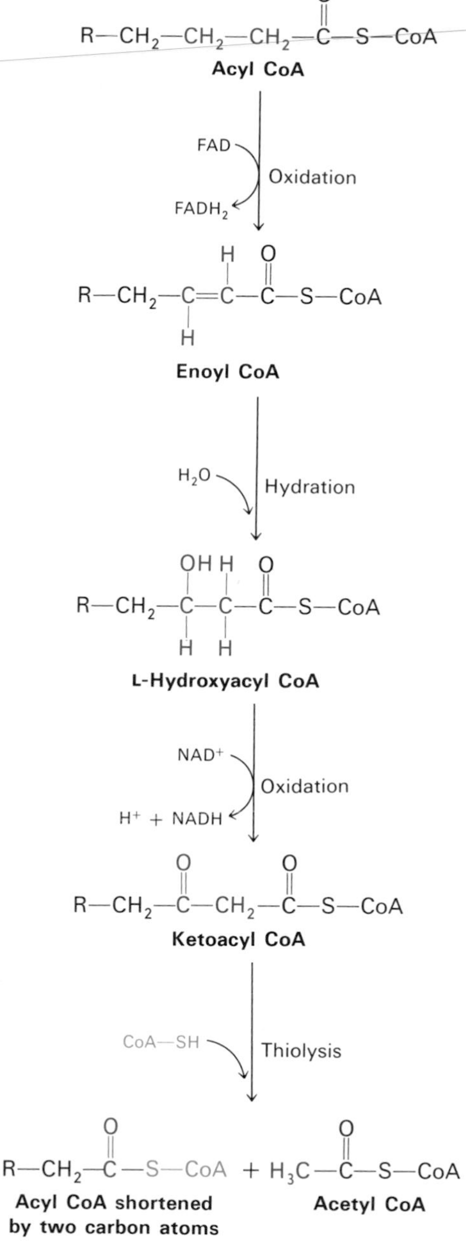

Figure 20-5
Reaction sequence in the degradation of fatty acids; oxidation, hydration, oxidation, and thiolysis.

These three reactions in each round of fatty acid degradation closely resemble the last steps in the citric acid cycle:

Acyl CoA	⟶	enol CoA	⟶	hydroxyacyl CoA	⟶	ketoacyl CoA
Succinate	⟶	fumarate	⟶	malate	⟶	oxaloacetate

The preceding reactions have oxidized the methylene group at C-3 to a keto group. The final step is the *cleavage* of 3-ketoacyl CoA by the thiol group of a second molecule of CoA, which yields acetyl CoA and an acyl CoA shortened by two carbon atoms. This thiolytic cleavage is catalyzed by β-ketothiolase.

$$\underset{(n \text{ carbons})}{\text{3-Ketoacyl CoA}} + \text{HS—CoA} \rightleftharpoons \text{acetyl CoA} + \underset{(n-2 \text{ carbons})}{\text{acyl CoA}}$$

The shortened acyl CoA then undergoes another cycle of oxidation, starting with the reaction catalyzed by acyl CoA dehydrogenase (Figure 20-6). β-Ketothiolase, hydroxyacyl dehydrogenase, and enoyl CoA hydratase have broad specificity with respect to the length of the acyl group.

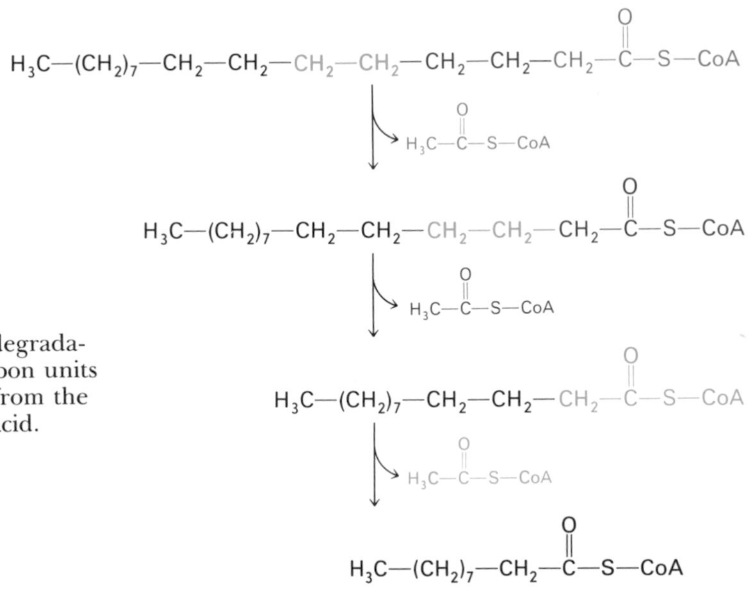

$$H_3C{-}(CH_2)_7{-}CH_2{-}CH_2{-}CH_2{-}CH_2{-}CH_2{-}CH_2{-}CH_2{-}\overset{O}{\overset{\|}{C}}{-}S{-}CoA$$

$$\downarrow \; \rightarrow H_3C{-}\overset{O}{\overset{\|}{C}}{-}S{-}CoA$$

$$H_3C{-}(CH_2)_7{-}CH_2{-}CH_2{-}CH_2{-}CH_2{-}CH_2{-}\overset{O}{\overset{\|}{C}}{-}S{-}CoA$$

$$\downarrow \; \rightarrow H_3C{-}\overset{O}{\overset{\|}{C}}{-}S{-}CoA$$

$$H_3C{-}(CH_2)_7{-}CH_2{-}CH_2{-}CH_2{-}\overset{O}{\overset{\|}{C}}{-}S{-}CoA$$

$$\downarrow \; \rightarrow H_3C{-}\overset{O}{\overset{\|}{C}}{-}S{-}CoA$$

$$H_3C{-}(CH_2)_7{-}CH_2{-}\overset{O}{\overset{\|}{C}}{-}S{-}CoA$$

Figure 20-6
First three rounds in the degradation of palmitate. Two-carbon units are sequentially removed from the carboxyl end of the fatty acid.

Table 20-2
Principal reactions in fatty acid oxidation

Step	*Reaction*	*Enzyme*
1	Fatty acid + CoA + ATP $\rightleftharpoons$ acyl CoA + AMP + PP_i	Acyl CoA synthetase (also called fatty acid thiokinase and fatty acid: CoA ligase [AMP])
2	Carnitine + acyl CoA $\rightleftharpoons$ acyl carnitine + CoA	Carnitine acyltransferase
3	Acyl CoA + E-FAD ⟶ *trans*-Δ^2-enoyl CoA + E-$FADH_2$	Acyl CoA dehydrogenases (several enzymes having different chain-length specificity)
4	*trans*-Δ^2-Enoyl CoA + H_2O $\rightleftharpoons$ L-3-hydroxyacyl CoA	Enoyl CoA hydratase (also called crotonase or 3-hydroxyacyl CoA hydrolyase)
5	L-3-Hydroxyacyl CoA + NAD^+ $\rightleftharpoons$ 3-ketoacyl CoA + NADH + H^+	L-3-Hydroxyacyl CoA dehydrogenase
6	3-Ketoacyl CoA + CoA $\rightleftharpoons$ acetyl CoA + acyl CoA (shortened by C_2)	β-Ketothiolase (also called thiolase)

THE COMPLETE OXIDATION OF PALMITATE YIELDS 129 ATP

We can calculate the energy yield derived from the oxidation of a fatty acid. In each reaction cycle, an acyl CoA is shortened by two carbons, and one $FADH_2$, NADH, and acetyl CoA are formed.

$$C_n\text{-acyl CoA} + \text{FAD} + \text{NAD}^+ + H_2O + \text{CoA} \longrightarrow C_{n-2}\text{-acyl CoA} + FADH_2 + \text{NADH} + \text{acetyl CoA} + H^+$$

The degradation of palmitoyl CoA (C_{16}-acyl CoA) requires seven reaction cycles. In the seventh cycle, the C_4-ketoacyl CoA is thiolyzed to two molecules of acetyl CoA. Hence, the stoichiometry of oxidation of palmitoyl CoA is

$$\text{Palmitoyl CoA} + 7\ \text{FAD} + 7\ \text{NAD}^+ + 7\ \text{CoA} + 7\ H_2O \longrightarrow 8\ \text{acetyl CoA} + 7\ FADH_2 + 7\ \text{NADH} + 7\ H^+$$

Three ATP are generated when each of these NADH is oxidized by the respiratory chain, whereas two ATP are formed for each $FADH_2$, because their electrons enter the chain at the level of ubiquinol. Recall that the oxidation of acetyl CoA by the citric acid cycle yields 12 ATP. Hence, the number of ATP formed in the oxidation of palmitoyl CoA is 14 from the 7 $FADH_2$, 21 from the 7 NADH, and 96 from the 8 molecules of acetyl CoA, which gives a total of 131. Two high-energy phosphate bonds are consumed in the activation of palmitate, in which ATP is split into AMP and 2 P_i. Thus, *the net yield from the complete oxidation of palmitate is 129 ATP*.

The efficiency of energy conservation in fatty acid oxidation can be estimated from the number of ATP formed and from the free energy of oxidation of palmitic acid to CO_2 and H_2O, as determined by calorimetry. The standard free energy of hydrolysis of 129 ATP is -940 kcal (129×-7.3 kcal). The standard free energy of oxidation of palmitic acid is -2340 kcal. Hence, *the efficiency of energy conservation in fatty acid oxidation under standard conditions is about 40%,* a value similar to those of glycolysis, the citric acid cycle, and oxidative phosphorylation.

AN ISOMERASE AND AN EPIMERASE ARE REQUIRED FOR THE OXIDATION OF UNSATURATED FATTY ACIDS

We turn now to the oxidation of unsaturated fatty acids. Many of the reactions are the same as those for saturated fatty acids. In fact, only two additional enzymes—an isomerase and an epimerase—are needed to degrade a wide range of unsaturated fatty acids.

Consider the oxidation of palmitoleate. This C_{16} unsaturated fatty acid, which has one double bond between C-9 and C-10, is activated and transported across the inner mitochondrial membrane in the same way as saturated fatty acids. Palmitoleoyl CoA then undergoes three cycles of degradation, which are carried out by the same enzymes as in the oxidation of saturated fatty acids. However, the *cis*-Δ^3-enoyl CoA formed in the third round is not a substrate for acyl CoA dehydrogenase. The presence of a double bond between C-3 and C-4 prevents the formation of another double bond between C-2 and C-3. This impasse is resolved by a new reaction that shifts the position and configuration of the *cis*-Δ^3 double bond. *An isomerase converts this double bond into a* trans-Δ^2 *double bond.* The subsequent reactions are those of the saturated fatty acid oxidation pathway, in which the *trans*-Δ^2-enoyl CoA is a regular substrate.

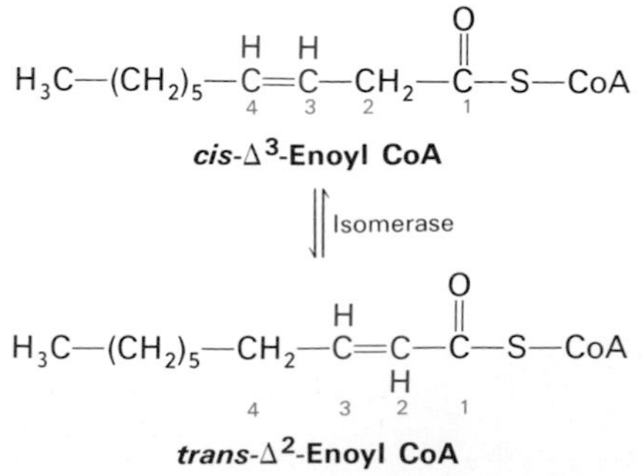

$H_3C{-}(CH_2)_5{-}CH{=}CH{-}CH_2{-}C({=}O){-}S{-}CoA$ (carbons numbered 4, 3, 2, 1 from the double bond to the carbonyl)

***cis*-Δ³-Enoyl CoA**

⇅ Isomerase

$H_3C{-}(CH_2)_5{-}CH_2{-}CH{=}CH{-}C({=}O){-}S{-}CoA$ (carbons numbered 4, 3, 2, 1)

***trans*-Δ²-Enoyl CoA**

A second accessory enzyme is needed for the oxidation of polyunsaturated fatty acids. The activated C_{18} unsaturated fatty acid with *cis*-Δ^6 and *cis*-Δ^9 double bonds undergoes two rounds of degradation by the saturated fatty acid oxidation pathway. Then *cis*-Δ^2,Δ^5-enoyl CoA is hydrated by enoyl CoA hydratase, the same enzyme that hydrates *trans*-Δ^2 double bonds in the saturated pathway. However, hydration of a *cis*-Δ^2 double bond yields the D-isomer of 3-hydroxyacyl CoA, which is not a substrate for L-3-hydroxyacyl CoA dehydrogenase. This hurdle is overcome by an *epimerase that inverts the configuration of the hydroxyl group at C-3*.

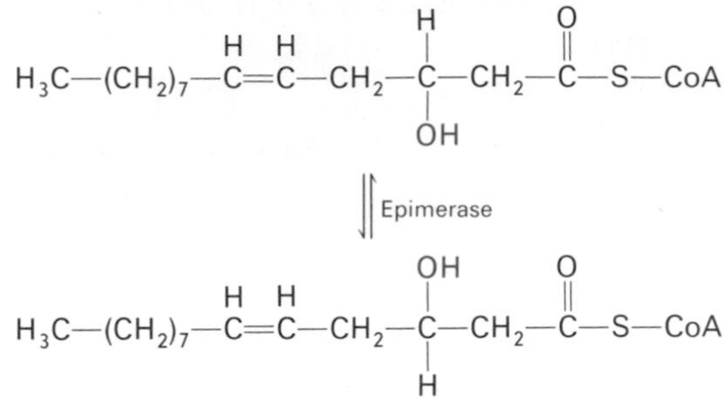

$$H_3C{-}(CH_2)_7{-}CH{=}CH{-}CH_2{-}C(H)(OH){-}CH_2{-}C({=}O){-}S{-}CoA$$

D-3-Hydroxy-*cis*-Δ⁵-enoyl CoA

Epimerase ⇅

$$H_3C{-}(CH_2)_7{-}CH{=}CH{-}CH_2{-}C(OH)(H){-}CH_2{-}C({=}O){-}S{-}CoA$$

L-3-Hydroxy-*cis*-Δ⁵-enoyl CoA

ODD-CHAIN FATTY ACIDS YIELD PROPIONYL COENZYME A IN THE FINAL THIOLYSIS STEP

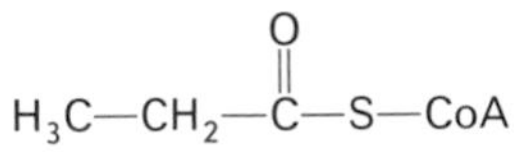

$$H_3C{-}CH_2{-}C({=}O){-}S{-}CoA$$

Propionyl CoA

Fatty acids having an odd number of carbon atoms are minor species. They are oxidized in the same way as fatty acids having an even number, except that propionyl CoA and acetyl CoA, rather than two molecules of acetyl CoA, are produced in the final round of degradation. The activated three-carbon unit in propionyl CoA enters the citric acid cycle after it is converted into succinyl CoA. The pathway from propionyl CoA to succinyl CoA will be discussed in the next chapter (p. 506) because propionyl CoA is also formed in the oxidation of several amino acids.

KETONE BODIES ARE FORMED IN THE LIVER FROM ACETYL COENZYME A IF FAT BREAKDOWN PREDOMINATES

The molecular basis of the adage that *fats burn in the flame of carbohydrates* is now evident. The acetyl CoA formed in fatty acid oxidation enters the citric acid cycle only if fat and carbohydrate degradation are appropriately balanced. However, if fat breakdown predominates, acetyl CoA in the liver undergoes a different fate. The reason is that the entry of acetyl CoA into the citric acid cycle depends on the availability of oxaloacetate for the formation of citrate, but the concentration of oxaloacetate is lowered if carbohydrate is unavailable or improperly utilized. In fasting or in diabetes, oxaloacetate is used to form glucose and is thus unavailable for condensation with acetyl CoA. Under these conditions, acetyl CoA is diverted to the formation of acetoacetate and D-3-hydroxybutyrate. Acetoacetate, D-3-hydroxybutyrate, and acetone are sometimes referred to as *ketone bodies*.

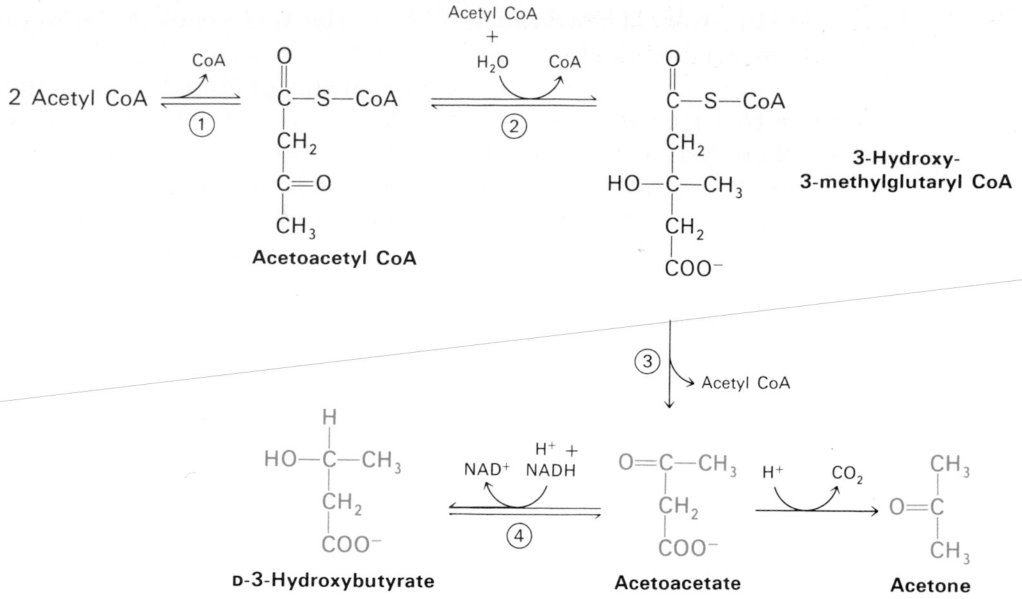

2 Acetyl CoA ⇌ (1, −CoA) Acetoacetyl CoA ⇌ (2, +Acetyl CoA + H_2O, −CoA) 3-Hydroxy-3-methylglutaryl CoA → (3, −Acetyl CoA) Acetoacetate

Acetoacetate ⇌ (4, NADH + H^+ → NAD^+) D-3-Hydroxybutyrate

Acetoacetate → (H^+, −CO_2) Acetone

Figure 20-7
Formation of acetoacetate, D-3-hydroxybutyrate, and acetone from acetyl CoA in the liver. Enzymes catalyzing these reactions are: (1) 3-ketothiolase, (2) hydroxymethylglutaryl CoA synthetase, (3) hydroxymethylglutaryl CoA cleavage enzyme, and (4) D-3-hydroxybutyrate dehydrogenase. Acetoacetate spontaneously decarboxylates to form acetone.

Acetoacetate is formed from acetyl CoA in three steps (Figure 20-7). Two molecules of acetyl CoA condense to form acetoacetyl CoA. This reaction, which is catalyzed by thiolase, is a reversal of the thiolysis step in the oxidation of fatty acids. Acetoacetyl CoA then reacts with acetyl CoA and water to give 3-hydroxy-3-methylglutaryl CoA (HMG-CoA) and CoA. The unfavorable equilibrium in the formation of acetoacetyl CoA is compensated for by this reaction, which has a favorable equilibrium due to the hydrolysis of a thioester linkage. 3-Hydroxy-3-methylglutaryl CoA is then cleaved to acetyl CoA and acetoacetate. The sum of these reactions is

$$\text{2 Acetyl CoA} + H_2O \longrightarrow \text{acetoacetate} + \text{2 CoA} + H^+$$

3-Hydroxybutyrate is formed by the reduction of acetoacetate in the mitochondrial matrix. The ratio of hydroxybutyrate to acetoacetate depends on the NADH/NAD^+ ratio inside mitochondria. Because it is a β-keto acid, acetoacetate also undergoes a slow, spontaneous decarboxylation to acetone. The odor of acetone may be detected in the breath of a person who has a high level of acetoacetate in the blood.

ACETOACETATE IS A MAJOR FUEL IN SOME TISSUES

The major site of production of acetoacetate and 3-hydroxybutyrate is the liver. These substances diffuse from the liver mitochondria into the blood and are transported to peripheral tissues. Until a few years ago, these ketone bodies were regarded as degradation products of little physiological value. However, the studies of George Cahill and others have revealed that these derivatives of acetyl CoA are important molecules in energy metabolism. *Acetoacetate and 3-hydroxybutyrate are normal fuels of respiration and are quantitatively important as sources of energy*. Indeed, heart muscle and the renal cortex use acetoacetate in preference to glucose. In contrast, glucose is the major fuel for the brain and red blood cells in well-nourished persons on a balanced diet. However, the brain adapts to the utilization of acetoacetate during starvation and

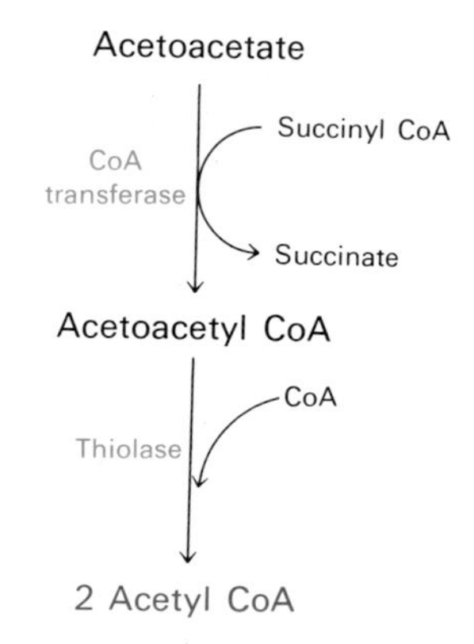

Figure 20-8
Utilization of acetoacetate as a fuel. Acetoacetate can be converted into two molecules of acetyl CoA, which then enter the citric acid cycle.

diabetes. In prolonged starvation, 75% of the fuel needs of the brain are met by acetoacetate.

Acetoacetate can be activated by the transfer of CoA from succinyl CoA in a reaction catalyzed by a specific CoA transferase. Acetoacetyl CoA is then cleaved by thiolase to yield two molecules of acetyl CoA, which can then enter the citric acid cycle. The liver can supply acetoacetate to other organs because it lacks this particular CoA transferase.

Acetoacetate can be regarded as a water-soluble, transportable form of acetyl units. Fatty acids are released by adipose tissue and converted into acetyl units by the liver, which then exports them as acetoacetate. As might be expected, acetoacetate also has a regulatory role. *High levels of acetoacetate in the blood signify an abundance of acetyl units and lead to a decrease in the rate of lipolysis in adipose tissue.*

ANIMALS CANNOT CONVERT FATTY ACIDS INTO GLUCOSE

It is important to note that *animals are unable to convert fatty acids into glucose.* Specifically, acetyl CoA cannot be converted into pyruvate or oxaloacetate in animals. The two carbon atoms of the acetyl group of acetyl CoA enter the citric acid cycle, but two carbon atoms leave the cycle in the decarboxylations catalyzed by isocitrate dehydrogenase and α-ketoglutarate dehydrogenase. Consequently, oxaloacetate is regenerated but it is not formed de novo when the acetyl unit of acetyl CoA is oxidized by the citric acid cycle. In contrast, plants have two additional enzymes enabling them to convert the carbon atoms of acetyl CoA into oxaloacetate (p. 388).

FATTY ACIDS ARE SYNTHESIZED AND DEGRADED BY DIFFERENT PATHWAYS

Fatty acid synthesis is not merely a reversal of the degradative pathway. Rather, it consists of a new set of reactions, which once again exemplifies the principle that *synthetic and degradative pathways in biological systems are usually distinct.* Some important features of the pathway for the biosynthesis of fatty acids are:

1. Synthesis takes place in the *cytosol,* in contrast with degradation, which occurs in the mitochondrial matrix.

2. Intermediates in fatty acid synthesis are covalently linked to the sulfhydryl groups of an *acyl carrier protein* (ACP), whereas intermediates in fatty acid breakdown are bonded to coenzyme A.

3. The enzymes of fatty acid synthesis in higher organisms are joined in a *single polypeptide chain* called *fatty acid synthase.* In contrast, the degradative enzymes do not seem to be associated.

4. The growing fatty acid chain is elongated by the *sequential addition of two-carbon units* derived from acetyl CoA. The activated donor of two-carbon units in the elongation step is *malonyl-ACP.* The elongation reaction is driven by the release of CO_2.

5. The reductant in fatty acid synthesis is *NADPH.*

6. Elongation by the fatty acid synthase complex stops upon formation of *palmitate* (C_{16}). Further elongation and the insertion of double bonds are carried out by other enzyme systems.

THE FORMATION OF MALONYL COENZYME A IS THE COMMITTED STEP IN FATTY ACID SYNTHESIS

Salih Wakil's finding that bicarbonate is required for fatty acid biosynthesis was an important clue in the elucidation of this process. In fact, fatty acid synthesis starts with the carboxylation of acetyl CoA to *malonyl CoA*. This irreversible reaction is the committed step in fatty acid synthesis.

$$H_3C{-}\overset{\overset{\large O}{\|}}{C}{-}S{-}CoA + ATP + HCO_3^- \longrightarrow {}^-O(O{=})C{-}CH_2{-}\overset{\overset{\large O}{\|}}{C}{-}S{-}CoA + ADP + P_i + H^+$$

Acetyl CoA **Malonyl CoA**

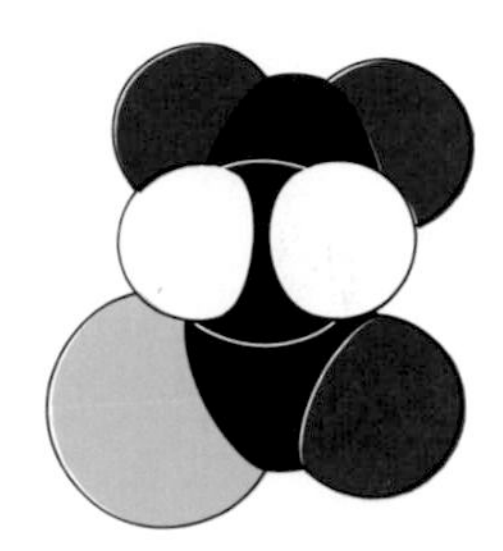

Molecular model of the malonyl thioester unit of malonyl CoA.

The synthesis of malonyl CoA is catalyzed by *acetyl CoA carboxylase*, which contains a biotin prosthetic group. The carboxyl group of biotin is covalently attached to the ϵ-amino group of a lysine residue, as in pyruvate carboxylase (p. 440). Another similarity between acetyl CoA carboxylase and pyruvate carboxylase is that acetyl CoA is carboxylated in two stages. First, a carboxybiotin intermediate is formed at the expense of an ATP. The activated CO_2 group in this intermediate is then transferred to acetyl CoA to form malonyl CoA.

$$\text{Biotin-enzyme} + \text{ATP} + HCO_3^- \rightleftharpoons CO_2 \sim \text{biotin-enzyme} + \text{ADP} + P_i$$

$$CO_2 \sim \text{biotin-enzyme} + \text{acetyl CoA} \longrightarrow \text{malonyl CoA} + \text{biotin-enzyme}$$

Substrates are bound to this enzyme and products are released in a specific sequence (Figure 20-9). Acetyl CoA carboxylase exemplifies a *ping-pong reaction mechanism* in which one or more products are released before all the substrates are bound.

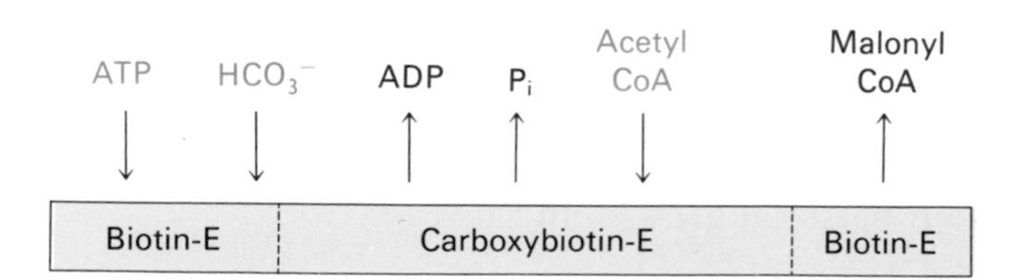

Figure 20-9
Reaction sequence of acetyl CoA carboxylase.

Acetyl CoA carboxylase from *E. coli* has been separated into subunits that catalyze partial reactions. Biotin is covalently attached to a small protein (22 kd) called the *biotin carboxyl carrier protein*. The carboxylation of the biotin unit in this carrier protein is catalyzed by *biotin carboxylase*, a second subunit. The third component of the system is a *transcarboxylase*, which catalyzes the transfer of the activated CO_2 unit from carboxybiotin to acetyl CoA. The length and flexibility of the link between biotin and its carrier protein enable the activated carboxyl group to move from one active site to another in the enzyme complex (Figure 20-10), as in pyruvate carboxylase (p. 441).

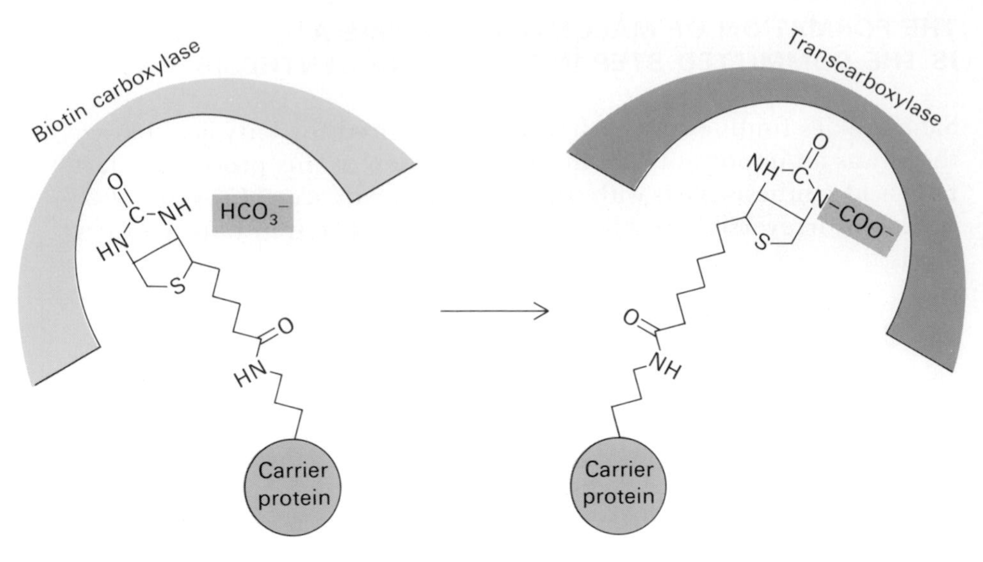

Figure 20-10
Schematic diagram showing the proposed movement of the biotin prosthetic group from the site where it acquires a carboxyl group from HCO_3^- to the site where it donates this group to acetyl CoA.

In eucaryotes, acetyl CoA carboxylase exists as an enzymatically inactive protomer (450 kd) or as an active, filamentous polymer (Figure 20-11). This interconversion is allosterically regulated, as might be expected because acetyl CoA carboxylase catalyzes the first committed step in fatty acid synthesis. *The key allosteric activator is citrate,* which shifts the enzyme equilibrium to the active filamentous form. The orientation of biotin with respect to its substrates is optimized in the filamentous form. In contrast, palmitoyl CoA shifts the equilibrium to the inactive protomer form. Thus, *palmitoyl CoA, the end product, inhibits the first committed step in the synthesis of fatty acids.* Acetyl CoA carboxylase in *E. coli* is controlled quite differently from the enzyme in eucaryotes. In bacteria, fatty acids are primarily precursors of phospholipids rather than of storage fuels, and so they are controlled differently. Citrate has no effect on the *E. coli* enzyme. Rather, the activity of its transcarboxylase component is regulated by guanine nucleotides that coordinate fatty acid synthesis with bacterial growth and division.

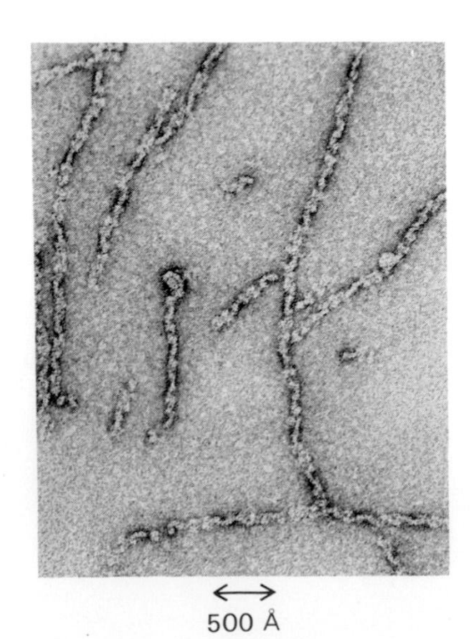

500 Å

Figure 20-11
Electron micrograph of the enzymatically active filamentous form of acetyl CoA carboxylase from chicken liver. [Courtesy of Dr. M. Daniel Lane.]

INTERMEDIATES IN FATTY ACID SYNTHESIS ARE ATTACHED TO AN ACYL CARRIER PROTEIN (ACP)

P. Roy Vagelos discovered that the intermediates in fatty acid synthesis in *E. coli* are linked to an acyl carrier protein. Specifically, they are linked to the sulfhydryl terminus of a phosphopantetheine group (Figure 20-12). In the degradation of fatty acids, this unit is part of CoA, whereas, in synthesis, it is attached to a serine residue of the ACP. This single polypeptide chain of 77 residues can be regarded as a giant prosthetic group, a "macro CoA."

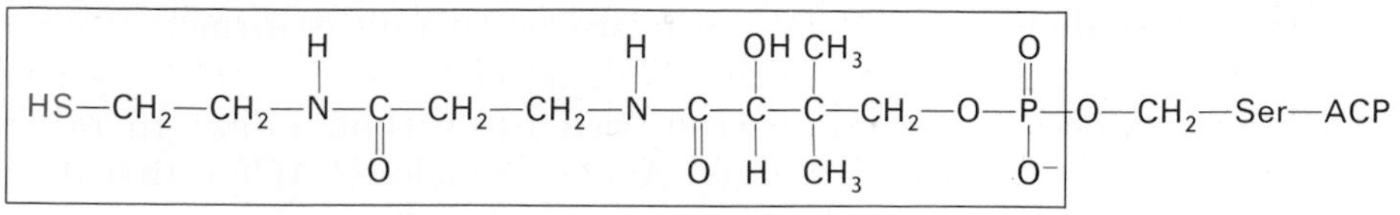

$HS{-}CH_2{-}CH_2{-}NH{-}CO{-}CH_2{-}CH_2{-}NH{-}CO{-}CH(OH){-}C(CH_3)_2{-}CH_2{-}O{-}PO_2^-{-}O{-}CH_2{-}Ser{-}ACP$

Phosphopantetheine prosthetic group of ACP

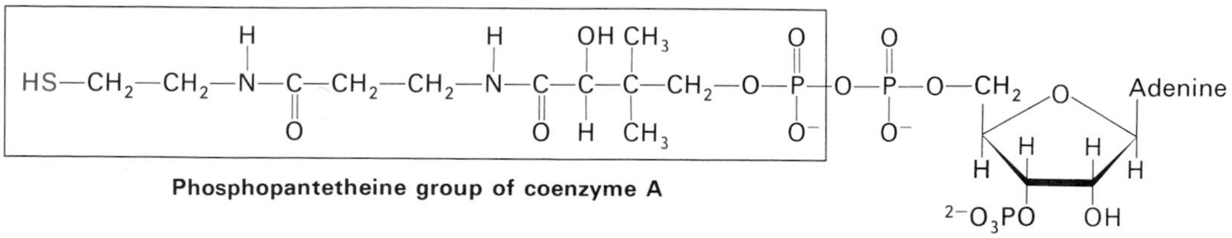

$HS{-}CH_2{-}CH_2{-}NH{-}CO{-}CH_2{-}CH_2{-}NH{-}CO{-}CH(OH){-}C(CH_3)_2{-}CH_2{-}O{-}PO_2^-{-}O{-}PO_2^-{-}O{-}CH_2$ (ribose bearing Adenine, OH, and $^{2-}O_3PO$)

Phosphopantetheine group of coenzyme A

Figure 20-12
Phosphopantetheine is the reactive unit of ACP and CoA.

THE ELONGATION CYCLE IN FATTY ACID SYNTHESIS

The enzyme system that catalyzes the synthesis of saturated long-chain fatty acids from acetyl CoA, malonyl CoA, and NADPH is called the *fatty acid synthase*. The constituent enzymes of bacterial fatty acid synthases are dissociated when the cells are disrupted. The availability of these isolated enzymes has facilitated the elucidation of the steps in fatty acid synthesis (Table 20-3). In fact, the reactions leading to fatty acid synthesis in higher organisms are very much like those of bacteria.

Table 20-3
Principal reactions in fatty acid synthesis

Step	*Reaction*	*Enzyme*
1	Acetyl CoA + HCO_3^- + ATP $\longrightarrow$ malonyl CoA + ADP + P_i + H^+	Acetyl CoA carboxylase
2	Acetyl CoA + ACP $\rightleftharpoons$ acetyl-ACP + CoA	Acetyl transacylase
3	Malonyl CoA + ACP $\rightleftharpoons$ malonyl-ACP + CoA	Malonyl transacylase
4	Acetyl-ACP + malonyl-ACP $\longrightarrow$ acetoacetyl-ACP + ACP + CO_2	Acyl-malonyl-ACP condensing enzyme
5	Acetoacetyl-ACP + NADPH + H^+ $\rightleftharpoons$ D-3-hydroxybutyryl-ACP + $NADP^+$	β-Ketoacyl-ACP-reductase
6	D-3-Hydroxybutyryl-ACP $\rightleftharpoons$ crotonyl-ACP + H_2O	3-Hydroxyacyl-ACP-dehydratase
7	Crotonyl-ACP + NADPH + H^+ $\longrightarrow$ butyryl-ACP + $NADP^+$	Enoyl-ACP reductase

The elongation phase of fatty acid synthesis starts with the formation of acetyl-ACP and malonyl-ACP. *Acetyl transacylase* and *malonyl transacylase* catalyze these reactions.

$$\text{Acetyl CoA} + \text{ACP} \rightleftharpoons \text{acetyl-ACP} + \text{CoA}$$

$$\text{Malonyl CoA} + \text{ACP} \rightleftharpoons \text{malonyl-ACP} + \text{CoA}$$

Malonyl transacylase is highly specific, whereas acetyl transacylase can transfer acyl groups other than the acetyl unit, though at a much slower rate. Fatty acids with an odd number of carbon atoms are synthesized starting with propionyl-ACP, which is formed from propionyl CoA by acetyl transacylase.

Acetyl-ACP and malonyl-ACP react to form acetoacetyl-ACP. This condensation reaction is catalyzed by the *acyl-malonyl-ACP condensing enzyme*.

$$\text{Acetyl-ACP} + \text{malonyl-ACP} \longrightarrow \text{acetoacetyl-ACP} + \text{ACP} + CO_2$$

In the condensation reaction, a four-carbon unit is formed from a two-carbon unit and a three-carbon unit, and CO_2 is released. Why isn't the four-carbon unit formed from two two-carbon units? In other words, why are the reactants acetyl-ACP and malonyl-ACP rather than two molecules of acetyl-ACP? The answer is that the equilibrium for the synthesis of acetoacetyl-ACP from two molecules of acetyl-ACP is highly unfavorable. In contrast, *the equilibrium is favorable if malonyl-ACP is a reactant because its decarboxylation contributes a substantial decrease in free energy*. In effect, the condensation reaction is driven by ATP, though ATP does not directly participate in the condensation reaction. Rather, ATP is used to carboxylate acetyl CoA to malonyl CoA. The free energy thus stored in malonyl CoA is released in the decarboxylation accompanying the formation of acetoacetyl-ACP. Although HCO_3^- is required for fatty acid synthesis, its carbon atom does not appear in the product. *Rather, all of the carbon atoms of fatty acids containing an even number are derived from acetyl CoA.*

The next three steps in fatty acid synthesis reduce the keto group at C-3 to a methylene group (Figure 20-13). First, acetoacetyl-ACP is reduced to D-3-hydroxybutyryl-ACP. This reaction differs from the corresponding one in fatty acid degradation in two respects: (1) the D- rather than the L-epimer is formed; and (2) NADPH is the reducing agent, whereas NAD^+ is the oxidizing agent in β-oxidation. This difference exemplifies the general principle that *NADPH is consumed in biosynthetic reactions, whereas NADH is generated in energy-yielding reactions*. Then D-3-hydroxybutyryl-ACP is *dehydrated* to form crotonyl-ACP, which is a

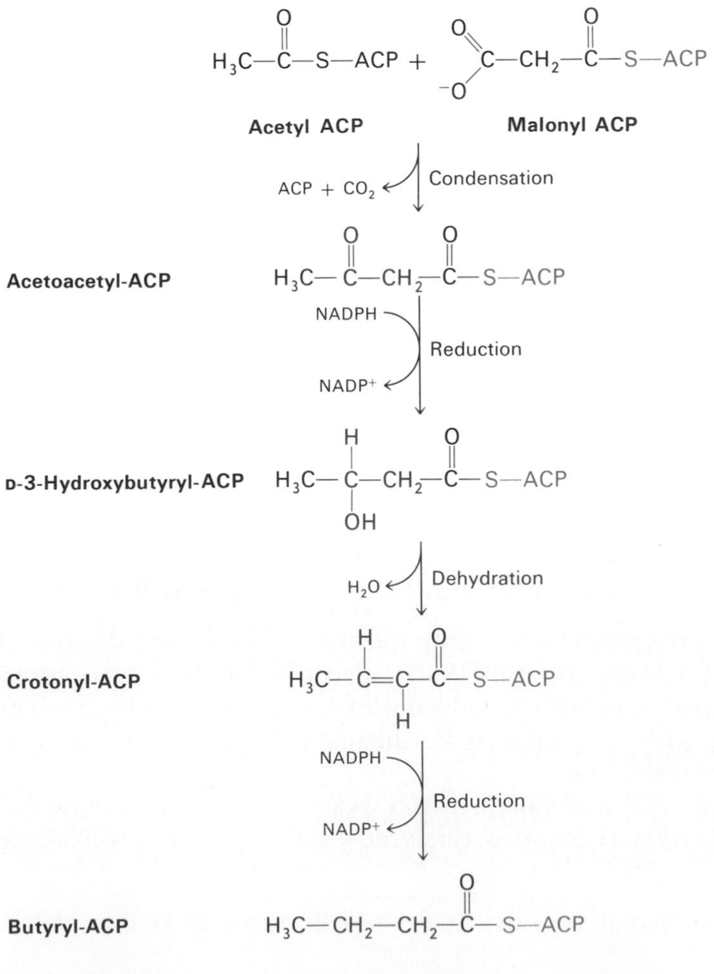

Acetyl ACP: $H_3C-C(=O)-S-ACP$ + Malonyl ACP: $^-O(O=)C-CH_2-C(=O)-S-ACP$

Condensation ($\rightarrow$ ACP + CO_2)

Acetoacetyl-ACP: $H_3C-C(=O)-CH_2-C(=O)-S-ACP$

Reduction (NADPH $\rightarrow$ $NADP^+$)

D-3-Hydroxybutyryl-ACP: $H_3C-CH(OH)-CH_2-C(=O)-S-ACP$

Dehydration ($\rightarrow$ H_2O)

Crotonyl-ACP: $H_3C-CH=CH-C(=O)-S-ACP$

Reduction (NADPH $\rightarrow$ $NADP^+$)

Butyryl-ACP: $H_3C-CH_2-CH_2-C(=O)-S-ACP$

Figure 20-13
Reaction sequence in the synthesis of fatty acids in *E. coli*: condensation, reduction, dehydration, and reduction. The intermediates shown here are produced in the first round of synthesis.

trans-Δ^2-enoyl-ACP. The final step in the cycle *reduces* crotonyl-ACP to butyryl-ACP. NADPH is again the reductant, whereas FAD is the oxidant in the corresponding reaction in β-oxidation. These last three reactions—a reduction, a dehydration, and a second reduction—convert acetoacetyl-ACP into butyryl-ACP, which completes the first elongation cycle.

In the second round of fatty acid synthesis, butyryl-ACP condenses with malonyl-ACP to form a C_6-β-ketoacyl-ACP. This reaction is like the one in the first round, in which acetyl-ACP condenses with malonyl-ACP to form a C_4-β-ketoacyl-ACP. Reduction, dehydration, and a second reduction convert the C_6-β-ketoacyl-ACP into a C_6-acyl-ACP, which is ready for a third round of elongation. The elongation cycles continue until C_{16}-acyl-ACP is formed. This intermediate is not a substrate for the condensing enzyme. Rather, it is hydrolyzed to yield palmitate and ACP.

STOICHIOMETRY OF FATTY ACID SYNTHESIS

The stoichiometry of the synthesis of palmitate is

$$\text{Acetyl CoA} + 7\ \text{malonyl CoA} + 14\ \text{NADPH} + 7\ \text{H}^+ \longrightarrow \text{palmitate} + 7\ \text{CO}_2 + 14\ \text{NADP}^+ + 8\ \text{CoA} + 6\ \text{H}_2\text{O}$$

The equation for the synthesis of the malonyl CoA used in the above reaction is

$$7\ \text{Acetyl CoA} + 7\ \text{CO}_2 + 7\ \text{ATP} \longrightarrow 7\ \text{malonyl CoA} + 7\ \text{ADP} + 7\ \text{P}_i + 7\ \text{H}^+$$

Hence, the overall stoichiometry for the synthesis of palmitate is

$$8\ \text{Acetyl CoA} + 7\ \text{ATP} + 14\ \text{NADPH} \longrightarrow \text{palmitate} + 14\ \text{NADP}^+ + 8\ \text{CoA} + 6\ \text{H}_2\text{O} + 7\ \text{ADP} + 7\ \text{P}_i$$

FATTY ACIDS ARE SYNTHESIZED IN EUCARYOTES BY A MULTIFUNCTIONAL ENZYME COMPLEX

The fatty acid synthases of eucaryotes, in contrast with those of *E. coli*, are well-defined *multienzyme complexes*. The one from yeast has a mass of 2200 kd and appears in electron micrographs as an ellipsoid with a length of 250 Å and a cross-sectional diameter of 210 Å (Figure 20-14).

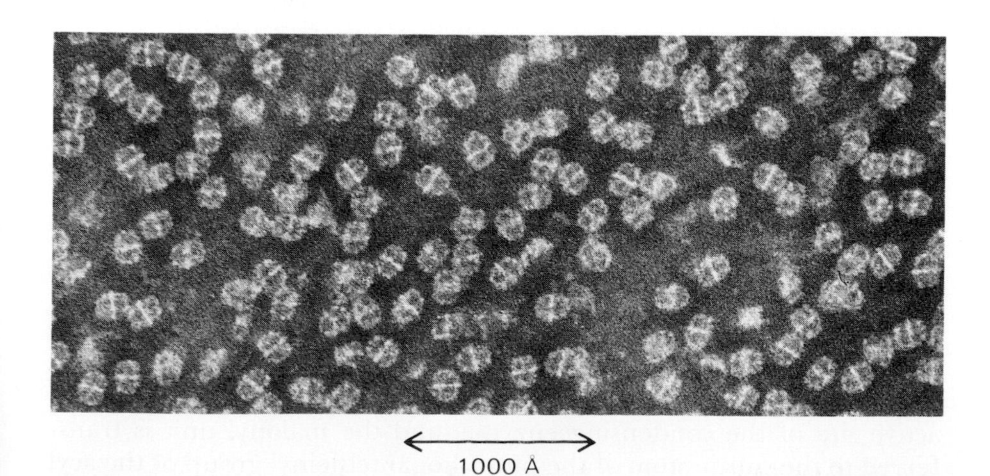

Figure 20-14
Electron micrograph of the fatty acid synthase complex from yeast. [Courtesy of Dr. Felix Wieland and Dr. Elmar A. Siess.]

It consists of just two kinds of polypeptide chains, and it has the subunit composition $\alpha_6\beta_6$. The α chain (185 kd) contains the acyl carrier protein, the condensing enzyme, and the β-ketoacyl reductase, whereas the β chain (175 kd) contains acetyl transacylase, malonyl transacylase, β-hydroxyacyl dehydratase, and enoyl reductase.

Mammalian fatty acid synthase is a dimer of identical subunits. Each chain is folded into three domains joined by flexible regions, as shown by proteolytic dissection experiments (Figure 20-15). *Domain 1, the substrate entry and condensation unit,* contains acetyl transferase, malonyl transferase, and β-ketoacyl synthase (condensing enzyme). *Domain 2, the reduction unit,* contains the acyl carrier protein, β-ketoacyl reductase, dehydratase, and enoyl reductase. *Domain 3, the palmitate release unit,* contains the thioesterase. It is interesting to note that many eucaryotic multienzyme complexes are multifunctional proteins, in which different enzymes are linked covalently in a single polypeptide chain. An advantage of this arrangement is that the synthesis of different enzymes is coordinated. Furthermore, a multienzyme complex consisting of covalently joined enzymes is more stable than one formed by noncovalent attractions. It seems likely that multifunctional enzymes such as fatty acid synthase arose in evolution by exon shuffling (p. 112).

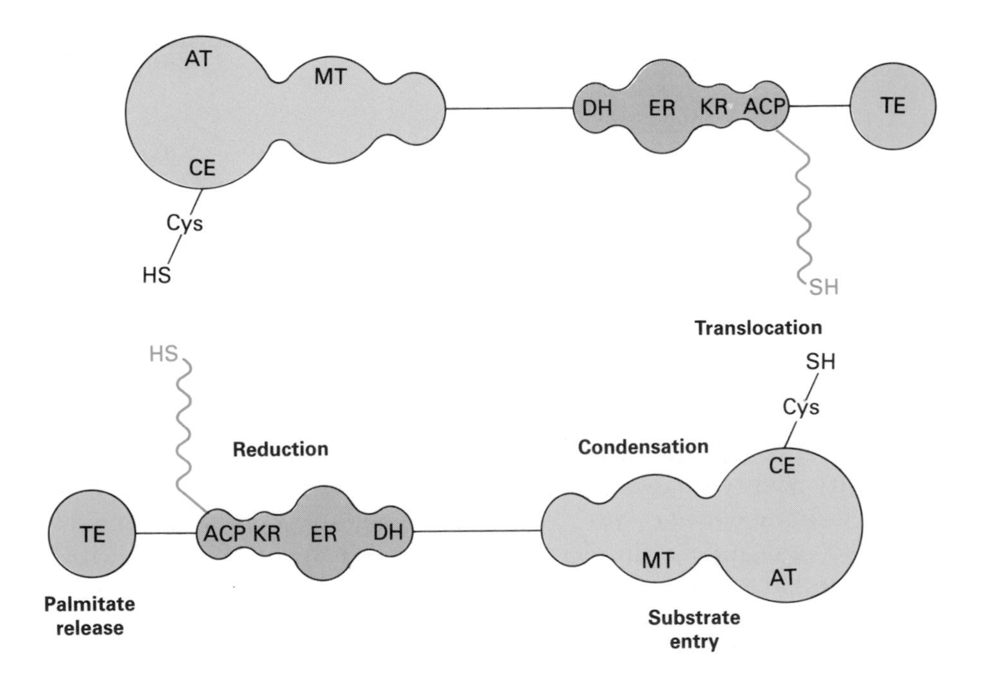

Figure 20-15
Schematic diagram of animal fatty acid synthase. The identical chains each contain three domains. Domain 1 (blue) contains acetyl transferase (AT), malonyl transferase (MT), and condensing enzyme (CE). Domain 2 (red) contains acyl carrier protein (ACP), β-ketoacyl reductase (KR), dehydratase (DH), and enoyl reductase (ER). Domain 3 (yellow) contains thioesterase (TE). The flexible phosphopantetheinyl group that carries the fatty acyl chain from one catalytic site to another is shown in green. [After Y. Tsukamoto, H. Wong, J. S. Mattick, and S. J. Wakil. *J. Biol. Chem.* 258(1983):15312.]

THE FLEXIBLE PHOSPHOPANTETHEINYL UNIT OF ACP CARRIES SUBSTRATE FROM ONE ACTIVE SITE TO ANOTHER

Fatty acid synthesis in animals begins with the attachment of the acetyl group of acetyl CoA to the oxygen atom in the side chain of serine in the active site of acetyl transferase. The malonyl group of malonyl CoA likewise becomes *O*-linked to the active site of malonyl transferase. These entry reactions take place in domain 1 of the synthase (Figure 20-15). The acetyl unit is then transferred to the cysteine sulfur in the active site of the condensing enzyme, and the malonyl unit is transferred to the sulfur atom of the phosphopantetheinyl group of the acyl carrier protein (ACP) of the *other* chain in the dimer. Domain 1 of each chain of this dimer interacts with domains 2 and 3 of the other chain. Thus, each of the two functional units of the synthase consists of domains formed by different chains. Indeed, the arenas of catalytic action are the interfaces between domains on opposite chains.

Elongation begins with the joining of the acetyl unit on the condensing enzyme to a two-carbon portion of the malonyl unit on ACP (Figure 20-16). CO_2 is released and an acetoacetyl-*S*-phosphopantetheinyl unit is formed on ACP. The active-site sulfhydryl on the condensing enzyme is restored. The acetoacetyl group is then delivered to three active sites in domain 2 of the opposite chain to reduce it to a butyryl unit. This saturated C_4 unit then migrates from the phosphopantetheinyl sulfur on ACP to the cysteine sulfur atom on the condensing enzyme. The synthase is now ready for another round of elongation. The butyryl unit on the condensing enzyme next becomes linked to a two-carbon part of the malonyl unit on ACP to form a C_6 unit on ACP, which undergoes reduction. Five more rounds of condensation and reduction produce a palmitoyl (C_{16}) chain on ACP, which is hydrolyzed to palmitate by the thioesterase on domain 3 of the opposite chain. The migration of the growing fatty acyl chain back and forth between ACP and the condensing enzyme in each round of elongation is noteworthy. Analogous translocations of growing peptide chains between two ribosomal sites take place in protein synthesis (p. 757).

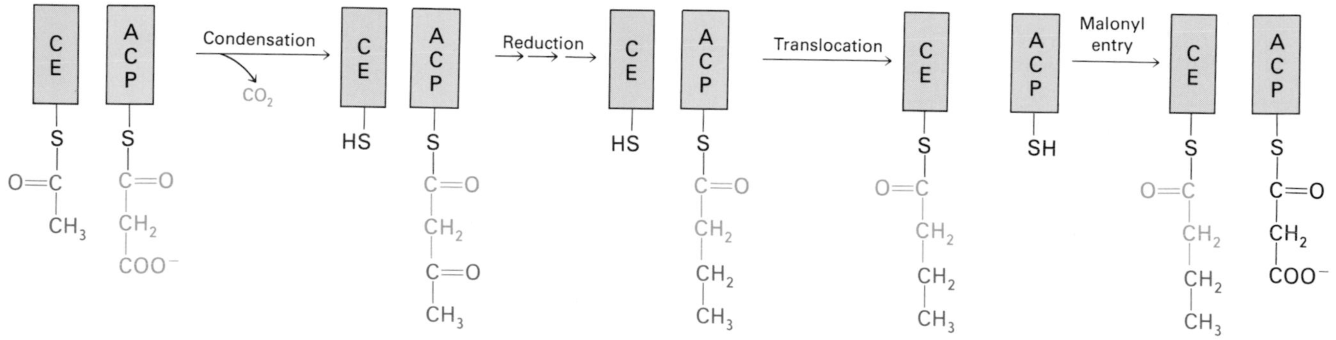

Figure 20-16
Translocations of the elongating fatty acyl chain between the cysteine sulfhydryl group of condensing enzyme (CE, blue) and the phosphopantetheine sulfhydryl group of acyl carrier protein (ACP, yellow).

The flexibility and 20-Å maximal length of the phosphopantetheinyl moiety are critical for the function of this multienzyme complex. The enzyme subunits need not undergo large structural rearrangements to interact with the substrate. Instead, the substrate is on a long, flexible arm that can reach each of the numerous active sites. Recall that biotin and lipoamide are also on long, flexible arms in their multienzyme complexes. The organization of the fatty acid synthases of yeast and higher organisms enhances the efficiency of the overall process because intermediates are directly transferred from one active site to the next. The reactants are not diluted in the cytosol. Moreover, they do not have to find each other by random diffusion. Another advantage of such a multienzyme complex is that covalently bound intermediates are sequestered and protected from competing reactions.

CITRATE CARRIES ACETYL GROUPS FROM MITOCHONDRIA TO THE CYTOSOL FOR FATTY ACID SYNTHESIS

The synthesis of palmitate requires the input of 8 molecules of acetyl CoA, 14 NADPH, and 7 ATP. Fatty acids are synthesized in the cytosol, whereas acetyl CoA is formed from pyruvate in mitochondria. Hence, acetyl CoA must be transferred from mitochondria to the cytosol. Mitochondria, however, are not readily permeable to acetyl CoA. Recall that carnitine carries only long-chain fatty acids. *The barrier to acetyl CoA is bypassed by citrate, which carries acetyl groups across the inner mitochondrial membrane.* Citrate is formed in the mitochondrial matrix by the condensation of acetyl CoA with oxaloacetate (Figure 20-17). When present at high levels, citrate is transported to the cytosol, where it is cleaved by *citrate lyase:*

Lyases—
Enzymes catalyzing the cleavage of C–C, C–O, or C–N bonds by elimination. A double bond is formed in these reactions.

$$\text{Citrate} + \text{ATP} + \text{CoA} + H_2O \longrightarrow \text{acetyl CoA} + \text{ADP} + P_i + \text{oxaloacetate}$$

Thus, acetyl CoA and oxaloacetate are transferred from mitochondria to the cytosol at the expense of an ATP.

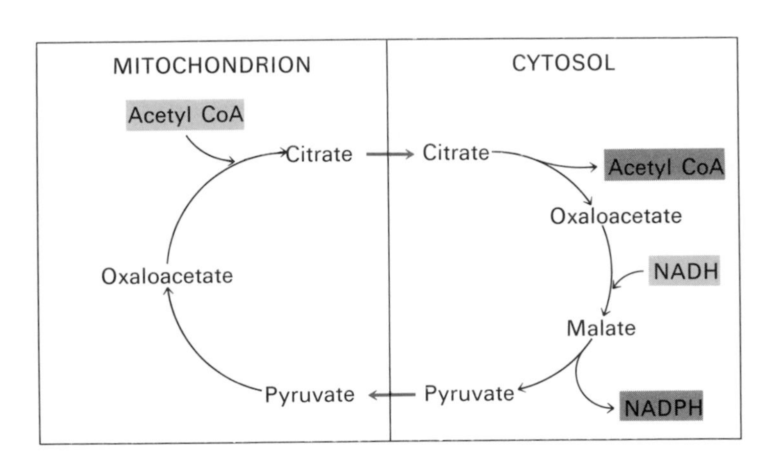

Figure 20-17
Acetyl CoA is transferred from mitochondria to the cytosol, and NADH is concomitantly converted into NADPH by this series of reactions.

SOURCES OF NADPH FOR FATTY ACID SYNTHESIS

Oxaloacetate formed in the transfer of acetyl groups to the cytosol must now be returned to the mitochondria. The inner mitochondrial membrane is impermeable to oxaloacetate. Hence, a series of bypass reactions are needed. Most important, these reactions generate much of the NADPH needed for fatty acid synthesis. First, oxaloacetate is reduced to malate by NADH. This reaction is catalyzed by a *malate dehydrogenase* in the cytosol.

$$\text{Oxaloacetate} + \text{NADH} + H^+ \rightleftharpoons \text{malate} + \text{NAD}^+$$

Second, malate is oxidatively decarboxylated by an *NADP⁺-linked malate enzyme* (also called *malic enzyme*). This reaction has not been mentioned before.

$$\text{Malate} + \text{NADP}^+ \longrightarrow \text{pyruvate} + CO_2 + \text{NADPH}$$

The pyruvate formed in this reaction readily diffuses into mitochondria, where it is carboxylated to oxaloacetate by pyruvate carboxylase.

$$\text{Pyruvate} + CO_2 + \text{ATP} + H_2O \longrightarrow \text{oxaloacetate} + \text{ADP} + P_i + 2\,H^+$$

The sum of these three reactions is

$$NADP^+ + NADH + ATP + H_2O \longrightarrow NADPH + NAD^+ + ADP + P_i + H^+$$

Thus, *one NADPH is generated for each acetyl CoA that is transferred from the mitochondria to the cytosol.* Hence, eight NADPH are formed when eight molecules of acetyl CoA are transferred to the cytosol for the synthesis of palmitate. *The additional six NADPH required for this process come from the pentose phosphate pathway* (p. 431).

ELONGATION AND UNSATURATION OF FATTY ACIDS ARE CARRIED OUT BY ACCESSORY ENZYME SYSTEMS

The major product of the fatty acid synthase is palmitate. In eucaryotes, longer fatty acids are formed by elongation reactions catalyzed by enzymes on the cytosolic face of the *endoplasmic reticulum membrane.* For study of these elongation reactions, the membrane is fragmented into closed vesicles called *microsomes.* As in the reactions leading to the synthesis of palmitate, prepared microsomal systems add two-carbon units sequentially to the carboxyl end of both saturated and unsaturated fatty acids. Malonyl CoA is the two-carbon donor in the elongation of fatty acyl CoAs. The decarboxylation of malonyl CoA drives the condensation of these units.

Microsomal systems also introduce double bonds into long-chain acyl CoAs. For example, in the conversion of stearoyl CoA into oleoyl CoA, a *cis*-Δ^9 double bond is inserted by an oxidase that employs *molecular oxygen* and *NADH* (or *NADPH*):

$$\text{Stearoyl CoA} + NADH + H^+ + O_2 \longrightarrow \text{oleoyl CoA} + NAD^+ + 2\ H_2O$$

This reaction is catalyzed by a complex of three membrane-bound enzymes: *NADH-cytochrome* b_5 *reductase*, *cytochrome* b_5, and a *desaturase* (Figure 20-18). First, electrons are transferred from NADH to the FAD

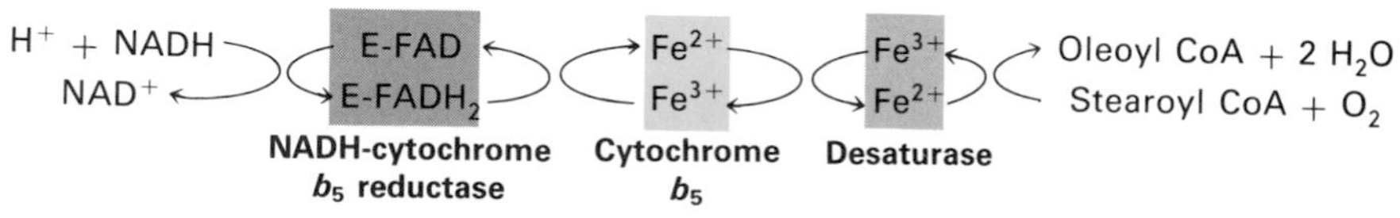

H^+ + NADH
NAD^+
E-FAD
E-$FADH_2$
Fe^{2+}
Fe^{3+}
Fe^{3+}
Fe^{2+}
Oleoyl CoA + 2 H_2O
Stearoyl CoA + O_2
NADH-cytochrome b_5 reductase
Cytochrome b_5
Desaturase

Figure 20-18
Electron-transport chain in the desaturation of fatty acids.

moiety of NADH-cytochrome b_5 reductase. The heme iron atom of cytochrome b_5 is then reduced to the ferrous form. The nonheme iron atom of the desaturase is subsequently converted to the Fe^{2+} state, which enables it to interact with O_2 and the saturated fatty acyl CoA substrate. A double bond is formed and two molecules of H_2O are released. Two electrons come from NADH and two from the single bond of the fatty acyl substrate.

A variety of unsaturated fatty acids can be formed from oleate by a combination of elongation and desaturation reactions. For example, oleate can be elongated to a 20:1 *cis*-Δ^{11} fatty acid. Alternatively, a second double bond can be inserted to yield an 18:2 *cis*-Δ^6, Δ^9 fatty acid. Similarly, palmitate (16:0) can be oxidized to palmitoleate (16:1 *cis*-Δ^9), which can then be elongated to *cis*-vaccenate (18:1 *cis*-Δ^{11}).

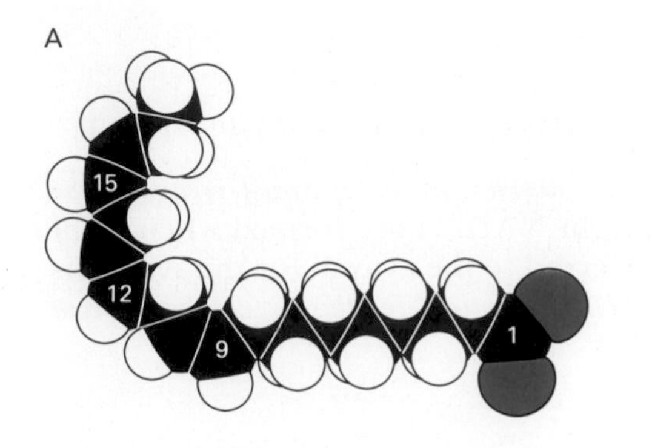

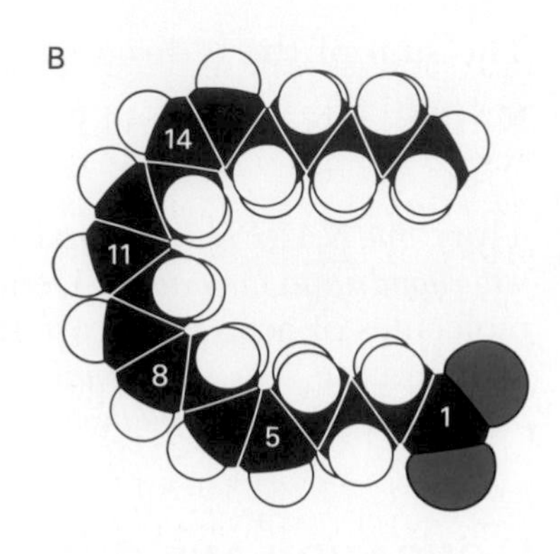

Figure 20-19
Models of polyunsaturated fatty acids. (A) Linolenate, a C_{18} fatty acid with three *cis* double bonds.
(B) Arachidonate, a C_{20} fatty acid with four *cis* double bonds.

Unsaturated fatty acids in mammals are derived from either palmitoleate (16:1), oleate (18:1), linoleate (18:2), or linolenate (18:3). The number of carbons from the ω-end of a derived unsaturated fatty acid to the nearest double bond identifies its precursor.

Precursor	*Formula*
Linolenate (ω-3)	CH_3—$(CH_2)_1$—CH═CH—R
Linoleate (ω-6)	CH_3—$(CH_2)_4$—CH═CH—R
Palmitoleate (ω-7)	CH_3—$(CH_2)_5$—CH═CH—R
Oleate (ω-9)	CH_3—$(CH_2)_7$—CH═CH—R

Mammals lack the enzymes to introduce double bonds at carbon atoms beyond C-9 in the fatty acid chain. Hence, mammals cannot synthesize linoleate (18:2 *cis*-Δ^9, Δ^{12}) and linolenate (18:3 *cis*-Δ^9, Δ^{12}, Δ^{15}). *Linoleate and linolenate are the two essential fatty acids.* The term "essential" means that they must be supplied in the diet because they are required by the organism and cannot be endogenously synthesized. Linoleate and linolenate furnished by the diet are the starting points for the synthesis of a variety of other unsaturated fatty acids. *Arachidonate*, a 20:4 fatty acid derived from linolenate, is in turn the source of many highly active signal molecules such as prostaglandin hormones (p. 991).

CONTROL OF FATTY ACID METABOLISM

The rate of fatty acid oxidation is largely determined by the availability of substrate. In starvation, the level of free fatty acids rises because adipose-cell lipase is stimulated by hormones such as epinephrine and glucagon (p. 458). The entry of fatty acyl CoAs into the mitochondrial matrix is also regulated. Malonyl CoA, which is present at a high level when fuel molecules are abundant, inhibits carnitine acyltransferase I. Hence, *fatty acyl CoAs do not have ready access to the mitochondrial matrix in times of plenty.* Moreover, two enzymes in the β-oxidation pathway are markedly inhibited when the energy charge is high. NADH inhibits 3-hydroxyacyl CoA dehydrogenase, and acetyl CoA inhibits thiolase.

The synthesis of fatty acids is maximal when carbohydrate is abundant and the level of fatty acids is low. Both short-term and long-term control mechanisms are important. The concentration of *citrate* in the cytosol is the most important short-term regulator of fatty acid synthesis. As was mentioned earlier, citrate stimulates acetyl CoA carboxylase, which catalyzes the formation of malonyl CoA, the committed step in fatty acid synthesis. The level of citrate is high when both acetyl CoA and ATP are abundant. Recall that isocitrate dehydrogenase is inhibited by a high energy charge (p. 390). Hence, *a high level of citrate indicates that two-carbon units and ATP are available for fatty acid synthesis.* The effect of citrate on acetyl CoA carboxylase is antagonized by *palmitoyl CoA,* which is abundant when there is an excess of fatty acids. Palmitoyl CoA also inhibits the translocase that transports citrate from mitochondria to the cytosol, as well as glucose 6-phosphate dehydrogenase, which generates NADPH.

Acetyl CoA carboxylase is also controlled by reversible phosphorylation. When the blood glucose level is low, glucagon is released. The binding of glucagon to receptors on the plasma membrane of liver cells triggers a cascade leading to the formation of cyclic AMP and the consequent activation of the phosphorylating enzyme protein kinase, important in the regulation of glycogen metabolism. Acetyl CoA carboxylase, like glycogen synthase (p. 462), is switched off by phosphorylation. Hence, fatty acid synthesis ceases when the blood glucose level is low. Insulin, a signal that fuels are abundant, has the reverse effect on acetyl CoA carboxylase and hence promotes fatty acid synthesis.

Long-term control is mediated by changes in the rates of synthesis and degradation of the enzymes participating in fatty acid synthesis. Animals that have fasted and are then fed high-carbohydrate, low-fat diets show marked increases in their amounts of acetyl CoA carboxylase and fatty acid synthase within a few days. This type of regulation is known as *adaptive control.*

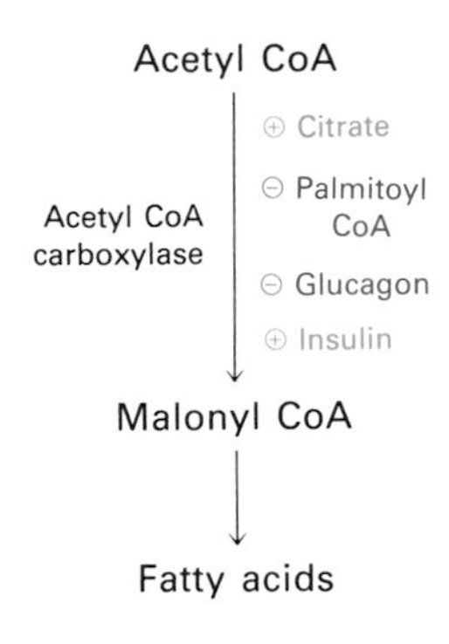

Figure 20-20
Acetyl CoA carboxylase is a key control site in fatty acid synthesis.

SUMMARY

Fatty acids are physiologically important both as components of phospholipids and glycolipids and as fuel molecules. They are stored in adipose tissue as triacylglycerols (neutral fat), which can be mobilized by the hydrolytic action of lipases that are under hormonal control. Fatty acids are activated to acyl CoAs, transported across the inner mitochondrial membrane by carnitine, and degraded in the mitochondrial matrix by a recurring sequence of four reactions: oxidation linked to FAD, hydration, oxidation linked to NAD^+, and thiolysis by CoA. The $FADH_2$ and NADH formed in the oxidation steps transfer their electrons to O_2 by means of the respiratory chain, whereas the acetyl CoA formed in the thiolysis step normally enters the citric acid cycle by condensing with oxaloacetate. When the concentration of oxaloacetate is insufficient, acetyl CoA gives rise to acetoacetate and 3-hydroxybutyrate, which are normal fuel molecules. In starvation and in diabetes, large amounts of acetoacetate, 3-hydroxybutyrate, and acetone (collectively known as ketone bodies) are formed in the liver and accumulate in the blood. Mammals are unable to convert fatty acids into glucose because they lack a pathway for the net production of oxaloacetate, pyruvate, or other gluconeogenic intermediates from acetyl CoA.

Fatty acids are synthesized in the cytosol by a different pathway from that of β-oxidation. A reaction cycle based on the formation and cleavage of citrate carries acetyl groups from mitochondria to the cytosol. NADPH needed for synthesis is generated by the pentose phosphate pathway and in the transfer of reducing equivalents from mitochondria by the malate-pyruvate shuttle. Synthesis starts with the carboxylation of acetyl CoA to malonyl CoA. This ATP-driven reaction is catalyzed by acetyl CoA carboxylase, a biotin-enzyme. Citrate allosterically stimulates this committed step in fatty acid synthesis. The intermediates in fatty acid synthesis are linked to an acyl carrier protein (ACP), specifically to the sulfhydryl terminus of its phosphopantetheine prosthetic group. Acetyl-ACP is formed from acetyl CoA, and malonyl-ACP is formed from malonyl CoA. Acetyl-ACP and malonyl-ACP condense to form acetoacetyl-ACP, a reaction driven by the release of CO_2 from the activated malonyl unit. This is followed by a reduction, a dehydration, and a second reduction. NADPH is the reductant in these steps. The butyryl-ACP formed in this way is ready for a second round of elongation, starting with the addition of a two-carbon unit from malonyl-ACP. Seven rounds of elongation yield palmitoyl-ACP, which is hydrolyzed to palmitate. The synthesis of palmitate requires eight molecules of acetyl CoA, fourteen NADPH, and seven ATP; seven HCO_3^- play a catalytic role. In higher organisms, the enzymes carrying out fatty acid synthesis are covalently linked in a multifunctional enzyme complex. The flexible phosphopantetheinyl unit of ACP carries the substrate from one active site to another in this complex.

Fatty acids are elongated and desaturated by enzyme systems in the endoplasmic reticulum membrane. Desaturation requires NADH and O_2 and is carried out by a complex consisting of a flavoprotein, a cytochrome, and a nonheme iron protein. Mammals lack the enzymes to introduce double bonds distal to C-9, and so they require linoleate and linolenate in their diets. Arachidonate, a key precursor of prostaglandins and other signal molecules, is derived from linolenate.

SELECTED READINGS

Where to start

Wakil, S. J., Stoops, J. K., and Joshi, V. C., 1983. Fatty acid synthesis and its regulation. *Ann. Rev. Biochem.* 52:537–579.

Lynen, F., 1972. The pathway from "activated acetic acid" to the terpenes and fatty acids. In *Nobel Lectures: Physiology or Medicine (1963–1970)*, pp. 103–138. American Elsevier (1973). [An account of the author's pioneering work on fatty acid degradation and synthesis and of the roles of acetyl CoA and biotin.]

Books

Vance, D. E., and Vance, J. E., (eds.), 1985. *Biochemistry of Lipids and Membranes.* Benjamin/Cummings. [Contains many excellent articles. Chapters 3, 4, 5, and 6 pertain directly to fatty acid metabolism.]

Boyer, P. D., (ed.), 1983. *The Enzymes* (3rd ed.), vol. 16: *Lipid Enzymology.* Academic Press. [This volume contains a wealth of information about a broad range of topics in lipid metabolism.]

Numa, S., (ed.), 1984. *Fatty Acid Metabolism and Its Regulation.* Elsevier. [A concise and highly readable volume emphasizing control of fatty acid metabolism in bacteria, plants, and animals.]

Newsholme, E. A., and Start, C., 1973. *Regulation in Metabolism.* Wiley. [Chapters 4 and 7 deal with the regulation of fat metabolism.]

Fatty acid oxidation

Foster, D. W., 1984. From glycogen to ketones—and back. *Diabetes* 33:1188–1199.

McGarry, J. D., and Foster, D. W., 1980. Regulation of hepatic fatty acid oxidation and ketone body production. *Ann. Rev. Biochem.* 49:395–420.

Garland, P. B., Shepherd, D., Nicholls, D. G., Yates, D. W., and Light, P. A., 1969. Interactions between fatty acid oxidation and the tricarboxylic acid cycle. *In* J. M. Lowenstein (ed.), *Citric Acid Cycle: Control and Compartmentation,* pp. 163–212. Dekker.

Schulz, H., 1987. Inhibitors of fatty acid oxidation. *Life Sci.* 40:1443–1449.

Carnitine

Borum, P. R., 1983. Carnitine. *Ann. Rev. Nutr.* 3:233–259.

Bremer, J., 1977. Carnitine and its role in fatty acid metabolism. *Trends Biochem. Sci.* 2:207–209.

Fritz, I. B., 1968. The metabolic consequences of the effects of carnitine on long-chain fatty acid oxidation. *In* F. C. Gran (ed.), *Symposium on Cellular Compartmentalization and Control of Fatty Acid Metabolism,* pp. 39–63. Academic Press.

Engel, W. K., Vick, N. A., Glueck, C. J., and Levy, R. I., 1970. A skeletal-muscle disorder associated with intermittent symptoms and a possible defect of lipid metabolism. *N. Engl. J. Med.* 282:697–704.

Fatty acid synthesis

Singh, N., Wakil, S. J., and Stoops, J. K., 1985. Yeast fatty acid synthase: structure to function relationship. *Biochemistry* 24:6598–6602.

Goodridge, A. G., 1986. Regulation of the gene for fatty acid synthase. *Fed. Proc.* 45:2399–2405.

Volpe, J. J., and Vagelos, P. R., 1976. Mechanism and regulation of biosynthesis of saturated fatty acids. *Physiol. Rev.* 56:339–417.

Bloch, K., and Vance, D., 1977. Control mechanisms in the synthesis of saturated fatty acids. *Ann. Rev. Biochem.* 46:263–298.

Lane, M. D., Moss, J., and Polakis, S. E., 1974. Acetyl coenzyme A carboxylase. *Curr. Top. Cell Regul.* 8:139–187.

Enzyme mechanisms

Wood, H. G., and Barden, R. E., 1977. Biotin enzymes. *Ann. Rev. Biochem.* 46:385–413.

Alberts, A. W., and Vagelos, P. R., 1972. Acyl-CoA carboxylases. *In* P. D. Boyer (ed.), *The Enzymes* (3rd ed.), vol. 6, pp. 37–82. Academic Press.

Vagelos, P. R., 1973. Acyl group transfer (acyl carrier protein). *In* P. D. Boyer (ed.), *The Enzymes* (3rd ed.), vol. 8, pp. 155–199.

Climent, I., and Rubio, V., 1986. ATPase activity of biotin carboxylase provides evidence for initial activation of HCO_3^- by ATP in the carboxylation of biotin. *Arch. Biochem. Biophys.* 251:465–470.

Singh, N., Wakil, S. J., and Stoops, J. K., 1985. The development and application of a novel chromophoric substrate for investigation of the mechanism of yeast fatty acid synthase. *Biochem. Biophys. Res. Commun.* 131:786–792.

PROBLEMS

1. Write a balanced equation for the conversion of glycerol into pyruvate. Which enzymes are required in addition to those of the glycolytic pathway?

2. Write a balanced equation for the conversion of stearate into acetoacetate.

3. Compare the following aspects of fatty acid oxidation and synthesis:
   (a) Site of the process.
   (b) Acyl carrier.
   (c) Reductants and oxidants.
   (d) Stereochemistry of the intermediates.
   (e) Direction of synthesis or degradation.
   (f) Organization of the enzyme system.

4. For each of the following unsaturated fatty acids, indicate whether the biosynthetic precursor in animals is palmitoleate, oleate, linoleate, or linolenate.
   (a) 18:1 *cis*-Δ^{11}
   (b) 18:3 *cis*-Δ^{6}, Δ^{9}, Δ^{12}
   (c) 20:2 *cis*-Δ^{11}, Δ^{14}
   (d) 20:3 *cis*-Δ^{5}, Δ^{8}, Δ^{11}
   (e) 22:1 *cis*-Δ^{13}
   (f) 22:6 *cis*-Δ^{4}, Δ^{7}, Δ^{10}, Δ^{13}, Δ^{16}, Δ^{19}

5. Consider a cell extract that actively synthesizes palmitate. Suppose that a fatty acid synthase in this preparation forms one palmitate in about five minutes. A large amount of malonyl CoA labeled with ^{14}C in each carbon of its malonyl unit is suddenly added to this system, and fatty acid synthesis is stopped a minute later by altering the pH. The fatty acids in the supernatant are analyzed for radioactivity. Which carbon atom of the palmitate formed by this system is more radioactive—C-1 or C-14?

6. Propose a reaction mechanism for the condensation of an acetyl unit with a malonyl unit to form an acetoacetyl unit in fatty acid synthesis.

7. Suppose that a promoter mutation leads to the overproduction of the cyclic AMP-dependent protein kinase in adipose cells. How would this mutation affect fatty acid metabolism?

8. Suppose that a mutation impaired the binding site on acetyl CoA carboxylase for the cyclic AMP-dependent protein kinase but left intact the catalytic site. What is a likely consequence of this mutation?

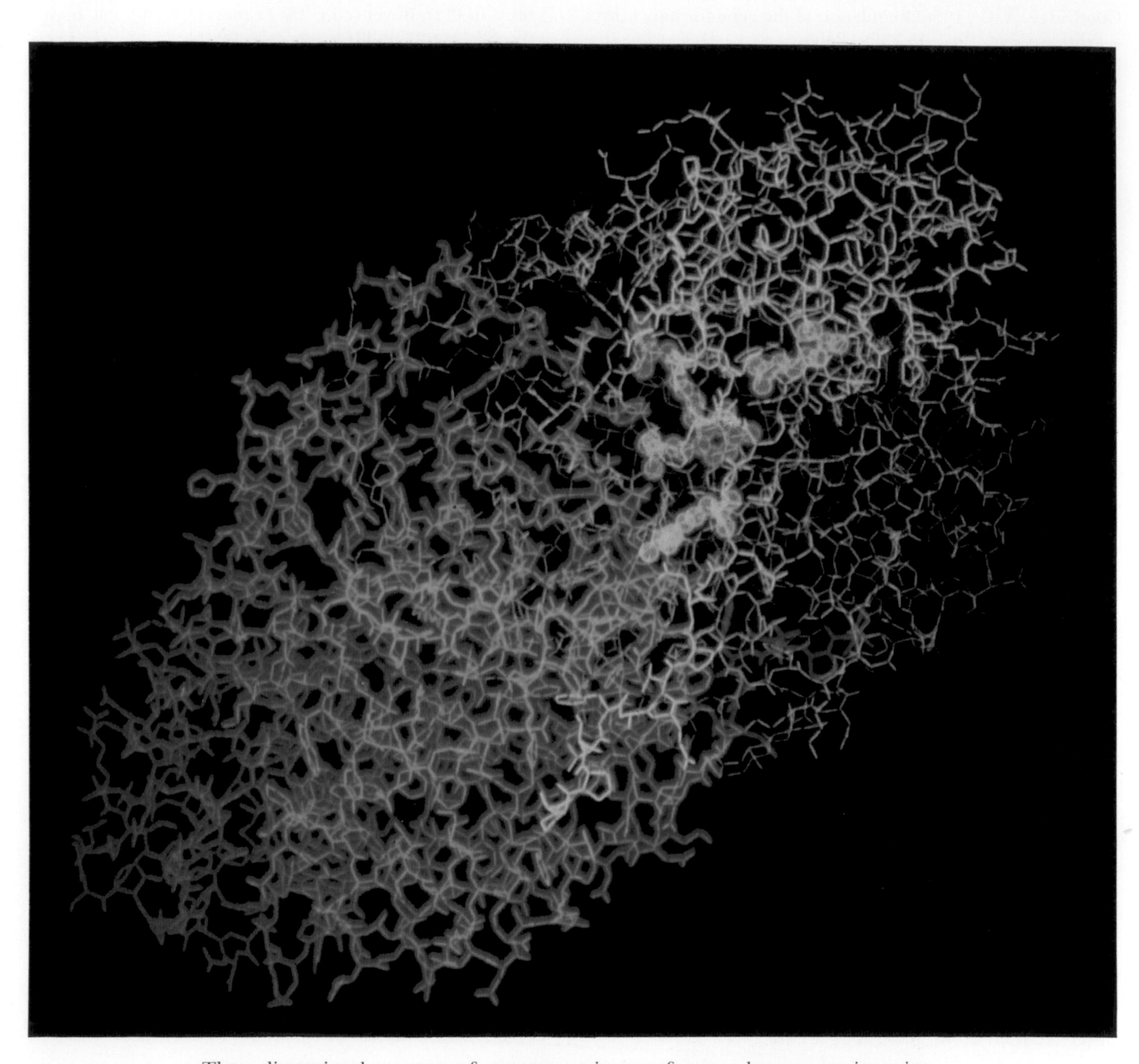

Three-dimensional structure of aspartate aminotransferase, a key enzyme in amino acid metabolism. The two identical subunits are shown in blue and green. A bound pyridoxal phosphate coenzyme and several key residues at the catalytic site are highlighted in orange. [Courtesy of Dr. Arthur Arnone, Dr. Craig Hyde, and Dr. David Metzler.]

CHAPTER 21

Amino Acid Degradation and the Urea Cycle

Amino acids in excess of those needed for the synthesis of proteins and other biomolecules cannot be stored, in contrast with fatty acids and glucose, nor are they excreted. Rather, surplus amino acids are used as metabolic fuel. *The α-amino group is removed and the resulting carbon skeleton is converted into a major metabolic intermediate.* Most of the amino groups of surplus amino acids are converted into urea, whereas their carbon skeletons are transformed into acetyl CoA, acetoacetyl CoA, pyruvate, or one of the intermediates of the citric acid cycle. Hence *fatty acids, ketone bodies, and glucose can be formed from amino acids.*

α-AMINO GROUPS ARE CONVERTED INTO AMMONIUM ION BY OXIDATIVE DEAMINATION OF GLUTAMATE

The major site of amino acid degradation in mammals is the liver. The fate of the α-amino group will be considered first, followed by that of the carbon skeleton. The α-amino group of many amino acids is transferred to α-ketoglutarate to form *glutamate,* which is then oxidatively deaminated to yield NH_4^+.

$$\underset{\text{Amino acid}}{{}^+H_3N-\overset{H}{\underset{COO^-}{C}}-R} \longrightarrow \underset{\text{Glutamate}}{{}^+H_3N-\overset{H}{\underset{COO^-}{C}}-CH_2-CH_2-COO^-} \longrightarrow NH_4^+$$

Aminotransferases catalyze the transfer of an α-amino group from an α-amino acid to an α-keto acid. These enzymes, also called *transami-*

$$H-\overset{NH_3^+}{\underset{R_1}{C}}-COO^- + \overset{O}{\underset{R_2}{C}}-COO^- \rightleftharpoons \overset{O}{\underset{R_1}{C}}-COO^- + H-\overset{NH_3^+}{\underset{R_2}{C}}-COO^-$$

nases, generally *funnel α-amino groups from a variety of amino acids to α-ketoglutarate for conversion into* NH_4^+. *Aspartate aminotransferase,* one of the most important of these enzymes, catalyzes the transfer of the amino group of aspartate to α-ketoglutarate.

$$\text{Aspartate} + \alpha\text{-ketoglutarate} \rightleftharpoons \text{oxaloacetate} + \text{glutamate}$$

Alanine aminotransferase, which is also prevalent in mammalian tissue, catalyzes the transfer of the amino group of alanine to α-ketoglutarate.

$$\text{Alanine} + \alpha\text{-ketoglutarate} \rightleftharpoons \text{pyruvate} + \text{glutamate}$$

Ammonium ion is formed from glutamate by oxidative deamination. This reaction is catalyzed by *glutamate dehydrogenase,* which is unusual in being able to utilize either NAD^+ or $NADP^+$.

$$\underset{\text{Glutamate}}{H\text{—}C(NH_3^+)(COO^-)\text{—}CH_2\text{—}CH_2\text{—}COO^-} + \underset{(\text{or } NADP^+)}{NAD^+} + H_2O \rightleftharpoons NH_4^+ + \underset{\alpha\text{-Ketoglutarate}}{O{=}C(COO^-)\text{—}CH_2\text{—}CH_2\text{—}COO^-} + \underset{(\text{or } NADPH)}{NADH} + H^+$$

The activity of glutamate dehydrogenase is allosterically regulated. The vertebrate enzyme consists of six identical subunits, which can polymerize further. Guanosine triphosphate (GTP) and adenosine triphosphate (ATP) are allosteric inhibitors, whereas guanosine diphosphate (GDP) and adenosine diphosphate (ADP) are allosteric activators. Hence, *a lowering of the energy charge accelerates the oxidation of amino acids.*

The sum of the reactions catalyzed by aminotransferases and glutamate dehydrogenase is

$$\alpha\text{-Amino acid} + \underset{(\text{or } NADP^+)}{NAD^+} + H_2O \rightleftharpoons \alpha\text{-keto acid} + NH_4^+ + \underset{(\text{or } NADPH)}{NADH} + H^+$$

In terrestrial vertebrates, NH_4^+ is converted into urea, which is excreted. The synthesis of urea will be discussed shortly.

α-Amino acid → α-Keto acid; α-Ketoglutarate ⇄ Glutamate; $NAD^+ + H_2O$ → NADH + NH_4^+ - - -> $H_2N\text{—}C({=}O)\text{—}NH_2$ **Urea**

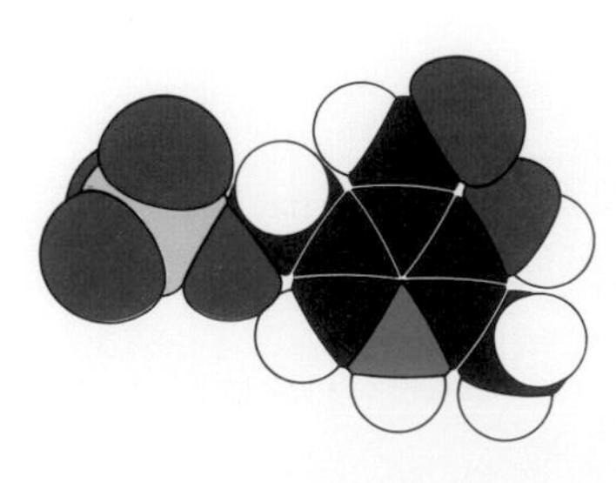

Figure 21-1
Space-filling model of pyridoxal 5-phosphate (PLP).

PYRIDOXAL PHOSPHATE FORMS SCHIFF-BASE INTERMEDIATES IN AMINOTRANSFERASES

The prosthetic group of all aminotransferases is *pyridoxal phosphate* (PLP), which is derived from *pyridoxine* (vitamin B_6). During transamination, pyridoxal phosphate is transiently converted into *pyridoxamine phosphate* (PMP).

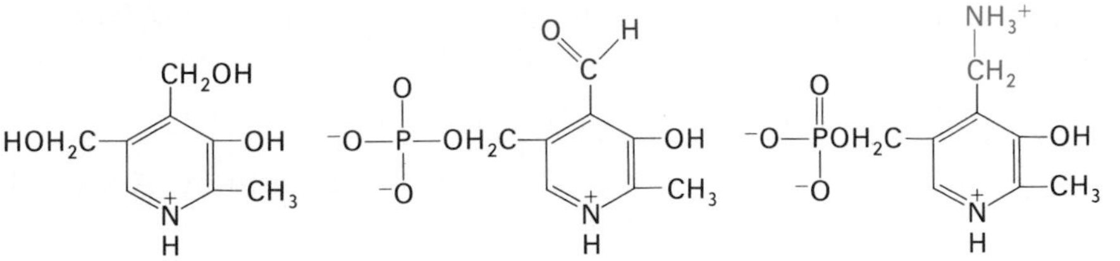

Pyridoxine (Vitamin B_6)

Pyridoxal phosphate (PLP)

Pyridoxamine phosphate (PMP)

PLP enzymes form covalent Schiff-base intermediates with their substrates. In the absence of substrate, the aldehyde group of PLP is in Schiff-base linkage with the *ϵ-amino group of a specific lysine residue at the active site*. A new Schiff-base linkage is formed on addition of an amino acid substrate. *The α-amino group of the amino acid substrate displaces the ϵ-amino group of the active-site lysine*. The amino acid–PLP Schiff base that is formed remains tightly bound to the enzyme by multiple noncovalent interactions.

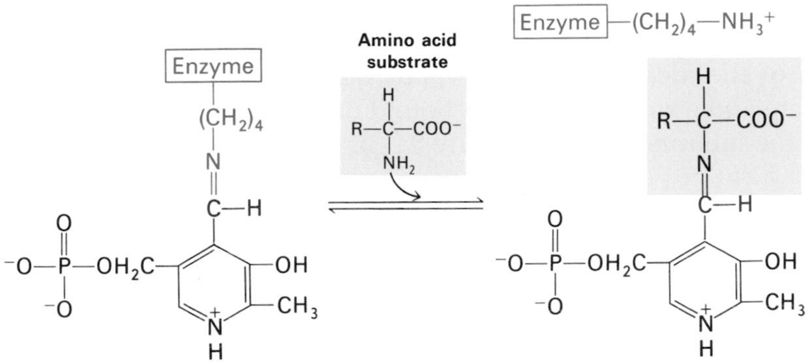

Amino acid substrate

Schiff base of PLP and enzyme

Schiff base of PLP and amino acid substrate

Esmond Snell and Alexander Braunstein proposed more than forty years ago a reaction mechanism for transamination that has proven to be generally valid. The Schiff base between the amino acid substrate and PLP, termed an *aldimine*, loses a proton from its α-carbon to form a *quinonoid* intermediate. Reprotonation yields a *ketimine*, which contains a double bond between N and C_α of the substrate. In contrast, the aldimine contains a double bond between N and the carbonyl C of PLP.

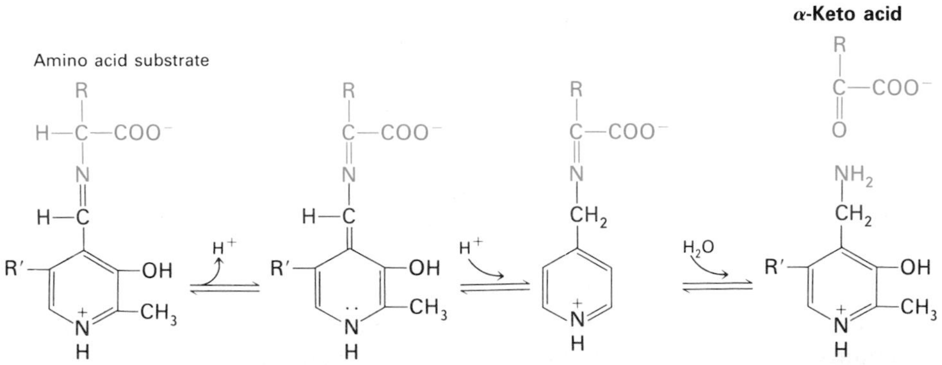

Amino acid substrate

α-Keto acid

Aldimine

Quinonoid intermediate

Ketimine

Pyridoxamine phosphate (PMP)

Figure 21-2
Proposed mechanism of transamination reactions.

The ketimine is then hydrolyzed to an α-keto acid and pyridoxamine phosphate. These steps comprise half of the transamination reaction.

$$\text{Amino acid}_1 + \text{E-PLP} \rightleftharpoons \alpha\text{-keto acid}_1 + \text{E-PMP}$$

The second half occurs by a reversal of the above pathway. A second α-keto acid reacts with the enzyme–pyridoxamine phosphate complex (E-PMP) to yield a second amino acid and regenerate the enzyme–pyridoxal phosphate complex (E-PLP).

$$\alpha\text{-Keto acid}_2 + \text{E-PMP} \rightleftharpoons \text{amino acid}_2 + \text{E-PLP}$$

The sum of these partial reactions is

$$\text{Amino acid}_1 + \alpha\text{-keto acid}_2 \rightleftharpoons \text{amino acid}_2 + \alpha\text{-keto acid}_1$$

THE ACTIVE SITE CLEFT OF ASPARTATE AMINOTRANSFERASE CLOSES WHEN SUBSTRATE FORMS A SCHIFF BASE LINKAGE

X-ray crystallographic studies of mitochondrial aspartate aminotransferase have provided detailed views of how PLP and substrate are bound and have confirmed much of the proposed catalytic mechanism. Each of the identical 45-kd subunits of this dimer consists of a large domain and a small one. PLP is bound to the large domain, in a pocket near the subunit interface (Figure 21-3). The pyridine nitrogen atom of PLP is hydrogen-bonded to an aspartate carboxylate, and the 2-methyl group is in van der Waals contact with several hydrophobic residues. The 3-hydroxyl group is ionized and hydrogen bonded to the phenolic OH of a tyrosine. The 5-phosphate group of PLP is hydrogen bonded to seven groups. Its two negative charges are balanced by an arginine side chain and by the positive dipole at the amino end of an α helix. As was postulated earlier on the basis of spectroscopic and chemical studies, the 4-aldehyde group of PLP has been shown by x-ray crystallography to be in Schiff base linkage with a lysine residue.

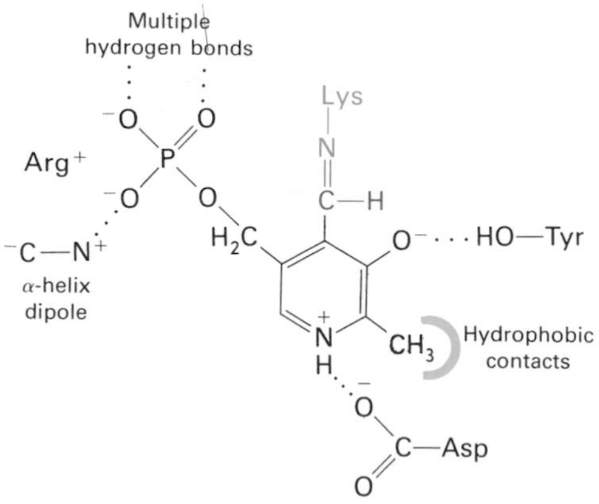

Figure 21-3
Mode of binding of PLP to aspartate aminotransferase. [After J. F. Kirsch, G. Eichele, G. F. Ford, M. G. Vincent, and J. N. Jansonius. *J. Mol. Biol.* 174(1984):510.]

Each of the two carboxylates of aspartate (or glutamate) forms a salt bridge with an arginine residue of the aminotransferase (Figure 21-4). These electrostatic interactions with guanido groups on different chains largely determine the substrate specificity of the enzyme. Moreover, they lead to a substantial movement of the small domain, which

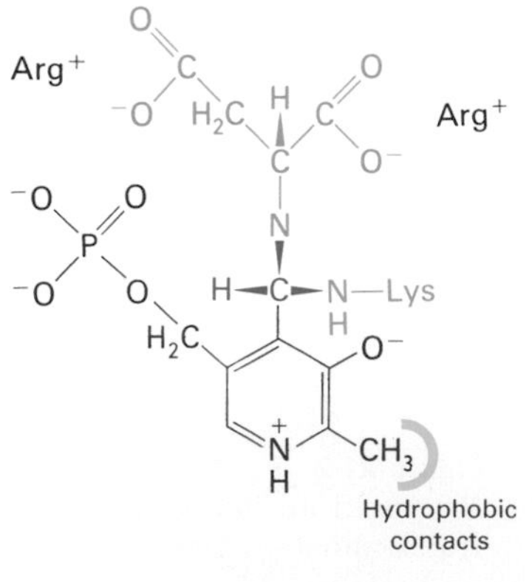

Figure 21-4
Mode of binding of aspartate in the tetrahedral intermediate of aspartate aminotransferase. [After J. F. Kirsch, G. Eichele, G. F. Ford, M. G. Vincent, and J. N. Jansonius. *J. Mol. Biol.* 174(1984):510.]

closes the active-site crevice. Recall that hexokinase (p. 365) and citrate synthase (p. 384) also undergo cleft closure during catalysis. The subsequent replacement of the ϵ-amino group of lysine by the α-amino group of the substrate induces a 30-degree tilting of the PLP ring, which promotes the subsequent conversions. Specifically, the C_α–H bond becomes nearly perpendicular to the plane containing the PLP ring. This orientation facilitates the release of the α-H to form the quinonoid intermediate (see Figure 21-2). Reprotonation gives the ketimine, which is hydrolyzed to give oxaloacetate, the α-keto acid. The lysine amino group that was initially in Schiff base linkage with PLP serves as a proton acceptor and donor in subsequent steps.

PYRIDOXAL PHOSPHATE, A HIGHLY VERSATILE COENZYME, CATALYZES MANY REACTIONS OF AMINO ACIDS

Transamination is just one of a wide range of amino acid transformations that are catalyzed by PLP enzymes. The other reactions at the α-carbon atom of amino acids are decarboxylations, deaminations, racemizations, and aldol cleavages (Figure 21-5). In addition, PLP enzymes catalyze elimination and replacement reactions at the β-carbon atom (e.g., tryptophan synthetase, p. 586) and the γ-carbon atom (e.g., cystathionase, p. 584) of amino acid substrates. The common features of PLP catalysis underlying these diverse reactions are: (1) A *Schiff base* is formed by the amino acid substrate (the amine component) and PLP (the carbonyl component). (2) The protonated form of PLP acts as an *electron sink* to stabilize catalytic intermediates that are negatively charged—the ring nitrogen of PLP attracts electrons from the amino acid substrate. In other words, PLP is an *electrophilic catalyst.* (3) The product Schiff base is then hydrolyzed.

How does an enzyme selectively break one of three bonds at the α-carbon atom of an amino acid substrate? An important principle is that *the bond being broken must be perpendicular to the π-orbitals of the electron sink.* In an aminotransferase, for example, this is accomplished by binding the amino acid substrate so that the C_α–H bond is perpendicular to the PLP ring. This means of choosing one of several possible catalytic outcomes is called *stereoelectronic control.*

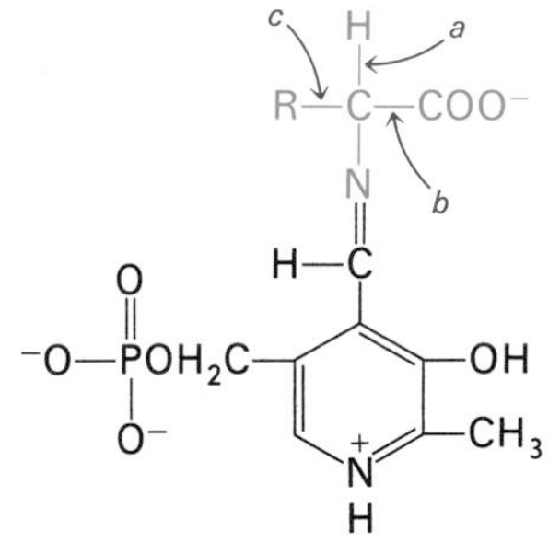

Figure 21-5
Pyridoxal phosphate enzymes labilize one of three bonds at the α-carbon atom of an amino acid substrate. For example, bond *a* is labilized by aminotransferases, bond *b* by decarboxylases, and bond *c* by aldolases (such as threonine aldolases). PLP enzymes also catalyze reactions at the β- and γ-carbon atoms of amino acids.

SERINE AND THREONINE CAN BE DIRECTLY DEAMINATED

The α-amino groups of serine and threonine can be directly converted into NH_4^+ because each of these amino acids contains a hydroxyl group attached to its β carbon atom. These direct deaminations are catalyzed by *serine dehydratase* and *threonine dehydratase,* in which PLP is the prosthetic group.

$$\text{Serine} \longrightarrow \text{pyruvate} + NH_4^+$$

$$\text{Threonine} \longrightarrow \alpha\text{-ketobutyrate} + NH_4^+$$

These enzymes are called dehydratases because dehydration precedes deamination. Serine loses a hydrogen atom from its α-carbon and a hydroxyl group from its β-carbon to yield aminoacrylate. This unstable compound reacts with H_2O to give pyruvate and NH_4^+.

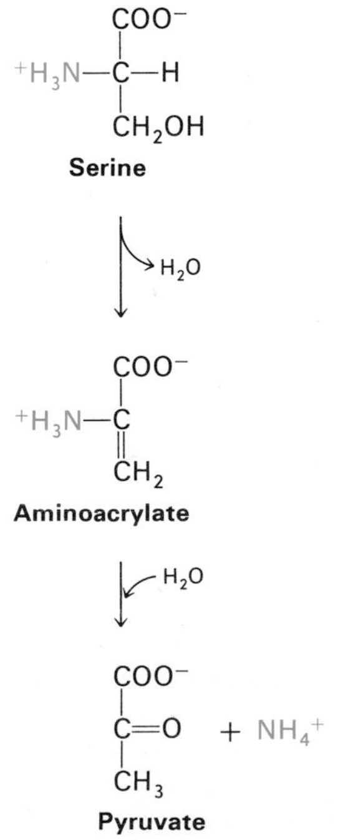

Serine → ($-H_2O$) → **Aminoacrylate** → ($+H_2O$) → **Pyruvate** $+ NH_4^+$

NH_4^+ IS CONVERTED INTO UREA IN MOST TERRESTRIAL VERTEBRATES AND THEN EXCRETED

Some of the NH_4^+ formed in the breakdown of amino acids is consumed in the biosynthesis of nitrogen compounds. In most terrestrial vertebrates, the excess NH_4^+ is converted into urea and then excreted. In birds and terrestrial reptiles, NH_4^+ is converted into uric acid for excretion, whereas, in many aquatic animals, NH_4^+ itself is excreted. These three classes of organisms are called *ureotelic, uricotelic,* and *ammonotelic*.

In terrestrial vertebrates, urea is synthesized by the *urea cycle* (Figure 21-6). This series of reactions was proposed by Hans Krebs and Kurt Henseleit (a medical student) in 1932, five years before the elucidation of the citric acid cycle. In fact, the urea cycle was the first cyclic metabolic pathway to be discovered. One of the nitrogen atoms of the urea synthesized by this pathway is transferred from an amino acid, aspartate. The other nitrogen atom and the carbon atom are derived from NH_4^+ and CO_2. *Ornithine* is the carrier of these carbon and nitrogen atoms.

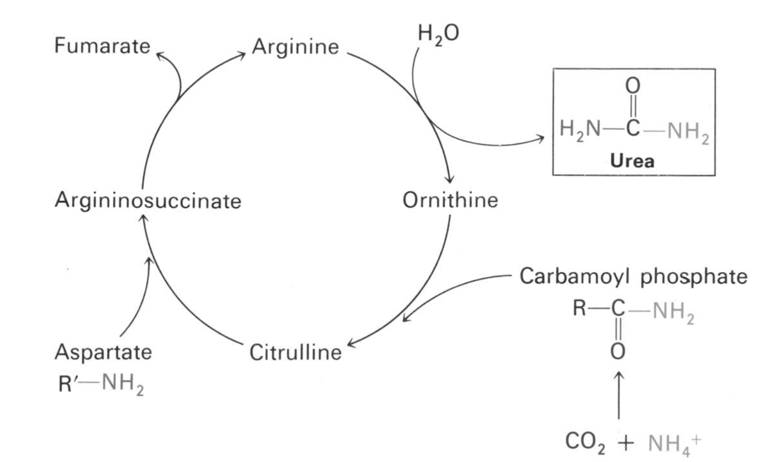

Figure 21-6
The urea cycle.

The immediate precursor of urea is *arginine,* which is hydrolyzed to urea and ornithine by *arginase.*

$$\underset{\textbf{Arginine}}{H_2N{-}C({=}NH_2^+){-}NH{-}CH_2{-}CH_2{-}CH_2{-}CH(NH_3^+){-}COO^-} \xrightarrow{H_2O} \underset{\textbf{Urea}}{H_2N{-}C({=}O){-}NH_2} + \underset{\textbf{Ornithine}}{NH_3^+{-}CH_2{-}CH_2{-}CH_2{-}CH(NH_3^+){-}COO^-}$$

The other reactions of the urea cycle lead to the synthesis of arginine from ornithine. First, a carbamoyl group is transferred to ornithine to form *citrulline,* in a reaction catalyzed by ornithine transcarbamoylase. The carbamoyl donor in this reaction is carbamoyl phosphate, which has a high transfer potential because of its anhydride bond.

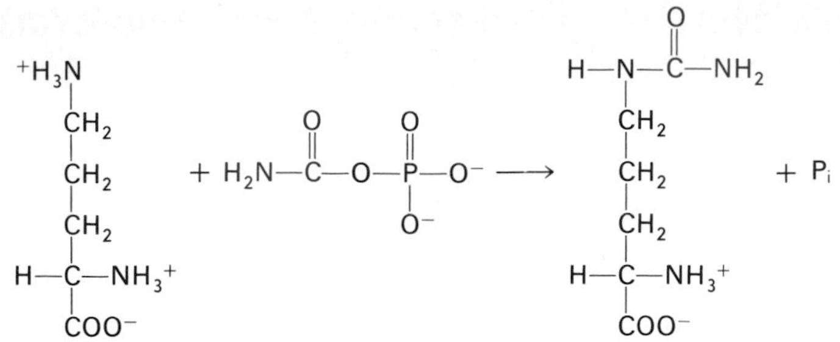

$$\text{Ornithine} + \text{Carbamoyl phosphate} \longrightarrow \text{Citrulline} + P_i$$

Ornithine　Carbamoyl phosphate　Citrulline

Argininosuccinate synthetase then catalyzes the condensation of citrulline and aspartate. This synthesis of *argininosuccinate* is driven by the cleavage of ATP into AMP and pyrophosphate and by the subsequent hydrolysis of pyrophosphate.

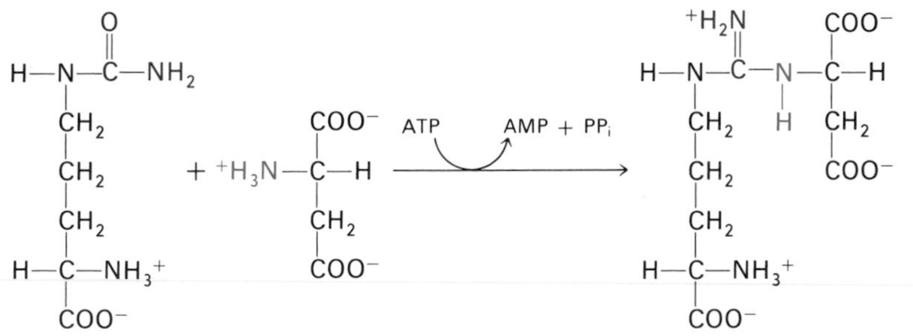

$$\text{Citrulline} + \text{Aspartate} \xrightarrow{ATP \;\; AMP + PP_i} \text{Argininosuccinate}$$

Citrulline　Aspartate　Argininosuccinate

Finally, argininosuccinase cleaves argininosuccinate into *arginine* and *fumarate*. Note that these reactions, which transfer the amino group of aspartate to form arginine, preserve the carbon skeleton of aspartate.

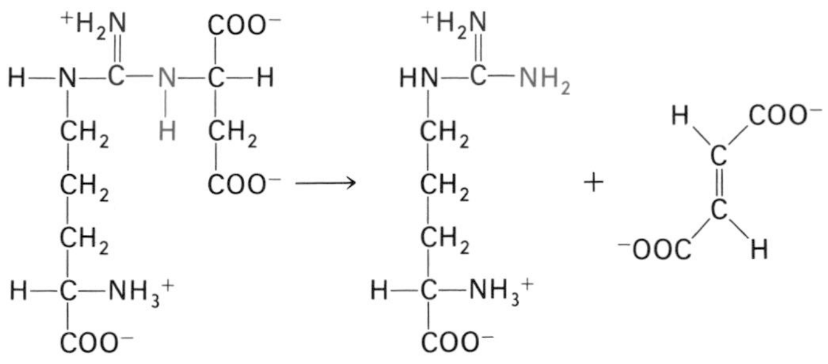

$$\text{Argininosuccinate} \longrightarrow \text{Arginine} + \text{Fumarate}$$

Argininosuccinate　Arginine　Fumarate

Carbamoyl phosphate is synthesized from NH_4^+, CO_2, ATP, and H_2O in a complex reaction that is catalyzed by *carbamoyl phosphate synthetase*. An unusual feature of this enzyme is that it requires *N*-acetylglutamate for activity.

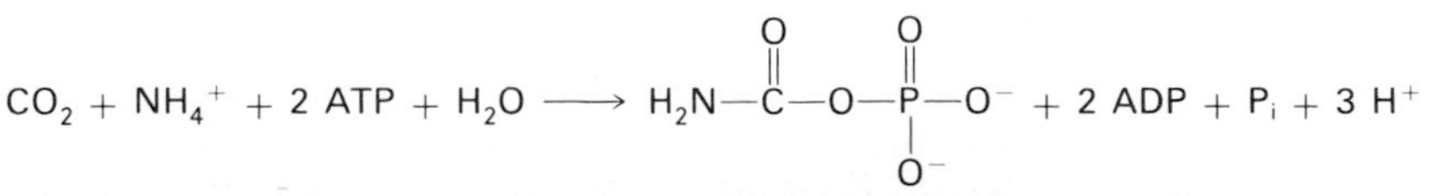

$$CO_2 + NH_4^+ + 2\ ATP + H_2O \longrightarrow H_2N{-}C(=O){-}O{-}P(=O)(O^-){-}O^- + 2\ ADP + P_i + 3\ H^+$$

Carbamoyl phosphate

The consumption of two molecules of ATP makes this synthesis of carbamoyl phosphate essentially irreversible.

THE UREA CYCLE IS LINKED TO THE CITRIC ACID CYCLE

The stoichiometry of urea synthesis is

$$CO_2 + NH_4^+ + 3\ ATP + \text{aspartate} + 2\ H_2O \longrightarrow \text{urea} + 2\ ADP + 2\ P_i + AMP + PP_i + \text{fumarate}$$

Pyrophosphate is rapidly hydrolyzed, and so four ~P are consumed in these reactions to synthesize one molecule of urea. *The synthesis of fumarate by the urea cycle is important because it links the urea cycle and the citric acid cycle* (Figure 21-7). Fumarate is hydrated to malate, which is in turn oxidized to oxaloacetate. Oxaloacetate has several possible fates: (1) transamination to aspartate; (2) conversion into glucose by the gluconeogenic pathway; (3) condensation with acetyl CoA to form citrate; or (4) conversion into pyruvate.

The compartmentation of the urea cycle and its associated reactions is also noteworthy. The formation of NH_4^+ by glutamate dehydrogenase, its incorporation into carbamoyl phosphate, and the subsequent synthesis of citrulline occur in the mitochondrial matrix. In contrast, the next three reactions of the urea cycle, which lead to the formation of urea, take place in the cytosol.

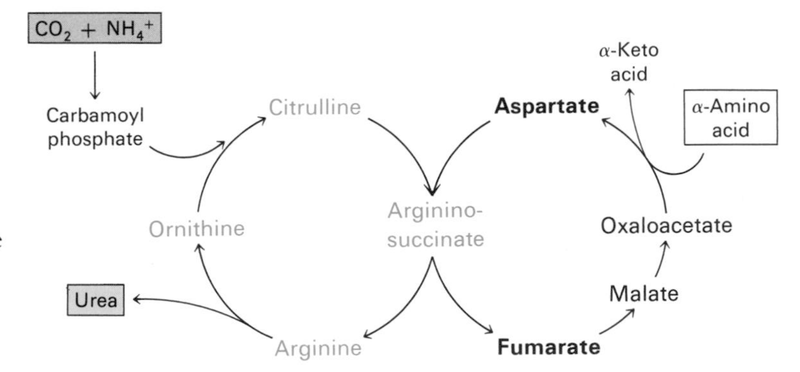

Figure 21-7
The urea cycle, the citric acid cycle, and the transamination of oxaloacetate are linked by fumarate and aspartate.

INHERITED ENZYMATIC DEFECTS OF THE UREA CYCLE CAUSE HYPERAMMONEMIA

High levels of NH_4^+ are toxic to humans. The synthesis of urea in the liver is the major route of removal of NH_4^+. A complete block of any of the steps of the urea cycle is probably fatal, because there is no known alternative pathway for the synthesis of urea. Inherited disorders caused by a partial block of each of the urea cycle reactions have been diagnosed. *The common condition is an elevated level of NH_4^+ in the blood (hyperammonemia).* A nearly total deficiency of any of the urea cycle enzymes results in coma and death shortly after birth. Partial deficiencies of these enzymes cause mental retardation, lethargy, and episodic vomiting. A low-protein diet leads to a lowering of the ammonium level in the blood and to clinical improvement in the milder forms of these inherited disorders.

Why are high levels of NH_4^+ toxic? A high concentration of ammonium ion shifts the equilibrium of the reaction catalyzed by glutamate dehydrogenase toward the formation of glutamate. NH_4^+ then reacts with glutamate to form glutamine (p. 577). Elevated levels of glutamine are found in the cerebrospinal fluid of patients with hyperammonemia and may lead directly to brain damage. This important question deserves further study.

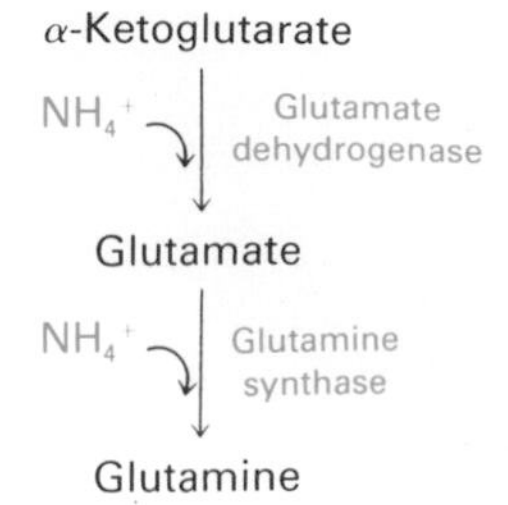

CARBON ATOMS OF DEGRADED AMINO ACIDS EMERGE IN MAJOR METABOLIC INTERMEDIATES

Thus far, we have considered a series of reactions that removes the α-amino group from amino acids and converts it into urea. We now turn to the fates of the remaining carbon skeletons. *The strategy of amino acid degradation is to form major metabolic intermediates that can be converted into glucose or be oxidized by the citric acid cycle.* In fact, the carbon skeletons of the diverse set of twenty amino acids are funneled into only seven molecules: *pyruvate, acetyl CoA, acetoacetyl CoA, α-ketoglutarate, succinyl CoA, fumarate, and oxaloacetate.* We see here a striking example of the remarkable economy of metabolic conversions.

Amino acids that are degraded to acetyl CoA or acetoacetyl CoA are termed *ketogenic* because they give rise to ketone bodies. In contrast, amino acids that are degraded to pyruvate, α-ketoglutarate, succinyl CoA, fumarate, or oxaloacetate are termed *glucogenic.* Net synthesis of glucose from these amino acids is feasible because these citric acid cycle intermediates and pyruvate can be converted into phosphoenolpyruvate and then into glucose (p. 438). Recall that mammals lack a pathway for the net synthesis of glucose from acetyl CoA or acetoacetyl CoA.

Of the basic set of twenty amino acids, only leucine and lysine are purely ketogenic. Isoleucine, phenylalanine, tryptophan, and tyrosine are both ketogenic and glucogenic. Some of their carbon atoms emerge in acetyl CoA or acetoacetyl CoA, whereas others appear in potential precursors of glucose. The other fourteen amino acids are classed as purely glucogenic. This classification is not universally accepted because different quantitative criteria are applied. Whether an amino acid is regarded as being glucogenic, ketogenic, or both depends partly on the eye of the beholder.

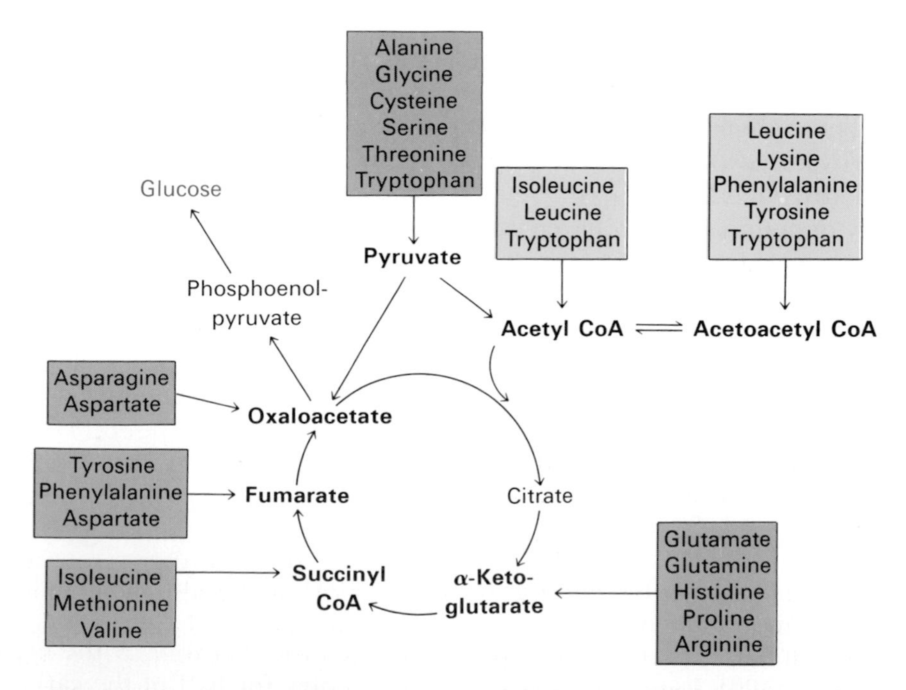

Figure 21-8
Fates of the carbon skeletons of amino acids. Glucogenic amino acids are shaded red, and ketogenic amino acids are shaded yellow.

THE C_3 FAMILY: ALANINE, SERINE, AND CYSTEINE ARE CONVERTED INTO PYRUVATE

Pyruvate is the entry point for the three-carbon amino acids: alanine, serine, and cysteine (Figure 21-9). The transamination of alanine directly yields pyruvate:

$$\text{Alanine} + \alpha\text{-ketoglutarate} \rightleftharpoons \text{pyruvate} + \text{glutamate}$$

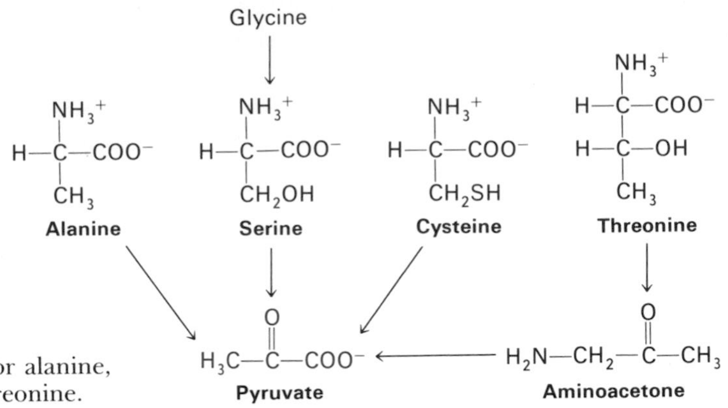

Figure 21-9
Pyruvate is the point of entry for alanine, serine, cysteine, glycine, and threonine.

As mentioned previously (p. 495), glutamate is then oxidatively deaminated, yielding NH_4^+ and regenerating α-ketoglutarate. The sum of these reactions is

$$\text{Alanine} + NAD^+ + H_2O \longrightarrow \text{pyruvate} + NH_4^+ + NADH + H^+$$

Another simple reaction in the degradation of amino acids is the *deamination of serine to pyruvate* by serine dehydratase (p. 499).

$$\text{Serine} \longrightarrow \text{pyruvate} + NH_4^+$$

Cysteine can be converted into pyruvate by several pathways, with its sulfur atom emerging in H_2S, SO_3^{2-}, or SCN^-.

The carbon atoms of three other amino acids can be converted into pyruvate. *Glycine* can be converted into serine by enzymatic addition of a hydroxymethyl group (p. 580). *Threonine* can give rise to pyruvate by way of aminoacetone. Three carbon atoms of *tryptophan* can emerge in alanine, which can be converted into pyruvate.

THE C_4 FAMILY: ASPARTATE AND ASPARAGINE ARE CONVERTED INTO OXALOACETATE

Aspartate, a four-carbon amino acid, is directly *transaminated to oxaloacetate,* a citric acid cycle intermediate:

$$\text{Aspartate} + \alpha\text{-ketoglutarate} \rightleftharpoons \text{oxaloacetate} + \text{glutamate}$$

Asparagine is hydrolyzed by *asparaginase* to NH_4^+ and aspartate, which is then transaminated.

Recall that aspartate can also be converted into *fumarate* by the urea cycle (p. 500). Fumarate is also a point of entry for half of the carbon atoms of tyrosine and phenylalanine, as will be discussed shortly.

THE C_5 FAMILY: SEVERAL AMINO ACIDS ARE CONVERTED INTO α-KETOGLUTARATE THROUGH GLUTAMATE

The carbon skeletons of several five-carbon amino acids enter the citric acid cycle at *α-ketoglutarate*. These amino acids are first converted into *glutamate*, which is then oxidatively deaminated by glutamate dehydrogenase to yield α-ketoglutarate (Figure 21-10).

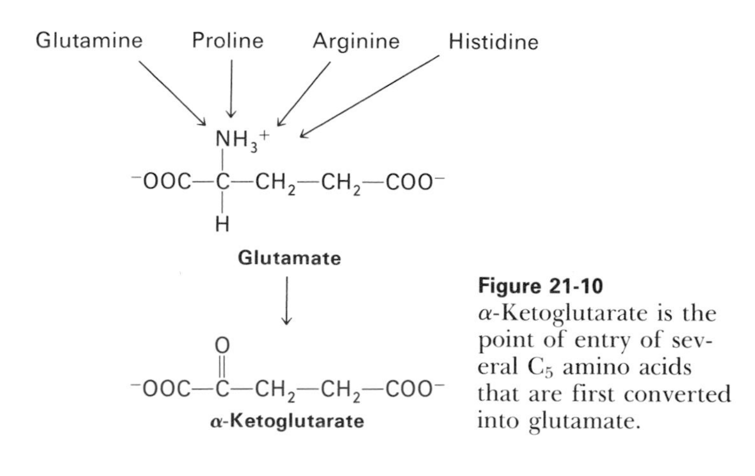

Figure 21-10
α-Ketoglutarate is the point of entry of several C_5 amino acids that are first converted into glutamate.

Histidine is converted into 4-imidazolone 5-propionate (Figure 21-11). The amide bond in the ring of this intermediate is hydrolyzed to the *N*-formimino derivative of glutamate, which is then converted into glutamate by transfer of its formimino group to tetrahydrofolate, a carrier of activated one-carbon units (see p. 580).

Glutamine is hydrolyzed to glutamate and NH_4^+ by glutaminase. *Proline* and *arginine* are converted into glutamate γ-semialdehyde, which is then oxidized to glutamate (Figure 21-12).

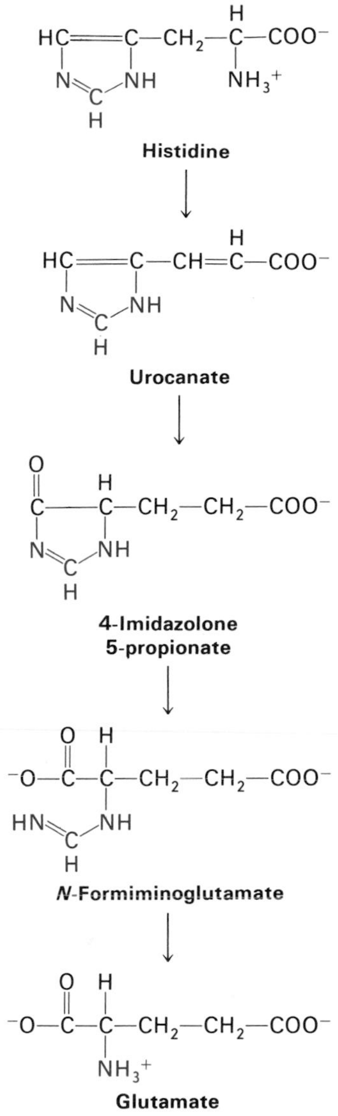

Histidine

Urocanate

4-Imidazolone 5-propionate

N-Formiminoglutamate

Glutamate

Figure 21-11
Conversion of histidine into glutamate.

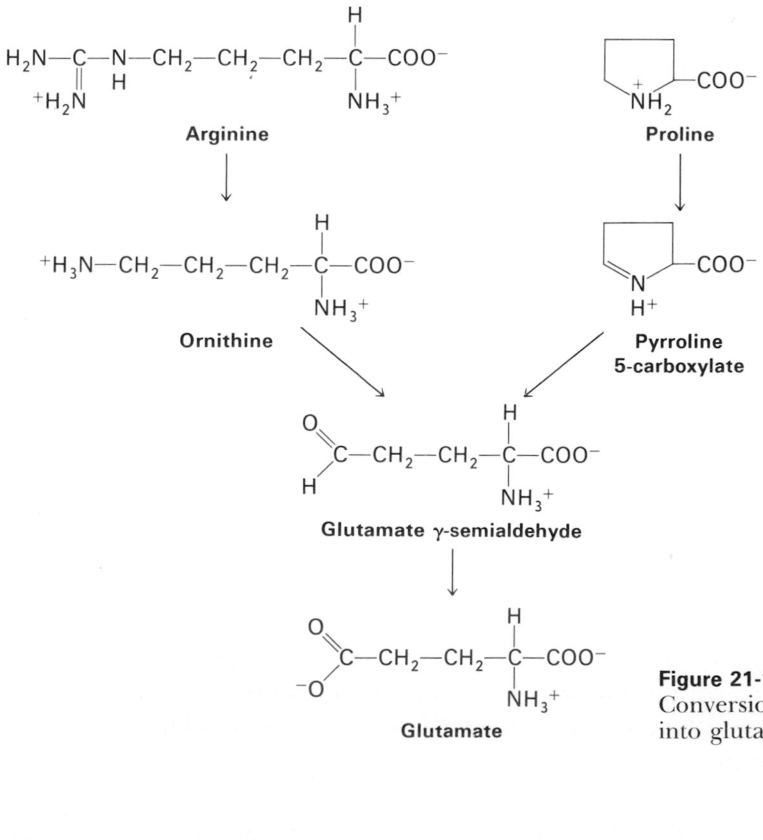

Arginine

Proline

Ornithine

Pyrroline 5-carboxylate

Glutamate γ-semialdehyde

Glutamate

Figure 21-12
Conversion of proline and arginine into glutamate.

SUCCINYL COENZYME A IS A POINT OF ENTRY FOR SOME AMINO ACIDS

Succinyl CoA is the point of entry for some of the carbon atoms of methionine, isoleucine, and valine. Methylmalonyl CoA is an intermediate in the breakdown of these three amino acids (Figure 21-13).

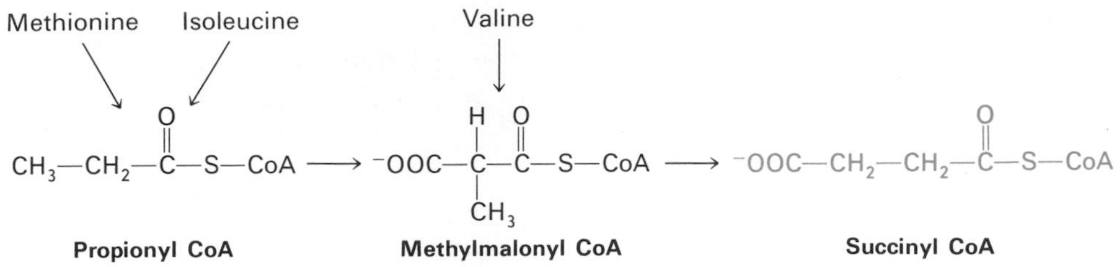

Methionine → Propionyl CoA; Isoleucine → Propionyl CoA; Valine → Methylmalonyl CoA

$$CH_3{-}CH_2{-}\overset{O}{\overset{\|}{C}}{-}S{-}CoA \longrightarrow {}^{-}OOC{-}\underset{CH_3}{\overset{H}{C}}{-}\overset{O}{\overset{\|}{C}}{-}S{-}CoA \longrightarrow {}^{-}OOC{-}CH_2{-}CH_2{-}\overset{O}{\overset{\|}{C}}{-}S{-}CoA$$

Propionyl CoA — Methylmalonyl CoA — Succinyl CoA

Figure 21-13
Conversion of methionine, isoleucine, and valine into succinyl CoA.

The pathway from propionyl CoA to succinyl CoA is especially interesting. Propionyl CoA is carboxylated at the expense of an ATP to yield the D-isomer of methylmalonyl CoA. This carboxylation reaction is catalyzed by *propionyl CoA carboxylase,* a biotin enzyme that has a catalytic mechanism like that of acetyl CoA carboxylase and pyruvate carboxylase. The D-isomer of methylmalonyl CoA is racemized to the L-isomer, which is the substrate for the mutase enzyme that converts it into succinyl CoA.

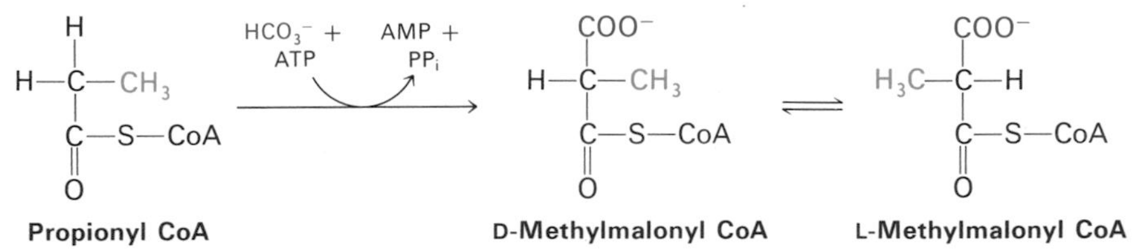

$$\text{H}{-}\overset{H}{C}{-}CH_3 \text{ (with } {-}C(=O){-}S{-}CoA) \xrightarrow[\text{ATP}\quad\quad PP_i]{HCO_3^- + \quad\quad AMP +} \text{H}{-}\overset{COO^-}{C}{-}CH_3 \text{ (with } {-}C(=O){-}S{-}CoA) \rightleftharpoons H_3C{-}\overset{COO^-}{C}{-}H \text{ (with } {-}C(=O){-}S{-}CoA)$$

Propionyl CoA — D-Methylmalonyl CoA — L-Methylmalonyl CoA

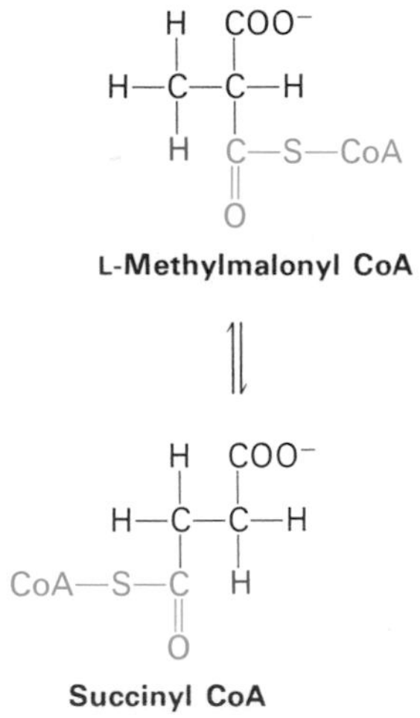

L-Methylmalonyl CoA ⇌ Succinyl CoA

Succinyl CoA is formed from L-methylmalonyl CoA by an *intramolecular rearrangement*. The —CO—S—CoA group migrates from C-2 to C-3 in exchange for a hydrogen atom. *This very unusual isomerization is catalyzed by methylmalonyl CoA mutase, one of the two mammalian enzymes known to contain a derivative of vitamin B_{12} as its coenzyme.*

This pathway from propionyl CoA to succinyl CoA also participates in the oxidation of *fatty acids that have an odd number of carbon atoms.* The final thiolytic cleavage of an odd-numbered acyl CoA yields acetyl CoA and propionyl CoA (p. 478). Hence, odd-carbon fatty acids are partly glucogenic; specifically, three of their carbon atoms can emerge in glucose.

THE COBALT ATOM OF VITAMIN B_{12} IS BONDED TO THE 5′-CARBON OF DEOXYADENOSINE IN COENZYME B_{12}

Cobalamin (vitamin B_{12}) has been a challenging problem in biochemistry and medicine since the discovery by George Minot and William Murphy in 1926 that pernicious anemia can be treated by feeding the patient large amounts of liver. Cobalamin was first purified in 1948; and it was crystallized then by Dorothy Hodgkin, who elucidated its complex three-dimensional structure in 1956. The core of cobalamin consists of a *corrin ring with a central cobalt atom* (Figure 21-14). The corrin ring, like a porphyrin, has *four pyrrole units.* Two of them (rings A and D) are directly bonded to each other, whereas the others are joined by methene bridges, as in porphyrins.

A cobalt atom is bonded to the four pyrrole nitrogens. *The fifth substituent* (below the corrin plane in Figure 21-15) is a derivative of *dimethylbenzimidazole* that contains ribose 3-phosphate and aminoisopropanol. One of the nitrogen atoms of dimethylbenzimidazole is linked to cobalt. The amino group of aminoisopropanol is in amide linkage with a side chain. The *sixth substituent* of the cobalt atom (located above the corrin plane in Figure 21-15) can be —CH_3, OH^-, or a 5′-deoxyadenosyl unit.

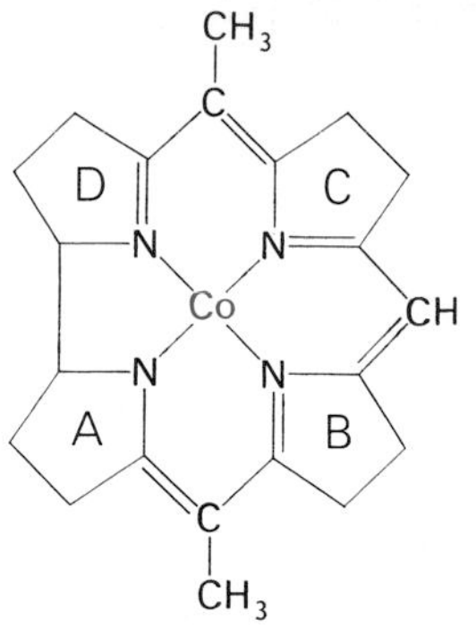

Figure 21-14
Corrin core of cobalamin (vitamin B_{12}). Substituents on the pyrroles and the other two ligands to cobalt are not shown in this diagram.

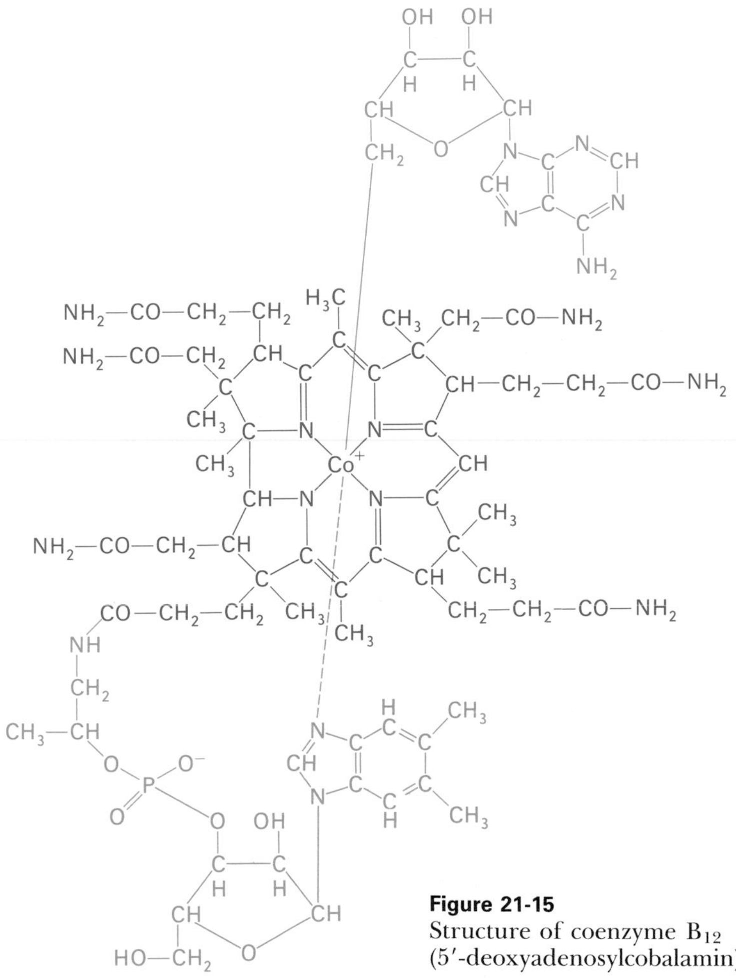

Figure 21-15
Structure of coenzyme B_{12} (5′-deoxyadenosylcobalamin).

The cobalt atom in cobalamin can have a +1, +2, or +3 oxidation state. The cobalt atom is in the +3 state in hydroxocobalamin (where OH^- occupies the sixth coordination site). This form, called B_{12a} (Co^{3+}), is reduced to a divalent state, called B_{12r} (Co^{2+}), by a flavoprotein reductase. The B_{12r} (Co^{2+}) form is reduced by a second flavoprotein reductase to B_{12s} (Co^+). NADH is the reductant in both reactions.

$$B_{12a}\ (Co^{3+}) \longrightarrow B_{12r}\ (Co^{2+}) \longrightarrow B_{12s}\ (Co^{+})$$

The B_{12s} form is the substrate for the final enzymatic reaction that yields the active coenzyme. Co^+ attacks the 5′-carbon atom of ATP and displaces the triphosphate group to form *5′-deoxyadenosylcobalamin,* also known as *coenzyme B_{12}* (Figure 21-16). This compound is remarkable in

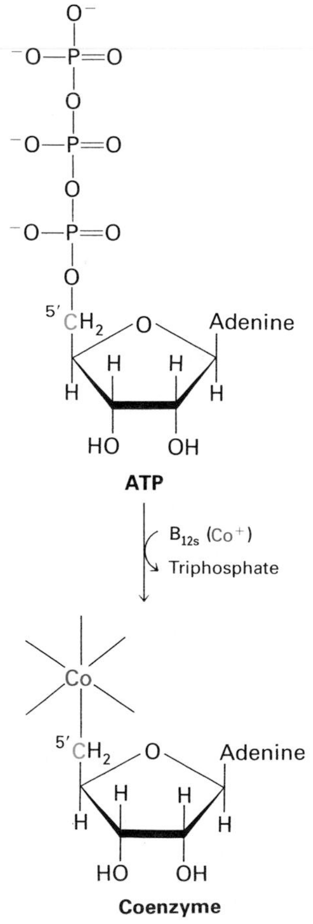

Figure 21-16
Formation of coenzyme B_{12} from cobalamin and ATP. The 5′ carbon of 5′-deoxyadenosine is coordinated to the cobalt atom in this coenzyme.

having a *carbon-metal bond*, the only one known in a biomolecule. Another unusual feature of this reaction is that the 5′-methylene carbon atom of ATP, rather than its α or β phosphorus atom, is the target of nucleophilic attack. The formation of *S*-adenosyl methionine (p. 582) is the only other biochemical reaction in which a nucleophile displaces the triphosphate group of ATP.

COENZYME B_{12} PROVIDES FREE RADICALS TO CATALYZE INTRAMOLECULAR MIGRATIONS INVOLVING HYDROGEN

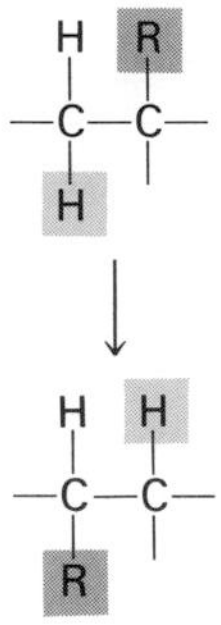

Cobalamin enzymes catalyze three types of reactions: (1) *intramolecular rearrangements*; (2) *methylations*, as in the synthesis of methionine (p. 583); and (3) *reduction of ribonucleotides to deoxyribonucleotides* (p. 610). The conversion of L-methylmalonyl CoA into succinyl CoA (an intramolecular rearrangement) and the formation of methionine by methylation of homocysteine are the only known reactions dependent on coenzyme B_{12} in mammals.

Figure 21-17
Rearrangement reaction catalyzed by cobalamin enzymes. The R group can be an amino group, a hydroxyl group, or a substituted carbon.

The rearrangement reactions catalyzed by coenzyme B_{12} are exchanges of two groups attached to adjacent carbon atoms (Figure 21-17). A hydrogen atom migrates from one carbon atom to the next, and an X group (such as the —CO—S—CoA group of methylmalonyl CoA) concomitantly moves in the reverse direction. The first step in these intramolecular rearrangements is the cleavage of the carbon–cobalt bond of 5′-deoxyadenosylcobalamin to form B_{12r} (Co^{2+}) and a 5′-deoxyadenosyl radical ($—CH_2\cdot$) (Figure 21-18). In this *homolytic cleavage reaction*, one electron of the Co–C bond stays with Co and the other with C, generating a free radical. In contrast, nearly all other cleavage reactions in biological systems are *heterolytic*—an electron pair is transferred to one of the two atoms that were bonded together.

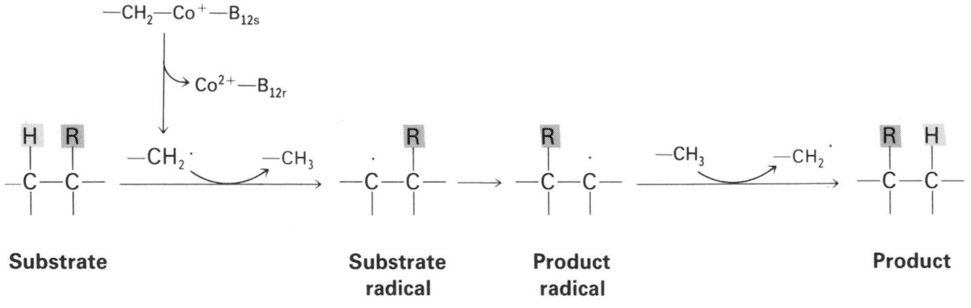

$—CH_2—Co^+—B_{12s}$ → $Co^{2+}—B_{12r}$

Substrate → Substrate radical → Product radical → Product

Figure 21-18
Coenzyme B_{12} provides the free radical that abstracts a hydrogen atom in rearrangement reactions.

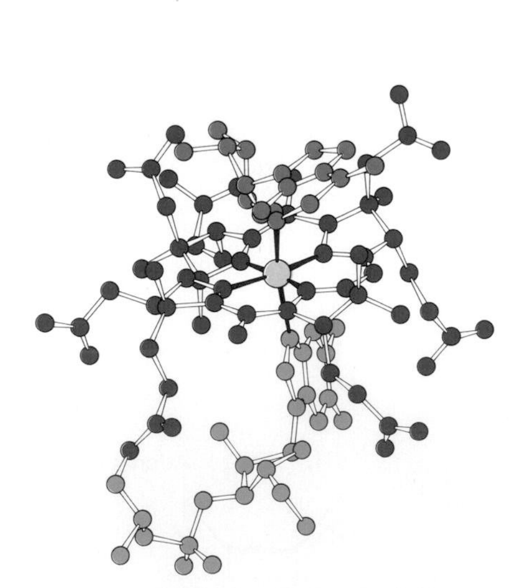

Figure 21-19
Model of coenzyme B_{12}. The cobalt atom is shown in yellow, the corrin unit in red, the 5′-deoxyadenosyl unit in blue, and the benzimidazole unit in green.

Why is this very unusual $—CH_2\cdot$ radical formed? This highly reactive species abstracts a *hydrogen atom* from the substrate to form 5′-deoxyadenosine ($—CH_3$) and a substrate radical. This sets the stage for the migration of X to the position formerly occupied by H on the neighboring carbon atom. Finally, the product radical abstracts a hydrogen atom from the 5′-methyl group to complete the rearrangement and return the deoxyadenosyl unit to the radical form. *The role of B_{12} in such intramolecular migrations is to serve as a source of free radicals for the abstraction of hydrogen atoms. A key property of coenzyme B_{12} is the weakness of its cobalt–carbon bond, whose facile cleavage generates a radical.* Steric crowding around the cobalt atom prevents the formation of a stronger bond, which would make the coenzyme a less effective catalyst.

ABSORPTION OF COBALAMIN IS IMPAIRED IN PERNICIOUS ANEMIA

Cobalamin is absorbed by a specialized transport system. The stomach secretes a glycoprotein called *intrinsic factor* (59 kd), which binds cobalamin in the intestinal lumen. This complex is subsequently bound by a *specific receptor in the lining of the ileum.* The complex of cobalamin and intrinsic factor is then dissociated by a *releasing factor* and actively transported across the ileal membrane into the bloodstream. *Pernicious anemia is caused by a deficiency of intrinsic factor, which leads to impaired absorption of cobalamin.* Consequently, the synthesis of purines and of thymine is impaired. This disease was originally treated by feeding patients large amounts of liver, a rich source of cobalamin, so that enough of the vitamin was absorbed even in the absence of intrinsic factor. The most reliable therapy is intramuscular injection of cobalamin at monthly intervals.

Animals and plants are unable to synthesize cobalamin. *This vitamin is unique in being synthesized only by microorganisms,* in particular, anaerobic bacteria. A normal person requires less than 10 μg of cobalamin per day. Nutritional deficiency of cobalamin is rare because this vitamin is found in virtually all animal tissues.

SEVERAL INHERITED DEFECTS OF METHYLMALONYL COENZYME A METABOLISM ARE KNOWN

Several *inherited disorders of methylmalonyl CoA metabolism* have been characterized. They usually become evident in the first year of life, when the striking symptom is *acidosis.* The pH of arterial blood is about 7.0, rather than the normal value of 7.4. *Large amounts of methylmalonate appear in the urine of patients who have these disorders.* A normal person excretes less than 5 mg of methylmalonate per day, whereas a patient with defective methylmalonyl CoA metabolism may excrete more than 1 g. About half of the patients with methylmalonic aciduria improve markedly when large doses of cobalamin are administered intramuscularly. The arterial blood pH returns to normal, and there is a marked decrease in the excretion of methylmalonate. These responsive patients usually have a defect in the transferase that catalyzes the synthesis of coenzyme B_{12} from B_{12s} and ATP.

$$B_{12s}\ (Co^{+}) \longrightarrow\!\!\!\!\!\!\!\!\!\!|\,|\quad \text{deoxyadenosylcobalamin}$$

In contrast, other patients with impaired methylmalonyl CoA metabolism do not respond to large doses of cobalamin. Some of them may have a defective methylmalonyl CoA mutase apoenzyme. This form of methylmalonic aciduria is frequently lethal.

LEUCINE IS DEGRADED TO ACETYL COENZYME A AND ACETOACETYL COENZYME A

As was mentioned earlier, leucine and lysine are the only purely ketogenic amino acids of the common set of twenty. First, leucine is transaminated to the corresponding α-keto acid, *α-ketoisocaproate*. This α-keto acid is then degraded by reactions akin to those occurring in the citric acid cycle and fatty acid oxidation. It is *oxidatively decarboxylated* to *isovaleryl CoA*. This reaction is analogous to the oxidative decarboxylation of pyruvate to acetyl CoA and of α-ketoglutarate to succinyl CoA.

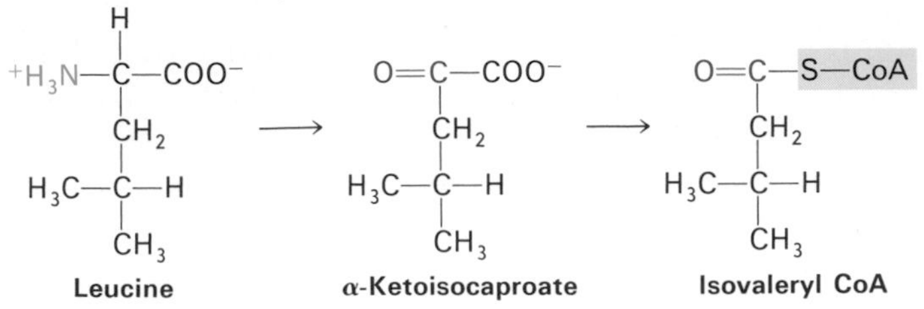

H
$^{+}H_3N$—C—COO^-
CH_2
H_3C—C—H
CH_3
Leucine

→

O=C—COO^-
CH_2
H_3C—C—H
CH_3
α-Ketoisocaproate

→

O=C—S—CoA
CH_2
H_3C—C—H
CH_3
Isovaleryl CoA

Isovaleryl CoA is *dehydrogenated* to yield *β-methylcrotonyl CoA*. This oxidation is catalyzed by isovaleryl CoA dehydrogenase, in which the hydrogen acceptor is FAD, as in the analogous reaction in fatty acid oxidation that is catalyzed by acyl CoA dehydrogenase. *β-Methylglutaconyl CoA* is formed by *carboxylation* of β-methylcrotonyl CoA at the expense of an ATP. As might be expected, the mechanism of carboxylation of β-methylcrotonyl CoA carboxylase is very similar to that of pyruvate carboxylase and acetyl CoA carboxylase. In fact, much of our present knowledge of the mechanism of biotin-dependent carboxylations comes from Feodor Lynen's pioneering work on this enzyme.

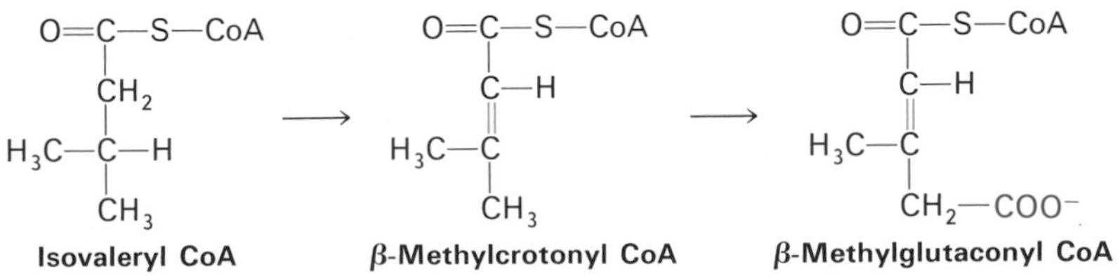

O=C—S—CoA
CH_2
H_3C—C—H
CH_3
Isovaleryl CoA

→

O=C—S—CoA
C—H
H_3C—C
CH_3
β-Methylcrotonyl CoA

→

O=C—S—CoA
C—H
H_3C—C
CH_2—COO^-
β-Methylglutaconyl CoA

β-Methylglutaconyl CoA is then *hydrated* to form *β-hydroxy-β-methylglutaryl CoA,* which is cleaved to *acetyl CoA* and *acetoacetate*. This reaction has already been discussed in regard to the formation of ketone bodies from fatty acids (p. 479).

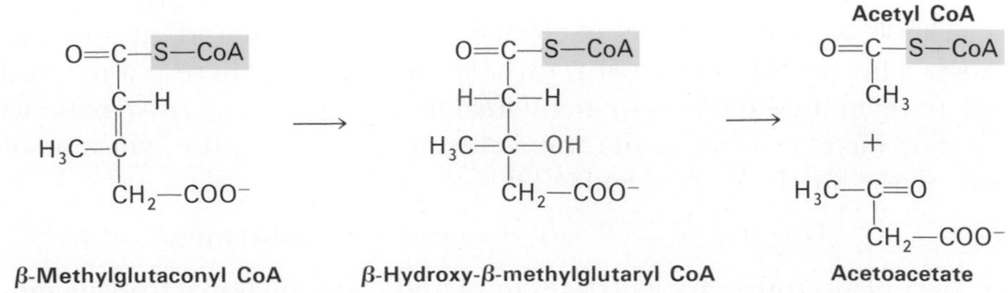

O=C—S—CoA
C—H
H_3C—C
CH_2—COO^-
β-Methylglutaconyl CoA

→

O=C—S—CoA
H—C—H
H_3C—C—OH
CH_2—COO^-
β-Hydroxy-β-methylglutaryl CoA

→

Acetyl CoA
O=C—S—CoA
CH_3
+
H_3C—C=O
CH_2—COO^-
Acetoacetate

It is interesting to note that many coenzymes participate in the degradation of leucine to acetyl CoA and acetoacetate: *PLP* in transamination; *TPP, lipoate, FAD, and NAD*$^+$ in oxidative decarboxylation; FAD again in dehydrogenation; and *biotin* in carboxylation. *Coenzyme A* is the acyl carrier in these reactions.

The degradative pathways of valine and isoleucine resemble that of leucine. All three amino acids are initially transaminated to the corresponding α-keto acid, which is then oxidatively decarboxylated to yield a derivative of CoA. The subsequent reactions are like those of fatty acid oxidation. Isoleucine yields acetyl CoA and propionyl CoA, whereas valine yields methylmalonyl CoA. There is an inborn error of metabolism that affects the degradation of valine, isoleucine, and leucine. In *maple syrup urine disease,* the oxidative decarboxylation of these three amino acids is blocked. The amounts of leucine, isoleucine, and valine in blood and urine are markedly elevated, which results in a corresponding increase in the α-keto acids derived from these amino acids. The urine of patients having this disease has the odor of maple syrup, hence the name of the disease. Maple syrup urine disease is usually fatal unless the patient is placed on a diet low in valine, isoleucine, and leucine early in life.

PHENYLALANINE AND TYROSINE ARE DEGRADED BY OXYGENASES TO ACETOACETATE AND FUMARATE

The pathway for the degradation of phenylalanine and tyrosine has some very interesting features. This series of reactions shows how *molecular oxygen is used to break an aromatic ring*. The first step is the hydroxylation of phenylalanine to tyrosine, a reaction catalyzed by *phenylalanine hydroxylase*. This enzyme is called a *monooxygenase* (also called a *mixed-function oxygenase*) because *one atom of* O_2 *appears in the product and the other in* H_2O.

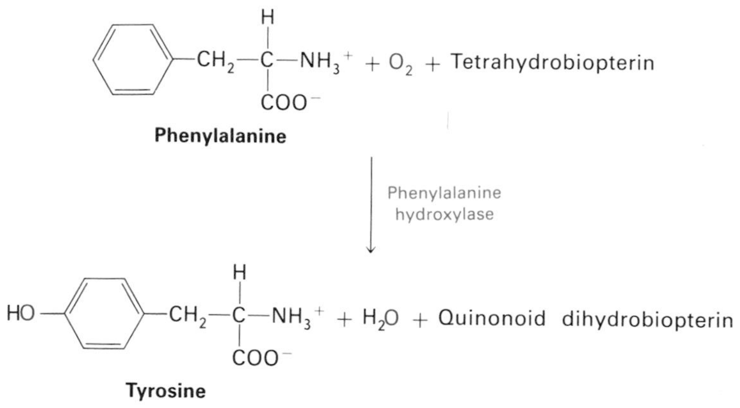

Phenylalanine + O_2 + Tetrahydrobiopterin

↓ Phenylalanine hydroxylase

Tyrosine + H_2O + Quinonoid dihydrobiopterin

The reductant here is *tetrahydrobiopterin,* an electron carrier that has not been previously discussed. Tetrahydrobiopterin is initially formed by reduction of dihydrobiopterin by NADPH, in a reaction catalyzed by dihydrofolate reductase (Figure 21-20). The quinonoid form of dihydrobiopterin produced in the hydroxylation of phenylalanine is reduced back to tetrahydrobiopterin by NADH in a reaction catalyzed by dihydropteridine reductase. The sum of the reactions catalyzed by phenylalanine hydroxylase and dihydropteridine reductase is

$$\text{Phenylalanine} + O_2 + \text{NADH} + H^+ \longrightarrow \text{tyrosine} + \text{NAD}^+ + H_2O$$

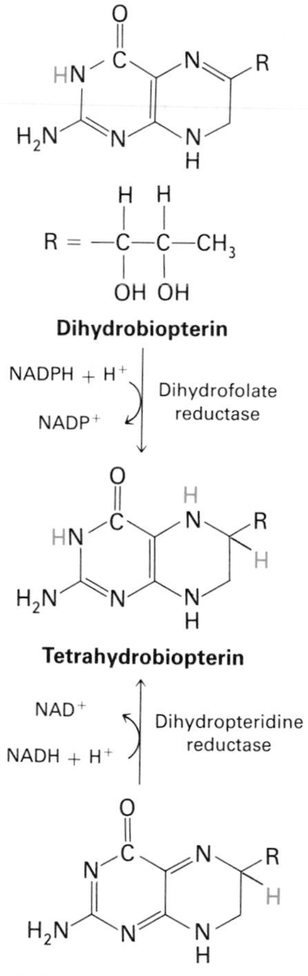

Dihydrobiopterin

$NADPH + H^+ \rightarrow NADP^+$ (Dihydrofolate reductase)

Tetrahydrobiopterin

$NADH + H^+ \rightarrow NAD^+$ (Dihydropteridine reductase)

Quinonoid dihydrobiopterin

Figure 21-20
Formation of tetrahydrobiopterin by reduction of either of two forms of dihydrobiopterin.

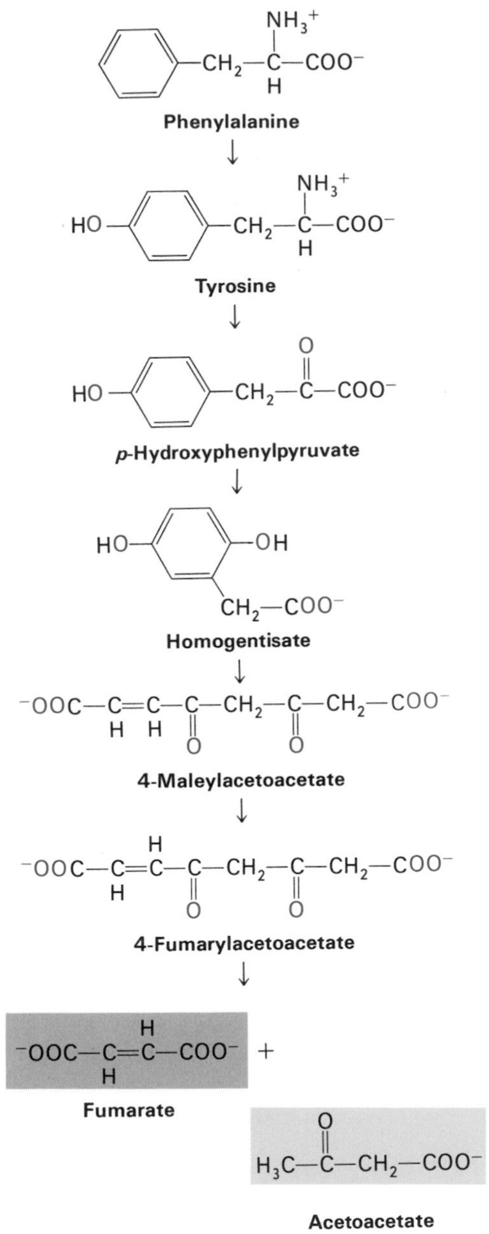

The next step is the transamination of tyrosine to p-*hydroxyphenylpyruvate* (Figure 21-21). This α-keto acid then reacts with O_2 to form *homogentisate*. The enzyme catalyzing this complex reaction, *p*-hydroxyphenylpyruvate hydroxylase, is called a *dioxygenase* because both atoms of O_2 become incorporated into the product, one on the ring and one in the carboxyl group. The aromatic ring of homogentisate is then cleaved by O_2, which yields 4-maleylacetoacetate. This reaction is catalyzed by homogentisate oxidase, another dioxygenase. 4-Maleylacetoacetate is then isomerized to *4-fumaryl acetoacetate* by an enzyme that uses glutathione as a cofactor. Finally, 4-fumarylacetoacetate is hydrolyzed to *fumarate* and *acetoacetate*.

We previously encountered a dioxygenase in the hydroxylation of proline (p. 263). Recall that α-ketoglutarate participates in this reaction and that one atom of O_2 emerges in hydroxyproline and the other in succinate. Prolyl hydroxylase and lysyl hydroxylase are *intermolecular dioxygenases* because they catalyze the incorporation of one atom of oxygen into each of *two* separate products. In contrast, *p*-hydroxyphenylpyruvate hydroxylase and homogentisate oxidase are *intramolecular dixoygenases* because both atoms of O_2 appear in the same product (Figure 21-22). The active sites of these enzymes contain iron that is not part of heme or an iron-sulfur cluster. *Nearly all cleavages of aromatic rings in biological systems are catalyzed by dioxygenases,* a class of enzymes discovered by Osamu Hayaishi.

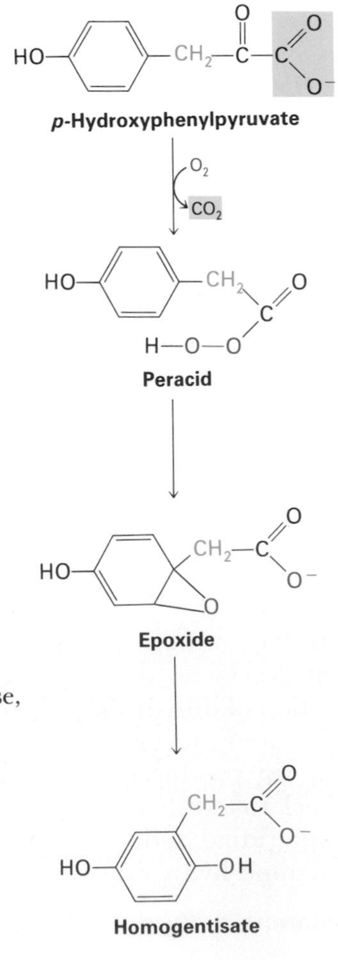

Phenylalanine

Tyrosine

p-Hydroxyphenylpyruvate

Homogentisate

4-Maleylacetoacetate

4-Fumarylacetoacetate

Fumarate

Acetoacetate

Figure 21-21
Pathway for the degradation of phenylalanine and tyrosine.

p-Hydroxyphenylpyruvate

Peracid

Epoxide

Homogentisate

Figure 21-22
Formation of homogentisate. *p*-Hydroxyphenylpyruvate hydroxylase, the enzyme catalyzing this reaction, is an intramolecular dixoygenase. Both atoms of O_2 emerge in homogentisate.

GARROD'S DISCOVERY OF INBORN ERRORS OF METABOLISM

Alcaptonuria is an inherited metabolic disorder caused by the absence of homogentisate oxidase. Homogentisate accumulates and is excreted in the urine, which turns dark on standing as homogentisate is oxidized and polymerized to a melaninlike substance. Alcaptonuria is a relatively benign condition, as described by Zacutus Lusitanus in 1649:

> The patient was a boy who passed black urine and who, at the age of fourteen years, was submitted to a drastic course of treatment which had for its aim the subduing of the fiery heat of his viscera, which was supposed to bring about the condition in question by charring and blackening his bile. Among the measures prescribed were bleedings, purgation, baths, a cold and watery diet and drugs galore. None of these had any obvious effect, and eventually the patient, who tired of the futile and superfluous therapy, resolved to let things take their natural course. None of the predicted evils ensued, he married, begat a large family, and lived a long and healthy life, always passing urine black as ink.

In 1902, Archibald Garrod showed that alcaptonuria is transmitted as a single recessive Mendelian trait. Furthermore, he recognized that homogentisate is a normal intermediate in the degradation of phenylalanine and tyrosine and that it accumulates in alcaptonuria because its degradation is blocked. He concluded that "the splitting of the benzene ring in normal metabolism is the work of a special enzyme, that in congenital alcaptonuria this enzyme is wanting." Garrod perceived the direct relationship between genes and enzymes, and he recognized the importance of chemical individuality. His book *Inborn Errors of Metabolism* was a most imaginative and important contribution to biology and medicine.

A BLOCK IN THE HYDROXYLATION OF PHENYLALANINE CAN LEAD TO SEVERE MENTAL RETARDATION

Phenylketonuria (PKU), an inborn error of phenylalanine metabolism, can have devastating effects, in contrast with alcaptonuria. Almost all untreated individuals with phenylketonuria are *severely mentally retarded.* In fact, about 1% of patients in mental institutions have phenylketonuria. The weight of the brain of these individuals is below normal, myelination of their nerves is defective, and their reflexes are hyperactive. The life expectancy of untreated phenylketonurics is drastically shortened. Half are dead by age twenty, and three-quarters by age thirty.

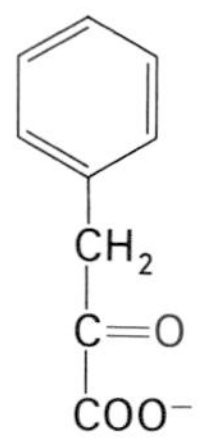

Phenylketonuria is caused by an *absence or deficiency of phenylalanine hydroxylase* or, more rarely, of its tetrahydrobiopterin cofactor. Phenylalanine cannot be converted into tyrosine, and so there is an accumulation of *phenylalanine in all body fluids.* Some minor fates of phenylalanine in normal persons become prominent in phenylketonurics. The most evident one is the transamination of phenylalanine to form *phenylpyruvate.* The disease acquired its name from the high levels of this phenylketone in urine. Phenyllactate, phenylacetate, and *o*-hydroxylphenylacetate are derived from phenylpyruvate. The α-amino group of glutamine forms an amide bond with the carboxyl group of phenylacetate, which yields phenylacetylglutamine.

CH_2
$C=O$
COO^-
Phenylpyruvate

There are many other derangements of amino acid metabolism associated with phenylketonuria, particularly among the aromatic compounds. Phenylketonu-

> *Melanin*—
> A black pigment in skin and hair. From the Greek word *melan,* meaning black. This polymeric pigment is formed in granules, called melanosomes, that are rich in *tyrosinase,* a monoxygenase.

rics have a lighter skin and hair color than their siblings. The hydroxylation of tyrosine is the first step in the formation of the pigment melanin. In phenylketonurics this reaction is competitively inhibited by the high levels of phenylalanine, and so less melanin is formed.

Phenylketonurics appear normal at birth but are severely defective by age one if untreated. The therapy for phenylketonuria is a *low phenylalanine diet.* The aim is to provide just enough phenylalanine to meet the needs for growth and replacement. Proteins that have an initially low content of phenylalanine, such as casein from milk, are hydrolyzed and phenylalanine is removed by adsorption. A low phenylalanine diet must be started very soon after birth to prevent irreversible brain damage. In one study, the average I.Q. of phenylketonurics treated within a few weeks after birth was 93; a control group of siblings treated starting at age one had an average I.Q. of 53. *However, the biochemical basis of mental retardation in untreated phenylketonuria is an enigma.*

Early diagnosis of phenylketonuria is essential and has been accomplished by mass screening programs. In past years, the urine of newborns was assayed by the addition of $FeCl_3$, which gives an olive green color in the presence of phenylpyruvate. The phenylalanine level in the blood is now the preferred diagnostic criterion because it is more reliable. Furthermore, the gene for human phenylalanine hydroxylase has recently been cloned, so that prenatal diagnosis of phenylketonuria is now feasible with DNA probes. The incidence of phenylketonuria is about 1 in 20,000 newborns. The disease is inherited as an *autosomal recessive.* Heterozygotes, which comprise about 1.5% of a typical population, appear normal. Carriers of the phenylketonuric gene have a reduced level of phenylalanine hydroxylase, as reflected in an increased level of phenylalanine in the blood. However, these criteria are not absolute, because the blood levels of phenylalanine in carriers and normal persons overlap to some extent. The measurement of the kinetics of disappearance of intravenously administered phenylalanine is a more definitive test for the carrier state. It should be noted that a high blood level of phenylalanine in a pregnant woman can result in abnormal development of the fetus. This is a striking example of maternal-fetal relationships at the molecular level.

SUMMARY

Surplus amino acids are used as metabolic fuel. The degradation of most surplus amino acids starts with the removal of their α-amino groups by transamination to an α-keto acid. Pyridoxal phosphate is the coenzyme in all aminotransferases and in many other enzymes catalyzing transformations of amino acids. The α-amino groups funnel into α-ketoglutarate to form glutamate, which is then oxidatively deaminated by glutamate dehydrogenase to give NH_4^+ and α-ketoglutarate. NAD^+ or $NADP^+$ is the electron acceptor in this reaction. In terrestrial vertebrates, NH_4^+ is converted into urea by the urea cycle. Urea is formed by the hydrolysis of arginine. The subsequent reactions of the urea cycle synthesize arginine from ornithine, the other product of the hydrolysis reaction. First, ornithine is carbamoylated to citrulline by carbamoyl phosphate. Citrulline then condenses with aspartate to form argininosuccinate, which is cleaved to arginine and fumarate. The carbon atom and one nitrogen atom of urea come from carbamoyl phosphate, which is synthesized from CO_2, NH_4^+, and ATP. The other nitrogen atom of urea comes from aspartate.

The carbon atoms of degraded amino acids are converted into pyruvate, acetyl CoA, acetoacetate, or an intermediate of the citric acid cycle. Most amino acids are purely glucogenic, two are purely ketogenic, and a few are both ketogenic and glucogenic. Alanine, serine, cysteine, glycine, and threonine are degraded to pyruvate. Asparagine and aspartate are converted into oxaloacetate. α-Ketoglutarate is the point of entry for glutamate and four amino acids (glutamine, histidine, proline, and arginine) that can be converted into glutamate. Succinyl CoA is the point of entry for some of the carbon atoms of three amino acids (methionine, isoleucine, and valine) that are degraded by way of methylmalonyl CoA. 5′-Deoxyadenosylcobalamin, a coenzyme formed from vitamin B_{12} and ATP, is the free-radical source in the isomerization of methylmalonyl CoA to succinyl CoA. Leucine is degraded to acetoacetyl CoA and acetyl CoA.

The aromatic rings of tyrosine and phenylalanine are degraded by oxygenases. Phenylalanine hydroxylase, a monooxygenase, uses tetrahydrobiopterin as the reductant. One of the oxygen atoms of O_2 emerges in tyrosine and the other in water. Absence or inactivity of this enzyme causes phenylketonuria; mental retardation results unless a low phenylalanine diet is started in infancy. Subsequent steps in the degradation of these aromatic amino acids are catalyzed by intramolecular dioxygenases that catalyze the insertion of both atoms of O_2 into a single product. Four of the carbon atoms of phenylalanine and tyrosine are converted into fumarate, and four emerge in acetoacetate.

SELECTED READINGS

Where to start

Halpern, J., 1985. Mechanisms of coenzyme B_{12}-dependent rearrangements. *Science* 227:869–875.

Books

Bender, D. A., 1985. *Amino Acid Metabolism* (2nd ed.). Wiley.

Christen, P., and Metzler, D. E., 1985. *Transaminases.* Wiley.

Meister, A., 1965. *Biochemistry of the Amino Acids* (2nd ed.), vols 1 and 2. Academic Press.

Grisolia, S., Báguena, R., and Mayor, F., (eds.), 1976. *The Urea Cycle.* Wiley.

Nozaki, M., Yamamoto, S., Ishimura, Y., Coon, M. J., Ernster, L., and Estabrook, R. W., (eds.), 1982. *Oxygenases and Oxygen Metabolism. A Symposium in Honor of Osamu Hayaishi.* Academic Press.

Reviews

Barker, H. A., 1981. Amino acid degradation by anaerobic bacteria. *Ann. Rev. Biochem.* 50:23–40.

Cooper, A. J. L., 1983. Biochemistry of sulfur-containing amino acids. *Ann. Rev. Biochem.* 52:187–222.

Nichol, C. A., Smith, G. K., and Duch, D. S., 1985. Biosynthesis and metabolism of tetrahydrobiopterin and molybdopterin. *Ann. Rev. Biochem.* 54:729–764.

Reaction mechanisms

Walsh, C., 1979. *Enzymatic Reaction Mechanisms.* W. H. Freeman. [Contains excellent accounts of the catalytic mechanisms of pyridoxal phosphate enzymes, cobalamin enzymes, and oxygenases.]

Kirsch, J. F., Eichele, G., Ford, G. C., Vincent, M. G., Jansonius, J. N., Gehring, H., and Christen, P., 1984. Mechanism of action of aspartate amino transferase proposed on the basis of its spatial structure. *J. Mol. Biol.* 174:497–525.

Arnone, A., Rogers, P. H., Hyde, C. C., Briley, P. D., Metzler, C. M., and Metzler, D. E., 1985. Cytosolic pig heart aspartate aminotransferase: the structure of the internal aldimine, external aldimine, ketimine and of the beta subform. *In* Christer, P., and Metzler, D. E., (eds.), *The Transaminases*, pp. 138–155. Wiley.

Snell, E. E., and DiMari, S. J., 1970. Schiff base intermediates in enzyme catalysis. *In* Boyer, P. D., (ed.), *The Enzymes* (3rd ed.), vol. 2, pp. 335–370. Academic Press.

Barker, H. A., 1972. Coenzyme B_{12}-dependent mutases causing carbon chain rearrangements. *In* Boyer, P. D., (ed.), *The Enzymes* (3rd ed.), vol. 6, pp. 509–537.

Abeles, R., and Dolphin, D., 1976. The vitamin B_{12} coenzyme. *Acc. Chem. Res.* 9:114–120.

Genetic diseases

Nyhan, W. L., (ed.), 1984. *Abnormalities in Amino Acid Metabolism in Clinical Medicine.* Appleton-Century-Crofts.

Stanbury, J. B., Wyngaarden, J. B., Fredrickson, D. S., Goldstein, J. L., and Brown, M. S., (eds.), 1983. *The Metabolic Basis of Inherited Disease* (5th ed.), McGraw-Hill. [Part 3, entitled "Disorders of Amino Acid Metabolism," contains many fine articles on this subject.]

Wellner, D., and Meister, A., 1981. A survey of inborn errors of amino acid metabolism and transport in man. *Ann. Rev. Biochem.* 50:911–968.

Ledley, F. D., Levy, H. L., and Woo, S. L. C., 1986. Molecular analysis of the inheritance of phenylketonuria and mild hyperphenylalanemia in families with both disorders. *New Engl. J. Med.* 314:1276–1280.

Batshaw, M. L., Brusilow, S., Waber, L., Blom, W., Brubakk, A. M., Burton, B. K., Cann, H. M., Kerr, D., Mamunes, P., Matalon, R., Myerberg, D., and Schafer, I. A., 1982. Treatment of inborn errors of urea synthesis. *New Engl. J. Med.* 306:1387–1392.

DiMagno, E. P., Lowe, J. E., Snodgrass, P. J., and Jones, J. D., 1986. Ornithine transcarbamylase deficiency: a cause of bizarre behavior in a man. *New Engl. J. Med.* 315:744–747.

Wilcken, D. E. L., Wilcken, B., Dudman, N. P. B., and Tyrrell, P. A., 1983. Homocystinuria—the effects of betaine in the treatment of patients not responsive to pyridoxine. *New Engl. J. Med.* 309:448–453.

Historical aspects and the process of discovery

Garrod, A. E., 1909. *Inborn Errors in Metabolism.* Oxford University Press (reprinted in 1963 with a supplement by H. Harris).

Childs, B., 1970. Sir Archibald Garrod's conception of chemical individuality: a modern appreciation. *New Engl. J. Med.* 282:71–78.

Holmes, F. L., 1980. Hans Krebs and the discovery of the ornithine cycle. *Fed. Proc.* 39:216–225.

PROBLEMS

1. Name the α-keto acid that is formed by transamination of each of the following amino acids:
   (a) Alanine. (d) Leucine.
   (b) Aspartate. (e) Phenylalanine.
   (c) Glutamate. (f) Tyrosine.

2. Write a balanced equation for the conversion of aspartate into glucose by way of oxaloacetate. Cite the coenzymes that participate in these steps.

3. Write a balanced equation for the conversion of aspartate into oxaloacetate by way of fumarate.

4. Consider the mechanism of the conversion of L-methylmalonyl CoA into succinyl CoA by L-methylmalonyl CoA mutase.
   (a) Design an experiment to distinguish between the migration of the COO^- group and that of the —CO—S—CoA group in this reaction.
   (b) What is the significance of the finding that no tritium is incorporated into succinyl CoA when the mutase reaction is carried out in tritiated water?

5. Pyridoxal phosphate stabilizes carbanionic intermediates by serving as an electron sink. Which other prosthetic group catalyzes reactions in this way?

6. Propose a role for the positively charged guanido nitrogen in the cleavage of argininosuccinate into arginine and fumarate.

7. Methylmalonyl mutase is incubated with deuterated methylmalonyl CoA. Coenzyme B_{12} extracted from mutase in this reaction mixture is found to contain deuterium in its 5′-methylene group. Account for the transfer of label from substrate to coenzyme.

8. Heterolytic cleavage of a C–H bond can yield two types of products. What are they?

CHAPTER 22

Photosynthesis

All free energy consumed by biological systems arises from solar energy that is trapped by the process of photosynthesis. The basic equation of photosynthesis is simple—indeed deceptively so:

$$H_2O + CO_2 \xrightarrow{\text{light}} (CH_2O) + O_2$$

In this equation, (CH_2O) represents carbohydrate, primarily sucrose and starch. The mechanism of photosynthesis is complex and requires the interplay of many proteins and small molecules. Photosynthesis in green plants takes place in *chloroplasts*. The energy conversion apparatus is an integral part of the *thylakoid membrane system* of these organelles (Figure 22-1). The first step in photosynthesis is the absorption of light by *chlorophyll*, a porphyrin with a coordinated magnesium ion. The resulting electronic excitation passes from one chlorophyll molecule to another in a light-harvesting complex until the excitation is trapped by a chlorophyll with special properties. At such a *reaction center*, the energy of the excited electron is converted into a separation of charge. In essence, *light is used to create reducing potential.*

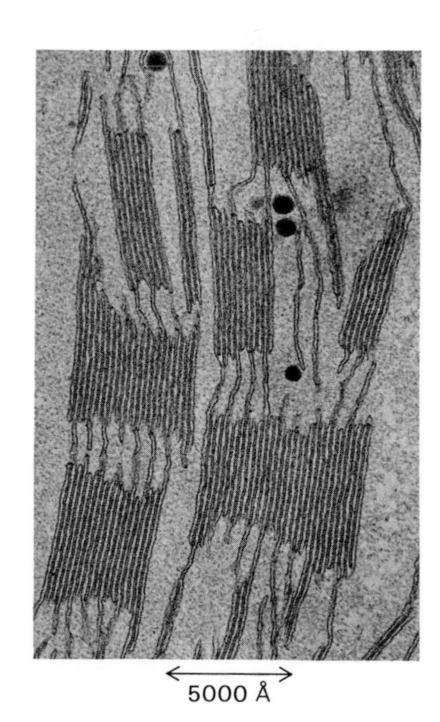

Figure 22-1
Electron micrograph of part of a chloroplast from a spinach leaf. The thylakoid membranes pile on top of each other to form grana. [Courtesy of Dr. Kenneth Miller.]

Photosynthesis in green plants is mediated by two kinds of light reactions. *Photosystem I* generates reducing power in the form of NADPH. *Photosystem II* transfers the electrons of water to a quinone and concomitantly evolves O_2. Electron flow between the photosystems generates a transmembrane proton gradient that is used to drive the synthesis of ATP, as in oxidative phosphorylation. Indeed, photosynthesis closely resembles oxidative phosphorylation in many ways. The principal difference between these energy transduction processes is in the sources of high-potential electrons. In oxidative phosphorylation, they come from the oxidation of fuels; in photosynthesis, they are produced by photoexcitation of chlorophyll. NADPH and ATP formed by the action of light then reduce CO_2 and convert it into *3-phosphoglycerate* by a series of dark reactions called the *Calvin cycle*, which occur in the stroma of chloroplasts. Hexoses are formed from 3-phosphoglycerate by the gluconeogenic pathway.

THE PRIMARY EVENTS OF PHOTOSYNTHESIS OCCUR IN THYLAKOID MEMBRANES

Chloroplasts, the organelles of photosynthesis, are typically 5 μm long. Like a mitochondrion, a chloroplast has an outer membrane and an inner membrane, with an intervening intermembrane space (Figure 22-2). The inner membrane surrounds a *stroma* containing soluble enzymes and membranous structures called *thylakoids,* which are flattened sacs. A pile of these sacs is called a granum. Different grana are connected by regions of thylakoid membrane called stroma lamellae. The thylakoid membranes separate the thylakoid space from the stroma space. Thus, chloroplasts have three different membranes (*outer, inner, and thylakoid membranes*) and three separate spaces (*intermembrane, stroma, and thylakoid spaces*). In developing chloroplasts, thylakoids arise from invaginations of the inner membrane, and so they are analogous to mitochondrial cristae.

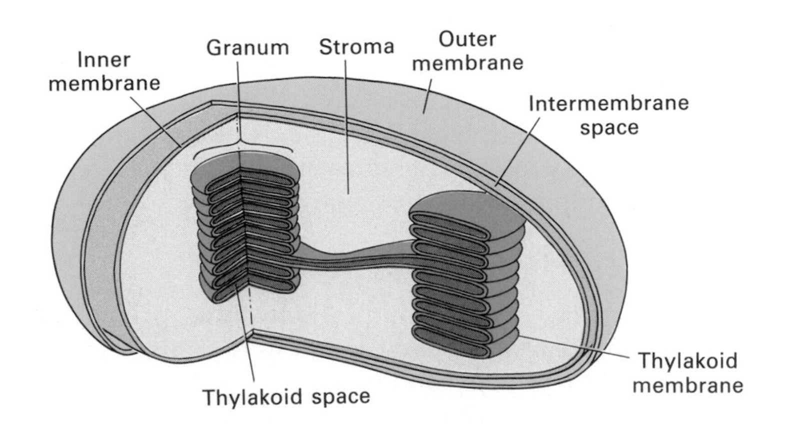

Figure 22-2
Diagram of a chloroplast. [After S. L. Wolfe, *Biology of the Cell*, p. 130. © 1972 by Wadsworth Publishing Company, Inc. Adapted by permission of the publisher.]

The thylakoid membranes contain the energy-transducing machinery: the light-harvesting proteins, reaction centers, electron-transport chains, and ATP synthase. They have nearly equal amounts of lipids and proteins. The lipid composition is highly distinctive: about 40% of the total lipids are *galactolipids* and 4% are *sulfolipids*, whereas only 10% are phospholipids. The thylakoid membrane, like the inner mitochondrial membrane, is impermeable to most molecules and ions. The outer membrane of a chloroplast, like that of a mitochondrion, is highly permeable to small molecules and ions. The stroma contains the soluble enzymes that utilize the NADPH and ATP synthesized by the thylakoids to convert CO_2 into sugar. Chloroplasts contain their own DNA and the machinery for replicating and expressing it. However, chloroplasts (like mitochondria) are not autonomous: they also contain proteins encoded by nuclear DNA.

Most of the basic equation of photosynthesis could have been written at the end of the eighteenth century. The production of oxygen in photosynthesis was discovered in 1780 by Joseph Priestley, an English chemist and nonconformist minister. He found that plants could "restore air which has been injured by the burning of candles." He placed a sprig of mint in an inverted glass jar in a vessel of water and found several days later "that the air would neither extinguish a candle, nor was it all inconvenient to a mouse which I put into it" (Figure 22-3).

The next major contribution to the elucidation of photosynthesis was made by Jan Ingenhousz, a Dutchman, who was court physician to the Austrian empress. Ingenhousz was a worldly man who liked to visit London. He once heard a discussion of Priestley's experiments on the restoration of air by plants and was so taken by it that he decided that he must do some experiments at the "earliest opportunity." This came six years later, when Ingenhousz rented a villa near London and spent a summer feverishly performing more than five hundred experiments. He discovered the role of light in photosynthesis:

Figure 22-3
Priestley's classic experiment on photosynthesis. [After E. I. Rabinowitch. Photosynthesis. Copyright © 1948 by Scientific American, Inc. All rights reserved.]

> I observed that plants not only have the faculty to correct bad air in six or ten days, by growing in it, as the experiments of Dr. Priestley indicate, but that they perform this important office in a complete manner in a few hours; that *this wonderful operation is by no means owing to the vegetation of the plant, but to the influence of light of the sun upon the plant.*

Similar experiments were being carried out in Geneva by Jean Senebier, a Swiss pastor. His distinctive contribution was to show that "fixed air"—namely, CO_2—is taken up in photosynthesis. The role of water in photosynthesis was demonstrated by Theodore de Saussure, also a Genevan. He showed that the sum of the weights of organic matter produced by plants and of the oxygen evolved is much more than the weight of CO_2 consumed. From Lavoisier's law of the conservation of mass, de Saussure concluded that another substance was utilized. The only inputs in his system were CO_2, water, and light. Hence, de Saussure concluded that the other reactant must be water.

The final contribution to the basic equation of photosynthesis came nearly a half-century later. Julius Robert Mayer, a German surgeon, discovered the law of conservation of energy in 1842. Mayer recognized that plants convert solar energy into chemical free energy:

> The plants take in one form of power, light; and produce another power, chemical difference.

The amount of energy stored by photosynthesis is enormous. More than 10^{17} kcal of free energy is stored annually by photosynthesis on earth, which corresponds to the assimilation of more than 10^{10} tons of carbon into carbohydrate and other forms of organic matter.

CHLOROPHYLLS TRAP SOLAR ENERGY

Mayer stated, "Nature has put itself the problem of how to catch in flight light streaming to the earth and to store the most elusive of all powers in rigid form." What is the mechanism of trapping this most elusive of all powers? The first step is the absorption of light by a photoreceptor molecule. The principal photoreceptor in the chloroplasts of green plants is *chlorophyll* a, a substituted tetrapyrrole (Figure 22-4). The four nitrogen atoms of the pyrroles are coordinated to a magnesium atom. Thus, chlorophyll is a *magnesium porphyrin,* whereas heme is an iron porphyrin. Another distinctive feature of chlorophyll is the presence of *phytol,* a highly hydrophobic 20-carbon alcohol, esterified to an acid side chain. *Chlorophyll* b differs from chlorophyll *a* in having a formyl group in place of a methyl group on one of its pyrroles.

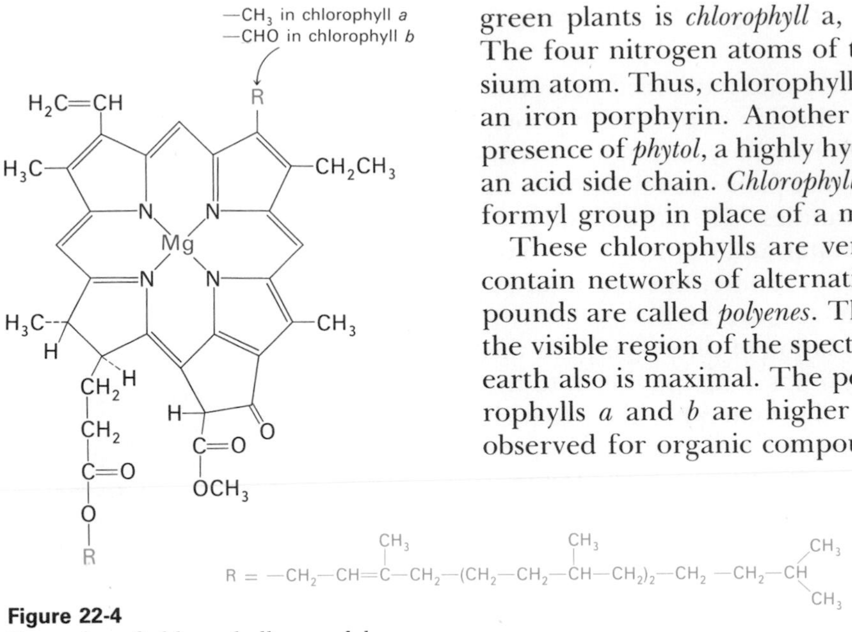

These chlorophylls are very effective photoreceptors because they contain networks of alternating single and double bonds. Such compounds are called *polyenes.* They have very strong absorption bands in the visible region of the spectrum, where the solar output reaching the earth also is maximal. The peak molar absorption coefficients of chlorophylls *a* and *b* are higher than 10^5 cm^{-1} M^{-1}, among the highest observed for organic compounds.

Figure 22-4
Formulas of chlorophylls *a* and *b*.

The absorption spectra of chlorophylls *a* and *b* are different (Figure 22-5). Light that is not appreciably absorbed by chlorophyll *a*—at 460 nm, for example—is captured by chlorophyll *b*, which has intense absorption at that wavelength. Thus, *these two kinds of chlorophyll complement each other in absorbing the incident sunlight.* The spectral region from 500 to 600 nm is only weakly absorbed by these chlorophylls, but this does not pose a problem for most green plants. Cyanobacteria (blue-green algae) and red algae, on the other hand, contain accessory light-harvesting pigments (p. 531) that enable them to efficiently utilize light not absorbed strongly by chlorophyll.

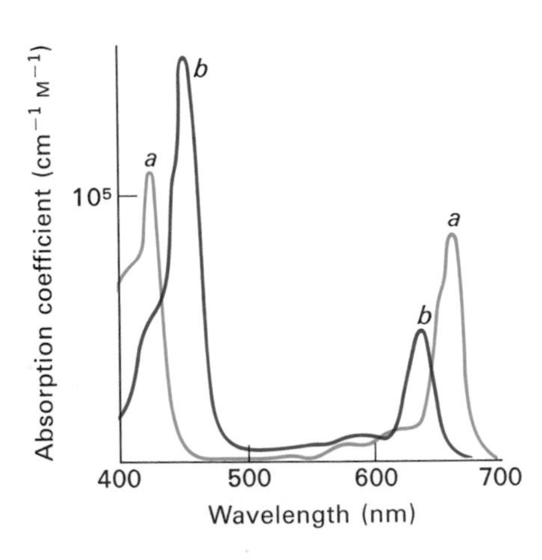

Figure 22-5
Absorption spectra of chlorophylls *a* and *b*.

PHOTONS ABSORBED BY MANY CHLOROPHYLLS FUNNEL INTO A REACTION CENTER

Measurements of the dependence of the rate of photosynthesis on the intensity of illumination show that it increases linearly at low intensities and reaches a saturating value at high intensities (Figure 22-6). A saturating value is observed in strong light because chemical reactions utilizing the absorbed photons become rate-limiting. This experiment provided the first intimation that *photosynthesis can be separated into light reactions and dark reactions.* As will be discussed shortly, the light reactions generate NADPH and ATP, whereas the dark reactions use these energy-rich molecules to reduce CO_2.

In 1932, Robert Emerson and William Arnold measured the oxygen yield of photosynthesis when *Chlorella* cells (unicellular green algae) were exposed to light flashes lasting a few microseconds. They expected to find that the yield per flash would increase with the flash intensity until each chlorophyll molecule absorbed a photon, which would then be used in dark reactions. Their experimental observation was entirely unexpected: a saturating light flash led to the production of only one molecule of O_2 per 2500 chlorophyll molecules.

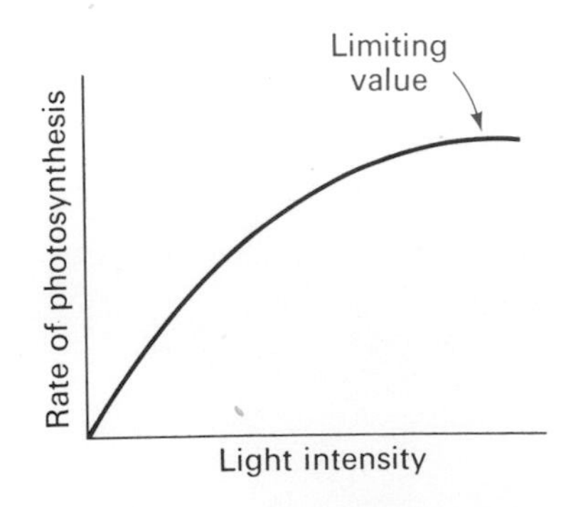

Figure 22-6
The rate of photosynthesis reaches a limiting value when the light intensity suffices to excite only a small fraction of the chlorophyll molecules.

This experiment led to the concept of the *photosynthetic unit.* Hans Gaffron proposed that light is absorbed by hundreds of chlorophyll molecules, which then transfer their excitation energy to a site at which chemical reactions occur (Figure 22-7). This site is called a *reaction center*. Thus, most chlorophyll molecules in the photosynthetic unit absorb light, but only a small proportion of them, those at reaction centers, mediate the transformation of light into chemical energy. The energy level of chlorophylls at the reaction center is lower than that of other chlorophylls, which enables the reaction center to trap the excitation (Figure 22-8). The transfer of energy by direct electromagnetic interactions between chlorophylls and then to the reaction center is very rapid, taking times measured in picoseconds (10^{-12} s).

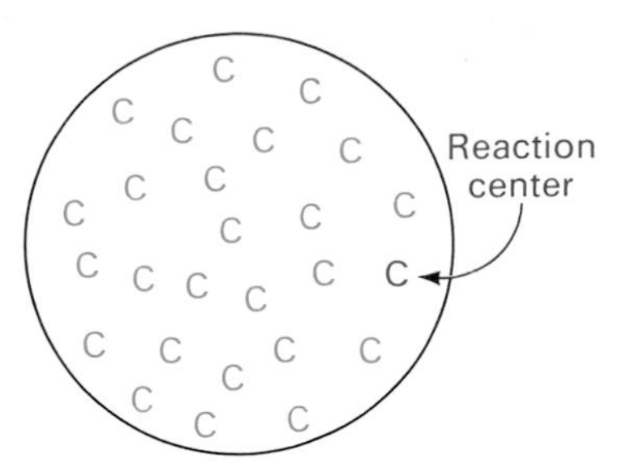

Figure 22-7
Diagram of a photosynthetic unit. The antenna chlorophyll molecules (denoted by green C's) transfer their excitation energy to a specialized chlorophyll at the reaction center (denoted by a red C).

O_2 EVOLVED IN PHOTOSYNTHESIS COMES FROM WATER

Let us turn now to the chemical changes in photosynthesis. The source of the oxygen evolved in green plants has important implications for the mechanism of photosynthesis. Comparative studies of photosynthesis in many organisms carried out as early as 1931 revealed the source. Some photosynthetic bacteria convert hydrogen sulfide into sulfur in the presence of light. Cornelis Van Niel perceived a common pattern in the overall reactions of photosynthesis carried out by green plants and by green sulfur bacteria:

$$CO_2 + 2\ H_2O \xrightarrow{\text{light}} (CH_2O) + O_2 + H_2O$$

$$CO_2 + 2\ H_2S \xrightarrow{\text{light}} (CH_2O) + 2\ S + H_2O$$

The sulfur that is formed by the photosynthetic bacteria is analogous to the oxygen that is evolved in plants. Van Niel proposed a general formula for photosynthesis:

$$\underset{\text{Hydrogen acceptor}}{CO_2} + \underset{\text{Hydrogen donor}}{2\ H_2A} \xrightarrow{\text{light}} \underset{\text{Reduced acceptor}}{(CH_2O)} + \underset{\text{Dehydrogenated donor}}{2\ A} + H_2O$$

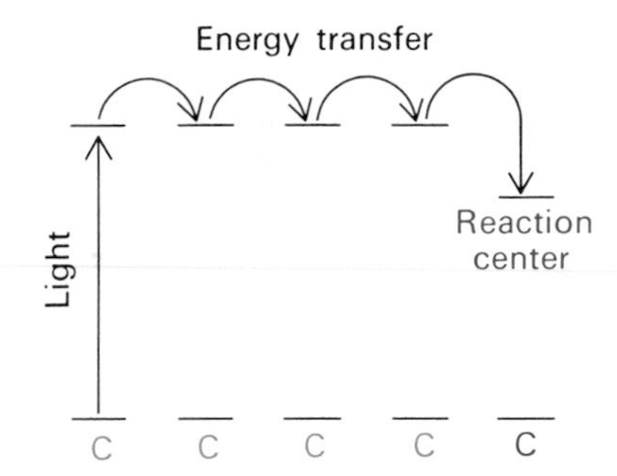

Figure 22-8
Diagram of the energy levels of the excited state of the antenna chlorophylls and of the reaction center.

The hydrogen donor H_2A is H_2O in green plants and H_2S in the photosynthetic sulfur bacteria. Thus, photosynthesis in plants could be formulated as a reaction in which CO_2 is reduced by hydrogen derived from water. Oxygen evolution would then be the necessary consequence of this dehydrogenation process. The essence of this view of photosynthesis is that *water is split by light.*

The availability in 1941 of a heavy isotope of oxygen, namely ^{18}O, made it feasible to test this concept directly. In fact, ^{18}O appeared in the evolved oxygen when photosynthesis was carried out in water enriched in this isotope. This result confirmed the proposal that the O_2 formed in photosynthesis comes from water.

$$H_2{}^{18}O + CO_2 \xrightarrow{\text{light}} (CH_2O) + {}^{18}O_2$$

HILL REACTION: ILLUMINATED CHLOROPLASTS EVOLVE O_2 AND REDUCE ELECTRON ACCEPTORS

In 1939, Robert Hill discovered that isolated chloroplasts evolve oxygen when they are illuminated in the presence of a suitable electron acceptor, such as ferricyanide. There is a concomitant reduction of ferricyanide to ferrocyanide. The Hill reaction is a landmark in the elucidation of the mechanism of photosynthesis for several reasons:

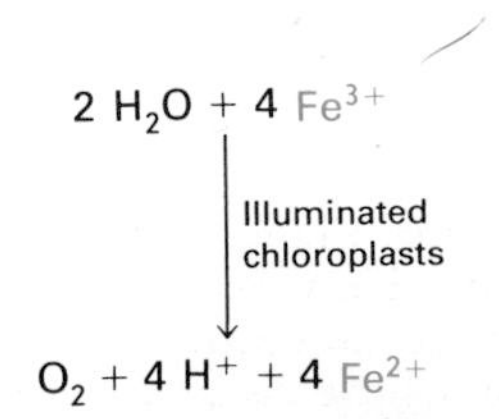

1. It dissected photosynthesis by showing that oxygen evolution can occur without the reduction of CO_2. Artificial electron acceptors such as ferricyanide can substitute for CO_2.

2. It confirmed that the evolved oxygen comes from water rather than from CO_2, because no CO_2 was present.

3. It showed that isolated chloroplasts can perform a significant partial reaction of photosynthesis.

4. It revealed that a primary event in photosynthesis is the *light-activated transfer of an electron from one substance to another against a chemical potential.* The reduction of ferric to ferrous ion by light is a conversion of light into chemical energy.

TWO LIGHT REACTIONS INTERACT IN PHOTOSYNTHESIS

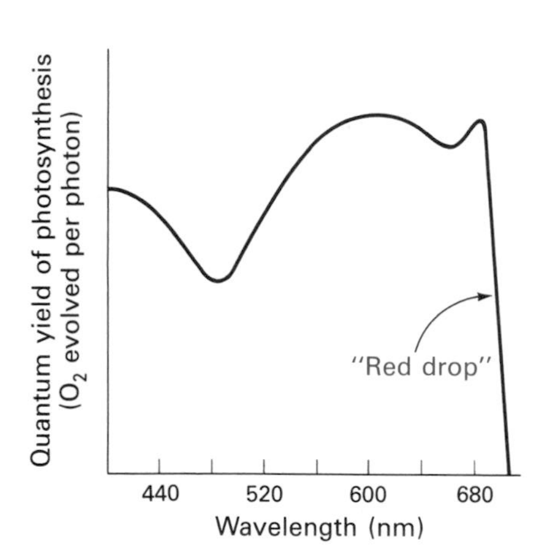

Figure 22-9
The quantum yield of photosynthesis drops abruptly when the excitation wavelength is greater than 680 nm.

Investigations of the dependence of the rate of photosynthesis on the wavelength of incident light led to the discovery that chloroplasts contain two different photosystems. The photosynthetic rate (i.e., the rate of O_2 evolution) divided by the number of quanta absorbed gives the relative quantum efficiency of the process. For a single kind of photoreceptor, the quantum efficiency is expected to be independent of wavelength over its entire absorption band. This is not the case in photosynthesis: the quantum efficiency of photosynthesis drops sharply at wavelengths longer than 680 nm, although chlorophyll still absorbs light in the range from 680 to 700 nm (Figure 22-9). However, the rate of photosynthesis using long-wavelength light can be enhanced by adding light of a shorter wavelength, such as 600 nm. The photosynthetic rate in the presence of both 600-nm and 700-nm light is greater than the sum of the rates when the two wavelengths are given separately. These observations, called the red drop and the enhancement phenomenon, led Emerson to propose that *photosynthesis requires the interaction of two light reactions: both of them can be driven by light of wavelength less than 680 nm, but only one of them by light of longer wavelength.*

PHOTOSYSTEMS I AND II HAVE COMPLEMENTARY ROLES

Indeed, photosynthesis by oxygen-evolving organisms depends on the interplay of two photosystems (Figure 22-10). Photosystem I, which can

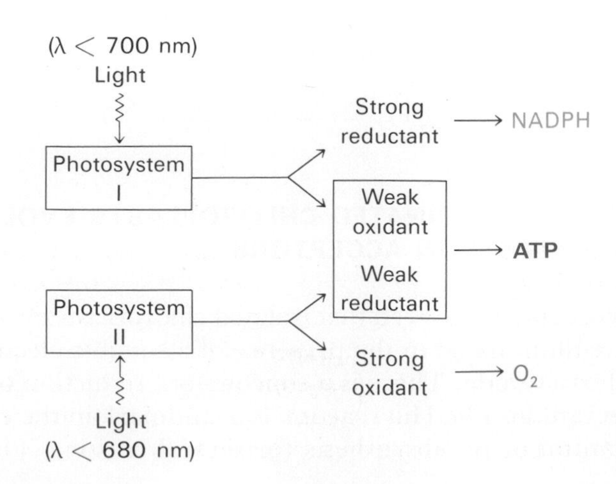

Figure 22-10
Interaction of photosystems I and II in photosynthesis by green plants.

be excited by light of wavelength shorter than 700 nm, generates a strong reductant that leads to the formation of NADPH. Photosystem II, which requires light of wavelength shorter than 680 nm, produces a strong oxidant that leads to the formation of O_2. In addition, photosystem I produces a weak oxidant, whereas photosystem II produces a weak reductant. The interaction of these species results in the generation of a transmembrane proton gradient and the subsequent formation of ATP. This part of photosynthesis, discovered by Daniel Arnon, is called photosynthetic phosphorylation or *photophosphorylation*.

PHOTOSYSTEM II TRANSFERS ELECTRONS FROM WATER TO PLASTOQUINONE

Photosystem II, a transmembrane assembly of more than ten polypeptide chains (>600 kd), catalyzes the light-driven transfer of electrons from water to *plastoquinone*. This electron acceptor closely resembles ubiquinone, a component of the electron-transport chain of mitochondria (p. 403). Plastoquinone cycles between an oxidized form (Q) and a reduced form (QH_2, plastoquinol). The intermediate in this two-electron reduction is a free-radical semiquinone ($QH\cdot$). The net reaction catalyzed by photosystem II is

$$2\ Q + 2\ H_2O \xrightarrow{\text{light}} O_2 + 2\ QH_2$$

The electrons in QH_2 are at a higher potential than those in H_2O. Recall that in oxidative phosphorylation electrons flow from ubiquinol to O_2 rather than in the reverse direction (p. 404). Photosystem II drives the reaction in the thermodynamically uphill direction by utilizing the free energy of light.

H_3C, H_3C, O, O, CH_3, $(CH_2—CH{=}C—CH_2)_n H$

n = 6 to 10

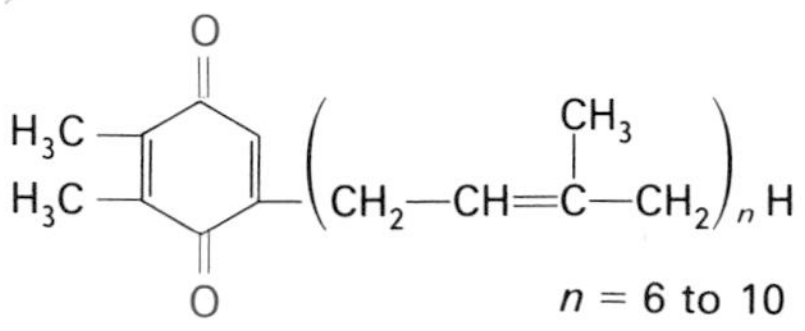

Plastoquinone (oxidized form, Q)

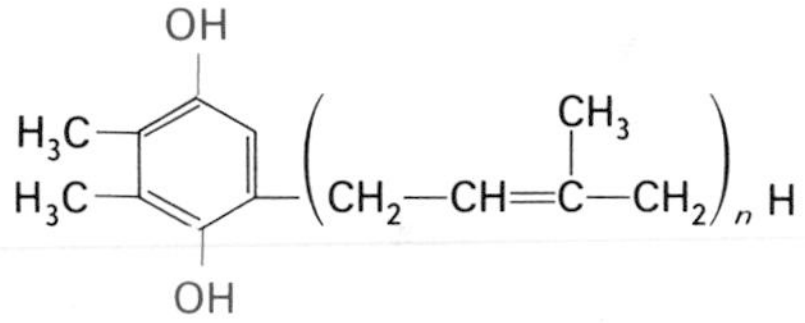

OH, OH, H_3C, H_3C, CH_3, $(CH_2—CH{=}C—CH_2)_n H$

Plastoquinol (reduced form, QH_2)

Photosystem II consists of *a light-harvesting complex, a core with a reaction center, and an oxygen-evolving complex*. The light-harvesting complex (LHC II) contains some 200 molecules of chlorophyll *a* and *b* bound to several polypeptide chains. The core contains an additional 50 molecules of bound chlorophyll *a*. Electronic excitation energy is funneled from these antenna chlorophylls to a reaction-center chlorophyll called *P680* (P stands for pigment and 680 for the wavelength, in nm, of maximal absorption). The excited state of this reaction center, $P680^*$, is a much stronger reductant than the ground state. Within picoseconds of excitation, an electron is transferred from $P680^*$ to bound *pheophytin* (Ph), a porphyrin identical to chlorophyll *a* except that it lacks Mg. The reaction center becomes a cation radical, $P680^+$. Much of the energy of the absorbed photon is conserved in this separation of charge.

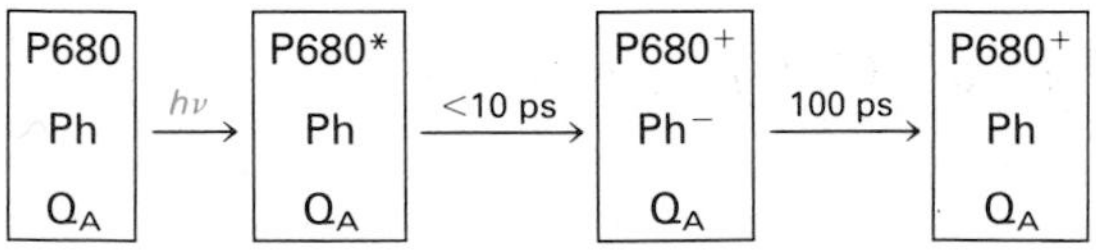

$$\begin{bmatrix} P680 \\ Ph \\ Q_A \end{bmatrix} \xrightarrow{h\nu} \begin{bmatrix} P680^* \\ Ph \\ Q_A \end{bmatrix} \xrightarrow{<10\text{ ps}} \begin{bmatrix} P680^+ \\ Ph^- \\ Q_A \end{bmatrix} \xrightarrow{100\text{ ps}} \begin{bmatrix} P680^+ \\ Ph \\ Q_A \end{bmatrix}$$

The electron then goes from reduced pheophytin to a plastoquinone bound to a protein site called Q_A, and finally to a second plastoquinone on site Q_B. Plastoquinone on Q_A alternately accepts an electron from reduced pheophytin and then donates it to Q_B; the quinone and radical anion on Q_A stay bound to the protein throughout this cycle. In contrast, plastoquinone on Q_B stays bound to a 32-kd protein on receiving the first electron but is released into the hydrophobic region of the membrane on accepting a second electron. At this point, *the energy of two photons has been safely stored in the reducing potential of* QH_2. As will be discussed shortly, QH_2 feeds its electrons into a proton-pumping electron-transport chain that is linked to photosystem I.

MANGANESE IONS PLAY A KEY ROLE IN EXTRACTING ELECTRONS FROM WATER TO FORM O_2

The $P680^+$ cation formed by photosystem II in the primary charge-separation step is a strong oxidant. $P680^+$ (through an intermediary called Z) extracts electrons from water, leading to the formation of O_2. The water-splitting enzyme, a constituent of photosystem II, contains a cluster of four manganese ions at its catalytic center. This Mn complex is oxidized by $P680^+$ to a sequence of five oxidation states (Figure 22-11). The lower oxidation states (S_0, S_1, and S_2) probably have a cubane-

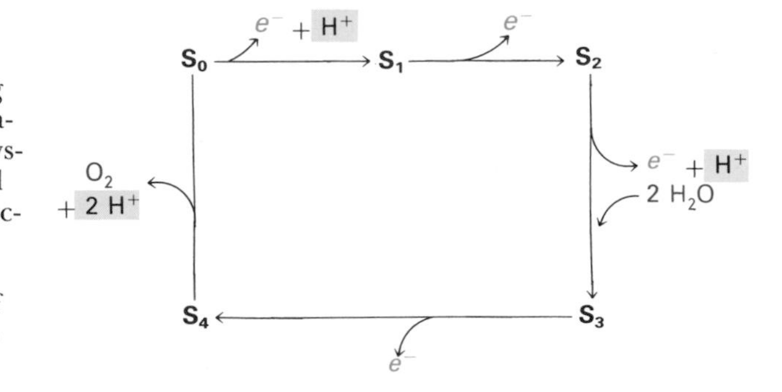

Figure 22-11
Charge-accumulator model for the splitting of water by the manganese center of photosystem II. The sequential withdrawal of four electrons by $P680^+$ drives the formation of O_2 from two molecules of H_2O. Four H^+ are released in each cycle.

like structure with four bound oxygens (Figure 22-12). S_2 loses an electron and a proton. It then acquires two water molecules and rearranges to form S_3, which has an adamantane-like structure with six bound oxygens. The loss of another electron gives S_4, which rearranges to form S_0 with the departure of O_2. The uptake of OH^- ions by S_2 and the release of O_2 by S_4 is triggered by electron deficiency in Mn ions resulting from electron transfer to $P680^+$. *This Mn cluster serves as a charge accumulator that enables* O_2 *to be formed without generating hazardous partly reduced intermediates.* Recall that cytochrome oxidase solved a similar problem by using an Fe^{2+}-Cu^+ center (p. 407).

The energetics and pathway of electron flow from H_2O to QH_2 are most simply depicted in terms of redox potentials (Figure 22-13). QH_2 lies at a higher potential (0.1 V) than H_2O (0.82 V), which means that QH_2 is a stronger reductant (see p. 400 to review redox potentials). This uphill transfer of electrons is achieved using the energy of photons absorbed by photosystem II. A 680-nm photon has an energy of 1.82 electron-volt (eV), which is more than sufficient to change the potential of an electron by 0.72 V (from 0.82 to 0.1 V) under standard conditions.

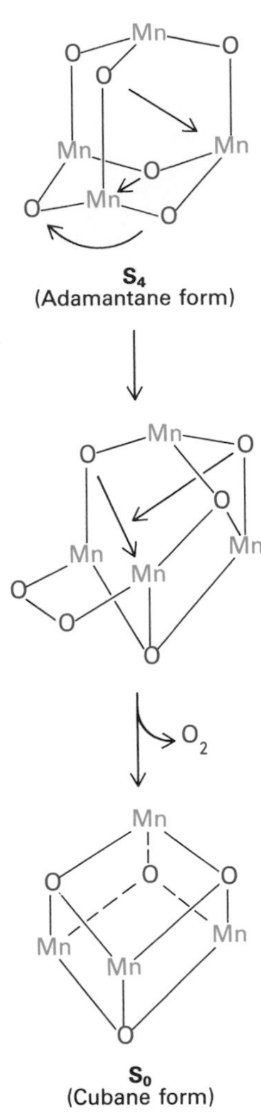

Figure 22-12
Proposed mechanism for the formation and release of O_2 from the S_4 state of the water-splitting enzyme of photosystem II. In this step, the manganese center undergoes a rearrangement from an adamantane-like structure to a cubane-like structure. [After G. W. Brudvig and R. H. Crabtree. *Proc. Nat. Acad. Sci.* 83(1986):4586.]

A PROTON GRADIENT IS FORMED AS ELECTRONS FLOW THROUGH CYTOCHROME *bf* FROM PHOTOSYSTEM II TO I

We turn now to the *cytochrome* bf *complex*, the next assembly mediating photosynthesis. Electrons flow through this complex from photosystem II to photosystem I. Cytochrome *bf* catalyzes the transfer of electrons

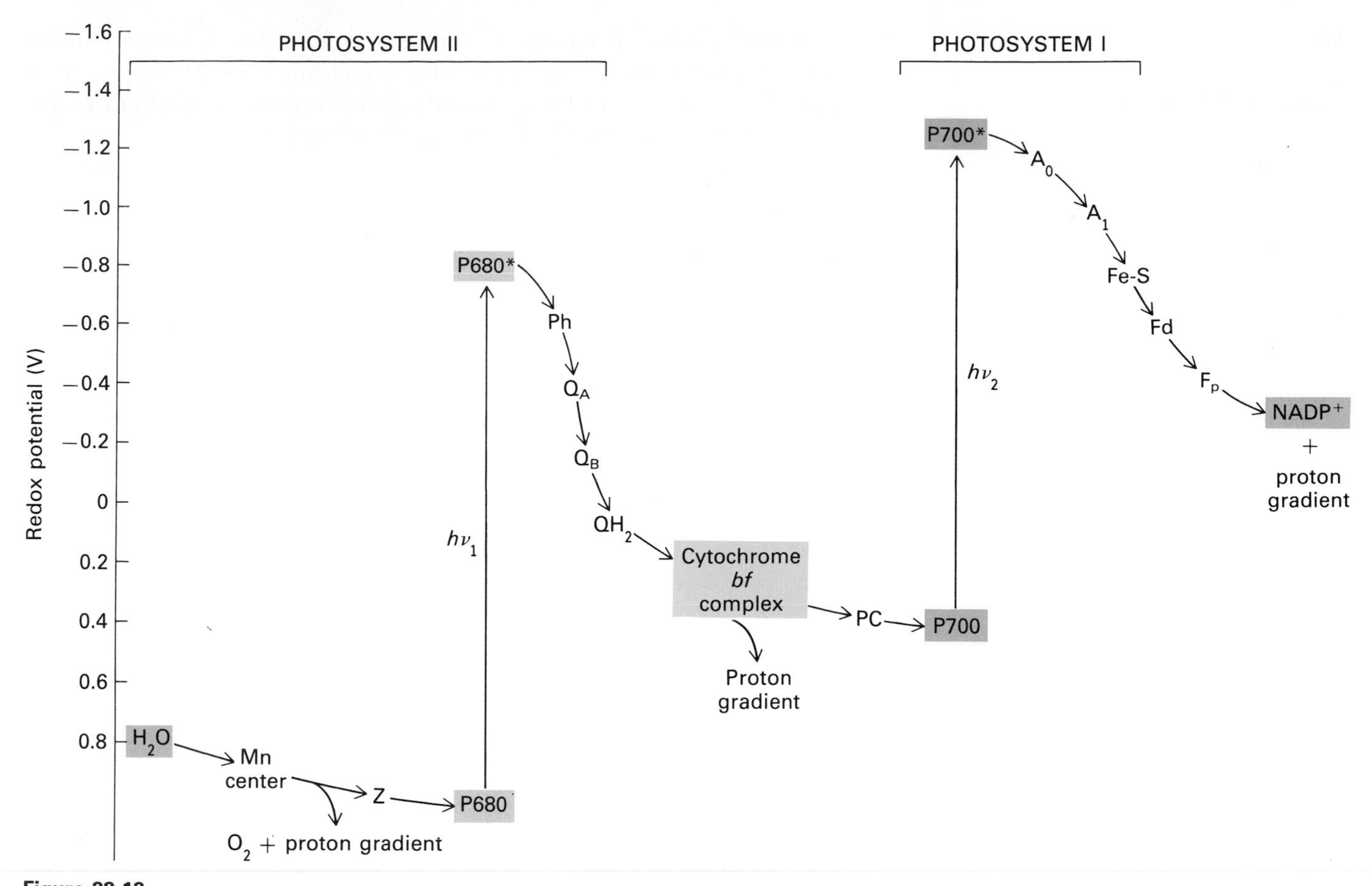

Figure 22-13
Pathway of electron flow from H_2O to $NADP^+$ in photosynthesis. This endergonic reaction is made possible by the absorption of light by photosystem II (P680) and photosystem I (P700). Reduced plastoquinone (QH_2) formed by photosystem II feeds electrons into the cytochrome *bf* complex. Reduced plastocyanin (PC) carries electrons to photosystem I, which generates reduced ferredoxin (Fd). This powerful reductant transfers its electrons to $NADP^+$ to form NADPH. A proton gradient across the thylakoid membrane (inside acidic) is formed when electrons flow through the cytochrome *bf* complex. The splitting of water and the reduction of $NADP^+$ on opposite sides of the thylakoid membrane also contribute to a proton gradient. Other abbreviations used: Z, intermediate between the Mn center and P680; Ph, pheophytin; Q_A and Q_B, plastoquinone-binding proteins; A_0 and A_1, acceptors of electrons from $P700^*$; Fp, flavoprotein (ferredoxin-$NADP^+$ reductase). [After R. E. Blankenship and R. C. Prince. *Trends Biochem. Sci.* 10(1985):383.]

from plastoquinol (QH_2) to plastocyanin (PC) and concomitantly pumps protons across the thylakoid membrane.

$$QH_2 + 2\ PC(Cu^{2+}) \longrightarrow Q + 2\ PC(Cu^{+}) + 2\ H^+$$

The cytochrome *bf* complex contains four subunits: a 34-kd cytochrome *f*, a 23-kd cytochrome b_{563} with two hemes, a 20-kd Fe-S protein, and a 17-kd polypeptide chain. This transmembrane complex closely resembles cytochrome reductase of mitochondria. Recall that cytochrome reductase pumps protons across the inner mitochondrial membrane in catalyzing the reduction of cytochrome *c*, a water-soluble protein, by ubiquinol (p. 404). The reaction mechanisms of these complexes also are similar. The Fe-S center participates directly in the reduction of plastocyanin, as it does in the reduction of cytochrome *c*.

The cytochrome *b* component is again a recycling device that enables a two-electron carrier (plastoquinol) to interact with a one-electron carrier (Fe-S center). These electron transfers drive the pumping of protons from the stroma to the inner thylakoid space.

Electrons flow from the Fe-S center of the cytochrome *bf* complex to *plastocyanin*, an 11-kd water-soluble protein. The redox center of plastocyanin consists of a *copper ion* coordinated to the side chains of a cysteine, a methionine, and two histidine residues (Figure 22-14). These ligands distort the coordination geometry from the planar array characteristic of low-molecular-weight Cu^{2+} complexes. The localized strain at the Cu atom facilitates electron transfer as Cu alternates between +1 and +2 oxidation states.

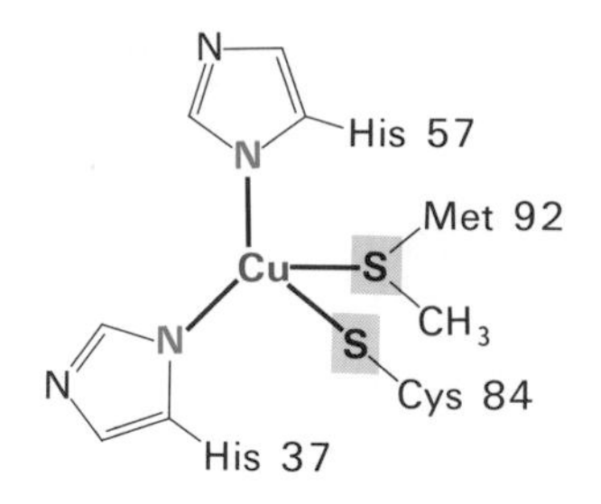

Figure 22-14
Structure of the coordinated copper ion in plastocyanin.

PHOTOSYSTEM I GENERATES NADPH BY FORMING REDUCED FERREDOXIN, A POWERFUL REDUCTANT

The actions of photosystem II and the cytochrome *bf* complex have resulted in the generation of a proton gradient across the thylakoid membrane and the reduction of plastocyanin. The next stage of photosynthesis is mediated by *photosystem I*, a transmembrane complex consisting of at least 13 polypeptide chains (>800 kd). Light is funneled from an accessory antenna protein (LHC I) containing 70 chlorophyll *a* and *b* molecules and from a core antenna with 130 chlorophyll *a* molecules to *P700*, the reaction center. As described for photosystem II, the primary event at the reaction center is a light-induced separation of charge. An electron is transferred from $P700^*$, the excited state of the reaction-center chlorophyll, to an acceptor chlorophyll called A_0 to form A_0^- and $P700^+$. A_0^- is a very powerful reductant ($E_0' = -1.1$ V)—indeed, the most reducing biomolecule known. Meanwhile, $P700^+$ captures an electron from reduced plastocyanin to return to P700, so that it can be excited once again.

The very high-potential electron of A_0^- is transferred to A_1 and then to a series of iron-sulfur centers within photosystem I. The final step is the reduction of *ferredoxin*, a 12-kd water-soluble protein containing a 2Fe-2S cluster. This reaction occurs on the stromal side of the thylakoid membrane. Thus, the net reaction catalyzed by photosystem I is

$$\mathrm{PC(Cu^+)} + \mathrm{ferredoxin_{oxidized}} \longrightarrow \mathrm{PC(Cu^{2+})} + \mathrm{ferredoxin_{reduced}}$$

$$2\ \mathrm{Ferredoxin_{reduced}} + \mathrm{H^+} + \mathrm{NADP^+} \xrightarrow{\text{Ferredoxin-NADP reductase}} 2\ \mathrm{Ferredoxin_{oxidized}} + \mathrm{NADPH}$$

The high-potential electrons of two molecules of ferredoxin are then transferred to $NADP^+$ to form NADPH. This reaction is catalyzed by *ferredoxin-NADP⁺ reductase*, a flavoprotein (Fp) with an FAD prosthetic group. The semiquinone form of the bound FAD is the intermediate in the convergence of two electrons from two molecules of reduced ferredoxin to one molecule of $NADP^+$.

This reaction occurs on the stromal side of the membrane. Hence, the uptake of a proton in the reduction of $NADP^+$ further contributes to the generation of a proton gradient across the thylakoid membrane, with the inside acidic.

The net reaction carried out by photosystem II, the cytochrome *bf* complex, and photosystem I is

$$2\ \mathrm{H_2O} + 2\ \mathrm{NADP^+} \xrightarrow{\text{light}} \mathrm{O_2} + 2\ \mathrm{NADPH} + 2\ \mathrm{H^+}$$

In essence, *light causes electrons to flow from H_2O to NADPH and leads to the generation of a proton-motive force* (Figure 22-13). This pathway is called the *Z-scheme of photosynthesis* because the redox diagram from P680 to $P700^*$ looks like a Z.

CYCLIC ELECTRON FLOW THROUGH PHOTOSYSTEM I LEADS TO THE PRODUCTION OF ATP INSTEAD OF NADPH

An alternative pathway for electrons arising from P700, the reaction center of photosystem I, contributes to the versatility of photosynthesis. The high-potential electron in ferredoxin can be transferred to the cytochrome *bf* complex rather than to $NADP^+$. This electron then flows back to the oxidized form of P700 through plastocyanin. This cyclic flow of electrons leads solely to proton pumping by the cytochrome *bf* complex. The proton gradient then drives the synthesis of ATP. In this process, called *cyclic photophosphorylation* (Figure 22-15), *ATP is generated without the concomitant formation of NADPH*. Photosystem II does not participate in cyclic photophosphorylation, and so O_2 is not formed from H_2O. Cyclic photophosphorylation takes place when $NADP^+$ is unavailable to accept electrons from reduced ferredoxin because of a very high ratio of NADPH to $NADP^+$.

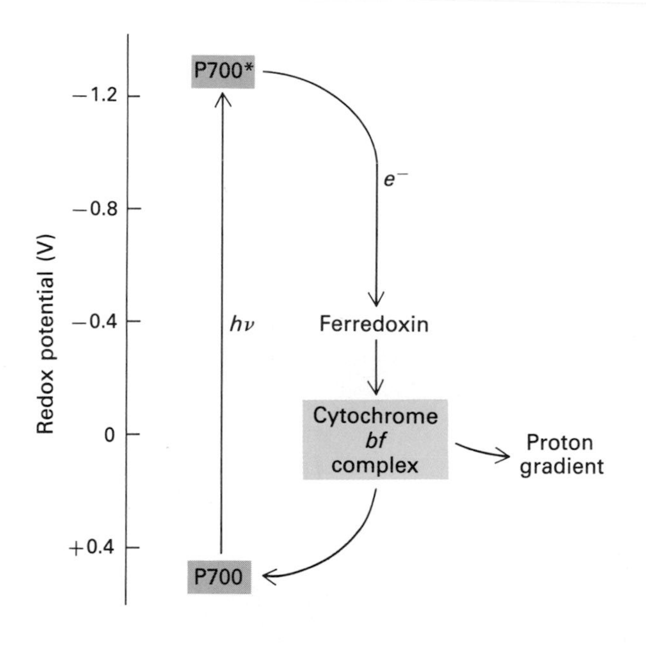

Figure 22-15
Electron flow in cyclic photophosphorylation. Electrons are transferred from $P700^*$, the excited reaction center of photosystem I, to ferredoxin and then to the cytochrome *bc* complex. Protons are pumped by this complex as electrons return to the reaction center through plastocyanin.

ATP SYNTHESIS IS DRIVEN BY A PROTON GRADIENT ACROSS THE THYLAKOID MEMBRANE

In 1966, André Jagendorf showed that chloroplasts synthesize ATP in the dark when an artificial pH gradient is imposed across the thylakoid membrane. To create this transient pH gradient, chloroplasts first were soaked in a pH 4 buffer for several hours. These chloroplasts were then rapidly mixed with a pH 8 buffer containing ADP and P_i. The pH of the stroma suddenly increased to 8, whereas the pH of the thylakoid space remained at 4. *A burst of ATP synthesis then accompanied the disappearance of the pH gradient across the thylakoid membrane* (Figure 22-16). This revealing experiment provided strong support for Mitchell's hypothesis that ATP synthesis is driven by a proton-motive force (p. 410).

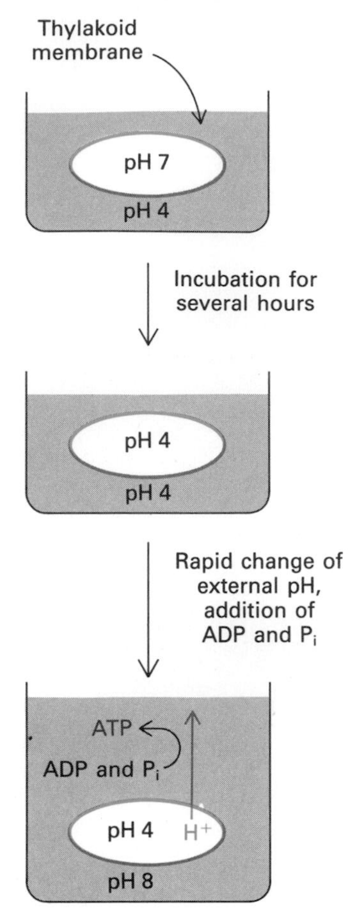

Figure 22-16
Synthesis of ATP by chloroplasts following the imposition of a pH gradient.

ATP SYNTHASE OF CHLOROPLASTS CLOSELY RESEMBLES THOSE OF BACTERIA AND MITOCHONDRIA

Indeed, the mechanism of ATP synthesis in chloroplasts is very similar to that in mitochondria. *ATP formation is driven by a proton-motive force in both photophosphorylation and oxidative phosphorylation.* Furthermore, the enzyme assembly catalyzing ATP formation in chloroplasts is very similar to that of mitochondria and bacteria. The *ATP synthase* of chloroplasts, also called the *CF_1-CF_0 complex* (C stands for chloroplast and F for factor), closely resembles the F_1-F_0 complex in oxidative phosphorylation (p. 414). CF_0, which consists of at least three kinds of subunits, conducts protons across the thylakoid membrane. The proton channel is probably formed by a hexamer of 8-kd chains. CF_1, like F_1, catalyzes the formation of ATP from ADP and P_i. The knobs on the external surface of thylakoid membranes are the CF_1 units of ATP synthase (Figure 22-17).

CF_1 has the subunit composition $\alpha_3\beta_3\gamma\delta\epsilon$. The α and β subunits contain the binding sites and catalytic sites for ATP and ADP. The δ subunit binds CF_1 to CF_0, and the γ subunit controls proton flow. The ϵ subunit inhibits the catalytic activity of the complex in the dark to block the wasteful hydrolysis of ATP.

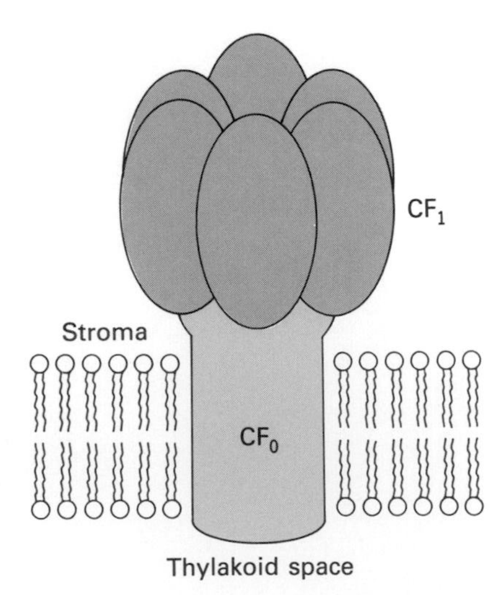

Figure 22-17
Schematic diagram of ATP synthase of chloroplasts. This assembly consists of a transmembrane CF_0 unit and a catalytic CF_1 unit on the stromal side of the thylakoid membrane.

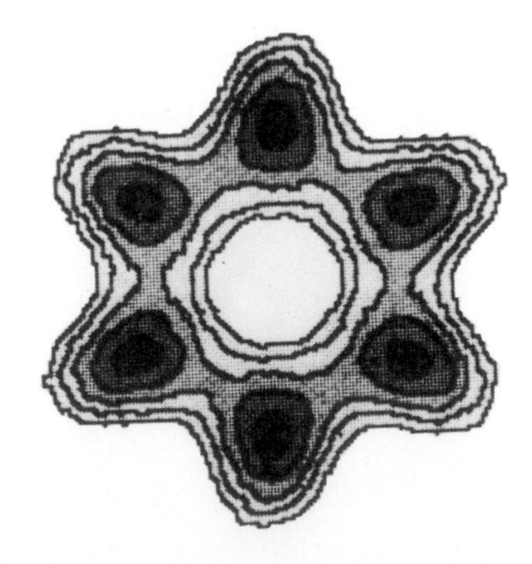

Figure 22-18
Two-dimensional image reconstruction of the CF_1 unit of chloroplast ATP synthase. This low-resolution image exhibits approximate six-fold symmetry, which implies that the three α and three β subunits are similar to each other. [From C. W. Akey, R. H. Crepeau, S. D. Dunn, R. E. McCarty, and S. J. Edelstein. *EMBO J.* 2(1983):1412.]

Electron transfer through the asymmetrically-oriented photosystems I and II and the cytochrome *bf* complex produces a large proton gradient across the thylakoid membrane (Figure 22-19). *The thylakoid space becomes markedly acidic, with the pH approaching 4. The light-induced transmembrane proton gradient is about 3.5 pH units.* As discussed previously (p. 411), the proton-motive force, Δp, consists of a pH-gradient contribution and a membrane-potential contribution. In chloroplasts, nearly all

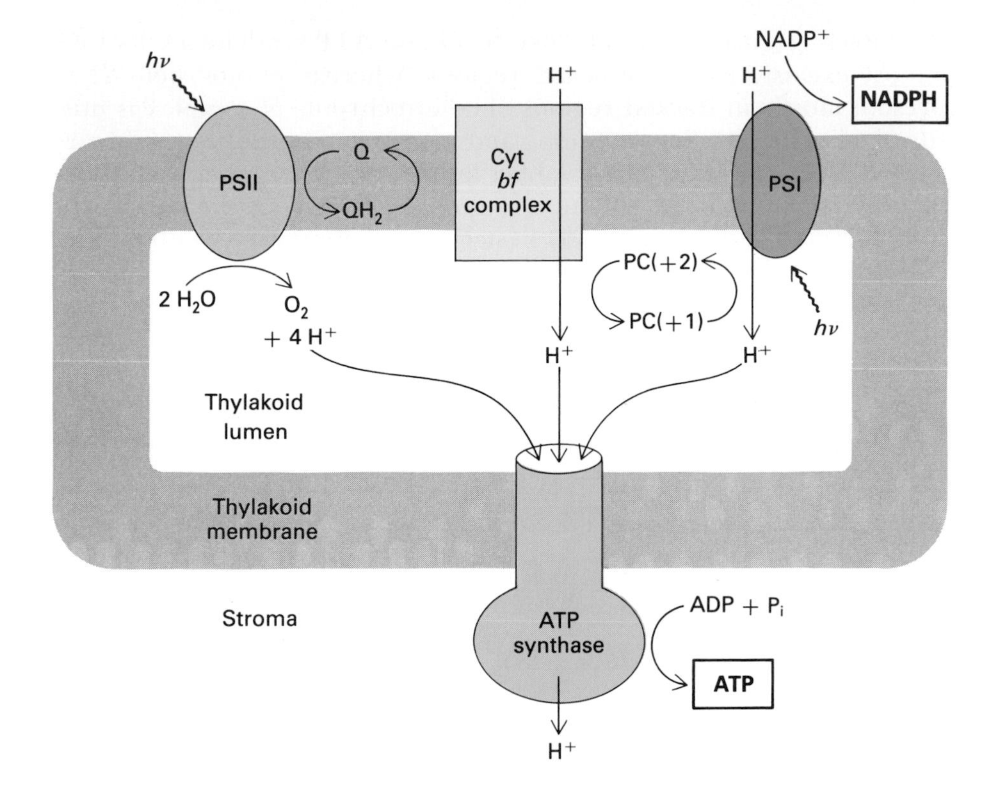

Figure 22-19
Vectorial arrangement of photosystems I and II, the cytochrome *bf* complex, and ATP synthase in the thylakoid membrane. Light-induced proton pumping makes the inner space acidic. The flow of protons through CF_0 to the stromal side leads to the synthesis of ATP by CF_1. NADPH is also formed on the stromal side. [After F. M. Harold, *The Vital Force: A Study of Bioenergetics* (W. H. Freeman, 1986), p. 271.]

of Δp arises from the pH gradient, whereas in mitochondria the contribution from the membrane potential is larger. The reason for this difference is that the thylakoid membrane is quite permeable to Cl^- and Mg^{2+}. The light-induced transfer of H^+ into the thylakoid space is accompanied by the transfer of either Cl^- in the same direction or Mg^{2+} (1 per 2 H^+) in the opposite direction. Consequently, electrical neutrality is maintained and no membrane potential is generated. A pH gradient of 3.5 units across the thylakoid membrane corresponds to a proton-motive force of 0.2 V or a ΔG of -4.8 kcal/mol. *About three protons flow through the CF_1-CF_0 complex per ATP synthesized, which corresponds to a free-energy input of 14.4 kcal per mole of ATP.* No ATP is synthesized if the pH gradient is less than two units, because the driving force is then too small.

CF_1 is on the stromal surface of the thylakoid membrane, and so the newly synthesized ATP is released into the stromal space. Likewise, NADPH formed by photosystem I is released into the stromal space. Thus, *ATP and NADPH, the products of the light reactions of photosynthesis, are appropriately positioned for the subsequent dark reactions, in which CO_2 is converted into carbohydrate.*

PHOTOSYSTEM I AND ATP SYNTHASE ARE LOCATED IN UNSTACKED THYLAKOID MEMBRANES

Thylakoid membranes of most plants are differentiated into *stacked* (appressed) and *unstacked* (nonappressed) regions (see Figures 22-1 and 22-2). Stacking increases the amount of thylakoid membrane in a given volume. Both regions surround a common internal thylakoid space, but only unstacked regions make direct contact with the chloroplast stroma. Stacked and unstacked regions differ in their content of photosynthetic assemblies (Figure 22-20). Photosystem I and ATP synthase are located almost exclusively in unstacked regions, whereas photosystem II is present mostly in stacked regions. The cytochrome *bf* complex is uniformly distributed. Plastoquinone and plastocyanin are the mobile carriers of electrons between assemblies located in different regions of the thylakoid membrane. A common internal thylakoid space enables protons liberated by photosystem II in stacked membranes to be utilized by ATP synthase molecules that are located far away in unstacked membranes.

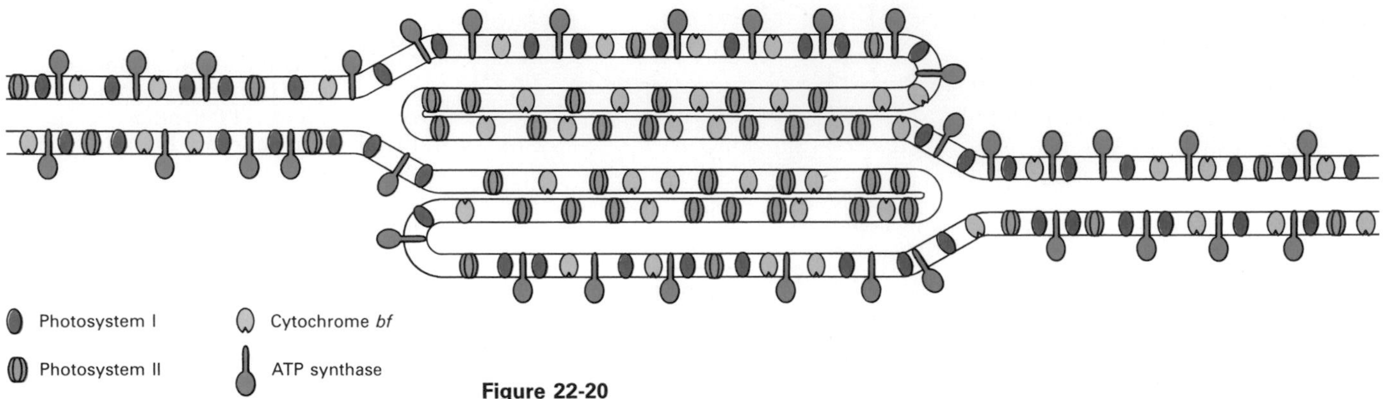

Figure 22-20
Photosynthetic assemblies are distributed differently in the stacked (appressed) and unstacked (nonappressed) regions of thylakoid membranes. [After a drawing kindly provided by Dr. Jan M. Anderson and Dr. Bertil Andersson.]

What is the functional significance of this lateral differentiation of the thylakoid membrane system? If both photosystems were present at high density in the same membrane region, a high proportion of photons absorbed by photosystem II would be transferred to photosystem I because the energy level of the excited state of II (P680*) relative to its ground state (P680) is higher than that of I (P700* relative to P700). A lateral separation of photosystems solves this problem by placing P680* more than 100 Å away from P700. The positioning of photosystem I in the unstacked membranes gives it direct access to the stroma for the reduction of $NADP^+$. It seems likely that ATP synthase is also located in the unstacked region to provide space for the large CF_1 globule and access to ADP. In contrast, the tight quarters of the appressed region pose no problem for photosystem II, which interacts with a small polar electron donor (H_2O) and a highly lipid-soluble electron carrier (plastoquinone).

The degree of stacking and the proportions of different photosynthetic assemblies are regulated in response to environmental variables such as the intensity and spectral character of incident light. The lateral distribution of light-harvesting complex II (LHC II) is controlled by reversible phosphorylation. At low light levels, LHC II is bound to photosystem II. At high light levels, a specific kinase is activated by plastoquinol, and phosphorylation of threonine side chains of LHC II leads to its release from photosystem II. The phosphorylated form of this light-harvesting unit diffuses freely in the thylakoid membrane and may become associated with photosystem I to increase its absorbance coefficient. Phosphorylation may also serve to tag LHC II for degradation.

PHYCOBILISOMES SERVE AS MOLECULAR LIGHT PIPES IN CYANOBACTERIA AND RED ALGAE

Little blue or red light reaches algae living at a depth of a meter or more in seawater because such light is absorbed by water and by chlorophyll molecules in organisms lying above. Cyanobacteria (blue-green algae) and red algae contain protein assemblies called *phycobilisomes* that enable them to harvest the green and yellow light that passes through. Phycobilisomes are bound to the outer face of thylakoid membranes, where they serve as light-absorbing antennae to funnel excitation energy into the reaction centers of photosystem II. They absorb maximally in the 470 to 650 nm region, in the valley between the blue and far-red absorption peaks of chlorophyll *a*. Phycobilisomes are very large assemblies (several million daltons) of many *phycobiliprotein* subunits, each containing many covalently attached *bilin* prosthetic groups, and linker polypeptides. Phycobilisomes contain hundreds of bilins. Phycocyanobilin and phycoerythrobilin are the two most common ones.

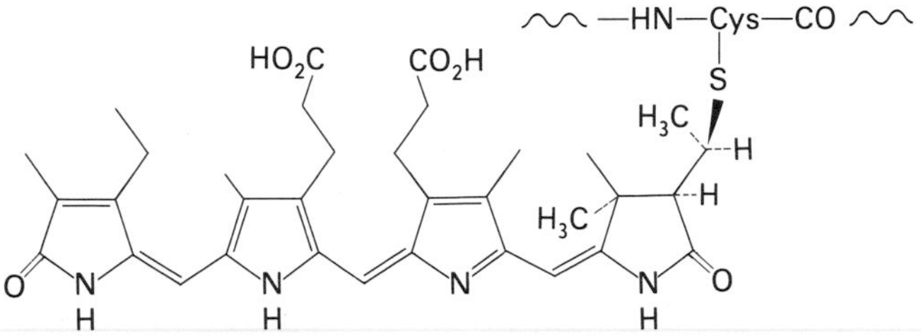

Peptide-linked phycocyanobilin

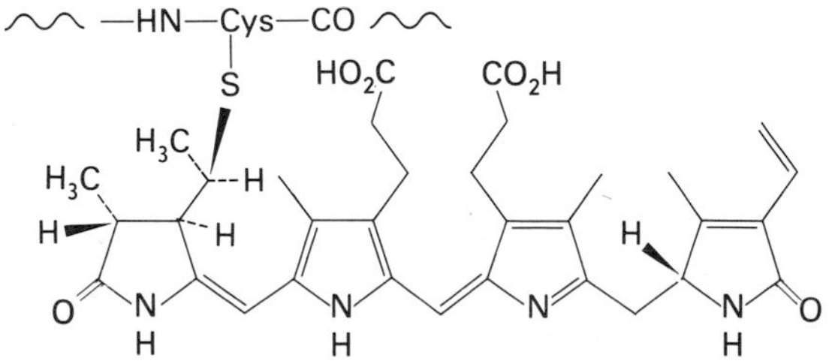

Peptide-linked phycoerythrobilin

Phycobilisomes absorb light over a broad spectral region because they contain several phycobiliproteins with different spectral properties. Light energy collected by phycobiliproteins absorbing maximally at shorter wavelengths is transferred to phycobiliproteins absorbing maximally at longer wavelengths and then to the reaction-center chlorophyll of photosystem II. The energy is transferred by a direct electromagnetic interaction that requires overlap of the absorption spectrum of the energy acceptor and the emission spectrum of the energy donor. A suitably matched donor and acceptor as much as 70 Å apart can transfer energy efficiently. In the phycobilisomes of blue-green algae, for example, excitation energy is transferred from one phycobiliprotein to another in the following sequence:

Phycoerythrin ⟶ phycocyanin ⟶ allophycocyanin ⟶ reaction center

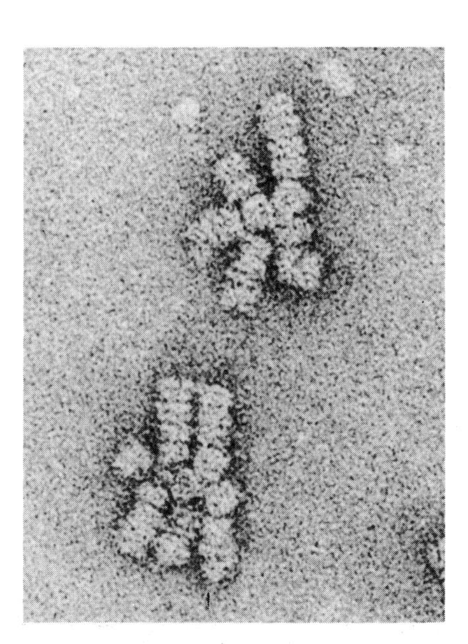

Figure 22-21
Electron micrograph of phycobilisomes from a cyanobacterium (*Synechocystis*). [Courtesy of Dr. Robley Williams and Dr. Alexander Glazer.]

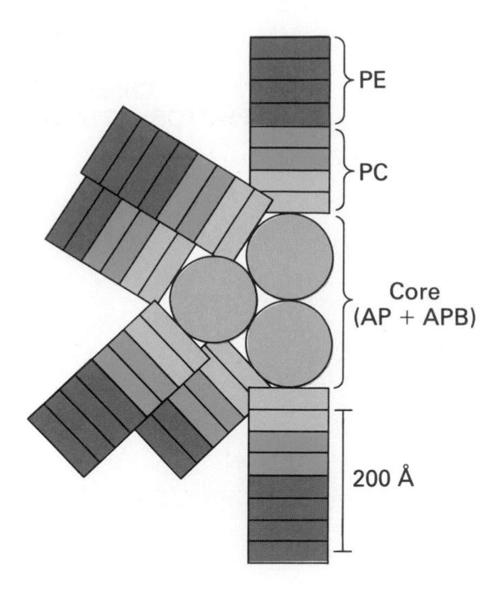

Figure 22-22
Schematic diagram of a phycobilisome from the cyanobacterium *Synechocystis* 6701. Rods containing phycoerythrin (PE) and phycocyanin (PC) emerge from a core made of allophycocyanin (AP) and allophycocyanin B (APB). The core region binds to the thylakoid membrane. [After a drawing kindly provided by Dr. Alexander Glazer.]

The geometrical arrangement of phycobiliproteins in phycobilisomes (Figure 22-22), as well as their spectral properties, contributes to the efficiency of energy transfer, which is greater than 95%. Excitation energy absorbed by phycoerythrin subunits at the periphery of these antennas appears at the reaction center in less than 100 ps. Phycobilisomes are elegantly designed light pipes that enable algae to occupy ecological niches that would not support organisms relying solely on chlorophyll for the trapping of light.

THE STRUCTURE OF A PHOTOSYNTHETIC REACTION CENTER HAS BEEN ELUCIDATED IN ATOMIC DETAIL

The photosynthetic reaction center of *Rhodopseudomonas viridis*, a purple sulfur bacterium, has recently been visualized in atomic detail from the x-ray crystallographic analysis of Johann Deisenhofer, Hartmut Michel, and Robert Huber. The reaction center consists of four polypeptides: L (31 kd), M (36 kd), and H (28 kd) subunits and a *c*-type cytochrome (Figure 22-23). The cytochrome lies on one side of the thylakoid membrane, and most of the H subunit on the other. The L and M subunits are very similar; each contains five transmembrane helices, in contrast to the H subunit, which has just one. Most of the side chains of these helical segments are hydrophobic.

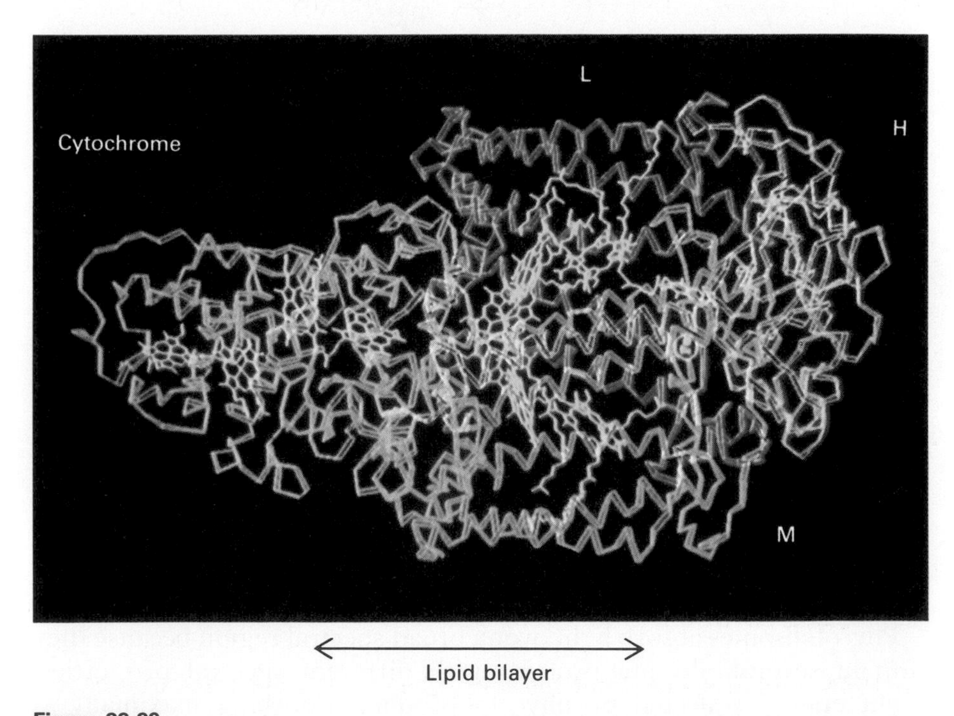

Figure 22-23
Three-dimensional structure of the photosynthetic reaction center of *Rhodopseudomonas viridis*, a purple sulfur bacterium. The reaction consists of four subunits: a cytochrome (green), L (orange), M (blue), and H (pink). The L and M subunits each contain five transmembrane helices and the H subunit contains one. [Courtesy of Dr. Hartmut Michel and Dr. Johann Deisenhofer.]

The cytochrome contains four covalently attached heme groups. Four bacteriochlorophyll *b* molecules (BChl-b), two bacteriopheophytin *b* molecules (BPh-b), two quinones (Q_A and Q_B), and a ferrous ion are associated noncovalently with the L and M subunits (Figure 22-24). Bacteriochlorophylls are similar to chlorophylls except for small modi-

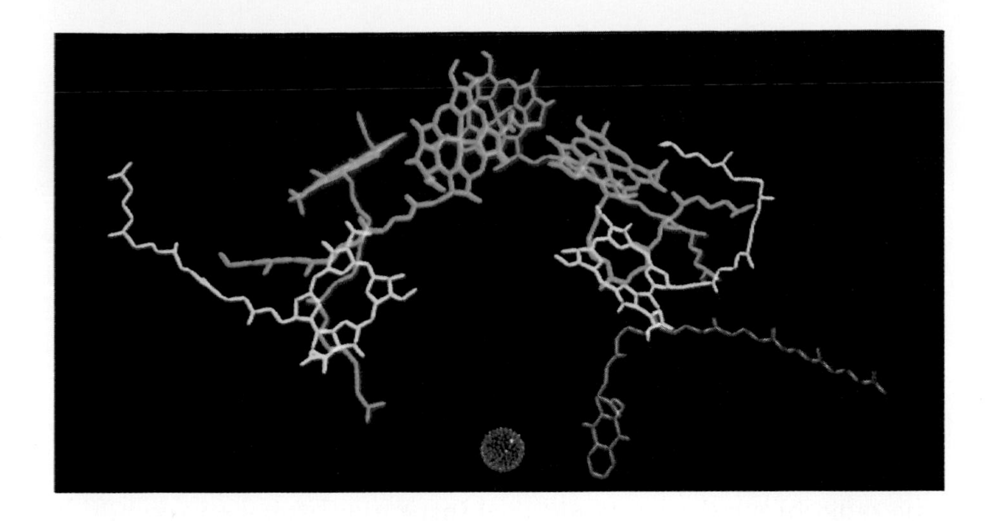

Figure 22-24
Prosthetic groups of the photosynthetic reaction center from *R. viridis*. The colors of the prosthetic groups are: hemes, blue; bacteriochlorophylls, green; bacteriopheophytins, yellow; quinone, red; and iron, orange.

fications that shift their absorption maxima to the near infrared, to wavelengths as long as 1000 nm. In *R. viridis*, the reaction center is a BChl-b dimer that absorbs maximally at 960 nm. Indeed, all known reaction centers are dimers of chlorophyll derivatives. Another recurring theme is that excitation of the reaction center leads to charge separation. In this case, an electron is transferred through a series of acceptors from the bacteriochlorophyll dimer to Q_B.

$$\underset{\text{(P960)}}{\text{BChl-b dimer}} \longrightarrow \text{BChl-b} \longrightarrow \text{BPh-b} \longrightarrow Q_A \longrightarrow Q_B$$

The hemes of the cytochrome subunit conduct electrons that return $P960^+$ to the ground state.

As in photosystem II of green plants, Q_A is very tightly bound, whereas Q_B is released as QH_2 on arrival of the second electron. Indeed, the reaction center of purple bacteria closely resembles that of photosystem II. The L subunit of purple bacteria and the 32-kd quinone-binding protein of plants are homologous. The elucidation of the three-dimensional structure of this bacterial reaction center is providing insight into the structure of integral membrane proteins, primary processes in photosynthesis, and evolutionary relations of photosynthetic assemblies.

THE PATH OF CARBON IN PHOTOSYNTHESIS WAS TRACED BY PULSE LABELING WITH RADIOACTIVE CO_2

In 1945, Melvin Calvin and his colleagues started a series of investigations that resulted in the elucidation of the dark reactions of photosynthesis. They used the unicellular green alga *Chlorella* in their work because it was easy to culture these organisms consistently. Their findings later proved to be pertinent to a wide variety of organisms ranging from photosynthetic bacteria to higher plants.

The aim of their work was to determine the pathway by which CO_2 becomes fixed into carbohydrate. The experimental strategy was to use radioactive carbon (^{14}C) as a tracer. $^{14}CO_2$ was injected into an illuminated suspension of algae that had been carrying out photosynthesis with normal CO_2. The algae were killed after a predetermined time by dropping the suspension into alcohol, which also stopped the enzymatic reactions.

The radioactive compounds in the algae were separated by two-dimensional paper chromatography. The paper chromatogram was then pressed against photographic film, which became black where the paper contained a radioactive spot. In his Nobel Lecture, Calvin noted that their primary data resided "in the number, position, and intensity—that is, radioactivity—of the blackened areas. The paper ordinarily does not print out the names of these compounds, unfortunately, and our principal chore for the succeeding ten years was to properly label those blacked areas on the film."

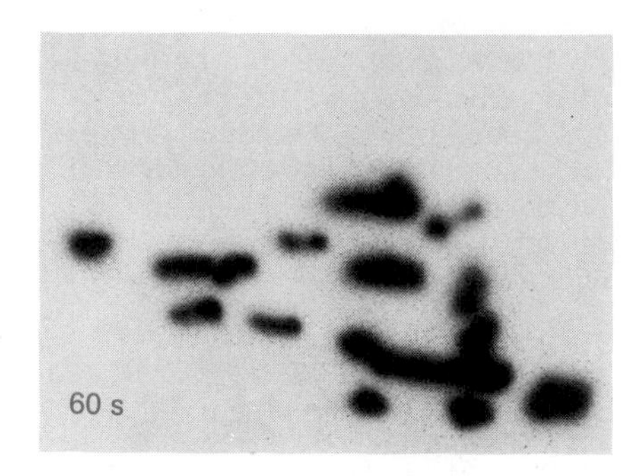

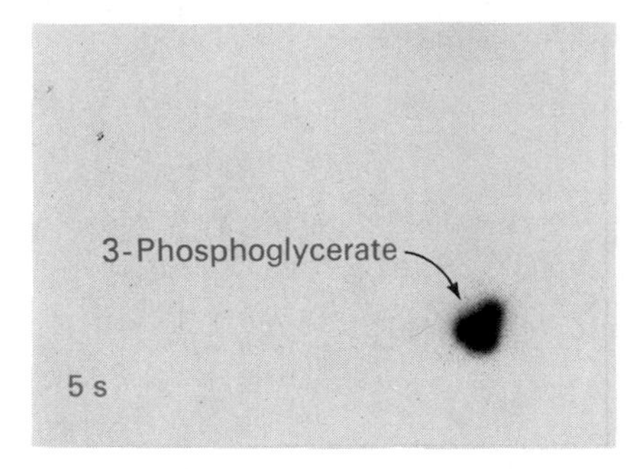

Figure 22-25
Radiochromatograms of illuminated suspensions of algae 5 s and 60 s after the injection of CO_2. [Courtesy of Dr. J. A. Bassham.]

CO_2 REACTS WITH RIBULOSE 1,5-BISPHOSPHATE TO FORM TWO MOLECULES OF 3-PHOSPHOGLYCERATE

The radiochromatogram of the algal suspension after sixty seconds of illumination was so complex (Figure 22-25) that it was not feasible to detect the earliest intermediate in the fixation of CO_2. However, the pattern after only five seconds of illumination was much simpler. In fact, there was just one prominent radioactive spot, which proved to be *3-phosphoglycerate.*

The formation of 3-phosphoglycerate as the first detectable radioactive intermediate suggested that a two-carbon compound is the acceptor for the CO_2. This proved not to be so. The actual reaction sequence is more complex:

$$C_5 \xrightarrow{CO_2} C_6 \xrightarrow{H_2O} C_3 + C_3$$

The CO_2 molecule condenses with ribulose 1,5-bisphosphate to form a transient six-carbon compound, which is rapidly hydrolyzed to two molecules of 3-phosphoglycerate (Figure 22-26). This highly exergonic reaction ($\Delta G°' = -12.4$ kcal/mol) is catalyzed by *ribulose 1,5-bisphosphate carboxylase* (also called *Rubisco*), an enzyme located on the stromal surface of thylakoid membranes. The enzyme in chloroplasts consists of

Ribulose 1,5-bisphosphate: $CH_2OPO_3^{2-}$–$C{=}O$–$H{-}C{-}OH$–$H{-}C{-}OH$–$CH_2OPO_3^{2-}$
→
Enediol intermediate: $CH_2OPO_3^{2-}$–$C{-}OH$ = $C{-}OH$–$H{-}C{-}OH$–$CH_2OPO_3^{2-}$
$\xrightarrow{CO_2,\ H^+}$
2′-Carboxy-3-keto-D-arabinitol 1,5-bisphosphate: $CH_2OPO_3^{2-}$–$HO{-}C{-}COO^-$–$C{=}O$–$H{-}C{-}OH$–$CH_2OPO_3^{2-}$
$\xrightarrow{H_2O}$
Hydrated intermediate: $CH_2OPO_3^{2-}$–$HO{-}C{-}COO^-$–$HO{-}C{-}OH$–$H{-}C{-}OH$–$CH_2OPO_3^{2-}$
→
Carbanion: $CH_2OPO_3^{2-}$–$HO{-}C^{\ominus}{-}COO^-$ $\xrightarrow{H^+}$ 3-Phosphoglycerate: $CH_2OPO_3^{2-}$–$H{-}C{-}OH$–COO^-
+
3-Phosphoglycerate: COO^-–$H{-}C{-}OH$–$CH_2OPO_3^{2-}$

Figure 22-26
The carbon-fixation reaction catalyzed by ribulose 1,5-bisphosphate carboxylase.

eight large subunits (L, 56 kd) and eight small ones (S, 14 kd). Each L chain contains a catalytic site and a regulatory site. The role of the S chain is unknown. This enzyme is very abundant in chloroplasts, comprising more than 16% of their total protein. In fact, Rubisco is probably the most abundant protein in the biosphere.

The first step in the reaction catalyzed by the carboxylase is the formation of an enediol intermediate, which then reacts with CO_2 to produce a six-carbon intermediate, 2′-carboxy-3-keto-D-arabinitol 1,5-bisphosphate. Hydration of this C_6 species yields a diol at C-3. Cleavage of a C–C bond gives a molecule of 3-phosphoglycerate and the carbanion of a second one. The second molecule of 3-phosphoglycerate is then formed by protonation of this carbanion.

The enzyme is converted into a catalytically active form by the addition of CO_2 to the ϵ-amino group of a specific lysine residue to form a *carbamate*. This negatively charged adduct then binds a divalent metal ion (Mg^{2+} or Mn^{2+}) to form a positively charged center. It seems likely that this enzyme-bound metal ion serves as an electron sink during catalysis.

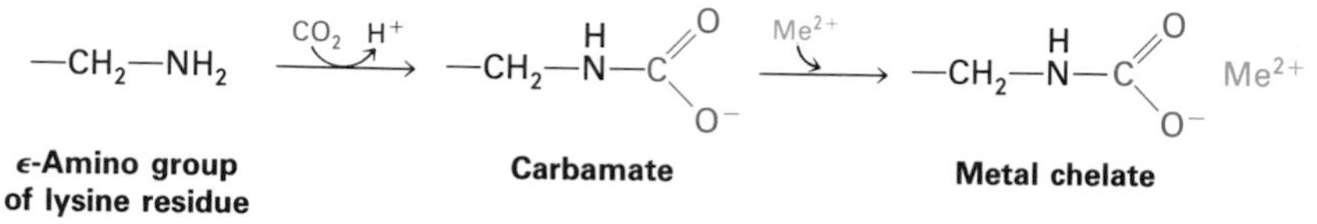

$$—CH_2—NH_2 \xrightarrow{CO_2,\ H^+} —CH_2—N(H)—C(=O)O^- \xrightarrow{Me^{2+}} —CH_2—N(H)—C(=O)O^-\ Me^{2+}$$

ε-Amino group of lysine residue | **Carbamate** | **Metal chelate**

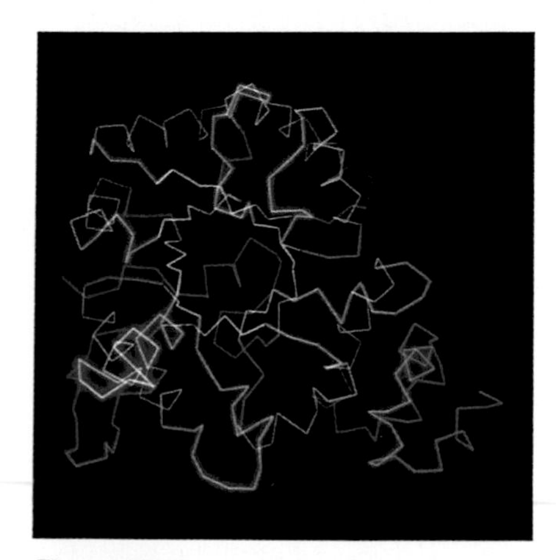

Figure 22-27
Structure of the catalytic domain of ribulose 1,5-bisphosphate carboxylase of a purple bacterium. The eight parallel β strands of the $\alpha\beta$ barrel are shown in yellow, the eight α helices in pink, and other parts of the main chain in blue. [From G. Schneider, Y. Lindquist, C.-I. Brändén, and G. Lorimer. *EMBO J.* 5(1986):3411.]

The three-dimensional structure of the carboxylase from *Rhodospirillum rubrum*, a purple bacterium, has recently been solved at high resolution. This enzyme is a dimer of identical subunits, which resemble the L chains of plants. The bacterial enzyme, too, is activated by the formation of a metal-binding carbamate and has an enzymatic mechanism like that of the chloroplast enzyme. Each subunit of the carboxylase is bipartite. The larger carboxyl-terminal domain containing the active site is shaped like a barrel. Eight parallel β strands forming the core of the barrel are surrounded by eight α helices (Figure 22-27). This motif is found in functionally and evolutionarily unrelated enzymes such as triose phosphate isomerase (p. 354) and pyruvate kinase. The active site of each is on the side of the carboxyl end of the barrel's core.

RIBULOSE 1,5-BISPHOSPHATE CARBOXYLASE ALSO CATALYZES A COMPETING OXYGENASE REACTION

Ribulose 1,5-bisphosphate carboxylase is also an *oxygenase*. It catalyzes the addition of O_2 to ribulose 1,5-bisphosphate to form *phosphoglycolate* and 3-phosphoglycerate (Figure 22-28). The oxygenase and carboxyl-

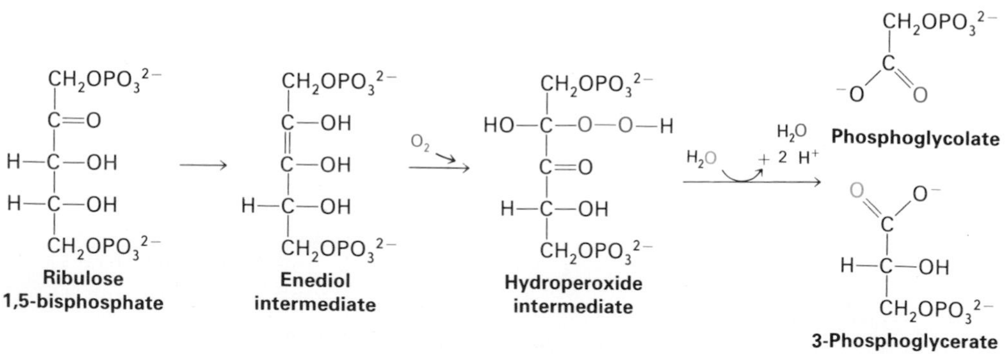

Ribulose 1,5-bisphosphate ($CH_2OPO_3^{2-}$–C=O–H–C–OH–H–C–OH–$CH_2OPO_3^{2-}$) → Enediol intermediate ($CH_2OPO_3^{2-}$–C–OH=C–OH–H–C–OH–$CH_2OPO_3^{2-}$) $\xrightarrow{O_2}$ Hydroperoxide intermediate ($CH_2OPO_3^{2-}$–HO–C–O–O–H–C=O–H–C–OH–$CH_2OPO_3^{2-}$) $\xrightarrow{H_2O \to H_2O + 2H^+}$ Phosphoglycolate ($CH_2OPO_3^{2-}$–C(=O)O$^-$) + 3-Phosphoglycerate ($^-$O(O=)C–H–C–OH–$CH_2OPO_3^{2-}$)

Figure 22-28
Oxygenase reaction catalyzed by ribulose 1,5-bisphosphate carboxylase.

Ribulose 1,5-bisphosphate

O_2 → 3-Phosphoglycerate

COO^-
$CH_2OPO_3^{2-}$
Phosphoglycolate

H_2O → P_i

COO^-
CH_2OH
Glycolate

O_2 → H_2O_2

COO^-
C
O H
Glyoxylate

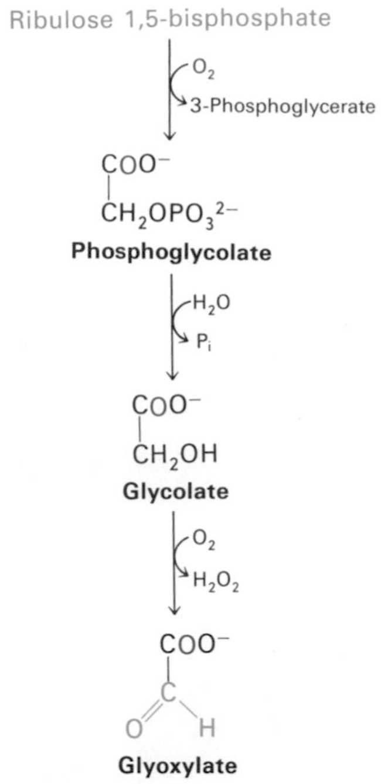

Figure 22-29
Formation and breakdown of glycolate.

ase reactions are carried out by the same active site and compete with each other. The rate of the carboxylase reaction is four times that of the oxygenase reaction under normal atmospheric conditions at 25°C; the stromal concentration of CO_2 is then 10 μM and that of O_2 is 250 μM. The oxygenase reaction, like the carboxylase reaction, requires that the same lysine be in the carbamate form with a bound divalent metal ion.

Phosphoglycolate is not a very versatile metabolite. A salvage pathway recovers part of its carbon skeleton (Figure 22-29). A specific phosphatase converts it into *glycolate*, which enters *peroxisomes* (also called *microbodies* (Figure 22-30). Glycolate is oxidized to *glyoxylate* by glycolate oxidase. The H_2O_2 produced in this reaction is cleaved by catalase to H_2O and O_2. Transamination of glyoxylate then yields *glycine*. In mitochondria, serine is formed from two molecules of glycine, with the release of CO_2 and NH_4^+.

This pathway serves to recycle three of the four carbon atoms of two molecules of glycolate. However, one of them is lost as CO_2, and one of two amino groups donated in transamination reactions is lost as NH_4^+. This process is called *photorespiration* because O_2 is consumed and CO_2 is released. Photorespiration is a seemingly wasteful process in that organic carbon is converted into CO_2 without the production of ATP or NADPH, or other evident gain. It appears to be a consequence of the imperfection of ribulose 1,5-bisphosphate carboxylase. Much effort is being devoted to engineering a carboxylase having less oxygenase activity than the naturally-occurring enzymes. The gene for the chloroplast enzyme has been cloned and expressed in *E. coli*, and site-specific mutagenesis studies are underway. The stakes are high because crop yields would increase substantially if photorespiration were prevented. It will be interesting to see whether nature's design of a CO_2-fixing enzyme can be improved upon in the laboratory.

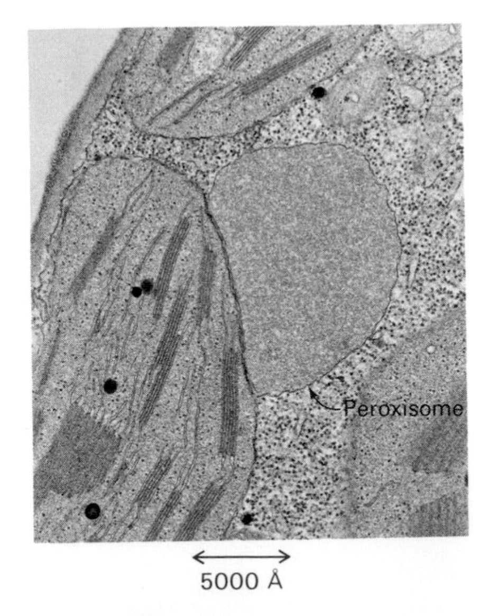

Figure 22-30
Electron micrograph of a peroxisome in a plant cell. [Courtesy of Dr. Sue Ellen Frederick.]

HEXOSE PHOSPHATES ARE MADE FROM PHOSPHOGLYCERATE AND RIBULOSE BISPHOSPHATE IS REGENERATED

The steps in the conversion of 3-phosphoglycerate into fructose 6-phosphate (Figure 22-31) are like those of the gluconeogenic pathway (p. 438), except that glyceraldehyde 3-phosphate dehydrogenase in chloroplasts is specific for NADPH, rather than NADH. These reactions bring CO_2 to the level of a hexose.

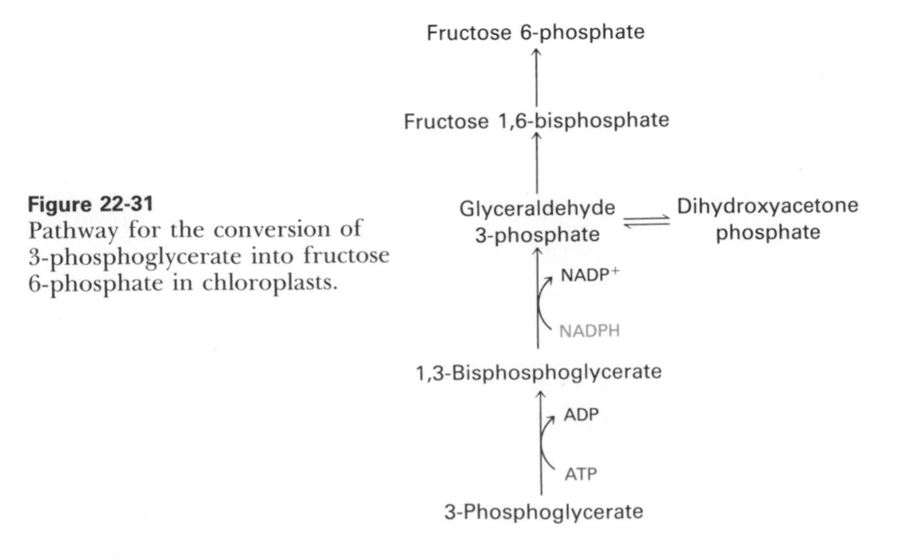

Figure 22-31
Pathway for the conversion of 3-phosphoglycerate into fructose 6-phosphate in chloroplasts.

The remaining task is to regenerate ribulose 1,5 bisphosphate, the acceptor of CO_2 in the first dark step. The problem is to construct a five-carbon sugar from six-carbon and three-carbon sugars. This is accomplished by the action of *transketolase* and *aldolase*. Transketolase also plays a role in the pentose phosphate pathway (p. 429). Recall that transketolase, a thiamine pyrophosphate (TPP) enzyme, transfers a two-carbon unit (CH_2OH—CO—) from a ketose to an aldose. Aldolase carries out an aldol condensation between dihydroxyacetone phosphate and an aldehyde. This enzyme is highly specific for dihydroxyacetone phosphate, but it accepts a wide variety of aldehydes. The specific dark reactions catalyzed by transketolase and aldolase are

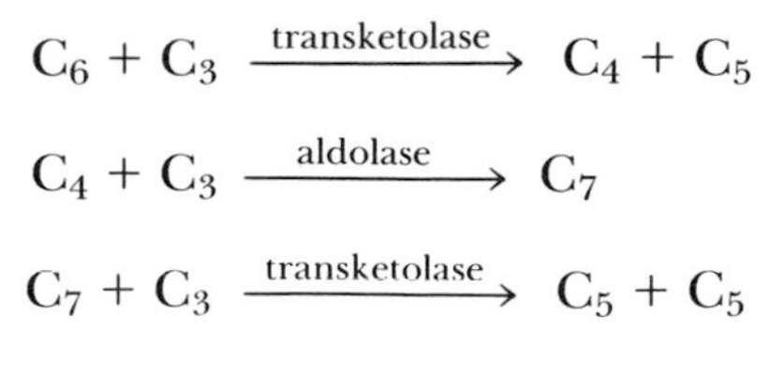

$$C_6 + C_3 \xrightarrow{\text{transketolase}} C_4 + C_5$$
$$C_4 + C_3 \xrightarrow{\text{aldolase}} C_7$$
$$C_7 + C_3 \xrightarrow{\text{transketolase}} C_5 + C_5$$

Fructose 6-phosphate + glyceraldehyde 3-phosphate $\xrightarrow{\text{transketolase}}$
xylulose 5-phosphate + erythrose 4-phosphate

Erythrose 4-phosphate + dihydroxyacetone phosphate $\xrightarrow{\text{aldolase}}$
sedoheptulose 1,7-bisphosphate

Sedoheptulose 7-phosphate + glyceraldehyde 3-phosphate $\xrightarrow{\text{transketolase}}$
ribose 5-phosphate + xylulose 5-phosphate

Four additional enzymes are needed for the dark reactions of photosynthesis. One, a *phosphatase*, hydrolyzes sedoheptulose 1,7-bisphosphate to sedoheptulose 7-phosphate. A second, *phosphopentose epimerase*, converts xylulose 5-phosphate into ribulose 5-phosphate. A third, *phosphopentose isomerase*, converts ribose 5-phosphate into ribulose 5-phosphate. Recall that the second and third enzymes also participate in the pentose phosphate pathway (p. 427). The sum of these reactions is

Fructose 6-phosphate + 2 glyceraldehyde 3-phosphate +
dihydroxyacetone phosphate $\longrightarrow$ 3 ribulose 5-phosphate

Finally, the fourth enzyme, *phosphoribulose kinase*, catalyzes the phosphorylation of ribulose 5-phosphate to regenerate ribulose 1,5-bisphosphate, the acceptor of CO_2.

Ribulose 5-phosphate + ATP $\longrightarrow$
ribulose 1,5-bisphosphate + ADP + H^+

This series of reactions is called the *Calvin cycle* (Figure 22-32).

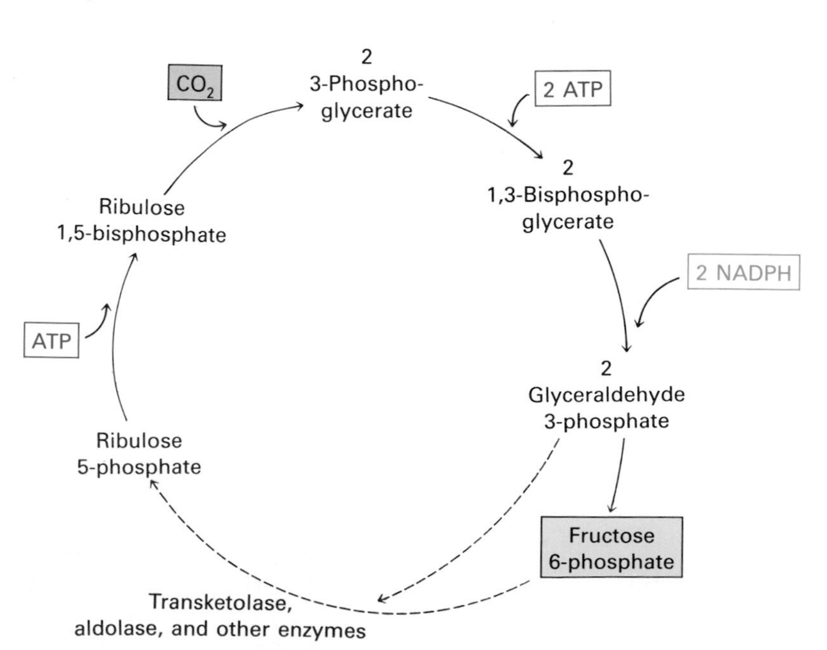

Figure 22-32
The Calvin cycle. The formation of ribulose 5-phosphate from three-carbon and six-carbon sugars is not explicitly shown in the diagram.

THREE ATP AND TWO NADPH ARE USED TO BRING CO_2 TO THE LEVEL OF A HEXOSE

What is the energy expenditure for synthesizing a hexose? Six rounds of the Calvin cycle are required, because one carbon atom is reduced in each (Figure 22-32). Twelve ATP are expended in phosphorylating twelve molecules of 3-phosphoglycerate to 1,3-bisphosphoglycerate, and twelve NADPH are consumed in reducing twelve molecules of 1,3-bisphosphoglycerate to glyceraldehyde 3-phosphate. An additional six ATP are spent in regenerating ribulose 1,5-bisphosphate.

We can now write a balanced equation for the net reaction of the Calvin cycle:

$$6\ CO_2 + 18\ ATP + 12\ NADPH + 12\ H_2O \longrightarrow C_6H_{12}O_6 + 18\ ADP + 18\ P_i + 12\ NADP^+ + 6\ H^+$$

Thus, three molecules of ATP and two of NADPH are consumed in converting CO_2 into a hexose such as glucose or fructose. The efficiency of photosynthesis can be estimated in the following way:

1. The $\Delta G^{\circ\prime}$ for the reduction of CO_2 to the level of hexose is +114 kcal/mol.

2. The reduction of $NADP^+$ is a two-electron process. Hence, the formation of two NADPH requires the pumping of four photons by photosystem I. The electrons given up by photosystem I are replenished by photosystem II, which needs to absorb an equal number of photons. Hence eight photons are needed to generate the required NADPH. The proton gradient generated in producing two NADPH is more than sufficient to drive the synthesis of three ATP.

3. A mole of 600-nm photons has an energy content of 47.6 kcal, and so the energy input of eight moles of photons is 381 kcal. Thus, the overall efficiency of photosynthesis under standard conditions is at least 114/381, or 30%.

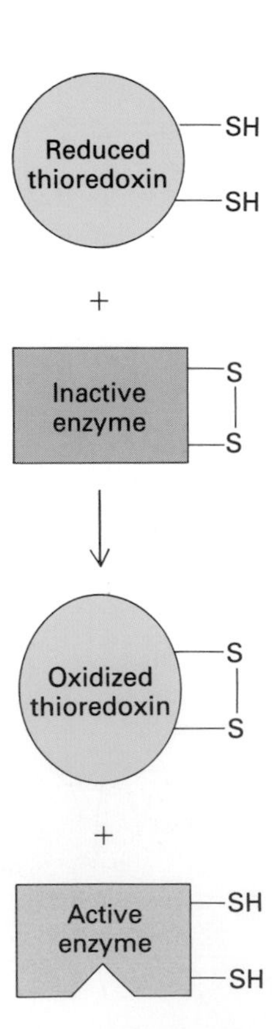

Figure 22-33
The reduced form of thioredoxin activates several enzymes in the Calvin cycle.

THIOREDOXIN PLAYS A KEY ROLE IN COORDINATING THE LIGHT AND DARK REACTIONS OF PHOTOSYNTHESIS

The Calvin cycle operates only when ATP and NADPH are produced by the light reactions of photosynthesis. Several enzymes of the cycle are activated by reduction of disulfide bridges. The reductant is *thioredoxin*, a 12-kd protein containing neighboring cysteine residues (Figure 22-33). These cysteines form a disulfide in oxidized thioredoxin. In chloroplasts, oxidized thioredoxin is reduced by ferredoxin. Thus, the activities of the light and dark reactions of photosynthesis are coordinated through the reducing potential of ferredoxin and then thioredoxin. The catalytic activity of phosphoribulose kinase, for example, increases 100-fold on illumination. We shall return to thioredoxin when we consider the reduction of ribonucleotides (p. 612).

The rate-limiting step in the Calvin cycle is the carboxylation of ribulose 1,5-bisphosphate to form two molecules of 3-phosphoglycerate. *The activity of ribulose 1,5-bisphosphate carboxylase increases markedly on illumination.* In the stroma, the pH increases from 7 to 8 and the level of Mg^{2+} increases. Both effects are consequences of proton-pumping into the thylakoid space. The activity of the carboxylase increases under these conditions because carbamate formation is favored at alkaline pH. CO_2 adds to the deprotonated ϵ-amino group of the regulatory lysine residue, and Mg^{2+} binds to the carbamate to form an adduct essential for catalysis (p. 535).

THE C_4 PATHWAY OF TROPICAL PLANTS ACCELERATES PHOTOSYNTHESIS BY CONCENTRATING CO_2

The oxygenase activity of ribulose 1,5-bisphosphate carboxylase increases more rapidly with temperature than does its carboxylase activity. How then do tropical plants, such as sugar cane, avoid very high rates of wasteful photorespiration? Their solution to this problem is to achieve a high local concentration of CO_2 at the site of the Calvin cycle in their photosynthetic cells. The first clue to the existence of a CO_2-transport mechanism came from studies showing that radioactivity from a pulse of $^{14}CO_2$ appeared initially in malate and aspartate, which are four-carbon compounds, rather than in 3-phosphoglycerate. The essence of this pathway, which was elucidated by M. D. Hatch and C. R. Slack, is that *C_4 compounds carry CO_2 from mesophyll cells, which are in contact with air, to bundle-sheath cells, which are the major sites of photosynthesis* (Figure 22-34). Decarboxylation of the C_4 compound in the bundle-sheath cell maintains a high concentration of CO_2 at the site of the Calvin cycle. The C_3 compound returns to the mesophyll cell for another round of carboxylation.

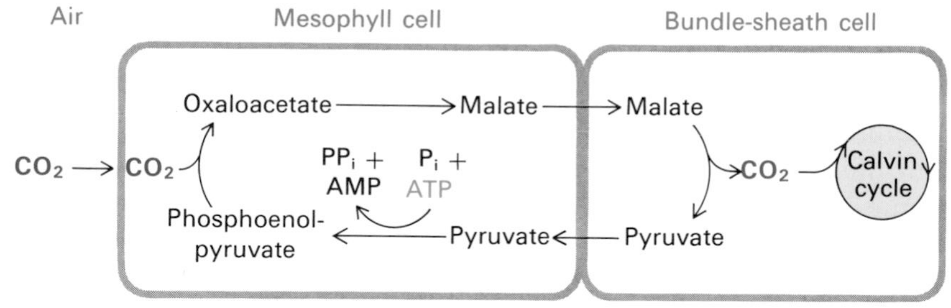

Figure 22-34
Schematic diagram of the essential features of the C_4 pathway. CO_2 is concentrated in bundle-sheath cells by the expenditure of ATP.

The C_4 pathway for the transport of CO_2 starts in the mesophyll cell with the condensation of CO_2 and phosphoenolpyruvate to form *oxaloacetate,* in a reaction catalyzed by phosphoenolpyruvate carboxylase. In some species, oxaloacetate is converted into *malate* by an $NADP^+$-linked malate dehydrogenase. Malate goes into the bundle-sheath cell and is decarboxylated within the chloroplasts by an $NADP^+$-linked malate enzyme. The released CO_2 enters the Calvin cycle in the usual way by condensing with ribulose 1,5-bisphosphate. Pyruvate formed in this decarboxylation reaction returns to the mesophyll cell. Finally, phosphoenolpyruvate is formed from pyruvate in an unusual reaction catalyzed by pyruvate-P_i dikinase (Figure 22-35).

$$\text{Pyruvate} + \text{ATP} + P_i \rightleftharpoons \text{phosphoenolpyruvate} + \text{AMP} + PP_i + H^+$$

ATP donates its γ-phosphoryl group to orthophosphate, and its β-phosphoryl group to a histidine residue of the enzyme. The phosphohistidine residue then reacts with pyruvate to form phosphoenolpyruvate. Why are two ~P consumed in this reaction rather than one? The reason for the phosphorylation of orthophosphate is that its subse-

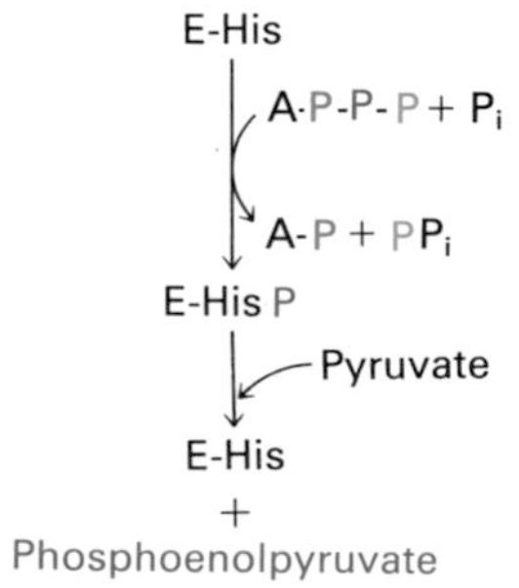

Figure 22-35
Reaction catalyzed by pyruvate-P_i dikinase (E-His). A-P-P-P denotes ATP, and A-P denotes AMP.

quent hydrolysis makes the overall reaction irreversible. The net reaction of this C_4 pathway is

$$CO_2 \text{ (in mesophyll cell)} + ATP + H_2O \longrightarrow CO_2 \text{ (in bundle-sheath cell)} + AMP + 2\ P_i + H^+$$

Thus, *two high-energy phosphate bonds are consumed in transporting CO_2 to the chloroplasts of the bundle-sheath cells.*

When the C_4 pathway and the Calvin cycle operate together, the net reaction is

$$6\ CO_2 + 30\ ATP + 12\ NADPH + 12\ H_2O \longrightarrow C_6H_{12}O_6 + 30\ ADP + 30\ P_i + 12\ NADP^+ + 18\ H^+$$

Note that 30 ATP are consumed per hexose formed when the C_4 pathway delivers CO_2 to the Calvin cycle, in contrast with 18 ATP per hexose in the absence of the C_4 pathway. The high concentration of CO_2 in the bundle-sheath cells of C_4 plants, which is due to the expenditure of the additional 12 ATP, is critical for their rapid photosynthetic rate, because CO_2 is limiting when light is abundant. A high CO_2 concentration also minimizes the energy loss caused by photorespiration.

Tropical plants with a C_4 pathway do little photorespiration because the high concentration of CO_2 in their bundle-sheath cells accelerates the carboxylase reaction relative to the oxygenase reaction. This effect is especially important at higher temperatures. The geographical distribution of plants having this pathway (C_4 plants) and those lacking it (C_3 plants) can now be understood in molecular terms. C_4 plants have the advantage in a hot environment and under high illumination, which accounts for their prevalence in the tropics. C_3 plants, which consume only 18 ATP per hexose formed in the absence of photorespiration (compared with 30 ATP for C_4 plants), are more efficient at temperatures of less than about 28°C, and so they predominate in temperate environments.

Ribulose 1,5-bisphosphate carboxylase emerged early in evolution, when the atmosphere was rich in CO_2 and almost devoid of O_2. The enzyme was not originally selected to operate in an environment rich in O_2 and almost devoid of CO_2. In essence, the C_4 pathway creates a microcosm in which CO_2 is abundant, which for photosynthesis is Paradise regained.

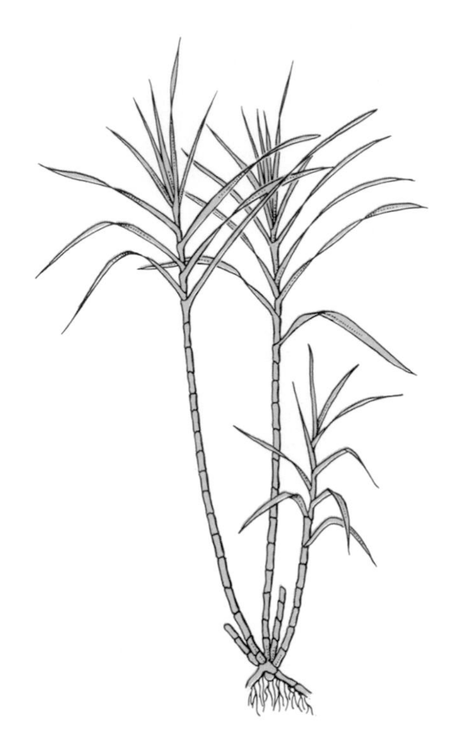

Figure 22-36
Sugar cane, a tropical plant, uses the C_4 pathway to concentrate CO_2.

SUMMARY

Photosynthesis in green plants is mediated by two photosystems located in the thylakoid membranes of chloroplasts. Illumination leads to (1) the generation of a transmembrane proton gradient for the formation of ATP, and (2) the creation of reducing power for the production of NADPH. Light absorbed by chlorophylls in the light-harvesting complex of photosystem II funnels into the reaction center P680. An electron is transferred from the excited P680* to pheophytin and then to plastoquinones attached to Q_A and Q_B to form reduced plastoquinone (QH_2). The reaction center regains electrons from water by the action of a manganese-containing protein, which causes the evolution of O_2. Thus, the net reaction catalyzed by photosystem II is the light-induced transfer of electrons from water to plastoquinone. The critical event at all photosynthetic reaction centers is the light-induced transfer of an electron to an acceptor against an electrochemical potential gradient.

Electrons from photosystem II flow to photosystem I through the cytochrome *bf* complex. This transmembrane complex pumps protons into the thylakoid space as electrons are transferred from QH_2 to plastocyanin, a water-soluble protein. Photosystem I mediates the light-activated transfer of electrons from plastocyanin to P700 and then to ferredoxin, a powerful reductant. Ferredoxin-NADP reductase, a flavoprotein located on the stromal side of the membrane, then catalyzes the formation of NADPH. Thus, the interplay of photosystems I and II leads to the transfer of electrons from H_2O to NADPH and the concomitant generation of a proton gradient for ATP synthesis. Alternatively, electrons from ferredoxin can flow back to photosystem I through the cytochrome *bf* complex; this mode of action of photosystem I, called cyclic photophosphorylation, leads to the generation of a proton gradient without the formation of NADPH. The ATP synthase of chloroplasts (also called CF_0-CF_1) closely resembles the ATP-synthesizing assemblies of bacteria and mitochondria (F_0-F_1). ATP synthesis is driven by the flow of protons from the thylakoid space through the CF_0 transmembrane channel into CF_1 on the stromal side of the membrane.

ATP and NADPH formed in the light reactions of photosynthesis are used to convert CO_2 into hexoses and other organic compounds. The dark phase of photosynthesis, called the Calvin cycle, starts with the reaction of CO_2 and ribulose 1,5-bisphosphate to form two molecules of *3-phosphoglycerate*. The steps in the conversion of 3-phosphoglycerate into fructose 6-phosphate and glucose 6-phosphate are like those of gluconeogenesis, except that glyceraldehyde 3-phosphate dehydrogenase in chloroplasts is specific for NADPH rather than NADH. Ribulose 1,5-bisphosphate is regenerated from fructose 6-phosphate, glyceraldehyde 3-phosphate, and dihydroxyacetone phosphate by a complex series of reactions. Several of the steps in the regeneration of ribulose 1,5-bisphosphate are like those of the pentose phosphate pathway. Reduced thioredoxin formed by transfer of electrons from ferredoxin activates enzymes of the Calvin cycle by reducing disulfide bridges. The light-induced increase in pH and Mg^{2+} level of the stroma are also important in stimulating the carboxylation of ribulose 1,5-bisphosphate. Three ATP and two NADPH are consumed for each CO_2 converted into a hexose. Four photons are absorbed by photosystem I and another four by photosystem II to generate two NADPH and a proton gradient sufficient to drive the synthesis of three ATP.

Ribulose 1,5-bisphosphate carboxylase also catalyzes a competing oxygenase reaction, which produces phosphoglycolate and 3-phosphoglycerate. The recycling of phosphoglycolate leads to the release of CO_2 and further consumption of O_2, in a process called photorespiration. This wasteful side reaction is minimized in tropical plants, which have an accessory pathway for concentrating CO_2 at the site of the Calvin cycle. This C_4 pathway enables tropical plants to take advantage of high light levels and minimize the oxygenation of ribulose 1,5-bisphosphate.

SELECTED READINGS

Where to start

Youvan, D. C., and Marrs, B. L., 1987. Molecular mechanisms of photosynthesis. *Sci. Amer.* 256:42–48.

Deisenhofer, J., Michel, H., and Huber, R., 1985. The structural basis of photosynthetic light reactions in bacteria. *Trends Biochem. Sci.* 10:243–248.

Hinkle, P. C., and McCarty, R. E., 1978. How cells make ATP. *Sci. Amer.* 238(3):104–123. [Contains a lucid discussion of photophosphorylation. Available as *Sci. Amer.* Offprint 1383.]

Miller, K. R., 1979. The photosynthetic membrane. *Sci. Amer.* 241(4):102–113. [Offprint 1448.]

Books

Staehelin, L. A., and Arntzen, C. J., (eds.), 1986. *Photosynthesis III. Photosynthetic Membranes and Light Harvesting Systems.* Springer-Verlag. [A valuable source of detailed articles on many facets of photosynthesis.]

Hoober, J. K., 1984. *Chloroplasts.* Plenum. [An excellent overview of the structure, function, and genetics of chloroplasts. Contains many informative and attractive illustrations.]

Halliwell, B., 1984. *Chloroplast Metabolism* (revised ed.). Oxford University Press. [Emphasizes carbon dioxide fixation, photorespiration, and membrane lipid biosynthesis.]

Clayton, R. K., 1980. *Photosynthesis: Physical Mechanisms and Chemical Patterns.* Cambridge University Press. [A lucid presentation of the physical basis of photosynthesis.]

Steinback, K. E., Bonitz, S., Arntzen, C. J., Bogorad, L., (eds.), 1985. *Molecular Biology of the Photosynthetic Apparatus.* Cold Spring Harbor Laboratory.

Harold, F. M., 1986. *The Vital Force: A Study of Bioenergetics.* W. H. Freeman. [See Chapter 8, "Harvesting the Light," for a perceptive overview of photosynthesis.]

Photosynthetic membrane topography

Anderson, J. M., 1986. Photoregulation of the composition, function, and structure of thylakoid membranes. *Ann. Rev. Plant Physiol.* 37:93–136.

Barber, J., 1985. Organization and dynamics of protein complexes within the chloroplast thylakoid membrane. *Biochem. Soc. Trans.* 14:1–4.

Miller, K. R., and Lyon, M. K., 1985. Do we really know why chloroplast membranes stack? *Trends Biochem. Sci.* 10:219–222.

Light-harvesting assemblies

Zuber, H., 1986. Structure of light-harvesting antenna complexes of photosynthetic bacteria, cyanobacteria, and red algae. *Trends Biochem. Sci.* 11:414–419.

Glazer, A. N., 1985. Light harvesting by phycobilisomes. *Ann. Rev. Biophys. Biophys. Chem.* 14:47–77.

Tronrud, D. E., Schmid, M. F., and Matthews, B. W., 1986. Structure and x-ray amino acid sequence of a bacteriochlorophyll *a* protein from *Prosthecochloris aestaurii* refined at 1.9 Å resolution. *J. Mol. Biol.* 188:443–454.

Reaction centers and the oxygen-evolving complex

Glazer, A. N., and Melis, A., 1987. Photochemical reaction centers: structure, organization, and function. *Ann. Rev. Plant Physiol.* 38:11–45

Deisenhofer, J., Epp, O., Miki, K., Huber, R., and Michel, H., 1985. Structure of the protein subunits in the photosynthetic reaction centre of *Rhodopseudomonas viridis* at 3 Å resolution. *Nature* 318:618–624.

Prince, R. C., 1986. Manganese at the active site of the chloroplast oxygen-evolving complex. *Trends Biochem. Sci.* 11:491–492.

Brudvig, G. W., and Crabtree, R. H., 1986. Mechanism for photosynthetic O_2 evolution. *Proc. Nat. Acad. Sci.* 83:4586–4588.

Blankenship, R. E., and Prince, R. C., 1985. Excited-state redox potentials and the Z scheme of photosynthesis. *Trends Biochem. Sci.* 10:382–383. [A concise and lucid statement of the redox properties of excited states.]

Electron carriers and proton pumps

Melandri, B. A., and Venturoli, G., 1984. Photosynthetic electron transfer. *In* Ernster, L., (ed.), *Bioenergetics*, pp. 95–148. Elsevier.

Cramer, W. A., Widger, W. R., Herrmann, R. G., and Trebst, A., 1985. Topography and function of thylakoid membrane proteins. *Trends Biochem. Sci.* 10:125–129.

Colman, P. M., Freeman, H. C., Guss, J. M., Murata, M., Norris, V. A., Ramshaw, J. A. M., and Venkatappa, M. P., 1978. X-ray crystal structure analysis of plastocyanin at 2.7 Å resolution. *Nature* 272:319–324.

ATP synthase

McCarty, R. E., and Moroney, J. V., 1985. Functions of the subunits and regulation of chloroplast coupling factor 1. *In* Martonosi, A., (ed.), *The Enzymes of Biological Membranes* (2nd ed.), vol. 4, pp. 383–413. Plenum.

Strotmann, H., and Bickel-Sandkötter, S., 1984. Structure, function, and regulation of chloroplast ATPase. *Ann. Rev. Plant Physiol.* 35:97–120.

Jagendorf, A. T., 1967. Acid-base transitions and phosphorylation by chloroplasts. *Fed. Proc.* 26:1361–1369.

Carbon dioxide fixation

Ellis, R. J., and Gray, J. C., (eds.), 1986. Ribulose bisphosphate carboxylase-oxygenase. *Phil. Trans. R. Soc. Lond.* B313:303–469. [Contains excellent articles on the catalytic mechanism, three-dimensional structure, evolution, and molecular genetics of this enzyme.]

Miziorko, H. M., and Lorimer, G. H., 1982. Ribulose-1,5-bisphosphate carboxylase-oxygenase. *Ann. Rev. Biochem.* 52:507–535.

Schneider, G., Lindqvist, Y., Brädén, C-I., and Lorimer, G., 1986. Three-dimensional structure of ribulose-1,5-bisphosphate carboxylase/oxygenase from *Rhodospirullum rubrum* at 2.9 Å resolution. *EMBO J.* 5:3409–3415.

Burnell, J. N., and Hatch, M. D., 1985. Light-dark modulation of leaf pyruvate,P_i dikinase. *Trends Biochem. Sci.* 10:288–291.

Ogren, W. L., 1984. Photorespiration: pathways, regulation, and modification. *Ann. Rev. Plant Physiol.* 35:415–442.

Historical aspects of photosynthesis

Arnon, D. I., 1987. Photosynthetic CO_2 assimilation by chloroplasts: assertion, refutation, discovery. *Trends Biochem. Sci.* 12:39–42.

Arnon, D. I., 1984. The discovery of photosynthetic phosphorylation. *Trends Biochem. Sci.* 9:258–262.

Bassham, J. A., 1962. The path of carbon in photosynthesis. *Sci. Amer.* 206(6):88–100. [Offprint 122.]

PROBLEMS

1. Calculate the $\Delta E'_0$ and $\Delta G^{\circ\prime}$ for the reduction of $NADP^+$ by ferredoxin. Use data given in Table 17-1 on p. 400.

2. Sedoheptulose 1,7-bisphosphate is an intermediate in the Calvin cycle but not in the pentose phosphate pathway. What is the enzymatic basis of this difference?

3. Suppose that an illuminated suspension of *Chlorella* was actively carrying out photosynthesis when the light was suddenly switched off. How would the levels of 3-phosphoglycerate and ribulose 1,5-bisphosphate change during the next minute?

4. Suppose that an illuminated suspension of *Chlorella* was actively carrying out photosynthesis in the presence of 1% CO_2 when the concentration of CO_2 was abruptly reduced to 0.003%. What effect would this have on the levels of 3-phosphoglycerate and ribulose 1,5-bisphosphate during the next minute?

5. Blue-green bacteria deprived of a source of nitrogen digest their least essential proteins. Which component of phycobilisomes is likely to be degraded first under starvation conditions?

6. The Van Niel equation for photosynthesis in higher plants is

$$6\ CO_2 + 12\ H_2O \longrightarrow C_6H_{12}O_6 + 6\ H_2O + 6\ O_2$$

   (a) H_2O appears on both sides of this equation. Is this merely a formalism or does it express an important aspect of the mechanism of photosynthesis?
   (b) The conventional equation for respiration is

$$C_6H_{12}O_6 + 6\ O_2 \longrightarrow 6\ CO_2 + 6\ H_2O$$

   Is the Van Niel equation (in reverse) more revealing regarding the mechanism of respiration? In other words, does the oxidation of glucose require the input of six molecules of H_2O?

7. Chloroplasts illuminated in the absence of ADP and P_i form ATP in a subsequent dark period on addition of ADP and P_i. The amount of ATP synthesized in the dark is markedly increased if pyridine is added before illumination. Why?

8. Dichlorophenyldimethylurea (DCMU), an herbicide, interferes with photophosphorylation and O_2 evolution. However, it does not block the Hill reaction. Propose a site for the inhibitory action of DCMU.

9. 2-Carboxyarabinitol 1,5-bisphosphate (CABP) has been useful in studies of ribulose 1,5-bisphosphate carboxylase.
   (a) Write the structural formula of CABP.
   (b) Which catalytic intermediate does it resemble?
   (c) Predict the effect of CABP on the carboxylase.

10. Write a balanced equation for the transamination of glyoxylate to yield glycine.

11. Consider the relation between the energy of a photon and its wavelength.
   (a) Some bacteria are able to harvest 1000-nm light. What is the energy (in kcal) of a mole (also called an einstein) of 1000-nm photons?
   (b) What is the maximum increase in redox potential that can be induced by a 1000-nm photon?
   (c) What is the minimum number of 1000-nm photons needed to form ATP from ADP and P_i? Assume a ΔG of 12 kcal/mol for the phosphorylation reaction.

On the facing page: Model of *S*-adenosylmethionine, a methyl donor in many biosyntheses.

PART IV

BIOSYNTHESIS OF MACROMOLECULAR PRECURSORS

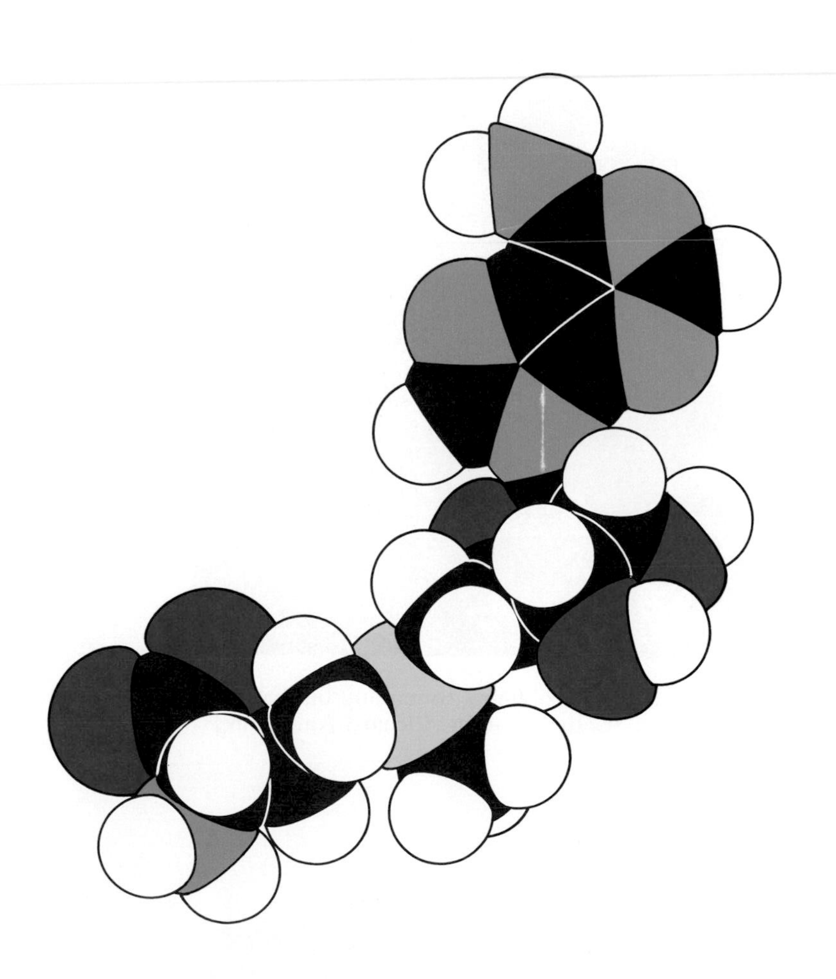

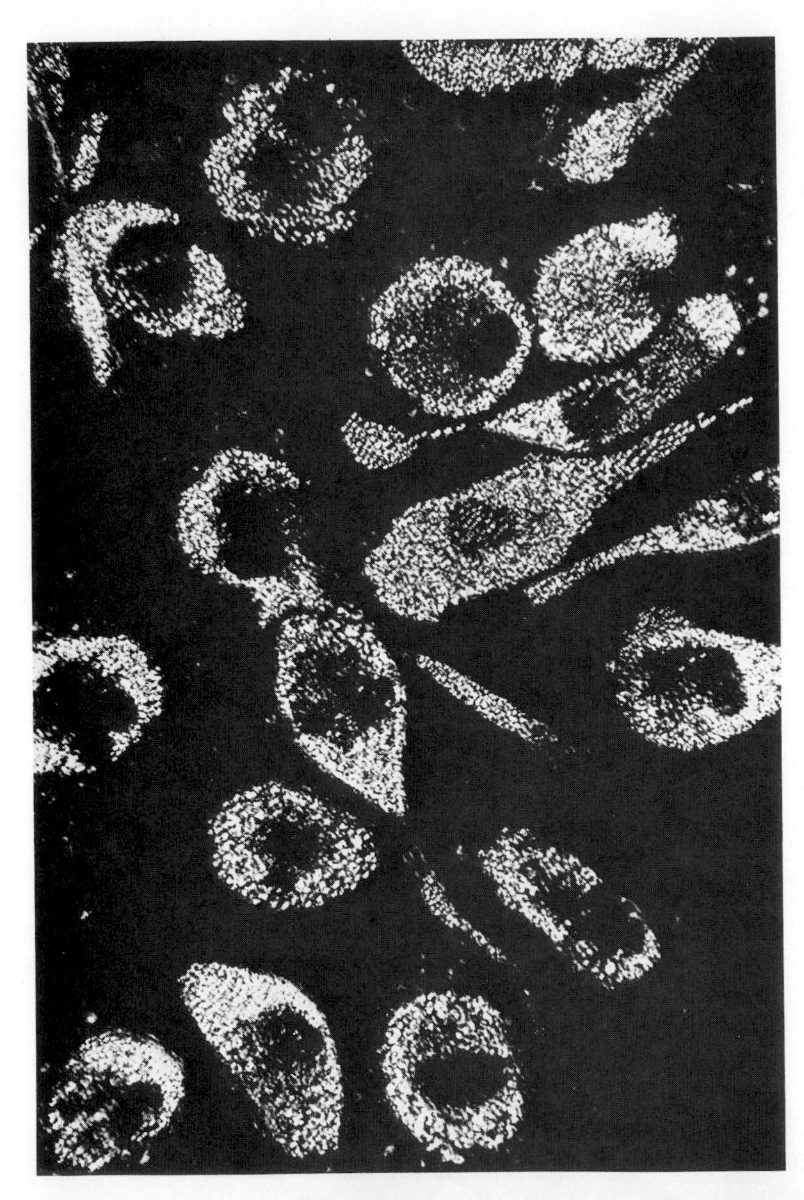

Polarized light micrograph of macrophages containing birefringent crystals of cholesterol esters. [Courtesy of Dr. Richard Anderson.]

CHAPTER 23

Biosynthesis of Membrane Lipids and Steroid Hormones

We turn now from the generation of metabolic energy to the biosynthesis of macromolecular precursors and related biomolecules. This chapter deals with the biosynthesis of three important components of biological membranes—phosphoglycerides, sphingolipids, and cholesterol (Chapter 12). The synthesis of triacylglycerols is also considered here because this pathway overlaps that of the phosphoglycerides; the formation of steroid hormones is presented because they are derived from cholesterol. The biosynthesis of cholesterol exemplifies a fundamental mechanism for the assembly of extended carbon skeletons from five-carbon units. The transport of cholesterol in blood by the low-density lipoprotein and its uptake by a specific receptor on the cell surface are discussed in some detail because they vividly illustrate a general mechanism for the entry of metabolites and signal molecules into cells. The absence of this receptor in familial hypercholesterolemia, a genetic disease, leads to markedly elevated cholesterol levels in the blood, cholesterol deposits on blood vessels, and heart attacks in childhood.

Anabolism—
Biosynthetic processes.

Catabolism—
Degradative processes.

Derived from the Greek words *ana*, up; *cata*, down; *ballein*, to throw.

PHOSPHATIDATE IS AN INTERMEDIATE IN THE SYNTHESIS OF PHOSPHOGLYCERIDES AND TRIACYLGLYCEROLS

Phosphatidate (diacylglycerol 3-phosphate) is a common intermediate in the synthesis of phosphoglycerides and triacylglycerols. The starting point is *glycerol 3-phosphate*, which is formed mainly by reduction of dihydroxyacetone phosphate and to a lesser extent by phosphorylation of glycerol. Glycerol 3-phosphate is acylated by acyl CoA to form *lysophosphatidate,* which is again acylated by acyl CoA to yield *phosphatidate.* These acylations are catalyzed by glycerol-phosphate acyl transferase. In most phosphatidates, the fatty acyl chain attached to C_1 is saturated, whereas the one attached to C_2 is unsaturated.

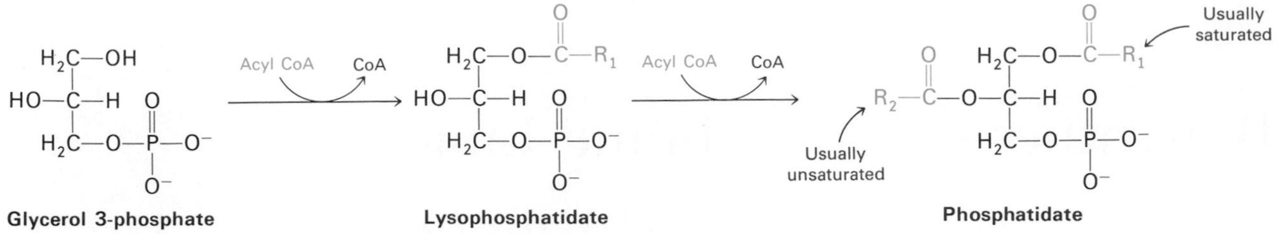

Glycerol 3-phosphate **Lysophosphatidate** **Phosphatidate**

The pathways diverge at phosphatidate. In the synthesis of triacylglycerols, phosphatidate is hydrolyzed by a specific phosphatase to give a *diacylglycerol.* This intermediate is acylated to a *triacylglycerol* in a reaction that is catalyzed by diglyceride acyl transferase. These enzymes are associated in a *triacylglycerol synthetase complex* that is bound to the endoplasmic reticulum membrane.

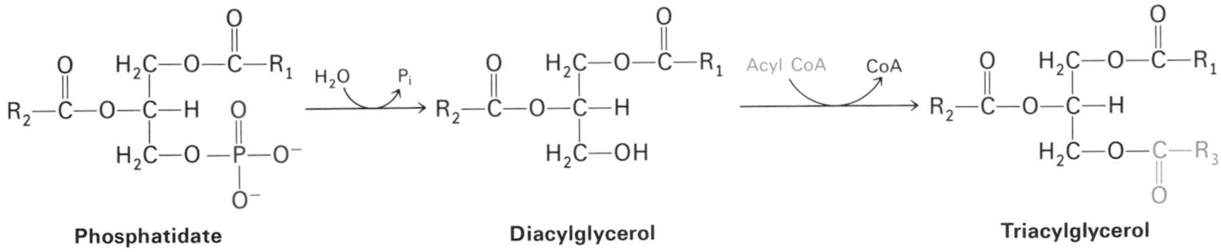

Phosphatidate **Diacylglycerol** **Triacylglycerol**

CDP-DIACYLGLYCEROL IS THE ACTIVATED INTERMEDIATE IN THE DE NOVO SYNTHESIS OF SOME PHOSPHOGLYCERIDES

There are several routes for the synthesis of phosphoglycerides. One de novo pathway starts with the formation of *cytidine diphosphodiacylglycerol* (CDP-diacylglycerol) from phosphatidate and cytidine triphosphate (CTP). This reaction, like many biosyntheses, is driven forward by the hydrolysis of pyrophosphate. Recall that PP_i is formed in the synthesis of UDP-glucose from glucose 1-phosphate and UTP (p. 456).

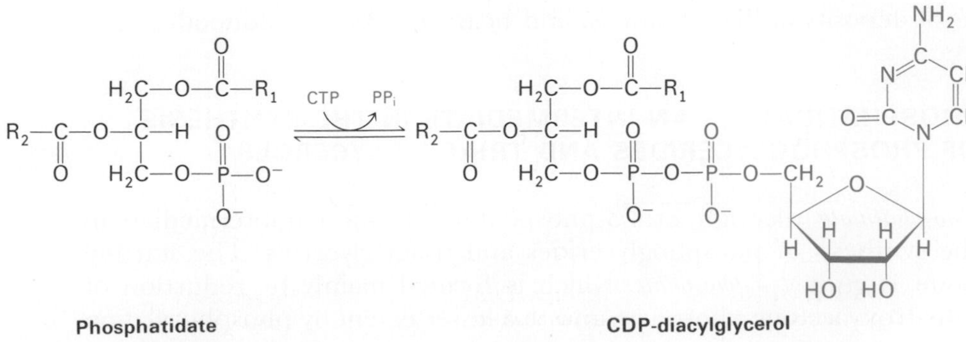

Phosphatidate **CDP-diacylglycerol**

The activated phosphatidyl unit then reacts with the hydroxyl group of a polar alcohol. If the alcohol is serine, the products are *phosphatidyl serine* and cytidine monophosphate (CMP).

$$\underset{\textbf{CDP-diacylglycerol}}{R_2{-}C({=}O){-}O{-}CH(CH_2{-}O{-}C({=}O){-}R_1){-}CH_2{-}O{-}P(O)(O^-){-}O{-}P(O)(O^-){-}O{-}\text{Cytidine}} + \underset{\textbf{Serine}}{HO{-}CH_2{-}CH(NH_3^+){-}COO^-} \rightleftharpoons$$

$$\underset{\textbf{Phosphatidyl serine}}{R_2{-}C({=}O){-}O{-}CH(CH_2{-}O{-}C({=}O){-}R_1){-}CH_2{-}O{-}P(O)(O^-){-}O{-}CH_2{-}CH(NH_3^+){-}COO^-} + \underset{\textbf{CMP}}{^-O{-}P(O)(O^-){-}O{-}\text{Cytidine}}$$

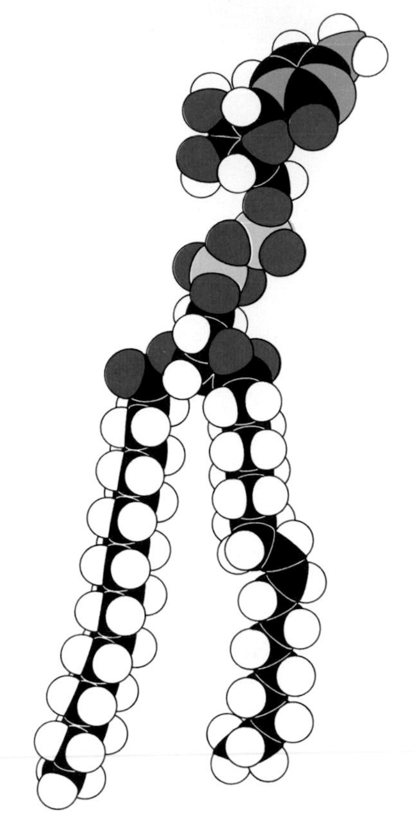

Figure 23-1
Model of a CDP-diacylglycerol, a key intermediate in the synthesis of phospholipids.

Likewise, *phosphatidyl inositol* is formed by the transfer of a diacylglycerol phosphate unit from CDP-diacylglycerol to inositol. Subsequent phosphorylations catalyzed by specific kinases lead to the synthesis of *phosphatidyl inositol 4,5-bisphosphate*, a key molecule in signal transduction. Hormonal stimuli activate a phospholipase that hydrolyzes this phospholipid into two intracellular messengers—diacylglycerol and inositol 1,4,5-trisphosphate. This important and ubiquitous cascade in eucaryotes will be discussed further in a later chapter (p. 986).

Thus, a cytidine nucleotide plays the same role in the synthesis of these phosphoglycerides as does a uridine nucleotide in the formation of glycogen (p. 456). In both biosyntheses, an activated intermediate (UDP-glucose or CDP-diacylglycerol) is formed from a phosphorylated substrate (glucose 1-phosphate or phosphatidate) and a nucleoside triphosphate (UTP or CTP). The activated intermediate then reacts with a hydroxyl group (the 4-OH terminus of glycogen or the hydroxyl side chain of serine).

$$\textbf{Inositol} \xrightarrow[\text{CMP}]{\text{CDP-diacylglycerol}} \textbf{Phosphatidyl inositol}$$

PHOSPHATIDYL ETHANOLAMINE AND PHOSPHATIDYL CHOLINE CAN BE FORMED FROM PHOSPHATIDYL SERINE

In bacteria, decarboxylation of phosphatidyl serine by a pyridoxal phosphate enzyme yields *phosphatidyl ethanolamine*. The amino group of this phosphoglyceride is then methylated three times to form *phosphatidyl choline*. *S-adenosylmethionine,* the methyl donor in this and many other methylations, will be discussed later (p. 582).

$$\underset{\textbf{Phosphatidyl serine}}{H{-}C(NH_3^+)(COO^-){-}CH_2{-}O{-}R} \xrightarrow{H^+ \;\; CO_2} \underset{\textbf{Phosphatidyl ethanolamine}}{H{-}C(NH_3^+)(H){-}CH_2{-}O{-}R} \xrightarrow{3 \sim CH_3 \;\; 3H^+} \underset{\textbf{Phosphatidyl choline}}{H{-}C(N^+(CH_3)_3)(H){-}CH_2{-}O{-}R}$$

PHOSPHOGLYCERIDES CAN ALSO BE SYNTHESIZED FROM A CDP-ALCOHOL INTERMEDIATE

In mammals, phosphatidyl choline is synthesized by a pathway that utilizes choline obtained from the diet (Figure 23-2). Choline is phosphorylated by ATP to *phosphorylcholine,* which then reacts with CTP to form *CDP-choline.* The phosphorylcholine unit of CDP-choline is then transferred to a diacylglycerol to form *phosphatidyl choline.* Note that the activated species in this pathway is the cytidine derivative of phosphorylcholine rather than of phosphatidate.

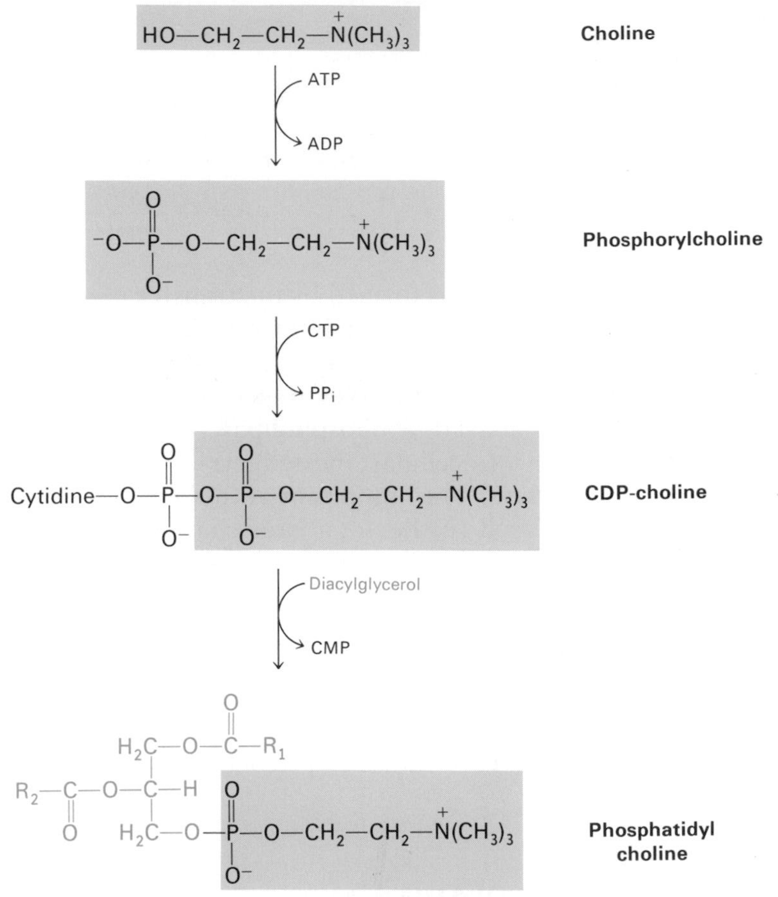

$HO{-}CH_2{-}CH_2{-}\overset{+}{N}(CH_3)_3$ **Choline**

↓ ATP → ADP

$^{-}O{-}P(=O)(O^{-}){-}O{-}CH_2{-}CH_2{-}\overset{+}{N}(CH_3)_3$ **Phosphorylcholine**

↓ CTP → PP_i

$Cytidine{-}O{-}P(=O)(O^{-}){-}O{-}P(=O)(O^{-}){-}O{-}CH_2{-}CH_2{-}\overset{+}{N}(CH_3)_3$ **CDP-choline**

↓ Diacylglycerol → CMP

$H_2C{-}O{-}C(=O){-}R_1$; $R_2{-}C(=O){-}O{-}C{-}H$; $H_2C{-}O{-}P(=O)(O^{-}){-}O{-}CH_2{-}CH_2{-}\overset{+}{N}(CH_3)_3$ **Phosphatidyl choline**

Figure 23-2
Synthesis of phosphatidyl choline.

Likewise, *phosphatidyl ethanolamine* can be synthesized from ethanolamine by forming a CDP-ethanolamine intermediate by analogous reactions. Alternatively, phosphatidyl ethanolamine can be formed from phosphatidyl serine by the enzyme-catalyzed exchange of ethanolamine for the serine moiety of the phospholipid.

PLASMALOGENS AND OTHER ETHER PHOSPHOLIPIDS ARE FORMED FROM DIHYDROXYACETONE PHOSPHATE

Some phospholipids contain an ether unit instead of an acyl unit at C_1. *Glyceryl ether phospholipids* are synthesized starting with dihydroxyacetone phosphate (Figure 23-3). Acylation by a fatty acyl CoA yields a 1-acyl derivative that exchanges with a long-chain alcohol to form an ether at C_1. The keto group at C_2 is reduced by NADPH, and the

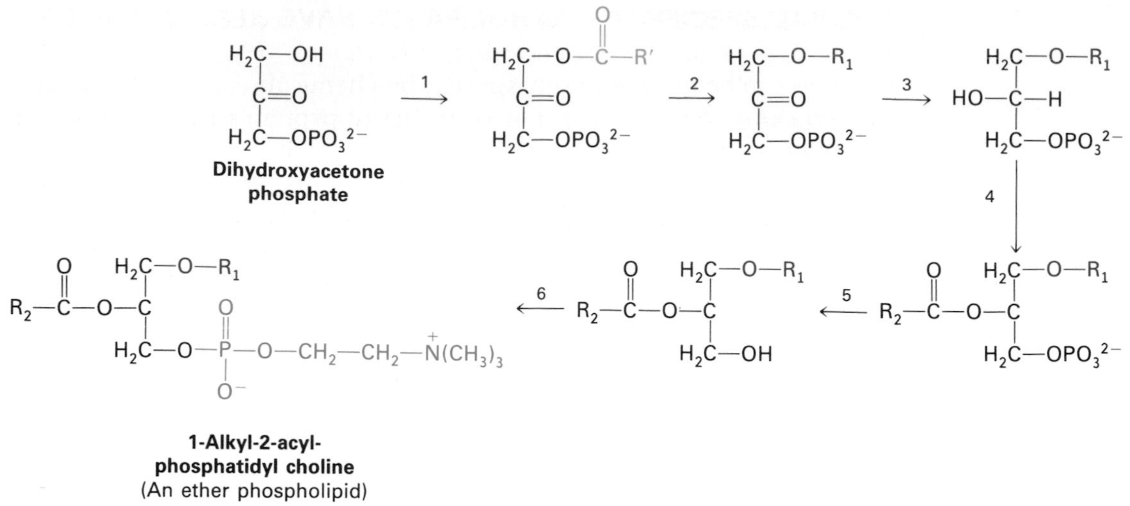

Dihydroxyacetone phosphate

1-Alkyl-2-acyl-phosphatidyl choline
(An ether phospholipid)

Figure 23-3
Synthesis of an ether phospholipid. The steps are (1) acylation by fatty acyl CoA, (2) exchange of an alcohol for the carboxylate moiety, (3) reduction by NADPH, (4) acylation by a second fatty acyl CoA, (5) hydrolysis of the phosphate ester, (6) transfer of a phosphocholine moiety.

resulting alcohol is acylated by a long-chain CoA. Removal of the 3-phosphate group yields 1-alkyl-2-acylglycerol, which reacts with CDP-choline to form the ether analog of phosphatidyl choline.

An ether phospholipid with striking activities has recently been identified. *Platelet-activating factor* is a 1-alkyl-2-acetyl ether analog of phosphatidyl choline. Even a very low concentration of this compound (0.1 nM) in the blood causes the aggregation of platelets and the dilation of blood vessels. The presence of an acetyl group rather than a long-chain acyl group at C_2 increases the water-solubility of this lipid, enabling it to function in an aqueous environment.

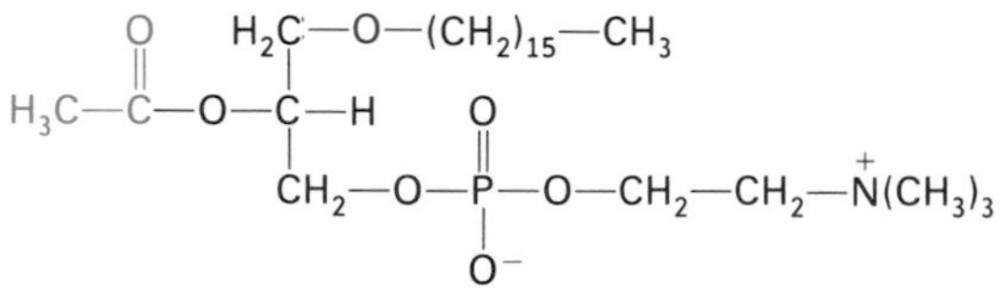

Platelet-activating factor

Plasmalogens are phospholipids containing an α,β-unsaturated ether at C_1. Phosphatid*al* choline, the plasmalogen corresponding to phosphatidyl choline, is formed by desaturation of a 1-alkyl precursor. The desaturase catalyzing this final step in the synthesis of a plasmalogen is a microsomal enzyme akin to the one that introduces double bonds into long-chain fatty acyl CoAs: O_2 and NADH are reactants, and cytochrome b_5 participates in catalysis (p. 489).

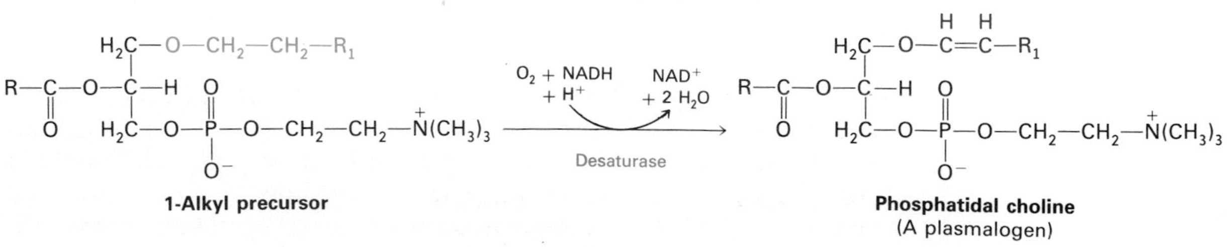

O_2 + NADH + H^+ → NAD^+ + 2 H_2O (Desaturase)

1-Alkyl precursor

Phosphatidal choline
(A plasmalogen)

SEVERAL SPECIFIC PHOSPHOLIPASES HAVE BEEN ISOLATED

Enzymes can be used as highly specific biochemical reagents. For example, trypsin is an invaluable tool in studies of protein structure. Specific phospholipases are likewise significant in studies of biological membranes and of their phospholipid constituents. Phospholipases are grouped according to their specificities. The bonds hydrolyzed by phospholipases A_1, A_2, C, and D are shown in Figure 23-4.

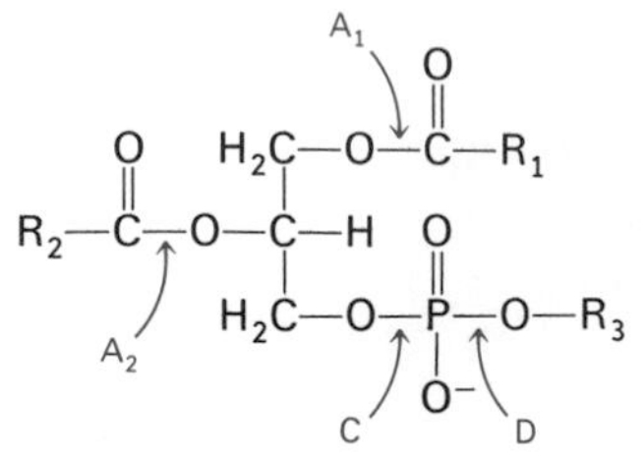

Phospholipases have two kinds of roles in nature. Many of them are *digestive enzymes* present in high concentration in intestinal juices, bacterial secretions, and venoms. *They also participate in enzymatic cascades that generate highly active lipids or transduce signals.* For example, phospholipase A_2 releases arachidonate, a precursor of prostaglandins (p. 991), and phospholipase C participates in the phosphoinositide cascade (p. 991).

Figure 23-4
Specificity of phospholipases.

SYNTHESIS OF CERAMIDE, THE BASIC STRUCTURAL UNIT OF SPHINGOLIPIDS

We turn now from phosphoglycerides to sphingolipids. The backbone of a sphingolipid is *sphingosine,* rather than glycerol. Palmitoyl CoA and serine condense to form dehydrosphinganine, which is then converted to sphingosine. The enzyme catalyzing this reaction requires pyridoxal phosphate.

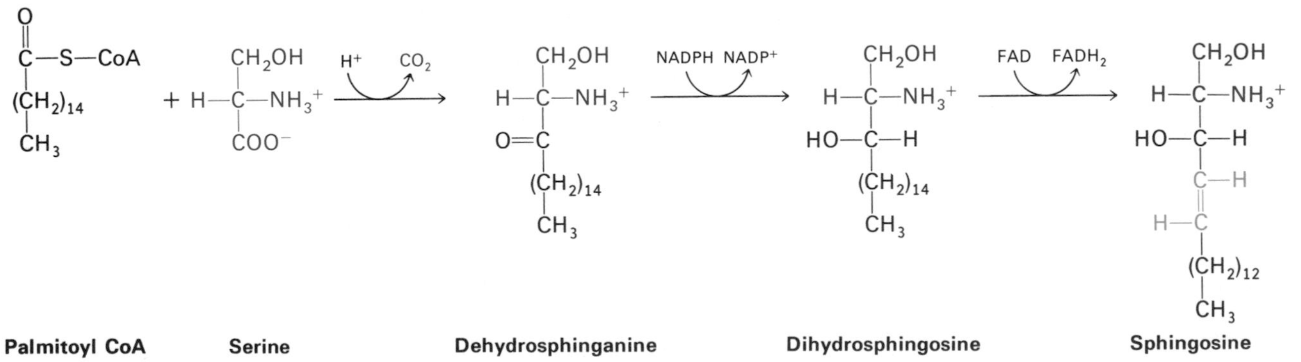

Palmitoyl CoA + **Serine** $\xrightarrow{H^+ \;\; CO_2}$ **Dehydrosphinganine** $\xrightarrow{NADPH \;\; NADP^+}$ **Dihydrosphingosine** $\xrightarrow{FAD \;\; FADH_2}$ **Sphingosine**

In all sphingolipids, the amino group of sphingosine is acylated (Figure 23-5): a long-chain acyl CoA reacts with sphingosine to form *ceramide* (*N*-acyl sphingosine). The terminal hydroxyl group also is substituted. In *sphingomyelin,* the substituent is phosphorylcholine, which comes

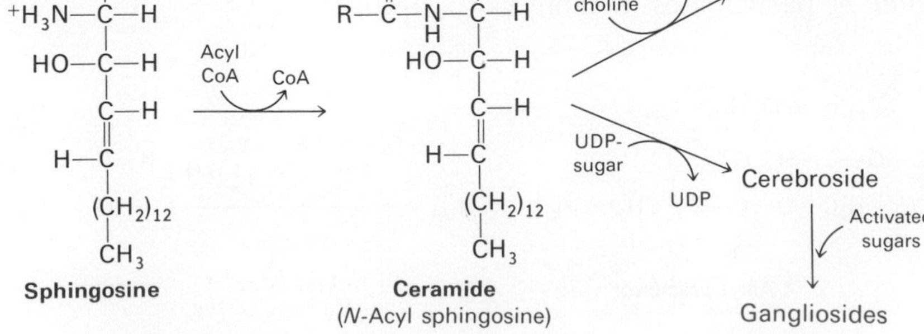

Sphingosine $\xrightarrow{\text{Acyl CoA} \;\; \text{CoA}}$ **Ceramide** (*N*-Acyl sphingosine) → (CDP-choline → CMP) Sphingomyelin; (UDP-sugar → UDP) Cerebroside → (Activated sugars) Gangliosides

Figure 23-5
Ceramide is the precursor of sphingomyelin and of gangliosides.

from CDP-choline. In a *cerebroside,* glucose or galactose is linked to the terminal hydroxyl group of ceramide. UDP-glucose or UDP-galactose is the sugar donor in the synthesis of cerebrosides. In a *ganglioside,* an oligosaccharide is linked to ceramide by a glucose residue (see Figure 23-6).

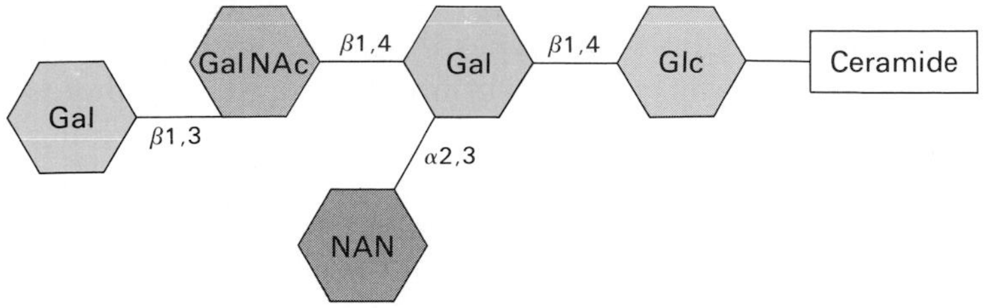

Figure 23-6
Structure of ganglioside G_{M1}. Abbreviations used: Gal, galactose; GalNAc, *N*-acetylgalactosamine; Glc, glucose; NAN, *N*-acetylneuraminate. The types of linkages between the sugars—such as β1,4—are noted at the bonds joining them.

GANGLIOSIDES ARE CARBOHYDRATE-RICH SPHINGOLIPIDS THAT CONTAIN ACIDIC SUGARS

In gangliosides, the most complex sphingolipids, an *oligosaccharide chain* containing at least one acidic sugar is attached to ceramide. The acidic sugar is N-*acetylneuraminate* or *N*-glycolylneuraminate. These acidic sugars are called *sialic acids.* Their 9-carbon backbones are synthesized from phosphoenolpyruvate (a C_3 unit) and *N*-acetylmannosamine 6-phosphate (a C_6 unit).

Gangliosides are synthesized by the ordered, step-by-step addition of sugar residues to ceramide. UDP-glucose, UDP-galactose, and UDP-*N*-acetylgalactosamine are activated donors of these sugar residues, as is the CMP derivative of *N*-acetyl neuraminate. CMP-*N*-acetylneuraminate is synthesized from CTP and *N*-acetylneuraminate. The structure of the resulting ganglioside is determined by the specificity of the glycosyl transferases in the cell. More than fifteen different gangliosides have been characterized. The structure of ganglioside G_{M1} is shown in Figure 23-6.

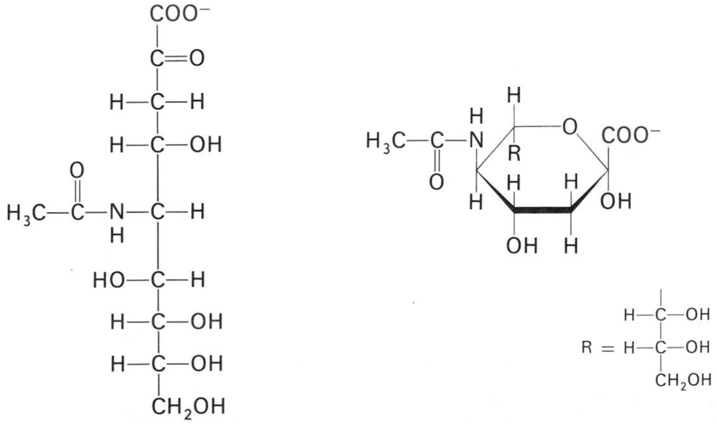

***N*-Acetylneuraminate**
(Open-chain form)

***N*-Acetylneuraminate**
(Pyranose form)

TAY-SACHS DISEASE: AN INHERITED DISORDER OF GANGLIOSIDE BREAKDOWN

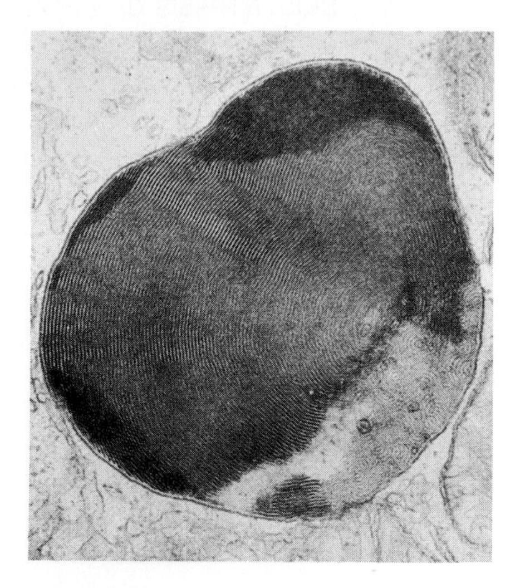

Figure 23-7
Electron micrograph of a lysosome containing an abnormal amount of lipid. [Courtesy of Dr. George Palade.]

Gangliosides are found in highest concentration in the nervous system, particularly in gray matter, where they constitute 6% of the lipids. Gangliosides are continually degraded by the sequential removal of their terminal sugars. The glycosyl hydrolases that catalyze these reactions are highly specific. Ganglioside breakdown occurs inside *lysosomes*. These organelles contain many types of degradative enzymes and are specialized for the orderly destruction of cellular components.

Disorders of ganglioside breakdown can have serious clinical consequences. In Tay-Sachs disease, the symptoms are usually evident before an affected infant is a year old. Weakness and retarded psychomotor development are typical early symptoms. Impaired vision and spasticity usually follow a few months later. By age two, the child is demented and blind. Tay-Sachs disease is usually fatal before age three. Striking pathological changes occur in the nervous system, where neurons become enormously swollen with lipid-filled lysosomes.

The ganglioside content of the brain of an infant with Tay-Sachs disease is greatly elevated. Specifically, *the concentration of ganglioside G_{M2} is many times higher than normal*. The abnormally high level of this ganglioside is caused by a deficiency of the enzyme that removes its terminal *N*-acetylgalactosamine residue. The missing or deficient enzyme is a specific *β*-N-*acetylhexosaminidase*.

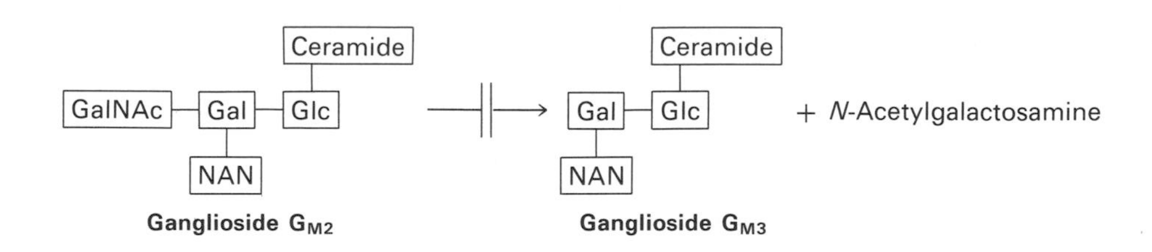

Tay-Sachs disease is inherited as an *autosomal recessive*. The carrier rate is 1/30 in Jewish Americans originally from eastern Europe and 1/300 in non-Jewish Americans. Consequently, the incidence of the disease is about 100 times higher in Jewish Americans. Tay-Sachs disease can be diagnosed during fetal development. Amniotic fluid is obtained by *amniocentesis* and assayed for *β*-*N*-acetylhexosaminidase activity.

CHOLESTEROL IS SYNTHESIZED FROM ACETYL COENZYME A

We now turn to the synthesis of cholesterol, a steroid that modulates the fluidity of eucaryotic membranes (p. 297). Cholesterol is also the precursor of steroid hormones such as progesterone, testosterone, estradiol, and cortisol. An early and important clue concerning the synthesis of cholesterol came from the work of Konrad Bloch in the 1940s. Acetate isotopically labeled in its carbon atoms was prepared and fed to rats. The cholesterol that was synthesized by these rats contained the isotopic label, which showed that acetate is a precursor of cholesterol. In fact, *all twenty-seven carbon atoms of cholesterol are derived from acetyl CoA*. Further insight came from studies in which cholesterol was synthe-

Figure 23-8
Model of cholesterol.

sized from acetate labeled in either its methyl or its carboxyl carbon. Degradation of cholesterol synthesized from such labeled acetate showed the origin of each atom of the cholesterol molecule (Figure 23-9). This pattern played a crucial role in the formation and testing of hypotheses concerning the pathway of cholesterol synthesis.

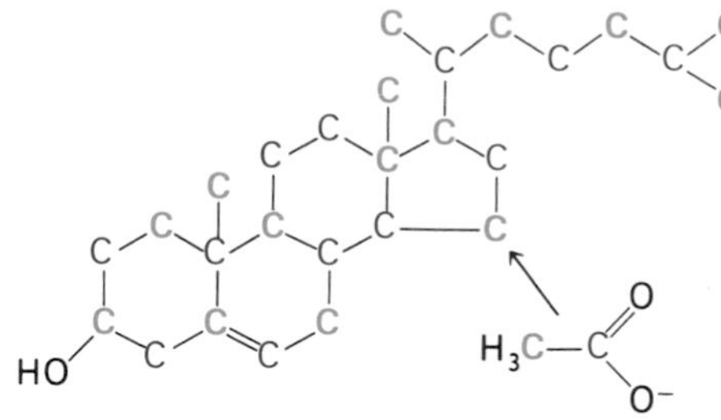

Figure 23-9
Isotope-labeling pattern of cholesterol synthesized from acetate labeled in its methyl (green) or carboxyl (red) carbon atom.

MEVALONATE AND SQUALENE ARE INTERMEDIATES IN THE SYNTHESIS OF CHOLESTEROL

The next major clue came from the discovery that *squalene*, a C_{30} hydrocarbon, is an intermediate in the synthesis of cholesterol. Squalene consists of six *isoprene* units.

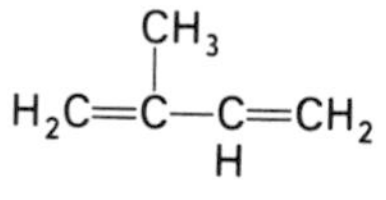

Isoprene

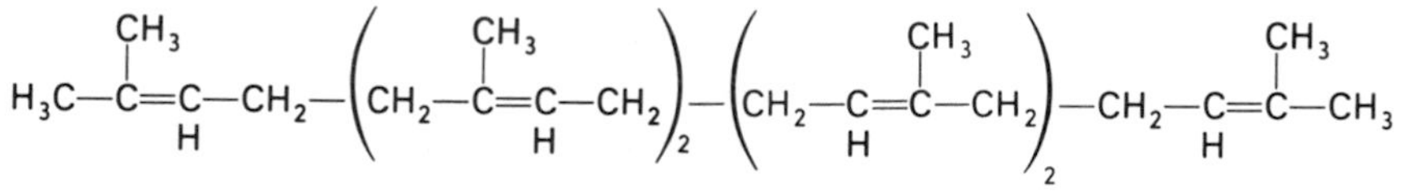

Squalene

This finding raised the question of how isoprene units are formed from acetate.

$$\underset{C_2}{\text{Acetate}} \longrightarrow \underset{C_5}{[\text{isoprene}]} \longrightarrow \underset{C_{30}}{\text{squalene}} \longrightarrow \underset{C_{27}}{\text{cholesterol}}$$

The answer came unexpectedly from unrelated studies of bacterial mutants. It was found that mevalonate could substitute for acetate to meet the nutritional needs of acetate-requiring mutants.

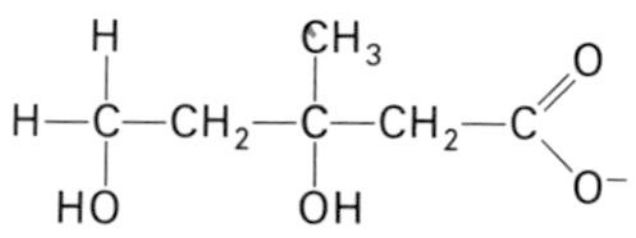

The discovery of mevalonate was a key step in the elucidation of the pathway of cholesterol biosynthesis, because it was soon recognized that this C_6 acid could decarboxylate to yield the postulated C_5 isoprene intermediate. Isotopic labeling studies then showed that mevalonate is indeed a precursor of squalene and that it could be formed from acetate. The activated isoprene intermediate proved to be *isopentenyl pyrophosphate*, which is formed by decarboxylation of a derivative of mevalonate. A pathway for the synthesis of cholesterol from acetate could then be outlined.

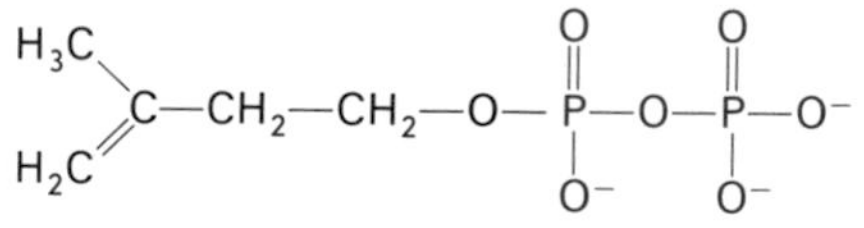

Mevalonate

Isopentenyl pyrophosphate

$$\underset{C_2}{\text{Acetate}} \longrightarrow \underset{C_6}{\text{mevalonate}} \longrightarrow$$

$$\underset{C_5}{\text{isopentenyl pyrophosphate}} \longrightarrow \underset{C_{30}}{\text{squalene}} \longrightarrow \underset{C_{27}}{\text{cholesterol}}$$

SYNTHESIS OF ISOPENTENYL PYROPHOSPHATE, AN ACTIVATED INTERMEDIATE IN CHOLESTEROL FORMATION

The first stage in the synthesis of cholesterol is the formation of isopentenyl pyrophosphate from acetyl CoA. This set of reactions starts with the formation of 3-hydroxy-3-methylglutaryl CoA from acetyl CoA and acetoacetyl CoA. One of the fates of 3-hydroxy-3-methylglutaryl CoA, its cleavage to acetyl CoA and acetoacetate, was discussed previously in regard to the formation of ketone bodies (p. 479). Alternatively, it can be reduced to mevalonate (Figure 23-10). 3-Hydroxy-3-methylglutaryl

Acetoacetyl CoA + Acetyl CoA + H_2O ⟶ 3-Hydroxy-3-methylglutaryl CoA

Cleavage enzyme in mitochondria ⟶ Acetyl CoA + Acetoacetate

Reductase in cytosol ⟶ Mevalonate ⟶ Cholesterol

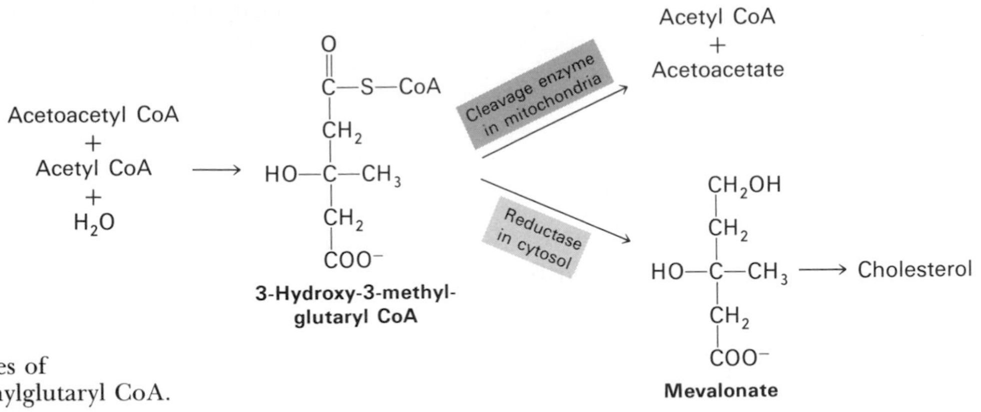

Figure 23-10
Synthesis and fates of 3-hydroxy-3-methylglutaryl CoA.

CoA is present both in the cytosol and in the mitochondria of liver cells. The mitochondrial pool of this intermediate is mainly a precursor of ketone bodies, whereas the cytoplasmic pool gives rise to mevalonate for the synthesis of cholesterol.

The synthesis of mevalonate is the committed step in cholesterol formation. The enzyme catalyzing this irreversible step, 3-hydroxy-3-methylglutaryl CoA reductase, is an important control site in cholesterol biosynthesis, as will be discussed shortly.

$$\text{3-Hydroxy-3-methylglutaryl CoA} + 2\ \text{NADPH} + 2\ \text{H}^+ \longrightarrow \text{mevalonate} + 2\ \text{NADP}^+ + \text{CoA}$$

Mevalonate is converted into *3-isopentenyl pyrophosphate* by three consecutive reactions involving ATP (Figure 23-11). In the last step, the release of CO_2 from 5-pyrophosphomevalonate occurs in concert with the hydrolysis of ATP to ADP and P_i.

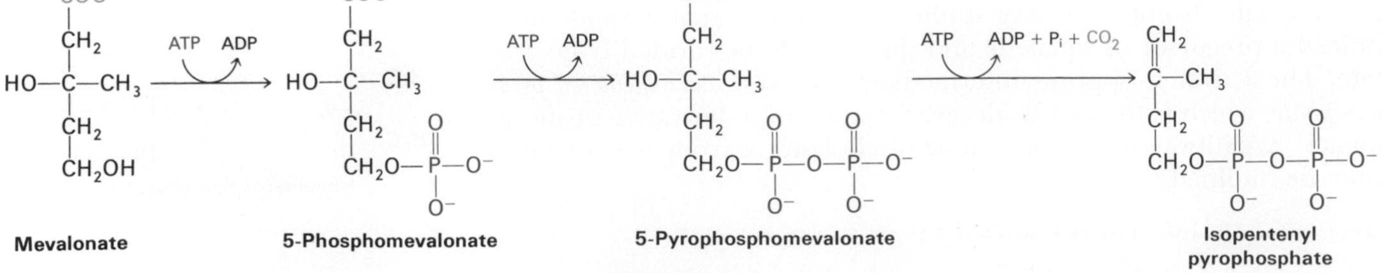

Mevalonate —(ATP → ADP)→ 5-Phosphomevalonate —(ATP → ADP)→ 5-Pyrophosphomevalonate —(ATP → ADP + P_i + CO_2)→ Isopentenyl pyrophosphate

Figure 23-11
Synthesis of isopentenyl pyrophosphate from mevalonate.

SYNTHESIS OF SQUALENE FROM ISOPENTENYL PYROPHOSPHATE

Squalene is synthesized from isopentenyl pyrophosphate by the reaction sequence

$$C_5 \longrightarrow C_{10} \longrightarrow C_{15} \longrightarrow C_{30}$$

This stage in the synthesis of cholesterol starts with the isomerization of *isopentenyl pyrophosphate* to *dimethylallyl pyrophosphate*.

$$(H_2C{=})(H_3C)C{-}CH_2{-}CH_2{-}O{-}P(O)(O^-){-}O{-}P(O)(O^-){-}O^- \rightleftharpoons (H_3C)_2C{=}CH{-}CH_2{-}O{-}P(O)(O^-){-}O{-}P(O)(O^-){-}O^-$$

Isopentenyl pyrophosphate **Dimethylallyl pyrophosphate**

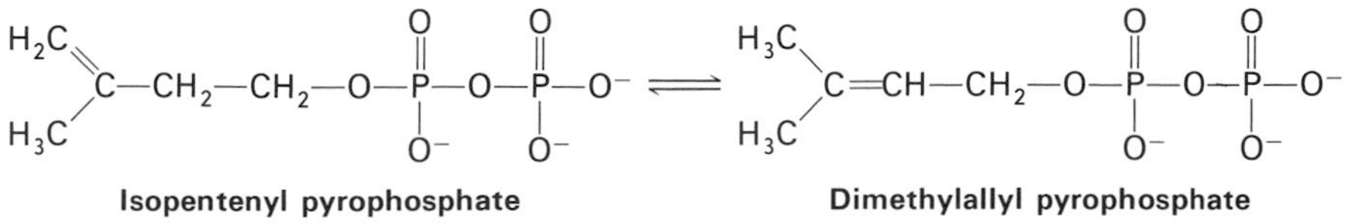

These isomeric C_5 units condense to form a C_{10} compound: an allylic carbonium ion formed from dimethylallyl pyrophosphate is attacked by isopentenyl pyrophosphate to form *geranyl pyrophosphate* (Figure 23-12).

Isopentenyl pyrophosphate

Allylic substrate $\xrightarrow{PP_i}$ **Allylic carbonium ion** (Resonance forms)

Condensation carbonium ion

$\downarrow H^+$

Geranyl (or farnesyl) pyrophosphate

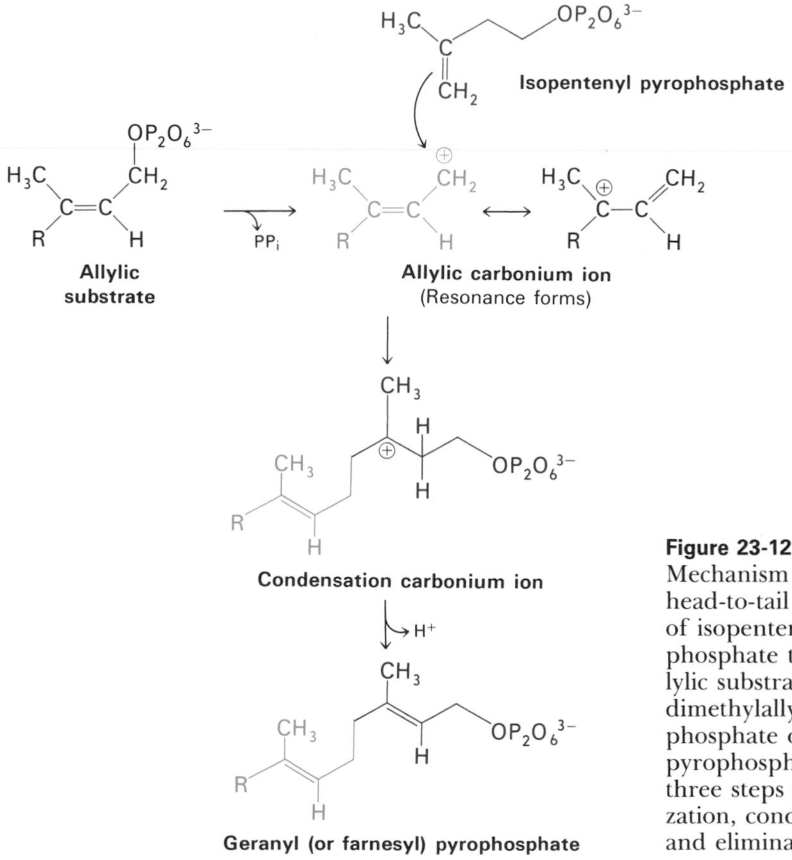

Figure 23-12
Mechanism for the head-to-tail joining of isopentenyl pyrophosphate to an allylic substrate (e.g., dimethylallyl pyrophosphate or geranyl pyrophosphate). The three steps are ionization, condensation, and elimination.

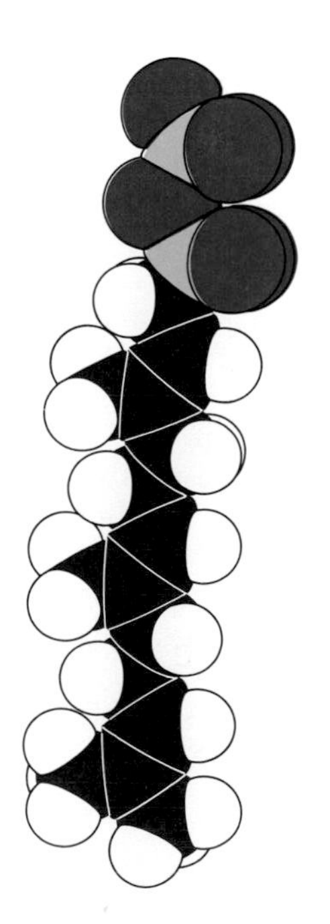

Figure 23-13
Model of farnesyl pyrophosphate.

The same kind of reaction occurs again: geranyl pyrophosphate is converted into an allylic carbonium ion, which is attacked by isopentenyl pyrophosphate. The resulting C_{15} compound is called *farnesyl pyrophosphate*.

The last step in the synthesis of *squalene* is a reductive condensation of two molecules of farnesyl pyrophosphate:

$$\underset{C_{15}}{\text{2 Farnesyl pyrophosphate}} + \text{NADPH} \longrightarrow \underset{C_{30}}{\text{squalene}} + 2\ PP_i + NADP^+ + H^+$$

The reactions leading from C_5 units to squalene, a C_{30} isoprenoid, are shown in Figure 23-14.

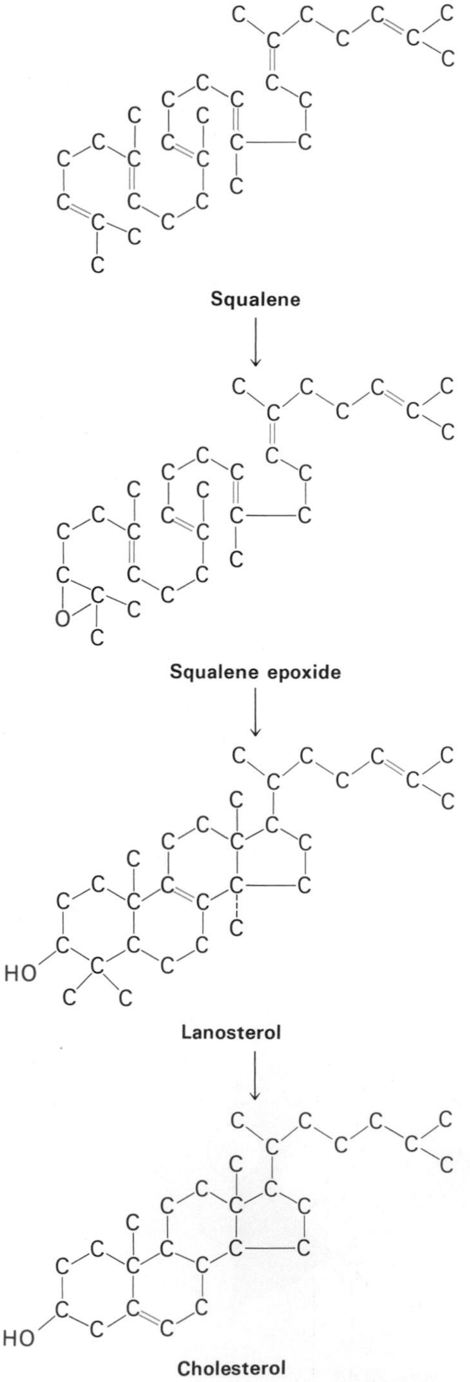

Squalene

Squalene epoxide

Lanosterol

Cholesterol

Figure 23-15
Synthesis of cholesterol from squalene.

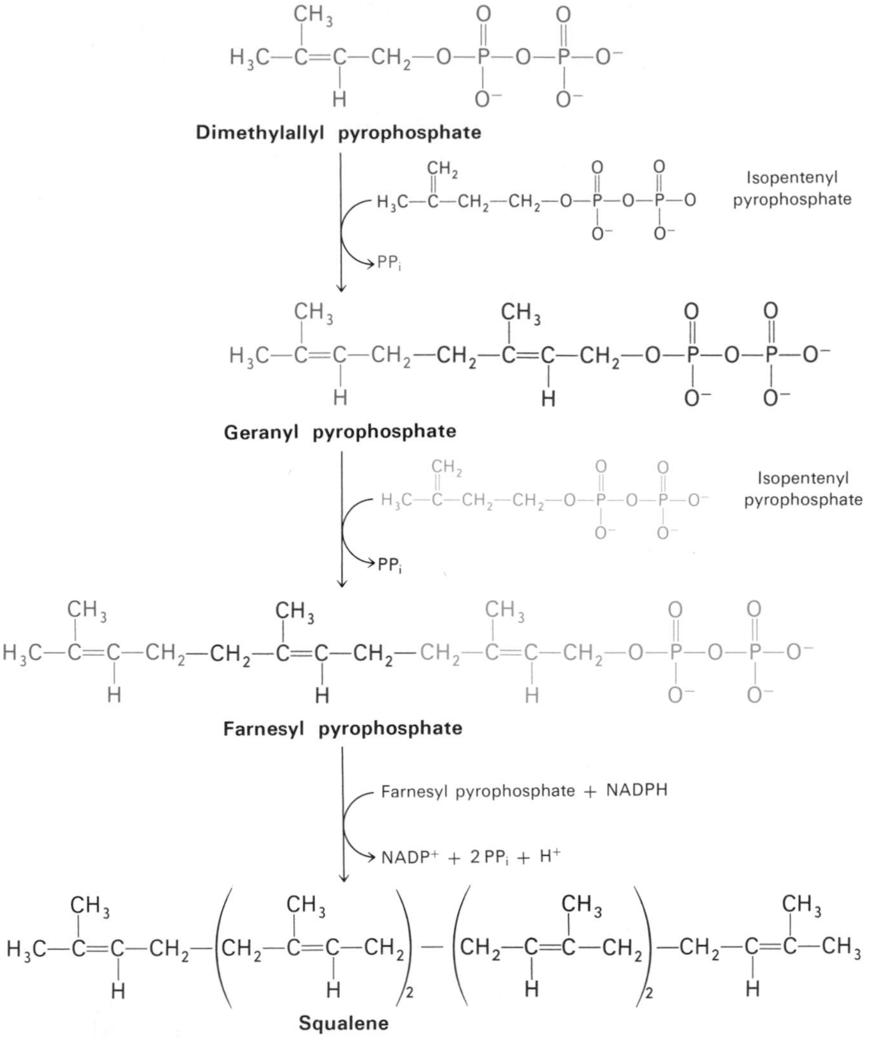

Dimethylallyl pyrophosphate

Isopentenyl pyrophosphate

PP_i

Geranyl pyrophosphate

Isopentenyl pyrophosphate

PP_i

Farnesyl pyrophosphate

Farnesyl pyrophosphate + NADPH

$NADP^+$ + 2 PP_i + H^+

Squalene

Figure 23-14
Synthesis of squalene from dimethylallyl pyrophosphate, an isomer of isopentenyl pyrophosphate.

SQUALENE EPOXIDE CYCLIZES TO LANOSTEROL, WHICH IS CONVERTED INTO CHOLESTEROL

The final stage of cholesterol biosynthesis starts with the cyclization of squalene (Figure 23-15). This stage, in contrast with the preceding ones, *requires molecular oxygen*. *Squalene epoxide,* the reactive intermediate, is formed in a reaction that uses O_2 and NADPH. Squalene epoxide is then cyclized to *lanosterol* by a cyclase. There is a concerted movement of electrons through four double bonds and a migration of two methyl groups in this remarkable closure. Finally, lanosterol is converted into *cholesterol* by the removal of three methyl groups, the reduction of one double bond by NADPH, and the migration of the other double bond. It is interesting to note that cholesterol evolved only after the earth's atmosphere became aerobic. Cholesterol is ubiquitous in eucaryotes but absent from most procaryotes.

BILE SALTS DERIVED FROM CHOLESTEROL FACILITATE THE DIGESTION OF LIPIDS

Bile salts are polar derivatives of cholesterol. These compounds are highly effective *detergents* because they contain both polar and nonpolar regions. Bile salts are synthesized in the liver, stored and concentrated in the gallbladder, and then released into the small intestine. Bile salts, the major constituent of bile, *solubilize dietary lipids*. The effective increase in the surface area of the lipids has two consequences: it promotes their hydrolysis by lipases and facilitates their absorption by the intestine. Bile salts are also the major breakdown products of cholesterol.

Cholesterol is converted into trihydroxycoprostanoate and then into *cholyl CoA*, the activated intermediate in the synthesis of most bile salts (Figure 23-16). The activated carboxyl carbon of cholyl CoA then reacts with the amino group of glycine to form *glycocholate*, or with the amino group of taurine ($H_2N—CH_2—CH_2—SO_3^-$) to form *taurocholate*. *Glycocholate is the major bile salt.*

"Cholesterol is the most highly decorated small molecule in biology. Thirteen Nobel Prizes have been awarded to scientists who devoted major parts of their careers to cholesterol. Ever since it was isolated from gallstones in 1784, cholesterol has exerted an almost hypnotic fascination for scientists from the most diverse areas of science and medicine. . . . Cholesterol is a Janus-faced molecule. The very property that makes it useful in cell membranes, namely its absolute insolubility in water, also makes it lethal."

MICHAEL BROWN AND JOSEPH GOLDSTEIN
Nobel Lectures (1985)
© The Nobel Foundation, 1985

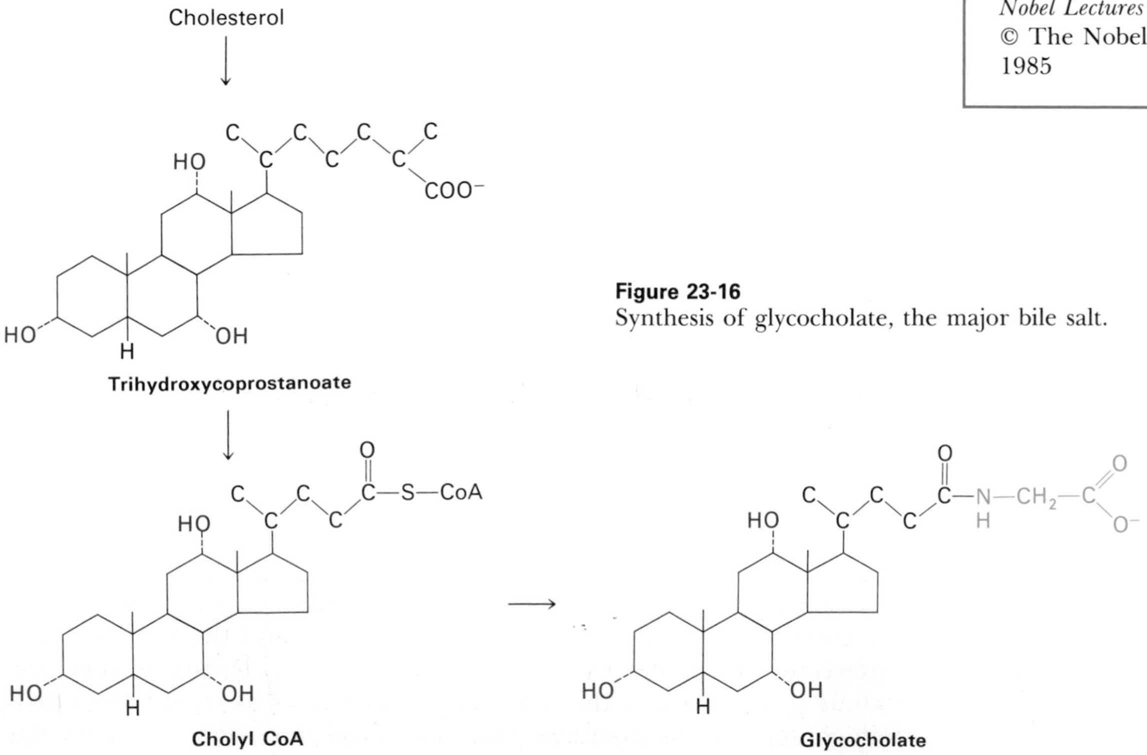

Cholesterol

↓

Trihydroxycoprostanoate

↓

Cholyl CoA → Glycocholate

Figure 23-16
Synthesis of glycocholate, the major bile salt.

CHOLESTEROL SYNTHESIS BY THE LIVER IS SUPPRESSED BY DIETARY CHOLESTEROL

Cholesterol can be obtained from the diet or it can be synthesized de novo. The major site of cholesterol synthesis in mammals is the liver. Appreciable amounts of cholesterol are also formed by the intestine. An adult on a low-cholesterol diet typically synthesizes about 800 mg of cholesterol per day. The rate of cholesterol formation by these organs is highly responsive to the amount of cholesterol absorbed from dietary sources. *This feedback regulation is mediated by changes in the activity of 3-hydroxy-3-methylglutaryl CoA reductase.* As discussed previously, this enzyme catalyzes the formation of mevalonate, which is the committed step in cholesterol biosynthesis. Dietary cholesterol suppresses the synthesis of the reductase in the liver and inactivates existing enzyme molecules.

CHOLESTEROL AND OTHER LIPIDS ARE TRANSPORTED TO SPECIFIC TARGETS BY LIPOPROTEINS

Cholesterol, triacylglycerols, and other lipids are transported in body fluids by lipoproteins classified according to increasing density (Table 23-1): *chylomicrons*, *chylomicron remnants*, *very low-density lipoproteins* (VLDL), *intermediate-density lipoproteins* (IDL), *low-density lipoproteins* (LDL), and *high-density lipoproteins* (HDL). A lipoprotein is a particle consisting of a core of hydrophobic lipids surrounded by a shell of polar lipids and apoproteins. Seven principal apoproteins—A-1, A-2, A-4, B-48, B-100, C, and E—have been isolated and characterized. They are synthesized and secreted by the liver and the intestine. These lipoproteins have two roles: *they solubilize highly hydrophobic lipids and they contain signals that regulate the movement of particular lipids into and out of specific target cells and tissues.*

Table 23-1
Properties of plasma lipoproteins

Lipoproteins	*Major core lipids*	*Apoproteins*	*Mechanism of lipid delivery*
Chylomicron	Dietary triacylglycerols	A-1, A-2, A-4, B-48	Hydrolysis by lipoprotein lipase
Chylomicron remnant	Dietary cholesterol esters	B-48, E	Receptor-mediated endocytosis by liver
Very-low-density lipoprotein (VLDL)	Endogenous triacylglycerols	B-100, C, E	Hydrolysis by lipoprotein lipase
Intermediate-density lipoprotein (IDL)	Endogenous cholesterol esters	B-100, E	Receptor-mediated endocytosis by liver and conversion to LDL
Low-density lipoprotein (LDL)	Endogenous cholesterol esters	B-100	Receptor-mediated endocytosis by liver and other tissues
High-density lipoprotein (HDL)	Endogenous cholesterol esters	A-1, A-2	Transfer of cholesterol esters to IDL and LDL

After M. S. Brown and J. L. Goldstein, *The Pharmacological Basis of Therapeutics*, 7th ed., (ed. by A. G. Gilman, L. S. Goodman, T. W. Rall, and F. Murad), Macmillan, 1985, p. 828.

Triacylglycerols, cholesterol, and other lipids obtained from the diet are carried from the intestine to adipose tissue and the liver by large *chylomicrons* (diameter of 80 to 500 nm). Their density is very low (<0.94 g/cm^3) because they are rich in triacylglycerols and have a protein content of less than 2%. Triacylglycerols in chylomicrons are hydrolyzed within a few minutes by lipases located in the capillaries of adipose and other peripheral tissues. The cholesterol-rich residue, known as *chylomicron remnants*, are taken up by the liver.

Triacylglycerols synthesized endogenously, in contrast with those obtained from the diet, are carried by *very-low-density lipoproteins* (VLDL) produced primarily by the liver (Figure 23-17). Triacylglycerols are released from them by the same lipase that acts on chylomicrons. The resulting remnants, which are rich in cholesterol esters, are called *intermediate-density lipoproteins* (IDL). These particles have two fates. Half of them are taken up by the liver, whereas the other half are converted into *low-density lipoprotein* (LDL), the major carrier of cholesterol in blood. LDL has a diameter of 22 nm and a mass of about three million daltons (Figure 23-18). It contains a core of about 1500 esterified cholesterol molecules; the most common fatty acyl chain in these esters is linoleate, a polyunsaturated fatty acid. This highly hydrophobic core is surrounded by a shell of phospholipids and unesterified cholesterols. The shell also contains a single copy of B-100, a very large protein (514 kd). The role of LDL is to transport cholesterol to peripheral tissues and regulate de novo cholesterol synthesis at these sites, as described below. A different purpose is served by *high-density lipoprotein* (HDL) (density >1.06 g/cm^3) which picks up cholesterol released into the plasma from dying cells and from membranes undergoing turnover. An acyl transferase in HDL esterifies these cholesterols, which are then rapidly shuttled to VLDL or LDL by a transfer protein.

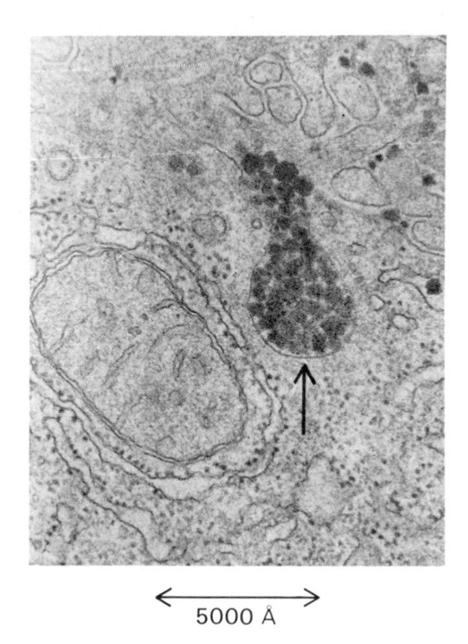

Figure 23-17
Electron micrograph of a part of a liver cell actively engaged in the synthesis and secretion of very low density lipoprotein (VLDL). The arrow points to a vesicle that is releasing its content of VLDL particles. [Courtesy of Dr. George Palade.]

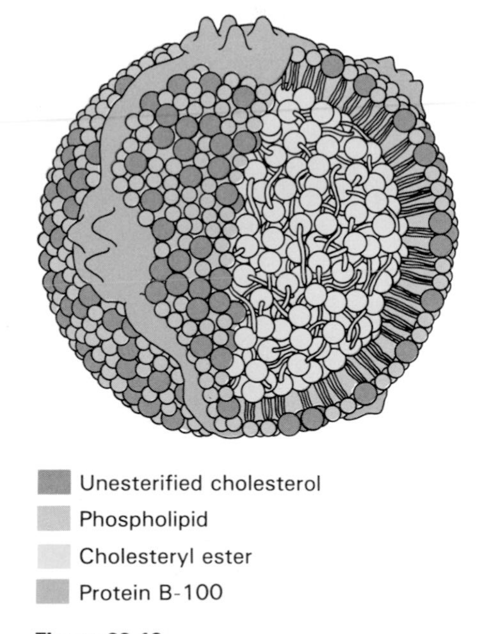

Figure 23-18
Schematic model of low-density lipoprotein (LDL). The conformation of the B-100 protein is not known. The diameter of the LDL particle is 22 nm (220 Å).

THE LOW-DENSITY-LIPOPROTEIN RECEPTOR PLAYS A KEY ROLE IN CONTROLLING CHOLESTEROL METABOLISM

Cholesterol, a component of all eucaryotic plasma membranes, is essential for the growth and viability of cells in higher organisms. However, high serum levels of cholesterol cause disease and death by contributing to the formation of atherosclerotic plaques in arteries throughout the

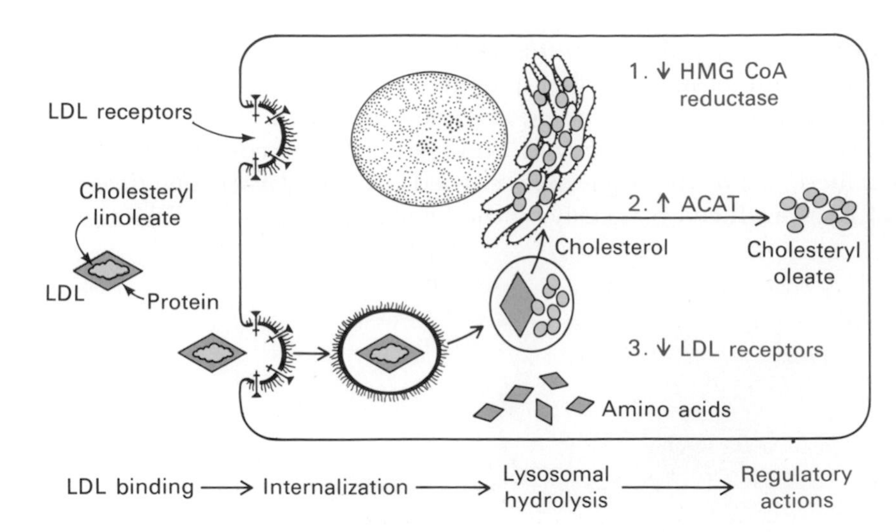

Figure 23-19
Steps in the low-density-lipoprotein pathway in cultured human fibroblasts. HMG CoA reductase denotes 3-hydroxy-3-methylglutaryl CoA reductase, and ACAT denotes acyl CoA:cholesterol acyltransferase. [After a drawing kindly provided by Dr. Michael Brown and Dr. Joseph Goldstein.]

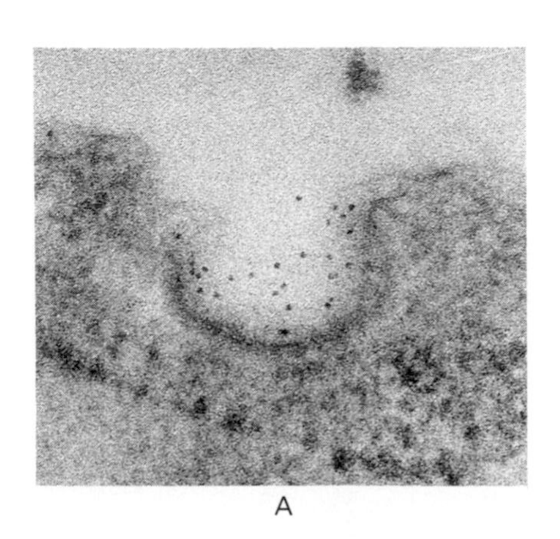

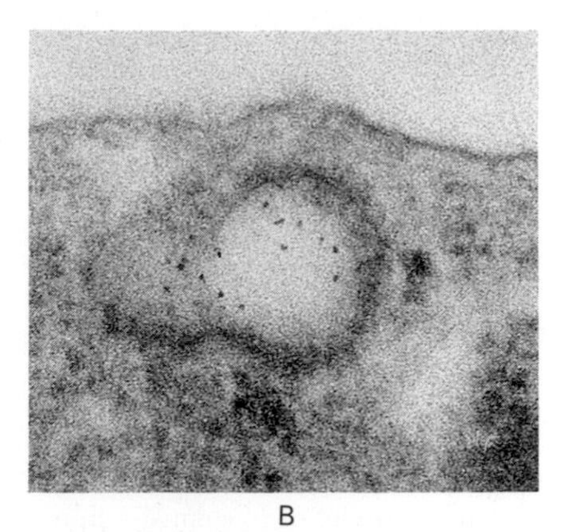

Figure 23-20
Endocytosis of LDL bound to its receptor on the surface of cultured human fibroblasts (LDL was visualized by conjugating it to ferritin): (A) electron micrograph showing LDL-ferritin (dark dots) bound to a coated-pit region of the cell surface; (B) this region invaginates and fuses to form an endocytic vesicle. [From R. G. W. Anderson, M. S. Brown, and J. L. Goldstein, *Cell* 10(1977): 351. MIT Press.]

body. It is evident that cholesterol metabolism must be precisely regulated. The mode of control in the liver, the primary site of cholesterol synthesis, has already been discussed: dietary cholesterol reduces the activity and amount of 3-hydroxy-3-methylglutaryl CoA reductase, the enzyme catalyzing the committed step. Studies of cultured human fibroblasts by Michael Brown and Joseph Goldstein have been sources of insight into the control of cholesterol metabolism in nonhepatic cells. In general, cells outside the liver and intestine obtain cholesterol from the plasma rather than by synthesizing it de novo. Specifically, *their primary source of cholesterol is the low-density lipoprotein.* The steps in the uptake of cholesterol by the *LDL pathway* (Figure 23-19) are:

1. The B-100 protein component of LDL binds to a specific receptor protein on the plasma membrane of nonhepatic cells. The receptors for LDL are localized in specialized regions called *coated pits,* which contain *clathrin* (p. 787).

2. The receptor-LDL complex is internalized by *endocytosis*—that is, the plasma membrane in the vicinity of the complex invaginates and then fuses to form an endocytic vesicle (Figure 23-20). Endocytosis will be discussed further in a later chapter (p. 787).

3. These vesicles, containing LDL, subsequently fuse with *lysosomes*, which carry a wide array of degradative enzymes. The protein component of the LDL is hydrolyzed to free amino acids. The cholesterol esters in the LDL are hydrolyzed by a lysosomal acid lipase. The LDL receptor itself usually returns unscathed to the plasma membrane. The round-trip time for a receptor is about 10 minutes; in its lifetime of about a day, it may bring many LDL particles into the cell.

4. *The released unesterified cholesterol can then be used for membrane biosynthesis.* Alternatively, it can be *reesterified for storage inside the cell.* In fact, free cholesterol activates acyl CoA:cholesterol acyl transferase (ACAT), the enzyme catalyzing this reaction. Reesterified cholesterol contains

mainly oleate and palmitoleate, which are monounsaturated fatty acids, in contrast with the cholesterol esters in LDL, which are rich in linoleate, a polyunsaturated fatty acid (Table 20-1).

The cholesterol content of cells having an active LDL pathway is regulated in two ways. First, the *released cholesterol suppresses the transcription of the gene for 3-hydroxy-3-methylglutaryl CoA reductase; hence, de novo synthesis of cholesterol is blocked.* Second, the LDL receptor is itself subject to feedback regulation. *When cholesterol is abundant inside the cell, new LDL receptors are not synthesized, and so the uptake of additional cholesterol from plasma LDL is blocked.*

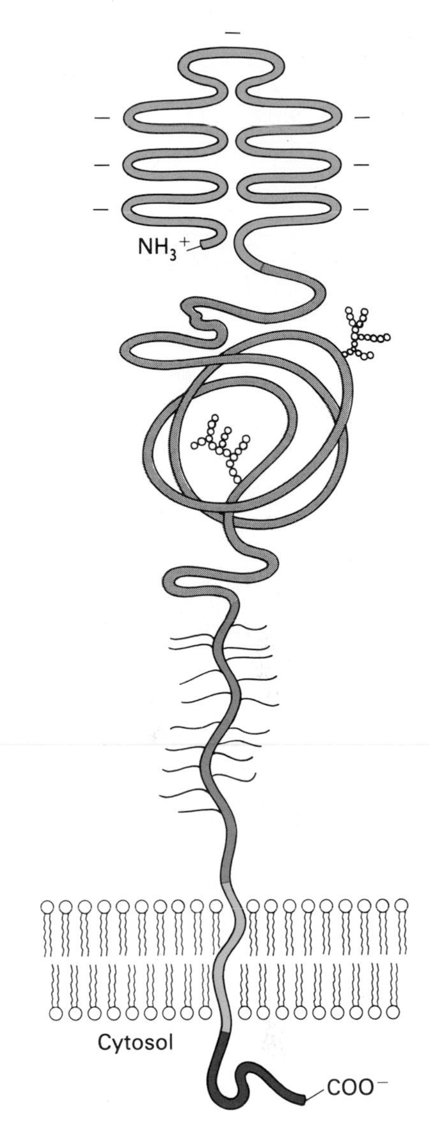

Figure 23-21
The LDL receptor consists of five domains with different functions: an LDL-binding domain, 292 residues (green); a domain bearing *N*-linked sugars, 350 residues (gray); a domain bearing *O*-linked sugars, 58 residues (blue); a membrane-spanning domain, 22 residues (yellow); and a cytosolic domain, 50 residues (red).

THE LDL RECEPTOR IS A TRANSMEMBRANE PROTEIN WITH FIVE DIFFERENT FUNCTIONAL DOMAINS

The cloning and sequencing of the cDNA for the human LDL receptor has revealed that this 115-kd protein consists of five domains (Figure 23-21). The amino-terminal domain of the mature receptor consists of a cysteine-rich sequence of 40 residues that is repeated, with some variation, seven times. A cluster of negatively charged side chains in this *LDL-binding domain* interacts with a positively charged site of the B-100 protein of the LDL. Protonation of the glutamate and aspartate side chains of the receptor in acidic endosomes (p. 789) leads to the release of LDL from its receptor. The second domain of the LDL receptor is homologous to part of the precursor of epidermal growth factor (EGF, p. 999). It contains two *N*-linked oligosaccharide chains. The third domain, which is very rich in serines and threonines, contains *O*-linked sugars. These oligosaccharides, like those in glycophorin (p. 303), may function as struts to keep the receptor extended from the membrane so that the amino-terminal domain is accessible to LDL. The fourth domain contains 22 hydrophobic residues that span the plasma membrane. This segment, which is probably α-helical, is just long enough to cross the hydrophobic part of the bilayer. The fifth domain, consisting of 50 residues, emerges on the cytosolic side of the membrane, where it controls the interaction of the receptor with coated pits and participates in endocytosis.

The gene for the LDL receptor is 45 kb long and consists of 18 exons, which correspond closely to structural units of the protein. Several of the cysteine-rich repeats of the LDL-binding domain are encoded by a single exon, which is also found in the gene for a component of the complement cascade (p. 899). The next eight exons are also found in the gene for the EGF precursor. It is as though a cassette had been taken from an ancestral gene and placed in the midst of the EGF precursor gene and the LDL receptor gene. *The LDL receptor is a striking example of a mosaic protein encoded by a gene that was assembled by exon shuffling.*

ABSENCE OF THE LDL RECEPTOR LEADS TO HYPERCHOLESTEROLEMIA AND ATHEROSCLEROSIS

The physiologic importance of the LDL receptor was revealed by Brown and Goldstein's pioneering studies of *familial hypercholesterolemia* (FH). The total concentration of cholesterol and LDL in the plasma is markedly elevated in this genetic disorder, which results from a mutation at a single autosomal locus. The cholesterol level in the plasma of homozygotes is typically 680 mg/dl, compared with 300 mg/dl in heterozygotes. A value of 175 mg/dl is regarded as desirable, but most Americans have higher levels. *In familial hypercholesterolemia, cholesterol is deposited in various tissues because of the high concentration of LDL-cholesterol in the plasma.* Nodules of cholesterol called xanthomas are prominent in skin and tendons. More harmful is the deposition of cholesterol in arterial plaques, which produce atherosclerosis and lead to heart attacks, strokes, and peripheral vascular disease. In fact, *most homozygotes die of coronary artery disease in childhood.* The disease in heterozygotes has a milder and more variable clinical course.

Figure 23-22
An atherosclerotic plaque (marked by the arrow) blocks most of the lumen of this blood vessel. The plaque is rich in cholesterol. [Courtery of Dr. Jeffrey Sklar.]

The molecular defect in most cases of familial hypercholesterolemia is an absence or deficiency of functional receptors for LDL. Homozygotes have almost no receptors for LDL, whereas heterozygotes have about half the normal number. Consequently, the entry of LDL into liver and other cells is impaired, leading to an increased plasma level of LDL. Furthermore, less IDL enters liver cells because IDL entry, too, is mediated by the LDL receptor (IDL, like LDL, contains the B-100 protein). Consequently, IDL stays in the blood longer in FH and more of it is converted into LDL than in normal people. All of the deleterious consequences of an absence or deficiency of the LDL receptor can be attributed to the ensuing elevated level of LDL cholesterol in the blood.

Several classes of FH mutations have been characterized: (1) No receptor is synthesized. (2) Receptors are synthesized but do not reach the plasma membrane because they lack signals for intracellular transport or do not fold properly. (3) Receptors reach the cell surface, but they fail to bind LDL normally because of a defect in the LDL-binding domain. (4) Receptors reach the cell surface and bind LDL, but they fail to cluster in coated pits because of a defect in their carboxyl-terminal region.

MEVINOLIN, AN INHIBITOR OF HMG CoA REDUCTASE, LEADS TO AN INCREASE IN THE NUMBER OF LDL RECEPTORS

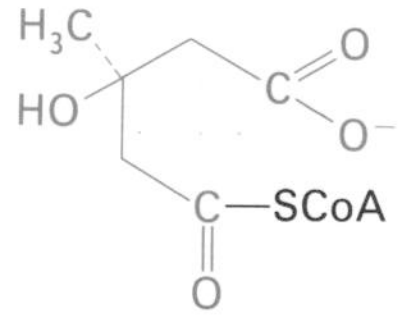

3-Hydroxy-3-methylglutaryl CoA

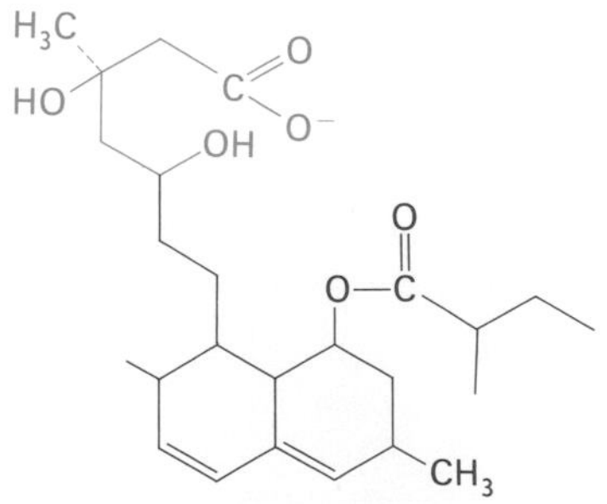

Mevinolin

Figure 23-23
Mevinolin, a potent competitive inhibitor of HMG CoA reductase, resembles 3-hydroxy-3-methylglutaryl CoA, the substrate.

Homozygous FH can be treated by transplanting a normal liver. A more generally applicable therapy is available for heterozygotes (1 in 500 persons). *The goal is to stimulate the single normal gene to produce more than the customary number of LDL receptors.* Studies of cultured fibroblasts showed that the production of LDL receptors is controlled by the cell's need for cholesterol. When cholesterol is required, the amount of mRNA for LDL receptor rises and more receptor is found on the cell surface. This state can be achieved by inhibiting the intestinal reabsorption of bile salts (which promote the absorption of dietary cholesterol, p. 559) and by blocking cholesterol synthesis. The reabsorption of bile is impeded by oral administration of positively charged polymers that bind negatively charged bile salts and are not themselves absorbed. Cholesterol synthesis can be effectively blocked by mevinolin (Figure 23-23), a potent competitive inhibitor (K_i of 1 nM) of HMG CoA reductase, the key enzyme in the biosynthetic pathway. The consequent increase in the number of LDL receptors on liver cells leads to a decrease in the LDL level in blood. Indeed, plasma cholesterol levels decrease by 50% in patients given both drugs. There is much interest in drugs that lower cholesterol levels because atherosclerosis is the leading cause of death in industrialized societies.

Carbon atoms in steroids are numbered as shown in Figure 23-24, for cholesterol. The rings in steroids are named A, B, C, and D. Cholesterol contains two angular methyl groups: the C-19 methyl group is attached to C-10, and the C-18 methyl group is attached to C-13. By convention, the C-18 and C-19 methyl groups of cholesterol are *above* the plane containing the four rings. A substituent that is above the plane is termed *β-oriented* and is shown by a *solid* bond. In contrast, a substituent that is below the plane is *α-oriented* and denoted by a *dashed* or dotted line.

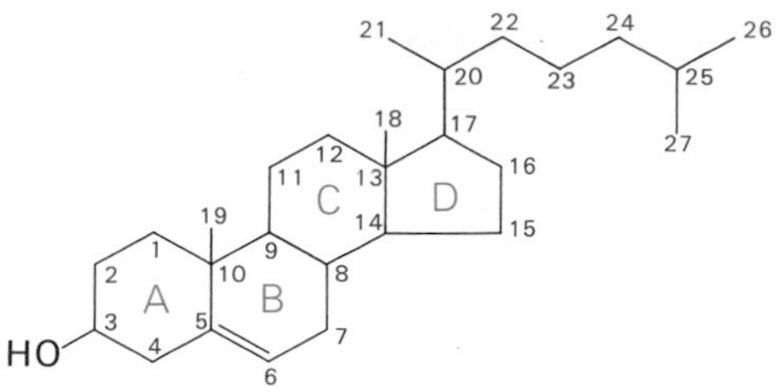

Figure 23-24
Numbering of the carbon atoms of cholesterol.

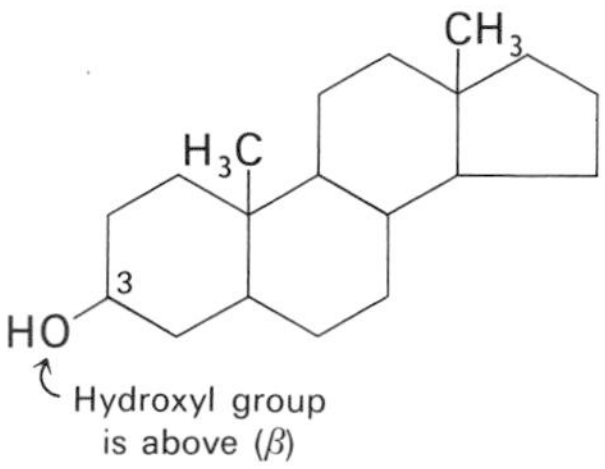

Hydroxyl group is above (β)
3β-Hydroxy

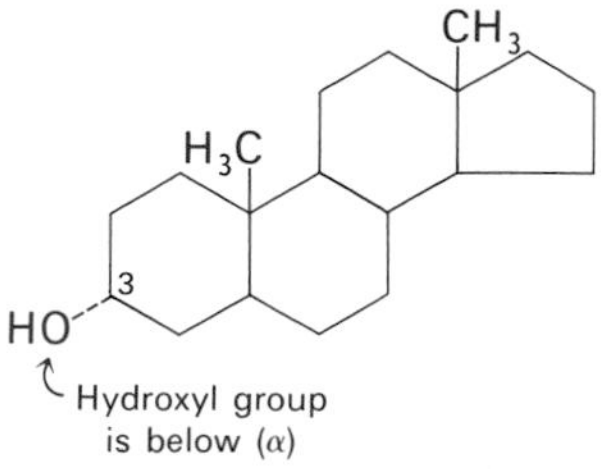

Hydroxyl group is below (α)
3α-Hydroxy

A hydrogen atom attached to C-5 can be α- or β-oriented. If this hydrogen atom is *α-oriented,* the A and B rings are fused in a *trans* conformation, whereas a *β-orientation* corresponds to a *cis* fusion. The absence of a symbol for the C-5 hydrogen atom implies a *trans* fusion. The C-5 hydrogen atom is α-oriented in all steroid hormones that contain a hydrogen atom in that position. In contrast, bile salts have a β-oriented hydrogen atom at C-5. Thus, *a* cis *fusion is characteristic of the bile salts, whereas a* trans *fusion is characteristic of all steroid hormones that possess a hydrogen atom at C-5*. A *trans* fusion yields a nearly planar structure, whereas a *cis* fusion gives a buckled structure.

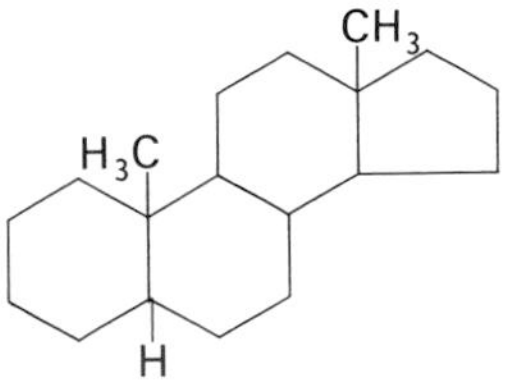

5β-Hydrogen
(A *cis* fusion)

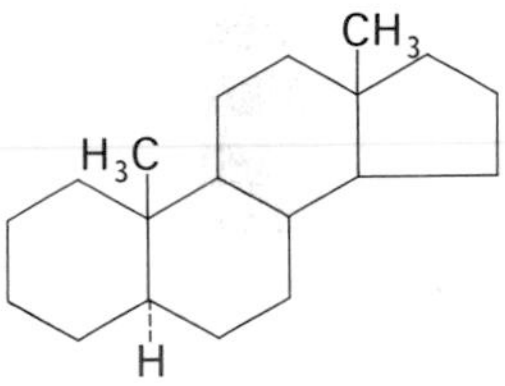

5α-Hydrogen
(A *trans* fusion)

STEROID HORMONES ARE DERIVED FROM CHOLESTEROL

Cholesterol is the precursor of the five major classes of steroid hormones: progestagens, glucocorticoids, mineralocorticoids, androgens, and estrogens (Figure 23-25). *Progesterone,* a *progestagen,* prepares the

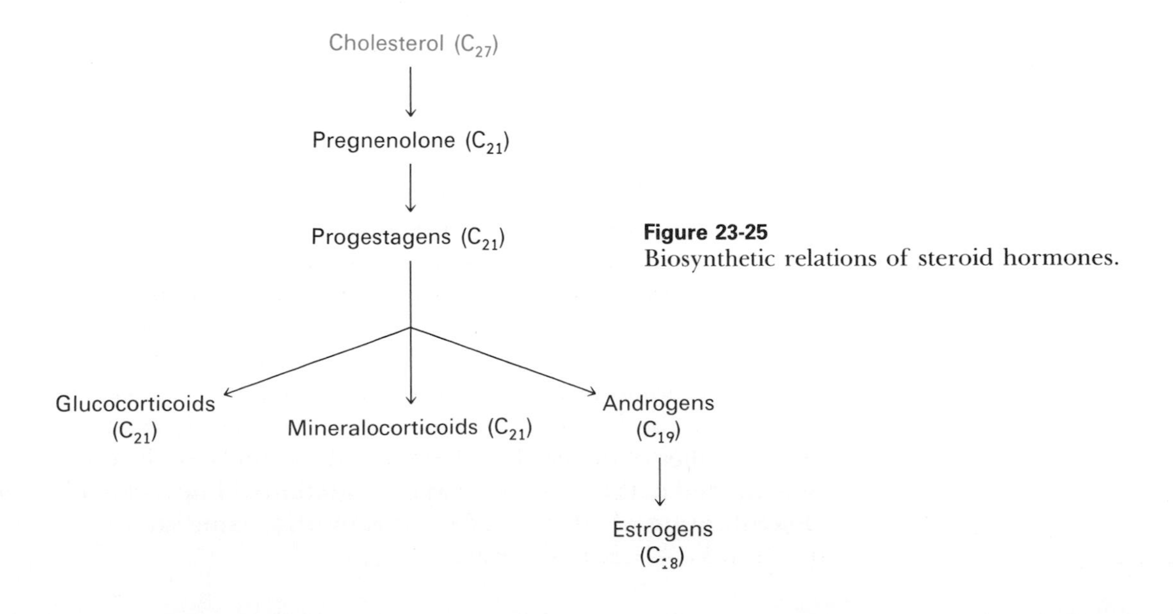

Figure 23-25
Biosynthetic relations of steroid hormones.

lining of the uterus for implantation of an ovum. Progesterone is also essential for the maintenance of pregnancy. *Androgens* (such as *testosterone*) are responsible for the development of male secondary sex characteristics, whereas *estrogens* (such as *estrone*) are required for the development of female secondary sex characteristics. Estrogens also participate in the ovarian cycle. *Glucocorticoids* (such as *cortisol*) promote gluconeogenesis and the formation of glycogen and enhance the degradation of fat and protein. *Mineralocorticoids* (primarily *aldosterone*) act on the distal tubules of the kidney to increase the reabsorption of Na^+ and the excretion of K^+ and H^+, which lead to an increase in blood volume and blood pressure. The major sites of synthesis of these classes of hormones are: progestagens, corpus luteum; estrogens, ovary; androgens, testis; glucocorticoids and mineralocorticoids, adrenal cortex.

STEROIDS ARE HYDROXYLATED BY MONOOXYGENASES THAT UTILIZE NADPH AND O_2

Hydroxylation reactions play a very important role in the synthesis of cholesterol from squalene and in the conversion of cholesterol into steroid hormones and bile salts. All of these hydroxylations require *NADPH and O_2*. The oxygen atom of the incorporated hydroxyl group comes from O_2 rather than from H_2O, as shown by the use of ^{18}O-labeled O_2 and H_2O. One oxygen atom of the O_2 molecule goes into the substrate, whereas the other is reduced to water. The enzymes catalyzing these reactions are called *monooxygenases* (or mixed-function oxygenases). Recall that a monooxygenase also participates in the hydroxylation of phenylalanine (p. 511).

$$RH + O_2 + NADPH + H^+ \longrightarrow ROH + H_2O + NADP^+$$

Hydroxylation requires the activation of oxygen. In the synthesis of steroid hormones and bile salts, activation is accomplished by P_{450}, a specialized cytochrome that absorbs light maximally at 450 nm when it is complexed with CO. Cytochrome P_{450} is the terminal component of an *electron-transport chain* in adrenal mitochondria and liver microsomes. The role of this assembly is hydroxylation rather than oxidative phosphorylation. NADPH transfers its high-potential electrons to a flavoprotein in this chain, which are then conveyed to *adrenodoxin,* a nonheme iron protein. Adrenodoxin transfers an electron to the oxidized form of cytochrome P_{450}. The reduced form of P_{450} then activates O_2.

The cytochrome P_{450} system is also important for the *detoxification of foreign substances* (xenobiotic compounds). For example, the hydroxylation of phenobarbital, a barbiturate, *increases its solubility* and *facilitates its excretion.* Likewise, polycyclic aromatic hydrocarbons are hydroxylated by the P_{450} system. The introduction of hydroxyl groups provides sites for conjugation with highly polar units (e.g., glucuronate or sulfate), which markedly increases the solubility of the modified aromatic molecule. However, the action of the P_{450} system is not always beneficial. Some of the *most powerful carcinogens are converted in vivo into a chemically reactive form.* This process of *metabolic activation* is usually carried out by the P_{450} system. P_{450} enzymes in mammals are encoded by eight families of genes. The earliest P_{450} proteins in evolution probably participated in the synthesis and degradation of fatty acids and steroids. Detoxification of foreign substances by P_{450} came later.

PREGNENOLONE IS FORMED FROM CHOLESTEROL BY CLEAVAGE OF ITS SIDE CHAIN

Steroid hormones contain 21 or fewer carbon atoms, whereas cholesterol contains 27. The first stage in the synthesis of steroid hormones is the removal of a C_6 unit from the side chain of cholesterol to form *pregnenolone*. The side chain of cholesterol is hydroxylated at C-20 and then at C-22, followed by the cleavage of the bond between C-20 and C-22. The last reaction is catalyzed by desmolase. All three reactions utilize NADPH and O_2.

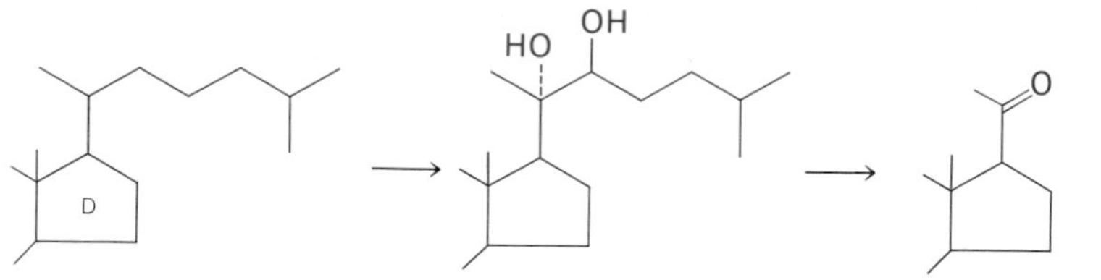

Cholesterol (Only ring D and the side chain are shown.) → **20α,22β-Dihydroxycholesterol** → **Pregnenolone**

Adrenocorticotropic hormone (ACTH, or corticotropin), a polypeptide synthesized by the anterior pituitary gland, stimulates the conversion of cholesterol into pregnenolone, the precursor of all steroid hormones.

SYNTHESIS OF PROGESTERONE AND CORTICOIDS

Progesterone is synthesized from pregnenolone in two steps. The 3-hydroxyl group of pregnenolone is oxidized to a 3-keto group, and the Δ^5 double bond is isomerized to a Δ^4 double bond (Figure 23-26).

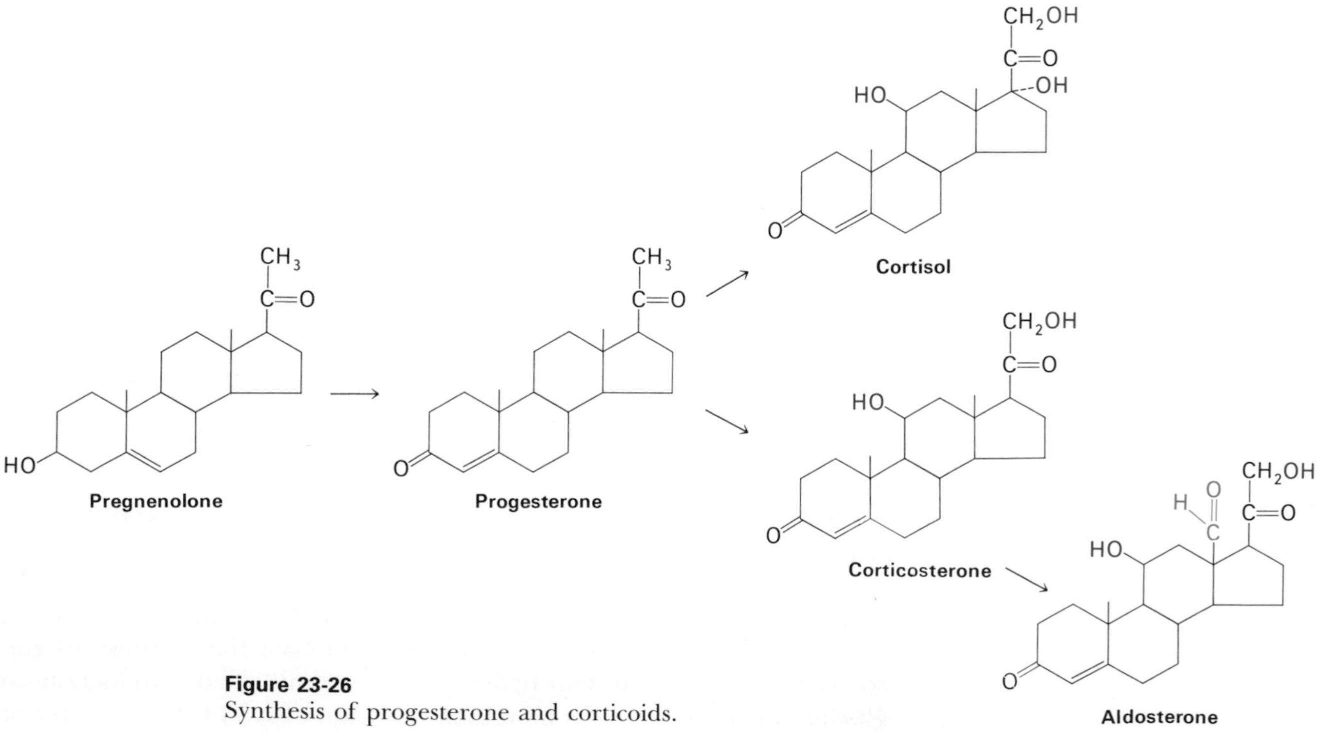

Pregnenolone → **Progesterone** → **Cortisol**; **Progesterone** → **Corticosterone** → **Aldosterone**

Figure 23-26
Synthesis of progesterone and corticoids.

Cortisol, the major glucocorticoid, is synthesized from progesterone by hydroxylations at C-17, C-21, and C-11; C-17 must be hydroxylated before C-21, whereas hydroxylation at C-11 may occur at any stage. The enzymes catalyzing these hydroxylations are highly specific, as shown by some inherited disorders of steroid metabolism. The initial step in the synthesis of *aldosterone,* the major mineralocorticoid, is the hydroxylation of progesterone at C-21. The resulting deoxycorticosterone is hydroxylated at C-11. The oxidation of the C-18 angular methyl group to an aldehyde then yields aldosterone.

SYNTHESIS OF ANDROGENS AND ESTROGENS

The synthesis of androgens (Figure 23-27) starts with the hydroxylation of progesterone at C-17. The side chain consisting of C-20 and C-21 is then cleaved to yield *androstenedione,* an androgen. *Testosterone,* another androgen, is formed by reduction of the 17-keto group of androstenedione. Androgens contain nineteen carbon atoms. Estrogens are synthesized from androgens by the loss of the C-19 angular methyl group and the formation of an aromatic A ring. These reactions require NADPH and O_2. *Estrone,* an estrogen, is derived from androstenedione, whereas *estradiol,* another estrogen, is formed from testosterone.

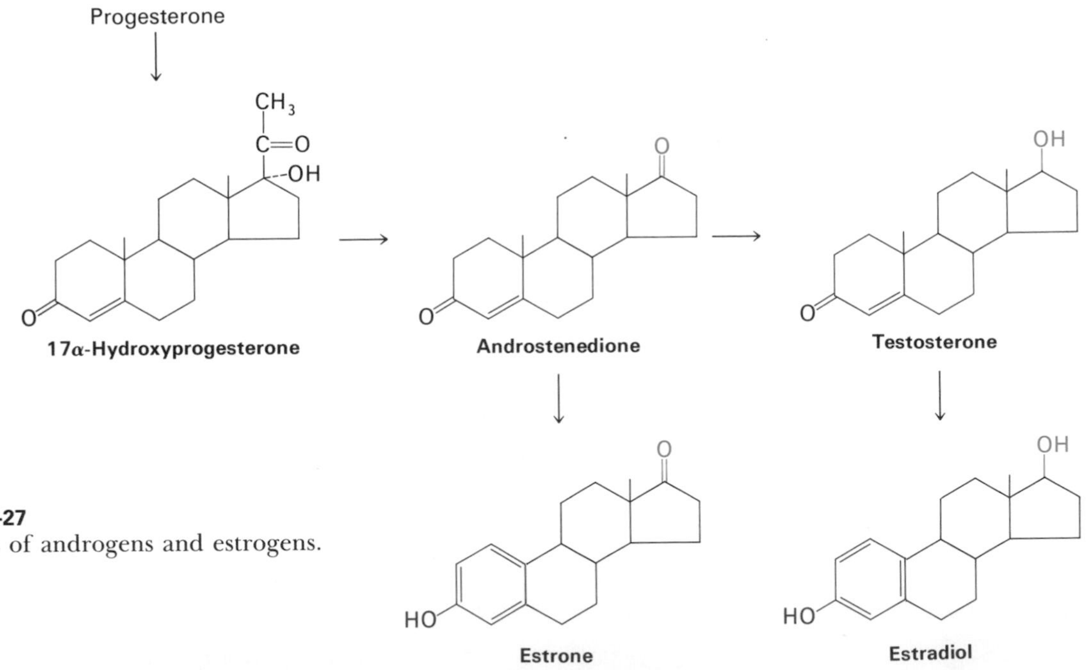

Progesterone

CH_3 C=O OH

17α-Hydroxyprogesterone

Androstenedione

Testosterone

Estrone

Estradiol

Figure 23-27
Synthesis of androgens and estrogens.

DEFICIENCY OF 21-HYDROXYLASE CAUSES VIRILIZATION AND ENLARGEMENT OF THE ADRENALS

The most common inherited disorder of steroid-hormone synthesis is a deficiency of *21-hydroxylase,* an enzyme needed for the synthesis of glucocorticoids and mineralocorticoids. The diminished production of glucocorticoids leads to an increased secretion of ACTH by the anterior pituitary gland. This response is an expression of a normal feedback mechanism that controls adrenal cortical activity. *The adrenal glands enlarge because of the high level of ACTH in the blood. ACTH, acting through cyclic AMP, stimulates the conversion of cholesterol to pregnenolone.* Consequently, the concentrations of progesterone and 17α-hydroxyprogesterone increase. In turn, there is a *marked increase in the amount of androgens,* because they are derived from 17α-hydroxyprogesterone. The striking clinical finding in 21-hydroxylase deficiency is *virilization* caused by high levels of androgens. In affected females, virilization is usually evident at birth. Androgens secreted during the development of the female fetus produce a masculinization of the external genitalia. In males, the sexual organs appear normal at birth. Sexual precocity becomes apparent several months later. There is accelerated growth and very early bone maturation so that short stature is the typical final result. About a third of the patients with 21-hydroxylase deficiency *persistently lose Na^+ in the urine.* They have very low levels of aldosterone, the principal mineralocorticoid. Loss of salt leads to dehydration and hypotension, which may lead to shock and sudden death.

Effective therapy is available for 21-hydroxylase deficiency. The administration of a glucocorticoid provides this needed hormone and concomitantly eliminates the excessive secretion of ACTH. Excessive formation of androgens is thereby stopped. A mineralocorticoid may also be given to patients who lose salt. Some of the symptoms of 21-hydroxylase deficiency are reversed if therapy is started in the first two years of life.

Several other inherited defects of steroid-hormone synthesis are known. The affected enzymes include 11-hydroxylase, 17-hydroxylase, 3β-dehydrogenase, and desmolase. All of these enzymatic lesions lead to a compensatory enlargement of the adrenal gland. Hence, the clinical term for this group of disorders is *congenital adrenal hyperplasia.* Like 21-hydroxylase deficiency, 11-hydroxylase deficiency is accompanied by virilization.

VITAMIN D IS DERIVED FROM CHOLESTEROL BY THE ACTION OF LIGHT

Cholesterol is also the precursor of vitamin D, which plays an essential role in the control of calcium and phosphorus metabolism. *7-Dehydrocholesterol (provitamin D_3) is photolyzed by ultraviolet light to previtamin D_3, which spontaneously isomerizes to vitamin D_3* (Figure 23-28). Vitamin D_3 (cholecalciferol) is converted into *calcitriol* (1,25-dihydroxycholecalciferol), the active hormone, by hydroxylation reactions occurring in the liver and kidneys. Vitamin D deficiency in childhood produces *rickets,* which is characterized by inadequate calcification of cartilage and bone. As described by Daniel Webster in 1645,

> . . . the whole bony structure is as flexible as softened wax, so that the flaccid and enervated legs can hardly support the superposed weight of the body; hence the tibia, giving way beneath the overpowering weight of the frame, bend inwards; and for the same reason the legs are drawn together at their tops; and the back, by reason of the bending of the spine, sticks out in a hump in the lumbar region . . . the patients in their weakness cannot (in the most severe stages of the disease) bear to sit upright, much less stand. . . .

Rickets was so common that Webster referred to it as the "Children's disease of the English." We now know that rickets was prevalent in these children because there was little sunlight for many months of the year.

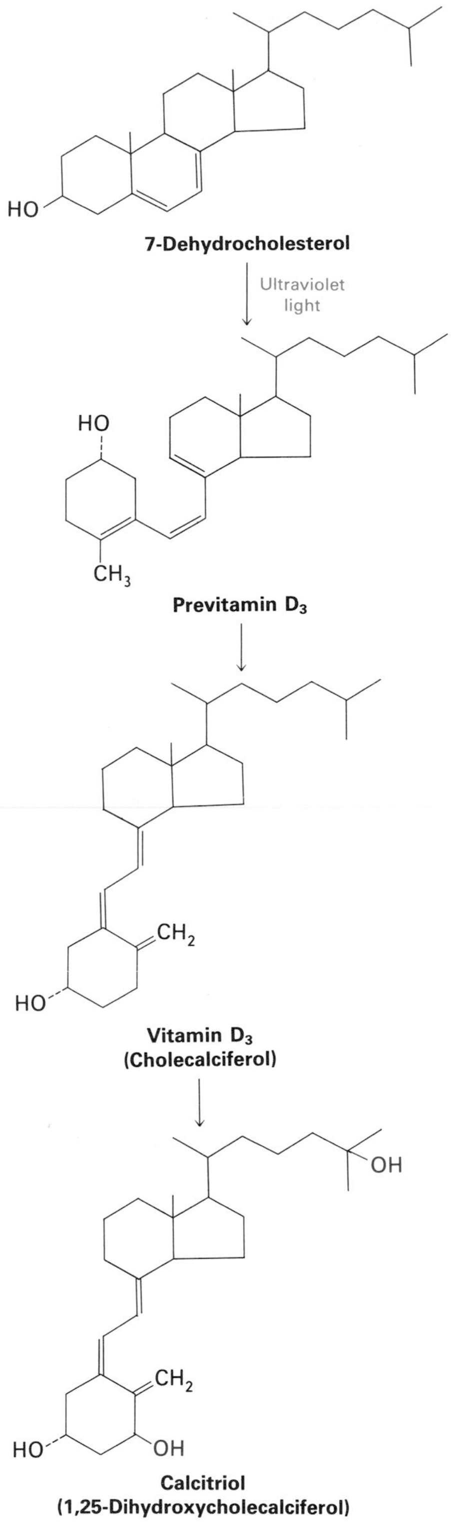

HO

7-Dehydrocholesterol

Ultraviolet light

HO

CH_3

Previtamin D_3

CH_2

HO

Vitamin D_3 (Cholecalciferol)

OH

CH_2

HO

OH

Calcitriol (1,25-Dihydroxycholecalciferol)

Figure 23-28
Conversion of 7-dehydrocholesterol into vitamin D_3 (cholecalciferol) and calcitriol, the active hormone. Vitamin D_2 (ergocalciferol) can be formed in a similar way starting with ergosterol, a plant sterol. Vitamin D_2 differs from D_3 in having a C-22–C-23 double bond and a C-24 methyl group.

Consequently, 7-dehydrocholesterol in the skin was not photolyzed to previtamin D_3. Furthermore, their diets provided little vitamin D because most naturally occurring foods have a low content of this vitamin. Fish-liver oils are a notable exception, and so cod-liver oil was used for many years as a rich source of vitamin D. Today, the most reliable dietary sources of vitamin D are fortified foods. In the United States, milk is fortified to a level of 400 international units per quart (10 μg per quart). The recommended daily intake of vitamin D is 400 international units, irrespective of age. In adults, vitamin D deficiency leads to softening and weakening of bones, a condition called *osteomalacia.* The occurrence of osteomalacia in Bedouin Arab women who are clothed so that only their eyes are exposed to sunlight is a striking reminder that vitamin D is needed by adults as well as by children.

> *Rickets*—
> From the Old English word *wrickken,* to twist.
>
> *Osteomalacia*—
> From the Greek words *osteon,* bone, and *malakia,* softness.

FIVE-CARBON UNITS ARE JOINED TO FORM A WIDE VARIETY OF BIOMOLECULES

The synthesis of squalene (C_{30}) from isopentenyl pyrophosphate (C_5) exemplifies a fundamental mechanism for the assembly of carbon skeletons of biomolecules. *A remarkable array of compounds are formed from isopentenyl pyrophosphate, the basic five-carbon building block.* The fragrances of many plants arise from volatile C_{10} and C_{15} compounds, which are called *terpenes.* For example, myrcene ($C_{10}H_{16}$) from bay leaves consists of two isoprene units, as does limonene ($C_{10}H_{15}$) from lemon oil (Figure 23-29). Zingiberene ($C_{15}H_{24}$), from the oil of ginger, is made up of three isoprene units. Some terpenes, such as geraniol from geraniums and menthol from peppermint oil, are alcohols, and others, such as citronellal, are aldehydes. *Natural rubber* is a linear polymer of *cis*-isoprene units.

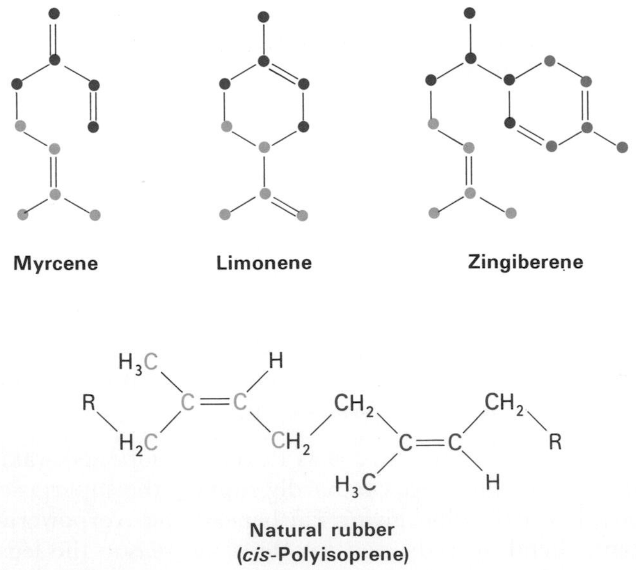

Myrcene **Limonene** **Zingiberene**

Natural rubber (*cis*-Polyisoprene)

Figure 23-29
Formulas for some isoprenoids. The five-carbon building blocks are shown in color.

Isoprenoid unit

Vitamin K_2

Isoprenoid unit

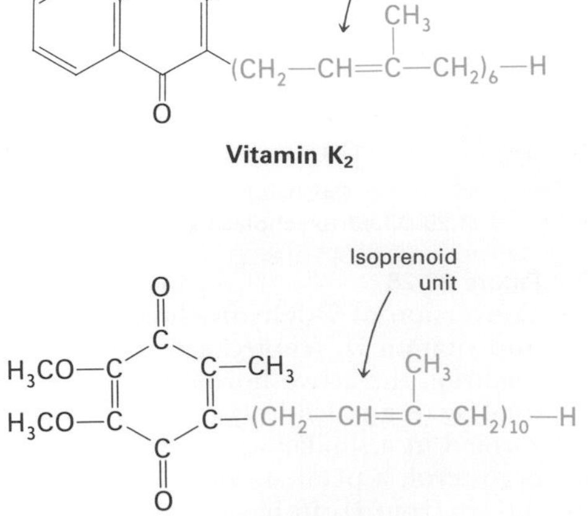

Ubiquinone (Coenzyme Q_{10})

We have already encountered several molecules that contain isoprenoid side chains. The C_{30} hydrocarbon *side chain of vitamin K_2,* a key molecule in clotting (p. 251), is built from six C_5 units. *Coenzyme Q_{10}* in the mitochondrial respiratory chain (p. 403) has a side chain made up of ten isoprene units. Yet another example is the *phytol side chain* of *chlorophyll* (p. 520), which is formed from four isoprene units.

Isoprenoids can delight by their color as well as by their fragrance. Indeed, isoprenoids can be regarded as the sensual molecules! The color of tomatoes and carrots comes from *carotenoids,* specifically from *lycopene* and *β-carotene,* respectively. These compounds absorb light because they contain extended networks of single and double bonds—that is, they are *polyenes.* Their C_{40} carbon skeletons are built by the successive addition of C_5 units to form *geranylgeranyl pyrophosphate, a* C_{20} *intermediate,* which then condenses tail-to-tail with another molecule of geranylgeranyl pyrophosphate. This biosynthetic pathway is like that of squalene except that C_{20} rather than C_{15} units are assembled and condensed.

> "Perfumes, colors, and sounds echo one another."
>
> CHARLES BAUDELAIRE
> *Correspondances*

$$C_5 \longrightarrow C_{10} \longrightarrow C_{15} \longrightarrow C_{30} \text{ (squalene)}$$
$$C_5 \longrightarrow C_{10} \longrightarrow C_{15} \longrightarrow C_{20} \longrightarrow C_{40} \text{ (phytoene)}$$

Phytoene, the C_{40} condensation product, is dehydrogenated to yield lycopene. Cyclization of both ends of lycopene gives β-carotene (Figure 23-30). Carotenoids serve as light-harvesting molecules in photosynthetic assemblies and also play a role in protecting procaryotes from the deleterious effects of light. Carotenoids are also essential for vision. β-Carotene is the precursor of retinal, the chromophore in all known visual pigments (p. 1029). *These examples illustrate the fundamental role of isopentenyl pyrophosphate in the assembly of extended carbon skeletons of biomolecules. It is also evident that isoprenoids are ubiquitous in nature and have diverse significant roles.*

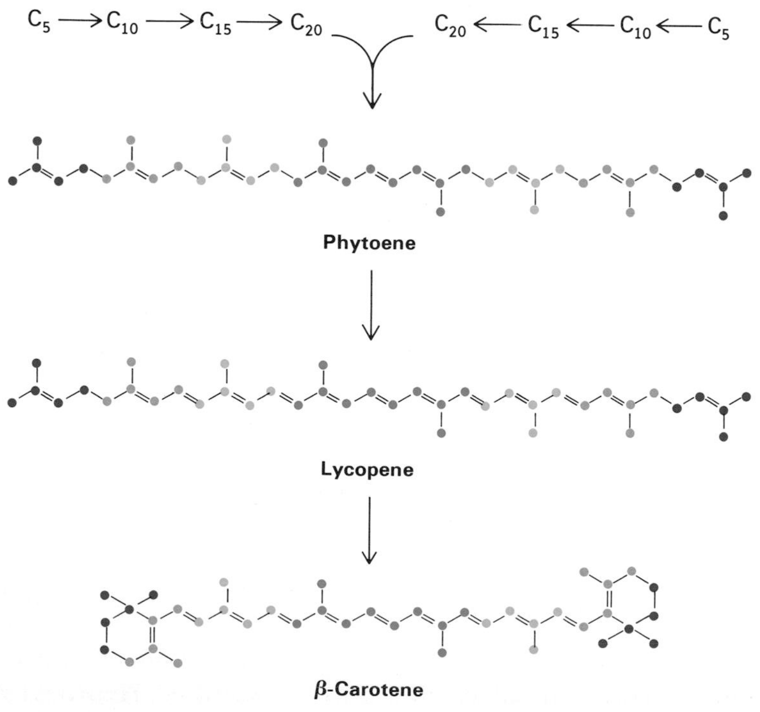

$C_5 \longrightarrow C_{10} \longrightarrow C_{15} \longrightarrow C_{20}$ $C_{20} \longleftarrow C_{15} \longleftarrow C_{10} \longleftarrow C_5$

Phytoene

Lycopene

β-Carotene

Figure 23-30
Synthesis of C_{40} carotenoids: phytoene, lycopene, and β-carotene.

SUMMARY

Phosphatidate, an intermediate in the synthesis of phosphoglycerides and triacylglycerols, is formed by successive acylations of glycerol 3-phosphate by acyl CoA. Hydrolysis of its phosphoryl group followed by acylation yields a triacylglycerol. CDP-diacylglycerol, the activated intermediate in the de novo synthesis of several phosphoglycerides, is formed from phosphatidate and CTP. The activated phosphatidyl unit is then transferred to the hydroxyl group of a polar alcohol, such as serine, to form phosphatidyl serine. In bacteria, decarboxylation of this phosphoglyceride yields phosphatidyl ethanolamine, which is methylated by *S*-adenosylmethionine to form phosphatidyl choline. In mammals, this phosphoglyceride is synthesized by a pathway that utilizes dietary choline. CDP-choline is the activated intermediate in this route. Sphingolipids are synthesized from ceramide, which is formed by the acylation of sphingosine. Gangliosides are sphingolipids that contain an oligosaccharide unit having at least one residue of *N*-acetylneuraminate or a related sialic acid. They are synthesized by the stepwise addition of activated sugars, such as UDP-glucose, to ceramide.

Cholesterol, a steroid component of eucaryotic membranes and a precursor of steroid hormones, is formed from acetyl CoA. The committed step in its synthesis is the formation of mevalonate from 3-hydroxy-3-methylglutaryl CoA (derived from acetyl CoA and acetoacetyl CoA). Mevalonate is converted into isopentenyl pyrophosphate (C_5), which condenses with its isomer, dimethylallyl pyrophosphate (C_5), to form geranyl pyrophosphate (C_{10}). Addition of a second molecule of isopentenyl pyrophosphate yields farnesyl pyrophosphate (C_{15}), which condenses with itself to form squalene (C_{30}). This intermediate cyclizes to lanosterol (C_{30}), which is modified to yield cholesterol (C_{27}). The synthesis of cholesterol by the liver is regulated by changes in the amount and activity of 3-hydroxy-3-methylglutaryl CoA reductase, the enzyme catalyzing the committed step in its biosynthesis.

Cholesterol and other lipids are transported in the blood to specific targets by a series of lipoproteins. Dietary lipids are carried by chylomicrons. Endogenously synthesized triacylglycerols are transported from the liver to adipose tissue by very-low-density lipoproteins (VLDL). After delivering its content of triacylglycerols, VLDL is converted into intermediate-density lipoprotein (IDL) and then into low-density lipoprotein (LDL). IDL and LDL carry cholesterol esters, primarily cholesterol linoleate. LDL is taken up by liver and peripheral tissue cells by receptor-mediated endocytosis. The LDL receptor, a protein spanning the plasma membrane of these target cells, binds LDL and mediates its entry into the cell. Absence of the LDL receptor in the homozygous form of familial hypercholesterolemia leads to a markedly elevated plasma level of LDL cholesterol, deposition of cholesterol on blood vessel walls, and heart attacks in childhood.

Five major classes of steroid hormones are derived from cholesterol: progestagens, glucocorticoids, mineralocorticoids, androgens, and estrogens. Hydroxylations by mixed-function oxidases that use NADPH and O_2 play an important role in the synthesis of steroid hormones and bile salts from cholesterol. Pregnenolone (C_{21}), a key intermediate in the synthesis of steroid hormones, is formed by scission of the side chain of cholesterol. Progesterone (C_{21}), synthesized from pregnenolone, is the precursor of cortisol and aldosterone. Hydroxylation of progesterone and cleavage of its side chain yields androstenedione, an androgen (C_{19}). Estrogens (C_{18}) are synthesized from androgens by the loss of an angular methyl group and the formation of an aromatic A ring. Vitamin D, which is important in the control of calcium and phosphorus metabolism, is formed from a derivative of cholesterol by the action of light. A remarkable array of biomolecules in addition to cholesterol and its derivatives are synthesized from isopentenyl pyrophosphate, the basic five-carbon building block. The hydrocarbon side chains of vitamin K_2, coenzyme Q_{10}, and chlorophyll are extended chains constructed from this activated C_5 unit.

SELECTED READINGS

Where to start

Brown, M. S., and Goldstein, J. L., 1986. A receptor-mediated pathway for cholesterol homeostasis. *Science* 232:34–47.

Brown, M. S., and Goldstein, J. L., 1984. How LDL receptors influence cholesterol and atherosclerosis. *Sci. Amer.* 251(5): 58–66.

Bloch, K., 1965. The biological synthesis of cholesterol. *Science* 150:19–28.

Books

Vance, D. E., and Vance, J. E., (eds.), 1985. *Biochemistry of Lipids and Membranes*. Benjamin/Cummings. [An excellent collection of articles on the metabolism, genetics, and assembly of lipids.]

Stanbury, J. B., Wyngaarden, J. B., Fredrickson, D. S., Goldstein, J. L., and Brown, M. S., (eds.), 1983. *The Metabolic Basis of Inherited Disease* (5th ed.). McGraw-Hill. [Contains excellent articles on genetic disorders of sphingolipid, lipoprotein, cholesterol, and steroid metabolism.]

Hawthorne, J. N., and Ansell, G. B., (eds.), 1982. *Phospholipids*. Elsevier.

DeLuca, H. F., (ed.), 1978. *The Fat-soluble Vitamins*. Plenum.

Nes, W. R., and McKean, M. L., 1977. *Biochemistry of Steroids and other Isopentenoids*. University Park Press.

Enzymes and reaction mechanisms

Walsh, C., 1979. *Enzymatic Reaction Mechanisms*. W. H. Freeman. [Contains excellent discussions of reaction mechanisms in the biosynthesis of cholesterol and other compounds derived from isopentenyl pyrophosphate.]

Hayaishi, O., (ed.), 1974. *Molecular Mechanisms of Oxygen Activation*. Academic Press.

Phospholipids, sphingolipids, and phosphoinositides

Hakomori, S., 1986. Glycosphingolipids. *Sci. Amer.* 254(5): 44–53. [An interesting presentation of changes in these lipids during cell differentiation and oncogenesis.]

Needleman, P., Turk, J., Jakschik, B. A., Morrison, A. R., and Lefkowith, J. B., 1986. Arachidonic acid metabolism. *Ann. Rev. Biochem.* 55:69.

Majerus, P., Connolly, T. M., Deckmyn, H., Ross, T. S., Bross, T. E., Ishii, H., Bansal, V. S., and Wilson, D. B., 1986. The metabolism of phosphoinositide-derived messenger molecules. *Science* 234:1519–1526.

Raetz, C. R. H., 1986. Molecular genetics of membrane phospholipid synthesis. *Ann. Rev. Genet.* 20:253–295.

Dennis, E. A., 1983. Phospholipases. *In* P. D. Boyer (ed.), *The Enzymes* (3rd ed.), vol. 16 pp. 307–353. Academic Press.

Hanahan, D. J., 1986. Platelet-activating factor: a biologically active phosphoglyceride. *Ann. Rev. Biochem.* 55:483–509.

Steroids

Bloch, K., 1983. Sterol structure and membrane function. *Crit. Rev. Biochem.* 14:47–92.

Schroepfer, G. J., Jr., 1982. Sterol biosynthesis. *Ann. Rev. Biochem.* 51:555–585.

DeLuca, H. F., and Schnoes, H. K., 1983. Vitamin D: recent advances. *Ann. Rev. Biochem.* 52:411–439.

Holick, M. F., and Clark, M. B., 1978. The photobiogenesis and metabolism of vitamin D. *Fed. Proc.* 37:2567–2574.

Nebert, D. W., and Gonzales, F. J., 1987. P450 genes: structure, evolution, and regulation. *Ann. Rev. Biochem.* 56:945–993.

Lipoproteins

Atkinson, D., and Small, D. M., 1986. Recombinant lipoproteins: implications for structure and assembly of native lipoproteins. *Ann. Rev. Biophys. Biophys. Chem.* 15:403.

Breslow, J. L., 1985. Human apolipoprotein molecular biology and genetic variation. *Ann. Rev. Biochem.* 54:699–727.

LDL and its receptor

Sudhof, T. C., Goldstein, J. L., Brown, M. S., and Russell, D. W., 1985. The LDL receptor gene: a mosaic of exons shared with different proteins. *Science* 228:815–822.

Goldstein, J. L., Brown, M. S., Anderson, R. G. W., Russell, D. W., and Schneider, W. J., 1985. Receptor-mediated endocytosis: concepts emerging from the LDL receptor system. *Ann. Rev. Cell Biol.* 1:1–39.

Yang, C.-Y., Chen, S.-H., Gianturco, S. H., Bradley, W. A., Sparrow, J. T., Tanimura, M., Li, W.-H., Sparrow, D. A., DeLoof, H., Rosseneu, M., Lee, F.-S., Gu, Z.-W., Gotto, A. M., Jr., and Chan, L., 1986. Sequence, structure, receptor-binding domains and internal repeats of human apolipoprotein B-100. *Nature* 323:738–742.

Goldstein, J. L., Kita, T., and Brown, M. S., 1983. Defective lipoprotein receptors and atherosclerosis. *New Engl. J. Med.* 309:288–296.

Knott, T. J., Pease, R. J., Powell, L. M., Wallis, S. C., Rall, S. C., Jr., Innerarity, T. L., Blackhart, B., Taylor, W. H., Marcel, Y., Milne, R., Johnson, D., Fuller, M., Lusis, A. J., McCarthy, B. J., Mahley, R. W., Levy-Wilson, B., and Scott, J., 1986. Complete protein sequence and identification of structural domains of human apolipoprotein B. *Nature* 323:734–738.

Ma, P. T., Gil, G., Sudhof, T. C., Bilheimer, D. W., Goldstein, J. L., and Brown, M. S., 1986. Mevinolin, an inhibitor of cholesterol synthesis, induces mRNA for low density lipoprotein receptor in livers of hamsters and rabbits. *Proc. Nat. Acad. Sci.* 83:8370–8374.

Ross, R., 1986. The pathogenesis of atherosclerosis—an update. *New Engl. J. Med.* 314:488–500.

PROBLEMS

1. Write a balanced equation for the synthesis of a triacylglycerol, starting from glycerol and fatty acids.

2. Write a balanced equation for the synthesis of phosphatidyl serine by the de novo pathway, starting from serine, glycerol, and fatty acids.

3. What is the activated reactant in each of these biosyntheses?
   (a) Phosphatidyl serine from serine.
   (b) Phosphatidyl ethanolamine from ethanolamine.
   (c) Ceramide from sphingosine.
   (d) Sphingomyelin from ceramide.
   (e) Cerebroside from ceramide.
   (f) Ganglioside G_{M1} from ganglioside G_{M2}.
   (g) Farnesyl pyrophosphate from geranyl pyrophosphate.

4. What is the distribution of isotopic labeling in cholesterol synthesized from each of these precursors?
   (a) Mevalonate labeled with ^{14}C in its carboxyl carbon atom.
   (b) Malonyl CoA labeled with ^{14}C in its carboxyl carbon atom.

CHAPTER 24

Biosynthesis of Amino Acids and Heme

This chapter deals with the biosynthesis of amino acids and some molecules derived from them. The flow of nitrogen into amino acids will be considered first. This process starts with the reduction of N_2 to NH_4^+ by nitrogen-fixing microorganisms. NH_4^+ is then assimilated into amino acids by way of glutamate and glutamine, the two pivotal molecules in nitrogen metabolism. Of the basic set of twenty amino acids, eleven are synthesized from intermediates of the citric acid cycle and other major metabolic pathways by quite simple reactions. We shall consider how they are formed and then examine the biosyntheses of aromatic amino acids and of histidine as examples of more complex pathways. In fact, humans must obtain the latter group of nine amino acids from their diets, and so they are called essential amino acids.

Two interesting carriers participate in amino acid metabolism: tetrahydrofolate, a highly versatile carrier of activated one-carbon units at three oxidation stages, and *S*-adenosylmethionine, the major methyl donor. The regulation of amino acid metabolism is another challenging area of inquiry. We shall take a look at glutamine synthetase—a highly responsive enzyme that controls the flow of nitrogen in bacteria. Many important biomolecules are derived from amino acids. For example, glutathione serves as a sulfhydryl buffer and transporter of amino acids. The final section of this chapter is concerned with the synthesis and degradation of heme.

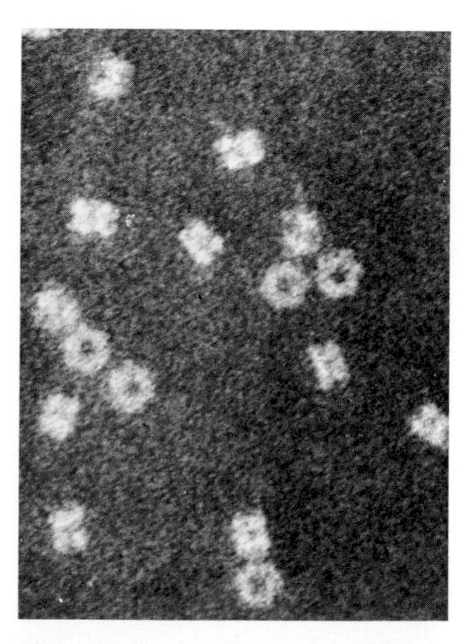

Figure 24-1
Electron micrograph of glutamine synthetase from *E. coli.* This enzyme plays a key role in nitrogen metabolism. [Courtesy of Dr. Earl Stadtman.]

MICROORGANISMS USE ATP AND A POWERFUL REDUCTANT TO CONVERT N_2 INTO NH_4^+

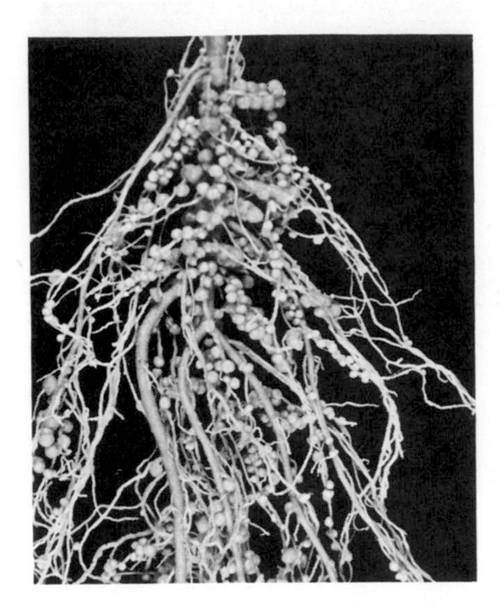

Figure 24-2
The nodules in the root system of the soybean are the sites of nitrogen fixation by *Rhizobium* bacteria. [Courtesy of Dr. Joe C. Burton, Nitragin Company, Inc.]

Nitrogen enters the synthesis of amino acids, purines, pyrimidines, and other biomolecules in reduced form, for example, as NH_4^+. Higher organisms are unable to convert N_2 into this form. Rather, this conversion—called *nitrogen fixation*—is carried out by bacteria and blue-green algae. Some of these microorganisms—namely, the symbiotic *Rhizobium* bacteria—invade the roots of leguminous plants and form root nodules, where nitrogen fixation takes place supplying both the bacteria and the plants (Figure 24-2). The amount of N_2 fixed by microorganisms has been estimated to be about 2×10^{11} kg per year.

The N≡N bond, which has a bond energy of 225 kcal/mol, is highly resistant to chemical attack. Indeed, Lavoisier named it "azote," meaning "without life," because it is quite unreactive. The industrial process for nitrogen fixation was devised by Fritz Haber in 1910 and is currently used in fertilizer factories:

$$N_2 + 3\ H_2 \rightleftharpoons 2\ NH_3$$

Fixation of N_2 is typically carried out over an iron catalyst at about 500°C and a pressure of 300 atm. It is not surprising, then, that the biological process of nitrogen fixation requires a complex enzyme. The *nitrogenase complex,* which carries out this process, consists of two kinds of protein components: a *reductase,* which provides electrons with high reducing power, and a *nitrogenase,* which uses these electrons to reduce N_2 to NH_4^+ (Figure 24-3). Each component is an *iron-sulfur protein,* in which iron is bonded to the sulfur atom of a cysteine residue and to inorganic sulfide (p. 403). The nitrogenase component of the complex also contains one or two *molybdenum* atoms, and so it has been known as the *MoFe protein.* It has the subunit structure $\alpha_2\beta_2$ and a mass of about 220 kd. The reductase (also called the *Fe protein*) consists of two identical polypeptides and has a mass of 65 kd. In the nitrogenase complex, one or two Fe proteins are associated with a MoFe protein.

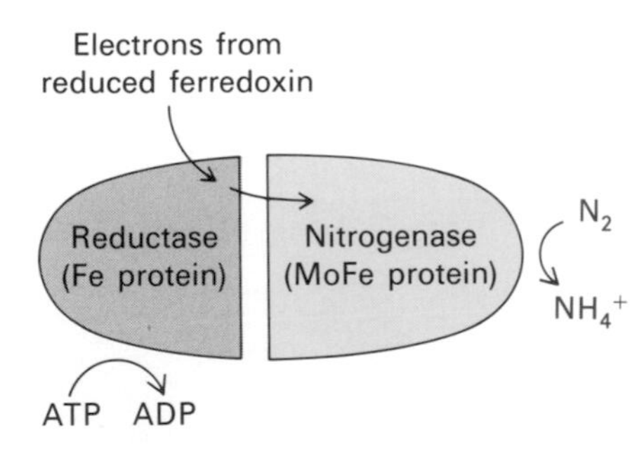

Figure 24-3
Schematic diagram of the nitrogenase complex. The reductase dissociates from the nitrogenase component before N_2 is converted into NH_4^+.

The conversion of N_2 into NH_4^+ by the nitrogenase complex *requires ATP and a powerful reductant.* In most nitrogen-fixing microorganisms, the source of high-potential electrons in this *six-electron reduction* is *reduced ferredoxin.* Recall that reduced ferredoxin is produced in chloroplasts by the action of photosystem I (p. 526). This electron carrier is generated by photosynthetic processes in some nitrogen-fixing bacteria, and by oxidative processes in others. The stoichiometry of the reaction catalyzed by the nitrogenase complex is

$$N_2 + 6\ e^- + 12\ ATP + 12\ H_2O \longrightarrow 2\ NH_4^+ + 12\ ADP + 12\ P_i + 4\ H^+$$

The following reaction sequence seems likely. First, reduced ferredoxin transfers its electrons to the reductase component of the complex. Second, ATP binds to the reductase and shifts the redox potential of the enzyme from −0.29 to −0.40 V by altering its conformation. This enhancement of its reducing power enables the reductase to transfer its electrons to the nitrogenase component. Third, electrons are transferred, ATP is hydrolyzed, and the reductase dissociates from the nitrogenase component. Finally, N_2 bound to the nitrogenase component of the complex is reduced to NH_4^+.

Sources of energy for the chemical production of ammonia by the Haber process are becoming scarcer and more costly, and so there is much current interest in enhancing nitrogen fixation by microorganisms. One potential approach is to insert the genes for nitrogen fixation into nonleguminous plants, such as cereals. A cluster of 18 genes (called the *nif* cluster) encode the proteins required for nitrogen fixation in *Klebsiella pneumoniae*. A difficulty that must be overcome is that the nitrogenase complex is exquisitely sensitive to inactivation by O_2. Leguminous plants maintain a very low concentration of free O_2 in their root nodules by binding O_2 to *leghemoglobin*. Another challenge that must be met in the formation of new nitrogen-fixing species is the requirement for a very high rate of ATP formation. In fact, nitrogen-fixing bacteria in the roots of a pea plant consume nearly a fifth of all the ATP generated by the plant. A complementary approach is to increase the rate of nitrogen fixation by blue-green algae, which generate their own ATP by photosynthesis and thus are not dependent on an energy-yielding symbiotic relation.

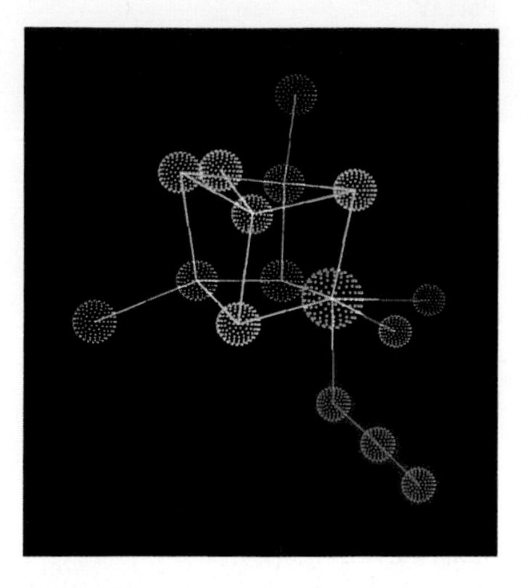

Figure 24-4
Structure of a synthetic molybdenum cluster. This model compound has some of the properties of the active site of nitrogenase. The roughly cubical complex consists of an $MoFe_3S_4$ core (Mo is shown in green, Fe in orange, and S in yellow). Thiolate ligands (also yellow) are bound to the irons. The two oxygen atoms (small red spheres) bound to Mo come from a substituted catechol. Azide ion (N_3^-, the three blue spheres), a substrate analog, is bound to Mo. The molybdenum cluster in the active site of the enzyme ($MoFe_{6-8}S_{8-10}$) is more complex than the one shown here. [Courtesy of Dr. Jeremy Berg.]

NH_4^+ IS ASSIMILATED INTO AMINO ACIDS BY WAY OF GLUTAMATE AND GLUTAMINE

The next step in the assimilation of nitrogen into biomolecules is the entry of NH_4^+ into amino acids. *Glutamate* and *glutamine* play pivotal roles in this regard. The α-amino group of most amino acids comes from the α-amino group of glutamate by transamination. Glutamine, the other major nitrogen donor, contributes its side-chain nitrogen in the biosynthesis of a wide range of important compounds.

Glutamate is synthesized from NH_4^+ and α-ketoglutarate, a citric acid cycle intermediate, by the action of *glutamate dehydrogenase*. This enzyme has already been encountered in the degradation of amino acids (p. 496). In the biosynthetic direction, NADPH is the reductant, whereas NAD^+ is the oxidant in the catabolic direction.

$$NH_4^+ + \alpha\text{-ketoglutarate} + NADPH + H^+ \rightleftharpoons \text{L-glutamate} + NADP^+ + H_2O$$

Ammonium ion is incorporated into glutamine by the action of *glutamine synthetase* on glutamate. This amidation is driven by the hydrolysis of ATP.

$$\underset{\textbf{Glutamate}}{{}^+H_3N-CH(COO^-)-CH_2-CH_2-COO^-} + NH_4^+ + ATP \longrightarrow \underset{\textbf{Glutamine}}{{}^+H_3N-CH(COO^-)-CH_2-CH_2-C(=O)-NH_2} + ADP + P_i + H^+$$

The regulation of glutamine synthetase plays a critical role in controlling nitrogen metabolism, as will be discussed shortly.

Glutamate dehydrogenase and glutamine synthetase are present in all organisms. Most procaryotes also contain *glutamate synthase*, which catalyzes the reductive amination of α-ketoglutarate. The nitrogen donor in this reaction is glutamine, and so two molecules of glutamate are formed.

$$\alpha\text{-Ketoglutarate} + \text{glutamine} + NADPH + H^+ \longrightarrow 2\ \text{glutamate} + NADP^+$$

When NH_4^+ is limiting, most of the glutamate is made by the sequential action of glutamine synthetase and glutamate synthase. The sum of these reactions is

$$NH_4^+ + \alpha\text{-ketoglutarate} + NADPH + ATP \longrightarrow \text{L-glutamate} + NADP^+ + ADP + P_i$$

Note that this stoichiometry differs from that of the glutamate dehydrogenase reaction in that an ATP is hydrolyzed. Why is this more expensive pathway sometimes used by *E. coli?* The answer is that the K_M of glutamate dehydrogenase for NH_4^+ is high (~1 mM), and so this enzyme is not saturated when NH_4^+ is limiting. In contrast, glutamine synthetase has very high affinity for NH_4^+.

AMINO ACIDS ARE MADE FROM INTERMEDIATES OF THE CITRIC ACID CYCLE AND OTHER MAJOR PATHWAYS

Table 24-1
Basic set of twenty amino acids

Nonessential	*Essential*
Alanine	Histidine
Arginine	Isoleucine
Asparagine	Leucine
Aspartate	Lysine
Cysteine	Methionine
Glutamate	Phenylalanine
Glutamine	Threonine
Glycine	Tryptophan
Proline	Valine
Serine	
Tyrosine	

Thus far, we have considered the conversion of N_2 into NH_4^+ and the assimilation of NH_4^+ into glutamate and glutamine. We turn now to the biosynthesis of the other amino acids. Bacteria such as *E. coli* can synthesize the entire basic set of twenty amino acids, whereas humans cannot make nine of them. The amino acids that must be supplied in the diet are called *essential,* whereas the others are termed *nonessential* (Table 24-1). *These designations refer to the needs of an organism under a particular set of conditions.* For example, enough arginine is synthesized by the urea cycle to meet the needs of an adult but perhaps not those of a growing child. A deficiency of even one amino acid results in a *negative nitrogen balance.* In this state, more protein is degraded than is synthesized, and so more nitrogen is excreted than is ingested.

The pathways for the biosynthesis of amino acids are diverse. However, they have an important common feature: *their carbon skeletons come from intermediates of glycolysis, the pentose phosphate pathway, or the citric acid cycle.* A further simplification is that there are only *six biosynthetic families* (Figure 24-5).

The nonessential amino acids are synthesized by quite simple reactions, whereas the pathways for the formation of the essential amino acids are quite complex. For example, the nonessential amino acids *alanine* and *aspartate* are synthesized in a single step from pyruvate and oxaloacetate, respectively. Each acquires its amino group from glutamate in a transamination reaction in which pyridoxal phosphate is the cofactor (p. 498):

$$\text{Pyruvate} + \text{glutamate} \rightleftharpoons \text{alanine} + \alpha\text{-ketoglutarate}$$

$$\text{Oxaloacetate} + \text{glutamate} \rightleftharpoons \text{aspartate} + \alpha\text{-ketoglutarate}$$

Asparagine is then synthesized by the amidation of aspartate:

$$\text{Aspartate} + NH_4^+ + ATP \longrightarrow \text{asparagine} + AMP + PP_i + H^+$$

In mammals, the nitrogen donor in the synthesis of asparagine is glutamine rather than NH_4^+, as in bacteria. Ammonia generated at the active site of the enzyme is directly transferred to bound aspartate. Recall that high levels of NH_4^+ are toxic to humans (p. 502).

Another one-step synthesis of a nonessential amino acid in mammals is the hydroxylation of phenylalanine (an essential amino acid) to *tyrosine.*

$$\text{Phenylalanine} + O_2 + NADPH + H^+ \longrightarrow \text{tyrosine} + NADP^+ + H_2O$$

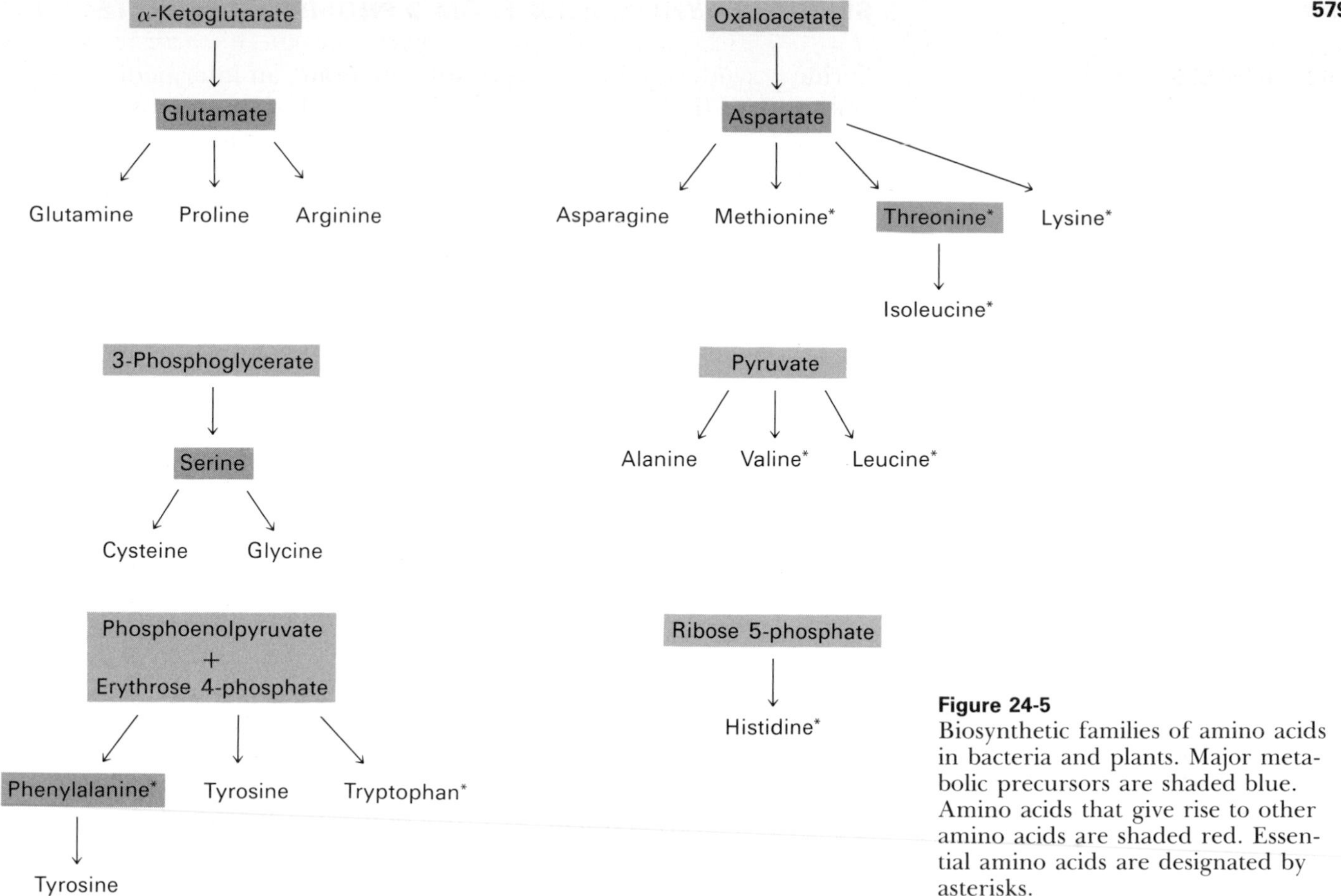

Figure 24-5
Biosynthetic families of amino acids in bacteria and plants. Major metabolic precursors are shaded blue. Amino acids that give rise to other amino acids are shaded red. Essential amino acids are designated by asterisks.

This reaction is catalyzed by phenylalanine hydroxylase, a monooxygenase discussed previously (p. 511). It is noteworthy that tyrosine is an essential amino acid in individuals lacking this enzyme.

GLUTAMATE IS THE PRECURSOR OF GLUTAMINE, PROLINE, AND ARGININE

The synthesis of glutamate by the reductive amination of α-ketoglutarate has already been discussed, as has the conversion of glutamate into *glutamine* (p. 577). Glutamate is the precursor of two other nonessential amino acids, *proline* and *arginine*. First, the γ-carboxyl group of glutamate reacts with ATP to form an acyl phosphate. This mixed anhydride is then reduced by NADPH to an aldehyde. Glutamic γ-semialdehyde cyclizes with a loss of H_2O to give Δ′-pyrroline-5-carboxylate, which is reduced by NADPH to proline. The semialdehyde also gives rise to ornithine, which is converted into arginine (p. 500).

$$\text{Glutamate} \xrightarrow[\text{NADPH} \;\to\; \text{NADP}^+]{\text{ATP} \;\to\; \text{ADP}} \text{Glutamic-}\gamma\text{-semialdehyde} \xrightarrow{-H_2O} \Delta'\text{-Pyrroline-5-carboxylate} \xrightarrow{\text{NADPH} \;\to\; \text{NADP}^+} \text{Proline}$$

Glutamate **Glutamic-γ-semialdehyde** **Δ′-Pyrroline-5-carboxylate** **Proline**

SERINE IS SYNTHESIZED FROM 3-PHOSPHOGLYCERATE

Serine is synthesized from 3-phosphoglycerate, an intermediate in glycolysis. The first step is an oxidation to 3-phosphohydroxypyruvate. This α-keto acid is transaminated to 3-phosphoserine, which is then hydrolyzed to serine.

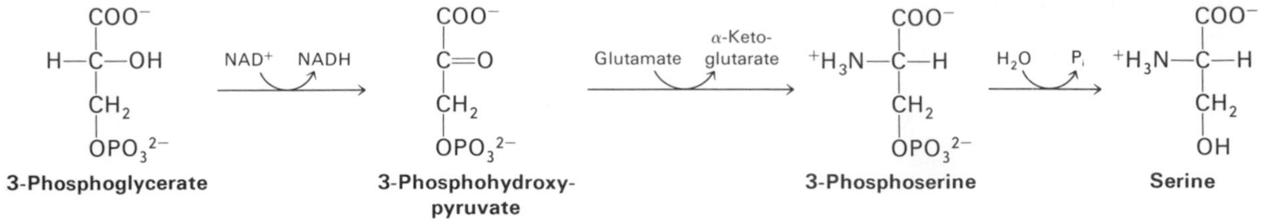

3-Phosphoglycerate $\xrightarrow{NAD^+ \rightarrow NADH}$ 3-Phosphohydroxypyruvate $\xrightarrow{\text{Glutamate} \rightarrow \alpha\text{-Ketoglutarate}}$ 3-Phosphoserine $\xrightarrow{H_2O \rightarrow P_i}$ Serine

Serine is the precursor of *glycine* and *cysteine*. In the formation of glycine, the side-chain β-carbon atom of serine is transferred to *tetrahydrofolate,* a carrier of one-carbon units that will be discussed shortly.

$$\text{Serine} + \text{tetrahydrofolate} \rightleftharpoons \text{glycine} + \text{methylenetetrahydrofolate} + H_2O$$

This conversion is catalyzed by *serine transhydroxymethylase,* a pyridoxal phosphate (PLP) enzyme. The bond between the α- and β-carbon atoms of serine is labilized by the formation of a Schiff base between serine and PLP. The β-carbon atom of serine is then transferred to tetrahydrofolate. Glycine can also be formed from CO_2, NH_4^+, and methylenetetrahydrofolate in a reaction catalyzed by *glycine synthase.* The conversion of serine into cysteine requires the substitution of a sulfur atom derived from methionine for the side-chain oxygen atom. This reaction sequence will be presented after one-carbon metabolism.

TETRAHYDROFOLATE CARRIES ACTIVATED ONE-CARBON UNITS AT SEVERAL OXIDATION LEVELS

Tetrahydrofolate (also called tetrahydropteroylglutamate), a highly versatile carrier of activated one-carbon units, consists of three groups: a substituted pteridine, *p*-aminobenzoate, and glutamate. Mammals are unable to synthesize a pteridine ring. They obtain tetrahydrofolate from their diets or from microorganisms in their intestinal tracts.

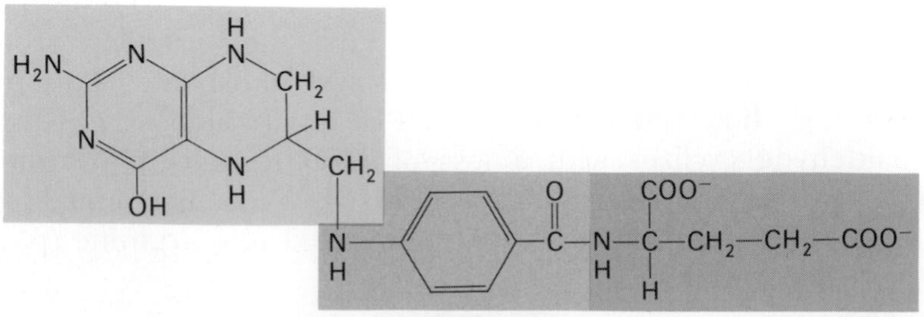

Tetrahydrofolate

Table 24-2
One-carbon groups carried by tetrahydrofolate

Oxidation state	Group	
Most reduced (= methanol)	$—CH_3$	Methyl
Intermediate (= formaldehyde)	$—CH_2—$	Methylene
Most oxidized (= formic acid)	—CHO	Formyl
	—CHNH	Formimino
	—CH=	Methenyl

The one-carbon group carried by tetrahydrofolate is bonded to its N-5 or N-10 nitrogen atom (denoted as N^5 and N^{10}) or to both. This unit can exist in three oxidation states (Table 24-2). The most reduced form carries a *methyl* group, whereas the intermediate form carries a *methylene* group. The most oxidized forms carry a *methenyl, formyl,* or *formimino* group. The most oxidized one-carbon unit, CO_2, is carried by biotin (p. 440) rather than by tetrahydrofolate.

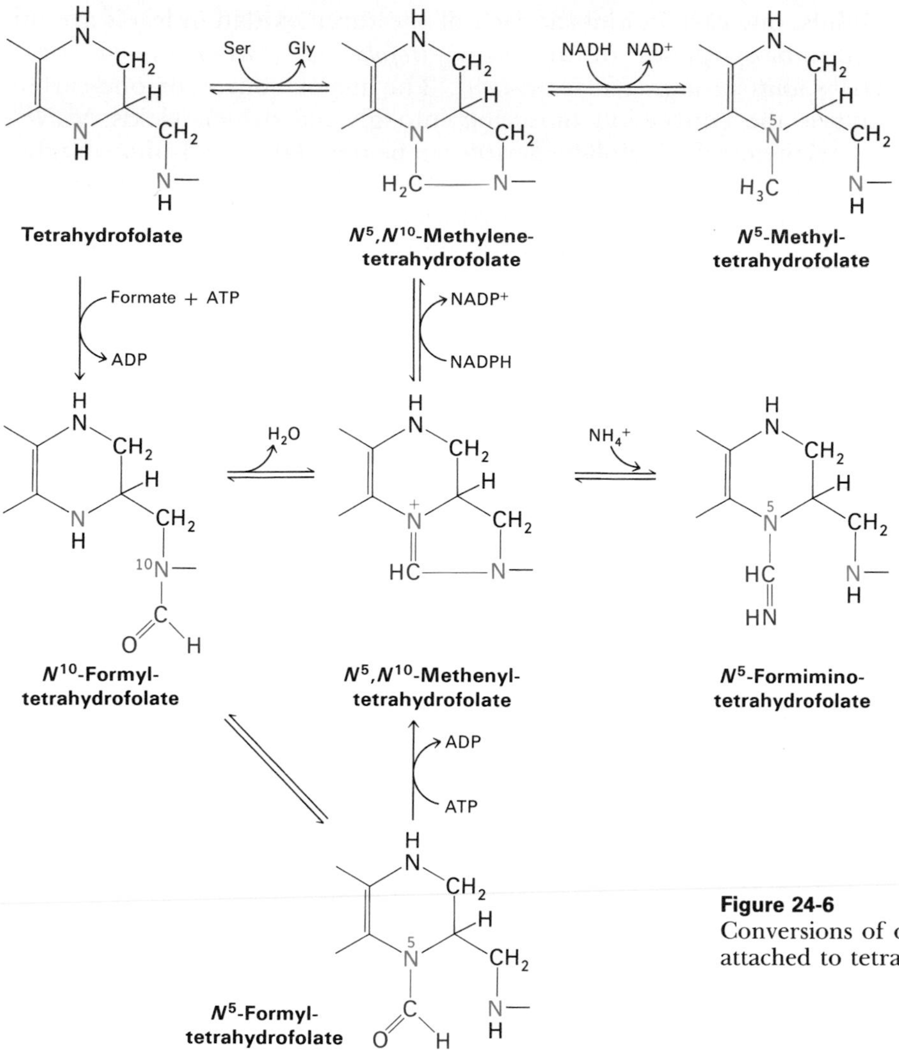

Tetrahydrofolate — Ser, Gly ⇌ **N^5,N^{10}-Methylenetetrahydrofolate** — NADH, NAD^+ ⇌ **N^5-Methyltetrahydrofolate**

Tetrahydrofolate — Formate + ATP → ADP → **N^{10}-Formyltetrahydrofolate**

N^5,N^{10}-Methylenetetrahydrofolate — $NADP^+$, NADPH ⇌ **N^5,N^{10}-Methenyltetrahydrofolate**

N^{10}-Formyltetrahydrofolate — H_2O ⇌ N^5,N^{10}-Methenyltetrahydrofolate — NH_4^+ ⇌ **N^5-Formiminotetrahydrofolate**

N^5-Formyltetrahydrofolate — ATP → ADP → N^5,N^{10}-Methenyltetrahydrofolate

Figure 24-6
Conversions of one-carbon units attached to tetrahydrofolate.

The one-carbon units carried by tetrahydrofolate are interconvertible (Figure 24-6). *N^5,N^{10}-Methylene*tetrahydrofolate can be reduced to *N^5-methyl*tetrahydrofolate or oxidized to *N^5,N^{10}-methenyl*tetrahydrofolate. *N^5,N^{10}-Methenyl*tetrahydrofolate can be converted into *N^5-formimino*tetrahydrofolate or *N^{10}-formyl*tetrahydrofolate, which are at the same oxidation level. N^{10}-Formyltetrahydrofolate can also be synthesized from tetrahydrofolate, formate and ATP:

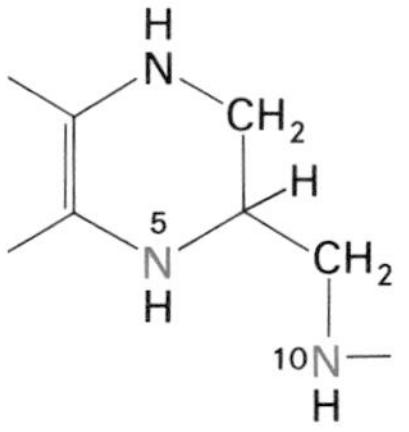

Reactive part of tetrahydrofolate

$$\text{Formate} + \text{ATP} + \text{tetrahydrofolate} \rightleftharpoons N^{10}\text{-formyltetrahydrofolate} + \text{ADP} + P_i$$

These tetrahydrofolate derivatives serve as donors of one-carbon units in a variety of biosyntheses. Methionine is regenerated from homocysteine by transfer of the methyl group of N^5-methyltetrahydrofolate, as will be discussed shortly. Some of the carbon atoms of *purines* are derived from the N^{10}-formyl derivatives of tetrahydrofolate. The methyl group of *thymine,* a pyrimidine, comes rom N^5,N^{10}-methylenetetrahydrofolate. This tetrahydrofolate derivative also donates a one-carbon unit in the synthesis of *glycine* from CO_2 and NH_4^+, in a reaction catalyzed by glycine synthase.

$$CO_2 + NH_4^+ + N^5,N^{10}\text{-methylenetetrahydrofolate} + \text{NADH} \rightleftharpoons \text{glycine} + \text{tetrahydrofolate} + NAD^+$$

Thus, one-carbon units at each of the three oxidation levels are utilized in biosyntheses. In turn, *tetrahydrofolate serves as an acceptor of one-carbon units in degradative reactions.* The major source of one-carbon units is the conversion of serine into glycine, which yields N^5,N^{10}-methylenetetrahydrofolate. Serine can be derived from 3-phosphoglycerate (p. 580), and so *this pathway enables one-carbon units to be formed de novo from carbohydrate.* The breakdown of *histidine* yields *N*-formiminoglutamate, which transfers its formimino group to tetrahydrofolate to form the N^5-derivative.

S-ADENOSYLMETHIONINE IS THE MAJOR DONOR OF METHYL GROUPS

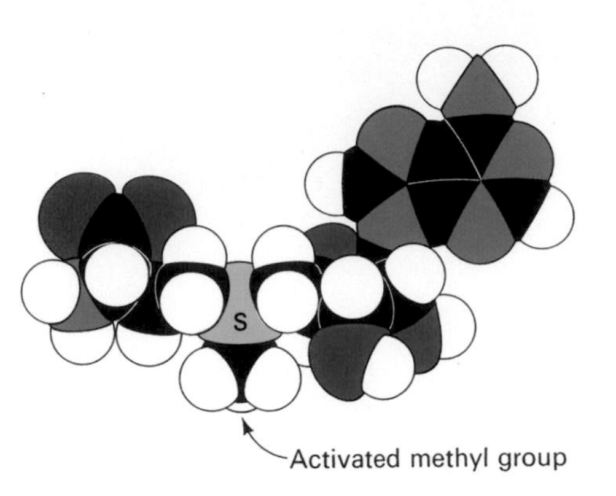

Figure 24-7
Model of *S*-adenosylmethionine.

Tetrahydrofolate can carry a methyl group on its N^5-atom, but its transfer potential is not sufficiently high for most biosynthetic methylations. Rather, the activated methyl donor is usually S-*adenosylmethionine,* which has already been encountered in the conversion of phosphatidyl ethanolamine into phosphatidyl choline (p. 549). *S*-Adenosylmethionine is synthesized by the transfer of an adenosyl group from ATP to the sulfur atom of methionine. The methyl group of the methionine unit is activated by the positive charge on the adjacent sulfur atom, which makes the molecule much more reactive than N^5-methyltetrahydrofolate.

$$\text{Methionine} + \text{ATP} \longrightarrow P_i + PP_i + S\text{-Adenosylmethionine}$$

Methionine (COO^-–H–C–NH_3^+–CH_2–CH_2–S–CH_3) **S-Adenosylmethionine**

The synthesis of *S*-adenosylmethionine is unusual in that the triphosphate group of ATP is split into pyrophosphate and orthophosphate. Pyrophosphate is then hydrolyzed. Thus, all of the phosphorus-oxygen bonds in ATP are split in this activation reaction, which markedly enhances the reactivity of the methyl group.

S-*Adenosylhomocysteine* is formed when the methyl group of *S*-adenosylmethionine is transferred to an acceptor such as phosphatidyl ethanolamine. *S*-Adenosylhomocysteine is then hydrolyzed to *homocysteine* and adenosine.

$$S\text{-Adenosylmethionine} \xrightarrow{R \;\to\; R\text{—}CH_3} S\text{-Adenosylhomocysteine} \xrightarrow{H_2O \;\to\; \text{Adenosine}} \text{Homocysteine}$$

S-Adenosylmethionine (H_3C—S^+—Adenosyl) **S-Adenosyl-homocysteine** (S—Adenosyl) **Homocysteine** (SH)

Methionine can be regenerated by the transfer of a methyl group to homocysteine from N^5-methyltetrahydrofolate, in a reaction catalyzed by *homocysteine methyltransferase*.

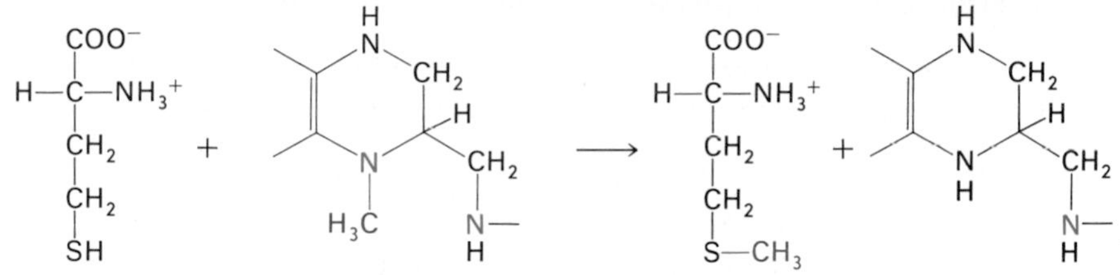

Homocysteine + N^5-Methyltetrahydrofolate ⟶ Methionine + Tetrahydrofolate

The coenzyme that mediates this transfer of a methyl group is *methylcobalamin* derived from vitamin B_{12}. In fact, this reaction and the rearrangement of L-methylmalonyl CoA to succinyl CoA (p. 506) are the only two known that are dependent on vitamin B_{12} in mammals. Alternatively, homocysteine can be methylated to methionine by donors such as *betaine,* an oxidation product of choline.

$(CH_3)_3\overset{+}{N}-CH_2-CH_2OH$
Choline

$(CH_3)_3\overset{+}{N}-CH_2-COO^-$
Betaine

These reactions constitute the *activated methyl cycle* (Figure 24-8). Methyl groups enter the cycle in the conversion of homocysteine into methionine and are then made highly reactive by the expenditure of 3 ~P. The high transfer potential of the methyl group in *S*-adenosylmethionine enables it to be transferred to a wide variety of acceptors, such as the amino group of the neurotransmitter norepinephrine (p. 1025) and a glutamate residue of a regulatory protein in chemotaxis (p. 1009).

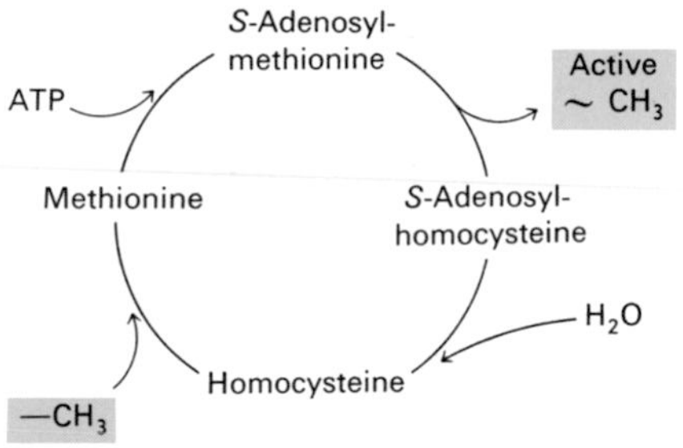

Figure 24-8
Activated methyl cycle. Humans derive methionine from the diet.

CYSTEINE IS SYNTHESIZED FROM SERINE AND HOMOCYSTEINE

In addition to being a precursor of methionine in the activated methyl cycle, homocysteine is an intermediate in the synthesis of cysteine. Serine and homocysteine condense to form *cystathionine* (Figure 24-9). This

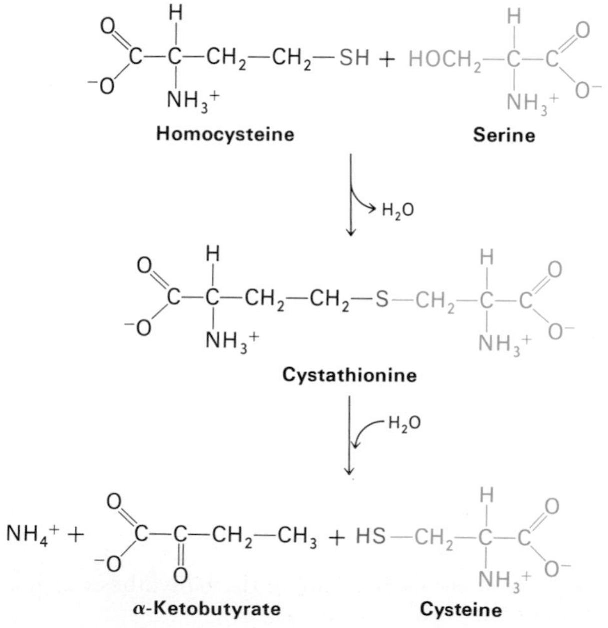

Figure 24-9
Synthesis of cysteine.

reaction is catalyzed by cystathionine synthase, a PLP enzyme. Cystathionine is then deaminated and cleaved to cysteine and α-ketobutyrate by *cystathioninase,* another PLP enzyme. The net reaction is

$$\text{Homocysteine} + \text{serine} \longrightarrow \text{cysteine} + \alpha\text{-ketobutyrate}$$

Note that the sulfur atom of cysteine is derived from homocysteine, whereas the carbon skeleton comes from serine.

SHIKIMATE AND CHORISMATE ARE INTERMEDIATES IN THE BIOSYNTHESIS OF AROMATIC AMINO ACIDS

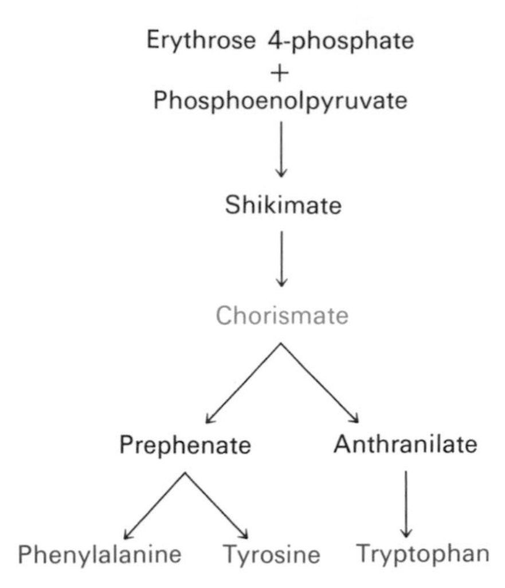

Figure 24-10
Pathway for the biosynthesis of aromatic amino acids in *E. coli.*

We turn now to the biosynthesis of essential amino acids, which are formed by much more complex routes than are the nonessential amino acids. Essential amino acids are synthesized by plants and microorganisms. The essential amino acids in the diet of humans are ultimately derived primarily from plants. Two bacterial pathways have been selected for discussion here—that of the aromatic amino acids and that of histidine—because they are well understood.

Phenylalanine, tyrosine, and tryptophan are synthesized by a common pathway in *E. coli* (Figure 24-10). The initial step is the condensation of phosphoenolpyruvate (a glycolytic intermediate) with erythrose 4-phosphate (a pentose phosphate pathway intermediate). The resulting C_7 open-chain sugar loses its phosphoryl group and cyclizes to 5-dehydroquinate. Dehydration then yields 5-dehydroshikimate, which is reduced by NADPH to *shikimate* (Figure 24-11). A second molecule of phosphoenolpyruvate then condenses with 5-phosphoshikimate to give an intermediate that loses its phosphoryl group, yielding *chorismate.*

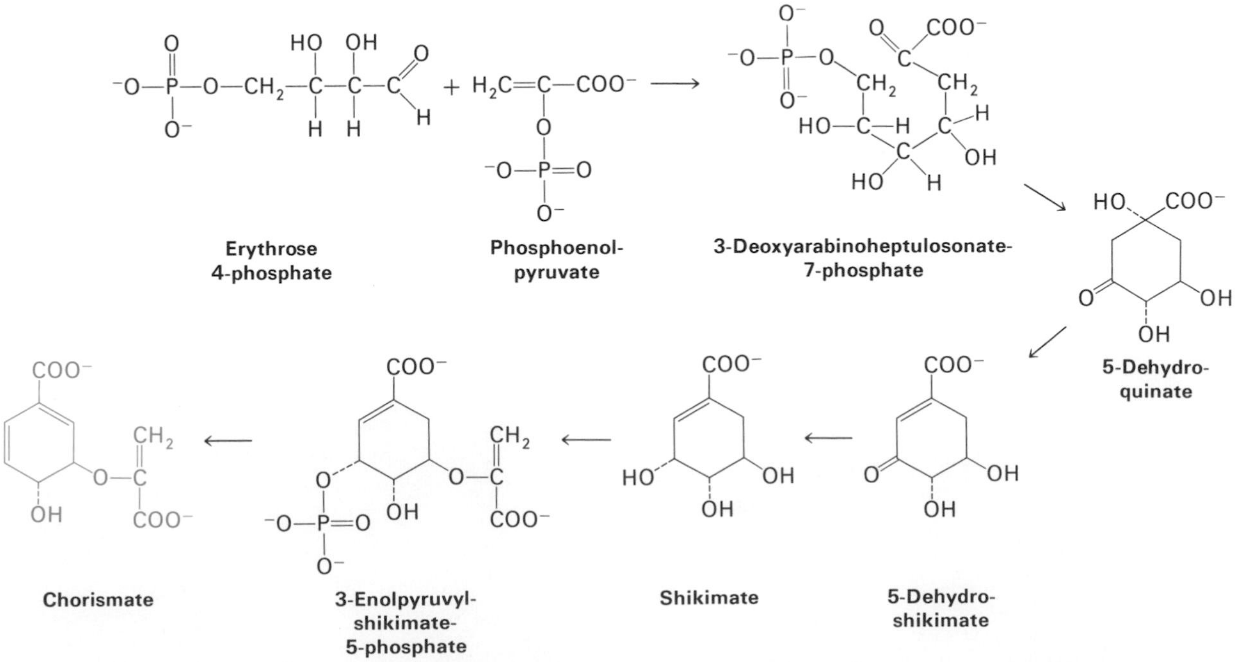

Erythrose 4-phosphate + Phosphoenolpyruvate ⟶ 3-Deoxyarabinoheptulosonate-7-phosphate ⟶ 5-Dehydroquinate ⟶ 5-Dehydroshikimate ⟶ Shikimate ⟶ 3-Enolpyruvylshikimate-5-phosphate ⟶ Chorismate

Figure 24-11
Synthesis of chorismate, an intermediate in the biosynthesis of phenylalanine, tyrosine, and tryptophan in *E. coli.*

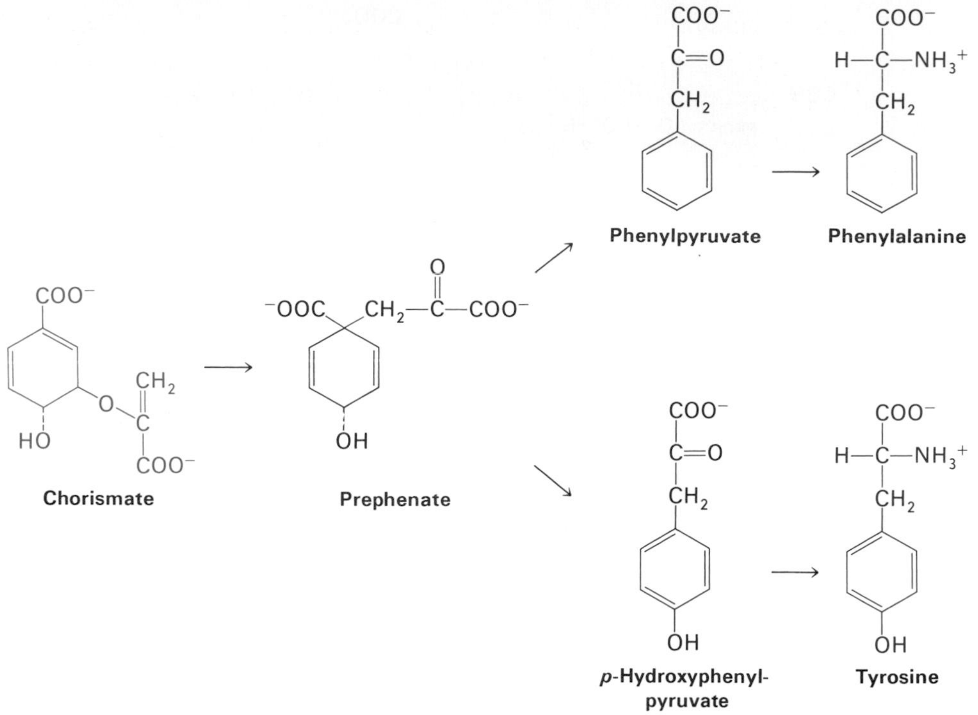

Phenylpyruvate Phenylalanine

Chorismate Prephenate

p-Hydroxyphenylpyruvate Tyrosine

Figure 24-12
Synthesis of phenylalanine and tyrosine from chorismate.

The pathway bifurcates at chorismate. Let us first follow the *prephenate branch* (Figure 24-12). A mutase converts chorismate into prephenate, the immediate precursor of the aromatic ring of phenylalanine and tyrosine. Dehydration and decarboxylation yield *phenylpyruvate.* Alternatively, prephenate can be oxidatively decarboxylated to yield p-*hydroxyphenylpyruvate.* These α-keto acids are then transaminated to yield *phenylalanine* and *tyrosine.*

The branch starting with *anthranilate* leads to the synthesis of *tryptophan.* Chorismate acquires an amino group from the side chain of glutamine to form anthranilate. In fact, *glutamine serves as an amino donor in many biosynthetic reactions.* Anthranilate then condenses with *phosphoribosylpyrophosphate* (PRPP), *an activated form of ribose phosphate.* PRPP is also a key intermediate in the synthesis of histidine, purine nucleotides, and pyrimidine nucleotides (p. 602). The C-1 atom of ribose 5-phosphate becomes bonded to the nitrogen atom of anthranilate in a reaction that is driven by the hydrolysis of pyrophosphate.

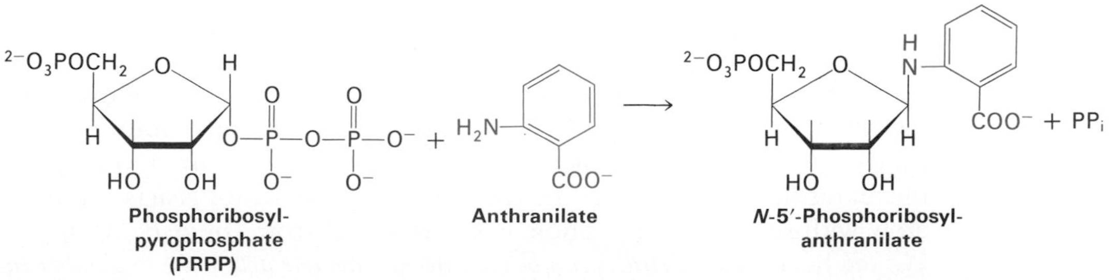

Phosphoribosylpyrophosphate (PRPP) + Anthranilate ⟶ *N*-5′-Phosphoribosylanthranilate + PP_i

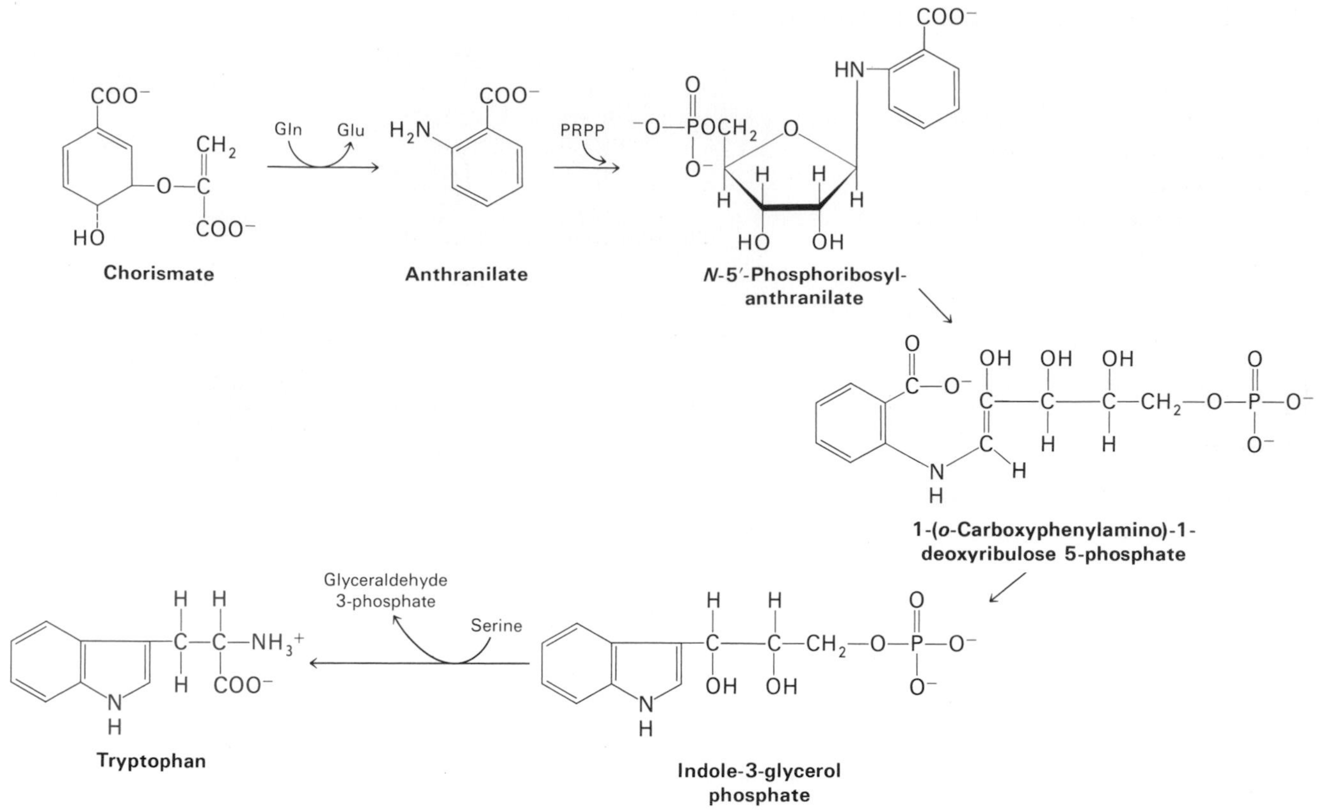

Figure 24-13
Synthesis of tryptophan from chorismate.

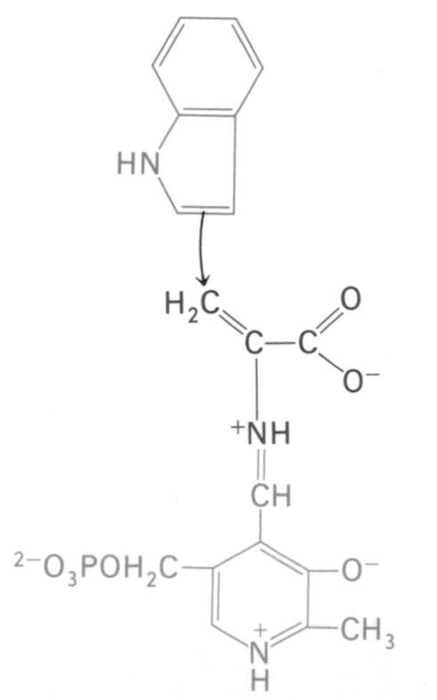

Figure 24-14
Intermediate in the synthesis of tryptophan. PLP (green) on a β chain forms a Schiff base with serine, which is then dehydrated to give the Schiff base of aminoacrylate (red). This enzyme-bound intermediate is attacked by indole, the product of the partial reaction catalyzed by the α subunit, to give tryptophan.

The ribose moiety of phosphoribosylanthranilate undergoes rearrangement (Figure 24-13) to yield 1-(*o*-carboxyphenylamino)-1-deoxyribulose 5-phosphate. This intermediate is dehydrated and decarboxylated to form *indole-3-glycerol phosphate*. Finally, indole-3-glycerol phosphate reacts with serine to form *tryptophan*. The glycerol phosphate side chain of indole-3-glycerol phosphate is replaced by the carbon skeleton and amino group of serine. This reaction is catalyzed by tryptophan synthetase.

Tryptophan synthetase of *E. coli* has the subunit structure $\alpha_2\beta_2$. The enzyme can be dissociated into two α subunits and a β_2 subunit. The isolated subunits catalyze partial reactions that lead to the synthesis of tryptophan:

$$\text{Indole-3-glycerol phosphate} \xrightarrow{\alpha \text{ subunit}} \text{indole} + \text{glyceraldehyde 3-phosphate}$$

$$\text{Indole} + \text{serine} \xrightarrow{\beta_2 \text{ subunit}} \text{tryptophan} + H_2O$$

Each active site on the β_2 subunit contains a PLP prosthetic group. *The catalytic properties of the α and β_2 subunits are markedly altered on the formation of the $\alpha_2\beta_2$ complex.* The rates of the partial reactions are more than ten times as great in the $\alpha_2\beta_2$ complex as in the isolated subunits. Furthermore, the $\alpha_2\beta_2$ complex synthesizes tryptophan by a concerted mechanism. Indole formed by the first partial reaction reacts immediately with serine, so that indole is not released from the $\alpha_2\beta_2$ complex. We see here that *substrates can be channeled from one active site to another in a multienzyme complex.*

HISTIDINE IS SYNTHESIZED FROM ATP, PRPP, AND GLUTAMINE

The pathway for histidine biosynthesis in *E. coli* and *Salmonella* contains many complex and novel features (Figure 24-15). The reaction sequence starts with the condensation of ATP with PRPP, in which N-1 of the purine ring becomes bonded to C-1 of the ribose unit of PRPP. In fact, five carbon atoms of histidine come from PRPP. The adenine unit

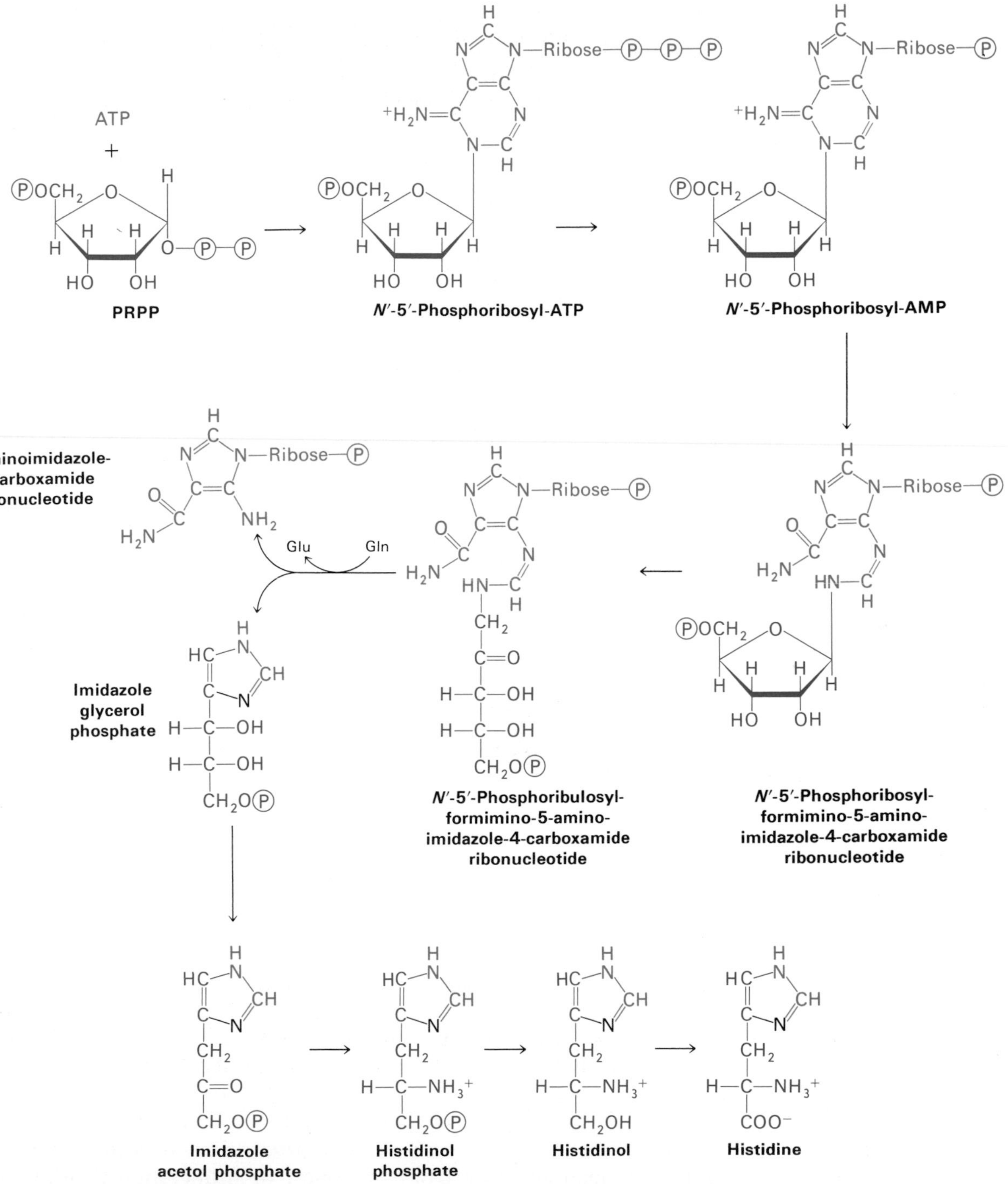

ATP + PRPP → *N*′-5′-Phosphoribosyl-ATP → *N*′-5′-Phosphoribosyl-AMP → *N*′-5′-Phosphoribosyl-formimino-5-amino-imidazole-4-carboxamide ribonucleotide → *N*′-5′-Phosphoribulosyl-formimino-5-amino-imidazole-4-carboxamide ribonucleotide → (Gln → Glu) 5-Aminoimidazole-4-carboxamide ribonucleotide + Imidazole glycerol phosphate → Imidazole acetol phosphate → Histidinol phosphate → Histidinol → Histidine

Figure 24-15
Pathway for the biosynthesis of histidine in *E. coli* and *Salmonella* (Ⓟ denotes a phosphoryl group).

of ATP provides a nitrogen and a carbon atom of the imidazole ring of histidine. The other nitrogen atom of the imidazole ring comes from the side chain of glutamine. A noteworthy aspect of this pathway is that 5-aminoimidazole-4-carboxamide ribonucleotide, which is produced in the cleavage reaction that forms the imidazole ring, is an intermediate in purine biosynthesis (p. 604). The synthesis of histidine from ATP and PRPP is a molecular relic of the early RNA world.

AMINO ACID BIOSYNTHESIS IS REGULATED BY FEEDBACK INHIBITION

The rate of synthesis of amino acids depends mainly on the *amounts* of the biosynthetic enzymes and on their *activities*. We shall now consider the control of enzymatic activity. The regulation of enzyme synthesis will be discussed in Chapter 32.

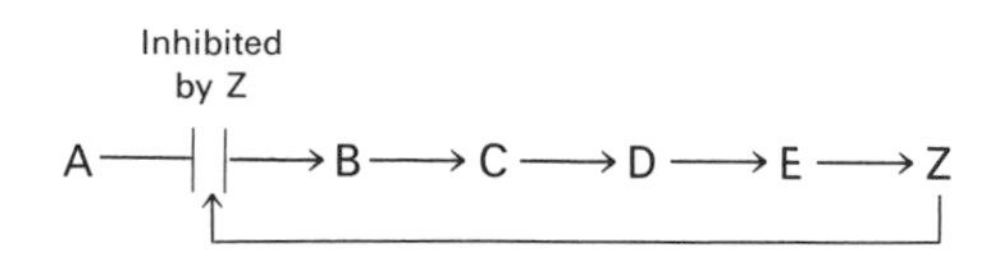

The first irreversible reaction in a biosynthetic pathway, called the committed step, is usually an important regulatory site. *The final product of the pathway* (*Z*) *often inhibits the enzyme that catalyzes the committed step* (*A* $\rightarrow$ B). This kind of control is essential for the conservation of building blocks and metabolic energy. The first example of this important principle of metabolic control appeared in studies of the biosynthesis of isoleucine in *E. coli*. The dehydration and deamination of threonine to α-ketobutyrate is the committed step. *Threonine deaminase,* the PLP enzyme that catalyzes this reaction, is allosterically inhibited by isoleucine.

$$\underset{\textbf{Threonine}}{H\text{—}\overset{NH_3^+}{\overset{|}{C}}\text{—}COO^- ;\ H\text{—}C\text{—}OH ;\ CH_3} \xrightarrow[\text{by isoleucine}]{\text{Inhibited}} \underset{\boldsymbol{\alpha}\textbf{-Ketobutyrate}}{O{=}C\text{—}COO^- ;\ CH_2 ;\ CH_3} \longrightarrow \longrightarrow \longrightarrow \longrightarrow \underset{\textbf{Isoleucine}}{H\text{—}\overset{NH_3^+}{\overset{|}{C}}\text{—}COO^- ;\ H\text{—}C\text{—}CH_3 ;\ CH_2 ;\ CH_3}$$

Likewise, tryptophan inhibits the enzyme complex that catalyzes the first two steps in the conversion of chorismate into tryptophan.

Consider a branched biosynthetic pathway in which Y and Z are the final products:

$$A \longrightarrow B \longrightarrow C \begin{cases} \nearrow D \longrightarrow E \longrightarrow Y \\ \searrow F \longrightarrow G \longrightarrow Z \end{cases}$$

Suppose that high levels of Y *or* Z completely inhibit the first common step (A $\rightarrow$ B). Then, high levels of Y would prevent the synthesis of Z even if there were a deficiency of Z. Such a regulatory scheme is obviously not optimal. In fact, several intricate feedback control mechanisms have been found in branched biosynthetic pathways:

1. *Sequential feedback control.* The first common step (A $\rightarrow$ B) is not inhibited directly by Y or Z. Rather, these final products inhibit the reactions leading away from the point of branching: Y inhibits the C $\rightarrow$ D step, and Z inhibits the C $\rightarrow$ F step. In turn, high levels of C inhibit the A $\rightarrow$ B step. Thus, the first common reaction is blocked only if both final products are present in excess.

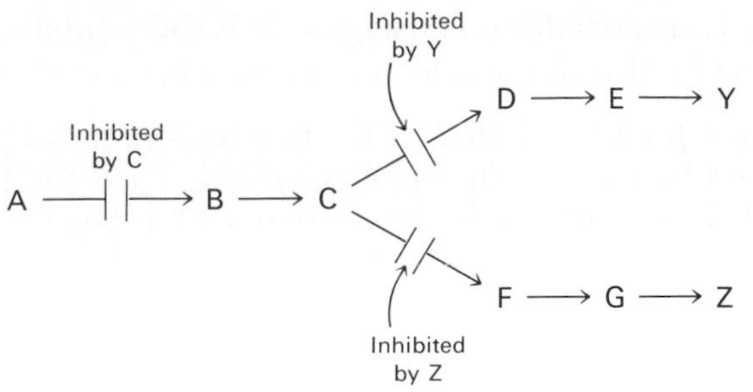

Inhibited by Y

D ⟶ E ⟶ Y

Inhibited by C

A ⟶ B ⟶ C

F ⟶ G ⟶ Z

Inhibited by Z

Sequential feedback control regulates the synthesis of aromatic amino acids in *Bacillus subtilis*. The first divergent steps in the synthesis of phenylalanine, tyrosine, and tryptophan are inhibited by their final products. If all three are present in excess, chorismate and prephenate accumulate. These branch-point intermediates in turn inhibit the first common step in the overall pathway, which is the condensation of phosphoenolpyruvate and erythrose 4-phosphate.

2. *Enzyme multiplicity*. The distinguishing feature of this mechanism is that the first common step (A → B) is catalyzed by two different enzymes. One of them is directly inhibited by Y, and the other by Z. Thus, both Y and Z must be present at high levels to prevent the conversion of A into B completely. In the rest of this control scheme, as in sequential feedback control, Y inhibits the C → D step and Z inhibits the C → F step.

Inhibited by Y

Inhibited by Y

D ⟶ E ⟶ Y

A B ⟶ C

Inhibited by Z

F ⟶ G ⟶ Z

Inhibited by Z

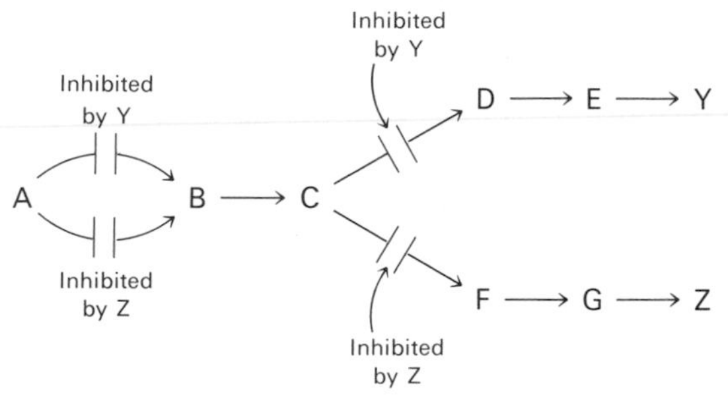

Differential inhibition of multiple enzymes controls a variety of biosynthetic pathways in microorganisms. In *E. coli*, the condensation of phosphoenolpyruvate and erythrose 4-phosphate is catalyzed by three different enzymes. One is inhibited by phenylalanine, another by tyrosine, and the third by tryptophan. Furthermore, there are two different mutases that convert chorismate into prephenate. One of them is inhibited by phenylalanine, the other by tyrosine.

3. *Concerted feedback control*. The first common step (A → B) is inhibited only if high levels of Y and Z are simultaneously present. A high level of either product alone does not inhibit the A → B step. As in the two control schemes just discussed, Y inhibits the C → D step and Z inhibits the C → F step.

Inhibited by Y

D ⟶ E ⟶ Y

Inhibited by (Y + Z)

A ⟶ B ⟶ C

F ⟶ G ⟶ Z

Inhibited by Z

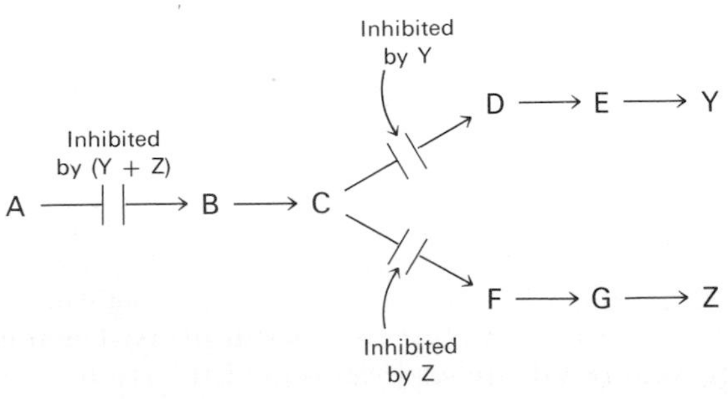

An example of concerted feedback control is the inhibition of bacterial aspartyl kinase by threonine and lysine, the final products.

4. *Cumulative feedback control.* The first common step ($A \rightarrow B$) is partly inhibited by each of the final products. Each final product acts independently of the others. Suppose that a high level of Y decreased the rate of the $A \rightarrow B$ step from 100 to 60 s^{-1} and that Z alone decreased the rate from 100 to 40 s^{-1}. Then, the rate of the $A \rightarrow B$ step in the presence of high levels of Y and Z would be 24 s^{-1} ($0.6 \times 0.4 \times 100\ s^{-1}$).

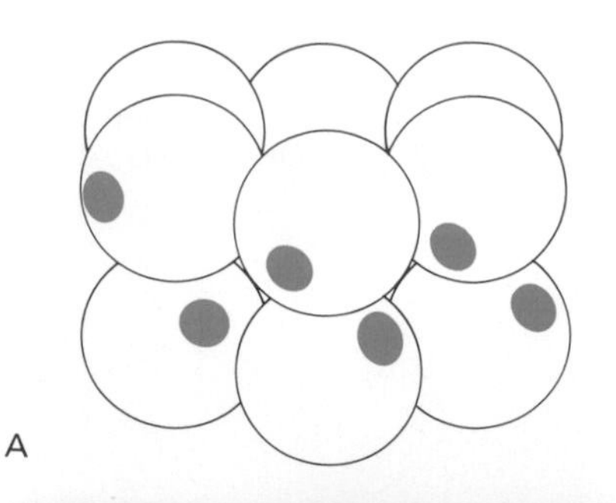

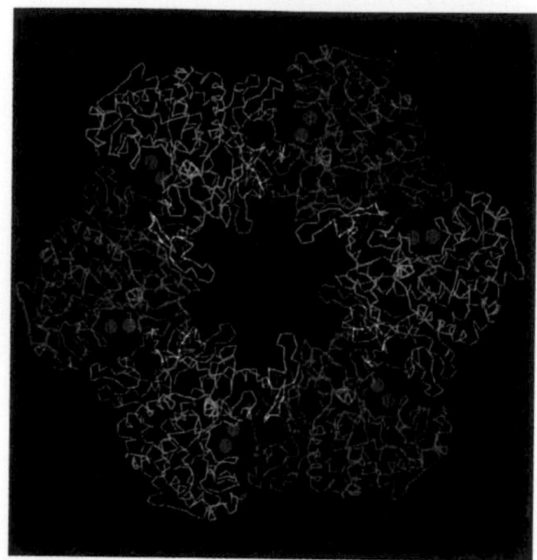

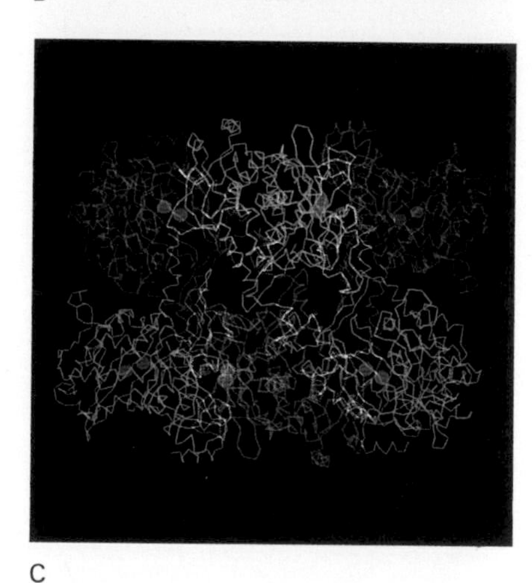

Figure 24-16
Three-dimensional structure of glutamine synthetase from *E. coli.* (A) Schematic diagram. The twelve identical subunits are arranged in two hexameric rings joined face to face. The adenylylation sites are marked. The outer diameter of the enzyme is 143 Å. (B) A view down the sixfold axis of one of the hexameric rings, (α-carbons only). Adjacent chains are shown in different colors. The two Mn^{2+} ions in each catalytic site are shown in blue. Catalytic sites are located at subunit interfaces. Adenylylation sites are more than 20 Å away from catalytic sites. (C) A view of the molecule along a twofold axis (perpendicular to that shown in B). [Courtesy of Dr. David Eisenberg.]

THE ACTIVITY OF GLUTAMINE SYNTHETASE IS MODULATED BY AN ENZYMATIC CASCADE

The regulation of glutamine synthetase in *E. coli* is a striking example of *cumulative feedback inhibition.* Recall that glutamine is synthesized from glutamate, NH_4^+, and ATP (p. 577). Glutamine synthetase consists of twelve identical 50-kd subunits arranged in two hexagonal rings that face each other (Figure 24-16). Earl Stadtman showed that this enzyme regulates the flow of nitrogen and hence plays a key role in controlling bacterial metabolism. The amide group of glutamine is a source of nitrogen in the biosyntheses of a variety of compounds, such as tryptophan, histidine, carbamoyl phosphate, glucosamine 6-phosphate, CTP, and AMP. Glutamine synthetase is cumulatively inhibited by each of these final products of glutamine metabolism, as well as by alanine and glycine. The enzymatic activity of glutamine synthetase is almost completely switched off when all eight final products are bound to the enzyme.

The activity of glutamine synthetase is also controlled by *reversible covalent modification*—the attachment of an *AMP unit* by a phosphodiester bond to the hydroxyl group of a specific tyrosine residue in each subunit (Figure 24-17). *This adenylylated enzyme is more susceptible to cumulative feedback inhibition than is the deadenylylated form.* The covalently attached AMP unit is removed from the adenylylated enzyme by phosphorolysis. An interesting feature of these reactions is that they are catalyzed by the same enzyme, *adenylyl transferase.* What determines whether it inserts or removes an AMP unit? The specificity of adenylyl transferase turns out to be controlled by a *regulatory protein* (designated P), which can exist in two forms, P_A and P_D. The complex of P_A and

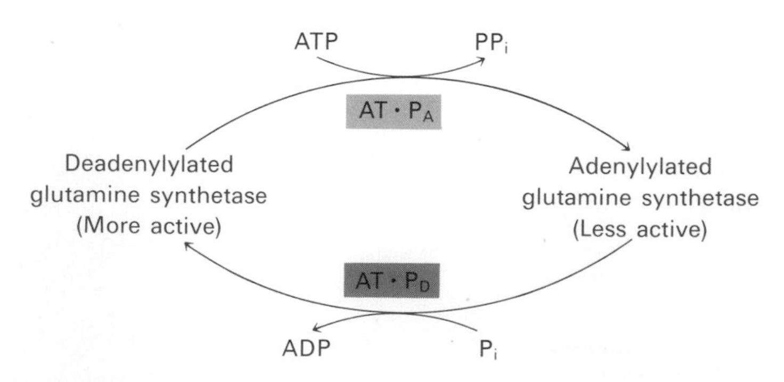

Figure 24-17
Control of the activity of glutamine synthetase by reversible covalent modification. Adenylylation is catalyzed by a complex of adenylyl transferase (AT) and one form of a regulatory protein (P_A). The same enzyme catalyzes deadenylylation when it is complexed with the other form (P_D) of the regulatory protein.

adenylyl transferase catalyzes the attachment of an AMP unit to glutamine synthetase, which reduces its activity. Conversely, the complex of P_D and adenylyl transferase removes AMP from the adenylylated enzyme. These opposing catalytic activities of adenylyl transferase are carried out by different catalytic sites on the same polypeptide chain—a motif encountered earlier in the control of the levels of fructose 2,6-bisphosphate (p. 361) and isocitrate dehydrogenase (p. 389).

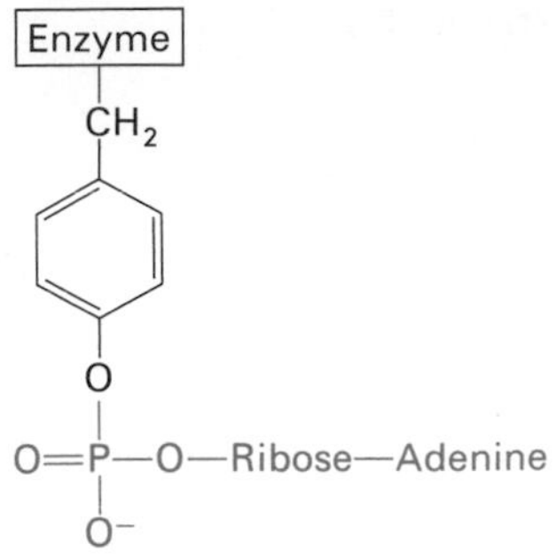

Adenylylated enzyme

This brings us to another level of reversible covalent modification. P_A is converted into P_D by the attachment of uridine monophosphate (UMP) (Figure 24-18). This reaction, which is catalyzed by *uridylyl transferase,* is stimulated by ATP and α-ketoglutarate, whereas it is inhibited by glutamine. In turn, the two UMP units on P_D are removed by hydrolysis, a reaction promoted by glutamate and inhibited by α-ketoglutarate. These opposing catalytic activities also are present on a single polypeptide chain and are controlled so that the enzyme does not simultaneously catalyze uridylylation and hydrolysis.

Why is an enzymatic cascade used to regulate glutamine synthetase? One advantage of a cascade is that it *amplifies signals,* as in blood clotting (p. 248) and the control of glycogen metabolism (p. 462). Another reason is that the *potential for allosteric control is markedly increased when each enzyme in the cascade is an independent target for regulation.* The integration of nitrogen metabolism in a cell requires that a large number of input signals be detected and processed. There are limits to what a single protein can accomplish on its own—even a molecule as sentient as glutamine synthetase! The evolution of a cascade provided many more regulatory sites and made possible a finer tuning of the flow of nitrogen in the cell.

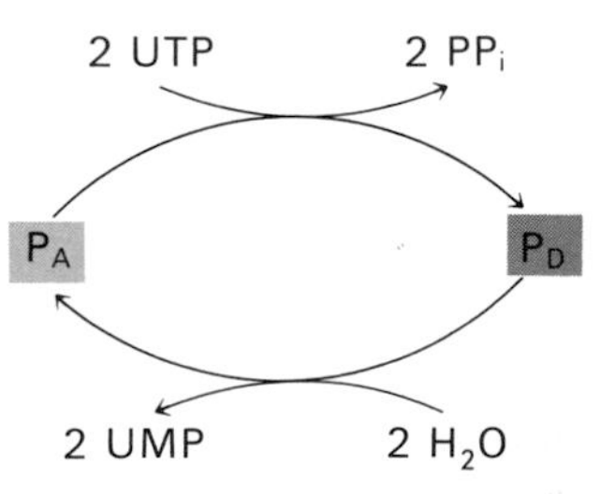

Figure 24-18
A higher level in the regulatory cascade of glutamine synthetase. P_A and P_D, the regulatory proteins that control the specificity of glutamine synthetase, are interconvertible. P_A is converted into P_D by uridylylation, which is reversed by hydrolysis. The enzymes catalyzing these reactions sense the concentrations of metabolic intermediates.

AMINO ACIDS ARE PRECURSORS OF A VARIETY OF BIOMOLECULES

Amino acids are the building blocks of proteins and peptides. They also serve as precursors of many kinds of small molecules that have important biological roles. Let us briefly survey some of the biomolecules that are derived from amino acids (Figure 24-19). *Purines* and *pyrimidines* are derived in part from amino acids. The biosynthesis of these precursors

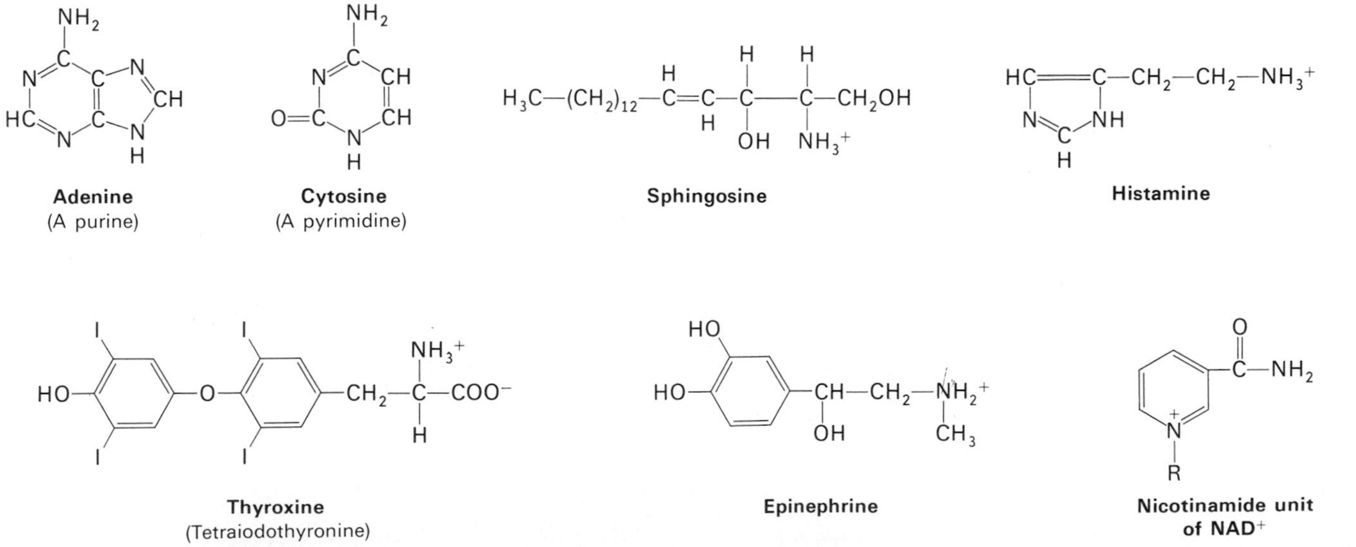

Adenine (A purine)

Cytosine (A pyrimidine)

Sphingosine

Histamine

Thyroxine (Tetraiodothyronine)

Epinephrine

Nicotinamide unit of NAD^+

Figure 24-19
Biomolecules derived from amino acids.

of DNA, RNA, and numerous coenzymes is discussed in detail in the next chapter. Six of the nine atoms of the purine rings and four of the six atoms of the pyrimidine ring are derived from amino acids. The reactive terminus of *sphingosine,* an intermediate in the synthesis of sphingolipids, comes from serine. *Histamine,* a potent vasodilator, is derived from histidine by decarboxylation. Tyrosine is a precursor of the hormones *thyroxine* (tetraiodothyronine) and *epinephrine* and of *melanin,* a polymeric pigment. The neurotransmitter 5-hydroxytryptamine (*serotonin*) and the *nicotinamide ring* of NAD^+ are synthesized from tryptophan. Glutamine contributes the amide group of the nicotinamide moiety.

GLUTATHIONE, A γ-GLUTAMYL PEPTIDE, SERVES AS A SULFHYDRYL BUFFER AND AMINO ACID TRANSPORTER

Glutathione, a tripeptide containing a sulfhydryl group, is a highly distinctive amino acid derivative with several important roles. For example, glutathione protects red cells from oxidative damage (p. 436). The first step in the synthesis of glutathione is the formation of a peptide linkage between the γ-carboxyl group of glutamate and the amino group of cysteine, in a reaction catalyzed by *γ-glutamylcysteine synthetase* (Figure 24-20). Formation of this peptide bond requires activation of the γ-carboxyl group, which is achieved by ATP. The resulting acylphosphate intermediate is then attacked by the amino group of cysteine. This reaction is feedback inhibited by glutathione. In the second step, which is catalyzed by *glutathione synthetase,* ATP activates the carboxyl group of cysteine to enable it to condense with the amino group of glycine.

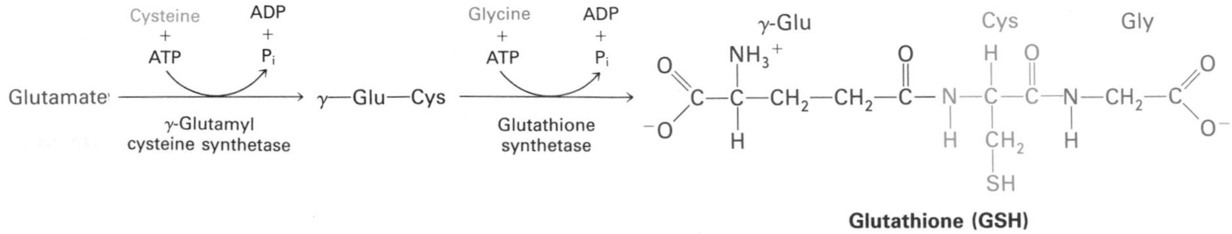

Glutamate + Cysteine + ATP → (γ-Glutamyl cysteine synthetase) → γ—Glu—Cys + ADP + P_i
γ—Glu—Cys + Glycine + ATP → (Glutathione synthetase) → Glutathione (GSH) + ADP + P_i

Glutathione (GSH)

Figure 24-20
Synthesis of glutathione.

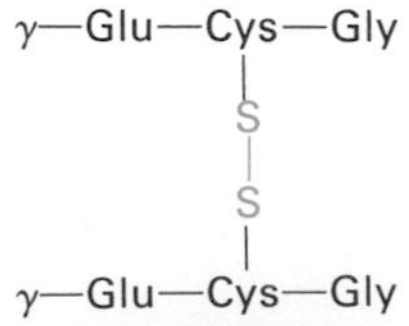

γ—Glu—Cys—Gly
S—S
γ—Glu—Cys—Gly
Oxidized glutathione (GSSG)

Glutathione, present at high levels (~5 mM) in animal cells, serves as a sulfhydryl buffer. It cycles between a reduced thiol form (GSH) and an oxidized form (GSSG) in which two tripeptides are linked by a disulfide bond. GSSG is reduced to GSH by glutathione reductase, a flavoprotein utilizing NADPH as the electron source (p. 437). The ratio of GSH to GSSG in most cells is greater than 500.

Glutathione plays a key role in detoxification by reacting with hydrogen peroxide and organic peroxides, the harmful byproducts of aerobic life.

$$2\ GSH + R\text{—}O\text{—}OH \longrightarrow GSSG + H_2O + ROH$$

Glutathione peroxidase, the enzyme catalyzing this reaction, is remarkable in having a covalently attached *selenium* (Se) atom. Its active site contains the selenium analog of cysteine, in which Se has replaced S (Figure

24-21). The selenolate (E-Se$^-$) form of this residue reduces the peroxide substrate to an alcohol and is in turn oxidized to selenenic acid (E-SeOH) (Figure 24-22). Glutathione now comes into action by forming a selenosulfide adduct (E-Se-S-G). A second glutathione then regenerates the active form of the enzyme by attacking the selenosulfide to form oxidized glutathione.

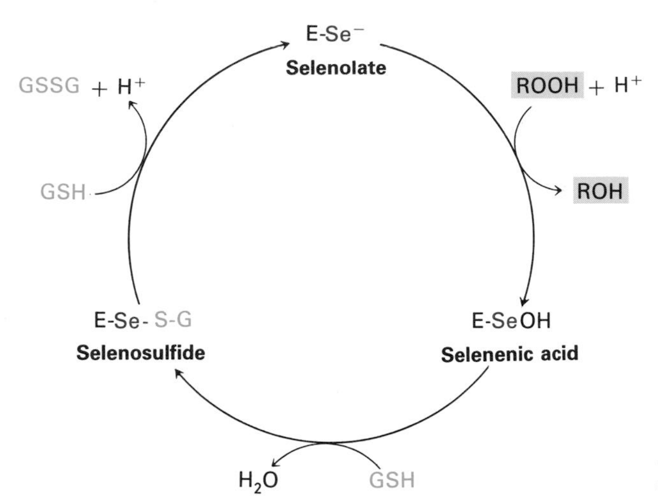

Figure 24-22
Proposed catalytic mechanism of glutathione peroxidase. [Based on O. Epp, R. Ladenstein, and A. Wendel. *Eur. J. Biochem.* 133(1983):51.]

Figure 24-21
Model of the selenocysteine residue in the active site of glutathione peroxidase. The selenium atom is shown in green.

Glutathione also participates in the transport of amino acids. A membrane-bound enzyme, *γ-glutamyl transpeptidase*, catalyzes the transfer of the γ-glutamyl group of glutathione to the α-amino group of an acceptor amino acid such as cystine or glutamine (Figure 24-23). The catalytic site of this enzyme is located on the extracellular side of the plasma membrane of kidney cells. Glutathione is translocated across the plasma membrane to undergo this reaction. γ-Glutamyl amino acids then are taken up by cells of other organs and cyclized to 5-oxoproline, and the transported amino acid is released. Finally, the peptide bond of 5-oxoproline is hydrolyzed in an ATP-requiring reaction that regenerates glutamate. Alton Meister showed that these reactions give rise to a *γ-glutamyl cycle* (Figure 24-24).

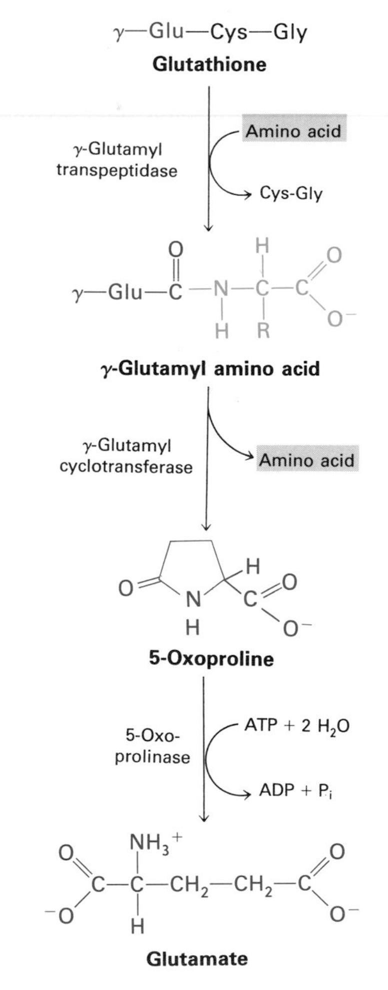

Figure 24-23
Glutathione is converted into glutamate in the transport of amino acids.

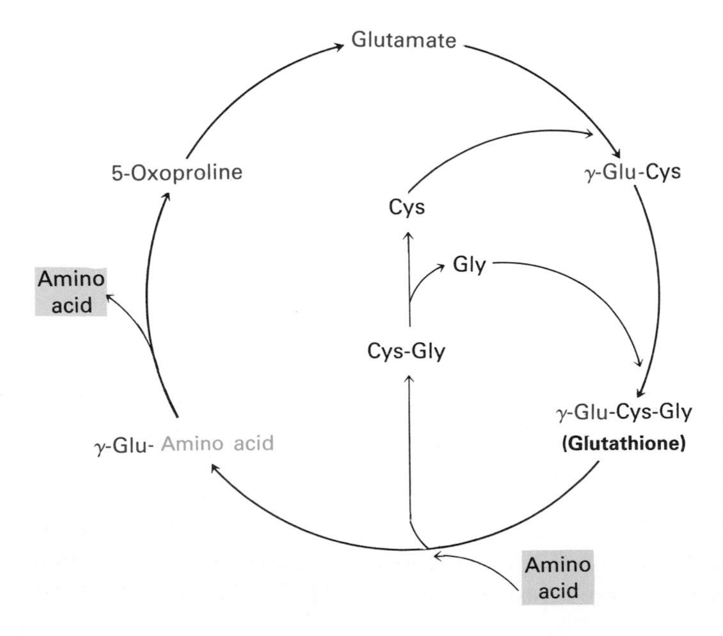

Figure 24-24
γ-Glutamyl cycle. An amino acid is transported from one cell to another as a γ-glutamyl derivative. All reactions of the cycle occur in the cytosol of cells except for the formation of the γ-glutamyl amino acid, which takes place on the extracellular face of the plasma membrane.

PORPHYRINS IN MAMMALS ARE SYNTHESIZED FROM GLYCINE AND SUCCINYL COENZYME A

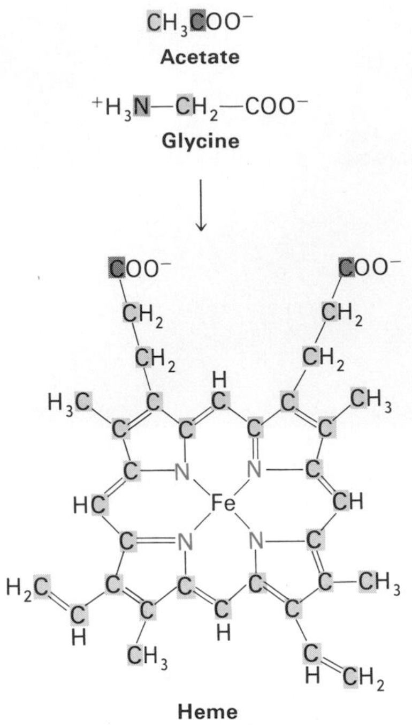

The involvement of an amino acid in the biosynthesis of the porphyrin rings of hemes and chlorophylls was first revealed by isotopic labeling experiments carried out by David Shemin and his colleagues. In 1945, they showed that the nitrogen atoms of heme were labeled after the feeding of ^{15}N-glycine to human subjects, whereas ^{15}N-glutamate resulted in very little labeling. Using carbon-14, which had just become available, they discovered that eight of the carbon atoms of heme in nucleated duck erythrocytes are derived from the α-carbon of glycine and none from the carboxyl carbon (Figure 24-25). Subsequent studies demonstrated that the other 26 carbon atoms of heme can arise from acetate. Moreover, the ^{14}C in methyl-labeled acetate emerged in 24 of these carbons, whereas the ^{14}C in carboxyl-labeled acetate appeared only in the other two. This highly distinctive labeling pattern led Shemin to propose that a heme precursor is formed by the condensation of glycine with an activated succinyl compound. In fact, *the first step in the biosynthesis of porphyrins in mammals is the condensation of glycine and succinyl CoA to form δ-aminolevulinate.*

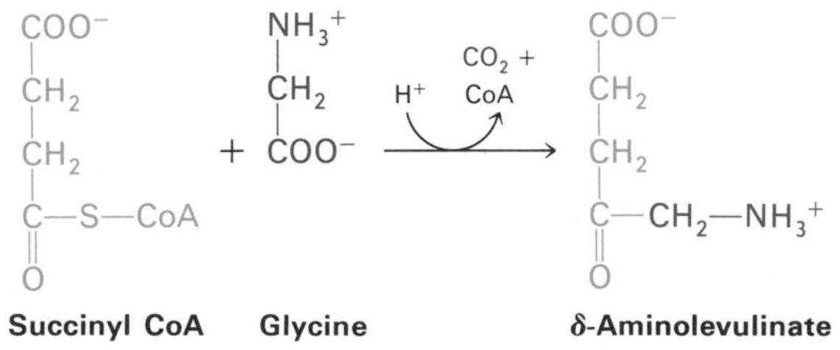

Figure 24-25
Labeling pattern of heme synthesized from glycine and acetate. The nitrogen atoms (blue) arise from the amino group of glycine. The origins of the carbon atoms are: yellow, from the α-carbon of glycine; green, mainly from the methyl carbon of acetate; and red, from the carboxyl carbon of acetate.

This reaction is catalyzed by δ-aminolevulinate synthase, a PLP enzyme in mitochondria. As might be expected, this committed step in the biosynthesis of porphyrins is regulated. Two molecules of δ-aminolevulinate then condense to form *porphobilinogen.* This dehydration reaction is catalyzed by δ-aminolevulinate dehydrase.

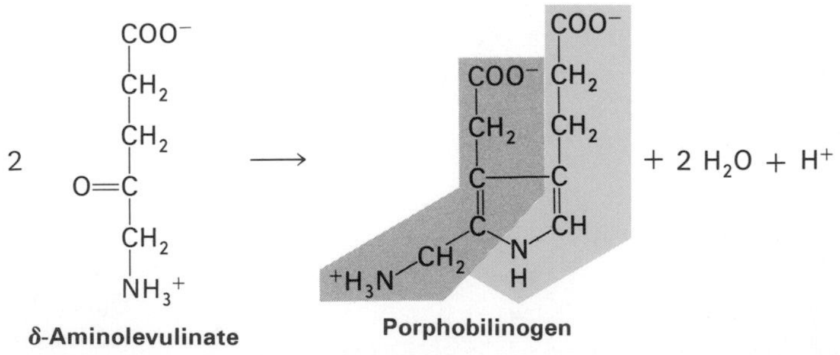

Four porphobilinogens condense head-to-tail to form a *linear tetrapyrrole,* which remains bound to the enzyme (Figure 24-26). An ammonium ion is released for each methylene bridge formed. This linear tetrapyrrole cyclizes by losing NH_4^+. The cyclic product is uroporphyrinogen III, which has an asymmetric arrangement of side chains. These reactions require a *synthase* and a *cosynthase.* In the presence of synthase alone, uroporphyrinogen I, the symmetric isomer, is produced. The cosynthetase is essential for isomerizing one of the pyrrole rings to yield asymmetric uroporphyrinogen III.

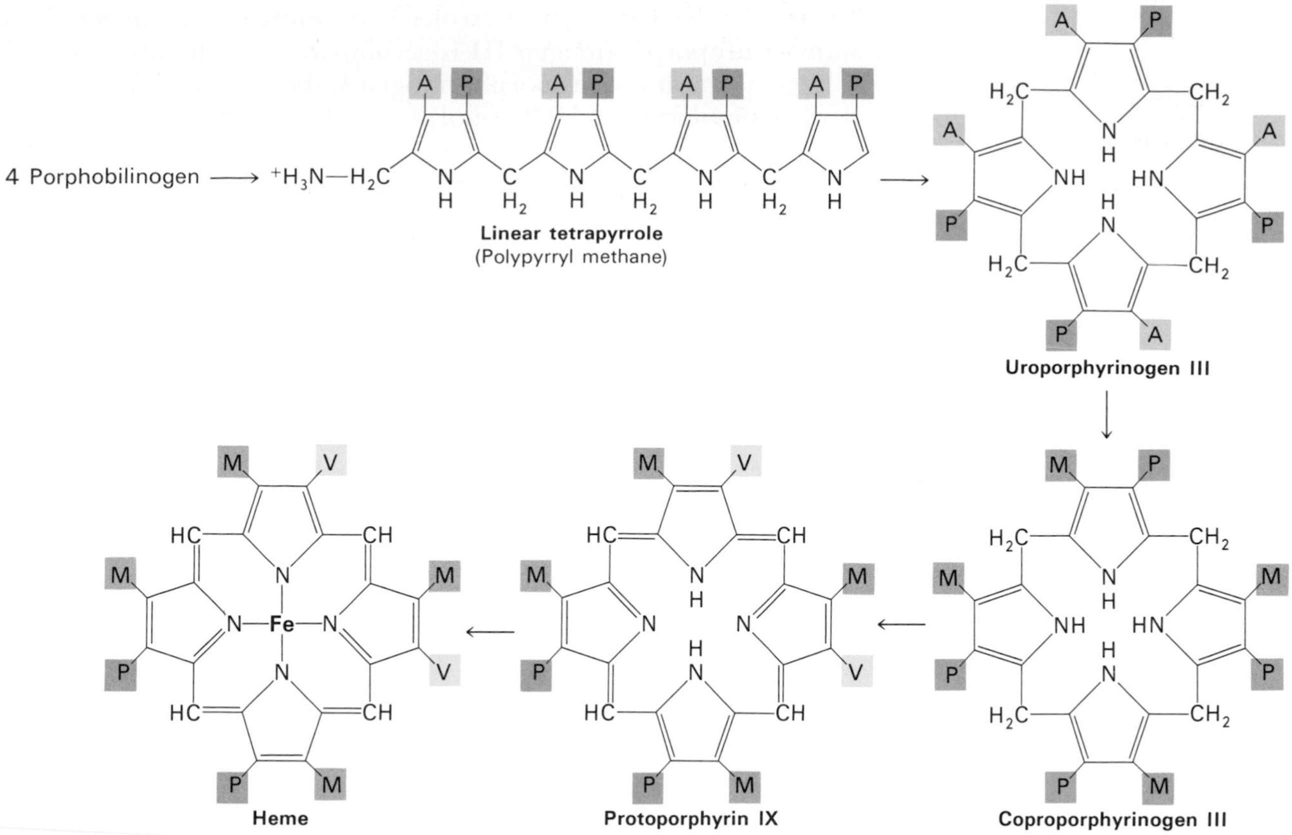

4 Porphobilinogen ⟶ $^{+}H_3N{-}H_2C{-}$ ... ⟶

Linear tetrapyrrole
(Polypyrryl methane)

Uroporphyrinogen III

Coproporphyrinogen III

Protoporphyrin IX

Heme

Figure 24-26
Pathway for the synthesis of heme from porphobilinogen. (Abbreviations: A, acetate; M, methyl; P, propionate; V, vinyl.)

The porphyrin skeleton is now formed. Subsequent reactions alter the side chains and the degree of saturation of the porphyrin ring (Figure 24-26). *Coproporphyrinogen III* is formed by decarboxylation of the acetate side chains. Desaturation of the porphyrin ring and conversion of two of the propionate side chains into vinyl groups yield *protoporphyrin IX*. Chelation of iron finally gives *heme*, the prosthetic group of proteins such as myoglobin, hemoglobin, catalase, peroxidase, and cytochrome *c*. The insertion of the *ferrous* form of iron is catalyzed by *ferrochelatase*. Iron is transported in the plasma by *transferrin*, a protein that binds two ferric ions, and stored in tissues inside molecules of *ferritin*. The large internal cavity (~80 Å diameter) of ferritin can hold as many as 4500 ferric ions.

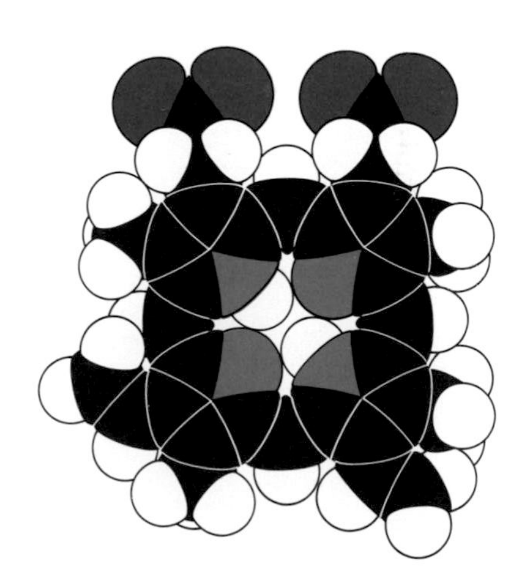

Figure 24-27
Model of protoporphyrin IX, the immediate precursor of heme.

Several factors regulating heme biosynthesis in animals have been elucidated. *δ-Aminolevulinate synthase, the enzyme that catalyzes the committed step in this pathway, is feedback inhibited by heme,* as is δ-aminolevulinate dehydrase and ferrochelatase. Regulation also occurs at the level of enzyme synthesis. *Heme represses the synthesis of δ-aminolevulinate synthase. The iron atom itself may be the active regulatory species.*

PORPHYRINS ACCUMULATE IN SOME INHERITED DISORDERS OF PORPHYRIN METABOLISM

Several inherited disorders of porphyrin metabolism are known. In *congenital erythropoietic porphyria,* there is a deficiency of uroporphyrinogen III synthase, the isomerase that yields the asymmetric isomer on cyclization of the linear tetrapyrrole. The synthesis of the required amount of uroporphyrinogen III is accompanied by the formation of very large quantities of uroporphyrinogen I, the symmetric isomer devoid of a physiologic role. Uroporphyrin I, coproporphyrin I, and other symmetric derivatives also accumulate. Erythrocytes are prematurely destroyed in this disease, which is transmitted as an autosomal recessive. The urine of patients having this disease is red because of the excretion of large amounts of uroporphyrin I. Their teeth exhibit a strong red fluorescence under ultraviolet light because of the deposition of porphyrins. Furthermore, their *skin is usually very sensitive to light.*

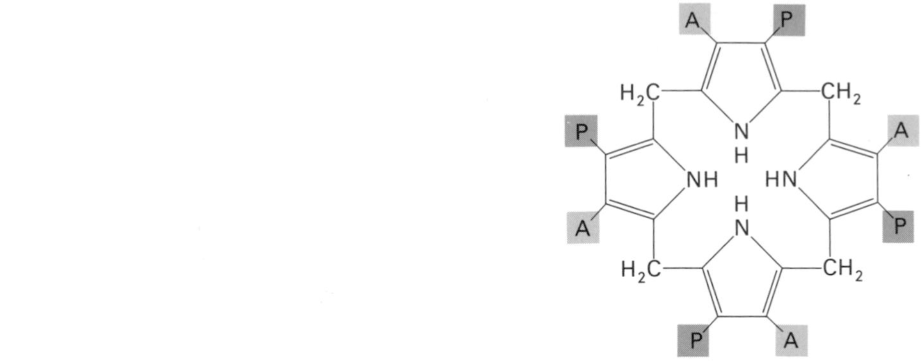

Uroporphyrinogen I

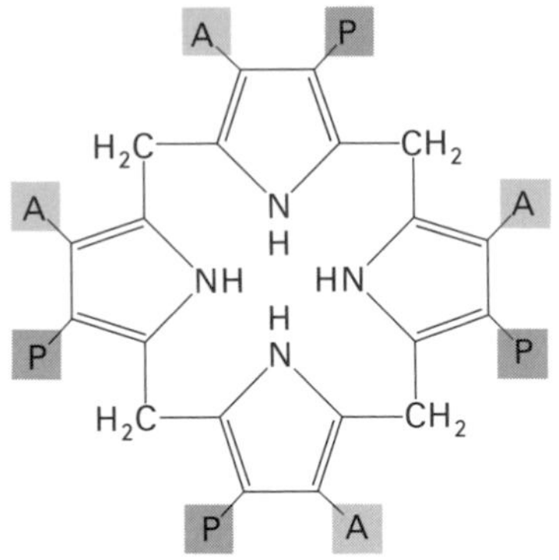

Uroporphyrinogen III

Acute intermittent porphyria is a quite different disease. The liver, rather than the red cells, is affected, and the skin is not typically photosensitive. The activity of uroporphyrinogen synthase is decreased in this disorder, and there is a compensatory increase in the level of δ-aminolevulinate synthase. Consequently, the concentrations of δ-aminolevulinate and porphobilinogen in the liver are increased, and so large amounts of these compounds are excreted in the urine. The disease is inherited as an *autosomal dominant.* The striking clinical symptoms are intermittent abdominal pain and neurologic disturbances. As its name implies, the disease is episodic in its clinical expression. Acute attacks are sometimes precipitated by drugs such as barbiturates and estrogens.

BILIVERDIN AND BILIRUBIN ARE INTERMEDIATES IN THE BREAKDOWN OF HEME

The normal human erythrocyte has a life span of about 120 days. Old cells are removed from the circulation and degraded by the spleen. The apoprotein of hemoglobin is hydrolyzed to its constituent amino acids. The first step in the degradation of the heme group to bilirubin (Figure 24-28) is the cleavage of its α-methene bridge to form *biliverdin,* a linear tetrapyrrole. This reaction is catalyzed by *heme oxygenase.* Two aspects of this reaction are noteworthy. First, this enzyme is a *monooxygenase:* O_2 and NADPH are required for the cleavage reaction. Second, a methene-bridge carbon is released as *carbon monoxide.* This endogenous production of CO posed a special problem in the evolution of oxygen carriers (p. 149). The central methene bridge of biliverdin is then reduced by biliverdin reductase to form *bilirubin.* Again, the reductant is NADPH. The changing color of a bruise is a highly visible indicator of these degradative reactions.

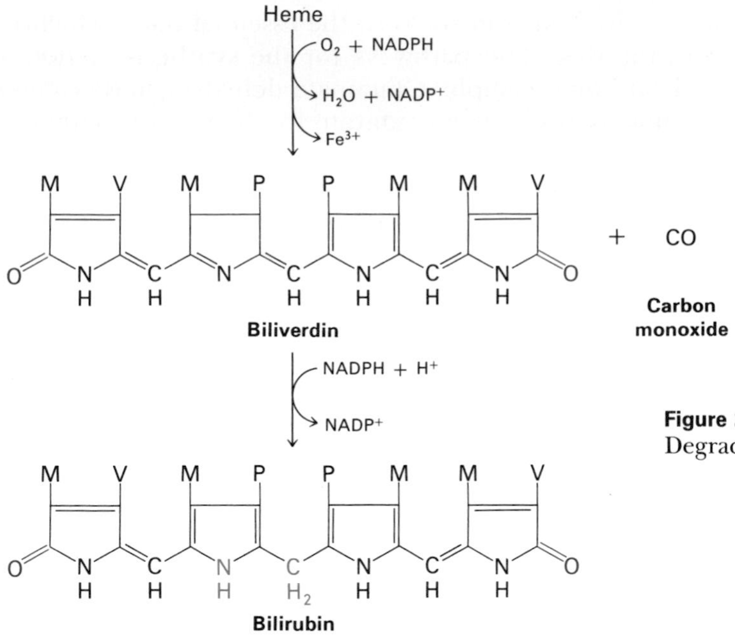

Heme

O_2 + NADPH

H_2O + $NADP^+$

Fe^{3+}

Biliverdin

+ CO

Carbon monoxide

NADPH + H^+

$NADP^+$

Bilirubin

Figure 24-28
Degradation of heme to bilirubin.

Bilirubin complexed to serum albumin is transported to the liver, where it is rendered more soluble by the attachment of sugar residues to its propionate side chains. The solubilizing sugar is *glucuronate,* which differs from glucose in having a COO^- group at C-6 rather than a CH_2OH group. The conjugate of bilirubin and two glucuronates, called *bilirubin diglucuronide,* is secreted into bile. *UDP-glucuronate, derived from the oxidation of UDP-glucose, is the activated intermediate in the synthesis of bilirubin diglucuronide.* Thus, the iron atom of heme is recycled, whereas the organic moiety is converted into a soluble, open-chain form that is excreted.

Bilirubin is much less soluble in aqueous media than is biliverdin. In reptiles and birds, the end product of heme catabolism is biliverdin rather than bilirubin. Why do mammals reduce biliverdin to bilirubin, a compound that poses a solubility problem? Recent studies have shown that *bilirubin is a very effective antioxidant,* whereas biliverdin is not. In scavenging two hydroperoxy radicals, bilirubin is oxidized to biliverdin, which is then rapidly reduced to again form bilirubin. On a molar basis, bilirubin bound to albumin is about a tenth as effective as ascorbate (vitamin C) in affording protection against water-soluble peroxides. Bilirubin, urate (p. 622), and ascorbate are the three principal antioxidants in plasma. In membranes, bilirubin is a highly potent antioxidant, rivaling vitamin E in this regard. These studies reveal that *the end product of a degradative pathway may be selected in evolution to exert a beneficial action.*

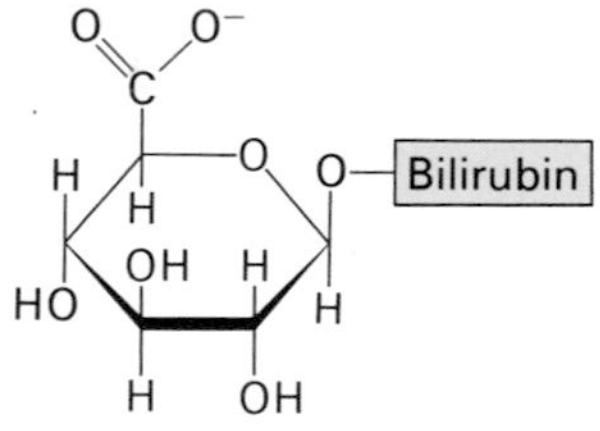

Bilirubin

A glucuronate unit in bilirubin diglucuronide

SUMMARY

Microorganisms use ATP and reduced ferredoxin, a powerful reductant, to convert N_2 into NH_4^+. Higher organisms consume NH_4^+ to synthesize amino acids, nucleotides, and other nitrogen-containing biomolecules. The major points of entry of NH_4^+ into metabolism are glutamine, glutamate, and carbamoyl phosphate. Humans can synthesize eleven of the basic set of twenty amino acids. These amino acids are called nonessential, in contrast with the essential ones, which must be supplied in the diet. The pathways for the synthesis of nonessential amino acids are quite simple. Glutamate dehydrogenase catalyzes the reductive amination of α-ketoglutarate to glutamate. Alanine and aspartate are synthesized by transamination of pyruvate and oxaloacetate, respectively. Glutamine is synthesized from NH_4^+ and glutamate, and asparagine is synthesized similarly. Proline and arginine are derived from glutamate. Serine, formed from 3-phosphoglycerate, is the precursor of glycine and cysteine. Tyrosine is synthesized by the hydroxylation of phenylalanine, an essential amino acid. The pathways for the biosynthesis of essential amino acids are much more complex than for the nonessential ones. Most of these pathways are regulated by feedback inhibition, in which the committed step is allosterically inhibited by the final product. The regulation of glutamine synthetase from *E. coli* provides a striking demonstration of cumulative feedback inhibition and of control by a cascade of reversible covalent modifications.

Tetrahydrofolate, a carrier of activated one-carbon units, plays an important role in the metabolism of amino acids and nucleotides. This coenzyme carries one-carbon units at three oxidation states, which are interconvertible: most reduced—methyl; intermediate—methylene; most oxidized—formyl, formimino, and methenyl. The major donor of activated methyl groups is *S*-adenosylmethionine, which is synthesized by the transfer of an adenosyl group from ATP to the sulfur atom of methionine. *S*-Adenosylhomocysteine is formed when the activated methyl group is transferred to an acceptor. It is hydrolyzed to adenosine and homocysteine, which is then methylated to methionine to complete the activated methyl cycle.

Amino acids are precursors of a variety of biomolecules. Glutathione (γ-Glu-Cys-Gly) serves as a sulfhydryl buffer, amino acid transporter, and detoxifying agent. Glutathione peroxidase, a selenoenzyme, catalyzes the reduction of hydrogen peroxide and organic peroxides by glutathione. Porphyrins are synthesized from glycine and succinyl CoA, which condense to give δ-aminolevulinate. This intermediate condenses with itself to form porphobilinogen. Four porphobilinogens combine to form a linear tetrapyrrole, which cyclizes to form uroporphyrinogen III. Oxidation and side-chain modifications lead to the synthesis of protoporphyrin IX, which acquires an iron atom to form heme. δ-Aminolevulinate synthase, the enzyme catalyzing the committed step in this pathway, is feedback inhibited by heme. This prosthetic group is degraded by a monooxygenase that converts it into biliverdin, a linear tetrapyrrole. Reduction of biliverdin by NADPH yields bilirubin, a potent scavenger of peroxides. Bilirubin is rendered soluble for excretion by the formation of the diglucuronide derivative.

SELECTED READINGS

Books

Bender, D. A., 1985. *Amino Acid Metabolism* (2nd ed.). Wiley.

Stanbury, J. B., Wyngaarden, J. B., Fredrickson, D. S., Goldstein, J. L., and Brown, M. S., (eds.), 1983. *The Metabolic Basis of Inherited Disease* (5th ed.), McGraw-Hill. [Contains many excellent articles on disorders of amino acid, porphyrin, and heme metabolism.]

Meister, A., 1965. *Biochemistry of the Amino Acids* (2nd ed.), vols. 1 and 2. Academic Press. [A comprehensive and authoritative treatise on amino acid metabolism.]

Evans, H. J., Bottomley, P. J., and Newton, W. E., (eds.), 1985. *Nitrogen Fixation Research Progress.* Martinus Nijhoff.

Blakley, R. L., and Benkovic, S. J., 1985. *Folates and Pterins,* vol. 2. Wiley. [This volume of an excellent series deals with the chemistry and biochemistry of folates and pterins. Includes several fine articles on phenylalanine hydroxylase.]

Blakley, R. L., 1969. *The Biochemistry of Folic Acid and Related Pteridines*. North-Holland.

Prusiner, S., and Stadtman, E. R., (eds.), 1973. *The Enzymes of Glutamine Metabolism*. Academic Press. [A valuable collection of articles on the role of glutamine in nitrogen metabolism.]

Nitrogen fixation

Orme-Johnson, W. H., 1985. Molecular basis of biological nitrogen fixation. *Ann. Rev. Biophys. Biophys. Chem.* 14:419–459.

Shah, V. K., Ugalde, R. A., Imperial, J., and Brill, W. J., 1984. Molybdenum in nitrogenase. *Ann. Rev. Biochem.* 53:231–257.

Emerich, D. W., Hageman, R. V., and Burris, R. H., 1981. Interactions of dinitrogenase and dinitrogenase reductase. *Advan. Enzymol.* 52:1–22.

One-carbon metabolism

Walsh, C., 1979. *Enzymatic Reaction Mechanisms*. W. H. Freeman. [Chapter 25 provides an excellent account of reaction mechanisms in one-carbon metabolism.]

Ridley, W. P., Dizikes, L. J., and Wood, J. M., 1977. Biomethylation of toxic elements in the environment. *Science* 197:329–332.

Amino acid biosynthesis and regulation

Cooper, A. J. L., 1983. Biochemistry of sulfur-containing amino acids. *Ann. Rev. Biochem.* 52:187–222.

Umbarger, H. E., 1978. Amino acid biosynthesis and its regulation. *Ann. Rev. Biochem.* 47:533–606. [An excellent review of amino acid biosynthesis in bacteria.]

Yanofsky, C., and Crawford, I. P., 1972. Tryptophan synthetase. *In* P. D. Boyer (ed.), *The Enzymes* (3rd ed.), vol. 7, pp. 1–31. Academic Press.

Glutamine synthetase

Stadtman, E. R., and Ginsburg, A., 1974. The glutamine synthetase of *Escherichia coli:* structure and control. *In* P. D. Boyer (ed.), *The Enzymes* (3rd ed.), vol. 10, pp. 755–807. Academic Press.

Almassy, R. J., Janson, C. A., Hamlin, R., Xuong, N.-H., and Eisenberg, D., 1986. Novel subunit-subunit interactions in the structure of glutamine synthetase. *Nature* 323:304–309.

Rhee, S. G., Park, R., Chock, P. B., and Stadtman, E. R., 1978. Allosteric regulation of monocyclic interconvertible enzyme cascade systems: use of *Escherichia coli* glutamine synthetase as an experimental model. *Proc. Nat. Acad. Sci.* 75:3138–3142.

Glutathione

Meister, A., and Anderson, M. E., 1983. Glutathione. *Ann. Rev. Biochem.* 52:711–760.

Meister, A., 1983. Selective modification of glutathione metabolism. *Science* 220:472–477.

Epp, O., Ladenstein, R., and Wendel, A., 1983. The refined structure of the selenoenzyme glutathione peroxidase at 0.2-nm resolution. *Eur. J. Biochem.* 133:51–69.

Biosynthesis of porphyrins, heme, and chlorophyll

Shemin, D., 1982. From glycine to heme. *In* Kaplan, N. O., and Robinson, A. (eds.), *From Cyclotrons to Cytochromes*, pp. 117–129. Academic Press. [A sensitive and reflective account of the early use of isotopes in unraveling this biosynthetic pathway.]

Smith, K. M., (ed.), 1976. *Porphyrins and Metalloporphyrins*. Elsevier.

Porra, R. J., and Meisch, H.-U., 1984. The biosynthesis of chlorophyll. *Trends Biochem. Sci.* 9:99–104.

Granick, S., and Beale, S. I., 1978. Hemes, chlorophylls, and related compounds: biosynthesis and metabolic regulation. *Advan. Enzymol.* 40:33–203. [An excellent review article.]

Antioxidant role of bilirubin

Stocker, R., Yamamoto, Y., McDonagh, A. F., Glazer, A. N., and Ames, B. N., 1987. Bilirubin is an antioxidant of possible physiologic importance. *Science* 235:1043–1046.

Stocker, R., Glazer, A. N., and Ames, B. N., 1987. Antioxidant activity of albumin-bound bilirubin. *Proc. Nat. Acad. Sci.* 84:5918–5922.

PROBLEMS

1. Write a balanced equation for the synthesis of alanine from glucose.

2. What are the intermediates in the flow of nitrogen from N_2 to heme?

3. Which derivative of folate is a reactant in each of the following conversions?
   (a) Glycine → serine.
   (b) Histidine → glutamate.
   (c) Homocysteine → methionine.

4. In the reaction catalyzed by glutamine synthetase, an oxygen atom is transferred from the side chain of glutamate to orthophosphate, as shown by ^{18}O-labeling studies. Propose an interpretation for this finding.

5. Isovaleric acidemia is an inherited disorder of leucine metabolism caused by a deficiency of isovaleryl-CoA dehydrogenase. Many infants having this disease die in the first month of life. It has recently been reported that the administration of large amounts of

glycine to two infants with this disease led to marked clinical improvement. What is the rationale for the use of glycine therapy?

6. Nitrogenase catalyzes the formation of HD in the presence of D_2, a strong reductant, and ATP. Furthermore, H–D exchange depends on the presence of N_2. Propose an intermediate in N_2 fixation that would account for these observations.

7. Blue-green algae form *heterocysts* when deprived of ammonia and nitrate. They lack nuclei and are attached to adjacent vegetative cells. Heterocysts have photosystem I activity but are entirely devoid of photosystem II activity. What is their role?

8. Most cytosolic proteins lack disulfide bonds, whereas extracellular proteins usually contain them. Why?

9. Methionine sulfoximine, a convulsion-causing agent isolated from some batches of bleached flour, is a potent inhibitor of glutamine synthetase and γ-glutamylcysteine synthetase. Propose a mechanism of action for this inhibitor.

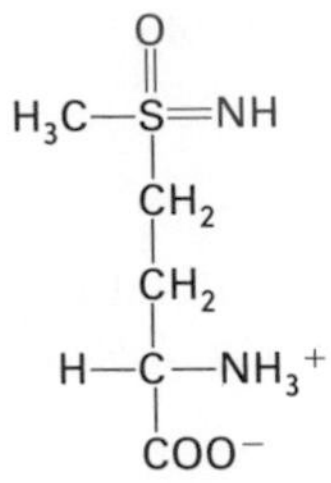

$$H_3C-S(=O)(=NH)-CH_2-CH_2-CH(NH_3^+)-COO^-$$

Methionine sulfoximine

10. An absence of glutathione synthetase leads to the massive excretion of 5-oxoproline in the urine. Why?

CHAPTER 25

Biosynthesis of Nucleotides

This chapter deals with the biosynthesis of nucleotides. These compounds play key roles in nearly all biochemical processes:

1. They are the *activated precursors of DNA and RNA.*

2. Nucleotide derivatives are *activated intermediates in many biosyntheses.* For example, UDP-glucose and CDP-diacylglycerol are precursors of glycogen and phosphoglycerides, respectively. *S*-Adenosylmethionine carries an activated methyl group.

3. ATP, an adenine nucleotide, is a *universal currency of energy* in biological systems. GTP powers many movements of macromolecules, such as the translocation of nascent peptide chains on ribosomes and the activation of signal-coupling proteins.

4. Adenine nucleotides are *components of three major coenzymes:* NAD^+, FAD, and CoA.

5. Nucleotides are *metabolic regulators.* Cyclic AMP is a ubiquitous mediator of the action of many hormones. Covalent modifications introduced by ATP alter the activities of some enzymes, as exemplified by the phosphorylation of glycogen synthase and the adenylylation of glutamine synthetase.

Inhibitors of nucleotide biosynthesis are being used as anti-cancer drugs. For example, *methotrexate* (an analog of folate) inhibits the formation of deoxythymidylate, a precursor of one of the four building blocks of DNA. *Azidothymidine* (AZT), an inhibitor of reverse transcription, is being used to treat AIDS.

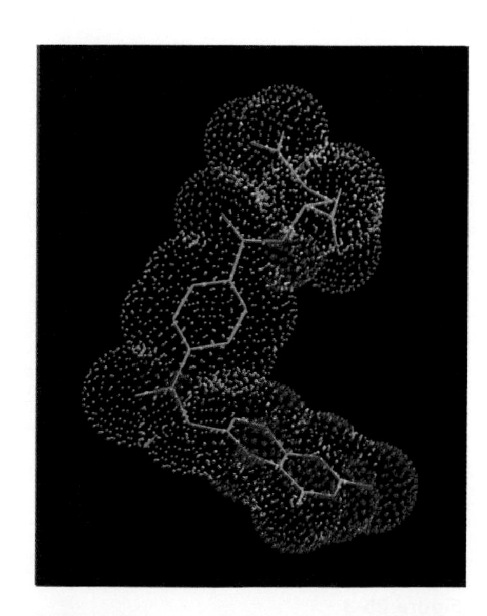

Figure 25-1
Model of methotrexate, a folate analog that is used to treat some cancers.

NOMENCLATURE OF BASES, NUCLEOSIDES, AND NUCLEOTIDES

The nomenclature of nucleotides and their constituent units was presented earlier (p. 72). Recall that a *nucleoside* consists of a purine or pyrimidine base linked to a pentose, and that a *nucleotide* is a phosphate ester of a nucleoside. The names of the major bases of RNA and DNA, and of their nucleoside and nucleotide derivatives, are given in Table 25-1.

Table 25-1
Nomenclature of bases, nucleosides, and nucleotides

Base	*Ribonucleoside*	*Ribonucleotide (5′-monophosphate)*
Adenine (A)	Adenosine	Adenylate (AMP)
Guanine (G)	Guanosine	Guanylate (GMP)
Uracil (U)	Uridine	Uridylate (UMP)
Cytosine (C)	Cytidine	Cytidylate (CMP)
Base	*Deoxyribonucleoside*	*Deoxyribonucleotide (5′-monophosphate)*
Adenine (A)	Deoxyadenosine	Deoxyadenylate (dAMP)
Guanine (G)	Deoxyguanosine	Deoxyguanylate (dGMP)
Thymine (T)	Deoxythymidine	Deoxythymidylate (dTMP)
Cytosine (C)	Deoxycytidine	Deoxycytidylate (dCMP)

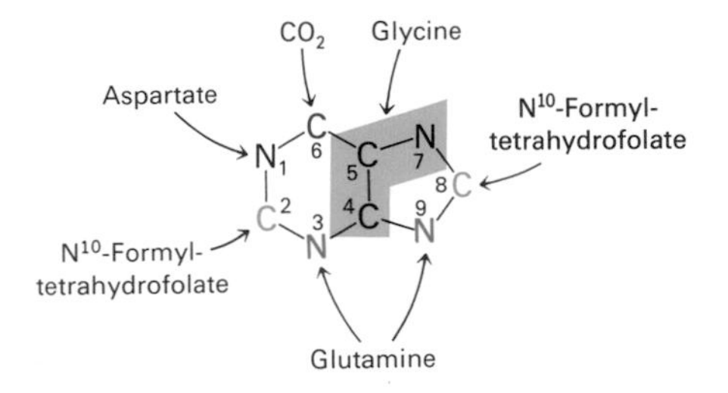

Figure 25-2
Origins of the atoms in the purine ring.

THE PURINE RING IS SYNTHESIZED FROM AMINO ACIDS, TETRAHYDROFOLATE DERIVATIVES, AND CO_2

Let us begin with the synthesis of purine nucleotides. The purine ring is assembled from a variety of precursors (Figure 25-2). *Glycine* provides C-4, C-5, and N-7. The N-1 atom comes from *aspartate*. The other two nitrogen atoms, N-3 and N-9, come from the amide group of the side chain of *glutamine*. Activated derivatives of *tetrahydrofolate* furnish C-2 and C-8, whereas CO_2 is the source of C-6.

PRPP IS THE DONOR OF THE RIBOSE PHOSPHATE UNIT OF NUCLEOTIDES

The pathway of purine biosynthesis was elucidated in the 1950s by John Buchanan and G. Robert Greenberg. The ribose phosphate portion of purine and pyrimidine nucleotides comes from *5-phosphoribosyl-1-pyrophosphate* (PRPP). We have already encountered PRPP in the synthesis of tryptophan (p. 585) and histidine (p. 587). PRPP is synthesized

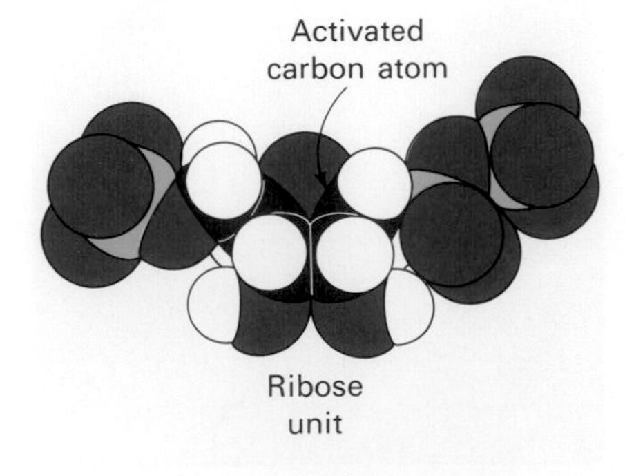

Figure 25-3
Model of 5-phosphoribosyl-1-pyrophosphate (PRPP), the activated donor of the sugar unit in the biosynthesis of nucleotides.

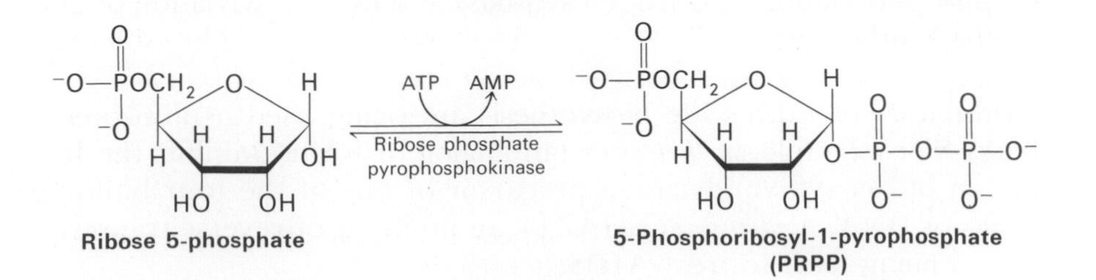

from ATP and ribose 5-phosphate, which is primarily formed by the pentose phosphate pathway (p. 428). The pyrophosphate group is transferred from ATP to C-1 of ribose 5-phosphate. PRPP has an α configuration at C-1.

THE PURINE RING IS ASSEMBLED ON RIBOSE PHOSPHATE

The committed step in the de novo synthesis of purine nucleotides is the formation of *5-phosphoribosylamine* from PRPP and glutamine. The amide group from the side chain of glutamine displaces the pyrophosphate group attached to C-1 of PRPP. The configuration at C-1 is inverted from α to β in this reaction. The resulting C–N glycosidic bond has the β configuration that is characteristic of naturally occurring nucleotides. This reaction is driven forward by the hydrolysis of pyrophosphate.

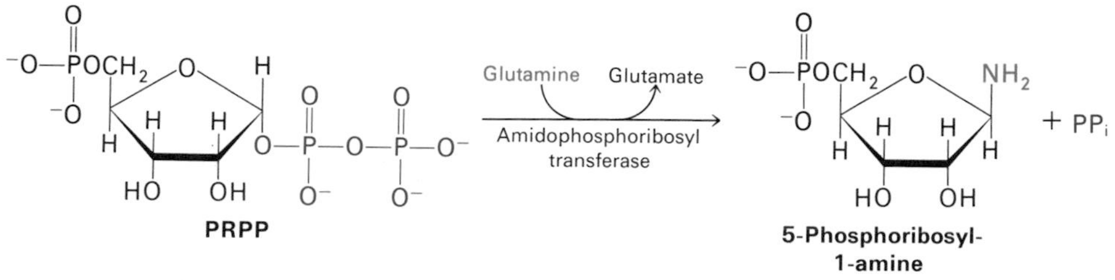

PRPP
Glutamine Glutamate
Amidophosphoribosyl transferase
5-Phosphoribosyl-1-amine
$+ PP_i$

Glycine joins phosphoribosylamine to yield *glycinamide ribonucleotide* (Figure 25-4). An ATP is consumed in the formation of an amide bond between the carboxyl group of glycine and the amino group of phosphoribosylamine. An acyl phosphate intermediate is formed, as in the biosyntheses of glutamine (p. 577), glutathione (p. 592), and almost all other amide bonds. The α-amino terminus of the glycine residue is then formylated by N^{10}-formyltetrahydrofolate to give *α-N-formylglycinamide ribonucleotide*. Folate derivatives serve as activated intermediates in several steps of the formation of purine and pyrimidine nucleotides. The amide group in this compound is then replaced by an amidine group. The nitrogen atom is donated by the side chain of glutamine in a reaction that consumes an ATP. *Formylglycinamidine ribonucleotide then undergoes ring closure to form 5-aminoimidazole ribonucleotide.* This intermediate contains the complete five-membered ring of the purine skeleton.

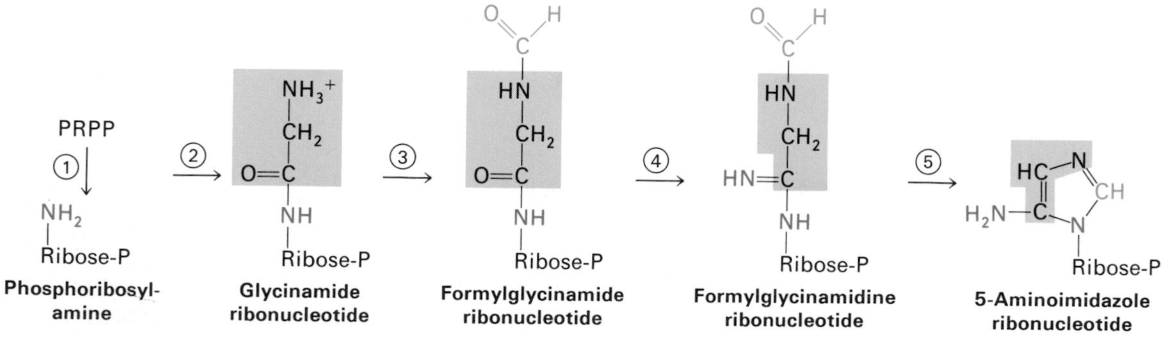

PRPP
Phosphoribosylamine
Glycinamide ribonucleotide
Formylglycinamide ribonucleotide
Formylglycinamidine ribonucleotide
5-Aminoimidazole ribonucleotide

Figure 25-4
First stage of purine biosynthesis: formation of 5-aminoimidazole ribonucleotide from PRPP. The essence of these reactions is (1) displacement of PP_i by the side-chain amino group of glutamine, (2) addition of glycine, (3) formylation by N^{10}-formyltetrahydrofolate, (4) transfer of a nitrogen atom from glutamine, and (5) dehydration and ring closure. The vertebrate enzymes catalyzing reactions 2, 3, and 5 are present on a single polypeptide chain.

The next phase in the synthesis of the purine skeleton, the formation of a six-membered ring, starts at this point (Figure 25-5). Three of the six atoms of this ring are already present in aminoimidazole ribonucleotide. The other three come from CO_2, aspartate, and formyltetrahydrofolate. The next carbon atom in the six-membered ring is introduced by the carboxylation of aminoimidazole ribonucleotide, yielding *5-aminoimidazole-4-carboxylate ribonucleotide*. This carboxylation reaction is unusual in not utilizing biotin.

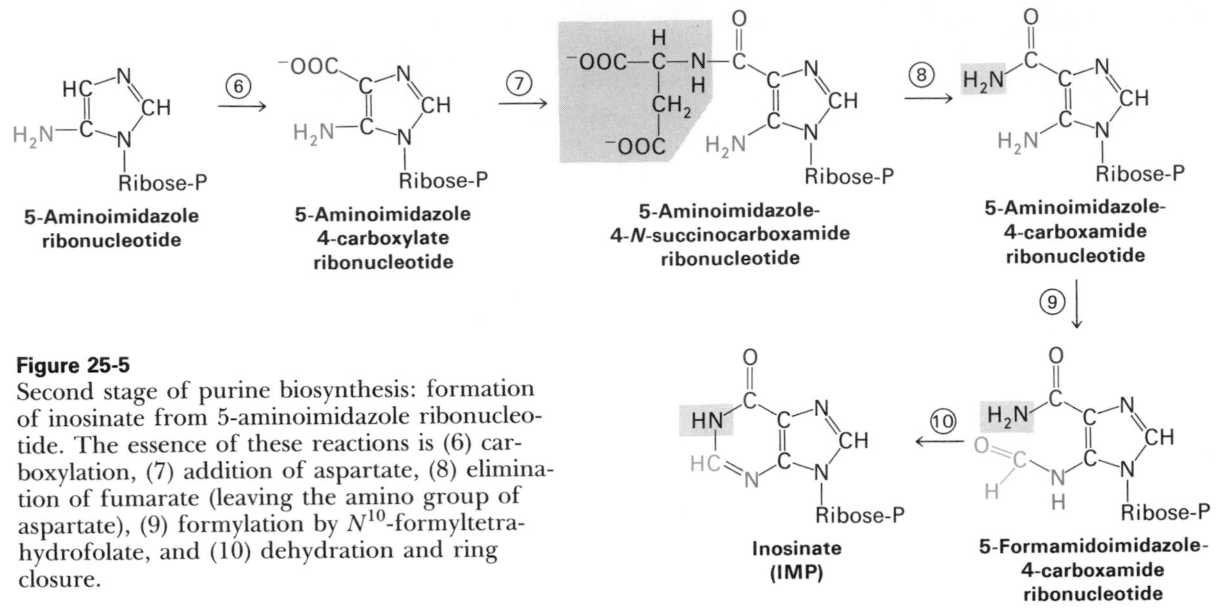

5-Aminoimidazole ribonucleotide → (6) → 5-Aminoimidazole 4-carboxylate ribonucleotide → (7) → 5-Aminoimidazole-4-*N*-succinocarboxamide ribonucleotide → (8) → 5-Aminoimidazole-4-carboxamide ribonucleotide → (9) → 5-Formamidoimidazole-4-carboxamide ribonucleotide → (10) → Inosinate (IMP)

Figure 25-5
Second stage of purine biosynthesis: formation of inosinate from 5-aminoimidazole ribonucleotide. The essence of these reactions is (6) carboxylation, (7) addition of aspartate, (8) elimination of fumarate (leaving the amino group of aspartate), (9) formylation by N^{10}-formyltetrahydrofolate, and (10) dehydration and ring closure.

The amino group of aspartate then reacts with the carboxyl group of this intermediate to form *5-aminoimidazole-4-*N*-succinocarboxamide ribonucleotide*. An ATP is consumed in the formation of this amide bond. The carbon skeleton of the aspartate moiety comes off as fumarate in the next reaction to give *5-aminoimidazole-4-carboxamide ribonucleotide*. Note that the result of these reactions is the conversion of a carboxylate into an amide. Thus, *aspartate contributes only its nitrogen atom to the purine ring*. The final atom of the purine ring is contributed by N^{10}-formyltetrahydrofolate. The resulting *5-formamidoimidazole-4-carboxamide ribonucleotide* undergoes dehydration and ring closure to form *inosinate* (IMP), which contains a complete purine ring. The purine base of inosinate is called *hypoxanthine*.

AMP AND GMP ARE FORMED FROM IMP

Inosinate is the precursor of AMP and GMP (Figure 25-6). *Adenylate* is synthesized from inosinate by the substitution of an amino group for the carbonyl oxygen atom at C-6. Again, the addition of aspartate followed by the elimination of fumarate contributes the amino group. GTP is the donor of a high-energy phosphate bond in the synthesis of the *adenylosuccinate* intermediate from inosinate and aspartate. The removal of fumarate from adenylosuccinate and from 5-aminoimidazole-4-*N*-succinocarboxamide ribonucleotide is catalyzed by the same enzyme.

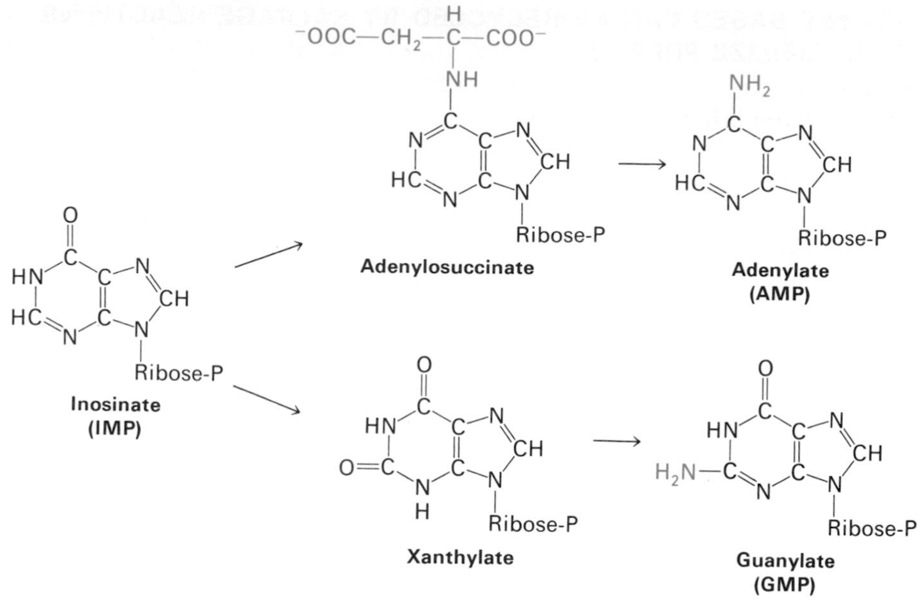

Inosinate (IMP)
Adenylosuccinate
Adenylate (AMP)
Xanthylate
Guanylate (GMP)

Figure 25-6
AMP and GMP are synthesized from IMP.

Guanylate (GMP) is synthesized by the oxidation of inosinate, followed by the insertion of an amino group at C-2. NAD^+ is the hydrogen acceptor in the oxidation of inosinate to xanthylate (XMP). The amino group in the side chain of glutamine is then transferred to xanthylate. Two high-energy phosphate bonds are consumed in this reaction, because ATP is cleaved into AMP and PP_i, which is subsequently hydrolyzed.

In the conversions of inosinate into adenylate and guanylate, a carbonyl oxygen atom is replaced by an amino group. A similar change occurs in the synthesis of formylglycinamidine ribonucleotide from its amide precursor (step 4 on p. 603), in the formation of CTP from UTP (p. 609), and in the conversion of citrulline into arginine in the urea cycle (p. 501). *The common mechanistic theme of these reactions is the conversion of the carbonyl oxygen into a derivative that can be readily displaced by an amino group.* The tautomeric form of the carbonyl group reacts with ATP (or GTP) to form a phosphoryl ester, which is then nucleophilically attacked by the nitrogen atom of an amine (Figure 25-7). Inorganic phosphate is then expelled from this tetrahedral adduct to complete the reaction. The attacking nitrogen can be from NH_3, the side-chain amide group of glutamine, or the α-amino group of aspartate. The leaving group in this class of reactions can be P_i, PP_i, or the AMP moiety. For example, PP_i is displaced by the amino group of glutamine in the synthesis of 5-phosphoribosyl-1-amine from PRPP (p. 603).

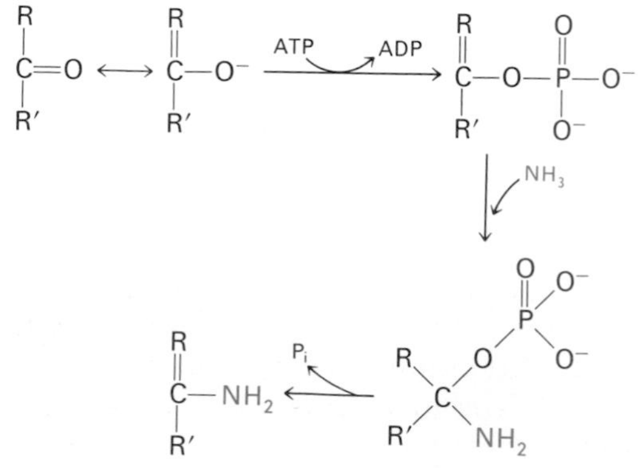

ATP ADP
NH_3
P_i

Figure 25-7
Reaction mechanism for the replacement of a carbonyl oxygen by an amino group.

PURINE BASES CAN BE RECYCLED BY SALVAGE REACTIONS THAT UTILIZE PRPP

Free purine bases are formed by the hydrolytic degradation of nucleic acids and nucleotides. Purine nucleotides can be synthesized from these preformed bases by a *salvage reaction,* which is simpler and much less costly than the reactions of the *de novo pathway* discussed above. In the salvage reaction, the ribose phosphate moiety of PRPP is transferred to the purine to form the corresponding nucleotide:

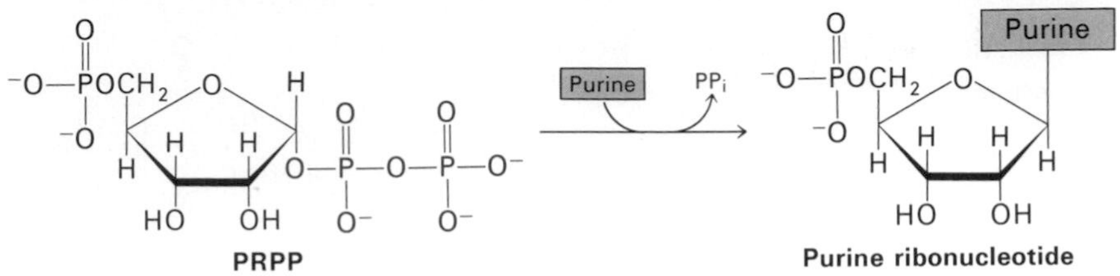

There are two salvage enzymes with different specificities. *Adenine phosphoribosyl transferase* catalyzes the formation of adenylate:

$$\text{Adenine} + \text{PRPP} \longrightarrow \text{adenylate} + \text{PP}_i$$

whereas *hypoxanthine-guanine phosphoribosyl transferase* catalyzes the formation of inosinate and guanylate:

$$\text{Hypoxanthine} + \text{PRPP} \longrightarrow \text{inosinate} + \text{PP}_i$$

$$\text{Guanine} + \text{PRPP} \longrightarrow \text{guanylate} + \text{PP}_i$$

The versatile and efficient use of the purine ring is also evident in the biosynthesis of histidine. The six-membered portion of the purine ring of ATP contributes part of the imidazole ring of histidine (p. 587). The rest of the purine skeleton is not discarded. Rather, it is conserved in 5-aminoimidazole-4-carboxamide ribonucleotide, an intermediate in the de novo pathway of purine biosynthesis.

AMP, GMP, AND IMP ARE FEEDBACK INHIBITORS OF PURINE NUCLEOTIDE BIOSYNTHESIS

The synthesis of purine nucleotides is controlled by feedback inhibition at several sites (Figure 25-8).

1. Feedback inhibition of *5-phosphoribosyl-1-pyrophosphate synthetase* by purine nucleotides regulates the level of PRPP. This synthetase is inhibited by AMP, GMP, and IMP.

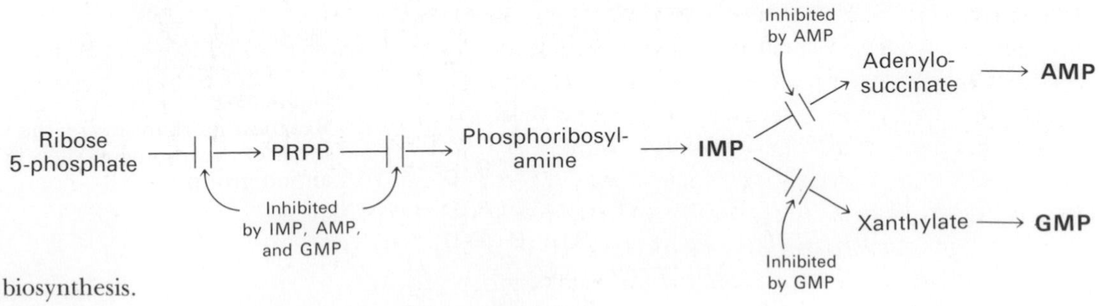

Figure 25-8
Control of purine biosynthesis.

2. The committed step in purine nucleotide biosynthesis is the conversion of PRPP into phosphoribosylamine by transfer of the side-chain amino group of glutamine. *Glutamine PRPP amidotransferase is feedback inhibited by many purine ribonucleotides.* It is noteworthy that AMP and GMP, the final products of the pathway, are synergistic in inhibiting this enzyme.

3. Inosinate is the branching point in the synthesis of AMP and GMP. *The reactions leading away from inosinate are sites of feedback inhibition.* AMP inhibits the conversion of inosinate into adenylosuccinate, its immediate precursor. Similarly, GMP inhibits the conversion of inosinate into xanthylate, its immediate precursor.

4. GTP is a substrate in the synthesis of AMP, whereas ATP is a substrate in the synthesis of GMP. This *reciprocal substrate relation* tends to balance the synthesis of adenine and guanine ribonucleotides.

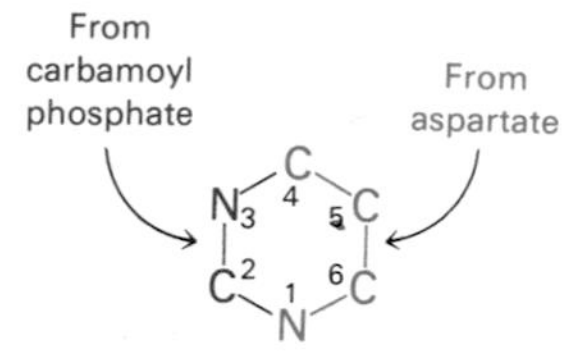

From carbamoyl phosphate
From aspartate

Figure 25-9
Origins of the atoms in the pyrimidine ring. C-2 and N-3 come from carbamoyl phosphate, whereas the other atoms of the ring come from aspartate.

THE PYRIMIDINE RING IS SYNTHESIZED FROM CARBAMOYL PHOSPHATE AND ASPARTATE

The pyrimidine ring is assembled first and then linked to ribose phosphate to form a pyrimidine nucleotide, in contrast with the reaction sequence in the de novo synthesis of purine nucleotides. However, PRPP is the donor of ribose phosphate in the synthesis of both purine and pyrimidine nucleotides. The precursors of the pyrimidine ring are carbamoyl phosphate and aspartate (Figure 25-9).

Pyrimidine biosynthesis starts with the formation of *carbamoyl phosphate,* which is also an intermediate in the synthesis of urea (p. 500). The synthesis of this activated carbamoyl donor is compartmentalized in eucaryotes. Carbamoyl phosphate used to synthesize pyrimidines is formed in the cytosol, whereas that used to make urea is formed in mitochondria, by a different carbamoyl phosphate synthetase (p. 501). Another noteworthy difference is that glutamine rather than NH_4^+ is the nitrogen donor in the cytosolic synthesis of carbamoyl phosphate. Also, *N*-acetylglutamate does not serve as an allosteric activator in the cytosolic synthesis.

$$\text{Glutamine} + 2\ \text{ATP} + HCO_3^- \longrightarrow \text{carbamoyl phosphate} + 2\ \text{ADP} + P_i + \text{glutamate}$$

The committed step in the biosynthesis of pyrimidines is the formation of N*-carbamoylaspartate from aspartate and carbamoyl phosphate.* This carbamoylation is catalyzed by *aspartate transcarbamoylase,* an especially interesting regulatory enzyme (p. 234).

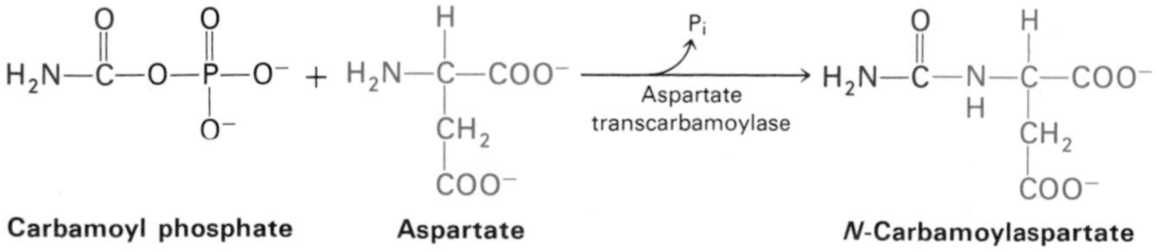

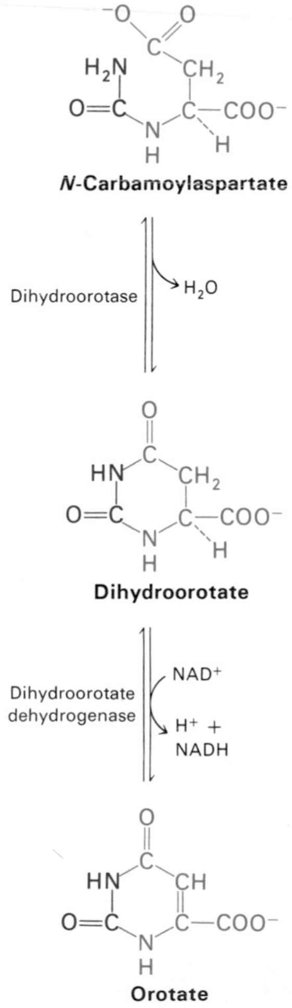

Carbamoyl phosphate + Aspartate —(P_i; Aspartate transcarbamoylase)→ *N*-Carbamoylaspartate

The pyrimidine ring is formed in the next reaction, in which carbamoylaspartate cyclizes with loss of water to yield *dihydroorotate. Orotate* is then formed by dehydrogenation of dihydroorotate.

N-Carbamoylaspartate —(Dihydroorotase; H_2O)→ **Dihydroorotate** —(Dihydroorotate dehydrogenase; NAD^+ → H^+ + NADH)→ **Orotate**

OROTATE ACQUIRES A RIBOSE PHOSPHATE MOIETY FROM PRPP

The next step in the synthesis of pyrimidine nucleotides is the *acquisition of a ribose phosphate group*. Orotate (a free pyrimidine) reacts with PRPP to form *orotidylate* (a pyrimidine nucleotide). This reaction, which is catalyzed by orotate phosphoribosyl transferase, is driven forward by the hydrolysis of pyrophosphate. Orotidylate is then decarboxylated to yield *uridylate* (UMP), a major pyrimidine nucleotide.

PRPP → PP_i (Orotate phosphoribosyl transferase); H^+ → CO_2 (Orotidylate decarboxylase)

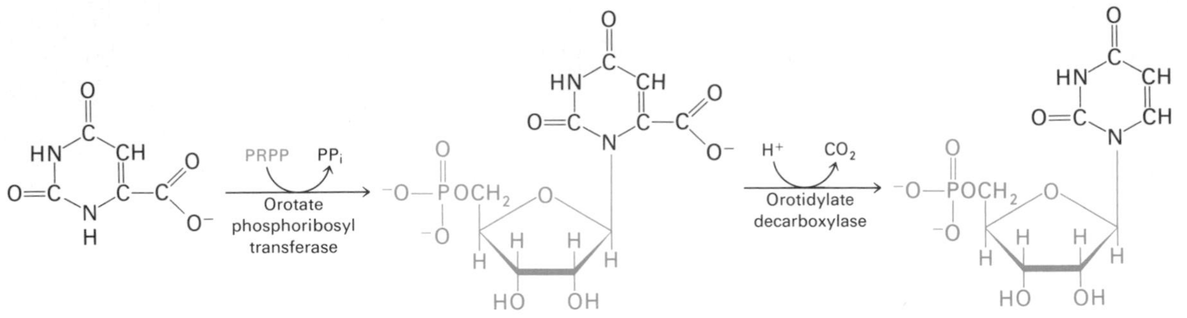

Orotate **Orotidylate** **Uridylate (UMP)**

PYRIMIDINE BIOSYNTHESIS IN HIGHER ORGANISMS IS CATALYZED BY MULTIFUNCTIONAL ENZYMES

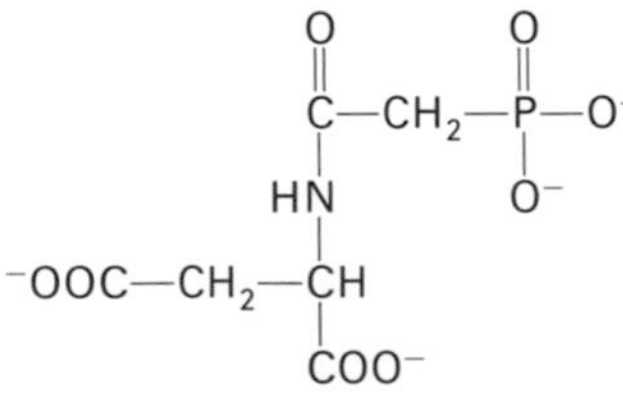

N-(Phosphonacetyl)-L-aspartate (PALA)

In *E. coli*, the six enzymes that synthesize UMP from simple precursors appear to be unassociated. In eucaryotes, by contrast, five of them are clustered in two complexes. One of these multifunctional enzymes was discovered when cultured mammalian cells were treated with N-*(phosphonacetyl)*-L-*aspartate* (PALA), a potent inhibitor of aspartate transcarbamoylase (ATCase). Recall that PALA binds tightly to ATCase (K_i = 10^{-8} M) because it resembles the transition state in catalysis (p. 237). The surviving cells overcame the inhibitory effect of PALA by synthesizing 100-fold more ATCase than do normal cells. The levels of carbamoyl phosphate synthetase and of dihydroorotase were also elevated 100-fold, whereas there was little change in the amounts of enzymes catalyzing subsequent steps in pyrimidine biosynthesis. These observations led to the finding that *carbamoyl phosphate synthetase, aspartate transcarbamoylase, and dihydroorotase are covalently joined on a single 240-kd polypeptide chain*. This multifunctional enzyme is called CAD.

Orotate phosphoribosyl transferase and orotidylate decarboxylase, the enzymes catalyzing the last two steps in pyrimidine biosynthesis, are also associated in eucaryotes. Each comprises a domain of a single 52-kd polypeptide. Indeed, *the covalent linkage of functionally related enzymes occurs often in eucaryotes*, in which such multifunctional enzymes probably evolved by exon shuffling. Another striking example is the mammalian fatty acid synthase, which contains seven enzymatic activities in two chains (p. 486). The clustering of enzymes catalyzing a reaction sequence has several potential advantages. Their synthesis is coordinated and their assembly into a coherent complex is easily assured. Side reactions are minimized as substrates are channeled from one catalytic site to the next. A covalently linked multifunctional complex is likely to be more stable than one formed by noncovalent interactions.

NUCLEOSIDE MONO-, DI-, AND TRIPHOSPHATES ARE INTERCONVERTIBLE

The active forms of nucleotides in biosyntheses and energy conversions are the diphosphates and triphosphates. Nucleoside monophosphates are converted by specific *nucleoside monophosphate kinases* that utilize ATP as the phosphoryl donor. For example, UMP is phosphorylated by *UMP kinase*.

$$\text{UMP} + \text{ATP} \rightleftharpoons \text{UDP} + \text{ADP}$$

AMP, ADP, and ATP are interconverted by *adenylate kinase* (also called myokinase). The equilibrium constants of these reactions are close to 1.

$$\text{AMP} + \text{ATP} \rightleftharpoons 2\ \text{ADP}$$

Nucleoside diphosphates and triphosphates are interconverted by *nucleoside diphosphate kinase,* an enzyme that has broad specificity, in contrast with the monophosphate kinases. In the following equation, X and Y can be any of several ribonucleosides or deoxyribonucleosides.

$$\text{XDP} + \text{YTP} \rightleftharpoons \text{XTP} + \text{YDP}$$

For example,

$$\text{UDP} + \text{ATP} \rightleftharpoons \text{UTP} + \text{ADP}$$

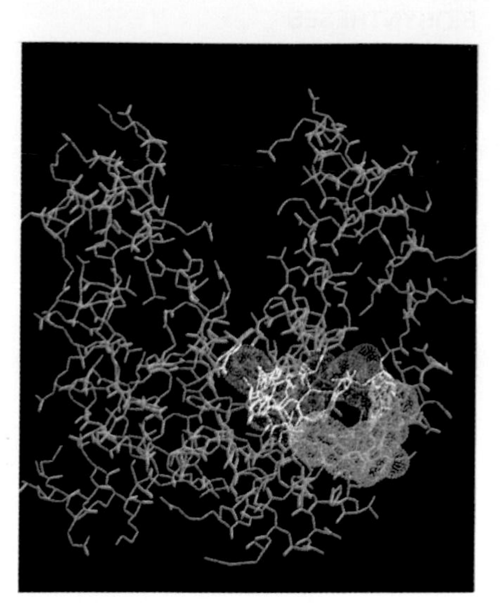

Figure 25-10
Model of adenylate kinase.

CTP IS FORMED BY AMINATION OF UTP

Cytidine triphosphate (CTP) is derived from uridine triphosphate (UTP), the other major pyrimidine ribonucleotide. The carbonyl oxygen at C-4 is replaced by an amino group. In mammals, the amide group of glutamine is the amino donor, whereas in *E. coli* NH_4^+ is used in this reaction. Mammals avoid having a high level of NH_4^+ in the plasma by generating it in situ from a donor such as glutamine. ATP is consumed in both amination reactions. As in the conversions of inosinate to AMP and GMP, an acyl phosphate intermediate is nucleophilically attacked by a nitrogen atom (p. 605).

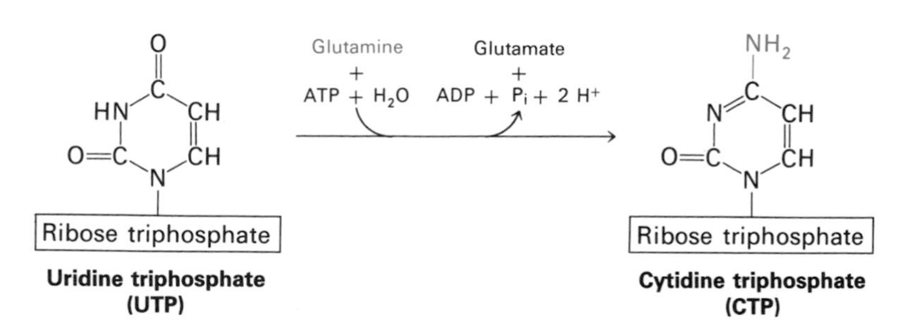

PYRIMIDINE NUCLEOTIDE BIOSYNTHESIS IN BACTERIA IS REGULATED BY FEEDBACK INHIBITION

The committed step in pyrimidine nucleotide biosynthesis in *E. coli* is the formation of *N*-carbamoylaspartate from aspartate and carbamoyl phosphate. *Aspartate transcarbamoylase (ATCase), the enzyme that catalyzes this reaction, is feedback inhibited by CTP, the final product in the pathway.* A second control site is *carbamoyl phosphate synthetase,* which is feedback inhibited by UMP (Figure 25-11). The structure and allosteric mechanism of *E. coli* ATCase were discussed in an earlier chapter (p. 234).

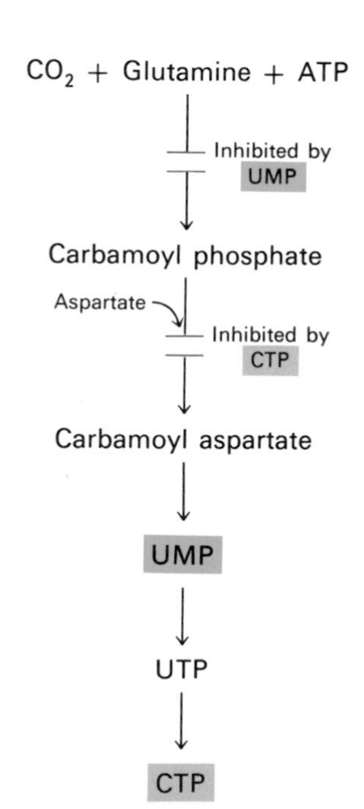

Figure 25-11
Control of pyrimidine biosynthesis in *E. coli.*

DEOXYRIBONUCLEOTIDES ARE SYNTHESIZED BY REDUCTION OF RIBONUCLEOSIDE DIPHOSPHATES

We turn now to the synthesis of deoxyribonucleotides. These precursors of DNA are formed by the reduction of ribonucleotides. The 2′-hydroxyl group on the ribose moiety is replaced by a hydrogen atom. In *E. coli* and in mammals, the substrates in this reaction are ribonucleoside *di*phosphates. The overall stoichiometry is

$$\text{Ribonucleoside diphosphate} \xrightarrow[\text{NADPH} + H^+ \;\; \text{NADP}^+ + H_2O]{} \text{Deoxyribonucleoside diphosphate}$$

Ribonucleoside diphosphate **Deoxyribonucleoside diphosphate**

The actual reaction mechanism is more complex than implied by this equation. Peter Reichard has shown that, in *E. coli*, electrons from NADPH are transferred to the substrate through a series of sulfhydryl groups. *Ribonucleotide reductase* catalyzes the final stage, which has the stoichiometry

$$\text{Ribonucleoside diphosphate} + R(SH)_2 \longrightarrow \text{Deoxyribonucleoside diphosphate} + R(S{-}S) + H_2O$$

This enzyme consists of two subunits, B1 (a 160-kd dimer) and B2 (a 78-kd dimer). The B1 subunit contains the binding sites for ribonucleotide substrates and allosteric effectors. B1 also contains sulfhydryls that serve as the immediate electron donors in the reduction of the ribose unit. B2 participates in catalysis by forming an unusual free radical, as described below. The B1 and B2 subunits together form the active sites of the enzyme (Figure 25-12).

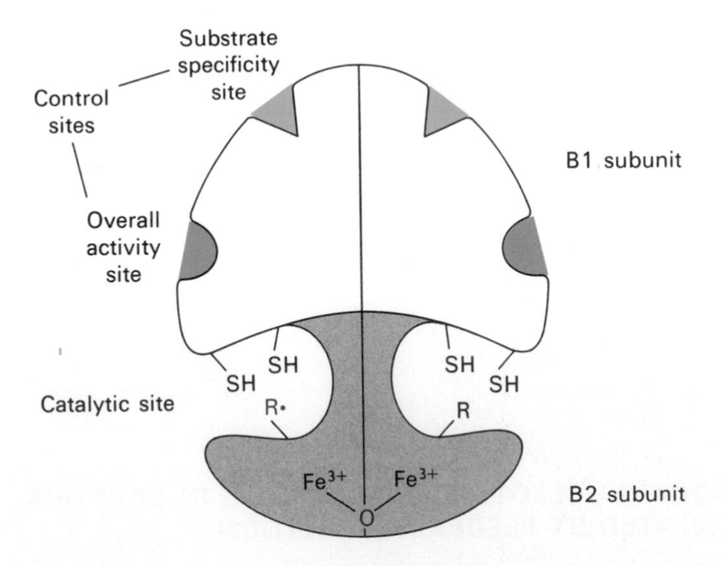

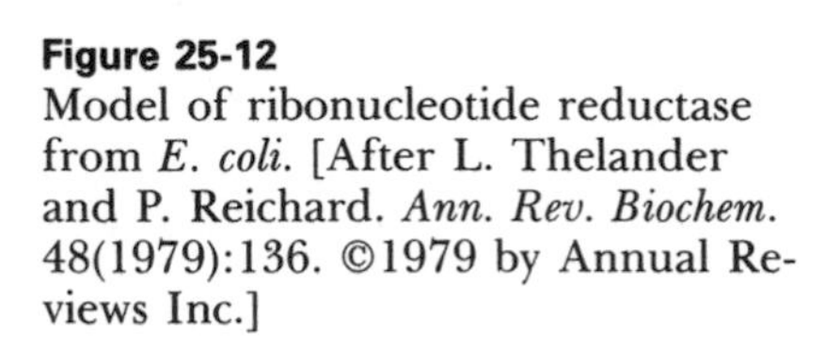

Figure 25-12
Model of ribonucleotide reductase from *E. coli*. [After L. Thelander and P. Reichard. *Ann. Rev. Biochem.* 48(1979):136. ©1979 by Annual Reviews Inc.]

RIBONUCLEOTIDE REDUCTASE CONTAINS A FREE RADICAL AT ITS ACTIVE SITE

The B2 subunit is very unusual in containing a stable organic free radical—specifically, a *tyrosyl radical cation*. This radical is essential for catalysis, as shown by the loss of enzymatic activity on addition of hydroxyurea, a quencher of free radicals. The radical in B2 is stabilized by an iron center consisting of two Fe^{3+} ions connected by an oxygen atom. B2 loses both its iron center and its radical when it is dialyzed in the presence of 8-hydroxyquinoline, a chelator of ferric ions. Active B2 is formed from the apoprotein and Fe^{3+} in the presence of O_2 and a thiol. Three proteins are needed to catalyze the regeneration of the radical and restore enzymatic activity. Interestingly, one of them is superoxide dismutase, an enzyme that catalyzes the removal of harmful superoxide anions (p. 422). O_2 is the electron acceptor in the one-electron oxidation of the phenolic ring of a tyrosine residue to the radical cation.

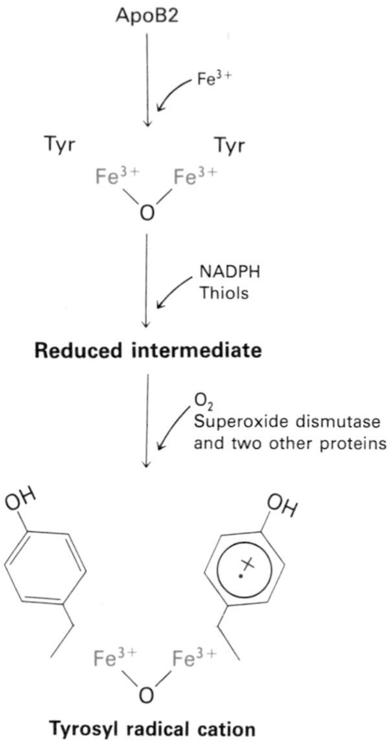

In the formation of a deoxyribonucleotide, the hydroxyl group bonded to C-2 of the ribose ring is stereospecifically replaced by H. No other hydrogen atoms are inserted or removed. The tyrosyl radical plays a catalytic rather than stoichiometric role in this reaction. The essence of the mechanism is a *transient transfer of radical properties from the enzyme to the substrate.* The tyrosyl radical abstracts a hydrogen atom from C-3 (Figure 25-13). The presence of the radical at C-3 promotes the ejection of OH^- from C-2. The sugar ring is a radical cation at this time. Deoxyribose is then formed by the reduction of C-2 by thiols on the B1 subunit. The hydrogen atom abstracted by the tyrosine radical is concomitantly returned to C-3.

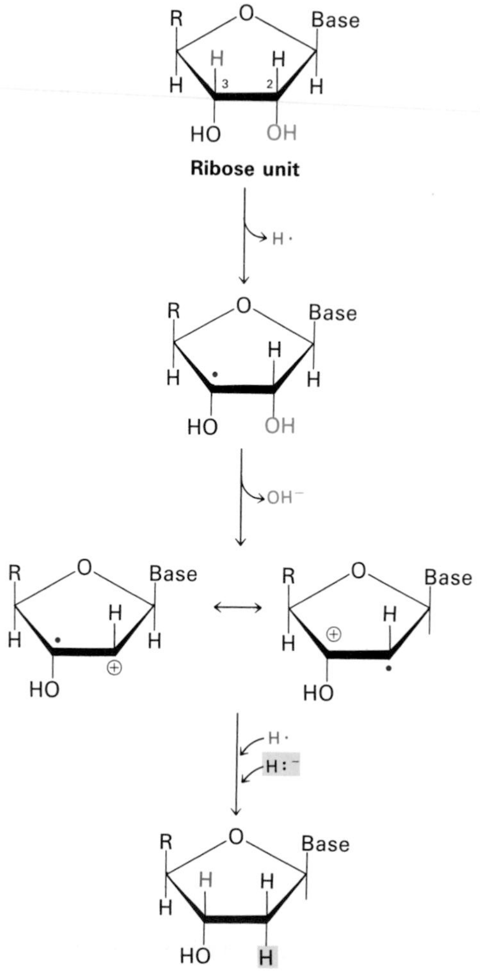

ApoB2 → (Fe^{3+}) → Tyr Tyr Fe^{3+}–O–Fe^{3+} → (NADPH, Thiols) → **Reduced intermediate** → (O_2, Superoxide dismutase and two other proteins) → **Tyrosyl radical cation**

Ribose unit

Figure 25-13
Proposed mechanism for the reduction of ribonucleotides by abstraction of a hydrogen atom and formation of free-radical intermediates. [After P. Reichard and A. Ehrenberg. *Science* 221(1983):514.]

The participation of a free radical in this reaction is reminiscent of the action of coenzyme B_{12} (p. 422), which also serves as a free-radical source. The tyrosine radical cation and the carbon–cobalt bond of coenzyme B_{12} are two means of achieving the same end—namely, the homolytic cleavage of a bond. Indeed, some bacterial ribonucleotide reductases (e.g., the one in *Lactobacillus leichmanni*) contain coenzyme B_{12} instead of an iron center and tyrosyl radical. Why do all known ribonucleotide reductases rely on a radical to achieve catalysis? The answer probably lies in the very demanding nature of the reaction. No change other than the exchange of —H for —OH is permitted in the conversion of a ribonucleotide into a deoxyribonucleotide.

THE SUBSTRATE SPECIFICITY AND CATALYTIC ACTIVITY OF RIBONUCLEOTIDE REDUCTASE ARE PRECISELY CONTROLLED

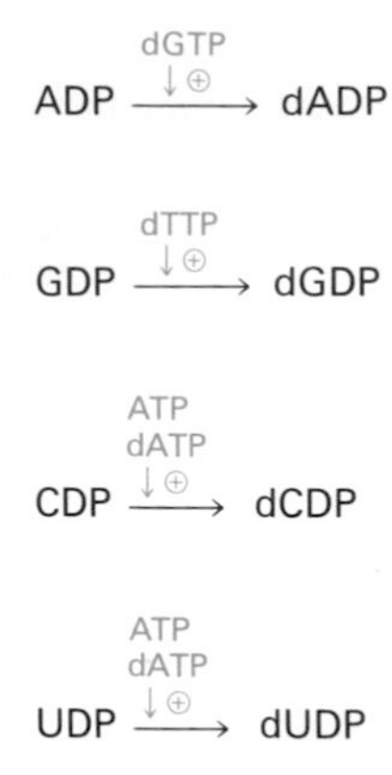

The reduction of ribonucleotide diphosphates is controlled by allosteric interactions. The B1 subunit of ribonucleotide reductase contains two types of allosteric sites: one of them controls the *overall activity* of the enzyme, whereas the other regulates *substrate specificity*. The overall catalytic activity of ribonucleotide reductase is diminished by the binding of dATP, which signals an abundance of deoxyribonucleotides. This feedback inhibition is reversed by the binding of ATP. The binding of dATP or ATP to the substrate-specificity control sites enhances the reduction of UDP and CDP, the pyrimidine nucleotides. The reduction of GDP is promoted by the binding of dTTP, which also inhibits the further reduction of pyrimidine ribonucleotides. The subsequent increase in the level of dGTP leads to a stimulation of ADP reduction. It is evident that ribonucleotide reductase has a variety of conformational states, each with different catalytic properties. This complex pattern of regulation provides the appropriate supply of the four deoxyribonucleotides needed for the synthesis of DNA.

THIOREDOXIN AND GLUTAREDOXIN CARRY ELECTRONS TO RIBONUCLEOTIDE REDUCTASE

How are electrons transferred from NADPH to the sulfhydryl groups at the catalytic site of ribonucleotide reductase? One carrier of reducing power is *thioredoxin,* a 12-kd protein with two exposed cysteine residues near each other (Figure 25-14). These sulfhydryls are oxidized to a

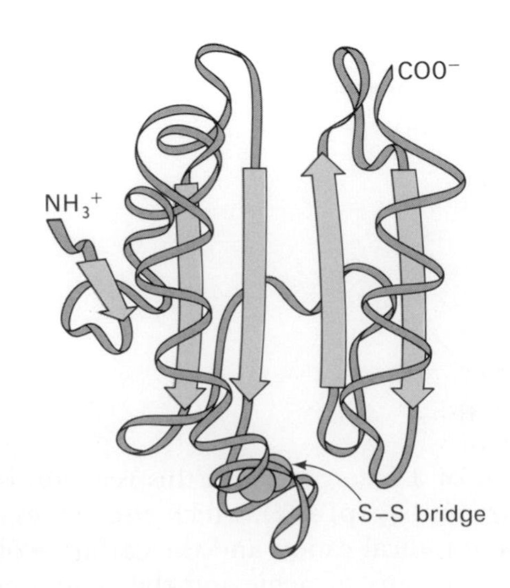

Figure 25-14
Schematic diagram of the main-chain conformation of oxidized thioredoxin from *E. coli.* The reactive disulfide is shown in yellow. [After a drawing kindly provided by Dr. Carl-Ivar Brändén.]

disulfide in the reaction catalyzed by ribonucleotide reductase. In turn, reduced thioredoxin is regenerated by electron flow from NADPH. This reaction is catalyzed by *thioredoxin reductase*, a flavoprotein. Electrons flow from NADPH to bound FAD of the reductase and then to the disulfide of oxidized thioredoxin.

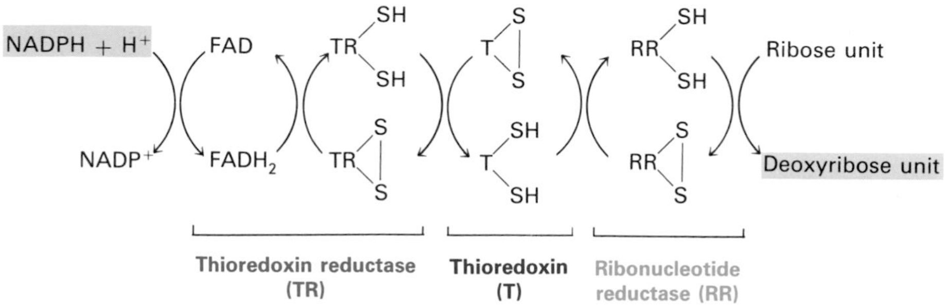

NADPH + H^+ → $NADP^+$; FAD → $FADH_2$; TR(SH)(SH) → TR(S–S); T(S–S) → T(SH)(SH); RR(SH)(SH) → RR(S–S); Ribose unit → Deoxyribose unit

Thioredoxin reductase (TR) | Thioredoxin (T) | Ribonucleotide reductase (RR)

Thioredoxin is an electron donor not only in the reduction of ribonucleotides. Recall that thioredoxin plays an important role in controlling the dark reactions of photosynthesis (p. 538). In chloroplasts, oxidized thioredoxin is reduced by ferredoxin. Thioredoxin regulates the activities of enzymes in many kinds of cells by reducing disulfide bonds.

Thioredoxin is not the only carrier of reducing power to ribonucleotide reductase. A mutant of *E. coli* totally devoid of thioredoxin was found to form deoxyribonucleotides. This surprising observation led to the isolation of a second carrier system. The electron donor in this mutant proved to be *glutathione,* a cysteine-containing tripeptide. As discussed previously (p. 436), *glutathione reductase* catalyzes the reduction of oxidized glutathione (the disulfide form) by NADPH. A flavin participates in this redox reaction, as in the one catalyzed by thioredoxin reductase. *Glutaredoxin,* a hitherto unknown protein, transfers the reducing power of glutathione to ribonucleotide reductase. Glutaredoxin and thioredoxin are homologous proteins. In particular, the protruding segment containing the thiol pair is very similar in these two proteins.

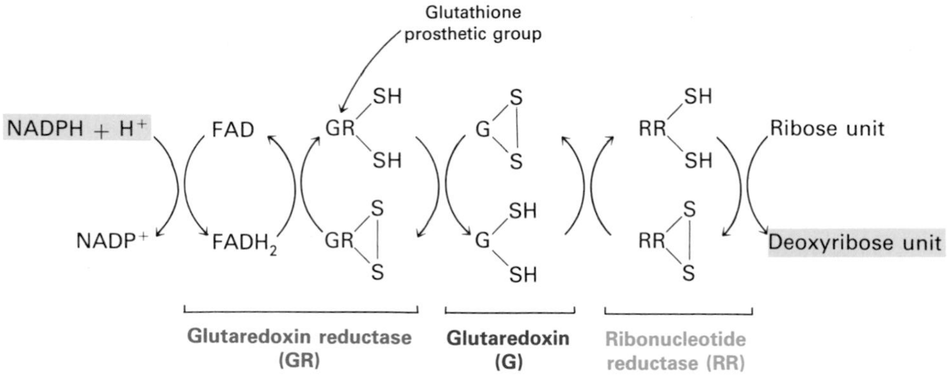

Glutathione prosthetic group

NADPH + H^+ → $NADP^+$; FAD → $FADH_2$; GR(SH)(SH) → GR(S–S); G(S–S) → G(SH)(SH); RR(SH)(SH) → RR(S–S); Ribose unit → Deoxyribose unit

Glutaredoxin reductase (GR) | Glutaredoxin (G) | Ribonucleotide reductase (RR)

DEOXYTHYMIDYLATE IS FORMED BY METHYLATION OF DEOXYURIDYLATE

Uracil is not a component of DNA. Rather, DNA contains *thymine,* the methylated analog of uracil. This finishing touch in the formation of thymine occurs at the level of the deoxyribonucleoside monophosphate: deoxyuridylate (dUMP) is methylated to deoxythymidylate (dTMP) by *thymidylate synthase*. The methyl donor in this reaction is a tetrahydrofolate derivative rather than *S*-adenosylmethionine. Specifically, the methyl carbon comes from N^5, N^{10}-methylenetetrahydrofolate. Note that the methyl group inserted into deoxyuridylate is more reduced than the methylene group in this tetrahydrofolate derivative. What is the source of electrons for this reduction? *The two electrons come from the tetrahydrofolate moiety itself in the form of a hydride ion (H:⁻), which is removed from the ring.* This hydrogen becomes part of the methyl group of dTMP. In this reaction, tetrahydrofolate is oxidized to dihydrofolate. Thus N^5, N^{10}-methylenetetrahydrofolate serves both as an *electron donor* and as a *one-carbon* donor in the methylation reaction (Figure 25-15).

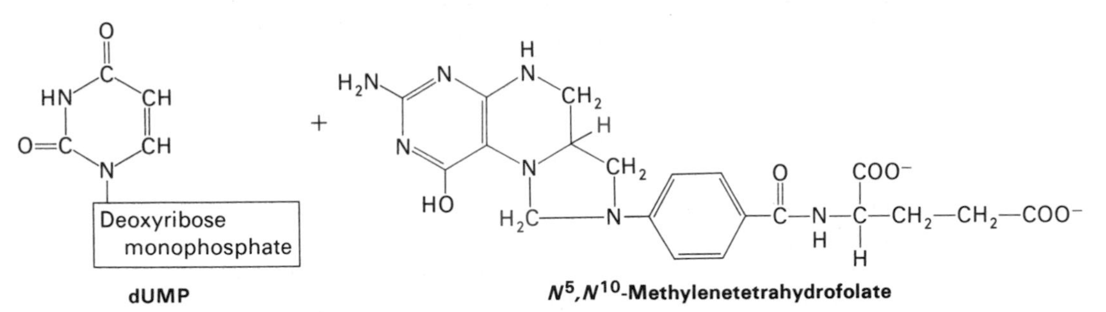

Figure 25-15
Synthesis of dTMP from dUMP.

Recall that one-carbon transfers are made by tetrahydrofolate rather than dihydrofolate. Hence, tetrahydrofolate must be regenerated. This is accomplished by *dihydrofolate reductase* using NADPH as the reductant.

$$\text{Dihydrofolate} + \text{NADPH} + \text{H}^+ \longrightarrow \text{tetrahydrofolate} + \text{NADP}^+$$

It is noteworthy that the deoxyribose and thymine units of DNA are formed by modification of ribonucleotides. In contrast, no known ribonucleotide is formed from a deoxyribonucleotide. These precursor-product relations strongly imply that ribonucleotides came first in evolution. *The reactions catalyzed by ribonucleotide reductase and thymidylate synthase are recapitulations of the transition from an RNA world to one in which DNA became the store of genetic information* (p. 113).

SEVERAL ANTICANCER DRUGS BLOCK THE SYNTHESIS OF DEOXYTHYMIDYLATE

Rapidly dividing cells require an abundant supply of deoxythymidylate for the synthesis of DNA. The vulnerability of these cells to the inhibition of dTMP synthesis has been exploited in cancer chemotherapy

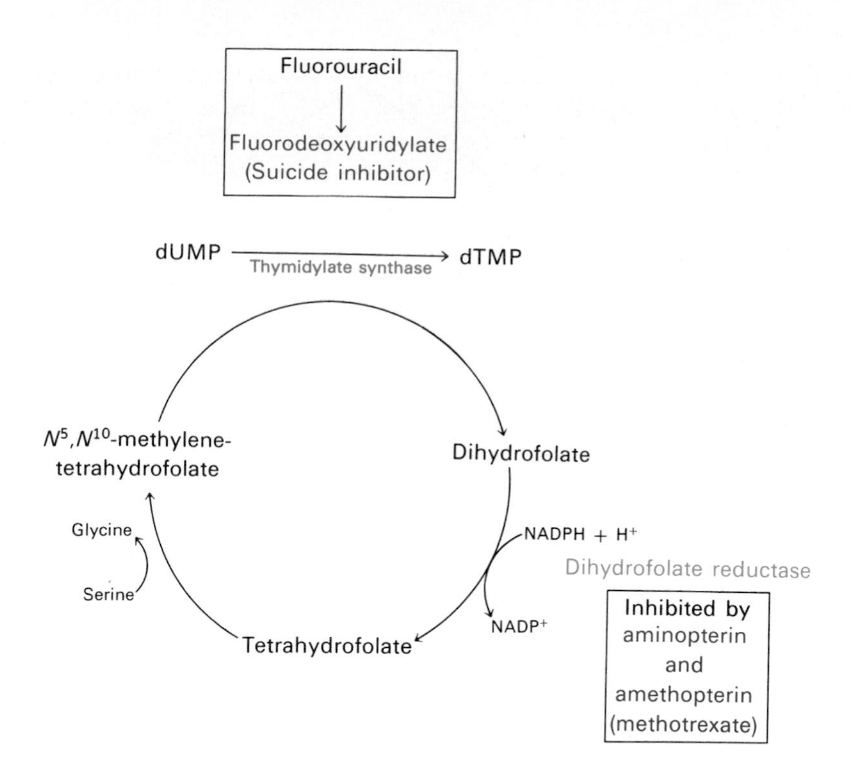

Figure 25-16
Thymidylate synthase and dihydrofolate reductase are target enzymes in cancer chemotherapy. Fluorodeoxyuridylate inhibits the methylation of dUMP. The folate analogs aminopterin and methotrexate block the regeneration of tetrahydrofolate.

(Figure 25-16). Thymidylate synthase and dihydrofolate reductase are choice target enzymes.

Fluorouracil (or fluorodeoxyuridine), a clinically useful anticancer drug, is converted in vivo into *fluorodeoxyuridylate* (F-dUMP). This analog of dUMP irreversibly inhibits thymidylate synthase after acting as a normal substrate through part of the catalytic cycle. First, a sulfhydryl group of the enzyme adds to C-6 of the bound F-dUMP. Methylenetetrahydrofolate then adds to C-5 of this intermediate. In the case of dUMP, a hydride ion of the folate is subsequently shifted to the methylene group, and a proton is taken away from C-5 of the bound nucleotide. However, F^+ cannot be abstracted from F-dUMP, by the enzyme, and so catalysis is blocked at the stage of the covalent complex formed by F-dUMP, methylenetetrahydrofolate, and the sulfhydryl group of the enzyme (Figure 25-17). We see here an example of *suicide inhibition*, in which an enzyme converts a substrate into a reactive inhibitor that immediately inactivates its catalytic activity.

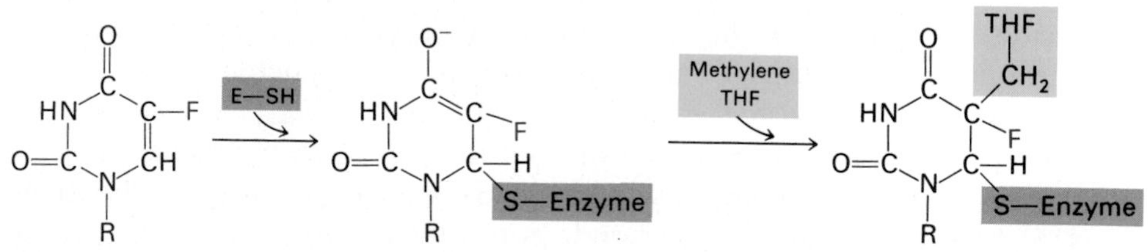

Figure 25-17
Thymidylate synthase is irreversibly inhibited by fluorodeoxyuridylate (F-dUMP). This analog forms a covalent complex with both a sulfhydryl residue of the enzyme (shown in blue) and methylenetetrahydrofolate (shown in yellow).

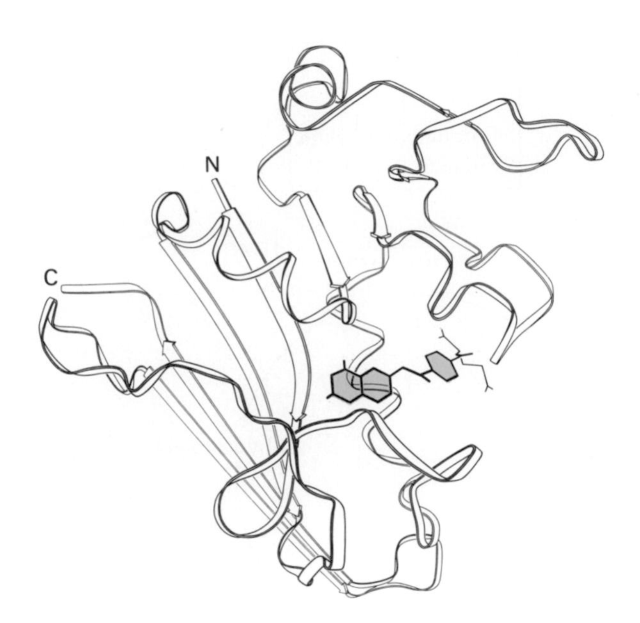

Figure 25-18
Three-dimensional structure of *E. coli* dihydrofolate reductase with a bound methotrexate. [Courtesy of Dr. Joseph Kraut].

The synthesis of dTMP can also be blocked by inhibiting the regeneration of tetrahydrofolate (Figure 25-18). Analogs of dihydrofolate, such as *aminopterin* and *methotrexate* (amethopterin) are potent competitive inhibitors ($K_i < 10^{-9}$ M) of dihydrofolate reductase. Methotrexate is a valuable drug in the treatment of many rapidly growing tumors, such as acute leukemia and choriocarcinoma. However, methotrexate is quite toxic because it kills rapidly replicating cells whether they are malignant or not. Stem cells in bone marrow, epithelial cells of the intestinal tract, and hair follicles are vulnerable to the action of this folate antagonist, accounting for many of its toxic side effects.

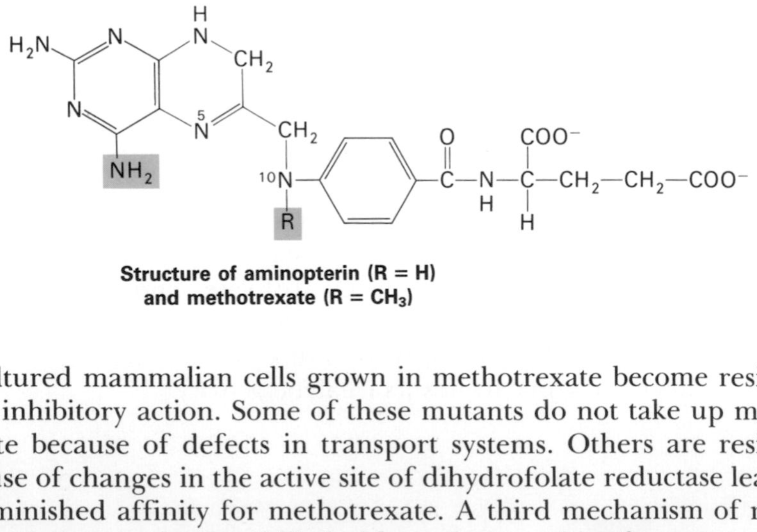

Structure of aminopterin (R = H) and methotrexate (R = CH_3)

Cultured mammalian cells grown in methotrexate become resistant to its inhibitory action. Some of these mutants do not take up methotrexate because of defects in transport systems. Others are resistant because of changes in the active site of dihydrofolate reductase leading to diminished affinity for methotrexate. A third mechanism of resistance is the *overproduction of dihydrofolate reductase arising from gene amplification* (p. 840). Robert Schimke showed that cells producing as much as 1000 times the normal level of enzyme are selected stepwise as the drug level in the medium is increased. Some overproducing cells contain hundreds of copies of the gene for dihydrofolate reductase. These studies of cultured cells strongly suggest that the dose of an anti-cancer drug should be sufficiently high to prevent the survival of mutants that become increasingly resistant by gene amplification.

The biosynthesis of *nicotinamide adenine dinucleotide* (NAD^+) starts with the formation of *nicotinate ribonucleotide* from nicotinate and PRPP. *Nicotinate* (also called *niacin*) is derived from tryptophan. Humans can synthesize the required amount of nicotinate if the supply of tryptophan in the diet is adequate. However, an exogenous supply of nicotinate is required if the dietary intake of tryptophan is low. A dietary deficiency of tryptophan and nicotinate can lead to *pellagra*, a disease characterized by dermatitis, diarrhea, and dementia. An endocrine tumor that consumes large amounts of tryptophan in synthesizing serotonin (5-hydroxytryptamine), a hormone and neurotransmitter, leads to pellagra-like symptoms.

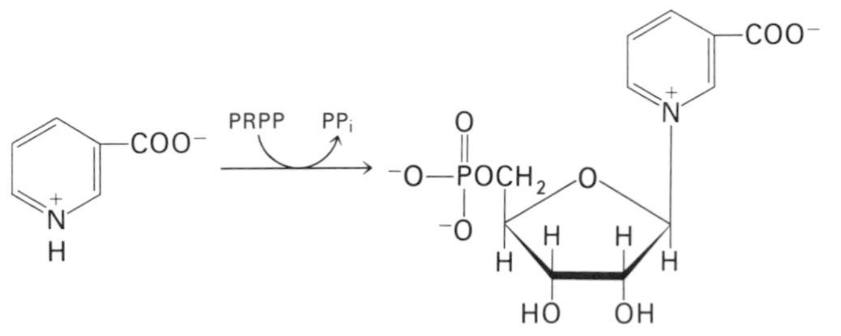

Nicotinate —(PRPP → PP_i)→ Nicotinate ribonucleotide

An AMP moiety is transferred from ATP to nicotinate ribonucleotide to form *desamido-NAD^+*. The final step is the transfer of the amide group of glutamine to the nicotinate carboxyl group to form NAD^+ (Figure 25-19). $NADP^+$ is derived from NAD^+ by phosphorylation of the 2′-hydroxyl group of the adenine ribose moiety. This transfer of a phosphoryl group from ATP is catalyzed by NAD^+ kinase.

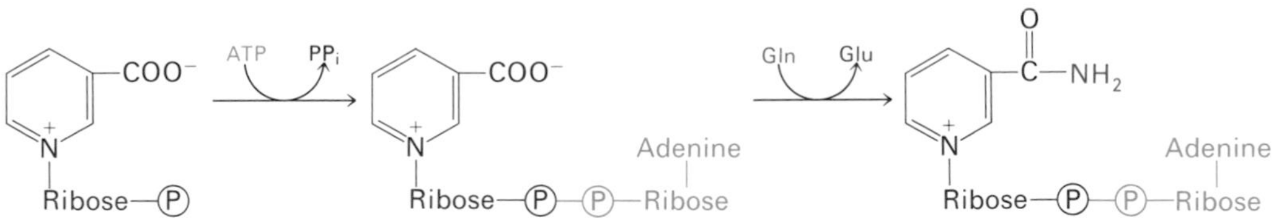

Nicotinate ribonucleotide —(ATP → PP_i)→ Desamido-NAD^+ —(Gln → Glu)→ NAD^+

Figure 25-19
Synthesis of NAD^+ from nicotinate ribonucleotide.

Flavin adenine dinucleotide (FAD) is synthesized from riboflavin and two molecules of ATP. Riboflavin is phosphorylated by ATP to give *riboflavin 5′-phosphate* (also called *flavin mononucleotide*). FAD is then formed by the transfer of an AMP moiety from a second molecule of ATP to riboflavin 5′-phosphate.

$$\text{Riboflavin} + \text{ATP} \longrightarrow \text{riboflavin 5'-phosphate} + \text{ADP}$$

$$\text{Riboflavin 5'-phosphate} + \text{ATP} \rightleftharpoons \text{flavin adenine dinucleotide} + PP_i$$

The *synthesis of coenzyme A* (CoA) in animals starts with the phosphorylation of *pantothenate* (Figure 22-20). Pantothenate is required in the diet of animals, whereas it is synthesized by plants and microorganisms. A peptide bond is formed between the carboxyl group of 4′-phos-

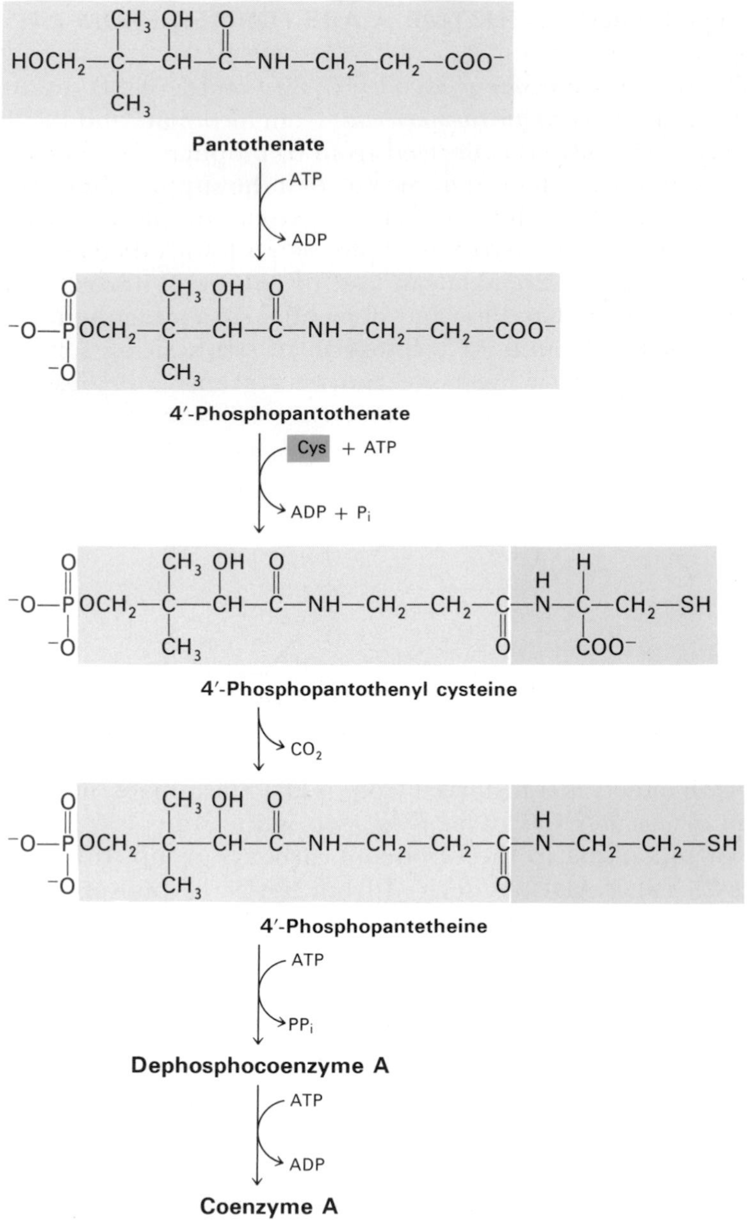

Pantothenate

$\downarrow$ ATP → ADP

4′-Phosphopantothenate

$\downarrow$ Cys + ATP → ADP + P_i

4′-Phosphopantothenyl cysteine

$\downarrow$ CO_2

4′-Phosphopantetheine

$\downarrow$ ATP → PP_i

Dephosphocoenzyme A

$\downarrow$ ATP → ADP

Coenzyme A

Figure 25-20
Synthesis of coenzyme A from pantothenate.

phopantothenate and the amino group of cysteine. The carboxyl group of the cysteine moiety is lost, which results in *4′-phosphopantotheine.* The AMP moiety of ATP is then transferred to this intermediate to form *dephosphocoenzyme A.* Finally, phosphorylation of its 3′-hydroxyl group yields coenzyme A.

A common feature of the biosyntheses of NAD^+, FAD, and CoA is the *transfer of the AMP moiety of ATP to the phosphate group of a phosphorylated intermediate.* The pyrophosphate formed in this reaction is hydrolyzed to orthophosphate. This is a recurring motif in biochemistry: *biosynthetic reactions are frequently driven by the hydrolysis of the released pyrophosphate.*

PURINES IN HUMANS ARE DEGRADED TO URATE

The nucleotides of a cell undergo continuous turnover. Nucleotides are hydrolytically degraded to nucleosides by *nucleotidases.* Phosphorolytic cleavage of nucleosides to free bases and ribose 1-phosphate (or deoxyribose 1-phosphate) is catalyzed by *nucleoside phosphorylases*. Ribose 1-phosphate is isomerized by *phosphoribomutase* to ribose 5-phosphate, a substrate in the synthesis of PRPP. Some of the bases are reused to form nucleotides by salvage pathways.

The pathway for the degradation of AMP (Figure 25-21) includes an additional step. AMP is deaminated to IMP by *adenylate deaminase*. The 5′-phosphate of IMP is hydrolytically removed and the glycosidic bond is then split by P_i to yield hypoxanthine, the free base. *Xanthine oxidase,* a molybdenum- and iron-containing flavoprotein, oxidizes hypoxanthine to *xanthine* and then to *urate* (Figure 25-22). Molecular oxygen, the oxidant in both reactions, is reduced to H_2O_2, which is decomposed to H_2O and O_2 by catalase. Xanthine is also an intermediate in the formation of urate from guanine. In humans, urate is the final product of purine degradation and is excreted in the urine.

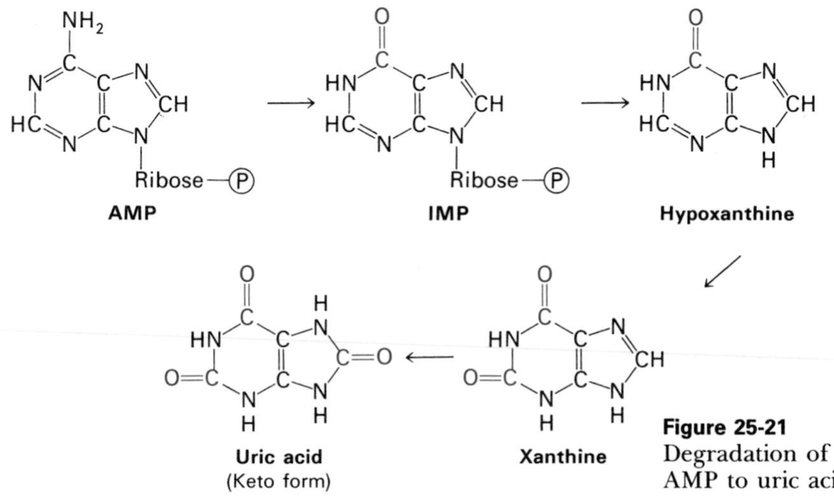

Figure 25-21
Degradation of AMP to uric acid.

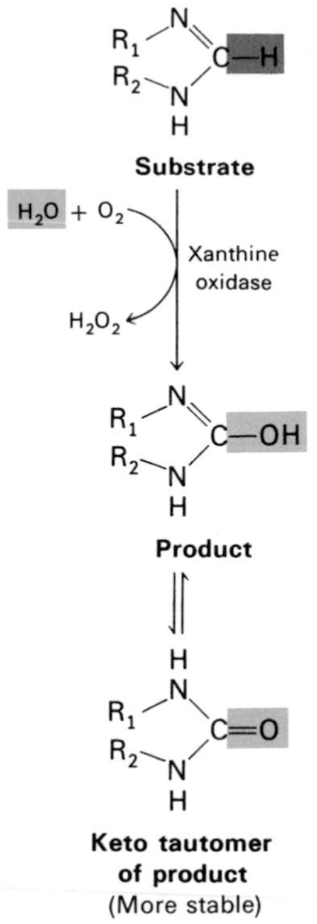

Figure 25-22
Oxidation reaction catalyzed by xanthine oxidase.

URATE IS FURTHER DEGRADED IN SOME ORGANISMS

The breakdown of purines proceeds further in some species (Figure 25-23). Mammals other than primates excrete *allantoin,* which is formed by oxidation of urate. Teleost fish excrete *allantoate,* which is formed by hydration of allantoin. The degradation proceeds a step further in amphibians and most fish. Allantoate is hydrolyzed to two molecules of

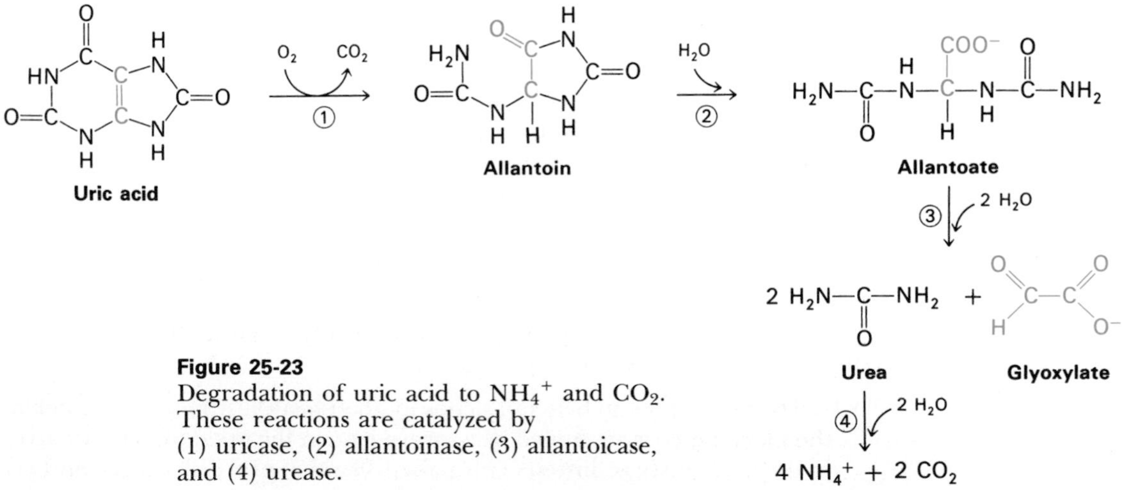

Figure 25-23
Degradation of uric acid to NH_4^+ and CO_2. These reactions are catalyzed by (1) uricase, (2) allantoinase, (3) allantoicase, and (4) urease.

urea and one of *glyoxylate*. Finally, some marine invertebrates hydrolyze urea to NH_4^+ and CO_2. It seems likely that the enzymes catalyzing these reactions were progressively lost in the evolution of primates.

BIRDS AND TERRESTRIAL REPTILES EXCRETE URATE INSTEAD OF UREA TO CONSERVE WATER

In terrestrial reptiles and birds, urea is not the final product of amino nitrogen metabolism. These animals synthesize purines from their excess amino nitrogen and then degrade these purines to urate. *This mode of eliminating nitrogen serves a vital function: the conservation of water. Urate is the vehicle for the excretion of amino nitrogen because of its very low solubility at acid pH.* The pK_a of the most acidic group in uric acid is 5.4. The acidic urine of birds and terrestrial reptiles consists of a paste of crystals of uric acid. Little water accompanies the excretion of these crystals, in contrast with the excretion of a comparable amount of highly soluble urea.

Uric acid (Keto form) ⇌ **Uric acid** (Enol form) ⇌ (pK 5.4, H^+) **Urate**

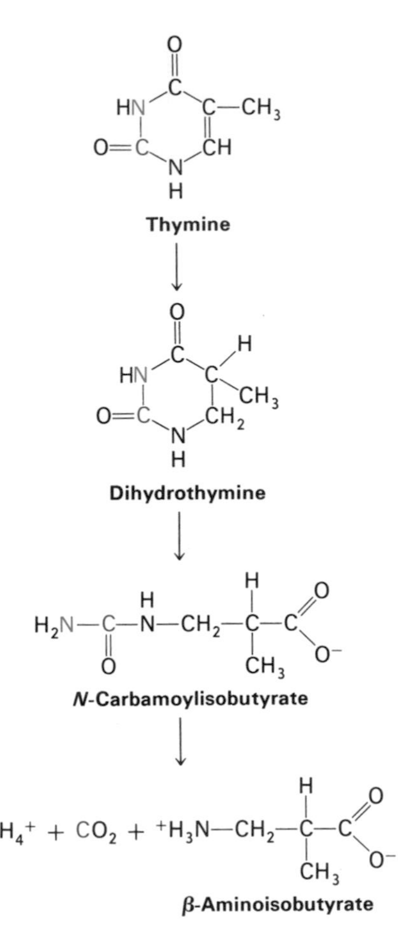

Figure 25-24
Degradation of thymine.

DEGRADATION OF PYRIMIDINES

The degradation of thymine (Figure 25-24) is illustrative of the breakdown of pyrimidines. Thymine is degraded to β-aminoisobutyrate, which is metabolized as though it were an α-amino acid. The amino group is removed by transamination to yield methylmalonate semialdehyde, which is converted into methylmalonyl CoA. The conversion of methylmalonyl CoA into succinyl CoA, the point of entry of several amino acid skeletons into the citric acid cycle, has already been discussed (p. 503). The nitrogen atoms of thymine are excreted in the form of urea.

HIGH SERUM LEVELS OF URATE CAUSES GOUT

Gout is a disease that affects the joints and kidneys. The major biochemical feature of gout is an *elevated level of urate in the serum*, a necessary but not sufficient condition for the production of the disease. Inflammation of the joints is triggered by the *precipitation of sodium urate crystals*. Kidney disease may also occur because of the deposition of urate crystals in that organ. Gout primarily affects adult males. A vivid description of an acute attack of gout was given by Thomas Sydenham, an outstanding seventeenth-century English physician, who himself was afflicted with this disease:

> The victim goes to bed and sleeps in good health. About two o'clock in the morning he is awakened by a severe pain in the great toe; more rarely in the heel, ankle or instep. This pain is like that of a dislocation, and yet

> the parts feel as if cold water were poured over them. Then follow chills and shivers, and a little fever. The pain, which was at first moderate, becomes more intense. With its intensity the chills and shivers increase. After a time this comes to its height, accommodating itself to the bones and ligaments of the tarsus and metatarsus. Now it is a violent stretching and tearing of the ligaments—now it is a gnawing pain and now a pressure and tightening. So exquisite and lively meanwhile is the feeling of the part affected, that it cannot bear the weight of the bedclothes nor the jar of a person walking in the room. The night is passed in torture, sleeplessness, turning of the part affected, and perpetual change of posture; the tossing about of the body being as incessant as the pain of the tortured joint, and being worse as the fit comes on.

The biochemical lesion in most cases of gout has not been elucidated. It seems likely that gout is an expression of a variety of inborn errors of metabolism in which *excessive production of urate* is a common finding. Some patients with this abnormality have a partial deficiency of *hypoxanthine-guanine phosphoribosyl transferase* (HGPRT), the enzyme catalyzing the salvage synthesis of IMP and GMP (p. 606).

$$\text{Hypoxanthine} + \text{PRPP} \xrightarrow{\text{HGPRT}} \text{IMP} + \text{PP}_i$$

$$\text{Guanine} + \text{PRPP} \xrightarrow{\text{HGPRT}} \text{GMP} + \text{PP}_i$$

Deficiency of HGPRT leads to reduced synthesis of GMP and IMP by the salvage pathway and an increased level of PRPP. There is a *marked acceleration of purine biosynthesis by the de novo pathway.* A few patients with gout have an *abnormally high level of active phosphoribosylpyrophosphate synthetase.* The allosteric control of the synthetase is impaired in these patients. This results in *excessive production of PRPP,* which in turn accelerates the rate of de novo synthesis of purines.

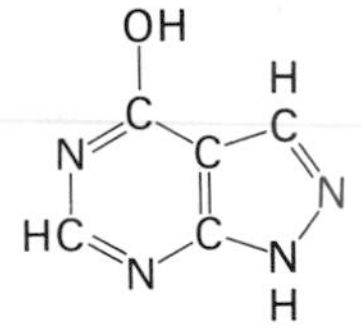

Allopurinol, an analog of hypoxanthine in which the N and C atoms at positions 7 and 8 are interchanged, is used to treat gout. The mechanism of action of allopurinol is very interesting: it acts *first as a substrate* and *then as an inhibitor* of xanthine oxidase. This enzyme hydroxylates allopurinol to *alloxanthine,* which then remains tightly bound to the active site. The molybdenum atom of xanthine oxidase is kept in the +4 oxidation state by the binding of alloxanthine instead of returning to the +6 oxidation state as it does in a normal catalytic cycle. We see here another example of *suicide inhibition.*

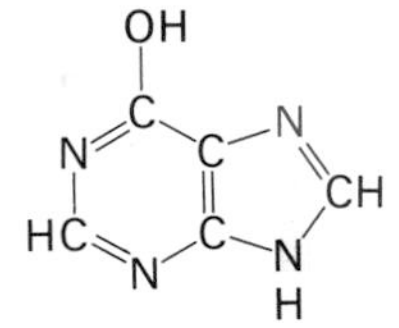

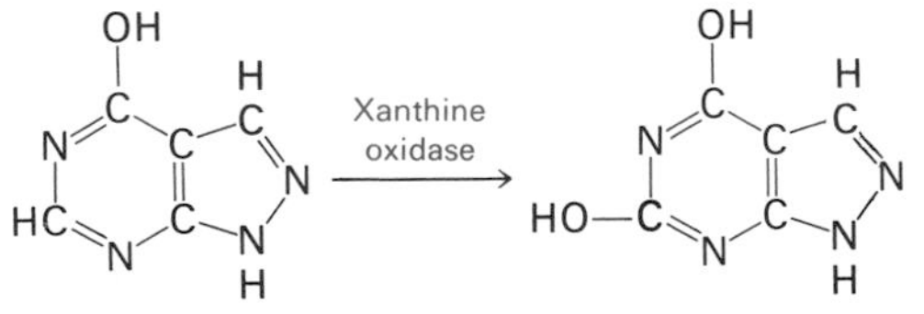

Allopurinol

Hypoxanthine

Xanthine oxidase

Allopurinol → **Alloxanthine**

The synthesis of urate from hypoxanthine and xanthine decreases soon after the administration of allopurinol.

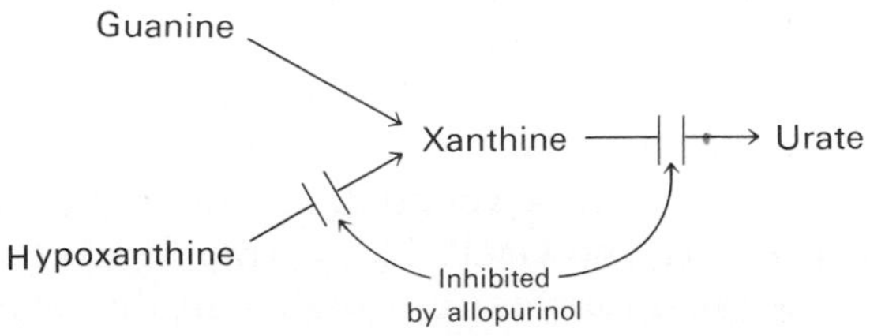

Guanine → Xanthine → Urate

Hypoxanthine → Xanthine

Inhibited by allopurinol

The serum concentrations of hypoxanthine and xanthine increase after administration of allopurinol, whereas that of urate drops. The formation of uric acid stones is virtually abolished by allopurinol, and there is some improvement in the arthritis. Also, *there is a decrease in the total rate of purine biosynthesis.* This inhibitory action of allopurinol depends on its reaction with PRPP to form the ribonucleotide. Consequently, the level of PRPP, the limiting substrate in the de novo synthesis of purines, is lowered. Furthermore, allopurinol ribonucleotide inhibits the conversion of PRPP into phosphoribosylamine by amidophosphoribosyltransferase.

URATE PLAYS A BENEFICIAL ROLE AS A POTENT ANTIOXIDANT

The average serum level of urate in humans is close to the solubility limit. In contrast, prosimians (such as lemurs) have ten-fold lower levels. A striking increase in urate level occurred in the evolution of primates. What is the selective advantage of a urate level so high that it teeters on the brink of gout in many people? It turns out that urate has a markedly beneficial action. Urate is a very efficient scavenger of highly reactive and harmful oxygen species—namely, hydroxyl radicals, superoxide anion, singlet oxygen, and oxygenated heme intermediates in high Fe valence states (+4 and +5). Indeed, urate is about as effective as ascorbate as an antioxidant. The increased level of urate in humans compared with prosimians and other lower primates may contribute significantly to the longer life span of humans and to the lower incidence of human cancer. We see in urate, as in bilirubin (p. 597), an expression of the principle that *some end products of degradative metabolic pathways play important roles as protective agents.*

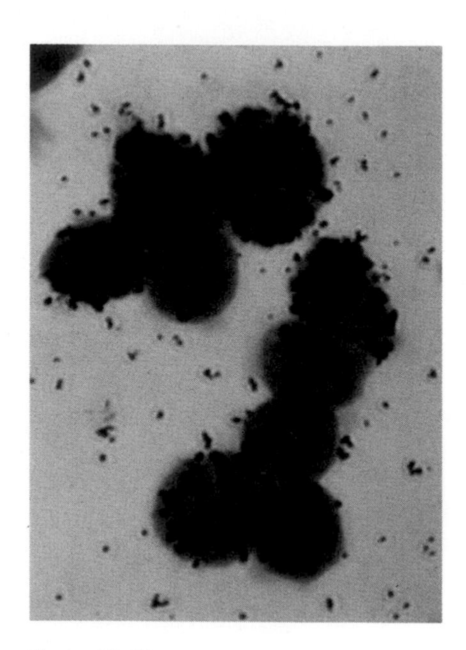

Figure 25-25
Mutant lymphoblastoid cells devoid of hypoxanthine-guanine phosphoribosyl transferase (HGPRT) were transfected with plasmids containing the HGPRT gene. The capacity of the transfected cells to synthesize purine ribonucleotides by the salvage pathway was measured by autoradiography following incubation of the cells in a medium containing tritiated hypoxanthine. A high concentration of grains in the photographic emulsion over a cell indicates a functional salvage pathway. About half of the transfected cells expressed the HGPRT gene. [Courtesy of Dr. Hiroshi Hayaka and Dr. Paul Berg.]

LESCH-NYHAN SYNDROME: SELF-MUTILATION, MENTAL RETARDATION, AND EXCESSIVE PRODUCTION OF URATE

A nearly total absence of hypoxanthine-guanine phosphoribosyltransferase has devastating consequences. The most striking expression of this inborn error of metabolism, called the *Lesch-Nyhan syndrome,* is *compulsive self-destructive behavior.* At age two or three, children with this disease begin to bite their fingers and lips. The tendency to self-mutilate is so extreme that it is necessary to protect these patients by such measures as wrapping their hands in gauze. Those afflicted also tend to be aggressive toward others. *Mental deficiency* and *spasticity* are other characteristics of the Lesch-Nyhan syndrome. Elevated levels of urate in the serum lead to the formation of stones early in life, followed by the symptoms of *gout* years later. The disease is inherited as a sex-linked recessive disorder.

The biochemical consequences of the virtual absence of hypoxanthine-guanine phosphoribosyltransferase are an *overproduction of urate* and an *elevated concentration of PRPP.* Also, there is a marked increase in the rate of purine biosynthesis by the de novo pathway. The relation between the absence of the transferase and the bizarre neurologic signs is an enigma. The brain may be very dependent on the salvage pathway for the synthesis of IMP and GMP. The normal level of hypoxanthine-guanine phosphoribosyltransferase is higher in the brain than in any other tissue. In contrast, the activity of the amidotransferase that catalyzes the committed step in the de novo pathway is rather low in the brain. Allopurinol is effective in diminishing urate synthesis in the Lesch-Nyhan syndrome. However, it has no effect on the rate of de novo synthesis of purines, and it fails to alleviate the neurologic expressions of the disease. Patients with the Lesch-Nyhan syndrome do not convert allopurinol into the ribonucleotide because they lack hypoxanthine-guanine phosphoribosyltransferase. Hence, the administration of allopurinol does not lower the level of PRPP in these individuals and so de novo purine synthesis is not diminished.

The Lesch-Nyhan syndrome demonstrates that the salvage pathway for the synthesis of IMP and GMP is not gratuitous. The salvage pathway evidently serves a critical role that is not yet fully understood. Furthermore, the interplay between the de novo and salvage pathways of purine synthesis remains to be elucidated. Moreover, *the Lesch-Nyhan syndrome reveals that abnormal behavior such as self-mutilation and extreme hostility can be caused by the absence of a single enzyme.* Psychiatry will no doubt benefit from a better understanding of the molecular basis of mental disorders.

SUMMARY

The purine ring is assembled from a variety of precursors: glutamine, glycine, aspartate, methylenetetrahydrofolate, N^{10}-formyltetrahydrofolate, and CO_2. The committed step in the de novo synthesis of purine nucleotides is the formation of 5-phosphoribosylamine from PRPP and glutamine. The purine ring is assembled on ribose phosphate. The addition of glycine, followed by formylation, amination, and ring closure, yields 5-aminoimidazole ribonucleotide. This intermediate contains the completed five-membered ring of the purine skeleton. The addition of CO_2, the nitrogen atom of aspartate, and a formyl group, followed by ring closure, yields inosinate (IMP), a purine ribonucleotide. AMP and GMP are formed from IMP. Purine ribonucleotides can also be synthesized by a salvage pathway in which a preformed base reacts directly with PRPP. Feedback inhibition of 5-phosphoribosyl-1-pyrophosphate synthetase and of glutamine-PRPP amidotransferase by purine nucleotides is important in regulating their biosynthesis.

The pyrimidine ring is assembled first and then linked to ribose phosphate to form a pyrimidine nucleotide, in contrast with the sequence in the de novo synthesis of purine nucleotides. PRPP is again the donor of the ribose phosphate moiety. The synthesis of the pyrimidine ring starts with the formation of carbamoylaspartate from carbamoyl phosphate and aspartate, a reaction catalyzed by aspartate transcarbamoylase. Dehydration, cyclization, and oxidation yield orotate, which reacts with PRPP to give orotidylate. Decarboxylation of this pyrimidine nucleotide yields UMP. CTP is then formed by amination of UTP. Pyrimidine biosynthesis in *E. coli* is regulated by feedback inhibition of aspartate transcarbamoylase, the enzyme that catalyzes the committed step. CTP inhibits and ATP stimulates this enzyme. In higher organisms, the first three enzymes of pyrimidine biosynthesis (called CAD) are present on a single polypeptide chain.

Deoxyribonucleotides, the precursors of DNA, are formed by the reduction of ribonucleoside diphosphates. These conversions are catalyzed by ribonucleotide reductase. Electrons are transferred from

NADPH to sulfhydryl groups at the active sites of this enzyme by thioredoxin or glutaredoxin. A tyrosine radical cation stabilized by an iron center participates in catalyzing the exchange of H for OH at C-2 of the sugar unit. dTMP is formed by methylation of dUMP. The one-carbon and electron donor in this reaction is N^5, N^{10}-methylenetetrahydrofolate, which is converted into dihydrofolate. Tetrahydrofolate is regenerated by the reduction of dihydrofolate by NADPH. Dihydrofolate reductase, which catalyzes this reaction, is inhibited by folate analogs such as aminopterin and methotrexate. These compounds are used as anti-cancer drugs.

Purines are degraded to urate in humans. Gout, a disease that affects joints and leads to arthritis, is associated with excessive production of urate. Allopurinol, a suicide inhibitor of xanthine oxidase, is used to treat gout. This analog of hypoxanthine blocks the formation of urate from hypoxanthine and xanthine. The Lesch-Nyhan syndrome, a genetic disease characterized by self-mutilation, mental deficiency, and gout, is caused by the absence of hypoxanthine-guanine phosphoribosyl transferase. This enzyme is essential for the synthesis of purine nucleotides by the salvage pathway.

SELECTED READINGS

Where to start

Reichard, P., and Ehrenberg, A., 1983. Ribonucleotide reductase—a radical enzyme. *Science* 221:514–519.

Villafranca, J. E., Howell, E. E., Voet, D. H., Strobel, M. S., Ogden, R. C., Abelson, J. N., and Kraut, J., 1983. Directed mutagenesis of dihydrofolate reductase. *Science* 222:782–788.

Hardy, L. W., Finer-Moore, J. S., Montfort, W. R., Jones, M. O., Santi, D. V., and Stroud, R. M., 1987. Atomic structure of thymidylate synthase: target for rational drug design. *Science* 235:448–455.

Ames, B. N., Cathcart, R., Schwiers, E., and Hochstein, P., 1981. Uric acid provides an antioxidant defense in humans against oxidant- and radical-caused aging and cancer: a hypothesis. *Proc. Nat. Acad. Sci.* 78:6858–6862.

Books on nucleotide metabolism

Kornberg, A., 1980. *DNA Replication.* W. H. Freeman. [Chapter 2 gives an excellent account of the biosynthesis of DNA precursors. An update is given in Chapter 2 of his *1982 Supplement to DNA Replication.*]

Much-Petersen, A., (ed.), 1983. *Metabolism of Nucleotides, Nucleosides, and Nucleobases in Microorganisms.* Academic Press.

Reese, C. B., (ed.), 1984. *Recent Aspects of the Chemistry of Nucleosides, Nucleotides, and Nucleic Acids.* Pergamon.

Walsh, C., 1979. *Enzymatic Reaction Mechanisms.* W. H. Freeman. [Several important reaction mechanisms in the biosynthesis of nucleotides are lucidly treated. See Chapter 5 for a discussion of amino transfers, Chapter 13 for xanthine oxidase, and Chapter 25 for one-carbon transfers.]

Hoffee, P. A., and Jones, M. E., (eds.), 1978. *Purine and Pyrimidine Nucleotide Metabolism.* Volume 51 of *Methods in Enzymology.* Academic Press.

Blakley, R. L., and Benkovic, S. J., 1985. *Folates and Pterins,* vols. 1 and 2. Wiley.

Blakley, R. L., 1969. *The Biochemistry of Folic Acid and Related Pteridines.* North-Holland.

Purine and pyrimidine biosynthesis

Daubner, S. C., Schrishsher, J. L., Schendel, F. J., Young, M., Henikoff, S., Patterson, D., Stubbe, J., and Benkovic, S. J., 1985. A multifunctional protein possessing glycinamide ribonucleotide synthetase, glycinamide ribonucleotide transformylase, and aminoimidazole ribonucleotide synthetase activities in de novo purine biosynthesis. *Biochemistry* 24:7059–7062.

Jones, M. E., 1980. Pyrimidine nucleotide biosynthesis in animals: genes, enzymes, and regulation of UMP biosynthesis. *Ann. Rev. Biochem.* 49:253–279.

Lee, L., Kelly, R. E., Pastra-Landis, S. C., and Evans, D. R., 1985. Oligomeric structure of the multifunctional protein CAD that initiates pyrimidine biosynthesis in mammalian cells. *Proc. Nat. Acad. Sci.* 82:6802–6806.

Coleman, P. F., Suttle, D. P., and Stark, G. R., 1977. Purification from hamster cells of the multifunctional protein that initiates de novo synthesis of pyrimidine nucleotides. *J. Biol. Chem.* 252:6379–6385.

Floyd, E. E., and Jones, M. E., 1985. Isolation and characterization of the orotidine 5′-monophosphate decarboxylase domain of the multifunctional protein uridine 5′-monophosphate synthase. *J. Biol. Chem.* 260:9443–9451.

Ribonucleotide reductase

Barlow, T., Eliasson, R., Platz, A., Reichard, P., and Sjoberg, B.-M., 1983. Enzymic modification of a tyrosine residue to a stable free radical in ribonucleotide reductase. *Proc. Nat. Acad. Sci.* 80:1492–1495.

Eliasson, R., Jornvall, H., and Reichard, P., 1986. Superoxide dismutase participates in the enzymatic formation of the tyrosine radical of ribonucleotide reductase from *Escherichia coli. Proc. Nat. Acad. Sci.* 83:2373–2377.

Thelander, L., and Reichard, P., 1979. Reduction of ribonucleotides. *Ann. Rev. Biochem.* 48:133–158.

Stubbe, J. A., 1983. Mechanism of B_{12}-dependent ribonucleotide reductase. *Mol. Cell Biochem.* 50:25–45.

Ashley, G. W., Harris, G., Stubbe, J., 1986. The mechanism of *Lactobacillus leichmannii* ribonucleotide reductase. Evidence for 3′ C-H bond cleavage and a unique role for coenzyme B_{12}. *J. Biol. Chem.* 261:3958–3964.

Thioredoxin and glutaredoxin

Holmgren, A., Soderberg, B.-O., Eklund, H., and Brändén, C.-I., 1975. Three-dimensional structure of *Escherichia coli* thioredoxin-S_2 to 2.8 Å resolution. *Proc. Nat. Acad. Sci.* 72:2305–2309.

Holmgren, A., 1985. Thioredoxin. *Ann. Rev. Biochem.* 54:237–271.

Eklund, H., Cambillau, C., Sjoberg, B. M., Holmgren, A., Jornvall, H., Hoog, J. O., and Brändén, C. I., 1984. Conformational and functional similarities between glutaredoxin and thioredoxins. *EMBO J.* 3:1443–1449.

Thymidylate synthase and dihydrofolate reductase

Howell, E. E., Villafranca, J. E., Warren, M. S., Oatley, S. J., and Kraut, J., 1986. Functional role of aspartic acid-27 in dihydrofolate reductase revealed by mutagenesis. *Science* 231:1123–1128.

Roth, B., 1986. Design of dihydrofolate reductase inhibitors from x-ray crystal structures. *Fed. Proc.* 45:2765–2772.

Schimke, R. T., 1986. Methotrexate resistance and gene amplification. Mechanisms and implications. *Cancer* 57:1912–1917.

Pagolotti, A. L., Jr., and Santi, D. V., 1977. The catalytic mechanism of thymidylate synthetase. *Bioorg. Chem.* 1:277–311.

Matthews, D. A., Alden, R. A., Bolin, J. T., Freer, S. T., Hamlin, R., Xuong, N., Kraut, J., Williams, M., Poe, M., and Hoogsteen, K., 1977. Dihydrofolate reductase: x-ray structure of the binary complex with methotrexate. *Science* 197:452–455.

Benkovic, S. J., 1980. On the mechanism of action of folate- and biopterin-requiring enzymes. *Ann. Rev. Biochem.* 49:227–251.

Uric acid

Kelley, W. N., and Weiner, I. M., (eds.), 1978. *Uric Acid.* Springer-Verlag. [A valuable collection of reviews on uric acid biochemistry and physiology, gout, and purine metabolism.]

Genetic diseases

Stanbury, J. B., Wyngaarden, J. B., Fredrickson, D. S., Goldstein, J. L., and Brown, M. S., (eds.), 1983. *The Metabolic Basis of Inherited Disease* (5th ed.). McGraw-Hill. [The chapters on inborn errors of purine and pyrimidine metabolism are excellent. See the articles on gout and the Lesch-Nyhan syndrome by J. B. Wyngaarden and W. B. Kelley, orotic aciduria by W. N. Kelly and L. H. Smith, Jr., and xanthinuria by J. B. Wyngaarden.]

Gruber, H. E., Finley, K. D., Hershberg, R. M., Katzman, S. S., Laikind, P. K., Seegmiller, J. E., Friedmann, T., Yee, J. K., and Jolly, D. J., 1985. Retroviral vector-mediated gene transfer into human progenitor cells. *Science* 230:1057–1061. [Transfer of the human HGPRT gene into human bone marrow cells.]

Wilson, J. M., Young, A. B., and Kelley, W. N., 1983. Hypoxanthine-guanine phosphoribosyltransferase deficiency. *New Engl. J. Med.* 309:900–910.

Seegmiller, J. E., 1985. Overview of possible relation of defects in purine metabolism to immune deficiency. *Ann. N. Y. Acad. Sci.* 451:9–19.

Seegmiller, J. E., 1980. *In* Bundy, P. K., and Rosenberg, L. E., (eds.), *Metabolic Control and Disease* (8th ed.), pp. 777–937. Saunders. [An excellent presentation of diseases of purine and pyrimidine metabolism.]

PROBLEMS

1. Write a balanced equation for the synthesis of PRPP from glucose via the oxidative branch of the pentose phosphate pathway.

2. Write a balanced equation for the synthesis of orotate from glutamine, CO_2, and aspartate.

3. What is the activated reactant in the biosynthesis of each of these compounds?
   (a) Phosphoribosylamine.
   (b) Carbamoylaspartate.
   (c) Orotidylate (from orotate).
   (d) Nicotinate ribonucleotide.
   (e) Phosphoribosyl anthranilate.

4. Amidotransferases are inhibited by the antibiotic azaserine (*O*-diazoacetyl-L-serine), which is an analog of glutamine.

   $$^{-}N{=}\overset{+}{N}{=}CH{-}\underset{\underset{O}{\|}}{C}{-}O{-}CH_2{-}\underset{\underset{NH_3^+}{|}}{CH}{-}COO^-$$

   Azaserine

   Which intermediates in purine biosynthesis would accumulate in cells treated with azaserine?

5. Write a balanced equation for the synthesis of dTMP from dUMP that is coupled to the conversion of serine into glycine.

6. Bacterial growth is inhibited by sulfanilamide and related sulfa drugs, and there is a concomitant accumulation of 5-aminoimidazole-4-carboxamide ribonucleotide. This inhibition is reversed by the addition of *p*-aminobenzoate.

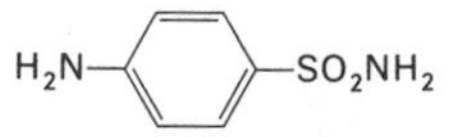

H_2N—⟨benzene⟩—SO_2NH_2

Sulfanilamide

Propose a mechanism for the inhibitory effect of sulfanilamide.

7. What are the major biosynthetic reactions that utilize PRPP?

8. Carbamoyl phosphate synthetase from *E. coli* consists of a small subunit (40 kd) and a large one (130 kd). The isolated small subunit has glutaminase activity, whereas the isolated large subunit can synthesize carbamoyl phosphate from NH_3 but not from glutamine. Furthermore, the isolated small subunit is an ATPase in the presence of bicarbonate. Propose a plausible reaction mechanism for the intact enzyme on the basis of the activities of its separated subunits.

9. In purine biosynthesis, aspartate becomes linked to the carboxylate group of an intermediate. In the next reaction, the carbon skeleton of the aspartate unit leaves in the form of fumarate. These reactions are reminiscent of a similar pair of reactions in the degradation of amino acids. Which ones?

10. Tyrosine 122 in the B2 subunit of ribonucleotide reductase has been implicated as the site of the free radical essential for enzymatic activity. Design an experiment to test this hypothesis.

11. The three-dimensional structure and enzymatic properties of two mutants of dihydrofolate reductase have been studied in detail. Aspartate 27 at the active site of the enzyme (Figure 25-26) was replaced by asparagine or serine. The structures of the Asn 27 mutant and the Ser 27 mutant were found to be nearly identical to the wild type. A water molecule serves as a bridge between Ser 27 and bound methotrexate. The binding constants for methotrexate at pH 7 were 0.07 nM for the wild type, 1.9 nM for the Asn 27 mutant, and 210 nM for the Ser 27 mutant.
   (a) Calculate the free energy of binding of methotrexate by these three enzymes. What is the decrease in binding energy resulting from each mutation?
   (b) The wild-type enzyme binds the protonated form of methotrexate. Which form of methotrexate is likely to be bound by each mutant? Sketch the likely interactions based on the above data.

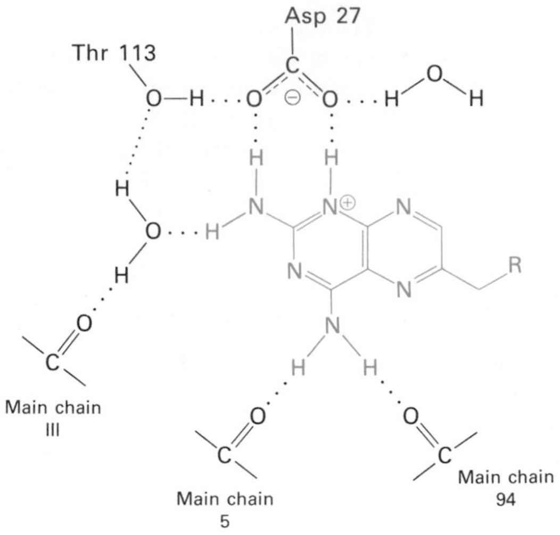

Figure 25-26
Mode of binding of methotrexate to native dihydrofolate reductase. [After E. H. Howell, J. E. Villafranca, M. S. Warren, S. J. Oatley, and J. Kraut. *Science* 231(1986):1125.]

12. Adenosine deaminase (ADA) catalyzes the irreversible deamination of adenosine and 2′-deoxyadenosine to inosine and 2′-deoxyinosine, respectively. Inherited defects of ADA lead to abnormalities in purine nucleoside metabolism that are selectively toxic to lymphocytes. They result in immune deficiency diseases.
   (a) The intracellular level of *S*-adenosylhomocysteine is elevated in ADA-deficient patients. Why?
   (b) Predict the major consequence of an elevated level of *S*-adenosylhomocysteine.

13. Mutant cells unable to synthesize nucleotides by salvage pathways are very useful tools in molecular and cell biology. Suppose that cell A lacks thymidine kinase, the enzyme catalyzing the phosphorylation of thymidine to thymidylate, and that cell B lacks hypoxanthine-guanine phosphoribosyl transferase.
   (a) Cell A and cell B do not grow and divide in a *HAT* medium containing *h*ypoxanthine, *a*minopterin or *a*methopterin (methotrexate), and *t*hymine. However, cell C formed by the fusion of A and B grows in this medium. Why?
   (b) Suppose that you wanted to introduce foreign genes into cell A. Devise a simple means of distinguishing between cells that have taken up foreign DNA and those that have not.

14. Allopurinol is often given to patients with acute leukemia who are being treated with anti-cancer drugs. Why is allopurinol used?

CHAPTER 26

Integration of Metabolism

How is the intricate network of reactions in metabolism coordinated to meet the needs of the whole organism? This chapter presents some of the principles underlying the integration of metabolism in mammals. It starts with a recapitulation of the strategy of metabolism and of recurring motifs in its regulation. The interplay of different pathways is then described in terms of the flow of molecules at three key crossroads: glucose 6-phosphate, pyruvate, and acetyl CoA. This view of major junctions is followed by a discussion of differences in the metabolic patterns of the brain, muscle, adipose tissue, and liver. The major hormonal regulators of fuel metabolism—insulin, glucagon, epinephrine, and norepinephrine—are considered next. We then turn to a most important aspect of the integration of metabolism—the control of the level of glucose in the blood. The last part of the chapter deals with the remarkable metabolic adaptations in prolonged starvation.

"To every thing there is a season, and a time to every purpose under the heaven:
A time to be born, and a time to die;
A time to plant and a time to pluck up that which is planted;
A time to kill, and a time to heal;
A time to break down, and a time to build up."

Ecclesiastes

STRATEGY OF METABOLISM: A RECAPITULATION

As was explained in Chapter 13, the basic strategy of metabolism is to form ATP, reducing power, and building blocks for biosyntheses. Let us briefly review these central themes:

1. *ATP is the universal currency of energy.* The high phosphoryl transfer potential of ATP enables it to serve as the energy source in muscle contraction, active transport, signal amplification, and biosyntheses. The hydrolysis of an ATP molecule changes the equilibrium ratio of products to reactants in a coupled reaction by a factor of about 10^8.

Hence, a thermodynamically unfavorable reaction sequence can be made highly favorable by coupling it to the hydrolysis of a sufficient number of ATP molecules. For example, three molecules of ATP are consumed in the conversion of mevalonate into isopentenyl pyrophosphate, the activated five-carbon unit in the synthesis of cholesterol.

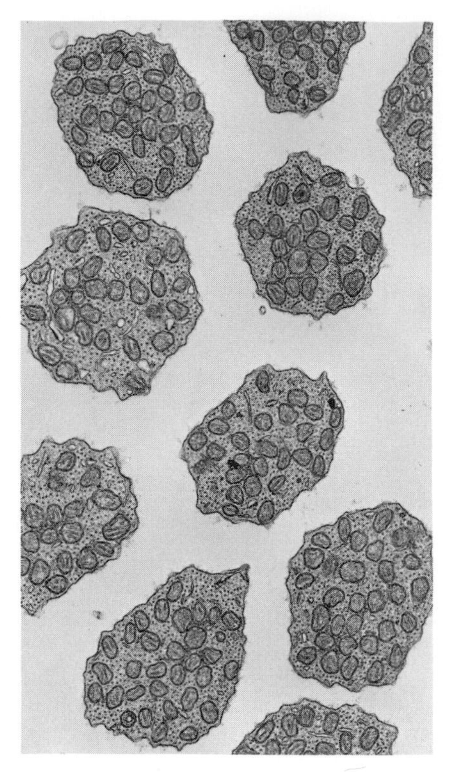

Figure 26-1
Electron micrograph showing numerous mitochondria in the inner segment of retinal rod cells. These photoreceptor cells generate large amounts of ATP and are highly dependent on a continuous supply of O_2. [Courtesy of Dr. Michael Hogan.]

2. *ATP is generated by the oxidation of fuel molecules such as glucose, fatty acids, and amino acids.* The common intermediate in most of these oxidations is acetyl CoA. The carbons of the acetyl unit are completely oxidized to CO_2 by the citric acid cycle with the concomitant formation of NADH and $FADH_2$. These electron carriers then transfer their high-potential electrons to the respiratory chain. The subsequent flow of electrons to O_2 leads to the pumping of protons across the inner mitochondrial membrane. This proton gradient is then used to synthesize ATP. Glycolysis also generates ATP, but the amount formed is much smaller than in oxidative phosphorylation. The oxidation of glucose to pyruvate yields only two ATP, whereas thirty-six (or thirty-eight) ATP are formed when glucose is completely oxidized to CO_2. However, glycolysis can proceed rapidly for a short time under anaerobic conditions, in contrast with oxidative phosphorylation, which requires a continuous supply of O_2.

3. *NADPH is the major electron donor in reductive biosyntheses.* In most biosyntheses, the products are more reduced than the precursors, and so reductive power is needed as well as ATP. The high-potential electrons required to drive these reactions are usually provided by NADPH. For example, in fatty acid biosynthesis, the keto group of an added two-carbon unit is reduced to a methylene group by the input of four electrons from two molecules of NADPH. The activation of O_2 by mixed-function oxygenases in hydroxylation reactions also illustrates the pervasive role of NADPH as a reductant. The pentose phosphate pathway supplies much of the required NADPH. This reductant is also formed by the tandem operation of the mitochondrial citrate-pyruvate shuttle and the malate enzyme in the cytosol.

4. *Biomolecules are constructed from a relatively small set of building blocks.* The highly diverse molecules of life are synthesized from a much smaller number of precursors. The metabolic pathways that generate ATP and NADPH also provide building blocks for the biosynthesis of more complex molecules. For example, dihydroxyacetone phosphate formed in glycolysis gives rise to the glycerol backbone of phosphatidyl choline and other phosphoglycerides. Phosphoenolpyruvate, another glycolytic intermediate, provides part of the carbon skeleton of the aromatic amino acids. Acetyl CoA, the common intermediate in the breakdown of most fuels, supplies a two-carbon unit in a wide variety of biosyntheses. Succinyl CoA, formed in the citric acid cycle, is one of the precursors of porphyrins. Ribose 5-phosphate, which is formed in addition to NADPH by the pentose phosphate pathway, is the source of the sugar unit of nucleotides. Many biosyntheses also require one-carbon units. Tetrahydrofolate is a source of these units at several oxidation levels. The formation of these derivatives and of *S*-adenosylmethionine, the major donor of methyl groups, is closely tied to amino acid metabolism. Thus, the central metabolic pathways have anabolic as well as catabolic roles.

5. *Biosynthetic and degradative pathways are almost always distinct.* For example, the pathway for the synthesis of fatty acids is different from

that of their degradation. Likewise, glycogen is synthesized and degraded by different sets of reactions. This separation enables both biosynthetic and degradative pathways to be thermodynamically favorable at all times. A biosynthetic pathway is made exergonic by coupling it to the hydrolysis of a sufficient number of ATP. For example, four more ~P are spent in converting pyruvate into glucose in gluconeogenesis than are gained in converting glucose into pyruvate in glycolysis. The four additional ~P assure that gluconeogenesis, like glycolysis, is highly exergonic under all cellular conditions. *The essential point is that the rates of metabolic pathways are governed more by the activities of key enzymes than by mass action.* The separation of biosynthetic and degradative pathways contributes greatly to the effectiveness of metabolic control.

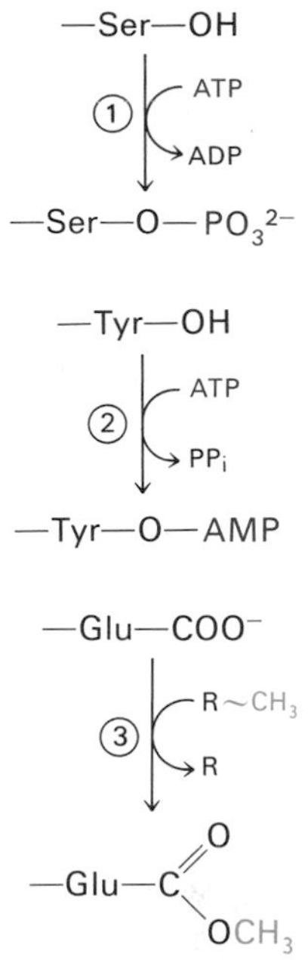

—Ser—OH

(1) ATP → ADP

—Ser—O—PO_3^{2-}

—Tyr—OH

(2) ATP → PP_i

—Tyr—O—AMP

—Glu—COO^-

(3) R~CH_3 → R

—Glu—C(=O)—OCH_3

Figure 26-2
Examples of reversible covalent modifications of proteins: (1) phosphorylation, (2) adenylylation, and (3) carboxymethylation.

RECURRING MOTIFS IN METABOLIC REGULATION

The complex network of reactions in a cell is precisely regulated and coordinated. Metabolism is controlled in several ways:

1. *Allosteric interactions.* The flow of molecules in most metabolic pathways is determined primarily by the amounts and activities of certain enzymes rather than by the amount of substrate available. Essentially irreversible reactions are potential control sites. The first irreversible reaction in a pathway (the committed step) is usually an important control element. Enzymes catalyzing committed steps are allosterically regulated, as exemplified by phosphofructokinase in glycolysis and acetyl CoA carboxylase in fatty acid synthesis. Subsequent irreversible reactions in a pathway also may be controlled. Allosteric interactions enable such enzymes to detect diverse signals and to integrate this information.

2. *Covalent modification.* Some regulatory enzymes are controlled by covalent modification in addition to allosteric interactions. For example, the catalytic activity of glycogen phosphorylase is enhanced by phosphorylation, whereas that of glycogen synthase is diminished. These covalent modifications are catalyzed by specific enzymes. Another example is glutamine synthetase, which is rendered less active by the covalent insertion of an AMP unit. Again, addition and removal of this modifying group is catalyzed by specific enzymes. Why is covalent modification used in addition to noncovalent allosteric control? Covalent modifications of key enzymes in metabolism are the final stage of amplifying cascades. Consequently, metabolic pathways can be rapidly switched on or off by very small triggering signals, as shown by the action of epinephrine in stimulating the breakdown of glycogen.

3. *Enzyme levels.* The amounts of enzymes, as well as their activities, are controlled. The rates of synthesis and degradation of some regulatory enzymes are subject to hormonal factors.

4. *Compartmentation.* The metabolic patterns of eucaryotic cells are markedly affected by the presence of compartments. Glycolysis, the pentose phosphate pathway, and fatty acid synthesis take place in the cytosol, whereas fatty acid oxidation, the citric acid cycle, and oxidative phosphorylation are carried out in mitochondria. Some processes, such as gluconeogenesis and urea synthesis, depend on the interplay of reactions that occur in both compartments. The fates of certain molecules

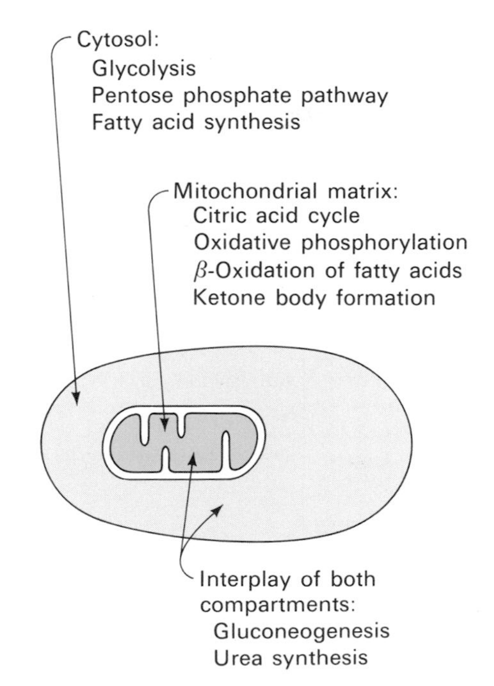

Figure 26-3
Compartmentation of the major pathways of metabolism.

depend on whether they are in the cytosol or in mitochondria, and so their flow across the inner mitochondrial membrane is often regulated. For example, fatty acids transported into mitochondria are rapidly degraded, in contrast with fatty acids in the cytosol, which are esterified or exported. Recall that long-chain fatty acids are transported into the mitochondrial matrix as esters of carnitine, a carrier that enables these molecules to traverse the inner mitochondrial membrane.

5. *Metabolic specializations of organs.* Regulation in higher eucaryotes is profoundly affected and enhanced by the existence of organs with different metabolic roles. These interactions will be discussed shortly.

MAJOR METABOLIC PATHWAYS AND CONTROL SITES

Let us review the roles of the major pathways of metabolism and the principal sites for their control:

> *Pasteur effect*—
> The inhibition of glycolysis by respiration, discovered by Louis Pasteur in studying fermentation by yeast. The consumption of carbohydrate is about sevenfold lower under aerobic conditions than under anaerobic ones. The inhibition of phosphofructokinase by citrate and ATP accounts for much of the Pasteur effect.

1. *Glycolysis.* This sequence of reactions in the cytosol converts glucose into two molecules of pyruvate with the concomitant generation of two ATP and two NADH. The NAD^+ consumed in the reaction catalyzed by glyceraldehyde 3-phosphate dehydrogenase must be regenerated for glycolysis to proceed. Under anaerobic conditions, as in highly active skeletal muscle, this is accomplished by the reduction of pyruvate to lactate. Alternatively, under aerobic conditions, NAD^+ is regenerated by the transfer of electrons from NADH to O_2 through the electron-transport chain. Glycolysis serves two main purposes: it degrades glucose to generate ATP and it provides carbon skeletons for biosyntheses. The rate of conversion of glucose into pyruvate is regulated to meet these dual needs.

Phosphofructokinase, which catalyzes the committed step in glycolysis, is the most important control site. A high level of ATP inhibits phosphofructokinase. This inhibitory effect is enhanced by citrate and reversed by AMP. Thus, the rate of glycolysis depends on the need for ATP, as signalled by the ATP/AMP ratio, and on the need for building blocks, as signalled by the level of citrate. *In liver, the most important regulator of phosphofructokinase activity is fructose 2,6-bisphosphate (F-2,6-BP).* Recall that the level of F-2,6-BP is determined by the activity of the kinase that forms it from fructose 6-phosphate and of the phosphatase that hydrolyzes the 2-phosphoryl group (p. 361). When the blood glucose level is low, a glucagon-triggered cascade leads to activation of this phosphatase and inhibition of this kinase in liver. The resulting decrease in the level of F-2,6-BP leads to the deactivation of phosphofructokinase, and hence glycolysis is slowed.

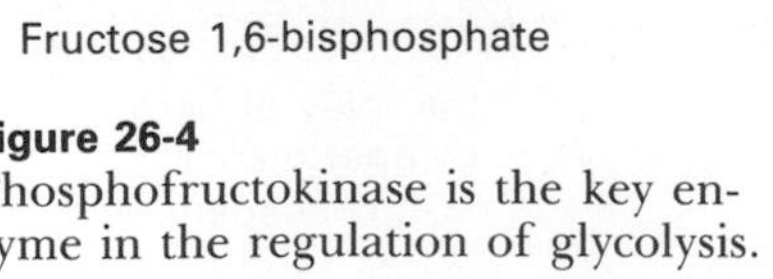

Fructose 6-phosphate

ATP → ADP

Phosphofructokinase
Activated by F-2,6-BP
Activated by AMP
Inhibited by ATP and citrate

Fructose 1,6-bisphosphate

Figure 26-4
Phosphofructokinase is the key enzyme in the regulation of glycolysis.

It is important to note that phosphofructokinase is controlled rather differently in muscle. Recent studies show that the muscle kinase catalyzing the synthesis of F-2,6-BP is stimulated, not inhibited, by cAMP-induced phosphorylation. Thus, epinephrine stimulates glycolysis in muscle but inhibits it in liver because of this key difference between the isozymes. The increased breakdown of liver glycogen induced by epinephrine serves to supply glucose to muscle, which rapidly consumes it to generate ATP for contractile activity.

2. *Citric acid cycle.* This common pathway for the oxidation of fuel molecules—carbohydrates, amino acids, and fatty acids—takes place inside mitochondria. Most fuels enter the cycle as acetyl CoA. The com-

plete oxidation of an acetyl unit generates one GTP, three NADH, and one $FADH_2$. The four pairs of electrons are then transferred to O_2 through the electron-transport chain, which results in the formation of a proton gradient that drives the synthesis of eleven ATP. NADH and $FADH_2$ are oxidized only if ADP is simultaneously phosphorylated to ATP. *This tight coupling, called respiratory control, ensures that the rate of the citric acid cycle matches the need for ATP.* An abundance of ATP also diminishes the activities of three enzymes in the cycle—citrate synthase, isocitrate dehydrogenase, and α-ketoglutarate dehydrogenase. The citric acid cycle also has an anabolic role. It provides intermediates for biosyntheses, such as succinyl CoA for the formation of porphyrins.

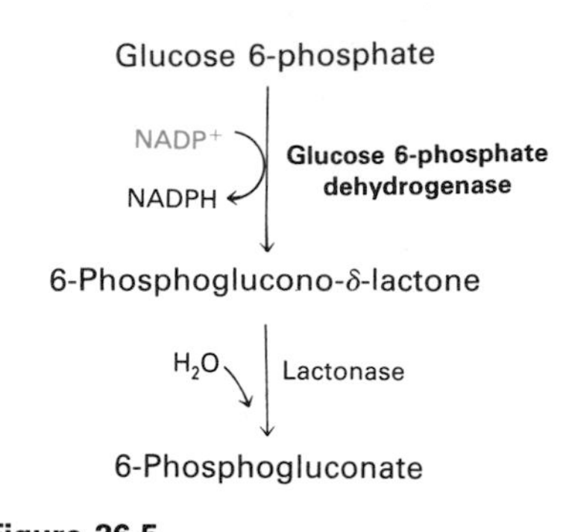

Figure 26-5
The dehydrogenation of glucose 6-phosphate is the committed step in the pentose phosphate pathway.

3. *Pentose phosphate pathway.* This series of reactions, which takes place in the cytosol, has two purposes: the generation of NADPH for reductive biosyntheses and the formation of ribose 5-phosphate for the synthesis of nucleotides. Two NADPH are generated in the conversion of glucose 6-phosphate into ribose 5-phosphate. The dehydrogenation of glucose 6-phosphate is the committed step in this pathway. This reaction is controlled by the level of $NADP^+$, the electron acceptor. The extra phosphoryl group in NADPH is a tag that distinguishes it from NADH. This differentiation makes it possible to have both a high $NADPH/NADP^+$ ratio and a high $NAD^+/NADH$ ratio in the same compartment. Consequently, reductive biosyntheses and glycolysis can proceed simultaneously at a high rate.

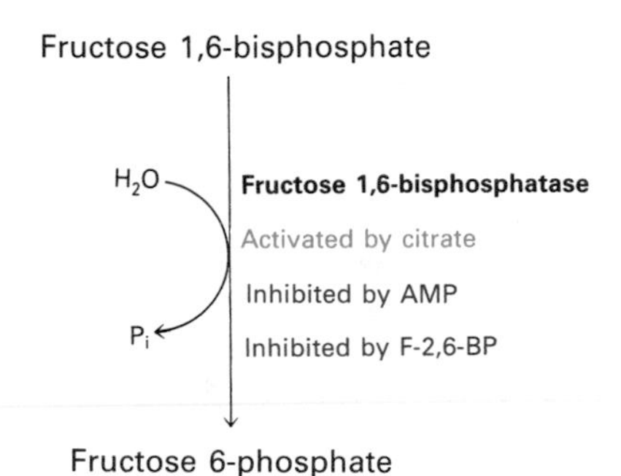

Figure 26-6
Fructose 1,6-bisphosphatase is the key control site in gluconeogenesis.

4. *Gluconeogenesis.* Glucose can be synthesized by the liver and kidneys from noncarbohydrate precursors such as lactate, glycerol, and amino acids. The major entry point of this pathway is pyruvate, which is carboxylated to oxaloacetate in mitochondria. Oxaloacetate is then decarboxylated and phosphorylated in the cytosol to form phosphoenolpyruvate. The other distinctive reactions of gluconeogenesis are two hydrolytic steps that bypass the irreversible reactions of glycolysis. *Gluconeogenesis and glycolysis are usually reciprocally regulated so that one pathway is quiescent while the other is highly active.* For example, AMP inhibits and citrate activates fructose 1,6-bisphosphatase, a key enzyme in gluconeogenesis, whereas these molecules have opposite effects on phosphofructokinase, the pacemaker of glycolysis. F-2,6-BP also coordinates these processes by inhibiting fructose 1,6-bisphosphatase. Hence, when glucose is abundant, the high level of F-2,6-BP inhibits gluconeogenesis and activates glycolysis.

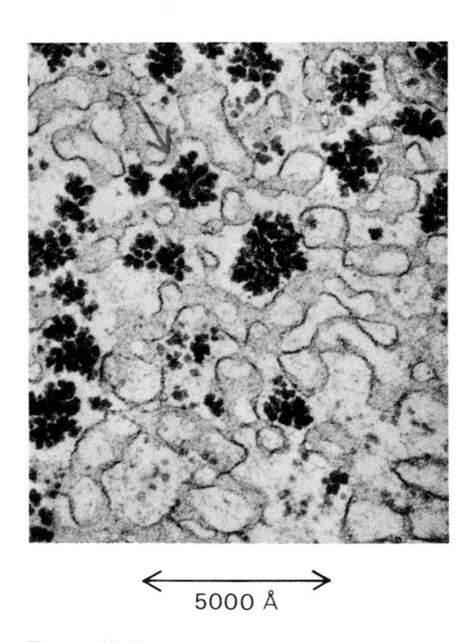

Figure 26-7
Electron micrograph of a portion of a liver cell showing glycogen particles. [Courtesy of Dr. George Palade.]

5. *Glycogen synthesis and degradation.* Glycogen, a readily mobilizable fuel store, is a branched polymer of glucose residues. The activated intermediate in its synthesis is UDP-glucose, which is formed from glucose 1-phosphate and UTP. Glycogen synthase catalyzes the transfer of glucose from UDP-glucose to the terminal hydroxyl residue of a growing strand. Glycogen is degraded by a different pathway. Phosphorylase catalyzes the cleavage of glycogen by orthophosphate to give glucose 1-phosphate. *Glycogen synthesis and degradation are coordinately controlled by a hormone-triggered amplifying cascade so that the synthase is inactive when phosphorylase is active, and vice versa.* These enzymes are regulated by phosphorylation and noncovalent allosteric interactions (p. 463).

6. *Fatty acid synthesis and degradation.* Fatty acids are synthesized in the cytosol by the addition of two-carbon units to a growing chain on an acyl carrier protein. Malonyl CoA, the activated intermediate, is formed

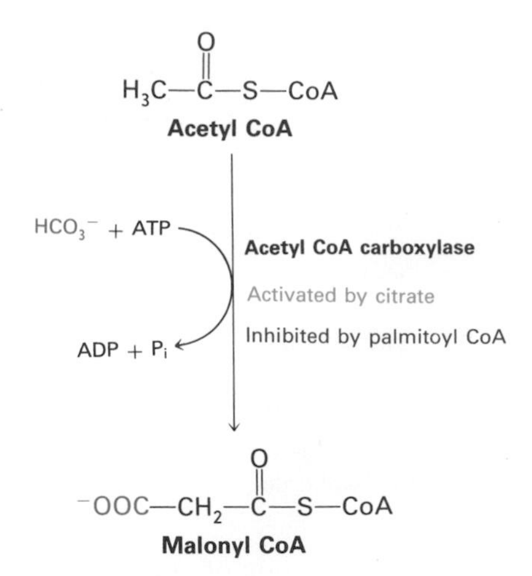

Figure 26-8
Acetyl CoA carboxylase is the key control site in fatty acid synthesis.

by the carboxylation of acetyl CoA. Acetyl groups are carried from mitochondria to the cytosol by the citrate-malate shuttle. *Citrate in the cytosol stimulates acetyl CoA carboxylase, the enzyme catalyzing the committed step. When ATP and acetyl CoA are abundant, the level of citrate increases, which accelerates the rate of fatty acid synthesis.* Fatty acids are degraded by a different pathway in a different compartment. They are degraded to acetyl CoA in the mitochondrial matrix by β-oxidation. The acetyl CoA then enters the citric acid cycle if the supply of oxaloacetate is sufficient. Alternatively, acetyl CoA can give rise to ketone bodies. The $FADH_2$ and NADH formed in the β-oxidation pathway transfer their electrons to O_2 through the electron-transport chain. Like the citric acid cycle, β-oxidation can continue only if NAD^+ and FAD are regenerated. *Hence, the rate of fatty acid degradation is also coupled to the need for ATP.*

KEY JUNCTIONS: GLUCOSE 6-PHOSPHATE, PYRUVATE, AND ACETYL CoA

The factors governing the flow of molecules in metabolism can be further understood by examining three key crossroads: glucose 6-phosphate, pyruvate, and acetyl CoA. Each of these molecules has several contrasting fates:

1. *Glucose 6-phosphate.* Glucose entering a cell is rapidly phosphorylated to glucose 6-phosphate, which can be stored as glycogen, degraded by way of pyruvate, or converted into ribose 5-phosphate (Figure 26-9). Glycogen is formed when glucose 6-phosphate and ATP are abundant. In contrast, glucose 6-phosphate flows into the glycolytic pathway when ATP or carbon skeletons for biosyntheses are required. Thus, the conversion of glucose 6-phosphate into pyruvate can be anabolic as well as catabolic. The third major fate of glucose 6-phosphate, to flow through the pentose phosphate pathway, provides NADPH for reductive biosyntheses and ribose 5-phosphate for the synthesis of nucleotides. The relative amounts of these two outputs can be adjusted over a very broad range by this highly flexible series of reactions, as discussed previously (p. 431). Glucose 6-phosphate can be formed by the mobilization of glycogen or it can be synthesized from pyruvate and glycogenic amino acids by the gluconeogenic pathway. As will be discussed shortly, a low level of glucose in the blood stimulates both glycogenolysis and gluconeogenesis in the *liver and kidneys. These organs are distinctive in possessing glucose 6-phosphatase, which enables glucose to be released into the blood.*

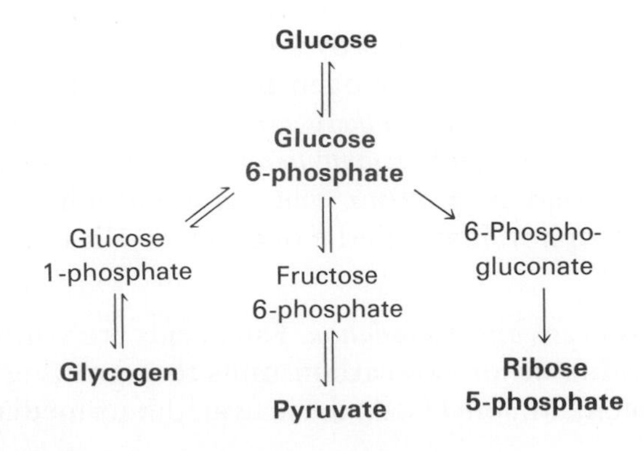

Figure 26-9
Metabolic fates of glucose 6-phosphate.

2. *Pyruvate.* This three-carbon α-keto acid is another major metabolic junction (Figure 26-10). Pyruvate is derived primarily from glucose 6-phosphate, alanine, and lactate (the reduced form of pyruvate). The facile reduction of pyruvate catalyzed by lactate dehydrogenase serves to regenerate NAD^+, enabling glycolysis to proceed transiently under anaerobic conditions. Lactate formed in active tissues such as contracting muscle is subsequently oxidized back to pyruvate, primarily in the liver. The essence of this interconversion is that it buys time and shifts part of the metabolic burden of active muscle to the liver. Another readily reversible reaction in the cytosol is the transamination of pyruvate, an α-keto acid, to alanine, the corresponding amino acid. Several amino acids can be synthesized from carbohydrate precursors by this route. Conversely, several amino acids can enter the central metabolic pathways in this way. Thus, transamination is a major link between amino acid and carbohydrate metabolism.

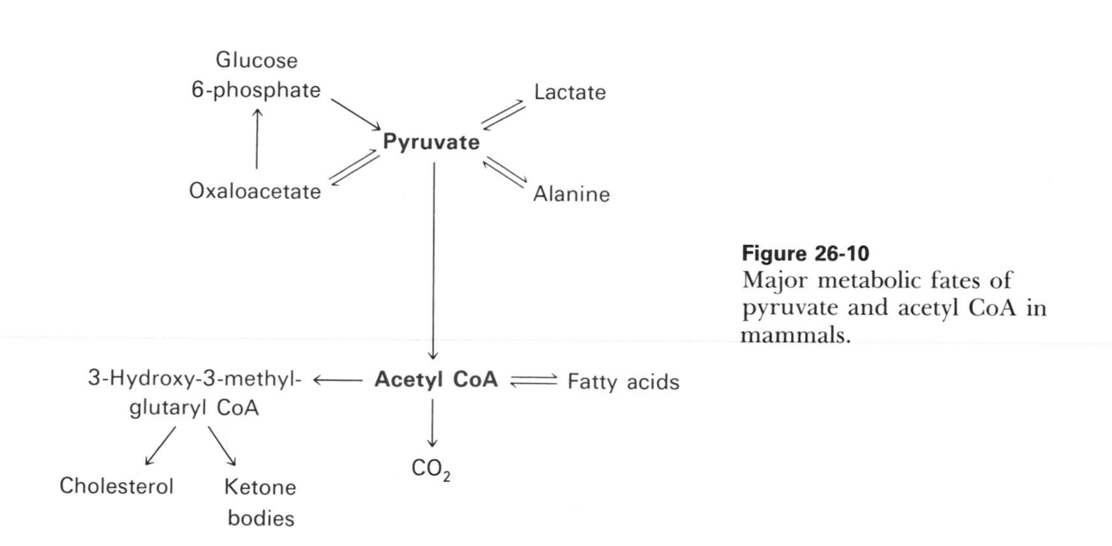

Figure 26-10
Major metabolic fates of pyruvate and acetyl CoA in mammals.

A third fate of pyruvate is its carboxylation to oxaloacetate inside mitochondria. This reaction and the subsequent conversion of oxaloacetate into phosphoenolpyruvate bypass an irreversible step of glycolysis and hence enable glucose to be synthesized from pyruvate. The carboxylation of pyruvate is also important for replenishing intermediates of the citric acid cycle. The activation of pyruvate carboxylase by acetyl CoA enhances the synthesis of oxaloacetate when the citric acid cycle is slowed by a paucity of this intermediate. On the other hand, oxaloacetate synthesized from pyruvate flows into the gluconeogenic pathway when the citric acid cycle is inhibited by an abundance of ATP.

A fourth fate of pyruvate is its oxidative decarboxylation to acetyl CoA. This irreversible reaction inside mitochondria is a decisive reaction in metabolism: it commits the carbon atoms of carbohydrates and amino acids to oxidation by the citric acid cycle or to the synthesis of lipids. The pyruvate dehydrogenase complex, which catalyzes this irreversible funneling, is stringently regulated by multiple allosteric interactions and covalent modifications. Pyruvate is rapidly converted into acetyl CoA only if ATP is needed or if two-carbon fragments are required for the synthesis of lipids.

3. *Acetyl CoA.* The major sources of this activated two-carbon unit are the oxidative decarboxylation of pyruvate and the β-oxidation of fatty

acids (Figure 26-10). Acetyl CoA is also derived from ketogenic amino acids. The fate of acetyl CoA, in contrast with many molecules in metabolism, is quite restricted. The acetyl unit can be completely oxidized to CO_2 by the citric acid cycle. Alternatively, 3-hydroxy-3-methylglutaryl CoA can be formed from three molecules of acetyl CoA. This six-carbon unit is a precursor of cholesterol and of ketone bodies, which are transport forms of acetyl units between liver and some peripheral tissues. A third major fate of acetyl CoA is its export to the cytosol in the form of citrate for the synthesis of fatty acids. It is important to stress again that acetyl CoA cannot be converted into pyruvate in mammals. Consequently, mammals are unable to convert lipid into carbohydrate.

METABOLIC PROFILES OF THE MAJOR ORGANS

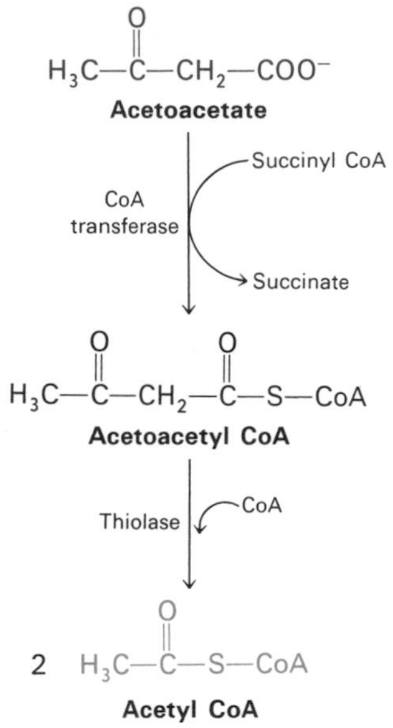

The metabolic patterns of the brain, muscle, adipose tissue, and the liver are strikingly different. Let us consider how these organs differ in their use of fuels to meet their energy needs:

1. *Brain. Glucose is virtually the sole fuel for the human brain, except during prolonged starvation.* The brain lacks fuel stores and hence requires a continuous supply of glucose, which enters freely at all times. It consumes about 120 g daily, which corresponds to an energy input of about 420 kcal. The brain accounts for some 60% of the utilization of glucose by the whole body in the resting state. *During starvation, ketone bodies (acetoacetate and 3-hydroxybutyrate, its reduced counterpart) partly replace glucose as fuel for the brain.* Acetoacetate is activated by the transfer of CoA from succinyl CoA to give acetoacetyl CoA (Figure 26-11). Cleavage by thiolase then yields two molecules of acetyl CoA, which enter the citric acid cycle. Fatty acids do not serve as fuel for the brain because they are bound to albumin in plasma and so they do not traverse the blood-brain barrier. In essence, *ketone bodies are transportable equivalents of fatty acids.* As will be discussed shortly, the shift in fuel usage from glucose to ketone bodies is critical in minimizing protein breakdown during starvation.

$$H_3C-\overset{O}{\overset{\|}{C}}-CH_2-COO^-$$
Acetoacetate

CoA transferase (Succinyl CoA → Succinate)

$$H_3C-\overset{O}{\overset{\|}{C}}-CH_2-\overset{O}{\overset{\|}{C}}-S-CoA$$
Acetoacetyl CoA

Thiolase (CoA)

$$2\ H_3C-\overset{O}{\overset{\|}{C}}-S-CoA$$
Acetyl CoA

Figure 26-11
Entry of ketone bodies into the citric acid cycle.

2. *Muscle. The major fuels for muscle are glucose, fatty acids, and ketone bodies.* Muscle differs from brain in having a large store of glycogen (1200 kcal). In fact, about three-fourths of all the glycogen in the body is stored in muscle (Table 26-1). The glycogen content of muscle after a

Table 26-1
Fuel reserves in a typical 70-kg man

	Available energy (kcal)		
Organ	*Glucose or glycogen*	*Triacylglycerols*	*Mobilizable proteins*
Blood	60	45	0
Liver	400	450	400
Brain	8	0	0
Muscle	1200	450	24,000
Adipose tissue	80	135,000	40

Source: After G. F. Cahill, Jr., *Clin. Endocrinol. Metab.* 5(1976):398.

meal can be as high as 1%. This glycogen is readily converted into glucose 6-phosphate for use within muscle cells. Muscle, like the brain, lacks glucose 6-phosphatase, and so it does not export glucose. *Rather, muscle retains glucose, its preferred fuel for bursts of activity.* In actively contracting skeletal muscle, the rate of glycolysis far exceeds that of the citric acid cycle. Much of the pyruvate formed under these conditions is reduced to lactate, which flows to the liver, where it is converted into glucose (Figure 26-12). These interchanges, known as the Cori cycle (p. 445), shift part of the metabolic burden of muscle to the liver. In addition, a large amount of alanine is formed in active muscle by the transamination of pyruvate. Alanine, like lactate, can be converted into glucose by the liver. The metabolic pattern of resting muscle is quite different. *In resting muscle, fatty acids are the major fuel.* Ketone bodies can also serve as fuel for heart muscle. In fact, heart muscle consumes acetoacetate in preference to glucose.

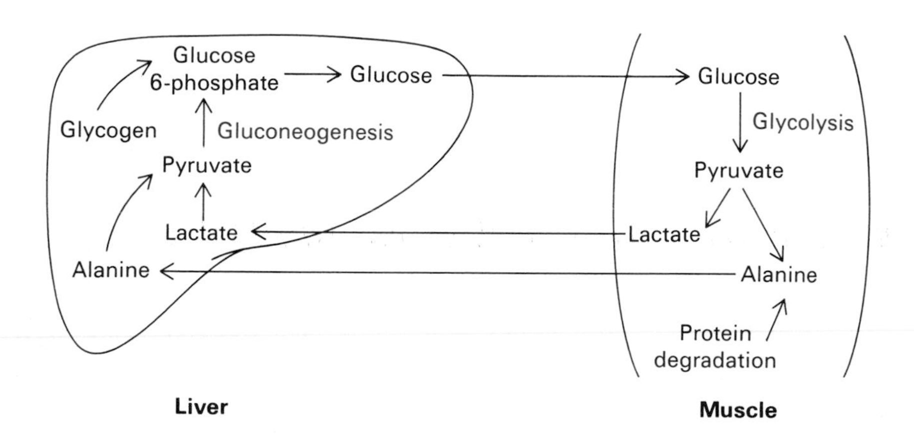

Figure 26-12
Metabolic interchanges between muscle and the liver.

3. *Adipose tissue.* The triacylglycerols stored in adipose tissue are an enormous reservoir of metabolic fuel. Their energy content is 135,000 kcal in a typical 70-kg man. Adipose tissue is specialized for the esterification of fatty acids and for their release from triacylglycerols. In humans, the liver is the major site of fatty acid synthesis, whereas the principal biosynthetic task of adipose tissue is to activate these fatty acids and transfer the resulting CoA derivatives to glycerol. Glycerol 3-phosphate, a key intermediate in this biosynthesis (p. 548), comes from the reduction of dihydroxyacetone phosphate, which is formed from glucose by the glycolytic pathway. Adipose cells are unable to phosphorylate endogeneous glycerol because they lack the kinase. Hence, *adipose cells need glucose for the synthesis of triacylglycerols.* Triacylglycerols are hydrolyzed to fatty acids and glycerol by lipases. The release of the first fatty acid from a triacylglycerol, the rate-limiting step, is catalyzed by a hormone-sensitive lipase that is reversibly phosphorylated. Cyclic AMP, the intracellular messenger in this hormone-triggered amplifying cascade, activates a protein kinase—a recurring theme in hormone action (p. 472). Triacylglycerols in adipose cells are continually being hydrolyzed and resynthesized. Glycerol derived from their hydrolysis is exported to the liver. Most of the fatty acids formed on hydrolysis are reesterified if glycerol 3-phosphate is abundant. In contrast, they are released into the plasma if glycerol 3-phosphate is scarce because of a paucity of glucose. Thus, the glucose level inside adipose cells is a major factor in determining whether fatty acids are released into the blood.

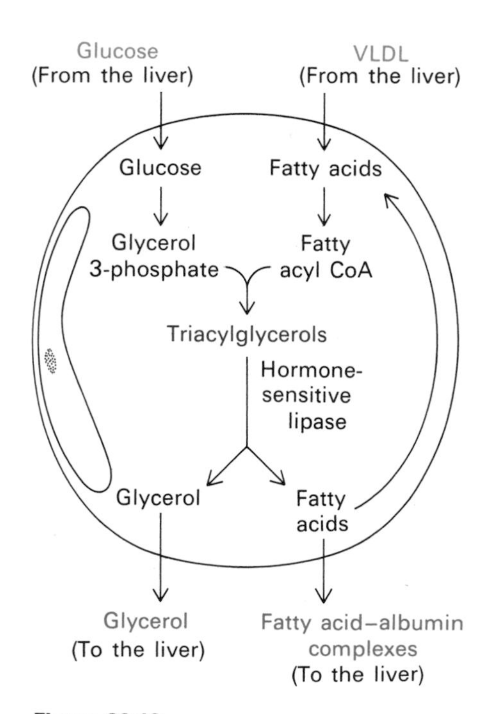

Figure 26-13
Synthesis and degradation of triacylglycerols by adipose tissue. Fatty acids are delivered to adipose cells in the form of very low density lipoproteins (VLDL).

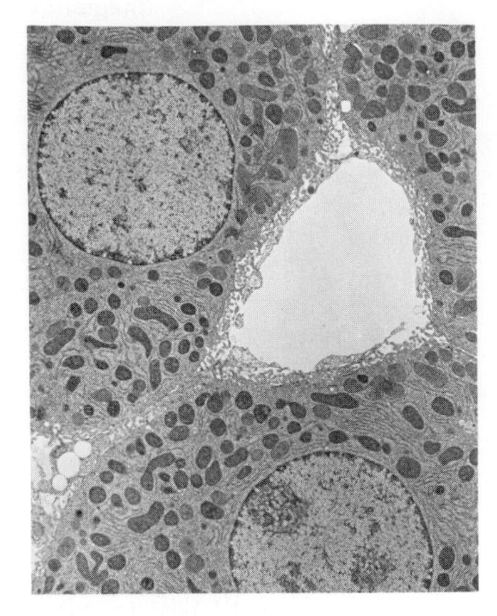

Figure 26-14
Electron micrograph of liver cells. The liver plays a key role in the integration of metabolism. [Courtesy of Dr. Ann Hubbard.]

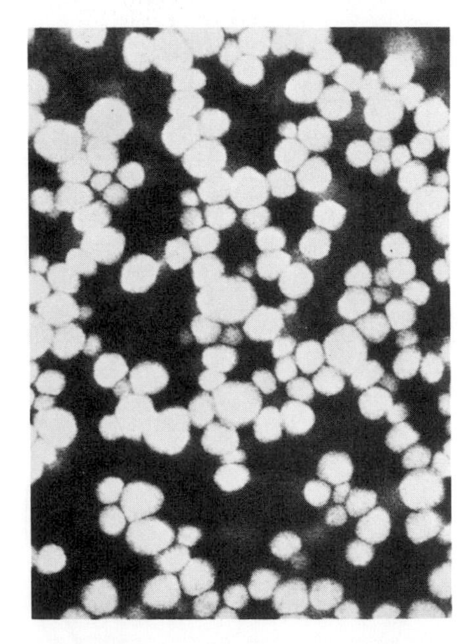

Figure 26-15
Electron micrograph of very low density lipoprotein particles (VLDL). These particles, which have diameters ranging between 300 and 800 Å, carry triacylglycerols from the liver to adipose tissue. [Courtesy of Dr. Robert Mahley.]

4. *Liver*. The metabolic activities of the liver are essential for providing fuel to the brain, muscle, and other peripheral organs. Most compounds absorbed by the intestine pass through the liver, which enables it to regulate the level of many metabolites in the blood. The liver can take up large amounts of glucose and convert it into glycogen. As much as 400 kcal can be stored in this way. The liver can release glucose into the blood by breaking down its store of glycogen and by carrying out gluconeogenesis. The main precursors of glucose are lactate and alanine from muscle, glycerol from adipose tissue, and glucogenic amino acids from the diet.

The liver also plays a central role in the regulation of lipid metabolism. When fuels are abundant, fatty acids are synthesized by the liver, esterified, and then secreted into the blood in the form of very low density lipoprotein (VLDL) (Figure 26-15). This plasma lipoprotein is the main source of fatty acids used by adipose tissue to synthesize triacylglycerols. However, in the fasting state, the liver converts fatty acids into ketone bodies. How does a liver cell choose between these contrasting paths? The selection is made according to whether the fatty acid enters the mitochondrial matrix. Recall that long-chain fatty acids traverse the inner mitochondrial membrane only if they are esterified to carnitine (p. 473). Carnitine acyl transferase I, which catalyzes the formation of acyl carnitine on the outer face of this membrane, is inhibited by malonyl CoA, the committed intermediate in the synthesis of fatty acids. *Thus, when long-chain fatty acids are being synthesized, they are prevented from entering the mitochondrial matrix, the compartment of β-oxidation and ketone-body formation. Instead, these fatty acids become incorporated into triacylglycerols and phospholipids.* In contrast, the level of malonyl CoA is low when fuels are scarce. Under these conditions, fatty acids liberated from adipose tissues enter the mitochondrial matrix for conversion into ketone bodies.

$$H_3C-\overset{CH_3}{\underset{CH_3}{\overset{|}{\underset{|}{N^+}}}}-CH_2-\overset{H}{\underset{O-C(=O)-R}{C}}-CH_2-C\overset{O}{\underset{O^-}{}}$$

Fatty acyl carnitine

How does the liver meet its own energy needs? Keto acids derived from the degradation of amino acids are preferred to glucose as the liver's own fuel. In fact, the main purpose of glycolysis in the liver is to form building blocks for biosyntheses. Furthermore, the liver cannot use acetoacetate as a fuel because it has little of the transferase needed for its activation to acetyl CoA. Thus, the liver eschews the fuels it exports to muscle and the brain—an altruistic organ indeed!

HORMONAL REGULATORS OF FUEL METABOLISM

Hormones play a key role in integrating metabolism. In particular, *insulin, glucagon, epinephrine, and norepinephrine* have large effects on the storage and mobilization of fuels and on related facets of metabolism:

1. *Insulin*. This 5.8-kd protein hormone (pp. 23 and 994), and glucagon are the most important regulators of fuel metabolism. The secre-

tion of insulin by the β cells of the pancreas is stimulated by glucose and the parasympathetic nervous system. *In essence, insulin signals the fed state: it stimulates the storage of fuels and the synthesis of proteins in a variety of ways.* This hormone promotes the dephosphorylation of key interconvertible enzymes. One consequence is to stimulate glycogen synthesis in both muscle and liver and to suppress gluconeogenesis by the liver. Insulin also accelerates glycolysis in the liver, which in turn increases the synthesis of fatty acids. The entry of glucose into muscle and adipose cells is promoted by insulin. The abundance of fatty acids and glucose in adipose tissue results in the synthesis and storage of triacylglycerols. The action of insulin also extends to amino acid and protein metabolism. Insulin promotes the uptake of branched-chain amino acids (valine, leucine, and isoleucine) by muscle, which favors a building up of muscle protein. Indeed, insulin has a general stimulating effect on protein synthesis. In addition, it inhibits the intracellular degradation of proteins.

2. *Glucagon.* This 3.5-kd polypeptide hormone (p. 458) is secreted by the α cells of the pancreas in response to a *low blood sugar level in the fasting state. The main target organ of glucagon is the liver.* Glucagon stimulates glycogen breakdown and inhibits glycogen synthesis by triggering the cAMP-mediated cascade that leads to the phosphorylation of phosphorylase and glycogen synthase (p. 463). Glucagon also inhibits fatty acid synthesis by diminishing the production of pyruvate and by lowering the activity of acetyl CoA carboxylase. In addition, glucagon stimulates gluconeogenesis and blocks glycolysis by lowering the level of F-2,6-BP. All known actions of glucagon are mediated by protein kinases that are activated by cyclic AMP. The net result of these actions of glucagon is to *markedly increase the release of glucose by the liver.* Likewise, glucagon raises the level of cyclic AMP in adipose cells, which activates a lipase that mobilizes triacylgylcerols.

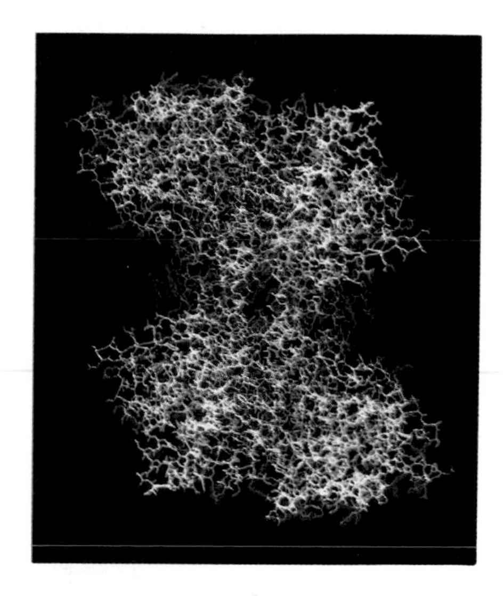

Model of phosphorylase *a*, a key enzyme in the regulation of the blood glucose level. [Courtesy of Dr. Robert Fletterick.]

3. *Epinephrine and norepinephrine.* These catecholamine hormones (p. 1025) are secreted by the adrenal medulla and sympathetic nerve endings in response to a *low blood glucose level.* Like glucagon, they stimulate the mobilization of glycogen and triacylglycerols by triggering the cAMP-mediated cascade. They differ from glucagon in that their glycogenolytic effect is greater in muscle than in the liver. Another action of the catecholamines is to inhibit the uptake of glucose by muscle. Instead, fatty acids released from adipose tissue are used as fuel. Epinephrine also stimulates the secretion of glucagon and inhibits the secretion of insulin. *Thus, the catecholamines increase the amount of glucose released into the blood by the liver and decrease the utilization of glucose by muscle.*

THE BLOOD GLUCOSE LEVEL IS BUFFERED BY THE LIVER

The blood glucose level in a typical person after an overnight fast is 80 mg/100 ml (4.4 mM). The blood glucose level during the day normally ranges from about 80 mg/100 ml before meals to about 120 mg/100 ml after meals. How is the blood glucose level maintained relatively constant despite large changes in the input and utilization of glucose? The major control elements have already been discussed, and so they need only be brought together here. The blood glucose level is controlled primarily by the action of the liver, which can take up or release large amounts of glucose in response to hormonal signals and the level of

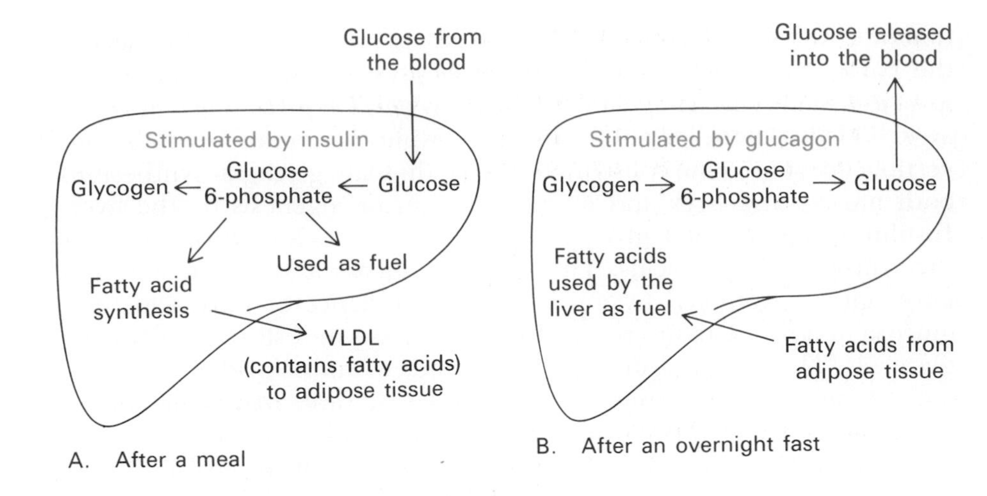

Figure 26-16
Control of the blood glucose level by the liver.

glucose itself (Figure 26-16). After a carbohydrate-containing meal, the increased concentration of glucose in the blood leads to a higher level of glucose 6-phosphate in the liver because only then do the catalytic sites of glucokinase become filled with glucose. Recall that glucokinase, in contrast with hexokinase, has a high K_M for glucose (~10 mM, compared with a fasting blood glucose level of 4.4 mM) and is not inhibited by glucose 6-phosphate. *Consequently, glucose 6-phosphate is formed more rapidly by the liver as the blood glucose level rises.*

The fate of glucose 6-phosphate is then largely controlled by the opposed effects of glucagon and insulin. Glucagon triggers the cAMP-mediated cascade (p. 463) that results in the breakdown of glycogen, whereas insulin antagonizes this action. *A high blood glucose level leads to a decreased secretion of glucagon and an increased secretion of insulin by the pancreas. Consequently, glycogen is rapidly synthesized when the blood glucose level is high.* These hormonal effects on glycogen synthesis and storage are reinforced by a direct action of glucose itself. As was discussed in an earlier chapter (p. 464), *phosphorylase* a *is a glucose sensor in addition to being the enzyme that cleaves glycogen.* When the glucose level is high, the binding of glucose to phosphorylase *a* renders it susceptible to the action of a phosphatase that converts it into phosphorylase *b*, which does not degrade glycogen. This conversion also releases the phosphatase, which enables it to activate glycogen synthase. *Thus, glucose allosterically shifts the glycogen system from a degradative to a synthetic mode.*

The high insulin level in the fed state also promotes *the entry of glucose into muscle and adipose tissue.* The synthesis of glycogen by muscle as well as by liver is stimulated by insulin. In fact, muscle can store about three times as much glycogen as the liver because of its larger mass. The entry of glucose into adipose tissue provides glycerol 3-phosphate for the synthesis of triacylglycerols.

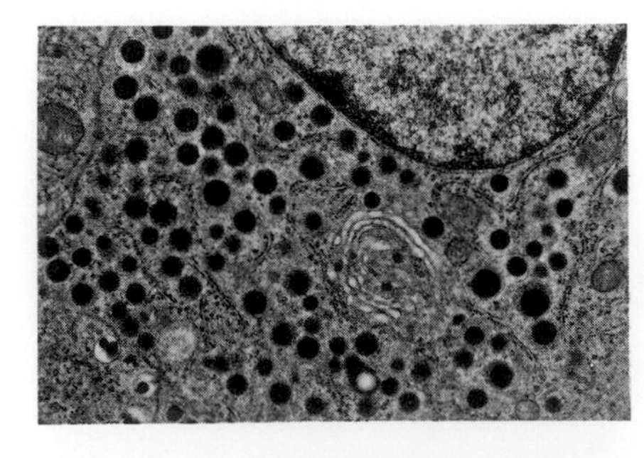

Figure 26-17
Electron micrograph of granules containing glucagon in an α-cell of the pancreas. [Courtesy of Dr. Arthur Like.]

The blood glucose level begins to drop several hours after a meal, which leads to decreased insulin secretion and increased glucagon secretion. The effects just described are then reversed. The activation of the cAMP-mediated cascade results in a higher level of phosphorylase *a* and a lower level of glycogen synthase *a*. The effect of hormones on this cascade is reinforced by the diminished binding of glucose to phosphorylase *a*, which makes it less susceptible to the hydrolytic action of the phosphatase. Instead, the phosphatase remains bound to phosphorylase *a* and so the synthase stays in the inactive phosphorylated form. Consequently, there is a rapid mobilization of glycogen. The large amount of glucose formed by the hydrolysis of glucose 6-phosphate derived from glycogen is then released from the liver into the blood. The diminished utilization of glucose by muscle and adipose tissue also contributes to the maintenance of the blood glucose level. The entry of glucose into muscle and adipose tissue decreases because of the low insulin level. Both muscle and liver use fatty acids as fuel when the blood glucose level drops. *Thus, the blood glucose level is kept higher than about 80 mg/100 ml by three major factors: the mobilization of glycogen and release of glucose by the liver, the release of fatty acids by adipose tissue, and the shift in the fuel used from glucose to fatty acids by muscle and the liver.*

METABOLIC ADAPTATIONS IN PROLONGED STARVATION MINIMIZE PROTEIN DEGRADATION

We turn now to metabolic adaptations in prolonged starvation. A typical well-nourished 70-kg man has fuel reserves of some 1600 kcal in glycogen, 24,000 kcal in mobilizable protein, and 135,000 kcal in triacylglycerols (see Table 26-1 on p. 634). The energy need for a 24-hour period ranges from about 1600 kcal in the basal state to 6000 kcal, depending on the extent of activity. Thus, stored fuels suffice to meet caloric needs in starvation for one to three months. However, the carbohydrate reserves are exhausted in only a day. Even under these conditions the blood glucose level is maintained above 50 mg/100 ml. The brain cannot tolerate appreciably lower glucose levels for even short periods. Hence, *the first priority of metabolism in starvation is to provide sufficient glucose to the brain and other tissues (such as red blood cells) that are absolutely dependent on this fuel.* However, precursors of glucose are not abundant. Most of the energy is stored in the fatty acyl moieties of triacylglycerols. Recall that fatty acids cannot be converted into glucose because acetyl CoA cannot be transformed into pyruvate (p. 480). The glycerol moiety of triacylglycerols can be converted into glucose, but only a limited amount is available. The only other potential source of glucose is amino acids derived from the breakdown of proteins. Muscle is the largest potential source of amino acids during starvation. However, survival for most animals depends on being able to move rapidly, which requires a large muscle mass. *Thus, the second priority of metabolism in starvation is to preserve protein. This is accomplished by shifting the fuel being used from glucose to fatty acids and ketone bodies* (Figure 26-18).

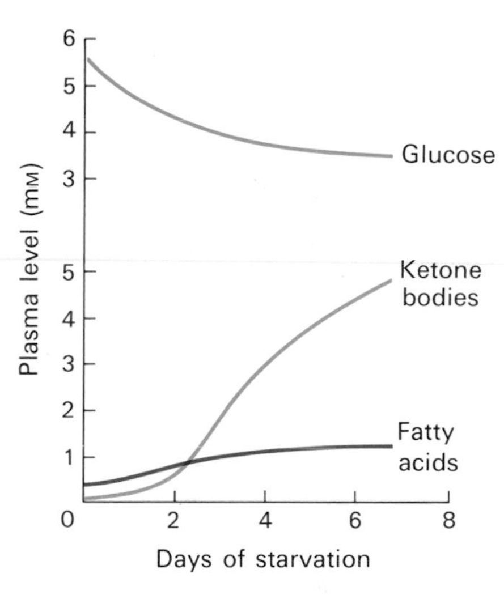

Figure 26-18
The plasma levels of fatty acids and ketone bodies increase in starvation, whereas that of glucose decreases.

The metabolic changes during the first day of starvation are like those after an overnight fast. The low blood sugar level leads to decreased secretion of insulin and increased secretion of glucagon. *The dominant metabolic processes are the mobilization of triacylglycerols in adipose tissue and gluconeogenesis by the liver. The liver obtains energy for its own needs by oxidizing fatty acids released from adipose tissue.* The concentrations of acetyl CoA and citrate consequently increase, which switches off glycolysis. The uptake of glucose by muscle is markedly diminished because of the low insulin level, whereas fatty acids enter freely. Consequently, *muscle also shifts from glucose to fatty acids for fuel.* The β-oxidation of fatty acids by muscle halts the conversion of pyruvate into acetyl CoA. Hence, pyruvate, lactate, and alanine are exported to the liver for conversion into glucose. Proteolysis of muscle protein provides some of these three-carbon precursors of glucose. Glycerol derived from the cleavage of triacylglycerols is another raw material for the synthesis of glucose by the liver.

The most important change after about three days of starvation is that large amounts of acetoacetate and 3-hydroxybutyrate (ketone bod-

2 Acetyl CoA

→ CoA

Acetoacetyl CoA

Acetyl CoA + H_2O

→ CoA

$$\begin{array}{c} O \\ \| \\ C-S-CoA \\ | \\ CH_2 \\ | \\ HO-C-CH_3 \\ | \\ CH_2 \\ | \\ COO^- \end{array}$$

3-Hydroxy-3-methylglutaryl CoA

→ Acetyl CoA

$$\begin{array}{c} O=C-CH_3 \\ | \\ CH_2 \\ | \\ COO^- \end{array}$$

Acetoacetate

NADH + H^+

→ NAD^+

$$\begin{array}{c} H \\ | \\ HO-C-CH_3 \\ | \\ CH_2 \\ | \\ COO^- \end{array}$$

D-3-Hydroxybutyrate

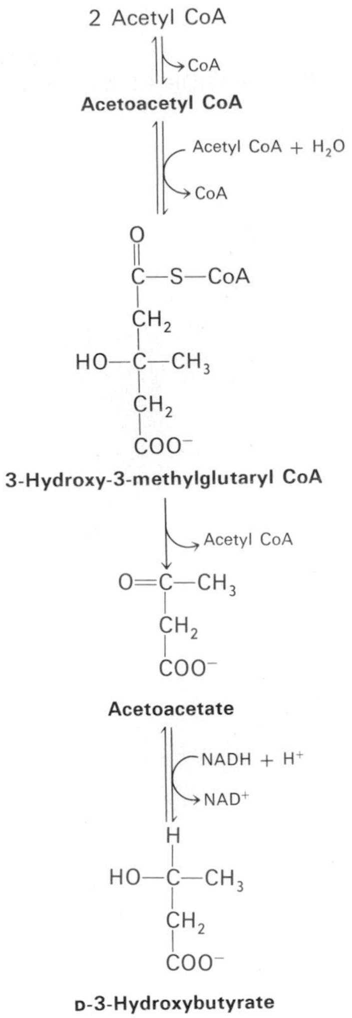

Figure 26-19
Synthesis of ketone bodies by the liver.

ies) are formed by the liver (Figure 26-19). Their synthesis from acetyl CoA increases markedly because the citric acid cycle is unable to oxidize all of the acetyl units generated by the degradation of fatty acids. Gluconeogenesis depletes the supply of oxaloacetate, which is essential for the entry of acetyl CoA into the citric acid cycle. Consequently, the liver produces large quantities of ketone bodies, which are released into the blood. At this time, the brain begins to consume appreciable amounts of acetoacetate in place of glucose. After three days of starvation, about a third of the energy needs of the brain are met by ketone bodies (Table 26-2). The heart also uses ketone bodies as fuel. These changes in fuel usage are accompanied by a rise in the level of ketone bodies in the plasma.

Table 26-2
Fuel metabolism in starvation

Fuel exchanges and consumption	*Amount formed or consumed in 24 hours (grams)*	
	3rd day	*40th day*
Fuel use by the brain		
Glucose	100	40
Ketone bodies	50	100
All other use of glucose	50	40
Fuel mobilization		
Adipose-tissue lipolysis	180	180
Muscle-protein degradation	75	20
Fuel output of the liver		
Glucose	150	80
Ketone bodies	150	150

After several weeks of starvation, ketone bodies become the major fuel of the brain. Only 40 g of glucose is needed per day for the brain, compared with about 120 g in the first day of starvation. *The effective conversion of fatty acids into ketone bodies by the liver and their use by the brain markedly diminishes the need for glucose. Hence, less muscle is degraded than in the first days of starvation.* The breakdown of 20 g of muscle compared with 75 g early in starvation is most important for survival. The duration of starvation compatible with life is mainly determined by the size of the triacylglycerol depot.

HUGE FAT DEPOTS ENABLE MIGRATORY BIRDS TO FLY LONG DISTANCES

Migratory birds provide a striking illustration of the biological value of triacylglycerols. Some small land birds fly in the autumn from their breeding grounds in New England to their wintering grounds in the West Indies and then back again in the spring. They fly nonstop over water for a distance of some 2400 kilometers. These birds sustain a velocity of 40 kilometers per hour for 60 hours. This remarkable feat is made possible by the existence of very large fat depots that are efficiently mobilized during the long flight. Birds that migrate for short distances or not at all are quite lean. They have fat indices of about 0.3. (The fat index is defined as the ratio of the dry weight of total body fat to that of nonfat.) In contrast, long-range migrants become moderately obese in preparing to travel over land and then become highly obese just before they begin their overseas journey. Indeed, their fat index is then close to 3. In the ruby-throated hummingbird, about 0.15 g of triacylglycerols is accumulated each day per gram of body weight. The comparable weight gain in a human would be 10 kg per day. The accumulated fat in migratory birds is stored under the skin, in the abdominal cavity, in muscle, and in the liver. About two-thirds of this fat store is consumed in the long flight over water. The transition to the use of fatty acids and ketone bodies as fuels must be very rapid because almost no protein is degraded during the 60-hour flight. Oxidation of fat also provides these birds with the water needed to make up for losses through respiratory passages. The high efficiency of triacylglycerols as a storage fuel is also noteworthy. Recall that triacylglycerols store six times as much energy as does glycogen because they are anhydrous and more highly reduced (p. 471). A migratory bird carrying the same amount of fuel in the form of glycogen would never get off the ground!

A ruby-throated hummingbird.

METABOLIC DERANGEMENTS IN DIABETES RESULT FROM RELATIVE INSULIN INSUFFICIENCY AND GLUCAGON EXCESS

We turn now to *diabetes mellitus*, a complex disease characterized by a grossly abnormal pattern of fuel usage—overproduction of glucose by the liver and underutilization by other organs. *Mellitus* refers to the high level of glucose in the urine. The incidence of diabetes mellitus (usually referred to simply as diabetes) in industrialized countries is about 1%. Indeed, diabetes is the most common serious metabolic disease in the world; it affects hundreds of millions.

Untreated diabetes is characterized by *abnormal glucose metabolism—the level of insulin is too low and that of glucagon is too high relative to the needs of the patient.* Because insulin is deficient, the entry of glucose into cells is impaired. The excessive level of glucagon relative to insulin leads to a decrease in the amount of F-2,6-BP in the liver. Hence, glycolysis is inhibited and gluconeogenesis is stimulated. The resulting drop in the level of malonyl CoA activates carnitine acyltransferase I. Consequently, fatty acyl CoA molecules are efficiently transported into the mitochondrial matrix for oxidation to acetoacetate. The elevated level of glucagon also leads to increased mobilization of triacylglycerols in adipose tissue. A striking feature of diabetes is the shift in fuel usage from carbohydrates to fats—glucose, more abundant than ever, is spurned.

Glucose is excreted in the urine when the blood glucose level exceeds the reabsorptive capacity of the renal tubules. Water accompanies the excreted glucose, and so an untreated diabetic in the acute phase of the disease is hungry and thirsty. The loss of glucose depletes the carbohydrate stores, which leads to the breakdown of fat and protein. The mobilization of fats leads to the formation of large amounts of acetyl CoA. However, much of the acetyl CoA cannot enter the citric acid cycle because there is insufficient oxaloacetate for the condensation step. Recall that mammals can synthesize oxaloacetate from pyruvate, a product of glycolysis, but not from acetyl CoA—instead, they generate ketone bodies. Most of the acid produced in normal metabolism is in the form of CO_2, which is readily excreted by the lungs. In contrast,

Diabetes—
Named for the excessive urination in the disease. Aretaeus, a Cappadocian physician of the second century A.D., wrote: "The epithet diabetes has been assigned to the disorder, being something like passing of water by a siphon." He perceptively characterized diabetes as "being a melting-down of the flesh and limbs into urine."

Mellitus—
From Latin, meaning sweetened with honey. Refers to the presence of sugar in the urine of patients having the disease.
Mellitus distinguishes this disease from diabetes *insipidus*, which is caused by impaired renal reabsorption of water.

ketone body acids cannot be excreted by the lungs. In high concentrations, they overwhelm the kidney's capacity to maintain acid-base balance. The untreated diabetic can go into a coma because of a lowered blood pH level and dehydration. *Accelerated ketone production leading to acidosis* is a characteristic of *insulin-dependent diabetes* (also known as type I or juvenile-onset diabetes). Some diabetics have a normal or even higher than normal level of insulin in their blood, but they are quite unresponsive to the hormone. This form of the disease, known as *non-insulin-dependent diabetes* (type II diabetes), typically arises later in life than does the insulin-dependent form.

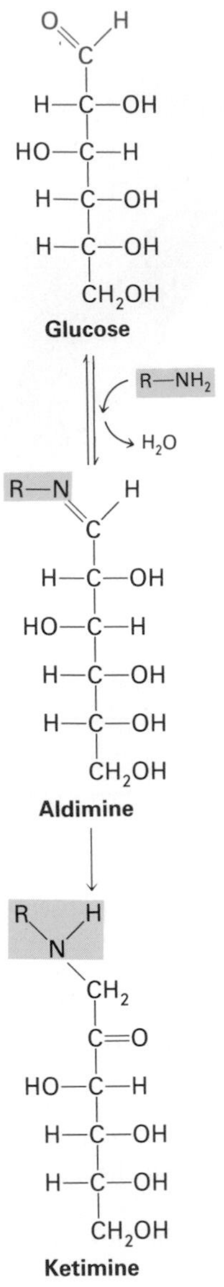

Glucose

Aldimine

Ketimine

Figure 26-20
Formation of hemoglobin A_{Ic} by the nonenzymatic addition of glucose (or glucose 6-phosphate) to an α-amino group. A stable adduct is formed by an Amadori rearrangement of the aldimine to a ketimine.

GLUCOSE REACTS WITH HEMOGLOBIN TO FORM A REVEALING INDICATOR OF BLOOD SUGAR LEVEL

One of the objectives in the treatment of diabetes is to lower the blood glucose level. A valuable indicator of blood sugar levels was unexpectedly found in a different field of inquiry—oxygen transport. During the 120-day lifetime of the red cell, glucose, glucose 6-phosphate, and other sugars react nonenzymatically to form stable conjugates with the α-amino group of the β chains of hemoglobin. The aldehyde group of the open-chain form of glucose condenses with this amino group to form a Schiff base (Figure 26-20). This reversible reaction is followed by a virtually irreversible Amadori rearrangement in which the double bond shifts to C-2 of the sugar to give a stable fructose derivative of hemoglobin, called hemoglobin A_{Ic}, having altered electrophoretic properties.

The red blood cells of all people contain a small proportion of hemoglobin A_{Ic}. The rate of its formation is proportional to the sugar level, and so diabetics have a higher proportion of hemoglobin A_{Ic} than do normal individuals (6 to 15% compared with 3 to 5%). *The level of hemoglobin A_{Ic} reveals the integral of the blood-sugar concentration over a period of several weeks.* Hence, measurements of hemoglobin A_{Ic} levels every several weeks are very useful in determining whether the blood glucose levels of diabetic patients were adequately controlled. Before the discovery of hemoglobin A_{Ic}, more frequent monitoring of the blood glucose levels of diabetics was necessary.

Hemoglobin A_{Ic} is also interesting as a model of how proteins can be damaged by high levels of reducing sugars. Many diabetics develop degenerative complications, such as atherosclerotic lesions in blood vessels, years after the onset of metabolic symptoms. One working hypothesis is that some of the late complications of diabetes arise from the covalent attachment of glucose to vulnerable proteins, as in hemoglobin A_{Ic}. Extensive clinical studies are in progress to learn whether the long-term implications of diabetes are consequences of hyperglycemia or of other factors. For the present, meticulous control of hyperglycemia, monitored by hemoglobin A_{Ic} assays, remains a key objective in the treatment of diabetes.

SUMMARY

The basic strategy of metabolism is to form ATP, reducing power, and building blocks for biosyntheses. This complex network of reactions is controlled by allosteric interactions and reversible covalent modifications of enzymes and changes in their amounts, by compartmentation, and by interactions between metabolically distinct organs. The enzyme catalyzing the committed step in a pathway is usually the most important control site, as exemplified by phosphofructokinase in glycolysis and acetyl CoA carboxylase in fatty acid synthesis. Opposing pathways such as gluconeogenesis and glycolysis are reciprocally regulated so that one pathway is usually quiescent while the other is highly active. Another pair of opposed reaction sequences, glycogen synthesis and degradation, are coordinately controlled by a hormone-triggered amplifying cascade that leads to the phosphorylation of glycogen synthase and phosphorylase. The role of compartmentation in control is illustrated by the contrasting fates of fatty acids in the cytosol and the mitochondrial matrix.

The metabolic patterns of brain, muscle, adipose tissue, and liver are very different. Glucose is essentially the sole fuel for the brain in a well-fed person. During starvation, ketone bodies (acetoacetate and 3-hydroxybutyrate) become the predominant fuel of the brain. Muscle uses glucose, fatty acids, and ketone bodies as fuel, and it synthesizes glycogen as a fuel reserve for its own needs. Adipose tissue is specialized for the synthesis, storage, and mobilization of triacylglycerols. The diverse metabolic activities of the liver support the other organs. The liver can rapidly mobilize glycogen and carry out gluconeogenesis to meet the glucose needs of other organs. It plays a central role in the regulation of lipid metabolism. When fuels are abundant, fatty acids are synthesized, esterified, and sent from the liver to adipose tissue in the form of very low density lipoprotein (VLDL). In the fasting state, however, fatty acids are converted into ketone bodies by the liver. The activities of these organs are integrated by hormones. Insulin signals the fed state: it stimulates the formation of glycogen and triacylglycerols and the synthesis of proteins. In contrast, glucagon signals a low blood glucose level: it stimulates glycogen breakdown and gluconeogenesis by the liver and triacylglycerol hydrolysis by adipose tissue. The effects of epinephrine and norepinephrine on fuels are like those of glucagon, except that muscle rather than the liver is their primary target.

The blood glucose level in a well-fed person typically ranges from 80 mg/100 ml to 120 mg/100 ml. After a meal, the rise in the blood glucose level leads to increased secretion of insulin and decreased secretion of glucagon. Consequently, glycogen is synthesized in muscle and the liver. The increased entry of glucose into adipose tissue provides glycerol 3-phosphate for the synthesis of triacylglycerols. These effects are reversed when the blood glucose level drops several hours later. Glucose is then formed by the degradation of glycogen and by the gluconeogenic pathway, and fatty acids are released by the hydrolysis of triacylglycerols. Liver and muscle then use fatty acids instead of glucose to meet their own energy needs so that glucose is conserved for use by the brain and other tissues that are highly dependent on it. The metabolic adaptations in starvation are designed to minimize protein degradation. Large amounts of ketone bodies are formed by the liver from fatty acids and released into the blood within a few days after the onset of starvation. After several weeks of starvation, ketone bodies become the major fuel of the brain. The diminished need for glucose decreases the rate of muscle breakdown, and so the likelihood of survival is enhanced.

Diabetes mellitus, the most common serious metabolic disease, is produced by an insufficiency of insulin and an excess of glucagon relative to the needs of the patient. Insulin deficiency impairs the entry of glucose into cells and its utilization. Excess glucagon enhances glucose formation by the liver. The elevated blood glucose level leads to the excretion of a large volume of glucose-rich urine. Triacylglycerols are mobilized and ketone bodies are formed to an abnormal extent. A striking feature of diabetes is the shift in fuel usage from carbohydrates to fats. Accelerated ketone body formation can lead to acidosis, coma, and death in untreated type I diabetics. Hemoglobin A_{Ic}, a conjugate formed by the nonenzymatic addition of glucose to terminal amino groups, is a valuable indicator of the blood sugar level over a period of weeks.

SELECTED READINGS

Where to start

Foster, D. W., 1984. From glycogen to ketones—and back. *Diabetes* 33:1188–1199. [A lucid account of how anabolism and catabolism are sensitively controlled by the insulin:glucagon ratio.]

Randle, P. J., 1986. Fuel selection in animals. *Biochem. Soc. Trans.* 14:799–806. [A fine presentation of the control of the pyruvate dehydrogenase complex and its significance in fuel selection.]

Hers, H.-G., Hue, L., and Van Schaftingen, E., 1983. Fructose 2,6-bisphosphate. *Trends Biochem. Sci.* 7:329–333.

Cerami, A., Vlassara, H., and Brownlee, M., 1987. Glucose and aging. *Sci. Amer.* 256(3):90–96. [An interesting account of deleterious consequences arising from the reactivity of glucose.]

Cahill, G. F., Jr., 1976. Starvation in man. *Clin. Endocrinol. Metab.* 5:397–415.

Books

Ochs, R. S., Hanson, R. W., and Hall, J., (eds.), 1985. *Metabolic Regulation.* Elsevier. [A collection of many interesting articles on metabolic control in procaryotes as well as higher organisms.]

Newsholme, E. A., and Start, C., 1973. *Regulation in Metabolism.* Wiley. [An excellent account of the control of carbohydrate and fat metabolism.]

Bondy, P. K., and Rosenberg, L. E., (eds.), 1980. *Metabolic Control and Disease.* Saunders.

Newsholme, E., and Leech, T., 1983. *The Runner: Energy and Endurance.* Fitness Books. [A fascinating presentation of the biochemical and physiological basis of energy utilization in running.]

Howald, H., and Poortmans, J. R., (eds.), 1975. *Metabolic Adaptation to Prolonged Physical Exercise.* Birkhauser Verlag, Basel.

Atkinson, D. E., 1977. *Cellular Energy Metabolism and Its Regulation.* Academic Press.

Carbohydrate metabolism

McGarry, J. D., Kuwajima, M., Newgard, C. B., and Foster, D. W., 1987. From dietary glucose to liver glycogen—the full circle round. *Ann. Rev. Nutrition* 7:51–73.

Nordlie, R. C., 1984. Fine tuning of blood glucose concentrations. *Trends Biochem. Sci.* 10:70–75.

Felig, P., 1980. Disorders of carbohydrate metabolism. *In* Bondy, P. K., and Rosenberg, L. E., (eds.), *Metabolic Control and Disease.* Saunders. [An excellent review of the control of carbohydrate metabolism and its relationship to protein and lipid metabolism.]

Esmann, V., (ed.), 1978. *Regulatory Mechanisms of Carbohydrate Metabolism.* Pergamon.

Newsholme, E. A., 1980. A possible metabolic basis for the control of body weight. *New Engl. J. Med.* 302:400–405. [Presentation of the hypothesis that obesity results from diminished cycling of substrate through pairs of opposed reactions, such as the ones catalyzed by phosphofructokinase and fructose 1,6-bisphosphatase.]

Felig, P., and Wahren, J., 1975. Fuel homeostasis in exercise. *New Engl. J. Med.* 293:1078–1084.

Cahill, G. F., Jr., and Owen, O. E., 1968. Some observations on carbohydrate metabolism in man. *In* Dickens, F., Randle, P. J., and Whelan, W. J., (eds.), *Carbohydrate Metabolism and Its Disorders,* vol. 1, pp. 497–522. Academic Press.

El-Maghrabi, M. R., Correia, J. J., Heil, P. J., Pate, T. M., Cobb, C. E., and Pilkis, S. J., 1986. Tissue distribution, immunoreactivity, and physical properties of 6-phosphofructo-2-kinase/fructose-2,6-bisphosphatase. *Proc. Nat. Acad. Sci.* 83:5005–5009. [The level of fructose 2,6-bisphosphate is controlled differently in liver and muscle.]

Kitakura, K., and Uyeda, K., 1987. The mechanism of activation of heart fructose 6-phosphate,2-kinase:fructose-2,6-bisphosphatase. *J. Biol. Chem.* 262:679–681. [Reports experiments showing that the enzyme in heart, in contrast to the one in liver, is activated by phosphorylation.]

Gluconeogenesis

Hers, H.-G., and Hue, L., 1983. Gluconeogenesis and related aspects of glycolysis. *Ann. Rev. Biochem.* 52:617–653.

Snell, K., 1979. Alanine as a gluconeogenic carrier. *Trends Biochem. Sci.* 4:124–128.

Hanson, R., and Mehlman, M., (eds.), 1976. *Gluconeogenesis.* Wiley.

Amino acid metabolism

Felig, P., 1975. Amino acid metabolism in man. *Ann. Rev. Biochem.* 44:933–955.

Dice, J. F., Walker, C. D., Byrne, B., and Cardiel, A., 1978. General characteristics of protein degradation in diabetes and starvation. *Proc. Nat. Acad. Sci.* 75:2093–2097.

Lipid metabolism

Foster, D. W., and McGarry, J. D., 1983. The metabolic derangements and treatment of diabetic ketoacidosis. *New Engl. J. Med.* 309:159–169.

McGarry, J. D., and Foster, D. W., 1980. Regulation of hepatic fatty acid oxidation and ketone body production. *Ann. Rev. Biochem.* 49:395–420.

Williamson, D. H., 1979. Recent developments in ketone-body metabolism. *Biochem. Soc. Trans.* 7:1313–1321.

Odum, E. P., 1965. Adipose tissue in migratory birds. *In* Renold, A. E., and Cahill, G. F., (eds.), *Handbook of Physiology,* section 5, *Adipose Tissue,* pp. 37–43. American Physiological Society.

Grey, N. J., Karl, I., and Kipnis, D. M., 1975. Physiologic mechanisms in the development of starvation ketosis in man. *Diabetes* 24:10–16.

Diabetes

Foster, D. W., 1983. Diabetes mellitus. *In* Stanbury, J. B., Wyngaarden, J. B., Fredrickson, D. S., Goldstein, J. L., and Brown, M. S., (eds.), *The Metabolic Basis of Inherited Disease* (5th ed.), pp. 99–117. McGraw-Hill.

Cerami, A., and Koenig, R. J., 1978. Hemoglobin A_{Ic} as a model for the development of the sequelae of diabetes mellitus. *Trends Biochem. Sci.* 3:73–75.

Cohen, M. P., 1986. *Diabetes and Protein Glycosylation.* Springer-Verlag.

PROBLEMS

1. What are the key enzymatic differences between liver, muscle, and brain that account for their differing utilization of metabolic fuels?

2. Predict the major consequence of each of the following enzymatic deficiencies:
   (a) Hexokinase in adipose tissue.
   (b) Glucose 6-phosphatase in liver.
   (c) Carnitine acyl transferase I in skeletal muscle.
   (d) Glucokinase in liver.
   (e) Thiolase in brain.
   (f) Phosphofructokinase 2 in liver.

3. Cerebrospinal fluid has a low content of albumin and other proteins compared with plasma.
   (a) What effect does this have on the concentration of fatty acids in the extracellular medium of the brain?
   (b) Propose a plausible reason for the selection by brain of glucose rather than fatty acids as the prime fuel.
   (c) How does the fuel preference of muscle complement that of the brain?

4. The rate of energy expenditure of a typical 70-kg person at rest is about 70 watts, like a light bulb.
   (a) Express this rate in kilojoules per second and in kilocalories per second.
   (b) How many electrons flow through the electron-transport chain of mitochondria per second under these conditions?
   (c) Estimate the corresponding rate of ATP production.
   (d) The total ATP content of the body is about 50 grams. Estimate how often an ATP molecule turns over in a person at rest.

5. During a 100-meter race, less than half a liter of O_2 is consumed by a champion sprinter. However, 10 liters would be used if the metabolism in this interval were entirely aerobic. How is ATP generated during the race?

6. The rate of energy expenditure in running a marathon is about twelve times that at rest. Consider the fuel reserves listed in Table 26-1. How long do the glucose and glycogen stores last during a marathon?

7. Ingesting large amounts of glucose before a marathon might seem to be a good way of increasing the fuel stores. However, experienced runners do not ingest glucose before a race. What is the biochemical reason for their avoidance of this potential fuel? [Hint: consider the effect of glucose ingestion on the level of insulin.]

On the facing page: Model of a pair of deoxyribonucleotide units of a DNA double helix. Adenine is hydrogen bonded to thymine in this base pair.

PART V

GENETIC INFORMATION

storage, transmission, and expression

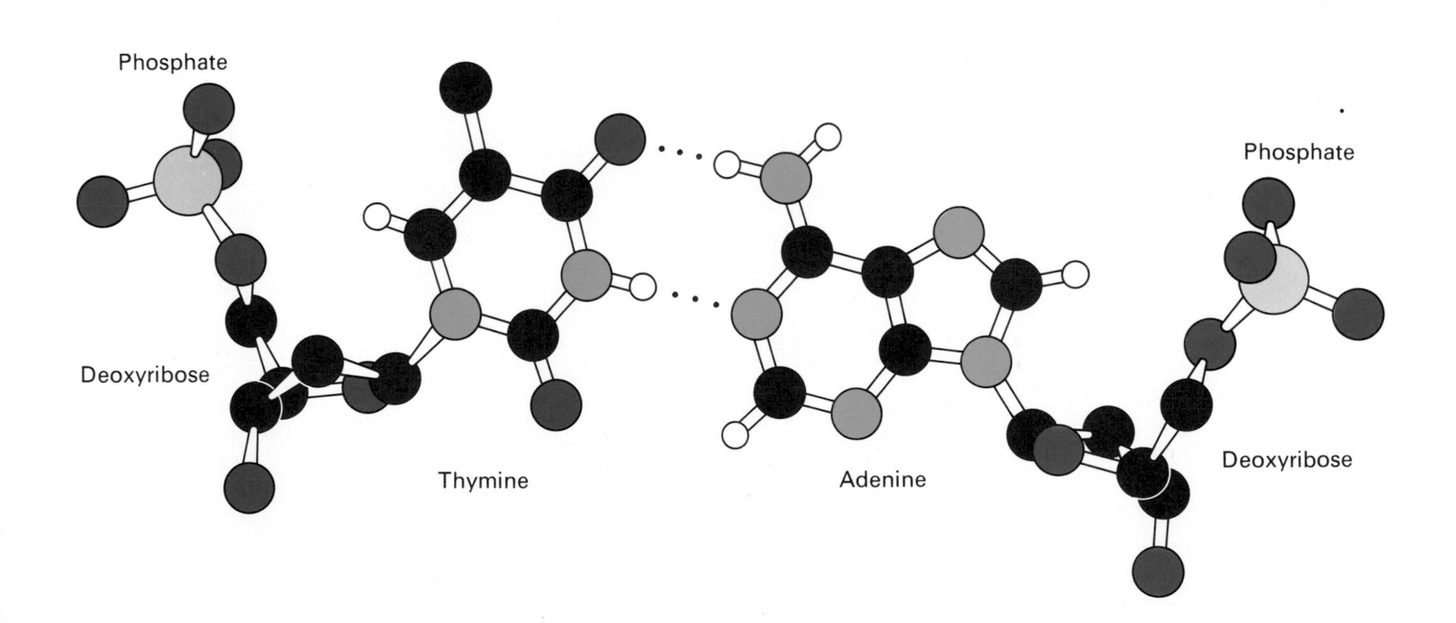

CHAPTER 27

DNA Structure, Replication, and Repair

The storage, transmission, and expression of genetic information is the central theme of this part of the book. The genetic role and structure of DNA were introduced in Part I (Chapter 4), as was the flow of genetic information (Chapter 5). Our consideration of this theme now resumes, enriched by a knowledge of proteins and metabolic transformations.

This chapter deals with the structure, replication, mutation, and repair of DNA. We shall focus on procaryotic systems because they have been intensively studied and exemplify general principles. More complex eucaryotic systems will be discussed in a later chapter (p. 823). The structure of DNA is dynamic. The Watson-Crick double helix can be bent, kinked, and unwound. It can also adopt different helical forms. X-ray analyses of crystals of DNA oligomers have provided insight into these conformational properties. Enzymes that cut and join DNA are considered next. The structure of DNA bound to a restriction endonuclease reveals how this exquisitely specific enzyme recognizes its target sequence. The next topic is the topology and supercoiling of DNA. Topoisomerases, a fascinating group of enzymes, catalyze the supercoiling of DNA molecules and their relaxation.

We shall then be ready to consider one of the most demanding and remarkable of all biological processes, the replication of DNA. How does DNA synthesis begin? How are parental strands unwound to serve as templates? How is very high fidelity replication achieved? A striking property of DNA polymerases is their built-in proofreading capability. The last part of this chapter deals with the molecular nature of mutations and their repair. Lesions in DNA are continually being repaired by multiple processes that use information in the intact strand to direct the correction of the damaged strand.

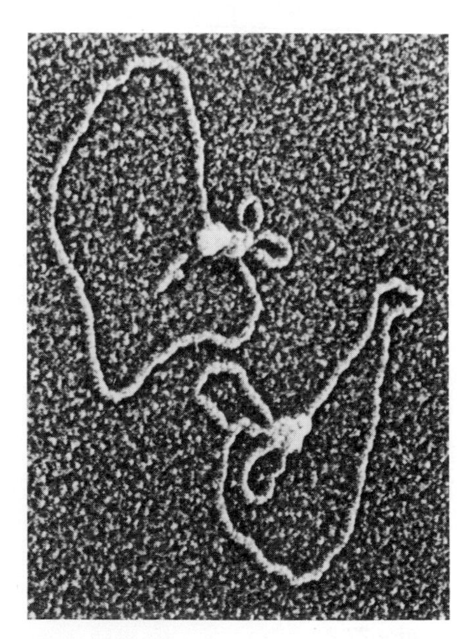

Figure 27-1
Electron micrograph showing the initiation of replication of ϕX174 viral DNA. Each of the two circular DNA molecules is bound to a primosome, a protein assembly that begins DNA replication. [Courtesy of Dr. Jack Griffith.]

DNA IS STRUCTURALLY DYNAMIC AND CAN ASSUME A VARIETY OF FORMS

The double-helical structure of DNA deduced by Watson and Crick profoundly influenced the course of biology because it immediately suggested how genetic information is stored and replicated. As was discussed before (p. 76), the essential features of their model are:

1. Two polynucleotide chains running in opposite directions coil around a common axis to form a right-handed double helix.

2. The purine and pyrimidine bases are on the inside of the helix, whereas the phosphate and deoxyribose units are on the outside.

3. Adenine (A) is paired with thymine (T), and guanine (G) with cytosine (C). AT base pairs are reinforced by two precisely directed hydrogen bonds, and GC base pairs by three such bonds. The double helix is also stabilized by interactions between stacked bases on the same strand.

The model proposed by Watson and Crick (known as the B-DNA helix) was based on x-ray diffraction patterns of DNA *fibers*, which provide information about properties of the double helix that are *averaged* over its constituent residues. Much more structural information can be obtained from x-ray analyses of DNA *crystals*. However, such studies had to await the development of techniques for synthesizing large amounts of DNA oligomers with defined base sequences (p. 123). X-ray analyses of these crystals at atomic resolution have revealed that DNA exhibits much more structural variability and diversity than formerly envisaged. A DNA backbone can be rotated about six bonds in each monomer, compared with two for polypeptide chains (p. 25). As will be discussed shortly, the puckering of the ribose ring and the orientation of the base with respect to the sugar are important structural determinants.

The x-ray analysis of a crystallized DNA dodecamer by Richard Dickerson and his coworkers revealed that its overall structure is very much like a Watson-Crick double helix (Figure 27-2). However, the dodecamer differed from the Watson-Crick model in not being uniform; there are rather large local deviations from the average structure. The Watson-Crick model has ten residues per complete turn, and so a residue is related to the next along a chain by a rotation of 36 degrees. In Dickerson's dodecamer, the rotation angles range from 28 to 42 degrees. Another deviation from the model is an opposite rotation of the two bases of a pair about their long axis (Figure 27-3). This

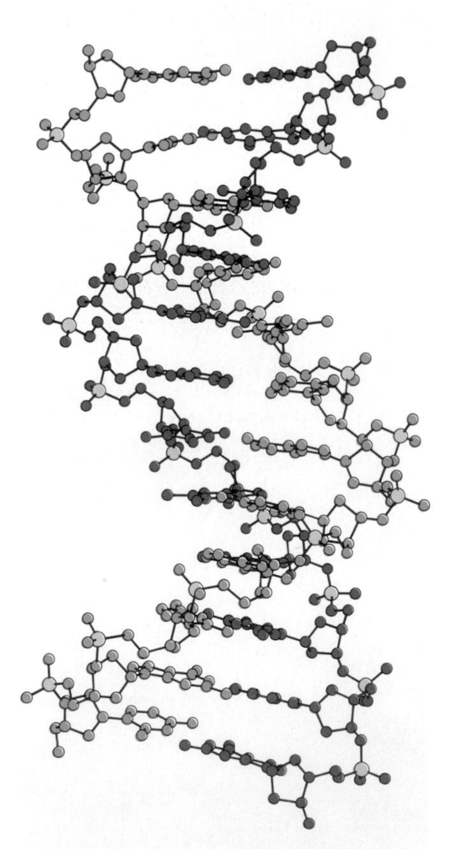

Figure 27-2
Structure of a 12-bp DNA molecule in the B form. One of the strands is shown in blue and the other in red; all phosphorus atoms are shown in yellow. Note the large local deviations from the classic Watson-Crick structure. This dynamic molecule can be bent, kinked, and unwound.

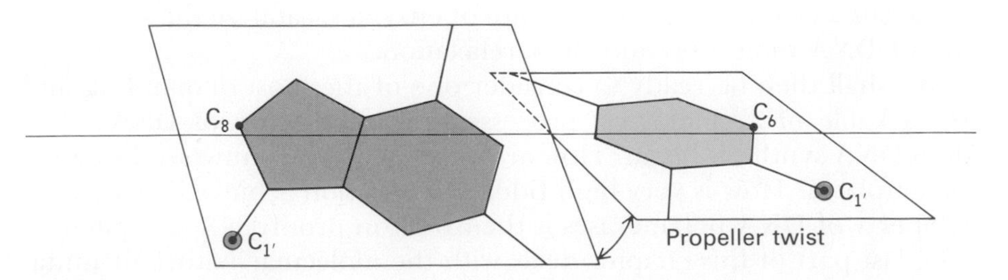

Figure 27-3
The bases of a DNA base pair are not perfectly coplanar. They are twisted with respect to each other, like the blades of a propeller.

structural feature, called *propeller twisting,* enhances the stacking of bases in each strand. Another source of variability is *base roll,* the tilting of base pairs relative to their neighbors. A key finding is that these and other local variations of the double helix depend on base sequence. A protein searching for a specific target sequence in DNA may sense its presence through its effect on the precise shape of the double helix.

One other feature of the B-DNA helix is noteworthy. *The helix can be smoothly bent into an arc or it can be supercoiled with rather little change in local structure.* This ease of deformation is biologically important because it enables circular DNA to be formed and allows DNA to be wrapped around proteins. The compacting of DNA to fit into a much smaller volume depends on its deformability. If DNA were constrained to be linear, it would not fit into a cell. DNA can also be *kinked*—that is, bent at discrete sites. Kinking can be induced by specific base sequences, such as a run of at least four adenine residues, or by the binding of a protein.

THE MAJOR AND MINOR GROOVES ARE LINED BY SEQUENCE-SPECIFIC HYDROGEN-BONDING GROUPS

A noteworthy feature of the B-DNA helix is the presence of two kinds of grooves, called the *major groove* (12 Å wide) and the *minor groove* (6 Å wide) (Figure 27-4). They arise because the glycosidic bonds of a base pair are not diametrically opposite each other. The minor groove contains the pyrimidine O-2 and the purine N-3 of the base pair, and the major groove is on the opposite side of the pair (Figure 27-5). The major groove is slightly deeper than the minor one (8.5 versus 7.5 Å). Each groove is lined by potential hydrogen-bond donor and acceptor atoms. In the minor groove, N-3 of adenine and guanine, and O-2 of thymine and cytosine, can serve as hydrogen acceptors, and the amino group attached to C-2 of guanine can be a hydrogen donor. Let us denote N-3 by n, O-2 by o, and an amino group hydrogen by h. Hence, the patterns of donors and acceptors in the minor groove are *no* (AT), *on* (TA), *nho* (GC), and *ohn* (CG). In the major groove, N-7 of guanine and adenine is a potential acceptor, as is O-4 of thymine and O-6 of guanine. The amino groups attached to C-6 of adenine and C-4 of cytosine can serve as hydrogen donors. Thus, the patterns exhibited by the major groove are *nho* (AT), *ohn* (TA), *noh* (GC), and *hon* (CG). Note that the major groove displays more distinctive features than does the minor groove. Also, the larger size of the major groove makes it more accessible for interactions with proteins that recognize specific DNA sequences.

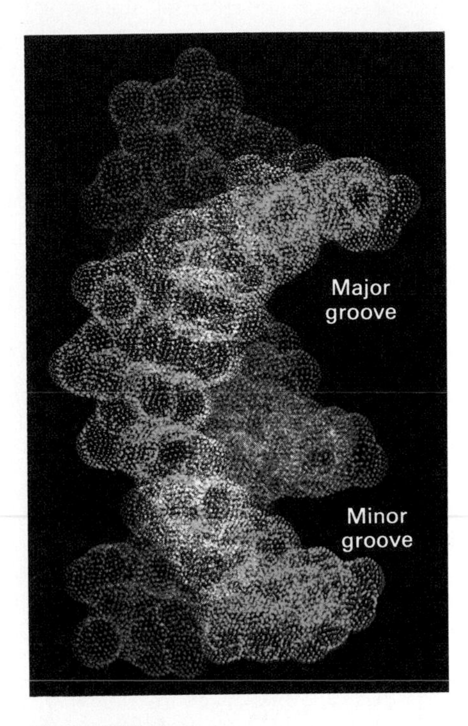

Figure 27-4
B-DNA contains a major groove and minor groove. This helix has been tilted to emphasize the markedly different sizes of these grooves.

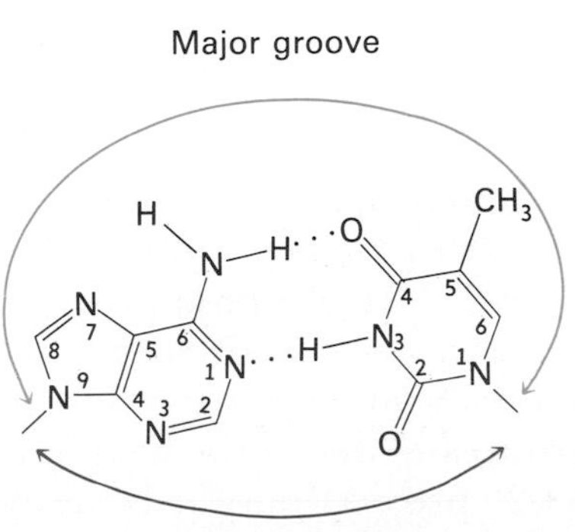

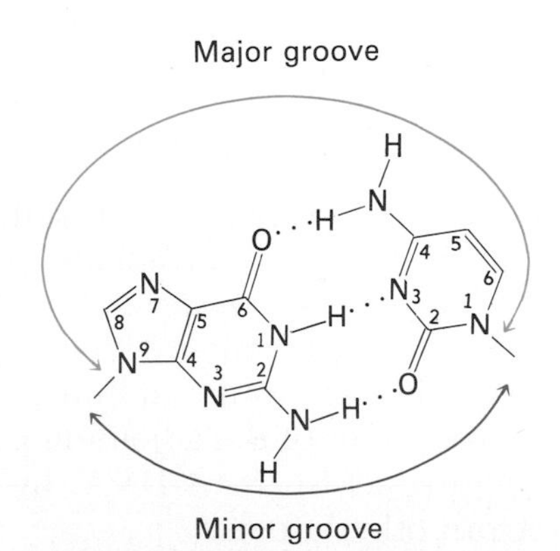

Figure 27-5
The AT and GC base pairs of DNA contain atoms that can form additional hydrogen bonds. The major and minor grooves are lined by different potential donors and acceptors.

THE 2′-OH OF RNA FITS IN AN A-DNA HELIX WITH TILTED BASE PAIRS BUT NOT IN A B-DNA HELIX

X-ray diffraction studies of dehydrated DNA fibers revealed a different form called *A-DNA*, which appears when the relative humidity is reduced below about 75%. A-DNA, like B-DNA, is a right-handed double helix made up of antiparallel strands held together by Watson-Crick base pairing. The A-helix is wider and shorter than the B-helix, and its base pairs are tilted rather than normal to the helix axis (Figure 27-6). Many of the structural differences between the two types of helices arise from different puckerings of the ribose units. As was discussed in an earlier chapter (p. 337), furanose rings can be puckered so that four atoms are nearly coplanar and the fifth is about 0.5 Å away from this plane. The structure formed when C-2′ is out of plane is called $C_{2'}$-*endo*, and the one formed when C-3′ is out of plane is called $C_{3'}$-*endo* (Figure 27-7). A-DNA has $C_{3'}$-*endo* sugar units, whereas B-DNA has $C_{2'}$-*endo* sugars. This rather small structural difference in the sugar units has amplified consequences. The $C_{3'}$-*endo* puckering leads to a 19-degree tilting of the base pairs away from the normal to the helix. Moreover, the minor groove nearly vanishes. The phosphate groups in the A-helix bind fewer H_2O molecules than do phosphates in B-DNA. Hence, dehydration favors the A-form.

Figure 27-6
Structure of the A-DNA helix.

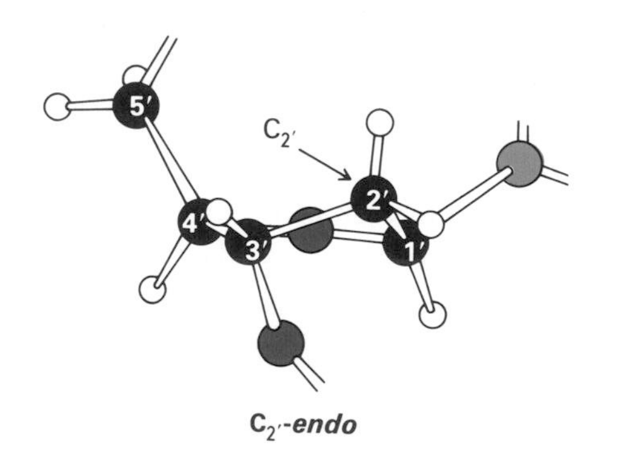

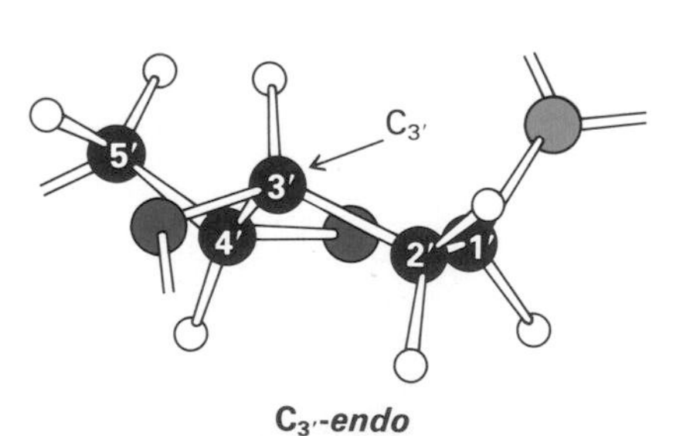

Figure 27-7
Sugar puckering markedly affects the orientation of the phosphodiester bridges and the glycosidic bond. $C_{2'}$-*endo* (found in B-DNA) and $C_{3'}$-*endo* (found in A-DNA) denote which atom of the ribose ring lies above the plane in the orientation shown here.

The A-type helix is not confined to dehydrated DNA. *Double-stranded regions of RNA (as in hairpins, p. 98) and RNA-DNA hybrids adopt a double-helical form very similar to that of A-DNA.* The 2′-OH of ribose prevents RNA from forming a Watson-Crick type helix because of steric hindrance. O-2′ cannot fit into a B-helix because it would come too close to three atoms of the adjoining phosphate group and one in the next base (Figure 27-8). In A-DNA, by contrast, O-2′ projects outwards, away from other atoms.

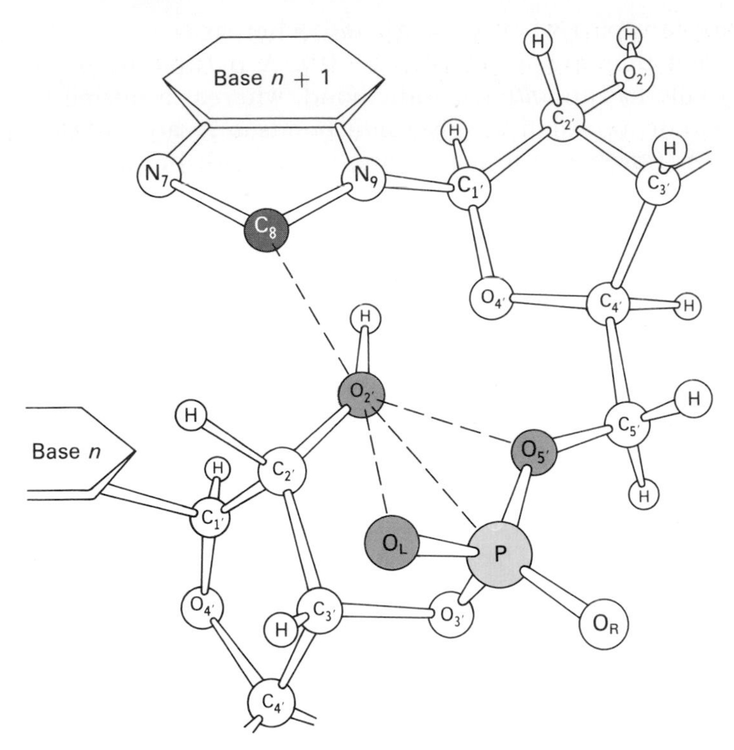

Figure 27-8
Ribose sugars do not fit into a helix of the B-DNA type because there is insufficient room for the 2′-oxygen atom. Four contacts would be closer than the allowed van der Waals distance. [After R. E. Dickerson. The DNA helix and how it is read. Copyright © 1983 by Scientific American, Inc. All rights reserved.]

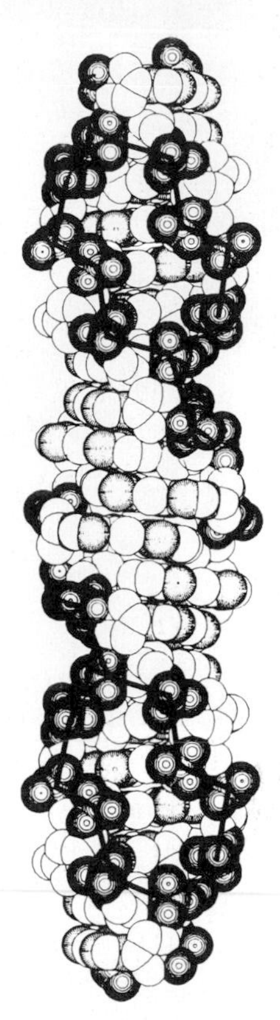

Z-DNA IS A LEFT-HANDED DOUBLE HELIX IN WHICH BACKBONE PHOSPHATES ZIGZAG

A third type of DNA helix was discovered by Alexander Rich and his associates when they solved the structure of CGCGCG. They found that this hexanucleotide forms a duplex of antiparallel strands held together by Watson-Crick base pairing, as expected. What was surprising, however, was that this double helix was *left-handed*, in contrast with the A and B helices. Another striking difference was that the phosphates in the backbone *zigzagged*; hence they called this new form *Z-DNA* (Figure 27-9). The zigzagging is a consequence of the fact that the repeating unit is a dinucleotide rather than a mononucleotide. A third difference is that Z-DNA contains only one deep helical groove.

How does such a different structure arise? Z-DNA is made possible by *sequences of alternating pyrimidines and purines*. This regularity allows the glycosidic bonds to alternate between *anti* and *syn*. The base and

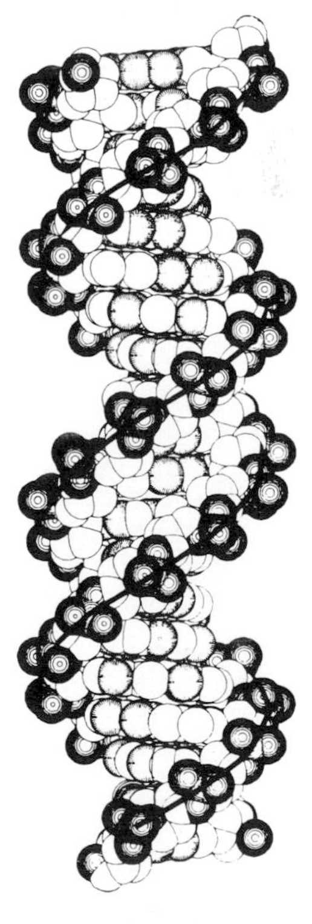

Figure 27-9
Comparison of Z-DNA, a left-handed helix, with B-DNA, a right-handed helix. Another difference is that the sugar-phosphate backbone zigzags in Z-DNA because the repeating unit is a dinucleotide. In contrast, the backbone coils smoothly in B-DNA, where the repeating unit is a mononucleotide. [Courtesy of Dr. Alexander Rich.]

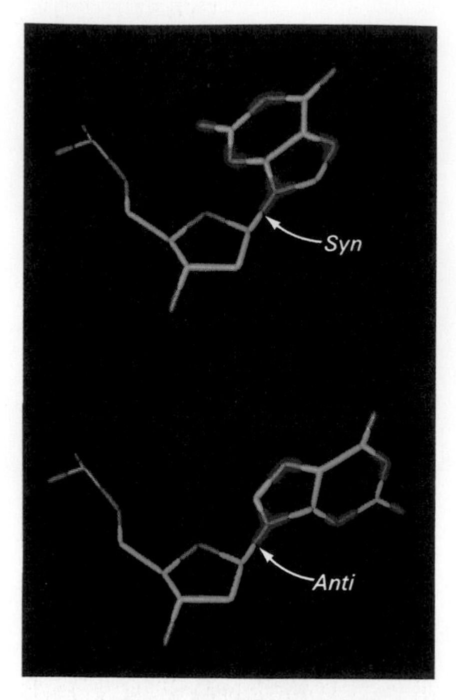

Figure 27-10
The glycosidic bond is *syn* when the base and sugar are on the same side, whereas it is *anti* when they are opposite one another.

sugar are far apart when they are *anti*, whereas they are next to each other when they are *syn* (Figure 27-10). A pyrimidine nucleotide by itself usually has an *anti* glycosidic bond, whereas a purine nucleotide can be *anti* or *syn*. All of the glycosidic bonds in A- and B-DNA are *anti*. In Z-DNA, however, the pyrimidine is *anti* and the purine is *syn*.

A stretch of Z-DNA can occur within B-DNA. The conversion of an alternating pyrimidine-purine sequence from B- to Z-DNA can be accomplished by flipping the base pairs 180 degrees and rotating the sugars of the purine residues (Figure 27-11). However, the formation of Z-DNA is usually thermodynamically unfavorable. Electrostatic repulsions between phosphate groups are stronger in Z- than in B-DNA because the closest groups on opposite strands are 8 Å rather than 12 Å apart. However, the transition to the Z-form is favored by methylation of cytosine at C-5, a modification that occurs in eucaryotes. A methyl group at this site in Z-DNA fills a small hole and promotes hydrophobic interactions; in B-DNA, this methyl group is surrounded by water, an unfavorable situation. The negative supercoiling of naturally occurring DNA molecules (p. 662) also promotes the transition from B- to Z-DNA. Indeed, antibodies specific for Z-DNA stain discrete regions of eucaryotic chromosomes. Also, eucaryotic DNAs contain many repeating pyrimidine-purine sequences (e.g., CACACA) that could form Z-DNA.

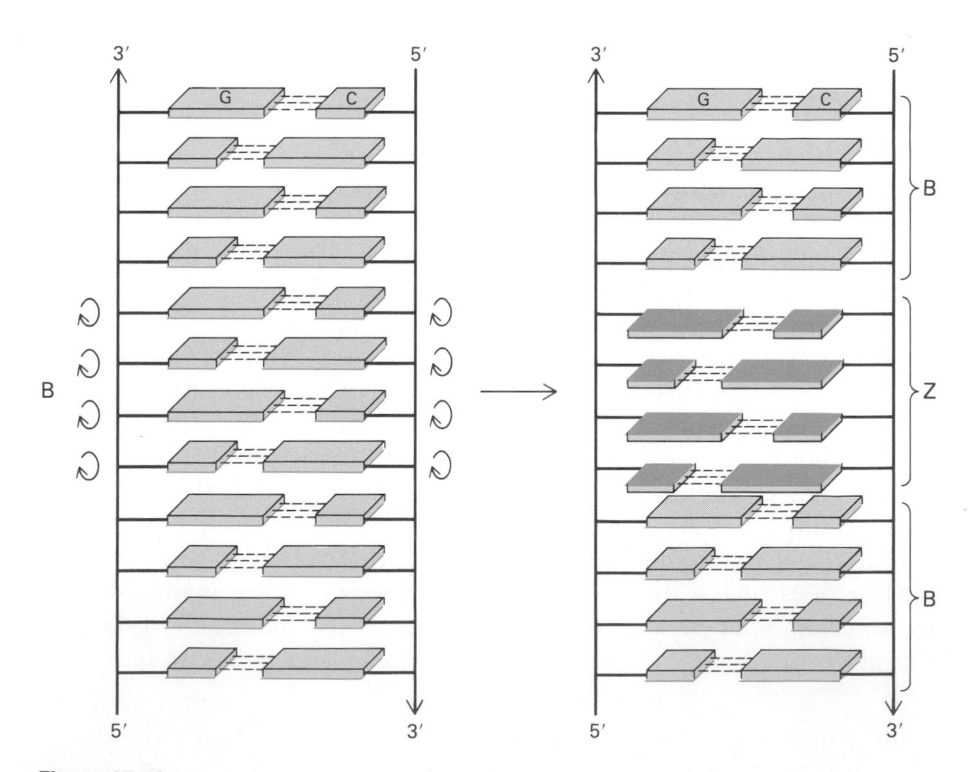

Figure 27-11
A segment of B-DNA can be converted into one of Z-DNA by a flipping of base pairs. The two sides of a base are colored light and dark. An alternating pyrimidine-purine sequence is required for the formation of Z-DNA. [After A. Rich, A. Nordheim, and A. H. J. Wang. The chemistry and biology of left-handed Z-DNA. *Ann. Rev. Biochem.* 53:791. Copyright © 1984 by Annual Reviews, Inc. All rights reserved.]

Most of the DNA in a bacterial or a eucaryotic genome is in the classic Watson-Crick B-DNA form. However, the discovery of Z-DNA, a form quite different from both A-DNA and B-DNA, emphasizes that DNA is a flexible, dynamic molecule. It will be very interesting to learn whether Z-DNA is biologically important. It would be surprising if nature did not exploit its unusual features.

Table 27-1
Comparison of A-, B-, and Z-DNA

	Helix type		
	A	*B*	*Z*
Shape	Broadest	Intermediate	Most elongated
Rise per base pair	2.3 Å	3.4 Å	3.8 Å
Helix diameter	25.5 Å	23.7 Å	18.4 Å
Screw sense	Right-handed	Right-handed	Left-handed
Glycosidic bond	*anti*	*anti*	*anti* for C, T *syn* for G
Base pairs per turn of helix	11	10.4	12
Pitch per turn of helix	24.6 Å	33.2 Å	45.6 Å
Tilt of base pairs from normal to helix axis	19°	1°	9°
Major groove	Narrow and very deep	Wide and quite deep	Flat
Minor groove	Very broad and shallow	Narrow and quite deep	Very narrow and deep

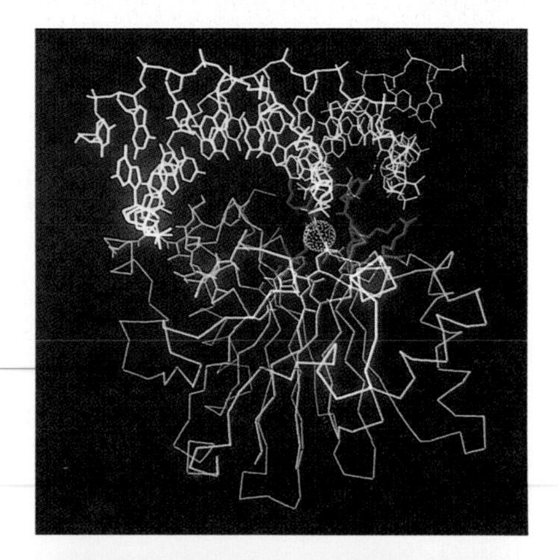

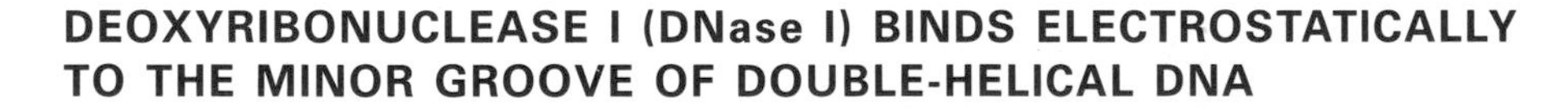

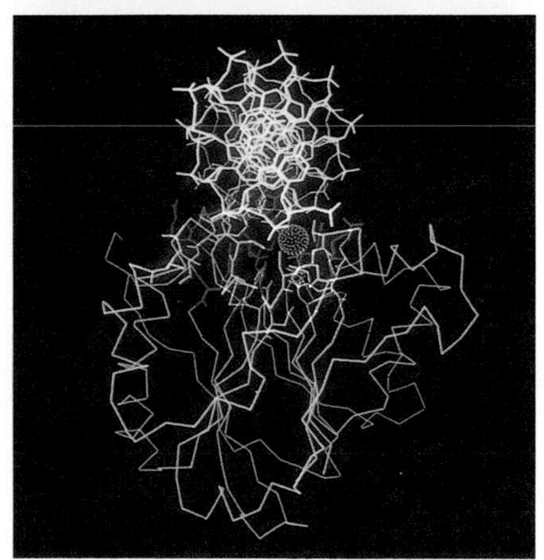

Figure 27-12
Interaction of DNA with DNase I inferred from models based on the crystal structures of the enzyme and its complexes with nucleotides. DNA is shown in yellow, the interacting groups of the enzyme in red, and the rest of the enzyme in blue. The sphere at the enzyme-DNA junction depicts the van der Waals radius of the essential divalent cation. The views shown here are related by a rotation of about 90 degrees. [Courtesy of Dr. Dietrich Suck and Dr. Christian Oefner.]

DEOXYRIBONUCLEASE I (DNase I) BINDS ELECTROSTATICALLY TO THE MINOR GROOVE OF DOUBLE-HELICAL DNA

We turn now to protein-DNA interactions, as exemplified by the recognition of DNA by nucleases. Deoxyribonuclease I (DNase I) from bovine pancreas is a digestive enzyme that degrades DNA rather nonspecifically to form short oligonucleotides having 5′-phosphate and 3′-hydroxyl termini. The crystal structure of this 30-kd endonuclease has recently been solved at high resolution (Figure 27-12). As expected, electrostatic interactions play a prominent role in binding the DNA substrate to the enzyme. A pentapeptide loop containing an arginine and a lysine residue fits tightly into the minor groove of DNA. These residues and five additional arginine and lysine side chains form salt bridges with phosphate groups of both strands flanking the minor groove. The enzyme interacts with an extensive region of DNA, about one turn of helix (34 Å). The matching of many positively charged groups of the enzyme with negatively charged groups of the duplex substrate accounts for the fact that double-helical DNA is hydrolyzed 10^4 times as rapidly as is single-stranded DNA. On the other hand, the enzyme does not interact intimately with the bases of DNA, so it does not discriminate nucleotide sequences well.

X-ray analyses of nucleotide-enzyme complexes suggest a plausible catalytic mechanism. A divalent cation is required for activity, as in all other enzymes catalyzing phosphoryl transfer. In the crystal, Ca^{2+} is bound close to the scissile phosphodiester bond; in a cell, Mg^{2+} may serve the same role. The other key groups are a bound water molecule, a histidine residue, and a glutamate residue (Figure 27-13). The water molecule becomes activated by the transfer of a proton to the histidine. The resulting positively charged imidazole side chain is stabilized electrostatically by the negatively charged glutamate side chain. This glutamate-histidine-water catalytic triad is reminiscent of the aspartate-histidine-serine triad in chymotrypsin and other serine proteases (p. 224). The hydroxide ion formed in this way attacks the phosphorus atom of the scissile bond to form a pentacovalent intermediate. Ca^{2+} (or Mg^{2+}) stabilizes this intermediate by interacting electrostatically with a negatively charged oxygen atom. Finally, the 3′-O group at the opposite corner of the trigonal bipyramid (p. 213) departs, leaving the 5′-phosphate group on the remaining end of the cleaved DNA strand.

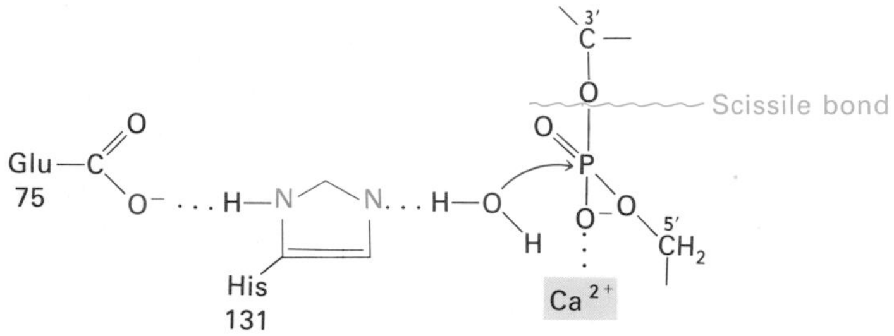

Figure 27-13
Catalytic triad in DNase I. The phosphorus atom of the scissile bond is attacked by hydroxide ion formed from water by the transfer of a proton to histidine. The resulting positively charged imidazole ring is stabilized by the negatively charged glutamate side chain.

THE EcoRI ENDONUCLEASE UNWINDS ITS PALINDROMIC TARGET AND FORMS MULTIPLE HYDROGEN BONDS

Restriction enzymes, in contrast with DNase I, cleave both strands of DNA at highly specific sites. These exquisitely precise scalpels made possible the DNA revolution in biology (p. 118). As will be discussed in a later chapter (p. 858), bacteria use restriction endonucleases as protective devices to degrade foreign DNA. Their own DNA evades cleavage because potential target sites are methylated. How do restriction endonucleases achieve very high specificity? The EcoRI endonuclease recognizes a hexanucleotide target sequence. This *palindromic* nucleotide sequence, like that of all known targets of restriction enzymes, possesses *twofold rotational symmetry*. Both strands are cleaved at identical sites symmetrically positioned with respect to the twofold axis.

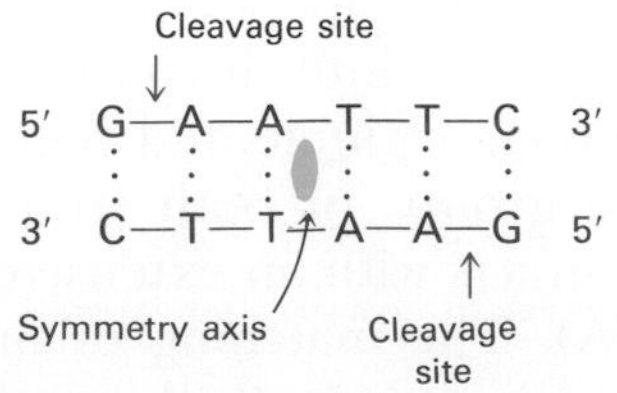

5′ G—A—A—T—T—C 3′
3′ C—T—T—A—A—G 5′

The enzyme recognizes a site with twofold symmetry and cleaves both strands symmetrically—hence, the enzyme, too, should be symmetrical. Indeed, it is. The EcoRI endonuclease is a dimer of identical subunits (31 kd each), and binds DNA so that the twofold axis of the DNA target site coincides with the twofold axis of the enzyme (Figure 27-14). Thus, *the symmetry of the enzyme matches the symmetry of its substrate.*

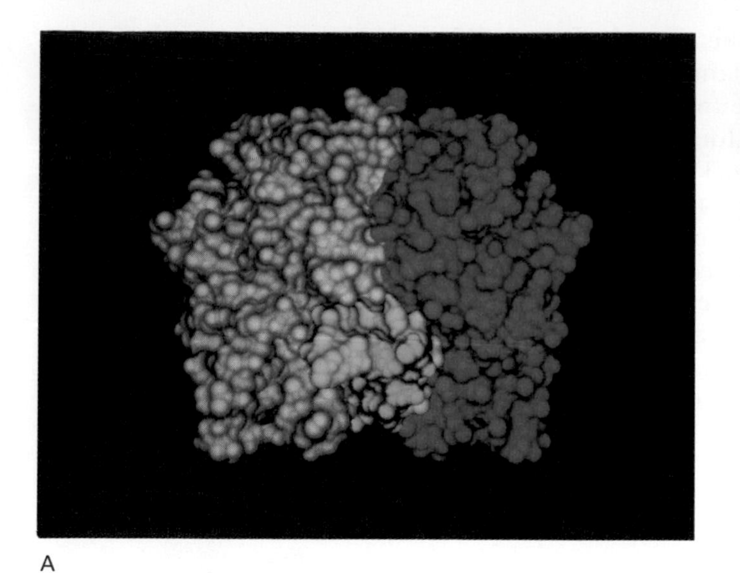

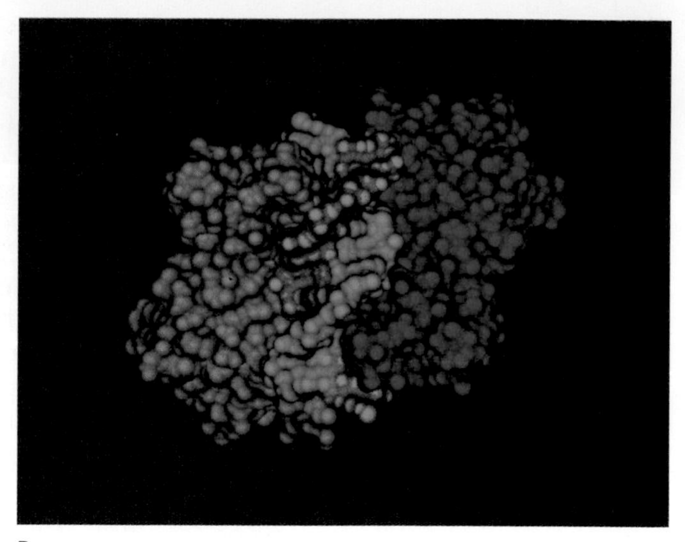

Figure 27-14
Structure of EcoRI endonuclease bound to its target DNA. The twofold symmetry of the enzyme matches that of its substrate. One subunit of the enzyme is shown in red and the other in orange; one strand of DNA is shown in green and the other in blue. View A looks down the DNA axis, and view B is in a perpendicular direction. [After J. A. McClarin, C. A. Frederick, B-C. Wang, P. Greene, H. W. Boyer, J. Grable, and J. M. Rosenberg. *Science* 234(1986):1530.]

How does the enzyme select GAATTC from all other hexanucleotides in DNA? A high degree of discrimination requires intimate contact between the enzyme and bases. This is accomplished by a kinking of the double helix at the center of the hexanucleotide target (Figure 27-15). This abrupt disruption of the double helix is called a *type I neokink*; the prefix *neo* denotes that the kink is imposed by an external agent (e.g., a drug or an enzyme). *The neokink unwinds the DNA helix by 25 degrees, widening the major groove by nearly 4 Å*. This expansion allows the entry of α helices into the major groove to sense the identity of the bases.

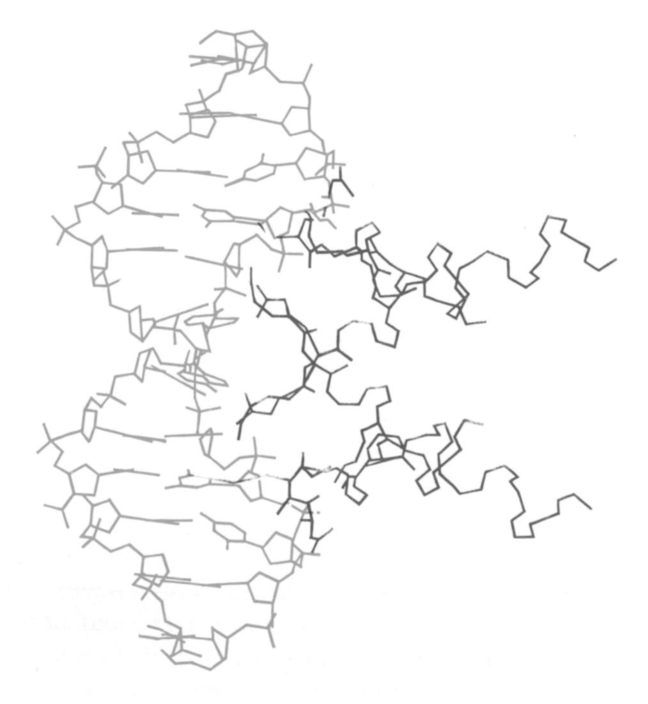

Figure 27-15
The site in DNA (green) that is recognized by EcoRI endonuclease (red) becomes kinked. Four helices of the enzyme enter the widened major groove of the palindromic target site. [Courtesy of Dr. John M. Rosenberg.]

One α helix bears an arginine side chain that forms two hydrogen bonds with a guanine of the target site (Figure 27-16). A second α helix of the same subunit contains an arginine and a glutamate residue that form a total of four hydrogen bonds with the adjacent adenines in the target sequence. The other subunit of the enzyme makes the same contacts with the symmetrical GAA on the other strand of DNA. Thus, *the very high specificity of this protein-DNA interaction is based on symmetry and the formation of 12 precisely directed hydrogen bonds.*

All four α helices are oriented so that their amino ends are in contact with DNA. This arrangement is significant because an α helix is dipolar; its amino end carries about half a positive charge, whereas the carboxyl end carries about half a negative charge (Figure 27-17).

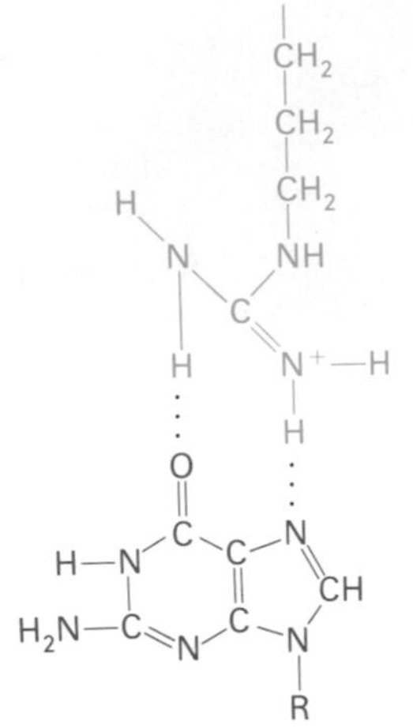

Figure 27-16
Hydrogen bonding between an arginine residue of the endonuclease and a guanine in the target site. The active site for cleavage of the backbone between G and A was not fully formed in the crystal because Mg^{2+} was omitted to obtain a stable enzyme-substrate complex.

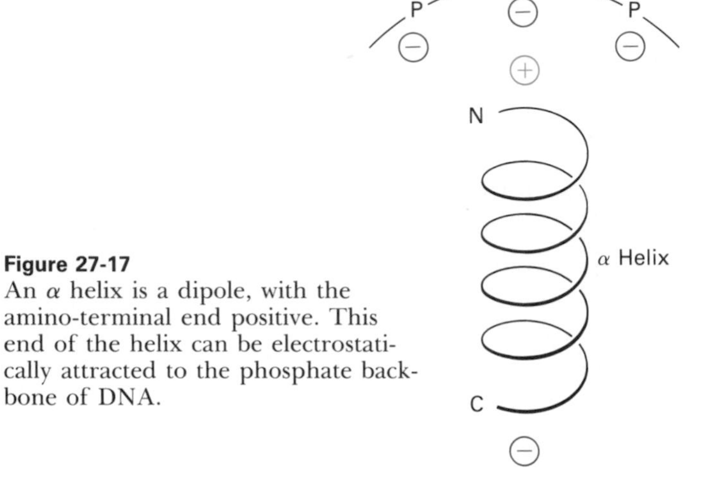

Figure 27-17
An α helix is a dipole, with the amino-terminal end positive. This end of the helix can be electrostatically attracted to the phosphate backbone of DNA.

Hence, *this bundle of four parallel α helices is electrostatically attracted to the phosphate groups of the DNA backbone.* We can picture the EcoRI endonuclease sliding along the major groove, transiently kinking the DNA, and searching for pairs of GAA related by a two-fold axis like that of the enzyme. It is noteworthy that precise recognition is achieved even though the bases of the target site remain fully paired with each other, as in a classic Watson-Crick duplex. This means that the enzyme can evaluate a hexanucleotide sequence without having to pry apart its base pairs, a process that would undoubtedly slow its scanning of foreign DNA. The mode of action of EcoRI endonuclease shows that *base pairs can be read with very high precision if the major groove of DNA is first widened.* The recognition of DNA sequences by other highly specific DNA-binding proteins will be discussed in a later chapter (p. 811).

DNA LIGASE JOINS ENDS OF DNA IN DUPLEX REGIONS

The finding of circular DNA pointed to the existence of an enzyme that connects the ends of DNA chains. In 1967, scientists in several laboratories simultaneously discovered *DNA ligase.* This enzyme catalyzes *the formation of a phosphodiester bond between the 3′-OH group at the end of one DNA chain and the 5′-phosphate group at the end of the other.* An energy source is required to drive this endergonic reaction. In *E. coli* and other

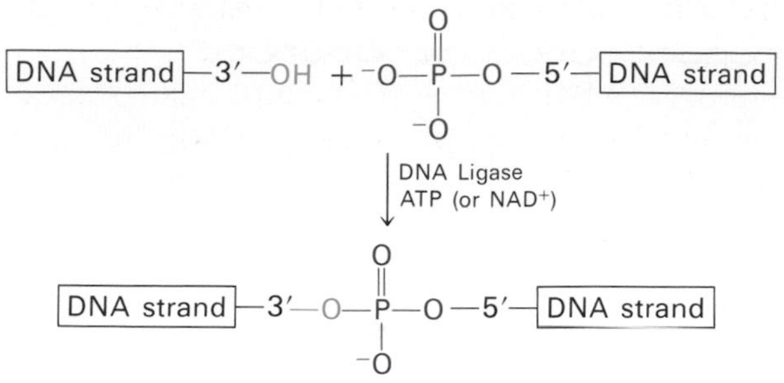

$$\boxed{\text{DNA strand}}\text{—3'—OH} + {}^{-}\text{O—}\overset{\overset{\text{O}}{\|}}{\underset{\underset{{}^{-}\text{O}}{|}}{\text{P}}}\text{—O—5'—}\boxed{\text{DNA strand}}$$

$$\downarrow \text{DNA Ligase, ATP (or NAD}^{+}\text{)}$$

$$\boxed{\text{DNA strand}}\text{—3'—O—}\overset{\overset{\text{O}}{\|}}{\underset{\underset{{}^{-}\text{O}}{|}}{\text{P}}}\text{—O—5'—}\boxed{\text{DNA strand}}$$

Figure 27-18
DNA ligase catalyzes the joining of two DNA chains that are part of a double-helical molecule.

bacteria, *NAD*$^+$ serves this role; in animal cells and bacteriophage, *ATP* is the energy source. As will be discussed later, this joining process is essential for the *normal synthesis of DNA, the repair of damaged DNA, and the splicing of DNA chains in genetic recombination.*

DNA ligase cannot link two molecules of single-stranded DNA. Rather, *the DNA chains joined by DNA ligase must belong to a double-helical molecule (or molecules).* The enzyme from *E. coli* ordinarily forms a phosphodiester bridge only if there are at least several base pairs near this link. Ligase encoded by T4 bacteriophage can link two blunt-ended double-helical fragments, a capability that is exploited in recombinant DNA technology.

Let us look at the mechanism of joining, which was elucidated by I. Robert Lehman. ATP (or NAD$^+$) reacts with DNA ligase to form a *covalent enzyme-AMP complex* (*enzyme-adenylate complex*) in which AMP is linked to the ϵ-amino group of a lysine residue of the enzyme through a phosphoamide bond (Figure 27-19). Pyrophosphate (or nicotinamide mononucleotide, NMN) is concomitantly released. The activated AMP moiety is then transferred from the lysine residue to the phosphate group at the 5′ terminus of a DNA chain, forming a *DNA-adenylate complex.* The final step is a nucleophilic attack by the 3′-OH group on this activated 5′-phosphorus atom. This sequence of reactions (Figure 27-20) is driven by the hydrolysis of pyrophosphate that was released in the formation of the enzyme-adenylate complex. Thus, two ~P are spent in constructing a phosphodiester bridge in the DNA backbone if ATP is the energy source and PP_i is hydrolyzed.

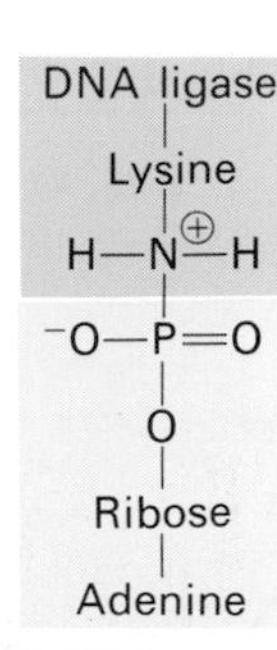

Figure 27-19
Covalent ligase-AMP intermediate formed by the reaction of NAD$^+$ with a lysine residue of the enzyme. The AMP moiety is activated in this intermediate.

$$\text{E} + \text{ATP (or NAD}^+) \rightleftharpoons \text{E-AMP} + \text{PP}_i \text{ (or NMN)}$$

$$\text{E-AMP} + \text{Ⓟ—5'-DNA} \rightleftharpoons \text{E} + \text{AMP—Ⓟ—5'-DNA}$$

$$\text{DNA-3'-OH} + \text{AMP—Ⓟ—5'-DNA} \rightleftharpoons \text{DNA-3'-O—Ⓟ—5'-DNA} + \text{AMP}$$

$$\text{DNA-3'-OH} + \text{Ⓟ—5'-DNA} + \text{ATP (or NAD}^+) \rightleftharpoons \text{DNA-3'-O—Ⓟ—5'-DNA} + \text{AMP} + \text{PP}_i \text{ (or NMN)}$$

Figure 27-20
Mechanism of the reaction catalyzed by DNA ligase.

THE LINKING NUMBER OF DNA, A TOPOLOGICAL PROPERTY, DETERMINES THE DEGREE OF SUPERCOILING

In 1963, Jerome Vinograd found that circular DNA from polyoma virus emerged in two bands when it was centrifuged. In pursuing this puzzle, he discovered an important property of circular DNA not possessed by linear DNA. Consider a 260-bp DNA duplex in the B-DNA

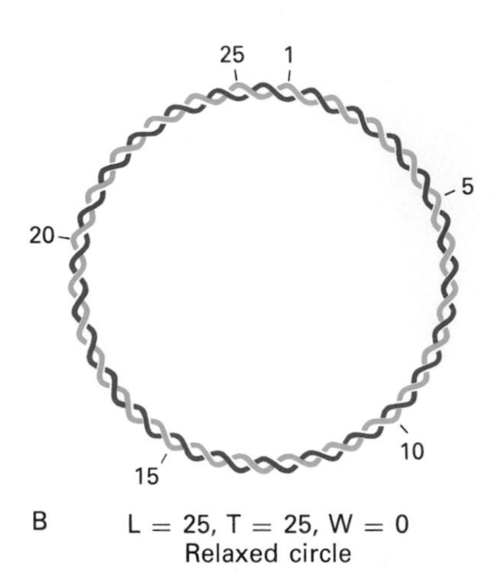

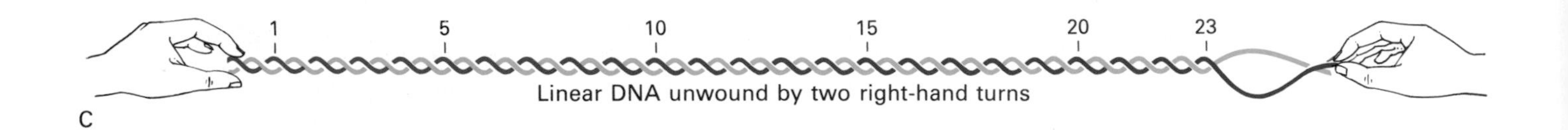

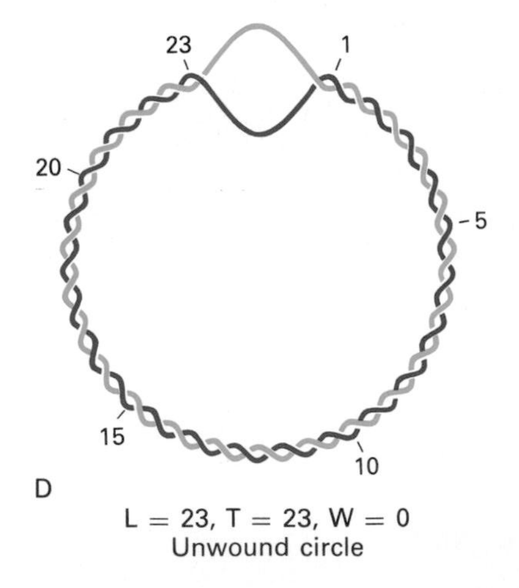

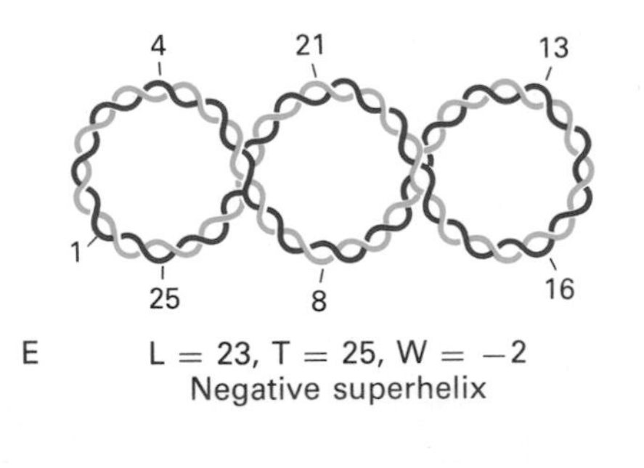

Figure 27-21
Definition of the linking number (L), twisting number (T), and writhing number (W) of a circular DNA molecule. [After W. Saenger. *Principles of Nucleic Acid Structure.* Springer-Verlag, 1984, p. 452.]

form (Figure 27-21A). The number of helical turns in this linear DNA is 25 (260/10.4). Now let us join the ends of this helix. This circular DNA is said to be *relaxed* (Figure 27-21B). A different circular DNA can be formed by unwinding the linear duplex by two turns before joining its ends (Figure 27-21C). The resulting circular DNA can fold into a structure containing 23 turns of B-helix and an unwound loop (Figure

27-21D). Alternatively, it can adopt a structure with 25 turns of B-helix and 2 turns of superhelix (Figure 27-21E). In fact, this *supercoiled* form is energetically favored over the one having an unwound loop.

Supercoiling markedly alters the overall form of DNA. *A supercoiled DNA molecule is more compact than a relaxed DNA molecule of the same length* (Figure 27-22). Hence, supercoiled DNA moves faster than relaxed DNA when they are centrifuged or electrophoresed. The rapidly sedimenting DNA in Vinograd's experiment was supercoiled, whereas the slowly sedimenting DNA was relaxed because one of its strands was nicked.

Our understanding of the conformation of DNA is enriched by concepts drawn from topology, a branch of mathematics dealing with structural properties that are unchanged by deformations such as stretching and bending. A key topological property of a circular DNA molecule is its *linking number (L),* which is defined as the number of times one strand of DNA winds around the other in the right-handed direction. For the relaxed DNA shown in Figure 27-21B, $L = 25$. For the unwound molecule shown in D and the supercoiled one shown in E, $L = 23$ because the linear duplex was unwound two complete turns before closure. Molecules differing only in linking number are topological isomers, or *topoisomers,* of one another. *Topoisomers of DNA can be interconverted only by cutting one or both DNA strands.*

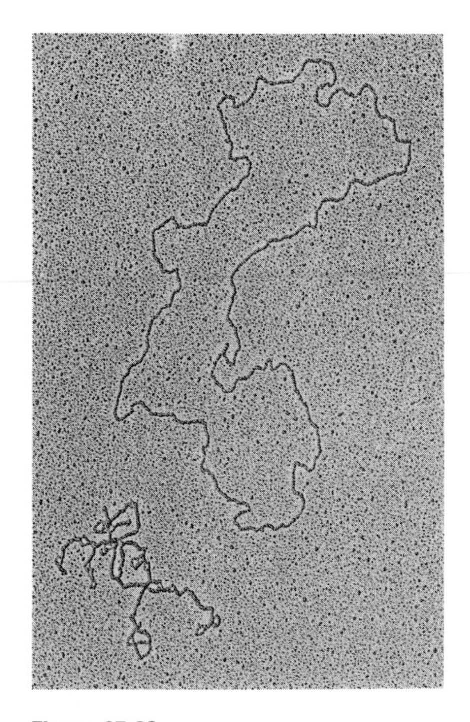

Figure 27-22
Electron micrograph of relaxed and negatively supercoiled DNA. [Courtesy of Dr. Jack Griffith.]

The unwound DNA and supercoiled DNA shown in parts D and E of Figure 27-21 have the same value of L but differ in the number of turns of Watson-Crick helix (denoted by T, for *twisting number*) and the number of turns of superhelix (denoted by W, for *writhing number*). Is there a relation between T and W? Indeed, there is. Topology tells us that the sum of T and W is simply equal to L.

$$L = T + W$$

T and W can be nonintegral, whereas L must be integral. The unwound circular DNA has $T = 23$ and $W = 0$, whereas the supercoiled DNA has $T = 25$ and $W = -2$. These forms can be interconverted without cleaving the DNA chain because they have the same value of L, namely, 23. The partitioning of L between the number of turns of Watson-Crick helix and of superhelix is determined by energetics. Supercoiled DNA is favored over unwound DNA because it contains more paired bases. In general, T has a value nearly equal to the number of base pairs divided by 10.4. Hence, *changes in the linking number of DNA usually lead to changes in the degree of supercoiling (the writhing number) rather than to changes in the number of Watson-Crick turns (the twisting number).* Topoisomers differing in L (and hence in W) by just 1 unit can readily be separated by agarose gel electrophoresis.

MOST NATURALLY OCCURRING DNA MOLECULES ARE NEGATIVELY SUPERCOILED

The degree of supercoiling of DNA can be expressed in terms of the *specific linking difference* (λ), which is independent of the length of the molecule.

$$\lambda = (L - L_0)/L_0$$

L_0 is the linking number of the relaxed circular DNA molecule. For the supercoiled DNA in Figure 27-21E, $L = 23$, $L_0 = 25$, and so $\lambda = -0.08$. This quantity is also known as the *superhelix density* because $L - L_0$ is usually about equal to W. The superhelix density of most natural DNA molecules lies between -0.03 and -0.09. The negative sign means that DNA superhelices in nature, as in this example, are left-handed. In other words, they arise from *underwinding* or *unwinding*. An important consequence is that negatively supercoiled DNA molecules are poised to be unwound. *Negative supercoiling prepares DNA for processes requiring the separation of strands: replication, recombination, and transcription.* Positive supercoiling would compact DNA as effectively as negative supercoiling does, but it would make strand separation more difficult. We see here a compelling reason for the selection of negative supercoiling.

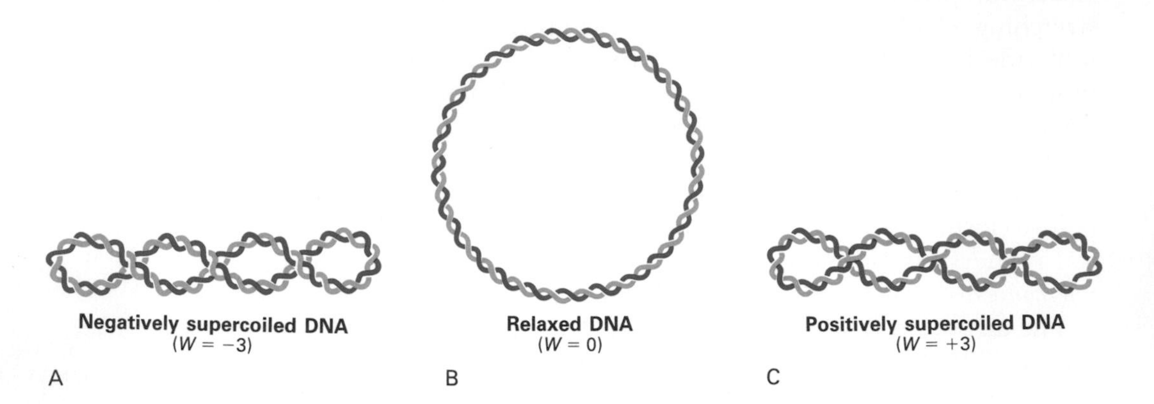

Figure 27-23
Schematic diagram of DNA molecules that differ in their supercoiling. (A) Negatively supercoiled ($W = -3$). (B) Relaxed ($W = 0$). (C) Positively supercoiled ($W = +3$). Most naturally occurring DNA molecules are negatively supercoiled.

TOPOISOMERASE I CATALYZES THE RELAXATION OF SUPERCOILED DNA

> "With thy sharp teeth this knot intrinsicate of life at once untie."
>
> WILLIAM SHAKESPEARE
> *Anthony and Cleopatra* (V, ii)

The interconversion of topoisomers of DNA is catalyzed by enzymes called *topoisomerases* that were discovered by James Wang and Martin Gellert. These enzymes alter the linking number of DNA by catalyzing a three-step process: (1) *cleavage* of one or both strands of DNA, (2) *passage* of a segment of DNA through this break, and (3) *resealing* of the

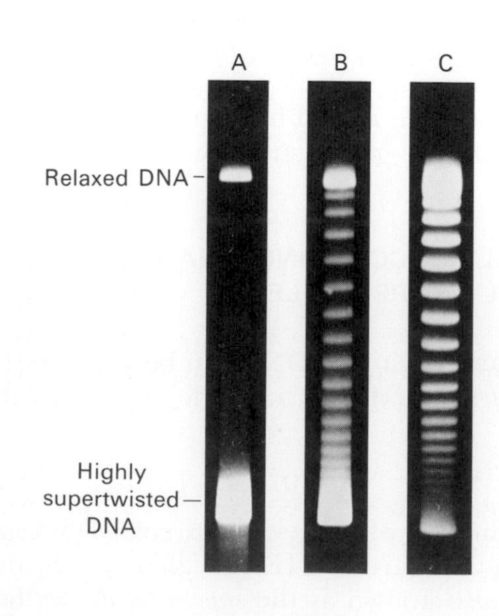

Figure 27-24
Gel patterns showing the relaxation of supercoiled SV40 viral DNA. Part A is a highly negatively supercoiled DNA. Incubation of the DNA with a topoisomerase for (B) 5 minutes and (C) 30 minutes leads to a series of bands that have less supercoiling. [From W. Keller. *Proc. Nat. Acad. Sci.* 72(1975):2553.]

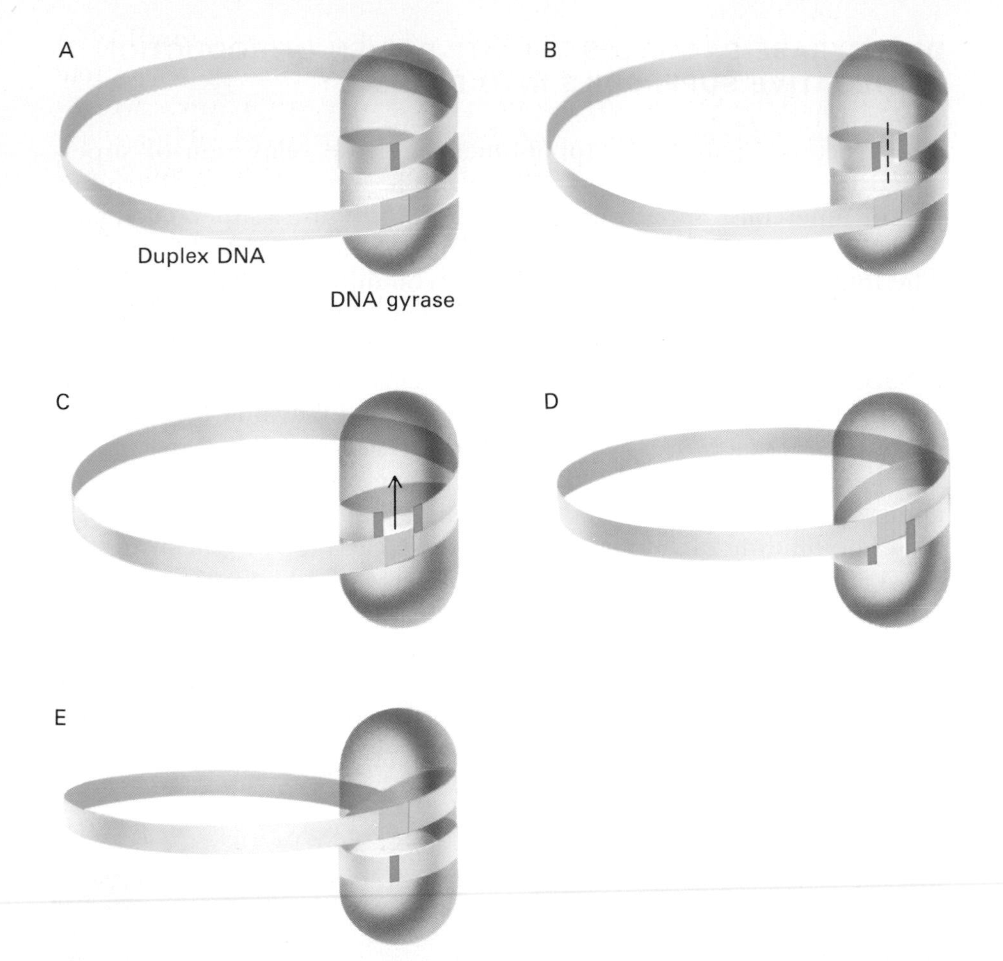

Figure 27-25
Topoisomerases catalyze changes in the linking number of DNA. DNA gyrase cleaves both strands of DNA, and passes a segment of double-helical DNA through this break. The severed strands are then rejoined. [After J. C. Wang. DNA topoisomerases. Copyright © 1982 by Scientific American, Inc. All rights reserved.]

DNA break (Figure 27-25). Type I topoisomerases cleave just one strand of DNA, whereas type II enzymes cleave both strands.

The type I topoisomerase of *E. coli* catalyzes the relaxation of negatively supercoiled DNA. One possibility a priori was that this enzyme acts first as a nuclease and then as a ligase. However, intact relaxed circles were formed in the absence of ATP, NAD^+, and other potential energy donors. This finding ruled out an initial hydrolytic event. It suggested instead that one end of the cleaved chain stays activated by being covalently attached to the enzyme. In fact, a covalent enzyme-DNA complex has been isolated from a mixture in which sodium dodecyl sulfate denatured the topoisomerase in the midst of its catalytic cycle. The 5′-phosphate moiety of a DNA strand was found to be linked to a tyrosine hydroxyl group of this 100-kd enzyme (Figure 27-26). The 3′-OH at the other end of the cleaved chain nucleophilically attacks this activated intermediate to restore the continuity of the circle. It is noteworthy that the chemical bonds in the substrate and product of this reaction are identical. The only role of this topoisomerase is to create a transient break that allows the passage of a segment of DNA. The linking number increases by +1 with each catalytic cycle.

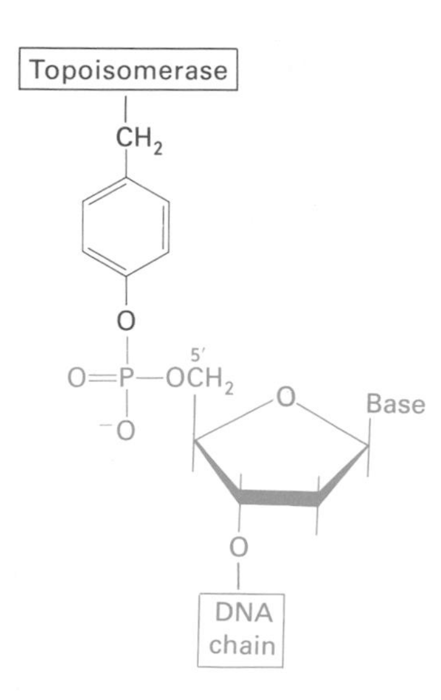

Figure 27-26
Covalent enzyme-substrate intermediate in the action of topoisomerases. The 5′-phosphate end of the cleaved DNA strand is covalently linked to the hydroxyl group of a specific tyrosine residue of the enzyme.

DNA GYRASE CATALYZES THE ATP-DRIVEN INTRODUCTION OF NEGATIVE SUPERCOILS INTO DNA

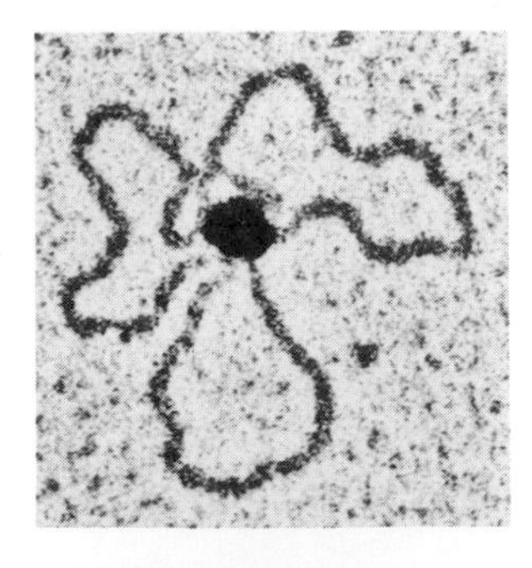

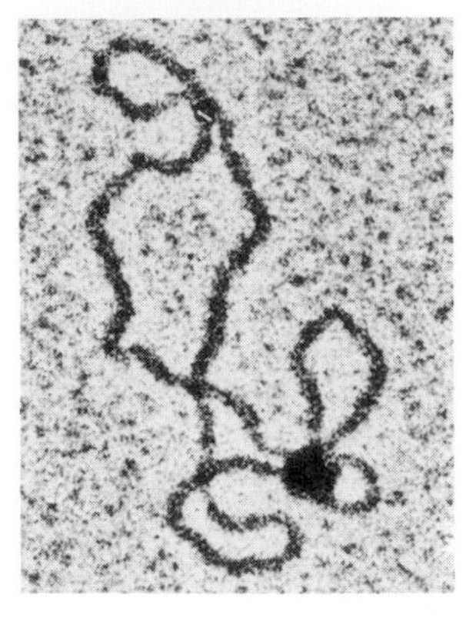

Figure 27-27
Electron micrographs showing DNA gyrase bound to negatively supercoiled DNA molecules.

The reaction catalyzed by topoisomerase I, the relaxation of supercoiled DNA, is thermodynamically downhill. What about the introduction of supercoils? Supercoiling costs energy because a supercoiled molecule, in contrast with its relaxed counterpart, is torsionally stressed. The introduction of an additional supercoil into a DNA molecule with $\lambda = -0.06$ costs about 9 kcal/mol. Thus, a good deal of energy is stored in the supercoiling of naturally occurring DNA molecules.

Supercoiling in *E. coli* is catalyzed by *DNA gyrase*, a topoisomerase consisting of two 105-kd A chains and two 95-kd B chains. *DNA gyrase is an energy-transducing device: it converts the free energy of ATP into the torsional energy of supercoiling.* The reaction begins with the wrapping of about 200 bp of DNA around the lozenge-shaped enzyme (Figure 27-27). The binding of ATP then triggers the cleavage of *both* strands of DNA; the cleavage site on one strand is staggered by four nucleotides from the site on the other strand. The 5′-phosphate terminus of each cleaved strand is linked to a specific tyrosine residue of an A subunit of the enzyme. The anchoring of the two ends of the cut DNA is essential for preventing their free rotation, which would quickly lead to the loss of supercoiling.

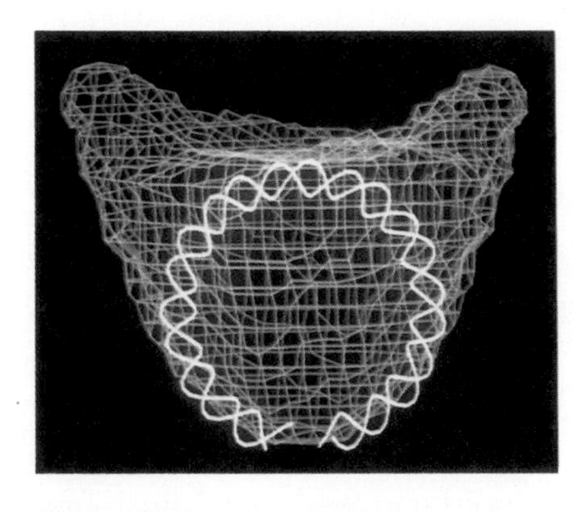

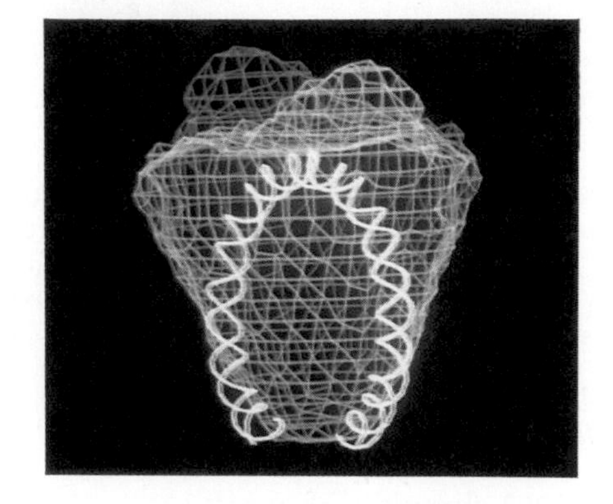

Figure 27-28
Model of DNA gyrase (blue) at low resolution. The bound DNA is depicted in yellow. [Courtesy of Dr. Stephen Harrison and Dr. James Wang.]

The next step is the passage of a segment of DNA through the ~35 Å gap between the fixed ends. This passage is *vectorial*—DNA gyrase allows facile movement only in the direction leading to negative supercoiling. The cleaved DNA ends then come together to reestablish the continuity of both strands of the duplex. Finally, hydrolysis of bound ATP leads to the release of the transported DNA segment, enabling the enzyme to begin another round of catalysis. DNA gyrase decreases the linking number of its substrate in steps of *two*, because duplex DNA passes through a break in *both* strands. About two negative supercoils are introduced per second. DNA gyrase can act repeatedly on the same DNA substrate without dissociating from it.

The degree of supercoiling of bacterial DNA is thus determined by the opposing actions of two enzymes. Negative supercoils are introduced by DNA gyrase and are removed by topoisomerase I. The amounts of these enzymes are regulated to maintain an appropriate degree of negative supercoiling. DNA gyrase is the target of several antibiotics. *Nalidixic acid*, which is used to treat urinary tract infections, interferes with the breakage and rejoining of DNA chains. *Novobiocin* blocks the binding of ATP to gyrase.

DISCOVERY OF DNA POLYMERASE I, THE FIRST TEMPLATE-DIRECTED ENZYME

We turn now to the molecular events in the replication of DNA. The search for an enzyme that synthesizes DNA was initiated by Arthur Kornberg and his associates in 1955. This search soon proved fruitful, largely because three inspired choices were made:

1. What are the *activated precursors* of DNA? They correctly deduced that *deoxyribonucleoside 5′-triphosphates* are the activated intermediates in DNA synthesis. This inference was based on two clues. First, pathways of purine and pyrimidine biosynthesis lead to nucleoside 5′-phosphates rather than to nucleoside 3′-phosphates (p. 603). Second, ATP is the activated intermediate in the synthesis of a pyrophosphate bond in coenzymes such as NAD^+, FAD, and CoA (p. 617).

2. What is the *criterion of DNA synthesis*? It was surmised that the net amount of DNA synthesized might be very small and so a sensitive assay was essential. This was accomplished by using *radioactive precursor nucleotides*. The incorporation of these precursors into DNA was detected by measuring the radioactivity of an *acid precipitate of the incubation mixture.* (DNA is precipitated by trichloroacetic acid, whereas precursor nucleotides stay in solution.)

3. Which *kinds of cells* should be analyzed? *E. coli* bacteria were used after initial experiments with animal-cell extracts were negative. *E. coli* was chosen because it has a generation time of only twenty minutes and can be harvested in large quantities. As expected, this bacterium is a choice source of enzymes that synthesize DNA.

An extract of *E. coli* was incubated with radioactive deoxythymidine. The level of radioactivity in this ^{14}C-labeled precursor was one million counts per minute. The acid precipitate from this incubation mixture gave just fifty counts. Only a few picomoles of DNA was synthesized, but it was a start. Kornberg wrote: "Although the amount of nucleotide incorporated into nucleic acid was miniscule, it was nonetheless significantly above the level of background noise. Through this tiny crack we tried to drive a wedge. The hammer was enzyme purification, a technique that had matured during the elucidation of alcoholic fermentation."

This new enzyme was named DNA polymerase. It is now called *DNA polymerase I* because other DNA polymerases have since been isolated. After a decade of effort in Kornberg's laboratory, DNA polymerase I was purified to homogeneity and characterized in detail. An appreciation of the magnitude of the task can be gained by noting that it took one hundred kilograms of *E. coli* cells to produce five hundred milligrams of pure enzyme. As will be discussed shortly, DNA polymerase I is not the enzyme that replicates most of the DNA in *E. coli*. However, it plays a critical role in replication and also in the repair of DNA. Moreover, it is the simplest DNA polymerase and the best understood. The study of this enzyme has been highly rewarding because it exemplifies many key principles of the action of both procaryotic and eucaryotic DNA polymerases.

DNA polymerase I, a 103-kd monomer, catalyzes the *step-by-step addition of deoxyribonucleotide units to the 3′ end of a DNA chain:*

$$(\text{DNA})_{n\text{ residues}} + \text{dNTP} \rightleftharpoons (\text{DNA})_{n+1} + \text{PP}_i$$

As was discussed in Chapter 4 (p. 84), DNA polymerase I requires *all four deoxyribonucleoside 5′-triphosphates—dATP, dGTP, dTTP, and dCTP—and* Mg^{2+} to synthesize DNA. The symbol dNTP denotes any deoxyribonucleoside 5′-triphosphate. The enzyme adds deoxyribonucleotides to the free 3′-OH of the chain undergoing elongation, *which proceeds in the 5′ → 3′ direction* (Figure 27-29). A *primer chain* with a free 3′-OH group is needed at the start. A *DNA template* containing a single-stranded region is also essential. The polymerase catalyzes the nucleophilic attack of the 3′-OH terminus of the primer on the innermost phosphorus atom of a dNTP. A phosphodiester bond is formed and pyrophosphate is released. The reaction is driven forward by the subsequent hydrolysis of PP_i by inorganic pyrophosphatase. DNA polymerase I is a *moderately processive* enzyme—it catalyzes multiple polymerization steps (~20) before dissociating from the template DNA.

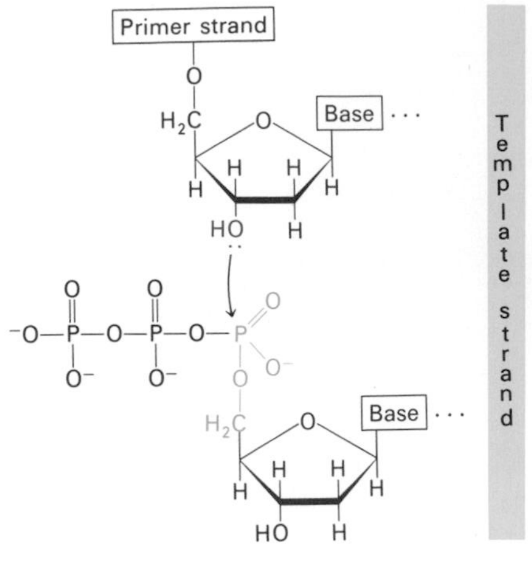

Figure 27-29
Chain-elongation reaction catalyzed by DNA polymerases.

A striking feature of the enzyme is that it takes instructions from its template. Indeed, DNA polymerase I was the first template-directed enzyme to be discovered. Polymerization is catalyzed by a single active site that can bind any of the four dNTPs. Which one binds depends on the corresponding base on the template strand. *The likelihood of binding and of making a phosphodiester bond is very low unless the incoming nucleotide forms a Watson-Crick base pair with the opposing nucleotide on the template.* The most compelling evidence for the high fidelity of this enzyme is that ϕX174 viral DNA replicated in vitro by the enzyme is fully infectious.

DNA POLYMERASE I IS ALSO A PROOFREADING 3′ → 5′ EXONUCLEASE

DNA polymerase I can catalyze the hydrolysis of DNA chains as well as their polymerization. The enzyme catalyzes the hydrolysis of nucleotides at the 3′-end of DNA chains. Thus, *DNA polymerase I is a 3′ → 5′ exonuclease.* To be removed, a nucleotide must have a free 3′-OH terminus, and it must not be part of a double helix. Experiments using synthetic DNA with a mismatched residue at the primer terminus revealed that the *3′ → 5′ nuclease activity has an editing function in polymerization.* The polymer shown in Figure 27-30 contains an unpaired C at the end of a sequence of T residues that form a double helix with a longer sequence of A residues. On addition of DNA polymerase I and dTTP, the unpaired C is first excised. dCMP is released and a 3′-OH group is exposed on the terminal T of the primer strand. Additional T are added only after removal of the unpaired C. In general, *DNA polymerase I removes mismatched residues at the primer terminus before proceeding with polymerization.*

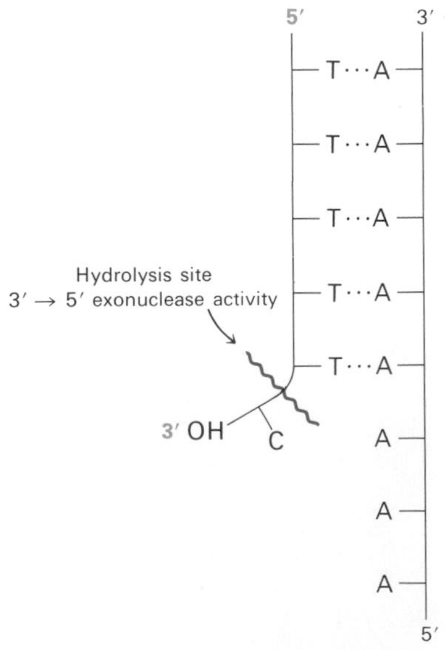

Figure 27-30
3′ → 5′ exonuclease action of DNA polymerase I.

This 3′ → 5′ exonuclease activity markedly enhances the accuracy of DNA replication by serving as a second test of the correctness of base pairing. Polymerization does not usually occur unless the base pair fits into a double helix. However, an error made at this stage is nearly always corrected before the next nucleotide is added. In effect, *DNA polymerase I examines the result of each polymerization it catalyzes before proceeding to the next.* The importance of exonuclease activity for the fidelity of replication is highlighted by some mutant T4 phage. Viruses encoding DNA polymerase with a reduced 3′ → 5′ exonuclease activity have a much higher mutation rate than normal. Conversely, phage with enhanced exonuclease activity have a decreased spontaneous mutation rate.

DNA POLYMERASE I IS ALSO AN ERROR-CORRECTING 5′ → 3′ EXONUCLEASE

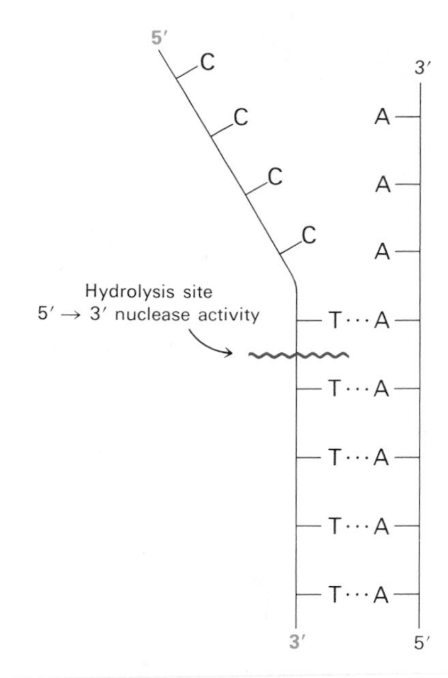

Figure 27-31
5′ → 3′ nuclease action of DNA polymerase I.

DNA polymerase I can also hydrolyze DNA starting from the 5′ end of a chain. This *5′ → 3′ nuclease activity* (Figure 27-31) is very different from the 3′ → 5′ exonuclease action discussed above. First, the cleaved bond must be in a double-helical region. Second, cleavage can occur at the terminal phosphodiester bond or at a bond several residues away from the 5′ terminus (which can bear a free hydroxyl group or be phosphorylated). Third, 5′ → 3′ exonuclease activity is enhanced by concomitant DNA synthesis. Fourth, the active site for exonuclease action is clearly separate from the active sites for polymerization and 3′ → 5′ hydrolysis. Thus, *DNA polymerase I contains three different active sites on a single polypeptide chain.*

The 5′ → 3′ exonuclease activity plays a key role in DNA replication by removing RNA primer, as will be discussed shortly (p. 675). Moreover, the 5′ → 3′ exonuclease complements the 3′ → 5′ exonuclease activity by correcting errors of a different type. For example, the 5′ → 3′ exonuclease participates in the excision of pyrimidine dimers formed by exposure of DNA to ultraviolet light.

DNA POLYMERASE I CONTAINS A DEEP CREVICE FOR BINDING DOUBLE-HELICAL DNA

This trifunctional enzyme can be split by proteases into a 36-kd small fragment with all of the original 5′ → 3′ exonuclease activity and a 67-kd large fragment (also called the Klenow fragment) with all of the polymerase and 3′ → 5′ exonuclease activities (Figure 27-32). X-ray crystallographic studies of the large fragment show that it is made up of two distinct domains. The small domain probably contains the 3′ → 5′ exonuclease site because it binds deoxyribonucleoside monophosphates, which inhibit nuclease but not polymerase activity. The phosphate group of dTMP binds next to a Zn^{2+} ion. The larger domain forms a structure with a deep cleft resembling a right hand holding a rod. The cleft is about 25 Å wide, 30 Å deep, and 50 Å long. The size and shape of the cleft suggests that it binds the duplex product in DNA

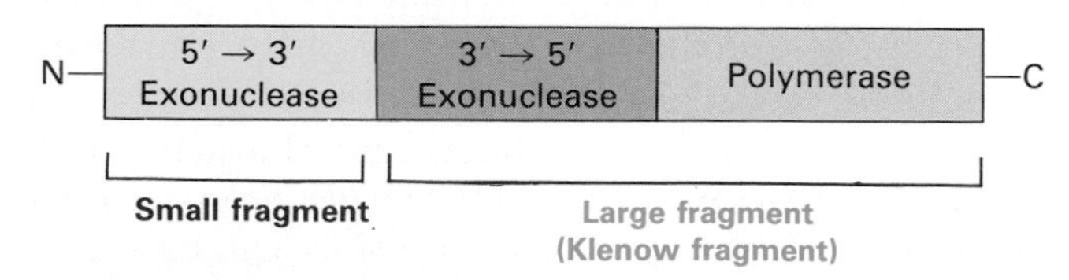

Figure 27-32
DNA polymerase I has three enzymatic activities in a single polypeptide chain.

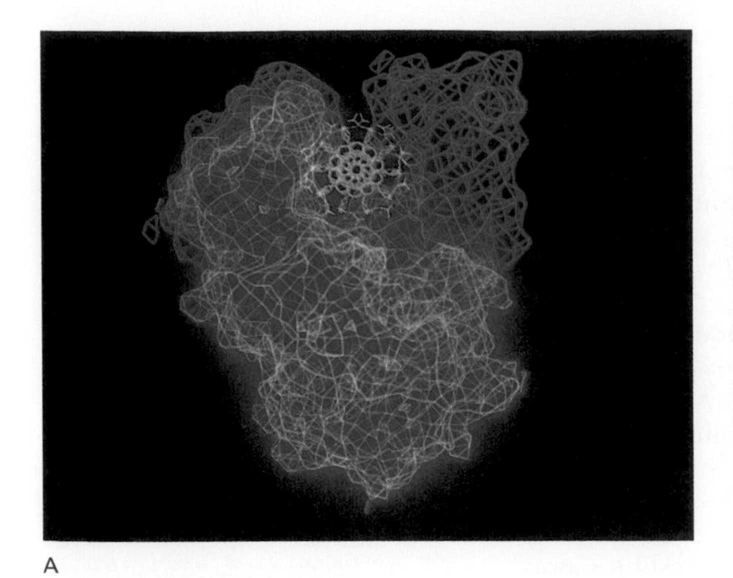

A

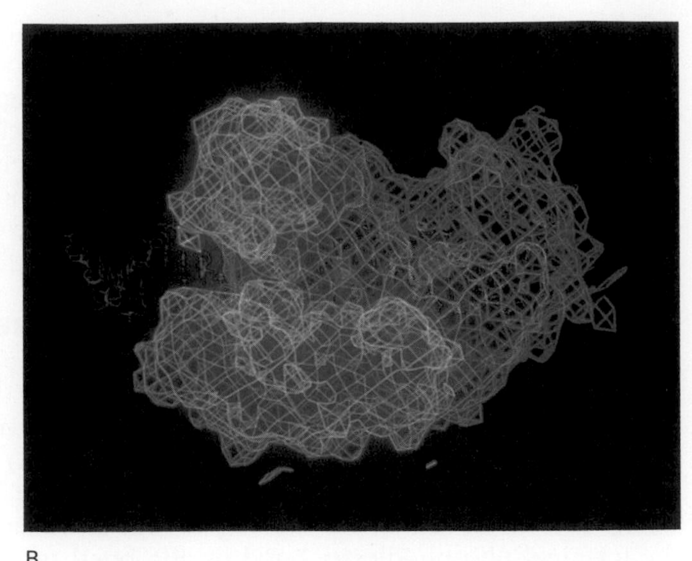

B

Figure 27-33
Model of the large fragment of DNA polymerase I (Klenow fragment) containing bound DNA. The enzyme is shown in blue and DNA in red. (A) View down the axis of the double helix, and (B) in a perpendicular direction. The location of DNA was determined by x-ray analysis. The location of DNA was inferred from the shape of the crevice in the enzyme and from chemical studies. [Courtesy of Dr. Art Perlo, Dr. Lorena Beese, and Dr. Thomas Steitz.]

synthesis (Figure 27-33). The processivity of DNA polymerase I may be due to closure of this cleft, which would slow the dissociation of DNA from the enzyme during polymerization. The exonuclease site on the small domain is some 25 Å away from the polymerase site on the large domain. An intriguing possibility is that the growing DNA chain stays in the polymerase site until a mismatched base pair is sensed, which would trigger its translocation to the $3' \rightarrow 5'$ exonuclease site for excision.

DISCOVERY OF DNA POLYMERASES II AND III

DNA polymerase I can synthesize and repair DNA in vitro. Does it perform these functions in vivo? This question is pertinent because an enzyme need not carry out the same reaction in vivo as it does in vitro. The reaction conditions inside a cell may be different and other enzymes may be present. In fact, *E. coli* contains two other DNA polymerases, named II and III, which were found some fifteen years after the discovery of DNA polymerase I. Why this lag? The reason is that the presence of polymerases II and III was masked by the high level of activity of polymerase I.

The breakthrough came in 1969, when Paula DeLucia and John Cairns isolated a mutant of *E. coli* that had about 1% of the normal polymerizing activity of DNA polymerase I. This mutant (named *polA1*) multiplied at the same rate as the parent strain. Furthermore, a variety of phages replicated as well in *polA1* as in the parent strain. By contrast, this mutant was much more readily killed by ultraviolet light than was the parent strain. DeLucia and Cairns inferred that DNA replication was normal in their *polA1* mutant, but that DNA repair was markedly impaired. They suggested that a polymerase different from DNA polymerase I might be essential for DNA synthesis.

Two such enzymes were soon isolated and characterized in several laboratories. *DNA polymerases II and III are like polymerase I in these respects:*

1. They catalyze a template-directed synthesis of DNA from deoxyribonucleoside 5′-triphosphate precursors.

2. A primer with a free 3′-OH group is required.
3. Synthesis is in the 5′ → 3′ direction.
4. They possess 3′ → 5′ exonuclease activity.

What are the roles of these polymerases in vivo? As will be discussed shortly, *a multisubunit assembly containing polymerase III synthesizes most of the new DNA, whereas polymerase I erases the primer and fills gaps*. The role of DNA polymerase II is not yet known. Biochemical and genetic studies have revealed that more than twenty proteins in addition to DNA polymerases I and III are required for DNA replication in *E. coli*. Before turning to the interplay of these proteins with DNA, let us take a view of replication at the level of the whole chromosome.

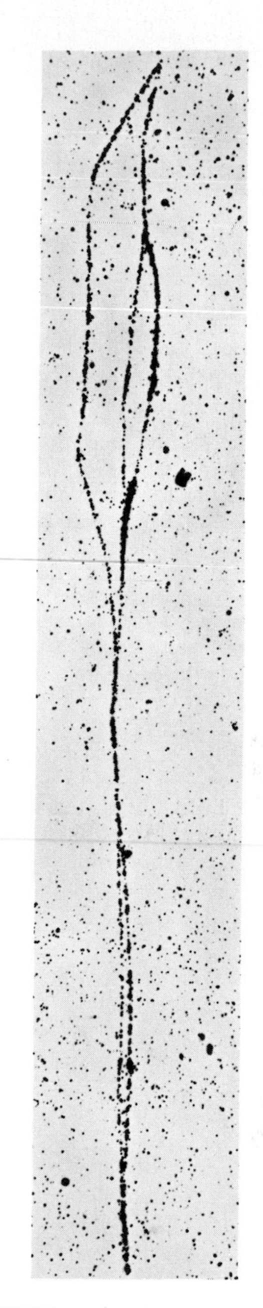

Figure 27-34
Autoradiograph of replicating DNA from *E. coli*. [Courtesy of Dr. John Cairns.]

PARENTAL DNA IS UNWOUND AND NEW DNA IS SYNTHESIZED AT REPLICATION FORKS

DNA in the midst of replication has been visualized by autoradiography and electron microscopy. Autoradiography has low resolution (~500 Å) but is attractive because only what is labeled is seen. DNA is made visible in autoradiographs by the incorporation of tritiated thymine or thymidine.

Autoradiographs and electron micrographs show the form of replicating DNA from *E. coli* to be a closed circle with an internal loop (Figure 27-34). Such forms are called theta structures because of their resemblance to the Greek letter θ. The theta structures show that the *DNA molecule maintains its circular form while it is being replicated.* The resolution of the technique does not suffice to display free ends. However, it is clear that long stretches of single-stranded DNA are absent. Thus, these pictures rule out a replication mechanism in which the parental DNA strands unwind completely before serving as templates for the synthesis of new DNA. Rather, *the synthesis of new DNA is closely coupled to the unwinding of parental DNA*. A site of simultaneous unwinding and synthesis is called a *replication fork*.

Figure 27-35
Schematic diagram of the circular chromosome of *E. coli* during replication. The green strands in this theta structure depict parental DNA, and the red strands represent newly synthesized DNA.

DNA REPLICATION STARTS AT A UNIQUE ORIGIN (*oriC*) AND PROCEEDS SEQUENTIALLY IN OPPOSITE DIRECTIONS

Does DNA replication in *E. coli* start anywhere in the chromosome or is there a specific initiation site? DNA replication is a rigorously controlled process, and so a specific initiation site seems much more likely a priori. In fact, DNA replication in *E. coli* starts at a unique origin, as shown by studies of the relative amounts of different genes during rapid DNA synthesis. Consider two genes, *a* and *b*. Suppose that *a* is located near the starting point of replication, whereas *b* is near the finish. Then, gene *a* will be replicated well ahead of *b*. In a rapidly growing culture, there will be nearly two *a* genes for each *b* gene. In contrast, if DNA replication starts randomly, there will be equal amounts of *a* and *b*.

The relative frequencies of a number of *E. coli* genes were determined by hybridization with complementary probes. The results of these experiments clearly showed that the relative gene frequency depends on map position (Figure 27-36) and revealed that:

Replicon—
A unit of DNA capable of being replicated. A replicon contains a unique origin and a unique termination site. An entire replicon is replicated once in a cell cycle. The whole chromosome of *E. coli* is a single replicon.

1. *Replication starts at a unique site.* This origin locus has since been identified as *oriC*, located at 83 minutes on the standard 100-minute genetic map of *E. coli.*

2. Replication proceeds simultaneously in opposite directions, at about the same velocity. In other words, *there are two replication forks: one moves clockwise, the other counterclockwise.*

3. The two replication forks meet at *tre* (near 31′ on the map), the *termination region*, which is opposite the origin of replication.

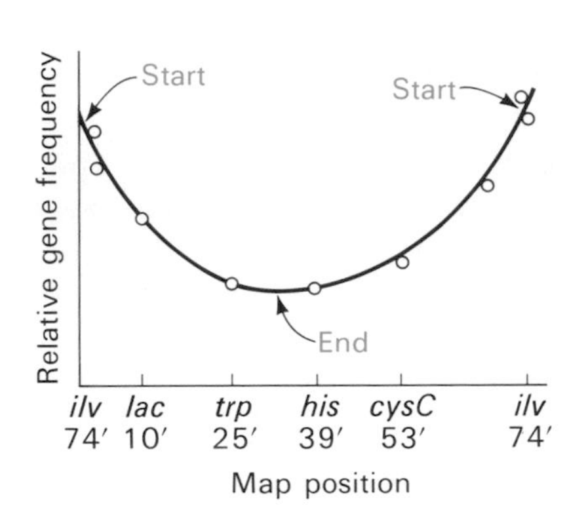

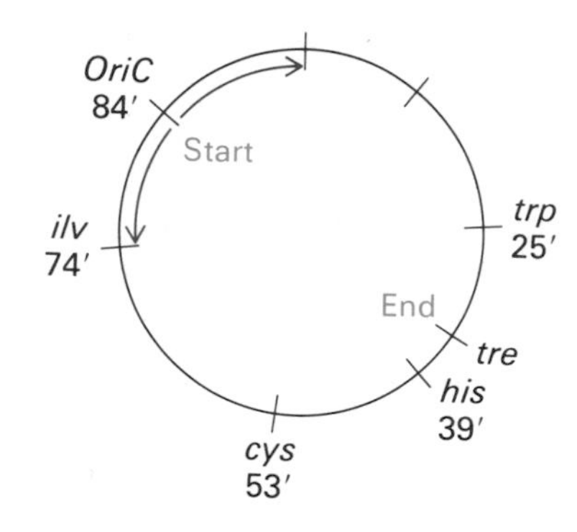

Figure 27-36
Relative amounts of different genes under conditions of rapid DNA synthesis in *E. coli* as a function of position on the genetic map.

Further evidence for the bidirectionality of replication of *E. coli* DNA comes from autoradiography. Replication was initiated in a medium containing *moderately* radioactive tritiated thymine. After a few minutes of incubation, the bacteria were transferred to a medium containing *highly* radioactive tritiated thymidine. Two levels of radioactivity were used to create two kinds of grain tracks in the autoradiographs: a low-density track for DNA synthesized early and a high-density track for DNA synthesized later. If replication were unidirectional, tracks with a low grain density at one end and a high grain density at the other end would be seen. On the other hand, if replication were bidirectional, the middle of a track would have a low density (Figure 27-37A). The autoradiographic patterns (Figure 27-37B) provided a vivid answer. All of the grain tracks were denser on both ends than in the middle, indicating that *replication of the* E. coli *chromosome is bidirectional.*

Unidirectional synthesis

Origin

Bidirectional synthesis

Origin

A

B

Figure 27-37
(A) Autoradiographic pattern expected for unidirectional replication and bidirectional replication when bacteria are transferred from a medium containing moderately radioactive thymine to one containing highly radioactive thymine. (B) Autoradiograph of *E. coli* DNA during replication (under conditions described in part A). The observed grain pattern indicates that replication is bidirectional. [From D. M. Prescott and P. L. Kuempel. *Proc. Nat. Acad. Sci.* 69(1972):2842.]

ONE STRAND OF DNA IS MADE IN FRAGMENTS AND THE OTHER STRAND IS SYNTHESIZED CONTINUOUSLY

At a replication fork, both strands of parental DNA serve as templates for the synthesis of new DNA. Recall that the parental strands are antiparallel (p. 76) and that DNA replication is semi-conservative (p. 79). Hence, the *overall* direction of DNA synthesis must be 5′ → 3′ for one daughter strand and 3′ → 5′ for the other (Figure 27-38). However, all known DNA polymerases synthesize DNA in the 5′ → 3′ direction but not in the 3′ → 5′ direction. How then does one of the daughter DNA strands *appear at low resolution* to grow in the 3′ → 5′ direction?

This dilemma was resolved by Reiji Okazaki, who found that a *significant proportion of newly synthesized DNA exists as small fragments.* These units of about a thousand nucleotides (called *Okazaki fragments*) are present briefly in the vicinity of the replication fork. As replication proceeds, these fragments become covalently joined by DNA ligase to form one of the daughter strands (Figure 27-39). The other new strand is synthesized continuously. The strand formed from Okazaki fragments is termed the *lagging strand*, whereas the one synthesized without interruption is the *leading strand*. Both the Okazaki fragments and the leading strand are synthesized in the 5′ → 3′ direction. *The discontinuous assembly of the lagging strand enables 5′ → 3′ polymerization at the nucleotide level to give rise to overall growth in the 3′ → 5′ direction.*

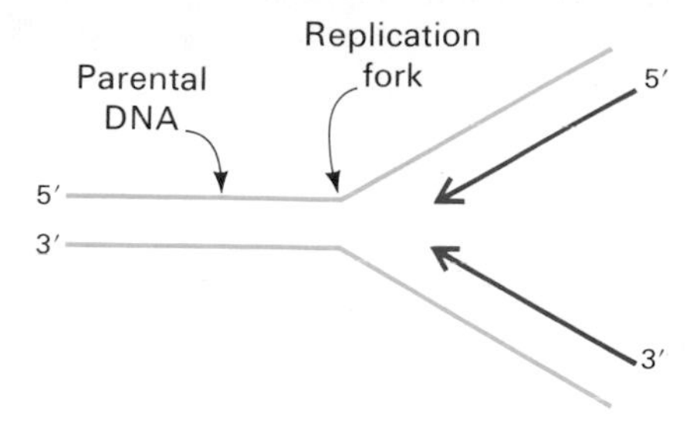

Figure 27-38
At *low resolution*, the *apparent* direction of DNA replication is 5′ → 3′ for one daughter strand and 3′ → 5′ for the other. Actually, both strands are synthesized in the 5′ → 3′ direction, as shown in Figure 27-39.

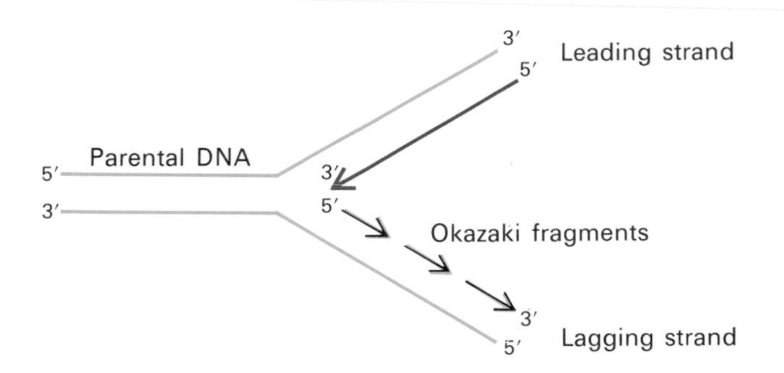

Figure 27-39
Schematic diagram of a replication fork. Both strands of DNA are synthesized in the 5′ → 3′ direction. The leading strand is synthesized continuously, whereas the lagging strand is synthesized in the form of short fragments (Okazaki fragments).

REPLICATION BEGINS WITH THE UNWINDING OF THE *oriC* (ORIGIN) SITE

How does DNA synthesis begin? Plasmids bearing *oriC*, the unique origin of replication in *E. coli*, have been invaluable in unraveling the molecular mechanism of this key process. The *oriC* locus is a sequence of 245 base pairs in the *E. coli* chromosome, a DNA molecule containing

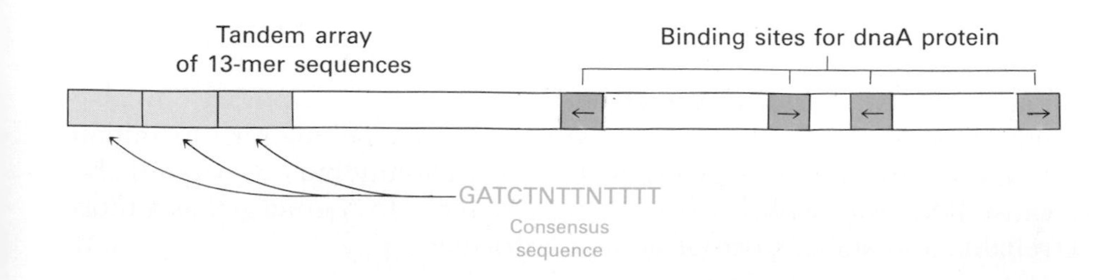

Figure 27-40
OriC, the origin of replication in *E. coli*, has a length of 245 bp. It contains a tandem array of three nearly identical 13-nucleotide sequences (green) and four binding sites (yellow) for dnaA protein. The orientations of the four dnaA sites are denoted by arrows.

4×10^6 base pairs. The origin has some unusual features (Figure 27-40). It contains a tandem array of three 13-mers with nearly identical sequences. Each begins with GATC, a sequence that appears 11 times in *oriC*, many times more often than would be expected on a random basis. Methylation of adenine in GATC may be important in controlling when replication begins. The timing of DNA replication is critical because it must respond to signals of increased cell mass and be coordinated with cell division.

The binding of the *dnaA protein* (named for its gene) to four sites on *oriC* initiates an intricate sequence of steps leading to the unwinding of the template DNA and the synthesis of a primer. The DNA must be negatively supercoiled to enable dnaA to bind. A complex of the dnaB and dnaC proteins joins dnaA to bend and open the double helix. *DnaB protein is a helicase: it catalyzes the ATP-driven unwinding of double-helical DNA.* The unwound portion of DNA is then stabilized by *single-strand-binding protein* (SSB), a tetramer of 19-kd subunits, which binds cooperatively to single-stranded DNA. The unwinding of DNA at the origin would lead to the positive supercoiling of the circular DNA if the linking number remained constant, and unwinding would soon stop. This is prevented by the compensatory action of *DNA gyrase* (p. 664), which introduces negative supercoils as it hydrolyzes ATP.

AN RNA PRIMER SYNTHESIZED BY PRIMASE ENABLES DNA SYNTHESIS TO BEGIN

The DNA template is now exposed, but new DNA cannot be synthesized until a primer is constructed. Recall that all known DNA polymerases require a primer with a free 3′-OH group for DNA synthesis. How is this primer formed? An important clue came from the observation that RNA synthesis is essential for the initiation of DNA synthesis. This finding, taken together with the fact that RNA polymerases can start chains de novo, suggested that RNA might prime the synthesis of DNA. Kornberg then found that *nascent DNA is covalently linked to a short stretch of RNA*.

In fact, *RNA primes the synthesis of DNA*. A specific RNA polymerase called *primase* joins the prepriming complex in a multisubunit assembly called the *primosome*. Primase synthesizes a short stretch of RNA (~5 nucleotides) that is complementary to one of the template DNA strands (Figure 27-41). This primer RNA is removed at the end of replication by the 5′ → 3′ exonuclease activity of DNA polymerase I. An RNA primer would be unnecessary if DNA polymerases could start chains de novo. However, such a property would be incompatible with the very high fidelity of DNA polymerases, which is due in part to their proofreading of nascent DNA. Recall that DNA polymerase I tests the correctness of the preceding base pair before forming a new phosphodiester bond. RNA polymerases can start chains de novo because they do not examine the preceding base pair. Consequently, their error rates are orders of magnitude higher than those of DNA polymerases. The ingenious solution that has evolved is to start DNA synthesis with a low-fidelity stretch of polynucleotide but mark it "temporary" by placing ribonucleotides in it. The use of an RNA polymerase to initiate DNA synthesis is also plausible from an evolutionary viewpoint, because RNA was probably present long before DNA emerged as a more reliable and stable store of genetic information.

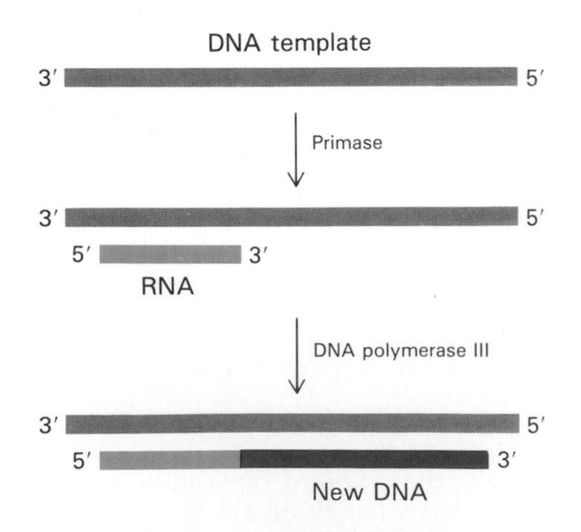

Figure 27-41
DNA replication is primed by a short stretch of RNA that is synthesized by primase, an RNA polymerase. The primer RNA is excised at a later stage of replication.

DNA POLYMERASE III HOLOENZYME, A HIGHLY PROCESSIVE AND PRECISE ENZYME, SYNTHESIZES MOST OF THE DNA

The stage is now set for the entry of *DNA polymerase III holoenzyme*. This multisubunit assembly (Table 27-2) is characterized by *very high processivity, catalytic potency, and fidelity*. The holoenzyme catalyzes the formation of many thousands of phosphodiester bonds before releasing its template, compared with only twenty for DNA polymerase I. The much lower degree of processivity of DNA polymerase I is well suited to its role as a gap filler (p. 675) and repair enzyme. DNA polymerase III holoenzyme, on the other hand, is designed to grasp its template and not let go until it has been completely replicated. A second distinctive feature of the holoenzyme is its catalytic prowess, 1000 nucleotides added per second compared with only 10 per second by DNA polymerase I. This accelerated catalysis is not accomplished at the cost of accuracy. The greater catalytic prowess of polymerase III is largely due to its processivity; no time is lost in repeatedly coming on and off the template.

Table 27-2
Composition of DNA polymerase III holoenzyme

Subunit	*Mass (kd)*	
α	140	Core
ϵ	27	Core
θ	10	Core
β	37	
τ	78	
γ	52	
δ	32	
δ'	32	

These striking features of DNA polymerase III do not come cheaply. The holoenzyme consists of at least eight kinds of polypeptide chains and has a mass of 800 kd, nearly an order of magnitude larger than that of single-chain DNA polymerase I. The roles of some of the subunits of the holoenzyme have been elucidated. The catalytic activity resides in the α subunit, and the $3' \rightarrow 5'$ exonuclease activity in the ϵ subunit. The α, ϵ, and θ subunits form a core that is catalytically active but not processive. The β subunit contributes to the high degree of processivity, probably by clamping the core of the enzyme onto its template. Indeed, many of the steps in DNA replication require the expenditure of ~P to sequentially bind and release reaction partners and to move the apparatus of replication along the template strand.

THE LEADING AND LAGGING STRANDS ARE SYNTHESIZED BY A DIMERIC DNA POLYMERASE III AT THE REPLICATION FORK

DNA polymerase III holoenzyme is now positioned at the replication fork, where it begins the synthesis of the leading strand using the RNA primer formed by primase (Figure 27-42). The duplex DNA ahead of

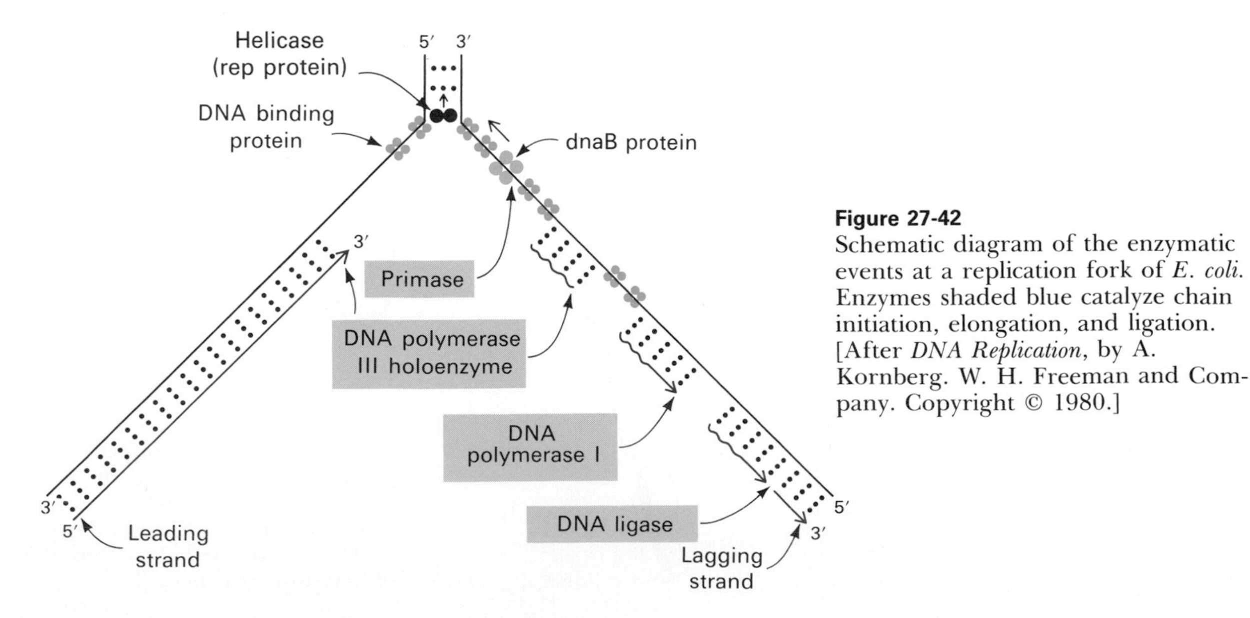

Figure 27-42
Schematic diagram of the enzymatic events at a replication fork of *E. coli*. Enzymes shaded blue catalyze chain initiation, elongation, and ligation. [After *DNA Replication*, by A. Kornberg. W. H. Freeman and Company. Copyright © 1980.]

Table 27-3
DNA-replication proteins of *E. coli*

Protein	*Role*	*Size (kd)*	*Molecules per cell*
dnaB protein	Begins unwinding the double helix	300	20
Primase	Synthesizes RNA primers	60	50
rep protein	Unwinds the double helix	65	50
SSB	Stabilizes single-stranded regions	74	300
DNA gyrase	Introduces negative supercoils	400	250
DNA polymerase III holoenzyme	Synthesizes DNA	~800	20
DNA polymerase I	Erases primer and fills gaps	103	300
DNA ligase	Joins the ends of DNA	74	300

the polymerase is unwound by a *helicase* (rep protein), which is driven by ATP, like the dnaB protein at the initiation site. Single-strand binding protein again keeps the unwound DNA extended and accessible so that both strands can serve as templates. The leading strand is synthesized continuously by the highly processive action of DNA polymerase III. DNA gyrase concurrently introduces negative supercoils to avert a topological crisis.

The mode of synthesis of the lagging strand is more complex. As was mentioned earlier, the lagging strand is synthesized in fragments so that $5' \rightarrow 3'$ polymerization leads to overall growth in the $3' \rightarrow 5'$ direction. *This may be accomplished by a looping of the template for the lagging strand (Figure 27-43). The lagging-strand template would then pass through*

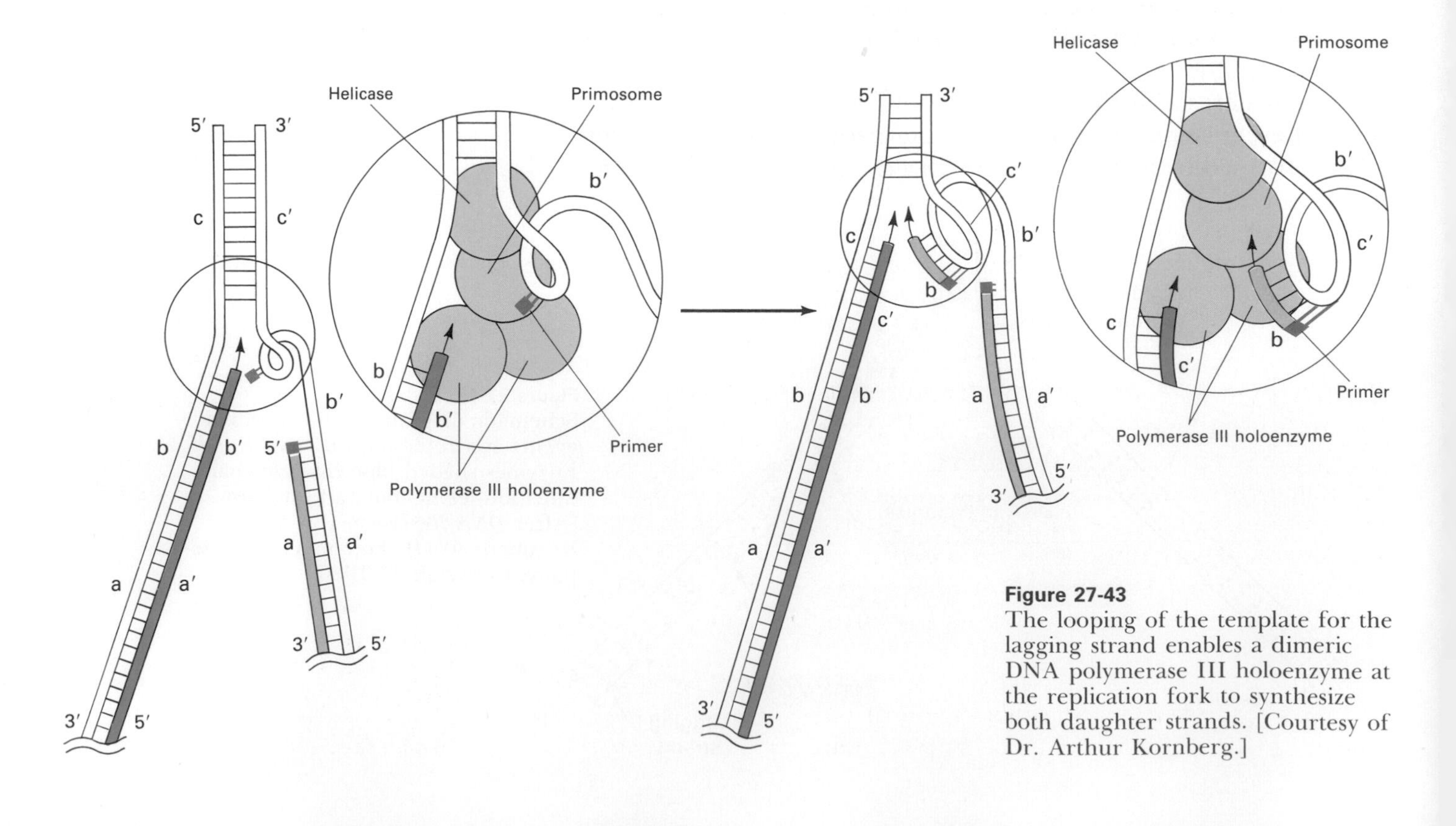

Figure 27-43
The looping of the template for the lagging strand enables a dimeric DNA polymerase III holoenzyme at the replication fork to synthesize both daughter strands. [Courtesy of Dr. Arthur Kornberg.]

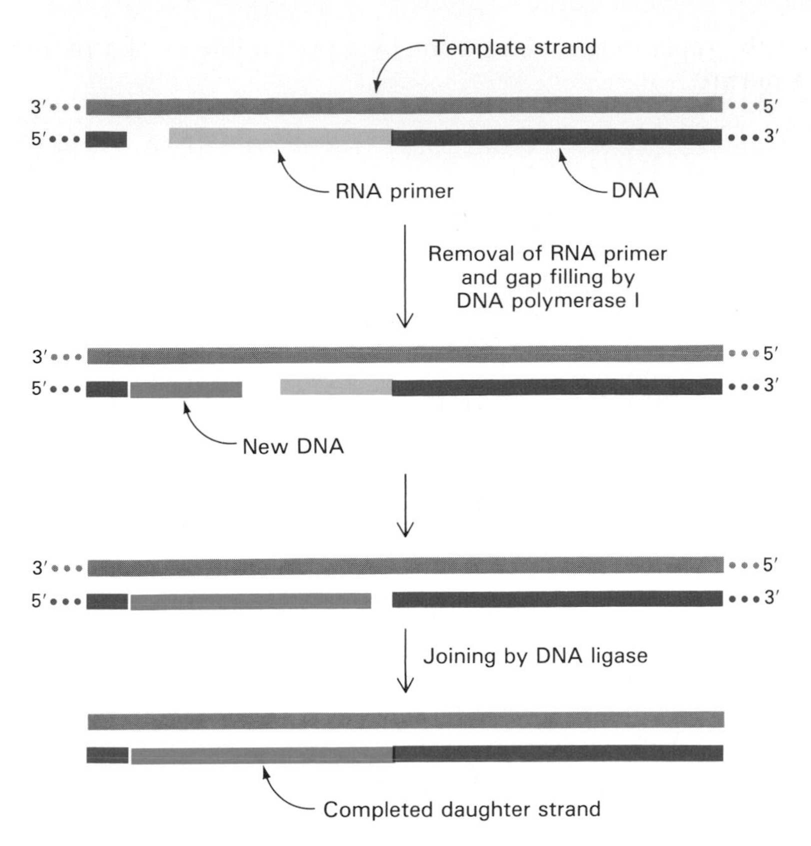

Figure 27-44
The RNA primer is removed by the 5′ → 3′ exonuclease activity of DNA polymerase I. The gap in the daughter strand is then filled by the polymerase action of this enzyme. DNA ligases catalyzes the final linking step.

the polymerase site in one subunit of a dimeric polymerase III in the same direction as the leading-strand template in the other subunit. DNA polymerase III would have to let go of the lagging strand template after about 1000 nucleotides have been added to the lagging strand. A new loop would then be formed, and primase would again synthesize a short stretch of RNA primer to initiate the formation of another Okazaki fragment. The gaps between fragments of the nascent lagging strand are filled by DNA polymerase I, which also uses its 5′ → 3′ exonuclease activity to remove the ribonucleotide primer lying ahead of the polymerase site (Figure 27-44). Finally, DNA ligase joins the fragments.

MUTATIONS ARE PRODUCED BY SEVERAL TYPES OF CHANGES IN THE BASE SEQUENCE OF DNA

Several types of mutations are known: (1) the *substitution* of one base pair for another; (2) the *deletion* of one or more base pairs; and (3) the *insertion* of one or more base pairs (Figure 27-45). The spontaneous mutation rate of T4 phage is about 10^{-7} per base per replication. *E. coli* and *Drosophila melanogaster* have much lower mutation rates, of the order of 10^{-10}

The substitution of one base pair for another is the most common type of mutation. A *transition* is the replacement of one purine by another purine or of one pyrimidine by another pyrimidine. In contrast, a *transver-*

There have been no problems in St. Louis created by pre-trial publicity because the newspapers have exorcised sound judgment and restraint in what they have printed.
—*St. Louis Globe-Democrat*

Substitution

In analyzing the pictures relayed by the Mariner 10 spacecraft en route to Mercury, the scientists said the planet becomes much more dramatic looking when viewed through an ultra-violent lens.
—*Boston Globe*

Insertion

"I can speak just as good nglish as you," Gorbulove corrected in a merry voice.
—*Seattle Times*

Deletion

Tomorrow: "Give Baby Time to Learn to Swallow Solid Food."
etaoin-oshrdlucmfwypvbgkq
—*Youngstown (Ohio) Vindicator*

Nonsense

Figure 27-45
Types of mutations, as illustrated by typographical errors. [After S. Benzer. *Harvey Lectures,* Ser. 56 (Academic Press, 1961), p. 3.]

sion is the replacement of a purine by a pyrimidine or of a pyrimidine by a purine.

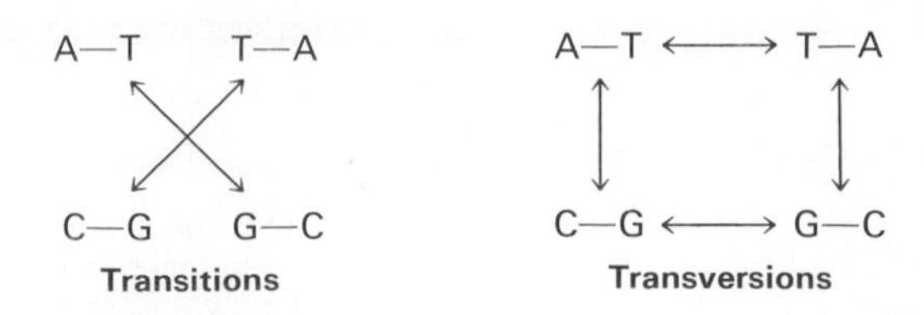

Cytosine

Rare tautomer of adenine

Figure 27-46
The rare tautomer of adenine pairs with cytosine instead of thymine. This tautomer is formed by the shift of a proton from the 6-amino group to N-1.

A mechanism for the spontaneous occurrence of transitions was suggested by Watson and Crick in their classic paper on the DNA double helix. They noted that some of the hydrogen atoms on each of the four bases can change their location to produce a tautomer. An *amino* group (—NH_2) can tautomerize to an *imino* form (=NH). Likewise, a *keto* group (—C=O) can tautomerize to an *enol* form (=C—OH). The fraction of each base in the form of these imino and enol tautomers is about 10^{-4}.

These transient tautomers can form nonstandard base pairs that fit into a double helix. For example, the imino tautomer of adenine can pair with cytosine (Figure 27-46). This A^*-C pairing (the asterisk denotes the imino tautomer) would allow C to become incorporated into a growing DNA strand where T was expected, so that it would lead to a mutation if left uncorrected. In the next round of replication, this adenine will probably retautomerize to the standard form, which pairs as usual with thymine, but the cytosine will pair with guanine. Hence, one of the daughter DNA molecules will contain a GC base pair in place of the normal AT base pair (Figure 27-47).

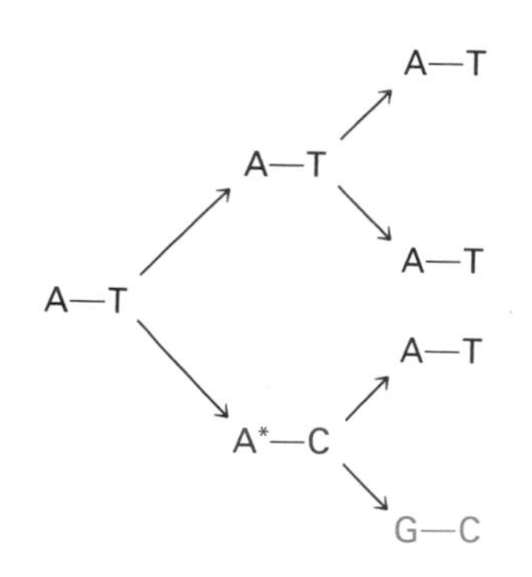

Figure 27-47
The pairing of the rare tautomer of adenine (A*) with cytosine leads to a GC base pair in the next generation.

BACTERIA DEFICIENT IN 3′ → 5′ EXONUCLEASE ACTIVITY HAVE ABNORMALLY HIGH MUTATION RATES

Most incorrect base pairs formed in the polymerization step do not become permanently incorporated into DNA. As was mentioned earlier, DNA polymerases proofread the outcome of a polymerization step before proceeding to the next one. An incorrectly paired nucleotide is nearly always removed. Indeed, about one in ten *correctly* paired nucleotides is removed by the 3′ → 5′ exonuclease activity of DNA polymerase III, a price paid for having a very vigilant editor. It has been estimated that the initial mispairing frequency is 10^{-4} and that 10^{-3} of the mispaired nucleotides are not removed by proofreading. The observed mutation frequency of 10^{-10} suggests that further repair processes after replication correct all but about 10^{-3} of these errors.

Some *E. coli* mutants with abnormally high mutation rates have a DNA polymerase III with an altered ϵ subunit. Their lowered 3′ → 5′ exonuclease activity allows a much higher proportion of mispaired nucleotides to evade excision. Conversely, mutants with a higher than normal ratio of exonuclease to polymerase activity have a lower than normal spontaneous mutation rate. The existence of mutator and antimutator strains strongly suggests that the mutation rate of a species has been optimized in the course of evolution. Too high a rate would lead to nonviable progeny, whereas too low a rate would diminish genetic diversity arising from point mutations, one of the sources of evolutionary change.

Base analogs such as 5-bromouracil and 2-aminopurine can be incorporated into DNA. They lead to transition mutations as a consequence of altered base pairing in a subsequent round of DNA replication (Figure 27-48). 5-*Bromouracil,* an analog of thymine, normally pairs with adenine. However, the proportion of the enol tautomer of 5-bromouracil is higher than that of thymine, probably because the bromine atom is much more electronegative than a methyl group at C-5. The enol form of 5-bromouracil pairs with guanine, which causes AT ↔ GC transitions. 2-*Aminopurine* normally pairs with thymine. In contrast with adenine, the normal tautomer of 2-aminopurine can form a single hydrogen bond with cytosine. Hence, 2-aminopurine can produce AT ↔ GC transitions.

Other mutagens act by chemically modifying the bases of DNA. For example, nitrous acid reacts with bases that contain amino groups. Adenine is oxidatively deaminated to hypoxanthine, cytosine to uracil, and guanine to xanthine. Hypoxanthine pairs with cytosine rather than with thymine (Figure 27-49). Uracil pairs with adenine rather than with guanine. Xanthine, like guanine, pairs with cytosine. Consequently, nitrous acid causes AT ↔ GC transitions. *Hydroxylamine* (NH_2OH) is a highly specific mutagen. It reacts almost exclusively with cytosine to give a derivative that pairs with adenine rather than with guanine. Hydroxylamine produces a unidirectional transition of CG to AT.

A different kind of mutation is produced by flat aromatic molecules such as the acridines. These compounds *intercalate* in DNA—that is, they slip in between adjacent base pairs in the DNA double helix (p. 715). Consequently, *they lead to the insertion or deletion of one or more base pairs.* The effect of such mutations is to alter the reading frame in translation, unless an integral multiple of three base pairs is inserted or deleted. In fact, it was the analysis of such mutants that revealed the triplet nature of the genetic code.

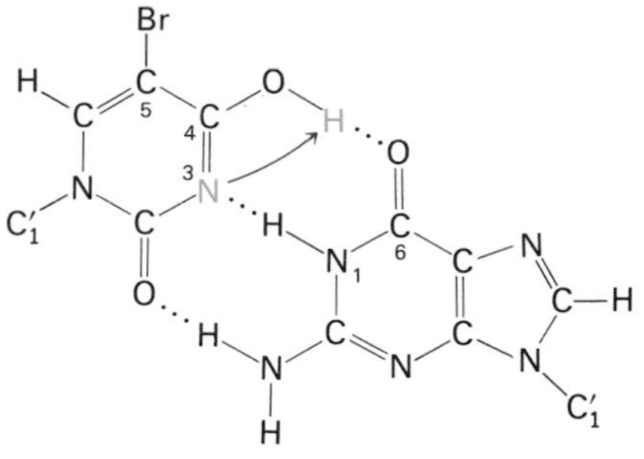

Minor tautomer of 5-bromouracil

Guanine

Figure 27-48
5-Bromouracil, an analog of thymine, occasionally pairs with guanine instead of adenine. The presence of bromine at C-5 increases the proportion of the rare tautomer formed by the shift of a proton from N-3 to the C-4 oxygen atom.

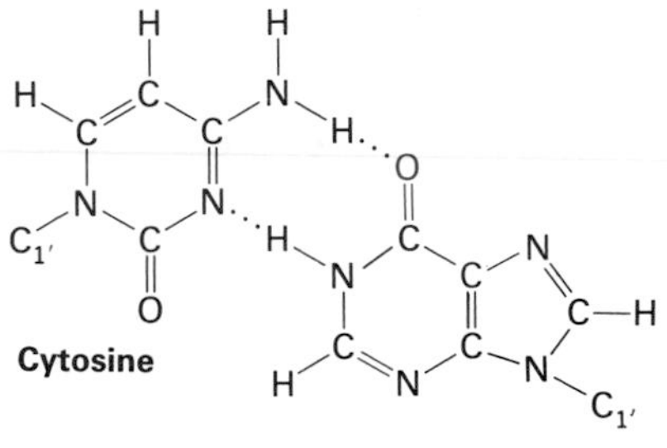

Cytosine

Hypoxanthine

Figure 27-49
Base pairing of hypoxanthine with cytosine. Hypoxanthine is formed by oxidative deamination of adenine. Hence, an AT base pair becomes a GC base pair in a later round of replication.

LESIONS IN DNA ARE CONTINUALLY BEING REPAIRED

DNA is damaged by a variety of chemical and physical agents, and so all cells possess mechanisms for repair. Bases can be altered or lost, phosphodiester bonds in the backbone can be broken, and strands can become covalently cross-linked. These lesions are produced by ionizing radiation, ultraviolet light, and a variety of chemicals. *Much of the damage sustained by DNA can be repaired because genetic information is stored in both strands of the double helix, so that information lost by one strand can be retrieved from the other.*

One of the best understood repair mechanisms is the excision of a *pyrimidine dimer* (Figure 27-50), *which is formed on exposure of DNA to ultraviolet light.* Adjacent pyrimidine residues on a DNA strand can become covalently linked under these conditions. Such a pyrimidine dimer cannot fit into a double helix, and so replication and gene expression are blocked until the lesion is removed. Three enzymatic activities are essential for this repair process in *E. coli* (Figure 27-51). First, an enzyme complex consisting of the proteins encoded by the *uvrABC* genes detects the distortion produced by the pyrimidine dimer. The

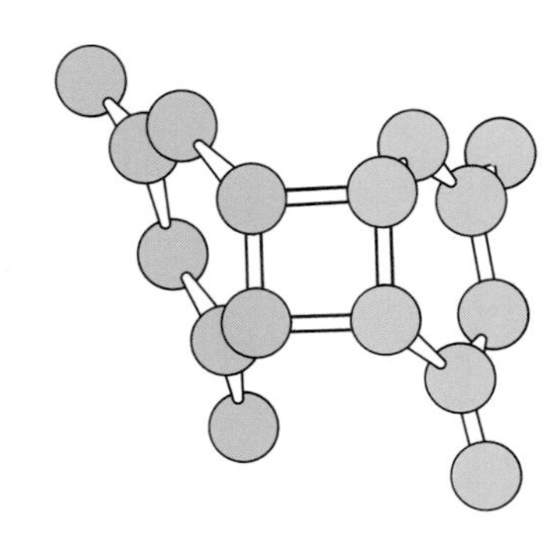

Figure 27-50
Model of a dimer of uracil produced by ultraviolet irradiation. A thymine dimer has nearly the same structure.

Thymine dimer

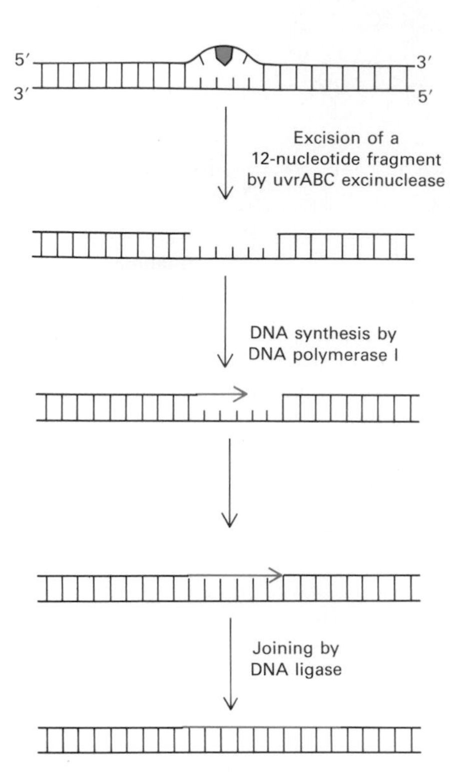

Figure 27-51
Repair of a region of DNA containing a thymine dimer by the sequential action of a specific endonuclease, a DNA polymerase, and a DNA ligase. The thymine dimer is shown in blue and the new region of DNA in red. [After P. C. Hanawalt, *Endeavor* 31(1972):83.]

uvrABC enzyme then cuts the damaged DNA strand at two sites, 8 nucleotides away from the dimer on the 5′ side and 4 nucleotides away on the 3′ side. The 12-residue oligonucleotide excised by this highly specific *excinuclease* then diffuses away. DNA polymerase I enters the gap to carry out repair synthesis. The 3′ end of the nicked strand is the primer, and the intact complementary strand is the template. Finally, the 3′ end of the newly synthesized stretch of DNA and the original portion of the DNA chain are joined by *DNA ligase*.

Alternatively, the pyrimidine dimer can be photochemically split. Nearly all cells contain a *photoreactivating enzyme* called *DNA photolyase*. The *E. coli* enzyme, a 35-kd protein, acquires an absorption band in the near ultraviolet and blue spectral region on binding to the distorted region of DNA. The absorption of a photon creates an excited state that cleaves the dimer into its original bases.

In the repair of DNA, it is essential to identify the normal and aberrant strands. This information is sometimes contained in the defect itself, if it is a modified base (e.g., a pyrimidine dimer). Consider, on the other hand, an AC base pair in a newly synthesized DNA molecule. Both bases are normal constituents of DNA. How does the repair machinery distinguish the parental strand (the authentic one) from the daughter strand (the one with the error)? *The tags that identify the strands are methyl groups on adenine residues in GATC sequences.* The parental strand is methylated at many of these sites but the new daughter strand has not yet been modified in this way, because methylation takes time. Hence, the mismatch-correction enzyme cuts the unmethylated strand to remove the wrong nucleotide and leaves the parental strand intact so that it can again serve as a correct template.

A SKIN CANCER IS CAUSED BY DEFECTIVE REPAIR OF DNA IN XERODERMA PIGMENTOSUM

Xeroderma pigmentosum is a rare skin disease in humans. It is genetically transmitted as an autosomal recessive trait. The skin in an affected homozygote is extremely sensitive to sunlight or ultraviolet light. In infancy, severe changes in the skin become evident and worsen with time. The skin becomes dry and there is a marked atrophy of the dermis. Keratoses appear, the eyelids become scarred, and the cornea ulcerates. Skin cancer usually develops at several sites. Many patients die before age thirty from metastases of these malignant skin tumors.

Ultraviolet light produces pyrimidine dimers in human DNA, as it does in *E. coli* DNA. Furthermore, the repair mechanisms seem to be similar. Studies of skin fibroblasts from patients with xeroderma pigmentosum have revealed a biochemical defect in one form of this disease. In normal fibroblasts, half of the pyrimidine dimers produced by ultraviolet radiation are excised in less than 24 hours. In contrast, almost no dimers are excised in this time interval in fibroblasts derived from patients with xeroderma pigmentosum. These studies show that *xeroderma pigmentosum can be produced by a defect in the excinuclease that hydrolyzes the DNA backbone near a pyrimidine dimer. The drastic clinical consequences of this enzymatic defect emphasize the critical importance of DNA repair processes.* The disease can also be caused by mutations in eight other genes, which attests to the complexity of the repair mechanism.

Xeroderma—
From the Greek words for dry skin.
The term was first used by F. Hebra and M. Kaposi in 1874 to describe the "parchment skin" and abnormal pigmentation seen in one of their patients.

DNA CONTAINS THYMINE INSTEAD OF URACIL TO PERMIT REPAIR OF DEAMINATED CYTOSINE

Cytosine in DNA spontaneously deaminates at a perceptible rate to form uracil. The deamination of cytosine is potentially mutagenic because uracil pairs with adenine, and so one of the daughter strands will contain an AU base pair rather than the original GC base pair.

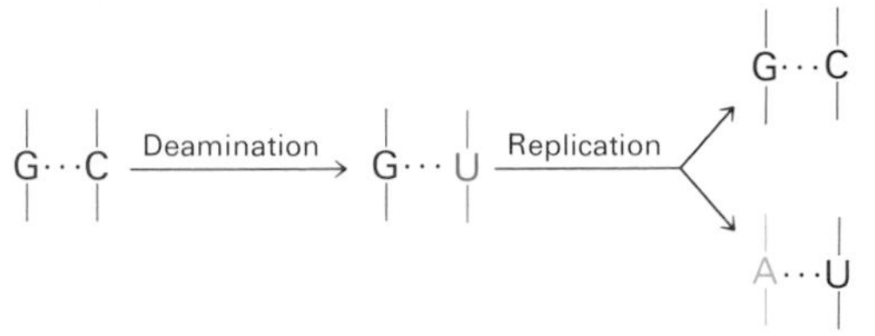

$$\text{G}\cdots\text{C} \xrightarrow{\text{Deamination}} \text{G}\cdots\text{U} \xrightarrow{\text{Replication}} \begin{cases} \text{G}\cdots\text{C} \\ \text{A}\cdots\text{U} \end{cases}$$

This mutation is prevented by a repair system that recognizes uracil to be foreign to DNA (Figure 27-52). First, a *uracil-DNA glycosidase* hydrolyzes the glycosidic bond between the uracil and deoxyribose moieties. At this stage, the DNA backbone is intact, but a base is missing. This hole is called an *AP site* because it is *ap*urinic (devoid of A or G) or *ap*yrimidinic (devoid of C or T). An *AP endonuclease* then recognizes this defect and nicks the backbone adjacent to the missing base. DNA polymerase I excises the residual deoxyribose phosphate unit and inserts cytosine, as dictated by the presence of guanine on the undamaged complementary strand. Finally, the repaired strand is sealed by DNA ligase.

The presence in DNA of thymine rather than uracil was an enigma for many years. Both bases pair with adenine. The only difference between them is a methyl group in thymine in place of the C-5 hydrogen in uracil. Why is a methylated base employed in DNA and not in RNA? Recall that the methylation of deoxyuridylate to form deoxythymidylate is energetically expensive (p. 614). The discovery of an active repair system to correct the deamination of cytosine provides a convincing solution to this puzzle. Uracil-DNA glycosidase does not remove thymine from DNA. *Thus, the methyl group on thymine is a tag that distinguishes it from deaminated cytosine.* If this tag were absent, uracil correctly in place would be indistinguishable from uracil formed by deamination. The defect would persist unnoticed, and so a GC base pair would necessarily be mutated to AU in one of the daughter DNA molecules. This mutation is prevented by a repair system that searches for uracil and leaves thymine alone. It seems likely that *thymine is used instead of uracil in DNA to enhance the fidelity of the genetic message.* In contrast, RNA is not repaired, and so uracil is used in RNA because it is a less expensive building block.

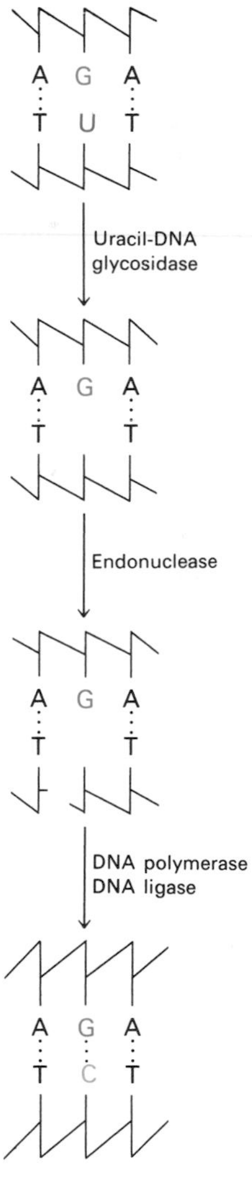

A G A
T U T
Uracil-DNA glycosidase
A G A
T T
Endonuclease
A G A
T T
DNA polymerase
DNA ligase
A G A
T C T

Figure 27-52
Uracil residues in DNA are excised and replaced with cytosine, the original base.

MANY POTENTIAL CARCINOGENS CAN BE DETECTED BY THEIR MUTAGENIC ACTION ON BACTERIA

Many human cancers are caused by exposure to toxic chemicals. These chemical carcinogens are usually mutagenic, which suggests that *damage to DNA is a fundamental event in both carcinogenesis and mutagenesis.* It is important to identify these compounds and ascertain their potency so that human exposure to them can be minimized. Bruce Ames has devised a simple and sensitive test for detecting chemical mutagens. A thin layer of agar containing about 10^9 bacteria of a specially constructed tester strain of *Salmonella* is placed on a petri dish. These bacteria are unable to grow in the absence of histidine because of a mutation in one of their genes for the biosynthesis of this amino acid. The addition of a mutagen to the center of a plate results in many new mutations. A small proportion of them reverse the original mutation so that histidine is synthesized. These *revertants* multiply in the absence of exogenous histidine and appear as discrete colonies after the plate is incubated at 37°C for two days (Figure 27-53). For example, 0.5 μg of 2-aminoanthracene gives 11,000 revertant colonies compared with only 30 spontaneous revertants in its absence. A series of concentrations of a chemical can readily be tested to generate a dose-response curve. These curves are usually linear, which suggests that there is no threshold concentration for mutagenesis.

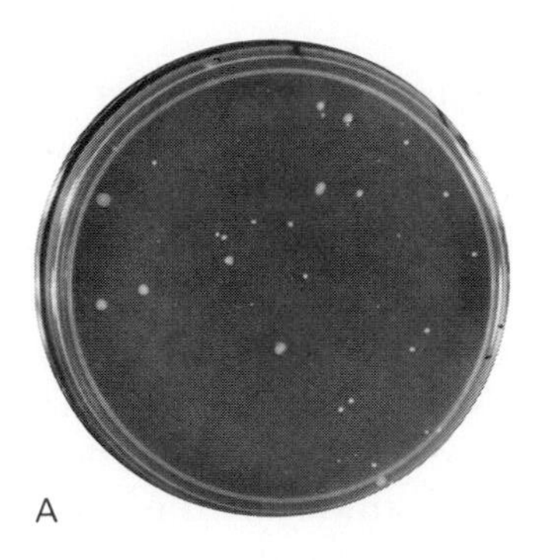

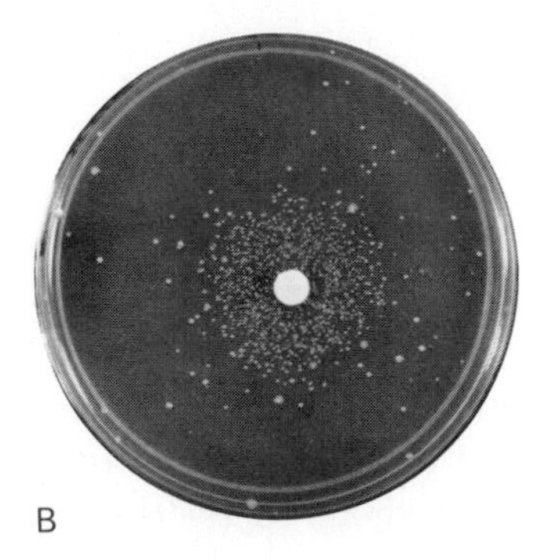

Figure 27-53
Salmonella test for mutagens: (A) petri plate containing about 10^9 bacteria that cannot synthesize histidine; (B) a plate containing a filter-paper disc with a mutagen, which produces a large number of revertants that can synthesize histidine. After two days, revertants appear as a ring of colonies around the disc. The small number of visible colonies in plate A are spontaneous revertants. [From B. N. Ames, J. McCann, and E. Yamasaki. *Mutation Res.* 31(1975):347.]

Some of the tester strains are responsive to *base-pair substitutions,* whereas others detect *deletions or additions of base pairs* (*frame shifts*). The sensitivity of these specially designed strains has been enhanced by genetically deleting their excision-repair systems. Also, the lipopolysaccharide barrier that normally coats the surface of *Salmonella* is incomplete in the tester strains, which facilitates the entry of potential mutagens. Another important feature of this detection system is a *mammalian liver homogenate.* Recall that some potential carcinogens are converted into their active forms by enzyme systems in the liver or other mammalian tissues (p. 566). Bacteria lack these enzymes, and so the test plate requires a few milligrams of a liver homogenate to activate this group of mutagens. For example, an activated compound with both a reactive side chain (enabling it to form a covalent bond with DNA) and an aromatic region (allowing it to intercalate) is far more mutagenic than one with just the aromatic region.

The *Salmonella* test is extensively used to help evaluate the mutagenic and carcinogenic risks of a large number of chemicals. This rapid and inexpensive bacterial assay for mutagenicity complements epidemiologic surveys and animal tests that are necessarily slower, more laborious, and far more expensive. The *Salmonella* test for mutagenicity is an outgrowth of studies of gene-protein relationships in bacteria. It is a striking example of how fundamental research in molecular biology can lead directly to important advances in public health.

SUMMARY

DNA is a structurally dynamic molecule that can exist in a variety of helical forms: A-DNA, B-DNA (the classic Watson-Crick helix), and Z-DNA. DNA can be bent, kinked, and unwound. In A-, B-, and Z-DNA, two antiparallel chains are held together by Watson-Crick base pairs and stacking interactions between bases in the same strand. The sugar-phosphate backbone is on the outside and the bases are inside the double helix. A- and B-DNA are right-handed helices in which the repeating unit is a mononucleotide. In B-DNA, the base pairs are nearly perpendicular to the helix axis, whereas in A-DNA they are tilted. Dehydration induces the transition from B- to A-DNA. The 2′-OH of ribose cannot fit into a B-helix because of steric hindrance. Double-helical RNA hairpins and RNA-DNA hybrids adopt the A-helix structure. Z-DNA is a left-handed helix in which the repeating unit is a dinucleotide. It can be formed in regions of DNA in which purines alternate repeatedly with pyrimidines, as in CGCG or CACA. The formation of Z-DNA is favored by methylation of cytosine and by negative supercoiling. Most of the DNA in a genome is in the Watson-Crick B-helical form. An important structural feature of the B helix is the presence of major and minor grooves, which display different potential hydrogen-bond acceptors and donors according to the base sequence.

DNase I, a quite nonspecific nuclease, and EcoRI endonuclease, a highly specific restriction enzyme, display contrasting modes of interaction with DNA. DNase binds electrostatically to the minor groove of duplex DNA. Its glutamate-histidine-water catalytic triad is reminiscent of the charge relay network of serine proteases. EcoRI endonuclease recognizes a specific palindromic hexanucleotide target; the twofold symmetry of the enzyme matches that of its substrate. Both strands are cleaved at symmetrically positioned identical sites. The binding of enzyme leads to the kinking of its target site. The resulting widening of the major groove allows the entry of four α-helical segments that form twelve precisely directed hydrogen bonds with bases in the target sequence.

A key topological property of DNA is its linking number (L), which is defined as the number of times one strand of DNA winds around the other in the right-hand direction. Molecules differing in linking number are topoisomers of each other and can be interconverted only by cutting one or both DNA strands; these reactions are catalyzed by topoisomerases. Changes in linking number can lead to changes in the number of turns of double helix (T, the twisting number) or in the number of turns of superhelix (W, the writhing number). DNA gyrase catalyzes the ATP-driven introduction of negative supercoils, which leads to the compaction of DNA and renders it more susceptible to unwinding. Supercoiled DNA is relaxed by topoisomerase I.

DNA polymerases are template-directed enzymes that catalyze the formation of phosphodiester bonds by the nucleophilic attack of a 3′-OH on the innermost phosphorus atom of a deoxyribonucleoside 5′-triphosphate. They cannot start chains de novo; a primer with a free 3′-OH is required. DNA polymerases proofread the nascent product;

their 3′ → 5′ exonuclease activity examines the outcome of each polymerization step. A mispaired nucleotide is excised before the next polymerization step. In *E. coli*, DNA polymerase I (a 103-kd single-chain protein) repairs DNA and participates in replication. Most of the DNA is synthesized by DNA polymerase III holoenzyme, a multisubunit assembly (800 kd) of eight kinds of polypeptide chains. Polymerase III is a highly processive enzyme with a very low error rate.

DNA replication in *E. coli* starts at a unique origin (*oriC*) and proceeds sequentially in opposite directions. More than twenty proteins are required for replication. DnaB protein, an ATP-driven helicase, unwinds the *oriC* region to create a replication fork. At this fork, both strands of parental DNA serve as templates for the synthesis of new DNA. DNA synthesis is primed by a short stretch of RNA formed by primase, an RNA polymerase. One strand of DNA (the leading strand) is synthesized continuously, whereas the other strand is synthesized discontinuously, in the form of 1-kb fragments (Okazaki fragments). Both new strands are formed by the action of dimeric DNA polymerase III holoenzyme at the replication fork. The discontinuous assembly of the lagging strand enables 5′ → 3′ polymerization at the atomic level to give rise to overall growth of this strand in the 3′ → 5′ direction. The RNA primer portion is hydrolyzed by the 5′ → 3′ nuclease activity of DNA polymerase I, which also fills gaps. Finally, nascent DNA fragments are joined by DNA ligase in a reaction driven by ATP or NAD^+. The unwinding of DNA at the replication fork is catalyzed by the rep protein, an ATP-driven helicase. DNA gyrase also plays a key role by introducing negative supercoils and allowing the daughter duplex to separate from the parental one.

Mutations are produced by mistakes in base pairing, covalent modification of bases, and the deletion and insertion of bases. The 3′ → 5′ exonuclease activity of DNA polymerases is critical in lowering the spontaneous mutation rate, which arises largely from mispairing by tautomeric bases. Lesions in DNA are continually being repaired. Multiple repair processes utilize information present in the intact strand to correct the damaged strand. For example, pyrimidine dimers formed by the action of ultraviolet light are excised by the uvrABC excinuclease, an enzyme that removes a 12-nucleotide region containing the dimer. Xeroderma pigmentosum, a genetically transmitted disease, is caused by defective repair of DNA; patients with this disease usually develop skin cancers. Damage to DNA is a fundamental event in both carcinogenesis and mutagenesis. Many potential carcinogens can be detected by their mutagenic action on bacteria.

SELECTED READINGS

Where to begin

Dickerson, R. E., 1983. The DNA helix and how it is read. *Sci. Amer.* 249(6):94–111.

Wang, J. C., 1982. DNA topoisomerases. *Sci. Amer.* 247(1):94–109.

Kornberg, A., 1984. DNA replication. *Trends Biochem. Sci.* 9:122–124.

Howard-Flanders, P., 1981. Inducible repair of DNA. *Sci. Amer.* 245(5):72–80.

Ames, B., 1983. Dietary carcinogens and anticarcinogens. *Science* 221:1256–1264.

Books

Kornberg, A., 1980. *DNA Replication*. W. H. Freeman. [An outstanding book, rich in information and insight. The *1982 Supplement to DNA Replication* is a valuable update.]

Saenger, W., 1984. *Principles of Nucleic Acid Structure.* Springer-Verlag. [A scholarly, well-illustrated account of fundamental principles. The stereochemistry of nucleotides and nucleic acids is lucidly presented.]

Friedberg, E. C., 1985. *DNA Repair.* W. H. Freeman. [A lucid and graphic presentation of molecular mechanisms of DNA repair in procaryotes and eucaryotes.]

Jurnak, F. A., and McPherson, A., (eds.), 1985. *Biological Macromolecules and Assemblies*, vol. 2. [Contains excellent articles on the structure of DNA and DNA-protein complexes.]

Kelly, T., and McMacken, R., (eds.), 1986. *Mechanisms of DNA Replication and Recombination.* Alan R. Liss.

Cold Spring Harbor Symposium on Quantitative Biology, 1982. *Structures of DNA,* vol. 47. Cold Spring Harbor Laboratory.

Kirkwood, T. B. L., Rosenberger, R. F., and Galas, D. J., (eds.), 1986. *Accuracy in Molecular Processes.* Chapman and Hall. [An interesting discussion of accuracy in the replication and expression of genetic information. Chapters 8 and 9 deal with the fidelity of DNA replication.]

DNA structure

Dickerson, R. E., Drew, H. R., Conner, B. N., Wing, R. M., Fratini, A. V., and Kopka, M. L., 1982. The anatomy of A-, B-, and Z-DNA. *Science* 216:475–485.

Rich, A., Nordheim, A., and Wang, A. H.-J., 1984. The chemistry and biology of left-handed Z-DNA. *Ann. Rev. Biochem.* 53:791–846.

Koo, H.-S., Wu, H.-M., and Crothers, D. M., 1986. DNA bending at adenine-thymine tracts. *Nature* 320:501–506.

Sherman, S. E., Gibson, D., Wang, A. H.-J., and Lippard, S. J., 1985. X-ray structure of the major adduct of the anticancer drug cisplatin with DNA: *cis*-[Pt(NH_3)$_2${d(pGpG)}]. *Science* 230:412–417.

Saenger, W., Hunter, W. N., and Kennard, O., 1986. DNA conformation is determined by economics in the hydration of phosphate groups. *Nature* 324:385–388.

Levene, S. D., and Crothers, D. M., 1986. Topological distributions and the torsional rigidity of DNA. A Monte Carlo study of DNA circles. *J. Mol. Biol.* 189:73–83.

Peck, L. J., Wang, J. C., Nordheim, A., and Rich, A., 1986. Rate of B to Z structural transition of supercoiled DNA. *J. Mol. Biol.* 190:125–127.

Nucleases and ligases

Suck, D., and Oefner, C., 1986. Structure of DNase I at 2.0 Å resolution suggests a mechanism for binding to and cutting DNA. *Nature* 321:620–625.

McClarin, J. A., Frederick, C. A., Wang, B.-C., Greene, P., Boyer, H. W., Grable, J., and Rosenberg, J. M., 1986. Structure of the DNA-EcoRI endonuclease recognition complex at 3 Å resolution. *Science* 234:1526–1541.

Lehman, I. R., 1974. DNA ligase: structure, mechanism, and function. *Science* 186:790–797.

Willis, A. E., and Lindahl, T., 1987. DNA ligase I deficiency in Bloom's syndrome. *Nature* 325:355–357.

DNA topology and topoisomerases

Wang, J. C., 1987. Recent studies of DNA topoisomerases. *Biochem. Biophys. Acta* 909:1–9.

Maxwell, A., and Gellert, M., 1986. Mechanistic aspects of DNA topoisomerases. *Advan. Protein Chem.* 38:69–107.

Wasserman, S. A., and Cozzarelli, N. R., 1986. Biochemical topology: applications to DNA recombination and replication. *Science* 232:951–960.

Wang, J. C., 1985. DNA topoisomerases. *Ann. Rev. Biochem.* 54:665–697.

Dean, F. B., and Cozzarelli, N. R., 1985. Mechanism of strand passage by *Escherichia coli* topoisomerase I. The role of the required nick in catenation and knotting of duplex DNA. *J. Biol. Chem.* 260:4984–4994.

Kirchhausen, T., Wang, J. C., and Harrison, S. C., 1985. DNA gyrase and its complexes with DNA: direct observation by electron microscopy. *Cell* 41:933–943.

Bauer, W. R., Crick, F. H. C., and White, J. H., 1980. Supercoiled DNA. *Sci. Amer.* 243(1):118–133. [A lucid presentation of topological principles.]

Crick, F. H. C., 1976. Linking numbers and nucleosomes. *Proc. Nat. Acad. Sci.* 73:2639–2643. [A lucid introduction to linkage, writhing, and twist of superhelical DNA.]

Gellert, M., 1981. DNA topoisomerases. *Ann. Rev. Biochem.* 50:879–910.

Menzel, R., and Gellert, M., 1983. Regulation of the genes for *E. coli.* DNA gyrase: homeostatic control of DNA supercoiling. *Cell* 34:105–113.

Cozzarelli, N. R., 1980. DNA gyrase and the supercoiling of DNA. *Science* 207:953–960.

DNA polymerases

Steitz, T. A., Freemont, P. S., Ollis, D. L., Joyce, C. M., and Grindley, J. M., 1986. Functional implications of the Klenow fragment structure. *Biochem. Soc. Trans.* 14:205–207.

Ollis, D. L., Brick, P., Hamlin, R., Xuong, N. G., and Steitz, T. A., 1985. Structure of large fragment of *Escherichia coli* DNA polymerase I complexed with dTMP. *Nature* 315:762–766.

Warwicker, J., Ollis, D., Richards, F. M., and Steitz, T. A., 1985. Electrostatic field of the large fragment of *Escherichia coli* DNA polymerase I. *J. Mol. Biol.* 186:645–649.

McHenry, C. S., 1985. DNA polymerase III holoenzyme of *Escherichia coli*: components and function of a true replicative complex. *Mol. Cell Biochem.* 66:71–85.

LaDuca, R. J., Crute, J. J., McHenry, C. S., and Bambara, R. A., 1986. The beta subunit of the *Escherichia coli* DNA polymerase III holoenzyme interacts functionally with the catalytic core in the absence of other subunits. *J. Biol. Chem.* 261:7550–7557.

Joyce, C. M., and Steitz, T. A., 1987. DNA polymerase I: from crystal structure to function via genetics. *Trends Biochem. Sci.* 12:288–292.

Baker, T. A., Funnell, B. E., and Kornberg, A., 1987. Helicase action of dnaB protein during replication from the *Escherichia coli* chromosomal origin in vitro. *J. Biol. Chem.* 262:6877–6855.

Echols, H., 1986. Multiple DNA-protein interactions governing high-precision DNA transactions. *Science* 233:1050–1056.

Kwon-Shin, O., Bodner, J. B., McHenry, C. S., and Bambara, R. A., 1987. Properties of initiation complexes formed between *Escherichia coli* DNA polymerase III holoenzyme and primed DNA in the absence of ATP. *J. Biol. Chem.* 262:2121–2130.

Origin and direction of DNA replication

Prescott, D. M., and Kuempel, P. L., 1972. Bidirectional replication of the chromosome in *E. coli. Proc. Nat. Acad. Sci.* 69:2842–2845.

Bird, R. E., Lourarn, J., Martuscelli, J., and Caro, L., 1972. Origin and sequence of chromosome replication in *E. coli. J. Mol. Biol.* 70:549–566.

Zyskind, J. W., and Smith, D. W., 1986. The bacterial origin of replication, *oriC. Cell* 46:489–490.

Events at the replication fork

Kornberg, A., 1983. Mechanism of replication of the *Escherichia coli* chromosome. *Eur. J. Biochem.* 137:377–382.

Chase, J. W., and Williams, K. R., 1986. Single-stranded DNA binding proteins required for DNA replication. *Ann. Rev. Biochem.* 55:103.

Alberts, B. M., 1984. The DNA enzymology of protein machines. *Cold Spring Harbor Symp. Quant. Biol.* 49:1–12. [Emphasizes the replication machinery used by T4 phage.]

DNA repair

Walker, G. C., 1985. Inducible DNA repair systems. *Ann. Rev. Biochem.* 4:425–457.

Sancar, G. B., and Sancar, A., 1987. Structure and function of DNA photolyases. *Trends Biochem. Sci.* 12:259–261.

Modrich, P., 1987. DNA mismatch correction. *Ann. Rev. Biochem.* 56:435–466.

Lindahl, T., 1982. DNA repair enzymes. *Ann. Rev. Biochem.* 51:61–87.

Sancar, G. B., and Rupp, W. D., 1983. A novel repair enzyme: uvrABC excision nuclease of *E. coli* cuts a DNA strand on both sides of the damaged region. *Cell* 33:249–260.

Yeung, A. T., Mattes, W. B., Oh, E. Y., and Grossman, L., 1983. Enzymatic properties of purified *Escherichia coli* uvrABC proteins. *Proc. Nat. Acad. Sci.* 80:6157–6161.

Hunter, W. N., Brown, T., Anand, N. N., and Kennard, O., 1986. Structure of an adenine-cytosine base pair in DNA and its implications for mismatch repair. *Nature* 320:552–555.

Sinha, N. K., 1987. Specificity and efficiency of editing of mismatches involved in the formation of base-substitution mutations by the $3' \rightarrow 5'$ exonuclease activity of phage T4 DNA polymerase. *Proc. Nat. Acad. Sci.* 84:915–919.

Hanawalt, P. C., Cooper, P. K., Ganesan, A. K., and Smith, C. A., 1979. DNA repair in bacteria and mammalian cells. *Ann. Rev. Biochem.* 48:783–836.

Lindahl, T., Ljungquist, S., Siegert, W., Nyberg B., and Sperens, B., 1977. DNA *N*-glycosidases. Properties of uracil-DNA glycosidase from *Escherichia coli. J. Biol. Chem.* 252:3286–3294.

Cleaver, J. E., and Karentz, D., 1986. DNA repair in man: regulation by a multigene family and association with human disease. *Bioessays* 6:122–127.

Cleaver, J. E., 1983. Xeroderma pigmentosum. *In* Stanbury, J. B., Wyngaarden, J. B., Fredrickson, D. S., Goldstein, J. L., and Brown, M. S., (eds.), *The Metabolic Basis of Inherited Disease* (5th ed.), pp. 1227–1248. McGraw-Hill.

Mutagenesis and carcinogenesis

Singer, B., and Kusmierek, J. T., 1982. Chemical mutagenesis. *Ann. Rev. Biochem.* 52:655–693.

Echols, H., Lu, C., and Burgers, P. M. J., 1983. Mutator strains of *Escherichia coli, mutD* and *dnaQ*, with defective exonucleolytic editing by DNA polymerase III holoenzyme. *Proc. Nat. Acad. Sci.* 80:2189–2192.

Hayes, W., 1968. *The Genetics of Bacteria and Their Viruses* (2nd ed.). Wiley. [Chapter 13, on the nature of mutations, is lucid and concise.]

Drake, J. W., and Baltz, R. H., 1976. The biochemistry of mutagenesis. *Ann. Rev. Biochem.* 45:11–38.

Ames, B. N., 1979. Identifying environmental chemicals causing mutations and cancer. *Science* 204:587–593.

Devoret, R., 1979. Bacterial tests for potential carcinogens. *Sci. Amer.* 241(2):40–49.

PROBLEMS

1. DNA polymerase I, DNA ligase, and topoisomerase I catalyze the formation of phosphodiester bonds. What is the activated intermediate in the linkage reaction catalyzed by each of these enzymes? What is the leaving group?

2. The catalytic rate of DNase I is maximal between pH 7.5 and 8.5. Why does the enzymatic rate decrease markedly in going to pH 6?

3. Would you expect DNase I to cleave Z-DNA? Why?

4. Whether the joining of two DNA chains by known DNA ligases is driven by NAD^+ or ATP depends on the species. Suppose that a new DNA ligase requiring a different energy donor is found. Propose a plausible substitute for NAD^+ or ATP in this reaction.

5. DNA ligase from *E. coli* relaxes supercoiled circular DNA in the presence of AMP but not in its absence. What is the mechanism of this reaction and why is it dependent on AMP?

6. *Sulfolobus acidocaldarius*, an archaebacterium found in acidic hot springs, contains a topoisomerase that catalyzes the ATP-driven introduction of positive supercoils into DNA. How might this enzyme be advantageous to this unusual bacterium?

7. Suppose that a turn of B-DNA in a circular DNA molecule with $L = 100$ and $W = -4$ becomes a turn of Z-DNA.
   (a) What are the values of L, T, and W following the transition?
   (b) How does this transition alter the electrophoretic mobility of this DNA in an agarose gel?
   (c) Transitions of B-DNA to Z-DNA are favored by negative supercoiling. Why?
   (d) Suppose that a protein binds a specific sequence of DNA in the B-DNA form but not in the Z-DNA form. How does this protein affect the dependence of the B-to-Z transition on the degree of supercoiling?

8. The transition from B-DNA to Z-DNA occurs over a small change in the superhelix density, which shows that the transition is highly cooperative.
   (a) Consider a DNA molecule at the midpoint of this transition. Are B- and Z-DNA regions frequently intermingled or are there long stretches of each?
   (b) What does this finding reveal about the energetics of forming a junction between the two kinds of helices?
   (c) Would you expect the transition from B- to A-DNA to be more or less cooperative than the one from B- to Z-DNA? Why?

9. AppNHp, the β,γ-imido analog of ATP, is hydrolyzed very slowly by most ATPases. The addition of AppNHp to DNA gyrase and circular DNA leads to the negative supercoiling of a single molecule of DNA per gyrase. DNA remains bound to gyrase in the presence of this analog. What does this finding reveal about the catalytic mechanism?

10. Suppose that the single-stranded RNA from tobacco mosaic virus was treated with a chemical mutagen, that mutants were obtained having serine or leucine instead of proline at a specific position, and that further treatment of these mutants with the same mutagen yielded phenylalanine at this position.

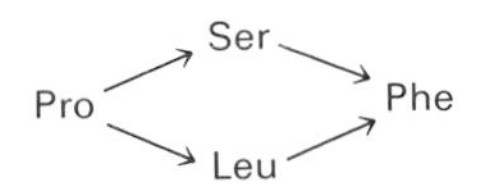

   (a) What are the plausible codon assignments for these four amino acids?
   (b) Was the mutagen 5-bromouracil, nitrous acid, or an acridine dye?

11. The amino acid sequences of part of lysozyme from wild-type T4 bacteriophage and a mutant are

Wild-type
-Thr-Lys-Ser-Pro-Ser-Leu-Asn-Ala-Ala-Lys-
-Thr-Lys-Val-His-His-Leu-Met-Ala-Ala-Lys-
Mutant

   (a) Could this mutant have arisen by the change of a single base pair in T4 DNA? If not, how might this mutant have been produced?
   (b) What is the base sequence of the mRNA that codes for the sequence of five amino acids in the wild-type phage that is different in the mutant?

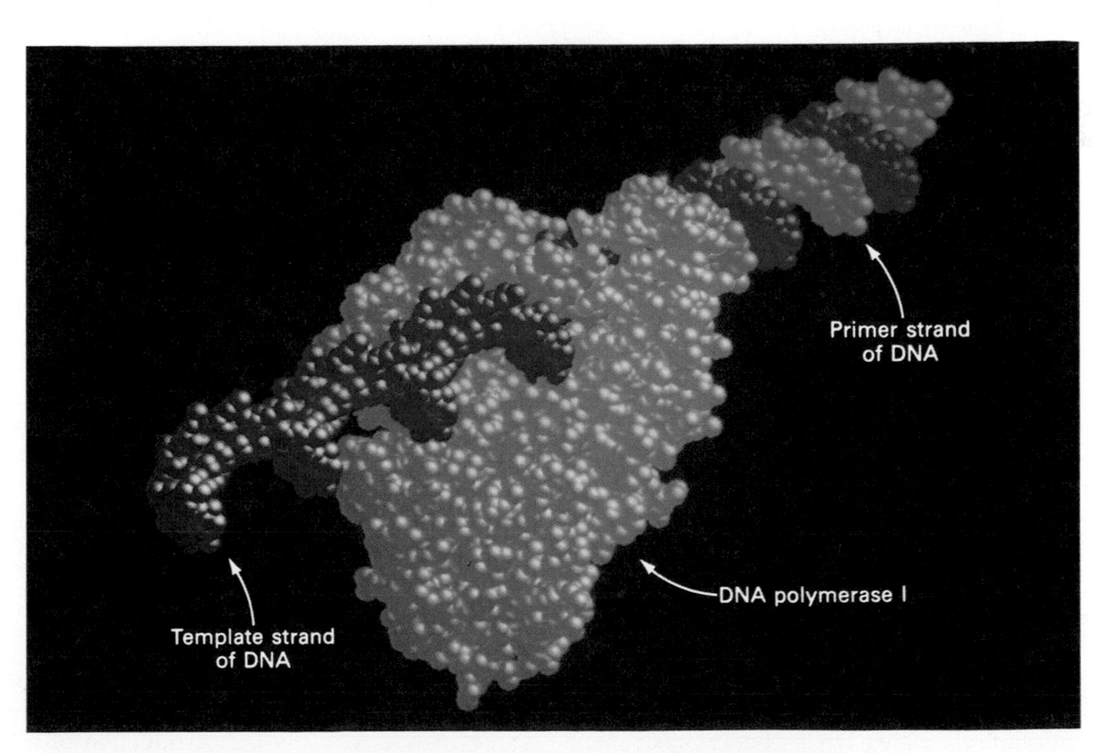

Model of DNA polymerase I containing bound DNA. [Courtesy of Dr. Art Perlo and Dr. Thomas Steitz.].

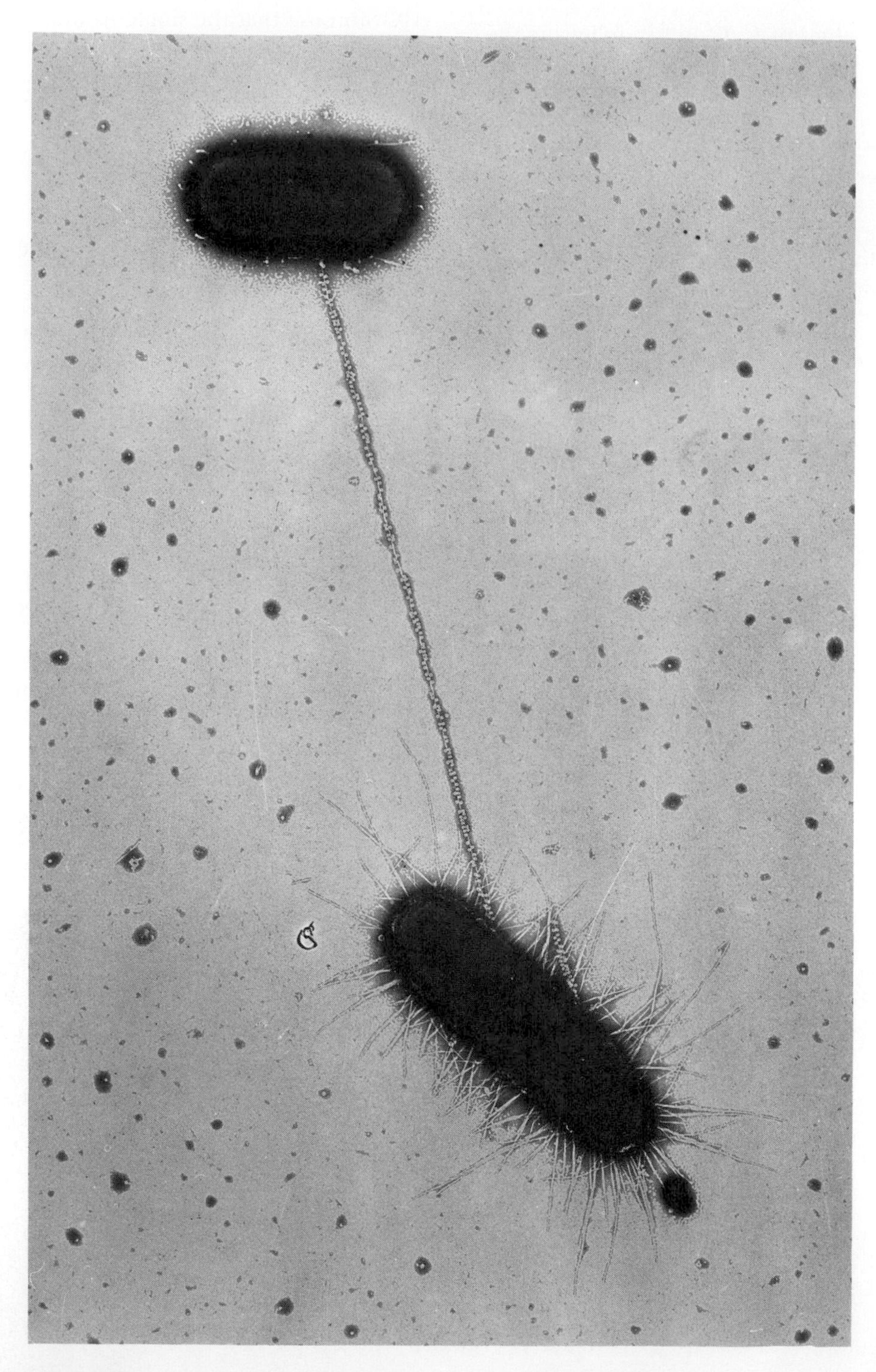

Figure 28-1
Transmission of genetic information from one *E. coli* cell to another. This electron micrograph shows two *E. coli* joined by a pilus during conjugation. DNA is transmitted through the pilus from the donor to the acceptor cell. [Courtesy of Dr. Charles Brinton and Dr. Judith Carnahan.]

CHAPTER 28

Gene Rearrangements: Recombination and Transposition

The theme of this chapter is the formation of new arrangements of genes in procaryotes by the movement of large blocks of DNA. In *general recombination*, parent DNA duplexes align at regions of sequence similarity (homology), and new DNA molecules are formed by the exchange of homologous segments. Intermediates in recombination have been isolated, and the actions of enzymes catalyzing the exchange of DNA strands have been delineated. *Transposition* is the movement of a gene from one chromosome to another or from one site to a different one on the same chromosome. In contrast with general recombination, transposition does not require extensive homology of nucleotide sequence. *Transposons* are mobile genetic elements that enable genes to move between nonhomologous sites in DNA. R-factor plasmids are a medically important group of transposons that make bacteria resistant to multiple antibiotics (Figure 28-2).

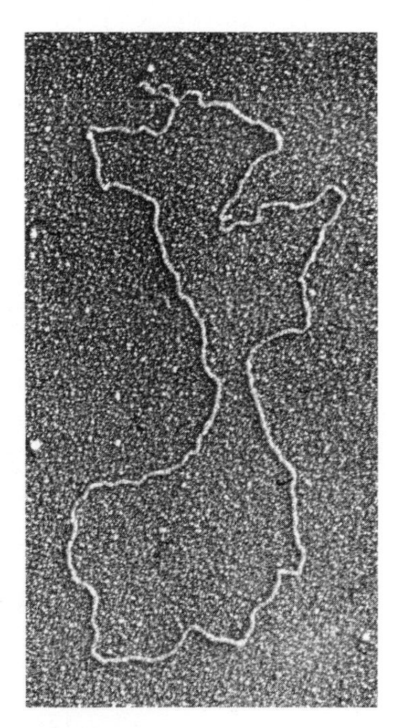

Figure 28-2
Electron micrograph of a circular DNA molecule containing genes that confer resistance to several antibiotics. Infectious drug resistance is a consequence of the transmission of such R (resistance) factor plasmids. [Courtesy of Dr. Stanley Cohen.]

General recombination plays a key role in the *repair of DNA*. When both strands of a duplex are damaged, information for repair must come from another DNA molecule. Recombination provides a template for the synthesis of DNA to fill the gap. The DNA of mutants incapable of undergoing recombination is highly vulnerable to irreversible damage by ultraviolet light and agents that react with DNA. Recombination and transposition also *generate new combinations of genes*. They are immensely important in evolution because they markedly enlarge the genetic repertoire from which natural selection chooses. A third function of gene rearrangements is to *regulate the expression of DNA*. Whether a gene is silent or active may depend on its location and orientation in a chromosome.

Site-specific recombination, the exchange of two specific (but not necessarily homologous) DNA sequences, will be discussed in later chapters. The integration of viral DNA into the *E. coli* chromosome (p. 860) and the inversion of a DNA segment that controls the expression of flagellar genes in *Salmonella* (p. 818) are examples of recombination at unique sites. We shall also encounter the strikingly mobile eucaryotic genes that encode antibodies (p. 905).

GENETIC RECOMBINATION IS MEDIATED BY THE BREAKAGE AND JOINING OF DNA STRANDS

In genetic recombination, a DNA molecule is formed with a base sequence derived partly from one parental DNA molecule and partly from another. Studies of *E. coli* infected with a mixture of ^{32}P-labeled and bromouracil-labeled T4 phages were sources of insight into the molecular nature of recombinant molecules. The buoyant density of bromouracil-labeled DNA is appreciably higher than that of ^{32}P-labeled DNA, and so the parental molecules could be separated from each other and from recombinants by centrifuging them in a CsCl gradient. Analyses of these gradients showed that recombinant DNA molecules containing both ^{32}P and bromouracil were formed after infection with this mixture of phages. The nature of these hybrids depended on whether DNA synthesis occurred during their formation. In the absence of DNA synthesis, the ^{32}P-DNA in the recombinant was not covalently joined to the bromouracil-DNA. The hybrid could be dissociated into light and heavy components by heating above the melting temperature of the duplex. The parental units in this hybrid are held together by base pairing, and so this intermediate is called a *lap-joint* (Figure 28-3). However, if DNA synthesis occurred, the gaps in the lap-joint intermediate were filled by DNA polymerase I and the ends were joined by DNA ligase. The resulting recombinant molecule could not be separated into light and heavy components because these pieces of DNA had become covalently linked. *These experiments revealed that single-stranded regions of DNA are intermediates in genetic recombination and suggested that enzymes are intimately involved in this process.*

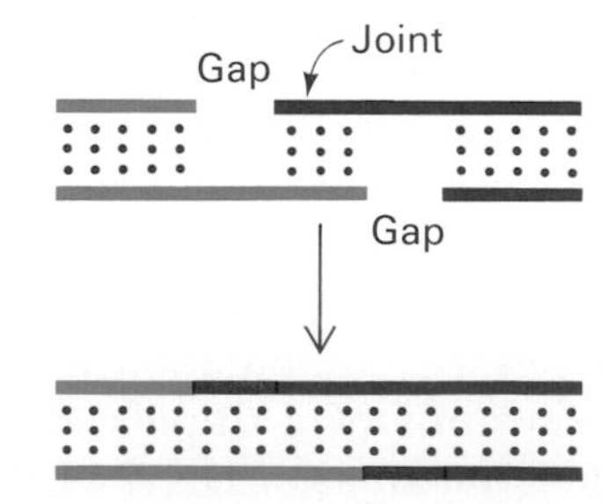

Figure 28-3
Sealing of a lap-joint intermediate in the recombination of T4 viral DNA in infected bacteria. The ^{32}P-labeled and bromouracil-labeled DNA are shown in different colors.

HOLLIDAY MODEL FOR GENERAL RECOMBINATION

In 1964, Robin Holliday proposed a model of general recombination to account for the products of meiosis in fungi. In his scheme (Figure 28-4), two homologous duplexes are aligned. A strand of one duplex and the corresponding strand of the other duplex are nicked by an endonuclease. An end of each nicked strand leaves its own duplex and

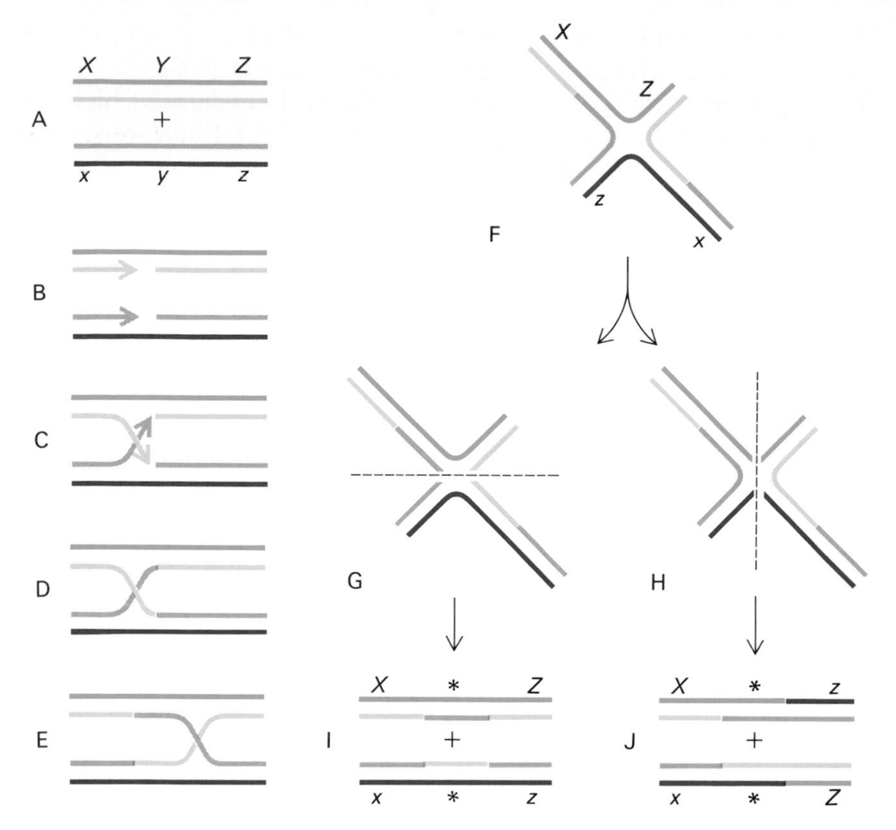

Figure 28-4
A model of genetic recombination proposed by Robin Holliday. One parental duplex is shown in green and the other in red. The darker strand of each duplex is the (+) strand. *X*, *Y*, and *Z* denote three genes; *x*, *y*, and *z* are alleles. The parental DNA molecules become covalently linked in parts C and D. In E, the exchange of strands has continued, so that the crossover point has moved. Part F is an alternative representation of part E. Note that part F can be cleaved along the horizontal or the vertical axis. Rejoining of strands in parts G and H gives two different sets of recombinants (I and J). Regions containing one strand from each parental duplex are labeled with an asterisk. [After H. Potter and D. Dressler. *Cold Spring Harbor Symp. Quant. Biol.* 43(1979):973.]

invades the other duplex. Strands from different duplexes are then joined to each other to form a recombination intermediate that is dynamic. Strand exchange can continue, allowing the crossover point between duplexes to move, a process called branch migration. This recombinational intermediate can be cleaved and rejoined in two different ways to form two kinds of recombinant products (G and H in Figure 28-4).

HOMOLOGOUS DNA STRANDS PAIR IN GENERAL RECOMBINATION TO FORM CHI INTERMEDIATES

The Holliday model has served as a stimulating conceptual framework for studies of molecular mechanisms of recombination. Intermediates in general recombination were visualized some fifteen years later by electron microscopic studies of plasmids in *E. coli.* Bacterial cells treated with chloramphenicol become filled with plasmids because this antibiotic inhibits replication of bacterial DNA but not that of plasmid DNA. These plasmids, like bacterial DNA, readily undergo recombination. When they were isolated from the treated cells, about a quarter of the plasmids were found to be dimers in the shape of a figure eight (Figure 28-5A), as predicted by the Holliday model (Figure 28-6).

These dimers were then cleaved by EcoRI endonuclease, which cuts the original plasmid at a unique site. If a figure-eight dimer were made of interlocked circles or if it were simply a twisted double-length circle, unit-size rods would be formed by this procedure. On the other hand, a structure with four arms resembling the Greek letter χ (chi) would be generated if a dimer consisted of two plasmid circles covalently joined at a single point. In fact, nearly all of the figure eights were converted into chi forms (Figure 28-5B), which strongly suggests that *figure-eight*

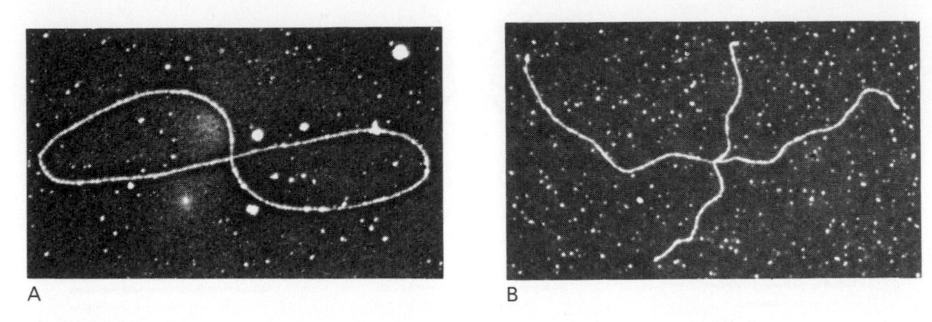

Figure 28-5
Electron micrographs of DNA molecules undergoing recombination: (A) a figure-eight intermediate consisting of two DNA molecules; (B) cleavage by a restriction endonuclease yields a chi form. [From H. Potter and D. Dressler. *Proc. Nat. Acad. Sci.* 76(1979):1086.]

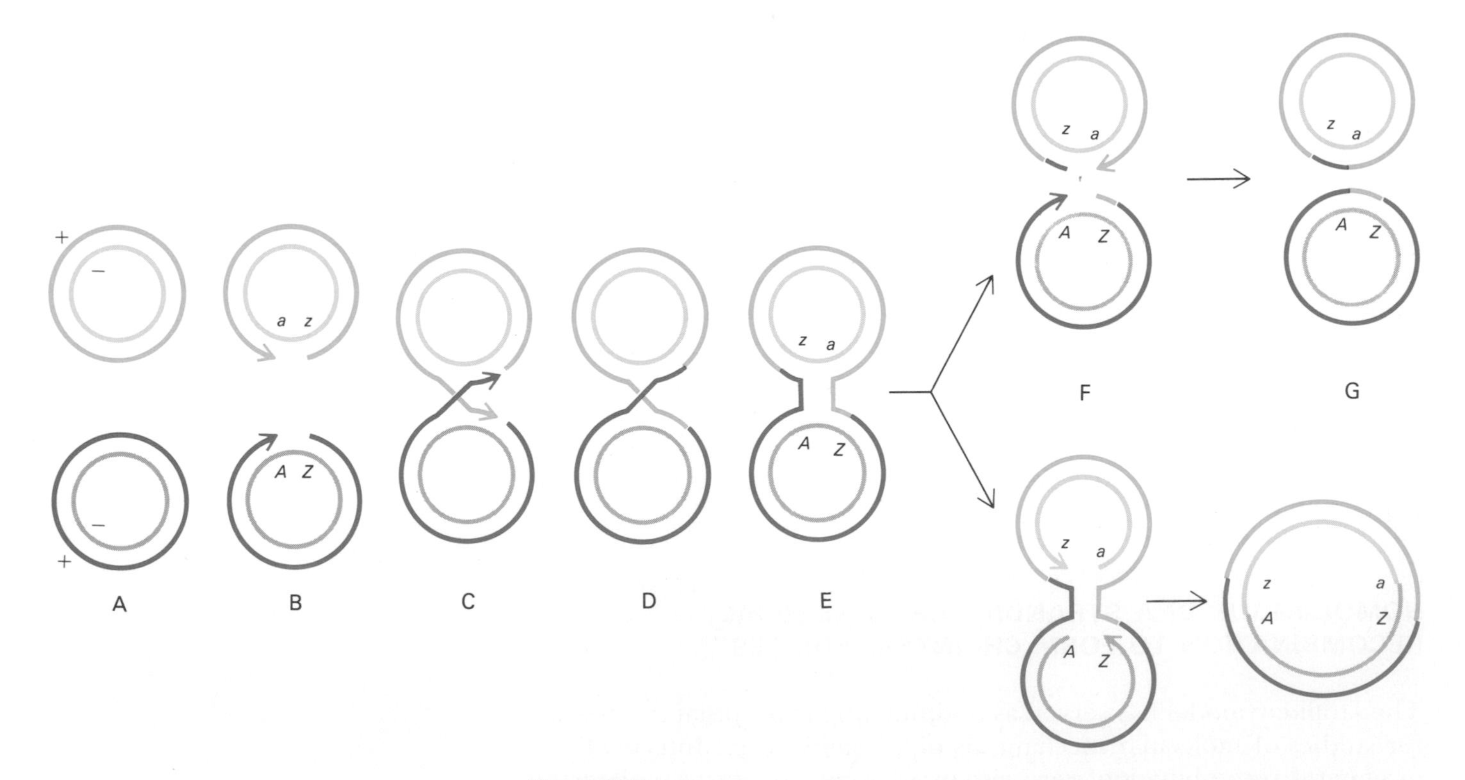

Figure 28-6
Recombination of circular DNA duplexes. The intermediate shown in (E) corresponds to the Holliday intermediate for the recombination of linear duplexes (F in Figure 28-4). Two kinds of products are formed, depending on how the Holliday intermediate is resolved. [After D. Dressler and H. Potter. Molecular Mechanisms in Genetic Recombination. *Ann. Rev. Biochem.* 51:733.]

forms are intermediates in recombination. This inference was supported by the finding that figure eights are not formed by *E. coli* mutants defective in recombination.

What is the nature of the crossover region of the two genomes in the figure-eight forms? The contact point in the chi forms always divides the structure into pairs of equal-length arms. This means that *the genomes are joined at a region of homology* (Figure 28-7). There would be no

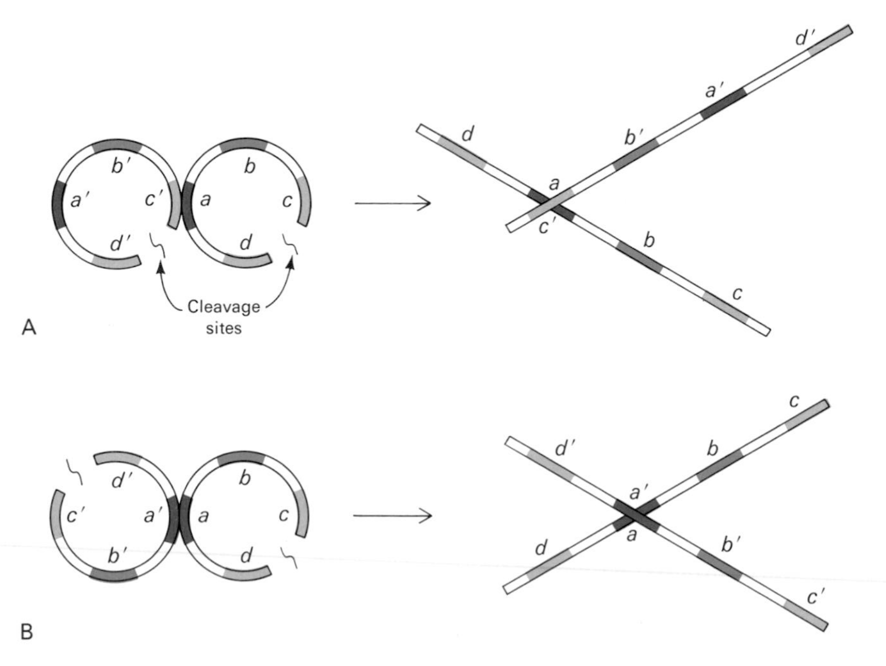

Figure 28-7
Expected cleavage patterns for two DNA molecules that are joined at (A) nonhomologous sites and (B) homologous sites. The observed symmetry of chi forms (as in Figure 28-5B) shows that DNA molecules in figure-eight forms are joined at homologous sites.

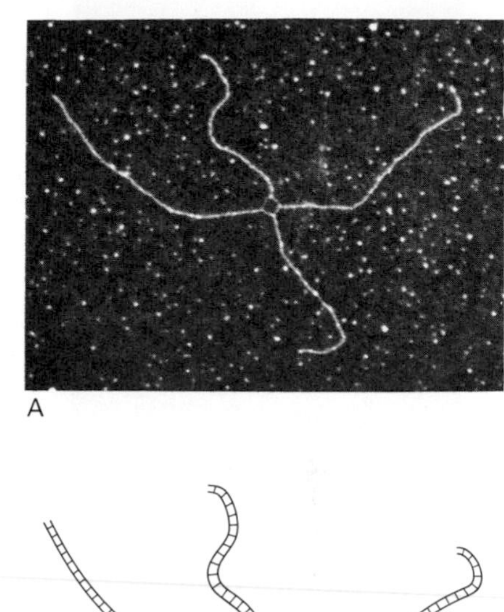

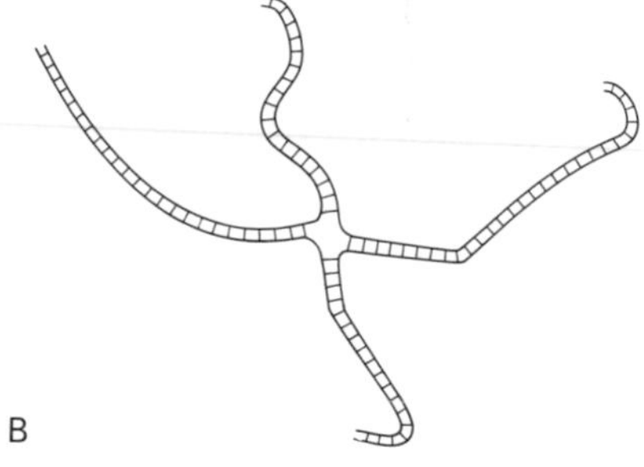

Figure 28-8
A. Electron micrograph of a chi form.
B. Interpretive diagram. The region of homology (which is rich in AT base pairs) was selectively denatured by formamide to show the strand connections in the crossover. [From H. Potter and D. Dressler. *Cold Spring Harbor Symp. Quant. Biol.* 43(1979):973.]

special relations between the sizes of the arms if the plasmids were joined at unrelated sequences. Furthermore, the contact point occurs with nearly equal probability along the entire plasmid, which shows that *pairing can occur at many locations.* The connections at the crossover region have been visualized by electron microscopy after being exposed by selective denaturation. *The four duplexes emerge from a ring of connecting single strands at the junction between the two genomes* (Figure 28-8).

THE recA PROTEIN CATALYZES THE ATP-DRIVEN EXCHANGE OF DNA STRANDS IN GENERAL RECOMBINATION

The process discussed thus far is called *general recombination* because exchange can occur between any pair of homologous sequences on the parental DNA molecules. A single-stranded region of DNA arises from cleavage of one of the strands in a duplex; it finds a homologous sequence in a second duplex, pairs with the complementary strand, and displaces the other one. In *E. coli,* general recombination depends on the *rec* genes. The *recA protein*, the product of the *recA* gene, catalyzes the strand-assimilation reaction, which is driven by the hydrolysis of ATP. The product, consisting of a duplex and a displaced single-

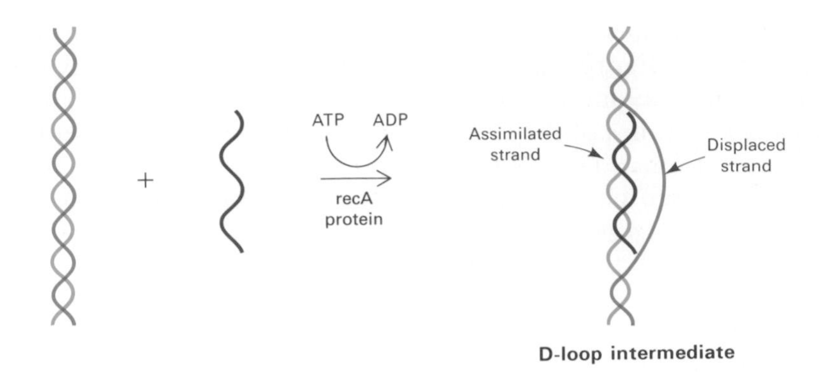

Figure 28-9
The pairing of a single-stranded DNA molecule (red) with the complementary strand (green) of a duplex is catalyzed by the recA protein. The resulting structure is called a D loop.

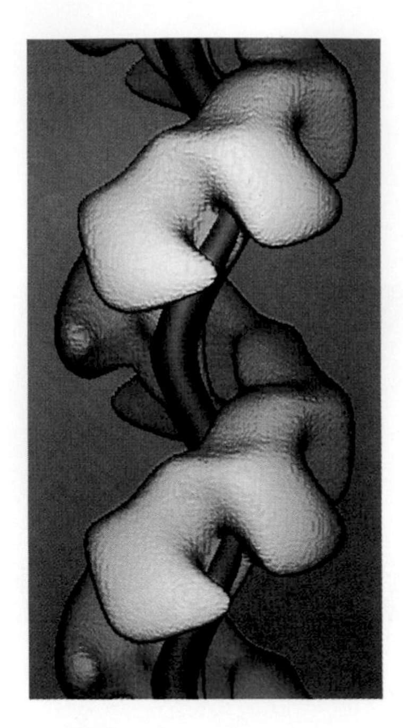

Figure 28-10
Low-resolution model of a recA filament, reconstructed from electron micrographs of filaments of recA bound to duplex DNA in the presence of ATPγS. The DNA (not shown in this model) is located inside the filament. [Courtesy of Dr. Edward Egelman and Dr. Andrzek Stasiak.]

stranded loop, has the shape of the letter D, and so it is called a *D-loop structure* (Figure 28-9). The central role of recA protein in general recombination is revealed by the finding that the frequency of recombination is 10^4-fold lower in *recA*$^-$ mutants than in the wild type.

RecA protein first binds to a region of single-stranded DNA (ssDNA). A filament of recA molecules forms around the ssDNA, with a stoichiometry of one 38-kd monomer per four nucleotides. *This recA-ssDNA filament then binds duplex DNA.* Both DNA molecules can be accommodated inside the filament, which has a diameter of about 100 Å (Figure 28-10). The presence of ssDNA markedly enhances the uptake of duplex DNA, a clear-cut example of induced fit. *RecA protein rapidly scans the duplex DNA for a sequence complementary to that of the ssDNA.* The duplex DNA is partly unwound to facilitate the reading of its base sequence; a turn of double helix in this complex contains 18.6 base pairs, compared with 10.4 for B-DNA. ssDNA and duplex DNA are probably next to each other in a single binding site of recA protein (Figure 28-11). When a complementary sequence is found, the duplex is further unwound to allow a switch in base pairing. *ssDNA now pairs with the complementary strand of the target duplex. Once started in this way, the exchange of strands continues, a process called branch migration.*

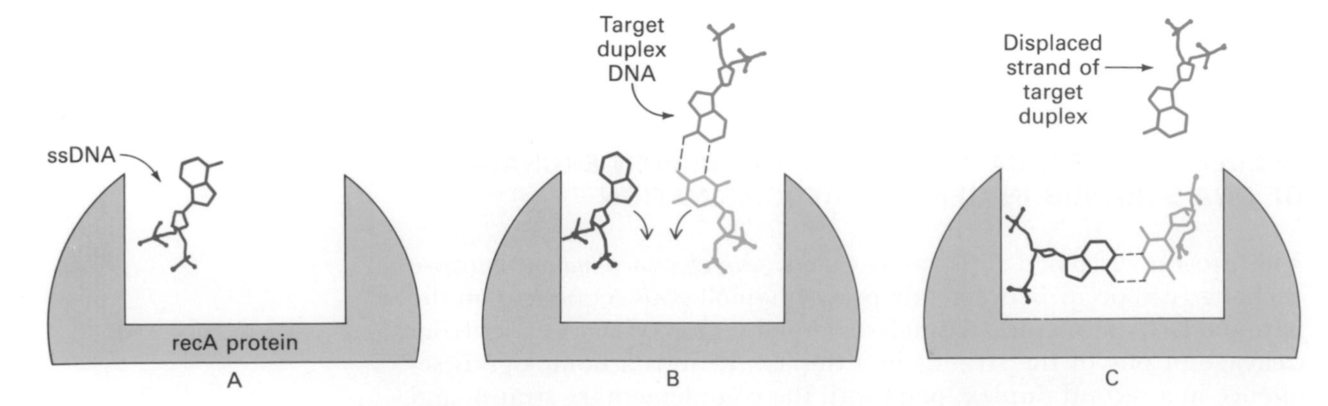

Figure 28-11
The binding of ssDNA to recA (A) leads to the binding of duplex DNA (B), which is scanned for a homologous region. The invading ssDNA becomes paired to the complementary strand of the target duplex (C). [After P. Howard-Flanders, S. C. West, and A. Stasiak. *Nature* 309(1984):217.]

The recombinational actions of recA protein are powered by the hydrolysis of ATP. The mechanism of free-energy transduction is not yet understood in molecular detail. However, several clues are known. ATP is required for the binding of DNA to recA protein. Hydrolysis of bound ATP enables recA protein to dissociate from DNA. A recA monomer in a filament can bind and release DNA as bound ATP is hydrolyzed. Such a reaction cycle is reminiscent of the action of DNA gyrase (p. 662). We shall encounter similar motifs in protein synthesis (p. 755) and muscle contraction (p. 931). Another clue is that the recA filament has polarity (see Figure 28-10). Moreover, it assembles unidirectionally ($5' \rightarrow 3'$) on ssDNA. These properties of recA enable it to catalyze unidirectional strand assimilation. The driving force for coherent movement of the recA filament and the invading ssDNA is the hydrolysis of bound ATP. Without an input of free energy, the invading ssDNA would take a few steps to the left and then to the right, an aimless random walk. Unidirectional, processive movement has another important function. It enables the invading DNA strand to bypass damaged regions of the target duplex that are not complementary to it. ATP hydrolysis also promotes branch migration by rotating a duplex DNA molecule around the filament (Figure 28-12). The mechanochemistry of recombination is an intriguing and challenging area of inquiry.

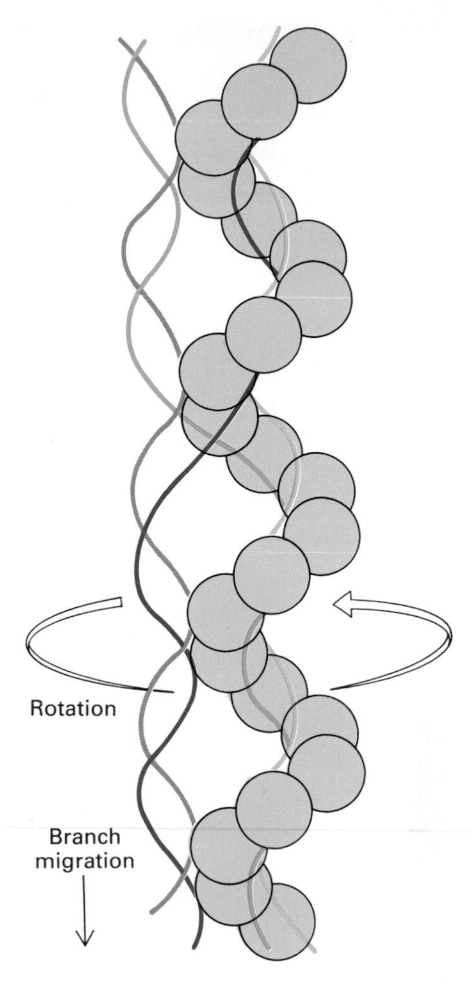

Figure 28-12
Model for branch migration, the movement of the crossover point between DNA duplexes. A recA filament containing one DNA molecule rotates a second DNA molecule joined to the first by a crossover junction. Branch migration is powered by the hydrolysis of ATP by recA. [After M. M. Cox and I. R. Lehman. Enzymes of General Recombination. *Ann. Rev. Biochem.* 56:252.]

THE UNWINDING AND ENDONUCLEASE ACTIONS OF recBCD GENERATE SINGLE-STRANDED DNA FOR RECOMBINATION

Recombination requires that the invading DNA be single stranded. How is ssDNA formed for this purpose? Genetic studies showed that the *recB*, *recC*, and *recD* genes are important in recombination. In fact, these genes encode proteins that form a 328-kd *recBCD complex*. RecB protein alone catalyzes the ATP-driven unwinding of duplex DNA, and recD protein by itself is a nuclease. The recBCD complex generates ssDNA from a duplex in an interesting way (Figure 28-13). The complex attaches to one end of a linear duplex and unwinds it. The hydro-

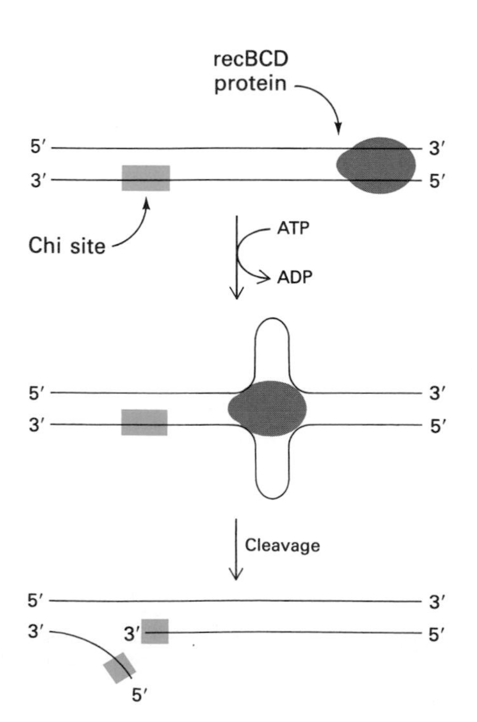

Figure 28-13
Model for the generation of single-stranded DNA by the recBCD complex (red), an ATP-driven helicase and nuclease. Two single-stranded loops are formed as the complex moves forward. The enzyme cleaves one of the strands when it encounters a chi sequence (green) to form single-stranded DNA with a free end. Recombination then takes place. [After A. Taylor and G. R. Smith. *Cell* 22(1980):447.]

lysis of ATP powers the unwinding, which proceeds at about 300 nucleotides per second. Rewinding of DNA proceeds at the back end of the complex, but slower than unwinding at the forward end. Hence, two single-stranded loops emerge from the complex, and they grow in size as the enzyme moves ahead on the DNA.

Cleavage occurs near a specific sequence called *chi*. The *E. coli* chromosome has about a thousand chi sites, about one per 4 kb. The sequence recognized is

$$5'\text{-GCTGGTGG-}3'$$

RecBCD cuts the backbone of this DNA strand between four and six nucleotides away from the 3′ end of this sequence. The enzyme complex continues to move forward and unwind the DNA to produce single-stranded DNA that can invade a homologous duplex (Figure 28-13).

Single-strand-binding protein (SSB) participates in general recombination as well as in DNA replication (p. 673). The cooperative binding of SSB to single-stranded DNA prevents the formation of hairpin loops. SSB stabilizes an unpaired DNA strand until recA protein forms a filament around it. *Topoisomerase I* (p. 660) also plays an important role in recombination by relieving torsional strain produced by the winding of one DNA molecule around another during branch migration.

DAMAGE TO DNA TRIGGERS AN SOS RESPONSE THAT IS INITIATED BY HYDROLYSIS OF A REPRESSOR PROTEIN

An *E. coli* cell normally contains several hundred copies of the recA protein. The level of this key enzyme in recombination increases a hundredfold following damage to DNA. Indeed, damage rapidly induces the synthesis of more than fifteen proteins mediating the repair of DNA. This switching on of the repair machinery is called the *SOS response*. Under normal conditions, repair proteins are present at low levels because the synthesis of their messenger RNAs is blocked by a repressor protein called *lexA*. The essence of the SOS response is the cleavage of the lexA protein. Single-stranded DNA arising from damaged DNA acts as a trigger by binding to recA protein (Figure 28-14). *The ssDNA-recA complex then hydrolyzes an Ala-Gly bond in lexA protein, which destroys its capacity to block the transcription of many genes involved in DNA repair. Thus, recA protein is multifunctional—it is a protease as well as a recombinase.* The synthesis of recA protein itself is unleashed in this way because expression of the recA gene is normally blocked by lexA re-

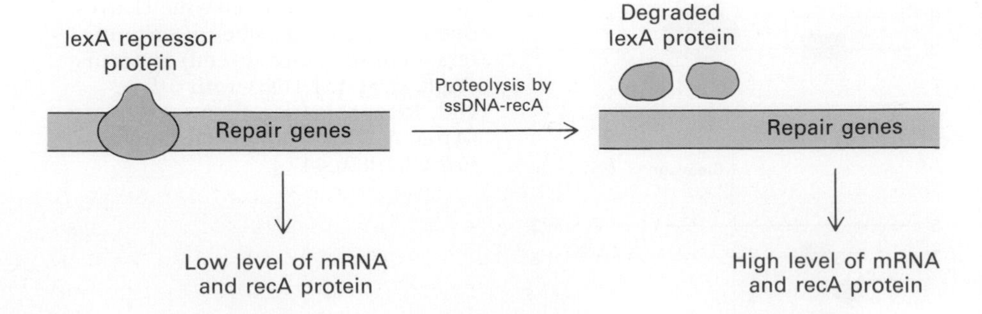

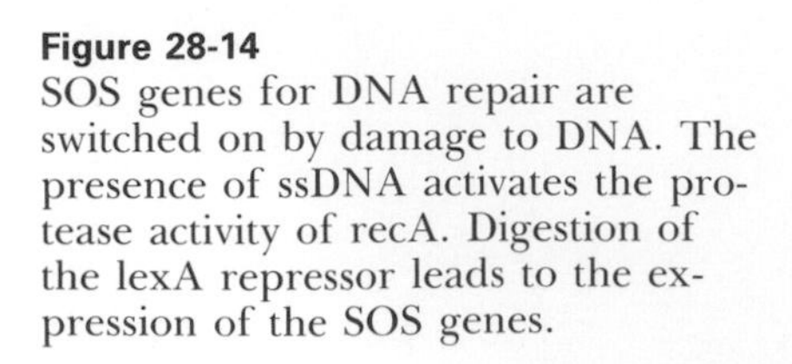
Figure 28-14
SOS genes for DNA repair are switched on by damage to DNA. The presence of ssDNA activates the protease activity of recA. Digestion of the lexA repressor leads to the expression of the SOS genes.

pressor. Destruction of this repressor also leads to increased synthesis of SSB and of the uvrABC complex, a nuclease that excises thymine dimers (p. 678).

BACTERIA CONTAIN PLASMIDS AND OTHER MOBILE GENETIC ELEMENTS

General recombination generates new combinations of specific *alleles,* but it does not readily change the arrangement of entire *loci.* In other words, *ABC′D′E′* can be readily formed by homologous recombination of *ABCDE* with *A′B′C′D′E′*, whereas *ABXYZCDE* and *ABE* cannot. These larger-scale genetic rearrangements are mediated by *mobile genetic elements,* which are also known as *transposable elements* (Table 28-1).

Allele—
Alternative forms of a gene that can occupy a particular chromosomal site.
From the Greek word meaning "of one another."
For example, the β^S sickle gene is an allele of the normal β^A gene. In contrast, the genes for the α and β chains of hemoglobin are not alleles of each other.

Table 28-1
Mobile genetic elements in *Escherichia coli*

Type	*Size (kb)*	*Characteristics*
Plasmids		
F (fertility) factor	93	Confers maleness; transmissible by conjugation.
F′ factors	>100	Carry *E. coli* genes in addition to F-factor genes.
R (resistance) factors	4 to 117	Carry drug-resistance genes; some also carry genes for conjugation.
Colicinogenic factors	6 to 141	Carry genes for colicin (toxin) production; some also carry genes for conjugation.
Lysogenic phages		
Lambda	48	A small proportion carry *E. coli* genes (*gal* or *bio*) in addition to viral genes.
Mu	38	All mu carry a small piece of the *E. coli* genome.
Insertion sequences (IS)	0.8 to 1.4	Genes flanked by a pair of IS are translocatable within a cell.

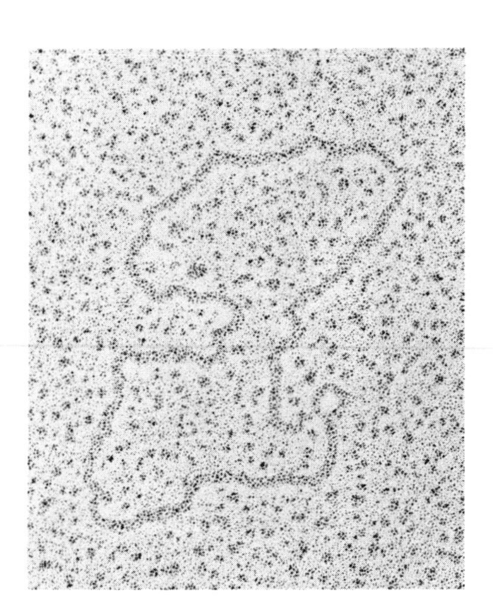

Figure 28-15
Electron micrograph of a small R-factor plasmid. [Courtesy of Dr. Jack Griffith.]

Plasmids are a major class of mobile genetic elements. These circular duplex DNA molecules (Figure 28-15) ranging in size from two to several hundred kilobases, carry genes for the inactivation of antibiotics (p. 127), the metabolism of natural products (p. 135), and the production of toxins. In essence, *plasmids are accessory chromosomes* that differ from the bacterial chromosome in being dispensable under certain conditions. *Plasmids are replicons; they contain their own origin of replication and hence can replicate autonomously.*

Plasmid pSC101	*E. coli*
5′	5′
G	G
A	A
T	T
C	C
T	T
A	A
T	T
T	T
C	T
T	A
T	T
T	T
T	T
3′	3′

***OriC* sequences**

THE F FACTOR ENABLES BACTERIA TO DONATE GENES TO RECIPIENTS BY CONJUGATION

Some plasmids enable bacteria to transfer genetic material to each other by forming a direct cell-cell contact. This process, called *conjugation,* was discovered in 1946 by Joshua Lederberg and Edward Tatum. During the conjugation of *E. coli,* one partner (the male) is the genetic donor and the other (the female) is the genetic recipient. Male bacteria contain special appendages called *sex pili* on their surfaces, whereas female cells contain receptor sites that bind pili. A male bacterium contains a plasmid called *F factor* (fertility factor), which contains genes for the formation of the sex pili and other components required for conjugation. A male and a female cell become joined by a pilus (see Figure 28-1), which retracts to draw the cells into direct contact for the passage of DNA (Figure 28-16). One strand of the F-factor plasmid gets nicked and the duplex unwinds (Figure 28-17). The 5′ end of the nicked strand enters the recipient cell and a complementary strand is synthesized to form a closed, circular duplex. *The presence of an F-factor plasmid in the recipient cell* (*originally* F^-) *converts it into a male* (F^+). On the other hand, a male cell can spontaneously lose the F factor and thereby revert to F^-.

Episome—
A genetic element that can exist as a separate piece of DNA or become integrated into the bacterial genome. Plasmids and F factors are episomes.

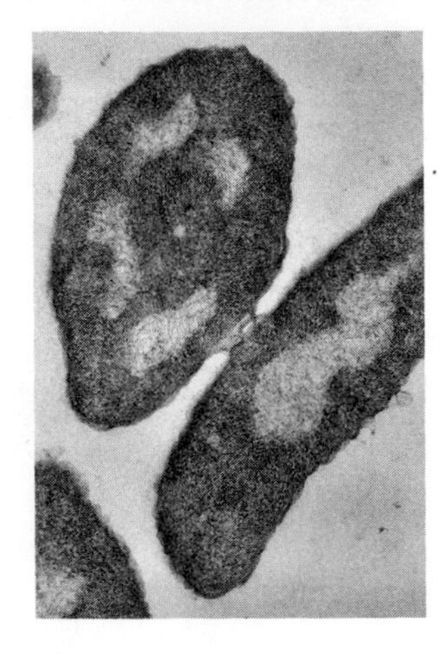

Figure 28-16
Electron micrograph of two *E. coli* in contact during conjugation. [Courtesy of Dr. Lucien Caro.]

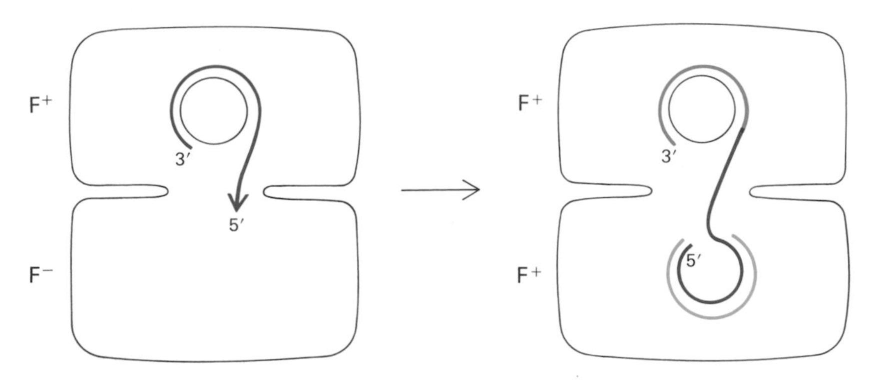

Figure 28-17
Proposed model for strand transfer in conjugation. The supercoiled F factor in the F^+ donor unwinds after being nicked. Transfer of one strand of the F factor to the F^- recipient is coupled to replication of that strand in the donor. A complementary strand is then synthesized in the recipient. [After G. J. Warren, A. J. Twigg, and D. J. Sherratt. *Nature* 274(1978):260.]

An F-factor plasmid can become integrated into the bacterial chromosome (Figure 28-18). Integration occurs by crossing-over at one of a number of sites on the bacterial chromosome. The frequency of integration is about 10^{-5} per generation. Bacteria harboring F factors in their chromosomes are called *Hfr cells* because they exhibit a high frequency of recombination. Hfr cells, like F^+ cells, are donors in conjugation. The difference between them is that *an Hfr cell donates the entire bacterial chromosome* (*including the integrated F factor*), whereas an F^+ cell donates only the F factor. The entire chromosome is transferred in about 90 minutes in an Hfr × F^- cross. The order of entry of the donated genes into the recipient depends on the site of integration of the F factor and on its orientation. Hence, it is feasible to ascertain the order of genes in the donor's chromosome by interrupting the conjugation at different times and determining which markers have been transferred. The donated chromosome can recombine with the recipient's chromosome. The frequency of recombination is highest for genes that first entered the recipient because they are present for the longest time. Thus, genetic maps can be constructed by determining the *time of entry* and the *frequency of recombination* of markers contributed by the donor.

An Hfr cell reverts to the F^+ state by the excision of the F factor from its chromosome. This reversal of the integration of the F factor also has a frequency of about 10^{-5} per generation. In a small proportion of the

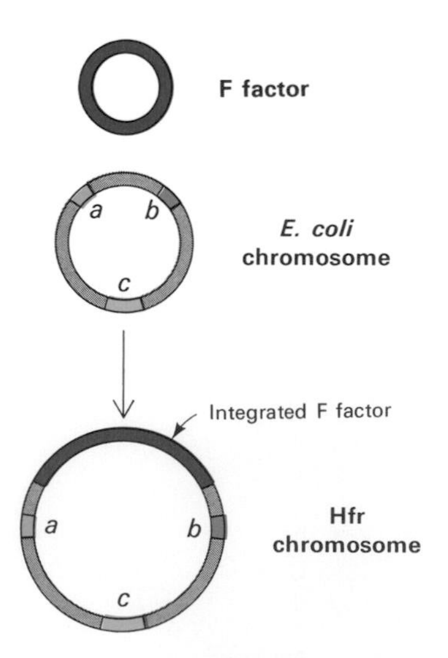

Figure 28-18
Schematic diagram showing the formation of an Hfr cell by integration of the F factor into the *E. coli* chromosome.

revertants, excision occurs at a site different from that of integration, which results in the *formation of a plasmid that contains chromosomal genes in addition to the F-factor genes* (Figure 28-19). Such a plasmid is called an *F′ factor,* the prime denoting the presence of chromosomal genes. Conjugation of an F′ cell with an F^- cell results in the transfer of these chromosomal genes from the donor to the recipient cell, which then becomes diploid for these genes.

Hfr cell

Abnormal excision

a b

F′ plasmid with host genes

c

E. coli **chromosome with deletion**

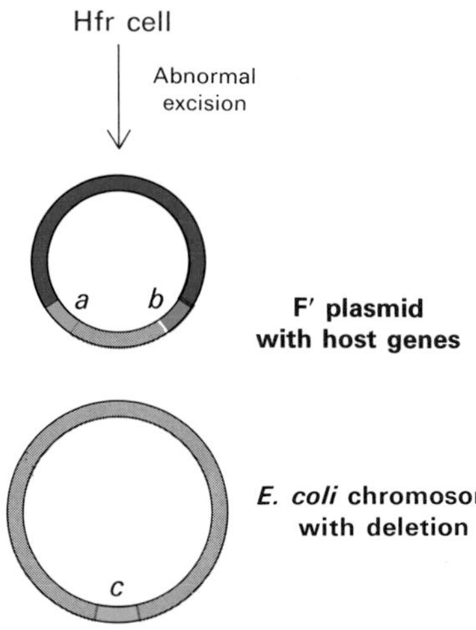

Figure 28-19
Abnormal excision leads to the formation of an F′ plasmid, which contains part of the *E. coli* chromosome.

Thus, bacteria have mechanisms for transferring sets of genes and even entire chromosomes from one cell to another. *The F factor can be regarded as a specially designed vector for the interchange of genetic material.* It is interesting to note that lysogenic phages can also mediate the exchange of host genes. For example, λ-phage DNA can be inserted between the *gal* and *bio* genes on the *E. coli* chromosome (p. 860). When progeny virions are formed, excision is usually but not invariably precise. In about 1 of 10^5 virions, λ DNA contains either the *gal* operon or the *bio* gene. Infection by such phages, which are called λgal or λbio, introduces these *E. coli* genes in addition to λ genes. A related phage, called φ80, inserts near the *trp* operon and can carry *trp* genes from one infected cell to another. Bacteriophage μ, which inserts nearly anywhere on the *E. coli* chromosome, invariably emerges with a piece of the bacterial chromosome. *These transducing phages, like the F factor, are mobile genetic elements that promote the exchange of bacterial genes and hence accelerate bacterial evolution.*

R-FACTOR PLASMIDS MAKE BACTERIA RESISTANT TO ANTIBIOTICS

A striking example of very rapid bacterial evolution took place in 1955 during an epidemic of bacterial dysentery. A strain of *Shigella dysenteriae* became simultaneously resistant to chloramphenicol, streptomycin, sulfanilamide, and tetracycline. Multiple-drug resistance of this kind is now common among many pathogenic microorganisms. The genes conferring resistance to multiple antibiotics are linked together on *R-factor* (*resistance factor*) *plasmids.* The larger of these plasmids contain a *resistance transfer factor* (RTF) in addition to several *r-genes* (Figure 28-20). The RTF region enables the plasmid to be transmitted to other bacteria by conjugation. In fact, the genes in the RTF region closely resemble those in F factors. The *r*-genes code for enzymes that inactivate specific drugs. R factors with an RTF region can be transmitted in mixed cultures even between different bacterial species. Thus, *multiple-drug resistance can be infectious.*

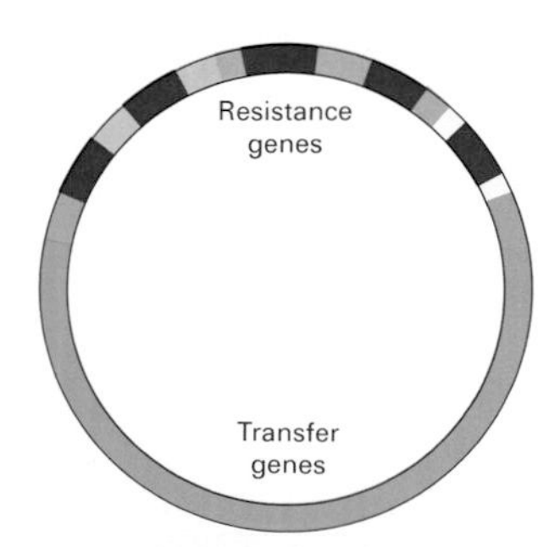

Figure 28-20
Schematic diagram of an R factor. The *RTF* genes (for conjugation and replication) are shown in green, and the *r*-genes (for drug resistance) in red. Insertion sequences are shown in yellow.

The small R-factor plasmids are devoid of an RTF region and usually confer resistance to a single antibiotic. For example, the 8.2-kb pSC101 plasmid carries a gene for tetracycline resistance, but it cannot be trans-

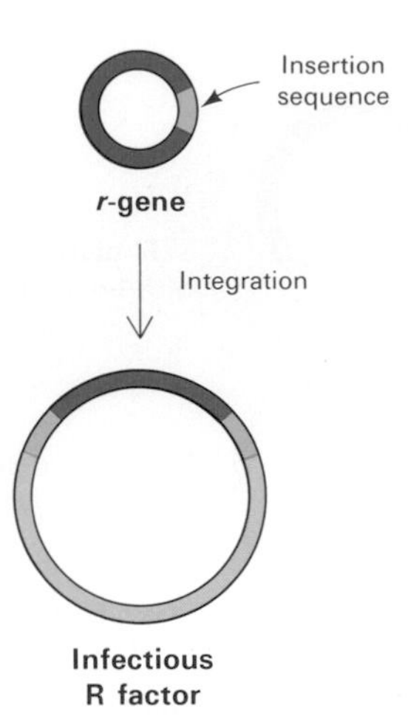

Figure 28-21
An infectious R factor is formed when an *r*-gene joins an RTF plasmid.

mitted by conjugation. However, this *r*-gene can join another plasmid with a different gene for drug resistance (Figure 28-21). A transmissible R plasmid is formed when these *r*-genes integrate into a plasmid that contains an RTF region. *Thus, complex R-factor plasmids are built from highly mobile modules that confer resistance to individual drugs.*

TRANSPOSONS ARE HIGHLY MOBILE GENETIC ELEMENTS

Mobile genetic elements such as R-factor plasmids are called *transposons.* What is the structural basis of their high mobility? The simplest transposons are *insertion sequences* (IS), which are typically about a kilobase long (Table 28-2). Some encode a *transposase,* and others encode a transposase and just one other protein essential for mobility. The sequences of twenty or so base pairs at opposite ends of an IS are very similar but their order is reversed. These boundary sequences are known as *inverted terminal repeats.* An IS can move to virtually any site on a chromosome. The recipient site becomes duplicated during transposition (Figure 28-22). A sequence of 4 to 12 base pairs that was present

Table 28-2
Properties of some transposons in *Escherichia coli*

Type	Length (bp)	Duplicate recipient length (bp)	Function (in addition to transposition)
Insertion sequences			
IS1	768	9	—
IS2	1,327	5	—
IS4	1,426	12	—
IS5	1,195	4	—
Complex transposons			
Tn3	4,957	5	Ampicillin resistance
Tn5	5,700	9	Kanamycin resistance
Tn2571	23,000	9	Resistance to multiple antibodies and to Hg

Sources: J. Cullum, in J. Scaife, D. Leach, and A. Galizzi, (eds.), *Genetics of Bacteria,* (Academic Press, 1985), p. 86; N. Kleckner, *Ann. Rev. Genet.* 15(1981):341.

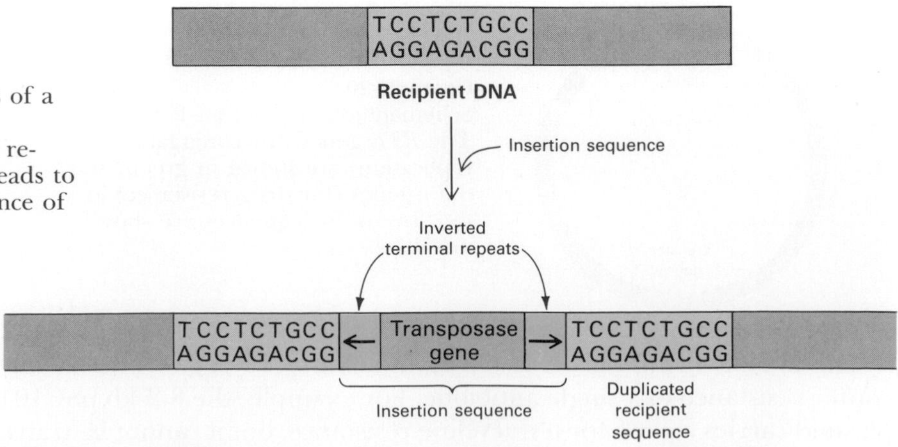

Figure 28-22
An insertion sequence consists of a gene for a transposase (red) bounded by inverted terminal repeats (yellow). Transposition leads to a duplication of a short sequence of the recipient DNA (blue).

once before transposition occurs twice after transposition, one on each side of the IS. The *duplicated recipient sequence* is a noninverted repeat. There is no homology between these flanking sequences and the terminal sequences of the IS. Furthermore, the products of *rec* genes do not participate in transposition. Rather, the inverted terminal repeats of an IS bind its own transposase, a highly specific nuclease that makes staggered cuts at both the donor site and recipient site. DNA polymerase I and DNA ligase are also essential for transposition, which occurs in bacteria at a frequency of the order of 10^{-6} per generation.

Most genes are inactivated by the transposition of an insertion sequence into their midst. However, each insertion sequence carries its own promoter for the transcription of its transposase gene. This promoter sometimes enhances the expression of a neighboring gene. Moreover, insertion sequences can cause deletions, inversions, and duplications of genes (Figure 28-23). The *E. coli* chromosome contains

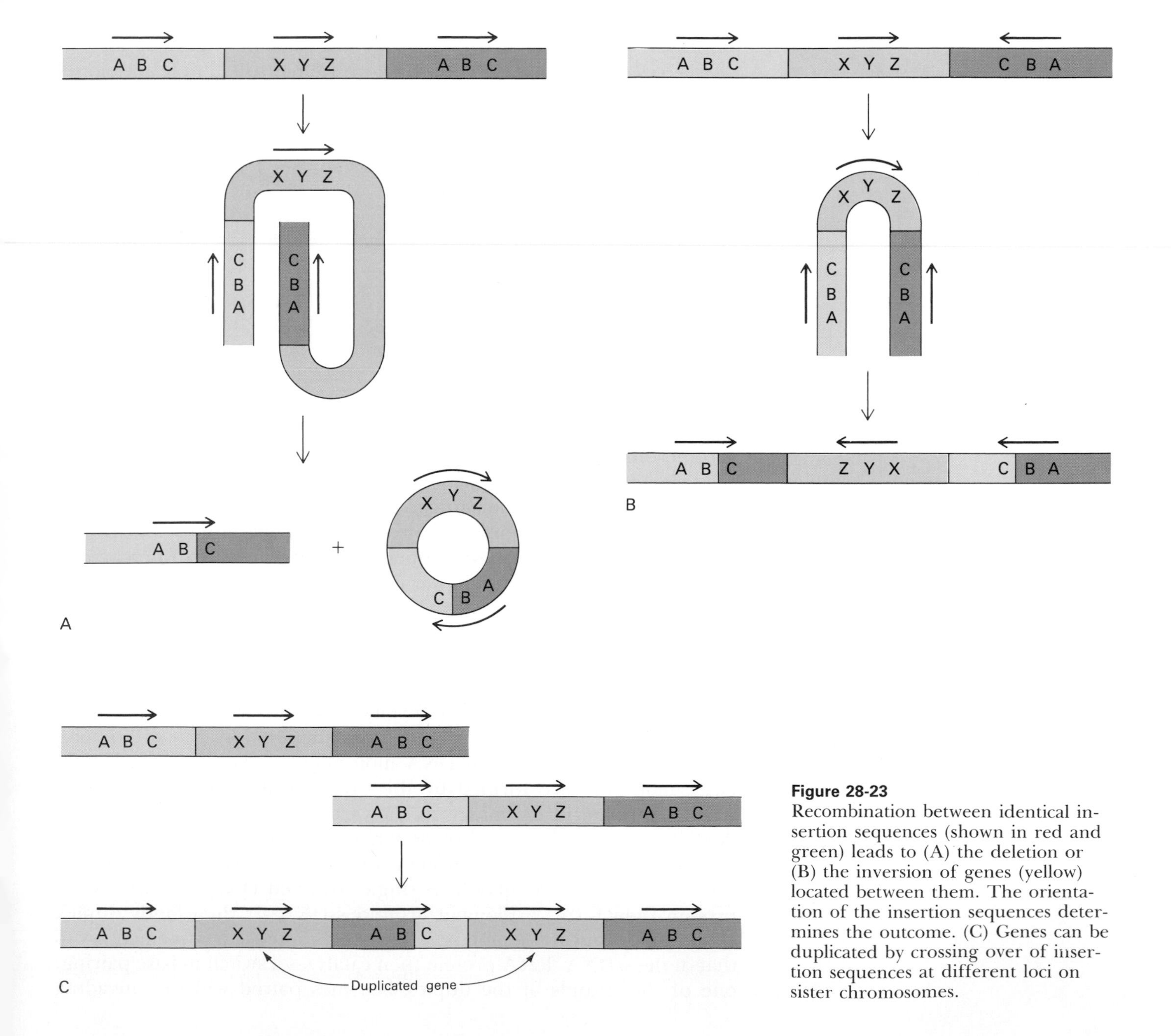

Figure 28-23
Recombination between identical insertion sequences (shown in red and green) leads to (A) the deletion or (B) the inversion of genes (yellow) located between them. The orientation of the insertion sequences determines the outcome. (C) Genes can be duplicated by crossing over of insertion sequences at different loci on sister chromosomes.

multiple copies of several insertion sequences. Recombination between two identical insertion sequences oriented the same way leads to the excision of genes between them. In contrast, the genes between oppositely oriented insertion sequences are inverted by a reciprocal crossover. The expression of a gene can be profoundly altered by inversion, as exemplified by a switching from one type of flagellar protein to another in *Salmonella* (p. 818). In eucaryotes, genes bounded by insertion sequences can be tandemly duplicated by unequal crossing over of sister chromosomes. Thus, the presence of transposons makes possible large-scale frequent rearrangements of the genome.

Complex transposons, in contrast with insertion sequences, contain genes mediating functions in addition to transposition. For example, Tn3 is a complex transposon that contains a gene for β-lactamase, an enzyme conferring resistance to the antibiotic ampicillin. It also contains genes for transposase and resolvase, an enzyme that catalyzes recombination between two insertion sequences (Figure 28-24). Tn3, like other transposons, contains inverted terminal repeats and induces duplication of the sequence at the recipient site. It seems likely that *complex transposons arose from cellular genes that became bounded by a pair of insertion sequences*. R-factor plasmids conferring resistance to multiple antibiotics probably evolved by the coming together of several transposons.

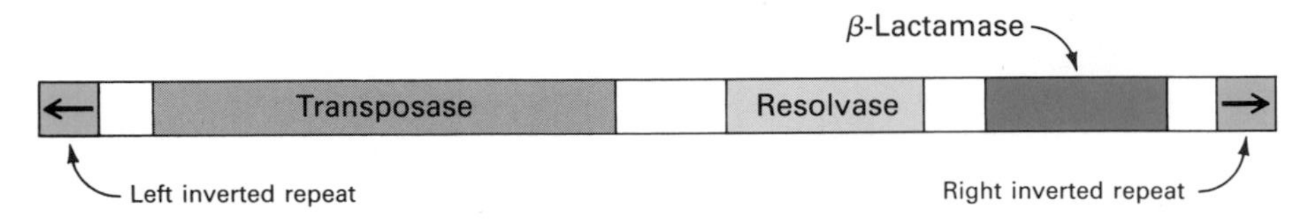

Figure 28-24
Tn3, a complex transposon, consists of genes for three proteins bounded by inverted terminal repeats.

Plasmids and phages can exchange blocks of genes with bacterial chromosomes and can recombine with each other. A tetracycline-resistance element has been shown to move from an R-factor plasmid to a *Salmonella* phage and then to the *Salmonella* chromosome, and from there to λ phage and then to the *trp* operon of *E. coli* and back to λ. This remarkable journey emphasizes the highly mobile nature of procaryotic genes, a theme that recurs in eucaryotes.

SUMMARY

New arrangements of genes are formed by the processes of recombination and transposition. In general recombination, any pair of homologous sequences on parental DNA molecules can be exchanged. New DNA molecules are formed by the breakage and reunion of DNA strands at points of homology, as shown by the formation of chi intermediates. In *E. coli,* general recombination is mediated by products of the *rec* genes. The recBCD protein complex, a helicase and sequence-specific endonuclease, generates single-stranded DNA (ssDNA). RecA protein then forms a filament around ssDNA. It then binds duplex DNA, partly unwinds it, and searches for a sequence complementary to that of the ssDNA. RecA protein then catalyzes a switch in base pairing; one of the strands of the duplex becomes paired with the invading ssDNA. The exchange of strands continues, a process called branch migration. These actions of recA protein, like the helicase activity of the recBCD complex, are powered by the hydrolysis of ATP.

General recombination plays a key role also in repairing DNA. Recombination provides a new template for the synthesis of DNA when both strands are damaged. Damage to DNA triggers an SOS response leading to increased synthesis of many proteins that mediate repair. The first step in this adaptive response is the proteolysis of lexA repressor by recA containing bound ssDNA, a signal that damage has occurred.

Genetic rearrangements of nonhomologous loci are mediated by mobile genetic elements called transposons (transposable elements). Plasmids, which are small circular DNA duplexes, are an important class of mobile genetic elements. These accessory chromosomes carry genes for the inactivation of antibiotics, metabolism of natural products, and production of toxins. Plasmids can replicate autonomously or become integrated into the host chromosome. The F-factor (fertility-factor) plasmid enables bacteria to donate genes to other bacteria by conjugation, a process mediated by direct physical contact between the donor (F^+) and recipient (F^-) cells. R-factor plasmids confer resistance to antibiotics. The larger of these plasmids contain a resistance transfer factor (RTF) that enables the plasmid to be transmitted to other bacteria by conjugation. Several genes that confer resistance to antibiotics can become linked to an RTF region. Hence, multiple-drug resistance can be infectious.

The simplest transposons are insertion sequences. They consist of a transposase and one other gene (or none) bounded by inverted terminal repeats. Transposons can move to nearly any site on a chromosome. Transposition is accompanied by duplication of a short sequence of the recipient site. Genes located between a pair of transposons that recombine can be inverted, deleted, inserted, and duplicated. Complex transposons contain genes mediating functions in addition to transposition. R-factor plasmids probably evolved by the coming together of several transposons.

SELECTED READINGS

Where to start

Stahl, F. W., 1987. Genetic recombination. *Sci. Amer.* 256(2):90–101.

Dressler, D., and Potter, H., 1982. Molecular mechanisms in genetic recombination. *Ann. Rev. Biochem.* 51:727–761. [The Holliday model is lucidly presented in this review.]

Cox, M. M., and Lehman, I. R., 1987. Enzymes of general recombination. *Ann. Rev. Biochem.* 56:229–262. [A perceptive presentation of the actions of the recA and recBCD proteins.]

Cohen, S. N., and Shapiro, J. A., 1980. Transposable genetic elements. *Sci. Amer.* 242(2):40–49.

Books

Cold Spring Harbor Laboratory, 1984. *Recombination at the DNA Level. Cold Spring Harbor Symp. Quant. Biol.* Volume 49. [A rich source of articles on many facets of recombination.]

Scaife, J., Leach, D., and Galizzi, A., (eds.), 1985. *Genetics of Bacteria.* Academic Press. [See chapters 7, 9, 11, 12, and 13 for excellent accounts of recombination and transposition.]

Shapiro, J. A., (ed.), 1983. *Mobile Genetic Elements.* Academic Press.

Stahl, F. W., 1979. *Genetic Recombination.* W. H. Freeman.

Suzuki, D. T., Griffiths, A. J. F., Miller, J. H., and Lewontin, R. C., 1986. *An Introduction to Genetic Analysis.* W. H. Freeman. [Chapters 16 and 17 give a concise account of recombination and transposition from a genetic viewpoint.]

Stanier, R. Y., Ingraham, J. L., Wheelis, M. L., and Painter, P. R., 1986. *The Microbial World* (5th ed.). Prentice-Hall. [Chapter 11 deals with genetic exchange and recombination in microorganisms.]

General recombination

Kowalczykowski, S. C., 1987. Mechanistic aspects of the DNA strand exchange activity of *E. coli* recA protein. *Trends Biochem. Sci.* 12:141–145.

Howard-Flanders, P., West, S. C., and Stasiak, A., 1984. Role of recA protein spiral filaments in genetic recombination. *Nature* 309:215–220.

Bryant, F. R., Taylor, A. R., and Lehman, I. R., 1985. Interaction of the recA protein of *E. coli* with single-stranded DNA. *J. Biol. Chem.* 260:1196–1202.

Honigberg, S. M., Rao, B. J., and Radding, C. M., 1986. Ability of recA protein to promote a search for rare sequences in duplex DNA. *Proc. Nat. Acad. Sci.* 83:9586–9590.

Potter, H., and Dressler, D., 1978. In vitro system from *Escherichia coli* that catalyzes generalized genetic recombination. *Proc. Nat. Acad. Sci.* 75:3698–3702.

McEntee, K., Weinstock, G. M., and Lehman, I. R., 1979. Initiation of general recombination catalyzed in vitro by the recA protein of *E. coli. Proc. Nat. Acad. Sci.* 76:2615–2619.

Cunningham, R. P., Shibata, T., DasGupta, C., and Radding, C. M., 1979. Single strands induce recA protein to unwind duplex DNA for homologous pairing. *Nature* 281:191–195.

Cox, M. M., and Lehman, I. R., 1981. recA protein of *E. coli* promotes branch migration, a kinetically distinct phase of DNA strand exchange. *Proc. Nat. Acad. Sci.* 78:3433–3437.

Egelman, E. H., and Stasiak, A., 1986. Structure of helical recA-DNA complexes. Complexes formed in the presence of ATP-gamma-S or ATP. *J. Mol. Biol.* 191:677–697.

DNA repair

Howard-Flanders, P., 1981. Inducible repair of DNA. *Sci. Amer.* 245(5):72–80.

Walker, G. C., 1985. Inducible DNA repair systems. *Ann. Rev. Biochem.* 54:425–457.

Friedberg, E. C., 1985. *DNA Repair*. W. H. Freeman. [SOS repair is discussed in Chapter 7.]

Transposition

Grindley, N. D. F., and Reed, R. R., 1985. Transpositional recombination in prokaryotes. *Ann. Rev. Biochem.* 54:863–896.

Derbyshire, K. M., and Grindley, N. D. F., 1986. Replicative and conservative transposition in bacteria. *Cell* 47:325–327.

Wasserman, S. A., and Cozzarelli, N. R., 1986. Biochemical topology: applications to DNA recombination and replication. *Science* 232:951–960. [A fascinating account of topological factors in site-specific recombination. DNA catenanes (linked rings) and knots are analyzed to reveal mechanisms of topoisomerases such as resolvase.]

Cohen, S. N., 1976. Transposable genetic elements and plasmid evolution. *Nature* 263:731–738.

Gill, R. E., Heffron, F., and Falkow, S., 1979. Identification of the protein encoded by the transposable element Tn3 which is required for its transposition. *Nature* 282:797–801.

Chou, J., Lemaux, P. G., Casadaban, M. J., and Cohen, S. N., 1979. Transposition protein of Tn3: identification and characterisation of an essential repressor-controlled gene product. *Nature* 282:801–806.

Boocock, M. R., Brown, J. L., and Sherratt, D. J., 1986. Structural and catalytic properties of specific complexes between Tn3 resolvase and the recombination site *res. Biochem. Soc. Trans.* 14:214–216.

Infectious drug resistance

Clewell, D. B., and Gawron-Burke, C., 1986. Conjugative transposons and the dissemination of antibiotic resistance in streptococci. *Ann. Rev. Microbiol.* 40:635–659.

Clowes, R. C., 1973. The molecule of infectious drug resistance. *Sci. Amer.* 228(4):18–27.

PROBLEMS

1. *E. coli* contains several ATP-driven proteins that unwind duplex DNA. Name them and cite their functions.

2. One of the strands of a 1040-bp duplex DNA is nicked. RecA protein is then added in an amount such that an average of 100 protein molecules are bound per DNA duplex. DNA ligase and NAD^+ are added to close the nicked circle. RecA protein is then removed.
   (a) Compare the linking number of the resulting DNA molecule with that of the corresponding relaxed duplex.
   (b) Compare the hydrodynamic properties of this DNA molecule with that of its relaxed counterpart.

3. The lexA protein represses its own synthesis. How does this control feature enable the SOS response to be reversed following repair of DNA?

4. ATPγS is a hydrolysis-resistant analog of ATP in which a nonesterified oxygen atom of the terminal phosphoryl group is replaced by a sulfur atom. Predict the consequences of using ATPγS instead of ATP in the reactions catalyzed by (a) recA protein, and (b) the recBCD complex.

CHAPTER 29

RNA Synthesis and Splicing

The flow of genetic information from DNA to RNA to protein was introduced in Chapter 5. An overview was given of the different kinds of RNA, the role of messenger RNA, the action of RNA polymerase, and the split nature of many eucaryotic genes. This chapter deals with how RNA is synthesized and spliced. We begin with transcription in *E. coli*, where it is best understood, and focus on three questions: What are the properties of promoters and how do they cause specific initiation? How do RNA polymerase, the DNA template, and the nascent RNA chain interact with each other? How is transcription terminated?

We shall then turn to transcription in eucaryotes, beginning with promoter sequences and the transcription-factor proteins that bind them. A distinctive feature of eucaryotic DNA templates is the presence of enhancer sequences that can stimulate initiation more than a thousand base pairs away. Primary transcripts in eucaryotes are extensively modified, as exemplified by the capping of the 5′ end of an mRNA precursor and the addition of a long poly A tail to its 3′ end. Most striking is the splicing of mRNA precursors, which is catalyzed by spliceosomes consisting of ribonucleoproteins (snRNPs). Other RNA molecules splice themselves, showing that RNA can have catalytic activity. This exciting finding enriches our understanding of molecular evolution and suggests that RNA components of molecular assemblies such as ribosomes may be catalytically active.

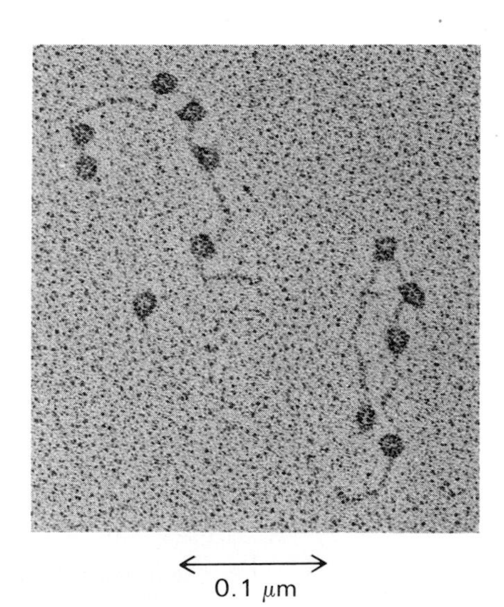

Figure 29-1
Electron micrograph of RNA polymerase holoenzyme molecules bound to several promoter sites on a fragment of T7 bacteriophage DNA. [From R. C. Williams. *Proc. Nat. Acad. Sci.* 74(1977):2313.]

RNA POLYMERASE FROM *E. coli* IS A MULTISUBUNIT ENZYME

RNA polymerase from *E. coli* is a very large (450 kd) and complex enzyme consisting of four kinds of subunits (Table 29-1). The subunit composition of the entire enzyme, called the *holoenzyme,* is $\alpha_2\beta\beta'\sigma$. As will be discussed shortly, the σ subunit finds a promoter site where transcription begins; it initiates RNA synthesis and then dissociates from the rest of the enzyme. RNA polymerase without this subunit is called the *core enzyme* ($\alpha_2\beta\beta'$); it contains the catalytic site. The β' subunit binds the DNA template, and the β subunit binds ribonucleoside triphosphate substrates.

Table 29-1
Subunits of RNA polymerase from *E. coli*

Subunit	*Number*	*Mass (kd)*	*Role*
α	2	37	Uncertain
β	1	151	Forms phosphodiester bonds
β'	1	155	Binds DNA template
σ^{70}	1	50	Recognizes promoter and initiates synthesis

The synthesis of RNA by *E. coli* RNA polymerase, like nearly all biological polymerization reactions, takes place in three stages: *initiation, elongation,* and *termination.* RNA polymerase performs multiple functions in this process:

1. It searches DNA for initiation sites. *E. coli* DNA has about 2000 promoter sites in its 4×10^6 base pairs.

2. It unwinds a short stretch of double-helical DNA to produce a single-stranded DNA template from which it takes instructions.

3. It selects the correct ribonucleoside triphosphate and catalyzes the formation of a phosphodiester bond. This process is repeated many times as the enzyme moves unidirectionally along the DNA template. RNA polymerase is totally processive—a transcript is synthesized from start to end by a single RNA polymerase molecule.

4. It detects termination signals that specify where a transcript ends.

5. It interacts with activator and repressor proteins that modulate the rate of transcription over a wide dynamic range. Gene expression is controlled mainly at the level of transcription, as will be discussed in detail in Chapters 32 and 33.

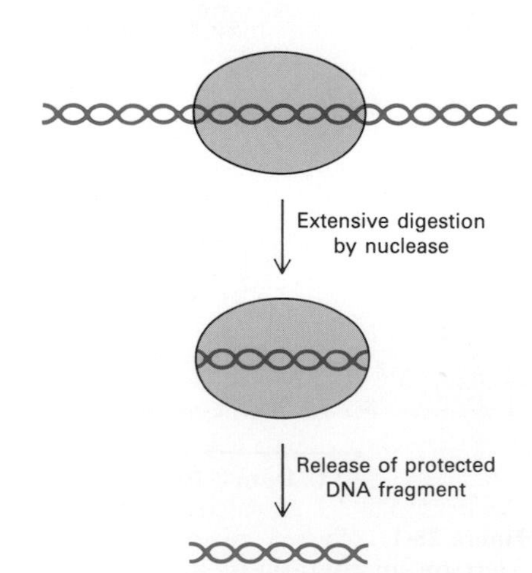

Figure 29-2
Binding sites on DNA for RNA polymerase and other proteins can be isolated by digesting DNA with nucleases. Fragments that are bound to protein are protected from digestion.

TRANSCRIPTION IS INITIATED AT PROMOTER SITES ON THE DNA TEMPLATE

Transcription starts at *promoters* on the DNA template. These initiation sites have been identified in several ways. One approach is to add RNA polymerase, and then a deoxyribonuclease, to DNA. In the absence of ribonucleoside triphosphates, RNA polymerase remains bound to promoters and protects them from digestion. Fragments of DNA about 60 bp long are produced after extensive digestion of the unprotected DNA (Figure 29-2). The position and sequence of the protected region can readily be identified by a complementary technique called *footprinting*. One end of a DNA chain of a specific DNA fragment (produced by the action of a restriction enzyme, for example) is first labeled with ^{32}P. RNA polymerase is added to the labeled DNA, and the complex is digested with DNase I just long enough to make an average of one cut in each chain. An aliquot without RNA polymerase is treated in the same way to serve as a control. The resulting DNA fragments are separated according to size by electrophoresis. The gel pattern (Figure 29-3) is highly revealing: a series of bands present in the control sample is absent from the aliquot containing RNA polymerase. These bands are missing because RNA polymerase shields DNA from cleavages that would give rise to the corresponding fragments.

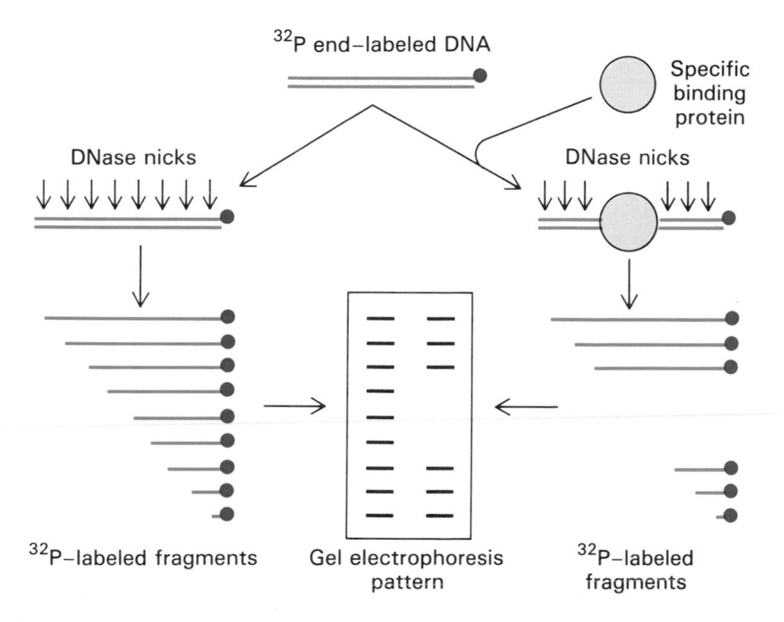

Figure 29-3
Footprinting technique. One end of a DNA chain is labeled with ^{32}P. This labeled DNA is then cut at a limited number of sites by DNase I. The same experiment is carried out in the presence of a protein that binds to specific sites on DNA. The bound protein protects a segment of DNA from the action of DNase I. Hence, certain fragments will be absent. The missing bands in the gel pattern identify the binding site on DNA.

The first nucleotide (the start site) of a DNA sequence is denoted as +1, the second one as +2; the nucleotide preceding the start site is denoted as −1. These designations refer to the coding (sense) strand, not the template (antisense) strand, of DNA; the coding strand has the same sequence as the RNA transcript except for T in place of U. A striking pattern emerges from the sequences of many procaryotic promoters. *Two common motifs are present on the 5′ side (upstream) of the start site.* They are known as the *−10 sequence* (Pribnow box) and the *−35 sequence* because they are centered at about 10 and 35 nucleotides upstream of the start site. These sequences are each 6 bp long. Their consensus sequences, deduced from analyses of more than a hundred promoters, are

−35 −10 +1
5′〰TTGACA〰〰〰〰〰TATAAT〰〰 Start site

mRNA starts here ↓

5′ −10 3′
(A) CG TATGTT GTGTGG A
(B) GC TATGGT TATTTC A
(C) G TTAACTA GTACGC A
(D) GT GATACT GAGCAC A
(E) GT TTTCAT GCCTCC A
TATAAA

Figure 29-4
Procaryotic promoter sequences showing homology in the −10 region. The sequences are from the (A) *lac*, (B) *gal*, and (C) *trp* operons of *E. coli*, (D) from λ phage, and (E) from ϕX174 phage. Homologies are shown in green and the −10 consensus sequence in red.

Promoters differ markedly in their efficacy. Strong promoters cause frequent initiation of transcription, as often as every two seconds in *E. coli*. In contrast, genes with very weak promoters are transcribed about once in ten minutes. The −10 and −35 regions of most strong promoters have sequences that correspond closely to the consensus ones, whereas weak promoters tend to have multiple substitutions at these sites. Indeed, mutation of a single base in either the −10 sequence or the −35 sequence can lead to a loss of promoter activity. The distance between these conserved sequences is also important; a separation of 17 nucleotides is optimal. The frequency of transcription of many genes is also markedly influenced by regulatory proteins that bind to specific sequences near promoter sites and interact with RNA polymerase (Chapters 32 and 33).

SIGMA SUBUNITS ENABLE RNA POLYMERASE TO RECOGNIZE PROMOTER SITES

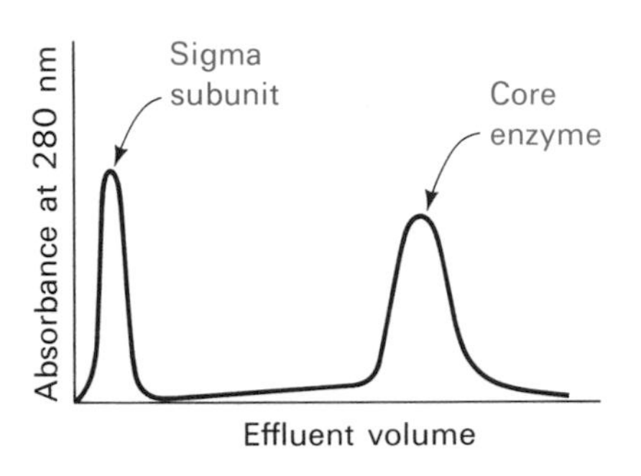

Figure 29-5
Resolution of RNA polymerase into the σ subunit and the core enzyme ($\alpha_2\beta\beta'$) on a phosphocellulose column.

The $\alpha_2\beta\beta'$ core of RNA polymerase is unable to start transcription at promoter sites. Rather, *the complete $\alpha_2\beta\beta'\sigma$ holoenzyme is essential for specific initiation.* The role of the sigma subunit was discovered in the following way. RNA polymerase purified by chromatography on a phosphocellulose column was found to be nearly inactive when assayed with T4 DNA as template but still quite active with calf-thymus DNA as template. In contrast, RNA polymerase purified by centrifugation on a glycerol gradient was highly active with both templates. This observation suggested that a factor might be missing from the phosphocellulose-purified RNA polymerase. This was indeed the case (Figure 29-5). The activity of the phosphocellulose-purified enzyme could be markedly enhanced by the addition of another fraction of the effluent of this column. This stimulatory fraction, which was the sigma (σ) factor, had no catalytic activity by itself. Subsequent experiments showed that the phosphocellulose-purified enzyme lacked σ subunit, whereas the enzyme purified on a glycerol gradient retained this subunit. Thus, the enzyme that was fully active with the T4 DNA template was the $\alpha_2\beta\beta'\sigma$ holoenzyme, whereas the $\alpha_2\beta\beta'$ core enzyme was ineffective in transcribing this DNA.

The addition of σ to the core enzyme led to the reconstitution of the fully active holoenzyme. The core enzyme was able to bind and transcribe calf-thymus DNA because this template contains many nicks and hence unpaired regions, but RNA synthesized in vitro by the core polymerase does not correspond to in vivo transcripts. For example, the core enzyme in vitro transcribes both strands of phage DNA templates, whereas the holoenzyme asymmetrically transcribes one strand, as it does in vivo.

The σ subunit contributes to specific initiation by *decreasing the affinity of RNA polymerase for general regions of DNA by a factor of 10^4. Sigma also enables RNA polymerase to recognize promoters.* The holoenzyme binds to duplex DNA and searches for a promoter site by forming transient hydrogen bonds with exposed hydrogen donor and acceptor groups of base pairs. The search is rapid for two reasons. First, RNA polymerase slides along the DNA template instead of repeatedly binding and dissociating from it. *The promoter site is encountered by a random walk in one dimension rather than in three dimensions.* The observed rate constant for the binding of RNA polymerase holoenzyme to promoter sequences is 10^{10} M^{-1} s^{-1}, more than 100 times as fast as could be accomplished by repeated encounters on and off the DNA. *Second, RNA polymerase can detect the −10 and −35 sequences of DNA without unwinding the double helix.* Recall that EcoRI endonuclease recognizes its palindromic target site without breaking base pairs (p. 657). A search requiring extensive DNA unwinding would be much slower than one exploring an essentially intact duplex.

PROMOTERS FOR HEAT-SHOCK GENES ARE RECOGNIZED BY A SPECIAL KIND OF SIGMA SUBUNIT

Recent studies have revealed that *E. coli* contains more than one kind of σ. The type that recognizes the consensus sequences described earlier is called σ^{70} because it has a mass of 70 kd. When the temperature is raised abruptly, *E. coli* responds by synthesizing a series of *heat-shock proteins* that enables it to adapt to this challenge. The promoters of heat-shock genes exhibit −10 sequences that differ markedly from the one for standard promoters (Figure 29-6). The −35 region of heat-

Heat-shock promoter 5′∼TNNCNCNCTTGAA∼CCCATNT∼3′
Standard promoter 5′∼TTGACA∼TATAAT∼3′

Figure 29-6
Comparison of the consensus sequences of heat-shock promoters (red) and standard promoters (blue) of *E. coli*.

shock promoters resembles that of general promoters but it also has some distinctive features. Holoenzyme containing a different sigma subunit, called σ^{32}, recognizes heat-shock promoters. The level of σ^{32} increases markedly when the temperature is raised rapidly, which leads to the coordinated synthesis of a series of protective proteins. A change of sigma subunit is also employed by *Bacillus subtilis* as a developmental device. Different σ factors control sporulation in this bacterium by determining which battery of genes is expressed. Likewise, SPO1, a phage that infects *B. subtilis*, encodes several new σ factors that lead to sequential expression of viral genes. These findings demonstrate that σ plays a key role in determining where RNA polymerase initiates transcription.

RNA POLYMERASE UNWINDS NEARLY TWO TURNS OF TEMPLATE DNA BEFORE INITIATING RNA SYNTHESIS

In searching for promoter sites, RNA polymerase is bound to double-helical DNA. However, duplex DNA cannot serve directly as the template for RNA synthesis because its bases are paired. A region of duplex DNA must be unpaired so that nucleotides on one of its strands become accessible for base pairing with incoming ribonucleoside triphosphates. The DNA template strand selects the correct ribonucleoside triphosphate by forming a Watson-Crick base pair with it (p. 76), as in DNA synthesis. What is the extent of unwinding of DNA by RNA polymerase? This question was answered by analyzing the effect of the enzyme on the supercoiling of a circular duplex DNA. Different amounts of RNA polymerase were added to the DNA. Topoisomerase I, an enzyme catalyzing the concerted nicking and resealing of duplex DNA (p. 662),

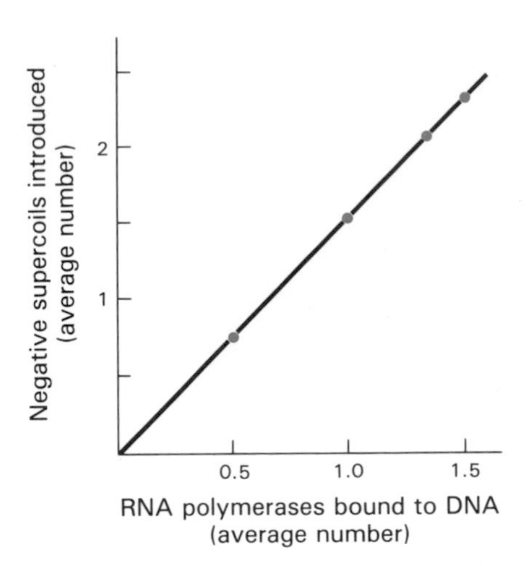

Figure 29-7
The unwinding of template DNA by RNA polymerase was measured by analyzing the change in linkage number produced by topoisomerase I in the presence of bound polymerase. [After H. B. Gamper and J. E. Hearst. *Cell* 29(1982):83.]

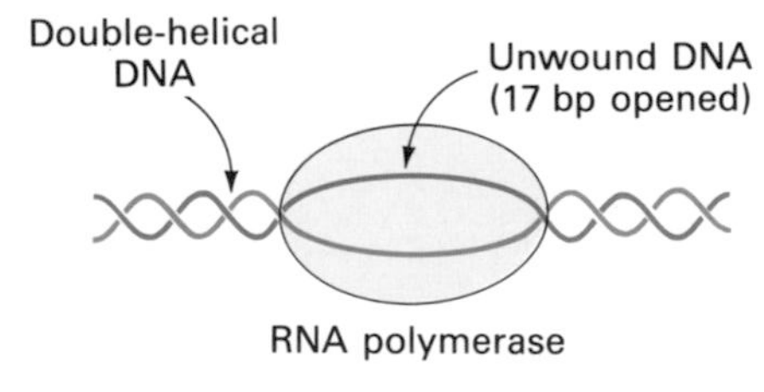

Figure 29-8
RNA polymerase unwinds about 17 base pairs of template DNA.

was then added to relax the portion of circular DNA not in contact with polymerase molecules. These DNA samples were analyzed by gel electrophoresis following removal of bound protein (Figure 29-7). *The degree of negative supercoiling increased in proportion to the number of RNA polymerases bound per template DNA, showing that the enzyme unwinds DNA. Each bound polymerase unwinds a 17-bp segment of DNA, which corresponds to 1.6 turns of B-DNA helix.*

Negative supercoiling of circular DNA templates favors the transcription of many genes because it facilitates unwinding (p. 662). Thus, the pumping of negative supercoils into DNA by DNA gyrase can increase the efficiency of promoters located at distant sites. However, not all promoters are stimulated by negative supercoiling. The promoter for DNA gyrase is a noteworthy exception. The rate of transcription of this gene is decreased by negative supercoiling, a nice feedback control assuring that DNA does not become excessively supercoiled. Negative supercoiling could decrease the efficiency of this promoter by changing the angle between the -10 and -35 regions.

The transition from the *closed promoter complex* (in which DNA is double-helical) to the *open promoter complex* (in which a DNA segment is unwound) is a key event in transcription. The stage is now set for the formation of the first phosphodiester bond of the new RNA chain.

RNA CHAINS START WITH pppG OR pppA

Most newly synthesized RNA chains carry a tag that reveals how RNA chains are started. The 5′ end of a new RNA chain is highly distinctive: *the molecule starts with either pppG or pppA.* In contrast with DNA synthesis, a primer is not needed. *RNA chains can be formed de novo.*

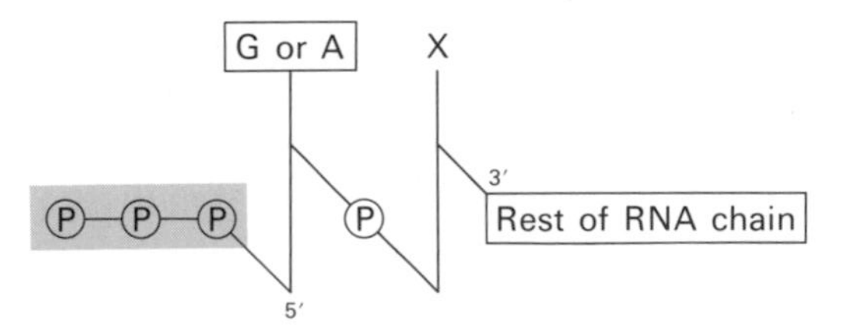

This telltale tag at the 5′ terminus was discovered in two ways. It was found that RNA chains incorporated ^{32}P when the incubation mixture contained ^{32}P γ-labeled ATP. Clearly, the incorporated label had to be in a terminal position because only the α phosphorus atom of a nucleoside triphosphate can become part of internal phosphodiester bridges in the RNA. Furthermore, alkaline hydrolysis of newly synthesized RNA yielded three kinds of products: nucleosides, nucleoside 2′- (or 3′-) monophosphates, and nucleoside tetraphosphates (Figure 29-9).

^{32}P γ-labeled ATP

Radioactive nucleoside tetraphosphate

Nucleoside 2′-(3′-) monophosphates

Nucleoside

Figure 29-9
Alkaline hydrolysis products from an RNA chain synthesized with ^{32}P γ-labeled ATP or GTP.

When ^{32}P γ-labeled ATP was used as a substrate, adenosine 3′-phosphate-5′-triphosphate was formed. The γ-phosphorus atom of this nucleoside tetraphosphate was labeled. The corresponding result was obtained when ^{32}P γ-labeled GTP was used. However, radioactivity was not incorporated when γ-labeled UTP or CTP were used. *Thus, a new RNA chain has a triphosphate group at its 5′ terminus. Other experiments showed that the 3′ terminus has a free hydroxyl group.*

RNA CHAINS ARE SYNTHESIZED IN THE 5′ → 3′ DIRECTION

Is RNA synthesized in the 5′ → 3′ direction or in the 3′ → 5′ direction? Two contrasting mechanisms of chain growth are depicted in Figure 29-10. In 5′ → 3′ growth, the triphosphate terminus is inserted at the start of the synthesis of a chain, whereas in 3′ → 5′ growth, the triphosphate terminus comes from the last residue that is incorporated. The kinetics of incorporation of radioactivity into RNA synthesized from ^{32}P γ-labeled GTP (or ATP) distinguished between these alternatives. When ^{32}P γ-labeled GTP was used as a substrate, the ratio of ^{32}P incorporation to total nucleotide incorporation was highest shortly after the components were mixed and then decreased progressively with time. The revealing finding was that the total radioactivity of the RNA already labeled was not reduced by the subsequent addition of a large excess of nonradioactive GTP to the incubation mixture. Thus, ^{32}P entered the RNA molecule at the start of its synthesis rather than at the last step. *Hence, the growth of an RNA chain is in the 5′ → 3′ direction, as in DNA synthesis.*

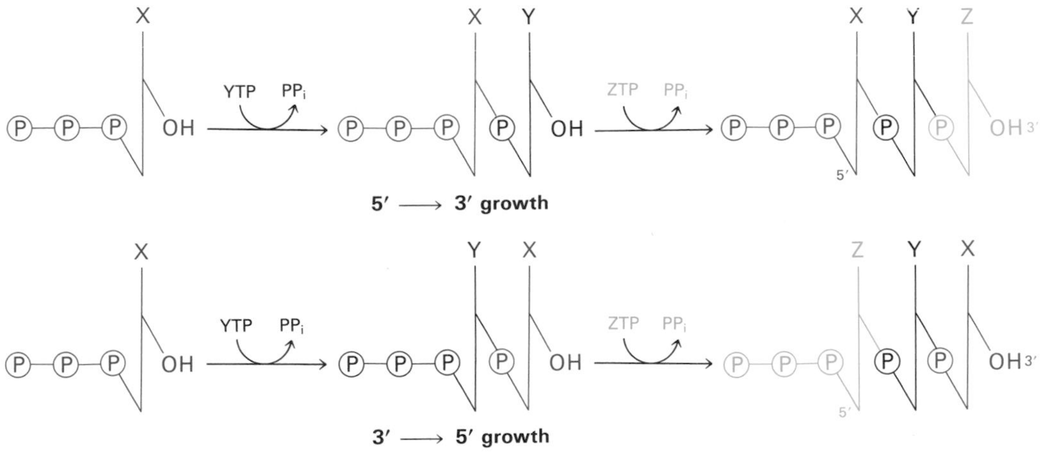

Figure 29-10
Location of ^{32}P expected for 5′ → 3′ growth and for 3′ → 5′ growth. The observed location of label shows that RNA chains are synthesized in the 5′ → 3′ direction.

ELONGATION TAKES PLACE AT TRANSCRIPTION BUBBLES THAT MOVE ALONG THE DNA TEMPLATE

The elongation phase of RNA synthesis begins after formation of the first phosphodiester bond. An important change is the loss of σ—the core enzyme without σ binds more strongly to the DNA template. Indeed, RNA polymerase stays bound to its template until a termination signal is reached. The region containing RNA polymerase, DNA, and nascent RNA is called a *transcription bubble* because it contains a locally melted "bubble" of DNA (Figure 29-11). The newly synthesized RNA forms a hybrid helix with the template DNA strand. This RNA-DNA helix is about 12 bp long, which corresponds to nearly one turn of an A-DNA type helix. The 3′-hydroxyl of the RNA in this hybrid helix is positioned so that it can attack the α phosphorus atom of an incoming ribonucleoside triphosphate. The core enzyme also contains a binding site for the other DNA strand. About 17 bp of DNA are unwound throughout the elongation phase, as in the initiation phase. The rate of elongation is about 50 nucleotides per second; the transcription bubble moves a distance of 170 Å in this time.

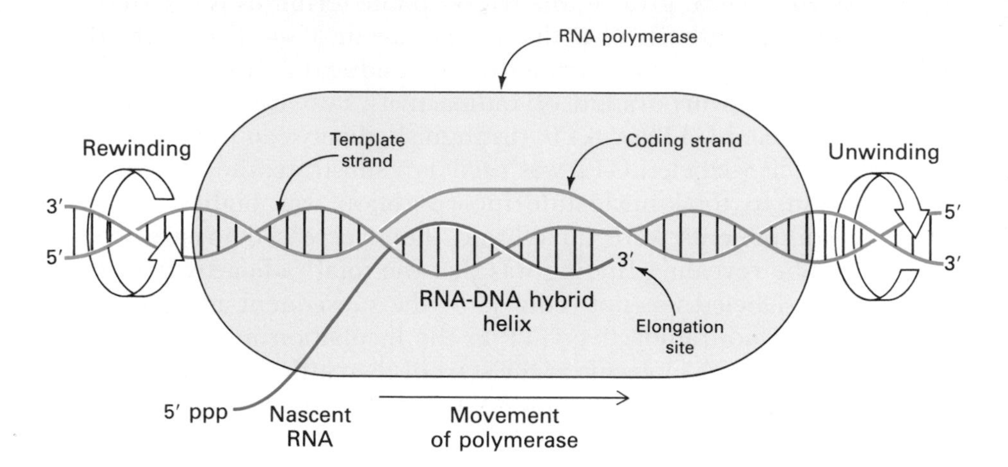

Figure 29-11
Model of a transcription bubble during elongation of the RNA transcript. The template DNA strand is shown in green, the coding DNA strand in blue, and the nascent RNA chain in red. Duplex DNA is unwound at the forward end of RNA polymerase and rewound at its rear end. The RNA-DNA hybrid helix rotates in synchrony.

The length of the RNA-DNA hybrid and of the unwound region of DNA stay constant as RNA polymerase moves along the DNA template. This finding indicates that DNA is rewound at the same rate at the rear of RNA polymerase as it is unwound at the front of the enzyme (Figure 29-11). The RNA-DNA hybrid must also rotate each time a nucleotide is added so that the 3′-OH end of the RNA stays at the catalytic site. The 12 bp length of the hybrid is just short of a full turn of helix to avoid a knotty problem. The RNA strand is bent sharply away from the template DNA strand just before a full turn is formed to prevent the 5′ end of the RNA from becoming intertwined with the DNA.

It is important to note that RNA polymerase lacks nuclease activity. *In contrast with DNA polymerase, it does not edit the nascent polynucleotide chain. Consequently, the fidelity of transcription is much lower than that of replication.* The error rate of RNA synthesis is of the order of one mistake per 10^4 or 10^5, which is about 10^5 times as high as that of DNA synthesis. The much lower fidelity of RNA synthesis can be tolerated because mistakes are not transmitted to progeny. For most genes, many RNA transcripts are synthesized each generation; a few defective transcripts are unlikely to be harmful.

AN RNA HAIRPIN FOLLOWED BY SEVERAL U RESIDUES CAUSES THE TERMINATION OF TRANSCRIPTION

In the termination phase of transcription, the formation of phosphodiester bonds ceases, the RNA-DNA hybrid dissociates, the melted region of DNA rewinds, and RNA polymerase releases the DNA. The termination of transcription is as precisely controlled as its initiation. The transcribed regions of DNA templates contain stop signals. The simplest one is a *palindromic GC-rich region followed by an AT-rich region* (Figure 29-12). *The RNA transcript of this DNA palindrome is self-complementary.* Hence, its bases can pair to form a hairpin structure with a stem and loop, a structure favored by its high content of GC residues (Figure 29-13). Recall that GC base pairs are more stable than AT pairs (p. 82). This stable hairpin is followed by a sequence of four or more U residues. The RNA transcript ends within or just after them.

How does this hairpin-oligo–U structure terminate transcription? First, it seems likely that RNA polymerase pauses when it encounters such a hairpin. Furthermore, the RNA-DNA hybrid helix produced after the hairpin is unstable because of its content of rU-dA base pairs, which are the weakest of the four kinds. Hence, *nascent RNA dissociates from the DNA template and then from the enzyme.* The solitary DNA template strand now rejoins its partner to reform the DNA duplex in the bubble region. The core enzyme (devoid of σ) has much less affinity for duplex DNA than for single-stranded DNA, and so the DNA is released. Sigma rejoins the core enzyme to form holoenzyme that can again search for a promoter site to initiate a new transcript.

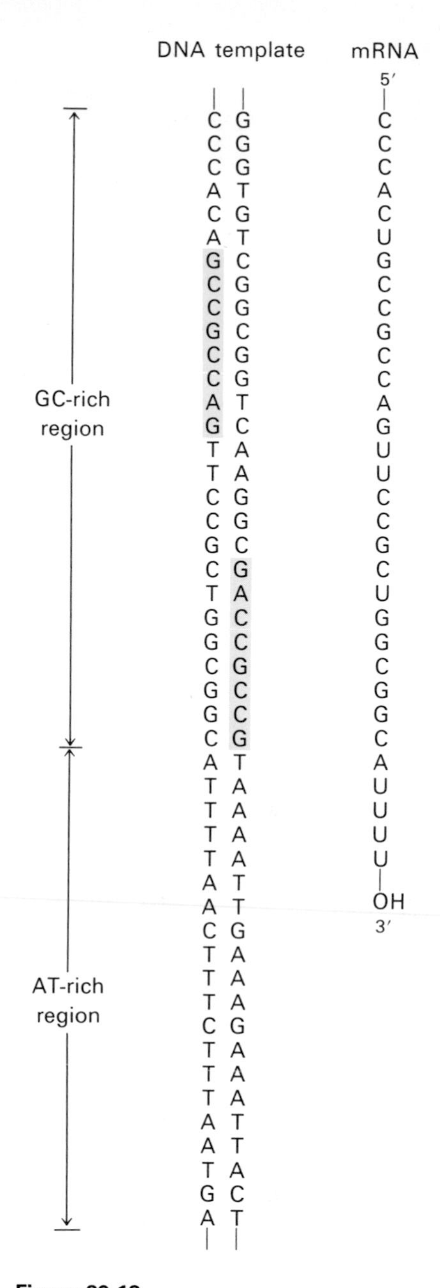

Figure 29-12
DNA sequence corresponding to the 3′ end of *trp* mRNA from *E. coli.* A twofold axis of symmetry (marked by the green symbol) relates the base pairs that are shaded yellow.

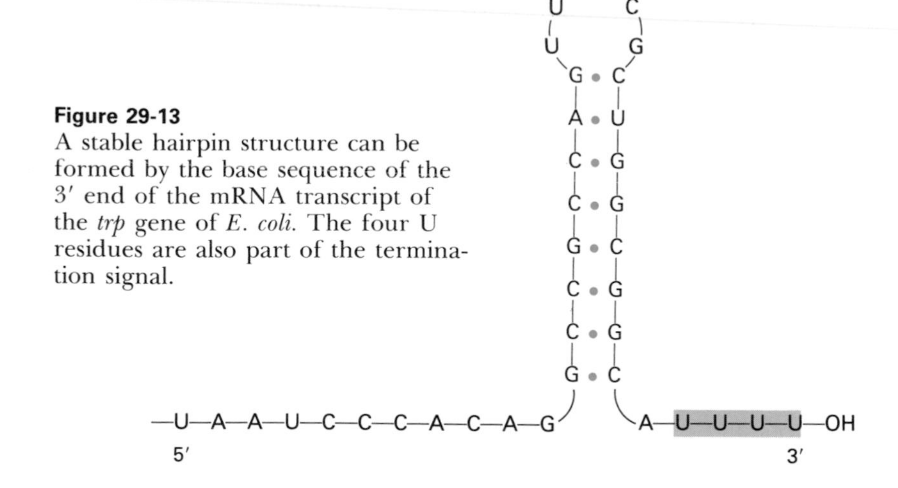

Figure 29-13
A stable hairpin structure can be formed by the base sequence of the 3′ end of the mRNA transcript of the *trp* gene of *E. coli.* The four U residues are also part of the termination signal.

RHO PROTEIN HELPS TERMINATE TRANSCRIPTION OF SOME GENES

RNA polymerase needs no help to terminate transcription at a hairpin followed by several U residues. At other sites, however, termination requires the participation of an additional factor. Evidence for it came from the finding that some RNA molecules synthesized in vitro by RNA polymerase alone are *longer* than those made in vivo. The missing factor, a protein, was isolated and named rho (ρ). For example, RNA synthesized from M13 filamentous phage DNA in the presence of ρ has a sedimentation coefficient of 10S, whereas RNA made in its absence is 23S. Additional information about the action of ρ was obtained by adding this termination factor to an incubation mixture at various times

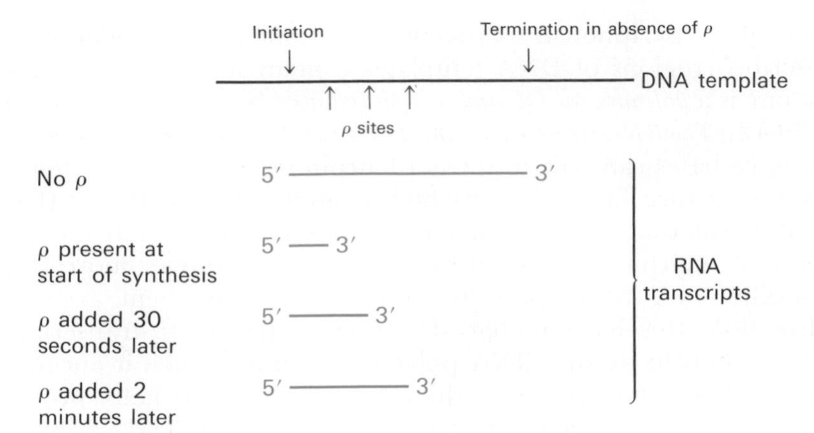

Figure 29-14
Effect of ρ protein on the size of RNA transcripts.

after the initiation of RNA synthesis (Figure 29-14). Species with sedimentation coefficients of 13S, 17S, and 23S were obtained when ρ was added a few seconds, two minutes, and ten minutes, respectively, after initiation. This result indicated that the template contains at least three termination sites that respond to ρ (yielding 10S, 13S, and 17S RNA) and one termination site that does not require rho (yielding 23S RNA). Thus, specific termination can occur in the absence of ρ. However, ρ detects additional termination signals that are not recognized by RNA polymerase alone.

How does ρ provoke termination of RNA synthesis? *A key clue is the finding that ρ hydrolyzes ATP in the presence of single-stranded RNA but not of DNA or duplex RNA.* Rho is a hexamer of 46-kd subunits that specifically bind single-stranded RNA; a stretch of 72 nucleotides is bound, 12 per subunit. Rho is brought into action by sequences located in the nascent RNA—the absence of a simple consensus sequence implies that ρ detects noncontiguous structural features. The ATPase activity of ρ enables it to move unidirectionally along nascent RNA toward the transcription bubble. It then breaks the RNA-DNA hybrid helix by pulling RNA away (Figure 29-15). *A common feature of ρ-independent and ρ-dependent termination is that the active signals lie in newly synthesized RNA rather than in the DNA template.*

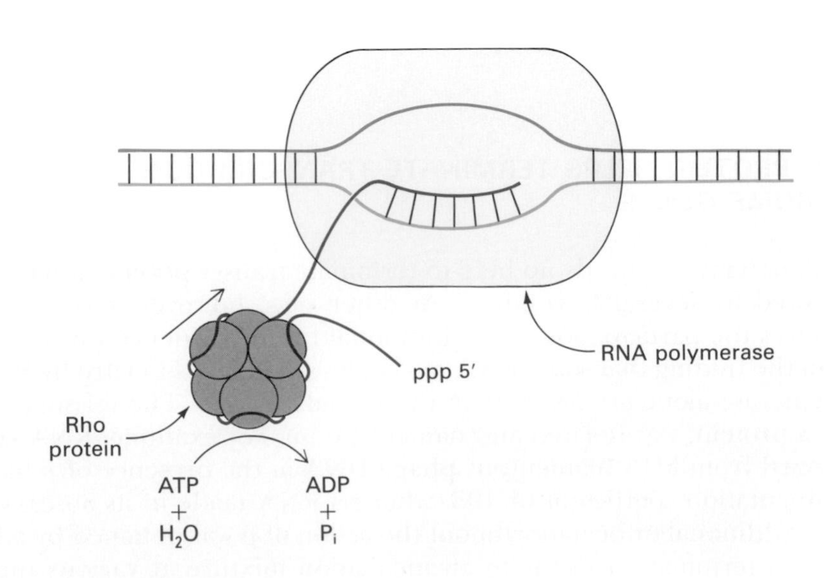

Figure 29-15
Model for the termination of transcription by ρ. This ATPase binds the nascent RNA chain and pulls it away from RNA polymerase and the DNA template.

Proteins in addition to ρ mediate and modulate termination. For example, the *nusA protein* enables RNA polymerase in *E. coli* to recognize a characteristic class of termination sites. As will be described later, bacteriophage λ synthesizes *antitermination proteins* that allow certain genes to be transcribed and expressed (p. 808). In *E. coli,* specialized termination signals called *attenuators* are regulated to meet the nutritional needs of the cell (p. 815).

PRECURSORS OF TRANSFER AND RIBOSOMAL RNA ARE CLEAVED AND CHEMICALLY MODIFIED AFTER TRANSCRIPTION

In procaryotes, mRNA molecules undergo little or no modification following synthesis by RNA polymerase. Indeed, many of them are translated while they are being transcribed. In contrast, *transfer RNA and ribosomal RNA molecules are generated by cleavage and modification of nascent RNA chains.* For example, in *E. coli,* three kinds of ribosomal RNA molecules and a transfer RNA molecule are excised from a single primary RNA transcript that also contains spacer regions (Figure 29-16). Other transcripts contain arrays of several kinds of tRNAs or of several copies of the same tRNA. The nucleases that cleave and trim these precursors of rRNA and tRNA are highly precise. Ribonuclease P, for example, generates the correct 5′-terminus of all tRNA molecules in *E. coli.* This interesting enzyme contains a catalytically active RNA molecule. Ribonuclease III excises 5S, 16S, and 23S rRNA precursors from the primary transcript by cleaving double-helical hairpin regions at specific sites.

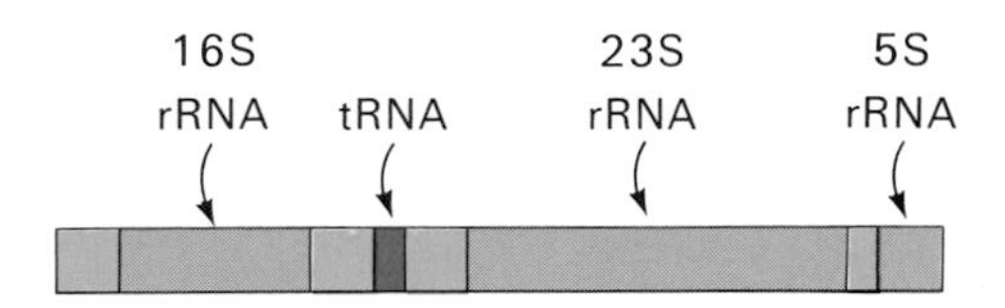

Figure 29-16
Cleavage of this primary transcript produces 5S, 16S, and 23S rRNA molecules and a tRNA molecule. Spacer regions are shown in yellow.

A second type of processing is the *addition of nucleotides to the termini of some RNA chains.* For example, CCA is added to the 3′ end of tRNA molecules that do not already possess this terminal sequence. *Modifications of bases and ribose units* of ribosomal RNAs are a third category. In procaryotes, some bases are methylated, whereas in eucaryotes the 2′-hydroxyl group of about one in a hundred ribose units is methylated. Unusual bases are found in all tRNA molecules (p. 740). They are formed by enzymatic modification of a standard ribonucleotide in a tRNA precursor. For example, *pseudouridylate* and *ribothymidylate* are formed by modification of uridylate residues after transcription. These modifications generate diversity, as if tRNAs were trying to emulate proteins.

2′-*O*-Methylribose

6-Dimethyladenine

Ribothymidine

Pseudouridine

In eucaryotes, tRNA precursors are converted into mature tRNAs by a series of alterations: cleavage of a 5′-leader sequence, splicing to remove an intron, replacement of the 3′-terminal UU by CCA, and modification of several bases. The processing of yeast tRNA precursors has

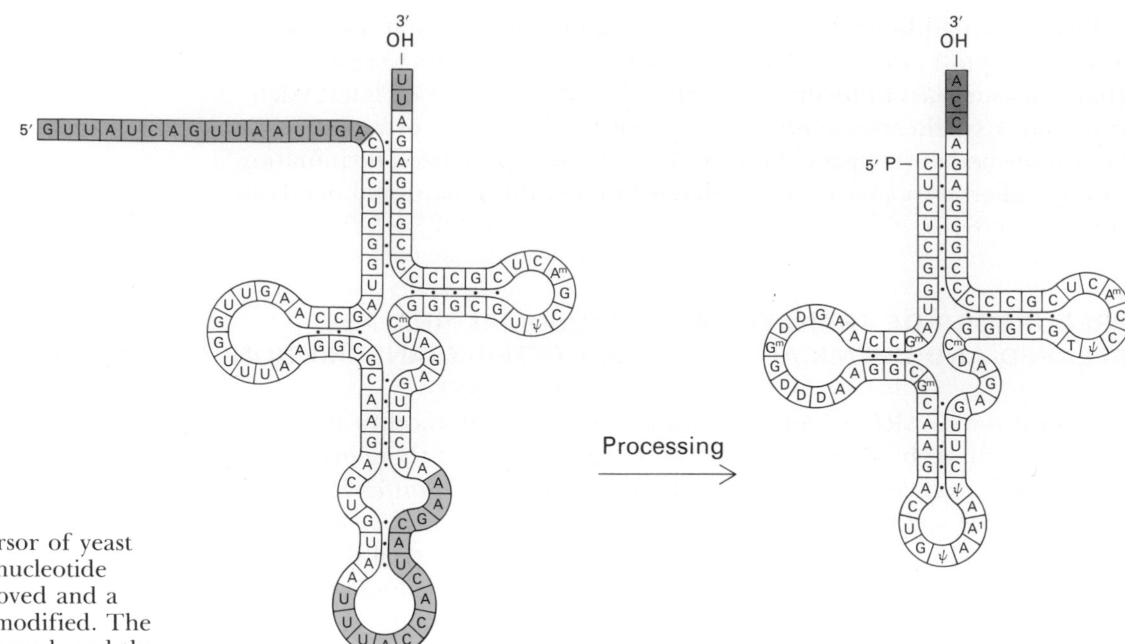

Figure 29-17
Processing of a precursor of yeast tyrosine tRNA. A 14-nucleotide intron (yellow) is removed and a number of bases are modified. The 5′ leader (green) is cleaved, and the 3′ end is changed to CCA (red).

been investigated in detail (Figure 29-17). The 14-nucleotide intron next to the anticodon is excised by an endonuclease that generates a 2′,3′-cyclic phosphate at the end of the upstream exon and a 5′-OH at the end of the downstream exon. These two halves of the tRNA molecule stay together during the subsequent steps leading to their ligation. The splicing of tRNAs is very similar in species that are evolutionarily far apart. Cloned genes for a yeast tRNA precursor were microinjected into *Xenopus oocytes* to determine whether a gene from a unicellular eucaryote can be transcribed and processed in an amphibian. The striking result is that the yeast gene is transcribed and properly processed in *Xenopus*, even though the genes for this tRNA are quite different in the two species. In particular, a 14-nucleotide intron was properly spliced out. Thus, the specificity of splicing enzymes has been conserved over a very long evolutionary period.

ANTIBIOTIC INHIBITORS OF TRANSCRIPTION: RIFAMYCIN AND ACTINOMYCIN

Antibiotics are interesting molecules because many of them are highly specific inhibitors of biological processes. Actinomycin and rifamycin are two antibiotics that inhibit transcription in quite different ways. Rifamycin, which is derived from a strain of *Streptomyces*, and rifampicin, a semisynthetic derivative, *specifically inhibit the initiation of RNA synthesis.* Rifampicin does not block the binding of RNA polymerase to the DNA template; rather, it interferes with the formation of the first phosphodiester bond in the RNA chain. In contrast, chain elongation is not appreciably affected. This high degree of selectivity makes rifampicin a very useful experimental tool. For example, it can be used to block the initiation of new RNA chains without interfering with the transcription of chains that are already being synthesized. The site of action of rifampicin is the *β subunit of RNA polymerase*. Some mutants (called *rif-r*) have an altered *β* chain and are resistant to rifampicin.

Rifamycin B

Actinomycin D, a polypeptide-containing antibiotic (Figure 29-18) from a different strain of *Streptomyces*, inhibits transcription by an entirely different mechanism. *Actinomycin D binds tightly and specifically to double-helical DNA and thereby prevents it from being an effective template for RNA synthesis.* It does not bind to single-stranded DNA or RNA, double-stranded RNA, or RNA-DNA hybrids. The binding of actinomycin to DNA is markedly enhanced by the presence of guanine residues. Spectroscopic and hydrodynamic studies of complexes of actinomycin D and DNA suggested that the phenoxazone ring of actinomycin slips in between neighboring base pairs in DNA. This mode of binding is called *intercalation.* At low concentrations, *actinomycin D inhibits transcription without appreciably affecting DNA replication or protein synthesis. Hence, actinomycin D has been extensively used as a highly specific inhibitor of the formation of new RNA in both procaryotic and eucaryotic cells.* Its inhibition of the growth of rapidly dividing cells makes it an effective therapeutic agent in the treatment of some cancers.

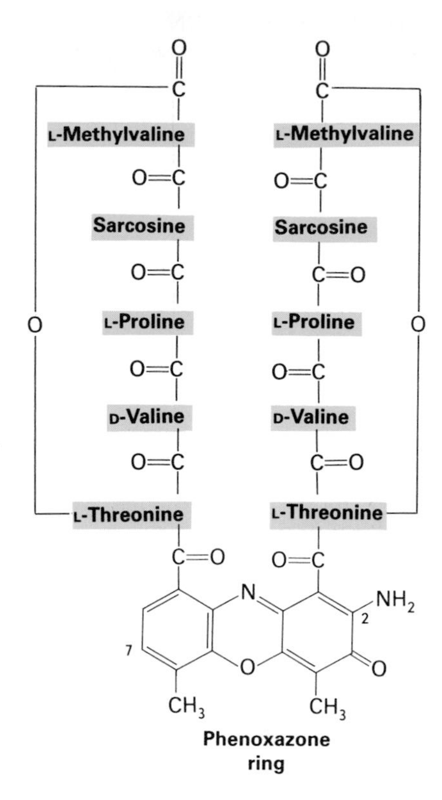

Figure 29-18
Structure of actinomycin D.

The structure of a crystalline complex of one actinomycin molecule and two deoxyguanosine molecules has been solved at atomic resolution. The phenoxazone ring of actinomycin is sandwiched between two guanine rings in this complex. One of the two cyclic peptides is located above the phenoxazone ring, the other below. Each peptide forms several hydrogen bonds with a guanine residue. Many favorable van der Waals interactions reinforce the binding of the antibiotic. A significant feature of this complex is that it is nearly symmetric. There is a twofold axis of symmetry along the line joining the central O and N atoms of the phenoxazone ring. The conformation of this complex suggested that *actinomycin recognizes the base sequence GpC in DNA.* If the sequence along one strand of DNA is 5′-GC-3′, the sequence of the complementary strand is 3′-CG-5′. Actinomycin intercalates between two GC base pairs in DNA and interacts with the G residues in much the same way as it does in its complex with deoxyguanosine (Figure 29-19). The cyclic peptide units are located in the minor groove of the DNA helix. Hence, *the symmetry of actinomycin matches the symmetry of a specific base sequence in DNA.* In fact, the symmetry of many DNA-binding proteins matches that of their palindromic target sites, as exemplified by EcoR1 endonuclease (p. 656) and repressor proteins (p. 811).

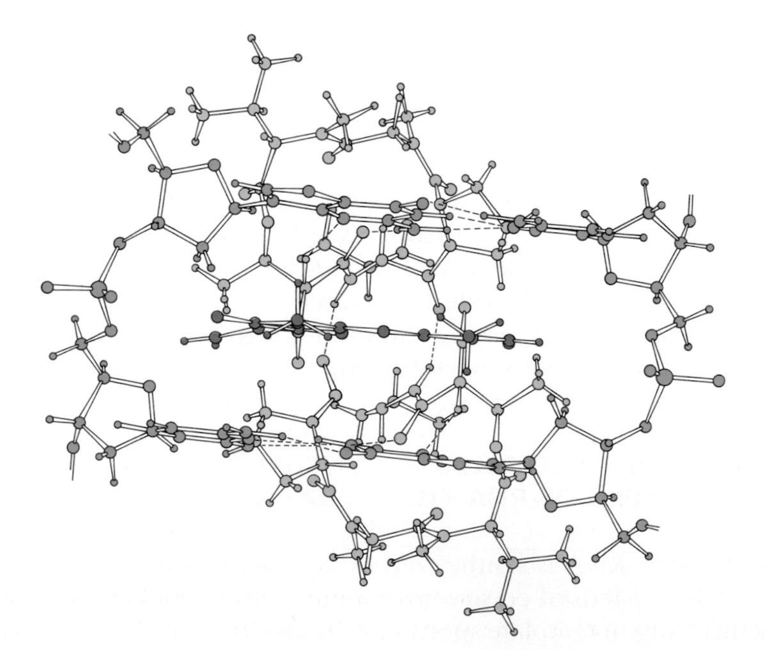

Figure 29-19
Proposed structure of the complex of actinomycin D and DNA. The phenoxazone ring of actinomycin D (shown in red) is intercalated between two GC base pairs of the DNA (shown in blue). The cyclic peptides of actinomycin D (shown in yellow) bind to the narrow groove of the DNA helix. There are several hydrogen bonds between actinomycin D and the adjacent guanines. The axis of symmetry of actinomycin coincides with the axis of symmetry of the two GC base pairs. [After a drawing kindly provided by Dr. Henry Sobell.]

TRANSCRIPTION AND TRANSLATION IN EUCARYOTES ARE SEPARATED IN SPACE AND TIME

We turn now to transcription in eucaryotes, a much more complex process than in procaryotes. Eucaryotes, by definition, contain a membrane-bounded nucleus, where transcription occurs. In contrast, translation takes place outside the nucleus. Thus, transcription and translation in eucaryotes occur in different cellular compartments, whereas in procaryotes they are closely coupled (Figure 29-20). Indeed, translation of bacterial mRNA begins while the transcript is still being synthesized. *The spatial and temporal separation of transcription and translation enables eucaryotes to regulate gene expression in much more intricate ways, contributing to the richness of eucaryotic form and function.*

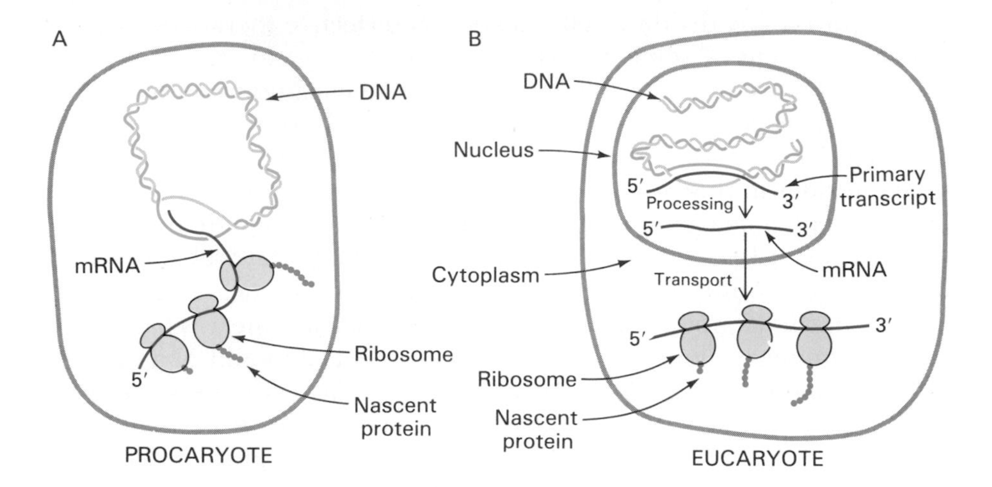

Figure 29-20
Transcription and translation are closely coupled in procaroytes, whereas they are spatially and temporally separate in eucaryotes. (A) In procaryotes, the primary transcript serves as mRNA and is used immediately as the template for protein synthesis. (B) In eucaryotes, mRNA precursors are processed and spliced in the nucleus before being transported to the cytosol. [After J. Darnell, H. Lodish, and D. Baltimore. *Molecular Cell Biology* (Scientific American Books, 1986), p. 270.]

A second major difference between procaryotes and eucaryotes is that *eucaryotes very extensively process nascent RNA destined to become mRNA.* Primary transcripts—the products of RNA polymerase action—acquire a cap at their 5′ end and a poly A tail at their 3′ end. Most important, *nearly all mRNA precursors in higher eucaryotes are spliced.* Introns are precisely excised from primary transcripts to form mature mRNAs with continuous messages. Some mRNAs are a tenth of the size of their precursors, which can be as large as 30 kb or more. The pattern of splicing of some primary transcripts is developmentally regulated to generate variations on a theme, such as the membrane-bound and secreted forms of antibody molecules. Alternative splicing enlarges the repertoire of proteins in eucaryotes.

Another noteworthy feature of eucaryotic gene expression is that a majority of transcripts are degraded within the nucleus. mRNA is derived from a class of precursors called *heterogeneous nuclear RNA (hnRNA).* However, *only about 20% of hnRNA molecules are converted into mRNA.* The others are hydrolyzed by nucleases within a few minutes of their formation. An interesting question is why eucaryotes indulge in what appears to be a very wasteful process.

RNA IN EUCARYOTIC CELLS IS SYNTHESIZED BY THREE TYPES OF RNA POLYMERASES

In procaryotes, RNA is synthesized by a single kind of polymerase. In contrast, the nucleus of eucaryotes contains three types of RNA polymerases differing in template specificity, localization, and susceptibility to

Table 29-2
Eucaryotic RNA polymerases

Type	*Localization*	*Cellular transcripts*	*Effect of α-amanitin*
I	Nucleolus	18S, 5.8S, and 28S rRNA	Insensitive
II	Nucleoplasm	mRNA precursors and hnRNA	Strongly inhibited
III	Nucleoplasm	tRNA and 5S rRNA	Inhibited by high concentrations

inhibitors (Table 29-2). RNA polymerase I is located in nucleoli, where it transcribes the tandem array of genes for 18S, 5.8S, and 28S ribosomal RNA (p. 837). The other ribosomal RNA molecule (5S rRNA, p. 844) and all of the transfer RNA molecules (p. 740), are synthesized by RNA polymerase III, which is located in the nucleoplasm rather than in nucleoli. The precursors of messenger RNA are synthesized by RNA polymerase II, which is also located in the nucleoplasm. Several small RNA molecules, such as the U1 snRNA of the splicing apparatus (p. 724), are also synthesized by polymerase II.

Eucaryotic RNA polymerases are like procaryotic ones in catalyzing the nucleophilic attack of the 3′-OH of the growing RNA chain on the α phosphorus atom of an incoming ribonucleoside triphosphate. Again, a primer is not needed, and synthesis proceeds in the $5' \rightarrow 3'$ direction according to instructions given by an antiparallel DNA template strand. Eucaryotic RNA polymerases also lack nuclease activity, which means that errors in nascent RNA are not corrected. The study of the subunit composition of eucaryotic polymerases and the roles of the subunits is at an early stage. These enzymes are large multisubunit complexes (>500 kd) consisting of two large subunits (~140 kd and ~200 kd) and several small ones.

More than a hundred deaths result each year in the world from the ingestion of poisonous mushrooms, particularly *Amanita phalloides*, which is also called the death cup or the destroying angel. One of the toxins in this mushroom is *α-amanitin*, a cyclic octapeptide that contains several unusual amino acids. α-Amanitin binds very tightly ($K = 10^{-8}$ M) to RNA polymerase II and thereby blocks the formation of precursors of mRNA. Polymerase III is inhibited by higher concentrations of α-amanitin (10^{-6} M), whereas polymerase I is insensitive to this toxin. α-Amanitin blocks the elongation phase of RNA synthesis.

Figure 29-21
Amanita phalloides, a poisonous mushroom that contains α-amanitin. This mushroom also contains phalloidin, a potent inhibitor of cytoskeletal function. [After G. Lincoff and D. H. Mitchel. *Toxic and Hallucinogenic Mushroom Poisoning* (Van Nostrand Reinhold, 1977), p. 30.]

α-Amanitin
(Inhibitor of RNA polymerase II)

EUCARYOTIC PROMOTERS CONTAIN A TATA BOX AND ADDITIONAL UPSTREAM SEQUENCES

5′

T_{82} A_7

A_{97} T_2

T_{93} A_7

A_{85} T_{10}

A_{63} T_{37}

A_{88} T_{10}

A_{50} T_{33}

3′

Figure 29-22
TATA box in eucaryotic promoters for mRNA precursors. Comparisons of the sequences of more than a hundred procaryotic promoters lead to the consensus shown in the left column. The second most frequently found bases are shown on the right. The subscript denotes the frequency (in percent) of the base at that position. The TATA box is centered at about −25.

Promoters for RNA polymerase II, like those for bacterial polymerases, are located on the 5′ side of the start site for transcription. Mutagenesis experiments and comparisons of many higher eucaryotic genes have demonstrated the importance of several upstream regions. The one closest to the start site, centered at about −25, is called the *TATA box* (Hogness box). The consensus sequence is a heptanucleotide of A and T residues (Figure 29-22). In yeast, a TATAAA sequence centered between −30 and −90 plays the same role. The TATA box is present in nearly all eucaryotic genes giving rise to mRNA. Note that *the TATA box of eucaryotes closely resembles the −10 sequence of procaryotes (TATAAT) but is farther from the start site.* Mutation of a single base in the TATA box markedly impairs promoter activity. Likewise, interchanges of A and T bases lead to loss of activity, showing that the precise sequence, not just a high content of AT pairs, is essential. Most TATA boxes are flanked by GC-rich sequences.

The TATA box is necessary but usually not sufficient for promoter activity. Additional elements are located between −40 and −110. Many promoters contain a *CAAT box* and some contain a *GC box* (Figure 29-23). Constitutive genes (genes that are continuously expressed rather than developmentally regulated) tend to have GC boxes in their promoters. The positions of these upstream sequences varies from one promoter to another, in contrast with the quite constant location of the −35 region in procaryotes. Another difference is that the CAAT box and the GC box can be effective when present on the template strand, unlike the −35 region.

5′ GGNCAATCT 3′
CAAT box

5′ GGGCGG 3′
GC box

Figure 29-23
Consensus sequences of CAAT and GC boxes of eucaryotic promoters for mRNA precursors. Most are located in the −40 to −110 region.

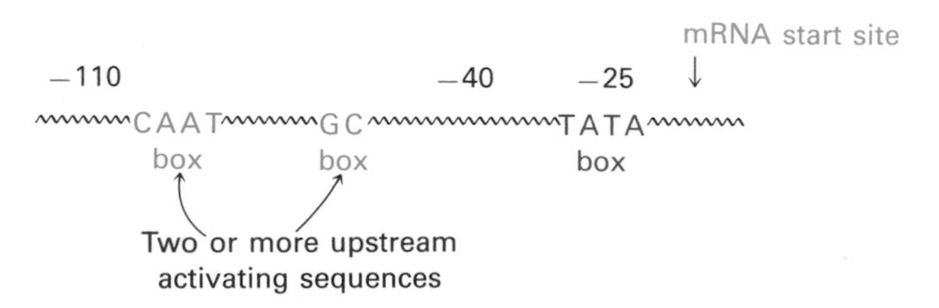

Figure 29-24
General features of promoters for mRNA precursors in higher eucaryotes.

SPECIFIC PROTEINS CALLED TRANSCRIPTION FACTORS INTERACT WITH EUCARYOTIC PROMOTERS

The diversity of upstream stimulatory sequences in eucaryotic genes and the variability of their position suggests that they may be recognized by different specific proteins rather than by a single polymerase. Indeed, this is the case. RNA polymerase II by itself cannot recognize promoter sites and begin transcription. However, the addition of cell extracts restores this ability. Several specific stimulatory proteins called *transcription factors* are being isolated and their sites of action are being identified by footprinting experiments. For example, *Sp1*, a 95-kd or 105-kd protein from mammalian cells, is required for transcription of genes whose promoters contain GC boxes. The duplex DNA of SV40 virus (a cancer-producing virus that infects monkey cells) contains five

GC boxes between 50 and 100 bp away from start sites on either side. DNase I digestion footprints showed that three GC boxes bind Sp1 tightly and two bind it weakly (Figure 29-25). The binding of a molecule of Sp1 protects about 20 bp (~68 Å) of DNA from digestion.

Several other transcription factors have recently been identified. *CTF* (CCAAT-binding transcription factor), a 60-kd protein from mammalian cells, binds to the CAAT box. The *B protein* from *Drosophila* stimulates transcription by binding to the TATA box. A *heat-shock transcription factor* (*HSTF*) is expressed in *Drosophila* following an abrupt increase in temperature. Multiple copies of HSTF bind to the promoter of heat-shock genes starting at a site 15 bp upstream from the TATA box (Fig. 29-25). HSTF differs from σ^{32}, a heat-shock protein of *E. coli* (p. 707), in binding directly to target sites in heat-shock promoters rather than first becoming associated with RNA polymerase. In yeast, *gal4 protein* stimulates the transcription of three genes encoding enzymes that metabolize galactose. Gal4 protein acts by binding to specific upstream sequences in the promoters of these genes.

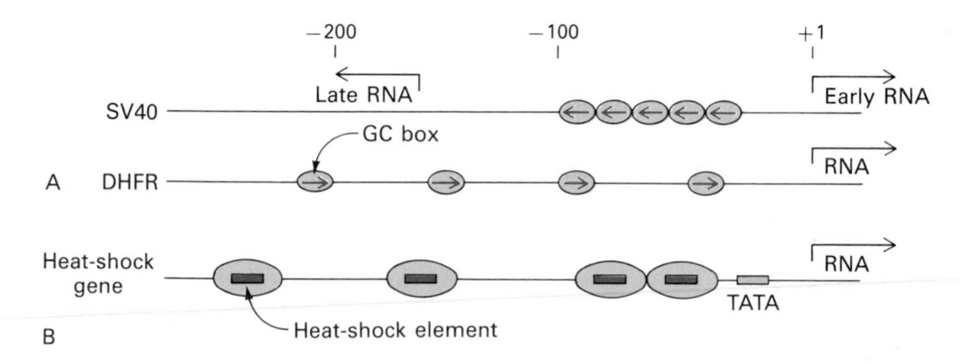

Figure 29-25
Multiple binding sites for transcription factors in eucaryotic promoters. (A) Binding of Sp1 to the dihydrofolate reductase (DHFR) promoter and the SV40 viral DNA promoter. (B) Binding of HSTF to a *Drosophila* heat-shock promoter. [After W. S. Dynan and R. Tjian. *Nature* 316(1985):774.]

ENHANCER SEQUENCES CAN STIMULATE TRANSCRIPTION AT START SITES THOUSANDS OF BASES AWAY

The activities of many promoters in higher eucaryotes are greatly increased by sequences called *enhancers* that have no promoter activity of their own. *Remarkably, enhancers can exert their stimulatory actions over distances of several thousand base pairs. They can be upstream, downstream, or even in the midst of a transcribed gene.* The SV40 viral enhancer, a tandem repeat of a 72-bp sequence, remains active when moved anywhere in the 5.2-kb circular viral genome. Moreover, enhancers are effective if they are located on either the coding or the template strand.

A particular enhancer is effective only in certain kinds of cells. The immunoglobulin enhancer functions in B-lymphocytes but not elsewhere. The action of enhancers requires proteins that are expressed only in some cells. This is best understood for the enhancer mediating the hormonal action of glucocorticoids. These steroids interact with a soluble receptor. The binding of the hormone-receptor complex to the glucocorticoid enhancer then leads to stimulation of transcription of a dis-

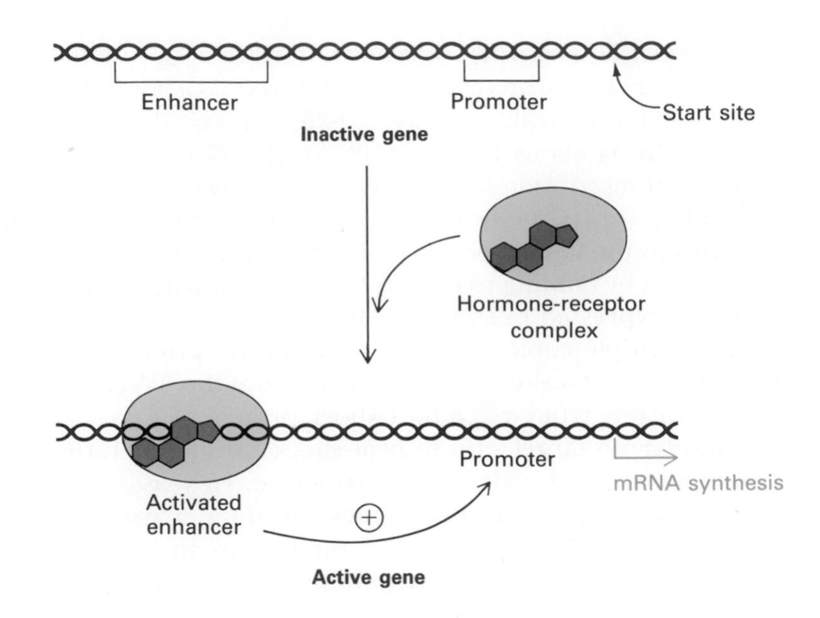

Figure 29-26
Enhancer sequences play a key role in mediating the action of steroid hormones. A glucocortocoid hormone binds to a soluble receptor protein. This complex binds to enhancer sequences, enabling them to stimulate the transcription of hormone-responsive genes.

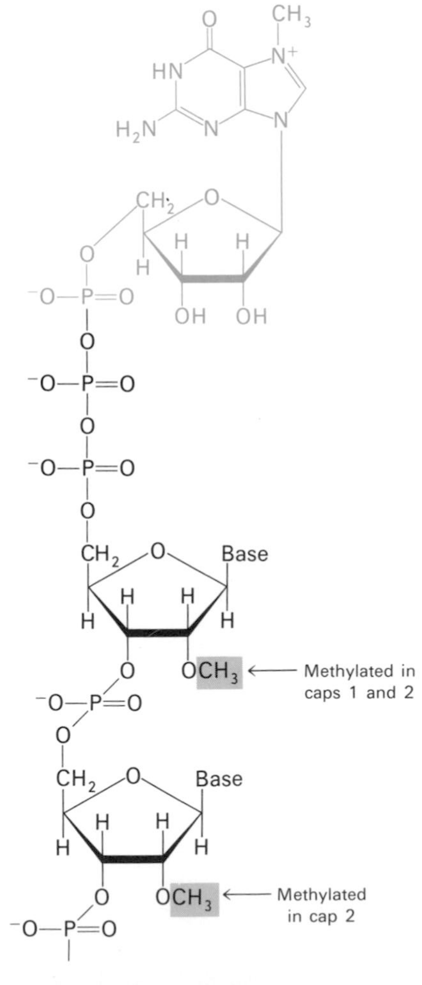

Figure 29-27
Structure of caps at the 5′ end of eucaryotic mRNAs. All caps contain 7-methylguanylate (shown in green) attached by a triphosphate linkage to the sugar at the 5′ end. None of the riboses are methylated in cap 0, one is methylated in cap 1, and 2 are methylated in cap 2.

tinctive set of genes (p. 873). As might be expected, the DNA of viruses infecting eucaryotic cells usually contains enhancers that are activated by host-cell proteins. For example, the DNA of mouse mammary tumor virus (MMTV) contains the glucocorticoid enhancer. Hence, MMTV thrives in cells that are normally stimulated by these steroids—namely, epithelial cells of the mammary gland—because they contain the hormone-receptor that activates the enhancer. The restricted host range of viruses is partly a consequence of their having tissue-specific and species-specific enhancers.

How do enhancers act over distances of several kilobases? An attractive hypothesis is that they serve as docking sites for the assembly of initiation complexes containing RNA polymerase II. DNA can be looped by the binding of regulatory proteins so that sites distant in the linear sequence are brought into proximity (p. 806). Detailed answers to this intriguing question are eagerly awaited. The discovery of enhancers and upstream activating sequences opens the door to understanding how genes are selectively expressed in eucaryotic cells. Transcription factors and other proteins that bind these regulatory sequences can be regarded as passwords that cooperatively open multiple locks, giving RNA polymerase access to specific genes.

mRNA PRECURSORS ACQUIRE 5′ CAPS DURING TRANSCRIPTION

As in procaryotes, transcription usually begins with A or G. However, the 5′ triphosphate end of the nascent RNA chain is almost immediately modified. A phosphate is released by hydrolysis. The diphosphate 5′ end then attacks the α phosphorus atom of GTP to form a very unusual 5′-5′ triphosphate linkage. This highly distinctive terminus is called a *cap*. The N-7 nitrogen of the terminal guanine is then methylated by *S*-adenosylmethionine to form *cap 0*. The adjacent riboses may be methylated to form *cap 1* and *cap 2* (Figure 29-27). Caps are important for subsequent splicing reactions. They also contribute to the stability of mRNAs by protecting their 5′ ends from phosphatases and nucleases. In addition, caps enhance the translation of mRNA by eucaryotic protein-synthesizing systems. Transfer RNA and ribosomal RNA molecules do not have caps.

A 3′ POLYADENYLATE TAIL IS ADDED TO MOST mRNA PRECURSORS AFTER CLEAVAGE BY AN ENDONUCLEASE

Rather little is known about eucaryotic termination signals. Some contain a hairpin followed by a series of U residues, as in procaryotic transcripts. *Most eucaryotic mRNA molecules contain a polyadenylate (poly A) tail at their 3′ end.* This poly A tail is not encoded by DNA. Indeed, the nucleotide preceding poly A is not the last nucleotide to be transcribed. Some primary transcripts contain hundreds of nucleotides beyond the 3′ end of the mature mRNA.

Eucaryotic primary transcripts are cleaved by a specific endonuclease that recognizes the sequence AAUAAA (Figure 29-28). Cleavage does not occur if this sequence or a segment of some 20 nucleotides on its 3′ side is deleted. Another clue is that some mature mRNAs contain internal AAUAAA sequences. Thus, AAUAAA is only part of the cleavage signal; its context is also important. After cleavage by the endonuclease, a *poly A polymerase* adds about 250 A residues to the 3′ end of the transcript; ATP is the donor in this reaction. This polyadenylate (poly A) tail then wraps around several copies of a 78-kd binding protein.

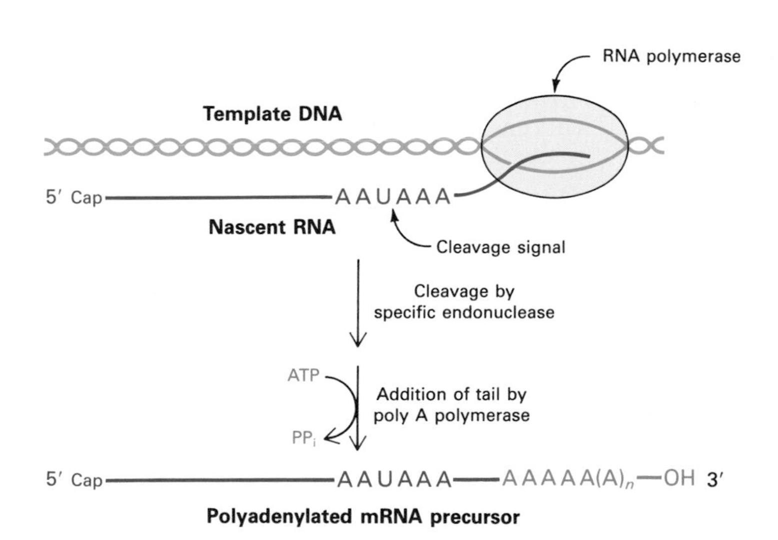

Figure 29-28
Cleavage and polyadenylation of a primary transcript. A specific endonuclease cleaves the RNA downstream of AAUAAA. Poly A polymerase then adds about 250 A residues.

The role of the poly A tail is an enigma. Its synthesis can be blocked by 3′-deoxyadenosine (cordycepin), which does not interfere with synthesis of the primary transcript. Messenger RNA devoid of a poly A tail can be transported out of the nucleus, and it serves as an effective template for protein synthesis. An important class of mRNAs, those encoding histone proteins that package DNA (p. 825), do not have a poly A terminus.

SPLICE SITES IN mRNA PRECURSORS ARE SPECIFIED BY SEQUENCES AT THE ENDS OF INTRONS

Introns are precisely spliced out of mRNA precursors. A one-nucleotide slippage in a splice point would shift the reading frame on the 3′ side of the splice to give an entirely different amino acid sequence. How does the splicing machinery process its targets with high fidelity? The base sequences of hundreds of intron-exon junctions of RNA transcripts are known, and they are revealing (Table 29-3). These sequences

Table 29-3
Base sequences of splice points in transcripts containing introns

Gene region	*Exon*		*Intron*		*Exon*
Ovalbumin intron 2	U A A G	G U G A G C	~~~	U U A C A G	G U U G
Ovalbumin intron 3	U C A G	G U A C A G	~~~	A U U C A G	U C U G
β-Globin intron 1	G C A G	G U U G G U	~~~	C C U U A G	G C U G
β-Globin intron 2	C A G G	G U G A G U	~~~	C C A C A G	U C U C
Immunoglobulin λ_1 intron 1	U C A G	G U C A G C	~~~	U U G C A G	G G G C
SV40 virus early T-antigen	U A A G	G U A A A U	~~~	U U U U A G	A U U C

for eucaryotes ranging in complexity from yeast to mammals have a common structural motif: *the base sequence of an intron begins with GU and ends with AG*. The consensus sequence at the 5′ end of vertebrate introns is AGGUAAGU (Figure 29-29). At the 3′ end of introns, the consensus sequence is a stretch of *ten pyrimidines* (U or C), followed by any base and then by C, and ending with the invariant AG. Introns also have an important internal site located between 20 and 50 nucleotides upstream of the 3′ splice site; it is called the *branch site* for reasons that will be evident shortly. In yeast, the branch site sequence is nearly always UACUAAC, whereas in mammals a variety of sequences are found.

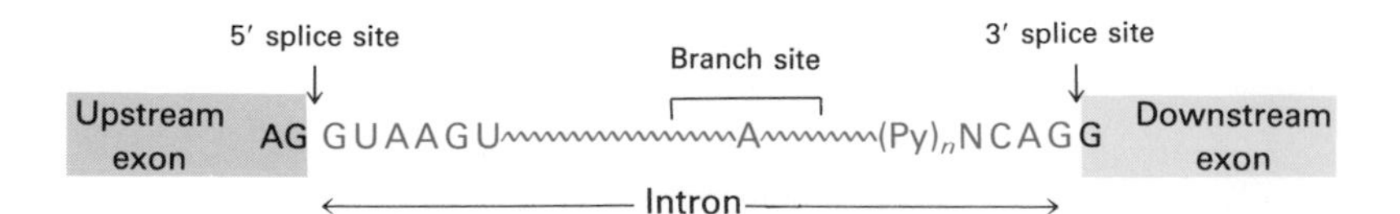

Figure 29-29
Splicing signals. Consensus sequences for the 5′ splice site and the 3′ splice site are shown.

Portions of introns other than the 5′ and 3′ splice sites and the branch site appear to be unimportant in determining where splicing occurs. Introns range in length from 50 to 10,000 nucleotides. Much of an intron can be deleted without altering the site or efficiency of splicing. Likewise, splicing is unaffected by the insertion of long stretches of DNA into the introns of genes. Moreover, chimeric introns crafted by recombinant DNA methods from the 5′ end of one intron and the 3′ end of a very different intron are properly spliced provided that the splice sites and branch site are unaltered. In contrast, mutations in each of these three critical regions lead to aberrant splicing. Mutations far from splice junctions can lead to abnormal splicing only if the change creates a consensus sequence for 5′ or 3′ splicing at a new location.

Some forms of thalassemia, a group of hereditary anemias characterized by defective synthesis of hemoglobin (p. 171), are caused by *aberrant splicing*. In one patient, a mutation of G to A twenty-two nucleotides away from the normal 3′ splice site of the first intron created a

Normal 3′ end of intron

Normal 5′ CCTATTGGTCTATTTTCCACCCTTAGGCTGCTG 3′

β-Thalassemia 5′ CCTATTAGTCTATTTTCCACCCTTAGGCTGCTG 3′

Figure 29-30
Mutation of a single base (G to A) in an intron of the β globin gene leads to thalassemia. This mutation probably generates a new 3′ splice site (blue) akin to the normal one (yellow) farther downstream. [After R. A. Spritz, P. Jagedeeswaran, P. V. Choudary, P. A. Biro, J. T. Elder, J. K. deRiel, J. L. Manley, M. L. Gefter, B. G. Forget, and S. M. Weissman. *Proc. Nat. Acad. Sci.* 78(1981):2457.]

new 3′ acceptor site (Figure 29-30). The resulting mRNA contains a series of codons not normally present. The seventh codon after the splice is a stop signal for protein synthesis, and so the aberrant protein ends prematurely.

LARIAT INTERMEDIATES ARE FORMED IN THE SPLICING OF mRNA PRECURSORS

How does the end of one exon in an mRNA precursor become linked to the beginning of the adjacent exon? Splicing begins with the cleavage of the phosphodiester bond between the upstream exon (exon 1) and the 5′ end of the intron (Figure 29-31). The attacking group in this reaction

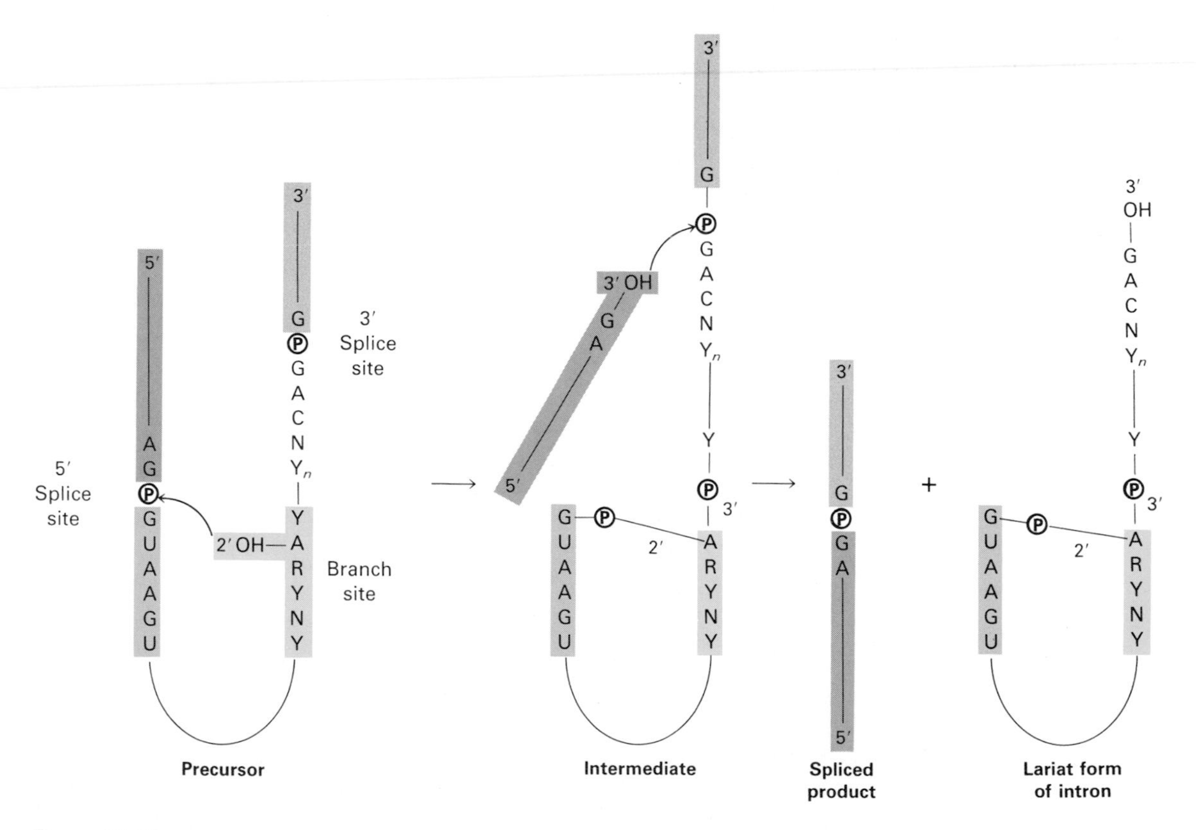

Figure 29-31
Mechanism of splicing of mRNA precursors in eucaryotic nuclei. The upstream (5′) exon is shown in red, the downstream (3′) exon in green, and the branch site in yellow. Y stands for a purine nucleotide, R for a pyrimidine nucleotide, and N for any nucleotide. The 5′-splice site is attacked by the 2′-OH of the branch-site adenosine. The 3′ splice site is then attacked by the newly formed 3′-OH of the upstream exon. The exons are joined, and the intron is released in the form of a lariat. [After P. A. Sharp. *Cell* 2(1985):3980.]

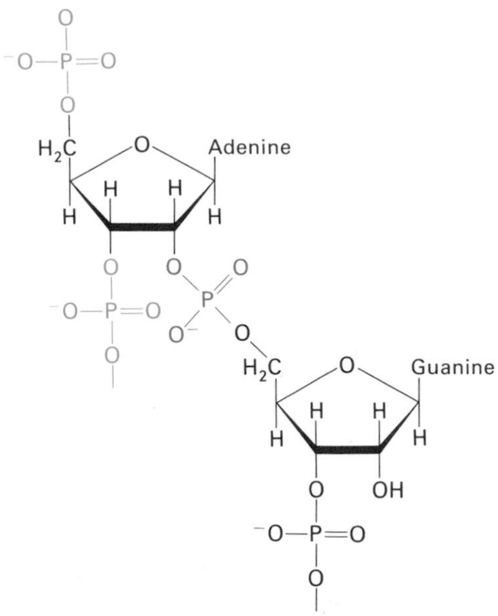

is the 2′-OH of an adenylate residue in the branch site. A 2′,5′-phosphodiester bond is formed between this A residue and the 5′-terminal phosphate of the intron. Note that this adenylate residue is also joined to two other nucleotides by normal 3′,5′-phosphodiester bonds (Figure 29-32). Hence a *branch* is generated at this site and a *lariat intermediate* is formed.

The 3′-OH terminus of exon 1 then attacks the phosphodiester bond between the intron and exon 2. Exons 1 and 2 become joined and the intron is released in lariat form. Note that splicing is accomplished by two *transesterification reactions* rather than by hydrolysis followed by ligation. The first reaction generates a free 3′-OH at the end of exon 1, and the second reaction links this group to the 5′-phosphate of exon 2. *The number of phosphodiester bonds stays the same during these steps.* Until the two exons are joined, the products of the first reaction are held together by a *spliceosome*, an assembly of ribonucleoproteins that recognizes the 5′ splice site, 3′ splice site, and branch site of mRNA precursors.

Figure 29-32
Structure of the branch point in the lariat intermediate in splicing. This adenylate residue is joined to three nucleotides by phosphodiester bonds. The new 2′ → 5′ phosphodiester bridge is shown in red and the usual 5′ → 3′ bridges in blue.

SMALL NUCLEAR RIBONUCLEOPROTEIN PARTICLES (snRNPs) BIND mRNA PRECURSORS TO FORM SPLICEOSOMES

The nucleus and cytosol contain many types of small RNA molecules with fewer than 300 nucleotides. They are called *snRNA* (small nuclear RNA) and *scRNA* (small cytoplasmic RNA). These RNA molecules are associated with specific proteins to form complexes called *snRNPs* (*s*mall *n*uclear *r*ibo*n*ucleoprotein *p*articles) and *scRNPs* (cytosolic particles); investigators often speak of them as "snurps" and "scurps". Spliceosomes are large (60S) complexes containing three kinds of snRNPs in addition to an mRNA precursor (Table 29-4). The key role of snRNPs in splicing was revealed by the finding that splicing is blocked by antibodies specific for these ribonucleoproteins. These antibodies were obtained from patients with systemic lupus erythematosus, an autoimmune disease characterized by the formation of antibodies against many nuclear components. Joints and most organ systems are affected in this disease.

Table 29-4
Small nuclear ribonucleoprotein particles (snRNPs) involved in the splicing of mRNA precursors

snRNP	*Size of snRNA (nucleotides)*	*Role*
U1	165	Binds the 5′ splice site
U2	185	Binds the branch site
U5	116	Binds the 3′ splice site
U4–U6	145 (U4) 106 (U6)	Assembles the spliceosome

In mammalian cells, U1 snRNP recognizes the 5′ splice site. In fact, U1 RNA contains a sequence that is complementary to this splice site, and the binding of U1 snRNP to an mRNA precursor protects a 15-nucleotide region at the 5′ splice site from digestion.

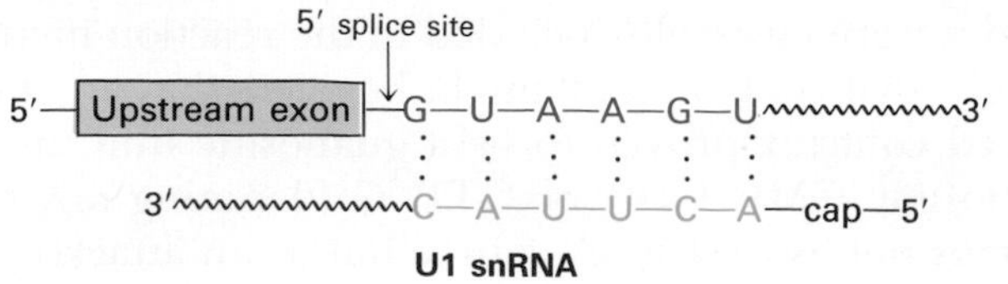

U2 snRNP binds the branch site and polypyrimidine tract in a reaction requiring ATP. The 3′ splice site is recognized by the U5 snRNP. A particle containing both U4 and U6 snRNA is also necessary for the assembly of the spliceosome. Yeast spliceosomes contain a 1.2-kb RNA molecule that has sequences homologous to mammalian U2, U4, U5, and U6 snRNA. This one RNA molecule mediates splicing functions carried out by several snRNAs in mammalian systems.

SELF-SPLICING RNA: THE DISCOVERY OF CATALYTIC RNA

Ribosomal RNA molecules in eucaryotes, like those of procaryotes, are formed by the cleavage of primary transcripts (p. 838). In *Tetrahymena* (a ciliated protozoan), a 414-nucleotide intron is then removed from a 6.4-kb precursor to yield the mature 26S rRNA molecule (Figure 29-33). Studies of this splicing reaction by Thomas Cech and his coworkers led to some very unexpected findings. They added cell extracts to the precursor RNA to determine the protein requirements for splicing. Surprisingly, they found that a control containing the precursor and nucleoside triphosphates but apparently devoid of protein underwent splicing. Could it be that an essential protein was not removed from the precursor RNA? This doubt was addressed by using recombinant-DNA methods to prepare the DNA corresponding to this 6.4-kb precursor (*E. coli*, the host cell, lacks this RNA molecule). Purified DNA encoding this precursor was then transcribed in vitro to yield a synthetic RNA substrate for the splicing reaction. The result was the same: in the presence of nucleotides, the RNA spliced itself to precisely excise the 414-nucleotide intron. *This remarkable experiment demonstrated that an RNA molecule can have highly specific catalytic activity and splice itself.*

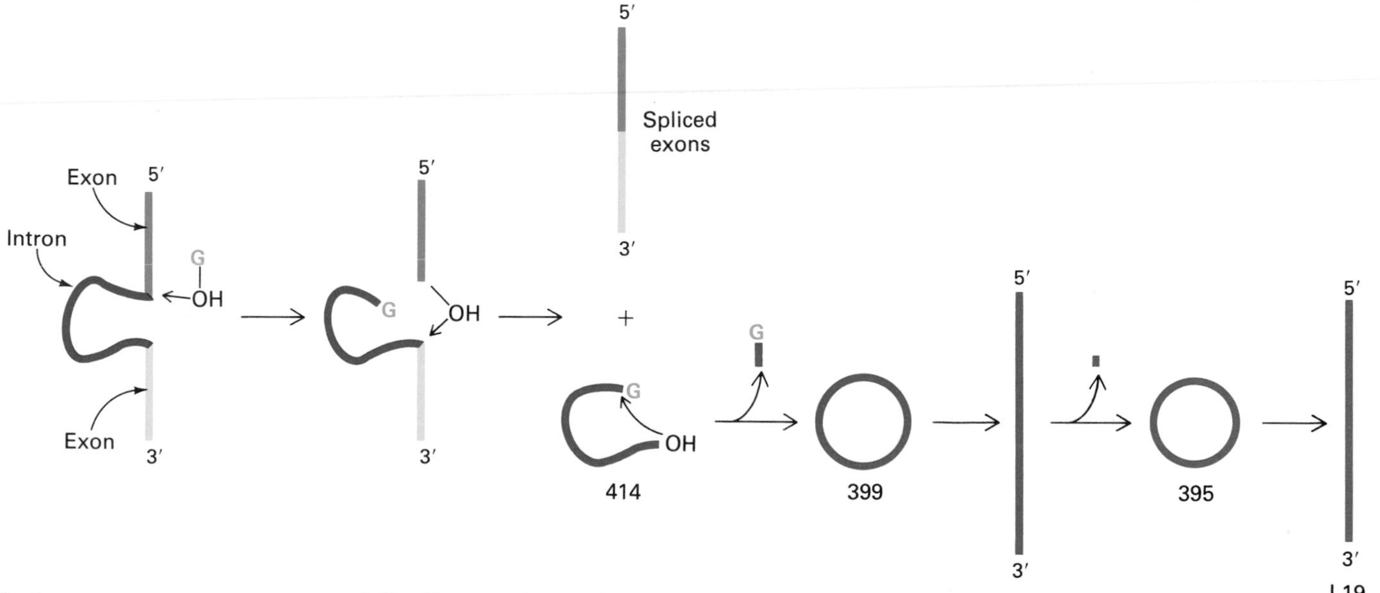

Figure 29-33
Self-splicing of a ribosomal RNA precursor from *Tetrahymena*. A 414-nucleotide intron (red) is released in the first splicing reaction in which guanosine or a guanyl nucleotide (green G) serves as a cofactor. This intron splices itself twice again to produce a linear RNA that has lost a total of 19 nucleotides. This L19 RNA is catalytically active. [After T. Cech. RNA as an enzyme. Copyright © 1986 by Scientific American, Inc. All rights reserved.]

Nucleotides were originally included in the reaction mixture because it was thought that ATP or GTP might be needed as an energy source. The required cofactor proved to be a guanosine unit, in the form of either guanosine, GMP, GDP, or GTP. G (denoting any one of these species) serves not as an energy source but as an attacking group that becomes transiently incorporated into the RNA (Figure 29-33). G binds to the RNA and then attacks the 5′ splice site to form a phosphodiester bond with the 5′ end of the intron. This transesterification reaction generates a 3′-OH at the end of the upstream exon. The 3′ splice site is then attacked by the newly formed 3′-OH group of the upstream exon. This second transesterification reaction joins the two exons and leads to the release of the 414-nucleotide intron.

Two more rounds of self-splicing take place. The 3′-OH of the intron attacks a phosphodiester bond near the 5′ end to form a circle and a 15-nucleotide fragment containing the G that was incorporated in the first step. This 399-nucleotide circle opens into a linear molecule, which cyclizes to lose a 4-nucleotide fragment. This circle opens into a linear RNA called L − 19 IVS (*l*inear minus 19 *i*ntervening *s*equence); we shall refer to it as L19 RNA.

Self-splicing depends on the structural integrity of the rRNA precursor. Much of the intron is needed for self-splicing. This molecule, like many RNAs, has a folded structure formed by many double-helical stems and loops (Figure 29-34). The folded RNA contains weak GU

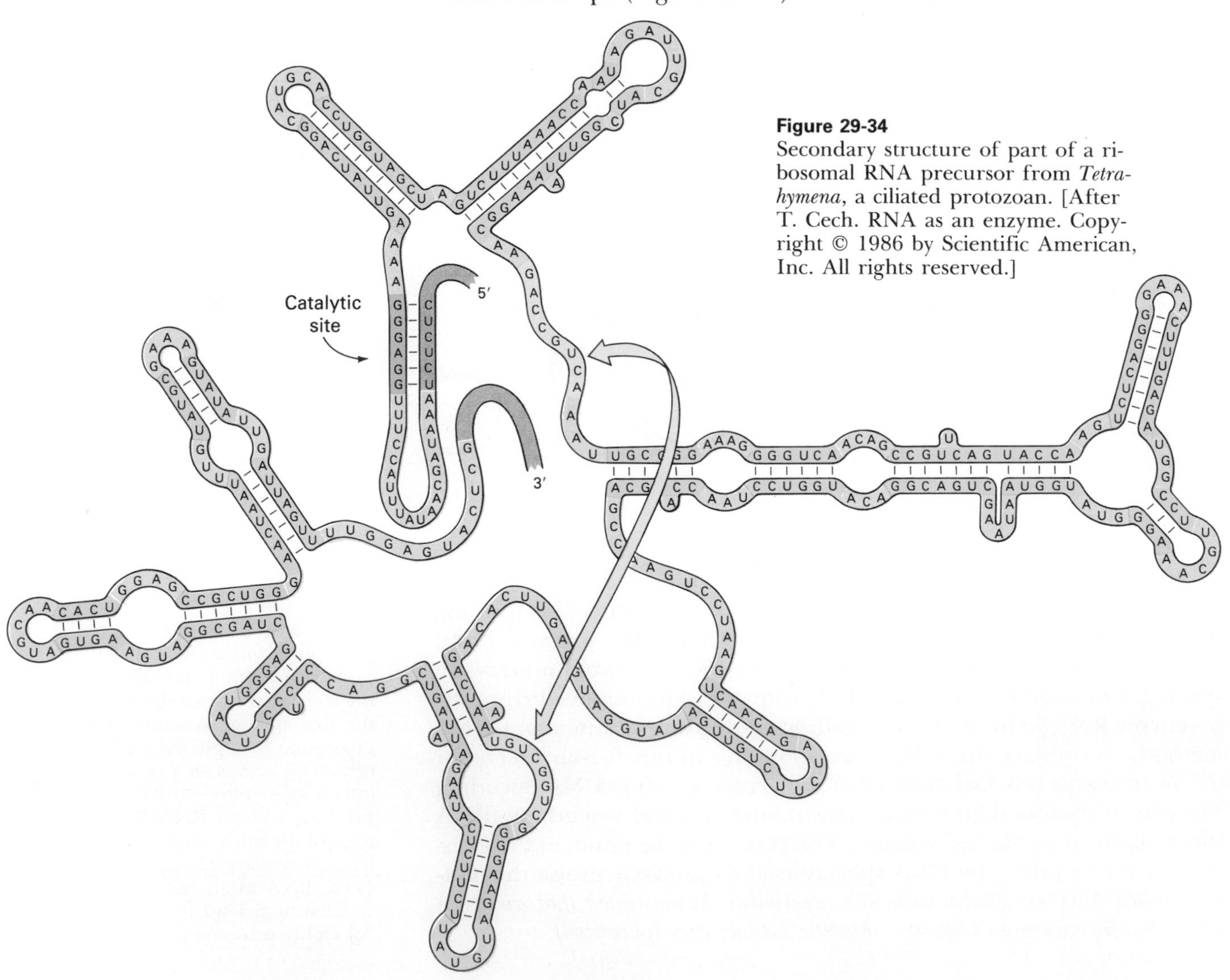

Figure 29-34
Secondary structure of part of a ribosomal RNA precursor from *Tetrahymena*, a ciliated protozoan. [After T. Cech. RNA as an enzyme. Copyright © 1986 by Scientific American, Inc. All rights reserved.]

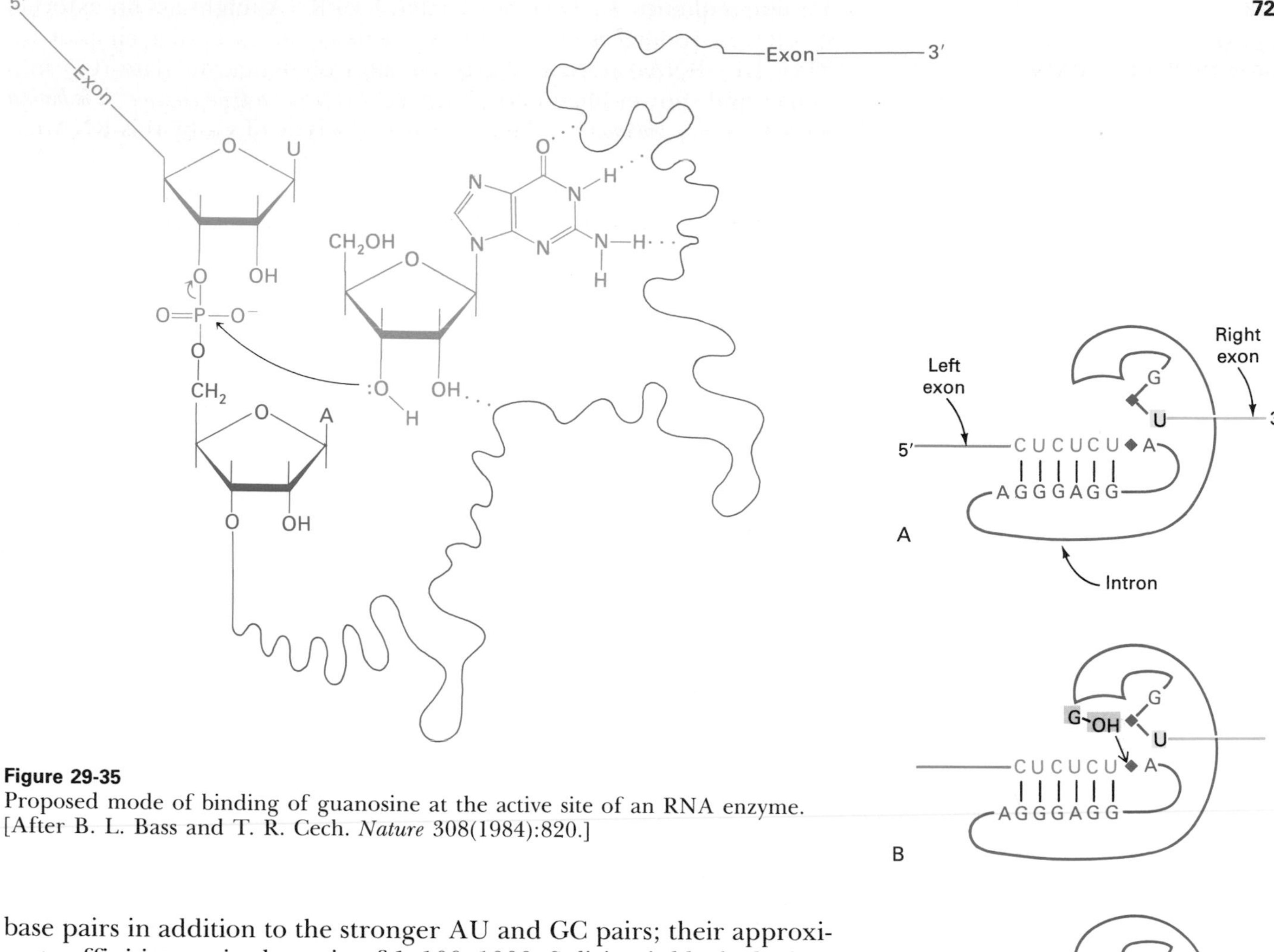

Figure 29-35
Proposed mode of binding of guanosine at the active site of an RNA enzyme. [After B. L. Bass and T. R. Cech. *Nature* 308(1984):820.]

base pairs in addition to the stronger AU and GC pairs; their approximate affinities are in the ratio of 1:100:1000. Splicing is blocked when secondary and tertiary structure are disrupted by the addition of denaturing agents such as dimethylformamide. Furthermore, the binding of G is saturable (K_M of 32 μM) and can be competitively inhibited. *These findings strongly imply that the precursor contains a specific binding pocket for G* (Figure 29-35).

Analysis of the base sequence suggested that the 5′ splice site is aligned with the catalytic residues by base pairing between a *pyrimidine-rich region* (CUCUCU) of the upstream exon and a *purine-rich guide sequence* (GGGAGG) within the intron (Figure 29-36). The intron brings together the guanosine cofactor and 5′ splice site so that the 3′-OH of G can nucleophilically attack the phosphorus atom at this splice site. Another part of the intron then holds the downstream exon in position for attack by the newly formed 3′-OH of the upstream exon. A phosphodiester bond is formed between the two exons and the intron is released as a linear molecule. Self-catalysis of bond formation and breakage in this rRNA precursor is highly stereospecific, like catalysis by protein enzymes.

The purine-rich guide sequence then binds another pyrimidine-rich sequence of the linear intron to form a circle and release a 15-nucleotide fragment. The circle opens, another pyrimidine-rich sequence binds to the guide sequence, and a new circle shortened by 4 nucleotides is produced. Finally, this 395-nucleotide intron opens to give L19 RNA, which is stable because no complementary pyrimidine-rich sequences are left. However, L19 RNA still possesses a G-binding site and

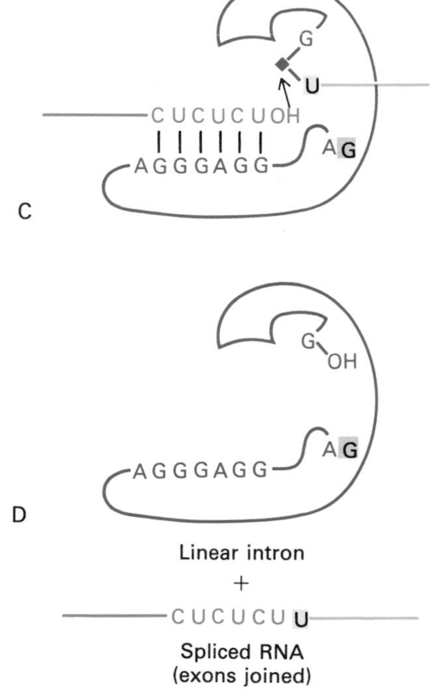

Figure 29-36
Postulated catalytic mechanism of the self-splicing intron from *Tetrahymena*. [After T. Cech. RNA as an enzyme. Copyright © 1986 by Scientific American, Inc. All rights reserved.]

a guide sequence. Cech reasoned that L19 RNA might act on external substrates. Indeed it does. As was discussed in an earlier chapter (p. 214), L19 RNA catalyzes the conversion of pentacytidylate (C_5) into longer and shorter oligomers. Thus, *L19 RNA is a true enzyme: it is both a nuclease and a polymerase*. The rate of hydrolysis of C_5 by this RNA enzyme (ribozyme) is about 10^{10} times the uncatalyzed rate, showing that RNA molecules can have great catalytic prowess. This discovery suggests that RNA at an early stage of evolution could have replicated itself without the participation of proteins.

SPLICEOSOME-CATALYZED SPLICING MAY BE EVOLUTIONARILY DERIVED FROM SELF-SPLICING

mRNA precursors in the mitochondria of yeast and fungi also undergo self-splicing, as do some RNA precursors in the chloroplasts of unicellular organisms such as *Chlamydomonas*. Self-splicing reactions can be classified according to the nature of the unit that attacks the upstream splice site. Group I self-splicing is mediated by a guanosine cofactor, as in *Tetrahymena*. The attacking moiety in group II splicing is the 2′-OH of a specific adenylate of the intron (Figure 29-37). Transfer RNA precursors are spliced by an entirely different mechanism, as was discussed earlier (p. 714).

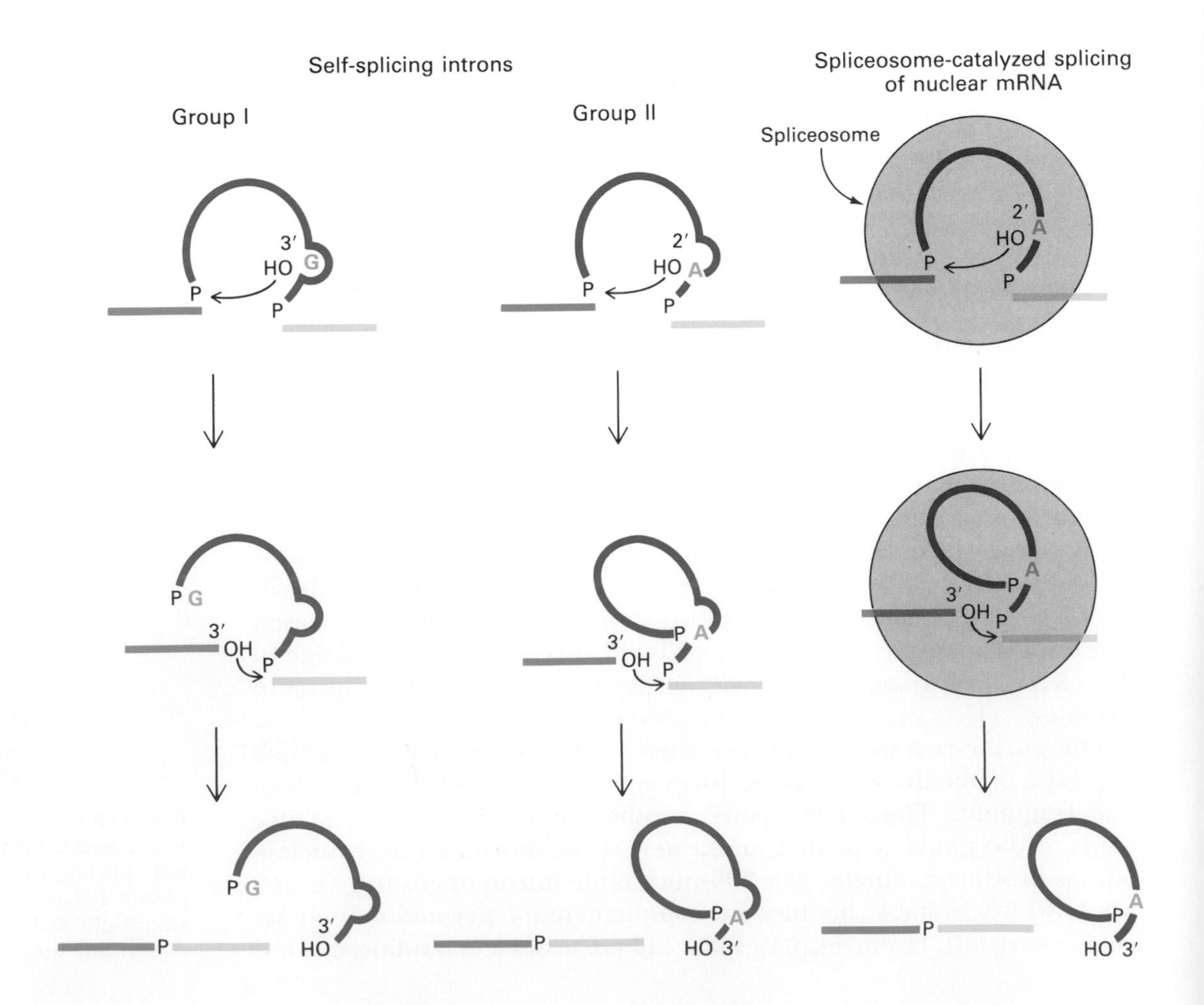

Figure 29-37
Comparison of self-splicing and spliceosome-catalyzed splicing. The exons being joined are shown in blue and yellow, and the attacking unit in green. The catalytic site is formed by the intron itself (red) in group I and II splicing. In contrast, the splicing of nuclear mRNA precursor is catalyzed by snRNPs in a spliceosome. [After P. A. Sharp. *Science* 235(1987):769.]

Group I and group II self-splicing resemble spliceosome-catalyzed splicing in two respects. First, the initial step is an attack by a ribose hydroxyl group on the 5′ splice site. The newly formed 3′-OH terminus of the upstream exon then attacks the 3′ splice site to form a phosphodiester bond with the downstream exon. Second, both reactions are transesterifications in which the phosphate moieties at each splice site are retained in the products. The number of phosphodiester bonds stays constant. Group II splicing is like spliceosome-catalyzed splicing of mRNA precursors in two additional ways. The first cleavage is carried out by a part of the intron itself (the 2′-OH of A) rather than by an external cofactor (G). Furthermore, the intron is released in the form of a lariat.

Phillip Sharp has proposed that spliceosome-catalyzed splicing of mRNA precursors evolved from RNA-catalyzed self-splicing. Group II splicing may well be an intermediate between group I splicing and that occurring in the nuclei of higher eucaryotes. *A major step in this transition was the transfer of catalytic power from the intron itself to other molecules.* The formation of spliceosomes gave introns a new freedom because they were no longer constrained to provide the catalytic center for splicing. Another advantage of external catalysts for splicing is that they can be more readily regulated. The complementarity of U1 snRNA to the 5′ splice site indicates that the RNA component of spliceosomes participates in recognizing splice sites. RNA is very well suited to interacting specifically with other RNA molecules because of the precision of base pairing. It will be very interesting to learn whether the catalytic groups at the active site of spliceosomes come from their snRNA or protein components. RNA in spliceosomes, ribosomes, and other ribonucleoprotein complexes may be catalytically active. Our eyes have been opened to more of the underlying continuity between the early RNA world and contemporary life.

SUMMARY

All cellular RNA molecules are synthesized by RNA polymerases according to instructions given by DNA templates. The activated monomer substrates are ribonucleoside triphosphates. The direction of RNA synthesis is $5' \rightarrow 3'$, as in DNA synthesis. RNA polymerases, unlike DNA polymerases, do not need a primer and do not possess proofreading nuclease activity. Another difference is that the DNA template is fully conserved in RNA synthesis, whereas it is semiconserved in DNA synthesis.

All RNA in *E. coli* is synthesized by one kind of RNA polymerase. The subunit composition of the 460-kd holoenzyme is $\alpha_2\beta\beta'\sigma$ and that of the core enzyme is $\alpha_2\beta\beta'$. Transcription is initiated at promoter sites consisting of two sequences, one centered near -10 and the other near -35, that is, 10 and 35 nucleotides away from the start site in the 5′ (upstream) direction. The consensus sequence of the -10 region is TATAAT. The sigma subunit enables the holoenzyme to recognize promoter sites. When the temperature is raised, *E. coli* expresses a special σ that selectively binds the distinctive promoter of heat-shock genes. Likewise, *Bacillus subtilis* switches σ subunits when it sporulates. The binding of RNA polymerase to a promoter leads to a local unwinding of the bound DNA, which exposes some 17 bases on the template strand and sets the stage for the formation of the first phosphodiester bond. RNA chains usually start with pppG or pppA. The sigma subunit dissociates from the holoenzyme following initiation of the new chain. Elongation takes place at transcription bubbles that move along the DNA template at a rate of about 50 nucleotides per second. The nascent RNA chain contains stop signals that end transcription. One stop signal is an RNA hairpin followed by several U residues. A different stop signal is read by rho protein, an ATPase. In *E. coli*, transfer RNA and ribosomal RNA precursors are cleaved and chemically modified after transcription, whereas mRNA is used unchanged as a template for protein synthesis.

Eucaryotes contain in their nuclei three kinds of RNA polymerases: type I makes ribosomal RNA precursors, II makes mRNA precursors, and III makes transfer RNA precursors. Promoters for RNA polymerase II are located on the 5′ side of the start site for transcription. They consist of a TATA box centered at about −25 and additional upstream sequences usually located between −40 and −110. They are recognized by proteins called transcription factors rather than by RNA polymerase II. The binding sites of several transcription factors have been delineated by footprinting, a technique that identifies regions of template DNA that are protected from digestion or chemical modification. The activity of many promoters is greatly increased by enhancer sequences that have no promoter activity of their own. Enhancers can act over distances of several kilobases, and they can be located either upstream or downstream of a gene. The 5′ ends of mRNA precursors become capped during transcription. A 3′ poly A tail is added to most mRNA precursors after the nascent chain is cleaved by an endonuclease.

Splicing of mRNA precursors is carried out by spliceosomes, which consist of small nuclear ribonucleoprotein particles (snRNPs). Splice sites are specified by sequences at ends of introns and by a branch site near their 3′ end. The 2′-OH of an A in the branch site attacks the 5′ splice site to form a lariat intermediate. The newly generated 3′-OH terminus of the upstream exon then attacks the 3′ splice site to become joined to the downstream exon. The number of phosphodiester bonds stays constant during these two transesterification reactions. Some RNA molecules, such as the 26S ribosomal RNA precursor from *Tetrahymena*, undergo self-splicing. The discovery of catalytic RNA has important implications concerning molecular evolution and raises the possibility that RNA components of ribonucleoproteins have enzymatic roles. Spliceosome-catalyzed splicing may have evolved from self-splicing.

SELECTED READINGS

Where to begin

Darnell, J. E., Jr., 1985. RNA. *Sci. Amer.* 253(4):68–78.

Darnell, J. E., Jr., 1983. The processing of RNA. *Sci. Amer.* 249(4):90–100.

Sharp, P. A., 1987. Splicing of messenger RNA precursors. *Science* 235:766–771.

Aebi, M., and Weissmann, C., 1987. Precision and orderliness in splicing. *Trends Genet.* 3:102–107.

Cech, T. R., 1986. RNA as an enzyme. *Sci. Amer.* 255(5):64–75.

Books

Lewin, B., 1987. *Genes* (3rd ed.). Wiley. [Chapters 9, 11, 23, and 24 give excellent accounts of RNA synthesis and processing.]

Watson, J. D., Hopkins, N. H., Roberts, J. W., Steitz, J. A., Weiner, A. M., 1987. *Molecular Biology of the Gene* (4th ed.). Benjamin/Cummings. [Chapters 13, 18, and 21, dealing with transcription and its control, show a wealth of experimental data.]

Darnell, J., Lodish, H., and Baltimore, D., 1986. *Molecular*

Cell Biology. Scientific American Books. [Chapters 8 and 9 deal with RNA synthesis and processing. A very interesting discussion of evolutionary aspects of splicing is given in Chapter 25.]

Losick, R., and Chamberlin, M., (eds.), 1976. *RNA Polymerase*. Cold Spring Harbor Laboratory. [Contains many excellent articles on this topic.]

RNA polymerases

Chamberlin, M., 1982. Bacterial DNA-dependent RNA polymerases. *In* Boyer, P. D., (ed.), *The Enzymes*, vol. 15, pp. 61–108. Academic Press.

Lewis, M. K., and Burgess, R. R., 1982. Eukaryotic RNA polymerases. *In* Boyer, P. D., (ed.), *The Enzymes*, vol. 15, pp. 109–153. Academic Press.

von Hippel, P. H., Bear, D. G., Morgan, W. D., and McSwiggen, J. A., 1984. Protein–nucleic acid interactions in transcription: a molecular analysis. *Ann. Rev. Biochem.* 53:389–446.

Travers, A., 1976. RNA polymerase specificity and the control of growth. *Nature* 263:641–646.

Burgess, R. R., Travers, A. A., Dunn, J. J., and Bautz, E. K. F., 1969. Factor stimulating transcription by RNA polymerase. *Nature* 221:43–46. [Discovery of the sigma factor.]

Sobell, H. M., 1974. How actinomycin binds to DNA. *Sci. Amer.* 231(2):82–91. [Mode of binding of an inhibitor of transcription.]

Promoters, enhancers, and initiation

McClure, W. R., 1985. Mechanism and control of transcription initiation in prokaryotes. *Ann. Rev. Biochem.* 54:171–204.

Park, C. S., Wu, F. Y.-H., and Wu, C.-W., 1982. Molecular mechanism of promoter selection in gene transcription. *J. Biol. Chem.* 257:6950–6956.

Johnson, W., Moran, C., Losick, R., 1983. Two RNA polymerase sigma factors from *Bacillus subtilis* discriminate between overlapping promoters for a developmentally regulated gene. *Nature* 302:800–804.

Grossman, A., Erickson, J., and Gross, C., 1984. The *htpR* gene product of *E. coli* is a sigma factor for heat-shock promoters. *Cell* 37:383–390.

Dynan, W. S., and Tjian, R., 1985. Control of eukaryotic messenger RNA synthesis by sequence-specific DNA-binding proteins. *Nature* 316:774–778.

Khoury, G., and Gruss, P., 1983. Enhancer elements. *Cell* 33:313–315.

Termination of transcription

Platt, T., 1986. Transcription termination and the regulation of gene expression. *Ann. Rev. Biochem.* 55:339–372.

Brennan, C. A., Dombroski, A. J., and Platt, T., 1987. Transcription termination factor rho is an RNA-DNA helicase. *Cell* 48:945–952.

Yager, T. D., and von Hippel, P. H., 1987. Transcript elongation and termination. *In* Neidhardt, F. (ed.), *Escherichia coli and Salmonella typhimurium: Cellular and Molecular Biology* (in preparation), American Society of Microbiology.

Birnstiel, M. L., Busslinger, M., and Strub, K., 1985. Transcription termination and 3′ processing: the end is in site! *Cell* 41:349–359.

Das, A., Merril, C., and Adhya, S., 1978. Interaction of RNA polymerase and rho in transcription termination: coupled ATPase. *Proc. Nat. Acad. Sci.* 75:4828–4832.

Splicing of mRNA precursors

Sharp, P. A., 1985. On the origin of RNA splicing and introns. *Cell* 42:397–400.

Padgett, R. A., Grabowski, P. J., Konarska, M. M., Seiler, S., and Sharp, P. A., 1986. Splicing of messenger RNA precursors. *Ann. Rev. Biochem.* 55:1119.

Brody, E., and Abelson, J., 1985. The "spliceosome": yeast pre-messenger RNA associates with a 40S complex in a splicing-dependent reaction. *Science* 228:963–967.

Steitz, J. A., 1987. The mammalian premessenger RNA splicing apparatus: a ribosome in pieces? *In* Inoue, M., and Dudock, B. S., (eds.), *Molecular Biology of RNA: New Perspectives* (in preparation).

Chabot, B., Black, D. L., LeMaster, D. M., and Steitz, J. A., 1985. The 3′ splice site of pre-messenger RNA is recognized by a small nuclear ribonucleoprotein. *Science* 230:1344–1349.

Berget, S. M., and Robberson, B. L., 1986. U1, U2, and U4/U6 small nuclear ribonucleoproteins are required for in vitro splicing but not polyadenylation. *Cell* 46:691–696.

Black, D. L., Chabot, B., and Steitz, J. A., 1985. U2 as well as U1 small nuclear ribonucleoproteins are involved in pre-mRNA splicing. *Cell* 42:737–750.

Grabowski, P. J., and Sharp, P. A., 1986. Affinity chromatography of splicing complexes: U2, U5, and U4 + U6 small nuclear ribonucleoprotein particles in the spliceosome. *Science* 233:1294–1299.

Krainer, A. R., and Maniatis, T., 1985. Multiple factors including the small nuclear ribonucleoproteins U1 and U2 are necessary for pre-mRNA splicing in vitro. *Cell* 42:725–736.

Padgett, R. A., Konarska, M. M., Grabowski, P. J., Hardy, S. F., and Sharp, P. A., 1984. Lariat RNAs as intermediates and products in the splicing of messenger RNA precursors. *Science* 225:898–903.

Reed, R., and Maniatis, T., 1986. A role for exon sequences and splice-site proximity in splice-site selection. *Cell* 46:681–690.

Ruskin, B., and Green, M. R., 1985. An RNA processing activity that debranches RNA lariats. *Science* 229:135–140.

Wolin, S.L., 1985. Small cytoplasmic ribonucleoproteins. *Trends Genet.* 1:201–204.

Self-splicing and RNA catalysis

Cech, T. R., 1987. The chemistry of self-splicing RNA and RNA enzymes. *Science* 236:1532–1539.

Cech, T. R., and Bass, B. L., 1986. Biological catalysis by RNA. *Ann. Rev. Biochem.* 55:599.

Zaug, A. J., Been, M. D., and Cech, T. R., 1986. The *Tetrahymena* ribozyme acts like an RNA restriction endonuclease. *Nature* 324:429–433.

Garriga, G., Lambowitz, A. M., Inoue, T., and Cech, T. R., 1986. Mechanism of recognition of the 5′ splice site in self-splicing group I introns. *Science* 322:86–89.

Szostak, J. W., 1986. Enzymatic activity of the conserved core of a group I self-splicing intron. *Nature* 322:83–86.

Waring, R. B., Towner, P., Minter, S. J., and Davies, R. W., 1986. Splice-site selection by a self-splicing RNA of *Tetrahymena*. *Nature* 321:133–139.

PROBLEMS

1. Heparin inhibits RNA polymerase by binding to its β' subunit. Why is heparin an effective inhibitor of transcription?

2. Sigma protein by itself does not bind to promoter sites. Predict the effect of a mutation enabling sigma to bind to the -10 and -35 regions in the absence of other subunits of RNA polymerase.

3. What would be the likely effect of a mutation that would prevent σ from dissociating from the RNA polymerase core?

4. What is the minimum length of time required for the synthesis by *E. coli* polymerase of an mRNA encoding a 100-kd protein?

5. How far apart are transcription bubbles on *E. coli* genes that are being transcribed at a maximal rate?

6. What is the amino acid sequence of the extra segment of protein synthesized in the thallasemic patient with a mutation leading to aberrant splicing (see Figure 29-30)? The reading frame after the splice site begins with TCT.

7. Another thalassemic patient had a mutation leading to the production of an mRNA for the β chain of hemoglobin that was 900 nucleotides longer than the normal one. The poly A tail of this mutant mRNA was located a few nucleotides after the only AAUAAA sequence in the additional sequence. Propose a mutation that would lead to the production of this altered mRNA.

Protein Synthesis

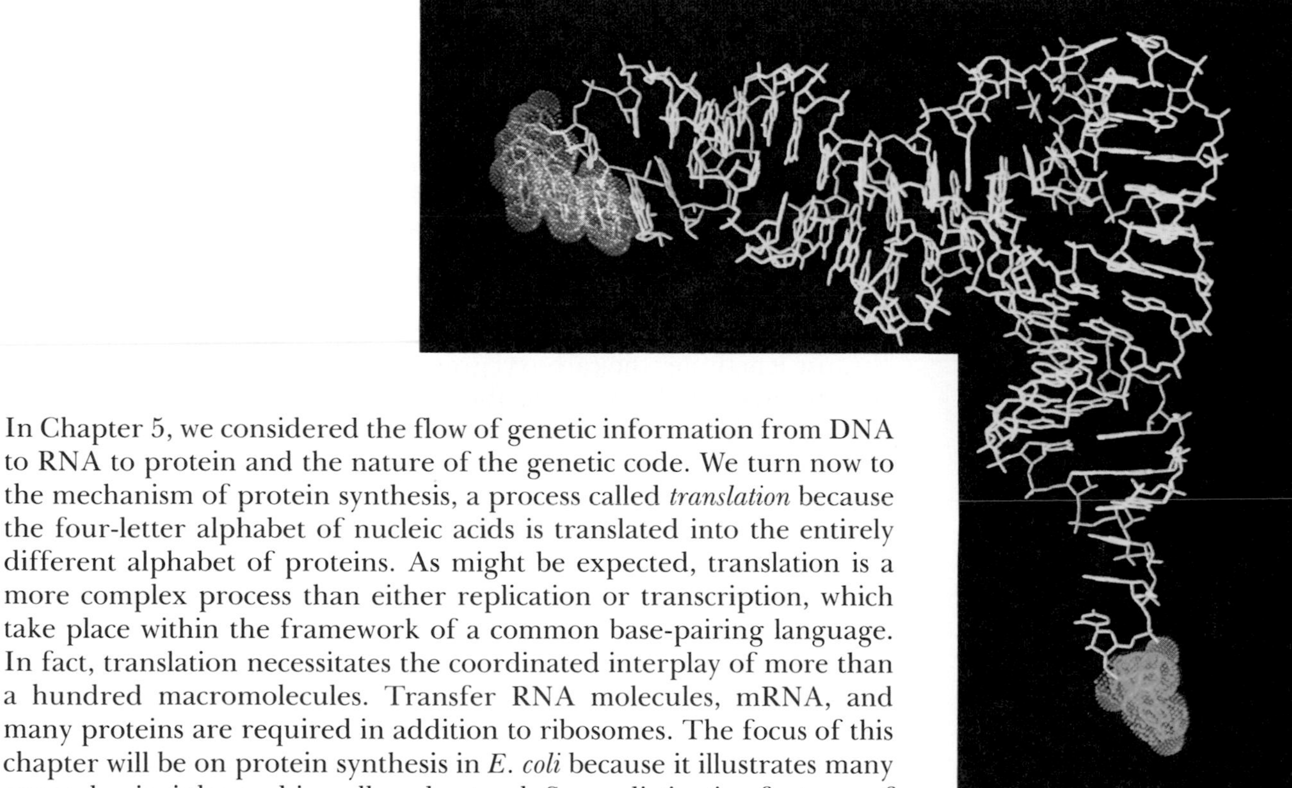

In Chapter 5, we considered the flow of genetic information from DNA to RNA to protein and the nature of the genetic code. We turn now to the mechanism of protein synthesis, a process called *translation* because the four-letter alphabet of nucleic acids is translated into the entirely different alphabet of proteins. As might be expected, translation is a more complex process than either replication or transcription, which take place within the framework of a common base-pairing language. In fact, translation necessitates the coordinated interplay of more than a hundred macromolecules. Transfer RNA molecules, mRNA, and many proteins are required in addition to ribosomes. The focus of this chapter will be on protein synthesis in *E. coli* because it illustrates many general principles and is well understood. Some distinctive features of protein synthesis in eucaryotes will also be presented.

Figure 30-1
Transfer RNAs, the central molecules of protein synthesis, serve as adaptors between the four-base language of nucleic acids and the twenty-amino-acid language of proteins. The van der Waals outline of the anticodon is shown in red and the outline of the attachment site for the amino acid in blue.

Let us take an overview of protein synthesis before examining it in some detail. A protein is synthesized in the amino-to-carboxyl direction by the sequential addition of amino acids to the carboxyl end of the growing peptide chain. The activated precursors are aminoacyl-tRNAs, in which the carboxyl group of an amino acid is joined to the 3′-OH of a transfer RNA (tRNA). The linking of an amino acid to its corresponding tRNA is catalyzed by an aminoacyl-tRNA synthetase. This activation reaction, which is analogous to the activation of fatty acids, is driven by ATP. For each amino acid, there is at least one kind of tRNA and activating enzyme.

Protein synthesis takes place in three stages: initiation, elongation, and termination. *Initiation* results in the binding of the initiator tRNA to the start signal of mRNA. The initiator tRNA occupies the P (peptidyl) site on a ribosome. *Elongation* starts with the binding of an aminoacyl-tRNA to the other tRNA-binding site on the ribosome, which is called the A (aminoacyl) site. A peptide bond then forms between the amino group of the incoming aminoacyl-tRNA and the carboxyl group of the formylmethionine carried by the initiator tRNA. The resulting dipeptidyl-tRNA then moves from the A site to the P site, and the initiator tRNA molecule leaves the ribosome. The binding of aminoacyl-tRNA, the movement of peptidyl-tRNA from the A site to the P site, and the associated movement of the mRNA to the next codon are powered by the hydrolysis of GTP. An aminoacyl-tRNA then binds to the vacant A site to start another round of elongation, which proceeds as described above. *Termination* occurs when a stop signal on the mRNA is read by a protein release factor, which leads to the release of the completed polypeptide chain from the ribosome.

AMINO ACIDS ARE ACTIVATED AND LINKED TO SPECIFIC TRANSFER RNAs BY SPECIFIC SYNTHETASES

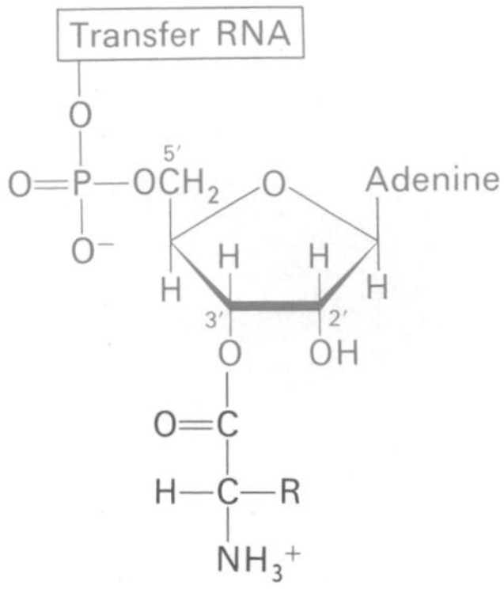

The formation of a peptide bond between the amino group of one amino acid and the carboxyl group of another is thermodynamically unfavorable. This barrier is overcome by activating the carboxyl group of the precursor amino acids. *The activated intermediates in protein synthesis are amino acid esters,* in which the carboxyl group of an amino acid is linked to either the 2′- or the 3′-hydroxyl group of the ribose unit at the 3′ end of tRNA. This activated intermediate is called an *aminoacyl-tRNA* (Figure 30-2).

Figure 30-2
An amino acid is esterified to the 2′- or the 3′-hydroxyl group of the terminal adenosine in an aminoacyl-tRNA.

The attachment of an amino acid to a tRNA is important not only because it activates the carboxyl group, but also because amino acids by themselves cannot recognize the codons on mRNA. Rather, amino acids must be carried to the ribosomes by specific tRNAs, which do recognize codons on mRNA and thereby act as adaptor molecules.

In 1957, Paul Zamecnik and Mahlon Hoagland discovered that the activation of amino acids and their subsequent linkage to tRNAs are catalyzed by specific *aminoacyl-tRNA synthetases,* which are also called *activating enzymes.* The first step is the formation of an *aminoacyl-adenylate* from an amino acid and ATP. This activated species is a mixed anhydride in which the carboxyl group of the amino acid is linked to the phosphoryl group of AMP; hence, it is also known as *aminoacyl-AMP*.

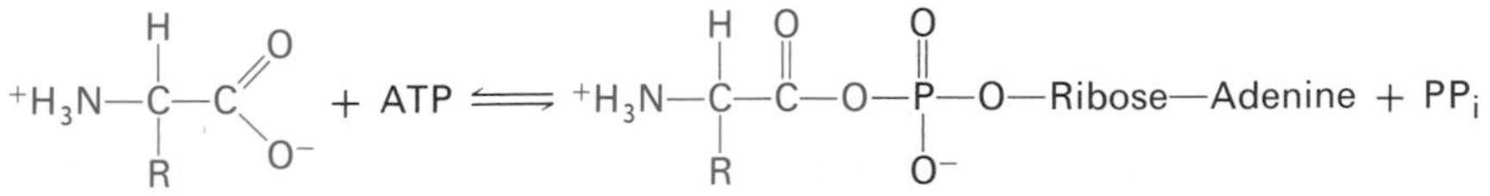

$$^{+}H_3N{-}CH(R){-}COO^{-} + \text{ATP} \rightleftharpoons \,^{+}H_3N{-}CH(R){-}C(=O){-}O{-}P(=O)(O^{-}){-}O{-}\text{Ribose}{-}\text{Adenine} + PP_i$$

Aminoacyl-adenylate (Aminoacyl-AMP)

The next step is the transfer of the aminoacyl group of aminoacyl-AMP to a tRNA molecule to form *aminoacyl-tRNA,* the activated intermediate in protein synthesis. For some tRNAs, the amino acid is transferred to the 2′-hydroxyl of the ribose unit and, for others, to the 3′-hydroxyl. The activated amino acid can migrate very rapidly between the 2′- and 3′-hydroxyl groups.

$$\text{Aminoacyl-AMP} + \text{tRNA} \rightleftharpoons \text{aminoacyl-tRNA} + \text{AMP}$$

The sum of these activation and transfer steps is

$$\text{Amino acid} + \text{ATP} + \text{tRNA} \rightleftharpoons \text{aminoacyl-tRNA} + \text{AMP} + \text{PP}_i$$

The $\Delta G^{\circ\prime}$ of this reaction is close to 0, because the free energy of hydrolysis of the ester bond of aminoacyl-tRNA is similar to that of the terminal phosphoryl group of ATP. What then drives the synthesis of aminoacyl-tRNA? As expected, the reaction is driven by the hydrolysis of pyrophosphate. The sum of these three reactions is highly exergonic:

$$\text{Amino acid} + \text{ATP} + \text{tRNA} + H_2O \longrightarrow \text{aminoacyl-tRNA} + \text{AMP} + 2\ \text{P}_i$$

Thus, *two high-energy phosphate bonds are consumed in the synthesis of an aminoacyl-tRNA*. One of them is consumed in forming the ester linkage of aminoacyl-tRNA, whereas the other is consumed in driving the reaction forward.

The activation and transfer steps for a particular amino acid are catalyzed by the same aminoacyl-tRNA synthetase. In fact, *the aminoacyl-AMP intermediate does not dissociate from the synthetase*. Rather, it is tightly bound to the active site of the enzyme by noncovalent interactions. The aminoacyl-AMP is normally a transient intermediate in the synthesis of aminoacyl-tRNA, but it is quite stable and readily isolated if tRNA is absent from the reaction mixture.

An acyl-adenylate intermediate has already been encountered in fatty acid activation (p. 473). In fact, Paul Berg first discovered this intermediate in fatty acid activation, and then recognized that it is also formed in amino acid activation. The major difference between these reactions is that the acceptor of the acyl group is CoA in the former and tRNA in the latter. The energetics of these biosyntheses are very similar: both are made irreversible by the hydrolysis of pyrophosphate.

$$\text{Amino acid}-\overset{\overset{\large O}{\|}}{C}-O-\text{tRNA}$$

$$\text{Fatty acid}-\overset{\overset{\large O}{\|}}{C}-S-\text{CoA}$$

At least one aminoacyl-tRNA synthetase exists for each amino acid. These enzymes differ markedly in size and subunit structure (Table 30-1).

Table 30-1
Properties of some aminoacyl-tRNA synthetases

Source	*Amino acid specificity*	*Mass (kd)*	*Subunit structure*
E. coli	Histidine	85	α_2
E. coli	Isoleucine	114	One chain
E. coli	Lysine	104	α_2
E. coli	Glycine	227	$\alpha_2\beta_2$
Yeast	Lysine	138	α_2
Yeast	Phenylalanine	270	$\alpha_2\beta_2$
Beef pancreas	Tryptophan	108	α_2

THE FIDELITY OF PROTEIN SYNTHESIS DEPENDS ON THE HIGH SPECIFICITY OF AMINOACYL-tRNA SYNTHETASES

Aminoacyl-tRNA synthetases are highly selective in their recognition of both the amino acid to be activated and the prospective tRNA acceptor. As will be discussed shortly, tRNA molecules that accept different amino acids have different base sequences, and so they can be readily distinguished by their synthetases. A much more demanding task for these enzymes is to discriminate between similar amino acids. For example, the only difference between isoleucine and valine is that isoleucine contains an additional methylene group (Figure 30-3). The additional binding energy contributed by this extra —CH_2— group favors the activation of isoleucine over valine by isoleucyl-tRNA synthetase by a factor of about 200. Even so, the concentration of valine in vivo is about five times that of isoleucine, and so valine would be mistakenly incorporated into proteins in place of isoleucine 1 in 40 times. However, the observed error frequency in vivo is only 1 in 3000, indicating that there must be a subsequent editing step to enhance fidelity.

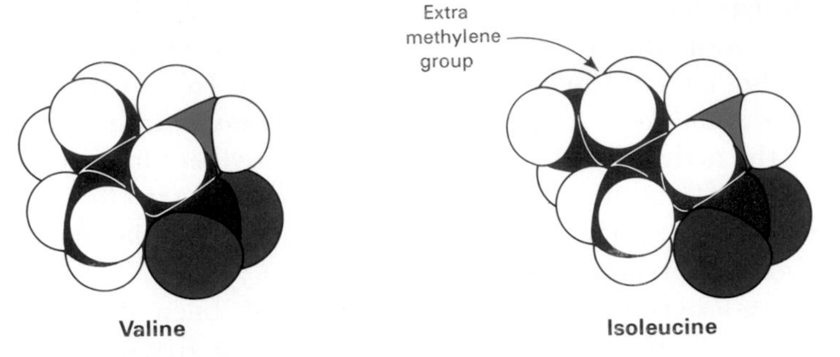

Figure 30-3
Molecular models of valine and isoleucine. The extra methylene group in isoleucine is marked. The synthetases specific for these amino acids are highly discerning.

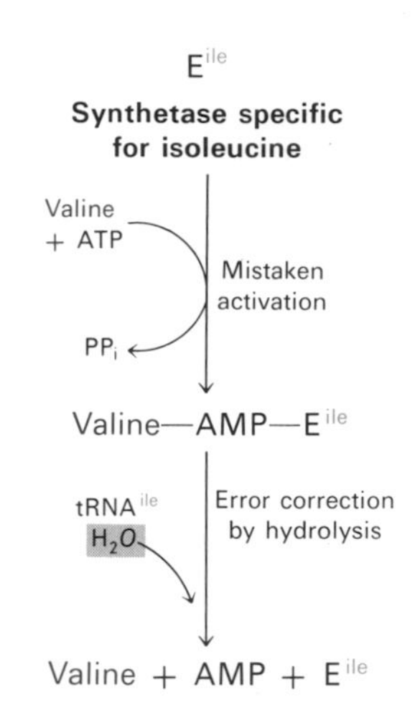

Figure 30-4
Proofreading by hydrolysis of an incorrect aminoacyl-AMP. The entry of isoleucyl tRNA induces hydrolysis of valyl-AMP.

In fact, *the synthetase corrects its own errors.* Valine that is mistakenly activated is not transferred to the tRNA specific for isoleucine. Instead, *this tRNA promotes the hydrolysis of valine-AMP and thereby prevents its erroneous incorporation into proteins* (Figure 30-4). Furthermore, hydrolysis frees the synthetase to activate and transfer isoleucine, the correct amino acid. How does the synthetase avoid hydrolyzing isoleucine-AMP, the correct intermediate? Most likely, the hydrolytic site is just large enough to accommodate valine-AMP but too small to allow the entry of isoleucine-AMP.

Most aminoacyl-tRNA synthetases contain hydrolytic sites in addition to synthetic sites. Complementary pairs of sites function as a *double sieve* to assure very high fidelity. The synthetic site rejects amino acids that are *larger* than the correct one because there is insufficient room for them, whereas the hydrolytic site destroys activated intermediates that are *smaller* than the correct species (Figure 30-5). Hydrolytic proofreading is central to the fidelity of many aminoacyl-tRNA synthetases, as it is to DNA polymerases (p. 666). However, a few synthetases achieve high accuracy without editing their covalently attached intermediates. For example, tyrosyl-tRNA synthetase has no difficulty discriminating between tyrosine and phenylalanine; the hydroxyl group on the tyrosine ring enables it to be bound to the enzyme 10^4 times as strongly as phenylalanine. Proofreading is costly in energy and time and hence is selected in the course of evolution only when fidelity must be enhanced beyond what can be obtained through an initial binding interaction.

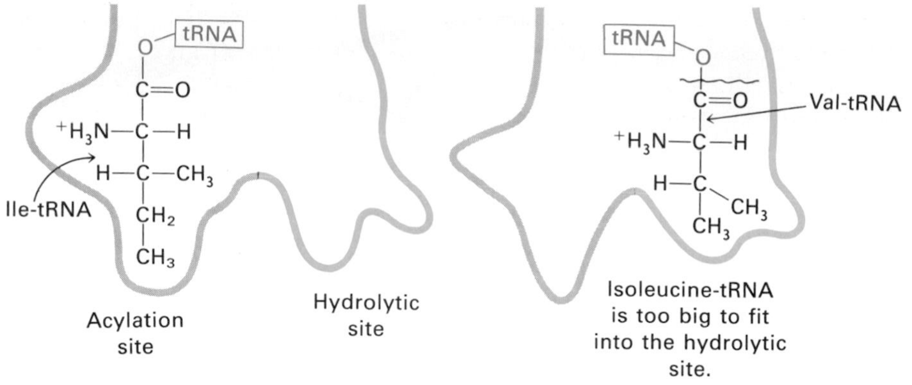

Figure 30-5
Double-sieve mechanism for rejecting amino acids that are either smaller or larger than the cognate one. The catalytic site for the joining of the amino acid to the AMP unit is the first sieve; the hydrolytic site for proofreading is the second sieve.

TYROSYL-AMP FORMATION IS GREATLY ACCELERATED BY THE BINDING OF γ-PHOSPHATE IN THE TRANSITION STATE

X-ray crystallographic and protein engineering studies have provided insight into the catalytic action of tyrosyl-tRNA synthetase, a dimer of 47-kd subunits. The amino-terminal 320 residues are needed for the activation reaction, whereas the carboxy-terminal 99 residues participate in the binding of tRNA and the formation of tyrosyl-tRNA. The crystal structure of the synthetase containing bound tyrosyl-AMP has been solved at high resolution. This intermediate is stable in the absence of the matching tRNA and is bound to the enzyme by some twelve hydrogen bonds (Figure 30-6).

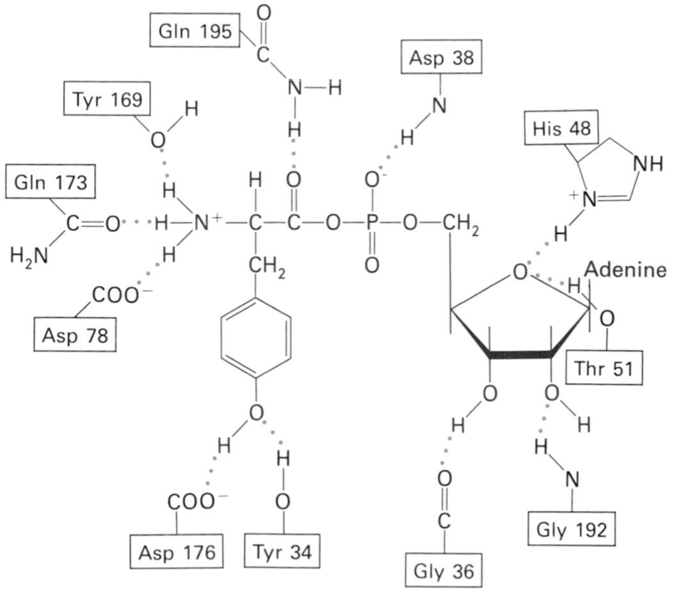

Figure 30-6
Tyrosyl-AMP (shown in red) is bound to its synthetase by multiple hydrogen bonds. [Courtesy of Dr. David Blow.]

The formation of tyrosyl-AMP from tyrosine and ATP leads to an *inversion* of configuration of the α-phosphorus atom. Hence, the reaction probably proceeds by an *in-line displacement* in which the tyrosyl carboxylate is the attacking nucleophile and pyrophosphate is the leaving group. The α-phosphorus atom in the transition state is pentacovalent and has the geometry of a trigonal bipyramid, as in the hydrolytic reaction catalyzed by ribonuclease (p. 212).

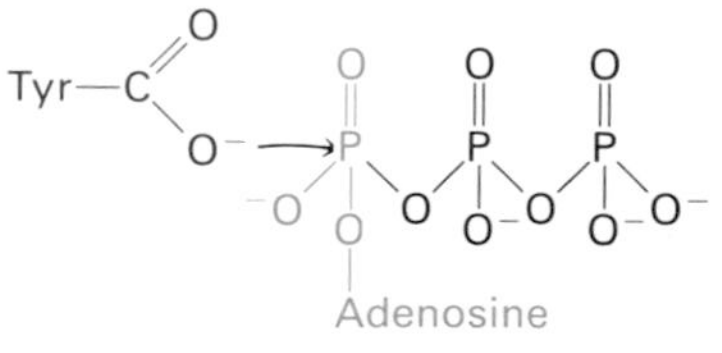

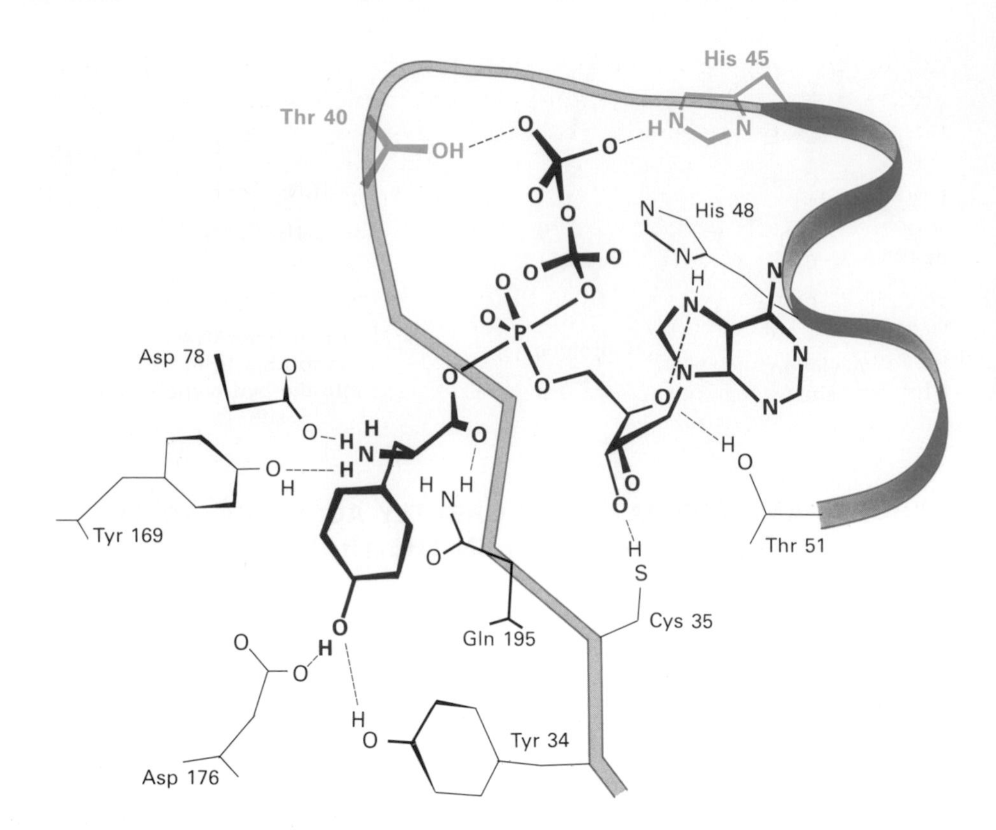

Figure 30-7
Proposed structure of the transition state in the formation of tyrosyl-AMP from tyrosine and ATP. The side chains of threonine 40 and histidine 45 (shown in green) play a critical role in catalysis by selectively binding the γ phosphate in the transition state. [After R. J. Leatherbarrow, A. R. Fersht, and G. Winter. *Proc. Nat. Acad. Sci.* 82(1985):7841.]

From this evidence, a plausible structure of the transition state has been deduced by model building (Figure 30-7). A key feature of this model is the hydrogen bonding of the γ-phosphate group to the side chains of threonine 40 and histidine 45. Do these residues enhance catalysis by selectively stabilizing the transition state? Their importance was assessed by applying site-specific mutagenesis to change histidine 45 to glycine, and threonine 40 to alanine, to eliminate the possibility of hydrogen bonding.

These engineered mutant proteins proved to be highly revealing (Table 30-2). The catalytic activity of the double mutant was less than that of the wild type by a factor of 3×10^{-6} but the binding affinities of ATP and tyrosine were altered relatively little. This experiment demonstrates that threonine 40 and histidine 45 are essential for catalysis but

Table 30-2
Properties of mutant tyrosyl-tRNA synthetase

				Dissociation constants (μM)	
Enzyme	*Critical residues*		k_{cat} (s^{-1})	*Tyrosine*	*ATP*
Wild type	Thr 40	His 45	38	12	4700
Mutant	Thr 40	Gly 45	0.16	10	1200
Mutant	Ala 40	His 45	0.0055	8	3800
Double mutant	Ala 40	Gly 45	0.00012	4.5	1100

After R. J. Leatherbarrow, A. R. Fersht, and G. Winter, *Proc. Nat. Acad. Sci.* 82(1985):7842.

not for the binding of the substrates (Figure 30-8). Apparently, they interact strongly with the γ-phosphate group in the transition state but not with the same group in the initial enzyme-substrate complex. It is the large change in position of the pyrophosphate unit accompanying the shift from a tetrahedral to a bipyramidal geometry that triggers this selective binding. Recall that *the essence of catalysis is selective stabilization of the transition state* (p. 184).

Which groups on the enzyme directly participate in making and breaking bonds? Probably none. The carboxylate group of tyrosine is an intrinsically effective nucleophile, ATP is already activated, and Mg^{2+}-pyrophosphate is a good leaving group. The enzyme accelerates catalysis by a factor of about 4×10^4 simply by bringing tyrosine and ATP together, and it gains another factor of 3×10^5 mainly by binding γ-phosphate in the transition state. It is noteworthy that the new interactions in the transition state are at some distance from the α-phosphorus atom, the reaction center. Thus, *catalysis can be delocalized*—what counts is selective binding of the transition state, however achieved.

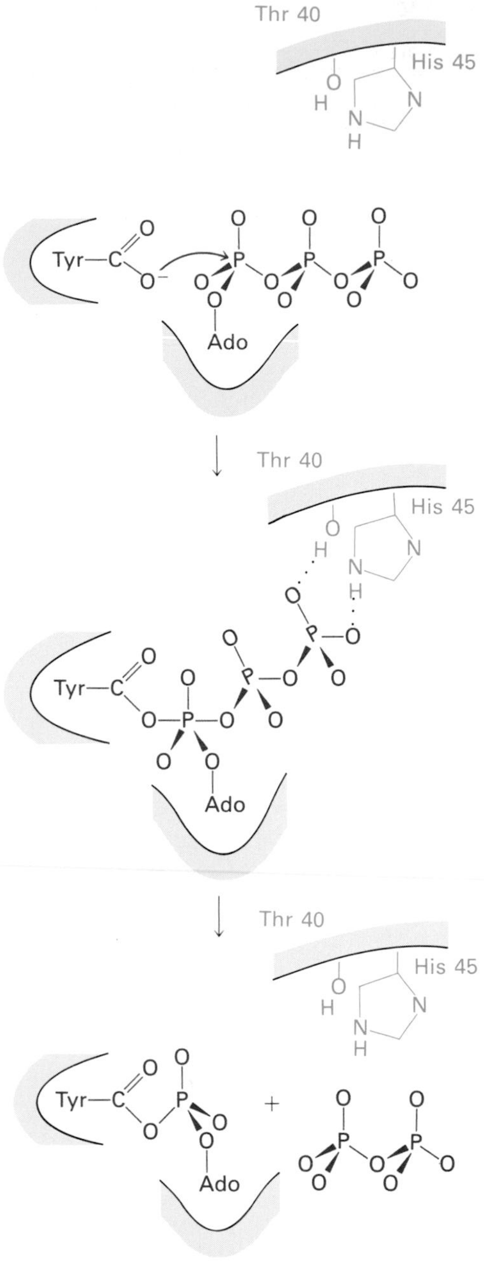

TRANSFER RNA MOLECULES HAVE A COMMON DESIGN

The base sequence of a transfer RNA molecule was first determined by Robert Holley in 1965, as the culmination of seven years of effort. Indeed, his study of yeast alanine tRNA provided the first complete sequence of any nucleic acid, as well as a source of insight into how tRNA functions. This adaptor molecule is a single chain of seventy-six ribonucleotides (Figure 30-9). The 5′ terminus is phosphorylated (pG),

Figure 30-8
Proposed mechanism for the formation of tyrosyl-AMP by tyrosyl-tRNA synthetase. The α-phosphorus atom of ATP is nucleophilically attacked by a carboxylate oxygen of tyrosine. The transition state is stabilized by hydrogen bonding of the γ phosphate to the side chains of a threonine and a histidine residue. Pyrophosphate bound to Mg^{2+} (not shown) leaves the pentacovalent transition state to give tyrosyl-AMP. [After R. J. Leatherbarrow, A. R. Fersht, and G. Winter. *Proc. Nat. Acad. Sci.* 82(1985):7841.]

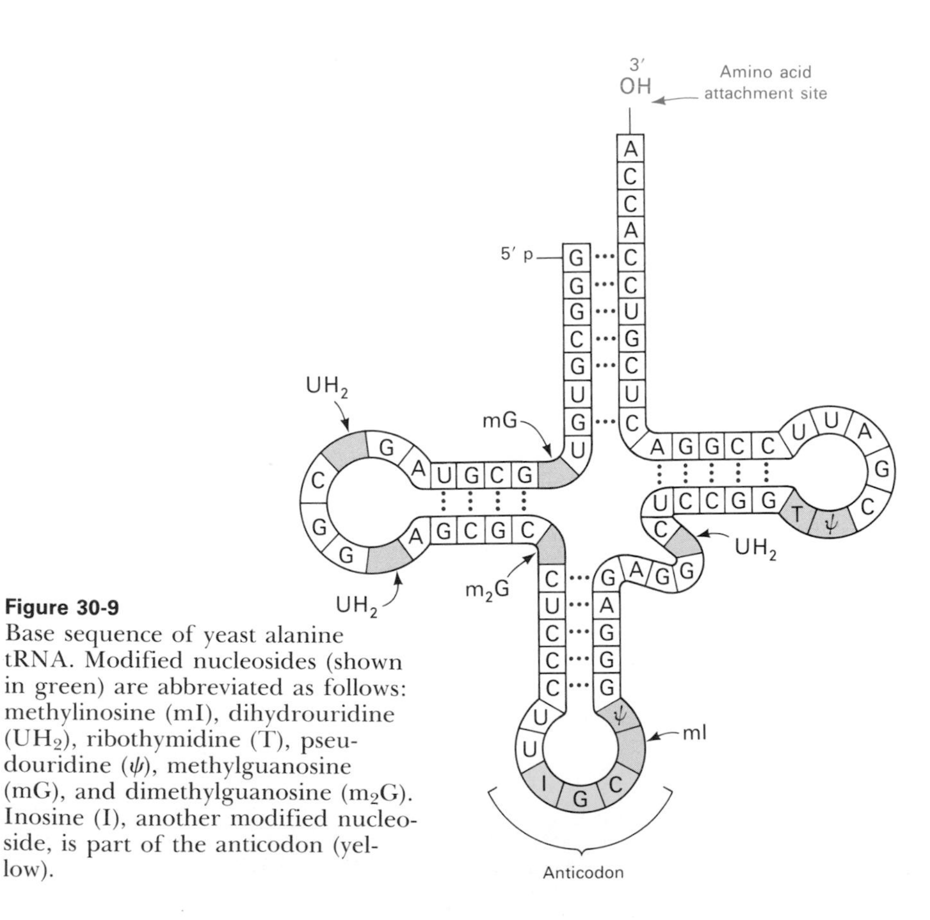

Figure 30-9
Base sequence of yeast alanine tRNA. Modified nucleosides (shown in green) are abbreviated as follows: methylinosine (mI), dihydrouridine (UH_2), ribothymidine (T), pseudouridine (ψ), methylguanosine (mG), and dimethylguanosine (m_2G). Inosine (I), another modified nucleoside, is part of the anticodon (yellow).

Anticodon

3′ —C—G—I— 5′

—G—C—C—

5′ 3′

Codon

whereas the 3′ terminus has a free hydroxyl group. A striking feature of this RNA is its high content of bases other than A, U, G, and C. Many unusual nucleosides are present: inosine, pseudouridine, dihydrouridine, ribothymidine, and methylated derivatives of guanosine and inosine. The *amino acid attachment site* is the 3′-hydroxyl group of the adenosine residue at the 3′ terminus of the molecule. The sequence IGC in the middle of the molecule is the *anticodon.* It is complementary to GCC, one of the codons for alanine.

The sequences of several other tRNA molecules were determined a short time later. More than one hundred sequences are now known. The striking finding is that all of them can be written in a cloverleaf pattern in which about half the residues are base paired. Hence, *tRNA molecules have many common structural features.* This finding is not unexpected, because all tRNA molecules must be able to interact in nearly the same way with ribosomes and mRNAs. Specifically, they must fit into the A and P sites on the ribosome and interact with the enzymatic site that catalyzes peptide-bond formation.

All known transfer RNA molecules share the following features (Figure 30-10):

1. They are single chains containing between *73 and 93 ribonucleotides* (about 25 kd) each.

NH_2, CH_3, H, O, Ribose

5-Methylcytidine

2. They contain *many unusual bases,* typically between 7 and 15 per molecule. Many of them are methylated or dimethylated derivatives of A, U, C, and G that are formed by enzymatic modification of a precursor tRNA (p. 713). Methylation prevents the formation of certain base pairs, thereby rendering some of the bases accessible for other interactions. Also, methylation imparts a hydrophobic character to some re-

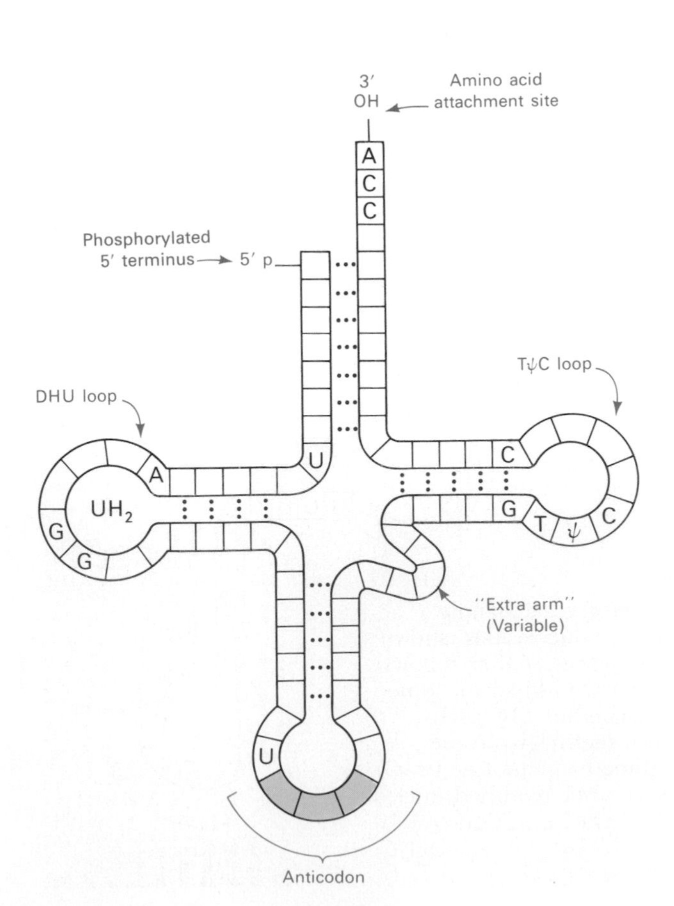

Figure 30-10
Common features of tRNA molecules.

gions of tRNAs, which may be important for their interaction with synthetases and ribosomal proteins. Other modifications alter codon recognition, as will be discussed shortly.

3. The 5′ end of tRNAs is phosphorylated. The 5′ terminal residue is usually pG.

4. The base sequence at the 3′ end of tRNAs is CCA. The activated amino acid is attached to the 3′-hydroxyl group of the terminal adenosine.

5. About half of the nucleotides in tRNAs are base paired to form double helices. Five groups of bases are not base paired: the 3′ *CCA terminal region;* the *TψC loop,* which acquired its name from the sequence ribothymine-pseudouracil-cytosine; the *"extra arm"*, which contains a variable number of residues; the *DHU loop,* which contains several dihydrouracil residues; and the *anticodon loop.*

Dihydrouridine

6. The *anticodon loop* consists of seven bases, with the following sequence:

5′ —Pyrimidine—Pyrimidine—X—Y—Z—Modified purine—Variable base— 3′

(X—Y—Z: Anticodon)

TRANSFER RNA IS AN L-SHAPED MOLECULE

The three-dimensional structure of a tRNA molecule was first solved in 1974 from the results of x-ray crystallographic studies carried out in the laboratories of Alexander Rich and Aaron Klug. Their independent analyses of yeast phenylalanine tRNA provided a wealth of structural information:

1. The molecule is *L-shaped* (Figure 30-11).

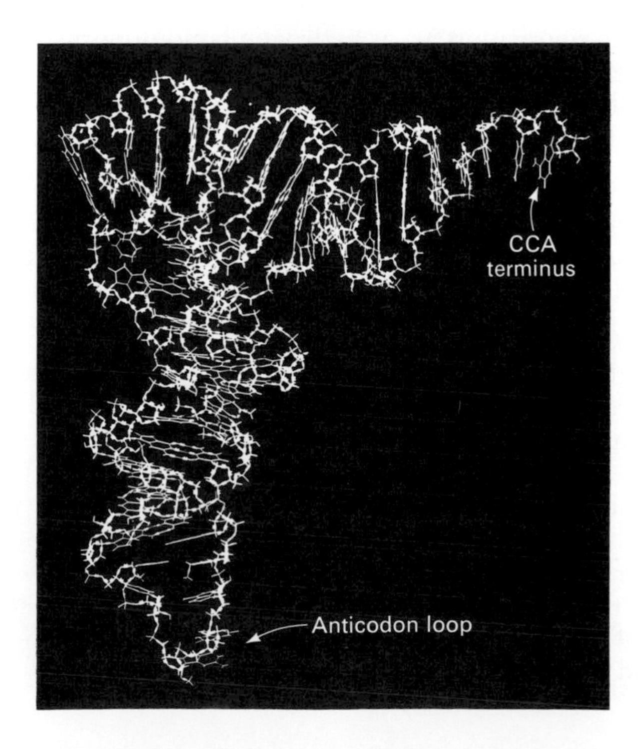

Figure 30-11
Photograph of a skeletal model of yeast phenylalanine tRNA based on an electron-density map at 3-Å resolution. [Courtesy of Dr. Sung-Hou Kim.]

2. There are *two segments of double helix*. Each of these helices contains about ten base pairs, which correspond to one turn of helix. The helical segments are perpendicular to each other, which gives the molecule its L-shape (Figure 30-12). The base pairing in the cloverleaf model, postulated on the basis of sequence studies, is correct.

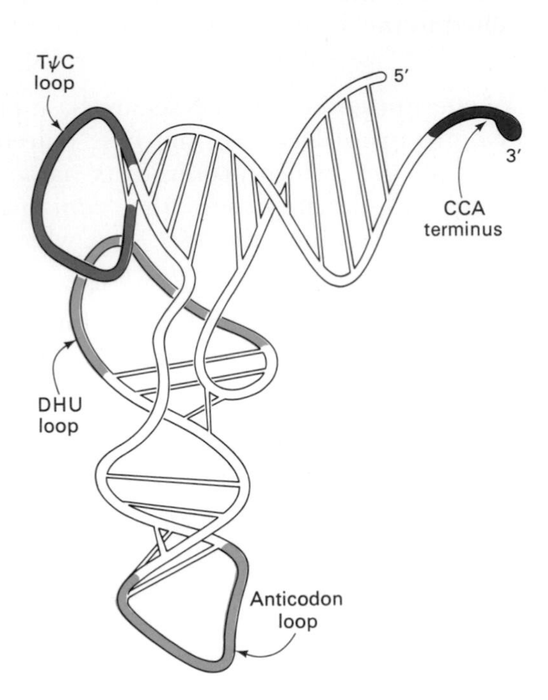

Figure 30-12
Schematic diagram of the three-dimensional structure of yeast phenylalanine tRNA. [After a drawing kindly provided by Dr. Sung-Hou Kim.]

3. Most of the bases in the nonhelical regions participate in unusual hydrogen-bonding interactions. These *tertiary interactions* are between bases that are not usually complementary (e.g., GG, AA, and AC). Moreover, the ribose-phosphate backbone interacts with some bases and even with another region of the backbone itself. The 2′—OH groups of the ribose units act as hydrogen bond donors or acceptors in many of these interactions. In addition, most bases are stacked. These hydrophobic interactions between adjacent aromatic rings play a major role in stabilizing the architecture of the molecule.

4. The CCA terminus containing the *amino acid attachment site* is at one end of the L. The other end of the L is the *anticodon loop*. Thus, *the amino acid in aminoacyl-tRNA is far from the anticodon* (about 80 Å). The DHU and TψC loops form the corner of the L.

5. The CCA terminus and the adjacent helical region do not interact strongly with the rest of the molecule. This part of the molecule may change its conformation during amino acid activation and also during protein synthesis on the ribosome.

Subsequent x-ray analyses of other procaryotic and eucaryotic tRNAs have shown that their molecular architecture follows the same plan as that of yeast phenylalanine tRNA. Efforts are now being made to achieve more demanding goals, such as solving the three-dimensional structure of tRNA-synthetase complexes and even of tRNA bound to mRNA and ribosomal proteins.

MULTIPLE TRANSFER RNA MOLECULES ARISE FROM THE CLEAVAGE OF A LARGE PRECURSOR BY RIBONUCLEASE P

The sixty genes for transfer RNA molecules in *E. coli* are clustered in twenty-five units that are transcribed into multimeric precursors. Some units encode ribosomal RNAs as well, whereas others contain only tRNAs. One of these transcripts is the precursor of seven tRNAs: one specific for leucine, two for internal methionines, and two each for two kinds of glutamine codons (Figure 30-13). The primary transcript of 950 nucleotides is cleaved by *ribonuclease P* (RNase P) on the 5′ side of the first nucleotide of each mature tRNA to be. *Ribonuclease D* (RNase D) then trims the exposed 3′ end of each until it reaches the CCA sequence, which becomes the 3′ terminus of the mature molecule.

Ribonuclease P is a ribonucleoprotein enzyme consisting of a 377-nucleotide M1 RNA molecule and a 20-kd protein. Sidney Altman discovered that the RNA component alone possesses enzymatic activity. In 60 mM Mg^{2+} (much higher than physiologic concentration), M1 RNA by itself recognizes target sites in primary transcripts and cleaves them at an appreciable rate. This experiment showed that M1 RNA interacts specifically with the substrate and possesses the catalytic groups. The role of the protein is subsidiary—it increases the hydrolytic rate and enables the reaction to occur at a much lower concentration of Mg^{2+}. We see here another example of *catalytically active RNA*.

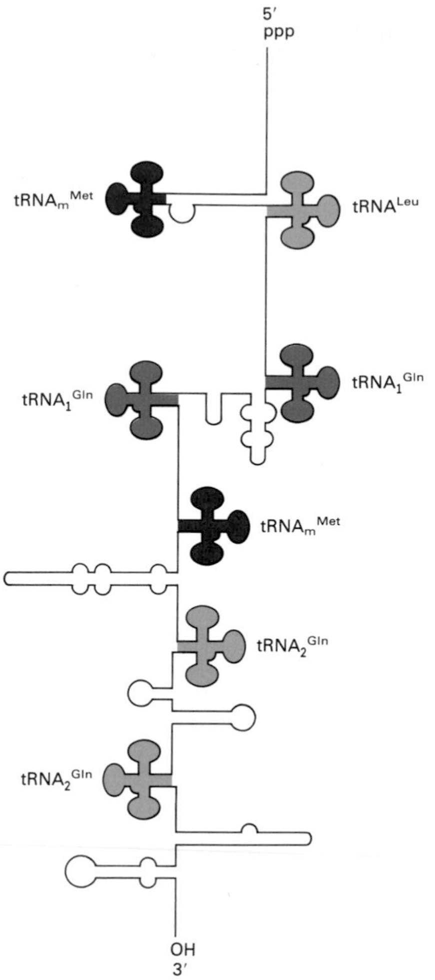

Figure 30-13
Seven tRNA molecules are formed by cleavage of this 950-nucleotide primary transcript. [After N. Nakajima, H. Ozeki, and Y. Shimura. *Cell* 23(1981):245.]

THE CODON IS RECOGNIZED BY THE ANTICODON OF tRNA RATHER THAN BY THE ACTIVATED AMINO ACID

It has already been mentioned that the anticodon on tRNA is the recognition site for the codon on mRNA and that recognition occurs by base pairing. Does the amino acid attached to the tRNA play a role in this recognition process? This question was answered in the following way. First, cysteine was attached to its cognate tRNA (denoted by $tRNA^{Cys}$). The attached cysteine unit was then converted into alanine by reacting Cys-$tRNA^{Cys}$ with Raney nickel, which removed the sulfur atom from the activated cysteine residue without affecting its linkage to tRNA. Thus, a *hybrid aminoacyl-tRNA* was produced in which alanine was covalently attached to a tRNA specific for cysteine.

$$\text{Cysteine} + tRNA^{Cys} \xrightarrow[\text{ATP} + H_2O \;\; \text{AMP} + 2\,P_i]{\text{Cysteine tRNA synthetase}} \underset{\textbf{Cys—tRNA}^{\textbf{Cys}}}{HS{-}CH_2{-}\overset{H}{\underset{NH_3^+}{C}}{-}\overset{O}{\overset{\|}{C}}{-}O{-}tRNA^{Cys}} \xrightarrow{\text{Raney nickel}} \underset{\textbf{Ala-tRNA}^{\textbf{Cys}}}{H{-}CH_2{-}\overset{H}{\underset{NH_3^+}{C}}{-}\overset{O}{\overset{\|}{C}}{-}O{-}tRNA^{Cys}}$$

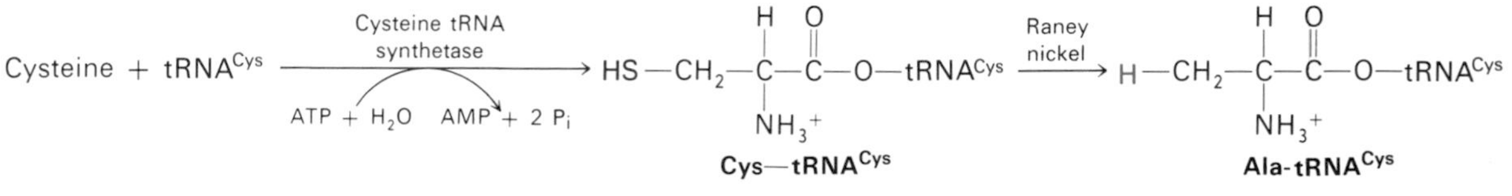

Does this hybrid tRNA recognize the codon for alanine or for cysteine? The answer came on adding the tRNA to a cell-free protein-synthesizing system. The template was a random copolymer of U and G in the ratio of 5:1, which normally leads to the incorporation of cysteine (UGU) but not of alanine (GCX). However, alanine was incorporated into a polypeptide when Ala-$tRNA^{Cys}$ was added to the incubation mixture, because it was attached to the tRNA specific for cysteine. The same result was obtained when mRNA for hemoglobin served as the template and ^{14}C alanyl-$tRNA^{Cys}$ was used as the hybrid aminoacyl-tRNA. The only radioactive tryptic peptide produced was one that normally contained cysteine but not alanine. On the other hand, peptides normally containing alanine but not cysteine were devoid of radioactivity. Thus, *the amino acid in aminoacyl-tRNA does not play a role in selecting a codon.*

SOME TRANSFER RNA MOLECULES RECOGNIZE MORE THAN ONE CODON BECAUSE OF WOBBLE IN BASE PAIRING

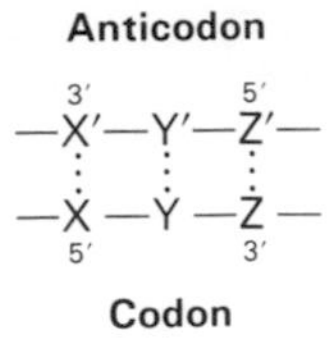

Anticodon

3′ 5′
—X′—Y′—Z′—
—X —Y —Z —
5′ 3′

Codon

What are the rules that govern the recognition of a codon by the anticodon of a tRNA? A simple hypothesis is that each of the bases of the codon forms a Watson-Crick type of base pair with a complementary base on the anticodon. The codon and anticodon would then be lined up in an antiparallel fashion. In the accompanying diagram at the left, the prime denotes the complementary base. Thus X and X′ would be either A and U (or U and A) or G and C (or C and G). A specific prediction of this model is that a particular anticodon can recognize only one codon.

The facts are otherwise. *Some pure tRNA molecules can recognize more than one codon.* For example, the yeast alanine tRNA studied by Holley binds to *three* codons: GCU, GCC, and GCA. The first two bases of these codons are the same, whereas the third is different. Could it be that the recognition of the third base of a codon is sometimes less discriminating than recognition of the other two? The pattern of degeneracy of the genetic code indicates that this might be so. XYU and XYC always code for the same amino acid, whereas XYA and XYG usually do. Crick surmised from these data that the steric criteria for pairing of the third base might be less stringent than for the other two. Models of various base pairs were built to determine which ones are similar to the standard AU and GC base pairs with regard to the distance and angle between the glycosidic bonds. Inosine was included in this study because it appeared in several anticodons. Assuming some steric freedom ("wobble") in the pairing of the third base of the codon, the combinations shown in Table 30-3 seemed plausible.

Table 30-3
Allowed pairings at the third base of the codon according to the wobble hypothesis

First base of anticodon	*Third base of codon*
C	G
A	U
U	A or G
G	U or C
I	U, C, or A

The wobble hypothesis is now firmly established. The anticodons of tRNAs of known sequence bind to the codons predicted by this hypothesis. For example, the anticodon of yeast alanine tRNA is IGC. This tRNA recognizes the codons GCU, GCC, and GCA. Recall that, by convention, nucleotide sequences are written in the 5′ → 3′ direction unless otherwise noted.

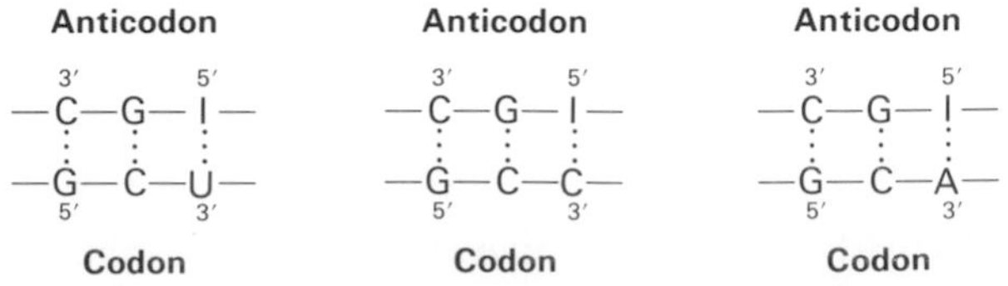

Anticodon
3′ 5′
—C—G—I—
—G—C—U—
5′ 3′
Codon

Anticodon
3′ 5′
—C—G—I—
—G—C—C—
5′ 3′
Codon

Anticodon
3′ 5′
—C—G—I—
—G—C—A—
5′ 3′
Codon

Thus, I pairs with U, C, or A, as predicted. Phenylalanine tRNA, which has the anticodon GAA, recognizes the codons UUU and UUC but not UUA and UUG.

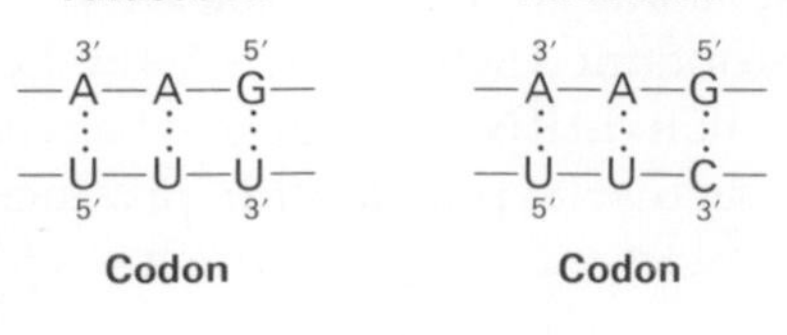

Anticodon
3′ 5′
—A—A—G—
—U—U—U—
5′ 3′
Codon

Anticodon
3′ 5′
—A—A—G—
—U—U—C—
5′ 3′
Codon

Thus, G pairs with either U or C in the third position of the codon, as predicted by the wobble hypothesis.

Two generalizations concerning the codon-anticodon interaction can be made:

1. The first two bases of a codon pair in the standard way. Recognition is precise. Hence, *codons that differ in either of their first two bases must be recognized by different tRNAs*. For example, both UUA and CUA code for leucine but are read by different tRNAs.

2. The first base of an anticodon determines whether a particular tRNA molecule reads one, two, or three kinds of codons: C or A (1 codon), U or G (2 codons), or I (3 codons). Thus, *part of the degeneracy of the genetic code arises from imprecision (wobble) in the pairing of the third base of the codon*. We see here a strong reason for the frequent appearance of inosine, one of the unusual nucleosides, in anticodons. Inosine maximizes the number of codons that can be read by a particular tRNA molecule (Figure 30-14).

Inosine-cytidine base pair

Inosine-adenosine base pair

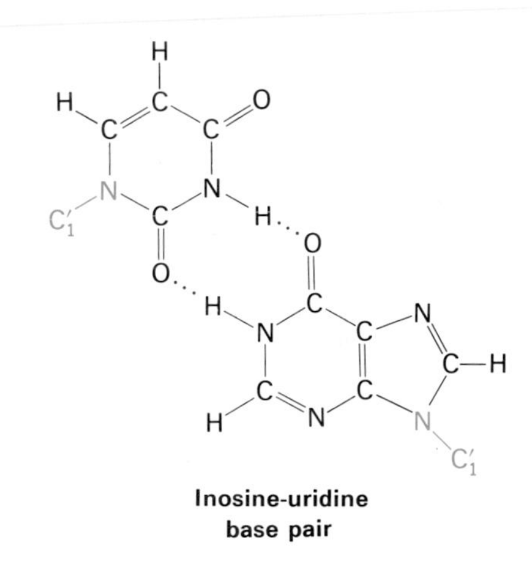

Inosine-uridine base pair

Figure 30-14
Inosine can base-pair with cytosine, adenine, or uracil because of wobble.

MUTANT TRANSFER RNA MOLECULES CAN SUPPRESS OTHER MUTATIONS

Further insight into the process of codon recognition came from studies of mutant tRNAs with an altered base in the anticodon. These spontaneous changes, the converse of the kind in which the amino acid attached to a tRNA is chemically modified in vitro, were discovered in an interesting way. Geneticists had known for many years that the deleterious effects of some mutations can be reversed by a second mutation. Suppose that an enzyme becomes inactive because of a mutation of an mRNA codon from GCU (alanine) to GAU (aspartate). What kind of mutations might restore the activity of this enzyme?

1. A reverse mutation of the same base, A → C, would give the original code word.

2. A different mutation, A → U, would give valine (GUU), which might result in partly or fully restored enzymatic activity.

3. A mutation at a different site in the same gene might reverse the effects of the first mutation. Such an altered protein would differ from the wild type at two amino acids.

4. A mutation in a *different gene* might overcome the deleterious effects of the first mutation. Such a mutation is called an *intergenic suppressor*.

The mode of action of intergenic suppressors was a puzzle for a long time. We now know from genetic and biochemical studies that *most of these suppressors act by altering the reading of mRNA*. For example, consider a mutation that leads to the premature appearance of the codon UAG, a termination signal, producing an incomplete polypeptide chain. Such mutations are called *nonsense mutations* because incomplete polypeptides are usually inactive. The UAG nonsense mutation can be suppressed by mutations in several different genes. One of these suppressors causes UAG to be read as a codon for tyrosine, which may lead to the synthesis

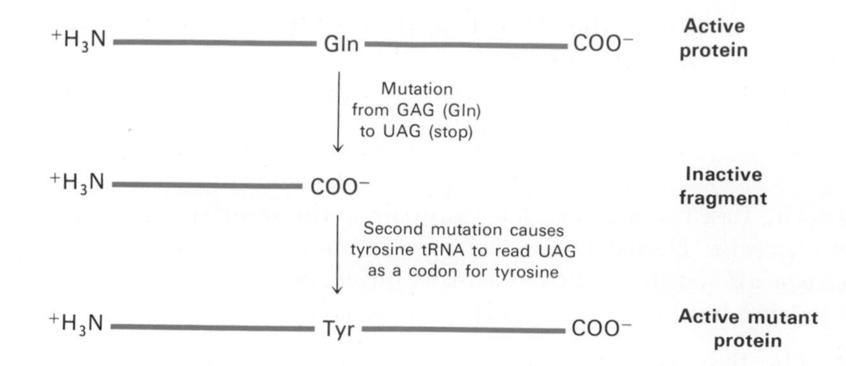

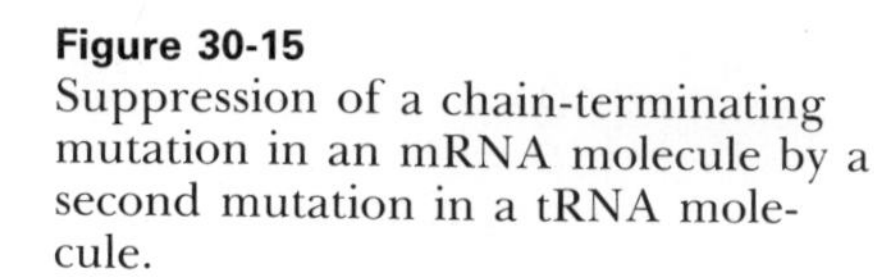

Figure 30-15
Suppression of a chain-terminating mutation in an mRNA molecule by a second mutation in a tRNA molecule.

of a functional protein rather than to an incomplete polypeptide chain (Figure 30-15).

Anticodon

3′ 5′
—A—U—C—
—U—A—G—
5′ 3′

Codon

Mutated base in the anticodon of a tyrosine tRNA (normally 3′ AUG 5′)

This chain-termination codon is read as tyrosine.

Why did this mutant tRNA insert tyrosine in response to the UAG codon? Tyrosine tRNA normally recognizes the codons UAC and UAU. This mutant tRNA proved to be identical with the normal tyrosine tRNA except for a *single base change in its anticodon*: GUA → CUA. The mutation G → C in the first base of its anticodon changed its recognition properties. The altered anticodon recognizes only UAG, as predicted by the wobble hypothesis.

This kind of suppressor mutation is more likely to be selected if the tRNA undergoing mutation is not essential. In other words, there must be another tRNA species that recognizes the same codons as does the one undergoing mutation. In fact, *E. coli* contains two different tRNAs that normally recognize both UAC and UAU. The minor species is the one that is altered in a suppressor mutation; its role under normal conditions is uncertain. The occurrence of this suppressor mutation raises a second question. If the mutation to UAG is read as tyrosine rather than as a stop signal, what happens to normal chain termination? Surprisingly, most polypeptide chains terminate normally in the suppressor mutant, possibly because the termination signal may be more than just the trinucleotide UAG. Indeed, some messages are known to end with two different stop codons. Also, suppression is not completely efficient.

Other kinds of mutant tRNAs have been identified. *Missense suppressors* alter the reading of mRNA so that one amino acid is inserted instead of another at certain codons (e.g., glycine instead of arginine). *Frameshift suppressors* are especially interesting. One of them has an extra base in its anticodon loop. Consequently, it reads *four* nucleotides as a codon rather than three. In particular, UUUC rather than UUU is read as the codon for phenylalanine. A mutation caused by the insertion of an extra base can be suppressed by this altered tRNA.

A A A G — Anticodon of suppressor tRNA

U U U C — Four nucleotides are read as a codon for phenylalanine.

RIBOSOMES ARE RIBONUCLEOPROTEIN PARTICLES (70S) CONSISTING OF A SMALL (30S) AND A LARGE (50S) SUBUNIT

Thus far, we have considered the synthetases that activate amino acids and couple them to transfer RNAs, which serve as adaptors between the two fundamental alphabets of biological systems. We turn now to ribosomes, the molecular machines that coordinate the interplay of tRNAs, mRNA, and proteins in the complex process of protein synthesis. An *E. coli* ribosome is a ribonucleoprotein particle with a mass of about 2700 kd, a diameter of approximately 200 Å, and a sedimentation coefficient of 70S (Figure 30-16). It can be dissociated into a *large subunit* (*50S*) and a *small subunit* (*30S*) (Figure 30-17). These subunits can be further split into their constituent proteins and RNAs. The 30S subunit contains 21 different proteins (labeled S1 to S21) and a 16S RNA molecule. The 50S subunit contains 34 different proteins (labeled L1 to L34) and two RNA molecules, a 23S and a 5S species. A ribosome contains one copy of each RNA molecule, two of L7 and L12, and one of each other protein. The twenty thousand ribosomes in a bacterial cell constitute nearly a fourth of its mass.

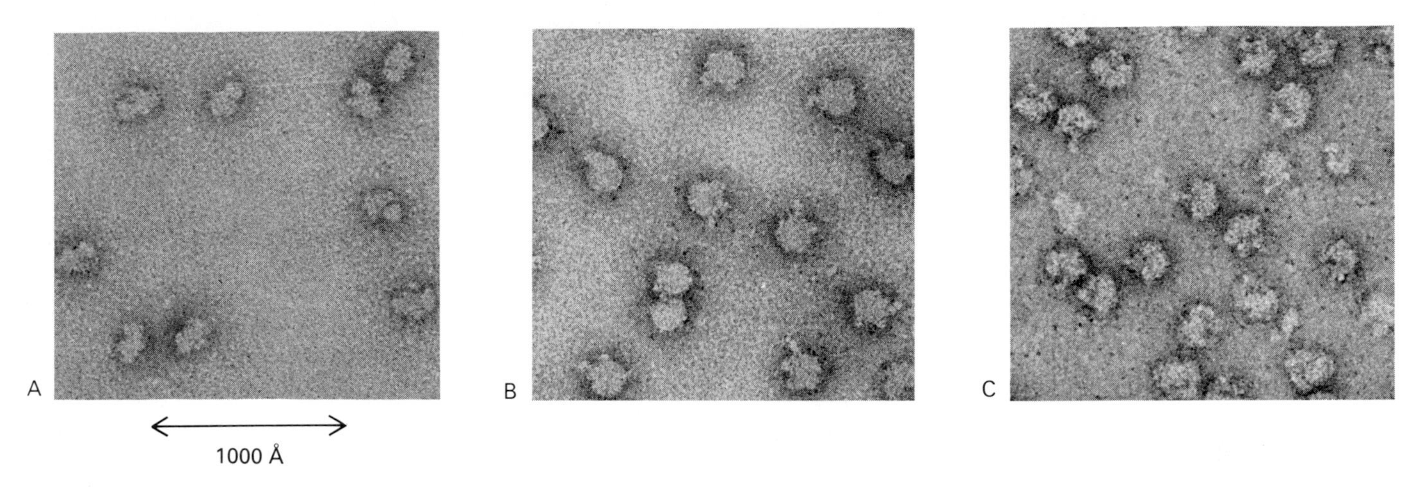

Figure 30-16
Electron micrographs of (A) 30S subunits, (B) 50S subunits, and (C) 70S ribosomes. [Courtesy of Dr. James Lake.]

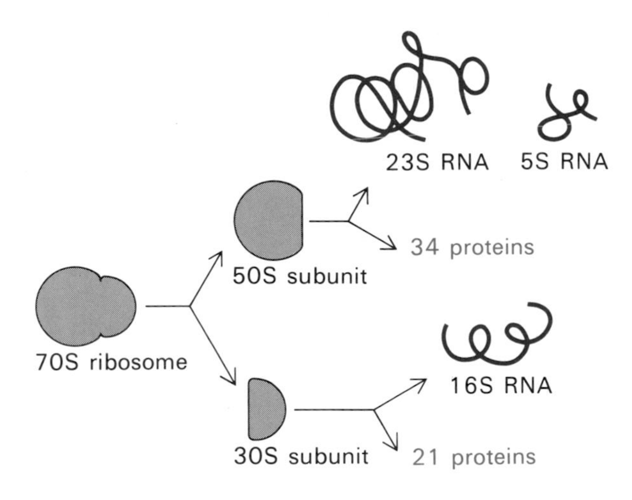

Figure 30-17
Ribosomes can be dissociated into fifty-five kinds of proteins and three RNA molecules.

RIBOSOMAL RNAs (5S, 16S, AND 23S rRNA) CONTAIN MANY BASE-PAIRED HELICAL REGIONS

The prefix *ribo* in the name *ribosome* is apt, for RNA constitutes nearly two-thirds of the mass of these large molecular assemblies. The three RNAs—5S, 16S, and 23S—are critical for ribosomal architecture and function. *A striking feature of these rRNAs is their folding into defined structures with many short duplex regions.* The base-pairing patterns of these molecules have been deduced by carrying out chemical modification and digestion experiments and by comparing nucleotide sequences of

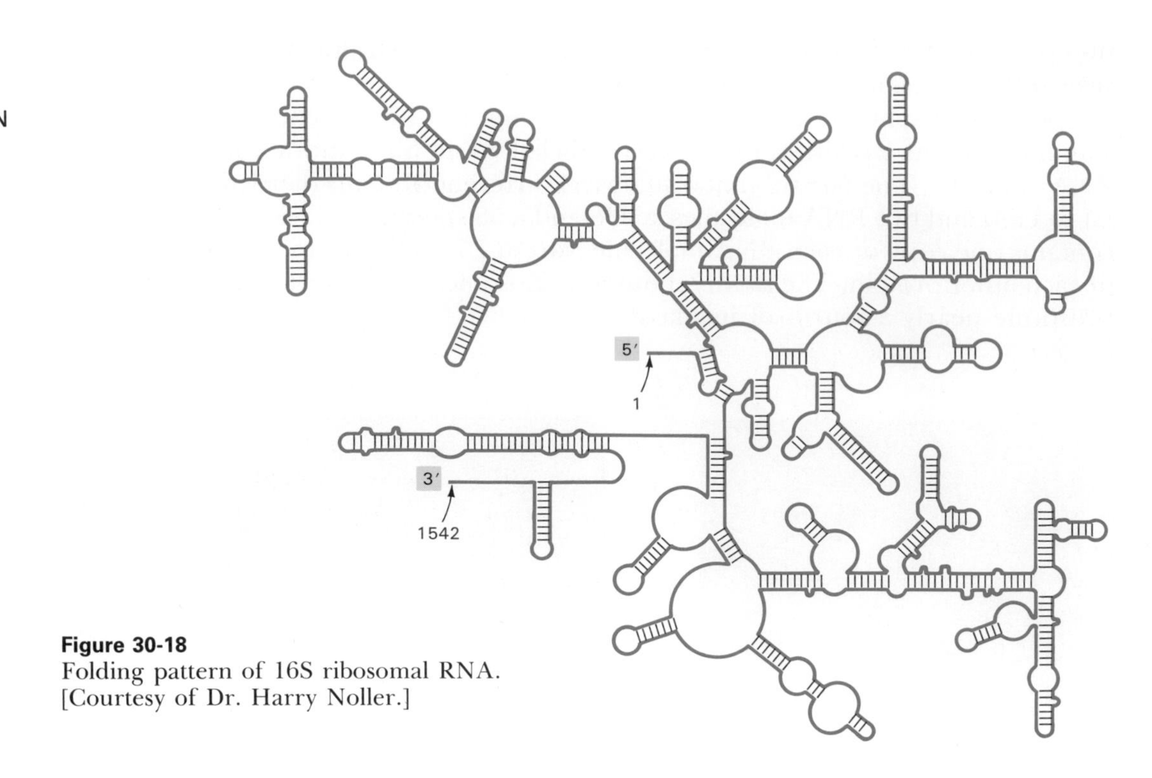

Figure 30-18
Folding pattern of 16S ribosomal RNA. [Courtesy of Dr. Harry Noller.]

many species to detect conserved features (Figure 30-18). The 3′ end of 16S RNA, a constituent of the 30S ribosomal subunit, plays a key role in selecting the start site on the mRNA template, as will be described shortly. One of the challenges now is to elucidate other specific roles of rRNAs.

Ribosomal RNA molecules are formed by cleavage of primary 30S transcripts containing a tandem array of 16S, 23S, and 5S units. These large precursors are encoded by seven different *rrn* operons. In each transcript, one or two tRNAs lie between 16S and 23S RNA and up to two tRNAs follow 5S RNA (Figure 30-19). The precursor RNA is cleaved by ribonuclease III (RNase III) into a pre-16S and a pre-23S fragment. Further processing takes place after the binding of ribosomal proteins to these fragments.

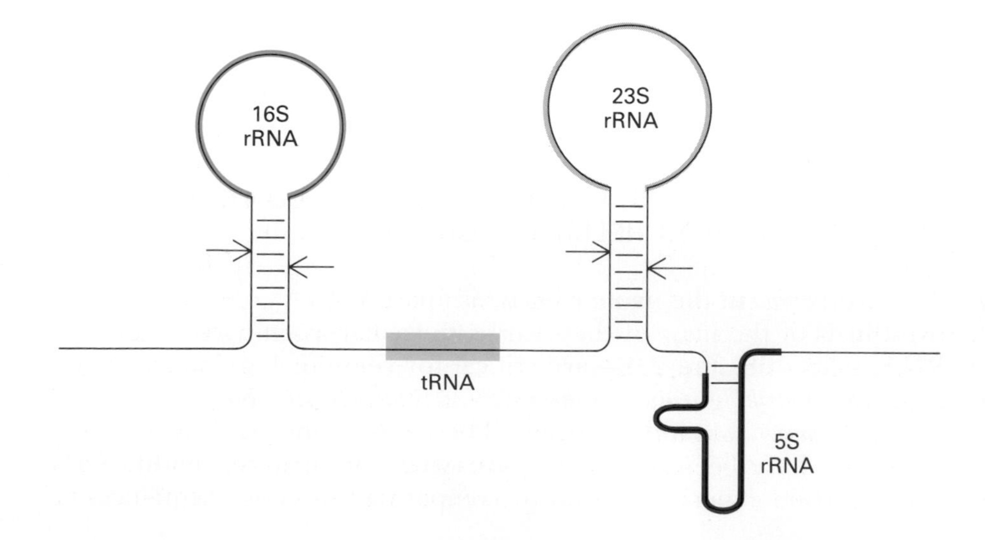

Figure 30-19
The three ribosomal RNA molecules are derived from primary transcripts that also contain at least one tRNA molecule. Arrows mark the sites of cleavage by RNase III.

RIBOSOMES CAN BE FORMED IN VITRO BY SELF-ASSEMBLY OF THEIR CONSTITUENT PROTEINS AND RNAs

The 30S ribosomal subunit can be reconstituted from a mixture of its 21 constituent proteins and 16S RNA. The spontaneous reassembly of these components to form a fully functional 30S subunit was first achieved by Masayasu Nomura in 1968. The 50S subunit was reconstituted several years later. The significance of these experiments is twofold. First, they demonstrate that all of the information needed for the correct assembly of this organelle is contained in the structure of its components. Nonribosomal factors are not needed. Thus, *the formation of a ribosome in vitro is a self-assembly process.* Second, reconstitution can be used to ascertain whether a particular component is essential for the assembly of the ribosome or for a specific function. For example, the component responsible for sensitivity to the antibiotic streptomycin was identified in this way (p. 759).

Studies of the reconstitution of the 30S subunit have shown that 16S RNA is essential for its assembly and function. The requirement is quite specific because 16S RNA from yeast, which folds similarly, cannot substitute for 16S RNA from *E. coli.* Most of the 21 proteins are also needed for assembly, indicating that *the 30S subunit is a cooperative functional entity.* The assembly process is ordered and proceeds in stages. For example, an intermediate containing fifteen proteins and 16S RNA can be trapped by reconstituting the particle at 0°C. The other six proteins join this intermediate to form a functional 30S subunit when the temperature is raised to 40°C. Likewise, an intermediate containing the two RNAs and some 20 of the 34 kinds of proteins of the large subunit has been trapped at low temperature.

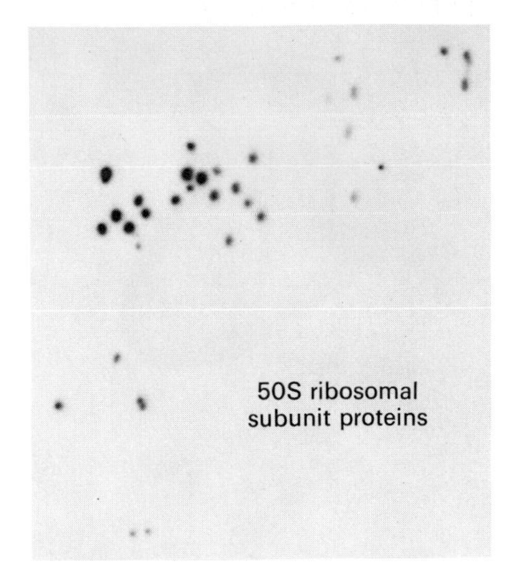

Figure 30-20
Separation of the proteins of the 50S ribosomal subunit by two-dimensional electrophoresis on a polyacrylamide gel. [Courtesy of Dr. Charles Cantor.]

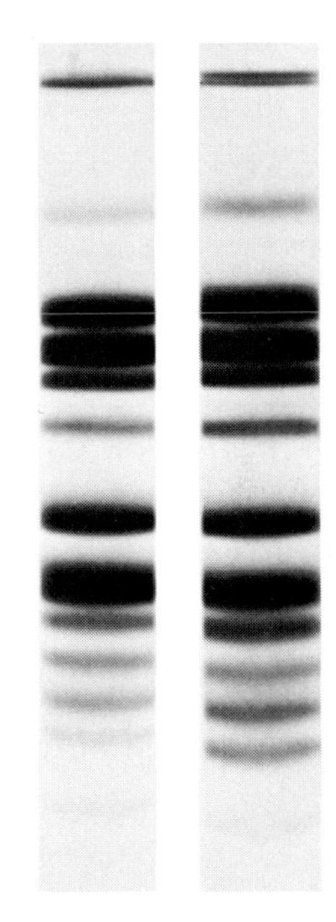

Figure 30-21
Polyacrylamide-gel electrophoresis patterns of the native 30S ribosomal subunit (left) and a reconstituted 30S subunit (right). All of the proteins present in the native subunit are present in the reconstituted one. [Courtesy of Dr. Masayasu Nomura.]

RIBOSOMAL ARCHITECTURE IS BEING MAPPED BY ELECTRON MICROSCOPY, NEUTRON DIFFRACTION, AND CROSSLINKING

Elucidating the relations between ribosomal structure and function is a formidable challenge because of their large size (megadaltons). Highly ordered three-dimensional crystals of ribosomes have not yet been obtained. Nevertheless, investigators are delineating the overall shape of the ribosome, its surface topography, and the location of its protein and RNA constituents. This impressive progress has been the result of the application of a wide range of techniques by many laboratories. The shapes of the ribosome and of its 30S and 50S subunits have been reconstructed from a large number of electron microscopic images. Immunoelectron microscopy using antibodies specific for particular proteins has revealed the identity of many surface features (Figure 30-22, on the following page). The mRNA binding site and the 3′ end of 16S RNA are situated on a platform located between the upper and lower parts of the 30S subunit. In a cleft formed by this platform and the upper third of the subunit are the two tRNA binding sites. The 50S subunit contains three protuberances. The peptidyl transferase site that catalyzes peptide-bond formation is located in the valley between two of them; a fingerlike projection contains the GTPase site that powers the movements of tRNAs and mRNA. The exit site for the growing polypeptide chain is on the opposite side of the 50S subunit.

The locations of all 21 proteins of the 30S subunit have been determined by neutron diffraction analyses of concentrated solutions. Neutrons rather than x-ray are used because neutrons are scattered very

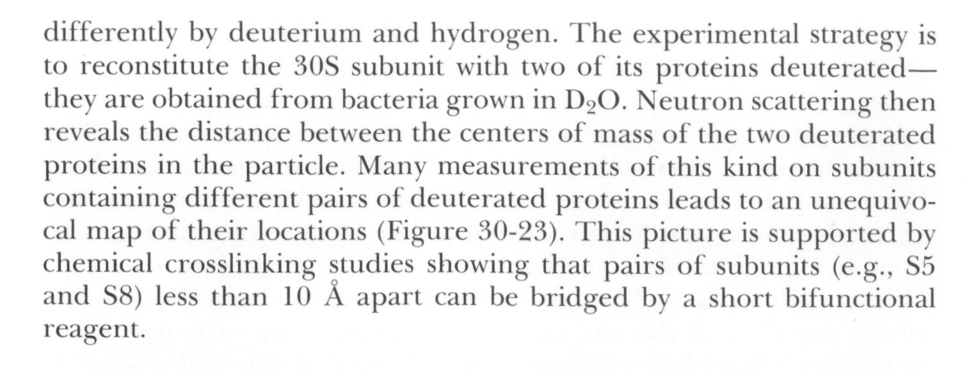

differently by deuterium and hydrogen. The experimental strategy is to reconstitute the 30S subunit with two of its proteins deuterated—they are obtained from bacteria grown in D_2O. Neutron scattering then reveals the distance between the centers of mass of the two deuterated proteins in the particle. Many measurements of this kind on subunits containing different pairs of deuterated proteins leads to an unequivocal map of their locations (Figure 30-23). This picture is supported by chemical crosslinking studies showing that pairs of subunits (e.g., S5 and S8) less than 10 Å apart can be bridged by a short bifunctional reagent.

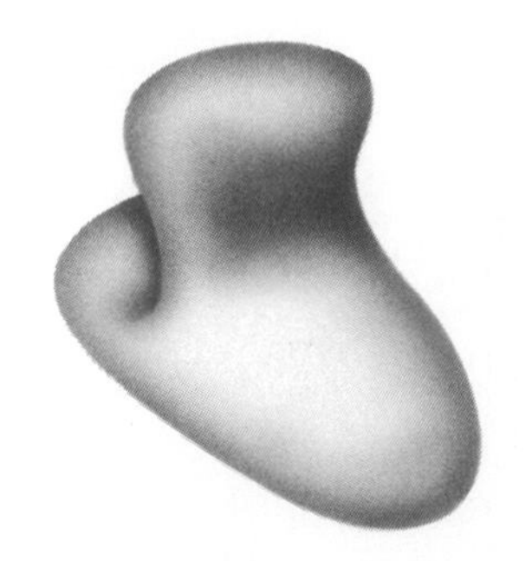

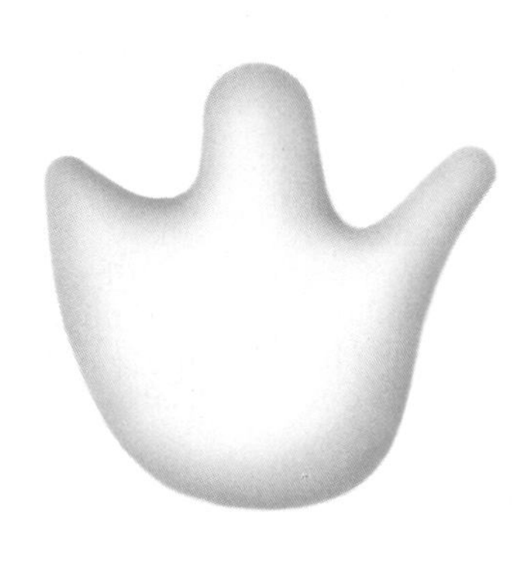

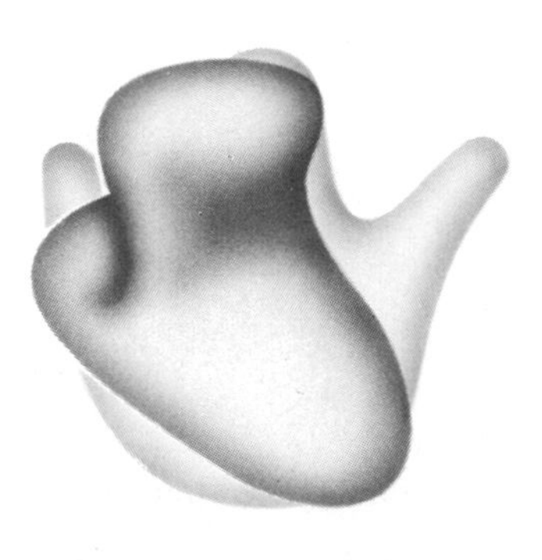

Figure 30-22
Shapes and surface topography of the 30S and 50S subunits and the intact 70S ribosome. [Courtesy of Dr. James Lake.]

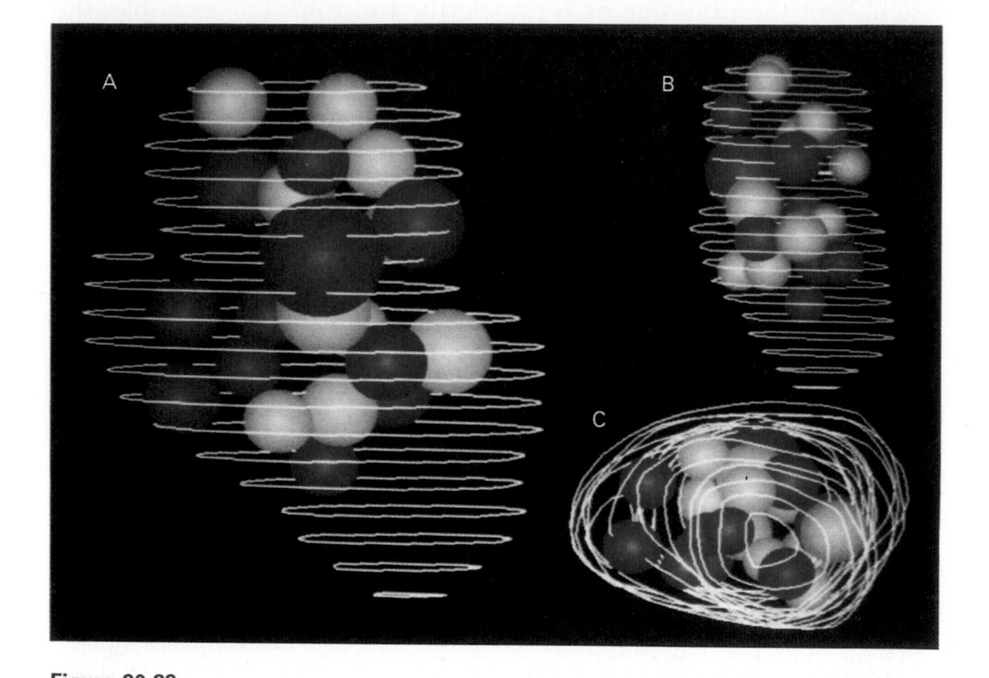

Figure 30-23
The location of all twenty-one proteins in the 30S ribosomal subunit has been mapped by neutron diffraction analyses of reconstituted ribsomes containing pairs of deuterated proteins. The white arcs depict the surface of the 30S particle, and the colored spheres show the positions of the proteins (the different colors are only to help distinguish individual spheres). If view A is considered the front of the subunit, then B is a view from the left and C is a view from the bottom. RNA occupies the unmarked volume of the particle. [Courtesy of Drs. M. S. Capel, D. M. Engelman, B. R. Freeborn, M. Kjeldgaard, J. A. Langer, V. Ramakrishnan, D. G. Schindler, D. K. Schneider, B. P. Schoenborn, I.-Y. Sillers, S. Yabuki, and P. B. Moore.]

PROTEINS ARE SYNTHESIZED IN THE AMINO-TO-CARBOXYL DIRECTION

One of the first questions asked about the mechanism of protein synthesis was whether proteins are synthesized in the amino-to-carboxyl direction or in the reverse direction. The pulse-labeling studies of Howard Dintzis provided a clear-cut answer. Reticulocytes that were actively synthesizing hemoglobin were exposed to ^{3}H-leucine. Completed hemoglobin was sampled frequently during a period shorter than required to synthesize a complete chain. Each sample was separated into α and β chains and then treated with trypsin. In the earliest samples, only peptides from the carboxyl ends were labeled. Later samples yielded labeled peptides closer and closer to the amino ends. Over all the samples, *a gradient of radioactivity increasing from the amino to the carboxyl end of each chain was found* (Figure 30-24). This would be expected if the amino part of the sampled chains was already synthesized prior to the addition of radioactive leucine. If the carboxyl end was synthesized last, radioactive label would appear there first, in chains that were almost complete when label was added to the medium. This experiment demonstrated that *the direction of chain growth is from the amino to the carboxyl end.*

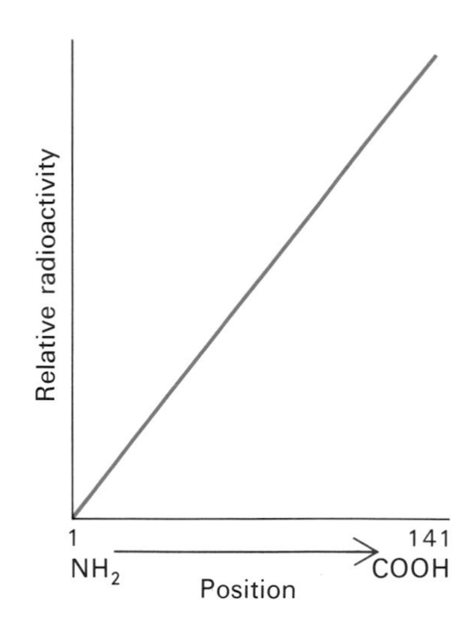

Figure 30-24
Distribution of ^{3}H-leucine in α chains of hemoglobin synthesized after exposure of reticulocytes to tritiated leucine. The higher radioactivity of the carboxyl ends relative to the amino ends indicates that the carboxyl end of each chain was synthesized last.

MESSENGER RNA IS TRANSLATED IN THE 5′ → 3′ DIRECTION

The direction of reading of mRNA was determined by using the synthetic polynucleotide

$$\overset{5'}{\text{A}}\text{—A—A—(A—A—A)}_n\text{— A—A—}\overset{3'}{\text{C}}$$

as the template in a cell-free protein-synthesizing system. AAA codes for lysine, whereas AAC codes for asparagine. The polypeptide product was

$$^{+}\text{H}_3\text{N—Lys—(Lys)}_n\text{— Asn—C}(=\text{O})\text{O}^-$$

Because asparagine was the carboxyl-terminal residue, the codon AAC was the last to be read. Hence, *the direction of translation is 5′ → 3′.*

If the direction of translation were opposite to that of transcription, only fully synthesized RNA could be translated. In contrast, if the directions are the same, mRNA can be translated while it is being synthesized. In fact, mRNA is synthesized also in the 5′ → 3′ direction (p. 709). In *E. coli,* almost no time is lost between transcription and translation. The 5′ end of mRNA interacts with ribosomes very soon after it is made (Figure 30-25). *An important feature of procaryotic gene expression is that translation and transcription are closely coupled in space and time.*

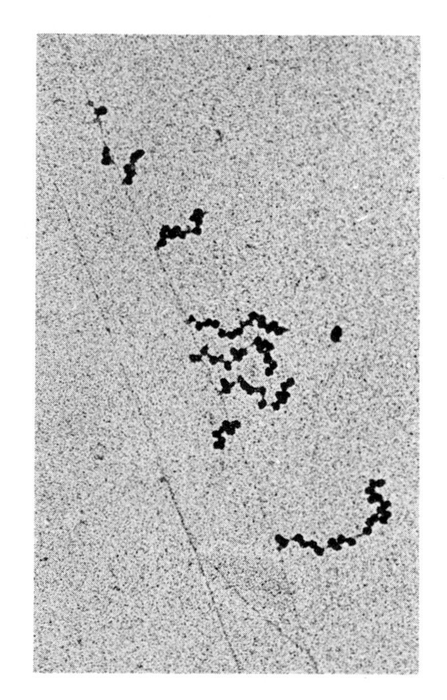

Figure 30-25
Transcription of a section of the DNA of *E. coli* and translation of the nascent mRNA. Only part of the chromosome is being transcribed [From O. L. Miller, Jr., Barbara A. Hamkalo, and C. A. Thomas, Jr. Visualization of bacterial genes in action. *Science* 169(1970):392.]

SEVERAL RIBOSOMES SIMULTANEOUSLY TRANSLATE A MESSENGER RNA MOLECULE

Many ribosomes can simultaneously translate an mRNA molecule. This markedly increases the efficiency of utilization of the mRNA. The group of ribosomes bound to an mRNA molecule is called a *polyribosome* or a *polysome.* The ribosomes in this unit operate independently, each synthesizing a complete polypeptide chain. The maximum density of ribosomes on mRNA is about one ribosome per eighty nucleotides. Polyribosomes synthesizing hemoglobin (which contains about 145 amino acids per chain, or 500 nucleotides per mRNA) typically consist of five ribosomes bound to an mRNA molecule. Ribosomes closest to the 5′

end of the messenger have the shortest polypeptide chains, whereas those nearest the 3′ end have almost finished chains (Figure 30-26). Ribosomes dissociate into 30S and 50S subunits after the polypeptide product is released.

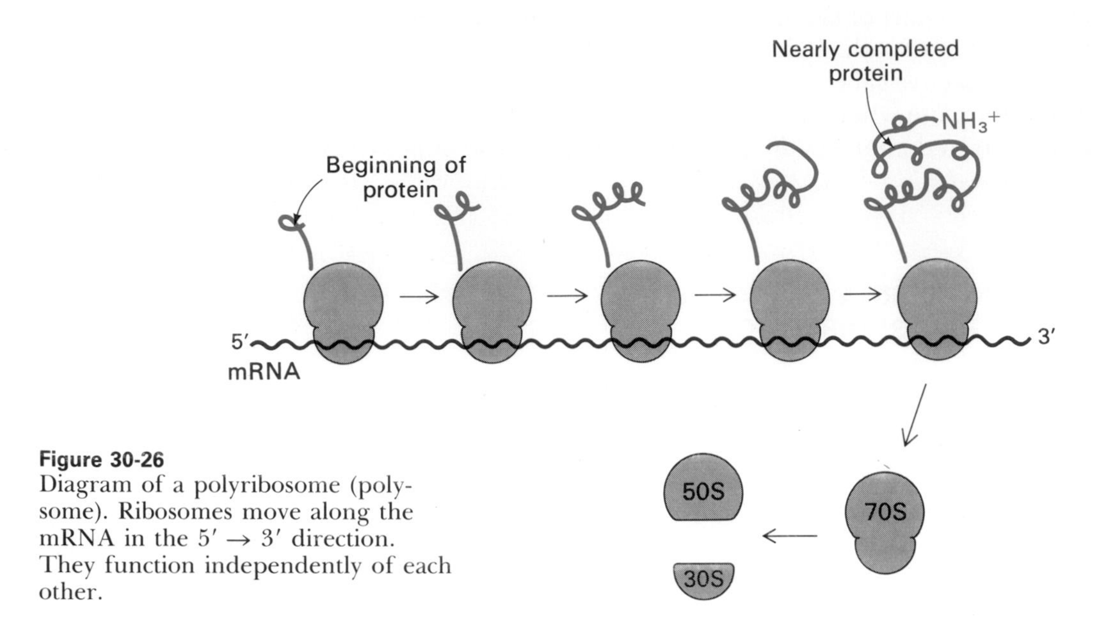

Figure 30-26
Diagram of a polyribosome (polysome). Ribosomes move along the mRNA in the 5′ → 3′ direction. They function independently of each other.

PROTEIN SYNTHESIS IN BACTERIA IS INITIATED BY FORMYLMETHIONYL TRANSFER RNA

tRNA$_f$

↓ Synthetase

$^+H_3N—CH(CH_2CH_2SCH_3)—C(=O)—O—tRNA_f$

Methionyl-tRNA$_f$ (Met-tRNA$_f$)

N^{10}-formyltetrahydrofolate → Tetrahydrofolate

↓ Transformylase

$HC(=O)—NH—CH(CH_2CH_2SCH_3)—C(=O)—O—tRNA_f$

Formylmethionyl-tRNA$_f$ (fMet-tRNA$_f$)

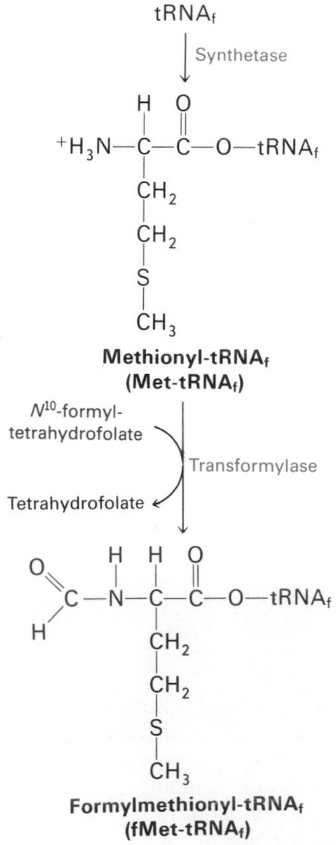

Figure 30-27
Formation of formylmethionyl-tRNA$_f$.

How does protein synthesis start? The simplest possibility a priori is that the entire mRNA is translated so that no special start signal is needed. However, the experimental fact is that translation does not begin immediately at the 5′ terminus of mRNA. Indeed, the first translated codon is nearly always more than 25 nucleotides away from the 5′ end. Furthermore, many mRNA molecules in procaryotes are *polycistronic*—that is, they code for two or more polypeptide chains. For example, a single mRNA molecule about 7000 nucleotides long specifies five enzymes in the biosynthetic pathway for tryptophan in *E. coli.* Each of these five proteins has its own start and stop signals on the mRNA. In fact, *all known bacterial mRNA molecules contain signals that define the beginning and end of each encoded polypeptide chain.*

A clue to the mechanism of chain initiation was the finding that nearly half of the amino-terminal residues of proteins in *E. coli* are methionine, yet this amino acid is uncommon at other positions of the polypeptide chain. Furthermore, the amino terminus of nascent proteins is usually modified, which suggests that a derivative of methionine participates in initiation. In fact, *protein synthesis in bacteria starts with formylmethionine* (fMet). A special tRNA brings formylmethionine to the ribosome to initiate protein synthesis. This *initiator tRNA* (abbreviated as *tRNA$_f$*) is different from the one that inserts methionine in internal positions (abbreviated as tRNA$_m$). The subscript *f* indicates that methionine attached to the initiator tRNA can be formylated, whereas it cannot be formylated when attached to tRNA$_m$.

Methionine is linked to these two kinds of tRNAs by the same aminoacyl-tRNA synthetase (Figure 30-27). A specific enzyme then formylates the amino group of methionine that is attached to $tRNA_f$. The activated formyl donor in this reaction is N^{10}-formyltetrahydrofolate. It is significant that free methionine and methionyl-$tRNA_m$ are not substrates for this transformylase.

THE START SIGNAL IS AUG (OR GUG) PRECEDED BY SEVERAL BASES THAT PAIR WITH 16S rRNA

The initiating codon in mRNA is AUG or, much less frequently, GUG. How is an initiating AUG or GUG distinguished from one that encodes an internal residue of a protein? The first step toward answering this question was the isolation of initiator regions from a number of mRNAs. This was accomplished by digesting mRNA-ribosome complexes (formed under conditions of chain initiation but not elongation) with pancreatic ribonuclease. In each case, a sequence of about thirty nucleotides was protected from digestion. As expected, each of these initiator regions displays an AUG (or GUG) codon (Figure 30-28). In addition, each initiator region contains a purine-rich sequence centered about ten nucleotides on the 5′ side of the initiator codon.

AGCACGAGGGGAAAUCUGAUGGAACGCUAC	*E. coli trpA*
UUUGGAUGGAGUGAAACGAUGGCGAUUGCA	*E. coli araB*
GGUAACCAGGUAACAACCAUGCGAGUGUUG	*E. coli thrA*
CAAUUCAGGGUGGUGAAUGUGAAACCAGUA	*E. coli lacI*
AAUCUUGGAGGCUUUUUUAUGGUUCGUUCU	φX174 phage A protein
UAACUAAGGAUGAAAUGCAUGUCUAAGACA	Qβ phage replicase
UCCUAGGAGGUUUGACCUAUGCGAGCUUUU	R17 phage A protein
AUGUACUAAGGAGGUUGUAUGGAACAACGC	λ phage *cro*

Pairs with 16S rRNA — Pairs with initiator tRNA

Figure 30-28
Sequences of initiation sites for protein synthesis in some bacterial and viral RNA molecules.

The role of this purine-rich region (called the Shine-Dalgarno sequence) became evident when the sequence of 16S RNA was elucidated. The 3′ end of this RNA component of the 30S subunit contains a sequence of several bases that is complementary to the purine-rich region in the initiator sites of mRNA. In fact, a complex of the 3′ end of 16S RNA and the initiator region of mRNA has been isolated from an enzymatic digest of the initiation complex (Figure 30-29). The sequences of more than seventy known initiator sites show that the number of base pairs between mRNA and 16S rRNA ranges from three to nine. *Thus, two kinds of interactions determine where protein synthesis starts: the pairing of mRNA bases with the 3′ end of 16S rRNA, and the pairing of the initiator codon on mRNA with the anticodon of fMet initiator tRNA.*

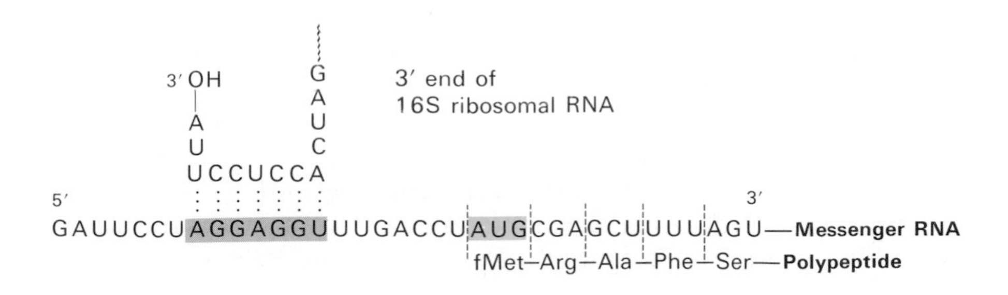

Figure 30-29
Base pairing between the purine-rich region (blue) in the initiator region of an mRNA and the 3′ end of 16S rRNA (red). The AUG codon (green) defines the start of the polypeptide chain. The mRNA shown here codes for the A protein of R17 phage. The purine-rich region is known as a Shine-Dalgarno sequence.

FORMATION OF THE 70S INITIATION COMPLEX PLACES FORMYLMETHIONYL-tRNA$_f$ IN THE P SITE OF THE RIBOSOME

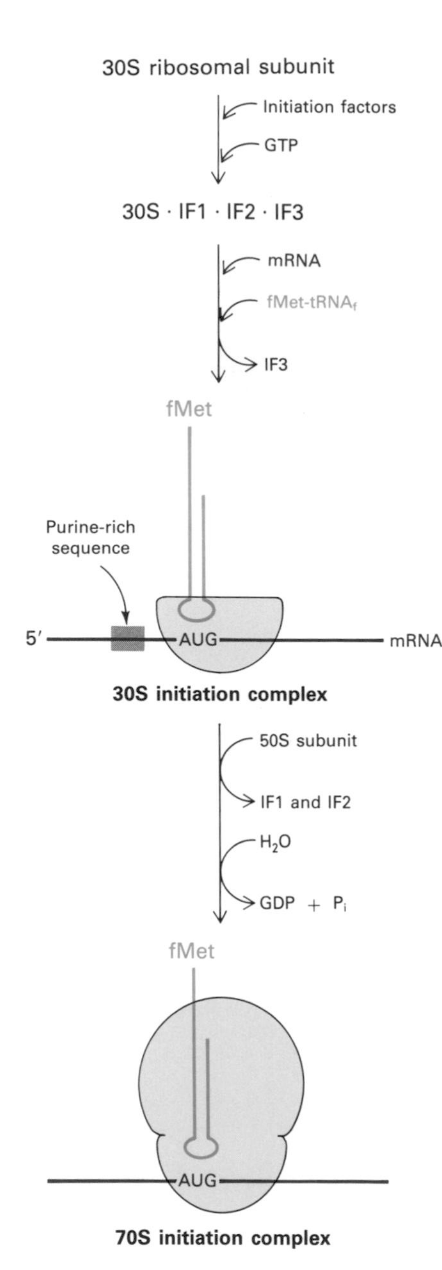

Figure 30-30
Initiation phase of protein synthesis: formation of a 30S initiation complex, followed by a 70S initiation complex.

How are mRNA and formylmethionyl-tRNA$_f$ brought to the ribosome to initiate protein synthesis? Three protein *initiation factors* (IF1, IF2, and IF3) are essential. The 30S ribosomal subunit first forms a complex with these three factors (Figure 30-30). The binding of GTP to IF2 enables mRNA and the initiator tRNA to join the complex as IF3 is released. fMet-tRNA$_f$ is specifically recognized by IF2, and the release of IF3 allows the 50S subunit to join the complex. Hydrolysis of GTP bound to IF2 on entry of the 50S subunit leads to the release of IF1 and IF2. The result is a *70S initiation complex.* The L7 and L12 proteins of the 50S subunit participate in GTP hydrolysis, which is essential for forming a productive initiation complex. L7 is identical to L12 except that its amino-terminus is acetylated. The L7/L12 tetramer, a fingerlike spike on the large subunit (see Figure 30-22, on p. 750) also participates in GTP hydrolysis during the elongation phase of protein synthesis.

The 70S initiation complex (Figure 30-30) is ready for the elongation phase of protein synthesis. *The fMet-tRNA$_f$ molecule occupies the P (peptidyl) site on the ribosome.* The other site for a tRNA molecule, the A (aminoacyl) site, is empty. The existence of distinct P and A sites was inferred from studies of puromycin, an antibiotic to be discussed shortly (p. 759). The important point now is that fMet-tRNA$_f$ is positioned so that its anticodon pairs with the initiating AUG (or GUG) codon on mRNA. *The reading frame is defined by this interaction and by the pairing of the adjoining purine-rich sequence to a pyrimidine-rich sequence in 16S RNA* (see Figures 30-28 and 30-29). How were poly U and other synthetic polypeptides without start signals translated in the studies leading to the elucidation of the genetic code (p. 101)? Fortunately, nonspecific initiation occurred because the concentration of Mg^{2+} in the reaction mixture was higher than it is in vivo.

THE GTP FORM OF ELONGATION FACTOR Tu DELIVERS AMINOACYL-tRNA TO THE A SITE OF THE RIBOSOME

The elongation cycle in protein synthesis begins with the insertion of an aminoacyl-tRNA into the empty A site on the ribosome. The particular species inserted depends on the mRNA codon that is positioned in the A site. The complementary aminoacyl-tRNA is delivered to the A site by a protein *elongation factor* called *EF-Tu.* EF-Tu, like IF2, contains a bound guanyl nucleotide and cycles between a GTP and a GDP form. After EF-Tu has positioned the aminoacyl-tRNA in the A site, GTP is hydrolyzed. The GDP form of EF-Tu then dissociates from the ribosome. *EF-Ts,* a second elongation factor, joins the EF-Tu complex and induces the dissociation of GDP. Finally, GTP binds to EF-Tu, and EF-Ts is concomitantly released. EF-Tu containing bound GTP is now ready to pick up another aminoacyl-tRNA and deliver it to the A site of the ribosome. We shall encounter similar GTP-GDP cycles in hormone action (p. 979) and visual excitation (p. 1033). Moreover, in muscle contraction, a similar cycle involving ATP and ADP takes place. The guanyl nucleotide site of EF-Tu (Figure 30-32) has been conserved over billions of years of evolution.

It is noteworthy that *EF-Tu does not interact with fMet-tRNA$_f$.* Hence, this initiator tRNA is not delivered to the A site. In contrast, Met-

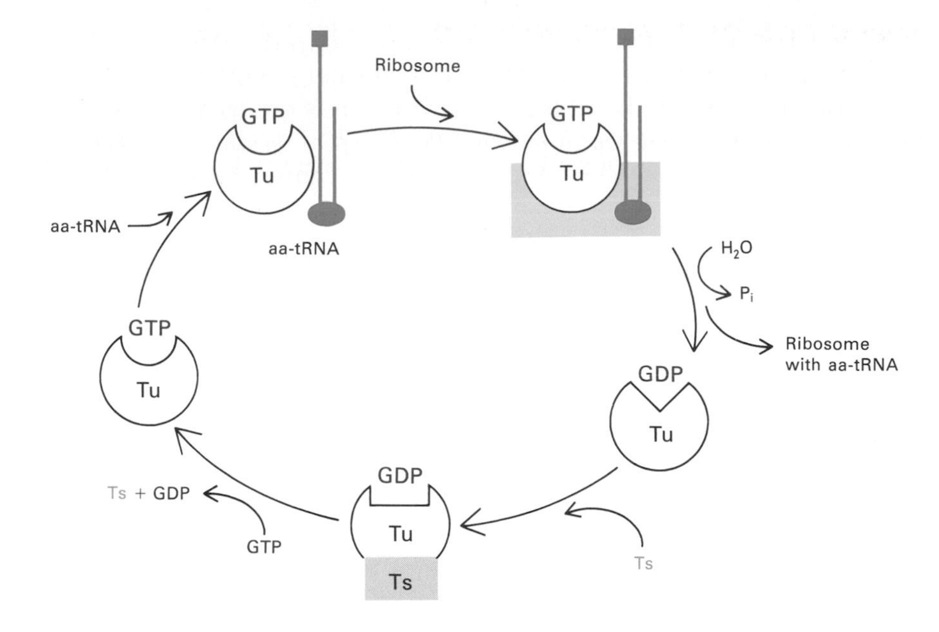

Figure 30-31
Reaction cycle of elongation factor Tu (EF-Tu).

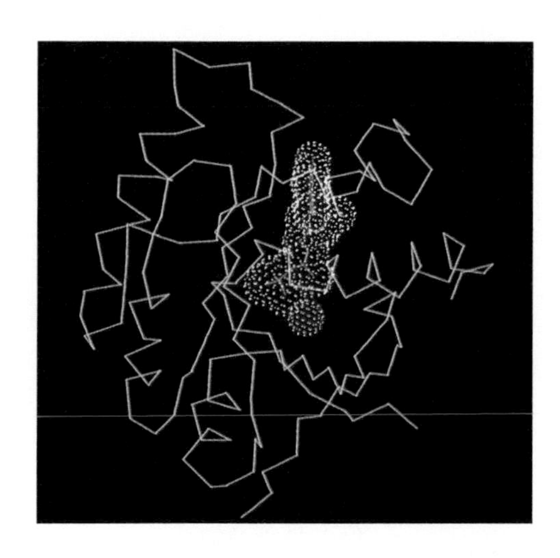

Figure 30-32
Structure of elongation factor Tu (EF-Tu) from *E. coli.* Bound GDP is shown in purple and Mg^{2+} in blue. [Courtesy of Dr. Fran Jurnak].

$tRNA_m$, like all other aminoacyl-tRNAs, does bind to EF-Tu. These findings account for the fact that *internal AUG codons are not read by the initiator tRNA.* Conversely, initiation factor 2 recognizes fMet-$tRNA_f$ but no other tRNA.

THE GTPase RATE OF EF-Tu SETS THE PACE OF PROTEIN SYNTHESIS AND DETERMINES ITS FIDELITY

The fidelity of protein synthesis depends on having the correct aminoacyl-tRNA in the A site when the peptide bond is formed. This is the point of no return, for cells lack a means of excising an incorrect amino acid residue once the peptide bond is made. As might be expected, the incoming aminoacyl-tRNA is carefully scrutinized to make sure that its anticodon is complementary to the mRNA codon occupying the A site. An aminoacyl-tRNA that can pair with two of the three bases of a codon may bind transiently but it will usually leave the A site before a peptide bond is formed. It takes time for the ribosome to decide whether the bound aminoacyl-tRNA is the right one. Protein synthesis would be inaccurate if a peptide bond formed immediately on arrival of an aminoacyl-tRNA. *The timer that determines the period of scrutiny is the GTPase site of EF-Tu* (Figure 30-33). A peptide bond cannot be formed until EF-Tu is released from the aminoacyl-tRNA; release requires that bound GTP be hydrolyzed to GDP. On average, several milliseconds pass before GTP is hydrolyzed, and another several milliseconds pass before EF-Tu-GDP leaves the ribosome. An incorrect aminoacyl-tRNA usually leaves the ribosome during one of these intervals, whereas the correct one stays bound. *The conformation of EF-Tu changes when it hydrolyzes GTP, and so the context of the codon-anticodon interaction is altered. The correct aminoacyl-tRNA interacts strongly with mRNA in both states, but an incorrect one does not.* In effect, the codon-anticodon interaction is scrutinized twice, in different ways, to achieve higher accuracy, just as proofreading of a manuscript by two readers for an hour each is much more effective than two hours of proofreading by one. This double-check mechanism for assuring that the right amino acid becomes attached to tRNA is another expression of the value of multiple views in testing authenticity.

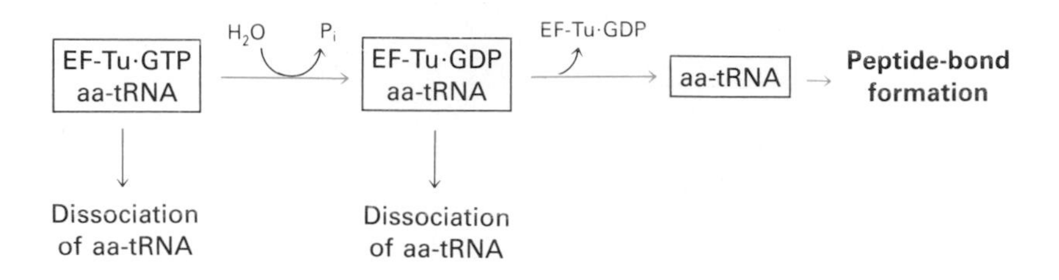

Figure 30-33
Proofreading of the aminoacyl-tRNA occupying the A site of the ribosome occurs both before and after hydrolysis of GTP bound to EF-Tu. Incorrect aminoacyl-tRNAs dissociate from the A site (steps shown in red), whereas the correct aminoacyl-tRNA stays bound and a peptide bond is formed (green arrows).

How accurate must aminoacyl-tRNA selection be? The probability p of forming a protein with no errors depends on n, the number of amino acid residues, and ϵ, the frequency of inserting a wrong amino acid:

$$p = (1 - \epsilon)^n$$

As Table 30-4 shows, an error frequency of 10^{-2} would be intolerable, even for quite small proteins. An ϵ of 10^{-3} would usually lead to the error-free synthesis of a 300-residue protein (~33 kd) but not of a 1000-residue protein (~110 kd). Thus, an error frequency of the order of no more than 10^{-4} per residue is needed to effectively produce the

Table 30-4
Accuracy of protein synthesis

Frequency of inserting an incorrect amino acid	*Probability of synthesizing an error-free protein*		
	Number of amino acid residues		
	100	*300*	*1000*
10^{-2}	0.364	0.049	0.000
10^{-3}	0.915	0.840	0.368
10^{-4}	0.990	0.970	0.905
10^{-5}	0.999	0.997	0.990

larger proteins. In fact, the observed values of ϵ are close to 10^{-4}. It is interesting to note that higher accuracy can be attained in vitro by using GTPγS, a GTP analog that is hydrolyzed by EF-Tu a thousand times less rapidly than is GTP. The longer residency time of EF-Tu on the ribosome enhances fidelity by giving more time for the departure of incorrect aminoacyl-tRNAs. However, higher accuracy achieved in this way carries a high kinetic price with little gain in the number of functional proteins synthesized. On the other hand, protein synthesis with an ϵ of 10^{-2} would be rapid but nearly all the products would be inactive. *An error frequency of about 10^{-4} per amino acid residue has been selected in the course of evolution to produce the greatest number of functional proteins in the shortest time.*

FORMATION OF A PEPTIDE BOND IS FOLLOWED BY TRANSLOCATION

We now have a complex in which aminoacyl-tRNA occupies the A site, and fMet-tRNA occupies the P site. EF-Tu has left the ribosome. The stage is set for the *formation of a peptide bond* (Figure 30-34). This reaction is catalyzed by *peptidyl transferase,* an enzymatic activity of the 50S subunit. The activated formylmethionine unit of fMet-$tRNA_f$ (in the P site) is transferred to the amino group of the aminoacyl-tRNA (in the A site) to form a dipeptidyl-tRNA. The attack of the amino group on the ester linkage to form a peptide bond is a thermodynamically favorable reaction; the free energy cost of forming a peptide bond was paid earlier in forming an aminoacyl-tRNA.

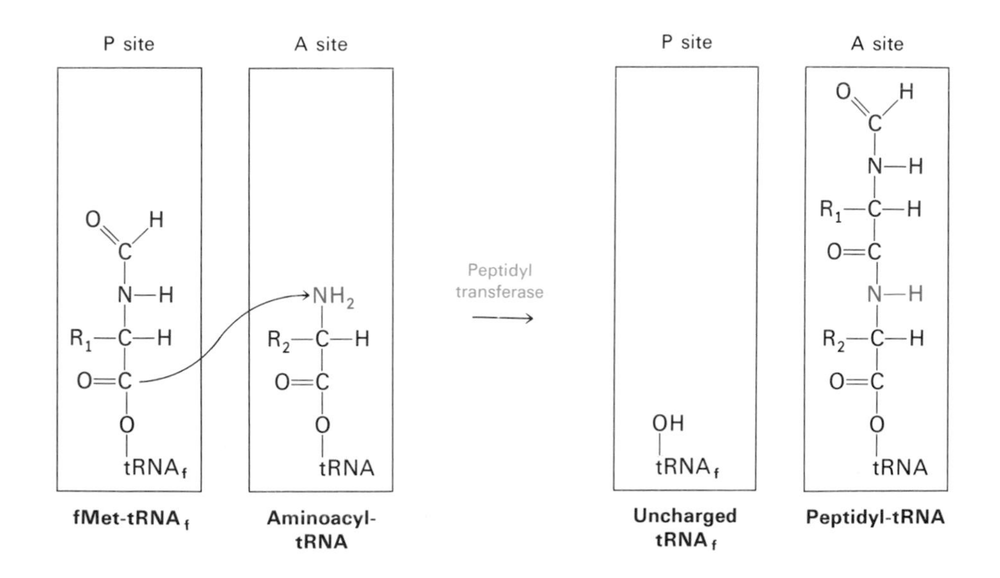

Figure 30-34
Peptide-bond formation is catalyzed by the peptidyl transferase site of the ribosome. The elongating chain is transferred to the amino group of the incoming aminoacyl-tRNA. This reaction proceeds by nucleophilic attack of the amino nitrogen atom of the aminoacyl-tRNA on the ester carbon atom of fMet-tRNA (or peptidyl-tRNA).

An uncharged $tRNA_f$ occupies the P site, whereas a dipeptidyl-tRNA occupies the A site, following the formation of a peptide bond. The next phase of the elongation cycle is *translocation.* Three movements occur: the uncharged tRNA leaves the P site, the peptidyl-tRNA moves from the A site to the P site, and mRNA moves a distance of three nucleotides. The result is that the next codon is positioned for reading by the incoming aminoacyl-tRNA. Translocation requires a third elongation factor, *EF-G* (also called *translocase*). EF-G, like IF2 and EF-Tu, cycles between a GTP and a GDP form. The GTP form of EF-G is the one that drives the translocation step. The subsequent hydrolysis of bound GTP enables EF-G to be released from the ribosome. After translocation, the A site is empty, ready to bind an aminoacyl-tRNA to start another round of elongation (Figure 30-35).

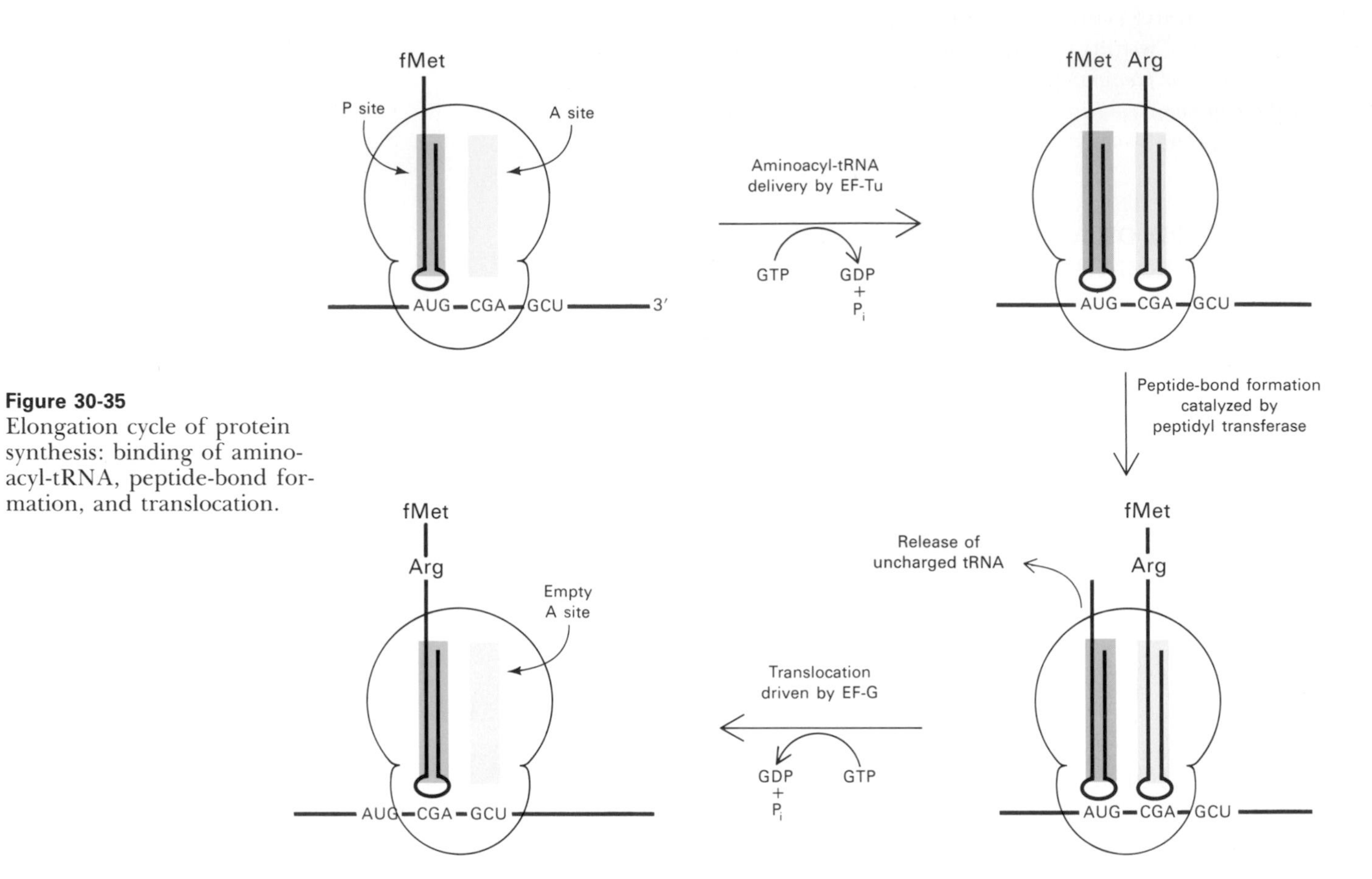

Figure 30-35
Elongation cycle of protein synthesis: binding of aminoacyl-tRNA, peptide-bond formation, and translocation.

PROTEIN SYNTHESIS IS TERMINATED BY RELEASE FACTORS THAT READ STOP CODONS

Aminoacyl-tRNA does not normally bind to the A site of a ribosome if the codon is UAA, UGA, or UAG. Normal cells do not contain tRNAs with anticodons complementary to these stop signals. Instead, the signals are recognized by release factors, which are proteins. One of these release factors, RF1, recognizes UAA or UAG. A second factor, RF2, recognizes UAA or UGA. Thus, proteins can recognize trinucleotide sequences with high specificity. The binding of a release factor to a termination codon in the A site somehow activates peptidyl transferase so that it hydrolyzes the bond between the polypeptide and the tRNA in the P site. The specificity of peptidyl transferase is altered by the release factor so that water rather than an amino group is the acceptor of the activated peptidyl moiety. The polypeptide chain leaves the ribosome, followed by tRNA and mRNA. Finally, the ribosome dissociates into 30S and 50S subunits as the prelude to the synthesis of another protein molecule. The binding of IF3 to the 30S subunit prevents it from joining the 50S subunit to form a dead-end 70S complex devoid of mRNA.

STREPTOMYCIN INHIBITS INITIATION AND INDUCES MISREADING OF MESSENGER RNA

We have seen that antibiotics are very useful tools in biochemistry because many have highly specific actions. For example, rifampicin is a potent inhibitor of the initiation of RNA synthesis in bacteria (p. 714). Many antibiotic inhibitors of protein synthesis are known, and the mechanisms of several have been elucidated (Table 30-5). Streptomycin, tetracycline, chloramphenicol, and erythromycin are highly potent antibacterial agents that act by blocking different steps in procaryotic protein synthesis.

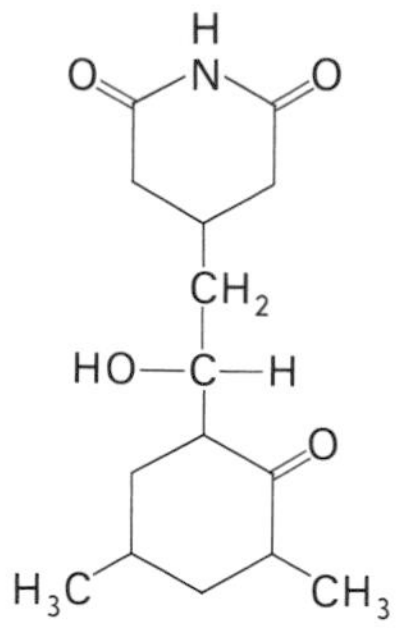

Cycloheximide

Table 30-5
Antibiotic inhibitors of protein synthesis

Antibiotic	*Action*
Streptomycin	Inhibits initiation and causes misreading of mRNA (procaryotes)
Tetracycline	Binds to the 30S subunit and inhibits binding of aminoacyl-tRNAs (procaryotes)
Chloramphenicol	Inhibits the peptidyl transferase activity of the 50S ribosomal subunit (procaryotes)
Cycloheximide	Inhibits the peptidyl transferase activity of the 60S ribosomal subunit (eucaryotes)
Erythromycin	Binds to the 50S subunit and inhibits translocation (procaryotes)
Puromycin	Causes premature chain termination by acting as an analog of aminoacyl-tRNA (procaryotes and eucaryotes)

Streptomycin, a highly basic trisaccharide, interferes with the binding of formylmethionyl-tRNA to ribosomes and thereby prevents the correct initiation of protein synthesis. Streptomycin also leads to a misreading of mRNA. For example, if poly U is the template, isoleucine (AUU) is incorporated in addition to phenylalanine (UUU). The site of action of streptomycin in the ribosome has been determined by forming hybrid ribosomes containing components from streptomycin-sensitive and streptomycin-resistant bacteria. These bacterial strains are related by mutation of a single gene. Hybrid ribosomes consisting of 50S subunits from resistant bacteria and 30S subunits from sensitive bacteria were found to be sensitive, whereas ribosomes prepared from the reverse combination were resistant. This experiment showed that the determinant of streptomycin sensitivity is located in the 30S subunit. A subsequent experiment demonstrated that a protein in the 30S subunit rather than its 16S RNA molecule specifies streptomycin sensitivity. Finally, an extensive series of experiments revealed that a single protein in the 30S subunit—namely, protein S12—is the determinant of streptomycin sensitivity.

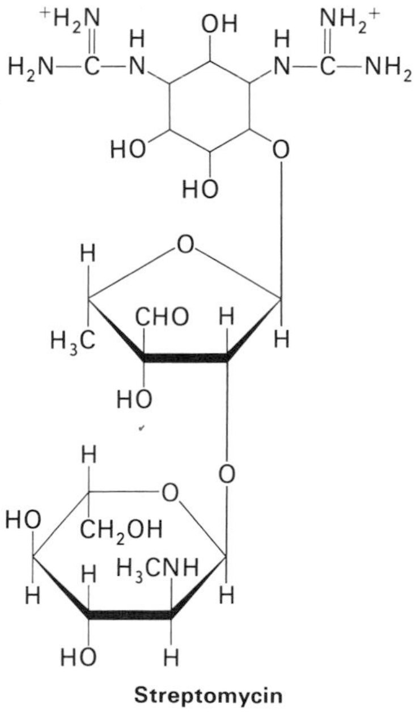

Streptomycin

PUROMYCIN CAUSES PREMATURE CHAIN TERMINATION BY MIMICKING AMINOACYL-TRANSFER RNA

The antibiotic puromycin inhibits protein synthesis by releasing nascent polypeptide chains before their synthesis is completed. Puromycin is an analog of the terminal aminoacyl-adenosine portion of aminoacyl-tRNA (Figure 30-36). It binds to the A site on the ribosome and inhibits the entry of aminoacyl-tRNA. Furthermore, puromycin contains an α-amino group. This amino group, like the one on aminoacyl-tRNA, forms a peptide bond with the carboxyl group of the growing peptide chain in a reaction that is catalyzed by peptidyl transferase. The product is a peptide having a covalently attached puromycin residue at its carboxyl end. Peptidyl puromycin then dissociates from the ribosome. Puromycin has been used to probe the functional state of ribosomes. In fact, the concept of A and P sites resulted from the use of puromycin to ascertain the location of peptidyl tRNA. When peptidyl tRNA is in the A site (before translocation), it cannot react with puromycin.

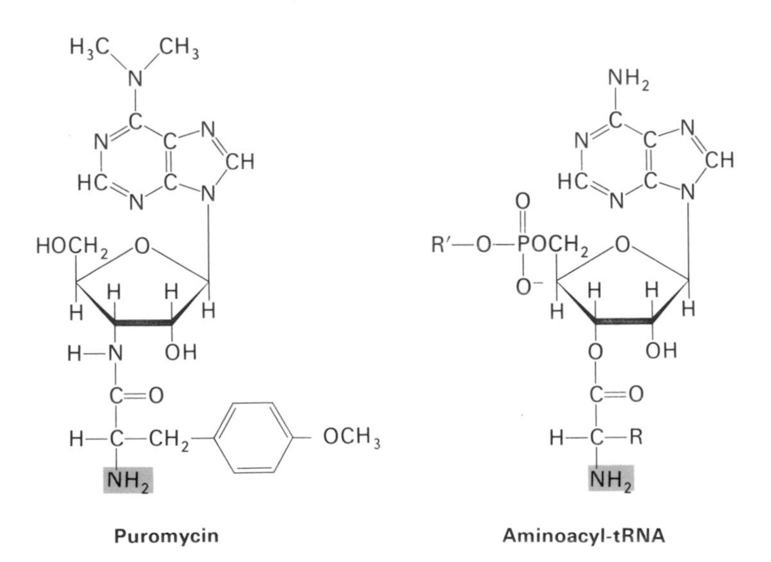

Figure 30-36
Puromycin resembles the aminoacyl terminus of an aminoacyl-tRNA. Its amino group becomes joined to the carboxyl group of the growing peptide chain to form an adduct that dissociates from the ribosome.

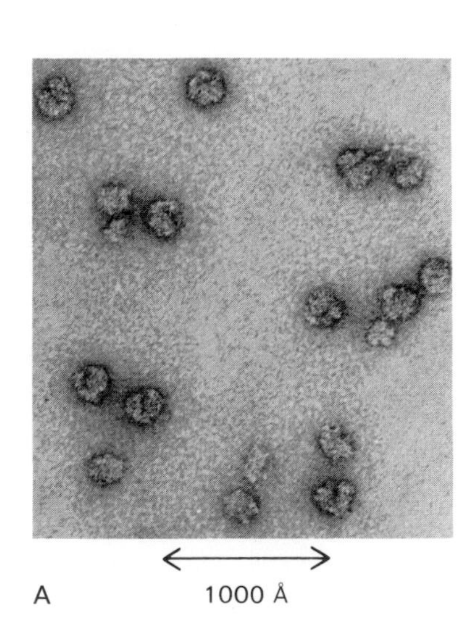

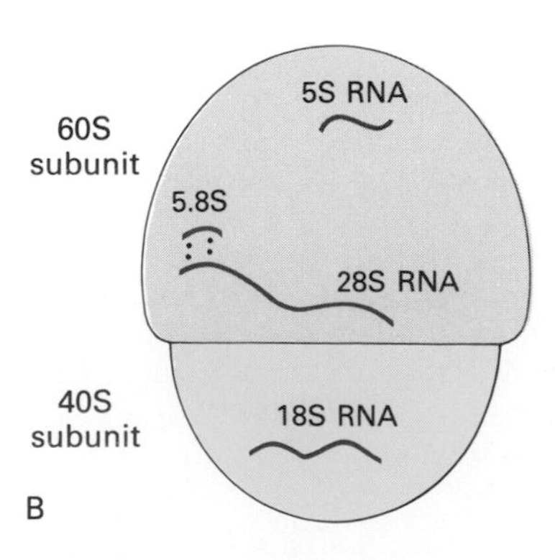

Figure 30-37
(A) Electron micrograph of eucaryotic ribosomes. [Courtesy of Dr. Miloslav Bublik.] (B) Schematic diagram of a eucaryotic ribosome.

EUCARYOTIC AND PROCARYOTIC PROTEIN SYNTHESIS HAVE MANY COMMON STRUCTURAL AND MECHANISTIC MOTIFS

The basic plan of protein synthesis in eucaryotes is similar to that in procaryotes. The major structural and mechanistic themes recur in higher forms. However, eucaryotic protein synthesis involves more protein components, and some steps are more intricate. Some similarities and differences are listed below:

1. *Ribosomes.* Eucaryotic ribosomes are larger. They consist of a 60S large subunit and a 40S small subunit, which come together to form an 80S particle having a mass of 4200 kd, compared with 2700 kd for the procaryotic 70S ribosome. The 40S subunit contains an 18S RNA that is homologous to the procaryotic 16S RNA. The 60S subunit contains three RNAs: the 5S and 28S RNAs are the counterparts of the procaryotic 5S and 23S molecules; its 5.8S RNA is unique to eucaryotes.

2. *Initiator tRNA*. In eucaryotes, the initiating amino acid is methionine rather than *N*-formyl methionine. However, as in procaryotes, a special tRNA participates in initiation. This aminoacyl-tRNA is called Met-$tRNA_f$ or Met-$tRNA_i$ (the subscript *f* indicates that it can be formylated in vitro, and *i* stands for initiation).

3. *Start signal*. The initiating codon in eucaryotes is always AUG. Eucaryotes, in contrast with procaryotes (p. 753), do not use a purine-rich sequence on the 5′ side to distinguish initiator AUGs from internal ones. Instead, the AUG nearest the 5′ end of mRNA is usually selected as the start site. 40S ribosomes attach to the cap at the 5′ end of eucaryotic mRNAs and search for an AUG codon by moving stepwise in the 3′ direction. This scanning process is driven by the hydrolysis of ATP. The anticodon of Met-$tRNA_i$ bound to the 40S subunit pairs with the AUG codon, signifying that the target has been found. A eucaryotic mRNA has only one start site and hence is the template for a single protein. In contrast, a procaryotic mRNA can have multiple start sites and be the template for the synthesis of several proteins.

4. *Initiation complexes*. Eucaryotes contain many more initiation factors than do procaryotes, and their interplay is much more intricate. Nine are known, and several consist of multiple subunits. The prefix *eIF* denotes a eucaryotic initiation factor. The GTP form of eIF2 (1000 kd) brings the initiator tRNA to the 40S subunit. Cap-binding proteins (CBPs) bind the cap of mRNA. They are joined by eIF3, which finds the AUG closest to the 5′ end. eIF4 is the ATP-driven engine in this search. eIF5 induces the release of eIF2 and eIF3 following the pairing of Met-$tRNA_i$ with the initiating AUG; eIF5 exerts these actions by triggering the hydrolysis of GTP bound to eIF2. Finally, the 60S subunit joins the complex of initiator tRNA, mRNA, and 40S subunit to form the 80S initiation complex.

5. *Elongation and termination factors*. Eucaryotic elongation factors EF1α and EF1$\beta\gamma$ are the counterparts of procaryotic EF-Tu and EF-Ts. The GTP form of EF1α delivers aminoacyl-tRNA to the A site of the ribosome, and EF1$\beta\gamma$ catalyzes the exchange of GTP for bound GDP. Eucaryotic EF2 mediates GTP-driven translocation in much the same way as does procaryotic EF-G. Termination in eucaryotes is carried out by a single release factor eRF, a GTP-driven protein, compared with two in procaryotes. Finally, eIF3, like its procaryotic counterpart (p. 754), prevents the reassociation of ribosomal subunits in the absence of an initiation complex.

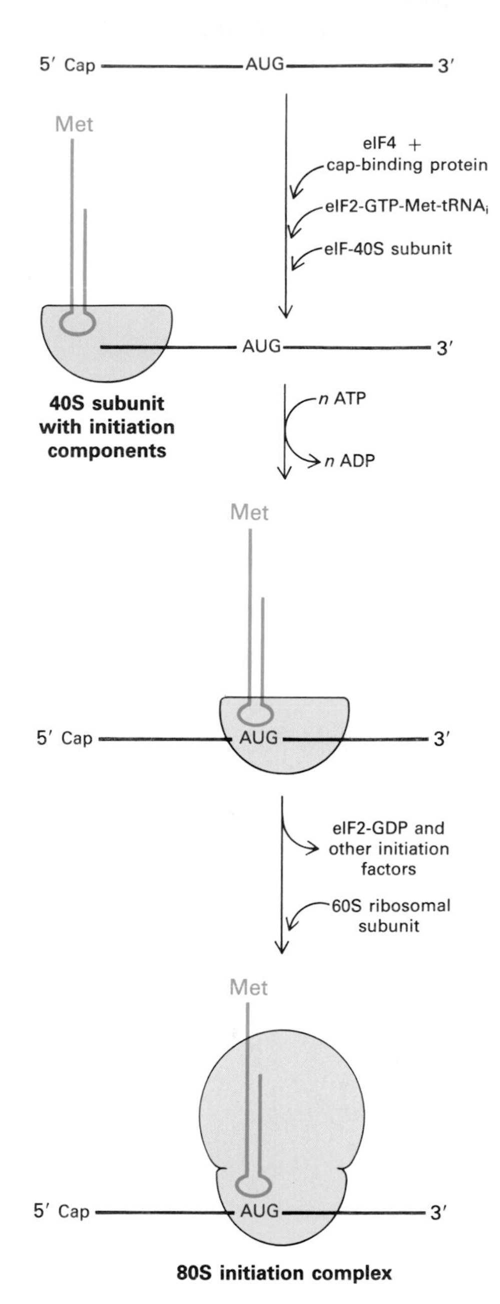

Figure 30-38
A 40S ribosomal subunit containing initiator tRNA and several initiation factors scans mRNA starting at the 5′ cap. The initiation factors dissociate when the first AUG codon is found. The 60S subunit then joins the 40S subunit to form an 80S initiation complex that is ready for the elongation phase.

TRANSLATION IN EUCARYOTES IS REGULATED BY PROTEIN KINASES THAT INACTIVATE AN INITIATION FACTOR

Extracts of reticulocytes synthesize hemoglobin subunits at a rapid rate until the supply of heme is depleted. In the absence of heme, protein synthesis halts because of the abrupt formation of an inhibitor of protein synthesis called *heme-controlled inhibitor* (HCI). The mechanism of action of HCI was an enigma until it was discovered that HCI is a kinase that phosphorylates the α subunit of eIF2, the initiation factor that brings Met-$tRNA_i$ to the 40S ribosomal subunit (see Figure 30-38). Phosphorylation blocks the recycling of eIF2 in the following way (Fig-

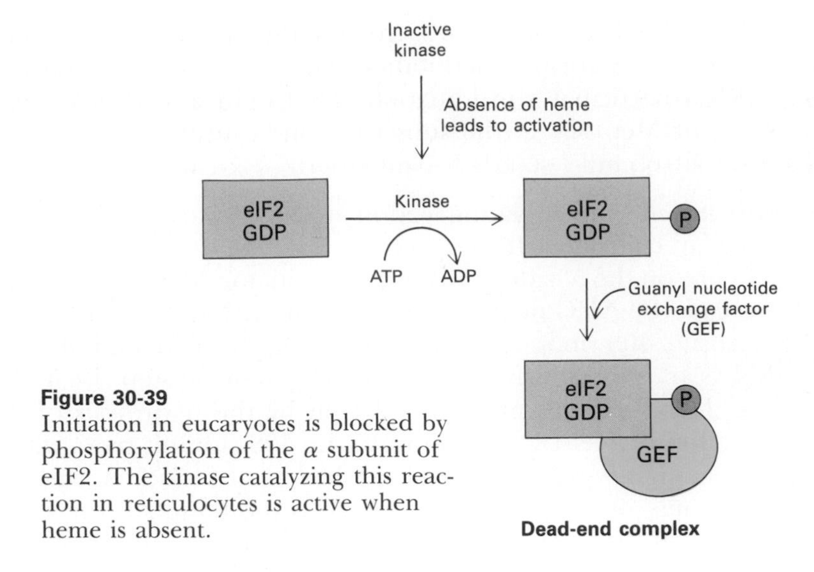

Figure 30-39
Initiation in eucaryotes is blocked by phosphorylation of the α subunit of eIF2. The kinase catalyzing this reaction in reticulocytes is active when heme is absent.

ure 30-39). eIF2 is in the GDP form when it is released from the 40S subunit on formation of the 80S initiation complex. Until it is converted into the GTP form, eIF2 cannot pick up a Met-tRNA$_i$ to initiate another round of protein synthesis. The needed exchange of GTP for GDP bound to eIF2 is catalyzed by a protein known as GEF (guanyl nucleotide exchange factor) or eIF2B. But if eIF2 is phosphorylated by the heme-controlled kinase, it has such high affinity for GEF that the complex is essentially irreversible. Hence, *phosphorylated eIF2 molecules are unable to initiate protein synthesis.* Moreover, they tie up GEF, which is present in lesser amounts than is eIF2. Consequently, the phosphorylation of only 30% of eIF2 leads to the complete cessation of protein synthesis. Phosphorylated eIF2 becomes functional again upon removal of its phosphate group by a specific phosphatase.

The physiologic role of the phosphorylation of eIF2 in reticulocytes is to coordinate the synthesis of heme and globin. When the level of heme is low, the synthesis of globins stops to prevent the formation of apohemoglobin, which is readily denatured. It seems likely that phosphorylation of eIF2 by different kinases is used to control translation in other cells too. The finding of an interferon-induced kinase that acts similarly in virus-infected cells (p. 881) suggests that this mechanism is widely used to control gene expression in eucaryotic cells.

DIPHTHERIA TOXIN BLOCKS PROTEIN SYNTHESIS IN EUCARYOTES BY INHIBITING TRANSLOCATION

Diphtheria was a major cause of death in childhood before the advent of effective immunization. The lethal effects of this disease are due mainly to a protein toxin produced by *Corynebacterium diphtheriae,* a bacterium that grows in the upper respiratory tract. The gene that encodes the toxin comes from a lysogenic phage that is harbored by some strains of *C. diphtheriae.* A few micrograms of diphtheria toxin is usually lethal in an unimmunized person because it inhibits protein synthesis. It consists of a single 61-kd chain with two disulfide bonds, and it is a bipartite protein. *The A domain catalyzes the covalent modification of a key component of the protein-synthesizing machinery, whereas the B domain enables the A domain to traverse the plasma membrane and enter the cytosol.* The binding of the B domain to a specific receptor (not yet identified) on the cell surface triggers the cleavage of the toxin into a 21-kd A fragment and a 40-kd B fragment, the reduction of the disulfide bonds between them, and the transfer of the A fragment into the cytosol.

A single A fragment of the toxin in the cytosol can kill a cell. Why is it so lethal? The target of the enzymatic activity of the A fragment is EF2, the elongation factor catalyzing translocation in eucaryotic protein synthesis. EF2 contains *diphthamide*, an unusual amino acid residue of unknown function that is formed by posttranslational modification of histidine. The A fragment of the toxin catalyzes the transfer of the adenosine diphosphate ribose unit of NAD^+ to a nitrogen atom of the diphthamide ring (Figure 30-40). *This ADP-ribosylation of a single side chain of EF2 blocks its capacity to carry out translocation of the growing polypeptide chain. Protein synthesis ceases, accounting for the remarkable toxicity of diphtheria toxin.* It is interesting to note that a variety of other toxins, such as cholera toxin (p. 793) and pertussis toxin also consist of a domain that binds a cell surface receptor and a catalytic domain that irreversibly modifies an important cell constituent.

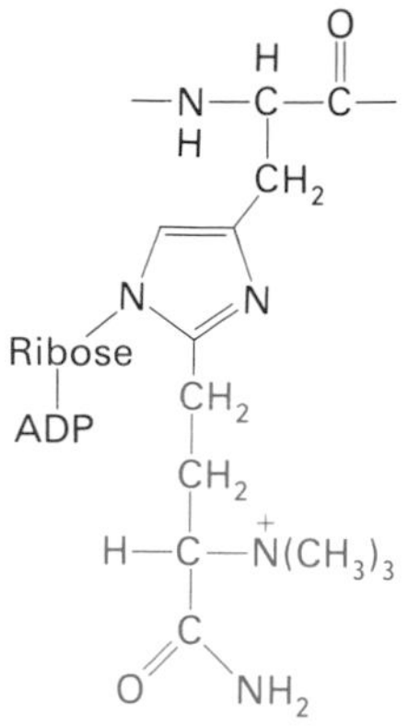

Figure 30-40
Diphtheria toxin blocks protein synthesis in eucaryotes by catalyzing the transfer of an ADP-ribose unit from NAD^+ to diphthamide, a modified amino acid residue on elongation factor 2 (translocase).

SUMMARY

Protein synthesis (translation) is mediated by the coordinated interplay of more than a hundred macromolecules, including mRNA, tRNAs, activating enzymes, and protein factors, in addition to ribosomes. Protein synthesis starts with the ATP-driven activation of amino acids by aminoacyl-tRNA synthetases (activating enzymes), which link the carboxyl group of an amino acid to the 2′- or 3′-hydroxyl group of the adenosine unit at the 3′ end of tRNA. There is at least one specific activating enzyme for each amino acid. Also, there is at least one specific tRNA for each amino acid. Transfer RNAs of different specificity have a common structural design. They are single RNA chains of about eighty nucleotides, some of which are modified after transcription. The base sequences of all known tRNAs can be written in a cloverleaf pattern in which about half of the nucleotides are base paired. X-ray crystallographic studies have shown that tRNA is L-shaped. The amino acid attachment site at the 3′-CCA terminus is at one end of the L, whereas the anticodon is 80 Å away at the other end. Messenger RNA recognizes the anticodon rather than the amino acid attached to a tRNA. A codon on mRNA forms base pairs with the anticodon of the tRNA. Some tRNAs are recognized by more than one codon because pairing of the third base of a codon is less discriminating than that of the other two (the wobble hypothesis).

Protein synthesis takes place on ribosomes, which consist of large and small subunits, each about two-thirds RNA and one-third protein by weight. In *E. coli*, the 70S ribosome (2700 kd) is made up of 30S and 50S subunits. Ribosomes can be formed in vitro by self-assembly of their constituent RNAs and proteins. Proteins are synthesized in the amino-to-carboxyl direction, and mRNA is translated in the 5′ → 3′ direction. In procaryotes, transcription and translation are closely coupled. Several ribosomes can simultaneously translate an mRNA.

Protein synthesis takes place in three phases: initiation, elongation, and termination. In procaryotes, mRNA, formylmethionyl-$tRNA_f$ (the special initiator tRNA), and a 30S ribosomal subunit come together to form a 30S initiation complex. Three protein initiation factors are essential for initiation. The start signal on mRNA is AUG (or GUG) preceded by a purine-rich sequence that can base-pair with 16S rRNA. A 50S ribosomal subunit then joins this complex to form a 70S initiation complex, in which fMet-$tRNA_f$ occupies the P (peptidyl) site of the ribosome. EF-Tu delivers the appropriate aminoacyl-tRNA to the A site (aminoacyl site) to begin the elongation phase. A peptide bond is then formed between the activated carboxyl group of the growing polypeptide and the amino group of the aminoacyl-tRNA. This reaction is catalyzed by the peptidyl transferase center of the large ribosomal subunit. The resulting uncharged tRNA leaves the P site, peptidyl-tRNA moves from the A to the P site, and mRNA moves a distance of three nucleotides. These translocations are mediated by EF-G (translocase). Protein synthesis is terminated by release factors, which recognize the termination codons UAA, UGA, and UAG and cause the hydrolysis of the bond between the polypeptide and tRNA. A GTP-GDP cycle drives the formation of the 70S initiation complex, the delivery of aminoacyl-tRNA, and translocation. Various steps in protein synthesis are specifically inhibited by toxins and antibiotics. Streptomycin inhibits initiation and induces misreading of mRNA, and puromycin causes premature chain termination.

The basic plan of protein synthesis in eucaryotes is similar to that of procaryotes, but there are some significant differences between them. Eucaryotic ribosomes (80S) consist of a 40S small subunit and a 60S large subunit. The AUG closest to the 5′ end of mRNA is nearly always the start site. The initiating amino acid is again methionine, but it is not formylated. The initiation of protein synthesis in eucaryotes is more complex than in procaryotes. eIF2, the eucaryotic initiation factor that delivers the initiator tRNA to the ribosome, is inactivated by phosphorylation. The kinases catalyzing this inactivation reaction are controlled. For example, a deficiency of heme in reticulocytes leads to the activation of this kinase and the consequent cessation of protein synthesis. Diphtheria toxin blocks protein synthesis in eucaryotes by catalyzing the ADP-ribosylation of EF2, the elongation factor mediating translocation.

SELECTED READINGS

Where to start

Lake, J. A., 1981. The ribosome. *Sci. Amer.* 245(2):84–97.

Nomura, M., 1984. The control of ribosome synthesis. *Sci. Amer.* 250(1):102–114.

Rich, A., and Kim, S. H., 1978. The three-dimensional structure of transfer RNA. *Sci. Amer.* 238(1):52–62. [Offprint 1377.]

Clark, B. F. C., and Marcker, K. A., 1968. How proteins start. *Sci. Amer.* 218(1):36–42. [Offprint 1092.]

Books

Spirin, A. S., 1986. *Ribosome Structure and Protein Synthesis.* Benjamin/Cummings.

Clark, B. F. C., and Petersen, H. F., (eds.), 1984. *Gene Expression. The Translational Step and Its Control.* Munksgaard, Copenhagen.

Bermek, E., (ed.), 1985. *Mechanisms of Protein Synthesis. Structure-Function Relations, Control Mechanisms, and Evolutionary Aspects.* Springer-Verlag.

Weissbach, H., and Pestka, S., (eds.), 1977. *Molecular Mechanisms of Protein Biosynthesis.* Academic Press.

Chambliss, G., Craven, G. R., Davies, J., Davis, K., Kahan, L., and Nomura, M., (eds.), 1980. *Ribosomes: Structure, Function, and Genetics.* University Park Press.

Cold Spring Harbor Laboratory, 1969. *Mechanisms of Protein Biosynthesis* (Cold Spring Harbor Symposium on Quantitative Biology, vol. 34).

Aminoacyl-tRNA synthetases

Schimmel, P., 1987. Aminoacyl tRNA synthetases: general scheme of structure-function relationships in the polypeptides and recognition of transfer RNAs. *Ann. Rev. Biochem.* 56:125–158.

Blow, D. M., and Brick, P., 1985. Aminoacyl-tRNA synthetases. *In* Jurnak, F. A., and McPherson, A., (eds.), *Biological Macromolecules and Assemblies,* vol. 2: *Nucleic Acids and Interactive Proteins,* pp. 441–469. Wiley-Interscience. [Presentation of x-ray crystallographic analyses.]

Fersht, A. R., Leatherbarrow, R. J., and Wells, T. N. C., 1986. Structure and reactivity of the tyrosyl-tRNA synthetase: the hydrogen bond in catalysis and specificity. *Phil. Trans. R. Soc. London* A317:305–320.

Bedouelle, H., and Winter, G., 1986. A model of synthetase/transfer RNA interaction as deduced by protein engineering. *Nature* 320:371–373.

Leatherbarrow, R. J., Fersht, A. R., and Winter, G., 1985. Transition-state stabilization in the mechanism of tyrosyl-tRNA synthetase revealed by protein engineering. *Proc. Nat. Acad. Sci.* 82:7840–7844.

Schimmel, P. R., and Söll, D., 1979. Amino acyl-tRNA synthetases: general features and recognition of transfer RNAs. *Ann. Rev. Biochem.* 48:601–648.

Ribosomes and ribosomal RNAs

Noller, H. F., 1984. Structure of ribosomal RNA. *Ann. Rev. Biochem.* 53:119–162.

Gutell, R. R., Weiser, B., Woese, C. R., and Noller, H. F., 1985. Comparative anatomy of 16S-like ribosomal RNA. *Prog. Nucleic Acid Res. Mol. Biol.* 32:154–216.

Thompson, J. F., and Hearst, J. E., 1983. Structure-function relations in *E. coli* 16S RNA. *Cell* 33:19–24.

Engelman, D. M., and Moore, P. B., 1976. Neutron scattering studies of the ribosome. *Sci. Amer.* 235:43–54.

Nomura, M., Gourse, R., and Baughman, G., 1984. Regulation of the synthesis of ribosomal components. *Ann. Rev. Biochem.* 53:75–117.

Wittmann, H. G., 1983. Architecture of procaryotic ribosomes. *Ann. Rev. Biochem.* 52:35–65.

Initiation, elongation, and termination

Dintzis, H. M., 1961. Assembly of the peptide chains of hemoglobin. *Proc. Nat. Acad. Sci.* 47:247–261. [These important labeling experiments showed that proteins are synthesized from the amino to the carboxyl end.]

Steitz, J. A., 1979. Genetic signals and nucleotide sequences in messenger RNA. *In* Goldberger, R. F., (ed.), *Biological Regulation and Development,* vol. 1, pp. 349–399. Plenum.

Shine, J., and Dalgarno, L., 1974. The 3′-terminal sequence of *Escherichia coli 16S* ribosomal RNA: complementarity to nonsense triplets and ribosome binding sites. *Proc. Nat. Acad. Sci.* 71:1342–1346.

Steitz, J. A., and Jakes, K., 1975. How ribosomes select initiator regions in mRNA: base pair formation between the 3′-terminus of 16S rRNA and the mRNA during initiation of protein synthesis in *Escherichia coli. Proc. Nat. Acad. Sci.* 72:4734–4738.

Maitra, U., Stringer, E. A., and Chaudhuri, A., 1982. Initiation factors in protein synthesis. *Ann. Rev. Biochem.* 51:869–900.

Jurnak, F., 1985. Structure of the GDP domain of EF-Tu and location of the amino acids homologous to *ras* oncogene proteins. *Science* 230:32–36.

Gupta, S. L., Waterson, J., Sopori, M., Weissman, S. M., and Lengyel, P., 1971. Movement of the ribosome along the messenger ribonucleic acid during protein synthesis. *Biochemistry* 10:4410–4421.

Kaziro, Y., 1978. The role of GTP in polypeptide chain elongation. *Biochim. Biophys. Acta* 505:95–127.

Caskey, T. H., 1980. Peptide chain termination. *Trends Biochem. Sci.* 5:234–237.

Transfer RNA

Schimmel, P., Söll, D., and Abelson, J., (eds.), 1979. *Transfer RNA.* Cold Spring Harbor Laboratory. [Part 1 deals with structure, properties, and recognition, and Part 2 with biosynthesis and genetic aspects of tRNA. An authoritative and valuable treatise.]

Björk, G. R., Ericson, J. U., Gustafsson, C. E. D., Hagervall, T. G., Jonssön, Y. H., and Wikström, P. M., 1987. Transfer RNA modification. *Ann. Rev. Biochem.* 56:263–288.

Nakajima, N., Ozeki, H., and Shimura, Y., 1981. Organization and structure of an *E. coli* tRNA operon containing seven tRNA genes. *Cell* 23:239–249.

Guerrier-Takada, C., and Altman, S., 1984. Catalytic activity of an RNA molecule prepared by transcription in vitro. *Science* 223:285–286. [Demonstration that the RNA moiety of RNase P, the enzyme that cleaves tRNA precursors, is catalytically active in the absence of protein.]

Kim, S.-H., 1978. Three-dimensional structure of transfer RNA and its functional implications. *Advan. Enzymol.* 46:279–315.

Jack, A., Ladner, J. E., and Klug, A., 1976. Crystallographic refinement of yeast phenylalanine transfer RNA at 2.5 Å resolution. *J. Mol. Biol.* 108:619–649.

Sussman, J. L., and Kim, S.-H., 1976. Three-dimensional structure of a transfer RNA in two crystal forms. *Science* 192:853–858.

Woo, N. H., Roe, B. A., and Rich, A., 1980. Three-dimensional structure of *Escherichia coli* initiator tRNA-fMet. *Nature* 286:346–351.

Temple, G. F., Dozy, A. M., Roy, K. L., and Kan, Y. W., 1982. Construction of a functional human suppressor tRNA gene: an approach to gene therapy for beta-thalassaemia. *Nature* 296:537–540.

Fidelity and proofreading

Kirkwood, T. B. L., Rosenberger, R. F., and Galas, D. J., (eds.), 1986. *Accuracy in Molecular Processes. Its Control and Relevance to Living Systems.* Chapman and Hall. [A fascinating book. Chapters 4, 5, and 6 deal with the fidelity of protein synthesis, and Chapter 11 with the kinetic costs of proofreading.]

Fersht, A., and Dingwall, C., 1979. Evidence for the double-sieve editing mechanism in protein biosynthesis. *Biochemistry* 18:2627–2631.

Fersht, A., 1985. *Enzyme Structure and Mechanism.* W. H. Freeman. [Chapter 13 on specificity and editing mechanisms is lucid and concise.]

Thompson, R. C., Dix, D. B., and Karim, A. M., 1986. The reaction of ribosomes with elongation factor Tu-GTP complexes. Aminoacyl-tRNA-independent reactions in the elongation cycle determine the accuracy of protein synthesis. *J. Biol. Chem.* 261:4868–4874.

Eucaryotic protein synthesis

Proud, C. G., 1986. Guanine nucleotides, protein phosphorylation, and the control of translation. *Trends Biochem. Sci.* 11:73–77.

Moldave, K., 1985. Eukaryotic protein synthesis. *Ann. Rev. Biochem.* 54:1109–1149.

Shatkin, A. J., 1985. mRNA cap binding proteins: essential factors for initiating translation. *Cell* 40:223–224.

Kozak, M., 1983. Comparison of initiation of protein synthesis in procaryotes, eucaryotes, and organelles. *Microbiol. Rev.* 47:1–45.

Antibiotics and toxins

Pestka, S., (1971). Inhibitors of ribosome function. *Ann. Rev. Microbiol.* 25:487–562.

Jimenez, A., 1976. Inhibitors of translation. *Trends Biochem. Sci.* 1:28–29.

Tai, P.-C., Wallace, B. J., and Davis, B. D., 1978. Streptomycin causes misreading of natural messenger by interacting with ribosomes after initiation. *Proc. Nat. Acad. Sci.* 75:275–279.

PROBLEMS

1. The formation of Ile-tRNA proceeds through an enzyme-bound Ile-AMP intermediate. Predict whether ^{32}P-labeled ATP is formed from $^{32}PP_i$ when each of the following sets of components is incubated with the specific activating enzyme:
   (a) ATP and $^{32}PP_i$.
   (b) tRNA, ATP, and $^{32}PP_i$.
   (c) Isoleucine, ATP, and $^{32}PP_i$.

2. Ribosomes were isolated from bacteria grown in a "heavy" medium (^{13}C and ^{15}N) and from bacteria grown in a "light" medium (^{12}C and ^{14}N). These 70S ribosomes were added to an in vitro system actively engaged in protein synthesis. An aliquot removed several hours later was analyzed by density-gradient centrifugation. How many bands of 70S ribosomes would you expect to see in the density gradient?

3. How many high-energy phosphate bonds are consumed in the synthesis of a 200-residue protein, starting from amino acids?

4. There are two basic mechanisms for the elongation of biomolecules (Figure 30-41). In type 1, the activating

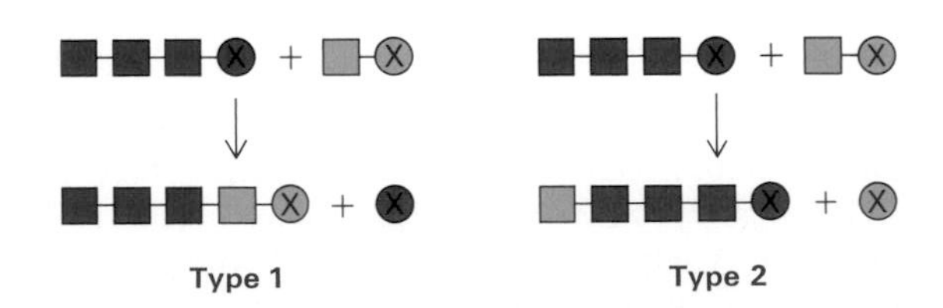

Figure 30-41
Two modes of elongation.

group (labeled X) is released from the growing chain. In type 2, the activating group is released from the incoming unit as it is added to the growing chain. Indicate whether each of the following biosyntheses occurs by means of a type 1 or a type 2 mechanism:
   (a) Glycogen synthesis.
   (b) Fatty acid synthesis.
   (c) $C_5 \rightarrow C_{10} \rightarrow C_{15}$ in cholesterol synthesis.
   (d) DNA synthesis.
   (e) RNA synthesis.
   (f) Protein synthesis.

5. Mutations that produce termination codons are called nonsense mutations. These mutations can be suppressed by altered tRNAs. For example, the UGA codon is translated into tryptophan by a mutant tRNA. What is the most likely base change in this mutant tRNA?

6. Design an affinity-labeling reagent for one of the tRNA binding sites in *E. coli* ribosomes. How would you synthesize such a reagent?

7. An mRNA transcript of a T7 phage gene contains the base sequence

   ↓
   5′-AACUGCACGAGGUAACACAAGAUGGCU-3′

   Predict the effect of a mutation that changes the G marked by an arrow to A.

8. What is the nucleophile in the reaction catalyzed by peptidyl transferase?

9. Fusidic acid prevents the dissociation of EF-G-GDP from ribosomes but does not block the entry of GTP into EF-G nor its hydrolysis to GDP. Suppose that Leu-Gly-tRNA occupies the P site. Fusidic acid and Trp-tRNA (the tRNA corresponding to the next codon) are added. At which step is protein synthesis blocked? What is the product and where is it situated on the ribosome?

10. Compare and contrast protein synthesis by ribosomes with peptide synthesis by the solid-phase method.

11. Compare the accuracy of (a) DNA replication, (b) RNA synthesis, and (c) protein synthesis. Which mechanisms are used to assure the fidelity of each of these processes?

CHAPTER 31

Protein Targeting

Nascent proteins contain signals that determine their ultimate destination. A newly synthesized protein in *E. coli*, for example, can stay in the cytosol, or it can be sent to the plasma membrane, the outer membrane, the space between them, or the extracellular medium. Eucaryotic cells can direct proteins to internal sites such as lysosomes, mitochondria, chloroplasts, and the nucleus. How is sorting accomplished? In eucaryotes, a key choice is made soon after the synthesis of a protein begins. A ribosome remains free in the cytosol unless it is directed to the endoplasmic reticulum (ER) by a signal sequence in the protein being synthesized. Nascent polypeptide chains formed by such membrane-bound ribosomes are translocated across the ER membrane. In the lumen of the ER, many of them are glycosylated and modified in other ways. They are then transported to the Golgi complex, where they are further modified. Finally, they are sorted for delivery to lysosomes, secretory vesicles, and the plasma membrane. The signals used to target eucaryotic proteins for transfer across the endoplasmic reticulum membrane are ancient—bacteria employ similar sequences to send proteins to their plasma membrane and to secrete them. Different strategies are used to send proteins synthesized by free ribosomes to the nucleus, mitochondria, and chloroplasts of eucaryotic cells.

A complementary aspect of protein targeting is the import of proteins into eucaryotic cells by *receptor-mediated endocytosis*. Proteins in the extracellular space bind to specific receptors in specialized regions of the plasma membrane called coated pits. Clathrin, a protein that forms polyhedral lattices, converts coated-pit regions into membrane-bound vesicles inside the cell. When these vesicles become acidified, the im-

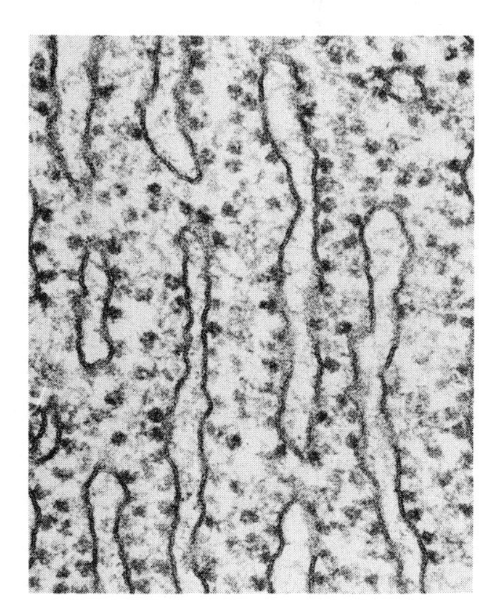

Figure 31-1
Electron micrograph of the rough endoplasmic reticulum. Ribosomes are bound to the cytosolic face of the membrane. [Courtesy of Dr. George Palade.]

ported proteins dissociate from their receptors. Many viruses and toxins enter cells by mimicking molecules that are normally imported by endocytosis.

The final section of this chapter deals with the *programmed destruction of proteins.* The half-life of a protein is very much influenced by the nature of its amino-terminal residue. Ubiquitin, a protein found in all eucaryotes, becomes conjugated to proteins marked for degradation.

RIBOSOMES BOUND TO THE ENDOPLASMIC RETICULUM MAKE SECRETORY AND MEMBRANE PROTEINS

"The choice of the pancreatic acinar cell, a very efficient protein producer, as the object of our studies reflected in part our training, and in part our environment . . . . Perhaps the most important factor in this selection was the appeal of the amazing organization of the pancreatic acinar cell, whose cytoplasm is packed with stacked endoplasmic reticulum cisternae studded with ribosomes. Its pictures had for me the effect of the song of a mermaid: irresistable and half transparent. Its meaning seemed to be buried only under a few years of work, and reasonable working hypotheses were already suggested by the structural organization itself."

George Palade
Nobel Lectures (1975)

In eucaryotic cells, some ribosomes are free in the cytosol, whereas others are bound to an extensive membrane system called the *endoplasmic reticulum* (ER), which comprises about half of the total membrane of a cell. The region that binds ribosomes is called the *rough ER* because of its studded appearance, in contrast with the smooth ER, which is devoid of ribosomes. Membrane-bound ribosomes synthesize three major classes of proteins: *lysosomal proteins, secretory proteins*—that is, proteins exported by the cell—and *proteins spanning the plasma membrane.* Indeed, virtually all integral membrane proteins of the cell, except those located in the membranes of mitochondria and chloroplasts, are formed by ribosomes bound to the ER. George Palade's studies of the mechanism of secretion of zymogens by the pancreatic acinar cell, a cell specialized for protein synthesis and export, opened a new area of inquiry—protein targeting—and delineated the pathway taken by secretory proteins.

Are there two classes of ribosomes—one free in the cytosol, the other membrane bound—or are all ribosomes intrinsically the same? Studies of the protein-synthesizing activities of ribosomes in cell-free systems answered this important question. Free ribosomes from the cytosol were isolated and then added to rough ER membranes that had been stripped of their ribosomes. This reconstituted system actively synthesized secretory proteins when presented with the appropriate mRNAs and other soluble factors. Likewise, ribosomes isolated from the rough ER were fully active in synthesizing proteins that are normally released into the cytosol. Furthermore, there were no detectable structural differences between free ribosomes and ribosomes derived from the rough ER. *Thus, membrane-bound ribosomes and free ribosomes are intrinsically the same. Whether a particular ribosome is free or attached to the rough ER depends only on the kind of protein it is making.*

SIGNAL SEQUENCES MARK PROTEINS FOR TRANSLOCATION ACROSS THE ENDOPLASMIC RETICULUM (ER) MEMBRANE

The attachment of an actively synthesizing ribosome to the ER membrane is a key event in translocating a protein across this membrane. What is it that directs the ribosome to the membrane? In 1970, David Sabatini and Günter Blobel postulated that *the signal for attachment is a sequence of amino acid residues near the amino-terminus of the nascent polypeptide chain.* This *signal hypothesis* was soon supported; Cesar Milstein and George Brownlee found that an immunoglobulin chain synthesized in vitro by free ribosomes contained an amino-terminal sequence of twenty residues that was absent from the mature protein synthesized in vivo. Blobel then found that the major secretory proteins of the pancreas contain amino-terminal extensions of about twenty residues when synthesized in vitro by free ribosomes. These *signal sequences* are absent from normally secreted proteins because they are cleaved by a *signal peptidase* on the lumenal side of the ER membrane.

The cleaved amino-terminal signal sequences of more than a hundred secretory proteins from a wide variety of eucaryotic species are now known (Figure 31-2). A well-defined consensus sequence, such as the TATA box guiding the initiation of transcription, is not evident. However, signal sequences do exhibit several common features: (1) They range in length from 13 to 36 residues. (2) The amino-terminal part of the signal contains at least one positively charged residue. (3) A highly hydrophobic stretch, typically 10 to 15 residues long, forms the center of the signal sequence. Alanine, leucine, valine, isoleucine, and phenylalanine are common in this region. A substitution of a charged residue for a nonpolar one in this hydrophobic core usually destroys the directing activity of the signal sequence. (4) The cleavage site at the carboxyl-terminal end is preceded by a sequence of about five residues that is more polar than the hydrophobic core. However, the residues one before (−1, on the amino-terminal side) and three before (−3) the cleavage site have small neutral side chains. Alanine is most common. Cysteine, threonine, serine, and glycine are also found at the −1 and −3 positions.

		Cleavage site
Human growth hormone	MATGSRTSLLLAFGLLCLPWLQEGSA	FPT
Human proinsulin	MALWMRLLPLLALLALWGPDPAAA	FVN
Bovine proalbumin	MKWVTFISLLLFSSAYS	RGV
Mouse antibody H chain	MKVLSLLYLLTAIPHIMS	DVQ
Chicken lysozyme	MRSLLILVLCFLPKLAALG	KVF
Bee promellitin	MKFLVNVALVFMVVYISYIYA	APE
Drosophila glue protein	MKLLVVAVIACMLIGFADPASG	CKD
Zea maize protein 19	MAAKIFCLIMLLGLSASAATA	SIF
Yeast invertase	MLLQAFLFLLAGFAAKISA	SMT
Human influenza virus A	MKAKLLVLLYAFVAG	DQI

Figure 31-2
Amino-terminal signal sequences of some eucaryotic secretory and plasma membrane proteins. The hydrophobic core (yellow) is preceded by basic residues (blue) and followed by a cleavage site (red) for signal peptidase. One-letter symbols for amino acids are defined in Table 2-2 on page 21.

Not all secretory and plasma membrane proteins contain an amino-terminal signal sequence that is cleaved following translocation across the ER membrane. Some proteins contain an *internal signal sequence* that serves the same role. The location of the internal signal sequence of ovalbumin, the major protein of hen egg-white, has been determined. By recombinant DNA techniques, different segments of the ovalbumin gene were deleted to find out which amino acid residues are essential for targeting. This analysis revealed that the sequence between residues 22 and 41 is critical for the transfer of nascent ovalbumin across the endoplasmic reticulum membrane.

A CYTOSOLIC PROTEIN CAN BE REDIRECTED TO THE ER BY JOINING A SIGNAL SEQUENCE TO ITS AMINO TERMINUS

Chimeric proteins formed by recombinant DNA techniques have provided insight into mechanisms of protein targeting. This approach has been used to determine whether a cleavable amino-terminal signal sequence contains all the information needed to direct the translocation of a protein across the ER membrane. The signal sequence in this experiment came from β-lactamase, an *E. coli* enzyme that hydrolyzes penicillin and other β-lactams. As it is synthesized, this enzyme is secreted into the periplasmic space between the outer and plasma membranes of the bacterium. In the process, a 23-residue signal sequence is cleaved from the polypeptide. This sequence and 5 adjoining residues of β-lactamase were joined to the amino terminus of the α chain of hemoglobin by forming a hybrid gene (Figure 31-3). α-Globin was chosen as the test protein because it normally stays in the cytosol and is highly hydrophilic.

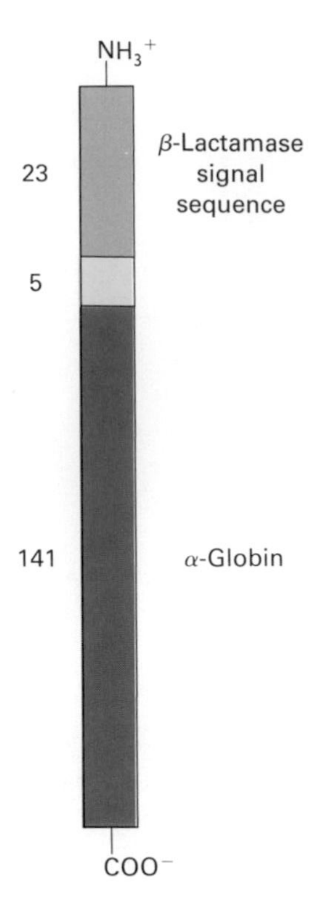

Figure 31-3
α-Globin (red) is redirected from the cytosol to the ER by joining a bacterial signal sequence (green) and five adjoining residues (yellow) to its amino terminus.

The mRNA for this chimeric protein was added to an in vitro protein-synthesizing system that can translocate secretory proteins across the ER membrane (Figure 31-4). It consists of microsomes, ribosomes, tRNAs, and other soluble factors required for protein synthesis and translocation. (Microsomes are closed vesicles formed by the self-sealing of fragments of the endoplasmic reticulum membrane.) Ribosomes synthesizing α-globin with a signal sequence became bound to the microsomes. Most striking, the nascent polypeptide chain traversed the ER membrane, the signal sequence was cleaved, and α-globin was sequestered inside the vesicle. Thus, *the addition of a signal sequence to α-globin converted it from a cytosolic protein into a secretory protein. This experiment also demonstrated that bacterial and eucaryotic signal sequences are functionally similar.*

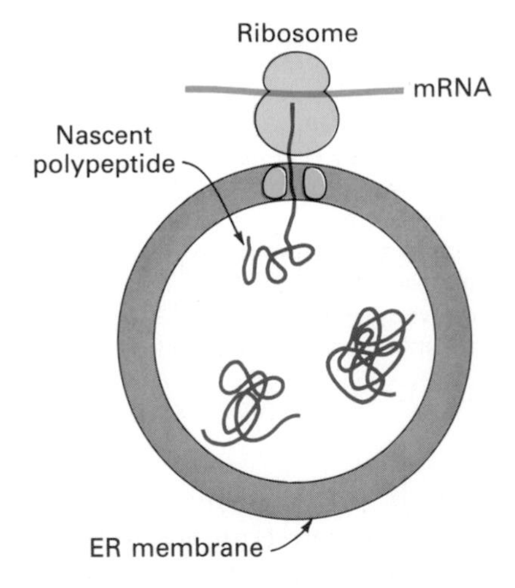

Figure 31-4
Membrane and secretory proteins can be synthesized and sequestered in vitro by using microsomal vesicles produced by fragmentation of the endoplasmic reticulum.

SIGNAL RECOGNITION PARTICLE (SRP) DETECTS SIGNAL SEQUENCES AND BRINGS RIBOSOMES TO THE ER MEMBRANE

How are signal sequences recognized and the proteins containing them translocated across the ER membrane? Blobel and Peter Walter showed that the adaptor coupling the protein-synthesizing machinery in the cytosol to the protein-translocating machinery in the ER membrane is a ribonucleoprotein called *signal recognition particle* (SRP). This 325-kd assembly consists of a 300-nucleotide RNA molecule (called 7SL RNA) and six different polypeptide chains. SRP binds tightly to ribosomes containing a nascent chain with a signal sequence but not to other ribosomes (Figure 31-5). This binding occurs soon after the emergence of the amino-terminal signal sequence from the ribosome. Elongation of the polypeptide chain is arrested or slowed while SRP is bound. The SRP-ribosome complex then diffuses to the ER membrane, where SRP binds to the *SRP receptor* (docking protein), a dimer of α (69 kd) and β

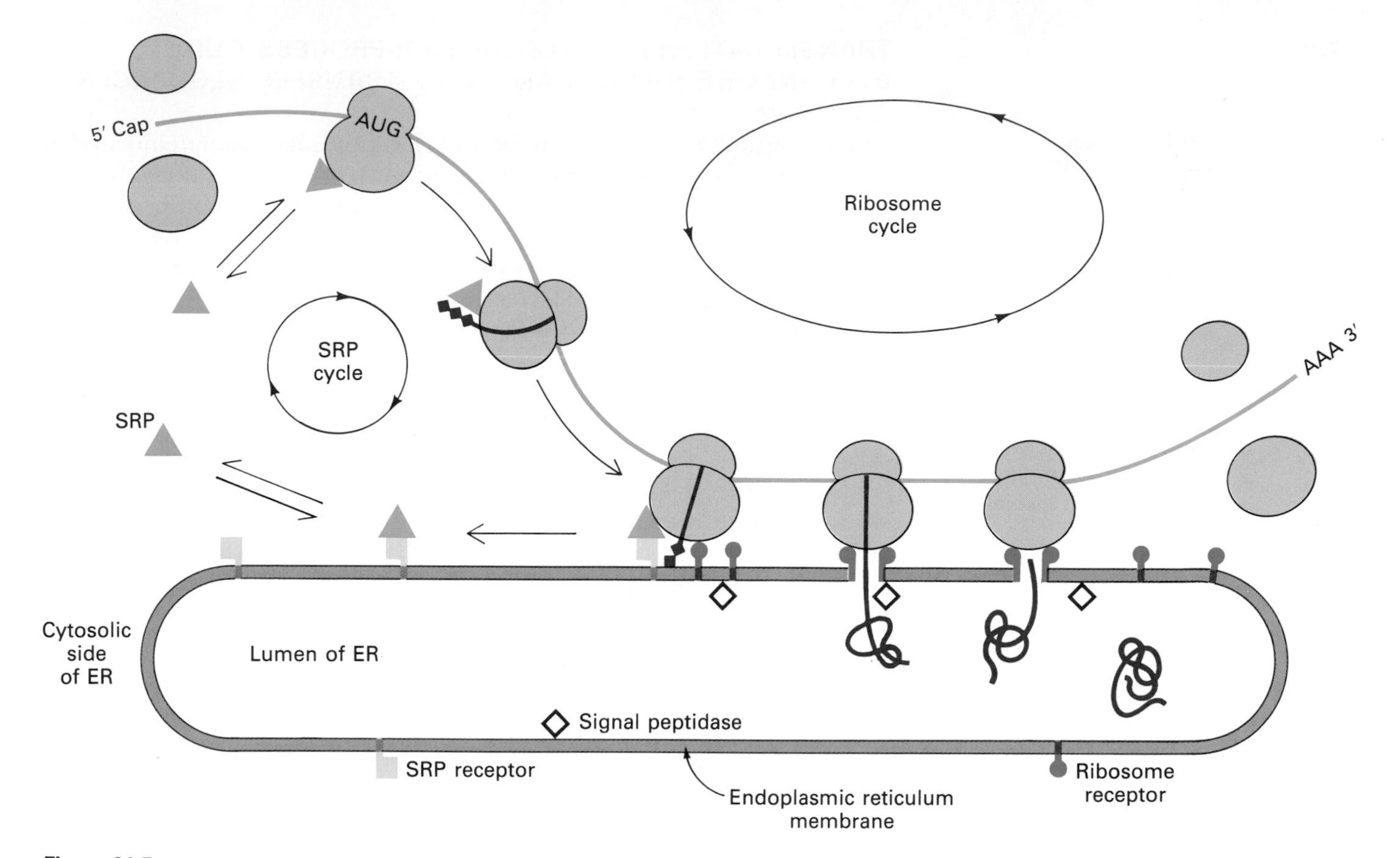

Figure 31-5
Signal recognition particle (SRP) participates in delivering ribosomes to the ER. The signal sequence of a nascent polypeptide chain is recognized by SRP. The complex consisting of SRP, the nascent chain, and the ribosome synthesizing it then binds to the SRP receptor in the ER membrane. The ribosome is then transferred to ribophorins and the translocation machinery, which actively threads the polypeptide chain across the ER membrane. SRP released from its receptor can then bind another signal sequence. [After P. Walter, R. Gilmore, and G. Blobel. *Cell* 38(1984):6.]

(30 kd) subunits. The ribosome containing the nascent polypeptide chain is delivered to the translocation machinery, which includes two integral membrane proteins called *ribophorin* I and II, and SRP is released into the cytosol. The nascent chain resumes elongation. *Thus, SRP acts catalytically to deliver ribosomes with a signal sequence to the ER membrane.*

SRP has the shape of a rod about 240 Å long and 50 Å wide. The structural backbone of SRP is its RNA component, which contains two sequences (called Alu sequences) that recur frequently in mammalian DNA (p. 835). The RNA binds six proteins (9, 14, 19, 54, 68, and 72 kd) to form a bipartite SRP. One domain of this highly elongated particle recognizes and binds the emerging signal sequence; affinity-labeling studies have shown that the 54-kd subunit of SRP is in contact with the signal sequence. The other domain of SRP sterically interferes with the entry of aminoacyl-tRNA and the peptidyl transferase step, and so elongation of the polypeptide chain is arrested. This arrest is relieved when SRP binds its receptor on the ER membrane. The cytosolic part of the α subunit of SRP receptor contains several highly positively charged loops that probably interact with the negatively charged RNA of SRP.

TRANSLOCATION IS AN ATP-DRIVEN PROCESS GUIDED BY SIGNAL SEQUENCES AND STOP-TRANSFER SEQUENCES

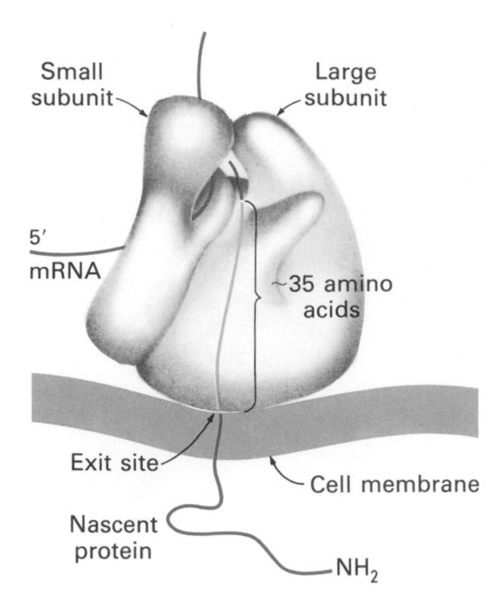

Figure 31-6
Diagram of a membrane-bound ribosome. [Courtesy of Dr. James Lake.]

The topography of membrane-bound ribosomes has been delineated at low resolution by immunoelectronmicroscopy and by image-reconstruction studies of crystalline arrays. The nascent polypeptide emerges from the ribosome at an exit site 150 Å away from the peptidyl transferase site. The most recently synthesized forty residues are located in a narrow tunnel through the large subunit of the ribosome. The exit site is situated on the face of the large subunit contacting the ER membrane.

What is the relation between elongation of the polypeptide chain and its translocation across the ER membrane? It was first thought that elongation drives translocation, that peptide synthesis pushes the nascent chain through a protein tunnel in the bilayer. However, recent studies have shown that some proteins, such as an 18-kd precursor of a yeast mating factor, can be translocated after they have been completely synthesized and released from the ribosome. Indeed, the glucose transporter of human erythrocytes, a large protein with 12 membrane-spanning α helices, can be slowly translocated in vitro after being fully synthesized. These post-translational translocations require the hydrolysis of ATP and the participation of membrane-bound and soluble proteins. Thus, *translocation and elongation are mechanistically separate processes.*

Why then do they usually occur simultaneously? The answer is that most fully synthesized proteins possessing signal sequences cannot be efficiently translocated, probably because they have folded up. A pair of interacting α helices, for example, probably cannot fit in the ribosomal tunnel. *Unfolded polypeptide chains are optimal substrates for translocation across the ER membrane.* The elongation arrest or slowing induced by the binding of SRP to ribosomes prevents premature folding on the cytosolic side of the membrane. The ribosome, too, participates in maintaining the translocatability of the nascent chain by keeping its most recently synthesized part fully stretched out in the narrow tunnel of the large subunit. The challenge now is to elucidate the components and mechanism of the membrane engine that grasps the nascent polypeptide chain and pulls it through the ER membrane. The ATP-driven processive movements of enzymes along DNA and RNA may be analogs of the unidirectional movement of polypeptide chains driven by membrane-bound motors.

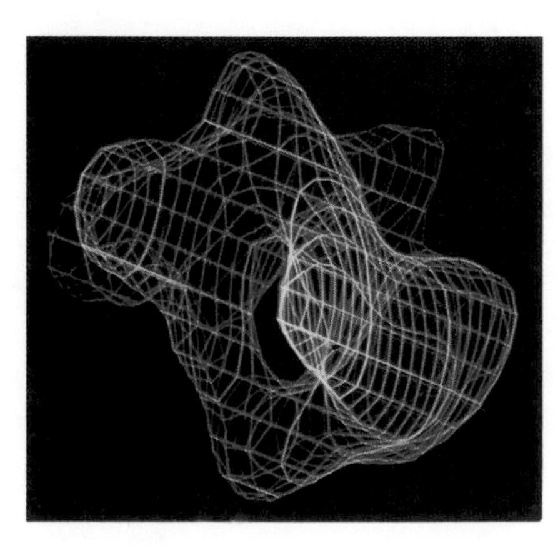

Figure 31-7
Tunnel in the large subunit of the *E. coli* ribosome. This image was obtained by analyses of electron micrographs of two-dimensional crystalline arrays. [From A. Yonath, K. R. Leonard, H. G. Wittmann. *Science* 236(1987):815.]

The translocation process for integral membrane proteins is more complex than for secretory and lysosomal proteins, which are threaded through in entirety. Some integral membrane proteins have one membrane-spanning helix, whereas others have several (Figure 31-8). Moreover, the amino and carboxyl termini can be on either side of the membrane, depending on the particular protein. How are these different topologies generated? The translocation machinery, like the Sorcerer's Apprentice, is unrelenting in its action unless stopped by a specific instruction, which in this case is a *stop-transfer sequence* (also called a membrane anchor sequence) on the nascent polypeptide. The appearance of a second signal sequence would initiate another round of translocation. The translocation machinery must also be able to thread polypeptide chains in the reverse direction. Much remains to be learned about the assembly and placement of membrane proteins, a fascinating area of inquiry.

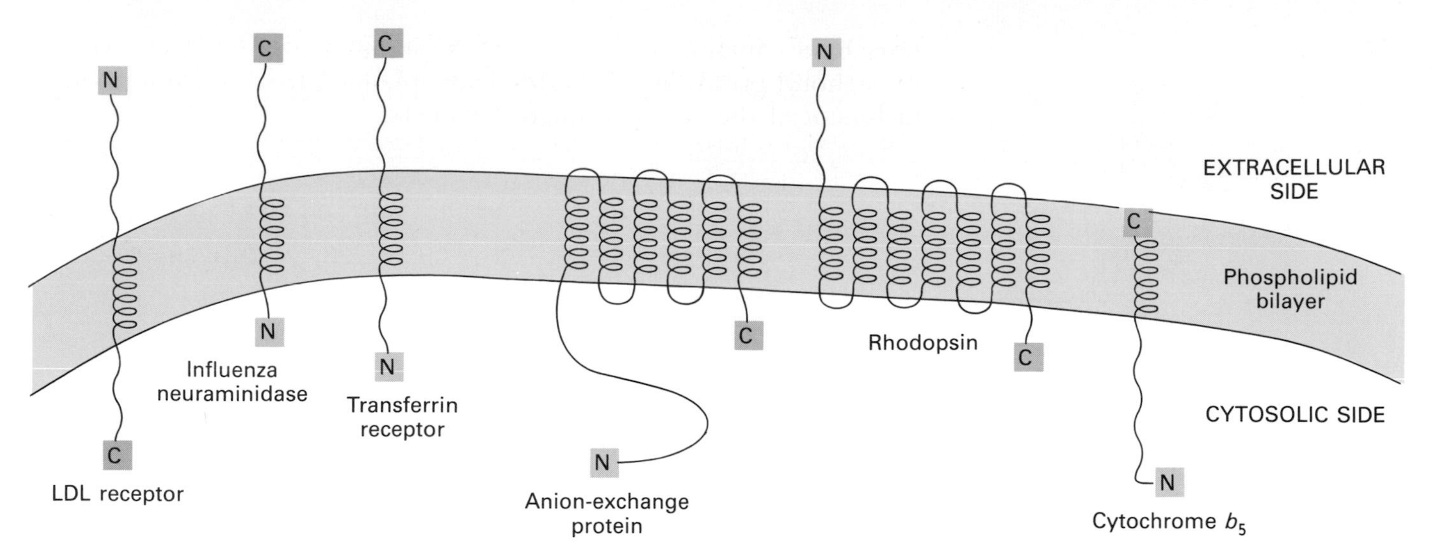

Figure 31-8
Different topological arrangements of integral membrane proteins. [After W. T. Wickner and H. F. Lodish. *Science* 230(1985):400.]

GLYCOPROTEINS ACQUIRE THEIR CORE SUGARS FROM DOLICHOL DONORS IN THE ENDOPLASMIC RETICULUM

After translocation, many proteins are modified in the lumen of the endoplasmic reticulum: amino-terminal signal sequences are cleaved by signal peptidase, disulfide bonds are formed, and many proteins are glycosylated. As was discussed earlier (p. 343), glycoproteins contain oligosaccharide units linked to either asparagine side chains by *N*-glycosidic bonds or to serine and threonine side chains by *O*-glycosidic bonds. *N*-linked oligosaccharides, though diverse, contain a common pentasaccharide core consisting of three mannose and two *N*-acetylglucosamine residues. This unifying motif expresses the mode of biosynthesis of such oligosaccharide units. A somewhat larger common oligosaccharide block is constructed on an activated lipid carrier, which transfers it to the growing polypeptide chain on the lumenal side of the ER membrane (Figure 31-9). The carrier is *dolichol phosphate,* a very

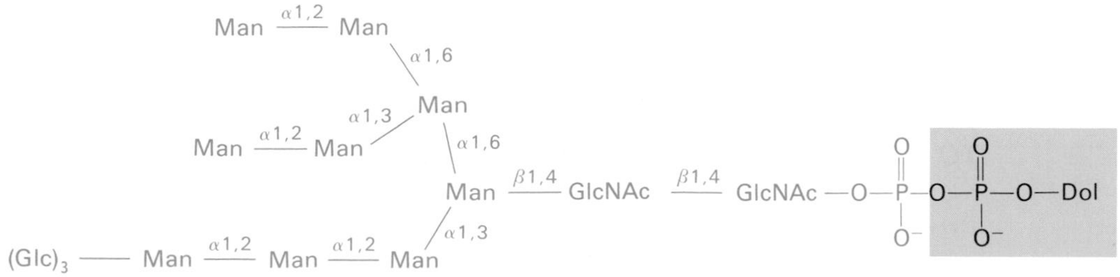

Man $\xrightarrow{\alpha 1,2}$ Man $\xrightarrow{\alpha 1,6}$ Man
Man $\xrightarrow{\alpha 1,2}$ Man $\xrightarrow{\alpha 1,3}$ Man
Man $\xrightarrow{\alpha 1,6}$ Man $\xrightarrow{\beta 1,4}$ GlcNAc $\xrightarrow{\beta 1,4}$ GlcNAc—O—P($=$O)(O$^-$)—O—P($=$O)(O$^-$)—O—Dol
$(Glc)_3$ —— Man $\xrightarrow{\alpha 1,2}$ Man $\xrightarrow{\alpha 1,2}$ Man $\xrightarrow{\alpha 1,3}$ Man

Figure 31-9
Structure of an activated core oligosaccharide (green). The carrier (yellow) is dolichol phosphate.

long lipid containing some twenty isoprene (C_5) units. The terminal phosphoryl group of this highly hydrophobic carrier is the site of attachment of the activated oligosaccharide.

$$^{-}O-\overset{\overset{\displaystyle O}{\|}}{\underset{\underset{\displaystyle O^{-}}{|}}{P}}-O-CH_2-CH_2-\overset{\overset{\displaystyle H}{|}}{\underset{\underset{\displaystyle CH_3}{|}}{C}}-CH_2-\left[CH_2-CH=\overset{\overset{\displaystyle CH_3}{|}}{C}-CH_2\right]_n-CH_2-CH=\overset{\overset{\displaystyle CH_3}{|}}{C}-CH_3$$

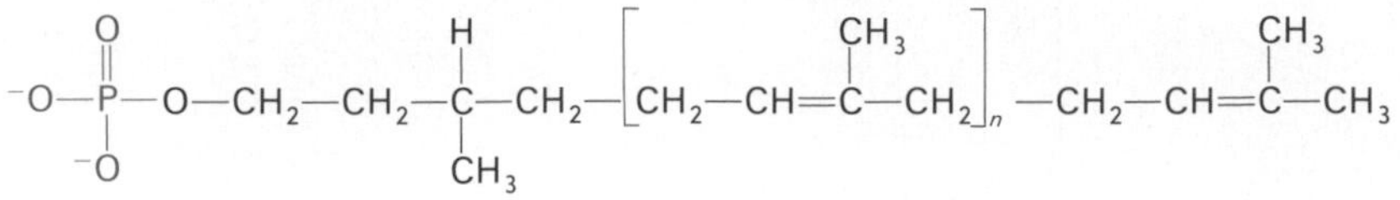

Dolichol phosphate
(n = 15 to 19)

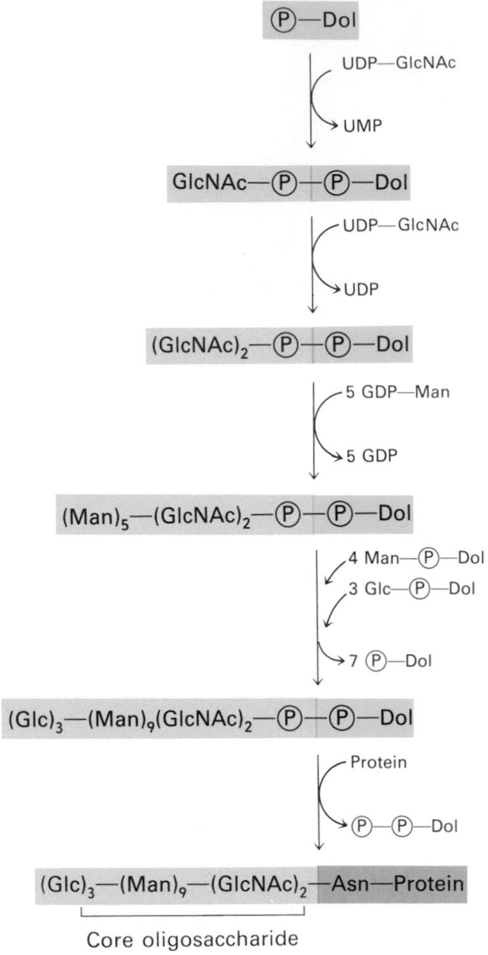

Ⓟ—Dol
↓ UDP—GlcNAc → UMP
GlcNAc—Ⓟ—Ⓟ—Dol
↓ UDP—GlcNAc → UDP
$(GlcNAc)_2$—Ⓟ—Ⓟ—Dol
↓ 5 GDP—Man → 5 GDP
$(Man)_5$—$(GlcNAc)_2$—Ⓟ—Ⓟ—Dol
↓ 4 Man—Ⓟ—Dol, 3 Glc—Ⓟ—Dol → 7 Ⓟ—Dol
$(Glc)_3$—$(Man)_9(GlcNAc)_2$—Ⓟ—Ⓟ—Dol
↓ Protein → Ⓟ—Ⓟ—Dol
$(Glc)_3$—$(Man)_9$—$(GlcNAc)_2$—Asn—Protein

Core oligosaccharide

Figure 31-10
An activated core oligosaccharide is built by the sequential addition of single sugar units. The block is then transferred to an asparagine side chain of a nascent protein in the lumen of the ER.

Dolichol phosphate acquires an oligosaccharide unit consisting of two *N*-acetylglucosamines, nine mannoses, and three glucoses by the sequential addition of monosaccharides (Figure 31-10). The activated sugar donors in these reactions are derivatives of UDP, GDP, and dolichol pyrophosphate. A series of specific transferases catalyzes the synthesis of the activated core oligosaccharide. Dolichol phosphate, located in the ER membrane with its phosphate terminus on the cytosolic face, acquires two *N*-acetylglucosamine residues and five mannoses from activated intermediates in the cytosol. Dolichol bearing this growing chain is then translocated across the membrane so that its sugar end is located on the lumenal side. It then acquires four mannose and three glucose residues from dolichol phosphate intermediates rather than from UDP or GDP. Dolichol glucose and dolichol mannose are regenerated on the cytosolic side of the membrane.

The fourteen sugar residues in this dolichol phosphate intermediate are then transferred en bloc to a specific asparagine residue of the growing polypeptide chain. The activated oligosaccharide and the specific transferase are located on the lumenal side of the ER, accounting for the fact that proteins in the cytosol are not glycosylated. An asparagine residue can accept the oligosaccharide only if it is part of an Asn-X-Ser or Asn-X-Thr sequence. However, not all asparagines in such sequences become glycolysated.

Dolichol pyrophosphate released in the transfer of the oligosaccharide to the protein is recycled to dolichol phosphate by the action of a phosphatase. This hydrolysis is blocked by *bacitracin,* an antibiotic. Another interesting antibiotic inhibitor of *N*-glycosylation is *tunicamycin,* a hydrophobic analog of UDP-*N*-acetylglucosamine. Tunicamycin blocks the addition of *N*-acetylglucosamine to dolichol phosphate, the first step in the formation of the core oligosaccharide.

The three glucose residues and one of the mannoses in the *N*-linked oligosaccharide are rapidly trimmed while the glycoprotein is still in the ER. Indeed, most glycoproteins are efficiently transported out of the ER only after removal of their glucoses.

TRANSFER VESICLES CARRY PROTEINS FROM THE ER TO THE GOLGI COMPLEX FOR GLYCOSYLATION AND SORTING

Proteins in the lumen of the ER and in the ER membrane are transported to the *Golgi complex,* which is a stack of flattened membranous sacs (Figure 31-11). The Golgi has two principal roles. First, *carbohydrate units of glycoproteins are altered and elaborated in the Golgi. O*-linked sugar units are fashioned there, and *N*-linked ones are modified in many

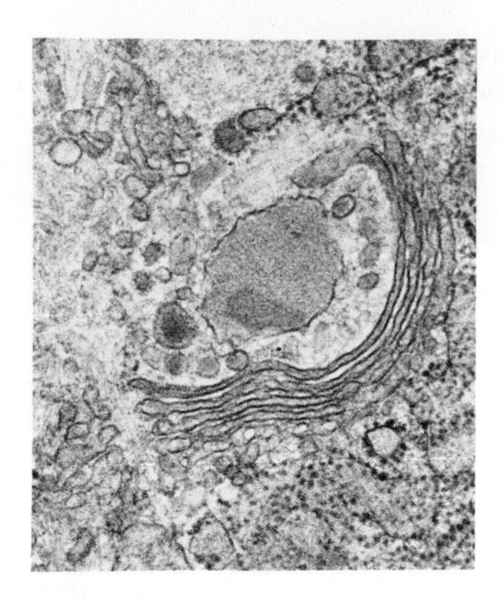

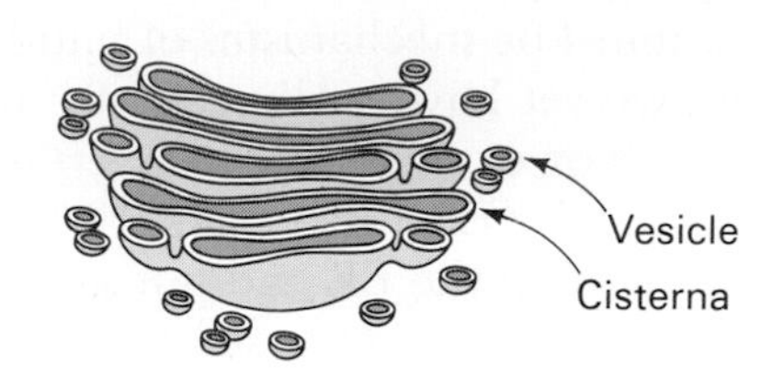

Figure 31-11
Electron micrograph (left) and schematic diagram (above) of the Golgi complex. [Micrograph courtesy of Lynne Mercer.]

different ways. Second, *the Golgi is the major sorting center of the cell.* It sends proteins to lysosomes, secretory granules, or the plasma membrane according to signals encoded by their three-dimensional structures.

The Golgi of a typical mammalian cell has about six membranous sacs (cisternae), and those of many plant cells have about twenty. The stack of cisternae is asymmetric. The Golgi is differentiated into (1) a *cis* compartment, the receiving end, which is closest to the ER; (2) *medial* compartments; and (3) a *trans* compartment, which exports proteins to a variety of destinations. These compartments contain different enzymes and mediate different functions. Proteins are delivered from the ER to the *cis* face of the Golgi by *transfer vesicles*, which have a diameter of about 500 Å. The lipid bilayer of these vesicles is surrounded by a bristle-like coat different from the clathrin coat of endocytic vesicles (p. 787). Vesicles bud from the ER and then fuse with the *cis* Golgi (Figure 31-12).

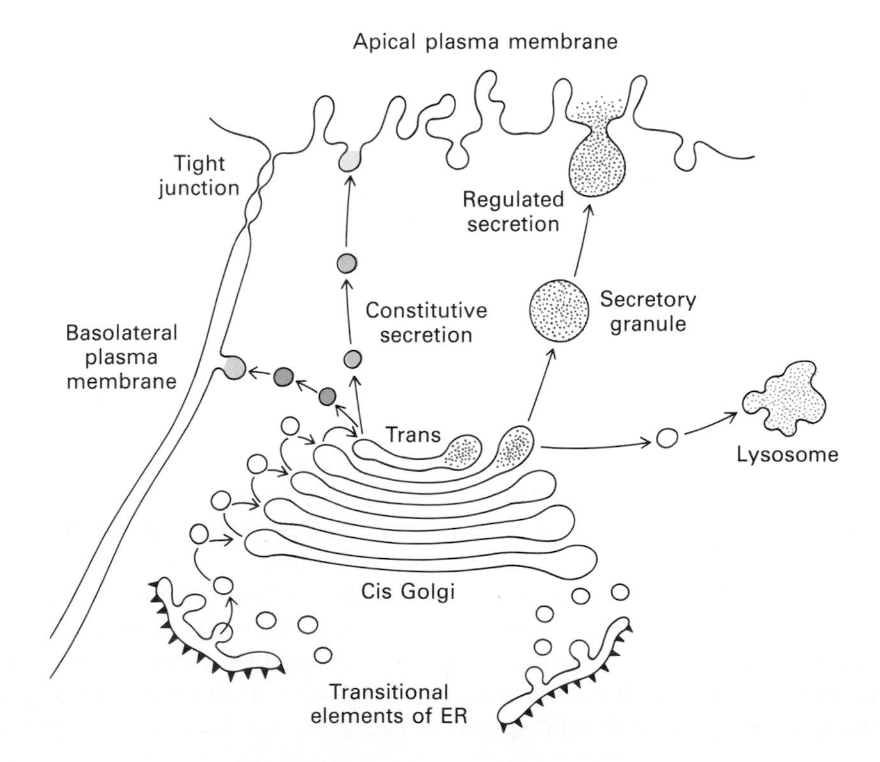

Figure 31-12
The Golgi complex is the sorting center in the targeting of proteins to lysosomes, secretory vesicles, and the plasma membrane. The *cis* face of the Golgi receives vesicles from the ER, and the *trans* face sends a different set of vesicles to different target sites. Proteins are also transferred by vesicles from one compartment of the Golgi to another. Polarized cells, such as those in epithelia, contain apical and basolateral regions separated by tight junctions that prevent their intermixture. Different proteins are targeted to these regions. [Courtesy of Dr. Marilyn Farquhar.]

Different vesicles transfer proteins from one Golgi compartment to another and then to lysosomes, secretory vesicles, and the plasma membrane. The mechanisms of budding and fusion are not understood, nor do we yet have a glimpse of how specificity is achieved in these transfers. For example, it is uncertain how integral membrane proteins essential for ER function, such as the SRP receptor and the ribophorins, are kept in the ER and not sent to the Golgi. However, it is evident that these transfer processes consume energy and are unidirectional. Moreover, *membrane asymmetry is preserved in these transport processes* (Figure 31-13). The lumenal side of membranes of transfer vesicles and the Golgi corresponds to the lumenal side of the ER membrane. However, when a vesicle ultimately fuses with the plasma membrane, its lumenal surface becomes the extracellular surface of the plasma membrane. Hence, the lumenal side of the ER membrane and of other organelle membranes corresponds to the external face of the plasma membrane. For this reason, the carbohydrate groups of glycoproteins in the plasma membrane are always on its extracellular surface (p. 298).

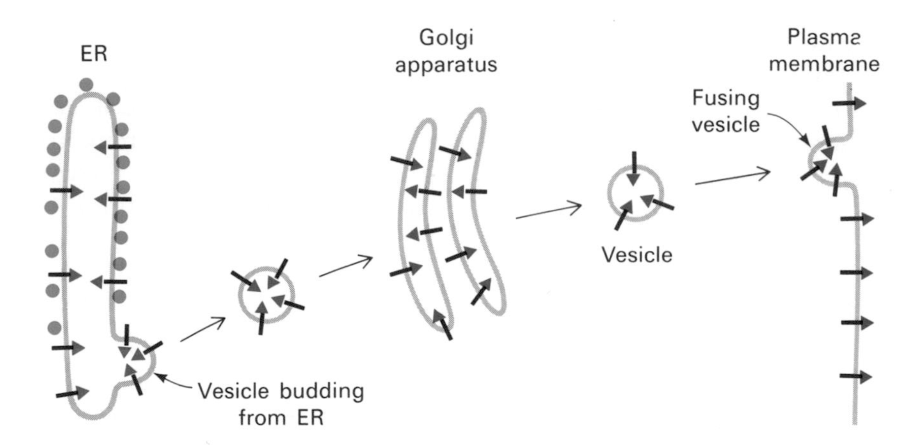

Figure 31-13
Vesicles transferred between membranes preserve their asymmetry. The lumenal face of the ER and other organelles corresponds to the extracellular face of the plasma membrane.

Carbohydrate units of glycoproteins are modified in each of the compartments of the Golgi (Figure 31-14). In the *cis* Golgi, a mannose residue is removed from the oligosaccharide chains of proteins destined for secretory vesicles or the plasma membrane. The carbohydrate units of glycoproteins targeted to the lumen of lysosomes are modified differently, as described below. In the *medial* Golgi of some cells, two more mannoses are removed, and an *N*-acetylglucosamine and a fucose residue are added. Finally, in the *trans* Golgi, galactose is added, followed by sialic acid, to form a complex oligosaccharide unit (p. 343). The structure of *N*-linked oligosaccharide units of glycoproteins is determined both by their conformation and by the glycosyl transferase composition of the Golgi in which they are processed. Carbohydrate processing in the Golgi is called *terminal glycosylation* to distinguish it from *core glycosylation,* which takes place in the ER.

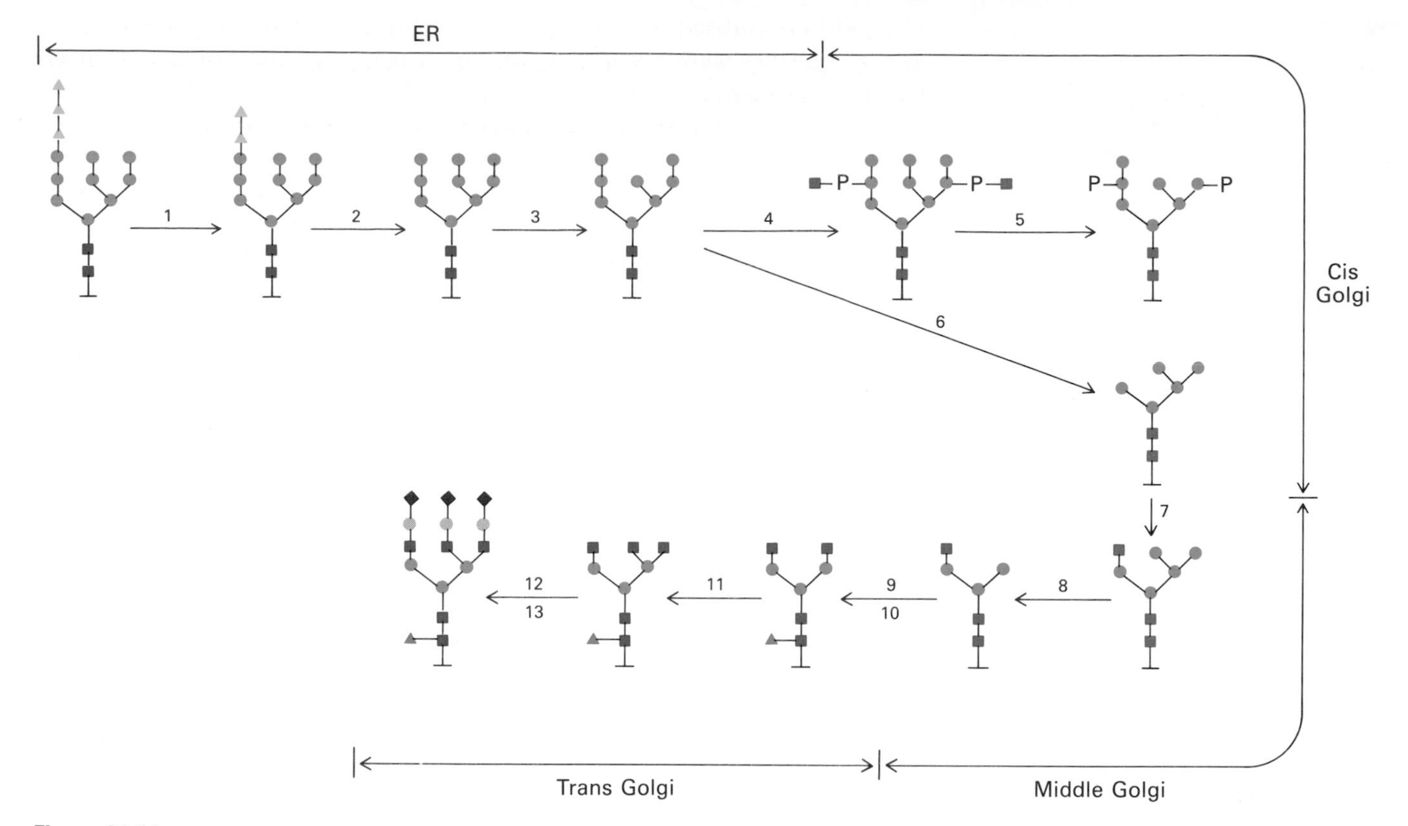

Figure 31-14
Asparagine-oligosaccharides are processed in the ER and the three compartments of the Golgi. The enzymes catalyzing these steps are: (1) glucosidase I, (2) glucosidase II, (3) ER α-1,2-mannosidase, (4) *N*-acetylglucosaminylphosphotransferase, (5) phosphodiester glycosidase, (6) Golgi mannosidase I, (7) GlcNAc transferase I, (8) mannosidase II, (9) GlcNAc transferase II, (10) fucosyl transferase, (11) GlcNAc transferase IV, (12) galactosyltransferase, (13) sialyltransferase. Steps 4 and 5 apply to lysosomal enzymes only. Symbols used: yellow triangles, glucose; red squares, GlcNAc; green circles, mannose; gray circles, galactose; blue triangles, fucose; black diamonds, sialic acid; P, phosphate. [After D. E. Goldberg and S. Kornfeld. *J. Biol. Chem.* 258(1983):3160.]

MANNOSE 6-PHOSPHATE IS THE MARKER THAT TARGETS LYSOSOMAL ENZYMES TO THEIR DESTINATION

The molecular marker that sends proteins from the Golgi to lysosomes has been identified. A key clue came from analyses of *I-cell disease* (also called mucolipidosis II), a lysosomal storage disease that is inherited as an autosomal recessive trait. Patients with I-cell disease have severe psychomotor retardation and skeletal deformities. Their lysosomes contain large inclusions of undigested glycosaminoglycans and glycolipids—hence the I in the name of the disease (Figure 31-15). These inclusions are present because at least eight acid hydrolases required for their degradation are missing from affected lysosomes. In contrast, very high levels of the enzymes are present in the blood and urine of these patients. Thus, active enzymes are synthesized but they are exported instead of being sequestered in lysosomes. In other words, *a whole series of enzymes is mislocated in I-cell disease.* Normally these enzymes contain mannose 6-phosphate residue, but in I-cell disease, this mannose is

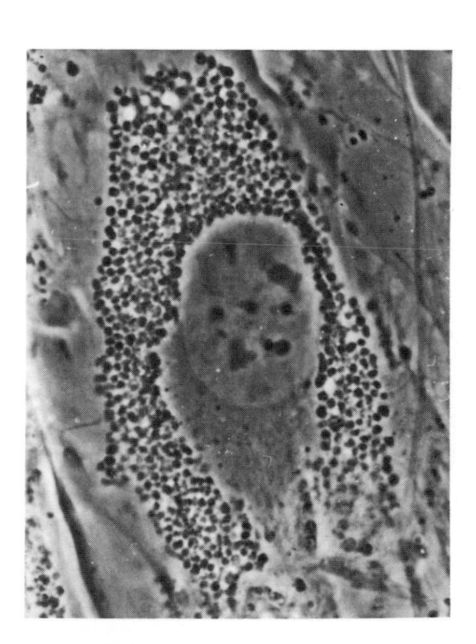

Figure 31-15
Phase-contrast micrograph of a cultured fibroblast from a patient with I-cell disease. [Courtesy of Dr. George H. Thomas.]

unmodified. Subsequent studies revealed that a *mannose 6-phosphate residue is in fact the marker that normally directs many hydrolytic enzymes from the Golgi to lysosomes.*

A glycoprotein destined for delivery to lysosomes acquires a mannose 6-phosphate marker in the *cis* Golgi (Figure 31-16). The activated phosphate donor is UDP-*N*-acetylglucosamine. A phosphotransferase catalyzes an attack by the 6-OH group of a mannose residue on the pyrophosphate bond of this activated donor. *N*-Acetylglucosamine becomes joined to the mannose by a phosphodiester bond. A phosphodiesterase then hydrolyzes this linkage to produce a mannose 6-phosphate residue in the core oligosaccharide. *I-cell patients are deficient in the phosphotransferase catalyzing the first step; the consequence is the mistargeting of eight essential enzymes.* This phosphotransferase is a highly discriminating enzyme. It does not act on a mannose residue in an isolated oligosaccharide unit nor on a peptide containing this unit. Rather, *the transferase recognizes a three-dimensional motif that is present only in glycoproteins addressed to lysosomes.*

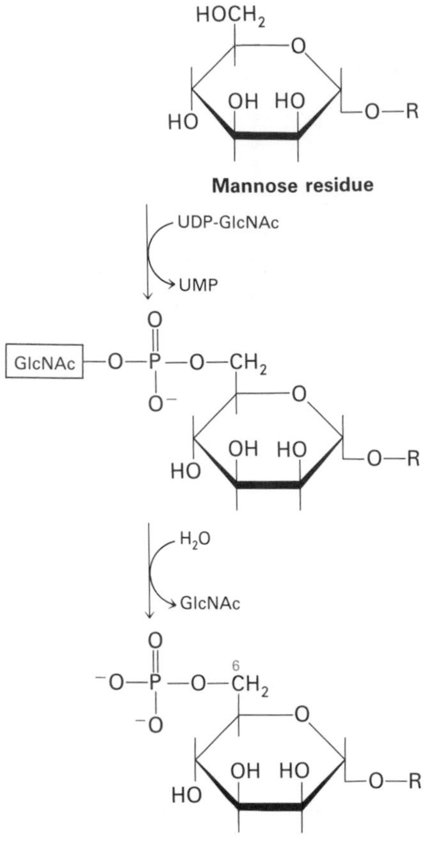

Mannose residue

UDP-GlcNAc

UMP

H_2O

GlcNAc

Mannose 6-phosphate residue

Figure 31-16
Formation of a mannose 6-phosphate marker for targeting to lysosomes. These reactions occur in the *cis* Golgi.

How does mannose 6-phosphate act as a marker for lysosomal targeting? The Golgi contains a receptor, an integral membrane protein, that specifically recognizes the mannose 6-phosphate unit and binds proteins marked by it. Vesicles containing this protein-receptor complex bud off the rims of the *trans* Golgi (see Figure 31-12). These carriers then fuse with sorting vesicles, which are more acidic (pH 5.0) than the Golgi because they contain proton pumps. The lowered pH leads to the dissociation of the marked glycoprotein from its receptor. In addition, a phosphatase removes the phosphate group from the mannose 6-phosphate moiety to further promote the release of the lysosomal enzyme into the vesicle. At this stage, vesicles containing lysosomal enzymes bud from the sorting vesicles and deliver their contents to lysosomes by fusion.

The membrane-bound mannose 6-phosphate receptor returns to the Golgi by a different set of vesicles. This receptor, like many others, is recycled so that it can be used many times. The importance of acidification in this cycle is evidenced by the finding that *lysosomal targeting is blocked by agents that make sorting vesicles less acidic.* NH_4Cl and chloroquine, for example, raise the pH of sorting vesicles and lead to the export of lysosomal enzymes from the cell rather than to lysosomes. The reason is that, if the sorting vesicles are not sufficiently acidic, the enzyme-receptor complex does not dissociate and so the receptor does not return to the Golgi. In the absence of receptors, newly formed glycoproteins containing mannose 6-phosphate continue along the secretory pathway and are exported from the cell.

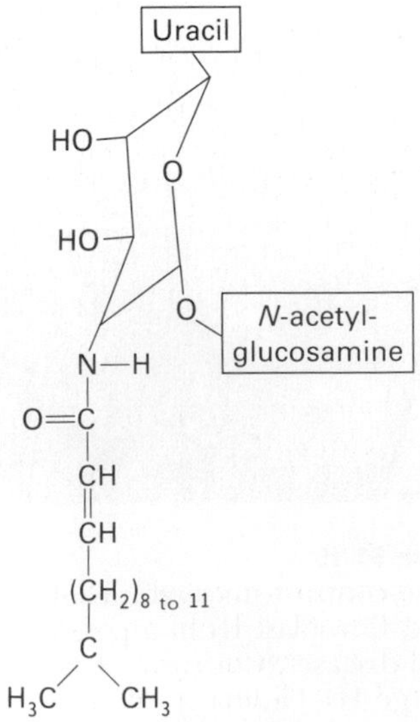

Uracil

N-acetyl-glucosamine

$(CH_2)_{8\text{ to }11}$

Tunicamycin

SECRETORY AND PLASMA-MEMBRANE PROTEINS ARE NOT TARGETED BY CARBOHYDRATE MARKERS

The finding that mannose 6-phosphate is the marker for targeting lysosomal enzymes raised the expectation that other carbohydrate units are used to direct proteins from the Golgi to other destinations. This hypothesis was tested by using inhibitors of glycosylation such as tunicamycin. As mentioned earlier, *tunicamycin* blocks *N*-linked glycosylation by inhibiting the addition of *N*-acetylglucosamine to dolichol phosphate, the first step in the formation of the core oligosaccharide (p. 774). Proteins transiting the Golgi in the presence of tunicamycin do not acquire any *N*-linked sugars. As anticipated, enzymes normally targeted to lysosomes do not reach their appointed destination in cells treated with tunicamycin. Instead, they are secreted. In contrast, *the targeting of most secretory and integral plasma-membrane proteins is essentially unaffected by treatment with tunicamycin.* This finding demonstrates that carbohydrate moieties do not *generally* serve as recognition markers in the sorting of these classes of proteins. Even so, the transport of some proteins is impeded if *N*- or *O*-linked glycosylations are inhibited. For example, *O*-linked glycosylation is important in enabling the low-density-lipoprotein receptor to reach the plasma membrane. These sugars do not serve as markers themselves, but they probably enable the protein to assume a conformation allowing it to be recognized by the sorting machinery.

Does the information for targeting reside in a short stretch of the amino acid sequence of secretory and integral plasma membrane proteins? Comparisons of their amino acid sequences do not reveal any equivalent of the cleavable signal sequence that directs many proteins to the ER. The information for targeting comes from three-dimensional motifs—*signal patches*—rather than linear markers akin to signal sequences. It is important to note that the proteins sorted by the Golgi are essentially fully folded, in contrast with the largely unfolded nascent chains scrutinized by the signal recognition particle.

What then are the roles of oligosaccharide groups of most glycoproteins? We are beginning to catch glimpses of why nature has developed an elaborate machinery for inserting and modifying carbohydrate units in many proteins. As discussed in an earlier chapter (p. 331), carbohydrates are highly polar, hydrophilic molecules. Hence, they increase the water-solubility of glycoproteins and assure that a protein region containing an attached oligosaccharide faces the aqueous medium. Second, carbohydrate units stabilize glycoproteins. For example, the oligosaccharide units of antibodies protect them from digestion by proteases. Third, carbohydrates can serve as markers for the uptake and subsequent destruction of proteins, as in the removal of aging plasma glycoproteins by the liver (p. 345). Fourth, carbohydrate processing by the ER and Golgi may determine the pace at which glycoproteins move through these organelles and help assure that enough time is spent to achieve high-fidelity sorting.

PROTEINS CAN BE TARGETED TO PARTICULAR PLASMA MEMBRANE REGIONS AND SECRETORY VESICLES

The question of how secretory and integral membrane proteins are targeted is made even more intriguing by the existence of multiple destinations in many cells. The plasma membrane of polarized cells, such as epithelial cells, is divided into an *apical* and a *basolateral* part. Proteins anchored in these regions are prevented from intermixing by barriers in the membrane called *tight junctions* (see Figure 31-12). An integral plasma membrane protein is sent to either the apical or basolateral region but not to both. The replication of membrane-enveloped viruses (p. 791) is a source of insight into this targeting process. Cells infected with *influenza virus* send the future viral coat protein from the Golgi to the *apical* plasma membrane, the site of budding in the formation of new virus particles. In contrast, the future protein coat of *vesicular stomatitis virus* (VSV) is sent to the *basolateral* plasma membrane (Figure 31-17). Vesicles that deliver proteins to the apical plasma membrane are different from those that move to the basolateral plasma membrane.

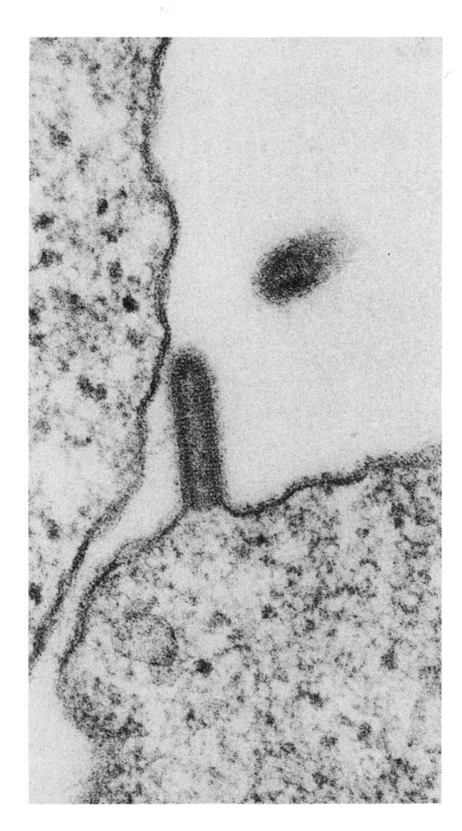

Figure 31-17
Emergence of a vesicular stomatitis virus particle by budding of the basolateral plasma membrane of an infected cell. [Courtesy of Dr. Carl-Henrik v. Bonsdorff.]

The distinction between these two kinds of vesicles is made either when they bud from the *trans* Golgi or shortly thereafter.

Many cells have more than one secretory pathway. Secretory proteins in these cells can be targeted to *constitutive vesicles* or *regulated vesicles*. *Constitutive secretion* is rapid and continuous. The released proteins are not appreciably concentrated nor do they remain in the cell for more than a few minutes. In contrast, *regulated secretion* is episodically triggered by specific signals. Proteins are stored in highly concentrated form (up to 200 times as high as in the ER), which often gives regulated vesicles an opaque appearance under the electron microscope. For example, precursors of digestive enzymes are stored in regulated vesicles of pancreatic acinar cells (Figure 31-18). Hormonal stimuli after a meal lead to the fusion of these vesicles with the apical plasma membrane and the release of their glycoproteins into the pancreatic duct. In contrast, components of the underlying extracellular matrix are continually released by constitutive vesicles fusing with the basolateral plasma membrane. Secretory proteins are targeted to constitutive or regulated vesicles in the *trans* Golgi or in a later compartment.

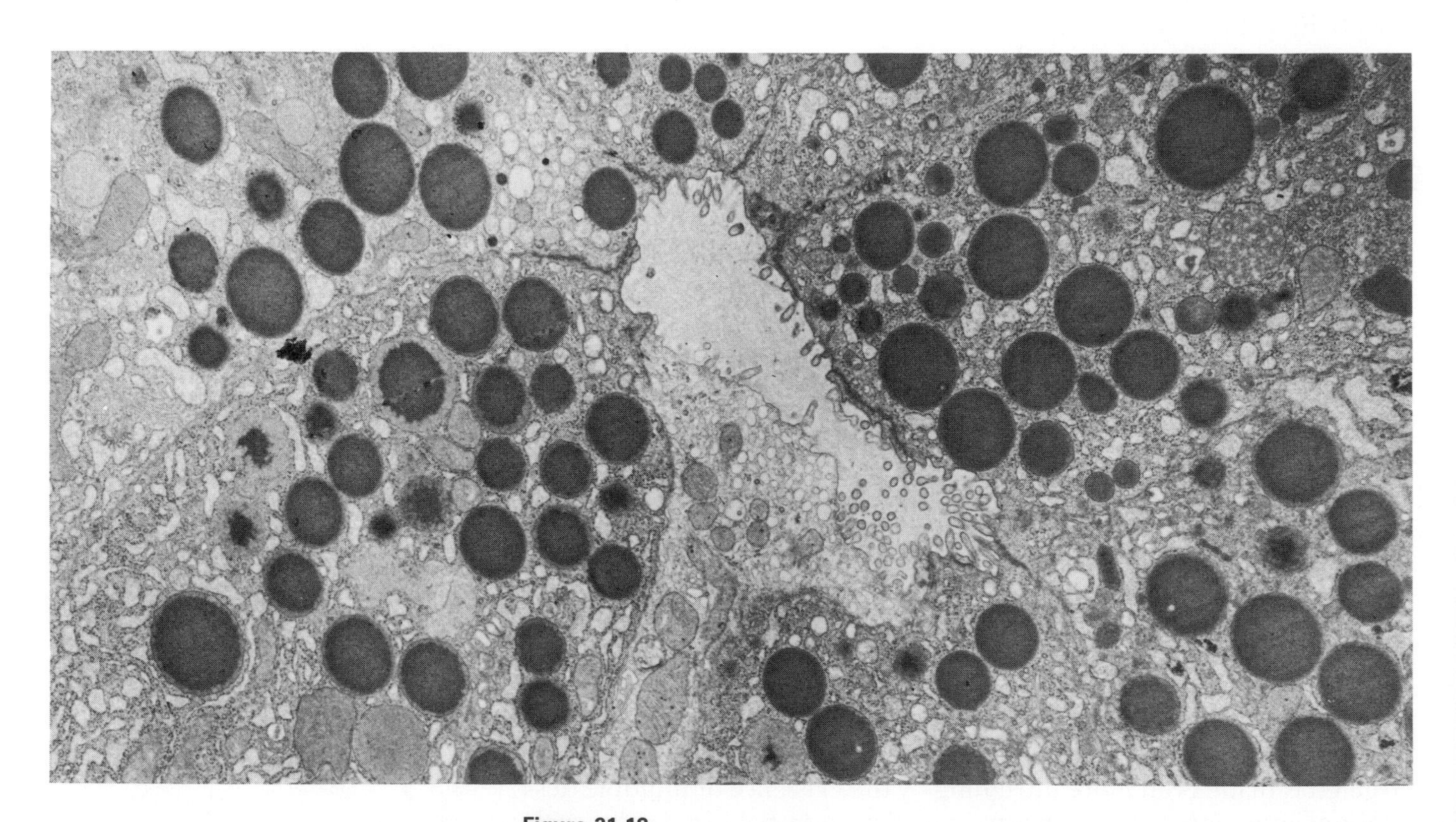

Figure 31-18
Electron micrograph of pancreatic acinar cells. The dense secretory vesicles contain zymogens, precursors of digestive enzymes. The zymogens are released into the lumen of an acinus (the clear area near the center) by fusion of the membrane of a secretory granule with the plasma membrane. [Courtesy of Dr. James Jamieson and Dr. George Palade.]

Targeting signals that send proteins to regulated vesicles have not been deciphered, but genetic engineering experiments with a line of cultured mouse pituitary cells have provided a tantalizing clue. Adrenocorticotropic hormone (ACTH) is stored in regulated vesicles of these cells and released by specific stimuli. Transfection of these cells with the cDNA for trypsinogen (a zymogen normally secreted by pancreatic acinar cells) leads to the synthesis of this precursor and its storage with ACTH in regulated vesicles. This finding strongly suggests that *endocrine cells* (which secrete hormones into the blood) and *exocrine cells* (which secrete digestive enzymes and other substances into ducts) *utilize similar signals to target their secretory proteins to regulated vesicles.*

Another informative finding is that thc addition of choloroquine redirects ACTH from its normal regulated secretory pathway to the constitutive pathway. Chloroquine interferes with the acidification of cellular compartments, and so it prevents the recycling of membrane receptors that recognize proteins destined for regulated vesicles. This experiment and others lead to the generalization that *soluble proteins in the lumen of the ER emerge in constitutive secretory vesicles and are exported unless special signals are present to direct them to other locations. Likewise, integral membrane proteins initially in the ER go to the plasma membrane unless they carry instructions to the contrary.* In other words, secretory vesicles and the plasma membrane are default destinations.

BACTERIA ALSO USE SIGNAL SEQUENCES TO TARGET PROTEINS

Protein targeting is not an invention of eucaryotes. Bacteria also target proteins to destinations encoded in their sequences. A gram-negative microorganism such as *E. coli* can send nascent proteins synthesized by ribosomes in the cytosol to the *plasma membrane* (inner membrane), the *outer membrane*, the *periplasmic space* between membranes, or (rarely) the *extracellular medium.* The targeting of periplasmic and outer-membrane proteins is best understood. Amino-terminal signal sequences (also called leader sequences) very much like those of eucaryotic secretory proteins direct bacterial proteins across the plasma and outer membranes (Figure 31-19), perhaps with the help of an assembly akin to the eucaryotic signal recognition particle. As in eucaryotes, the signal sequence is usually cleaved by a signal peptidase. Another similarity is that polypeptide chain elongation and translocation usually occur at about the same time but are not mechanistically coupled. Translocation across the plasma membrane of procaryotes is driven by the proton-motive force rather than by the phosphoryl potential of ATP or another energy-rich molecule. Proteins encoded by the *sec* (secretory) genes may be components of the translocation machinery.

H_3^+N-Met-Lys-Ala-Thr-Lys-Leu-
-Val-Leu-Gly-Ala-Val-Ile-
-Leu-Gly-Ser-Thr-Leu-Leu-
-Ala-Gly⌇Cys-Ser-Ser-Asn
Cleavage
Site

Figure 31-19
Amino-terminal sequence of prolipoprotein, the precursor of the most abundant outer membrane protein of *E. coli.* This signal sequence contains many hydrophobic residues (yellow) preceded by two basic residues (blue).

Some components of protein targeting appear to be highly conserved in evolution. The signal sequence of a nascent procaryotic secretory protein

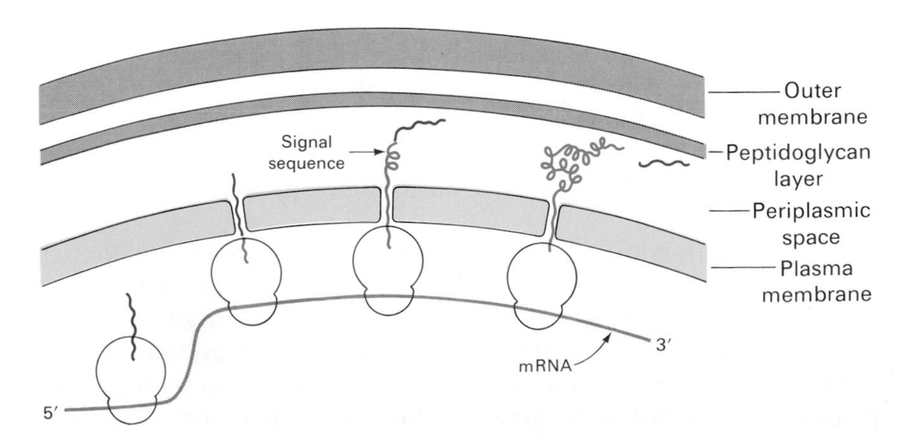

Figure 31-20
Procaryotic proteins destined for locations other than the cytosol are synthesized by ribosomes bound to the plasma membrane. A signal sequence (red) on the nascent chain directs the ribosome to the plasma membrane and enables it to be translocated. As in eucaryotes, translocation is not mechanistically coupled to chain elongation. The translocation machinery is not depicted in this schematic diagram.

(β-lactamase) that was synthesized on plant ribosomes was correctly recognized by mammalian SRP. The specificities of bacterial and eucaryotic signal peptidases also are very similar. It will be interesting to learn whether other molecular components of recognition and translocation are homologous.

MOST MITOCHONDRIAL PROTEINS ARE SYNTHESIZED IN THE CYTOSOL AND IMPORTED INTO THE ORGANELLE

Mitochondrial DNA encodes all of the RNA but only a few of the proteins in mitochondria. Most mitochondrial proteins are encoded by nuclear DNA and synthesized in the cytosol by free ribosomes. How do these proteins reach their mitochondrial destinations? The problem is intriguing because mitochondrial proteins reside in four locations: the outer membrane, the inner membrane, the intermembrane space, and the matrix. Gottfried Schatz discovered that *amino-terminal sequences of imported mitochondrial proteins specify their ultimate location in mitochondria.* The entry of a protein into a mitochondrion requires the presence of a specific sequence that is recognized by a receptor in the outer membrane of the organelle. This *mitochondrial entry sequence*, which differs from the endoplasmic reticulum signal sequence, is rich in positively charged residues and in serines and threonines (Figure 31-21). A protein remains in the *outer membrane* if this sequence is followed by a *membrane anchoring sequence* and a *second positively charged sequence*. For example, the information needed to target and anchor the major (70 kd) outer-membrane protein resides in its amino-terminal sequence of 41 residues, which is not cleaved.

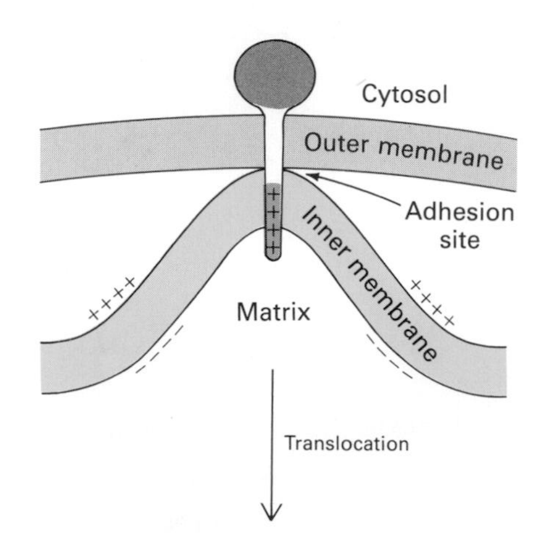

Figure 31-21
Mitochondrial entry sequence. This sequence is recognized by receptors on the external face of the outer membrane of mitochondria and leads to the import of the protein bearing it. Hydrophobic residues are shown in yellow, basic ones in blue, and serine and threonine in red.

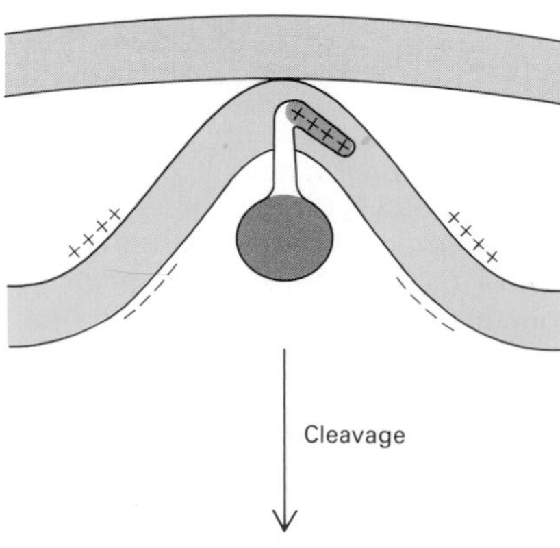

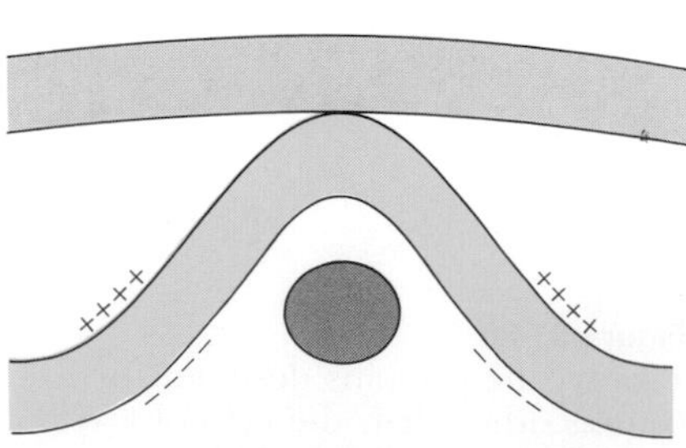

Figure 31-22
Proposed mechanism for the delivery of a protein from the cytosol to the mitochondrial matrix. A transmembrane potential across the inner membrane is essential for transport. [After E. C. Hurt and A. P. G. M. van Loon. *Trends Biochem. Sci.* 11(1986):206.]

The targeting of mitochondrial proteins to the other three locations is more complex. The inner membrane participates in their translocation. Indeed, *a proton-motive force across this membrane is required for their transport, whereas none is needed for the import of outer-membrane proteins.* The addition of uncouplers (such as dinitrophenol) that dissipate the proton-motive force blocks their movement from the outer membrane to interior targets. A second difference is that nearly all proteins delivered to the inner membrane, the matrix, and the intermembrane space are proteolytically cleaved, whereas outer-membrane proteins stay intact. The cleaved amino-terminal targeting sequences are called *presequences.*

A protein destined to be located in the *matrix* enters a mitochondrion by the binding of its presequence to a receptor in the outer membrane (Figure 31-22A). It then traverses both the outer and inner membranes at an adhesion site, where these membranes come together. The precursor protein, now in the matrix, is cleaved by a soluble metalloprotease in the matrix to produce the mature protein.

Inner-membrane proteins and *intermembrane proteins*, like matrix proteins, are transported into mitochondria at adhesion sites. But only the amino-terminal region of the cytochrome b_2 precursor crosses the inner membrane; it is cleaved on the matrix side by the metalloprotease (Figure 31-22B). Heme binds to the remaining polypeptide chain, and a second protease then splits the chain on the external face of the inner membrane to release a soluble cytochrome b_2 from the membrane anchor. *Thus, a protein is placed in the intermembrane space by anchoring its precursor in the inner membrane and then severing this membrane stalk.* An inner-membrane protein, such as cytochrome c_1, is fashioned similarly, except that its precursor contains three transmembrane segments. Two of them stay in the membrane after proteolytic removal of the amino-terminal membrane anchor.

CYTOSOLIC PROTEINS CAN BE REDIRECTED TO MITOCHONDRIA BY ATTACHMENT OF PRESEQUENCES

The strongest evidence for the targeting of mitochondrial proteins by discrete amino-terminal sequences comes from studies of genetically engineered chimeric proteins. *Dihydrofolate reductase (DHFR), a cytosolic enzyme, can be redirected into mitochondria by attaching a mitochondrial presequence to its amino-terminus.* For example, attachment of the 61-residue presequence from cytochrome c_1 causes the import of DHFR into the *intermembrane space*, the physiologic site of cytochrome c_1 (Figure 31-23).

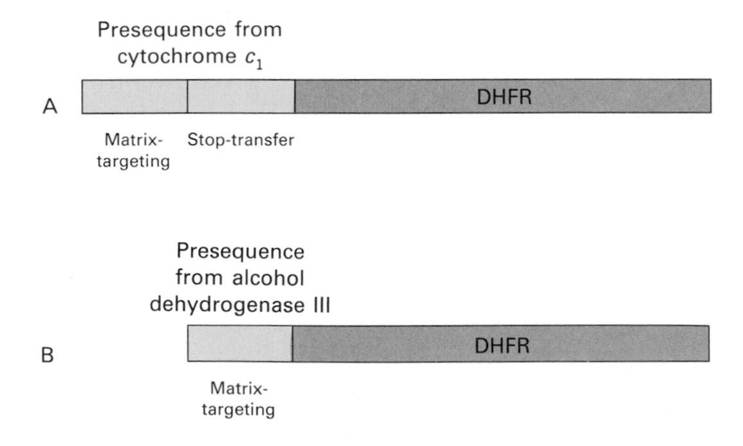

Figure 31-23
Dihydrofolate reductase (DHFR), a cytosolic enzyme, can be redirected to (A) the intermembrane space of mitochondria or (B) the mitochondrial matrix by adding mitochondrial presequences to its amino-terminus.

Alternatively, DHFR can be placed in the *mitochondrial matrix* by linking the 27-residue presequence of a matrix enzyme (an alcohol dehydrogenase isozyme) to the amino-terminus of DHFR. The localization of a third chimeric protein also proved to be illuminating. When only the first 35 residues of the 61-residue cytochrome c_1 presequence were joined to DHFR, the enzyme appeared in the matrix rather than in the intermembrane space. This experiment showed that residues 36 to 61 of the presequence contain a *stop-transfer* sequence that prevents inner-membrane and intermembrane proteins from proceeding into the matrix. It seems likely that the stretch of 19 uncharged residues beginning at position 36 is the membrane anchor that stops translocation into the matrix.

The import of DHFR linked to a mitochondrial presequence is inhibited by methotrexate, a folate antagonist that binds very tightly to DHFR (p. 616), but it does not block the transport of authentic mitochondrial proteins. This folate antagonist also does not interfere with the binding of the presequence-DHFR chimeric protein to receptors on the outer mitochondrial membrane. Methotrexate probably prevents transport of the chimeric protein into the mitochondrial matrix by interfering with the unfolding of DHFR. These results strongly suggest that *proteins must be partly or completely unfolded during their translocation across membranes.*

CHLOROPLASTS ALSO IMPORT MOST OF THEIR PROTEINS AND SORT THEM ACCORDING TO THEIR PRESEQUENCES

Most proteins of chloroplasts, like those of mitochondria, are synthesized by cytosolic ribosomes and imported. Proteins imported into chloroplasts have six potential destinations: the outer membrane, the inner membrane, the intermembrane space, the stroma, the thylakoid membrane, and the thylakoid lumen. The existence of thylakoid membranes separate from the inner membrane gives rise to two more locations than are present in mitochondria. As in mitochondria, targeting is largely achieved by *amino-terminal presequences*. Chloroplast presequences (also called transit sequences) resemble mitochondrial presequences in being positively charged and rich in serine and threonine residues. Indeed, a chloroplast presequence derived from the small subunit of ribulose bisphosphate carboxylase, a plant enzyme, effectively targets a cytosolic protein to yeast mitochondria in vitro. It will be interesting to learn why the enzyme in vivo is imported into chloroplasts only.

The presequences of chloroplast proteins targeted to the stroma and the thylakoids are cleaved during the transport process. Transport is powered by ATP hydrolysis rather than by proton-motive force. The presequences of proteins destined for the thylakoid lumen appear to contain two signals (Figure 31-24). The amino-terminal signal leads to

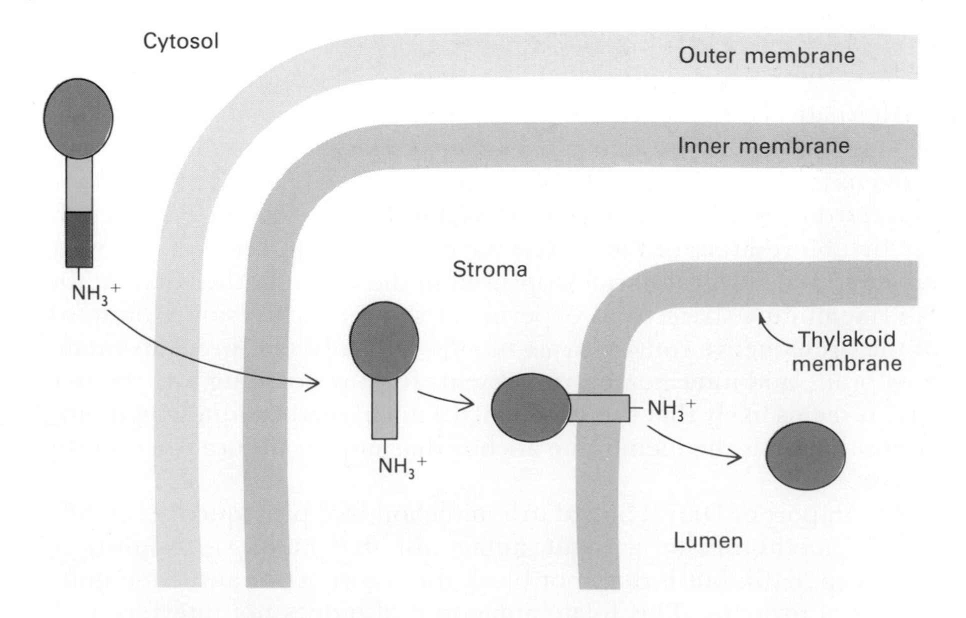

Figure 31-24
Plastocyanin is targeted to the thylakoid lumen of chloroplasts by the sequential action of two amino-terminal sequences. The first one enables it to enter the stroma and the second (exposed by proteolysis) enables it to cross the thylakoid membrane. [After S. Smeckens, C. Bauerle, J. Hageman, K. Keegstra, and P. Weisbeck. *Cell* 46(1986):373.]

the import of the precursor protein into the stroma of the chloroplast. This part of the presequence is cleaved in the stroma or enroute to it, exposing a second signal that directs the translocation of the modified precursor across the thylakoid membrane. A protein targeted to the stroma lacks this second signal. This working hypothesis is supported by the finding that plastocyanin (normally located in the thylakoid lumen, p. 529) is redirected to the stroma when its own presequence is replaced with that of ferredoxin (a stromal protein).

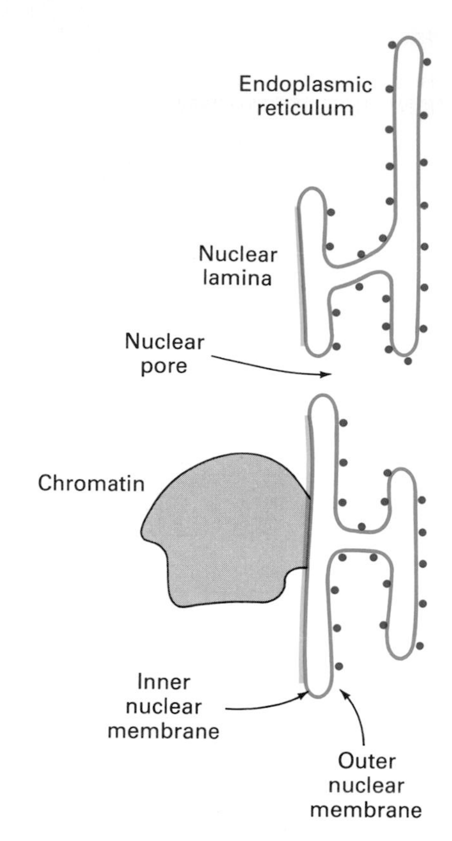

Figure 31-25
Diagram of a nuclear envelope. The outer nuclear membrane is continuous with the endoplasmic reticulum.

NUCLEAR LOCALIZATION SIGNALS ENABLE PROTEINS TO RAPIDLY ENTER THE NUCLEUS THROUGH NUCLEAR PORES

The nucleus of eucaryotic cells is surrounded by a *nuclear envelope* consisting of two membranes—an *inner nuclear membrane* and an *outer nuclear membrane*—separated by a perinuclear space (Figure 31-25). The outer nuclear membrane is continuous with the rough endoplasmic reticulum, and the perinuclear space is continuous with the lumen of the ER. All of the proteins of the cell nucleus are synthesized in the cytosol by free ribosomes. How do DNA polymerases, RNA polymerases, histones (p. 825), and other nuclear proteins traverse the nuclear envelope? This selective barrier contains openings, called *nuclear pores* (Figure 31-26), having a diameter of about 70 Å. Small proteins (~15 kd) such as histones enter readily, whereas large proteins (>~90 kd) are excluded unless they contain specific signals. Such signals can also accelerate the entry of small proteins.

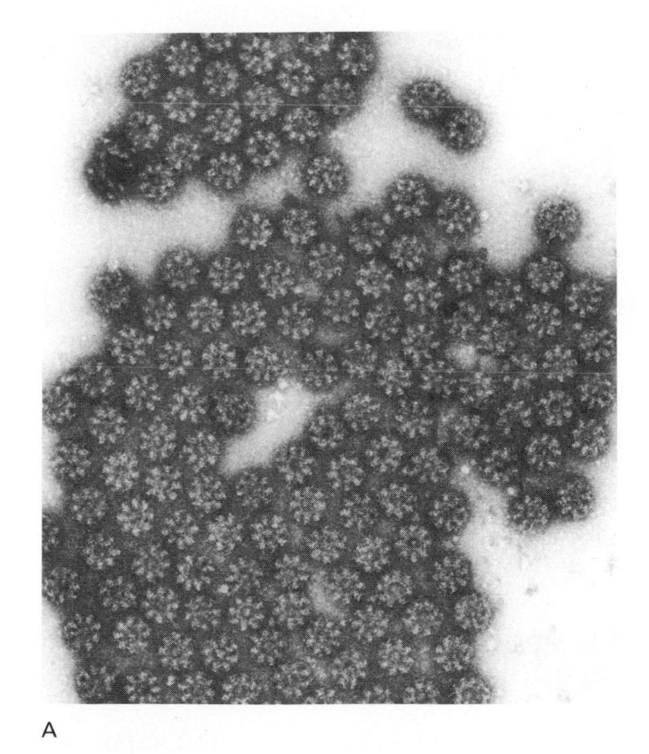

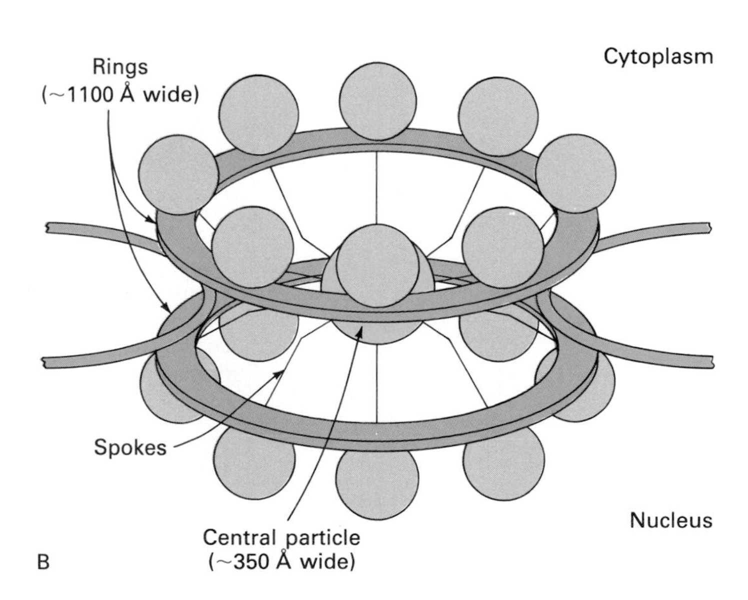

Figure 31-26
(A) Electron micrograph of a preparation of nuclear pores from *Xenopus laevis* oocytes. A nuclear pore is a very large assembly (50 megadaltons) with eightfold symmetry. (B) Model of the structure of a nuclear pore. [Courtesy of Dr. Nigel Unwin and Dr. Ronald Milligan.]

The T antigen of SV40 virus is a 92-kd protein that regulates transcription and replication of viral DNA. Studies of the accumulation of T antigen in nuclei have shown that the transport of this large protein into the nucleus depends on the presence of a *nuclear localization sequence* containing five consecutive positively charged residues:

-Pro-Lys-Lys-Lys-Arg-Lys-Val-
128

A change of a single amino acid residue can render this sequence inactive. For example, T antigen containing threonine or asparagine in place of lysine at residue 128 stays in the cytosol.

Can a protein that normally resides in the cytosol be brought into the nucleus by attaching to it this viral heptapeptide sequence? A test was carried out on pyruvate kinase, a tetramer of 58-kd subunits that normally resides in the cytosol. By recombinant DNA techniques, cDNA encoding the targeting sequence was joined to cDNA encoding the kinase. A plasmid containing this chimeric cDNA was then microinjected into mammalian cells. Pyruvate kinase containing this SV40 nuclear location sequence was found almost exclusively in the cell nucleus (Figure 31-27). In contrast, pyruvate kinase bearing an altered nuclear localization sequence (threonine instead of lysine at position 128) stayed in the cytosol. The joining of this viral heptapeptide to a variety of other cytosolic proteins led in each case to their localization in the cell nucleus. Hence, *a short peptide can direct a protein into the nucleus.* Nuclear localization signals, in contrast with most other targeting sequences discussed thus far, are not cleaved.

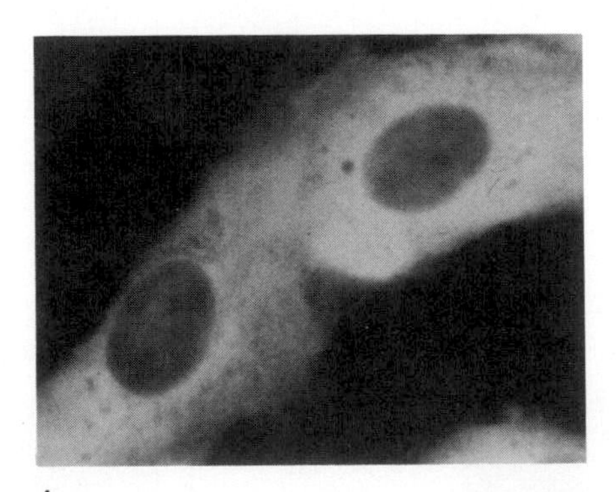

A

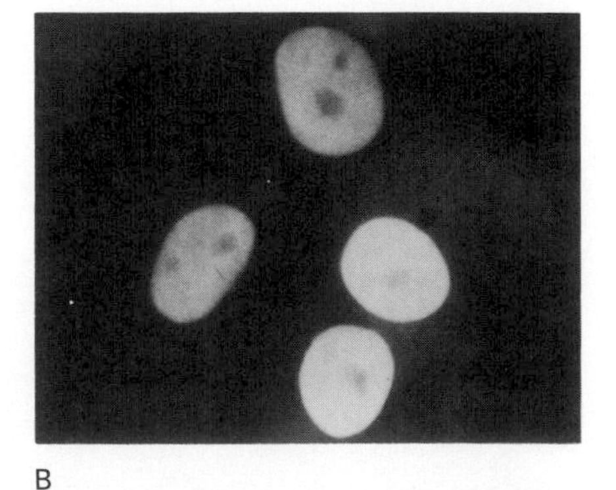

B

Figure 31-27
Localization of (A) unmodified pyruvate kinase, and (B) pyruvate kinase containing at its amino terminus a nuclear localization signal derived from the T antigen of SV40 virus. Pyruvate kinase was visualized by fluorescence microscopy after staining with a specific antibody. [From W. D. Richardson, B. L. Roberts, and A. E. Smith. *Cell* 44(1986):79.]

SPECIFIC PROTEINS ARE IMPORTED INTO CELLS BY RECEPTOR-MEDIATED ENDOCYTOSIS

We turn now to a different facet of protein targeting—the import of specific proteins into a cell by their binding to receptors in the plasma membrane and their inclusion into vesicles. This process of *receptor-mediated endocytosis* has broad biological significance. First, it is a means of *delivering essential metabolites to cells.* As was discussed earlier (p. 561), low-density lipoprotein (LDL) carrying cholesterol is taken up by the

LDL receptor in the plasma membrane and internalized. Likewise, the complexes of vitamin B_{12} bound to transcobalamin II and of iron bound to transferrin are recognized by cell-surface receptors and imported. Second, *endocytosis modulates responses to many protein hormones and growth factors.* Insulin, epidermal growth factor, and nerve growth factor bound to their receptors are taken into the cell and degraded together with them. Endocytosis removes these hormones from the circulation and makes the cell temporarily less responsive to them by decreasing the number of receptors. Third, *proteins targeted for destruction are taken up and delivered to lysosomes for digestion.* For example, phagocytic cells have receptors enabling them to take up antigen-antibody complexes. Liver cells take up and destroy senescent plasma glycoproteins that are marked by galactose residues exposed by the loss of sialic acid residues from the termini of their oligosaccharide chains. Fourth, *receptor-mediated endocytosis is exploited by many viruses and toxins to gain entry into cells.* The ingenious mode of entry and departure of Semliki Forest virus will be considered shortly. Fifth, *disorders of receptor-mediated uptake can lead to disease,* as exemplified by some forms of familial hypercholesterolemia.

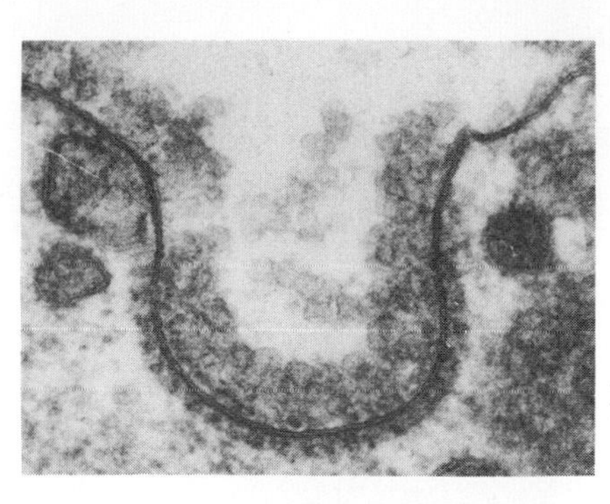

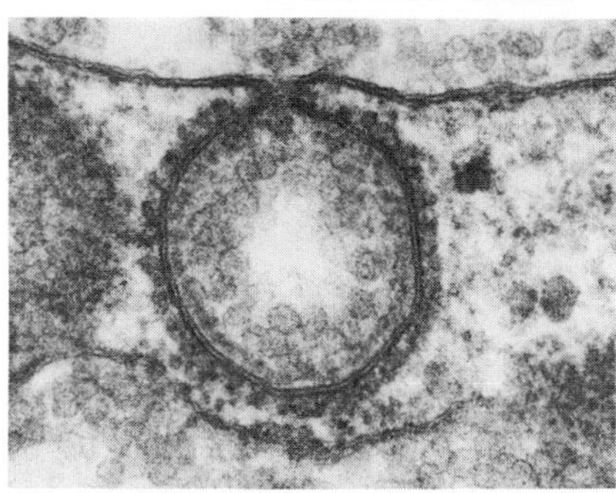

Figure 31-28
Receptor-mediated endocytosis takes place at coated pits in the plasma membrane. These electron micrographs show the uptake of vitellogenin, a lipoprotein, by hen oocytes. [Courtesy of Dr. M. M. Perry and Dr. A. A. Gilbert.]

CLATHRIN PARTICIPATES IN ENDOCYTOSIS BY FORMING A POLYHEDRAL LATTICE AROUND COATED PITS

Most cell-surface receptors mediating endocytosis are *transmembrane glycoproteins.* They have a large extracellular domain, one or two transmembrane helices, and a small cytosolic region, as exemplified by the receptors for transferrin and asialoglycoproteins (Figure 31-29). Many

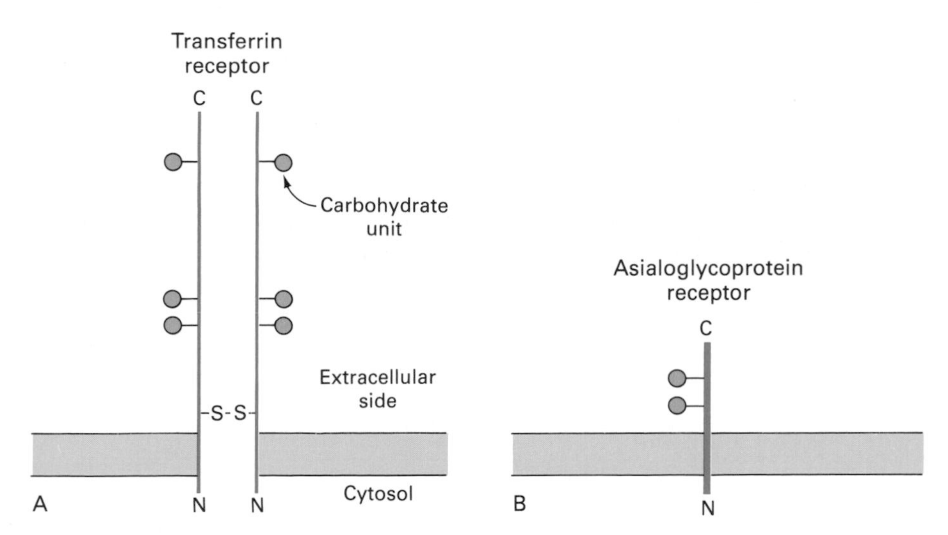

Figure 31-29
Diagram of the (A) transferrin receptor and (B) the asialoglycoprotein receptor.

of these receptors are located in specialized regions of the plasma membrane called *coated pits.* The cytosolic side of these indentations has a thick coat of *clathrin,* a protein designed to form lattices around membranous vesicles (Figure 31-30). Coated pits occupy about 2% of the surface of a typical animal cell. A number of receptors (such as those for LDL, transferrin, asialoglycoproteins, and insulin) congregate in

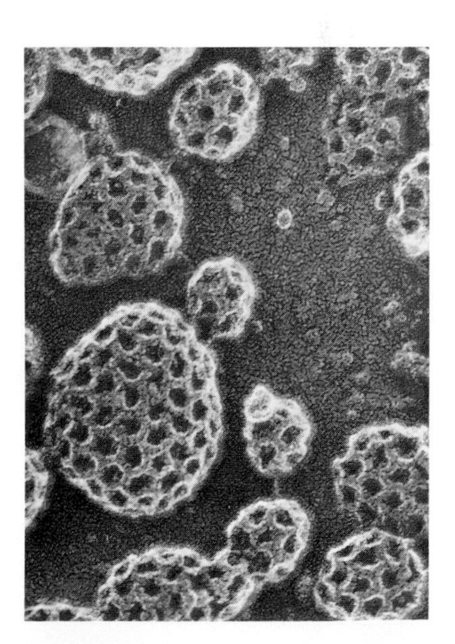

Figure 31-30
Electron micrograph of clathrin lattices on the cytosolic face of a plasma membrane. [Courtesy of Dr. John Heuser.]

coated pits whether or not ligand is bound. Others (such as the receptor for epidermal growth factor) cluster there after binding their cognate protein.

Receptor-mediated endocytosis begins with the invagination of a coated pit (see Figure 31-28). Clathrin forms a lattice around the coated pit, excising it from the plasma membrane to form a *coated vesicle*, which has a diameter of about 800 Å. The coated vesicle rapidly loses its clathrin shell to become an *endosome* (also called a *receptosome*). Endosomes fuse with each other to form vesicles having diameters ranging from 2000 to 6000 Å. The conversion of a coated pit into an endosome takes about twenty seconds. An important characteristic of endosomes is their acidity, which dissociates most protein-receptor complexes, enabling the partners to have different fates. Hence, *sorting decisions are made in endosomes.*

What is the structural basis of clathrin's propensity to form closed polyhedral lattices? Clathrin is a trimer of three heavy chains (H, 180 kd) and three light chains (L, ~35 kd). The $(HL)_3$ clathrin unit (8S, 650 kd), obtained by solubilizing coats at alkaline pH in the absence of divalent cations, has a striking appearance in electron micrographs (Figure 31-31). It is a three-legged structure, a *triskelion.* The carboxyl-termini

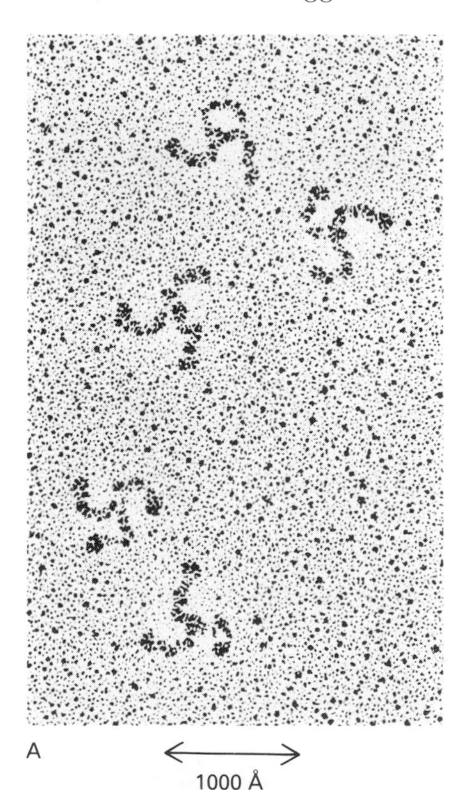

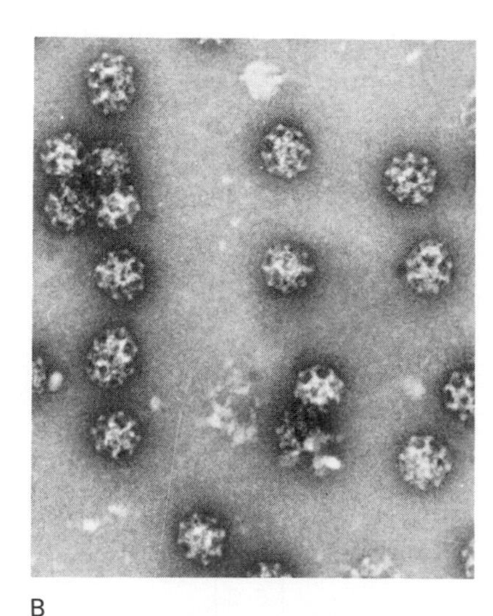

Figure 31-31
Electron micrographs showing (A) triskelions, the building blocks of the polyhedral shell surrounding (B) coated vesicles. [Courtesy of Dr. Daniel Branton (part A) and Barbara Pearse (part B).]

of the three heavy chains, each about 500 Å long, come together at a vertex (Figure 31-32A). A bend in the heavy chain divides it into a proximal unit, closest to the vertex, and a distal unit. Each of the light chains is aligned with the proximal unit of a heavy chain. Purified trimers spontaneously reassemble into closed shells when dialyzed into a Mg^{2+}-containing buffer at pH 6. These closed polyhedra are made of pentagons and hexagons; an icosahedral shell cannot be constructed from pentagons alone or hexagons alone (p. 861). *A single edge of a polyhedron is made of parts of four triskelions, two proximal units and two distal ones* (Figure 31-32B). The flexibility of a triskelion is important in enabling it to fit into a pentagon or a hexagon.

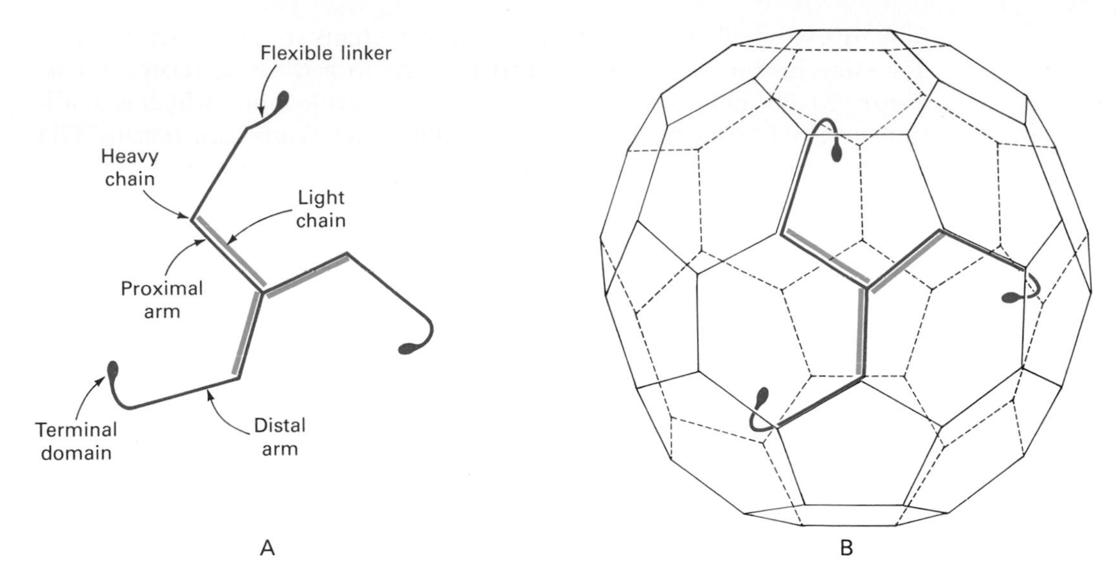

Figure 31-32
Diagram of (A) a triskelion molecule, and (B) its position in the polyhedral lattice formed by the association of triskelions. [Courtesy of Dr. Stephen Harrison.]

A variety of open and closed lattices of different sizes are formed by the assembly of triskelions alone in vitro. Additional proteins participate in vivo. A 180-kd *clathrin assembly protein* (different from the H chain) promotes the association of triskelions into regular polyhedrons closely resembling those of coated vesicles. Several proteins, having masses between 100 kd and 110 kd, participate in the binding of clathrin to the membrane of coated vesicles. The release of clathrin from coated vesicles is also a key step in endocytosis because the presence of a protein shell around the vesicle prevents it from sending imported proteins and receptors to their destinations. Clathrin is recycled by an *ATP-driven uncoating enzyme*. This 70-kd protein hydrolyzes three ATP for each triskelion removed from the polyhedral lattice. It is evident that lattice formation through noncovalent binding interactions is thermodynamically favorable, whereas its disruption requires an input of free energy.

ENDOCYTOSED PROTEINS AND RECEPTORS ARE SORTED IN ACIDIC ENDOSOMES

The pathway from coated pits to coated vesicles to endosomes is common to all proteins that have undergone endocytosis. The fates of internalized proteins and receptors then diverge. Acidification of endosomes by ATP-driven proton pumps (p. 958) leads to the dissociation of protein-receptor complexes, a necessary prelude to their sorting and targeting. The pathway taken by transferrin and its receptor illustrates one of four potential outcomes. *Transferrin* transports iron from sites of absorption and storage to sites of utilization. Two Fe^{3+} ions are bound per 77-kd protein, which contains two similar domains. The protein devoid of iron is called *apotransferrin*. The binding of Fe^{3+} to the protein involves HCO_3^- and a tyrosine side chain in the anionic form. Transferrin, but not apotransferrin, binds to a dimeric receptor (see Figure 31-29) in coated pits. On reaching an endosome, which is acidified to a pH between 5 and 6, Fe^{3+} dissociates from transferrin. The acidity lowers the affinity of transferrin for Fe^{3+} more than a millionfold. *Acidification leads to the release of Fe^{3+} by protonating bicarbonate, the phenolate —O^-, and other groups that contribute to binding.* However, apotransferrin remains bound to the receptor (Figure 31-33).

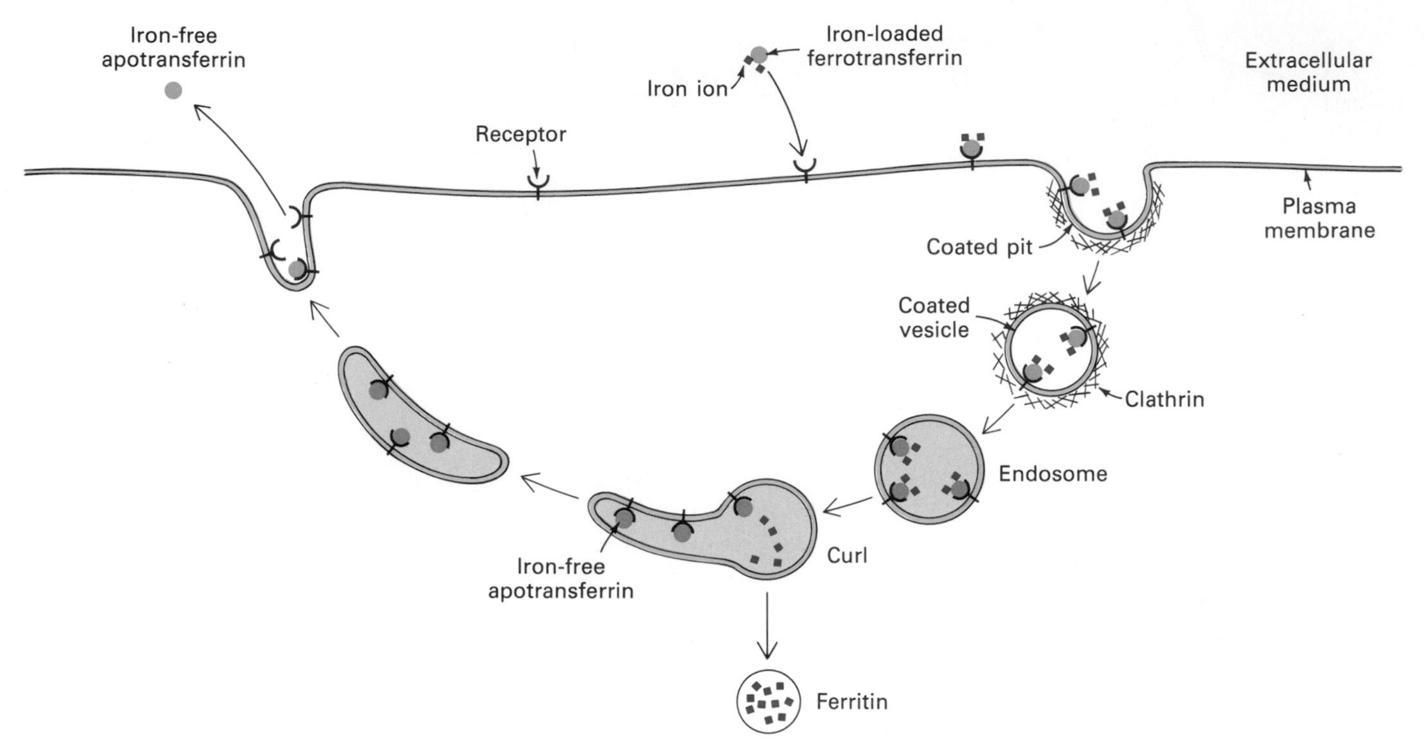

Figure 31-33
Endocytic pathway for transferrin. Iron is released in acidic endosomes. Apotransferrin and the receptor are recycled. [After A. Dautry-Varsat and H. F. Lodish. How receptors bring proteins and particles into cells. Copyright © 1984 by Scientific American, Inc. All rights reserved.]

The endosome then fuses with a tubular vesicle called CURL (*c*ompartment of *u*ncoupling of *r*eceptor and *l*igand), in which sorting takes place: a portion of the vesicle containing apotransferrin bound to the receptor pinches off and is directed to the plasma membrane, whereas the remaining Fe^{3+} is stored in ferritin in the cytosol. *When the pinched-off vesicle fuses with the plasma membrane, apotransferrin is released from the receptor because of the sudden shift in pH.* Apotransferrin has little affinity for the receptor at pH 7.4. Thus, pH changes are used twice to drive the transport cycle, first to release iron from transferrin in the endosome, and then to discharge apotransferrin into the extracellular fluid. Both the carrier of iron and the receptor are recycled with little loss. The cycle takes about sixteen minutes—four minutes for the binding of transferrin, five minutes for transport to endosomes, and seven minutes for the return of the iron carrier and the receptor to the cell surface. A liver cell can take up some 20,000 iron atoms per minute in this way.

The recycling of both apotransferrin and the transferrin receptor illustrates one mode of receptor-mediated endocytosis. Three other

Table 31-1
Four modes of receptor-mediated endocytosis

	Fate of		
Mode	*Receptor*	*Protein*	*Examples*
1	Recycled	Recycled	Transferrin, major histocompatibility proteins
2	Recycled	Degraded	Low-density lipoprotein, transcobalamin II
3	Degraded	Degraded	Epidermal growth factor, immune complexes
4	Transported	Transported	Maternal immunoglobulin G, immunoglobulin A

modes of operation are known (Table 31-1). The LDL pathway (p. 562) exemplifies a mode in which the receptor recycles but the endocytosed protein is sent to lysosomes for digestion. The pH of lysosomes (between 4.5 and 5) is lower than that of endosomes. Lysosomal proteases are activated by this acidic environment, which also serves to unfold the protein destined for destruction. In a third mode of endocytosis, both the receptor and transported protein are sent to lysosomes for digestion. Receptor-mediated endocytosis is also used to convey a protein from one surface of a polarized cell to the opposite one. For example, intestinal epithelial cells of newborns take up antibodies obtained from their mother's milk. The endocytic vesicle containing the maternal antibody bound to a receptor is directed to the other side of the cell (facing blood vessels), where it fuses with the plasma membrane. How is the maternal antibody released? While in the endocytic vesicle, the receptor is cleaved, so that the transported antibody emerges from the cell with a piece of the receptor called the *secretory component*. Thus, the receptor is sacrificed in carrying an antibody molecule from one face of the plasma membrane to the opposite one.

The acid-induced dissociation of protein-receptor complexes in endocytosis can be blocked by the addition of permeant bases. For example, NH_4Cl enters cells and internal compartments as NH_3, which readily crosses bilayer membranes. NH_3 acquires a proton inside the compartment, which raises the pH there. Organic bases such as chloroquine act in the same way to equilibrate the pH of endosomes and lysosomes with that of the cytosol. The pH of these compartments can also be raised by adding ionophores (p. 957) that exchange lumenal H^+ for cytosolic K^+. All of these agents block the release of endocytosed proteins, the recycling of receptors, and the action of degradative enzymes in lysosomes.

Chloroquine

MANY VIRUSES AND TOXINS ENTER CELLS BY RECEPTOR-MEDIATED ENDOCYTOSIS

Membrane-enveloped viruses exploit endocytic pathways to enter cells and exocytic pathways to leave them (Figure 31-34). Studies of the life cycle of *Semliki Forest virus* (SFV) have provided insight into how membrane-enveloped viruses take advantage of the protein-targeting capabilities of host cells. SFV, an RNA virus (p. 868) that infects mosquitoes, was named for a rain forest in Uganda. It is closely related to the virus causing yellow fever in humans. The 12.7-kb RNA molecule of SFV is

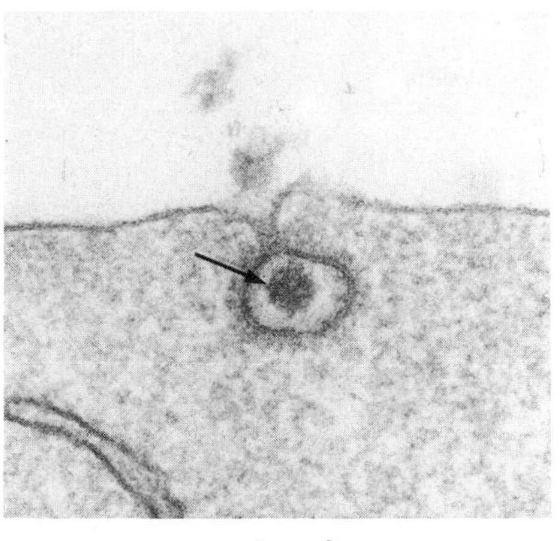

100 Å

Figure 31-34
Electron micrograph showing the entry of Semliki Forest virus into an infected cell. [Courtesy of Dr. Kai Simons.]

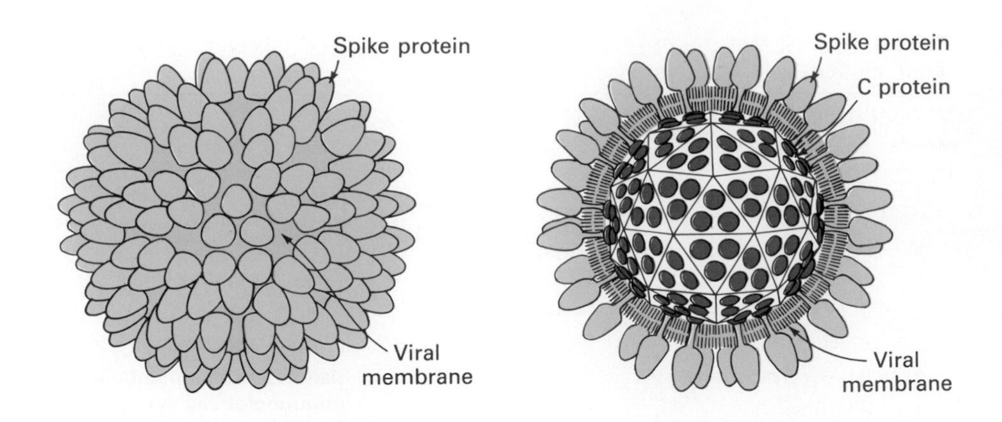

Figure 31-35
Diagram of the structure of Semliki Forest virus (SFV), a membrane-enveloped virus. [After K. Simons, H. Garoff, and A. Helenius. How an animal virus gets into and out of its host cell. Copyright © 1982 by Scientific American, Inc. All rights reserved.]

surrounded by 180 copies of C protein (Figure 31-35). This nucleoprotein capsid is enveloped by a lipid bilayer membrane containing 180 spike glycoproteins, each consisting of three subunits. The E1 and E2 subunits each include a large external domain, a single membrane-spanning helix, and a short carboxyl-terminal region that makes contact with the C protein of the nucleocapsid. The small E3 protein is bound to the external domain of E2; these subunits arise by cleavage of a common precursor.

Electron micrographs of infected cells show virus particles in coated pits, coated vesicles, and endosomes. It is evident that SFV enters susceptible cells by binding to receptors in coated pits that are then endocytosed (Figure 31-36). *The acidic environment of the endosome leads to a conformational change that is necessary for infection.* The E1 spike protein changes shape and triggers the fusion of the viral membrane with the membrane of the endosome. Hence, the nucleocapsid is released into the cytosol. Inhibitors of acidification prevent infection by blocking this release step.

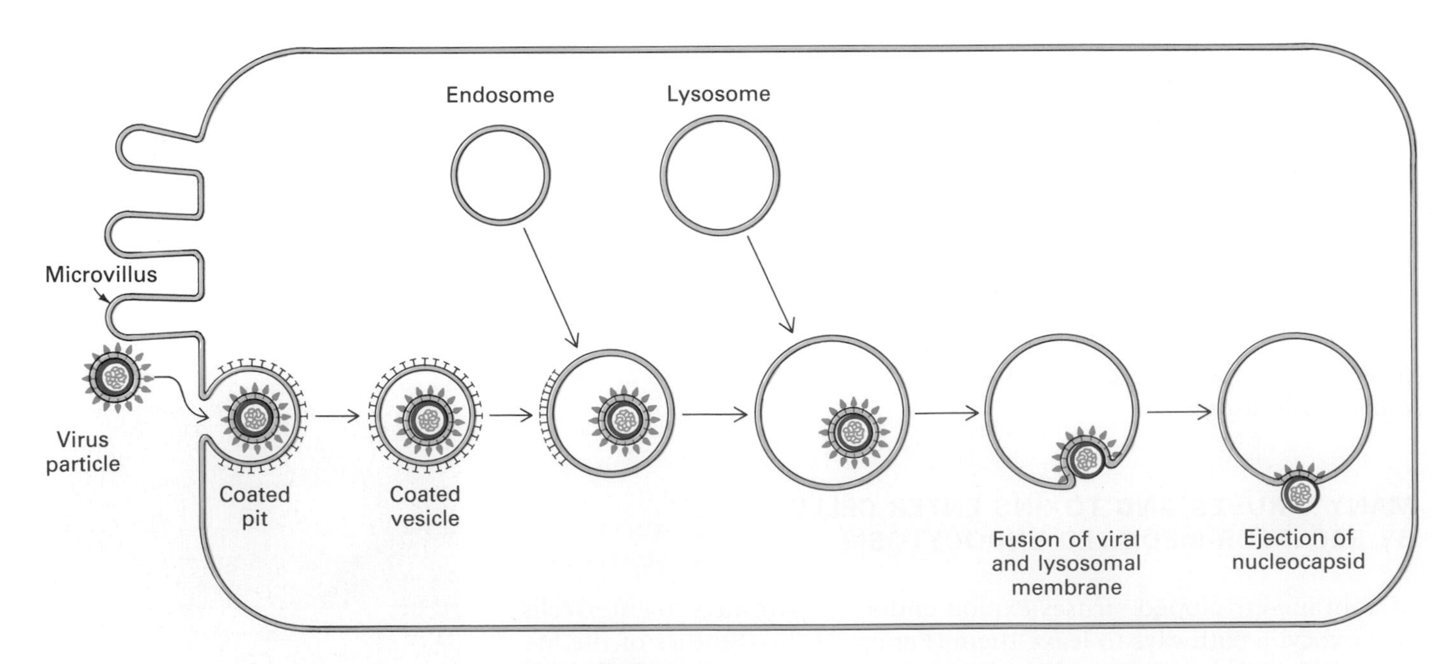

Figure 31-36
Entry of Semliki Forest virus into a susceptible cell by receptor-mediated endocytosis. The acidic milieu of the endosome triggers the fusion of membranes and release of the viral nucleocapsid. [After K. Simons, H. Garoff, and A. Helenius. How an animal virus gets into and out of its host cell. Copyright © 1982 by Scientific American, Inc. All rights reserved.]

The viral RNA is then translated and replicated. The C protein is synthesized on free ribosomes. In contrast, the precursors of spike protein contain signal sequences, which direct ribosomes synthesizing them to the endoplasmic reticulum. These viral glycoproteins are processed by the Golgi and sent to the plasma membrane, where they cluster (Figure 31-37). Newly-synthesized nucleocapsids consisting of RNA and C protein interact with these viral spike proteins. The plasma membrane curves around the nucleocapsid as successively more bonds are made between spikes and C proteins. Finally, when all 180 C proteins are bound to spikes, the membrane coat around the virus breaks off from the plasma membrane, releasing the virus particle into the extracellular space. As will be discussed in a later chapter (p. 869), influenza virus, also a membrane-enveloped virus, uses similar mechanisms to enter and leave cells.

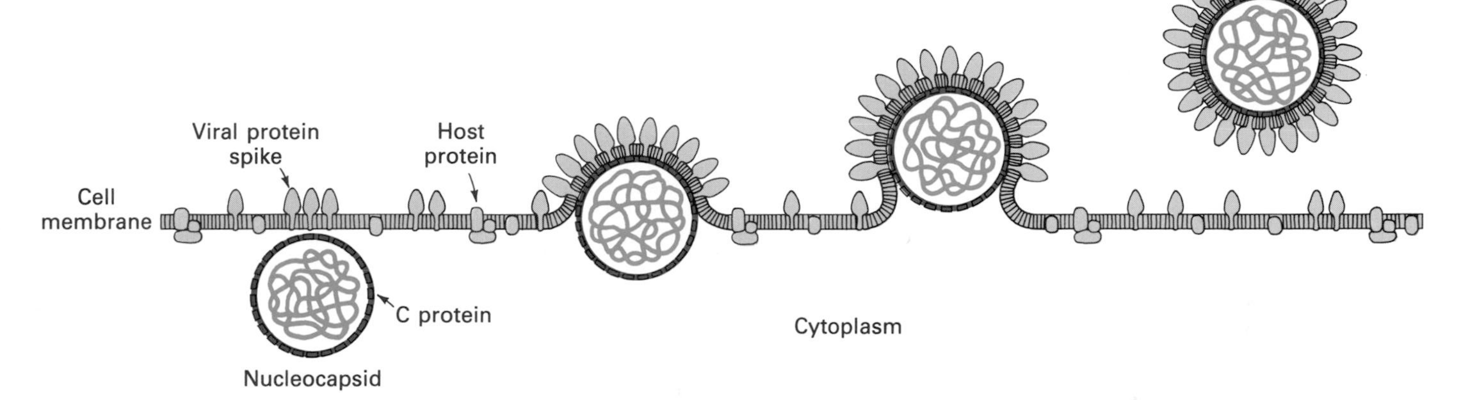

Figure 31-37
Membrane assembly and release of newly formed Semliki Forest virus particles. [After K. Simons, H. Garoff, and A. Helenius. How an animal virus gets into and out of its host cell. Copyright © 1982 by Scientific American, Inc. All rights reserved.]

Many toxins enter cells by receptor-mediated pathways. Diphtheria toxin, for example, is a 61-kd protein having two domains that is internalized after binding of the B domain to a cell-surface receptor. The two domains are separated in an endosome by cleavage of a peptide bond and reduction of an interchain disulfide bond. The A fragment inhibits protein synthesis by catalyzing the ADP-ribosylation of diphthamide, a modified histidine residue of elongation factor 2 (p. 763). Acidification of the endosome leads to the release of the A fragment into the cytosol, perhaps by triggering the insertion of hydrophobic segments of the B fragment into the endosomal membrane. *Cholera toxin* also consists of a *catalytic unit* (an A chain) and a *membrane penetration unit* (five B chains). Its A chain catalyzes the ADP-ribosylation of a key signal-coupling protein, leading to the persistent activation of adenylate cyclase (p. 982). The pentameric B unit binds to ganglioside GM_1, the cell-surface receptor for this toxin, and enables the A chain to enter the cytosol.

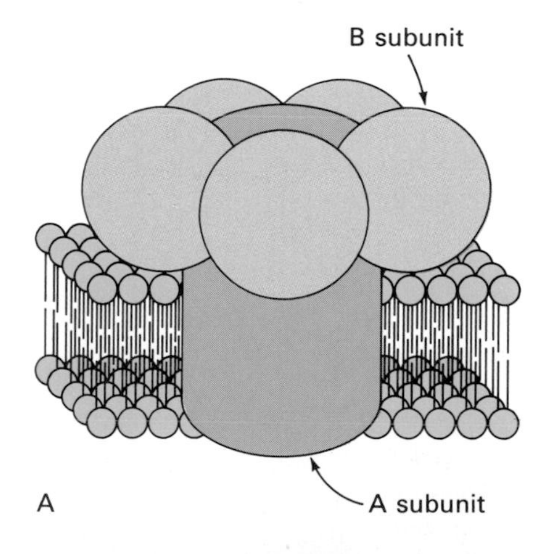

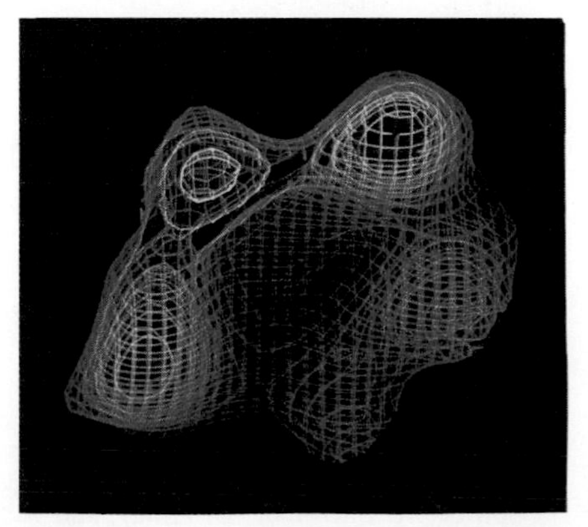

Figure 31-38
Structure of cholera toxin. (A) Schematic diagram of the intact toxin molecule. (B) Image of the intact toxin bound to GM_1 gangliosides in the membrane. This image was obtained by analyses of electron micrographs of crystalline arrays of membrane-bound toxin. [Courtesy of Dr. Hans Ribi and Dr. Roger Kornberg.]

UBIQUITIN TAGS PROTEINS FOR DESTRUCTION

Proteins are also targeted for destruction. They differ markedly in their half-lives. Some are turned over very rapidly, whereas others are very stable. Most enzymes that are important in metabolic regulation have short lives, enabling their concentration and hence activity to be rapidly changed. Controlled degradation is also important in removing abnormal proteins. A significant proportion of newly synthesized protein molecules are defective because of errors in translation. Moreover, proteins undergo oxidative damage and are altered in other ways with the passage of time. Cells have mechanisms for detecting and removing damaged proteins.

Ubiquitin, a small (8.5 kd) protein present in all eucaryotic cells, plays an important role in tagging proteins for destruction. This protein is highly conserved in evolution: yeast and human ubiquitin differ at only 3 of 76 residues. The carboxyl-terminal glycine of ubiquitin becomes covalently attached to the ϵ-amino group of lysine residues of proteins destined to be degraded. The energy for the formation of these *isopeptide bonds* comes from ATP. Three enzymes (E_1, E_2, and E_3) participate in the conjugation of ubiquitin to proteins. First, the terminal carboxylate group of ubiquitin becomes linked to a sulfhydryl group of E_1 by a thioester bond (Figure 31-39). This ATP-driven reaction (Figure 31-39) is reminiscent of fatty acid activation (p. 473) and amino acid activation (p. 734). Activated ubiquitin is then shuttled to a sulfhydryl of E_2. Finally, E_3 catalyzes the transfer of ubiquitin from E_2 to the target protein. A protein tagged for destruction usually acquires several molecules of ubiquitin. It is not yet evident how the presence of ubiquitin on a protein induces proteolysis.

What determines whether a protein becomes ubiquitinated? One signal turns out to be unexpectedly simple. *The half-life of a cytosolic protein is determined to a large extent by its amino-terminal residue* (Table 31-2). A yeast protein with methionine at its amino terminus typically has a

Table 31-2
Dependence of half-lives of cytosolic proteins on the nature of their amino-terminal residue

Amino-terminal residue	*Half-life*
Stabilizing	
Methionine	
Glycine	
Alanine	>20 hours
Serine	
Threonine	
Valine	
Destabilizing	
Isoleucine	~30 minutes
Glutamate	
Tyrosine	~10 minutes
Glutamine	
Proline	~7 minutes
Highly destabilizing	
Leucine	
Phenylalanine	~3 minutes
Aspartate	
Lysine	
Arginine	~2 minutes

Source: A. Bachmair, D. Finley, and A. Varshavsky, *Science* 234(1986):179.

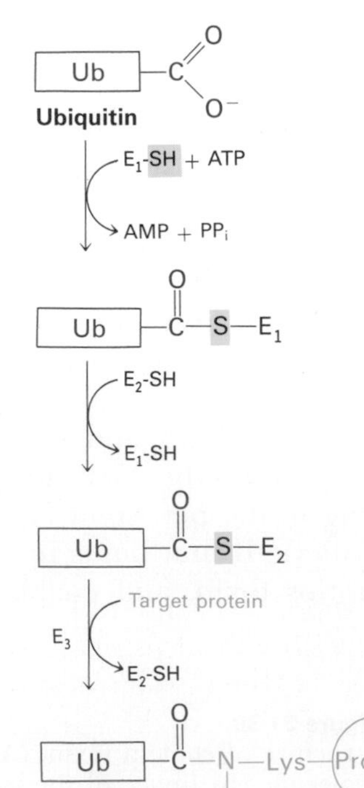

Figure 31-39
Activation and attachment of ubiquitin to a protein targeted for degradation.

half-life of more than twenty hours, whereas one with arginine at this position has a half-life of about 2 minutes. Highly destabilizing amino-terminal residues such as arginine and aspartate favor rapid ubiquitination, whereas stabilizing residues such as methionine and serine do not. The E_3 enzyme in the conjugation reaction may be the reader of amino-terminal residues. It is interesting to note that *the ordering of amino-terminal residues according to whether they are stabilizing, destabilizing, or highly destabilizing is similar for bacteria, yeast, and mammals. This signal determining the half-life of a protein has been retained over several billion years of evolution.* The elucidation of other signals controlling protein degradation is eagerly awaited.

SUMMARY

Proteins contain signals that determine their ultimate destination. The synthesis of all proteins (except those encoded by mitochondrial and chloroplast DNA) begins on free ribosomes in the cytosol. In eucaryotes, protein synthesis continues in the cytosol unless the nascent chain contains a hydrophobic signal sequence that directs the ribosome to the endoplasmic reticulum (ER). Signal recognition particle (SRP), a ribonucleoprotein assembly, recognizes signal sequences and brings ribosomes bearing them to the ER. The nascent chain is then translocated across the ER membrane by an ATP-driven motor. Translocation and chain elongation are mechanistically separate processes that usually occur at the same time. Translocation continues until a stop-transfer sequence is encountered. Multiple signal sequences and stop-transfer sequences give rise to proteins that weave back and forth across the bilayer. Most signal sequences are cleaved after translocation. Bacterial signal sequences are like those of eucaryotes. Indeed, a cytosolic protein can be directed to the ER by placing a bacterial signal sequence at its amino-terminus.

Glycoproteins acquire the core of their *N*-linked sugars from a dolichol pyrophosphate donor. A block of fourteen sugar residues is transferred from this highly hydrophobic carrier to asparagine side chains. Three glucose residues and a mannose residue are rapidly trimmed while the glycoprotein is still in the ER. Transfer vesicles carry proteins from the ER to the Golgi complex for further modification of carbohydrate units and for sorting. The Golgi consists of a stack of membranous sacs that are differentiated into *cis*, *medial*, and *trans* compartments. The *cis* compartment is closest to the ER and receives vesicles from it. A different set of vesicles transfers proteins from the *cis* to the *medial* and then to the *trans* compartment of the Golgi. Carbohydrate units of glycoproteins are modified in each of these compartments. Enzymes destined to be delivered to lysosomes contain a conformational motif that leads to the addition of a mannose 6-phosphate unit. This phosphorylated sugar is recognized by a membrane-bound receptor that brings glycoproteins to lysosomes. Sorting takes place primarily in the *trans* Golgi and in vesicles derived from it.

Most mitochondrial proteins are encoded by nuclear DNA and are synthesized by free ribosomes in the cytosol. They contain amino-terminal sequences (called mitochondrial presequences) that direct their transport into mitochondria and specify whether they are to remain in the outer membrane or be translocated to the inner membrane, the intermembrane space, or the matrix. Transport to locations other than the outer membrane is driven by the proton-motive force across the inner membrane. A cytosolic protein can be directed to the mitochondrial matrix by joining a mitochondrial presequence to its amino terminus. Chloroplasts also import most of their proteins and send them to six sites according to the nature of their presequences (also called transit sequences). The import of chloroplast proteins to interior sites is driven by the hydrolysis of ATP. Most mitochondrial and chloroplast presequences are cleaved after import. A different series of signals targets proteins to the cell nucleus by enabling them to traverse nuclear pores.

Specific proteins are imported into eucaryotic cells by receptor-mediated endocytosis. Proteins bound to transmembrane receptors cluster in coated pit regions of the plasma membrane. Clathrin participates in endocytosis by forming a polyhedral lattice around the coated pit. The resulting coated vesicle rapidly loses its clathrin shell by the action of an uncoating ATPase. The vesicle then becomes an endosome (receptosome). The acidity of endosomes (pH 5 to 6) induces dissociation of most protein-receptor complexes and leads to their sorting. Many viruses and toxins enter cells by receptor-mediated endocytosis. Acid-induced conformational changes permit their release from endosomes into the cytosol.

Proteins are also targeted for destruction. Ubiquitin, a protein present in all eucaryotes and highly conserved in evolution, becomes covalently linked to proteins destined to be degraded. The conjugation of the carboxyl-terminus of ubiquitin to ϵ-amino groups of proteins involves activation by ATP and recognition of a marker. The rate of ubiquitination of cytosolic proteins is determined to a large extent by the nature of their amino-terminus. Methionine, for example, leads to a long half-life, whereas arginine leads to a very short half-life.

SELECTED READINGS

Where to start

Palade, G., 1975. Intracellular aspects of the process of protein synthesis. *Science* 189:347–358. [This Nobel Lecture gives a lucid and beautiful account of the mechanism of secretion of zymogens by pancreatic acinar cells.]

Pfeffer, S. R., and Rothman, J. E., 1987. Biosynthetic protein transport by the endoplasmic reticulum and Golgi. *Ann. Rev. Biochem.* 56:829–852.

Wickner, W. T., and Lodish, H. F., 1985. Multiple mechanisms of protein insertion into and across membranes. *Science* 230:400–407.

Walter, P., Gilmore, R., and Blobel, G., 1984. Protein translocation across the endoplasmic reticulum. *Cell* 38:5–8.

Goldstein, J. L., Brown, M. S., Anderson, R. G. W., Russell, D. W., and Schneider, W. J., 1985. Receptor-mediated endocytosis. *Ann. Rev. Cell Biol.* 1:1–39.

General reviews

Garoff, H., 1985. Using recombinant DNA techniques to study protein targeting in the eucaryotic cell. *Ann. Rev. Cell Biol.* 1:403–445.

Schekman, R., 1985. Protein localization and membrane traffic. *Ann. Rev. Cell Biol.* 1:115–143.

Sabatini, D. D., Kreibich, G., Morimoto, T., and Adesnik, M., 1982. Mechanisms for the incorporation of proteins in membranes and organelles. *J. Cell Biol.* 92:1–22.

Kelly, R. B., 1985. Pathways of protein secretion in eukaryotes. *Science* 230:25–32.

Signal sequences and signal recognition

Perara, E., Rothman, R. E., and Lingappa, V. R., 1986. Uncoupling translocation from translation: implications for transport of proteins across membranes. *Science* 232:348–352.

Milligan, R. A., and Unwin, P. N. T., 1986. Location of exit channel for nascent protein in 80S ribosome. *Nature* 319:693–695. [Electron microscopic analysis and three-dimensional image reconstruction of crystalline arrays of membrane-bound ribosomes.]

Friedlander, M., and Blobel, G., 1985. Bovine opsin has more than one signal sequence. *Nature* 318:338–343.

Kurzchalia, T. V., Wiedmann, M., Girshovich, A. S., Bochkareva, E. S., Bielka, H., and Rapoport, T. A., 1986. The signal sequence of nascent preprolactin interacts with the 54K polypeptide of the signal recognition particle. *Nature* 320:634–636.

Lauffer, L., Garcia, P. D., Harkins, R. N., Coussens, L., Ullrich, A., and Walter, P., 1985. Topology of signal recognition particle receptor in endoplasmic reticulum membrane. *Nature* 318:334–338.

Golgi complex

Farquhar, M. G., 1985. Progress in unraveling pathways of Golgi traffic. *Ann. Rev. Cell Biol.* 1:447–488.

Rothman, J. E., 1985. The compartmental organization of the Golgi apparatus. *Sci. Amer.* 253(3):74–89.

Griffiths, G., and Simons, K., 1986. The trans Golgi network: sorting at the exit site of the Golgi complex. *Science* 234:438–443.

Dunphy, W. G., and Rothman, J. E., 1985. Compartmental organization of the Golgi stack. *Cell* 42:13–21.

von Figura, K., and Hasilik, A., 1986. Lysosomal enzymes and their receptors. *Ann. Rev. Biochem.* 55:167–193.

Targeting to mitochondria and chloroplasts

Hurt, E. C., and van Loon, A. P. G. M., 1986. How proteins find mitochondria and intramitochondrial compartments. *Trends Biochem. Sci.* 11:204–207.

Schmidt, G. W., and Mishkind, M. L., 1986. The transport of proteins into chloroplasts. *Ann. Rev. Biochem.* 55:879–912.

van Loon, A. P. G. M., Brandli, A. W., and Schatz, G., 1986. The presequences of two imported mitochondrial proteins contain information for intracellular and intramitochondrial sorting. *Cell* 44:801–812.

Schleyer, M., and Neupert, W., 1985. Transport of proteins into mitochondria: translocational intermediates spanning contact sites between outer and inner membranes. *Cell* 43:339–350.

Eilers, M., and Schatz, G., 1986. Binding of a specific ligand inhibits import of a purified precursor protein into mitochondria. *Nature* 322:228–232.

Chen, W.-J., and Douglas, M., 1987. Phosphodiester bond cleavage outside mitochondria is required for the completion of protein import into the mitochondrial matrix. *Cell* 49:651–658.

Targeting to nuclei

Dingwall, C., 1985. The accumulation of proteins in the nucleus. *Trends Biochem. Sci.* 10:64–66.

De Robertis, E. M., 1983. Nucleocytoplasmic segregation of proteins and RNAs. *Cell* 32:1021–1025.

Kalderon, D., Roberts, B. L., Richardson, W. D., and Smith, A. E., 1984. A short amino acid sequence able to specify nuclear location. *Cell* 39:499–509.

Richardson, W. D., Roberts, B. L., and Smith, A. E., 1986. Nuclear location signals in polyoma virus large-T. *Cell* 44:77–85.

Targeting in bacteria

Benson, S. A., Hall, M. N., and Silhavy, T. J., 1985. Genetic analysis of protein export in *Escherichia coli* K12. *Ann. Rev. Biochem.* 54:101–134.

Randall, L. L., and Hardy, S. J. S., 1984. Export of protein in bacteria. *Microbiol. Rev.* 48:290–298.

Coleman, J., Inukai, M., and Inouye, M., 1985. Dual functions of the signal peptide in protein transfer across the membrane. *Cell* 43:351–360.

Davis, N. G., and Model, P., 1985. An artificial anchor domain: hydrophobicity suffices to stop transfer. *Cell* 41:607–614.

Kuhn, A., Wickner, W., and Kreil, G., 1986. The cytoplasmic carboxy terminus of M13 procoat is required for the membrane insertion of its central domain. *Nature* 322:335–339.

Muller, M., and Blobel, G., 1984. In vitro translocation of bacterial proteins across the plasma membrane of *Escherichia coli. Proc. Nat. Acad. Sci.* 81:7421–7425.

Endocytosis

Dautry-Varsat, A., and Lodish, H. F., 1984. How receptors bring proteins and particles into cells. *Sci. Amer.* 250(5):52–58.

Pastan, I., and Willingham, M. C., (eds.), 1985. *Endocytosis.* Plenum. [Contains excellent articles on topics such as receptor-mediated endocytosis, clathrin, asialoglycoprotein receptor, and toxins.]

Pearse, B. M. F., and Crowther, R. A., 1987. Structure and assembly of coated vesicles. *Ann. Rev. Biophys. Biophys. Chem.* 16:49–68.

Bretscher, M. S., and Pearse, B. M. F., 1984. Coated pits in action. *Cell* 38:3–4.

Mellman, I., Fuchs, R., and Helenius, A., 1986. Acidification of the endocytic and exocytic pathways. *Ann. Rev. Biochem.* 55:663–700.

Harrison, S. C., and Kirschhausen, T., 1983. Clathrin, cages, and coated vesicles. *Cell* 33:650–652.

Ahle, S., and Ungewickell, E., 1986. Purification and properties of a new clathrin assembly protein. *EMBO J.* 5:3143–3149.

Rothman, J. E., and Schmid, S. L., 1986. Enzymatic recycling of clathrin from coated vesicles. *Cell* 46:5–9.

Bretscher, M. S., 1984. Endocytosis: relation to capping and cell locomotion. *Science* 224:681–686.

Entry of viruses and toxins

Simons, K., Garoff, H., and Helenius, A., 1982. How an animal virus gets into and out of its host cell. *Sci. Amer.* 246(2):58–66.

Ward, W. H. J., 1987. Diphtheria toxin: a novel cytocidal enzyme. *Trends Biochem. Sci.* 12:28–31.

Neville, Jr., D. M., and Hudson, T. H., 1986. Transmembrane transport of diphtheria toxin, related toxins, and colicins. *Ann. Rev. Biochem.* 55:195–224.

Allured, V. S., Collier, R. J., Carroll, S. F., and McKay, D. B., 1986. Structure of exotoxin A of *Pseudomonas aeruginosa* at 3.0-Å resolution. *Proc. Nat. Acad. Sci.* 83:1320–1324.

Protein degradation

Ciechanover, A., 1987. Regulation of the ubiquitin-mediated proteolytic pathway: role of the substrate α-NH_2 group and of transfer RNA. *J. Cell. Biochem.* 34:81–100.

Waxman, L., and Goldberg, A. L., 1986. Selectivity of intracellular proteolysis: protein substrates activate the ATP-dependent protease La. *Science* 232:500–503.

Bachmair, A., Finley, D., and Varshavsky, A., 1986. In vivo half-life of a protein is a function of its amino-terminal residue. *Science* 234:179–186.

PROBLEMS

1. What are the distinctive features of each of the following targeting sequences?
   (a) Eucaryotic signal directing a nascent protein to the ER.
   (b) Procaryotic signal directing a nascent protein to the plasma membrane.
   (c) Signal directing a protein in the Golgi to lysosomes.
   (d) Signal directing a membrane protein in the Golgi to the plasma membrane.
   (e) Signal directing a protein in the cytosol to mitochondria.
   (f) Signal targeting a cytosolic protein for rapid destruction.

2. A mutant LDL receptor is found to be uniformly distributed in the plasma membrane rather than being concentrated in coated pit regions. The mutant binds LDL normally but is not internalized. Which region of the receptor is likely to be altered?

3. Cells derived from patients with I-cell disease take up lysosomal enzymes purified from normal cells. The added enzymes appear in the lysosomes of these cells.
   (a) What is the likely pathway for the transport of these enzymes from the extracellular medium to the lysosomes of I-cells?
   (b) This experiment led to the hypothesis that lysosomal enzymes are normally secreted and taken up by this route. Which sugar phosphate would you add to the extracellular medium to test this hypothesis?

4. Synthetic oligonucleotides were used to construct a variety of mitochondrial presequences to determine the requirements for effective targeting signals. These sequences were fused to the amino terminus of yeast cytochrome oxidase subunit IV after its own presequence had been removed. The sequences studied were:
   (a) MLSLRQSIRFFKPATRTLCSSRYLL-
   (b) MLSRLSLRLLSRLSLRLLSRYLL-
   (c) MLSSLLRLRSLSLRLRLSRYLL-
   (d) MLSRQSQTQQGQRSKSQQARYLL-

   The synthetic presequences (b) and (c) were nearly as effective as the wild-type sequence (a) in targeting the protein to the inner mitochondrial membrane. In contrast, presequence (d) was ineffective. What do these experiments suggest about the nature of the recognition process?

5. Membrane-bound immunoglobin and its secreted counterpart differ in the carboxyl-terminal regions of their heavy chains. As was noted in an earlier chapter (p. 113), these variations on a theme arise from alternative splicing of mRNA.
   (a) Suppose that the transmembrane sequence of the antibody is placed by recombinant DNA methods at the amino-terminus of a cytosolic protein. Where is this chimeric protein likely to be located?
   (b) Suppose that this transmembrane sequence is placed at the carboxyl-terminus of chymotrypsinogen. Where is this chimeric protein likely to be located?

6. Chloroplasts and mitochondria are somewhat similar in how they import proteins. However, they differ in the energy source used to drive translocation to interior sites. Propose a reason why chloroplasts do not use the same energy source as do mitochondria for importing proteins.

7. A new virus is believed to gain entry into the cytosol by receptor-mediated endocytosis. Propose a simple experiment to test this hypothesis.

8. The amino-terminal residue of secreted eucaryotic proteins is usually different from that of cytosolic proteins. A majority of secretory proteins have leucine, phenylalanine, aspartate, lysine, or arginine as their amino-terminal residue. In contrast, these residues are rarely found in cytosolic proteins. What is a potential benefit of having different amino-terminal residues for these two classes of proteins?

CHAPTER 32

Control of Gene Expression in Procaryotes

Bacteria are highly versatile and responsive organisms because they sense the levels of many metabolites and regulate their metabolic patterns accordingly by a wide variety of mechanisms. As was discussed in earlier chapters, the activities of many key proteins are controlled by allosteric changes and reversible covalent modifications. Equally critical in determining the pattern of cellular processes is the control of gene expression. An *E. coli* cell contains only ten or so copies of scarce proteins and as many as 10^5 copies of highly abundant ones. Moreover, the rate of synthesis of some proteins varies over a 1000-fold range in response to the supply of nutrients or to environmental challenges.

Gene activity is regulated primarily at the level of transcription. In bacteria, many genes are clustered in units called *operons*. The coordinate transcription of genes in an operon is blocked by *repressor proteins* and activated by stimulatory proteins. We shall focus on the lactose and tryptophan operons of *E. coli* because much is known about the molecular basis of their control. Bacteriophage lambda is another rewarding object of inquiry because it illustrates how a choice is made between alternative developmental pathways. The elucidation of the three-dimensional structure of several control proteins has revealed a recurring motif in protein-DNA interactions (Figure 32-1). The successful exploration of the regulation of gene expression in procaryotes has inspired studies of more complex eucaryotic systems, the theme of the next chapter.

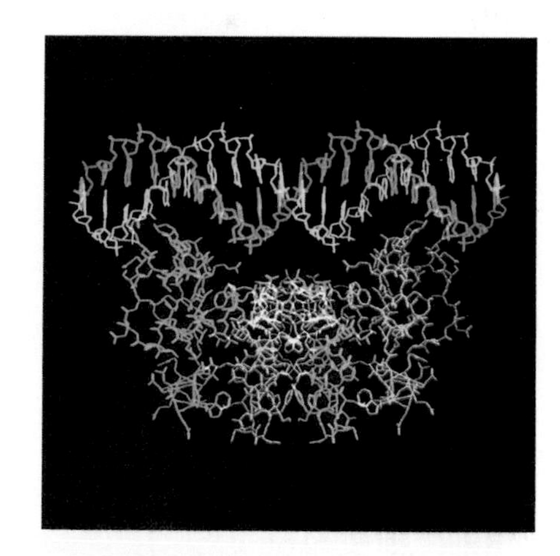

Figure 32-1
Three-dimensional structure of a repressor-DNA complex. Binding of the tryptophan repressor protein to a specific site on DNA silences the expression of a series of genes for the biosynthesis of tryptophan. The structure of the repressor was solved at atomic resolution; its interaction with DNA is inferred from model building. [Based on coordinates kindly provided by Dr. Paul Sigler.]

β-GALACTOSIDASE IS AN INDUCIBLE ENZYME

E. coli can use lactose as its sole source of carbon. An essential enzyme in the metabolism of this disaccharide is *β-galactosidase,* which hydrolyzes lactose to galactose and glucose (Figure 32-2). An *E. coli* cell growing on lactose contains several thousand molecules of β-galactosidase. In contrast, the number of β-galactosidase molecules per cell is fewer than ten if *E. coli* is grown on other sources of carbon, such as glucose or glycerol. The presence of lactose in a culture medium induces a large increase in the amount of β-galactosidase in *E. coli* by eliciting the synthesis of new enzyme molecules rather than by activating a proenzyme (Figure 32-3). Hence, β-galactosidase is an *inducible enzyme.* Two other proteins are synthesized in concert with β-galactosidase—namely, *galactoside permease* and *thiogalactoside transacetylase.* The permease is required for the transport of lactose across the bacterial cell membrane. The transacetylase is not essential for lactose metabolism; its physiologic role is uncertain. In vitro, it catalyzes the transfer of an acetyl group from acetyl CoA to the C-6 hydroxyl group of a thiogalactoside.

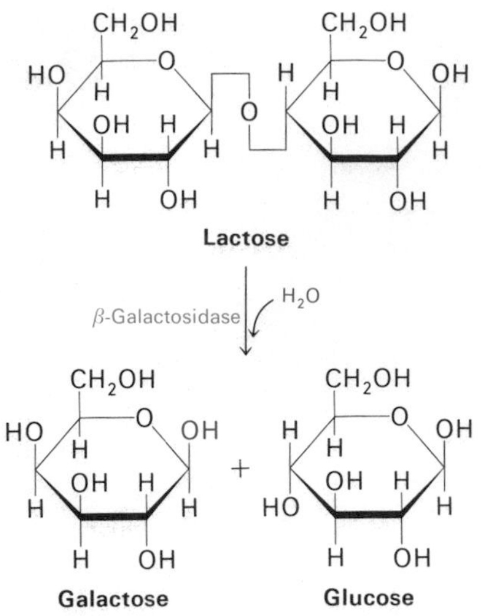

Figure 32-2
Lactose is hydrolyzed by β-galactosidase.

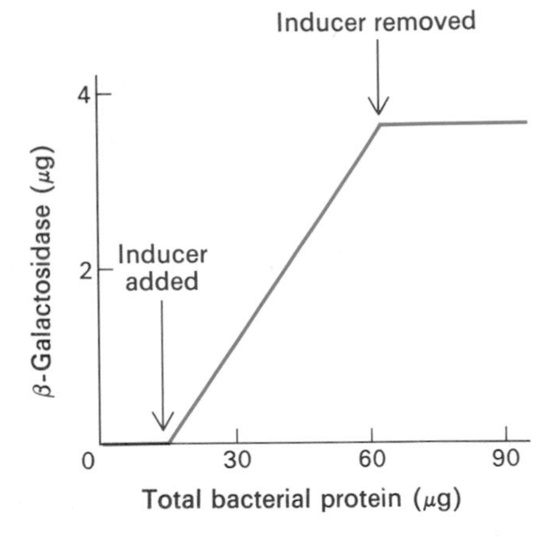

Figure 32-3
The increase in the amount of β-galactosidase parallels the increase in the number of cells in a growing culture of *E. coli.* The slope of this plot indicates that 6.6% of the protein synthesized is β-galactosidase.

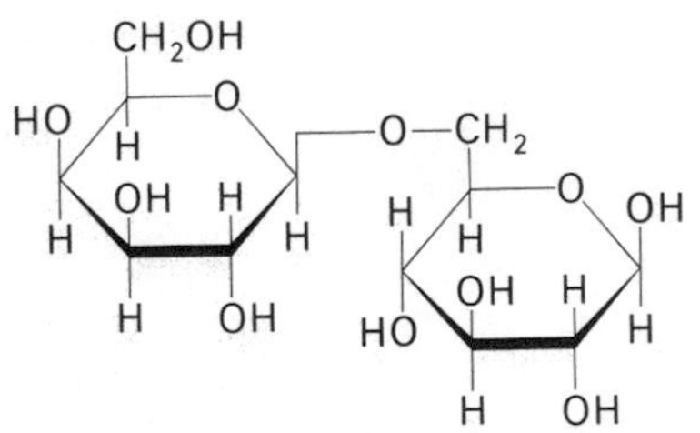

Within an *E. coli* cell, the physiologic inducer is *allolactose,* which is formed from lactose by transglycosylation. The synthesis of allolactose is catalyzed by the few β-galactosidase molecules that are present prior to induction. Studies of synthetic inducers showed that some β-galactosides are inducers without being substrates of β-galactosidase, whereas other compounds are substrates without being inducers. For example, *isopropylthiogalactoside* (IPTG) is a nonmetabolizable inducer.

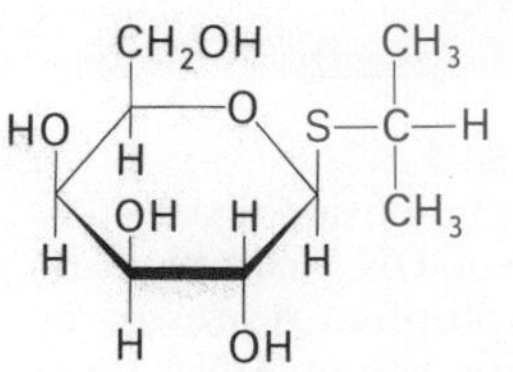

DISCOVERY OF A REGULATORY GENE

An important clue concerning the nature of the induction process was the finding that the amounts of the permease and the transacetylase increased in direct proportion to that of β-galactosidase for all inducers tested. Further insight came from studies of mutants, which showed that β-galactosidase, the permease, and the transacetylase are encoded by three contiguous genes, called *z, y,* and *a,* respectively. Mutants defective in only one of these proteins were isolated. For example, $z^- y^+ a^+$ denotes a mutant lacking β-galactosidase but having normal amounts of the permease and the transacetylase. A most interesting class of mutants affecting all three proteins was then isolated. These *constitutive mutants* synthesize large amounts of β-galactosidase, the permease, and the transacetylase whether or not inducer is present. Francois Jacob and Jacques Monod deduced that *the rate of synthesis of these three proteins is governed by a common element that is different from the genes specifying their structures.* The gene for this common regulatory element was named *i*. Wild-type inducible bacteria have the genotype $i^+z^+y^+a^+$, whereas the constitutive lactose mutants have the genotype $i^-z^+y^+a^+$.

How does the i^+ gene affect the rate of synthesis of the proteins encoded by the *z*, *y*, and *a* genes? The simplest hypothesis was that the i^+ gene determines the synthesis of a cytoplasmic substance called a *repressor*, which is missing or inactive in i^-. This idea was tested in an ingenious series of genetic experiments involving partly diploid bacteria that contained two sets of genes for the lactose region. One set was on the bacterial chromosome whereas the other was on an F′ sex factor (p. 696), an *episome*, introduced by conjugation. For example, an i^+z^-/Fi^-z^+ diploid was isolated. In this diploid, i^+z^- is on the chromosome, whereas i^-z^+ is on the episome. Is this diploid inducible or constitutive for β-galactosidase? In other words, does i^+ on the bacterial chromosome repress the expression of z^+ on the episome? The experimental result was clear-cut: *the diploid was inducible rather than constitutive.* The same result was obtained for the diploid i^-z^+/Fi^+z^-. Hence, *a diffusible repressor is specified by the* i$^+$ *gene.* A diffusible repressor is an example of a *trans-acting factor*, one that can be encoded by a locus on a DNA molecule different from the one containing its target.

AN OPERON IS A COORDINATED UNIT OF GENE EXPRESSION

These experiments led Jacob and Monod to propose the *operon model* for the regulation of protein synthesis. The genetic elements of this model are a *regulator gene*, an *operator site*, and a set of *structural genes* (Figure 32-4). The regulator gene produces a *repressor* that can interact

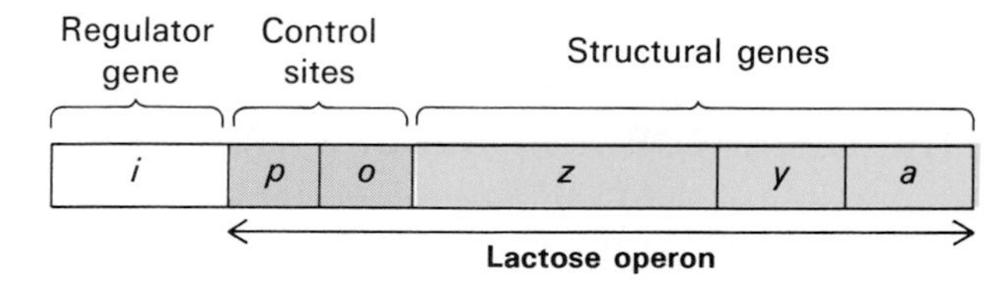

Figure 32-4
Map of the lactose operon and its regulator gene. (This map is not drawn to scale: the *p* and *o* sites are actually much smaller than the other genes.)

with the operator. Subsequent work revealed that the repressor is a protein. The operator is adjacent to the structural genes it controls. The binding of the repressor to the operator prevents the transcription of the structural genes. The operator and its associated structural genes are called an *operon*. For the lactose operon, the *i* gene is the regulator gene, *o* is the operator, and the *z*, *y*, and *a* genes are the structural genes. The operon also contains a *promoter site* (denoted by *p*) for the binding of RNA polymerase. This site for the initiation of transcription is next to the operator. An *inducer* such as IPTG binds to the repressor,

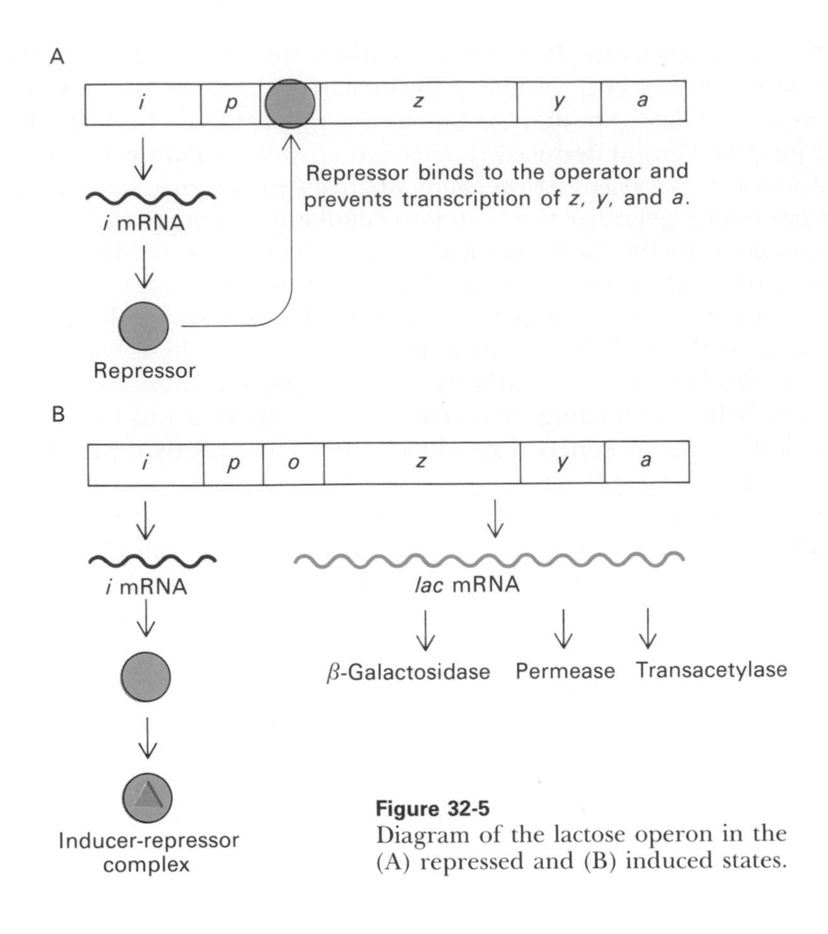

Figure 32-5
Diagram of the lactose operon in the (A) repressed and (B) induced states.

which prevents it from interacting with the operator. The *z*, *y*, and *a* genes can then be transcribed to give a single mRNA molecule that codes for all three proteins (Figure 32-5). An mRNA molecule coding for more than one protein is known as a *polygenic (or polycistronic) transcript*.

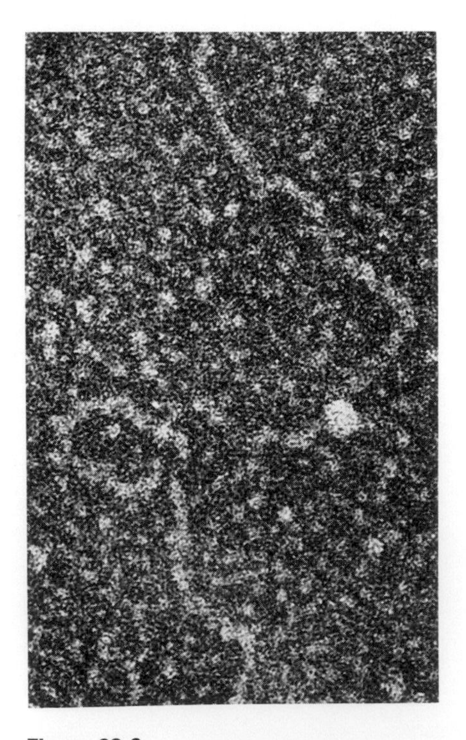

Figure 32-6
Electron micrograph of the *lac* repressor bound to DNA containing the *lac* operator. [Courtesy of Dr. Jack Griffith.]

Lac REPRESSOR PROTEIN IN THE ABSENCE OF INDUCER BINDS TO THE OPERATOR AND BLOCKS TRANSCRIPTION

The repressor of the lactose operon (called the *lac* repressor) was isolated on the basis of its binding of IPTG. Walter Gilbert and Benno Müller-Hill showed that the *lac* repressor is a protein that binds to DNA carrying the *lac* operon but not to other DNA molecules. As predicted, IPTG prevents the binding of *lac* repressor to *lac* operator DNA. A wild-type *E. coli* cell contains only about ten molecules of the *lac* repressor, only 0.001% of the total protein, so that it was difficult to purify the repressor from these cells. However, much larger amounts of the *lac* repressor were made in mutants having a more efficient promoter for the *i* gene. The amount of *lac* repressor was increased further by infecting the bacteria with transducing phages that carry the *lac* region. Such an infected *E. coli* cell contains about 20,000 repressors (about 2% of the total protein), which made it a choice starting material for the purification of *lac* repressor.

The repressor is a tetramer of identical 37-kd subunits, each with a binding site for the inducer. The dissociation constant for the IPTG tetramer is about 10^{-6} M. The repressor binds very tightly and rapidly to the operator. The dissociation constant of the repressor-operator complex is about 10^{-13} M. The rate constant for association ($\sim10^{10}$ M^{-1} s^{-1}) is strikingly high, indicating that the repressor finds the operator site by diffusing along a DNA molecule (a one-dimensional search) rather than by encountering it from the aqueous medium (a three-dimensional search). Recall that RNA polymerase finds promoter sites in a similar way (p. 706). The *lac* repressor binds 4×10^6 times as strongly to its operator as to other sites on the chromosome. This high degree of selectivity is necessary because the genome contains a vast excess (1.6×10^5) of competing sites.

THE *lac* OPERATOR HAS A SYMMETRIC BASE SEQUENCE

The availability of pure *lac* repressor made it feasible to isolate the *lac* operator and determine its base sequence. Gilbert and his associates sonicated the DNA of a phage carrying the *lac* region into fragments that were approximately 1000 base pairs long. The *lac* repressor was added to the mixture of fragments, which was then filtered through a cellulose nitrate membrane. DNA fragments without bound *lac* repressor passed through this filter, whereas DNA-repressor complexes bound tightly to it. The bound DNA was released by adding IPTG. These released DNA fragments were then treated with pancreatic deoxyribonuclease in the presence of repressor, which protected the operator region from digestion (Figure 32-7).

The base sequence of this region is very interesting: a total of 28 base pairs are related by a twofold axis of symmetry (Figure 32-8), just as the subunits of *lac* repressor are. Thus, *the symmetry of the repressor molecule matches that of its target site in DNA*. Recall that this recognition principle is also utilized in the binding of EcoRI endonuclease to its cleavage site (p. 656). *Symmetry matching is a recurring motif in protein-DNA interactions.*

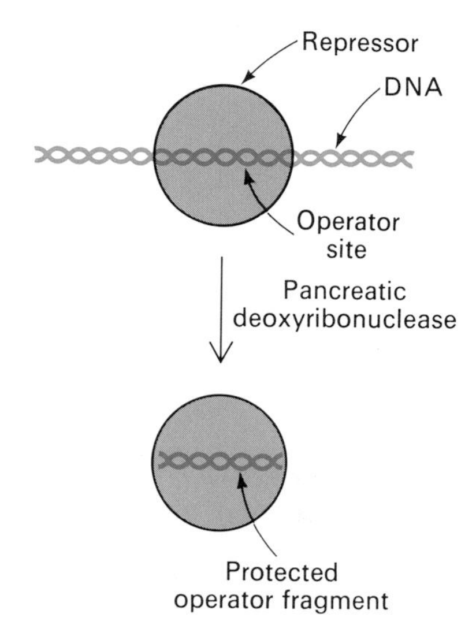

Figure 32-7
The *lac* repressor protects the *lac* operator from digestion by pancreatic deoxyribonuclease.

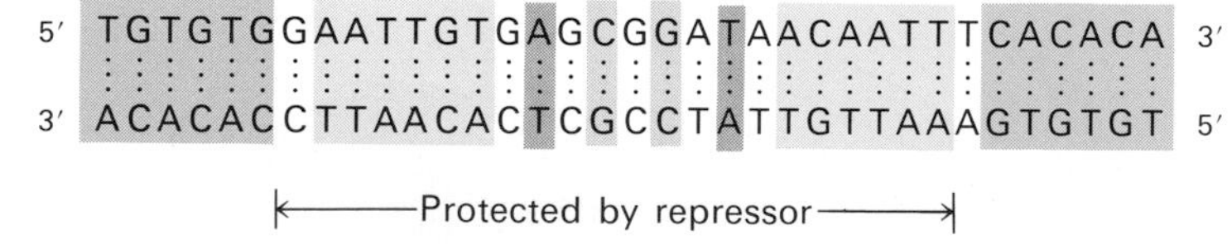

Figure 32-8
Nucleotide sequence of the *lac* operator. The symmetrically related regions are shown in matching colors.

INDUCIBLE CATABOLIC OPERONS ARE GLOBALLY REGULATED BY CAP PROTEIN CONTAINING BOUND CYCLIC AMP

It has long been known that *E. coli* grown on glucose, a preferred energy source, have very low levels of catabolic enzymes, such as β-galactosidase, galactokinase, arabinose isomerase, and tryptophanase. Clearly, it would be wasteful to synthesize these enzymes when glucose is abundant. The molecular basis of this inhibitory effect of glucose, called *catabolite repression*, has been elucidated. A key clue was the observation that glucose lowers the concentration of cyclic AMP in *E. coli*. It was then found that exogenous cyclic AMP can relieve the repression exerted by glucose. Subsequent biochemical and genetic studies revealed that *cyclic AMP stimulates the initiation of transcription of many inducible operons.* It is interesting to note that cyclic AMP serves as a hunger signal both in bacteria and in mammals. Recall that a low level of blood sugar stimulates the secretion of glucagon, which leads to elevated cyclic AMP levels inside hormone-sensitive cells (p. 458).

In mammalian cells, cyclic AMP (cAMP) acts by stimulating a protein kinase that phosphorylates many target proteins, such as those controlling glycogen synthesis and breakdown (p. 462). The action of cAMP in bacteria is very different. cAMP binds to CAP (the *catabolite gene activator protein*), a dimer of identical 22-kd subunits. Proteolytic digestion experiments have shown that each subunit contains a DNA-binding domain and a cAMP binding domain. *The complex of CAP and cAMP, but not CAP alone, stimulates transcription by binding to certain promoter sites.* In the *lac* operon, CAP binds next to the site for RNA polymerase, as shown by nuclease digestion studies. Specifically, CAP protects nucleotides −87 to −49 from digestion, whereas RNA polymerase protects nucleotides −48 to +5 (Figure 32-9). In this numbering system, the first transcribed nucleotide is +1. CAP exhibits twofold symmetry (Figure 32-10) that matches that of its DNA binding site.

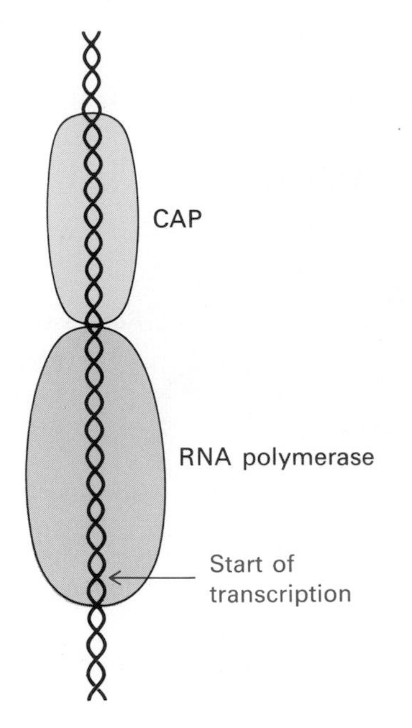

Figure 32-9
Schematic diagram of CAP and RNA polymerase on the DNA template. The locations of these proteins were inferred from nuclease digestion studies.

CAP stimulates the initiation of *lac* mRNA synthesis by a factor of 50. How? The contiguous and nonoverlapping arrangement of the binding sites for CAP and RNA polymerase suggested that *the binding of CAP to DNA creates an additional interaction site for RNA polymerase.* Indeed, the binding of RNA polymerase to the promoter is enhanced by its energetically favorable contacts with bound CAP. In contrast, the *lac* repressor binds to nucleotides −3 to +21, which significantly overlaps the RNA polymerase site. Hence, *the repressor sterically interferes with the binding of RNA polymerase.*

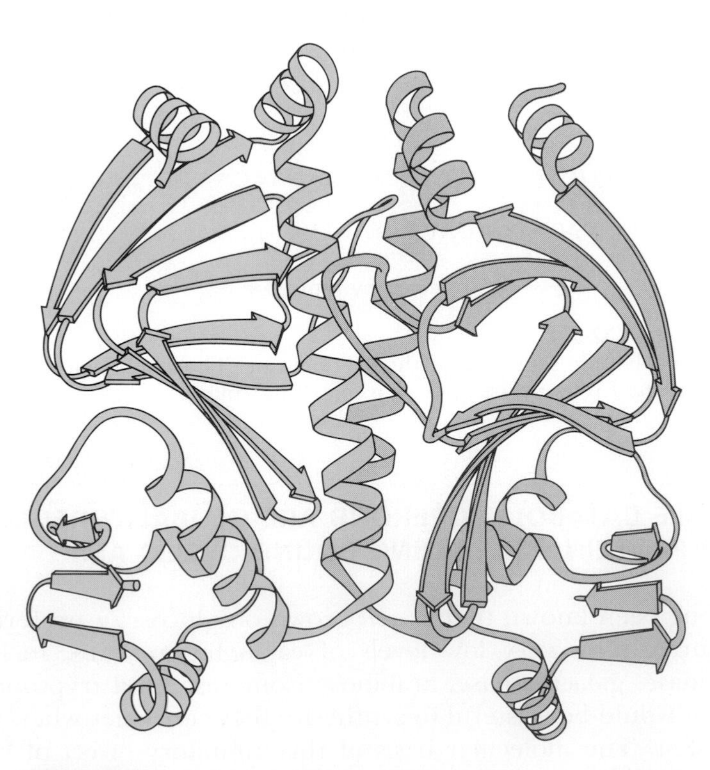

Figure 32-10
Three-dimensional structure of CAP protein. Each subunit of the dimer contains a binding site for a cyclic AMP molecule. [Courtesy of Dr. Jane Richardson, based on coordinates kindly provided by Dr. Thomas Steitz.]

The cAMP-CAP complex probably acts similarly at other inducible catabolic operons. All contain the sequence TGTGA upstream of the promoter site. Another common feature is that their −35 and −10 sequences differ markedly from the consensus sequence of strong promoters (Figure 32-11). Evolution has probably weakened these promoters to make their operons dependent on a helper protein for efficient initiation of transcription. Thus, *inducible catabolic operons are under dual control.* A high level of expression requires the simultaneous presence of cAMP and a specific inducer, such as a galactoside for the *lac* operon. The specific inducer acts on a single operon, whereas the cAMP-CAP complex affects many.

The operons controlled in concert by cyclic AMP are members of a *global regulatory circuit.* Another such network is made up of the heat-shock genes that are activated by a distinctive sigma subunit of RNA polymerase induced by a rise in temperature (p. 707). Yet another example of global control is the induction of SOS genes that encode DNA repair enzymes following damage to DNA (p. 694). Each of these higher-level regulatory systems is specifically triggered and has built-in mechanisms for returning to the unstimulated state.

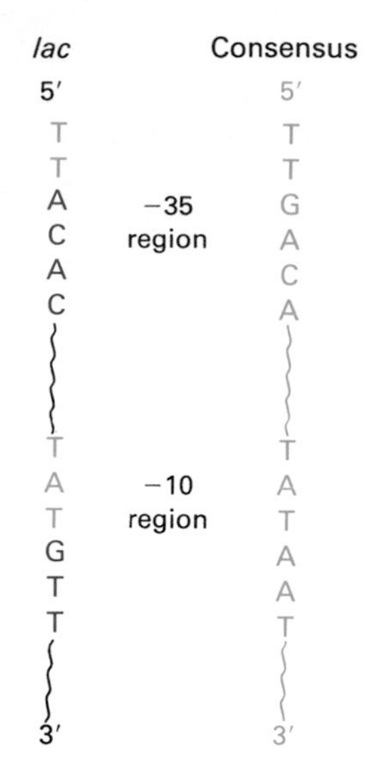

Figure 32-11
The base sequence of *lac* promoter deviates considerably from the consensus sequence (green) of strong promoters. cAMP-CAP is required for optimal expression of the *lac* operon.

DIFFERENT FORMS OF THE SAME PROTEIN ACTIVATE AND INHIBIT TRANSCRIPTION OF THE ARABINOSE OPERON

Bacteria can use arabinose as a fuel by converting it into xylulose 5-phosphate, a pentose phosphate pathway intermediate (p. 429), by the sequential action of arabinose isomerase, ribulokinase, and ribulose 5-phosphate epimerase (Figure 32-12). These enzymes are encoded by the *araA, araB,* and *araD* genes, respectively, which all belong to the arabinose operon. The operon also contains a gene (*araC*) for a regulatory C protein, two operator sites (*ara*O_1 and *ara*O_2), and a promoter (*araI*) (Figure 32-13).

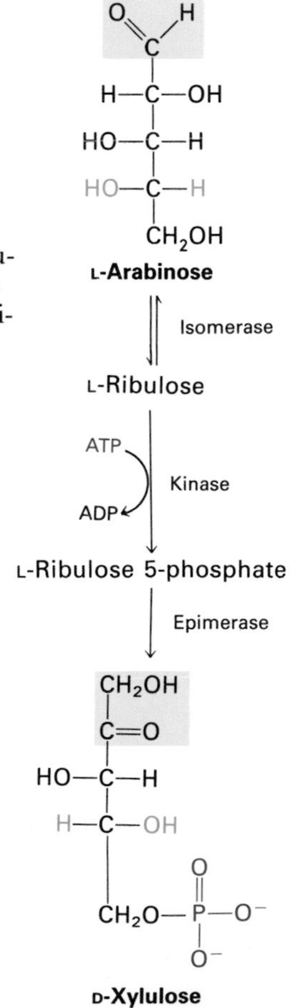

Figure 32-12
The conversion of arabinose to xylulose 5-phosphate is catalyzed by the three enzymes encoded by the arabinose operon.

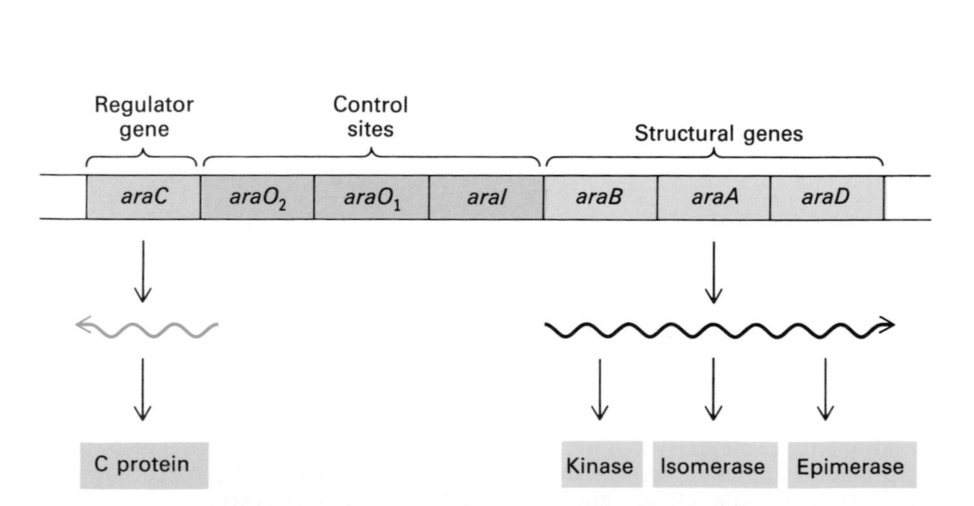

Figure 32-13
Map of the arabinose operon and its regulator gene.

The arabinose operon, like other inducible catabolic operons, is under dual control. Two signals—the cAMP-CAP complex and arabinose bound to the C protein—are required for efficient transcription. A distinctive feature of this operon is that the C protein also serves as a negative regulator. The direction of transcription of *araC* is opposite to that of *araBAD* and is controlled by the O_1 rather than the O_2 operator site. mRNA for C protein is formed when the level of this protein and of cAMP-CAP is low (Figure 32-14A). When C protein is abundant and cAMP-CAP is not, transcription of the *C* gene stops because a C protein binds to *ara*O_1. Thus, *the synthesis of C protein is autoregulated.* The binding of a second molecule of C to *ara*O_2 blocks the synthesis of *BAD* mRNA by forming a DNA loop and also binding to *araI*, which is adjacent to the promoter for the *BAD* genes (Figure 32-14B). This loop is not formed when cAMP-CAP is abundant and arabinose is bound to the C protein, which alters its conformation. Instead, the presence of these positive regulatory factors enables RNA polymerase to bind to the promoter site for the *BAD genes* and to transcribe them (Figure 32-14C). In this case, mRNA for C protein is not formed because *ara*O_1 is occupied by C protein.

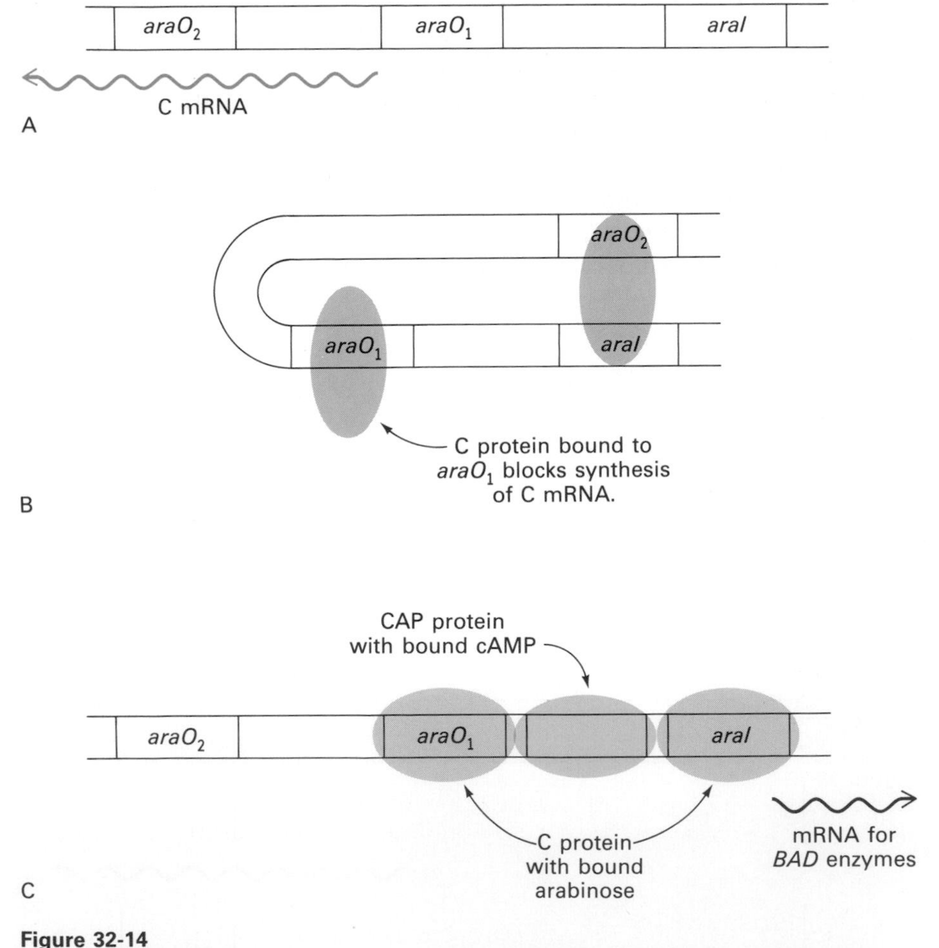

Figure 32-14
Three states of the arabinose operon. (A) mRNA for C protein is formed at low levels of this protein. (B) No mRNA is formed when the cAMP level is low and C protein is abundant, irrespective of the level of arabinose. (C) When both cyclic AMP and arabinose are abundant, mRNA for enzymes catabolizing arabinose is formed.

The arabinose operon illustrates several general principles of gene regulation. First, a protein can regulate its own synthesis by repressing the transcription of its gene. Second, the binding of a signal molecule to a protein can switch it from being an inhibitor of transcription to being an activator. These alternative conformations bind to different regulatory sites on DNA. Third, protein-binding regulatory sites on DNA need not be contiguous with the genes controlled by them. The arabinose operon provides a concrete example of how transcription can be modulated by a site at some distance from the transcribed gene. Fourth, the changes induced by signal molecules are readily reversed. The system responds continuously and rapidly to variations in the levels of metabolites.

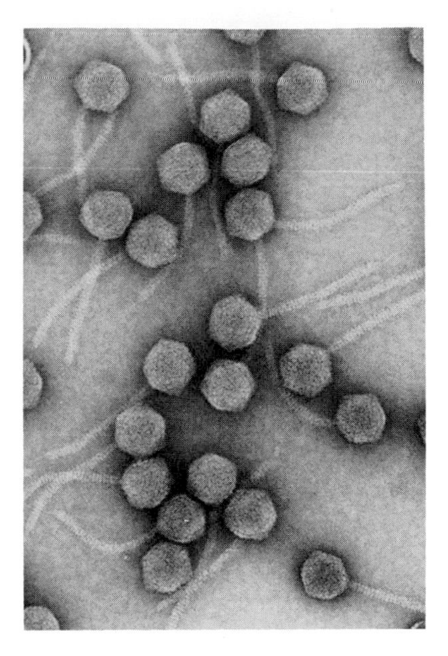

Figure 32-15
Electron micrograph of λ phages. [Courtesy of Dr. A. Dale Kaiser.]

REPRESSORS AND ACTIVATORS OF TRANSCRIPTION DETERMINE THE DEVELOPMENT OF TEMPERATE PHAGE

We turn now to the role of repressors and activators of transcription in regulating the life cycle of *lambda* (λ) *bacteriophage.* The mature virus particle consists of a linear double-helical DNA molecule (48 kb) surrounded by a protein coat. After it has infected a bacterium, two developmental pathways are open to this virus: destroy the bacterium or join it (hence the name *temperate*; see Figure 6-15 on p. 128). In the *lytic pathway,* the viral functions are fully expressed, leading to the lysis of the bacterium and the production of about 100 progeny virus particles. Alternatively, λ can enter the *lysogenic pathway,* in which its DNA becomes covalently inserted into the host-cell DNA at a specific site. This recombination process, involving a circular λ DNA molecule, will be discussed later (p. 860). Most of the phage functions are switched off when its DNA is integrated in the host DNA. The viral DNA in this state is called a *prophage;* a host cell containing a prophage is called a *lysogenic bacterium.* The prophage replicates as part of the host chromosome in a lysogenic bacterium, usually for many generations. The lytic functions of λ are dormant but not lost in the lysogenic state (Figure 32-16). A variety of agents that interfere with DNA replication in the host induce the prophage to undergo lytic development.

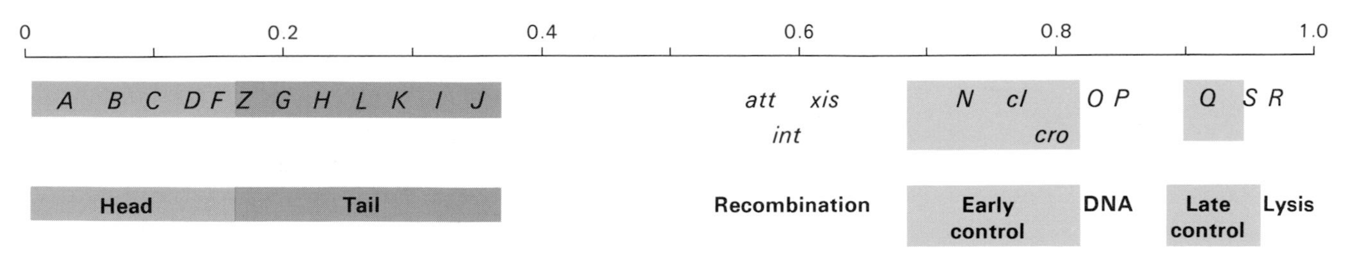

Figure 32-16
Genetic map of λ phage. Only some of its genes are shown here. The linear duplex DNA is converted into the circular form after it enters the bacterial cell.

Let us first consider the pattern of gene expression of λ in the lytic pathway. The goal of producing a large number of progeny is achieved by the *sequential transcription of viral genes.* Proteins needed for DNA replication and recombination are made first, followed by the synthesis of the head and tail proteins of the virus particle and of the proteins required for lysis of the host cell. Timing is critical; premature destruction of the host would of course be disadvantageous to the virus. There are three stages of gene expression in the lytic pathway: immediate-early, delayed-early, and late (Figure 32-17).

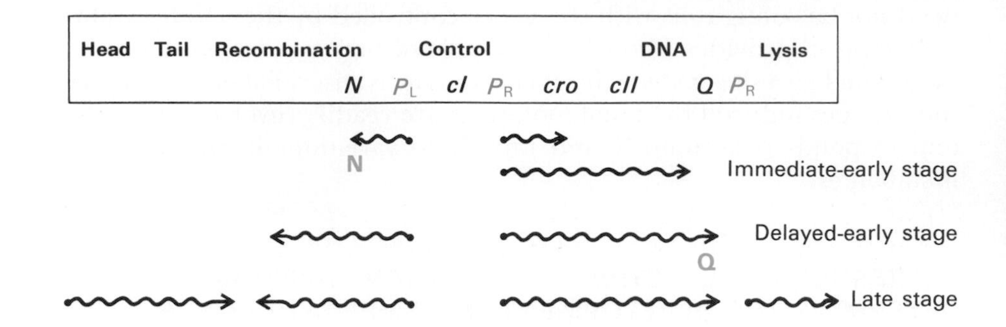

Figure 32-17
Three stages of transcription in the lytic growth of λ phage. The N protein formed in the immediate-early stage activates the delayed-early stage. In turn, the Q protein is formed, which activates the late stage. [After H. Echols. *The Bacteria* 8(1979):502.]

In the *immediate-early stage,* RNA synthesis starts at two promoter sites, P_L and P_R. One of these transcripts is the message for the N protein, which has a critical regulatory role. In the absence of N protein, the immediate-early transcripts end at either of two termination sites. The N protein antagonizes the termination of transcription at these sites and thereby permits further expression of λ genes. Thus, the N protein turns on the *delayed-early stage.* Proteins necessary for the replication of λ DNA and for recombination are made at this time. In addition, the *Q* gene is transcribed in the delayed-early stage. The Q protein is another critical regulator of gene expression in λ; it is required for the *late stage.* The genes for proteins that compose the head and tail of the phage and those that lyse the host are transcribed in the late stage. The Q protein, like the N protein, antagonizes the termination of transcription. *In short, the sequential regulation of lytic development is effected by two positive regulatory proteins, encoded by genes* N *and* Q, *which act by allowing transcription to proceed beyond several termination sites.*

Now let us consider lysogeny, the alternative developmental pathway for λ. The three stages of the lysogenic cycle are *establishment, maintenance,* and *release.* The establishment of the prophage state requires the integration of the viral DNA into the host DNA and the inactivation of the lytic functions of the virus. These processes are complex and not fully understood. The maintenance of the prophage state is much simpler. A. Dale Kaiser showed that *only the* cI *gene is expressed in the prophage.* This gene codes for the λ repressor, which binds to two operator regions, O_L and O_R (Figure 32-18).

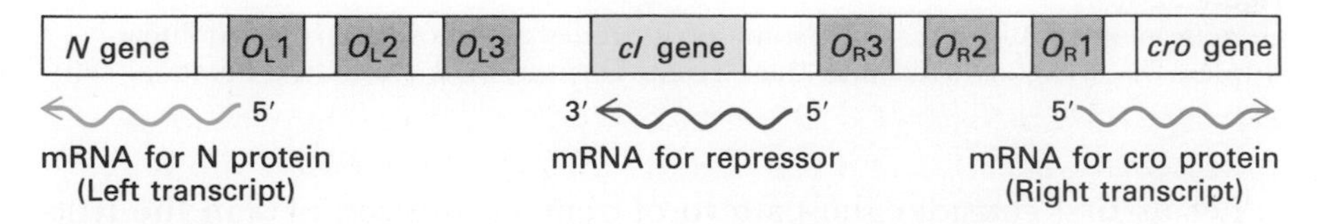

Figure 32-18
Diagram of the O_L and O_R operator regions and the adjacent genes. O_L1 and O_R1 have the highest affinity for the λ repressor. The repressor gene is *cI*. The left transcript starts with the *N* gene, whereas the right transcript starts with the *cro* gene.

Binding of the λ repressor to O_L directly prevents the leftward transcription of the immediate-early genes. In particular, the N protein is not synthesized and so the lytic pathway is blocked. By binding to O_R, the λ repressor prevents the rightward expression of the *cro* and *Q* genes. Indeed, the binding of the λ repressor to O_L and O_R silences the whole λ genome except for the *cI* gene, which codes for the λ repressor. As will be discussed shortly, the λ repressor itself controls the *cI* gene, thereby regulating its own level. Inactivation of the λ repressor enables lytic genes to be transcribed. The prophage is then excised from the host chromosome and the lytic functions are expressed.

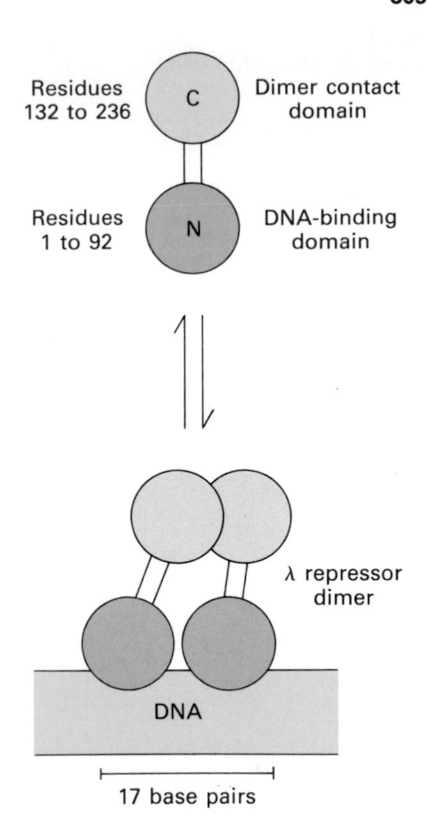

Figure 32-19
Schematic diagram of the domain structure of λ repressor.

TWO OPERATORS IN LAMBDA CONTAIN A SERIES OF BINDING SITES FOR THE REPRESSOR

The λ repressor, a dimer of identical 26-kd subunits, has been purified and studied in detail by Mark Ptashne. Each chain contains an amino-terminal domain that binds DNA and a carboxyl-terminal domain that interacts with its counterpart to hold the dimer together (Figure 32-19). The three-dimensional structure of the repressor and how it binds to an operator will be discussed shortly (p. 812). Two operator regions, O_L and O_R, are recognized by the same λ repressor. The *cI* gene that codes for the repressor is located between O_L and O_R (see Figure 32-18). Each of these operators contains three binding sites for the λ repressor. Nuclease-digestion studies have shown that the binding sites are seventeen base pairs long and that they are separated from each other by AT-rich regions that are from three to seven base pairs in length. The base sequences of these operator sites are similar but not identical (Figure 32-20), which accounts for their differing affinities for the λ repressor. The six operator sites, like the *lac* operator, exhibit partial twofold symmetry. Again, the symmetry of the binding sites in DNA matches that of the dimeric repressor molecule.

The strongest binding site for repressor in both O_L and O_R is the one closest to the start of the first structural gene in the operon—O_L1 and O_R1, respectively. The promoter site for the *N* gene lies within O_L1 and the promoter site for the *cro* gene lies within O_R1. As in the lactose and arabinose operons, the binding of repressor to these operators blocks the binding of RNA polymerase to the corresponding promoter and so transcription of the *N* gene (left transcript) and of the *cro* gene (right transcript) is not initiated. In contrast, transcription of the *cI* gene itself is not blocked by the binding of the repressor to these high-affinity sites.

O_L1	TATCACCGCCAGTGGTA
O_R1	TATCACCGCCAGAGGTA
O_L2	TATCTCTGGCGGTGTTG
O_L3	TATCACCGCAGATGGTT
O_R2	TAACACCGTGCGTGTTG
O_R3	TATCACCGCAAGGGATA

Figure 32-20
Sequences of one of the strands of the six binding sites for λ repressor. The base lying on the twofold axis is shown in red. Bases obeying twofold symmetry are shown in blue. The affinity for λ repressor is greatest for O_L1 and least for O_R3, the reverse of the affinities for cro protein.

THE LAMBDA REPRESSOR REGULATES ITS OWN SYNTHESIS

The number of λ repressor molecules in a lysogenized *E. coli* cell is precisely regulated. Too little repressor, even briefly, would put the cell on the lytic pathway. On the other hand, too much repressor would make it difficult for the phage to emerge should the bacterium become inhospitable. What controls the level of the repressor? The solution is elegantly simple: the repressor regulates its own synthesis. The different affinities of O_R1, O_R2, and O_R3 for repressor are at the heart of its autoregulation. O_R1 has the highest intrinsic affinity for the repressor. At low levels of repressor, the O_R1 site is filled, which blocks the tran-

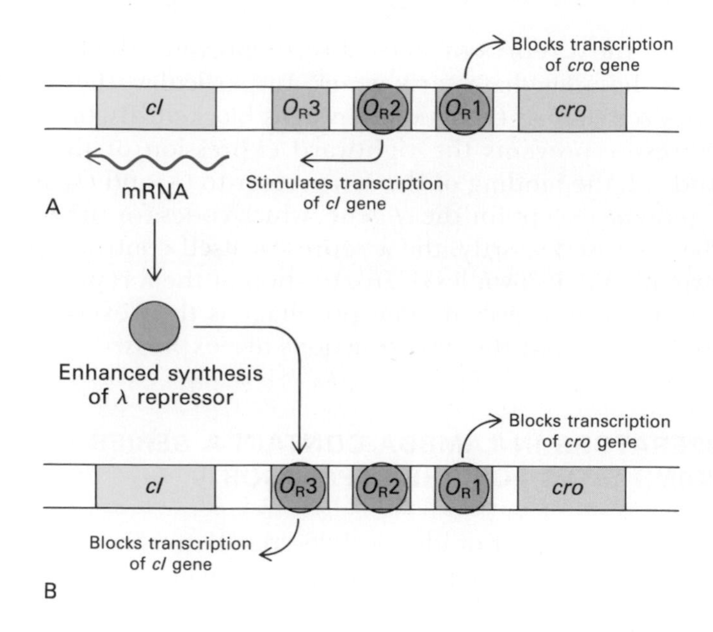

Figure 32-21
Self-regulation of the level of λ repressor: (A) the repressor binds to O_R1 and then to O_R2 when its level is low. Occupancy of O_R2 stimulates transcription of *cI*. (B) As the level of the λ repressor increases, it binds to O_R3, which inhibits further transcription of *cI*.

scription of the *cro* gene and others lying to the right (Figure 32-21). A repressor molecule bound to O_R1 favors the binding of another one to O_R2, which enhances the transcription of the *cI* gene and leads to the production of more repressor. As the concentration of repressor increases, O_R3 becomes filled. However, the occupancy of this site, the closest to the *cI* gene, blocks transcription of the gene. Thus, expression of the *cI* gene is self-regulated.

LYSOGENY IS TERMINATED BY PROTEOLYSIS OF LAMBDA REPRESSOR AND SYNTHESIS OF CRO PROTEIN

The feedback circuit just described will tend to maintain the level of repressor so that the rest of the phage genome is not expressed. How then does the phage ever emerge from lysogeny? The critical trigger is a reduction in the number of λ repressor molecules just sufficient to enable the *cro* gene to be transcribed. The completed cro protein (a dimer of 7-kd subunits) then binds to O_R3 to prevent the transcription of the *cI* gene (Figure 32-22). The important point is that O_R3 has a higher affinity for the cro protein than does O_R1. Thus, *low levels of the cro protein repress the synthesis of λ repressor without turning off the synthesis of the cro protein itself*. The λ repressor is now prevented from regaining the upper hand. At this point, a chain of events leading to lysis is irreversibly set in motion.

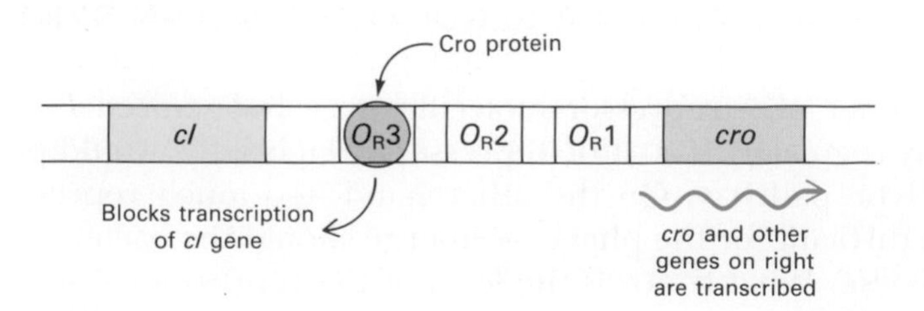

Figure 32-22
Cro protein at low levels blocks the expression of *cI* by binding first to O_R3. mRNA for Cro is formed because site O_R1 is empty.

The transition from lysogeny to lysis is induced by damage to DNA arising from ultraviolet radiation or chemical agents. What is the link between these events? *The key step is the proteolytic destruction of λ repressor by bacterial recA protein* (Figure 32-23). Recall that recA protein catalyzes the exchange of DNA strands in general recombination (p. 691). RecA protein is also a protease when bound to single-stranded DNA, a signal that DNA has been damaged. This protease activity is important in activating the SOS response leading to the synthesis of more than fifteen proteins that take part in DNA repair. Activated recA protein degrades the *lexA* repressor protein that inhibits transcription of SOS genes by binding to their operators. Bacteriophage λ takes advantage of this triggered protease activity to switch to its lytic mode so that it can flee from what is no longer a safe haven. The subtle behavior of λ arises from the remarkable interplay of just a few proteins and operator sites. The economy and elegance of its regulatory circuits are striking.

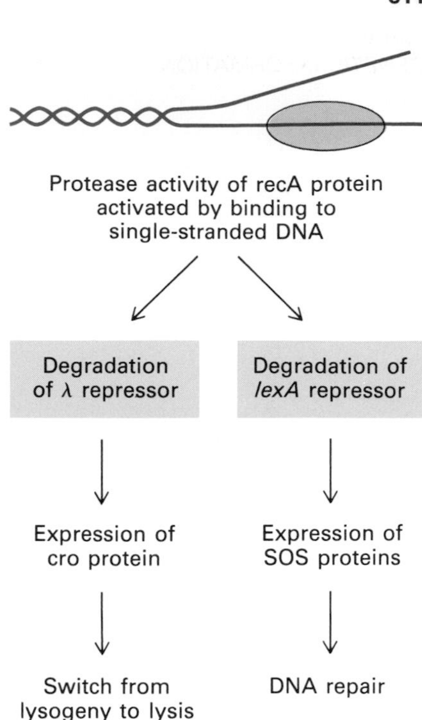

Figure 32-23
Proteolysis of λ repressor by recA protein bound to single-stranded DNA leads to the termination of lysogeny and the induction of the lytic pathway of this phage.

A HELIX-TURN-HELIX MOTIF MEDIATES THE BINDING OF MANY REGULATORY PROTEINS TO CONTROL SITES IN DNA

How do repressors and activators bind to DNA and recognize their target sequences? The elucidation of the three-dimensional structure of cro protein (Cro) by Brian Matthews and his coworkers immediately suggested how it functions. The 66-residue monomer contains three α helices and three antiparallel β sheets (Figure 32-24). The dimer is held

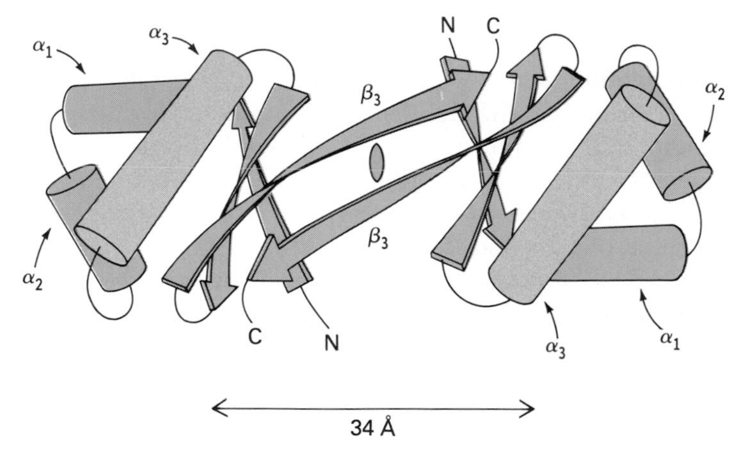

Figure 32-24
Three-dimensional structure of cro protein, a dimer of identical subunits. [Courtesy of Dr. Brian Matthews.]

together by the association of two antiparallel β strands; the twofold axis of symmetry lies between them. Helix 3 in one subunit is separated from its counterpart in the other subunit by 34 Å, the same distance as the separation between neighboring major grooves of a Watson-Crick double helix. Furthermore, an α helix (including its side chains) has a diameter of about 12 Å, just the right size to fit into the major groove of DNA, which is about 12 Å wide and 7 Å deep. Another important structural feature is the match between the twofold symmetry of Cro and the approximate symmetry of the 17-base pair operator sites recognized by it. Thus, it seemed likely that Cro fits into DNA by placing the third helix of each subunit into adjoining major grooves of DNA (Fig-

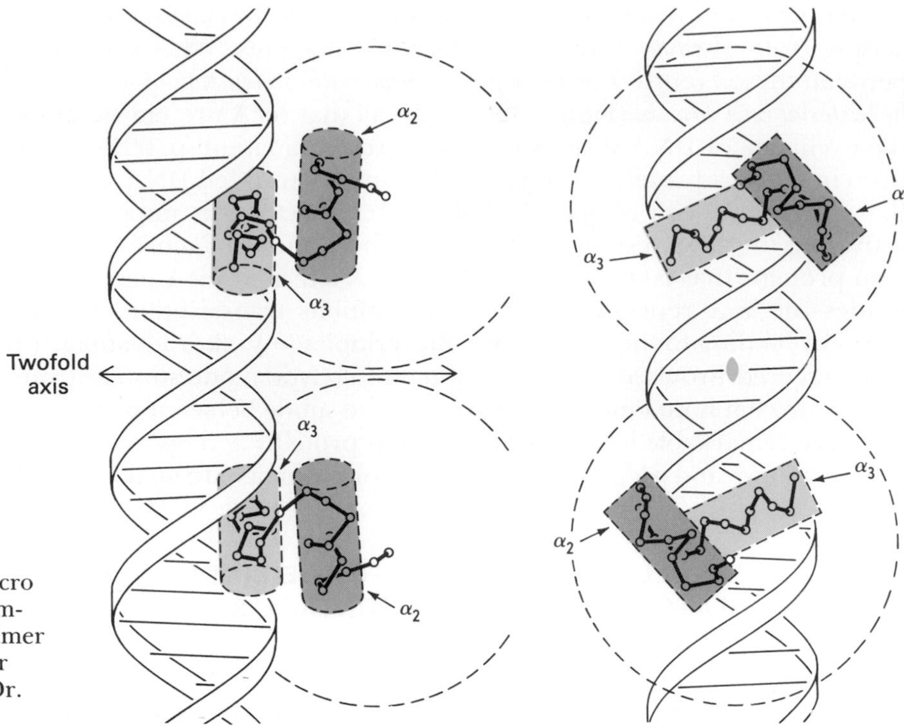

Figure 32-25
Two views of the interaction of cro protein with DNA. A pair of symmetry-related α helices of the dimer fit neatly into neighboring major grooves of DNA. [Courtesy of Dr. Brian Matthews.]

ure 32-25). Specificity would arise from hydrogen bonding between the side chains on the exposed side of this helix and exposed portions of base pairs in the major groove. For example, a glutamine side chain could form two hydrogen bonds with adenine. Van der Waals contacts between complementary DNA and protein surfaces also contribute to specificity. An α helix is the right size and shape to readily interact with five base pairs in the major groove, which would be sufficient to establish the specificity of Cro and other DNA-binding proteins. This proposal is supported by analyses of mutant operators and genetically engineered variants of Cro that do not form strong Cro-DNA complexes.

Many other DNA-binding proteins contain two α helical recognition units related by a twofold axis of symmetry and separated by 34 Å, the pitch of the DNA helix. The structures of Cro, the DNA-binding domain of λ repressor, and CAP protein are compared in Figure 32-26. A

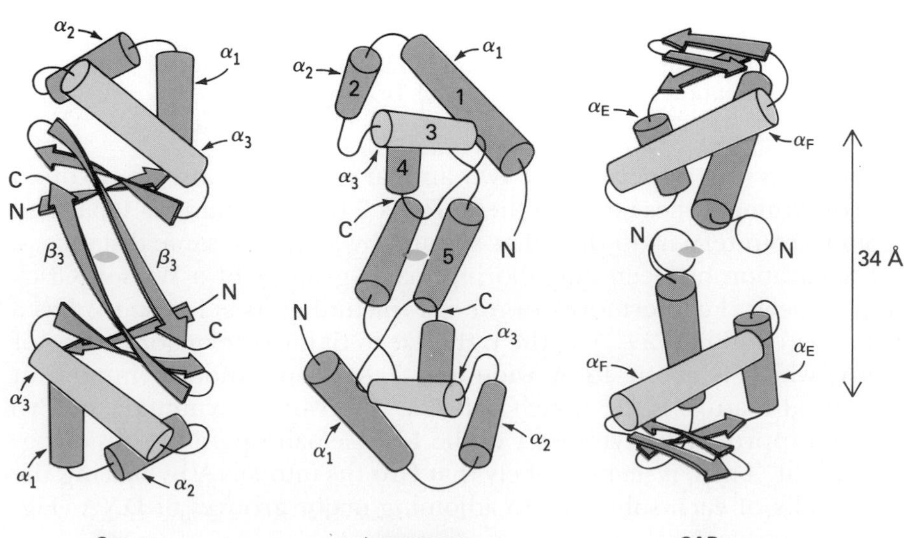

Figure 32-26
Comparison of the three-dimensional structures of cro protein, λ repressor, and CAP protein. The recognition helices that bind to the major grooves of DNA are shown in green.

common element of these proteins and others that bind to specific sites in DNA is a *helix-turn-helix* motif. The first helix (the recognition helix) contains side chains that interact with base pairs in the major groove of DNA; the second helix stabilizes this motif by interacting hydrophobically with it.

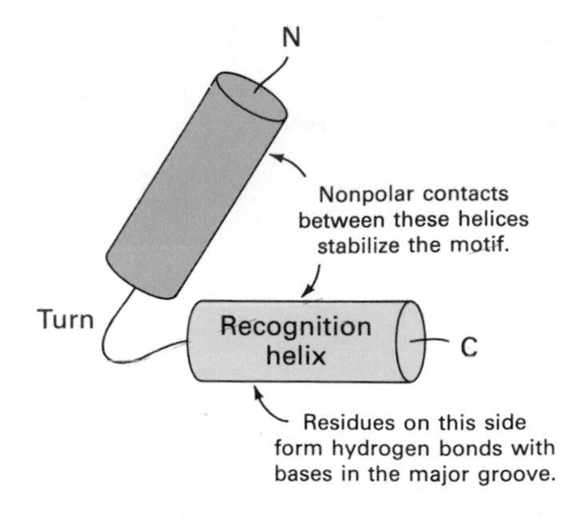

Figure 32-27
Helix-turn-helix motif of DNA-binding proteins. The recognition helix is shown in green. These dimeric proteins contain two such units separated by 34 Å, the pitch of the DNA helix.

THE SPECIFICITY OF A REPRESSOR CAN BE CHANGED BY ALTERING ITS RECOGNITION HELIX

The three-dimensional structure of a repressor-operator complex has recently been solved. The repressor from phage 434, a close relative of λ, also exhibits the helix-turn-helix motif. Moreover, as predicted, its recognition helix lies in the major groove of DNA, which is in the B-DNA form (Figure 32-28). Repressors and other regulatory proteins probably search for target sequences by binding to B-DNA and scanning its major groove. Most regulatory proteins recognize B-DNA or a slight variation of it because they can slide along the DNA while searching for the specific target sequence. The search would be much slower for a protein that cannot bind B-DNA.

Can the specificity of a repressor be changed by altering only its recognition helix? Five amino acid residues on the outside of this helix in the 434 repressor were replaced with those occupying the same position in the P22 repressor encoded by a phage that infects *Salmonella*. This helix-swapping experiment changed the specificity of the chimeric protein to that of the P22 repressor, as assayed in vitro and in vivo. Thus, *residues on the external face of the recognition helix of a repressor determine its DNA-binding specificity*. It will be interesting to learn whether a small set of rules—a recognition code—relates amino acid residues on the recognition helix to the base sequence of the target site.

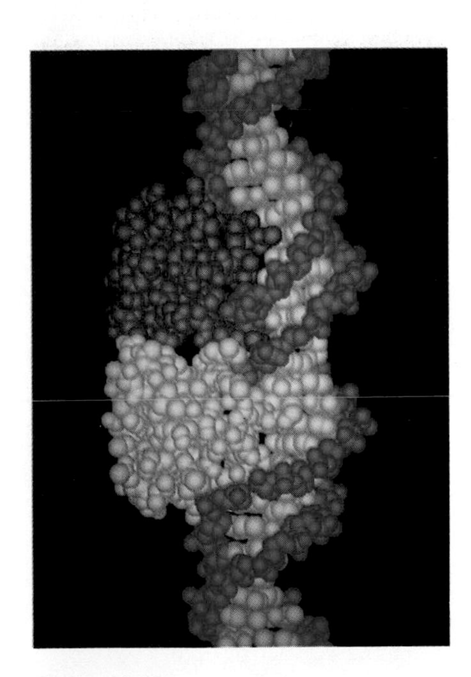

Figure 32-28
Structure of phage 434 repressor bound to DNA. One subunit of the dimeric repressor is shown in red, and the other in yellow. The DNA backbone is shown in dark blue, and the bases in light blue. [Courtesy of Dr. Stephen Harrison and Howard Holley.]

TRANSCRIPTION OF THE *trp* OPERON IS BLOCKED BY REPRESSOR CONTAINING BOUND TRYPTOPHAN

The 7-kb mRNA transcript of this operon codes for the five enzymes that convert chorismate into tryptophan (p. 586). These five proteins are synthesized sequentially, coordinately, and in equimolar amounts by translation of this polygenic (polycistronic) *trp* mRNA. Translation begins prior to the completion of transcription. *Trp* mRNA is synthesized in about four minutes and then rapidly degraded. The short lifetime of *trp* mRNA, which is only about three minutes, enables bacteria to respond quickly to their changing needs for tryptophan. In fact, *E. coli* can vary the rate of production of its biosynthetic enzymes for tryptophan over a 700-fold range.

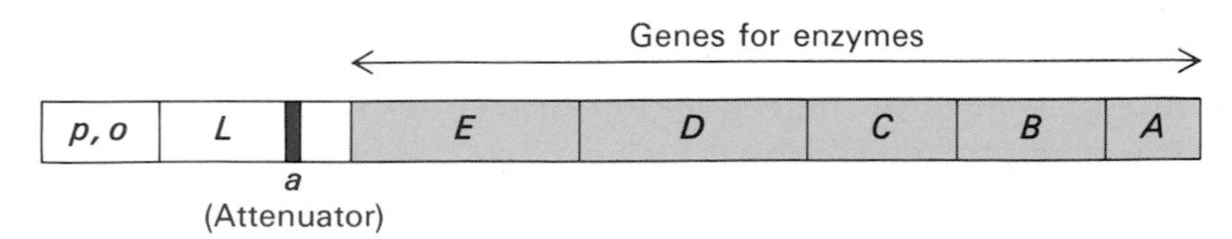

Figure 32-29
Diagram of the *trp* operon showing the promoter (*p*), operator (*o*), and attenuator (*a*) control sites and the genes for the leader sequence (*L*) and enzymes of the tryptophan pathway (*E, D, C, B,* and *A*).

How is this regulation accomplished? One level of control is achieved by the interaction of a specific repressor with the *trp* operator site on the DNA. The *trp* repressor is a dimeric 58-kd protein encoded by the *trpR* gene, which is far from the *trp* operon. *A complex of this repressor and tryptophan binds tightly to the operator, whereas the repressor alone does not.* In other words, tryptophan itself is a *corepressor.* The target for the tryptophan-repressor complex is a DNA sequence with twofold symmetry (Figure 32-30). Again, symmetry plays an important role in the interaction of a protein with DNA. This operator site overlaps the promoter site for the initiation of transcription. Hence, *binding of the* trp *repressor to the operator prevents RNA polymerase from binding to the* trp *promoter, so that the* trp *genes are not transcribed.*

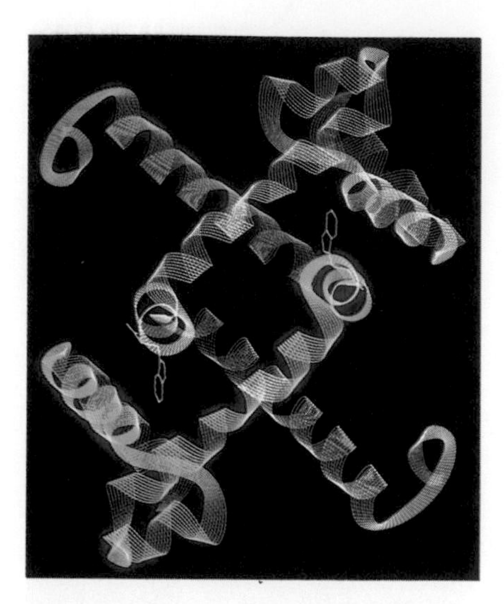

Figure 32-31
Drawing of the main chains of the *trp* repressor protein. The two subunits are extensively interlocked. [Courtesy of Dr. Jane Richardson. Based on atomic coordinates kindly provided by Dr. Paul Sigler.]

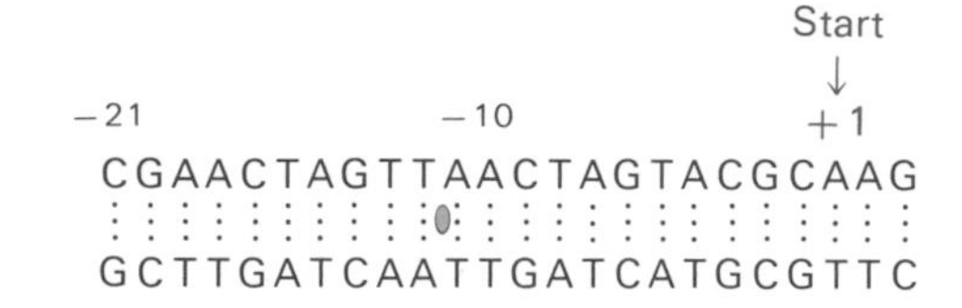

Figure 32-30
Base sequence of the *trp* operator. The twofold axis of symmetry is denoted by the green symbol. The base pair labeled +1 is the start of the transcribed part of the operon.

The *trp* repressor consists of two extensively interlocked subunits (Figure 32-31). The dimer contains three domains: a central core, formed by the amino-terminal half of both subunits, and two flexible DNA-reading heads, each formed from the carboxyl-terminal half of a subunit. Helices D and E form a helix-turn-helix unit akin to those of λ repressor, Cro, and CAP. The central core serves as a spacer between the reading heads, 26 Å apart in repressor without bound tryptophan. X-ray crystallographic studies of the repressor protein with and without bound tryptophan have revealed how it acts as a corepressor. *Tryptophan induces the formation of a DNA-binding surface by altering the conformation of the helix-turn-helix unit to enable it to bind to the major groove.* Moreover, the binding of tryptophan increases the distance between subunits by 8 Å, so that they fit readily into adjacent major grooves of DNA (see Figure 32-1).

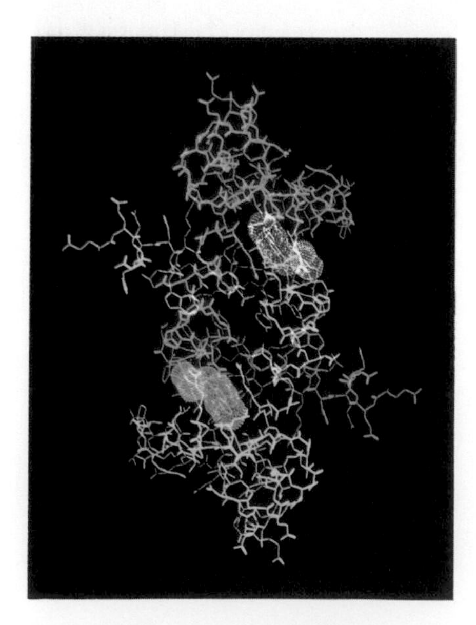

Figure 32-32
Model of *trp* repressor. One subunit is shown in blue, the other in yellow. The bound tryptophans are shown in green and red. [Courtesy of Dr. Paul Sigler.]

DISCOVERY OF ATTENUATION, A KEY MEANS OF CONTROLLING OPERONS FOR AMINO ACID BIOSYNTHESIS

A new means of controlling gene expression was discovered by Charles Yanofsky and his colleagues as a result of their studies of the tryptophan operon. For some time it was thought that end-product inhibition of the catalytic activity of the first enzyme complex in the tryptophan pathway (p. 586) and repressor-mediated inhibition of transcription accounted for most of the regulation of tryptophan biosynthesis. This view was abruptly altered by the unexpected finding that certain mutants with *deletions* between the operator and the gene for the first enzyme (*trpE*) in the operon showed *increased* production of *trp* mRNA. Furthermore, analysis of the 5′ end of *trp* mRNA revealed the presence of a *leader sequence* of 162 nucleotides before the initiation codon of *trpE*. It was then found that deletion mutants with enhanced *trp* mRNA levels mapped in this leader region, some 30 to 60 nucleotides before the start of *trpE*. The next striking observation was that nonmutants produced a transcript consisting of only the first 130 nucleotides of the leader when the tryptophan level was high, but produced a 7000-nucleotide *trp* mRNA including the entire leader sequence when tryptophan was scarce. Hence, Yanofsky concluded that transcription of the *trp* operon must be regulated by a *controlled termination site,* called an *attenuator,* that is located between the operator and the gene for the first enzyme in the pathway. This physiologically regulated termination site, like the stop signals at the ends of some operons (p. 711), contains a GC-rich sequence followed by an AT-rich one. Each of these regions in the attenuator exhibits a twofold axis of symmetry (Figure 32-33). Moreover, the terminated leader transcript ends with a series of U residues.

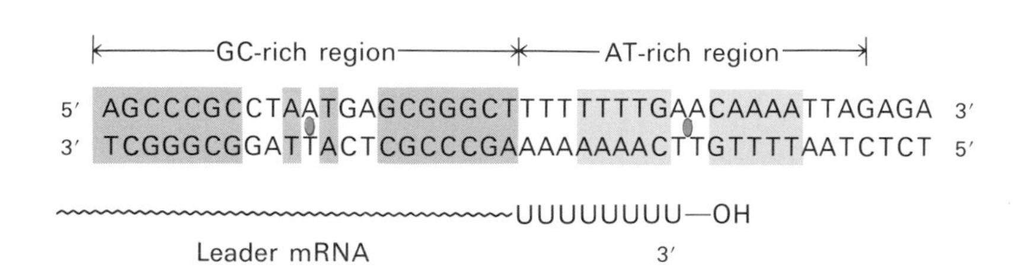

Figure 32-33
Base sequence of the *trp* attenuator site. Base pairs in the GC-rich region related by a twofold axis of symmetry are shown in blue, and those in the AT-rich region are shown in yellow.

ATTENUATION IS MEDIATED BY THE COUPLING OF TRANSCRIPTION AND TRANSLATION

The *attenuator site* complements the operator site in regulating transcription of *trp* genes. When tryptophan is plentiful, initiation of transcription is blocked by the binding of the tryptophan-repressor complex to the operator. As the level of tryptophan in the cell decreases, repression is lifted and transcription begins. However, some of the RNA polymerase molecules dissociate from the template at the attenuator site, whereas others continue to synthesize the entire *trp* message. *The proportion of RNA polymerase molecules that proceed past the attenuator site increases as tryptophan becomes scarcer.*

How does the attenuator site in the *trp* operon sense the level of tryptophan in the cell? An important clue was the finding that part of the leader mRNA is translated. The presence of tryptophan residues at positions 10 and 11 of the fourteen-residue leader polypeptide (Figure 32-34) is highly significant. When tryptophan is abundant, this complete polypeptide is synthesized. *However, when tryptophan is scarce, the ribosome stalls at the tandem UGG codons because of a paucity of tryptophanyl tRNA. The stalled ribosome somehow alters the structure of the mRNA so that the RNA polymerase transcribing it proceeds beyond the attenuator site.* A key aspect of this mechanism is that translation and transcription are closely coupled. The ribosome translating the *trp* leader mRNA follows closely behind the RNA polymerase molecule that is transcribing the DNA template. The stalled ribosome switches the secondary structure of the mRNA from a base-paired arrangement that favors termination of

- Lys - Ala - Ile - Phe - Val - Leu - Lys - Gly - Trp - Trp - Arg - Thr - Ser - Stop

~AUG AAA GCA AUU UUC GUA CUG AAA GGU UGG UGG CGC ACU UCC UGA~

Figure 32-34
Amino acid sequence of the *trp* leader peptide and the base sequence of the corresponding leader mRNA.

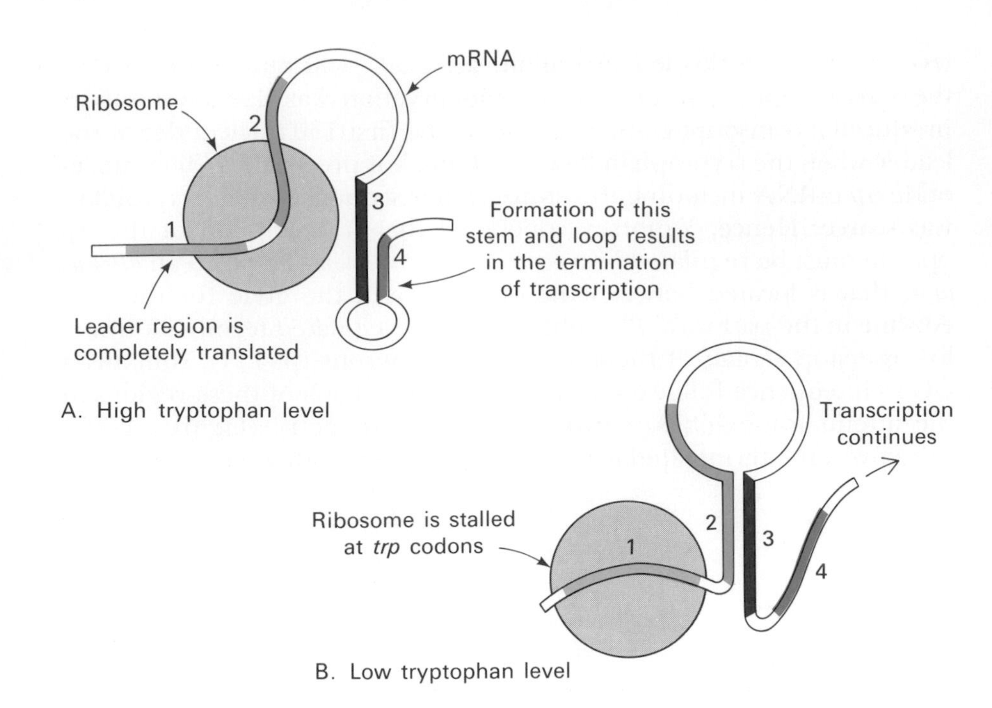

Figure 32-35
Model for attenuation in the *E. coli trp* operon. When tryptophan is abundant (A), the leader region (segment 1) of the *trp* mRNA is fully translated. Segment 2 enters the ribosome, which enables segments 3 and 4 to base-pair. This base-paired region somehow signals RNA polymerase to terminate transcription. In contrast, when tryptophan is scarce (B), the ribosome is stalled at the *trp* codons of segment 1. Segment 2 interacts with 3 instead of being drawn into the ribosome and so segments 3 and 4 cannot pair. Consequently, transcription continues. [After D. L. Oxender, G. Zurawski, and C. Yanofsky. *Proc. Nat. Acad. Sci.* 76(1979):5524.]

transcription to a very different one that allows RNA polymerase to read through the attenuator site (Figure 32-35). RNA molecules, like proteins, can adopt alternative conformations that are regulated with far-reaching physiological consequences.

THE ATTENUATOR SITE IN THE HISTIDINE OPERON CONTAINS SEVEN HISTIDINE CODONS IN A ROW

Several other operons for the biosynthesis of amino acids in *E. coli* are now known to have attenuator sites. The leader peptide of each contains an abundance of amino acid residues of the kind synthesized by the operon. For example, the threonine operon encodes enzymes that synthesize both threonine and isoleucine; the leader peptide contains eight threonines and four isoleucines in a sequence of twenty-one residues (Figure 32-36). Seven phenylalanines are present in the fifteen-residue leader of the phenylalanine operon. Even more striking, the leader peptide of the histidine operon contains *seven histidine residues in a row.* Clearly, these leader mRNAs are designed to sense the level of the amino acid synthesized by the encoded proteins. If the corresponding charged tRNA is scarce, translation of the leader is arrested. As in

A

Met- Lys - Arg - Ile - Ser - Thr - Thr - Ile - Thr - Thr - Thr - Ile - Thr - Ile - Thr - Thr -

5′ AUG AAA CGC AUU AGC ACC ACC AUU ACC ACC ACC AUC ACC AUU ACC ACA 3′

B

Met- Lys - His - Ile - Pro - Phe - Phe - Phe - Ala - Phe - Phe - Phe - Thr - Phe - Pro - Stop

5′ AUG AAA CAC AUA CCG UUU UUC UUC GCA UUC UUU UUU ACC UCC CCC UGA 3′

C

Met- Thr - Arg - Val - Gln - Phe - Lys - His - His - His - His - His - His - His - Pro - Asp -

5′ AUG ACA CGC GUU CAA UUU AAA CAC CAC CAU CAU CAC CAU CAU CCU GAC 3′

Figure 32-36
Amino acid sequence of the leader peptide and base sequence of the corresponding portion of mRNA from the (A) threonine operon, (B) phenylalanine operon, and (C) histidine operon.

the *trp* operon, a stalled ribosome facilitates switching to an mRNA conformation that allows RNA polymerase to read through the attenuator site. The presence of seven consecutive codons for histidine in the histidine operon leader mRNA markedly heightens the sensitivity of this detection system. Indeed, the histidine operon is controlled solely by attenuation.

FREE RIBOSOMAL PROTEINS REPRESS THE TRANSLATION OF mRNA ENCODING THEM

How is the synthesis of more than fifty kinds of ribosomal proteins coordinated? Their genes are located in more than twenty operons, yet their synthesis is precisely balanced. Tight control is essential, for ribosomal components make up some 40% of the dry weight of a bacterial cell. Control of ribosomal protein synthesis is exerted primarily at the level of *translation* rather than of transcription, in contrast with that of most other proteins. At least one protein encoded by each ribosomal protein operon serves as a *translational repressor* (Figure 32-37). It binds to mRNA near the initiation site for its own synthesis and blocks the synthesis of several proteins encoded by that polygenic message. This binding to mRNA does not interfere with the assembly of ribosomes because ribosomal proteins bind more tightly to ribosomal RNA than to mRNA. Translational repression occurs only when the production of ribosomal proteins runs ahead of that of ribosomal RNAs or when these proteins are not formed in equimolar amounts.

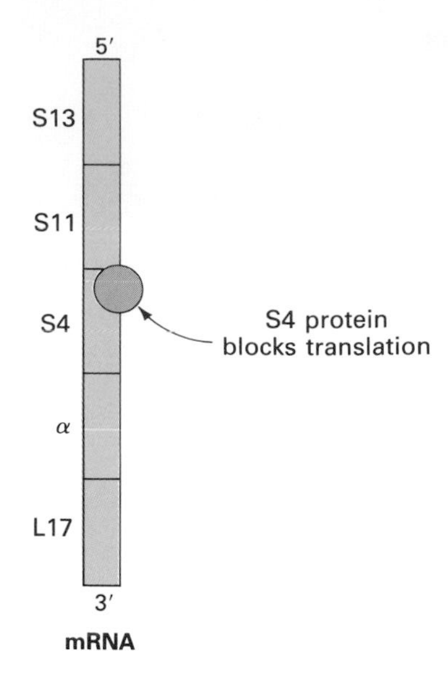

Figure 32-37
Feedback control by translational repression. A ribosomal protein (S4) represses the translation of an mRNA encoding components of the small (S) and large (L) ribosomal subunits and the α subunit of RNA polymerase.

The synthesis of ribosomal RNAs and transfer RNAs is also coordinated with the synthesis of other proteins. When amino acids are scarce, the synthesis of these RNAs stops. This adaptation to adversity, called the *stringent response*, is mediated by an unusual nucleotide, originally called *magic spot* because of its mysterious emergence in chromatograms of certain cell extracts. The new compound formed under conditions of amino acid deprivation turned out to be *guanosine tetraphosphate (ppGpp)*. It is formed by the transfer of a diphosphate unit from ATP to the 3′-OH of GDP. This reaction requires *uncharged* tRNA located in the A site of a ribosome and a protein called *stringent factor* that binds to the ribosome. ppGpp is formed only in cells starved of amino acids because the A site is occupied by an uncharged tRNA only if its aminoacyl counterpart is absent. Once formed, ppGpp inhibits the initiation of transcription of operons encoding ribosomal RNAs and slows the elongation of many other transcripts. Its precise mechanism of action is an intriguing problem.

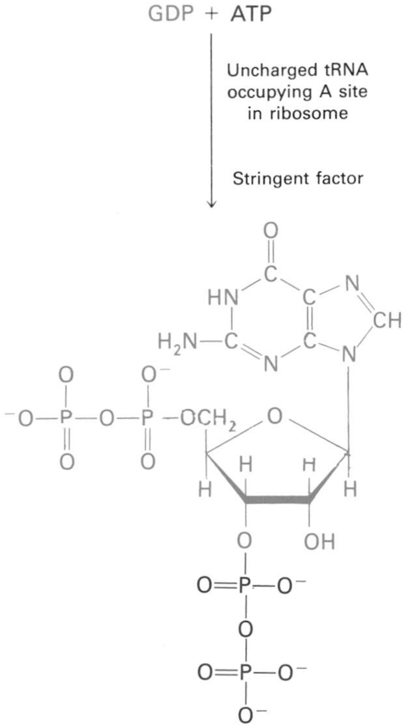

GDP + ATP

Uncharged tRNA occupying A site in ribosome

Stringent factor

ppGpp (Guanosine 5′-diphosphate 3′-diphosphate)

Figure 32-38
ppGpp is synthesized from ATP and GDP during amino acid starvation. This signal molecule blocks the formation of ribosomal and transfer RNAs in the stringent response.

DNA INVERSIONS LEAD TO ALTERNATING EXPRESSION OF A PAIR OF FLAGELLAR GENES

Salmonella, a bacterium living in the intestine of mammals, swims by rotating flagella that emerge from its surface (p. 1006). Flagella are thin helical protein filaments built from many 53-kd flagellin subunits. *Salmonella* contains two genes for flagellins and expresses only one of them at a given time. The type expressed changes spontaneously an average of once in a thousand divisions. This switching between flagellins H1 and H2, called *phase variation*, helps the bacterium evade the immune response of its host. To be effective, the change must be complete. In phase 2, for example, the presence in the flagellum of even a few H1 molecules amidst many H2 would render this motile appendage vulnerable to inactivation by antibodies directed against H1, the type produced in phase 1. The switch must be both *absolute* and *reversible*.

These two objectives are met by an ingeniously simple means of switching back and forth between two genes (Figure 32-39). The mRNA for H2 is also the template for a repressor that silences the H1 gene. This finding explains phase 2, but how then can H1 be expressed to the exclusion of H2? In phase 1, the promoter for H2 and the repressor is ineffective because the DNA segment containing it has been inverted. In this phase, the promoter is in the wrong place and points in the wrong direction for expression of the genes for H2 and the repressor. The H1 gene containing its own promoter is now expressed because the repressor is absent.

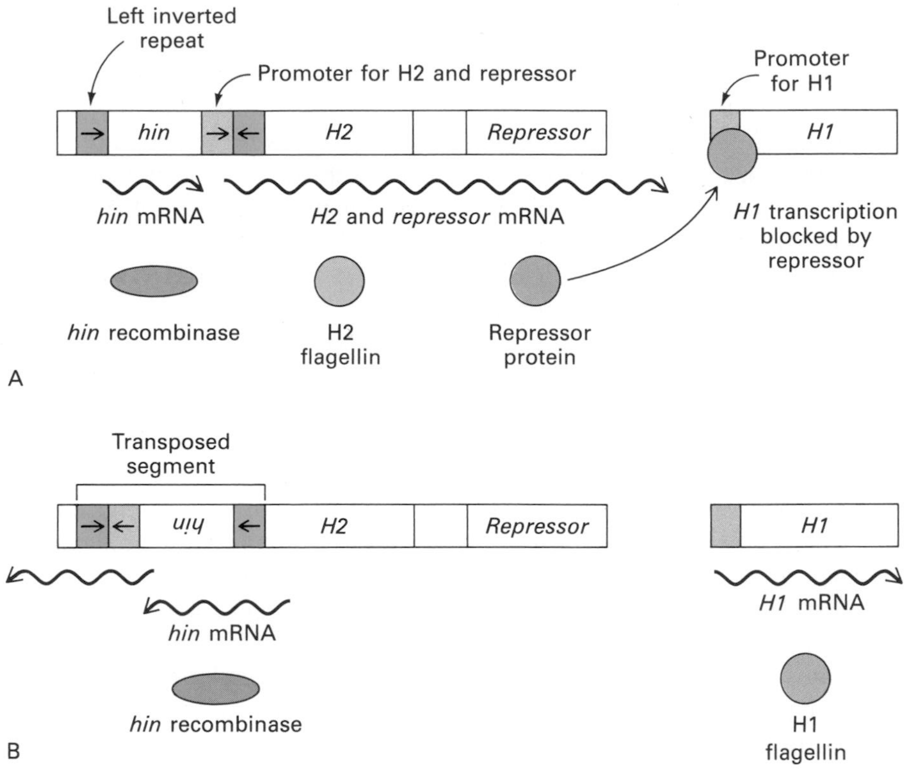

Figure 32-39
Phase variation in *Salmonella*. Flagellins H1 and H2 are expressed in a mutually exclusive manner. (A) In phase 2, the H1 gene is silenced by a repressor protein formed along with the H2 protein. (B) In phase 1, inversion of a DNA segment catalyzed by a recombinase (Hin) encoded by it leads to the loss of the promoter for H2 and the repressor. H1 is then expressed. Further inversions switch the system back and forth between these phases.

What controls the inversion of the DNA segment bearing the promoter for H2 and the repressor? The ends of this 970-bp segment are a pair of 14-bp inverted terminal repeats; the middle of it encodes a recombinase, called Hin, that catalyzes recombination between the repeats. Recall that recombination between oppositely oriented sites leads to the inversion of the DNA segment located between them (p. 699). Thus, four genes—those for H1, H2, the repressor, and Hin—constitute a *flip-flop circuit that changes an exposed bacterial protein at a frequency determined by the activity of the recombinase*. Eucaryotes, too, use gene rearrangements to regulate their expression, as will be discussed in the next chapter.

SUMMARY

Procaryotes regulate the amounts of proteins synthesized in a variety of ways. The expression of most genes is regulated primarily at the level of transcription rather than translation. Many genes are clustered into operons, which are units of coordinated genetic expression. An operon consists of control sites (an operator and a promoter) and a set of structural genes. In addition, regulator genes encode proteins that interact with the operator and promoter sites to stimulate or inhibit transcription. In the absence of a galactoside inducer, *lac* repressor protein binds to the operator and blocks transcription of the genes encoding β-galactosidase and two other proteins of the lactose operon. The twofold symmetry of the repressor protein matches the nearly palindromic base sequence of the *lac* operator. Symmetry matching is an important general feature of protein-DNA interactions. Binding of an inducer to the *lac* repressor displaces it from the operator. RNA polymerase can then move through the operator to transcribe the *lac* operon.

Cyclic AMP, a hunger signal, stimulates the transcription of many catabolic operons by binding to the catabolite gene activator protein (CAP). The binding of cAMP-CAP to a specific site in the promoter region of these inducible catabolic operons enhances the binding of RNA polymerase and the initiation of transcription. Full expression of the *lac* operon requires both a galactoside inducer and cAMP, which is formed when glucose is scarce. Different forms of a control protein activate and inhibit transcription of the arabinose operon, another member of this global regulatory circuit controlled by cAMP.

Repressors and activators of transcription determine whether λ phage follows the lysogenic or lytic pathway. The λ repressor prevents the expression of all viral genes except for the one (*cI*) encoding the repressor itself. The expression of the repressor is autoregulated; high concentrations block the transcription of *cI*. Damage to host DNA leads to the stimulation of the protease activity of recA protein, which degrades λ repressor as well as the *lexA* repressor (the controller of the SOS genes for DNA repair). The consequent expression of cro protein silences further formation of λ repressor and commits the virus to the lytic pathway. The switch between the lysogenic and lytic pathways is mediated by the interplay of λ repressor and cro protein with six binding sites in the O_R and O_L operator sites of the viral DNA.

The λ repressor, cro protein, CAP, *trp* repressor, and many other DNA-binding proteins are dimers of a common helix-turn-helix motif. One of the α helices of this recurring unit detects specific sequences in DNA by binding to the major groove. The angular relation and distance between the two DNA-reading heads of these dimeric control proteins enables them to fit into adjoining major grooves of B-DNA. The binding of inducer or corepressor changes the form of these DNA-binding units and their spatial relationship.

The tryptophan operon and several others for the biosynthesis of amino acids are controlled by attenuation, a mechanism based on the coupling of transcription and translation. Transcription terminates at an attenuator site preceding the first structural gene if the amino acid end product is abundant. Attenuation is mediated by translation of a leader mRNA. A ribosome stalled by the absence of an aminoacyl-tRNA alters the structure of mRNA so that RNA polymerase transcribes the operon beyond the attenuator site.

Purely translational control is evident in the regulation of ribosomal protein synthesis. Free ribosomal proteins repress the translation of mRNAs encoding them by binding to initiation sites. The synthesis of ribosomal RNAs is inhibited by ppGpp, a signal molecule formed when amino acids are scarce.

A specialized mechanism of controlling gene expression is used by *Salmonella* to alternately express H1 or H2 flagellin in a mutually exclusive manner. This reversible switching is mediated by the inversion of a DNA segment containing the promoter for *H2* and the gene for a recombinase catalyzing this inversion.

SELECTED READINGS

Where to start

Ptashne, M., Johnson, A. D., and Pabo, C. O., 1982. A genetic switch in a bacterial virus. *Sci. Amer.* 247(5):128–140.

Maniatis, T., and Ptashne, M., 1976. A DNA operator-repressor system. *Sci. Amer.* 234(1):64–76.

Yanofsky, C., 1981. Attenuation in the control of expression of bacterial operons. *Nature* 289:751–758.

Pabo, C. O., and Sauer, R. T., 1984. Protein-DNA recognition. *Ann. Rev. Biochem.* 53:293–321. [X-ray crystallographic studies of repressor-operator interactions are lucidly reviewed in this article.]

Books

Beckwith, J., Davies, J., and Gallant, J., (eds.), 1983. *Gene Function in Prokaryotes.* Cold Spring Harbor Laboratory. [Contains excellent reviews of the control of gene expression in procaryotes. The chapters on attenuation, global control systems, translational repression, and recombinational regulation are especially pertinent.]

Ptashne, M., 1986. *A Genetic Switch. Gene Control and Phage λ.* Cell Press and Blackwell Scientific Publications. [A superb account of how λ phage switches between two developmental pathways.]

Schleif, R., 1986. *Genetics and Molecular Biology.* Addison-Wesley. [Chapters 12, 13, and 14 provide an excellent account of the experimental basis of major concepts of gene regulation in procaryotes.]

Jurnak, F. A., and McPherson, A., (eds.), 1985. *Biological Macromolecules and Assemblies. Volume 2: Nucleic Acids and Interactive Proteins.* Wiley-Interscience. [Contains articles on DNA-binding proteins such as λ repressor and Cro protein.]

Miller, J. H., and Reznikoff, W. S., (eds.), 1978. *The Operon.* Cold Spring Harbor Laboratory. [An excellent collection of articles about regulation in *E. coli* and λ phage. The lactose, tryptophan, arabinose, histidine, and galactose operons are discussed in detail.]

General reviews

Gottesman, S., 1984. Bacterial regulation: global regulatory networks. *Ann. Rev. Genet.* 18:415–442.

Takeda, Y., Ohlendorf, D. H., Anderson, W. F., and Matthews, B. W., 1983. DNA-binding proteins. *Science* 221:1020–1026.

Ptashne, M., 1986. Gene regulation by proteins acting nearby and at a distance. *Nature* 322:697–701.

Lactose operon

Gilbert, W., and Müller-Hill, B., 1966. Isolation of the *lac* repressor. *Proc. Nat. Acad. Sci.* 56:1891–1898.

Dickson, R., Abelson, J., Barnes, W., and Reznikoff, W., 1975. Genetic regulation: the *lac* control region. *Science* 187:27–35.

McKay, D. B., Pickover, C. A., and Steitz, T. A., 1982. *Escherichia coli lac* repressor is elongated with its operator DNA binding domains located at both ends. *J. Mol. Biol.* 156:175–183.

Cyclic AMP and catabolite repression

de Crombrugghe, B., Busby, S., and Buc, H., 1984. Cyclic AMP receptor protein: role in transcription activation. *Science* 224:831–838.

Pastan, I., and Adhya, S., 1976. Cyclic adenosine 3′,5′-monophosphate in *Escherichia coli. Bacteriol. Rev.* 40:527–551.

McKay, D. B., Weber, I. T., and Steitz, T. A., 1982. Structure of catabolite gene activator protein at 2.9 Å resolution. *J. Biol. Chem.* 257:9518–9524.

Weber, I. T., Takio, K., Titani, K., and Steitz, T. A., 1982. The cAMP-binding domains of the regulatory subunit of cAMP-dependent protein kinase and the catabolite gene activator protein are homologous. *Proc. Nat. Acad. Sci.* 79:7679–7683.

Arabinose operon

Dunn, T. M., Hahn, S., Ogden, S., and Schleif, R. F., 1984. An operator at −280 base pairs that is required for repression of araBAD operon promoter: addition of DNA helical turns between the operator and promoter cyclically hinders repression. *Proc. Nat. Acad. Sci.* 81:5017–5020.

Wilcox, G., Meuris, P., Bass, P., and Englesberg, E., 1974. Regulation of the arabinose operon in vitro. *J. Biol. Chem.* 249:2946–2952.

Hirsh, J., and Schleif, R., 1976. Electron microscopy of gene regulation: the L-arabinose operon. *Proc. Nat. Acad. Sci.* 73:1518–1522.

Control of transcription in λ phage

Johnson, A. D., Poteete, A. R., Lauer, G., Sauer, R. T., Ackers, G. K., and Ptashne, M., 1981. Lambda repressor and cro—components of an efficient molecular switch. *Nature* 294:217–223.

Griffith, J., Hochschild, A., and Ptashne, M., 1986. DNA loops induced by cooperative binding of λ repressor. *Nature* 322:750–752.

Reichardt, L., and Kaiser, A. D., 1971. Control of λ repressor synthesis. *Proc. Nat. Acad. Sci.* 68:2185–2189.

Structure of λ repressor and cro protein

Ohlendorf, D. H., Anderson, W. F., Lewis, M., Pabo, C. O., and Matthews, B. W., 1983. Comparison of the structures of cro and λ repressor proteins from bacteriophage λ. *J. Mol. Biol.* 169:757–769.

Anderson, J. E., Ptashne, M., and Harrison, S. C., 1985. A phage repressor-operator complex at 7 Å resolution. *Nature* 316:596–601.

Wharton, R. P., and Ptashne, M., 1985. Changing the binding specificity of a repressor by redesigning an alpha-helix. *Nature* 316:601–605.

Wharton, J. E., Ptashne, M., and Harrison, S. C., 1987. Structure of the repressor-operator complex of bacteriophage 434. *Nature* 326:846–852.

Tryptophan operon

Schevitz, R. W., Otwinowski, Z., Joachimiak, A., Lawson, C. L., and Sigler, P. B., 1985. The three-dimensional structure of *trp* repressor. *Nature* 317:782–786.

Zhang, R.-g., Joachimiak, A., Lawson, C. L., Schevitz, R. W., Otwinowski, Z., and Sigler, P. B., 1987. The crystal structure of *trp* aporepressor at 1.8 Å shows how binding tryptophan enhances DNA affinity. *Nature* 327:591–597.

Platt, T., 1978. Regulation of gene expression in the tryptophan operon of *Escherichia coli. In* Miller, J. H., and Reznikoff, W. S., (eds.), *The Operon,* pp. 213–302. Cold Spring Harbor Laboratory. [A lucid introduction to the *trp* operon.]

Oxender, D. L., Zurawski, G., and Yanofsky, C., 1979. Attenuation in the *Escherichia coli* tryptophan operon: the role of RNA secondary structure involving the Trp codon region. *Proc. Nat. Acad. Sci.* 76:5524–5528.

Landick, R., Carey, J., and Yanofsky, C., 1985. Translation activates the paused transcription complex and restores transcription of the trp operon leader region. *Proc. Nat. Acad. Sci.* 82:4663–4667.

Control of ribosomal protein and RNA synthesis

Nomura, M., Gourse, R., and Baughman, G., 1984. Regulation of the synthesis of ribosomes and ribosomal components. *Ann. Rev. Biochem.* 53:75–117.

Stephens, J. C., Artz, S. W., and Ames, B. N., 1975. Guanosine 5′-diphosphate 3′-diphosphate (ppGpp): positive effector for histidine operon transcription and general signal for amino acid deficiency. *Proc. Nat. Acad. Sci.* 72:4389–4393.

DNA inversion and phase variation

Zeig, J., Silverman, M., Hilmen, M., and Simon, M., 1978. Recombinational switch for gene expression. *Science* 196:170–172.

Discovery of operons and control elements

Ptashne, M., and Gilbert, W., 1970. Genetic repressors. *Sci. Amer.* 222(6):36–44. [Available as *Sci. Amer.* Offprint 1179.]

Jacob, F., and Monod, J., 1961. Genetic regulatory mechanisms in the synthesis of proteins. *J. Mol. Biol.* 3:318–356. [The operon model and the concept of messenger RNA were proposed in this superb paper.]

Lwoff, A., and Ullmann, A., (eds.), 1979. *Origins of Molecular Biology. A Tribute to Jacques Monod.* Academic Press. [A wonderful collection of essays about the discovery of the operon concept and the scientists involved in this research. A book to be savored.]

PROBLEMS

1. What is likely to be the effect of each of these mutations?
   (a) Deletion of the *lac* regulator gene.
   (b) Deletion of the *trp* regulator gene.
   (c) Deletion of the *ara* regulator gene.
   (d) Deletion of the *cI* gene of λ.
   (e) Deletion of the *N* gene of λ.

2. Superrepressed mutants (i^s) of the *lac* operon behave as noninducible mutants. The i^s gene is dominant over i^+ in partial diploids. What is the molecular nature of this mutation likely to be?

3. A mutant of *E. coli* synthesizes large amounts of β-galactosidase whether or not inducer is present. A partial diploid formed from this mutant and $Fi^+o^+z^-$ also synthesizes large amounts of β-galactosidase whether or not inducer is present. What kind of mutation might give these results?

4. A mutant unable to grow on galactose, lactose, arabinose, and several other carbon sources is isolated from a wild-type culture. The cyclic AMP level in this mutant is normal. What kind of mutation might give these results?

5. An *E. coli* cell bearing a λ prophage is immune to lytic infection by λ. Why?

6. Transcription of *cI* can be initiated at p_{RE}, the promoter for repressor establishment, or at p_{RM}, the promoter for repressor maintenance. The p_{RM} transcript begins at its 5′ end with an AUG codon for the λ repressor, whereas in the p_{RE} transcript this initiation codon is preceded by a sequence that is complementary to the 3′ end of 16S rRNA. Initiation at the p_{RE} site requires phage-encoded proteins that are not expressed in the lysogenic state.
   (a) Which transcript will be translated more efficiently?
   (b) What is the likely physiological significance of this difference?

7. Predict the consequence of a mutation that increases by a factor of 100 the binding affinity of *lac* repressor for *lac* operator without changing the binding affinity for nonspecific sites on DNA.

8. The binding of most repressors to operator sites is much faster at low ionic strength than at high. In contrast, the affinity of the operator for the repressor is not markedly affected by ionic strength. Why?

9. Crystals of normal *trp* aporepressor shatter on addition of tryptophan. However, crystals of a mutant aporepressor are unaffected by the addition of tryptophan. The unit cell dimensions of this crystal are virtually identical to those of a crystal of normal repressor containing bound tryptophan. Predict the physiologic activity of this mutant repressor protein.

10. A novel mechanism for controlling the amount of termination factor RF2 has recently been discovered. A truncated polypeptide ending at residue 25 or a complete protein ending at residue 340 is produced from the mRNA for RF2. The amino acid sequence in the vicinity of residue 25 and the corresponding base sequence are:

    -Gly-Tyr-Leu-Asp-Tyr-Asp-
     23 24 25 26 27 28

    5′-GGGUAUCUUUGACUACGAC-3′

    Propose a regulatory mechanism based on these data.

CHAPTER 33

Eucaryotic Chromosomes and Gene Expression

Eucaryotic chromosomes are larger and have a higher degree of structural organization than those of procaryotes. The genomes of yeast, fruit flies, and humans contain about 4, 40, and 1000 times as much DNA as the genome of *E. coli*. This abundance endows eucaryotes with potentialities that are absent from procaryotes and poses additional challenges in replication and expression. This chapter focuses on three aspects of eucaryotic chromosomes and gene expression: (1) How is eucaryotic DNA packaged and replicated? Basic proteins called *histones* play a key role in the compaction of DNA, which is wound around histones to form *nucleosomes*, like beads on a string. The entire DNA-protein complex is called *chromatin*. (2) How are genes arranged in a eucaryotic chromosome? A striking feature of the genomes of higher eucaryotes is the *abundance of repeated sequences and the small proportion of protein-coding sequences*. (3) How is gene expression controlled in eucaryotes? As in procaryotes, control is exerted primarily at the level of transcription. Transcriptionally active regions of chromatin are hypersensitive to digestion by DNase I, and, in general, fewer of their cytosines are methylated. The expression of particular genes in these accessible regions of chromosomes depends on transcriptional factors. Two examples having general significance are presented here. The transcription of genes for 5S RNA depends on the binding of a protein containing multiple *metal-binding fingers* that fit into the grooves of DNA. Many genes controlling development in insects and vertebrates contain a recurring motif called the *homeo box* that encodes a DNA-binding domain.

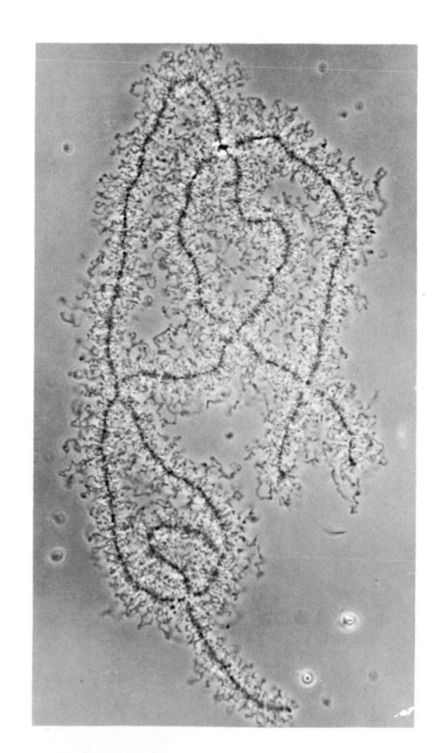

Figure 33-1
Phase-contrast light micrograph of a lampbrush chromosome from an oocyte. [Courtesy of Dr. Joseph Gall.]

A EUCARYOTIC CHROMOSOME CONTAINS A SINGLE LINEAR MOLECULE OF DOUBLE-HELICAL DNA

How is DNA packed into chromosomes? This important question was difficult to answer for many years because very large DNA molecules are exquisitely sensitive to degradation by shearing forces. Bruno Zimm circumvented this problem by using the *viscoelastic technique,* which provides a measure of the size of the *largest DNA molecules in a mixture.* DNA molecules are stretched out by flow and then allowed to recoil to their normal form. The half-time for recoiling depends on the mass of the molecule. Cells from fruit flies were lysed in the measuring chamber to avoid the breakage of DNA that would be caused by transferring samples. Nucleases were inactivated by incubating the sample in detergent at 65°C. In addition, pronase was added to digest proteins bound to DNA.

The observed mass of the largest DNA molecules in this mixture was 41×10^6 kd, which agreed closely with the known value of the DNA content of the largest chromosome in *Drosophila melanogaster,* which is 43×10^6 kd. Excellent agreement also was obtained for a translocation mutant of the largest chromosome, which has an extra piece of DNA, giving it a calculated DNA content of 59×10^6 kd. The measured viscoelastic mass of this DNA molecule was 58×10^6 kd. Radioautographs of *D. melanogaster* DNA (Figure 33-2) confirm the existence of very long DNA molecules. These studies reveal that *a* Drosophila *chromosome contains a single uninterrupted molecule of DNA.* Furthermore, this DNA molecule is *linear and unbranched.* A mass of 43×10^6 kd corresponds to 76×10^6 base pairs (76 Mb). Thus, the largest *Drosophila* chromosome is about twenty times as large as the *E. coli* chromosome (Table 33-1).

Megabase (Mb)—
A length of DNA consisting of 10^6 base pairs (if double-stranded) or 10^6 bases (if single-stranded).

1 Mb = 10^3 kb = 10^6 bases

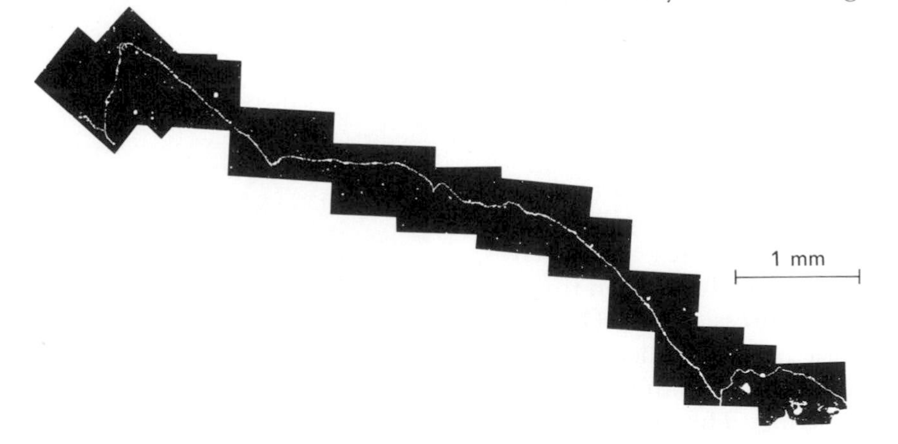

Figure 33-2
Radioautograph of a DNA molecule from *Drosophila melanogaster.* The contour length of this DNA is 1.2 cm. [From R. Kavenoff, L. C. Klotz, and B. H. Zimm. *Cold Spring Harbor Symp. Quant. Biol.* 38(1974):4.]

Table 33-1
DNA contents of several procaryotic and eucaryotic genomes

Organism	*Number of base pairs*	*DNA length (mm)*	*Number of chromosomes*
E. coli	4×10^6	1.4	1
Yeast (*S. cerevisiae)*	1.4×10^7	4.6	16
Fruit fly (*D. melanogaster)*	1.7×10^8	56	4
Human	3.9×10^9	990	23

Note: The values given are for haploid genomes.

Baker's yeast (*Saccharomyces cerevisiae*) contains sixteen chromosomes ranging in size from 0.2 to 2.2 Mb. These DNA molecules can be separated by *pulsed field electrophoresis* (Figure 33-3). Electric fields oriented 120 degrees apart in the plane of the agarose gel are applied in an alternating sequence. This new technique makes it possible to analyze and separate much larger DNA molecules than by using a constant electric field in a single direction. Megabase DNA molecules, such as entire yeast chromosomes, can now be resolved.

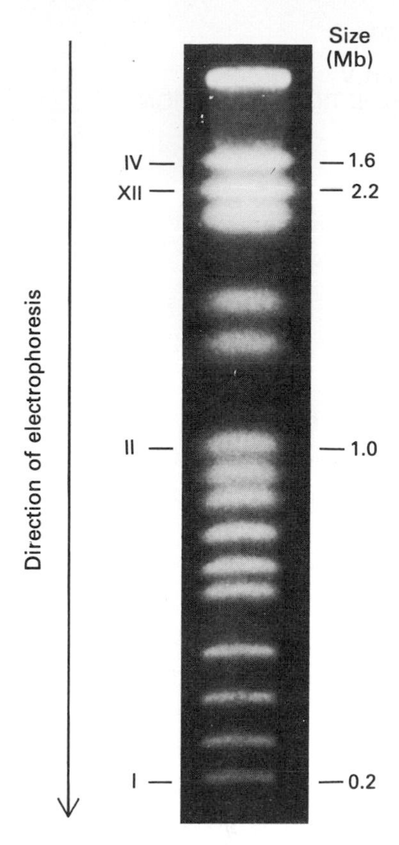

Figure 33-3
Separation of yeast chromosomes by pulsed field electrophoresis, which resolves megabase DNA molecules. The locations of chromosomes I, II, IV, and XII were determined by Southern blotting. [From G. Chu, D. Vollrath, and R. W. Davis. *Science* 234(1986):1583.]

EUCARYOTIC DNA IS TIGHTLY BOUND TO BASIC PROTEINS CALLED HISTONES

The DNA in eucaryotic chromosomes is not bare. Rather, eucaryotic DNA is tightly bound to a group of small basic proteins called *histones*. In fact, histones constitute about half of the mass of eucaryotic chromosomes, the other half being DNA. This nucleoprotein chromosomal material is termed *chromatin*. Histones can be dissociated from DNA by treating chromatin with salt or dilute acid. The resulting mixture can then be fractionated by ion-exchange chromatography. The five types of histones are called H1, H2A, H2B, H3, and H4. They range in mass from about 11 kd to 21 kd (Table 33-2). A striking feature of histones is their *high content of positively charged side chains:* about one in four residues is either lysine or arginine.

Table 33-2
Types of histones

Type	*Lys/Arg ratio*	*Number of residues*	*Mass (kd)*	*Location*
H1	20.0	215	21.0	Linker
H2A	1.25	129	14.5	Core
H2B	2.5	125	13.8	Core
H3	0.72	135	15.3	Core
H4	0.79	102	11.3	Core

Each type of histone can exist in a variety of forms because of *posttranslational modifications* of certain side chains. For example, lysine 16 in H4 is often acetylated. Histones can also be methylated, ADP-ribosylated, and phosphorylated. The modulation of the charge, hydrogen-bonding capabilities, and shape of histones by these reversible covalent modifications may be important in packaging DNA and in regulating its availability for replication and transcription.

Ser-Gly-Arg-Gly-Lys-Gly-Gly-Lys-Gly-Leu- 10
Gly-Lys-Gly-Gly-Ala-Lys-Arg-His-Arg-Lys- 20
Val-Leu-Arg-Asp-Asn-Ile-Gln-Gly-Ile-Thr- 30
Lys-Pro-Ala-Ile-Arg-Arg-Leu-Ala-Arg-Arg- 40
Gly-Gly-Val-Lys-Arg-Ile-Ser-Gly-Leu-Ile- 50
Tyr-Glu-Glu-Thr-Arg-Gly-Val-Leu-Lys-Val- 60
Phe-Leu-Glu-Asn-Val-Ile-Arg-Asp-Ala-Val- 70
Thr-Tyr-Thr-Glu-His-Ala-Lys-Arg-Lys-Thr- 80
Val-Thr-Ala-Met-Asp-Val-Val-Tyr-Ala-Leu- 90
Lys-Arg-Gln-Gly-Arg-Thr-Leu-Tyr-Gly-Phe- 100
Gly-Gly 102

Figure 33-4
Amino acid sequence of histone H4 from calf thymus. Several residues are modified. The α-amino and the ϵ-amino group of Lys 16 are acetylated. The ϵ-amino group of Lys 20 is methylated or dimethylated. Histone H4 from pea seedlings has the same amino acid sequence except that residue 60 is isoleucine and residue 77 is arginine.

THE AMINO ACID SEQUENCES OF H3 AND H4 ARE NEARLY THE SAME IN ALL PLANTS AND ANIMALS

Emil Smith and Robert DeLange showed that the amino acid sequences of H4 from pea seedlings and calf thymus differ at only two sites out of 102 residues. The changes at these two sites are quite small: valine in place of isoleucine, and lysine in place of arginine. Thus, *the amino acid sequence of H4 has remained nearly constant in the 1.2 × 10⁹ years since the divergence of plants and animals.* Likewise, H3 has changed little throughout this very long evolutionary period. The amino acid sequences of H3 from pea seedlings and calf thymus differ at just four positions.

It is interesting to compare the rate of change of these histones with those of other proteins in the course of evolution. A useful index is the *unit evolutionary period*, the time in which the sequence has changed by 1% after the divergence of two evolutionary lines. The times for H3 and H4 of 300 million years and 600 million years are much longer than those of nearly all other proteins studied thus far. For example, the unit evolutionary period is 20 million years for cytochrome *c*, 6 million years for hemoglobin, and 1 million years for fibrinopeptides. The remarkably conserved structure of H3 and H4 strongly suggests that they play a critical role that was established early in the evolution of eucaryotes and has remained nearly invariant since then.

NUCLEOSOMES ARE THE REPEATING UNITS OF CHROMATIN

How do histones combine with DNA to form the chromatin fiber? In 1974, Roger Kornberg proposed on the basis of several lines of evidence that *chromatin is made up of repeating units, each consisting of 200 base pairs of DNA and of two each of H2A, H2B, H3, and H4*. These repeating units are now known as *nucleosomes*. Most of the DNA is wound around the outside of a core of histones. The remainder of the DNA, called the linker, joins adjacent nucleosomes and contributes to the flexibility of the chromatin fiber. Thus, *a chromatin fiber is a flexibly jointed chain of nucleosomes*, rather like beads on a string.

This model for the structure of chromatin is supported by a wide range of experimental findings:

1. *Electron microscopy*. Linear arrays of 100-Å-diameter spheres that are connected by a thin strand are seen in electron micrographs of chromatin (Figure 33-5). The degree of extension of the chromatin fiber depends on the procedure used to prepare the specimen for electron microscopy. Other preparative methods yield electron micrographs in which the 100-Å spheres are more compactly arranged. Thus, electron microscopy directly supports the notion that chromatin is a chain of roughly spherical particles that are separated by flexible regions.

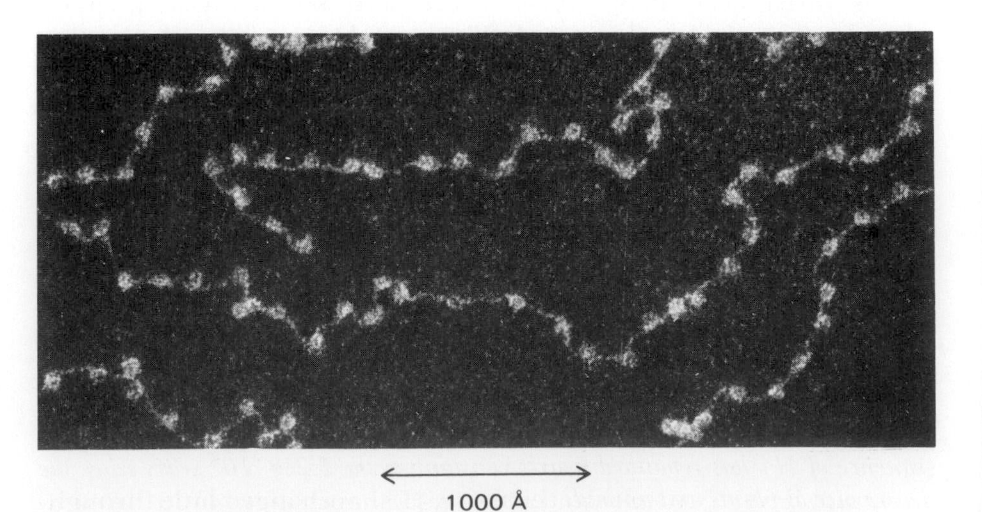

Figure 33-5
Electron micrograph of chromatin. The beadlike particles have diameters of nearly 100 Å. [Courtesy of Dr. Ada Olins and Dr. Donald Olins.]

2. *X-ray and neutron diffraction.* A 100-Å repetition is also seen in x-ray diffraction patterns of chromatin fibers. Neutron diffraction studies indicate that the DNA is located near the outside of the nucleosome.

3. *Nuclease digestion.* Free DNA in solution can be cleaved at any of its phosphodiester bonds by pancreatic deoxyribonuclease I (DNase I) or micrococcal nuclease. In contrast, DNA in chromatin is protected from digestion except at a few sites. The digestion pattern of chromatin is striking in its simplicity: it consists of a ladder of discrete bands (Figure 33-6). The DNA contents of these fragments are multiples of a basic

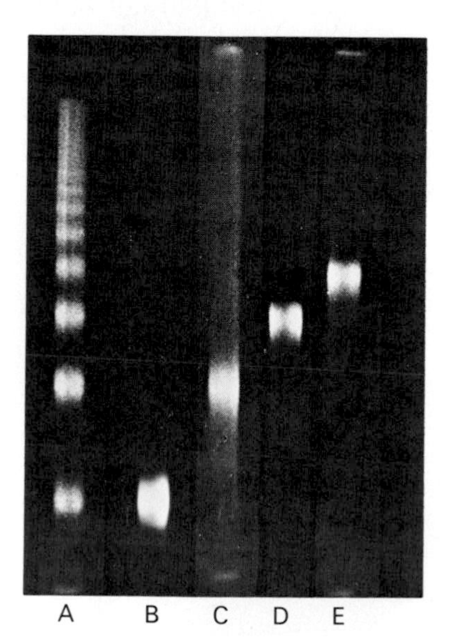

Figure 33-6
Gel-electrophoresis patterns of defined lengths of chromatin produced by limited micrococcal nuclease digestion. Part A shows an unfractionated digest. Fractionation by centrifugation on a sucrose gradient yielded a (B) monomer, (C) dimer, (D) trimer, and (E) tetramer. [From J. T. Finch, M. Noll, and R. D. Kornberg. *Proc. Nat. Acad. Sci.* 72(1975):3321.]

unit of about 200 base pairs. Electron micrographs show that the number of spherical particles in a fragment of chromatin is equal to the number of 200 base-pair units (Figure 33-7). For example, a fragment with 600 base pairs of DNA consists of three 100-Å-diameter particles. Thus, a bead seen in an electron micrograph corresponds to a nucleosome defined by nuclease digestion.

4. *Reconstitution.* A chromatin-like fiber can be formed in vitro by adding histones to DNA from adenovirus or simian virus 40 (SV40). The amount of DNA associated with a nucleosome in these reconstituted systems is close to 200 base pairs. Furthermore, equimolar amounts of H2A, H2B, H3, and H4 are required to form nucleosomes. Characteristic beads are not formed if any of the four histones is absent from the reconstitution mixture. X-ray diffraction studies also showed that H2A, H2B, H3, and H4 must be added to DNA to regain the diffraction pattern of chromatin. In contrast, H1 is not required, which fits the finding that H1 is not an integral part of the nucleosome.

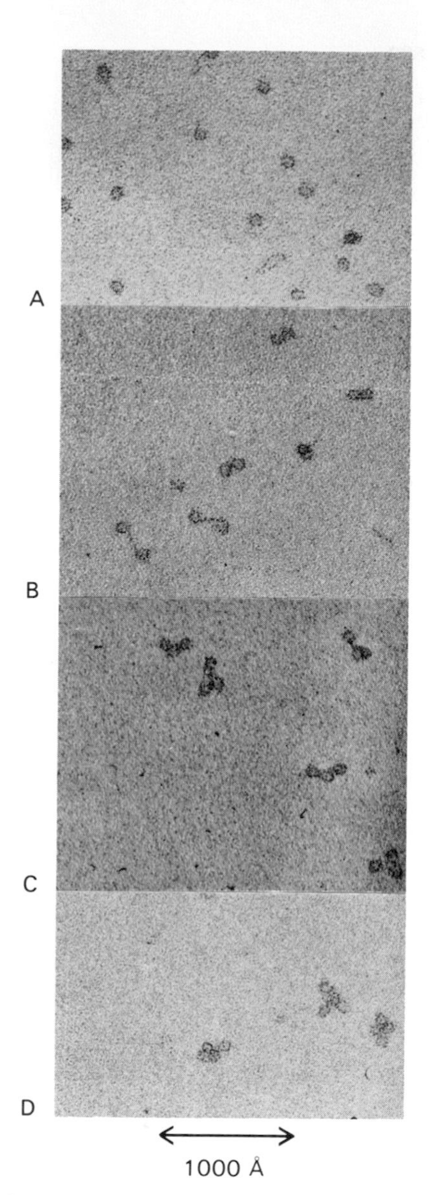

Figure 33-7
Electron micrographs of (A) monomer, (B) dimer, (C) trimer, and (D) tetramer nucleosomes produced as described in Figure 33-6. [From J. T. Finch, M. Noll, and R. D. Kornberg. *Proc. Nat. Acad. Sci.* 72(1975):3321.]

Table 33-3
DNA content of nucleosomes

Cell type	*Number of base pairs*
Yeast	165
HeLa	183
Rat bone marrow	192
Rat liver	196
Rat kidney	196
Chick oviduct	196
Chicken erythrocyte	207
Sea urchin gastrula	218
Sea urchin sperm	241

A NUCLEOSOME CORE CONSISTS OF 140 BASE PAIRS OF DNA WOUND AROUND A HISTONE OCTAMER

The DNA content of nucleosomes from different organisms and cell types ranges from about 160 to 240 base pairs (Table 33-3). What is the structural basis of these differences? Again, nucleases have proved to be highly informative probes. Nucleosomes can be further digested with micrococcal nuclease to yield a *core particle* that contains 140 base pairs of DNA, irrespective of the initial DNA content of the particular nucleosome. The nucleosome core is probably nearly the same in all eucaryotes. *This core consists of 140 base pairs of DNA bound to a histone octamer (two each of H2A, H2B, H3, and H4).*

The crystallization of nucleosome cores (Figure 33-8) demonstrated that the particles are quite homogeneous and opened the door to a deeper understanding of how eucaryotic DNA is packaged. X-ray and electron microscopic analyses of these crystals by Aaron Klug and John Finch revealed that the core is a flat particle of dimensions 110 × 110 × 55 Å and that it is made up of two layers. The 140 base pairs of DNA are wound on the outside of the core to form 1¾ turns of a left-handed supercoil with a pitch of about 28 Å. A model of the nucleosome based on low-resolution (20 Å) data is shown in Figure 33-9. Further analysis suggested that a dimer of histones H3 and H4 occupy the center of the nucleosome, and that one each of H2A and H2B are at either end (Figure 33-10).

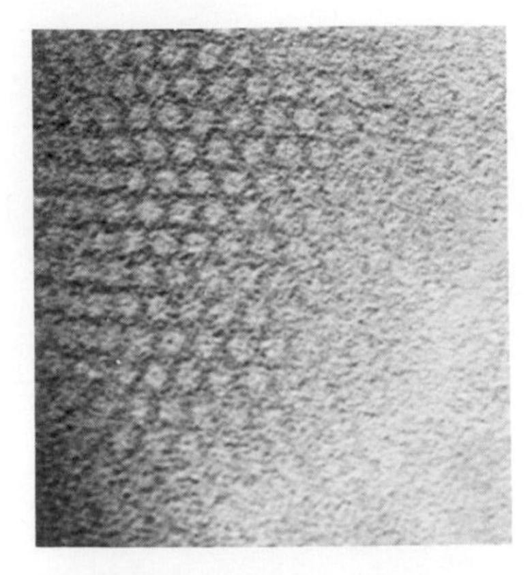

Figure 33-8
Electron micrograph of a crystal of nucleosome cores. Centers of adjacent nucleosome cores in this hexagonal array are 100 Å apart. [From J. T. Finch, L. C. Lutter, D. Rhodes, R. S. Brown, B. Rushton, M. Levitt, and A. Klug. *Nature* 269(1977):31.]

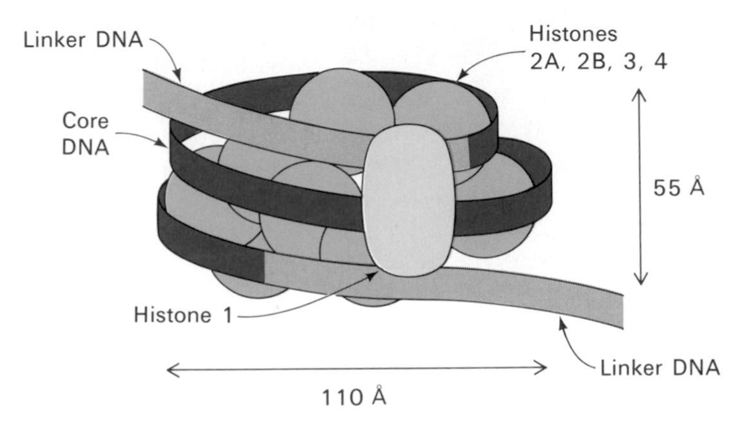

Figure 33-9
Schematic diagram of a region of chromatin containing a nucleosome. The DNA double helix (shown in red) is wound around an octamer of histones (two molecules each of 2A, 2B, 3, and 4, shown in blue). Histone 1 (shown in yellow) binds to the outside of this core particle and to the linker DNA. [After A. Kornberg, *DNA Replication*. W. H. Freeman (1980), p. 294.]

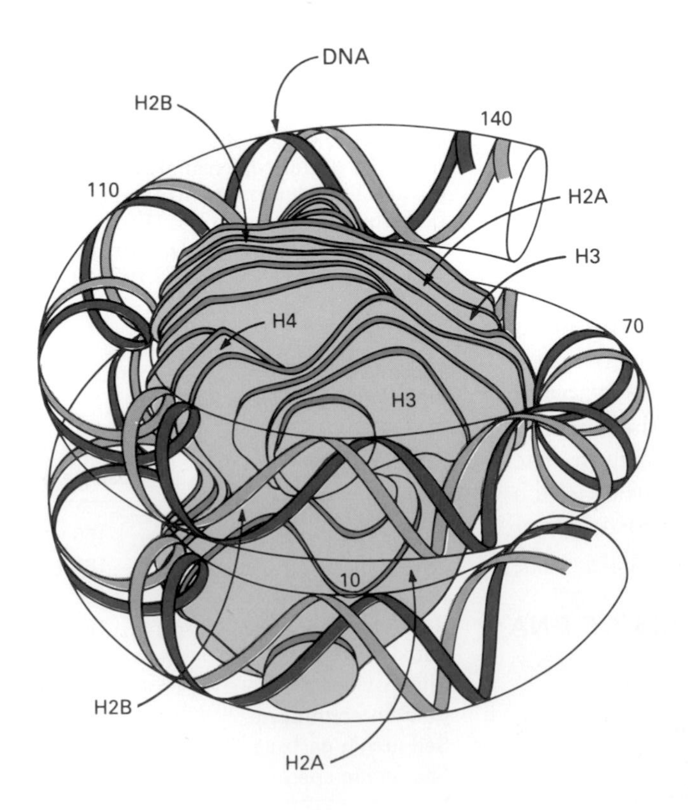

Figure 33-10
Model of a nucleosome core showing DNA wound in a left-handed superhelix around the histone octamer. The locations of the four kinds of histone subunits are marked. [After R. Kornberg and A. Klug, The nucleosome. Copyright © 1981 by Scientific American, Inc. All rights reserved.]

The structure of the nucleosome core determined at higher resolution (7 Å) shows that this model is essentially correct and provides a wealth of additional information. DNA-histone contacts are made on nearly every turn of the DNA double helix and are confined to the inner surface of the superhelix. The histones do not embrace the DNA nor do they protrude between the turns of the superhelix. The most substantial contacts are made between the H3-H4 tetramer and the central part of the DNA double helix. α-Helical rods projecting from the H3 dimer fit snugly into the minor grooves of DNA, on either side of the twofold symmetry axis of the nucleosome (Figure 33-11). The histones can interact with most DNA sequences, in keeping with their role as a device for packaging DNA.

The H2A-H2B dimers are attached to each exposed end of the H3-H4 tetramer. Their binding to the last half turns of the superhelix further stabilizes the nucleosome core. The disassembly of a nucleosome, as in DNA replication, is probably initiated by the dissociation or pulling away of an H2A-H2B dimer.

Histone H1 plays a key role at the next level of chromosome structure by serving as a bridge between different nucleosomes. H1 is located on the outside of the nucleosome, near the linker DNA, where it interacts with H2A subunits of the core. It is released from nucleosomes when the DNA per particle is trimmed from 160 to 140 base pairs. H1 differs also in its stoichiometry of one per nucleosome compared with two for the other histones. It is noteworthy that several types of H1 have been found, in contrast with the constancy of the other histones. Moreover, H1 is phosphorylated just before mitosis, and is dephosphorylated following mitosis, suggesting that this covalent modification regulates its capacity to make DNA compact.

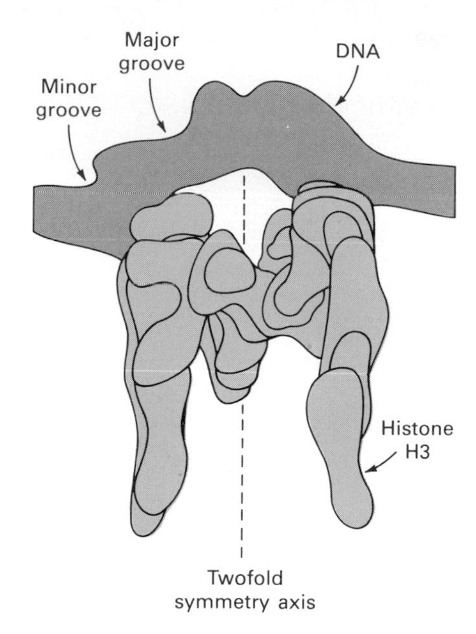

Figure 33-11
The H3 dimer interacts intimately with two minor grooves of DNA. The twofold axis is marked. [After T. J. Richmond, J. T. Finch, B. Rushton, D. Rhodes, and A. Klug. *Nature* 311(1984):535.]

NUCLEOSOMES ARE THE FIRST STAGE IN THE CONDENSATION OF DNA

The winding of DNA around a nucleosome core contributes to the packing of DNA by decreasing its linear extent. A 200-base-pair stretch of DNA would have a length of about 680 Å in solution. In contrast, this amount of DNA fits within the 100-Å diameter of the nucleosome. Thus, the *packing ratio* (degree of condensation) of the nucleosome is about 7. How does this value compare with the degree of condensation of DNA in chromosomes? Human metaphase chromosomes, which are highly condensed, contain a total of 7.8×10^9 base pairs, corresponding to a contour length of 2.6 m. This DNA is packed into forty-six cylinders whose total length is 200 μm. Thus, *the packing ratio of DNA in these metaphase chromosomes is about 10^4. DNA in interphase nuclei, where the chromatin is more dispersed, has a packing ratio of about 10^2 to 10^3*. Clearly, the nucleosome is just the first step in the compaction of DNA.

What is the next level of organization of DNA? One possibility is that the nucleosomes themselves form a helical array. For example, a postulated solenoidal model of chromatin with a 360-Å diameter (Figure 33-12) would have a packing ratio of about 40. The folding of such solenoids into loops would provide additional condensation. It seems likely that a series of nonhistone proteins stabilizes higher-order structures of chromosomes. For example, metaphase chromosomes stripped of histones display a central protein scaffold that is surrounded by

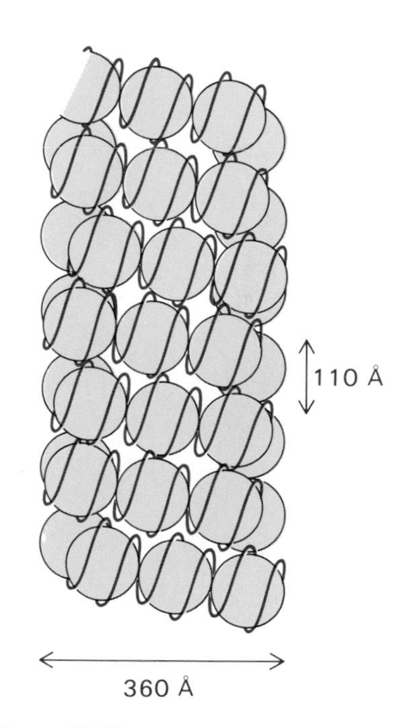

Figure 33-12
A proposed solenoidal model of chromatin consisting of six nucleosomes (shown in blue) per turn of helix. The DNA double helix (shown in red) is wound around each nucleosome. [After J. T. Finch and A. Klug. *Proc. Nat. Acad. Sci.* 73(1976):1900.]

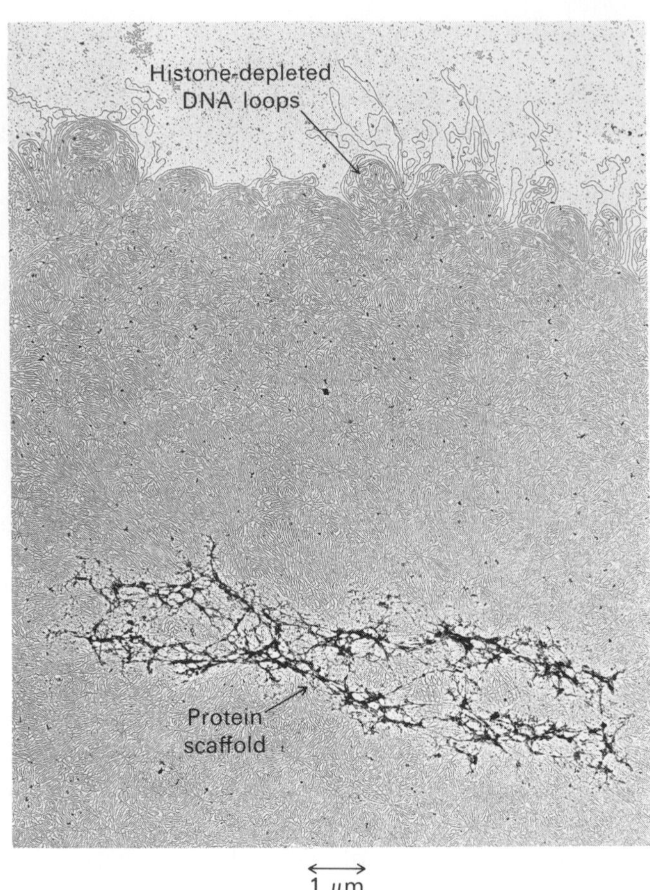

Figure 33-13
Electron micrograph showing histone-depleted DNA attached to a central protein scaffold. Histones were removed from metaphase chromosomes of HeLa cells by treating them with polyanions. [Courtesy of Dr. Ulrich Laemmli.]

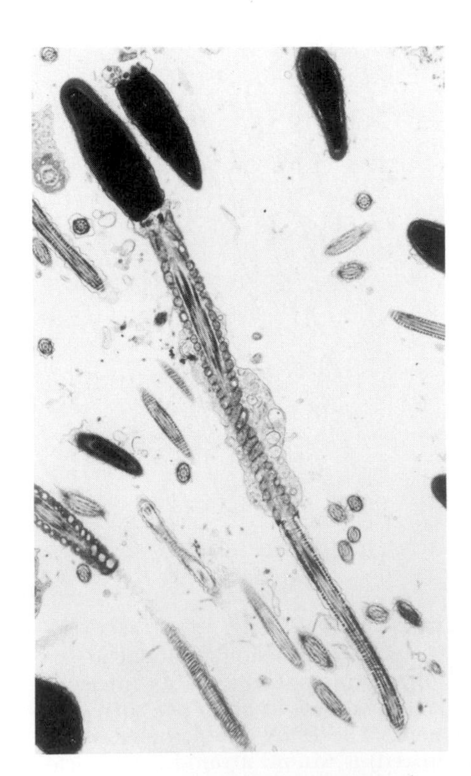

Electron micrograph of a sperm. The dense head of the sperm contains highly compacted DNA. [Courtesy of Lynne Mercer.]

many very long loops of DNA (Figure 33-13). These DNA loops are probably bound to the scaffold. The presence of topoisomerase II in the scaffold makes it likely that changes in supercoiling are important in altering the architecture and accessibility of large segments of DNA in mitosis and meiosis.

The most compacted DNA is found in sperm heads, in which histones are replaced by *protamines*, a series of arginine-rich proteins that become highly α-helical on binding to DNA. The α helices of protamine probably lie in the major grooves of DNA, where they neutralize the negatively charged phosphate backbone and so enable DNA duplexes to pack tightly together.

EUCARYOTIC DNA IS REPLICATED BIDIRECTIONALLY FROM MANY ORIGINS

Eucaryotic DNA, like all other DNA molecules, is *replicated semiconservatively*. Electron microscopic studies have also shown that eucaryotic DNA is replicated *bidirectionally from many origins*. The use of many initiation points is necessary for rapid replication because of the great length of eucaryotic DNAs. Consider, for example, a 62-Mb *Drosophila* chromosome. A DNA replication fork in *Drosophila* moves at the rate of 2.6 kb/min (0.16 Mb/hr), compared with about 16 kb/min in *E. coli*. The replication of this *Drosophila* chromosome would take more than sixteen days if there were only one origin for replication. The actual replication time of less than three minutes is achieved by *the cooperative action of more than 6000 replication forks per DNA molecule.* A DNA molecule from the cleavage nuclei of *Drosophila* exhibits a serial array of replicated regions, or "eye forms" (Figure 33-14). The activation of each initiation point generates two diverging replication forks. The eye forms expanding in both directions merge to form the two daughter DNA molecules (Figure 33-15). An eye form within an eye form has not been observed, indicating that an origin cannot be reactivated until after the entire DNA molecule is replicated.

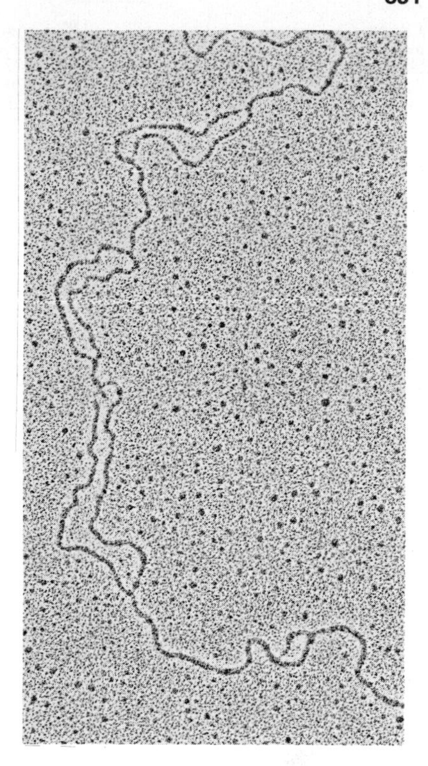

Figure 33-14
Electron micrograph of replicating chromosomal DNA from cleavage nuclei of *Drosophila*. The eye forms are the newly replicated regions. [Courtesy of Dr. David Hogness.]

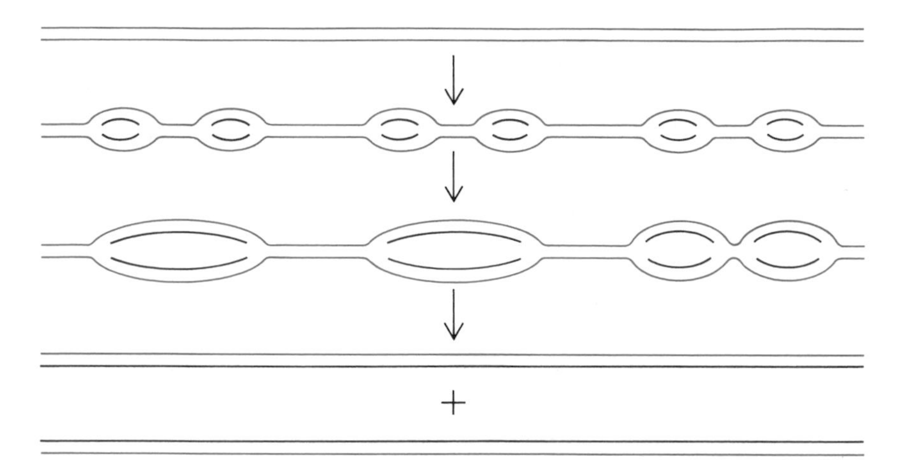

Figure 33-15
Schematic diagram of the replication of a eucaryotic chromosome. The parental DNA is shown in blue, and the newly replicated DNA in red.

A yeast chromosome contains about 400 initiation sites for DNA replication. These origins of replication share an 11-bp consensus sequence called *ars* for *autonomous replication sequence.* Insertion of an *ars* sequence into a bacterial plasmid enables it to replicate autonomously in yeast cells.

Eucaryotic cells contain several types of DNA polymerases (Table 33-4). The α polymerase plays a major role in chromosome replication, whereas the β enzyme participates in the repair of DNA. The amount of the α polymerase increases more than tenfold when quiescent cells begin to divide rapidly. The α polymerase is a multisubunit complex, whereas the β enzyme is a single polypeptide chain. The γ polymerase is responsible for the replication of mitochondrial DNA. These enzymes, like procaryotic DNA polymerases, use deoxyribonucleoside triphosphates as activated intermediates and carry out template-directed elongation of a primer in the $5' \rightarrow 3'$ direction. The α, β, and γ polymerases lack nuclease activities. It seems likely that proofreading is per-

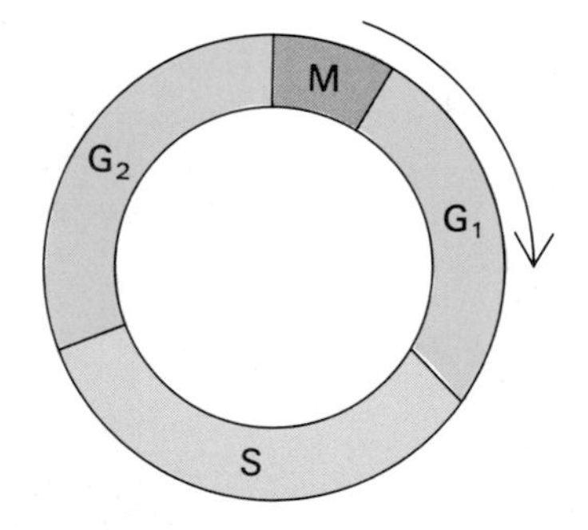

Figure 33-16
Phases in the life cycle of a eucaryotic cell: M (mitosis); G_1 (gap 1, prior to DNA synthesis); S (period of DNA synthesis); and G_2 (gap 2, between DNA synthesis and mitosis). The length of these periods depends on the cell type and conditions of growth. Mitosis is usually the shortest phase.

Table 33-4
Eucaryotic DNA polymerases

Type	*Location*	*Major role*	*Mass (kd)*
α	Nucleus	Replication of nuclear DNA	140
β	Nucleus	Repair of nuclear DNA	40
γ	Mitochondria	Replication of mitochondrial DNA	150

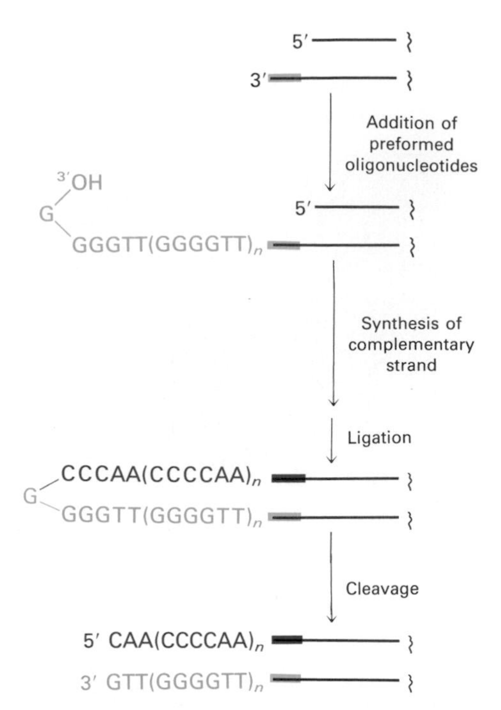

Figure 33-17
Completion of the replication of a linear duplex DNA.

formed by nucleases that are associated with these polymerases. A recently discovered eucaryotic δ polymerase possesses a 3′ exonuclease activity with editing capabilities akin to those of procaryotic polymerases. The α polymerase can be specifically inhibited by *aphidocolin*, a fungal diterpenoid.

THE ENDS OF CHROMOSOMES (TELOMERES) ARE REPLICATED BY ADDING BLOCKS OF PREFORMED OLIGONUCLEOTIDES

The linearity of eucaryotic chromosomes poses a problem not encountered with circular DNA molecules such as those of procaryotes. Eucaryotic DNA polymerases, like procaryotic ones, are unable to synthesize in the 3′ → 5′ direction or start chains de novo, and so erasure of the RNA primers leaves the 5′ ends of nascent daughter DNA strands incomplete (Figure 33-17). How are the 5′ ends finished? First, preformed oligonucleotide blocks are added to the overhanging 3′ ends of newly synthesized daughter strands. In the ciliated protozoan *Tetrahymena*, some fifty copies of 5′-TTGGGG-3′ are added to each 3′-OH end. These extensions then loop around so that their 3′-OH serves as a primer for the synthesis of complementary 3′-AACCCC-5′ sequences. DNA ligase joins each of these lengthened 3′ extensions to the 5′ ends, and a nuclease cleaves the unpaired loops to fashion a linear duplex with flush ends. Thus, *Tetrahymena* completes the replication of its linear chromosome by transiently joining the two strands at each end of the double helix.

> "There's a divinity that
> shapes our ends,
> Rough-hew them how we will."
>
> WILLIAM SHAKESPEARE (1601)
> *Hamlet* (V,ii)

NEW HISTONES FORM NEW NUCLEOSOMES ON THE LAGGING DAUGHTER DNA DUPLEX

At replication forks, the overall direction of DNA synthesis is 5′ → 3′ for one daughter strand and 3′ → 5′ for the other. As in the replication of procaryotic DNA (p. 671), this is accomplished by having one strand, the leading strand, synthesized continuously, and the other, the lagging strand, discontinuously. How are old and new histones distributed? This question was answered by carrying out DNA synthesis in vitro in the presence of cycloheximide, an inhibitor of protein synthesis. Under these conditions, DNA synthesis continues for about 15 minutes. Half of the newly synthesized DNA is completely degraded by DNase I, whereas the other half is split into fragments containing 200 base pairs. This experiment, as well as density-labeling studies, suggested that parental histones are associated with one of the daughter DNA duplexes, the other being bare because of the absence of new histones. This interpretation is directly supported by electron micrographs showing that one of the daughter duplexes at a replication fork is beaded, whereas the other is naked (Figure 33-18). In other words, *parental histones segregate conservatively during replication.*

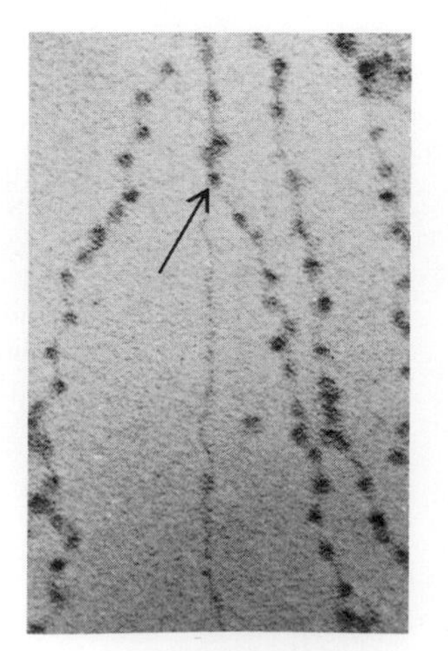

Figure 33-18
An asymmetric replication fork generated by DNA synthesis in the presence of cycloheximide, which blocks the formation of new histones. One of the daughter duplexes is beaded, whereas the other is bare, which shows that parental histones are bound to only one daughter duplex. [From D. Riley and H. Weinstraub. *Proc. Nat. Acad. Sci.* 76(1979):331.]

This arrangement indicates that histones do not dissociate from DNA during replication. In fact, *old histones stay with the DNA duplex containing the leading strand, whereas new histones assemble on the DNA duplex containing the lagging strand.* A likely reason for this difference between the daughter DNA molecules is that histones bind much more strongly to double-stranded than to single-stranded DNA. Old histones probably do not follow the lagging duplex because it contains single-stranded regions prior to the joining of its Okazaki fragments.

1 μm

Figure 33-19
Electron micrograph of a mitochondrial DNA molecule containing two genomes joined head-to-tail to form a circle. Replication of this DNA molecule has just started. The arrows point to two loops 180 degrees apart on the circle. These displacement loops (D loops) contain newly synthesized DNA. The thinner line at each loop is a displaced single-stranded region of parental DNA. [Courtesy of Dr. David Clayton.]

MITOCHONDRIA AND CHLOROPLASTS CONTAIN THEIR OWN DNA

Not all of the genetic information of eucaryotic cells is encoded by their nuclear chromosomal DNA. Genetic studies of yeast led to the discovery of a mitochondrial genome distinct from the nuclear one. In 1949, Boris Ephrussi found that some mutants of baker's yeast are unable to carry out oxidative phosphorylation. These respiration-deficient mutants grow slowly by fermentation. They are called *petites* because they form very small colonies. Genetic analyses led to the surprising finding that petite mutations segregate independently of the nuclear genes, which suggested that mitochondria have their own genomes. Several years later, mitochondria were found to contain DNA. Furthermore, mitochondrial DNA from a petite strain has a different buoyant density from that of wild-type yeast, which shows that a large part of the mitochondrial genome is altered in this mutant. It was subsequently found that chloroplasts in photosynthetic eucaryotes likewise contain DNA that is replicated, transcribed, and translated. In fact, chloroplast DNA is larger and more complex than mitochondrial DNA.

Human mitochondrial DNA is a double-helical circle containing 16,569 base pairs, which have been sequenced in entirety. This genome encodes 13 proteins, 22 tRNAs, and 2 rRNAs (Figure 33-20). About

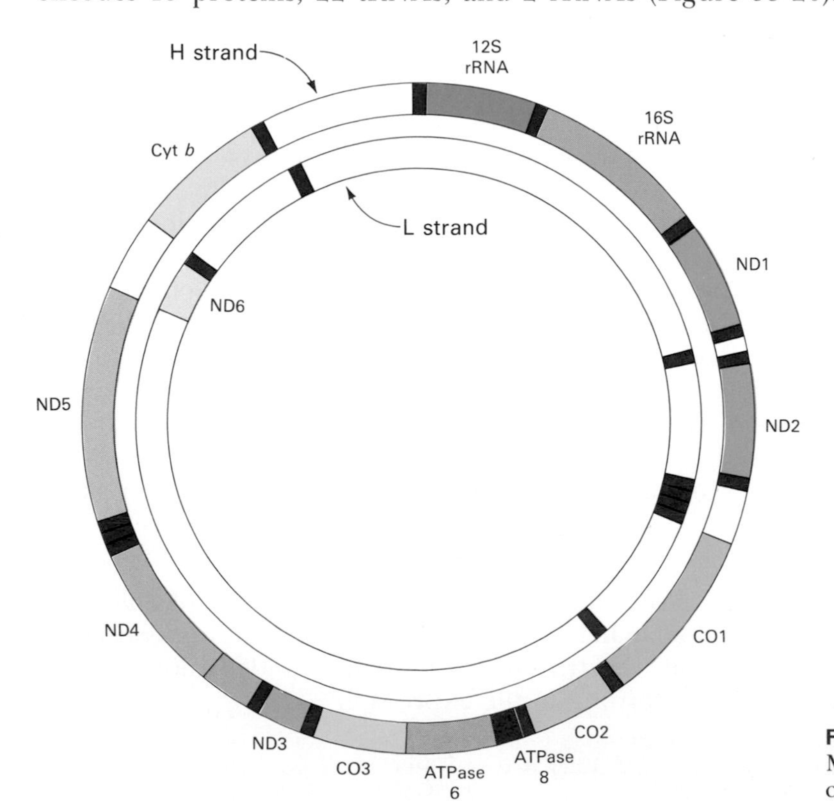

Figure 33-20
Map of the 16,569-bp human mitochondrial genome.

60% of the protein-coding capacity of this DNA is used to specify seven subunits of NADH-Q reductase, the first of three proton-pumping complexes in the inner mitochondrial membrane (p. 402). This DNA molecule also encodes a cytochrome reductase subunit, three cytochrome oxidase subunits, and two ATP synthase subunits. *A striking feature of the human mitochondrial genome is its extreme economy.* Nearly every base pair in human mitochondrial DNA encodes a protein or RNA product, and some even do double duty in that the last base of one gene serves as the first one of the next gene. The H strand (the denser of the two DNA strands) encodes all of the RNA and protein products except for the one protein and fourteen tRNAs that are specified by the L strand. A single primary transcript arises from the H strand, and another one from the L strand. These transcripts are cleaved on either side of tRNA sequences.

Another remarkable feature of gene expression in human mitochondria is the use of only 22 kinds of tRNAs, compared with 61 in the cytosol. Some mitochondrial tRNAs read four kinds of codons. As was mentioned in an earlier chapter (p. 108), mitochondria have a distinctive genetic code. Four codons have nonstandard meanings: AGA and AGG, which normally encode arginine, are stop codons; AUA codes for methionine instead of isoleucine; and UGA codes for tryptophan rather than being a stop signal.

EUCARYOTIC DNA CONTAINS MANY REPEATED BASE SEQUENCES

Studies of the kinetics of reassociation of thermally denatured DNA by Roy Britten and his associates have revealed that eucaryotic DNA, in contrast with procaryotic DNA, contains many repeated base sequences. In these experiments, DNA was sheared into small fragments and then denatured by heating the solution above the melting temperature (T_m) of the DNA. This solution of single-stranded DNA was then cooled to about 25°C below the T_m, which is optimal for the reassociation of complementary strands to form double-helical DNA. The kinetics of reassociation can be measured in a variety of ways. One technique is to follow the absorbance of the solution at 260 nm (p. 82). At this wavelength, the absorbance coefficient of double-stranded DNA is about 40% less than that of single-stranded DNA; this decrease is called *hypochromicity*. Another experimental approach is based on the fact that *double-stranded DNA binds to hydroxyapatite (calcium phosphate) columns, whereas single-stranded DNA passes through.* An attractive feature of this technique is that large amounts of DNA can be fractionated according to their rates of reassociation.

The observed kinetics of reassociation of DNA from *E. coli* or T4 phage followed the time course expected for the bimolecular reaction

$$S + S' \xrightarrow{k} D$$

in which S and S′ are complementary single-stranded molecules, D is the reassociated double helix, and k is the rate constant for association. For such a reaction, the fraction f of single-stranded molecules decreases with time according to the expression

$$f = \frac{1}{1 + kC_0t}$$

in which C_0 is the total concentration of DNA (expressed in moles of nucleotides per liter) and t is time (in seconds). For a particular DNA and set of experimental conditions (e.g., ionic strength, temperature, size of the DNA fragments), f depends only on C_0t, the product of DNA concentration and time. It is convenient to depict the kinetics of reassociation by plotting f versus the logarithm of C_0t. Such a *C_0t curve* has a sigmoidal shape (Figure 33-21).

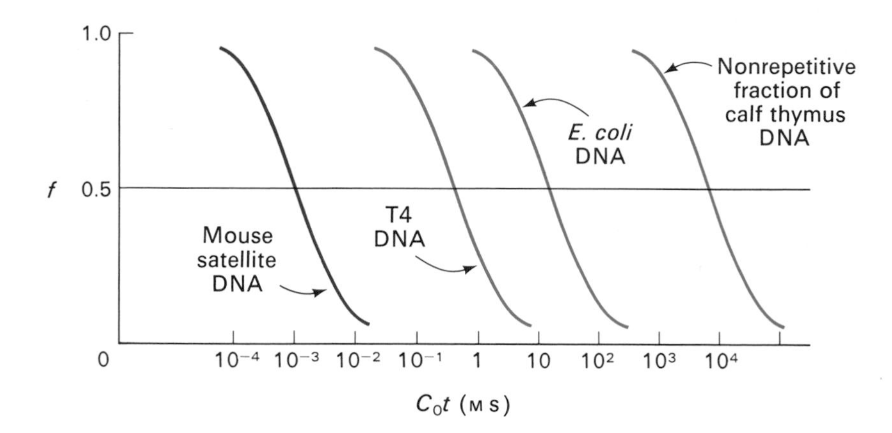

Figure 33-21
These C_0t curves depict the kinetics of reassociation of several kinds of thermally denatured DNA. The fraction of single-stranded molecules is plotted as a function of C_0t. The rapid reassociation of mouse satellite DNA shows that it contains very many repeated sequences. [After R. J. Britten and D. E. Kohne. *Science* 161(1968):530.]

An informative index of this plot is $C_0t_{0.5}$, the value of C_0t corresponding to reassociation of half of the DNA ($f = 0.5$). For *E. coli* DNA, $C_0t_{0.5}$ is about 15 M s, whereas for T4 phage, $C_0t_{0.5}$ is 0.3 M s. These values indicate that *E. coli* DNA reassociates about thirty times as slowly as T4 phage DNA. The reason is that *E. coli* DNA is larger and so the number of kinds of fragments in a sheared sample is greater than for T4 DNA. Hence, *the concentration of complementary fragments is lower in a solution of sheared* E. coli *DNA compared with one of T4 DNA* (containing the same concentration of nucleotides) *and thus its rate of reassociation is slower.* Studies of a variety of procaryotic DNA molecules showed that $C_0t_{0.5}$ is directly related to the size of the genome.

An unexpected result was obtained when mouse DNA was studied in this way. Because mammalian genomes are about three orders of magnitude larger than that of *E. coli*, a $C_0t_{0.5}$ value of the order of 10^4 M s seemed likely. A 10^{-4} M solution of DNA having such a $C_0t_{0.5}$ value would take 10^8 s (about 3 years) to reassociate halfway. It came as a surprise to find that a certain 10% fraction of the mouse DNA was half-reassociated in a few seconds. Indeed, *this fraction of mouse DNA reassociates more rapidly than even the smallest viral DNAs, which indicates that it contains many repeated sequences.* An analysis of the C_0t curve suggested that this fraction of mouse DNA contains *on the order of a million copies of a repeating sequence of about 300 base pairs.* About 20% of the mouse DNA renatured at an intermediate rate, pointing to the presence of moderately repetitive sequences. The other 70% renatured very slowly, suggesting that it consists of sequences that are unique or nearly so.

THE HUMAN GENOME CONTAINS NEARLY A MILLION COPIES OF VERY SIMILAR ALU SEQUENCES

Alu sequences—
A family of 300-bp sequences occurring nearly a million times in the human genome. Many of these sequences contain a target site for the *AluI* restriction endonuclease; hence their name.

Indeed, *all higher eucaryotes have an abundance of repetitive DNA.* For example, more than 30% of human DNA consists of sequences repeated at least twenty times. *Alu* sequences are especially abundant. These 300-bp sequences recur nearly a million times in the human genome and hence account for about 7% of the DNA. Alu sequences are dispersed throughout the genome; most 20-kb fragments of human DNA contain an Alu sequence. The degree of sequence identity between any two members of this family is about 85%. An intriguing finding is that *Alu sequences are similar to 7SL RNA, the small cytoplasmic RNA molecule in signal recognition particle* (Figure 33-22). Recall that this assembly plays a key role in delivering membrane and secretory proteins to the endoplasmic reticulum for translocation (p. 771). Alu sequences may have arisen by reverse transcription of 7SL RNA and integration of this cDNA into the genome.

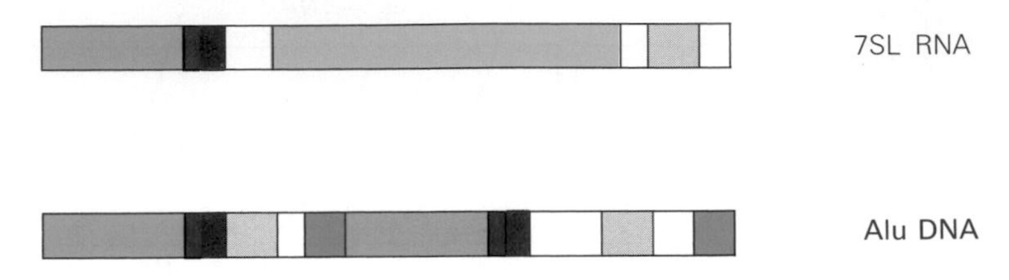

Figure 33-22
Alu sequences in human DNA are similar to portions of 7SL RNA, an abundant cytosolic RNA that is a component of signal recognition particle and participates in protein targeting. The distinctive central sequence of 7SL RNA is shown in yellow, and the A-rich regions of Alu sequences in blue. Human Alu sequences are head-to-tail dimers of similar sequences. [After E. Ullu and C. Tschudi. *Nature* 312(1984):171.]

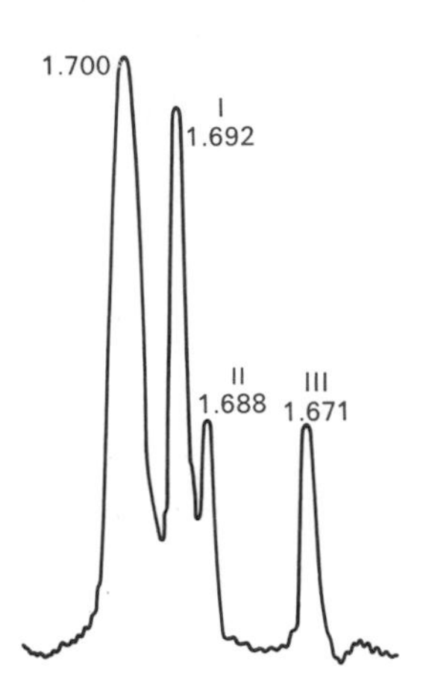

Figure 33-23
Three prominent bands of satellite DNA (ρ = 1.692, 1.688, and 1.671) are evident in this sedimentation pattern. DNA from *Drosophila virilis* was centrifuged to equilibrium in neutral CsCl. [After J. G. Gall, E. H. Cohen, and D. D. Atherton. *Cold Spring Harbor Symp. Quant. Biol.* 38(1974):417.]

CENTROMERES CONTAIN HIGHLY REPETITIVE DNA THAT CAN BE SEPARATED FROM OTHER CHROMOSOMAL DNA

Many highly repetitive DNAs can be isolated by density-gradient centrifugation because they have distinctive buoyant densities (Figure 33-23). For example, the sedimentation pattern of fragmented *Drosophila virilis* DNA exhibits a main band and three less-dense satellite bands, each consisting exclusively of highly repetitive DNA. In fact, the DNA in each band is a heptanucleotide sequence:

Satellite I	Satellite II	Satellite III
5′—ACAAACT—3′	5′—ATAAACT—3′	5′—ACAAATT—3′
3′—TGTTTGA—5′	3′—TATTTGA—5′	3′—TGTTTAA—5′

The chromosomal localization of this *satellite DNA* fraction has been determined by *in situ hybridization,* a technique devised by Joseph Gall and Mary Lou Pardue. Cells were immobilized under a thin layer of agar and treated with alkali to denature the DNA. This preparation was then incubated with tritium-labeled RNA that had been transcribed in vitro from purified mouse satellite DNA. Hybrids of this radioactive RNA and chromosomal regions containing satellite DNA were detected by autoradiography. A clear-cut result was obtained: *mouse satellite DNA is confined to centromeric regions.* (In mitosis, the daughter DNA molecules are joined to each other and to the mitotic spindle at a specialized chromosomal site called the *centromere.*) These satellite sequences (also called *simple sequence DNA*) do not encode proteins. Rather, they probably help to align chromosomes during mitosis and meiosis.

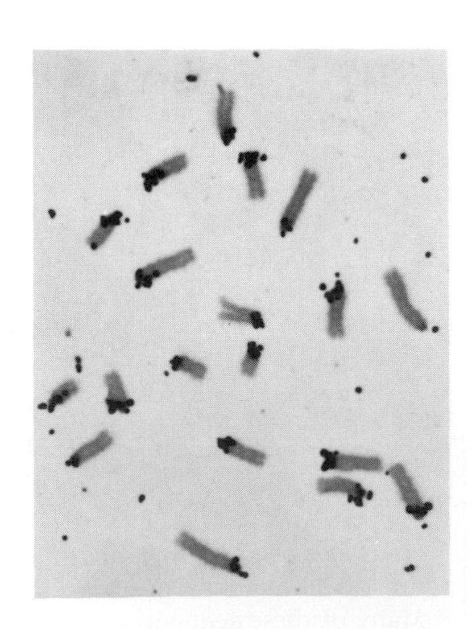

Figure 33-24
Autoradiograph of a mouse cell showing the localization of the satellite DNA. [Courtesy of Dr. Joseph Gall.]

GENES FOR RIBOSOMAL RNAs ARE TANDEMLY REPEATED SEVERAL HUNDRED TIMES

The genes coding for ribosomal RNA molecules are highly distinctive in two respects. First, they are *tandemly repeated.* Nearly all eucaryotes have more than 100 copies of these genes. Second, most rRNA genes are located in specialized regions of chromosomes that are associated with *nucleoli.* Mutants lacking nucleoli synthesize very little rRNA and hence are not viable. Much is known about these genes, largely because of the work of Max Birnstiel, Donald Brown, Oscar Miller, and their associates. The genes for the four kinds of ribosomal RNA—18S, 5.8S, 28S, and 5S rRNA—have been purified from the DNA of the African clawed toads *Xenopus laevis* and *Xenopus mulleri.* These creatures were chosen because their oocytes contain especially large amounts of these genes.

The genes for 18S, 5.8S, and 28S rRNA are clustered and tandemly repeated. In situ hybridization studies showed that these genes are located in nucleoli. Repeated clusters of them are separated by spacers that are not transcribed (Figures 33-26 and 33-27). Somatic cells of toads contain about 500 copies of this repeating unit, all in a tandem array. In oogenesis, these genes are selectively replicated several thousand times to form some 2×10^6 copies. Indeed, the genes for these rRNA molecules comprise nearly 75% of the total DNA in the oocyte. The amplified DNA is present as extrachromosomal circles that are bound to the very many new nucleoli. This *selective gene amplification* enables the oocyte to amass the 10^{12} ribosomes needed for the very rapid protein synthesis that occurs during cell cleavages. In the absence of gene amplification, it would take several centuries to assemble 10^{12} ribosomes!

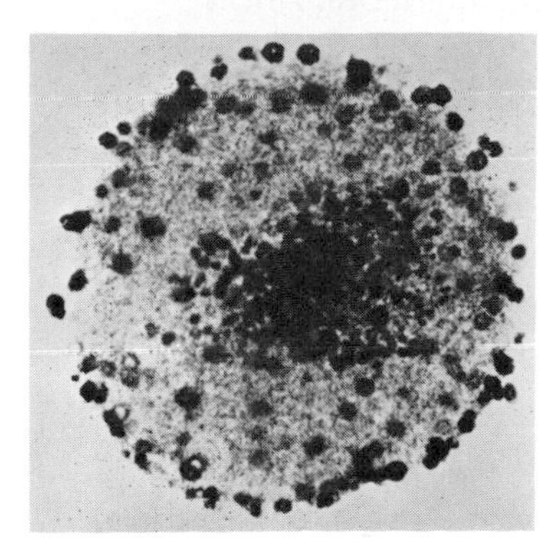

Figure 33-25
Micrograph of an isolated nucleus from a *Xenopus* oocyte stained to show the hundreds of nucleoli that are formed in the amplification of the ribosomal RNA genes. [From D. D. Brown and I. B. Dawid. *Science* 160(1968):272.]

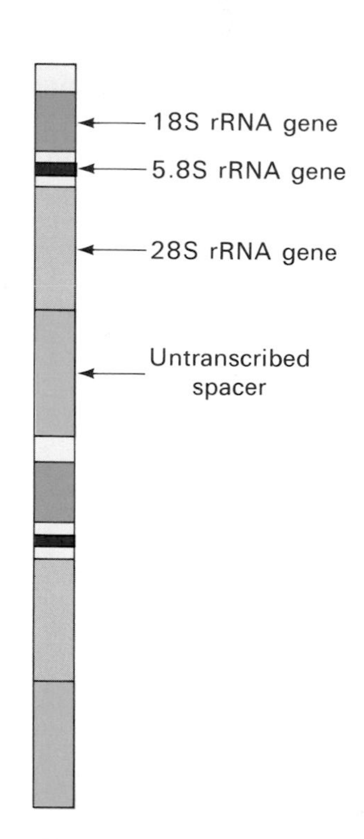

Figure 33-26
Organization of the genes for the 40S precursor of 18S, 5.8S, and 28S rRNAs in *Xenopus*. The tandemly repeated genes are separated by untranscribed spacers (yellow). The repeating unit is about 13 kb long.

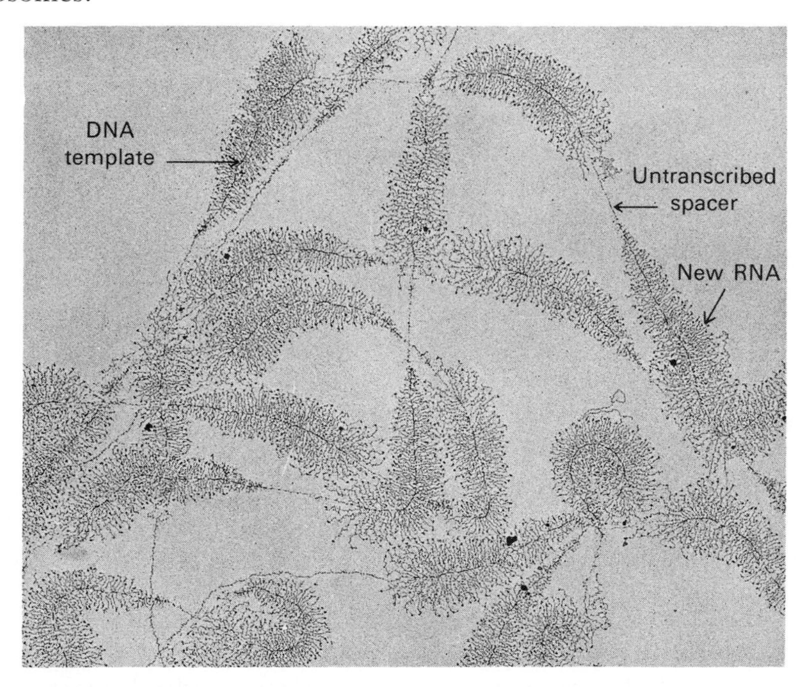

Figure 33-27
The tandem array of genes for 18S, 5.8S, and 28S rRNA is evident in this electron micrograph of nucleolar DNA. The dense axial fiber is DNA. The fine lateral fibers are newly synthesized RNA molecules with bound proteins. The tip of each arrowhead of the RNA molecules corresponds to an initiation point for transcription. The bare regions between arrowheads are the untranscribed spacers. [From O. L. Miller, Jr., and B. R. Beatty. Portrait of a gene. *J. Cell Physiol.* 74(supp. 1, 1969):225.]

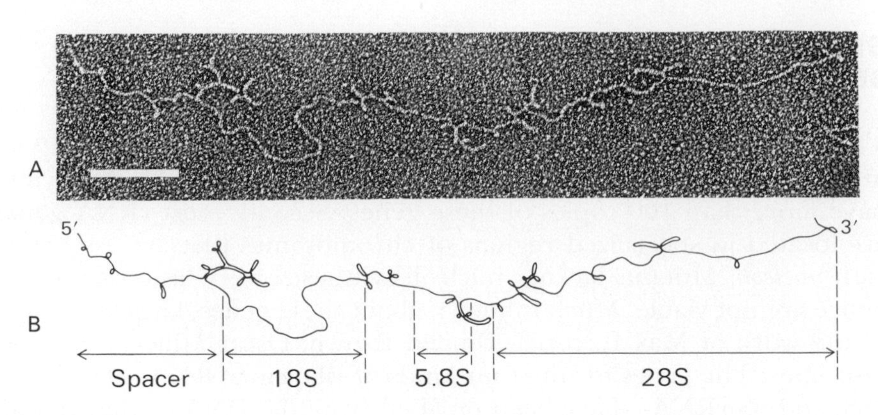

Figure 33-28
The 45S precursor of ribosomal RNA in HeLa cells has a highly distinctive pattern of hairpin loops when spread and examined by electron microscopy. This pattern makes it feasible to map the linear arrangement of 28S and 18S RNA molecules derived from this precursor. Part A is an electron micrograph of the 45S precursor (bar = 2000 Å). Part B is a tracing of the molecule shown in part A. [From P. K. Wellauer and I. B. Dawid. *Proc. Nat. Acad. Sci.* 70(1973):2828.].

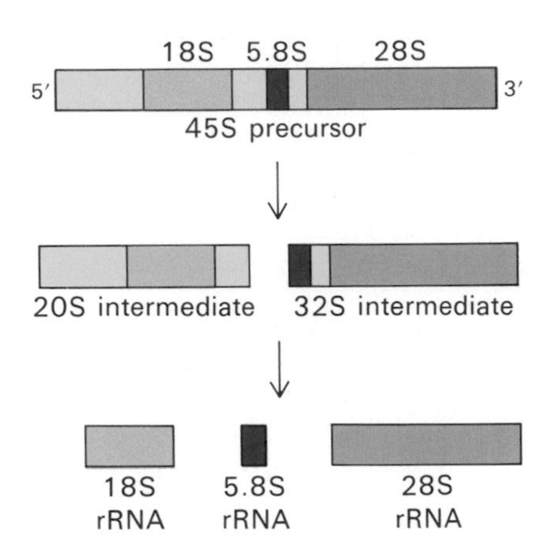

Figure 33-29
Formation of mammalian ribosomal RNAs from a primary transcript. The regions shown in yellow are removed.

The cluster of three rRNA genes is transcribed by RNA polymerase I in nucleoli to give a 45S RNA (Figure 33-28). This primary transcript is enzymatically modified and cleaved to yield the mature 18S, 5.8S, and 28S rRNAs (Figure 33-29). About a hundred nucleotides are methylated, nearly all on the 2′-hydroxyl groups of their ribose units. The highly specific methylation sites are conserved in the rRNAs of eucaryotes as distant as yeast and fruit flies. In addition, more than a hundred uridines in the precursor RNA are isomerized to pseudouridines. Many ribosomal proteins become associated with these RNAs and their precursors during the maturation process. Indeed, interaction of the RNA with ribosomal proteins may render certain sites susceptible to the action of nucleases.

The ribosomes of oocytes and somatic cells also contain 5S rRNA molecules that differ at eight of their 120 nucleotides. A *Xenopus* cell contains 400 copies of the somatic-type 5S gene or 20,000 copies of the oocyte-type 5S gene. They are tandemly repeated at the ends of most of the chromosomes. Again, untranscribed spacers separate the copies. Indeed, the spacer is several times as long as the gene.

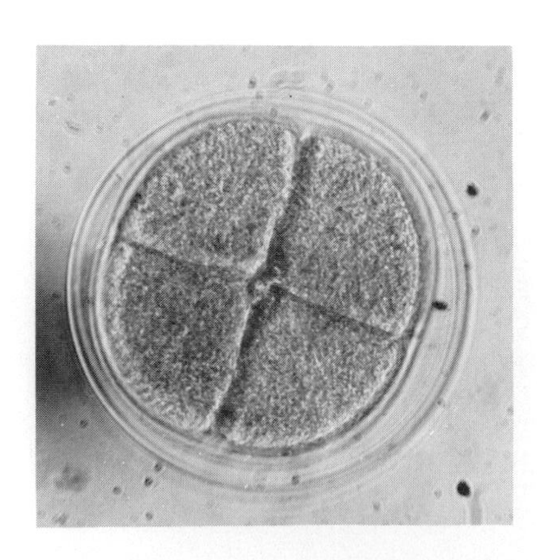

Figure 33-30
Light micrograph of a sea urchin embryo at the four-cell stage. [Courtesy of Dr. Annamma Spudich.]

HISTONE GENES ARE CLUSTERED AND TANDEMLY REPEATED MANY TIMES

What is the arrangement of genes that code for proteins? The histone genes were among the first to be isolated and characterized because of the abundance of histone mRNA in rapidly dividing sea urchin embryos. These marine invertebrates develop from a zygote to a 1000-cell blastula in about 10 hours, and so large amounts of histones must be synthesized for the assembly of new chromatin. Indeed, more than a quarter of the proteins synthesized during early embryogenesis are histones. Moreover, about 70% of the messengers synthesized at this stage are histone mRNAs, and so it was relatively easy to isolate them. The kinetics of hybridization of histone mRNA with sea urchin DNA was then measured to determine the number of copies of histone genes. *The rate of hybridization was several hundred times as fast as would be given by single-copy sequences, showing that histone genes are highly reiterated.* The number of copies of the histone genes ranges from 300 to 1000 in several species of sea urchin. In other organisms, the number is lower (Table 33-5). Yeast, for example, have only two copies of each histone gene. The number of copies of histone genes in an organism is correlated with its need to rapidly synthesize histone mRNAs.

Table 33-5
Reiteration frequency of the histone genes

Species	*Number of copies*
Sea urchin	300–1000
Fruit fly *(Drosophila melanogaster)*	110
Clawed toad *(Xenopus laevis)*	20–50
Mouse	10–20
Chicken	10
Human	30–40

The genes for the five major histones in the sea urchin are clustered in a basic 7-kb repeating unit (Figure 33-31). The five coding regions of this repeating unit alternate with five spacers. *This cluster of five histone genes is tandemly repeated many times.* The repetitions are very similar but not identical, which fits the finding that H1, H2A, and H2B are groups of closely related proteins rather than unique species. Different sets of histone genes are expressed in different tissues and at different times in development. The genes for the five kinds of histones are also clustered together and tandemly repeated in the fruit fly (Figure 33-31). However, in some vertebrates, histone genes are dispersed, indicating that clustering is not essential for coordinated gene expression.

Two additional features of histone gene expression are noteworthy. First, *histone genes lack introns.* Continuous genes are rare in vertebrates. Second, *histone mRNAs do not have poly A tails.* The absence of introns and poly A tails may enable histone mRNAs to be formed and transported to the cytosol very rapidly.

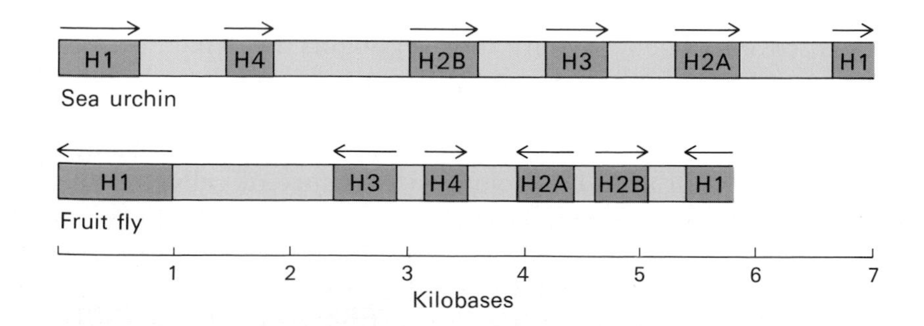

Figure 33-31
Map of the histone gene cluster in a sea urchin (*Strongylocentrotus purpuratus*) and a fruit fly (*Drosophila melanogaster*). Coding regions are shown in blue and spacers in yellow. The arrows denote the direction of transcription.

MANY MAJOR PROTEINS ARE ENCODED BY SINGLE-COPY GENES

We have seen that the genes for ribosomal RNAs and histones are reiterated many times. Are there a large number of copies of other genes that code for abundant products? Donald Brown explored this question before the advent of recombinant DNA technology by isolating mRNA for silk fibroin from the silkworm *Bombyx mori.* This system was chosen because the mRNA for silk fibroin has distinctive chemical characteristics. It is the template for repeating amino acid sequences that are rich in glycine. Furthermore, very large amounts of silk fibroin are synthesized by a single type of giant cell at a particular stage of larval development. Fibroin mRNA was purified on the basis of its large size (9.1 kb) and its high content of G compared with that of rRNA and other RNA molecules. Hybridization of purified fibroin mRNA with genomic DNA then revealed that *there is only one gene for silk fibroin per haploid genome.*

Figure 33-32
A silkworm. [Courtesy of Dr. Karen Sprague.]

This result has important implications concerning gene expression and differentiation in eucaryotes. It shows that large amounts of a particular protein can be synthesized even if there is only one copy of its gene. *The single gene for silk fibroin is the template for the synthesis of 10^4 mRNA molecules, which are stable for days. Each mRNA is the template for the synthesis of 10^5 proteins. Thus, a single gene is sufficient for the synthesis of 10^9 protein molecules.*

The reiterated genes for ribosomal RNAs and histones are exceptions rather than the common pattern. *The single gene found for silk fibroin is much more typical of genes that code for proteins, even abundant ones.* For example, reticulocytes contain one or just a few copies of the genes for the subunits of hemoglobin (p. 841). Likewise, large amounts of ovalbumin, the major protein in egg white, are synthesized by the oviduct of a laying hen, which contains only one copy of this gene per haploid genome.

SINGLE-COPY GENES CAN BECOME HIGHLY AMPLIFIED UNDER SELECTIVE PRESSURE

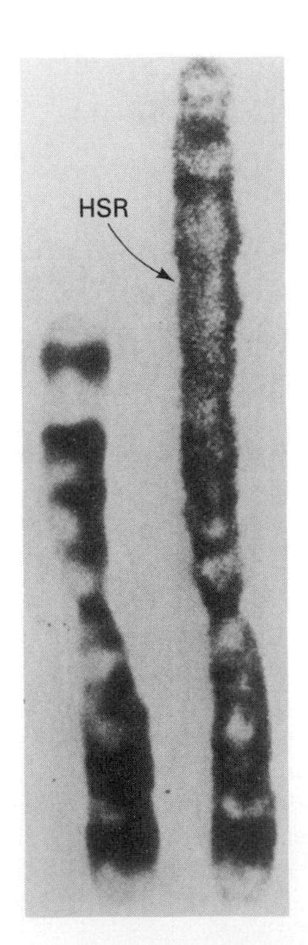

Micrograph of an abnormally long chromosome (right) that harbors some 300 copies of the DHFR gene in a homogeneously-staining region (HSR). The normal Chinese hamster chromosome 2 bearing a single copy of the DHFR gene is shown on the left. [Courtesy of Dr. Lawrence Chasin.]

Single-copy genes can become highly reiterated as a result of *gene amplification and selection.* For example, cells containing hundreds of copies of the gene for dihydrofolate reductase (DHFR), a key enzyme in the synthesis of deoxythymidylate, have been obtained by exposing them to methotrexate (p. 616). The rate of *spontaneous* duplication of the DHFR gene is about 10^{-3} per division. Methotrexate, an inhibitor of DHFR, selects for cells that contain multiple copies of this gene. In this way, a chromosomal region can acquire a tandem array of DHFR or other genes. The size of the repeating unit is typically 1 Mb, which is much larger than the genes undergoing amplification.

The discovery of gene amplification highlights the dynamic nature of the genome. Gene duplications occur frequently, and they can be stabilized by selective pressures. The work on dihydrofolate reductase has also influenced the strategy of cancer chemotherapy by revealing that drug resistance can arise by prolonged exposure of cells to sublethal doses of an inhibitory agent.

HEMOGLOBIN GENES IN TWO CLUSTERS ARE ARRANGED IN THE ORDER OF EXPRESSION DURING DEVELOPMENT

The study of hemoglobin has contributed a great deal to our understanding of protein structure and function (Chapter 7). Likewise, investigations of the genes for hemoglobin and their mode of expression have been sources of insight into eucaryotic gene function in general. For example, studies of the hybridization of β-globin mRNA with the

β-globin gene contributed to the discovery that eucaryotic genes are interrupted by introns (p. 110).

As was discussed earlier (p. 150), different hemoglobins are synthesized in development: embryonic hemoglobins are followed by fetal hemoglobin and then by adult hemoglobins. Each of these hemoglobins consists of two α-typc and two β-type chains. *The α-type genes are clustered on one chromosome, and the β-type genes on another* (Figure 33-33A). In the α cluster, the gene for the embryonic zeta (ζ) chain precedes two genes for α chains, which are components of both fetal and adult hemoglobins. In the β cluster, the gene for the embryonic epsilon ϵ chain is followed by two genes for fetal γ chains, and then by the genes for the adult δ and β chains. *Thus, the sequence of the human globin genes matches the order in which they are expressed during development.* It will be interesting to learn why these genes are clustered in developmental sequences. One possibility is that the $\zeta \rightarrow \alpha$ switch and the $\epsilon \rightarrow \gamma \rightarrow \delta,\beta$ switch depend on the proximity and ordering of these genes.

The linkage of these genes also reflects their evolutionary history. Differences between the DNA sequences of these genes suggest that an ancestral hemoglobin gene duplicated and diverged to give α and β globin genes

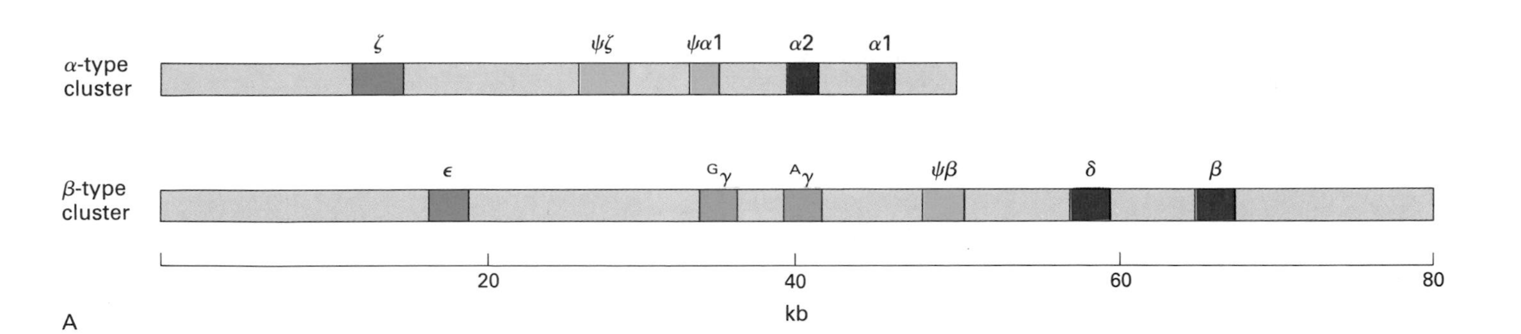

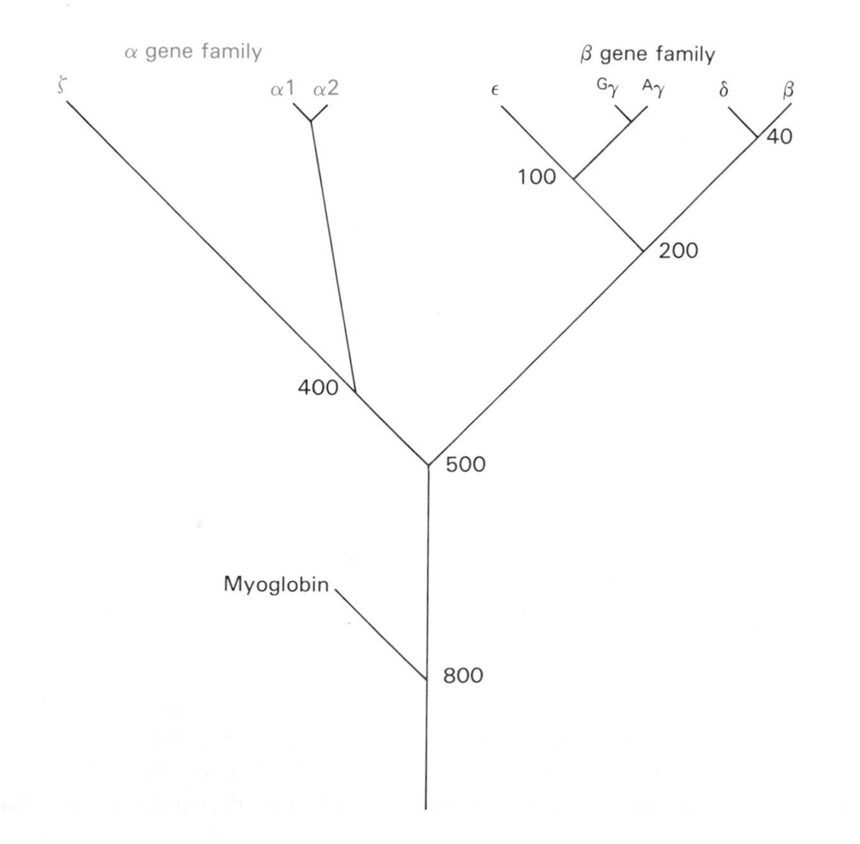

Figure 33-33
(A) Map of the gene clusters encoding the α-type and β-type chains of human hemoglobin. Embryonic genes are shown in blue, fetal genes in green, adult genes in red, and pseudogenes in yellow. Spacer regions between genes are shown in gray. Each of these genes consists of three exons and two introns (not shown). (B) Evolutionary tree of the globin genes deduced from DNA and amino acid sequence differences. The numbers shown at branch points are the estimated times of divergence in millions of years ago. [Part B is after A. J. Jeffreys, S. Harris, P. A. Barrie, D. Wood, A. Blanchetot, and S. M. Adams. In *Evolution from Molecules to Men*, D. S. Bendall, ed. (Cambridge University Press, 1983), p. 176.]

some 500 million years ago, early in the evolution of vertebrates. Myoglobin differs from both the α and β subunits of hemoglobin more than they differ from each other, indicating that myoglobin diverged before the α and β genes arose. Mammals, reptiles, birds, amphibians, and bony fish have distinct α and β subunits, whereas primitive vertebrates have only a single kind of hemoglobin subunit. The α and β genes later duplicated and diversified to give the two series of genes now present in humans (Figure 33-33B). Evolutionary trees deduced from DNA and protein sequences have been highly informative. *Biomolecules are indeed documents of evolutionary history*, as proposed by Emile Zuckerkandl and Linus Pauling in 1965.

Both the α and β gene clusters contain *pseudogenes*. These sequences are homologous to adjoining genes but they do not code for functional products. The prefix ψ denotes a pseudogene; $\psi\alpha1$, for example, is a pseudogene resembling the $\alpha1$ gene. Several structural features of this pseudogene account for its inactivity: (1) it lacks the 5′ consensus sequence for splicing; (2) the polyadenylation signal has mutated from AATAAA to AATGAA; (3) the normal ATG initiation codon is replaced by GTG; (4) a 20-nucleotide deletion starting at amino residue 38 generates three stop codons that prevent completion of the polypeptide chain. Why are pseudogenes present? Most of them are relics of genes that arose by duplication but have drifted in sequence and become inactive. We see in eucaryotic genomes both the mighty and the fallen.

Table 33-6
Classes of eucaryotic DNA

Class
Protein-coding genes
Single-copy genes
Duplicated genes
RNA-coding genes
Most are tandemly duplicated
Pseudogenes
Repetitive DNA
Simple-sequence DNA (as in satellite DNA)
Dispersed repetitive DNA (includes mobile genetic elements)
Spacer DNA of unknown function

After J. Darnell, H. Lodish, and D. Baltimore. *Molecular Cell Biology* (Scientific American Books, 1986), p. 408.

ONLY A SMALL PART OF A MAMMALIAN GENOME CODES FOR PROTEINS

The functional globin genes in the α and β clusters are quite far from one another. The clusters occupy 35 and 60 kb, which corresponds to about 12 kb per gene, but the amount of DNA needed to encode a ~150-residue globin chain is only 0.45 kb. Thus, less than 4% of these clusters codes for protein. Furthermore, the globin genes are far from the nearest protein-coding DNA. *Perhaps only 2% of mammalian DNA actually codes for proteins.* Thus, mammalian DNA is strikingly different from bacterial DNA, which is very compact and largely dedicated to encoding proteins. *The human genome is about a thousand times as large as the* E. coli *genome, but the number of genes is more likely to be fiftyfold rather than a thousandfold greater.* One of the great challenges in molecular biology today is to unravel the function of the vast majority of DNA, of the more than three billion base pairs in the human genome that do not encode protein and RNA products.

TRANSCRIPTIONALLY ACTIVE REGIONS OF CHROMOSOMES ARE UNDERMETHYLATED AND HYPERSENSITIVE TO DNase I

Gene expression in eucaryotes, as in procaryotes, is controlled to a large extent by the pattern of transcription. Some of the earliest evidence came from studies of the chromosomes of developing insects. The giant *polytene chromosomes* from *Drosophila* salivary glands contain more than a thousand aligned replicated DNA molecules that have not separated. Each polytene chromosome has a characteristic series of bands (chromomeres) that can be seen under a light microscope. The *Drosophila* genome contains about 5000 bands, each containing, on average, fewer than 10 genes. In the development of a larva into a pupa, certain bands become transiently enlarged (puffed) because the DNA in that region is converted from a condensed form into a dispersed one (Figure 33-34).

Puffs correspond to transcriptionally active regions. A striking finding is that puffs can be induced in isolated salivary glands by *ecdysone*, an insect steroid hormone. Specific bands enlarge and then contract in a precise temporal sequence, and there are concomitant changes in the population of mRNAs being synthesized. Puffing is necessary but not sufficient for gene activation. Some bands can be puffed either by developmental signals or by heat shock, but the batteries of genes activated by these stimuli are different.

The underlying molecular events in gene activation are being elucidated. The susceptibility of DNA to DNase I is not uniform. DNA packaged in nucleosomes is digested by low concentrations of this enzyme into fragments that are multiples of 200 base pairs, like those produced by digestion with micrococcal nuclease (see Figure 33-6 on p. 827). In contrast, *transcriptionally active regions of DNA are digested by DNase I into much smaller and irregular pieces.* These susceptible regions are rather large. For example, in the globin genes they extend some 7 kb beyond the 5′ and 3′ boundaries of the α and β clusters. Enhanced susceptibility to digestion implies that these regions have a lower density of nucleosomes or that their nucleosomes have an altered structure. Covalent modifications of histones, such as phosphorylation and ubiquitination, may be important in this regard. Moreover, some sites become exquisitely sensitive to cleavage by DNase I. Most of these *hypersensitive sites* are located at the 5′ ends of transcribed genes. Naked DNA is not hypersensitive to digestion by DNase I. Rather, hypersensitivity is probably due to the binding of specific regulatory proteins. Hypersensitivity of the globin genes and others is associated with the binding of two nonhistone proteins, HMG 14 and 17. *HMG* stands for *high mobility group*; these proteins are small and highly charged.

Enhanced sensitivity and hypersensitivity to digestion are tissue-specific and developmentally regulated. For example, the globin genes in the precursors of erythroid cells are insensitive to DNase I in 20-hour-old chicken embryos. However, when hemoglobin synthesis begins at 35 hours, these genes become sensitive to digestion. In contrast, globin genes in the brain are resistant to DNase I throughout development and in adulthood. Hypersensitivity is also correlated with active expression of specific genes in an array. The 5′ ends of γ, δ, and β genes in fetal erythroid cells are hypersensitive to DNase I, whereas only the δ and β genes are hypersensitive in adult bone marrow cells.

The degree of methylation of cytosines in DNA is also correlated with gene activity. C-5 of cytosine can be methylated by specific methyltransferases (Figure 33-35). About 70% of CG sequences in mammalian DNA are methylated. Methylation patterns can be assessed by comparing Southern blots of DNA cleaved by a pair of restriction endonucleases. HpaII cleaves CCGG but not $C^{m}CGG$ sequences (^{m}C denotes a methylated cytosine), whereas MspI cleaves both. *Studies using these enzymes have shown that globin genes are extensively methylated in nonerythroid cells but much less so in erythroid cells.* The γ genes, for example, contain a CCGG at position −54, between the TATA box and an upstream promoter site. Methylation of a cytosine in this sequence places a protruding methyl group in the major groove of DNA, which could interfere with the binding of a factor that stimulates transcription. Methylation also favors the transition from B-DNA to Z-DNA (p. 653). The importance of methylation is reinforced by the finding that methylation of

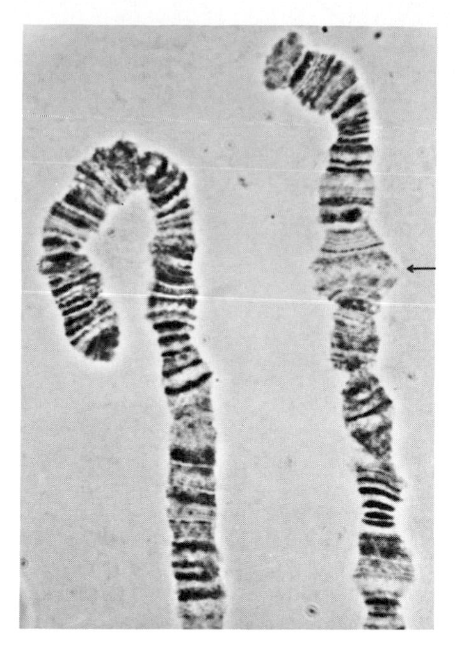

Figure 33-34
Puff formation in a polytene chromosome during development. The arrow marks a prominent puff in a *Drosophila* chromosome. [Courtesy of Dr. Joseph Gall.]

Cytosine

5-Methylcytosine

5-Azacytosine

Figure 33-35
Formulas of cytosine, 5-methylcytosine, and 5-azacytosine (the pyrimidine base of 5-azacytidine). Transcriptionally inactive regions of DNA have a high content of 5-methyl cytosine. 5-Azacytosine incorporated into DNA cannot be methylated at N-5 by the methyltransferase.

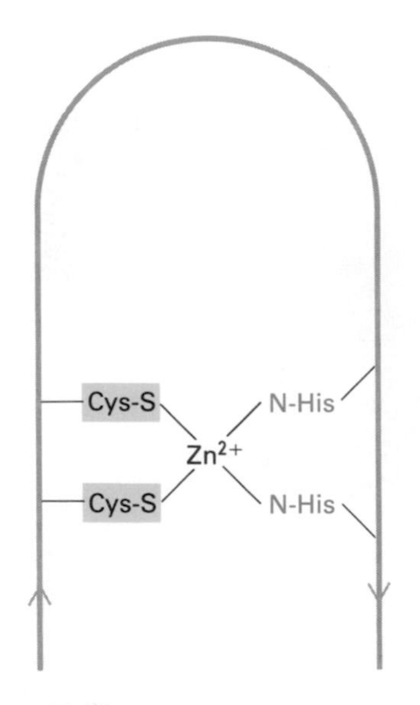

Figure 33-36
Structure of a metal-binding finger domain. A zinc ion is tetrahedrally coordinated to two cysteine and two histidine residues.

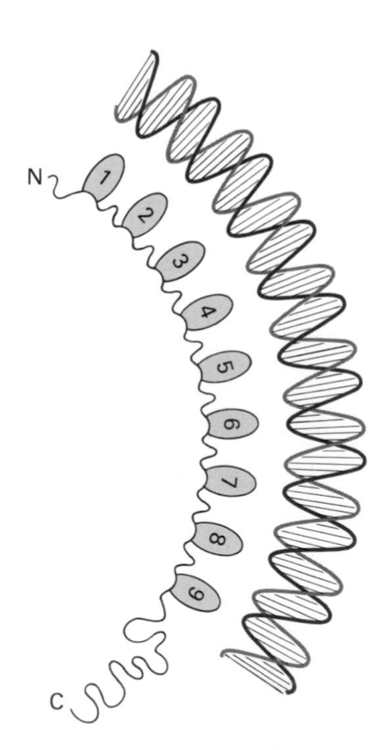

Figure 33-37
Domain structure of transcription factor IIIA (TFIIIA). The metal-binding fingers interact with the internal control region of 5S rRNA genes to stimulate their transcription. The carboxyl-terminal 10 kd of this 40-kd protein participates in forming a complex that contains RNA polymerase III and other transcription factors. [After J. Miller, A. D. McLachlan, and A. Klug. *EMBO J.* 4(1985):1613.]

recombinant DNA containing a globin gene blocks its expression in transfected cells. Specifically, methylation of the −760 to +100 region, but not other parts of the gene, interferes with transcription.

Further evidence for the significance of methylation in gene regulation comes from studies of the effects of incorporating *5-azacytidine* (5-azaC) into DNA. The substitution of N-5 for C-5 (Figure 33-35) in this analog prevents methylation and induces differentiation—that is, activation of genes that would otherwise be quiescent. *In general, active mammalian genes are less methylated than inactive ones.* However, it should be noted that methylation of DNA is not a universal regulatory device in higher eucaryotes. *Drosophila* DNA is not methylated at all.

TRANSCRIPTION FACTOR IIIA CONTAINS METAL-BINDING FINGERS THAT ACTIVATE GENES FOR 5S RIBOSOMAL RNA

The control of eucaryotic genes by specific promoters called transcription factors was introduced in an earlier chapter (p. 718). For example, heat-shock genes are activated by a specific transcription factor that binds to a promoter site upstream of the TATA box. A different mode of transcriptional control has been found for the genes encoding 5S ribosomal RNA in *Xenopus*. 5S ribosomal RNA and transfer RNA genes are transcribed by RNA polymerase III, whereas protein-coding genes are transcribed by RNA polymerase II. A sequence essential for transcription of 5S genes was identified by deleting parts of the gene and ascertaining the effect on the rate of transcription. The surprising finding was that deletions upstream of the start site and those downstream of the termination site had little effect on the number of transcripts formed. Rather, a 50-nucleotide region in the *center* of the gene proved to be critical for efficient transcription. This regulatory site is termed an *internal control region*.

The internal control region is recognized by a 40-kd protein called *transcription factor IIIA* (TFIIIA). One molecule of TFIIIA binds to a 5S RNA gene to form a complex that sequentially binds TFIIIC, TFIIIB, and RNA polymerase III. Factors B and C also serve to initiate transcription of tRNA genes, whereas TFIIIA is specific for 5S RNA genes. *A molecule of TFIIIA stays bound to the gene during transcription.* TFIIIA interacts more intimately with the coding strand of DNA than with the template strand.

TFIIIA has a remarkable structure. It contains nine similar domains, each about 30 residues long. Each of these repeats contains two cysteines and two histidines in identical locations. Moreover, the protein contains about nine zinc ions. X-ray absorption experiments have shown that each zinc is tetrahedrally coordinated to two cysteine sulfur atoms and two histidine nitrogen atoms (Figure 33-36). *The 30-residue repeating unit is probably folded into an elongated metal-binding domain having the shape of a finger. Each finger binds to about five base pairs in DNA, which corresponds to half a turn of the double helix.* Nine fingers would encompass about 45 base pairs, nearly the length of the internal control region (Figure 33-37). The tip of a finger unit could fit into either the major or minor groove of DNA. An advantage of a many-fingered protein is that it can remain bound to an internal control region during transcription. As RNA polymerase traverses the gene, several fingers of TFIIIA could release the template DNA strand, while others remain attached to the duplex or to the coding strand.

The organization of the TFIIIA gene matches the domain structure of the encoded protein. The first six exons in the gene correspond to the first six fingers of the protein. The last exon encodes the 10-kd carboxyl-terminal domain, which interacts with other transcription factors and RNA polymerase III. The structure of the gene supports the notion that *TFIIIA evolved from a primordial gene encoding a 30-residue metal-binding finger.* These finger domains have recently been found in a wide variety of RNA-binding and DNA-binding proteins. For example, the glucocorticoid receptor, a protein that binds a steroid hormone and controls the transcription of genes sensitive to it, contains metal-binding finger domains (p. 1001).

THE HOMEO BOX IS A RECURRING MOTIF IN GENES CONTROLLING DEVELOPMENT IN INSECTS AND VERTEBRATES

Insect bodies are divided into segments that are formed early in embryonic development. *Drosophila* contains three thoracic segments and eight abdominal ones (Figure 33-38). Cells of different segments usually retain their identity and do not intermix. However, certain mutations, called *homeotic mutations*, transform one body part into another. The Antennapedia (Antp) mutation, for example, changes an antenna into a leg. It can be inferred that the normal $Antp^{+}$ gene determines that legs are formed in the right place. In general, *homeotic genes control the fundamental architectural plan of the embryo.* The two main clusters of homeotic genes are the Antennapedia complex (ANT-C), which regulates the head and anterior thoracic segments, and the Bithorax complex (BX-C), which regulates the posterior thoracic segments and abdominal segments.

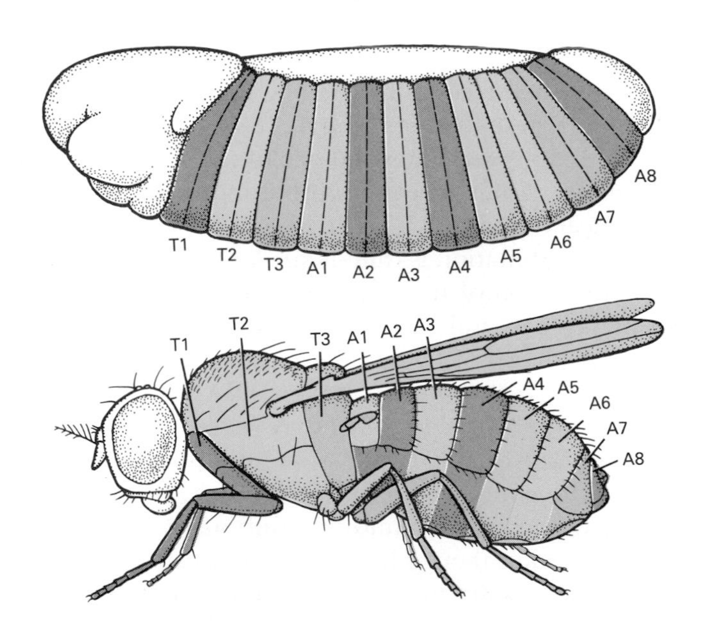

Figure 33-38
Segmentation pattern of a *Drosophila* embryo and adult fly. Segmentation is controlled by homeotic genes.

These large clusters of homeotic genes have been explored by David Hogness and Walter Gehring using the technique of chromosome walking (p. 131). ANT-C is longer than 100 kb, and BX-C is longer than 310 kb. Hybridization studies revealed that many genes in these clusters contain a common sequence. The recurring motif proved to be a 180-bp sequence, which was called the *homeo box*. It encodes a 60-residue protein *homeo domain* (Figure 33-39). The high content of basic residues suggested that the homeo domain binds to DNA. Indeed, *proteins containing the homeo domain are localized in the nucleus and bind to specific DNA sequences.* The homeo domain may contain the helix-turn-helix motif that is present in procaryotic DNA-binding proteins such as Cro and lambda repressor (p. 811).

```
A  SKRGRTAYTRPQLVELEKEFHFNRYLMRPRRVEMANLLCLTERQIKIWFQNRRMKYKKDN
B  RKRGRQTYTRYQTLELEKEFHFNRYLTRRRRIEIAHVLCLTERQIKIWFQNRRMKWKKEN
C  RKRGRQTYTRYQTLELEKEFHFNRYLTRRRRIEIAHALCLTERQIKIWFQNRRMKWKKEN
```

Figure 33-39
The homeo domain is a recurring motif in genes controlling development. Sequences from mammals, amphibians, and insects are highly homologous. The sequences compared here are from (A) the mouse *MO*-10 gene, (B) the frog *MM*3 gene, and (C) the *Drosophila Antp* gene. The nine-residue sequence shaded green is virtually invariant in more than ten homeo domains sequenced thus far.

The homeo box is not confined to insects. It is also present in the genes of amphibians and vertebrates, including humans. This finding raises the intriguing possibility that the underlying processes in the development of insects and mammals are more similar than was once thought. The molecular circuitry of development is now being delineated. We are on the threshold of understanding how the one-dimensional information contained in DNA is converted into the intricate and beautiful three-dimensional form of multicellular organisms.

SUMMARY

A eucaryotic chromosome contains a single linear molecule of double-helical DNA. The largest chromosome of *Drosophila* contains 76×10^6 base pairs (76 Mb), nineteen times the size of the entire *E. coli* genome. The human genome contains nearly a thousand times as much DNA as that of *E. coli*. Eucaryotic DNA is tightly bound to basic proteins called histones, and the combination is called chromatin. The chromatin fiber is a flexibly jointed chain of nucleosomes. The core of these repeating units consists of 140 base pairs of DNA wound around the outside of a histone octamer made up of two each of histones H2A, H2B, H3, and H4. Nucleosome cores are joined by linker DNA, typically 60 base pairs long, which binds one molecule of H1. A sevenfold reduction in the length of DNA is achieved by organization into nucleosomes, which are the first stage in the condensation of DNA. Eucaryotic DNA is replicated semiconservatively and semidiscontinuously from several thousand origins. A large number of initiation points are necessary because of the great length of eucaryotic DNA. Telomeres (the ends of chromosomes) in *Tetrahymena* are replicated by the addition of preformed oligonucleotide blocks followed by transient circularization. Old histones stay with the DNA duplex containing the leading strand, whereas new histones assemble on the DNA duplex containing the lagging strand.

Mitochondria and chloroplasts contain their own DNA, which encodes a small proportion of the proteins in these organelles. The human mitochondrial genome is a 16.6-kb double-helical circular DNA molecule. Nearly all of the DNA is used to encode proteins, tRNAs, and rRNAs. Several codons in mitochondria have meanings different from those in the cytosol.

The kinetics of reassociation of thermally denatured DNA revealed that eucaryotic DNA contains many repeated base sequences. The human genome, for example, contains nearly a million copies of highly homologous 300-bp Alu sequences. They are dispersed through the genome and may serve as initiation sites in DNA replication. A different class of highly repetitive DNA, called satellite DNA, is localized at centromeres. In *Drosophila*, this DNA consists of a heptanucleotide sequence repeated more than ten thousand times. The genes for 5.8S, 18S, and 28S rRNA are clustered and tandemly repeated many times. These ribosomal RNA genes are located in nucleoli. The genes for histones are also clustered and tandemly repeated many times. In contrast, other abundant proteins, such as silk fibroin, hemoglobin, and ovalbumin, are encoded by genes that are present once (or a few times) per genome. Single-copy genes, such as the one for dihydrofolate reductase, can become highly amplified under selective pressure, as when cells are exposed to an inhibitor of the gene product.

Our understanding of the fundamental architecture of eucaryotic genomes and the regulation of their expression is at an early stage. Only a small proportion (perhaps as little as 2%) of the human genome codes for proteins. The function of most of the DNA is an enigma. Intensive study of a number of gene clusters, such as the ones encoding the α-type and β-type genes of hemoglobin, are proving to be highly rewarding. Control is mediated largely at the level of transcription. Actively transcribed regions of DNA are more susceptible to digestion by DNase I than are inactive regions, and fewer of their cytosine bases are methylated. These differences are tissue-specific and developmentally regulated.

Specific genes or batteries of genes are activated by proteins called transcription factors, which bind to DNA upstream or downstream of the coding sequences. The genes for 5S rRNA, however, are regulated by the binding of TFIIIA to an internal control region in the middle of the gene. TFIIIA contains nine metal-binding domains that have a finger-like shape. The tips of these fingers bind to the grooves of DNA in a sequence-specific manner. Metal-binding fingers have been found in a wide variety of proteins that bind nucleic acids. The homeo box is another recurring motif in eucaryotic gene regulation. The fundamental architectural plan of the *Drosophila* embryo is determined by homeotic genes, which specify the segmentation pattern. Many of these genes contain the homeo box, a sequence encoding a 60-residue DNA-binding domain. The presence of the homeo box in the genes of both amphibians and mammals points to an underlying unity in the molecular basis of development.

SELECTED READINGS

Where to start

Kornberg, R. D., and Klug, A., 1981. The nucleosome. *Sci. Amer.* 244(2):52–64.

Gehring, W. J., 1987. Homeo boxes in the study of development. *Science* 236:1245–1252.

Grivell, L. A., 1983. Mitochondrial DNA. *Sci. Amer.* 248(3):78–89.

Books

Darnell, J., Lodish, H., and Baltimore, D., 1986. *Molecular Cell Biology*. Scientific American Books.

Watson, J. D., Hopkins, N. H., Roberts, J. W., Steitz, J. A., Weiner, A. M., 1987. *Molecular Biology of the Gene* (4th ed.). Benjamin/Cummings.

Lewin, B., 1987. *Genes* (3rd ed.). Wiley.

Cold Spring Harbor Symposia, 1985. *Molecular Biology of Development.* Cold Spring Harbor Symposia on Quantitative Biology, vol. 50.

Nucleosomes and chromatin

Richmond, T. J., Finch, J. T., Rushton, B., Rhodes, D., and Klug, A., 1984. Structure of the nucleosome core particle at 7 Å resolution. *Nature* 311:532–537.

Burlingame, R. W., Love, W. E., Wang, B-C., Hamlin, R., Xuong, N-h., and Moudrianakis, E. N., 1985. Crystallographic structure of the octameric histone core of the nucleosome at a resolution of 3.3 Å. *Science* 228:546–553. [Also see *Science* 229:1109–1113 for a discussion of these crystallographic studies.]

Kornberg, R. D., 1974. Chromatin structure: a repeating unit of histones and DNA. *Science* 184:868–871.

Cold Spring Harbor Symposia, 1978. *Chromatin.* Cold Spring Harbor Symposia on Quantitative Biology, vol. 42.

Pedersen, D. S., Thoma, F., and Simpson, R. T., 1986. Core particle, fiber, and transcriptionally active chromatin structure. *Ann. Rev. Cell Biol.* 2:117–147.

Sperling, R., and Wachtel, E. J., 1981. The histones. *Advan. Protein Chem.* 29:85–133.

Mitochondrial DNA

Borst, P., and Grivell, L. A., 1981. Small is beautiful—portrait of a mitochondrial genome. *Nature* 290:443–444.

Anderson, S., Bankier, A. T., Barrell, B. G., de Bruijn, M. H. L., Coulson, A. R., Drouin, J., Eperon, I. C., Nierlich, D. P., Roe, B. A., Sanger, F., Schreier, P. H., Smith, A. J. H., Staden, R., and Young, I. G., 1981. Sequence and organization of the human mitochondrial genome. *Nature* 290:457–465.

DNA replication in eucaryotes

Kornberg, A., 1980. *DNA Replication.* W. H. Freeman. [Chapter 6 deals with eucaryotic DNA polymerases and replication.]

DePamphilis, M. L., and Wassarman, P. M., 1980. Replication of eucaryotic chromosomes: a close-up of the replication fork. *Ann. Rev. Biochem.* 49:627–666.

Challberg, M. D., and Kelly, T. J., 1982. Eukaryotic DNA replication: viral and plasmid model systems. *Ann. Rev. Biochem.* 51:901–934.

Kriegstein, H. J., and Hogness, D. S., 1974. The mechanism of DNA replication in *Drosophila* chromosomes: structure of replication forks and evidence for bidirectionality. *Proc. Nat. Acad. Sci.* 71:135–139.

Repetitive DNA sequences

Jelinek, W. R., and Schmid, C. W., 1982. Repetitive sequences in eukaryotic DNA and their expression. *Ann. Rev. Biochem.* 51:813–844.

Singer, M. F., and Skowronski, J., 1985. Making sense out of LINES: long interspersed repeat sequences in mammalian genomes. *Trends Biochem. Sci.* 10:119–122.

Pardue, M. L., and Gall, J. G., 1970. Chromosomal localization of mouse satellite DNA. *Science* 168:1356–1358.

Gene amplification

Schimke, R. T., 1984. Gene amplification in cultured animal cells. *Cell* 37:705–713.

Stark, G. R., and Wahl, G. M., 1984. Gene amplification. *Ann. Rev. Biochem.* 53:447–491.

Centromeres, telomeres, and nuclear matrix

Blackburn, E. H., and Szostak, J. W., 1984. The molecular structure of centromeres and telomeres. *Ann. Rev. Biochem.* 53:163–194.

Carbon, J., 1984. Yeast centromeres: structure and function. *Cell* 37:351–353.

Newport, J. W., and Forbes, D. J., 1987. The nucleus: structure, function, and dynamics. *Ann. Rev. Biochem.* 56:535–566.

Nelson, W. G., Pienta, K. J., Barrack, E. R., and Coffey, D. S., 1986. The role of the nuclear matrix in the organization and function of DNA. *Ann. Rev. Biophys. Biophys. Chem.* 15:457.

Hemoglobin gene clusters

Karlsson, S., and Nienhuis, A. W., 1985. Developmental regulation of human globin genes. *Ann. Rev. Biochem.* 54:1071–1108.

Bunn, H. F., and Forget, B. G., 1986. *Hemoglobin: Molecular, Genetic, and Clinical Aspects.* Saunders. [Chapter 7 of this excellent book deals with the structure and expression of globin genes.]

DNase I sensitivity and methylation

Elgin, S. C. R., 1984. Anatomy of hypersensitive sites. *Nature* 309:213–214.

Bird, A. P., 1986. CpG-rich islands and the function of DNA methylation. *Nature* 321:209–213.

Hutchinson, N., and Weintraub, H., 1985. Localization of DNAse I-sensitive sequences to specific regions of interphase nuclei. *Cell* 43:471–482.

Metal-binding fingers in regulatory proteins

Brown, D. D., 1984. The role of stable complexes that repress and activate eucaryotic genes. *Cell* 37:359–365. [Discusses the interaction of TFIIIA with 5S RNA genes.]

Miller, J., McLachlan, A. D., and Klug, A., 1985. Repetitive zinc-binding domains in the protein transcription factor IIIA from *Xenopus* oocytes. *EMBO J.* 4:1609–1614.

Tso, J. Y., van den Berg, D. J., and Korn, L. J., 1986. Structure of the gene for *Xenopus* transcription factor IIIA. *Nucl. Acids Res.* 14:2187–2200.

Berg, J. M., 1986. Potential metal-binding domains in nucleic acid binding proteins. *Science* 232:485–487.

Rhodes, D., and Klug, A., 1986. An underlying repeat in some transcriptional control sequences corresponding to half a double helical turn of DNA. *Cell* 46:123–132.

Chowdhury, K., Deutsch, U., and Gruss, P., 1987. A multigene family encoding several finger structures is present and differentially active in mammalian genomes. *Cell* 48:771–778.

Homeo box and homeotic genes

Gehring, W. J., 1985. The molecular basis of development. *Sci. Amer.* 253(4):152–162.

Bender, W., Akam, M., Karch, F., Beachy, P. A., Peifer, M., Spierer, P., Lewis, E. B., and Hogness, D. S., 1983. Molecular genetics of the bithorax complex in *Drosophila melanogaster. Science* 221:23–29.

Gehring, W. J., 1985. The homeo box: a key to the understanding of development? *Cell* 40:3–5.

Laughon, A., and Scott, M. P., 1984. Sequence of a *Drosophila* segmentation gene: protein structure homology with DNA-binding proteins. *Nature* 310:25–31.

Desplan, C., Theis, J., and O'Farrell, P. H., 1985. The *Drosophila* developmental gene, *engrailed*, encodes a sequence-specific DNA binding activity. *Nature* 318:630–635.

Manley, J. L., and Levine, M. S., 1985. The homeo box and mammalian development. *Cell* 43:1–2.

Molecular evolution

Bendall, D. S., (ed.), 1983. *Evolution from Molecules to Men.* Cambridge University Press. [Contains many stimulating articles on molecular evolution.]

Doolittle, R. F., 1986. *Of Urfs and Orfs: A Primer on How to Analyze Derived Amino Acid Sequences.* University Science Books. [A concise and engaging account. Many interesting examples of molecular evolution are presented.]

Wilson, A. C., 1985. The molecular basis of evolution. *Sci. Amer.* 253(4):164–173.

Dickerson, R. E., and Geis, I., 1983. *Hemoglobin: Structure, Function, Evolution, and Pathology.* Benjamin/Cummings. [A fascinating account of hemoglobin evolution is given in Chapter 3.]

Zuckerkandl, E., and Pauling, L., 1965. Molecules as documents of evolutionary history. *J. Theor. Biol.* 8:357–366.

PROBLEMS

1. Compare procaryotic and eucaryotic gene expression in regard to:
   (a) Degree of coupling of transcription and translation.
   (b) Number of gene products on a primary transcript.
   (c) Number of proteins arising from the translation of a primary transcript.
   (d) Density of coding sequences in DNA.
   (e) Organization of genes into operons.

2. Synthetic yeast chromosomes can be constructed by the joining of three kinds of DNA elements. What are they?

3. The error rate of the mitochondrial DNA polymerase is much higher than that of the nuclear DNA polymerase. Why can lower fidelity be tolerated in the mitochondrial enzyme?

4. In the course of evolution, genes encoding mitochondrial proteins probably moved from the mitochondrial to the nuclear genome. Human mitochondrial DNA, for example, is fivefold smaller and simpler than yeast mitochondrial DNA. Different sets of mitochondrial proteins are encoded by mitochondrial DNA in the two species. Hence, there seems to be no structural reason why a mitochondrial protein must be synthesized within the organelle rather than be imported. Do you expect the human mitochondrial genome to be even shorter ten million years from now? Which aspect of mitochondrial gene expression would impede the transfer of a mitochondrial gene to the nuclear genome?

5. Design an experiment showing that DNA is coiled around histones to form a left-handed rather than a right-handed superhelix.

6. Yeast contain internal $C_{1-3}A$ sequences that are like those found in telomeres. It is also known that chromosomes that have lost their ends can be repaired. How might the internal $C_{1-3}A$ sequences be used to fix the damage?

7. Transcription factor IIIA (TFIIIA) binds to 5S RNA as well as to the internal control region of the gene encoding it. Why? How might this binding of the gene product be used to regulate expression of the 5S genes?

8. Baker's yeast (*S. cerevisiae*) interconverts between haploid and diploid states. These conversions occur by *mating* and *sporulation*. In mating, two haploid spores fuse to form a diploid. In sporulation, haploid spores are formed by meiosis of diploids. Cells cannot mate unless they are of opposite mating types, called *a* and α. Yeast strains can switch back and forth between different mating types (sexes), which are controlled by the *MAT* locus (mating type locus). Strains carrying the *HO* allele change their mating type as often as once a generation, whereas those harboring the recessive *ho* allele switch mating types at a frequency of 10^{-6} per generation.
   (a) Mating type switching in yeast is reminiscent of a reversible change in gene expression in procaryotes discussed in an earlier chapter. What is the analogous procaryotic switch? How is it achieved?
   (b) Analysis of the yeast mating type switch showed that it proceeds by a different mechanism from the alternation seen in the procaryote. The active *MAT* gene is flanked by an unexpressed locus *HML* on the left, and another unexpressed locus *HMR* on the right. When *a* occupies *MAT*, a silent *a* occupies *HMR* and a silent α occupies *HML*. The switching

of mating type is mediated by the donation of genetic information (called a *casette*). What are two contrasting means of bringing information from *HML* or *HMR* into *MAT*?

9. Sleeping sickness, an often fatal neurological disease, is caused by the African trypanosome. This protozoan has a life cycle that alternates between tsetse flies and mammalian hosts. A key to the trypanosome's success is its ability to circumvent the mammalian immune system. A trypanosome changes its surface coat every two weeks by expressing a different *variable surface glycoprotein* (VSG). VSGs are encoded by clusters of DNA sequences called *basic copies* located in the interior of a chromosome. However, basic-copy genes are not transcribed. Moreover, a basic-copy gene is incomplete. Expression requires the duplication of a basic-copy gene and its translocation to a telomere to join a gene for the 3′ end of the transcript. Even this assembled gene, called an *expression-linked copy*, does not contain all the information present in the mRNA for the glycoprotein. The 35-nucleotide sequence at the 5′ end of the mRNA comes from sequences upstream in the telomere that are controlled by separate promoters. Propose a mechanism for the formation of the complete mRNA.

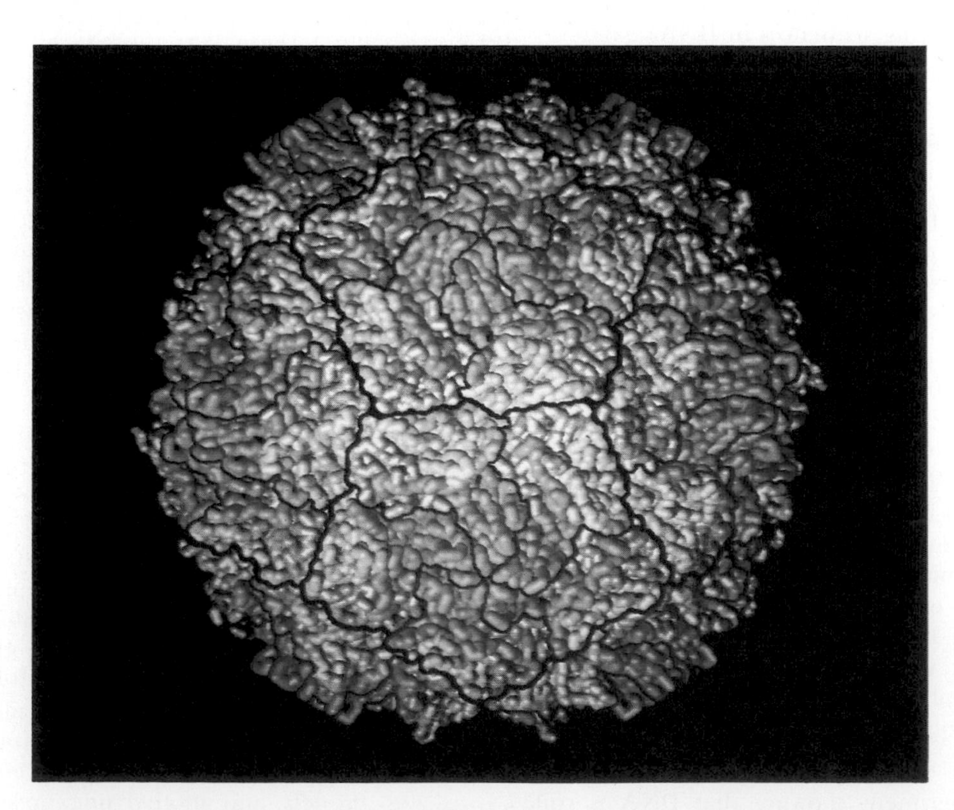

Structure of poliovirus. The shell of the virus particle is made of 60 copies of each of four proteins. Three of them are located on the surface. The RNA genome is located inside this 310-Å particle. [Courtesy of Dr. Arthur J. Olson, Research Institute of Scripps Clinic. © 1987.]

CHAPTER 34

Viruses and Oncogenes

Viruses are packets of infectious nucleic acid surrounded by protective coats. They are the most efficient of the self-reproducing intracellular parasites. Viruses are unable to generate metabolic energy or to synthesize proteins. They also differ from cells in having either DNA or RNA, but not both. The nucleic acid is single stranded in some viruses, and double stranded in others. Viruses range in complexity from Qβ, an RNA phage having only 4 genes, to the poxviruses, which have about 250 genes. The complete extracellular form of a virus is called a *virion* (or virus particle). In a virion, the viral nucleic acid is covered by a protein *capsid,* which protects it from enzymatic attack and mechanical breakage and delivers it to a susceptible host. In some of the more complex animal viruses, the capsid itself is surrounded by an *envelope* containing membrane lipids and glycoproteins.

The study of viruses has greatly advanced biochemistry and molecular biology, as exemplified by the discovery of messenger RNA (p. 94). Viruses continue to attract great interest for several reasons. First, *viral multiplication is a model for cellular development* because it requires the sequential expression of genes and the assembly of macromolecules into highly ordered structures. The relatively small number of viral genes, their rapid rate of replication, and the feasibility of genetic analysis enhance the value of viruses as models. Second, *viruses cause life-threatening diseases* such as smallpox, polio, influenza, and AIDS. Third, the study of viruses has provided insight into *evolutionary processes and host-parasite relations.* Fourth, viruses can produce cancers in susceptible animals. The cancer-producing potential of an RNA tumor virus comes from a single viral gene—an *oncogene.* All known oncogenes are altered forms of normal cellular genes. *Oncogenes are revealing the molecular circuitry that controls normal growth and development.*

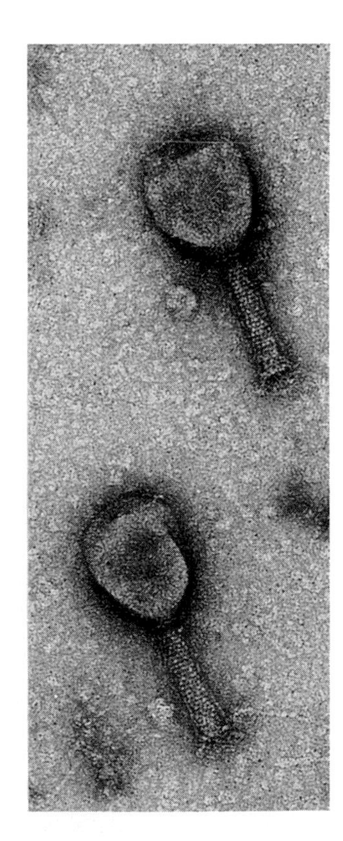

Figure 34-1
Electron micrograph of T4 virions from an infected cell. [Courtesy of Dr. Jonathan King, Dr. Yoshiko Kikuchi, and Dr. Elaine Lenk.]

Table 34-1
Types of viruses

Nucleic acid	*Representative virus*	*Approximate number of genes*
Single-stranded DNA	φX174 phage	5
Double-stranded DNA	Polyoma virus	6
	Adenovirus 2	30
	T4 phage	150
	Vaccinia poxvirus	240
Single-stranded RNA	Qβ phage	4
	Rous sarcoma virus	4
	Tobacco mosaic virus	6
	Poliovirus	8
	Influenza virus	12
Double-stranded RNA	Reovirus	22

Source: B. D. Davis, R. Dulbecco, H. N. Eisen, H. S. Ginsberg, and W. B. Wood, Jr. *Microbiology* (Harper & Row, 1973).

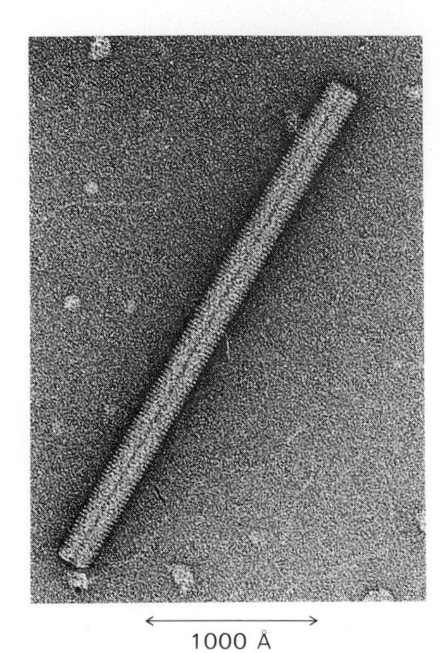

Figure 34-2
Electron micrograph of a tobacco mosaic virus (TMV) particle. [Courtesy of Dr. Robley Williams.]

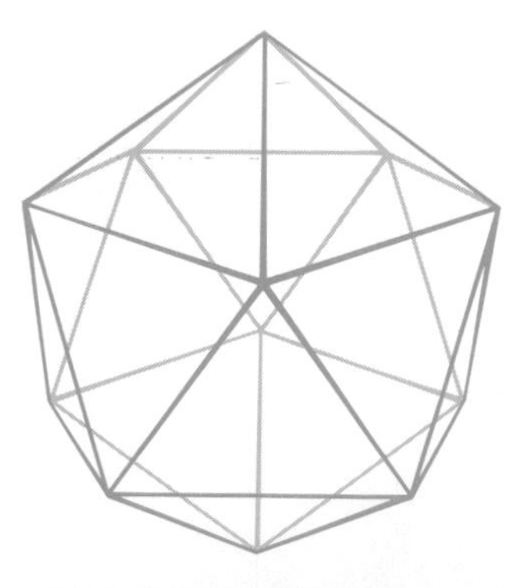

Figure 34-3
Model of an icosahedron.

A VIRAL COAT CONSISTS OF MANY COPIES OF ONE OR A FEW KINDS OF PROTEINS

The total number of amino acids in the coat of a virus always exceeds the number of nucleotides in its genome. For example, the protein coat of a tobacco mosaic virus (TMV) particle contains about 340,000 amino acid residues, whereas its RNA contains only about 6400 nucleotides. In 1957, Francis Crick and James Watson noted that the protein coat of a virus cannot be one large molecule or an assembly of very many different small proteins because the amount of viral nucleic acid is much too small to code for such a large number of amino acid residues. On the other hand, the protein coat must be large enough to cover all the nucleic acid. Viruses circumvent their genetic poverty by having coats made up of a *large number of one or a few kinds of protein subunits.* For example, the coat of TMV consists of 2130 identical protein subunits (each containing 158 residues).

A viral coat can be built from identical subunits in only a limited number of ways. Stability is achieved by forming a maximum number of bonds and by using the same kinds of contacts again and again. Given these requirements, the resulting structure must be symmetric. Two kinds of arrangements of the protein coat are most likely: *a cylindrical shell having helical symmetry and a spherical shell having icosahedral symmetry* (Figure 34-3). In fact, *all small viruses are either rods or spheres* (or a combination of these shapes). The rules governing the structure of spherical viruses were deduced by Donald Caspar and Aaron Klug, who received some of their inspiration from the architectural designs of Buckminster Fuller.

SELF-ASSEMBLY OF TOBACCO MOSAIC VIRUS (TMV)

The simplest and best understood viral assembly process is that of TMV. This rod-shaped virus is 3000 Å long and 180 Å in diameter (Figure 34-4) and has a mass of about 40,000 kd. The 2130 identical subunits in the protein coat are closely packed in a helical array around a single-stranded RNA molecule consisting of 6390 nucleotides. The RNA is deeply buried in the protein, which renders it insusceptible to attack by ribonucleases. Each protein subunit interacts with three nucleotides. The dissociated RNA is labile, whereas intact TMV remains infective for decades.

TMV can be dissociated into its protein and RNA components by agents such as concentrated acetic acid. In 1955, Heinz Fraenkel-Conrat and Robley Williams showed that the *dissociated coat subunits and RNA of TMV spontaneously reassemble under suitable conditions into virus particles that are indistinguishable from the original TMV in structure and infectivity*. This was the first known example of the self-assembly of an active biological structure. The formation of hybrid virus particles from RNA of one strain and protein of another revealed that all of the genetic information of TMV is contained in its RNA (p. 87).

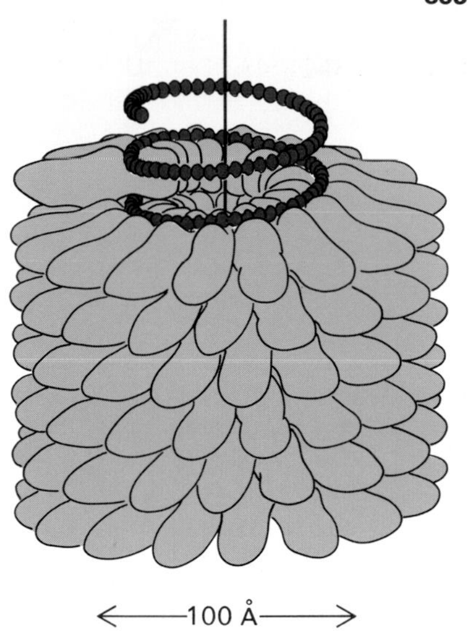

Figure 34-4
Model of a part of TMV, showing the helical array of protein subunits around a single-stranded RNA molecule. [After A. Klug and D. L. D. Caspar. *Adv. Virus Res.* 7(1960):274.]

PROTEIN DISCS ADD TO A LOOP OF RNA IN THE ASSEMBLY OF TMV

A priori, the simplest mechanism for the assembly of TMV would be the step-by-step addition of single protein subunits to the RNA. The difficulty with such a mechanism, however, is that nucleation would be very slow. About 17 coat subunits would have to add to a flexible RNA molecule before the complex could close on itself by forming a turn of helix and thereby acquire stability. This problem can be solved by adding a complex of many subunits to the RNA rather than one subunit at a time. In fact, the coat protein alone readily forms a *two-layered disc* consisting of 34 subunits. Each layer of a disc is a ring of 17 subunits, which is nearly the same as the number of subunits ($16\frac{1}{3}$) in a turn of the TMV helix. Klug and his associates solved the three-dimensional structure of the disc (Figure 34-5) and showed that it is a key intermediate in the assembly of TMV. *A critical property of the disc is that its subunits can slide over each other to form a two-turn helix, called a lockwasher* (Figure 34-6).

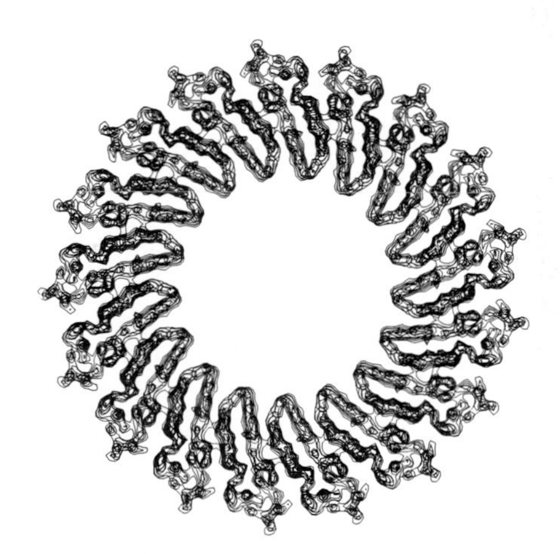

Figure 34-5
Electron-density map of a TMV protein disc. A slice 6 Å thick is shown here. [Courtesy of Dr. Aaron Klug.]

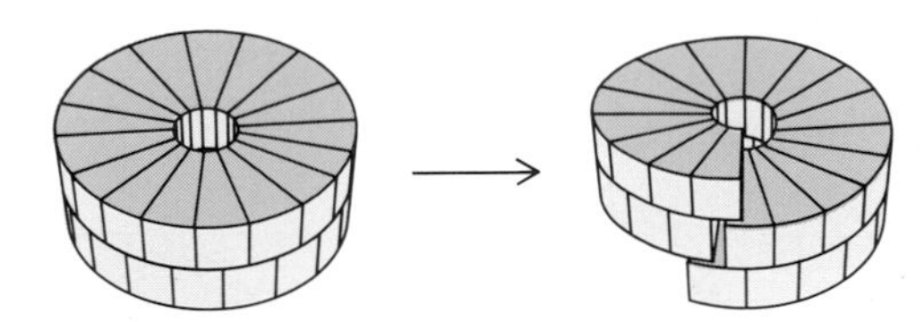

Figure 34-6
Schematic diagram showing the conversion of a TMV protein disc into the helical lock-washer form. [After A. Klug. *Fed. Proc.* 31(1972):40.]

A disc interacts much more rapidly with TMV RNA than with foreign RNAs. Hence, it seemed likely that TMV RNA contains a base sequence that is specifically recognized by the disc and serves to initiate assembly. This initiation region was isolated by adding a few discs to TMV RNA to coat the initiation region and then digesting the rest of the RNA with a nuclease. The protected fragment contains a common core of about 65 nucleotides that binds very tightly and specifically to discs. The *base sequence of this initiation region strongly suggests that it forms a*

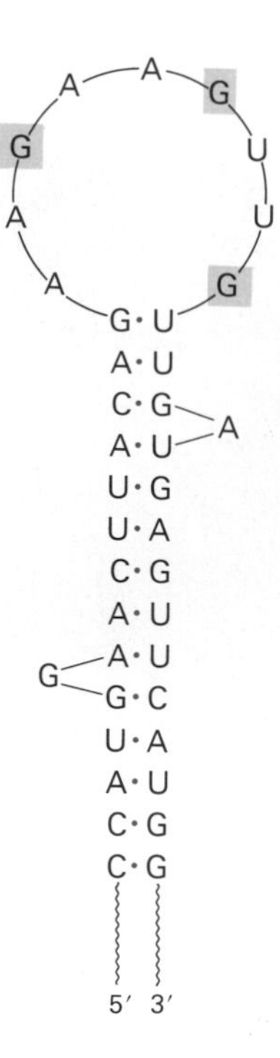

Figure 34-7
Initiation region in TMV RNA for the assembly of the virus particle.

hairpin structure with a base-paired stem and a loop (Figure 34-7). Most interesting, the loop contains G at every third position. This repeating triplet pattern matches the stoichiometry of three nucleotides per coat subunit in the virus. Hence, it seems likely that the loop binds the first disc to start the assembly of the virus.

Surprisingly, the initiation loop is far from either end of the RNA. Indeed, this starting point for assembly is about 5300 nucleotides from the 5′ end and about 1000 nucleotides from the 3′ end of the RNA. Another unexpected finding is that two tails of RNA emerge from the *same* end of a growing TMV particle (Figure 34-8). The length of the 3′ tail is rather constant throughout most of the assembly process, whereas the 5′ tail becomes shorter as the virus particle becomes longer.

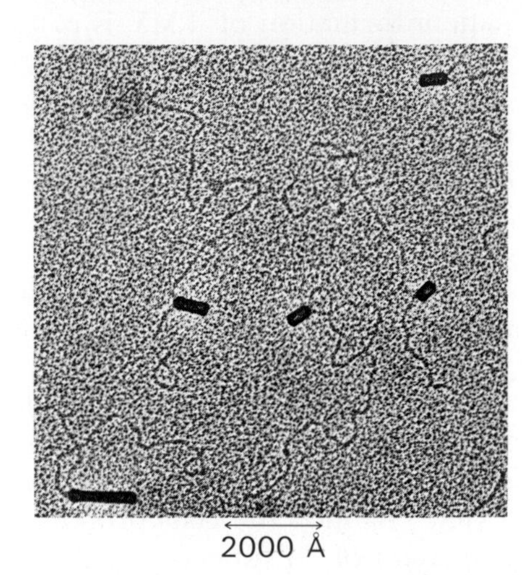

Figure 34-8
Electron micrograph of partly reconstituted TMV particles. Two RNA tails emerge from each growing virion. [From G. Lebeurier, A. Nicholaieff, and K. E. Richards. *Proc. Nat. Acad. Sci.* 74(1977):150.]

A plausible model for the formation of TMV particles is shown in Figure 34-9. Assembly starts with the insertion of an initiation loop into the central hole of a two-layered protein disc. The loop binds to the first turn of the disc, and the adjoining base-paired stem opens. This interaction transforms the disc into the helical lock washer form and traps the viral RNA. The viral helix has now been started. Another disc then

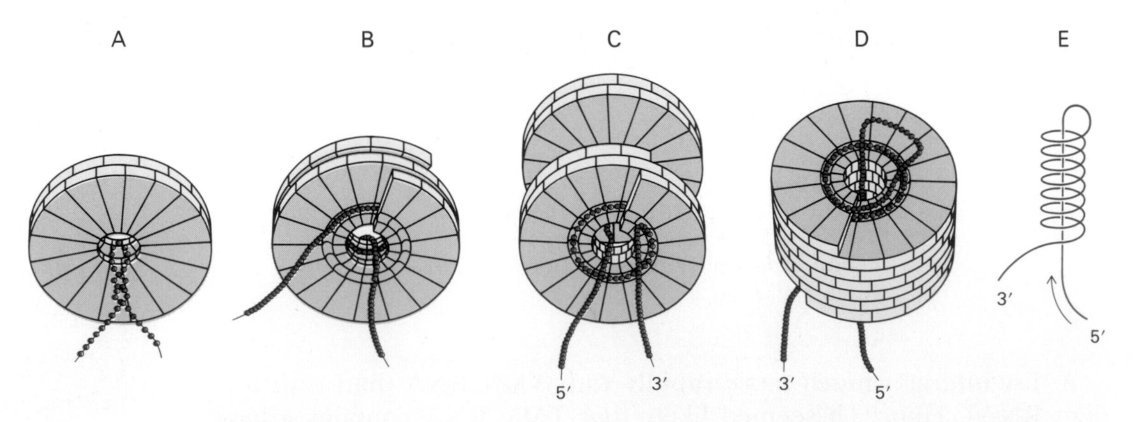

Figure 34-9
Model for the assembly of TMV: (A) the initiation region of an RNA loops into the central hole of a protein disc and transforms it into (B) the helical lock-washer form; (C) additional discs add to the looped end of the RNA; (D) one of the RNA tails is continually pulled through the central hole to interact with incoming discs; (E) schematic diagram of RNA in a partly assembled virus. The direction of RNA movement is denoted by the arrow. [After P. J. G. Butler and A. Klug. The assembly of a virus. Copyright © 1978 by Scientific American, Inc. All rights reserved.]

adds to a newly formed loop of RNA that protrudes from the central hole. The new loop is formed after the addition of each disc by the drawing up of the 5′ tail through the central hole of the growing virus particle. Finally, the 3′ tail is coated with protein in a way not yet determined.

The *two-layered disc markedly enhances the specificity of coating, in addition to assuring rapid nucleation.* A disc can bind a sequence of many nucleotides, whereas a single subunit can interact with only three. Consequently, the disc is much more discriminating than a single subunit could be in selecting TMV RNA instead of host mRNA for coating. Another important feature of discs is that, under physiological conditions, they do not form helices devoid of RNA. Two closely spaced carboxyl groups in each subunit play a critical role in this regard. At neutral pH, both carboxyl groups are ionized in the helix form, but only one is ionized in the disc. This favors the disc form because of electrostatic repulsion between the carboxylates in the helix form. The binding of RNA to the helix form provides enough free energy to overcome the electrostatic repulsion of the carboxylates. Thus, *the carboxylates are a negative switch to prevent the formation of a helix devoid of RNA.*

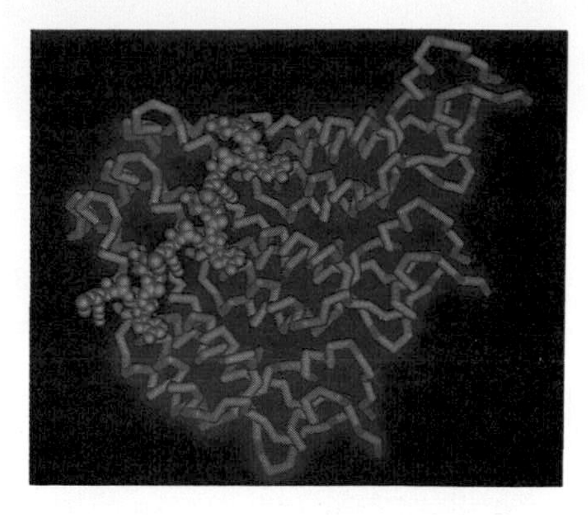

Model of TMV showing RNA (red) bound to three protein subunits (blue). [Courtesy of Dr. Gerald Stubbs.]

T4 PHAGE REDIRECTS BIOSYNTHESES IN INFECTED BACTERIAL CELLS

The T4 bacteriophage is a much more complex virus than TMV. Its double-stranded DNA contains about 165 genes, compared with 6 in TMV. However, much is known about the structure, multiplication, and assembly of T4 because it has been subjected to intensive genetic and biochemical analysis. The T4 virion consists of a *head,* a *tail,* and six *tail fibers* (Figure 34-10). The DNA molecule is tightly packed inside an icosahedral protein coat to form the head of the virus. The tail is made up of two coaxial hollow tubes that are connected to the head by a short neck. In the tail, a contractile sheath surrounds a central core through which the DNA is injected into the bacterial host. The tail terminates in a baseplate that has six short spikes and gives off six long, slender fibers.

Table 34-2
Genes in T4 phage

Type	*Number*
Metabolic, essential	22
Metabolic, nonessential	60
Particle assembly, structural proteins	40
Particle assembly, other proteins	13
Total number of identified genes	135

Source: W. B. Wood and H. R. Revel, *Bacterial Rev.* 40(1976):860.

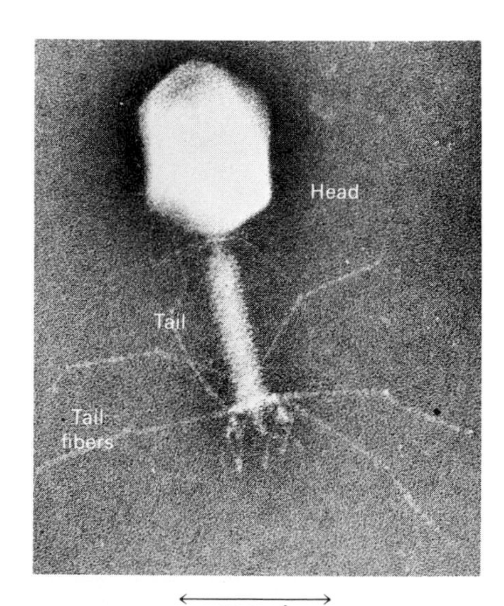

Figure 34-10
Electron micrograph of a T4 phage. [From R. C. Williams and H. W. Fisher. *An Electron Micrographic Atlas of Viruses*, 1974. Courtesy of Charles C Thomas, Publisher, Springfield, Illinois.]

The tips of the tail fibers bind to a specific site on the outer membrane of *E. coli.* An ATP-driven contraction of the tail sheath pulls the phage head toward the baseplate and tail fibers, which causes the central core to penetrate the cell wall but not the cell membrane. The naked T4 DNA then penetrates the cell membrane. *A few minutes later, all synthesis of cellular DNA, RNA, and protein stops, and the synthesis of viral*

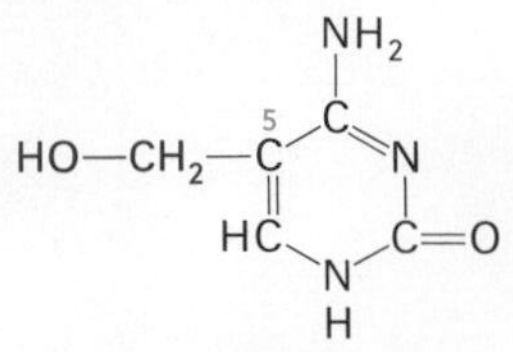

5-Hydroxymethylcytosine

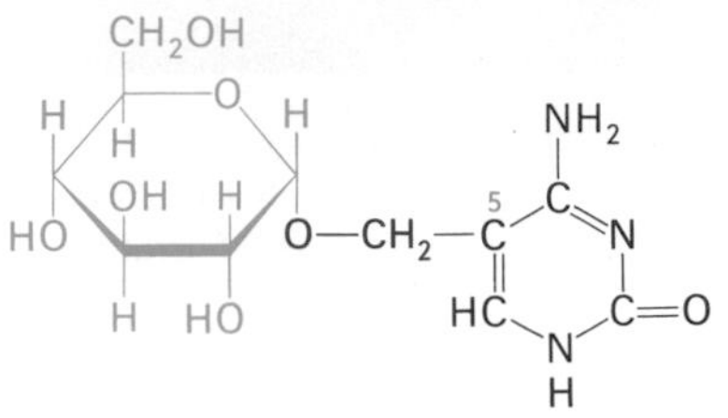

α-Glucosylated derivative of 5-hydroxymethylcytosine

macromolecules begins. The infecting virus commandeers the biosynthetic machinery of the cell and substitutes its genes for bacterial ones.

T4 DNA contains three sets of genes that are transcribed at different times after infection: *immediate-early, delayed-early,* and *late.* The early genes are transcribed and translated before T4 DNA is replicated. Some proteins coded by these genes are responsible for switching off the synthesis of cellular macromolecules. *Soon after infection, the host-cell DNA is degraded by a deoxyribonuclease that is specified by a T4 early gene. This enzyme does not hydrolyze T4 DNA because it lacks clusters of cytosines.* T4 DNA contains 5-hydroxymethylcytosine (HMC) instead of cytosine. Furthermore, some HMC residues in T4 DNA are *glucosylated.*

The DNA of T4 bacteriophage contains these derivatives of cytosine because of the action of several phage-specified enzymes that are synthesized in the early phase of infection. A pyrophosphatase hydrolyzes dCTP and dCDP to dCMP to prevent the incorporation of dCTP into T4 DNA. A hydroxymethylase then converts dCMP into 5-hydroxymethylcytidylate and a kinase phosphorylates it to the diphosphate. A bacterial kinase phosphorylates it to 5-hydroxymethyl dCTP for incorporation into DNA. Finally, some hydroxymethylcytosines in DNA are glucosylated by yet another viral-encoded enzyme.

Synthesis of the *late proteins* is coupled to the replication of T4 DNA. The capsid proteins and a lysozyme are formed at this stage. The lysozyme digests the bacterial cell wall and causes its rupture when the assembly of progeny virions has been completed. Some twenty minutes after infection, about two hundred new virus particles emerge.

SCAFFOLD PROTEINS AND PROTEASES PARTICIPATE IN THE ORDERED ASSEMBLY OF T4 PHAGE

The assembly of the T4 virion is a far more intricate process than that of TMV. The T4 capsid has a much higher degree of structural organization and contains some forty different proteins. An additional thirteen proteins participate in the construction of this virus. Insight into the mechanism of assembly of T4 has come from the combined use of genetic, biochemical, and electron-microscopic techniques. The studies of William Wood and Robert Edgar on mutants of T4 that are defective in assembly have revealed that:

1. *Three major pathways in the construction of the virus lead independently to the head, the tail, and the tail fibers* (Figure 34-11). A block in the formation of one of these components does not affect the synthesis of the other two.

2. *A strict sequential order* is followed in each pathway. All of the capsid proteins are synthesized simultaneously during the latter half of the infectious cycle. Hence, the sequential nature of the assembly process is imposed by structural features of the intermediates themselves. *None of the associations in a pathway occur at a significant rate unless the preceding ones have taken place.* Each step produces conformational changes that direct the next one. Part of the binding energy at each step is used to lower the activation energy for the formation of the next complex and thereby increase its rate.

3. The head and tail must be completed before they can combine. Completed tail fibers then attach to the baseplate. Again, the strict sequential order assures that only complete virus particles are formed.

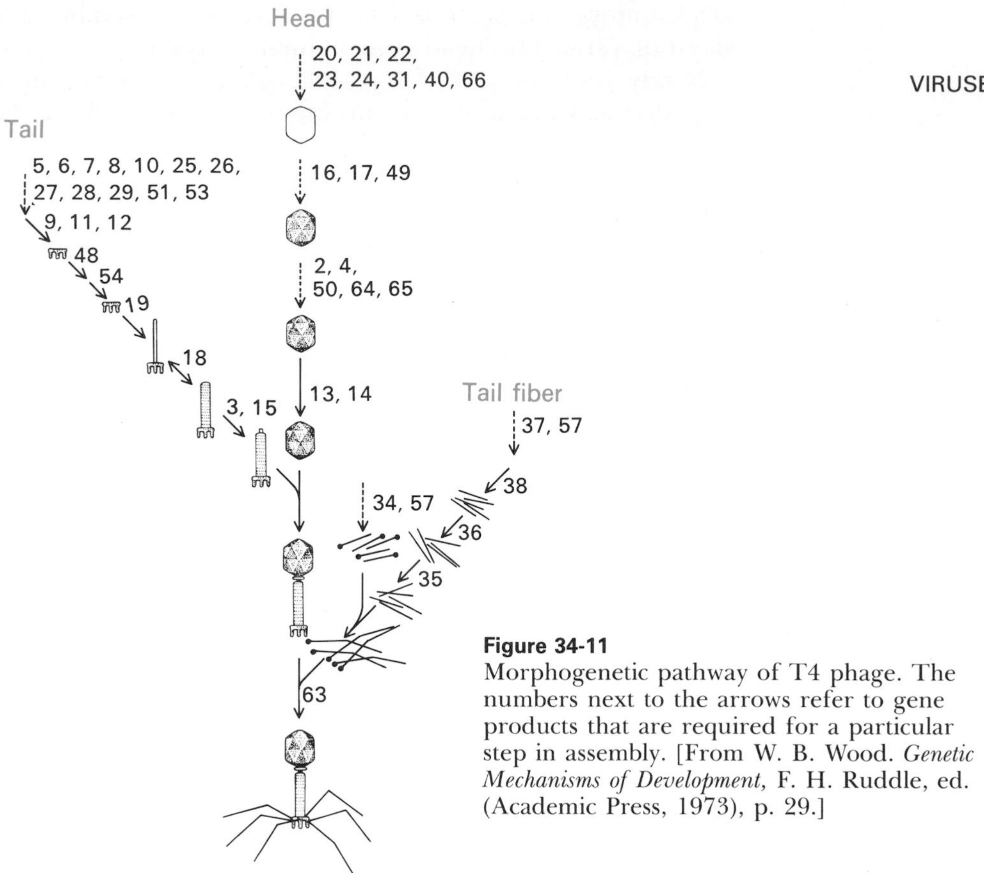

Figure 34-11
Morphogenetic pathway of T4 phage. The numbers next to the arrows refer to gene products that are required for a particular step in assembly. [From W. B. Wood. *Genetic Mechanisms of Development,* F. H. Ruddle, ed. (Academic Press, 1973), p. 29.]

The construction of T4 virions does not occur by self-assembly alone. Rather, *scaffold proteins* and *proteases* play essential roles at certain stages of the assembly process. For example, three proteins that do not become part of the finished structure are needed for the formation of the central plug in the baseplate of the tail. These scaffold proteins serve as transient templates to promote the association of the constituents of the plug. Proteases have a prominent role in the assembly of the head. The 45-kd major protein of the head, called *gp 23** (gp stands for gene product), is derived from gp 23, the 55-kd precursor. Cleavage takes place when the head is partly assembled, which suggests that it may be the trigger for the entry of DNA. Three other head proteins are also known to be cleaved during assembly. *Thus, T4 is constructed by a combination of self-assembly, scaffold-assisted assembly, and enzyme-directed assembly.*

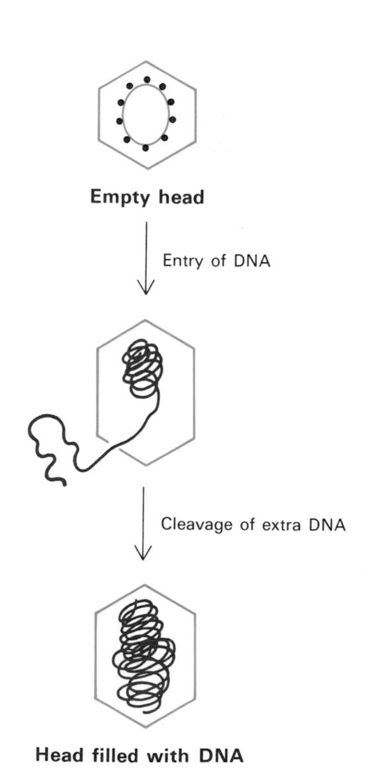

Figure 34-12
Schematic diagram of the packing of DNA in T4 head assembly.

T4 DNA IS INSERTED INTO A PREFORMED HEAD

How is a DNA molecule formed and packaged inside a phage head? The problem is formidable: the DNA has a contour length of 56 μm, yet it must fit inside a head only 0.1 μm long. Furthermore, the volume of the DNA (1.8×10^{-4} μm^3) is not much less than that of the head (2.5×10^{-4} μm^3). A priori, the DNA could be introduced into a preformed head or, alternatively, a head could be assembled around a core of condensed DNA. The isolation of empty phage heads capable of packaging DNA strongly suggested that *T4 DNA is inserted into preformed heads* (Figure 34-12). Indeed, several proteins in the head are cleaved as

DNA enters and is wound into a structure resembling an elongated skein of yarn. The head concomitantly expands.

Newly replicated T4 DNA is concatameric—the entire genome is repeated several times (see problem 2, on page 885). *This long DNA duplex containing multiple genomes is cleaved when a unit-length genome has entered the head.* A phage-encoded nuclease cleaves the DNA when the head is filled rather than on recognizing a specific sequence. Virion DNA is packed so that it can be rapidly injected into a bacterium on the next round of infection.

BACTERIAL RESTRICTION ENDONUCLEASES RECOGNIZE UNMETHYLATED TARGET SITES IN FOREIGN DNA MOLECULES

T4 phage has enzymatic devices for specifically degrading the DNA of its host. Likewise, bacteria contain enzymes called *restriction endonucleases* that cleave foreign DNA molecules such as viral DNAs. The discovery of these protective enzymes followed from the observation that phage propagated in one strain of bacteria (such as *E. coli* B) grew poorly in a second strain (such as *E. coli* K) and vice versa. These phage were said to be *restricted* on encountering a strain different from the one in which they had been cultivated. However, a small proportion (about 10^{-5}) of the phage evaded restriction and grew well in their new host. They then lost the capacity to grow in their former host. Werner Arber showed that restriction is caused by degradation of phage DNA and that viral DNA not restricted carries a host-specific tag.

What is the nature of this tag? *Methylation of a base in specific palindromic target sequences protects the bacterium's own DNA from cleavage by its restriction endonuclease.* The essence of this surveillance system is the possession by the bacterium of a methylase and a restriction endonuclease that recognize the same base sequence (Figure 34-13). Methylation of these specific sites is the password denoting that the DNA belongs to the bacterium and is not an intruder. Newly synthesized DNA methylated on only one strand is also protected from digestion. Methylases then convert hemimethylated DNA into fully methylated DNA before the next round of replication. Nearly all foreign DNA molecules containing unmethylated target sites are rapidly degraded by the endonuclease. However, this highly effective mechanism for distinguishing between self and nonself is not perfect. A small proportion of invading

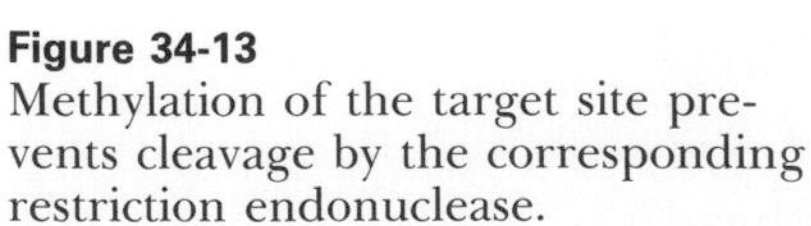
Figure 34-13
Methylation of the target site prevents cleavage by the corresponding restriction endonuclease.

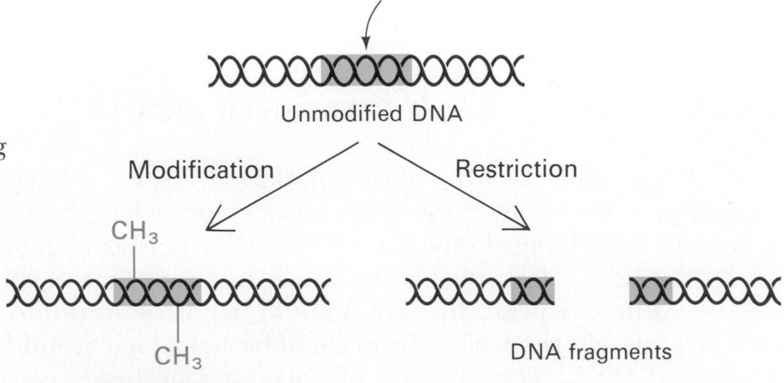

DNA molecules acquire a full complement of methyl groups before they can be cut by the restriction enzyme; the bacterial cell can no longer identify them as being foreign.

Three types of restriction-modification systems are found in bacteria. We have already considered type II restriction endonucleases, which are indispensable tools for analyzing DNA and in constructing and cloning new DNA molecules. The discovery of these exquisitely specific enzymes made possible the recombinant DNA revolution (p. 118). Most of these endonucleases recognize a specific sequence of four to six base pairs and hydrolyze a single phosphodiester bond in each strand of this target. These target sites are *palindromic*—that is, they possess a twofold rotational symmetry axis (Figure 34-14). The symmetry of a target site is matched by that of the corresponding restriction enzyme. The structure and mode of action of the EcoRI endonuclease was discussed in an earlier chapter (p. 657). The methylase, a different enzyme, recognizes the same DNA sequence as does the endonuclease. *S*-Adenosylmethionine is the methyl donor in the modification of DNA by methylases.

The methylase and nuclease activities of type I and III systems are combined in large multisubunit complexes. For example, the *E. coli* B and K type I complexes consist of three kinds of polypeptide chains. The α chain has the endonuclease activity, the β chain has the methylase activity, and the γ chain has the recognition site for DNA. Type I enzymes cleave unmodified DNA at random sites a kilobase or more from the recognition site, and type III enzymes cut the backbone some 25 base pairs from the recognition site. Type I and III enzymes are processive; their movement along a DNA strand is powered by the hydrolysis of ATP.

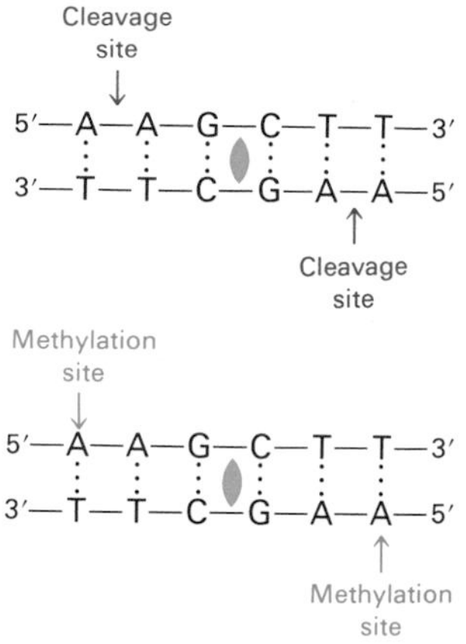

Cleavage site
5′—A—A—G—C—T—T—3′
3′—T—T—C—G—A—A—5′
Cleavage site

Methylation site
5′—A—A—G—C—T—T—3′
3′—T—T—C—G—A—A—5′
Methylation site

Figure 34-14
Specificity of the HindIII restriction endonuclease from *Hemophilus influenzae* and of the corresponding methylase. The green symbol denotes the twofold axis of symmetry.

LYSOGENIC PHAGES CAN INSERT THEIR DNA INTO THE HOST-CELL DNA

Some bacteriophages have a choice of life styles: they can multiply and then lyse an infected cell (*lytic pathway*), or their DNA can join the infected cell, retaining the capacity for multiplication and lysis (*lysogenic pathway*). Viruses that do not always kill their hosts are called *temperate* or *moderate*. The best-understood temperate virus is *lambda* (λ), whose control of transcription was discussed previously (p. 807). Recall that the λ repressor binds to two sets of operator sites (the O_L and O_R regions) and that it regulates its own synthesis.

The DNA in the λ virion is a linear 48-kb duplex. The 5′ ends of this DNA molecule are single-stranded sequences of twelve nucleotides.

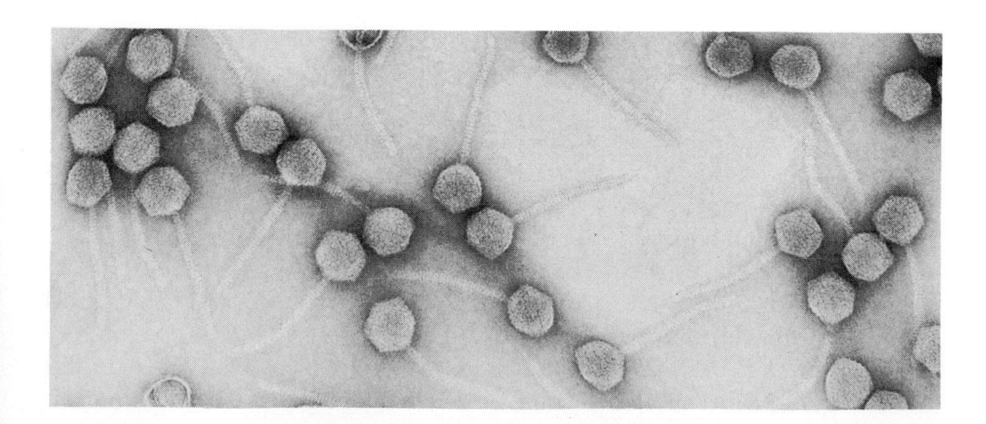

Figure 34-15
Electron micrograph of λ phage. [Courtesy of Dr. A. Dale Kaiser.]

These sequences are called *cohesive ends* because they are mutually complementary and can base-pair to each other. In fact, the cohesive ends come together after infection of a bacterium. The 5′ phosphate terminus of each strand is then adjacent to the 3′ hydroxyl end of the same strand. A host DNA ligase seals these two gaps to yield a *circular λ DNA molecule* (Figure 34-16).

This circular λ DNA molecule can be replicated by the λ proteins acting with the DNA replication machinery of the host. Alternatively, the λ DNA circle can be inserted by enzymes into the bacterial chromosome by recombination between specific loci of 15 base pairs on the λ and the *E. coli* DNA. This event is called *site-specific recombination* to contrast it with general recombination (p. 688).

The attachment site for λ in *E. coli* DNA, called *attB*, is located between *galE* and *bioA*, genes in the galactose and biotin operons. The base sequence of *attB* is symbolized by B-O-B′ (B for bacterial). The specific attachment site on λ, called *attP*, is located next to the genes *int* (for integrate) and *xis* (for excise). The base sequence of *attP* is symbolized by P-O-P′ (P for phage). O denotes the common core sequences in the phage and bacterial segments that recombine. Integrase, the protein encoded by the *int* gene, recognizes the P-O-P′ sequence in the phage DNA and the B-O-B′ sequence in *E. coli* DNA. *IHF* (integration host factor), a bacterial protein, is also required. It is interesting to note that the DNA must be negatively supercoiled to undergo recombination. *Integrase then catalyzes a reciprocal transfer of DNA chains: P joins B′ and B joins P′*. This scheme (Figures 34-17 and 34-18) was originally proposed by Allan Campbell on the basis of genetic evidence.

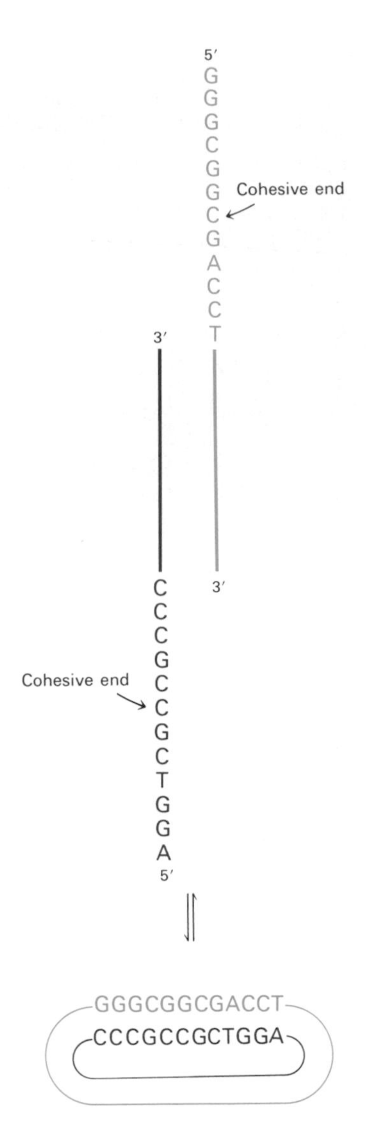

Figure 34-16
Conversion of linear λ DNA into the circular form.

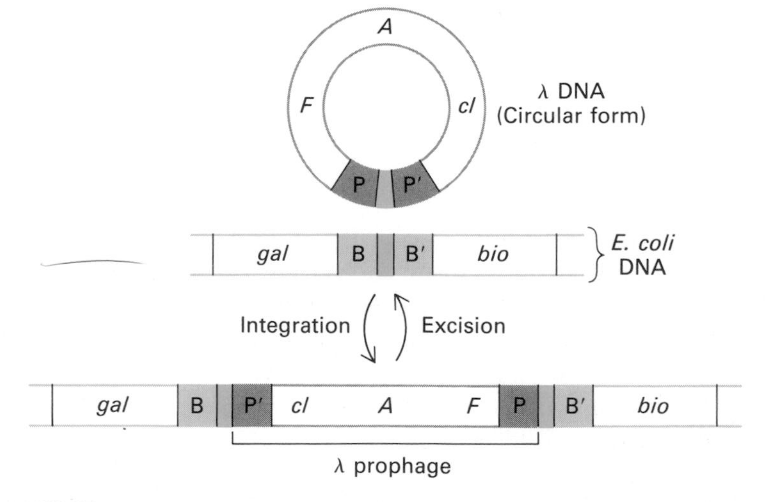

Figure 34-17
Schematic diagram of reciprocal recombination of λ DNA and *E. coli* DNA.

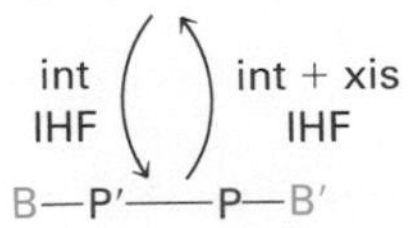

B—B′ + P—P′

int / IHF ⇅ int + xis / IHF

B—P′——P—B′

Figure 34-18
Integration and excision of λ DNA. *Int* denotes integrase and *xis* denotes excisionase. IHF (integration host factor) is a host protein, whereas *int* and *xis* are encoded by λ.

The λ DNA is now part of the *E. coli* DNA molecule. This form of λ is called the *prophage,* and the *E. coli* cell containing the prophage is called a *lysogenic bacterium.* The prophage is stable in the absence of *excisionase*, an enzyme encoded by the *xis* gene. Transcription of *xis* is blocked by the λ repressor (p. 809). When repression is released, excisionase and integrase together catalyze the breaking of the B-P′ and P-B′ sequences, and again a transfer occurs (Figure 34-18): P rejoins P′ and B rejoins B′, which recreates a circular molecule of λ DNA and a nonlysogenic *E. coli* chromosome. A key feature of this recombination system is that integrase alone cannot recognize the two new sequences at the ends of the prophage (B-P′ and P-B′), which makes the prophage DNA stable. Thus, insertion occurs when integrase alone is present, whereas excision occurs when both integrase and excisionase are present. Hence, integration and excision are ultimately controlled by the level of the λ repressor and cro protein (p. 810).

THE FLEXIBILITY OF THE COAT PROTEIN OF TBSV ENABLES IT TO FORM A LARGE ICOSAHEDRAL SHELL

Tomato bushy stunt virus (TBSV), a spherical virus (Figure 34-19), illustrates a principle of virus construction different from those of tobacco mosaic virus, T4 phage, and λ phage. TBSV consists of a single molecule of RNA (4800 nucleotides) surrounded by a shell of 180 identical protein subunits (each 41 kd). How are these coat subunits arranged? The highest possible symmetry of an isometric shell, found in an icosahedron, is 60-fold (Figure 34-20A). In other words, no more than 60 identical subunits can pack together in a *precisely* symmetric way to form a spherical shell. However, TBSV and a number of other spherical viruses contain 180 identical subunits. *An advantage of building a shell from 180 instead of 60 subunits of the same size is that more nucleic acid can be packaged in a larger virion. This is achieved not by abandoning symmetry but by relaxing it slightly* (Figure 34-20B). The high-resolution x-ray analysis of the structure of TBSV by Stephen Harrison shows that its chemically identical coat subunits belong to three sets (A,B,C), each consisting of 60 proteins that obey strict icosahedral symmetry. However, members of different sets are related to each other slightly differently, and so they are said to be *quasi-equivalent.*

What is the physical basis of quasi-equivalence? X-ray analyses showed that each subunit consists of an S domain that forms part of the surface shell, a P domain that projects outward, an amino-terminal arm that goes inward, and an R domain that is internal. (Figure 34-21). The

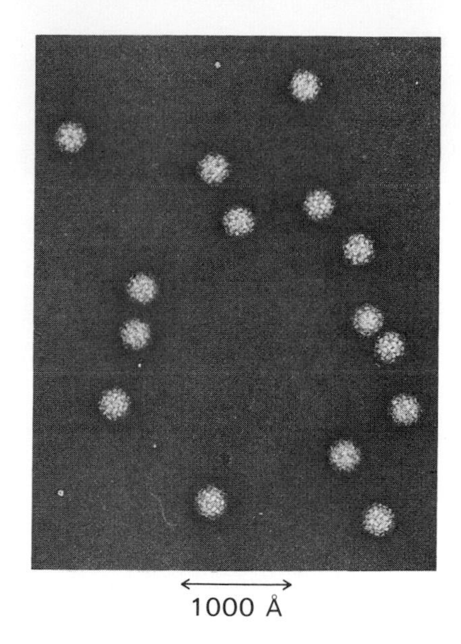

Figure 34-19
Electron micrograph of tomato bushy stunt virus. [Courtesy of Dr. Robley Williams.]

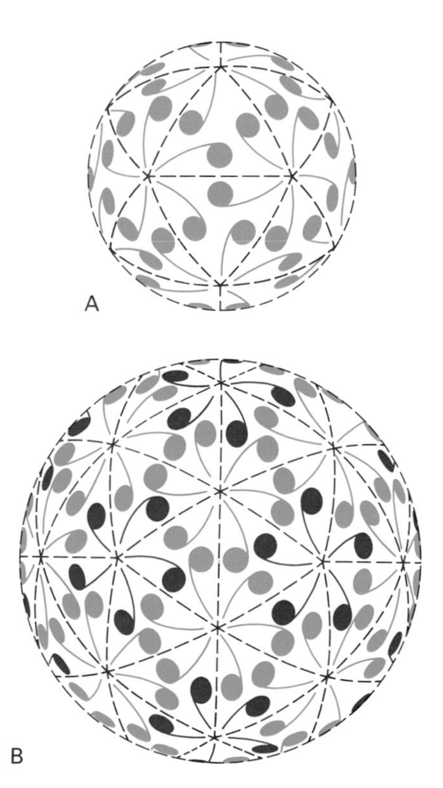

Figure 34-20
Icosahedral surface lattices showing the packing of (A) 60 strictly equivalent subunits and (B) 180 quasi-equivalent subunits. Note that all of the tail-to-tail contacts in part A are in rings of 5, whereas some of these contacts in part B are in rings of 5 and others in rings of 6. [After S. C. Harrison. *Trends Biochem. Sci.* 3(1978):4.]

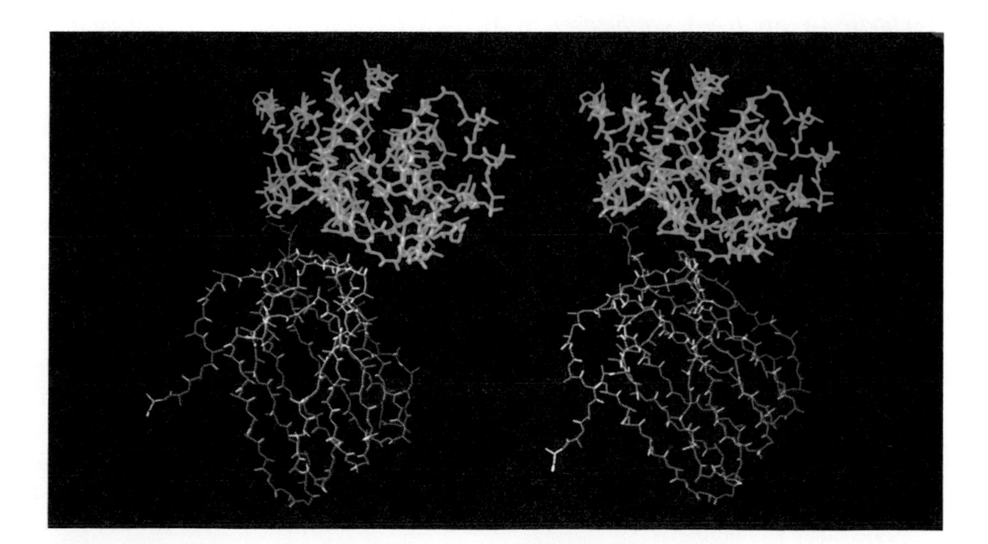

Figure 34-21
The coat protein of tomato bushy stunt virus can adopt three different conformations because of the presence of a hinge between its P (red) and S (green) domains. Two of these conformations are compared here (left and right views). The carboxyl-terminal residue is shown in white. The amino-terminal region (not shown) is disordered in one of the conformations and ordered in the other. [Based on a diagram kindly provided by Dr. Stephen Harrison.]

P and S domains have nearly the same structure in all of the subunits. In contrast, the angle between P and S domains is very different in the C subunits from that in the A and B subunits. The P and S domains are joined by a *hinge* that has an angular range of at least 20 degrees. Another difference is that part of the amino-terminal arm has a well-defined structure in the C subunits, whereas it is highly disordered in the others. *These modes of flexibility enable subunits belonging to different sets to interact in a nearly identical manner, which is necessary for self-assembly.* On the other hand, the RNA strand trapped inside does not appear to adopt a well-defined shape. It does interact with the flexible amino-terminal arm and the R domain of the coat subunit, which penetrate into the interior. TBSV undergoes a reversible expansion at pH values above 7.0 on removal of its bound Ca^{2+} ions. This conformational transition, caused by electrostatic repulsion of aspartate side chains at subunit interfaces, may be important in allowing RNA to enter and leave the particle.

Southern bean mosaic virus (SBMV), a spherical RNA virus that infects legumes, is also built of 180 identical subunits. Indeed, the folding of its polypeptide chains resembles that of TBSV. SBMV contains R and S domains joined by a flexible hinge but lacks the P domain of TBSV. The positively charged R domains of both viruses serve to neutralize the negative charge of the RNA housed inside the virion. The wedge shape of their S domains is well suited to forming icosahedral shells.

STRATEGIES FOR THE REPLICATION OF RNA VIRUSES

We turn now from strategies used to form viral structures to those employed in their replication. DNA viruses recruit many host proteins in the replication and expression of their genomes. In the case of RNA viruses, a special problem arises because uninfected host cells lack enzymes for synthesizing RNA according to the instructions of an RNA template. Consequently, RNA viruses must contain genetic information for the synthesis of an *RNA-directed RNA polymerase* (also called an *RNA replicase* or an *RNA synthetase*) or for an *RNA-directed DNA polymerase* (also called a *reverse transcriptase*). It is informative to classify RNA viruses according to the relation between their virion RNA and mRNA. By convention, mRNA is defined as (+) RNA and its complement as (−) RNA. Four pathways of replication and transcription of RNA viruses are known (Figure 34-22). Class 1 viruses (e.g., poliovirus) are *positive-strand RNA viruses.* They synthesize (−) RNA, which then serves as the template for the formation of (+) mRNA. Class 2 viruses (e.g., rabies virus) are *negative-strand RNA viruses* in which virion (−) RNA is the template for the synthesis of (+) mRNA. Class 3 viruses (e.g., reovi-

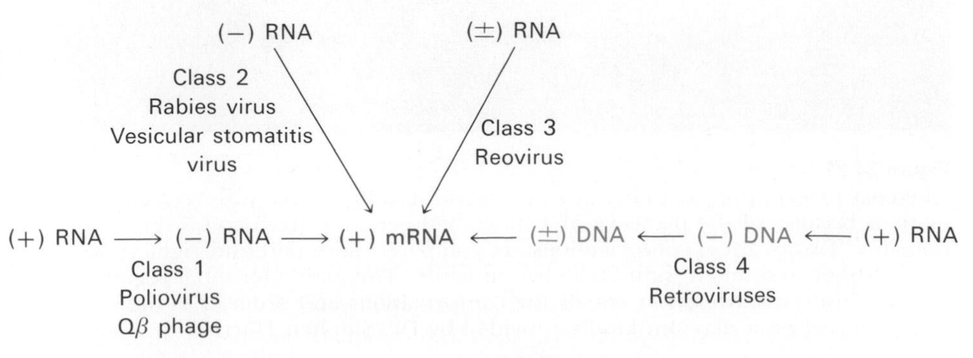

Figure 34-22
Modes of gene expression of RNA viruses. [After D. Baltimore. *Bacteriol. Rev.* 35(1971):326.]

rus) are *double-stranded RNA viruses* in which the virion (±) RNA directs the asymmetric synthesis of (+) mRNA. Class 4 viruses, the most unusual, are *retroviruses* (e.g., Rous sarcoma virus). They express the genetic information in their virion (+) RNA through a DNA intermediate that serves as the template for the synthesis of (+) RNA. Thus, the flow of information in retroviruses is from RNA to DNA and then back to RNA (p. 87).

SMALL RNA PHAGES CONTAIN OVERLAPPING GENES

RNA phages such as R17 (MS2, F2) and Qβ are very simple Class 1 viruses that infect *E. coli.* They have a regular polyhedral shape and a diameter of about 200 Å. The capsid of these closely related phages contains 180 molecules of the coat protein (14 kd) and 1 molecule of the A (maturation) protein (38 kd). The single-stranded (+) RNA also codes for a subunit of the replicase. In 1976, Walter Fiers and his coworkers determined the complete nucleotide sequence of MS2 RNA (3569 nucleotides), the first entire sequence of a viral genome. This nucleotide sequence provided a wealth of information on how these simple RNA viruses store and express genetic information. It was thought for some time that they contain only three genes. However, the finding of a phage mutant that formed normal virions but was unable to lyse the host cell led to a search for another virus-encoded protein. Indeed, these RNA phages do contain a fourth gene, which specifies a protein needed to lyse the bacterial host. The gene for this lysis protein overlaps the genes for the coat protein and replicase subunit (Figure 34-23). *These small RNA phages use overlapping genes to pack more information into their small genomes.* The (+) RNA molecule of the virion serves both as a messenger for the synthesis of the four proteins and as a template for the synthesis of (−) RNA. Then, (−) RNA is used as the template for producing many copies of (+) RNA.

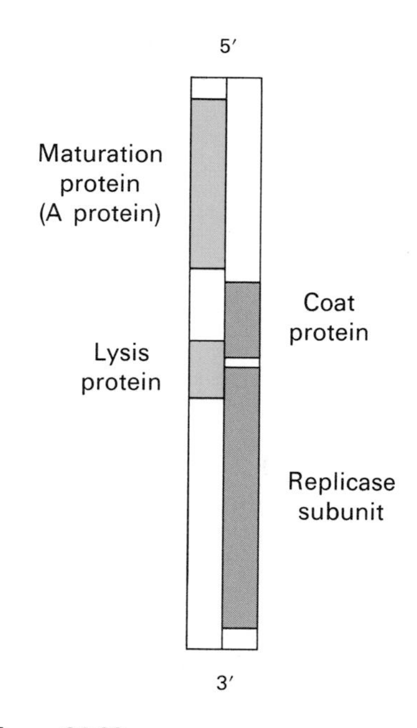

Figure 34-23
Overlapping genes in the RNA of R2 phage and the closely related MS2 and F2 phages. One of the reading frames is shown in yellow and the other in blue.

The replicase that synthesizes the (+) and (−) strands of phage RNA is noteworthy in two respects. First, it is highly specific for phage RNA. Hence, host RNA molecules do not compete with phage RNA for replication. Second, the replicase consists of four subunits, of which only one is encoded by Qβ RNA. *The other three subunits of the replicase are host proteins that have been recruited by the phage to serve its own needs.* Two of them are elongation factors for protein synthesis, EF-Tu and EF-Ts, and the third is a protein that is normally bound to the 30S ribosomal subunit. Thus, Qβ creates a highly specific enzyme very economically.

Regulatory mechanisms assure that replication and translation are appropriately timed. The (+) RNA is both a messenger for protein synthesis and a template for the synthesis of (−) RNA. It would be undesirable for both processes to occur simultaneously on the same (+) RNA because ribosomes travelling in the 5′ to 3′ direction would collide with the replicase moving in the 3′ to 5′ direction. This does not occur because Qβ replicase strongly inhibits the binding of ribosomes to (+) RNA until an adequate number of (−) RNA have been synthesized.

The four phage proteins are synthesized in different amounts. The coat protein, which is synthesized throughout infection, is the major product of translation. One reason is that ribosomes bind much more strongly to the initiation site for the coat protein than to the others on the (+) RNA. Furthermore, the coat protein inhibits translation of the replicase gene by blocking its initiation site. Thus, *the coat protein is a specific translational repressor.* The A protein is translated from unfinished nascent

(+) RNA molecules because its initiation site is blocked by base pairing in finished RNA molecules, whose secondary structure permits only small amounts of the A protein to be made.

DARWINIAN EVOLUTION OF PHAGE RNA OUTSIDE A CELL

The purification of Qβ RNA and Qβ replicase free from contaminating nucleases enabled Sol Spiegelman to examine evolutionary events outside a living cell. One question asked was: *What happens to RNA molecules if the only demand made on them is that they multiply as rapidly as possible?* Molecules of Qβ RNA, Qβ replicase, and ribonucleoside triphosphates were incubated for twenty minutes, an incubation period that facilitates the selection of mutant RNA molecules that are rapidly replicated. An aliquot was diluted into a fresh standard reaction mixture containing only Qβ replicase and ribonucleoside triphosphates. A series of seventy-five such transfers was carried out, and the RNA products were analyzed. The length of incubation was reduced periodically to select the first RNA molecules to be completed. *The striking outcome was that the RNA molecules after seventy-five transfers ("generations") were only 17% as long as the original Qβ RNA.* In this in vitro environment, the viral genes encoding proteins could be shed. This mutant RNA was replicated fifteen times as rapidly as the wild type. The only requirements imposed by selection were (1) that the 3′ end of the RNA contain the initiation sequence recognized by Qβ replicase, and (2) that the secondary structure of the RNA not impede the movement of the enzyme. An even shorter mutant containing only 220 of the original 4220 nucleotides was obtained by further selection.

The cloning of single Qβ RNA molecules by Charles Weissman has revealed that the wild-type sequence is found in only a small fraction of the RNA molecules; most sequences deviate in one or two positions from the wild type. The rate of mutation of Qβ virus is about 3×10^{-4}

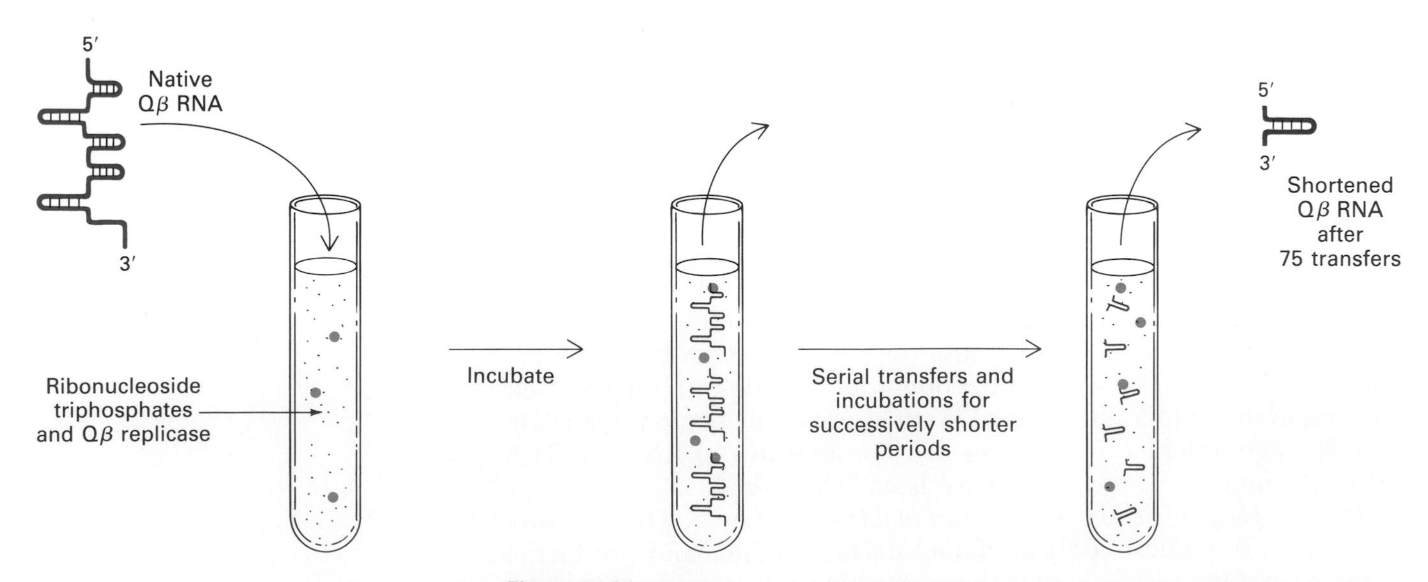

Figure 34-24
Molecular evolution in a test tube. Qβ RNA was added to a medium containing the replicase and nucleoside triphosphates. An aliquot was transferred to another tube after a short incubation to select for the first RNA molecules to be completed. After 75 transfers, the mutant RNA was only 17% as long as the wild type.

per nucleotide per generation. How is a wild-type sequence perpetuated in the face of this high mutation rate? The answer is that mutants revert to the wild type after many generations because the wild type has a significant selective advantage. The 4220-nucleotide genome of $Q\beta$ is close to the maximum possible for a virus having a mutation rate of 3×10^{-4}. Indeed, *all RNA viruses have high mutation rates and hence their replicative units are usually not much larger than 10 kb.* DNA viruses can have larger, more stable genomes than RNA viruses because DNA polymerases, in contrast with RNA replicases and polymerases, proofread their nascent products and correct their mistakes.

VIROIDS, THE SIMPLEST PLANT PATHOGENS, ARE VERY SMALL CIRCULAR RNA MOLECULES RESEMBLING INTRONS

Viroids are single-stranded circular RNA molecules about 300 nucleotides long that infect plants. Indeed, viroids are the smallest pathogens. They lack a coat and do not appear to encode any proteins. Theodor Diener proposed in 1979 that *viroids may be escaped introns.* This hypothesis is supported by sequence comparisons showing that viroids resemble group I introns, a class of introns found in genes encoding nuclear ribosomal RNA, mitochondrial mRNA and ribosomal RNA, and chloroplast tRNA. The self-splicing intron of the ciliated protozoan *Tetrahymena* (p. 725) belongs to this class. Viroids contain a 16-base sequence that is conserved in these introns. Several other sequences are common to viroids and introns of this class, suggesting that they have a common secondary and tertiary structure. It will be very interesting to learn whether the intron-like structure of viroids enables them to cause plant diseases such as cadang-cadang, which has killed more than 30 million coconut palms in the Philippines.

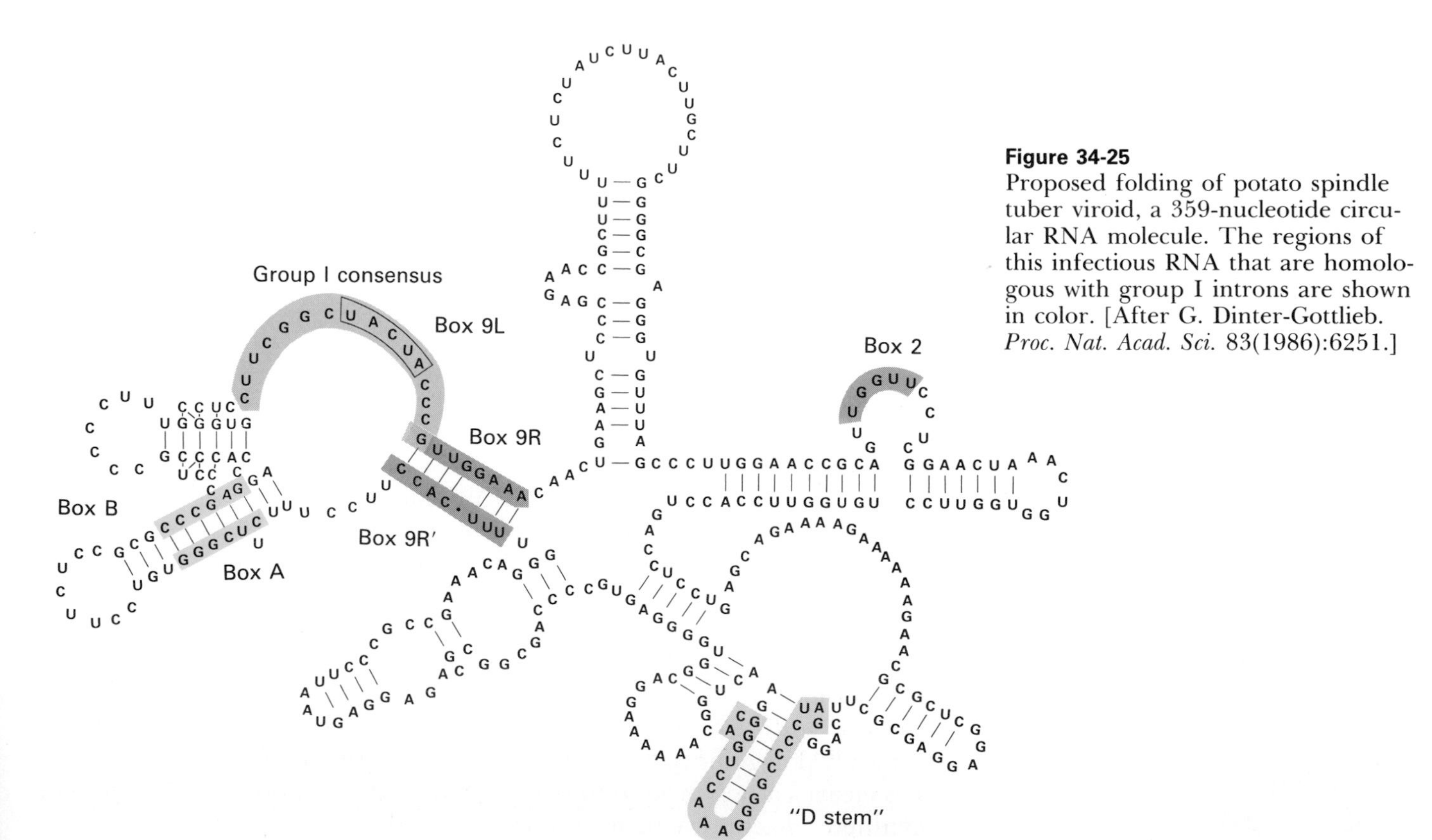

Figure 34-25
Proposed folding of potato spindle tuber viroid, a 359-nucleotide circular RNA molecule. The regions of this infectious RNA that are homologous with group I introns are shown in color. [After G. Dinter-Gottlieb. *Proc. Nat. Acad. Sci.* 83(1986):6251.]

POLIOVIRUS PROTEINS ARE FORMED BY MULTIPLE CLEAVAGES OF A GIANT PRECURSOR

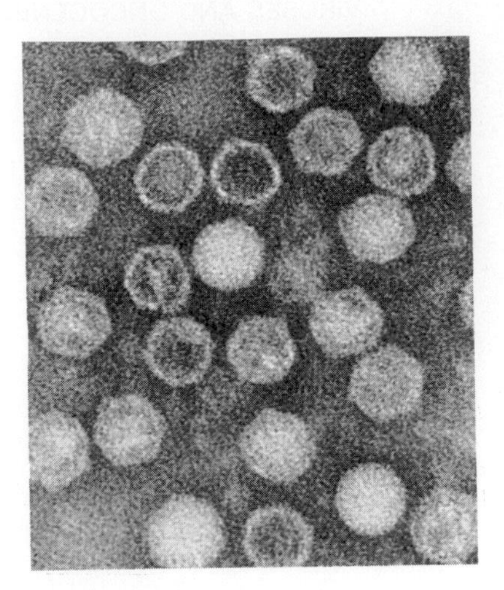

Figure 34-26
Electron micrograph of poliovirus particles. [Courtesy of Dr. John Finch.]

Poliovirus consists of a 7.5-kb single-stranded (+) RNA in an icosahedral capsid. This virion RNA molecule, like those of other Class 1 RNA viruses, acts as a messenger on entering the cytoplasm of its host cell. It is translated by host ribosomes to give capsid proteins and a special RNA polymerase that takes instructions from an RNA template (an *RNA replicase*). The virion (+) RNA then directs this RNA replicase to synthesize (−) strands (Figure 34-26). In turn, the (−) RNA acts as the template for the synthesis of many (+) strands, which serve as messenger or are encapsidated to form new virions.

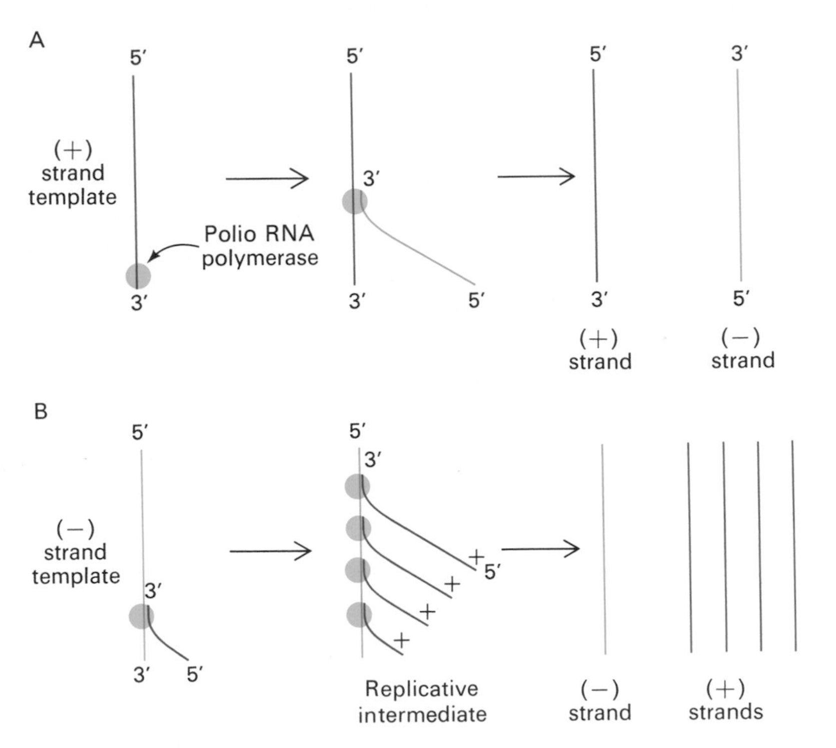

Figure 34-27
Replication of poliovirus RNA.

A striking feature of the expression of poliovirus genes is that the virion (+) RNA is the messenger for the synthesis of a continuous polypeptide chain of more than two thousand residues. David Baltimore found that *this giant polypeptide is cleaved into multiple proteins* (Figure 34-28). Four coat proteins, an RNA replicase, a protease, and several other proteins arise from this very large precursor polypeptide. The first cleavage, an autocatalytic one, occurs while the nascent chain is being synthesized on the ribosome.

Why does poliovirus synthesize its proteins by sequential cleavages? It appears to have no choice. Recall that *in eucaryotic cells an mRNA molecule can be translated into only one polypeptide chain* (p. 761). In contrast, procaryotic mRNAs and viral mRNAs expressed in procaryotic cells are typically polygenic (e.g., *lac* mRNA and Qβ viral mRNA). Thus, poliovirus cleaves a polyprotein to overcome a limitation imposed by the biosynthetic machinery of its animal host cell.

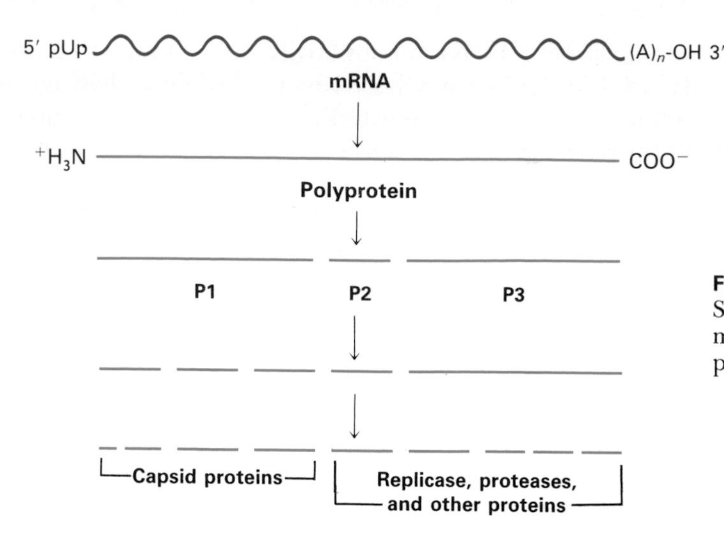

Figure 34-28
Synthesis of poliovirus proteins by multiple cleavages of a giant polypeptide precursor.

The three-dimensional structure of poliovirus has recently been solved at high resolution. The icosahedral shell consists of 60 copies of each of four proteins, called VP1, VP2, VP3, and VP4 (VP stands for virion protein). Cleavage of P1 (one of the three polypeptides initially formed by cleavage of the polyprotein) gives VP0, VP3, and VP1, which assemble into pentamers (Figure 34-29). The pentamers come together

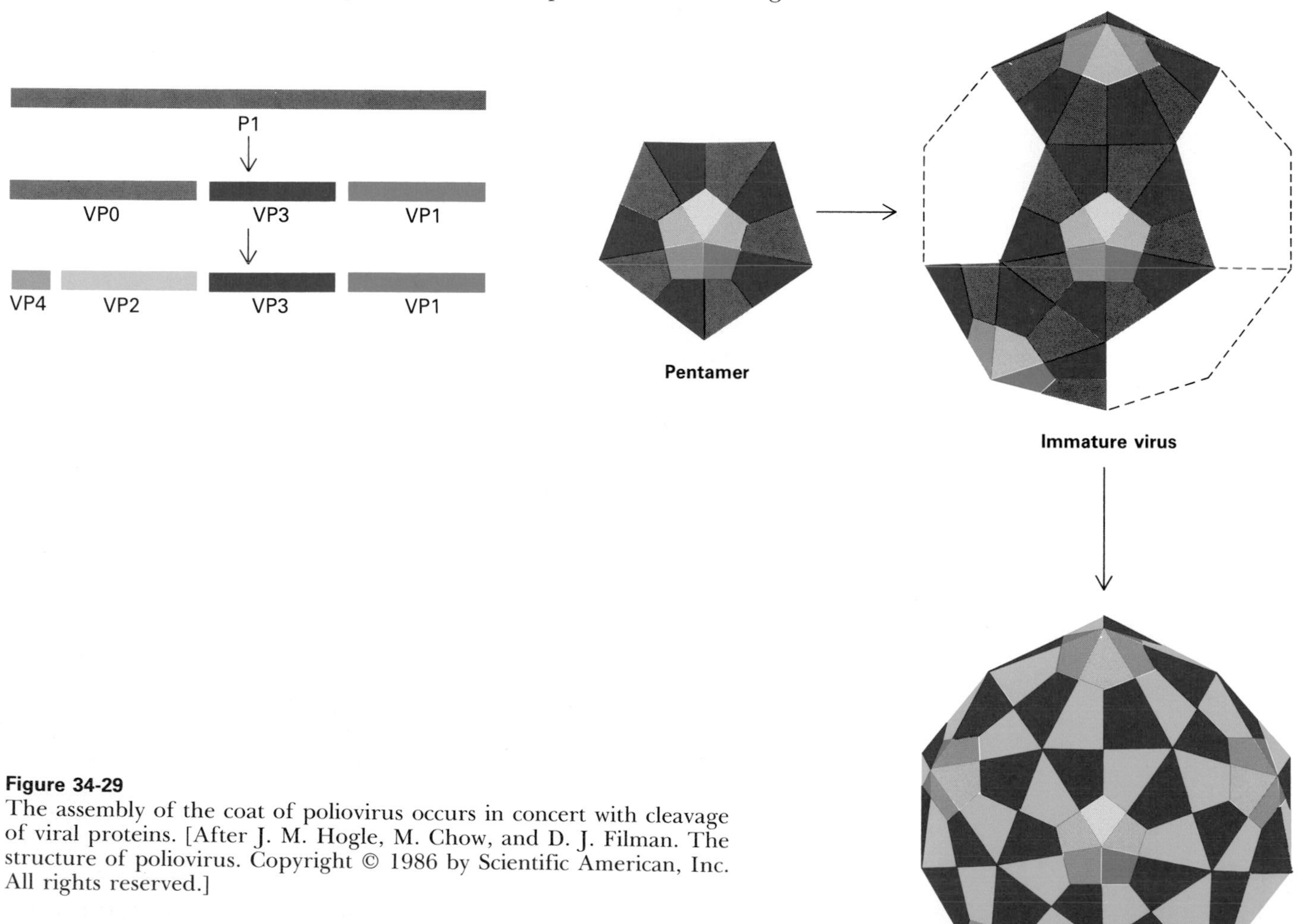

Figure 34-29
The assembly of the coat of poliovirus occurs in concert with cleavage of viral proteins. [After J. M. Hogle, M. Chow, and D. J. Filman. The structure of poliovirus. Copyright © 1986 by Scientific American, Inc. All rights reserved.]

Picornaviruses—
A group of small *(pico)* RNA *(rna)* viruses. One single-stranded (+) RNA molecule is surrounded by an icosahedral protein shell (270 Å diameter). Poliovirus, rhinovirus (which causes the common cold), and the virus that causes foot-and-mouth disease in cattle are picornaviruses.

Rhabdoviruses—
Bullet-shaped viruses, such as vesicular stomatitis virus and rabies virus. From *rhabdo,* the Greek word for rod.

to form the viral shell; VP0 is concomitantly cleaved into VP2 and VP4, which makes the assembly process irreversible. This cleavage is catalyzed by VP0 itself and may be assisted by RNA, which is located inside the assembling particle.

The common cold is caused by *human rhinovirus 14,* a picornavirus that is structurally similar to poliovirus. It also has a capsid of 60 copies each of four proteins, which arise from a polyprotein by sequential cleavages. Moreover, the packing arrangements of the poliovirus and rhinovirus coat proteins are reminiscent of those in tomato bushy stunt virus and other icosahedral plant viruses. The recurring structural motifs seen in these small spherical viruses may reflect a common evolutionary origin.

VESICULAR STOMATITIS VIRUS (VSV) TRANSCRIBES FIVE MONOCISTRONIC mRNAs FROM ITS GENOMIC RNA

Vesicular stomatitis virus (VSV), a mild pathogen of cattle, and rabies virus display a second mode of gene expression. Their virions contain a single-stranded (−) RNA molecule, which does not serve as messenger. *Hence the first step in its expression is the synthesis of (+) mRNA. Uninfected cells lack an RNA replicase, and so these viruses must carry this enzyme in their virions and deliver it to the infected cell.* In fact, two of the five proteins in the VSV virion carry out RNA replication. The 200-kd L (large) and 45-kd NS (nonstructural) proteins, though present in small amounts, are essential for infectivity. The genomic RNA is complexed to many copies of N, the 50-kd nucleocapsid protein, which is the major protein in the virion.

VSV is enveloped by a lipid bilayer, which is acquired from the plasma membrane of the host in the process of budding out of the cell (Figure 34-30). VSV, like Semliki Forest virus (p. 792), uses normal endocytic and exocytic pathways to enter and leave epithelial cells on their basolateral surfaces. The spikes in the membrane of VSV come from the G (glycoprotein) protein, a 65-kd species encoded by the virus. The 29-kd matrix protein, called M, lies between the envelope and nucleocapsid. *These five VSV proteins are formed by the translation of five (+) mRNAs rather than by cleavage of a polyprotein.* The same RNA replicase also synthesizes a long (+) RNA containing all of the genetic information of the virus. This complete (+) RNA in turn serves as the template for the synthesis of (−) RNA, which is packaged to form new virions.

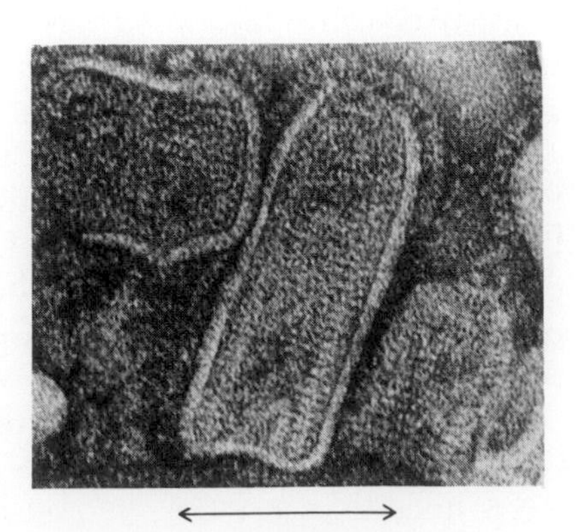

Figure 34-30
Electron micrograph of vesicular stomatitis virus. [From R. C. Williams and H. W. Fisher. *An Electron Micrographic Atlas of Viruses,* 1974. Courtesy of Charles C Thomas, Publisher, Springfield, Illinois.]

THE HEMAGGLUTININ IN THE MEMBRANE OF INFLUENZA VIRUS ENABLES IT TO ENTER SUSCEPTIBLE CELLS

Influenza virus, like VSV, is a membrane-enveloped virus with a (−) RNA genome. Its ten genes are located on eight separate segments of single-stranded RNA, ranging in size from 890 to 2341 nucleotides. The membrane envelope of this 1000-Å diameter virus particle (Figure 34-31) contains many copies of three proteins: HA (*hemagglutinin*), NA (*neuraminidase*), and M (*membrane protein*). HA and NA are transmembrane glycoproteins, whereas M lies on the inner surface of the membrane envelope. The spikes seen in electron micrographs are trimers of HA and tetramers of NA. The hemagglutinin molecule enables the virus to enter susceptible cells, such as those lining the respiratory tract. HA has affinity for sialic acid residues on the surface of these cells, enabling the virus to bind to the cells and enter by receptor-mediated endocytosis (p. 786). HA also fuses the viral membrane with that of the endosome to release the viral RNA into the cytosol. The neuraminidase cleaves terminal sialic acid residues, which may help prevent entrapment of the virus in mucus secretions that are rich in sialic acid.

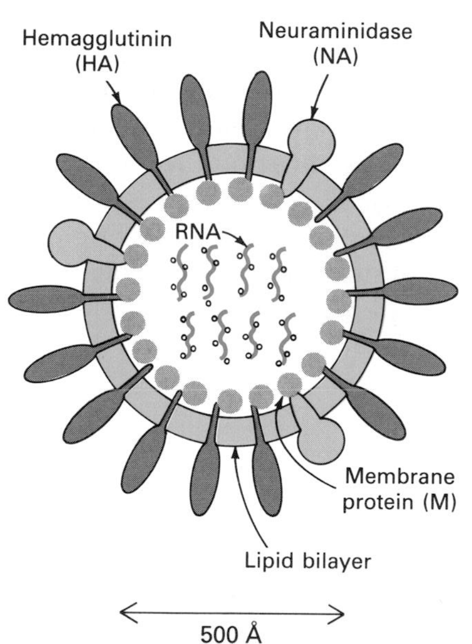

Figure 34-31
Schematic diagram of an influenza A virus.

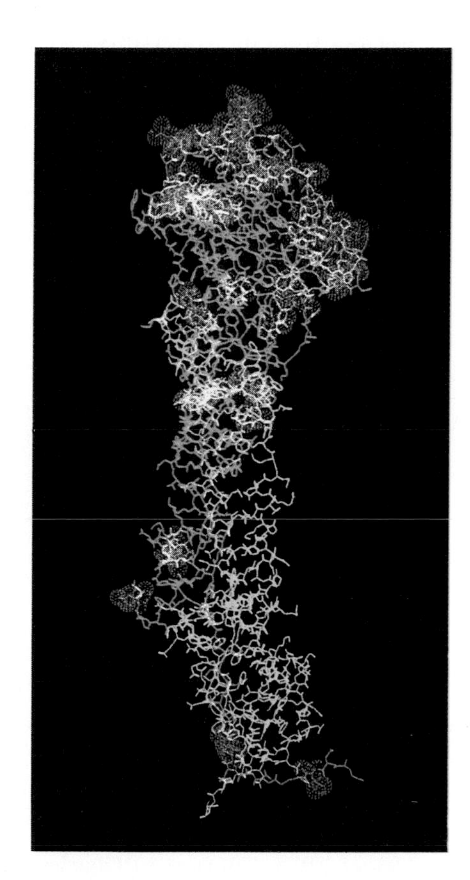

Figure 34-32
Three-dimensional structure of the hemagglutinin protein of influenza A virus. A soluble fragment was crystallized containing all of the HA_1 chain (blue) and the portion of the HA_2 chain (yellow) that extends outside the viral membrane. The red dots depict the van der Waals surface of residues in the HA_1 chain that mutated in the interval from 1969 and 1983. Most of these changes are located in the portion of the protein (top) farthest from the membrane; this domain binds sialic acid units on the surface of target cells.

Not until newly assembled virus has left the host cell is the 62-kd hemagglutinin cleaved into two chains, HA_1 and HA_2, which are held together by a disulfide bond and numerous noncovalent interactions. The three-dimensional structure of a soluble fragment containing a trimer of the entire HA_1 and most of HA_2 has been determined by Don Wiley (Figure 34-32). The protein is highly elongated, extending some 135 Å out of the viral membrane. The binding site for sialic acid, formed by HA_1, is located near the tip of the protein, far from the viral lipid bilayer. The highly hydrophobic amino-terminal region of HA_2 is responsible for the fusion of the viral membrane envelope with the endosome membrane. This *fusion sequence* is activated by the low pH of the endosome.

Antibodies formed against hemagglutinin effectively block the infection by the virus. However, this protection is not permanent because RNA encoding the HA_1 region of influenza virus changes at a rapid rate. New virus strains emerge often because of genetic reassortment of viral RNA segments in infected cells. Influenza virus infects birds, pigs, and horses, in addition to humans. The incorporation of an HA gene from an animal source into a viral genome optimized for infection in humans can lead to a highly virulent new strain. *The segmented character of the viral genome markedly enlarges the viral repertoire of surface appearances, thereby lessening the effectiveness of immune surveillance.* The 1919 pandemic, which caused 20 million deaths, was a striking expression of the capacity of the influenza virus to evade immune responses.

THE GENOME OF REOVIRUS CONSISTS OF TEN DIFFERENT DOUBLE-STRANDED RNA MOLECULES

> *Reovirus*—
> A double-stranded RNA virus isolated from the respiratory and intestinal tracts of humans and other mammals. The prefix *reo* is an acronym for respiratory enteric orphan, a virus in search of a disease.

Reovirus, a double-stranded RNA virus that infects mammalian cells, exemplifies a third type of viral genetic system. The core of the virion contains *ten different double-stranded* (±) *RNA molecules* that are associated with proteins. On entering the host, the virion loses its outer icosahedral shell, which is made up of three kinds of proteins. Removal of this shell activates an RNA polymerase that is carried in the core of the virion. This RNA-directed polymerase transcribes the ten (±) RNA molecules in entirety, so that the (+) mRNAs formed have the same length as the genome segments. However, the (±) RNA template is transcribed *asymmetrically* and *conservatively*—that is, only (+) RNA is formed and the original (±) RNA duplex is not disrupted. The 5′ ends of these mRNAs are capped by enzymes in the core and then extruded through channels in the core (Figure 34-34). Thus, the core is a highly organized assembly for the synthesis of mRNA. *Each of the ten mRNAs is translated into one protein.* In the assembly of new virus, a complete set of ten (+) RNA molecules then joins some viral proteins to form a precore on which ten (−) strands are synthesized.

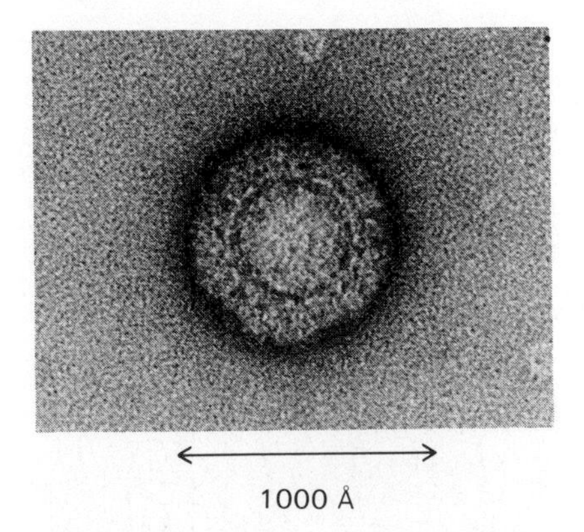

Figure 34-33
Electron micrograph of reovirus. [From R. C. Williams and H. W. Fisher. *An Electron Micrographic Atlas of Viruses*, 1974. Courtesy of Charles C Thomas, Publisher, Springfield, Illinois.]

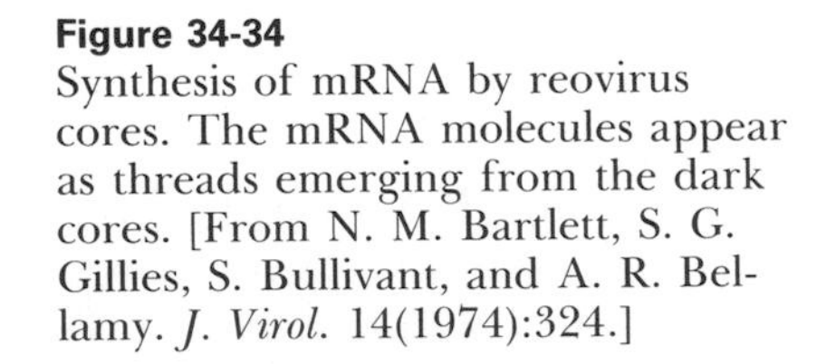

Figure 34-34
Synthesis of mRNA by reovirus cores. The mRNA molecules appear as threads emerging from the dark cores. [From N. M. Bartlett, S. G. Gillies, S. Bullivant, and A. R. Bellamy. *J. Virol.* 14(1974):324.]

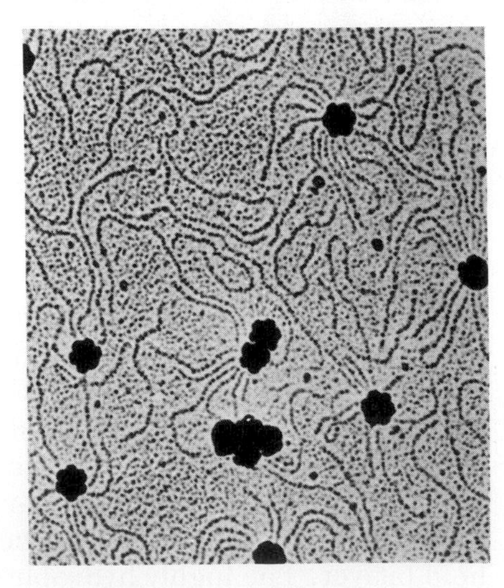

RETROVIRUSES AND SOME DNA VIRUSES CAN INDUCE CANCER IN SUSCEPTIBLE HOSTS

In 1909, a poultry farmer in upstate New York observed a tumor in a hen and brought the sick fowl to Peyton Rous, an investigator at the Rockefeller Institute. Rous prepared a cell-free filtrate from this connective-tissue tumor and injected it into normal chickens. The striking result was that the recipients developed highly malignant tumors of the same kind, called *sarcomas.* Rous also found that the tumor-causing agent in the filtrate, now known as *Rous sarcoma virus* (RSV) or *avian sarcoma virus* (ASV) (Figure 34-35), could be propagated by serial passage through chickens. Rous received the Nobel Prize fifty-five years later in recognition of this seminal discovery.

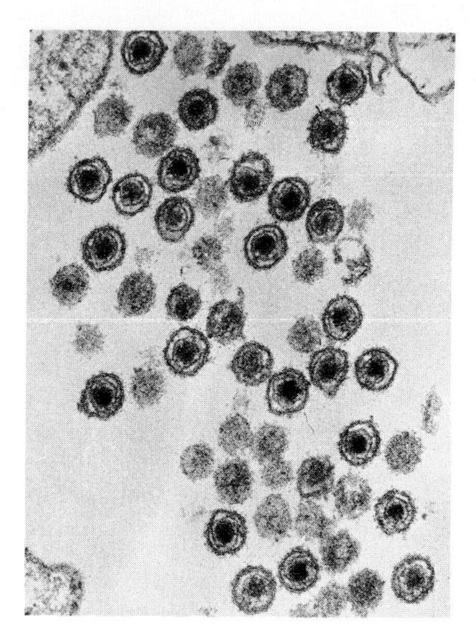

Figure 34-35
Electron micrograph of avian sarcoma virus (Rous sarcoma virus), a retrovirus. The densely stained virions are near the surface of an infected chicken cell. [Courtesy of Dr. Samuel Dales.]

Avian sarcoma virus is a member of a group of *RNA tumor viruses* (oncogenic RNA viruses). These Class 4 viruses contain (+) RNA in their virions and propagate through a double-helical DNA intermediate. Hence, they are known as *retroviruses.* In fact, retroviruses are the only cancer-producing RNA viruses. Malignant tumors can also be induced by a variety of DNA viruses. *Simian virus 40 (SV40)* and *polyoma virus,* which belong to the papovavirus group, are the most intensively studied *oncogenic DNA viruses* (Figure 34-36).

Tissue-culture systems have been developed for the study of cancer at the molecular level. Appropriate animal cells become permanently *transformed*—that is, rendered cancerlike—by infecting them with an oncogenic virus. Transformed cells differ from normal cells in their growth patterns and in the nature of their cell surfaces (Table 34-3).

Table 34-3
Altered properties of cells transformed by DNA or RNA tumor viruses

Growth patterns
Form tumors when injected into susceptible animals
Grow to much higher densities
Growth becomes unoriented and cells detach from the substratum
Protease activators are secreted into the medium, making the cells more invasive
Requirement for serum factors decreases
Surface properties
New viral-specific antigens appear
Fibronectin, an external surface protein, disappears
Ganglioside content decreases
Fetal antigens become exposed
Rate of transport of nutrients increases
Agglutinability by plant lectins increases
Signs of the presence of tumor virus
Virus DNA sequences are present
Virus mRNAs are present
Virus-specific antigens are detectable

Source: J. Tooze, ed., *The Molecular Biology of Tumour Viruses* (Cold Spring Harbor Laboratory, 1973), p. 351.

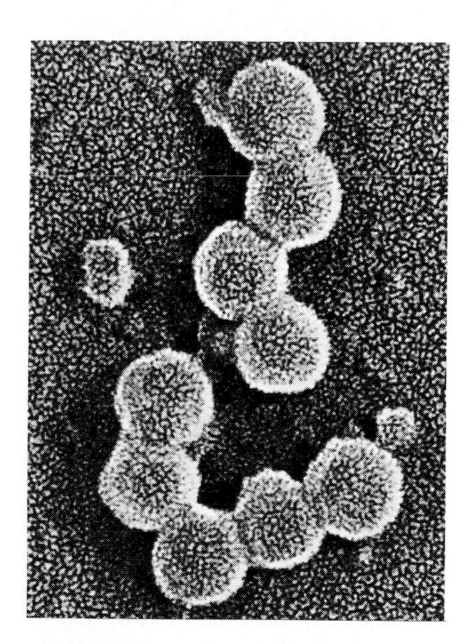

Figure 34-36
Electron micrograph of SV40, a DNA tumor virus. [Courtesy of Dr. Jack Griffith.]

The most striking change is that *transformed cells grow continuously and chaotically,* without regard for their neighbors. Furthermore, transformed cells contain *viral-specific DNA that is integrated into the host genome,* which accounts for the fact that transformation is a heritable alteration. The abnormal characteristics of transformed cells are perpetuated in cultures derived from transformed colonies. Furthermore, some *transformed cells from tissue culture grow into a cancer when they are injected in sufficient number into an appropriate host.*

SV40 AND POLYOMA VIRUS CAN PRODUCTIVELY INFECT OR TRANSFORM HOST CELLS

Papovaviruses—
A group of DNA viruses named after three prototypes: papilloma, polyoma, and vacuolating (SV40) viruses.

SV40 and polyoma virus contain a small circular DNA duplex inside an icosahedral shell. In some cells (called *permissive hosts*), these viruses go through a *lytic cycle,* which results in the production of many new virions (Figure 34-37). Productively infected cells are killed by these viruses. In other types of cells (called *nonpermissive hosts*), some of the steps in viral expression are blocked for unknown reasons. No progeny virus are formed but a small proportion of the cells, of the order of 1 in 10^5, are *transformed* following integration of the viral DNA into the host genome.

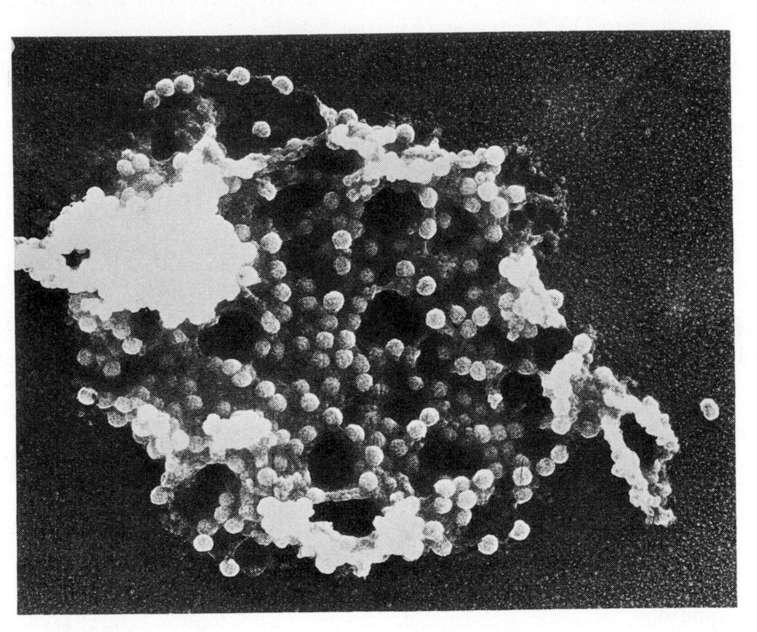

Figure 34-37
Electron micrograph of a fragment of the nuclear membrane of an SV40-infected cell showing nuclear pores and many imbedded virions. [Courtesy of Dr. Jack Griffith.]

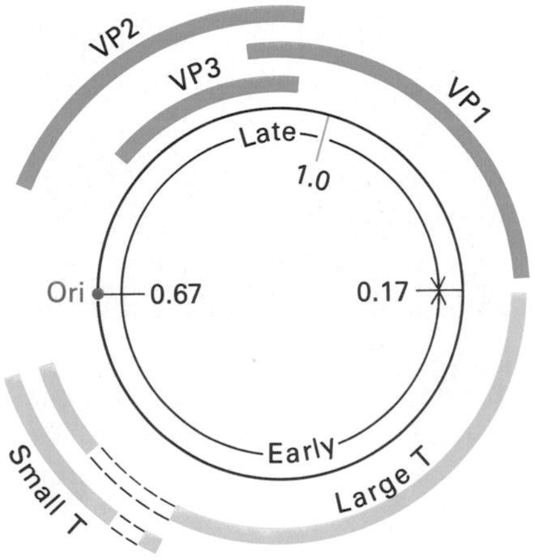

VP2
VP3
VP1
Late
1.0
Ori
0.67
0.17
Early
Small T
Large T

Figure 34-38
Genetic map of SV40 DNA, which contains 5243 base pairs. The early region (transcribed counterclockwise) is shown in yellow, and the late region (transcribed clockwise) in blue. The origin of replication is marked in red. [After W. Fiers, R. Contreras, G. Haegeman, R. Rogiers, A. van de Voorde, H. van Heuverswyn, J. van Herreweghe, G. Volckaert, and M. Ysebaert. *Nature* 273(1978):113.]

The determination of the sequence of the 5243 base pairs of SV40 DNA opened the door to a deeper understanding of how this virus expresses its genes and transforms cells. Half of the DNA is expressed early in infection, whereas the other half is expressed later, concomitant with the synthesis of viral DNA (Figure 34-38). In fact, the origin of DNA replication corresponds to the common starting point of the early and late regions, which are transcribed in opposite directions. The *early region* codes for two proteins, called large T (81 kd) and small T (20 kd). Large T stimulates the replication of viral DNA by binding to three sites at the origin. Noteworthy features of the origin of replication are the presence of a twofold axis of symmetry and an adjoining AT-rich sequence.

5′ C A G A G G C C G A G G C G G C C T C G G C C T C T G C A T A A A T A A A A A A A A T T A 3′

3′ G T C T C C G G C T C C G C C G G A G C C G G A G A C G T A T T T A T T T T T T T T A A T 5′

Table 34-4
Proteins encoded by SV40

Protein	Mass (kd)	mRNA	Role
Large T antigen	81	Early	Necessary for initiation of DNA replication and for transformation
Small T antigen	20	Early	Unknown
VP1	40	Late	Major capsid protein
VP2	39	Late	Minor capsid protein
VP3	27	Late	Minor capsid protein

SV40 DNA also carries two consecutive 72-bp enhancer sequences near the start site for transcription.

Large T and small T arise from differential splicing. A small intron is removed in forming mRNA for small T and a larger one is taken out for large T. Large T is a larger protein because its mRNA does not contain the stop codon that terminates translation of small t. Transcription of the *late region* leads to the synthesis of three capsid proteins: VP1, VP2, and VP3. Again, diversity is achieved by differential splicing of the primary transcript. The amino acid sequence of VP3 is identical to the carboxyl-terminal 70% of VP2. Furthermore, an overlap region of 22 nucleotides is read in one frame for VP2 and VP3 and in a different frame for VP1. Thus, *the limited amount of genetic information in SV40 is fully exploited.* SV40 is also frugal in not synthesizing its own proteins to package its DNA. Rather, host-cell histones become bound to the newly synthesized viral DNA (Figure 34-39). This supercoiled complex is then encapsidated by VP1, VP2, and VP3. Finally, progeny virions are released by lysis of the host cell, which is killed in the process.

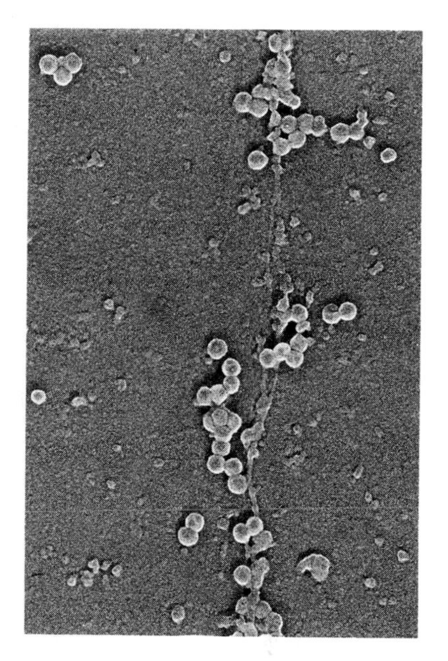

Figure 34-39
Electron micrograph of SV40 tumor virus particles developing in association with the host chromosome. [Courtesy of Dr. Jack Griffith.]

In nonpermissive cells, the early region of SV40 is expressed, but DNA replication and transcription of the late region do not occur. *A small fraction of these cells become transformed by the integration of the SV40 genome into the host DNA.* In contrast with λ phage, SV40 DNA does not integrate at specific sites on either the viral or host DNA. Injection of transformed cells into susceptible animals leads to the rapid production of tumors. Large T is required for transformation and the maintenance of the transformed state. Studies of mutants have shown that the transforming action of large T is independent of its capacity to bind to the origin and stimulate replication. Large T has other stimulatory actions, but its specific role in inducing conversion to the cancerous state is an enigma. Much more is known about the oncogenic mechanisms of RNA tumor viruses, to which we now turn.

RETROVIRUSES CONTAIN A REVERSE TRANSCRIPTASE THAT SYNTHESIZES DOUBLE-HELICAL DNA FROM (+) RNA

Retroviruses, the other class of tumor-producing viruses, contain a (+) RNA genome inside an icosahedral shell. This spherical nucleoprotein core is surrounded by an envelope consisting of viral-encoded glycoprotein molecules in a lipid bilayer derived from the plasma membrane of the host. Retroviruses are typically 1000 Å in diameter.

In 1964, Howard Temin observed that infection by RNA tumor viruses such as avian sarcoma virus is blocked by inhibitors of DNA synthesis. Methotrexate, 5-fluorodeoxyuridine, and cytosine arabinoside block viral replication if present during the first twelve hours following the introduction of the virus. This finding suggested that *DNA synthesis is required for the growth of RNA tumor viruses.* Furthermore, the production of progeny virus particles is inhibited by actinomycin D, which is known to block the synthesis of RNA from DNA templates (p. 715). Hence, *transcription of DNA seemed to be essential for the multiplication of RNA tumor viruses.* These unexpected results led Temin to propose that *a DNA provirus is an intermediate in the replication and oncogenic action of RNA tumor viruses.*

$$\text{RNA tumor virus} \longrightarrow \text{DNA provirus} \longrightarrow \text{RNA tumor virus}$$

Temin's bold hypothesis that genetic information can flow from RNA to DNA was initially greeted with little enthusiasm by most investigators. It required the existence of a then unknown enzyme—one that synthesizes DNA according to instructions given by an RNA template (an *RNA-directed DNA polymerase*). In 1970, Temin and Baltimore independently discovered such an enzyme, which is called *reverse transcriptase,* in the virions of some RNA tumor viruses. All RNA tumor viruses subsequently studied have been found to contain a reverse transcriptase in their virions, and so they became known as *retroviruses.* Not all retroviruses induce cancer; human immunodeficiency virus, for example, causes AIDS.

The life cycle of a typical retrovirus starts when infecting virions bind to specific receptors on the surface of the host and enter the cell. The viral (+) RNA is uncoated in the cytosol. Reverse transcriptase brought in by the virus particle then synthesizes both the (−) and (+) strands of DNA and digests the viral (+) RNA. *Thus, reverse transcriptase carries out three kinds of reactions: RNA-directed DNA synthesis, hydrolysis of RNA, and DNA-directed DNA synthesis.*

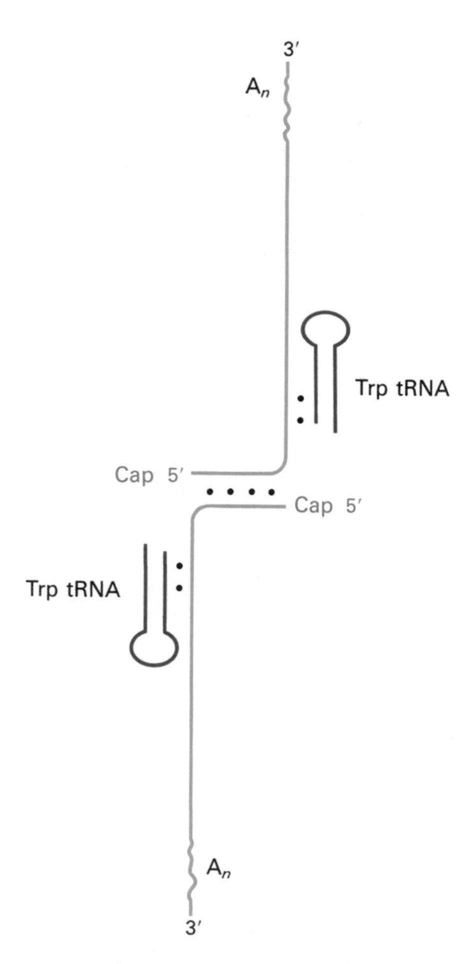

Figure 34-40
Schematic diagram of the genome of avian sarcoma virus. Two identical (+) RNA molecules are noncovalently bonded. The 5′ ends are capped, and the 3′ ends are poly A tails. A tRNA molecule that serves as primer is base paired to each RNA molecule.

Reverse transcriptase, like other DNA polymerases, synthesizes DNA in the 5′ to 3′ direction and is unable to initiate chains de novo. How then is viral DNA synthesis primed? Initiation is accomplished very economically: the (+) RNA viral genome contains a noncovalently bound transfer RNA (trp tRNA in avian sarcoma virus) that was acquired from the host during the preceding round of infection (Figure 34-40). *The 3′ OH of this base-paired tRNA acts as the primer for DNA synthesis.* How is the entire (+) RNA strand replicated? Recall that a special problem is encountered in the replication of any *linear* DNA, in that the 5′ ends of daughter strands are initially incomplete (p. 832). Retroviruses have met this challenge in an ingenious way. Their genomic (+) RNA contains the same sequence (called R) at the 5′ and 3′ ends. This terminal redundancy plays a critical role in the synthesis of duplex DNA, as shown in Figure 34-41. Nascent DNA strands undergo two shifts in base pairing before a complete duplex is formed. The resulting double-helical intermediate contains identical ends called *LTR* (long terminal repeats). LTRs are rich in signals for integration and transcription.

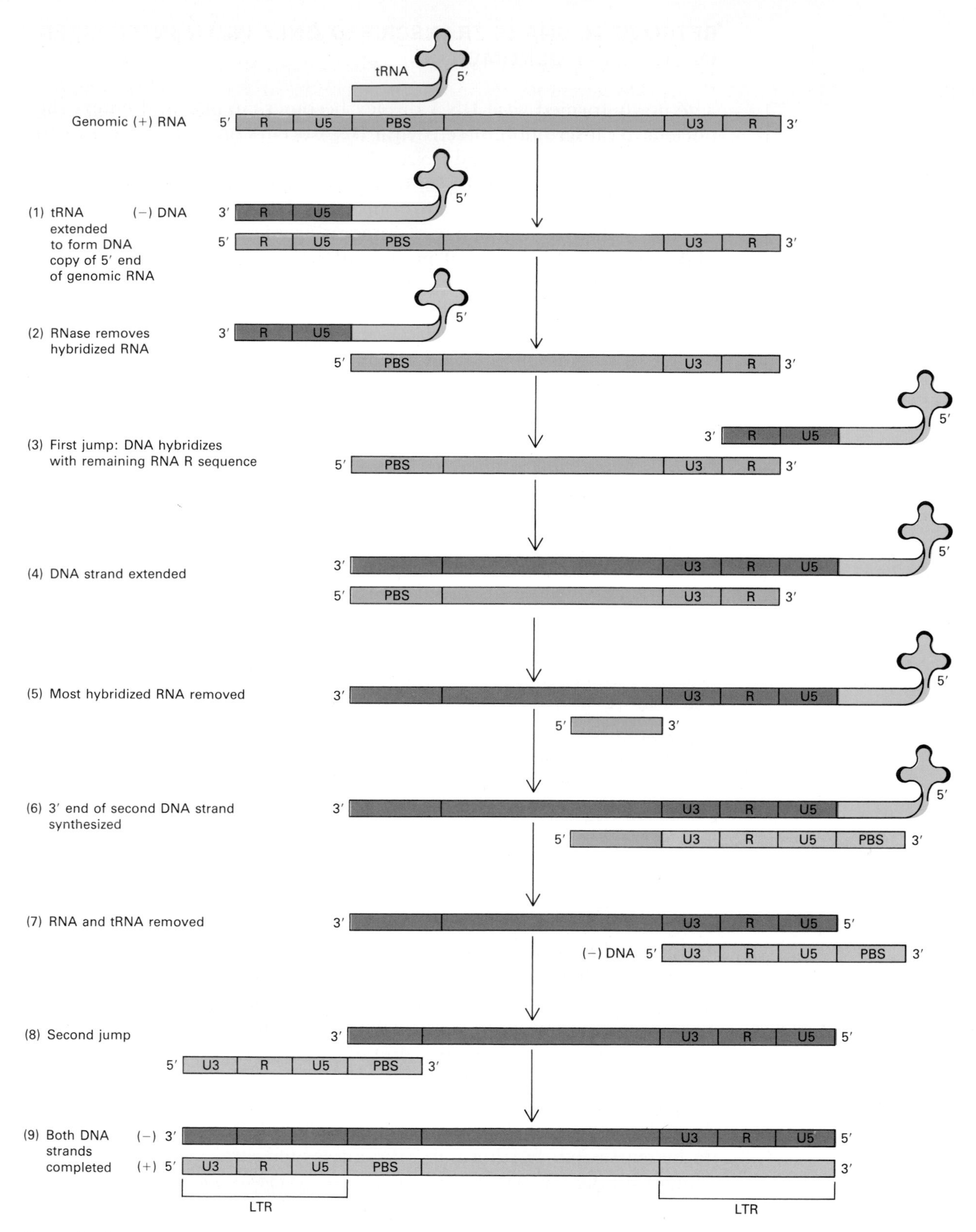

Figure 34-41
Conversion of the genomic RNA of a retrovirus into a double-stranded proviral DNA. (+) RNA is shown in blue, (−) DNA in red, (+) DNA in green, and the tRNA primer in yellow. PBS is the primer binding site, U5 and U3 are different sequences, and R is a repeated sequence. LTR denotes the long terminal repeat at the ends of the proviral DNA. [After J. Darnell, H. Lodish, and D. Baltimore. *Molecular Cell Biology* (Scientific American Books, 1986), p. 1052.]

RETROVIRAL DNA IS TRANSCRIBED ONLY WHEN INTEGRATED IN THE HOST GENOME

> "The study of biology is partly an exercise in natural esthetics. We derive much of our pleasure as biologists from the continuing realization of how economical, elegant, and intelligent are the accidents of evolution that have been maintained by selection. A virologist is among the luckiest of biologists because he can see into his chosen pet down to the details of all of its molecules. The virologist sees how an extreme parasite functions using just the most fundamental aspects of biological behavior . . ."
>
> David Baltimore
> Nobel Lecture (1976)
>

The newly formed viral DNA duplex becomes circular and enters the nucleus. Transcription of retroviral DNA occurs only after it has been integrated into the host-cell DNA. Thus, *integration is an obligatory step in the life cycle of retroviruses.* In contrast, integration and productive infection are alternative pathways for oncogenic DNA viruses. Another difference is that the frequency of integration of retroviral DNA is very high, as might be expected in view of its central role in productive infection. Integration occurs at TCAG sites in the host genome. Four to six bases at the host integration site are duplicated, as in the movement of transposons (p. 698).

The 10-kb genome of avian sarcoma virus contains four genes (Figure 34-42). Three of them—*gag, pol,* and *env*—are essential for productive infection. The *gag* gene specifies a 76-kd polyprotein that is cleaved to form the four proteins of the viral core. *Pol* codes for the reverse transcriptase, which consists of an α and a β subunit. The 65-kd α subunit is a proteolytic fragment of the 90-kd β chain. *Env* specifies the glycoprotein of the viral envelope, which is important in the attachment of the virus to the surface of the host. The fourth gene, *src* (for sarcoma), is not needed for viral multiplication but is required for transformation, as will be discussed shortly.

R	U5	PBS	*gag*	*pol*	*env*	*src*	U3	R

Figure 34-42
Genetic map of avian sarcoma virus. The genome is 10 kb long. The abbreviations are the same as those used in Figure 34-41.

LTRs carry multiple signals—enhancers, promoters, and polyadenylation sites—for the expression of retroviral genes. Three mRNAs are formed: an unspliced mRNA for both *gag* and *pol,* and spliced *env* mRNA and *src* mRNA. The *gag-pol* mRNA is usually translated into the *gag* protein, but a termination signal is infrequently ignored to form a *gag-pol* protein. Genomic RNA (the unspliced primary transcript) and viral proteins migrate to the plasma membrane and become incorporated into it. A portion of the altered membrane then buds to form a new virus particle. Thus, *productive infection by retroviruses, in contrast with oncogenic DNA viruses, is not lytic. Most retroviruses do not usually kill their hosts.* Their DNA stays in the genome of the infected cell and continues to be expressed. Furthermore, the integrated viral DNA is replicated along with the host DNA, and so *the viral genome is inherited by the daughter cells.*

ONCOGENES IN RETROVIRUSES ARE DERIVED FROM NORMAL CELLULAR GENES

Studies of *temperature-sensitive mutants* of avian sarcoma virus provided insight into the mechanism of transformation. Some mutants multiplied normally at high temperature but did not transform their hosts, whereas they did both at low temperature. Furthermore, fibroblasts transformed by these mutants at low temperature reverted to normal when the temperature was raised; they again became transformed when the temperature was lowered. Analyses of these mutants revealed that only one viral gene—the *src* gene—was involved in transformation. The *src* gene acquired its name from its capacity to direct the synthesis of a *sar*coma-producing protein. *The* src *protein is a 60-kd kinase that phosphorylates tyrosine residues on proteins.* Immunocytochemical labeling studies have shown that the *src* protein is bound to the cytosolic surface of the plasma membrane. It is not yet known how phosphorylation of target proteins by *src* protein induces a cancerous state.

Does the viral *src* gene have a cellular counterpart? Hybridization studies carried out by Michael Bishop and Harold Varmus revealed that normal chicken cells contain a *src* gene that is very similar to the viral one. The cellular gene is called c-*src* to distinguish it from the viral v-*src* gene. How is it known that c-*src* is truly a cellular gene and not a viral one that became incorporated into the cellular genome? The c-*src* gene is interrupted by several introns, whereas the v-*src* gene is continuous. Furthermore, the c-*src* gene occupies the same chromosomal location in all cells of an organism. It seems likely that a precursor of avian sarcoma virus acquired c-*src* by unequal crossing over and that this gene mutated to become v-*src*.

Cancer-producing genes such as v-*src* are called *oncogenes. Analyses of more than twenty retroviral oncogenes have shown that they are closely related to normal cellular genes.* They can be grouped in five classes (Table 34-5):

1. Many of them, like the *src* protein, are *tyrosine kinases*. As will be discussed in a subsequent chapter (p. 997), tyrosine kinases participate in the transmission of growth control signals.

2. Some oncogene products act as *growth factors*. The product of the v-*sis* oncogene of simian sarcoma virus is similar to platelet-derived growth factor, a potent stimulator of growth of fibroblasts, smooth muscle cells, and glial cells.

3. Other oncogenes encode *receptors for growth factors*. For example *erbA* encodes a receptor for thyroid hormone, and *erbB* specifies a truncated form of the receptor for epidermal growth factor. The erbB protein also has tyrosine kinase activity.

Sarcoma—
Malignant tumor derived from mesoderm. From the Greek words *sarkos*, meaning flesh, and *oma*, meaning tumor. Leukemias and lymphomas are types of sarcomas.

Carcinoma—
Malignant tumor derived from ectoderm or endoderm. From the Greek word *karkinos*, meaning crab or cancer.

Table 34-5
Retroviral oncogenes

Class	*Oncogene*	*Retrovirus*
Tyrosine kinases	*abl*	Abelson murine leukemia virus
	erbB	Avian erythroblastosis virus
	src	Avian sarcoma virus (Rous sarcoma virus)
Growth factors	*sis*	Simian sarcoma virus
Growth factor receptors	*erbB*	Avian erythroblastosis virus
Guanyl-nucleotide-binding proteins	Ha-*ras*	Harvey murine sarcoma virus
	Ki-*ras*	Kirsten murine sarcoma virus
Nuclear proteins	*fos*	FBJ osteosarcoma virus
	myb	Avian myeloblastosis virus
	myc	Avian myelocytomatosis virus

4. Viral *ras* genes encode 21-kd *guanyl-nucleotide-binding proteins. Ras* proteins cycle between an active GTP state and an inactive GDP state. They become highly oncogenic as a result of single amino acid mutations that inhibit their GTPase activity. The mutant protein in the GTP form is persistently activated.

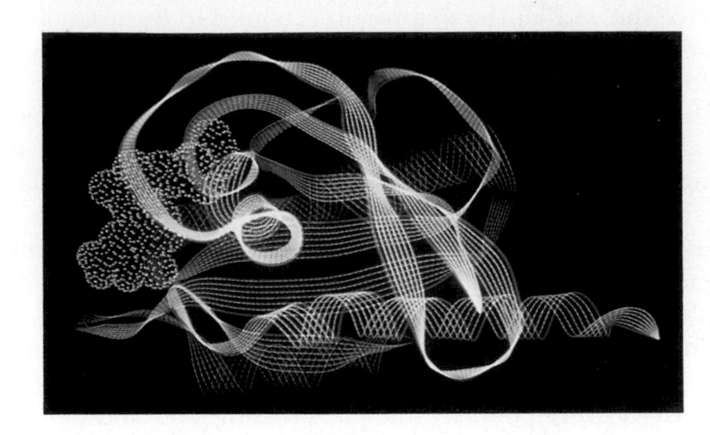

Three-dimensional structure of the ras protein. The bound GDP is shown in green. [Courtesy of Dr. Sung-hou Kim.]

5. Several oncogene products are *nuclear proteins*. For example, the *myb* gene of avian myeloblastosis virus, which causes leukemia, encodes a protein that binds to the nuclear matrix. These proteins probably exert their oncogenic action by altering transcription.

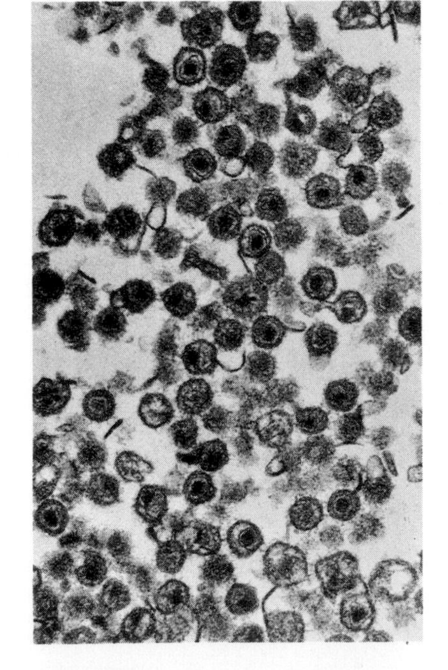

Figure 34-43
Electron micrograph of particles of avian myeloblastosis virus. Injection of this RNA tumor virus into newborn chicks produces a fatal leukemia within three weeks. The v-*myb* oncogene of this virus encodes a protein that is bound to the nuclear matrix of transformed cells. [Courtesy of Dr. Ursula Heine, National Cancer Institute.]

Retroviruses do not appear to be a major cause of human cancers. However, their oncogenic actions in animals have been sources of insight into the molecular basis of cancer. *Retroviruses transform susceptible cells by producing excessive quantities of certain key proteins in growth control, or by forming altered proteins that can no longer be controlled.* Analysis of oncogenes has revealed that many of them encode proteins with key roles in controlling normal growth and development. It also shows that *mutation, duplication, or translocation of normal cellular genes involved in growth control can lead to cancer.* A cellular gene that can become cancer-producing is called a *proto-oncogene*.

ACQUIRED IMMUNE DEFICIENCY SYNDROME (AIDS) IS CAUSED BY A RETROVIRUS

In 1981, the first cases of a new disease called *acquired immune deficiency syndrome* (AIDS) were recognized. The victims died of rare infections because their immune systems were crippled. The cause was identified two years later by Luc Montagnier and Robert Gallo. AIDS is produced by *human immunodeficiency virus* (HIV). This retrovirus is related to human T-cell lymphotrophic virus I (HTLV-I), the cause of a rare leukemia. The host cell for HIV is the T4 lymphocyte, a vital contributor to the immune response; this lymphocyte was named for an antigen present on its surface. T4 lymphocytes serve as helper and inducer cells in the action of B-lymphocytes and other T-lymphocytes (p. 912). The HIV virion, like the avian sarcoma virion, is about 1000 Å in diameter. It is enveloped by a lipid bilayer membrane containing two glycoproteins: gp41 spans the membrane and is disulfide-bonded to gp120,

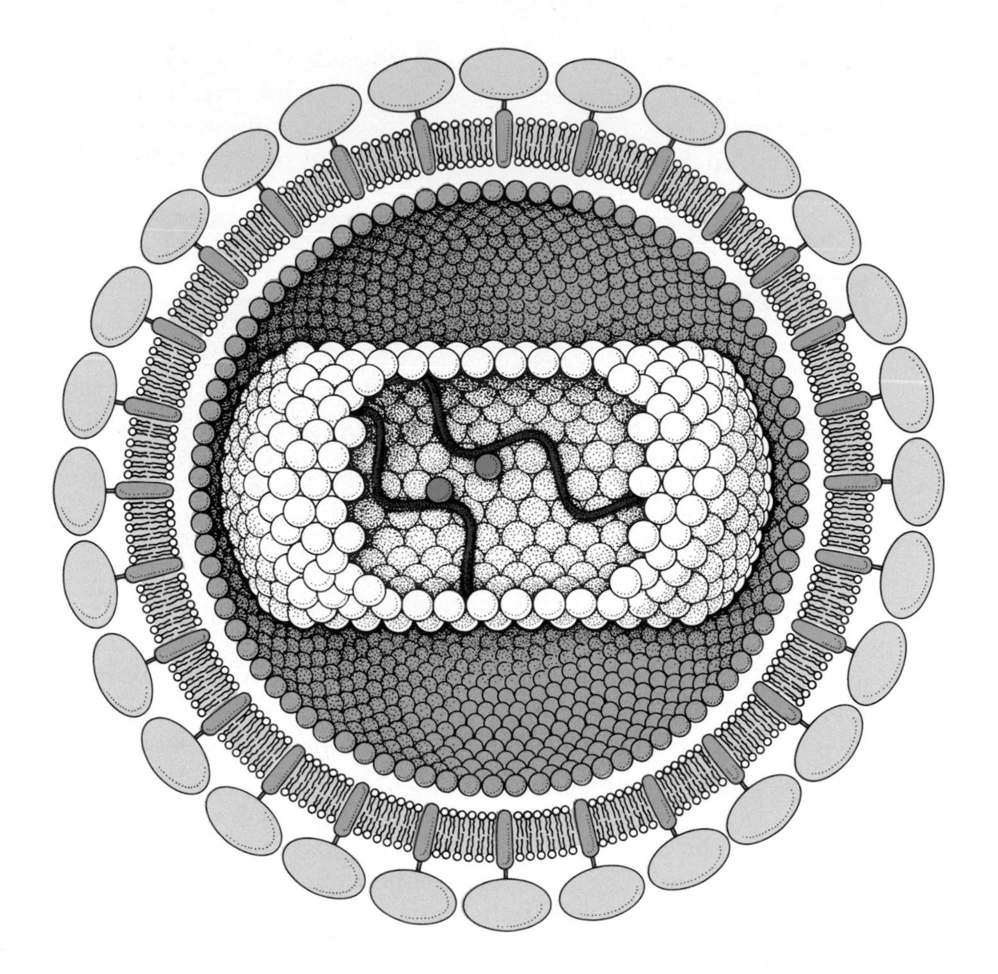

Figure 34-44
Schematic diagram of human immunodeficiency virus (HIV), the cause of AIDS. The membrane envelope glycoproteins gp41 and gp120 are shown in dark and light green. The core of the virus contains two kinds of protein subunits, p18 (orange) and p24 (white), an RNA genome (red), and several molecules of reverse transcriptase (blue). [After R. C. Gallo. The AIDS virus. Copyright © 1987 by Scientific American, Inc. All rights reserved.]

which is located on the external face (Figure 34-44). The core of the virus contains two copies of the RNA genome and associated tRNAs, and several molecules of reverse transcriptase. They are surrounded by many copies of two proteins called p18 and p24. HIV enters T4 lymphocytes by the interaction of gp120 in the viral envelope with a plasma membrane receptor called CD4. The two membranes fuse and the viral core is released directly into the cytosol (Figure 34-45).

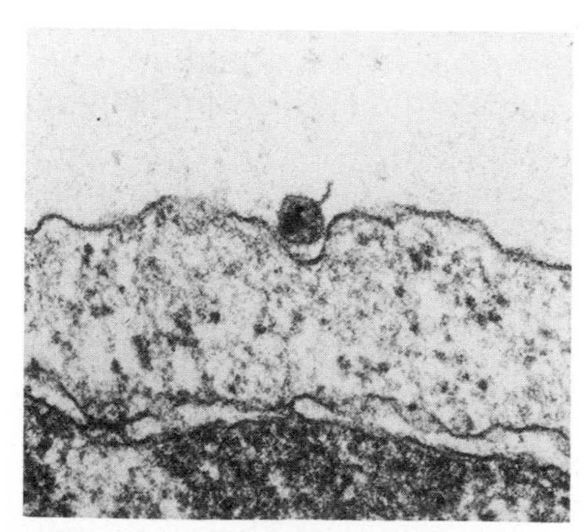

A

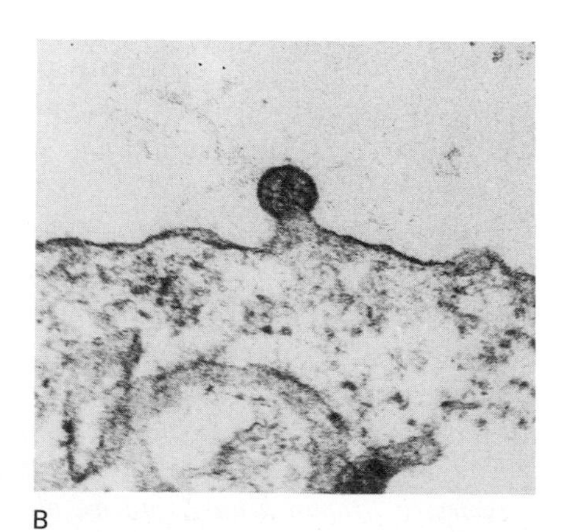

B

Figure 34-45
Entry of human immunodeficiency virus into a precursor of a T-lymphocyte. The viral membrane fuses with the plasma membrane of the target cell. [After B. S. Stein, S. D. Gowda, J. D. Lifson, R. C. Penhallow, K. G. Bensch, and E. G. Engleman. *Cell* 49(1987): 664.]

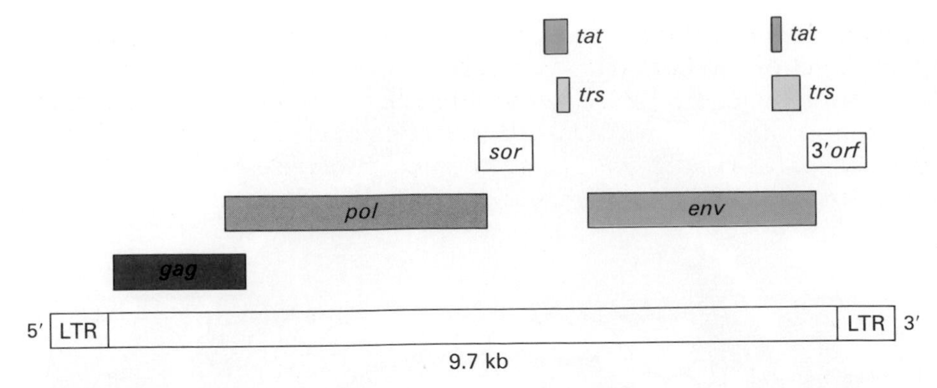

Figure 34-46
Genetic map of the DNA form of the genome of human immunodeficiency virus. [After R. Gallo. The AIDS virus. Copyright © 1987 by Scientific American, Inc. All rights reserved.]

On AIDS—
"It is a modern plague: the first great pandemic of the second half of the 20th century. . . . Does this terrible tale have a moral? Yes. In the past two decades one of the fondest boasts of medical science has been the conquest of infectious disease, at least in the wealthy countries of the industrialized world. The advent of retroviruses with the capacity to cause extraordinarily complex and devastating disease has exposed the claim for what it was: hubris. Nature is never truly conquered. The human retroviruses and their intricate interrelation with the human cell are but one example of that fact. Indeed, perhaps conquest is the wrong metaphor to describe our relation to nature, which not only surrounds but in the deepest sense also constitutes our being."

Robert Gallo (1987)

The 9.7-kb RNA genome of HIV is similar in overall organization to that of avian sarcoma virus but is more complex. The *gag*, *pol*, and *env* genes are bounded by LTRs (Figure 34-46). HIV also contains four unusual genes called *sor*, *trs*, *tat*, and *3′orf*. Cleavage of polyproteins and alternative splicing of mRNA lead to the formation of numerous proteins from this small genome. The *tat* protein stimulates the formation of viral mRNAs by binding to promoter sites and may also increase their translation. The *trs* protein affects splicing to increase the expression of *gag*, *pol*, *env*, and *sor* genes. The subtle interplay of these proteins makes HIV a formidable pathogen. Moreover, *env* undergoes mutation at a very rapid rate. The amino acid sequence of gp120 in virus particles obtained from different patients or even the same patient at different times exhibits a high degree of variability. Preparing an effective vaccine against HIV is likely to be much more difficult than for viruses that are genetically stable, such as smallpox and poliovirus.

INTERFERONS, A FAMILY OF PROTEINS INDUCED BY VIRAL INFECTION, BLOCK PROTEIN SYNTHESIS

The resistance of animal cells to many viruses is markedly enhanced by *interferons*, a family of small proteins. *They are synthesized and secreted by vertebrate cells following a virus infection.* Double-stranded RNA molecules are particularly effective in stimulating the formation of interferon. Secreted interferons bind to the plasma membrane of other cells in the organism and induce an antiviral state in them. These cells acquire resistance to a broad spectrum of viruses. In contrast, immunity conferred by an antibody is highly specific. Interferons are very potent—as little as 10^{-11} M can have a significant antiviral effect.

Three classes of interferons have been isolated and purified: α interferon (originally called leukocyte interferon), β interferon (fibroblast interferon), and γ interferon (immune interferon). These proteins range in size from 15 kd to 35 kd. The cloning of several interferon genes and their expression in *E. coli* has provided abundant amounts of protein for mechanistic studies and clinical trials. Interferons are currently being evaluated as antiviral and anticancer agents.

Interferons lead to an antiviral state by stimulating the production of two enzymes: a protein kinase and an oligoadenylate synthetase. These enzymes become active in the presence of *double-stranded RNA* (Figure 34-47). The target of the kinase is the α subunit of eIF2, the initiation factor that brings initiator tRNA to the 40S ribosomal subunit. Recall that phosphorylation of the α blocks the recycling of eIF2 and hence inhibits translation (p. 762). In particular, the phosphorylation of only 30% of eIF2 leads to the complete cessation of protein synthesis.

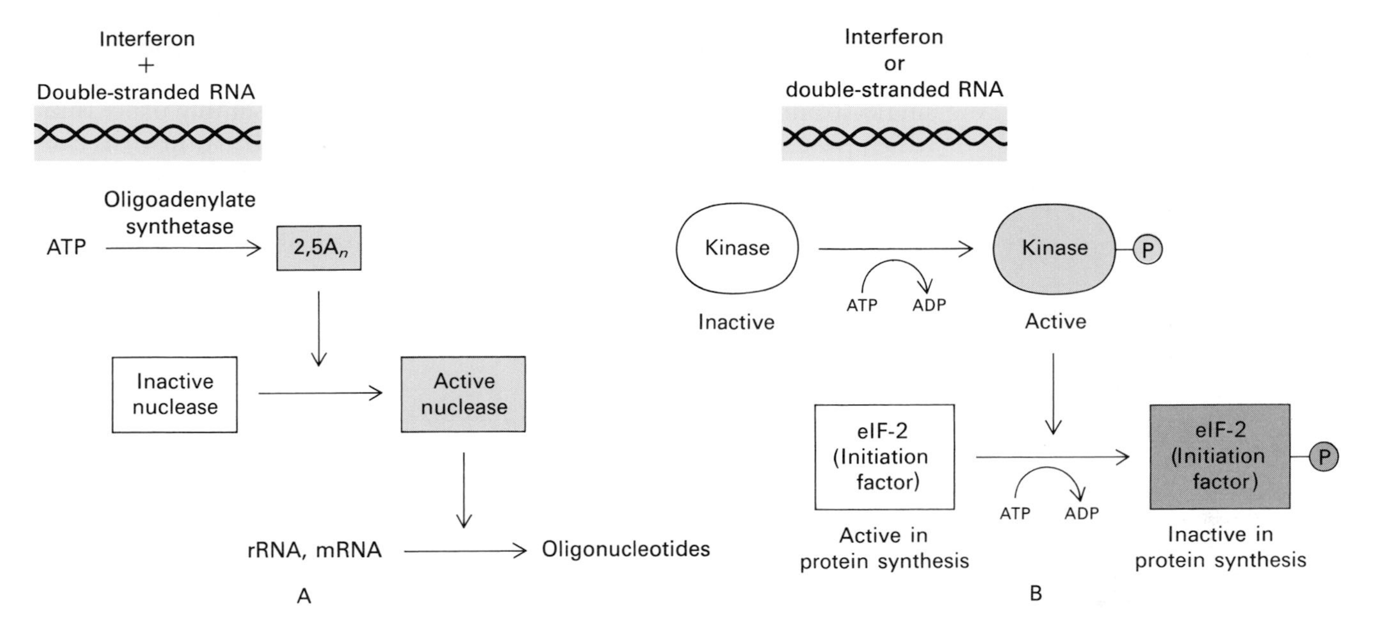

Figure 34-47
Double-stranded RNA (A) stimulates degradation of mRNA and rRNA and (B) inhibits the initiation of protein synthesis in interferon-treated cells. [After P. J. Farrell, G. C. Sen, M. F. Dubois, L. Ratner, E. Slattery, and P. Lengyel. *Proc. Nat. Acad. Sci.* 75(1978):5896.]

The second enzyme activated by interferon and double-stranded RNA is *2′,5′-oligoadenylate synthetase*. This enzyme catalyzes the formation of oligoadenylates joined by 2′,5′ rather than by the usual 3′,5′ phosphodiester bonds. The 2′-oxygen atom of ATP nucleophilically attacks the α-phosphorus atom of another ATP molecule to begin the synthesis of these oligoadenylates (denoted as 2,5-A_n), which range in length from two to about fifteen residues. 2,5-A_n then activates an endoribonuclease called *RNase L* that hydrolyzes mRNAs and rRNAs at —UpXpY— sequences to yield —UpXp and Y—. Activation of RNase L is reversed by hydrolysis of 2,5-A_n by a specific phosphatase that is not controlled by interferon or double-stranded RNA. It is interesting to note that 2,5-A_n, like many other signal molecules (p. 1027), is formed in one step from a molecule that is central in metabolism.

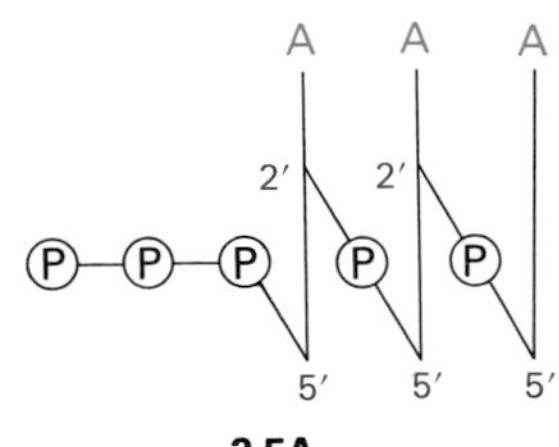

2,5A$_n$
(Endonuclease activator)

These two actions of interferon, in concert with double-stranded RNA, block cell growth and proliferation. The activated kinase interferes with the formation of new polypeptide chains, and the activated oligoadenylate synthetase leads to the destruction of mRNA templates and rRNA components of the protein-synthesizing machinery. It will be interesting to learn whether interferon plays a key role in normal cell growth and development in addition to serving as an important line of defense against viral infection.

SUMMARY

Viruses are packets of infectious nucleic acid surrounded by a protective coat. They contain either DNA or RNA, which can be single stranded or double stranded. The simplest viruses have only 4 genes, whereas the most complex ones have about 250 genes. Most viral coats consist of a large number of one or a few kinds of protein subunits because viruses have a very limited amount of genetic information. The two principal arrangements are a cylindrical shell with helical symmetry and a spherical shell with icosahedral symmetry. Tobacco mosaic virus, which consists of a helical array of 2130 identical subunits around a single-stranded RNA molecule, is formed by a self-assembly process in which protein discs add to a loop of RNA. Tomato bushy stunt virus can form a large icosahedral shell because the domains of its coat protein are hinged, allowing them to pack quasi-equivalently. The conformation of the coat proteins of poliovirus and rhinovirus (the common cold virus) and their packing arrangements are similar to those of tomato bushy stunt virus. Vesicular stomatitis virus and influenza virus acquire their lipid bilayer envelope by budding from the plasma membrane of infected host cells.

The assembly of T4 phage is a much more complex process that involves approximately fifty proteins. Three major pathways lead independently to the head, the tail, and the tail fibers. There is a stringent sequential order in the formation of T4, which occurs by a combination of self-assembly and enzyme-directed assembly. Scaffold proteins are also important in the assembly of T4. Proteolytic cleavages are also critical in the assembly of icosahedral viruses such as poliovirus.

Cellular enzymes cannot synthesize RNA from an RNA template. Hence, a common feature of RNA viruses (except for retroviruses) is their specification of an RNA-directed RNA polymerase. The RNA in the virion of some RNA viruses (e.g., Qβ phage) serves as a messenger on infection. For other RNA viruses (e.g., reovirus), the genome RNA must first be transcribed following infection. This is accomplished by an enzyme that is carried in the virion. A variety of strategies are employed by RNA viruses to synthesize proteins. Poliovirus proteins are formed by cleavage of a giant precursor. Reovirus contains a separate chromosome for each protein it synthesizes, whereas vesicular stomatitis virus transcribes several monogenic mRNAs from a single genomic RNA.

Not all viruses destroy their hosts. The temperate phage λ can assume a dormant prophage form in which its DNA is integrated into the bacterial chromosome by reciprocal recombination. This prophage retains the capacity for multiplication and lysis, which occur when it is excised from the host DNA. Likewise, the DNA tumor viruses SV40 and polyoma can multiply and lyse their hosts or their DNA can become integrated into the host-cell DNA. Retroviruses can propagate only by integrating into the host-cell DNA. These viruses contain a reverse transcriptase that synthesizes duplex DNA from genomic RNA. Productive infection by retroviruses, in contrast with oncogenic DNA viruses, is typically not lytic. Retroviruses, like other membrane-enveloped viruses, enter a cell by fusion of the lipid bilayer of the virus with the plasma membrane of the infected cell.

Cells harboring SV40 DNA or retroviral DNA in their genomes have abnormal growth patterns and cell surfaces. They cause cancers when injected into susceptible animals. These viruses are attractive models for the study of the molecular basis of cancer because they contain few

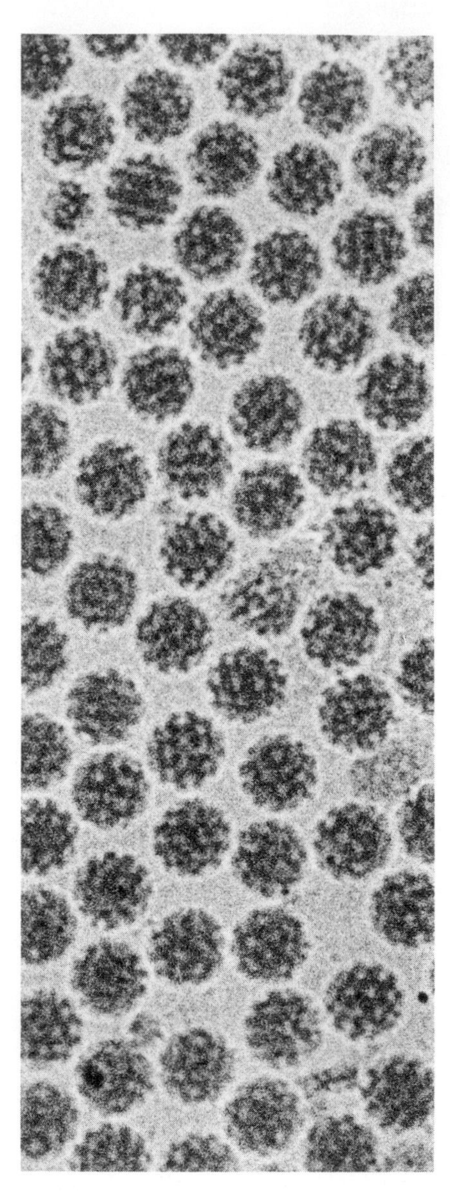

Electron micrograph of Semliki Forest virions. [Courtesy of Dr. Stephen Fuller.]

genes. Indeed, the cancer-producing action of numerous retroviruses has been shown to reside in a single gene called an oncogene. The *src* oncogene of avian sarcoma virus (Rous sarcoma virus) encodes a 60-kd tyrosine kinase. Several other oncogenes specify other tyrosine kinases. The other classes of oncogene products are growth factors, growth factor receptors, guanyl-nucleotide-binding proteins, and nuclear proteins. Oncogenes act by forming excessive amounts of proteins that stimulate growth and by encoding proteins that have lost the capacity to be regulated. These cancer-producing genes of retroviruses are derived from normal cellular genes that control growth and development. Viral oncogenes provide insight into normal growth control processes. Moreover, they reveal that cancer can be directly caused by the mutation, duplication, or translocation of cellular genes called proto-oncogenes.

Acquired immune deficiency syndrome (AIDS) is caused by human immunodeficiency virus (HIV). This retrovirus cripples the immune system by destroying T4 lymphocytes. The RNA genome of HIV contains *gag*, *pol*, and *env* genes and also encodes several small proteins that stimulate transcription and translation. The presence of these stimulatory factors and the rapid variation of its gp120 envelope protein make HIV a formidable pathogen.

The resistance of animal cells to many viruses is markedly enhanced by interferon, a protein synthesized and secreted by cells following a viral infection. Interferon stimulates the production of two enzymes, a protein kinase and an oligoadenylate synthetase, which are then activated by double-stranded RNA. The kinase blocks protein synthesis by phosphorylating the α subunit of eIF2, an initiation factor. Oligoadenylates (2,5-A_n) activate RNase L, which degrades mRNAs and rRNAs.

SELECTED READINGS

Where to start

Butler, P. J. G., and Klug, A., 1978. The assembly of a virus. *Sci. Amer.* 239(5):62–69.

Hogle, J. M., Chow, M., and Filman, D. J., 1987. The structure of poliovirus. *Sci. Amer.* 256(3):42–49.

Simons, K., Garoff, H., and Helenius, A., 1982. How an animal virus gets into and out of its host cell. *Sci. Amer.* 246(2):58–66.

Varmus, H. E., 1987. Reverse transcription. *Sci. Amer.* 257(3):56–64.

Bishop, J. M., 1987. The molecular genetics of cancer. *Science* 235:305–311.

Gallo, R. C., 1987. The AIDS virus. *Sci. Amer.* 256(1):46–56.

Books on viruses and oncogenes

Fields, B. N., and Knipe, D. M., (eds.), 1986. *Fundamental Virology.* Raven Press. [Contains many informative articles on animal viruses.]

Luria, S. E., Darnell, J. E., Jr., Baltimore, D., and Campbell, A., 1978. *General Virology* (3rd ed.). Wiley. [A highly readable and interesting introduction to viruses.]

Botchan, M., Grodzicker, T., and Sharp, P. A., (eds.), 1986. *DNA Tumor Viruses.* Cold Spring Harbor Laboratory.

Weiss, R., (ed.), 1982. *RNA Tumor Viruses.* Cold Spring Harbor Laboratory.

Williams, R. C., and Fisher, H. W., 1974. *An Electron Micrographic Atlas of Viruses.* Thomas. [Contains superb electron micrographs and a very interesting accompanying text.]

Virus structure

Jurnak, F. A., and McPherson, A., (eds.), 1984. *Biological Macromolecules and Assemblies. Volume 1: Virus Structure.* Wiley. [Contains excellent articles on the structure of several viruses discussed in this chapter.]

Hogle, J. M., Chow, M., and Filman, D. J., 1985. Three-dimensional structure of poliovirus at 2.9 Å resolution. *Science* 229:1358–1365.

Smith, T. J., Kremer, M. J., Luo, M., Vriend, G., Arnold, E., Kamer, G., Rossmann, M. G., McKinlay, M. A., Diana, G. D., and Otto, M. J., 1986. The site of attachment in human rhinovirus 14 for antiviral agents that inhibit uncoating. *Science* 233:1286–1293.

Erickson, J. W., Silva, A. M., Murthy, M. R. N., Fita, I., and Rossmann, M. G., 1985. The structure of a T =1 icosahedral empty particle from Southern bean mosaic virus. *Science* 229:625–629.

Bloomer, A. C., Champness, J. N., Bricogne, G., Staden, R., and Klug, A., 1978. Protein disk of tobacco mosaic virus at 2.8 Å resolution showing the interactions within and between subunits. *Nature* 276:362–368.

Namba, K., and Stubbs, G., 1986. Structure of tobacco mosaic virus at 3.6 Å resolution: implications for assembly. *Science* 231:1401–1406.

Harrison, S. C., 1984. Multiple modes of subunit association in the structures of simple viruses. *Trends Biochem. Sci.* 9:345–351.

Crick, F. H. C., and Watson, J. D., 1957. Virus structure: general principles. *In* Wolstenholme, G. E. W., (ed.), *Ciba Foundation Symposium on the Nature of Viruses,* pp. 5–13.

Caspar, D. L. D., and Klug, A., 1962. Physical principles in the construction of regular viruses. *Cold Spring Harbor Symp. Quant. Biol.* 27:1–24. [The basic theory of the structure of spherical viruses is presented in this classic paper.]

Virus assembly

Lebeurier, G., Nicolaieff, A., and Richards, K. E., 1977. Inside-out model for self-assembly of tobacco mosaic virus. *Proc. Nat. Acad. Sci.* 74:149–153.

Butler, P. J. G., Finch, J. T., and Zimmern, D., 1977. Configuration of tobacco mosaic virus RNA during virus assembly. *Nature* 265:217–219.

Wood, W. B., and King, J., 1979. Genetic control of complex bacteriophage assembly. *In* Fraenkel-Conrat, H., and Wagner, R. R., (eds.), *Comprehensive Virology,* vol. 13, pp. 581–633. Plenum.

Lamb, R. A., and Choppin, P. W., 1983. The gene structure and replication of influenza virus. *Ann. Rev. Biochem.* 52:467–506.

Arnold, E., Luo, M., Vriend, G., Rossmann, M. G., Palmenberg, A. C., Parks, G. D., Nicklin, M. J. H., and Wimmer, E., 1987. Implications of the picornavirus capsid structure for polyprotein processing. *Proc. Nat. Acad. Sci.* 84:21–25.

Viruses and evolution

Mills, D. R., Peterson, R. L., and Spiegelman, S., 1967. An extracellular Darwinian experiment with a self-duplicating nucleic acid molecule. *Proc. Nat. Acad. Sci.* 58:217–224.

Batschelet, E., Domingo, E., and Weissmann, C., 1976. The proportion of revertant and mutant phage in a growing population, as a function of mutation and growth rate. *Gene* 1:27–32.

Eigen, M., 1983. Self-replication and molecular evolution. *In* Bendall, D. S., (ed.), *Evolution from Molecules to Men,* pp. 105–130. Cambridge University Press.

Viroids

Dinter-Gottlieb, G., 1986. Viroids and virusoids are related to group I introns. *Proc. Nat. Acad. Sci.* 83:6250–6254.

Riesner, D., and Gross, H. J., 1985. Viroids. *Ann. Rev. Biochem.* 54:531–564.

Marmamorosch, K., 1987. The curse of cadang-cadang. *Natural History* 96(7):20–22. [A fascinating account of the epidemiology of a disease caused by a viroid.]

Restriction and modification of DNA

Smith, D. H., 1979. Nucleotide sequence specificity of restriction endonucleases. *Science* 205:455–462.

Arber, W., 1979. Promotion and limitation of genetic exchange. *Science* 205:361–365.

Lysogeny

Campbell, A. M., 1976. How viruses insert their DNA into the DNA of the host cell. *Sci. Amer.* 235(6):102–113.

Sadowski, P., 1986. Site-specific recombinases: changing partners and doing the twist. *J. Bacteriol.* 165:341–347.

Lwoff, A., 1966. The prophage and I. *In* Cairns, J., Stent, G. S., and Watson, J. D., (eds.), *Phage and the Origins of Molecular Biology,* pp. 88–99. Cold Spring Harbor Laboratory. [A delightful account of the discovery of the basis of lysogeny.]

Tumor viruses

Rous, P., 1911. A sarcoma of the fowl transmissible by an agent separable from the tumor cells. *J. Exp. Med.* 13:397–411.

Baltimore, D., 1976. Viruses, polymerases, and cancer. *Science* 192:632–636.

Temin, H., 1976. The DNA provirus hypothesis: the establishment and implications of RNA-directed DNA synthesis. *Science* 192:1075–1080.

Dulbecco, R., 1976. From the molecular biology of oncogenic DNA viruses to cancer. *Science* 192:437–440.

Oncogenes and cancer

Weinberg, R. A., 1983. A molecular basis of cancer. *Sci. Amer.* 249(5):126–142.

Hunter, T., 1984. The proteins of oncogenes. *Sci. Amer.* 251(2):70–79.

Barbacid, M., 1987. *Ras* genes. *Ann. Rev. Biochem.* 56:779–828.

Croce, C. M., and Klein, G., 1985. Chromosome translocations and human cancer. *Sci. Amer.* 252(3):54–60.

Bishop, J. M., 1985. Viral oncogenes. *Cell* 42:23–38.

Doolittle, R. F., Hunkapiller, M. W., Hood, L. E., Devare, S. G., Robbins, K. C., Aaronson, S. A., and Antoniades, H. N., 1983. Simian sarcoma *onc* gene, v-*sis*, is derived from the gene (or genes) encoding a platelet-derived growth factor. *Science* 221:275–277.

Kingston, R. E., Baldwin, A. S., and Sharp, P. A., 1985. Transcription control by oncogenes. *Cell* 41:3–5.

AIDS and human immunodeficiency virus

Chen, I. S. Y., 1986. Regulation of AIDS virus expression. *Cell* 47:1–2.

Mitsuya, H., and Broder, S., 1987. Strategies for antiviral therapy in AIDS. *Nature* 325:773–778.

Coffin, J. M., 1986. Genetic variation in AIDS viruses. *Cell* 46:1–4.

Rabson, A. B., and Martin, M. A., 1985. Molecular organization of the AIDS retrovirus. *Cell* 40:477–480.

Laurence, J., 1985. The immune system in AIDS. *Sci. Amer.* 253(6):84–93.

Maddon, P. J., Dalgleish, A. G., McDougal, J. S., Clapham, P. R., Weiss, R. A., and Axel, R., 1986. The T4 gene encodes the AIDS virus receptor and is expressed in the immune system and the brain. *Cell* 47:333–348.

Interferons

Pestka, S., Langer, J. A., Zoon, K. C., and Samuel, C. E., 1987. Interferons and their actions. *Ann. Rev. Biochem.* 56:727–778.

Lengyel, P., 1982. Biochemistry of interferons and their actions. *Ann. Rev. Biochem.* 51:251–282.

Farrell, P. J., Sen, G. C., Dubois, M. F., Ratner, L., Slattery, E., and Lengyel, P., 1978. Interferon action: two distinct pathways for inhibition of protein synthesis by double-stranded RNA. *Proc. Nat. Acad. Sci.* 75:5893–5897.

PROBLEMS

1. The cloning and sequencing of the entire genome of human immunodeficiency virus, the cause of AIDS, has enhanced our understanding of this pathogen. In particular, the nucleotide sequence led to the finding of the *tat* and *trs* proteins, which stimulate transcription and translation. How can base sequence information be used to predict the existence of hitherto unknown proteins? What experiments can be carried out to isolate such proteins?

2. A special problem is encountered in replicating T4 DNA and other linear DNA molecules because their 5′ ends are incomplete. The mechanism used by eucaryotic replication systems to circumvent this difficulty was discussed in the previous chapter (p. 832). A different solution is used by T4 phage. Its linear DNA molecule has *redundant ends*—that is, the base sequence at the left end of the DNA is precisely repeated at the right end:

   5′ A B C D E——V W X Y Z A B C 3′
   3′ a b c d e——v w x y z a b c 5′

   How might redundant ends be used to make possible the replication of the entire chromosome?

3. ϕX174, like a number of small viruses, contains overlapping genes to maximize its information content. The peptides encoded by the two reading frames are

   Leu-Asp-Phe-Val -Gly -Tyr -Pro -Arg-Phe-

   Trp-Thr-Leu-Trp-Asp-Thr-Leu-Ala -Phe-

   (a) What is the sequence of the corresponding mRNA?
   (b) A mutation changes proline in the first peptide to leucine. What is the concomitant change in the second peptide?
   (c) Why are extensive overlapping sequences not generally used to encode proteins?

4. What would be the effect of a translocation placing the T4 lysozyme gene under the control of a promoter for the immediate-early set of genes?

5. Turnip yellow mosaic virus is similar in many respects to tomato bushy stunt virus and Southern bean mosaic virus. However, its coat protein lacks an R domain. Instead, the virus particle is rich in polyamines such as spermidine. What is the function of these polyamines?

6. Influenza virus agglutinates erythrocytes. Why? How might you specifically inhibit agglutination in vitro?

7. The localization of viral nucleic acid in infected cells depends on the type of virus.
   (a) The genome of all DNA viruses but only some RNA viruses is found in the nucleus of infected cells. Why?
   (b) Influenza virus contains ten genes on its eight RNA segments. Passage of viral RNA through the nucleus is necessary for the synthesis of viral proteins. Why?

8. Human immunodeficiency virus can infect brain cells as well as cells of the immune system.
   (a) Propose a mechanism accounting for its entry into cells other than T4 lymphocytes.
   (b) How would you test your hypothesis?

9. What would be a selective advantage of RNA-assisted proteolytic cleavage in the assembly of an icosahedral RNA virus?

On the facing page: A retinal rod cell can be triggered by a single photon. This micrograph shows a piece of retina from which the outer segment of a rod cell has been drawn into a suction pipet. The rod cell is excited by light (green) shining through a slit. The change in ionic current across the plasma membrane induced by a single photon can be measured. The molecular events underlying this highly amplified response have been elucidated. [Courtesy of Dr. Denis Baylor.]

PART VI

MOLECULAR PHYSIOLOGY

interaction of information, conformation, and metabolism in physiological processes

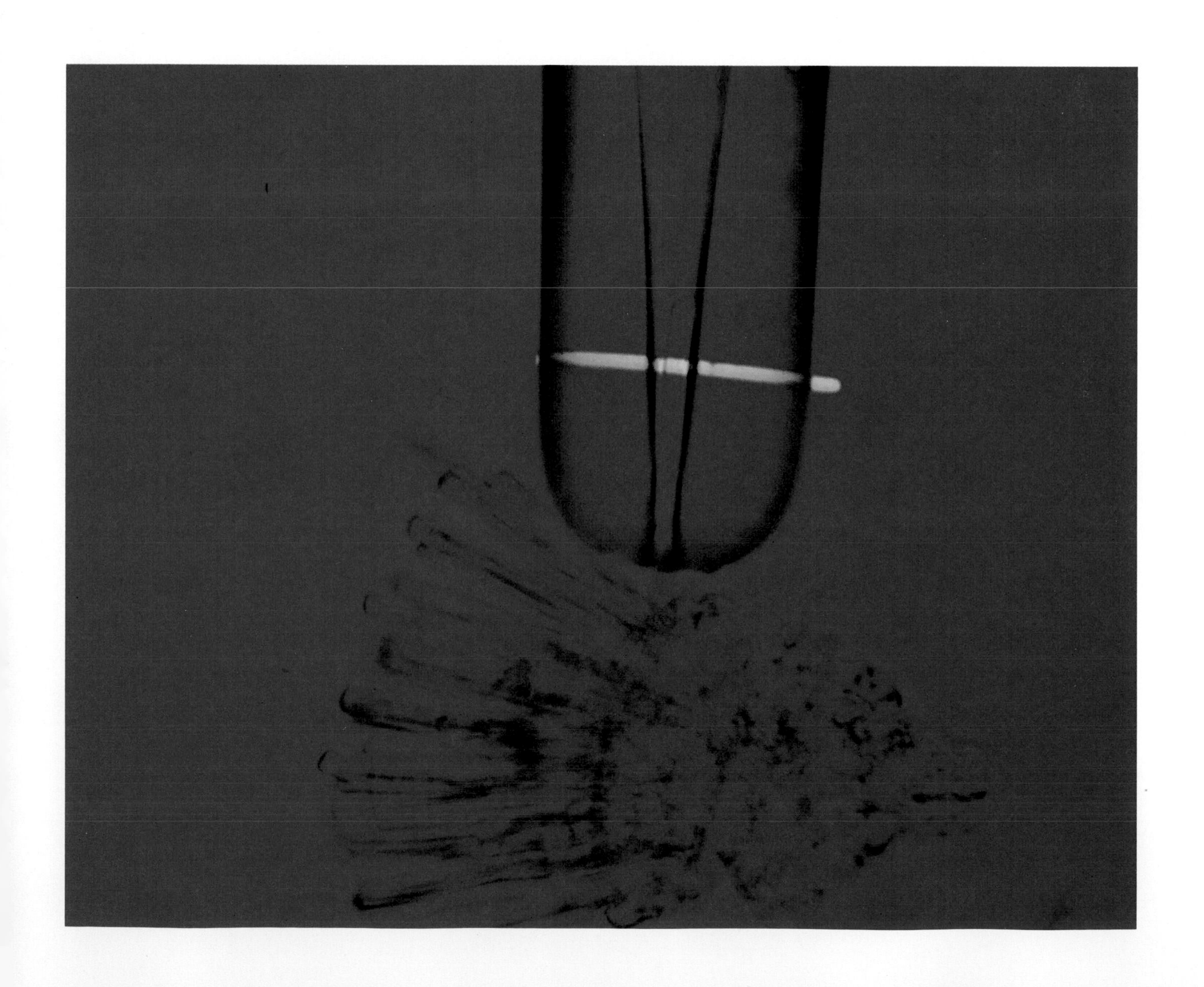

CHAPTER 35

Molecular Immunology

The vertebrate immune system is a vast network of molecules and cells having a single goal: *to distinguish between self and nonself.* Its primary function is to protect vertebrates against microorganisms—viruses, bacteria, and parasites. It continuously scans countless numbers of molecular units to decide which ones are foreign and initiate their destruction. The immune system learns from experience and remembers its encounters. Its hallmarks are *specificity, adaptation, and memory.* The immune system uses two different but related strategies. The recognition elements of the *humoral immune response* are soluble proteins called *antibodies*, which are produced by *plasma cells.* In the *cellular immune response, T lymphocytes* kill cells that display foreign motifs on their surface. They also stimulate the humoral response by helping B cells, the precursors of plasma cells.

The molecular basis of these remarkable processes is rapidly being unraveled. Immunology is deepening our understanding of disease processes and providing new diagnostic and therapeutic approaches. Monoclonal antibodies, for example, serve as highly specific detectors of pathogenic microorganisms, tissue damage, and cancer. Many chronic diseases are likely to be autoimmune disorders, in which the immune systems turns against self. Immunology is also providing insight into how cell development is triggered and how cells stimulate and suppress one another. The proteins and genes mediating these interactions are being identified and mapped, and their molecular architecture is coming into view. It is a time of ferment and excitement.

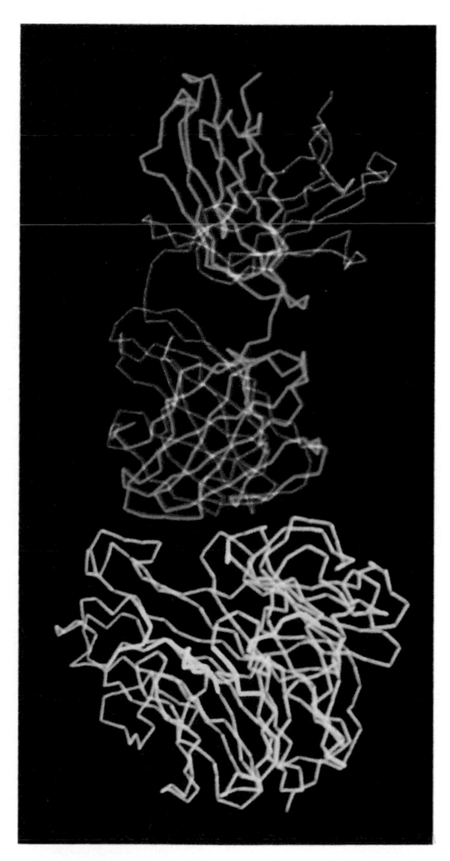

Figure 35-1
Structure of an antigen-antibody complex. Neuraminidase (yellow), an enzyme on the surface of influenza virus, is bound to a fragment of an antibody molecule (purple and blue). [Courtesy of Dr. Peter Colman.]

SPECIFIC ANTIBODIES ARE SYNTHESIZED FOLLOWING EXPOSURE TO AN ANTIGEN

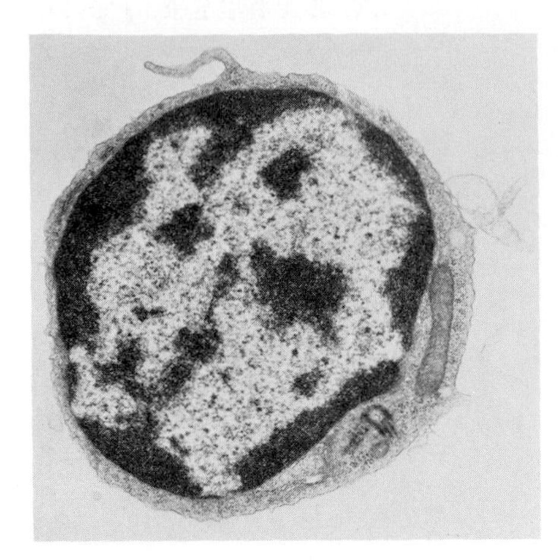

Figure 35-2
Electron micrograph of a B lymphocyte. [Courtesy of Lynne Mercer.]

Antibodies (or *immunoglobulins*) are proteins synthesized by an animal in response to the presence of a foreign substance. They are secreted by *plasma cells,* which are derived from *B lymphocytes* (*B cells*). These soluble proteins are the recognition elements of the *humoral immune response* (*humor* is the Latin word for liquid). Each antibody has specific affinity for the foreign material that stimulated its synthesis. A foreign macromolecule capable of eliciting antibody formation is called an *antigen* (or *immunogen*). Proteins, polysaccharides, and nucleic acids are usually effective antigens. The specific affinity of an antibody is not for the entire macromolecular antigen, but for a particular site on it called the *antigenic determinant* (or *epitope*).

Most small foreign molecules do not stimulate antibody formation. However, they can elicit the formation of specific antibody if they are attached to macromolecules. The macromolecule is then the *carrier* of the attached chemical group, which is called a *haptenic determinant.* The small foreign molecule by itself is called a *hapten*. Antibodies elicited by attached haptens will bind unattached haptens as well.

Animals can make specific antibodies against virtually any foreign chemical group. The dinitrophenyl group (DNP) is particularly effective in eliciting antibody formation and therefore has been used extensively as a haptenic determinant. Antibodies specific for DNP (termed anti-DNP antibody) can be obtained in the following way:

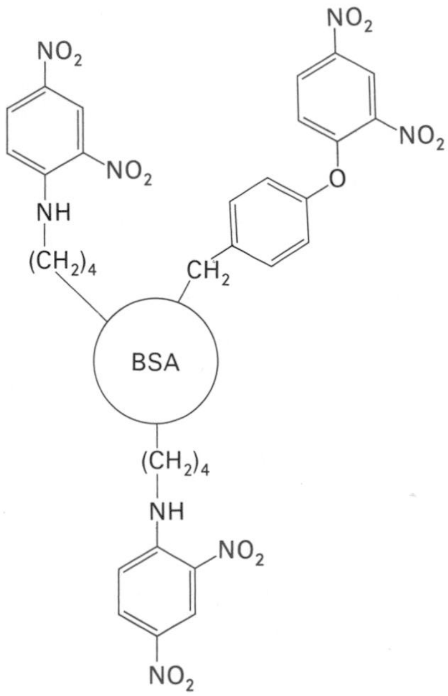

1. DNP groups are covalently attached to a carrier protein such as bovine serum albumin (BSA) by reaction of fluorodinitrobenzene with lysine and other nucleophilic side chains (Figure 35-3).

2. DNP-BSA, the immunogen (antigen), is injected into a rabbit. The level of anti-DNP antibody in the serum of the rabbit starts to rise a few days later (Figure 35-4). These *early antibody molecules* belong to the *immunoglobulin M* (IgM) class and have a mass of nearly 1000 kd.

Figure 35-3
Dinitrophenylated bovine serum albumin (DNP-BSA) is an effective immunogen.

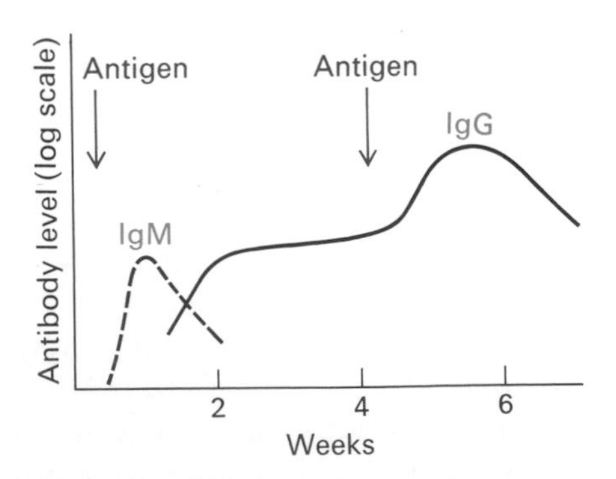

Figure 35-4
Kinetics of the appearance of immunoglobulins M and G (IgM and IgG) in the serum following immunization.

3. Approximately ten days after the injection of immunogen, the amount of immunoglobulin M decreases and there is a concurrent increase in the amount of anti-DNP antibody of a different class, called *immunoglobulin G* (IgG), which has a mass of 150 kd.

4. The level of anti-DNP antibody of the immunoglobulin G class reaches a plateau approximately three weeks after the injection of im-

munogen. A booster dose of DNP-BSA given at that time produces a further increase in the level of anti-DNP antibody in the rabbit's serum.

5. Blood is drawn from the immunized rabbit. The separated serum (called an *antiserum* because it is obtained after immunization) may contain as much as 1 mg/ml of anti-DNP antibody. Nearly all of it is of the immunoglobulin G class, the principal one in serum.

6. The next step is to separate anti-DNP antibody from antibodies of other specificities and from other serum proteins. The distinguishing feature of anti-DNP antibody is its very high affinity for DNP. Consequently, it can be separated by *affinity chromatography* using a column consisting of dinitrophenyl groups covalently attached to an insoluble carbohydrate matrix.

ANTIBODIES ARE FORMED BY SELECTION RATHER THAN BY INSTRUCTION

Enzymes acquired their specificities over millions of years of evolution. A specific antibody appears in the serum of an animal only a few weeks after exposure to any foreign determinant. How is a specific antibody formed in such a short time? The *instructive theory,* which was proposed by Linus Pauling in 1940, postulated that the antigen acts as a template that directs the folding of the nascent antibody chain. In this model, antibody molecules of a given amino acid sequence have the potential for forming combining sites of many different specificities. The particular one formed depends on the antigen present at the time of folding. A contrasting hypothesis, called the *selective theory* (or clonal selection theory), was proposed and developed by Macfarlane Burnet, Niels Jerne, David Talmage, and Joshua Lederberg in the 1950s. It postulated that the *combining site of an antibody molecule is completely determined before it ever encounters antigen.* In this theory, the antigen affects only the *amount* of specific antibody produced.

The instructive theory predicted that an antibody will lose its specificity if it is unfolded and then refolded in the absence of antigen. This prediction contrasted with experimental results obtained in Christian Anfinsen's laboratory on the refolding of ribonuclease. Recall that unfolded ribonuclease spontaneously assumes the three-dimensional structure, specificity, and catalytic activity of the native enzyme upon removal of the denaturing agent (p. 32). A substrate need not be present during the refolding of ribonuclease. This important experiment on ribonuclease provided the impetus for devising a similar test for antibodies (Figure 35-5). An antibody fragment (called F_{ab}) containing the antigen-binding site of anti-DNP antibody was unfolded in 7 M guanidine hydrochloride. Denatured F_{ab} had no detectable affinity for DNP haptens. Hydrodynamic and optical measurements showed that its conformation approached that of a random coil. On removal of guanidine hydrochloride by dialysis, *the chain refolded in the absence of hapten into a compact unit that regained specific affinity for DNP.* This important finding agreed with the selective theory but contradicted a critical prediction of the instructive theory. The selective theory was further supported by the finding that *antibody-producing cells can synthesize large amounts of specific antibody in the complete absence of the corresponding antigen.*

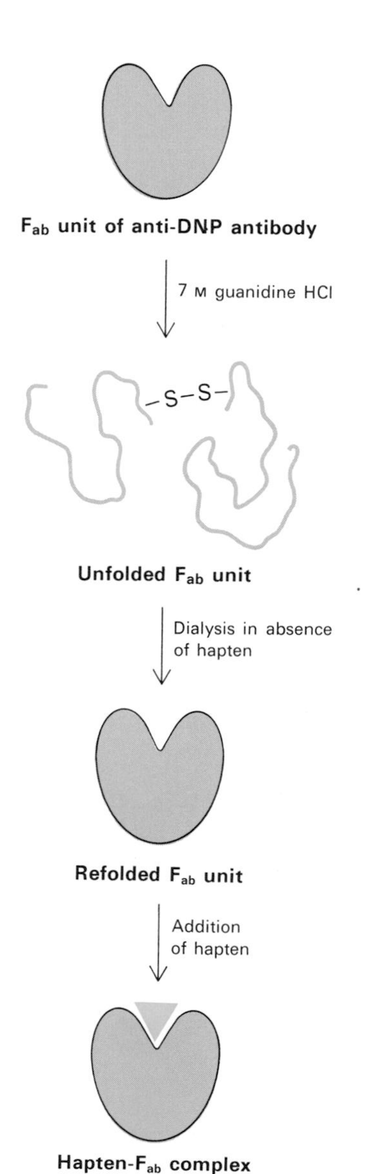

Figure 35-5
The F_{ab} fragment of anti-DNP antibody binds DNP after being unfolded and refolded in the absence of this hapten. This experiment showed that specificity is inherent in the amino acid sequence.

The *clonal selection theory* provides a unifying view of the immune response. The essential features of this visionary hypothesis, now firmly established, are:

1. Each antibody-producing cell makes antibody of a single kind. *The commitment to synthesize a particular kind of antibody is made before the cell ever encounters antigen.*

2. Each cell has a distinctive base sequence in its DNA that determines the amino acid sequence of its immunoglobulin chains. The specificity of an antibody is determined by its amino acid sequence.

3. As it begins to mature, each cell produces small amounts of specific antibody, which appears on its surface. An immature cell is killed if it encounters the antigen corresponding to its antibody. Such antibody-producing cells are eliminated during fetal life. Hence, an animal does not usually make antibody against its own macromolecules—it is *self-tolerant.* In contrast, *a mature cell is stimulated if it encounters antigen. Large amounts of antibody are then synthesized and the cell is triggered to divide.* The descendants of this cell are a *clone.* They have the same genetic makeup as the cell that was initially triggered by its encounter with antigen, and therefore all of them make antibody of the same specificity.

4. A clone tends to persist after the disappearance of the antigen. These cells retain the capacity to be stimulated by antigen if it reappears, which provides for *immunological memory* and a quick response to reinfection.

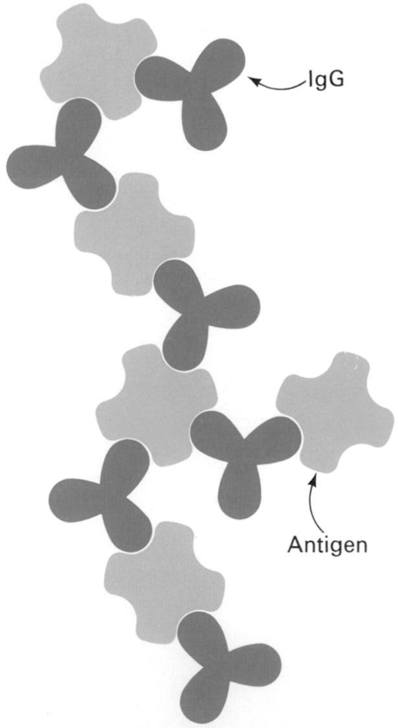

IgG

Antigen

Figure 35-6
Schematic diagram of a lattice formed by the cross-linking of IgG (shown in blue) and an antigen (shown in yellow).

IMMUNOGLOBULIN G CAN BE CLEAVED INTO ANTIGEN-BINDING AND EFFECTOR FRAGMENTS

Immunoglobulin G has a mass of 150 kd. A fruitful approach in studying such a large protein is to split it into fragments that retain activity. In 1959, Rodney Porter showed that immunoglobulin G can be cleaved into three active 50-kd fragments by the limited proteolytic action of papain.

$$\underset{(150\text{ kd})}{\text{IgG}} \xrightarrow{\text{papain}} \underset{(50\text{ kd each})}{2\ \text{F}_{\text{ab}}} + \underset{(50\text{ kd})}{\text{F}_{\text{c}}}$$

Two of these fragments bind antigen. They are called F_{ab} (*ab* stands for antigen-binding, *F* for fragment). *Each F_{ab} contains one combining site for hapten or antigen,* and it has the same binding affinity for hapten as does the whole molecule. However, *F_{ab} does not form a precipitate with antigen* because it is univalent (i.e., contains only one combining site), in contrast with an intact immunoglobulin G antibody molecule, which contains two *identical* antigen-binding sites. Immunoglobulin G can form a precipitate with an antigen that contains more than one antigenic determinant. An extended lattice is then formed, in which each antibody molecule cross-links two antigens, and vice versa (Figure 35-6). The amount of precipitate formed is largest when antibody and antigen are present in equivalent amounts.

The other fragment (also 50 kd), called F_c because it crystallizes readily, does not bind antigen, but it has other important biological activities. F_c mediates *effector functions,* such as *complement fixation.* The binding of an antibody to an antigenic determinant on the surface of a foreign cell often leads to a cascade of reactions that lyses the cell. This sequel of the combination of antibody and antigen is mediated by a group of proteins known collectively as complement (p. 899). The binding of complement proteins to the F_c units of antigen-antibody complexes triggers the cascade.

IMMUNOGLOBULIN G CONSISTS OF PAIRS OF LIGHT (L) AND HEAVY (H) CHAINS

In 1959, too, Gerald Edelman found that immunoglobulin G consists of two kinds of polypeptide chains. The disulfide bonds of the molecule were reduced by mercaptoethanol, and noncovalent interactions between the constituent chains were disrupted by 6 M urea. Chromatographic analysis of the resulting mixture revealed that immunoglobulin G consists of 25- and 50-kd polypeptide chains, which are called *light* (L) and *heavy* (H) chains. Rodney Porter then found milder conditions that made it possible to reconstitute immunoglobulin G from two H and two L chains. He proposed that immunoglobulin G has the subunit structure L_2H_2. Each L chain is bonded to an H chain by a disulfide bond, and the H chains are bonded to each other by at least one disulfide bond (Figure 35-7).

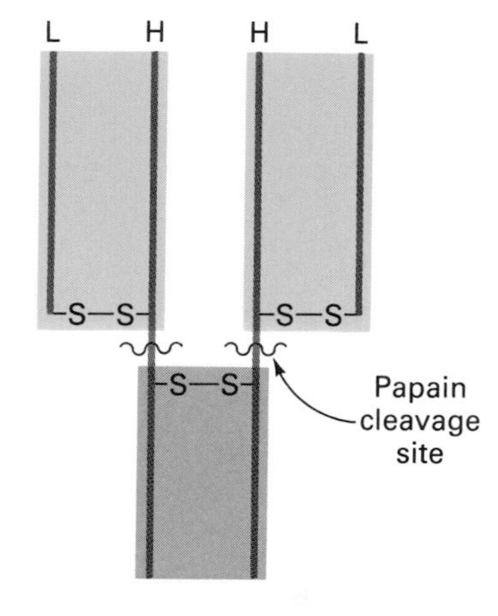

Figure 35-7
The subunit structure of IgG is L_2H_2. The F_{ab} units obtained by digestion with papain are shaded blue, whereas the F_c unit is shaded red.

The fragments produced by papain digestion were then related to this model. Papain cleaves the H chains on the carboxyl-terminal side of the disulfide bond that links the L and H chains. Thus, F_{ab} consists of the entire L chain and the amino-terminal half of the H chain, whereas F_c consists of the carboxyl-terminal halves of both H chains.

IMMUNOGLOBULIN G IS A FLEXIBLE Y-SHAPED MOLECULE

Electron-microscopic studies by Robin Valentine and Michael Green showed that immunoglobulin G has the shape of the letter Y and that the hapten binds near the ends of the F_{ab} units. Furthermore, the F_c and the two F_{ab} units of the intact antibody are joined by a hinge that allows rapid variation in the angle between the F_{ab} units through a wide range (Figure 35-8). This kind of mobility, called *segmental flexibility*, enhances the formation of antibody-antigen complexes by enabling

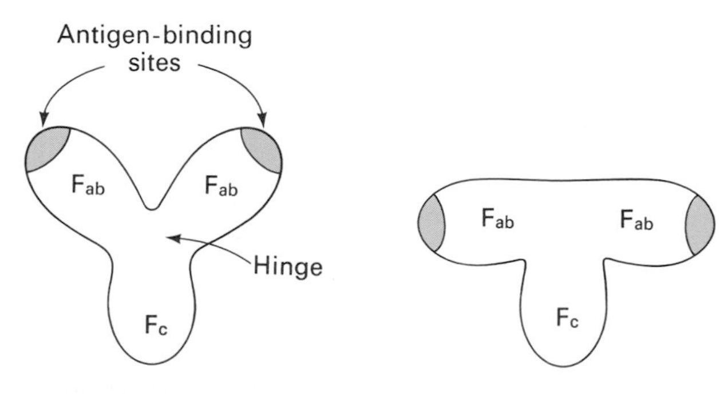

Figure 35-8
Immunoglobulin G has the shape of the letter Y. It contains a hinge that gives it segmental flexibility.

both combining sites on an antibody to bind a multivalent antigen. The distance between combining sites at the tips of the F_{ab} units can be adjusted to match the distance between specific determinants on the antigen (e.g., a virus with many repeating subunits). The affinity of an antibody for a multivalent antigen is typically 10^4 times as high when both combining sites are used as when only one binds the antigen. Segmental flexibility may also play a role in the transmission of information from the F_{ab} units to the F_c unit.

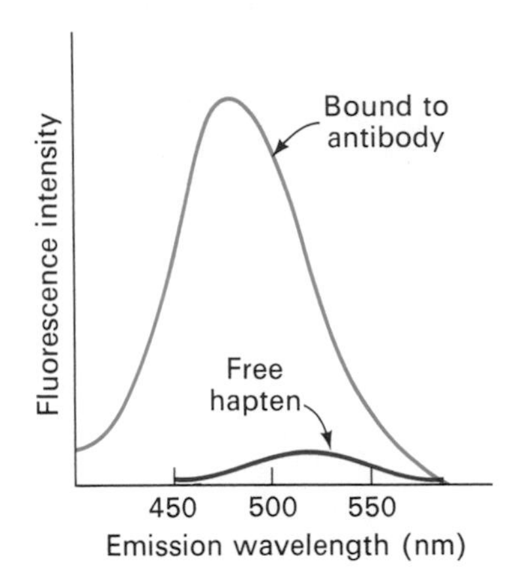

Figure 35-9
The fluorescence properties of ϵ-dansyl-lysine change markedly when this hapten binds to antidansyl antibody. The decrease in the wavelength of the emission maximum and the increase in quantum yield indicate that the hapten is bound to a nonpolar site.

THE COMBINING SITES OF ANTIBODIES ARE LIKE THE ACTIVE SITES OF ENZYMES

The *combining sites* of antibodies (i.e., their antigen-binding sites) resemble the active sites of enzymes in several ways:

1. The binding constants for haptens have been determined by equilibrium dialysis and spectroscopic techniques. For example, the binding of a colored hapten such as a DNP derivative quenches the fluorescence of the tryptophan residues of the antibody. The extent of quenching is a measure of the saturation of the antibody combining sites. Alternatively, the fluorescence of some haptens increases when they bind to antibodies (Figure 35-9). *Binding constants for most haptens range from* $K = 10^{-4}$ *to* 10^{-10} M. Thus, the standard free energies of binding typically fall between about −6 and −15 kcal/mol, which is the same range as for enzyme-substrate and enzyme-coenzyme complexes. Furthermore, *the binding forces in hapten-antibody complexes are like those in enzyme-substrate complexes.* Weak, noncovalent interactions of the electrostatic, hydrogen bond, and van der Waals types combine to give strong and specific binding.

2. An antibody specific for dextran, a polysaccharide of glucose residues, was tested for its binding of glucose oligomers. The full binding affinity was obtained when the oligomer contained six glucose residues. An interesting comparison here is with lysozyme, which also has space for six sugar residues in its active-site cleft. Thus, *the binding site of this antidextran antibody is about 25 Å long.*

3. The spectroscopic properties of some haptens provide information about the polarity of the antibody combining site. For example, certain naphthalenes exhibit a weak yellow fluorescence when they are in a highly polar environment (such as water) and an intense blue fluorescence when they are in a markedly nonpolar environment (such as hexane). An intense blue fluorescence is observed when such a naphthalene hapten is bound to specific antibody, indicating that water is largely excluded from the combining site. This site is a *nonpolar niche*; it enhances the binding of antigens by excluding water, which competes for hydrogen bonds and attenuates electrostatic interactions (p. 9).

4. The fit of hapten is rather precise. *Specificity is high but certainly not absolute.* The hapten is firmly held to the combining site, so that it has little rotational freedom. The rate constant for the binding of a variety of haptens is about 10^8 M^{-1} s^{-1}. This high rate constant indicates that the binding step is a diffusion-controlled reaction. The binding of hapten does not lead to large-scale conformational rearrangements.

MYELOMA ANTIBODIES ARE HOMOGENEOUS, WHEREAS MOST NORMAL ANTIBODIES ARE HETEROGENEOUS

Antibodies differ from enzymes in a very important way. *Most naturally occurring antibodies of a given specificity, such as anti-DNP antibody, are not a single molecular species.* Analyses of the binding of DNP haptens to a preparation of anti-DNP antibody reveal a wide range of binding affinities. Some antibody molecules bind DNP with a K of 10^{-6} M, whereas others have a K of 10^{-10} M. In contrast, most enzymes have a single binding constant for a particular substrate or coenzyme. Furthermore, a large number of bands are evident when anti-DNP antibody or other antibodies of a particular specificity are subjected to isoelectric focusing. In contrast, an enzyme displays a single electrophoretic band or a few discrete bands (e.g., the isozymes of lactate dehydrogenase).

Antibodies produced by a single cell are homogeneous. However, *antibody molecules having a common specificity are normally heterogeneous because they are products of many different antibody-producing cells.* This heterogeneity was a serious impediment to elucidating the molecular basis of antibody action. The hurdle was overcome by taking advantage of *multiple myeloma,* a malignant disorder of antibody-producing cells. In this cancer, a single transformed lymphocyte or plasma cell divides uncontrollably. Consequently, a very large number of cells of a single kind are produced. They are a *clone* because they are descended from the same cell and have identical properties. Large amounts of immunoglobulin of a single kind are secreted by these tumors. *Myeloma immunoglobulins have a normal structure and are typical of normal immunoglobulins, but they are homogeneous.* Myelomas also occur in mice. These tumors can be transplanted to other mice, where they proliferate. Furthermore, these antibody-producing tumors synthesize the same kind of homogeneous antibody generation after generation. As will be discussed shortly, the analysis of myeloma immunoglobulins contributed to a major advance in our understanding of antibody specificity.

MONOCLONAL ANTIBODIES WITH DESIGNED SPECIFICITIES CAN READILY BE PREPARED

Myeloma immunoglobulins are advantageous in being homogeneous but their corresponding antigens are usually not known. In 1975, Cesar Milstein and Georges Köhler discovered that *large amounts of homogeneous antibody of nearly any desired specificity can be obtained by fusing an antibody-producing cell with a myeloma cell.* A mouse is immunized with an antigen, and its spleen is removed several weeks later (Figure 35-10). A mixture of lymphocytes and plasma cells from this spleen is fused in vitro with myeloma cells by exposing them to polyethylene glycol, a polymer that induces cell fusion. A mutant myeloma cell line lacking hypoxanthine-guanosine phosphoribosyl transferase (HGPRT) is used to enable hybrids to be selected with ease. Recall that HGPRT catalyzes the synthesis of inosinate (a precursor of AMP and GMP) in the salvage pathway (p. 606). The cells are grown in a medium containing *h*ypoxanthine, *a*minopterin (methotrexate), and *t*hymine (called HAT medium) to kill unfused myeloma cells. The role of aminopterin in this medium is to block de novo synthesis of nucleotides (p. 616). Hypoxanthine cannot be used by unfused myeloma cells because they lack HGPRT.

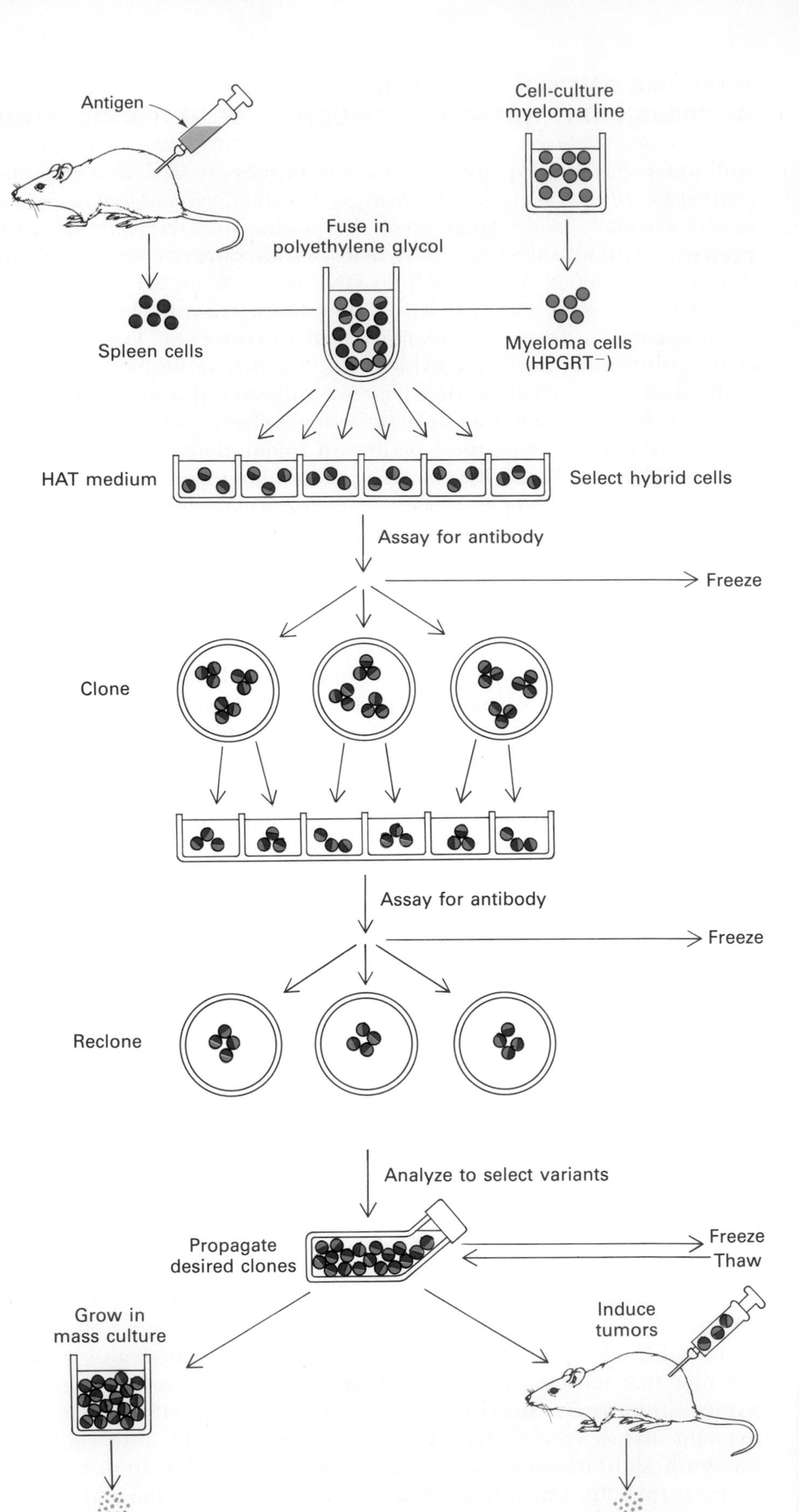

Figure 35-10
Preparation of monoclonal antibodies. Hybridoma cells formed by fusion of antibody-producing cells and myeloma cells are selected by growth in HAT medium. These hybrid cells are screened to determine which ones produce antibody of the desired specificity. [After C. Milstein. Monoclonal antibodies. Copyright © 1980 by Scientific American, Inc. All rights reserved.]

Spleen cells contain HGPRT, but they die in tissue culture because they are not able to proliferate in vitro. Hybrid cells, in contrast, survive because they have the neoplastic character of their myeloma cell parent and contain HGPRT genes of their spleen cell parent. They are called *hybridoma cells*.

Hybridoma cells and their progeny indefinitely produce large amounts of the homogeneous antibody specified by the parent cell from the spleen. They are grown in wells in a tissue-culture plate (Figure 35-10). Supernatants from these wells are screened for the presence of antibody molecules specific for the antigen of interest. The cells in positive wells are cloned and screened again to obtain hybridomas of a single kind. These cells can be grown in bottles containing culture medium or they can be injected into mice to induce myelomas that secrete the desired antibody. Alternatively, the cells can be frozen and stored for long periods.

The hybridoma method of producing *monoclonal antibodies* has opened new vistas in biology and medicine. *Large amounts of homogeneous antibodies with tailor-made specificities can readily be prepared. They provide insight into relations between antibody structure and specificity. Moreover, monoclonal antibodies can serve as precise analytical and preparative reagents.* For example, a pure antibody can be obtained against an antigen that has not yet been isolated. This feature is especially advantageous in studies of cell-surface proteins. Subsets of lymphocytes can be isolated by fluorescence-activated cell sorting using monoclonal antibodies that are specific for distinctive protein markers on these cells. Monoclonal antibodies attached to solid supports can be used as affinity columns to purify scarce proteins. This method has been used to purify interferon 5000-fold from a crude mixture. *Clinical laboratories are using monoclonal antibodies in many assays.* For example, the detection in blood of isozymes that are normally localized in the heart points to a myocardial infarction. Monoclonal antibodies are also being evaluated for use as therapeutic agents, as in the treatment of cancer.

Fluorescence-activated cell sorting—A technique (also called FACS) for analyzing and sorting cells according to their content of fluorescent molecules. A stream of cells is divided into droplets, each containing no more than one cell. A charge is imposed on the surface of the droplets to enable them to be deflected. The droplets are passed one at a time through a focused laser beam, which excites the fluorescence of probe molecules inside the cells or on their surface. Droplets are deflected into different test tubes according to the intensity and spectral distribution of their fluorescence. For example, killer T cells can readily be sorted by tagging them with a fluorescent-labeled antibody specific for a distinctive marker on their surface.

L AND H CHAINS CONSIST OF A VARIABLE (V) AND A CONSTANT (C) REGION

An abnormal protein appears in large amounts in the urine of many patients who have multiple myeloma. In 1847, this protein attracted the attention of Henry Bence-Jones because of its unusual solubility properties: it precipitates on being heated to 50°C and becomes soluble again on boiling. The *Bence-Jones protein* is now known to be a *dimer of the L chains* of a patient's myeloma immunoglobulin.

The first complete amino acid sequences of myeloma L chains, which contain about 214 amino acids, were determined in 1965. These studies showed that Bence-Jones proteins from different patients have different amino acid sequences. Most striking, these differences in sequence are confined to the amino-terminal half of the polypeptide chain. Each Bence-Jones protein studied has a unique amino acid sequence from positions 1 to 108. In contrast, the sequences of many of these proteins are the same starting at position 109. Thus, *the L chain consists of a variable region (residues 1 to 108) and a constant region (residues 109 to 214)* (Figure 35-11). This remarkable finding had no precedent.

Not all L chains have the same constant region. In fact, there are two types of L chains, called kappa (κ) and lambda (λ), which exhibit many

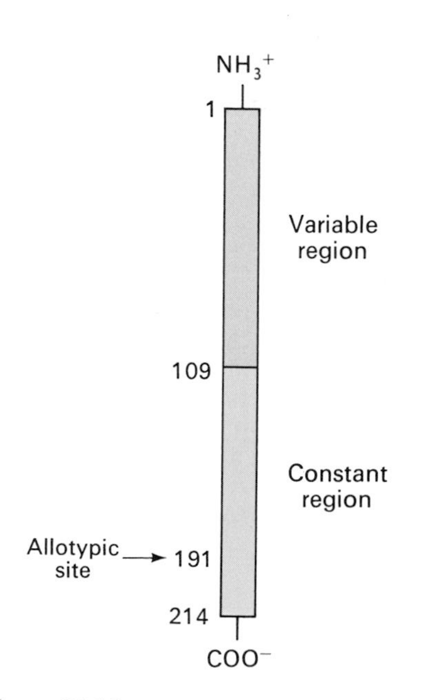

Figure 35-11
The light chain of an immunoglobulin consists of a variable region and a constant region.

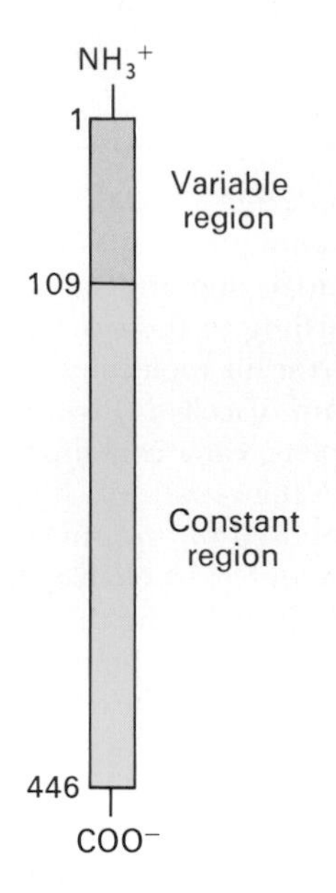

Figure 35-12
The heavy chain of IgG also consists of a variable region and a constant region.

similarities. For example, each consists of 214 residues and contains an amino-terminal variable half and a carboxyl-terminal constant half. Each half of the κ and λ chains contains an intrachain disulfide loop. Both types have a carboxyl-terminal cysteine residue, which, in an immunoglobulin molecule, forms a disulfide bond with the H chain. Furthermore, the κ and λ chains have identical amino acid residues at 40% of the positions, which is nearly the same degree of identity exhibited by the α and β chains of human hemoglobin.

All κ chains have the same constant region except for residue 191, which is either leucine or valine. The λ chains also exhibit variability only at position 191, which is either lysine or arginine. For each type of chain, this *allotypic* difference is inherited in a classic Mendelian manner.

The H chain consists of 446 amino acid residues. A comparison of the sequences of H chains from different myeloma immunoglobulins shows that all of the differences in sequence are located in the 108 residues at the amino-terminal end. Thus, *the heavy chain, like the light chain, consists of a variable region* and *a constant region.* The variable region of the heavy chain has the same length as that of the light chain, whereas the constant region is about three times as long (Figure 35-12).

VARIABLE REGIONS OF L AND H CHAINS FORM ANTIGEN-BINDING SITES

As mentioned earlier, the antigen-binding fragment F_{ab} consists of the entire L chain and part of the H chain. What are the relative contributions of the two chains to binding activity? An early clue came from a comparison of the amino acid sequences of many myeloma proteins, which revealed that the variable regions of the L and H chains are not uniformly variable. Rather, three segments in the L chain and four in the H chain display far more variability than do other residues in the variable regions of these chains (Figure 35-13). In 1970, Elvin Kabat proposed that these *hypervariable segments* form the antigen-binding site and that antibody specificity is determined by the nature of their amino acid residues.

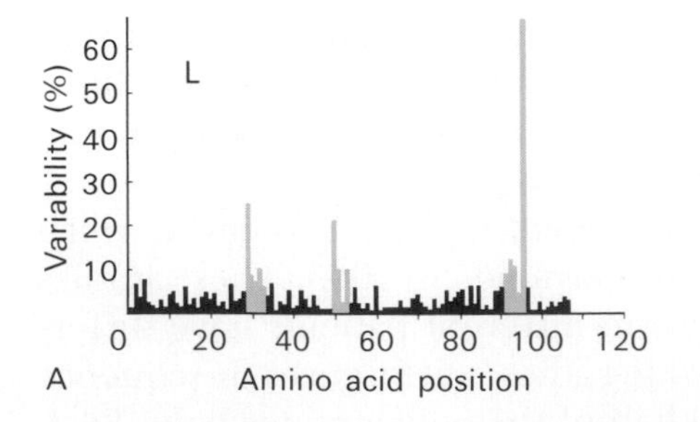

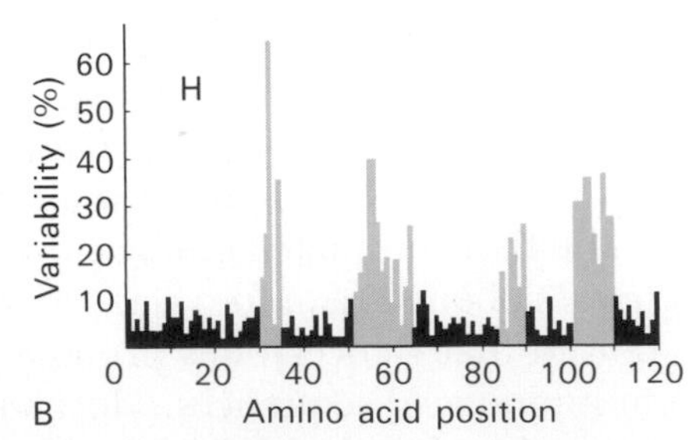

Figure 35-13
Hypervariable regions in (A) light and (B) heavy chains. The degree of variability in amino acid sequence is plotted versus amino acid position. [After J. D. Capra and A. B. Edmundson. The antibody combining site. Copyright © 1976 by Scientific American, Inc. All rights reserved.]

Affinity-labeling studies were then carried out to test this hypothesis. Haptens containing reactive groups can be readily synthesized. The reactive hapten binds specifically to the antibody combining site and then forms a covalent bond with a suitable neighboring residue on the protein. For example, anti-DNP antibody can be labeled at its combining sites by a nitrophenyldiazonium hapten. The diazonium group is highly reactive and can form a diazo bond with tyrosine, histidine, and lysine residues. Antibody modified in this way contains the label in both its L and its H chains. In fact, the labeled residues are located in the hypervariable regions of these chains. Thus, *hypervariable residues of both the L and the H chains form the antigen-binding site.* X-ray crystallographic analyses have provided the most direct evidence in support of this conclusion, as will be discussed shortly. Hypervariable regions are also called *complementarity-determining regions* (CDRs), because they determine antibody specificity.

$O_2N-C_6H_4-N_2^+$

***p*-Nitrophenyldiazonium hapten**

CONSTANT REGIONS MEDIATE EFFECTOR FUNCTIONS SUCH AS THE TRIGGERING OF THE COMPLEMENT CASCADE

The amino-terminal parts of the light and heavy chains contain the variable regions, which are responsible for the binding of antigen. Differences in amino acid sequence yield combining sites that have different specificities. The rest of each chain contains the constant region, which mediates effector functions such as the binding of complement and the transfer of antibodies across the placental membrane. The same effector functions are mediated by antibodies of very different specificity. Thus, *antibodies consist of regions of unique sequence that confer antigen-binding specificity and regions of constant sequence that mediate common effector functions.* A striking structural feature of immunoglobulins is that the variable and constant regions are sharply demarcated in the amino acid sequences of the light and heavy chains.

The complement system lyses microorganisms and infected cells by forming holes in their plasma membrane. More than fifteen soluble proteins in the plasma participate in this precisely regulated cascade. The triggering of lysis by antibody-antigen complexes is mediated by the *classical pathway*, beginning with the activation of C1, the first component of the pathway. C1 consists of a recognition unit *q* and two zymogens, *r* and *s*. C1q is a hexamer of three kinds of chains (A, B, and C), each about 23 kd. The high proline content of C1q suggested that it has an unusual structure. Indeed, each ABC trimer forms a triple-stranded helix of the kind seen in collagen (p. 264). The overall structure of C1q resembles a bunch of six tulips (Figure 35-14). Each of the six globular heads contains a binding site for the F_c moiety of an antibody molecule. C1 is not activated by an antibody molecule alone or by a single one complexed with an antigen. Rather, a group of several F_c units serves as the trigger when they are brought together by the formation of a multivalent antibody-antigen complex.

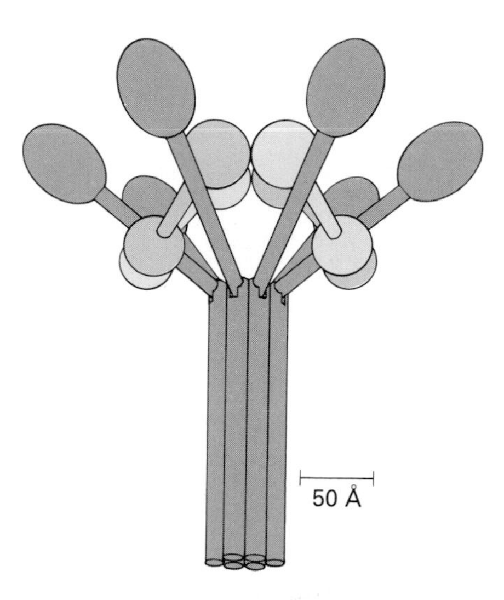

Figure 35-14
Model of the C1 component of complement. C1q is shown in red, C1r in blue, and C1s in green. The globular heads of C1q emerge from long stems consisting of triple-stranded helices of the type found in collagen. The binding of the F_c units of antibody-antigen complexes to the globular heads of C1q triggers the complement cascade by activating C1r and then C1s. [After G. J. Arlaud, M. G. Colomb, and J. Gagnon. *Immunology Today* 8(1987):1076.]

C1r and C1s are located in a cavity formed by the heads and stems of C1q (Figure 35-14). The simultaneous binding of several F_c units to the globular heads of C1q converts C1r from a zymogen into a protease. This activation step, in which C1r undergoes an internal cleavage, is reminiscent of the activation of pepsinogen by acid (p. 246). It seems likely that the binding of F_c units to C1q pulls an inhibitory segment out from C1r. Activated C1r in turn splits a peptide bond in C1s to convert it into an active protease. C4, C2, C3, and C5 are sequentially cleaved. C5b, a fragment of C5, then joins C6, C7, and C8 to penetrate the membrane bearing the antigen. Finally, *the binding of some sixteen molecules of C9 to this "bridgehead" produces large pores in the membrane, which cause the lysis and destruction of the target cell* (Figure 35-15).

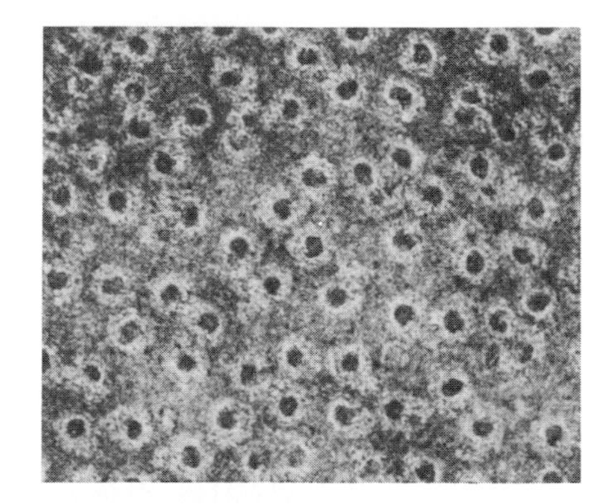

Figure 35-15
Electron micrograph of an erythrocyte membrane that was lysed by complement. The 100-Å-diameter pores are polymers of C9, the final component of the cascade. [Courtesy of Dr. Eckhard Podack.]

IMMUNOGLOBULIN DOMAINS HAVE SIMILAR SEQUENCES AND THREE-DIMENSIONAL STRUCTURES

The complete amino acid sequence of an immunoglobulin molecule was elucidated by Gerald Edelman in 1968. A most interesting feature of this sequence is the periodic location of intrachain disulfide bonds in both the light and the heavy chains (Figure 35-16). The variable region of the light chain (V_L) is similar in sequence to the variable region of the heavy chain (V_H). In addition, the constant region of the heavy chain (C_H) consists of equal thirds (C_H1, C_H2, and C_H3) that are similar in sequence. Furthermore, the constant region of the light chain (C_L) closely resembles the three domains of the constant region of the heavy chain.

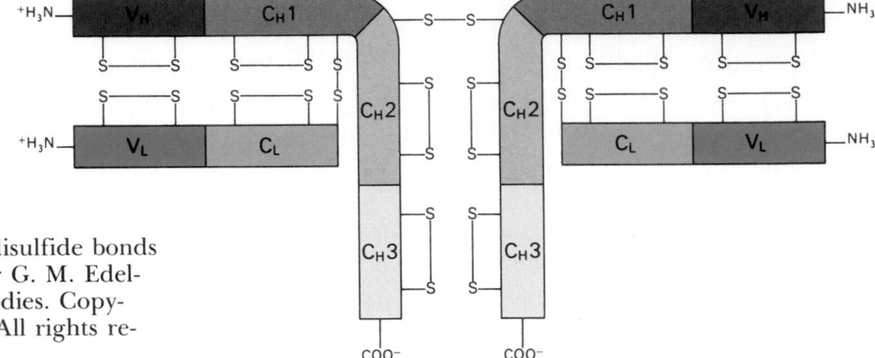

Figure 35-16
Structure of IgG, showing the pattern of disulfide bonds and the regions of similar sequence. [After G. M. Edelman. The structure and function of antibodies. Copyright © 1970 by Scientific American, Inc. All rights reserved.]

The similarities suggested that immunoglobulins are folded into compact domains, each consisting of a similar sequence containing at least one region that serves a distinctive molecular function. It also seemed likely that these domains would resemble each other in tertiary structure because they have similar amino acid sequences. X-ray crystallographic studies have verified the domain hypothesis. The three-dimensional structure of the $F_{ab'}$ fragment (which is nearly the same as the F_{ab} fragment) of a human immunoglobulin has been solved at 2.0-Å resolution by Roberto Poljak. This fragment consists of a *tetrahedral array of four globular subunits—V_L, V_H, C_L, and C_H1—which are strikingly similar in three-dimensional structure* (Figure 35-17).

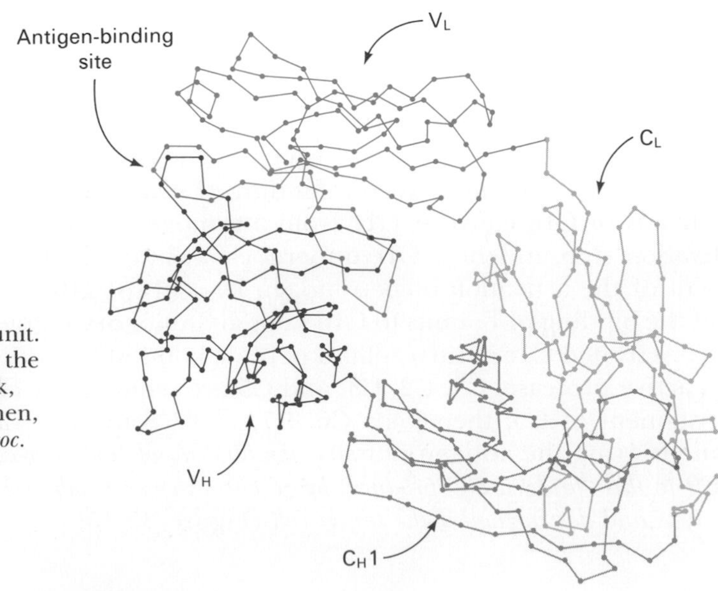

Figure 35-17
An α-carbon diagram of the $F_{ab'}$ unit. The L chain is shown in blue and the H chain in red. [From R. L. Poljak, L. M. Amzel, H. P. Avey, B. L. Chen, R. P. Phizackerley, and F. Saul. *Proc. Nat. Acad. Sci.* 70(1973):3306.]

A common structural feature of these domains is the presence of two broad sheets of antiparallel β-strands (Figure 35-18). Many hydrophobic side chains are tightly packed between these sheets, which are joined by a disulfide bond. This recurring structural motif is called the *immunoglobulin fold*.

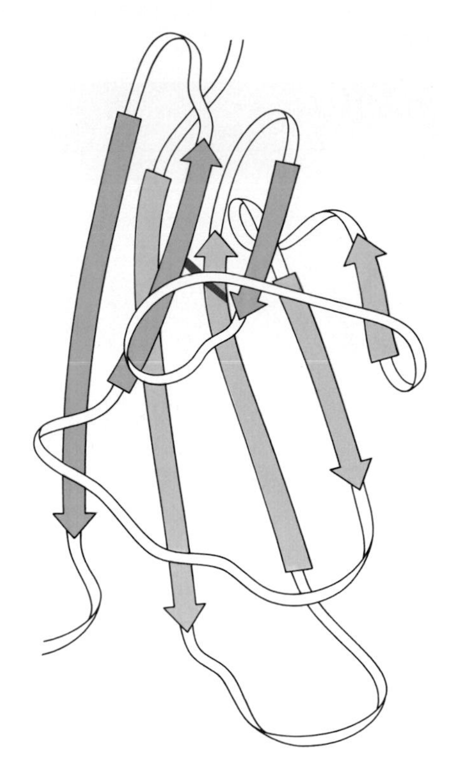

Figure 35-18
The basic structural motif of domains in immunoglobulins consists of two sheets of antiparallel β strands (shown in green and yellow). A conserved disulfide bond (shown in red) bridges the two sheets. [After J. D. Capra and A. B. Edmundson. The antibody combining site. Copyright © 1976 by Scientific American, Inc. All rights reserved.]

The three-dimensional structure of an intact immunoglobulin G molecule has been solved by David Davies. A 6-Å electron-density map shows that this antibody molecule is T-shaped (Figure 35-19). The domains in the F_c unit, like those in the F_{ab} units, display the characteristic immunoglobulin fold. Comparison of the relative orientations of adjacent domains in a variety of immunoglobulins and their fragments shows that the polypeptide chain between domains is quite flexible. The structure of the F_c unit has been solved at high resolution by Robert

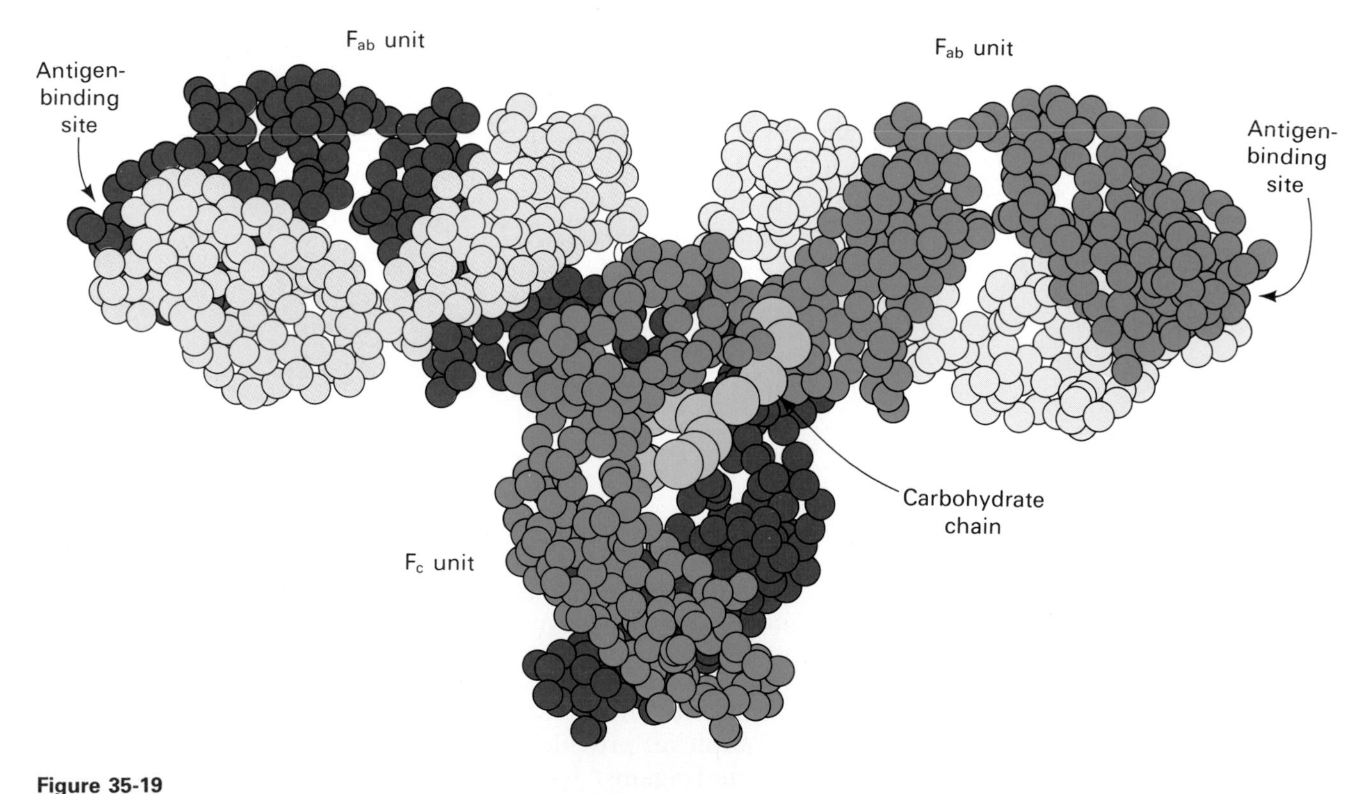

Figure 35-19
Schematic drawing of the three-dimensional structure of an IgG molecule. Each amino acid residue is represented by a small sphere. One of the H chains is shown in dark red, the other in dark blue. One of the L chains is shown in light red, the other in light blue. A carbohydrate unit attached to a C_H2 domain is shown in yellow. [After E. W. Silverton, M. A. Navia, and D. R. Davies. *Proc. Nat. Acad. Sci.* 74(1977):5142.]

Huber (Figure 35-20). The C_H2 domains differ from the others in that they interact through asparagine-linked carbohydrate units rather than through their polypeptide chains.

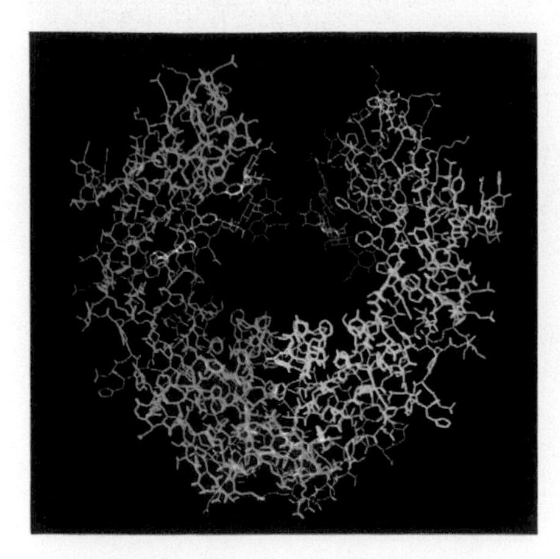

Figure 35-20
Three-dimensional structure of the F_c unit of IgG. The oligosaccharide units attached to the C_H2 domains are shown in red, and the polypeptide chains in blue and yellow.

X-RAY ANALYSES HAVE REVEALED HOW ANTIBODIES BIND HAPTENS AND ANTIGENS

X-ray crystallographic studies of haptens and antigens bound to F_{ab} provide insight into the nature of antibody combining sites and the structural basis of their specificity. As predicted from comparisons of amino acid sequences and affinity-labeling studies, *the combining site for antigen is formed by residues from the hypervariable segments (complementarity-determining regions) of both the L and the H chains.* The mode of binding of a vitamin K_1 derivative exemplifies this important structural feature (Figure 35-21).

Likewise, phosphorylcholine binds to a cavity that is lined with residues from five hypervariable segments, two from the L chain and three from the H chain (Figure 35-22). This hapten interacts most strongly with residues from the H chain. The positively charged trimethylammonium group of phosphorylcholine is buried inside the wedge-shaped cavity, where it interacts electrostatically with two negatively charged glutamate side chains. The negatively charged phosphate group of the hapten binds to the positively charged guanidinium group of an arginine residue at the mouth of the crevice. The phosphate group is also hydrogen bonded to the hydroxyl group of a tyrosine residue and to the amino group of the arginine side chain. Numerous van der Waals interactions, such as those made by a tryptophan side chain (see Figure 35-22), also stabilize this complex.

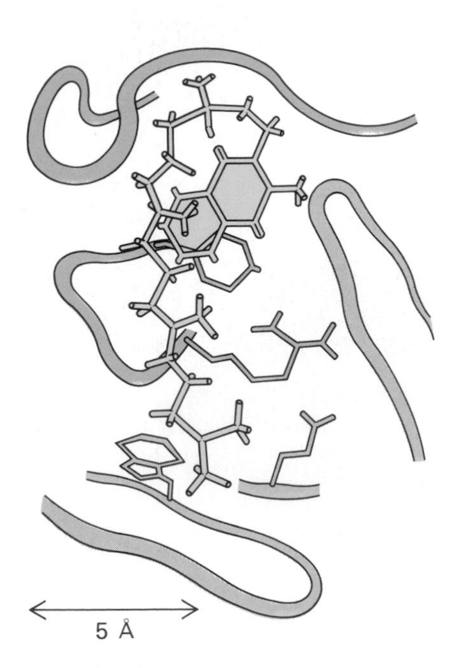

Figure 35-21
Mode of binding of a derivative of vitamin K_1 (shown in yellow) to an antibody combining site. Amino acid residues from the hypervariable segments of the H chain are shown in red, and those from the L chain in blue. [After I. M. Amzel, R. Poljak, F. Saul, J. Varga, and F. Richards. *Proc. Nat. Acad. Sci.* 71(1974):1427.]

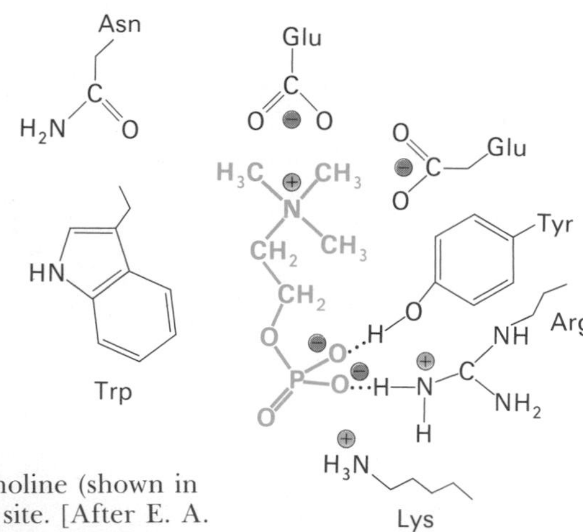

Figure 35-22
Mode of binding of phosphorylcholine (shown in green) to an antibody combining site. [After E. A. Padlan, D. R. Davies, S. Rudikoff, and M. Potter. *Immunochemistry* 13(1976):945.]

How do large antigens interact with antibodies? Two recently solved antigen-F_{ab} complexes provide highly informative views. A monoclonal antibody directed against lysozyme binds two polypeptide segments that are widely separated in the primary structure, residues 18 to 27 and 116 to 129. Each of the complementarity-determining regions of the L and H chains makes contact with this epitope. Twelve hydrogen bonds and numerous van der Waals interactions hold together the contacting surfaces, which are rather flat. Lysozyme and the F_{ab} fragment

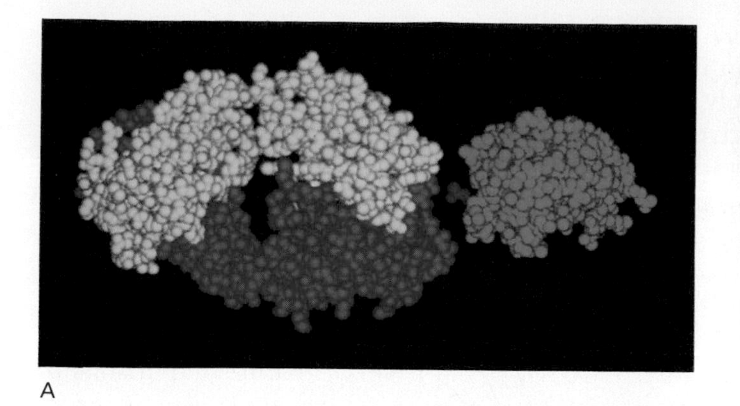

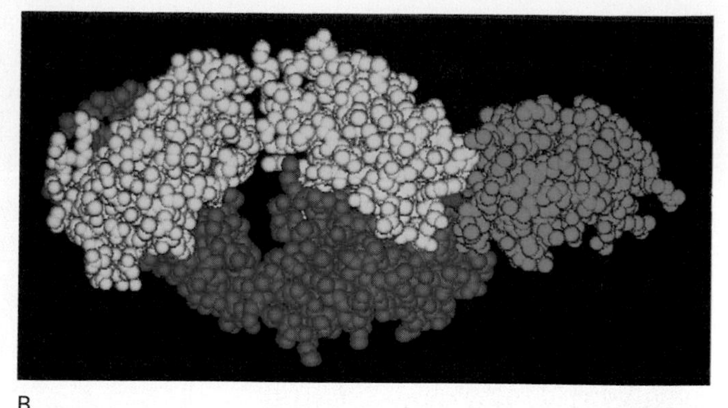

Figure 35-23
Structure of lysozyme bound to an F_{ab} molecule. The H chain is shown in blue, the L chain in yellow, and lysozyme in green. The antibody-antigen complex is shown separated (A) and together (B). The side chain of a glutamine residue (shown in red) of lysozyme penetrates deeply into the antibody. [From A. G. Amit, R. A. Mariuzza, S. E. V. Phillips, and R. J. Poljak. *Science* 233(1986)749.]

undergo little change in structure on forming this complex (Figure 35-23), which has a dissociation constant of 2×10^{-8} M. The lysozyme-antibody complex can be regarded as a lock-and-key combination, an association of two nearly rigid bodies. In contrast, the structure of neuraminidase, an enzyme in the membrane envelope of influenza virus (p. 869), changes significantly when bound to F_{ab} (Figure 35-24). Furthermore, the V_L and V_H domains of the antibody slide 3 Å past each other on binding this antigen. The combination has the character of a handshake, rather than the insertion of a key into a lock. An intriguing question is whether antigen-induced conformational changes are important in triggering effector functions such as the complement cascade.

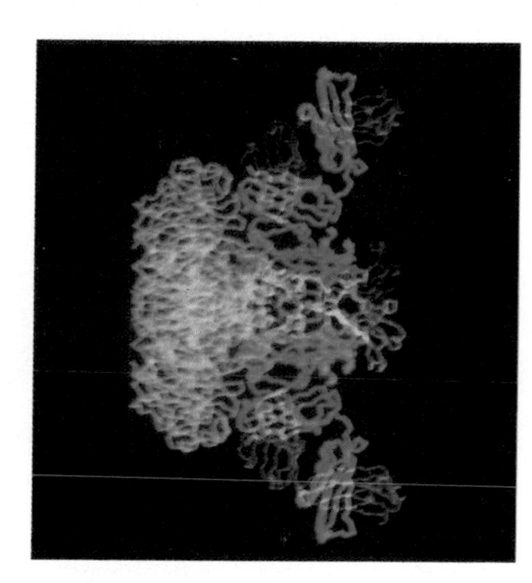

Figure 35-24
Structure of four F_{ab} molecules bound to a tetrameric neuraminidase from influenza virus. The H chains are shown in orange, L chains in blue, and neuraminidase in green. [From P. M. Colman, W. G. Laver, J. N. Varghese, A. T. Baker, P. A. Tulloch, G. M. Air, and R. G. Webster. *Nature* 326(1987):361.]

DIFFERENT CLASSES OF IMMUNOGLOBULINS HAVE DISTINCT BIOLOGICAL ACTIVITIES

Thus far, we have focused on immunoglobulin G. There are in fact five classes of immunoglobulins (Table 35-1), each consisting of heavy and light chains. The constant regions of the heavy chains differ from one class to another, whereas those of the light chains are the same, either λ

Table 35-1
Properties of immunoglobulin classes

Class	*Serum concentration (mg/ml)*	*Mass (kd)*	*Sedimentation coefficient(s)*	*Light chains*	*Heavy chains*	*Chain structure*
IgG	12	150	7	κ or λ	γ	$\kappa_2\gamma_2$ or $\lambda_2\gamma_2$
IgA	3	180–500	7, 10, 13	κ or λ	α	$(\kappa_2\alpha_2)_n$ or $(\lambda_2\alpha_2)_n$
IgM	1	950	18–20	κ or λ	μ	$(\kappa_2\mu_2)_5$ or $(\lambda_2\mu_2)_5$
IgD	0.1	175	7	κ or λ	δ	$\kappa_2\delta_2$ or $\lambda_2\delta_2$
IgE	0.001	200	8	κ or λ	ϵ	$\kappa_2\epsilon_2$ or $\lambda_2\epsilon_2$

Note: n = 1, 2, or 3. IgM and oligomers of IgA also contain J chains that join immunoglobulin molecules. IgA in secretions has an additional secretory piece.

or κ. The heavy chains in immunoglobulin G are called γ chains, whereas those in immunoglobulins A, M, D, and E are called α, μ, δ, and ϵ, respectively. These different heavy chains, variations on a fundamental theme, give the five classes of immunoglobulins distinct biological characteristics.

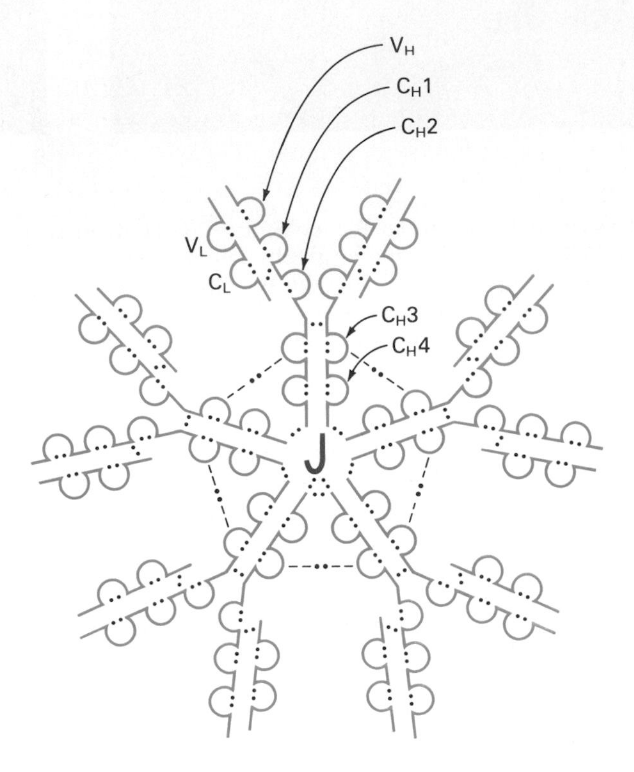

Figure 35-25
Schematic diagram of IgM, a pentamer of L_2H_2 units (blue) joined by disulfide bonds (dots) to a J chain (red).

As mentioned previously, *immunoglobulin M* (IgM) is the first class of antibodies to appear in the serum after injection of an antigen, whereas *immunoglobulin G* (IgG) is the principal antibody in the serum. IgG is also known as γ-globulin. Subclasses of IgG (e.g., IgGl and IgG2) differ in their degree of segmental flexibility and their capacity to trigger complement fixation and other effector functions. *Immunoglobulin A* (IgA) is the major class of antibodies in external secretions, such as saliva, tears, bronchial mucus, and intestinal mucus. Thus, IgA serves as a first line of defense against bacterial and viral antigens. IgA is transported across epithelial cells from the blood side to the extracellular side by a receptor called *secretory component*. The benefits conferred by immunoglobulin D (IgD) are not yet known. *Immunoglobulin E* (IgE) is important in conferring protection against parasites. A price paid for having IgE is allergic reactions mediated by this immunoglobulin class.

VARIABLE AND CONSTANT REGIONS ARE ENCODED BY SEPARATE GENES THAT BECOME JOINED

How are immunoglobulin genes organized and expressed? The discovery of distinct variable and constant regions in the L and H chains raised the possibility that immunoglobulin genes, as well as their polypeptide products, have an unusual architecture. As mentioned earlier, the constant regions of κ chains are identical except at position 191, which is either leucine or valine. Pedigree analyses showed that these two variants (allotypes) are inherited in a classical Mendelian manner, which strongly implied that there is only one gene for the constant region of these chains. In contrast, genetic analyses showed that their variable regions are encoded by multiple genes. In 1965, William Dreyer and Claude Bennett proposed that multiple *V* genes are separate from a single *C* gene in the germ line. According to their model, one of these *V* genes becomes joined to the *C* gene during differentiation of the antibody-producing cell. It was known then that DNA can be spliced, as in the integration of λ phage DNA.

A critical test of this novel hypothesis had to await the isolation of pure immunoglobulin mRNA and the development of techniques for analyzing complex mammalian genomes. The use of restriction enzymes to split large chromosomes into specific pieces of DNA that can be readily analyzed was particularly important in this regard. The distribution of *V* and *C* genes on these restriction fragments could then be assayed by hybridizing them with mRNA segments specific for either the variable or the constant region. In 1976, Susumu Tonegawa found that V *and* C *genes are far apart in embryonic (germ-line) DNA but are closely associated in the DNA of antibody-producing cells. Thus, immunoglobulin genes are translocated during the differentiation of lymphocytes* (Figure 35-26).

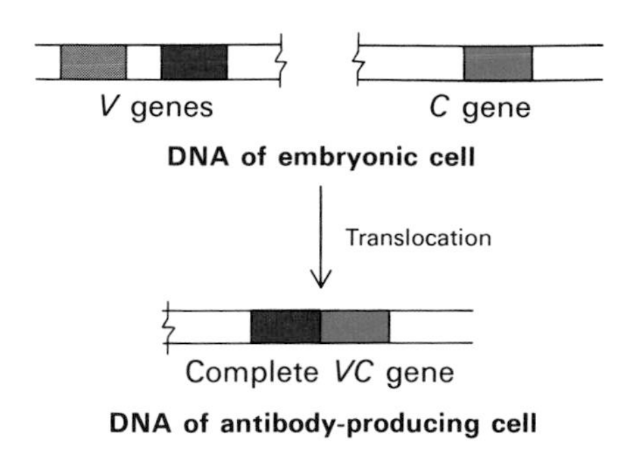

Figure 35-26
A *V* gene is translocated near a *C* gene in the differentiation of an antibody-producing cell.

SEVERAL HUNDRED GENES ENCODE THE VARIABLE REGIONS OF L AND H CHAINS

An animal can synthesize large amounts of specific antibody against virtually any foreign determinant within a few weeks after being exposed to it. The number of different kinds of antibodies that can be made by an animal is very large, at least a million. We have seen that antibody specificity is determined by the amino acid sequences of the variable regions of both light and heavy chains. This brings us to the key question: *How are different variable-region sequences generated?* More specifically, we can ask:

1. *When* is diversity generated? During the life of an individual animal (somatic) or in the course of evolution (genetic)?

2. *How* is diversity generated? By a random evolutionary process, somatic recombination, or somatic hypermutation?

The recombinant DNA revolution provided the means to obtain definitive answers to these questions. A key finding is that mice have *several hundred genes for the variable regions of κ light chains* (V_κ) *and of heavy chains* (V_H). For 300 V_κ genes and 300 V_H genes, a maximum of 9×10^4 different specificities could be generated by the combination (300 × 300) of L and H chains. This calculated upper limit of 9×10^4 falls far short of the number of different kinds of antibodies that an animal can produce ($>10^6$). The disparity is even greater for mouse λ light chains, which are encoded by only two V_λ genes. In short, *the number of variable-region genes in the germ line is simply too small to be the sole source of antibody diversity. Clearly, some of this diversity must be generated during the lifetime of an animal in the differentiation of its lymphocytes.*

J (JOINING) GENES AND *D* (DIVERSITY) GENES INCREASE ANTIBODY DIVERSITY

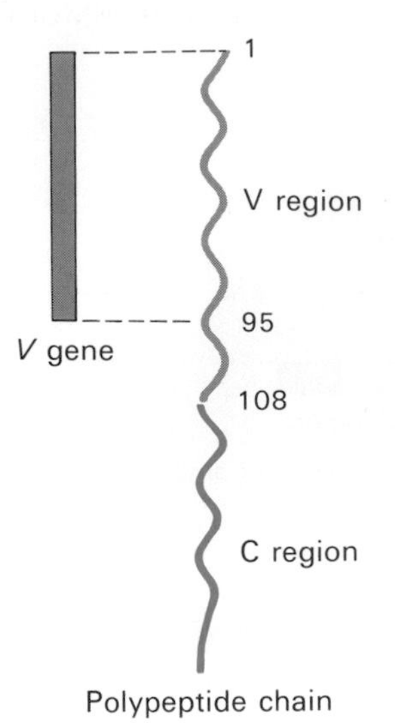

Figure 35-27
A *V* gene isolated from embryonic cells is foreshortened. It does not encode the last thirteen amino acid residues of the variable (V) region of the polypeptide chain.

The next advance in the elucidation of how diversity is generated came from the determination of the base sequences of cloned immunoglobulin genes from embryonic cells and myeloma cells. These highly informative studies were carried out by Tonegawa, Philip Leder, and Leroy Hood. The first surprise was that V *genes in embryonic cells do not encode the entire variable region of L and H chains.* The germ-line *V* gene stops at the codon for amino acid residue 95 rather than 108, which is the end of the variable region of the polypeptide chain (Figure 35-27).

Where is the DNA that encodes the last thirteen residues of the V region? For light chains in embryonic cells, this stretch of DNA is located in an unexpected place: near the *C* gene. It is called the *J* gene because it joins the *V* and *C* genes in a differentiated cell. In fact, *a tandem array of several* J *genes is located near the* C *gene in embryonic cells* (Figure 35-28). In the differentiation of an antibody-producing cell, a *V* gene is translocated to a site near the *C* gene by intrachromosomal recombination. Specifically, *a* V *gene becomes spliced to a* J *gene to form a complete gene for the variable region.* The *J* genes are important contributors to antibody diversity because they encode part of the last hypervariable segment. In forming a continuous V_κ gene, any of several hundred *V* genes can become linked to any of five *J* genes. For example, 1250 kinds of continuous *V* genes can be formed by the joining of 250 incomplete *V* genes and 5 *J* genes. Thus, *somatic recombination of these gene segments amplifies the diversity already present in the germ line.*

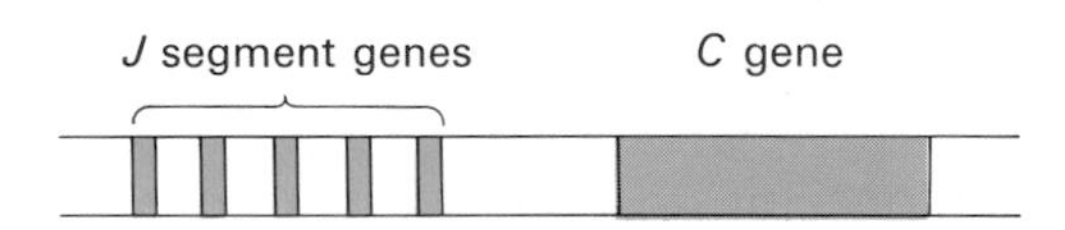

Figure 35-28
A tandem array of *J* genes, which encode part of the last hypervariable segment of the V region of the light chain, is located near the *C* gene.

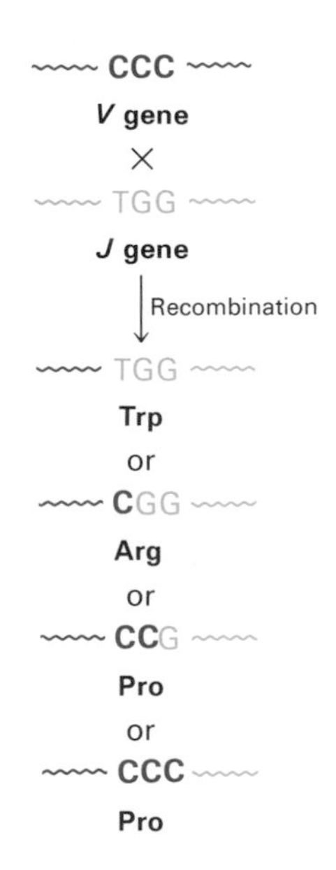

Figure 35-29
Imprecision in the exact site of splicing of a *V* gene to a *J* gene gives rise to additional diversity.

The other surprising finding was that the set of five *J* genes yielded more than four amino acid sequences for the corresponding region of the light chain. Analyses of amino acid and base sequence data then showed that recombination of *V* and *J* genes is not absolutely precise. Recombination between these genes can occur at one of several bases near the codon for residue 95 (Figure 35-29). Thus, *additional diversity is generated by allowing* V *and* J *genes to become spliced in different joining frames.* The immune system clearly takes delight in sweet disorder!

The variable region of heavy chains is encoded by yet another gene segment, called *D* for diversity. Some fifteen *D* segments lie between hundreds of V_H and five J_H segments (Figure 35-30). A *D* segment joins

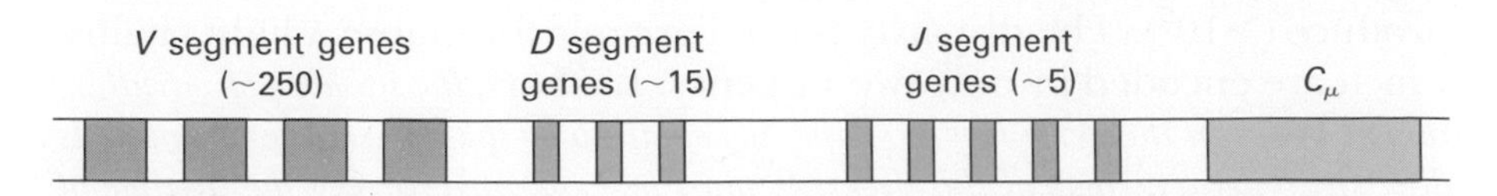

Figure 35-30
The variable region of the heavy chain is encoded by *V*, *D*, and *J* segment genes.

a J_H segment; a V_H segment then becomes linked to DJ_H. The third complementarity-determining region (CDR) of the heavy chain is encoded mainly by a D segment. More kinds of antigen-binding patches and clefts can be formed by the H than by the L chain because it is encoded by three rather than two gene segments. Moreover, the third CDR of the heavy chain is diversified by the action of *terminal deoxyribonucleotidyl transferase*, a special polymerase that does not use a template. This enzyme inserts extra nucleotides between V_H and D.

The site-specific recombination of V, D, and J genes is mediated by enzymes that recognize sequences flanking these gene segments. The recognition sequence on the 3′ side consists of a conserved palindromic heptamer separated by a nonconserved 12-bp spacer from a conserved, AT-rich nonamer (Figure 35-31). The one on the 5′ side of a segment undergoing recombination consists of a conserved AT-rich nonamer separated by an unconserved 23-bp spacer from a conserved palindromic heptamer. Recombination can occur only between the 12-bp and 23-bp types but not between two 12-bp types or two 23-bp types. For example, V_H segments and J_H segments are flanked by 23-bp types on both their 5′ and 3′ sides. Hence, V_H and J_H cannot recombine with each other, nor amongst themselves. Rather, they recombine with D segments, which are flanked on both the 5′ and 3′ sides by recognition sequences of the 12-bp type.

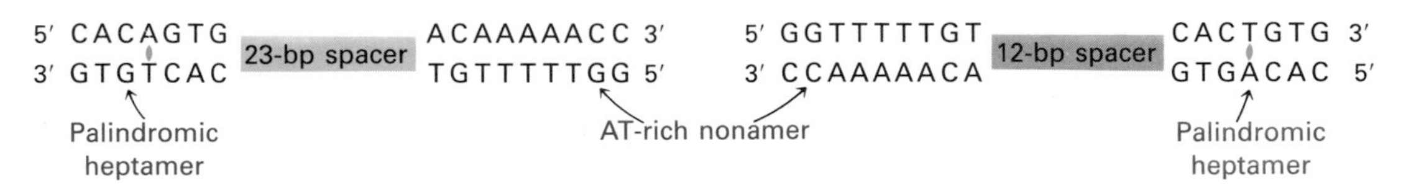

Figure 35-31
Recognition sites for the recombination of V, D, and J segment genes. Sites are of two classes: one with a 12-bp spacer (red) and the other with a 23-bp spacer (blue). Recombination can occur only between different classes of sites.

Lymphocytes undergo *allelic exclusion*: only one member of an allelic pair of genes is expressed by a particular cell. Analyses of restriction fragments have shown that an incomplete V gene becomes correctly joined to a J gene on only one of a pair of homologous chromosomes. Only the properly recombined immunoglobulin gene is expressed. Hence, *all of the antigen-binding sites produced by a single cell are the same.*

mRNAs FOR L AND H CHAINS ARE FORMED BY THE SPLICING OF PRIMARY TRANSCRIPTS

The mRNA for a κ light chain contains about 1250 bases (Figure 35-32). Like other eucaryotic mRNAs, it has a poly A tail at its 3′ end and untranslated sequences at both its 5′ and its 3′ ends. A *leader sequence*

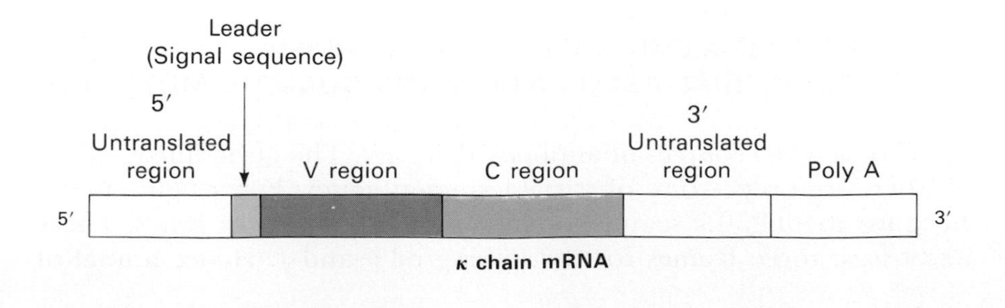

Figure 35-32
Structure of the mRNA for an L chain.

H_3^+N-Met-Arg-Ala-Pro-Ala-
-Gln-Ile-Phe-Gly-Phe-
-Leu-Leu-Leu-Leu-Phe-
-Pro-Gly-Thr-Arg-Cys-
Cleavage Site

Figure 35-33
Signal sequence of a nascent L chain. Hydrophobic residues are shown in yellow.

near the 5′ end of the κ-chain mRNA codes for the hydrophobic amino-terminal region of the nascent chain. This *signal sequence* (Figure 35-33) is recognized by the signal recognition particle, which directs the ribosome to the translocation machinery in the endoplasmic reticulum membrane (p. 768). Signal peptidase cleaves the signal sequence after it has been translocated to the lumenal side of the endoplasmic reticulum. The variable and constant regions of the light chain are encoded by a contiguous sequence on the mRNA that immediately follows the leader sequence.

The primary transcript that gives rise to this mRNA contains two introns (Figure 35-34). One of them separates the leader sequence from the start of the variable-region sequence, and the other separates the distal end of the joining-region sequence from the start of the constant-region sequence. *These introns are removed in the processing of the primary transcript.* It is interesting to note that the *J* gene carries information for two kinds of splicing, one at the DNA level for the fusion of *V* and *J*, and the other at the RNA level for the joining of *J* and *C*.

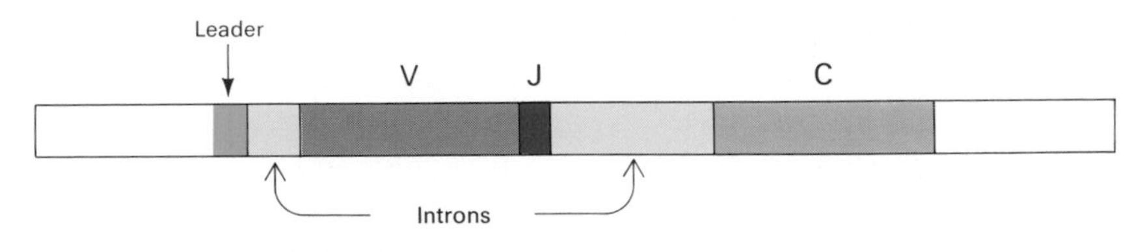

Figure 35-34
Primary transcript of an L chain mRNA.

What is the structure of the gene for the heavy chain? Recall that an H chain consists of four domains—V_H, C_H1, C_H2, and C_H3—and a hinge between C_H1 and C_H2. Gene-mapping studies have revealed that these five components are encoded by separate DNA segments (Figure 35-35). Thus, *the domain construction of immunoglobulins is an expression of the underlying architecture of their genes.*

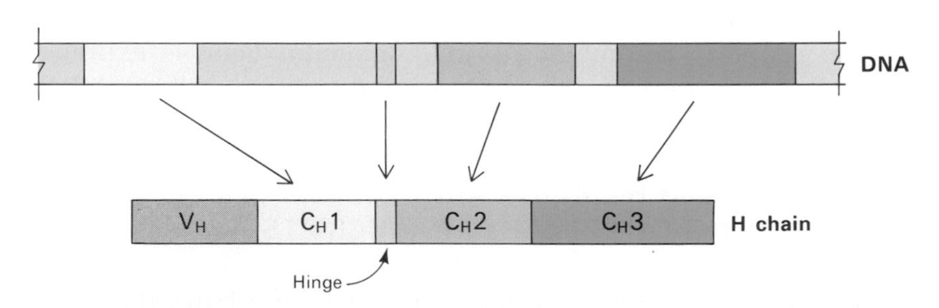

Figure 35-35
The three domains of the constant (C) region of a heavy chain and the hinge region are encoded by separate gene segments.

MORE THAN 10^8 ANTIBODIES CAN BE FORMED BY COMBINATORIAL ASSOCIATION AND SOMATIC MUTATION

Let us review the sources of antibody diversity. The germ line contains a rather large repertoire of variable-region genes. For κ light chains, there are about 250 *V* segment genes and four *J* segment genes. There are at least three frames for the joining of *V* and *J*. Hence a total of

$250 \times 4 \times 3$, or 3000, kinds of complete V_κ genes can be formed by the combinations of V and J. A larger number of heavy chain genes can be formed because of the existence of D segments. For 250 V, 15 D, and 5 J gene segments that can be joined in three frames, the number of complete V_H genes that can be formed is 56,250. The association of 3000 kinds of L chains with 56,250 kinds of H chains would yield 1.7×10^8 different antibody specificities. This calculated number is sufficiently large to account for the remarkable range of antibodies that can be synthesized by an animal.

As mentioned earlier, there are far fewer V_λ genes than V_κ genes. In fact, mice seem to have only two V_λ genes. However, many more V_λ amino acid sequences are known. It seems likely that somatic mutation gives rise to much of the diversity of λ light chains. Likewise, somatic mutation increases the diversity of heavy chains. *Thus, nature draws on each of the three proposed sources of diversity—a germ-line repertoire, somatic recombination, and somatic mutation—to form the rich variety of antibodies that protect an organism from foreign incursions.*

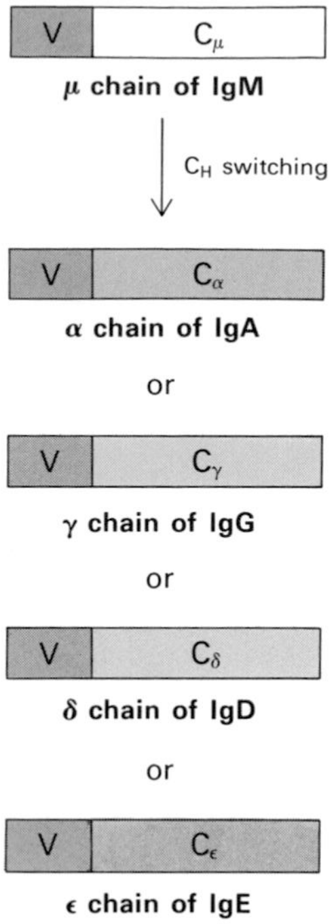

V | C_μ

μ chain of IgM

C_H switching

V | C_α

α chain of IgA

or

V | C_γ

γ chain of IgG

or

V | C_δ

δ chain of IgD

or

V | C_ϵ

ϵ chain of IgE

Figure 35-36
Synthesis of different classes of immunoglobulins. The V_H region is first associated with C_μ and then with another C region to form an H chain of a different class.

DIFFERENT CLASSES OF ANTIBODIES ARE FORMED BY THE HOPPING OF V_H GENES

As mentioned earlier, there are five classes of immunoglobulins. An antibody-producing cell first synthesizes IgM and then makes either IgG, IgA, IgD, or IgE of the *same specificity*. In this switch, the light chain is unchanged. Furthermore, the variable region of the heavy chain stays the same. *Only the constant region of the heavy chain changes,* and so this step in the differentiation of an antibody-producing cell is called *class switching* (Figure 35-36).

In embryonic mouse cells, the genes for the constant regions of all kinds of heavy chains—called C_μ, C_δ, C_γ, C_ϵ, and C_α—are next to each other (Figure 35-37). There are eight in all, including four genes for the constant regions of γ chains, which agrees nicely with genetic analyses showing that there are four subclasses of IgG. A tandem array of J genes for the last hypervariable segment of the variable region is located next to the C_μ gene. *A complete gene for the heavy chains of IgM antibody is formed by the translocation of a V_H gene segment to a DJ_H gene segment* (Figure 35-38).

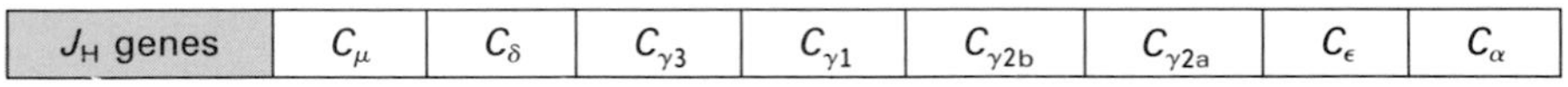

J_H genes | C_μ | C_δ | $C_{\gamma 3}$ | $C_{\gamma 1}$ | $C_{\gamma 2b}$ | $C_{\gamma 2a}$ | C_ϵ | C_α

Figure 35-37
The genes for the constant regions of μ, δ, γ, ϵ, and α chains are next to each other.

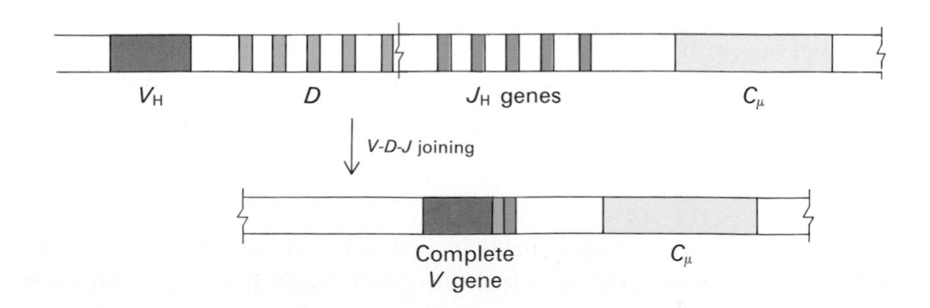

Figure 35-38
The joining of V_H and DJ_H creates a functional gene for a μ chain.

This joining brings V_H, D, J_H, and C_μ together to form a functional gene. The introns between the leader and start of the variable region, between the end of J_H and the start of C_μ, and within C_μ are excised in the processing of the primary transcript to form mRNA for the μ chain.

Analyses of restriction digests of DNA from embryonic and myeloma cells showed that C_H switching takes place in DNA instead of RNA. *In the switch from IgM to IgA, for example, the* V_HDJ_H *gene moves from a site near the* C_μ *gene to one near the* C_α *gene* (Figure 35-39). In this recombination, the genes from C_μ to C_α loop out and are lost. Class switches occur by genetic recombination between long switching sequences consisting of many repeated elements. The translocation of the entire V_HDJ_H gene accounts for the fact that the antigen-combining specificity of IgA produced by a particular cell is the same as that of IgM synthesized at an earlier stage of its development. It is not known how a cell chooses one of several C_H genes as the target in this translocation; the switching sequences preceding these genes do differ. *The biological significance of* C_H *switching is that a whole recognition domain (the variable domain) is shifted from the early constant region* (C_μ) *to one of several other constant regions that mediate different effector functions.*

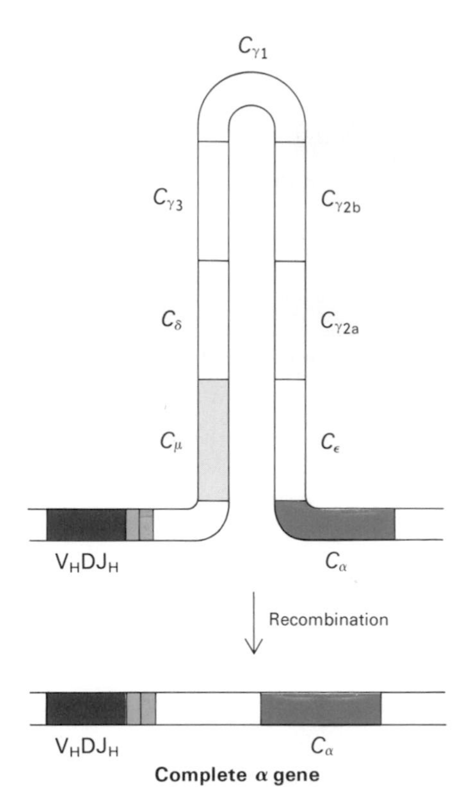

Figure 35-39
Structural basis of C_H switching. The V_HDJ_H gene moves from its position near C_μ to one near C_α by intrachromosomal recombination.

B LYMPHOCYTES ARE ACTIVATED BY THE BINDING OF ANTIGEN TO TRANSMEMBRANE IMMUNOGLOBULINS

B lymphocytes, the precursors of plasma cells, are triggered to divide and proliferate by the binding of antigen to receptors on the cell surface. In fact, *these receptors for antigen are immunoglobulin molecules that span the plasma membrane.* The membrane-bound antibodies of a particular cell have precisely the same specificity as the soluble antibodies that will be secreted by the cell when it is activated. How is this achieved? Let us consider IgM, the first antibody to appear on the cell surface and to be secreted. The membrane-bound (μ_m) and soluble (μ_s) forms of IgM differ only in the carboxyl-terminal region of their H chain. μ_m contains a hydrophobic sequence that anchors it in the plasma membrane (Figure 35-40). The carboxyl end of the H chains of μ_m lies in the cytosol, where it transmits the activation signal that results from the binding of antigen to the part of the molecule on the external side of the membrane. In contrast, μ_s is soluble in aqueous media and is secreted because it lacks a hydrophobic tail. The rest of the H chain and the entire L chain of μ_m and μ_s are the same. Hence, *the membrane-bound form and its soluble counterpart have the same antigen-binding specificity.*

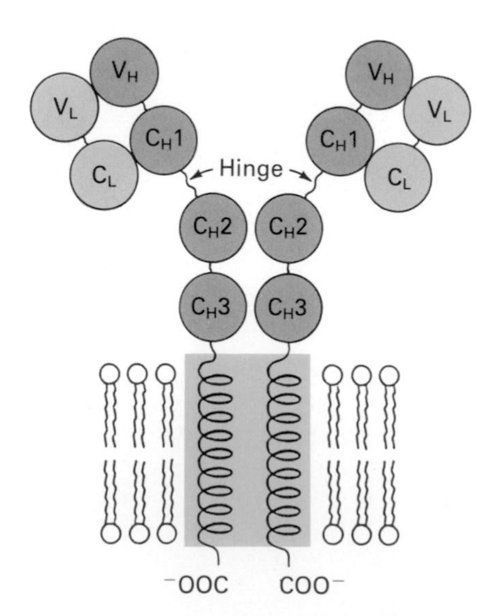

Figure 35-40
Domain structure of a membrane-bound IgG. The membrane-bound form of IgM is similar except that the short hinge of IgG is replaced by a larger domain.

How are μ_m and μ_s formed? Analyses of the gene for the μ chain and mRNAs for the membrane-bound and soluble forms have revealed that they are generated by *alternative splicing* (Figure 35-41). The assembled gene contains eight exons and encodes two potential polyadenylation sites (AAUAAA, p. 721). The first polyadenylation site is located on the 5′ side of the last two exons, the ones that encode the transmembrane anchor. Usage of this site leads to a primary transcript that includes only six exons; splicing then gives the mRNA for μ_s. A longer primary transcript containing all eight exons is formed when the more distal polyadenylation site is used. Splicing of this transcript leads to the inclusion of the last two exons and the excision of a small portion of the sixth exon. Prior to antigen stimulation, the second polyadenylation site is used, leading to the formation of μ_m. We see here a clear-cut example of how alternative splicing is used in cell differentiation to switch between closely related forms of a protein. Alternative splicing is also used

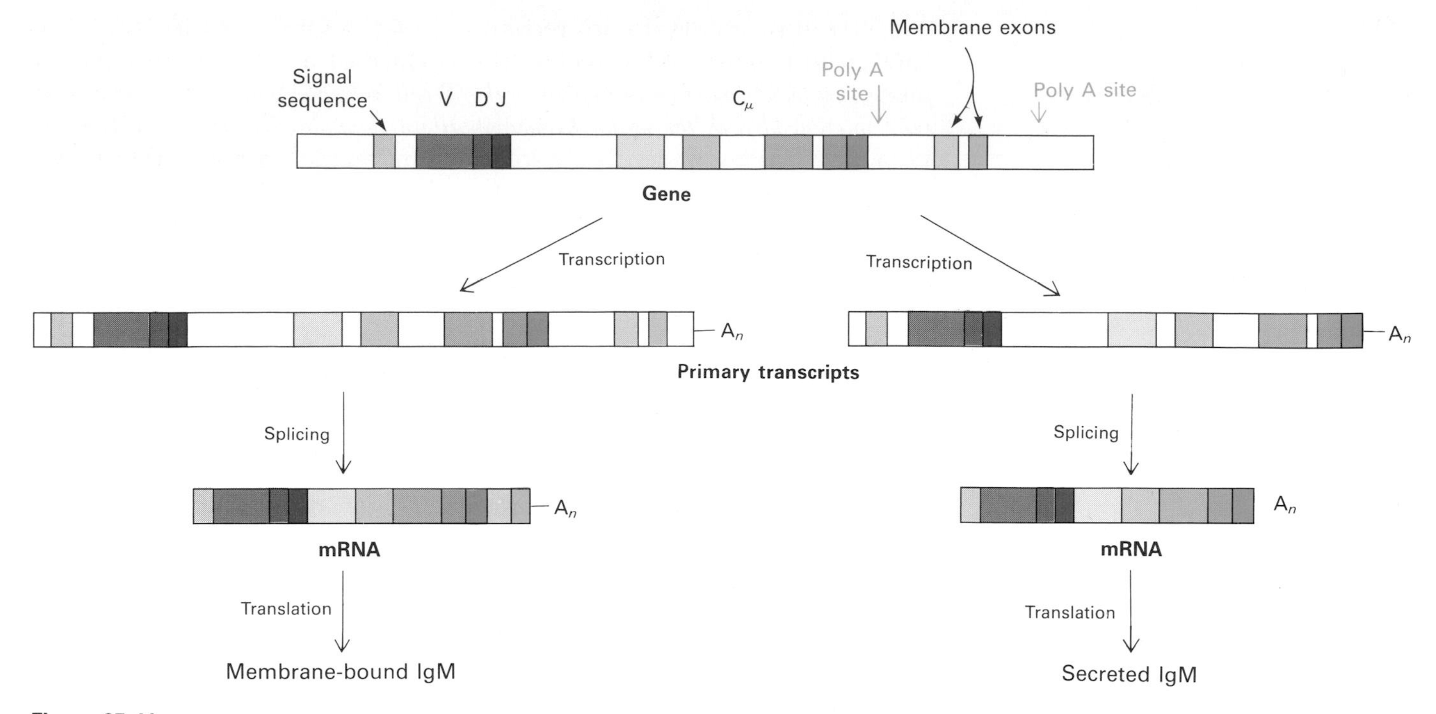

Figure 35-41
The membrane-bound and secreted forms of IgM are formed by the use of different polyadenylation sites and alternative splicing.

to generate the membrane-bound and secreted forms of several other classes of immunoglobulins (IgD, IgE, and IgG).

B lymphocytes are activated by the binding of multivalent antigens to membrane-bound IgM. The molecular events linking antigen binding to cell division and differentiation of B lymphocytes are largely unknown. The activation of phospholipase C, a membrane-bound enzyme that hydrolyzes phosphatidylinositol trisphosphate, may be a key step. The role of this enzyme in signal transduction will be discussed in a subsequent chapter (p. 985). Some stimulated B lymphocytes become *plasma cells*, which are specialized for the synthesis and secretion of soluble antibodies (Figure 35-42). Others become long-lived *memory B cells*, which can be triggered at a later time to differentiate into plasma cells.

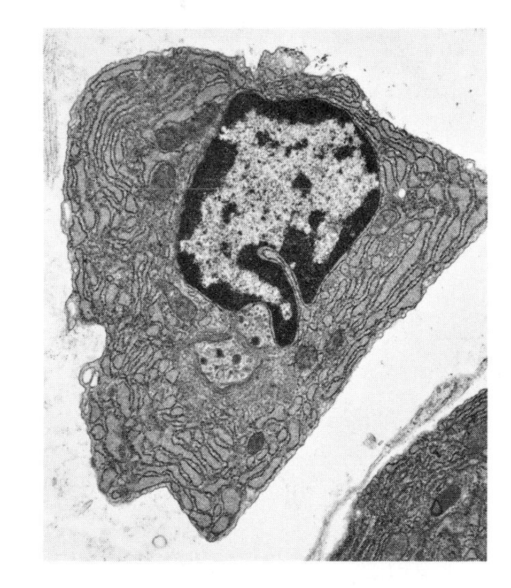

Figure 35-42
Electron micrograph of a plasma cell. The rough endoplasmic reticulum of this secretory cell is highly developed. [Courtesy of Lynne Mercer.]

T CELLS KILL INFECTED CELLS AND MODULATE THE ACTION OF B CELLS

We turn now to T cells, which mediate the cellular immune response. The humoral immune response is most effective in combating bacteria and viruses in extracellular media. The cellular immune response, in contrast, destroys virus-infected cells, parasites, and cancer cells. The surface of T cells contains transmembrane proteins called *T cell receptors* that recognize foreign molecules on the surface of other cells. As will be discussed shortly, T cell receptors are antibody-like proteins. However, they differ from antibodies in several ways:

1. T cell receptors are not secreted. *The immune response mounted by T cells is triggered by receptors located on their surface.* In contrast, the humoral immune response is mediated by soluble antibody molecules.

2. T cell receptors do not recognize isolated foreign molecules. The foreign unit must be located on the surface of a cell. Specifically, *the foreign molecule must be presented to the T cell by a particular membrane protein, one encoded by the major histocompatibility complex (MHC).* Antibodies recognize foreign molecules in solution and on membranes irrespective of their context.

3. Most T cell receptors recognize relatively short peptides, less than about ten residues in length. These epitopes typically do not have a well-defined structure in solution. The conformational determinants detected by antibodies, on the other hand consist of residues that are brought together by the folding of a long polypeptide chain. In essence, *antibodies recognize native macromolecules, whereas T cell receptors recognize fragments derived from them.*

Lymphokines—
Peptides and proteins secreted by stimulated T cells. These hormone-like molecules direct the movements and activities of other cells. Some examples are:

Interleukin-2 (IL-2)
γ Interferon (γ-IFN)
Colony-stimulating factor (CSF)
Macrophage chemotactic factor
Lymphocyte growth factors

The suffix *-kine* comes from the Greek word *kinesis,* meaning movement.

T cells are of several kinds. *Cytotoxic T cells,* also called *killer T cells,* destroy cells displaying a foreign epitope bound to an MHC protein. The binding of this foreign motif to the T cell receptor causes the cell to secrete granules containing *perforin.* This 70-kd protein lyses the target cell by polymerizing to form transmembrane pores 100-Å wide (Figure 35-43). Perforin is similar to C9, the final component in the complement cascade (p. 899). Other classes of T cells modulate the responses of B cells. *Helper T cells* bind B cells and stimulate their proliferation by secreting B-cell growth factor. For stimulation to occur, the foreign epitope recognized by the T cell must be on the surface of the B cell.

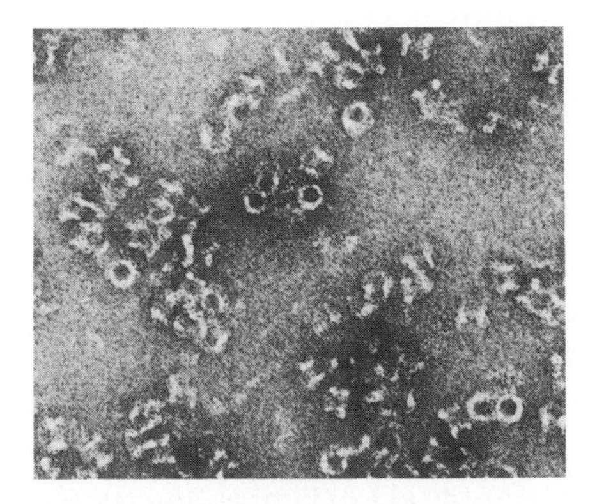

Figure 35-43
Electron micrograph showing pores in the membrane of a cell that was lysed by a cytotoxic (killer) T cell. The pores are formed by the polymerization of perforin, a protein akin to the C9 component of complement. [Courtesy of Dr. Eckhard Podack.]

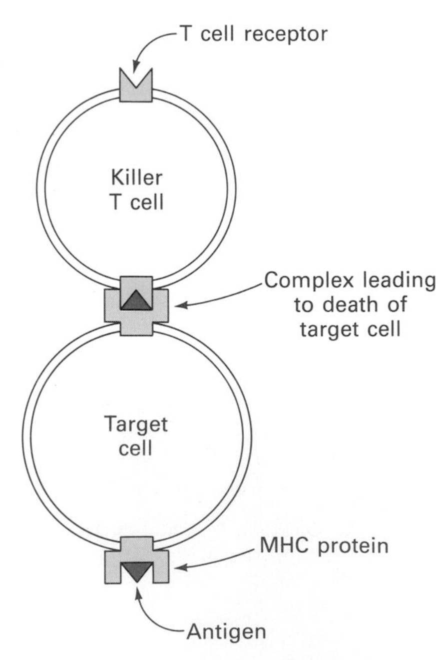

The T-cell receptor (yellow) binds antigen (red) that is presented by an MHC protein (green) on the infected target cell.

ANTIGENS ARE PRESENTED TO T CELLS BY TWO CLASSES OF MAJOR HISTOCOMPATIBILITY COMPLEX (MHC) PROTEINS

The proteins that present antigens to T cells were discovered by their role in *transplantation rejection.* A tissue transplanted from one person to another, or from one wild mouse to another, is usually rejected by the immune system. In contrast, tissues transplanted from one identical twin to another, or between mice of an inbred strain, are accepted. Genetic analyses revealed that rejection occurs when tissues are transplanted between individuals having different genes in a chromosomal region called the *major histocompatibility complex* (MHC). This large region (~3000 kb) encodes three classes of transmembrane proteins. Class I proteins present foreign epitopes to cytotoxic T cells, class II proteins present them to helper T cells, and class III proteins are components of the complement cascade. Class I proteins are present on nearly all cells, whereas class II proteins are confined to cells of the immune system and phagocytes. MHC proteins are highly variable.

Multiple gene loci encode class I and II proteins, and many alleles of each are present in members of a species. Hence, the likelihood that two unrelated individuals have identical class I and II proteins is very small ($<10^{-4}$), accounting for transplantation rejection unless there is close matching of genotypes.

Class I proteins consist of a 44-kd α chain noncovalently bound to a 12-kd polypeptide called β_2-microglobulin. The α chain has three extracellular domains—α_1, α_2, and α_3—a transmembrane segment, and a cytosolic tail (Figure 35-44). Differences between class I proteins are located mainly in the α_1 and α_2 domains. The α_3 domain, which interacts with a constant β_2-microglobulin, is largely conserved. Class II proteins have a similar overall plan. They consist of a 35-kd α chain and a noncovalently bound 29-kd β chain. Each contains two extracellular domains, a transmembrane segment, and a short cytosolic tail (Figure 35-45).

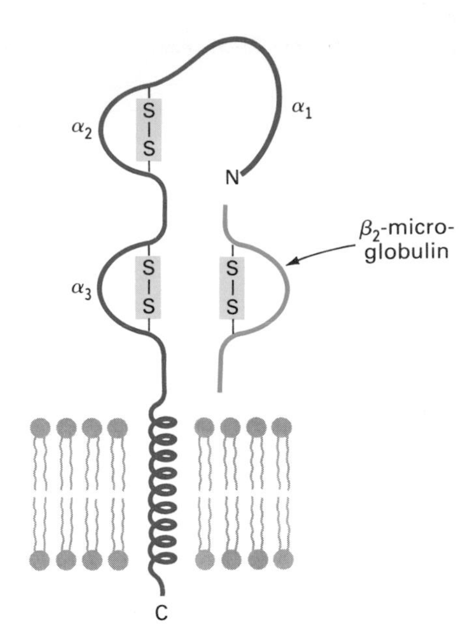

Figure 35-44
Schematic diagram of the domain structure of a class I major histocompatibility complex (MHC) protein.

Class I and II proteins bind peptides derived from foreign proteins rather than the intact proteins themselves. The foreign protein is internalized and digested in lysosomes. Some of the resulting peptides are transported to the plasma membrane in association with either a class I or II protein. Only peptides having an appropriate sequence, called an *aggretope*, can bind to a particular MHC protein. For example, peptides capable of binding to E^d, a class II protein, have the consensus sequence

Leu / Val	Glu	× ×	Arg / Lys	Arg / Lys	× ×	Ala / Val	× ×	Lys

An aggretope sequence

where X denotes a variable residue. An intriguing finding is that this consensus sequence is present in the variable region of the MHC molecule capable of binding it. In other words, the MHC protein contains an *internal aggretope*. Thus, *peptides capable of being presented by an MHC protein are homologous to a portion of the MHC protein itself.*

The three-dimensional structure of a human MHC class I protein (HLA-A2) has recently been solved by Don Wiley, Jack Strominger, and their coworkers (Figure 35-46). The antigen-binding site is formed by

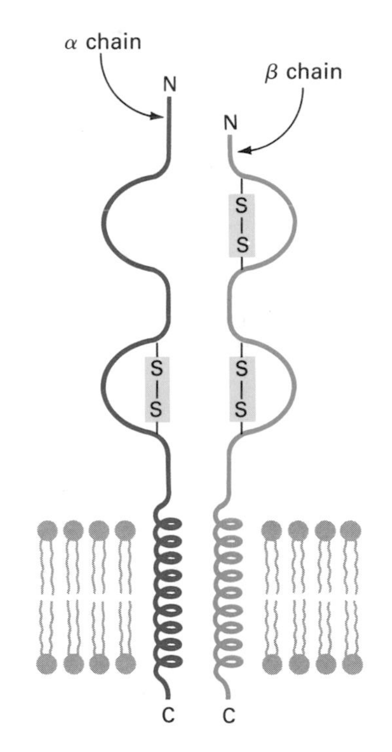

Figure 35-45
Schematic diagram of the domain structure of a class II MHC protein.

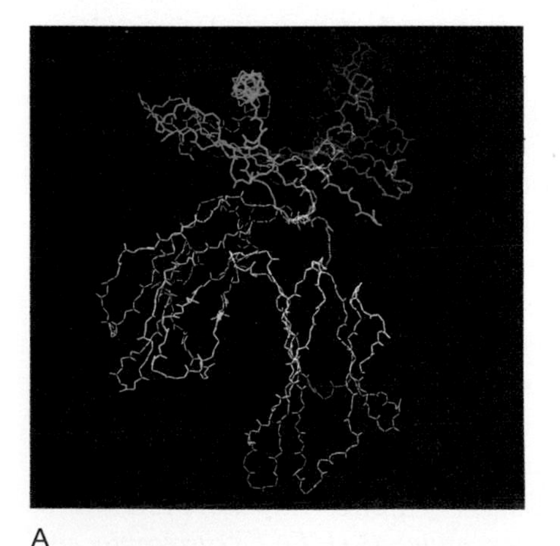

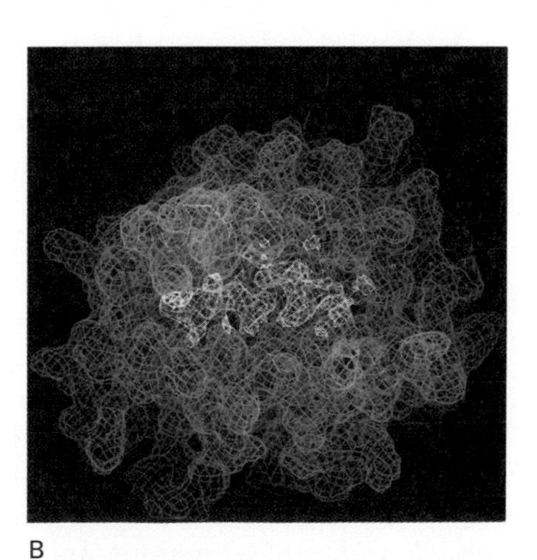

Figure 35-46
Three-dimensional structure of a human class I MHC protein (HLA-A2). (A) The main chain. The α_1 domain is shown in green, α_2 in red, and α_3 in blue. The β_2-microglobulin chain is shown in yellow. (B) An unidentified peptide (orange) occupies the antigen-binding site (blue) formed by residues from the α_1 and α_2 domains. The polygons depict van der Waals surfaces of the binding site and the bound peptide. [Courtesy of Dr. Pamela Bjorkman and Dr. Don Wiley.]

the α_1 and α_2 domains, which are very similar to one another and different from antibody domains. The binding site is a deep groove, 25 Å long and 10 Å wide, which runs between two long helices of α_1 and α_2. The α_3 and β_2-microglobulin domains resemble antibody domains but are paired differently.

Why are MHC proteins so highly variable? Their diversity makes possible the presentation of a wide range of peptides to T cells. A *particular* class I or class II molecule may not be able to bind any of the peptides formed by hydrolysis of the proteins from a particular virus because they lack the appropriate aggretope consensus sequence. The existence of multiple loci and alleles for these MHC proteins makes it likely that at least one of them will have a matching internal aggretope sequence. An effective T cell response can then be mounted against the infected cell.

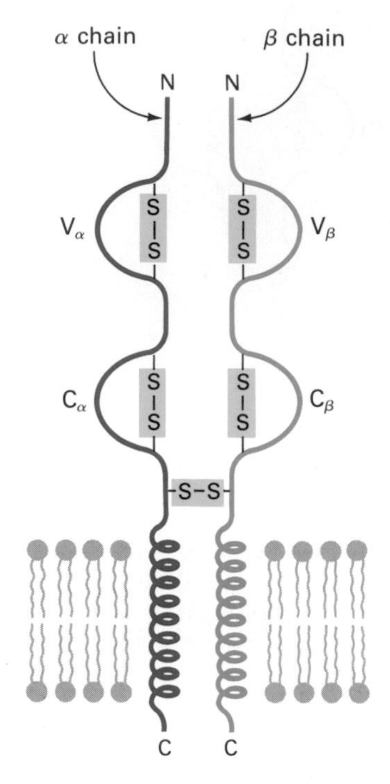

Figure 35-47
Schematic diagram of a T cell receptor. Each polypeptide chain of this disulfide-linked heterodimer consists of a variable and a constant region.

T-CELL RECEPTORS ARE ANTIBODY-LIKE PROTEINS CONTAINING VARIABLE AND CONSTANT REGIONS

The T cell receptor eluded investigators until 1983. It was first identified on a T cell hybridoma bearing a receptor for a peptide from ovalbumin (a hen egg-white protein) and a *d*-strain Class II MHC protein. Monoclonal antibodies directed against the large number of proteins in these T cells were screened to find those specific for the combining site of the T cell receptor. The assay was based on the expectation that some of these antibodies would block the capacity of ovalbumin peptides to trigger T cell activation and hence lead to a decrease in the amount of lymphokines secreted. The preparation of several such antibodies led to the identification and isolation of the T cell receptor, which consists of a 43-kd α chain (T_α) joined by a disulfide bond to a 43-kd β chain (T_β) (Figure 35-47). Both chains span the plasma membrane and have a short carboxyl-terminal region on the cytosolic side.

The cloning and sequencing of the receptor genes followed several months later. The strategy used was based on three assumptions: (1) The T cell receptor genes should be expressed in T cells but not in B cells. (2) The mRNAs for the receptor should be located on membrane-bound polysomes because the receptor is a membrane protein. (3) By analogy with immunoglobulins, the T cell receptor genes should rearrange during development to generate diversity. cDNAs were formed from membrane-bound mRNAs, which constitutes about 3% of the total mRNA. Sequences also expressed by B cells were removed by hybridization of the cDNAs with B cell mRNAs. This technique of *subtractive hybridization* removed most of the cDNAs. The remaining 2%, corresponding to about 250 sequences, were screened to determine whether any arose from genomic DNA that underwent rearrangement. Indeed, one clone with this property was found, the one containing the gene for the α chain of the T cell receptor.

The cloning and sequencing of many T cell receptor genes has revealed that T_α and T_β, like immunoglobulin L and H chains, consist of *variable* and *constant* regions. Indeed, *these domains of the T cell receptor are homologous in sequence and three-dimensional structure to the V and C domains of immunoglobulins.* Hypervariable sequences are present in the V regions of T_α and T_β, suggesting that they form the binding site for the foreign epitope. Moreover, the genetic architecture of these homologous proteins is similar to that of immunoglobulins (Figure 35-48). The variable region of T_α is encoded by about 100 V segment genes and 50

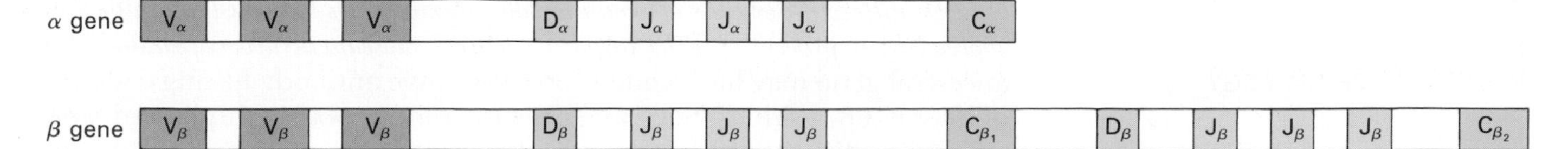

Figure 35-48
The T cell receptor is encoded by multiple *V*, *D*, and *J* gene segments. *C* genes follow *J* segment genes.

J (joining) segment genes. T_β is encoded by two *D* (diversity) gene segments in addition to about 30 *V* and 12 *J* segments. Again, diversity of component genes and different reading frames increase the number of kinds of proteins formed. *At least 10^7 different specificities could arise from combinations of this repertoire of genes. Thus, T cell receptors, like immunoglobulins, can recognize a very large number of different epitopes.*

How do T cells recognize their targets? A working hypothesis is depicted in Figure 35-49. The foreign peptide is bound to an MHC Class I or II protein on the target cell. This *combined epitope* is recognized by the variable regions of the T cell receptor. The surface of the T cell contains other molecules that bind MHC proteins and other markers on the target cell. One of the challenges now is to understand how the T cell is activated by binding the combined epitope. The elucidation of the three-dimensional structure of the T cell receptor and its complexes is eagerly awaited.

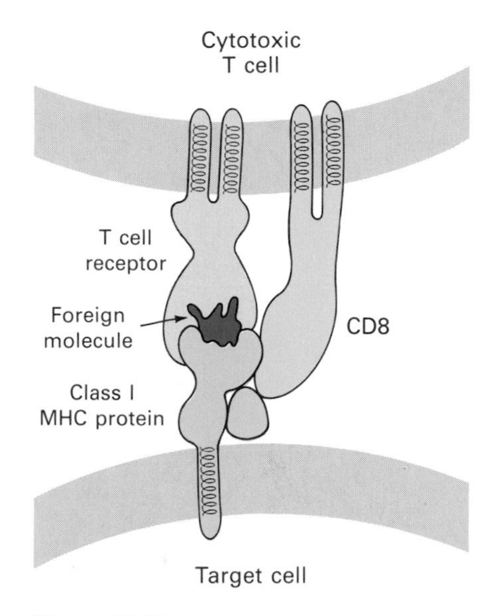

Figure 35-49
Model of antigen recognition by a T cell. A class I MHC protein (blue) presents a foreign molecule (red) to a T cell receptor (green). The T cell contains other cell-surface proteins that bind markers on target cells. CD8 (yellow), which is present on cytotoxic T cells, binds to class I MHC proteins. [After J. Goverman, T. Hunkapiller, and L. Hood. *Cell* 45:475–484.]

As was mentioned earlier, T cell receptors recognize peptides displayed on cell surfaces, whereas antibodies recognize intact proteins in solution or on cell surfaces. What is the advantage of having two different recognition strategies? One possibility is that specificity is enhanced by examining proteins from different vantage points to determine whether they are self or foreign. A virus particle may evade detection by antibodies by frequently changing its coat or camouflaging it. In contrast, it seems much less likely that the many peptides derived from its digestion in an infected cell will escape the notice of the T cell system. Multiple MHC proteins present foreign peptides in different ways to assure their recognition by T cells. *The immune system exploits endocytosis, lysosomal degradation, and receptor recycling* (p. 790) *to present peptide components of antigens for scrutiny by the T cell system.*

The interplay of B and T cells also helps to prevent the mounting of an immune response to an organism's own proteins. All recognition processes have a finite error rate. However, mistakes can be minimized by viewing different facets of an object. A B cell is triggered by the binding of what appears to be a foreign antigen, but a full response requires confirmation by a helper T cell. The T cell receptor must agree with the judgment of the membrane-bound immunoglobulin on the B cell to elicit a full antibody response against a foreign protein.

THE IMMUNOGLOBULIN GENE SUPERFAMILY ENCODES MANY PROTEINS THAT MEDIATE CELL-CELL INTERACTIONS

The structural homologies between domains in immunoglobulins have important evolutionary implications. The similarities in the amino acid sequences of chymotrypsin, trypsin, elastase, and thrombin indicate that they evolved from a common ancestor (p. 227). Likewise, *homologies between different domains of immunoglobulin molecules suggest that antibodies evolved by a process of gene duplication and subsequent diversification.* An ancestral gene may have coded for a primitive antibody having a length of about 108 residues. Such an ancestral gene probably duplicated, and the copies diverged to give rise to different kinds of variable and constant regions. Allen Edmundson has proposed that a dimer of L chains served as a primitive antibody (Figure 35-50).

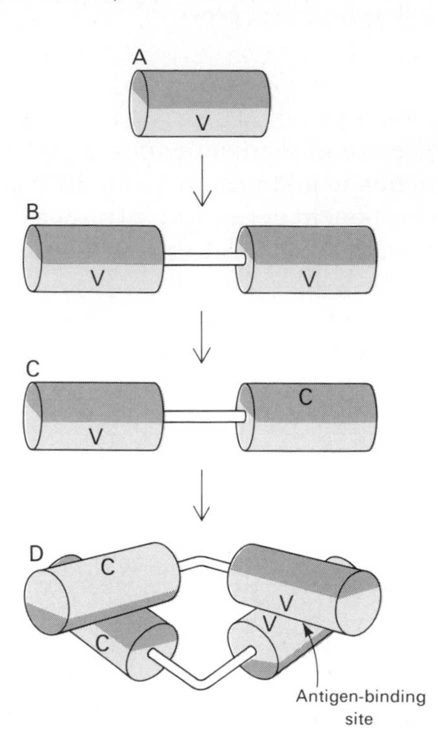

Figure 35-50
Proposed stages in the evolution of a primitive antibody. The gene coding for (A) a single polypeptide chain with about 108 residues duplicated to give (B) two identical domains joined by a short linear polypeptide chain. Divergent evolution of the duplicated genes altered the amino acid compositions of the two domains to give (C) variable and constant regions rotated with respect to each other. Association of these molecules yielded (D) a primitive F_{ab} unit with a combining site for antigen. [From J. D. Capra and A. B. Edmundson. The antibody combining site. Copyright © 1976 by Scientific American, Inc. All rights reserved.]

The immunoglobulin fold (see Figure 35-18, on p. 901), the repeating motif, is found not only in antibody molecules. It is found in MHC class I and class II proteins, T cell receptors, and many other proteins that participate in cell-cell recognition (Figure 35-51). For example, CD4 and CD8 are cell-surface proteins on helper and cytotoxic T cells

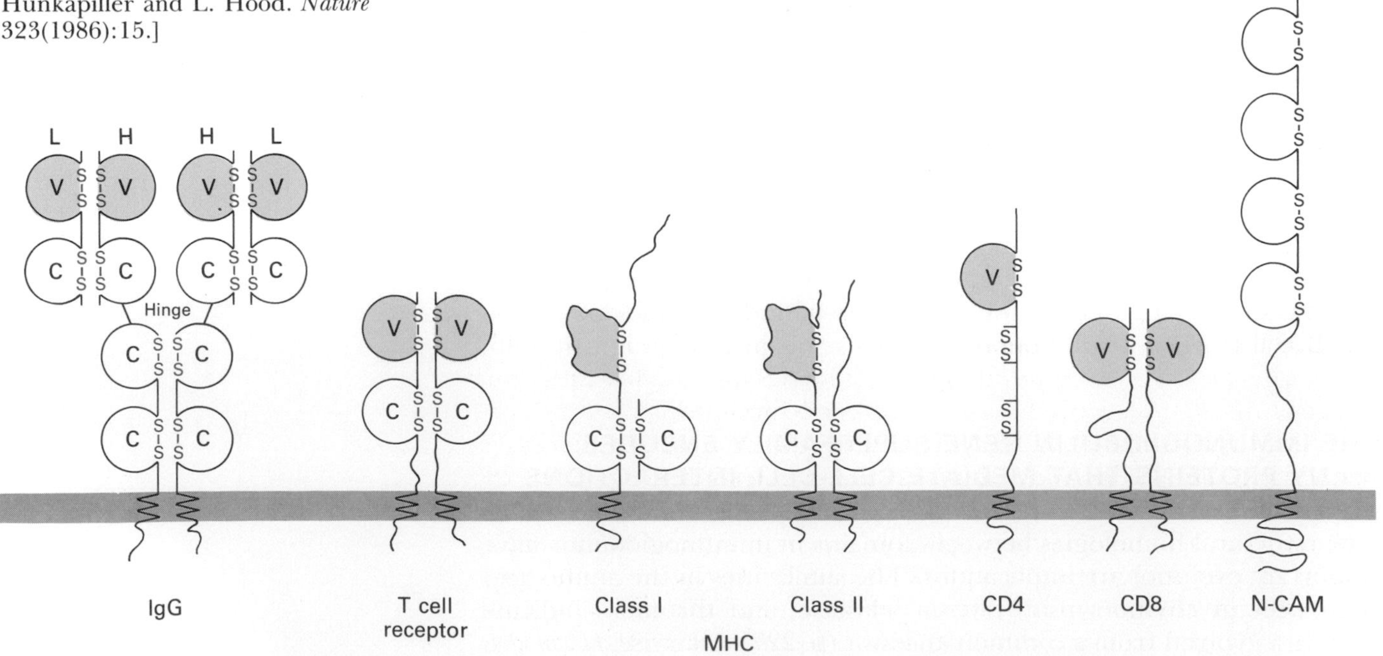

Figure 35-51
Domain structure of several members of the immunoglobulin gene superfamily. Variable domains are shown in green. [After T. Hunkapiller and L. Hood. *Nature* 323(1986):15.]

that recognize class II and I MHC proteins. The immunoglobulin fold also occurs in a developmentally regulated neuronal cell-adhesion molecule, N-CAM. This motif may play key roles in other recognition molecules important in the development of vertebrates.

It is useful to distinguish between two kinds of gene families. A *multigene family* is a group of homologous genes with similar functions, such as those encoding the different classes of immunoglobulins or the various serine proteases. A *superfamily* is a set of multigene families and single-copy genes that are related by sequence and have a common ancestor but divergent functions. Thus, antibodies and MHC molecules are members of the immunoglobulin superfamily, but play different roles in the immune response. N-CAM has diverged further in function, but its sequence, cell-surface location, and capacity to engage in recognition events point to its membership in the immunoglobulin superfamily.

> *On immunity and the plague of Athens—*
> All speculation as to its origins and its causes, if causes can be found adequate to produce so great a disturbance, I leave to other writers . . . for myself, I shall simply set down its nature, and explain the symptoms by which it may be recognized by the student, if it should ever break out again. This I can the better do, as I had the disease myself, and watched its operation in the case of others. . . .
> Yet it was with those who had recovered from the disease that the sick and the dying found most compassion. These knew what it was from experience, and had now no fear for themselves; *for the same man was never attacked twice*—never at least fatally. And such persons not only received the congratulations of others, but themselves also, in the elation of the moment, *half entertained the vain hope that they were for the future safe from any disease whatsoever*. . . .
>
> THUCYDIDES (460–400 B.C.)
> *The History of the Peloponnesian War*

THE IMMUNE RESPONSE REVEALS THE POWER OF SELECTIVE MECHANISMS

The essence of the clonal selection theory, as expressed by Burnet, is that

> No combining site is in any evolutionary sense adapted to a particular antigenic determinant. The pattern of the combining site is there and if it happens to fit, in the sense that the affinity of adsorption to a given antigenic determinant is above a certain value, immunologically significant reaction will be initiated.

The selection of a preexisting pattern that is randomly generated is not a new theme in biology. Indeed, it is the heart of Darwin's theory of evolution. It is interesting to note that instructive theories have preceded selective theories: Lamarck's before Darwin's, antigen-directed folding before clonal selection.

Jerne has suggested that we should consider the possibility that selective mechanisms play a role in the operations of the nervous system, such as memory. In fact, the instructive and selective theories of learning have been stated long ago. Locke held that the brain could be likened to white paper, devoid of all characteristics, on which experience paints with almost endless variety. In contrast, Socrates asserted that "all learning consists of being reminded of what is preexisting in the brain." I hope that this chapter has seemed familiar to you!

SUMMARY

Antibodies are populations of protein molecules (immunoglobulins) that are synthesized by an animal in response to a foreign macromolecule, called an antigen or immunogen. Antibodies have high affinity for antigens that induced their formation. Small foreign molecules (haptens) elicit the formation of specific antibody if they are attached to a macromolecule. Antibody synthesis occurs by selection rather than by instruction. An antigen binds to the surface of lymphocytes already committed to making antibodies specific for that antigen. The combination of antigen and surface receptors triggers cell division and the synthesis of large amounts of specific antibody. Antibodies directed against a specific determinant are usually heterogeneous, because they are the products of many antibody-producing cells. The antibodies produced by a single cell or by a clone are homogeneous. Monoclonal antibodies are formed by hybridoma cells obtained by fusing an immortal myeloma tumor cell with a lymphocyte that produces an antibody having the desired specificity.

Five classes of antibodies are formed. Immunoglobulin G (IgG) is the principal one in serum, but IgM is the first class to appear following exposure to an antigen. IgA is the major class in external secretions, and IgE protects against parasites; the role of IgD is not known. Antibodies consist of light (L) and heavy (H) chains. Immunoglobulin G, which has the subunit structure L_2H_2, contains two binding sites for antigen. Immunoglobulin G can be enzymatically cleaved into two F_{ab} fragments, which bind antigen but do not form a precipitate, and one F_c fragment, which mediates effector functions such as the binding of complement. A comparison of the amino acid sequences of myeloma immunoglobulins has revealed that L and H chains consist of a variable (V) region (an amino-terminal sequence of about 108 residues) and a constant (C) region. Antigen-binding sites are formed by residues from the hypervariable segments of the variable regions of both the L and the H chains. Antibody molecules are folded into compact domains of about 108 residues that have homologous sequences. The constant-region domains, which mediate effector functions, probably evolved by duplication and diversification of an antigen-binding (variable) domain. Antigen-antibody complexes trigger the complement cascade, leading to the lysis of target cells.

Variable and constant regions are encoded by separate genes that become joined during cell differentiation. There are several hundred genes for the V regions of L and H chains. A complete gene for the V region of a light chain is formed by the recombination of an incomplete *V* segment gene with one of several *J* (joining) segment genes encoding part of the last hypervariable segment. A tandem array of *J* genes is located near the *C* gene. Heavy-chain V region genes are formed by the recombination of *V*, *D* (diversity), and *J* segment genes. Introns are removed from the primary transcript. The resulting mRNA codes for an amino-terminal hydrophobic signal sequence that is cleaved from the nascent chain as well as for the V and C regions. More than 10^8 different specificities can be generated by combining different *V*, *D*, and *J* gene segments, somatic mutation of segments, and combining different L and H chains. Different classes of antibodies are formed by the translocation of a V_H gene from the *C* gene of one class to that of another. The membrane-bound and soluble counterparts of immunoglobulins are formed by the usage of different polyadenylation sites and alternative splicing. The binding of antigen to membrane-bound immunoglobulins triggers the proliferation of B cells.

The cellular immune response mounted by T cells is triggered by the binding of antigenic fragments on target cell membranes to T cell receptors. The foreign molecule must be presented to the T cell by a protein encoded by the major histocompatibility complex (MHC). MHC class I proteins present antigens to cytotoxic (killer) T cells, and class II proteins present them to helper T cells that stimulate B cells. T cell receptors are transmembrane antibody-like proteins containing variable and constant regions. They usually recognize peptides produced by digestion of foreign proteins. Immunoglobulins, MHC proteins, and T cell receptors are members of the immunoglobulin superfamily. This group of genes encodes proteins that contain the immunoglobulin fold and mediate cell-cell recognition processes.

SELECTED READINGS

Where to start

Tonegawa, S., 1985. The molecules of the immune system. *Sci. Amer.* 253(4):122–131.

Leder, P., 1982. The genetics of antibody diversity. *Sci. Amer.* 246(5):102–115.

Ada, G. L., and Nossal, G., 1987. The clonal selection theory. *Sci. Amer.* 257(2):62–69.

Capra, J. D., and Edmundson, A. B., 1977. The antibody combining site. *Sci. Amer.* 236(1):50–59.

Milstein, C., 1980. Monoclonal antibodies. *Sci. Amer.* 243(4):66–74.

Books

Hood, L. E., Weissman, I. L., Wood, W. B., and Wilson, J. H., 1984. *Immunology* (2nd ed.). Benjamin. [A lucid, highly readable account of molecular and cellular immunology.]

Nisinoff, A., 1985. *Introduction to Molecular Immunology* (2nd ed.). Sinauer. [A concise and perceptive introduction.]

Weir, D. M., (ed.), 1986. *Handbook of Experimental Immunology.* Oxford University Press.

Paul, W. E., (ed.), 1984. *Fundamental Immunology.* Raven Press.

Annual Review of Immunology. [Contains excellent review articles on many facets of molecular and cellular immunology. Volume 1 was published in 1983.]

Immunoglobulin structure

Davies, D. R., and Metzger, H., 1983. Structural basis of antibody function. *Ann. Rev. Immunol.* 1:87–117.

Huber, R., 1980. Spatial structure of immunoglobulin molecules. *Klin. Wochenschr.* 58:1217–1231.

Marquart, M., Deisenhofer, J., Huber, R., and Palm, W., 1980. Crystallographic refinement and atomic models of the intact immunoglobulin molecule Kol and its antigen-binding fragment at 3.0 Å and 1.9 Å resolution. *J. Mol. Biol.* 141:369–391.

Silverton, E. W., Navia, M. A., and Davies, D. R., 1977. Three-dimensional structure of an intact human immunoglobulin. *Proc. Nat. Acad. Sci.* 74:5140–5144.

Valentine, R. C., and Green, N. M., 1967. Electron microscopy of an antibody-hapten complex. *J. Mol. Biol.* 27:615–617.

Antigen-antibody complexes

Mariuzza, R. A., Phillips, S. E. V., and Poljak, R. J., 1987. The structural basis of antigen-antibody recognition. *Ann. Rev. Biophys. Biophys. Chem.* 16:139–159.

Amit, A. G., Mariuzza, R. A., Phillips, S. E. V., and Poljak, R. J., 1986. Three-dimensional structure of an antigen-antibody complex at 2.8 Å resolution. *Science* 233:747–753.

Colman, P. M., Laver, W. G., Varghese, J. N., Baker, A. T., Tulloch, P. A., Air, G. M., Webster, R. G., 1987. Three-dimensional structure of a complex of antibody with influenza virus neuraminidase. *Nature* 326:356–363.

Effector functions of immunoglobulins

Reid, K. B., 1986. Activation and control of the complement system. *Essays Biochem.* 22:27–68.

Müller-Eberhard, H. J., 1986. The membrane attack complex of complement. *Ann. Rev. Immunol.* 4:503–528.

Reid, K. B. M., and Porter, R. R., 1981. The proteolytic activation systems of complement. *Ann. Rev. Biochem.* 50:433–464.

Metzger, H., Alcaraz, G., Hohman, R., Kinet, J. P., Pribluda, V., Quarto, R., 1986. The receptor with high affinity for immunoglobulin E. *Ann. Rev. Immunol.* 4:419–470.

Generation of antibody diversity

Honjo, T., and Habu, S., 1985. Origin of immune diversity: genetic variation and selection. *Ann. Rev. Biochem.* 54:803–830.

Berek, C., Griffiths, G. M., and Milstein, C., 1985. Molecular events during maturation of the immune response to oxazolone. *Nature* 316:412–418.

Shimizu, A., and Honjo, T., 1984. Immunoglobulin class switching. *Cell* 36:801–803.

Tonegawa, S., 1983. Somatic generation of antibody diversity. *Nature* 302:575–581.

Sakano, H., Hüppi, K., Heinrich, G., and Tonegawa, S., 1979. Sequences at the somatic recombination sites of immunoglobulin light-chain genes. *Nature* 280:288–294.

Kataoka, T., Kawakami, T., Takahashi, N., and Honjo, T., 1980. Rearrangement of immunoglobulin γ1-chain gene and mechanism for heavy-chain class switch. *Proc. Nat. Acad. Sci.* 77:919–923.

Major histocompatibility complex

Bjorkman, P. J., Saper, M. A., Samraoui, B., Bennett, W. S., Strominger, J. L., and Wiley, D. C., 1987. Structure of the human class I histocompatibility antigen, HLA-A2. *Nature* 329:506–512.

Widera, G., 1986. Molecular biology of the H-2 histocompatibility complex. *Science* 233:437–443.

Becker, J. W., and Reeke, G. N., Jr., 1985. Three-dimensional structure of β_2-microglobulin. *Proc. Nat. Acad. Sci.* 82:4225–4229.

T cell receptors

Allison, J. P., and Lanier, L. L., 1987. The structure, function, and serology of the T-cell antigen receptor complex. *Ann. Rev. Immunol.* 5:503–540.

Hedrick, S. M., Cohen, D. I., Nielsen, E. A., and Davis, M. M., 1984. Isolation of cDNA clones encoding T cell-specific membrane-associated proteins. *Nature* 308:149–152.

Hood, L., Kronenberg, M., and Hunkapiller, T., 1985. T cell antigen receptors and the immunoglobulin supergene family. *Cell* 40:225–229.

Novotny, J., Tonegawa, S., Saito, H., Kranz, D. M., and Eisen, H. N., 1986. Secondary, tertiary, and quaternary structures of the T-cell-specific immunoglobulin-like polypeptide chains. *Proc. Nat. Acad. Sci.* 83:742–746.

Fink, P. J., Matis, L. A., McElligott, D. L., Bookman, M., and Hedrick, S. M., 1986. Correlations between T-cell specificity and the structure of the antigen receptor. *Nature* 321:219–226.

Goverman, J., Hunkapiller, T., and Hood, L., 1986. A speculative view of the multicomponent nature of T cell antigen recognition. *Cell* 45:475–484.

Hunkapiller, T., and Hood, L., 1986. The growing immunoglobulin gene superfamily. *Nature* 323:15–16.

Discovery of major concepts

Porter, R. R., 1973. Structural studies of immunoglobulins. *Science* 180:713–716.

Edelman, G. M., 1973. Antibody structure and molecular immunology. *Science* 180:830–840.

Kohler, G., 1986. Derivation and diversification of monoclonal antibodies. *Science* 233:1281–1286.

Milstein, C., 1986. From antibody structure to immunological diversification of immune response. *Science* 231:1261–1268.

Jerne, N. K., 1967. Antibodies and learning: selection versus instruction. *In* Quarton, G. C., Melnechuk, T., and Schmitt, F. O., (eds.), *The Neurosciences: A Study Program.* Rockefeller University Press.

Jerne, N. K., 1973. The immune system. *Sci. Amer.* 229(1):52–60.

CHAPTER 36

Muscle Contraction and Cell Motility

Coordinated movement is central to life. This chapter considers three ATP-driven motility systems in eucaryotes. *Muscle contraction* is mediated by the sliding of interdigitating *myosin* and *actin* filaments. The beating of *cilia* and *flagella* depends on the interplay of a different pair of proteins—*dynein* and *tubulin*. Microtubules made of tubulins form the mitotic spindle, which organizes the movement of chromosomes in cell division. They also serve as tracks for the movement of vesicles and organelles within cells, as in the transport of secretory vesicles along the axons of neurons. *Kinesin* is one of several proteins mediating the *movement of vesicles on microtubules*. The binding of ATP to myosin, dynein, and kinesin induces conformational transitions in these proteins; these changes are reversed by the hydrolysis of bound ATP and the release of ADP and P_i. How do ATP-driven conformational cycles in these molecular engines lead to coordinated motion? The concerted use of a wide range of experimental approaches drawn from protein chemistry, structural biology, and molecular genetics is providing answers to the fundamental question of how chemical-bond energy is transduced into mechanical energy. We begin with the structural basis of contraction in vertebrate striated muscle, the best understood of these processes.

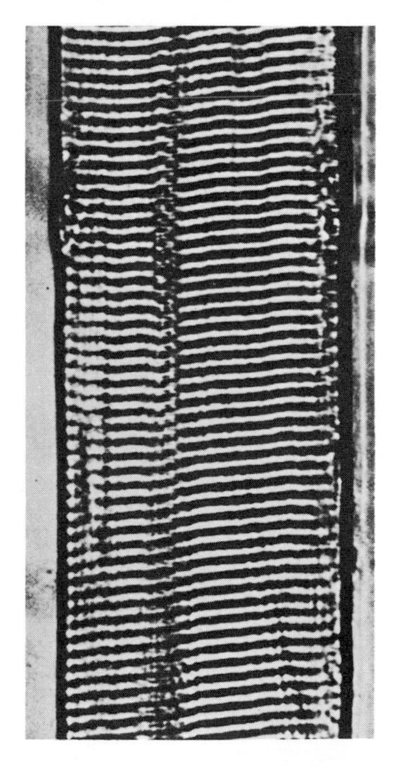

Figure 36-1
Phase-contrast light micrograph of a skeletal muscle fiber about 50 μm in diameter. The A bands are dark and the I bands are light. [Courtesy of Dr. Hugh Huxley.]

MUSCLE IS MADE UP OF INTERACTING THICK AND THIN PROTEIN FILAMENTS

Vertebrate muscle that is under voluntary control has a *striated* appearance when examined under the light microscope (Figure 36-1). It consists of multinucleate cells that are bounded by an electrically excitable plasma membrane called the *sarcolemma*. A muscle cell contains many parallel *myofibrils,* each about 1 μm in diameter. The myofibrils are immersed in a cytosol (called the sarcoplasm) that is rich in glycogen, ATP, creatine phosphate, and glycolytic enzymes. Many regularly spaced mitochondria are found in cells of red muscle, which are capable of sustained contractile activity.

Figure 36-2
Electron micrograph of a longitudinal section of a skeletal muscle fiber, showing about a half dozen myofibrils extending from upper left to lower right. [Courtesy of Dr. Hugh Huxley.]

An electron micrograph of a longitudinal section of a myofibril displays a wealth of structural detail (Figures 36-2 and 36-3). The functional unit, called a *sarcomere,* repeats every 2.3 μm (23,000 Å) along the fibril axis. A dark *A band* and a light *I band* alternate regularly. The central region of the A band, termed the *H zone,* is less dense than the rest of the band. A dark *M line* is found in the middle of the H zone. The I band is bisected by a very dense, narrow *Z line.*

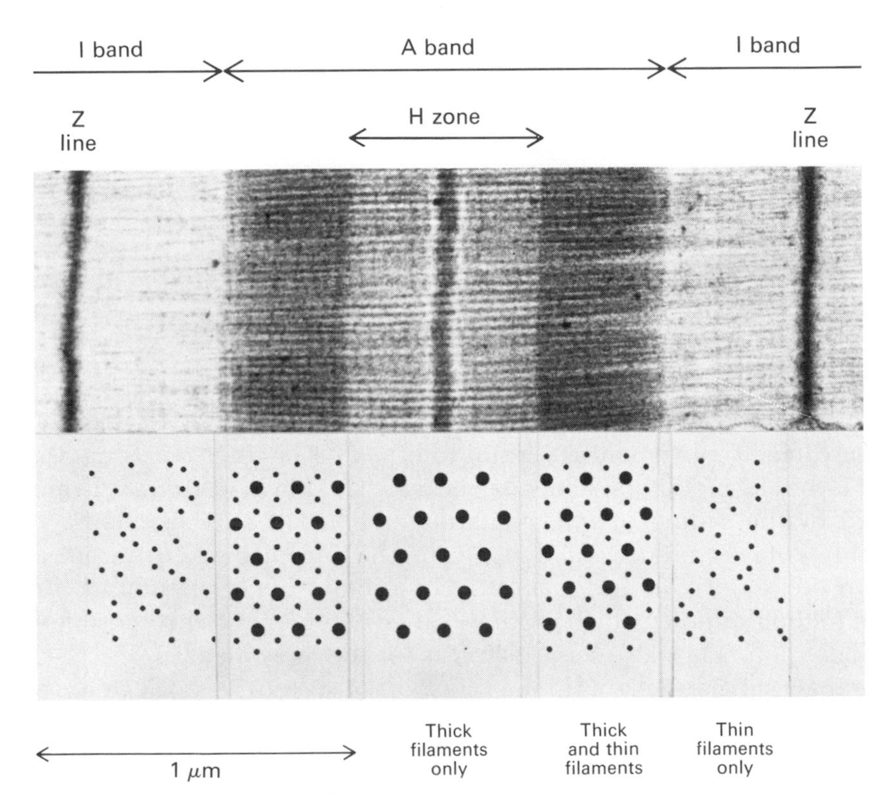

Figure 36-3
Electron micrograph of a longitudinal section of a skeletal muscle myofibril, showing a single sarcomere. Schematic diagrams of cross sections are shown below the micrograph. [Courtesy of Dr. Hugh Huxley.]

The underlying molecular plan of a sarcomere is revealed by electron micrographs of cross sections of a myofibril, which show that there are *two kinds of interacting protein filaments.* The *thick filaments* have diameters of about 150 Å, whereas the *thin filaments* have diameters of about 70 Å. The thick filament is primarily *myosin.* The thin filament contains *actin, tropomyosin,* and *troponin.* α-Actinin is present in the Z line, and an M-protein is located in the M line. *Titin,* composed of two extraordinarily long proteins (1.2 and 1.4 megadaltons), and *nebulin* (600 kd) form a flexible mesh around the filaments.

The I band consists of only thin filaments, whereas only thick filaments are found in the H zone of the A band. Both types of filaments are present in the other parts of the A band, where a regular hexagonal array is evident in cross sections: Each thin filament has three neighboring thick filaments and each thick filament is encircled by six thin filaments (Figure 36-3). The thick and thin filaments interact by *cross-bridges,* which are domains of myosin molecules. In muscle depleted of ATP, cross-bridges emerge at regular intervals from the thick filaments and bridge a gap of 130 Å between the surfaces of thick and thin filaments (Figures 36-4 and 36-5). In fact, the contractile force of muscle is generated by the interaction of myosin cross-bridges with actin units in thin filaments.

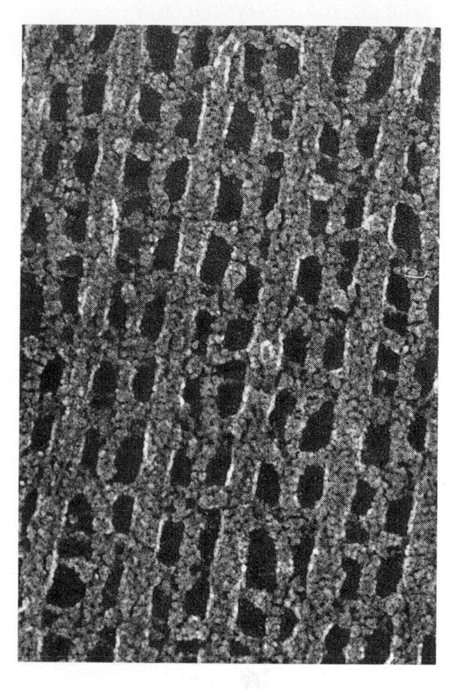

Figure 36-4
Electron micrograph showing cross-bridges between thick and thin filaments. [Courtesy of Dr. John Heuser.]

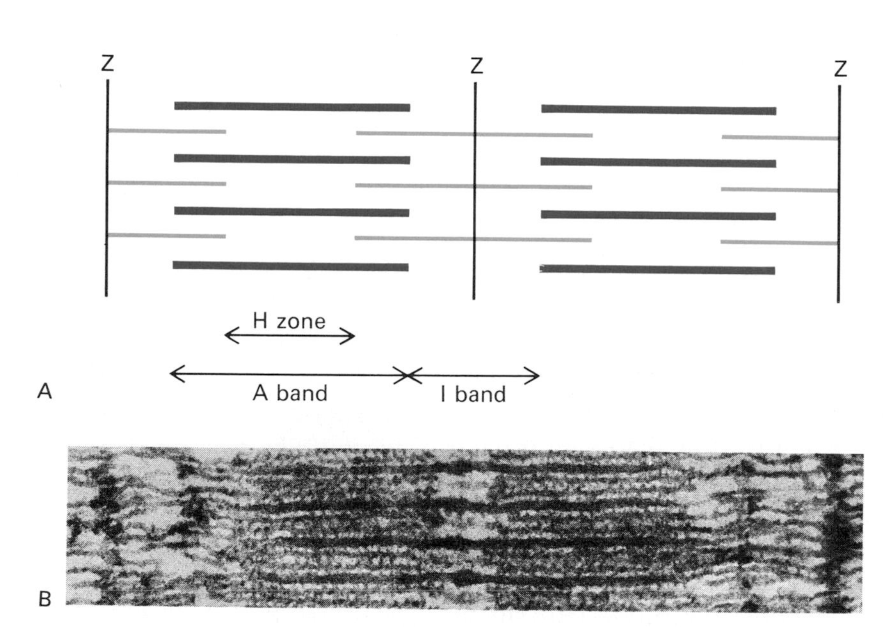

Figure 36-5
(A) Schematic diagram of the structure of striated muscle, showing overlapping arrays of thick and thin filaments. (B) Electron micrograph of a very thin longitudinal section. [Courtesy of Dr. Hugh Huxley.]

THICK AND THIN FILAMENTS SLIDE PAST EACH OTHER IN MUSCLE CONTRACTION

Muscle shortens by as much as a third of its original length as it contracts. What causes this shortening? In the 1950s, two groups of investigators independently proposed a *sliding-filament model* on the basis of x-ray, light-microscopic, and electron-microscopic studies. The essential features of this model (Figure 36-6), which was proposed by Andrew Huxley and Ralph Niedergerke and by Hugh Huxley and Jean Hanson, are:

1. The lengths of the thick and thin filaments do not change during muscle contraction.

2. Instead, the length of the sarcomere decreases during contraction because the two types of filaments overlap more. *The thick and thin filaments slide past each other in contraction.*

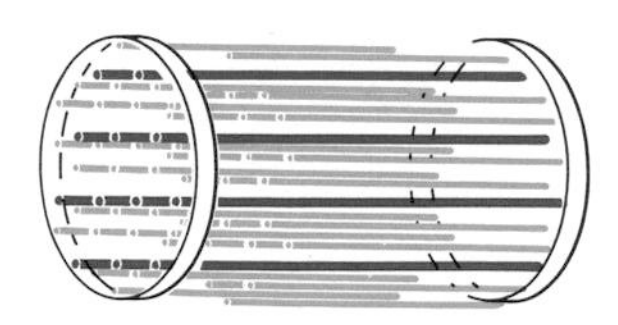

Figure 36-6
Sliding filament model. [From H. E. Huxley. The mechanism of muscular contraction. Copyright © 1965 by Scientific American, Inc. All rights reserved.]

3. The force of contraction is generated by a process that actively moves one type of filament past the neighboring filaments of the other type.

This model is supported by measurements of the lengths of the A and I bands and the H zone in stretched, resting, and contracted muscle. The length of the A band is constant, which means that the thick filaments do not change size. The distance between the Z line and the adjacent edge of the H zone is also constant, which indicates that the thin filaments do not change size. In contrast, the size of the H zone and also of the I band decreases on contraction, because the thick and thin filaments overlap more.

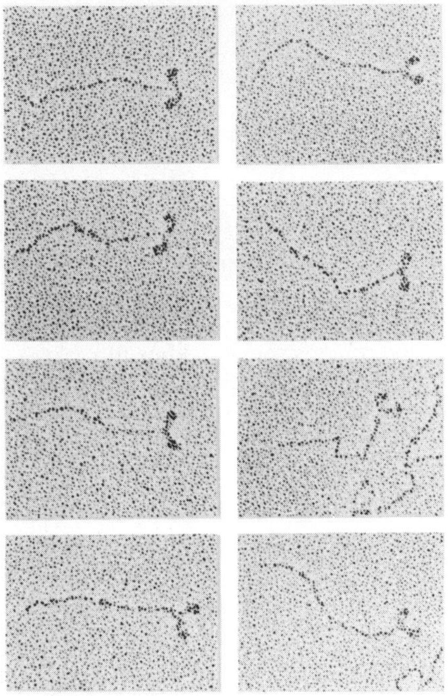

MYOSIN FORMS THICK FILAMENTS, HYDROLYZES ATP, AND BINDS ACTIN

Figure 36-7
Electron micrographs of myosin molecules. [Courtesy of Dr. Paula Flicker, Dr. Theo Walliman, and Dr. Peter Vibert.]

Myosin has three important biological activities. First, myosin molecules spontaneously assemble into filaments in solutions of physiologic ionic strength and pH. In fact, the *thick filament* consists mainly of myosin molecules. Second, myosin is an enzyme. Vladimir Engelhardt and Militsa Lyubimova discovered in 1939 that myosin is an *ATPase.*

$$ATP + H_2O \longrightarrow ADP + P_i + H^+$$

This reaction provides the free energy that drives muscle contraction. Third, myosin binds to the polymerized form of *actin,* the major constituent of the thin filament. Indeed, this interaction is critical for the generation of the force that moves the thick and thin filaments past each other.

Myosin is a very large molecule (540 kd) made of six polypeptide chains: two identical heavy chains (each 230 kd) and four light chains (each about 20 kd). Electron micrographs show that the molecule consists of a double-headed globular region joined to a very long rod (Figures 36-7 and 36-8). The rod is a two-stranded α-helical coiled coil formed by the heavy chains. In each head, two different light chains are bound to the heavy chain.

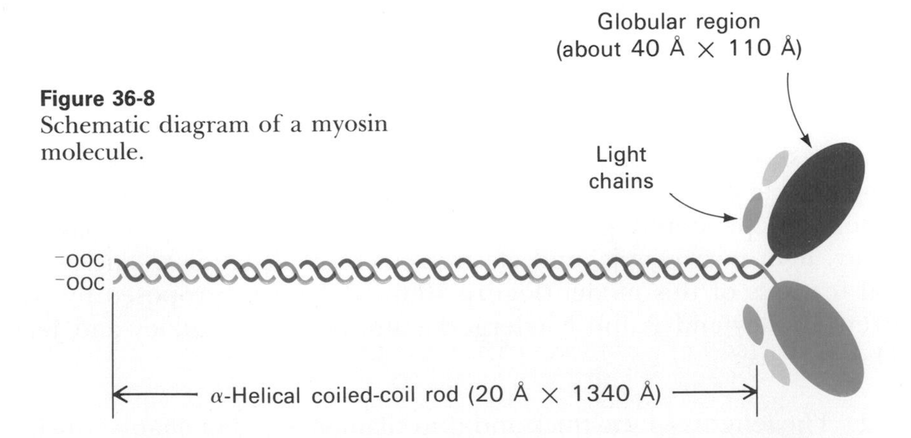

Figure 36-8
Schematic diagram of a myosin molecule.

The amino acid sequences of myosins of many species are known from studies of these proteins and their cloned genes. The myosins of species as different as *Dictyostelium discoideum* (a slime mold), *Caenorhabditis elegans* (a nematode), chicken, and rabbit have many common features. A prominent one is the demarcation of the heavy chain into

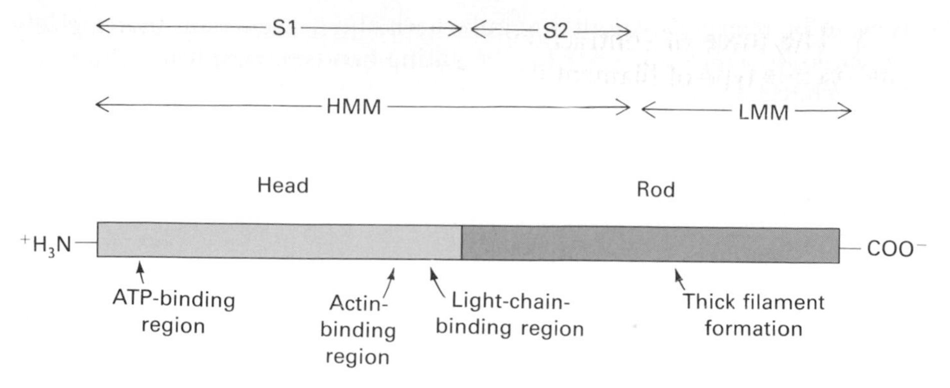

Figure 36-9
Functions of different regions of the myosin molecule. [Courtesy of Dr. James Spudich.]

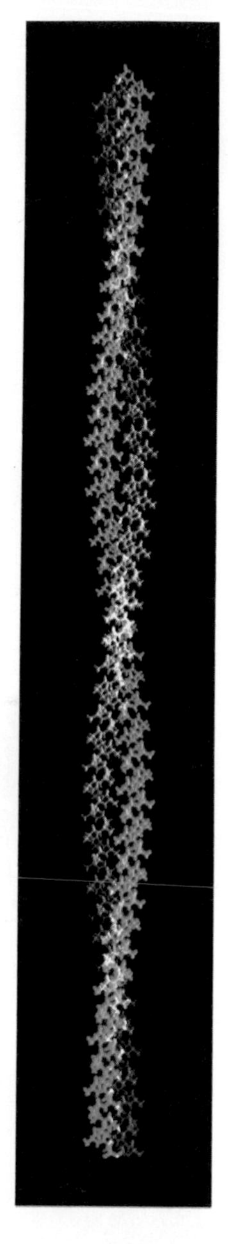

an *amino-terminal globular head* of about 820 residues (~95 kd), and a *carboxyl-terminal tail* of some 1300 residues (~145 kd) (Figure 36-9). All known myosins share the sequence:

Gly-Glu-Ser-Gly-Ala-Gly-Lys-Thr

which is similar to the sequences found in the active sites of other ATPases. The lysine in this conserved sequence binds the α-phosphate of ATP.

The highly distinctive amino acid sequence of the myosin tail gives rise to its coiled-coil structure. Its uninterrupted α-helical structure is favored by the absence of proline over a span of more than a thousand residues and by the abundance of helix formers such as leucine, alanine, and glutamate (p. 37). The two α-helical strands are in register, pointing the same way. Their axes are about 10 Å apart, enabling the side chains of the two strands to interact intimately to reinforce the helical structure (a solitary α helix in water is usually unstable). The resulting superhelical structure is called an *α-helical coiled coil* (Figure 36-10A). Regularities in the amino acid sequence promote these interactions. The tail consists of *repeating 7-residue units* (*abcdefg*) in which *a* and *d* are usually hydrophobic (Figure 36-10B). Residues *a* and *d* form a zig-zag pattern of knobs and holes that interlock with those of the other strand to form a tight-fitting hydrophobic core. In contrast, residues *b*, *c*, and *f*, which are located on the periphery of the coiled coil,

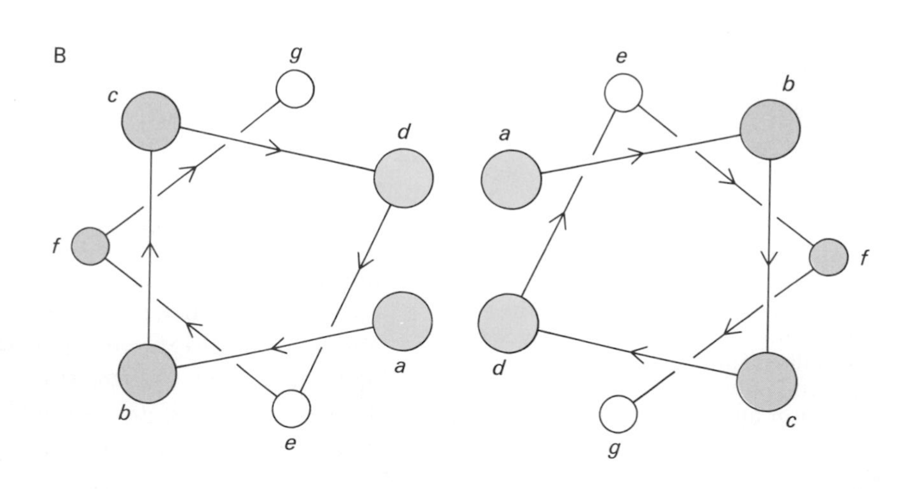

Figure 36-10
(A) Model of a two-stranded α-helical coiled coil. (B) Axial view showing the positions of side chains. Residues *a* and *d* (yellow) of each strand pack tightly to form a hydrophobic core. Residues *b*, *c*, and *f* (green) on the periphery tend to be charged.]

tend to be charged. Another regularity is the presence of *repeating 28-residue units*, which gives rise to alternating bands of positively and negatively charged side chains on the outside of the coiled coil. Yet another regularity, the 196-residue repeating unit, governs the packing of myosin molecules in the thick filament and gives it a structure that repeats every 143 Å. Thus, *the amino acid sequence of the myosin tail leads to a hierarchy of reinforcing interactions culminating in the highly ordered array of the thick filament.*

MYOSIN CAN BE CLEAVED INTO ACTIVE FRAGMENTS: LIGHT AND HEAVY MEROMYOSIN (LMM AND HMM)

Myosin can be enzymatically split into fragments that retain some key activities of the intact molecule. The fruitfulness of such an experimental approach was evident in the analysis of immunoglobulins (p. 892). In fact, the cleavage of myosin into active fragments came first. In 1953, Andrew Szent-Györgyi showed that myosin is split by trypsin into two fragments, called *light meromyosin* and *heavy meromyosin* (Figure 36-11). Light meromyosin (LMM), like myosin, forms filaments. However, it lacks ATPase activity and does not combine with actin. Electron microscopy and x-ray diffraction have shown that *LMM is a two-stranded α-helical rod for its entire length of 850 Å*. Heavy meromyosin (HMM) has entirely different properties. It catalyzes the hydrolysis of ATP and binds to actin but does not form filaments. HMM is the force-generating unit in muscle contraction. It consists of a short rod attached to two globular domains, the myosin heads. HMM can be split further into two globular subfragments (called S1) and one rod-shaped subfragment (called S2). *Each S1 fragment contains an ATPase site and a binding site for actin.*

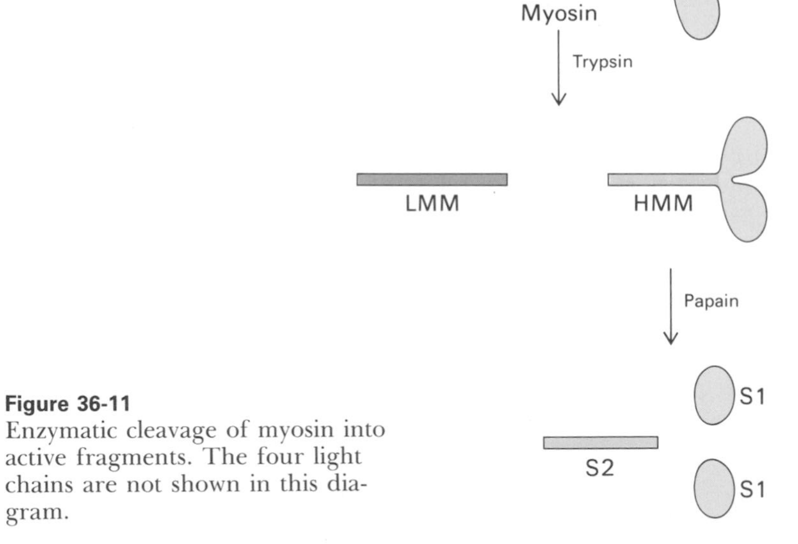

Figure 36-11
Enzymatic cleavage of myosin into active fragments. The four light chains are not shown in this diagram.

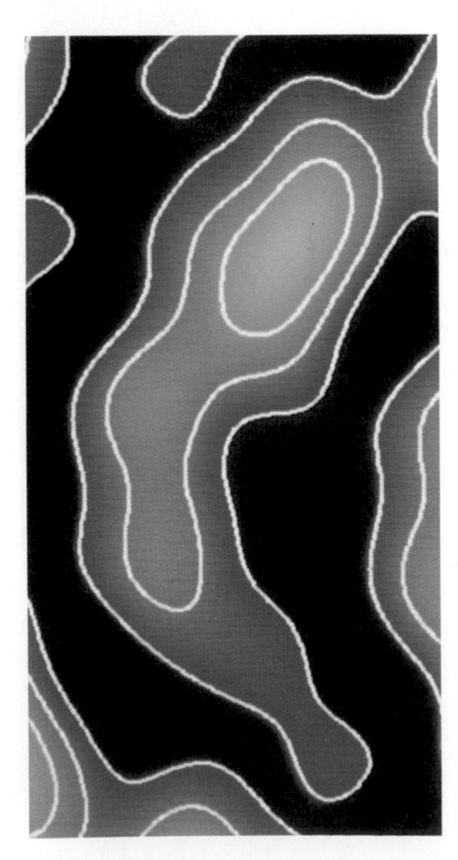

Figure 36-12
Structure of the S1 head of myosin. This image is a projection of the electron density determined by x-ray crystallographic analysis at low resolution. [Courtesy of Dr. Donald Winkelmann and Dr. Ivan Rayment.]

ACTIN FORMS FILAMENTS THAT BIND MYOSIN

Actin, a ubiquitous protein in eucaryotes, is the major constituent of thin filaments. In solutions of low ionic strength, actin is a 42-kd monomer called *G-actin* because of its globular shape. X-ray crystallographic analyses have shown that actin consists of two globular domains and has dimensions of about 30 × 40 × 70 Å (Figure 36-13). As the ionic strength is increased to the physiologic level, G-actin polymerizes into a fibrous form called *F-actin,* which closely resembles thin filaments. An F-actin fiber looks in electron micrographs like two strings of beads wound around each other (Figure 36-14). X-ray diffraction patterns show that *F-actin is a helix of actin monomers.* The helix has a diameter of about 90 Å. The structure repeats at intervals of about 360 Å along the helix axis (Figure 36-15).

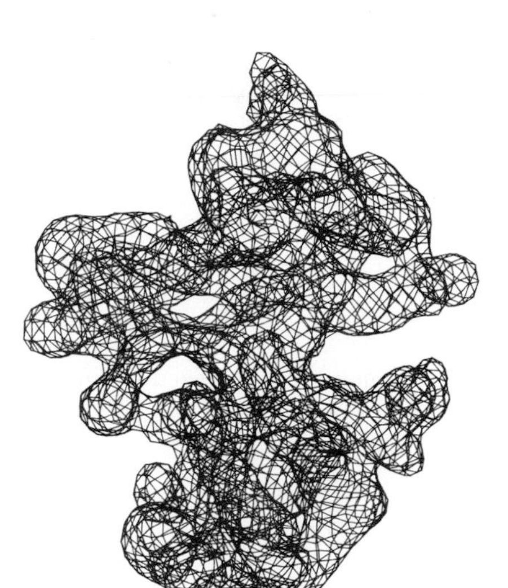

Figure 36-13
Structure of monomeric actin determined by x-ray crystallographic analysis of a complex of actin and DNase I. [Courtesy of Dr. Wolfgang Kabsch.]

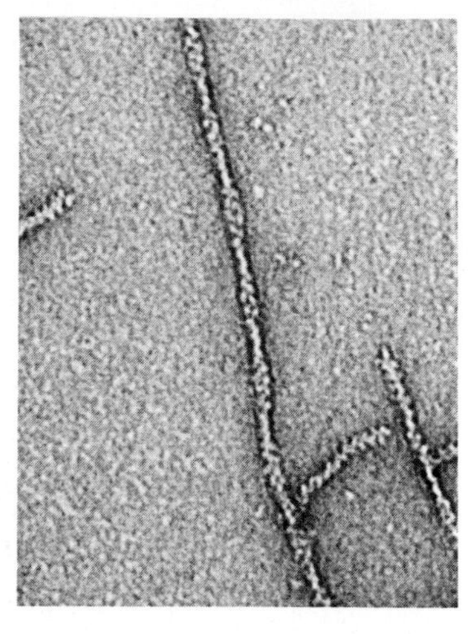

Figure 36-14
Electron micrograph of purified F-actin filaments. [Courtesy of Dr. James Spudich.]

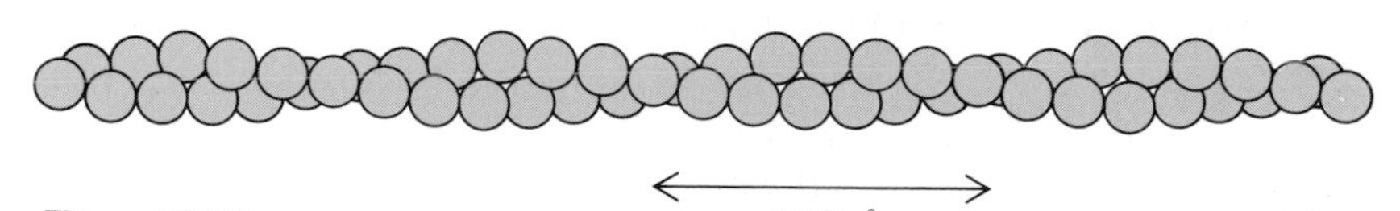

Figure 36-15
Schematic diagram of the helical array of actin units in F-actin. Each actin monomer is represented by two spheres. The directionality of the filament is not depicted here. [After a diagram kindly provided by Dr. James Spudich.]

MUSCLE CONTRACTION IS PRODUCED BY THE INTERACTION OF ACTIN AND MYOSIN

A complex called *actomyosin* is formed when a solution of actin is added to one of myosin. Complex formation is accompanied by a large increase in the viscosity of the solution. In the 1940s, Albert Szent-Györgyi showed that this increase is reversed by the addition of ATP. This observation revealed that *ATP dissociates actomyosin* into actin and myosin. Szent-Györgyi also prepared threads of actomyosin in which the molecules were oriented by flow. A striking result was obtained when the threads were immersed in a solution containing ATP, K^+, and Mg^{2+}. The *actomyosin threads contracted,* whereas threads formed from myosin alone did not. *These incisive experiments suggested that the force of muscle contraction arises from the interplay of myosin, actin, and ATP.*

THE THICK AND THIN FILAMENTS HAVE DIRECTION

The cyclic formation and dissociation of the complex between myosin and actin produces coherent motion because these molecules are components of highly ordered assemblies. The organization of myosin molecules in the thick filament was elucidated by Hugh Huxley. A *dissociated thick filament* has a diameter of 160 Å and a length of 1.5 μm (15,000 Å). *Cross-bridges* emerge in a regular helical array at intervals of 143 Å along the filament axis; recall that amino acid sequence regularities in the tail of myosin give rise to this period (p. 926). Midway along the length of a thick filament is a 1500-Å *bare region* devoid of projecting cross-bridges (Figure 36-16).

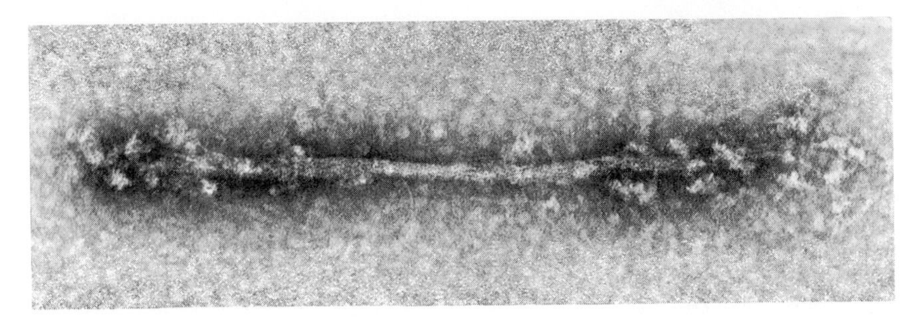

Figure 36-16
Electron micrograph of a reconstituted thick filament. The projections on either side of the bare zone are the cross-bridges. [Courtesy of Dr. Hugh Huxley.]

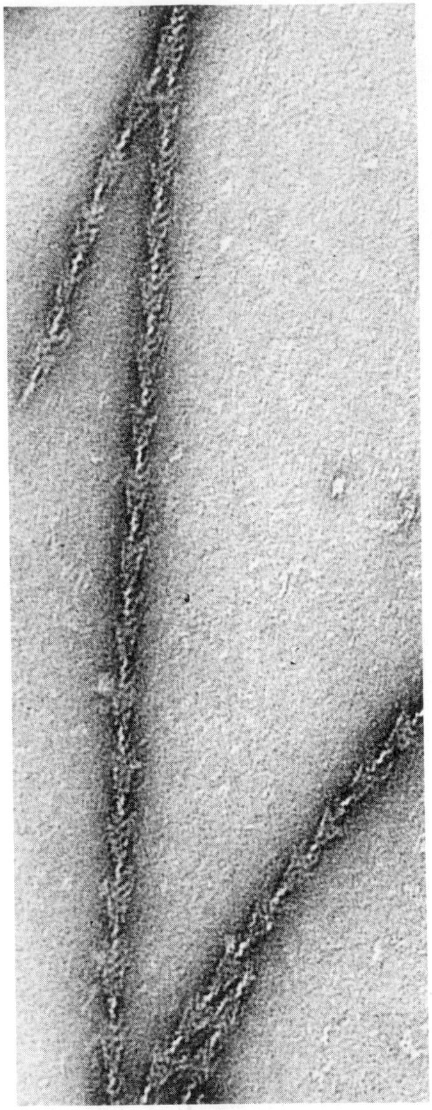

The same structural features are evident in *synthetic thick filaments*, which are produced by lowering the ionic strength of a solution of myosin. The shortest synthetic filament is about 3000 Å long and contains a central bare region 1500 Å long, about the length of the myosin tail. The thick filament grows by the addition of molecules parallel to those already assembled. *The myosin molecules on one side of the bare zone point one way, whereas those on the other side point in the opposite direction. Thus, a thick filament is inherently bipolar.*

Thin filaments also have direction. If myosin (or HMM or S1) is added to unattached thin filaments or to F-actin, an arrowhead pattern appears in electron micrographs (Figure 36-17). These structures are picturesquely called *decorated filaments*. The arrows on a decorated filament always point the same way for its entire length. Thus, *thin filaments have inherent directionality.*

Figure 36-17
Electron micrograph of filaments of F-actin decorated with the S1 moiety of HMM. The "arrowheads" all point the same way. [Courtesy of Dr. James Spudich.]

THE POLARITY OF THICK AND THIN FILAMENTS REVERSES IN THE MIDDLE OF A SARCOMERE

In muscle depleted of ATP, the cross-bridges decorate the thin filament so that the arrows on all filaments point away from the Z line. Thus, *all the thin filaments on one side of a Z line have the same orientation, whereas those on the opposite side have the reverse polarity.* The structural polarity of both the thick and the thin filaments is crucial for coherent motion. The sliding force developed by the interaction of individual actin and myosin units adds up because the interacting units have the same relative orientation. Furthermore, *the absolute direction of the sites reverses half-way between Z lines* (Figure 36-18). Consequently, the two thin filaments that bind the cross-bridges of a thick filament are drawn toward each other, which shortens the distance between Z lines (Figure 36-19).

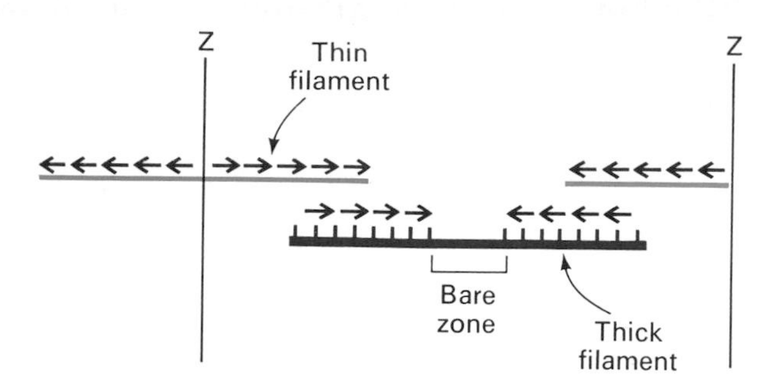

Figure 36-18
The polarity of the thick and thin filaments reverses halfway between the Z lines.

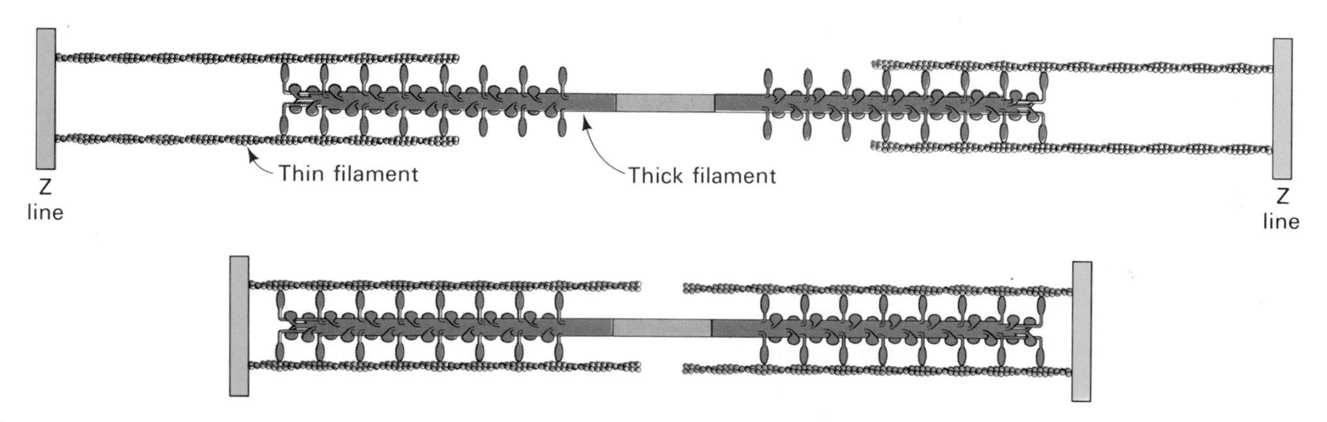

Figure 36-19
Schematic diagram showing the interaction of thick and thin filaments in skeletal muscle contraction. [After a diagram kindly provided by Dr. James Spudich.]

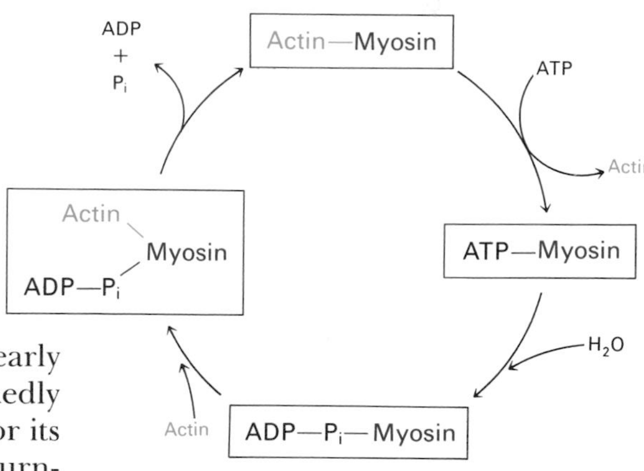

Figure 36-20
The hydrolysis of ATP drives the cyclic association and dissociation of actin and myosin.

THE ATPase ACTIVITY OF MYOSIN IS MARKEDLY INCREASED BY ACTIN

We turn now to the generation of contractile force. A revealing early clue was the finding that the ATPase activity of myosin is markedly enhanced by F-actin. Indeed, Albert Szent-Györgyi named actin for its capacity to activate ATP hydrolysis by myosin. Actin increases the turnover number of myosin 200-fold, from 0.05 s^{-1} to 10 s^{-1}. ATP bound to myosin is rapidly hydrolyzed, but ADP and P_i are slow to leave. Edwin Taylor proposed a model in which actin increases the turnover number of myosin by binding to the myosin-ADP-P_i complex and accelerating the release of products (Figure 36-20). Actomyosin then binds ATP, which leads to the dissociation of actin and myosin. The resulting ATP-myosin complex is ready for another round of catalysis. These reactions, like those of all known ATPases, require Mg^{2+}.

The interaction of the S1 heads of myosin with actin has been visualized by electron microscopy of thin filaments decorated with S1. The image shown in Figure 36-21 was obtained by averaging data from many micrographs of unstained filaments that were frozen in thin aqueous films at low temperature. An advantage of this technique is that the native structure, rather than the distribution of electron-dense stains, is seen. The actin monomers in these decorated thin filaments consist of two domains; the line joining them is nearly perpendicular to the filament axis. Tropomyosin, which participates in controlling contraction (p. 933), runs longitudinally along the filament. The S1 heads in these decorated filaments are helically arrayed on its periphery, Each S1 binds the outer domain of a single actin unit. The other tip of S1 is at least 130 Å away from the actin-binding site. These links between S1 and actin are formed in the absence of ATP. The geometry probably corresponds to the end of a power stroke in vivo, just before the binding of ATP.

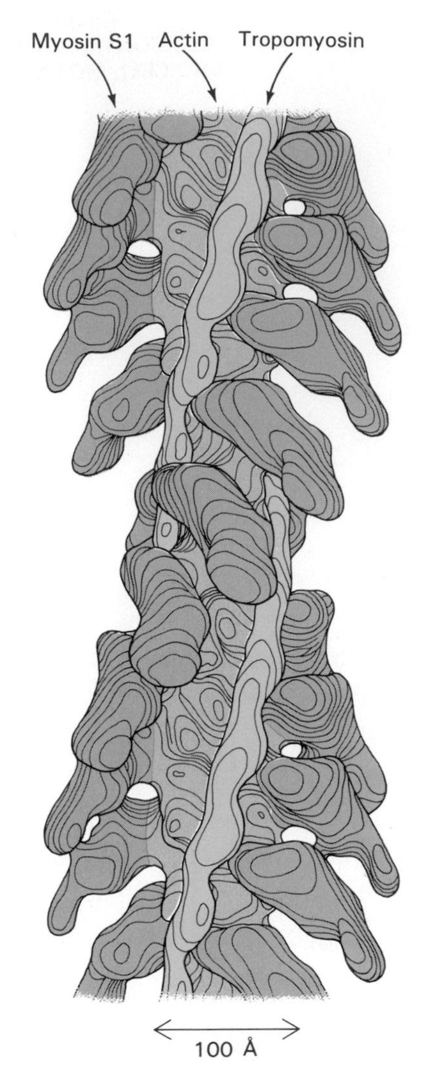

Figure 36-21
Structure of a reconstituted thin filament decorated with the S1 fragment of myosin. This image, obtained by three-dimensional analyses of electron micrographs of unstained frozen filaments, probably corresponds to the structure of the intact system at the end of the power stroke. The location of the troponin complex is not defined by this analysis. The pitch of the helical array is 370 Å. [After a drawing kindly provided by Dr. Ronald Milligan and Dr. Paula Flicker.]

THE POWER STROKE IN CONTRACTION IS DRIVEN BY CONFORMATIONAL CHANGES IN THE MYOSIN S1 HEAD

The contractile force is generated by the cyclic formation and dissociation of complexes between actin and the S1 heads of myosin. Repeated cycles of attachment, pulling, and detachment take place during a single contraction. Ideally, one should have images of each state of the intact system at atomic resolution to elucidate how the cycle is driven. We are far from attaining this objective for two reasons. First, three-dimensional crystals of intact contractile assemblies have not been obtained because of the size and complexity of the assemblies. Second, the interactions of myosin and actin are asynchronous; at any moment, S1 heads are in different states instead of in step. What we have today are glimpses of different facets of the contractile system provided by ingenious use of biochemical, electron microscopic, spectroscopic, and x-ray diffraction methods. They do not add up to a definitive model for the conversion of chemical-bond energy into mechanical energy. However, they do provide valuable clues that can be pieced together into a working hypothesis.

A plausible proposed mechanism for the generation of force is shown in Figure 36-22. S1 heads in resting muscle are detached from thin filaments (see part A); ADP and P_i, the hydrolysis products, are bound. When muscle is stimulated, S1 heads reach out from the thick filament and attach to actin units on thin filaments (Figure 36-22B). P_i leaves this complex of actin with myosin-ADP-P_i, and the S1 head of myosin concomitantly changes structure. This change in orientation of S1 relative to actin causes the *power stroke* of muscle contraction—the thin filament is pulled a distance of some 75 Å. The dissociation of ADP at the end of the power stroke and the subsequent binding of ATP reverses the conformational change in S1 and leads to the very rapid release of S1 from actin (Figure 36-22D). Finally, the bound ATP is hydrolyzed by the free myosin head, which resets it for the next interaction with the thin filament (Figure 36-22E).

The myosin molecule contains two kinds of hinges that enable its S1 heads to attach and detach from actin and to change their orientation when bound. One type of hinge is located between each S1 head and the S2 rod, and the other is positioned between S2 and the LMM unit of

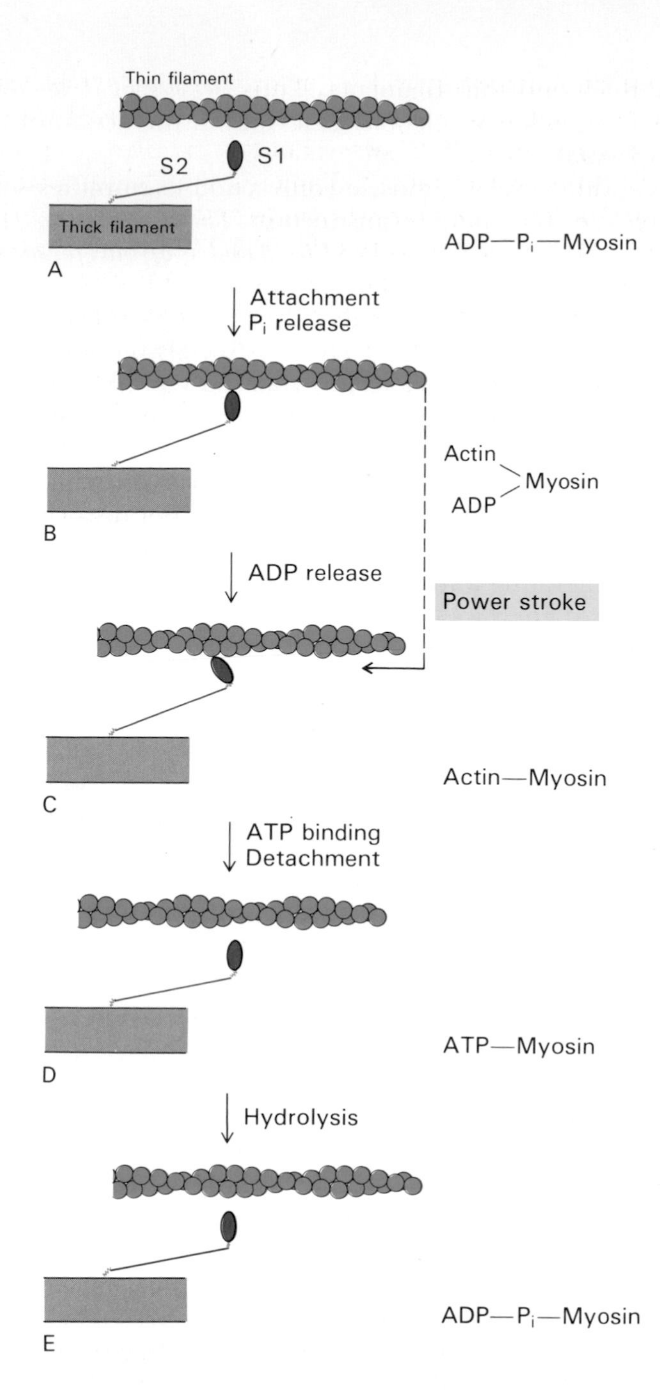

Figure 36-22
Proposed mechanism for the generation of force by the interaction of an S1 unit of a myosin filament with an actin filament. In the power stroke, the thick filament moves relative to the thin filament when S1 undergoes conformational changes initiated by the release of P_i.

myosin (Figure 36-23). These hinges are flexible regions of polypeptide chain, vulnerable to cleavage by proteolytic enzymes. Indeed, the enzymatic splitting of myosin into LMM, S2, and S1 pieces expresses the fact that *myosin is constructed of domains joined by hinges.* The role of the S2 domain is to transmit tension from S1 bound to a thin filament to the LMM domain that forms part of the thick filament. The hinge between S1 and S2 enables S1 to interact with actin in one orientation when ADP is bound and in a different one when ADP is released. The other hinge, between S2 and LMM, allows considerable variation in the position of S1 relative to the thick filament, which permits S1 to interact precisely with actin. Hence, the system works over a wide range of lateral spac-

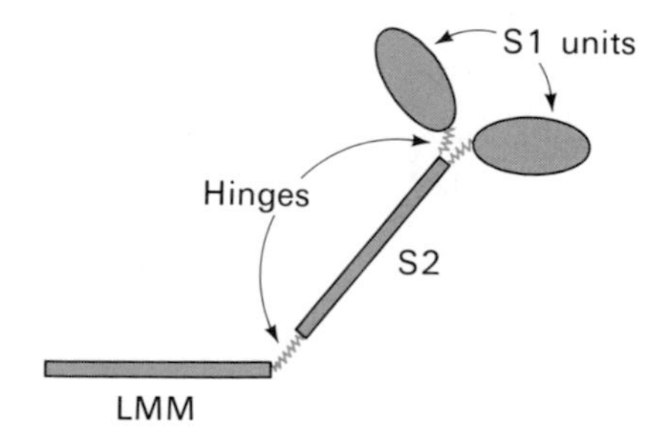

Figure 36-23
Two kinds of hinges in myosin—one between S1 and S2 and the other between S2 and the LMM region—are important in allowing the S1 unit to alter its contacts with actin during the power stroke.

ings of the thick and thin filaments. Thus, *segmental flexibility* plays a critical role in muscle contraction, as it does in the action of antibody molecules (p. 893).

The cross-bridge cycle in muscle contraction exemplifies some general principles of free-energy transduction. *The essence of the process is a cyclic change in the shape of the myosin S1 head and its affinity for actin with an ATP-ADP cycle at its catalytic site.* The binding of ATP to the catalytic site, near the center of S1, leads to the detachment of S1 from actin at a site at least 60 Å away. S1 is designed so that it strongly binds either ATP or actin but not both. Given a choice, S1 markedly prefers ATP to actin. The bound ATP is then hydrolyzed. The weak binding of actin to the resulting myosin-ADP-P_i complex leads to the release of P_i, which further changes the conformation of myosin and markedly increases its affinity for actin. A key point is that ADP does not dissociate immediately. It stays on myosin for about 20 μs, enabling S1 and actin to interact strongly, generating a force between the two sets of filaments. The large drop in free energy (~7 kcal/mol) in this step, with the associated structural changes, enables it to be the power stroke. The subsequent release of ADP allows ATP to bind, inducing the very rapid release of actin from the S1 head. *Thus, the free energy of ATP relative to that of ADP and P_i is used by myosin in two steps and for two purposes: (1) to dissociate actin from myosin, and (2) to release P_i so that myosin can bind strongly to actin, generating force.*

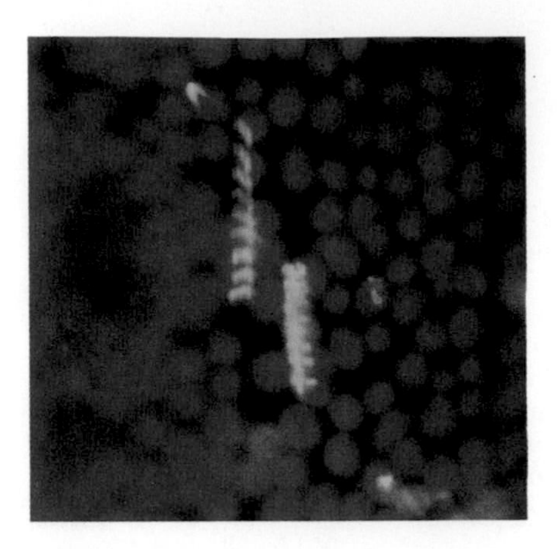

Figure 36-24
Movement of myosin-coated beads on actin cables. Yellow-fluorescent beads coated with myosin were added to an opened *Nitella* cell. On addition of ATP, the beads moved on actin cables lying above the chloroplasts, which fluoresce red. A series of exposures taken at intervals of 1 second indicate that the velocity of movement was 3 μm/s. [Courtesy of Dr. James Spudich and Dr. Michael Sheetz.]

MYOSIN-COATED BEADS MOVE UNIDIRECTIONALLY ON ORIENTED ACTIN CABLES

One of the goals of biochemical studies is to reconstitute biological function in vitro using purified components. We have previously discussed reconstituted systems that carry out processes such as light-driven ATP synthesis (p. 411) and DNA replication (p. 673). An assay system for direct observation of movements produced by interactions between myosin and actin has recently been devised by Michael Sheetz and James Spudich. They made use of *Nitella*, a giant alga that contains cables of actin fixed to the cytoplasmic face of rows of chloroplasts. A key feature of these cables, which mediate cytoplasmic streaming, is that actin filaments in them point in the same direction. The striking finding is *myosin-coated beads move unidirectionally along these actin cables in the presence of ATP* (Figure 36-24), in the same direction as the cytosol in cytoplasmic streaming. Beads coated with skeletal muscle myosin move with a speed of 5 μm/s, which is approximately the speed of contraction of intact sarcomeres in vivo. It has been estimated that 25 myosin heads on each bead give rise to this movement. In essence, myosin molecules "walk" on actin cables (Figure 36-25). Alternatively, actin filaments can move on a glass surface containing bound myosin molecules if ATP is present.

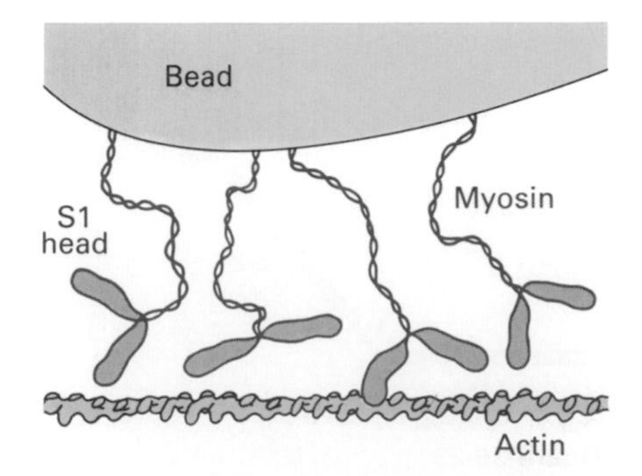

Figure 36-25
Schematic diagram of an in vitro motility system. Myosin molecules attached to beads "walk" unidirectionally along actin filaments as ATP is hydrolyzed. [After a drawing kindly provided by Dr. James Spudich.]

The velocity of movement depends on the type of myosin used in the assay, not on the type of actin. Smooth muscle myosin moves at 0.4 μm/s and *Dictyostelium* myosin at 1 μm/s, considerably slower than skeletal muscle myosin, which is also the fastest in vivo. These in vitro motility assay systems have also provided valuable information concerning the region of myosin essential for movement along actin. Heavy meromyosin moves at nearly the same velocity as does intact myosin, indicating that the distal 850 Å of the myosin tail does not participate in the generation of contractile force. In fact, actin filaments move on films coated with S1. Thus, no portion of the α-helical tail of myosin is essential for movement. *The force of muscle contraction is generated within the S1 head.* Site-specific mutants of myosin may reveal the parts of S1 that are crucial for generating this force.

TROPONIN AND TROPOMYOSIN MEDIATE THE REGULATION OF MUSCLE CONTRACTION BY CALCIUM ION

The physiologic regulator of muscle contraction is Ca^{2+}. In the resting (relaxed) state of skeletal muscle, Ca^{2+} is sequestered in the sarcoplasmic reticulum, a specialized form of the endoplasmic reticulum, by an active transport system (p. 955). This ATP-driven pump lowers the concentration of Ca^{2+} in the cytosol to less than 1 μM. A nerve impulse leads to the release of Ca^{2+} from the sacs of the sarcoplasmic reticulum, which raises the cytosolic concentration to about 10 μM and leads to muscle contraction. Setsuro Ebashi discovered that the effect of Ca^{2+} on the interaction of actin and myosin is mediated by *tropomyosin* and the *troponin complex,* which are located in the thin filament and constitute about a third of its mass. Tropomyosin is a two-stranded α-helical rod. This highly elongated 70-kd protein is aligned nearly parallel to the long axis of the thin filament (Figure 36-26). Troponin is a complex of three polypeptide chains: TnC (18 kd), TnI (24 kd), and TnT (37 kd). TnC binds calcium ions, TnI binds to actin, and TnT binds to tropomyosin. The troponin complex is located in the thin filaments at intervals of 385 Å, a period set by the length of a tropomyosin unit. Each troponin complex regulates the interactions of some seven actin units.

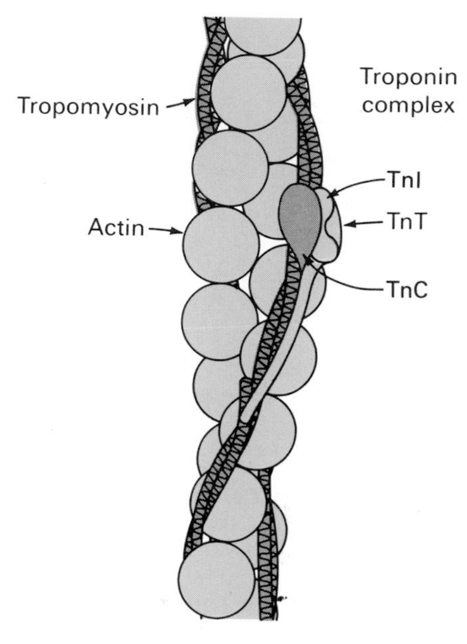

Figure 36-26
Model of a thin filament. The troponin complex consists of three components: TnI (yellow), TnC (gray), and TnT (green). In relaxed muscle (low Ca^{2+}), tropomyosin prevents actin from interacting with myosin S1 units to generate force. [After C. Cohen. Protein switch of muscle contraction. Copyright © 1975 by Scientific American, Inc. All rights reserved. The arrangement of the troponin complex is based on a drawing kindly provided by Dr. Larry Smillie.]

Troponin C, the calcium-sensing component of the complex, consists of two homologous domains. An amino-terminal and a carboxyl-terminal globular domain are connected by a long nine-turn α-helix (Figure 36-27). Each domain contains two binding sites for Ca^{2+}. The ones in

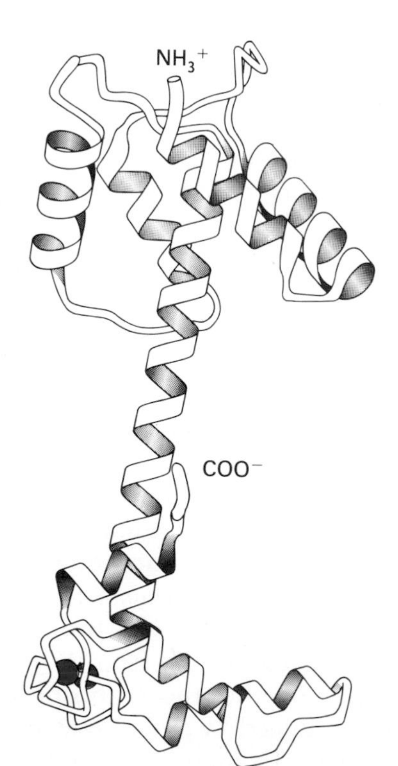

Figure 36-27
Structure of troponin C. The amino-terminal regulatory domain (top) is separated from a homologous carboxyl-terminal domain (bottom) by a long α helix. The Ca^{2+}-binding sites in the regulatory domain are unoccupied in this crystal form. The two sites with high affinity for Ca^{2+} or Mg^{2+} (red) in the carboxyl-terminal domain are filled. [After O. Herzberg and M. N. G. James. *Nature* 313(1985):655.]

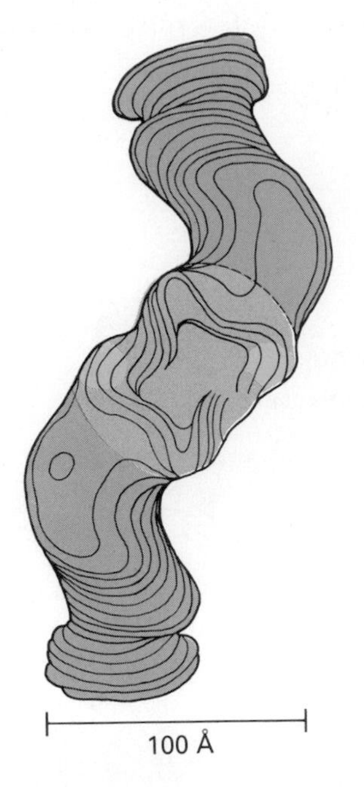

Figure 36-28
Axial view down a reconstituted thin filament showing the inferred positions of tropomyosin in the contracting state. The view here is perpendicular to the one shown in Figure 36-21. Tropomyosin is shown in yellow, actin in blue, and S1 in red. Nerve excitation causes the sarcoplasmic reticulum to release Ca^{2+}, which binds to TnC. The resulting conformational change is transmitted to TnT and then to tropomyosin. The shift in the position of tropomyosin enables actin to interact with myosin. [Courtesy of Dr. Ronald Milligan and Dr. Paula Flicker.]

the carboxyl-terminal domain have high affinity for Ca^{2+} ($K_d = 0.1$ μM), whereas those in the amino-terminal domain have low affinity ($K_d = 10$ μM). TnC is a member of a family of calcium-binding proteins that have important regulatory roles. The structure of TnC closely resembles that of calmodulin, a ubiquitous calcium-sensing protein (p. 988).

In resting muscle, the high-affinity sites are occupied by Ca^{2+} but the low-affinity ones are empty. When Ca^{2+} is released from the sarcoplasmic reticulum, it occupies the low-affinity sites, which changes the conformation of the amino-terminal domain, making it more similar to the carboxyl-terminal domain. This conformational change in TnC is transmitted to the other components of the troponin complex and then to tropomyosin. It seems likely that TnT, which has a highly elongated shape, controls the position of tropomyosin on the thin filament, near the interface between actin and the S1 head of myosin (Figure 36-28). *A rather small shift in the position of tropomyosin could markedly alter the binding of actin to S1 or the capacity of this complex to undergo structural changes in the cross-bridge cycle.*

In fact, time-resolved x-ray diffraction experiments have revealed that tropomyosin moves before S1 heads do when skeletal muscle is activated by a nerve impulse. The half-time for tropomyosin movement was 17 ms, compared with 25 ms for the movement of myosin and 45 ms for the generation of tension. Further evidence for the essentiality of the troponin-tropomyosin couple in the regulation of contraction comes from in vitro motility assays. The movement of myosin-coated beads on actin cables is insensitive to Ca^{2+} when only these two proteins are present. The addition of troponin and tropomyosin makes the beads respond to Ca^{2+}. Movement is blocked if Ca^{2+} is absent and restored by the addition of micromolar levels of Ca^{2+}. Thus, a reconstituted system containing only four proteins—myosin, actin, troponin, and tropomyosin—exhibits calcium-regulated movement. It is evident that *Ca^{2+} controls muscle contraction by an allosteric mechanism in which the flow of information is*

$$Ca^{2+} \longrightarrow \text{troponin} \longrightarrow \text{tropomyosin} \longrightarrow \text{actin} \longrightarrow \text{myosin}$$

SMOOTH MUSCLE CONTRACTS WHEN ITS LIGHT CHAINS BECOME PHOSPHORYLATED BY A Ca^{2+}-ACTIVATED KINASE

Myosin from vertebrate striated muscle contains three kinds of light chains, named LC1 (22 kd), LC2 (18 kd), and LC3 (16 kd). Each S1 head contains one LC2 and an LC1 or an LC3. LC2 (also known as the regulatory light chain) is phosphorylated by a Ca^{2+}-stimulated light-chain kinase and dephosphorylated by phosphatase C. Phosphorylation of LC2 doubles the actin-stimulated ATPase activity of myosin; the physiologic significance of this change is uncertain.

Smooth muscle, in contrast with skeletal muscle, is not regulated by a troponin-tropomyosin mechanism. Instead, vertebrate smooth muscle contraction is controlled by the degree of phosphorylation of its light chains. *Phosphorylation of smooth muscle LC2 leads to contraction, and dephosphorylation results in relaxation* (Figure 36-29). Beads coated with phosphorylated smooth-muscle myosin move on actin cables from *Nitella* at a velocity of 0.15 μm/s, but at less than 0.01 μ/s with the dephosphorylated form. Smooth-muscle contraction is induced by an increase in the cytosolic concentration of calcium ion. The effect is mediated through Ca^{2+}-calmodulin, which stimulates myosin light-chain kinase (MLCK). Phosphoryl groups introduced by MLCK are removed from myosin by a Ca^{2+}-independent phosphatase. Contraction of smooth muscle is more sustained than that of skeletal muscle because this dephosphorylation is slower than the lowering of the cytosolic Ca^{2+} level by active transport of the ion into the sarcoplasmic reticulum.

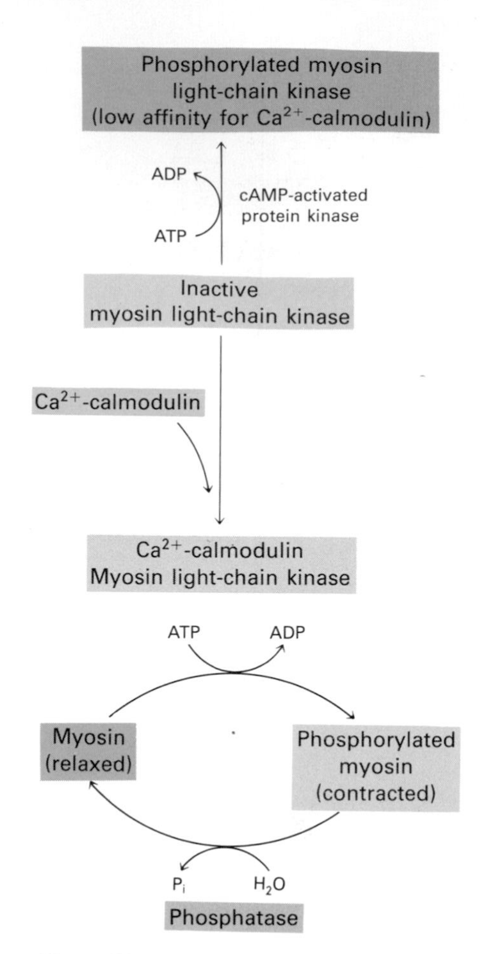

Figure 36-29
The contraction of vertebrate smooth muscle is regulated by phosphorylation of myosin. Ca^{2+} bound to calmodulin activates a specific kinase that phosphorylates the light chains of myosin, leading to contraction. A cascade mediated by cyclic AMP leads to the relaxation of smooth muscle by phosphorylating this light-chain kinase, which lowers its affinity for Ca^{2+}-calmodulin.

Smooth muscle activity is modulated by epinephrine, which acts as a relaxant in cells containing receptors of the β type (p. 980). Epinephrine triggers the adenylate cyclase cascade in these cells (p. 462). The resulting increase in the intracellular level of cAMP leads to the activation of protein kinase, which phosphorylates MLCK in addition to other proteins (e.g., phosphorylase kinase). Phosphorylation of MLCK lowers its affinity for Ca^{2+}-calmodulin and hence stabilizes its inactive form. Consequently, epinephrine lowers the proportion of phosphorylated myosin in the muscle, which then relaxes.

Molluscan smooth muscle is activated by the binding of Ca^{2+} directly to LC2. In the absence of Ca^{2+}, LC2 blocks the interaction of this myosin with actin. It is interesting to note that the LC2 chains of both molluscs and vertebrates are members of the calmodulin-troponin family of calcium-sensing proteins. Indeed, vertebrate LC2 substituted for molluscan LC2 enables molluscan smooth muscle to be activated by Ca^{2+}. The light chains of molluscan myosin, like those of vertebrate muscle, bind near the junction of S1 and the rod region of myosin.

CREATINE PHOSPHATE IS A RESERVOIR OF ~P

The amount of ATP in muscle suffices to sustain contractile activity for less than a second. Vertebrate muscle contains a reservoir of high-potential phosphoryl groups in the form of *creatine phosphate* (phosphocreatine). The phosphoryl transfer potential of creatine phosphate is higher than that of ATP (p. 319). *Creatine kinase catalyzes the reversible transfer of a phosphoryl group from creatine phosphate to ADP to form ATP.*

$$\text{Creatine phosphate} + \text{ADP} + \text{H}^+ \rightleftharpoons \text{ATP} + \text{creatine}$$

```
   -O        NH                                 -O        NH                    H
    |        ||                                  |        ||                    |
-O—P—N—C—N—CH2—COO-                         -O—P—N—C—N—(CH2)3—C—COO-
    ||  |      |                                 ||  |       |                  |
    O   H      CH3                               O   H       H                  NH3+
```

Creatine phosphate **Arginine phosphate**

At pH 7, the free energy of hydrolysis of creatine phosphate is −10.3 kcal/mol, compared with −7.3 kcal/mol for ATP. Hence, the standard free-energy change in forming ATP from creatine phosphate is −3 kcal/mol, which corresponds to an equilibrium constant of 162.

$$K = \frac{[\text{ATP}][\text{creatine}]}{[\text{ADP}][\text{creatine phosphate}]} = 10^{3/1.36} = 162$$

In resting muscle, typical concentrations of these metabolites are: [ATP] = 4 mM, [ADP] = 0.013 mM, [creatine phosphate] = 25 mM, and [creatine] = 13 mM. *The abundance of creatine phosphate and its high phos-*

phoryl transfer potential relative to that of ATP makes it a highly effective ~P buffer. Creatine phosphate maintains a high concentration of ATP during periods of muscular exertion. Indeed, creatine phosphate is the major source of ~P for a runner during the first four seconds of a 100-meter sprint.

In active muscle, the ATP level decreases little until most of the creatine phosphate is consumed. However, the levels of ADP and P_i do rise rapidly following the onset of contractile activity. Likewise, the level of AMP rises by the action of adenylate kinase (myokinase).

$$2\ \text{ADP} \rightleftharpoons \text{ATP} + \text{AMP}$$

The reduced energy charge of active muscle stimulates glycogen breakdown, glycolysis, the citric acid cycle, and oxidative phosphorylation. The relative contributions of these processes to the generation of ATP depend on the type of muscle. Red muscle, which derives its color from myoglobin and the cytochromes of the respiratory chain, has a much more aerobic metabolism than white muscle.

Some invertebrates use *arginine phosphate* (phosphoarginine) to store high-potential phosphoryl groups in muscle. Arginine phosphate, like creatine phosphate, contains a *phosphoguanido group*; these compounds are *phosphagens*. The transfer of the phosphoryl group of arginine phosphate to ADP is catalyzed by *arginine phosphokinase*.

$$\text{Arginine phosphate} + \text{ADP} + \text{H}^+ \rightleftharpoons \text{arginine} + \text{ATP}$$

NMR studies of intact molluscan muscles have shown that arginine phosphate, like creatine phosphate, buffers the ATP level during contraction (Figure 36-30).

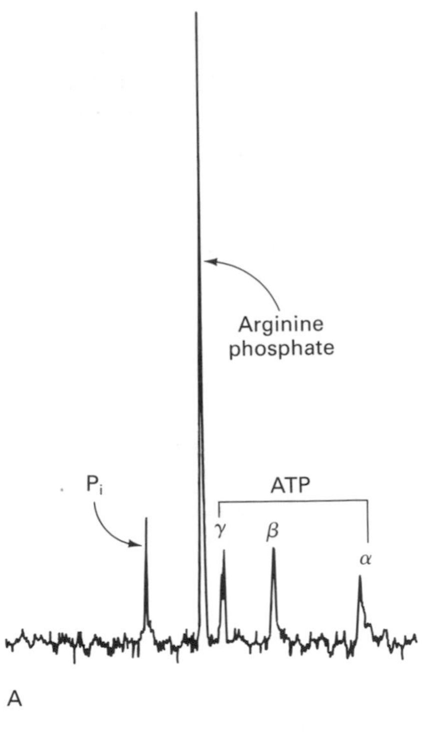

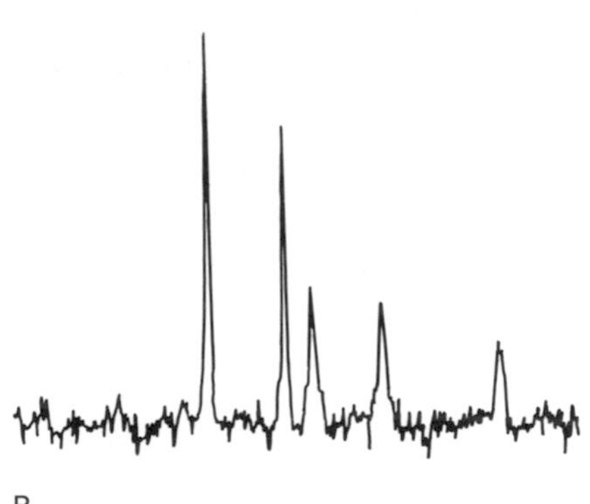

Chemical shift

Figure 36-30
^{31}P NMR spectra of a molluscan (scallop) muscle (A) before and (B) after contractile activity. The level of arginine phosphate drops and that of P_i rises. The level of ATP changes little during this period. [Courtesy of Dr. W. Ross Ellington.]

ACTIN AND MYOSIN HAVE CONTRACTILE ROLES IN NEARLY ALL EUCARYOTIC CELLS

It has long been known that many nonmuscle cells can move and change shape. The migration of cells in the development of embryos, the movement of macrophages to injured tissues, and the retraction of clots by blood platelets vividly exemplify the universality of cell motility. Some aspects of cell movement can be studied in tissue cultures. For example, many kinds of cultured cells have sheetlike extensions called *lamellipodia,* which slowly change their shape. They resemble ruffles on a dress in a slight breeze, and so they are known as *ruffled edges,* or borders (Figure 36-31). A ruffled edge projects forward a distance of a

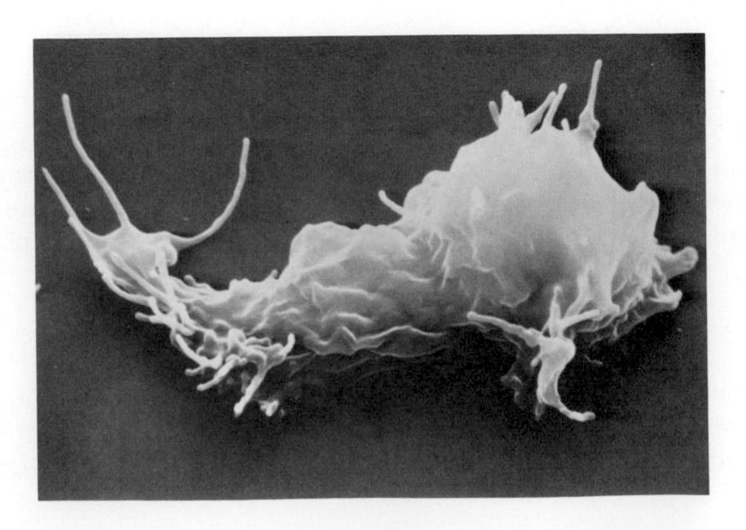

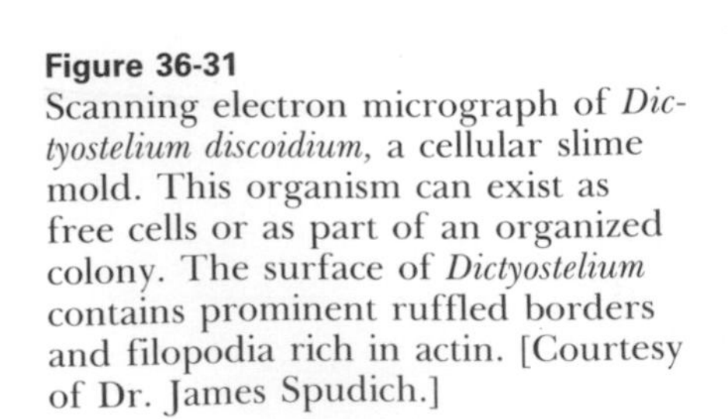

Figure 36-31
Scanning electron micrograph of *Dictyostelium discoidium,* a cellular slime mold. This organism can exist as free cells or as part of an organized colony. The surface of *Dictyostelium* contains prominent ruffled borders and filopodia rich in actin. [Courtesy of Dr. James Spudich.]

few micrometers and may adhere to the surface. If it does, the rear edge of the cell is retracted, which results in an overall movement of the cell.

What is the molecular basis of cell movement? An early clue came from the work of Ariel Loewy on *Physarum polycephalum,* a plasmodial slime mold, which contains a streaming mass of cytoplasm. Extracts made at high ionic strength had properties like that of actomyosin from striated muscle (p. 927). The addition of ATP produced a rapid drop in viscosity, which was followed by a slow rise accompanying the hydrolysis of ATP. Several years later, Sadashi Hatano and Fumio Oosawa found that this slime mold contains a large amount of actin that is very similar to muscle actin. It forms thin filaments and interacts with myosin. Most interesting, *slime-mold actin* reacts with S1 heads from *vertebrate skeletal muscle* to give decorated filaments like those formed from vertebrate actin and myosin. In fact, slime-mold actin differs from rabbit-muscle actin at only 17 of 375 amino acid residues. Thus, *actin is a highly conserved, ancient protein of eucaryotes.* Likewise, myosin has been isolated from this slime mold and from many other eucaryotic cells. Myosins derived from different species differ considerably in their sequences and properties. *Myosin is less highly conserved in evolution than is actin.* For example, myosins from many nonmuscle cells do not readily form thick filaments of the kind seen in skeletal muscle. However, nonmuscle myosins do form short bipolar filaments.

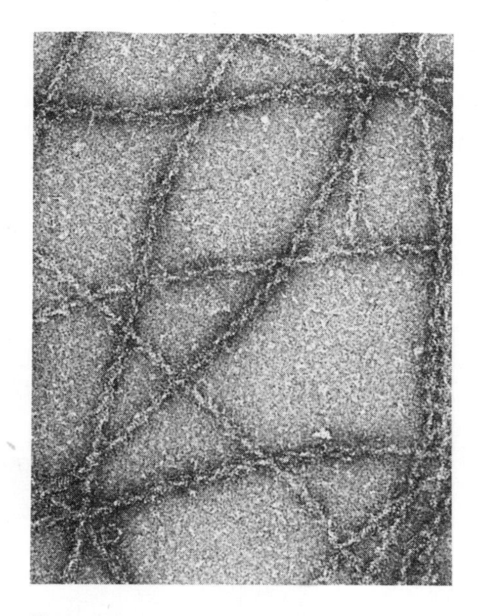

Figure 36-32
Electron micrograph of F-actin from *Dictyostelium* decorated with *Dictyostelium* myosin S1 heads. The arrowhead pattern is like that of skeletal-muscle F-actin decorated with muscle S1. [Courtesy of Dr. James Spudich.]

Most eucaryotic cells are rich in actin, which typically makes up 10% of the total cell protein. Indeed, actin is the most abundant protein in many kinds of cells. Nonmuscle cells typically contain ten- to a hundredfold less myosin than actin. Hence, the ratio of actin to myosin is much higher in nonmuscle cells than in muscle. Some of the actin in nonmuscle cells is polymerized into *microfilaments,* which resemble the thin filaments of muscle. *These 70-Å-diameter filaments are evident in electron micrographs of nearly all eucaryotic cells.* Regular arrays of arrowheads are seen when microfilaments are decorated with S1 heads, which shows that they are made up of actin units and can interact with myosin (Figure 36-32).

Genetic studies have provided insight into the roles of actin and myosin in nonmuscle cells. *Gene disruption* is a generally applicable means of interfering with a particular gene and ascertaining its consequences. For example, the yeast *Saccharomyces cerevisiae* contains a single gene for actin per haploid genome. Homologous recombination with a plasmid containing only part of the actin gene gave rise to transformed yeast cells containing two incomplete actin genes (Figure 36-33). This mutation behaved as a recessive lethal, indicating that the *actin gene encodes a function essential for the growth of yeast cells.* Analyses of temperature-sensitive mutants have shown that yeast cells lacking a functional actin gene are unable to bud.

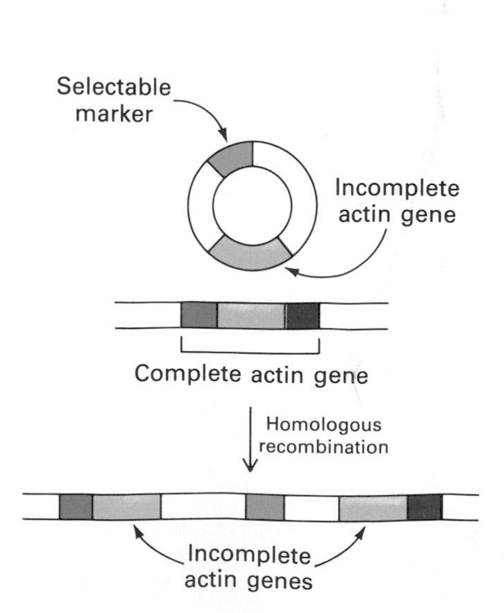

Figure 36-33
Disruption of the actin gene in yeast by homologous recombination with a plasmid gene for an incomplete actin chain. The incomplete chain is shown in yellow, and the additional flanking segments in the complete gene are shown in red and blue. The plasmid also contains a marker gene that enables recombinants to be selected. [After D. Shortle, J. E. Haber, and D. Botstein. *Science* 217(1982):371.]

Gene disruption has recently been used to explore the function of myosin in *Dictyostelium*. Transfection of slime mold cells with a plasmid encoding the HMM moiety of myosin led to the disruption of the sole gene for the heavy chain of myosin; mutant cells express HMM but not a complete heavy chain. The formation of large, multinucleate cells (Figure 36-34) shows that they undergo *karyokinesis* (chromosomal segregation) but not *cytokinesis* (cell division). Even without functional myosin, they exhibit many forms of cell movement, including membrane ruffling, phagocytosis, and chemotaxis. Similar results were obtained by transfection with a plasmid containing the myosin gene in the wrong orientation relative to a promoter. Such a vector encodes an *antisense mRNA* (the complement of mRNA), which blocks the translation of normal cellular myosin mRNA by forming a duplex with it. The precise roles of actin, myosin, and other contractile proteins in vivo can now be determined using these powerful molecular genetic approaches in concert with biochemical and structural methods.

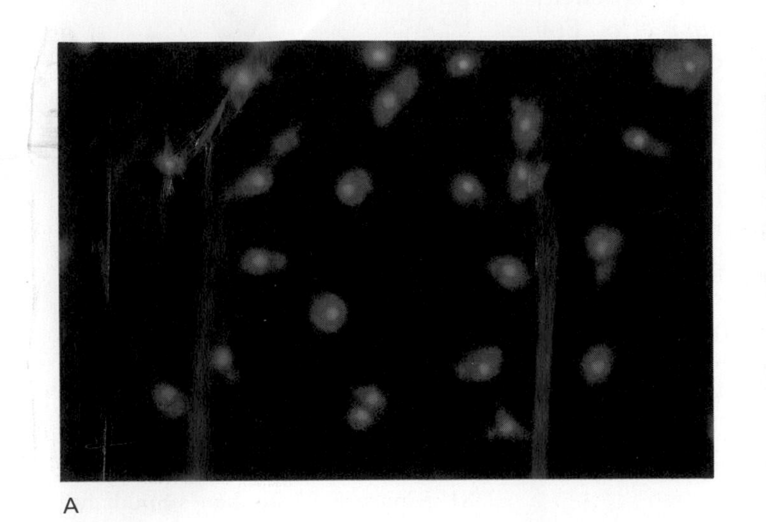

A

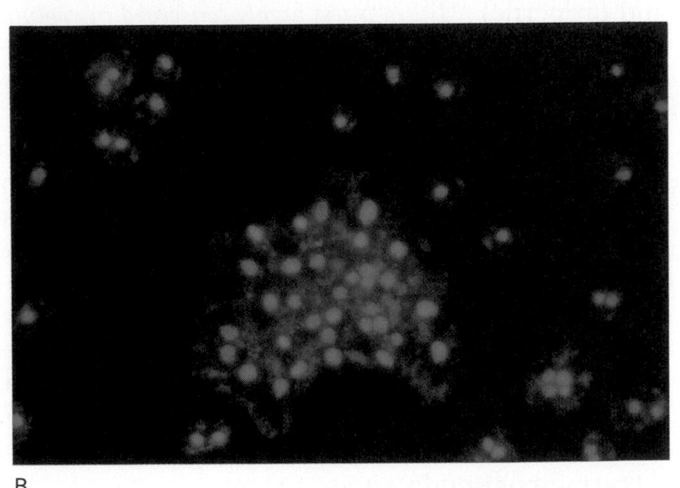

B

Figure 36-34
Fluorescence micrographs of (A) normal and (B) multinucleate *Dictyostelium* cells. The disruption of the myosin heavy-chain gene by homologous recombination prevented cell division. Otherwise, mitosis was essentially normal in these *hmm* mutants. [Courtesy of Dr. Arturo De Lozanne.]

MICROFILAMENTS, INTERMEDIATE FILAMENTS, AND MICROTUBULES FORM DYNAMIC CYTOSKELETONS

Eucaryotic cells have an internal scaffolding called the *cytoskeleton* that gives them their distinctive shapes. The cytoskeleton also enables cells to transport vesicles, undergo changes in shape, and migrate. This dynamic structure is formed by three classes of filamentous assemblies: *microfilaments*, *intermediate filaments*, and *microtubules*. As mentioned earlier, microfilaments (70 Å diameter) are made of actin. Intermediate filaments (70 to 110 Å diameter) and microtubules (about 300 Å diameter) are formed from other proteins. The cellular distribution of microfilaments can be displayed by *immunofluorescence microscopy*. Antibodies specific for actin are covalently labeled with a fluorescent tag such as

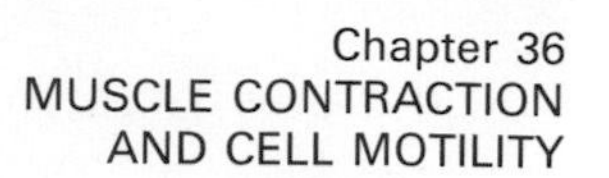

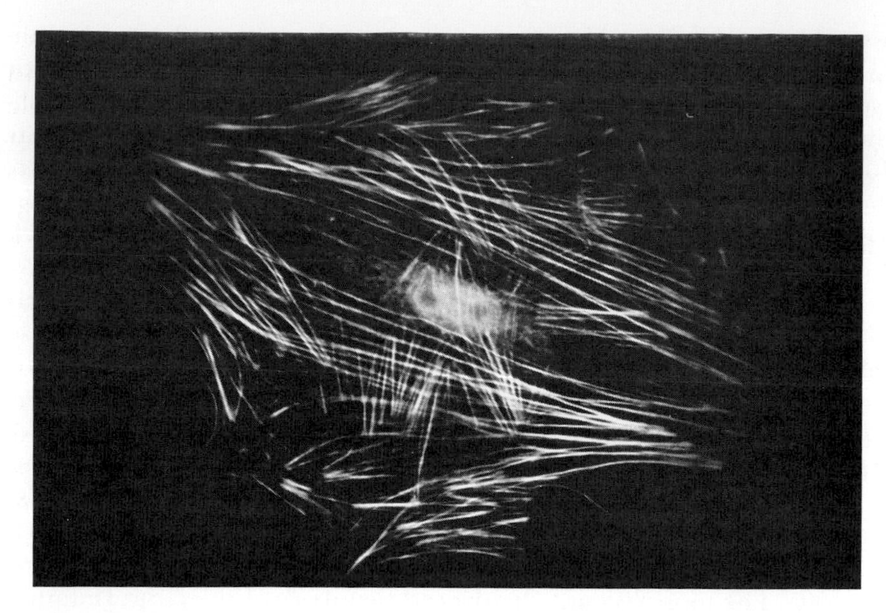

Figure 36-35
Immunofluorescence micrograph of a resting cell in a culture dish that was stained with a fluorescent antibody to actin. Long bundles of microfilaments are evident. [Courtesy of Dr. Elias Lazarides.]

fluorescein and then added to cultured cells. Fluorescence micrographs of these cells show a remarkable array of filaments, some spanning the entire length of the cell (Figure 36-35).

Actin filaments in nonmuscle cells are transient structures, continually dissolving and reforming. They are elongated by the stepwise addition of actin monomers to exposed ends. Polymerization is a cooperative process; filaments can form when the concentration of actin exceeds a value known as the *critical concentration*. Actin contains a single bound nucleotide, either ATP or ADP, and catalyzes the slow hydrolysis of bound ATP to ADP. ATP-actin has a stronger tendency to polymerize than does ADP-actin; their critical concentrations are about 0.1 and 2 μM under cellular conditions. The typical concentration of nonfilamentous actin in a nonmuscle cell is 200 μM, much higher than the critical concentration. However, a significant proportion of the actin is not polymerized because of the presence of control proteins that bind actin. Indeed, more than fifty actin-binding proteins have been identified (Table 36-1). Some bind actin monomers, whereas others interact with the ends of filaments or along their sides. Yet another category are proteins that crosslink actin filaments to each other. In addition, some proteins crosslink actin filaments to other structures. For example, actin filaments interact with spectrin in the membrane skeleton that enables erythrocytes to resist strong shearing forces (p. 306).

Cytochalasin B, an alkaloid from fungi, alters the shape of eucaryotic cells and inhibits many of their movements. For example, the ruffling of fibroblasts, the outgrowth of axons from ganglia, the retraction of blood clots by platelets, and the division of fertilized sea urchin eggs are inhibited by this alkaloid. Electron-microscopic studies suggested that cytochalasin acts on microfilaments because they disappear from cells treated with this alkaloid. In fact, *cytochalasin interferes with the assembly of actin filaments by capping one of their ends*. The importance of continual

Table 36-1
Actin-binding proteins

Proteins that bind actin monomers
Profilin
DNase I
Barbed-end capping proteins
Fragmin
Gelsolin
Severin
Villin
Pointed-end capping proteins
Acumentin
β-Actinin
Spectrin and band 4.1
Proteins that bundle and crosslink actin filaments
Fascin
Fimbrin
Villin
Filamin
Spectrin
Fodrin

Cytochalasin B

assembly and disassembly of microfilaments in cell motility is also underscored by studies of the mechanism of action of *phalloidin*. This toxic cyclic peptide comes from *Amanita phalloides,* a highly poisonous mushroom that also contains α-amanitin, an inhibitor of RNA polymerase (p. 703). Phalloidin, like cytochalasin, blocks cell movements that require the turnover of microfilaments. *Phalloidin binds to actin units in microfilaments and prevents their depolymerization.* Thus, microfilaments are in effect locked by phalloidin.

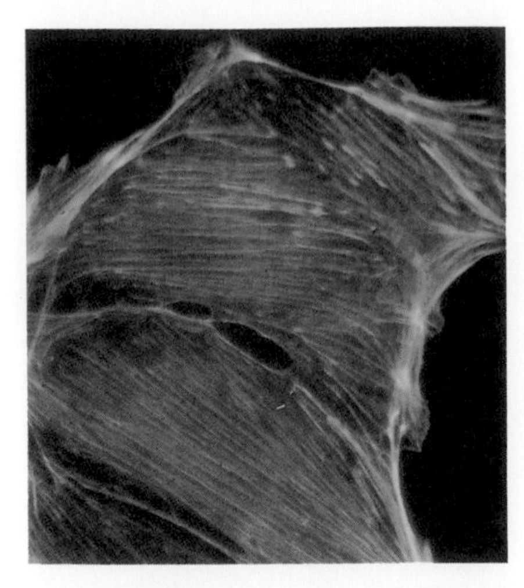

Figure 36-36
Fluorescence micrograph of microfilaments stained with fluorescent-labeled phalloidin. [Courtesy of Dr. Watt Webb and Dr. William Carley.]

Animal cells, but not other eucaryotic cells, also contain intermediate filaments, which range in diameter from 70 to 110 Å. *Intermediate filament (IF) proteins are encoded by a large multigene family.* The common structural motif of IF proteins is a *two- or three-stranded α-helical coiled-coil core* some 300 residues long. The regions on either side of this conserved core are diverse. *Keratins*, one of several types of IF, contribute to the mechanical stability of sheets of epithelial cells. Keratins are also major proteins of skin and hair. Neurons contain *neurofilaments*, muscle cells contain *desmin* filaments, and many kinds of cells contain *vimentin filaments*. The most recently discovered members of the large IF family are *lamins*. They form the nuclear lamina, a fibrous meshwork that lies on the nuclear side of endoplasmic reticulum membrane surrounding the nucleus.

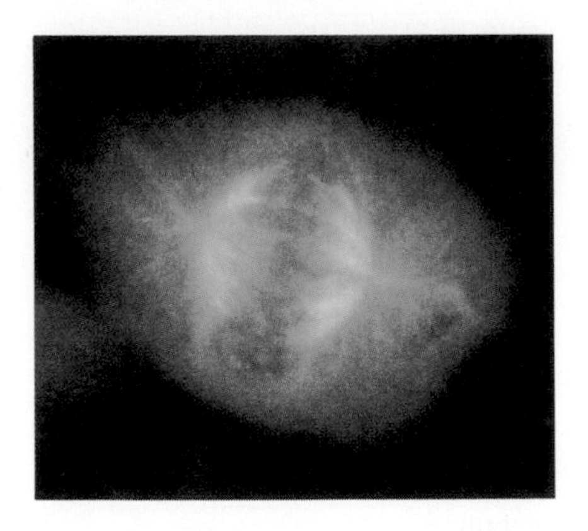

Figure 36-37
Fluorescence micrograph of microtubules in the mitotic spindle of a mouse fibroblast. A fluorescent-labeled antitubulin antibody was used to display these filaments. [Courtesy of Dr. Marc Kirschner.]

THE BEATING OF CILIA AND FLAGELLA IS PRODUCED BY THE DYNEIN-INDUCED SLIDING OF MICROTUBULES

We turn now to *microtubules*, the third class of cytoskeletal elements. Microtubules have multiple architectural and contractile roles in nearly all eucaryotic cells. They are hollow cylindrical structures built from two kinds of similar 50-kd subunits, α- and *β-tubulin*. Their outer diameter of about 300 Å clearly distinguishes them from microfilaments and intermediate filaments. The rigid wall of a microtubule is made of a helical array of alternating α- and β-tubulin subunits (Figure 36-38). A

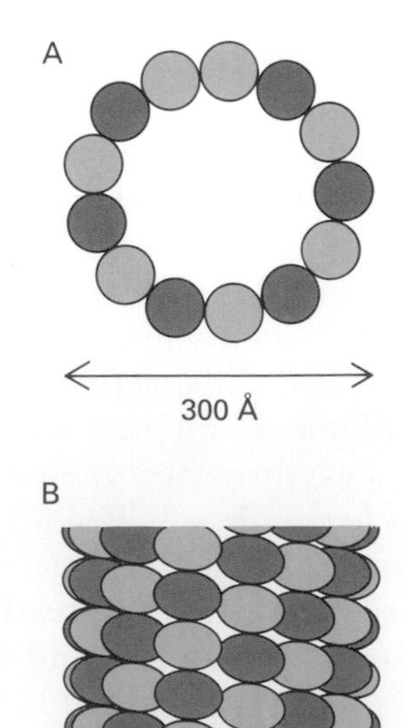

Figure 36-38
Schematic diagram showing the helical pattern of tubulin subunits in a microtubule: (A) cross-sectional view showing the arrangement of thirteen protofilaments; (B) surface lattice of α and β subunits. [After J. A. Snyder and J. R. McIntosh. *Ann. Rev. Biochem.* 45(1976):706.]

microtubule can be regarded as being made of thirteen protofilaments that run parallel to its long axis (Figure 36-39).

Microtubules are major components of eucaryotic cilia (short) and flagella (long). These hairlike organelles protrude from the surface of many cells. Cilia act as oars to move a stream of liquid parallel to a stationary cell's surface. For example, the coordinated beating of cilia on cells lining the respiratory passages serves to sweep out foreign particles. Free cells such as sperm and protozoa are propelled by either cilia or flagella. Electron-microscopic studies have revealed that cilia and flagella from nearly all eucaryotes have the same fundamental design: a bundle of fibers called an *axoneme* is surrounded by a membrane that is continuous with the plasma membrane. In fact, the fibers in an axoneme are microtubules. A peripheral group of nine pairs of microtubules surrounds two singlet microtubules (Figure 36-40). This recurring motif is known as a *9 + 2 array.*

A schematic diagram of the structure of an axoneme is shown in Figure 36-41. Each of the nine outer doublets looks like a figure eight, 370 Å × 250 Å. The smaller of the pair, *subfiber A*, is joined to a central

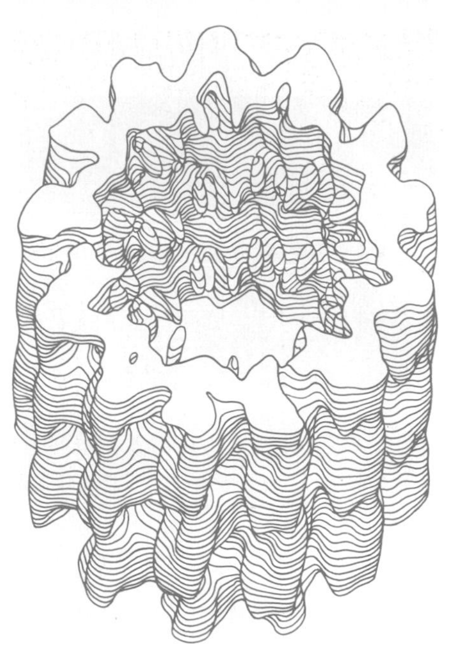

Figure 36-39
Image of a microtubule obtained from x-ray diffraction analysis of fibers. The resolution of the analysis was 18 Å. [After a drawing kindly provided by Dr. Lorena Beese, Dr. Gerald Stubbs, and Dr. Carolyn Cohen.]

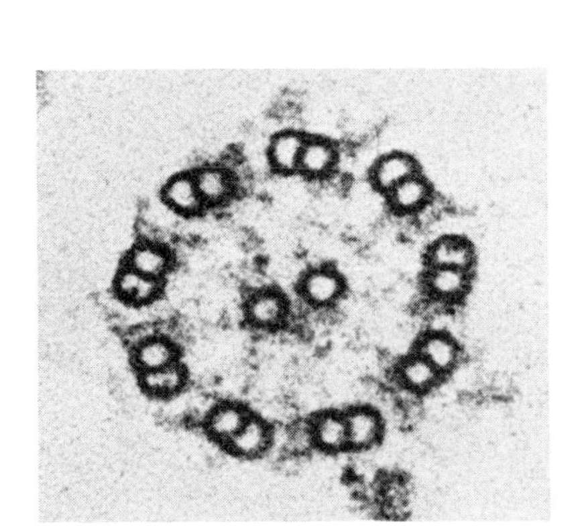

Figure 36-40
Electron micrograph of a cross section of a flagellar axoneme. Nine outer microtubule doublets surround two singlets. [Courtesy of Dr. Joel Rosenbaum.]

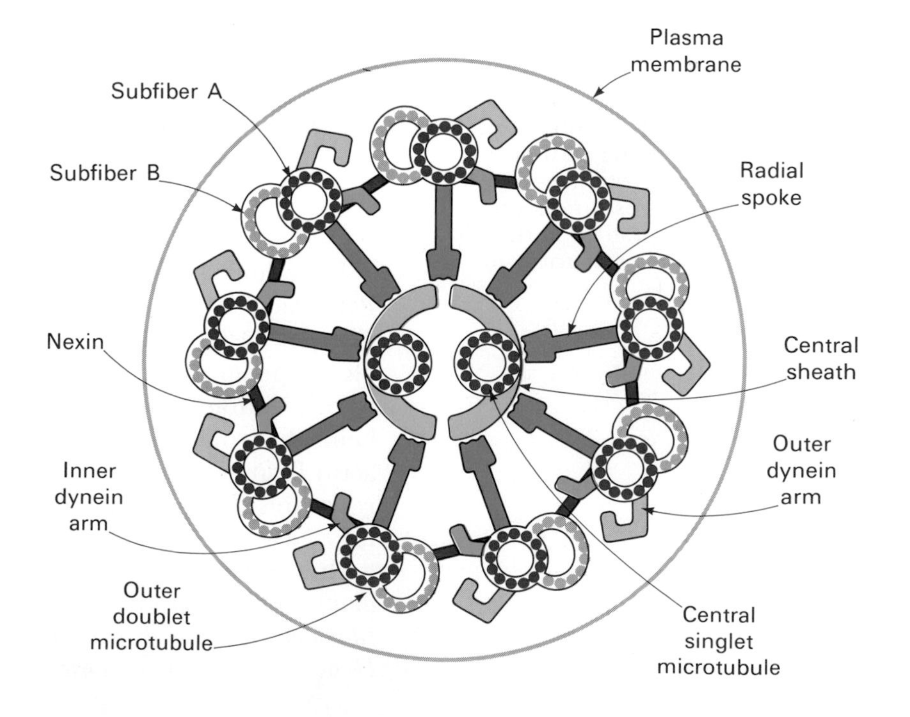

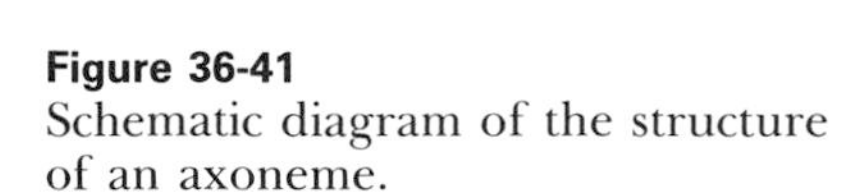

Figure 36-41
Schematic diagram of the structure of an axoneme.

sheath of the cilium by radial spokes. Microtubule doublets are also held together by *nexin* links. Two *arms* emerge from each subfiber A. All the arms in a given cilium point in the same direction. Treatment of cilia with detergent and then with salt at high concentration removes the plasma membrane surrounding them and solubilizes an ATPase called *dynein.* The outer fibers retain their ninefold cylindrical arrangement but are devoid of arms. They can be restored by the addition of dynein under suitable ionic conditions. Hence, *the arms of subfibers A are dyneins with ATPase activity.*

Dynein is a very large protein (~1500 kd) containing three heads and stalks (Figure 36-43). The ATPase sites are located in the heads, which act as crossbridges, much like the S1 heads of myosin. The ATPase cycle of dynein closely resembles that of myosin. The binding of ATP to dynein releases it from the B subfiber. The hydrolysis of bound ATP yields dynein-ADP-P_i, which binds again to the B subfiber. The association of dynein with the B subfiber induces the release of P_i, as occurs in muscle when S1 attaches to actin. The power stroke probably occurs in this step.

Figure 36-42
Freeze-fracture electron micrograph of microtubules in the motile axostyle (central supporting rod) of a protozoan. The microtubules are linked by oblique cross-bridges that resemble the dynein arms in the cilia of higher organisms. [Courtesy of Dr. John Heuser.]

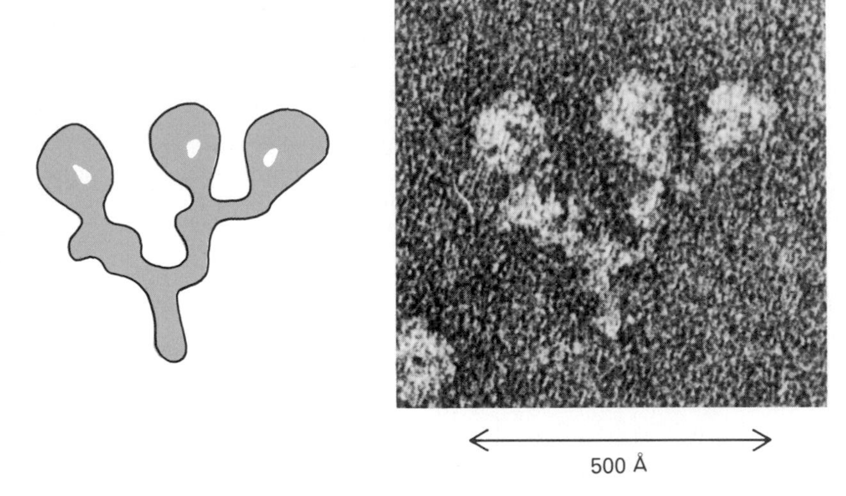

Figure 36-43
Electron micrograph and interpretive diagram of a dynein molecule. The three heads of this ATPase interact with an adjacent microtubule to generate force. [Courtesy of Dr. Yoko Yano Toyoshima.]

How does the ATPase cycle of dynein cause cilia and flagella to beat? Peter Satir and Ian Gibbons showed that the *outer doublets of the axoneme slide past each other to produce bending.* The force between adjacent doublets is generated by *dynein cross-bridges.* The dynein arms on subfiber A of one doublet walk along subfiber B of an adjacent doublet as ATP is being hydrolyzed, as in the movement of myosin cross-bridges on an actin filament in skeletal muscle. In an intact cilium, the radial spokes resist this sliding motion, which instead is converted into a local bending. Nexin, a highly extensible protein, keeps adjacent doublets together during the sliding process. *An important difference between muscle and cilia is that myosin forms bipolar filaments that crosslink antiparallel actin filaments, whereas dynein drives the sliding of parallel adjacent microtubules. Hence, the sarcomere shortens but the axoneme bends.*

Defective cilia have been found in a group of patients with chronic pulmonary disorders. Bjorn Afzelius showed that the cilia in their respiratory tracts are immotile. Males with this genetic defect are also infertile because their sperm are unable to move. This disease, called the *immotile-cilia syndrome*, arises from a variety of molecular lesions. The most common one is the absence of outer and inner dynein arms. Other defects producing this disease are the absence of spokes, nexin links, and central microtubules. An intriguing finding is that many patients with immotile cilia have a complete left-right reversal of internal organs (situs inversus), which suggests that microtubular arrays play a critical role in establishing left-right asymmetry early in embryogenesis.

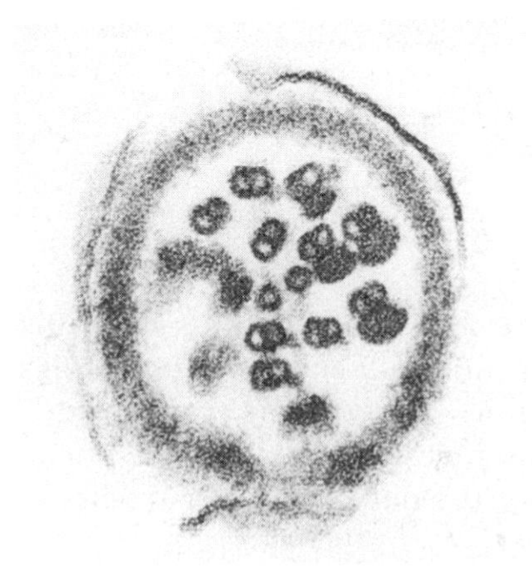

Figure 36-44
Electron micrograph of a cilium from a patient with the immotile-cilia syndrome. The arrangement of microtubules in this cilium is disordered. [Courtesy of Dr. Bjorn Afzelius.]

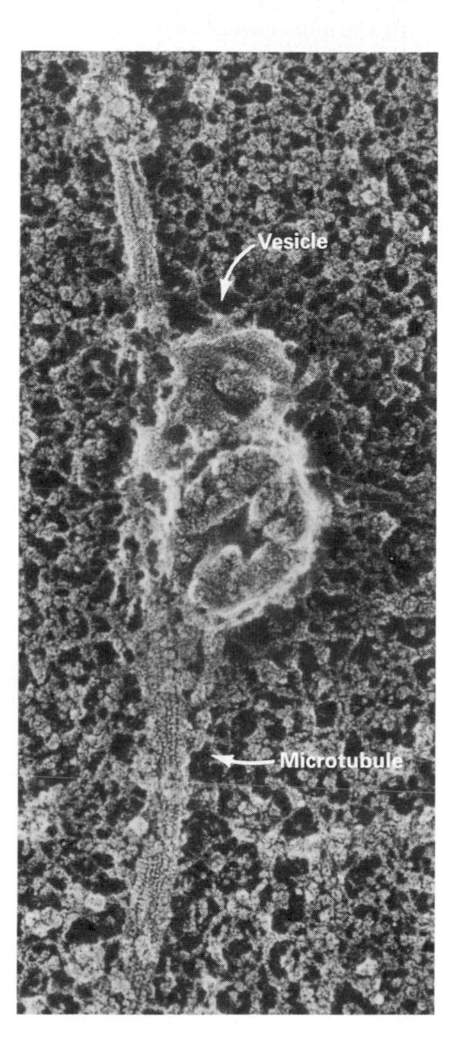

Figure 36-45
Freeze-fracture electron micrograph of a vesicle on a microtubule. [Courtesy of Dr. Ronald Vale.]

KINESIN MOVES VESICLES AND ORGANELLES UNIDIRECTIONALLY ALONG MICROTUBULE TRACKS

Microtubules also participate in the movements of vesicles and organelles, which are found in nearly all eucaryotic cells. The development of video-enhanced contrast microscopy has made it feasible to view the motions of 300-Å vesicles in vivo (Figure 36-45). Vesicle transport is most striking in neurons, where it occurs rapidly over long distances. Vesicles move from the cell body to the nerve terminal at rates of up to 5 μm/s, enabling them to traverse a distance of a meter in about a day. It was found that vesicles also move rapidly in extruded axoplasm (the cytoplasm of nerve axons). *The tracks for these movements are single microtubules.*

The development of a reconstituted in vitro system that transports vesicles led to the discovery of a new ATP-driven molecular engine. Vesicles and organelles are transported by *kinesin*, a large water-soluble protein consisting of two 110-kd subunits and one 65- or 70-kd subunit. Kinesin is a highly elongated protein, about 1000 Å long. One end of kinesin binds to a specific receptor in the membrane of a vesicle, and the other end binds to a microtubule (Figure 36-46). Observations of the movements of asymmetric organelles, such as mitochondria, suggest that they walk rather than roll along microtubule tracks. Five kinesin molecules per organelle would provide more than enough force to give a velocity of 2 μm/s. Organelles and vesicles containing kinesin move from the minus end of a microtubule (the one at the centrosome,

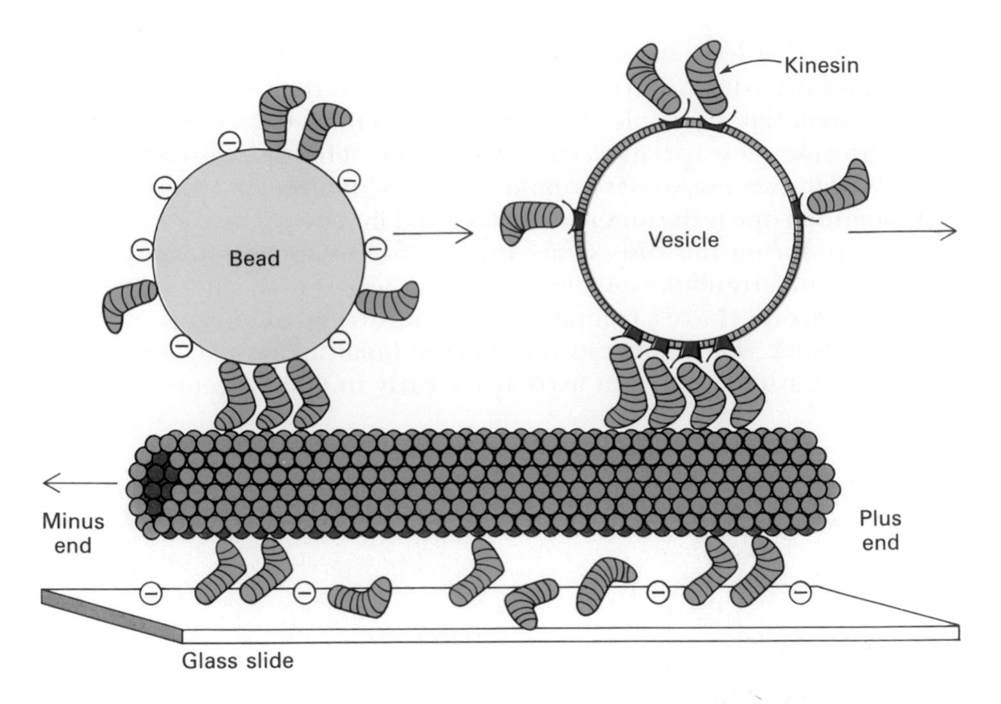

Figure 36-46
Schematic diagram showing movements mediated by kinesin, an ATP-driven motor, on microtubule tracks. Vesicles and beads containing kinesin molecules on their surface move toward the plus end of microtubules. Kinesin is bound to vesicles by receptors (shown in red) in their membrane. Conversely, microtubules can move on glass slides containing bound kinesin. [After a drawing kindly provided by Dr. Ronald Vale.]

where microtubule elongation begins; see p. 945) to the plus end; this usually produces movement from the center of a cell to its periphery (anterograde transport). A different engine (a retrograde transporter), yet to be isolated, drives movements in the reverse direction. The polarity of microtubules serves as a compass for the navigation of particles in the cytoplasm.

Beads coated with kinesin move in the same direction as do vesicles (Figure 36-46). Kinesin bound to glass coverslips also drives the movement of microtubule filaments. These actions of kinesin are like those of the S1 heads of myosin (p. 931). The ATPase activity of kinesin is stimulated 50-fold by the addition of microtubules, just as the ATPase activities of myosin and dynein are enhanced by actin and microfilaments. *ATP hydrolysis by kinesin is tightly coupled to its binding to a microtubule and to the generation of force.* Kinesin differs from myosin and dynein in that ATP promotes its binding to, rather than its release from, its partner protein. ATP is hydrolyzed by kinesin while it is attached to a microtubule; hydrolysis enables kinesin to be released so that it can take another step in the walk along the microtubule. In this respect, kinesin resembles elongation factor Tu, which stays attached to aminoacyl-tRNA until its bound nucleoside triphosphate is hydrolyzed (p. 755). *A motor can have either higher or lower affinity for its partner protein when bound ATP has been hydrolyzed; in either case, the key to energy transduction is the large change in affinity of the pair of proteins during the ATP-ADP cycle.*

THE RAPID GTP-DRIVEN ASSEMBLY AND DISASSEMBLY OF MICROTUBULES IS CENTRAL TO MORPHOGENESIS

Microtubules are also important in determining the shapes of cells and in separating daughter chromosomes in mitosis (Figure 36-47). Microtubules in cells are formed by the addition of α and β tubulin molecules to preexisiting filaments or nucleation centers. De novo polymerization is very slow because multiple subunits must come together simultaneously to form the thirteen-protofilament helix. Instead, microtubules in cells radiate from *centrosomes*, which lie on the periphery of centrioles, and from the *poles* of mitotic spindles. These sites of the initiation of microtubule growth are called *microtubule-organizing centers* (MTOC). The minus end of a microtubule is the one bound to an MTOC, and the plus end is the free one. *Colchicine,* an alkaloid from the autumn crocus, inhibits cell processes that depend on functioning microtubules by blocking their polymerization. For example, dividing cells are arrested at metaphase by the action of colchicine because microtubules are essential for moving chromosomes. Colchicine also inhibits the movement of vesicles along microtubule tracks; hence, many secretory processes are blocked by this compound. Colchicine has been used for several centuries to treat acute attacks of gout.

H_3CO H_3CO H_3CO N—C—CH_3 H O O H_3CO

Colchicine

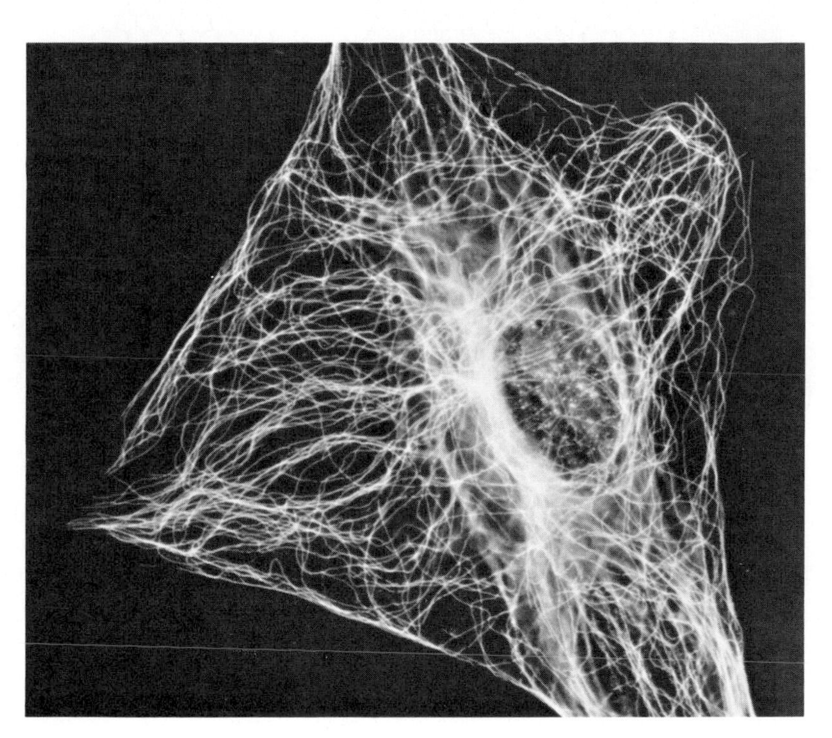

Figure 36-47
Immunofluorescence micrograph showing the distribution of microtubules in a fibroblast. [Courtesy of Dr. Klaus Weber.]

Most microtubules undergo rapid assembly and disassembly. Fluorescent tubulins injected into cells become incorporated into mitotic spindles in about 15 seconds, and into other microtubules in several minutes. The hydrolysis of GTP bound to tubulin is at the heart of the rapid turnover of microtubules. The critical concentration for the formation of microtubules is much lower for GTP-tubulin than for GDP-tubulin. GTP-tubulin adds to the plus end of microtubules. The bound nucleotide is hydrolyzed seconds later; tubulin is a GTPase with a low turnover number. A GDP-tubulin unit located within a microtubule stays there but one situated at an exposed end dissociates from the filament. Hence, microtubules become longer when the rate of addition of GTP-tubulin to the plus end is faster than the rate of hydrolysis of bound GTP. Marc Kirschner and Tim Mitchison found that some microtubules in a population of filaments lengthen while others simultaneously shorten. This property, called *dynamic instability,* arises from random fluctuations in whether the plus end contains GTP-tubulin or GDP-tubulin (Figure 36-48).

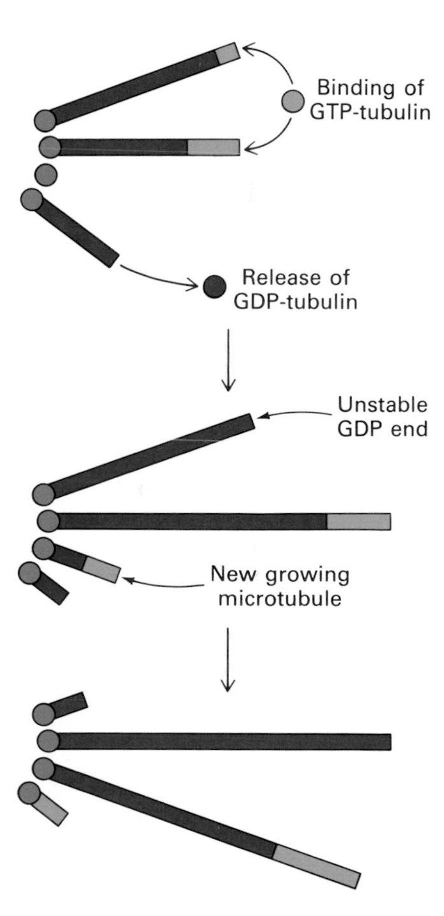

Figure 36-48
Some microtubules grow and others simultaneously regress because their ends are different. Microtubules with GDP-tubulin (red) at their ends are unstable, in contrast with those having GTP-tubulin (green) at their ends. Tubulin has an inherent GTPase activity that converts bound GTP to GDP. Nucleation sites, which bind the minus ends of microtubules, are shown in blue.

The dynamic instability of microtubules is exploited in cell development. The formation of the mitotic spindle is an instructive example. Microtubules join each pole of the mitotic spindle to kinetochores that are located at the centromeres of daughter chromosomes (Figure 36-49). The nucleation sites at the two poles do not send out microtubules precisely aimed at kinetochore targets. Instead, hundreds of randomly oriented microtubules emanate from the poles. Microtubules reaching kinetochores are stabilized, whereas the other microtubules fall apart because they have an exposed plus end. Thus, *dynamic instability produces a large repertoire of structures; those engaged in constructive interactions become stabilized.* We see here, as in the operations of the immune system (p. 891), *the importance of selection in generating pattern out of random variation.*

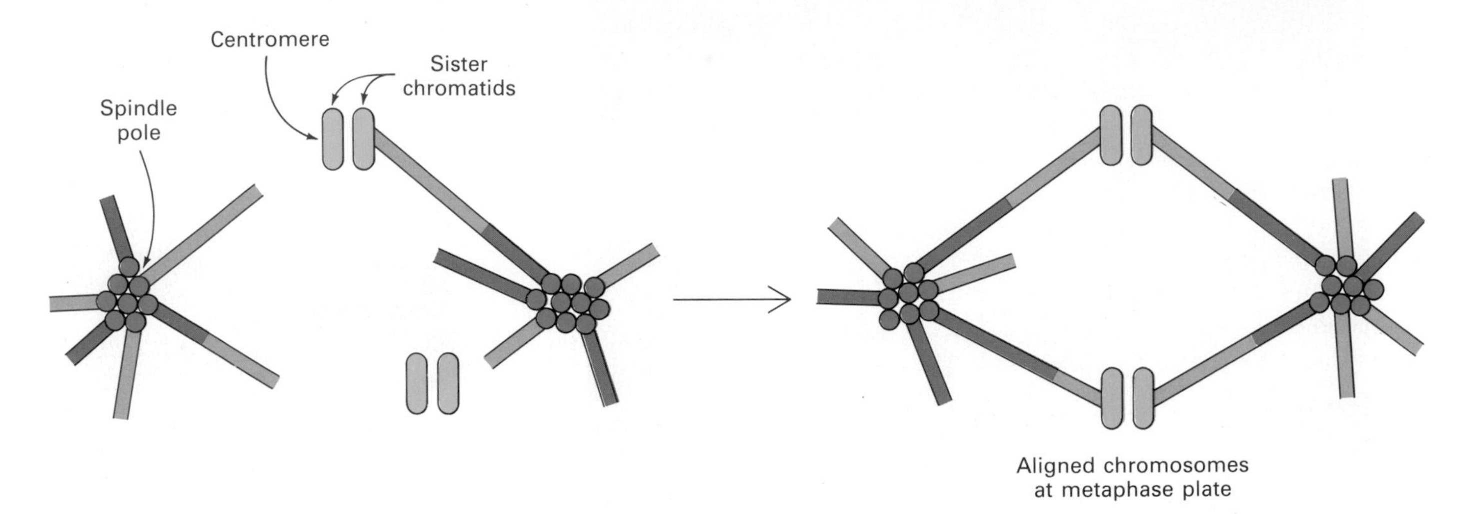

Figure 36-49
The dynamic instability of microtubules is exploited in the formation of the mitotic spindle. Microtubule growth in many directions is nucleated from a pole of the spindle (blue). Microtubules reaching kinetochores (yellow) at the centromeres of chromosomes are stabilized. The other microtubules regress when spontaneous fluctuations give rise to GDP ends (red) as a result of hydrolysis of bound GTP (green). [After M. Kirschner and T. Mitchison. *Cell* 45(1986):339.]

SUMMARY

Vertebrate striated muscle consists of two kinds of interacting protein filaments. The thick filaments contain myosin, whereas the thin filaments contain actin, tropomyosin, and troponin. The hydrolysis of ATP bound to myosin drives the sliding of these filaments past each other. Myosin is a large protein (540 kd) composed of two heavy chains and four light chains. The heavy chains are folded to form two S1 heads joined to a double-stranded α-helical rod; two light chains are bound to each S1 head. The S1 heads and part of the rod form cross-bridges that interact with actin to generate the contractile force. The remainder of the myosin molecule forms the backbone of the thick filament. Actin is a globular protein (42 kd) that polymerizes to form thin filaments. Both thin and thick filaments have direction, which reverses half-way between the Z lines. ATP binding and hydrolysis drives the cyclic formation and dissociation of complexes between the myosin cross-bridges of thick filaments and the actin units of thin filaments, which shortens the distance between Z lines. The power stroke of muscle contraction comes from conformational changes within the S1 head, which alter its affinity for actin. Hinges between domains of myosin are important in enabling S1 heads to reversibly attach and detach from actin and to change their orientation when bound.

The contraction of skeletal muscle is regulated by Ca^{2+}. The interaction of actin and myosin is inhibited by the troponin complex and tropomyosin when the level of Ca^{2+} is low. Nerve excitation triggers the release of Ca^{2+} from the sarcoplasmic reticulum. The binding of Ca^{2+} to troponin C, a member of a family of calcium-sensing proteins, alters the interaction of tropomyosin with actin to allow myosin to bind actin and generate the contractile force. Vertebrate smooth muscle contracts when its light chains become phosphorylated by a kinase that is activated by Ca^{2+}-calmodulin. Molluscan smooth muscle is directly activated by the binding of Ca^{2+} to the light chains of myosin.

Actin and myosin are ancient proteins, as shown by their presence in even the simplest eucaryotes, such as yeast and slime molds. In fact, these proteins have contractile roles in nearly all eucaryotic cells. Actin, which is particularly abundant, forms microfilaments 70 Å in diameter. They participate in a wide range of cellular movements, as exemplified by cell migration in development, clot retraction by blood platelets, and the movement of macrophages to injured tissues. Cytochalasin and phalloidin inhibit cell movements requiring the turnover of actin filaments. The dynamic cytoskeletons of eucaryotic cells also contain intermediate filaments (~100 Å in diameter) and microtubules (~300 Å in diameter). Intermediate filaments are formed from proteins that have a common α-helical coiled-coil core but are otherwise diverse.

Microtubules have multiple architectural and contractile roles in nearly all eucaryotic cells. These 300-Å-diameter hollow fibers are built from globular α and β tubulin subunits. Colchicine inhibits movements mediated by microtubules by blocking their polymerization. A eucaryotic cilium or flagellum contains nine microtubule doublets around two singlets. The outer doublets are cross-linked by dynein, an ATPase. The dynein-induced sliding of adjacent microtubule doublets produces a local bending of the cilium and causes it to beat. Microtubules also serve as tracks for the movements of vesicles and organelles. These transport processes are carried out by kinesin, a different ATP-driven motor. The rapid GTP-driven assembly and disassembly of microtubules is central to their actions in cell development, as exemplified by the formation of the mitotic spindle. Microtubules spontaneously disassemble if they do not reach a target that stabilizes their ends.

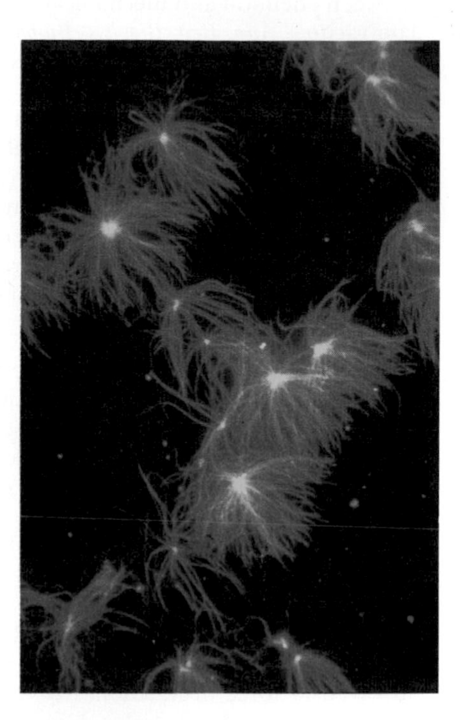

Figure 36-50
Fluorescence micrograph showing the in vivo nucleation of microtubules from centrosomes. [Courtesy of Dr. Tim Mitchison and Dr. Marc Kirschner.]

SELECTED READINGS

Where to start

Eisenberg, E., and Hill, T. L., 1985. Muscle contraction and free energy transduction in biological systems. *Science* 227:999–1006.

Weber, K., and Osborn, M., 1985. The molecules of the cell matrix. *Sci. Amer.* 253(4):110–120.

Dustin, P., 1980. Microtubules. *Sci. Amer.* 243(2):66–76.

Vale, R. D., Scholey, J. M., and Sheetz, M. P., 1986. Kinesin: possible roles for a new microtubule motor. *Trends Biochem. Sci.* 11:464–468.

Kirschner, M., and Mitchison, T., 1986. Beyond self-assembly: from microtubules to morphogenesis. *Cell* 45:329–342.

Books

Bagshaw, C. R., 1982. *Muscle Contraction*. Chapman and Hall. [A concise and informative introduction.]

Squire, J. M., 1986. *Muscle: Design, Diversity, and Disease*. Benjamin/Cummings.

Pollack, G. H., and Sugi, H., (eds.), 1984. *Contractile Mechanisms in Muscle*. Plenum.

Hatano, S., Ishikawa, H., and Sato, H., (eds.), 1979. *Cell Motility: Molecules and Organization*. University Park Press.

Myosin and actin

Warrick, H. M., and Spudich, J. A., 1987. Myosin: structure and function in cell motility. *Ann. Rev. Cell Biol.* 3:379–422.

McLachlan, A. D., 1984. Structural implications of the myosin amino acid sequence. *Ann. Rev. Biophys. Bioeng.* 13:167–189.

Milligan, R. A., and Flicker, P. F., 1987. Structural relationships of actin, myosin, and tropomyosin revealed by cryo-electron microscopy. *J. Cell. Biol.* 105:29–39.

Kabsch, W., Mannherz, H. G., and Suck, D., 1985. Three-dimensional structure of the complex of actin and DNase I at 4.5 Å resolution. *EMBO J.* 4:2113–2118.

Generation of contractile force

Hibberd, M. G., and Trentham, D. R., 1986. Relationships between chemical and mechanical events during muscular contraction. *Ann. Rev. Biophys. Biophys. Chem.* 15:119.

Cooke, R., 1986. The mechanism of muscle contraction. *CRC Crit. Rev. Biochem.* 21:53–118.

Thomas, D. D., 1987. Spectroscopic probes of muscle cross-bridge rotation. *Ann. Rev. Physiol.* 49:691–709.

Kress, M., Huxley, H. E., Faruqi, A. R., and Hendrix, J., 1986. Structural changes during activation of frog muscle studied by time-resolved x-ray diffraction. *J. Mol. Biol.* 188:325–342.

In vitro motility systems

Sheetz, M. P., and Spudich, J. A., 1983. Movement of myosin-coated fluorescent beads on actin cables in vitro. *Nature* 303:31–35.

Honda, H., Nagashima, H., and Asakura, S., 1986. Directional movement of F-actin in vitro. *J. Mol. Biol.* 191:131–133.

Kron, S. J., and Spudich, J. A., 1986. Fluorescent actin filaments move on myosin fixed to a glass surface. *Proc. Nat. Acad. Sci.* 83:6272–6276.

Hynes, T. R., Block, S. M., White, B. T., and Spudich, J. A., 1987. Movement of myosin fragments in vitro: domains involved in force production. *Cell* 48:953–963.

Control of muscle contraction

Cohen, C., 1975. The protein switch of muscle contraction. *Sci. Amer.* 233(5):36–45.

Sundaralingam, M., Bergstrom, R., Strasburg, G., Rao, S. T., Roychowdhury, P., Greaser, M., and Wang, B. C., 1985. Molecular structure of troponin C from chicken skeletal muscle at 3-angstrom resolution. *Science* 227:945–948.

Herzberg, O., Moult, J., and James, M. N. G., 1985. Calcium binding to skeletal muscle troponin C and the regulation of muscle contraction. *Ciba Found. Symp.* 122:120–144.

Reinach, F. C., Nagai, K., and Kendrick-Jones, J., 1986. Site-directed mutagenesis of the regulatory light-chain Ca^{2+}/Mg^{2+} binding site and its role in hybrid myosins. *Nature* 322:80–83.

White, S. P., Cohen, C., and Phillips, G. N., Jr., 1987. Structure of co-crystals of tropomyosin and troponin. *Nature* 325:826–828.

Actin in nonmuscle cells

Pollard, T. D., and Cooper, J. A., 1986. Actin and actin-binding proteins. A critical evaluation of mechanisms and functions. *Ann. Rev. Biochem.* 55:987.

Stossel, T. P., Chaponnier, C., Ezzell, R. M., Hartwig, J. H., Janmey, P. A., Kwiatkowski, D. J., Lind, S. E., Smith, D. B., Southwick, F. S., Yin, H. L., and Zaner, K. S., 1985. Non-muscle actin-binding proteins. *Ann. Rev. Cell Biol.* 1:353–402.

Weeds, A., 1982. Actin-binding proteins—regulators of cell architecture and motility. *Nature* 296:811–816.

Microtubules

Johnson, K. A., 1985. Pathway of the microtubule-dynein ATPase and the structure of dynein: a comparison with actomyosin. *Ann. Rev. Biophys. Biophys. Chem.* 14:161–188.

Goodenough, U. W., and Heuser, J. E., 1985. Outer and inner dynein arms of cilia and flagella. *Cell* 41:341–342.

Afzelius, B., 1985. The immotile-cilia syndrome: a microtubule-associated defect. *CRC Crit. Rev. Biochem.* 19:63–87.

Vale, R. D., Schnapp, B. J., Reese, T. S., and Sheetz, M. P., 1985. Organelle, bead, and microtubule translocations promoted by soluble factors from the squid giant axon. *Cell* 40:559–569.

Schnapp, B. J., Vale, R. D., Sheetz, M. P., and Reese, T. S., 1985. Single microtubules from squid axoplasm support bidirectional movement of organelles. *Cell* 40:455–462.

Horio, T., and Hotani, H., 1986. Visualization of the dynamic instability of individual microtubules by dark-field microscopy. *Nature* 321:605–607.

Intermediate filaments

Steinert, P. M., and Parry, D. A. D., 1985. Intermediate filaments. *Ann. Rev. Cell Biol.* 1:41–65.

Weber, K., and Geisler, N., 1985. Intermediate filaments: structural conservation and divergence. *Ann. N. Y. Acad. Sci.* 455:649–668.

Wang, K., 1986. Sarcomere-associated cytoskeletal lattices in striated muscle. *Cell Muscle Motility* 6:315–369.

Development of major concepts

Huxley, H. E., 1965. The mechanism of muscular contraction. *Sci. Amer.* 213(6):18–27.

Huxley, H. E., 1971. The structural basis of muscle contraction. *Proc. Roy. Soc. London* (B) 178:131–149.

Huxley, A. F., 1975. The origin of force in skeletal muscle. *Ciba Found. Symp.* 31:271–290.

Summers, K. E., and Gibbons, I. R., 1971. ATP-induced sliding of tubules in trypsin-treated flagella of sea-urchin sperm. *Proc. Nat. Acad. Sci.* 68:3092–3096. [These ingenious experiments revealed that the bending waves of normal flagella are the result of ATP-induced shearing forces between adjacent doublet-tubules.]

CHAPTER 37

Membrane Transport

The permeability of biological membranes is highly selective. The flow of molecules and ions between a cell and its environment is precisely regulated by specific transport systems. These systems have several important roles:

1. They *regulate cell volume* and *maintain the intracellular pH and ionic composition within a narrow range* to provide a favorable environment for enzyme activity.

2. They extract and *concentrate metabolic fuels and building blocks* from the environment and *extrude toxic substances.*

3. They generate ionic gradients that are essential for the *excitability* of nerve and muscle.

The molecular mechanisms of many transport processes are now being elucidated. This chapter considers several bacterial and animal transport systems that carry ions, sugars, and amino acids across biological membranes. Transport antibiotics produced by microorganisms are also discussed because analyses of their structures shed light on how transport systems discriminate between similar ions such as Na^+ and K^+. The last part of this chapter deals with channels that connect the interiors of apposed cells. These conduits (Figure 37-1) play an important role in intercellular communication and transfer of metabolites.

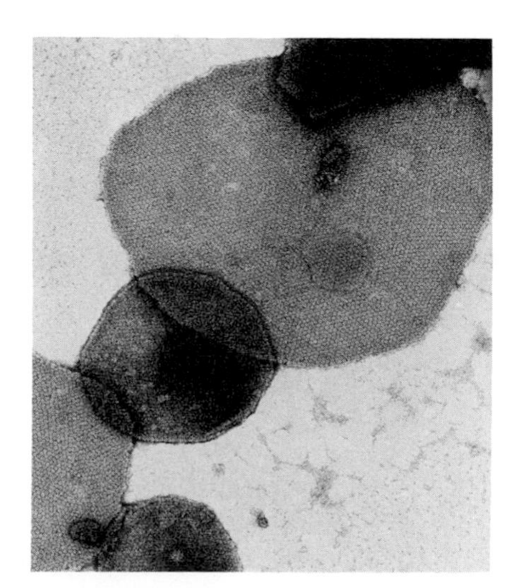

Figure 37-1
Electron micrograph of negatively stained junctions from liver cells. These 15-Å-diameter channels allow ions and small molecules to flow between the interiors of adjoining cells. [From E. L. Hertzberg and N. B. Gilula, *J. Biol. Chem.* 254(1979):2143.]

DISTINCTION BETWEEN PASSIVE AND ACTIVE TRANSPORT

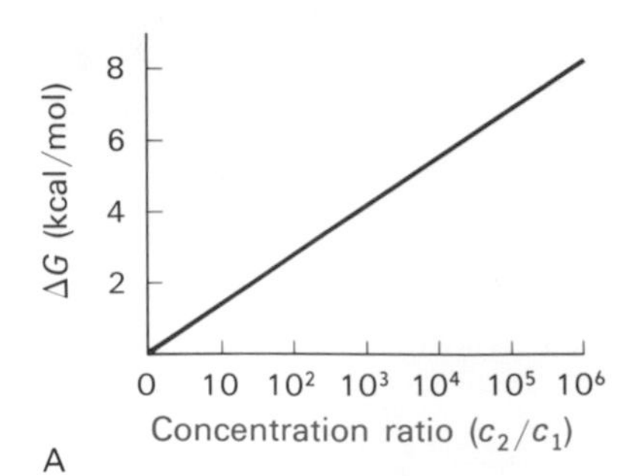

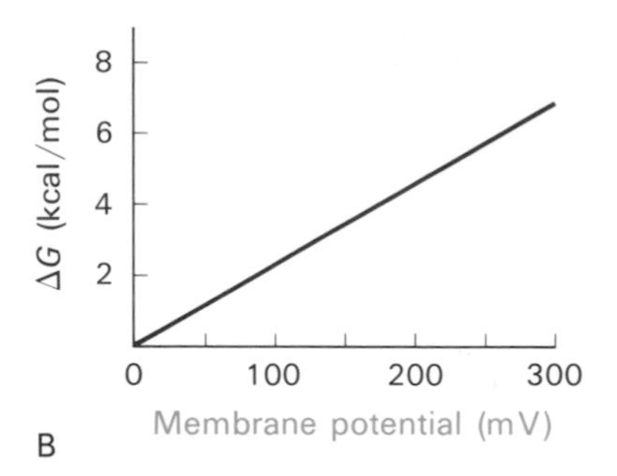

Figure 37-2
Free-energy change in transporting (A) an uncharged solute from a compartment at concentration c_1 to one at c_2 and (B) a singly charged species across a membrane to the side having the same charge as that of the transported ion. Note that the free-energy change imposed by a membrane potential of 59 mV is equivalent to that imposed by a concentration ratio of 10 for a singly charged ion at 25°C.

Whether a transport process is passive or active depends on the change in free energy of the transported species. Consider an uncharged solute molecule. The free-energy change in transporting this species from side 1 where it is present at a concentration of c_1 to side 2 where it is present at concentration c_2 is

$$\Delta G = RT \log_e \frac{c_2}{c_1} = 2.303\ RT \log_{10} \frac{c_2}{c_1}$$

For a charged species, the electrical potential across the membrane must also be considered. The sum of the concentration and electrical terms is called the *electrochemical potential.* The free-energy change is then given by

$$\Delta G = RT \log_e \frac{c_2}{c_1} + ZF\ \Delta V$$

in which Z is the electrical charge of the transported species, ΔV is the potential in volts across the membrane, and F is the faraday (23.062 kcal V^{-1} mol^{-1}).

A transport process must be active when ΔG is positive, whereas it can be passive when ΔG is negative. *Active transport requires a coupled input of free energy, whereas passive transport can occur spontaneously.* For example, consider the transport of an uncharged molecule from $c_1 = 10^{-3}$ mM to $c_2 = 10^{-1}$ mM.

$$\begin{aligned} \Delta G &= 2.303\ RT \log_{10} \frac{10^{-1}}{10^{-3}} \\ &= 2.303 \times 1.99 \times 298 \times 2 \\ &= +2.7 \text{ kcal/mol} \end{aligned}$$

At 25°C (298 K), ΔG is +2.7 kcal/mol, indicating that this transport process is active and hence requires an input of free energy. It could be driven, for example, by the hydrolysis of ATP, which yields −7.3 kcal/mol under standard conditions.

DISCOVERY OF THE ATP-DRIVEN ACTIVE-TRANSPORT SYSTEM FOR SODIUM AND POTASSIUM IONS

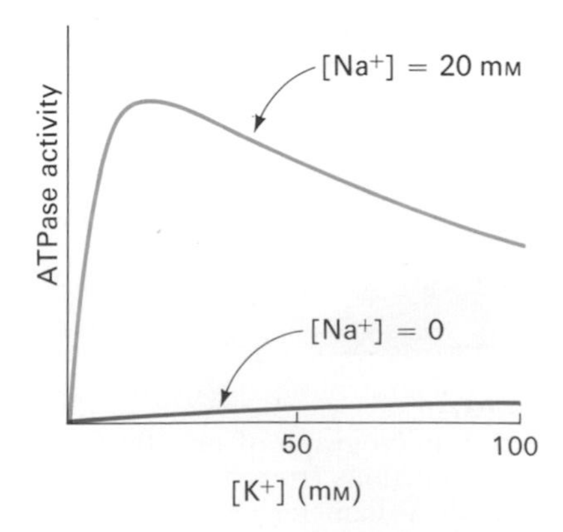

Figure 37-3
The Na^+-K^+ ATPase, a component of the Na^+-K^+ pump, hydrolyzes ATP only if both Na^+ and K^+ are present, in addition to Mg^{2+}.

Most animal cells have a high concentration of K^+ and a low concentration of Na^+ relative to the external medium. These ionic gradients are generated by a specific transport system that is called the *Na^+-K^+ pump* because the movement of these ions is linked. The active transport of Na^+ and K^+ is of great physiological significance. Indeed, more than a third of the ATP consumed by a resting animal is used to pump these ions. The Na^+-K^+ gradient in animal cells controls cell volume, renders nerve and muscle cells electrically excitable, and drives the active transport of sugars and amino acids (p. 958).

In 1957, Jens Skou discovered an enzyme that hydrolyzes ATP only if Na^+ and K^+ are present in addition to Mg^{2+}, which is required by all ATPases (Figure 37-3). Hence, this enzyme was named the *Na^+-K^+ ATPase.*

$$\text{ATP} + \text{H}_2\text{O} \xrightarrow{\text{Na}^+,\ \text{K}^+,\ \text{Mg}^{2+}} \text{ADP} + \text{P}_i + \text{H}^+$$

Skou proposed that *the Na^+-K^+ ATPase is an integral part of the Na^+-K^+ pump and that the splitting of ATP provides the energy needed for the active transport of these cations*. His hypothesis is supported by several lines of evidence:

1. The Na^+-K^+ ATPase is present wherever Na^+ and K^+ are actively transported. The level of enzymatic activity is quantitatively correlated with the extent of ion transport. For example, nerve cells are rich in both the Na^+-K^+ ATPase and the pump, whereas red blood cells have low levels of both.

2. Both the Na^+-K^+ ATPase and the pump are tightly associated with the plasma membrane.

3. Both the Na^+-K^+ ATPase and the pump are oriented in the same way in the plasma membrane.

4. Variations in the concentration of Na^+ and K^+ have parallel effects on the ATPase activity and on the rate of transport of these ions.

5. Both the Na^+-K^+ ATPase and the pump are specifically inhibited by cardiotonic steroids. The concentration of steroids causing half-maximal inhibition is the same for both processes.

BINDING SITES FOR IONS, ATP, AND INHIBITORS ARE ASYMMETRICALLY POSITIONED

Studies of the Na^+-K^+ pump in erythrocyte ghosts have revealed the orientation of this ATPase and pump. In a hypotonic salt solution, an erythrocyte swells and holes appear in its membrane. The inside of the swollen erythrocyte equilibrates with the external medium, so that hemoglobin diffuses out to leave a pale cell (hence, a ghost). If the external medium is then made isotonic, the membrane again becomes a permeability barrier. Thus, the molecular and ionic composition inside a ghost can be controlled by resealing it in an appropriate medium. Transport and enzymatic studies of such erythrocyte ghosts have shown that the Na^+-K^+ pump is oriented in the following way (Figure 37-4):

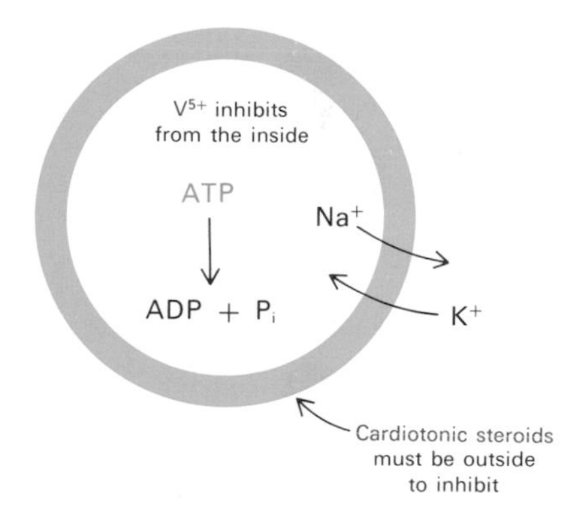

Figure 37-4
The Na^+-K^+ pump is oriented in the plasma membrane.

1. Na^+ must be inside, whereas K^+ must be outside to activate the ATPase and to be transported across the membrane.

2. ATP is an effective substrate for the ATPase and the pump only when this energy source is located inside the cell.

3. Cardiotonic steroids inhibit the pump and the ATPase only when they are located outside the cell.

4. Vanadate inhibits the pump and the ATPase only when it is located inside the cell.

ATP TRANSIENTLY PHOSPHORYLATES THE Na^+-K^+ PUMP

How does ATP drive the active transport of Na^+ and K^+? An important clue was the finding that the ATPase is phosphorylated by ATP only when both Na^+ and Mg^{2+} are present. The site of phosphorylation is the side chain of a specific aspartate residue. This phosphorylated intermediate (E-P) is then hydrolyzed if K^+ is present. Phosphorylation

β-Aspartyl phosphate
(E-P intermediate)

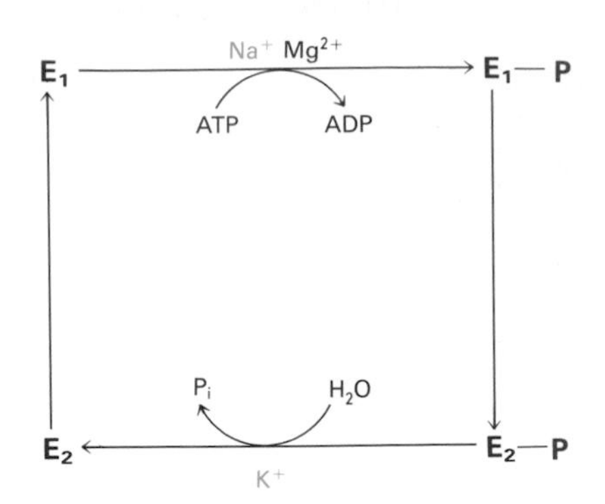

Figure 37-5
Enzymatic cycle of the Na^+-K^+ ATPase.

does not require K^+, whereas dephosphorylation does not require Na^+ or Mg^{2+}.

$$E + ATP \xrightleftharpoons{Na^+, Mg^{2+}} E\text{—}P + ADP$$

$$E\text{—}P + H_2O \xrightarrow{K^+} E + P_i$$

Na^+-dependent phosphorylation and K^+-dependent dephosphorylation are not the only critical reactions. The pump also interconverts between at least two different conformations, denoted by E_1 and E_2. Thus, at least four conformational states—E_1, E_1-P, E_2-P, and E_2—participate in the transport of Na^+ and K^+ and concomitant hydrolysis of ATP (Figure 37-5). *Three Na^+ and two K^+ are transported per ATP hydrolyzed.* Hence, the pump generates an electric current across the plasma membrane. In other words, the Na^+-K^+ATPase pump is *electrogenic*. The maximal turnover number of the ATPase is about 100 s^{-1}.

The ATPase is inhibited by nanomolar concentrations of vanadate ion (VO_4^{3-}). Vanadate is an analog of the transition state in phosphoryl transfer reactions because it readily forms a pentacovalent bipyramidal structure (Figure 37-6) like that of a phosphate ester undergoing hydrolysis (p. 213).

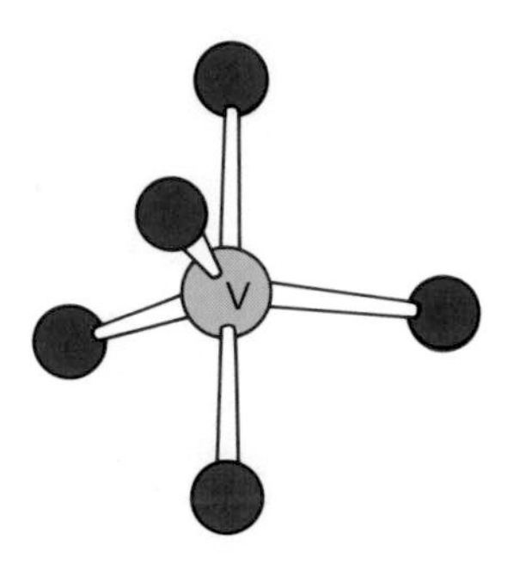

Figure 37-6
Postulated structure of a vanadate ion bound to an ATPase. The bipyramidal array of oxygen atoms around this ion is like that around a phosphorus atom during hydrolysis of a phosphoryl group.

ION TRANSPORT AND ATP HYDROLYSIS ARE TIGHTLY COUPLED

An important feature of the pump is that *ATP is not hydrolyzed unless Na^+ and K^+ are transported.* In other words, the system is coupled so that the energy stored in ATP is not dissipated. Tight coupling is a general characteristic of biological assemblies that mediate energy conversion. Recall that electrons do not normally flow through the mitochondrial electron-transport chain unless ATP is concomitantly generated (p. 421). The coupling of ATP hydrolysis and muscle contraction (p. 927) also illustrates this principle. The Na^+-K^+ pump can be reversed to *synthesize* ATP by exposing it to steep ionic gradients. This is achieved by incubating red cells in a medium containing a much higher than normal concentration of Na^+ and a much lower than normal concentration of K^+.

BOTH SUBUNITS OF THE Na^+-K^+ PUMP SPAN THE PLASMA MEMBRANE

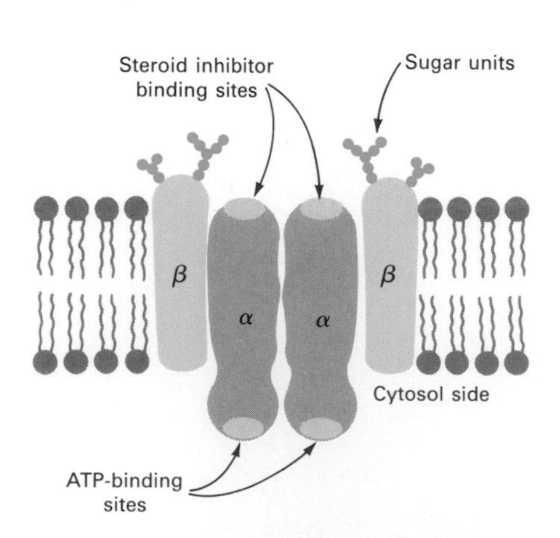

Figure 37-7
Schematic diagram of the subunit structure and orientation of the Na^+-K^+ pump.

The pump consists of two kinds of subunits, α (112 kd) and β (~40 kd), most likely associated in the membrane as an $\alpha_2\beta_2$ tetramer (Figure 37-7). The genes for both chains have been cloned and sequenced. Hydrophobicity plots (p. 304) suggest that the α chain contains at least seven transmembrane α helices. Much of the α chain, including its ATPase site, is located on the cytosolic side of the membrane. The small portion of α on the extracellular side contains the binding site for cardiotonic steroid inhibitors. The β chain probably contains four transmembrane helices. Much of β, including its three *N*-linked oligosaccharide chains, lies on the extracellular side of the membrane. The β chain does not appear to be essential for ATPase or transport activity. Chemical cross-links can readily be formed between α subunits or between an α and a β subunit but not between β subunits, suggesting that α chains are in contact and that the β chains are far apart.

A MODEL FOR THE MECHANISM OF THE Na^+-K^+ PUMP

How does the phosphorylation and dephosphorylation of the ATPase result in the transport of Na^+ and K^+ across the membrane? Not enough is known about the three-dimensional structure of the pump to formulate a detailed mechanism of how it acts. Nevertheless, it is instructive to consider a simple model for a pump that was proposed in 1966 by Oleg Jardetzky. In this model, a protein must fulfill three structural conditions to function as a pump:

1. It must contain a cavity large enough to admit a small molecule or ion.

2. It must be able to assume two conformations, such that the cavity is open to the inside in one form and to the outside in the other.

3. The affinity for the transported species must be different in the two conformations.

Let us apply this model to Na^+ and K^+ transport (Figure 37-8). The two conformations are the E_1 and E_2 forms described earlier. It is assumed that (1) the ion-binding cavity in E_1 faces the inside of the cell, whereas in E_2 the cavity faces the outside, and that (2) E_1 has a high affinity for Na^+, whereas E_2 has a high affinity for K^+. The model also makes use of two established facts: (1) Na^+ triggers phosphorylation, whereas K^+ triggers dephosphorylation; and (2) E_2 is stabilized by phosphorylation, whereas E_1 is stabilized by dephosphorylation.

A conceptual reaction cycle for the operation of the pump can be formulated on the basis of these findings and assumptions. The binding of Na^+ inside triggers the phosphorylation of E by ATP (steps 1 and 2 in Figure 37-8). Phosphorylation switches the conformation to E_2, everting the ion-binding sites so that they now face the outside (step 3). E_2-P has low affinity for Na^+, which is released outside (steps 4 and 5). In contrast, E_2-P has high affinity for K^+, which binds on the outside (step 6). The binding of K^+ triggers the dephosphorylation of E_2-P (step 7). The enzyme devoid of a covalently attached phosphoryl group is not stable in the E_2 form. It everts to E_1 (step 8), which has low

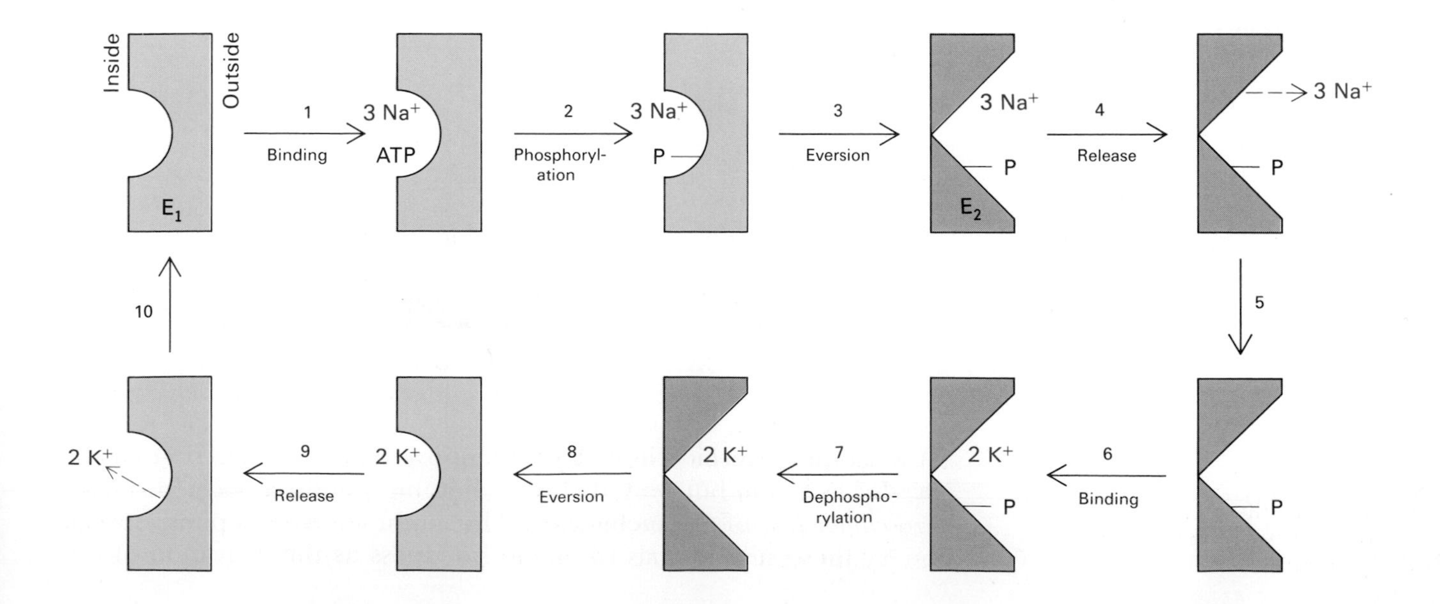

Figure 37-8
Schematic diagram of a proposed mechanism for the Na^+-K^+ pump. The upper sequence of reactions depicts the extrusion of three Na^+ ions, whereas the lower reactions show the entry of two K^+ ions. The E_1 (yellow) and E_2 (blue) forms are shown here as having very different conformations. The actual conformational differences may be quite small.

affinity for K^+. Hence, K^+ is released inside (steps 9 and 10) to complete the cycle. The actual cycle is undoubtedly more complex; for example, ATP accelerates steps 9 and 10.

In Figure 37-8, E_1 and E_2 are depicted as having very different shapes. However, it must be stressed that the conformational differences between them need not be large. A shift of a few atoms a distance of 2 Å might suffice to alter the relative affinities of the cavity for Na^+ and K^+ and change its orientation. There is ample precedent for assuming that phosphorylation can readily produce changes of this magnitude. Recall the effect of phosphorylation on the properties of glycogen phosphorylase and synthetase (p. 463) and the change in oxygen affinity elicited by the noncovalent binding of bisphosphoglycerate to hemoglobin (p. 156).

The change in free energy accompanying the transport of Na^+ and K^+ can be calculated. Suppose that the concentration of Na^+ outside and inside the cell is 143 and 14 mM, and that of K^+ is 4 and 157 mM. At a membrane potential of −50 mV, the free energy change in transporting 3 moles of Na^+ out and 2 moles of K^+ into the cell is +9.9 kcal. The hydrolysis of a single ATP per transport cycle provides sufficient free energy, about −12 kcal/mol under cellular conditions, to drive the uphill transport of Na^+ and K^+.

CARDIOTONIC STEROIDS SPECIFICALLY INHIBIT THE Na^+-K^+ PUMP

Certain steroids derived from plants are potent inhibitors of the Na^+-K^+ ATPase and pump. The inhibition of both processes is half-maximal at an inhibitor concentration of the order of 10^{-8} M. Digitoxigenin and ouabain are members of this class of inhibitors, which are known as *cardiotonic steroids* because of their profound effects on the heart (Figure 37-9). Inhibition requires the presence of a five- or six-membered unsaturated lactone ring in the β configuration at C-17. Also, a hydroxyl group at C-14 and a *cis* fusion of rings C and D are essential. Ouabain and some other cardiotonic steroids contain a sugar residue attached to C-3, but this sugar does not contribute to inhibition.

Figure 37-9
Cardiotonic steroids such as digitoxigenin and ouabain inhibit the Na^+-K^+ pump.

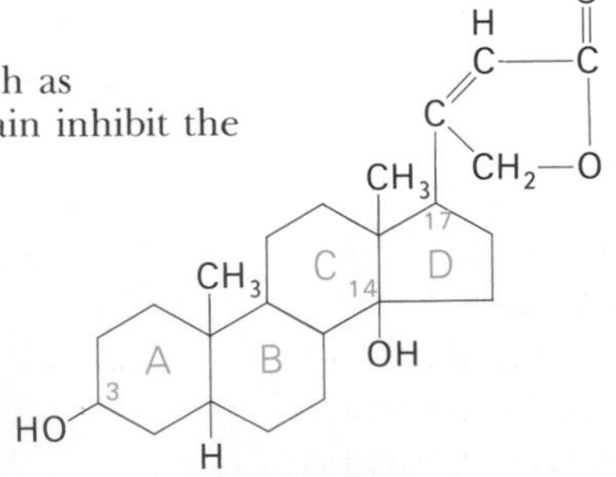

Digitoxigenin

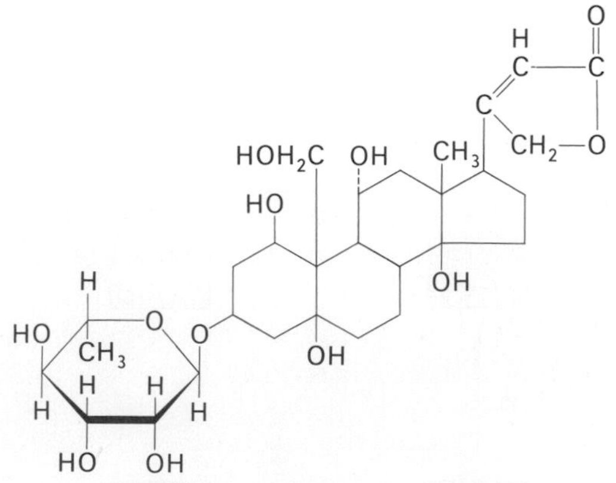

Ouabain

$$E\text{—}P + H_2O \xrightarrow{K^+} E + P_i$$

Inhibited by cardiotonic steroids

Cardiotonic steroids inhibit the dephosphorylation reaction of the Na^+-K^+ ATPase, but only if the cardiotonic steroid is located on the *extracellular face* of the membrane. Thus, inhibition of dephosphorylation by these steroids has the same sidedness as the activation of dephosphorylation by K^+. Cardiotonic steroids stabilize the E_2-P form of the pump.

Cardiotonic steroids such as *digitalis* are of great clinical significance. Digitalis increases the force of contraction of heart muscle, which makes it a choice drug in the treatment of congestive heart failure. Inhibition of the Na^+-K^+ pump by digitalis leads to a higher level of Na^+ inside the cell. The diminished Na^+ gradient results in slower extrusion of Ca^{2+} by the sodium-calcium exchanger (p. 959). The subsequent increase in the intracellular level of Ca^{2+} enhances the contractility of cardiac muscle. It is interesting to note that digitalis was effectively used long before the discovery of the Na^+-K^+ ATPase. In 1785, William Withering, a physician and botanist, published "An Account of the Foxglove and Some of its Medicinal Uses." He described how he first learned about the use of digitalis to cure congestive heart failure:

> In the year 1775, my opinion was asked concerning a family receipt for the cure of the dropsy. I was told that it had long been kept a secret by an old woman in Shropshire, who had sometimes made cures after the more regular practitioners had failed. . . . This medicine was composed of twenty or more different herbs; but it was not very difficult for one conversant in these subjects to perceive that the active herb could be no other than the foxglove. . . . It has a power over the motion of the heart to a degree yet unobserved in any other medicine, and this power may be converted to salutary ends.

Figure 37-10
A foxglove plant. [From A. Krochmal and C. Krochmal. *A Guide to the Medicinal Plants of the United States* (Quadrangle Books, 1973), p. 243. Courtesy of Dr. James Hardin.]

CALCIUM IONS ARE PUMPED OUT OF THE CYTOSOL BY A RELATED ATPase

Calcium ion plays an important role in the regulation of muscle contraction. Skeletal muscle contains an intricate network of membrane-bound tubules and vesicles. This membrane system, called the *sarcoplasmic reticulum*, regulates the Ca^{2+} concentration surrounding the contractile fibers of the muscle (Figure 37-11). In resting muscle, Ca^{2+} is pumped into the sarcoplasmic reticulum so that the Ca^{2+} concentration around the muscle fibers is very low. Excitation of the sarcoplasmic-reticulum membrane by a nerve impulse leads to a sudden release of large amounts of Ca^{2+}, which triggers muscle contraction through troponin and tropomyosin (p. 933). Likewise, Ca^{2+} is an intracellular signal in the action of many hormones.

The cytosolic level of Ca^{2+} typically increases from 0.1 μM to 10 μM on stimulation. Ca^{2+} rapidly enters the cytosol through specific channels (p. 987). In muscle, the Ca^{2+} level in the cytosol is lowered during the recovery phase by the action of an ATP-driven Ca^{2+} pump in the sarcoplasmic reticulum membrane. Muscle, like nearly all eucaryotic cells, also contains Ca^{2+} transporters in the plasma membrane, but the Ca^{2+}-ATPase of the sarcoplasmic reticulum of skeletal muscle has been a choice object of inquiry because of its abundance. It constitutes more

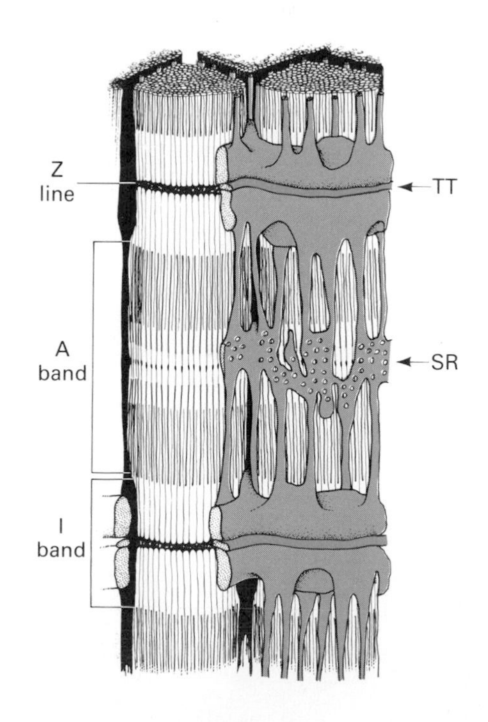

Figure 37-11
Schematic diagram of the sarcoplasmic reticulum. Excitation of the transverse tubule (TT) leads to the release of Ca^{2+} from the sarcoplasmic reticulum (SR). Their relation to the Z line, A band, and I band of adjoining skeletal muscle (p. 922) is shown. [After L. D. Peachey. *J. Cell Biol.* 25(1965):222.]

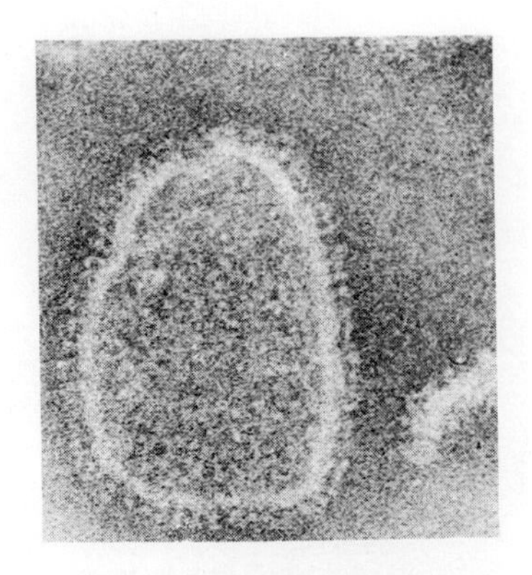

Figure 37-12
Membrane vesicle formed from purified Ca^{2+}-ATPase. The globular particles on the surface of the membrane are part of the ATPase molecule, which extends across the membrane. [From P. S. Stewart and D. H. MacLennan. *J. Biol. Chem.* 249(1974):987.]

than 80% of the integral membrane protein and takes up a third of the membrane surface area (Figure 37-12). The density of this pump in the sarcoplasmic reticulum is about 20,000 per μm^2, amongst the highest of any membrane protein.

Analysis of the cloned gene for the Ca^{2+}-ATPase (110 kd) has revealed that its sequence is similar to that of the α subunit of the Na^+-K^+ ATPase. Hydrophobicity plots suggest that these pumps contain the same pattern of transmembrane helices and that much of the polypeptide chain is located on the cytosolic side of the membrane. Moreover, *the Ca^{2+}-ATPase is mechanistically similar to the Na^+-K^+ ATPase:* a specific aspartate residue is transiently phosphorylated by ATP in a Ca^{2+}-dependent reaction; this phosphorylated intermediate is then hydrolyzed.

$$\text{E} + \text{ATP} \xrightleftharpoons{Ca^{2+},\ Mg^{2+}} \text{E—P} + \text{ADP}$$

$$\text{E—P} + H_2O \xrightarrow{Mg^{2+}} \text{E} + P_i$$

A cycle of conformational changes driven by phosphorylation and dephosphorylation transports two Ca^{2+} for each ATP hydrolyzed (Figure 37-13). The cycle begins with the binding of two Ca^{2+} ions to the E_1 form of the pump, in which the binding cavity faces the cytosol. E_1 containing Ca^{2+} binds ATP. Phosphorylation changes the conformational state to E_2, which everts the Ca^{2+} binding sites so that they face the lumen of the sarcoplasmic reticulum. The bound Ca^{2+} are then released because E_2-P has low affinity for them. E_2-P devoid of bound ions is then hydrolyzed to give E_2, which is unstable and everts to E_1 to complete the cycle.

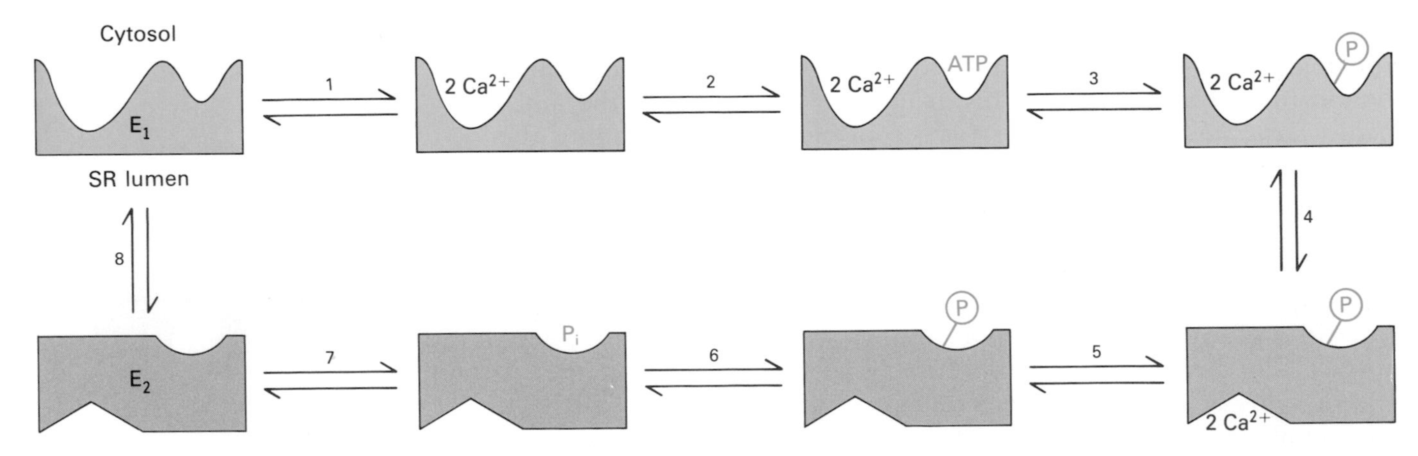

Figure 37-13
Conformational cycle of the calcium pump in the sarcoplasmic reticulum (SR) membrane.

This Ca^{2+}-ATPase cycle resembles the Na^+-K^+ ATPase cycle in several ways. (1) The enzyme exists in two major conformations, E_1 and E_2. (2) Phosphorylation favors E_2, and dephosphorylation favors E_1. (3) The ion-binding sites evert and their affinity for Ca^{2+} changes more than a thousandfold in the transition from E_1 to E_2. (4) An aspartyl phosphate E-P intermediate is formed. A ten-residue sequence containing this aspartate is identical in the two pumps. (5) Vanadate inhibits Ca^{2+} transport and ATPase activity by stabilizing the E_2 form. (6) The transition from E_2 to E_1 is accelerated by ATP.

A functional Ca^{2+} pump has been reconstituted from purified Ca^{2+} ATPase and phospholipids. The Ca^{2+} ATPase was isolated from sarcoplasmic-reticulum membranes after it had been solubilized by cholate, a detergent (p. 300). This solubilized enzyme was added to phospholipids obtained from soybeans. Membrane vesicles formed spontaneously when the detergent was removed by dialysis. *These reconstituted vesicles pumped in Ca^{2+} at a rapid rate (30 s^{-1}) when provided with ATP and Mg^{2+}. The pump is tightly coupled so that ATP is hydrolyzed only when Ca^{2+} is transported.* A striking feature of the pump is that it can readily be reversed. *Calcium gradients can drive the synthesis of ATP.* Vesicles loaded with Ca^{2+} are placed in a medium containing P_i, Mg^{2+}, and EGTA (a chelator of calcium). Under these conditions, E_2-P is formed, but there is no flow of calcium out of the vesicle as might be expected from the reversal of reactions 5 and 4 in Figure 37-13. The addition of ADP to the external medium then leads to calcium efflux and the synthesis of ATP. In fact, one ATP is formed per two Ca^{2+} transported (Figure 37-14). Thus, the Ca^{2+}-ATPase is a reversible molecular machine; it can transduce phosphoryl potential into an ion gradient, or vice versa.

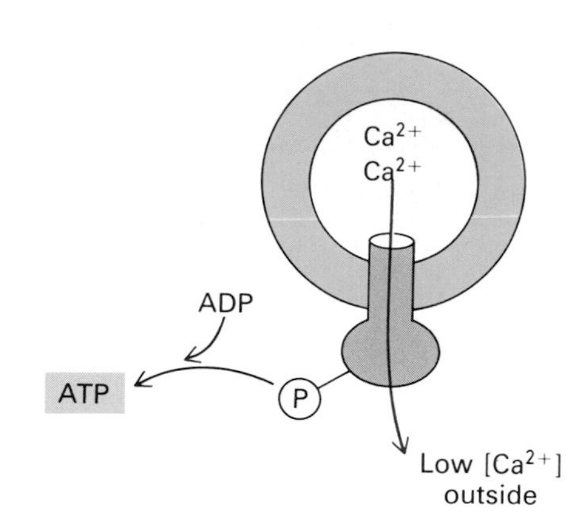

Figure 37-14
The calcium pump can be reversed in vitro to synthesize ATP.

The high affinity of this ATPase for Ca^{2+} ($K \sim 10^{-7}$ M) enables it to pump Ca^{2+} from the cytosol (where $[Ca^{2+}] < 10^{-5}$ M) into the sarcoplasmic reticulum (where $[Ca^{2+}] \sim 10^{-2}$ M). Much of the Ca^{2+} inside the lumen of the sarcoplasmic reticulum is bound to *calsequestrin*, a 45-kd anionic protein containing some 40 low-affinity binding sites for Ca^{2+}. The rate of calcium transport is controlled by *phospholamban*. Phosphorylation of this cytosolic protein by a cyclic-AMP-stimulated kinase or by a Ca^{2+}-calmodulin-stimulated kinase leads to the activation of the Ca^{2+}-ATPase. Thus, the level of Ca^{2+} in the cytosol is feedback-regulated.

THREE CLASSES OF ION-MOTIVE ATPases ARE PRESENT IN MOST EUCARYOTIC CELLS

The secretion of acid into the stomach is carried out by the H^+-K^+ ATPase, which generates a striking millionfold proton gradient. This 114-kd protein catalyzes the electroneutral exchange of H^+ for K^+. It is homologous in sequence to the Na^+-K^+ ATPase and the Ca^{2+} ATPase and has a similar pattern of transmembrane helices. The H^+-K^+ ATPase, like the other two transporters, undergoes phosphorylation during its catalytic cycle and interconverts between two major conformational states. The structural and mechanistic motifs shared by these mammalian ATP-driven ion pumps are ancient ones. The K^+-ATPase of *E. coli*, which pumps K^+ into the bacterial cell, has a very similar sequence in the vicinity of the aspartate residue that is transiently phosphorylated in the transport cycle (Figure 37-15).

A AKRMARKNCLVKNLEAVETLGSTSTICSDKTGTLT
B TRRMAKKNAIVRSLPSVETLGCTSVICSDKTGTLT
C AKRLASKNCVVKNLEAVETLGSTSVICSDKTGTLT
D MSRMLGANVIATSGRAVEAAGDVDVLLLDKTGTIT

Phosphorylation site

Figure 37-15
The amino acid sequence in the vicinity of the aspartate at the active site of P-type ATPases is highly conserved. The sequences of (A) sheep kidney Na^+-K^+ ATPase, (B) rabbit cardiac Ca^{2+}-ATPase, (C) rat gastric H^+-K^+ ATPase, and (D) *E. coli* K^+-ATPase are compared. Sequence identities are shown in red.

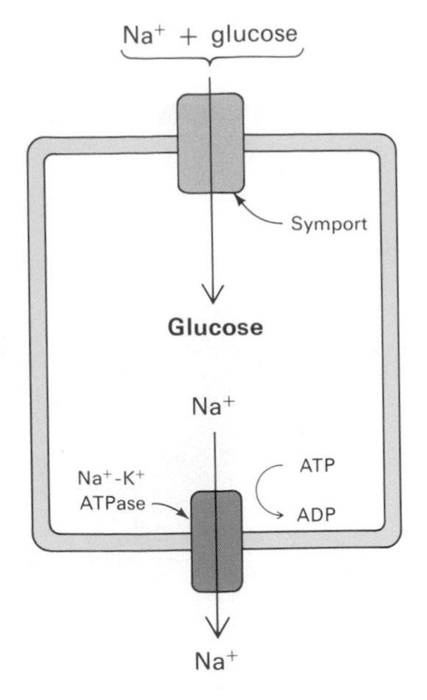

Figure 37-16
A Na^+ gradient drives the active transport of glucose. This symport system is present in the plasma membrane of intestinal cells and kidney cells.

These enzymes are members of a class of transporters called *P-type ion-motive ATPases*. The *P* in the name of this family of transporters denotes a key shared feature, a phosphorylated aspartate intermediate. They have masses of about 100 kd, cycle between E_1 and E_2 forms, and are inhibited by vanadate. P-type ATPases transport a variety of cations. *V-type ATPases*, in contrast, transport only H^+. The *V* denotes that they are found in the membranes surrounding the *vacuoles* of fungi and yeast. Related ATPases are present in the membranes of vesicles involved in endocytosis and exocytosis. They play key roles in lysosomes, endosomes, clathrin-coated vesicles, secretory vesicles, and the Golgi apparatus, in which acidification is important for protein targeting (p. 789). V-type enzymes are large (>400 kd) and consist of multiple subunits; ATP hydrolysis and proton translocation are carried out by different subunits that are conformationally coupled. We have already considered *F-type ATPases*, the third class. They are made of a membrane-spanning F_0 unit and an ATP-binding F_1 catalytic unit (p. 413). Despite their name, the physiologic role of F-type ATPases is to synthesize ATP using a proton gradient, rather than to carry out the reverse reaction. Most eucaryotic cells contain all three types of ATPases. Procaryotes contain ATPases of the P and F type but not of the V type. Vacuolar ATP-driven proton pumps are an invention of eucaryotes related to their distinctive capacity to carry out exocytosis and endocytosis.

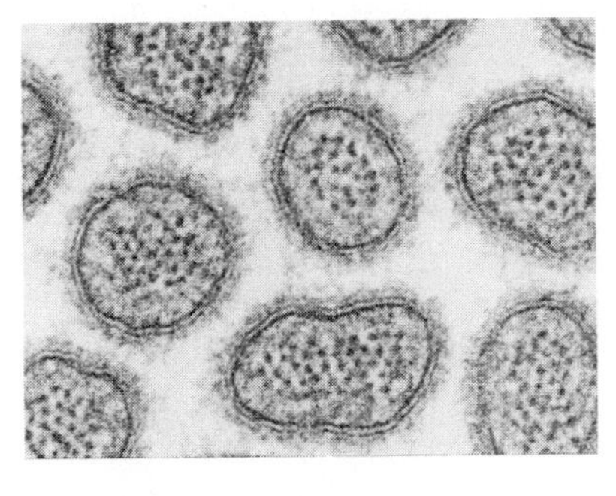

Figure 37-17
Electron micrograph of a cross section of intestinal microvilli, which greatly increase the surface area for transport. [Courtesy of Dr. George Palade.]

THE FLOW OF Na^+ DRIVES THE ACTIVE TRANSPORT OF SUGARS AND AMINO ACIDS INTO ANIMAL CELLS

Many transport processes are not directly driven by the hydrolysis of ATP. Instead, they are coupled to the flow of an ion down its electrochemical gradient. For example, *glucose is pumped into some animal cells by the simultaneous entry of Na^+*. In fact, sodium ion and glucose bind to a specific transport protein and enter together. A concerted movement of two species is called *cotransport*. A *symport* carries two species in the same direction, whereas an *antiport* carries them in opposite directions. The rate and extent of glucose transport depend on the Na^+ gradient across the plasma membrane. Sodium ions entering the cell in the company of glucose are pumped out again by the Na^+-K^+ ATPase (Figure 37-16). It should be noted that glucose enters many cells by facilitated diffusion through the glucose transporter rather than by cotransport with Na^+. The subsequent phosphorylation of glucose assures its retention by the cell.

Symports driven by Na^+ are widely used by animal cells to accumulate amino acids. Some cells, such as the brush border of intestinal microvilli, are also rich in symports for the active transport of sugars. The small intestine has a specialized Na^+-driven symport for the uphill transport of Cl^-. Sodium ions are also the driving force in antiports that extrude calcium ions from a variety of cells (Figure 37-18). The

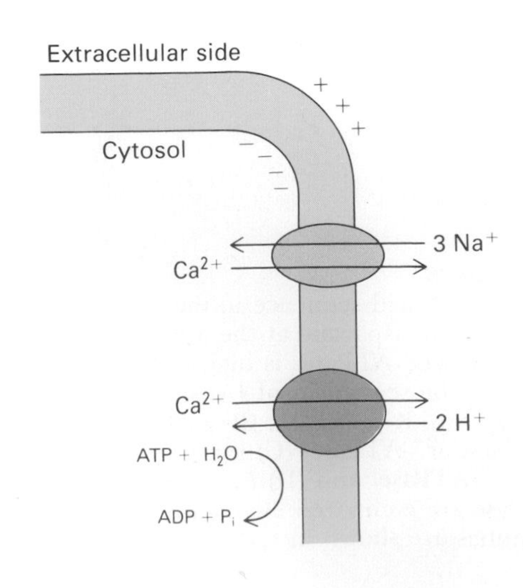

Figure 37-18
The Na^+-Ca^{2+} antiporter (green) works in parallel with the Ca^{2+}-ATPase (red) to maintain a low cytosolic level of Ca^{2+}. Extrusion of Ca^{2+} by the antiporter is driven by the sodium-motive force across the plasma membrane.

sodium-calcium exchanger sends one Ca^{2+} ion out of the cell in exchange for the entry of three Na^+ ions. This antiporter is important in maintaining a low cytosolic calcium level in many cells. *Most symports and antiports in animal cells are driven by Na^+ gradients that are generated by the Na^+-K^+ ATPase.*

THE FLOW OF PROTONS DRIVES MANY BACTERIAL TRANSPORT PROCESSES

Symports and antiports are ancient molecular machines. Many bacterial transport systems are driven by the flow of protons across the plasma membrane. The best-understood symport system in bacteria is the one for lactose in *E. coli* (Figure 37-19). This denizen of the mammalian lower intestinal tract has evolved a highly efficient mechanism for concentrating lactose. Eugene Kennedy found that this disaccharide is actively transported by *lactose permease*, a 47-kd integral membrane protein that is encoded by the *y* gene of the *lac* operon (p. 800). It comprises about 4% of the membrane protein of induced cells. A hydrophobicity analysis of the amino acid sequence of lactose permease suggests that it contains twelve transmembrane α helices; the optical rotatory properties of the permease also point to a high α helix content.

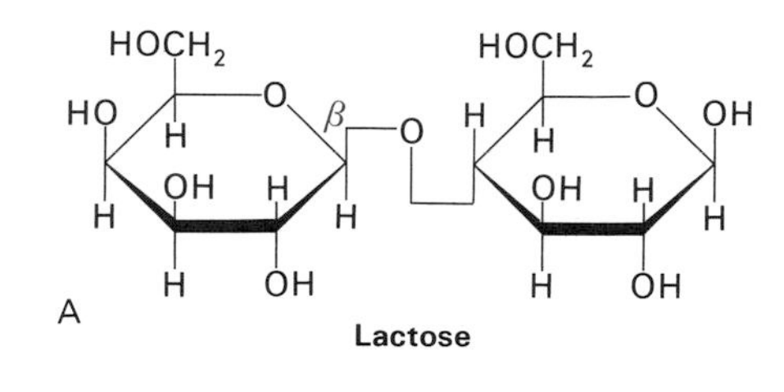

Figure 37-19
Lactose permease transports β-galactosides such as (A) lactose and (B) isopropylthiogalactoside. Thiogalactosides have been useful in studies of this system because they are transported by lactose permease but are not hydrolyzed by β-galactosidase.

Studies of vesicles formed from bacterial membranes have contributed to an understanding of the transport mechanism of lactose permease. Such vesicles are attractive for transport studies because they are much simpler than whole bacteria. They contain the machinery for oxidative phosphorylation as well as other membrane-bound proteins but are devoid of the cytoplasmic constituents of the intact cell. Vesicles alone do not accumulate lactose but can be stimulated to do so by the addition of substrates that can transfer their high-potential electrons to the respiratory chain. Alternatively, an imposed pH gradient, with the outside acid, leads to the accumulation of lactose in the vesicles. A membrane potential, inside negative, generated by a K^+ gradient also causes the pumping of lactose. *These results revealed that the driving force for the active transport of lactose is the proton-motive force across the plasma membrane.*

Reconstituted vesicles containing only purified lactose permease in a phosphatidyl choline bilayer are fully active in transporting lactose. Entry of a lactose molecule is coupled to the simultaneous movement of a proton into the cell; the permease contains a specific proton-binding site in addition to a β-galactoside binding site. Under physiological conditions, the proton gradient needed for this active transport is generated by the flow of electrons from high-potential donors (such as NADH) through the respiratory chain. This *lactose-proton symport* exemplifies the "transduction of energy by proticity," a unifying concept developed by Peter Mitchell (p. 423). Ribose, arabinose, proline, several amino acids, and succinate are also transported into *E. coli* by proton symports.

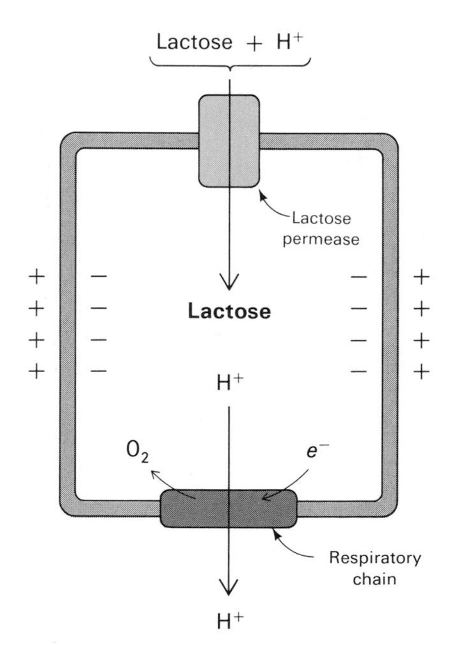

Figure 37-20
A proton gradient drives the active transport of some sugars and amino acids into bacteria. This proton gradient is generated by the flow of electrons through the respiratory chain.

THE ACTIVE TRANSPORT OF SOME SUGARS IS COUPLED TO THEIR PHOSPHORYLATION

Symports are not the only kinds of pumps for the active transport of sugars. *Some bacteria accumulate sugars by coupling their entry to phosphorylation.* For example, glucose is converted into glucose 6-phosphate during its transport into many bacteria. The distinctive feature of this kind of

Table 37-1
Sugars transported by the phosphotransferase system in *E. coli*

Glucose
Fructose
Mannose
N-Acetylglucosamine
Mannitol
Sorbitol
Galactitol
Lactose

transport, called *group translocation,* is that the *solute is changed* during transport. The cell membrane is impermeable to sugar phosphates, and so they accumulate inside the bacterium.

The best-understood group-translocation processes are the ones carried out by the *phosphotransferase system* (*PTS*), which was discovered by Saul Roseman. An unusual feature of this system is that *phosphoenolpyruvate, rather than ATP or another nucleoside triphosphate, is the phosphoryl donor.* The overall reaction carried out by the phosphotransferase system is

$$\text{Sugar}_{\text{outside}} + \text{phosphoenolpyruvate} \xrightarrow[\text{Mg}^{2+}]{\text{PTS}} \text{sugar-phosphate}_{\text{inside}} + \text{pyruvate}$$

Four proteins participate in this group translocation: HPr, enzyme I, enzyme II, and enzyme III. Enzyme II, an integral membrane protein, forms the transmembrane channel and catalyzes the phosphorylation of the sugar. The phosphoryl group of phosphoenolpyruvate is not transferred directly to the sugar. Rather, it is transferred from phosphoenolpyruvate to enzyme I and then to a specific histidine residue of HPr, a small heat-stable protein (Figure 37-21). This phosphohistidine

Phosphohistidine

intermediate has a high group-transfer potential, intermediate between those of phosphoenolpyruvate and ATP. The phosphorylated form of HPr then transfers its phosphoryl group to enzyme III, a peripheral membrane protein that interacts with enzyme II, the actual channel. The final step is the transfer of a phosphoryl group from enzyme III to the sugar that is being translocated (Figure 37-22). A functional enzyme complex has been reconstituted from these four purified proteins.

Figure 37-21
Flow of phosphoryl groups from phosphoenolpyruvate to the sugar that is translocated by the phosphotransferase system.

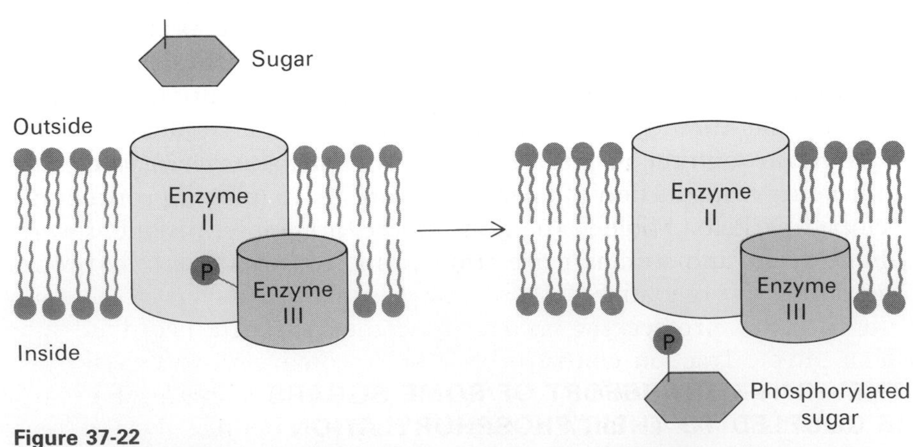

Figure 37-22
Proposed mechanism of group translocation by the phosphotransferase system. Enzyme II catalyzes phosphoryl transfer from enzyme III to the sugar substrate, which is concomitantly translocated across the membrane. Some members of the PTS family contain a second integral membrane protein (called II-A) instead of enzyme III, a peripheral membrane protein. II-A is then the phosphoryl donor and II-B is the sugar translocator.

Some of the proteins in the phosphotransferase system are general, whereas others are specific. HPr and enzyme I, which are soluble proteins in the cytosol, participate in the transport of all sugars that are translocated by this system. In contrast, enzymes II and III are specific for particular sugars. For example, there are different enzymes II and III for the transport of glucose, lactose, and fructose. Genetic studies have confirmed these conclusions. Mutants defective in HPr or enzyme I are unable to transport many different sugars, whereas mutants defective in an enzyme II or III are unable to transport a particular sugar. In the translocation of hexitols such as galactitol, enzyme III does not participate; instead, a phosphoryl group is directly transferred from HPr to the sugar.

Why is the phosphotransferase system so complex compared with transporters such as lactose permease? A likely reason is that PTS has regulatory roles in addition to mediating the transport of sugars. An abundance of any of the sugars taken up by PTS strongly inhibits the active transport of many sugars, including some that are brought in by other systems. The increased level of fructose 1,6-bisphosphate activates a kinase that specifically phosphorylates a serine residue on HPr. This modification prevents it from serving as a phosphoryl donor in all of the transport processes mediated by PTS. Also, a high concentration of a sugar translocated by PTS leads to the decreased production of cyclic AMP (Figure 37-23). The transcription of several inducible operons then stops. Recall that the expression of inducible catabolic operons such as the *lac* and *ara* operons is markedly enhanced by the binding of a complex of cyclic AMP and the CAP protein to their promoter sites (p. 803). Thus, *the phosphotransferase system regulates the acquisition of carbon sources.*

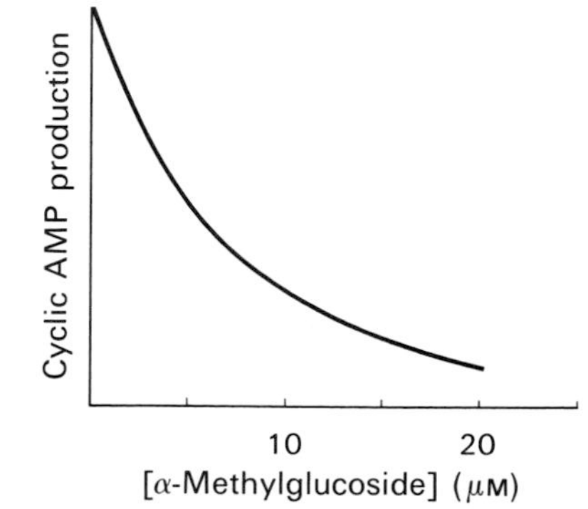

Figure 37-23
α-Methylglucoside, a sugar transported by the phosphotransferase system, inhibits cyclic AMP production.

PHOTOISOMERIZATION OF RETINAL IN BACTERIORHODOPSIN GENERATES A PROTON GRADIENT IN HALOBACTERIA

In photosynthesis, light induces the separation of charge; the downhill flow of the resulting high-potential electrons then generates a proton gradient (p. 523). A much simpler and more direct proton-pumping process takes place in *Halobacterium halobium.* This archaebacterium requires high concentrations of NaCl for growth, the optimum being 4.3 M NaCl (ordinary seawater contains 0.6 M NaCl). Halobacteria are found in natural salt lakes and in areas where seawater is evaporated to produce salt, such as the enclosed shallows in San Francisco Bay. Walther Stoeckenius separated the cell membrane into yellow, red, and purple fractions. The yellow fraction mainly contains the walls of gas vacuoles, which enable the bacteria to adjust their depth in the water. The red fraction contains the respiratory chain and other enzymatic machinery for oxidative phosphorylation, in addition to a red screening pigment that protects the bacteria from the lethal effects of blue light. The purple fraction consists of a 26-kd protein and forty associated lipids. This *purple-membrane protein* was named *bacteriorhodopsin* because it contains *retinal* as its light-absorbing group. Retinal, derived from β-carotene (an isoprenoid, p. 571), was already known to be the chromophore of *rhodopsin,* the photoreceptor protein in animal eyes (p. 1030). Retinal is joined to the ϵ-amino group of a specific lysine residue of bacteriorhodopsin by a Schiff base linkage (Figure 37-24). As was discussed in an earlier chapter (p. 308), electron microscopic analyses of

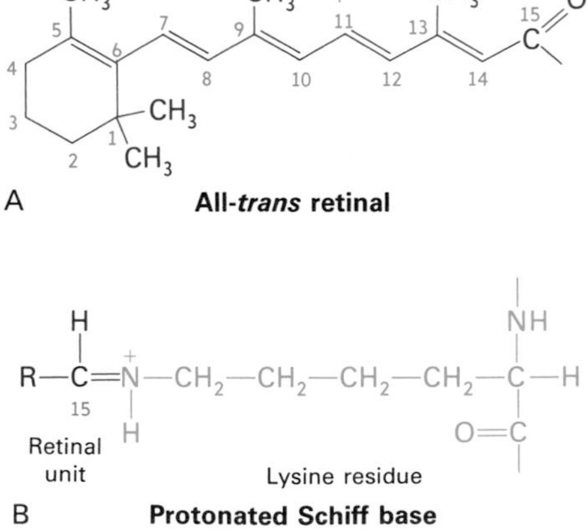

Figure 37-24
The light-absorbing group in bacteriorhodopsin is (A) all-*trans* retinal joined to a specific lysine residue by (B) a protonated Schiff base linkage.

two-dimensional crystals of bacteriorhodopsin revealed that it contains seven tightly packed transmembrane helices.

In the presence of O_2, halobacteria carry out oxidative phosphorylation to generate ATP. When oxygen is scarce, they switch to a photosynthetic mode. In 1973, Stoeckenius and Dieter Oesterhelt discovered that *bacteriorhodopsin is a light-driven proton pump.* Two protons per absorbed photon are translocated across the cell membrane from the cytosol to the outside. A cycle of conformational changes (Figure 37-25) is initiated by the absorption of a photon by bacteriorhodopsin, denoted as BR_{570} because its absorption spectrum peaks at 570 nm, which gives it a purple color. Light isomerizes the all-*trans* retinal group of BR_{570} to the 13-*cis* form of K_{610}, a transient red-absorbing intermediate. This conformational transition occurs very rapidly, in picoseconds. A series of intermediates with distinctive absorption spectra (L_{550}, M_{412}, N_{520}, and O_{640}) then appear in the microsecond to millisecond time range. Another key event in the photocycle is the deprotonation of the Schiff base in the steps leading to M_{412}. The Schiff base is reprotonated and the retinal group isomerizes to the all-*trans* form in the subsequent steps, bringing bacteriorhodopsin back to the BR_{570} form. In bright light, several hundred H^+ are pumped per second by each bacteriorhodopsin molecule. The proton-motive force then drives the synthesis of ATP and powers other ion pumps. The goal now is to learn how light-driven conformational changes in retinal lead to the vectorial movement of protons through bacteriorhodopsin, the simplest known energy-transducing protein.

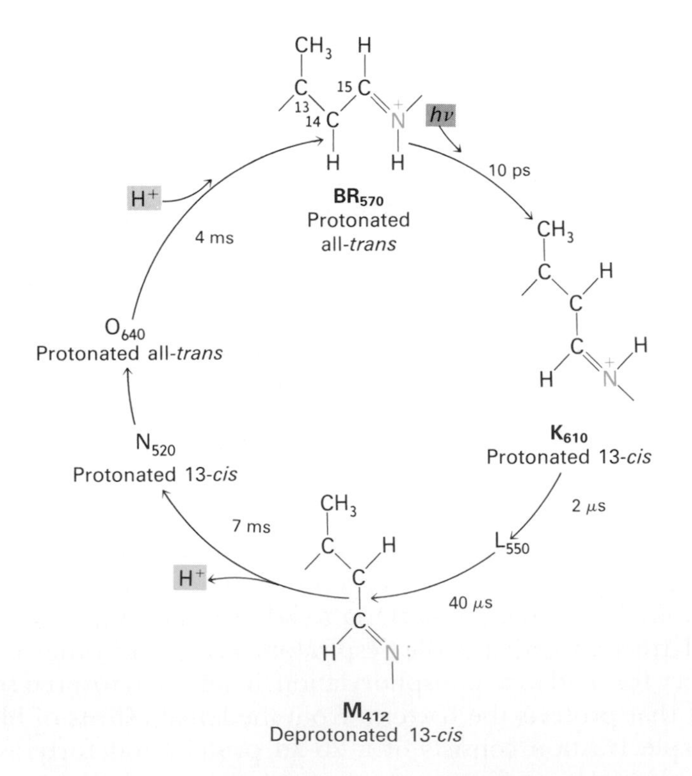

Figure 37-25
Photocycle of bacteriorhodopsin. The primary energy-transducing step is the photoisomerization of retinal from the all-*trans* to the 13-*cis* form. Some 30 kcal/mol of the 60 kcal/mol energy content of the absorbed photon is conserved in K_{610}, the first photoproduct.

Halobacteria contain another ion transporter that is driven by the photoisomerization of bound retinal. *Halorhodopsin*, a 27-kd protein, pumps chloride ions from the outside to the inside of the cell. Like bacteriorhodopsin, halorhodopsin contains an all-*trans* retinal chromophore joined to a specific lysine residue by a protonated Schiff base linkage. The hydrophobicity profiles of these retinal-containing proteins are very similar, suggesting that halorhodopsin also contains seven transmembrane helices. Moreover, its amino acid sequence is similar to that of bacteriorhodopsin. *These transporters are variations of a fundamental motif that couples the cis-trans isomerization of retinal to the directed transmembrane movement of ions.* They have different binding sites for ions but a similar framework for converting light into the vectorial movement of atoms. A third member of this family of retinal proteins is *sensory rhodopsin.* This photoreceptor protein triggers a transduction cascade leading to the movement of halobacteria toward red light and away from blue light.

This concludes our consideration of active transport processes. We have seen that they can be driven by a variety of energy sources—ATP, H^+ gradients, Na^+ gradients, light, and high-potential electrons (p. 401) (Table 37-2). *All known active transport processes in biological systems are mediated by asymmetrically oriented transmembrane proteins. They undergo a cycle of conformational transitions that simultaneously change the orientation and affinity of the binding site for the transported species.* A small portion of the transport protein is transiently everted during each transport cycle.

Table 37-2
Sources of free energy in active transport

Free-energy source	*Transport system*
ATP	**P-type** Na^+-K^+ ATPase of animal cells Ca^{2+}-ATPase of animal cells H^+-K^+ ATPase of stomach K^+-ATPase of *E. coli*
ATP	**V-type** Proton pumps in endocytic vesicles
ATP	**F-type** F_0-F_1 H^+-ATPase of mitochondria
Phosphoenolpyruvate	Phosphotransferase system of *E. coli*
Proton motive force	Lactose permease of *E. coli*
Sodium motive force	Sugar transporter of animal cells Ca^{2+}-Na^+ antiporter of animal cells
Electron motive force	Proton-pumping assemblies in oxidative phosphorylation (e.g., cytochrome oxidase) and photophosphorylation (e.g., photosystem I)
Light	Bacteriorhodopsin and halorhodopsin of *Halobacterium halobium*

THE ANION-EXCHANGE PROTEIN OF ERYTHROCYTES EXCHANGES CHLORIDE FOR BICARBONATE

We turn now to passive transport processes, in which transported species move down their electrochemical gradient. Erythrocytes contain a 103-kd integral membrane protein that mediates the exchange of HCO_3^- for Cl^-. This *anion exchange protein* (also called the anion channel or the band 3 protein) is critical in transporting carbon dioxide from respiring tissues to the lung (p. 305). The exchanger consists of two domains. The hydrophilic amino-terminal domain serves as an anchor for the cytoskeleton. The hydrophobic carboxyl-terminal domain, which spans the erythrocyte membrane at least twelve times, transports anions. A key feature of this transporter is the *obligatory nature of the exchange process.* The efflux of an anion (e.g., HCO_3^-) is very tightly coupled to the entry of another anion (Cl^-). In other words, the protein is an *antiporter*. This one-for-one exchange means that the transport process is electrically neutral and hence is electrically silent. The biological significance of electroneutrality is that the membrane potential is not perturbed by bicarbonate-chloride exchange.

Anion exchange cannot occur simply by diffusion of ions through a pore that selects anions and is simultaneously accessible from both sides. A minimal model (Figure 37-26) must have four states: *e* is the empty transporter with the anion-binding site on the extracellular side; *c* is the empty transporter with the binding site facing the cytosol; *ea* is the transporter containing a bound anion, with the binding site facing the extracellular side; and *ca* is the transporter containing a bound anion, with the binding site facing the cytosol. A critical feature of the action of this transporter is that the $ea \rightleftharpoons ca$ interconversion (exchange) is much faster than the $e \rightleftharpoons c$ interconversion (slippage). The exchange transport rate is $4 \times 10^4\ s^{-1}$, compared with less than 1 s^{-1} for slippage between the two empty states. *Exchange is obligatory because the anion-binding site of the protein does not evert at an appreciable rate in the absence of a bound anion.* In effect, anions catalyze the interconversion of the inward and outward facing forms of the binding site. NMR studies have shown that the transported anion binds to an arginine residue that is alternately exposed to opposite sides of the membrane. Further evidence for the essentiality of arginine is that 1,2-cyclohexadione, which reacts specifically with this residue, blocks anion exchange.

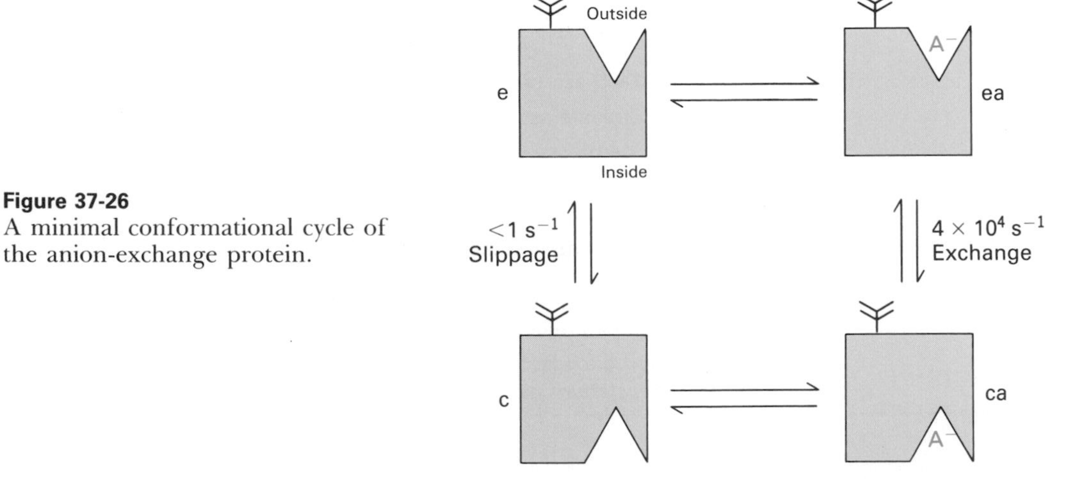

Figure 37-26
A minimal conformational cycle of the anion-exchange protein.

TRANSPORT ANTIBIOTICS INCREASE THE IONIC PERMEABILITY OF MEMBRANES

Some microorganisms synthesize compounds that make membranes permeable to certain ions. These small molecules, called *transport antibiotics,* are valuable experimental tools, in addition to being sources of insight into the mechanism of ion binding. For example, *valinomycin* interferes with oxidative phosphorylation in mitochondria by making them permeable to K^+. The result is that mitochondria use the energy generated by electron transport to accumulate K^+ rather than to make

ATP. Valinomycin is a repeating cyclic molecule made of four kinds of residues (A, B, C, and D) taken three times (Figure 37-27). The four kinds of residues are alternately joined by ester and peptide bonds.

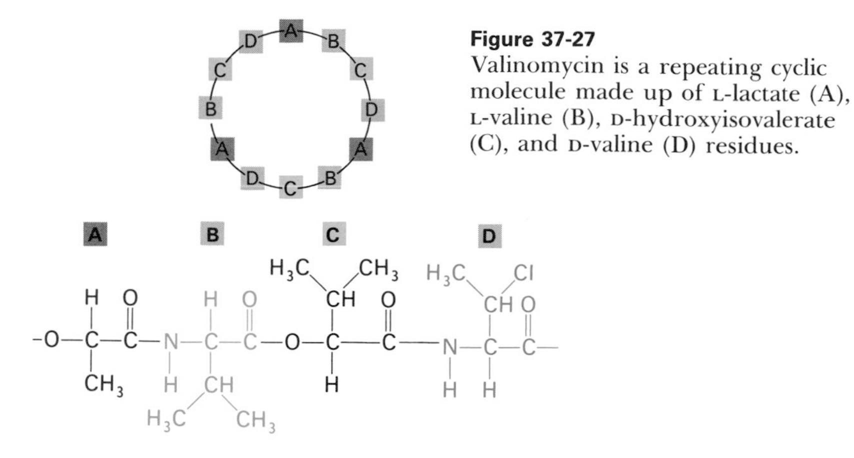

Figure 37-27
Valinomycin is a repeating cyclic molecule made up of L-lactate (A), L-valine (B), D-hydroxyisovalerate (C), and D-valine (D) residues.

Gramicidin A is another transport antibiotic that has been studied extensively (Figure 37-28). Gramicidin is an open-chain polypeptide consisting of fifteen amino acid residues. Two noteworthy features of its structure are the alternation of L- and D-residues and the modification of the amino and carboxyl termini. As will be discussed shortly, gramicidin A and valinomycin transport ions in quite different ways.

Ion transport by these antibiotics is readily studied in well-defined model systems such as lipid bilayer vesicles and planar bilayer membranes (p. 291). Vesicles containing a radioactive ion such as $^{42}K^+$ are prepared by sonicating membranes in the presence of this ion and then removing untrapped $^{42}K^+$ by gel filtration. The rate of efflux of this radioactive ion is then measured in the presence and absence of a transport antibiotic. Another way to obtain information about the ionic permeability of a bilayer membrane is to measure electrical properties such as the membrane potential and conductance (the inverse of resistance). For example, the conductance of a bilayer membrane to K^+ increases more than 10^4-fold in the presence of 10^{-7} M valinomycin or 10^{-9} M gramicidin.

O=C(H)–
HN-Val(L)-Gly-Ala(L)-Leu(D)-Ala(L)-5
Val(D)-Val(L)-Val(D)-Trp(L)-Leu(D)-10
Trp(L)-Leu(D)-Trp(L)-Leu(D)-Trp(L)-C(=O)-N(H)-CH_2-CH_2-OH

Figure 37-28
Structure of gramicidin A.

TRANSPORT ANTIBIOTICS ARE CARRIERS OR CHANNEL FORMERS

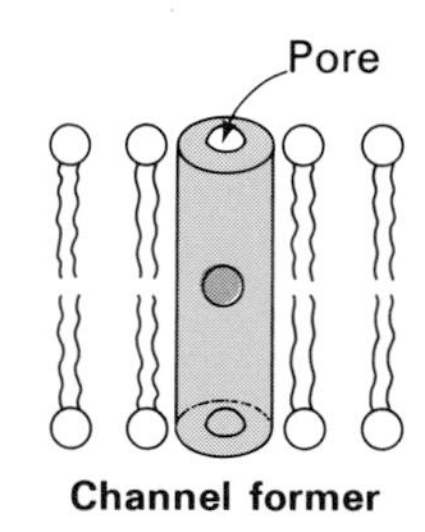

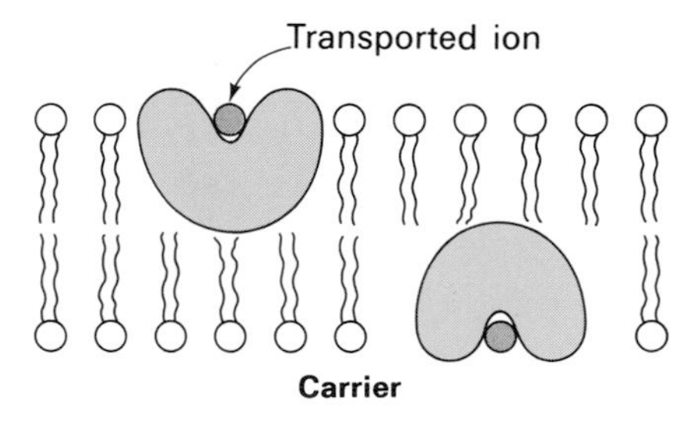

Figure 37-29
Schematic diagram comparing a channel-forming with a carrier transport antibiotic. All known transport proteins are of the channel-forming type.

Transport antibiotics make membranes permeable to ions in two contrasting ways (Figure 37-29). Some of these antibiotics (e.g., gramicidin A) form channels that traverse the membrane. Ions enter such a channel at one surface of the membrane and diffuse through it to the other side of the membrane. The channel former itself need not move for ion transport to occur. The other group of antibiotics (e.g., valinomycin) functions by carrying ions through the hydrocarbon region of the membrane. Diffusion of these transport antibiotics across the membrane is essential for their activities.

Carriers and *channel formers* can be experimentally distinguished in the following way. The ionic conductance of a synthetic lipid bilayer membrane containing the transport antibiotic is measured as a function of temperature near the transition temperature of the lipid. This is the

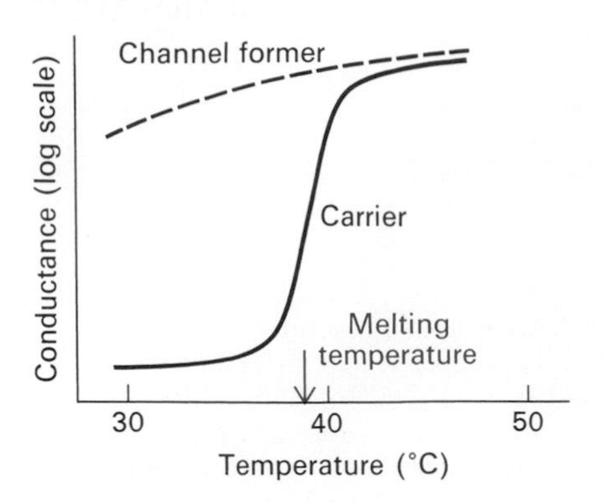

Figure 37-30
Comparison of the temperature dependence of the conductance of a lipid bilayer membrane containing a channel-forming and a carrier transport antibiotic.

temperature at which the hydrocarbon interior of the membrane changes between an essentially frozen state and a highly fluid one. A channel former need not diffuse in order to mediate ion transport across the membrane. Hence, freezing of the hydrocarbon core should have little effect on its ion-transport activity. In contrast, freezing should markedly diminish the efficiency of a carrier, because a carrier must diffuse through the hydrocarbon region to be active. The experimental data for valinomycin and gramicidin A distinguish between these mechanisms (Figure 37-30). The conductance of a bilayer membrane containing valinomycin increases more than a thousandfold as the membrane is melted. In contrast, the transport activity of gramicidin A is relatively unaffected by the melting of the membrane.

It is important to note that *all known naturally occurring transport systems are channels.* Integral membrane proteins cannot serve as diffusional carriers because they do not flip-flop (p. 295). Binding sites within them undergo eversion during the transport cycle, but the proteins themselves do not rotate across the membrane because it is energetically very expensive to remove their numerous polar groups from solution.

DONUT-SHAPED CARRIER ANTIBIOTICS BIND IONS IN THEIR CENTRAL CAVITIES

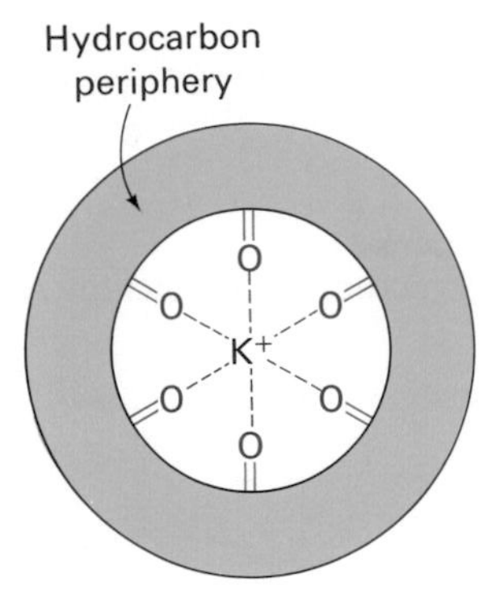

Figure 37-31
Schematic diagram of the chelation of K^+ by oxygen atoms of a carrier transport antibiotic.

Ion-carrying transport antibiotics have some common structural features, as shown by x-ray crystallographic and spectroscopic studies. The carriers studied thus far have a shape like that of a donut. *A single metal ion is coordinated to several oxygen atoms that surround a central cavity.* The number of oxygen atoms that bind the metal ion is typically six or eight. *The periphery of the carrier consists of hydrocarbon groups* (Figure 37-31).

The roles of the central oxygen atoms and the hydrocarbon exterior are evident. In an aqueous medium, a metal ion such as K^+ binds several water molecules through their oxygen atoms. A carrier competes with water for binding the ion by chelating it to several appropriately arranged oxygen atoms in its central cavity. The hydrocarbon periphery makes the ion-carrier complex soluble in the lipid interior of the membranes. In essence, *these antibiotics catalyze the transport of ions across membranes by making them soluble in lipid.*

The structures of valinomycin and its complex with K^+ are shown in Figure 37-32. The K^+ ion is coordinated to six oxygen atoms, which are

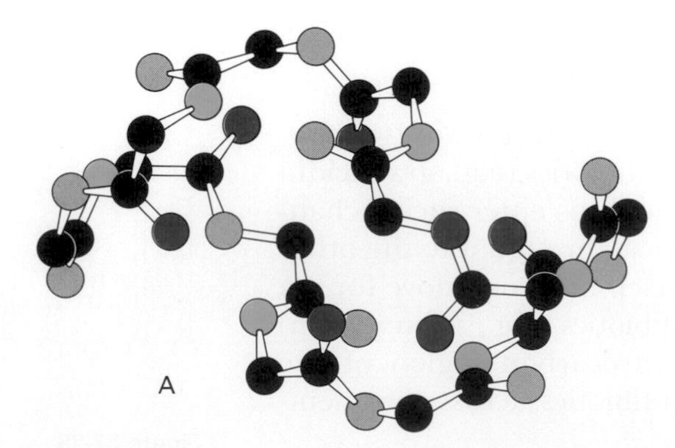

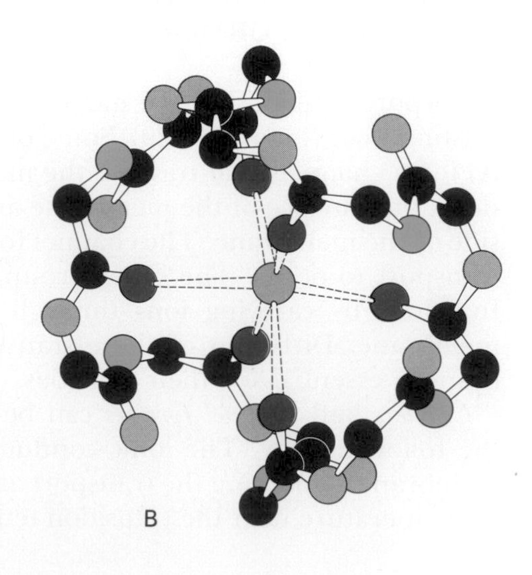

Figure 37-32
Models of (A) valinomycin and (B) its complex with K^+. The conformation of the antibiotic changes upon binding K^+. [After atomic coordinates kindly provided by Dr. William Duax (for valinomycin) and by Dr. Larry Steinrauf (for the valinomycin-K^+ complex).]

arranged octahedrally around the center of the molecule. These oxygen atoms come from the carbonyl groups of the six valine residues in the antibiotic. The methyl and isopropyl side chains constitute the hydrocarbon periphery of valinomycin.

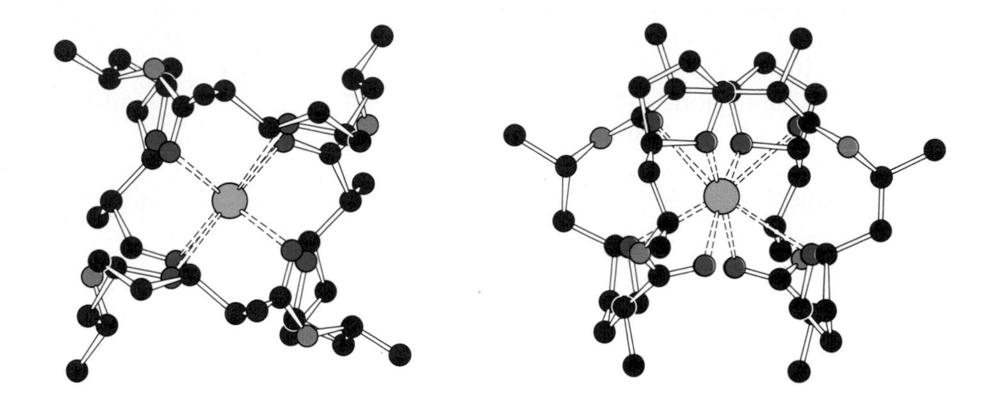

Two views of a model of the K^+ complex of nonactin. The ion is bound to eight oxygen atoms in the center of the molecule. Four are carbonyl oxygen atoms and the other four are ether oxygens.

VALINOMYCIN BINDS K^+ A THOUSAND TIMES AS STRONGLY AS Na^+

How do ion transporters distinguish between such related ions as Na^+ and K^+? Valinomycin binds K^+ about a thousand times as strongly as Na^+. This degree of discrimination is even higher than that achieved by the Na^+-K^+ ATPase, and so it is instructive to examine its basis. Spectroscopic studies indicate that the structure of the Na^+ complex of valinomycin closely resembles that of the K^+ complex. In particular, these complexes have similar bond distances between the central cation and coordinating oxygen atoms. Why then does valinomycin bind K^+ much more tightly than Na^+? The reason for this selectivity is that water has less attraction for K^+ than for Na^+. The free energy of solvation of Na^+ is 17 kcal/mol more favorable than that of K^+ (Table 37-3). *Thus, valinomycin binds K^+ more tightly because it is energetically more costly to pull Na^+ away from water.*

Table 37-3
Properties of alkali cations

Ion	*Atomic number*	*Ionic radius (Å)*	*Hydration free energy (kcal/mol)*
Li^+	3	0.60	−98
Na^+	11	0.95	−72
K^+	19	1.33	−55
Rb^+	37	1.48	−51
Cs^+	55	1.69	−47

A transporter that markedly prefers Na^+ has not yet been analyzed in atomic detail, but a key aspect of its structure can be surmised. Such a transporter must take advantage of the smaller ionic radius of Na^+ (r = 0.95 Å) compared with that of K^+ (r = 1.33 Å). Negatively charged oxygen atoms can approach Na^+ more closely than K^+, and so the higher cost of removing Na^+ from water could be overcome. *The prediction then is that Na^+-specific sites have a high density of negative charge and a geometry that allows Na^+ to fit snugly in such a cluster. In contrast, K^+-specific sites are expected to have a lower density of negative charge and a geometry less conducive to the formation of very short bonds between the cation and coordinating groups.*

The flexibility of the valinomycin molecule (see Figure 37-32) is also noteworthy. Chelation of K^+ is a step-by-step process in which water molecules in the hydration shell are successively displaced by oxygen atoms of the antibiotic. The activation barrier for binding is small because new bonds are being made as old ones are being broken. Likewise, the energy of activation for release, the reverse process, is small. Hence, valinomycin picks up and unloads K^+ many times a second. The importance of flexibility is evident here, as in enzyme action.

THE FLOW OF IONS THROUGH A SINGLE CHANNEL IN A MEMBRANE CAN BE DETECTED

The conductance of a planar bilayer membrane containing a very small amount of gramicidin A is not constant. Denis Haydon discovered that the conductance of this membrane to Na^+ fluctuates with time in a quantized manner (Figure 37-33). These steps in conductance arise from the spontaneous opening and closing of gramicidin channels. A single channel stays open for about a second. This channel is highly permeable to monovalent cations but not to divalent cations or anions. *In fact, more than 10^7 ions can traverse a single channel in a second.* This transport rate is only a factor of ten less than that of diffusion through pure water. In contrast, the maximal transport rate of diffusional carriers is less than 10^3 ions per second because rotation and translation of these transporters in the membrane takes at least a millisecond.

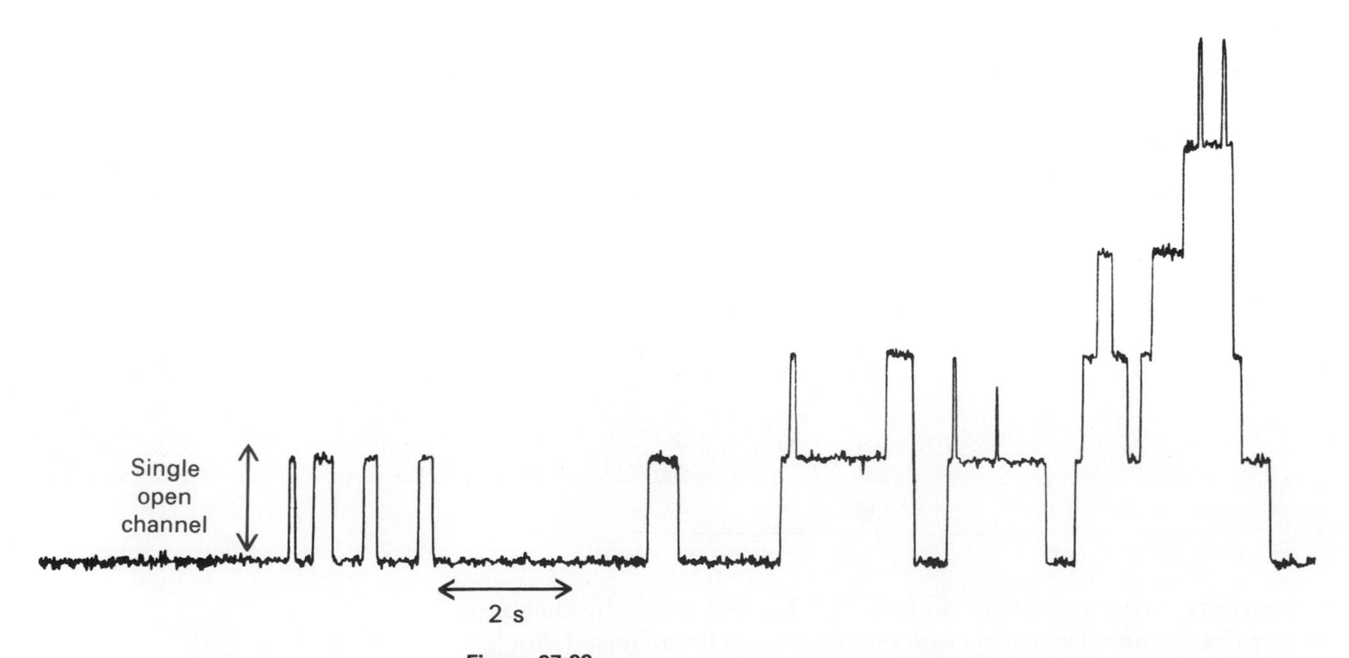

Figure 37-33
The conductance of a lipid bilayer membrane containing a few molecules of gramicidin A fluctuates in a step-by-step manner. The smallest step in conductance arises from a single open channel. [Courtesy of Dr. Olaf Andersen.]

Spectroscopic and x-ray-crystallographic studies have revealed that a transmembrane channel is formed from two molecules of gramicidin A (Figure 37-34). In the membrane, this conducting helical dimer is in equilibrium with nonconducting monomers. In fact, the steps in conductance in Figure 37-33 correspond to the formation and dissociation of dimers. The dimer contains a 4-Å-diameter aqueous channel that is surrounded by polar peptide groups. In contrast, the hydrophobic side chains are at the periphery of the channel, in contact with hydrocarbon chains of membrane phospholipids. The carbonyl groups surrounding the aqueous pore transiently coordinate the cation as it passes through the channel. The alternation of L and D residues places all the side chains on the outside of the cylinder, leaving a central hole.

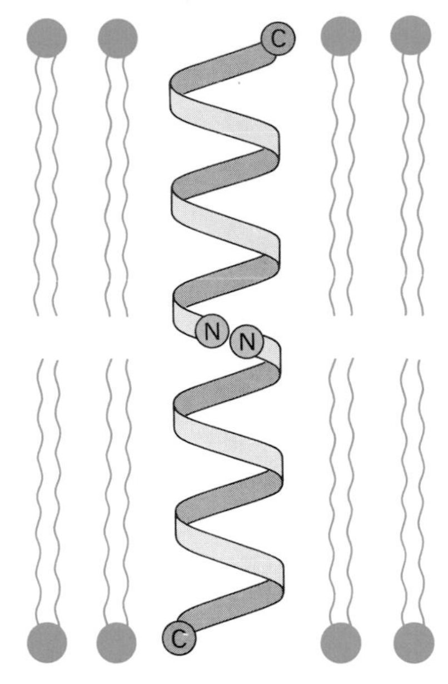

Figure 37-34
Schematic diagram of a gramicidin-A channel formed by the association of two polypeptides at their *N*-formyl ends. Each chain is folded into a β helix, which resembles a rolled-up β pleated sheet. This model was proposed by Dan Urry. [After S. Weinstein, B. A. Wallace, E. R. Blout, J. S. Morrow, and W. Veatch. *Proc. Nat. Acad. Sci.* 76(1979):4230.]

GAP JUNCTIONS ALLOW IONS AND SMALL MOLECULES TO FLOW BETWEEN COMMUNICATING CELLS

Large aqueous channels for passive transport are present in many procaryotic and eucaryotic biological membranes. The one best understood in eucaryotes is the *gap junction,* which is also known as the *cell-to-cell channel* because it serves as a passageway between the interiors of contiguous cells. Gap junctions, which were discovered by Jean-Paul Revel and Morris Karnovsky, are clustered in discrete regions of the plasma membranes of apposed cells. Electron micrographs of sheets of gap junctions show them tightly packed in a regular hexagonal array (Figure 37-35). A 20 Å central hole, the lumen of the channel, is prominent; the centers of neighboring channels are about 85 Å apart. The density of channels in a gap junction is very high, about 28,000 per μm^2. A tangential view (Figure 37-36) shows that they span the intervening space, or gap, between apposed cells (hence, the name gap junction). The width of the gap between the cytosols of the two cells is about 35 Å.

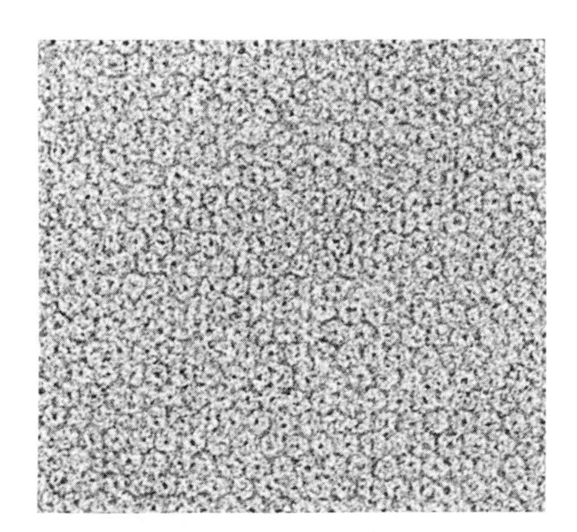

Figure 37-35
Electron micrograph of a sheet of isolated gap junctions. The cylindrical connexons are organized in a hexagonal lattice with a unit-cell length of 85 Å. The densely stained central hole has a diameter of about 20 Å. [Courtesy of Dr. Nigel Unwin and Dr. Guido Zampighi.]

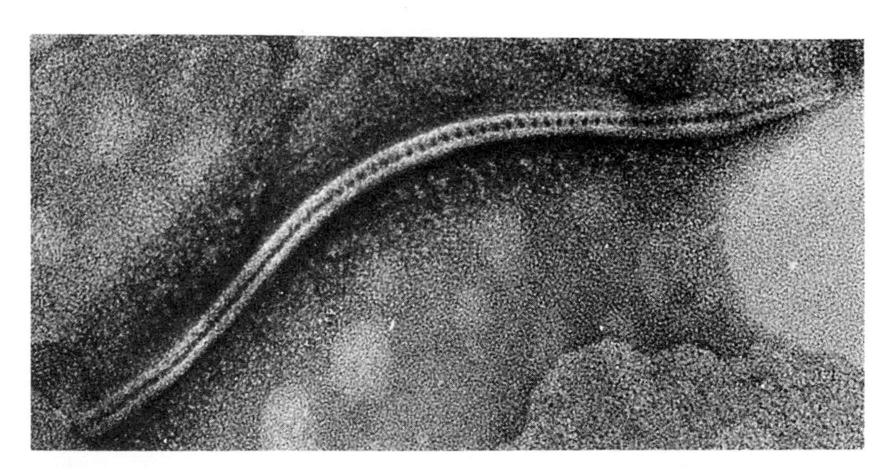

Figure 37-36
Electron micrograph of a tangential view of apposed cell membranes that are joined by gap junctions. [From E. L. Hertzberg and N. B. Gilula, *J. Biol. Chem.* 254(1979):2143.]

In 1964, Werner Loewenstein showed that small hydrophilic molecules as well as ions can pass through gap junctions. In fact, the bore size of the junctions was determined by microinjecting a series of fluorescent molecules into cells and observing their passage into adjoining cells. Loewenstein found that all polar molecules with a mass of less than about 1 kd can readily pass through these cell-to-cell channels. Thus, *inorganic ions and most metabolites (e.g., sugars, amino acids, and nucleotides) can flow between the interiors of cells joined by gap junctions.* In contrast, proteins, nucleic acids, and polysaccharides are too large to traverse these channels. *Gap junctions are important for intercellular communication.* Cells in some excitable tissues, such as heart muscle, are coupled by the rapid flow of ions through these junctions, which assure a rapid and synchronous response to stimuli. Gap junctions are also essential for the nourishment of cells that are distant from blood vessels, as in lens and bone. Moreover, communicating channels are important in development and differentiation. For example, a pregnant uterus is transformed from a quiescent protector of the fetus to a forceful ejector at the onset of labor; the formation of functional gap junctions at that time creates a syncytium of muscle cells that contract in synchrony.

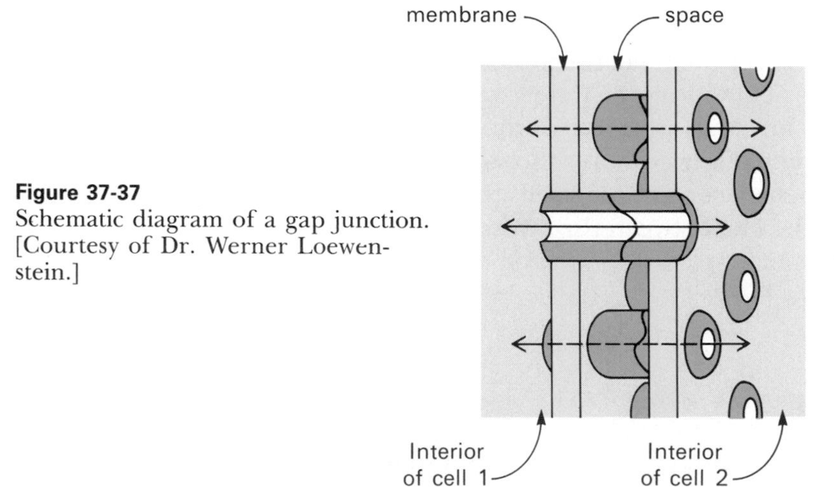

Figure 37-37
Schematic diagram of a gap junction. [Courtesy of Dr. Werner Loewenstein.]

A cell-to-cell channel is made of twelve molecules of *connexin*, a 32-kd transmembrane protein. Six connexins form a half-channel, called a *connexon* or *hemichannel*. Two connexons join end to end in the intercellular space to form a functional channel between the communicating cells. Cell-to-cell channels differ from other membrane channels in three respects: (1) they traverse *two* membranes rather than one; (2) they connect cytosol to cytosol, rather than to the extracellular space or the lumen of an organelle; and (3) the connexons forming a channel are synthesized by different cells. Gap junctions form readily when cells are brought together. A cell-to-cell channel, once formed, tends to stay open for seconds to minutes. They are closed by high concentrations of calcium ion (~mM) and by a lowering of pH. *The closing of gap junctions by Ca^{2+} and H^+ serves to seal normal cells from traumatized or dying neighbors.* The Hill coefficient for Ca^{2+}-induced closure is 3 and that of H^+-

induced closure is 4.5, indicating that channel closing is a cooperative event. Gap junctions are also controlled by membrane potential and by hormone-induced phosphorylations.

The subunits of a connexon are cylindrical, about 25 Å in diameter and 75 Å long. Each protrudes from the bilayer only slightly on the cytoplasmic side (7 Å) compared with the extracellular side (17 Å). The central hole is about 20 Å wide in the extracellular region but is narrower within the membrane. On exposure to Ca^{2+}, the subunits of each connexon rotate and slide so that the long axis of each becomes more aligned with the axis of the channel (Figure 37-38). The change in tilt, of only 7.5 degrees, leads to a quite large displacement—about 9 Å—at the cytoplasmic end of each subunit. Hence, the channel-lining faces of a pair of subunits on opposite sides of a channel move 18 Å toward each other, which closes the channel. A noteworthy feature of this mechanism is that switching is achieved without gross distortion of the individual connexin chains. In both the open and closed states, the polar faces of the subunits remain exposed to water, and the nonpolar ones to the hydrophobic core of the bilayer. Switching is facile because the free energies of the open and closed states do not differ markedly. A relatively small conformational change due to the binding of Ca^{2+} at one end of a subunit is amplified into a much larger one at the other end by the cooperative sliding and tilting of subunits. This allosteric mechanism is probably generally used to control the activity of channels.

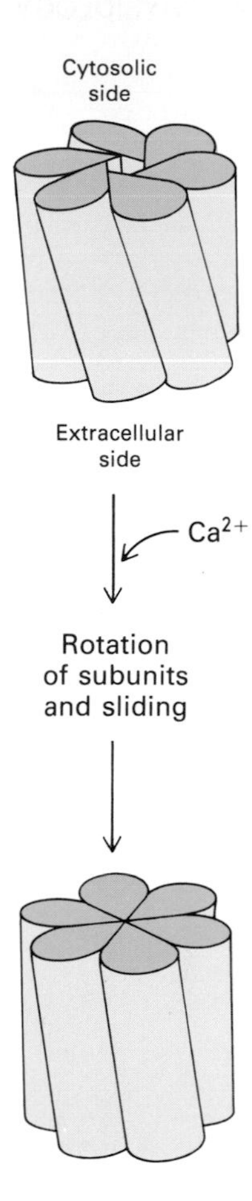

Figure 37-38
Model for the closure of gap junctions by Ca^{2+}. [After a drawing kindly provided by Dr. Nigel Unwin and Dr. Guido Zampighi.]

SUMMARY

Molecules and ions are transported across biological membranes by oriented transmembrane proteins that form channels. A transport process is passive if ΔG for the transported species is negative, whereas it is active if ΔG is positive. The free-energy change depends on the concentration ratio of the transported species and on the membrane potential if it is charged. Active transport requires an input of free energy. Membrane transporters are driven by ATP, phosphoenolpyruvate, proton gradients, Na^+ gradients, light, and high-potential electrons.

The most ubiquitous transport system in animal cells is the Na^+-K^+ pump, which hydrolyzes a molecule of ATP to pump out three Na^+ ions and pump in two K^+ ions. When Na^+ is bound to the cytosolic side of the pump, ATP phosphorylates an aspartate side chain of the α subunit. Phosphorylation everts the binding site and releases Na^+ on the extracellular side. The binding of K^+ induces the hydrolysis of the attached phosphoryl group, which everts the binding site to its initial orientation and releases K^+ inside the cell. Cardiotonic steroids such as digitalis bind to an external site and inhibit dephosphorylation. Calcium ion, which plays an important role in the control of muscle contraction, is transported by a related ATPase located in the sarcoplasmic-reticulum membrane. An aspartate residue is phosphorylated during the transport cycle, as in the Na^+-K^+ pump. Both the Ca^{2+} and the Na^+-K^+ transport systems have been reconstituted from purified ATPases and phospholipids. These transporters are P-type pumps, characterized by a phosphorylated intermediate. The two other classes of ion-motive ATPases are the V-type (found in the membranes of vacuoles and vesicles) and the F-type (the F_0-F_1 ATPases).

Some transport systems are driven by ionic gradients rather than by the hydrolysis of ATP. For example, the active transport of glucose and amino acids into some animal cells requires the concerted movement of Na^+, a process called cotransport. Na^+ and glucose enter together on a specific symport. In turn, the Na^+ gradient for this coupled entry is maintained by the Na^+-K^+ pump. Bacteria generally use H^+ instead of Na^+ to drive symports and antiports. For example, the active transport of lactose into *E. coli* by lactose permease is coupled to the movement of a proton into the bacterium. The driving force is the proton-motive force generated by the respiratory chain. A different type of transport process in bacteria is group translocation, in which a solute is changed during its transport. The phosphotransferase system phosphorylates several sugars (e.g., glucose to glucose 6-phosphate) as it brings them into the cell. Phosphoenolpyruvate is the phosphoryl donor in this process, which is mediated by three kinds of enzymes and a small phosphoryl-carrier protein (HPr).

Halobacteria contain bacteriorhodopsin, the simplest known light-driven proton pump. This 26-kd protein contains a covalently attached retinal chromophore that cycles between an all-*trans* and a 13-*cis* form. Photoisomerization of all-*trans* retinal is followed by the deprotonation of the Schiff base linkage and the transport of two H^+ from the inside to the outside of the cell. Light also drives the active transport of Cl^- into the cell, a process mediated by halorhodopsin, a related retinal-containing protein.

Transport antibiotics make membranes permeable to certain ions by serving as carriers (e.g., valinomycin) or as channel formers (e.g., gramicidin A). Carrier antibiotics are donut-shaped molecules that bind a single metal ion in a central cavity. The hydrocarbon periphery enables the complex to traverse the hydrocarbon interior of the membrane. Channel-forming antibiotics create aqueous pores that span membranes.

Procaryotic and eucaryotic cells also contain aqueous channels that allow ions and small polar molecules to passively diffuse across membranes. Many contiguous cells in higher organisms are joined by gap junctions. These 20-Å-diameter cell-to-cell channels enable ions and most metabolites (e.g., monosaccharides, amino acids, and nucleotides) to flow from the interior of one cell to an adjoining one. These channels are important for intercellular communication and the transfer of metabolites. A channel is formed from two connexons, one from each apposed cell. A connexon consists of six identical 32-kd subunits. Closure of gap junctions by high concentrations of Ca^{2+} is achieved by a rotation and tilting of subunits, which occlude the aqueous pore.

SELECTED READINGS

Where to start

Pedersen, P. L., and Carafoli, E., 1987. Ion motive ATPases. *Trends Biochem. Sci.* 12:146–150 and 12:186–189. [A lucid overview of ATP-driven ion pumps emphasizing common motifs.]

Jay, D., and Cantley, L., 1986. Structural aspects of the red cell anion exchange protein. *Ann. Rev. Biochem.* 55:511–538.

Stoeckenius, W., 1985. The rhodopsin-like pigments of halobacteria: light-energy and signal transducers in an archaebacterium. *Trends Biochem. Sci.* 10:483–486.

Books

Tonomura, Y., 1986. *Energy-transducing ATPases—Structure and Kinetics*. Cambridge University Press. [A concise and highly informative presentation of the catalytic and energy-transducing mechanisms of ATPases involved in cell motility, oxidative phosphorylation, and membrane transport.]

Harold, F. M., 1986. *The Vital Force: A Study of Bioenergetics.* W. H. Freeman. [A highly readable account of energy transduction in membrane transport is given in chapters 5, 9, and 10.]

Martonosi, A. N., (ed.), 1985. *The Enzymes of Biological Membranes* (2nd ed.). Plenum. [Volume 3 contains many excellent articles on procaryotic and eucaryotic membrane transporters.]

Poste, G., and Crooke, S. T., (eds.), 1986. *New Insights into Cell and Membrane Transport Processes.* Plenum.

Senenza, G., and Kinne, R., 1985. *Membrane Transport Driven by Ion Gradients. Ann. N. Y. Acad. Sci.* vol. 456.

Andreoli, T. E., Hoffman, J. F., and Fanestil, D. D., (eds.), 1986. *Physiology of Membrane Disorders* (2nd ed.). Plenum.

Conformational transitions and energetics

Tanford, C., 1983. Mechanism of free energy coupling in active transport. *Ann. Rev. Biochem.* 52:379–409.

Jencks, W. P., 1982. Rules and the economics of energy balance in coupled vectorial processes. *In* Martonosi, A. N., (ed.), *Membranes and Transport,* vol. 1, pp. 515–520. Plenum.

Eisenberg, E., and Hill, T. L., 1985. Muscle contraction and free energy transduction in biological systems. *Science* 227:999–1006

Jardetzky, O., 1966. Simple allosteric model for membrane pumps. *Nature* 211:969.

Sodium-potassium pump

Shull, G. E., Schwartz, A., and Lingrel, J. B., 1985. Amino-acid sequence of the catalytic subunit of the ($Na^+ + K^+$) ATPase deduced from a complementary DNA. *Nature* 316:691–695.

Skou, J. C., and Norby, J. G., (eds.), 1979. *Na,K-ATPase: Structure and Kinetics.* Academic Press.

Sweadner, K. J., and Goldin, S. M., 1980. Active transport of sodium and potassium ions: mechanism, function, and regulation. *New Engl. J. Med.* 302:777–783.

Cantley, L. C., 1981. Structure and mechanism of the (Na-K)-ATPase. *Curr. Top. Bioenerg.* 11:201–237.

Estes, J. W., and White, P. D., 1965. William Withering and the purple foxglove. *Sci. Amer.* 212(6):110–117. [An interesting historical account of digitalis.]

Calcium transport

Carafoli, E., 1987. Intracellular calcium homeostasis. *Ann. Rev. Biochem.* 56:395–433.

MacLennan, D. H., Brandl, C. J., Korczak, B., and Green, N. M., 1985. Amino-acid sequence of a $Ca^{2+} + Mg^{2+}$-dependent ATPase from rabbit muscle sarcoplasmic reticulum, deduced from its complementary DNA sequence. *Nature* 316:696–700.

Pickart, C. M., and Jencks, W. P., 1984. Energetics of the calcium-transporting ATPase. *J. Biol. Chem.* 259:1629–1643.

deMeis, L., and Vianna, A. L., 1979. Energy interconversion by the Ca^{2+}-dependent ATPase of the sarcoplasmic reticulum. *Ann. Rev. Biochem.* 48:275–292.

Racker, E., 1972. Reconstitution of a calcium pump with phospholipids and a purified Ca^{2+}-adenosine triphosphatase from sarcoplasmic reticulum. *J. Biol. Chem.* 247:8198–8200.

Sugar-ion cotransport

Wright, J. K., Seckler, R., and Overath, P., 1986. Molecular aspects of sugar:ion cotransport. *Ann. Rev. Biochem.* 55:225–248.

Semenza, G., Kessler, M., Hosang, M., Weber, J., and Schmidt, U., 1984. Biochemistry of the Na^+,D-glucose cotransporter of the small-intestinal brush-border membrane. *Biochim. Biophys. Acta* 779:343–379.

Bacterial transport processes

Kaback, H. R., 1986. Active transport in *Escherichia coli*: passage to permease. *Ann. Rev. Biophys. Biophys. Chem.* 15:279–319.

Ferro-Luzzi Ames, G., 1986. Bacterial periplasmic transport systems: structure, mechanism, and evolution. *Ann. Rev. Biochem.* 55:397–425.

Saier, M. H., Jr., 1987. *Enzymes in Metabolic Pathways: A Comparative Study of Mechanism, Structure, Evolution, and Control.* Harper & Row. [Chapter 5 contains an interesting discussion of sugar transport in bacteria.]

Bouma, C. L., Meadow, N. D., Stover, E. W., and Roseman, S., 1987. II-B^{Glc}, a glucose receptor of the bacterial phosphotransferase system: molecular cloning of the *pts*G and purification of the receptor from an overproducing strain of *Escherichia coli. Proc. Nat. Acad. Sci.* 84:930–934.

Bacteriorhodopsin and halorhodopsin

Nassal, M., Mogi, T., Karnik, S. S., and Khorana, H. G., 1987. Structure-function studies on bacteriorhodopsin. III. Total synthesis of a gene for bacterio-opsin and its expression in *Escherichia coli. J. Biol. Chem.* 262:9264–9270.

Stoeckenius, W., 1976. The purple membrane of salt-loving bacteria. *Sci. Amer.* 234(6):38–46.

Stoeckenius, W., and Bogomolni, R. A., 1982. Bacteriorhodopsin and related pigments of halobacteria. *Ann. Rev. Biochem.* 52:587–616.

Smith, S. O., Hornung, I., van der Steen, R., Pardoen, J. A., Braiman, M. S., Lugtenburg, J., and Mathies, R. A., 1986. Are C_{14}-C_{15} single bond isomerizations of the retinal chromophore involved in the proton-pumping mechanism of bacteriorhodopsin? *Proc. Nat. Acad. Sci.* 83:967–971. [Reports resonance Raman studies showing that the primary step is a C_{13}-C_{14} *trans-cis* isomerization.]

Lanyi, J. K., 1986. Halorhodopsin: a light-driven chloride ion pump. *Ann. Rev. Biophys. Biophys. Chem.* 15:11–28.

Blanck, A., and Oesterhelt, D., 1987. The halo-opsin gene II. Sequence, primary structure of halorhodopsin and comparison with bacteriorhodopsin. *EMBO J.* 6:265–273.

Anion exchange protein

Kopito, R. R., and Lodish, H. F., 1985. Primary structure and transmembrane orientation of the murine anion exchange protein. *Nature* 316:234–238.

Falke, J. J., Pace, R. J., and Chan, S. I., 1984. Chloride binding to the anion transport binding sites of band 3. A ^{35}Cl NMR study. *J. Biol. Chem.* 259:6472–6480.

Cell-cell channels

Unwin, P. N. T., and Ennis, P. D., 1984. Two configurations of a channel-forming membrane protein. *Nature* 307:609–613.

Young, J. D., Cohn, Z. A., and Gilula, N. B., 1987. Functional assembly of gap junction conductance in lipid bilayers: demonstration that the major 27 kd protein forms the junctional channel. *Cell* 48:733–743.

Paul, D. L., 1986. Molecular cloning of cDNA for rat liver gap junction protein. *J. Cell Biol.* 103:123–134.

Loewenstein, W. R., 1981. The cell-to-cell channel. *Fed. Proc.* 37:2645–2650.

Spray, D. C., Stern, J. H., Harris, A. L., and Bennett, M. V. L., 1982. Gap junctional conductance: comparison of sensitivities to H^+ and Ca^{2+} ions. *Proc. Nat. Acad. Sci.* 79:441–445.

Staehelin, L. A., and Hull, B. E., 1978. Junctions between living cells. *Sci. Amer.* 238(5):140–152.

Transport antibiotics

Dobler, M., 1981. *Ionophores and Their Structures.* Wiley-Interscience. [A well-illustrated account of transport antibiotics and other ionophores.]

Ovchinnikov, Y. A., 1979. Physico-chemical basis of ion transport through biological membranes: ionophores and ion channels. *Eur. J. Biochem.* 94:321–336.

Laüger, P., 1980. Kinetic properties of ion carriers and channels. *J. Memb. Biol.* 57:163–178.

Urban, B. W., Hladky, S. B., and Haydon, D. A., 1978. The kinetics of ion movements in the gramicidin channel. *Fed. Proc.* 37:2628–2632.

Krasne, S., Eisenman, G., and Szabo, G., 1971. Freezing and melting of lipid bilayers and the mode of action of nonactin, valinomycin, and gramicidin. *Science* 174:412–415.

Anderson, O. S., 1984. Gramicidin channels. *Ann. Rev. Physiol.* 46:531–548.

CHAPTER 38

Hormone Action

Hormones are chemical messengers that coordinate the activities of different cells in multicellular organisms. The term *hormone* (from Greek, meaning to spur on) was first used in 1904 by William Bayliss and Ernest Starling to describe the action of secretin, a molecule secreted by the duodenum that stimulates the flow of pancreatic juice. Several very fruitful concepts emerged from their work: (1) hormones are molecules synthesized by specific tissues (*glands*); (2) they are secreted directly into the blood, which carries them to their sites of action; and (3) they specifically alter the activities of responsive tissues (target organs or *target cells*).

Hormones are chemically diverse. Some hormones, such as epinephrine and thyroxine, are *small molecules derived from amino acids.* Others, such as oxytocin, insulin, thyroid-stimulating hormone, and many growth factors are *polypeptides* or *proteins.* A third group, made from cholesterol, are the *steroid hormones.* The most recently discovered group are the *eicosanoids*, which are formed from arachidonate, a C_{20}-polyunsaturated fatty acid.

Recent research has revealed several recurring motifs in hormone action. This chapter will focus on four general mechanisms. Most hormones bind to cell surface receptors and trigger cascades of enzymatic reactions. The *adenylate cyclase cascade,* leading to an increased level of cyclic AMP and the activation of a protein kinase, is a major transduction pathway. The *phosphoinositide cascade* is another generally used pathway of hormone action: the hydrolysis of a membrane phospholipid produces two intracellular messengers—inositol trisphosphate opens calcium channels, and diacylglycerol activates a protein kinase.

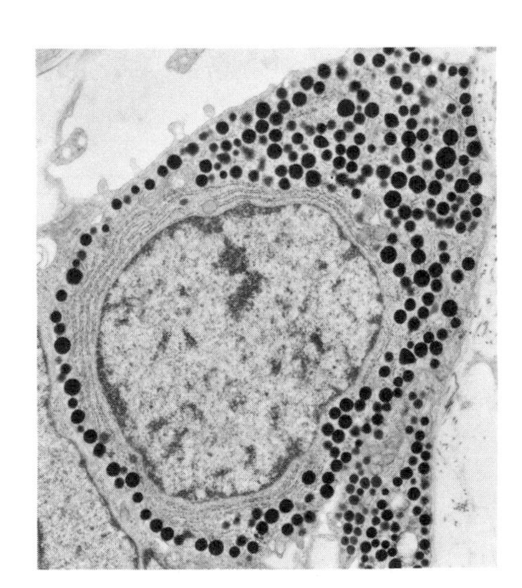

Figure 38-1
Electron micrograph of a somatotrope cell in the pituitary gland. Growth hormone (somatotropin) is stored in the prominent dense granules. [Courtesy of Lynne Mercer.]

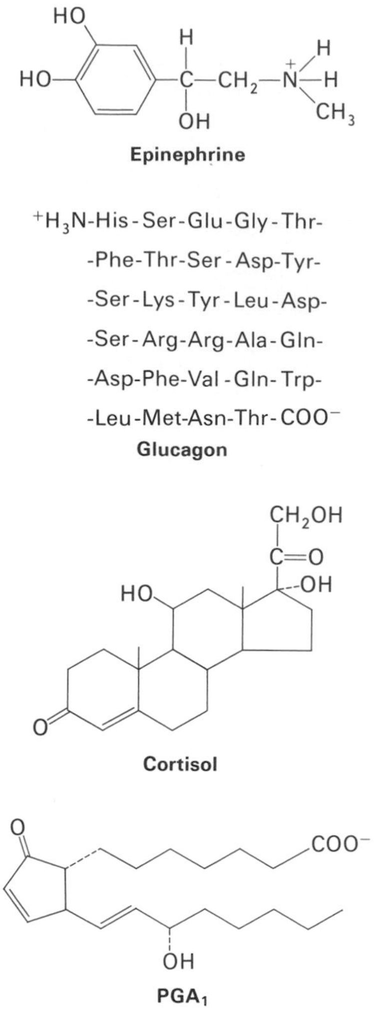

Insulin and many growth factors act by stimulating the *tyrosine kinase* activity of their receptors. Moreover, several oncogene products have tyrosine kinase activity or trigger this cascade. Steroid and thyroid hormones act quite differently. They enter cells and bind to receptors that are transported to the nucleus, where they bind to specific sites on DNA. These hormone-receptor complexes are *transcriptional enhancers.*

DISCOVERY OF CYCLIC AMP, A MEDIATOR OF THE ACTION OF MANY HORMONES

A major breakthrough by Earl Sutherland in the elucidation of the mechanism of action of hormones grew out of studies started in the 1950s. The initial aim was to determine how epinephrine and glucagon elicit the breakdown of glycogen by the liver. Sutherland chose this system because the effects of these hormones are large and rapid. Another advantage is that liver slices are readily prepared in large quantity. Furthermore, much was already known about the biochemistry of glycogen breakdown (Chapter 19). In fact, Sutherland began these studies in the laboratory of Carl Cori and Gerty Cori.

The first step was to identify the hormone-sensitive reaction. Measurements of labeled intermediates formed on incubating liver slices with $^{32}P_i$ showed that the reaction catalyzed by phosphorylase, rather than those catalyzed by phosphoglucomutase or glucose 6-phosphatase, was rate-limiting in the production of glucose. Moreover, its activity was markedly increased by epinephrine or glucagon. Sutherland then found an enzyme that catalyzed the inactivation of active phosphorylase. This deactivating enzyme proved to be a phosphatase, suggesting that phosphorylase is activated by phosphorylation. Indeed, the rate of incorporation of ^{32}P into phosphorylase was found to be increased by glucagon and epinephrine, in direct proportion to their acceleration of glycogen breakdown. These studies revealed that *phosphorylase is activated by phosphorylation and inactivated by dephosphorylation.* It was the first example of *enzyme regulation by covalent modification.*

The activation of phosphorylase by hormones in a preparation of broken liver cells was then studied. The exciting finding was that epinephrine or glucagon activated phosphorylase, as they did in liver slices. *This observation of hormone action in a cell-free homogenate is a landmark in biochemistry.* Specific hormone effects had not previously been observed in cell-free systems. Cell-fractionation experiments then showed that the soluble component alone (which contained phosphorylase) was unresponsive to hormone; the particulate fraction (which contained the plasma membrane) was also required. The working hypothesis at this point was that the binding of hormone to the plasma membrane led to the production of a substance that stimulated the phosphorylation of phosphorylase. Indeed, *a heat-stable activator was produced when the particulate fraction was incubated with hormone and nucleoside triphosphates.*

The next challenge was to identify this heat-stable intermediate, of which only small amounts were available. Chemical analysis showed that it was an adenine ribonucleotide, but its properties were unusual. Sutherland wrote Leon Heppel about this molecule in the hope that he might be able to help in elucidating its structure. At the same time, David Lipkin wrote Heppel describing a new nucleotide that was produced by treating ATP with barium hydroxide. Heppel surmised that Lipkin and Sutherland were studying the same molecule, and he put

Epinephrine

^+H_3N-His-Ser-Glu-Gly-Thr-
-Phe-Thr-Ser-Asp-Tyr-
-Ser-Lys-Tyr-Leu-Asp-
-Ser-Arg-Arg-Ala-Gln-
-Asp-Phe-Val-Gln-Trp-
-Leu-Met-Asn-Thr-COO^-

Glucagon

Cortisol

PGA_1

Figure 38-2
Examples of four chemical classes of hormones: epinephrine, an amino acid derivative; glucagon, a polypeptide; cortisol, a steroid; and prostaglandin PGA_1, a fatty acid derivative.

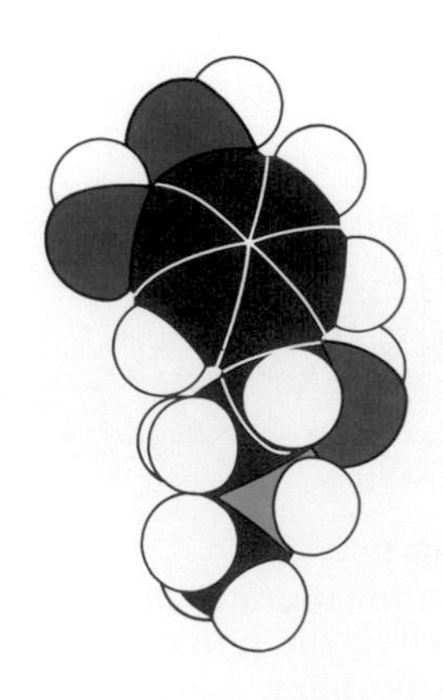

Figure 38-3
Model of epinephrine.

them in touch with each other. Indeed, both investigators were studying the same molecule, which turned out to be adenosine 3′,5′-monophosphate, now commonly referred to as *cyclic AMP* or *cAMP*. Another dividend of this chance encounter was that large amounts of cyclic AMP could be prepared for biochemical studies. Furthermore, the laboratory synthesis of cyclic AMP from ATP and barium hydroxide suggested a plausible route for its biosynthesis.

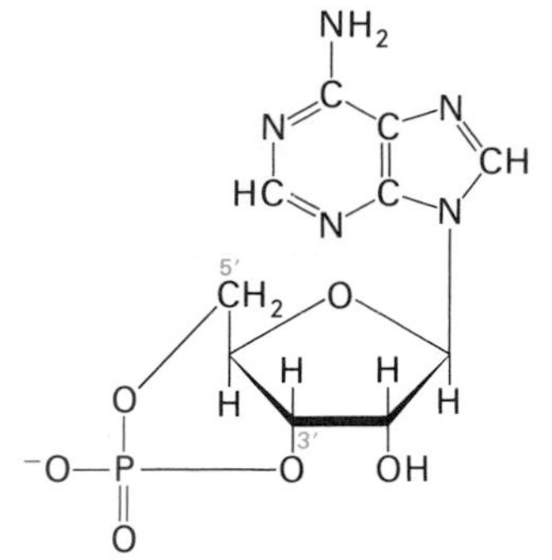

Adenosine 3′,5′-monophosphate (Cyclic AMP)

CYCLIC AMP IS SYNTHESIZED BY ADENYLATE CYCLASE AND DEGRADED BY A PHOSPHODIESTERASE

Cyclic AMP is formed from ATP by the action of *adenylate cyclase*, an integral membrane protein.

$$\text{ATP} \xrightarrow{Mg^{2+}} \text{cyclic AMP} + \text{PP}_i + \text{H}^+$$

This reaction is slightly endergonic, having a $\Delta G^{\circ\prime}$ of about 1.6 kcal/mol. It is driven in the direction of the synthesis of cyclic AMP by the subsequent hydrolysis of pyrophosphate. Cyclic AMP is degraded by a specific *phosphodiesterase*, which hydrolyzes it to AMP.

$$\text{Cyclic AMP} + H_2O \xrightarrow{Mg^{2+}} \text{AMP} + \text{H}^+$$

This reaction is highly exergonic, having a $\Delta G^{\circ\prime}$ of about −12 kcal/mol, yet cyclic AMP is a very stable compound in the absence of the phosphodiesterase. We see here a nice example of kinetic stability in the face of thermodynamic instability. This favorable property of phosphate esters gave them a selective advantage early in evolution, which accounts for their universal presence in nearly all biochemical pathways.

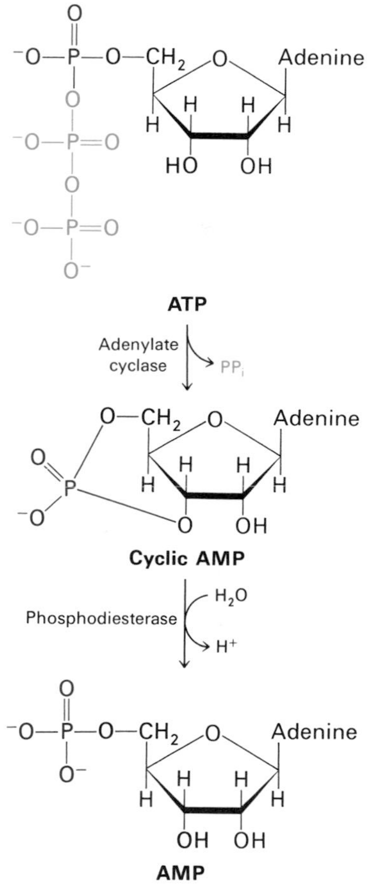

Figure 38-4
Enzyme-catalyzed synthesis and degradation of cyclic AMP.

CYCLIC AMP IS A SECOND MESSENGER IN THE ACTION OF MANY HORMONES

Sutherland's work led to the concept that cyclic AMP is a *second messenger* in the action of some hormones. The first messenger is the hormone itself. The essential features of this concept are:

1. The plasma membrane contains receptors for hormones.

2. The combination of a hormone with its specific receptor leads to the activation of adenylate cyclase, which is also bound to the plasma membrane.

3. The increased activity of adenylate cyclase increases the level of cyclic AMP in the cytosol.

4. Cyclic AMP then acts inside the cell to alter the rate of one or more processes.

An important feature of the second-messenger model is that the hormone need not enter the cell. Its impact is made at the cell membrane. The biological effects of the hormone are mediated inside the cell by cyclic AMP rather than by the hormone itself. This model makes several predictions that can be experimentally tested:

1. Adenylate cyclase in a target cell should be stimulated by hormones affecting that cell. Cells that do not show a characteristic biologi-

cal response to a hormone should not increase their cyclase level in its presence.

2. The change in concentration of cyclic AMP in a target cell should precede or occur at the same time as the response to hormonal stimulation. Variations in hormone levels should be matched by variations in the concentration of cyclic AMP.

3. Inhibitors of the phosphodiesterase, such as theophylline and caffeine, should act synergistically with hormones that use cyclic AMP as a second messenger.

4. The biological effects of a hormone should be mimicked by the addition of cyclic AMP or a related compound to the target cells. (In practice, cyclic AMP cannot be readily used in this way because it penetrates cells poorly. However, less-polar derivatives of cyclic AMP, such as dibutyryl cyclic AMP, do enter cells and are active.)

Caffeine
(1,3,7-Trimethylxanthine)

Theophylline
(1,3-Dimethylxanthine)

Experiments based on these criteria have revealed that *cyclic AMP is a second messenger for many hormones in addition to epinephrine and glucagon* (Table 38-1). Cyclic AMP affects a very wide range of cellular processes. For example, it enhances the degradation of storage fuels, increases the secretion of acid by the gastric mucosa, leads to the dispersion of melanin pigment granules, and diminishes the aggregation of blood platelets.

A GUANYL-NUCLEOTIDE-BINDING PROTEIN (G-PROTEIN) COUPLES HORMONE RECEPTORS TO ADENYLATE CYCLASE

Table 38-1
Hormones using cyclic AMP as a second messenger

Hormone
Calcitonin
Chorionic gonadotropin
Corticotropin
Epinephrine
Follicle-stimulating hormone
Glucagon
Luteinizing hormone
Lipotropin
Melanocyte-stimulating hormone
Norepinephrine
Parathyroid hormone
Thyroid-stimulating hormone
Vasopressin

How does the binding of a hormone such as epinephrine or glucagon to a specific receptor lead to the activation of adenylate cyclase? A key clue was Martin Rodbell's finding that GTP in addition to hormone is essential for activation. Equally revealing was the observation that hormone stimulates GTP hydrolysis. These findings led to the discovery that *a guanyl-nucleotide-binding protein is an intermediary in the activation process.* This signal-coupling protein was named *G protein* (G stands for guanyl nucleotide). The hormone-receptor complex does not directly stimulate adenylate cyclase. Rather, the activated receptor stimulates the G protein, which carries the excitation signal to adenylate cyclase. This signal-transduction process takes place in the plasma membrane.

How does the G protein control adenylate cyclase and what is the role of GTP in this process? Alfred Gilman has shown that the G protein is a peripheral membrane protein consisting of α (45 kd), β (35 kd), and γ (7 kd) subunits. It interconverts between a GDP form and a GTP form (Figure 38-5). The GTP form activates adenylate cyclase, whereas the GDP form does not. In the absence of hormone, nearly all of the G protein is in the inactive GDP form. *The binding of hormone to the receptor triggers the exchange of GTP for bound GDP:* the hormone-receptor complex (but not the unoccupied receptor) binds to the G protein, induces the release of bound GDP, and allows GTP to enter. The α subunit bearing GTP (G_α-GTP) dissociates from the $\beta\gamma$ subunit. Adenylate cyclase is then activated by G_α-GTP. Thus, *the flow of information is from the hormone-receptor complex to the G protein and then to adenylate cyclase* (Figure 38-6). The key point is that the hormone-receptor complex catalyzes the activation of the G protein. *Many G_α-GTP are formed for each bound hormone, giving an amplified response.*

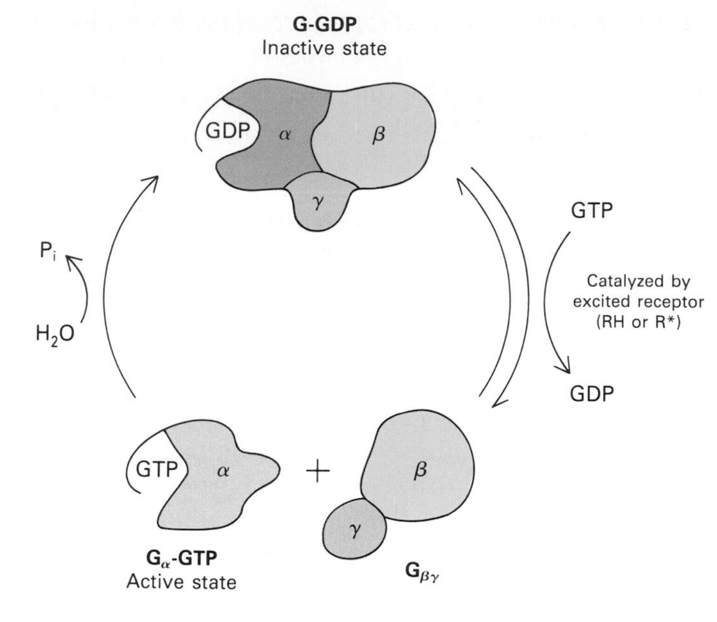

Figure 38-5
G proteins interconvert between an inactive GDP form and an active GTP form. The exchange of GTP for bound GDP is catalyzed by the hormone-receptor complex. G_α-GTP activates the effector protein. Hydrolysis of bound GTP brings the G protein back to the inactive state. The cycle is driven by the phosphoryl potential of GTP.

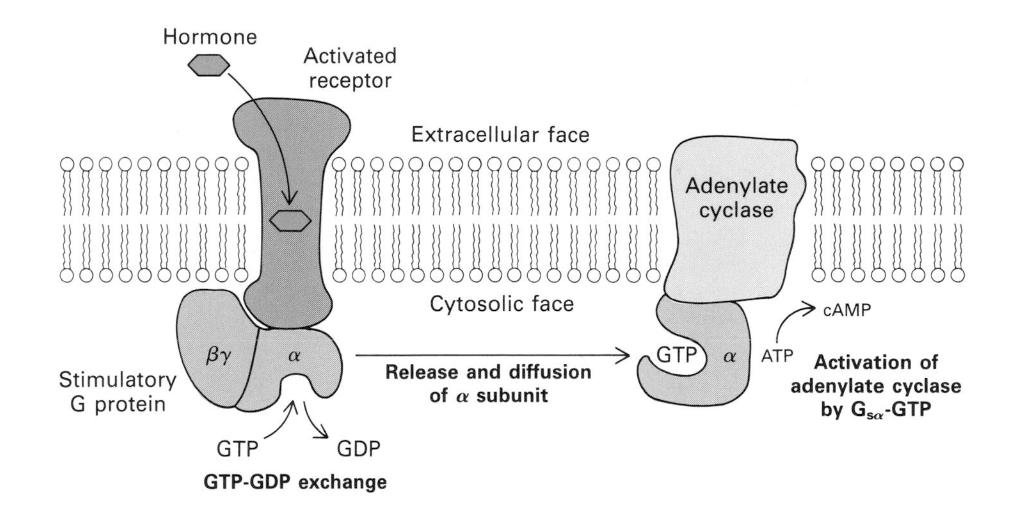

Figure 38-6
The activation of adenylate cyclase by the binding of a hormone to its specific receptor is mediated by G_s, the stimulatory G protein. A single hormone-receptor complex catalyzes the formation of many molecules of G_s. Hydrolysis of GTP bound to the α subunit of G_s terminates the activation of adenylate cyclase.

How is the activation of adenylate cyclase switched off? The G protein possesses yet another property that enables it to serve as the information-carrying intermediate between hormone receptors and adenylate cyclase. GTP bound to the α subunit of the G protein is slowly hydrolyzed to GDP. In other words, *the G protein is a GTPase. Thus, this regulatory protein has a built-in device for deactivation.* The proportion of G protein in the GTP state, and hence of adenylate cyclase in the active form, depends on the rate of exchange of GTP for GDP compared with the rate of hydrolysis of bound GTP. The rate of GTP-GDP exchange in the absence of hormone is very low because the uncatalyzed reaction has a very large activation barrier. Consequently, nearly all of the G protein is in the GDP form and nearly all of the adenylate cyclase, in turn, is inactive. The binding of hormone to the receptor leads to the amplified formation of G_α-GTP, which rapidly activates adenylate cy-

> *Adrenergic receptors*—
> Receptors on the plasma membrane that are triggered by epinephrine (adrenaline) and related catecholamines. A series of receptors were differentiated according to the relative potencies of agonists (hormones) and antagonists (inhibitors).
>
> α_1-Adrenergic receptors are coupled to the phosphoinositide cascade, whereas β_1- and β_2-receptors are coupled by G_s to the adenylate cyclase cascade.

clase by binding to it. *The hydrolysis of bound GTP by G_α closes this hormone-triggered cycle.* As will be discussed shortly, G proteins participate in many signal-transduction processes. The one activating adenylate cyclase is called the *stimulatory G protein* (G_s).

THE SEVEN-HELIX MOTIF OF THE β-ADRENERGIC RECEPTOR IS A RECURRING THEME IN G PROTEIN CASCADES

Epinephrine (also called adrenaline) triggers the adenylate cyclase cascade by binding to the *β-adrenergic receptor*, a 64-kd protein spanning the plasma membrane of target cells. The purification of this receptor led to the cloning and sequencing of its gene. The hydrophobicity profile of the protein showed that it contains seven transmembrane helices, like other receptors that are coupled to G proteins (Figure 38-7). This *seven-helix motif* is found, for example, in rhodopsin (the photoreceptor

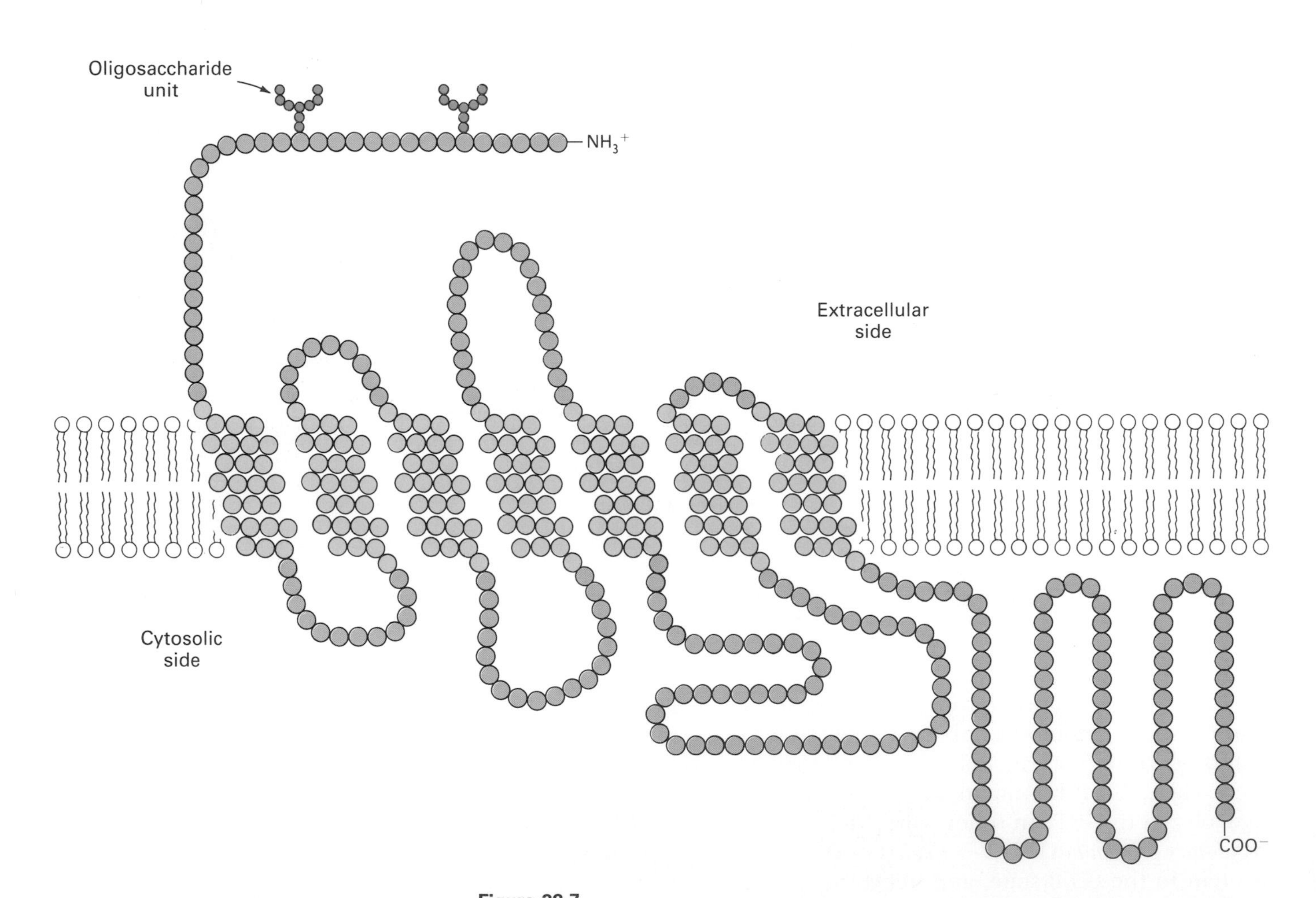

Figure 38-7
Proposed folding of the β-adrenergic receptor. This seven-helix motif is a common feature of transmembrane receptors that activate G proteins. Transmembrane helices are shown in yellow. Two *N*-linked oligosaccharide units (green) are located on the extracellular side of the plasma membrane. A loop on the cytosolic side participates in activating G proteins. Phosphorylation of multiple serine residues in the carboxyl-terminal tail prevents the receptor from interacting with the G protein. [After H. G. Dohlman, M. G. Caron, and R. J. Lefkowitz. *Biochemistry* 26(1987):2660.]

protein in retinal rod cells, p. 1030) and in the muscarinic acetylcholine receptor (which controls K^+ channels in heart cells). The amino-terminal region of these receptors, containing *N*-linked oligosaccharides, lies on the extracellular side of the membrane, and the carboxyl-terminal region resides on the cytosolic side. The binding site for epinephrine is located in a pocket formed by transmembrane helices.

The activation of G protein by the β-adrenergic receptor does not depend solely on whether hormone is bound. Receptor molecules exposed to a constant level of epinephrine for an extended period no longer catalyze GTP-GDP exchange effectively. This changing relation between stimulus and response, a general feature of sensory systems, is called *desensitization* or *adaptation*. Signal-transducing systems are designed to respond to *changes* in the concentration of stimuli rather than to their absolute concentration. Adaptation is advantageous because it enables receptors to operate over a wide range of background concentration of stimuli. The carboxyl-terminal region of the β-adrenergic receptor plays a key role in adaptation. Serine residues in this region are phosphorylated by a specific kinase that acts on the hormone-receptor complex, but not on the receptor alone. *Phosphorylation at multiple sites prevents the hormone-receptor complex from catalyzing GTP-GDP exchange and thereby blocks signal transmission* (Figure 38-8). Sensitivity is restored by the removal of the attached phosphates by a phosphatase. We shall encounter this regulatory device again when we consider the molecular mechanism of visual excitation (p. 1033).

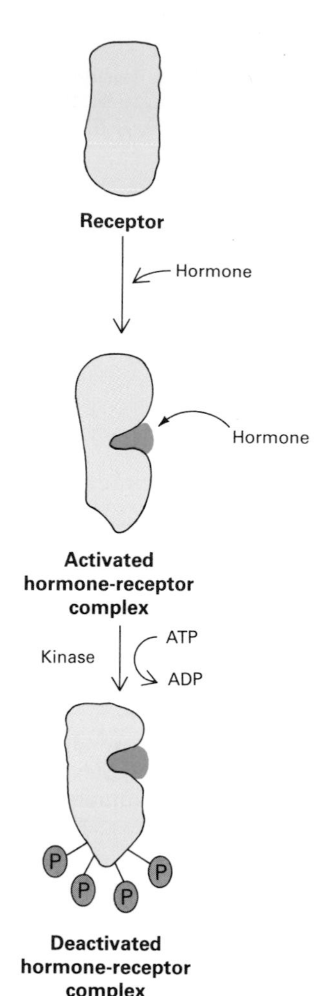

Figure 38-8
The β-adrenergic receptor is deactivated by phosphorylation. The hormone-receptor complex, but not the receptor alone, is multiply phosphorylated by a specific kinase. Dephosphorylation restores the capacity of the receptor to activate the G protein.

CYCLIC AMP ACTIVATES A PROTEIN KINASE BY RELEASING ITS REGULATORY SUBUNITS

How does cyclic AMP influence so many cellular processes? Is there a common denominator for its diverse effects? Indeed there is, and again the answer has come from studies of the control of glycogen metabolism, which has been described in a lighter vein as the metabolic birthplace of cyclic AMP. Edwin Krebs and Donal Walsh discovered that *cyclic AMP activates a protein kinase*. In skeletal muscle, this kinase phosphorylates both glycogen synthase (rendering it inactive) and phosphorylase kinase (rendering it active). In this way, cyclic AMP stimulates glycogen breakdown and stops glycogen synthesis (p. 462). A similar mechanism operates in the liver. In fact, *all known effects of cyclic AMP in eucaryotic cells result from the activation of protein kinases.* These protein kinases, which are activated by cyclic AMP concentrations of the order

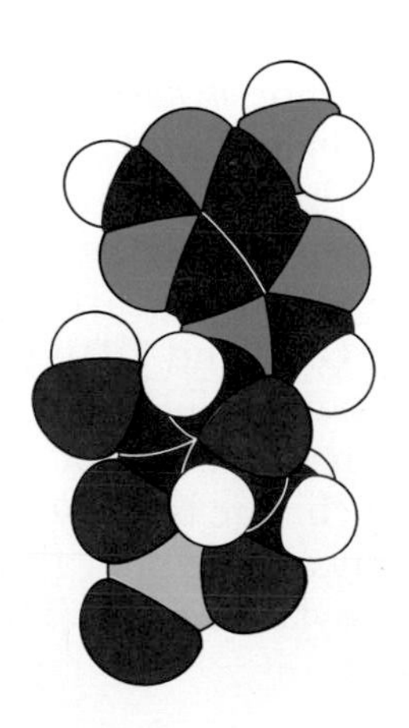

Model of cyclic AMP.

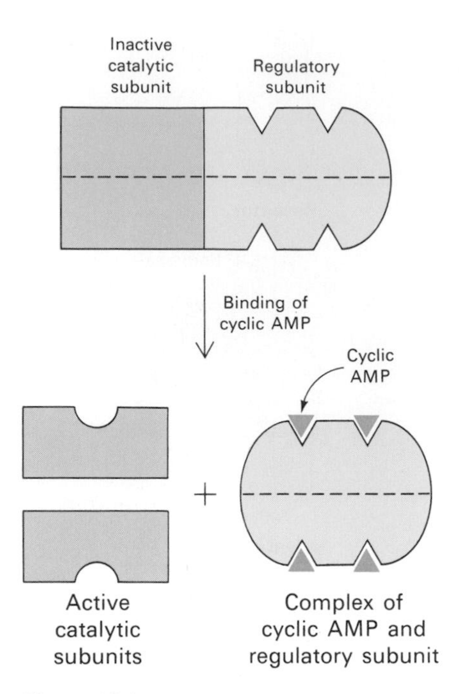

Figure 38-9
Cyclic AMP activates protein kinases by dissociating the complex of regulatory and catalytic subunits.

of 10^{-8} M, modulate the activities of target proteins by phosphorylating serine residues in Arg-Arg-X-Ser-X sequences (X denotes any residue).

The mechanism of activation of the cyclic-AMP-controlled kinase is interesting. The enzyme in muscle consists of two kinds of subunits: a 49-kd regulatory (R) subunit, which can bind cyclic AMP, and a 38-kd catalytic (C) subunit. In the absence of cyclic AMP, the regulatory and catalytic subunits form an R_2C_2 complex that is enzymatically inactive. The binding of two molecules of cyclic AMP to each of the regulatory subunits leads to the dissociation of the R_2C_2 complex into an R_2 subunit and two C subunits. These free catalytic subunits are then enzymatically active. Thus, *the binding of cyclic AMP to the regulatory subunit relieves its inhibition of the catalytic subunit. Cyclic AMP acts as an allosteric effector.* The presence of distinct regulatory and catalytic subunits in the protein kinase is reminiscent of aspartate transcarbamoylase (p. 235).

The concentration of many hormones in the blood is of the order of 10^{-10} M. The adenylate cyclase cascade greatly amplifies such weak hormonal signals. Three steps in the cascade provide a high degree of amplification: (1) Each hormone-receptor complex catalyzes the formation of many G_α-GTP; (2) Many molecules of cyclic AMP are formed by an activated adenylate cyclase; and (3) Each cAMP-activated protein kinase can alter the activity of many molecules of each target protein.

CHOLERA TOXIN STIMULATES ADENYLATE CYCLASE BY INHIBITING THE GTPase ACTIVITY OF G_s

The direct participation of cyclic AMP in a disease process has been clearly established in cholera. This potentially lethal disease is caused by *Vibrio cholerae,* a gram-negative bacterium. The striking clinical feature of the disease is a massive diarrhea. Several liters of body fluid may be lost within a few hours, which leads to shock and death if the fluids are not replaced. The diarrhea is caused by a toxin secreted by these bacteria rather than by the action of the bacteria themselves.

Cholera toxin, an 87-kd protein, consists of an A subunit (an A_1 peptide linked by a disulfide bond to an A_2 peptide) and five B subunits. It enters intestinal mucosal cells by interacting with a G_{M1} ganglioside (p. 553) on the cell surface. This carbohydrate-rich sphingolipid is recognized by the B chains of the toxin. After gaining entry, the 23-kd A_1 chain covalently modifies G_s, the G protein that stimulates adenylate cyclase. Specifically, A_1 catalyzes the transfer of an ADP-ribose unit from NAD^+ to a specific arginine side chain of the α subunit of G_s (Figure 38-10). *This ADP-ribosylation of G_s blocks its capacity to hydrolyze bound GTP to GDP,* and so impairs the built-in deactivation device (p. 979). *The G protein is locked in the active form. Hence, adenylate cyclase stays persistently activated in the absence of hormone.* The level of cyclic AMP becomes abnormally high, which stimulates the active transport of ions and leads to a very large efflux of Na^+ and water into the gut.

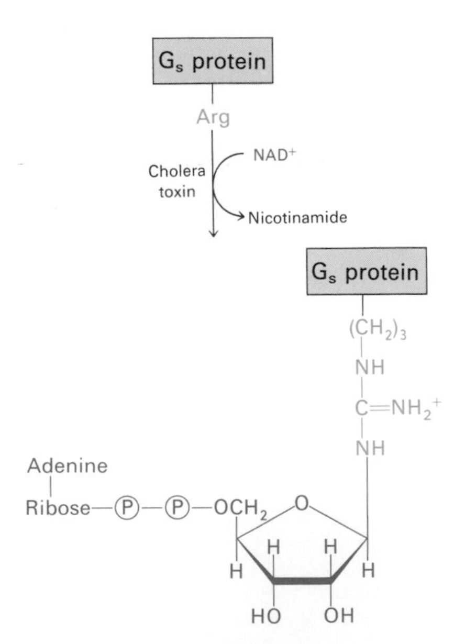

Figure 38-10
Cholera toxin catalyzes the ADP-ribosylation of the G_s protein, a regulator of the activity of adenylate cyclase.

Impaired GTPase activity in a regulatory protein can also lead to cancer. The v-*ras* oncogene of murine sarcoma viruses encodes a 21-kd protein that cycles between GTP and GDP forms. This viral oncogene product differs from its normal cellular counterpart in having diminished GTPase activity, which keeps it in a persistently activated form. Indeed, a single amino acid change (glycine to valine) converts the normal cellular gene into an oncogenic one. In yeast, ras proteins are involved in the control of adenylate cyclase. They have a different, as yet unknown, function in mammals. The oncogenic action of v-*ras* and mutant c-*ras* reveals that a guanyl-nucleotide-binding protein participates in growth control.

CYCLIC AMP IS AN ANCIENT HUNGER SIGNAL

Cyclic AMP has a regulatory role in bacteria, too, where it stimulates the transcription of certain genes (p. 673). It is evident that *cyclic AMP has a long evolutionary history as a regulatory molecule.* In bacteria, cyclic AMP is a hunger signal. It signifies an absence of glucose and leads to the synthesis of enzymes that can exploit other energy sources. In some mammalian cells, such as liver and muscle, cyclic AMP retains its ancient role as a hunger signal. However, it acts by stimulating a protein kinase rather than by enhancing the transcription of certain genes. Another difference is that cyclic AMP in higher organisms mediates intercellular signaling rather than intracellular signaling. The role of cyclic AMP in the life cycle of the slime mold *Dictyostelium discoideum,* a simple eucaryote, is especially interesting. When food is abundant, *Dictyostelium* exists as independent cells. When food become scarce, cyclic AMP is secreted by the free-living amoebae. Cyclic AMP serves as a chemoattractant that leads to the aggregation of *Dictyostelium* into a slug and to major changes in gene expression.

Why was cyclic AMP chosen in the course of evolution to be a signal molecule? Three factors seem important:

1. Cyclic AMP is derived from ATP, a ubiquitous molecule, in a simple reaction driven forward by the subsequent hydrolysis of pyrophosphate.

2. Though derived from a molecule that is at the center of metabolic transformations, cyclic AMP itself is not on a major metabolic pathway. It is used only as an integrator of metabolism, not as a biosynthetic precursor or intermediate in energy production. Hence, its concentration can be independently controlled. Furthermore, it is stable unless hydrolyzed by a specific phosphodiesterase.

3. The numerous functional groups of cyclic AMP enable it to bind tightly and specifically to receptor proteins, such as the regulatory subunit of the protein kinase in muscle, and elicit allosteric effects.

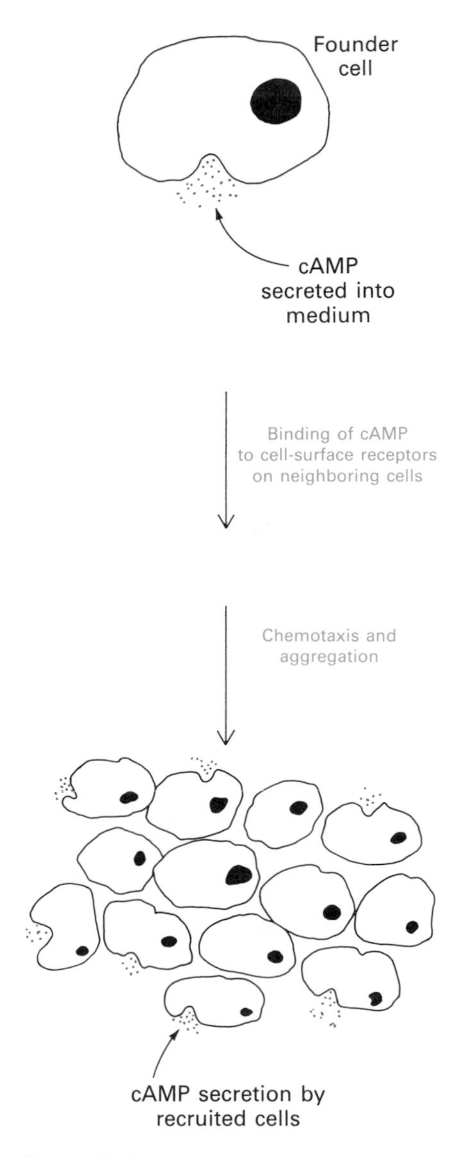

Figure 38-11
Cyclic AMP secreted by a *Dictyostelium* cell triggers the aggregation of many cells into a slug.

A FAMILY OF G PROTEINS TRANSDUCES MANY HORMONAL AND SENSORY STIMULI

Certain hormones, such as opiates and α_2-adrenergic amines, lower cyclic AMP levels in target cells and reduce adenylate cyclase activity in membrane preparations derived from them. These agents stimulate GTP hydrolysis as they inhibit adenylate cyclase. The protein mediating their inhibitory effect was identified as a result of studies aimed at a different goal, the understanding of the mode of action of the toxin produced by *Bordetella pertussis,* the bacterium that causes whooping cough. Pertussis toxin blocks the inhibition of adenylate cyclase by catalyzing the covalent modification of what proved to be the *inhibitory G protein* (G_i).

Like G_s, G_i consists of α, β, and γ subunits. Indeed, the β and γ subunits of these proteins are identical, and their α subunits are similar. The actions of these two G proteins are linked by their common β and γ subunits. The hormone-triggered exchange of GTP for GDP in unmodified G_i splits it into α and $\beta\gamma$ subunits. *The released $\beta\gamma$ subunits then bind the α subunit of G_s, reversing its activation of adenylate cyclase.* This inhibitory mechanism is effective because the cell contains much more G_i than G_s. *Pertussis toxin* catalyzes the ADP-ribosylation of a specific cysteine side chain on the α subunit of G_i. This covalent modification locks G_i in the GDP form.

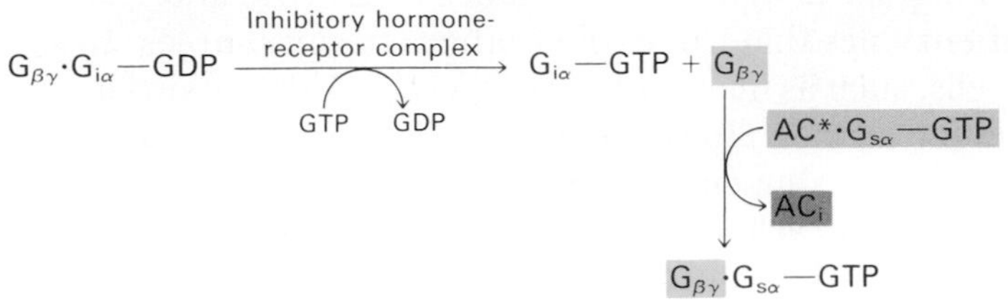

$$G_{\beta\gamma}\cdot G_{i\alpha}\text{—GDP} \xrightarrow[\text{GTP} \curvearrowright \text{GDP}]{\text{Inhibitory hormone-receptor complex}} G_{i\alpha}\text{—GTP} + G_{\beta\gamma}$$

$$G_{\beta\gamma} + \text{AC}^{*}\cdot G_{s\alpha}\text{—GTP} \longrightarrow \text{AC}_i + G_{\beta\gamma}\cdot G_{s\alpha}\text{—GTP}$$

G_s and G_i are members of a large family of signal-coupling proteins (Table 38-2). The brain is rich in G_o, a G protein whose function is not yet known (*o* stands for other). G proteins have been implicated in controlling phospholipase C, an effector enzyme in the phosphoinositide cascade (p. 985). They are involved in regulating the activity of K^+ channels in cardiac cells. The best understood G protein is *transducin*, which plays a key role in visual excitation (p. 1033). G proteins also participate in leukocyte chemotaxis, olfaction, and simple forms of learning in invertebrates. This family has many common structural and functional motifs:

1. *G proteins are intermediaries between activated receptors and effectors.* They are heterotrimers of α, β, and γ subunits. The α subunit contains the guanyl nucleotide binding site and carries the excitation signal to the effector protein (an enzyme or channel). The β and γ subunits are required for the interaction of α with the activated receptor.

2. They interconvert between an inactive GDP form and an active GTP form. The role of the activated receptor is to catalyze the exchange of GTP for GDP, leading the G protein to release G_α-GTP. Hydrolysis of bound GTP brings the system back to the inactive state.

Table 38-2
Physiological processes mediated by G proteins

Stimulus	*Receptor*	*G protein*	*Effector*	*Physiological response*
Epinephrine	β-Adrenergic receptor	G_s	Adenylate cyclase	Glycogen breakdown
Serotonin	Serotonin receptor	G_s	Adenylate cyclase	Behavioral sensitization and learning in *Aplysia*
Light	Rhodopsin	Transducin	cGMP phosphodiesterase	Visual excitation
IgE-antigen complexes	Mast cell IgE receptor	G_{PLC}	Phospholipase C	Secretion
f-Met peptide	Chemotactic receptor	G_{PLC}	Phospholipase C	Chemotaxis
Acetylcholine	Muscarinic receptor	G_K	Potassium channel	Slowing of pacemaker activity

Note: G_{PLC} and G_K refer to as yet unidentified G proteins in these cascades.
After L. Stryer and H. R. Bourne. *Ann. Rev. Cell Biol.* 2(1986):393.

3. G proteins are permanently activated by hydrolysis-resistant analogs of GTP, such as GppNHp and GTPγS. These analogs stay bound to the α subunit and keep it in the GTP state. Likewise, AlF_4^- persistently activates G proteins. AlF_4^- has nearly the same shape as PO_4^{2-}; it activates by binding next to GDP, where it mimics the γ-phosphate group of GTP.

Guanosine—P(=O)(O⁻)—O—P(=O)(O⁻)—N(H)—P(=O)(O⁻)—O⁻

Guanosine 5′-[β,γ-imido]triphosphate (GppNHp)

Guanosine—P(=O)(O⁻)—O—P(=O)(O⁻)—O—P(=S)(O⁻)—O⁻

Guanosine 5′-*O*-thiotriphosphate (GTPγS)

4. Nearly all known G proteins are ADP-ribosylated by cholera toxin or pertussis toxin.

5. G proteins are activated by receptors that are integral membrane proteins. These receptors share a seven-helix motif (p. 980). Phosphorylation of the carboxyl-terminal region of the receptors blocks their activation of G proteins. Further evidence for the homology of these receptors is the finding that some receptors can activate several G proteins.

The structural and mechanistic similarities of G proteins imply that they arose by duplication and divergence of a common ancestral gene. Sequence similarities between procaryotic elongation factor Tu (p. 755) and mammalian G proteins suggest that the controlled binding and release of macromolecules coupled to GTP-GDP exchange and hydrolysis was perfected early in evolution and elaborated over several billion years. This elegant molecular device now serves a wide variety of cellular functions, including the translocation of macromolecules in protein synthesis, the regulation of cell proliferation, and the transduction of hormonal signals and sensory stimuli.

RECEPTOR-TRIGGERED HYDROLYSIS OF PHOSPHATIDYL INOSITOL BISPHOSPHATE GENERATES TWO MESSENGERS

We turn now to the phosphoinositide cascade, which evokes a wide variety of responses in many kinds of cells (Table 38-3). The phosphoinositide cascade, like the adenylate cyclase cascade, converts extracellular signals into intracellular ones. The intracellular messengers formed by activation of this pathway arise from *phosphatidyl inositol 4,5-bisphosphate* (PIP_2), a phospholipid in the plasma membrane (Figure 38-12). PIP_2 is formed by phosphorylation of phosphatidyl inositol (PI), a membrane constituent synthesized from CDP-diacylglycerol (p. 549) and inositol. The binding of a hormone such as serotonin to a cell-surface receptor leads to the activation of *phosphoinositidase* (also called phospholipase C or polyphosphoinositide phosphodiesterase). This membrane-bound enzyme hydrolyzes the phosphodiester bond linking the phosphorylated inositol unit to the acylated glycerol moiety. Two messengers—*inositol 1,4,5-trisphosphate* (IP_3) and *diacylglycerol*—are formed by the cleavage of PIP_2. This hydrolysis is markedly stimulated

Table 38-3
Effects mediated by the phosphoinositide cascade

Glycogenolysis in liver cells
Histamine secretion by mast cells
Serotonin release by blood platelets
Aggregation of blood platelets
Insulin secretion by pancreatic islet cells
Epinephrine secretion by adrenal chromaffin cells
Smooth muscle contraction
Visual transduction in invertebrate photoreceptors

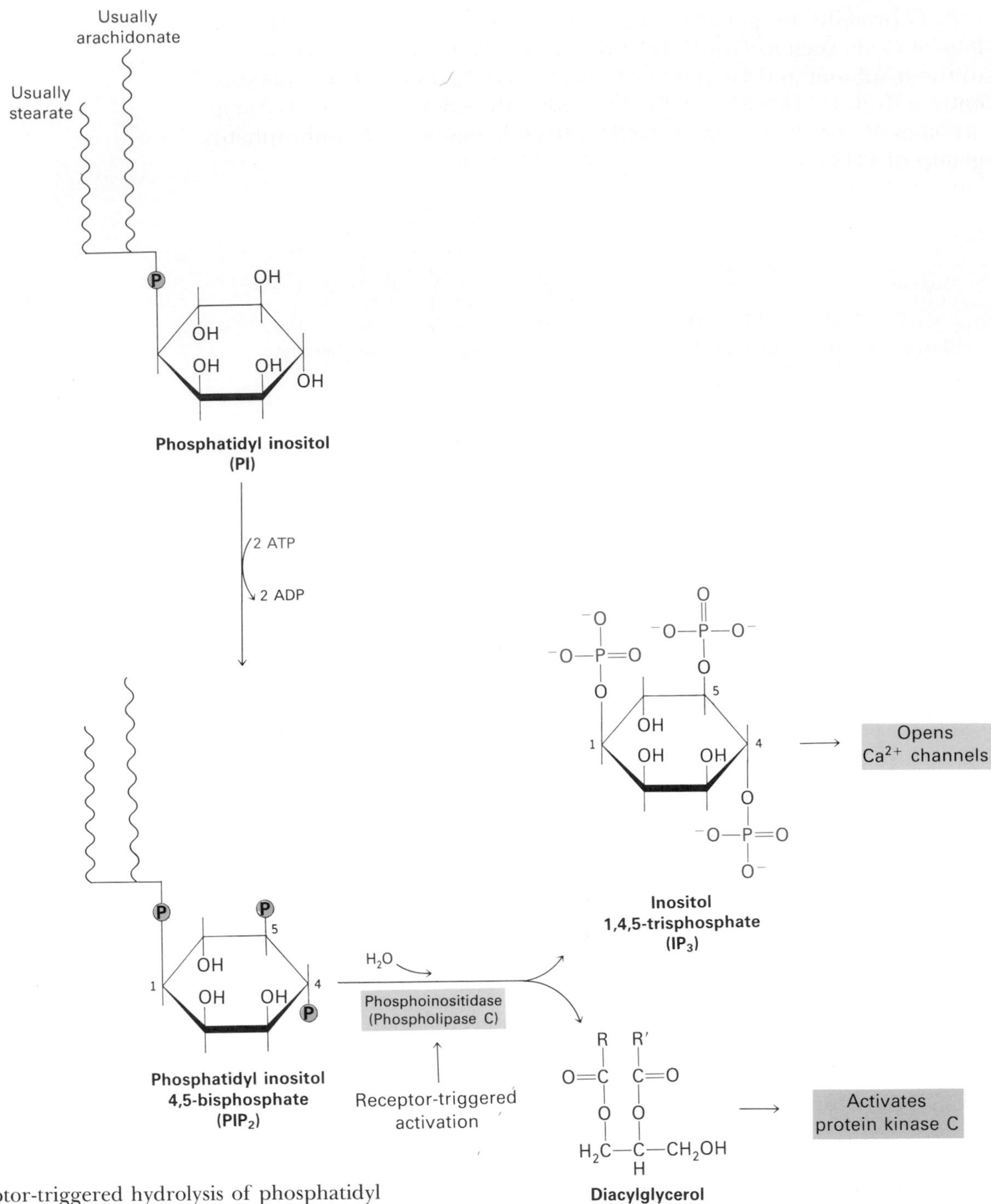

Figure 38-12
Synthesis and receptor-triggered hydrolysis of phosphatidyl inositol 4,5-bisphosphate (PIP_2). Two messengers are formed.

by hydrolysis-resistant analogs of GTP and by AlF_4^-, suggesting that a G-protein carries the excitation signal from the activated receptor to phosphoinositidase. The inhibition of this cascade by pertussis toxin also points to the participation of a G protein.

IP_3 is a short-lived messenger, lasting only a few seconds. It can be converted to inositol by the sequential action of three phosphatases. The removal of the 5-phosphate group terminates its messenger role. Inositol 1,4-bisphosphate is hydrolyzed to inositol 4-phosphate and then to inositol. Alternatively, IP_3 can be phosphorylated to inositol 1,3,4,5-tetrakisphosphate, which is then hydrolyzed to inositol 1,3,4-trisphosphate, a different isomer from the one formed by cleavage of

PIP_2. This 1,3,4-isomer is converted to inositol by successive dephosphorylations. The phosphatase that acts on inositol 1,3,4-phosphate is inhibited by millimolar levels of Li^+. It is interesting to note that lithium ion is widely used to treat manic-depressive disorders. Li^+ may exert its therapeutic action by inhibiting the recycling of 1,3,4-trisphosphate. Some of the compounds formed in the conversion of IP_3 to inositol may have messenger roles.

Diacylglycerol can be phosphorylated to phosphatidic acid, which reacts with CTP to form CDP-diacylglycerol. Alternatively, diacylglycerol can be hydrolyzed to glycerol and its constituent fatty acids. Arachidonate, the C_{20}-polyunsaturated fatty acid that usually occupies the 2-position on the glycerol moiety of PIP_2, is the precursor of a series of eicosanoid hormones (p. 991). Thus, the phosphoinositide pathway gives rise to many molecules that have signaling roles.

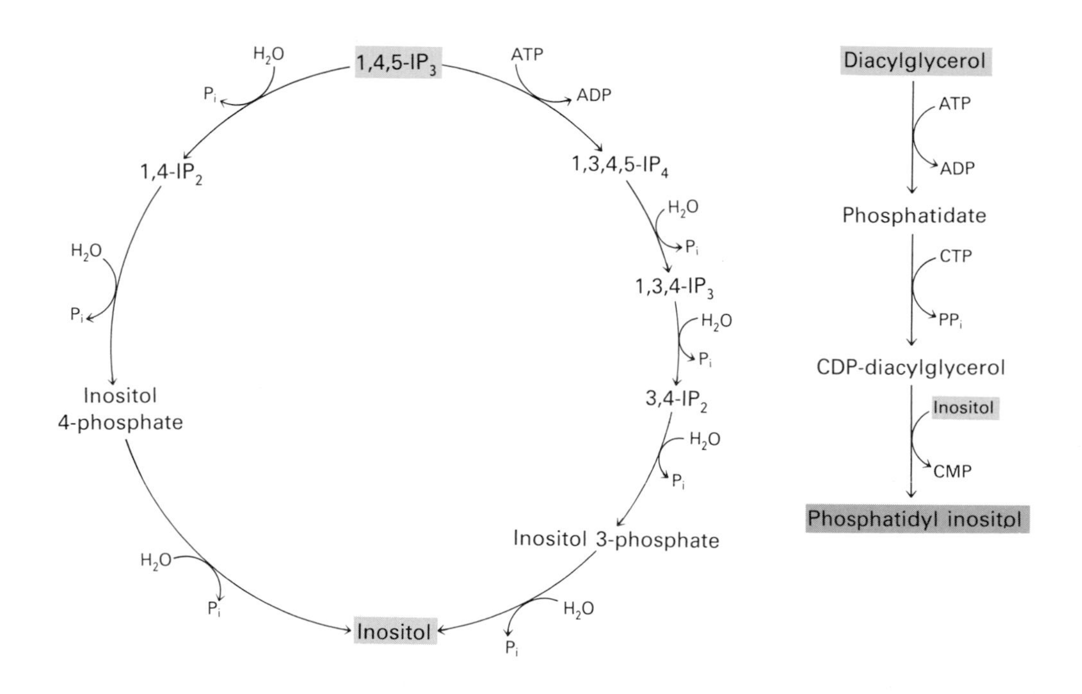

Figure 38-13
Recycling of inositol trisphosphate and diacylglycerol (DAG).

IP_3 RELEASES CALCIUM ION INTO THE CYTOSOL AND DIACYLGLYCEROL ACTIVATES PROTEIN KINASE C

The action of IP_3 has been investigated by microinjecting it into cells or by adding it to cells whose plasma membrane has been made permeable. Michael Berridge found that *IP_3 causes the rapid release of Ca^{2+} from intracellular stores—the endoplasmic reticulum and, in smooth muscle cells, the sarcoplasmic reticulum.* The elevated level of Ca^{2+} in the cytosol then triggers processes such smooth muscle contraction (p. 934), glycogen breakdown (p. 461), and exocytosis. Indeed, the injection of IP_3 into *Xenopus* oocytes suffices to activate many of the early events of fertilization, such as the release of granules and the opening of ion channels in the plasma membrane. Submicromolar levels of IP_3 mobilize Ca^{2+} from intracellular stores by directly opening calcium channels in the membrane of the endoplasmic reticulum and the sarcoplasmic reticulum.

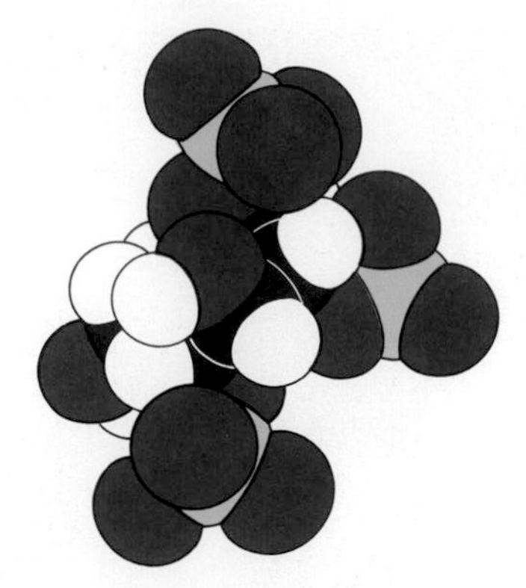

Figure 38-14
Model of inositol 1,4,5-trisphosphate (IP_3).

A phorbol ester

Diacylglycerol, the other messenger formed by the receptor-triggered hydrolysis of PIP_2, activates protein kinase C. This 77-kd enzyme phosphorylates serine and threonine residues in many target proteins. For example, the phosphorylation of glycogen synthase by protein kinase C stops the synthesis of glycogen. This action of protein kinase C nicely complements the IP_3-induced increase in the activity of glycogen phosphorylase, which is mediated by an increase in the cytosolic level of Ca^{2+}. Indeed, most effects of diacylglycerol and IP_3 are synergistic. Yasutomi Nishizuka found that protein kinase C is enzymatically active only in the presence of Ca^{2+} and phosphatidyl serine. *Diacylglycerol greatly increases the affinity of protein kinase C for Ca^{2+} and thereby renders it active at physiological levels of this ion.* The inactive enzyme is located mainly in the cytosol, whereas the active form is membrane-bound. Protein kinase C contains a catalytic domain and a regulatory domain; the binding of diacylglycerol reverses the inhibition imposed by the regulatory portion of the enzyme.

The importance of protein kinase C in controlling cell division and proliferation is revealed by the action of *phorbol esters*. These polycyclic alcohol derivatives from croton oil are carcinogenic; they are known as *tumor promoters*. Phorbol esters activate protein kinase C because they resemble diacylglycerol. The activation is persistent because phorbol esters, unlike diacylglycerol, are not readily degraded.

CALMODULIN BELONGS TO A FAMILY OF CALCIUM SENSORS CONTAINING HELIX-LOOP-HELIX SITES (EF HANDS)

Calcium ion is an intracellular messenger in many signal-transducing pathways, as exemplified by the phosphoinositide cascade and the regulation of muscle contraction. Why has nature chosen this ion to mediate so many signaling processes? *The intracellular level of Ca^{2+} must be kept low because phosphate esters are highly abundant and calcium phosphates are quite insoluble.* All cells have transport systems for the extrusion of Ca^{2+}. The cytosolic level of Ca^{2+} in unexcited cells is typically 0.1 μM, several orders of magnitude less than the concentration in the extracellular milieu. This steep gradient presents cells with an opportunity: *the cytosolic Ca^{2+} concentration can be abruptly raised for signaling purposes by transiently opening calcium channels in the plasma membrane or in an intracellular membrane.*

A second property of Ca^{2+} that makes it a highly suitable intracellular messenger is that it can bind tightly to proteins. Negatively charged oxygens (from the side chains of glutamate and aspartate) and uncharged oxygens (main-chain carbonyls) bind well to Ca^{2+}. *The capacity of Ca^{2+} to be coordinated to multiple ligands—six to eight oxygen atoms—enables it to cross-link different segments of a protein and induce large conformational changes. Furthermore, the binding of Ca^{2+} can be highly selective.* Mg^{2+}, a potential competitor, can be spurned because it does not have appreciable affinity for uncharged oxygen atoms. Another important difference between these ions is that Mg^{2+} prefers to form small and symmetric coordination shells, whereas Ca^{2+} can form asymmetric complexes having a larger radius. Thus, Ca^{2+} is well suited for binding to irregularly shaped crevices in proteins and can be selected over Mg^{2+} even when the latter is a thousandfold more abundant.

X-ray crystallographic studies of calcium-binding proteins have provided informative views of how these sensors function. Parvalbumin, a

12-kd protein in carp muscle, contains two similar Ca^{2+}-binding sites, which are formed by a helix, a loop, and another helix. Eight oxygen atoms are coordinated to each Ca^{2+}: six carboxylate oxygens of three aspartates and a glutamate, a main-chain carbonyl oxygen, and an oxygen of a bound water molecule (Figure 38-15). One of the sites is formed by helices E and F of this protein, which are positioned like the forefinger and thumb of the right hand (Figure 38-16). The Ca^{2+}-binding site is formed by a loop between these helices. Robert Kretsinger named this structural motif the *EF hand* and proposed that the two Ca^{2+}-binding sites of parvalbumin arose by duplication of a primordial gene encoding a calcium-binding loop. He noted that the amino acid sequences of parvalbumin and troponin C (p. 933) are similar and suggested that troponin C also contains EF hands. *Subsequent x-ray analyses have shown that the EF hand is a recurring motif in troponin C and other calcium-binding proteins.*

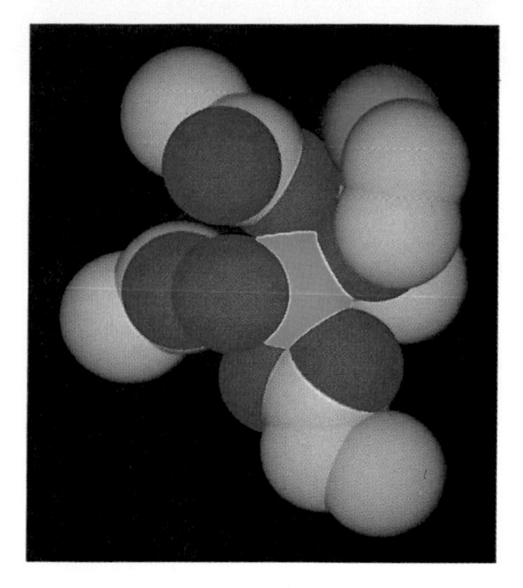

Figure 38-15
Mode of binding of Ca^{2+} (green) to a cluster of eight oxygen atoms (red) at one of the two calcium-binding sites of parvalbumin. Four of the oxygens come from aspartate side chains, two from a glutamate side chain, one from a main-chain carbonyl group, and one from water.

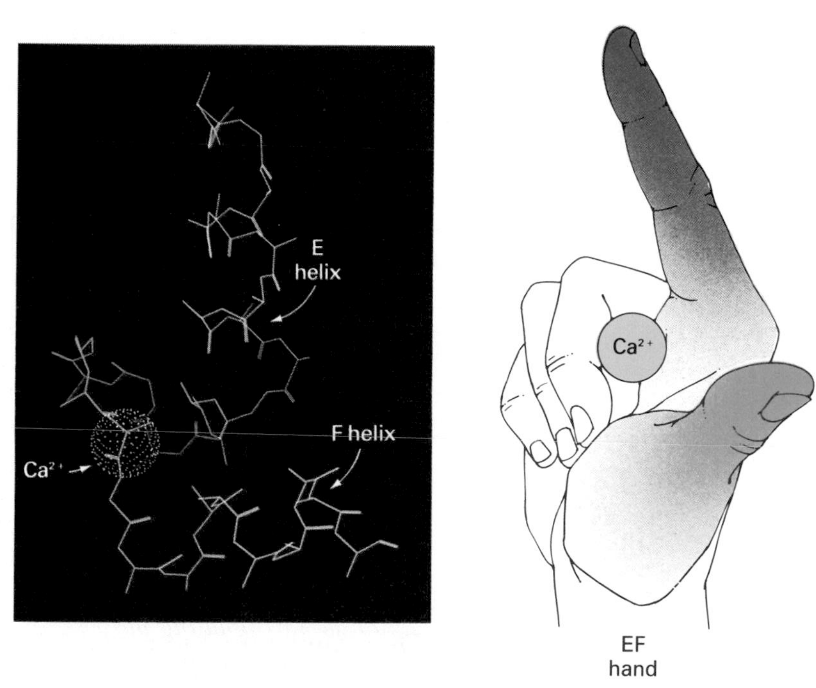

Figure 38-16
The binding sites for Ca^{2+} in many calcium-sensing proteins are formed by a helix-loop-helix unit. This recurring motif is called an EF hand.

Calmodulin, a 17-kd member of this family of proteins, serves as calcium detector in nearly all eucaryotic cells. It consists of two similar globular lobes joined by a long α helix (Figure 38-17). Each lobe contains two Ca^{2+}-binding sites; EF-hand units that are 11 Å apart. Thus, calmodulin is built from repeating modules based on the EF-hand motif. The sites in one lobe have high affinity for Ca^{2+}, and those in the other lobe have low affinity for the ion. The binding of Ca^{2+} to the latter two sites activates calmodulin and enables it to stimulate a variety of enzymes. The conformational changes induced by Ca^{2+} have not yet been directly visualized. Spectroscopic studies and model building suggest that the binding of Ca^{2+} to the loop between helices E and F causes

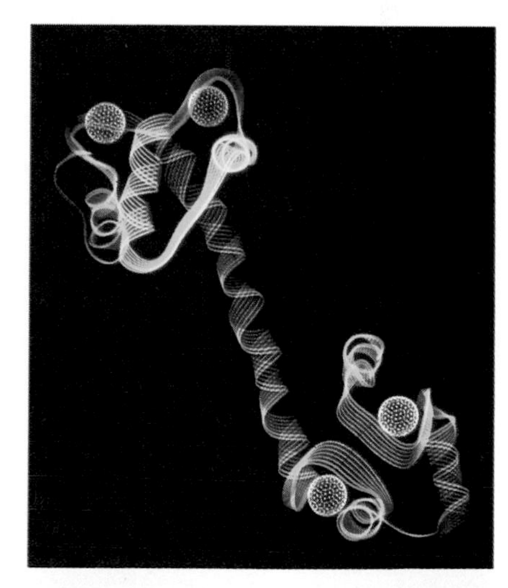

Figure 38-17
Schematic diagram of the structure of calmodulin. The four bound calcium ions are shown in blue. [Courtesy of Dr. Y. S. Babu and Dr. W. J. Cook.]

each helix to rotate about its axis and to shift location (Figure 38-18). Such a change could switch calmodulin into a form having high affinity for target proteins.

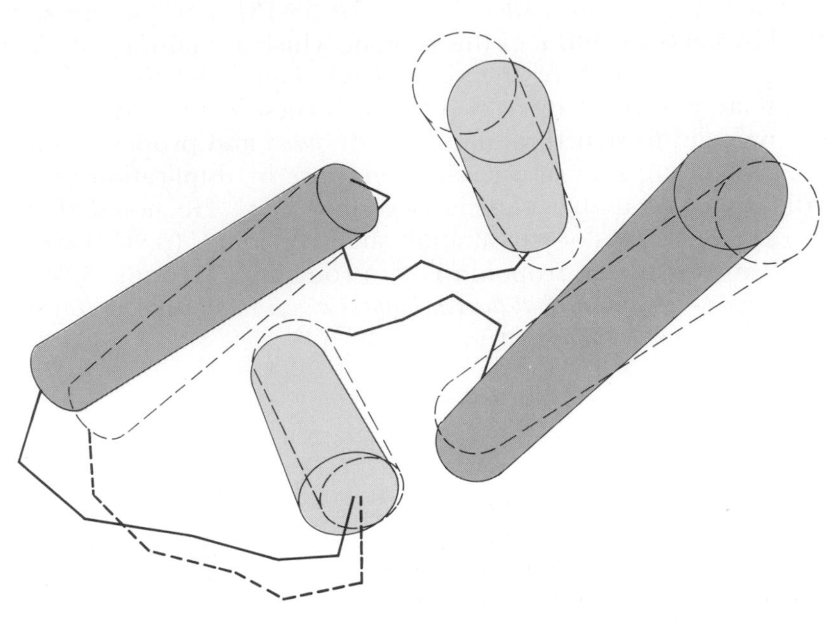

Figure 38-18
Postulated conformational change induced by the binding of Ca^{2+} to a pair of helix-loop-helix units (EF hands). The dotted lines depict the structure proposed to be present before the binding of Ca^{2+}. [After B. A. Levine, D. C. Dalgarno, M. P. Esnouf, R. E. Klevit, and R. J. P. Williams. *Ciba Found. Symp.* 83(1982):81.]

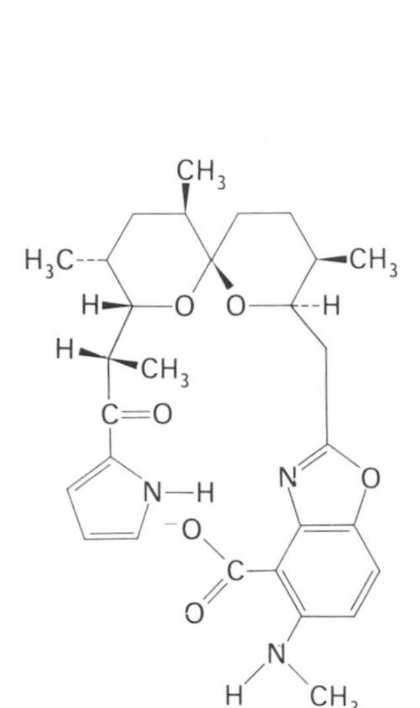

Figure 38-19
Structural formula of A23187, a calcium ionophore.

CALCIUM IONOPHORES, BUFFERS, AND INDICATORS ARE VALUABLE EXPERIMENTAL TOOLS

Our understanding of the role of calcium in cellular processes has been greatly enhanced by the use of calcium-specific reagents. *Ionophores* such as *A23187* can traverse a lipid bilayer because they have a hydrophobic periphery (Figure 38-19). They can be used to introduce Ca^{2+} into cells and organelles. Many physiologic responses that are normally triggered by the binding of hormones to cell-surface receptors can also be elicited by using calcium ionophores to raise the cytosolic calcium level. Conversely, the concentration of unbound calcium in a cell can be made very low (nanomolar or less) by introducing a calcium-specific chelator such as *EGTA* (Figure 38-21). A process postulated to be mediated by a rise in the cytosolic level of Ca^{2+} should be blocked by the free calcium level with a chelator.

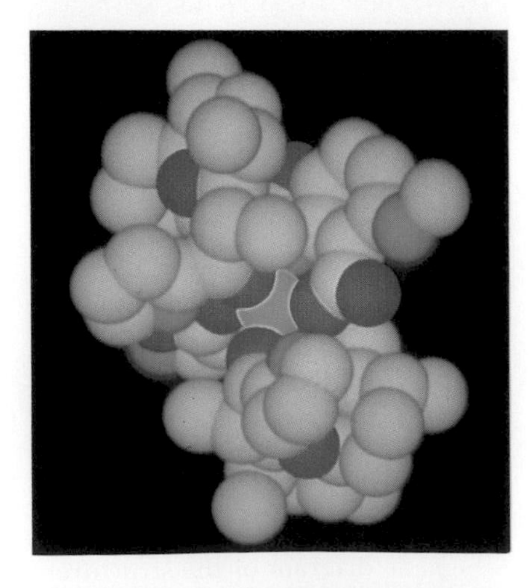

Figure 38-20
Mode of binding of Ca^{2+} (green) to two molecules of A23187, an ionophore used to introduce calcium ion into cells.

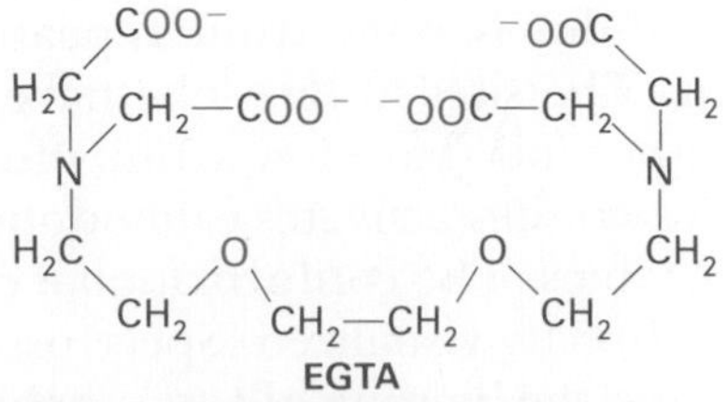

Figure 38-21
EGTA specifically binds Ca^{2+} with high affinity.

EGTA
(Ethylene glycol bis(β-aminoethyl ether)-*N,N,N',N'*-tetraacetate)

The concentration of unbound Ca^{2+} in cells can be monitored by fluorescent probes. The fluorescence intensity of *Fura-2*, a polycyclic chelator, changes when it complexes Ca^{2+} (Figure 38-22). Calcium concentrations in the nanomolar to micromolar range can also be measured using *aequorin*, a protein produced by a luminous jellyfish (*Aequorea forskalea*). This bioluminescent protein contains a fluorescent chromophore that emits light on binding Ca^{2+}; a total of 10^{10} photons are emitted by 0.1 pmol (about 2 ng) of aequorin injected into a cell. The rate of emission depends on the concentration of Ca^{2+}.

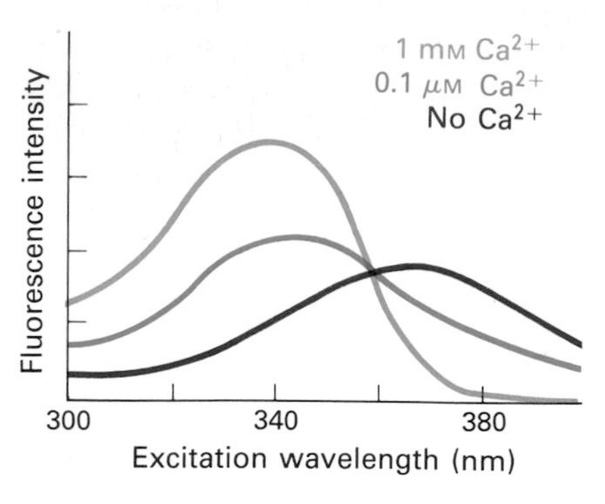

Figure 38-22
The binding of Ca^{2+} to Fura-2, a calcium chelater, alters its absorption and fluorescence properties. This fluorescent indicator can be used to measure free Ca^{2+} levels ranging from 10 nM to 1 μM. [After G. Grynkiewicz, M. Poenie, and R. Y. Tsien. *J. Biol. Chem.* 260(1985):3440.]

EICOSANOID HORMONES ARE DERIVED FROM POLYUNSATURATED FATTY ACIDS

We now turn to the prostaglandins, a group of fatty acids that affect a wide variety of physiological processes. These compounds were discovered in the 1930s but became prominent only in the 1960s, largely because of the pioneering work of Sune Bergstrom. *A prostaglandin is a twenty-carbon fatty acid containing a five-carbon ring.* They were first found in secretions of the prostate gland. Prostaglandins are derived from *arachidonate* (a $C_{20:4}$ fatty acid) and other polyunsaturated fatty acids (Figure 38-23). A series of prostaglandins are fashioned by reductases and isomerases. The major classes are designated PGA through PGI; a subscript denotes the number of carbon-carbon double bonds outside the ring. Prostaglandins with two double bonds, such as PGE_2, are derived from arachidonate; the other two double bonds of this precursor are lost in forming a five-membered ring. *Thromboxanes* are related compounds containing a six-membered ether ring. Alternatively, arachidonate can be converted into *leukotrienes* by the action of *lipoxygenase*. These compounds, first found in leukocytes, contain three conjugated double bonds—hence, the name. Prostaglandins, thromboxanes, and leukotrienes are called *eicosanoids* because they contain twenty carbon atoms (*eikosi* is the Greek word for twenty).

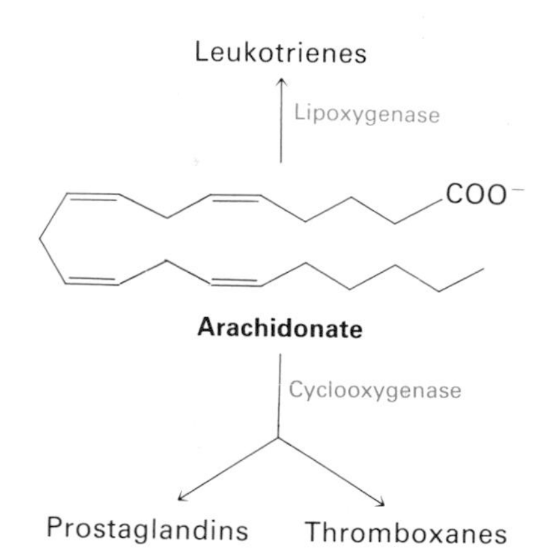

Figure 38-23
Arachidonate is the major precursor of eicosanoid hormones. Cyclooxygenase catalyzes the first step in a pathway leading to prostaglandins and thromboxanes. Lipoxygenase catalyzes the initial step in a pathway leading to leukotrienes.

OH OH COO⁻

Leukotriene B_4

O COO⁻ OH

Prostaglandin A_2

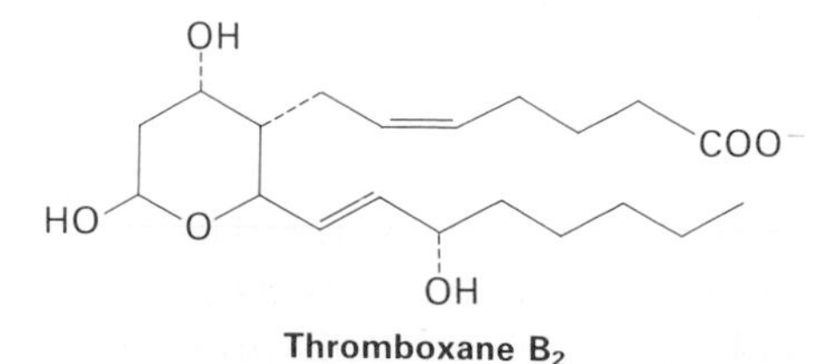

Figure 38-24
Structures of eicosanoids.

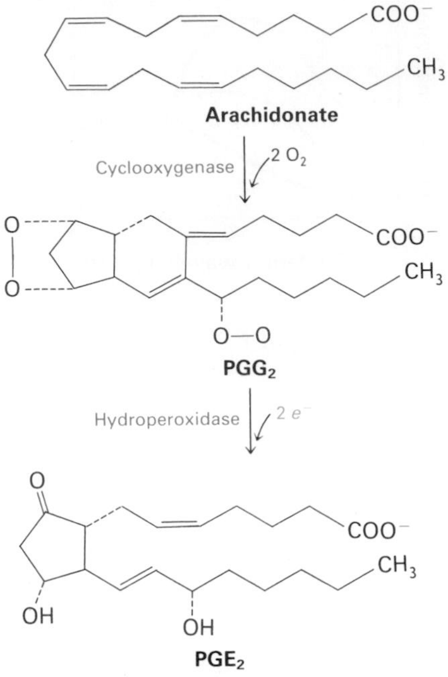

Prostaglandins and other eicosanoids are *local hormones* because they are short-lived. They alter the activities of the cells in which they are synthesized and of adjoining cells. The nature of these effects may vary from one type of cell to another, in contrast with the more uniform actions of global hormones such as insulin and glucagon. The mechanism of action of prostaglandin PGE_1 in adipose tissue has been studied in detail. Lipolysis is stimulated by hormones such as epinephrine, glucagon, corticotropin, and thyroid-stimulating hormone. PGE_1 at a concentration of 10^{-8} M strongly inhibits the lipolytic effects of these hormones. A related finding is that PGE_1 prevents the rise of the intracellular level of cyclic AMP that is elicited by these hormones. However, PGE_1 does not inhibit lipolysis caused by the addition of dibutyryl cyclic AMP, an analog that readily traverses the plasma membrane. Hence, *PGE_1 inhibits adenylate cyclase in fat cells.* In some other cells, prostaglandins *stimulate* adenylate cyclase. Other effects of prostaglandins are the stimulation of inflammation, the regulation of blood flow to particular organs, the control of ion transport across membranes, and the modulation of synaptic transmission.

Figure 38-25
Synthesis of PGE_2 from arachidonate. Prostaglandin synthase, the enzyme catalyzing these reactions, contains cyclooxygenase and hydroperoxidase components.

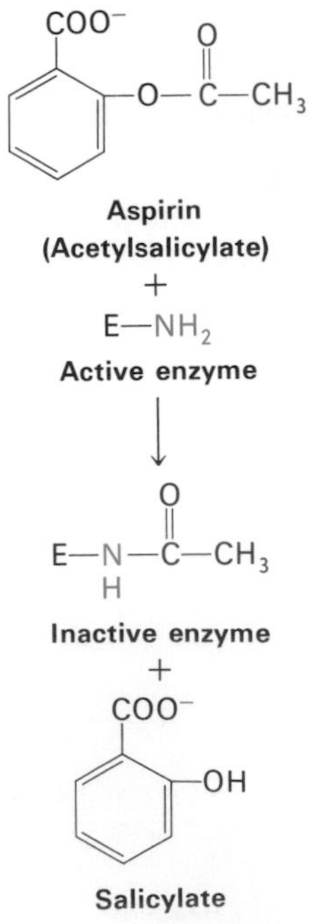

ASPIRIN INHIBITS THE SYNTHESIS OF PROSTAGLANDINS BY ACETYLATING THE CYCLOOXYGENASE

The immediate precursors of prostaglandins are synthesized in membranes from C_{20} fatty acids that contain at least three double bonds. These polyunsaturated fatty acids are required in the diets of mammals (p. 490). The prostaglandin precursors are released from membrane phospholipids by phospholipases. For example, the biosynthesis of PGE_2 starts at arachidonate, which has four double bonds. A cyclopentane ring is formed and four oxygen atoms are introduced by the *cyclooxygenase* component of *prostaglandin synthase*. All of the oxygen atoms introduced into prostaglandin come from molecular oxygen (Figure 38-25). The *hydroperoxidase* component of the synthase catalyzes a two-electron reduction of the 15-hydroperoxy group to a 15-hydroxyl group. The heme-containing dioxygenase catalyzing these reactions is bound to the smooth endoplasmic reticulum.

John Vane discovered that *aspirin inhibits the biosynthesis of prostaglandins by inactivating prostaglandin synthase.* Specifically, aspirin (acetylsalicylate) inhibits the cyclooxygenase activity of this enzyme by acetylating the terminal amino group of this subunit (Figure 38-26). Prostaglandins enhance inflammatory effects, whereas aspirin diminishes them. Aspirin is a potent anti-inflammatory agent because it blocks the first step in the synthesis of prostaglandins.

Figure 38-26
Aspirin inactivates the cyclooxygenase component of prostaglandin synthase.

ENDORPHINS ARE BRAIN PEPTIDES THAT ACT LIKE OPIATES

Opiates such as morphine have been used for centuries to relieve pain. In 1680, Thomas Sydenham wrote, "Among the remedies which it has pleased Almighty God to give to man to relieve his sufferings, none is so universal and so efficacious as opium." Why do the brains of vertebrates contain receptors for alkaloids derived from the juice of poppy seeds? Neuropharmacologists surmised that opiate receptors serve to detect endogenous regulators of pain perception. According to this view, morphine exerts its pharmacological effects by mimicking molecules that are normally present in the bodies of vertebrates. The break-

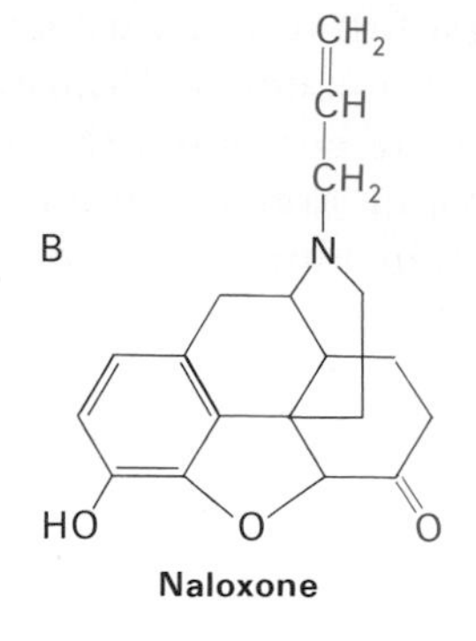

Figure 38-27
Structures of morphine, an opiate, and naloxone, an antagonist of morphine.

through came in 1975, when John Hughes isolated two peptides with opiatelike activity from pig brains. These related pentapeptides, called *methionine enkephalin* and *leucine enkephalin,* are abundant in certain nerve terminals (Figure 38-28). It seems likely that they participate in the integration of sensory information pertaining to pain.

A year later, Roger Guillemin isolated longer peptides, called *endorphins,* from the intermediate lobe of the pituitary gland. They are about as potent as morphine in relieving pain. Laboratory animals respond in a remarkable way to the injection of endorphins into the ventricles of the brain. For example, β-endorphin induces a profound analgesia of the whole body for several hours. Body temperature is lowered during this interval. Moreover, the animal becomes stuporous and assumes a stretched-out posture. These effects of endorphins disappear within a few hours and the animal again behaves normally. Another striking finding is that the actions of endorphins are reversed a few seconds after administering *naloxone* (see Figure 38-27), a known antagonist of morphine. The behavioral effects induced by the endorphins suggest that these peptides may normally participate in regulating emotional responses. We see here a new and promising area of neurobiology and neuropsychiatry.

$^{+}H_3N$-Tyr-Gly-Gly-Phe-Met-COO^-
Methionine enkephalin
(Met-enkephalin)

$^{+}H_3N$-Tyr-Gly-Gly-Phe-Leu-COO^-
Leucine enkephalin
(Leu-enkephalin)

$^{+}H_3N$-Tyr-Gly-Gly-Phe-Met-
-Thr-Ser-Glu-Lys-Ser-
-Gln-Thr-Pro-Leu-Val-
-Thr-Leu-Phe-Lys-Asn-
-Ala-Ile-Val-Lys-Asn-
-Ala-His-Lys-Lys-Gly-
-Gln-COO^-
β-Endorphin

Figure 38-28
Amino acid sequences of methionine enkephalin, leucine enkephalin, and β-endorphin. The common tetrapeptide sequence is shown in blue.

CLEAVAGE OF PRO-OPIOCORTIN YIELDS SEVERAL PEPTIDE HORMONES

β-Endorphin has the same sequence as the carboxyl-terminal region of *β-lipotropin,* a hormone isolated from the pituitary gland by Choh Li. In fact, β-endorphin is formed in vivo by the proteolytic cleavage of β-lipotropin contained in storage granules. An even larger protein encompassing the sequences of both corticotropin and β-lipotropin was then discovered. This 29-kd prohormone is called *pro-opiocortin* because it is the precursor of an opiate hormone and of corticotropin (Figure 38-29). *Corticotropin* (also called adrenocorticotropin, or ACTH) pro-

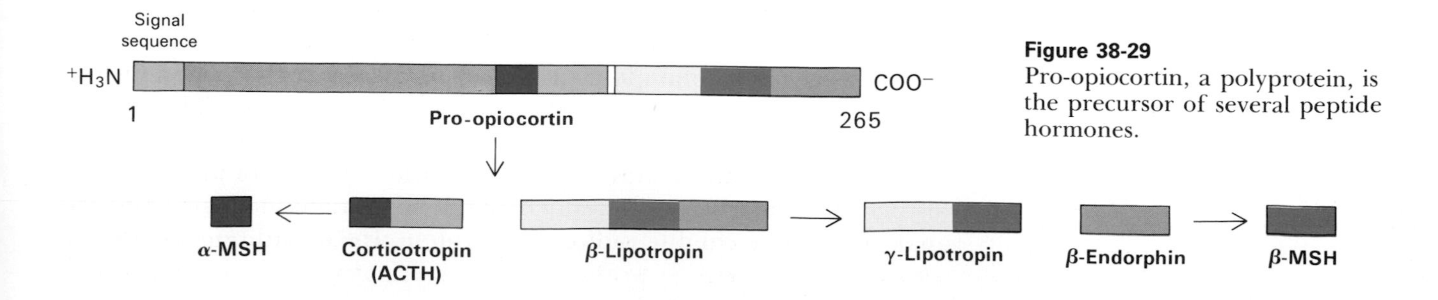

Figure 38-29
Pro-opiocortin, a polyprotein, is the precursor of several peptide hormones.

motes the growth of the adrenal cortex and stimulates it to synthesize a set of steroid hormones. Pro-opiocortin also gives rise to two melanocyte-stimulating hormones (MSH). One of them, α-MSH, is a fragment of corticotropin, and the other, β-MSH, comes from β-lipotropin (Figure 38-29). The amino-terminal half of pro-opiocortin may be a source of yet other hormones. Clearly, *pro-opiocortin is a cornucopia of peptide hormones.* This prohormone is made of four similar regions, which probably evolved by successive gene duplications.

The junctions between prospective active hormones in pro-opiocortin contain *pairs of basic residues* (Lys-Arg, Arg-Arg, or Lys-Lys). It is interesting to note that pairs of basic residues are also present at the boundaries of the connecting peptide in proinsulin (p. 995) and proparathyroid hormone. Pairs of basic residues mark cleavage sites in many other prohormones, such as the precursors of enkephalins, glucagon, and vasopressin.

INSULIN STIMULATES ANABOLIC PROCESSES AND INHIBITS CATABOLIC PROCESSES

We now turn to insulin, a polypeptide hormone that plays a key role in the integration of fuel metabolism, as was discussed in an earlier chapter (p. 636). Interest in insulin has been heightened by the discovery that its receptor has tyrosine kinase activity when the hormone is bound. As will be discussed shortly, the stimulation of tyrosine kinase activity is a recurring mechanism in the actions of numerous growth factors and oncogene products.

Insulin promotes anabolic processes and inhibits catabolic ones in muscle, liver, and adipose tissue. Specifically, insulin increases the rate of synthesis of glycogen, fatty acids, and proteins. By stimulating glycolysis, insulin leads to the formation of building blocks for the synthesis of macromolecules. An important action of the hormone is that *it promotes the entry of glucose, some other sugars, and amino acids into muscle and fat cells.* Hence, the level of glucose in the blood is lowered by insulin (called the *hypoglycemic effect*). Insulin inhibits catabolic processes such as the breakdown of glycogen and fat. It also decreases gluconeogenesis by lowering the level of enzymes such as pyruvate carboxylase and fructose 1,6-bisphosphatase. Many of the effects of insulin are the opposite of those elicited by epinephrine and glucagon. In essence, *epinephrine and glucagon signal that glucose is scarce, whereas insulin signals that glucose is abundant.*

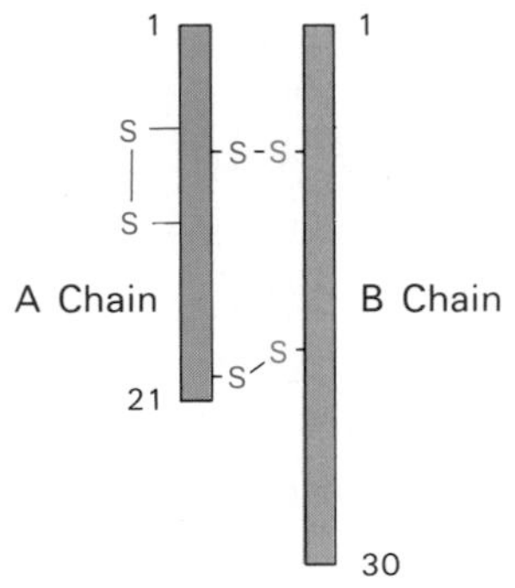

Figure 38-30
Chain structure and disulfide pairing in insulin.

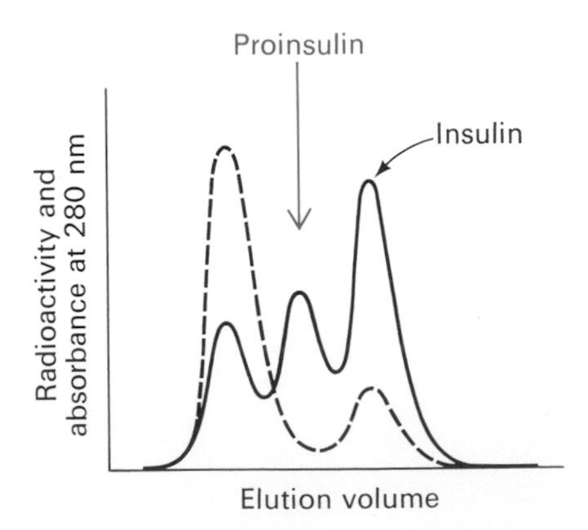

Figure 38-31
The appearance of a highly radioactive protein following pulse-labeling of an islet-cell adenoma led to the discovery of proinsulin. (The solid line refers to radioactivity; the dashed line to absorbance at 280 nm.)

PREPROINSULIN AND PROINSULIN ARE THE PRECURSORS OF THE ACTIVE HORMONE

In 1953, Frederick Sanger showed that bovine insulin consists of two chains—an A chain of 21 residues and a B chain of 30 residues—which are covalently joined by two disulfide links (Figure 38-30). The same pattern exists in insulin molecules from many different species, including humans. How is this two-chain protein synthesized? The answer came from Donald Steiner's study of an islet-cell adenoma of the pancreas. This rare human tumor produces large amounts of insulin. Slices of this pancreatic tumor were incubated with tritiated leucine and then analyzed. A new, highly radioactive protein was found (Figure 38-31).

Subsequent studies revealed that this new protein, called *proinsulin,* is the *biosynthetic precursor of insulin.*

Proinsulin is a single polypeptide chain containing a sequence of about thirty residues that is absent from insulin (Figure 38-32). This *connecting peptide* (*C-peptide*) joins the carboxyl end of the B chain and the amino terminus of the A chain of the future insulin molecule. Proinsulin is not the earliest form of the hormone. The nascent polypeptide chain, called *preproinsulin,* contains an additional nineteen residues at its amino terminus. This hydrophobic stretch is the signal sequence that directs the nascent chain to the endoplasmic reticulum (p. 769). Preproinsulin is converted into proinsulin in the lumen of this membrane, very soon after the polypeptide chain traverses it. Proinsulin is transported to the Golgi complex and then to secretory granules, where the connecting peptide is proteolyzed. The connecting peptides in proinsulins from different species have common structural features. They all contain Arg-Arg at their amino end and Lys-Arg at their carboxyl end. A proteolytic enzyme similar to trypsin hydrolyzes the polypeptide chain at these positively charged residues. The insulin molecules in mature storage granules are secreted when the membrane of a granule fuses with the plasma membrane of the cell.

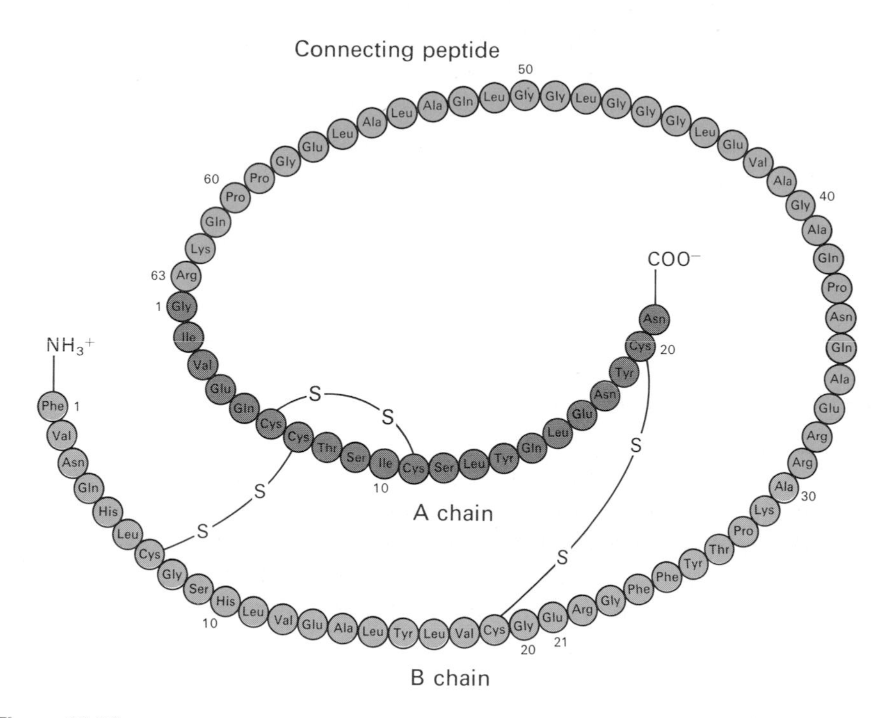

Figure 38-32
Amino acid sequence of porcine proinsulin: the A chain of insulin is shown in red, the B chain in blue, and the connecting peptide (C-peptide) in yellow. [After R. E. Chance, R. M. Ellis, and W. W. Bromer. *Science* 161(1968):165. Copyright 1968 by the American Association for the Advancement of Science.]

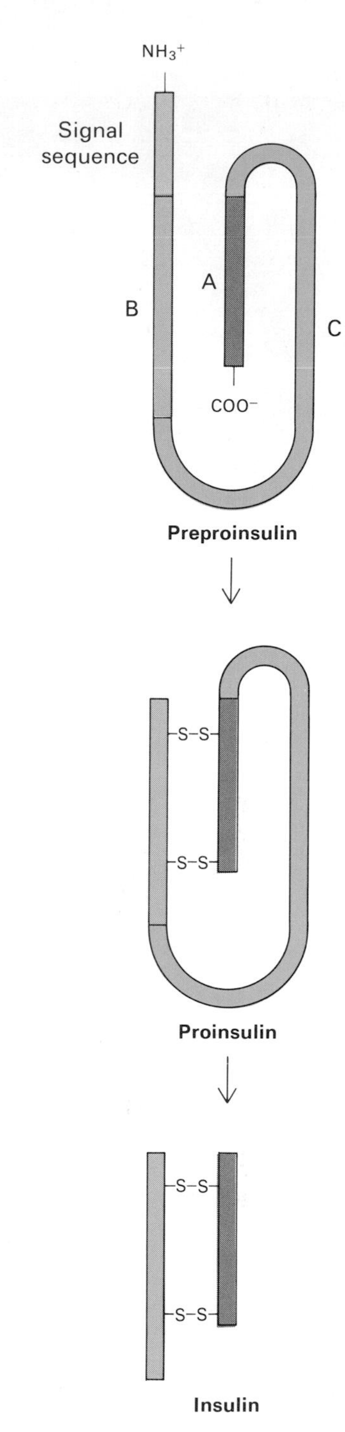

Figure 38-33
Enzymatic conversion of preproinsulin into proinsulin and then into insulin.

X-ray crystallographic analyses carried out in the laboratory of Dorothy Crowfoot Hodgkin have revealed the three-dimensional structure of porcine insulin at a resolution of 1.9 Å. This accomplishment was completed some thirty-six years after Hodgkin obtained the first x-ray diffraction pattern of a protein crystal (pepsin), when she was a graduate student in the laboratory of John Bernal. In the intervening years,

Hodgkin solved the structures of such biologically important molecules as cholesterol, penicillin, and vitamin B_{12}. Insulin has a compact three-dimensional structure (Figure 38-34). Only the amino and carboxyl termini of the B chain extend away from the rest of the protein. The A chain is nestled between these extended arms of the B chain. A nonpolar core is formed by buried aliphatic side chains from both chains. Insulin is also stabilized by several salt links, hydrogen bonds between groups on the A and B chains, and two interchain disulfide bonds.

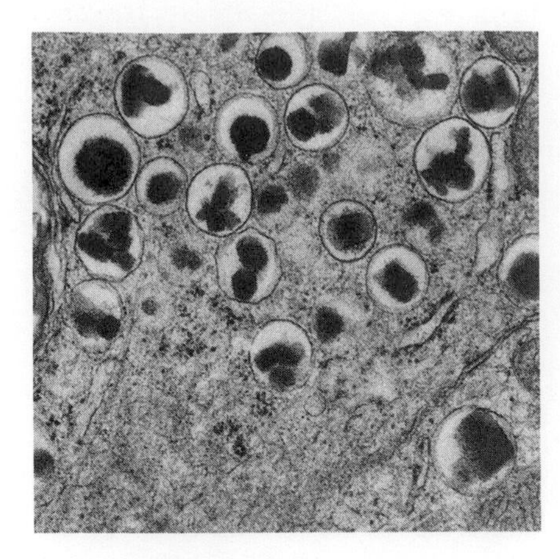

Electron micrograph of storage granules containing insulin in a β cell of the pancreas. Small crystals of insulin form within these granules because they contain high concentrations of the hormone. [Courtesy of Dr. Arthur Like.]

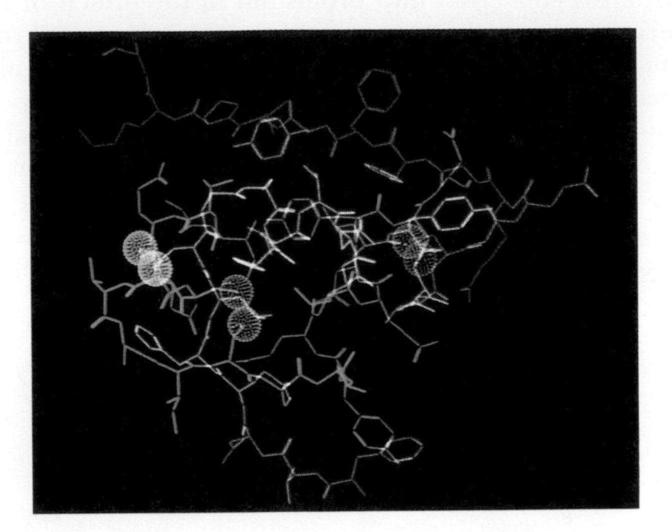

Figure 38-34
Three-dimensional structure of insulin. The A chain is shown in red, and the B chain in blue. Pairs of yellow spheres represent disulfide bonds.

INSULIN SWITCHES ON THE TYROSINE KINASE ACTIVITY OF ITS RECEPTOR

Insulin acts by binding to receptors in the plasma membrane of target cells. It binds very tightly—the dissociation constant of the hormone-receptor complex is 0.1 nM. The high affinity of the receptor for insulin is essential because the insulin level in the blood is low, of the order of 0.1 nM. The first purification of the receptor was difficult because it is present at low density, about one receptor per square micrometer of plasma membrane of fat cells, which corresponds to one receptor per 10^6 phospholipid molecules. The receptor was solubilized by the addition of a nonionic detergent to a plasma membrane preparation, and purified by affinity chromatography on a column containing covalently attached insulin. Pedro Cuatrecasas achieved a 250,000-fold purification of the insulin receptor in this way. A milligram of pure receptor could be obtained from 500 g of protein in the homogenate derived from the livers of two hundred rats.

The sequencing of tryptic peptides derived from purified insulin receptor led to the synthesis of oligonucleotide probes and the cloning of its cDNA. The insulin receptor is an integral membrane glycoprotein consisting of two α (135 kd) and two β (95 kd) chains joined by three disulfide bonds (Figure 38-35). The α chains are on the extracellular

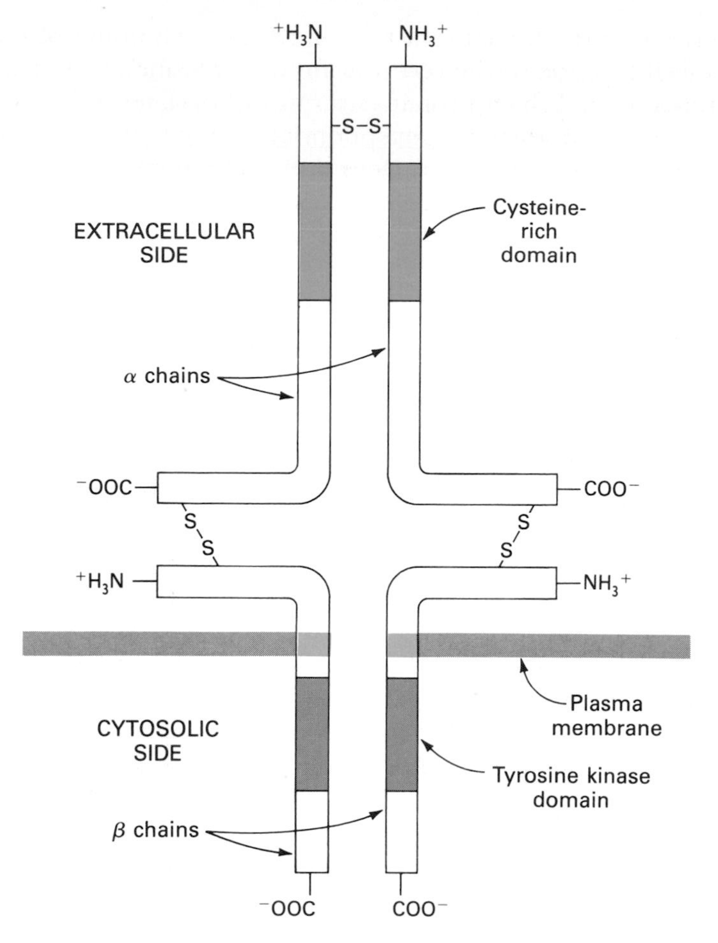

Figure 38-35
The insulin receptor consists of two α and two β chains. The binding sites for insulin are on the extracellular side of the plasma membrane, and the tyrosine kinase domains are on the cytosolic side.

side of the membrane, whereas the β chains traverse the membrane. Each $\alpha\beta$ unit of the receptor is derived from a single-chain precursor of 1382 residues. The precursor begins with a signal sequence (which is cleaved), followed by the α subunit sequence, a highly basic tetrapeptide, and then the β subunit sequence. The Arg-Lys-Arg-Arg sequence separating the α and β subunits is recognized by a processing protease.

The activated insulin receptor is an enzyme—it catalyzes the phosphorylation of tyrosine residues in target proteins. The tyrosine kinase domains of the receptor are located on the cytosolic portion of the β chains. The binding sites for insulin are on the α chains, on the extracellular side of the membrane. The binding of insulin switches on the tyrosine kinase activity of the receptor. The activated receptor then phosphorylates two tyrosine residues on the same chain as the catalytic site. (Figure 38-36). Autophosphorylation increases its capacity to phosphorylate tyrosine residues on target proteins. Another key consequence of autophosphorylation is that the kinase activity is switched on even if insulin dissociates from the receptor. (Figure 38-36). Phosphorylation of serine and threonine residues by protein kinase A (activated by the cylic AMP cascade) and protein kinase C (activated by the phosphoinositide cascade) decreases the tyrosine kinase activity of the receptor. Many intriguing

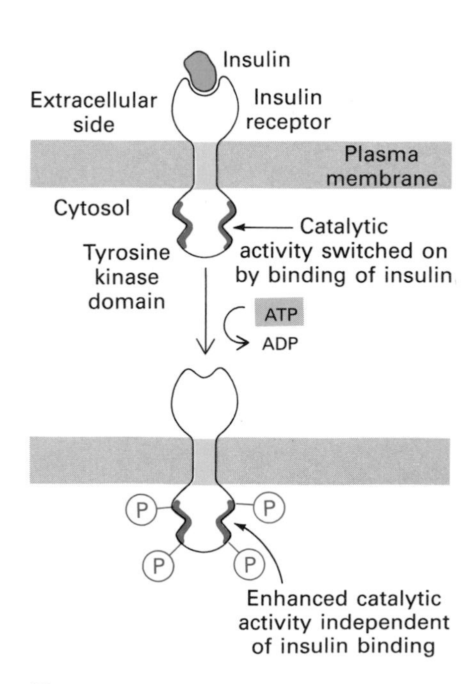

Figure 38-36
Activation of the tyrosine kinase activity of the insulin receptor by autophosphorylation.

questions are raised by these facts: How does the binding of insulin to the extracellular side of the receptor induce tyrosine kinase activity on the cytosolic side? Which proteins are phosphorylated by activated receptor, and how does their phosphorylation have multiple metabolic and growth-promoting effects? Does insulin do more than induce tyrosine kinase activity? How do the cyclic AMP cascade, the phosphoinositide cascade, and the tyrosine cascade interact?

NERVE GROWTH FACTOR AND EPIDERMAL GROWTH FACTOR STIMULATE THE PROLIFERATION OF TARGET CELLS

How is the growth of eucaryotic cells controlled? The isolation of growth-factor proteins that specifically stimulate target cells is a significant start in answering this challenging and important question. In 1954, Rita Levi-Montalcini discovered that *nerve growth factor (NGF) plays a critical role in the development of sympathetic neurons and of certain sensory neurons in vertebrates.* NGF stimulates these cells to divide and differentiate. A tumor that synthesized large amounts of NGF was instrumental in these pioneering studies. Levi-Montalcini went to Brazil to carry out tissue-culture studies of this tumor; she carried in her handbag two mice bearing transplants of mouse sarcomas rich in NGF activity. She wrote that "the tumor had given a first hint of its existence in St. Louis but it was in Rio de Janeiro that it revealed itself, and it did so in a theatrical and grand way, as if spurred by the bright atmosphere of that explosive and exuberant manifestation of life that is the Carnival in Rio." The outgrowth of neurites from a ganglion in culture is a sensitive assay for this growth factor (Figure 38-37). The biologically active NGF molecule consists of two identical 13-kd polypeptide chains. This β dimer is stored in the submaxillary gland, where it is synthesized, in a complex having the subunit structure $\alpha_2\beta\gamma_2$. The γ subunit is a protease, and α is a protease inhibitor.

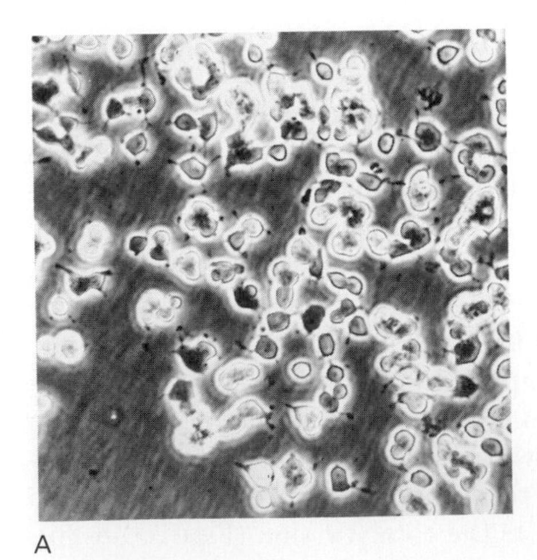

A

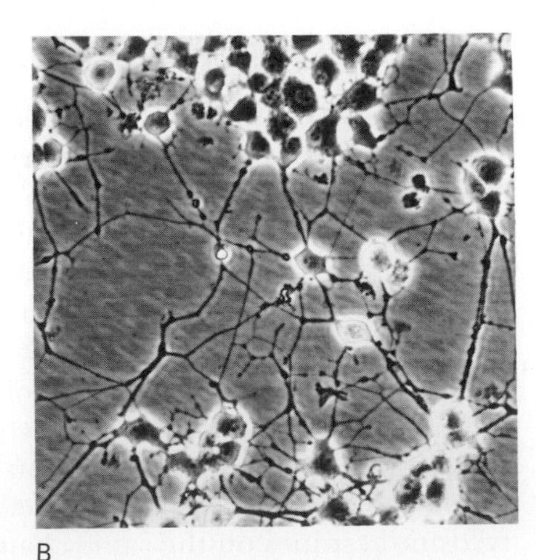

B

Figure 38-37
Nerve growth factor (NGF) induces neurites to grow out of cultured nerve cells: (A) no NGF added; (B) with NGF. [Courtesy of Bruce Yankner, Christiana Richter-Landsberg, and Dr. Eric Shooter.]

Epidermal growth factor (EGF) was discovered serendipitously by Stanley Cohen while he was studying nerve growth factor. He found that the injection of crude preparations of NGF into newborn mice led to precocious eyelid opening and tooth eruption. These unexpected effects pointed to existence of a new growth factor, which turned out to be EGF, a 6-kd polypeptide that stimulates the growth of epidermal and epithelial cells (Figure 38-38). This 53-residue chain is produced by the cleavage of a 1168-residue *EGF precursor,* a transmembrane protein. This very large precursor may be the source of other active polypeptides. The presence of pairs of basic residues (Lys-Arg, Lys-Lys, and Arg-Arg) is reminiscent of pro-opiocortin.

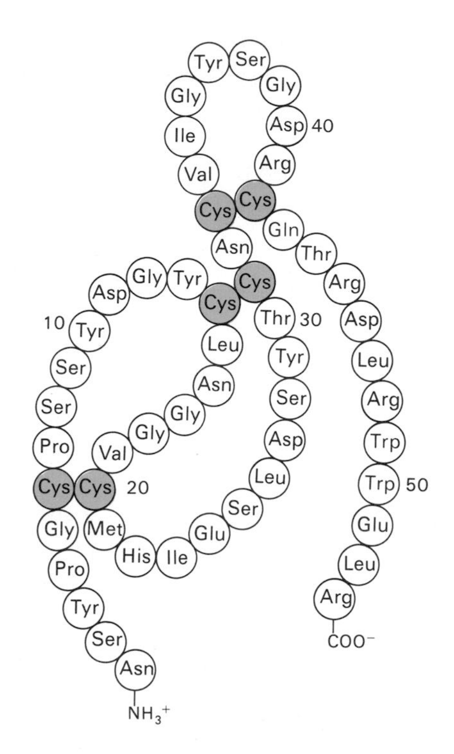

Figure 38-38
Amino acid sequence of epidermal growth factor (EGF). [After C. R. Savage, Jr., J. H. Hash, and S. Cohen. *J. Biol. Chem.* 248(1973):7669.]

THE RECEPTOR FOR EPIDERMAL GROWTH FACTOR IS A MEMBER OF THE TYROSINE KINASE FAMILY

Epidermal growth factor, like insulin, binds tightly to a receptor on the surface of a cell. The A431 cell line, derived from a human epidermoid carcinoma, is a rich source of receptor. A431 cells contain 2×10^6 receptors, compared with 10^4 to 10^5 receptors in other epidermal cells. The abundance of receptor in A431 cells is a consequence of gene amplification. The availability of this cell line has facilitated studies of the EGF receptor and has led to the cloning of its gene. The EGF receptor is a single polypeptide chain (175 kd) that spans the plasma membrane (Figure 38-39). The cytosolic side, like that of the insulin receptor, contains a *tyrosine kinase domain* and tyrosine residues that become autophosphorylated. The amino-terminal portion of the chain, located on the extracellular side of the membrane, contains the binding site for EGF. This segment of the receptor has two similar *cysteine-rich domains*; many of the cysteines are disulfide bonded. The segment also contains a large number of potential glycosylation sites.

The tyrosine kinase activity of the EGF receptor, like that of the insulin receptor, is activated by the binding of growth factor to the extracellular portion. Again, kinase activity is stimulated by autophosphorylation of the receptor. Is the binding of growth factor signaled across the membrane portion of both receptors in the same way? Axel Ullrich answered this question by synthesizing a gene that encoded a *chimeric receptor*—the

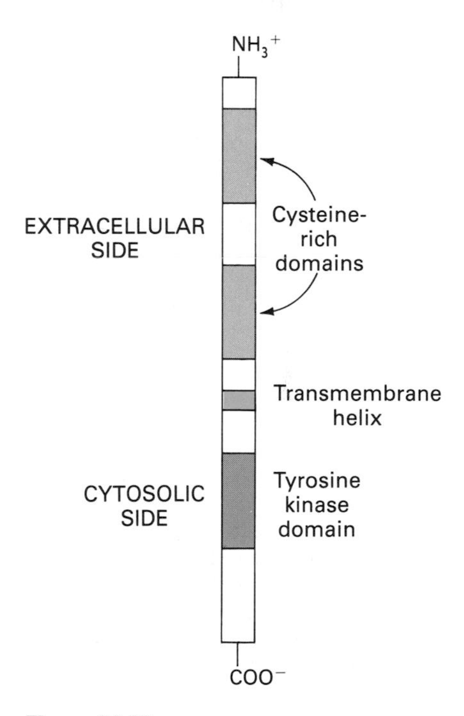

Figure 38-39
The epidermal growth factor receptor (EGF receptor) is a transmembrane protein with an extracellular hormone-binding site and a cytosolic tyrosine kinase domain.

extracellular portion came from the insulin receptor, and the membrane-spanning and cytosolic parts from the EGF receptor (Figure 38-40). The striking result was that the binding of insulin induced tyrosine kinase activity, as evidenced by rapid autophosphorylation. *Hence, the insulin receptor and the EGF receptor use a common mechanism of signal transmission across the plasma membrane.* It will be interesting to learn how transmembrane signaling is accomplished. One possibility is that the binding of insulin causes a rotation or sliding of the transmembrane helices, which would change the conformation of the tyrosine kinase domain on the other side.

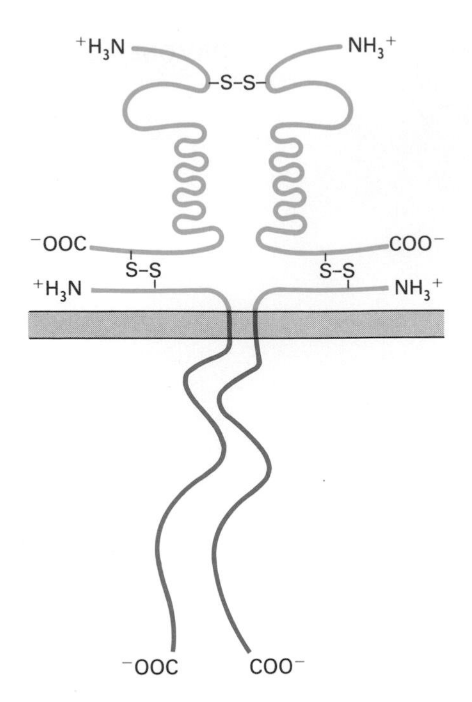

Figure 38-40
A chimeric receptor consisting of a unit from the insulin receptor (green) and the EGF receptor (red). The binding of insulin stimulates the tyrosine kinase activity of this chimeric molecule and leads to its autophosphorylation. [After A. Ullrich, H. Riedel, Y. Yarden, L. Coussens, A. Gray, T. Dull, J. Schlessinger, M. D. Waterfield, and P. J. Parker. *Cold Spring Harbor Symp. Quant. Biol.* 51(1986):715.]

SEVERAL ONCOGENES ENCODE GROWTH FACTORS OR RECEPTORS HAVING TYROSINE KINASE ACTIVITY

Tumor virologists anticipated that the study of oncogenes (p. 876) would converge with the study of growth control in normal cells. This hope has been realized. The retroviral oncogene v-*sis* encodes one of the subunits of platelet-derived growth factor (PDGF). The viral gene product probably stimulates cell replication by inducing the tyrosine kinase activity of the PDGF receptor, an integral membrane glycoprotein. The proteins encoded by the v-*src* gene of avian sarcoma virus (Rous sarcoma virus) and by the c-*src* gene are tyrosine kinases. Indeed, many cellular and viral oncogenes encode tyrosine kinases, as exemplified by the c and v forms of *yes*, *fgr*, *fes*, *abl*, and *ros*, which are either peripheral or integral membrane proteins. An especially interesting pair of oncogenes are c-*erb*-B and v-*erb*-B (from avian erythroblastosis virus). *The viral form of* erb-*B encodes a truncated receptor for epidermal growth factor* (Figure 38-41). The transmembrane and the cytosolic portions of the receptor are present in this viral oncogene product, but nearly all of the amino-terminal half of the EGF receptor is missing. *The tyrosine kinase activity of the v-*erb-*B protein is switched on permanently.* The loss of regulatory control leads to oncogenesis.

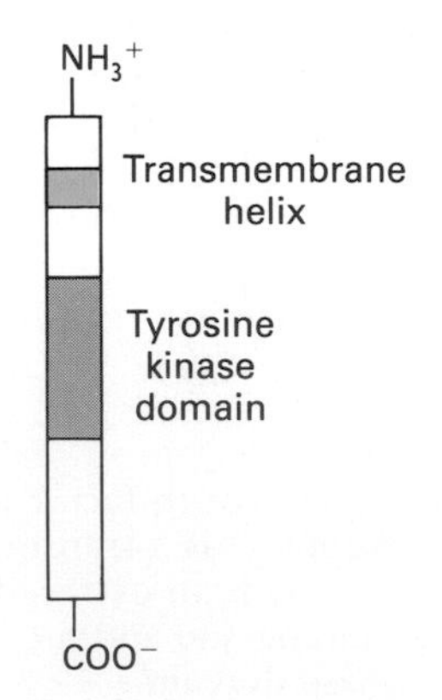

Figure 38-41
The viral oncogene v-*erb*-B encodes a truncated form of the EGF receptor. This oncogene product has a permanently stimulated tyrosine kinase activity.

STEROID HORMONES AND THYROID HORMONES ACTIVATE SPECIFIC GENES BY BINDING TO ENHANCER PROTEINS

The primary effect of steroid hormones such as estradiol, progesterone, and cortisone is on *gene expression* rather than on enzyme activities or transport processes. These hormones, in contrast with epinephrine, must enter target cells to exert their effects. Furthermore, their primary site of action is in the *cell nucleus* rather than on the plasma membrane. The full impact of these steroids is achieved in hours rather than minutes because their biological effects depend on the *synthesis of new proteins*. Actinomycin D inhibits the action of these steroid hormones,

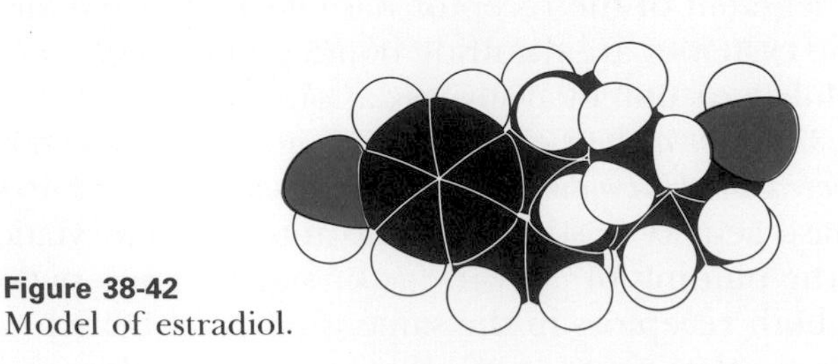

Figure 38-42
Model of estradiol.

indicating that the *synthesis of new messenger RNA is required for their action.* The first step in the stimulation of uterine growth by 17β-estradiol is its binding to a specific receptor in the cytoplasm of the uterine cell. The binding is very tight ($K \sim 10^{-9}$ M). The hormone-receptor complex then migrates to the nucleus of the cell, where it binds to specific sites on DNA. Other steroid hormones act similarly.

17β-Estradiol

Receptors for estrogens, progesterone, and glucocorticoids have been purified by taking advantage of their high affinity for hormone. Antibodies specific for these receptors were then used to screen proteins expressed by a library of cDNA clones (p. 132). The sequencing and expression of these cDNAs has been highly rewarding. It revealed that *steroid hormone receptors contain a DNA-binding region rich in cysteine, arginine, and lysine residues. Most interesting, this region has Cys-X_2-Cys sequences that probably form metal-binding fingers.* Such fingers are well suited for binding specific sequences in DNA, as has been shown for transcription factor III, a regulatory protein that activates expression of the 5S ribosomal RNA gene (p. 844). Indeed, *steroid hormones bound to their receptors act as enhancers of transcription.* Mutants of the glucocortocoid receptor containing insertions that disrupt the Cys-X_2-Cys motif bind hormone but do not activate transcription. The hormone-binding regions of these steroid hormone receptors have similar sequences, which indicates that they arose from a common ancestral receptor.

Thyroxine
(L-3,5,3′,5′-Tetraiodothyronine)

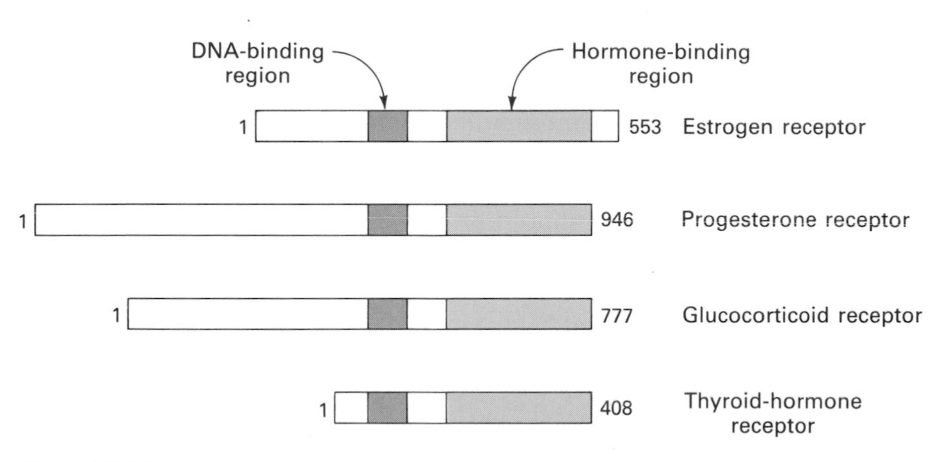

Figure 38-43
Receptors for steroid hormones and thyroid hormones contain similar DNA-binding domains (red) and different hormone-binding domains (blue).

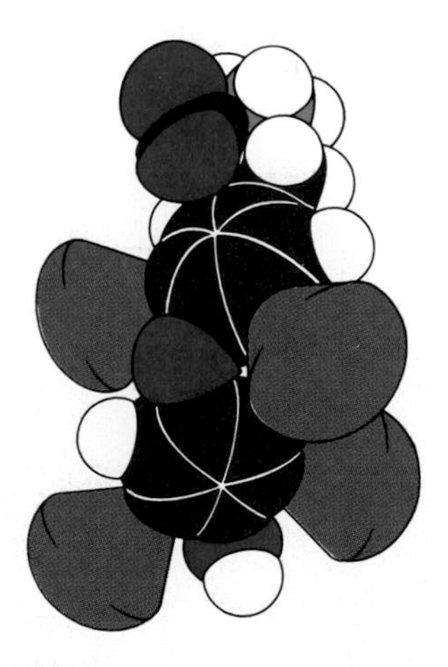

Figure 38-44
Model of thyroxine. The four iodine atoms of this thyroid hormone are shown in purple. Thyroxine is formed by the iodination and joining of tyrosine residues in thyroglobulin. Proteolysis of this precursor gives rise to thyroxine.

Further insight into hormone receptors acting on DNA has come from studies of the v-*erb*-A oncogene of avian erythroblastosis virus and c-*erb*-A, its cellular counterpart. The c-*erb*-A gene encodes a receptor for thyroid hormones. The gene product binds thyroid hormones such such as *thyroxine* (tetraiodothyronine, T_4) (Figure 38-44). The receptor contains a DNA-binding domain with putative metal-binding fingers, like those of the steroid receptors. This unexpected finding has revealed *an underlying unity in the action of thyroid hormones and steroid hormones. Both classes of receptor-hormone complexes are transcriptional enhancers.*

SUMMARY

Hormones are diverse molecules that integrate the activities of different cells in multicellular organisms. Cyclic AMP plays a critical role in the action of many hormones (e.g., epinephrine and glucagon) by serving as a second messenger inside target cells. The binding of such a hormone to a receptor protein in the plasma membrane triggers the exchange of GTP for GDP bound to a G protein. The GTP form of the G protein then activates adenylate cyclase. Activation is reversed by hydrolysis of bound GTP, which brings the G protein back to the inactive GDP form. A family of G proteins transduces many hormonal and sensory stimuli; G_s (stimulatory G protein) is the one that activates adenylate cyclase. Cyclic AMP formed by stimulated adenylate cyclase then activates a protein kinase by releasing its regulatory subunits. The free catalytic subunits of the kinase phosphorylate serine and threonine residues in target proteins. For example, the phosphorylation of glycogen synthase and phosphorylase kinase results in decreased synthesis and increased degradation of glycogen. This reaction cascade greatly amplifies the initial hormonal stimulus. The ADP-ribosylation of G_s by cholera toxin blocks its GTPase activity, leading to a persistent activation of adenylate cyclase. Elevated levels of cyclic AMP in intestinal epithelial cells produce a large efflux of Na^+ and water into the gut.

The phosphoinositide cascade also evokes a wide variety of responses in many kinds of cell. The binding of hormone to a cell surface receptor activates a G protein, which switches on the catalytic activity of phosphoinositidase (phospholipase C). The hydrolysis of phosphatidyl inositol 4,5-bisphosphate (PIP_2) generates two messengers. Inositol trisphosphate (IP_3) opens calcium channels in the endoplasmic and sarcoplasmic reticulum membrane, leading to an elevated level of Ca^{2+} in the cytosol. Diacylglycerol activates protein kinase C, which phosphorylates serine and threonine residues in target proteins. Intracellular Ca^{2+} acts by binding to calmodulin and other calcium sensors; they contain a recurring helix-loop-helix motif (EF hand).

Arachidonate, a C_{20} polyunsaturated fatty acid, is the precursor of eicosanoids; many of these compounds are local hormones. Prostaglandins and thromboxanes are formed from arachidonate by the action of a cyclooxygenase, whereas leukotrienes are formed by the action of a lipoxygenase. Aspirin inhibits the synthesis of prostaglandins by covalently modifying the cyclooxygenase.

Insulin promotes anabolic processes (e.g., the synthesis of glycogen, fatty acids, and proteins) and inhibits catabolic ones (e.g., the breakdown of glycogen and fat). The signal sequence of preproinsulin is cleaved to form proinsulin, which is further cleaved to generate insulin, a two-chain protein linked by disulfides. Insulin binds tightly to a receptor in the plasma membrane of target cells. The insulin receptor, an $\alpha_2\beta_2$ transmembrane protein, is a tyrosine kinase. The binding of insulin switches on this catalytic activity. The receptor phosphorylates tyrosine residues on itself, which further stimulates its kinase activity. It seems likely that the multiple effects of insulin are mediated by the phosphorylation of tyrosine residues in target proteins. Epidermal growth factor (EGF) stimulates target cells in a similar way. The EGF receptor, like the insulin receptor, contains an extracellular hormone-binding domain that is rich in cysteine and a cytosolic tyrosine kinase domain. The importance of tyrosine kinases in growth control is highlighted by the finding that many oncogenes act through tyrosine kinase cascades.

Endorphins and enkephalins are brain peptides that act like opiates such as morphine. β-Endorphin is derived from pro-opiocortin, a prohormone that is the source of other potent peptides such as corticotropin (ACTH), β-lipotropin, and melanocyte-stimulating hormones (α- and β-MSH). The junctions between prospective hormones in pro-opiocortin, as in proinsulin, contain pairs of basic residues.

The primary effect of steroid hormones (such as estrogens, progesterone, and glucocorticoids) and thyroid hormones is on gene expression. These hormones, in contrast with those of other classes, must enter a cell to exert their actions, which take place in the nucleus. The receptors for these hormones are DNA-binding proteins that possess metal-binding fingers. The binding of a steroid or thyroid hormone to its receptor converts it into a transcriptional enhancer; specific genes are then expressed.

SELECTED READINGS

Where to start

Berridge, M. J., 1985. The molecular basis of communication within the cell. *Sci. Amer.* 253(4):142–152.

Snyder, S. H., 1985. The molecular basis of communication between cells. *Sci. Amer.* 253(4):132–140.

Stryer, L., and Bourne, H. R., 1986. G proteins: a family of signal transducers. *Ann. Rev. Cell Biol.* 2:391–419.

Nishizuka, Y., 1986. Studies and perspectives of protein kinase C. *Science* 233:305–312.

Carafoli, E., and Penniston, J. T., 1985. The calcium signal. *Sci. Amer.* 253(5):70–78.

Books

Greenspan, F. S., and Forsham, P. H., (eds.) 1986. *Basic and Clinical Endocrinology* (2nd ed.). Lange. [A concise introduction that emphasizes human endocrinology and disorders.]

Levitzki, A., 1984. *Receptors: A Quantitative Approach.* Benjamin/Cummings.

Bradshaw, R. A., and Gill, G. N., (eds.), 1983. *Evolution of Hormone-Receptor Systems.* Liss.

Chadwick, C. M., and Garrod, D. R., (eds.), 1986. *Hormones, Receptors, and Cellular Interactions in Plants.* Cambridge University Press.

Adenylate cyclase cascade and G proteins

Gilman, A. G., 1987. G proteins: transducers of receptor-generated signals. *Ann. Rev. Biochem.* 56:615–649.

Schramm, M., and Selinger, Z., 1984. Message transmission: receptor controlled adenylate cyclase system. *Science* 225:1350–1356.

Feder, D., Im, M.-J., Klein, H. W., Hekman, M., Holzhofer, A., Dees, C., Levitzki, A., Helmreich, E. J. M., and Pfeuffer, T., 1986. Reconstitution of beta$_1$-adrenoceptor-dependent adenylate cyclase from purified components. *EMBO J.* 5:1509–1514.

Brandt, D. R., Asano, T., Pedersen, S. E., and Ross, E. M., 1983. Reconstitution of catecholamine-stimulated GTPase activity. *Biochemistry* 22:4357–4362.

Dohlman, H. G., Caron, M. G., and Lefkowitz, R. J., 1987. A family of receptors coupled to guanine nucleotide regulatory proteins. *Biochemistry* 26:2657–2664.

Benovic, J. L., Strasser, R. H., Caron, M. G., and Lefkowitz, R. J., 1986. β-Adrenergic receptor kinase: identification of a novel protein kinase that phosphorylates the agonist-occupied form of the receptor. *Proc. Nat. Acad. Sci.* 83:2797–2801.

Gerisch, G., 1987. Cyclic AMP and other signals controlling cell development and differentiation in *Dictyostelium. Ann. Rev. Biochem.* 56:853–879.

Phosphoinositide cascade

Berridge, M. J., 1987. Inositol trisphosphate and diacylglycerol: two interacting second messengers. *Ann. Rev. Biochem.* 56:159–193.

Majerus, P. W., Connolly, T. M., Deckmyn, H., Ross, T. S., Bross, T. E., Ishii, H., Bansal, V. S., and Wilson, D. B., 1986. The metabolism of phosphoinositide-derived messenger molecules. *Science* 234:1519–1526.

Woods, N. M., Cuthbertson, S. R., and Cobbold, P. H., 1986. Repetitive transient rises in cytoplasmic free calcium in hormone-stimulated hepatocytes. *Nature* 319:600–602.

Hokin, L. E., 1985. Receptors and phosphoinositide-generated second messengers. *Ann. Rev. Biochem.* 54:205–235.

Irvine, R. F., Letcher, A. J., Heslop, J. P., and Berridge, M. J., 1986. The inositol tris/tetrakisphosphate pathway—demonstration of Ins(1,4,5)P_3 3-kinase activity in animal tissues. *Nature* 320:631–634.

Wakelam, M. J. O., Davies, S. A., Houslay, M. D., McKay, I., Marshall, C. J., and Hall, A., 1986. Normal $p21^{N\text{-}ras}$ couples bombesin and other growth factor receptors to inositol phosphate production. *Nature* 323:173–176.

Bell, R. M., 1986. Protein kinase C activation by diacylglycerol second messengers. *Cell* 45:631–632.

Parthasarathy, R., and Eisenberg, F., Jr., 1986. The inositol phospholipids: a stereochemical view of biological activity. *Biochem. J.* 235:313–322.

Eicosanoid hormones

Hammarstrom, S., 1983. Leukotrienes. *Ann. Rev. Biochem.* 52:355–377.

Smith, W. L., and Borgeat, P., 1985. The eicosanoids: prostaglandins, thromboxanes, leukotrienes, and hydroxyeicosaenoic acids. *In* Vance, D. E., and Vance, J. E., (eds.), *Biochemistry of Lipids and Membranes*, pp. 325–360. Benjamin/Cummings.

Roth, G. J., and Siok, C. J., 1978. Acetylation of the NH_2-terminal serine of prostaglandin synthetase by aspirin. *J. Biol. Chem.* 253:3782–3784.

Vane, J. R., 1971. Inhibition of prostaglandin synthesis as a mechanism of action for aspirin-like drugs. *Nature New Biol.* 231:232–235.

Insulin and other growth factors

James, R., and Bradshaw, R. A., 1984. Polypeptide growth factors. *Ann. Rev. Biochem.* 53:259–292.

Kurachi, K., Davie, E. W., Strydom, D. J., Riordan, J. F., and Vallee, B. L., 1985. Sequence of the cDNA and gene for angiogenin, a human angiogenesis factor. *Biochemistry* 24:5494–5499.

Ross, R., Raines, E. W., and Bowen-Pope, D. F., 1986. The biology of platelet-derived growth factor. *Cell* 46:155–169.

Stoscheck, C. M., and King, Jr., L. E., 1986. Functional and structural characteristics of EGF and its receptor and their relationship to transforming proteins. *J. Cellul. Biochem.* 31:135–152.

Levi-Montalcini, R., and Calissano, P., 1979. The nerve-growth factor. *Sci. Amer.* 240(6):68–77.

Orci, L., Ravazzola, M., Storch, M.-J., Anderson, R. G. W., Vassalli, J.-D., and Perrelet, A., 1987. Proteolytic maturation of insulin is a post-Golgi event which occurs in acidifying clathrin-coated secretory vesicles. *Cell* 49:865–868.

Growth factor receptors and tyrosine kinases

Carpenter, G., 1987. Receptors for epidermal growth factor and other polypeptide mitogens. *Ann. Rev. Biochem.* 56:881–914.

Ullrich, A., Coussens, L., Hayflick, J. S., Dull, T. J., Gray, A., Tam, A. W., Lee, J., Yarden, Y., Libermann, T. A., Schlessinger, J., Diwnward, J., Mayes, E. L. V., Whittle, N., Waterfield, M. D., and Seeburg, P. H., 1984. Human epidermal growth factor cDNA sequence and aberrant expression in A431 epidermoid carcinoma cells. *Nature* 309:418–425.

Ebina, Y., Ellis, L., Jarnagin, K., Edery, M., Graf, L., Clauser, E., Ou, J., Masiarz, F., Kan, Y. W., Goldfine, I. D., Roth, R. A., and Rutter, W. J., 1985. The human insulin receptor cDNA: the structural basis for hormone-activated transmembrane signalling. *Cell* 40:747–758.

White, M. F., Maron, R., and Kahn, C. R., 1985. Insulin rapidly stimulates tyrosine phosphorylation of a M_r-185,000 protein in intact cells. *Nature* 318:183–186.

Riedel, H., Dull, D. J., Schlesinger, J., and Ullrich, A., 1986. A chimaeric receptor allows insulin to stimulate tyrosine kinase activity of epidermal growth factor receptor. *Nature* 324:68–70.

Chao, M. V., Bothwell, M. A., Ross, A. H., Koprowski, H., Lanahan, A. A., Buck, C. R., and Sehgal, A., 1986. Gene transfer and molecular cloning of the human NGF receptor. *Science* 232:518–521.

Steroid receptors

Weinberger, C., Giguerre, V., Hollenberg, S., Rosenfeld, M. G., and Evans, R. M., 1986. Human steroid receptors and *erbA* proto-oncogene products: members of a new superfamily of enhancer binding proteins. *Cold Spring Harbor Symp. Quant. Biol.* 51:759–772.

Giguere, V., Hollenberg, S. M., Rosenfeld, M. G., and Evans, R. M., 1986. Functional domains of the human glucocorticoid receptor. *Cell* 46:645–652.

Green, S., Walter, P., Kumar, V., Krust, A., Bornert, J.-M., Argos, P., and Chambon, P., 1986. Human oestrogen receptor cDNA: sequence, expression and homology to v-*erb*-A. *Nature* 320:134–139.

Hollenberg, S. M., Weinberger, C., Ong, E. S., Cerelli, G., Oro, A., Lebo, R., Thompson, E. B., Rosenfeld, M. G., and Evans, R. M., 1985. Primary structure and expression of a functional human glucocorticoid receptor cDNA. *Nature* 318:635–641.

Sap, J., Munoz, A., Damm, K., Goldberg, Y., Ghysdael, J., Leutz, A., Beug, H., and Vennström, B., 1986. The c-*erb*-A protein is a high-affinity receptor for thyroid hormone. *Nature* 324:635–640.

McDonnell, D. P., Mangelsdorf, D. J., Pike, J. W., Haussler, M. R., and O'Malley, B. W., 1987. Molecular cloning of complementary DNA encoding the avian receptor for vitamin D. *Science* 235:1214–1217.

Neuropeptides

Douglass, J., Civelli, O., and Herbert, E., 1984. Polyprotein gene expression: generation of diversity of neuroendocrine peptides. *Ann. Rev. Biochem.* 53:665–715.

Synder, S. H., 1977. Opiate receptors and internal opiates. *Sci. Amer.* 236(3):44–56.

Development of major concepts

Sutherland, E. W., 1972. Studies on the mechanism of hormone action. *Science* 177:401–408.

Guillemin, R., 1978. Peptides in the brain: the new endocrinology of the neuron. *Science* 202:390–402.

Samuelsson, B., 1983. Leukotrienes: mediators of immediate hypersensitivity reactions and inflammation. *Science* 220:568–575.

Cohen, S., 1986. Epidermal growth factor. *Biosci. Rep.* 6:1017–1028.

Levi-Montalcini, R., 1987. The nerve growth factor: thirty-five years later. *EMBO J.* 6:1145–1154.

CHAPTER 39

Excitable Membranes and Sensory Systems

Many cell membranes can be excited by specific chemical or physical stimuli. The responses of a nerve-axon membrane to an electrical stimulus, of a synapse to a transmitter substance, of a retinal rod cell to light, and of motile cells to attractant molecules are mediated by excitable assemblies in membranes. These processes and others carried out by excitable assemblies have common features:

1. The stimulus is detected by a highly specific protein receptor, which is an *integral component of the excitable membrane.*

2. The specific stimulus elicits a *conformational change* in the receptor. As a result, the permeability of the membrane or the activity of a membrane-bound enzyme changes. Many of the responses are *highly amplified.*

3. The conformational change in the receptor and the resulting alterations in function are *reversible.* There are mechanisms that take the receptor back to its resting state and restore its excitability.

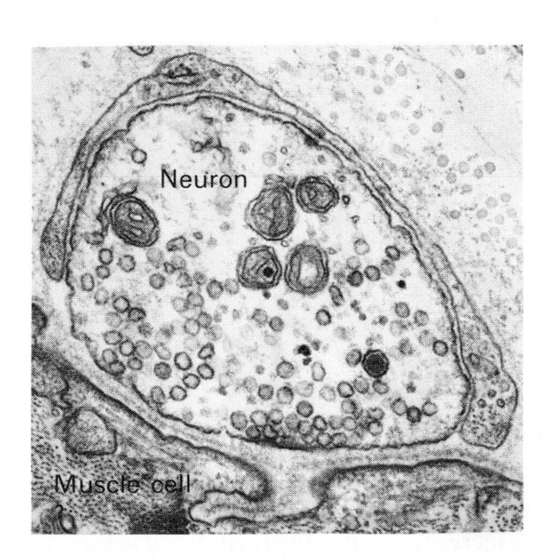

Figure 39-1
Electron micrograph of a synapse. [Courtesy of Dr. U. Jack McMahan.]

This chapter deals with four types of excitable assemblies. Bacterial chemotaxis will be considered first. The movement of bacteria toward nutrients and away from noxious substances begins with detection by chemoreceptors on the cell surface. A sensory processing system then sends signals to flagellar motors that determine whether a bacterium swims smoothly in a straight line or abruptly alters its course. We then turn to higher eucaryotes and consider the sodium channel in nerve-axon membranes. This voltage-sensitive gate participates in the generation of action potentials, the fundamental mode of signal transmission in the nervous system. A different means of controlling the activity of a channel is illustrated by the acetylcholine receptor, a chemically regulated channel that enables nerve impulses to be transmitted across synapses. The interplay of receptors, enzymes, and channels in mediating visual excitation is the final topic of this chapter. The photoisomerization of rhodopsin, the photoreceptor protein of retinal rod cells, triggers a cyclic nucleotide cascade that leads to the generation of a nerve signal. A remarkable feature of this system is its exquisite sensitivity—a rod cell can be excited by a single photon.

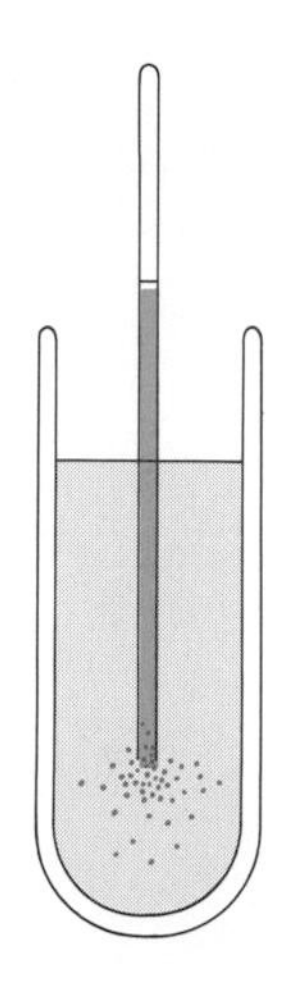

Figure 39-2
Chemotaxis of bacteria to a capillary containing an attractant such as glucose.

CHEMORECEPTORS ON BACTERIA DETECT SPECIFIC MOLECULES AND SEND SIGNALS TO FLAGELLA

In the late nineteenth century, Wilhelm Pfeffer, a German botanist, showed that motile bacteria cluster near the mouth of a thin capillary tube containing an attractant such as a sugar (Figure 39-2). In contrast, they move away from the tube if it contains a repellent, such as a harmful substance or a bacterial excretory product. This directed movement of bacteria toward some specific substances and away from others is called *chemotaxis*. In the 1960s, Julius Adler began to investigate the molecular basis of bacterial chemotaxis. Biochemical, genetic, and structural analyses carried out by him and numerous other investigators have revealed much about this process. Chemotaxis begins with the detection of chemicals by specific *chemoreceptors*. Some of these receptors are soluble proteins located in the periplasmic space, whereas others are proteins that span the plasma membrane. Twenty different chemoreceptors—some for attractants and others for repellents—have been identified. Information from these sensors is transmitted to a *processing system* that analyzes and integrates many stimuli. This sensory processing system then sends signals to the *motors* that drive the *flagella*. These signals determine whether a bacterium continues to swim smoothly in a straight line or abruptly alters its course.

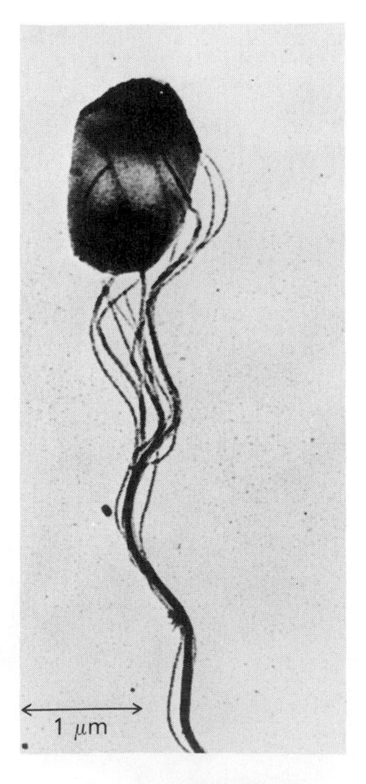

Figure 39-3
Electron micrograph of *S. typhimurium* showing flagella in a bundle. [Courtesy of Dr. Daniel Koshland, Jr.]

BACTERIAL FLAGELLA ARE ROTATED BY REVERSIBLE MOTORS AT THEIR BASES

Bacteria swim by rotating flagella that emerge from their surfaces (Figure 39-3). These thin helical filaments are built from 53-kd *flagellin* subunits. An *E. coli* or *S. typhimurium* bacterium has about six flagella, which are 150 Å in diameter and 10 μm long. Bacterial flagella are much smaller and far less complex than eucaryotic flagella and cilia (p. 941). A bacterial flagellum is an extracellular appendage. It cannot actively bend because it lacks a contractile apparatus. Instead, it is rotated by a motor located at the junction of the flagellum and cell envelope. The isolation of flagella with the basal structure still attached has contributed to an understanding of these assemblies. The flagellum itself consists of a *filament*, a *hook*, and a *rod*. In *E. coli*, a basal structure consists of four rings mounted on the rod. The outer pair of rings is attached to the outer membrane; one side of the inner pair of rings is attached to the plasma membrane. This basal body (Figure 39-4) anchors the flagellum to the cell envelope and actively drives its rotation.

The motor is powered by the proton-motive force across the plasma membrane rather than by the hydrolysis of ATP. In fact, the angular velocity of rotation is directly proportional to the proton-motive force. The maximum ve-

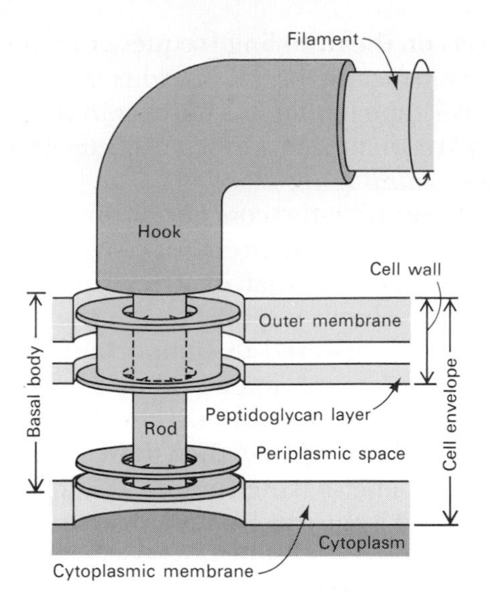

Figure 39-4
Basal body of the flagellum of *E. coli.* This motor is encoded by about thirty-five genes. [After J. Adler. The sensing of chemicals by bacteria. Copyright © 1976 by Scientific American, Inc. All rights reserved.]

locity is about 60 revolutions per second. A torque is generated by the movement of protons through multiple (~16) force-generating units in the inner pair of rings. About a thousand protons move across per revolution. The efficiency of this energy-conversion process is high, about 50%. An intriguing feature of this motor is that it can rotate clockwise or counterclockwise.

An *E. coli* bacterium typically swims smoothly in nearly a straight line for about a second. In this interval, it travels a distance of about 30 μm, which is equal to about fifteen body lengths. The bacterium then abruptly alters its course by tumbling (Figure 39-5). The average change in direction is about 60 degrees. What determines whether a bacterium swims smoothly or tumbles? When the flagella rotate counterclockwise, the helical filaments form a coherent bundle and the cell swims smoothly. When the flagella rotate clockwise, in contrast, the bundle flies apart because the screwsense of the helical flagella does not match the direction of rotation. Each flagellum then pulls in a different direction, and so the cell tumbles.

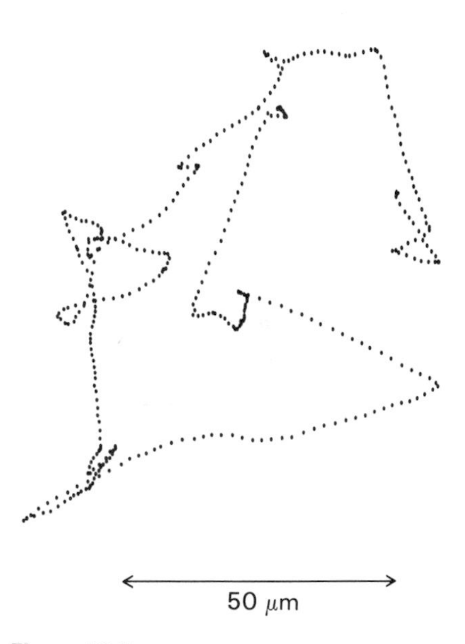

Figure 39-5
A projection of the track of an *E. coli* bacterium obtained with a microscope that automatically follows its motion in three dimensions. The points show the locations of the bacterium at 80 ms intervals. [From H. C. Berg, *Nature* 254(1975):390.]

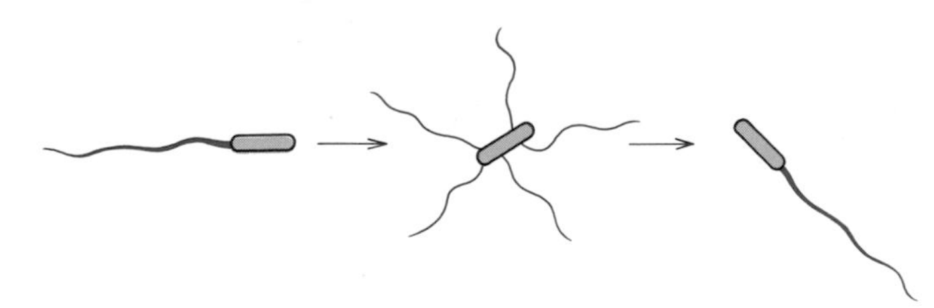

Figure 39-6
Tumbling is caused by an abrupt reversal of the flagellar motor, which disperses the flagellar bundle. A second reversal of the motor restores smooth swimming, but (most likely) in a different direction. [After a drawing kindly provided by Dr. Daniel Koshland, Jr.]

BACTERIA DETECT TEMPORAL GRADIENTS RATHER THAN INSTANTANEOUS SPATIAL GRADIENTS

When a bacterium moves toward an increasing concentration of an attractant, tumbling becomes less frequent. In contrast, when it moves away from an attractant, tumbling becomes more frequent. Repellents have opposite effects on the tumbling frequency. Thus, if a bacterium is moving toward an attractant or away from a repellent it will swim smoothly for a longer time than if it is moving in the opposite direction. Tumbling serves to reorient a misdirected bacterium. *The regulation of tumbling frequency is central to chemotaxis.*

Does a bacterium compare the concentration of an attractant at one end of the cell with that at the other end or does it compare the concentration of attractant now with that of a few moments ago? In other words, is the sensing mechanism spatial or temporal? Daniel Koshland, Jr., and Robert Macnab answered this important question by carrying out an ingeniously simple experiment. They rapidly mixed a suspension of bacteria in a medium devoid of an attractant with a solution containing an attractant, and they then observed the tumbling frequency of the bacteria. The striking finding was that tumbling was suppressed within a second after mixing. The bacteria swam for long distances in a straight line in this rapidly mixed solution, which was devoid of a spatial gradient. Hence, this experiment demonstrated that *bacteria possess a temporal sensing mechanism.* In other words, a bacterium detects a spatial gradient of attractant not by comparing the concentration at its head and tail, but by *traveling through space and comparing its observations through time.* In essence, *bacterial chemotaxis is a biased random walk.* Purposeful behavior arises from the *selection* of random movements.

METHYL-ACCEPTING CHEMOTAXIS PROTEINS TRANSMIT CHEMOTACTIC SIGNALS ACROSS THE PLASMA MEMBRANE

Chemotaxis begins with the binding of chemicals to specific *chemosensors* on the cell surface. Information concerning the presence of attractants and repellents is then transmitted across the plasma membrane by *transducers.* A *tumble regulator* is postulated to carry signals from these transducers to the *switch* in the flagellar motor that determines whether it rotates clockwise or counterclockwise.

Transmembrane signaling is mediated by transducers known as *methyl-accepting chemotaxis proteins* (MCPs) because they become reversibly methylated, as will be discussed shortly. Repellents and some attractants bind directly to MCPs, whereas other attractants bind to soluble chemosensors, which then interact with MCPs (Figure 39-7). Four

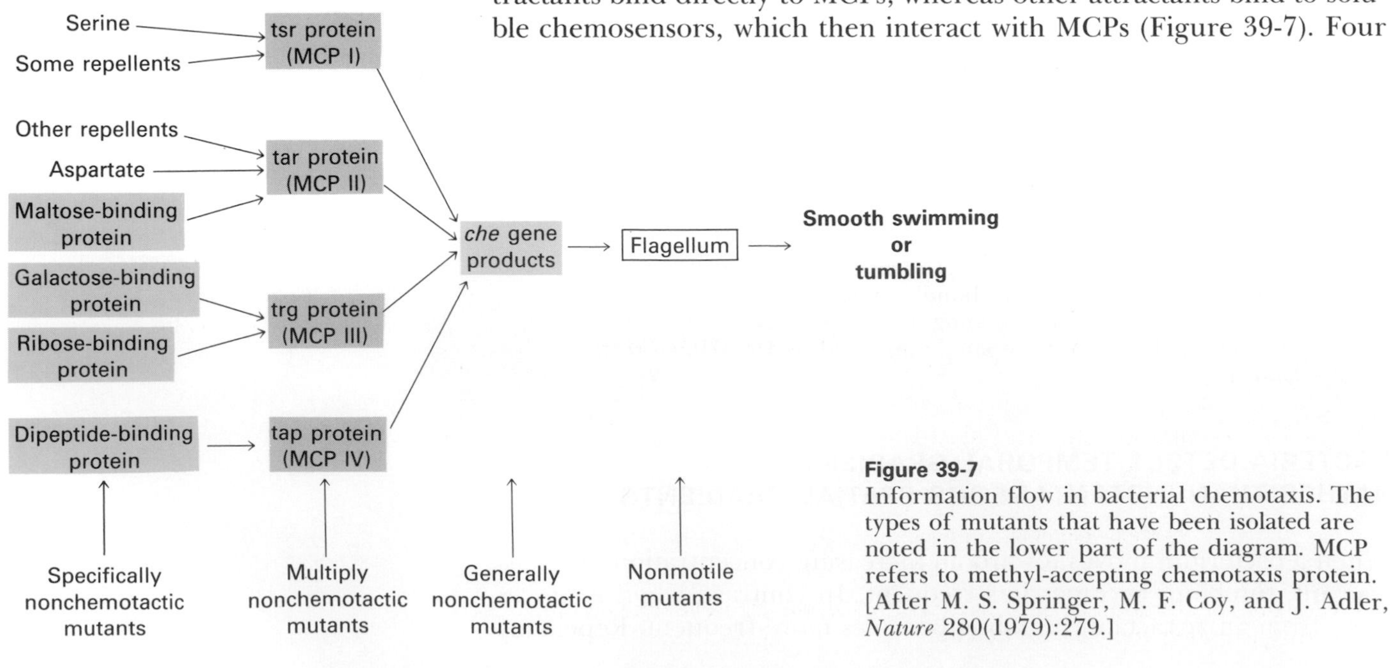

Figure 39-7
Information flow in bacterial chemotaxis. The types of mutants that have been isolated are noted in the lower part of the diagram. MCP refers to methyl-accepting chemotaxis protein. [After M. S. Springer, M. F. Coy, and J. Adler, *Nature* 280(1979):279.]

homologous MCPs are known; they are encoded by the *tsr* (I), *tar* (II), *trg* (III), and *tap* (IV) genes. For example, aspartate binds directly to MCP II. Maltose, on the other hand, binds to a maltose-binding protein that is present in the periplasmic space; this complex then binds to MCP II. The maltose-binding protein has another role: it delivers the disaccharide to a membrane transporter. Indeed, *each of the soluble chemosensors is also a component of a membrane transport system for a nutrient.*

Each of the four members of the MCP family of transducers has a mass of about 60 kd and contains four regions: (1) a periplasmic domain that binds attractant or repellent (or an attractant-protein complex); (2) a transmembrane segment consisting of two α helices; (3) a cytosolic region that interacts with the tumble regulator; and (4) a cytosolic region that can be reversibly methylated (Figure 39-8). The cytosolic domains of MCPs are highly conserved because they interact with the same components of the information-processing system. For example, a 48-residue sequence of the cytosolic domain is the same in MCPs I and II. In contrast, their periplasmic domains are divergent because they bind different attractants, repellents, and chemosensor proteins. A chimeric MCP formed from the amino-terminal half of II and the carboxyl-terminal half of I is triggered by attractants that are normally sensed by II. Hence, it seems likely that different MCPs transmit signals across the membrane in essentially the same way.

Several glutamate side chains on each MCP are reversibly methylated by a *methyltransferase* that uses *S*-adenosylmethionine as the activated donor (Figure 39-9). This enzyme methylates the second glutamate in Glu-Glu-X-X-Ala-Ser and Glu-Glu-X-X-Ala-Thr sequences in α helices of these MCPs. These γ-methyl groups on MCPs are removed by the hydrolytic action of a *methylesterase*. The extent of methylation of an MCP is increased by attractants and decreased by repellents. A methylated MCP is less readily triggerd by attractants and more readily triggered by repellents than is an unmethylated MCP. Reversible methylation enables the bacterial chemotactic apparatus, like the visual system (p. 1032) and other sensory systems, to respond to *changes* in the intensity of stimuli rather than to their absolute level. This process of adjustment is called *adaptation*.

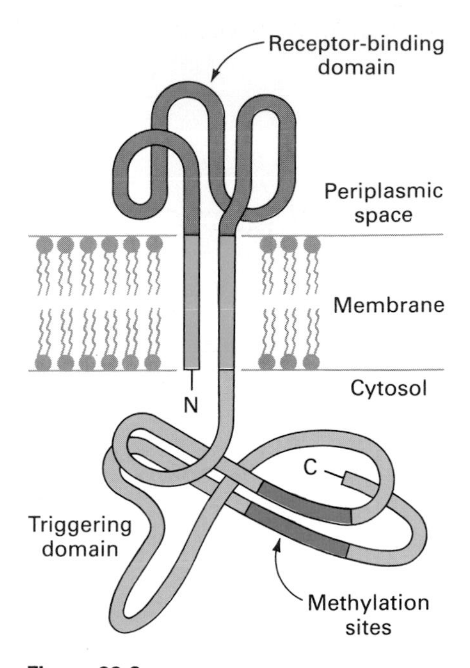

Figure 39-8
Schematic diagram of a methyl-accepting chemotaxis protein (MCP). This receptor transmits chemotactic signals across the plasma membrane. Repellents, attractants, and chemosensor proteins bind to the periplasmic domain. The cytosolic domain communicates with a central processing system that controls the direction of flagellar rotation. Reversible methylation of the cytosolic domain is important in adaptation. [After D. E. Koshland, Jr., A. F. Russo, and N. I. Gutterson. *Cold Spring Harbor Symp. Quant. Biol.* 48(1983):805.]

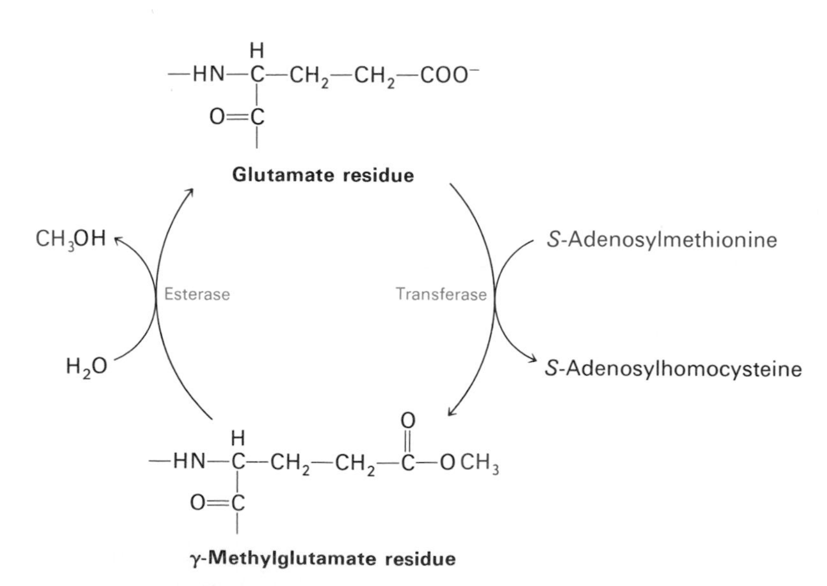

Figure 39-9
Reversible methylation of glutamate residues in the cytosolic domain of methyl-accepting chemotaxis proteins.

The output of a single flagellar motor can be monitored by tethering the flagellar filament to a glass surface. Bacteria grown so that they contain an average of one flagellum are treated with antiflagellin antibody, which adheres to a glass coverslip. When the filament is held fixed, the motor spins the entire bacterial cell body. The alternation between clockwise (CW) and counterclockwise (CCW) rotations can readily be observed under a microscope and automatically recorded by an optoelectronic device that tracks the image of the spinning cell. The probability of spinning CCW, which normally results in smooth swimming is increased by exposing the bacterium to a transient increase in the concentration of an attractant such as aspartate (Figure 39-10A). Conversely, a pulse of a repellent such as benzoate markedly increases the probability of spinning CW, which normally leads to tumbling (Figure 39-10B). The effect of a sustained increase in the level of an attractant is also very revealing. The probability of spinning CCW increases within a second of the stepwise increase in concentration but returns to the basal value shortly thereafter (Figure 39-10C). In other words, the system has adapted to the higher level of attractant; it now has higher expectations.

In essence, *a bacterium decides whether or not to tumble by comparing the concentrations of attractants and repellents sensed in the past second with those encountered in the three seconds before.* The decision is made by changing the probabilities of clockwise and counterclockwise rotation. Signals are highly amplified by the chemotactic system. The binding of an attractant to one receptor molecule of 600 present on the cell surface transiently increases the probability of CCW rotation by 0.1.

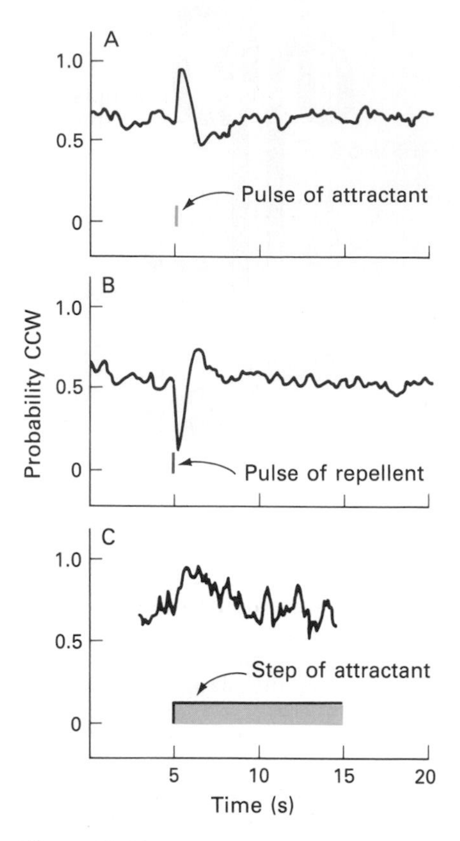

Figure 39-10
Biasing of the probability of counterclockwise (CCW) rotation of bacterial flagella by changes in the concentration of attractants and repellents. The responses of many tethered cells were averaged. (A) Response to a pulse of attractant. (B) Response to a pulse of repellent. (C) Response to a sustained increase in the concentration of an attractant. The sensory system adapts within a few seconds. [After S. M. Block, J. E. Segall, and H. C. Berg. *Cell* 31(1982):215.]

A CENTRAL PROCESSING SYSTEM ENCODED BY *che* GENES MEDIATES THE SWITCHING OF FLAGELLAR ROTATION

Genetic analyses revealed that MCPs determine the direction of flagellar rotation through a common processing system that is encoded by the *che* genes. Eight *che* loci have been identified: *A, B, C, D, W, R, Y* and *Z*. Mutations in these genes impair chemotaxis to all attractants and repellents, in contrast with mutations in an MCP, which blocks the response to a subset of these agents (see Figure 39-7). Gene *cheR* encodes the methyltransferase, and *cheB* encodes the methylesterase. Mutants lacking both *cheR* and *cheB* are still able to carry out chemotaxis but only over a limited range of concentrations of attractants and repellents. Adaptation is severely impaired in these double mutants, showing that *reversible methylation is central to adaptation.* The loss of only *cheB* leads to persistent tumbling; such a bacterium keeps changing directions and executes a truly random walk that never gets anywhere. In contrast, the loss of *cheR* leads to persistent smooth swimming; such a bacterium is rather like Candide, eternally optimistic, convinced that it is heading in just the right direction.

The motor rotates exclusively CW in mutants that express *cheY* at much higher than normal levels. Flagella of bacterial envelopes entirely lacking the cytosolic *che* gene products rotate exclusively CCW. The addition of micromolar amounts of CheY (14 kd) to these envelopes leads to CW rotation. Hence, *it seems likely that the direction of flagellar rotation is controlled by the cheY protein.* CheZ antagonizes the effect of CheY but does not itself switch the motor. CheA and CheW are also needed for reversible rotation of the flagellar motor; they may mediate switching by CheY. The picture that emerges from such genetic and biochemical studies is that an activated form of CheY promotes CW rotation (tumbling). Activation of CheY requires ATP. Whether an MCP activates CheY depends on three factors (Figure 39-11). The binding of a repellent to an MCP and methylation of an MCP increase the probability of activation of CheY. In contrast, the binding of an attractant to an MCP decreases the likelihood that it will activate CheY. It is evident that *MCPs themselves are miniature integrators of information as well as transmembrane signalers*. The challenge now is to determine the molecular nature of the active and inactive states of CheY.

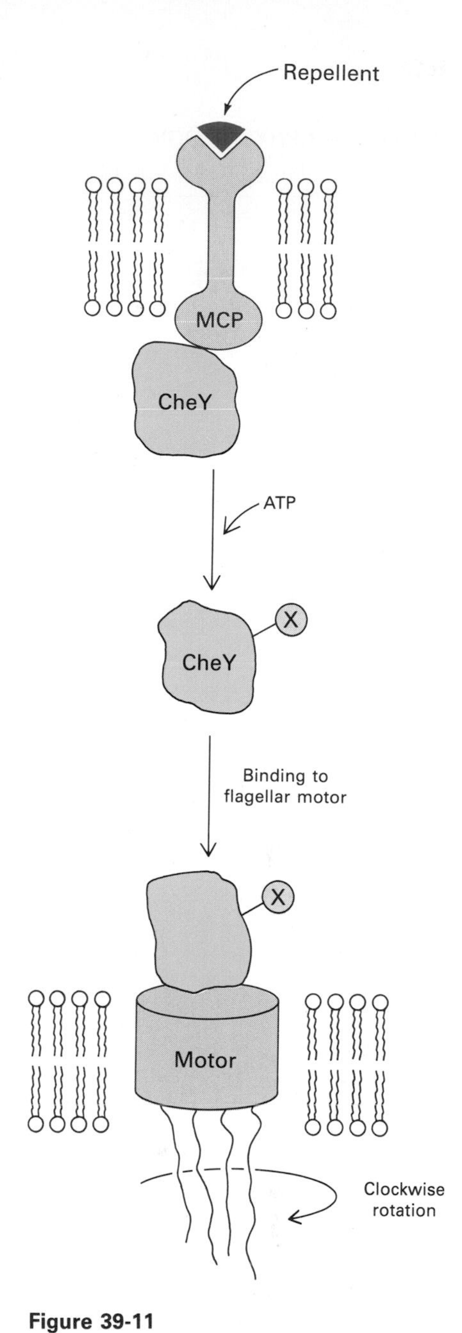

Figure 39-11
Proposed mechanism for the coupling of methyl-accepting chemotaxis proteins (MCPs) to the flagellar motor. An MCP containing a bound repellent activates the cheY protein in an ATP-dependent reaction. A labile active form of CheY (marked X) is produced. This activated form of CheY then diffuses to the motor and induces it to rotate clockwise (tumbling). This CW signal is transient because of the rapid deactivation of CheY by CheZ.

ACTION POTENTIALS ARE MEDIATED BY TRANSIENT CHANGES IN Na^+ AND K^+ PERMEABILITY

We turn now to excitable membranes and sensory systems in higher eucaryotes with highly differentiated nervous systems. A nerve impulse is an electrical signal produced by the flow of ions across the plasma membrane of a neuron. The interior of a neuron, like that of most other cells, has a high concentration of K^+ and a low one of Na^+. These ionic gradients are generated by the Na^+-K^+ ATPase pump (p. 950). In the resting state, the membrane of an axon is much more permeable to K^+ than to Na^+, and so the membrane potential is largely determined by the ratio of the internal to the external concentration of K^+ (Figure 39-12A). The membrane potential of −60 mV in unstimulated axons is near the −75-mV level (the K^+ equilibrium potential) that would be given by a membrane permeable to only K^+. A nerve impulse or *action potential* is generated when the membrane potential is depolarized beyond a critical threshold value (i.e., from −60 to −40 mV). The membrane potential becomes positive within about a millisecond and attains a value of about +30 mV before turning negative again. This amplified depolarization is propagated along the nerve terminal. The giant axons of squids played an important role in the discovery of the nature of the action potential. Electrodes can be readily inserted into these unusually large axons, which have a diameter of about a millimeter, and so squid axons have been favorite objects of inquiry.

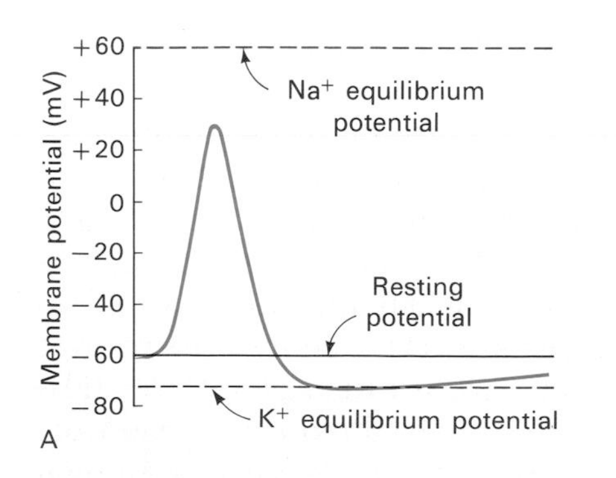

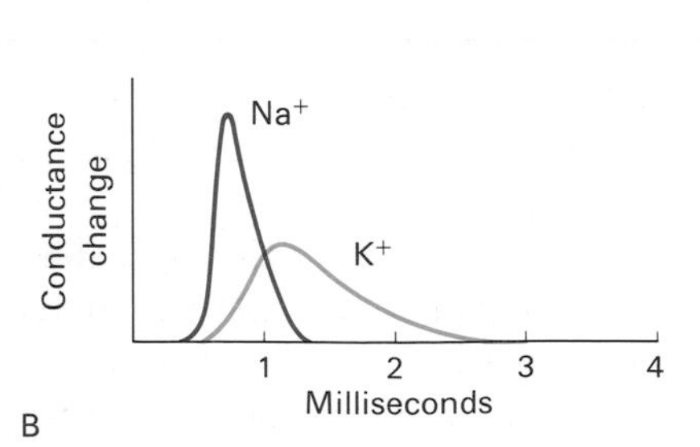

Figure 39-12
Depolarization of an axon membrane results in an action potential. Time course of (A) the change in membrane potential and (B) the change in Na^+ and K^+ conductances.

What is the mechanism of the action potential? Alan Hodgkin and Andrew Huxley carried out ingenious studies that revealed that *the action potential arises from large, transient changes in the permeability of the axon membrane to Na^+ and K^+ ions* (Figure 39-12B). The conductance of the membrane to Na^+ changes first. Depolarization of the membrane beyond the threshold level leads to an opening of Na^+ channels. *The sodium channel is gated by membrane voltage rather than by a small molecule or macromolecule.* Sodium ions begin to flow into the cell because of the large electrochemical gradient across the plasma membrane. The entry of Na^+ further depolarizes the membrane, and so more gates for Na^+ are opened. This positive feedback between depolarization and Na^+ entry leads to a very rapid and large change in membrane potential, from about −60 mV to +30 mV in a millisecond. The entry of Na^+ stops at about +30 mV because this is the Na^+ equilibrium potential. In other words, the thermodynamic driving force for the entry of Na^+ vanishes when this potential is attained.

Na^+ channels spontaneously close and the K^+ gates begin to open at about this time (Figure 39-12B). Consequently, potassium ions flow outward and so the membrane potential returns to a negative value. At about 2 ms, the membrane potential is −75 mV, the K^+ equilibrium potential. The resting level of −60 mV is restored in a few milliseconds as the K^+ conductance decreases to the value characteristic of the unstimulated state. It is important to stress that a very small proportion of the sodium and potassium ions in a nerve cell, of the order of one in a million, flows across the plasma membrane during the action potential. In other words, only a tiny fraction of the Na^+-K^+ gradient is dissipated by a single nerve impulse. Clearly, the action potential is a very efficient means of signaling over large distances.

THE NARROW PORE SIZE OF THE SODIUM CHANNEL MAKES IT MUCH MORE PERMEABLE TO Na^+ THAN TO K^+

Table 39-1
Relative permeabilities of the sodium and potassium channels in axon membranes

Ion	*Na^+ channel*	*K^+ channel*
Li^+	0.93	<0.01
Na^+	1.00	<0.01
K^+	0.09	1.00
Rb^+	<0.01	0.91
Cs^+	<0.01	<0.08
NH_4^+	0.16	0.13
$OHNH_3^+$	0.94	<0.03
$H_2NNH_3^+$	0.59	<0.03
$H_3CNH_3^+$	<0.01	<0.02

The Na^+ channel favors the passage of Na^+ over K^+ by a factor of 11. How is this selectivity achieved? A definitive answer to this question will have to await a high-resolution analysis of the structure of the channel. However, electrophysiological studies of the relative permeabilities of alkali cations and organic cations provide some clues. The dependence of permeability on ionic size (Table 39-1) indicates that the channel is narrow. Ions having diameters greater than 5 Å are excluded.

Conductance is not governed by size alone. For example, methylamine ($H_3CNH_3^+$) has nearly the same dimensions as hydrazine ($H_2NNH_3^+$) and hydroxylamine ($HONH_3^+$), yet it is much less permeable. A likely reason for this difference is that the methyl group of methylamine, in contrast with the amino group of hydrazine or the hydroxyl group of hydroxylamine, cannot form a hydrogen bond with an oxygen atom in the channel. Hence, methylamine does not fit. Another significant finding is that the conductance of the Na^+ channel to all permeant cations decreases markedly when the pH is lowered. In fact, the relative permeability follows a titration curve for an acid with a pK of 5.2, which suggests that the active form of the channel contains a negatively charged carboxylate group. Thus, *the Na^+ channel selects for Na^+ by providing a negatively charged site with a small radius.* A K^+ ion cannot readily pass through this region because it is larger than Na^+ (Figure 39-13). The ionic radius of K^+ is 1.33 Å, compared with 0.95 Å for Na^+.

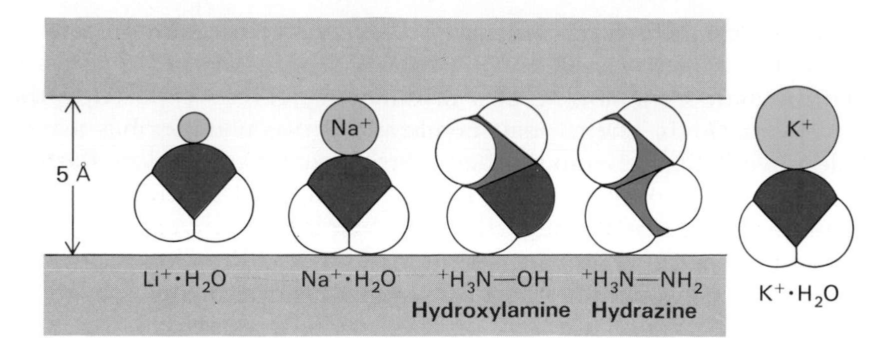

Figure 39-13
The ionic selectivity of the sodium channel partly depends on steric factors. Na^+ and Li^+, together with a water molecule, fit in the channel, as do hydroxylamine and hydrazine. In contrast, K^+ with a water molecule is too large. [After R. D. Keynes. Ion-channels in the nerve-cell membrane. © 1979 by Scientific American, Inc. All rights reserved.]

TETRODOTOXIN AND SAXITOXIN BLOCK SODIUM CHANNELS IN NERVE-AXON MEMBRANES

Tetrodotoxin, a highly potent poison from the puffer (fugu) fish, blocks the conduction of nerve impulses along axons and in excitable membranes of nerve fibers, which leads to respiratory paralysis. The lethal dose for a mouse is about 10 ng. Tetrodotoxin is a very useful probe because it is highly specific. It binds very tightly ($K \sim 10^{-9}$ M) to the Na^+ channel and blocks the flow of sodium ions but has no effect on the K^+ channel. *Saxitoxin,* which is produced by a marine dinoflagellate, acts in the same way. Shellfish, particularly clams and mussels, feeding on such dinoflagellates are poisonous. Indeed, a small mussel may contain enough saxitoxin to kill fifty humans! A common feature of tetrodotoxin and saxitoxin is the presence of a guanido group (Figure 39-14). This positively charged group of the toxin interacts with a negatively charged carboxylate at the mouth of the channel on the extracellular side of the membrane. These toxins obstruct the conductance pathway.

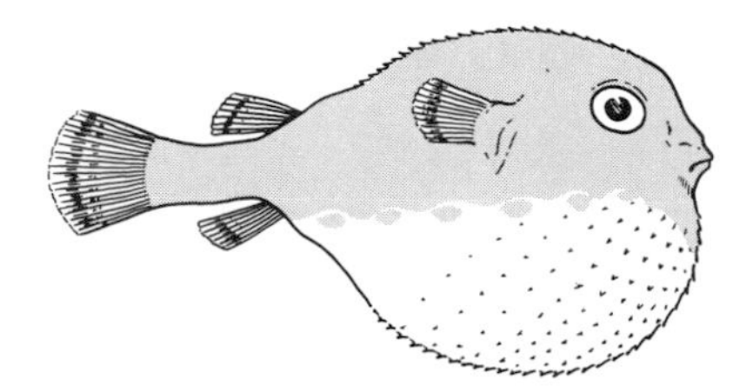

A puffer fish, regarded as a culinary delicacy in Japan.

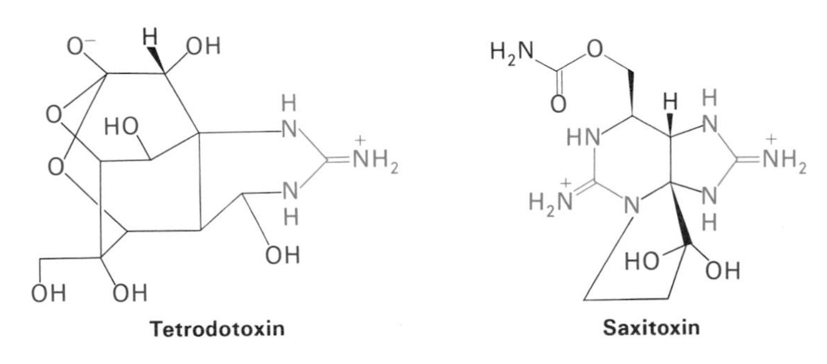

Figure 39-14
Poisons of the Na^+ channel. Tetrodotoxin has one guanido group (blue), whereas saxitoxin has two.

Tetrodotoxin and saxitoxin are very useful probes because of their specificity and high affinity for the Na^+ channel. For example, the densities of Na^+ channels in a variety of excitable membranes have been determined by measuring the binding of highly radioactive tetrodotoxin. *Unmyelinated nerve fibers, which are devoid of an insulating sheath of myelin, typically have low densities of Na^+ channels, of the order of twenty per square micrometer.* Na^+ channels in these axonal membranes are about 2000 Å apart. *In contrast, myelinated nerve fibers have a very high density of*

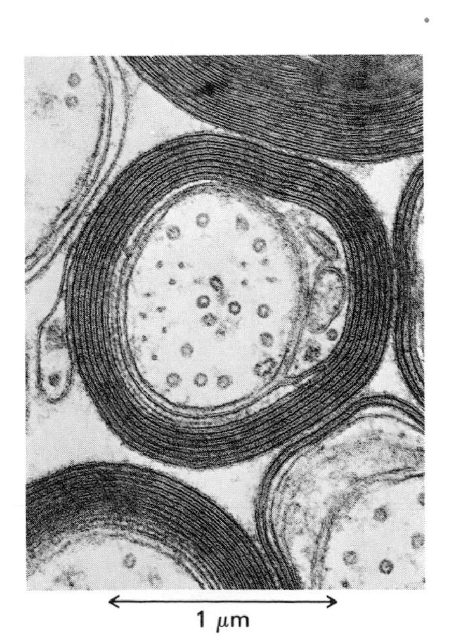

Electron micrograph of a myelinated axon from the spinal cord. Myelin, a membranous wrapping around the axon, acts as an insulating layer. Myelinated nerves have a much higher conduction velocity than unmyelinated ones of the same diameter. [Courtesy of Dr. Cedric Raine.]

channels, of the order of 10^4 per μm^2, in specialized regions called nodes of Ranvier. These nodes, spaced at intervals of 2 mm, are the only sites at which the axonal membrane of a myelinated nerve is exposed to the extracellular fluid. The axonal membrane between nodes has a very low density of channels and does not participate in conduction. Rather, the action potential jumps from node to node, and so the impulse is transmitted more rapidly and efficiently than in an unmyelinated fiber. The presence of 10^4 channels per μm^2 at a node of Ranvier means that about half of its membrane is occupied by Na^+ channels.

PURIFIED SODIUM CHANNELS RECONSTITUTED IN LIPID BILAYERS ARE FUNCTIONALLY ACTIVE

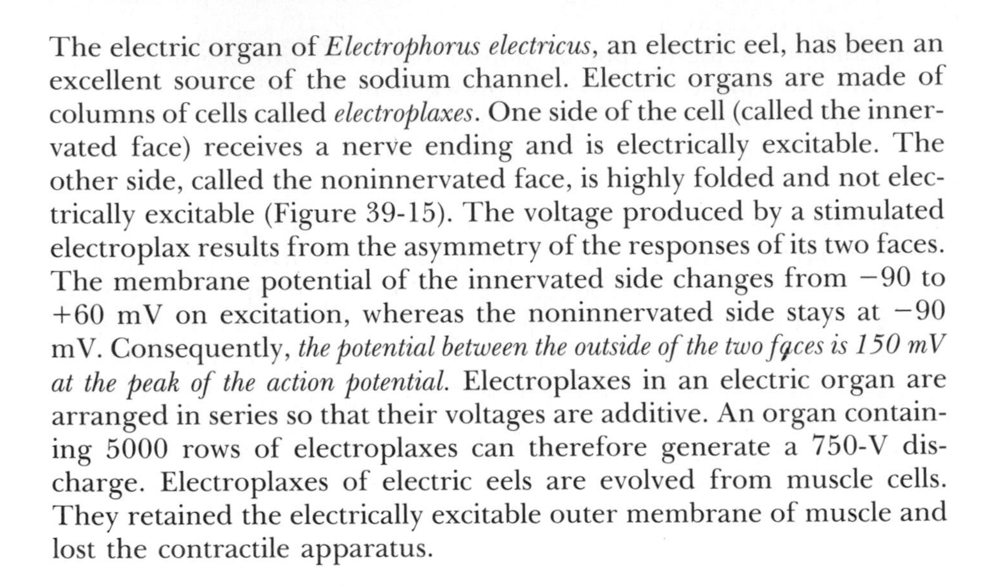

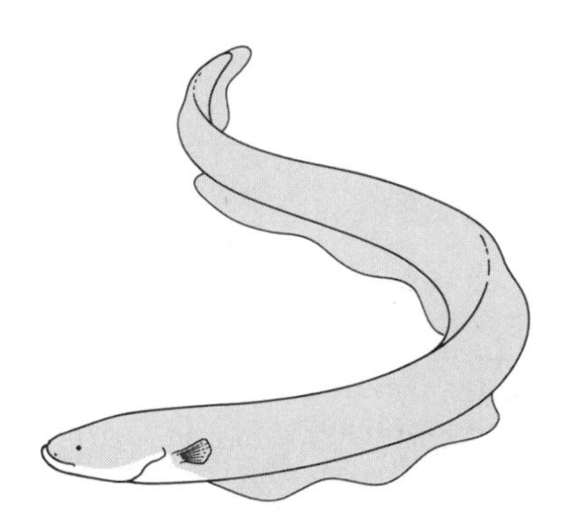

Electrophorus electricus, the electric eel.

The electric organ of *Electrophorus electricus*, an electric eel, has been an excellent source of the sodium channel. Electric organs are made of columns of cells called *electroplaxes*. One side of the cell (called the innervated face) receives a nerve ending and is electrically excitable. The other side, called the noninnervated face, is highly folded and not electrically excitable (Figure 39-15). The voltage produced by a stimulated electroplax results from the asymmetry of the responses of its two faces. The membrane potential of the innervated side changes from −90 to +60 mV on excitation, whereas the noninnervated side stays at −90 mV. Consequently, *the potential between the outside of the two faces is 150 mV at the peak of the action potential.* Electroplaxes in an electric organ are arranged in series so that their voltages are additive. An organ containing 5000 rows of electroplaxes can therefore generate a 750-V discharge. Electroplaxes of electric eels are evolved from muscle cells. They retained the electrically excitable outer membrane of muscle and lost the contractile apparatus.

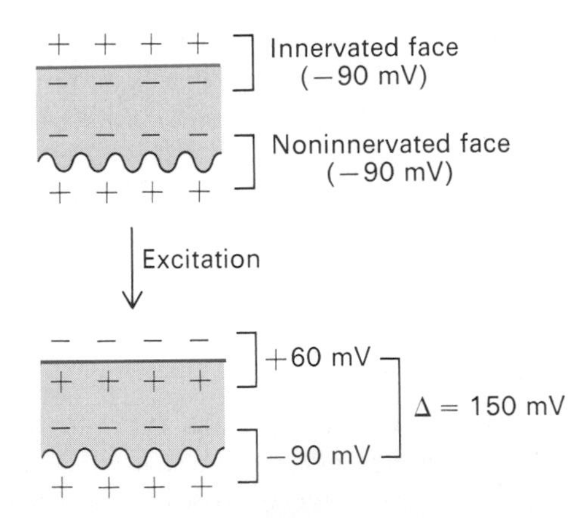

Figure 39-15
Generation of a voltage by an electroplax cell of the electric eel.

The specific binding of tetrodotoxin has also been exploited in purifying the Na^+ channel. Integral membrane proteins of electrically excitable membranes were solubilized by adding a nonionic detergent (Triton X-100) to intact membranes. The Na^+ channel was separated from other proteins by ion-exchange chromatography, followed by affinity chromatography on an immobilized lectin column and sucrose density sedimentation. The Na^+ channel from rat brain consists of a 270-kd α chain, a 39-kd β_1 chain, and a 37-kd β_2 chain. The sodium channel from eel electric organ is a single 270-kd polypeptide chain.

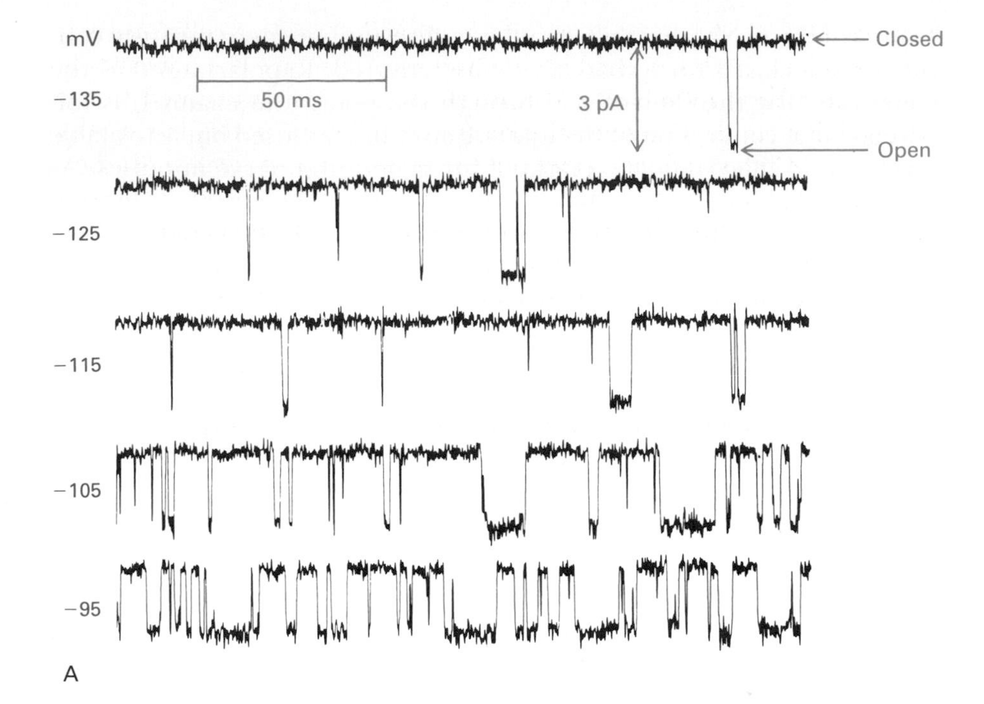

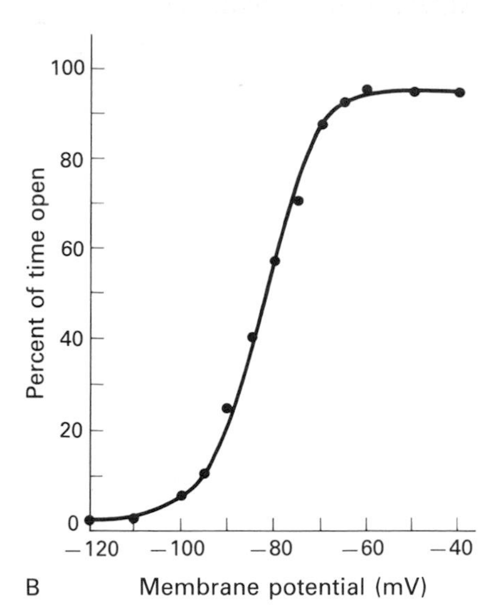

Figure 39-16
Voltage-dependence of reconstituted sodium channels. (A) Conductance recordings showing transitions between the closed and open states of the channel at several membrane voltages. (B) Dependence of the fraction of open channels on the applied membrane voltage. [Courtesy of Dr. Mauricio Montal.]

Purified Na^+ channel reconstituted in lipid bilayer membranes has conductance properties that are similar to those of the channel in native membranes. The reconstituted channels are *Na^+-selective*, favoring Na^+ over K^+ by a factor of 8. Channel opening is highly *voltage-dependent* (Figure 39-16), indicating that the transition between closed and open states involves movements of charges partly or completely through the lipid bilayer. The equilibrium constant K for such a voltage-dependent transition is given by

$$K = \frac{[\text{Closed}]}{[\text{Open}]} = e^{-zF(V-V_0)/RT}$$

where V is the membrane voltage, V_0 is the voltage at which half the channels are open, z is the number of charges that move across the membrane in the transition, F is the Faraday constant, R is the universal gas constant, and T is the absolute temperature. The fraction f of open channels is then given by

$$f_{\text{open}} = \frac{1}{1 + e^{-zF(V-V_0)/RT}}$$

The observed dependence of f on V indicates that z is 4 for the reconstituted channel, compared with values of 4 to 6 for native ones.

The reconstituted channels are blocked by tetrodotoxin applied to the extracellular side. They also undergo *inactivation*, a key property of native channels. Na^+ channels have multiple conformational states (Figure 39-17). The open form induced by depolarization of the mem-

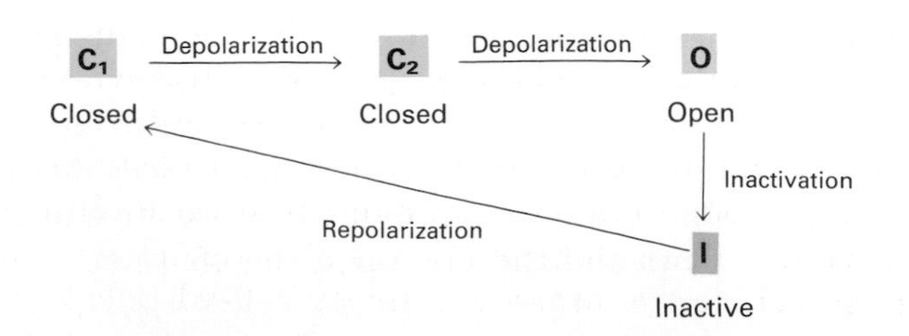

Figure 39-17
Multiple conformational states of the Na^+ channel.

brane conducts Na^+ for only about a millisecond; it spontaneously converts into a closed form that cannot be activated. Repolarization of the membrane (by the flow of K^+ through the potassium channel) is followed by the conversion of the inactive state into a closed but activatable form. Rapid inactivation is essential for generating nerve impulses. As in music, the rests are as important as the notes. *Veratridine* (a steroid from plants of the lily family) and *batrachotoxin* (a steroidal alkaloid from the skin of a poisonous Columbian tree frog) block inactivation by binding specifically to the open state of the channel.

THE SODIUM CHANNEL PROTEIN CONTAINS FOUR REPEATING UNITS THAT SPAN THE BILAYER AND FORM A PORE

Shosaku Numa and coworkers have cloned and sequenced the cDNAs for several Na^+ channels—one from the electric organ of eel and two different ones from rat brain. These studies have shown that the amino acid sequence of the Na^+ channel has been conserved over a long evolutionary period. Most interesting, they revealed that *the channel contains four internal repeats having similar amino acid sequences.* (Figure 39-18).

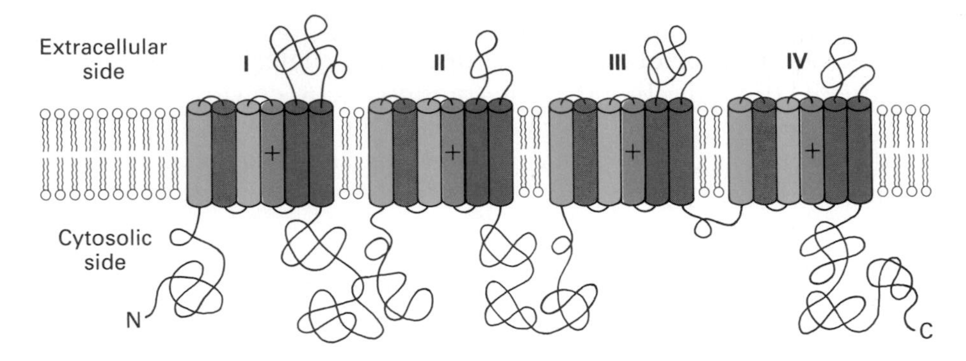

Figure 39-18
The Na^+ channel is a single polypeptide chain with four repeating units (I–IV). Each repeat probably folds into six transmembrane helices. The one labeled + contains many lysine and arginine residues and is likely to be involved in voltage-gating. [After M. Noda, T. Ikeda, T. Kayono, H. Suzuki, H. Takeshima, M. Kurasaki, H. Takahashi, and S. Numa. *Nature* 320(1986):188.]

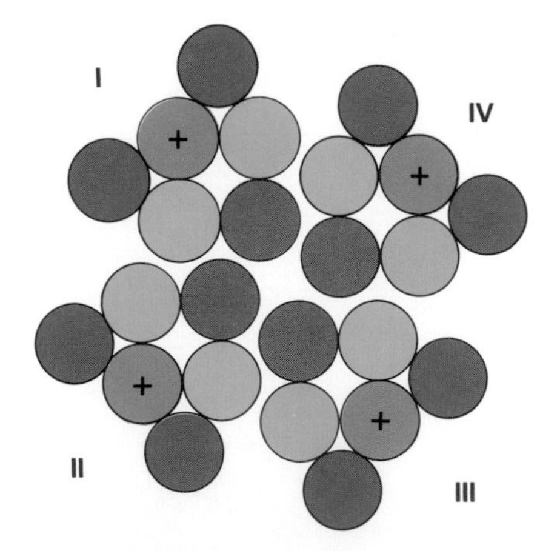

Figure 39-19
Proposed arrangement of the transmembrane segments of the Na^+ channel. The view shown here is perpendicular to the one shown in Figure 39-18. [After M. Noda, T. Ikeda, T. Kayono, H. Suzuki, H. Takeshima, M. Kurasaki, H. Takahashi, and S. Numa. *Nature* 320(1986):188.]

They probably arose by gene duplication and divergence. Hydrophobicity profiles indicate that each homology unit contains five hydrophobic segments (S1, S2, S3, S5, and S6). Each unit also contains a highly positively charged S4 segment; arginine or lysine residues are present at nearly every third residue. Numa has proposed that segments S1–S6 are membrane-spanning α helices and that the pore is formed by the walls of four S2 helices (Figure 39-19). The S4 segments probably participate in making the channel responsive to membrane voltage. The positively charged side chains on the S4 helices are likely to be paired with negatively charged side chains from other helices. Membrane depolarization could shift the position of one helix relative to the other in each of four such pairs and thereby open the channel.

ACETYLCHOLINE IS A NEUROTRANSMITTER AT MANY SYNAPSES AND AT NERVE-MUSCLE JUNCTIONS

Nerve cells interact with other nerve cells at junctions called *synapses* (Figure 39-20). Nerve impulses are communicated across most synapses by chemical transmitters, which are small, diffusible molecules such as acetylcholine and norepinephrine. Acetylcholine is also the transmitter at motor end plates (neuromuscular junctions), which are the junctions between nerve and striated muscle. The *presynaptic membrane* of a *cholinergic synapse*—that is, one that uses acetylcholine as the neurotransmitter—is separated from the *postsynaptic membrane* by a gap of about 500 Å, called the *synaptic cleft.* The end of the presynaptic axon is filled with *synaptic vesicles* containing acetylcholine. The arrival of a nerve impulse leads to the release of acetylcholine into the cleft. The acetylcholine molecules then diffuse to the postsynaptic membrane, where they combine with specific receptor molecules. This produces a *depolarization of the postsynaptic membrane,* which is propagated along the electrically excitable membrane of the second nerve cell. The acetylcholine messenger is *hydrolyzed by acetylcholinesterase,* and the polarization of the postsynaptic membrane is restored.

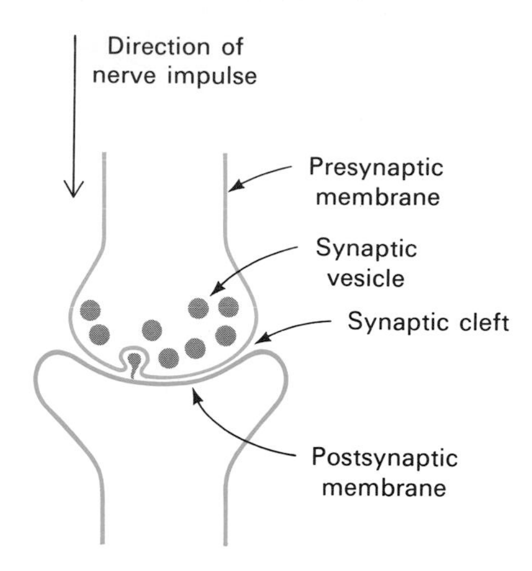

Figure 39-20
Schematic diagram of a cholinergic synapse.

Acetylcholine is synthesized near the presynaptic end of axons by the transfer of an acetyl group from acetyl CoA to choline. This reaction is catalyzed by *choline acetyltransferase* (choline acetylase). Some of the acetylcholine is taken up by synaptic vesicles, typically 400 Å in diameter, whereas the remainder stays in the cytosol. The study of synaptic function has been facilitated by the isolation of *synaptosomes* from homogenates of nerve tissue. Synaptosomes are presynaptic endings that have resealed after being pinched off. They consist of mitochondria, cytosol, and synaptic vesicles surrounded by an intact presynaptic membrane.

$$OH{-}CH_2{-}CH_2{-}\overset{+}{N}{-}(CH_3)_3$$
Choline

$$H_3C{-}\overset{O}{\overset{\|}{C}}{-}S{-}CoA \text{ (Acetyl CoA)} \longrightarrow HS{-}CoA$$

$$H_3C{-}\overset{O}{\overset{\|}{C}}{-}O{-}CH_2{-}CH_2{-}\overset{+}{N}{-}(CH_3)_3$$
Acetylcholine

ACETYLCHOLINE IS RELEASED IN PACKETS

Studies by Bernard Katz of nerve-impulse transmission at neuromuscular junctions revealed that acetylcholine is released from the presynaptic membrane in the form of packets containing some 10^4 molecules. The evidence for *quantal release of acetylcholine* comes from an analysis of the membrane potential of the motor end plate, which shows spontaneous electrical activity even when the associated nerve is not stimulated. Depolarizing pulses that are 0.5 mV in amplitude and last about 20 msec occur intermittently. These *miniature end-plate potentials* occur randomly with a probability that stays constant for long periods. A minia-

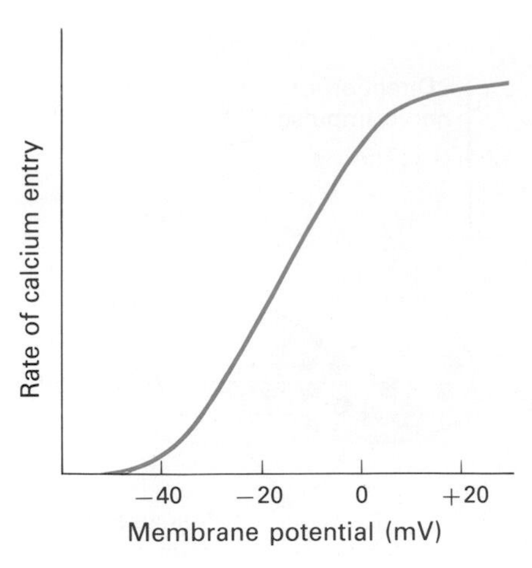

Figure 39-21
Depolarization of the presynaptic membrane opens calcium channels. The elevated level of Ca^{2+} in the cytosol then promotes the release of vesicles containing acetylcholine. Vesicle release is an electrically-controlled form of secretion.

Acetylcholine receptors—
Acetylcholine triggers two classes of membrane receptors that were distinguished pharmacologically by the effects of nicotine and muscarine, which are alkaloids.

The *nicotinic acetylcholine receptor* is a cation-specific channel in postsynaptic membranes.

The *muscarinic acetylcholine receptor* is a seven-helix transmembrane protein that triggers G proteins. It is homologous with the β-adrenergic receptor and rhodopsin. In cardiac cells, this receptor controls a K^+ channel through the intermediacy of a G protein.

ture end-plate potential is caused by the spontaneous release of a single synaptic vesicle. The full depolarization of the end plate elicited by an action potential results from the synchronous release of up to three hundred packets of acetylcholine in less than a millisecond. *The number of packets released depends on the potential of the presynaptic membrane.* In other words, *the release of acetylcholine is an electrically controlled form of secretion.* Release of acetylcholine requires the presence of Ca^{2+} in the extracellular fluid. The depolarization of the presynaptic membrane leads to the entry of this Ca^{2+} through voltage-sensitive calcium channels. The elevated level of Ca^{2+} in the cytosol promotes the fusion of the synaptic vesicle membrane and the presynaptic membrane.

ACETYLCHOLINE OPENS CATION GATES IN THE POSTSYNAPTIC MEMBRANE

The resting potential of a postsynaptic membrane or of a motor end plate is about −75 mV. The interaction of acetylcholine with specific receptors changes the ionic permeabilities of these membranes (Figure 39-22). *The conductance of both Na^+ and K^+ increases markedly within 0.1 msec, and there is a large inward current of Na^+ and a smaller outward current of K^+.* The inward Na^+ current depolarizes the postsynaptic membrane and triggers an action potential in the adjacent axon or muscle membrane. *Acetylcholine opens a single kind of cation channel, which is almost equally permeable to Na^+ and K^+.* The influx of Na^+ is much larger than the efflux of K^+ because the electrochemical gradient across the membrane is steeper for Na^+. This change in ion permeability is mediated by the *nicotinic acetylcholine receptor*; for simplicity, we shall refer to it as the acetylcholine receptor. The *muscarinic acetylcholine receptor* has a different design and function.

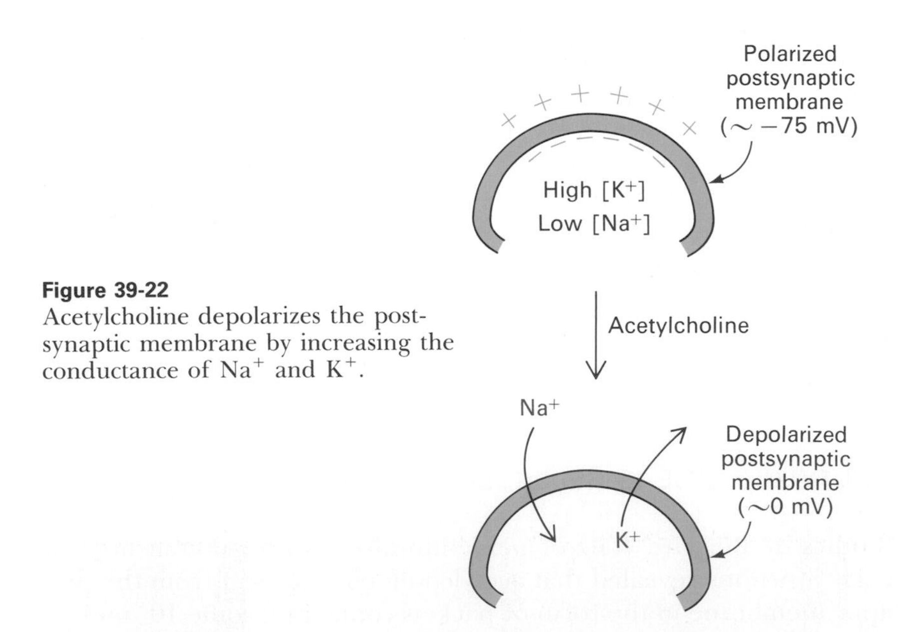

Figure 39-22
Acetylcholine depolarizes the postsynaptic membrane by increasing the conductance of Na^+ and K^+.

PATCH-CLAMP CONDUCTANCE MEASUREMENTS REVEAL THE ACTIVITIES OF SINGLE CHANNELS

The study of ion channels has been revolutionized by the *patch-clamp technique*, which was introduced by Erwin Neher and Bert Sakmann in 1976. A clean glass pipet with a tip diameter of about 1 μm is pressed

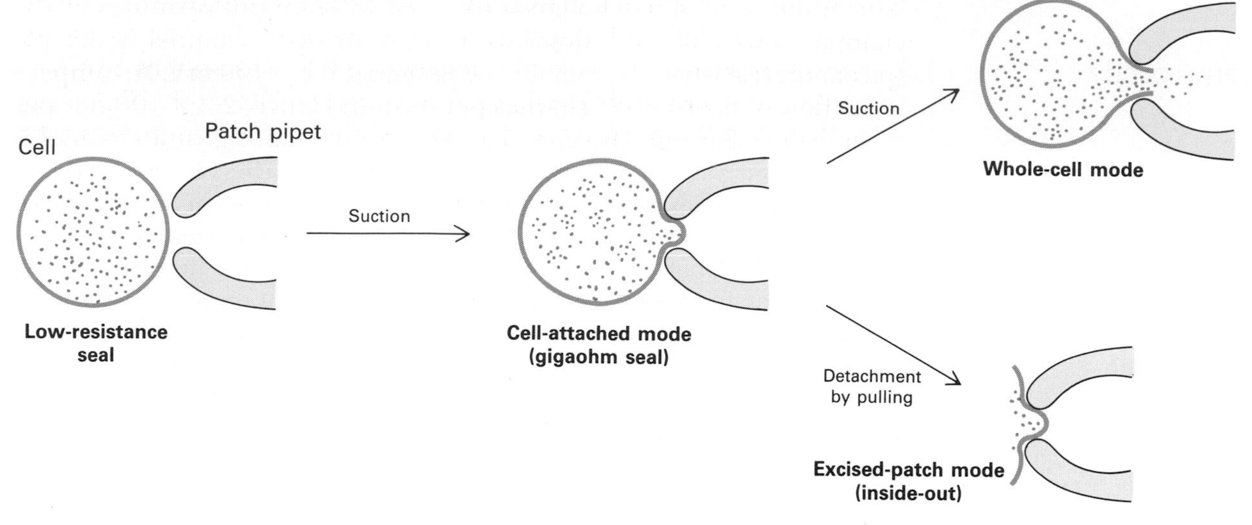

Figure 39-23
The patch-clamp technique for monitoring channel activity is highly versatile. A high-resistance seal (gigaseal) is formed between the pipet and a small patch of plasma membrane. This configuration is called *cell-attached.* The breaking of the membrane patch by increased suction produces a low-resistance pathway between the pipet and interior of the cell. The activity of the channels in the entire plasma membrane can be monitored in this *whole-cell* mode. To prepare a membrane in the *excised-patch mode,* the pipet is pulled away from the cell. A piece of plasma membrane with its cytosolic side now facing the medium is monitored by the patch pipet.

against an intact cell to form a seal. Slight suction leads to the formation of a very tight seal so that the resistance between the inside of the pipet and the bathing solution is many gigaohms (one gigaohm is equal to 10^9 ohms). Thus, a gigaohm seal (called a gigaseal) ensures that an electric current flowing through the pipet is identical with the current flowing through the membrane covered by the pipet. The gigaseal makes possible high-resolution current measurements while a known voltage is applied across the membrane. In fact, patch clamping increased the precision of such measurements a hundredfold. The flow of ions through a single channel and transitions between different states of a channel can now be directly monitored with a time resolution of microseconds.

Patch-clamp recordings of the acetylcholine receptor channel in the postsynaptic membrane of skeletal muscle are shown in Figure 39-24.

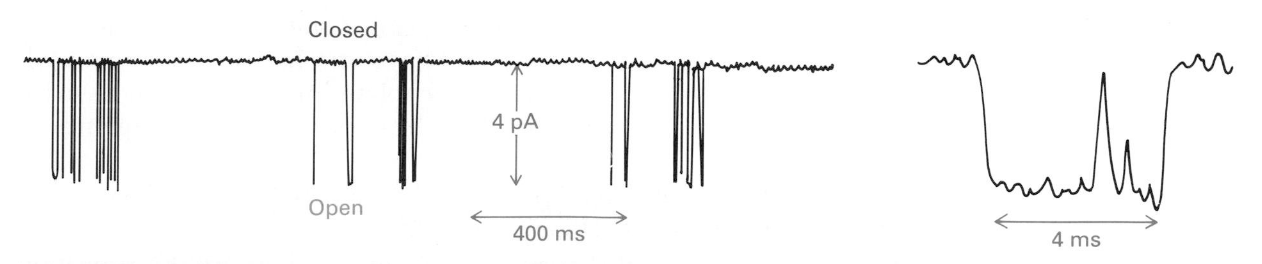

Figure 39-24
Patch-clamp recordings showing the conductance of an acetylcholine receptor channel in the presence of acetylcholine. The channel undergoes frequent transitions between open and closed states.

The addition of acetylcholine is followed by transient openings of the channel. The current i flowing through an open channel is 3.5 pA (picoamperes) when the membrane potential V is -100 mV. An ampere is the flow of 6.24×10^{18} charges per second. Hence, 2.2×10^{7} ions per second flow through the channel. The conductance g of a channel is equal to i/V; g is expressed in units of siemens (the reciprocal of an ohm), i in amperes, and V in volts. A current of 3.5 pA at a potential of 100 mV corresponds to a conductance of 35 pS (picosiemens).

The acetylcholine receptor channel contains two noninteracting binding sites for acetylcholine. The channel can open only if both are occupied by acetylcholine. Records of single-channel activity such as the one shown in Figure 39-24 indicate that openings occur in groups called bursts lasting a few hundred microseconds. Each burst is separated from the next burst by a much longer shut period. The gaps within a burst are tens of microseconds long. The simplest interpretation of these data is that the liganded receptor exists in at least two states, a closed A_2R form and an open A_2R^* form. A gap occurs when the open A_2R^* state makes a transition to the closed A_2R state. The channel opens again when A_2R switches back to A_2R^*. This flickering is faster than the dissociation of acetylcholine from A_2R to form AR or R, which cannot open until acetylcholine binds again.

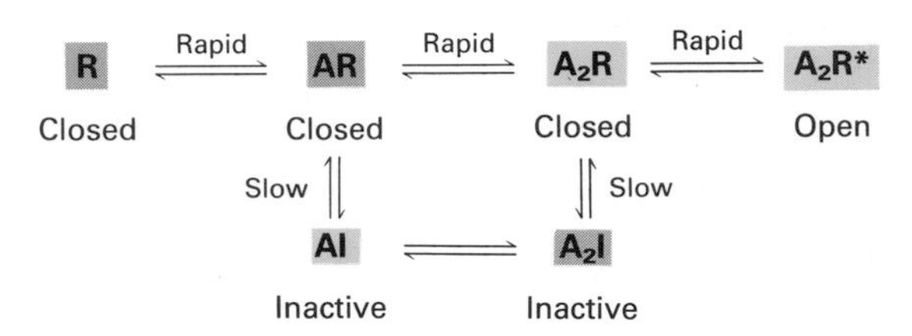

Figure 39-25
Multiple conformational states of the acetylcholine receptor. The abbreviations used are: A, acetylcholine; R, closed but activatable form of the receptor; R*, open state; I, inactive state.

After a short period of activity, the liganded receptor becomes inactive, just as the sodium channel becomes inactive after it is opened by depolarization. Rapid turnoff is a characteristic of nearly all signal transduction systems. This *desensitization* is reversible. The receptor recovers its capacity to open after the acetylcholine concentration is lowered from 0.3 mM to 10 nM by the action of acetylcholinesterase (p. 1022). The kinetics of desensitization and recovery have been studied by George Hess, who has carried out millisecond kinetic studies of cation entry into vesicles containing the receptor. These measurements point to the existence of two additional closed states, AI and A_2I, which are more stable than AR and A_2R and have very high affinity for acetylcholine. Nevertheless, when the acetylcholine concentration in a synaptic cleft is suddenly increased by the release of vesicles, the sequence of transitions is:

$$\text{R} \longrightarrow \text{AR} \longrightarrow \text{A}_2\text{R} \longrightarrow \text{A}_2\text{R}^*$$

The channel goes to the open A_2R^* form first because the transition from A_2R to A_2I occurs in hundreds of milliseconds, whereas the other steps take place in about a millisecond or less. Thus, *the channel is designed to open and then automatically shut following a pulse of acetylcholine.*

THE ACETYLCHOLINE RECEPTOR CHANNEL IS FORMED BY FIVE HOMOLOGOUS TRANSMEMBRANE POLYPEPTIDES

The *electric organ* of *Torpedo*, an electric fish, has been a choice source of the acetylcholine receptor. The electroplaxes of this electric organ are very rich in cholinergic postsynaptic membranes. Another exotic biological material has been invaluable in the isolation of acetylcholine receptors. Neurotoxins from snakes, such as *α-bungarotoxin*, which comes from the venom of a Formosan snake, and *cobratoxin*, block neuromuscular transmission. These small basic proteins (7 kd) bind specifically to acetylcholine receptors. They can be made highly radioactive by iodinating them with ^{125}I or by forming a Schiff base with pyridoxal phosphate and then reducing it with ^{3}H-borohydride. Labeled cobratoxin binds tightly to the acetylcholine receptor with a dissociation constant of 10^{-9} M. Most important, it does not bind appreciably to other macromolecules in the postsynaptic membrane. Thus, the acetylcholine receptor can be specifically marked with a radioactive label.

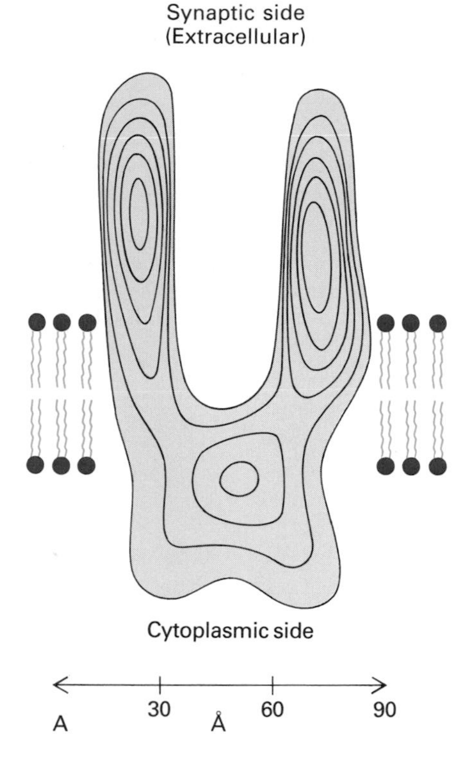

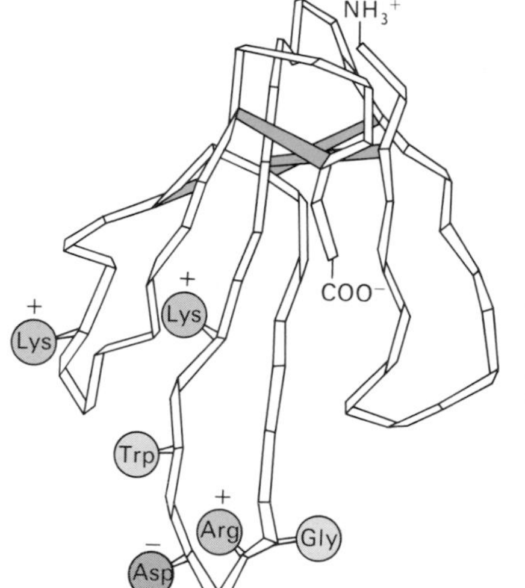

Figure 39-26
Three-dimensional structure of a neurotoxin inhibitor of the acetylcholine receptor. This toxin comes from a sea snake. [Courtesy of Dr. Demetrius Tsernoglou and Dr. Gregory Petsko.]

The acetylcholine receptor of the electric organ has been solubilized by adding a nonionic detergent to a postsynaptic membrane preparation and purified by affinity chromatography on a column containing covalently attached cobratoxin. This 268-kd receptor is a pentamer of four kinds of subunits, $\alpha_2\beta\gamma\delta$. Each α chain contains a binding site for acetylcholine. The cloning and sequencing of the cDNAs for these subunits (50–58 kd) has shown that they have similar sequences. Each chain contains four predominantly hydrophobic segments that span the membrane. The wall of the pore is probably formed by hydrophilic residues contributed by these transmembrane segments. It seems likely that the genes for the four kinds of subunits arose by duplication of an ancestral gene and divergence.

The structure of the acetylcholine receptor from the electric organ of *Torpedo* has recently been elucidated at 25 Å resolution by Nigel Unwin. His three-dimensional electron crystallographic analysis shows that the five subunits of the receptor are arranged regularly around its central axis (Figure 39-27). The structure has approximate *pentagonal symmetry*, in harmony with the homologous character of the five constituent subunits. The receptor is cylindrical with a mean diameter of about 65Å. It

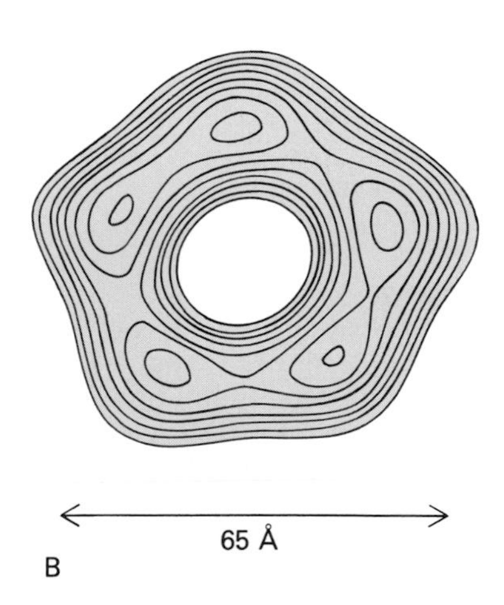

Figure 39-27
Structure of the acetylcholine receptor channel determined by three-dimensional image analysis of electron micrographs of crystalline arrays. (A) Longitudinal view of the channel. (B) Axial view from the synaptic side of the membrane (top of part A). The pore diameter at the synaptic mouth of the channel is 22 Å. [After A. Brisson and P. N. T. Unwin. *Nature* 315(1985):476.]

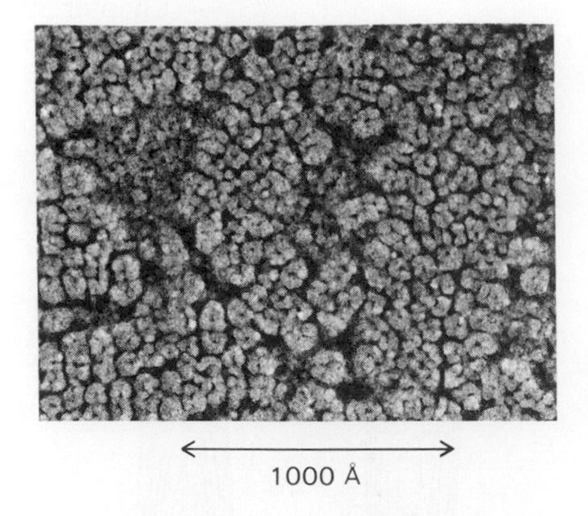

Electron micrograph of acetylcholine receptors in a postsynaptic membrane. [Courtesy of Dr. John Heuser and Dr. Shelly Salpeter.]

consists of five rod-shaped units, each about 140 Å long. The receptor protrudes about 70 Å on the synaptic side of the membrane and about 30 Å on the cytosolic side. The lumen of the channel is very wide, 22 Å, on the synaptic surface of the receptor. It narrows markedly near the middle of the bilayer and cannot be resolved at the cytoplasmic face; this form of the receptor is the closed but activatable state. Images of the open and desensitized forms, and higher-resolution views, are eagerly awaited.

Xenopus OOCYTES EXPRESS MICROINJECTED mRNAs ENCODING SUBUNITS OF THE ACETYLCHOLINE RECEPTOR

The cloning of cDNAs for the subunits of the acetylcholine receptor has led to new ways of exploring how this channel functions. mRNAs transcribed in vitro from these cDNAs are microinjected into *Xenopus* oocytes, which lack their own acetylcholine receptor (Figure 39-28). Polypeptide chains translated from these injected mRNAs are targeted to the plasma membrane of the oocyte. Patch-clamp measurements then reveal whether functional acetylcholine receptors are formed. Numa has found that fully active channels are formed only when all four kinds of subunits are expressed. Partly active channels are assembled if either the γ or δ chain is missing; one of these chains probably substitutes for the other in forming a pentameric channel. The oocyte expression system is also being used to test the effects of site-specific mutants. It opens new avenues to understanding relations between the structure and function of ion channels.

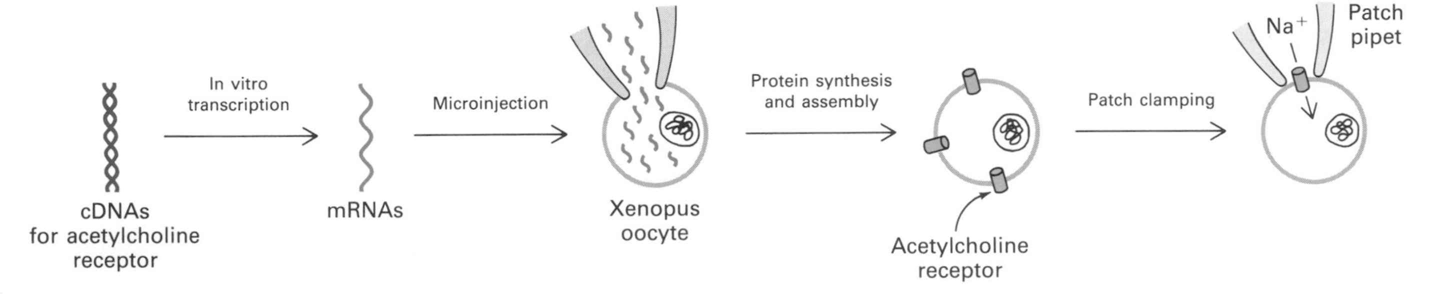

Figure 39-28
Xenopus oocytes can be used to express mRNAs derived from cDNAs. The steps in expression are in vitro transcription of the cDNA, microinjection of the resulting mRNA, translation of the mRNA by the oocyte. Microinjection of the mRNAs for the four kinds of subunits of the acetylcholine receptor results in the formation of functional channels, as evidenced by patch-clamping studies of membrane conductance.

ACETYLCHOLINE IS RAPIDLY HYDROLYZED AND THE END PLATE REPOLARIZES

The depolarizing signal must be switched off to restore the excitability of the postsynaptic membrane. Acetylcholine is hydrolyzed to acetate and choline by *acetylcholinesterase,* an enzyme discovered by David Nachmansohn in 1938. The permeability properties of the postsynaptic membrane are restored, and the membrane becomes repolarized.

$$\underset{\text{Acetylcholine}}{H_3C{-}\overset{O}{\overset{\|}{C}}{-}O{-}CH_2{-}CH_2{-}\overset{+}{N}{-}(CH_3)_3} + H_2O \rightleftharpoons \underset{\text{Acetate}}{H_3C{-}C\genfrac{}{}{0pt}{}{\nearrow O}{\searrow O^-}} + \underset{\text{Choline}}{HO{-}CH_2{-}CH_2{-}\overset{+}{N}{-}(CH_3)_3} + H^+$$

Acetylcholinesterase is located in the synaptic cleft, where it is bound to a network of collagen and glycosaminoglycans derived from the postsynaptic cell. This 260-kd enzyme, which has an $\alpha_2\beta_2$ structure, can be readily separated from the acetylcholine receptor. A striking feature of acetylcholinesterase is its very high turnover number of 25,000 s^{-1}, which means that it cleaves an acetylcholine molecule in 40 μs. Indeed, acetylcholinesterase is a diffusion-controlled enzyme (p. 191), as indicated by its k_{cat}/K_M value of 2×10^8 M^{-1} s^{-1}. The high turnover number of the enzyme is essential for the rapid restoration of the polarized state of the postsynaptic membrane. Synapses can transmit 1000 impulses per second only if the membrane recovers its polarization within a fraction of a millisecond.

The catalytic mechanism of acetylcholinesterase resembles that of chymotrypsin. Acetylcholine reacts with a specific serine residue at the active site of acetylcholinesterase to form a covalent acetyl-enzyme intermediate, and choline is released. The acetyl-enzyme intermediate then reacts with water to form acetate and regenerate the free enzyme (Figure 39-29).

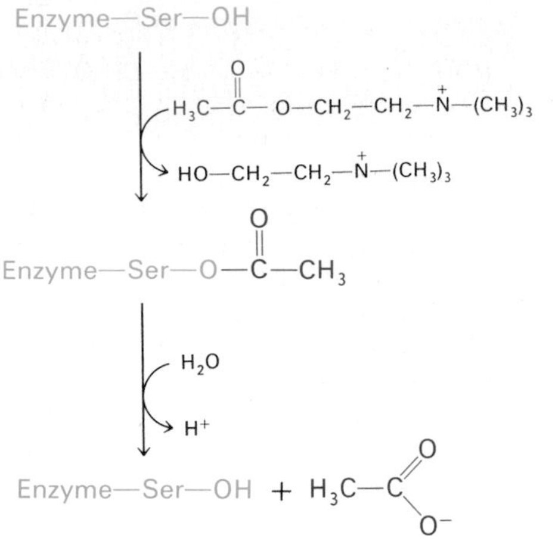

Figure 39-29
Catalytic mechanism of acetylcholinesterase.

ACETYLCHOLINESTERASE INHIBITORS ARE USED AS DRUGS AND POISONS

The therapeutic and toxic properties of acetylcholinesterase inhibitors are of considerable practical importance. *Physostigmine* (also called eserine) is an alkaloid derived from the Calabar bean, which was once used as an ordeal poison in witchcraft trials. Physostigmine and related inhibitors such as neostigmine are *carbamoyl esters* (Figure 39-30). They inhibit acetylcholinesterase by forming a covalent intermediate that is hydrolyzed very slowly. The serine residue at the active site becomes carbamoylated. *This carbamoyl-enzyme intermediate is subsequently hydrolyzed at a very slow rate, in contrast with the acetyl-enzyme intermediate normally formed when acetylcholine is the substrate.* Thus, the active site of the enzyme is effectively blocked. Neostigmine is used to treat glaucoma, an eye disease characterized by abnormally high intraocular pressure. The therapeutic rationale is that neostigmine inhibits acetylcholinesterase and thereby enhances the effects of acetylcholine.

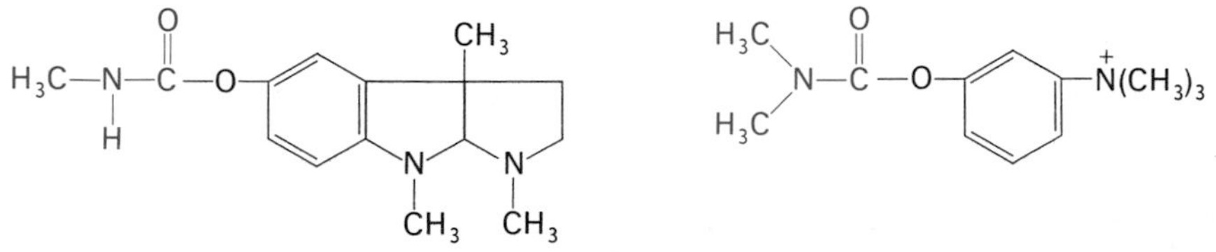

Physostigmine

Neostigmine

Figure 39-30
Physostigmine and neostigmine inhibit acetylcholinesterase by carbamoylating the serine in the active site.

Even more potent inhibitors are the *organic fluorophosphates,* such as diisopropyl phosphofluoridate (DIPF). These compounds react with acetylcholinesterase to form *very stable covalent phosphoryl-enzyme complexes* (Figure 39-31). The phosphoryl group becomes bonded to the active-site serine, as in serine proteases that have reacted with DIPF (p. 223). Many organic phosphate compounds have been synthesized for

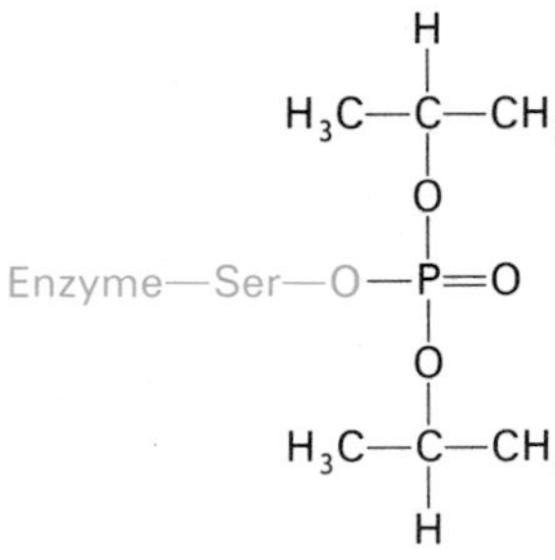

DIPF-inhibited enzyme

Figure 39-31
DIPF-inhibited acetylcholinesterase.

use as *agricultural insecticides* or as *nerve gases* for chemical warfare (Figure 39-32). These compounds kill by causing respiratory paralysis. Tabun and sarin are among the most toxic of fluorophosphates. Parathion has been widely used as an agricultural insecticide.

The number of acetylcholinesterase molecules in the end plate of mouse-diaphragm muscle has been counted using radioactive DIPF as a label. The density is 12,000 μm^{-2}, only a factor of 2 less than the density of acetylcholine receptors. Thus, *the postsynaptic membrane is very densely packed with both acetylcholinesterase and the acetylcholine receptor.* Only a small fraction of the acetylcholinesterase is required for the transmission of nerve impulses at low frequencies. However, most of the enzyme molecules must be active to sustain transmission at high rates of firing.

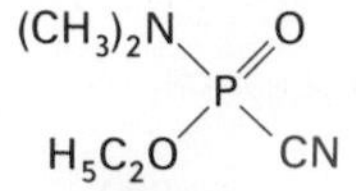

Tabun 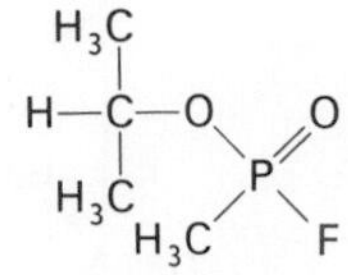 **Sarin** 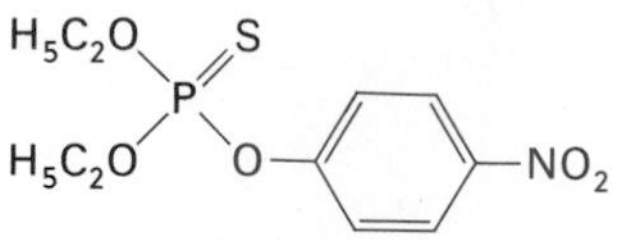 **Parathion**

Figure 39-32
Organic phosphate inhibitors of acetylcholinesterase.

INHIBITORS OF THE ACETYLCHOLINE RECEPTOR

Neuromuscular transmission can also be impaired by *compounds that act directly on the acetylcholine receptor. Curare* has been used for centuries by South American Indians. Soon after Columbus returned, d'Anghera wrote in his *De Orbe Novo* that "the natives poisoned their arrows with the juice of a death-dealing herb. . . ." One of the active components of curare is d-*tubocurarine* (Figure 39-33). *Tubocurarine inhibits the depolarization of the end plate by competing with acetylcholine for binding to the acetylcholine receptor.* α-Bungarotoxin and cobratoxin have a similar action.

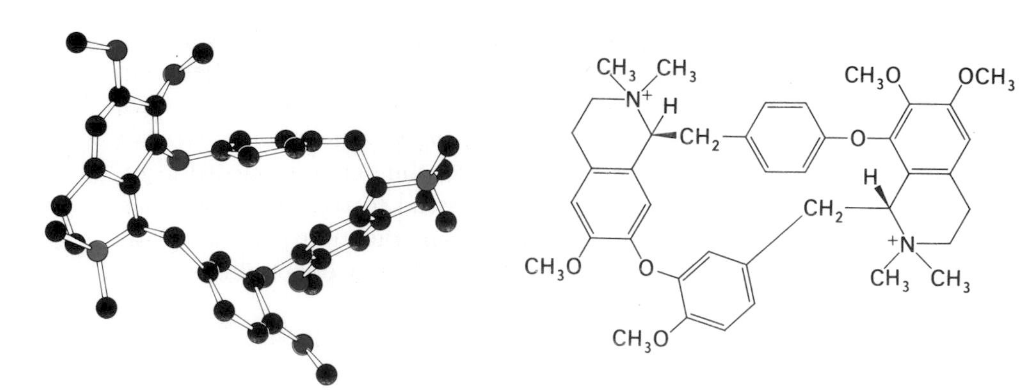

Figure 39-33
Formula and model of *d*-tubocurarine.

In contrast, compounds such as *decamethonium* and *succinylcholine* bind to the receptor and cause a *persistent depolarization* of the end plate. Decamethonium is not hydrolyzed by acetylcholinesterase, and succinylcholine is slowly hydrolyzed by this enzyme.

$$(CH_3)_3{-}\overset{+}{N}{-}(CH_2)_{10}{-}\overset{+}{N}{-}(CH_3)_3$$

Decamethonium

$$(CH_3)_3{-}\overset{+}{N}{-}CH_2{-}CH_2{-}O{-}\overset{O}{\overset{\|}{C}}{-}CH_2{-}CH_2{-}\overset{O}{\overset{\|}{C}}{-}O{-}CH_2{-}CH_2{-}\overset{+}{N}{-}(CH_3)_3$$

Succinylcholine

Succinylcholine is also hydrolyzed by less specific cholinesterases in the plasma and in the liver; they are called *plasma cholinesterase* or *pseudocholinesterase* to distinguish them from acetylcholinesterase in the postsynaptic membrane. Succinylcholine is used to produce muscular relaxation in surgical procedures. An attractive feature of succinylcholine is that neuromuscular transmission resumes soon after drug infusion is stopped. In a small proportion of patients, however, muscular relaxation and respiratory paralysis persist for many hours. In such patients, succinylcholine is hydrolyzed very slowly because their plasma cholinesterase has little affinity for succinylcholine. Sensitivity to succinylcholine, like sensitivity to pamaquine (p. 436), is an example of a genetically determined drug idiosyncrasy.

Rabbits exhibit muscle weakness and fatigability several weeks after being immunized with purified acetylcholine receptor from the electric organ of *Torpedo*. The reason for their symptoms is that they produce antibodies that react with acetylcholine receptors in their own neuromuscular junctions. The resulting decrease in the number of functional receptors leads to defective neuromuscular transmission. The symptoms displayed by these rabbits are very much like those of *myasthenia gravis*, a disease of humans. Indeed, the sera of myasthenic patients contain antibody directed against their own acetylcholine receptors. Thus, myasthenia gravis is an *autoimmune disorder*, in which self becomes the target of the immune system. Neostigmine is used in the treatment of this disease. This acetylcholinesterase inhibitor leads to an elevated level of acetylcholine in the synaptic cleft, which enables myasthenic muscle to respond to repetitive stimuli.

EPINEPHRINE AND DOPAMINE BELONG TO A FAMILY OF CATECHOLAMINE TRANSMITTERS

Many neurotransmitters in addition to acetylcholine have been identified. An established neurotransmitter meets several criteria. First, microinjection of the proposed transmitter into the synaptic cleft must elicit the same response as does excitation of the presynaptic nerve. Second, the presynaptic nerve terminals must be rich in this substance. The isolation of synaptic vesicles containing the putative transmitter is the strongest evidence in this regard. Third, the presynaptic nerve must release the postulated transmitter at the right time and in a quantity sufficient to act on the postsynaptic nerve.

Several catecholamines meet these criteria. For example, norepinephrine is the transmitter at smooth-muscle junctions that are innervated by sympathetic nerve fibers, whereas acetylcholine is the transmitter at parasympathetic junctions. Epinephrine and dopamine are two other catecholamine transmitters. In fact, these catecholamines are synthesized from tyrosine in sympathetic-nerve terminals and in the adrenal gland (Figure 39-34). The first step, which is rate limiting, is the hydroxylation of tyrosine to form 3,4-dihydroxyphenylalanine (*dopa*). This reaction is catalyzed by tyrosine hydroxylase, which is like phenylalanine hydroxylase. Molecular oxygen is activated by tetrahydrobiopterin, the cofactor. The second step is the decarboxylation of dopa by dopa decarboxylase, a pyridoxal phosphate enzyme, to yield 3,4-dihydroxyphenylethylamine (*dopamine*). Dopa and inhibitors of the decarboxylase are used in the treatment of Parkinson's disease, a neurological disorder characterized by tremor, rigidity, and disturbances of posture. Dopamine is then hydroxylated to *norepinephrine* by a copper-containing hydroxylase. Finally, *epinephrine* is formed by the methylation of norepinephrine by a transmethylase that utilizes *S*-adenosylmethionine.

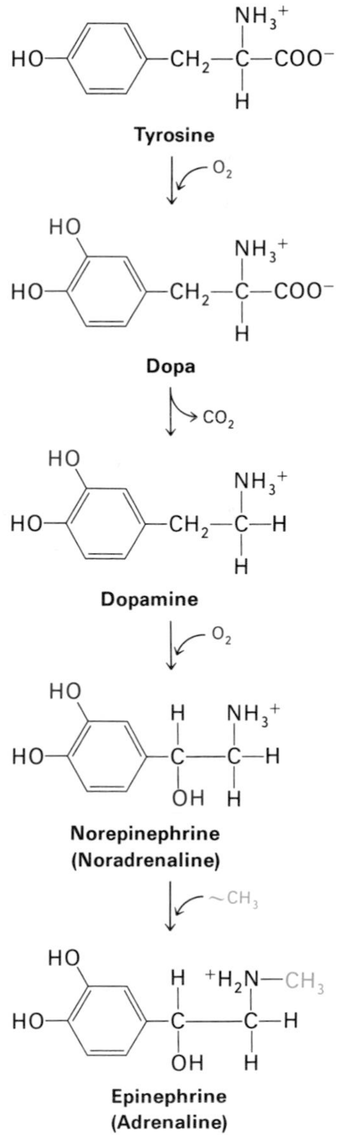

Tyrosine

O_2

Dopa

CO_2

Dopamine

O_2

Norepinephrine (Noradrenaline)

$\sim CH_3$

Epinephrine (Adrenaline)

Figure 39-34
Pathway for the synthesis of catecholamine neurotransmitters.

Catecholamine neurotransmitters are inactivated by methylation of the 3-hydroxyl group of the catechol ring. This reaction is catalyzed by *catechol-O-methyltransferase,* which uses *S*-adenosylmethionine as the methyl donor. Alternatively, these neurotransmitters can be inactivated by oxidative removal of their amino group by *monoamine oxidase* (Figure 39-35).

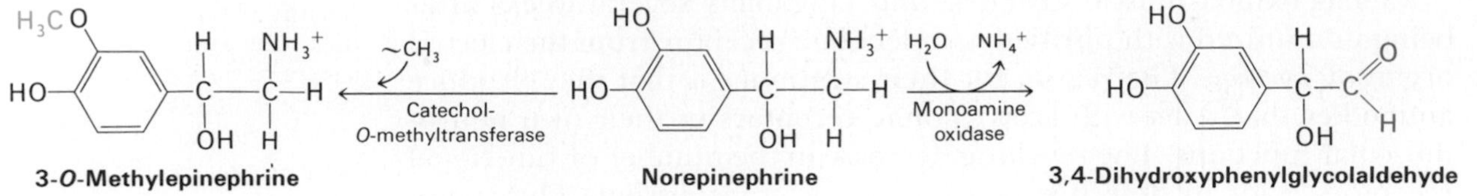

3-*O*-Methylepinephrine ← Catechol-*O*-methyltransferase ($\sim CH_3$) — Norepinephrine → Monoamine oxidase (H_2O, NH_4^+) → 3,4-Dihydroxyphenylglycolaldehyde

Figure 39-35
Inactivation of norepinephrine.

GLYCINE AND γ-AMINOBUTYRATE (GABA) OPEN CHLORIDE CHANNELS

Glycine and γ-aminobutyrate (also called γ-aminobutyric acid or GABA) increase the permeability of postsynaptic membranes to chloride ion. An increase in chloride conductance leads to membrane *hyperpolarization,* which increases the threshold for the triggering of an action potential. Glycine and GABA serve as *inhibitory transmitters* in the central nervous system. γ-Aminobutyrate is formed by the decarboxylation of glutamate, in a reaction catalyzed by glutamate decarboxylase (Figure 39-36). As might be expected, pyridoxal phosphate is the prosthetic group of this decarboxylase. γ-Aminobutyrate is inactivated by transamination to succinate semialdehyde, which is then oxidized to succinate.

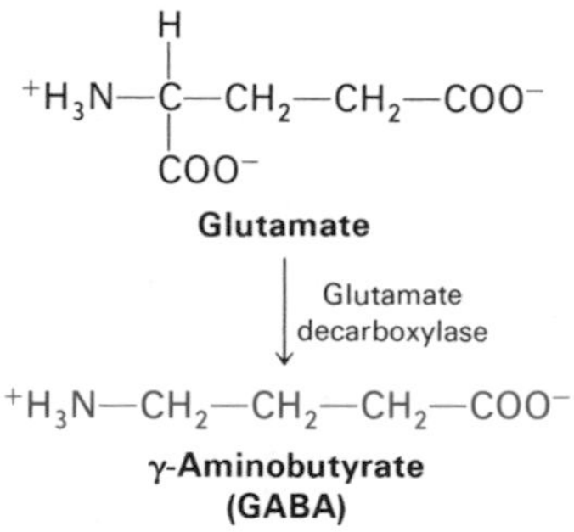

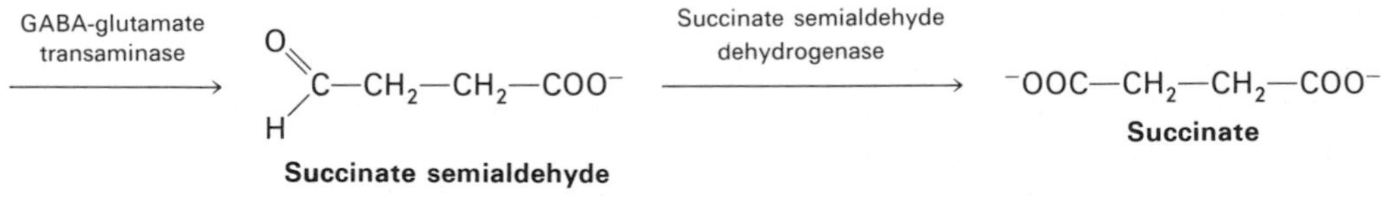

Glutamate: $^{+}H_3N-CH(COO^-)-CH_2-CH_2-COO^-$
↓ Glutamate decarboxylase
γ-Aminobutyrate (GABA): $^{+}H_3N-CH_2-CH_2-CH_2-COO^-$
→ GABA-glutamate transaminase →
Succinate semialdehyde: $OHC-CH_2-CH_2-COO^-$
→ Succinate semialdehyde dehydrogenase →
Succinate: $^{-}OOC-CH_2-CH_2-COO^-$

Figure 39-36
Synthesis and inactivation of γ-aminobutyrate (GABA).

The GABA receptor from brain is an $\alpha_2\beta_2$ tetramer. The 53-kd α subunit contains a binding site for barbiturates and other drugs affecting the central nervous system. The 57-kd β subunit binds GABA. The cDNAs for these chains have recently been cloned and sequenced. Hydrophobicity profiles suggest that each subunit contains four membrane-spanning helices (Figure 39-37). The 6-Å diameter pore is formed by the coming together of at least one helix from each of the four subunits. The clustering of positively charged side chains at each end of the pore makes the channel highly selective for anions. The glycine receptor has a similar structure. Its strychnine-binding subunit closely resembles the subunits of the GABA receptor. Furthermore, this subunit of the glycine-activated channel and both subunits of the GABA receptor exhibit 25% sequence identify with the nicotinic acetylcholine receptor. Thus, *a superfamily of ion channels is emerging.*

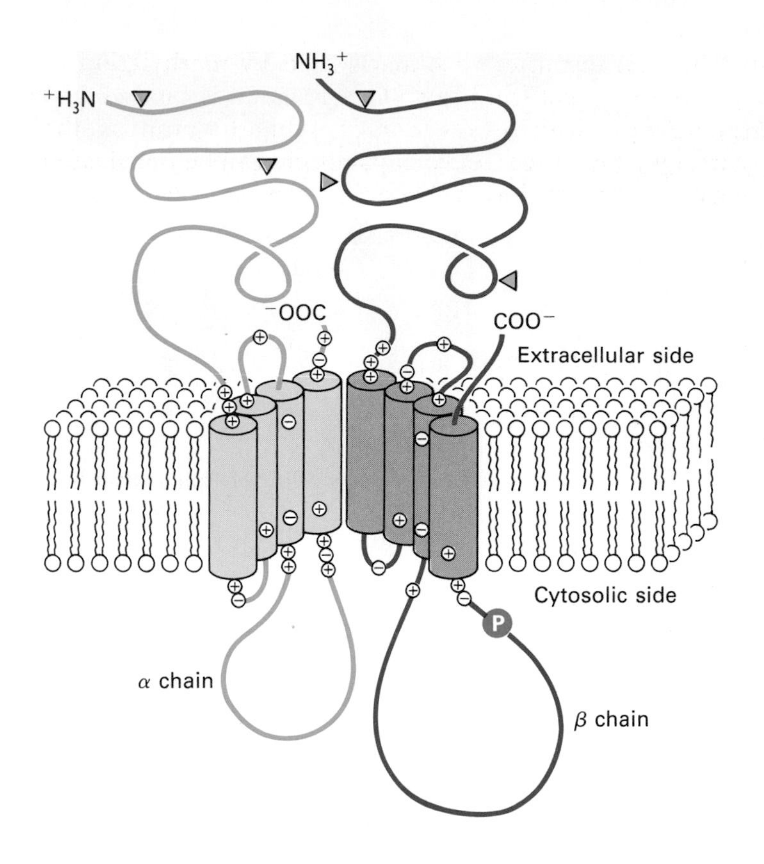

Figure 39-37
Proposed model of the GABA receptor. One α chain and one β chain of the $\alpha_2\beta_2$ receptor is shown. Oligosaccharide units are shown in yellow. A potential site for phosphorylation by the cyclic-AMP-dependent protein kinase is shown in blue (marked P). [After P. R. Schofield, M. G. Darlison, N. Fujita, D. R. Burt, F. Stephenson, H. Rodriguez, L. M. Rhee, J. Ramachandran, V. Reale, T. A. Glencorse, P. H. Seeburg, and E. A. Barnard. *Nature* 328(1987):226.]

MANY SIGNAL MOLECULES ARE FORMED IN A SINGLE STEP FROM A MAJOR METABOLITE

The synthesis of GABA by the decarboxylation of glutamate exemplifies a recurring mechanistic motif: *many signal molecules are formed in a single step from a major metabolic intermediate* (Table 39-2). Recall that cyclic AMP, a second messenger in the action of many hormones (p. 976), is formed by the cyclization of ATP. As discussed earlier (p. 1025), dopa and several other neurotransmitters arise from the degradation of tyrosine. Another recurring feature of signal molecules is the irreversibility of their synthesis and degradation. Hence, their concentration can be precisely controlled. Major metabolites such as glycine can serve as neurotransmitters outside cells but not as intracellular messengers because their concentrations cannot be adjusted over a wide range without disrupting the economy of the cell. Most signal molecules are by-products of metabolism that have been recruited in the course of evolution for loftier roles. Some await discovery, as highlighted by the recent finding of fructose 2,6-bisphosphate, a key regulator of metabolism formed by phosphorylation of a major glycolytic intermediate (p. 360).

Table 39-2
Signal molecules formed in one step from a major metabolic intermediate

Precursor	*Signal molecule*
ATP	Cyclic AMP
ATP	2′,5′-Oligoadenylates
GTP	Cyclic GMP
GTP	Guanosine tetraphosphate
Glutamate	GABA
Tyrosine	Dopa
Choline	Acetylcholine
Arachidonate	Prostaglandin PGE_2
Fructose 6-phosphate	Fructose 2,6-bisphosphate
1,3-Bisphosphoglycerate	2,3-Bisphosphoglycerate

A RETINAL ROD CELL CAN BE EXCITED BY A SINGLE PHOTON

We now turn to excitable receptors that are activated by *light.* Vertebrates have two kinds of photoreceptor cells, called rods and cones because of their distinctive shapes. *Cones* function in bright light and are responsible for color vision, whereas *rods* function in dim light but do not perceive color. A human retina contains about three million cones

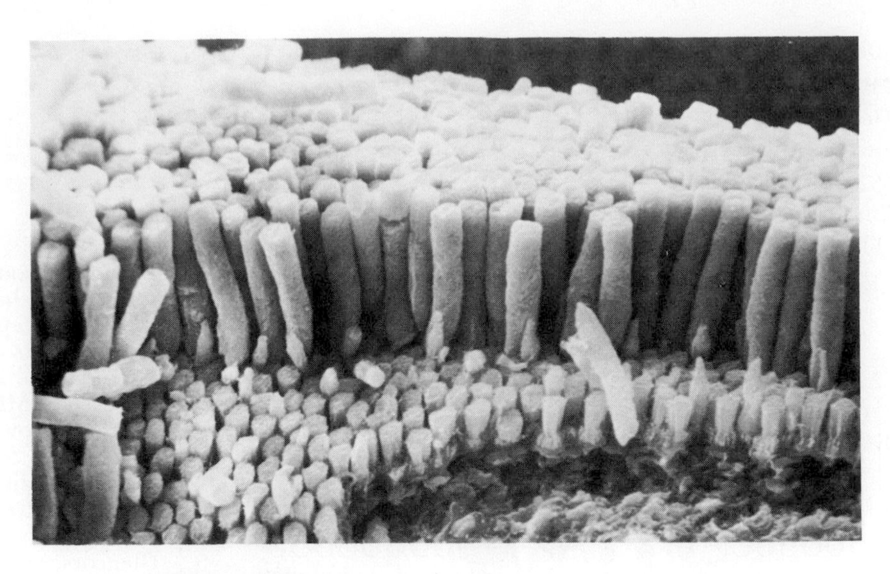

Figure 39-38
Scanning electron micrograph of retinal rod cells. [Courtesy of Dr. Deric Bownds.]

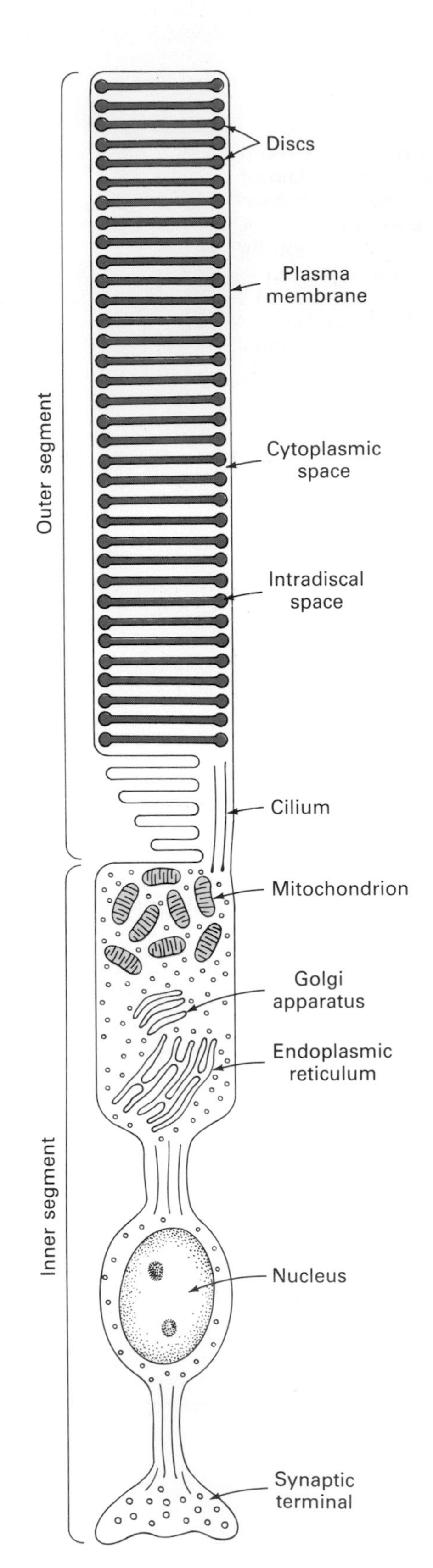

Figure 39-39
Schematic diagram of a retinal rod cell.

and a hundred million rods. These photoreceptor cells *convert light into atomic motion and then into a nerve impulse.* Rods and cones form synapses with bipolar cells, which in turn interact with other nerve cells in the retina. The electrical signals generated by the photoreceptors are processed by an intricate array of nerve cells within the retina and then transmitted to the brain by the fibers of the optic nerve. Thus, the retina has a dual function: to transform light into nerve impulses and to integrate visual information.

In 1938, Selig Hecht discovered through psychophysical studies that *a human rod cell can be excited by a single photon.* Let us explore the molecular basis of this exquisite sensitivity. Rods are slender, elongated structures; in humans, they have a diameter of 1 μm and a length of 40 μm. The major functions of a rod cell are highly compartmentalized (Figure 39-39). *The outer segment of a rod is specialized for photoreception.* It contains a stack of about 1000 *discs* (Figure 39-40), which are closed, flattened sacs about 160 Å thick. These membranous structures are densely

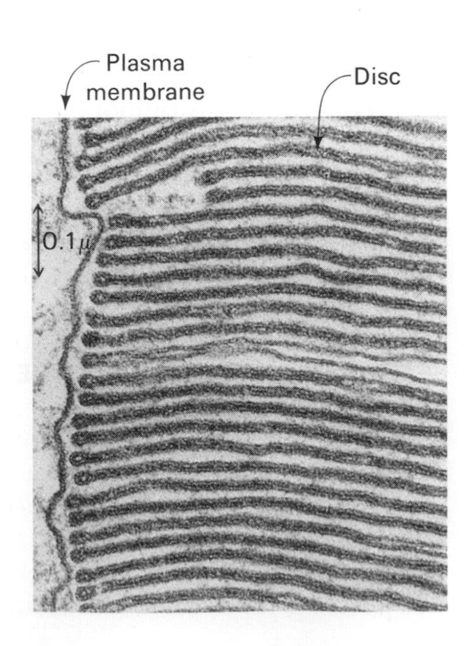

Figure 39-40
Electron micrograph of a rod outer segment, showing the stack of discs. [From J. E. Dowling. The organization of vertebrate visual receptors. In *Molecular Organization and Biological Function,* J. M. Allen, ed. (Harper & Row, 1967).]

packed with photoreceptor molecules. The disc membranes are separate from the plasma membrane of the outer segment. A slender immotile cilium joins the outer segment and the *inner segment,* which is rich in mitochondria and ribosomes. The inner segment generates ATP at a very rapid rate and is highly active in synthesizing proteins. The discs in the outer segment have a life of a month and are continually renewed. The inner segment is contiguous with the nucleus, which is next to the *synaptic body*. Many synaptic vesicles are present in the synaptic body, which forms a synapse with a bipolar neuron.

RHODOPSIN, THE PHOTORECEPTOR PROTEIN OF RODS, CONTAINS AN ATTACHED 11-*cis*-RETINAL CHROMOPHORE

Light must be absorbed to stimulate a photoreceptor cell. In addition, the light-absorbing group (called a *chromophore*) must undergo a conformational change after it has absorbed a photon. The photosensitive molecule in the discs of rod cells is *rhodopsin,* which consists of *opsin,* a protein, and *11*-cis-*retinal,* a prosthetic group (Figure 39-41).

Opsin, like other proteins devoid of prosthetic groups, does not itself absorb visible light. The color of rhodopsin and its responsiveness to light depend on the presence of 11-*cis*-retinal, which is a very effective chromophore. 11-*cis*-Retinal gives rhodopsin a broad absorption band in the visible region of the spectrum with a peak at 500 nm, which nicely matches the solar output. The intensity of the visible absorption band of rhodopsin is also noteworthy. The absorption coefficient of rhodopsin at 500 nm is 40,000 cm^{-1} M^{-1}, a high value (Figure 39-42). The integrated absorption strength of the visible absorption band of rhodopsin approaches the maximum value attainable by organic compounds. 11-*cis*-Retinal has these favorable chromophoric properties because it is a *polyene*. Its six alternating single and double bonds constitute a long, unsaturated electron network. Recall that retinal also serves as the chromophore in bacteriorhodopsin (p. 308) and halorhodopsin (p. 962).

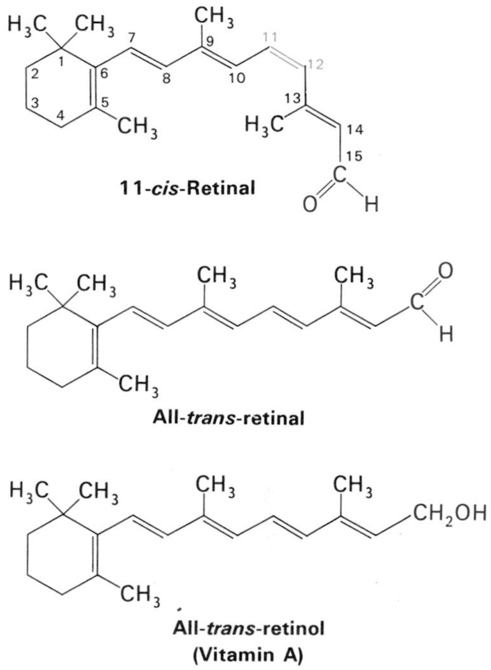

11-*cis*-Retinal

All-*trans*-retinal

All-*trans*-retinol (Vitamin A)

Figure 39-41
Structure of 11-*cis*-retinal, all-*trans*-retinal, and all-*trans*-retinol (vitamin A).

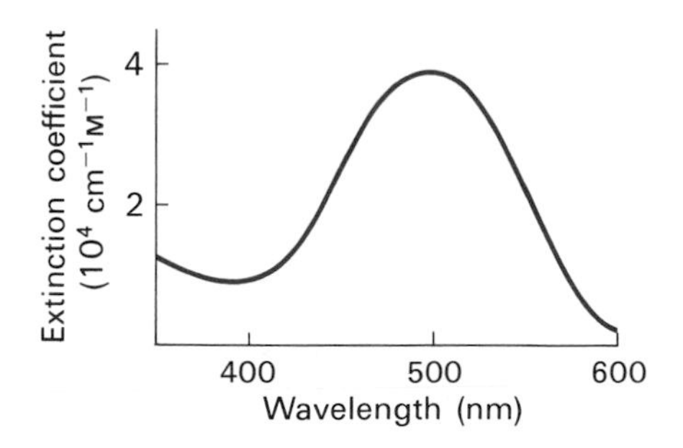

Figure 39-42
Absorption spectrum of rhodopsin.

11-*cis*-Retinal is attached to rhodopsin by a *Schiff-base linkage*. The aldehyde group of 11-*cis*-retinal is linked to the ϵ-amino group of a specific lysine residue of opsin. The spectral properties of rhodopsin indicate that the Schiff base is protonated.

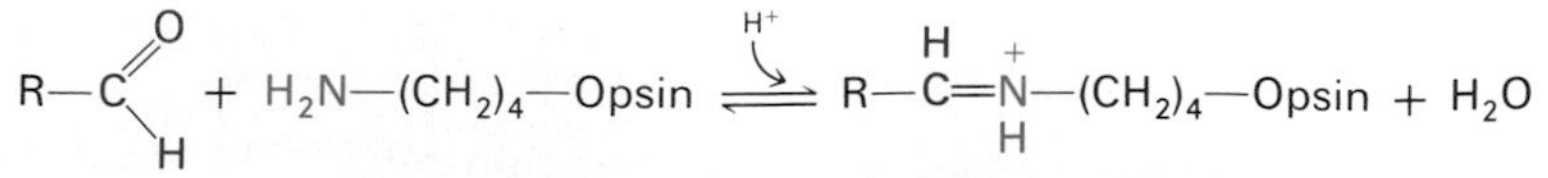

$$R{-}CHO + H_2N{-}(CH_2)_4{-}\text{Opsin} \underset{}{\overset{H^+}{\rightleftharpoons}} R{-}CH{=}\overset{+}{N}H{-}(CH_2)_4{-}\text{Opsin} + H_2O$$

11-*cis*-Retinal **Lysine side chain** **Protonated Schiff base**

The precursor of 11-*cis*-retinal is *all*-trans-*retinol* (vitamin A), which cannot be synthesized de novo by mammals. All-*trans*-retinol (Figure 39-41) is converted into 11-*cis*-retinal in two steps. The alcohol group is oxidized to an aldehyde by *retinol dehydrogenase*, which uses $NADP^+$ as the electron acceptor. Then, the double bond between C-11 and C-12 is isomerized from a *trans* to a *cis* configuration by *retinal isomerase*. A deficiency of vitamin A leads to *night blindness* and eventually to the deterioration of the outer segments of rods.

RHODOPSIN IS A MEMBER OF A FAMILY OF RECEPTORS CONTAINING SEVEN TRANSMEMBRANE HELICES

Rhodopsin is a 40-kd integral membrane protein containing seven transmembrane helices (Figure 39-43). The amino terminus is on the intradiscal side of the membrane, and the carboxyl terminus is on the cytosolic side. About half of the protein is within the hydrophobic core of the membrane, a quarter on the intradiscal side, and the other quarter on the cytosolic side. Rhodopsin contains two *N*-linked oligosaccharide units in its amino terminal region. The 11-*cis*-retinal chromophore lies in a pocket of the protein, near the center of the membrane, with its long axis nearly parallel to the plane of the membrane. The cytosolic side contains a region that transmits the excitation signal to an enzymatic cascade. An overlapping region participates in deactivating the excited receptor. The phosphorylation of serine and threonine residues in the carboxyl-terminal tail is a step in deactivation. The seven-helix pattern of rhodopsin recurs in other membrane receptors. The other members of this family are the photoreceptor proteins of cone cells (p. 1036), the β-adrenergic receptor (p. 980), and the muscarinic acetylcholine receptor.

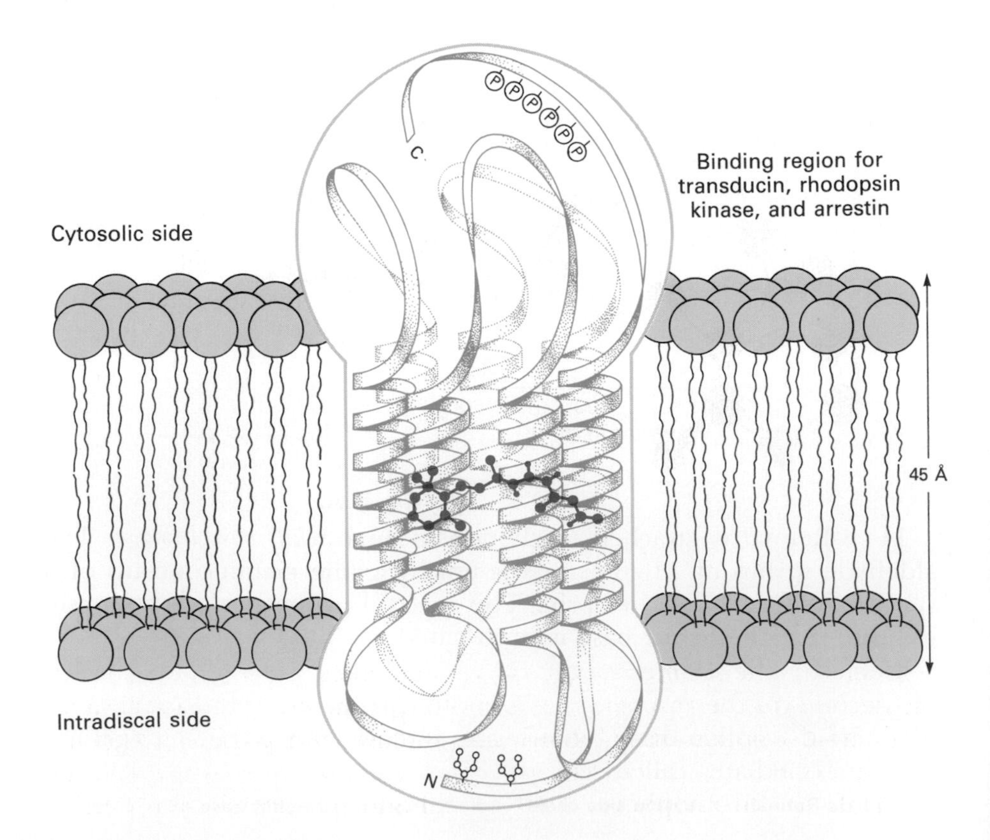

Figure 39-43
Model of rhodopsin. The seven-helix motif of this transmembrane receptor occurs in other sensory receptors and in hormone receptors. The 11-*cis*-retinal chromophore is located near the center of the bilayer. The amino-terminal segment, located on the intradiscal side of the membrane, contains two *N*-linked oligosaccharide units. The cytosolic domain transmits the excitation signal to transducin. Phosphorylation of multiple serines and threonines (marked P) in the carboxyl-terminal tail deactivates photoexcited rhodopsin. [After E. A. Dratz and P. A. Hargrave. *Trends Biochem. Sci.* 8(1983):128, and L. Stryer. *Sci. Amer.* 255(7):42.]

Rhodopsin, like nearly all other integral membrane proteins, is synthesized by ribosomes that are attached to the endoplasmic reticulum. The newly synthesized protein is transported to the Golgi apparatus and then to the plasma membrane. New discs are formed at the base of the outer segment by invagination of the plasma membrane, which accounts for the fact that the sugar units of rhodopsin are located inside the discs, whereas they face the extracellular space while on the plasma membrane (Figure 39-44). The intradiscal space, like the interior of other membrane-bounded compartments, is topologically equivalent to the extracellular space.

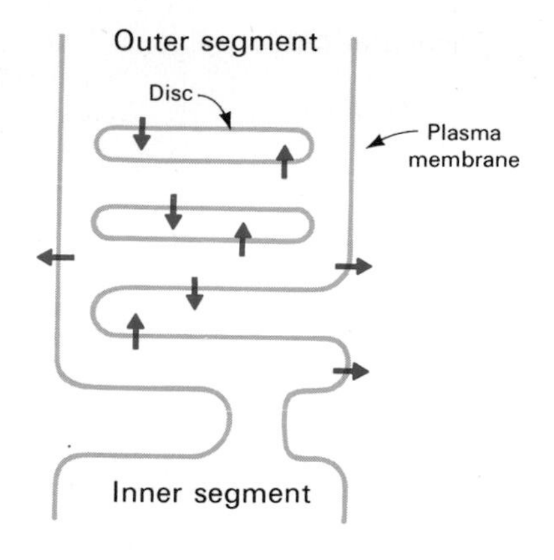

Figure 39-44
Formation of discs by invagination of the plasma membrane. The points of the arrows mark the location of the oligosaccharide units of rhodopsin.

THE PRIMARY EVENT IN VISUAL EXCITATION IS THE ISOMERIZATION OF 11-*cis*-RETINAL

In 1958, George Wald and his coworkers showed that *light isomerizes the 11-*cis*-retinal group of rhodopsin to all-*trans*-retinal.* This isomerization, the initial event in visual excitation, markedly alters the geometry of retinal (Figure 39-45). The Schiff-base linkage of retinal moves approximately 5 Å in relation to the ring portion of the chromophore. In essence, *a photon has been converted into atomic motion.*

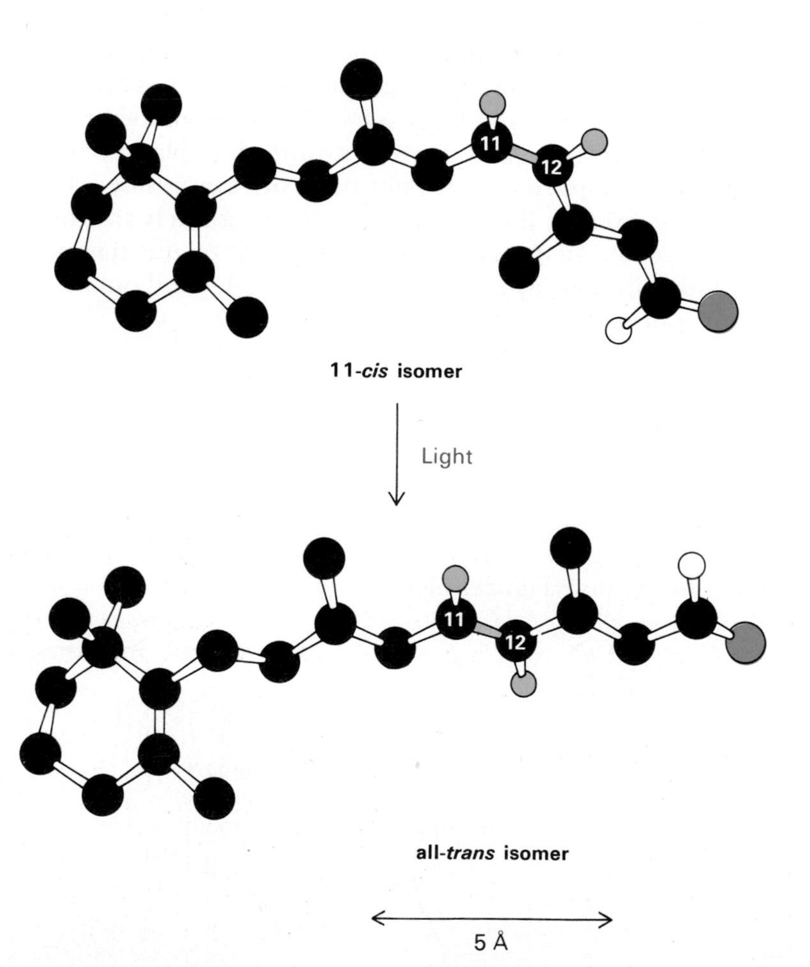

Figure 39-45
The primary event in visual excitation is the isomerization of the 11-*cis* isomer of the Schiff base of retinal to the all-*trans* form. The double bond between C11 and C12 is shown in green; hydrogen atoms attached to these carbons are shown in yellow.

Much of the isomerization of retinal takes place within a few picoseconds of the absorption of a photon, as shown by the appearance of a new absorption band following an intense laser pulse. This photolytic intermediate, called *bathorhodopsin,* contains a strained all-*trans* form of the chromophore. Both retinal and the protein continue to

Rhodopsin (500 nm)
ps ⇣ Light
Bathorhodopsin (543 nm)
ns ↓
Lumirhodopsin (497 nm)
μs ↓
Metarhodopsin I (480 nm)
ms ↓
Metarhodopsin II (380 nm)
s ↓ (R*)
Opsin + All-*trans*-retinal (380 nm)

Figure 39-46
Intermediates in the photolysis of rhodopsin. The wavelength of the absorption maximum of each species and the time constant of each transition are given.

change their conformations, as reflected in the formation of a series of transient intermediates with distinctive spectral properties (Figure 39-46). The Schiff-base linkage becomes deprotonated in the transition from metarhodopsin I to II, which takes about a millisecond. Metarhodopsin II, called *photoexcited rhodopsin,* triggers an enzymatic cascade, as will be discussed shortly. The unprotonated Schiff base in metarhodopsin II is hydrolyzed in about a minute to yield opsin and all-*trans*-retinal, which diffuses away from the protein because it does not fit into the binding site for the 11-*cis* isomer. All-*trans*-retinal is isomerized in the dark to 11-*cis*-retinal, which associates with opsin to regenerate rhodopsin.

LIGHT HYPERPOLARIZES THE PLASMA MEMBRANE BY CLOSING CATION-SPECIFIC CHANNELS

The *cis-trans* isomerization of retinal and the consequent conformational changes in rhodopsin are the primary events in visual excitation. An important later step in the generation of a nerve impulse has been elucidated by electrophysiological studies of intact retinas. The plasma membrane of a rod cell contains cation-specific channels that are open in the dark. Sodium ions rapidly flow into the outer segment in the dark because the electrochemical gradient for Na^+ is large. This gradient is maintained by Na^+-K^+ ATPase pumps located in the inner segment. *Light blocks these cation-specific channels in the outer segment.* Consequently, the influx of Na^+ decreases, and the plasma membrane becomes hyperpolarized—more negative on the inside. This *light-induced hyperpolarization* (Figure 39-47) is then passively transmitted by the plasma membrane from the outer segment to the synaptic body.

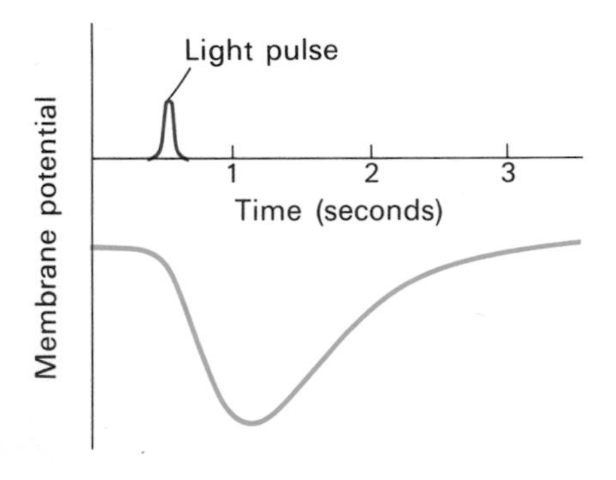

Figure 39-47
Light hyperpolarizes the plasma membrane of a retinal rod cell.

The speed and intensity of hyperpolarization depend on the intensity of the light pulse and on the level of steady background illumination. The response to a single photon takes place in about a second, whereas an intense pulse hyperpolarizes the plasma membrane in a few milliseconds. The rod cell does not have action potentials. Instead, its response to light is graded. The signal sent from the outer segment to the synaptic body depends on the number of absorbed photons. The hyperpolarization of a fully sensitive dark-adapted rod is half maximal when only 30 photons are absorbed by an outer segment containing 40×10^6 rhodopsin molecules (Figure 39-48). *A single photon absorbed by a dark-adapted rod closes hundreds of cation-specific channels and leads to a hyperpolarization of about 1 mV*, which is sensed by the synapse and conveyed to other neurons of the retina.

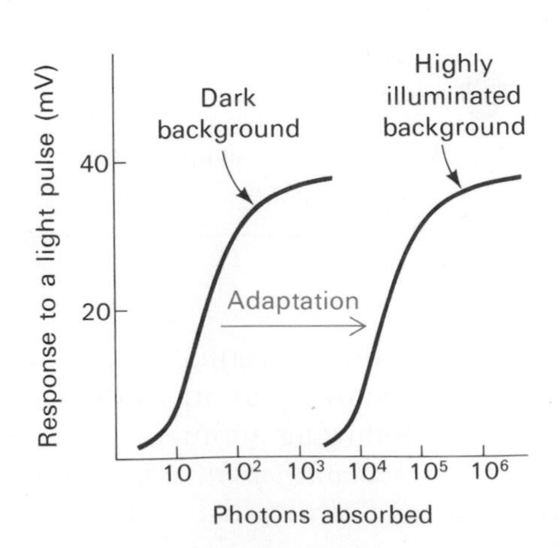

Figure 39-48
The sensitivity of a retinal rod cell to a light pulse depends on the background light level.

Exquisite sensitivity is not the only remarkable property of the rod cell. The response of this photodetector to a light pulse depends on the background light level. Many more photons are needed to excite a constantly illuminated rod than to excite one in the dark (Figure 39-48). This property, called *adaptation*, enables the rod cell to perceive contrast over a ten-thousand-fold range of background light intensity.

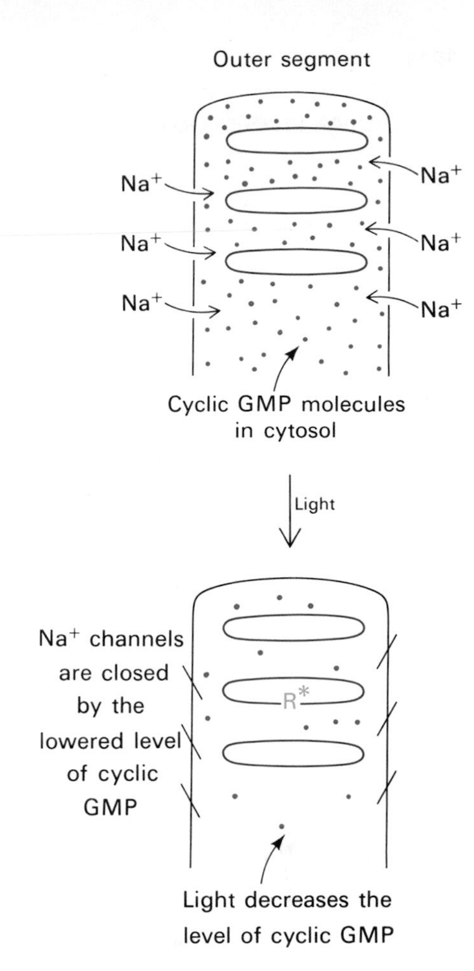

Figure 39-49
Cyclic GMP is the internal transmitter in visual excitation.

PHOTOEXCITED RHODOPSIN TRIGGERS A CASCADE LEADING TO THE HIGHLY AMPLIFIED HYDROLYSIS OF CYCLIC GMP

The change in permeability of the plasma membrane to Na^+ and the consequent hyperpolarization are highly amplified responses of the outer segment. The flow of more than a *million* sodium ions is blocked by the absorption of a *single* photon by a dark-adapted rod. What is the mechanism of this remarkable amplification? Photoexcited rhodopsin triggers an enzymatic cascade resulting in the hydrolysis of cyclic GMP (Figure 39-49). In outline, the flow of information in visual excitation is from *photoexcited rhodopsin* (R*) to *transducin* (T-GTP) to *activated phosphodiesterase* (PDE*), which hydrolyzes cyclic GMP. The light-induced drop in the concentration of cyclic GMP then closes the cation-specific channels in the plasma membrane.

$$\text{R} \xrightarrow{\text{light}} \text{R}^* \longrightarrow \text{T}_\alpha\text{-GTP} \longrightarrow \text{PDE}^* \begin{cases} \text{cGMP} \longrightarrow \text{Open channels} \\ \text{5}'\text{-GMP} \quad \text{Closed channels} \end{cases}$$

Transducin, the signal-coupling protein in visual excitation, is a member of the G-protein family (p. 984). This peripheral membrane protein interconverts between an inactive GDP state and an active GTP state. It consists of α (39 kd), β (36 kd), and γ (8 kd) subunits; the guanyl-nucleotide-binding site is on the α subunit. Photoexcited rhodopsin activates transducin by forming a complex with it and catalyzing the exchange of GTP for bound GDP (Figure 39-50). The binding of GTP to transducin leads to the release of R* for another round of catalysis.

The conformational change induced by the binding of GTP also in-

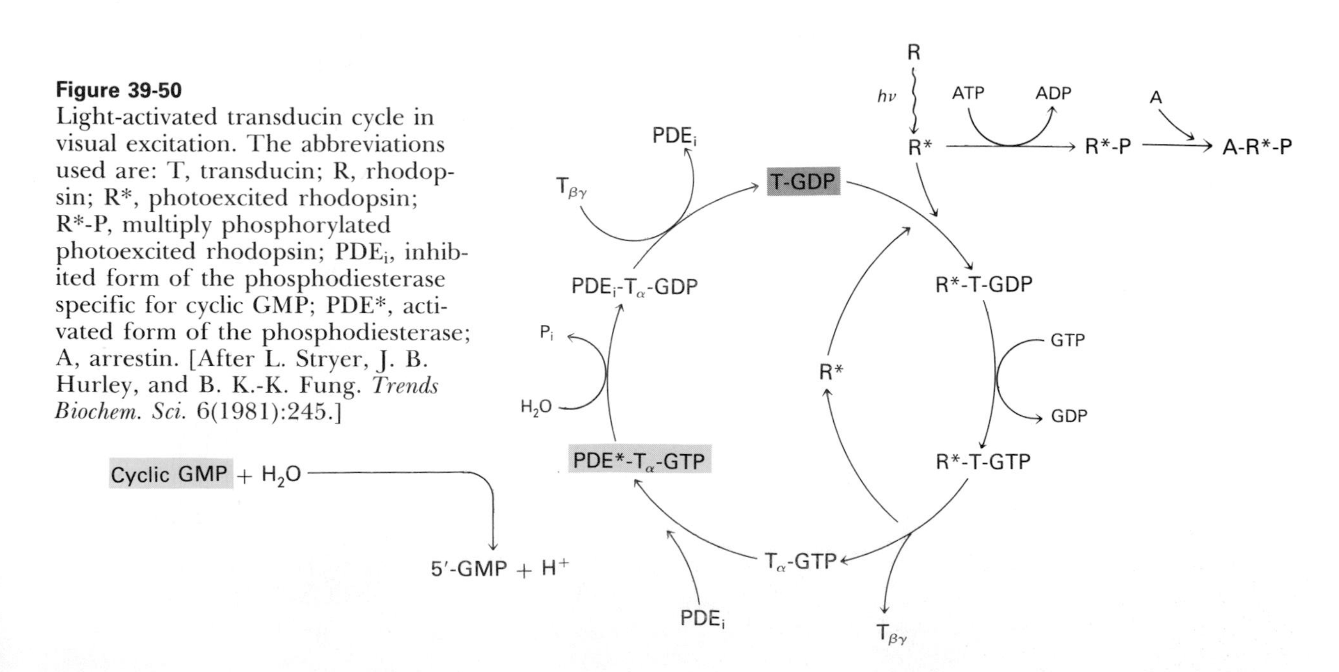

Figure 39-50
Light-activated transducin cycle in visual excitation. The abbreviations used are: T, transducin; R, rhodopsin; R*, photoexcited rhodopsin; R*-P, multiply phosphorylated photoexcited rhodopsin; PDE_i, inhibited form of the phosphodiesterase specific for cyclic GMP; PDE*, activated form of the phosphodiesterase; A, arrestin. [After L. Stryer, J. B. Hurley, and B. K.-K. Fung. *Trends Biochem. Sci.* 6(1981):245.]

duces the dissociation of T_α-GTP, from $T_{\beta\gamma}$. T_α-GTP, the active form of transducin, then activates the phosphodiesterase, which very rapidly hydrolyzes cyclic GMP.

$$\text{Cyclic GMP} + H_2O \xrightarrow{\text{phosphodiesterase}} \text{GMP} + H^+$$

The phosphodiesterase is deactivated by the hydrolysis of GTP bound to T_α. This subunit of transducin has a built-in *GTPase activity*. The return to the dark state also requires that R* be deactivated so that it does not continue to catalyze the activation of transducin. *Rhodopsin kinase* catalyzes the phosphorylation of R* at multiple sites in its carboxyl-terminal region. Phosphorylated R* then binds *arrestin*, an inhibitory protein that blocks the binding of transducin. The restoration of the dark state also requires the synthesis of cyclic GMP, which is catalyzed by *guanylate cyclase*.

$$\text{GTP} \xrightarrow{\text{guanylate cyclase}} \text{cyclic GMP} + PP_i$$

This reaction is driven forward by the hydrolysis of pyrophosphate.

Several features of this light-triggered cyclic GMP cascade are noteworthy:

1. A single molecule of photoexcited rhodopsin catalyzes the activation of about five hundred molecules of transducin. *The formation of activated transducin is the first stage of amplification in visual excitation.*

2. The catalytic activity of the phosphodiesterase is blocked in the dark by its inhibitory subunit. T_α-GTP activates the enzyme by pulling

Figure 39-51
Schematic diagram of the cyclic GMP cascade of vision. [After L. Stryer. *Cold Spring Harbor Symp. Quant. Biol.* 48(1983):841.]

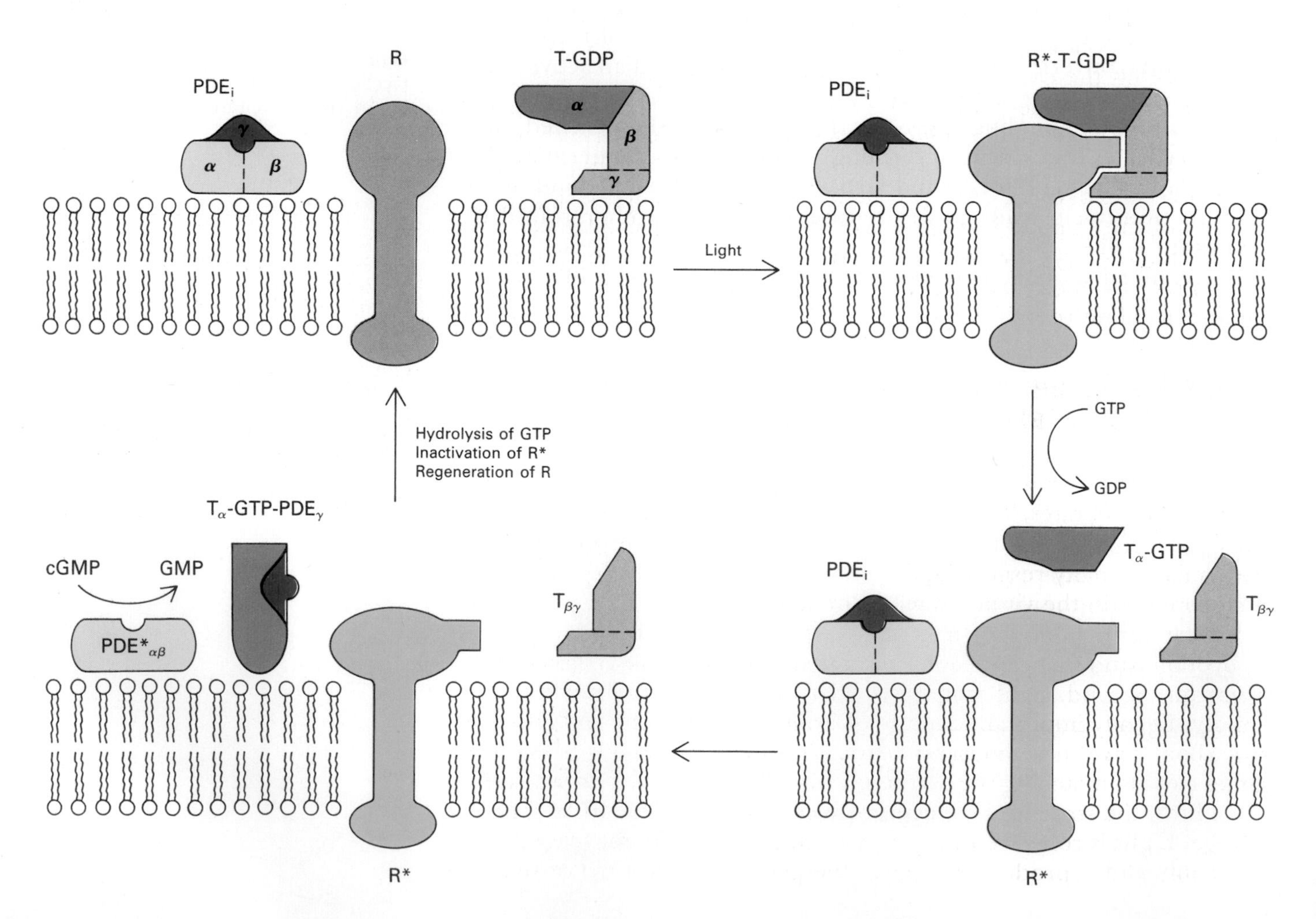

this inhibitory subunit away (Figure 39-51). The activated enzyme has great catalytic prowess. It has a turnover number of 4200 s^{-1} and a k_{cat}/K_M value of 6×10^7 M^{-1} s^{-1}, near the diffusion-controlled limit. *The hydrolysis of cyclic GMP by the phosphodiesterase is the second stage of amplification.*

3. The transducin cycle is powered by the hydrolysis of GTP. *We see here the use of ~P to achieve amplification.*

4. *This cyclic GMP cascade is like the cyclic AMP cascade that mediates the action of hormones such as epinephrine.* The structure and mechanism of action of transducin is like that of the stimulatory G protein (p. 979). Both signal-coupling proteins are activated by receptor proteins with the seven-helix motif. Their GTP-GDP cycles are also very similar. They switch on effector enzymes by relieving constraints imposed by inhibitory domains or subunits.

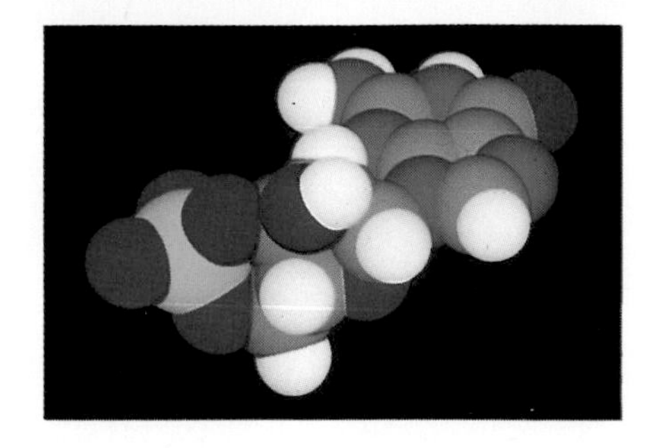

Model of cyclic GMP.

THE LIGHT-SENSITIVE CHANNEL OF RETINAL RODS IS GATED BY THE BINDING OF CYCLIC GMP

How does the light-triggered hydrolysis of cyclic GMP lead to the closure of cation-specific channels in the plasma membrane? The answer has come from patch-clamp studies. A patch electrode was applied to the outer segment of a rod and a gigaseal was formed. The pipet was then pulled away from the cell. A piece of plasma membrane adhered to the pipet. What had been the cytosolic side of the membrane in the intact cell now faced the external solution. The channels in this membrane were opened by micromolar concentrations of cyclic GMP but not other nucleotides tested. This experiment showed that the effect of cyclic GMP on the channel is direct rather than mediated by covalent modification or by the binding of a cytosolic protein. The opening of the channel by cyclic GMP is highly cooperative. The Hill coefficient (p. 155) is 3.0, indicating channel opening requires the binding of at least three molecules of cyclic GMP (Figure 39-52). The high degree of cooperativity of the channel increases its sensitivity to small changes in the concentration of cyclic GMP. The channel is likely to be multimeric, with its cation-conducting pore located along its symmetry axis, as in the acetylcholine receptor channel.

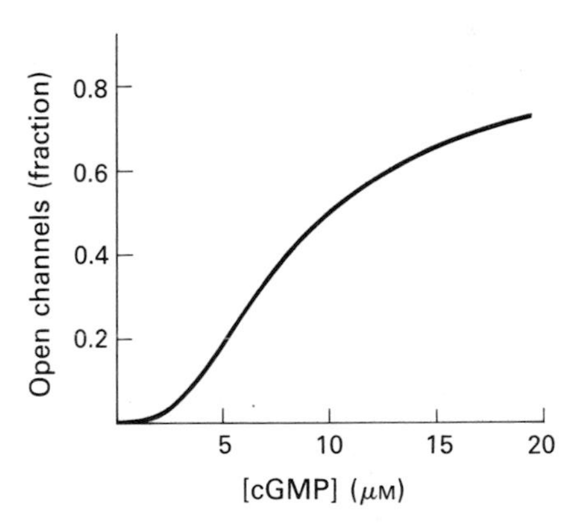

Figure 39-52
Cooperative opening of the cation-specific channel of retinal rod cells by cyclic GMP. The slope of the corresponding Hill plot is 3.0, indicating that an open channel contains at least three bound molecules of cyclic GMP. [Based on data in A. L. Zimmerman and D. A. Baylor. *Nature* 321(1986):70.]

11-*cis*-RETINAL, RHODOPSIN, AND TRANSDUCIN ALSO MEDIATE VISUAL EXCITATION IN INVERTEBRATES

Image-forming eyes are found in only three phyla: molluscs, arthropods, and vertebrates. Though the three kinds of eyes are anatomically quite different, the visual transduction mechanisms employed by them are similar in many respects at the molecular level. 11-*cis*-Retinal is the chromophore in the visual pigments of all three phyla. What is so special about this light-absorbing group? First, 11-*cis*-retinal has an intense absorption band that can readily be shifted into the visible region of the spectrum. Second, light isomerizes 11-*cis*-retinal very efficiently and rapidly. Equally important, its rate of isomerization in the dark is very low, about once in a thousand years. No other compound in nature comes close to matching the extremely high signal-to-noise ratio of retinal. Third, the structural change produced by isomerization of retinal is large. Light is converted into atomic motion of sufficient magnitude to reliably and reproducibly trigger the generation of a nerve signal.

Fourth, retinal is derived from β-carotene (p. 571), a precursor having a very broad biological distribution. For example, carotenoids protect bacteria from the harmful effects of light. The splitting of β-carotene by O_2 directly yields all-*trans*-retinal.

Recent biochemical and molecular genetic studies have revealed that the rhodopsins of *Drosophila* (an arthropod) and octopus (a mollusc) are homologous with those of vertebrates. They have the same seven-helix motif. Furthermore, guanyl-nucleotide-binding proteins like transducin are found in these invertebrates. Indeed, octopus rhodopsin can trigger mammalian transducin. It is evident that the rhodopsin-transducin couple is ancient, at least 700 million years old. Both proteins may be much older. Rhodopsin is present in the eye spot of *Chlamydomonas*, a flagellated photosynthetic eucaryote, and G proteins occur in yeast. It will be interesting to learn where rhodopsin and transducin first came together to mediate visual excitation.

COLOR VISION IS MEDIATED BY THREE KINDS OF PHOTORECEPTOR CELLS

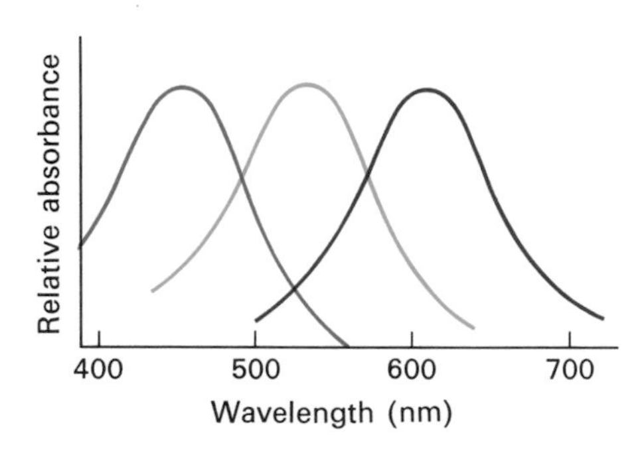

Figure 39-53
Absorption spectra of the three receptors mediating color vision.

The remarkable Thomas Young was a practicing physician, a professor of physics, and a distinguished Egyptologist. He proved the wave nature of light and deciphered many of the hieroglyphs of the Rosetta Stone. In 1802, Young proposed that color vision is mediated by *three fundamental receptors.* Spectrophotometric studies of intact retinas carried out more than a century and a half later revealed that there are indeed *three types of cone cells: blue-, green-, and red-absorbing.* The absorption spectra of their three photoreceptor pigments have been measured by illuminating cones with a beam of light having a diameter of only 1 μm (Figure 39-53). The responses of different cone cells to monochromatic light of different wavelengths have shown that they belong to three groups: some are excited maximally by blue light, others by green light, and the rest by red light. In goldfish, the absorption maxima of the three color receptors are at 455, 530, and 625 nm, and that of rhodopsin is at 500 nm.

*The chromophore in all three kinds of cones is 11-*cis*-retinal.* A protonated Schiff base with 11-*cis*-retinal in the absence of a protein has an absorption maximum at 440 nm. Groups on opsin tune the position of the spectral maximum of bound 11-*cis*-retinal. The chromophoric properties of retinal can readily be shifted more than a hundred nanometers by placing dipoles and charged groups in close proximity and by twisting the polyene chain. The dependence of the absorption properties of this chromophore on its protein environment exemplifies a general principle: *the properties of a prosthetic group are modulated by its interaction with the protein.* Another example is the heme group, which functions as an oxygen carrier in hemoglobin, an electron carrier in cytochrome *c*, and a catalyst in peroxidase.

The photoreceptor proteins of cone cells have not yet been isolated. However, their genes have recently been cloned and sequenced. As predicted by psychophysical and electrophysiological studies, humans have three genes for cone pigments—two closely linked genes on the X chromosome encode the green- and red-absorbing photoreceptor proteins, and one on an autosome encodes the blue-absorbing protein. About half of the amino acid sequence of each cone pigment is identical to that of rhodopsin. The red- and green-absorbing pigments differ from one another at only 15 of 348 residues (Figure 39-54). It is evident

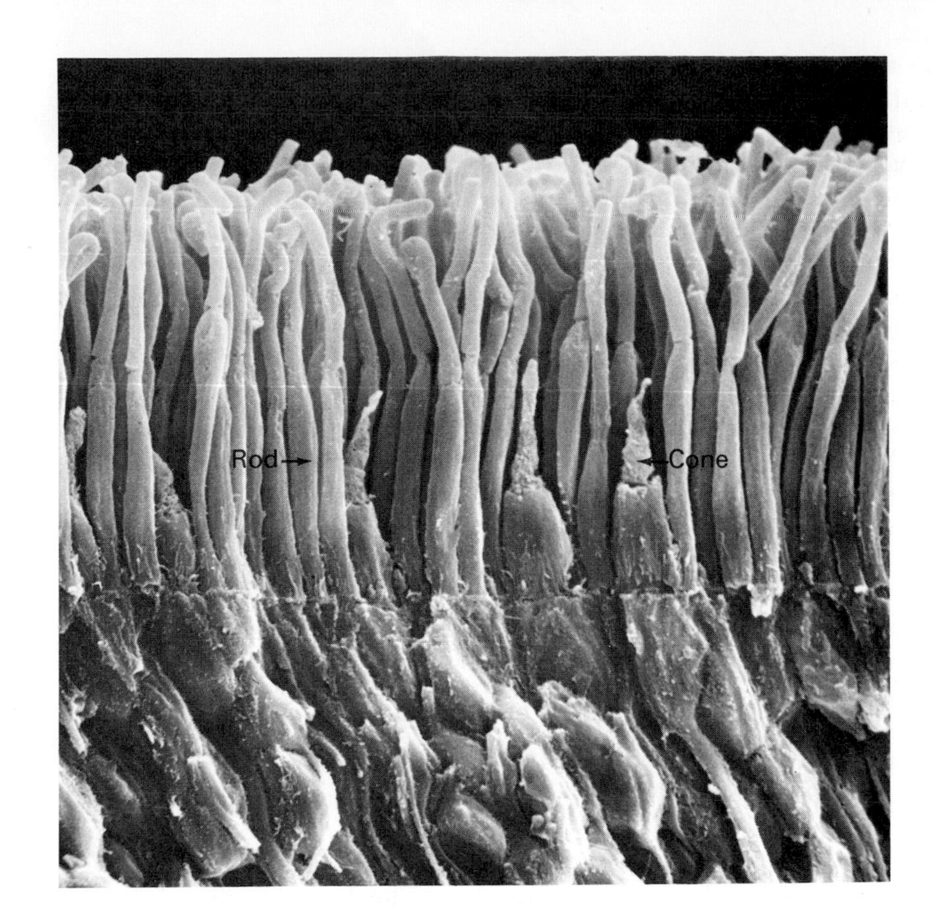

Scanning electron micrograph of rod and cone cells in the photoreceptor layer of the retina. [Courtesy of Dr. William Miller.]

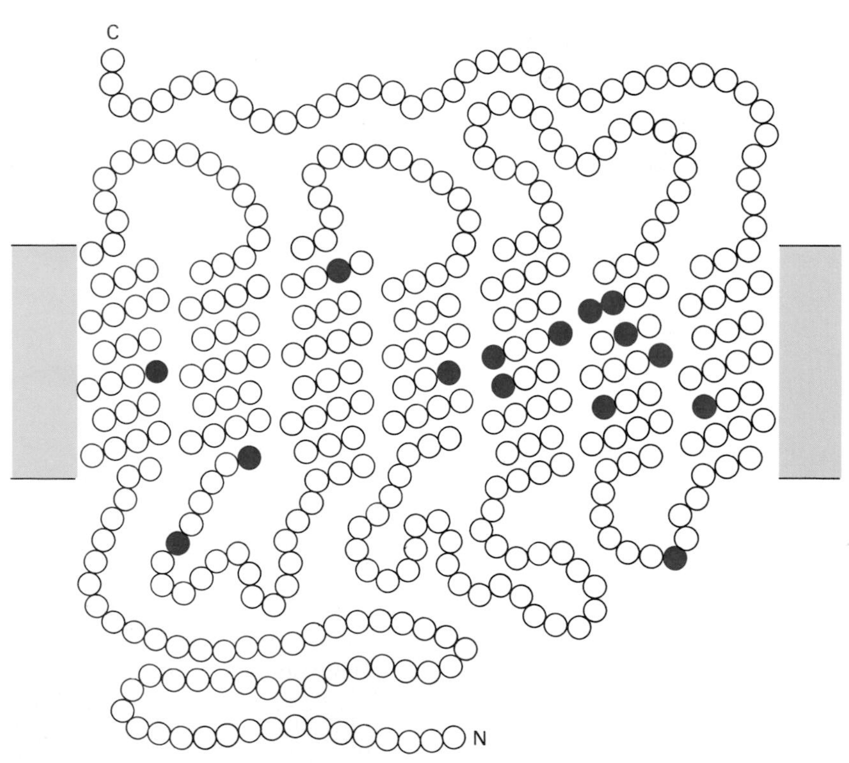

Figure 39-54
The amino acid sequences of the green- and red-absorbing photoreceptor proteins of cone cells are very similar. Open circles denote identical residues and colored circles mark residues that are different. These receptors probably fold into a seven-helix motif like that of rhodopsin. [Courtesy of Dr. Jeremy Nathans and Dr. David Hogness.]

that the cone pigments, like rhodopsin, fold into a seven-helix motif. Furthermore, the transduction mechanism of cones is like that of rods. The transducin, phosphodiesterase, and cyclic GMP channel of cones are very similar to their counterparts in rods.

The genes for the cone pigments and rhodopsin probably arose from a common ancestral gene that underwent duplication and divergence. Amino acid sequence differences between rod and cone pigments suggest that blue cone pigment came first (Figure 39-55). It then gave rise to a long wavelength pigment that diverged to give the green and red pigments. It is interesting to note that New World monkeys have only two cone pigments (a blue one and a long wavelength one), whereas Old World monkeys and humans have three cone pigments. Old World and New World monkeys separated about 30 million years ago. Thus, the red and green cone pigments of primates came into being rather recently.

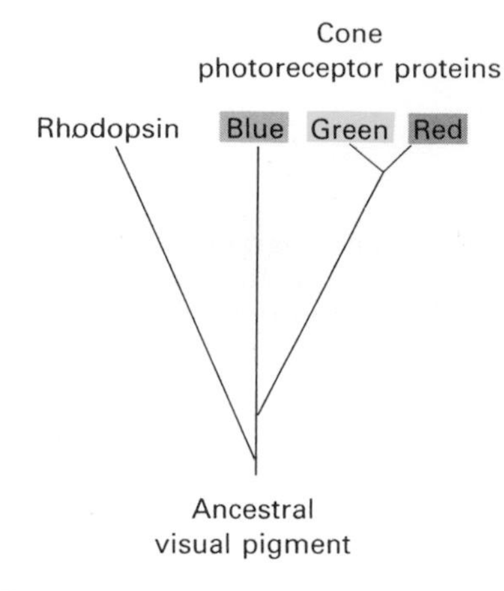

Figure 39-55
Proposed evolutionary relations of human visual pigments.

Most forms of color blindness are caused by a sex-linked recessive mutation. About 1% of men are red-blind, and 2% are green-blind. Spectral measurements of intact eyes have shown that they lack either the red-absorbing or the green-absorbing photoreceptor molecule or that they have an altered pigment with a shifted absorption spectrum. Thus, *color blindness is caused by an absence or defect of one kind of cone opsin.* Recent analyses of the DNA of color-blind males have revealed that they lack the green gene or contain a gene that is a hybrid of the green and red ones (Figure 39-56). The hybrid gene probably encodes a protein having spectral properties intermediate between those of the normal green- and red-absorbing pigments. The high incidence of red-green color blindness arises from recombination between these very similar, closely linked genes. Unequal crossing over leads, in some recombinants, to the loss of the green gene. Are there recombinants with multiple green genes? Indeed, some people harbor two or even three tandemly arrayed green genes. Some of their progeny millions of years from now may have four-color vision!

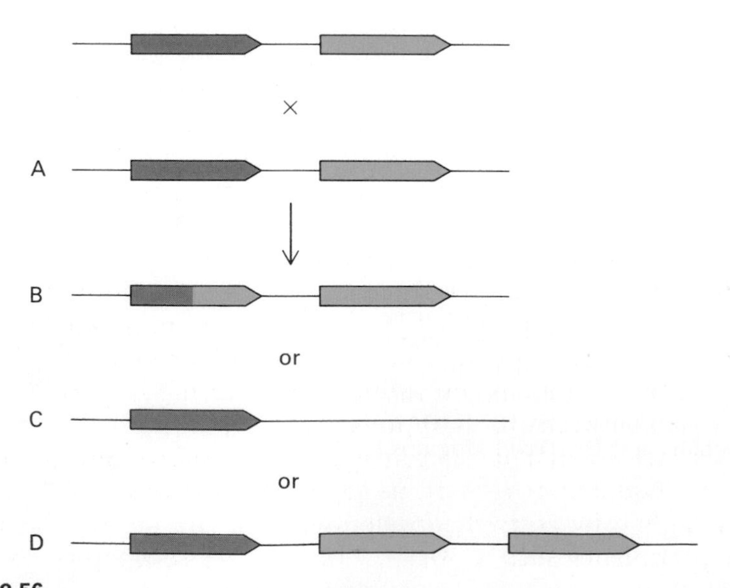

Figure 39-56
(A) The genes for the green-absorbing and red-absorbing visual pigments of human cones are next to each other on a normal X chromosome. Recombination between these genes can lead to (B) the generation of a hybrid gene, (C) the loss of a gene, or (D) the duplication of a gene. [Courtesy of Dr. Jeremy Nathans.]

SUMMARY

Motile bacteria can move toward attractants (such as glucose) and away from repellents (such as fatty acids). The sensors for chemotaxis are located in the periplasmic space or the plasma membrane. Bacterial flagella are rotated by motors that are powered by the flow of protons across the plasma membrane. Bacteria swim smoothly for about a second and then tumble when the direction of rotation of their flagella abruptly changes from counterclockwise to clockwise. Tumbling is suppressed when a bacterium moves toward an attractant or away from a repellent. Bacteria detect temporal rather than instantaneous spatial gradients. Methyl-accepting chemotaxis proteins (MCPs) transmit chemotactic signals across the plasma membrane. The MCPs themselves are chemoreceptors for repellents and some attractants. They also interact with periplasmic proteins (such as the ribose-binding protein) that bind attractants. A central processing system encoded by *che* genes mediates the switching of the direction of flagellar rotation. The activated form of the cheY protein promotes tumbling. Reversible methylation of the MCPs enables the system to respond to changes in the intensity of stimuli rather than to their absolute level, a process called adaptation.

Action potentials in nerve-axon membranes are mediated by transient changes in Na^+ and K^+ permeability. Both the Na^+ and the K^+ channels are gated by transmembrane voltage rather than by chemical signals. Tetrodotoxin and saxitoxin specifically block the Na^+ channel. These inhibitors have been invaluable in purifying the Na^+ channel, which is a single 270-kd polypeptide chain, in some cases associated with smaller chains. The channel contains four repeating transmembrane units. Its narrow pore size (~5 Å) makes it much more permeable to Na^+ than to K^+. Depolarization opens the channel by moving the equivalent of at least four charges across the bilayer. The open state, which lasts about a millisecond, converts into an inactive closed state, which reverts on repolarization to a closed but activatable state.

Nerve impulses are communicated across most synapses by chemical transmitters. Acetylcholine, the neurotransmitter at many synapses and at motor end plates, is synthesized from acetyl CoA and choline. Acetylcholine molecules are packaged in synaptic vesicles, which fuse with the presynaptic membrane upon arrival of a nerve impulse. Released acetylcholine diffuses across the synaptic cleft and combines with specific protein receptors on the postsynaptic membrane. The consequent increase in cation permeability leads to the influx of Na^+ and the depolarization of the membrane. The high affinity of the acetylcholine receptor for α-bungarotoxin has facilitated its purification. The acetylcholine receptor channel is formed by five homologous transmembrane subunits. The binding of acetylcholine to each of its two α subunits transiently opens the cation-conducting pore. Patch-clamp conductance studies have shown that 2×10^7 ions per second flow through an open channel, which spontaneously converts into an inactive nonconducting form. The postsynaptic membrane repolarizes following the hydrolysis of acetylcholine by acetylcholinesterase. This enzyme is inhibited by organic phosphates such as DIPF, which causes respiratory paralysis. Catecholamines (epinephrine, norepinephrine, and dopamine), glycine, and γ-aminobutyrate (GABA) also serve as neurotransmitters. GABA and glycine inhibit the generation of action potentials in the central nervous system by opening chloride channels.

A retinal rod cell can be excited by a single photon. The outer segment of a rod contains a stack of about a thousand discs, which are closed bilayer membranes densely packed with molecules of rhodopsin, a transmembrane protein. The chromophore of rhodopsin, the photoreceptor protein, is 11-*cis*-retinal, which is derived from all-*trans*-retinol (vitamin A). 11-*cis*-Retinal forms a Schiff-base linkage with a specific lysine residue of opsin. Rhodopsin is a member of the seven-helix family of membrane receptors that includes the β-adrenergic receptor. The primary event in visual excitation is the isomerization of 11-*cis*-retinal to the all-*trans* form. Photoexcited rhodopsin (R*) triggers an enzymatic cascade leading to the hydrolysis of cyclic GMP and the consequent closure of cation-specific channels in the plasma membrane. The blockage of the influx of more than 10^6 Na^+ ions by a single photon results in the hyperpolarization of the plasma membrane, which is communicated to the synapse. A single R* activates hundreds of molecules of transducin by catalyzing the exchange of GTP for bound GDP. The GTP form of transducin then activates a phosphodiesterase specific for cyclic GMP by relieving an inhibitory constraint. Hydrolysis of bound GTP by the built-in GTPase activity of transducin brings it back to the inactive GDP state and deactivates the phosphodiesterase. Color vision is mediated by three kinds of photoreceptors, each containing 11-*cis*-retinal. In fact, the photoreceptor proteins of cones are akin to rhodopsin. Moreover, the cyclic GMP cascade mediating visual excitation in cones is similar to that of rods. This amplification cascade of vertebrate photoreceptor cells closely resembles the cyclic AMP cascade mediating the action of hormones such as epinephrine and glucagon.

SELECTED READINGS

Where to start

Adler, J., 1976. The sensing of chemicals by bacteria. *Sci. Amer.* 234(4):40–47.

Berg, H., 1975. How bacteria swim. *Sci. Amer.* 233(2):36–44.

Numa, S., 1986. Molecular basis for the function of ionic channels. *Biochem. Soc. Symp.* 52:119–143.

Snyder, S. H., 1985. The molecular basis of communication between cells. *Sci. Amer.* 253(4):132–140.

Stryer, L., 1987. The molecules of visual excitation. *Sci. Amer.* 255(7):42–50.

Books

Hille, B., 1984. *Ionic Channels of Excitable Membranes*. Sinauer. [An outstanding presentation of concepts, experimental methods, and channel properties. A book to savor.]

Miller, C., 1986. *Ion Channel Reconstitution*. Plenum. [Contains many excellent articles on the purification and reconstitution of channels. The sodium channel and the acetylcholine receptor are discussed in detail.]

Sakmann, B., and Neher, E., (eds.), 1983. *Single-channel Recording*. Plenum. [Many interesting applications of the patch-clamp technique are discussed in this important volume.]

Kuffler, S. W., Nicholls, J. G., and Martin, A. R., 1984. *From Neuron to Brain* (2nd ed.). Sinauer. [Part II of this fine book contains an excellent account of mechanisms for neuronal signalling.]

Cold Spring Harbor Symposium, 1983. *Molecular Neurobiology. Cold Spring Harbor Symp. Quant. Biol.* Volume 48. [Contains many excellent articles on membrane channels and sensory systems.]

Bacterial chemotaxis

Hazelbauer, G. L., and Harayama, S., 1983. Sensory transduction in bacterial chemotaxis. *Int. Rev. Cytol.* 81:33–70.

McNab, R. M., and Aizawa, S.-I., 1984. Bacterial motility and the flagellar motor. *Ann. Rev. Biophys. Bioeng.* 13:51–83.

Koshland, D. E., Jr., Russo, A. F., and Gutterson, N. I., 1983. Information processing in a sensory system. *Cold Spring Harbor Symp. Quant. Biol.* 48:805–810.

Berg, H. C., and Purcell, E. M., 1977. Physics of chemoreception. *Biophys. J.* 20:193–219. [A beautiful paper, rich in ideas and insight. Also see H. C. Berg's *Random Walks in Biology* (Princeton, 1983).]

Segall, J. E., Block, S. M., and Berg, H. C., 1986. Temporal comparisons in bacterial chemotaxis. *Proc. Nat. Acad. Sci.* 83:8987–8991.

Macnab, R., and Koshland, D. E., Jr., 1972. The gradient-sensing mechanism in bacterial chemotaxis. *Proc. Nat. Acad. Sci.* 69:2509–2512. [These experiments revealed that bacteria detect a temporal gradient rather than an instantaneous spatial one.]

Khan, S., and Macnab, R. M., 1980. Proton chemical potential, proton electrical potential and bacterial motility. *J. Mol. Biol.* 138:599–614.

Meister, M., Lowe, G., and Berg, H. C., 1987. The proton flux through the bacterial flagellar motor. *Cell* 49:643–650.

Krikos, A., Conley, M. P., Boyd, A., Berg, H. C., and Simon, M. I., 1985. Chimeric chemosensory transducers of *Escherichia coli*. *Proc. Nat. Acad. Sci.* 82:1326–1330.

Bogonez, E., and Koshland, D. E., Jr., 1985. Solubilization of a vectorial transmembrane receptor in functional form: aspartate receptor of chemotaxis. *Proc. Nat. Acad. Sci.* 82:4891–4895.

Wolfe, A. J., Conley, M. P., Kramer, T. J., and Berg, H. C., 1987. Reconstitution of signaling in bacterial chemotaxis. *J. Bacteriol.* 169:1878–1885.

Ravid, S., Matsumura, P., and Eisenbach, M., 1986. Restoration of flagellar clockwise rotation in bacterial envelopes by insertion of the chemotaxis protein cheY. *Proc. Nat. Acad. Sci.* 83:7157–7161.

Voltage-sensitive sodium channel

Catterall, W. A., 1986. Molecular properties of voltage-sensitive sodium channels. *Ann. Rev. Biochem.* 55:953–985.

Kao, C. Y., and Levinson, S. R., (eds.), 1986. *Tetrodotoxin, Saxitoxin, and the Molecular Biology of the Sodium Channel. Ann. N. Y. Acad. Sci.* Volume 479.

Noda, M., Ikeda, T., Suzuki, H., Takeshima, H., Takahashi, T., Kuno, M., and Numa, S., 1986. Expression of functional sodium channels from cloned cDNA. *Nature* 322:826–828.

Hartshorne, R. P., Keller, B. U., Talvenheimo, J. A., Catterall, W. A., and Montal, M., 1985. Functional reconstitution of the purified brain sodium channel in planar lipid bilayers. *Proc. Nat. Acad. Sci.* 82:240–244.

Keynes, R. D., 1979. Ion channels in the nerve-cell membrane. *Sci. Amer.* 240(3):126–135.

Acetylcholine receptor channel

Stroud, R. M., and Finer-Moore, J., 1985. Acetylcholine receptor structure, function, and evolution. *Ann. Rev. Cell Biol.* 1:317–351.

Brisson, A., and Unwin, P. N. T., 1985. Quaternary structure of the acetylcholine receptor. *Nature* 315:474–477.

Colquhoun, D., and Sakmann, B, 1981. Fluctuations in the microsecond time range of the current through single acetylcholine receptor ion channels. *Nature* 294:464–466.

Sakmann, B., Methfessel, C., Mishina, M., Takahashi, T., Takai, T., Kurasaki, M., Fukuda, K., and Numa, S., 1985. Role of acetylcholine receptor subunits in gating of the channel. *Nature* 318:538–543.

Hess, G. P., Kolb, H. A., Lauger, P., Schoffeniels, E., and Schwarze, W., 1984. Acetylcholine receptor (from *Electrophorus electricus*): a comparison of single-channel current recordings and chemical kinetic measurements. *Proc. Nat. Acad. Sci.* 81:5281–5285.

Huganir, R. L., Delcour, A. H., Greengard, P., and Hess, G. P., 1986. Phosphorylation of the nicotinic acetylcholine receptor regulates its rate of desensitization. *Nature* 321:774–776.

Lester, H. A., 1977. The response to acetylcholine. *Sci. Amer.* 236(2):106–118.

GABA and glycine receptor channels

Schofield, P. R., Darlison, M. G., Fujita, N., Burt, D. R., Stephenson, F. A., Rodriguez, H., Rhee, L. M., Ramachandran, J., Reale, V., Glencorse, T. A., Seeburg, P. H., and Barnard, E. A., 1987. Sequence and functional expression of the $GABA_A$ receptor shows a ligand-gated receptor super-family. *Nature* 328:221–227.

Grenningloh, G., Rienitz, A., Schmitt, B., Methfessel, C., Zensen, M., Beyreuther, K., Gundelfinger, E. D., and Betz, H., 1987. The strychnine-binding subunit of the glycine receptor shows homology with nicotinic acetylcholine receptors. *Nature* 328:215–220.

Visual excitation

Schnapf, J. L., and Baylor, D. A., 1987. How photoreceptor cells respond to light. *Sci. Amer.* 256(4):40–47. [A lucid account of the electrophysiology of rods and cones.]

Stryer, L., 1986. Cyclic GMP cascade of vision. *Ann. Rev. Neurosci.* 9:87–119.

Chabre, M., 1985. Trigger and amplification mechanisms in visual phototransduction. *Ann. Rev. Biophys. Biophys. Chem.* 14:331–360.

Fesenko, E. E., Kolesnikov, S. S., and Lyubarsky, A. L., 1985. Induction by cyclic GMP of cationic conductance in plasma membrane of retinal rod outer segments. *Nature* 313:310–313.

Ovchinnikov, Yu. A., 1982. Rhodopsin and bacteriorhodopsin: structure-function relationships. *FEBS Lett.* 148:179–191.

Miller, W. H., (ed.), 1981. *Molecular Mechanisms of Photoreceptor Transduction.* Academic Press.

Fung, B.-K., Hurley, J. B., and Stryer, L., 1981. Flow of information in the light-triggered cyclic nucleotide cascade of vision. *Proc. Nat. Acad. Sci.* 78:152–156.

Liebman, P. A., and Pugh, E. N., Jr., 1979. The control of phosphodiesterase in rod disk membranes: kinetics, possible mechanisms, and significance for vision. *Vision Res.* 19:375–380.

Wald, G., 1968. The molecular basis of visual excitation. *Nature* 219:800–807. [This Nobel lecture contains an interesting account of the discovery of the primary event in vision.]

Color vision

Nathans, J., 1987. Molecular biology of visual pigments. *Ann. Rev. Neurosci.* 10:163–194.

Nathans, J., Thomas, D., and Hogness, D. S., 1986. Molecular genetics of human color vision: the genes encoding blue, green, and red pigments. *Science* 232:193–202.

MacNichol, E. F., Jr., 1964. Three-pigment color vision. *Sci. Amer.* 211(6):48–56.

Rushton, W. A. H., 1975. Visual pigments and color blindness. *Sci. Amer.* 232(3):64–74.

Appendixes
Answers to Problems
Index

APPENDIX 

Physical Constants and Conversion of Units

Values of physical constants

Physical constant	*Symbol*	*Value*
Atomic mass unit (dalton)	amu	1.660×10^{-24} g
Avogadro's number	N	6.022×10^{23} mol^{-1}
Boltzmann's constant	k	1.381×10^{-23}J deg^{-1} 3.298×10^{-24} cal deg^{-1}
Electron volt	eV	1.602×10^{-19} J 3.828×10^{-20} cal
Faraday constant	F	9.649×10^{4} C mol^{-1} 2.306×10^{4} cal volt^{-1} eq^{-1}
Curie	Ci	3.70×10^{10} disintegrations s^{-1}
Gas constant	R	8.315 J mol^{-1} deg^{-1} 1.987 cal mol^{-1} deg^{-1}
Planck's constant	h	6.626×10^{-34} J s 1.584×10^{-34} cal s
Speed of light in a vacuum	c	2.998×10^{10} cm s^{-1}

Abbreviations: C, coulomb; cal, calorie; cm, centimeter; deg, degree Kelvin; eq, equivalent; g, gram; J, joule; mol, mole; s, second.

Mathematical constants

$\pi = 3.14159$

$e = 2.71828$

$\log_e x = 2.303 \log_{10} x$

Conversion factors

Physical quantity	*Equivalent*
Length	$1\text{ cm} = 10^{-2}\text{ m} = 10\text{ mm} = 10^4\ \mu\text{m} = 10^7\text{ nm}$
	$1\text{ cm} = 10^8\text{ Å} = 0.3937\text{ inch}$
Mass	$1\text{ g} = 10^{-3}\text{ kg} = 10^3\text{ mg} = 10^6\ \mu\text{g}$
	$1\text{ g} = 3.527 \times 10^{-2}$ ounce (avoirdupoir)
Volume	$1\text{ cm}^3 = 10^{-6}\text{ m}^3 = 10^3\text{ mm}^3$
	$1\text{ ml} = 1\text{ cm}^3 = 10^{-3}\text{ l} = 10^3\ \mu\text{l}$
	$1\text{ cm}^3 = 6.1 \times 10^{-2}\text{ in}^3 = 3.53 \times 10^{-5}\text{ ft}^3$
Temperature	$\text{K} = {}^\circ\text{C} + 273.15$
	${}^\circ\text{C} = (5/9)({}^\circ\text{F} - 32)$
Energy	$1\text{ J} = 10^7\text{ erg} = 0.239\text{ cal} = 1\text{ watt s}$
Pressure	$1\text{ torr} = 1\text{ mm Hg } (0^\circ\text{C})$
	$= 1.333 \times 10^2\text{ newton/m}^2$
	$= 1.333 \times 10^2$ pascal
	$= 1.316 \times 10^{-3}$ atmospheres

Standard prefixes

Prefix	*Symbol*	*Factor*
kilo	k	10^3
hecto	h	10^2
deca	da	10^1
deci	d	10^{-1}
centi	c	10^{-2}
milli	m	10^{-3}
micro	μ	10^{-6}
nano	n	10^{-9}
pico	p	10^{-12}

APPENDIX B

Atomic Numbers and Weights of the Elements

Element	Symbol	Atomic number	Atomic weight
Actinium	Ac	89	227.03
Aluminum	Al	13	26.98
Americium	Am	95	243.06
Antimony	Sb	51	121.75
Argon	Ar	18	39.95
Arsenic	As	33	74.92
Astatine	At	85	210.99
Barium	Ba	56	137.34
Berkelium	Bk	97	247.07
Beryllium	Be	4	9.01
Bismuth	Bi	83	208.98
Boron	B	5	10.81
Bromine	Br	35	79.90
Cadmium	Cd	48	112.40
Calcium	Ca	20	40.08
Californium	Cf	98	249.07
Carbon	C	6	12.01
Cerium	Ce	58	140.12
Cesium	Cs	55	132.91
Chlorine	Cl	17	35.45
Chromium	Cr	24	52.00
Cobalt	Co	27	58.93
Copper	Cu	29	63.55
Curium	Cm	96	245.07
Dysprosium	Dy	66	162.50
Einsteinium	Es	99	254.09
Erbium	Er	68	167.26
Europium	Eu	63	151.96
Fermium	Fm	100	252.08
Fluorine	F	9	18.99
Francium	Fr	87	223.02
Gadolinium	Gd	64	157.25
Gallium	Ga	31	69.72
Germanium	Ge	32	72.59
Gold	Au	79	196.97
Hafnium	Hf	72	178.49
Helium	He	2	4.00
Holmium	Ho	67	164.93
Hydrogen	H	1	1.01
Indium	In	49	114.82
Iodine	I	53	126.90
Iridium	Ir	77	192.22
Iron	Fe	26	55.85
Khurchatovium	Kh	104	260
Krypton	Kr	36	83.80
Lanthanum	La	57	138.91
Lawrencium	Lr	103	256
Lead	Pb	82	207.20
Lithium	Li	3	6.94
Lutetium	Lu	71	174.97
Magnesium	Mg	12	24.31
Manganese	Mn	25	54.94
Mendelevium	Md	101	255.09
Mercury	Hg	80	200.59
Molybdenum	Mo	42	95.94
Neodymium	Nd	60	144.24
Neon	Ne	10	20.18
Neptunium	Np	93	237.05
Nickel	Ni	28	58.71
Niobium	Nb	41	92.91
Nitrogen	N	7	14.01
Nobelium	No	102	255
Osmium	Os	76	190.20
Oxygen	O	8	16.00
Palladium	Pd	46	106.40
Phosphorus	P	15	30.97
Platinum	Pt	78	195.09
Plutonium	Pu	94	242.06
Polonium	Po	84	208.98
Potassium	K	19	39.10
Praseodymium	Pr	59	140.91
Promethium	Pm	61	145
Protactinium	Pa	91	231.04
Radium	Ra	88	226.03
Radon	Rn	86	222.02
Rhenium	Re	75	186.20
Rhodium	Rh	45	102.91
Rubidium	Rb	37	85.47
Ruthenium	Ru	44	101.07
Samarium	Sm	62	150.40
Scandium	Sc	21	44.96
Selenium	Se	34	78.96
Silicon	Si	14	28.09
Silver	Ag	47	107.87
Sodium	Na	11	22.99
Strontium	Sr	38	87.62
Sulfur	S	16	32.06
Tantalum	Ta	73	180.95
Technetium	Tc	43	98.91
Tellurium	Te	52	127.60
Terbium	Tb	65	158.93
Thallium	Tl	81	204.37
Thorium	Th	90	232.04
Thulium	Tm	69	168.93
Tin	Sn	50	118.69
Titanium	Ti	22	47.90
Tungsten	W	74	183.85
Uranium	U	92	238.03
Vanadium	V	23	50.94
Xenon	Xe	54	131.30
Ytterbium	Yb	70	173.04
Yttrium	Y	39	88.91
Zinc	Zn	30	65.37
Zirconium	Zr	40	91.22

APPENDIX C

p*K*′ Values of Some Acids

Acid	p*K*′ *(at 25°C)*
Acetic acid	4.76
Acetoacetic acid	3.58
Ammonium ion	9.25
Ascorbic acid, pK'_1	4.10
pK'_2	11.79
Benzoic acid	4.20
n-Butyric acid	4.81
Cacodylic acid	6.19
Carbonic acid, pK'_1	6.35
pK'_2	10.33
Citric acid, pK'_1	3.14
pK'_2	4.77
pK'_3	6.39
Ethylammonium ion	10.81
Formic acid	3.75
Glycine, pK'_1	2.35
pK'_2	9.78
Imidazolium ion	6.95

Acid	p*K*′ *(at 25°C)*
Lactic acid	3.86
Maleic acid, pK'_1	1.83
pK'_2	6.07
Malic acid, pK'_1	3.40
pK'_2	5.11
Phenol	9.89
Phosphoric acid, pK'_1	2.12
pK'_2	7.21
pK'_3	12.67
Pyridinium ion	5.25
Pyrophosphoric acid, pK'_1	0.85
pK'_2	1.49
pK'_3	5.77
pK'_4	8.22
Succinic acid, pK'_1	4.21
pK'_2	5.64
Trimethylammonium ion	9.79
Tris (hydroxymethyl) aminomethane	8.08
Water	14.0

APPENDIX D

Standard Bond Lengths

Bond	*Structure*	*Length (Å)*
C—H	R_2CH_2	1.07
	Aromatic	1.08
	RCH_3	1.10
C—C	Hydrocarbon	1.54
	Aromatic	1.40
C=C	Ethylene	1.33
C≡C	Acetylene	1.20
C—N	RNH_2	1.47
	O=C—N	1.34
C—O	Alcohol	1.43
	Ester	1.36
C=O	Aldehyde	1.22
	Amide	1.24
C—S	R_2S	1.82
N—H	Amide	0.99
O—H	Alcohol	0.97
O—O	O_2	1.21
P—O	Ester	1.56
S—H	Thiol	1.33
S—S	Disulfide	2.05

Answers to Problems

Chapter 2

1. 477 Å (318 residues per strand, 1.5 Å per residue).
2. The methyl group attached to the β carbon of isoleucine sterically interferes with α-helix formation. In leucine, this methyl group is attached to the γ carbon atom, which is farther from the main chain and hence does not interfere.
3. The first mutation destroys activity because valine occupies more space than alanine, and so the protein must take a different shape. The second mutation restores activity because of a compensatory reduction of volume; glycine is smaller than isoleucine.
4. The native conformation of insulin is not the thermodynamically most stable form. Indeed, insulin is formed from proinsulin, a single-chain precursor (p. 994).
5. A segment of the main chain of the protease could hydrogen bond to the main chain of the substrate to form an extended parallel or antiparallel pair of β strands.

Chapter 3

1. (a) Phenyl isothiocyanate.
   (b) Dansyl chloride or dabsyl chloride.
   (c) Urea; β-mercaptoethanol to reduce disulfides.
   (d) Chymotrypsin.
   (e) CNBr.
   (f) Trypsin.
2. 0.01, 0.1, 1, 10, and 100.
3. Each amino acid residue, except the carboxyl-terminal one, gives rise to a hydrazide on reacting with hydrazine. The carboxyl-terminal residue can be identified because it yields a free amino acid.
4. (a) Approximately +1.
   (b) Two peptides.
5. The *S*-aminoethylcysteine side chain resembles that of lysine. The only difference is a sulfur atom in place of a methylene group.
6. A 1 mg/ml solution of myoglobin (17.8 kd) corresponds to 5.62×10^{-5} M. The absorbance of a 1-cm path length is 0.84, which corresponds to an I_0/I ratio of 6.96. Hence 14.4% of the incident light is transmitted.
7. Tropomyosin is rod shaped, whereas hemoglobin is approximately spherical.
8. 50 kd.
9. Reduction of disulfide bonds by dithiothreitol makes the protein less compact, so that it migrates less rapidly.
10. The positions of disulfide bonds can be determined

by diagonal electrophoresis (p. 57). The disulfide pairing is unaltered by the mutation if the off-diagonal peptides formed from the native and mutant proteins are the same.

11. Electrostatic repulsion between positively charged ϵ-amino groups prevents α-helix formation at pH 7. At pH 10, the side chains become deprotonated, allowing α-helix formation.
12. Poly-L-glutamate is a random coil at pH 7 and becomes α helical below pH 4.5 because the γ-carboxylate groups become protonated.
13. Glycine has the smallest side chain of any amino acid. Its smallness often is critical in allowing polypeptide chains to make tight turns or to approach one another closely.
14. Affinity chromatography on a column containing a covalently attached analog of vasopressin would be an effective purification method. The first step would be to solubilize the receptor by adding a nondenaturing detergent to a membrane preparation. The solubilized receptor would then be added to the affinity column, which would be washed to remove proteins that do not have high affinity for it. The receptor would be eluted by adding vasopressin.
15. The addition of substrate makes the enzyme less compact, which increases its frictional coefficient. The partial specific volume is unlikely to be altered.
16. The centrifugal force F_c corresponds to Ez. The centrifugal field $\omega^2 r$ is analogous to E, and the effective mass m' is analogous to z.
17. (a) A compact particle sediments more rapidly than does an extended one of the same mass and partial specific volume.
    (b) Shape does not affect the position of the molecule in a sedimentation equilibrium experiment.
18. Large molecules emerge first from a gel-filtration column because a smaller volume is accessible to them; they cannot enter the beads. In contrast, a continuous polymer framework impedes the movement of large molecules in gel electrophoresis, causing them to migrate less rapidly than small molecules.
19. Different monoclonal antibodies recognize different determinants (epitopes). Suppose that one antibody is specific for epitope *a* and the other for epitope *b*. The Western blot suggests that the 23-kd protein contains *a* and *b*, the 57-kd protein contains only *a*, and the 69-kd protein contains only *b*.
20. The amino-terminus of myoglobin is flexible. The existence of many different conformations prevents it being seen in the electron density map.
21. A fluorescent-labeled derivative of a bacterial degradation product (e.g., a formylmethionyl peptide) would bind to cells containing the receptor of interest.

Chapter 4

1. (a) TTGATC; (b) GTTCGA; (c) ACGCGT; and (d) ATGGTA.
2. (a) [T] + [C] = 0.46.
   (b) [T] = 0.30, [C] = 0.24, and [A] + [G] = 0.46.
3. 5.88×10^3 base pairs.
4. After 1.0 generation, one-half of the molecules would be ^{15}N-^{15}N, the other half ^{14}N-^{14}N. After 2.0 generations, one-quarter of the molecules would be ^{15}N-^{15}N, the other three-quarters ^{14}N-^{14}N. Hybrid ^{14}N-^{15}N molecules would not be observed in conservative replication.
5. The DNA renatured when the heat-killed pneumococci were cooled before they were injected into mice.
6. Non-competent strains may not be able to take up DNA. Alternatively, they may have potent deoxyribonucleases, or they may not be able to integrate fragments of DNA into their genome.
7. In the Hershey-Chase experiment, ^{35}S-labeled T2 viral proteins did not become incorporated into infected cells. The labeled viral proteins were found in the supernatant when infected cells were centrifuged. In contrast, M13 proteins become imbedded in the inner membrane of infected cells; they would appear in the pellet rather than the supernatant after centrifugation. Hershey and Chase would not have been able to separate M13 into genetic and nongenetic parts, as they did for T2.
8. Tritiated thymine.
9. dATP, dGTP, dCTP, and dTTP labeled with ^{32}P in the innermost (α) phosphorus atom.
10. Molecules (a) and (b) would not lead to DNA synthesis because they lack a 3′-OH group (a primer). Molecule (d) has a free 3′-OH at one end of each strand but no template strand beyond. Only (c) would lead to DNA synthesis.
11. A deoxythymidylate oligonucleotide should be used as the primer. The poly rA template specifies the incorporation of dT; hence radioactive dTTP should be used in the assay.
12. The ribonuclease serves to degrade the RNA strand, a necessary step in forming duplex DNA from the RNA-DNA hybrid.
13. (a) Treat one aliquot of the sample with ribonuclease and another with deoxyribonuclease. Test these nuclease-treated samples for infectivity.
    (b) An essential protein enzyme carried by the virus particle seems unlikely because the phenol-treated material was infectious. The nucleic acid contains all the information needed for its replication.
14. Deamination changes the original GC base pair into a GU pair. After one round of replication, one daughter duplex will contain a GC pair, and the other duplex an AU pair. After two rounds of replication, there would be two GC pairs, one AU pair, and one AT pair.

Chapter 5

1. (a) Deoxyribonucleoside triphosphates versus ribonucleoside triphosphates.
   (b) $5' \rightarrow 3'$ for both.
   (c) Semiconserved for DNA polymerase I, conserved for RNA polymerase.

(d) DNA polymerase I needs a primer, whereas RNA polymerase does not.
2. 5′-UAACGGUACGAU-3′
3. The 2′-OH group in RNA acts as an intramolecular catalyst. In the alkaline hydrolysis of RNA, it forms a 2′-3′ cyclic intermediate.
4. Cordycepin terminates RNA synthesis. An RNA chain containing cordycepin lacks a 3′-OH group.
5. Leu-Pro-Ser-Asp-Trp-Met.
6. Poly (Leu-Leu-Thr-Tyr).
7. (a) A codon for lysine cannot be changed to one for aspartate by the mutation of a single nucleotide.
   (b) Arg, Asn, Gln, Glu, Ile, Met, or Thr.
8. A peptide terminating with Lys (UGA is a stop codon); -Asn-Glu-; and -Met-Arg-.
9. Highly abundant amino acid residues have the most codons (e.g., Leu and Ser each have six), whereas the least abundant ones have the fewest (Met and Trp each have only one). Degeneracy allows (a) variation in base composition and (b) decreases the likelihood that a substitution of a base will change the encoded amino acid. If the degeneracy were equally distributed, each of the twenty amino acids would have three codons. Benefits (a) and (b) are maximized by assigning more codons to prevalent amino acids than to less frequently used ones.
10. Phe-Cys-His-Val-Ala-Ala.
11. GUG and GUC are likely to be used more by the alga from the hot springs to increase the melting temperature of its DNA (see p. 82).
12. The genetic code is degenerate. Eighteen of the twenty amino acids are specified by more than one codon. Hence, many nucleotide changes (especially in the third base of a codon) do not alter the nature of the encoded amino acid.
13. (a) Green-red color blindness.
    (b) Recombination between similar genes can lead to gene duplication. The duplicate of an essential gene can undergo mutation and diversification without deleterious consequences. New genes nearly always arise by diversification of duplicated genes.

Chapter 6

1. 5′-GGCATAC-3′.

2.

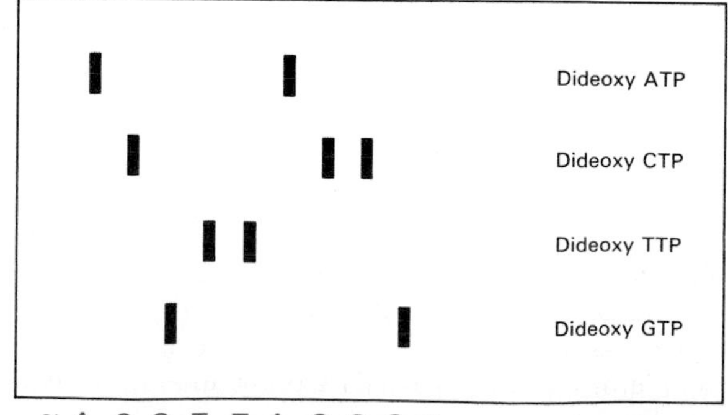

3. Ovalbumin cDNA should be used. *E. coli* lacks the machinery to splice the primary transcript arising from genomic DNA.
4. (a) No, because most human genes are much longer than 4 kb. One would obtain fragments containing only a small part of a complete gene.
   (b) No, chromosome walking depends on having *overlapping* fragments. Exhaustive digestion with a restriction enzyme produces nonoverlapping, short fragments.
5. Southern blotting of an MstII digest would distinguish between the normal and mutant genes. The loss of a restriction site would lead to the replacement of two fragments on the Southern blot by a single longer fragment (see p. 169). Such a finding would not prove that GTG replaced GAG; other sequence changes at the restriction site could yield the same result.
6. Cech replicated the recombinant DNA plasmid in *E. coli*, and then transcribed the DNA in vitro using bacterial RNA polymerase. He then found that this RNA underwent self-splicing in vitro in the complete absence of any proteins from *Tetrahymena*.
7. (a) Only one of the foreign DNA strands encodes an mRNA specifying a functional protein.
   (b) One could analyze RF molecules from the original single virus infection by restriction enzyme mapping. Asymmetric restriction sites in the foreign DNA fragments and phage chromosome would yield different gel patterns according to the orientation of the foreign DNA with the M13 DNA.
   (c) Single-strand circles from two clones containing the same foreign DNA strand will not hybridize to one another. In contrast, the circles will hybridize to form a duplex if the foreign DNA in them are complements of one another. S1 nuclease digests unpaired DNA strands but not duplex DNA.
8. Knowledge of the amino acid sequence is essential. It would be helpful to know which bonds are highly susceptible to proteolysis, and which residues are critical for the biological function of the peptide.

Chapter 7

1. (a) 2.96×10^{-11} g.
   (b) 2.71×10^{8} molecules.
   (c) No. There would be 3.22×10^{8} hemoglobin molecules in a red cell if they were packed in a cubic crystalline array. Hence, the actual packing density is about 84% of the maximum possible.
2. 2.65 g (or 4.75×10^{-2} moles) of Fe.
3. (a) In humans, 1.44×10^{-2} g (4.49×10^{-4} moles) of O_2 per kg of muscle. In sperm whale, 0.144 g (4.49×10^{-3} moles) of O_2 per kg.
   (b) 128.
4. (a) $k_{off} = k_{on}K = 20\ s^{-1}$.
   (b) Mean duration is 0.05 s (the reciprocal of k_{off}).
5. (a) Increased, (b) decreased, (c) decreased, and (d) increased oxygen affinity.

6. Inositol hexaphosphate.
7. The pK is (a) lowered, (b) raised, and (c) raised.
8. (a) Yes. $K_{AB} = K_{BA}(K_B/K_A) = 2 \times 10^{-5}$ M.
   (b) The presence of A enhances the binding of B; hence, the presence of B enhances the binding of A.
9. Carbon monoxide bound to one heme alters the oxygen affinity of the other hemes in the same hemoglobin molecule. Specifically, CO increases the oxygen affinity of hemoglobin and thereby decreases the amount of O_2 released in actively metabolizing tissues. Carbon monoxide stabilizes the quaternary structure characteristic of oxyhemoglobin. In other words, CO mimics O_2 as an allosteric effector.
10. (a) For maximal transport, $K = 10^{-5}$ M. In general, maximal transport is achieved when $K = ([L_A][L_B])^{0.5}$.
    (b) For maximal transport, $P_{50} = 44.7$ torrs, which is considerably higher than the physiological value of 26 torrs. However, it must be stressed that this calculation ignores cooperative binding and the Bohr effect.
11. (a) Lys or Arg at position 6.
    (b) GAG (Glu) to AAG (Lys).
    (c) This mutant hemoglobin moves more rapidly toward the anode than does Hb A and Hb S because it is more positively charged.
12. No. The target site for MstII is CCTNAGG, which encodes Pro-Glu-Glu in hemoglobin A. Other mutations of this heptanucleotide sequence could lead to the loss of the 1.3-kb fragment in the Southern blot—for example, a mutation of CCTGAGG to CTTGAGG (Pro-Glu-Glu to Leu-Glu-Glu).
13. Mutations in the α gene affect all three hemoglobins because their subunit structures are $\alpha_2\beta_2$, $\alpha_2\delta_2$, and $\alpha_2\gamma_2$. Mutations in the β, δ, or γ genes affect only one of them.
14. Deoxy Hb A contains a complementary site, and so it can add on to a fiber of deoxy Hb S. The fiber cannot then grow further because the terminal deoxy Hb A molecule lacks a sticky patch.
15. Carbamoylation of hemoglobin increases its oxygen affinity. Oxygenated Hb S does not sickle.
16. (a and b) Protons are released from hemoglobin because of increased oxygenation.

Chapter 8

1. (a) 31.1 μmoles.
   (b) 0.05 μmoles.
   (c) 622 s^{-1}.
2. Yes. $K_M = 5.2 \times 10^{-6}$ M.
   (b) $V_{max} = 6.84 \times 10^{-10}$ moles/min.
   (c) 337 s^{-1}.
3. Penicillinase, like glycopeptide transpeptidase, forms an acyl-enzyme intermediate with its substrate but transfers it to water rather than to the terminal glycine of the pentaglycine bridge.
4. (a) In the absence of inhibitor, V_{max} is 47.6 μmole/min and K_M is 1.1×10^{-5} M. In the presence of inhibitor, V_{max} is the same, and the apparent K_M is 3.1×10^{-5} M.
   (b) Competitive.
   (c) 1.1×10^{-3} M.
   (d) f_{ES} is 0.243 and f_{EI} is 0.488.
   (e) f_{ES} is 0.73 in the absence of inhibitor and 0.49 in the presence of 2×10^{-3} M inhibitor. The ratio of these values, 1.49, is the same as the ratio of the reaction velocities under these conditions.
5. (a) V_{max} is 9.5 μmole/min. K_M is 1.1×10^{-5} M, the same as without inhibitor.
   (b) Noncompetitive.
   (c) 2.5×10^{-5} M.
   (d) 0.73, in the presence or absence of this noncompetitive inhibitor.
6. (a) $V = V_{max} - (V/[S])\,K_M$.
   (b) Slope $= -K_M$, y-intercept $= V_{max}$, x-intercept $= V_{max}/K_M$.
   (c)

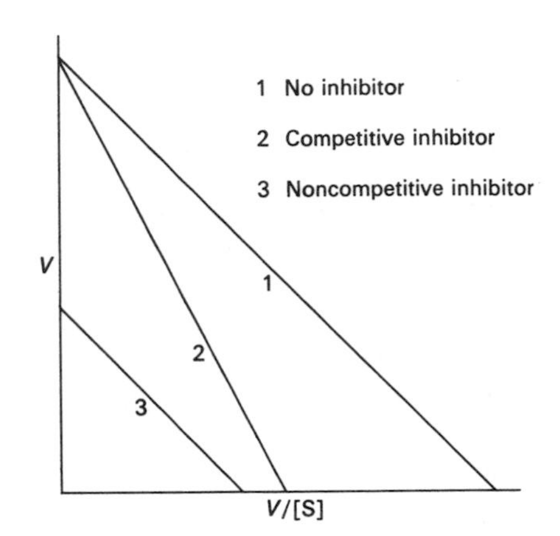

7. Potential hydrogen-bond donors at pH 7 are the side chains of the following residues: arginine, asparagine, glutamine, histidine, lysine, serine, threonine, tryptophan, and tyrosine.
8. The rates of utilization of A and B are given by

$$V_A = \left(\frac{k_3}{K_M}\right)_A [E][A]$$

and

$$V_B = \left(\frac{k_3}{K_M}\right)_B [E][B]$$

Hence, the ratio of these rates is

$$V_A/V_B = \left(\frac{k_3}{K_M}\right)_A [A] \Big/ \left(\frac{k_3}{K_M}\right)_B [B]$$

Thus, an enzyme discriminates between competing substrates on the basis of their values of k_3/K_M rather than of K_M alone.

Chapter 9

1. The fastest is (b) and the slowest is (a).
2. (a)B-C, (b) A-B or E-F, and (c) A-B-C (one sugar residue does not interact with the enzyme, thus avoiding site D, which is energetically unfavorable).

3. The ^{18}O emerges in the C-4 hydroxyl of di-NAG (residues E-F).
4. This analog lacks a bulky substituent at C-5 and so it can probably bind to site D without being strained. Consequently, the binding of residue D of this analog is likely to be energetically favorable, whereas the binding of residue D of tetra-NAG costs free energy. See P. van Eikeren and D. M. Chipman, *J. Am. Chem. Soc.* 94(1972):4788.
5. (a) In oxymyoglobin, Fe is bonded to five nitrogens and one oxygen. In carboxypeptidase A, Zn is bonded to two nitrogens and two oxygens.
   (b) In oxymyoglobin, one of the nitrogens bonded to Fe comes from the proximal histidine residue, whereas the other four come from the heme. The oxygen atom linked to Fe is that of O_2. In carboxypeptidase A, the two nitrogen atoms coordinated to Zn come from histidine residues. One of the oxygen ligands is from a glutamate side chain, the other from a water molecule.
   (c) Aspartate, cysteine, and methionine.
6. (a) Yes.
   (b) Histidine 119 in ribonuclease A, histidine 57 in chymotrypsin, glutamic acid 35 in lysozyme, and a zinc-bound water molecule in carboxypeptidase A.
7. (a) Tosyl-L-lysine chloromethyl ketone (TLCK).
   (b) First, determine whether substrates protect trypsin from inactivation by TLCK and, second, ascertain whether the D-isomer of TLCK inactivates trypsin.
8. (a) Serine.
   (b) A hemiacetal between the aldehyde of the inhibitor and the hydroxyl group of the active site serine.
9. The boron atom becomes bonded to the oxygen atom of the active-site serine. This tetrahedral intermediate has a geometry similar to that of the transition state.
10. Precise positioning of catalytic residues and substrates, geometrical strain (distortion of the substrate), electronic strain, and desolvation of the substrate.
11. The zinc ion polarizes the carbonyl group of the scissile bond to make its carbon atom more positively charged. Zn^{2+} also enhances the nucleophilicity of its bound water molecule. Glutamate pulls a proton away from the zinc-bound water molecule, which makes it a stronger nucleophile. See M. A. Holmes and B. W. Matthews, *Biochemistry* 20(1981):6912.

Chapter 10

1. The protonated form of histidine probably stabilizes the negatively charged carbonyl oxygen atom of the scissile bond in the transition state. Deprotonation would lead to a loss of activity. Hence, the rate is expected to be half-maximal at a pH of about 6.5 (the p*K* of an unperturbed histidine side chain in a protein) and decrease as the pH is raised.
2. (a) 100. By linkage (p. 175), the change in [R]/[T] is the same as the ratio of affinities of the two forms.
   (b) 100. The binding of four substrate molecules changes the [R]/[T] by a factor of $100^4 = 10^8$. The ratio in the absence of substrate is 10^{-6}. Hence, the ratio in the fully liganded molecule is $10^8 \times 10^{-6} = 10^2$.
3. Activation is independent of zymogen concentration because the reaction is intramolecular (p. 246).
4. Add blood from the second patient to a sample from the first. If the mixture clots, the second patient has a defect different from that of the first. This type of assay is called a complementation test.
5. Activated factor X remains bound to blood platelet membranes, which accelerates its activation of prothrombin.
6. Antithrombin III is a very slowly hydrolyzed substrate of thrombin. Hence, its interaction with thrombin requires a fully formed active site on the enzyme.
7. Residues *a* and *d* are located in the interior of an α-helical coiled coil, near the axis of the superhelix. Hydrophobic interactions between these side chains contribute to the stability of the coiled coil.
8. Replace methionine 358 with leucine, which occupies nearly the same volume and is hydrophobic (p. 248).

Chapter 11

1. (a) Every third residue in each strand of a collagen triple helix must be glycine because there is insufficient space for a larger residue.
   (b) Poly (Gly-Pro-Gly) melts at a lower temperature than poly (Gly-Pro-Pro).
   (c) No. Glycine does not occupy every third position.
2. (a and b)

```
       OH         OH          OH
       |          |           |
-Gly-Leu-Pro-Gly-Pro-Pro-Gly-Ala-Pro-Gly
        ↑
         \
          Susceptible
          peptide bond
```

3. (a) Disulfides.
   (b) None.
   (c) Peptide bonds between specific glutamine and lysine side chains.
   (d) Aldol cross-link and hydroxypyridinium cross-link.
   (e) Lysinonorleucine and desmosine.
4. The decarboxylation of α-ketoglutarate is half of the physiological reaction. It seems likely that α-ketoglutarate is first attacked by oxygen to form a peroxy acid, which then reacts with the prolyl substrate. See D. F. Counts, G. J. Cardinale, and S. Udenfriend, *Proc. Nat. Acad. Sci.* 75(1978):2145, for a discussion of this experiment and its mechanistic implications.
5. (a) The aldehyde is more reactive. For example, it can form a hemiacetal with the hydroxyl group of a serine at an active site.

(b) A tetrahedral intermediate, like the one formed in catalysis by chymotrypsin (p. 225), seems likely.

6. The peptide Arg-Gly-Asp-Ser competitively inhibits the binding of one of the modules of fibronectin to integrin, a protein that spans the plasma membrane of fibroblasts.
7. The three-quarter-length and one-quarter-length fragments of collagen formed by the action of tissue collagenases are likely to be more stable than normal. Collagen resorption will probably be impeded.
8. This mutation will probably mimic vitamin C deficiency. Defective hydroxylation of proline will lead to collagen having a lower than normal melting temperature. The abnormal collagen cannot properly form fibers; skin lesions and blood vessel fragility are likely consequences. The clinical picture will resemble scurvy.

Chapter 12

1. 2.86×10^6 molecules.
2. Cyclopropane rings interfere with the orderly packing of hydrocarbon chains, and so they increase membrane fluidity.
3. 2×10^{-7} cm, 6.32×10^{-6} cm, and 2×10^{-4} cm.
4. The radius of this molecule is 3.08×10^{-7} cm and its diffusion coefficient is 7.37×10^{-9} cm²/s. The average distances traversed are 1.72×10^{-7} cm in 1 μs, 5.42×10^{-6} in 1 ms, and 1.72×10^{-4} cm in 1 s.
5. The initial decrease in the amplitude of the paramagnetic resonance spectrum results from the reduction of spin-labeled phosphatidyl cholines in the outer leaflet of the bilayer. Ascorbate does not traverse the membrane under these experimental conditions, and so it does not reduce the phospholipids in the inner leaflet. The slow decay of the residual spectrum is due to the reduction of phospholipids that have flipped over to the outer leaflet of the bilayer.

Chapter 13

1. Reactions (a) and (c), to the left; reactions (b) and (d), to the right.
2. None whatsoever.
3. (a) $\Delta G^{\circ\prime} = +7.5$ kcal/mol and $K'_{eq} = 3.06 \times 10^{-6}$.
   (b) 3.28×10^4.
4. $\Delta G^{\circ\prime} = 1.7$ kcal/mol. The equilibrium ratio is 17.8.
5. (a) +0.2 kcal/mol.
   (b) −7.8 kcal/mol. The hydrolysis of PP_i drives the reaction toward the formation of acetyl CoA.
6. (a) $\Delta G^\circ = 2.303\, RT\, pK$.
   (b) 6.53 kcal/mol at 25°C.
7. An ADP unit (or a closely related derivative, in the case of CoA).
8. The activated form of sulfate in most organisms is 3′-phosphoadenosine 5′-phosphosulfate. See P. W. Robbins and F. Lipmann, *J. Biol. Chem.* 229(1957):837.
9. (a) 310 Hz.
   (b) Both the protonation and deprotonation rates must be faster than 310 s^{-1}.
   (c) The p*K*′ for the equilibrium of $H_2PO_4^{2-}$ and HPO_4^- is 7.21 (see Appendix C). Hence, the dissociation constant *K* is 6.16×10^{-8} M. The rate constant for association k_{on} is equal to k_{off}/K. Because k_{off} is greater than 310 s^{-1}, k_{on} must be greater than 5×10^9 $M^{-1}\,s^{-1}$.

Chapter 14

1. (a) Aldose-ketose; (b) epimers; (c) aldose-ketose; (d) anomers; (e) aldose-ketose; and (f) epimers.
2. Aldoses are converted into aldonic acids; the aldehyde group of the sugar is oxidized to a carboxylate.
3. Glucose is reactive because of the presence of an aldehyde group in its open-chain form. The aldehyde group slowly condenses with amino groups to form Schiff-base adducts (see p. 642).
4. A pyranoside reacts with two molecules of periodate; formate is one of the products. A furanoside reacts with only one molecule of periodate; formate is not formed.
5. From methanol.
6. (a) β-D-Mannose; (b) β-D-galactose; (c) β-D-fructose; (d) β-D-glucosamine.
7. The trisaccharide itself should be a competitive inhibitor of cell adhesion if the trisaccharide unit of the glycoprotein is critical for the interaction.

Chapter 15

1. Glucose is reactive because its open-chain form contains an aldehyde group (see p. 642).
2. (a) The label is in the methyl carbon of pyruvate.
   (b) 5 mCi/mM. The specific activity is halved because the number of moles of product (pyruvate) is twice that of the labeled substrate (glucose).
3. (a) $\Delta G^{\circ\prime}$ is −29.5 kcal/mol for the reaction

$$\text{Glucose} + 2\,P_i + 2\,\text{ADP} \longrightarrow 2\,\text{lactate} + 2\,\text{ATP}$$

   (b) $\Delta G' = -27.2$ kcal/mol.
4. 3.06×10^{-5}.
5. The equilibrium concentrations of fructose 1,6-bisphosphate, dihydroxyacetone phosphate, and glyceraldehyde 3-phosphate are 7.76×10^{-4} M, 2.24×10^{-4} M, and 2.24×10^{-4} M, respectively.
6. All three carbon atoms of 2,3-BPG are ^{14}C-labeled. The phosphorus atom attached to the C-2 hydroxyl is ^{32}P-labeled.
7. Hexokinase has a low ATPase activity in the absence of a sugar because it is in a catalytically inactive conformation (p. 271). The addition of xylose closes the cleft between the two lobes of the enzyme. However, xylose lacks a hydroxymethyl group, and so it cannot be phosphorylated. Instead, a water molecule at the site normally occupied by the C-6 hydroxymethyl group acts as the phosphoryl acceptor from ATP.

8. (a) 2,3-Bisphosphoglycerate (BPG) lowers the oxygen affinity of hemoglobin (p. 157). The rate of synthesis of 2,3-BPG is controlled by the level of 1,3-BPG, a glycolytic intermediate.
   (b) The lowered level of glycolytic intermediates leads to less 2,3-BPG, and hence, a higher oxygen affinity.
   (c) Glycolytic intermediates are present at a higher than normal level. The level of 2,3-BPG is increased, which makes the oxygen affinity lower than normal.
9. (a) The fructose 1-phosphate pathway (p. 357) forms glyceraldehyde 3-phosphate. Phosphofructokinase, a key control enzyme, is bypassed. Furthermore, fructose 1-phosphate stimulates pyruvate kinase.
   (b) The rapid, unregulated production of lactate can lead to metabolic acidosis.
10. The catalytic site of an activated enzyme molecule contains a phosphorylated serine residue. This phosphoryl group is transferred to either substrate to form a glucose 1,6-bisphosphate intermediate (p. 454). The phosphoryl group on the enzyme is slowly lost by hydrolysis; it is regenerated by phosphoryl transfer from 1,6-glucose bisphosphate.
11. The metal ion serves as an electron sink, as does the protonated Schiff base in animal aldolases.
12. EDTA removes the metal ion from the catalytic site of procaryotic aldolases, whereas sodium borohydride reduces the Schiff base intermediate in catalysis by animal aldolases.

Chapter 16

1. (a) After one round of the citric acid cycle, the label emerges in C-2 and C-3 of oxaloacetate.
   (b) After one round of the citric acid cycle, the label emerges in C-1 and C-4 of oxaloacetate.
   (c) The label emerges in CO_2 in the formation of acetyl CoA from pyruvate.
   (d and e) Same fate as in (a).
2. No, because two carbon atoms are lost in the two decarboxylation steps of the cycle. Hence, there is no *net* synthesis of oxaloacetate.
3. 0.90, 0.03, and 0.07.
4. −9.8 kcal/mol.
5. The coenzyme stereospecificity of glyceraldehyde 3-phosphate dehydrogenase is the opposite of that of alcohol dehydrogenase (type B versus type A, respectively.)
6. Thiamine thiazolone pyrophosphate is a transition-state analog. The sulfur-containing ring of this analog is uncharged, and so it closely resembles the transition state of the normal coenzyme in thiamine-catalyzed reactions (e.g., the uncharged resonance form of hydroxyethyl-TPP, p. 380). See J. A. Gutowski and G. E. Lienhard, *J. Biol. Chem.* 251(1976):2863, for a discussion of this analog.
7. The ratio of malate to oxaloacetate must be greater than 1.75×10^4 for oxaloacetate to be formed.
8. Methane is first oxidized by a monooxygenase to methanol; NADH is the reductant. Methanol is then oxidized to formaldehyde; PQQ, a novel quinone, is the electron acceptor in this step. Formaldehyde is oxidized to formic acid, which is in turn oxidized to CO_2. NADH is formed in each of these steps. About 5 ATP are formed (3 ATP from NADH and 2 ATP from $PQQH_2$). See G. Gottschalk, *Bacterial Metabolism* (2nd ed.) (Springer-Verlag, 1986), p. 163.
9. The enolate anion of acetyl CoA attacks the carbonyl carbon atom of glyoxylate to form a C–C bond. This reaction is like the condensation of oxaloacetate with the enolate anion of acetyl CoA (p. 383). Glyoxylate contains a hydrogen atom in place of the $—CH_2COO^-$ of oxaloacetate; the reactions are otherwise nearly identical.

Chapter 17

1. (a) 15, (b) 2, (c) 38, (d) 16, (e) 36, and (f) 19.
2. (a) $\Delta E'_0$ is +1.05 V and $\Delta G°'$ is −48.4 kcal/mol for the reaction

$$2\ G—SH + \tfrac{1}{2}\ O_2 \rightleftharpoons G—S—S—G + H_2O$$

   (b) $\Delta E'_0$ = +0.09 V and $\Delta G^{0\prime}$ is −4.15 kcal/mol.
3. (a) Blocks electron transport and proton pumping at site 3.
   (b) Blocks electron transport and ATP synthesis by inhibiting the exchange of ATP and ADP across the inner mitochondrial membrane.
   (c) Blocks electron transport and proton pumping at site 1.
   (d) Blocks ATP synthesis without inhibiting electron transport by dissipating the proton gradient.
   (e) Blocks electron transport and proton pumping at site 3.
   (f) Blocks electron transport and proton pumping at site 2.
4. Oligomycin inhibits ATP formation by interfering with the utilization of the proton gradient. It does not block electron transport.
5. $\Delta G°'$ is +16.1 kcal/mol for oxidation by NAD^+ and +1.4 kcal/mol for oxidation by FAD. The oxidation of succinate by NAD^+ is not thermodynamically feasible.
6. Cyanide can be lethal because it binds to the ferric form of cytochrome ($a + a_3$) and thereby inhibits oxidative phosphorylation. Nitrite converts ferrohemoglobin to ferrihemoglobin, which also binds cyanide. Thus, ferrihemoglobin competes with cytochrome ($a + a_3$) for cyanide. This competition is therapeutically effective because the amount of ferrihemoglobin that can be formed without impairing oxygen transport is much greater than the amount of cytochrome ($a + a_3$).
7. The available free energy from the translocation of 2, 3, and 4 protons is −9.23, −13.8, and −18.5 kcal, respectively. The free energy consumed in synthesizing a mole of ATP under standard conditions is 7.3 kcal. Hence, the residual free energy of −1.93, −6.5, and

-11.2 kcal can drive the synthesis of ATP until the $[ATP]/[ADP][P_i]$ ratio is 26.2, 6.51×10^4, and 1.62×10^8, respectively. Suspensions of isolated mitochondria synthesize ATP until this ratio is greater than 10^4, which shows that the number of protons translocated per ATP synthesized is at least three.

8. Biochemists use E_0', the value at pH 7, whereas chemists use E_0, the value in 1 N H^+. The prime denotes that pH 7 is the standard state.
9. Such a defect (called Luft's syndrome) was found in a thirty-eight-year-old woman who was incapable of performing prolonged physical work. Her basal metabolic rate was more than twice normal, but her thyroid function was normal. A muscle biopsy showed that her mitochondria were highly variable and atypical in structure. Biochemical studies then revealed that oxidation and phosphorylation were not tightly coupled in these mitochondria. In this patient, much of the energy of fuel molecules was converted into heat rather than ATP. See R. Luft, D. Ikkos, G. Palmieri, L. Ernster, and B. Afzelius. *J. Clin. Invest.* 41(1962):1776.
10. The absolute configuration of thiophosphate is opposite to that of ATP in the reaction catalyzed by ATP synthase. This result is consistent with an in-line phosphoryl transfer reaction occurring in a single step. The retention of configuration in the Ca^{2+}-ATPase reaction points to two phosphoryl transfer reactions—inversion by the first, and a return to the starting configuration by the second. The Ca^{2+}-ATPase reaction proceeds by a phosphorylated enzyme intermediate. See M. R. Webb, C. Grubmeyer, H. S. Penefsky, and D. R. Trentham, *J. Biol. Chem.* 255(1980):255.
11. Dicylohexylcarbodiimide reacts readily with carboxyl groups, as was discussed earlier in regard to its use in peptide synthesis (p. 65). Hence, the most likely targets are aspartate and glutamate side chains. In fact, aspartate 61 of subunit c of *E. coli* F_0 is specifically modified by this reagent. Conversion of this aspartate to an asparagine by site-specific mutagenesis also eliminates proton conduction. See A. E. Senior, *Biochim. Biophys Acta* 726(1983):81.

Chapter 18

1. (a) 5 Glucose 6-phosphate + ATP $\longrightarrow$ 6 ribose 5-phosphate + ADP + H^+

   (b) Glucose 6-phosphate + 12 $NADP^+$ + 7 H_2O $\longrightarrow$ 6 CO_2 + 12 NADPH + 12 H^+ + P_i
2. The label emerges at C-5 of ribulose 5-phosphate.
3. Oxidative decarboxylation of isocitrate to α-ketoglutarate. A β-keto acid intermediate is formed in both reactions.
4. C-1 and C-3 of fructose 6-phosphate are labeled, whereas erythrose 4-phosphate is not labeled.
5. Reactions b and e would be blocked.
6. Form a Schiff base between a ketose substrate and transaldolase, reduce it with tritiated $NaBH_4$, and fingerprint the labeled enzyme.
7. $\Delta E_0'$ for the reduction of glutathione by NADPH is +0.09 V. Hence, $\Delta G^{\circ\prime}$ is -4.15 kcal/mol, which corresponds to an equilibrium constant of 1126. The required $[NADPH]/[NADP^+]$ ratio is 8.9×10^{-5}.
8. The lactate level in the maternal circulation, and hence in the fetal circulation, increases during pregnancy because the mother is carrying a growing fetus with its own metabolic demands. The shift to H_4 in the fetal heart enables the fetus to use lactate as a fuel. Consequently, there is less need for gluconeogenesis by the liver and kidneys of the mother.
9. Fructose 2,6-bisphosphate, present at a high concentration when glucose is abundant, normally inhibits gluconeogenesis by blocking fructose 1,6-bisphosphatase. In this genetic disorder, the phosphatase is active irrespective of the glucose level. Hence, substrate cycling is increased, which generates heat. The level of fructose 1,6-bisphosphate is consequently lower than normal. Less pyruvate is formed, resulting in less acetyl CoA. In turn, less ATP is formed by the citric acid cycle and oxidative phosphorylation.

Chapter 19

1. Galactose + ATP + UTP + H_2O + glycogen$_n$ $\longrightarrow$ glycogen$_{n+1}$ + ADP + UDP + 2 P_i + H^+
2. Fructose + 2 ATP + 2 H_2O $\longrightarrow$ glucose + 2 ADP + 2 P_i
3. There is a deficiency of the branching enzyme.
4. The concentration of glucose 6-phosphate is elevated in von Gierke's disease. Consequently, the phosphorylated D form of glycogen synthetase is active.
5. Glucose is an allosteric inhibitor of phosphorylase *a*. Hence, crystals grown in its presence are in the T state. The addition of glucose 1-phosphate, a substrate, shifts the R $\rightleftharpoons$ T equilibrium toward the R state. The conformational differences between these states are sufficiently large that the crystal shatters unless it is stabilized by chemical cross-links. The shattering of a crystal caused by an allosteric transition was first observed by Haurowitz in the oxygenation of crystals of deoxyhemoglobin.
6. H. G. Hers [*Ann. Rev. Biochem.* 45(1976):167] suggested that these kinetics would ensure a lag in the dephosphorylation of subunit B, which would allow glycogen to be degraded before phosphorylase kinase is inactivated by its phosphatase.
7. (a) The control of glycogen phosphorylase and synthase will be impaired. Specifically, epinephrine will not trigger the breakdown of glycogen and the cessation of glycogen synthesis.

   (b) Calmodulin mediates the activation of phosphorylase kinase by elevated Ca^{2+} during muscle contraction. Hence, glycogen will not be degraded in concert with contraction.

(c) Protein phosphatase 1 will be continually active. Hence, the level of phosphorylase *b* will be higher than normal, and glycogen will be less readily degraded.

Chapter 20

1. (a) Glycerol + 2 NAD^+ + P_i + ADP $\longrightarrow$ pyruvate + ATP + H_2O + 2 NADH + H^+
   (b) Glycerol kinase and glycerol phosphate dehydrogenase.
2. Stearate + ATP + 13½ H_2O + 8 FAD + 8 NAD^+ $\longrightarrow$ 4½ acetoacetate + 12½ H^+ + 8 $FADH_2$ + 8 NADH + AMP + 2 P_i
3. (a) Oxidation in mitochondria, synthesis in the cytosol.
   (b) Acetyl CoA in oxidation, acyl carrier protein for synthesis.
   (c) FAD and NAD^+ in oxidation, NADPH for synthesis.
   (d) L-isomer of 3-hydroxyacyl CoA in oxidation, D-isomer in synthesis.
   (e) Carboxyl to methyl in oxidation, methyl to carboxyl in synthesis.
   (f) The enzymes of fatty acid synthesis, but not those of oxidation, are organized in a multienzyme complex.
4. (a) Palmitoleate, (b) linoleate, (c) linoleate, (d) oleate, (e) oleate, and (f) linolenate.
5. C-1 is more radioactive (see p. 751 for a discussion of this experimental approach in the elucidation of the direction of synthesis of a polypeptide chain).
6. The enolate anion of one thioester attacks the carbonyl carbon atom of the other thioester to form a C–C bond.
7. Adipose cell lipase is activated by phosphorylation. Hence, overproduction of the cAMP-activated kinase will lead to accelerated breakdown of triacylglycerols and depletion of fat stores.
8. When the blood glucose level is low, acetyl CoA carboxylase is switched off by phosphorylation. Impaired phosphorylation will lead to persistent activation of the carboxylase. Malonyl CoA will be synthesized even when glucose is scarce.

Chapter 21

1. (a) Pyruvate, (b) oxaloacetate, (c) α-ketoglutarate, (d) α-ketoisocaproate, (e) phenylpyruvate, and (f) hydroxyphenylpyruvate.
2. Aspartate + α-ketoglutarate + GTP + ATP + 2 H_2O + NADH + H^+ $\longrightarrow$ ½ glucose + glutamate + CO_2 + ADP + GDP + NAD^+ + 2 P_i
3. Aspartate + CO_2 + $NH_4{}^+$ + 3 ATP + NAD^+ + 4 H_2O $\longrightarrow$ oxaloacetate + urea + 2 ADP + 4 P_i + AMP + NADH + H^+
4. (a) Label the methyl carbon atom of L-methylmalonyl CoA with ^{14}C. Determine the location of ^{14}C in succinyl CoA. The group transferred is the one bonded to the labeled carbon atom.
   (b) The proton that is abstracted from the methyl group of L-methylmalonyl CoA is directly transferred to the adjacent carbon atom.
5. Thiamine pyrophosphate.
6. It acts as an electron sink. See C. Walsh, *Enzymatic Reaction Mechanisms,* (W. H. Freeman, 1979), p. 178.
7. Deuterium is abstracted by the radical form of the coenzyme. The methyl group rotates before hydrogen is returned to the product radical.
8. A carbanion or a carbonium ion.

Chapter 22

1. $\Delta E'_0 = +0.11$ V and $\Delta G^{\circ\prime} = -5.1$ kcal/mol.
2. Aldolase participates in the Calvin cycle, whereas transaldolase participates in the pentose phosphate pathway.
3. The concentration of 3-phosphoglycerate would increase, whereas that of ribulose 1,5-bisphosphate would decrease.
4. The concentration of 3-phosphoglycerate would decrease, whereas that of ribulose 1,5-bisphosphate would increase.
5. Phycoerythrin, the most peripheral protein in the phycobilisome.
6. (a) It expresses a key aspect of photosynthesis—namely, that water is split by light. The evolved oxygen in photosynthesis comes from water.
   (b) The Van Niel equation for respiration expresses the fact that the combustion of glucose requires the input of six molecules of H_2O. For an interesting discussion, see G. Wald, On the nature of cellular respiration, *Current Aspects of Biochemical Energetics,* N. O. Kaplan and E. P. Kennedy, eds. (Academic Press, 1966), pp. 27–37.
7. The addition of pyridine increases the proton storage capacity of the thylakoid space. More pumped protons can then flow through the ATP-synthesizing complex in the dark. For a discussion of this experiment, see M. Avron, *Ann. Rev. Biochem.* 46(1977):145.
8. DCMU inhibits electron transfer between Q and plastoquinone in the link between photosystems II and I. O_2 evolution can occur in the presence of DCMU if an artificial electron acceptor such as ferricyanide can accept electrons from Q.
9. CABP resembles the addition compound formed in the reaction of CO_2 and ribulose 1,5-bisphosphate

(p. 534). As predicted, CABP is a potent inhibitor of the enzyme.

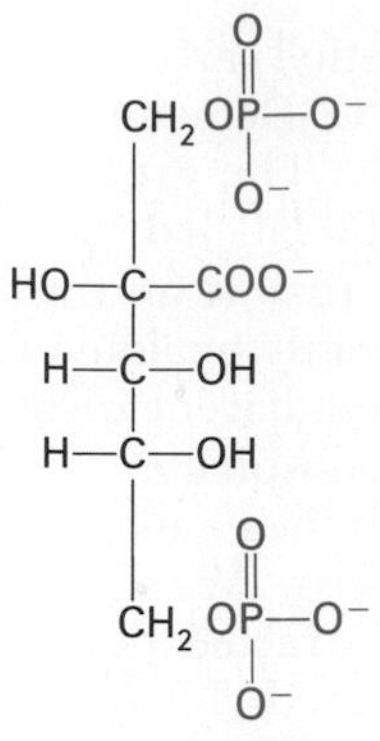

```
          O
          ‖
    CH2 OP—O⁻
    |     |
    |     O⁻
HO—C—COO⁻
 H—C—OH
 H—C—OH
    |     O
    |     ‖
    CH2 OP—O⁻
          |
          O⁻
```

2-Carboxyarabinitol 1,5-bisphosphate (CABP)

10. Aspartate + glyoxylate ⟶ oxaloacetate + glycine
11. (a) 28.7 kcal/einstein (or 120 kJ/einstein).
    (b) 1.24 V.
    (c) One 1000-nm photon has the free energy content of 2.39 ATP. A minimum of 0.42 photon is needed to drive the synthesis of an ATP.

Chapter 23

1. Glycerol + 4 ATP + 3 fatty acids + 4 H_2O ⟶ triacylglycerol + ADP + 3 AMP + 7 P_i + 4 H^+
2. Glycerol + 3 ATP + 2 fatty acids + 2 H_2O + CTP + serine ⟶ phosphatidyl serine + CMP + ADP + 2 AMP + 6 P_i + 3 H^+
3. (a) CDP-diacylglycerol, (b) CDP-ethanolamine, (c) acyl CoA, (d) CDP-choline, (e) UDP-glucose or UDP-galactose, (f) UDP-galactose, and (g) geranyl pyrophosphate.
4. (a and b) None, because the label is lost as CO_2.

Chapter 24

1. Glucose + 2 ADP + 2 P_i + 2 NAD^+ + 2 glutamate ⟶ 2 alanine + 2 α-ketoglutarate + 2 ATP + 2 NADH + H^+
2. N_2 ⟶ NH_4^+ ⟶ glutamate ⟶ serine ⟶ glycine ⟶ δ-aminolevulinate ⟶ porphobilinogen ⟶ heme
3. (a) Tetrahydrofolate, (b) tetrahydrofolate, and (c) N^5-methyltetrahydrofolate.
4. γ-Glutamyl phosphate may be a reaction intermediate.
5. The administration of glycine led to the formation of isovalerylglycine. This water-soluble conjugate, in contrast with isovaleric acid, is excreted very rapidly by the kidneys. See R. M. Cohn, M. Yudkoff, R. Rothman, and S. Segal, *New Engl. J. Med.* 299(1978):996.
6. H-D exchange points to the existence of a diimide intermediate. See W. A. Bulen, *Proc. Int. Symp. N_2 Fixation* (Washington State University Press, 1976).

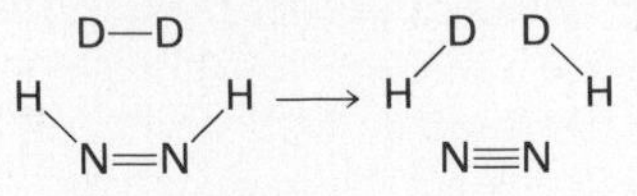

```
    D—D              D   D
 H       H   ⟶   H/      \H
  \N=N/             N≡N
```

7. They carry out nitrogen fixation. The absence of photosystem II provides an environment in which O_2 is not produced. Recall that the nitrogenase is very rapidly inactivated by O_2. See R. Y. Stanier, J. L. Ingraham, M. L. Wheelis, and P. R. Painter, *The Microbial World,* 5th ed. (Prentice-Hall, 1986), pp. 356–359, for a discussion of heterocysts.
8. The cytosol is a reducing environment, whereas the extracellular milieu is an oxidizing environment. Glutathione is the major sulfhydryl buffer in the cytosol.
9. The sulfoximine moiety, erroneously recognized as the γ-carboxylate of glutamate, is enzymatically phosphorylated to form methionine sulfoximine phosphate. This phosphorylated product binds very tightly to glutamine synthetase. See R. Rando, *Accts. Chem. Res.* 8(1975):281.
10. Glutathione normally feedback-inhibits γ-glutamylcysteine synthetase, the enzyme catalyzing the first step in its biosynthesis. The absence of glutathione leads to high levels of γ-glutamylcysteine, which is converted into 5-oxoproline by γ-glutamyl cyclotransferase.

Chapter 25

1. Glucose + 2 ATP + 2 $NADP^+$ + H_2O ⟶ PRPP + CO_2 + ADP + AMP + 2 NADPH + H^+
2. Glutamine + aspartate + CO_2 + 2 ATP + NAD^+ ⟶ orotate + 2 ADP + 2 P_i + glutamate + NADH + H^+
3. (a, c, d, and e) PRPP, (b) carbamoyl phosphate.
4. PRPP and formylglycinamide ribonucleotide.
5. dUMP + serine + NADPH + H^+ ⟶ dTMP + $NADP^+$ + glycine
6. There is a deficiency of N^{10}-formyltetrahydrofolate. Sulfanilamide inhibits the synthesis of folate by acting as an analog of *p*-aminobenzoate, one of the precursors of folate.
7. PRPP is the activated intermediate in the synthesis of (a) phosphoribosylamine in the de novo pathway of purine formation, (b) purine nucleotides from free bases by the salvage pathway, (c) orotidylate in the formation of pyrimidines, (d) nicotinate ribonucleotide, (e) phosphoribosyl-ATP in the pathway leading to histidine, and (f) phosphoribosyl-anthranilate in the pathway leading to tryptophan.
8. It seems likely that glutamine yields ammonia as a

result of the catalytic action of the small subunit. The nascent ammonia would then react with an activated form of CO_2 that is formed by the large subunit. The bicarbonate-dependent ATPase activity implies that this activated species is carbonyl phosphate. Reaction of this carbonic-phosphoric mixed anhydride with NH_3 yields carbamate, which would then react with ATP to give carbamoyl phosphate. For a discussion of this enzymatic mechanism, see C. Walsh, *Enzymatic Reaction Mechanisms* (W. H. Freeman, 1979), pp. 150–154.

9. Analogous reactions occur in the urea cycle (p. 501)—from citrulline to arginosuccinate, and then to arginine.
10. Tyrosine 122 was converted into phenylalanine 122 by site-specific mutagenesis. This mutant enzyme has the same size, iron content, and iron-sensitive absorption spectrum as the wild type but is totally devoid of enzymatic activity. This experiment by A. Larsson and B. M. Sjoberg (*EMBO J.* 5(1986):2037) provides strong evidence that tyrosine 122 is the site of the free radical in ribonucleotide reductase.
11. (a) At 25°C, the free energies of binding are −13.8 (wild type), −11.9 (Asn 27), and −9.1 (Ser 27) kcal/mol. The loss in binding energy is 1.9 and 4.7 kcal/mol.
    (b) The mutants bind the unprotonated form of methotrexate. The side chain amide of asparagine 27 donates a proton to the N-1 nitrogen of methotrexate to form a hydrogen bond. The other side chain hydrogen atom is hydrogen-bonded to the oxygen atom of a water molecule. In the serine 27 mutant, the oxygen atom of a water molecule accepts a proton from the serine OH group, and a hydrogen atom of water is donated to the N-1 nitrogen of methotrexate.

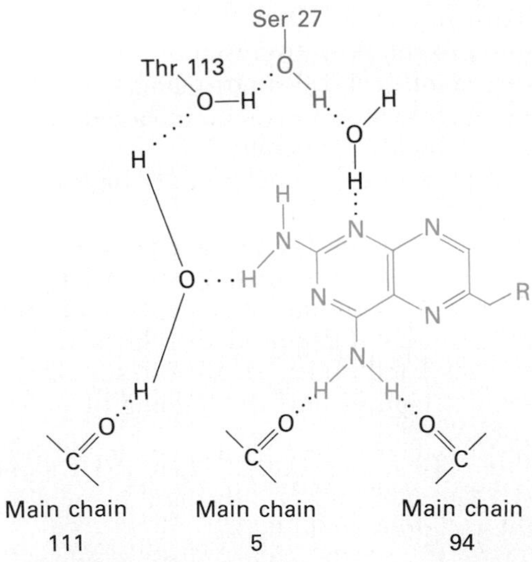

Mode of binding of methotrexate to the serine 27 mutant of dihydrofolate reductase. [After E. H. Howell, J. E. Villafranca, M. S. Warren, S. J. Oatley, and J. Kraut. *Science* 231(1986):1125.]

12. (a) *S*-Adenosylhomocysteine hydrolase activity is markedly diminished in ADA-deficient patients because of reversible inhibition by adenosine and suicide inactivation by 2′-deoxyadenosine.
    (b) The activated methyl cycle (p. 583) is blocked in these patients. *S*-Adenosylhomocysteine is a potent inhibitor of methyl transfer reactions involving *S*-adenosylmethionine. For a discussion of adenosine deaminase deficiency, see N. M. Kredich and M. S. Hershfield in Stanbury, J. B., Wyngaarden, J. B., Fredrickson, D. S., Goldstein, J. L., and Brown, M. S., (eds.), *The Metabolic Basis of Inherited Disease* (5th ed., McGraw-Hill, 1983), pp. 1157–1183.
13. (a) Cell A cannot grow in a *HAT* medium because it cannot synthesize dTMP either from thymidine or dUMP. Cell B cannot grow in this medium because it cannot synthesize purines either by the de novo pathway or the salvage pathway. Cell C can grow in a *HAT* medium because it contains active thymidine kinase (enabling it to phosphorylate thymidine to dTMP) and hypoxanthine-guanine phosphoribosyl transferase (enabling it to synthesize purines from hypoxanthine by the salvage pathway).
    (b) Transform cell A with a plasmid containing foreign genes of interest and a functional thymidine kinase gene. The only cells that will grow in a *HAT* medium are those that have acquired a thymidylate kinase gene; nearly all of these transformed cells will also contain the other genes on the plasmid.
14. These patients have a high level of urate because of the breakdown of nucleic acids. Allopurinol prevents the formation of kidney stones and blocks other deleterious consequences of hyperuricemia by preventing the formation of urate (p. 621).

Chapter 26

1. Liver contains glucose 6-phosphatase, whereas muscle and brain do not. Hence, muscle and brain, in contrast with liver, do not release glucose. Another key enzymatic difference is that liver has little of the transferase needed to activate acetoacetate to acetoacetyl CoA. Consequently, acetoacetate and 3-hydroxybutyrate are exported by the liver for use by heart muscle, skeletal muscle, and brain.
2. (a) Adipose cells normally convert glucose to glycerol 3-phosphate for the formation of triacylglycerols. A deficiency of hexokinase would interfere with the synthesis of triacylglycerols.
   (b) A deficiency of glucose 6-phosphatase would block the export of glucose from liver following glycogenolysis. This disorder (called von Gierke's disease) is characterized by an abnormally high content of glycogen in the liver and a low blood glucose level.
   (c) A deficiency of carnitine acyltransferase I impairs the oxidation of long-chain fatty acids. Fasting and exercise precipitate muscle cramps in these individuals.

(d) Glucokinase enables the liver to phosphorylate glucose even in the presence of a high level of glucose 6-phosphate. A deficiency of glucokinase would interfere with the synthesis of glycogen.

(e) Thiolase catalyzes the formation of two molecules of acetyl CoA from acetoacetyl CoA and CoA. A deficiency of thiolase would interfere with the utilization of acetoacetate as a fuel when the blood sugar level is low.

(f) Phosphofructokinase will be less active than normal because of the lowered level of F-2,6-BP. Hence, glycolysis will be much slower than normal.

3. (a) A high proportion of fatty acids in the blood are bound to albumin. Cerebrospinal fluid has a low content of fatty acids because it has little albumin.

(b) Glucose is highly hydrophilic and soluble in aqueous media, in contrast with fatty acids, which must be carried by transport proteins such as albumin. Micelles of fatty acids would disrupt membrane structure.

(c) Fatty acids, not glucose, are the major fuel of resting muscle.

4. (a) A watt is equal to 1 joule per second (0.239 calorie per second). Hence, 70 watts is equivalent to 0.07 kJ/s or 0.017 kcal/s.

(b) A watt is a current of one ampere across a potential of one volt. For simplicity, let us assume that all of the electron flow is from NADH to O_2 (a potential drop of 1.14 V). Hence, the current is 61.4 amperes, which corresponds to 3.86×10^{20} electrons per second (1 ampere = 1 coulomb/s = 6.28×10^{18} charges/s).

(c) Three ATP are formed per NADH oxidized (two electrons). Hence, one ATP is formed per 0.67 electrons transferred. A flow of 3.86×10^{20} electrons per second therefore leads to the generation of 5.8×10^{20} ATP per second or 0.96 mmole per second.

(d) The molecular weight of ATP is 507. The total body content of ATP of 50 grams is equal to 0.099 mole. Hence, ATP turns over about once per hundred seconds when the body is at rest.

5. The store of ATP at rest is used in a half second. Creatine phosphate is the major source of ~P during the first four seconds of the sprint. Glycolysis provides most of the additional ATP that is consumed.

6. The 1748 kcal available in glucose or glycogen would be consumed in 8740 seconds, or 146 minutes, if the rate of energy expenditure during the marathon were 0.2 kcal/s (12 times the basal level of 0.017 kcal/s given in problem 4).

7. A high blood glucose level would trigger the secretion of insulin, which would stimulate the synthesis of glycogen and triacylglycerols. A high insulin level would impede the mobilization of fuel reserves during the marathon.

Chapter 27

1. DNA polymerase I uses deoxyribonucleoside triphosphates; pyrophosphate is the leaving group. DNA ligase uses a DNA-adenylate (AMP joined to the 5′-phosphate) as a reaction partner; AMP is the leaving group. Topoisomerase I uses a DNA-tyrosyl intermediate (5′-phosphate linked to the phenolic OH); the tyrosine residue of the enzyme is the leaving group.

2. At pH 6, the histidine residue at the active site is likely to be protonated and hence unable to accept a proton to activate the bound water molecule.

3. DNase I does not cleave Z-DNA because it lacks the minor groove of B-DNA. The arginine and lysine side chains at the active site are not sterically complementary to phosphate groups in Z-DNA.

4. FAD, CoA, and $NADP^+$ are plausible alternatives.

5. DNA ligase relaxes supercoiled DNA by catalyzing the cleavage of a phosphodiester bond in a DNA strand. The attacking group is AMP, which becomes attached to the 5′-phosphoryl group at the site of scission. AMP is required because this reaction is the reverse of the final step in the joining of pieces of DNA (see Figure 27-20 on p. 659).

6. Positive supercoiling resists the unwinding of DNA. The melting temperature of DNA increases in going from negatively supercoiled to relaxed to positively supercoiled DNA. Positive supercoiling is probably an adaptation to high temperature.

7. (a) The twisting number changes by −2 when a turn of right-handed B-DNA ($T = +1$) switches into a turn of left-handed Z-DNA ($T = -1$). Because L stays the same, W changes by +2. Hence, $L = 100$, $T = 102$, and $W = -2$ after the transition.

(b) The DNA becomes less supercoiled and hence less compact. It will move more slowly.

(c) The torsional energy stored in a supercoiled DNA molecule is proportional to the square of the superhelix density. Free energy is released when the degree of negative supercoiling decreases. The endergonic B-to-Z transition is driven by the accompanying exergonic decrease in negative supercoiling.

(d) This protein stabilizes B-DNA. Hence, the midpoint of the B-to-Z transition will occur at higher degree of negative supercoiling.

8. (a) Long stretches of each occur because the transition is highly cooperative.

(b) B-Z junctions are energetically highly unfavorable.

(c) A–B transitions are less cooperative than B–Z transitions because the helix stays right-handed at an A-B junction but not at a B-Z junction.

9. ATP hydrolysis is required to release DNA gyrase after it has acted on its DNA substrate. Negative supercoiling requires only the binding of ATP, not its hydrolysis.

10. (a) Pro (CCC), Ser (UCC), Leu (CUC) and Phe (UUC). Alternatively, the last base of each of these codons could be U.

(b) These C → U mutations were produced by nitrous acid.

11. (a) No, it was produced by the deletion of the first base in the sequence shown below and the insertion of another base at the end of this sequence.

(b) -AGUCCAUCACUUAAU-

Chapter 28

1. The dnaB protein and rep protein participate in DNA replication. The recB component of the recBCD complex generates single-stranded DNA in general recombination. The recA protein binds to single-stranded DNA and catalyzes its invasion of duplex DNA and a switch in base pairing.
2. (a) The recA-DNA complex contains 18.6 base pairs per turn, compared with 10.4 for B-DNA. Hence, the linking number is 56, compared with 100 for the relaxed circle formed in the absence of recA protein.
   (b) The DNA molecule after removal of recA protein is highly supercoiled and hence much more compact than its relaxed counterpart.
3. Proteolysis of the lexA protein at the onset of the SOS response relieves the repression of synthesis of lexA mRNA. Increased formation of lexA protein will terminate the SOS response.
4. (a) In the presence of ATPγS, recA protein binds to DNA but its dissociation is blocked because this analog is not readily hydrolyzed. Strand exchange does not take place.
   (b) RecB protein binds to DNA in the presence of this analog but its release is blocked. Consequently, single-stranded DNA is not formed.

Chapter 29

1. Heparin, a glycosaminoglycan (p. 276), is highly anionic. Its negative charges, like the phosphodiester bridges of DNA templates, bind to lysine and arginine residues of β'.
2. This mutant sigma would competitively inhibit the binding of holoenzyme and prevent the specific initiation of RNA chains at promoter sites.
3. The core enzyme without sigma binds more tightly to the DNA template than does the holoenzyme. The retention of sigma after chain initiation would make the mutant RNA polymerase less processive. Hence, RNA synthesis would be much slower than normal.
4. A 100-kd protein contains about 910 residues, which are encoded by 2730 nucleotides. At a maximal transcription rate of 50 nucleotides per second, the protein would be synthesized in 54.6 seconds.
5. Initiation at strong promoters occurs every two seconds. In this interval, 100 nucleotides are transcribed. Hence, centers of transcription bubbles are 340 Å apart.
6. Ser-Ile-Phe-His-Pro-Stop.
7. A change from U to C in the recognition sequence AAUAAA for the endonuclease caused this defect in a thalasemic patient. Cleavage occurred at the AAUAAA 900 nucleotides downstream from this mutant AACAAA site. See S. H. Orkin, T-C. Cheng, S. E. Antonarakis, and H. H. Kazazian, Jr., *EMBO J.* 4(1985):453.

Chapter 30

1. (a) No, (b) no, and (c) yes.
2. Four bands: light, heavy, a hybrid of light 30S and heavy 50S, and a hybrid of heavy 30S and light 50S.
3. About 799 high-energy phosphate bonds are consumed—400 to activate the 200 amino acids, 1 for initiation, and 398 to form 199 peptide bonds.
4. (b, c, and f) Type 1; (a, d, and e) type 2.
5. The simplest hypothesis is that the CCA anticodon of a tryptophan tRNA has mutated to UCA, which is complementary to UGA. However, analysis of this altered tRNA produces a surprise. Its anticodon is unaltered. Rather, there is a substitution of A for G at position 24. Thus, a residue far from the anticodon in the linear base sequence can influence the fidelity of codon recognition.
6. One approach is to synthesize a tRNA charged with a reactive amino acid analog. For example, bromoacetyl-phenylalanyl-tRNA is an affinity-labeling reagent for the P site of *E. coli* ribosomes. See H. Oen, M. Pellegrini, D. Eilat, and C. R. Cantor, *Proc. Nat. Acad. Sci.* 70(1973):2799.
7. The sequence GAGGU is complementary to a sequence of five bases at the 3′ end of 16S rRNA and is located several bases on the 5′ side of an AUG codon. Hence this region is a start signal for protein synthesis. The replacement of G by A would be expected to weaken the interaction of this mRNA with the 16S rRNA and thereby diminish its effectiveness as an initiation signal. In fact, this mutation results in a tenfold decrease in the rate of synthesis of the protein specified by this mRNA. See J. J. Dunn, E. Buzash-Pollert, and F. W. Studier, *Proc. Nat. Acad. Sci.* 75(1978):2741, for a discussion of this informative mutant.
8. The nitrogen atom of the deprotonated α-amino group of aminoacyl-tRNA is the nucleophile in peptide-bond formation.
9. Leu-Gly-Trp-tRNA occupies the P site because peptide bond formation and translocation have occurred. The entry of the next aminoacyl-tRNA into the A site is blocked because fusidic acid prevents EF-G-GDP from dissociating.
10. Proteins are synthesized from the amino to the carboxyl end on ribosomes, and in the reverse direction in the solid-phase method. The activated intermediate in ribosomal synthesis is an aminoacyl-tRNA; in the solid-phase method, the adduct of the amino acid and dicyclohexylcarbodiimide.
11. The error rates of DNA, RNA, and protein synthesis are of the order of 10^{-10}, 10^{-5}, and 10^{-4} per nucleotide (or amino acid) incorporated. The fidelity of all three processes depends on the precision of base pairing to the DNA or mRNA template. No error correction occurs in RNA synthesis. In contrast, the fidelity of DNA synthesis is markedly increased by the 3′ → 5′ proofreading nuclease activity and by post-replicative repair. In protein synthesis, the mischarging of some tRNAs is corrected by the hydrolytic action of the aminoacyl-tRNA synthetase. Proofreading also takes place when aminoacyl-tRNA occupies the A site on the ribosome; the GTPase activity of EF-Tu sets the pace of this final stage of editing.

Chapter 31

1. (a) Cleavable amino-terminal signal sequences are usually 13 to 36 residues long and contain a highly hydrophobic central region 10 to 15 residues long. The amino-terminal part has at least one basic residue. The cleavage site is preceded (at −1 and −3) by small neutral residues.
   (b) Procaryotic signal sequences are similar to eucaryotic ones. In addition, a stop-transfer sequence is needed to keep the protein in the plasma membrane.
   (c) Mannose 6-phosphate residues direct proteins to lysosomes.
   (d) Integral membrane proteins initially in the endoplasmic reticulum go to the plasma membrane unless they carry instructions to the contrary. No specific signal is needed.
   (e) An amino-terminal sequence containing positively charged residues, serine, and threonine, in addition to hydrophobic residues.
   (f) An amino-terminal arginine, aspartate, leucine, lysine, or phenylalanine.
2. The cytosolic portion of the receptor, which enables it to interact with coated pits, is likely to be altered.
3. (a) Endocytosis mediated by a receptor specific for mannose 6-phosphate residues. Endocytic vesicles containing the added lysosomal enzymes then fuse with lysosomes.
   (b) The addition of mannose 6-phosphate to the extracellular medium should prevent lysosomal enzymes from reaching their destination if the normal pathway involves secretion outside the cell and import back into the cell. However, mannose 6-phosphate does not inhibit normal lysosomal targeting, showing that lysosomal enzymes reach their destination without leaving the cell.
4. A specific targeting sequence is not required. Rather, amino-terminal sequences are scanned for overall features such as the presence of hydrophobic, basic, and hydroxyl amino acids. Sequence (d) is ineffective because it contains too many polar residues (glutamines) in place of hydrophobic residues (such as leucine and phenylalanine).
5. (a) The chimeric protein will probably be found in the cytosol. The transmembrane sequence of a membrane-bound immunoglobulin functions as a stop-transfer sequence and not as a signal sequence.
   (b) The chimeric protein will probably be found in the plasma membrane because chymotrypsinogen is synthesized with a signal sequence.
6. Chloroplasts contain three kinds of membranes (outer, inner, and thylakoid), whereas mitochondria contain only two (outer and inner). The inner membrane of chloroplasts, in contrast with that of mitochondria, is not energized. Hence, chloroplasts use ATP instead of a proton-motive force as the energy source for importing proteins.
7. The addition of a permeant base such as chloroquine will prevent acidification of endosomes, which should block the entry of virus into the cytosol.
8. Secretory proteins that are erroneously targeted to the cytosol will be rapidly degraded because their amino termini mark them for rapid destruction (Table 31-2, on p. 794). Rapid elimination of mistargeted secretory proteins is important because some of them (e.g., trypsinogen) could wreak havoc in the cytosol.

Chapter 32

1. (a) The *lac* repressor is missing in an i^- mutant. Hence, this mutant is constitutive for the proteins of the *lac* operon.
   (b) This mutant is constitutive for the proteins of the *trp* operon because the *trp* repressor is missing.
   (c) The arabinose operon is not expressed in this mutant because the P2 form of the *araC* protein is needed to activate transcription.
   (d) This mutant is lytic but not lysogenic because it cannot synthesize the λ repressor.
   (e) This mutant is lysogenic but not lytic because it cannot synthesize the N protein, a positive control factor in transcription.
2. One possibility is that an i^s mutant produces an altered *lac* repressor that has almost no affinity for inducer but normal affinity for the operator. Such a *lac* repressor would bind to the operator and block transcription even in the presence of inducer.
3. This mutant has an altered *lac* operator that fails to bind the repressor. Such a mutator is called O^c (operator constitutive).
4. The cyclic AMP binding protein (CAP) is probably defective or absent in this mutant.
5. An *E. coli* cell bearing a λ prophage contains λ repressor molecules, which also block the transcription of the immediate-early genes of other λ viruses.
6. (a) Translation of the p_{RE} transcript is from five to ten times as rapid as that of the p_{RM} transcript because it contains the full protein synthesis initiation signal (as discussed on p. 753).
   (b) The more effective translation of the p_{RE} transcript provides a burst of λ repressor molecules needed to establish the lysogenic state. See M. Ptashe, K. Backman, Z. Humagun, A. Jeffrey, R. Maurer, B. Meyer, and R. T. Sauer, *Science* 194(1976):156, for a discussion.
7. The rate constant for association of wild-type *lac* repressor with the operator site is near the diffusion-controlled limit. Hence, a 100-fold increase in binding affinity implies that the mutant dissociates from the operator site about 100-fold more slowly than does the wild-type. The lag between addition of inducer and the initiation of transcription of the *lac* operon will be much longer in the mutant than in the wild type.
8. A repressor finds its target site by first binding to a nonspecific site anywhere in the DNA molecule and then diffusing along the DNA to reach the specific site. The repressor binds less tightly to nonspecific DNA at low ionic strength than at high ionic strength

because of increased electrostatic repulsion. In contrast, binding to the operator is nearly independent of ionic strength because the interaction is mediated by hydrogen-bond and van der Waals interactions with the bases of the operator site.

9. The mutant protein acts as a repressor even when tryptophan is not bound. Expression of the *trp* operon is permanently repressed in this mutant bacterium.
10. When RF2 is present, UGA is read as a stop codon, leading to the formation of a truncated 25-residue polypeptide.

- Gly - Tyr - Leu - Stop
GGG UAU CUU UGA   Read by RF2

When the level of RF2 is very low, termination at UGA does not occur. Instead, a shift of the reading frame occurs, resulting in continued protein synthesis with Asp 26 as the next residue. See W. Craigen, R. Cook, W. Tate, and C. Caskey. *Proc. Nat. Acad. Sci.* 82(1985):3619.

- Gly - Tyr - Leu - Asp - Tyr - Asp -
GGG UAU CUU UGAC UAC GAC

Frameshift in absence of RF2. U is skipped. GAC is first codon of new reading frame.

Chapter 33

1. (a) In procaryotes, translation begins while transcription is still in progress. In eucaryotes, these processes are separate in space and time.
   (b) Eucaryotic transcripts are monogenic, whereas procaryotic transcripts are typically polygenic.
   (c) In procaryotes, several proteins can be specified by a single primary transcript, one for each gene encoding the mRNA. In eucaryotes, a primary transcript encoded by a single gene can give rise to multiple proteins through alternative splicing. Cleavage of a polyprotein can also yield multiple proteins in eucaryotes.
   (d) Most of procaryotic DNA encodes proteins and functional RNAs. In contrast, most of eucaryotic DNA does not encode functional macromolecules.
   (e) Most procaryotic genes are clustered in operons. Eucaryotes do not have operons.
2. Centromeres, telomeres, and *ars* (autonomous replicating sequences serving as origins of replication) are needed to fashion a synthetic yeast chromosome. See A. Murray and J. W. Szostak, *Nature* 305(1983):189.
3. The mitochondrial genome is much smaller than the nuclear genome. Furthermore, a cell contains a large number of mitochondria. A substantial proportion of them, say 5%, could be nonfunctional without injuring the cell.
4. Mitochondrial DNA is evolving at a very rapid rate (see R. L. Cann, M. Stoneking, and A. C. Wilson. *Nature* 325(1986):31.). It is possible but not likely that a human mitochondrial gene will be transferred to the nuclear genome in the next ten million years. The existence of different genetic codes in the mitochondrion and the cytosol is a formidable barrier to gene transfer. In particular, UGA specifies tryptophan in mitochondrial proteins but is a stop signal in cytosolic protein synthesis.
5. One experimental approach is to add histones to linear duplex DNA and then form closed circles using DNA ligase. The histones are then removed from the supercoiled circles. Topoisomerase I is added to an aliquot to form the relaxed counterpart. The melting temperature of the supercoiled DNA is compared with that of the relaxed circle. A left-handed superhelix has a lower melting temperature than the relaxed circle, whereas a right-handed superhelix has a higher melting temperature. Recall that left-handed supercoiled DNA (negatively supercoiled; *W* is negative) is poised to be unwound (p. 660).
6. The broken end invades internal C_{1-3} sequences that are complementary. The intact strand serves as a template in repair synthesis.
7. 5S RNA has the same base sequence (except for U in place of T) as the coding strand of its gene, the one that strongly binds TFIIIA. Hence, 5S RNA competes with its gene for the binding of TFIIIA. When 5S RNA is abundant, little TFIIIA is bound to the internal control region, and so transcription is slowed. The binding of TFIIIA to 5S RNA as well as the internal control region of the gene is the basis of a feedback loop that regulates the amount of 5S RNA.
8. (a) DNA inversions lead to alternating expression of a pair of flagellar genes (H1 and H2) in *Salmonella*. A flip-flop circuit changes an exposed bacterial protein at a frequency determined by the activity of a recombinase (p. 818).
   (b) One possible mechanism for changing *MAT* is to transfer *HML* or *HMR* into the *MAT* site. Alternatively, *HML* or *HMR* can serve as a template for the synthesis of a new *MAT* gene. In fact, *HML* and *HMR* are preserved at their original sites; they serve as templates without being physically removed from their original loci. See J. Rine, R. Jensen, D. Hagen, L. Blair, and I. Herskowitz. *Cold Spring Harbor Symp. Quant. Biol.* 45(1981):951.
9. One possibility is that the transcript of the 35-nucleotide sequence primes the transcription of the rest of the other segment of the gene. Alternatively, the two RNAs could be joined by an unusual *trans* (intermolecular) splicing reaction.

Chapter 34

1. The nucleotide sequence is searched for an open reading frame—a sequence that has no internal stop

codons between the codon for the initiating methionine and the stop codon after the carboxyl-terminal residue. Peptides about ten residues long corresponding to portions of the deduced amino acid sequence are synthesized by the solid-phase method and used as immunogens. Antibodies against these peptides then serve as specific reagents for identifying the encoded protein. A Western blot would be a good way to start. The antibodies could also be used to immunoprecipitate the protein and to isolate it by affinity chromatography. See R. F. Doolittle, *Of Urfs and Orfs: A Primer on How to Analyze Derived Amino Acid Sequences* (University Books, 1986) for a concise and interesting discussion.

2. Long concatamers are formed by the association of complementary single-stranded tails of nascent duplexes. Complementarity arises from the redundancy in base sequence of the ends. The 3′ end of one duplex in the concatamer then serves as the primer to fill the gap at the 5′ end of the adjoining duplex in the repeating chain of new DNA molecules.

```
 3'   5'
a|    |A
b|    |B
c|    |C
x|    |X
y|    |Y
      |Z
a|    |A
b|    |B
c|
d|    |D
e|    |E
f|    |F
a|    |A
b|    |B
 5'   3'
```

3. (a) 5′-UGGACUUUGUGGGAUACCCUCGCUUU-3′
   (b) The second leucine in the sequence becomes phenylalanine.
   (c) Each sequence is highly constrained by the other.
4. The infected cell would be lysed before progeny virions are fully synthesized and assembled.
5. Polyamines neutralize the negatively charged RNA inside the virion. See S. S. Cohen and F. P. McCormick, *Adv. Virus Res.* 24(1979):331–387.
6. Influenza hemagluttinin (HA) binds sialic acid (*N*-acetylneuraminate) residues on the surface of cells in the respiratory tract. This binding enables influenza virus to be endocytosed into susceptible cells. Erythrocytes, like cells of the respiratory tract, contain sialic acid residues on their cell-surface glycoproteins. The binding of influenza virus to erythrocytes in vitro can be inhibited by adding glycopeptides containing sialic acid.
7. (a) DNA viruses use the replication machinery of their eucaryotic hosts. Hence, their genomes must pass through the nucleus of infected cells. DNA viruses also exploit the transcriptional and splicing apparatus of their hosts. Some RNA viruses can replicate and form mRNA in the cytosol because they encode their own transcriptase and carry it in the virion. The genomes of other RNA viruses (e.g., retroviruses) must pass through the nucleus.
   (b) Influenza virus exploits the splicing apparatus of the host cell to form ten mRNAs from its eight genomic segments.
8. (a) The CD4 cell-surface receptor is present on brain cells as well as T4 lymphocytes.
   (b) Prepare an antibody to the CD4 protein of T4 lymphocytes, and use immunofluorescence microscopy to determine whether this receptor is present on other kinds of cells.
9. RNA-assisted cleavage assures that virus particles are not formed of a protein coat containing no RNA. A closed icosahedral shell is formed only after RNA is packaged inside.

INDEX

References to structural formulas are given in **boldface** type.

COMMON ABBREVIATIONS IN BIOCHEMISTRY

A	adenine
ACP	acyl carrier protein
ADP	adenosine diphosphate
Ala	alanine
AMP	adenosine monophosphate
cAMP	cyclic AMP
cGMP	cyclic GMP
Arg	arginine
Asn	asparagine
Asp	aspartate
ATP	adenosine triphosphate
ATPase	adenosine triphosphatase
C	cytosine
CDP	cytidine diphosphate
CMP	cytidine monophosphate
CTP	cytidine triphosphate
CoA	coenzyme A
CoQ	coenzyme Q (ubiquinone)
cyclic AMP	adenosine 3′,5′-cyclic monophosphate
cyclic GMP	guanosine 3′,5′-cyclic monophosphate
Cys	cysteine
cyt	cytochrome
d	2′-deoxyribo
DNA	deoxyribonucleic acid
cDNA	complementary DNA
DNase	deoxyribonuclease
EcoRI	EcoRI restriction endonuclease
FAD	flavin adenine dinucleotide (oxidized form)
$FADH_2$	flavin adenine dinucleotide (reduced form)
fMET	formylmethionine
FMN	flavin mononucleotide (oxidized form)
$FMNH_2$	flavin mononucleotide (reduced form)
G	guanine
Gln	glutamine
Glu	glutamate
Gly	glycine
GDP	guanosine diphosphate
GMP	guanosine monophosphate
GSH	reduced glutathione
GSSG	oxidized glutathione
GTP	guanosine triphosphate
GTPase	guanosine triphosphatase
Hb	hemoglobin
HDL	high-density lipoprotein
HGPRT	hypoxanthine-guanine phosphoribosyl transferase
His	histidine
Hyp	hydroxyproline
IgG	immunoglobulin G
Ile	isoleucine
IP_3	inositol trisphosphate
ITP	inosine triphosphate
LDL	low-density lipoprotein
Leu	leucine
Lys	lysine
Met	methionine
NAD^+	nicotinamide adenine dinucleotide (oxidized form)
NADH	nicotinamide adenine dinucleotide (reduced form)
$NADP^+$	nicotinamide adenine dinucleotide phosphate (oxidized form)
NADPH	nicotinamide adenine dinucleotide phosphate (reduced form)
PFK	phosphofructokinase
Phe	phenylalanine
P_i	inorganic orthophosphate
PLP	pyridoxal phosphate
PP_i	inorganic pyrophosphate
Pro	proline
PRPP	phosphoribosylpyrophosphate
Q	ubiquinone (or plastoquinone)
QH_2	ubiquinol (or plastoquinol)
RNA	ribonucleic acid
mRNA	messenger RNA
rRNA	ribosomal RNA
scRNA	small cytoplasmic RNA
snRNA	small nuclear RNA
tRNA	transfer RNA
RNase	ribonuclease
Rubisco	ribulose 1,5-bisphosphate carboxylase
Ser	serine
T	thymine
Thr	threonine
TPP	thiamine pyrophosphate
Trp	tryptophan
TTP	thymidine triphosphate
Tyr	tyrosine
U	uracil
UDP	uridine diphosphate
UDP-galactose	uridine diphosphate galactose
UDP-glucose	uridine diphosphate glucose
UMP	uridine monophosphate
UTP	uridine triphosphate
Val	valine
VLDL	very low-density lipoprotein